The Greek Alphabet

Alpha	A	α	Iota	I	ι	Rho	P	ρ
Beta	B	β	Kappa	K	κ	Sigma	Σ	σ
Gamma	Γ	γ	Lambda	Λ	λ	Tau	T	τ
Delta	Δ	δ	Mu	M	μ	Upsilon	Υ	υ
Epsilon	E	ϵ	Nu	N	ν	Phi	Φ	ϕ
Zeta	Z	ζ	Xi	Ξ	ξ	Chi	X	χ
Eta	H	η	Omicron	O	o	Psi	Ψ	ψ
Theta	Θ	θ	Pi	Π	π	Omega	Ω	ω

Abbreviations for Units

A	ampere
Å	angstrom (10^{-10} m)
atm	atmosphere
Btu	British thermal unit
Bq	becquerel
C	coulomb
°C	degree Celsius
cal	calorie
Ci	curie
cm	centimeter
dyn	dyne
eV	electron volt
°F	degree Fahrenheit
fm	femtometer, fermi (10^{-15} m)
ft	foot
Gm	gigameter (10^9 m)
G	gauss
Gy	gray
g	gram
H	henry
h	hour
Hz	hertz
in	inch
J	joule
K	kelvin
kg	kilogram
km	kilometer
keV	kilo-electron volts
lb	pound
L	liter
m	meter
MeV	mega-electron volts
Mm	megameter (10^6 m)
mi	mile
min	minute
mm	millimeter
ms	millisecond
N	newton
nm	nanometer (10^{-9} m)
pt	pint
qt	quart
rev	revolution
R	roentgen
Sv	seivert
s	second
T	tesla
u	unified mass unit
V	volt
W	watt
Wb	weber
y	year
yd	yard
μm	micrometer (10^{-6} m)
μs	microsecond
μC	microcoulomb
Ω	ohm

Physics For Scientists and Engineers

Third Edition

Physics

For Scientists and Engineers

Paul A. Tipler

Worth Publishers

For Claudia

Physics for Scientists and Engineers, Third Edition

Paul A. Tipler

Printed in the United States of America

Library of Congress Catalog Card Number: 89-52165

ISBN: 0-87901-430-x

Printing: 5 4 3 2 1 Year: 95 94 93 92 91

Development Editors: Valerie Neal and Steven Tenney

Design: Malcolm Grear Designers

Art Director: George Touloumes

Production Editor: Elizabeth Mastalski

Production Supervisor: Sarah Segal

Layout: Patricia Lawson

Photographs: Steven Tenney, John Schultz of PAR/NYC, and Lana Berkovich

Line Art: York Graphic Services and Demetrios Zangos

Composition: York Graphic Services

Printing and binding: R. R. Donnelley and Sons

Cover: Supersonic Candlelight. A stroboscopic color schlieren or shadow picture taken at one-third microsecond exposure shows a supersonic bullet passing through the hot air rising above a candle. Schlieren pictures make visible the regions of nonuniform density in air. Estate of Harold E. Edgerton/ Courtesy of Palm Press.

Illustration credits begin on p. **IC-1** which constitute an extension of the copyright page.

Worth Publishers

33 Irving Place

New York, NY 10003

Preface

This third edition of *Physics*, now titled *Physics for Scientists and Engineers*, is a textbook for the standard two- or three-semester introductory physics course for engineering and science majors. It is assumed that the student has taken, or is concurrently taking, calculus. The book is divided into six parts: Mechanics, Oscillations and Waves, Thermodynamics, Electricity and Magnetism, Optics, and Modern Physics. It is available both in the standard version of 35 chapters, including two on modern physics (Relativity and The Origins of Quantum Theory), and an extended version of 42 chapters (available in one volume or two), which has seven additional chapters on modern physics (Quantum Mechanics, Atoms, Molecules, Solids, Nuclei, Elementary Particles, and Astrophysics and Cosmology), of which six are new.

SI units are used throughout the text. Except for a few problems on the conversion of force units (for example, pounds to newtons) in Chapter 4 and in the discussion of *R* factors for insulation materials in Chapter 16, all the worked examples, exercises, and problems are given in SI units.

My primary goals in writing this text have been

1. To provide a balanced introduction to the most important concepts and phenomena in classical and modern physics in a way that reflects the beauty and excitement of physics and also provides a solid foundation for further study.
2. To present physics in a logical and coherent manner that is interesting and accessible to all students.
3. To help students build self-confidence in their understanding of physics and in their problem-solving skills.
4. To stimulate students by exposing them to some of the many contemporary applications of and developments in physics in everyday life, in today's technology, and in the study of the cosmos.

Some of the features of this new edition are described below.

Streamlining and Consolidation

Every chapter has been extensively revised to make the presentation more concrete and to provide clear, logical, and succinct introductions to the central ideas of physics. For example,

> Work and energy are treated in a single chapter (Chapter 6) that discusses both the application of the work–energy theorem and conservation of mechanical energy to the solving of mechanics problems, and the generalized law of the conservation of energy. The difficulties that arise in the application of the work–energy theorem to work done on extended objects is discussed fully, as is the conversion of internal energy to mechanical energy. The concept of pseudowork is presented in Chapter 7 after the center of mass has been introduced. The treatment of escape speed has been moved to Chapter 10 (Gravity).

The discussion of rotational motion is now covered in a single chapter (Chapter 8) and has been reorganized so that angular momentum and rolling bodies can be discussed before the full vector treatment of rotation.

The chapter on oscillations (Chapter 12) has been moved so that it immediately precedes the chapters on mechanical waves. Care has been taken in the discussion of resonance to match the equations in it to those in Chapter 28 (Alternating-Current Circuits) describing resonance in *LRC* circuits. Many of the concepts of waves are introduced in Chapter 13 (Waves on Strings), with Chapter 14 being devoted entirely to sound waves. Reflection, refraction, interference, and diffraction are discussed qualitatively as applying to all types of waves, but all detailed calculations are deferred to the chapters on optics (Chapters 30 through 33).

The thermodynamics unit has been rewritten and is now organized into three chapters (Chapter 15 through 17). The introduction to temperature has been made more physical, and the section on the kinetic-theory interpretation of temperature has been expanded by the inclusion of the Maxwell–Boltzmann distribution.

The unit on electricity and magnetism (Chapters 18 through 29) has been extensively revised with particular attention to enhancing its accessibility. Electric charge, electric fields due to discrete charges, and the behavior of point charges and dipoles in an electric field are covered in Chapter 18 (Electric Fields I). In Chapter 19 (Electric Fields II), the fields due to continuous charge distributions are calculated using Coulomb's law, Gauss's law, or both. Expanded coverage of the classical model of conduction, together with a qualitative discussion of contemporary theory in Chapter 22 (Electric Current), complements the more detailed introduction to the band theory of solids in Chapter 39 of the extended edition.

In Chapter 25 (Sources of the Magnetic Field), the Biot–Savart law is given in terms of the field of a moving charge as well as that of a current element. The field is then calculated for a current loop, a solenoid, and a straight wire before Ampère's law is discussed. The material on magnetic flux has been moved to Chapter 26 (Magnetic Induction), the coverage of the magnetization of a bar magnet has been moved to Chapter 27 (Magnetism in Matter), and the discussion of Maxwell's displacement current has been moved to Chapter 29 (Maxwell's Equations and Electromagnetic Waves).

In the optics unit (Chapters 30 through 33), the material on the eye, microscopes, and telescopes is now in a short, new chapter on optical instruments (Chapter 32), which also includes a new section on the camera. A discussion of rainbows has been added to Chapter 30 (Light). The diffraction of light is first covered qualitatively and is then given a full mathematical treatment using phasors in Chapter 33 (Interference and Diffraction).

Because of this streamlining and consolidation, the length of the book has not increased significantly despite the substantial number of new worked examples, problems, photographs, and figures.

Modern Physics and Contemporary Applications

In addition to the separate chapters on modern physics (two in the standard version and nine in the extended version), the coverage of modern and applied physics has been greatly expanded throughout this edition. A section on numerical methods to support computer-based problem solving has been written for Chapter 5 (Newton's Laws II). A set of computer problems that supplement the end-of-chapter problems is available.

A solid foundation for the study of modern physics has been provided in the presentation of classical physics. For example, wave packets and group velocity are introduced in Chapter 14 (Sound) so that these concepts will be available for the study of electron waves in Chapter 36 (Quantum Mechanics) in the extended edition. Similarly, the Maxwell–Boltzmann energy distribution is given in Chapter 15 (Temperature) so that it can be used in the discussion of the electrical properties of solids in Chapter 39, and vacuum-tube diodes and triodes are discussed in Chapter 28 (Alternating-Current Circuits) to provide a basis for the treatment of semiconductor devices, also in Chapter 39.

New to this edition are hundreds of photographs, most of them in color, that complement and amplify the text and provide effective examples of contemporary applications of physics.

Seventeen guest essays, of which fourteen are new, have been written for the enjoyment and enlightenment of students and instructors. These include:

Chapter 1 Hans Christian von Baeyer, *The Fermi Solution* **p. 11**
7 Ralph A. Llewellyn, *Discovery of the Neutrino* **p. 218**
10 A.P. French, *Isaac Newton (1642–1727)* **p. 322**
11 Robert G. Hunt, *Aerodynamics of Bicycling* **p. 356**
12 James S. Walker, *Chaos: An Orderly Disorder* **p. 397**
14 Jack L. Flinner, *Seismic Waves* **p. 472**
16 Jerrold H. Krenz, *The Earth's Energy Balance and Global Warming* **p. 553**
17 Albert A. Bartlett, *Exponential Growth and Decay* **p. 586**
20 Richard Zallen, *Electrostatics and Xerography* **p. 682**
22 Elizabeth Pfleg Nickles, *Conduction in Nerve Cells* **p. 740**
26 Syun-Ichi Akasofu, *The Aurora* **p. 866**
28 John Dentler, *Electric Motors* **p. 929**
29 C.W.F. Everitt, *James Clerk Maxwell (1831–1879)* **p. 964**
30 Robert Greenler, *Beyond the (Visible) Rainbow* **p. 1004**
36 Ellen D. Williams, *Scanning Tunneling Microscopy*
37 David J. Wineland, *Atom Trapping*
39 Samuel J. Williamson, *Superconducting Quantum Interference Devices (SQUID)*

Examples, Exercises, and Problems

The understanding of physics and the development of problem-solving skills are enhanced by the extensive and integrated use of examples, in-text exercises, and graded sets of problems. Of the 391 worked examples in the standard 35-chapter version, 247 are new. Nearly all of the examples are numerical and have been written to ensure correspondence between the examples and the end-of-chapter problems (especially those at the intermediate level). Worked examples are often paired with a numerical in-text exercise (with its answer given immediately) that asks the student to perform a simple calculation to extend the example and reinforce understanding. Many of the simple, single-step examples in the second edition have been converted into in-text exercises, and many of the nonnumerical examples have been converted into text material under an appropriate subhead. Despite these conversions, the total number of worked examples has been increased by about 50 percent.

Problems at the end of each chapter are grouped into three levels of difficulty. Level I problems are relatively easy, single-step problems. They are keyed to the appropriate sections of the chapter so that the student can quickly find help if needed. Level II problems require a more sophisticated understanding and are not divided by section. Great effort has been invested in providing an extensive selection of these intermediate level problems. Level III problems are the most challenging and will be of value for more advanced students. A typical chapter has about 35 problems at Level I, 20 at Level II, and 12 at Level III. The number of intermediate level problems has been increased by more than 40 percent.

Mathematics

The teaching of introductory physics is complicated by the wide range of student backgrounds and abilities. Some have had both physics and calculus courses in high school; others have had no high school physics and are just beginning their first calculus course. To accommodate this diversity of backgrounds, a general review of mathematics (including algebra, geometry, trigonometry, complex numbers, and calculus) is provided in Appendix A.

In the text itself, the mathematical sophistication required increases gradually so that those students who are studying calculus concurrently will not be at a disadvantage. New mathematical methods and concepts are introduced as dictated by the physics, with the introduction of a new method or concept immediately followed by an example. The derivative and integral are presented in Chapter 2 (Motion in One Dimension) as an intrinsic component of the discussion of instantaneous velocity and displacement. Vector dot products are introduced in Chapter 6, where they are needed for the discussion of work and energy, and cross products are introduced in Chapter 8 in conjunction with the discussion of torque and rotational motion. A few sections that are more mathematically sophisticated are optional and are located at the ends of chapters, where they do not interrupt the flow of ideas and can easily be omitted.

The gradual increase in mathematical level through the text, combined with the relatively easy in-text exercises and numerous Level I problems, should help the least prepared students to build confidence. At the same time, the broad spectrum of complexity in the problem sets, culminating in the Level III problems, ensures that all students can be challenged and stimulated at a level appropriate to their abilities.

Ease of Review

Several pedagogical features will aid students as they review the material just covered. Important equations, laws, and tables are highlighted by a color screen. Margin heads are provided for quick reference. Key terms are introduced in boldface type, defined in the text, and listed in the review section of each chapter. Thought questions immediately follow some sections within each chapter. These include routine questions that can be easily answered from the preceding text as well as open-ended questions that can serve as a basis for classroom discussion.

Each chapter concludes with a summary, which lists the important laws and results that were discussed in the chapter, along with the equations that will be most useful in solving problems. Following each chapter is a list of suggested further readings, a review section, and the graded sets of problems. The review section contains a list of learning objectives, a list of key terms that the student should be able to identify and define, and a set of true–false questions.

Use of Color

Color has been used to improve the pedagogical effectiveness of the illustrations, most of which have been redrawn to enhance their clarity and dimensionality. Students will appreciate, for example, the way color is used to distinguish between vectors and their components, between force and velocity vectors in the same diagram, or between conductor surfaces and gaussian surfaces. The conventions used in the illustrations, such as the placement of force vectors in a diagram, have been chosen with pedagogical effectiveness as the primary goal.

Comparison with the Second Edition

Those instructors familiar with my previous edition should be reassured that the features that contributed to its usefulness have not been diminished in this extensive revision. All of the many changes have been made to enhance and complement what was already sound. A few of the changes are given here to expedite familiarization with this edition for those who have taught from the second edition:

The list of learning objectives, formerly at the beginning of each chapter, now appears in the review section following the summary.

The reviews of algebra and trigonometry have been moved from Chapter 1 to Appendix A.

Relative velocity is presented in Chapter 3 (Motion in Two and Three Dimensions).

Each of Newton's three laws of motion is discussed in a separate section in Chapter 4 (Newton's Laws I). The basic interactions are then listed with a brief qualitative discussion. In Chapter 4, applications are limited to problems involving a single particle. A brief discussion of rolling friction has been included in Chapter 5 (Newton's Laws II).

Static equilibrium is now treated separately in a new chapter (Chapter 9).

Discussions of the Maxwell–Boltzmann distribution and the heat pump have been added to the thermodynamics unit, while the Claussius inequality and the proof of the existence of entropy have been deleted.

Chapter 29 (Maxwell's Equations and Electromagnetic Waves) now includes material on Maxwell's displacement current, energy and momentum in electromagnetic waves, and the electromagnetic spectrum.

The Lorentz transformation is treated early in Chapter 34 and is used to simplify the discussion of clock synchronization and simultaneity.

A number of deletions have been made to streamline the text or to make space for the coverage of contemporary topics. These include reaction thresholds, the vector nature of angular displacement, gravitational field lines, the buoyant force in accelerated reference frames, the capacitance of an isolated sphere, the betatron, the Wheatstone Bridge, and the magnetic vector **H**.

Acknowledgements

Many people have contributed to this edition. I would like to thank everyone who used the first or second edition and offered comments and suggestions. Your letters have been valuable and welcome.

Ralph Llewellyn (University of Central Florida) wrote the exciting and informative chapter on astrophysics and cosmology (Chapter 42) in the extended edition.

Many new and interesting end-of-chapter problems were provided by Howard Miles (Washington State University), Lawrence B. Golden and James Klein (Pennsylvania State University), Robert Rundel (Mississippi State University), and William E. Fasnacht (United States Naval Academy). Among them, they have also provided one of the independently worked sets of solutions for all of the problems in the text.

Lawrence Golden and James Klein also wrote the Computer Investigations supplement.

James Walker (Washington State University) prepared the answers listing at the end of the text, produced the elegant solutions that are published in the accompanying complete *Solutions Manual* (for instructors), and also offered many valuable suggestions for improving and clarifying the end-of-chapter problems.

Robin Macqueen (University of British Columbia) contributed the Suggestions for Further Reading for each chapter.

The accuracy of the numerical calculations in the examples and exercises has been expertly checked by Robert Weidman (Michigan Technological University), Chelcie Liu (City College of San Francisco), and Edward Brown (Manhattan College). Professor Brown also offered many helpful suggestions in his reviews of the end-of-chapter problems.

Gene Mosca (United States Naval Academy) has applied his profound understanding of the teaching of physics to the revision of the *Study Guide* originally written by Granvil C. Kyker.

David Mills (College of the Redwoods) has produced a test bank of about 3000 conceptual and numerical multiple-choice questions.

The *Instructors Resource Manual* has been prepared by Robert Allen (Inver Hills Community College), John Davis (University of Washington), John Risley (North Carolina State University), James Walker (Washington State University), Vicki Williams (Pennsylvania State University), and myself. It provides a comprehensive selection of demonstrations, listings of commercially available and public-domain software programs suitable for use in teaching introductory physics, a film and video guide, suggested homework assignments, critical-thinking questions, and a complete answers listing for all end-of-chapter problems.

Roger Clapp (University of South Florida), Manuel Gómez-Rodríguez (University of Puerto Rico, Río Piedras), John Russell (Southeastern Massachusetts University), and Jim Smith (University of Illinois, Champaign–Urbana) reviewed the entire second edition; John Russell also reviewed parts of the second draft manuscript, and Jim Smith reviewed the entire first draft for the third edition. Their substantive and constructive comments and suggestions were a continual source of inspiration over the many months of this project.

Many other instructors have provided extensive and invaluable reviews. They have all made a deeply appreciated and fundamental contribution to the quality of this revision, and I therefore wish to thank:

Edward Brown, *Manhattan College*

James Brown, *The Colorado School of Mines*

Walter Borst, *Texas Technological University*

Christopher Cameron, *University of Southern Mississippi*

Bob Coakley, *University of Southern Maine*

Andrew Coates, *University College, London, United Kingdom*

Miles Dresser, *Washington State University*

Allin Gould, *John Abbott College C.E.G.E.P. Canada*

Dennis Hall, *University of Rochester*

Grant Hart, *Brigham Young University*

Jerold Izatt, *University of Alabama*

Alvin Jenkins, *North Carolina State University*

Lorella Jones, *University of Illinois, Champaign–Urbana*

Michael Kambour, *Miami-Dade Junior College*

Patrick Kenealy, *California State University at Long Beach*

Doug Kurtze, *Clarkson University*

Lui Lam, *San Jose State University*

Chelcie Liu, *City College of San Francisco*

Robert Luke, *Boise State University*

Stefan Machlup, *Case Western Reserve University*

Eric Matthews, *Wake Forest University*

Konrad Mauersberger, *University of Minnesota, Minneapolis*

Duncan Moore, *University of Rochester*

Gene Mosca, *United States Naval Academy*

Elizabeth Nickles, *Albany College of Pharmacy*

Harry Otteson, *Utah State University*

Jack Overley, *University of Oregon*

Larry Panek, *Widener University*

Malcolm Perry, *Cambridge University, United Kingdom*

Brooke Pridmore, *Clayton State College*

Arthur Quinton, *University of Massachusetts, Amherst*

John Risley, *North Carolina State University*

Robert Rundel, *Mississippi State University*

Michael Simon, *Housatonic Community College*

S. Sivananthan, *University of Illinois at Chicago*

Richard Smith, *Montana State University*

Larry Sorenson, *University of Washington*

Thor Stromberg, *New Mexico State University*

Colin Thomson, *Queens University, Canada*

Gianfranco Vidali, *Syracuse University*

Brian Watson, *St. Lawrence University*

Robert Weidman, *Michigan Technological University*

Stan Williams, *Iowa State University*

Thad Zaleskiewicz, *University of Pittsburgh, Greensburg*

George Zimmerman, *Boston University*

Finally, I would like to thank everyone at Worth Publishers for their help and encouragement, and particularly Steven Tenney, Valerie Neal, Betsy Mastalski, Wendy Schechter, Anne Vinnicombe, George Touloumes, and Sarah Segal.

Berkeley, California
February 1991

Paul Tipler

Supplements

For Students

Study Guide, Volumes 1 and 2
Student's Solutions Manual
Computer Investigations

For Instructors

Solutions Manual, Volumes 1 and 2
Test Bank, Volumes 1 and 2
Computerized Test-Generation System (IBM and Macintosh)
Instructor's Resource Manual
Transparencies, Volumes 1 and 2

About the Author

Paul Tipler was born in the small farming town of Antigo, Wisconsin, in 1933. He graduated from high school in Oshkosh, Wisconsin, where his father was Superintendent of the Public Schools. He received his BS at Purdue University in 1955 and his PhD at the University of Illinois in 1962, where he studied the structure of nuclei. He taught for one year at Wesleyan University in Connecticut while writing his thesis. He then moved to Oakland University in Michigan, where he was one of the original members of the physics department, playing a major role in developing the physics curriculum. During the next 20 years, he taught nearly all the physics courses and wrote the first and second editions of his widely used textbooks *Modern Physics* (1969, 1978) and *Physics* (1976, 1982). In 1982 he moved to Berkeley, California, where he now resides and where he wrote *College Physics* (1987). In addition to physics, his interests include music, hiking, and camping, and he is an accomplished jazz pianist and poker player.

Contents in Brief

Volume 1 Chapter 1 Systems of Measurement 1

Part 1 Mechanics 19

Chapter 2 Motion in One Dimension 20
3 Motion in Two and Three Dimensions 47
4 Newton's Laws I 77
5 Newton's Laws II 107
6 Work and Energy 135
7 Systems of Particles and Conservation of Momentum 182
8 Rotation 227
9 Static Equilibrium of a Rigid Body 275
10 Gravity 295
11 Mechanics of Solids and Fluids 331

Part 2 Oscillations and Waves 367

Chapter 12 Oscillations 368
13 Waves on a String 409
14 Sound 439

Part 3 Thermodynamics 485

Chapter 15 Temperature 486
16 Heat and the First Law of Thermodynamics 517
17 The Availability of Energy 563

Appendix A Review of Mathematics AP-1
B SI Units AP-21
C Numerical Data AP-22
D Conversion Factors AP-25
E Periodic Table of the Elements AP-27
Illustration Credits IC-1
Answers to True–False and Odd-Numbered Problems A-1
Index I-1

Volume 2 Part 4 Electricity and Magnetism 597

Chapter 18 The Electric Field I: Discrete Charge Distributions 598
19 The Electric Field II: Continuous Charge Distributions 624
20 Electric Potential 656
21 Capacitance, Dielectrics, and Electrostatic Energy 690
22 Electric Current 716
23 Direct Current Circuits 749
24 The Magnetic Field 781
25 Sources of the Magnetic Field 811
26 Magnetic Induction 840
27 Magnetism in Matter 878
28 Alternating-Current Circuits 898
29 Maxwell's Equations and Electromagnetic Waves 943

Part 5 Optics 973

Chapter 30 Light 974
31 Geometrical Optics 1012
32 Optical Instruments 1041
33 Interference and Diffraction 1061

Part 6 Modern Physics 1099

Chapter 34 Relativity 1100
35 The Origins of Quantum Theory 1145
36 Quantum Mechanics
37 Atoms
38 Molecules
39 Solids
40 Nuclei
41 Elementary Particles
42 Astrophysics and Cosmology

Appendix A Review of Mathematics AP-1
B SI Units AP-21
C Numerical Data AP-22
D Conversion Factors AP-25
E Periodic Table of the Elements AP-27
Illustration Credits IC-1
Answers to True–False and Odd-Numbered Problems A-1
Index I-1

Contents

Chapter 1

Systems of Measurement 1

1-1 Units **2**

1-2 Conversion of Units **4**

1-3 Dimensions of Physical Quantities **5**

1-4 Scientific Notation **6**

1-5 Significant Figures and Orders of Magnitude **7**

Summary **10**

Essay Hans Christian von Baeyer, *The Fermi Solution* **11**

Suggestions for Further Reading, Review, Problems **15**

Part **1** Mechanics **19**

Chapter 2

Motion in One Dimension 20

2-1 Speed, Displacement, and Velocity **21**

2-2 Instantaneous Velocity **24**

2-3 Acceleration **28**

2-4 Motion with Constant Acceleration **30**

2-5 Integration **37**

Summary **39**

Suggestions for Further Reading, Review, Problems **40**

Chapter 3

Motion in Two and Three Dimensions 47

3-1 The Displacement Vector and Vector Addition **48**

3-2 Addition of Vectors by Components **50**

3-3 Unit Vectors and Multiplication of Vectors by Scalars **53**

3-4 The Velocity Vector **54**

3-5 The Acceleration Vector **56**

3-6 Relative Velocity **57**

3-7 Projectile Motion **58**

3-8 Circular Motion **65**

Summary **69**

Suggestions for Further Reading, Review, Problems **70**

Chapter 4

Newton's Laws I 77

4-1 Newton's First Law: The Law of Inertia **78**

4-2 Force, Mass, and Newton's Second Law **80**

4-3 The Force Due to Gravity: Weight **83**

4-4 Newton's Third Law **86**

4-5 Forces in Nature **88**

4-6 Applications to Problem Solving **92**

Summary **99**

Suggestions for Further Reading, Review, Problems **100**

Chapter 5

Newton's Laws II 107

5-1 Friction **107**

5-2 Drag Forces **116**

5-3 Problems with Two or More Objects **117**

5-4 Pseudoforces **121**

5-5 Numerical Methods **124**

Summary **127**

Suggestions for Further Reading, Review, Problems **128**

Chapter 6

Work and Energy 135

6-1 Work and Kinetic Energy: Motion in One Dimension with Constant Forces **136**

6-2 Work Done by a Variable Force **140**

6-3 Work and Energy in Three Dimensions and the Dot Product **143**

6-4 Work and Energy for Particle Systems: Potential Energy **148**

6-5 Potential Energy and Equilibrium in One Dimension **152**

6-6 The Conservation of Mechanical Energy **155**

6-7 The Generalized Work–Energy Theorem **161**

6-8 The Conservation of Energy **166**

6-9 Power **169**

Summary **171**

Suggestions for Further Reading, Review, Problems **173**

Chapter 7

Systems of Particles and Conservation of Momentum 182

7-1 The Center of Mass **183**

7-2 Motion of the Center of Mass of a System **187**

7-3 The Conservation of Momentum **190**

7-4 The Center-of-Mass Reference Frame **194**

7-5 Kinetic Energy of a System of Particles **195**

7-6 Collisions in One Dimension **198**

7-7 Collisions in Three Dimensions **206**
7-8 Impulse and Time Average of a Force **210**
7-9 Jet Propulsion **213**
Summary **216**
Essay Ralph A. Llewellyn, *Discovery of the Neutrino* **218**
Suggestions for Further Reading, Review, Problems **221**

Chapter 8

Rotation 227
8-1 Angular Velocity and Angular Acceleration **228**
8-2 Torque and Moment of Inertia **231**
8-3 Rotational Kinetic Energy **236**
8-4 Calculating the Moment of Inertia **238**
8-5 Angular Momentum **242**
8-6 Rolling Objects **249**
8-7 The Vector Nature of Rotation and the Cross Product **255**
8-8 Motion of a Gyroscope **260**
8-9 Static and Dynamic Imbalance **262**
Summary **263**
Suggestions for Further Reading, Review, Problems **265**

Chapter 9

Static Equilibrium of a Rigid Body 275
9-1 Conditions for Equilibrium **275**
9-2 The Center of Gravity **278**
9-3 Some Examples of Static Equilibrium **280**
9-4 Couples **284**
9-5 Stability of Equilibrium **285**
Summary **287**
Suggestions for Further Reading, Review, Problems **288**

Chapter 10

Gravity 295
10-1 Kepler's Laws **295**
10-2 Newton's Law of Gravity **299**
10-3 Measurement of the Universal Gravitational Constant **305**
10-4 Gravitational and Inertial Mass **307**
10-5 Escaping the Earth **308**
10-6 Potential Energy, Total Energy, and Orbits **313**
10-7 The Gravitational Field of a Spherical Shell and a Solid Sphere **316**
Summary **320**
Essay A. P. French, *Isaac Newton (1642–1727)* **322**
Suggestions for Further Reading, Review, Problems **326**

Chapter 11

Mechanics of Solids and Fluids 331

11-1 Density **331**
11-2 Stress and Strain **333**
11-3 Pressure in a Fluid **336**
11-4 Buoyancy and Archimedes' Principle **340**
11-5 Surface Tension and Capillarity **344**
11-6 Fluids in Motion and Bernoulli's Equation **346**
11-7 Viscous Flow **351**
Summary **354**
Essay Robert G. Hunt, *Aerodynamics of Bicycling* **356**
Suggestions for Further Reading, Review, Problems **360**

Part 2 Oscillations and Waves 367

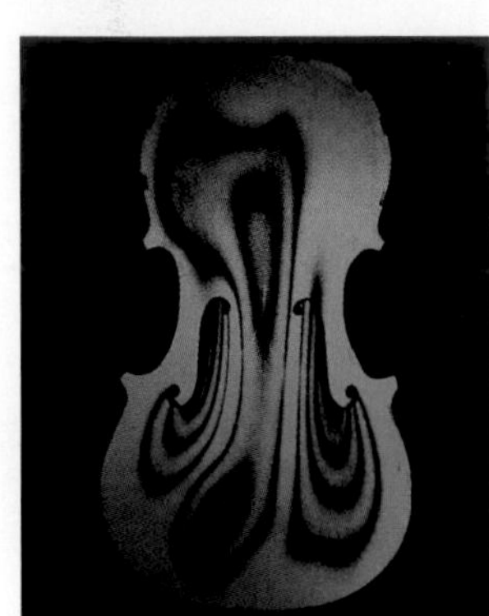

Chapter 12

Oscillations 368

12-1 Simple Harmonic Motion: A Mass on a Spring **369**
12-2 Simple Harmonic Motion and Circular Motion **376**
12-3 Energy in Simple Harmonic Motion **377**
12-4 An Object on a Vertical Spring **380**
12-5 Pendulums **382**
12-6 General Motion near Equilibrium **386**
12-7 Damped Oscillations **388**
12-8 Driven Oscillations and Resonance **392**
Summary **395**
Essay James S. Walker, *Chaos: An Orderly Disorder* **397**
Suggestions for Further Reading, Review, Problems **403**

Chapter 13

Waves on a String 409

13-1 Wave Pulses **410**
13-2 Speed of Waves **414**
13-3 Harmonic Waves **416**
13-4 Energy Transmitted by Waves **417**
13-5 Superposition and Interference of Harmonic Waves **419**
13-6 Standing Waves **420**
13-7 The Superposition of Standing Waves **427**
13-8 The Wave Equation (Optional) **429**
Summary **431**
Suggestions for Further Reading, Review, Problems **433**

Chapter 14

Sound 439

14-1 Speed of Sound Waves **440**
14-2 Harmonic Sound Waves **442**
14-3 Waves in Three Dimensions: Intensity **445**
14-4 Interference: Beats **449**
14-5 Standing Sound Waves **454**
14-6 Harmonic Analysis and Synthesis **458**
14-7 Wave Packets and Dispersion **460**
14-8 Reflection, Refraction, and Diffraction **461**
14-9 The Doppler Effect **464**
Summary **469**
Essay Jack L. Flinner, *Seismic Waves* **472**
Suggestions for Further Reading, Review, Problems **477**

Part **3** Thermodynamics **485**

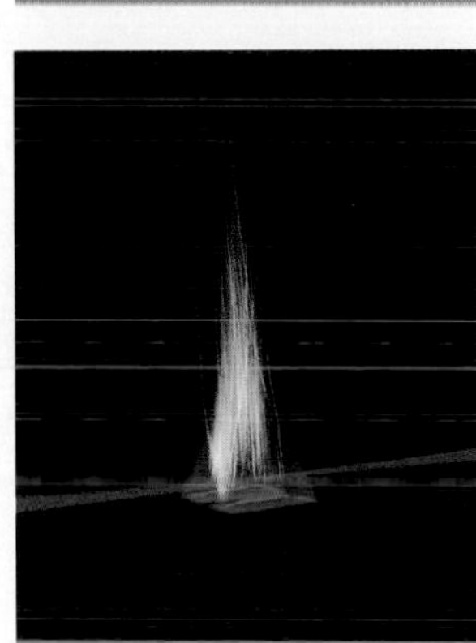

Chapter 15

Temperature 486

15-1 The Celsius and Fahrenheit Temperature Scales **487**
15-2 Gas Thermometers and the Absolute Temperature Scale **490**
15-3 Thermal Expansion **493**
15-4 The Ideal-Gas Law **496**
15-5 The Molecular Interpretation of Temperature: The Kinetic Theory of Gases **500**
15-6 The van der Waals Equation and Liquid–Vapor Isotherms **506**
15-7 Phase Diagrams **508**
Summary **511**
Suggestions for Further Reading, Review, Problems **512**

Chapter 16

Heat and the First Law of Thermodynamics 517

16-1 Heat Capacity and Specific Heat **518**
16-2 Change of Phase and Latent Heat **522**
16-3 The Transfer of Thermal Energy **525**
16-4 The First Law of Thermodynamics **534**
16-5 The Internal Energy of an Ideal Gas **537**
16-6 Work and the *PV* Diagram for a Gas **539**
16-7 Heat Capacities and the Equipartition Theorem **542**
16-8 Quasi-Static Adiabatic Expansion of a Gas **547**
Summary **550**

Essay Jerrold H. Krenz, *The Earth's Energy Balance and Global Warming* **553**
Suggestions for Further Reading, Review, Problems **557**

Chapter 17
The Availability of Energy 563
17-1 Heat Engines and the Second Law of Thermodynamics **564**
17-2 Refrigerators and the Second Law of Thermodynamics **568**
17-3 Equivalence of the Heat-Engine and Refrigerator Statements **569**
17-4 The Carnot Engine **569**
17-5 The Heat Pump **575**
17-6 Entropy and Disorder **576**
17-7 Entropy and Probability **582**
Summary **584**
Essay Albert A. Bartlett, *Exponential Growth and Decay* **586**
Suggestions for Further Reading, Review, Problems **590**

Part 4 Electricity and Magnetism 597

Chapter 18
The Electric Field I: Discrete Charge Distributions 598
18-1 Electric Charge **599**
18-2 Conductors and Insulators and Charging by Induction **601**
18-3 Coulomb's Law **603**
18-4 The Electric Field **607**
18-5 Electric-Field Lines **612**
18-6 Motion of Point Charges in Electric Fields **614**
18-7 Electric Dipoles in Electric Fields **616**
Summary **618**
Suggestions for Further Reading, Review, Problems **619**

Chapter 19
The Electric Field II: Continuous Charge Distributions 624
19-1 Calculation of the Electric Field from Coulomb's Law **625**
19-2 Gauss's Law **631**
19-3 Calculation of the Electric Field from Gauss's Law **635**
19-4 Charge and Field at Conductor Surfaces **644**
19-5 Mathematical Derivation of Gauss's Law (Optional) **649**
Summary **650**
Suggestions for Further Reading, Review, Problems **651**

Chapter 20

Electric Potential 656

20-1 Electric Potential and Potential Difference **657**
20-2 Potential Due to a System of Point Charges **660**
20-3 Electrostatic Potential Energy **664**
20-4 Calculation of Electric Potential for Continuous Charge Distributions **665**
20-5 Electric Field and Potential **671**
20-6 Equipotential Surfaces, Charge Sharing, and Dielectric Breakdown **674**
Summary **680**
Essay: Richard Zallen, *Electrostatics and Xerography* **682**
Suggestions for Further Reading, Review, Problems **685**

Chapter 21

Capacitance, Dielectrics and Electrostatic Energy 690

21-1 The Parallel-Plate Capacitor **691**
21-2 The Cylindrical Capacitor **693**
21-3 Dielectrics **694**
21-4 The Storage of Electrical Energy **698**
21-5 Combinations of Capacitors **702**
Summary **708**
Suggestions for Further Reading, Review, Problems **709**

Chapter 22

Electric Current 716

22-1 Current and Motion of Charges **716**
22-2 Resistance and Ohm's Law **720**
22-3 Energy in Electric Circuits **725**
22-4 Combinations of Resistors **731**
22-5 A Microscopic Picture of Conduction **735**
Summary **738**
Essay: Elizabeth Pflegl Nickles, *Conduction in Nerve Cells* **740**
Suggestions for Further Reading, Review, Problems **744**

Chapter 23

Direct-Current Circuits 749

23-1 Kirchhoff's Rules **750**
23-2 *RC* Circuits **760**
23-3 Ammeters, Voltmeters, and Ohmmeters **767**
Summary **772**
Suggestions for Further Reading, Review, Problems **773**

Chapter 24

The Magnetic Field 781

24-1 The Force Exerted by a Magnetic Field **782**
24-2 Motion of a Point Charge in a Magnetic Field **787**
24-3 Torques on Current Loops and Magnets **798**
24-4 The Hall Effect **801**
Summary **804**
Suggestions for Further Reading, Review, Problems **806**

Chapter 25

Sources of the Magnetic Field 811

25-1 The Magnetic Field of Moving Point Charges **812**
25-2 The Magnetic Field of Currents: The Biot–Savart Law **815**
25-3 Definition of the Ampere **826**
25-4 Ampère's Law **828**
Summary **832**
Suggestions for Further Reading, Review, Problems **833**

Chapter 26

Magnetic Induction 840

26-1 Magnetic Flux **841**
26-2 Induced Emf and Faraday's Law **843**
26-3 Lenz's Law **846**
26-4 Motional Emf **848**
26-5 Eddy Currents **852**
26-6 Generators and Motors **854**
26-7 Inductance **857**
26-8 *LR* Circuits **859**
26-9 Magnetic Energy **862**
Summary **864**
Essay: Syun-Ichi Akasofu, *The Aurora* **866**
Suggestions for Further Reading, Review, Problems **871**

Chapter 27

Magnetism in Matter 878

27-1 Magnetization and Magnetic Susceptibility **879**
27-2 Atomic Magnetic Moments **882**
27-3 Paramagnetism **884**
27-4 Ferromagnetism **886**
27-5 Diamagnetism **890**
Summary **892**
Suggestions for Further Reading, Review, Problems **894**

Chapter 28

Alternating-Current Circuits 898

28-1 Alternating Current in a Resistor **899**

28-2 Alternating Current in Inductors and Capacitors **902**

28-3 Phasors **907**

28-4 *LC* and *LCR* Circuits without a Generator **908**

28-5 *LCR* Circuits with a Generator **912**

28-6 The Transformer **920**

28-7 Rectification and Amplification **923**

Summary **927**

Essay: John Dentler, *Electric Motors* **929**

Suggestions for Further Reading, Review, Problems **937**

Chapter 29

Maxwell's Equations and Electromagnetic Waves 943

29-1 Maxwell's Displacement Current **944**

29-2 Maxwell's Equations **947**

29-3 The Wave Equation for Electromagnetic Waves (Optional) **947**

29-4 Energy and Momentum in an Electromagnetic Wave **953**

29-5 The Electromagnetic Spectrum **957**

Summary **962**

Essay: C.W.F. Everitt, *James Clerk Maxwell (1831–1879)* **964**

Suggestions for Further Reading, Review, Problems **967**

Part 5 Optics 973

Chapter 30

Light 974

30-1 The Speed of Light **976**

30-2 The Propagation of Light: Huygens' Principle **981**

30-3 Reflection **982**

30-4 Refraction **985**

30-5 Fermat's Principle **995**

30-6 Polarization **997**

Summary **1003**

Essay: Robert Greenler, *Beyond the (Visible) Rainbow* **1004**

Suggestions for Further Reading, Review, Problems **1007**

Chapter 31

Geometrical Optics 1012

31-1 Plane Mirrors **1012**

31-2 Spherical Mirrors **1015**

31-3 Images Formed by Refraction **1022**

31-4 Thin Lenses **1025**
31-5 Aberrations **1034**
Summary **1035**
Suggestions for Further Reading, Review, Problems **1036**

Chapter 32

Optical Instruments 1041

32-1 The Eye **1041**
32-2 The Simple Magnifier **1046**
32-3 The Camera **1047**
32-4 The Compound Microscope **1050**
32-5 The Telescope **1052**
Summary **1056**
Suggestions for Further Reading, Review, Problems **1058**

Chapter 33

Interference and Diffraction 1061

33-1 Phase Difference and Coherence **1061**
33-2 Interference in Thin Films **1063**
33-3 The Michelson Interferometer **1066**
33-4 The Two-Slit Interference Pattern **1068**
33-5 The Addition of Harmonic Waves Using Phasors **1071**
33-6 Interference Pattern of Three or More Equally Spaced Sources **1073**
33-7 Diffraction Pattern of a Single Slit **1076**
33-8 Interference–Diffraction Pattern of Two Slits **1080**
33-9 Fraunhofer and Fresnel Diffraction **1082**
33-10 Diffraction and Resolution **1083**
33-11 Diffraction Gratings **1085**
Summary **1090**
Suggestions for Further Reading, Review, Problems **1091**

Part **6** Modern Physics **1099**

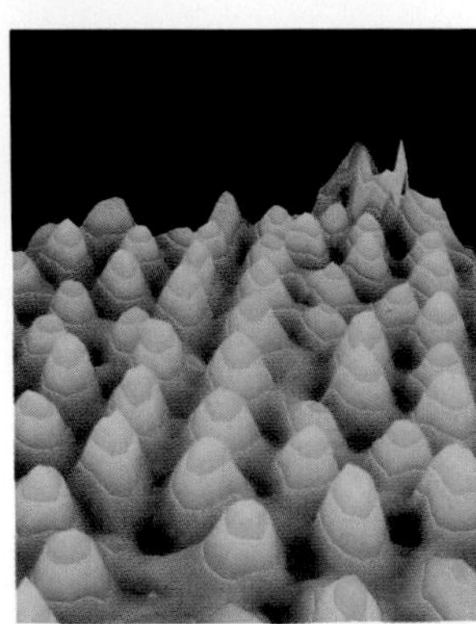

Chapter 34

Relativity 1100

34-1 Newtonian Relativity **1101**
34-2 The Michelson–Morley Experiment **1103**
34-3 Einstein's Postulates **1106**
34-4 The Lorentz Transformation **1107**
34-5 Clock Synchronization and Simultaneity **1113**

34-6 The Doppler Effect **1118**
34-7 The Twin Paradox **1119**
34-8 The Velocity Transformation **1122**
34-9 Relativistic Momentum **1124**
34-10 Relativistic Energy **1126**
34-11 General Relativity **1132**
Summary **1135**
Suggestions for Further Reading, Review, Problems **1136**

Chapter 35

The Origins of Quantum Theory 1145

35-1 The Origin of the Quantum Constant: Blackbody Radiation **1147**
35-2 The Photoelectric Effect **1149**
35-3 X Rays **1152**
35-4 Compton Scattering **1154**
35-5 Quantization of Atomic Energies: The Bohr Model **1156**
35-6 Electron Waves and Quantum Theory **1161**
Summary **1166**
Suggestions for Further Reading, Review, Problems **1167**

Appendix

A Review of Mathematics **AP-1**
B SI Units **AP-21**
C Numerical Data **AP-22**
D Conversion Factors **AP-25**
E Periodic Table of the Elements **AP-27**

Illustration Credits IC-1

Answers to True–False and Odd-Numbered Problems A-1

Index I-1

Physics For Scientists and Engineers

Chapter 1

Systems of Measurement

Tycho Brahe (1546–1601) with his large brass quadrant for measuring the position of planets and stars.

We have always been curious about the world around us. Since the beginning of recorded thought, we have sought ways to impose order on the bewildering diversity of the events we observe. This search for order has taken a variety of forms: one is religion, another is art, and a third is science. Although the term "science" has its origins in the Latin verb meaning "to know," science has come to mean not merely knowledge but specifically knowledge of the natural world. Most important, this knowledge is organized in a systematic and rational way.

We generally think of science as being divided into several separate but interrelated fields. Biology, for example, is the study of living organisms. Chemistry deals with the interaction of elements and compounds. Geology is the study of the earth. Astronomy is the study of the solar system, the stars, and galaxies, and the universe as a whole. Physics deals with matter and energy, with the principles that govern the motion of particles and waves, with the interactions of particles, and with the properties of molecules, atoms, and atomic nuclei, and of larger-scale systems such as gases, liquids, and solids. Some consider physics to be the most fundamental science because it is the basis of all the other fields of science.

1-1 Units

The laws of physics express relationships between physical quantities, such as length, time, force, energy, and temperature. Thus, the ability to define such quantities precisely and measure them accurately is a requisite of physics. The measurement of any physical quantity involves comparing it with some precisely defined unit value of the quantity. For example, to measure the distance between two points, we compare that distance with a standard unit of distance, such as the meter. The statement that a certain distance is "25 meters" means that it is 25 times the length of the unit meter. That is, a standard meterstick fits into the distance 25 times. It is important to include the unit "meter" along with the number "25" in expressing a distance because there are other units of distance, such as feet or miles, in common use. To say that a distance is "25" is meaningless. The magnitude of any physical quantity must include both a number and a unit.

All physical quantities can be expressed in terms of a small number of fundamental units. For example, speed is expressed in terms of a unit of length and a unit of time, such as meters per second or miles per hour. Many of the quantities that we will be studying, such as force, momentum, work, energy, and power, can be expressed in terms of three fundamental quantities—length, time, and mass. The choice of standard units for these fundamental quantities determines a system of units. The system used universally in the scientific community is the **International System (SI).*** In the SI the standard unit of length is the meter, the standard unit of time is the second, and the standard unit of mass is the kilogram.

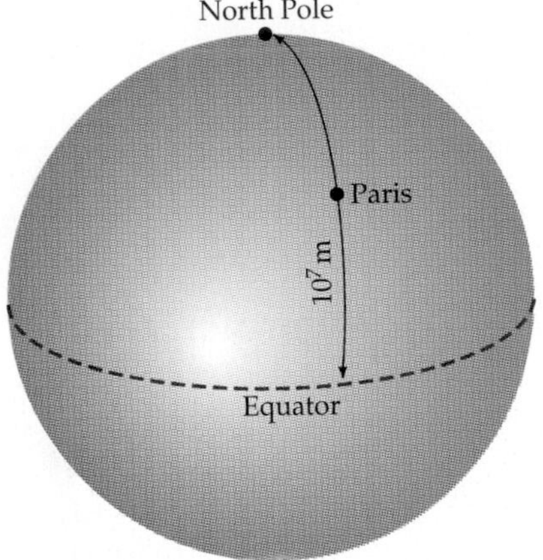

Figure 1-1 The meter was originally chosen so that the distance from the equator to the North Pole along the meridian through Paris would be 10^7 m.

The standard unit of length, the meter (abbreviated m), was originally indicated by two scratches on a bar made of a platinum-iridium alloy kept at the International Bureau of Weights and Measures in Sèvres, France. This length was chosen so that the distance between the equator and the North Pole along the meridian through Paris would be 10 million meters (Figure 1-1). The standard meter is now defined in terms of the speed of light as the distance traveled by light in empty space during a time of 1/299,729,458 second. (This makes the speed of light exactly 299,792,458 m/s.) The standard meter is used to construct secondary standards that are used to calibrate measuring rods throughout the world.

The standard unit of time, the second (s), was originally defined in terms of the rotation of the earth as $\frac{1}{60} \times \frac{1}{60} \times \frac{1}{24}$ of the mean solar day. The second is now defined in terms of light. All atoms, after absorbing energy, emit light with wavelengths and frequencies that are characteristic of the particular element. There is a particular frequency and wavelength associated with each energy transition within the atom. As far as we know, these frequencies remain constant. The second is defined so that the frequency of the light from a certain transition in cesium is 9,192,631,770 cycles per second. With these definitions, the fundamental units of length and time are accessible to laboratories throughout the world.

The standard unit of mass, the kilogram (kg), which equals 1000 grams (g), is defined to be the mass of a standard body, also kept at Sèvres. We discuss the concept of mass in detail in Chapter 4. We will see that the weight of an object at a given point on earth is proportional to its mass. Thus, masses of ordinary size can be compared by weighing them. A duplicate of the standard 1-kg body is kept at the National Bureau of Standards in Gaithersburg, Maryland.

In our study of thermodynamics and electricity, we will need three more fundamental physical units: the unit of temperature, the kelvin (K) (formerly the degree Kelvin); the unit for the amount of a substance, the mole (mol);

*SI stands for Système International. Complete definitions of the SI units are given in Appendix B.

(a)

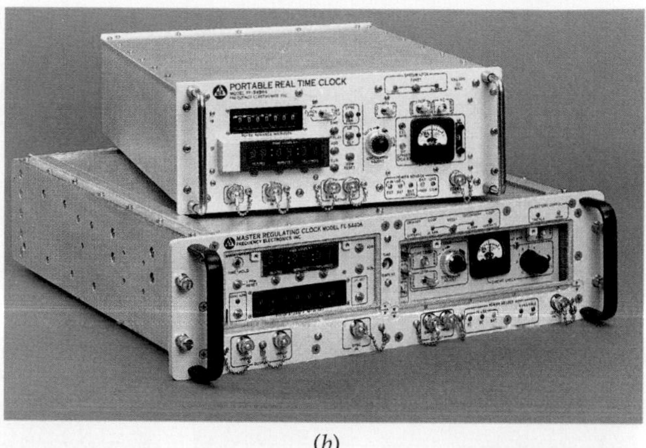

(b)

(*a*) Water clock used to measure time intervals in the 13th century. (*b*) Modern portable cesium clock.

and the unit of current, the ampere (A). There is another fundamental unit, the candela (cd), for luminous intensity, which we will have no occasion to use in this book. These seven fundamental units—the meter (m), the second (s), the kilogram (kg), the kelvin (K), the ampere (A), the mole (mol), and the candela (cd)—constitute the SI.

The units of every physical quantity can be expressed in terms of the fundamental SI units. Some frequently used combinations are given special names. For example, the SI unit of force, kg·m/s^2, is called a newton (N). Similarly, the SI unit of power, kg·m^2/s^3 = N·m/s, is called a watt (W).

Prefixes for common multiples and submultiples of SI units are listed in Table 1-1. These multiples are all powers of 10, and so such a system is called a **decimal system.** The decimal system based on the meter is called the **metric system.** The prefixes can be applied to any SI unit; for example, 0.001 second is 1 millisecond (ms); 1,000,000 watts is 1 megawatt (MW).

Another decimal system still in use but gradually being replaced by the SI is the cgs system, based on the centimeter, gram, and second. The centimeter is defined as 0.01 m. The gram is now defined as 0.001 kg. Originally, the gram was defined as the mass of one cubic centimeter of water. (The kilogram was then the mass of 1000 cubic centimeters or one liter of water.)

In another system of units used in the United States, the **U.S. customary system,** a unit of force, the pound, is chosen to be a fundamental unit. The pound is defined in terms of the gravitational attraction of the earth at a particular place for a standard body. The unit of mass is then defined in terms of the pound. (We will see in Chapter 4 that mass is a better choice for a fundamental unit than force because mass is an intrinsic property of an object that is independent of its location relative to the earth.) The fundamental unit of length in this system is the foot (ft). The foot is defined as exactly one-third of a yard (yd), which is now defined in terms of the meter:

$$1 \text{ yd} = 0.9144 \text{ m} \tag{1-1}$$

$$1 \text{ ft} = \tfrac{1}{3} \text{ yd} = 0.3048 \text{ m} \tag{1-2}$$

This makes the inch (in) exactly 2.54 cm. The fundamental unit of time is the second, which is defined as it is in the SI. This system is not a decimal system. It is less convenient than the SI or other decimal systems because common multiples of its units are not powers of 10. For example, 1 yd = 3 ft and 1 ft = 12 in. Relationships between the units in the U.S. customary system and those in the SI are given in Appendix D.

Table 1-1 **Prefixes for Powers of 10**

Multiple	Prefix	Abbreviation
10^{18}	exa	E
10^{15}	peta	P
10^{12}	tera	T
10^{9}	giga	G
10^{6}	mega	M
10^{3}	kilo	k
10^{2}	hecto†	h
10^{1}	deka†	da
10^{-1}	deci†	d
10^{-2}	centi†	c
10^{-3}	milli	m
10^{-6}	micro	μ
10^{-9}	nano	n
10^{-12}	pico	p
10^{-15}	femto	f
10^{-18}	atto	a

†The prefixes hecto (h), deka (da), and deci (d) are not powers of 10^3 or 10^{-3} and are rarely used. The other prefix which is not a power of 10^3 or 10^{-3} is centi (c), now used only with the meter, as in 1 cm = 10^{-2} m.

(a)

(b)

(*a*) Laser beams from the Macdonald Observatory used to measure the distance to the moon. The distance can be measured to within a few centimeters by measuring the time for the beam to go to the moon and return after reflecting off a corner mirror (*b*) placed on the moon by the Apollo 14 astronauts.

Questions

1. What are the advantages and disadvantages of using the length of your arm for a standard length?
2. A certain clock is known to be consistently 10 percent fast compared with the standard cesium clock. A second clock varies in a random way by 1 percent. Which clock would make a more useful secondary standard for a laboratory? Why?

1-2 Conversion of Units

We have said that the magnitude of a physical quantity must include both a number and a unit. When such quantities are added, subtracted, multiplied, or divided in an algebraic equation, the unit can be treated like any other algebraic quantity. For example, suppose you wish to find the distance traveled in 3 hours (h) by a car moving at a constant rate of 80 kilometers per hour (km/h). The distance is the product of the speed v and the time t:

$$x = vt = \frac{80\text{ km}}{\cancel{\text{h}}} \times 3\cancel{\text{h}} = 240\text{ km}$$

We cancel the unit of time, the hours, just as we would any algebraic quantity to obtain the distance in the proper unit of length, the kilometer. This method of treating units makes it easy to convert from one unit to another. Suppose we want to convert our answer of 240 km to miles (mi). We use the fact that

$$1\text{ mi} = 1.61\text{ km}$$

If we divide each side of this equation by 1.61 km, we obtain

$$\frac{1\text{ mi}}{1.61\text{ km}} = 1$$

Since any quantity can be multiplied by 1 without changing its value, we can now change 240 km to miles by multiplying by the factor (1 mi)/(1.61 km):

$$240\ \text{km} = 240\ \cancel{\text{km}} \times \frac{1\ \text{mi}}{1.61\ \cancel{\text{km}}} = 149\ \text{mi}$$

The factor (1 mi)/(1.61 km) is called a **conversion factor.** All conversion factors have a value of 1 and are used to convert a quantity expressed in one unit of measure into its equivalent in another unit of measure. By writing out the units explicitly and canceling them, we need not think about whether we multiply by 1.61 or divide by 1.61 to change kilometers to miles because the units that remain tell us whether we have chosen the correct or incorrect factor.

Example 1-1

What is the equivalent of 90 km/h in meters per second and in miles per hour?

We use the facts that 1000 m = 1 km, 60 s = 1 min, and 60 min = 1 h to convert kilometers per hour to meters per second. We multiply the quantity 90 km/h by a set of conversion factors each having the value 1 so that the value of the speed is not changed:

$$\frac{90\ \cancel{\text{km}}}{\cancel{\text{h}}} \times \frac{1000\ \text{m}}{1\ \cancel{\text{km}}} \times \frac{1\ \cancel{\text{h}}}{60\ \cancel{\text{min}}} \times \frac{1\ \cancel{\text{min}}}{60\ \text{s}} = 25\ \text{m/s}$$

To convert this speed into miles per hour, we use the conversion factor (1 mi)/(1.61 km) = 1.

$$\frac{90\ \cancel{\text{km}}}{\text{h}} \times \frac{1\ \text{mi}}{1.61\ \cancel{\text{km}}} = 55.9\ \text{mi/h}$$

Exercise

What is the equivalent of 65 mi/h in meters per second?
(Answer: 29.1 m/s)

1-3 Dimensions of Physical Quantities

The area of a surface is found by multiplying one length by another. For example, the area of a rectangle of sides 2 m and 3 m is $A = (2\ \text{m})(3\ \text{m}) = 6\ \text{m}^2$. The unit of area is square meters. Because area is the product of two lengths, it is said to have the dimensions of length times length or length squared, often written L^2. The idea of dimensions is easily extended to nongeometric quantities. For example, speed is said to have the dimensions of length divided by time, or L/T. The dimensions of other quantities, such as force or energy, are written in terms of the fundamental quantities of length, time, and mass.

Adding two physical quantities makes sense only if the quantities have the same dimensions. For example, we cannot add an area to a speed and obtain a meaningful sum. If we have an equation like

$$A = B + C$$

the quantities A, B, and C must all have the same dimensions. The addition of B and C also requires that the dimensions of these quantities be expressed in the same units. For example, if B is an area of 500 in^2 and C is an area of 4 ft^2, we must either convert B into square feet or C into square inches in order to find the sum of the two areas.

We can often find mistakes in a calculation by checking the dimensions or units of the quantities in the calculation. Suppose, for example, we mistakenly use the formula $A = 2\pi r$ for the area of a circle. We can see immediately that this cannot be correct because the right side of the equation, $2\pi r$, has dimensions of length whereas area should have dimensions of length squared. For another example, consider this formula for distance x:

$$x = vt + \tfrac{1}{2}at$$

where t is the time, v is the speed, and a is the acceleration, which (as we will see) has dimensions of L/T^2. We can see by looking at the dimensions of each quantity that this formula cannot be correct. Since x has dimensions of length, each term on the right side of the equation must also have dimensions of length. The term vt has dimensions of length, but the dimensions of $\tfrac{1}{2}at$ are $(L/T^2)T = L/T$. Since this term does not have the correct dimensions, the formula cannot be correct. Dimensional consistency is a necessary but not a sufficient condition for an equation to be correct. An equation can have the correct dimensions in each term and still not describe any physical situation.

1-4 Scientific Notation

Handling very large or very small numbers is simplified by using scientific notation. In this notation, a number is written as a product of a number between 1 and 10 and a power of 10, such as 10^2 (= 100) or 10^3 (= 1000). For example, the number 12,000,000 is written 1.2×10^7; the distance from the earth to the sun, about 150,000,000,000 m, is written 1.5×10^{11} m. The number 11 in 10^{11} is called the **exponent.** For numbers smaller than 1, the exponent is negative. For example, $0.1 = 10^{-1}$ and $0.0001 = 10^{-4}$. The diameter of a virus, which is about 0.00000001 m, is written 1×10^{-8} m.

In multiplication, exponents are added; in division, they are subtracted. Some simple examples follow that show the use of these rules.

$$10^2 \times 10^3 = 100 \times 1000 = 100{,}000 = 10^{2+3} = 10^5$$

Similarly,

$$\frac{10^2}{10^3} = \frac{100}{1000} = \frac{1}{10} = 10^{2-3} = 10^{-1}$$

In scientific notation, 10^0 is defined to be 1. We can see this as follows. Suppose we divide 1000 by itself. We have

$$\frac{1000}{1000} = \frac{10^3}{10^3} = 10^{3-3} = 10^0 = 1$$

Example 1-2

Using scientific notation, compute (*a*) 120 × 6000 and (*b*) 3,000,000/0.00015.

(*a*) $(1.20 \times 10^2)(6.00 \times 10^3) = (1.20)(6.00) \times 10^{2+3} = 7.20 \times 10^5$

(*b*) $\dfrac{3.00 \times 10^6}{1.50 \times 10^{-4}} = \dfrac{3.00}{1.50} \times 10^{6-(-4)} = 2.00 \times 10^{10}$

The two minus signs in (*b*) arise because −4 is subtracted from 6.

Exercise

Using scientific notation, calculate $(2.50 \times 10^7)(1.90 \times 10^{-3})$. (Answer: 4.75×10^4)

In raising a power to another power, the exponents are multiplied. For example,

$$(10^2)^4 = 10^2 \times 10^2 \times 10^2 \times 10^2 = 10^8$$

Example 1-3

A liter is the volume of a cube that is 10 cm by 10 cm by 10 cm. Find the volume of a liter in cubic centimeters and in cubic meters.

The volume V of a cube of side L is L^3.

$$V = L^3 = (10\text{ cm})(10\text{ cm})(10\text{ cm}) = 10^3\text{ cm}^3$$

To convert to cubic meters (m^3), we use $1\text{ cm} = 10^{-2}\text{ m}$,

$$10^3\text{ cm}^3 = 10^3\text{ cm}^3 \times \left(\frac{10^{-2}\text{ m}}{1\text{ cm}}\right)^3 = 10^3\text{ cm}^3 \times \frac{10^{-6}\text{ m}^3}{1\text{ cm}^3} = 10^{-3}\text{ m}^3$$

Note that raising the conversion factor (which equals 1) to the third power does not change its value and enables us to cancel the units.

Care must be taken when adding or subtracting numbers that are written in scientific notation when the exponents don't match. Consider, for example,

$$(1.200 \times 10^2) + (8 \times 10^{-1}) = 120.0 + 0.8 = 120.8$$

To find the sum without first converting both numbers into ordinary decimal form, it is sufficient to rewrite either of the numbers so that its power of 10 is the same as that of the other. For example, we can find the sum by writing $1.200 \times 10^2 = 1200 \times 10^{-1}$ and then adding:

$$(1200 \times 10^{-1}) + (8 \times 10^{-1}) = 1208 \times 10^{-1} = 120.8$$

If the exponents are very different, then one of the numbers is much smaller than the other and can often be neglected in addition or subtraction. For example,

$$(2 \times 10^6) + (9 \times 10^{-3}) = 2{,}000{,}000 + 0.009$$

$$= 2{,}000{,}000.009 \approx 2 \times 10^6$$

where the symbol $\approx$ means "is approximately equal to."

1-5 Significant Figures and Orders of Magnitude

Many of the numbers in science are the result of measurement and are therefore known only within the limits of some experimental uncertainty. The magnitude of the uncertainty depends on the skill of the experimenter and on the apparatus used and often can only be estimated. A rough indication of the uncertainty in a measurement is implied by the number of digits used. For example, if we say that a table is 2.50 m long, we are implying that its length is probably between 2.495 m and 2.505 m. That is, we know the length to within about $\pm 0.005\text{ m} = \pm 0.5\text{ cm}$ of the stated length. If we used a meterstick with millimeter markings and measured the table length carefully, we might estimate that we could measure the length to ± 0.5 mm rather than ± 0.5 cm. We would indicate this precision by using four digits, such as 2.503 m, to give the length. A reliably known digit (other than a zero used to locate the decimal point) is called a **significant figure.** The number 2.50 has three significant figures; 2.503 has four. The number 0.00103 has three significant figures. (The first three zeros are not significant figures but merely locate the decimal point.) In scientific notation, this number is expressed as 1.03×10^{-3}. A common student error, particularly since the advent of hand calculators, is to carry many more digits in an answer than are warranted. Suppose, for example, that you measure the area of a circular playing field by pacing off the radius and using the formula $A = \pi r^2$. If you estimate the radius to be 8 m by pacing and use a 10-digit calculator to compute the area, you obtain $\pi(8\text{ m})^2 = 201.0619298\text{ m}^2$. The digits after the decimal point are misleading about the accuracy with which you know the area. Since you found the radius by pacing, you might expect that your measurement is accurate to only about 0.5 m. That is, the radius could be as great as 8.5 m or as small as 7.5 m. If the radius is 8.5 m, the area is $\pi(8.5\text{ m})^2 = 226.9800692\text{ m}^2$, whereas if the radius is 7.5 m, the area is $\pi(7.5\text{ m})^2 = 176.714587\text{ m}^2$. A general rule to follow when multiplying or dividing several numbers is

> The number of significant figures in the result of multiplication or division is no greater than the least number of significant figures in any of the factors.

In the previous example, the radius of the playing field is known to only one significant figure, so the area is also known only to one significant figure. It should be written as $2 \times 10^2\text{ m}^2$, which implies that the area is somewhere between 150 m^2 and 250 m^2.

The precision of the sum or difference of two measurements is only as good as the precision of the least precise of the two measurements. A general rule to follow is

> The result of the addition or subtraction of two numbers has no significant figures beyond the last decimal place where both of the original numbers had significant figures.

Example 1-4

Find the sum of 1.040 and 0.2134.

The first number, 1.040, has only three significant figures beyond the decimal point, whereas the second, 0.2134, has four. According to our rule, the sum can have only three significant figures beyond the decimal point. We thus obtain

$$1.040 + 0.2134 = 1.253$$

Exercise

Apply the appropriate rule for determining the number of significant figures to calculate the following: (*a*) 1.58×0.03, (*b*) $1.4 + 2.53$, and (*c*) $(2.34 \times 10^2) + 4.93$. [Answers: (*a*) 0.05, (*b*) 3.9, (*c*) 2.39×10^2]

In a textbook, it is cumbersome to write every number with the proper number of significant figures. Most examples and exercises will be done with data to three (or sometimes four) significant figures, but occasionally we will say, for example, that a table top is 3 ft by 8 ft rather than taking the time and space to say it is 3.00 ft by 8.00 ft. Any data you see in an example or exercise can be assumed to be known to three significant figures unless otherwise indicated.

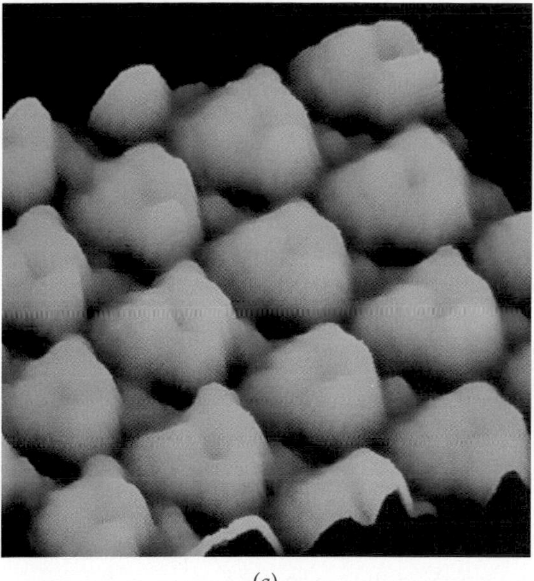
(a)

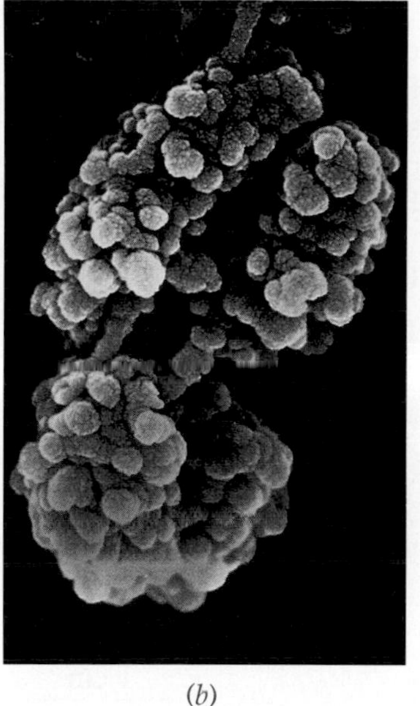
(b)

(c)

(d)

(e)

(a) Benzene molecules of the order of 10^{-10} m diameter, as seen in a scanning electron microscope. (b) Chromosomes measuring of the order of 10^{-6} m across, as seen in a scanning electron microscope. (c) Distances familiar in our everyday macroscopic world. The height of the girl is of the order of 10^{0} m and that of the mountain is of the order of 10^{4} m. (d) Earth with a diameter of the order of 10^{7} m, as seen from space in the Apollo Mission. (e) Andromeda galaxy with diameter of the order of 10^{21} m.

In doing rough calculations or comparisons, we sometimes will round off a number to the nearest power of 10. Such a number is called an **order of magnitude.** For example, the height of a small insect, say an ant, might be 8×10^{-4} m $\approx 10^{-3}$ m. We would say that the order of magnitude of the height of an ant is 10^{-3} m. Similarly, though the height of most people is about 2 m, we might round that off and say that the order of magnitude of the height of a person is 10^{0} m. By this we do not mean to imply that a typical height is really 1 m but that it is closer to 1 m than to 10 m or to $10^{-1} = 0.1$ m. We might say that a typical human being is 3 orders of magnitude taller than a typical ant, meaning that the ratio of their heights is about 1000 to 1. An order of magnitude does not provide any digits that are reliably known. It may be thought of as having no significant figures.

Tables 1-2 to 1-4 give some typical order-of-magnitude values for some lengths, masses, and time intervals encountered in physics.

Table 1-2 **Order of Magnitude of Some Lengths**

Length	m
Radius of proton	10^{-15}
Radius of atom	10^{-10}
Radius of virus	10^{-7}
Radius of giant amoeba	10^{-4}
Radius of walnut	10^{-2}
Height of human being	10^{0}
Height of highest mountains	10^{4}
Radius of earth	10^{7}
Radius of sun	10^{9}
Earth-sun distance	10^{11}
Radius of solar system	10^{13}
Distance to nearest star	10^{16}
Radius of Milky Way galaxy	10^{21}
Radius of visible universe	10^{26}

Table 1-3 **Order of Magnitude of Some Masses**

Mass	kg
Electron	10^{-30}
Proton	10^{-27}
Amino acid	10^{-25}
Hemoglobin	10^{-22}
Flu virus	10^{-19}
Giant amoeba	10^{-8}
Raindrop	10^{-6}
Ant	10^{-2}
Human being	10^{2}
Saturn 5 rocket	10^{6}
Pyramid	10^{10}
Earth	10^{24}
Sun	10^{30}
Milky Way galaxy	10^{41}
Universe	10^{52}

Table 1-4 Order of Magnitude of Some Time Intervals

Interval	s
Time for light to cross nucleus	10^{-23}
Period of visible-light radiation	10^{-15}
Period of microwaves	10^{-10}
Half-life of muon	10^{-6}
Period of highest audible sound	10^{-4}
Period of human heartbeat	10^{0}
Half-life of free neutron	10^{3}
Period of earth's rotation (day)	10^{5}
Period of revolution of earth (year)	10^{7}
Lifetime of a human	10^{9}
Half-life of plutonium 239	10^{12}
Lifetime of a mountain range	10^{15}
Age of earth	10^{17}
Age of universe	10^{18}

In many cases, the order of magnitude of a quantity can be estimated using reasonable assumptions and simple calculations. The physicist Enrico Fermi was a master at computing approximate answers to questions that seemed impossible to answer because of limited available information. Such questions are often called **Fermi questions.** The following is an example of a Fermi question.

Example 1-5

What thickness of rubber tread is worn off an automobile tire as it travels 1 km (0.6 mi)?

We assume the tread thickness of a new tire is 1 cm. This may be off by a factor of 2 or so, but 1 mm is certainly too small and 10 cm is too large. Since tires have to be replaced after about 60,000 km (about 37,000 mi), we will assume that the tread is completely worn off after 60,000 km. In other words, the rate of wear is 1 cm of tire per 60,000 km of travel. The thickness that is worn per kilometer is therefore

$$\frac{1\text{ cm}}{60{,}000\text{ km}} = 1.7 \times 10^{-5}\text{ cm/km} \approx 0.2\ \mu\text{m/km}$$

Summary

1. The magnitudes of physical quantities (for example, length, time, force, and energy) are expressed as a number times a unit.
2. The fundamental units in the International System (SI) are the meter (m), the second (s), the kilogram (kg), the kelvin (K), the ampere (A), the mole (mol), and the candela (cd). Every physical quantity can be expressed in terms of these fundamental units.
3. Units in equations are treated just like any other algebraic quantity.
4. The two sides of an equation must have the same dimensions.
5. Conversion factors, which are always equal to 1, provide a convenient method for converting from one unit to another.
6. Very small and very large numbers are most easily handled when written as a number between 1 and 10 times a power of 10. This is called scientific notation. When multiplying two numbers, the exponents are added; when dividing, the exponents are subtracted. When a number containing an exponent is itself raised to an exponent, the exponents are multiplied.
7. The number of significant figures in the result of multiplication or division is no greater than the least number of significant figures in any of the factors. The result of the addition or subtraction of two numbers has no significant figures beyond the last decimal place where both of the original numbers had significant figures.
8. A number rounded to the nearest power of 10 is called an order of magnitude. The order of magnitude of a quantity can often be estimated using reasonable assumptions and simple calculations.

The Fermi Solution

Hans Christian von Baeyer
The College of William and Mary

At twenty-nine minutes past five, on a Monday morning in July 1945, the world's first atom bomb exploded in the desert sixty miles northwest of Alamogordo, New Mexico. Forty seconds later, the blast's shock wave reached the base camp, where scientists stood in stunned contemplation of the historic spectacle. The first person to stir was the Italian-American physicist Enrico Fermi, who was on hand to witness the culmination of a project he had helped begin.

Before the bomb detonated, Fermi tore a sheet of notebook paper into small bits. Then, as he felt the first quiver of the shock wave spreading outward through the still air, he released the shreds above his head. They fluttered down and away from the mushroom cloud growing on the horizon, landing about two and half yards behind him. After a brief mental calculation, Fermi announced that the bomb's energy had been equivalent to that produced by ten thousand tons of TNT. Sophisticated instruments also were at the site. Analyses of their readings of the shock wave's velocity and pressure, which took several weeks to complete, confirmed Fermi's instant estimate. (It is not known for sure how Fermi did it, but what he did was probably this: by measuring the speed with which air was pushed out by the explosion, he estimated the total kinetic energy dissipated by the atmosphere, and then divided that by the amount of energy liberated by a ton of TNT.)

Hans Christian von Baeyer is a professor of physics at the College of William and Mary in Williamsburg, Virginia. After attending high school in Canada, he graduated from Columbia University and earned his Ph.D. in physics at Vanderbilt University. His research specialty is elementary particle theory, with particular emphasis on classical relativistic mechanics—a problem that should have been solved 70 years ago but wasn't. He is a fellow of the American Physical Society, and a recipient of the American Institute of Physics Science Writing Award. This essay originally appeared in his bimonthly column entitled "PHYSIKA" that appears in the journal *The Sciences*. Professor von Baeyer likes to run in 10-km races and finishes in the top half of the field.

The bomb-test team was impressed, but not surprised, by this brilliant bit of scientific improvisation. Enrico Fermi's genius was known throughout the world of physics. In 1938, he won a Nobel Prize for his work in elementary particle physics, and, four years later, in Chicago, produced the first sustained nuclear chain reaction, thereby ushering in the age of atomic weapons and commercial nuclear power. No other physicist of his generation, and no one since, had been at once a masterful experimentalist and a leading theoretician. In miniature, the bits of paper and the analysis of their motion exemplified this unique combination of gifts.

Like all virtuosos, Fermi had a distinctive style. His approach to physics brooked no opposition; it simply never occurred to him that he might fail to find the solution to a problem. His scientific papers and books reveal a disdain for embellishment—a preference for the most direct, rather than the most intellectually elegant, route to an answer. When he reached the limits of his cleverness, Fermi completed a task by brute force.

To illustrate this approach, imagine that a physicist must determine the volume of an irregular object—say, the earth, which is slightly pear-shaped. Without some kind of formula, one might feel stymied; there are several ways of getting one. Consulting a mathematician is one way, but finding one with enough knowledge and interest to be of help is usually difficult. Searching through the mathematical literature is another, though time-consuming and probably fruitless, because the ideal shapes that interest mathematicians often do not match those of the irregular objects found in nature. Finally the physicist could consider setting aside all other research in order to derive the formula from basic mathematical principles, but, of course, anyone who wanted to devote vast amounts of time to theoretical geometry wouldn't have become a physicist.

Alternatively, the physicist could do what Fermi would have done—compute the volume *numerically*. Instead of relying on a formula, this could per-

Continued

Enrico Fermi (1901–1954).

haps involve mentally dividing the planet into a number of small cubes, each with a volume easily determined by multiplying the length times the width times the height, and then adding together the answers to these more tractable problems (increasing the number of these cubes will increase the accuracy of the calculation). This method yields only an approximate solution, but because it does not depend on any unknown formulas, it is sure to produce the desired result; this is what mattered to Fermi. With the introduction, after the Second World War, of computers and, later, pocket calculators, numerical computation has become standard procedure in physics.

The technique of dividing difficult problems into small, manageable ones applies to many problems besides those amenable to numerical computation. Fermi excelled at this rough-and-ready method. To pass it on to his students, he developed a type of question that has become associated with his name. A Fermi problem has a characteristic profile. Upon first hearing it, one doesn't have even the remotest notion what the answer might be. And one feels certain that too little information exists to find a solution. Yet, when the problem is broken down into subproblems, each one answerable without the help of experts or reference books, an estimate can be made, either mentally or on the back of an envelope, that comes remarkably close to the exact solution.

Suppose, for example, that one wants to determine the earth's circumference without looking it up. If one knows that New York and Los Angeles are separated by about three thousand miles and that the time difference between the two coasts is three hours. Three hours corresponds to one-eighth of a day, and a day is the time it takes the planet to complete one revolution. Therefore, the circumference of the earth can be estimated to be eight times three thousand, or twenty-four thousand miles. At the equator, the earth's circumference is in fact 24,902 miles. In John Milton's words:

> so easy it seemed
> Once found, which yet unfound most
> would have thought
> Impossible.

Fermi problems might seem to resemble the brainteasers that appear among the back pages of airline magazines and other popular publications (given three containers that hold eight, five, and three liters, respectively, how do you measure out a single liter?), but the two genres differ significantly. The answer to a Fermi problem, in contrast to that of a brainteaser, cannot be verified by logical deduction alone and is always approximate. (To determine precisely the circumference of the earth, it is necessary that the planet actually be measured.) Then, too, solving a Fermi problem requires a knowledge of facts not mentioned in the statement of the problem. (In contrast, the decanting puzzle above contains all the information necessary for its solution.)

These differences mean that Fermi problems are more closely tied to the physical world than are mathematical puzzles, which rarely have anything practical to offer physicists. By the same token, Fermi problems are reminiscent of the ordinary dilemmas that nonphysicists encounter every day of their lives. Indeed, Fermi problems, and the way they are solved, are not only essential to the practice of physics but also valuable for the lesson they give in the art of living.

How many piano tuners are there in Chicago? The whimsical nature of this question, the improbability that anyone knows the answer, and the fact that Fermi posed it to his classes at the University of Chicago have elevated it to the status of legend. There is no standard solution (that's exactly the point), but anyone can make assumptions that quickly lead to an approximate answer. Here is one way: If the population of metropolitan Chicago is three million, an average family consists of four people, and one-third of all families own pianos, there are 250,000 pianos in the city. If each piano is tuned every ten years, there are 25,000 tunings a year. If each tuner can service 4 pianos a day, 250 days a year, for a total of 1000 tunings a year, there must be about 25 piano tuners in the city. The answer is not exact; it could be as low as 10 or as high as 50. But, as the yellow pages of the telephone directory attest, it is definitely in the ballpark.

Fermi's intent was to show that although at the outset even the answer's order of magnitude is unknown, one can proceed on the basis of different assumptions and still arrive at estimates that fall within the range of the answer. The reason is that, in any string of calculations, errors tend to cancel out one another. Every sixth, rather than third, family might be assumed to own a piano, but then pianos might just as well be tuned every five, rather than ten, years. It is just as improbable that all of one's errors will be underestimates (or overestimates) as it is that all the throws in a series of coin tosses will be heads (or tails). The law of probabilities dictates that deviations from the correct assumptions will tend to compensate for one another, so the final results will converge toward the right number. An essential caveat is that one must be cautious that no possible source of bias favors deviations in one direction over another.

Of course, the Fermi problems that physicists face deal with atoms and molecules more often than pianos. To answer them, one needs to commit to memory a few basic magnitudes, such as the approximate diameter of a typical atom or the number of molecules in a thimbleful of water. Equipped with such facts, one can estimate, in a fashion similar to that given in Example 1-5, the amount by which a car tire is worn down in one revolution during normal driving conditions. Assume that the tread is about 1 cm thick and that it wears off in 60,000 km of driving. If 1 cm is divided by the number of revolutions made by a typical tire, with its typical circumference, in 60,000 km, the answer is roughly one molecular diameter.

Another Fermi problem illustrates the immensity of the number of atoms and molecules that surround us. The task is to prove a theorem called "Caesar's Last Breath," which states that with every breath you take, you inhale a single molecule of the air that Julius Caesar exhaled as he died. Behind the statement lurks a qualification and several assumptions. The qualification is that, since the theorem is based on averages, it can only be approximately true. If you take three or four breaths without encountering one of Caesar's air molecules, don't be disappointed; later you may inhale several at once. One of the assumptions is that over the last two millennia no air molecules have been added to the atmosphere or been taken out of circulation by becoming attached to other molecules, which isn't true. Also taken for granted is that Caesar's last breath has had a chance to mingle evenly throughout the entire atmosphere, which isn't quite realistic either. But if both assumptions *were* tenable, the theorem would be true and could also be applied to the dying gasp of Attila, Socrates, or Jesus.

The point of the parable is that, although the earth's atmosphere is vast, the number of molecules exhaled in a single human breath is also vast. In order to prove the theorem, you estimate the total volume of the atmosphere, and then divide it by the volume of air in your lungs. The answer is the number of lungsful of air in the atmosphere, and this turns out to be approximately equal to the estimated number of molecules in Caesar's last breath. Both are roughly equal to one-tenth of Avogadro's number, 6×10^{23}, a number that physicists carry around in their heads.

More momentous Fermi problems might concern energy policy (the number of solar cells required to produce a certain amount of electricity), environmental quality (the amount of acid rain caused annually by coal consumption in the United States), or the arms race. Prudent physicists—those who want to avoid false leads and dead ends—operate according to a long-standing principle: Never start a lengthy calculation until you know the range of values within which the answer is likely to fall (and, equally important, the range within which the answer is *un*likely to fall). These physicists attack every problem as if it were a Fermi problem, estimating the order of magnitude of the result before engaging in an investigation.

Physicists also use Fermi problems to communicate with one another. Whether gathering in university hallways, convention-center lobbies, or French restaurants, as they describe, say, new experiments

Continued

or discuss unfamiliar subjects, they often first survey the lay of the land, staking out, in a numerical way, the perimeter of the problem at hand. Those accustomed to tackling Fermi problems approach the experiment or subject as if it were their own, demonstrating their understanding of the underlying physical principles by performing rough calculations. If the conversation turns to a new particle accelerator, for example, physicists might estimate the strength of the magnets it requires; if the subject is the structure of a novel crystal, they could calculate the spacing between its atoms. The goal is to arrive at a reasonable answer using the least effort. It is this spirit of independence, which he himself possessed in ample measure, that Fermi sought to instill by posing his unconventional problems.

Questions about atom bombs, piano tuners, automobile tires, particle accelerators, and crystal structure have little in common. But the manner in which they are answered is the same in every case and can be applied to questions outside the realm of physics. Whether the problem concerns cooking, automobile repair, or personal relationships, there are two basic types of responses. The fainthearted turn to authority—reference books, bosses, expert consultants, physicians, or ministers—whereas the independent of mind delve into that private store of common sense and limited factual knowledge that everyone possesses, make reasonable assumptions, and derive their own, admittedly approximate, solutions. To be sure, it would be foolish to practice neurosurgery at home, but mundane challenges, such as chili from scratch, replacing a water pump, or resolving a family quarrel, can often be sorted out with nothing more than logic, common sense, and patience.

Not everyone trusts these informal approaches. For example, some people were probably skeptical of Fermi's analysis of a two-billion-dollar bomb test that he carried out with the aid of a handful of confetti. Such an attitude demonstrates less, perhaps, about their knowledge of the problem than about their attitude toward life.

Enrico Fermi and Niels Bohr on a stroll.

Ultimately, the value of dealing with the problems of science, or everyday life, the way Fermi did lies in the rewards one gains for making independent discoveries and inventions. It doesn't matter whether the discovery is as momentous as the determination of the yield of an atom bomb or as insignificant as an estimate of the number of piano tuners in a midwestern city. Looking up the answer, or letting someone else find it, actually impoverishes one; it robs one of the pleasure and pride that accompany creativity and deprives one of an experience that, more than anything else in life, bolsters self-confidence. Self-confidence, in turn, is the essential prerequisite for solving Fermi problems. Thus, approaching personal dilemmas as Fermi problems can become, by a kind of chain reaction, a habit that enriches life.

Suggestions for Further Reading

McMahon, Thomas A., and John Tyler Bonner: "The Physics of Dimensions," Chapter 3 of *On Size and Life,* Scientific American Books, New York, 1983.

Submarines, airplanes, and stringed instruments are used as examples to illustrate the principles of dimensional analysis. This prepares the reader for later chapters in which the importance of scale to living things is considered.

Root-Bernstein, Robert S.: "Setting the Stage for Discovery: Breakthroughs Depend on More Than Luck," *The Sciences,* May/June 1988, p. 26.

The author's view of the nature of scientific discovery, illustrated with examples from the life sciences and astronomy.

von Baeyer, Hans Christian: "New Measures of Matter," *The Sciences,* May/June 1988, p. 2.

How the atom is replacing man as the measure of all things: a personal view of the significance of the redefinition of the second and meter.

von Baeyer, Hans Christian: "How Fermi Would Have Fixed It," *The Sciences,* September/October 1988, p. 2.

Enrico Fermi's approach to obtaining approximate solutions, illustrated with entertaining examples.

von Baeyer, Hans Christian: "The Inverse Problem," *The Sciences,* November/December 1985, p. 13.

Two approaches to deducing general principles from observations in physics: working backwards versus guessing and working forwards.

von Baeyer, Hans Christian: "Rainbows, Whirlpools, and Clouds: A Field Guide to the Physics of a Summer's Day," *The Sciences,* July/August 1984, p. 24.

An introduction to physics as a way of thinking about and investigating the world around you.

Review

A. Objectives: After studying this chapter you should:

1. Be able to define the units of length, time, and mass.

2. Know what is meant by SI units, U.S. customary units, and cgs units.

3. Know what conversion factors are, and be able to use them to convert from one kind of unit to another.

4. Know what is meant by the dimensions of a quantity.

B. Define, explain, or otherwise identify the following:

Units
SI units
Metric system
U.S. customary system
Conversion factor
Dimensions
Significant figure

C. True or false: If the statement is true, explain why it is true. If it is false, give a counterexample, that is, an example that contradicts the statement.

1. Two quantities to be added must have the same dimensions.

2. Two quantities to be multiplied must have the same dimensions.

3. All conversion factors have the value 1.

Problems

Level I

1-1 Units

1. Write the following using the prefixes listed in Table 1-1 and the abbreviations listed on the inside cover. For example, 10,000 meters = 10 km. (*a*) 1,000,000 watts; (*b*) 0.002 gram; (*c*) 3×10^{-6} meter; (*d*) 30,000 seconds.

2. Write the following without using prefixes: (*a*) 40 μW; (*b*) 4 ns; (*c*) 3 MW; (*d*) 25 km.

3. Write out the following (which are not SI units) without using any abbreviations. For example, 10^3 meters = 1 kilometer. (*a*) 10^{-12} boo; (*b*) 10^9 low; (*c*) 10^{-6} scope; (*d*) 10^{-18} boy; (*e*) 10^6 phone; (*f*) 10^{-9} goat; (*g*) 10^{12} bull.

4. In the following equations, the distance x is in meters, the time t is in seconds, and the velocity v is in meters per second. What are the SI units of the constants C_1 and C_2? (*a*) $x = C_1 + C_2t$; (*b*) $x = \frac{1}{2}C_1t^2$; (*c*) $v^2 = 2C_1x$; (*d*) $x = C_1 \cos C_2t$; (*e*) $v = C_1e^{-C_2t}$. *Hint:* The arguments of trigonometric functions and exponentials must be dimensionless. The "argument" of $\cos\theta$ is θ and that of e^x is x.

5. If x is in feet, t is in seconds, and v is in feet per second, what are the units of the constants C_1 and C_2 in each part of Problem 4?

1-2 Conversion of Units

6. From the original definition of the meter in terms of the distance from the equator to the North Pole, find in meters (*a*) the circumference and (*b*) the radius of the earth. (*c*) Convert your answers to (*a*) and (*b*) from meters into miles.

7. Complete the following: (*a*) 100 km/h = ______ mi/h; (*b*) 60 cm = ______ in; (*c*) 100 yd = ______ m.

8. In the following, x is in meters, t is in seconds, v is in meters per second, and the acceleration a is in meters per second squared. Find the SI units of each combination: (*a*) v^2/x; (*b*) $\sqrt{x/a}$; (*c*) $\frac{1}{2}at^2$.

9. Find the conversion factor to convert from miles per hour into kilometers per hour.

10. (*a*) Find the number of seconds in a year. (*b*) If one could count $1 per second, how many years would it take to count 1 billion dollars (1 billion = 10^9)? (*c*) If one could count one molecule per second, how many years would it take to count the molecules in a mole? (The number of molecules in a mole is Avogadro's number, $N_A = 6.02 \times 10^{23}$.)

11. Complete the following: (*a*) 1.296×10^5 km/h^2 = ______ km/h · s; (*b*) 1.296×10^5 km/h^2 = ______ m/s^2; (*c*) 60 mi/h = ______ ft/s; (*d*) 60 mi/h = ______ m/s.

12. There are 1.057 quarts in a liter and 4 quarts in a gallon. (*a*) How many liters are there in a gallon? (*b*) A barrel equals 42 gallons. How many cubic meters are there in a barrel?

13. There are 640 acres in a square mile. How many square meters are there in one acre?

14. Sometimes a conversion factor can be derived from the knowledge of a constant in two different systems. (*a*) The speed of light in a vacuum is 186,000 mi/s = 3×10^8 m/s. Use this fact to find the number of kilometers in a mile. (*b*) The weight of 1 ft^3 of water is 62.4 lb. Use this and the fact that 1 cm^3 of water has a mass of 1 g to find the weight in pounds of a 1-kg mass.

1-3 Dimensions of Physical Qualities

15. What are the dimensions of the constants in each part of Problem 4?

16. The SI unit of force, the kilogram-meter per second squared (kg·m/s^2), is called the newton (N). Find the dimensions and the SI units of the constant G in Newton's law of gravitation $F = Gm_1m_2/r^2$.

17. An object on the end of a string moves in a circle. The force exerted by the string depends on the mass of the object, its speed, and the radius of the circle. What combination of these variables gives the correct dimensions (ML/T^2) for the force?

18. Kepler's third law relates the period of a planet to its radius r, the constant G in Newton's law of gravitation ($F = Gm_1m_2/r^2$), and the mass of the sun M_s. What combination of these factors gives the correct dimensions for the period of a planet?

1-4 Scientific Notation

19. Express each of the following as a decimal number without using power-of-10 notation: (*a*) 3×10^4; (*b*) 6.2×10^{-3}; (*c*) 4×10^{-6}; (*d*) 2.17×10^5.

20. Write the following in scientific notation in watts, seconds, and meters: (*a*) 3.1 GW = ______ W; (*b*) 10 pm = ______ m; (*c*) 2.3 ks = ______ s; (*d*) 4 μs = ______ s.

21. Calculate the following, round off to the correct number of significant figures, and express your result in scientific notation: (*a*) $(1.14)(9.99 \times 10^4)$; (*b*) $(2.78 \times 10^{-8}) - (5.31 \times 10^{-9})$; (*c*) $12\pi/(4.56 \times 10^{-3})$; (*d*) $27.6 + (5.99 \times 10^2)$.

22. Calculate the following, round off to the correct number of significant figures, and express your result in scientific notation: (*a*) (200.9)(569.3); (*b*) $(0.000000513)(62.3 \times 10^7)$; (*c*) $28{,}401 + (5.78 \times 10^4)$; (*d*) $63.25/(4.17 \times 10^{-3})$.

23. A cell membrane has a thickness of about 7 nm. How many cell membranes would it take to make a stack 1 in high?

24. Calculate the following, round off to the correct number of significant figures, and express your result in scientific notation: (*a*) $(2.00 \times 10^4)(6.10 \times 10^{-2})$; (*b*) $(3.141592)(4.00 \times 10^5)$; (*c*) $(2.32 \times 10^3)(1.16 \times 10^8)$; (*d*) $(5.14 \times 10^3) + (2.78 \times 10^2)$; (*e*) $(1.99 \times 10^2) + (9.99 \times 10^{-5})$.

25. Perform the following calculations and round off the answers to the correct number of significant figures: (*a*) $3.141592654 \times (23.2)^2$; (*b*) $2 \times 3.141592654 \times 0.76$; (*c*) $\frac{4}{3}\pi \times (1.1)^3$; (*d*) $(2.0)^5/(3.141592654)$.

26. The sun has a mass of 1.99×10^{30} kg. The sun is mostly hydrogen, with only a small amount of heavier elements. The hydrogen atom has a mass of 1.67×10^{-27} kg. Estimate the number of hydrogen atoms in the sun.

Level II

27. The angle subtended by the moon's diameter at a point on the earth is about 0.524° (Figure 1-2). Use this and the fact that the moon is about 384 Mm away to find the diameter of the moon. (The angle θ subtended by the moon is approximately D/r_m, where D is the diameter of the moon and r_m is the distance to the moon.)

Figure 1-2 Problem 27.

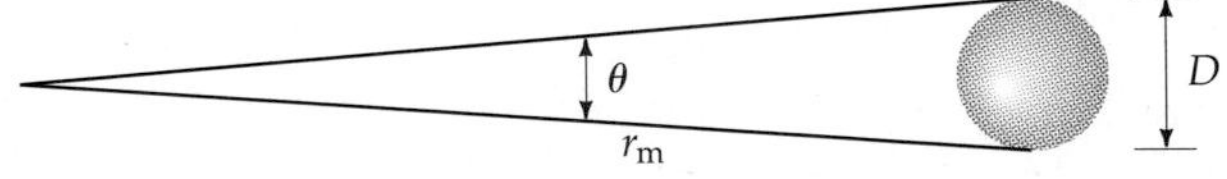

28. The astronomical unit is defined as the mean distance from the earth to the sun, namely, 1.496×10^{11} m. The parsec is the radial length at which one astronomical unit of arc length subtends an angle of 1 second (Figure 1-3). The lightyear is the distance that light travels in one year. (*a*) How many parsecs are there in one astronomical unit? (*b*) How many meters are there in a parsec? (*c*) How many meters are there in a lightyear? (*d*) How many astronomical units are there in a lightyear? (*e*) How many lightyears are there in a parsec?

Figure 1-3 Problem 28.

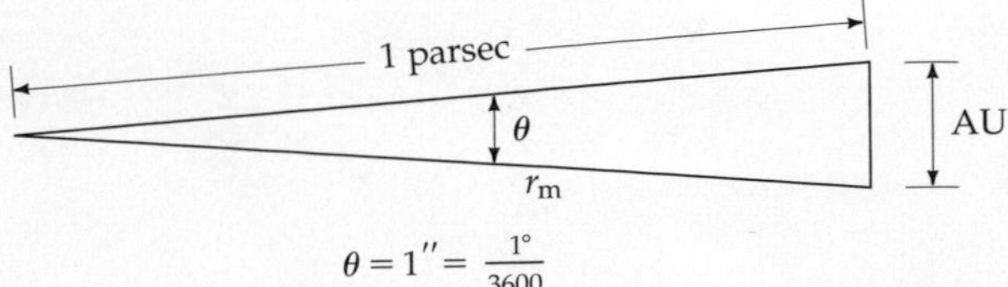

29. The United States imports 6 million barrels of oil per day. This imported oil provides about one-fourth of our total energy needs. A barrel fills a drum that stands about 1 m high. (*a*) If the drums are laid end to end, what is the length in kilometers of the oil imported each day? (*b*) The largest tankers hold about a quarter million barrels. How many tanker loads per year would supply our imported oil? (*c*) If oil costs $20 a barrel, how much do we spend for imported oil each year?

30. Every year the United States generates 160 million tons of municipal solid waste and a grand total of 10 billion tons of solid waste of all kinds. If one allows one cubic meter of volume per ton of waste, how many square miles of area at an average height of 10 m is needed for landfills each year?

31. An iron nucleus has a radius of 5.4×10^{-15} m and a mass of 9.3×10^{-26} kg. (*a*) What is its mass per unit volume in kilograms per cubic meter? (*b*) If the earth has the same mass per unit volume, what would its radius be? (The mass of the earth is 5.98×10^{24} kg.)

32. If the universe is to stop expanding and begin contracting, its average density must be at least 6×10^{-27} kg/m^3. (*a*) How many electrons are needed in a cubic meter to produce the critical density? (*b*) How many protons per cubic meter would produce the critical density? (The mass of an electron and of a proton are listed on the inside cover.)

33. Observational estimates of the density of the universe yield an average of about 2×10^{-28} kg/m^3. (*a*) If a 100-kg football player had his mass uniformly spread out in a sphere to match the estimate for the average density of the universe, what would be the radius of the sphere? (*b*) Compare this radius with the earth–moon distance of 3.84×10^8 m.

34. Beer and soft drinks are sold in aluminum cans. The mass of a typical can is about 0.018 kg. (*a*) Estimate the number of aluminum cans used in the United States in one year. (*b*) Estimate the total mass of aluminum in a year's consumption from these cans. (*c*) Recycled aluminum sells for about $1 per kilogram. How much is this year's accumulation of aluminum cans worth?

Level III

35. The following table gives experimental results for the measurement of the period of motion T of an object of mass m suspended on a spring versus the mass of the object. These data are consistent with a simple equation expressing T as a function of m of the form $T = Cm^n$, where C and n are constants and n is not necessarily an integer. (*a*) Find n and C. (There are several ways to do this. One way is to guess the value of n and check by plotting T versus m^n on graph paper. If your guess is right, the plot will be a straight line. Another way is to plot log T versus log m. The slope of the straight line on this plot is n.) (*b*) Which data points deviate the most from a straight-line plot of T versus m^n?

Mass m, kg	0.10	0.20	0.40	0.50	0.75	1.00	1.50
Period T, s	0.56	0.83	1.05	1.28	1.55	1.75	2.22

36. The table that follows gives the period T and orbit radius r for the motions of four satellites orbiting a dense, heavy asteroid.

Period T, y	0.44	1.61	3.88	7.89
Radius r, Gm	0.088	0.208	0.374	0.600

(*a*) These data can be fitted by the formula $T = Cr^n$. Find C and n. (*b*) A fifth satellite is discovered to have a period of 6.20 y. Find the radius for the orbit of this satellite, which fits the same formula.

37. The period T of a simple pendulum depends on the length L of the pendulum and the acceleration of gravity g (dimensions L/T^2). (*a*) Find a simple combination of L and g that has the dimensions of time. (*b*) Check the dependence of the period T on the length L by measuring the period (the time for a complete to-and-fro swing) of a pendulum for two different values of L. (*c*) The correct formula relating T to L and g involves a constant that is a multiple of π and cannot be obtained by the dimensional analysis of part (*a*). It can be found by experiment as in part (*b*) if g is known. Using the value $g = 9.81$ m/s^2 and your experimental results from part (*b*), find the formula relating T to L and g.

38. A projectile fired at an angle of 45° travels a total distance R, called the range, which depends only on the initial speed v and the acceleration of gravity g (dimensions L/T^2). Using dimensional analysis, find how R depends on the speed and on g.

39. A ball thrown horizontally from a height H with speed v travels a total horizontal distance R. (*a*) Do you expect R to increase or decrease with increasing H? With increasing v? (*b*) From dimensional analysis, find a possible dependence of R on H, v, and g.

40. The earth's atmosphere pushes down on the surface of the earth with a force of 14.7 lb for each square inch of the earth's surface. What is the weight in pounds of the earth's atmosphere? (The radius of the earth is about 6370 km.)

41. Each binary digit is termed a bit. A series of bits grouped together is called a word. An eight-bit word is called a byte. Suppose a computer disk has a capacity of 100 megabytes. (*a*) How many bits can be stored on the disk? (*b*) Estimate the number of typical books that can be stored on the disk.

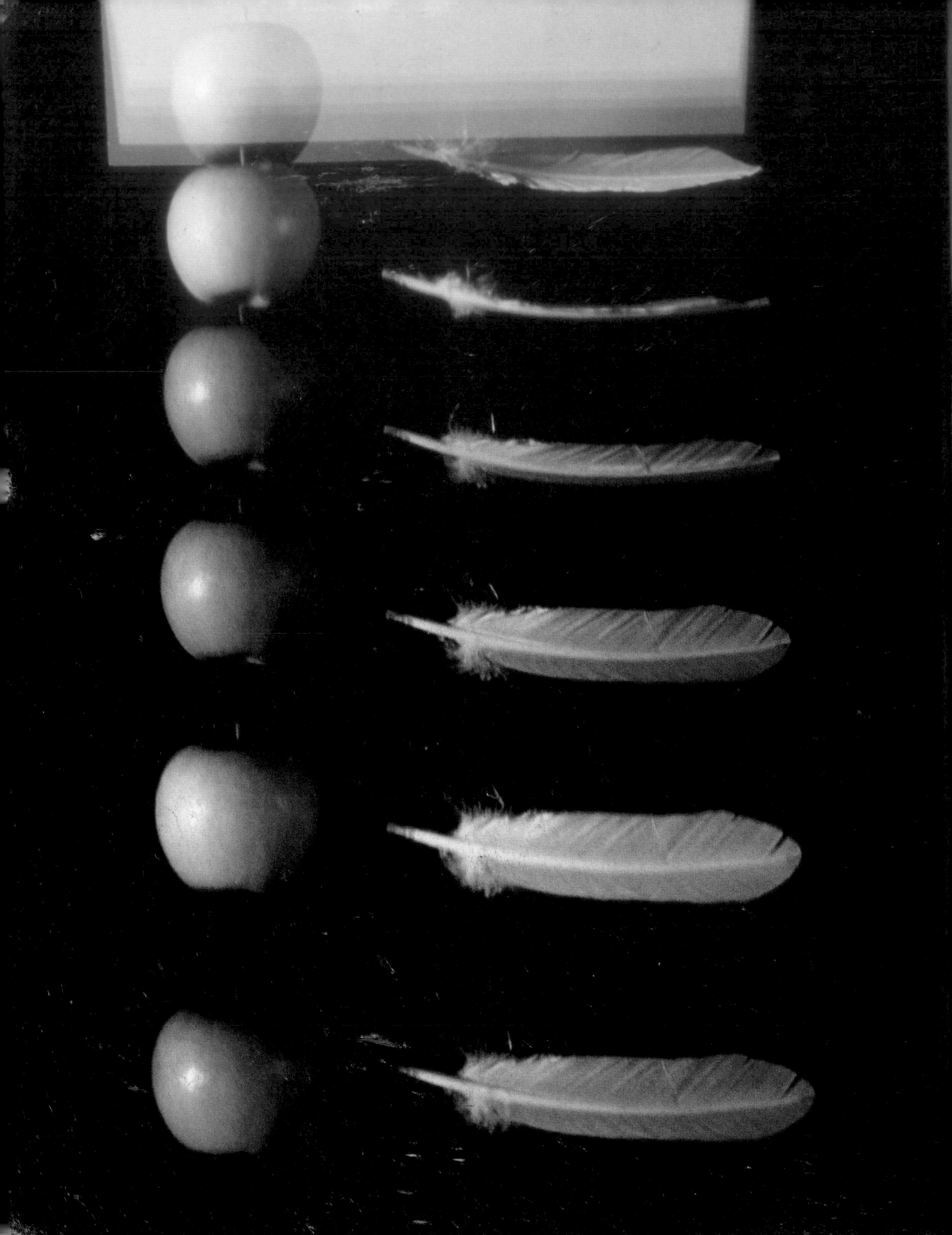

Part 1 Mechanics

Modern version of Galileo's legendary experiment in which he dropped a heavy ball and a light ball from the Leaning Tower of Pisa. In this demonstration, a feather and an apple fall together in a large vacuum chamber with the same acceleration due to gravity. The light bar at the top of the photograph is a portion of a trap door, which initially held the apple and feather at rest, and then dropped them simultaneously.

Chapter 2

Motion in One Dimension

This jet-fuel powered car on the Bonneville Salt Flats deploys its chute to slow it down after a qualifying run of 243.3 km/h.

The description of the motion of objects is an important part of the description of the physical universe. Indeed, it was central to the development of science from Aristotle to Galileo. The laws of how things fall were developed long before Newton described why things fall. One of the earliest scientific puzzles concerned the apparent motion of the sun across the sky and the seasonal motion of the planets and stars. A great triumph of newtonian mechanics was the discovery that the motion of the earth and the other planets around the sun could be explained in terms of a force of attraction between the sun and the planets.

In this chapter and the next, we will be concerned with the description of motion (kinematics) without worrying about its causes. (We will consider the causes of motion in Chapter 4 when we study Newton's laws.) We will confine our discussion in this chapter to motion in one dimension, that is, motion along a straight line. A simple example of one-dimensional motion is a car moving along a flat, straight, narrow road. For such restricted motion, there are only two possible directions, which we distinguish by designating one positive and the other negative.

To simplify our discussion of motion, we will start with objects whose position can be described by locating one point. Such an object is called a

particle. One tends to think of a particle as a very small object, but actually no size limit is implied by the word "particle." For example, it is sometimes convenient to consider the earth to be a particle moving around the sun in a nearly circular path. (Certainly, when viewed from a distant planet or a distant galaxy, the earth looks like a point.) In such cases, we are interested only in the motion of the center of the earth, so we ignore the size of the earth and its rotation. In some astronomical problems, the entire solar system or even a whole galaxy is sometimes treated as a particle. When we are analyzing the rotation or the internal structure of an object, we can no longer treat it as a single particle. But our study of particle motion is useful even in these cases because any object, no matter how complex, can be treated as a collection or "system" of particles.

2-1 Speed, Displacement, and Velocity

We are all familiar with the concept of speed. We define the **average speed** of a particle as the ratio of the total distance traveled to the total time taken:

$$\text{Average speed} = \frac{\text{total distance}}{\text{total time}}$$

The SI unit of average speed is meters per second (m/s), and the U.S. customary unit is feet per second (ft/s). The familiar, everyday unit of speed in the United States is miles per hour (mi/h). Internationally, the more common unit is kilometers per hour (km/h). If you drive 200 km in 5 h, your average speed is (200 km)/(5 h) = 40 km/h. The average speed tells you nothing about the details of the trip. You may have driven at a steady rate of 40 km/h for the whole 5 h, or you may have driven faster part of the time and slower the rest of the time, or you may have stopped for an hour and then driven at varying rates during the other 4 h.

The concept of velocity is similar to that of speed but differs because it includes the *direction* of motion. To understand this concept, we first introduce the idea of displacement. Let us set up a coordinate system by choosing some reference point on a line for the origin O. To every other point on the line we assign a number x that indicates how far the point is from the origin. The value of x depends on the unit (feet, meters, or whatever) chosen to measure the distance. The sign of x depends on its position relative to the origin O. The usual convention we choose is that points to the right of the origin are given positive values and those to the left, negative values.

Figure 2-1 shows a car (which we can treat as a particle) that is at position x_1 at some time t_1 and at position x_2 at time t_2. The change in the position of the particle, $x_2 - x_1$, is called the **displacement** of the particle. It is customary to use the Greek letter Δ (capital delta) to indicate the change in quantity. Thus, the change in x is written Δx:

$$\Delta x = x_2 - x_1 \tag{2-1}$$

Displacement defined

The notation Δx (read "delta x") stands for a single quantity, the change in x. It is not a product of Δ and x any more than $\cos\theta$ is a product of cos and θ.

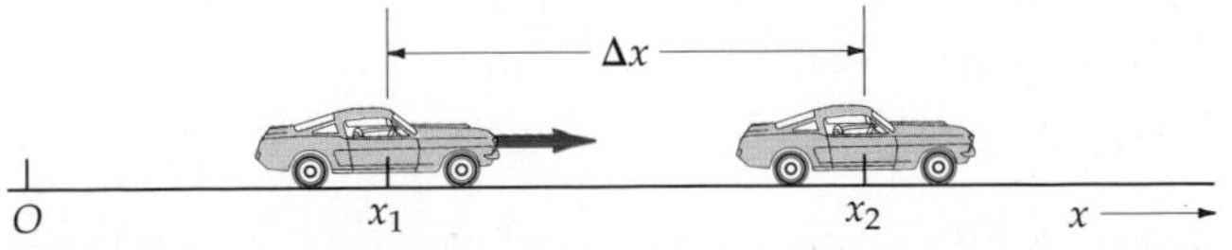

Figure 2-1 When the car travels from point x_1 to point x_2, its displacement is $\Delta x = x_2 - x_1$.

Velocity is the rate at which the position changes. The **average velocity** of the particle is defined as the ratio of the displacement Δx to the time interval $\Delta t = t_2 - t_1$:

Average velocity defined

$$v_{av} = \frac{\Delta x}{\Delta t} = \frac{x_2 - x_1}{t_2 - t_1} \tag{2-2}$$

Note that the displacement and the average velocity may be either positive or negative, depending on whether x_2 is greater or less than x_1. A positive value indicates motion to the right and a negative value indicates motion to the left.

Example 2-1

A snail is at $x_1 = 18$ mm at $t_1 = 2$ s and is later found at $x_2 = 14$ mm at $t_2 = 7$ s. Find the displacement and the average velocity of the snail for this time interval.

By the definition, the displacement of the snail is

$$\Delta x = x_2 - x_1 = 14 \text{ mm} - 18 \text{ mm} = -4 \text{ mm}$$

and the average velocity is

$$v_{av} = \frac{\Delta x}{\Delta t} = \frac{x_2 - x_1}{t_2 - t_1} = \frac{14 \text{ mm} - 18 \text{ mm}}{7 \text{ s} - 2 \text{ s}} = \frac{-4 \text{ mm}}{5 \text{ s}} = -0.8 \text{ mm/s}$$

The displacement and average velocity are negative, indicating that the snail moved to the left.

Note that the unit millimeters per second is included as part of the answer for the average velocity found in Example 2-1. Since there are many other possible choices for units of length (for example, feet, inches, miles, lightyears) and time (for example, hours, days, years), it is essential to include the unit with a numerical answer. The statement "the average velocity of a particle is −3" is meaningless.

Example 2-2

How far does a car go in 5 min if its average velocity is 80 km/h during this time interval?

We are interested in the displacement during a time interval of 5 min. From Equation 2-2, the displacement Δx is given by

$$\Delta x = v_{av}\,\Delta t$$

Since the time is given in minutes and the average velocity in kilometers per hour, we must convert either the time to hours or the average velocity to kilometers per minute. Doing the latter, we obtain

$$v_{av} = \frac{80 \text{ km}}{1 \text{ h}} \times \frac{1 \text{ h}}{60 \text{ min}} = \frac{4 \text{ km}}{3 \text{ min}}$$

Then

$$\Delta x = \frac{4 \text{ km}}{3 \text{ min}} \times 5 \text{ min} = 6.67 \text{ km}$$

Exercise

A runner runs with an average velocity of 0.25 km/min along a straight line. How long does it take to run a distance of 10 km? (Answer: 40 min)

Example 2-3

A runner runs 100 m in 12 s, then turns around and jogs 50 m back toward the starting point in 30 s. What are her average *speed* and average *velocity* for the total trip?

The total distance traveled is 100 m + 50 m = 150 m, and the total time taken is 42 s. The average speed is therefore (150 m)/(42 s) = 3.57 m/s. Note that this is not the average of her running and jogging speeds because she ran for 12 s but jogged for 30 s. To find the average velocity, we first find the total displacement. If her starting point x_1 is taken to be 0, then her final point x_2 is 50 m, so her total displacement is $x_2 - x_1 = 50$ m. The average velocity is then

$$v_{av} = \frac{\Delta x}{\Delta t} = \frac{50 \text{ m}}{42 \text{ s}} = +1.19 \text{ m/s}$$

Again, this is not the average of her running velocity (+8.33 m/s) and her jogging velocity (−1.67 m/s) because the times are different.

Figure 2-2 shows a graph of x versus t for some arbitrary motion along the x axis. Each point on the curve has an x value, which is the location of the particle at a particular time, and a t value, which is the time the particle was at that location. On the graph, we have drawn a straight line between the position labeled P_1 and the position labeled P_2. The displacement $\Delta x = x_2 - x_1$ and the time interval $\Delta t = t_2 - t_1$ between these points are indicated. The line between P_1 and P_2 is the hypotenuse of the triangle having sides Δx and Δt. The ratio $\Delta x/\Delta t$ is called the **slope** of this straight line. In geometric terms, it is a measure of the steepness of the straight line in the graph. For a given interval Δt, the steeper the line, the greater the value of $\Delta x/\Delta t$. Since the slope of this line is just the average velocity for the interval Δt, we have a geometric representation of average velocity.

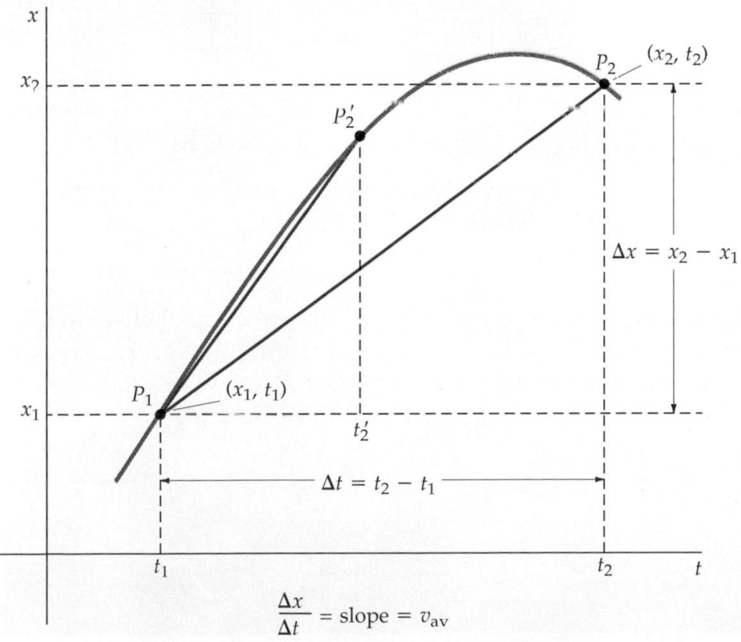

Figure 2-2 Graph of x versus t for a particle moving in one dimension. The points P_1 and P_2 are connected by a straight line. The average velocity is the slope of this line, $\Delta x/\Delta t$, which depends on the time interval, as indicated by the fact that the line from P_1 to P_2' has a greater slope than the line from P_1 to P_2.

The average velocity is the slope of the straight line connecting the points (x_1, t_1) and (x_2, t_2).

Unless the velocity is constant, the average velocity will depend on the time interval on which it is based. For example, in Figure 2-2, if we chose a smaller time interval by choosing a time t_2' closer to t_1, the average velocity would be greater, as indicated by the greater steepness of the line connecting points P_1 and P_2'.

Questions

1. What sense, if any, does the following statement make? "The average velocity of the car at 9 A.M. was 60 km/h."
2. Is it possible for the average velocity for some interval to be zero even though the average velocity for a shorter interval included in the first interval is not zero? Explain.
3. What is the approximate average velocity of the race cars during the Indianapolis 500?

2-2 Instantaneous Velocity

At first glance, to define the velocity of a particle at a single instant, that is, at a specific time, might seem impossible. At a time t_1, the particle is at a single point x_1. If it is at a single point, how can it be moving? On the other hand, if it is not moving, shouldn't it stay at the same point? This is an age-old paradox, which can be resolved when we realize that, to observe motion and thus define it, we must look at the position of the object at more than one time. It is then possible to define the velocity at an instant by using a limiting process.

The fastest running animal, the cheetah, can reach speeds of greater than 100 km/h.

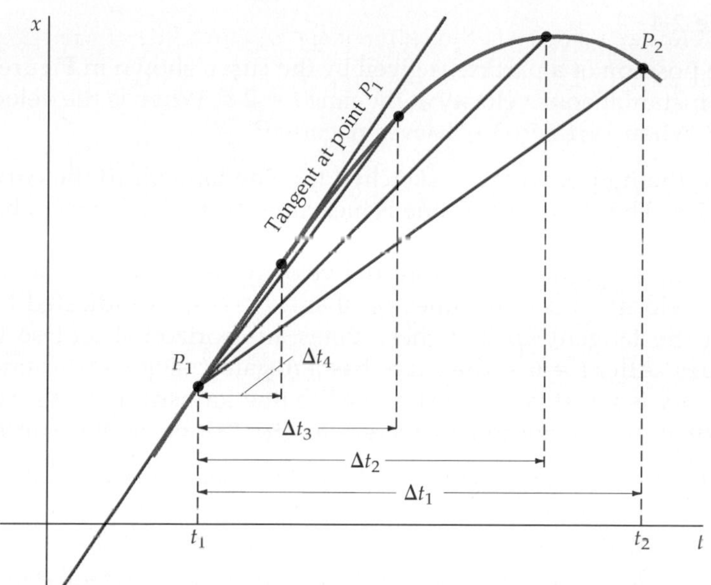

Figure 2-3 Graph of x versus t from Figure 2-2. As the time interval beginning at t_1 is decreased, the average velocity for that interval approaches the slope of the line tangent to the curve at time t_1. The instantaneous velocity is defined as the slope of this line.

Figure 2-3 is the same x-versus-t curve as Figure 2-2, showing a sequence of time intervals, Δt_1, Δt_2, Δt_3, . . . , each smaller than the previous one. For each time interval Δt, the average velocity is the slope of the straight line appropriate for that interval. The figure shows that, as the time interval becomes smaller, the straight lines get steeper, but they never incline more than the tangent to the curve at point t_1. We define the slope of this tangent as the **instantaneous velocity** at the time t_1.

The instantaneous velocity at some point is the slope of the line tangent to the x-versus-t curve at that point.

It is important to realize that the displacement Δx depends on the time interval Δt. As Δt approaches zero, Δx does too (as can be seen from Figure 2-3), and the ratio $\Delta x/\Delta t$ approaches the slope of the line tangent to the curve. Since the slope of the tangent is the limit of the ratio $\Delta x/\Delta t$ as Δt approaches zero, we can restate our definition as follows:

The instantaneous velocity is the limit of the ratio $\Delta x/\Delta t$ as Δt approaches zero.

$$v = \lim_{\Delta t \to 0} \frac{\Delta x}{\Delta t} = \begin{array}{c}\text{slope of the line tangent}\\ \text{to the } x\text{-versus-}t \text{ curve}\end{array} \qquad \text{2-3}$$

Instantaneous velocity defined

This limit is called the **derivative** of x with respect to t. In usual calculus notation, the derivative is written dx/dt:

$$v = \lim_{\Delta t \to 0} \frac{\Delta x}{\Delta t} = \frac{dx}{dt} \qquad \text{2-4}$$

This slope may be positive (x is increasing) or negative (x is decreasing); correspondingly, the instantaneous velocity may be positive or negative in one-dimensional motion. The magnitude of the instantaneous velocity is called the **instantaneous speed.**

Example 2-4

The position of a particle is given by the curve shown in Figure 2-4. Find the instantaneous velocity at the time $t = 2$ s. When is the velocity greatest? When is it zero? Is it ever negative?

In the figure, we have sketched the line tangent to the curve at time $t = 2$ s. The slope of this line is measured from the figure to be (4.5 m)/(3 s) = 1.5 m/s. Thus $v = 1.5$ m/s at time $t = 2$ s. According to the figure, the slope (and therefore the velocity) is greatest at about $t = 4$ s. The velocity is zero at times $t = 0$ and $t = 6$ s, as indicated by the fact that the tangent lines at these times are horizontal and so have zero slopes. After $t = 6$ s, the curve has a negative slope, indicating that the velocity is negative. (The slope of the line tangent to a curve at a given point is often referred to as merely as the "slope of the curve" at that point.)

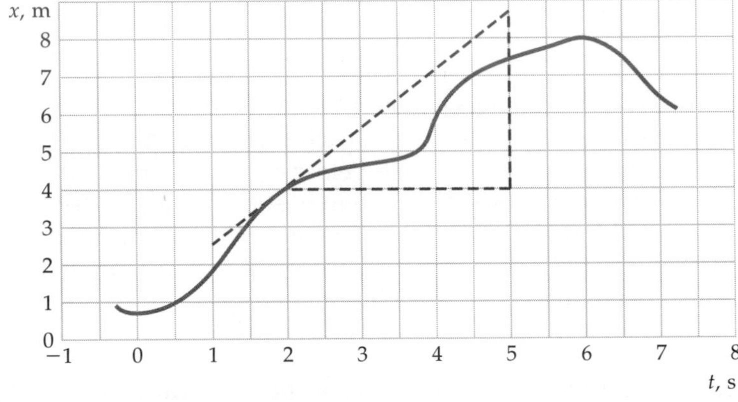

Figure 2-4 Graph of x versus t for Example 2-4. The instantaneous velocity at time $t = 2$ s can be found by measuring the slope of the line tangent to the curve at that time.

Example 2-5

The position of a stone dropped from rest from a cliff is given by $x = 5t^2$, where x is in meters measured downward from the original position at $t = 0$, and t is in seconds. Find the velocity at any time. (We omit explicit indication of the units to simplify the notation.)

The curve corresponding to $x = 5t^2$ is shown in Figure 2-5. Tangent lines have been drawn for three different times, t_1, t_2, and t_3. The slopes of these lines differ. As time passes, the slope of the curve increases, indicating that the instantaneous velocity is increasing with time. We can compute the velocity at some time t by computing the derivative dx/dt directly from the definition in Equation 2-3. At time t, the position is

$$x(t) = 5t^2$$

At a later time, $t + \Delta t$, the position is $x(t + \Delta t)$, which is given by

$$x(t + \Delta t) = 5(t + \Delta t)^2 = 5[t^2 + 2t\ \Delta t + (\Delta t)^2]$$
$$= 5t^2 + 10t\ \Delta t + 5(\Delta t)^2$$

The displacement for this time interval is thus

$$\Delta x = x(t + \Delta t) - x(t) = [5t^2 + 10t\ \Delta t + 5(\Delta t)^2] - 5t^2$$
$$= 10t\ \Delta t + 5(\Delta t)^2$$

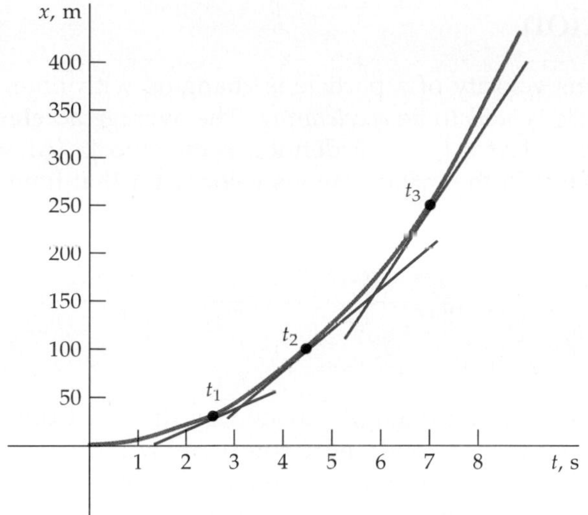

Figure 2-5 Graph of the function $x(t) = 5t^2$ for Example 2-5. Tangent lines are drawn at times t_1, t_2, and t_3. The slopes of these tangent lines increase steadily from t_1 to t_2 to t_3, indicating that the instantaneous velocity increases steadily with time.

The average velocity for this time interval is

$$v_{av} = \frac{\Delta x}{\Delta t} = \frac{10t\ \Delta t + 5(\Delta t)^2}{\Delta t} = 10t + 5\ \Delta t$$

As we consider shorter and shorter time intervals, Δt approaches zero, so the second term, $5\ \Delta t$, approaches zero while the first term, $10t$, remains unaltered. The instantaneous velocity at time t is thus

$$v = \lim_{\Delta t \to 0} \frac{\Delta x}{\Delta t} = 10t$$

For this example, the instantaneous velocity is proportional to the time.

Notice that if we set Δt in Equation 2-3 directly equal to 0, then Δx would also be 0 and the ratio $\Delta x/\Delta t$ would not be defined. However, by starting with an equation that represents how x varies with t, we have calculated exactly the limit of $\Delta x/\Delta t$ as Δt approaches zero.

It is instructive to examine the limiting process numerically by computing the average velocity for smaller and smaller time intervals. Table 2-1 gives the average velocity for Example 2-5 for $t = 2$ s and various time intervals Δt, each smaller than the previous one. The table shows that for very small time intervals, the average velocity is very nearly equal to the instantaneous velocity, which is 20 m/s. The difference between $\Delta x/\Delta t$ and $\lim_{\Delta t \to 0} (\Delta x/\Delta t)$ can be made arbitrarily small by choosing a sufficiently small Δt.

It is important to distinguish carefully between average velocity and instantaneous velocity. By custom, the word "velocity" alone is assumed to mean instantaneous velocity.

Table 2-1 **Displacement and Average Velocity for Various Time Intervals Δt Beginning at $t = 2$ s for the Function $x = 5t^2$**

Δt, s	Δx, m	$\Delta x/\Delta t$, m/s
1.00	25	25
0.50	11.25	22.5
0.20	4.20	21.0
0.10	2.05	20.5
0.05	1.0125	20.25
0.01	0.2005	20.05
0.005	0.100125	20.025
0.001	0.020005	20.005
0.0001	0.00200005	20.0005

Questions

4. If the instantaneous velocity does not change, will the average velocities for different intervals differ?

5. If $v_{av} = 0$ for some time interval Δt, must the instantaneous velocity v be zero at some point in the interval? Support your answer by sketching a possible x-versus-t curve that has $\Delta x = 0$ for some interval Δt.

2-3 Acceleration

When the instantaneous velocity of a particle is changing with time, as in Example 2-5, the particle is said to be *accelerating*. The **average acceleration** for a particular time interval $\Delta t = t_2 - t_1$ is defined as the ratio $\Delta v/\Delta t$, where $\Delta v = v_2 - v_1$ is the change in the instantaneous velocity for that time interval:

Average acceleration defined

$$a_{av} = \frac{\Delta v}{\Delta t} \tag{2-5}$$

The dimensions of acceleration are length divided by $(\text{time})^2$. Convenient units are meters per second per second, more compactly written meters per second squared (m/s^2), or feet per second squared (ft/s^2). For example, if we say that a particle is accelerating at 5 m/s^2, we mean that, if it begins at rest, after 1 s it is traveling with a velocity of 5 m/s, after 2 s it is traveling with a velocity of 10 m/s, after 3 s it is traveling with a velocity of 15 m/s, and so on.

The **instantaneous acceleration** is the limit of the ratio $\Delta v/\Delta t$ as Δt approaches zero. If we plot the velocity versus time, the instantaneous acceleration at time t is defined as the slope of the line tangent to the curve at that time:

Instantaneous acceleration defined

$$a = \lim_{\Delta t \to 0} \frac{\Delta v}{\Delta t} = \begin{array}{c}\text{slope of the line tangent}\\ \text{to the } v\text{-versus-}t \text{ curve}\end{array} \tag{2-6}$$

The acceleration is thus the derivative of the velocity with respect to time. The calculus notation for this derivative is dv/dt. Since velocity is the derivative of the position x with respect to t, the acceleration is the second derivative of x with respect to t, which is usually written d^2x/dt^2. We can see the reason for this notation by writing the acceleration as dv/dt and replacing v with dx/dt:

$$a = \frac{dv}{dt} = \frac{d(dx/dt)}{dt} = \frac{d^2x}{dt^2} \tag{2-7}$$

If the velocity is constant, the acceleration is zero because $\Delta v = 0$ for all time intervals. In this case, the slope of the corresponding x-versus-t curve does not change. In Example 2-5 we found that for the position function $x = (5\ m/s^2)t^2$, the velocity increases linearly with the time according to $v = (10\ m/s^2)t$. In this case, the acceleration is constant and equal to 10 m/s^2, which is the slope of the corresponding v-versus-t curve.

We will see in Chapter 4 that the acceleration is directly proportional to the net force on the particle.

Example 2-6

A fast car can accelerate from 0 to 90 km/h in 5 s. What is the average acceleration during this period? Compare this with the free-fall acceleration due to gravity, which is 9.81 m/s^2.

From Equation 2-5, we have for the average acceleration

$$a_{av} = \frac{\Delta v}{\Delta t} = \frac{90\ \text{km/h}}{5\ \text{s}} = 18\ \text{km/h·s}$$

To compare this with the acceleration due to gravity, we convert to meters per second squared using 1 h = 3600 s = 3.6 ks. Then

$$\frac{18\ \text{km}}{\text{h}\cdot\text{s}} \times \frac{1\ \text{h}}{3.6\ \text{ks}} = 5\ \text{m/s}^2$$

This is about half the free-fall acceleration due to gravity.

Exercise

A car is traveling at 45 km/h at time $t = 0$. It accelerates at a constant rate of 10 km/h·s. How fast is it traveling at $t = 2$ s? (Answer: 65 km/h)

Example 2-7

The position of a particle is given by $x = Ct^3$, where C is a constant with the unit m/s^3. Find the velocity and acceleration as functions of time.

As in Example 2-5, we can compute the velocity by computing the derivative dx/dt directly from its definition as $\lim_{\Delta t \to 0} (\Delta x/\Delta t)$ (Equation 2-3). At some time t the position is $x(t) = Ct^3$. At a later time $t + \Delta t$, the position is

$$x(t + \Delta t) = C(t + \Delta t)^3 = Ct^3 + 3Ct^2\,\Delta t + 3Ct(\Delta t)^2 + C(\Delta t)^3$$

The displacement is thus

$$\begin{aligned}\Delta x = x(t + \Delta t) - x(t) &= C(t + \Delta t)^3 - Ct^3 \\ &= Ct^3 + 3Ct^2\,\Delta t + 3Ct(\Delta t)^2 + C(\Delta t)^3 - Ct^3 \\ &= 3Ct^2\,\Delta t + 3Ct(\Delta t)^2 + C(\Delta t)^3\end{aligned}$$

The average velocity for this time interval is

$$v_{\text{av}} = \frac{\Delta x}{\Delta t} = 3Ct^2 + 3Ct(\Delta t) + C(\Delta t)^2$$

As $\Delta t \to 0$, the two terms containing Δt approach zero and the term $3Ct^2$ remains unchanged. The instantaneous velocity at time t is thus

$$v = \lim_{\Delta t \to 0} \frac{\Delta x}{\Delta t} = 3Ct^2$$

We find the acceleration by repeating the process and finding the derivative of v with respect to t, which is the second derivative of x with respect to t. We omit some of the algebra in this derivation because it is similar to that just shown. The change in velocity for the time interval from t to $t + \Delta t$ is

$$\Delta v = 3C(t + \Delta t)^2 - 3Ct^2 = 6Ct\,\Delta t + 3C(\Delta t)^2$$

We divide this expression by Δt to obtain the average acceleration for this interval:

$$a_{\text{av}} = \frac{\Delta v}{\Delta t} = 6Ct + 3C\,\Delta t$$

The instantaneous acceleration is thus

$$a = \lim_{\Delta t \to 0} \frac{\Delta v}{\Delta t} = 6Ct$$

In this example the acceleration is not constant but increases with time.

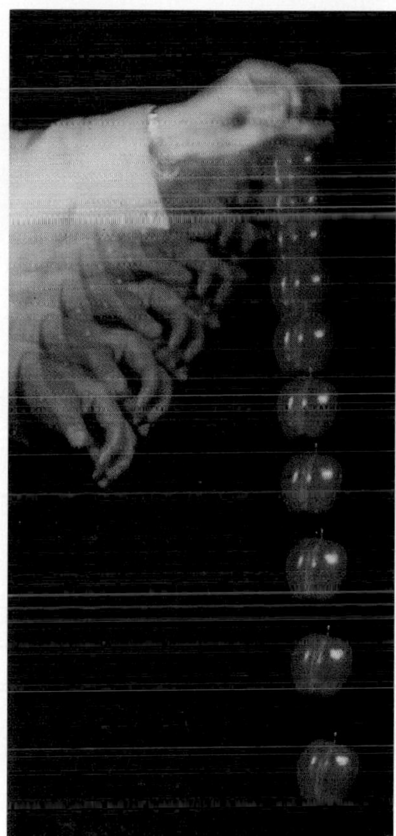

Strobe photograph of a falling apple at 60 flashes per second. The acceleration of the apple is indicated by the greater spacing of the images at the bottom.

In the examples so far, we have computed derivatives directly from the definition by taking the appropriate limit explicitly. It is useful to examine various properties of the derivative and to develop rules that allow us to calculate the derivatives of many functions quickly without applying the definition each time. Appendix Table A-4 contains a list of such rules, followed by brief discussions of their origins. We will use these rules without comment and leave detailed study of them to your calculus course.

Rule 7 in Appendix Table A-4 is used so often that we will repeat it here. If x is a simple power function of t, such as

$$x = Ct^n$$

where C and n are any constants, the derivative of x with respect to t is given by

$$\frac{dx}{dt} = \frac{d}{dt}(Ct^n) = Cnt^{n-1} \qquad 2\text{-}8$$

We have already seen applications of this rule. For example, for the position function of Example 2-7, $x = Ct^3$, we found $v = dx/dt = 3Ct^2$ and $a = dv/dt = 6Ct$.

"It goes from zero to 60 in about 3 seconds."

Questions

6. Give an example of a motion for which the velocity is negative but the acceleration is positive; for example, sketch a graph of v versus t.
7. Give an example of a motion for which both the acceleration and the velocity are negative.
8. Is it possible for a body to have zero velocity and nonzero acceleration?

2-4 Motion with Constant Acceleration

The motion of a particle that has constant acceleration is common in nature. For example, near the surface of the earth all unsupported objects fall vertically with constant acceleration due to gravity provided that air resistance can be neglected. The acceleration due to gravity is designated by g and has the approximate value

$$g = 9.81 \text{ m/s}^2 = 32.2 \text{ ft/s}^2$$

A constant acceleration means that the slope of the v-versus-t curve is constant; that is, the velocity varies linearly with time. If the velocity is v_0 at time $t = 0$, its value v at a later time t is given by

Constant acceleration, v versus t

$$v = v_0 + at \qquad 2\text{-}9$$

If the particle starts at x_0 at time $t = 0$ and its position is x at time t, the displacement $\Delta x = x - x_0$ is given by

$$\Delta x = v_{av}t$$

(This is the same as Equation 2-2, with t replacing Δt because we have chosen the initial value of t to be zero.) For constant acceleration, the velocity varies linearly with time and the average velocity is the mean value of the initial and final velocities, as shown in Figure 2-6.* If v_0 is the initial velocity and v the final velocity, the average velocity is $\frac{1}{2}(v_0 + v)$.

$$v_{av} = \tfrac{1}{2}(v_0 + v) \qquad \text{2-10}$$

Constant acceleration, v_{av}

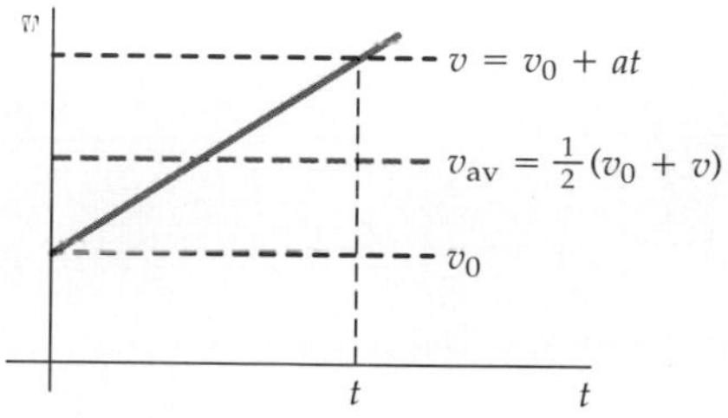

v

$v = v_0 + at$

$v_{av} = \frac{1}{2}(v_0 + v)$

v_0

t t

Figure 2-6 Average velocity for constant acceleration.

The displacement is then

$$\Delta x = v_{av}t = \tfrac{1}{2}(v_0 + v)t \qquad \text{2-11}$$

We can eliminate v by substituting $v = v_0 + at$ from Equation 2-9:

$$\Delta x = \tfrac{1}{2}(v_0 + v)t = \tfrac{1}{2}(v_0 + v_0 + at)t = v_0t + \tfrac{1}{2}at^2 \qquad \text{2-12}a$$

The position function is thus

$$x = x_0 + v_0t + \tfrac{1}{2}at^2 \qquad \text{2-12}b$$

Constant acceleration, x versus t

We can eliminate the time t from Equations 2-9 and 2-12a and obtain a relation among the displacement Δx, the acceleration a, and the initial and final velocities. Solving Equation 2-9 for $t = (v - v_0)/a$ and substituting this result into Equation 2-12a, we obtain

$$\Delta x = v_0t + \tfrac{1}{2}at^2 = v_0\left(\frac{v - v_0}{a}\right) + \tfrac{1}{2}a\left(\frac{v - v_0}{a}\right)^2$$

Multiplying each side by a and writing out the terms, we obtain

$$a\,\Delta x = v_0v - v_0^2 + \tfrac{1}{2}v^2 - v_0v + \tfrac{1}{2}v_0^2 = \tfrac{1}{2}v^2 - \tfrac{1}{2}v_0^2$$

or

$$v^2 = v_0^2 + 2a\,\Delta x \qquad \text{2-13}$$

Constant acceleration, v versus x

Equation 2-13 is useful, for example, if we want to find the final velocity of a ball dropped from rest from some height x and we are not interested in the time the fall takes.

We will now look at some examples of constant-acceleration problems. The first step in solving such a problem is to choose a convenient coordinate system. When possible, we choose the origin to be the location of the particle at $t = 0$ so that $x_0 = 0$. The choice of the positive direction for the x axis determines the positive direction for the velocity and acceleration. Although this choice is arbitrary, a judicious choice for the positive direction can make the solution of a problem easier. For example, if we have a problem in which we drop a ball from some height, it is easiest to choose the downward direction as positive. Then the acceleration is positive (since the acceleration due

*This is not true if the acceleration is not constant.

to gravity is always downward), and the velocity is also positive since the ball moves only downward. On the other hand, if we throw a ball upwards, it is usually more convenient to choose the upward direction as positive. The acceleration is then negative, and the velocity is positive when the ball is moving upward and negative when it is moving downward.

The next step in solving a constant-acceleration problem is to list the information given in terms of an equation. For example, we would write $v_0 = 0$ if the initial velocity is given as zero. Next, list the quantities to be found; for example, $v = ?$ if the final velocity is asked for. Then choose the equation (from Equations 2-9 to 2-13) that contains the given and unknown quantities, and solve for the unknown quantity. When possible, it is always a good idea to solve a problem in two different ways to check your solution. Finally, you should check your answer to see if it is reasonable. For example, if a ball is thrown upward and you find that it travels 4500 m before reaching its highest point, you have probably made a numerical error.

Example 2-8

A ball is thrown upward with an initial velocity of 30 m/s. If its acceleration is 10 m/s^2 downward, how long does it take to reach its highest point, and what is the distance to the highest point? (In this example, we approximate the acceleration of gravity to be 10 m/s^2 to simplify the calculations.)

We choose the origin at the initial position of the ball and take the upward direction to be positive. The given quantities are $v_0 = 30$ m/s and $a = -10$ m/s^2. The acceleration is negative because it is in the downward direction. As the ball moves upward (v is positive), the velocity decreases from its initial value until it is zero. When the velocity is zero, the ball is at its highest point. It then begins to fall, and the velocity becomes negative, indicating that the ball is moving downward. We wish to find the time $t = ?$ when the velocity v is zero. Equation 2-9 contains the unknown time t in terms of the given quantities v_0, a, and v:

$$v = v_0 + at$$

$$0 = 30\text{ m/s} + (-10\text{ m/s}^2)t$$

$$t = \frac{30\text{ m/s}}{10\text{ m/s}^2} = 3.0\text{ s}$$

Note that the units work out correctly.

We can find the distance traveled from Equation 2-11. Since the initial velocity is +30 m/s and the final velocity is 0, the average velocity for the upward motion is 15 m/s. The distance traveled is then

$$\Delta x = v_{av}t = (15\text{ m/s})(3.0\text{ s}) = 45\text{ m}$$

We can also find Δx from Equation 2-12*a*, but the calculation is somewhat more difficult and there is more of a chance of making a mistake. We will use this method to check our result, however.

$$\begin{aligned}\Delta x &= v_0 t + \tfrac{1}{2}at^2\\ &= (30\text{ m/s})(3.0\text{ s}) + \tfrac{1}{2}(-10\text{ m/s}^2)(3.0\text{ s})^2\\ &= +90\text{ m} - 45\text{ m}\\ &= 45\text{ m}\end{aligned}$$

Example 2-9

What is the total time the ball in Example 2-8 is in the air?

We could guess the answer to be 6 s since, by symmetry, if it takes 3.0 s to rise 45 m, it should take the same time to fall 45 m. This is correct. We can also find the time from Equation 2-12*a* by setting $\Delta x = 0$.

$$\Delta x = v_0 t + \tfrac{1}{2}at^2 = 0$$

Factoring, we obtain

$$t(v_0 + \tfrac{1}{2}at) = 0$$

The two solutions are $t = 0$ and

$$t = -\frac{2v_0}{a} = -\frac{2(30 \text{ m/s})}{-10 \text{ m/s}^2} = 6 \text{ s}$$

The solution $t = 0$ corresponds to the initial condition that the ball was at x_0 at $t = 0$.

Figures 2-7*a* and *b* show *x*-versus-*t* and *v*-versus-*t* curves for the ball in Examples 2-8 and 2-9. Note that at the time 3.0 s, the velocity of the ball is zero but the slope of the *v*-versus-*t* curve is not. The slope of the *v*-versus-*t* curve has the value -10 m/s^2 at this time and at all other times because the acceleration is constant. Note that when the ball is moving upward, its velocity is positive and decreasing so its speed is decreasing. When it is moving downward, its velocity is negative and decreasing, so its speed is increasing.

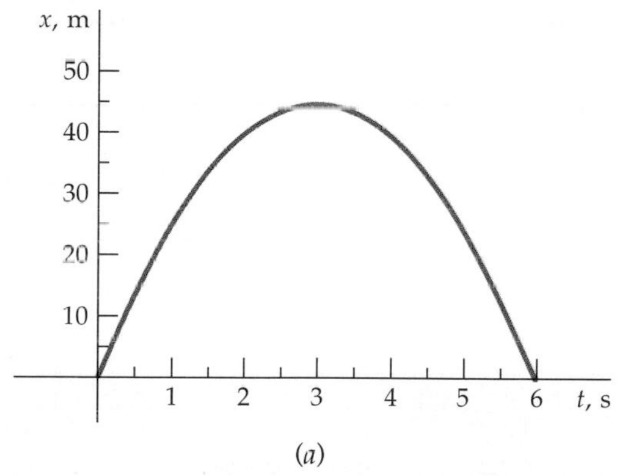

(*a*)

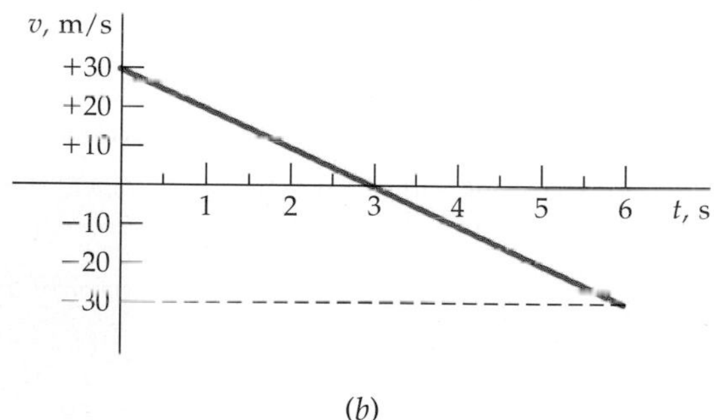

(*b*)

Figure 2-7 (*a*) Graph of x versus t for the ball thrown into the air in Examples 2-8 and 2-9. The curve is a parabola, $x = (30 \text{ m/s})t - \frac{1}{2}(10 \text{ m/s}^2)t^2$. (*b*) Graph of v versus t for the same ball. The velocity decreases steadily from its initial value of 30 m/s to its final value of -30 m/s just before the ball hits the ground. At time $t = 3.0$ s, when the ball is at its highest point, the velocity is zero, but its rate of change is -10 m/s^2, the same as at any other time.

Exercise

A car accelerates from rest at a constant rate of 8 m/s^2. (*a*) How fast is it going after 10 s? (*b*) How far has it gone after 10 s? (*c*) What is its average velocity for the interval $t = 0$ to $t = 10$ s? [Answers: (*a*) 80 m/s, (*b*) 400 m, (*c*) 40 m/s]

Example 2-10

A car traveling at 30 m/s (about 67 mi/h) brakes to a stop. If the acceleration is -5 m/s^2, how far does the car travel before stopping? This distance is called the **stopping distance.**

In this example, we choose the original direction of motion to be positive. The stopping distance will then also be positive, but the acceleration will be negative. (An acceleration that causes the speed to decrease is sometimes called *deceleration*.) In this problem we are given the initial velocity $v_0 = 30$ m/s, the final velocity $v = 0$, the acceleration $a = -5$ m/s^2, and we wish to find the distance traveled $\Delta x = ?$. We do not need to know the time it takes for the car to stop, so Equation 2-13 is the most convenient. Setting $v = 0$ in Equation 2-13, we have

$$v^2 = v_0^2 + 2a\,\Delta x = 0$$

$$0 = (30 \text{ m/s})^2 + 2(-5 \text{ m/s}^2)\,\Delta x$$

$$\Delta x = 90 \text{ m}$$

Note that this is a considerable distance. The force that produces this deceleration is that of friction between the car tires and the road. On wet pavement or gravel, the frictional force is smaller and the magnitude of the acceleration is less than 5 m/s^2, giving an even greater stopping distance.

Example 2-11

What is the stopping distance under the same conditions as Example 2-10 if the car is initially traveling at 15 m/s?

From Equation 2-13 with $v = 0$, we see that the stopping distance is proportional to the square of the initial speed. If we double the speed, the stopping distance is increased by a factor of 4. Similarly, if we halve the initial speed, the stopping distance is reduced by a factor of 4. The stopping distance at 15 m/s is therefore one-fourth that at 30 m/s, or $\frac{1}{4}(90 \text{ m}) = 22.5$ m.

Sometimes, even when the acceleration is not constant, valuable insight can be gained about the motion of an object by assuming that the constant-acceleration formulas still apply to it and analyzing what happens in such an ideal case.

(*a*)

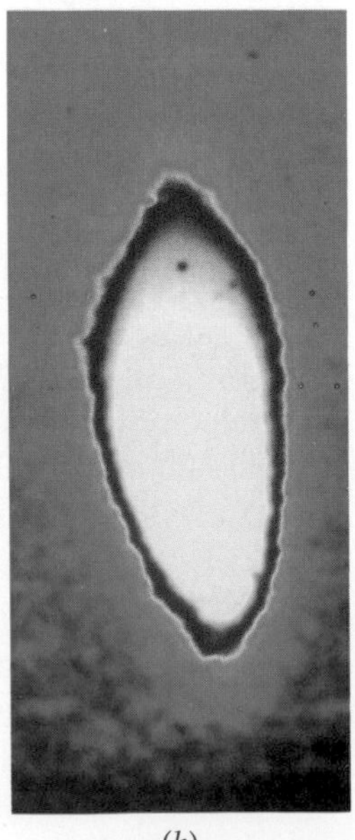

(*b*)

(*a*) The two-mile long linear accelerator at Stanford University is used to accelerate electrons and positrons in a straight line to nearly the speed of light.
(*b*) Cross section of the electron beam as shown on a television screen monitor.

Example 2-12

A car traveling 100 km/h (about 62 mi/h) crashes into a concrete wall, which does not move. How long does it take the car to come to rest, and what is its acceleration?

In this example, it is not accurate to treat the car as a particle because different parts of it have different accelerations. Moreover, the accelerations are not constant. Nevertheless, let us assume that the crashing of the car can be described by the constant acceleration of a point particle. We need more information in order to find either the stopping time or the acceleration. The information missing is the stopping distance. We can estimate this from our practical knowledge. The center of the car certainly moves less than half the length of the car. (Otherwise the car would be flattened.) A reasonable estimate for the stopping distance is probably between 0.5 and 1.0 m. Let us use 0.75 m as our estimate. We can then find the time it takes the car to stop from $\Delta x = v_{av}\,\Delta t$, with $v_{av} = \frac{1}{2}v_0 = 50$ km/h $= 14$ m/s. (Since we are making estimates, two significant figures are sufficient.) Then

$$\Delta t = \frac{\Delta x}{v_{av}} = \frac{0.75\ \text{m}}{14\ \text{m/s}} = 0.054\ \text{s}$$

Since the car is brought from $v_0 = 100$ km/h $= 28$ m/s to rest in this time, the acceleration is

$$a = \frac{\Delta v}{\Delta t} = \frac{0 - 28\ \text{m/s}}{0.054\ \text{s}} = -520\ \text{m/s}^2$$

To get a feeling for the magnitude of this acceleration, we note that it is more than 50 times the acceleration due to gravity.

Example 2-13

A car is traveling 80 km/h in a school zone. A police car starts from rest just as the speeder passes it and accelerates at a constant rate of 8 km/h·s. (*a*) When does the police car catch the speeding car? (*b*) How fast is the police car traveling when it catches the speeder?

This problem is more difficult because there are two moving objects. We take the origin to be at the original position of both cars, with the positive direction in the direction of motion, and define $t = 0$ to be when the speeder passes the police car.

(*a*) Since the car is traveling with constant velocity, its position x_c is given by Equation 2-12*b*, with $x_0 = 0$ and $a = 0$:

$$x_c = v_0 t = (80\text{ km/h})t$$

The position of x_p of the police car is given by

$$x_p = \tfrac{1}{2}at^2 = \tfrac{1}{2}(8\text{ km/h·s})t^2$$

We find the time when the two cars are at the same position by setting $x_c = x_p$ and solving for t:

$$(80\text{ km/h})t = \tfrac{1}{2}(8\text{ km/h·s})t^2$$

or

$$t(4t\text{ km/h·s} - 80\text{ km/h}) = 0$$

The two solutions are

$$t = 0$$

corresponding to the initial conditions and

$$t = \frac{80\text{ km/h}}{4\text{ km/h·s}} = 20\text{ s}$$

The police car thus catches the speeder at time $t = 20$ s.

(*b*) The velocity of the police car is given by Equation 2-9, with $v_0 = 0$:

$$v_p = at = (8\text{ km/h·s})t$$

At $t = 20$ s, the velocity of the police car is

$$v_p = (8\text{ km/h·s})(20\text{ s}) = 160\text{ km/h}$$

At this time, the speed of the police car is twice that of the speeder. This must be true because the average velocity of the police car is half its final velocity, and since both cars cover the same distance in the same time, they must have equal average velocities. Figure 2-8 shows x-versus-t curves for the two cars.

Questions

9. Two boys standing on a bridge throw two rocks straight down into the water below. They throw the rocks at the same time, but one hits the water before the other. How can this be if the rocks have the same acceleration?

10. A ball is thrown straight up. What is its velocity at the top of its flight? What is its acceleration at that point?

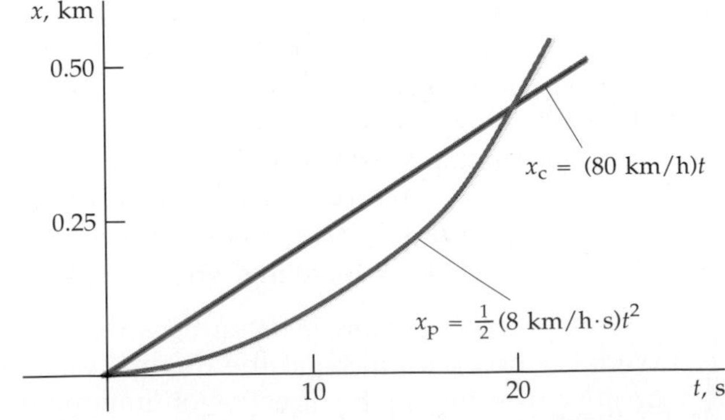

Figure 2-8 Graphs of x versus t for the speeding car (x_c) and the police car (x_p) for Example 2-13. The curve for the police car has a zero slope at $t = 0$ because the police car starts from rest. It catches the speeding car at $t = 20$ s. Note that at that time the slope of $x_p(t)$ is greater than that of $x_c(t)$. At $t = 20$ s, the police car is traveling twice as fast as the speeding car.

2-5 Integration

We have learned how to obtain the velocity and acceleration functions from a given position function by differentiation. The inverse problem is to find the position function x given the velocity v or the acceleration a. To do this we use a procedure called **integration.** Consider, for example, the problem of finding the velocity and position from a given acceleration. If we know the acceleration as a function of time, we can find the velocity by finding the function $v(t)$ whose derivative is the acceleration. For example, if the acceleration is constant,

$$\frac{dv}{dt} = a$$

the velocity is the function of time that when differentiated equals this constant. One such function is

$$v = at$$

This is not the most general expression for v that will satisfy the relation $dv/dt = a$, however. In particular, we can add any constant to at without changing the value of its derivative with respect to time. Calling this constant v_0, we have

$$v = at + v_0$$

The constant v_0 is the initial velocity. The position function x is that function whose derivative is the velocity:

$$\frac{dx}{dt} = v = v_0 + at$$

We can treat each term separately. The function whose derivative is a constant v_0 is v_0t plus any constant. The function whose derivative is at is $\frac{1}{2}at^2$ plus any constant. (This result is easily verified by differentiation.) Combining these results and writing x_0 for the combined arbitrary constants, we have for the position function

$$x = x_0 + v_0t + \tfrac{1}{2}at^2$$

Whenever we find a function from its derivative, we must include an arbitrary constant in the general function. Since we go through the integration process twice to find $x(t)$ from the acceleration, two constants arise. These constants are usually determined from the velocity and position at some given time, which is usually chosen to be $t = 0$. They are therefore called the **initial conditions.** The problem "given $a(t)$, find $x(t)$" is called the **initial-value problem.** The solution depends on the form of the function $a(t)$ and on the values of v and x at some particular time. This problem is particularly important in physics because the acceleration of a particle is determined by the forces acting on it. Thus, if we know the forces acting on a particle and the position and velocity of the particle at some particular time, we can find its position uniquely at all other times.

The problem of integration is related to the problem of finding the area under a curve. Consider the case of a constant velocity v_0. The change in position Δx during some time interval Δt is just the velocity times the time interval:

$$\Delta x = v_0\,\Delta t$$

This is the area under the v-versus-t curve, as can be seen in Figure 2-9. This

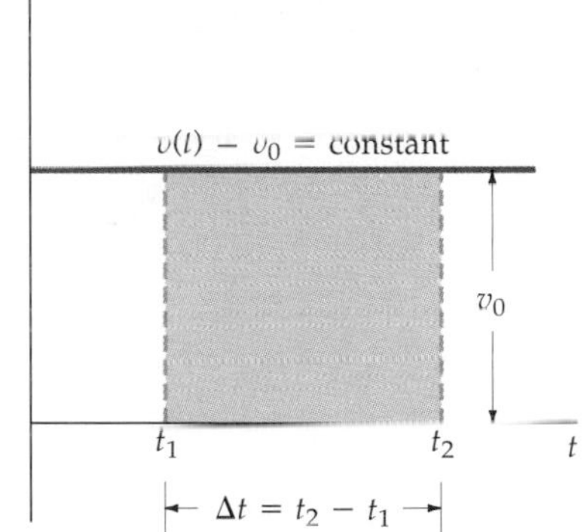

Figure 2-9 The displacement for the interval Δt equals the area under the velocity-versus-time curve for that interval. For $v(t) = v_0 =$ constant, the displacement equals the area of the rectangle shown.

graphical interpretation of the displacement as the area under the v-versus-t curve is true both for cases in which the velocity is constant and for cases in which it varies, as in Figure 2-10. In this case, the area under the curve can be approximated by first dividing the time interval into a number of small intervals, Δt_1, Δt_2, and so on, and drawing a set of rectangular

Figure 2-10 Graph of a general $v(t)$-versus-t curve. The displacement for the interval Δt_i is approximately $v_i\ \Delta t_i$, which is indicated by the shaded rectangular area. The total displacement from t_1 to t_2 is the area under the curve for this interval, which can be approximated by summing the areas of the rectangles.

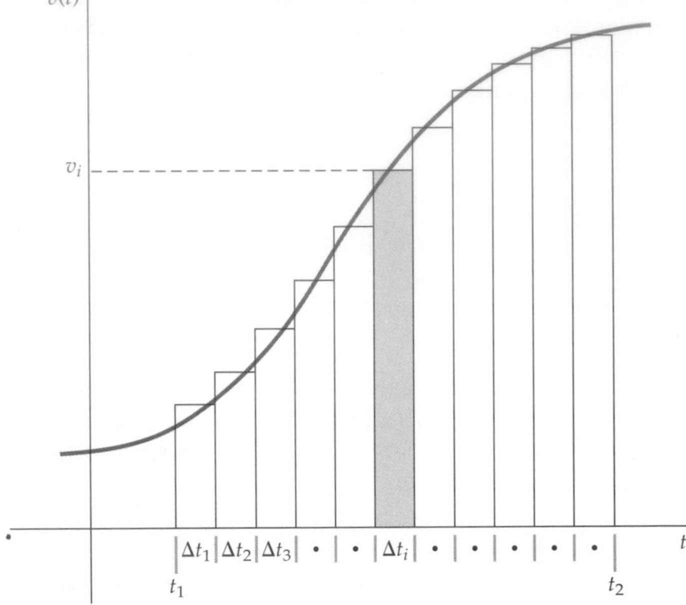

areas. The area of the shaded rectangle is $v_i\ \Delta t_i$, which is approximately equal to the displacement Δx_i of the particle during the interval Δt_i. We can make the approximation as good as we wish by taking enough rectangles and making each Δt very small. The sum of the rectangular areas is therefore approximately the sum of the displacements during the time intervals and is approximately equal to the total displacement from time t_1 to t_2.

Mathematically, we write this as

$$\Delta x \approx \sum_i v_i\ \Delta t_i$$

where the Greek letter Σ (capital sigma) stands for "sum." For the limit of smaller and smaller time intervals, this sum equals the area under the curve, which equals the displacement. This limit is called an **integral** and is written

$$\Delta x = \lim_{\Delta t_i \to 0} \sum_i v_i\ \Delta t_i = \int_{t_1}^{t_2} v\ dt \qquad \text{2-14}$$

It is helpful to think of the integral sign ∫ as an elongated S indicating a sum. The limits t_1 and t_2 indicate the initial and final values of the variable t.

The change in velocity for some time interval can similarly be interpreted as the area under the a-versus-t curve for that interval. This is written

$$\Delta v = \lim_{\Delta t_i \to 0} \sum_i a_i\ \Delta t_i = \int_{t_1}^{t_2} a\ dt \qquad \text{2-15}$$

The average velocity has a simple geometric interpretation in terms of the area under a curve. Consider the v-versus-t curve in Figure 2-11. The displacement Δx during the time interval $\Delta t = t_2 - t_1$ is indicated by the shaded area. By the definition of the average velocity for this interval (Equation 2-2), the displacement is the product of v_{av} and Δt:

$$\Delta x = v_{av}\,\Delta t$$

The average velocity is indicated in Figure 2-11 by the horizontal line, which is drawn such that the area under it from t_1 to t_2 equals the area under the actual v-versus-t curve.

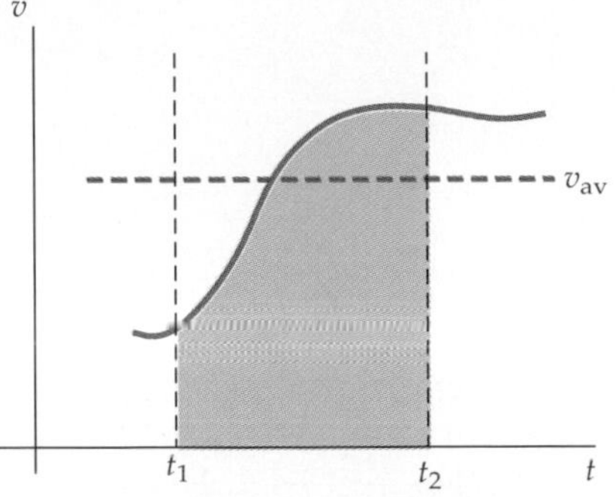

Figure 2-11 Geometric interpretation of average velocity. By definition, $v_{av} = \Delta x/\Delta t$. Thus, the rectangular area $v_{av}(t_2 - t_1)$ must equal the displacement in the time interval $t_2 - t_1$. It follows that the rectangular area $v_{av}(t_2 - t_1)$ and the shaded area under the curve must be equal.

Example 2-14

A particle moves from rest with constant acceleration a. Show by calculating the area under the v-versus-t curve that the average velocity for some time interval beginning at $t = 0$ equals half the final velocity.

Figure 2-12 shows the v-versus-t curve for this problem. The displacement from $t = 0$ to some final time t is indicated by the shaded area. The area of this triangle is $\frac{1}{2}v_f\,\Delta t$, where v_f is the final velocity. The displacement is

$$\Delta x = \tfrac{1}{2}v_f\,\Delta t = v_{av}\,\Delta t$$

The average velocity is thus half the final velocity for this time interval.

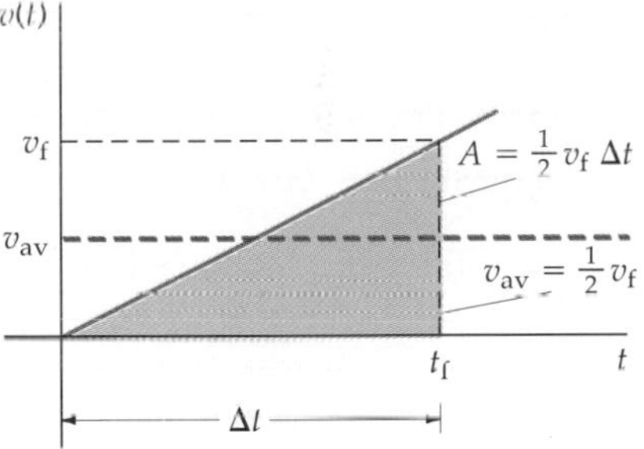

Figure 2-12 Proof that $v_{av} = \frac{1}{2}v_f$ for a particle starting from rest and moving with constant acceleration. The displacement equals the area under the curve, which equals $\frac{1}{2}v_f\,\Delta t$. It also equals $v_{av}\,\Delta t$, so $v_{av} = \frac{1}{2}v_f$.

Summary

1. Average velocity is the ratio of the displacement Δx to the time interval Δt:

$$v_{av} = \frac{\Delta x}{\Delta t}$$

2. The instantaneous velocity v is the limit of this ratio as the time interval approaches zero. This is the derivative of x with respect to t:

$$v = \lim_{\Delta t \to 0} \frac{\Delta x}{\Delta t} = \frac{dx}{dt}$$

The instantaneous velocity is represented graphically as the slope of the x-versus-t curve. In one dimension, both average and instantaneous velocity can be either positive or negative. The magnitude of the instantaneous velocity is called the speed.

3. The average acceleration is the ratio of the change in velocity Δv to the time interval Δt:

$$a_{av} = \frac{\Delta v}{\Delta t}$$

The instantaneous acceleration is the limit of this ratio as the time interval approaches zero. Instantaneous acceleration is the derivative of v with respect to t, which is the second derivative of x with respect to t:

$$a = \frac{dv}{dt} = \frac{d^2x}{dt^2}$$

The instantaneous acceleration is represented graphically as the slope of the v-versus-t curve.

4. In the special case of constant acceleration, the following formulas hold:

$$v = v_0 + at$$

$$\Delta x = v_{av}t = \tfrac{1}{2}(v_0 + v)t$$

$$\Delta x = x - x_0 = v_0 t + \tfrac{1}{2}at^2$$

$$v^2 = v_0^2 + 2a\,\Delta x$$

A common example of motion with constant acceleration is the motion of an object near the surface of the earth in free fall under the influence of gravity. In this case, the acceleration of the body is directed downward and has the magnitude $g = 9.81\ \text{m/s}^2 = 32.2\ \text{ft/s}^2$.

5. The displacement is represented graphically as the area under the v-versus-t curve. This area is the integral of v over time from some initial time t_1 to some final time t_2 and is written

$$\Delta x = \lim_{\Delta x_i \to 0} \sum_i v_i\,\Delta t_i = \int_{t_1}^{t_2} v\,dt$$

Similarly, the change in velocity for some time is represented graphically as the area under the a-versus-t curve.

Suggestions for Further Reading

Drake, Stillman: "Galileo's Discovery of the Law of Free Fall," *Scientific American*, May 1973, p. 84.

The author's reexamination of unpublished manuscripts clarifies the development of Galileo's ideas about motion.

Magie, W. F.: *A Source Book in Physics*, McGraw-Hill, New York and London, 1935.

This useful book contains short life histories and extracts from the important works of the great physicists from the time of Galileo to the year 1900.

U.S. Department of Transportation, National Highway Traffic Safety Administration: "Acceleration and Passing Ability: A Comparison of Acceleration and Passing Ability for 1974 Passenger Cars and Motorcycles," *Consumer Aid Series*, vol. 1, part 3, 1974.

A tabulation of the time and distance required for vehicles sold in the United States to pass a 55-foot-long truck.

U.S. Department of Transportation, National Highway Traffic Safety Administration: "Brakes: A Comparison of Braking Performance for 1974 Passenger Cars and Motorcycles," *Consumer Aid Series*, vol. 4, part 1, 1974.

A tabulation of stopping distance from 60 mi/h, under specified conditions, for vehicles sold in the United States.

Review

A. Objectives: After studying this chapter you should:

1. Be able to define displacement, velocity, and acceleration.

2. Be able to distinguish between velocity and speed.

3. Be able to calculate the instantaneous velocity from a graph of position versus time.

4. Be able to state the important equations relating displacement, velocity, acceleration, and time that apply when the acceleration is constant, and use them to work problems.

5. Be able to calculate the displacement of a particle from the v-versus-t curve and the change in the velocity of a particle from the a-versus-t curve by finding the areas under the appropriate curves.

B. Define, explain, or otherwise identify:

Particle
Speed
Displacement
Average velocity
Instantaneous velocity
Slope
Derivative
Average acceleration
Instantaneous acceleration
Initial-value problem
Integral

C. True or false: If the statement is true, explain why it is true. If it is false, give a counterexample.

1. The equation $\Delta x = v_0 t + \frac{1}{2}at^2$ is true for all motion in one dimension.

2. If the velocity is zero at an instant, the acceleration must also be zero at that instant.

3. If the acceleration is zero, the body cannot be moving.

4. If the acceleration is zero, the x-versus-t curve is a straight line.

5. The equation $\Delta x = v_{av}\,\Delta t$ holds for all motion in one dimension.

Problems

Unless instructed otherwise, use $g = 9.81\ m/s^2 = 32.2\ ft/s^2$ for the acceleration due to gravity.

Level I

2-1 Speed, Displacement, and Velocity

1. A runner runs 2 km in 5 min and then takes 10 min to walk back to the starting point. (*a*) What is the average velocity for the first 5 min? (*b*) What is the average velocity for the time spent walking? (*c*) What is the average velocity for the total trip? (*d*) What is the average speed for the total trip?

2. Work Problem 1 if the runner walked only halfway back in 10 min and then stopped.

3. A particle is at $x = +5$ m at $t = 0$, $x = -7$ m at $t = 6$ s, and $x = +2$ m at $t = 10$ s. Find the average velocity of the particle during the intervals (*a*) $t = 0$ to $t = 6$ s, (*b*) $t = 6$ s to $t = 10$ s, and (*c*) $t = 0$ to $t = 10$ s.

4. A driver begins a 200-km trip at noon. (*a*) She drives nonstop and arrives at her destination at 5:30 P.M. Calculate her average velocity for the trip. (*b*) She drives for 3 h, rests for $\frac{1}{2}$ h, and continues driving, arriving at 5:30 P.M. Calculate her average velocity. (*c*) After resting 2 h, she drives back home, taking 6 h for the return trip. What is her average velocity for the total round trip? (*d*) What is her displacement?

5. A car travels in a straight line with an average velocity of 80 km/h for 2.5 h and then with an average velocity of 40 km/h for 1.5 h. (*a*) What is the total displacement for the 4-h trip? (*b*) What is the average velocity for the total trip?

6. As you drive into Kentucky, you notice that the mile marker on the interstate is 325. You drive straight through to mile marker 0 on the other side of the state in 6 h. You are then struck by a desire for fried chicken, so you turn around and drive back 25 mi for a snack. This return takes 30 min. (*a*) What was your average speed in miles per hour for the 350-mi trip? (*b*) What was your average velocity for the trip? (*c*) What was your average velocity for the last 25 mi?

7. (*a*) How long does it take a supersonic jet flying at 2.4 times the speed of sound to fly across the Atlantic, which is about 5500 km wide? Take the speed of sound to be 350 m/s. (*b*) How long does it take a subsonic jet flying at 0.9 times the speed of sound to make the same trip? Assuming it takes 2 h at each end of the trip to get to or from the airport and to check or pick up your baggage, what is your average speed from your home to your final destination for (*c*) the supersonic jet and (*d*) the subsonic jet?

8. As you drive along the freeway, your speedometer quits working. You measure your speed by measuring the time it takes to travel between mile markers. (*a*) How many seconds should elapse between mile markers if your average speed is 55 mi/h? (*b*) What is your average speed in miles per hour if the time between mile markers is 45 s?

9. Light travels at a speed of $c = 3 \times 10^8$ m/s. (*a*) How long does it take light to travel from the sun to the earth, a distance of 1.5×10^{11} m? (*b*) How long does it take light to travel from the moon to the earth, a distance of 3.84×10^8 m? (*c*) A lightyear is a unit of distance equal to that traveled by light in 1 year. Find the equivalent distance of 1 lightyear in kilometers and in miles.

10. The nearest star, Proxima Centauri, is 4.1×10^{13} km away. (*a*) How much time is required for a light signal sent from the earth to reach Proxima Centauri? (*b*) How many years would it take a spaceship traveling with a speed of $10^{-4}\,c$ to reach the nearest star? (See Problem 9.)

11. A car making a 100-km journey covers the first 50 km at 40 km/h. How fast must it cover the second 50 km to average 50 km/h?

2-2 Instantaneous Velocity

12. For each of the four graphs of x versus t in Figure 2-13, indicate whether (*a*) the velocity at time t_2 is greater than, less than, or equal to the velocity at time t_1 and (*b*) the speed at time t_2 is greater than, less than, or equal to the speed at time t_1.

Figure 2-13 Graphs of x versus t for Problem 12.

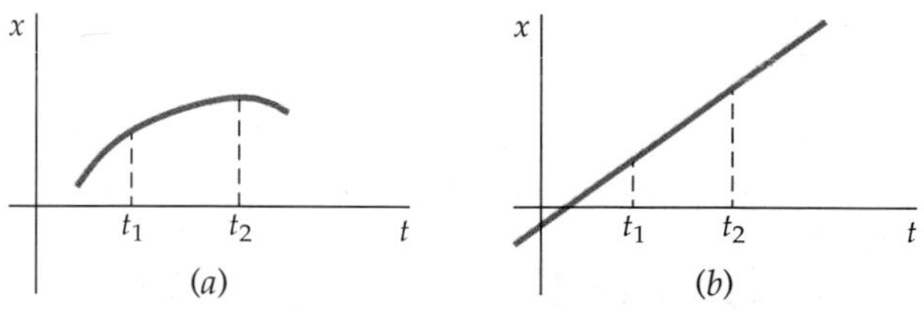

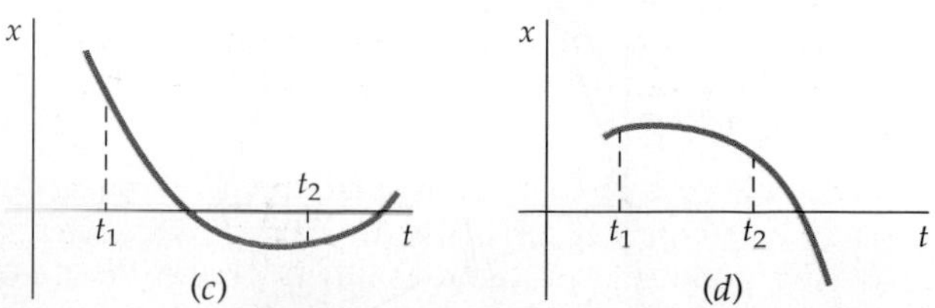

13. From the graph of x versus t in Figure 2-14, (*a*) find the average velocity between the times $t = 0$ and $t = 2$ s. (*b*) Find the instantaneous velocity at $t = 2$ s by measuring the slope of the tangent line indicated.

Figure 2-14 Graph of x versus t with a tangent line drawn at $t = 2$ s for Problem 13.

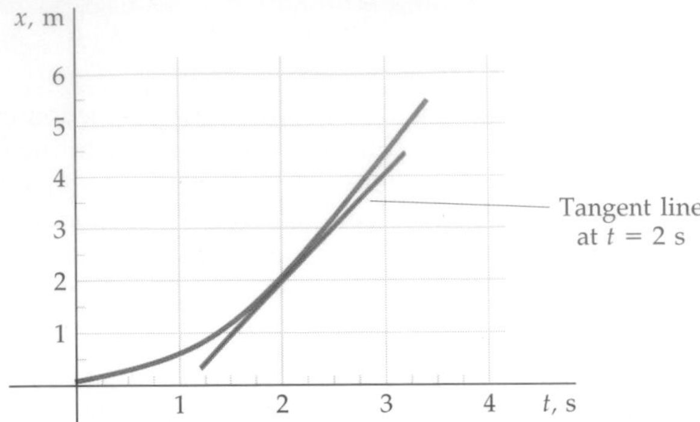

14. For the graph of x versus t in Figure 2-15, find the average velocity for the time intervals $\Delta t = t_2 - 0.75$ s when t_2 is 1.75, 1.5, 1.25, and 1.0 s. What is the instantaneous velocity at $t = 0.75$ s?

Figure 2-15 Graph of x versus t for Problem 14.

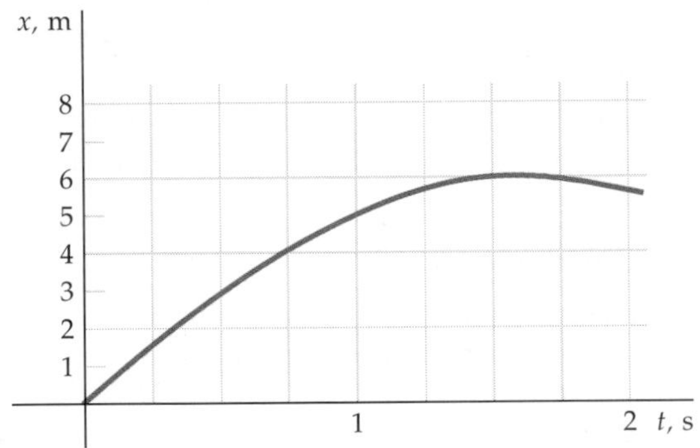

15. For the graph of x versus t shown in Figure 2-16, (*a*) find the average velocity for the interval $t = 1$ s to $t = 5$ s. (*b*) Find the instantaneous velocity at $t = 4$ s. (*c*) At what time is the velocity of the particle zero?

Figure 2-16 Graph of x versus t with a tangent line drawn at $t = 4$ s for Problem 15.

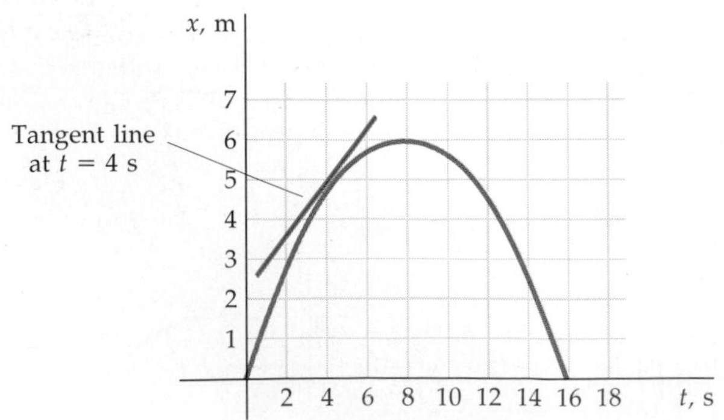

2-3 Acceleration

16. A car is traveling at 45 km/h at time $t = 0$. It accelerates at a constant rate of 10 km/h·s. (*a*) How fast is the car going at $t = 1$ s? At $t = 2$ s? (*b*) What is its speed at a general time t?

17. At $t = 5$ s an object is traveling at 5 m/s. At $t = 8$ s its velocity is -1 m/s. Find the average acceleration for this interval.

18. State whether the acceleration is positive, negative, or zero for each of the position functions $x(t)$ in Figure 2-17.

Figure 2-17 Graphs of x versus t for Problem 18.

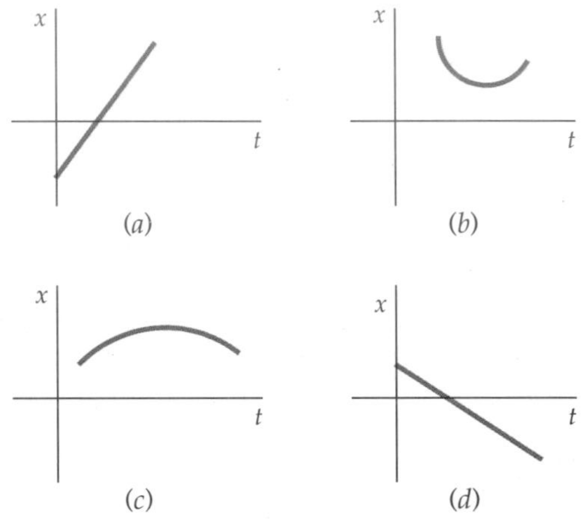

2-4 Motion with Constant Acceleration

19. A car accelerates from rest at a constant rate of 8 m/s^2. (*a*) How fast is it going after 10 s? (*b*) How far has it gone after 10 s? (*c*) What is its average velocity for the interval $t = 0$ to $t = 10$ s?

20. An object with initial velocity of 5 m/s has a constant acceleration of 2 m/s^2. When its speed is 15 m/s, how far has it traveled?

21. An object with constant acceleration has velocity $v = 10$ m/s when it is at $x = 6$ m and $v = 15$ m/s when it is at $x = 10$ m. What is its acceleration?

22. An object has constant acceleration $a = 4$ m/s^2. Its velocity is 1 m/s at $t = 0$, when it is at $x = 7$ m. How fast is it moving when it is at $x = 8$ m? At what time is this?

23. How long does it take for a particle to travel 100 m if it begins from rest and accelerates at 10 m/s^2? What is the velocity when it has traveled 100 m? What is the average velocity for this time?

24. A ball is thrown upward with initial velocity of 20 m/s. (*a*) How long is the ball in the air? (*b*) What is the greatest height reached by the ball? (*c*) When is the ball 15 m above the ground?

25. Suppose a rifle fires a bullet straight up with a muzzle speed of 300 m/s. Neglecting air friction, how high will the bullet rise?

26. The minimum distance for a controlled stop, starting from 98 km/h, for a certain car is 50 m for level braking.

Find the acceleration, assuming it to be constant, and express your answer as a fraction of the free-fall acceleration due to gravity. How long does it take to stop?

27. A world-land-speed car can decelerate at about 1*g*. (That is, the magnitude of *a* is *g*.) (*a*) How long does it take the car to come to a stop from a record-setting run of 885 km/h? (*b*) How far does it travel while stopping?

2-5 Integration

28. The velocity of a particle is given by $v = 6t + 3$, where t is in seconds and v is in meters per second. (*a*) Sketch $v(t)$ versus t, and find the area under the curve for the interval $t = 0$ to $t = 5$ s. (*b*) Find the general position function $x(t)$. Use it to calculate the displacement during the interval $t = 0$ to $t = 5$ s.

29. The velocity of a particle in meters per second is given by $v = 7 - 4t$, where t is in seconds. (*a*) Sketch $v(t)$ versus t, and find the area between the curve and the t axis from $t = 2$ s to $t = 6$ s. (*b*) Find the position function $x(t)$ by integration, and use it to find the displacement during the interval $t = 2$ s to $t = 6$ s. (*c*) What is the average velocity for this interval?

30. Figure 2-18 shows the velocity of a particle versus time. (*a*) What is the magnitude in meters of the area of the rectangle indicated? (*b*) Find the approximate displacement of the particle for the 1-s intervals beginning at $t = 1$ s and $t = 2$ s. (*c*) What is the approximate average velocity for the interval from $t = 1$ s to $t = 3$ s?

Figure 2-18 Graph of v versus t for Problems 30 and 41.

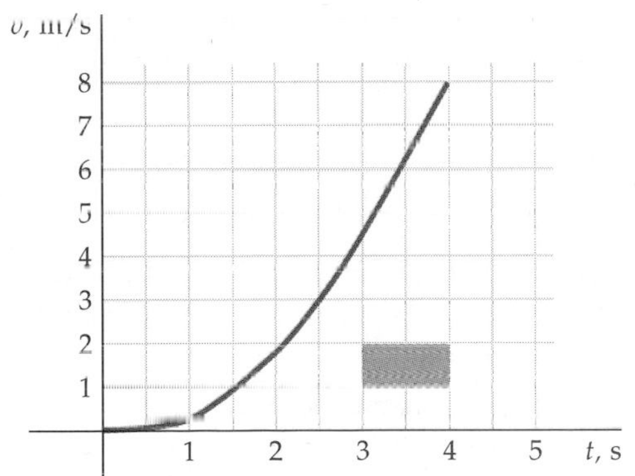

Level II

31. Figure 2-19 shows the position of a particle versus time. Find the average velocity for the time intervals *a*, *b*, *c*, and *d* indicated in the figure.

32. The position of a particle depends on the time according to $x = (1\ \text{m/s}^2)t^2 - (5\ \text{m/s})t + 1$ m. (*a*) Find the displacement and average velocity for the interval $t = 3$ s to $t = 4$ s. (*b*) Find a general formula for the displacement for the time interval from t to $t + \Delta t$. (*c*) Find the instantaneous velocity for any time t.

33. The height of a certain projectile is related to time by $y = -5(t - 5)^2 + 125$, where y is in meters and t is in seconds. (*a*) Sketch y versus t for $t = 0$ to $t = 10$ s. (*b*) Find the average velocity for each of the 1-s time intervals between integral time values from $t = 0$ to $t = 10$ s. Sketch v_{av} versus t. (*c*) Find the instantaneous velocity as a function of time.

34. A particle moves with velocity given by $v = 8t - 7$, where v is in meters per second and t is in seconds. (*a*) Find the average acceleration for the 1-s intervals beginning at $t = 3$ s and $t = 4$ s. (*b*) Sketch v versus t. What is the instantaneous acceleration at any time?

35. The position of a particle versus time is given by

t, s	0	1	2	3	4	5	6	7	8	9	10	11
x, m	0	5	15	45	65	70	60	−30	−50	−50	−55	−55

Plot x versus t, and draw a smooth curve $x(t)$. Indicate the times or time intervals for which (*a*) the speed is greatest, (*b*) the speed is least, (*c*) the speed is zero, (*d*) the speed is constant, (*e*) the acceleration is positive, and (*f*) the acceleration is negative.

36. The position of an object is related to time by $x = At^2 - Bt + C$, where $A = 8\ \text{m/s}^2$, $B = 6$ m/s, and $C = 4$ m. Find the instantaneous velocity and acceleration as functions of time.

Figure 2-19 Graph of x versus t for Problem 31.

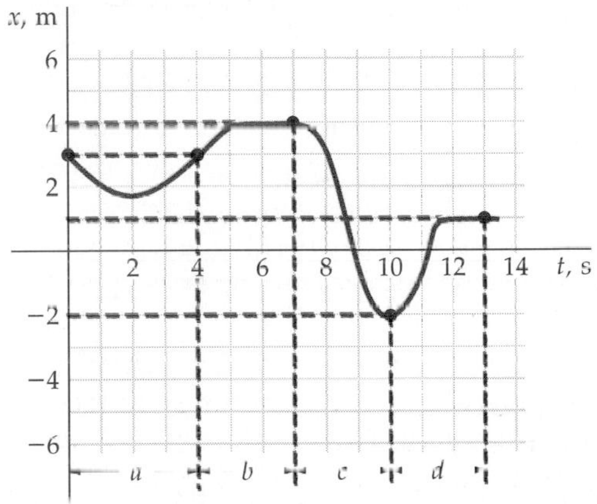

37. A ball is dropped from a height of 3 m and rebounds from the floor to a height of 2 m. (*a*) What is the velocity of the ball just as it reaches the floor? (*b*) What is the velocity just as it leaves the floor? (*c*) If it is in contact with the floor for 0.02 s, what are the magnitude and direction of its average acceleration during this interval?

38. The velocity of a particle in meters per second is given by $v = 7t + 5$, where t is in seconds. Find the general position function $x(t)$.

39. The acceleration of a certain rocket is given by $a = Ct$, where C is a constant. (*a*) Find the general position function $x(t)$. (*b*) Find the position and velocity at $t = 5$ s if $x = 0$ and $v = 0$ at $t = 0$ and $C = 3\ \text{m/s}^3$.

40. Figure 2-20 shows the acceleration of a particle versus time. (*a*) What is the magnitude of the area of the rectangle indicated? (*b*) The particle starts from rest at $t = 0$. Find the velocity at $t = 1$, 2, and 3 s by counting the rectangles under the curve. (*c*) Sketch the curve $v(t)$ versus t from your results for part (*b*), and estimate how far the particle travels in the interval $t = 0$ to $t = 3$ s.

41. The equation of the curve shown in Figure 2-18 is $v = 0.5t^2$ m/s. Find the displacement of the particle for the interval $t = 1$ s to $t = 3$ s by integration, and compare this answer with your answer for Problem 30. Is the average velocity equal to the mean of the initial and final velocities for this case?

42. A car traveling at a constant speed of 20 m/s passes an intersection at time $t = 0$, and 5 s later another car passes the same intersection traveling 30 m/s in the same direction. (*a*) Sketch the position functions $x_1(t)$ and $x_2(t)$ for the two cars. (*b*) Find when the second car overtakes the first. (*c*) How far from the intersection have the cars traveled when this happens?

43. Hare and Tortoise begin a 10-km race at time $t = 0$. Hare runs at 4 m/s and quickly outdistances Tortoise, who runs at 1 m/s (about 10 times faster than a tortoise can actually run but convenient for this problem). After running 5 min, Hare stops and falls asleep. His nap lasts 135 min. He awakes and again runs at 4 m/s but loses the race. Plot x-versus-t curves for Hare and Tortoise on the same axes. At what time does Tortoise pass Hare? How far behind is Hare when Tortoise crosses the finish line? How long can Hare nap and still win the race?

44. A particle moves with a constant acceleration of 3 m/s^2. At $t = 4$ s, it is at $x = 100$ m. At $t = 6$ s, it has a velocity $v = 15$ m/s. Find its position at $t = 6$ s.

45. Figure 2-21 shows the position of a car plotted as function of time. At which of the times t_0 to t_7 is (*a*) the velocity negative, (*b*) the velocity positive, (*c*) the velocity zero, (*d*) the acceleration negative, (*e*) the acceleration positive, and (*f*) the acceleration zero?

46. It has been found that galaxies are moving away from the earth at a speed that is proportional to their distance from the earth. This discovery is known as Hubble's law. The speed of a galaxy a distance r away is given by $v = Hr$, where H is Hubble's constant, which equals 1.58×10^{-18} s^{-1}. What is the speed of a galaxy (*a*) 5×10^{22} m from earth and (*b*) 2×10^{25} m from earth? (*c*) If each of these galaxies has traveled with constant speed, how long ago were they both located at the same place as the earth?

47. For each of the graphs in Figure 2-22, indicate (*a*) at what times the acceleration of the object is positive, negative, and zero; (*b*) at what times the acceleration is constant; and (*c*) at what times the instantaneous velocity is zero.

48. In the Blackhawk landslide in California, a mass of rock and mud fell 460 m down a mountain and then traveled 8 km across a level plain on a cushion of compressed air. Assume that the mud dropped with the free-fall acceleration due to gravity and then slid horizontally with constant deceleration. (*a*) How long did the mud take to drop the 460 m? (*b*) How fast was it traveling when it reached the bottom? (*c*) How long did the mud take to slide the 8 km horizontally?

Figure 2-20 Graph of a versus t for Problem 40.

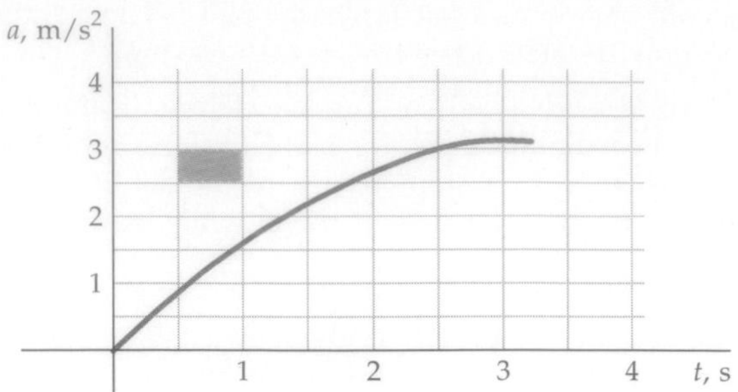

Figure 2-21 Graph of the position of a car versus time for Problem 45.

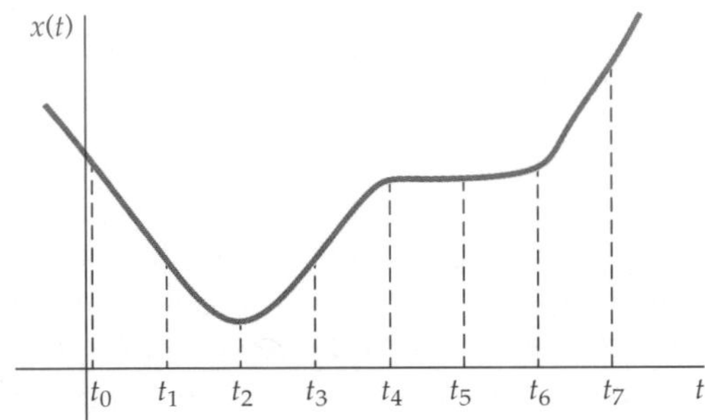

Figure 2-22 Graphs of (*a*) v versus t and (*b*) x versus t for Problem 47.

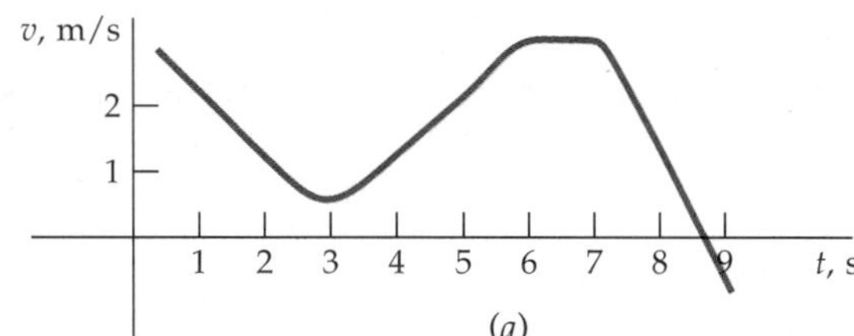

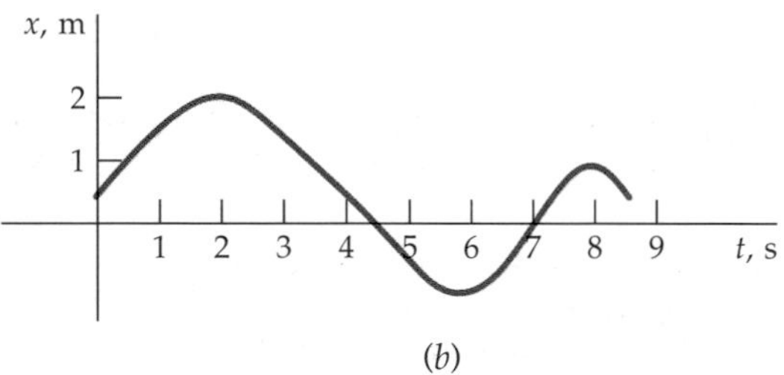

49. A police car chases a speeder who is traveling at 125 km/h. The maximum speed of the police car is 190 km/h. The police car travels from rest with a constant acceleration of 8 km/h·s until its speed reaches 190 km/h. It then moves at constant speed. (*a*) When does the police car catch the speeder if it starts just as the speeder passes? (*b*) How far has each car traveled? (*c*) Sketch $x(t)$ for each car.

50. When the police car of Problem 49 is traveling at 190 km/h and is 100 m behind the speeder (traveling at 125 km/h), the speeder sees the police car and slams on his brakes, locking the wheels. (*a*) Assuming that each car can brake at 6 m/s^2 and that the driver of the police car brakes as soon as she sees the brake lights of the speeder, that is, with no reaction time, show that the cars collide. (*b*) At what time after the speeder applies his brakes do they collide? (*c*) The interval between the time the police officer sees the speeder's brake lights and the time she puts on her own brakes is called her *reaction time T*. Discuss how reaction time affects this problem.

51. A glider moves along an air track with constant acceleration a. It is projected from the start of the track ($x = 0$) with an initial velocity v_0. At time $t = 8$ s, it is at $x =$ 100 cm and is moving along the track at velocity $v =$ -15 cm/s. Find the initial speed v_0 and the acceleration a.

52. A load of bricks is being lifted by a crane at the steady velocity of 5 m/s when one brick falls off 6 m above the ground. Describe the motion of the free brick by sketching $x(t)$. (*a*) What is the greatest height the brick reaches above the ground? (*b*) How long does it take to reach the ground? (*c*) What is its speed just before it hits the ground?

53. A student running at 9 m/s is 40 m behind Eddy when Eddy starts from rest on his motor bike with an acceleration of 0.9 m/s^2. (*a*) How long does it take for the student to catch Eddy? (*b*) What is the time interval during which the student remains ahead of Eddy?

54. An automobile accelerates from rest at 2 m/s^2 for 20 s. The speed is then held constant for 20 s, after which there is an acceleration of -3 m/s^2 until the automobile stops. What is the total distance traveled?

55. A flower pot falls from a ledge of an apartment building. A person in an apartment below, who just happens to have a stopwatch handy, notices that it takes 0.2 s for the pot to fall past his window, which is 4 m high. How far above the top of the window is the ledge from which the pot fell?

56. A pilot jumps out of a burning plane without a parachute. She reaches a speed of 120 km/h prior to impact. The pilot can survive a $35g$ deceleration. She falls onto a stack of loose hay. Assuming uniform deceleration to the bottom of the haystack, how high a stack of hay is needed for the pilot to barely survive?

57. A bolt comes loose from the bottom of an elevator that is moving upward at a speed of 6 m/s. The bolt reaches the bottom of the elevator shaft in 3 s. (*a*) How high up was the elevator when the bolt came loose? (*b*) What is the speed of the bolt when it hits the bottom of the shaft?

58. An archer fires an arrow that produces a loud noise when it strikes its target. The average velocity of the arrow is 150 m/s. The archer hears the impact exactly 1 s after firing the arrow. If the velocity of sound is 340 m/s, how far away is the target?

Level III

59. Sketch a single v-versus-t curve in which there are points or segments for which (*a*) the acceleration is zero and constant while the velocity is not zero; (*b*) the acceleration is zero but not constant; (*c*) the velocity and acceleration are both positive; (*d*) the velocity and acceleration are both negative; (*e*) the velocity is positive and the acceleration is negative; (*f*) the velocity is negative and the acceleration is positive; and (*g*) the velocity is zero, but the acceleration is not.

60. When a car traveling at speed v_1 rounds a corner, the driver sees another car traveling at a slower speed v_2 a distance d ahead. (*a*) If the maximum acceleration the driver's brakes can provide is a, show that the distance d must be greater than $(v_1 - v_2)^2/2a$ if a collision is to be avoided. (*b*) Evaluate this distance for $v_1 = 90$ km/h, $v_2 =$ 45 km/h, and $a = 6$ m/s^2. (*c*) Estimate or measure your reaction time, and calculate the effect it would have on the distance found in part (*b*).

61. A passenger is running at his maximum velocity of 8 m/s to catch a train. When he is a distance d from the nearest entry to the train, the train starts from rest with a constant acceleration $a = 1.0$ m/s^2 away from him. (*a*) If $d = 30$ m and the passenger keeps running, will he be able to jump onto the train? (*b*) Sketch the position function $x(t)$ for the train, choosing $x = 0$ at $t = 0$. On the same graph sketch $x(t)$ for the passenger for various initial separation distances d, including $d = 30$ m and the critical separation distance d_c such that he just catches the train. (*c*) For the critical separation distance d_c, what is the speed of the train when the passenger catches it? What is the train's average speed for the time interval from $t = 0$ until he catches it? What is the value of d_c?

62. A train starts from a station with a constant acceleration of 0.40 m/s^2. A passenger arrives at the track 6.0 s after the end of the train left the very same point. What is the slowest constant speed at which she can run and catch the train? Sketch curves for the motion of the passenger and the train as functions of time.

63. Ball A is dropped from the top of a building at the same instant that ball B is thrown vertically upward from the ground. When the balls collide, they are moving in opposite directions, and the speed of A is twice the speed of B. At what fraction of the height of the building does the collision occur?

64. Solve Problem 63 if the collision occurs when the balls are moving in the same direction and the speed of A is 4 times that of B.

65. Figure 2-23 is a graph of v versus t for a particle moving along a straight line. The position of the particle at time $t = 0$ is $x_0 = 5$ m. (*a*) Find x for various times t by counting squares, and sketch x versus t. (*b*) Sketch the acceleration a versus t.

Figure 2-23 Graph of v versus t for Problem 65.

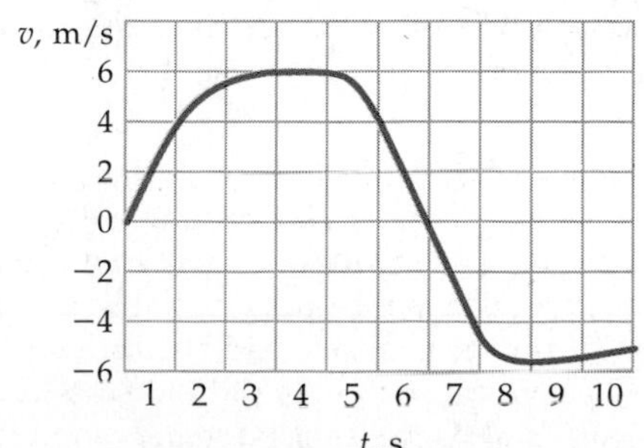

66. Figure 2-24 shows a plot of x versus t for a body moving along a straight line. Sketch rough graphs of v versus t and a versus t for this motion.

Figure 2-24 Graph of x versus t for Problem 66.

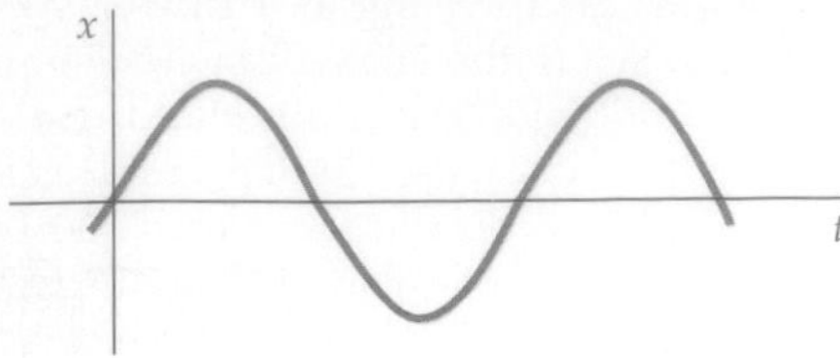

67. The position of a body oscillating on a spring is given by $x = A \sin \omega t$, where A and ω are constants and have the values $A = 5$ cm and $\omega = 0.175\ \mathrm{s}^{-1}$. Sketch x versus t for times from $t = 0$ to $t = 36$ s. (*a*) Measure the slope of your graph at $t = 0$ to find the velocity at this time. (*b*) Calculate the average velocity for a series of intervals beginning at $t = 0$ and ending at $t = 6$, 3, 2, 1, 0.5, and 0.25 s. (*c*) Compute dx/dt and find the velocity at time $t = 0$.

68. A car has a maximum acceleration a, which remains constant to high speeds, and it has a maximum deceleration $2a$. The car has to travel a short distance L, starting and ending at rest, in the minimum time T. (The distance is so short that the car can never reach top speed.) After what fraction of L should the driver move her foot from the gas pedal to the brake, and what fraction of the time for the trip has elapsed at this point?

69. Table 2-2 lists some world track records for short sprints. A simple model of sprinting assumes that a sprinter starts from rest, accelerates with constant acceleration a for a short time T, and then runs with constant speed $v_0 = aT$. According to this model, for times t greater than T, the distance x varies linearly with time. (*a*) Make a graph of distance x versus time t from the data in the table. (*b*) Set up an equation for x versus T according to the simple model described, and show that for $t > T$, x can be written $x = v_0(t - \frac{1}{2}T)$. (*c*) Connect the points on your graph with a straight line, and determine the slope and the intercept of the line with the time axis. From the fact that the slope is v_0 and the intercept is $\frac{1}{2}v_0T$, compute the acceleration a. (*d*) The record for $x = 200$ m is 19.5 s. Discuss the applicability of this simple model to races of 200 m or more.

x		
yd	m	t, s
50		5.1
	50	5.5
60		5.9
	60	6.5
100		9.1
	100	9.9

70. A Sprint ABM missile can accelerate at $100g$. If an ICBM is detected at an altitude of 100 km moving straight down at a constant speed of 3×10^4 km/h and the ABM missile is launched to intercept it, at what time and altitude will the interception take place? *Note:* You can neglect the acceleration of gravity in this problem. Why?

71. The acceleration of a particle falling under the influence of gravity and a resistive force, such as air resistance, is given by

$$a = \frac{dv}{dt} = g - bv$$

where g is the free-fall acceleration due to gravity and b is a constant that depends on the mass and shape of the particle and on the medium. Suppose the particle begins with zero velocity at time $t = 0$. (*a*) Discuss qualitatively how the speed v varies with time from your knowledge of the rate of change dv/dt given by this equation. What is the value of the velocity when the acceleration is zero? This is called the *terminal speed*. (*b*) Sketch the solution $v(t)$ versus t without solving the equation. This can be done as follows: At $t = 0$, v is zero and the slope is g. Sketch a straight-line segment, neglecting any change in slope for a short time interval. At the end of the interval the velocity is not zero, so the slope is less than g. Sketch another straight-line segment with a smaller slope. Continue until the slope is zero and the velocity equals the terminal velocity.

72. Suppose acceleration is a function of x, where $a(x) = 2x\ \mathrm{m/s^2}$. (*a*) If the velocity at $x = 1$ m is zero, what is the speed at $x = 3$ m? (*b*) How long does it take to travel from $x = 1$ m to $x = 3$ m?

73. Suppose that a particle moves in a straight line such that, at any time t, its position, velocity, and acceleration all have the same numerical value. Give the position x as a function of time.

74. A typical automobile has a maximum deceleration of about $7\ \mathrm{m/s^2}$; and the typical reaction time to engage the brakes is 0.50 s. A school board sets the speed limit in a school zone to meet the condition that all cars should be able to stop in a distance of 4 m. (*a*) What maximum speed should be allowed for a typical automobile? (*b*) What fraction of the 4 m is due to the reaction time?

75. An object moving in a straight line doubles its velocity each second for the first 10 s. Let the initial speed be 2 m/s. (*a*) Sketch a smooth function $v(t)$ that gives the velocity. (*b*) What is the average velocity over the first 10 s?

76. Suppose a fast-moving object encounters resistance such that its velocity is reduced by one-half for each second that elapses. Let the initial velocity be 1000 m/s. (*a*) Sketch a smooth function $v(t)$ that gives the velocity. (*b*) What is the average velocity over the first 10 s?

Chapter 3

Motion in Two and Three Dimensions

Illuminated fountains, St. Louis Missouri. The jets follow parabolic paths like those followed by projectiles.

We now extend our description of the motion of a particle to two and three dimensions. In these cases, displacement, velocity, and acceleration are quantities that have direction in space as well as magnitude. We call such quantities **vectors**. In future chapters, we will encounter many other vector quantities, such as force, momentum, and the electric field. Quantities that have magnitude only and no associated direction, such as distance, speed, mass, or temperature, are called **scalars.**

In this chapter we will investigate the properties of vectors in general and those of displacement, velocity, and acceleration in particular. Many of the interesting features of motion in three dimensions are also present in motion in two dimensions. Since two-dimensional motion is more easily illustrated on paper or a blackboard, most of our examples will be limited to it. Two important special cases, projectile motion and circular motion, will be discussed in detail.

3-1 The Displacement Vector and Vector Addition

If you ask someone where the post office is and she says it is 10 blocks away, you would likely ask in what direction before you set off to find it. It makes quite a difference whether the post office is 10 blocks east, or 10 blocks north, or 6 blocks west and 8 blocks south (and therefore 14 blocks walking but 10 blocks "as the crow flies"). The quantity that gives the straight-line distance and the direction from one point in space to another is a directed line segment called the **displacement vector.** We use displacement vectors to illustrate results for vectors in general because, by definition, **vectors** *are quantities with magnitude and direction that add and subtract like displacements.* A vector is represented graphically by an arrow whose direction is the same as the direction of the vector and whose length is proportional to the magnitude of the vector.

Figure 3-1 shows the path of a particle that moves from point P_1 to a second point P_2 and then to a third point P_3. The displacement from point P_1 to point P_2 is represented by the arrow **A**. Note that the displacement **A** does not depend on the path taken by the particle as it moves from P_1 to P_2 but only on the endpoints P_1 and P_2. A second displacement from P_2 to P_3 is indicated by the arrow **B**. The resultant displacement from P_1 to P_3 is represented by the arrow **C**. The resultant displacement vector **C** is the sum of the two successive displacements **A** and **B**:

$$\mathbf{C} = \mathbf{A} + \mathbf{B} \qquad 3\text{-}1$$

Any two vectors (whose units are the same) may be added graphically in this way, by placing the tail of one arrow at the head of the other.

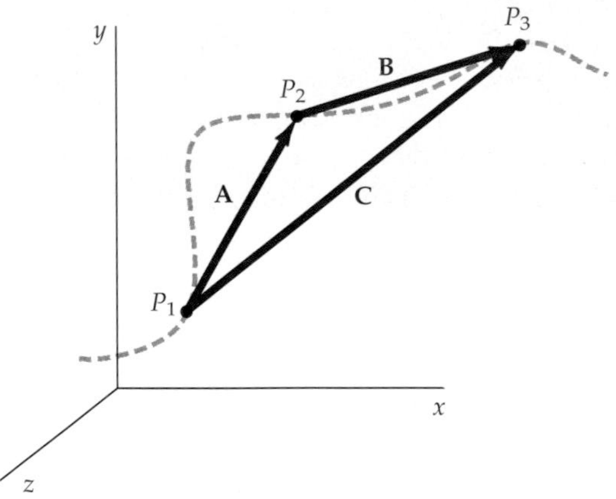

Figure 3-1 The addition of vectors. The displacement **C** is equivalent to the successive displacements **A** and **B**; that is, **C** = **A** + **B**.

The resultant vector is then from the tail of the first vector to the head of the second.

We denote vector quantities with boldface type, as in **A**. We will use this convention throughout the text to distinguish vectors from scalar quantities, which are denoted with italic type. (In handwriting, we indicate a vector by drawing an arrow over the symbol as in $\vec{A}$.) The magnitude of a vector **A** is written |**A**| or simply *A*. Ordinarily, the magnitude of a vector has physical units. For example, a displacement vector has a magnitude that can be expressed in feet, meters, or any other unit of distance.

Note that the sum of the magnitudes of **A** and **B** does not equal the magnitude of **C** unless **A** and **B** are in the same direction. That is, **C** = **A** + **B** does not imply that $C = A + B$.

Example 3-1

A man walks 3 km east and then 4 km north. What is his resultant displacement?

The two displacements and the resultant displacement are shown in Figure 3-2. Since these three vectors form a right triangle, we can find the magnitude of the resultant displacement using the pythagorean theorem, which states

$$C^2 = A^2 + B^2 = (3\text{ km})^2 + (4\text{ km})^2 = 25\text{ km}^2$$

$$C = \sqrt{25\text{ km}^2} = 5\text{ km}$$

To describe the resultant displacement, we need to give its direction as well as its magnitude. If θ is the angle between the east axis and the resultant displacement, we have from Figure 3-2

$$\tan\theta = \frac{4\text{ km}}{3\text{ km}} = 1.33$$

We find θ from trigonometric tables or from a calculator that has trigonometric functions:

$$\theta = \tan^{-1} 1.33 = 53.1°$$

The resultant displacement is therefore 5 km at 53.1° north of east.

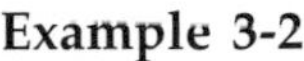

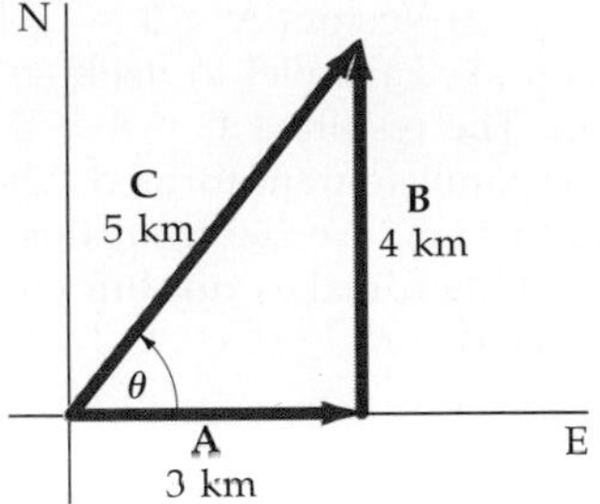

Figure 3-2 Displacement vectors for Example 3-1. The magnitude of the resultant displacement vector **C** can be found from the pythagorean theorem.

Example 3-2

A man walks 3 km east and then 4 km at 60° north of east. What is his resultant displacement?

The displacement vectors for this example are shown in Figure 3-3. In this case, the triangle formed by the three vectors is not a right triangle, so we cannot use the pythagorean theorem to find the resultant displacement. In the next section, we will learn how to find the resultant vector for a case like this using vector components. For now, we find the resultant graphically by drawing each of the displacements to scale and measuring the resultant displacement with a ruler. For example, if we draw the first displacement vector 3 cm long and the second 4 cm long, we find the resultant vector to be about 6 cm long. Thus, the magnitude of the resultant displacement is 6 km. The angle between the resultant displacement and the east direction can be measured with a protractor. It is about 35°.

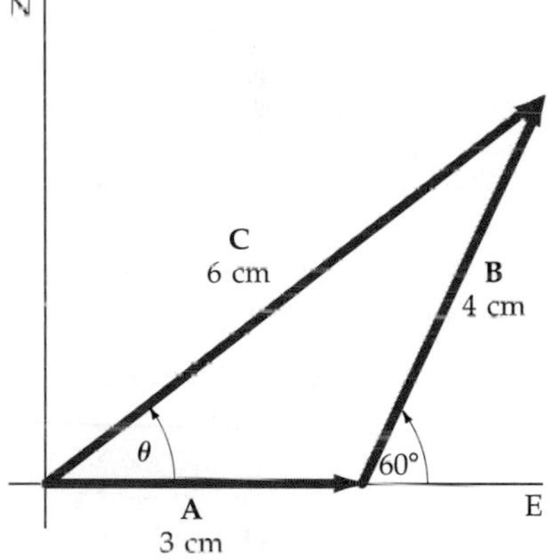

Figure 3-3 Displacement vectors for Example 3-2. Since **A** and **B** are not perpendicular, the pythagorean theorem cannot be used to find the magnitude of **C**. Instead, C can be found graphically.

Since vectors are defined only by their magnitude and direction, two vectors are equal if they have the same magnitude and the same direction, independent of their points of origin. Graphically, two vectors are equal if they have the same length and are parallel to one another. Thus, all the vectors in Figure 3-4 are equal. Figure 3-5*a* shows two vectors **A** and **B**

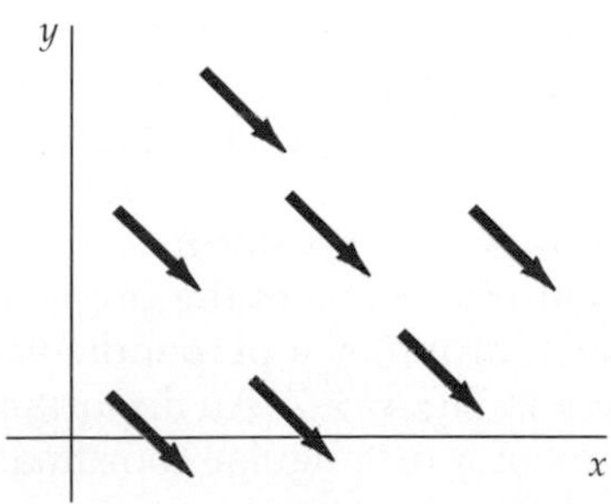

Figure 3-4 Vectors are equal if their magnitudes and directions are the same. All the vectors in this figure are equal.

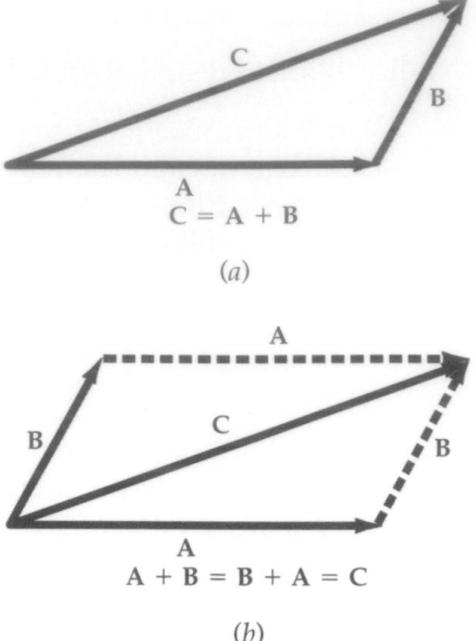

Figure 3-5 (*a*) Vectors **A** + **B** = **C**. In (*b*) the vector **B** has been moved parallel to itself so that **A** and **B** are tail to tail. The resultant **C** = **A** + **B** is along the diagonal of the parallelogram formed when **A** and **B** are placed tail to tail. We see from this figure that the order of addition makes no difference; that is, **A** + **B** is the same as **B** + **A**.

whose sum is **C**. In Figure 3-5*b*, we have moved vector **B** parallel to itself so that it originates at the same point as **A**. The resultant vector **C** lies along the diagonal of the parallelogram formed by **A** and **B**. The graphical addition of two vectors by placing them tail to tail and finding the diagonal of the parallelogram formed is known as the **parallelogram method of vector addition**. From Figure 3-5*b*, we can see that it makes no difference in which order we add two vectors, that is, **A** + **B** = **B** + **A**.

We can subtract vector **B** from vector **A**, as shown in Figure 3-6*a*, by simply adding −**B**, which has the same magnitude as **B** but points in the opposite direction. The result is **C** = **A** + (−**B**) = **A** − **B**. Another method of subtraction, illustrated in Figure 3-6*b*, is to draw the two vectors **A** and **B** tail to tail and then note that the vector **C** = **A** − **B** is that vector that must be added to **B** to obtain the resultant vector **A**.

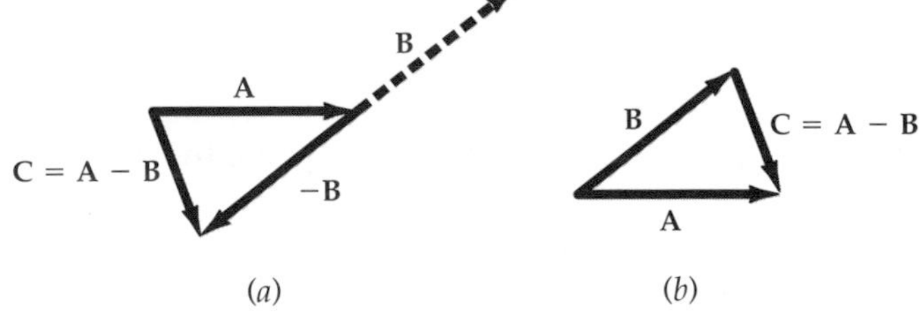

Figure 3-6 Subtraction of vectors. (*a*) Here **C** = **A** − **B** is found by adding −**B** to **A**. (*b*) An alternative method of finding **A** − **B** is to find the vector **C** that when added to **B** gives the vector **A**.

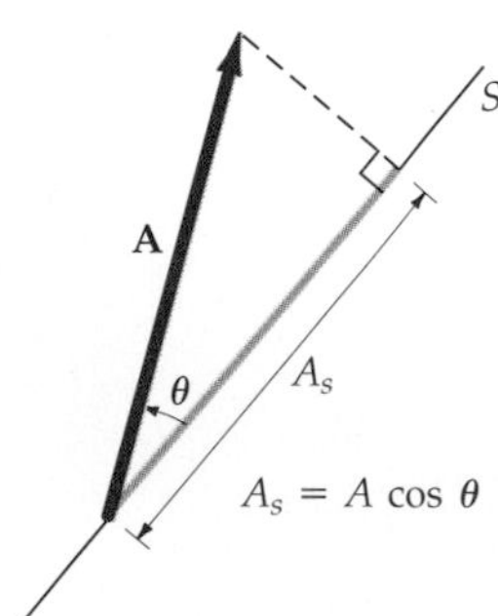

Figure 3-7 The component A_s of a vector **A** along a line in space is found by dropping the perpendicular from the head of the vector to the line.

Questions

1. Can the displacement of a particle have a magnitude that is less than the distance traveled by the particle along its path? Can its magnitude be more than the distance traveled? Explain.
2. Give an example in which the distance traveled is a significant amount yet the corresponding displacement is zero.

3-2 Addition of Vectors by Components

We add or subtract vectors analytically by first breaking down the vectors into their components. The component of a vector is the projection of that vector on a line in space that is found by dropping a perpendicular from the head of the vector to the line, as shown in Figure 3-7. An important example is the projection of a vector on an axis of a rectangular coordinate system.

Such a projection is called a **rectangular component** of the vector. Figure 3-8 shows a vector **A** lying in the xy plane. It has the rectangular components A_x and A_y. In general, the components can be either positive or negative. For example, if **A** points in the negative x direction, A_x is negative. If θ is the angle between vector **A** and the x axis, we see from the figure that

$$\tan\theta = \frac{A_y}{A_x} \quad \text{3-2}a$$

$$\sin\theta = \frac{A_y}{A} \quad \text{3-2}b$$

$$\cos\theta = \frac{A_x}{A} \quad \text{3-2}c$$

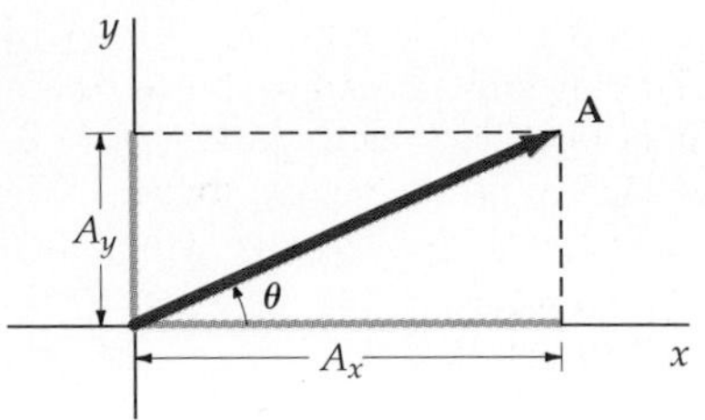

Figure 3-8 The rectangular components of the vector **A** are related to the magnitude A and the angle θ by $A_x = A\cos\theta$ and $A_y = A\sin\theta$.

where A is the magnitude of **A**. We can therefore determine the components of **A** analytically from the magnitude A and the angle θ by

$$A_x = A\cos\theta \quad \text{3-3}$$

and

$$A_y = A\sin\theta \quad \text{3-4}$$

Conversely, if we know the components A_x and A_y, we can find the angle θ from Equation 3-2a and the magnitude A from the pythagorean theorem:

$$A = \sqrt{A_x^2 + A_y^2} \quad \text{3-5}$$

Figure 3-9 illustrates the use of components in the addition of two vectors **A** and **B** that lie in the xy plane. The rectangular components of each vector and those of the sum **C** = **A** + **B** are shown in the figure. We can see from the figure that **C** = **A** + **B** implies both

$$C_x = A_x + B_x \quad \text{3-6}a$$

and

$$C_y = A_y + B_y \quad \text{3-6}b$$

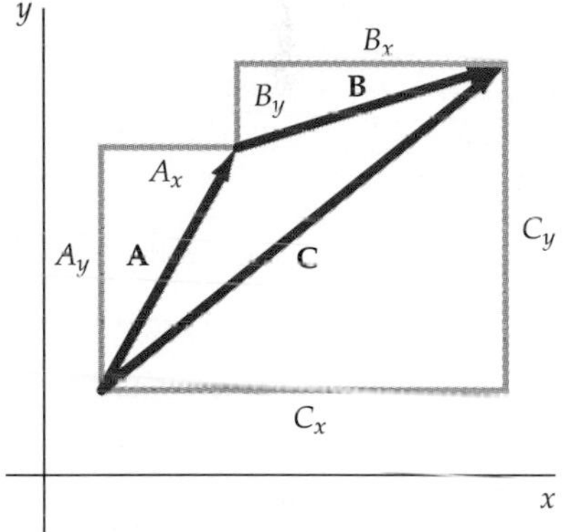

Figure 3-9 The x and y components of the vectors **A**, **B**, and **C** = **A** + **B**. We see from this figure that $C_x = A_x + B_x$ and $C_y = A_y + B_y$.

Exercise

A car travels 20 km at an angle of 30° north of west. Assuming the x axis points east and the y axis points north as in Figure 3-10, find the x and y components of the displacement vector of the car.
(Answers. $A_x = -17.3$ km, $A_y = +10$ km)

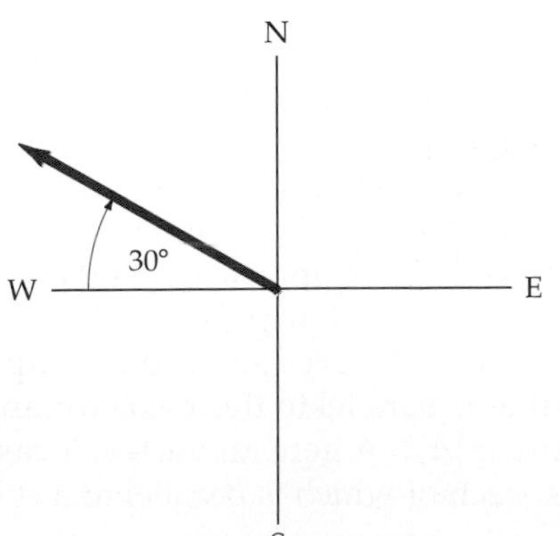

Figure 3-10 Geometry for the exercise.

Example 3-3

Work Example 3-2 using vector components.

In this example, a man first walked east 3 km. If **A** represents his displacement and we take the x axis to be in the easterly direction, the components of **A** are

$$A_x = 3 \text{ km}$$

and

$$A_y = 0$$

He next walked 4 km at 60° north of east. The vector **B** representing this displacement has the components

$$B_x = (4 \text{ km}) \cos 60° = (4 \text{ km})(0.5) = 2 \text{ km}$$

and

$$B_y = (4 \text{ km}) \sin 60° = (4 \text{ km})(0.866) = 3.46 \text{ km}$$

The components of the resultant displacement are thus

$$C_x = A_x + B_x = 3 \text{ km} + 2 \text{ km} = 5 \text{ km}$$

and

$$C_y = A_y + B_y = 0 + 3.46 \text{ km} = 3.46 \text{ km}$$

We obtain the magnitude of the resultant displacement **C** from the pythagorean theorem:

$$C^2 = C_x^2 + C_y^2 = (5 \text{ km})^2 + (3.46 \text{ km})^2 = 37.0 \text{ km}^2$$

So

$$C = \sqrt{37.0 \text{ km}^2} = 6.1 \text{ km}$$

The angle between **C** and the x axis is found from

$$\tan \theta = \frac{C_y}{C_x} = \frac{3.46 \text{ km}}{5 \text{ km}} = 0.692$$

The angle θ is

$$\theta = \tan^{-1} 0.692 = 34.7°$$

This agrees with our results for Example 3-2 within the accuracy of our measurement in that example.

3-3 Unit Vectors and Multiplication of Vectors by Scalars

A vector **A** may be multiplied by a scalar s. The result is the vector $\mathbf{B} = s\mathbf{A}$, which points in the direction of **A** and has a magnitude of sA. The dimensions of **B** are the dimensions of s multiplied by the dimensions of **A**.

We can express a vector conveniently in terms of its components by making use of unit vectors. A **unit vector** is a dimensionless vector that is defined to have the magnitude of 1 and points in some specified direction. For example, let **i**, **j**, and **k** be unit vectors that point in the x, y, and z directions, respectively. The vector $A_x\mathbf{i}$ is the product of the component A_x and the unit vector **i**. It is a vector that is parallel to the x axis (or antiparallel if A_x is negative) and has a magnitude $|A_x|$. A general vector **A** can thus be written as the sum of three vectors, each of which is parallel to a coordinate:

$$\mathbf{A} = A_x\mathbf{i} + A_y\mathbf{j} + A_z\mathbf{k} \qquad 3\text{-}7$$

The vector sum of Equation 3-7 is illustrated in Figure 3-11. The addition of two vectors **A** and **B** can be written in terms of unit vectors as

$$\mathbf{A} + \mathbf{B} = (A_x\mathbf{i} + A_y\mathbf{j} + A_z\mathbf{k}) + (B_x\mathbf{i} + B_y\mathbf{j} + B_z\mathbf{k})$$
$$= (A_x + B_x)\mathbf{i} + (A_y + B_y)\mathbf{j} + (A_z + B_z)\mathbf{k} \qquad 3\text{-}8$$

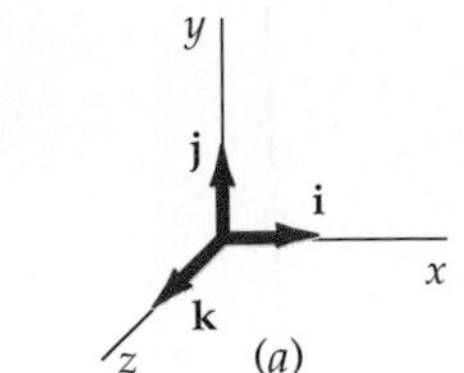

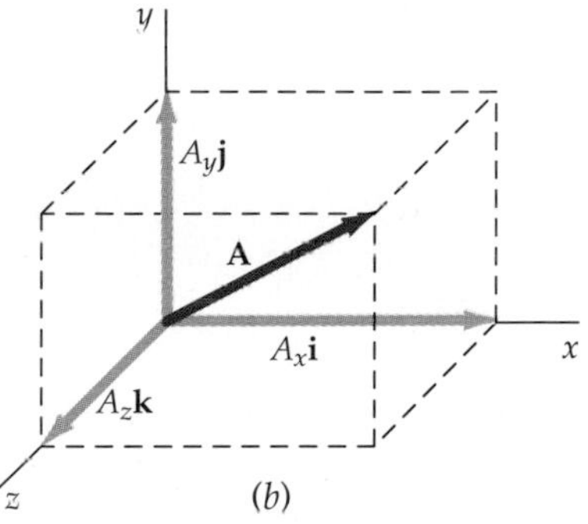

Figure 3-11 (*a*) The unit vectors **i**, **j**, and **k** in a rectangular coordinate system. (*b*) The vector **A** can be written in terms of the unit vectors as $\mathbf{A} = A_x\mathbf{i} + A_y\mathbf{j} + A_z\mathbf{k}$.

Exercise

Given the two vectors $\mathbf{A} = (4\text{ m})\mathbf{i} + (3\text{ m})\mathbf{j}$ and $\mathbf{B} = (2\text{ m})\mathbf{i} - (3\text{ m})\mathbf{j}$, find (*a*) A, (*b*) B, (*c*) $\mathbf{A} + \mathbf{B}$, and (*d*) $\mathbf{A} - \mathbf{B}$. [Answers: (*a*) $A = 5$ m, (*b*) $B = 3.61$ m, (*c*) $\mathbf{A} + \mathbf{B} = (6\text{ m})\mathbf{i}$, (*d*) $\mathbf{A} - \mathbf{B} = (2\text{ m})\mathbf{i} + (6\text{ m})\mathbf{j}$]

The general properties of vectors that we have discussed are summarized in Table 3-1.

Questions

3. How would you subtract two vectors using the component method?
4. Can a component of a vector have a magnitude greater than the magnitude of the vector? Under what circumstances can a component of a vector have a magnitude equal to the magnitude of the vector?
5. Can a vector be equal to zero and still have one or more components not equal to zero?
6. Are the components of $\mathbf{C} = \mathbf{A} + \mathbf{B}$ necessarily larger than the corresponding components of either **A** or **B**?

Table 3-1 **Properties of Vectors**

Property	Explanation	Figure	Component representation
Equality	$\mathbf{A} = \mathbf{B}$ if $\lvert\mathbf{A}\rvert = \lvert\mathbf{B}\rvert$ and their directions are the same	A B	$A_x = B_x$ $A_y = B_y$ $A_z = B_z$
Addition	$\mathbf{C} = \mathbf{A} + \mathbf{B}$	C B A	$C_x = A_x + B_x$ $C_y = A_y + B_y$ $C_z = A_z + B_z$
Negative of a vector	$\mathbf{A} = -\mathbf{B}$ if $\lvert\mathbf{B}\rvert = \lvert\mathbf{A}\rvert$ and their directions are opposite	A B	$A_x = -B_x$ $A_y = -B_y$ $A_z = -B_z$
Subtraction	$\mathbf{C} = \mathbf{A} - \mathbf{B}$	B A C −B	$C_x = A_x - B_x$ $C_y = A_y - B_y$ $C_z = A_z - B_z$
Multiplication by a scalar	$\mathbf{B} = s\mathbf{A}$ if $\lvert\mathbf{B}\rvert = s\lvert\mathbf{A}\rvert$ and the direction of **B** is the same as the direction of **A**	B A sA	$B_x = sA_x$ $B_y = sA_y$ $B_z = sA_z$

3-4 The Velocity Vector

Suppose you are driving a car at 50 km/h as indicated by the speedometer reading and are headed south as indicated by a compass. The speedometer gives the speed, magnitude of the velocity, and the compass gives the direction of the velocity. The instantaneous-velocity vector is a vector that points in the direction of motion and has a magnitude equal to the speed of the car. It equals the rate of change of the displacement vector.

Figure 3-12 shows a particle moving along some curve in space. (The curve represents the physical path or trajectory actually followed by the particle. It should not be confused with the x-versus-t plots in the previous chapter.) We define the position of the particle by means of its vector displacement from the origin O. This displacement is called the **position vector r**. If the particle is at point (x, y), the position vector is

$$\mathbf{r} = x\mathbf{i} + y\mathbf{j} \tag{3-9}$$

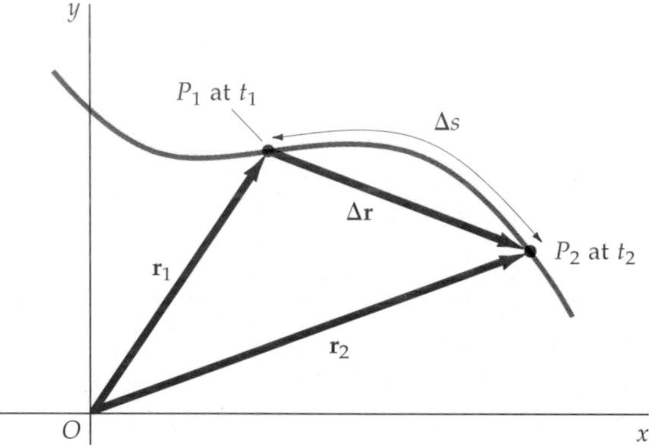

Figure 3-12 A particle moving along some arbitrary curve in space, with position vectors $\mathbf{r}_1$ and $\mathbf{r}_2$ at two different times t_1 and t_2. The displacement vector $\Delta\mathbf{r}$ is the difference in the position vectors, $\Delta\mathbf{r} = \mathbf{r}_2 - \mathbf{r}_1$.

At some time t_1, the particle is at point P_1. This is characterized by the position vector $\mathbf{r}_1$ drawn from the origin to P_1. At some later time t_2, it is at point P_2 and has the position vector $\mathbf{r}_2$. The displacement vector is the change in the position vector:

$$\Delta\mathbf{r} = \mathbf{r}_2 - \mathbf{r}_1 \tag{3-10}$$

(This is analogous to our one-dimensional definition in Chapter 2, where the displacement is the change in the position coordinate x.) The new position vector $\mathbf{r}_2$ is the sum of the original position vector $\mathbf{r}_1$ and the displacement vector $\Delta\mathbf{r}$, as shown in the figure. The ratio of the displacement vector to the time interval $\Delta t = t_2 - t_1$ is the **average-velocity vector**:

Average-velocity vector defined

$$\mathbf{v}_{\text{av}} = \frac{\Delta\mathbf{r}}{\Delta t} \tag{3-11}$$

We note from Figure 3-12 that the magnitude of the displacement vector is not equal to the actual distance traveled Δs as measured along the curve. It is, in fact, less than this distance (unless the particle travels in a straight line between points P_1 and P_2). However, if we consider smaller and smaller time intervals, as indicated in Figure 3-13, the magnitude of the displacement approaches the actual distance traveled by the particle along the curve, and the direction of $\Delta\mathbf{r}$ approaches the direction of the line tangent to the

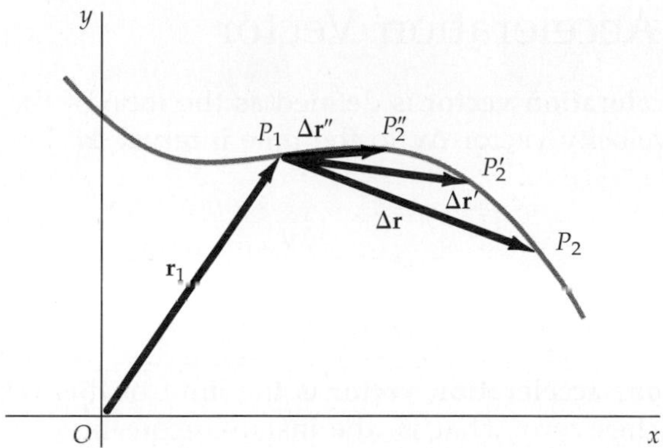

Figure 3-13 As smaller and smaller time intervals are considered, the magnitude of the displacement vector approaches the actual distance traveled along the curve, and the direction of the displacement vector approaches the direction of the line that is tangent to the curve at point P_1.

curve at point P_1. We define the **instantaneous-velocity vector** as the limit of the average-velocity vector as the time interval Δt approaches zero:

$$\mathbf{v} = \lim_{\Delta t \to 0} \frac{\Delta \mathbf{r}}{\Delta t} = \frac{d\mathbf{r}}{dt} \qquad 3\text{-}12$$

Instantaneous-velocity vector defined

The instantaneous-velocity vector is thus the derivative of the position vector with respect to time. Its direction is along the line tangent to the curve traveled by the particle in space. It therefore points in the direction of motion of the particle. The magnitude of the instantaneous velocity is the speed ds/dt, where s is the distance traveled measured along the curve.

To calculate the derivative in Equation 3-12, we must express the position vector in terms of its components as in Equation 3-9:

$$\Delta \mathbf{r} = \mathbf{r}_2 - \mathbf{r}_1 = (x_2 - x_1)\mathbf{i} + (y_2 - y_1)\mathbf{j}$$

$$\mathbf{v} = \lim_{\Delta t \to 0} \frac{\Delta \mathbf{r}}{\Delta t} = \lim_{\Delta t \to 0} \frac{\Delta x\,\mathbf{i} + \Delta y\,\mathbf{j}}{\Delta t} = \lim_{\Delta t \to 0} \left(\frac{\Delta x\,\mathbf{i}}{\Delta t}\right) + \left(\frac{\Delta y\,\mathbf{j}}{\Delta t}\right)$$

or

$$\mathbf{v} = \frac{dx}{dt}\mathbf{i} + \frac{dy}{dt}\mathbf{j}$$

Example 3-4

A sailboat has initial coordinates $(x_1, y_1) = (100 \text{ m}, 200 \text{ m})$. Two minutes later, it has the coordinates $(x_2, y_2) = (120 \text{ m}, 210 \text{ m})$. What are the components, magnitude, and direction of its average velocity for this 2.00-min interval?

$$v_{x,\text{av}} = \frac{x_2 - x_1}{\Delta t} = \frac{120 - 100 \text{ m}}{2.00 \text{ min}} = 10.0 \text{ m/min}$$

$$v_{y,\text{av}} = \frac{y_2 - y_1}{\Delta t} = \frac{210 - 200 \text{ m}}{2.00 \text{ min}} = 5.0 \text{ m/min}$$

$$v_{\text{av}} = \sqrt{(v_{x,\text{av}})^2 + (v_{y,\text{av}})^2} = \sqrt{10.0^2 + 5.0^2} = \sqrt{125} = 11.2 \text{ m/min}$$

$$\tan\theta = \frac{v_{y,\text{av}}}{v_{x,\text{av}}} = \frac{5.0 \text{ m/min}}{10.0 \text{ m/min}} = 0.500$$

$$\theta = \tan^{-1} 0.500 = 26.6°$$

3-5 The Acceleration Vector

The **average-acceleration vector** is defined as the ratio of the change in the instantaneous-velocity vector $\Delta\mathbf{v}$ to the time interval Δt:

Average-acceleration vector defined

$$\mathbf{a}_{av} = \frac{\Delta\mathbf{v}}{\Delta t} \qquad 3\text{-}13$$

The **instantaneous-acceleration vector** is the limit of this ratio as the time interval approaches zero. That is, the instantaneous-acceleration vector is the derivative of the velocity vector with respect to time:

Instantaneous-acceleration vector defined

$$\mathbf{a} = \lim_{\Delta t \to 0} \frac{\Delta\mathbf{v}}{\Delta t} = \frac{d\mathbf{v}}{dt} \qquad 3\text{-}14$$

To calculate the instantaneous acceleration, we express $\mathbf{v}$ in terms of its rectangular coordinates:

$$\mathbf{v} = v_x\mathbf{i} + v_y\mathbf{j} = \frac{dx}{dt}\mathbf{i} + \frac{dy}{dt}\mathbf{j}$$

Then

$$\mathbf{a} = \frac{dv_x}{dt}\mathbf{i} + \frac{dv_y}{dt}\mathbf{j} = \frac{d^2x}{dt^2}\mathbf{i} + \frac{d^2y}{dt^2}\mathbf{j}$$

It is particularly important to note that the velocity vector may be changing in magnitude, direction, or both. If the velocity vector is changing in any way, the particle is accelerating. We are perhaps most familiar with acceleration in which the velocity changes in magnitude, that is, in which the speed changes. However, a particle can be traveling with constant speed and still be accelerating if the direction of the velocity vector is changing. A particularly important example of this is circular motion, which will be discussed in Section 3-8. This acceleration is just as real as when the speed is changing.

Example 3-5

A car is traveling east at 60 km/h. It rounds a curve, and 5 s later it is traveling north at 60 km/h. Find the average acceleration of the car.

Figure 3-14 shows the initial and final velocity vectors $\mathbf{v}_1 =$ (60 km/h)$\mathbf{i}$ and $\mathbf{v}_2 =$ (60 km/h)$\mathbf{j}$. The change in velocity

$$\Delta\mathbf{v} = \mathbf{v}_2 - \mathbf{v}_1 = (60\text{ km/h})\mathbf{j} - (60\text{ km/h})\mathbf{i}$$

is also shown in the figure. (Note that we have drawn the figure such that $\mathbf{v}_1 + \Delta\mathbf{v} = \mathbf{v}_2$.) The average acceleration is given by

$$\mathbf{a}_{av} = \frac{\Delta\mathbf{v}}{\Delta t} = \frac{(60\text{ km/h})\mathbf{j} - (60\text{ km/h})\mathbf{i}}{5\text{ s}}$$
$$= (12\text{ km/h·s})\mathbf{j} - (12\text{ km/h·s})\mathbf{i}$$

The magnitude of the average acceleration is

$$a = \sqrt{a_x^2 + a_y^2} = \sqrt{(-12\text{ km/h·s})^2 + (12\text{ km/h·s})^2} = 17.0\text{ km/h·s}$$

Note that in this example, the car accelerates even though the speed does not change.

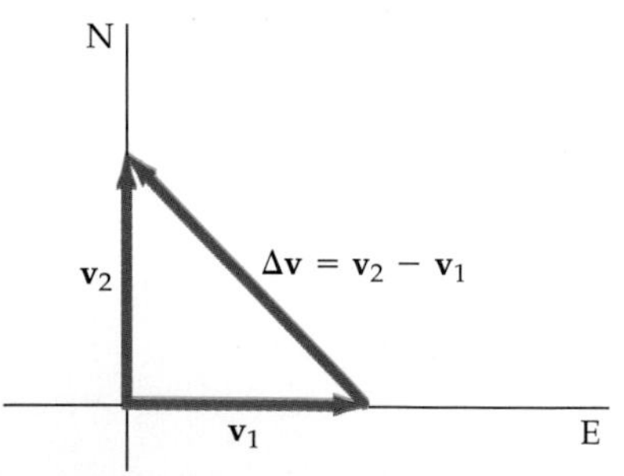

Figure 3-14 Velocity vectors for Example 3-5.

Questions

7. For an arbitrary motion of a given particle, does the direction of the velocity vector have any particular relation to the direction of the position vector?
8. Give examples in which the directions of the velocity and position vectors are (*a*) opposite, (*b*) the same, and (*c*) mutually perpendicular.
9. How is it possible for a particle moving at constant speed to be accelerating? Can a particle with constant velocity be accelerating at the same time?
10. Is it possible for a particle to round a curve without accelerating?
11. Can the velocity vector change direction without changing magnitude? If so, give an example.

3-6 Relative Velocity

The velocity of an object is sometimes measured relative to a coordinate system that is itself moving relative to another coordinate system. For example, suppose that a person is walking on a railroad car with velocity $\mathbf{v}_{pc}$ relative to the car while the car is moving with velocity $\mathbf{v}_{cg}$ relative to the ground, as shown in Figure 3-15. The velocity of the person relative to the ground $\mathbf{v}_{pg}$ is the sum of these two velocities:

$$\mathbf{v}_{pg} = \mathbf{v}_{pc} + \mathbf{v}_{cg} \qquad (3\text{-}15)$$

Relative velocity

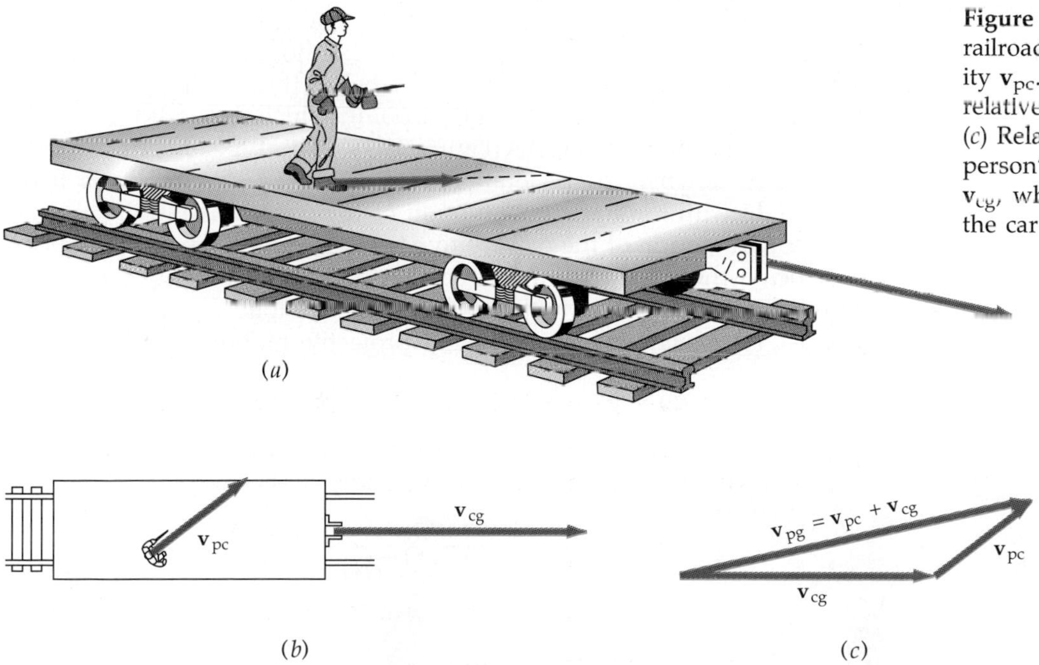

(*a*)

$\mathbf{v}_{pc}$ $\mathbf{v}_{cg}$

(*b*)

$\mathbf{v}_{pg} = \mathbf{v}_{pc} + \mathbf{v}_{cg}$ $\mathbf{v}_{pc}$ $\mathbf{v}_{cg}$

(*c*)

Figure 3-15 (*a*) Relative to the railroad car, the person has velocity $\mathbf{v}_{pc}$. (*b*) The velocity of the car relative to the ground is $\mathbf{v}_{cg}$. (*c*) Relative to the ground, the person's velocity is $\mathbf{v}_{pg} = \mathbf{v}_{pc} + \mathbf{v}_{cg}$, where $\mathbf{v}_{cg}$ is the velocity of the car relative to the ground.

The addition of relative velocities is done in the same way as the addition of displacements, either graphically by placing the tail of one velocity vector at the head of the other or analytically using vector components.

Example 3-6

A river flows from west to east with a speed of 3 m/s. A boy swims north across the river with a speed of 2 m/s relative to the water. What is the velocity of the boy relative to the shore?

Figure 3-16 shows the velocity vectors for this problem. The velocity of the boy relative to the shore is the vector sum of the velocity of the boy relative to the water $\mathbf{v}_{bw}$ and the velocity of the water relative to the shore $\mathbf{v}_{ws}$, as shown in the figure. The magnitude of this velocity is

$$v = \sqrt{v_{bw}^2 + v_{ws}^2} = \sqrt{(2\text{ m/s})^2 + (3\text{ m/s})^2}$$

$$= \sqrt{13\text{ m}^2/\text{s}^2} = 3.61\text{ m/s}$$

The direction is at an angle θ to the shore:

$$\tan\theta = \frac{v_{bw}}{v_{ws}} = \frac{2\text{ m/s}}{3\text{ m/s}} = 0.667$$

$$\theta = \tan^{-1} 0.667 = 33.7°$$

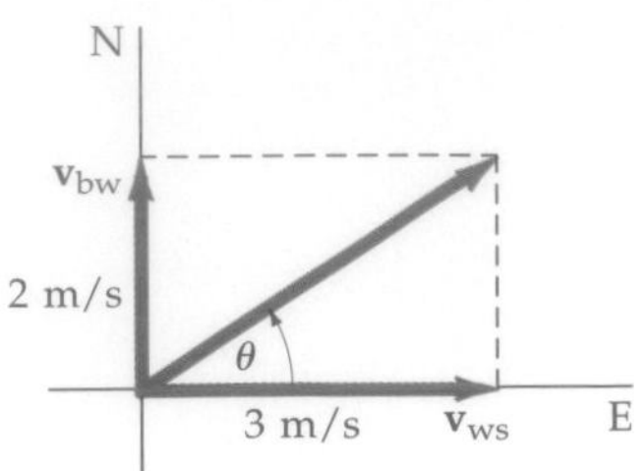

Figure 3-16 Velocity vectors for Example 3-6.

3-7 Projectile Motion

An interesting application of motion in two dimensions is that of a projectile, which is a body that is launched into the air and is then allowed to move freely. The motion of a projectile is complicated by air resistance, the rotation of the earth, and variations in the acceleration due to gravity. For simplicity, we will neglect these complications. The projectile then has a constant acceleration directed vertically downward with a magnitude $g = 9.81\text{ m/s}^2 = 32.2\text{ ft/s}^2$. In projectile motion, the horizontal and vertical components of the motion are independent. For example, consider a ball that is thrown from a platform car that is moving horizontally with constant velocity, as illustrated in Figure 3-17. If the ball is thrown straight upwards relative to the car, it moves to its highest point, which depends on its initial vertical velocity, and then returns (Figure 3-17*a*). This type of vertical motion with constant downward acceleration was discussed in Chapter 2. It has nothing to do with the horizontal motion of the car relative to the ground. The horizontal motion of the ball relative to the ground is motion with constant velocity, the velocity of the car. It has nothing to do with the vertical motion of the ball. Relative to the ground, the ball follows a parabolic path (Figure 3-17*b*), which is characteristic of projectile motion.

Figure 3-17 (*a*) A ball is thrown vertically into the air by a person riding on a cart. Relative to the cart, the ball moves vertically upwards and then falls back to its original position. (*b*) Relative to the ground, the ball has an initial horizontal velocity equal to the velocity of the cart and therefore moves along a parabolic path.

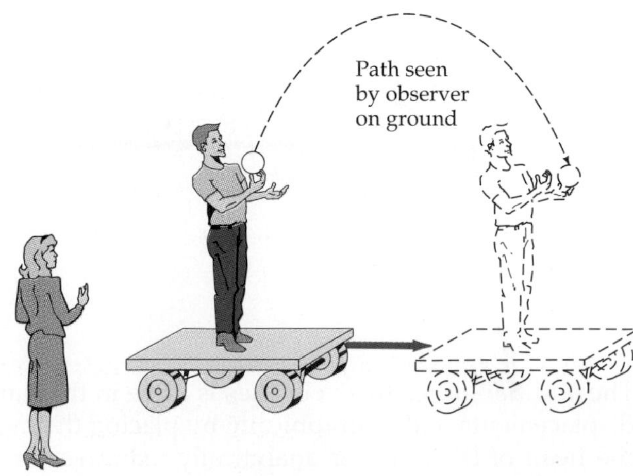

Consider a particle that is launched with some initial velocity that has both vertical and horizontal components relative to a fixed origin. If we take the y axis to be vertical with the positive direction upward and the x axis to be horizontal with the positive direction in the direction of the original horizontal component of the projectile's velocity, we have for the acceleration

$$a_y = -g \qquad \text{3-16}a$$

and

$$a_x = 0 \qquad \text{3-16}b$$

Suppose we launch a projectile from the origin with an initial speed v_0 at an angle of θ with the horizontal axis (Figure 3-18). The initial velocity then has the components

$$v_{0x} = v_0 \cos \theta \qquad \text{3-17}a$$

$$v_{0y} = v_0 \sin \theta \qquad \text{3-17}b$$

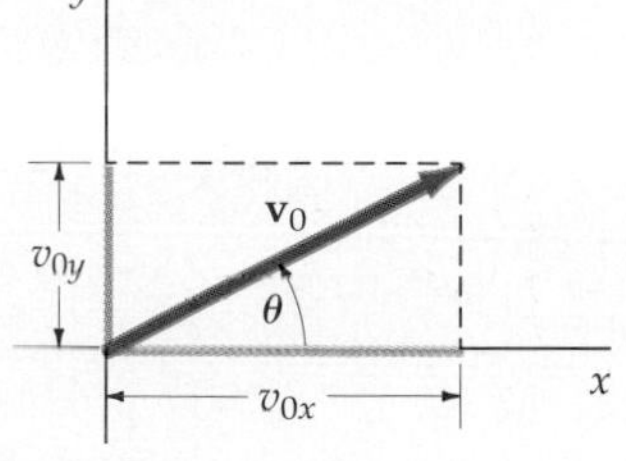

Figure 3-18 The components of the initial velocity of a projectile are $v_{0x} = v_0 \cos \theta$ and $v_{0y} = v_0 \sin \theta$, where θ is the angle between $\mathbf{v}_0$ and the horizontal x axis.

Since there is no horizontal acceleration, the x component of the velocity is constant:

$$v_x = v_{0x} \qquad \text{3-18}a$$

Projectile motion

The y component varies with time according to

$$v_y = v_{0y} - gt \qquad \text{3-18}b$$

Projectile motion

(This is the same as Equation 2-9 with $a = -g$.) The components of the displacement of the projectile are

$$\Delta x = v_{0x}t \qquad \text{3-19}a$$

$$\Delta y = v_{0y}t - \tfrac{1}{2}gt^2 \qquad \text{3-19}b$$

Projectile motion

(Compare this equation for y with Equation 2-12a.)

Example 3-7

A ball is thrown into the air with an initial velocity of 50 m/s at 37° to the horizontal. Find the total time the ball is in the air and the total horizontal distance it travels using the approximation $g = 10\ \text{m/s}^2$.

The components of the initial velocity vector are

$$v_{0x} = (50\ \text{m/s}) \cos 37° = 40\ \text{m/s}$$

$$v_{0y} = (50\ \text{m/s}) \sin 37° = 30\ \text{m/s}$$

The total time the ball is in the air can be found from Equation 3-19b by setting $\Delta y = 0$ and solving for t:

$$\Delta y = v_{0y}t - \tfrac{1}{2}gt^2 = t(v_{0y} - \tfrac{1}{2}gt) = 0$$

This has the solutions $t = 0$, corresponding to the initial conditions, and

$$t = \frac{2v_{0y}}{g} = \frac{2\ (30\ \text{m/s})}{10\ \text{m/s}^2} = 6\ \text{s}$$

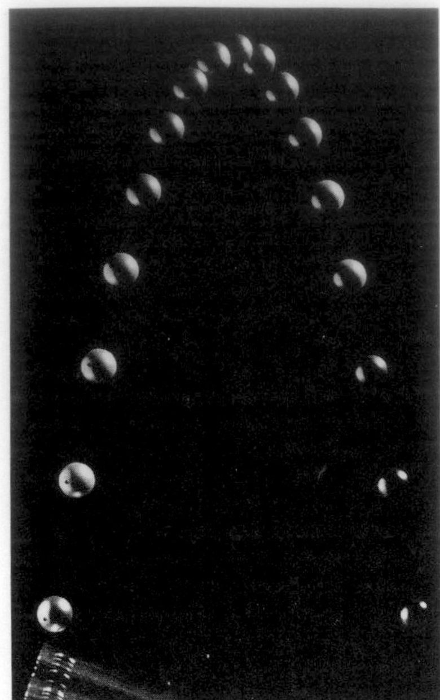

Multiflash photograph of a ball thrown into the air. The position of the ball is recorded at approximately 0.43-s intervals.

This is the same result we found in Example 2-9, in which the vertical acceleration and initial vertical velocity of the ball were the same as in this example. As in that example, the total time the ball is in the air is twice the time t_1 it takes the ball to reach its highest point, which can be found by setting $v_y = 0$ in Equation 3-18*b*:

$$v_y = v_{0y} - gt_1 = 0$$

$$t_1 = \frac{v_{0y}}{g} = \frac{30\text{ m/s}}{10\text{ m/s}^2} = 3\text{ s}$$

Since the ball moves horizontally with a constant velocity of 40 m/s, the total horizontal distance traveled is

$$\Delta x = v_{0x}t = 40\text{ m/s} \times 6\text{ s} = 240\text{ m}$$

This distance is called the **range** of the projectile.

Figure 3-19 shows the height y versus the time t for this example. This curve is identical to that in Figure 2-7*a* for Examples 2-8 and 2-9 because the vertical accelerations and vertical velocities are the same for all three examples. Since the projectile moves 40 m horizontally during each second, we can interpret this curve as a graph of y versus x if we change the horizontal axis from a time scale to a distance scale by multiplying the time values by 40 m/s. The curve of y versus x is a parabola.

The general equation for the path $y(x)$ can be obtained from Equations 3-19*a* and *b* by eliminating the variable t between these equations. Choosing $x_0 = y_0 = 0$ and using $t = x/v_{0x}$ in Equation 3-19*b*, we obtain

$$y = v_{0y}\left(\frac{x}{v_{0x}}\right) - \frac{1}{2}g\left(\frac{x}{v_{0x}}\right)^2$$

or

$$y = \left(\frac{v_{0y}}{v_{0x}}\right)x - \frac{1}{2}\left(\frac{g}{v_{0x}^2}\right)x^2 \qquad 3\text{-}20$$

This equation is of the form $y = ax + bx^2$, which is the equation for a parabola that passes through the origin. Figure 3-20 shows the path of a projectile with the velocity vector and its components indicated at several points.

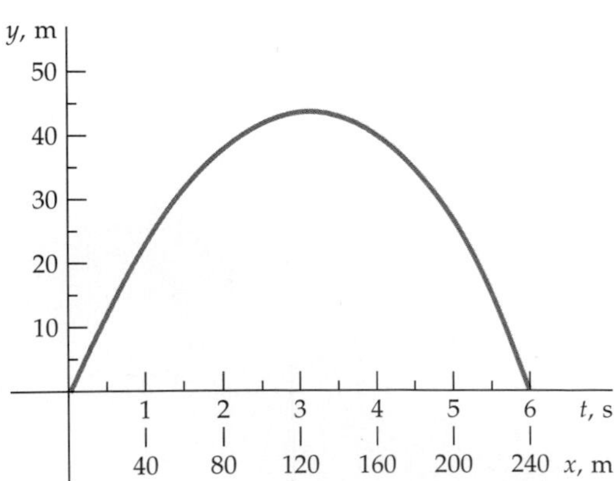

Figure 3-19 Graph of y versus t and y versus x for the ball in Example 3-7. The time scale can be converted into a horizontal-distance scale by multiplying each time by 40 m/s because x is related to t by $x = (40\text{ m/s})t$.

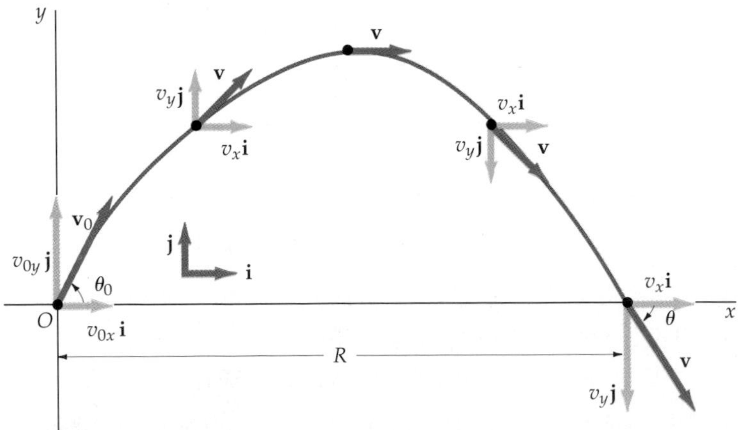

Figure 3-20 Path of a projectile with the velocity vector and its rectangular components shown at several points. The total horizontal distance traveled is the range R.

Exercise

A bullet is fired horizontally from a height of 2 m with an initial speed of 200 m/s. How far does it travel before striking the ground? Use $g = 9.81\ \text{m/s}^2$. (Answer: 129 m)

In Example 3-7, we found the range of the projectile by finding the total time the projectile was in the air from the vertical component of the velocity and then using this time and the horizontal component of the velocity to find the horizontal distance traveled. This method works whether or not the projectile lands at the same elevation as that from which it is projected.

Example 3-8

The ball in Example 3-7 is projected with the same initial velocity but from a cliff that is 55 m above a plane below (Figure 3-21). Where does the ball land?

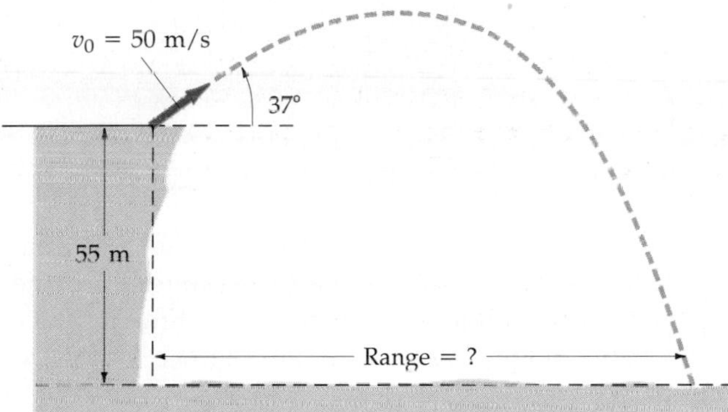

Figure 3-21 A ball thrown from a cliff in Example 3-8. The range is found by first finding the total time the ball is in the air and then multiplying it by the x component of the ball's velocity.

In this case, the time it takes for the ball to reach its maximum height is again 3 s, but the time of fall is longer because the ball falls a greater distance. We first find the maximum height the ball reaches, as we did in Example 2-8. The vertical component of velocity is initially 30 m/s and at the ball's greatest height it is zero. The average upward velocity for the 3 s it takes the ball to reach its maximum height is therefore 15 m/s, and the maximum height is $y = (15\ \text{m/s})(3\ \text{s}) = 45$ m. After reaching this maximum height, the ball falls 45 m back to the level of its projection plus 55 m to the plane below, so the total distance of fall is 45 m + 55 m = 100 m. The time for the ball to fall from "rest" (the vertical component of velocity is zero at the top of the path) a distance of 100 m is found from the constant-acceleration formula:

$$\Delta y = v_{0y}t - \tfrac{1}{2}gt^2$$

$$-100\ \text{m} = 0 - \tfrac{1}{2}(10\ \text{m/s}^2)t^2$$

or

$$t = \sqrt{20\ \text{s}^2} = 4.5\ \text{s}$$

The total time the ball is in the air is thus 3 s + 4.5 s = 7.5 s. The horizontal distance covered in this time is

$$x = v_{0x}t = (40\ \text{m/s})(7.5\ \text{s}) = 300\ \text{m}$$

We could have used Equation 3-19*b* to find the time directly without considering the problem in two parts. Since the ball lands 55 m below its starting point, we set $\Delta y = -55$ m. The time is then found from

$$\Delta y = v_{0y}t - \tfrac{1}{2}gt^2$$

$$-55\ \text{m} = (30\ \text{m/s})t - \tfrac{1}{2}(10\ \text{m/s}^2)t^2$$

There are two solutions for this quadratic equation, $t = -1.5$ s and $t = +7.5$ s. The negative time corresponds to the time of 1.5 s it would take for the ball to reach its starting point at $y = 0$ if it were projected from $y = -55$ m with an initial speed such that the component of its vertical velocity component would be 30 m/s when it reached $y = 0$, as shown in Figure 3-22.

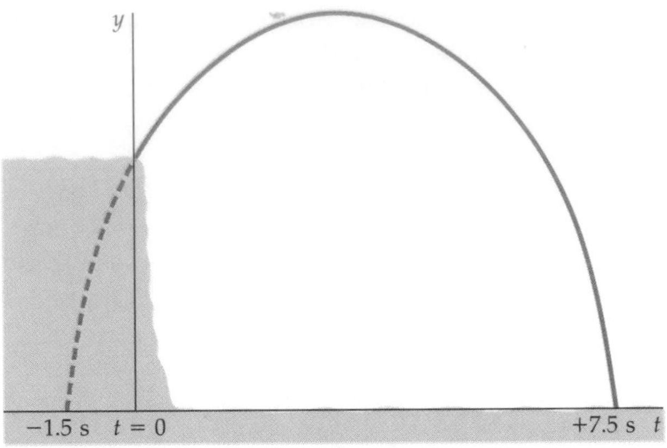

Figure 3-22 The solution $t = -1.5$ s, found by setting $\Delta y = -55$ m in Example 3-8, describes the situation that would occur if the ball were projected from $y = -55$ m such that it reaches $y = 0$ at time $t = 0$. This mathematical solution is discarded because it does not describe the given problem.

For the special case in which the initial and final elevations are equal, we can derive a general formula for the range of a projectile in terms of its initial speed and the angle of projection. The time it takes the projectile to reach its maximum height is found by setting the vertical component of its velocity equal to zero:

$$v_y = v_{0y} - gt = 0$$

or

$$t = \frac{v_{0y}}{g}$$

The range R is then the distance traveled in twice this time:

$$R = 2v_{0x}\left(\frac{v_{0y}}{g}\right) = \frac{2v_{0x}v_{0y}}{g}$$

In terms of the initial speed v_0 and the angle of projection θ, the range is

$$R = \frac{2(v_0 \cos\theta)(v_0 \sin\theta)}{g} = \frac{2v_0^2 \sin\theta \cos\theta}{g}$$

This can be further simplified by using the trigonometric identity for the sine of twice an angle (Appendix A):

$$\sin 2\theta = 2 \sin\theta \cos\theta$$

We then have

Range of a projectile

$$R = \frac{v_0^2}{g} \sin 2\theta \qquad 3\text{-}21$$

Since the maximum value of $\sin 2\theta$ is 1 when $2\theta = 90°$ or $\theta = 45°$, the range is a maximum and equal to v_0^2/g when $\theta = 45°$.

Equation 3-21 is useful for finding the range for projectile problems when the initial and final elevations are the same. Note, however, that the formula could not be used in Example 3-8 because the elevations were not the same. More important, Equation 3-21 is useful because we can learn something from it about the dependence of the range in the initial angle of projection. For example, we have just seen that the range is maximum when the angle of projection is 45°.

We note that the horizontal distance traveled is the product of the initial horizontal component of the velocity v_{0x} and the time the projectile is in the air, which is in turn proportional to v_{0y}. The maximum range occurs when the horizontal and vertical components are equal, which means that the angle of projection is 45°. In some practical applications, other considerations are important. For example, in the shot put, the initial and final elevations are not equal because the ball is projected from an initial height of about 2 m rather than from ground level, which is where the ball lands. The extra initial height increases the time the shot is in the air. In this case, the range is maximum when v_{0x} is somewhat greater than v_{0y}, that is, when the angle of projection is somewhat smaller than 45° (Figure 3-23). Studies of the form of the best shot putters show that the maximum range occurs with an angle of projection of about 42°. With artillery shells, air resistance must be taken into account to predict the range accurately. Air resistance reduces the range for a given angle of projection. It also decreases the optimum angle of projection slightly.

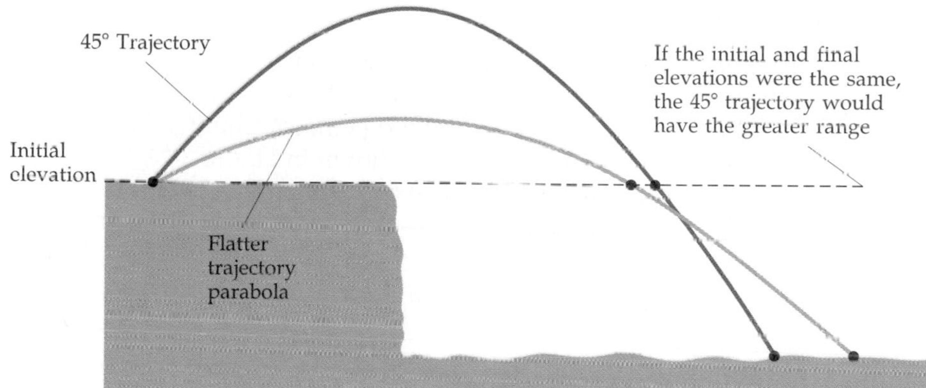

Figure 3-23 If a projectile lands at an elevation lower than the elevation of projection, the range is maximum when the projection angle is slightly less than 45°.

According to our analysis of projectile motion, an object dropped from a height h above the ground will hit the ground in the same time as one projected horizontally from the same height. In each case, the distance the object falls is given by $y = \frac{1}{2}gt^2$ (measuring y downward from the original height). This remarkable fact can easily be demonstrated. It was first commented upon during the Renaissance by Galileo Galilei (1564–1642), the first person to give the modern, quantitative description of projectile motion we have discussed. In the following, Galileo illustrates the validity of treating the horizontal and vertical components of a projectile's motion as independent motions:*

> Let the ship be motionless and the fall of the stone from the mast take two pulse beats. Then cause the ship to move, and drop the same stone from the same place; from what has been said, it will take two pulse beats to arrive at the deck. In these two pulse beats, the ship will have gone, say, twenty yards so that the natural motion of the stone will have been a diagonal line much longer than the

Galileo Galilei (1564-1642)

*Galileo Galilei, *Dialogue Concerning Two Chief World Systems—Ptolemaic and Copernican*, pp. 154–155, trans. Stillman Drake, University of California Press, Berkeley, 1953; reprinted by permission of The Regents of the University of California.

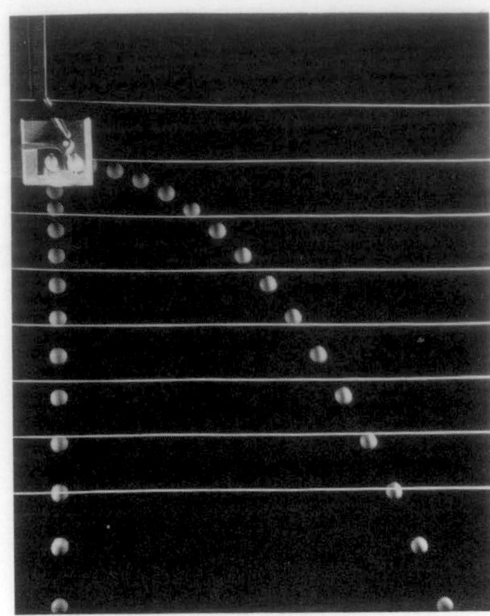

Comparison of a ball dropped with one projected horizontally. The vertical position is independent of the horizontal motion.

first straight and perpendicular one, which was merely the length of the mast; nevertheless, it will have traversed this distance in the same time. Now assuming the ship to be speeded up still more, so that the stone in falling must follow a diagonal line very much longer still than the other, eventually the velocity of the ship may be increased by any amount while the falling rock will describe always longer and longer diagonals, and still pass over them in the same two pulse beats. Similarly, if a perfectly level cannon on a tower were fired parallel to the horizon, it would not matter whether a small charge or a great one was put in, so that the ball would fall a thousand yards away, four thousand, or six thousand, or ten thousand or more; all these shots would require equal times and each time would be equal to that which the ball would have taken in going from the mouth of the cannon to the ground if it were allowed to fall straight down without any other impulse.

Example 3-9

A park ranger with a tranquilizer dart gun wishes to shoot a monkey hanging from a branch of a tree. The ranger aims right at the monkey, not realizing that the dart will follow a parabolic path and thus fall below the monkey. The monkey, however, seeing the dart leave the gun, lets go of the branch and drops out of the tree, expecting to avoid the dart. Show that the monkey will be hit regardless of the initial speed of the dart so long as the speed of the dart is great enough for it to travel the horizontal distance to the tree before hitting the ground and the monkey drops the instant the dart is fired.

This situation is sometimes demonstrated using a target suspended by an electromagnet. When the dart leaves the gun, the circuit is broken and the target falls. Let the horizontal distance to the tree be x and the original height of the monkey be h, as shown in Figure 3-24. The dart is projected at an angle given by $\tan\theta = h/x$. If there were no gravity, the dart would reach the height h in the time t it takes for it to travel the horizontal distance x:

$$y = v_{0y}t = h$$

$$t = \frac{x}{v_{0x}}$$

However, because of gravity, the dart has a vertical acceleration downward. In time $t = x/v_{0x}$, the dart reaches a height y given by

$$y = v_{0y}t - \tfrac{1}{2}gt^2 = h - \tfrac{1}{2}gt^2$$

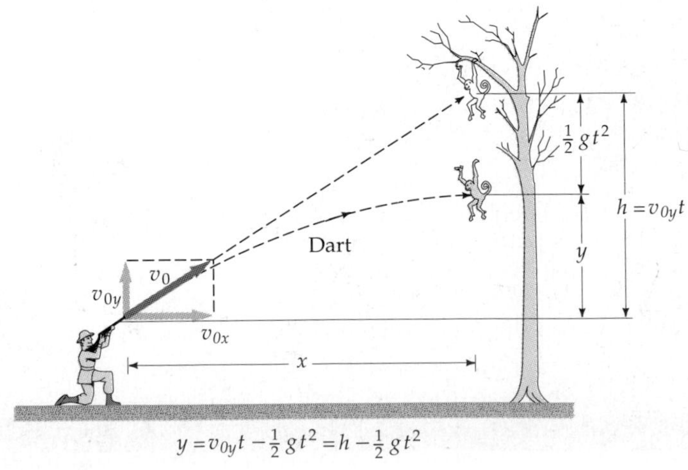

Figure 3-24 The monkey and dart from Example 3-9. The height of the dart at any time is $y = v_{0y}t - \frac{1}{2}gt^2$. The height of the monkey at any time is $h - \frac{1}{2}gt^2$, which is the same as that of the dart because $h = v_{0y}t$. The dart will therefore always hit the monkey if the monkey drops the instant the dart is fired.

This is lower than h by $\frac{1}{2}gt^2$, which is exactly the distance the monkey falls in the same time. In the usual lecture demonstration, the initial speed of the dart is varied so that for a large v_0 the target is hit very near its original height and for a small v_0 it is hit just before it reaches the floor.

Questions

12. What is the acceleration of a projectile at the top of its flight?

13. Can the velocity of an object change direction while its acceleration is constant in both magnitude and direction? If so, give an example.

3-8 Circular Motion

Circular motion is common in nature and in our everyday experience. The earth revolves in a nearly circular orbit around the sun; the moon, around the earth. Wheels rotate in circles, cars travel in circular arcs as they round corners, and so on. In this section, we consider a particle moving in a circle with constant speed. In everyday usage, we might say that since the speed is constant the particle is not accelerating. But we have defined acceleration as the rate of change of the velocity vector, and when a particle moves in a circle its velocity vector is continually changing direction.

Newton was one of the first to recognize the importance of circular motion. He showed that when a particle moves with constant speed v in a circle of radius r, it has an acceleration with a magnitude of v^2/r directed toward the center of the circle. This acceleration is called **centripetal acceleration.**

Consider a satellite that is moving in a circular orbit around the earth. Why doesn't the satellite fall toward the earth? The answer is not that there is no force of gravity acting on the satellite. At 200 km above the surface of the earth, the gravitational force acting on a satellite is about 94 percent of what it would be were the satellite at the earth's surface. The satellite does "fall" toward the earth, but because of its tangential velocity, it continually

The rings of Saturn as photographed by Voyager 1 in 1980. The rings consist of ice or ice-covered rock in circular orbits close to the planet.

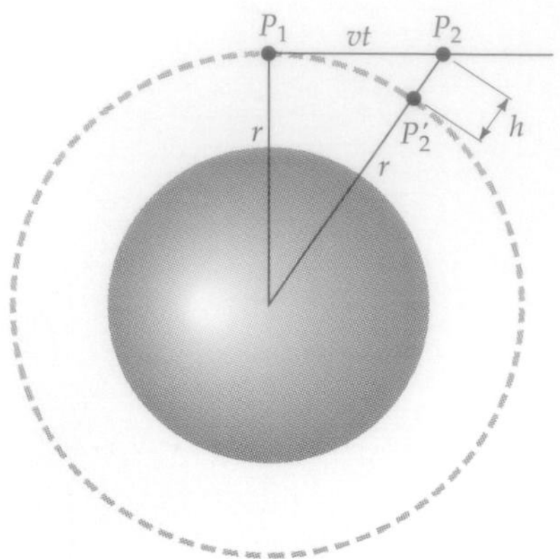

Figure 3-25 Satellite moving with speed v in a circular orbit of radius r about the earth. If the satellite did not accelerate toward the earth, it would move in a straight line from point P_1 to P_2 in time t. Because of its acceleration, it falls a distance h in this time. For small t, $h = \frac{1}{2}(v^2/r)t^2 = \frac{1}{2}at^2$.

misses the earth. To see this, consider Figure 3-25. If the satellite were not accelerating, it would move from point P_1 to P_2 in some time t. Instead, it arrives at point P_2' on its circular orbit. Thus, in a sense, the satellite "falls" the distance h shown. If we take the time t to be very small, the points P_2 and P_2' are approximately on a radial line, as shown in the figure, and we can use the approximation that h is much smaller than the orbit radius r. (From the figure, we can see that the smaller we make the time t, the shorter the distance vt and the smaller h will be for any given orbital radius r.) We can then calculate h from the right triangle with the sides vt, r, and $r + h$. Since $r + h$ is the hypotenuse of the right triangle, the pythagorean theorem gives

$$(r + h)^2 = (vt)^2 + r^2$$

$$r^2 + 2hr + h^2 = v^2t^2 + r^2$$

or

$$h(2r + h) = v^2t^2$$

For very short times t, h will be much less than r, so we can neglect h compared with $2r$ for the term in the parentheses. Thus,

$$2rh \approx v^2t^2$$

or

$$h \approx \frac{1}{2}\left(\frac{v^2}{r}\right)t^2$$

Comparing this with the constant-acceleration expression $h = \frac{1}{2}at^2$, we see that the magnitude of the acceleration of the satellite is

Centripetal acceleration

$$a = \frac{v^2}{r} \tag{3-22}$$

The direction of the acceleration is inward toward the center of the circle.

We can show that this result holds in general for circular motion with constant speed by considering the position and velocity vectors (Figure 3-26). Here the initial velocity vector $\mathbf{v}_1$ is perpendicular to the initial position vector $\mathbf{r}_1$. A short time later, the velocity is $\mathbf{v}_2$, which is perpendicular to $\mathbf{r}_2$. The angle between the velocity vectors $\Delta\theta$ is the same as that between the position vectors because both the position and velocity vectors must move through equal angles to remain mutually perpendicular. If we take the time interval to be very small, the magnitude of the displacement $|\Delta\mathbf{r}|$ is approximately equal to the distance traveled along the arc Δs. The average acceleration is the ratio of the change in velocity $\Delta\mathbf{v} = \mathbf{v}_2 - \mathbf{v}_1$ to the time interval Δt. From the figure we see that for very small Δt, the change in velocity (and therefore the average acceleration) is approximately perpendicular to the velocity vectors and is directed toward the center of the circle. We can find the magnitude of the average acceleration from the angle $\Delta\theta$ in the similar triangles in Figure 3-26. We have

$$\Delta\theta = \frac{\Delta s}{r} \approx \frac{|\Delta\mathbf{v}|}{v}$$

where r is the radius of the circle and v is the speed of the particle. If we now use the fact that the distance is given by $\Delta s = v\,\Delta t$, we have

$$\Delta\theta = \frac{v\,\Delta t}{r} \approx \frac{|\Delta\mathbf{v}|}{v}$$

or

$$\frac{|\Delta \mathbf{v}|}{\Delta t} \approx \frac{v^2}{r}$$

It is often convenient to describe the motion of a particle moving in a circle with constant speed in terms of the time required for one complete revolution, called the period T. If the radius of the circle is r, the particle travels a distance $2\pi r$ during one period, so its speed is related to the radius and the period by

$$v = \frac{2\pi r}{T} \quad \text{3-23}$$

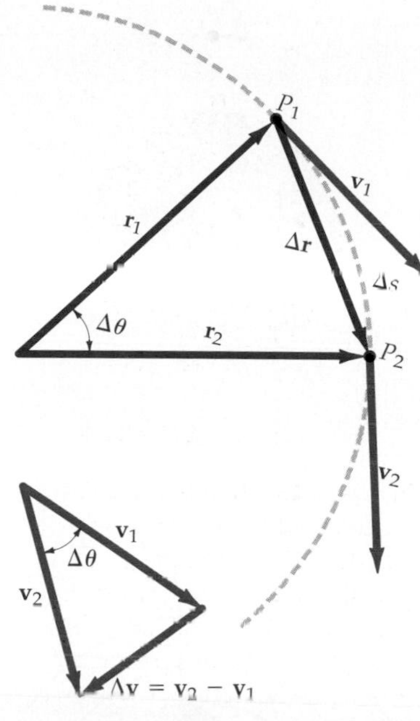

Figure 3-26 Position and velocity vectors for a particle moving in a circle. The angle $\Delta\theta$ between $\mathbf{v}_1$ and $\mathbf{v}_2$ is the same as the angle between $\mathbf{r}_1$ and $\mathbf{r}_2$. For very small time intervals, the change in velocity $\Delta \mathbf{v}$ is approximately perpendicular to $\mathbf{v}$ and points inward toward the center of the circle.

Example 3-10

A tether ball moves in a horizontal circle of radius 2 m. It makes one revolution in 3 s. Find its acceleration.

To find the acceleration we must first find the speed, which is given by Equation 3-23:

$$v = \frac{2\pi r}{T} = \frac{2\pi(2\text{ m})}{3\text{ s}} = 4.19\text{ m/s}$$

The magnitude of the ball's acceleration is then

$$a = \frac{v^2}{r} = \frac{(4.19\text{ m/s})^2}{2\text{ m}} = 8.78\text{ m/s}^2$$

This is about 0.9 times the acceleration due to gravity g. The direction of the acceleration is radially inward, towards the center of the circle.

Example 3-11

A car rounds a curve of radius 30 m. If the maximum centripetal acceleration that can be provided by friction is 5 m/s^2, what is the maximum speed of the car in kilometers per hour?

From Equation 3-22, we have

$$\frac{v^2}{r} = a$$

$$v_{\text{max}} = \sqrt{ra_{\text{max}}}$$

$$= \sqrt{(30\text{ m})(5\text{ m/s}^2)} = 12.2\text{ m/s}$$

To convert this speed to kilometers per hour, we use 1 h = 3600 s and 1 km = 1000 m:

$$12.2\,\frac{\text{m}}{\text{s}} \times \frac{3600\text{ s}}{1\text{ h}} \times \frac{1\text{ km}}{1000\text{ m}} = 44\text{ km/h}$$

Exercise

A car rounds a curve of radius 40 m at 48 km/h. What is its centripetal acceleration? (Answer: 4.44 m/s^2)

Example 3-12

A satellite moves at constant speed in a circular orbit about the center of the earth near the surface of the earth. If its acceleration is $9.81\ \mathrm{m/s^2}$, what is its speed, and how long does it take to make one complete revolution?

This acceleration is the same as that for any body falling freely near the surface of the earth. We take the radius of the earth, about 6370 km, to be the approximate radius of the orbit. (For actual satellites that are put into orbit a few hundred kilometers above the earth's surface, the orbital radius will be slightly greater, so the acceleration will be slightly less than $9.81\ \mathrm{m/s^2}$ because of the decrease in the gravitational force with distance from the center of the earth.) The speed of the satellite can be found from Equation 3-22:

$$v^2 = rg = (6370\ \mathrm{km})(9.81\ \mathrm{m/s^2})$$

$$v = 7.91\ \mathrm{km/s}$$

The period is related to the speed by Equation 3-23:

$$T = \frac{2\pi r}{v} = \frac{2\pi(6370\ \mathrm{km})}{7.91\ \mathrm{km/s}} = 5060\ \mathrm{s} = 84.3\ \mathrm{min}$$

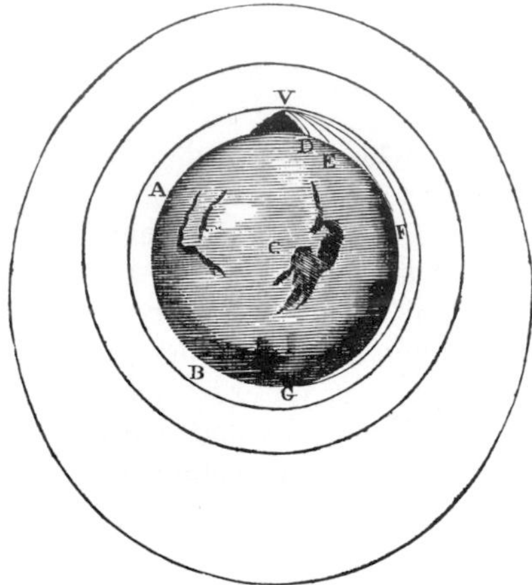

Figure 3-27 Drawing from Newton's *System of the World,* published in 1728, illustrating the connection between projectile motion and satellite motion.

Figure 3-27 is a drawing from Newton's *System of the World* that illustrates the connection between projectile motion and satellite motion. Newton's description of it reads*

> That by means of centripetal forces the planets may be retained in certain . . . orbits, we may easily understand, if we consider the motions of projectiles; for a stone that is projected is by the pressure of its own weight forced out of the rectilinear path, which by the initial projection alone it should have pursued, and made to describe a curved line in the air; and through that crooked way is at last brought down to the ground; and the greater the velocity is with which it is projected, the farther it goes before it falls to the earth. We may therefore suppose the velocity to be so increased, that it would describe an arc of 1, 2, 5, 10, 100, 1000 miles before it arrived at the earth, till at last, exceeding the limits of the earth, it should pass into space without touching it.

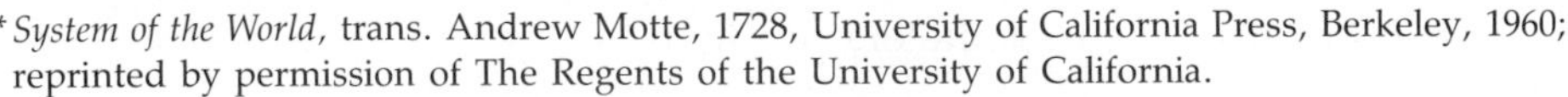

* *System of the World,* trans. Andrew Motte, 1728, University of California Press, Berkeley, 1960; reprinted by permission of The Regents of the University of California.

The space shuttle orbits freely after disengagement from the booster rockets shown at the left.

If a particle moves in a circle with a speed that is varying, there is a component of acceleration that is tangent to the circle as well as the centripetal acceleration inward. The tangential component of the acceleration is simply the rate of change of the speed, whereas the radially inward component has the magnitude v^2/r. For any general motion along a curve, we can treat a portion of the curve as an arc of a circle (Figure 3-28). The particle then has centripetal acceleration v^2/r toward the center of curvature, and if the speed is changing, it has tangential acceleration of magnitude dv/dt.

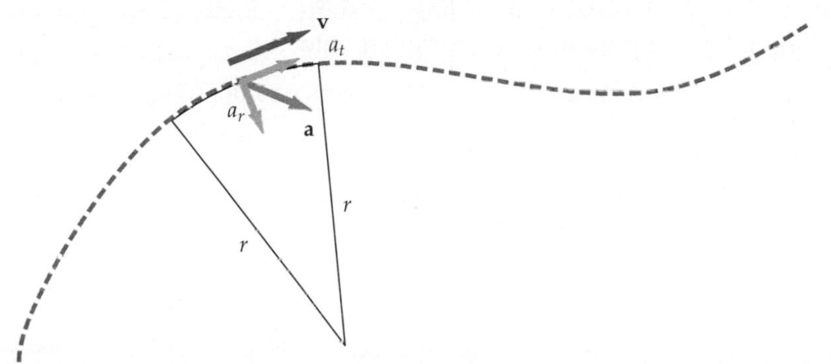

Figure 3-28 A particle moving along an arbitrary curve can be considered to be moving in a circular arc during a small time interval. The instantaneous-acceleration vector has a component a_r of magnitude v^2/r toward the center of curvature of the arc and a component a_t of magnitude dv/dt that is tangential to the curve in the direction of motion.

Summary

1. Quantities that have both magnitude and direction, such as displacement, velocity, and acceleration, are vector quantities.

2. Vectors can be added graphically by placing the tail of one vector at the head of the other and drawing the resultant vector from the tail of the first to the head of the second. Subtracting a vector **B** is the same as adding $-\mathbf{B}$, where $-\mathbf{B}$ is a vector with a magnitude equal to B but in the opposite direction.

3. Vectors can be added analytically by first finding the components of the vectors given by

$$A_x = A \cos \theta$$

$$A_y = A \sin \theta$$

where θ is the angle between **A** and the x axis. The x component of the resultant vector is the sum of the x components of the individual vectors, and the y component is the sum of the y components of the individual vectors.

4. The position vector **r** points from an arbitrary origin to a particle's position. In the time interval Δt, **r** changes by $\Delta\mathbf{r}$. The velocity vector **v** is the rate of change of the position vector. Its magnitude is the speed, and it points in the direction of motion, tangent to the curve along which the particle is traveling. The instantaneous-velocity vector is given by

$$\mathbf{v} = \lim_{\Delta t \to 0} \frac{\Delta \mathbf{r}}{\Delta t} = \frac{d\mathbf{r}}{dt}$$

5. The acceleration vector is the rate of change of the velocity vector. The instantaneous-acceleration vector is given by

$$\mathbf{a} = \lim_{\Delta t \to 0} \frac{\Delta \mathbf{v}}{\Delta t} = \frac{d\mathbf{v}}{dt}$$

A particle is accelerating if its velocity vector is changing in magnitude or direction or both.

6. If a particle moves with velocity $\mathbf{v}_{pA}$ relative to a coordinate system A, which is in turn moving with velocity $\mathbf{v}_{AB}$ relative to another coordinate system B, the velocity of the particle relative to B is

$$\mathbf{v}_{pB} = \mathbf{v}_{pA} + \mathbf{v}_{AB}$$

7. In projectile motion, the horizontal and vertical motions are independent. The horizontal motion has constant velocity that is equal to the horizontal component of the original velocity:

$$v_x = v_{0x} = v_0 \cos \theta$$

$$\Delta x = v_{0x} t$$

The vertical motion is the same as motion in one dimension with constant acceleration due to gravity g downward:

$$v_y = v_{0y} - gt$$

$$\Delta y = v_{0y} t - \tfrac{1}{2} g t^2$$

The total distance traveled by the projectile, called the range R, is found by first finding the total time the projectile is in the air and then multiplying this time by the constant horizontal component of the velocity. For the special case in which the initial and final elevations are the same, the range is related to the projection angle θ by

$$R = \frac{v_0^2}{g} \sin 2\theta$$

and is maximum at $\theta = 45°$.

8. When an object moves in a circle with constant speed, it is accelerating because its velocity is changing in direction. This acceleration is called centripetal acceleration, and it points toward the center of the circle. The magnitude of the centripetal acceleration is

$$a = \frac{v^2}{r}$$

where v is the speed and r is the radius of the circle.

Suggestions for Further Reading

Drake, Stillman, and James MacLachan: "Galileo's Discovery of the Parabolic Trajectory," *Scientific American*, March 1975, p. 102.

Galileo knew that a falling body with a horizontal component of velocity describes a parabola 30 years before he published the fact, according to unpublished manuscripts. (See also Chapter 2 references.)

Review

A. Objectives: After studying this chapter, you should

1. Be able to add and subtract vectors graphically.

2. Be able to obtain the components of vectors and use them to add and subtract vectors.

3. Be able to express arbitrary vectors in terms of unit vectors.

4. Be able to combine relative velocities to express the velocity of a particle relative to a coordinate system that is itself moving relative to another coordinate system.

5. Know that in projectile motion the horizontal and vertical motions are independent, and be able to use this fact in working projectile problems.

6. Know that when a particle moves in a circle with constant speed it has centripetal acceleration of magnitude v^2/r directed towards the center of the circle.

B. Define, explain, or otherwise identify:

Vectors
Scalars
Displacement vector
Parallelogram method of vector addition
Rectangular component of a vector
Unit vector
Position vector
Velocity vector
Acceleration vector
Relative velocity
Range
Centripetal acceleration

C. True or false: If the statement is true, explain why it is true. If it is false, give a counterexample.

1. The magnitude of the sum of two vectors must be greater than the magnitude of either vector.

2. The instantaneous-velocity vector is always in the direction of motion.

3. The instantaneous-acceleration vector is always in the direction of motion.

4. If the speed is constant, the acceleration must be zero.

5. If the acceleration is zero, the speed must be constant.

6. It is impossible to go around a curve without acceleration.

7. The time required for a bullet fired horizontally to reach the ground is the same as if it were dropped from rest from the same height.

Problems

Level I

3-1 The Displacement Vector and Vector Addition

1. A bear walks northeast for 10 m and then east for 10 m. Show each displacement graphically, and find the resultant displacement vector.

2. (*a*) A man walks along a circular arc from the position $x = 5$ m, $y = 0$ to a final position $x = 0$, $y = 5$ m. What is his displacement? (*b*) A second man walks from the same initial position along the x axis to the origin and then along the y axis to $y = 5$ m and $x = 0$. What is his displacement?

3. A circular path has a radius of 10 m. An xy coordinate system is established such that the center of the circle is on the positive y axis and the circle passes through the origin. A woman starts at the origin and walks around the path at a steady speed, returning to the origin exactly 1 min after she started. (*a*) Find the magnitude and direction of her displacement from the origin 15, 30, 45, and 60 s after she starts. (*b*) Find the magnitude and direction of her displacement for each of the four successive 15-s intervals of her walk. (*c*) How is her displacement for the first 15 s related to that for the second 15 s? (*d*) How is her displacement for the second 15-s interval related to that for the last 15-s interval?

4. For the two vectors **A** and **B** shown in Figure 3-29, find the following graphically: (*a*) **A** + **B**, (*b*) **A** − **B**, (*c*) 2**A** + **B**, (*d*) **B** − **A**, (*e*) 2**B** − **A**.

Figure 3-29 Problems 4 and 9.

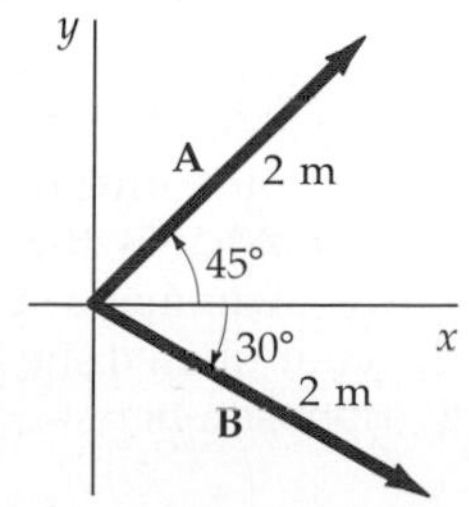

5. A scout walks 2 km due east from camp, then turns left and walks 2 km along the arc of a circle centered at the campsite, and finally walks 1 km directly toward camp. (*a*) How far is the scout from camp? (*b*) What is the direction of the scout's position as measured from the campsite? (*c*) What is the ratio of the final displacement to the total distance walked?

3-2 Addition of Vectors by Components

6. A hiker sets off at 8 A.M. on level terrain. At 9 A.M. she is 2 km due east of her starting point. At 10 A.M. she is 1 km northwest of where she was at 9 A.M. At 11 A.M. she is 3 km due north of where she was at 10 A.M. (*a*) Make a drawing showing these successive displacements as vectors, with the tail of each new vector being at the head of the previous one. What are the magnitudes and directions of these displacements? (Specify the direction of a vector by giving the angle it makes with the eastward direction.) (*b*) What are the north and east components of these displacements? (*c*) How far is the hiker from her starting point at 11 A.M.? In what direction? (*d*) Draw the three displacement vectors to scale and add them graphically. Do these successive straight lines represent the actual path the hiker followed? Is the distance she walked the sum of the lengths of the three displacement vectors?

7. Find the rectangular components of the vectors that lie in the xy plane, have a magnitude of A, and make an angle θ with the x axis, as shown in Figure 3-30, for the following values of A and θ: (*a*) $A = 10$ m, $\theta = 30°$; (*b*) $A = 5$ m, $\theta = 45°$; (*c*) $A = 7$ km, $\theta = 60°$; (*d*) $A = 5$ km, $\theta = 90°$; (*e*) $A = 15$ km/s, $\theta = 150°$; (*f*) $A = 10$ m/s, $\theta = 240°$; and (*g*) $A = 8$ m/s^2, $\theta = 270°$.

Figure 3-30 Problem 7.

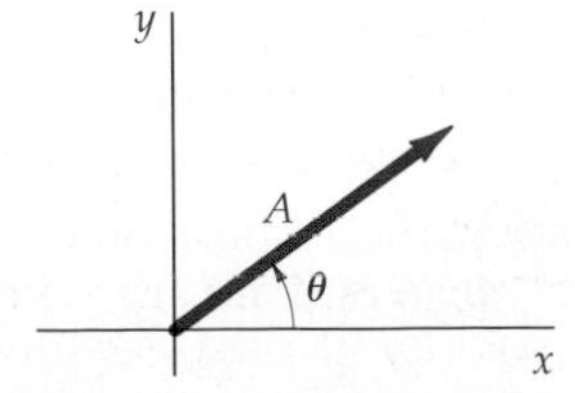

8. A plane is inclined at an angle of 30° with the horizontal. Choose the x axis pointing down the slope of the plane and the y axis perpendicular to the plane. Find the x and y components of the acceleration of gravity, which has the magnitude 9.81 m/s^2 and points vertically down.

9. The displacement vectors **A** and **B** shown in Figure 3-29 both have a magnitude of 2 m. (*a*) Find their x and y components. (*b*) Find the components, magnitude, and direction of the sum **A** + **B**. (*c*) Find the components, magnitude, and direction of the difference **A** − **B**.

3-3 Unit Vectors and Multiplication of Vectors by Scalars

10. Find the magnitude and direction of the following vectors: (*a*) **A** = 5**i** + 3**j**, (*b*) **B** = 10**i** − 7**j**, and (*c*) **C** = −2**i** − 3**j** + 4**k**.

11. Find the magnitude and direction of **A**, **B**, and **A** + **B** for (*a*) **A** = −4**i** − 7**j**, **B** = 3**i** − 2**j**, and (*b*) **A** = 1**i** − 4**j**, **B** = 2**i** + 6**j**.

12. A cube of side 2 m has its faces parallel to the coordinate planes with one corner at the origin. A fly begins at the origin and walks along three edges until it is at the far corner. Write the displacement vector of the fly using the unit vectors **i**, **j**, and **k**, and find the magnitude of this displacement.

13. Describe the following vectors using the unit vectors **i** and **j**: (*a*) a velocity of 10 m/s at an angle of elevation of 60°; (*b*) a vector **A** of magnitude $A = 5$ m and $\theta = 225°$; and (*c*) a displacement from the origin to the point $x = 14$ m, $y = -6$ m.

14. For the vector **A** = 3**i** + 4**j**, find any three other vectors **B** that also lie in the xy plane and have the property that $A = B$ but **A** ≠ **B**. Write these vectors in terms of their components and show them graphically.

15. Two vectors **A** and **B** lie in the xy plane. Under what conditions does the ratio A/B equal A_x/B_x?

16. If **A** = 5**i** − 4**j** and **B** = −7.5**i** + 6**j**, write an equation relating **A** to **B**.

3-4 The Velocity Vector

17. A vector **A**(t) has a constant magnitude but is changing direction in a uniform way. Draw the vectors **A**(t + Δt) and **A**(t) for a small time interval Δt, and find the difference Δ**A** = **A**($t + \Delta t$) − **A**(t) graphically. How is the direction of Δ**A** related to **A** for small time intervals?

18. Consider the path of a particle as it moves in space. (*a*) How is the velocity vector related geometrically to the path of the particle? (*b*) Sketch a curved path, and draw the velocity vector for the particle for several positions along the path.

19. A stationary radar operator determines that a ship is 10 km south of him. An hour later the same ship is 20 km southeast of him. If the ship moved at constant speed and always in the same direction, what was its velocity during this time?

20. A particle's position coordinates (x, y) are (2 m, 3 m) at $t = 0$; (6 m, 7 m) at $t = 2$ s; and (13 m, 14 m) at $t = 5$ s. (*a*) Find $\mathbf{v}_{av}$ from $t = 0$ to $t = 2$ s. (*b*) Find $\mathbf{v}_{av}$ from $t = 0$ to $t = 5$ s.

21. The position vector of a particle is given by **r** = 5t**i** + 10t**j**, where t is in seconds and **r** is in meters. (*a*) Draw the path of the particle in the xy plane. (*b*) Find **v** in component form and find its magnitude.

3-5 The Acceleration Vector

22. A ball is thrown directly upward. Consider the 2-s time interval $\Delta t = t_2 - t_1$, where t_1 is 1 s before the ball reaches its highest point and t_2 is 1 s after it reaches its highest point. Find (*a*) the change in speed, (*b*) the change in velocity, and (*c*) the average acceleration for this time interval.

23. Figure 3-31 shows the path of an automobile that is made up of segments of straight lines and arcs of circles. The automobile starts from rest at point A. After it reaches point B, it travels at constant speed until it reaches point E. It comes to rest at point F. (*a*) At the middle of each segment (AB, BC, CD, DE, and EF), what is the direction of the velocity vector? (*b*) At which of these points does the automobile have an acceleration? In those cases, what is the direction of the acceleration? (*c*) How do the magnitudes of the acceleration compare for segments BC and DE?

Figure 3-31 Problem 23.

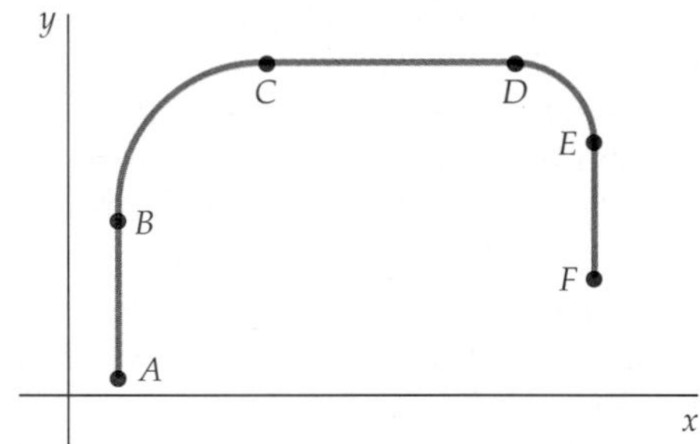

24. Initially, a particle is moving due west with a speed of 40 m/s, and 5 s later it is moving north with a speed of 30 m/s. (*a*) What was the change in the magnitude of the particle's velocity during this time? (*b*) What was the change in the direction of the velocity? (*c*) What are the magnitude and direction of Δ**v** for this interval? (*d*) What are the magnitude and direction of $\mathbf{a}_{av}$ for this interval?

25. At $t = 0$ a particle located at the origin has a speed of 40 m/s at $\theta = 45°$. At $t = 3$ s the particle is at $x = 100$ m and $y = 80$ m with speed of 30 m/s at $\theta = 50°$. Calculate (*a*) the average velocity and (*b*) the average acceleration of the particle during this interval.

3-6 Relative Velocity

26. A swimmer heads directly across a river, swimming at 1.6 m/s relative to still water. She arrives at a point 40 m downstream from the point directly across the river, which is 80 m wide. (*a*) What is the speed of the river current? (*b*) What is the swimmer's speed relative to the shore? (*c*) In what direction should she head so as to arrive at the point directly opposite her starting point?

27. A plane flies at a speed of 250 km/h relative to still air. There is a wind blowing at 80 km/h in the northeast direction (that is, at 45° to the east of north). (*a*) In what direction should the plane head so as to fly due north? (*b*) What is the speed of the plane relative to the ground?

28. A missile fired from a jet takes off from the wing with an acceleration of 500 m/s^2 that lasts for 3 s. The jet's velocity is 500 m/s due east. (*a*) What are the position and velocity of the missile as observed by the jet pilot 2 s after it is fired? (*b*) What are the position and velocity of the missile as observed by a person on the ground 2 s after it is fired? (Neglect the acceleration due to gravity.)

3-7 Projectile Motion

29. A bullet is fired horizontally with an initial velocity of 245 m/s. The gun is 1.5 m above the ground. How long is the bullet in the air?

30. A supersonic transport is flying horizontally at an altitude of 20 km with a speed of 2500 km/h when an engine falls off. (*a*) How long does it take the engine to hit the ground? (*b*) How far horizontally is the engine from where it fell off when it hits the ground? (*c*) How far is the engine from the aircraft (assuming it continues to fly as if nothing had happened) when the engine hits the ground? Neglect air resistance.

31. A cannon is elevated at an angle of 45°. It fires a ball with a speed of 300 m/s. (*a*) What height does the ball reach? (*b*) How long is the ball in the air? (*c*) What is the horizontal range?

32. A projectile is launched with speed v_0 at an angle of θ_0 with the horizontal. Find an expression for the maximum height it reaches above its starting point in terms of v_0, θ_0, and g.

33. A projectile is fired with an initial velocity of 30 m/s at 60° to the horizontal. At the projectile's highest point, what is its velocity? Its acceleration?

34. A pitcher throws a fastball at 140 km/h toward home plate, which is 18.4 m away. Neglecting air resistance (not a good idea if you are the batter), find how far the ball drops because of gravity by the time it reaches home plate.

3-8 Circular Motion

35. A particle travels in a circular path of radius 5 m with a constant speed of 15 m/s. What is the magnitude of its acceleration?

36. An airplane pilot pulls out of a dive by following the arc of a circle whose radius is 300 m. At the bottom of the circle, where her speed is 180 km/h, what are the direction and magnitude of her acceleration?

37. An object travels with a constant speed v in a circular path of radius r. (*a*) If v is doubled, how is the acceleration a affected? (*b*) If r is doubled, how is a affected? (*c*) Why is it impossible for an object to travel around a perfectly sharp angular turn?

38. In Figure 3-32 the particles are traveling counterclockwise in circles of radius 5 m with speeds that may be varying. The acceleration vectors are indicated at certain times. Find the values of v and dv/dt at each of these three times.

39. A boy whirls a ball on a string in a horizontal circle of radius 1 m. How many revolutions per minute must the ball make if its acceleration toward the center of the circle is to have the same magnitude as the acceleration due to gravity?

Level II

40. A particle moves in a circle of radius 4 cm. It takes 8 s to make a complete trip. Draw the path of the particle to scale, and indicate the particles's positions at 1-s intervals. Draw the displacement vectors for these 1-s intervals. These vectors also indicate the average-velocity vectors for these intervals. Find graphically the change in the average velocity $\Delta\mathbf{v}$ for two consecutive 1-s intervals. Compare $\Delta\mathbf{v}/\Delta t$, measured in this way, with the instantaneous acceleration computed from $a_r = v^2/r$.

41. A particle moves in the xy plane with constant acceleration. At $t = 0$ the particle is at $\mathbf{r} = 4\text{ m}\mathbf{i} + 3\text{ m}\mathbf{j}$. At $t = 2$ s the particle has moved to $\mathbf{r} = 10\text{ m}\mathbf{i} - 2\text{ m}\mathbf{j}$, and its velocity has changed to $\mathbf{v} = 5\text{ m/s}\mathbf{i} - 6\text{ m/s}\mathbf{j}$. (*a*) What is the acceleration of the particle? (*b*) What is the velocity of the particle as a function of time? (*c*) What is the position vector of the particle as a function of time?

42. A particle moves in an xy plane with constant acceleration. At time zero, the particle is at $x = 4$ m, $y = 3$ m. The acceleration is given by the vector $\mathbf{a} = 4\text{ m/s}^2\mathbf{i} + 3\text{ m/s}^2\mathbf{j}$. The velocity vector initially is $\mathbf{v} = 2\text{ m/s}\mathbf{i} - 9\text{ m/s}\mathbf{j}$. (*a*) Find the velocity vector at $t = 2$ s. (*b*) Find the position vector at $t = 4$ s. Give its magnitude and direction.

Figure 3-32 Problem 38.

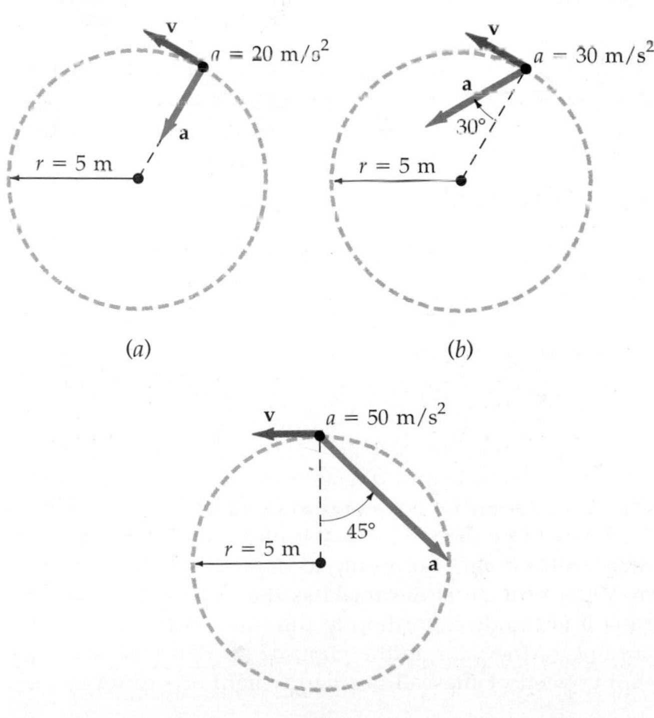

Figure 3-33 Problem 43.

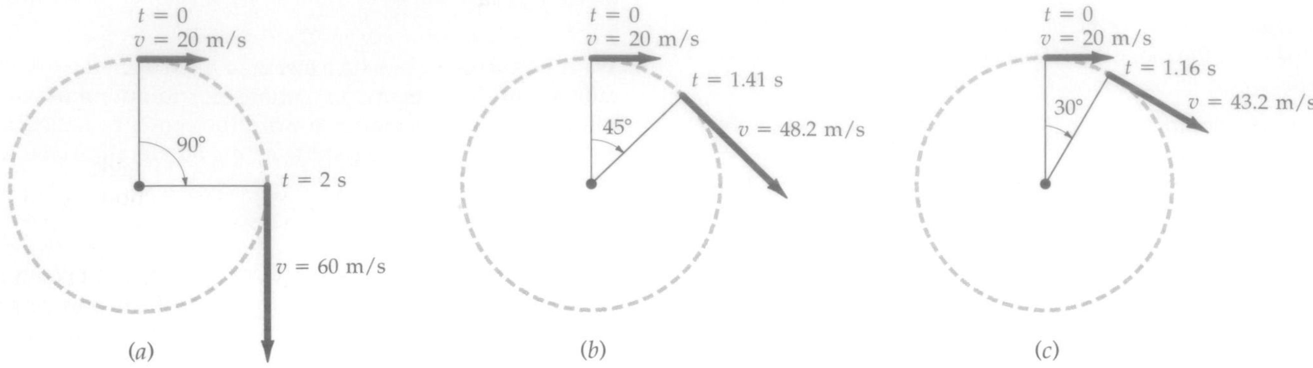

43. In parts (*a*) to (*c*) of Figure 3-33, particles are traveling in circular paths with varying speeds. The velocity vectors are shown. Find the magnitude of the average-acceleration vector between the two given positions in each case.

44. A particle travels with constant speed in a circular path of radius 5 m around the origin. It begins at $t = 0$ at $x = 5$ m, $y = 0$ and takes 100 s to make a complete revolution. (*a*) What is the speed of the particle? (*b*) Give the magnitude and direction of the position vector **r** at the times $t = 50$ s, $t = 25$ s, $t = 10$ s, and $t = 0$. (*c*) Find the magnitude of $\mathbf{v}_{av}$ and show its direction graphically for each of the following time intervals: $t = 0$ to $t = 50$ s, $t = 0$ to $t = 25$ s, and $t = 0$ to $t = 10$ s. (*d*) How does $\mathbf{v}_{av}$ for the interval $t = 0$ to $t = 10$ s compare with the instantaneous velocity at $t = 0$?

45. A particle has a position vector given by $\mathbf{r} = 30t\mathbf{i} + (40t - 5t^2)\mathbf{j}$, where **r** is in meters and t in seconds. Find the instantaneous-velocity and instantaneous-acceleration vectors as functions of time t.

46. A particle has a constant acceleration of $\mathbf{a} = 6\mathbf{i} + 4\mathbf{j}$ m/s^2. At time $t = 0$, the velocity is zero and the position vector is $\mathbf{r}_0 = (10\text{ m})\mathbf{i}$. (*a*) Find the velocity and position vectors at any time t. (*b*) Find the equation of the particle's path in the xy plane, and sketch the path.

47. A stone thrown horizontally from the top of a tower hits the ground at a distance of 18 m from the base of the tower. (*a*) Find the speed at which the stone was thrown if the tower is 24 m high. (*b*) Find the speed of the stone just before it hits the ground.

48. A freight train is moving at a constant speed of 10 m/s. A man standing on a flatcar throws a ball into the air and catches it as it falls. Relative to the flatcar, the initial velocity of the ball is 15 m/s straight up. (*a*) What are the magnitude and direction of the initial velocity of the ball as seen by a second man standing next to the track? (*b*) How long is the ball in the air according to the man on the train? According to the man on the ground? (*c*) What horizontal distance has the ball traveled by the time it is caught according to the man on the train? According to the man on the ground? (*d*) What is the minimum speed of the ball during its flight according to the man on the train? According to the man on the ground? (*e*) What is the acceleration of the ball according to the man on the train? According to the man on the ground?

49. A projectile is fired into the air from the top of a 200-m cliff above a valley (Figure 3-34). Its initial velocity is 60 m/s at 60° with the horizontal. Neglecting air resistance, where does the projectile land?

Figure 3-34 Problem 49.

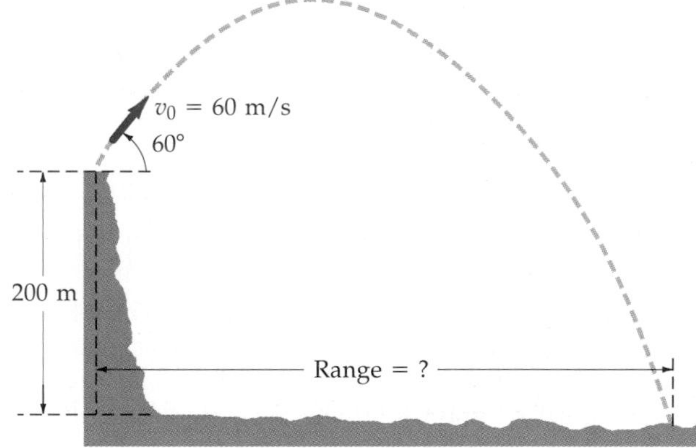

50. A car is traveling down a highway at 25 m/s. Just as the car crosses a perpendicularly intersecting crossroad, the passenger throws out a can at a 45° angle of elevation in a plane perpendicular to the motion of the car. The initial speed of the can relative to the car is 10 m/s. It is released at a height of 1.2 m above the road. (*a*) Write the initial velocity of the can (relative to the road) in terms of the unit vectors **i**, **j**, and **k**. (*b*) Where does the can land?

51. Car A is traveling east at 20 m/s. As car A crosses the intersection shown in Figure 3-35, car B starts from rest 40 m north of the intersection and moves south with a constant acceleration of 2 m/s^2. (*a*) What is the position of B relative to A 6 s after A crosses the intersection? (*b*) What is the velocity of B relative to A for $t = 6$ s? (*c*) What is the acceleration of B relative to A for $t = 6$ s?

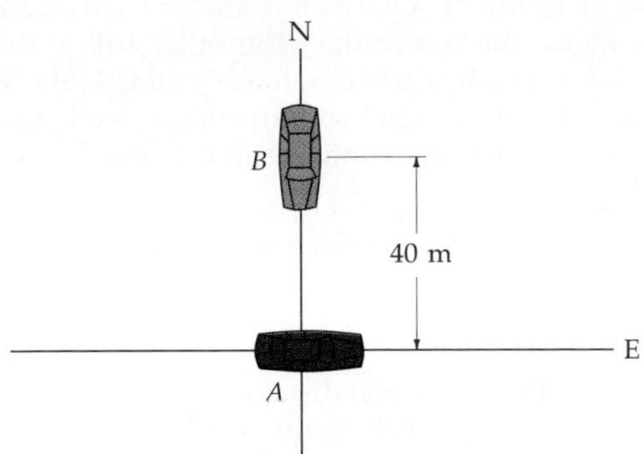

Figure 3-35 Problem 51.

52. An object on the equator has an acceleration toward the center of the earth because of the earth's rotation and an acceleration toward the sun because of the earth's motion along its orbit. Calculate the magnitudes of both accelerations, and express them as fractions of the free-fall acceleration due to gravity g.

53. A baseball is thrown toward a player with an initial speed of 20 m/s at an angle of 45° with the horizontal. At the moment the ball is thrown, the player is 50 m from the thrower. At what speed and in what direction must he run to catch the ball at the same height at which it was released?

54. Find the angle of projection such that the maximum height of a projectile is equal to its horizontal range.

55. A trail bike comes to a ditch. A ramp with an incline of 10° has been built so that the bike can jump the ditch. If the bike needs to jump a horizontal distance of 7 m to clear the ditch, how fast must it be going when it leaves the ramp?

56. The position of a particle is given by the vector

$$\mathbf{r} = -10 \text{ m} \cos \omega t \, \mathbf{i} + 10 \text{ m} \sin \omega t \, \mathbf{j}$$

where $\omega = 2 \text{ s}^{-1}$. (*a*) Show that the motion is a circle. What is the radius of the circle? Does the particle move clockwise or counterclockwise around the circle? (*b*) What is the speed of the particle? (*c*) What is the time for one complete revolution?

57. A workman on the roof of a house drops his hammer, which slides down the roof at a constant speed of 4 m/s. The roof makes an angle of 30° with the horizontal, and its lowest point is 10 m from the ground. What is the horizontal distance traveled by the hammer after it leaves the roof of the house before it hits the ground?

58. A motorboat begins on a course heading toward an island that is 1 km east and 3 km north from the boat's initial position. After traveling 45 min, the driver discovers that the boat is due east of the island. He reverses his course, and 45 min later the boat is 6 km east of its starting position. (*a*) What is the velocity of the water current? (*b*) What is the velocity of the boat relative to the water for the first 45 min? (*c*) What is the velocity of the boat relative to the island for the first 45 min?

Level III

59. A gun shoots bullets that leave the muzzle at 250 m/s. If a bullet is to hit a target 100 m away at the level of the muzzle, the gun must be aimed at a point above the target. How far above the target is this point? (Neglect air resistance.)

60. Galileo showed that, if air resistance is neglected, the ranges for projectiles whose angles of projection exceed or fall short of 45° by the same amount are equal. Prove this.

Projectile paths for various projection angles (taken from a book on physics by Petrus van Musschenbroek in 1762).

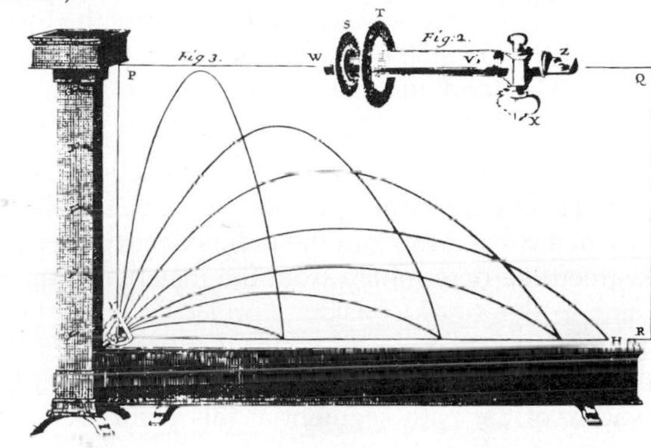

61. A particle moves clockwise in a circle of radius 1 m with its center at $(x, y) = (1 \text{ m}, 0)$. The particle starts from rest at the origin at time $t = 0$. Its speed increases at the constant rate of $(\pi/2) \text{ m/s}^2$. (*a*) How long does it take the particle to travel halfway around the circle? (*b*) What is its speed at that time? (*c*) What is the direction of its velocity at that time? (*d*) What is its radial acceleration then? Its tangential acceleration? (*e*) What are the magnitude and direction of the total acceleration halfway around the circle?

62. A boy stands 4 m from a vertical wall and throws a ball (Figure 3-36). The ball leaves the boy's hand at 2 m above the ground with initial velocity $\mathbf{v} = (10\mathbf{i} + 10\mathbf{j})$ m/s. When the ball hits the wall, the horizontal component of its velocity is reversed and the vertical component remains unchanged. Where does the ball hit the ground?

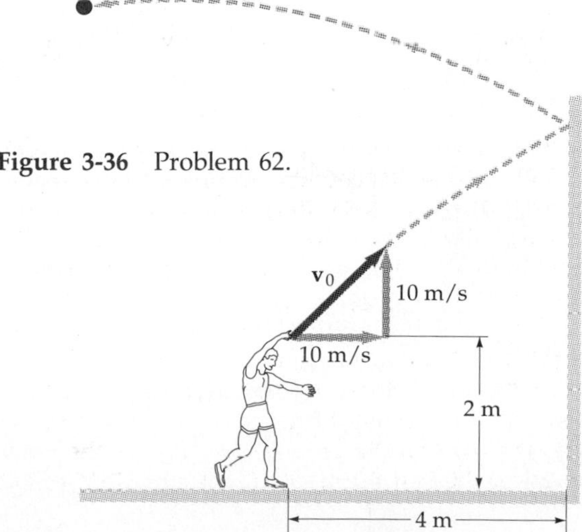

Figure 3-36 Problem 62.

63. A baseball just clears a 3-m wall that is 120 m from home plate. If it leaves the bat at 45° and 1.2 m above the ground, what must its initial velocity be (again, making the unrealistic assumption that air resistance can be ignored)?

64. A baseball is struck by a bat, and 3 s later it is caught 30 m away. (*a*) If the baseball was 1 m above the ground when it was struck and caught, what was the greatest height it reached above the ground? (*b*) What were the horizontal and vertical components of its velocity when it was struck? (*c*) What was its speed when it was caught? (*d*) At what angle with the horizontal did it leave the bat? (Neglect air resistance.)

65. A hockey puck struck at ice level just clears the top of a glass wall that is 2.80 m high. The flight time to this point is 0.650 s, and the horizontal distance is 12.0 m. Find (*a*) the initial speed of the puck and (*b*) the maximum height it will reach.

66. For short time intervals any path can be considered to be an arc of a circle. How can the radius of curvature of a path segment be determined from the instantaneous velocity and acceleration? Consider a projectile at the top of its path. Indicate the velocity vector just before and just after this point. Is the speed changing? What is the radius of curvature of the path segment at this point?

67. The position of a particle as a function of time is

$$\mathbf{r} = 4 \text{ m} \sin 2\pi t \, \mathbf{i} + 4 \text{ m} \cos 2\pi t \, \mathbf{j}$$

where t is in seconds. (*a*) Show that the path of this particle is a circle of radius 4 m with its center at the origin. (*b*) Compute the velocity vector. Show that $v_x/v_y = -y/x$. (*c*) Compute the acceleration vector and show that it is in the radial direction and has the magnitude v^2/r.

68. The position of a particle is given by

$$\mathbf{r} = 3 \text{ m} \sin 2\pi t \, \mathbf{i} + 2 \text{ m} \cos 2\pi t \, \mathbf{j}$$

where t is in seconds. (*a*) Plot the path of the particle in the xy plane. (*b*) Find the velocity vector. (*c*) Find the acceleration vector and show that its direction is along **r**; that is, it is radial. (*d*) Find the times for which the speed is a maximum or minimum.

69. In Figure 3-37, a trail bike takes off from a ramp at angle θ to clear a ditch of width x and land on the other side, which is elevated at a height H. (*a*) For a given angle θ and distance x, what is the upper limit for H such that the bike has any chance of making the jump? (*b*) For H less than this upper limit, what is the minimum takeoff speed v_0 necessary for a successful jump? (Neglect the size of the bike, and assume that covering a horizontal distance x and a vertical distance H is sufficient to clear the ditch.)

70. Two balls are thrown with equal speeds from the top of a cliff of height H. One ball is thrown upward at an angle α above the horizontal. The other ball is thrown downward at an angle β below the horizontal. Show that each ball strikes the ground with the same speed, and find that speed in terms of H and the initial speed v_0.

71. A baseball player hits a baseball that lands in the stands 24 m above the playing field. The ball lands with a velocity of 50 m/s at an angle of 35° below the horizontal. (*a*) If the player hits the ball 1 m above the playing field, what was the velocity of the ball upon leaving the bat? (*b*) What was the horizontal distance traveled by the ball? (*c*) How long was the ball in the air? (Neglect air resistance.)

72. A steel ball is dropped from a point with (x, y) coordinates of (8 m, 16 m). At the same time, another ball is launched from the origin with a speed of 20 m/s at an angle of 30°. (*a*) Find the minimum distance of separation of the two balls. (*b*) At what time does this separation occur? (*c*) Give the coordinates of the two balls for the minimum separation.

73. The position vector of a particle moving in the xy plane is given by

$$\mathbf{r} = (10t + 5 \cos 2t) \, \mathbf{i} + (5 - 5 \sin 2t) \, \mathbf{j}$$

where **r** is in meters and t is in seconds. (*a*) What is the velocity of the particle? (*b*) The position vector describes a particle on the rim of a rolling wheel. Sketch the locus of points that the particle traces out as the wheel rolls along. Find the x values for which the particle is in contact with the surface upon which the wheel rolls. (*c*) What is the acceleration of the particle? (*d*) At what times is the particle not moving, and how is the particle on the rim related to the horizontal surface at these times?

Figure 3-37 Problem 69.

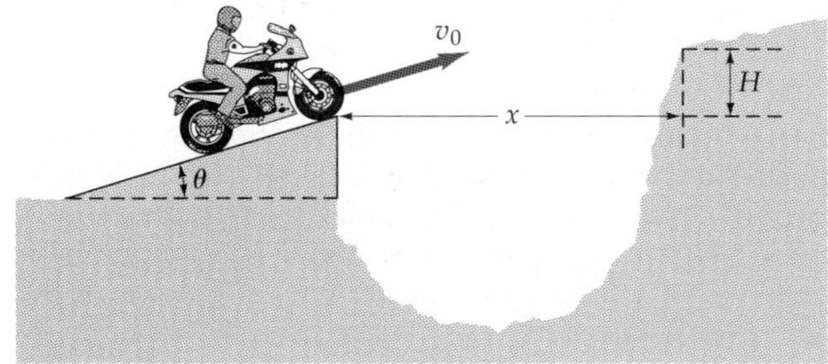

Chapter 4

Newton's Laws I

Isaac Newton (1642-1727)

Classical, or newtonian, mechanics is a theory of motion based on the ideas of mass and force and the laws connecting these physical concepts to the kinematic quantities—displacement, velocity, and acceleration—discussed in the preceding chapters. All phenomena in classical mechanics can be described using just three simple laws called Newton's laws of motion. Newton's laws relate the acceleration of an object to its mass and the forces acting on it. In this chapter, we will examine each of Newton's laws in detail, and we will carefully define the concepts of force and mass. We will see how Newton's laws can be applied to simple problems in which a single object is acted on by forces that are constant in magnitude. In Chapter 5, we will examine more general applications of Newton's laws.

A modern version of Newton's laws is

> Law 1. An object continues in its initial state of rest or motion with uniform velocity unless it is acted on by an unbalanced, or net external, force. (The net force acting on an object, also called the resultant force, is the vector sum of all the forces acting on it: $\mathbf{F}_{\text{net}} = \Sigma\mathbf{F}$.)

Newton's Laws

Law 2. The acceleration of an object is inversely proportional to its mass and directly proportional to the net external force acting on it:

$$\mathbf{a} = \frac{\mathbf{F}_{\text{net}}}{m}$$

or

$$\mathbf{F}_{\text{net}} = m\mathbf{a}$$

Law 3. Forces always occur in pairs. If object A exerts a force on object B, an equal but opposite force is exerted by object B on object A.

4-1 Newton's First Law: The Law of Inertia

Newton's first law states that an object at rest or moving with constant velocity will remain at rest or will continue moving with constant velocity unless it is acted on by a net external force. This tendency is described by saying that the object has **inertia.** Accordingly, Newton's first law is often called the **law of inertia.** Prior to Galileo, it was generally thought that a force, such as a push or a pull, was necessary to keep an object moving with constant velocity. In everyday experience, if a book is pushed along a table and then let go, it slides for a while and then stops. Galileo, and later Newton, recognized that in such a situation the book is not free from external forces because friction is present. If we polish the surface of the table, the book slides farther, and the decrease in its velocity in a given time is smaller. If we support the book on a thin cushion of air (this is possible with an air table), the book will glide for a considerable time and distance with almost no perceptible change in its velocity.

Galileo studied motion by doing experiments in which he rolled balls down and up inclined planes. He found, for example, that when a ball rolled down an incline its speed increased by equal amounts in equal time intervals; that is, the acceleration of the ball was constant. Similarly, when a ball rolled up an incline, its speed decreased by equal amounts in equal time intervals. Figure 4-1 shows a ball rolling down one incline and up another.

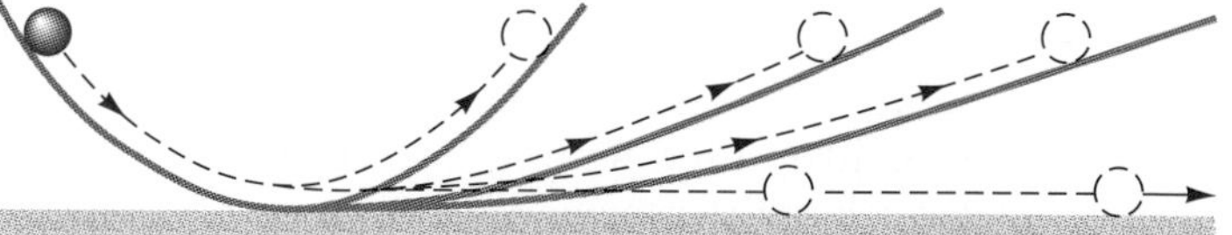

Figure 4-1 Galileo's experiment with balls rolling down and up inclined planes. The ball rolls up the second incline to nearly its original height no matter what the angle of incline. As the angle is decreased, the ball rolls farther and farther.

The ball rolls up the second incline to nearly the same height from which it started, regardless of the relative slopes of the inclines. As the slope of the second incline is reduced, the ball rolls farther and farther. Galileo reasoned that, if he could eliminate the effects of friction, a ball rolling on a horizontal plane would roll forever with no change in speed. Newton stated these results as his first law of motion.

Note that Newton's first law makes no distinction between an object at rest and one that is moving with constant velocity. The question of whether an object is at rest or is moving with constant velocity depends on the refer-

Midair refueling. Each plane is nearly at rest relative to the other plane, although they are both moving with a very large velocity relative to the earth.

Friction is greatly reduced by a cushion of air that supports the hovercraft.

ence frame in which the object is observed. Consider a book resting on an air table in a railroad boxcar. In a coordinate system whose origin O' is attached to the boxcar (Figure 4-2*a*), the book is at rest. The coordinate system provides a **reference frame** for measuring the position, velocity, and acceleration of the book. We will call the reference frame attached to the boxcar S'. Now suppose the boxcar is moving along the track to the right with speed v. The book is placed on the air table so that it is again at rest relative to the boxcar. The speed v of the boxcar is measured relative to a second coordinate system whose origin O is attached to the track. In the reference frame of the track, which we will call S, the book is moving to the right with speed v (Figure 4-2*b*). According to Newton's first law, the book will continue to move with constant velocity in reference frame S or will remain at rest in reference frame S' unless it is acted on by a net force.

Newton's first law does not hold in all reference frames. Suppose reference frame S'' is attached to a boxcar that is accelerating along the track with an acceleration a relative to the track. Let us suppose that the boxcar starts from rest at time $t = 0$ and at that time we place a book on an air table in the

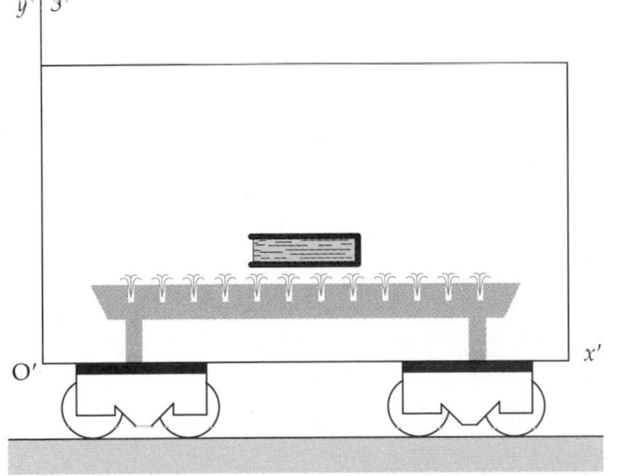

(*a*)

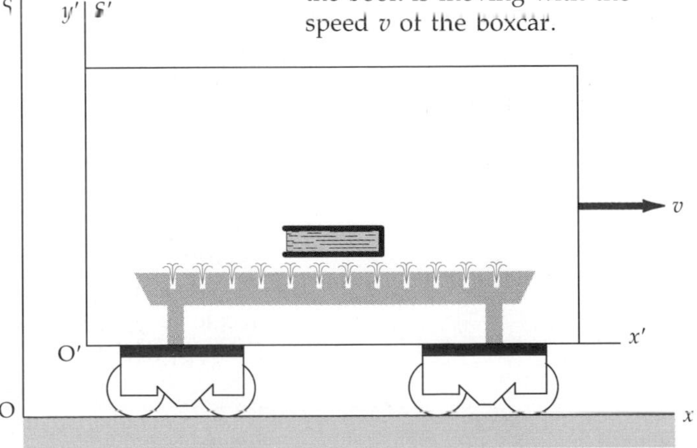

(*b*)

Figure 4-2 (*a*) A book is at rest on an air table in a boxcar in frame S'. (*b*) Relative to frame S, the book is moving with the speed v of the boxcar.

car. The book will remain at rest relative to the track, but it will accelerate backward with acceleration $-a$ relative to the boxcar. Thus, relative to the boxcar, the book has a horizontal acceleration with no horizontal force. If the book is to remain at rest relative to the boxcar, a horizontal force must be exerted on the book (Figure 4-3). Newton's first law does not hold in this reference frame.

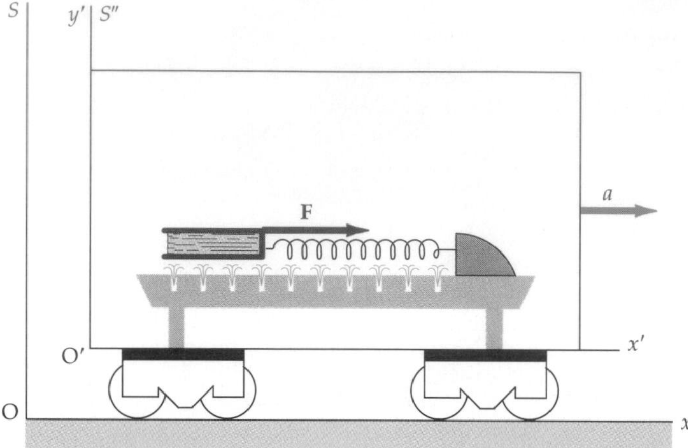

Figure 4-3 A horizontal force is exerted on the book by the attached spring to give the book an acceleration a relative to frame S. Relative to frame S'' attached to the boxcar, the book is at rest even though there is a net force acting on it. Newton's first law does not apply in the reference frame S'' that is accelerating.

A reference frame in which Newton's first law holds is called an **inertial reference frame.** Any reference frame that is moving with constant velocity relative to an inertial reference frame is itself an inertial reference frame. A reference frame attached to the surface of the earth is not quite an inertial reference frame because of the small acceleration of the surface of the earth (relative to the center of the earth) due to the rotation of the earth, and because of the small centripetal acceleration of the earth itself due to its revolution around the sun. However, these accelerations are of the order of $0.01\ \text{m/s}^2$ or less, so to a good approximation, a reference frame attached to the surface of the earth is an inertial reference frame.

4-2 Force, Mass, and Newton's Second Law

Newton's first and second laws may be considered to be a definition of force. A **force** is an influence on an object that causes the object to change its velocity, that is, to accelerate. The direction of the force is the direction of the acceleration it causes if it is the only force acting on the object. The magnitude of the force is the product of the mass of the object and the magnitude of the acceleration produced by the force. (We will define the mass of an object shortly.) This definition of force is in accord with our intuitive idea of a force as a push or pull like that exerted by our muscles. It has been found experimentally that when two or more forces act on the same object, the acceleration of the object is the same as if it were acted on by a single force equal to the vector sum of the individual forces. That is, forces combine as vectors.

Mass is an intrinsic property of an object that measures its resistance to acceleration. The ratio of any two masses can be defined as follows. If a force F is applied to an object of mass m_1, producing acceleration a_1, then

$$F = m_1 a_1$$

If the same force is applied to a second object of mass m_2, producing acceleration a_2, then

$$F = m_2 a_2$$

Combining these equations, we have

$$F = m_1a_1 = m_2a_2$$

or

$$\frac{m_2}{m_1} = \frac{a_1}{a_2} \qquad 4\text{-}1$$

Mass defined

Thus the ratio of the masses of any two objects is defined by applying the same force to each and comparing their accelerations. This definition is in agreement with our intuitive idea of mass. For example, if one object is more "massive" than another according to our everyday use of this term, we will find that a given force produces a smaller acceleration on the more massive object. Experimentally, we find that the ratio of the accelerations a_2/a_1 produced by the same force acting on two objects is independent of the magnitude or direction of the force. It is also independent of the kind of force used; that is, it does not matter whether the force is due to springs, the pull of gravity, electric or magnetic attraction or repulsion, and so forth. We also find that if mass m_2 is found to be twice mass m_1 by direct comparison and if a third mass m_3 is found to be 4 times the mass m_1, then m_3 will be twice the mass of m_2 when the two masses are compared directly. We can therefore set up a mass scale by choosing one particular object to be the standard object and assigning it a mass of 1 unit. The international standard object is a cylinder of a platinum alloy carefully preserved at the International Bureau of Weights and Measures at Sèvres, France. The mass of the standard object is 1 **kilogram,** the SI unit of mass.* The standard object can be used to produce secondary standards by direct comparison, and the mass of any other object can then be found by comparing the acceleration produced on it by a given force with the acceleration produced on one of the secondary standards. The mass of an object is an intrinsic property of the object that does not depend on the location of the object. That is, the mass of an object remains the same whether the object is on the earth, on the moon, or in outer space.

The force required to produce an acceleration of 1 m/s^2 on the standard object is defined to be 1 **newton** (N). Similarly, the force that produces an acceleration of 2 m/s^2 on the standard object is defined to be 2 N.

Example 4-1

A given force produces an acceleration of 5 m/s^2 on the standard object. When the same force is applied to a second object, it produces an acceleration of 15 m/s^2. What is the mass of the second object, and what is the magnitude of the force?

Since the acceleration of the second object is 3 times that of the standard object under the influence of the same force, the mass of the second object is $\frac{1}{3}$ that of the standard object or 0.33 kg. The magnitude of the force is $F = (1\ \text{kg})(5\ \text{m/s}^2) = 5$ N.

Note that we have defined the concepts of force and mass so that Newton's second law follows directly from these definitions:

$$\mathbf{F}_{\text{net}} = m\mathbf{a} \qquad 4\text{-}2$$

Newton's second law

*As mentioned in Chapter 1, the mass of the standard object, 1 kg, was originally intended to be equal to the mass of 1000 cm^3 = 1 liter of water.

Newton's second law provides the link between the dynamic quantities of force and mass and the kinematic quantities of acceleration, velocity, and displacement discussed in Chapters 2 and 3. It is useful because it allows us to describe a wide variety of physical phenomena using just a few, relatively simple force laws. For example, with the addition of Newton's law of gravitational attraction between two bodies, we can describe such phenomena as the motion of the moon, the orbits of the planets around the sun, the orbits of artificial satellites, the variation in the acceleration of gravity g with altitude and with latitude, the variations in the acceleration of gravity due to the presence of mineral deposits, and the paths of ballistic missiles.

Example 4-2

A 4-kg block is at rest at $t = 0$. A single constant horizontal force F_x acts on the block. At $t = 3$ s the block has moved 2.25 m. Find the force F_x.

Since the net force acting on the block is constant, the acceleration of the block is constant and can be found from the constant acceleration formulas from Chapter 2. Using Equation 2-12 with $v_0 = 0$, we have

$$\Delta x = \tfrac{1}{2}at^2$$

$$a = \frac{2\Delta x}{t^2} = \frac{2(2.25\text{ m})}{(3\text{ s})^2} = 0.500\text{ m/s}^2$$

The force is therefore

$$F_x = ma = (4\text{ kg})(0.500\text{ m/s}^2) = 2.00\text{ N}$$

Example 4-3

A particle of mass 0.4 kg is subjected to two forces, $\mathbf{F}_1 = 2\text{ N }\mathbf{i} - 4\text{ N }\mathbf{j}$ and $\mathbf{F}_2 = -2.6\text{ N }\mathbf{i} + 5\text{ N }\mathbf{j}$. If the particle starts from rest at the origin at $t = 0$, find its position and velocity at $t = 1.6$ s.

The net force acting on the particle is the vector sum of the two given forces:

$$\mathbf{F}_{\text{net}} = \mathbf{F}_1 + \mathbf{F}_2 = (2\text{ N }\mathbf{i} - 4\text{ N }\mathbf{j}) + (-2.6\text{ N }\mathbf{i} + 5\text{ N }\mathbf{j})$$
$$= -0.6\text{ N }\mathbf{i} + 1.0\text{ N }\mathbf{j}$$

The acceleration of the particle is given by Newton's second law:

$$\mathbf{a} = \frac{\mathbf{F}_{\text{net}}}{m} = \frac{-0.6\text{ N }\mathbf{i} + 1.0\text{ N }\mathbf{j}}{0.4\text{ kg}} = -1.5\text{ m/s}^2\,\mathbf{i} + 2.5\text{ m/s}^2\,\mathbf{j}$$

or

$$a_x = -1.5\text{ m/s}^2 \quad\text{and}\quad a_y = 2.5\text{ m/s}^2$$

Since the particle starts from rest at the origin at $t = 0$, its x and y coordinates at $t = 1.6$ s are given by

$$x = \tfrac{1}{2}a_x t^2 = \tfrac{1}{2}(-1.5\text{ m/s}^2)(1.6\text{ s})^2 = -1.92\text{ m}$$

and

$$y = \tfrac{1}{2}a_y t^2 = \tfrac{1}{2}(2.5\text{ m/s}^2)(1.6\text{ s})^2 = 3.20\text{ m}$$

The x and y components of the particle's velocity at $t = 1.6$ s are

$$v_x = a_x t = (-1.5\text{ m/s}^2)(1.6\text{ s}) = -2.40\text{ m/s}$$

and

$$v_y = a_y t = (2.5\ \text{m/s}^2)(1.6\ \text{s}) = 4.00\ \text{m/s}$$

In vector notation, the velocity of the particle at $t = 1.6$ s is

$$\mathbf{v} = -2.40\ \text{m/s}\ \mathbf{i} + 4.00\ \text{m/s}\ \mathbf{j}$$

Exercise

A force of 3 N produces an acceleration of 2 m/s^2 on an object of unknown mass. (*a*) What is the mass of the object? (*b*) If the force is increased to 4 N, what is the acceleration? [Answers: (*a*) 1.5 kg (*b*) 2.67 m/s^2]

Questions

1. If an object has no acceleration, can you conclude that no forces are acting on it?
2. If only a single force acts on an object, must the object accelerate? Can it ever have zero velocity?
3. Is there an unbalanced force acting when (*a*) an object moves at constant speed in a circle, (*b*) an object that is moving in a straight line slows down, or (*c*) an object moves at constant speed in a straight line?
4. Is it possible for an object to round a curve without any force acting on it?
5. If a single known force acts on an object, can you tell in which direction the object will move from this information alone?
6. If several forces of different magnitudes and directions are applied to an object initially at rest, how can you predict the direction in which it will move?
7. Can you judge the mass of an object by its size? If A is twice as big as B, does that mean that $m_A = 2m_B$?
8. Can the mass of an object be negative?

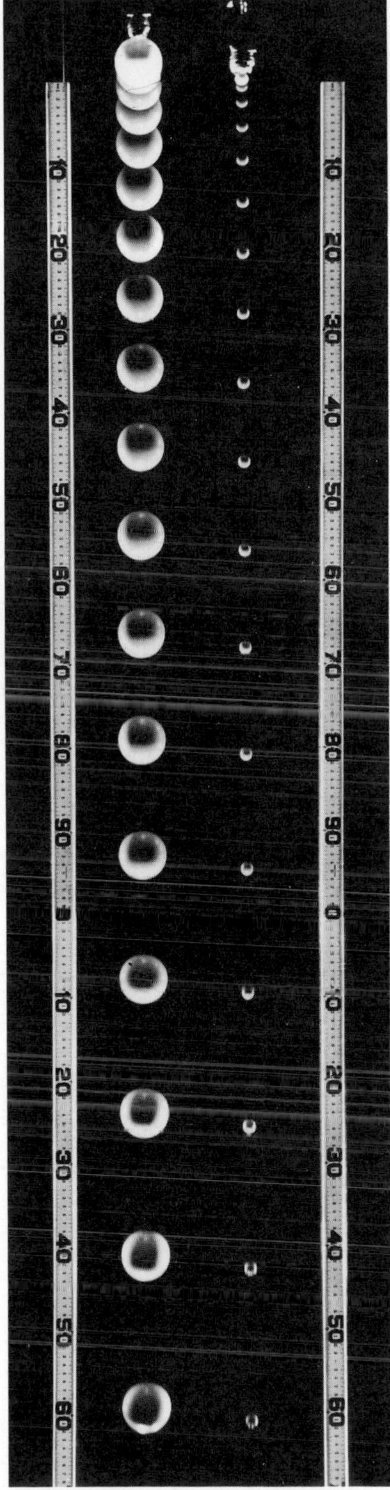

When air resistance can be neglected, balls of different mass fall with the same acceleration due to gravity.

4-3 The Force Due to Gravity: Weight

The force most common in our everyday experience is the force of the gravitational attraction of the earth for an object. This force is called the **weight** w of the object. If we drop an object near the surface of the earth and neglect air resistance so that the only force acting on the object is the force due to gravity (this situation is called **free fall**), the object accelerates toward the earth with acceleration of 9.81 m/s^2. At any given point in space, this acceleration is the same for all objects, independent of their masses. We call the value of this acceleration **g**. From Newton's second law, we can write the gravitational force $\mathbf{F}_g$ on an object of mass m as

$$\mathbf{F}_g = m\mathbf{a}$$

Using $\mathbf{a} = \mathbf{g}$ and writing $\mathbf{w}$ for the gravitational force, we have

$$\mathbf{w} = m\mathbf{g} \qquad 4\text{-}3$$

Since **g** is the same for all objects at a given point, we can conclude that the weight of an object is proportional to its mass. The vector **g** in Equation 4-3 is the force per unit mass exerted by the earth on any object and is called the

gravitational field of the earth. (We will discuss gravitational fields in Chapter 10.) It is equal to the free-fall acceleration experienced by an object when the only force acting on it is the gravitational force of the earth. Near the surface of the earth g has the value

$$g = 9.81 \text{ N/kg} = 9.81 \text{ m/s}^2$$

Careful measurements of g at various places show that it does not have the same value everywhere. *The force of attraction of the earth for an object varies with location.* In particular, at points above the surface of the earth, the force due to gravity varies inversely with the square of the distance of the object from the center of the earth. Thus, an object weighs slightly less at very high altitudes than it does at sea level. The gravitational field also varies slightly with latitude because the earth is not exactly spherical but is slightly flattened at the poles. Thus weight, unlike mass, is not an intrinsic property of an object; that is, it is not a property of the object itself.

Although the weight of an object varies from place to place because of changes in g, this variation is too small to be noticed in most practical applications. Thus, in our everyday experience, the weight of an object *appears* to be as much a constant characteristic of the object as its mass.

Near the surface of the moon, the gravitational attraction of the earth on an object is much less than that of the moon. The force exerted on the object by the moon is usually called the weight of the object when it is near the moon. Note again that the mass of an object is the same whether it is on the earth, on the moon, or somewhere in space. Mass is a property of the object itself, whereas weight depends on the nature and distance of other objects that exert gravitational forces on the object. An example should help clarify the difference between mass and weight. Suppose you take a heavy ball such as a bowling ball to the moon. Since the ball weighs only about one-sixth as much on the moon as on earth, lifting the ball is much easier on the moon. But throwing the ball with some given horizontal velocity requires the same force on the moon as on the earth. That is, the same force is required to produce a given acceleration of the ball on the moon as on the earth because the mass of the ball is the same. The same force would also be required to produce the same acceleration of the ball in free space, far from any gravitational field.

Weightlessness in a space capsule. The astronauts are accelerating toward the earth in free fall with the acceleration of gravity.

Since, at any particular location, the weight of an object is proportional to its mass, we can conveniently compare the mass of one object with that of another by comparing their weights as long as we determine the weights at the same location.

Our sensation of our weight usually comes from other forces that balance it. For example, sitting on a chair, we feel the force exerted by the chair that balances our weight and thereby prevents us from falling to the floor. When we stand on a spring scale, our feet feel the force exerted on us by the scale. The scale is calibrated to read the force it must exert (by the compression of its springs) to balance our weight. The force that balances our weight is called our **apparent weight.** It is the apparent weight that is given by a spring scale. If there is no force to balance your weight, as in free fall, your apparent weight is zero. This condition, called **weightlessness,** is experienced by astronauts in orbiting satellites. Consider a satellite in a circular orbit near the surface of the earth with a centripetal acceleration v^2/r, where r is the orbital radius and v is the speed. The only force acting on the satellite is its weight. Thus it is in free fall with the acceleration of gravity. An astronaut in the satellite is also in free fall. The only force acting on the astronaut is his or her weight, which produces the acceleration $g = v^2/r$. Since there is no force balancing the force of gravity, the astronaut's apparent weight is zero.

Units of Force and Mass

The SI unit of mass is the kilogram. Like the second and the meter, the kilogram is a fundamental unit in SI. The unit of force, the newton, and the units for other quantities that we will study, such as momentum and energy, are derived from these three units. Since 1 N produces an acceleration of 1 m/s^2 when it acts on 1 kg, we have from $F = ma$

$$1\text{ N} = 1\text{ kg}\cdot\text{m/s}^2 \qquad 4\text{-}4$$

Although we will generally use SI units in this book, we need to know another system, the U.S. customary system, which is based on the foot, the second, and a force unit (the pound) and is still used in the United States. The U.S. customary system differs from SI in that a unit of force rather than a unit of mass has been chosen as a fundamental unit. The **pound** was originally defined as the weight of a particular standard object at a particular location. It is now defined as 4.448222 N. (This is the weight of an object of mass 0.45359237 kg at a point where g has the value 9.80665 m/s^2 = 32.1740 ft/s^2.) Rounding to three places, we have

$$1\text{ lb} \approx 4.45\text{ N}$$

Since 1 kg weighs 9.81 N, its weight in pounds is

$$9.81\text{ N} \times \frac{1\text{ lb}}{4.45\text{ N}} = 2.20\text{ lb}$$ *Weight of 1 kg*

The unit of mass in the U.S. customary system is that mass that will be given an acceleration of one foot per second squared when a force of one pound is applied to it. This unit, called a *slug*, is the mass of an object that weighs 32.2 pounds. We will not need to use this unit. Instead, when working problems in this system, we substitute w/g for the mass m, where w is the weight in pounds and g is the acceleration due to gravity in feet per second per second (feet per second squared).

Example 4-4

The net force acting on a 10.0-lb object is 3.00 lb. What is its acceleration?

The acceleration is the force divided by the mass:

$$a = \frac{F}{m} = \frac{F}{w/g} = \frac{3.00\text{ lb}}{(10.0\text{ lb})/(32.2\text{ ft/s}^2)} = 9.66\text{ ft/s}^2$$

Exercise

What force is needed to give an acceleration of 3 ft/s^2 to a 5-lb block? (Answer: 0.466 lb)

Questions

9. Suppose an object was sent far out in space, away from galaxies, stars, or other bodies. How would its mass change? Its weight?
10. How would an astronaut in a condition of apparent weightlessness be aware of his or her mass?
11. Under what circumstances would your apparent weight be greater than your true weight?
12. What is your weight in newtons?
13. What is your mass in kilograms?
14. What would your weight be in pounds on the moon, where objects fall freely with an acceleration of about 5.33 ft/s^2?

4-4 Newton's Third Law

Newton's third law is sometimes called the **law of interaction.** It describes an important property of forces, that they always occur in pairs. If a force is exerted on some object *A*, there must be another object *B* exerting the force. Furthermore, if *B* exerts a force on *A*, then *A* must exert a force on *B* that is equal in magnitude and opposite in direction. For example, the earth exerts a gravitational force $\mathbf{F}_g$ on a projectile, causing it to accelerate toward the earth. According to the third law, the projectile in turn exerts a force on the earth equal in magnitude and opposite in direction. Thus, the projectile exerts a force $\mathbf{F}'_g = -\mathbf{F}_g$ on the earth toward the projectile. If this were the only force acting on the earth, the earth would accelerate toward the projectile. Since the earth has a very large mass, the acceleration it experiences due to this force is negligible and unobserved.

In discussions of Newton's third law, the words "action" and "reaction" are frequently used. If the force exerted on object *A* is called the action of object *B* upon *A*, then the force *A* exerts back on *B* is called the reaction of *A* upon *B*. It does not matter which force in such a pair is called the action and which the reaction. The important points are that forces always occur in action–reaction pairs, and that the reaction force is equal in magnitude and opposite in direction to the action force.

Note that the action and reaction forces can never balance each other because they act on *different objects*. This is illustrated in Figure 4-4, which shows two action–reaction pairs for a block resting on a table. The force acting downward on the block is the weight $\mathbf{w}$ due to the attraction of the earth. An equal and opposite force $\mathbf{w}' = -\mathbf{w}$ is exerted by the block on the earth. These forces are an action–reaction pair. If they were the only forces present, the block would accelerate downward because it would have only a single force acting on it. However, the table in contact with the block exerts an upward force $\mathbf{F}_n$ on it. This force balances the weight of the block. The block also exerts a force $\mathbf{F}'_n = -\mathbf{F}_n$ downward on the table. The forces $\mathbf{F}_n$ and $\mathbf{F}'_n$ are also an action–reaction pair.

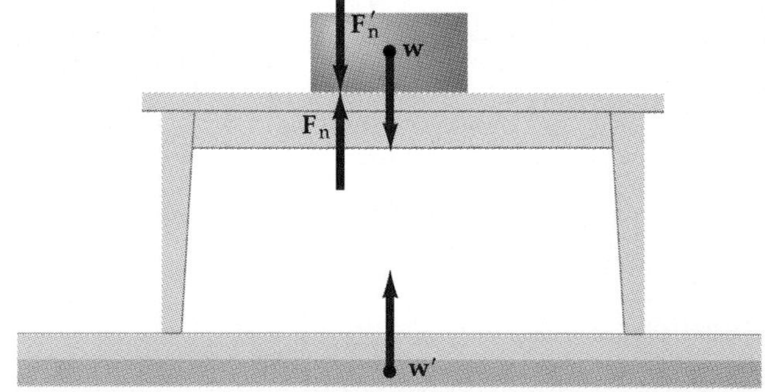

Figure 4-4 Action–reaction forces. The weight $\mathbf{w}$ is the force exerted on the block by the earth. The equal and opposite reaction force exerted on the earth by the block is $\mathbf{w}' = -\mathbf{w}$. Similarly, the table exerts a force $\mathbf{F}_n$ on the block, and the block exerts an equal and opposite force $\mathbf{F}'_n$ on the table. Action–reaction forces are exerted on different objects and therefore cannot balance.

Example 4-5

A horse refuses to pull a cart. The horse reasons, "according to Newton's third law, whatever force I exert on the cart, the cart will exert an equal and opposite force on me, so the net force will be zero and I will have no chance of accelerating the cart." What is wrong with this reasoning?

Figure 4-5 is a sketch of a horse pulling the cart. Since we are interested in the motion of the cart, we have enclosed it with a dashed line and have indicated the forces acting on it. The force exerted by the horse is labeled **T**. It is exerted by the horse on the harness. (Since the harness is attached to the cart, we are considering it part of the cart.) Other forces acting on the cart are its weight **w**, the vertical support force of the ground $\mathbf{F}_n$, and the horizontal force exerted by the ground, labeled **f** (for friction). The vertical forces **w** and $\mathbf{F}_n$ balance. (We know this because we know the cart does not accelerate vertically.) The horizontal forces are **T** to the right and **f** to the left. The cart will accelerate if **T** is greater than **f**. Note that the reaction force to **T**, which we call **T′**, is exerted on the horse, not on the cart. It has no effect on the motion of the cart, but it does affect the motion of the horse. If the horse is to accelerate to the right, there must be a force **F** (to the right) exerted by the ground on the horse's feet that is greater than **T′**. This example illustrates the importance of drawing a simple diagram when solving mechanics problems. Had the horse drawn a simple diagram, he would have seen that he need only push back hard against the ground so that the ground would push him forward.

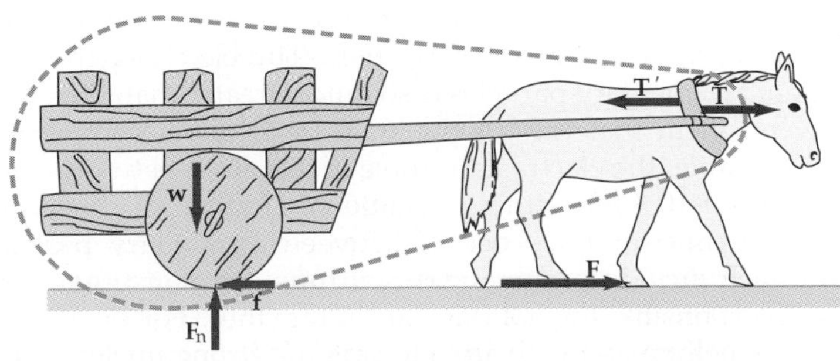

Figure 4-5 Horse pulling a cart. The cart will accelerate to the right if the force **T** exerted on it by the horse is greater than the frictional force **f** exerted on the cart by the ground. The force **T′** is equal and opposite to **T**, but because it is exerted on the *horse*, it has no effect on the motion of the cart.

(*a*)

(*b*)

(*a*) The tides are caused by the gravitational force exerted by the moon and the sun. Mont-Saint-Michel, France is an island when the tide is in. (*b*) These lightning bolts at the Kitt Peak National Observatory are the result of the electromagnetic force. (*c*) This hydrogen bomb explosion illustrates the strong nuclear force. (*d*) The weak interaction between muons and electrons is illustrated by this false-color cloud chamber photograph showing the path of a cosmic ray muon (green) and an electron (red) that is knocked out of an atom.

4-5 Forces in Nature

Fundamental Forces

All the different forces observed in nature can be explained in terms of four basic interactions that occur between elementary particles:

1. The gravitational force
2. The electromagnetic force
3. The strong nuclear force (also called the hadronic force)
4. The weak nuclear force

The gravitational force between the earth and an object near the earth's surface is the weight of the object. The gravitational force exerted by the sun on the earth and the other planets is responsible for keeping the planets in their orbits around the sun. Similarly, the gravitational force exerted by the earth on the moon keeps the moon in its nearly circular orbit around the earth. The gravitational forces exerted by the moon and the sun on the oceans of the earth are responsible for the tides. We will study the gravitational force in detail in Chapter 10.

The electromagnetic force includes both the electric and the magnetic forces. A familiar example of the electric force is the attraction between small bits of paper and a comb that has been electrified by being rubbed across hair. Although the familiar magnetic force between a magnet and iron objects seems to be very different from the electric force, the magnetic force arises when electric charges are in motion. The electromagnetic force between charged elementary particles is so much greater than the gravitational force between them that the gravitational force can nearly always be ignored. For example, the electrostatic force of repulsion between two protons is of the order of 10^{36} times the gravitational attraction between them.

The strong nuclear force occurs between elementary particles called hadrons, which include protons and neutrons, the constituents of atomic nuclei. It is responsible for holding nuclei together. For example, the two protons in the helium atom are attracted via the strong nuclear force, which more than balances the electrostatic repulsion of the protons. However, the strong nuclear force has a very short range. It decreases rapidly with separa-

(c)

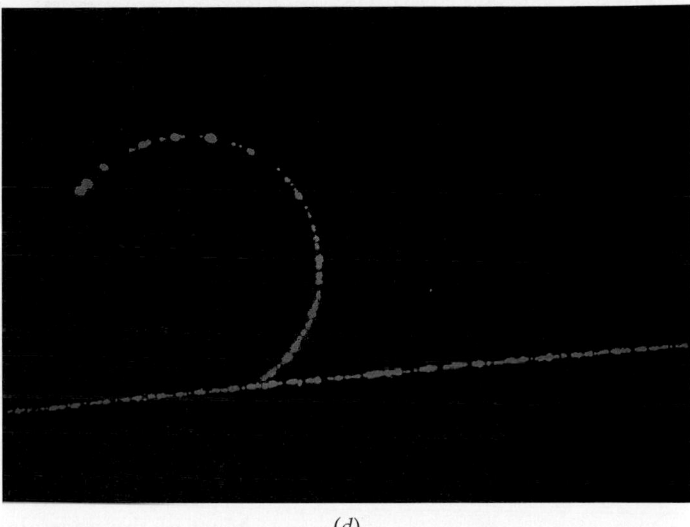

(d)

tion of the particles and is negligible when they are a few nuclear diameters apart. The weak nuclear force, which also has a short range, occurs between electrons and protons or neutrons. It is responsible for a certain kind of radioactive decay called beta decay.

The fundamental forces act between particles that are separated in space. This concept is referred to as **action at a distance.** Newton perceived action at a distance as a flaw in his theory of gravitation, but he avoided giving any other hypothesis. Indeed, in 1692 he wrote the following:*

> It is inconceivable that inanimate, brute matter should, without the mediation of something else, which is not material, operate upon, and affect other matter without mutual contact, as it must be if gravitation, in the sense of Epicurus, be essential and inherent in it. And this is one reason why I desired you would not ascribe innate gravity to me. That gravity should be innate, inherent, and essential to matter, so that one body may act upon another at a distance through a vacuum, without the mediation of anything else, by and through which their action and force may be conveyed from one to another, is to me so great an absurdity that I believe no man who has in philosophical matters a competent faculty of thinking can ever fall into it.

Today, we treat the problem of action at a distance by introducing the concept of a **field.** For example, we can consider the attraction of the earth by the sun in two steps. The sun creates a condition in space that we call the gravitational field. This field produces a force on the earth. The field is thus the intermediary agent. Similarly, the earth produces a gravitational field that exerts a force on the sun. If the earth moves to a new position, the field of the earth is changed. This change is not propagated through space instantly but rather with the speed $c = 3 \times 10^8$ m/s, which is also the speed of light. If we can neglect the time it takes for propagation of the field, we can ignore this intermediary agent and treat the gravitational forces as if they were exerted by the sun and the earth directly on each other. For example, during the 8 min it takes for propagation of the gravitational field from the earth to the sun, the earth moves through only a small fraction of its total orbit around the sun.

*Isaac Newton, Third Letter to Bentley (Feb. 25, 1692), R. and J. Dodsley, London, 1756.

Contact forces exerted by one object on another produce deformations in the objects that are often not visible. Here, the forces exerted by the C-clamp on the plastic block produce stress patterns in the block that are made visible by polarized light.

Contact Forces

Most of the everyday forces we observe on macroscopic objects are contact forces exerted by springs, strings, and surfaces in direct contact with the object. The forces are the result of molecular forces exerted by the molecules of one object on those of another. These molecular forces are themselves complicated manifestations of the basic electromagnetic force.

A spring made by winding a stiff wire into a helix is a familiar device. The force exerted by the spring when it is compressed or extended is the result of complicated intermolecular forces in the spring, but an empirical description of the macroscopic behavior of the spring is sufficient for most applications. If the spring is compressed or extended and released, it returns to its original, or natural, length, provided the displacement is not too great. There is a limit to such displacements beyond which the spring does not return to its original length but remains permanently deformed. If we allow only displacements below this limit, we can calibrate the extension or compression Δx in terms of the force needed to produce the extension or compression. It has been found experimentally that, for small Δx, the force exerted by the spring is approximately proportional to Δx and in the opposite direction. This relationship, known as **Hooke's law,** can be written

$$F_x = -k(x - x_0) = -k\,\Delta x \qquad \text{4-5}$$

where the constant k is called the **force constant** of the spring. The distance x is the coordinate of the free end of the spring or of any object attached to that end of the spring. The constant x_0 is the value of this coordinate when the spring is unstretched from its equilibrium position. There is a negative sign in Equation 4-5 because, if the spring is stretched (Δx is positive), the force F_x is negative, whereas if the spring is compressed (Δx is negative), F_x is positive (Figure 4-6). Such a force is called a **restoring force** because it tends to restore the spring to its initial configuration.

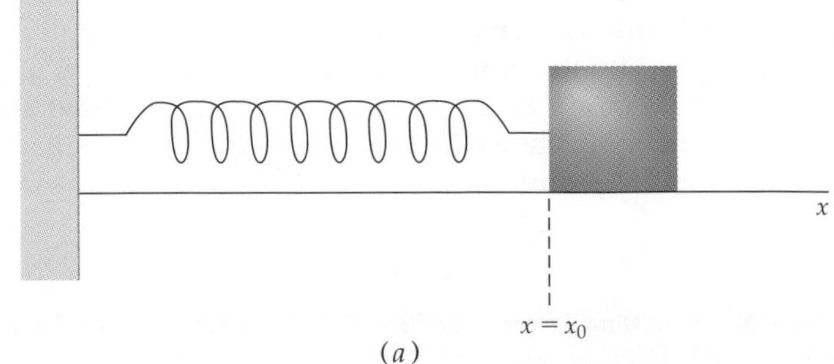

Figure 4-6 A horizontal spring attached to a block. (*a*) When the spring is unstretched, it exerts no force on the block. (*b*) When the spring is stretched such that Δx is positive, it exerts a force on the block of magnitude $k\,\Delta x$ in the negative x direction. (*c*) When the spring is compressed such that Δx is negative, the spring exerts a force on the block of magnitude $k\,\Delta x$ in the positive x direction.

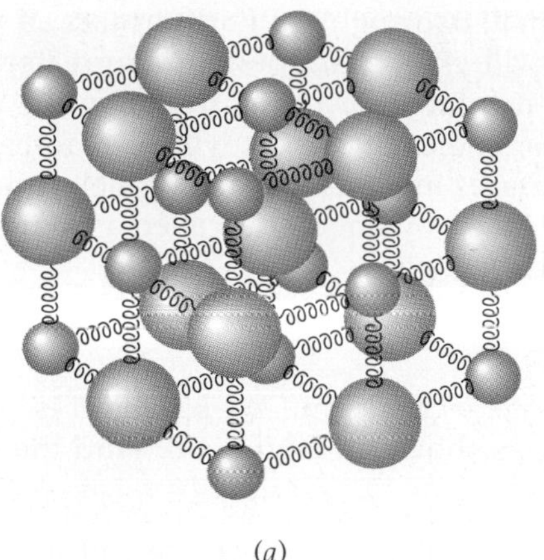

(a)

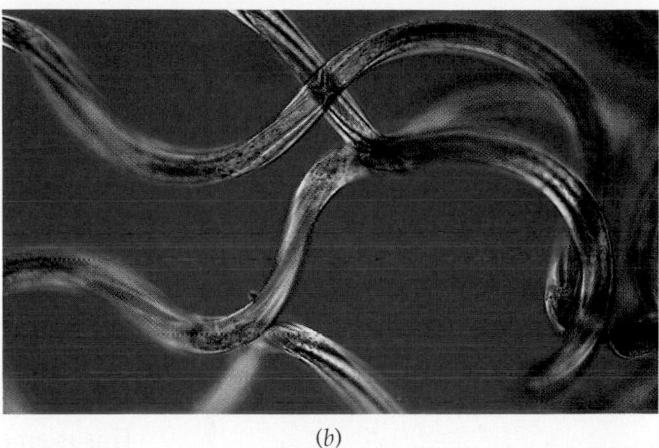

(b)

Figure 4-7 (a) Model of a solid as consisting of atoms connected to each other by springs. (b) The elasticity of nylon arises from the shape and cross linking of its fibers shown here under polarized light.

The force exerted by a spring is similar to that exerted by one atom on another in a molecule or in a solid in the sense that, for small displacements from equilibrium, the restoring force is proportional to the displacement. It is often useful to visualize the atoms in a molecule or solid as being connected by springs (Figure 4-7). For example, if we slightly increase the separation of the atoms in a molecule and release them, we would expect the atoms to oscillate back and forth as if they were two masses connected by a spring.

If we pull on a flexible string, the string stretches slightly and pulls back with an equal but opposite force (unless the string breaks). We can think of a string as a spring with such a large force constant that the extension of the string is negligible. Because the string is flexible, however, we cannot exert a force of compression on it. When we push on a string, it merely flexes or bends.

When two bodies are in contact with each other, they exert forces on each other due to the interaction of the molecules of one object with those of the other. Consider a block resting on a horizontal table. The weight of the block pulls the block downward, pressing it against the table. Because the molecules in the table have a great resistance to compression, the table exerts a force upward on the block perpendicular, or normal, to the surface. Such a force is called a **normal force.** (The word *normal* means perpendicular.) Careful measurement would show that a supporting surface always bends slightly in response to a load, but this compression is not noticeable to the naked eye. Since the table exerts an upward force on the block, the block must exert an equal force downward on the table. Note that the normal force exerted by one surface on another can vary over a wide range of values. For example, unless the block is so heavy that the table breaks, the table will exert an upward support force on the block exactly equal to the weight of the

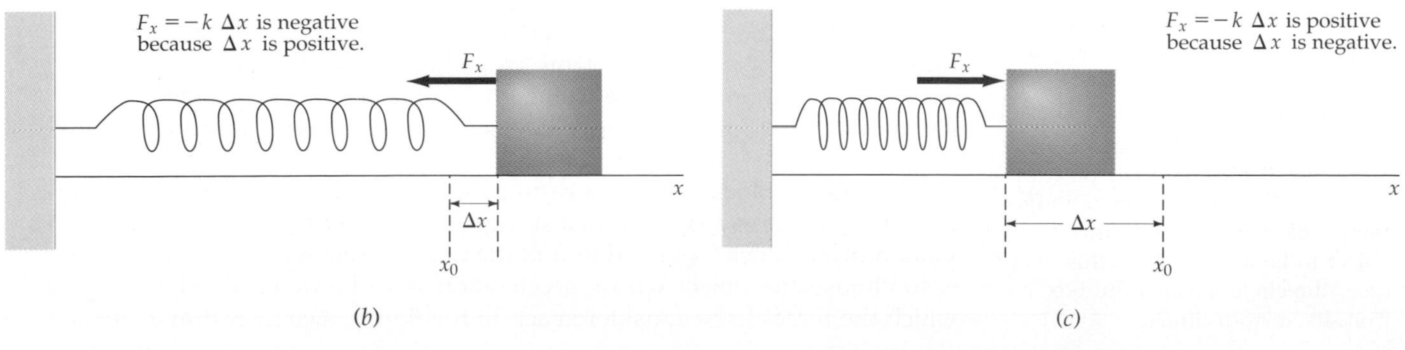

(b)

(c)

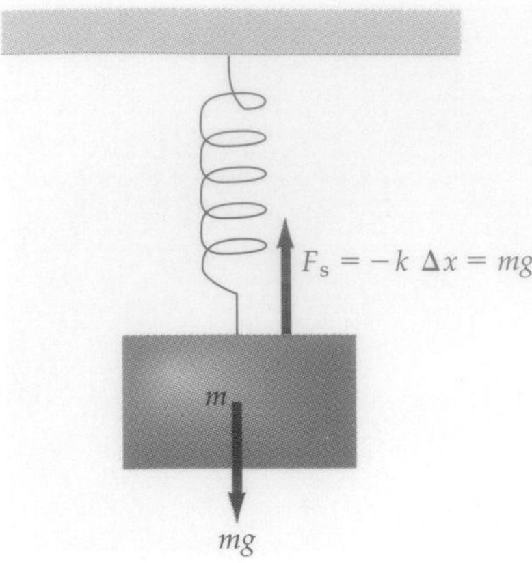

Figure 4-8 An object suspended from a spring. The weight of the object $\mathbf{w} = m\mathbf{g}$ is balanced by the upward force $\mathbf{F}_s$, exerted by the spring.

block no matter how large or small its weight is. Furthermore, if you press down on the block, the table will exert a support force greater than the weight of the block to prevent it from accelerating downward.

Under certain circumstances, bodies in contact will exert forces on each other that are parallel to the surfaces in contact. The parallel component of the contact force exerted by one object on another is called a **frictional force.** We will study friction in the next chapter.

Example 4-6

A spring has a force constant $k = 300$ N/m. A 4-kg object is suspended motionless from the spring, as shown in Figure 4-8. Find the extension of the spring.

This is an example of static equilibrium, which we will study in more detail in Chapter 9. Since the block is not accelerating, the net force acting on it must be zero. There are two forces acting on the block: its weight mg downward and the force of the spring F_s upward. If we take the downward direction to be positive, the weight is positive and the extension of the spring Δx is also positive. The force the spring exerts is upward and is therefore negative. Setting the sum of the forces on the block equal to zero, we obtain

$$mg + (-k\ \Delta x) = 0$$

Using $m = 4$ kg, $k = 300$ N/m, and $g = 9.81\ \text{m/s}^2 = 9.81$ N/kg, we obtain

$$(4\ \text{kg})(9.81\ \text{N/kg}) - (300\ \text{N/m})(\Delta x) = 0$$

or

$$\Delta x = \frac{(4\ \text{kg})(9.81\ \text{N/kg})}{300\ \text{N/m}} = 0.131\ \text{m}$$

Exercise

A spring of force constant 400 N/m is attached to a 3-kg block that rests on an air track so that friction is negligible. What extension of the spring is needed to give the block an acceleration of 4 m/s^2?
(Answer: 3.0 cm)

4-6 Applications to Problem Solving

Newton's laws can be used to solve a variety of mechanics problems. For instance, they allow us to determine the acceleration of a particle from a knowledge of all the forces acting on it, or to determine the forces acting on a particle given its acceleration. In this section, we will examine the application of Newton's laws to some simple problems involving motion under forces that are constant in magnitude. Careful study of these simple examples will make you aware of the content of newtonian mechanics and how it can be applied. Practical problems are generally more complex than these examples, but the methods of solving them are natural extensions of those illustrated here. Some more general examples of the applications of Newton's laws will be discussed in Chapter 5.

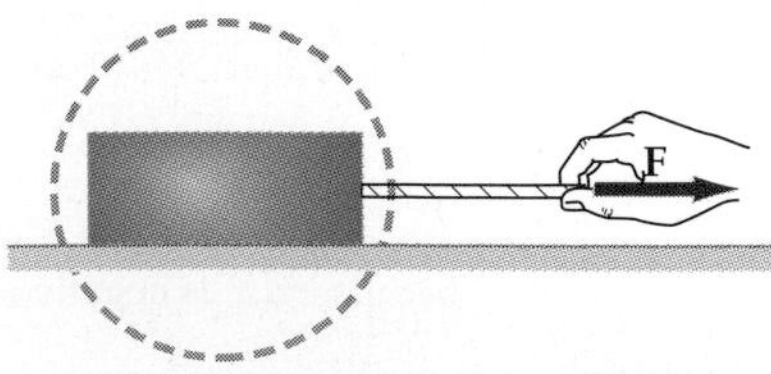

Figure 4-9 A block on a frictionless horizontal surface with a horizontal force exerted on it by a string. The first step in solving the problem is to isolate the object to be analyzed. In this case, the circle isolates the box from its surroundings.

Consider a block of mass m resting on a frictionless horizontal table and pulled with a force $\mathbf{F}$ applied to a string, as shown in Figure 4-9. To find the motion of the block, we need to find the net force acting on it. The first step is to choose the object whose acceleration is to be determined and upon which the forces to be considered act. In the figure, a circle is drawn around the block to help us mentally isolate it from its surroundings. We then look for all the possible forces that act on the chosen object. Such forces may arise from the contact of the object with its surroundings, or they may be action-at-a-distance forces such as gravity.

Three significant external forces act on the block in this example. They are shown in the diagram in Figure 4-10. Such a diagram is called a **free-body diagram.** The three forces are

1. The weight of the block **w**.
2. The contact force $\mathbf{F}_n$ exerted by the table. Since we are assuming the table to be frictionless, the contact force is perpendicular to the table.
3. The contact force **T** exerted by the string.

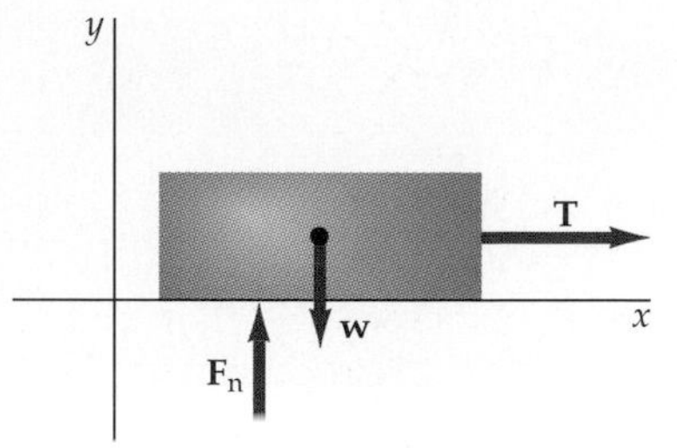

Figure 4-10 A free-body diagram for the block in Figure 4-9. The three significant forces acting on the box are the force exerted by the earth **w**, the normal force $\mathbf{F}_n$, and the force exerted by the string **T**.

A convenient coordinate system is also indicated in Figure 4-10. Note that the normal force $\mathbf{F}_n$ and the weight **w** are drawn with equal magnitudes. We know these forces have equal magnitudes because the block does not accelerate vertically. Since the resultant force is in the x direction and has magnitude T, Newton's second law gives

$$T = ma_x$$

The force **F** exerted by the hand on the string equals the force **T** exerted by the string on the block. We can see this by considering the free-body diagram for the string shown in Figure 4-11. The force **T′** is the force the block exerts on the string. It is equal in magnitude and opposite the direction of the force **T** exerted by the string on the block. (We have neglected the weight of the string. In actual fact, the string will sag slightly and the forces **F** and **T′** will have small vertical components, but they are so small that we can safely ignore them in this problem). Assuming the string remains taut, it must have the same acceleration as the block. If m_s is the mass of the string, applying Newton's second law to the string gives us

$$F - T' = m_s a_x$$

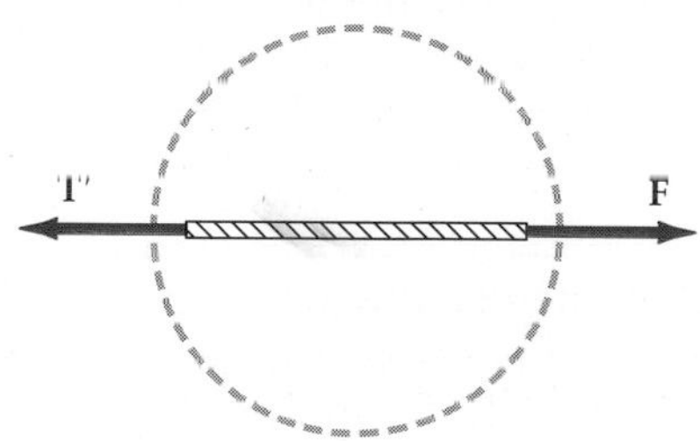

Figure 4-11 Free-body diagram for the string in Figure 4-9. If the string is light enough that its mass can be neglected, the forces **F** and **T′** are equal in magnitude.

If the string is light enough so that we can neglect its mass, we have

$$F - T' = m_s a_x \approx 0$$

Since **T′** and **T** have the same magnitudes, **F** and **T** are equal.

Figure 4-12 shows a small segment of the string of mass Δm_s. The forces acting on this segment are $\mathbf{T}_1$ exerted by that part of the string to the right of the segment and $\mathbf{T}_2$ exerted by that part of the string to the left of the segment. Because the string has negligible mass, these two forces are equal in magnitude, $T_1 = T_2 = T$. The force T is called the **tension** in the string. Each segment of the string exerts a force T on each of its neighbors. These forces act along the string, so a light string connecting two points has a tension that has a constant magnitude throughout. This result also holds for a string that passes over a frictionless peg or a pulley of negligible mass as long as there are no tangential forces acting on the string between the two points considered.

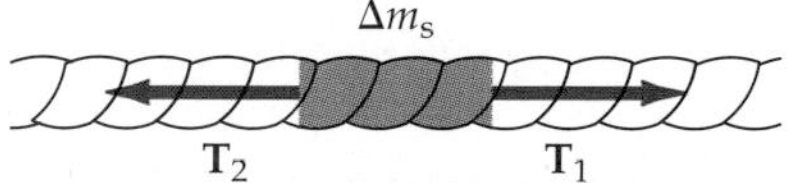

Figure 4-12 Free-body diagram for a segment of the string of mass Δm_s. If the mass of the segment is negligible, the force $\mathbf{T}_1$ exerted by that part of the string to the right of the segment is equal in magnitude to the force $\mathbf{T}_2$ exerted by that part of the string to the left of the segment. The tension T is the magnitude of the force each segment exerts on each adjacent segment.

Even in this simple example, both kinds of applications of Newton's laws are used: The horizontal acceleration is found in terms of the given force **F**, and the vertical force $\mathbf{F}_n$ exerted by the table is found from the fact that the block remains on the table and thus $a_y = 0$. Conditions on the motion of an object, such as the requirement that the block remain on the table, are called **constraints.**

According to Newton's third law, forces always act in pairs. Figure 4-10 shows only those forces that act on the block. Figure 4-13 shows the reaction forces to those in Figure 4-10. These are the gravitational force **w′** exerted by the block on the earth, the force $\mathbf{F}_n'$ exerted by the block on the table, and the force **T′** exerted by the block on the string. Since these forces are not exerted on the block they have nothing to do with its motion. We can therefore omit them in applying Newton's second law to the motion of the block.

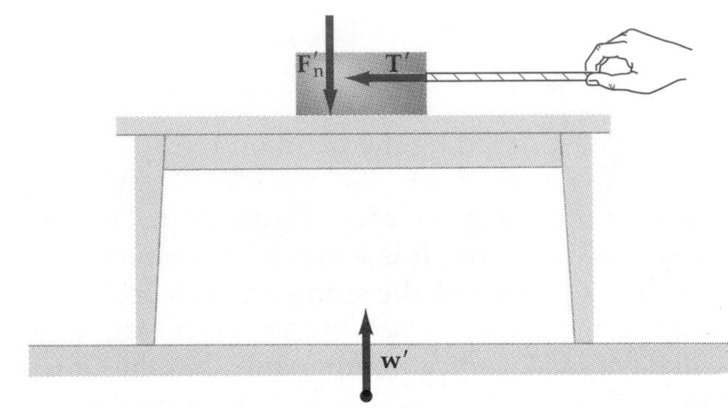

Figure 4-13 The reaction forces corresponding to the three forces shown in Figure 4-10. Note that these forces *do not* act on the block. **T′** acts on the string, $\mathbf{F}_n'$ acts on the table, and **w′** acts on the earth.

This simple example illustrates a general method of attack for solving problems using Newton's laws. Basically, this method consists of the following steps:

Solving problems using Newton's laws

1. Draw a neat diagram.
2. Isolate the object (particle) of interest, and draw a free-body diagram, showing each external force that acts on the object. If there is more than one object of interest in the problem, draw a separate free-body diagram for each.
3. Choose a convenient coordinate system for each object, and apply Newton's second law $\mathbf{F}_{\text{net}} = m\mathbf{a}$ in component form.
4. Solve the resulting equations for the unknowns using whatever additional information is available. The unknowns may include the masses, the components of the accelerations, or the components of some of the forces.
5. Finally, inspect your results carefully, checking to see whether they correspond to reasonable expectations. Particularly valuable is the determination of the predictions of your solution when variables are assigned extreme values. In this way you can check your work for errors.

We will now look at some examples.

Example 4-7

Find the acceleration of a block of mass m that slides down a fixed, frictionless surface inclined at an angle θ to the horizontal.

There are only two forces acting on the block, the weight $\mathbf{w}$ and the normal force $\mathbf{F}_n$ exerted by the incline (see Figure 4-14). (For real surfaces, there would be a force of friction parallel to the incline, but here we are assuming an ideal frictionless surface.) Since the two forces do not act along the same line, they cannot add to zero, so the block must accelerate. The acceleration is along the incline. This is another example of a constraint. For this problem, it is convenient to choose a coordinate frame with one axis parallel to the incline and the other perpendicular to it, as shown in the figure. Then the acceleration has only one component, a_x. For this choice, $\mathbf{F}_n$ is in the y direction, and the weight $\mathbf{w}$ has the components

$$w_x = w \sin \theta = mg \sin \theta$$
$$w_y = -w \cos \theta = -mg \cos \theta \qquad 4\text{-}6$$

where m is the mass of the block and g is the acceleration due to gravity (Figure 4-15). The net force in the y direction is $F_n - mg \cos \theta$. From Newton's second law and the fact that $a_y = 0$,

$$\sum F_y = F_n - mg \cos \theta = ma_y = 0$$

and thus

$$F_n = mg \cos \theta \qquad 4\text{-}7$$

Similarly, for the x components,

$$\sum F_x = mg \sin \theta = ma_x$$
$$a_x = g \sin \theta \qquad 4\text{-}8$$

The acceleration down the incline is constant and equal to $g \sin \theta$. It is useful to check our results at the extreme values of inclination, $\theta = 0$ and $\theta = 90°$. At $\theta = 0$, the surface is horizontal. The weight has only a y component, which is balanced by the normal force $\mathbf{F}_n$. The acceleration is zero: $a_x = g \sin 0° = 0$. At the opposite extreme, $\theta = 90°$, the incline is vertical. Then the weight has only an x component along the incline, and the normal force is zero: $F_n = mg \cos 90° = 0$. The acceleration is $a_x = g \sin 90° = g$. That is, the block is in free fall.

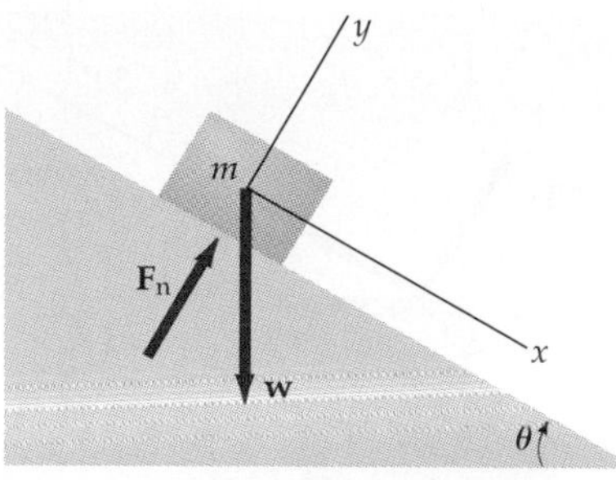

Figure 4-14 Forces acting on a block of mass m on a frictionless incline. It is convenient to choose the x axis parallel to the incline.

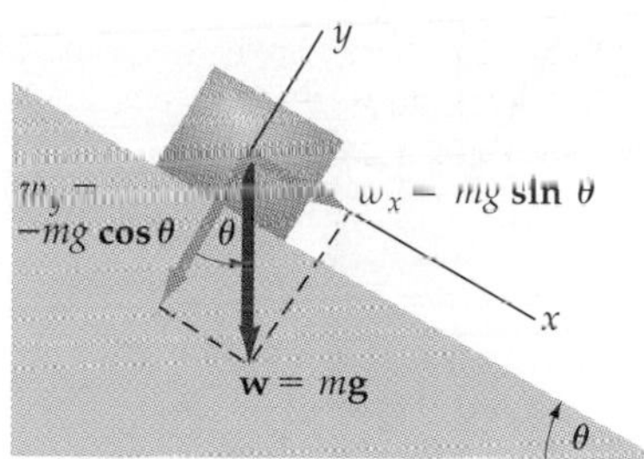

Figure 4-15 The weight of the block can be replaced by its components along the x and y axes. Since the weight vector is perpendicular to the horizontal and the negative y axis is perpendicular to the incline, the angle between these two lines is the same as the angle θ of the incline. The y component of the weight is thus $-mg \cos \theta$, and the x component is $mg \sin \theta$.

When friction is neglected, the body slides down the incline with acceleration $g \sin \theta$.

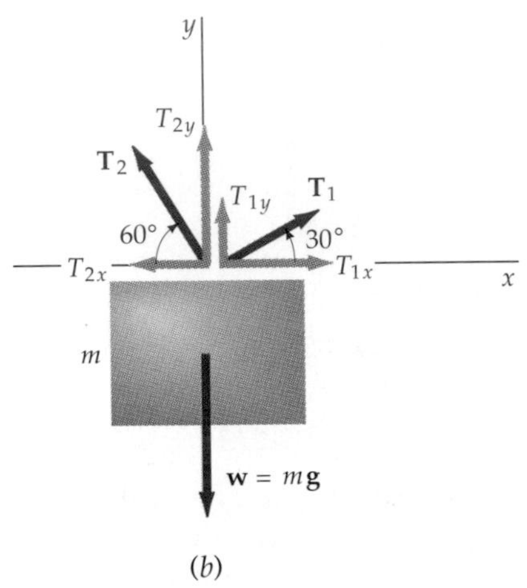

Figure 4-16 (*a*) Picture supported by two wires in Example 4-8. (*b*) Free-body diagram for the picture with the forces resolved into their x and y components.

Example 4-8

A picture weighing 8 N is supported by two wires of tension $\mathbf{T}_1$ and $\mathbf{T}_2$ as shown in Figure 4-16*a*. Find the tension in the wires.

Since the picture does not accelerate, the net force acting on it must be zero. The three forces acting on the picture—its weight $m\mathbf{g}$, the tension $\mathbf{T}_1$ in one wire, and the tension $\mathbf{T}_2$ in the other—must therefore sum to zero. Figure 4-16*b* shows the free-body diagram for the picture with the forces resolved into their horizontal and vertical components. Since the weight has only a vertical component mg downward, the horizontal components of the tensions $\mathbf{T}_1$ and $\mathbf{T}_2$ must balance each other, and the vertical components of the tensions must balance the weight:

$$\sum F_x = T_1 \cos 30° - T_2 \cos 60° = 0$$

$$\sum F_y = T_1 \sin 30° + T_2 \sin 60° - mg = 0$$

Using $\cos 30° = \sqrt{3}/2 = \sin 60°$ and $\sin 30° = \frac{1}{2} = \cos 60°$ and solving for the tensions, we obtain

$$T_1 = \tfrac{1}{2}mg = 4 \text{ N}$$

$$T_2 = \sqrt{3}T_1 = \frac{\sqrt{3}}{2}\, mg = 6.93 \text{ N}$$

We will now consider two examples of the application of Newton's laws to objects moving in circles. As discussed in Chapter 3, when a particle moves with a speed v in a circle of radius r, it has an acceleration with a magnitude of v^2/r directed toward the center of the circle whether or not the speed is changing. If the speed is changing, there is also a component of acceleration tangential to the circle and equal to the rate of change of the speed.

As with any acceleration, there must be a net force in the direction of the centripetal acceleration to produce it. This net force is called the **centripetal force.** It is important to understand that the centripetal force is not a new kind of force that we have not yet studied. It is merely a name for the net inward force that must be present to provide the centripetal acceleration needed for circular motion. Whenever an object moves in a circle, there must be a net force to provide its centripetal acceleration just as there must be a net force to provide linear acceleration. The centripetal force may be due to a string, a spring, or some other type of contact force such as a normal force or friction; it may be an action-at-a-distance force such as a gravitational force; or it may be a combination of any of these familiar forces.

Example 4-9

A tetherball of mass m is suspended from a rope of length L and travels at a constant speed v in a horizontal circle of radius r. The rope makes an angle θ given by $\sin \theta = r/L$, as shown in Figure 4-17. Find the tension in the rope and the speed of the ball.

The two forces acting on the ball are the weight $m\mathbf{g}$, which acts vertically downward, and the tension $\mathbf{T}$, which acts along the rope. In this problem, we know that the acceleration is horizontal, directed toward the center of the circle, and of magnitude v^2/r. Thus, the vertical component of the tension must balance the weight mg. The horizontal component of the tension is the centripetal force. The vertical and horizontal components of $\mathbf{F}_{\text{net}} = m\mathbf{a}$ therefore give

$$T \cos \theta - mg = 0$$

or

$$T \cos \theta = mg$$

and

$$T \sin \theta = ma = \frac{mv^2}{r}$$

The tension is found directly from the first equation since θ is given. We can find the speed v in terms of the known quantities r and θ by dividing one equation by the other to eliminate T. We thereby obtain

$$\frac{T \sin \theta}{T \cos \theta} = \frac{mv^2/r}{mg}$$

$$\frac{\sin \theta}{\cos \theta} = \tan \theta = \frac{v^2}{rg}$$

so

$$v = \sqrt{rg \tan \theta}$$

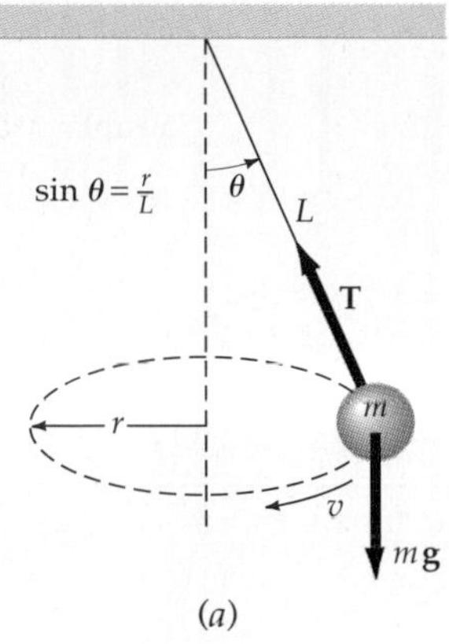

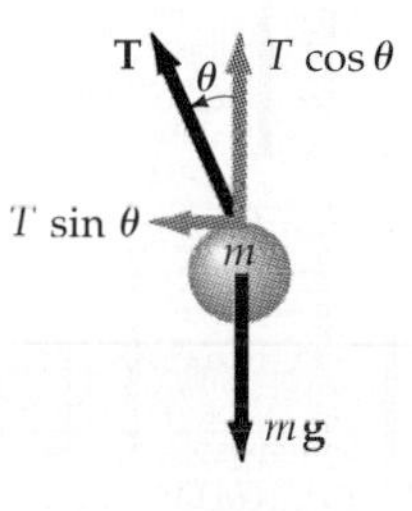

Figure 4-17 (*a*) Tetherball moving in a horizontal circle. (*b*) Free-body diagram for the ball with the force **T** resolved into its vertical and horizontal components.

Example 4-10

A pail of water is whirled in a vertical circle of radius r. If its speed is v_t at the top of the circle, find the force exerted on the water by the pail at the top of the circle. Find also the minimum value of v_t for which the water will remain in the pail.

The forces on the water at the top of the circle are shown in Figure 4-18. They are the force of gravity $m\mathbf{g}$ and the force $\mathbf{F}_p$ exerted by the pail. Both these forces act downward. The acceleration, which is toward the center of the circle, is also downward at this point. Newton's second law gives

$$F_p + mg = m\frac{v_t^2}{r}$$

The force exerted by the pail is therefore

$$F_p = m\frac{v_t^2}{r} - mg$$

Note that both the force of gravity and the contact normal force exerted by the pail contribute to the necessary centripetal force.

If we increase the speed of the pail, the bottom of the pail will exert a larger force on the water to keep it moving in a circle. If we decrease the speed, F_p will decrease. Since the pail cannot exert an upward force on the water, the minimum speed the water can have at the top of the circle occurs when $F_p = 0$. Then,

$$mg = m\frac{v_{t,min}^2}{r}$$

or

$$v_{t,min} = \sqrt{rg} \qquad 4\text{-}9$$

When the water is moving at this minimum speed, its acceleration at the top of the circle is **g**, the free-fall acceleration of gravity, and the only force acting on the water is the gravitational attraction of the earth, the weight mg.

If you wish to try this, you might want to know how fast you need to move the pail so that you will not get wet. A reasonable value for r is

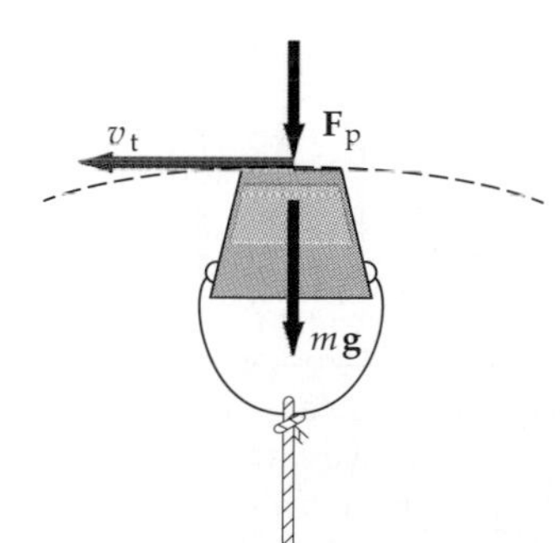

Figure 4-18 Pail of water whirled in a vertical circle. At the top of the circle the forces acting on the water are its weight $m\mathbf{g}$ and the force exerted by the pail $\mathbf{F}_p$. Both of these forces act downward at this point, toward the center of the circle.

1 m, assuming that your arm is 70 cm long and that there is another 30 cm from your hand to the water. The minimum speed at the top of the circle is then

$$v_{\text{t,min}} = \sqrt{rg} = \sqrt{(1\text{ m})(9.81\text{ m/s}^2)} = 3.13\text{ m/s}$$

If you rotate the pail at constant speed, the maximum period of revolution can be found using Equation 3-23 from Chapter 3:

$$T = \frac{2\pi r}{v} = \frac{(2)(3.14)(1\text{ m})}{3.13\text{ m/s}} \approx 2\text{ s}$$

Our final example is an application of Newton's laws to an object that is at rest relative to an elevator that is itself accelerating.

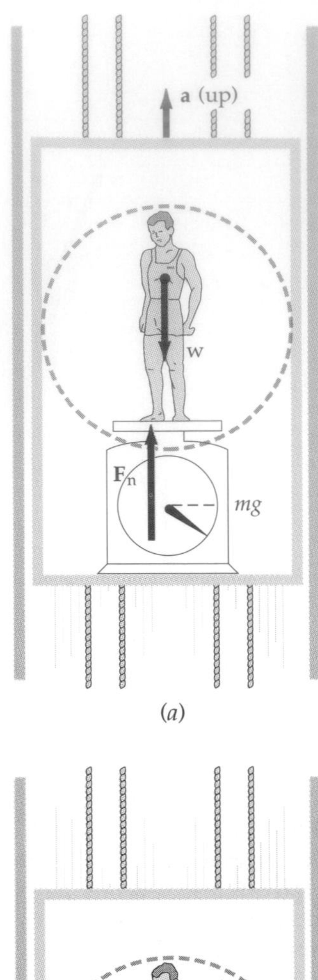

(a)

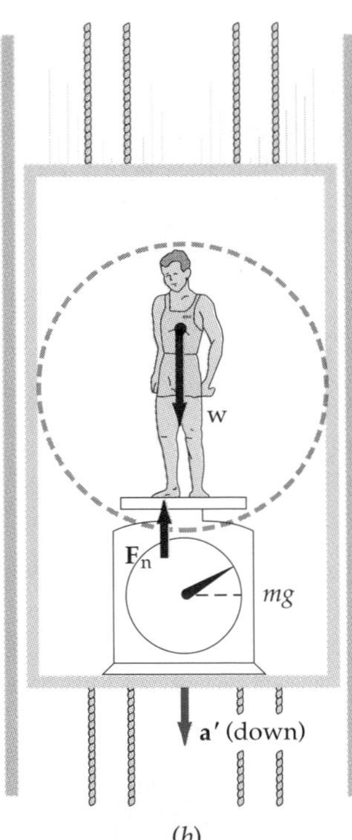

(b)

Figure 4-19 Man on a scale in an accelerating elevator. The scale indicates the man's *apparent weight*, which equals the magnitude of $\mathbf{F}_n$. (*a*) When the acceleration is upward, his apparent weight is greater than *mg*. (*b*) When the acceleration is downward, his apparent weight is less than *mg*.

Example 4-11

A man stands on a scale fastened to the floor of an elevator, as illustrated in Figure 4-19. What does the scale read when the elevator is accelerating (*a*) up and (*b*) down?

(*a*) Since the man is at rest relative to the elevator, he is also accelerating up. The forces acting on the man are $\mathbf{F}_n$ upward, exerted by the scale platform on which he stands, and **w** downward, the force of gravity. The net force is $F_n - w$ upward in the direction of the acceleration **a**. Newton's second law gives

$$F_n - w = ma$$

or

$$F_n = w + ma = mg + ma \tag{4-10}$$

The force $\mathbf{F}'_n$ exerted by the man on the scale determines the scale reading, which is his apparent weight. Since $\mathbf{F}'_n$ and $\mathbf{F}_n$ are an action–reaction pair, they are equal in magnitude. Thus, when the elevator accelerates up, the apparent weight of the man is greater than his true weight by the amount *ma*.

(*b*) For the case of the elevator accelerating downward, let us call the acceleration **a′**. In this case, the net force must be downward, implying that the weight *mg* is greater than F_n. Choosing down to be the positive direction, Newton's second law gives

$$w - F_n = ma'$$

or

$$F_n = w - ma' = mg - ma' \tag{4-11}$$

Again, the scale reading, or apparent weight, equals F_n in magnitude. In this case, the apparent weight is less than *mg*. If $a' = g$, as it would if the elevator were in free fall, the man would be apparently weightless. What if the downward acceleration of the elevator is greater than *g*? (For this to happen something in addition to gravity would have to pull down on the elevator.) Assuming that the surface of the scale is not sticky, the scale cannot exert a downward force on the man. Since the downward force on the man cannot be greater than *w*, the scale will soon leave the man. The man will have the acceleration *g*, which is less than that of the elevator, so he will eventually hit the ceiling of the elevator. Then, if the ceiling is strong enough, it can provide the force downward necessary to give him the acceleration a'.

Exercise

An elevator is descending to the ground floor and comes to a stop with an acceleration of magnitude 4 m/s^2. If your mass is 70 kg and you are standing on a scale in the elevator, what does the scale read as the elevator is stopping? (Answer: 967 N)

Questions

15. A picture is supported by two wires as in Example 4-8. Do you expect the tension to be greater or less in the wire that is more nearly vertical?
16. A weight is hung on a wire that was originally stretched horizontally. Can the wire remain horizontal? Explain.
17. What effect does the velocity of the elevator have on the apparent weight of the man in Example 4-11?

Summary

1. The fundamental relations of classical mechanics are contained in Newton's laws of motion:

   Law 1. An object continues in its initial state of rest or motion with uniform velocity unless it is acted on by an unbalanced, or net external, force.

   Law 2. The acceleration of an object is inversely proportional to its mass and directly proportional to the net external force acting on it:

   $$\mathbf{a} = \frac{\mathbf{F}_{\text{net}}}{m}$$

   or

   $$\mathbf{F}_{\text{net}} = m\mathbf{a}$$

   Law 3. Forces always occur in pairs. If object A exerts a force on object B, an equal but opposite force is exerted by object B on object A.

2. A reference frame in which Newton's laws hold is called an inertial reference frame. Any reference frame that is moving with constant velocity relative to an inertial reference frame is itself an inertial reference frame. A reference frame that is accelerating relative to an inertial frame is not an inertial reference frame. A reference frame attached to the earth is approximately an inertial reference frame.

3. Force is defined in terms of the acceleration it produces on a given object. A force of 1 newton (N) is that force that produces an acceleration of 1 m/s^2 on a standard object with a mass of 1 kilogram (kg).

4. Mass is an intrinsic property of an object that measures its resistance to acceleration. The mass of one object can be compared to that of another by applying an equal force to each object and measuring their accelerations. The ratio of the masses of the objects is then equal to the inverse ratio of the accelerations of the objects produced by the same force:

   $$\frac{m_1}{m_2} = \frac{a_2}{a_1}$$

   The mass of an object does not depend on the location of the object.

5. The weight **w** of an object is the force of gravitational attraction between the object and the earth. It is proportional to the mass m of the object and the gravitational field **g**, which also equals the free-fall acceleration of gravity:

$$\mathbf{w} = m\mathbf{g}$$

Weight is not an intrinsic property of an object. It does depend on the location of the object because **g** does.

6. All the forces observed in nature can be explained in terms of four basic interactions:

   1. The gravitational force
   2. The electromagnetic force
   3. The strong nuclear force (also called the hadronic force)
   4. The weak nuclear force

   The everyday forces we observe between macroscopic objects, such as the contact forces of support and friction and those exerted by springs and strings, are due to molecular forces that arise from the basic electromagnetic force.

7. The general method of attack for solving a problem using Newton's laws includes the following steps:

   1. Draw a neat diagram.
   2. Isolate the object (particle) of interest and draw a free-body diagram, showing each external force that acts on the object. Draw a separate free-body diagram for each object of interest.
   3. Choose a convenient coordinate system for each object, and apply Newton's second law in component form.
   4. Solve the resulting equations for the unknowns using whatever additional information is available.
   5. Check your results to see if they are reasonable. Examine your solutions when the variables are assigned extreme values.

Suggestions for Further Reading

Cohen, I. Bernard: "Isaac Newton," *Scientific American*, December 1955, p. 73.

This is a short biographical sketch of Newton.

Diggins, John Patrick: "Science and the American Experiment: How Newton's Laws Shaped the Constitution," *The Sciences*, November/December 1987, p. 28.

Newton's laws had social consequences!

Feld, Michael S., Ronald E. McNair, and Stephen R. Wilk: "The Physics of Karate," *Scientific American*, April 1979, p. 150.

An analysis of the forces that a karate expert can apply with bare hands to a block of wood or concrete.

Shamos, Morris H.: "The Laws of Motion—Isaac Newton," in *Great Experiments in Physics*, Henry Holt and Company, New York, 1959. Reprinted by Dover, 1987.

Newton's own account of the laws of motion, with editorial annotations for clarity and a biographical sketch.

Walker, Jearl: "The Amateur Scientist: In Judo and Aikido Application of the Physics of Forces Makes the Weak Equal to the Strong," *Scientific American*, July 1980, p. 150.

Many diagrams help to make this an entertaining and instructive exposition of physical principles employed in two martial arts.

Weaver, Jefferson Hane: "Newton's World," in *The World of Physics: A Small Library of the Literature of Physics from Antiquity to the Present*, Simon and Schuster, New York, 1987.

Contains selections from Newton's writings, including "Mathematical Principles of Natural Philosophy" and "Rules of Reasoning in Philosophy."

Review

A. Objectives: After studying this chapter, you should

1. Be able to discuss the definitions of force and mass and state Newton's laws of motion.

2. Be able to define an inertial reference frame.

3. Be able to distinguish between mass and weight.

4. Be familiar with the following units and know how they are defined: kilogram, newton, pound.

5. Be able to distinguish between action–reaction pairs of forces that act on different bodies and balancing forces that act on the same body.

6. Be able to apply Newton's laws in a systematic way to solve mechanics problems.

B. Define, explain, or otherwise identify:

Newton's laws	Weight	Fundamental forces	Force constant
Law of inertia	Apparent weight	Action at a distance	Normal force
Inertia	Free fall	Field	Free-body diagram
Reference frame	Weightlessness	Contact forces	Constraints
Inertial reference frame	Newton	Hooke's law	Centripetal force
Force	Pound		
Mass	Law of interaction		
Kilogram	Action and reaction forces		

C. Write out the steps in the general method of attack for solving problems using Newton's laws of motion.

D. True or false: If the statement is true, explain why it is true. If it is false give a counterexample.

1. If there are no forces acting on an object, the body will not accelerate.

2. If an object is not accelerating, there must be no forces acting on it.

3. The motion of an object is always in the direction of the resultant force.

4. Action–reaction forces never act on the same object.

5. The mass of an object depends on its location.

6. Action equals reaction only if the objects are not accelerating.

Problems

Level I

4-1 Newton's First Law: The Law of Inertia

There are no problems for this section.

4-2 Force, Mass, and Newton's Second Law

1. An object experiences an acceleration of 4 m/s^2 when a certain force F_0 acts on it. (*a*) What is its acceleration when the force is doubled? (*b*) A second object experiences an acceleration of 8 m/s^2 under the influence of the force F_0. What is the ratio of the masses of the two objects? (*c*) If the two objects are tied together, what acceleration will the force F_0 produce?

2. An object is pulled in a straight line along a level, frictionless surface with a constant force. The increase in its speed in a 10-s interval is 5 km/h. When a second constant force is applied in the same direction in addition to the first force, the speed increases by 15 km/h in a 10-s interval. How do the magnitudes of the two forces compare?

3. A force F_0 causes an acceleration of 5 m/s^2 when it acts on an object of mass m. Find the acceleration of the same object when it is acted on by the forces shown in Figure 4-20*a* and *b*.

Figure 4-20 Forces acting on an object for Problem 3.

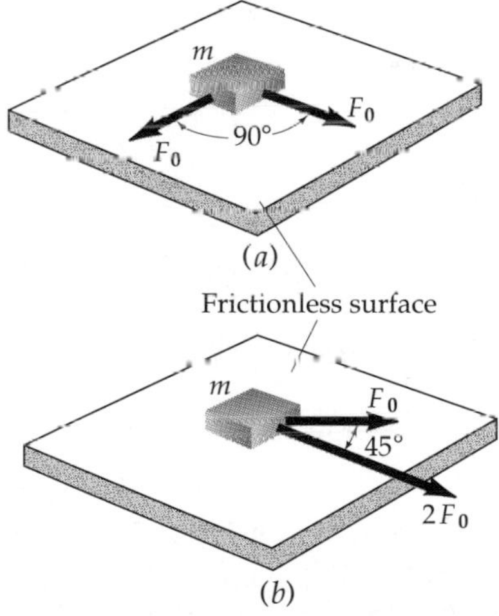

4. A certain force F_0 gives an object an acceleration of 6×10^6 m/s^2. Another force gives the same object an acceleration of 9×10^6 m/s^2. (*a*) What is the magnitude of the second force? What is the acceleration of the object (*b*) if the two forces act together on the object in the same direction, (*c*) if the two forces act in opposite directions on the object, and (*d*) if the two forces are perpendicular to each other?

5. A certain force applied to a particle of mass m_1 gives it an acceleration of 20 m/s^2. The same force applied to a particle of mass m_2 gives it an acceleration of 30 m/s^2. If the two particles are tied together and the same force is applied to the combination, find the acceleration.

6. A force of 15 N is applied to an object of mass m. The object moves in a straight line, with its speed increasing by 10 m/s every 2 s. Find the mass m.

7. A 5-kg object is pulled along a frictionless horizontal surface by a horizontal force of 10 N. (*a*) If the object is at rest at $t = 0$, how fast is it moving after 3 s? (*b*) How far does it travel from $t = 0$ to $t = 3$ s?

8. A force $\mathbf{F} = (6\mathbf{i} - 3\mathbf{j})$ N acts on an object of mass of 2 kg. Find the acceleration $\mathbf{a}$. What is the magnitude a?

9. A single force of 10 N acts on a particle of mass m. The particle starts from rest and travels in a straight line a distance of 18 m in 6 s. Find its mass m.

10. In order to drag a 100-kg log along the ground at constant velocity, you have to pull on it with a horizontal force of 300 N. (*a*) What is the resistive force exerted by the ground? (*b*) What force must you exert if you want to give the log an acceleration of 2 m/s^2?

11. The graph in Figure 4-21 shows a plot of v_x versus t for an object of mass 10 kg moving in a straight line. Make a plot of the net force acting on the object as a function of time.

Figure 4-21 Graph of velocity v_x versus time t for Problem 11.

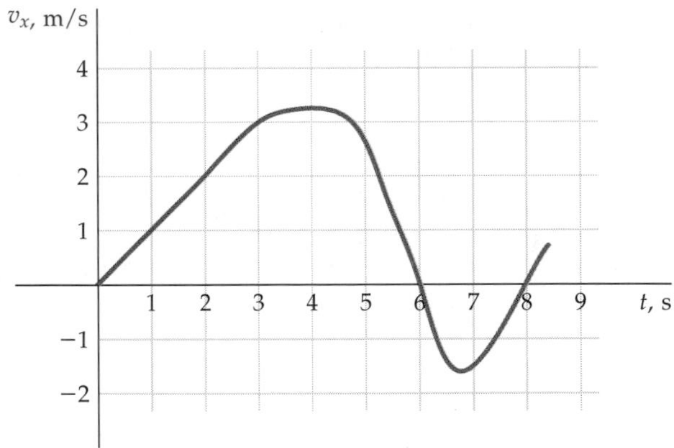

12. Figure 4-22 shows the position x versus time t of a particle moving in one dimension. During what time intervals is there a net force acting on the particle? Give the direction (+ or −) of the net force during these time intervals.

13. Figure 4-23 shows the path taken by an automobile. It consists of straight lines and arcs of circles. The automobile starts from rest at point A and accelerates until it reaches point B. It then proceeds at constant speed until it reaches point E. From point E on it slows down, coming to rest at point F. What is the direction of the net force, if there is any, on the automobile at the midpoint of each section of the path?

4-3 The Force Due to Gravity: Weight

14. Find the weight of a 50-kg girl in (*a*) newtons and (*b*) pounds.

15. Find the mass of a 175-lb man in (*a*) kilograms and (*b*) grams.

16. Find the weight of a 50-g object in newtons.

17. The gravitational force exerted by the earth on an object at height h above the earth's surface can be written

$$F = mg\frac{R_E^2}{(R_E + h)^2}$$

where R_E is the radius of the earth (about 6370 km) and g is the acceleration due to gravity at the earth's surface. (*a*) Find the weight of an 80-kg woman in newtons and in pounds at the earth's surface. (*b*) Find the weight of the woman (in newtons and in pounds) at a height of 300 km above the earth's surface. (*c*) What is the mass of this woman at this altitude?

4-4 Newton's Third Law

18. A 2-kg object hangs at rest from a string attached to the ceiling. (*a*) Draw a diagram showing the forces acting on the object and indicate each reaction force. (*b*) Do the same for the forces acting on the string.

19. A box slides down a frictionless inclined plane. Draw a diagram showing the forces acting on the box. For each force in your diagram, indicate the reaction force.

Figure 4-22 Graph of position x versus time t for Problem 12.

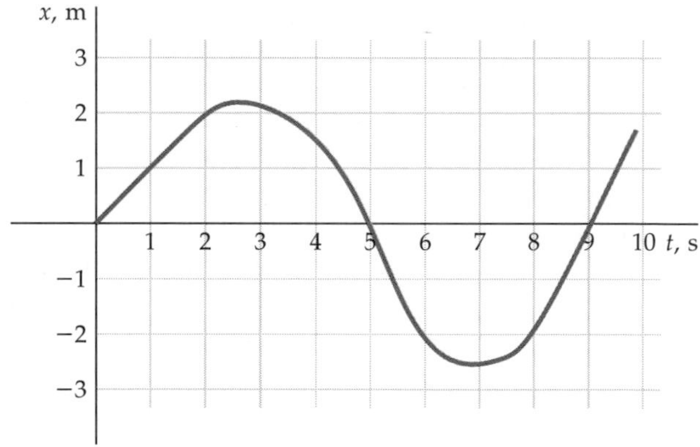

Figure 4-23 Path taken by automobile in Problem 13.

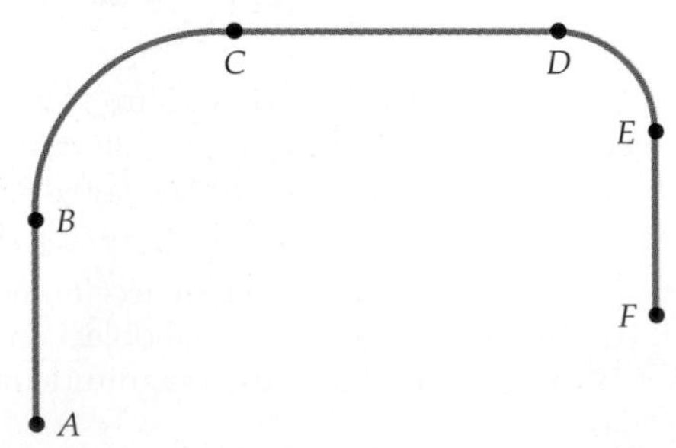

20. A 3-kg box rests on a 7-kg box that rests on a horizontal table. (*a*) Draw a diagram showing the forces acting on the 3-kg box and indicate each reaction force. (*b*) Do the same for the forces acting on the 7-kg box.

4-5 Forces in Nature

21. The moon is 3.84×10^8 m from the earth and the star Sirius is about 8.25×10^{16} m from the earth. (*a*) How long does it take for the gravitational field of the earth to propagate to the moon? (*b*) How long does it take for the gravitational field of the earth to propagate to the star Sirius?

22. A vertical spring of force constant 400 N/m is attached to a 10-kg block resting on a horizontal surface so that the spring exerts an upward force on the block. The spring is stretched by 10 cm. (*a*) What force does the spring exert on the block? (*b*) What is the force that the surface exerts on the block?

23. A 6-kg box slides on a frictionless horizontal surface. It is pulled by a horizontal spring with force constant of 800 N/m. If the spring is stretched 4 cm from its equilibrium position, what is the acceleration of the box?

4-6 Applications to Problem Solving

24. A 10-kg object is subjected to the two forces, $\mathbf{F}_1$ and $\mathbf{F}_2$, as shown in Figure 4-24. (*a*) Find the acceleration **a** of the object. (*b*) A third force $\mathbf{F}_3$ is applied so that the object is in static equilibrium. Find $\mathbf{F}_3$.

Figure 4-24 Problem 24.

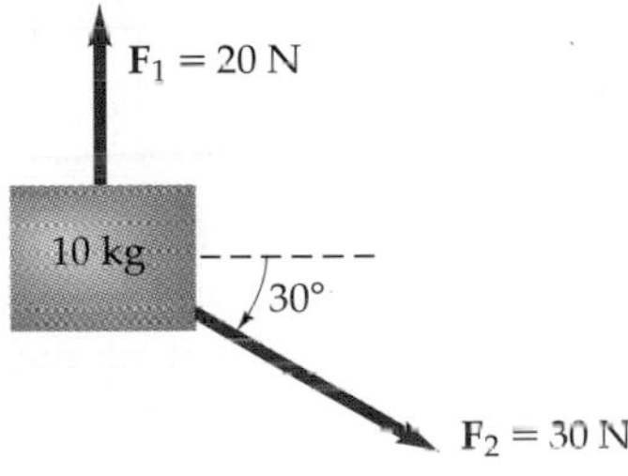

25. In Figure 4-25, the objects are attached to spring balances calibrated in newtons. Give the readings of the balances in each case, assuming that the strings are massless and the incline is frictionless.

26. A vertical force **T** is exerted on a 5-kg body near the surface of the earth, as shown in Figure 4-26. Find the acceleration of the body if (*a*) $T = 5$ N, (*b*) $T = 10$ N, and (*c*) $T = 100$ N.

Figure 4-26 Problem 26.

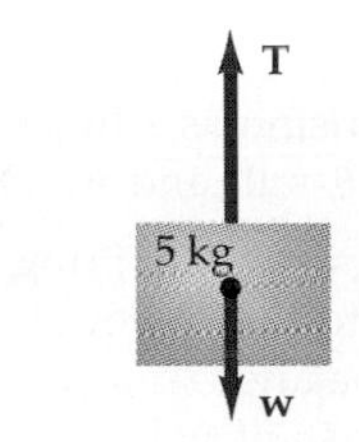

27. A 3-kg body is acted on by a single force **F** perpendicular to the velocity of the body. The body travels in a circle of radius 2 m. It makes one complete revolution every 3 s. (*a*) What is the magnitude of the acceleration? (*b*) What is the magnitude of **F**?

28. A woman holding a 10-kg block by a cord designed to withstand a tension of 150 N steps into an elevator. When the elevator starts up, the cord breaks. What was the minimum acceleration of the elevator?

29. A 60-kg girl weighs herself by standing on a scale in an elevator. What does the scale read when (*a*) the elevator is descending at a constant rate of 10 m/s, (*b*) the elevator is accelerating downward at 2 m/s^2, and (*c*) the elevator is ascending at 10 m/s but its speed is decreasing by 2 m/s in each second?

Level II

30. A 4-kg object is subjected to two forces, $\mathbf{F}_1 = 2\text{ N}\mathbf{i} - 3\text{ N}\mathbf{j}$ and $\mathbf{F}_2 = 4\text{ N}\mathbf{i} + 11\text{ N}\mathbf{j}$. The object is at rest at the origin at time $t = 0$. (*a*) What is the object's acceleration? (*b*) What is its velocity at time $t = 3$ s? (*c*) Where is the object at time $t = 3$ s?

Figure 4-25 Problem 25.

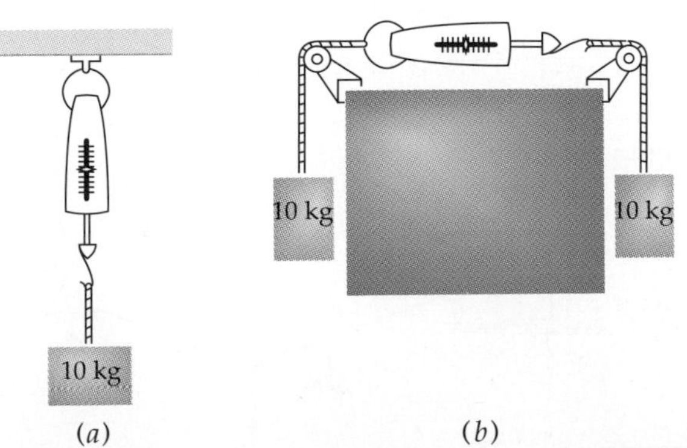

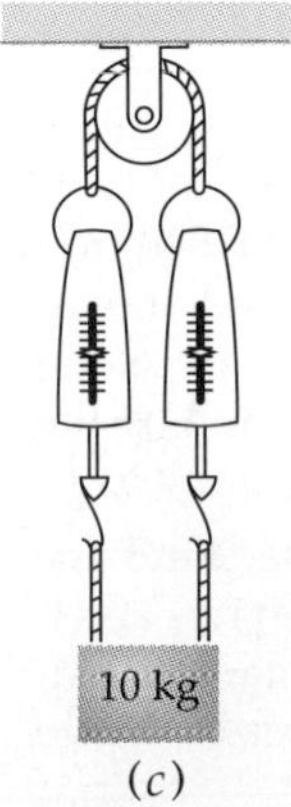

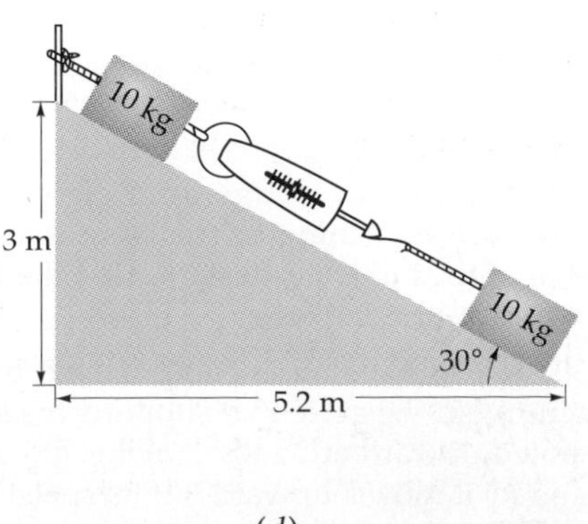

31. A 2-kg picture is hung by two wires of equal length. Each makes an angle of θ with the horizontal, as shown in Figure 4-27. (*a*) Find the tension T for general values of θ and weight w of the picture. For what angle θ is T the least? The greatest? (*b*) If $\theta = 30°$, find the tension in the wires.

32. A box is held in position by a cable along a smooth incline (Figure 4-28). (*a*) If $\theta = 60°$ and $m = 50$ kg, find the tension in the cable and the normal force exerted by the incline. (*b*) Find the tension as a function of θ and m, and check your result for $\theta = 0°$ and $\theta = 90°$.

33. A 2-kg block hangs from a spring balance (calibrated in newtons) attached to the ceiling of an elevator (Figure 4-29). What does the balance read (*a*) when the elevator is moving up with a constant velocity of 30 m/s, (*b*) when the elevator is moving down with a constant velocity of 30 m/s, and (*c*) when the elevator is accelerating upward at 10 m/s^2? (*d*) From $t = 0$ to $t = 2$ s, the elevator moves upward at 10 m/s. Its velocity is then reduced uniformly to zero in the next 2 s, so that it is at rest at $t = 4$ s. Describe the reading on the balance during the interval $t = 0$ to $t = 4$ s.

Figure 4-29 Problem 33.

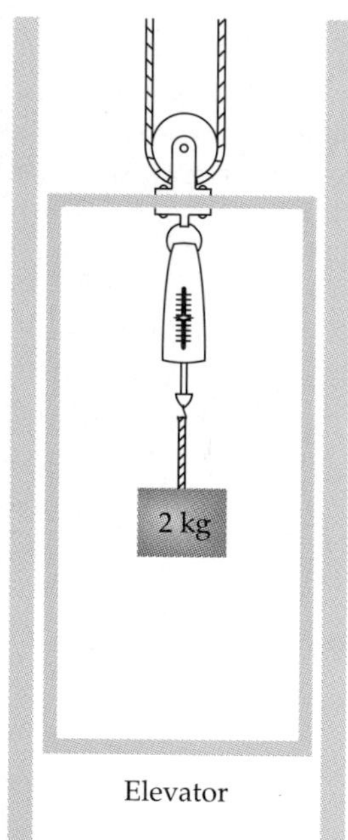

34. A bullet of mass 1.8×10^{-3} kg moving at 500 m/s embeds itself in a large fixed piece of wood and travels 6 cm before coming to rest. Assuming that the deceleration of the bullet is constant, find the force exerted by the wood on the bullet.

35. A 1000-kg load is being moved by a crane. Find the tension in the cable that supports the load as (*a*) it is accelerated upward at 2 m/s^2, (*b*) it is lifted at constant speed, and (*c*) it moves upward but its speed decreases by 2 m/s in each second.

Figure 4-27 Problem 31.

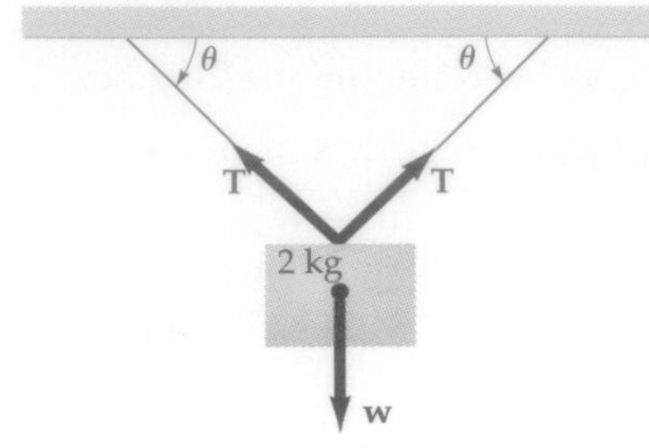

Figure 4-28 Problem 32.

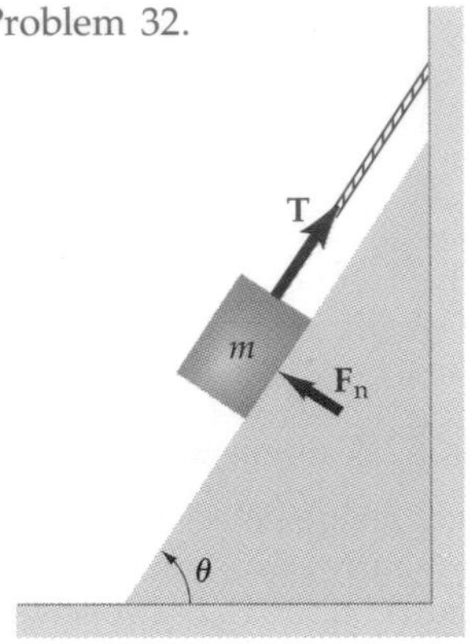

36. The acceleration a versus spring length L observed when a 0.5-kg mass is pulled along a frictionless table by a single spring is shown in the following table:

L, cm	4	5	6	7	8	9	10	11	12	13	14
a, m/s^2	0	2.0	3.8	5.6	7.4	9.2	11.2	12.8	14.0	14.6	14.6

(*a*) Make a plot of the force exerted by the spring versus length L. (*b*) If the spring is extended to 12.5 cm, what force does it exert? (*c*) By how much is the spring extended when the mass suspended from it is at rest near sea level, where $g = 9.81$ N/kg?

37. Your car is stuck in a mudhole. You are alone, but you have a long, strong rope. Having studied physics, you tie the rope tautly to a tree and pull on it sideways, as shown in Figure 4-30. (*a*) Find the force exerted by the rope on the car when the angle θ is 3° and you are pulling with a force of 400 N but the car does not move. (*b*) How strong must the rope be if it takes a force of 600 N to move the car when $\theta = 3°$?

Figure 4-30 Problem 37.

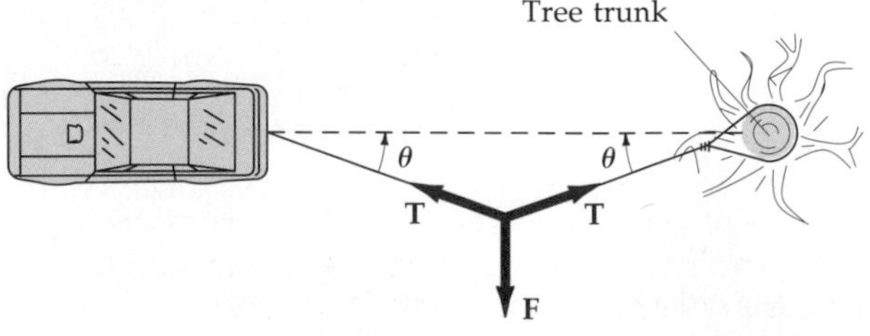

38. A rifle bullet of mass 9 g starts from rest and moves 0.6 m in the rifle barrel. The speed of the bullet as it leaves the barrel is 1200 m/s. Find the force exerted on the bullet assuming it to be constant while the bullet is in the barrel.

39. A car traveling 90 km/h comes to a sudden stop to avoid an accident. Fortunately the driver is wearing a seat belt. Using reasonable values for the mass of the driver and the time it takes the car to come to a stop, estimate the force (assuming it to be constant) exerted on the driver by the seat belt.

40. A man stands on a scale in an elevator that has an upward acceleration a. The scale reads 960 N. When he picks up a 20-kg box, the scale reads 1200 N. Find the mass of the man, his weight, and the acceleration a.

41. A horizontal force of 100 N pushes a 12-kg block up a frictionless incline that makes an angle of 25° with the horizontal. (*a*) What is the normal force that the incline exerts on the block? (*b*) What is the acceleration of the block?

42. In a tug-of-war, two boys pull on a rope, each trying to pull the other over a line midway between their original positions. According to Newton's third law, the forces exerted by each boy on the other are equal and opposite. Show with a force diagram how one boy can win.

43. A man swings his child in a circle of radius 0.75 m as shown in Figure 4-31. If the mass of the child is 25 kg and the child makes one revolution in 1.5 s, what is the magnitude and direction of the force that must be exerted by the man on the child?

Figure 4-31 Problem 43.

Level III

44. A box of mass m_1 is pulled by a force **F** exerted at the end of a rope that has a much smaller mass m_2, as shown in Figure 4-32. The box slides along a smooth horizontal surface. (*a*) Find the acceleration of the rope and box assuming them to be one object. (*b*) What is the net force acting on the rope? (*c*) Find the tension in the rope at the point where it is attached to the box. (*d*) The diagram in Figure 4-32, with the rope horizontal, is not quite correct for this situation. Correct the diagram, and state how this correction affects your solution.

Figure 4-32 Problem 44.

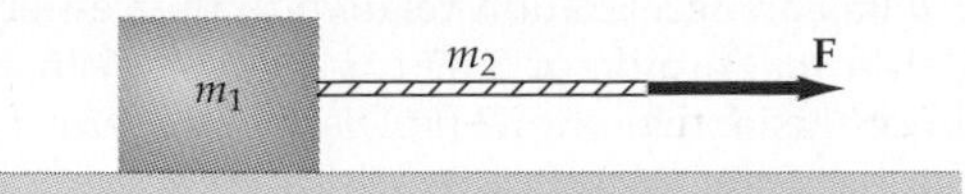

45. A 2-kg body rests on a smooth wedge that has an inclination of 60° (Figure 4-33) and an acceleration a to the right such that the mass remains stationary relative to the wedge. (*a*) Find a. (*b*) What would happen if the wedge were given a greater acceleration?

Figure 4-33 Problem 45.

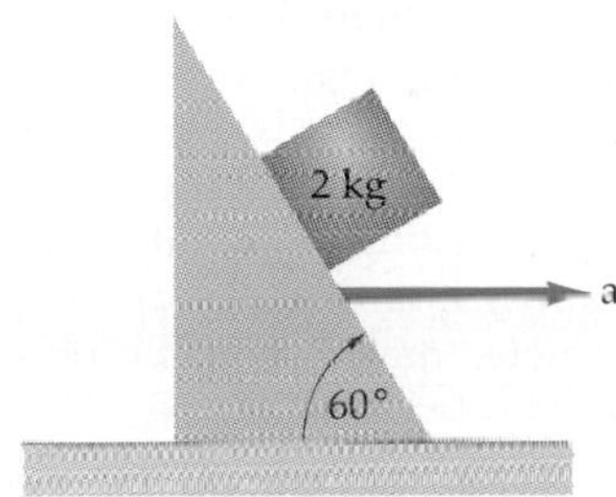

46. A student has to escape from his girlfriend's dormitory through a window that is 15.0 m above the ground. He has a heavy 20-m-long rope, but it will break when the tension exceeds 360 N and he weighs 600 N. The student will be injured if he hits the ground with a speed greater than 10 m/s. (*a*) Show that he cannot safely slide down the rope. (*b*) Find a strategy using the rope that will permit the student to reach the ground safely.

47. A 60-kg house painter stands on a 15-kg aluminum platform. A rope attached to the platform and passing over an overhead pulley allows the painter to raise herself and the platform (Figure 4-34). (*a*) To get started, she accelerates herself and the platform at a rate of 0.8 m/s^2. With what force must she pull on the rope? (*b*) When her speed reaches 1 m/s, she pulls such that she and the platform go up at a constant speed. What force must she exert on the rope? (Ignore the mass of the rope.)

Figure 4-34 Problem 47.

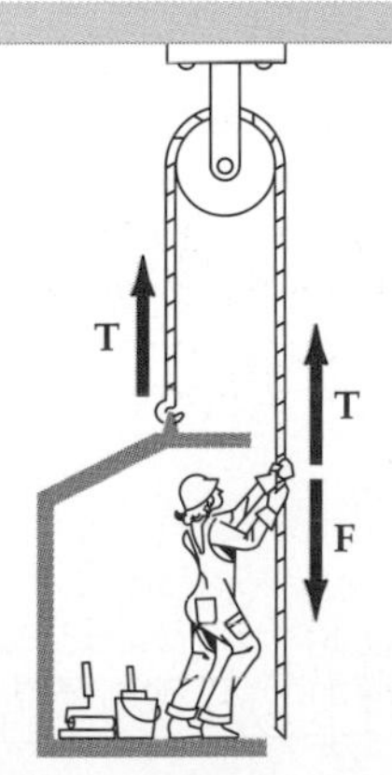

48. (*a*) Show that a point on the surface of the earth at latitude θ has an acceleration relative to the center of the earth with a magnitude of $3.37 \cos \theta$ cm/s^2. What is the direction of this acceleration? (*b*) Discuss the effect of this acceleration on the apparent weight of an object near the surface of the earth. (*c*) The free-fall acceleration of an object at sea level measured *relative to the earth's surface* is 9.78 m/s^2 at the equator and 9.81 m/s^2 at latitude $\theta = 45°$. What are the values of the gravitational field g at these points?

49. A simple accelerometer can be made by suspending a small object from a string attached to a fixed point in an accelerating object—to the ceiling of a passenger car, for example. When there is an acceleration, the object will deflect and the string will make some angle with the vertical. (*a*) How is the direction in which the suspended object deflects related to the direction of the acceleration? (*b*) Show that the acceleration a is related to the angle θ the string makes by $a = g \tan \theta$. (*c*) Suppose the accelerometer is attached to the ceiling of an automobile that brakes to rest from 50 km/h in a distance of 60 m. What angle will the accelerometer make? Will the object swing forward or backward?

50. A 65-kg girl weighs herself by standing on a scale mounted on a skateboard that is rolling down an incline as shown in Figure 4-35. Assume there is no friction so that the force exerted by the incline on the skateboard is perpendicular to the incline. What is the reading on the scale if $\theta = 30°$?

51. A small bead with a mass of 100 g slides along a semicircular wire with a radius of 10 cm that rotates about a vertical axis at a rate of 2 revolutions per second as shown in Figure 4-36. Find the values of θ for which the bead will remain stationary with respect to the rotating wire.

Figure 4-35 Problem 50.

Figure 4-36 Problem 51.

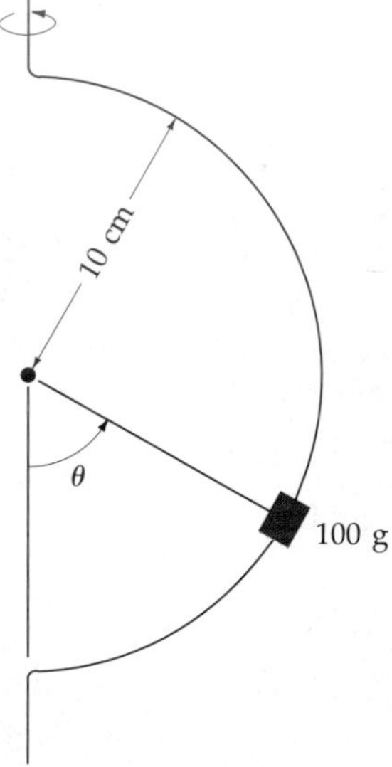

Chapter 5

Newton's Laws II

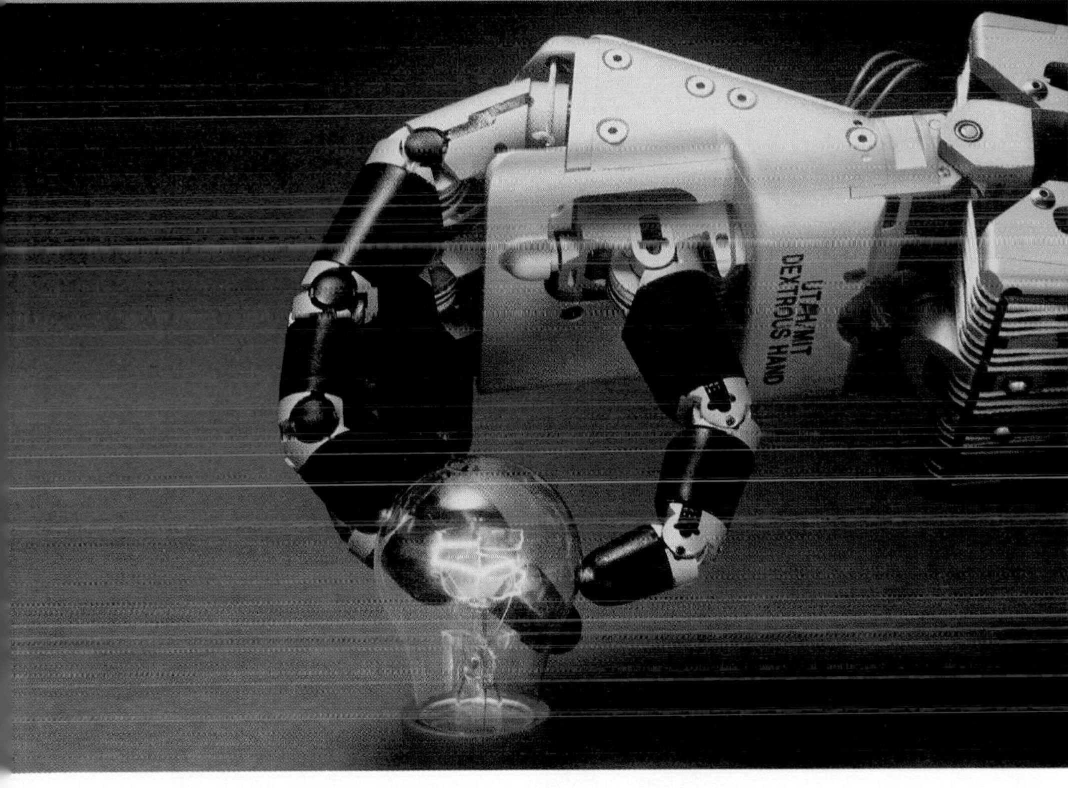

The Utah/MIT Dextrous Hand is a tendon-operated, multidegree-of-freedom dextrous hand that has multichannel touch sensing capability.

In this chapter, we give examples of the application of Newton's laws to problems that involve frictional forces and problems that involve two or more objects. We will also briefly discuss the motion of an object under the influence of drag forces, which are not constant but depend on the velocity of the object.

The method of problem solving discussed in Section 4-6 applies to many of the problems in this chapter, too. The steps involved should be reread and learned well. In particular, you should develop the habit of approaching a problem by first drawing a picture and then indicating the important forces that act on each object in a free-body diagram.

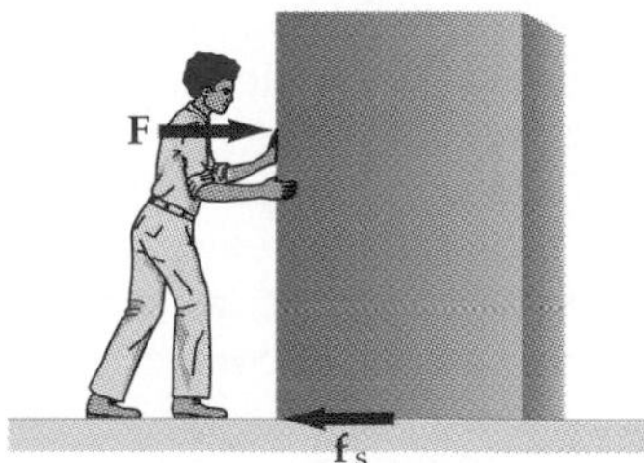

Figure 5-1 When you try to push a large box across the floor, friction opposes the motion. The floor exerts a force of static friction that balances the applied force unless the applied force is greater than the maximum possible force of static friction.

5-1 Friction

If you push on a large box resting on the floor with a small horizontal force, it is possible that the box won't move at all. The reason is that the floor is exerting a horizontal force called the force of **static friction** $\mathbf{f}_s$, which balances the force you are exerting (Figure 5-1). This frictional force is due to the bonding of the molecules of the box and the floor at those places where the

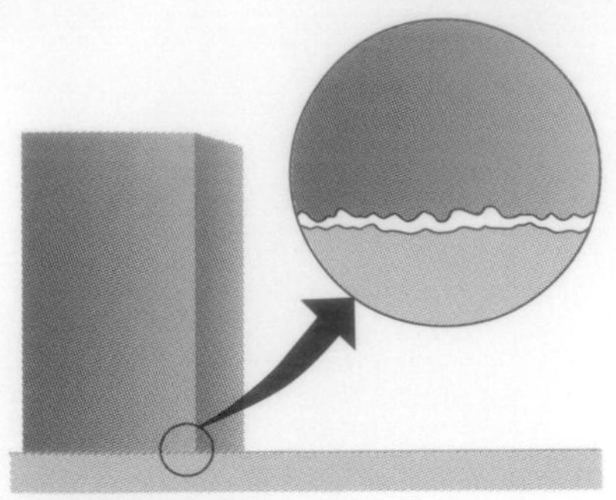

Figure 5-2 The microscopic area of contact between a box and a table is only a small fraction of the macroscopic area of contact. This fraction is proportional to the normal force exerted between the surfaces.

surfaces are in very close contact. It is in the opposite direction of the applied force. The force of static friction is somewhat like a support force in that it can adjust from zero to some maximum force $\mathbf{f}_{s,max}$, depending on how hard you push. If you push hard enough, the box will slide across the floor. When the box is sliding, molecular bonds are continually being formed and ruptured and small pieces of the surfaces are being broken off. The result is a force of **kinetic friction** $\mathbf{f}_k$ (also called sliding friction) that opposes the motion. To keep the box sliding with constant velocity, you must exert a force equal in magnitude and opposite in direction to this force of kinetic friction.

We will consider static friction first. We might expect the maximum force of static friction to be proportional to the area of contact between the two surfaces. However, it has been shown experimentally that, to a good approximation, this force is independent of the area of contact and is simply proportional to the normal force exerted by one of the surfaces on the other. Figure 5-2, which includes an enlarged view of the contact between a box and a table, shows that the actual microscopic area of contact where the molecules can bond together is just a small fraction of the total macroscopic area of contact. Static friction is a complicated phenomenon that is not completely understood, even today. The following, however, is a possible model that is consistent both with our intuition and with empirical results. The maximum force of static friction is proportional to the microscopic area of contact, but this microscopic area is proportional to the total macroscopic area A and to the normal force per unit area F_n/A exerted between the surfaces. The product of A and F_n/A is thus independent of the total macroscopic area A.

Consider, for example, a 1-kg box with a side area of 60 cm^2 and an end area of 20 cm^2. When the box is on its side on a table, a small fraction of the total 60 cm^2 is actually in microscopic contact with the table. When the box is placed on end, the fraction of the total area actually in microscopic contact is increased by a factor of 3 because the normal force per unit area is 3 times as great. However, since the area of the end is one-third that of the side, the actual microscopic area of contact is unchanged. Thus, the maximum force of static friction $f_{s,max}$ is proportional to the normal force between the surfaces:

$$f_{s,max} = \mu_s F_n \tag{5-1}$$

where μ_s is called the **coefficient of static friction.** It depends on the nature of the surfaces of the box and the table. If we exert a horizontal force smaller than $f_{s,max}$ on the box, the frictional force will just balance this horizontal force. In general, we can write

$$f_s \leq \mu_s F_n \tag{5-2}$$

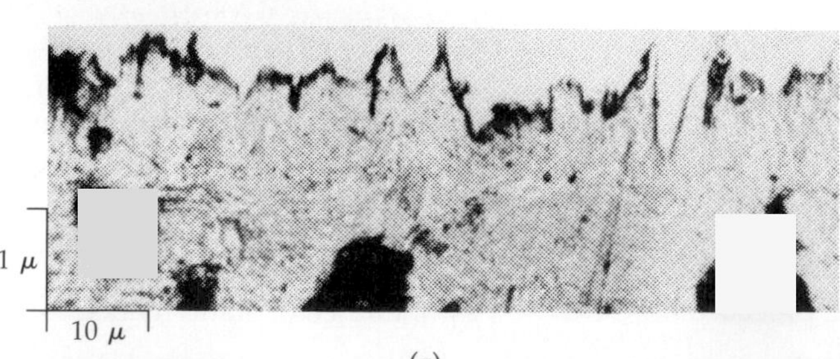

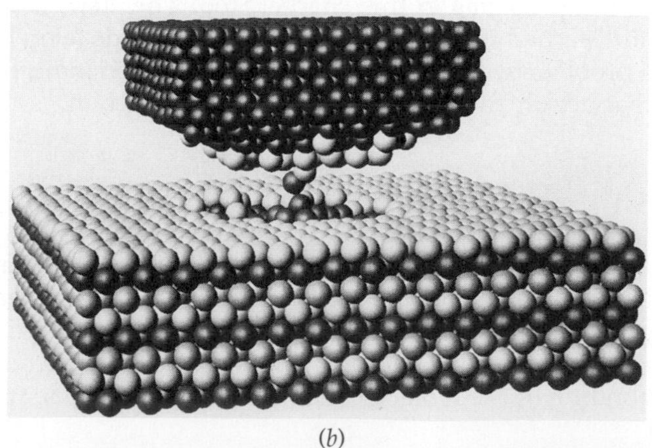

(*a*) (*b*)

(*a*) Magnified section of a polished steel surface showing surface irregularities. The irregularities are about 5×10^{-5} cm high, corresponding to about several thousand atomic diameters. (*b*) Computer depiction derived from microscope data. It shows gold atoms (bottom) adhering to a nickel tip (top), as the tip is lifted after being in contact with the gold surface.

The force of kinetic friction is opposite the direction of motion. Like static friction, kinetic friction is a complicated and incompletely understood phenomenon. The **coefficient of kinetic friction** μ_k is defined as the ratio of the magnitudes of the kinetic frictional force f_k and the normal force F_n. Then

$$f_k = \mu_k F_n \tag{5-3}$$

Experimentally, it has been found that:

1. μ_k is less than μ_s.
2. μ_k depends on the relative speed of the surfaces; but for speeds in the range from about 1 cm/s to several meters per second, μ_k is approximately constant.
3. μ_k (like μ_s) depends on the nature of the surfaces in contact, but it is independent of the (macroscopic) area of contact.

We will neglect any variation in μ_k with speed and assume that it is a constant that depends only on the nature of the surfaces in contact.

Figure 5-3 shows the frictional force exerted on the box by the table as a function of the applied force. The force of friction balances the applied force until the applied force equals $\mu_s F_n$, at which point the box begins to slide. Then the frictional force is constant and equal to $\mu_k F_n$.

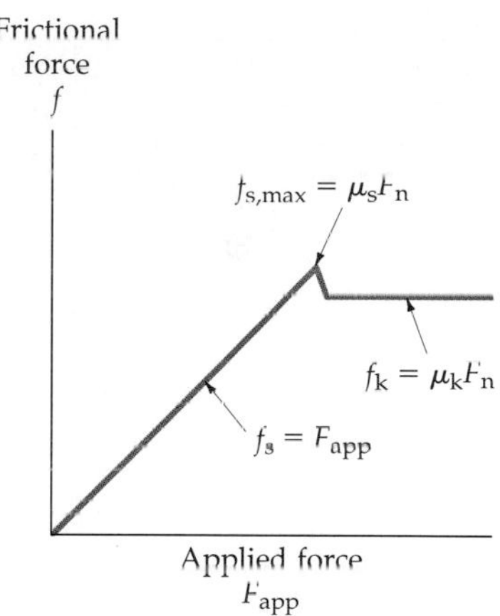

Figure 5-3 Graph of the force of friction acting on a box versus the force applied to the box. When the applied force exceeds the maximum force of static friction $\mu_s F_n$, the box slides and the friction is kinetic friction.

We can measure μ_s and μ_k for two surfaces simply by placing a block on a plane surface and inclining the plane until the block begins to slide. Let θ_c be the critical angle at which sliding starts. For angles of inclination less than this, the block is in static equilibrium under the influence of its weight $m\mathbf{g}$, the normal force $\mathbf{F}_n$, and the force of static friction $\mathbf{f}_s$ (see Figure 5-4). Choosing the x axis parallel to the plane and the y axis perpendicular to the plane, we have

$$\sum F_y = F_n - mg \cos \theta = 0$$

and

$$\sum F_x = mg \sin \theta - f_s = 0$$

We can eliminate the weight mg from these two equations by solving the first equation for $mg = F_n/\cos \theta$ and substituting this into the second equation:

$$f_s = mg \sin \theta = \frac{F_n}{\cos \theta} \sin \theta = F_n \tan \theta$$

At the critical angle θ_c, the force of static friction equals its maximum value, and we can replace f_s by $\mu_s F_n$. Then

$$\mu_s = \tan \theta_c \tag{5-4}$$

Thus, the coefficient of static friction equals the tangent of the angle of inclination at which the block just begins to slide.

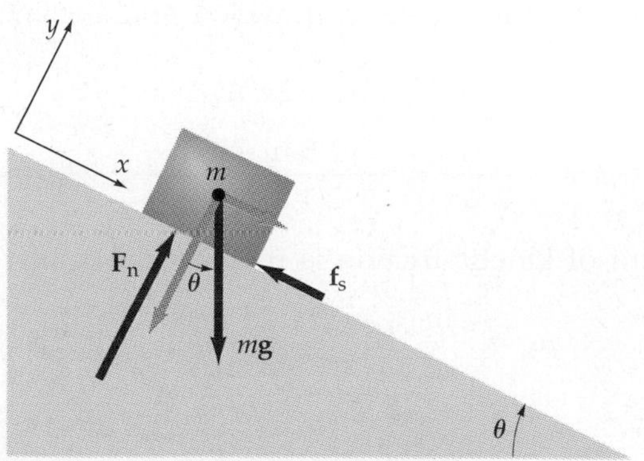

Figure 5-4 Forces acting on a block on a rough inclined plane. At angles less than the critical angle θ_c, the frictional force balances the component $mg \sin \theta$ directed down the incline. At angles greater than θ_c the block slides down the incline. The critical angle is related to the coefficient of static friction by $\tan \theta_c = \mu_s$.

At angles greater than θ_c, the block slides down the incline with an acceleration a_x. In this case, the frictional force is $\mu_k F_n$, and we have

$$F_x = mg \sin \theta - \mu_k F_n = ma_x$$

Substituting $mg \cos \theta$ for F_n, we obtain for the acceleration

$$a_x = g(\sin \theta - \mu_k \cos \theta)$$

By measuring the acceleration a_x, we can determine μ_k for the two surfaces. Table 5-1 lists some approximate values of μ_s and μ_k for various surfaces.

Table 5-1 **Approximate Values of Frictional Coefficients**

Materials	μ_s	μ_k
Steel on steel	0.7	0.6
Brass on steel	0.5	0.4
Copper on cast iron	1.1	0.3
Glass on glass	0.9	0.4
Teflon on Teflon	0.04	0.04
Teflon on steel	0.04	0.04
Rubber on concrete (dry)	1.0	0.80
Rubber on concrete (wet)	0.30	0.25
Waxed ski on snow (0°C)	0.1	0.05

Exercise

The coefficient of static friction between a car's tires and the road on a particular day is 0.7. What is the steepest angle of inclination of the road for which the car can be parked with its wheels locked and not slide down the hill? (Answer: 35°)

Example 5-1

A box slides along a horizontal floor with an initial speed of 2.5 m/s. It comes to rest after sliding 1.4 m. Find the coefficient of kinetic friction.

The weight of the box mg is balanced by the normal force of the floor $F_n = mg$. The only horizontal force acting on the box is the force of kinetic friction, which acts opposite to the direction of motion. If we assume the velocity to be in the positive direction, the force of friction is

$$f = -\mu_k F_n = -\mu_k mg$$

The acceleration is then

$$a = \frac{f}{m} = -\mu_k g$$

Since the acceleration is constant, we may use Equation 2-13 to relate the acceleration to the total distance traveled and the initial velocity. We have

$$v^2 = v_0^2 + 2a\,\Delta x = 0$$

$$a = -\frac{v_0^2}{2\,\Delta x} = -\frac{(2.5\text{ m/s})^2}{2(1.4\text{ m})} = -2.23\text{ m/s}^2$$

The coefficient of kinetic friction is then found from

$$\mu_k = -\frac{a}{g} = -\frac{-2.23\text{ m/s}^2}{9.81\text{ m/s}^2} = 0.228$$

Example 5-2

Two children are pulled on a sled over snow-covered ground. The sled is pulled by a rope that makes an angle of 40° with the horizontal as shown in Figure 5-5. The children have a combined mass of 45 kg and the sled has a mass of 5 kg. The coefficients of static and kinetic friction are $\mu_s = 0.2$ and $\mu_k = 0.15$. Find the frictional force exerted by the ground on the sled and the acceleration of the children and sled if the tension in the rope is (*a*) 100 N and (*b*) 140 N.

Figure 5-5 Man pulling children on a sled for Example 5-2.

(*a*) The vertical and horizontal components of the tension in the rope are

$$T_y = T \sin 40° = (100 \text{ N})(0.643) = 64.3 \text{ N}$$

and

$$T_x = T \cos 40° = (100 \text{ N})(0.766) = 76.6 \text{ N}$$

The vertical forces acting on the sled are its weight acting downward, the normal force exerted by the ground acting upward, and the upward component of the tension in the rope. Since there is no vertical acceleration, the net vertical force must be zero:

$$\sum F_y = F_n + T_y - mg = 0$$

The normal force is therefore

$$F_n = mg - T_y = (50 \text{ kg})(9.81 \text{ m/s}^2) - 64.3 \text{ N}$$
$$= 490 \text{ N} - 64.3 \text{ N} = 426 \text{ N}$$

The maximum possible force of static friction is

$$f_{s,\max} = \mu_s F_n = 0.2(426 \text{ N}) = 85.2 \text{ N}$$

Since the applied horizontal force $T_x = 76.6$ N does not exceed the maximum possible force of static friction, the sled remains at rest. The frictional force is therefore 76.6 N to the left to balance the horizontal force applied to the right.

There are two important points to note about this example: (1) The normal force is not equal to the weight of the children and the sled because the vertical component of the tension helps lift the sled off the ground. (2) The force of static friction is not equal to $\mu_s F_n$; it is less than this maximum possible limiting value.

(*b*) When the tension is increased to 140 N, its vertical and horizontal components are

$$T_y = (140\text{ N})(\sin 40°) = 90.0\text{ N}$$

and

$$T_x = (140\text{ N})(\cos 40°) = 107\text{ N}$$

The normal force is then

$$F_n = mg - T_y = 490\text{ N} - 90.0\text{ N} = 400\text{ N}$$

so the maximum possible force of static friction is

$$f_{s,max} = \mu_s F_n = 0.2(400\text{ N}) = 80.0\text{ N}$$

Since this maximum force of static friction is less than the applied horizontal force, the sled will slide. The frictional force on the sled will thus be due to kinetic friction and will have the value

$$f_k = \mu_k F_n = 0.15(400\text{ N}) = 60.0\text{ N}$$

The sum of the horizontal forces (taking the rightward direction in the figure to be positive) is

$$\sum F_x = T_x - f_k = 107\text{ N} - 60.0\text{ N} = 47.0\text{ N}$$

and the acceleration of the sled is

$$a_x = \frac{\Sigma F_x}{m} = \frac{47.0\text{ N}}{50\text{ kg}} = 0.940\text{ m/s}^2$$

When a car is accelerating along a straight road, the unbalanced force that provides the acceleration is due to friction between the tires and the road. Figure 5-6 shows the forces acting on a car that is just starting to move from rest. The weight of the car is balanced by the normal force exerted on the tires. To start the car moving, the engine delivers a torque to the drive shaft that makes the wheels rotate. (We will discuss torque in Chapter 8.) If the road were frictionless, the wheels would merely spin, with the surface of the tire in contact with the road moving backward. If there is friction, and if the torque delivered by the engine is not too great, the tire will not slip on the road because of static friction. This frictional force exerted by the road on the car is in the forward direction, and it provides the acceleration needed for the car to start moving forward. If each wheel rolls without slipping, the tire surface in contact with the road is at rest relative to the road. Although the tire surface in contact with the road is moving backward with speed v relative to the axle, the axle is moving forward with speed v relative to the road as the car moves forward (Figure 5-7). The friction between the road and the tire is therefore static friction. If the engine torque is too great, the wheels will spin, in which case the surface of each tire in contact with the road will slip backward relative to the road. Then the force that accelerates the car is the force of kinetic friction, which is less than the force of static friction. We can see from this example that if we are stuck on

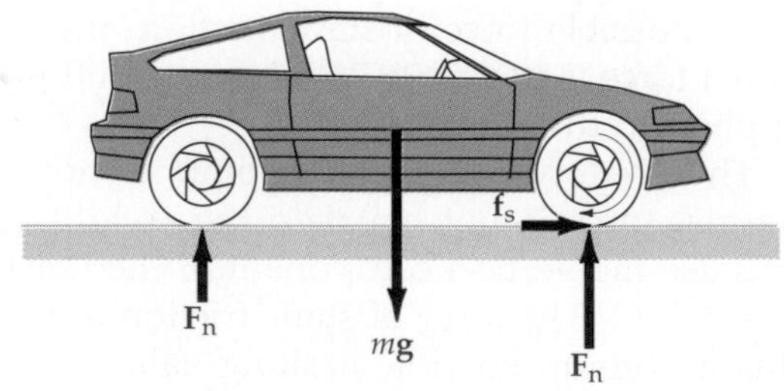

Figure 5-6 Forces acting on a car. As the engine turns the front wheels (for front-wheel drive), the force of static friction $\mathbf{f}_s$ prevents the tires from slipping on the road, causing the car to roll forward. If the wheels spin, the smaller force of kinetic friction pushes the car forward. (The normal forces $\mathbf{F}_n$ are not generally equal on the front and rear tires.)

ice or snow, our chances of getting free are better if we use a light touch on the accelerator pedal of the car so that the wheels do not slip.

When braking a car to a stop, the force exerted by the road on the tires may be either static friction or kinetic friction, depending on how the brakes are applied. If the brakes are applied so hard that the wheels lock, the tires will slide along the road and the stopping force will be that of kinetic friction. However, if the brakes are applied gently so that no slipping occurs between the tires and the road, the stopping force will be that of static friction.

When a wheel rolls *at constant speed* along a horizontal road without slipping, there is neither static nor kinetic friction. However, it is observed that a small force is required to keep a wheel rolling with constant velocity. As the wheel rolls, the surfaces of the wheel and road in contact must be continually peeled apart. In addition, each surface is deformed slightly so that in effect, the wheel is rolling uphill (Figure 5-8). We can define a **coefficient of rolling friction** μ_r as the ratio of the force needed to keep a wheel rolling at constant velocity on a level surface to the normal force exerted by the surface on the wheel. Typical values of μ_r are 0.01 to 0.02 for rubber tires on concrete, and 0.001 to 0.002 for steel wheels on steel rails.

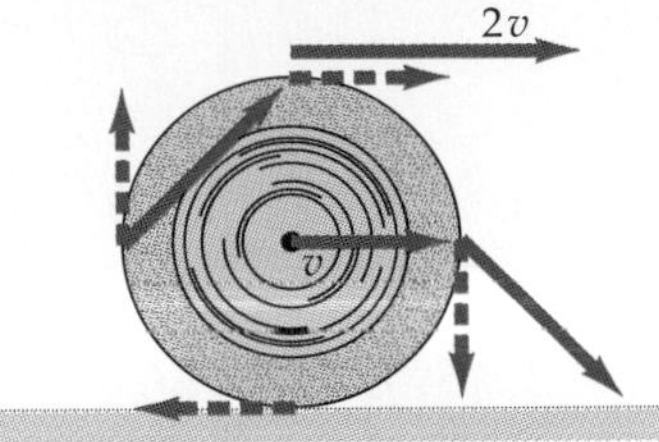

Figure 5-7 When a wheel rolls without slipping, each point on the rim has a velocity of magnitude v relative to the center of the wheel, where v is the speed of the center of the wheel relative to the ground. The velocity of the point on the tire in contact with the ground is zero relative to the ground. In this figure, the dashed lines represent velocities relative to the center of the wheel, and the solid lines represent velocities relative to the ground.

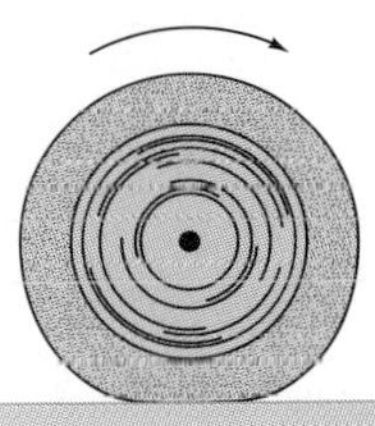

Figure 5-8 Rolling friction. When a wheel rests on a horizontal surface, both surfaces are deformed. As the wheel rolls, its center is raised slightly, just as if the wheel were rolling up a small hill.

Example 5-3

A car is traveling at 30 m/s along a horizontal road. The coefficients of friction between the road and the tires are $\mu_s = 0.5$ and $\mu_k = 0.3$. How far does the car travel before stopping if (*a*) the car is braked firmly so that the wheels almost slip and (*b*) the car is braked hard so that the wheels lock?

(*a*) Figure 5-9 shows the forces on the car as it is stopping. Since there is no vertical acceleration, the weight of the car mg is balanced by the normal force F_n exerted by the road:

$$\sum F_y = F_n - mg = 0$$

so

$$F_n = mg$$

Since the wheels do not slip, the horizontal force exerted by the road is that of static friction.* Taking the direction of motion as the positive direction (to the right in the figure), Newton's second law gives

$$\sum F_x = -\mu_s F_n = ma_x$$

$$a_x = -\frac{\mu_s F_n}{m} = -\frac{\mu_s mg}{m} = -\mu_s g$$

$$= -(0.5)(9.81 \text{ m/s}^2) = -4.90 \text{ m/s}^2$$

*For simplicity, we assume that the weight of the car is distributed equally over all four tires and that the brakes are applied to each wheel so that the total frictional force is μ_s times the total normal force.

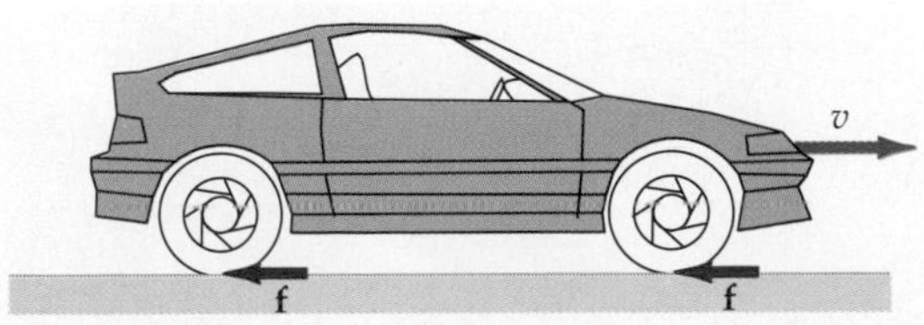

Figure 5-9 The force that stops a car when it brakes is the force of friction exerted by the road on the tires. If the brakes are applied gently so that the tires do not slip, the force is that of static friction. If the tires slip, the force is that of kinetic friction.

Since the acceleration is constant, we can use Equation 2-13 to find the stopping distance:

$$v^2 = v_0^2 + 2a\ \Delta x$$

$$0 = (30\ \text{m/s})^2 + 2(-4.90\ \text{m/s}^2)\ \Delta x$$

$$\Delta x = 91.8\ \text{m}$$

(*b*) When the wheels lock, the force exerted by the road on the car is that of kinetic friction. Using reasoning similar to that in part (*a*), we obtain for the acceleration

$$a_x = -\mu_k g = -(0.3)(9.81\ \text{m/s}^2) = -2.94\ \text{m/s}^2$$

The stopping distance is then

$$\Delta x = \frac{-v_0^2}{2a} = \frac{-(30\ \text{m/s})^2}{2(-2.94\ \text{m/s}^2)} = 153\ \text{m}$$

Exercise

What must the coefficient of static friction be between the road and the tires of a four-wheel-drive car if the car is to accelerate from rest to 25 m/s in 8 s? (Answer: 0.319)

When a car rounds a curve on a horizontal road, the centripetal force is provided by the force of friction exerted by the road on the tires of the car. If the car does not slide radially, the friction is static friction.

Example 5-4

A car travels on a horizontal road in a circle of radius 30 m. If the coefficient of static friction is $\mu_s = 0.6$, how fast can the car travel without slipping?

Figure 5-10 shows the forces acting on the car. The normal force $\mathbf{F}_n$ balances the downward force due to gravity $m\mathbf{g}$. The only horizontal force is due to friction. Its maximum value is $f_{s,max} = \mu_s F_n = \mu_s mg$. In this case, the frictional force is the centripetal force. The maximum speed of the car v_{max} occurs when the frictional force equals its maximum value. From Newton's second law

$$f_{s,max} = \mu_s mg = m\frac{v_{max}^2}{r}$$

or

$$v_{max} = \sqrt{\mu_s gr} = \sqrt{(0.6)(9.81\ \text{m/s}^2)(30\ \text{m})} = 13.3\ \text{m/s}$$

This speed is approximately equal to 47.8 km/h = 29.7 mi/h. If the car travels at a speed greater than 13.3 m/s, the force of static friction will not be great enough to provide the acceleration needed for the car to travel in a circle. The car will slide out away from the center of the circle; that is, it will tend to travel in a straight line.

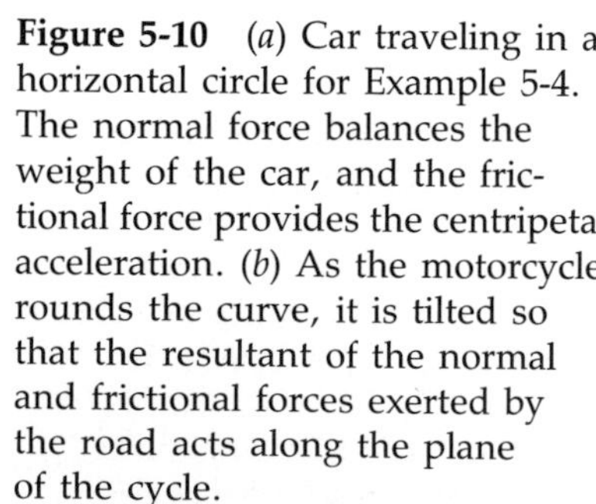

Figure 5-10 (*a*) Car traveling in a horizontal circle for Example 5-4. The normal force balances the weight of the car, and the frictional force provides the centripetal acceleration. (*b*) As the motorcycle rounds the curve, it is tilted so that the resultant of the normal and frictional forces exerted by the road acts along the plane of the cycle.

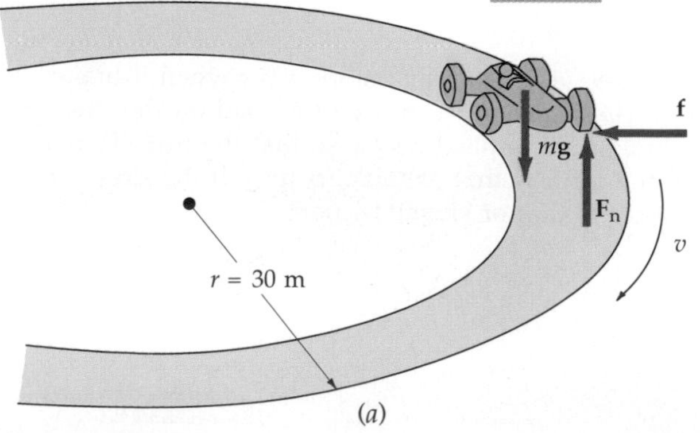

(*a*)

(*b*)

Exercise

If a car is to travel at 60 km/h around a curve whose radius is 40 m, what is the least value for the coefficient of static friction such that the car will not slide? (Answer: 0.71)

If a curved road is not horizontal but banked, the normal force of the road will have a component directed inward toward the center of the circle that will contribute to the centripetal force. The banking angle can be such that, for a given speed, no friction is needed for a car to make the curve.

Cars on a race track that is banked so that the normal force exerted by the track on a car has a horizontal component toward the center of the circle to provide the centripetal force.

Example 5-5

A curve of radius 30 m is banked at an angle θ as shown in Figure 5-11. Find θ for which a car can round the curve at 40 km/h even if the road is frictionless.

In this example, the normal force of the road acting on the car has a component directed inward toward the center of the circle that provides the centripetal force. From the figure, we can see that the angle between the normal force F_n and the vertical is θ, the same as the banking angle. The vertical component of the normal force $F_y = F_n \cos\theta$ must balance the weight of the car:

$$F_y = F_n \cos\theta = mg$$

The horizontal component of the normal force $F_x = F_n \sin\theta$ provides the centripetal force:

$$F_x = F_n \sin\theta = \frac{mv^2}{r}$$

If we divide this equation by the first, we can eliminate m and F_n and obtain an equation relating θ to the speed v and the radius r:

$$\frac{F_n \sin\theta}{F_n \cos\theta} = \tan\theta = \frac{v^2}{rg}$$

Substituting in $v = 40$ km/h $= 11.1$ m/s, $r = 30$ m, and $g = 9.81$ m/s^2, we obtain

$$\tan\theta = \frac{(11.1 \text{ m/s})^2}{(30 \text{ m})(9.81 \text{ m/s}^2)} = 0.419$$

$$\theta = 22.7^\circ$$

Questions

1. Various objects lie on the floor of a truck. If the truck accelerates, what force acts on the objects to cause them to accelerate?
2. Any object resting on the floor of a truck will slide if the acceleration of the truck is too great. How does the critical acceleration at which a light object slips compare with that at which a much heavier object slips?

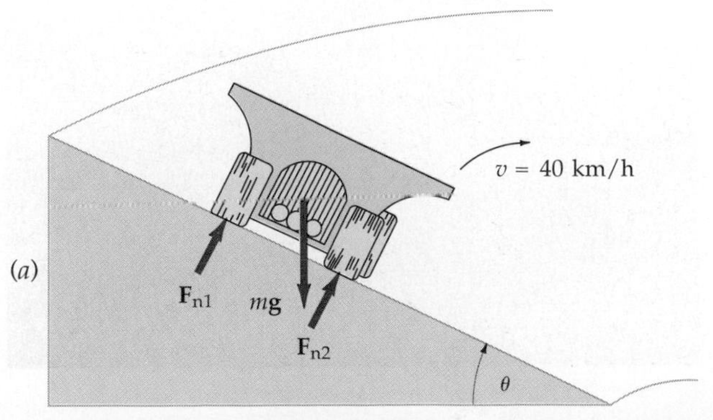

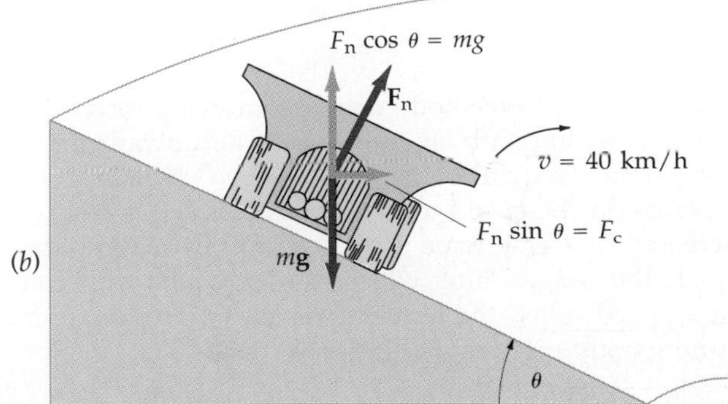

Figure 5-11 (*a*) Car on banked curve for Example 5-5. The forces exerted by the road on the car are represented by $\mathbf{F}_{n1}$ and $\mathbf{F}_{n2}$. (*b*) The force diagram for the car. The total normal force has the component $F_n \sin\theta$ directed towards the center of the curve, which contributes to the centripetal acceleration of the car.

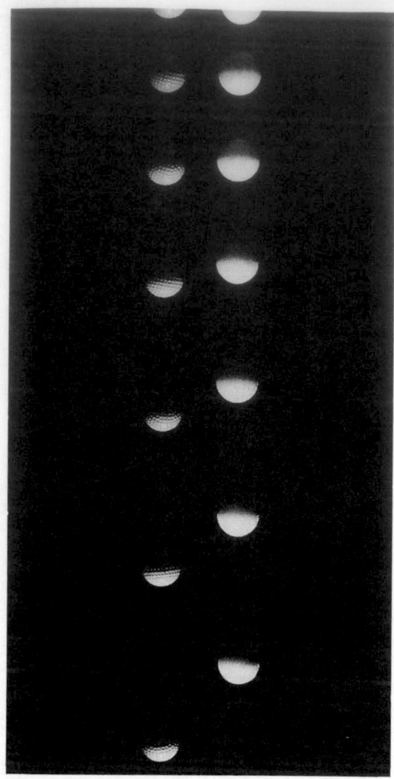

A golf ball and styrofoam ball falling in air. The air resistance is negligible for the heavier golf ball, which falls with essentially constant acceleration. The styrofoam ball reaches terminal velocity quickly, as indicated by the nearly equal spacing at the bottom.

5-2 Drag Forces

When an object moves through a fluid such as air or water, the fluid exerts a retarding or **drag force** that tends to reduce the speed of the object. This drag force depends on the shape of the object, on the properties of the fluid, and on the speed of the object relative to the fluid. Like the force of friction, the drag force is very complicated. Unlike ordinary friction, the drag force increases as the speed of the object increases. For small speeds, the drag force is approximately proportional to the speed of the object; for higher speeds, it is more nearly proportional to the square of the speed.

Consider an object dropped from rest and falling under the influences of the force of gravity, which we assume to be constant, and a drag force of magnitude bv^n, where b and n are constants. We then have a constant downward force mg and an upward force bv^n (Figure 5-12). If we take the downward direction to be positive, we obtain from Newton's second law

$$F_{\text{net}} = mg - bv^n = ma \qquad 5\text{-}5$$

At $t = 0$, when the object is dropped, the speed is zero, so the retarding force is zero and the acceleration is g downward. As the speed of the object increases, the drag force increases and the acceleration is less than g. Eventually, the speed is great enough for the drag force bv^n to equal the force of gravity mg, so the acceleration is zero. The object then continues moving at a constant speed v_t, called its **terminal speed.** Setting the acceleration a in Equation 5-5 equal to zero, we obtain

$$bv_t^n = mg$$

Solving for the terminal speed, we get

$$v_t = \left(\frac{mg}{b}\right)^{1/n} \qquad 5\text{-}6$$

The larger the constant b, the smaller the terminal speed. The constant b depends on the shape of the object. A parachute is designed to make b large so that the terminal speed will be small. On the other hand, cars are designed to make b small to reduce the effect of wind resistance.

For a sky diver (with parachute closed), the terminal speed is about 60 m/s = 216 km/h. When the parachute is opened, the drag force is greater than the force of gravity and the sky diver has an upward acceleration while falling; that is, the downward speed of the sky diver decreases. The drag force then decreases until a new terminal speed, about 20 km/h, is reached.

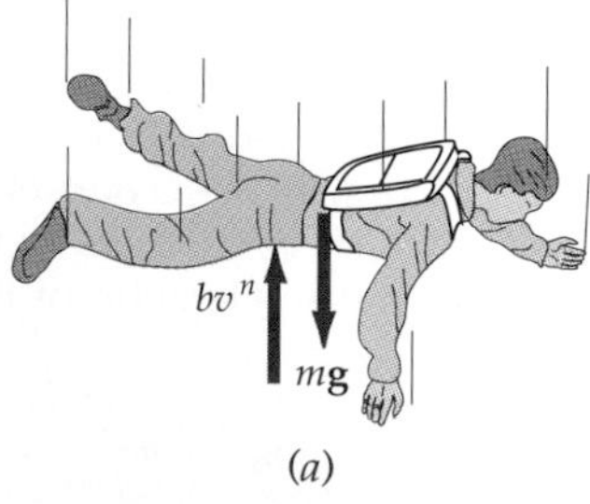

(*a*)

(*b*)

Figure 5-12 (*a*) Free-body diagram showing forces on an object falling with air resistance. The upward force is the drag force due to air resistance bv^n, which depends on the speed v of the body. As the speed increases, the drag force increases until its magnitude equals the weight, after which the body falls with constant speed called the terminal speed. (*b*) Use of a parachute greatly reduces the terminal speed.

(*a*) The space shuttle Columbia just after landing at Dryden Flight Research Center, Edwards, California on Nov. 14, 1981. The sides of the fuselage show scorching marks from the tremendous heat generated by air resistance as the shuttle fell through the atmosphere.

(*b*) As it falls through the atmosphere, a meteor burns making it visible in the night sky.

Example 5-6

A sky diver of mass 64 kg reaches a terminal speed of 180 km/h with her arms and legs outspread. (*a*) What is the magnitude of the upward drag force on the sky diver? (*b*) If the drag force is equal to bv^2, what is the value of b?

(*a*) Since the sky diver is moving with constant velocity, the downward force on her, her weight mg, must be balanced by the upward drag force F_d:

$$F_d = mg = (64 \text{ kg})(9.81 \text{ N/kg}) = 628 \text{ N}$$

(*b*) Setting bv^2 equal to 628 N and using $v = 180$ km/h = 50 m/s, we obtain

$$bv^2 = 628 \text{ N}$$

$$b = \frac{628 \text{ N}}{(50 \text{ m/s})^2} = 0.251 \text{ N·s}^2/\text{m}^2 = 0.251 \text{ kg/m}$$

where we have used 1 N = 1 kg·m/s^2.

Question

3. How would you expect the value of b for air resistance to depend on the density of air?

5-3 Problems with Two or More Objects

In many mechanics problems, two or more objects are in contact or are connected by a string or spring. Such problems can be solved by treating each object separately. For each object, a free-body diagram is drawn and Newton's second law is applied. The resulting equations are then solved simultaneously for the unknown forces or accelerations. For a system of two objects, the force exerted by the first object on the second must be equal and opposite to that exerted by the second object on the first, as stated in Newton's third law.

Example 5-7

A block hangs by a massless string that passes over a frictionless peg* and is connected to another block on a frictionless table. Find the acceleration of each block and the tension in the string.

Figure 5-13 shows the important elements of this problem. The string tensions T_1 and T_2 have equal magnitudes because the string is assumed to be massless, and since the peg is assumed to be frictionless, there are no tangential forces acting on the string. Because the block on the table has no vertical acceleration, the vertical forces $\mathbf{F}_n$ and $m_1\mathbf{g}$ must balance. If a_1 is the horizontal acceleration of m_1, Newton's second law gives

$$T = m_1 a_1 \qquad 5\text{-}7$$

where $T = T_1 = T_2$ is the tension in the string. The acceleration of the hanging block is vertically downward. The forces acting on it are its weight $m_2\mathbf{g}$ downward and the tension $\mathbf{T}_2$ upward. If we take the downward direction to be positive for the acceleration a_2 of this block, Newton's second law gives

$$m_2 g - T = m_2 a_2 \qquad 5\text{-}8$$

If the connecting string does not stretch or become slack, each block must move at the same speed. This constraint allows us to simplify Equations 5-7 and 5-8. Since the speeds of the blocks are the same, the accelerations a_1 and a_2 must be equal in magnitude (but not in direction). Let us call this magnitude a. We then have

$$T = m_1 a \qquad 5\text{-}9$$

$$m_2 g - T = m_2 a \qquad 5\text{-}10$$

We solve these two equations for the two unknowns T and a by first eliminating one of the unknowns. For example, we can eliminate T by substituting $m_1 a$ from Equation 5-9 for T in Equation 5-10 (or by simply adding the two equations). We then obtain

$$m_2 g - m_1 a = m_2 a$$

or

$$a = \frac{m_2}{m_1 + m_2}\, g \qquad 5\text{-}11$$

Note that the result for the magnitude of a is the same as that for a mass $m = m_1 + m_2$ acted on by a force $m_2 g$. The result for a can be substituted

*In some laboratory experiments, the string is replaced by a tape that passes over a curved channel that supports the tape on a cushion of air blown through tiny holes in the channel. In others, the "frictionless" peg is a pulley. If the pulley has negligible mass, it acts the same as a frictionless peg. We will learn how to handle cases involving a pulley with mass in Chapter 8.

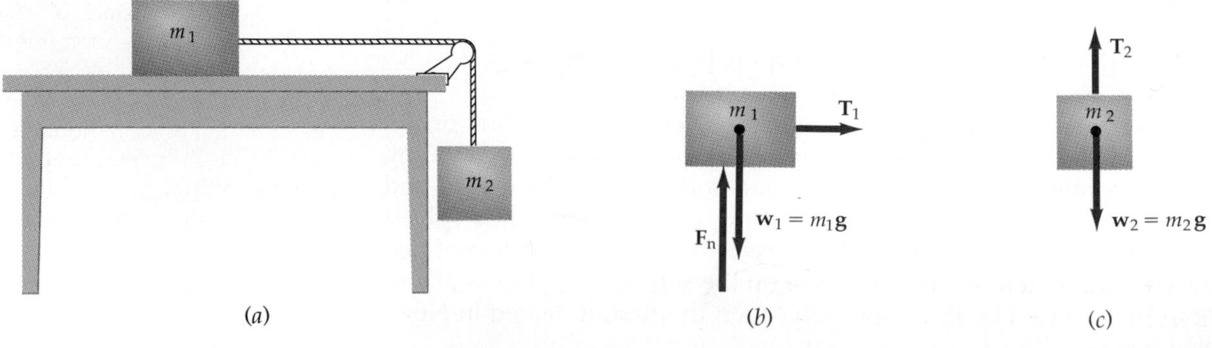

Figure 5-13 (*a*) The two blocks of Example 5-7. (*b*) Free-body diagram for the block on the table. (*c*) Free-body diagram for the hanging block.

into Equation 5-9 to find T:

$$T = \frac{m_1 m_2}{m_1 + m_2} g \qquad 5\text{-}12$$

If the block on the table is very much lighter than the one hanging, we would expect the hanging block to fall with the acceleration of gravity g and the tension T to be zero. Setting $m_1 = 0$ in Equations 5-11 and 5-12, we obtain $a = g$ and $T = 0$. At the other extreme, if m_1 is very much heavier than m_2, we would expect essentially no acceleration. If $m_1 \gg m_2$, the denominators in Equations 5-11 and 5-12 are approximately equal to m_1, and we obtain

$$a \approx \frac{m_2}{m_1} g \approx 0 \qquad \text{for } m_1 \gg m_2$$

and

$$T \approx \frac{m_1 m_2}{m_1} g = m_2 g \qquad \text{for } m_1 \gg m_2$$

Exercise

(*a*) Find the acceleration of the blocks in Example 5-6 if the masses are $m_1 = 2$ kg and $m_2 = 8$ kg. (*b*) Find the acceleration if these two masses are interchanged. [Answer: (*a*) $a = 0.8g = 7.85\ \text{m/s}^2$, (*b*) $a = 0.2g = 1.96\ \text{m/s}^2$]

Example 5-8

A block of mass m_1 sits on a second block of mass m_2, which in turn sits on a frictionless horizontal table as shown in Figure 5-14*a*. A force **F** is applied to the bottom block as shown in the figure. The coefficients of static and sliding friction between the blocks are μ_s and μ_k. (*a*) Find the maximum value of F for which the blocks will not slide relative to each other. (*b*) Find the acceleration of each block when F is greater than this value.

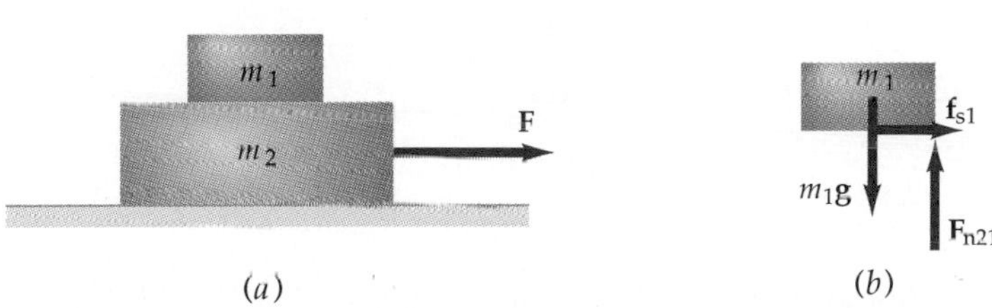

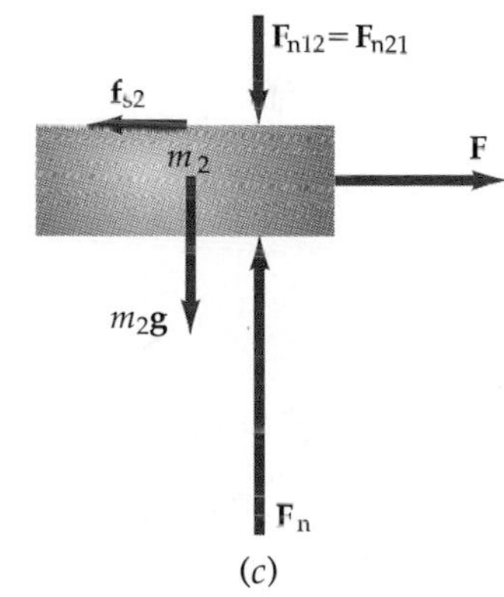

Figure 5-14 (*a*) A small block of mass m_1 on top of a large block of mass m_2 for Example 5-8. (*b*) Free-body diagram for the small block. The only horizontal force is that of friction exerted by the larger block. (*c*) Free-body diagram for the large block.

(*a*) The free-body diagrams for each block are shown in Figure 5-14*b* and *c*. The forces acting on the top block are its weight $m_1\mathbf{g}$ downward, the normal contact force $\mathbf{F}_{n21}$ upward exerted by the lower block, and the force of static friction $\mathbf{f}_{s1}$ exerted to the right by the lower block. Since there is no vertical acceleration, the vertical forces must balance. For the top block, Newton's second law gives

$$\sum F_y = F_{n21} - m_1 g = 0 \qquad 5\text{-}13a$$

and

$$\sum F_x = f_{s1} = m_1 a_1 \qquad 5\text{-}13b$$

The forces acting on the bottom block are its weight $m_2\mathbf{g}$ downward, the downward normal contact force $\mathbf{F}_{n12} = -\mathbf{F}_{n21}$ exerted by the upper block, the upward normal contact force $\mathbf{F}_n$ exerted by the table, the force of static friction $\mathbf{f}_{s2}$ exerted to the left by the upper block, and the applied force $\mathbf{F}$ to the right. Again, the vertical forces must balance. For the lower block, Newton's second law gives

$$\sum F_y = F_n - F_{n12} - m_2 g = 0 \qquad \text{5-14}a$$

and

$$\sum F_x = F - f_{s2} = m_2 a_2 \qquad \text{5-14}b$$

Since f_{s1} and f_{s2} are action–reaction forces, they are equal in magnitude, $f_{s1} = f_{s2} = f_s$. The blocks are moving together so they have the same acceleration. Thus, $a_1 = a_2 = a$. We can then eliminate f_s by adding Equations 5-13*b* and 5-14*b* and obtaining

$$F = (m_1 + m_2)a \qquad \text{5-15}$$

The greater the applied force F, the greater the acceleration a. But the acceleration of the upper block is limited because the force of static friction f_{s1} is limited. Solving Equation 5-13*b* for a_1, we obtain

$$a_1 = \frac{f_{s1}}{m_1} \leq \frac{\mu_s F_{n21}}{m_1}$$

But from Equation 5-13*a*, $F_{n21} = m_1 g$. Then

$$a \leq \frac{\mu_s m_1 g}{m_1}$$

or

$$a \leq \mu_s g$$

When a_1 has its maximum value $\mu_s g$, the force F has the maximum value for which the blocks will not slide relative to each other. Substituting $\mu_s g$ for a in Equation 5-15, we obtain

$$F_{max} = (m_1 + m_2)\mu_s g \qquad \text{5-16}$$

(*b*) When F is greater than this value, the blocks slide relative to each other, and the accelerations of the blocks are not equal. From Newton's second law, the horizontal component of the net force acting on the upper block is

$$f_k = m_1 a_1$$

Using $f_k = \mu_k F_{n21} = \mu_k m_1 g$, we obtain

$$\mu_k m_1 g = m_1 a_1$$

or

$$a_1 = \mu_k g \qquad \text{5-17}$$

Similarly, the horizontal component of the net force acting on the lower block is

$$F - f_k = m_2 a_2$$

$$F - \mu_k m_1 g = m_2 a_2$$

$$a_2 = \frac{F - \mu_k m_1 g}{m_2} \qquad \text{5-18}$$

Note that in part (*a*) we could find the acceleration of the two objects moving together by treating them as a single object of mass $m_1 + m_2$ moving under the net force $\mathbf{F}$. The vertical forces balance, and the horizontal frictional forces are equal and opposite forces that are internal to the system. However, when the upper block slides on the lower block, the accelerations are not the same, and each block must be treated separately.

5-4 Pseudoforces

Newton's laws hold only in inertial reference frames. When the acceleration of any object is measured relative to a reference frame that is itself accelerating relative to an inertial frame, the net force on the object does not equal the mass of the object times its acceleration as measured in the accelerating frame. In some cases, the object will be at rest relative to the noninertial frame even though there is obviously a net force acting on it. In other cases, the object has no forces acting on it even though it is accelerating relative to the noninertial frame. If we wish to use Newton's second law, $\mathbf{F}_{net} = m\mathbf{a}$, in such an accelerating reference frame, we must introduce fictitious forces, called **pseudoforces,** that depend on the acceleration of the reference frame. These fictitious forces are not exerted by any agent. They are introduced merely to make $\mathbf{F}_{net} = m\mathbf{a}$ work when the acceleration $\mathbf{a}$ is measured relative to the noninertial frame. However, to observers in the noninertial frame, the pseudoforces appear to be as real as other forces.

Let us first consider a railroad boxcar that is moving in a straight line along a horizontal track with a constant acceleration $\mathbf{a}_c$ relative to the track, which we assume to be in an inertial reference frame. If we drop an object in the car, it does not fall straight down but rather toward the back of the car. Relative to the car, the object has a vertical acceleration $\mathbf{g}$ and a horizontal acceleration $-\mathbf{a}_c$ (Figure 5-15). Similarly, if an object is placed on a smooth table in the car such that the net force acting on it is zero, it accelerates toward the back of the car relative to an observer in the car. Of course, from the point of view of an observer on the tracks in an inertial frame, the object does not accelerate. Instead, the car and table accelerate out from under the object. We can use Newton's second law in the reference frame of the car if we introduce a pseudoforce $\mathbf{F}_p = -m\mathbf{a}_c$ that acts on any object of mass m.

Consider, for example, a lamp hanging by a cord from the ceiling of the car. The acceleration of the lamp and the forces acting on it relative to both the inertial and the noninertial reference frames are shown in Figure 5-16.

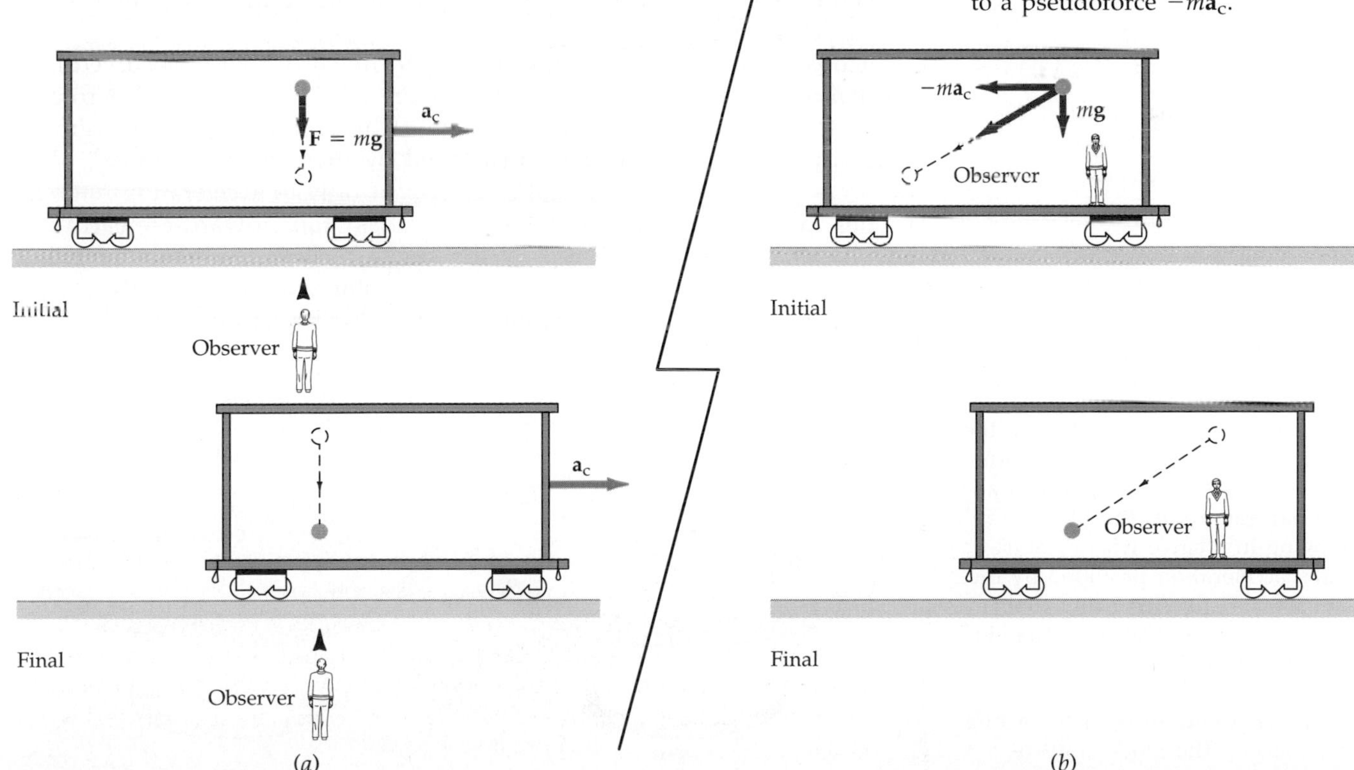

Figure 5-15 An object is dropped inside a railroad boxcar that has an initial velocity of zero and a constant acceleration $\mathbf{a}_c$ to the right. (*a*) An observer on the ground in an inertial reference frame sees the ball fall straight downward. (*b*) An observer in the accelerating car sees the ball fall downward and toward the back of the car. He attributes the backward acceleration of the ball to a pseudoforce $-m\mathbf{a}_c$.

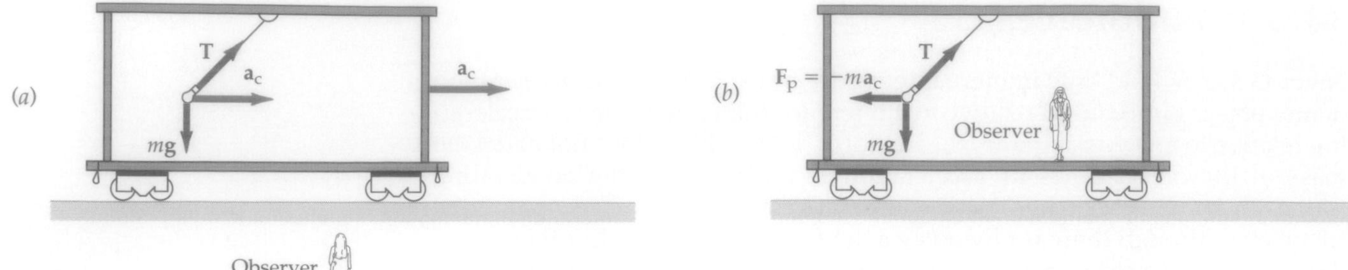

Figure 5-16 A lamp is hung by a cord from the roof of an accelerating railroad boxcar. (*a*) According to an observer in an inertial reference frame, the lamp accelerates to the right because there is a net force acting on it to the right, the horizontal component of the tension **T**. (*b*) In the accelerating frame, the lamp appears to be at rest and not accelerating. The force acting on it to the right is balanced by the pseudoforce $-m\mathbf{a}_c$ that must be introduced in this frame so that Newton's second law can be used.

The vertical component of the tension in the cord equals the weight of the lamp according to each observer. In the inertial frame of the track, the lamp is accelerating. This acceleration is provided by the net force due to the horizontal component of the tension in the cord. In the frame of the car, the lamp is at rest and therefore has no acceleration. This is explained by the fact that the horizontal component of the tension balances the pseudoforce $\mathbf{F}_p = -m\mathbf{a}_c$ seen to act on all objects in the car by an observer in the car.

Figure 5-17 shows another noninertial frame, a rotating platform. Each point on the platform is moving in a circle and therefore has centripetal acceleration. Thus, a frame attached to the platform is a noninertial frame. In the figure, a block at rest relative to the platform is attached to the center post by a string. According to an observer in an inertial frame, the block is moving in a circle with a speed v, so it is accelerating toward the center of the circle. This centripetal acceleration v^2/r is provided by the unbalanced force of the tension T in the string. However, according to an observer on the platform, the block is at rest and so is not accelerating. In order to use $\mathbf{F}_{net} = m\mathbf{a}$, this observer must introduce a pseudoforce with a magnitude of mv^2/r that is acting on the object radially outward to balance the string tension. This fictitious outward force, called the **centrifugal force,** appears quite real to the observer on the platform. If he wants to stand "at rest" on the platform, an inward force of this magnitude must be exerted on him (by the floor) to "balance" the outward centrifugal force. It is important to understand that the centrifugal force is a pseudoforce that exists *only in a rotating reference frame.*

Consider a satellite near the surface of the earth that is observed in an inertial frame attached to the earth. (We are neglecting the earth's rotation here.) It is incorrect to say that the satellite does not fall because the gravitational attraction of the earth "is balanced by the centrifugal force." Pseudoforces such as the centrifugal force appear only in accelerating reference frames. In the earth's frame, the satellite does "fall" toward the earth with an acceleration v^2/r produced by the single, unbalanced force of gravity acting on it. However, an observer in the satellite who considers the satellite to be at rest can use $\mathbf{F}_{net} = m\mathbf{a}$ only by introducing an outward centrifugal force to balance the force of gravity.

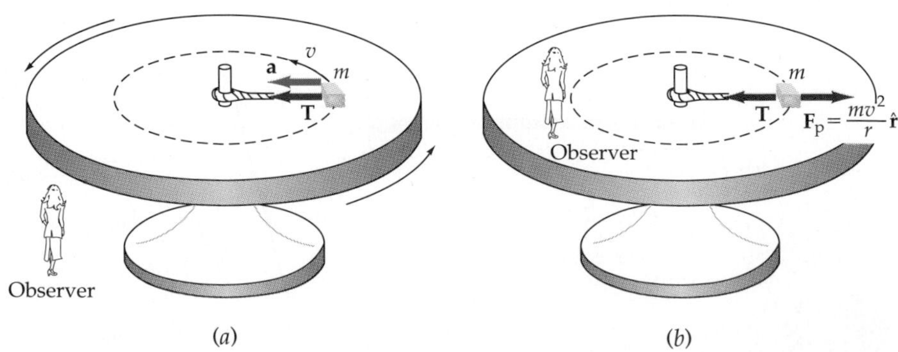

Figure 5-17 A block is tied to the center post of a rotating platform by a string. (*a*) An observer in an inertial frame sees the block moving in a circle with a centripetal acceleration provided by the net force **T**. (*b*) According to an observer on the platform, the block is not accelerating. Therefore, Newton's second law can be used only if a pseudoforce mv^2/r acting outward on the block is introduced to balance the tension.

Large centrifuge used for research at Sandia National Laboratories.

A second pseudoforce, which depends on the velocity of a particle, must be introduced in a rotating frame in order to use $\mathbf{F}_{\text{net}} = m\mathbf{a}$ in that frame. Called the **Coriolis force,** this force is perpendicular to the velocity of the particle (relative to the rotating frame) and causes a sideways deflection. Consider two boys standing along a radial line on a rotating platform and playing catch (Figure 5-18). If the ball is thrown outward along the radial line, an observer in an inertial frame will see the ball travel in a straight line and miss the receiver because the receiver is moving to the left (Figure 5-18*a*). The path of the ball relative to the rotating platform is shown in Figure 5-18*b*. To reach the receiver, the ball must be thrown to the left of the receiver to take into account this sideways deflection.

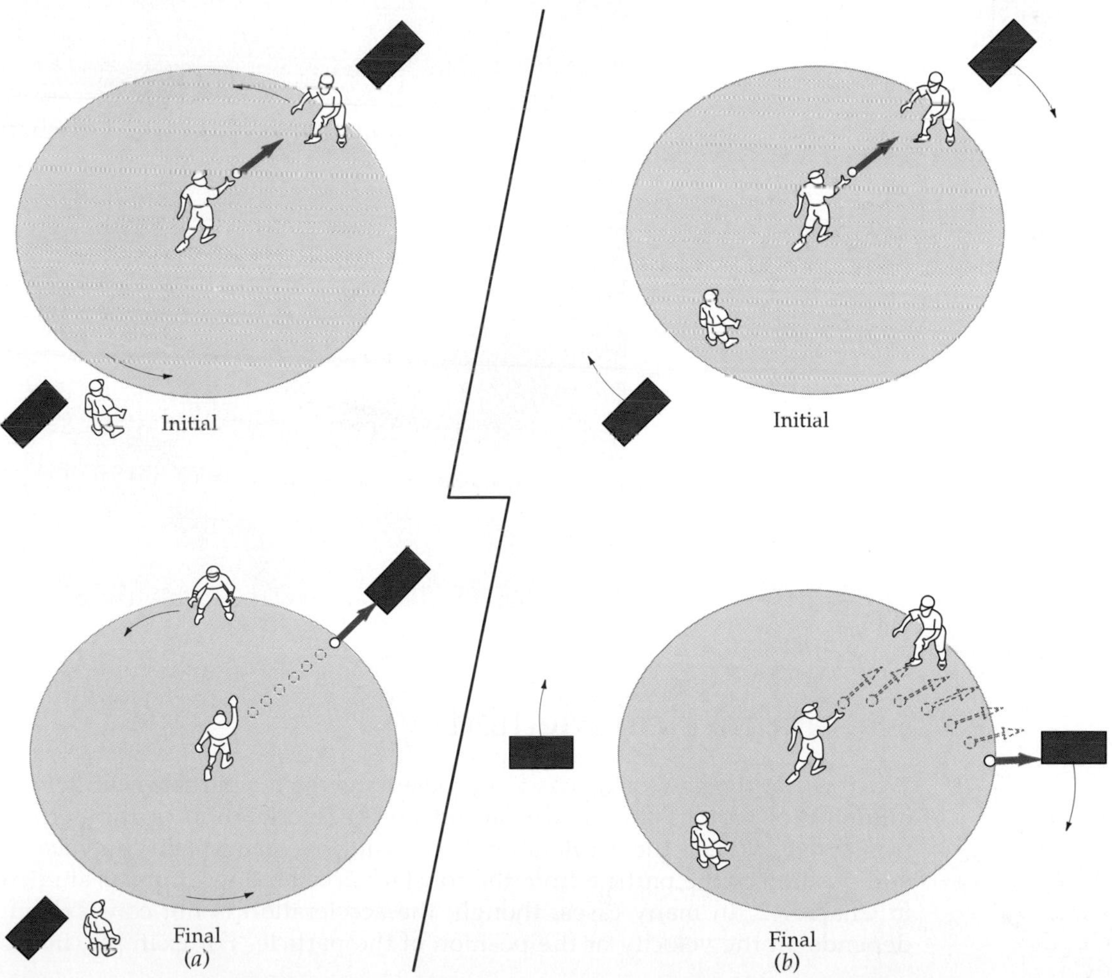

Figure 5-18 A boy at the center of a rotating platform throws a ball toward a friend on the edge of the platform. (*a*) In an inertial frame, the ball travels in a straight line and misses the receiver because the receiver moves with the platform. (*b*) In the rotating frame of the platform, the receiver is at rest, and the ball deflects to the right. The pseudoforce that deflects the ball from a straight line in this frame is called the Coriolis force.

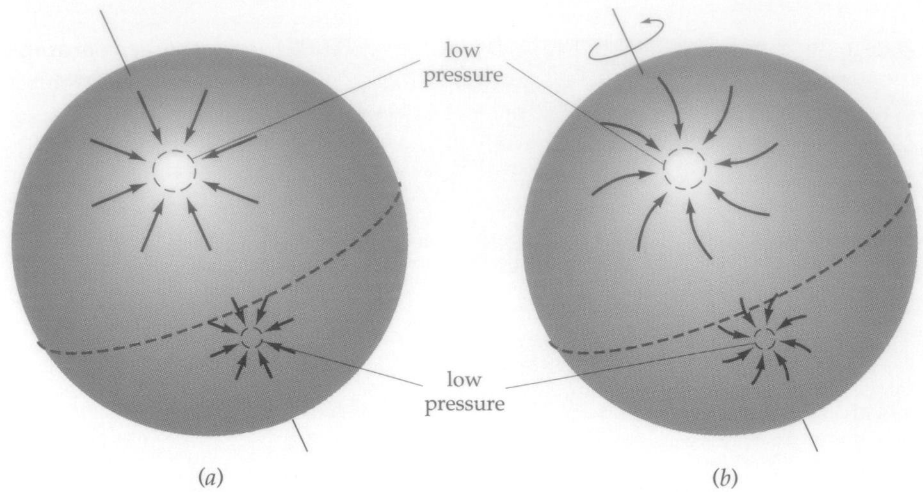

Figure 5-19 (*a*) If the earth were not rotating, the winds would blow inward toward a low-pressure center. (*b*) Because the earth does rotate, the Coriolis force deflects the winds to the right in the northern hemisphere, setting up a counterclockwise pattern. (In the southern hemisphere, the pattern is clockwise.)

These two types of pseudoforce for a rotating reference frame—the centrifugal force and Coriolis force—have direct application to reference frames that are attached to the earth because of the earth's rotation. In particular, Coriolis forces are important for understanding the weather. For example, these forces are responsible for cyclones being counterclockwise in the northern hemisphere and clockwise in the southern hemisphere, when viewed from above (see Figure 5-19).

Typhoon Pat photographed over the Western Pacific by Shuttle Discovery astronauts. Shown in the photo is the counterclockwise rotation of the typhoon due to the Coriolis effect.

5-5 Numerical Methods

When we apply Newton's laws to a problem in which a particle is under the influence of various forces, we can obtain the acceleration of the particle from $\mathbf{a} = \mathbf{F}_{\text{net}}/m$. If the acceleration is constant, we can obtain the velocity and position of the particle from the constant-acceleration formulas studied in Chapter 2. In many cases, though, the acceleration is not constant but depends on the velocity or the position of the particle. For example, in the motion of a particle under the influence of a drag force, the acceleration depends on the velocity of the particle. In such cases, an analytical determination of the velocity and position of the particle may be very difficult or even impossible. However, it is always possible to find the velocity and position approximately by **numerical integration.** For problems that can be solved analytically, the analytical solution has an advantage over numerical integration in that it gives the general features of the solution, whereas numerical integration applies only to a specific case. However, numerical integration has the advantage that it can be applied to any problem no matter how complex.

By definition, the average acceleration of a particle in one dimension is

$$a_{av} = \frac{\Delta v}{\Delta t}$$

and the average velocity is

$$v_{av} = \frac{\Delta x}{\Delta t}$$

The change in the velocity for a time interval Δt is then

$$\Delta v = a_{av}\, \Delta t$$

and the displacement during this interval is

$$\Delta x = v_{av}\, \Delta t$$

Numerical integration is based on the assumption that if the time interval Δt is very small, the average values of the acceleration and velocity can be replaced by particular values chosen at some time during the interval. In **Euler's method,** a_{av} and v_{av} are replaced by the values of the acceleration and velocity at the beginning of the time interval.* The acceleration and velocity are then assumed to be constant during that time interval. Let x_0, v_0, and a_0 be the initial values of the position, velocity, and acceleration at time $t = t_0$. At some later time $t_1 = t_0 + \Delta t$, the velocity is approximately

$$v_1 = v_0 + a_0\, \Delta t$$

and the position is approximately

$$x_1 = x_0 + v_0\, \Delta t$$

We now use the new values v_1 and x_1 to compute the acceleration a_1 for the next interval, and we then use this acceleration to compute the new values of v and x:

$$v_2 = v_1 + a_1\, \Delta t$$

and

$$x_2 = x_1 + v_1\, \Delta t$$

In general, if x_i, v_i, and a_i are the values of the position, velocity, and acceleration at some time t_i, the velocity and position at time $t_{i+1} = t_i + \Delta t$ are approximately

$$v_{i+1} = v_i + a_i\, \Delta t \qquad \text{5-19}a$$

Euler's Method

$$x_{i+1} = x_i + v_i\, \Delta t \qquad \text{5-19}b$$

*Other methods of numerical integration are more accurate but less simple to use. For example, the accuracy is improved if a_{av} and v_{av} are replaced by the values that occur at the midpoint of the interval rather than those at the beginning of the interval.

To find the velocity and position at some time t, we therefore divide the time interval $t - t_0$ into a large number of smaller intervals Δt and apply Equations 5-19a and b beginning at the initial time t_0. This involves a large number of simple, repetitive calculations that are easily done on a computer.

To illustrate the use of numerical methods, let us consider a problem in which a sky diver is dropped from rest at some height under the influences of gravity and a drag force that is proportional to the square of the speed. We will find the velocity v and the distance x traveled as functions of time. The equation describing the motion of an object of mass m dropped from rest is Equation 5-5 with $n = 2$:

$$F_{\text{net}} = mg - bv^2 = ma$$

The acceleration is thus

$$a = g - \left(\frac{b}{m}\right)v^2 \tag{5-20}$$

It is convenient to write the constant b/m in terms of the terminal speed v_t. Setting $a = 0$ in Equation 5-20, we obtain

$$0 = g - \left(\frac{b}{m}\right)v_t^2$$

$$\frac{b}{m} = \frac{g}{v_t^2}$$

Substituting g/v_t^2 for b/m in Equation 5-20 gives

$$a = g\left(1 - \frac{v^2}{v_t^2}\right) \tag{5-21}$$

To solve Equation 5-21 numerically, we need numerical values for g and v_t. A reasonable terminal speed for a sky diver is 60 m/s. Using this and $g = 9.81\ \text{m/s}^2$, we obtain

$$a = 9.81\left(1 - \frac{v^2}{3600}\right) \tag{5-22}$$

We have omitted the units in this equation. Since we are using SI, the unit for v is meters per second, and the unit for x is meters. If we choose $x_0 = 0$ for the initial position, the initial values are $x_0 = 0$, $v_0 = 0$, and $a_0 = g = 9.81$. To find the velocity v and position x after some time, say $t = 20$ s, we divide the time interval $0 < t < 20$ s into many small intervals Δt and apply Equations 5-19a and b. We do this by writing a computer program or by using a computer spreadsheet. Figure 5-20 shows graphs of v versus t and x versus t based on data found using a spreadsheet with $\Delta t = 0.5$ s. At $t = 20$ s, the computed values are $v = 59.97$ m/s and $x = 957.5$ m.

But how accurate are our computations? We can estimate the accuracy by running the program again using a smaller time interval. If we use $\Delta t = 0.25$ s, one-half of the value we originally used, we obtain $v = 59.92$ m/s and $x = 952.0$ m at $t = 20$ s. The difference in v is about 0.1 percent and that in x is about 0.5 percent. These are our estimates of the accuracy of the original computations.

Since the difference between the value of a_{av} for some time interval Δt and the value of a_i at the beginning of the interval becomes smaller as the time interval becomes smaller, we might expect that it would be better to use very small time intervals, say $\Delta t = 0.000000001$ s. But there are two reasons for not using very small time intervals. First, the smaller the time interval, the larger the number of calculations that are required and the longer the program takes to run. Second, the computer keeps only a fixed number of digits at each step of the calculation, so that at each step there is a round-off

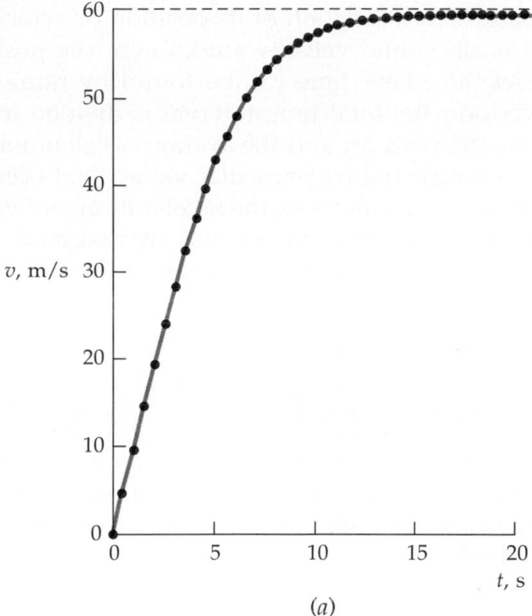

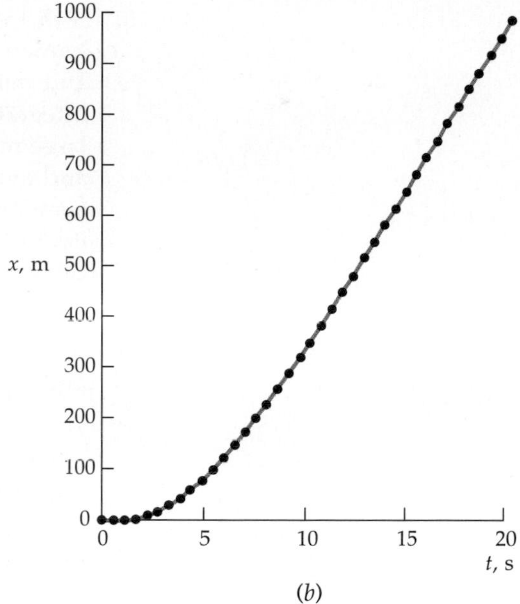

Figure 5-20 (*a*) Graph of v versus t for the sky diver found by numerical integration using $\Delta t = 0.5$ s. The horizontal dashed line is the terminal speed $v_t = 60$ m/s. (*b*) Graph of x versus t using $\Delta t = 0.5$ s.

error. These round-off errors add up. The larger the number of calculations, the more significant the total round-off errors become. When we first decrease the time interval, the accuracy improves because a_i more nearly approximates a_{av} for the interval. However, as the time interval is decreased further, the round-off errors build up and the accuracy of the computation decreases. A good rule of thumb to follow is to use no more than about 10^4 or 10^5 time intervals for the typical numerical integration.

Summary

1. When two bodies are in contact, they can exert frictional forces on each other. These frictional forces are parallel to the surfaces of the bodies at the points of contact. When the surfaces are at rest relative to each other, the frictional force is static friction, which can vary anywhere from 0 to its maximum value $\mu_s F_n$, where F_n is the normal force of contact and μ_s is the coefficient of static friction. When the surfaces are moving relative to each other, the frictional force is kinetic or sliding friction and has the value $\mu_k F_n$, where μ_k is the coefficient of kinetic friction. The coefficient of kinetic friction is slightly less than the coefficient of static friction.

2. When a body moves through a fluid such as air or water, the body experiences a drag force that opposes its motion. The drag force increases with increasing speed. If the body is dropped from rest, its speed increases until the drag force equals the force of gravity, after which it moves with a constant speed called its terminal speed. The terminal speed depends on the shape of the body and on the medium through which it falls.

3. In applying Newton's laws to problems with two or more objects, a free-body diagram should be drawn for each object. $\mathbf{F}_{net} = m\mathbf{a}$ should then be applied to each object separately.

4. Newton's laws do not hold in accelerating reference frames. However, they can be applied if certain fictitious forces or pseudoforces that depend on the acceleration of the frame are introduced.

5. If the acceleration of a particle as a function of its position or velocity is known and if the initial position and velocity are known, the position and velocity of the particle at a later time can be found by numerical integration. In this procedure, the total time interval is divided into a large number of small time intervals Δt, and the average acceleration a_{av} and average velocity v_{av} are replaced by particular values that occur at some time in the interval. In Euler's method, the acceleration and velocity are found at the beginning of a time interval and are assumed to be constant over that interval. The velocity and position at the end of an interval are found from

$$v_{i+1} = v_i + a_i \, \Delta t$$

and

$$x_{i+1} = x_i + v_i \, \Delta t$$

where x_i, v_i, and a_i are the values of the position, velocity, and acceleration at the beginning of the interval. The accuracy of the results obtained by numerical integration can be estimated by repeating the computation using a smaller time interval.

Suggestions for Further Reading

Armstrong, H. L.: "How Dry Friction Really Behaves," *American Journal of Physics,* vol. 53, no. 9, 1958, p. 910.

This one-page article discusses the dependence of friction on velocity and points out that the change from static to kinetic friction is not instantaneous but takes place gradually without discontinuity.

Brancazio, Peter J.: "Trajectory of a Fly Ball," *The Physics Teacher,* January 1985, p. 20.

This article shows why drag forces cannot be neglected in any realistic analysis of the motion of a baseball.

Gross, A. C., C. R. Kyle, and D. J. Malewicki: "The Aerodynamics of Human-Powered Land Vehicles," *Scientific American,* December 1983, p. 142.

Drag forces play an important role in determining the maximum velocity a bicycle can attain. By applying the principles of aerodynamics, a bicycle that can travel 60 miles per hour on level ground can be constructed.

von Baeyer, Hans Christian: "Physika: Einstein at the Ex," *The Sciences,* May/June 1985, p. 10.

An account of the author's "first conscious encounter with a fictitious force" (or pseudoforce) at the carnival. Complements the article "Thinking About Physics While Scared to Death (on a Falling Roller Coaster)" by Jearl Walker (see Suggestions for Further Reading, Chapter 8).

Review

A. Objectives: After studying this chapter you should:

1. Know that the maximum force of static friction and the force of kinetic friction are proportional to the normal force between the surfaces involved.

2. Be able to apply Newton's laws to problems involving frictional forces.

3. Be able to apply Newton's laws to problems involving two or more objects.

4. Be able to discuss qualitatively motion with a velocity-dependent drag force.

5. Be able to discuss qualitatively the use of Newton's laws in accelerating reference frames using pseudoforces.

6. Be able to apply Euler's method of numerical integration to solve problems in which the acceleration of a particle depends on its velocity or position.

B. Define, explain, or otherwise identify:

Static friction
Kinetic friction
Coefficient of static friction
Coefficient of kinetic friction
Coefficient of rolling friction
Drag force
Terminal speed
Pseudoforces
Centrifugal force
Coriolis force
Numerical integration
Euler's method

C. True or false: If the statement is true, explain why it is true. If it is false, give a counterexample.

1. The force of static friction always equals $\mu_s F_n$.

2. Friction always opposes the motion of an object.

3. The terminal speed of an object depends on its shape.

4. Pseudoforces exist only in noninertial reference frames.

5. The accuracy of numerical integration always increases when smaller time intervals are used.

Problems

Level I

5-1 Friction

1. A box weighing 600 N is pushed along a horizontal floor at constant velocity with a force of 250 N parallel to the floor. What is the coefficient of kinetic friction between the box and the floor?

2. The coefficient of static friction between the tires of a car and a horizontal road is $\mu_s = 0.6$. If the net force on the car is the force of static friction exerted by the road, (*a*) what is the maximum acceleration of the car? (*b*) What is the least distance in which the car can stop if it is initially traveling at 30 m/s?

3. The force that accelerates a car along a flat road is the frictional force exerted by the road on the car's tires. (*a*) Explain why the acceleration can be greater when the wheels do not spin. (*b*) If a car is to accelerate from 0 to 90 km/h in 12 s at constant acceleration, what is the minimum coefficient of friction needed between the road and tires?

4. A chair is sliding across a polished floor with an initial speed of 3 m/s. It comes to rest after sliding 2 m. What is the coefficient of kinetic friction between the floor and chair?

5. An 800-N box rests on a plane surface inclined at 30° to the horizontal. A physics student finds that she can prevent the box from sliding if she pushes on it with a force of at least 200 N parallel to the surface. (*a*) What is the coefficient of static friction between the box and the surface? (*b*) What is the greatest force that can be applied to the box parallel to the incline before the box slides up the incline?

6. A 5-kg block is held at rest against a vertical wall by a horizontal force of 100 N. (*a*) What is the frictional force exerted by the wall on the block? (*b*) What is the minimum horizontal force needed to prevent the block from falling if the coefficient of friction between the wall and the block is $\mu_s = 0.40$?

7. On a snowy day with the temperature near the freezing point, the coefficient of static friction between the tires and an icy road is 0.08. What is the maximum incline that a vehicle with four-wheel drive can climb with zero acceleration?

8. A 50-kg box that is resting on a level floor must be moved. The coefficient of static friction between the box and the floor is 0.6. One method to move the box is to push down on the box at an angle θ with the horizontal. Another method is to pull up on the box at an angle θ with the horizontal. (*a*) Explain why one method is better than the other. (*b*) Calculate the force necessary to move the box by each method if $\theta = 30°$ and compare these results with those for $\theta = 0°$.

5-2 Drag Forces

9. What are the dimensions and SI units of the constant b in the retarding force bv^n (*a*) if $n = 1$, and (*b*) if $n = 2$?

10. A small pollution particle drops to earth in still air with a terminal speed of 0.3 mm/s. The particle has a mass of 10^{-10} g and a retarding force of the form bv. What is the value of b?

11. A Ping-Pong ball has a mass of 2.3 g and a terminal speed of 9 m/s. The retarding force is of the form bv^2. What is the value of b?

12. A sky diver of mass 60 kg can slow herself to a constant speed of 90 km/h by adjusting her form. (*a*) What is the magnitude of the upward drag force on the sky diver? (*b*) If the drag force is equal to bv^2 what is the value of b?

13. Newton showed that the air resistance on a falling object with a circular cross section should be approximately $\frac{1}{2}\rho\pi r^2 v^2$, where $\rho = 1.2$ kg/m^3 is the density of air. Find the terminal speed for a 56-kg sky diver, assuming that his cross-sectional area is equivalent to that of a disk of radius 0.30 m.

5-3 Problems with Two or More Objects

14. A hand pushes two bodies on a frictionless horizontal surface, as shown in Figure 5-21. The masses of the bodies are 2 and 1 kg. The hand exerts a force of 5 N on the 2-kg body. (*a*) What is the acceleration of the system? (*b*) What is the acceleration of the 1-kg body? What force is exerted on it? What is the origin of this force? (*c*) Show all the forces acting on the 2-kg body. What is the net force acting on this body?

Figure 5-21 Problem 14.

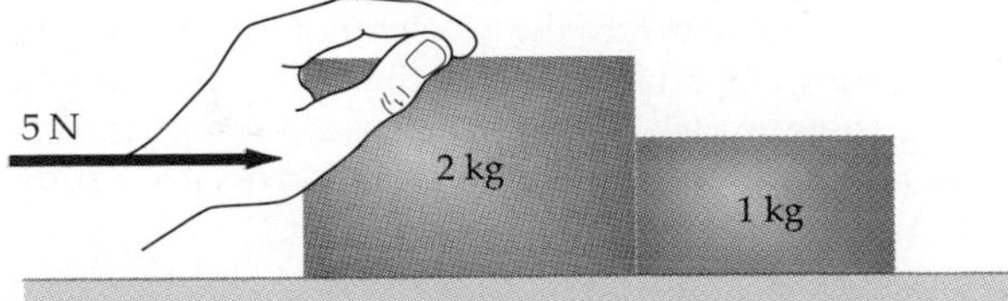

15. Two objects are connected by a massless string as shown in Figure 5-22. The incline and peg are frictionless. Find the acceleration of the objects and the tension in the string for (*a*) $\theta = 30°$ and $m_1 = m_2 = 5$ kg, and (*b*) for general values of θ, m_1, and m_2.

Figure 5-22 Problem 15.

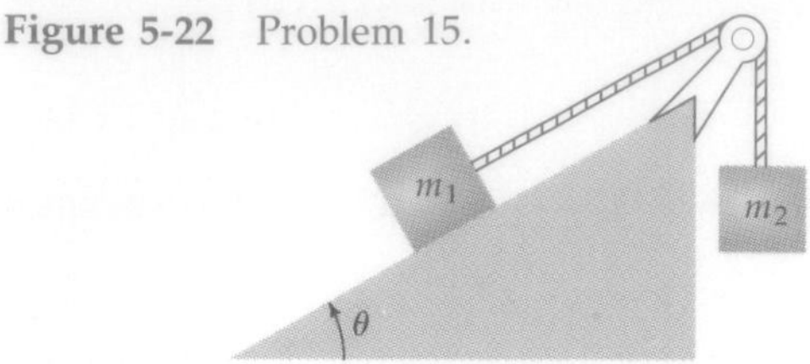

16. A 3-kg box rests on a horizontal table. It is attached to a 2-kg box by a light string, as shown in Figure 5-23. (*a*) What is the minimum coefficient of static friction such that the objects remain at rest? (*b*) If the coefficient of static friction is less than that found in part (*a*) and the coefficient of kinetic friction between the box and the table is 0.3, find the time for the 2-kg mass to fall 2 m to the floor if the system starts from rest.

Figure 5-23 Problem 16.

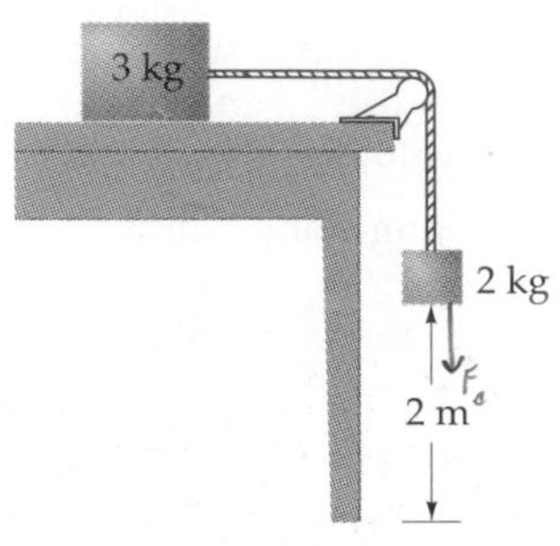

17. A block of mass m_1 is attached to a cord of length L_1, which is fixed at one end. The mass moves in a horizontal circle supported by a frictionless table. A second block of mass m_2 is attached to the first by a cord of length L_2 and also moves in a circle, as shown in Figure 5-24. If the period of the motion is T, find the tension in each cord.

Figure 5-24 Problem 17.

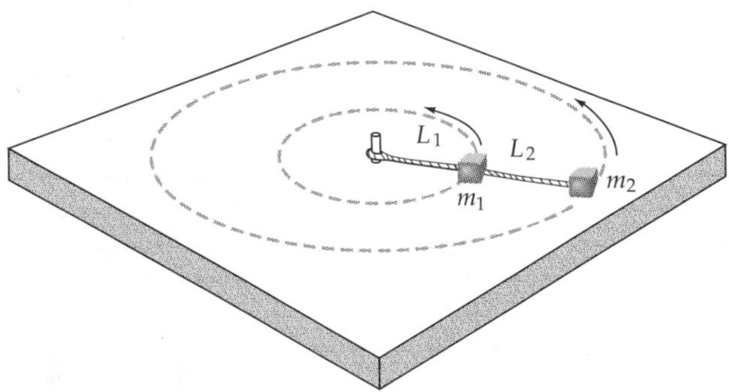

18. The apparatus shown in Figure 5-25, called *Atwood's machine*, is used to measure the acceleration due to gravity g by measuring the acceleration of the bodies. Assuming the string to be massless and the peg to be frictionless, show that the magnitude of the acceleration of either body and the tension in the string are

$$a = \frac{m_1 - m_2}{m_1 + m_2} g \quad \text{and} \quad T = \frac{2m_1 m_2 g}{m_1 + m_2}$$

Figure 5-25 Problems 18 and 50.

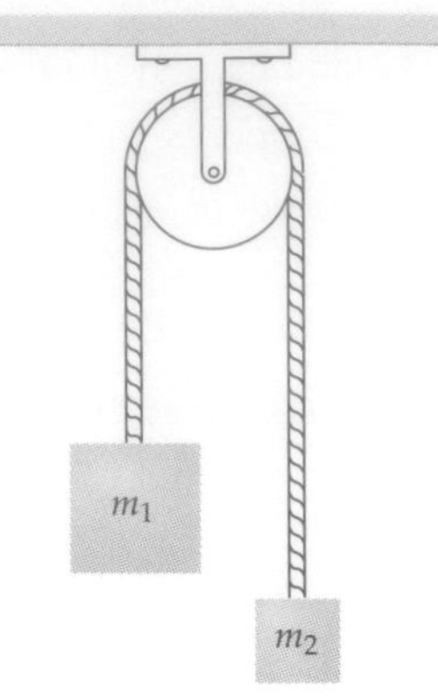

5-4 Pseudoforces

In Problems 19 through 23 the situations described take place in a boxcar that has an initial velocity $v = 0$ and an acceleration $\mathbf{a} = (5\ \text{m/s}^2)\ \mathbf{i}$ (Figure 5-26). Work the exercises in the frame of the boxcar using pseudoforces and in an inertial frame using only real forces.

Figure 5-26 Boxcar initially at rest with a constant acceleration $\mathbf{a} = (5\ \text{m/s}^2)\ \mathbf{i}$ to the right for Problems 19 through 23.

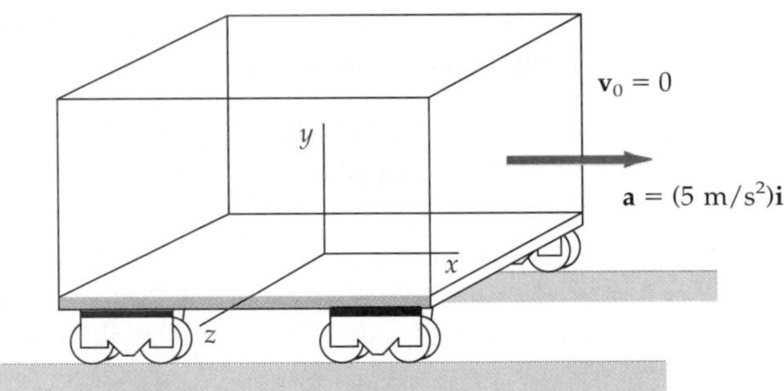

19. A 2-kg object slides along the frictionless floor of the boxcar with an initial velocity of (10 m/s) **i** relative to the boxcar. (*a*) Describe the motion of the object. (*b*) When does the object reach its original position relative to the boxcar?

20. A 2-kg object slides along the frictionless floor of the boxcar with an initial transverse velocity of (10 m/s) **k** relative to the boxcar. Describe the motion.

21. A 2-kg object slides along the floor of the boxcar, which is now assumed to be rough, with a coefficient of sliding friction of 0.3. The initial velocity of the object is (10 m/s) **i** relative to the boxcar. Describe the motion of the object, assuming that the coefficient of static friction is 0.6.

22. A 2-kg object is suspended from the ceiling of the boxcar by a massless, unstretchable string. (*a*) What angle does the string make with the vertical? (*b*) Indicate all the forces acting on the object in each frame of reference.

23. A 6-kg object is attached to the ceiling of the boxcar by a massless spring of force constant 1000 N/m. By how much is the spring stretched?

5-5 Numerical Methods

24. An object falls under the influence of gravity and a drag force $F_d = -bv$. (*a*) Show that the acceleration of the object can be written

$$a = g\left(1 - \frac{v}{v_t}\right)$$

where $v_t = mg/b$ is the terminal speed. (*b*) Solve numerically to obtain a graph of v versus t and a graph of x versus t for $v_t = 60$ m/s.

Level II

25. An 800-kg car rolls down a very long, 6° grade. The drag force for motion of the car has the form $F_d = 100\text{ N} + (1.2\text{ N}\cdot\text{s}^2/\text{m}^2)v^2$. What is the terminal velocity for the car rolling down this grade?

26. An 8-kg box and a 10-kg box that are attached by a rope that passes over a frictionless pulley slide on frictionless inclines, as shown in Figure 5-27. (*a*) Find the acceleration of the boxes and the tension in the rope. (*b*) The two boxes are replaced by two others of mass m_1 and mass m_2 such that there is no acceleration. Find whatever information you can about the mass of these two new boxes.

Figure 5-27 Problem 26.

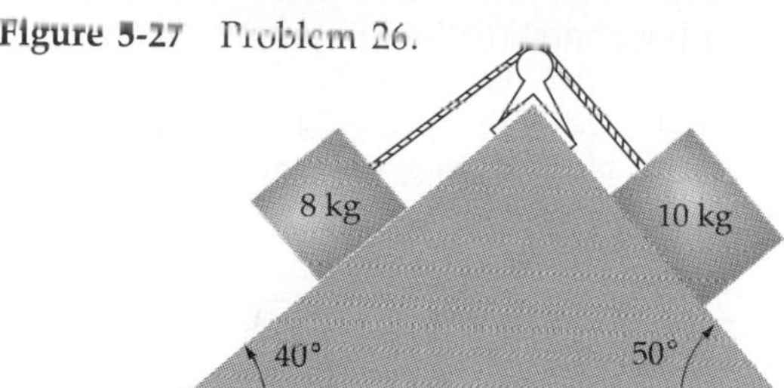

27. The coefficient of static friction between the floor of a truck and a box resting on it is 0.30. The truck is traveling at 80 km/h. What is the least distance in which the truck can stop if the box is not to slide?

28. Two 100-kg boxes are dragged along a frictionless surface with a constant acceleration of 1 m/s^2, as shown in Figure 5-28. Each rope has a mass of 1 kg. Find the force **F** and the tension in the ropes at points *A*, *B*, and *C*.

Figure 5-28 Problem 28.

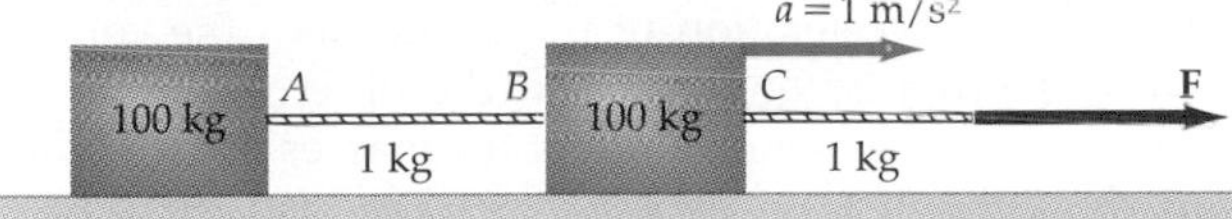

29. One end of a heavy rope of length 5 m and mass 4 kg is attached to the bottom of a 6-kg block on a frictionless horizontal table. At the other end of the rope, a constant force of 100 N is applied. (*a*) What is the acceleration of the system? (*b*) Give the tension in the rope as a function of position along the rope.

30. A block is on an incline whose angle can be varied. The angle is gradually increased from 0°. At 30°, the block starts to slide down the incline. It slides 3 m in 2 s. Calculate the coefficients of static and kinetic friction between the block and incline.

31. A platform scale calibrated in newtons is placed on the bed of a truck driven at a constant speed of 14 m/s. A box weighing 500 N is placed on the scale. Find the reading on the scale if (*a*) the truck passes over the crest of a hill with a radius of curvature 100 m, and (*b*) the truck passes through the bottom of a dip with a radius of curvature of 80 m.

32. A 100-g disk sits on a horizontally rotating turntable. The turntable makes one revolution each second. The disk is located 10 cm from the axis of rotation of the turntable. (*a*) What is the frictional force acting on the disk? (*b*) The disk will slide off the turntable if it is located at a radius larger than 16 cm from the axis of rotation. What is the coefficient of static friction?

33. In an amusement-park ride, riders stand against the wall of a spinning cylinder. The floor falls away and the riders are held up by friction. If the radius of the cylinder is 4 m, find the minimum number of revolutions per minute necessary when the coefficient of friction between a rider and the wall is 0.4.

34. Show with a force diagram how a motorcycle can travel in a circle on the inside vertical wall of a hollow cylinder. Assume reasonable parameters (coefficient of friction, radius of the circle, mass of the motorcycle, or whatever is required), and calculate the minimum speed needed.

35. The coefficient of friction between box *A* and the cart in Figure 5-29 is 0.6. The box has a mass of 2 kg. (*a*) Find the minimum acceleration *a* of the cart and box if the box is not to fall. (*b*) What is the magnitude of the frictional force in this case? (*c*) If the acceleration is greater than this minimum, will the frictional force be greater than in part (*b*)? Explain. (*d*) Show that for a box of any mass, the box will not fall if the acceleration is $a \geq g/\mu_s$, where μ_s is the coefficient of static friction.

Figure 5-29 Problem 35.

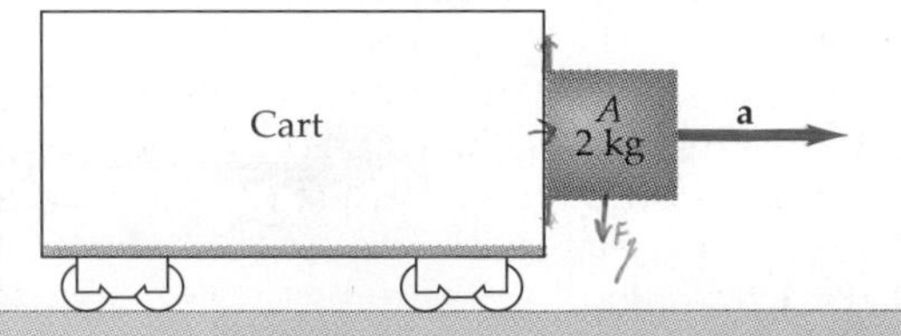

36. (*a*) A parachute creates enough air resistance to hold an 80-kg sky diver to a constant downward speed of

6.0 m/s after it has been open for a while. Assuming the force of air resistance is given by $f = bv^2$, calculate b for this case. (*b*) A sky diver drops in a free fall until his speed is 60 m/s before opening his parachute. If it opens instantaneously, calculate the initial upward force exerted by the parachute on the sky diver moving at 60 m/s. Explain why it is important that the parachute takes a few seconds to open.

37. A box of mass m rests on a horizontal table. The box is pulled by a massless rope with a force **F** at an angle θ as shown in Figure 5-30. The coefficient of static friction is 0.6. The minimum value of the force needed to move the box depends on the angle θ. (*a*) Discuss qualitatively how you would expect this force to depend on θ. (*b*) Compute the force for the angles $\theta = 0$, 10, 20, 30, 40, 50, and 60°, and make a plot of F versus θ for $mg = 400$ N. From your plot, at what angle is it most efficient to apply the force to move the box?

Figure 5-30 Problems 37 and 49.

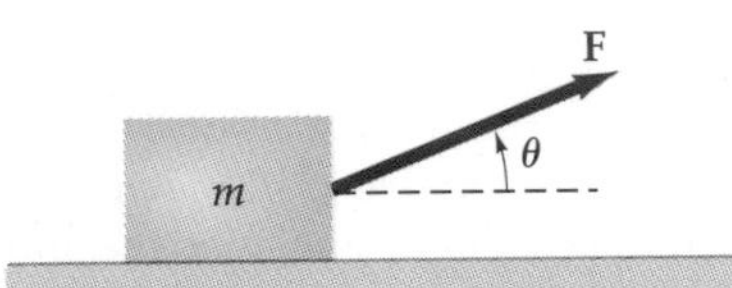

38. (*a*) Estimate the maximum acceleration of a car if 40 percent of its weight is on the tires that are powered by the engine, and if the coefficient of static friction is 0.7. (*b*) What is the shortest possible time in which this car can achieve a speed of 100 km/h? Assume the engine has unlimited power.

39. A road on a college campus has a sharp, level curve with a radius of 30 m that has a speed limit of 30 km/h. A serious accident occurs at this curve. Upon investigating the accident, a physics professor discovers that the curve has become covered with small, gravel-sized particles that act like miniature ball bearings and reduce the effective coefficient of static friction between the average tire and road from 0.7 to 0.2. (*a*) What is the maximum safe speed one can drive on this curve under normal conditions (with no gravel)? (*b*) What is the maximum safe speed one can drive on the curve when the road is covered with the gravel?

40. Consider the block of mass m_2 in Figure 5-31 to be a variable mass that can be adjusted until the block of mass m_1 is on the verge of sliding on the table. (*a*) If the critical mass m_2 is 5 kg and mass m_1 is 7 kg, what is the coefficient of static friction between the table and the block? (*b*) With a slight nudge, the system accelerates at 1 m/s^2. What is the coefficient of kinetic friction between the table and the block?

Figure 5-31 Problem 40.

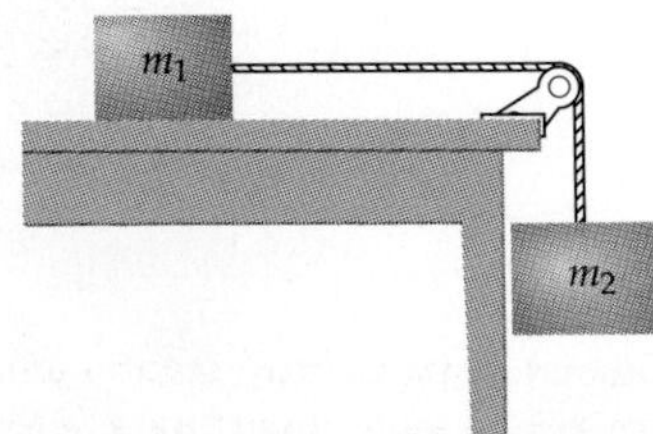

41. A 100-kg mass is pulled along a frictionless surface by a force **F** such that its acceleration is 6 m/s^2 (see Figure 5-32). A 20-kg mass slides along the top of the 100-kg mass and has an acceleration of 4 m/s^2. (It thus slides back relative to the 100-kg mass.) (*a*) What is the frictional force exerted by the 100-kg mass on the 20-kg mass? (*b*) What is the net force acting on the 100-kg mass? What is the force F? (*c*) After the 20-kg mass falls off the 100-kg mass, what is the acceleration of the 100-kg mass?

Figure 5-32 Problem 41.

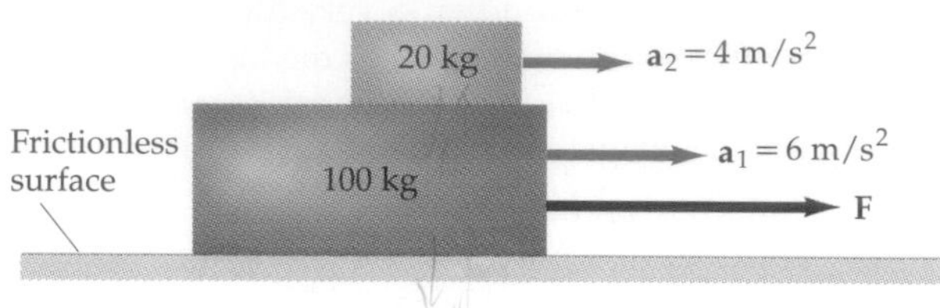

42. A 60-kg block slides along the top of a 100-kg block with an acceleration of 3 m/s^2 when a horizontal force **F** of 320 N is applied, as shown in Figure 5-33. The 100-kg block sits on a horizontal frictionless surface, but there is friction between the two blocks. (*a*) Find the coefficient of kinetic friction between the blocks. (*b*) Find the acceleration of the 100-kg block during the time that the 60-kg block maintains contact.

Figure 5-33 Problem 42.

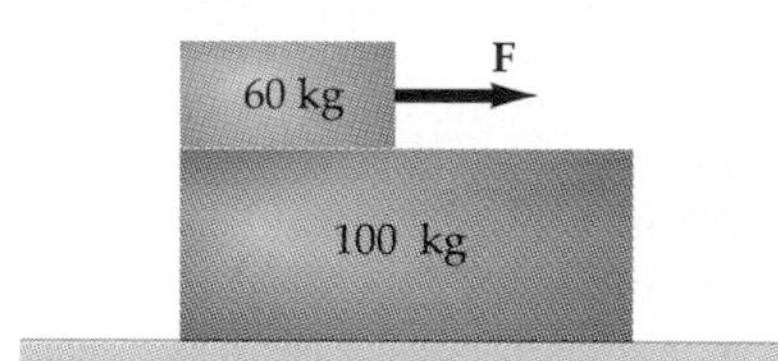

43. Two bodies of mass m_1 and m_2 rest on a horizontal frictionless table, as shown in Figure 5-34. A force **F** is applied as shown. (*a*) If $m_1 = 2$ kg, $m_2 = 4$ kg, and $F = 3$ N, find the acceleration of the bodies and the contact force F_c exerted by one body on the other. (*b*) Find the contact force for general values of the masses of the bodies and show that if $m_2 = nm_1$, $F_c = nF/(n + 1)$.

Figure 5-34 Problem 43.

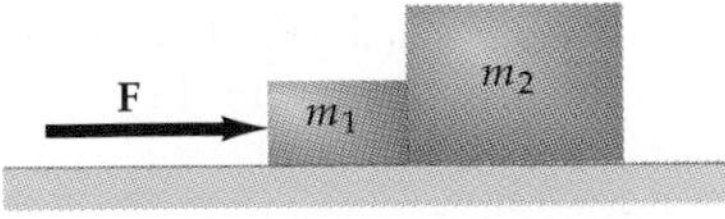

44. Two climbers on an icy (frictionless) slope, tied together by a 30-m rope, are in the predicament shown in

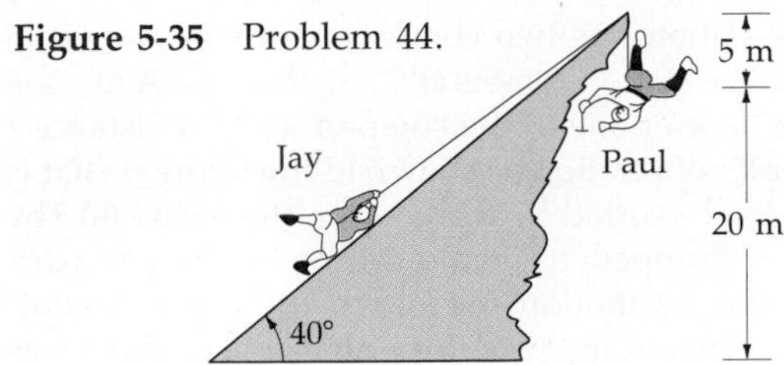

Figure 5-35 Problem 44.

Figure 5-35. At time $t = 0$ the speed of each is zero, but the top climber, Paul (mass 52 kg), has taken one step too many and his friend Jay (mass 74 kg) has dropped his pick. (*a*) Find the tension in the rope as Paul falls and his speed just before he hits the ground. (*b*) If Paul unties his rope after hitting the ground, find Jay's speed as he hits the ground. Assume the rope's length to be 30 m.

45. A 4-kg block resting on a 30° incline is attached to a second block of mass m by a cord that passes over a smooth peg, as shown in Figure 5-36. The coefficient of static friction between the block and the incline is 0.4. (*a*) Find the range of possible values for m for which the system will be in static equilibrium. (*b*) If $m = 1$ kg, the system will be in static equilibrium. What is the frictional force on the 4-kg block in this case?

Figure 5-36 Problem 45.

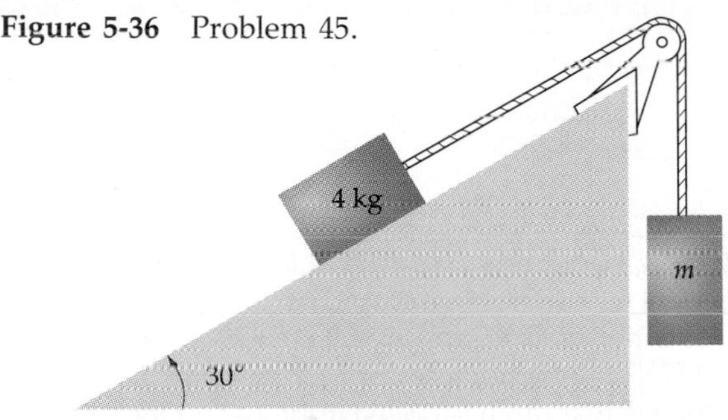

46. An automobile is going up a grade of 15° at a speed of 30 m/s. The coefficient of static friction between the tires and the road is 0.7. (*a*) What minimum distance does it take to stop the car? (*b*) What minimum distance would it take if the car were going down the grade?

Level III

47. A 2-kg block sits on a 4-kg block that is resting on a frictionless table (Figure 5-37). The coefficients of friction between the blocks are $\mu_s = 0.3$ and $\mu_k = 0.2$. (*a*) What is the maximum force F that can be applied if the 2-kg block is not to slide on the 4-kg block? (*b*) If F has half this value, find the acceleration of each block and the force of friction acting on each block. (*c*) If F is twice the value found in (*a*), find the acceleration of each block.

Figure 5-37 Problem 47.

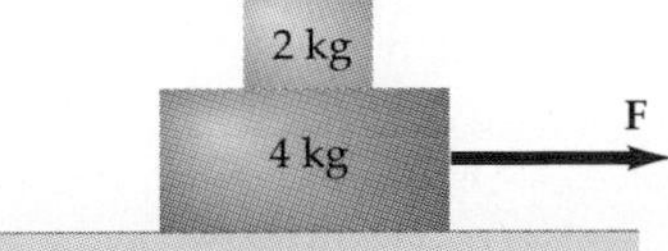

48. In Figure 5-38, the mass $m_2 = 10$ kg slides on a frictionless table. The coefficients of static and kinetic friction between m_2 and $m_1 = 5$ kg are $\mu_s = 0.6$ and $\mu_k = 0.4$. (*a*) What is the maximum acceleration of m_1? (*b*) What is the maximum value of m_3 if m_1 moves with m_2 without slipping? (*c*) If $m_3 = 30$ kg, find the acceleration of each body and the tension in the string.

Figure 5-38 Problem 48.

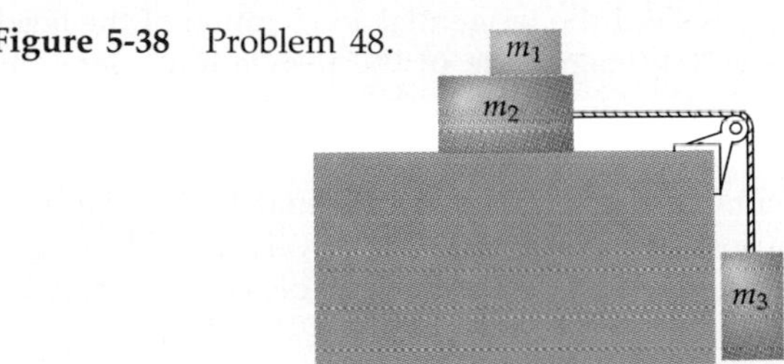

49. A box of mass m rests on a horizontal table. The coefficient of static friction is μ_s. A force **F** is applied at an angle θ as shown in Figure 5-30. (*a*) Find the force F needed to move the box as a function of angle θ. (*b*) At the angle θ for which this force is minimum, the slope $dF/d\theta$ of the curve F versus θ is zero. Compute $dF/d\theta$ and show that this derivative is zero at the angle θ that obeys $\tan\theta = \mu_s$. Compare this general result with that obtained in Problem 37.

50. Using the Atwood's machine illustrated in Figure 5-25 (Problem 18), the acceleration of gravity g is determined by measuring the time t for the larger body of mass m_1 to fall a distance L. (*a*) Find an expression for g in terms of m_1, m_2, L, and t. (*b*) Show that if there is a small error in the time measurement dt, it will lead to an error in the determination of g by an amount dg given by $dg/g = -2\,dt/t$. If m_1 is 1 kg, find the value of m_2 such that g can be measured with an accuracy of ±5 percent with a watch that is accurate to 0.1 s if $L = 3$ m. Assume that the only important uncertainty in the measurement is the time of fall.

51. A student rides a bicycle on a horizontal surface in a circle with a radius of 20 m. The bicycle and student are leaning inward at an angle of 15° with respect to the vertical. (*a*) What is the speed of the student? (*b*) If the frictional force is half its maximum value, what is the coefficient of static friction?

52. An object falls under the influence of gravity and a drag force $F_d = -bv$, as in Problem 24. (*a*) Show that the acceleration of the object can be written

$$a = \frac{dv}{dt} = g - \frac{b}{m}v$$

(*b*) Rearrange this equation to obtain

$$\frac{dv}{g - (b/m)v} = dt$$

(*c*) Integrate this equation to obtain the exact solution

$$v = \frac{mg}{b}(1 - e^{-bt/m}) = v_t(1 - e^{-gt/v_t})$$

(*d*) Plot v versus t for $v_t = 60$ m/s, and compare this with the numerical solution obtained in Problem 24.

53. Consider a bead of mass m that is free to move on a thin, circular wire of radius r. The bead is given an initial speed v_0, and there is a coefficient of sliding friction μ_k. The experiment is performed in a spacecraft drifting in space. Find the speed of the bead at any subsequent time t.

54. In Problem 53, (*a*) find the centripetal acceleration of the bead. (*b*) Find the tangential acceleration of the bead. (*c*) What is the magnitude of the resultant acceleration?

55. A curve of radius 30 m is banked so that a car traveling 40 km/h can round it even if the road is so icy that the coefficient of static friction is approximately zero. Find the range of speeds at which a car can travel around this curve without skidding if the coefficient of static friction between the road and the tires is 0.3.

56. Small spherical particles experience a viscous drag force given by Stokes' law, which is given by $F_d = 6\pi\eta r v$, where r is the radius of the particle, v is its speed, and η is the dynamic viscosity of the air or fluid medium. (*a*) Estimate the terminal speed of a spherical pollution particle of radius 10^{-5} m and density of 2000 kg/m^3. Assume that the air is still and that η is 1.8×10^{-5} N · s/m^2. (*b*) Estimate the time it takes for such a particle to fall from a 100 m smokestack.

57. Figure 5-39 shows a 20-kg block sliding on a 10-kg block. All surfaces are frictionless. Find the acceleration of each block and the tension in the string that connects the blocks.

Figure 5-39 Problem 57.

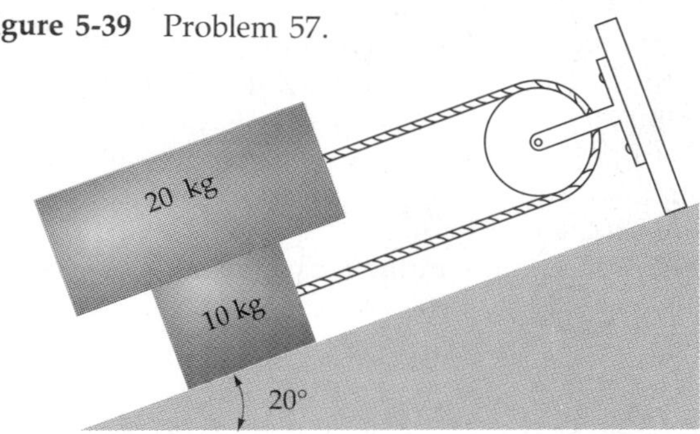

58. A 20-kg block with a pulley attached slides along a frictionless table. It is connected by a massless string to a 5-kg block, as shown in Figure 5-40. Find the acceleration of each block and the tension in the connecting string.

Figure 5-40 Problem 58.

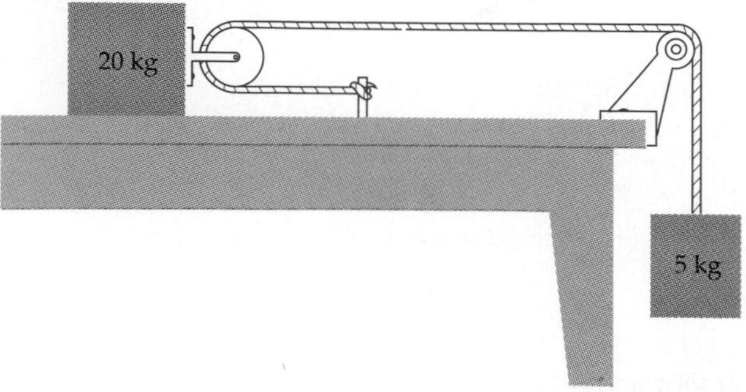

59. A space station has two compartments, as shown in Figure 5-41. The station rotates at B rev/min. (*a*) A mass m rests on the floor of one of the compartments a distance r from the center of rotation as shown in the figure. What is the normal force exerted by the floor on the mass? (*b*) The mass is now dropped from the ceiling of the compartment. Describe its motion relative to the compartment. What forces (including pseudoforces) act on the mass while it is falling? (*c*) Explain qualitatively why the mass falls to the floor from the point of view of an inertial reference frame in which there are no forces acting on the mass.

Figure 5-41 Problem 59.

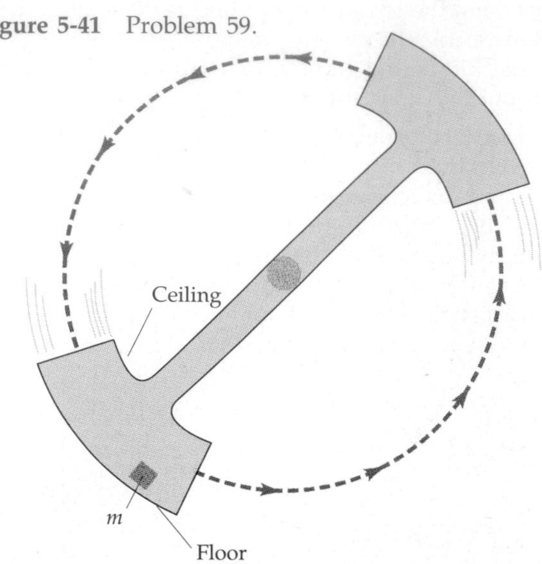

60. A 10-kg block rests on a 5-kg bracket as shown in Figure 5-42. The coefficients of friction between the 10-kg block and the bracket are $\mu_s = 0.40$ and $\mu_k = 0.30$. The bracket sits on a frictionless surface. (*a*) What is the maximum force F that can be applied if the 10-kg block is not to slide on the bracket? (*b*) What is the corresponding acceleration of the bracket?

Figure 5-42 Problem 60.

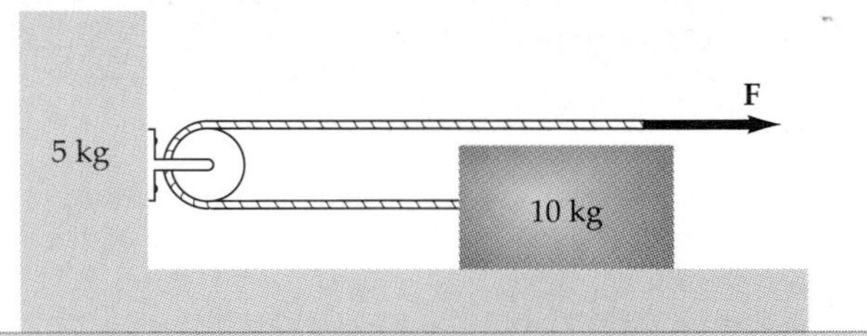

Chapter 6

Work and Energy

The action of a pole vaulter demonstrates several kinds of energy. First he transforms the internal chemical energy of his body into kinetic energy as he runs. Some of this kinetic energy is then converted into elastic potential energy, as shown in the deformation of the pole; and the rest is eventually converted into gravitational potential energy, which in turn is converted into kinetic energy as he drops. Mechanical energy is finally converted into heat energy when he reaches the ground.

Work and energy are among the most important concepts in physics, and they play important roles in our everyday life, as well. In physics, **work** is given a precise definition that differs from our everyday usage. Work is done on an object by a force only when the point of application of the force moves through a distance and there is a component of the force along the line of motion. Thus, when you exert a force on a sled and thereby pull it across the snow, you do work on the sled. However, if the sled were immobilized (for example, if it were chained to a tree) and you exert the same force on it, no work would be done on the sled because the point of application of the force does not move through a distance.

Closely associated with the concept of work is the concept of **energy,** which is the capacity to do work. When work is done by one system on another, energy is transferred between the two systems. For example, when you pull a sled, the work you do goes partly into the energy of motion of the sled, called its kinetic energy, and partly into the thermal energy that arises from the friction between the sled and the snow. At the same time, the internal chemical energy of your body decreases as you pull the sled. The net result is the transfer of the internal chemical energy of your body to the external kinetic energy of the sled plus thermal energy. One of the most important principles in science is the conservation of energy: The total energy of a system and its surroundings does not change. When the energy of a system decreases, there is always a corresponding increase in the energy of its surroundings or of another system.

There are many forms of energy. Kinetic energy is associated with the motion of an object. Potential energy is stored energy associated with the configuration of a system, such as the separation distance between some object and the earth. Thermal energy is associated with the motion of the molecules within a system and is closely connected with the temperature of the system.

In this chapter, we will explore the concepts of work, kinetic energy, and potential energy, and we will see how to use the law of conservation of energy to solve various problems. Later, when we study thermodynamics, we will look at heat energy, which is energy that is transferred due to differences in temperature, and we will explore the internal molecular energy of systems in more depth.

6-1 Work and Kinetic Energy: Motion in One Dimension with Constant Forces

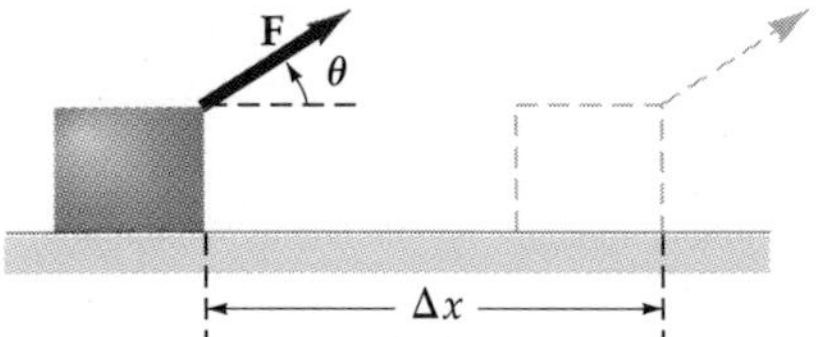

Figure 6-1 When a constant force **F** acts on a particle at an angle θ through a distance Δx, the work done on the particle is $F \cos \theta \, \Delta x = F_x \, \Delta x$.

We define the work done by a force on an object as the product of the force and the displacement of the point on which the force acts. If the force and the displacement are in different directions, then only the component of the force in the direction of the displacement does work. For simplicity, we will first consider the special case of constant forces and motion in one dimension. We will also assume that the object can be considered to be a particle. This eliminates such complications as the deformation of the object, in which case the displacement is different for different parts of the object. For a particle, the displacement of the point of application of the force is the same as the displacement of the particle.

If a force **F** makes an angle θ with the displacement Δx, as in Figure 6-1, the work done is

Work by a constant force

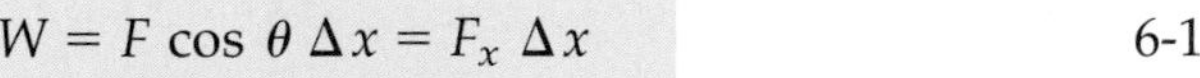

$$W = F \cos \theta \, \Delta x = F_x \, \Delta x \qquad 6\text{-}1$$

Work is a scalar quantity that is positive if Δx and F_x have the same signs and negative if they have opposite signs. The dimensions of work are those of force times distance. The SI unit of work and energy is the **joule** (J), which equals the product of a newton and a meter:

$$1 \text{ J} = 1 \text{ N·m} \qquad 6\text{-}2$$

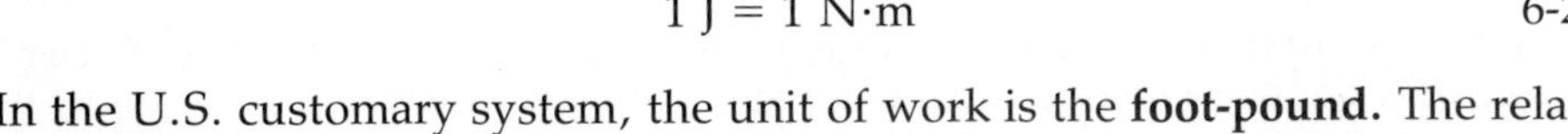

In the U.S. customary system, the unit of work is the **foot-pound.** The relation between these units of work is easily found using the relations between pounds and newtons and between meters and feet. The result is

$$1 \text{ J} = 0.738 \text{ ft·lb} \qquad 6\text{-}3$$

Exercise

A force of 12 N is exerted on a box at an angle of $\theta = 20°$, as in Figure 6-1. How much work is done by the force as the box moves along the table a distance of 3 m? (Answer: 33.8 J)

Figure 6-2 shows a boy pulling a sled across the snow by exerting a force **F** at an angle θ to the horizontal. For this discussion, we will assume that friction can be neglected. The other forces acting on the sled are then its weight mg and a vertical support force $\mathbf{F}_n$. The work done by the weight is zero because this force is perpendicular to the direction of motion. Similarly the support force $\mathbf{F}_n$ is perpendicular to the direction of motion and therefore does no work. The only force that does work on the sled is the horizontal force $F \cos \theta$. When there are several forces that do work on a particle, the total work is found by computing the work done by each force and summing. For a particle, the same result is obtained by summing all the forces to find the net force and then computing the work done by the net force. This is true because the displacement of the point of application is the same for each force and is equal to the displacement of the particle. In this case, the vertical upward force $F \sin \theta + F_n$ must equal the weight mg because the sled does not accelerate vertically. The net force acting on the sled is thus the horizontal force $F \cos \theta$.

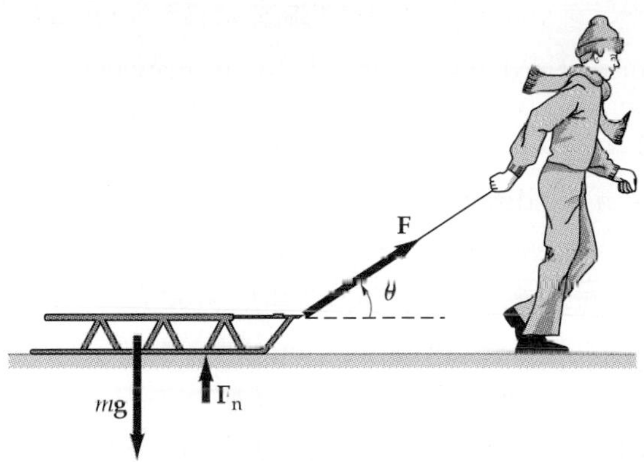

Figure 6-2 Boy pulling a sled. The work done by the boy as he pulls the sled a distance Δx is $F \cos \theta \, \Delta x$.

There is an important relation between the total work done on a particle and the initial and final speeds of the particle. If F_x is the net force acting on a particle, Newton's second law gives

$$F_x = ma_x$$

For a constant force, the acceleration is constant, and we can relate the distance the particle moves to its initial and final speed by using the constant-acceleration formula (Equation 2-13). If the initial speed is v_i and the final speed is v_f, we have

$$v_f^2 = v_i^2 + 2a_x \, \Delta x \qquad 6\text{-}4$$

Since the work done by the net force equals the total work done on the particle,

$$W_{total} = F_x \, \Delta x = ma_x \, \Delta x$$

Substituting $\frac{1}{2}(v_f^2 - v_i^2)$ for $a_x \, \Delta x$ we obtain

$$W_{total} = \tfrac{1}{2}mv_f^2 - \tfrac{1}{2}mv_i^2 \qquad 6\text{-}5$$

The quantity $\frac{1}{2}mv^2$ is called the **kinetic energy** K of the particle. It is a scalar quantity that depends on the particle's mass and speed:

$$K = \tfrac{1}{2}mv^2 \qquad 6\text{-}6$$

Kinetic energy defined

The quantity on the right side of Equation 6-5 is the change in the kinetic energy of the particle, that is, the kinetic energy $\frac{1}{2}mv_f^2$ at the end of the interval minus the kinetic energy $\frac{1}{2}mv_i^2$ at the beginning of the interval. The total work done on the particle is therefore equal to the change in the kinetic energy of the particle:

Work–energy theorem

$$W_{\text{total}} = \Delta K = \tfrac{1}{2}mv_f^2 - \tfrac{1}{2}mv_i^2 \qquad 6\text{-}7$$

The total work done on a particle equals the change in the kinetic energy of the particle.

This result is known as the **work–energy theorem.** It holds whether the net force is constant or not, as we will see in the next section.

Exercise

A girl of mass 50 kg is running at 3.5 m/s. What is her kinetic energy? (Answer: 306 J)

Example 6-1

A 4-kg box is raised from rest a distance of 3 m by an upward applied force of 60 N. Find (*a*) the work done by the applied force, (*b*) the work done by gravity, and (*c*) the final speed of the box.

(*a*) The applied force is in the direction of motion ($\theta = 0°$), so the work done by it is positive:

$$W_{\text{app}} = F_{\text{app}} \cos 0° \,\Delta y = (60\text{ N})(1)(3\text{ m}) = 180\text{ J}$$

(*b*) The force of gravity is opposite the direction of motion ($\theta = 180°$), so the work done by gravity is negative:

$$W_g = mg \cos 180° \,\Delta y = (4\text{ kg})(9.81\text{ N/kg})(-1)(3\text{ m}) = -118\text{ J}$$

The total work done on the box is thus $W_{\text{total}} = 180\text{ J} - 118\text{ J} = 62\text{ J}$. Applying the work–energy theorem with $v_i = 0$, we obtain

$$W_{\text{total}} = 62\text{ J} = \tfrac{1}{2}mv_f^2 - \tfrac{1}{2}mv_i^2 = \tfrac{1}{2}mv_f^2$$

(*c*) The final speed of the box is then

$$v_f = \sqrt{\frac{2W_{\text{total}}}{m}} = \sqrt{\frac{2(62\text{ J})}{4\text{ kg}}} = 5.57\text{ m/s}$$

Example 6-2

If the mass of the sled in Figure 6-2 is 5 kg and the boy exerts a force of 12 N at 30°, find the work done by the boy and the final speed of the sled after it moves 3 m, assuming that it starts from rest and there is no friction.

The forces are shown in the figure. The vertical forces are the force due to gravity,

$$mg = (5\text{ kg})(9.81\text{ N/kg}) = 49.0\text{ N}$$

the upward component of the force exerted by the rope,

$$F_y = (12\text{ N}) \sin 30° = 6\text{ N}$$

and the vertical support force exerted by the ground, which equals $mg - F_y$ because there is no vertical acceleration,

$$F_n = mg - F_y = 49.0 \text{ N} - 6 \text{ N} = 43.0 \text{ N}$$

The only horizontal force is

$$F_x = (12 \text{ N}) \cos 30° = 10.4 \text{ N}$$

The total work done on the sled is the product of the component of **F** in the direction of motion (10.4 N) and the distance traveled (3 m):

$$W_{\text{total}} = F_x\, \Delta x = (10.4 \text{ N})(3 \text{ m}) = 31.2 \text{ J}$$

The total work done on the sled equals the change in its kinetic energy. If the sled starts from rest, its kinetic energy after traveling 3 m is therefore 31.2 J. Its speed can be found from

$$W_{\text{total}} = \Delta K = \tfrac{1}{2}mv_f^2 - \tfrac{1}{2}mv_i^2 = \tfrac{1}{2}mv_f^2 - 0$$

$$\tfrac{1}{2}mv_f^2 = K = 31.2 \text{ J}$$

$$v_f = \sqrt{\frac{2K}{m}} = \sqrt{\frac{2(31.2 \text{ J})}{5 \text{ kg}}} = 3.53 \text{ m/s}$$

Atlas does no work holding up the earth because there is no motion.

We could have found the final speed in Examples 6-1 and 6-2 by first finding the acceleration of the box or sled and then using the constant-acceleration formulas. Setting the total work equal to the change in kinetic energy provides an alternative method to using Newton's laws of motion to solve mechanics problems. When the forces are not constant, this method is often much easier than using Newton's laws.

We have seen that for work to be done on a particle, the particle must move through a distance. But what if we are holding a weight at a fixed position a distance h off the floor, as the man in Figure 6-3 is doing? In everyday usage, we might say that it takes work to do this, but using our scientific definition, no work is done by a force acting on a stationary object. We could eliminate the effort of holding the weight merely by tying the rope to some other object, in which case the weight would be supported with no help from us. Although no external work is done when we hold a weight in a fixed position, we do become tired from this activity. While we hold the weight, nerve impulses continually trigger contractions of muscle fibers inside our body. As the muscle fibers contract and relax, work is done on the molecular level inside our body. In this process, some of our internal chemical energy is converted into thermal energy.

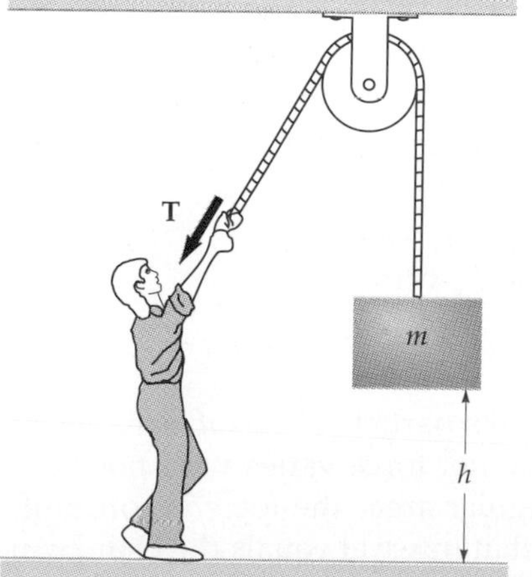

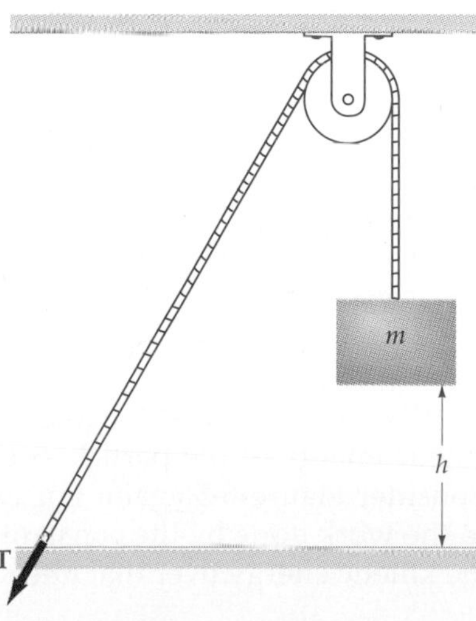

Figure 6-3 No external work is done by the woman holding the weight at a fixed position. The same task could be accomplished by tying the rope to a fixed point.

Questions

1. A heavy box is to be moved from the top of one table to the top of another table of the same height on the other side of the room. Is work required to do this?
2. To get out of bed in the morning, do you have to do work?
3. By what factor does the kinetic energy of a car change when its speed is doubled?
4. An object moves in a circle at constant speed. Does the force that accounts for its acceleration do work on it? Explain.
5. Is it possible to exert a force that does work on an object without increasing the kinetic energy of the object? If so, give an example.

6-2 Work Done by a Variable Force

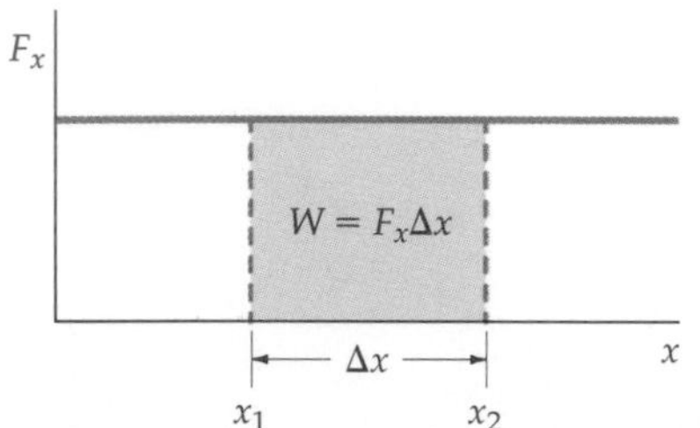

Figure 6-4 The work done by a constant force is represented graphically as the area under the F_x-versus-x curve.

In the previous section, we defined the work done by a constant force F_x as $W = F_x\ \Delta x$. In Figure 6-4, we have plotted a constant force F_x as a function of position x. The work done on a particle whose displacement is Δx is represented by the area under the force-versus-position curve, which is indicated by the shading in the figure. However, many forces encountered in nature are not constant but vary with position. For example, when you stretch a spring, the force exerted by the spring is proportional to the amount the spring is stretched. Similarly, the gravitational force exerted by the earth on a spaceship varies inversely with the square of the distance from the spaceship to the center of the earth. We can use the graphical representation of work as the area under a force-versus-position curve to extend our definition of work to cases in which the force varies with position.

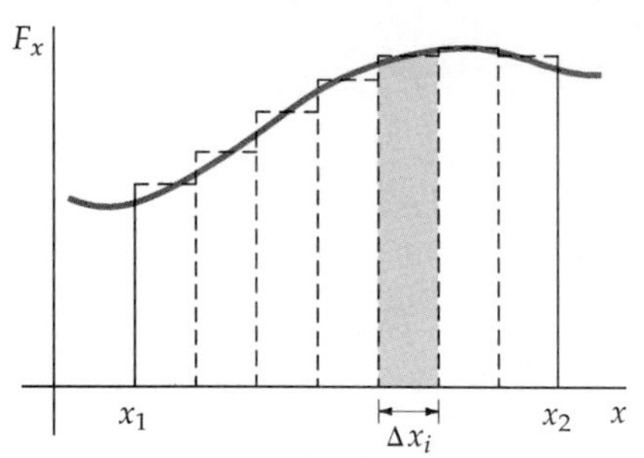

Figure 6-5 A variable force can be approximated by a series of constant forces acting over small intervals. The work done by one of these constant forces is the rectangular area beneath that force. The total work done by the variable force from x_1 to x_2 is the total area under the curve for this interval.

Figure 6-5 shows a variable force F_x as a function of position x. In the figure, we have divided the interval from x_1 to x_2 into a set of smaller intervals Δx_i. If each interval is small enough, we can approximate the variable force by a series of constant forces, as shown in the figure. For each interval, the work done by the constant force is the area of the rectangle beneath the force. The sum of these rectangular areas is the sum of the work done by the set of constant forces that approximates the varying force. As can be seen in the figure, this area is approximately equal to the area under the curve. In the limit of infinitesimally small Δx_i, the sum of the areas of the rectangles equals the area under the curve. We can therefore define the work done by a variable force to be the area under the F_x-versus-x curve for that force. (Note that this is the same procedure we used in Chapter 2 when we found the displacement Δx to be the area under the v-versus-t curve.)

$$W = \lim_{\Delta x_i \to 0} \sum_i F_x\ \Delta x_i = \text{area under the } F_x\text{-versus-}x \text{ curve} \qquad 6\text{-}8$$

This limit is the integral of F_x over x. Thus, the work done by a variable force F_x acting on a particle as it moves from point x_1 to point x_2 is

Work by a variable force

$$W = \int_{x_1}^{x_2} F_x\ dx = \text{area under the } F_x\text{-versus-}x \text{ curve} \qquad 6\text{-}9$$

We now show that the total work done on a particle equals the change in kinetic energy of the particle even when the net force varies with position. Consider Figure 6-5 again. For each rectangular area, the force is constant, so the work done by the constant force for that interval equals the change in the kinetic energy over that interval. When we sum over all the intervals, we sum the changes in kinetic energy for all the intervals, which gives us the total or net change in kinetic energy for the entire interval from x_1 to x_2. The total area under the curve therefore equals the sum of the changes in the kinetic energy for each interval Δx_i, which in turn equals the net change in the kinetic energy for the complete interval, $\Delta K = \frac{1}{2}mv_2^2 - \frac{1}{2}mv_1^2$. This result is the same as Equation 6-7, provided that W_{total} is interpreted to be the integral given by Equation 6-9.

Example 6-3

A 4-kg block on a frictionless table (Figure 6-6*a*) is attached to a horizontal spring that obeys Hooke's law and exerts a force $F_x = -kx$, where x is measured from the equilibrium position of the block and the force constant $k = 400$ N/m. The spring is compressed to $x_1 = -5$ cm. Find (*a*) the work done by the spring on the block as the block moves from $x_1 = -5$ cm to its equilibrium position $x_2 = 0$ and (*b*) the speed of the block at $x_2 = 0$.

(*a*) Figure 6-6*b* is a graph of this force versus distance. The work done on the block as it moves from x_1 to $x_2 = 0$ equals the triangular area under the F_x-versus-x curve between these limits, which is shaded in the figure. The area of this triangle is one-half its base times its height. The base is 5 cm = 0.05 m, and the height is the value of the force at x_1, which is

$$F_x = -kx = -(400\ \text{N/m})(-0.05\ \text{m}) = +20\ \text{N}$$

The work done is thus

$$W = \int_{-0.05\ \text{m}}^{0} F_x\, dx = \tfrac{1}{2}(0.05\ \text{m})(20\ \text{N}) = 0.500\ \text{N·m} = 0.500\ \text{J}$$

The work is positive because the force is in the direction of motion. This is indicated in the figure by the fact that the area is above the x axis.

We can also compute this work using integral calculus:

$$W = \int_{x_1}^{0} F_x\, dx = \int_{x_1}^{0} -kx\, dx = -k\int_{x_1}^{0} x\, dx$$

$$= -\tfrac{1}{2}kx^2\Big|_{x_1}^{0} = \tfrac{1}{2}kx_1^2 = \tfrac{1}{2}(400\ \text{N/m})(0.05\ \text{m})^2 = 0.500\ \text{J}$$

(*b*) We find the speed of the block from the fact that its kinetic energy is 0.500 J:

$$K = \tfrac{1}{2}mv^2 = 0.500\ \text{J}$$

The speed is then

$$v = \sqrt{\frac{2K}{m}} = \sqrt{\frac{2(0.500\ \text{J})}{4\ \text{kg}}} = 0.50\ \text{m/s}$$

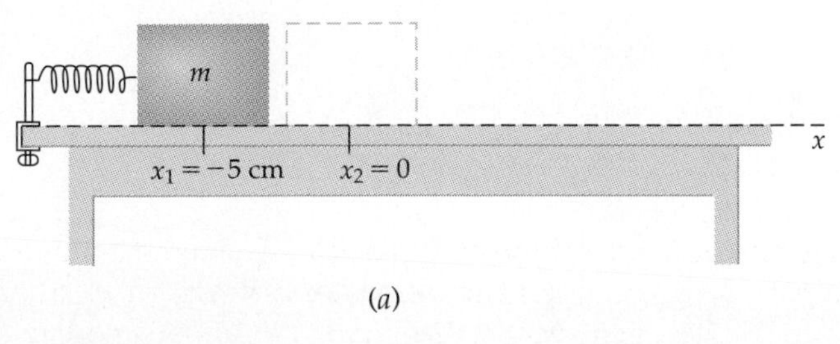

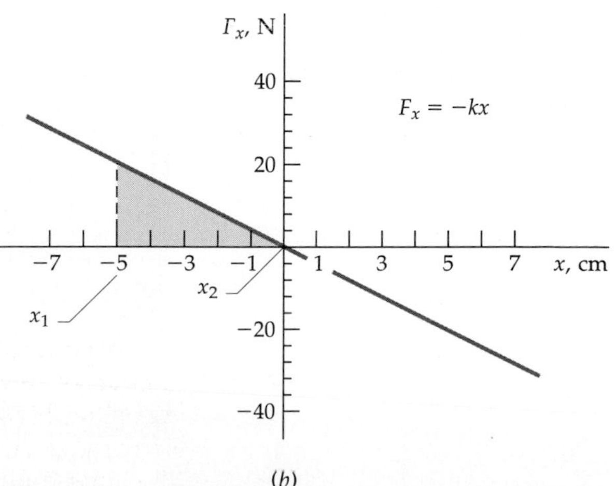

Figure 6-6 (*a*) Block attached to a spring for Example 6-3. The spring is compressed from equilibrium and released. (*b*) Plot of the force F_x exerted by the spring versus x. The work done by the spring as the block moves from x_1 to x_2 is the shaded area.

Note that we could not have found this result by first finding the acceleration and then using the constant-acceleration equations. The force varies with position, so the acceleration also varies.

Since the force exerted on the block by the spring is $-kx$ whether x is positive or negative, the work done by a spring on a block as the block moves from some initial compression or extension x to its equilibrium position $x = 0$ is given by

$$W = \int_x^0 -kx\, dx = \tfrac{1}{2}kx^2 \qquad \text{6-10}$$

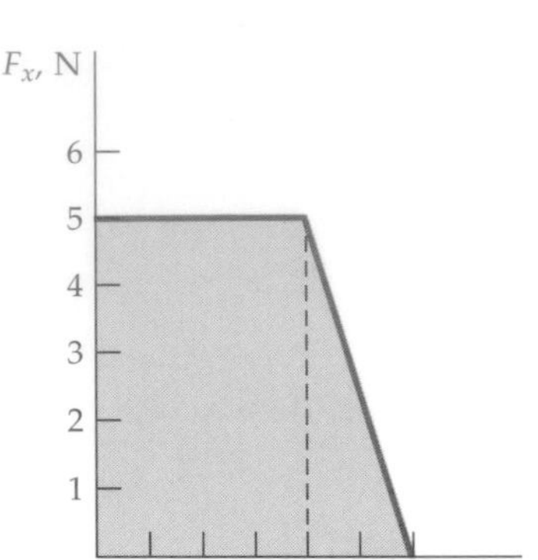

Figure 6-7 Variable force for Example 6-4. The work done equals the shaded area.

Example 6-4

A force F_x varies with x as shown in Figure 6-7. Find the work done by the force on a particle as the particle moves from $x = 0$ to $x = 6$ m.

We find the work done by calculating the area under the F_x-versus-x curve. From 0 to 4 m, the force is constant and the area is that of a rectangle, (5 N)(4 m) = 20 J. From 4 m to 6 m, the force is decreasing at a constant rate and the area is that of the triangle shown in the figure: $\frac{1}{2}$(5 N)(2 m) = 5 J. The total area is 20 J + 5 J = 25 J, which equals the work done by the force.

Exercise

The force shown in Figure 6-7 is the only force that acts on a particle of mass 3 kg. If the particle starts from rest at $x = 0$, how fast is it moving when it reaches $x = 6$ m? (Answer: 4.08 m/s)

Question

6. How does the work required to stretch a spring 2 cm compare with that required to stretch it 1 cm?

Wind farm at Altamount pass in California uses hundreds of windmills to convert wind energy into electrical energy.

6-3 Work and Energy in Three Dimensions and the Dot Product

The work done by a force is defined to be the product of the displacement of the point of application of the force and the component of the force in the direction of the displacement. The component of the force in the direction of the displacement is important because it changes the speed of the particle. Figure 6-8 shows a particle of mass m that is acted on by a single force $\mathbf{F}$ as it moves along an arbitrary curve in space. We consider a small displacement along the curve. The force $\mathbf{F}$ makes an angle ϕ with the displacement. By Newton's second law, the tangential component of this force F_s is related to the rate of change of speed of the particle, $F_s = m\,dv/dt$, and the perpendicular component is related to the centripetal acceleration, $F_\perp = mv^2/r$, where r is the radius of curvature of the path at this point. The perpendicular component does not affect the speed of the particle. Instead, it changes the direction of the particle's velocity. The work done by the force on the particle for a small displacement Δs is

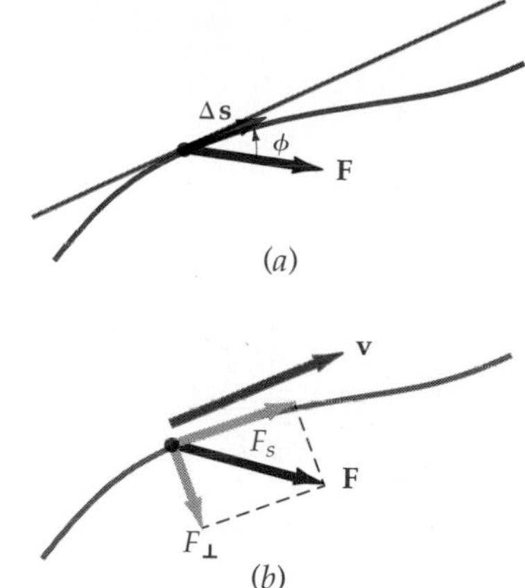

Figure 6-8 (*a*) A particle moving along an arbitrary curve in space. (*b*) The perpendicular component of the force $F_\perp$ changes the direction of the particle's motion but not its speed. The tangential component F_s changes the particle's speed but not its direction. F_s equals the mass m times the tangential acceleration dv/dt. Only this component does work.

$$\Delta W = F_s\,\Delta s$$

To find the work done by the force as the particle moves along the curve from point 1 to point 2, we compute the product $F_s\,\Delta s$ for each element of the path and sum. In the limit of smaller and smaller displacement elements, this sum becomes an integral:

$$W = \int_{s_1}^{s_2} F_s\,ds$$

As in the case of one-dimensional motion, we can show that the total work done on the particle equals the change in its kinetic energy. From Newton's second law, we have

$$F_s = m\,\frac{dv}{dt}$$

If we think of the speed as a function of the distance s measured along the curve, we can apply the chain rule for derivatives (rule 3 in Table A-4 of Appendix A):

$$\frac{dv}{dt} = \frac{dv}{ds}\frac{ds}{dt} = v\,\frac{dv}{ds}$$

where we have used the fact that ds/dt is the speed v. The work done by the net force, which for a particle is the total work, is thus

$$W_{\text{total}} = \int_{s_1}^{s_2} F_s\,ds = \int_{s_1}^{s_2} m\,\frac{dv}{dt}\,ds = \int_{s_1}^{s_2} mv\,\frac{dv}{ds}\,ds = \int_{v_1}^{v_2} mv\,dv$$

or

$$W_{\text{total}} = \int_{s_1}^{s_2} F_s\,ds = \tfrac{1}{2}mv_2^2 - \tfrac{1}{2}mv_1^2 \qquad 6\text{-}11$$

Work–energy theorem in three dimensions

Equation 6-11, along with its one-dimensional counterpart, Equation 6-7, follows directly from the definition of work and from Newton's second law of motion.

The Dot Product

If ϕ is the angle between the force **F** acting on a particle traveling along a curve in space and a small displacement $d\mathbf{s}$ along that curve, the component of the force parallel to $d\mathbf{s}$ is $F_s = F \cos \phi$. The work done by the force over the small displacement is then

$$dW = F_s\, ds = (F \cos \phi)\, ds$$

This type of scalar combination of two vectors and the cosine of the angle between them occurs often in physics, and is called the **scalar product** of the vectors. The scalar product of two general vectors **A** and **B** is written $\mathbf{A}\cdot\mathbf{B}$ and defined as

Dot product defined

$$\mathbf{A}\cdot\mathbf{B} = AB \cos \phi \qquad 6\text{-}12$$

where ϕ is the angle between **A** and **B**. Because of the way it is written, the scalar product is also called the **dot product.** The dot product $\mathbf{A}\cdot\mathbf{B}$ can be thought of as the product of A and $B \cos \phi$, which is the component of **B** in the direction of **A**, or alternately as the product of B and $A \cos \phi$, which is the component of **A** in the direction of **B** (Figure 6-9).

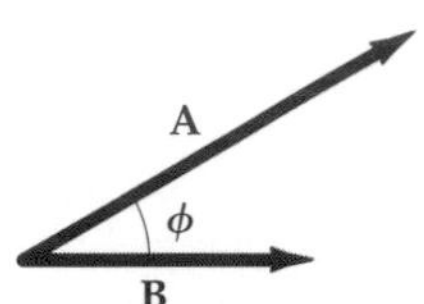

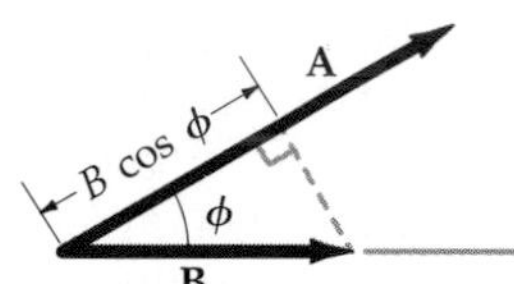

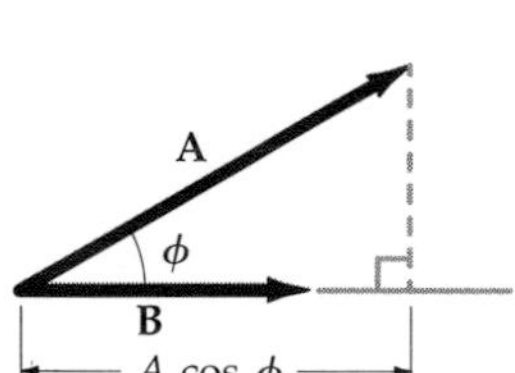

Figure 6-9 Geometric representation of the dot product $\mathbf{A}\cdot\mathbf{B}$. We can think of this as the product of A times $B \cos \phi$ or B times $A \cos \phi$.

If $\mathbf{A}\cdot\mathbf{B} = 0$, then $\mathbf{A} = 0$ or $\mathbf{B} = 0$ or **A** and **B** are perpendicular. If **A** and **B** are perpendicular, their dot product is zero because $\phi = 90°$ and $\cos 90° = 0$. If **A** and **B** are parallel, the dot product is just the product of their magnitudes. The dot product of a vector with itself is the square of the magnitude of the vector:

$$\mathbf{A}\cdot\mathbf{A} = A^2$$

It follows from the definition of the dot product that it is independent of the order of multiplication; that is, $\mathbf{A}\cdot\mathbf{B} = \mathbf{B}\cdot\mathbf{A}$. This is known as the *commutative rule of multiplication*. The dot product also obeys the *distributive rule of multiplication:*

$$(\mathbf{A} + \mathbf{B})\cdot\mathbf{C} = \mathbf{A}\cdot\mathbf{C} + \mathbf{B}\cdot\mathbf{C}$$

We sometimes wish to write the dot product in terms of the rectangular components of the two vectors. Consider the dot product of the vectors $\mathbf{A} = A_x\mathbf{i} + A_y\mathbf{j} + A_z\mathbf{k}$ and $\mathbf{B} = B_x\mathbf{i} + B_y\mathbf{j} + B_z\mathbf{k}$. Since the unit vectors **i**, **j**, and **k** are mutually perpendicular, the dot product of two different unit vectors is zero; that is, $\mathbf{i}\cdot\mathbf{j} = \mathbf{i}\cdot\mathbf{k} = \mathbf{j}\cdot\mathbf{k} = 0$. In addition, the dot product of a unit vector with itself is 1; that is, $\mathbf{i}\cdot\mathbf{i} = \mathbf{j}\cdot\mathbf{j} = \mathbf{k}\cdot\mathbf{k} = 1$. Thus, for the dot product $\mathbf{A}\cdot\mathbf{B}$, we have

$$\begin{aligned}\mathbf{A}\cdot\mathbf{B} &= (A_x\mathbf{i} + A_y\mathbf{j} + A_z\mathbf{k})\cdot(B_x\mathbf{i} + B_y\mathbf{j} + B_z\mathbf{k}) \\ &= A_xB_x + A_yB_y + A_zB_z \qquad 6\text{-}13\end{aligned}$$

In terms of the dot-product notation, the work dW done by a force **F** during a small displacement $d\mathbf{s}$ is written

$$dW = F \cos \phi\, ds = \mathbf{F}\cdot d\mathbf{s} \qquad 6\text{-}14$$

and the work done on the particle as it moves from point 1 to point 2 is written

Work—general definition

$$W = \int_{s_1}^{s_2} \mathbf{F}\cdot d\mathbf{s} \qquad 6\text{-}15$$

When several forces $\mathbf{F}_i$ act on a particle whose displacement is $d\mathbf{s}$, the total work is

$$\Delta W_{\text{total}} = \mathbf{F}_1 \cdot d\mathbf{s} + \mathbf{F}_2 \cdot d\mathbf{s} + \cdots = \sum_i \mathbf{F}_i \cdot d\mathbf{s} = \left(\sum_i \mathbf{F}_i\right) \cdot d\mathbf{s}$$

or

$$\Delta W_{\text{total}} = \mathbf{F}_{\text{net}} \cdot d\mathbf{s} \qquad 6\text{-}16$$

Example 6-5

A particle is given a displacement $\Delta\mathbf{s} = 2\text{ m }\mathbf{i} - 5\text{ m }\mathbf{j}$ along a straight line. During the displacement, a constant force $\mathbf{F} = 3\text{ N }\mathbf{i} + 4\text{ N }\mathbf{j}$ acts on the particle. Find the work done by the force and the component of the force in the direction of the displacement.

The work done by the force is

$$W = \mathbf{F} \cdot \Delta\mathbf{s} = (3\text{ N }\mathbf{i} + 4\text{ N }\mathbf{j}) \cdot (2\text{ m }\mathbf{i} - 5\text{ m }\mathbf{j})$$
$$= 6\text{ N·m} - 20\text{ N·m} = -14\text{ N·m}$$

In terms of the magnitudes of the force and displacement, the work is

$$W = \mathbf{F} \cdot \Delta\mathbf{s} = F \cos\phi\, \Delta s$$

The component of the force in the direction of the displacement is $F \cos\phi$, which equals the work divided by the magnitude of the displacement. We find Δs from

$$\Delta\mathbf{s} \cdot \Delta\mathbf{s} = (\Delta s)^2 = (2\text{ m }\mathbf{i} - 5\text{ m }\mathbf{j}) \cdot (2\text{ m }\mathbf{i} - 5\text{ m }\mathbf{j})$$
$$= 4\text{ m}^2 + 25\text{ m}^2 = 29\text{ m}^2$$

so $\Delta s = \sqrt{29}$ m. The component of $\mathbf{F}$ in the direction of the displacement is therefore

$$F \cos\phi = \frac{W}{\Delta s} = \frac{-14\text{ N·m}}{\sqrt{29}\text{ m}} = -2.60\text{ N}$$

The component of the force in the direction of the displacement is negative so the work done is negative.

Example 6-6

Find the angle between the vectors $\mathbf{A} = (3\text{ m})\mathbf{i} + (2\text{ m})\mathbf{j}$ and $\mathbf{B} = (4\text{ m})\mathbf{i} - (3\text{ m})\mathbf{j}$.

We find the angle ϕ between the vectors from the definition of the dot product:

$$\mathbf{A} \cdot \mathbf{B} = AB \cos\phi$$

$$\cos\phi = \frac{\mathbf{A} \cdot \mathbf{B}}{AB}$$

The dot product of these vectors is

$$\mathbf{A} \cdot \mathbf{B} = A_xB_x + A_yB_y = (3\text{ m})(4\text{ m}) + (2\text{ m})(-3\text{ m})$$
$$= 12\text{ m}^2 - 6\text{ m}^2 = 6\text{ m}^2$$

The magnitudes of the vectors are obtained from

$$\mathbf{A} \cdot \mathbf{A} = A^2 = A_x^2 + A_y^2$$
$$= (2\text{ m})^2 + (3\text{ m})^2 = 13\text{ m}^2$$
$$A = \sqrt{13}\text{ m}$$

and

$$\mathbf{B}\cdot\mathbf{B} = B^2 = B_x^2 + B_y^2$$
$$= (4\text{ m})^2 + (-3\text{ m})^2 = 25\text{ m}^2$$
$$B = 5\text{ m}$$

Then, substituting these values into the equation for $\cos\phi$, we get

$$\cos\phi = \frac{\mathbf{A}\cdot\mathbf{B}}{AB} = \frac{6\text{ m}^2}{(\sqrt{13}\text{ m})(5\text{ m})} = 0.333$$
$$\phi = 70°$$

Exercise

Find $\mathbf{A}\cdot\mathbf{B}$ for (*a*) $\mathbf{A} = (3\text{ m})\mathbf{i} + (4\text{ m})\mathbf{j}$ and $\mathbf{B} = (5\text{ m})\mathbf{i} + (8\text{ m})\mathbf{j}$ and (*b*) $\mathbf{A} = (2\text{ m/s})\mathbf{i} + (6\text{ m/s})\mathbf{j}$ and $\mathbf{B} = (5\text{ m/s})\mathbf{i} - (3\text{ m/s})\mathbf{j}$. [Answers: (*a*) 47 m^2, (*b*) $-8\text{ m}^2/\text{s}^2$]

Example 6-7

A skier of mass m skis down a frictionless hill that has a constant angle of inclination θ as shown in Figure 6-10. The skier starts from rest at a height h. Considering the skier to be a particle, find the work done on the skier by all the forces and the speed of the skier at the bottom of the hill.

The forces acting on the skier are the force of gravity $m\mathbf{g}$ and the contact force $\mathbf{F}_n$ exerted by the snow, as shown in Figure 6-10*a*. Since the force $\mathbf{F}_n$ is perpendicular to the hill and to the motion of the skier, it has no component in the direction of motion of the skier and therefore does no work on the skier. The only force that does work is the weight $m\mathbf{g}$. The angle between this force and the displacement is $\phi = 90° - \theta$ (Figure 6-10*b*). The component of the weight in the direction of motion is therefore $mg\cos\phi = mg\cos(90° - \theta) = mg\sin\theta$. For a displacement $\Delta\mathbf{s}$ of the skier down the incline, the force of gravity does work

$$\Delta W = m\mathbf{g}\cdot\Delta\mathbf{s} = (mg\cos\phi)\,\Delta s = (mg\sin\theta)\,\Delta s$$

The total work done by gravity when the skier moves a distance s down the incline is

$$W = \sum \Delta W = \sum (mg\sin\theta)\,\Delta s$$

Since the force exerted by the earth is constant, $\Sigma(mg\sin\theta)\,\Delta s = (mg\sin\theta)\,\Sigma\,\Delta s = (mg\sin\theta)s$. We can see from Figure 6-10*a* that the

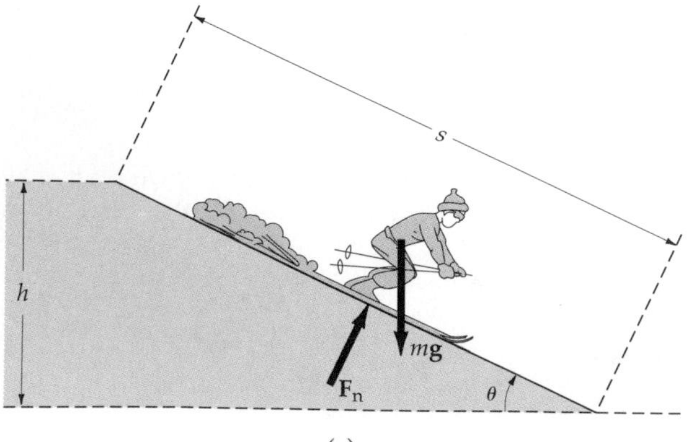

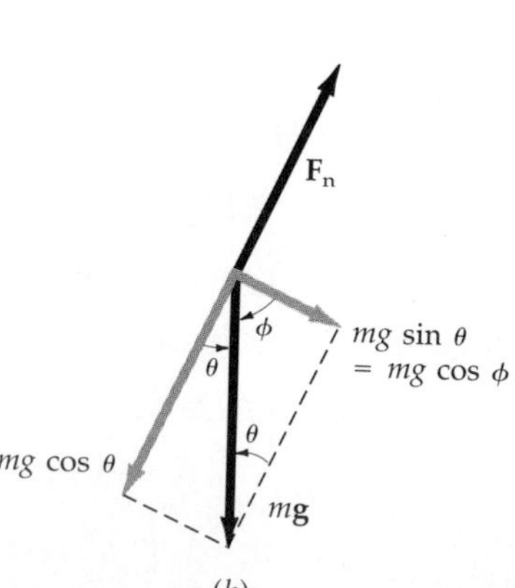

Figure 6-10 (*a*) Skier skiing down a hill of constant slope for Example 6-7. (*b*) Free-body diagram for the skier. The net force is $mg\sin\theta$, which is the component of the weight in the direction of the displacement Δs.

total distance s measured along the incline is related to the initial height h by $\sin\theta = h/s$ or $h = s\sin\theta$, so the work done by gravity is

$$W = (mg\sin\theta)s = mgh$$

Since this is the total work done by all the forces acting on the skier, it equals the change in the kinetic energy of the skier:

$$W_{\text{total}} = mgh = \Delta K = \tfrac{1}{2}mv^2 - 0$$

The speed of the skier at the bottom of the hill is thus given by

$$v = \sqrt{2gh}$$

This result is the same as if the skier had dropped the total distance h in free fall. If the angle θ were changed to θ', the skier would travel a different distance $s' = h/\sin\theta'$ to drop the same vertical distance h, but the component of $m\mathbf{g}$ parallel to the motion, $mg\sin\theta'$, would also change, making the work done, $(mg\sin\theta')s' = mgh$, the same. Thus, the work done by gravity on the skier is mgh, independent of the angle of incline of the hill.

The results of Example 6-7 can be generalized. Consider a skier who is skiing down a hill of arbitrary shape as shown in Figure 6-11. The small displacement $\Delta\mathbf{s}$ is tangent to the hill. The work done by gravity for this displacement is $(mg\cos\phi)\,\Delta s$, where ϕ is the angle between the displacement and the downward force of gravity. The quantity $\Delta s\cos\phi$ is just Δh, the vertical distance dropped over Δs. As the skier skis down the hill, the angle ϕ varies, but for each displacement $\Delta\mathbf{s}$, the work done is $\Delta W = m\mathbf{g}\cdot\Delta\mathbf{s} = mg\cos\phi\,\Delta s = mg\,\Delta h$. Thus, the total work done by gravity is mgh, where h is the total vertical distance the skier descends. If the hill is frictionless, the weight of the skier is the only force that does work. In this case, the speed of the skier after descending a vertical distance h is obtained from

$$\tfrac{1}{2}mv^2 - \tfrac{1}{2}mv_0^2 = mgh$$

where v_0 is the initial speed. This result applies only if the hill is frictionless. If there is friction between the skis and the snow, the frictional force will do work on the skier, which will be negative because the frictional force is in the direction opposite the motion. The work done on the skier by the frictional force depends on the length and shape of the hill and the coefficient of kinetic friction.

Questions

7. Suppose there is a net force acting on a particle but it does no work. Can the particle be moving in a straight line?
8. How are the vectors $\mathbf{A}$ and $\mathbf{B}$ related if $\mathbf{A}\cdot\mathbf{B} = -AB$?

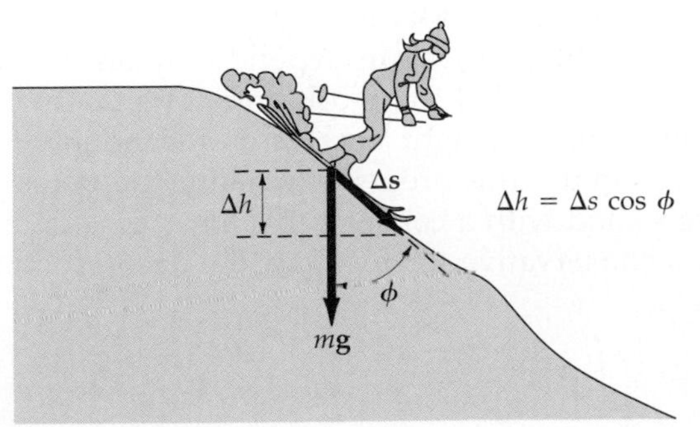

Figure 6-11 Skier skiing down a hill of arbitrary shape. The work done by the earth for the displacement $\Delta\mathbf{s}$ is $(mg\cos\phi)\,\Delta s = mg\,\Delta h$, where Δh is the vertical component of the displacement.

The ski lift does work on the skiers, which is stored as potential energy.

6-4 Work and Energy for Particle Systems: Potential Energy

When we have a system consisting of two or more particles, we must be careful when we calculate the work done by a force acting on a part of the system because the displacement of the point of application of the force is not always equal to the displacement of the entire system. In fact, different parts of a system may have different displacements.

In many cases, the work done *on a system* does not result in a change in the kinetic energy of the system but is instead stored as **potential energy.** Consider a skier of mass m going up a hill on a ski lift that raises the skier at a negligible speed to a height h. The lift does work on the skier of the amount mgh, independent of the angle of incline of the lift. The kinetic energy of the skier does not change because the gravitational force of the earth on the skier does work of the amount $-mgh$, so the total work done on the skier is zero. Let us now consider the earth and the skier (excluding the lift) to be a system consisting of two particles. The work done by gravity on the skier is internal to this system. The total work done *on this system* is the work mgh done by the lift, which we are considering to be external to our system. This work is stored as potential energy mgh of the earth–skier system. When the skier skis down a frictionless slope, this potential energy is converted into kinetic energy of the system, which is just the kinetic energy of the skier since the motion of the earth is negligible. Note that when the skier is lifted to the top of the hill, the work done by gravity is negative and the potential energy of the system increases, whereas when the skier skis down the hill, the work done by gravity is positive and the potential energy of the system decreases. The total work done by gravity when the skier moves to the top of the hill and back down to her original position is zero. The force of gravity is called a **conservative force.** In general

> A force is conservative if the total work it does on a particle is zero when the particle moves around any closed path returning to its initial position.

It follows that

> The work done by a conservative force on a particle is independent of how the particle moves from one point to another.

This is illustrated in Figure 6-12 on facing page, which shows three possible paths connecting point 1 and point 2. If the work done by a conservative force as the particle moves from 1 to 2 along one of the paths is W, it must be $-W$ when the particle returns along any of the paths because the total round-trip work must be zero. Since the work done by a conservative force on a particle does not depend on the path, it can depend only on the endpoints 1 and 2. (For example, we saw that the work done by gravity on a skier when the skier skies down a hill of height h is mgh, independent of the angle of incline of the hill.) We can use this property to define the potential-energy function U that is associated with a conservative force. It is defined such that the work done by a conservative force equals the decrease in the potential-energy function:

$$W = \int \mathbf{F}\cdot d\mathbf{s} = -\Delta U$$

or

$$\Delta U = U_2 - U_1 = -W = -\int_{s_1}^{s_2} \mathbf{F} \cdot d\mathbf{s} \qquad \text{6-17}a$$

Potential-energy function defined

For infinitesimal displacement, we have

$$dU = -\mathbf{F} \cdot d\mathbf{s} \qquad \text{6-17}b$$

The potential-energy function of a system depends on the configuration of the system. For the earth–skier system, it depends on the distance of the skier from the center of the earth. Since only a change in the potential energy is defined, the value of the function U for a system having any particular configuration is not specified by the definition. We are free to choose U to be zero at some convenient reference point. The potential energy at any other point is then the difference between the potential energy at that point and that at the reference point. For example, if the gravitational potential energy of the earth–skier system is chosen to be zero when the skier is at the bottom of the hill, its value when the skier is at a height h above that level is mgh. Or we could choose the potential energy to be zero when the skier is at sea level, in which case its value at any other point would be mgy, where y is measured from sea level.

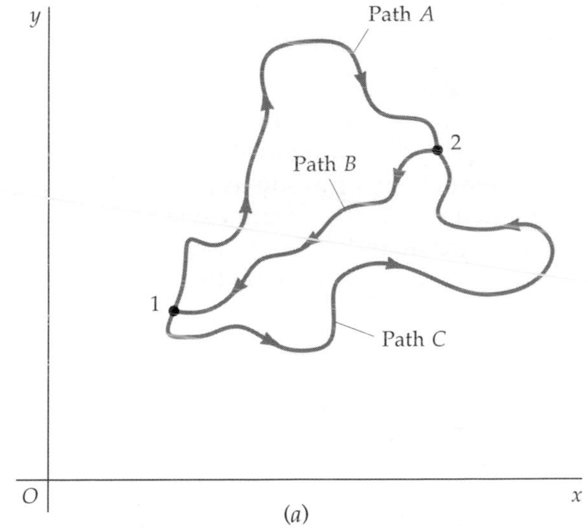

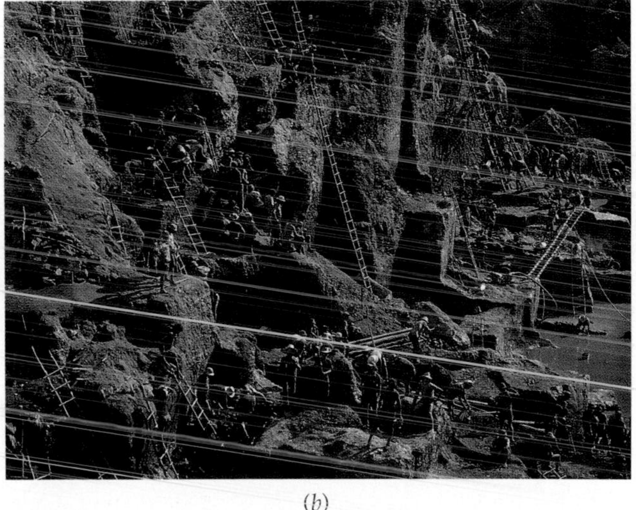

Figure 6-12 (*a*) Three paths in space connecting the points 1 and 2. If the work done by a conservative force along path A from 1 to 2 is W, the work done on the return trip along path B must be $-W$ because the round-trip work is zero. It follows that the work done as a particle goes from point 1 to 2 is the same along any path connecting the two points. (*b*) Gold mine in Brazil. The work done by the earth on a miner as he descends to the bottom is the same no matter which ladder he uses.

We can calculate the potential-energy function associated with the gravitational force near the surface of the earth from Equation 6-17*b*. For the force $\mathbf{F} = -mg\mathbf{j}$, we have

$$dU = -\mathbf{F} \cdot d\mathbf{s} = -(-mg\mathbf{j}) \cdot (dx\,\mathbf{i} + dy\,\mathbf{j} + dz\,\mathbf{k}) = +\,mg\,dy$$

Integrating, we obtain

$$U = U_0 + mgy \qquad \text{6-18}$$

Gravitational potential energy near the earth's surface

where U_0 is the value of the potential energy at $y = 0$.

Exercise

A 55-kg girl is on the third floor of a building, which is 8 m above the ground floor. What is the potential energy U of the girl–earth system if (*a*) U is chosen to be zero on the ground floor and (*b*) U is chosen to be zero on the second floor, which is 4 m above the ground floor? [Answer: (*a*) 4.32 kJ, (*b*) 2.16 kJ]

Example 6-8

A bottle of mass 0.350 kg falls from rest from a shelf that is 1.75 m above the floor. Find the original potential energy of the bottle–earth system relative to the floor and its kinetic energy just before it hits the floor.

If we choose the potential energy of the bottle–earth system to be zero when the bottle is on the floor, the potential energy when the bottle is at $y = 1.75$ m is

$$\begin{aligned} U &= mgy \\ &= (0.350 \text{ kg})(9.81 \text{ N/kg})(1.75 \text{ m}) \\ &= 6.01 \text{ J} \end{aligned}$$

As the bottle falls, the earth does work on the bottle. The total work done by the earth as the bottle falls a distance $y = 1.75$ m is $W = mgy =$ 6.01 J. The kinetic energy of the bottle just before it hits the floor is thus 6.01 J. In this example, the original potential energy of the bottle–earth system is converted into kinetic energy of the bottle.

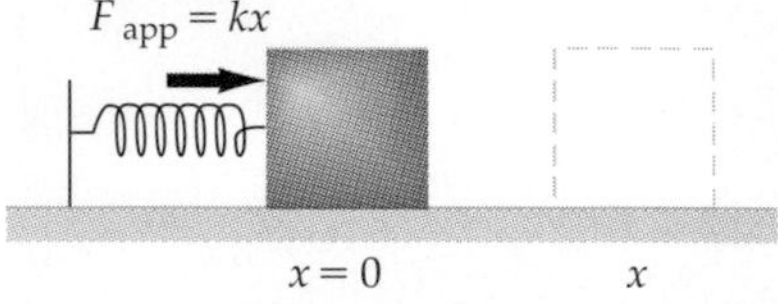

Figure 6-13 A mass attached to a spring. To stretch the spring, a force $F_{app} = +kx$ must be applied to the block.

Another example of a conservative force is that exerted by a spring. Figure 6-13 shows a system consisting of a spring and a block. The block is at rest on a frictionless table at $x = 0$ with the spring unstretched. Let us now slowly push the block with a force equal and opposite to that exerted by the spring so that the spring is extended. Since the spring exerts a force $F_{sb} = -kx$ on the block, we must exert an equal and opposite force $F_{app} = +kx$ to push the block without acceleration. The work done by the applied force is

$$W_{app} = \int_0^x kx\, dx = \tfrac{1}{2}kx^2$$

The spring exerts a force on the block opposite the direction of the displacement, so it does negative work on the block:

$$W_{sb} = \int_0^x -kx\, dx = -\tfrac{1}{2}kx^2$$

Since the total work done on the block is zero, its kinetic energy does not change. The block exerts a force on the spring F_{bs}, which is the reaction force to the force F_{sb} exerted by the spring on the block:

$$F_{bs} = -F_{sb} = +kx$$

The displacement of the point of application of force F_{bs} is the same as the displacement of the block, so the block does work on the spring equal in magnitude to W_{sb}:

$$W_{bs} = +\tfrac{1}{2}kx^2$$

We cannot assign a value for the displacement of the spring because different parts of the spring have different displacements. The other force exerted on the spring is that of the wall to which it is fixed. The total work done on the *block–spring system* is W_{app}. This work is stored as potential energy of the block–spring system. When the block is released, the spring does positive work on the block as the potential energy of the system is converted into kinetic energy. The round-trip work done on the block by the spring when the block moves from its original position out to some position x and then back is zero. The force of the spring is thus a conservative force. We can calculate the potential-energy function associated with this force from Equation 6-17*b*. We have

$$dU = -\mathbf{F}\cdot d\mathbf{s} = -F_x\, dx = -(-kx)\, dx = +kx\, dx$$

Then

$$U = \tfrac{1}{2}kx^2 + U_0$$

where U_0 is the potential energy when $x = 0$, that is, when the spring is unstretched. Choosing U_0 to be zero, we have

$$U = \tfrac{1}{2}kx^2 \qquad \text{6-19}$$

Potential energy of a spring

Exercise

A 3-kg block is hung vertically from a spring with a force constant of 600 N/m. (*a*) By how much is the spring stretched when the block is in equilibrium? (*b*) How much potential energy is stored in the spring–block system? [Answer: (*a*) 4.9 cm, (*b*) 0.72 J]

Although potential energy is associated with a system of particles, we sometimes have systems in which only one particle moves. For example, in the earth–skier system, the earth's motion is negligible. Since the configuration of the system is expressed in terms of the position of the skier relative to some chosen reference position, we sometimes speak of the potential energy of this system as the potential energy of the skier. Similarly, we may refer to the potential energy of the block–spring system as the potential energy of the block.

Not all forces are conservative. An example of a nonconservative force is a force, such as a push or a pull, applied by a human agent. Consider a box on a rough horizontal table. Suppose you push the box around some closed path so that the box ends up at its original position. The work that you do is not generally zero. It depends on how great a force you decide to exert on the box. Since the work done by such an applied force is not zero when the box completes a round trip, no potential-energy function can be defined for this force. The work done on the box by the force of kinetic friction is also not zero. Since the force of kinetic friction is always opposite the direction of motion, the total round-trip work done by kinetic friction will be negative. The force of kinetic friction is therefore not a conservative force, and no potential-energy function can be defined for it.

Question

9. When you climb a mountain, is the work done on you by gravity different if you take a short, steep trail instead of a long, gentle trail? If not, why do you find one trail easier than the other?

6-5 Potential Energy and Equilibrium in One Dimension

The potential-energy function U associated with a conservative force $\mathbf{F} = F_x\mathbf{i}$ is related to the force by Equation 6-17*b*:

$$dU = -\mathbf{F}\cdot d\mathbf{s} = -F_x\, dx$$

The force is therefore the negative derivative of the potential-energy function:

$$F_x = -\frac{dU}{dx} \qquad 6\text{-}20$$

We can illustrate this general relation for the block–spring system by differentiating the function $U = \frac{1}{2}kx^2$. We obtain

$$F_x = -\frac{dU}{dx} = -\frac{d}{dx}\left(\tfrac{1}{2}kx^2\right) = -kx$$

Figure 6-14 shows a plot of the potential-energy function $U = \frac{1}{2}kx^2$ versus the displacement x for a block–spring system. The derivative of this function is represented graphically as the slope of the tangent line to the curve. The force is thus equal to the negative of the slope of the curve. When x is positive, the slope is positive and the force F_x is negative. When x is negative the slope is negative and the force F_x is positive. In either case, the force is in the direction such as to accelerate the block toward lower potential energy.

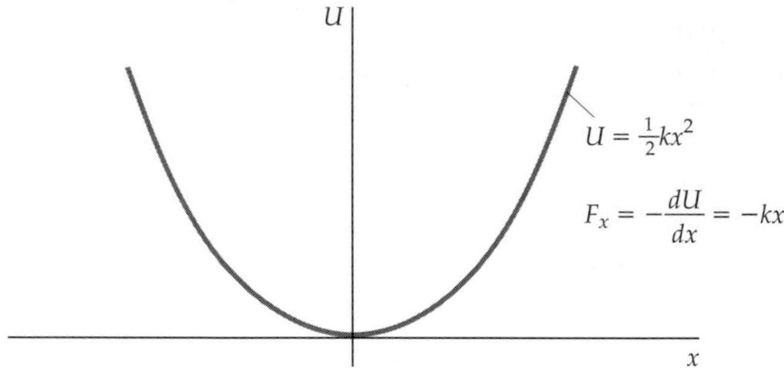

Figure 6-14 Plot of the potential-energy function U versus displacement x for the object on a spring. The force $F_x = -dU/dx$ is the negative slope of this curve. It is zero at the minimum in the curve, which occurs at $x = 0$. For positive values of x the force is negative, and for negative values of x the force is positive. A minimum in a potential-energy curve is a point of stable equilibrium because a displacement to either side of this point results in a force directed toward the equilibrium position.

At the position on the potential-energy curve where the slope is zero, the force is zero and, if F_x is the only force acting on it, the block is in equilibrium; that is, the block will remain at rest in this position if it is placed there at rest.

Condition for equilibrium

A particle is in equilibrium if the net force acting on it is zero.

This occurs at $x = 0$, where the spring is unstretched. For x slightly greater than zero, the force $F_x = -dU/dx$ is negative. Thus, if the block is displaced slightly in the positive x direction, the spring force will accelerate the block back toward its equilibrium position at $x = 0$. Similarly, if x is negative, F_x is positive. If the block is displaced in the negative x direction, the spring force accelerates the block in the positive x direction back toward its equilibrium position. This equilibrium is called **stable equilibrium.**

In stable equilibrium a small displacement results in a restoring force that accelerates the particle back toward its equilibrium position.

Figure 6-15 shows a potential-energy curve with a maximum rather than a minimum at $x = 0$. Such a curve could represent the potential energy of a skier at the top of a hill between two valleys. For this curve, when x is positive, the slope is negative and the force F_x is positive, and when x is negative, the slope is positive and the force F_x is negative. Again, the force is in the direction such as to accelerate the particle toward lower potential energy.

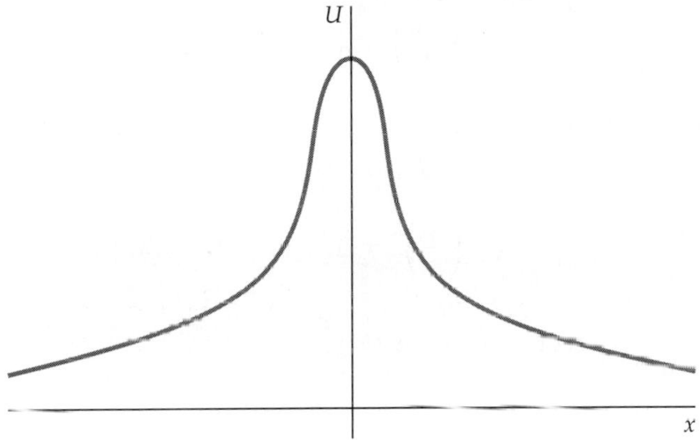

Figure 6-15 For this potential-energy curve, $F_x = -dU/dx$ is zero at $x = 0$, but this equilibrium is unstable because a displacement in either direction results in a force directed away from the equilibrium position.

At $x = 0$, $dU/dx = -F_x = 0$, so a particle placed at this point is in equilibrium. However, this equilibrium is **unstable equilibrium.** For x slightly greater than 0, the force F_x is positive. Thus, if a particle in equilibrium at $x = 0$ is given a small positive displacement x, it will experience a positive force F_x that will accelerate it to the right away from its equilibrium position. Similarly, for x slightly less than 0, F_x is negative. If the particle is displaced slightly to the left of its equilibrium position, it will accelerate to the left, away from its equilibrium position.

In unstable equilibrium a small displacement results in a force that accelerates the particle away from its equilibrium position.

Both maximum and minimum points on the potential-energy curve are positions of equilibrium. At minimum points the equilibrium is stable, and at maximum points it is unstable.

A conservative force always tends to accelerate a particle toward lower potential energy.

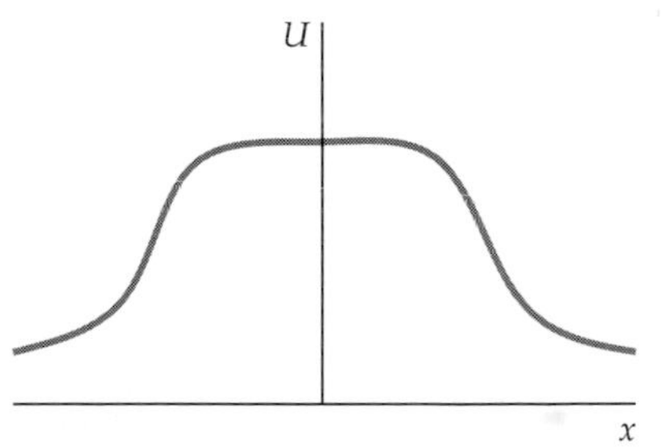

Figure 6-16 Plot of a potential-energy function U for which $x = 0$ is a position of neutral equilibrium. The force $F_x = -dU/dx$ is zero at $x = 0$ and at neighboring points. If a particle is displaced in either direction from $x = 0$, it experiences no force, and so remains in equilibrium.

Figure 6-16 shows a potential-energy curve that is flat in the region near $x = 0$. If a particle is placed at the point $x = 0$, it will experience no force so the particle will be in equilibrium at that point. If the particle is displaced slightly in either direction, the force will again be zero. This is an example of **neutral equilibrium.**

In neutral equilibrium a small displacement results in zero force so the particle is again in equilibrium.

Example 6-9

The force between two atoms in a diatomic molecule can be represented approximately by the potential-energy function

$$U = U_0\left[\left(\frac{a}{x}\right)^{12} - 2\left(\frac{a}{x}\right)^{6}\right] \qquad \text{6-21}$$

where U_0 and a are constants. (*a*) At what value of x is the potential energy zero? (*b*) Find the force F_x. (*c*) At what value of x is the potential energy a minimum? What is the value of this minimum potential energy?

(*a*) Setting $U = 0$ and solving for x, we obtain

$$U_0\left[\left(\frac{a}{x}\right)^{12} - 2\left(\frac{a}{x}\right)^{6}\right] = 0$$

$$\left(\frac{a}{x}\right)^{12} = 2\left(\frac{a}{x}\right)^{6}$$

$$\left(\frac{a}{x}\right)^{6} = 2$$

$$x = \frac{a}{2^{1/6}}$$

(*b*) We find the force from Equation 6-20. We have

$$U = U_0\left[\left(\frac{a}{x}\right)^{12} - 2\left(\frac{a}{x}\right)^{6}\right] = U_0(a^{12}x^{-12} - 2a^6x^{-6})$$

$$F_x = -\frac{dU}{dx} = -U_0[a^{12}(-12x^{-13}) - 2a^6(-6x^{-7})]$$

$$= \frac{12U_0}{a}\left[\left(\frac{a}{x}\right)^{13} - \left(\frac{a}{x}\right)^{7}\right]$$

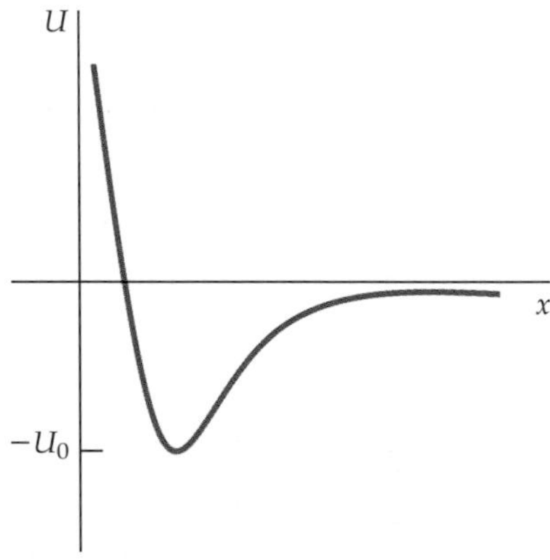

Figure 6-17 Plot of potential energy U versus separation x for a diatomic molecule. The minimum potential energy is $-U_0$, where U_0 is approximately equal to the magnitude of the energy needed to separate the atoms. The total energy of the atoms in a diatomic molecule in its lowest energy state is somewhat greater than $-U_0$.)

(*c*) When the potential energy is minimum, the force is zero. We see from the above equation that $F_x = 0$ when $x = a$. Setting $x = a$ in Equation 6-21, we obtain $U = -U_0$ for the minimum value of the potential energy. Figure 6-17 shows a plot of U versus x. The energy U_0 is approximately equal to the energy needed to separate the atoms, which is called the dissociation energy.

6-6 The Conservation of Mechanical Energy

The negative sign in the definition of the potential-energy function in Equation 6-17 is introduced so that the work done by a conservative force on a particle equals the decrease in the potential energy of the system. Let us consider a system in which work is done on only one of the particles such as a skier–earth system. If a conservative force is the only force that does work on the particle, the work done by the force equals the decrease in the potential energy of the system and also equals the increase in the kinetic energy of the particle (which is the increase in the kinetic energy of the system):

$$W_{\text{total}} = \int \mathbf{F}\cdot d\mathbf{s} = -\Delta U = +\Delta K \qquad 6\text{-}22$$

Hence,

$$\Delta K + \Delta U = \Delta(K + U) = 0 \qquad 6\text{-}23$$

The sum of the kinetic energy and the potential energy of the system is called the **total mechanical energy** E:

$$E = K + U \qquad 6\text{-}24$$

If only conservative forces do work, Equation 6-23 states that the change in the total mechanical energy is zero. Thus, the total mechanical energy remains constant during the motion of the particle.

$$E = K + U = \text{constant} \qquad 6\text{-}25$$

Conservation of mechanical energy

This is known as the **law of conservation of mechanical energy** and is the origin of the expression "conservative force."

We can extend our discussion to systems in which work is done on more than one particle. Figure 6-18 shows a system consisting of two blocks of mass m_1 and m_2 resting on a frictionless table connected by a spring of negligible mass. The spring is stretched by applying a force to each mass as shown in Figure 6-18*b*. The work done by the applied forces to stretch the spring by an amount x is stored as potential energy $\Delta U = \frac{1}{2}kx^2$ in the system.

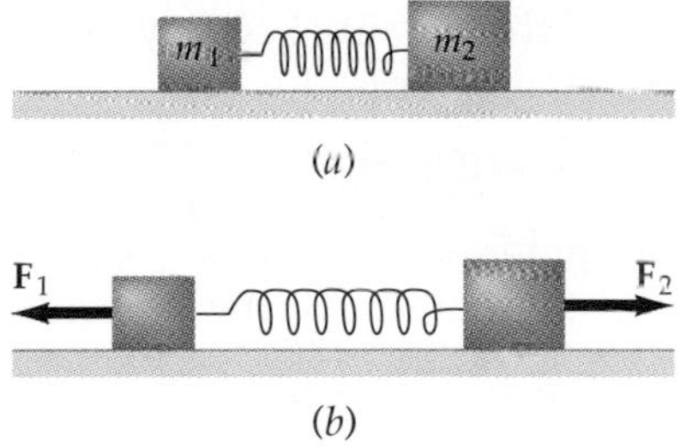

Figure 6-18 (*a*) A system consisting of two blocks connected by a spring. (*b*) The work done by the forces applied to stretch the spring equals the change in the potential energy of the system.

When the blocks are released, the only force that does work on each block is the conservative force exerted by the spring. The work done on each block equals the change in the kinetic energy of that block. The total work done by the spring therefore equals the change in the total kinetic energy of the two blocks. But this total work also equals the (negative) change in the potential energy of the system. Thus, Equations 6-22 through 6-25 also apply to this system, where K is the total kinetic energy of the system, U is the total potential energy of the system, and $E = K + U$ is the total mechanical energy of the system.

Applications of the Conservation of Mechanical Energy

Consider a problem in which a single object of mass m moves in one dimension under the influence of a conservative force. Writing $\frac{1}{2}mv^2$ for the kinetic energy and $U(x)$ for the potential energy in Equation 6-25, we obtain

$$\tfrac{1}{2}mv^2 + U(x) = E \qquad \text{6-26}a$$

If the total (constant) energy E is known, Equation 6-26a allows the determination of the speed v as a function of x. We have

$$v = \sqrt{\frac{2[E - U(x)]}{m}} \qquad \text{6-26}b$$

In problems in which we are not concerned with the time variable t, Equation 6-26 is often much easier to use than is Newton's second law. For example, consider the simple problem of a skier skiing down a hill starting from rest at an initial height h above the bottom of a hill. If we choose the potential energy of the skier–earth system to be zero at the bottom of the hill, the original potential energy is mgh. This equals the total energy because the initial kinetic energy is zero. At any other height y above the bottom of the hill, the potential energy is mgy. The speed v at any height y is then given by the conservation of mechanical energy:

$$\tfrac{1}{2}mv^2 + mgy = E = mgh$$
$$v = \sqrt{2g(h - y)}$$

The waterfall in this 1961 lithograph by the Swiss artist M. C. Escher violates the law of conservation of energy. As the water falls, part of its potential energy is converted into the kinetic energy of the waterwheel. How does the water get back to the top of the waterfall?

Example 6-10

A pendulum consists of a bob of mass m attached to a string of length L. It is pulled aside so that the string makes an angle θ_0 with the vertical and is released from rest. Find the speed v at the bottom of the swing and the tension in the string at that time.

The two forces acting on the bob (neglecting air resistance) are the force of gravity $m\mathbf{g}$, which is conservative, and the tension $\mathbf{T}$, which is perpendicular to the motion and therefore does no work. The mechanical energy of the bob–earth system is therefore conserved in this problem.

Let us choose the gravitational potential energy to be zero at the bottom of the swing. Initially, the bob is at a height h above the bottom and at rest. Its kinetic energy is zero and the potential energy of the system is mgh. The initial total energy of the system is thus

$$E_i = K_i + U_i = 0 + mgh$$

As the pendulum swings down, potential energy is converted into kinetic energy. The final energy at the bottom of the swing is therefore

$$E_f = K_f + U_f = \tfrac{1}{2}mv^2 + 0 = \tfrac{1}{2}mv^2$$

Conservation of energy then gives

$$E_f = E_i$$
$$\tfrac{1}{2}mv^2 = mgh$$

To find the speed in terms of the initial angle θ_0, we must relate h to θ_0. According to Figure 6-19, the distance h is related to θ_0 and the length of the pendulum L by

$$h = L - L\cos\theta_0 = L(1 - \cos\theta_0)$$

Thus, the speed at the bottom is found from

$$v^2 = 2gh = 2gL(1 - \cos\theta_0)$$

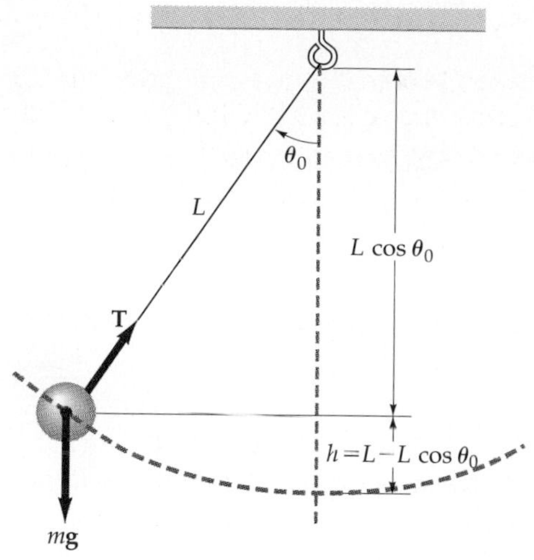

Figure 6-19 Simple pendulum for Example 6-10. The tension is perpendicular to the motion and does no work. The speed of the bob at the bottom is found from the conservation of energy $\frac{1}{2}mv^2 = mgh$, where the initial height h above the bottom is related to the initial angle θ_0 by $h = L - L \cos \theta_0$.

To find the tension at the bottom, we use Newton's second law. (One of the things that makes physics challenging is that we must apply everything we have learned previously as well as the material studied in this chapter.) At the bottom of the swing, the forces on the bob are the weight $m\mathbf{g}$ acting down and the tension $\mathbf{T}$ acting up. Since the bob is moving in a circle of radius L with a speed v, it has a centripetal acceleration v^2/L. This acceleration is directed toward the center of the circle, which is upward at this point. Newton's second law thus gives

$$T - mg = \frac{mv^2}{L} = 2mg(1 - \cos \theta_0)$$

$$T = mg + 2mg(1 - \cos \theta_0)$$

Note that if the bob is released from $\theta_0 = 90°$, the tension at the bottom is 3 times the force of gravity mg.

The speed of the bob at the bottom could also be found using Newton's laws, but the solution is difficult and requires calculus because the acceleration tangential to the curve varies with the angle θ and therefore with time, so the constant-acceleration formulas do not apply. (See Problem 88.)

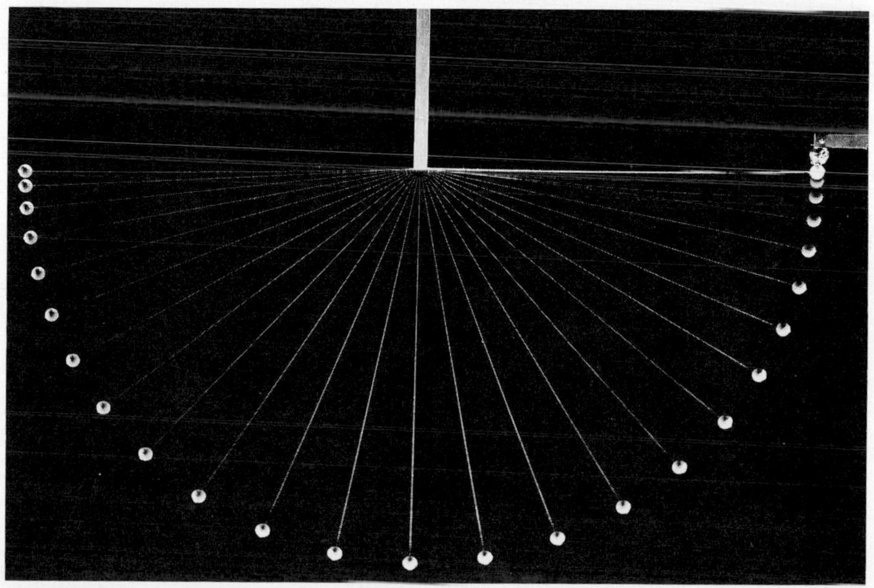

Multiflash photograph of a simple pendulum. As the bob descends, gravitational potential energy is converted into kinetic energy, and the speed increases as indicated by the increased spacing of the recorded positions. The speed decreases as the bob moves up, and the kinetic energy is changed into potential energy.

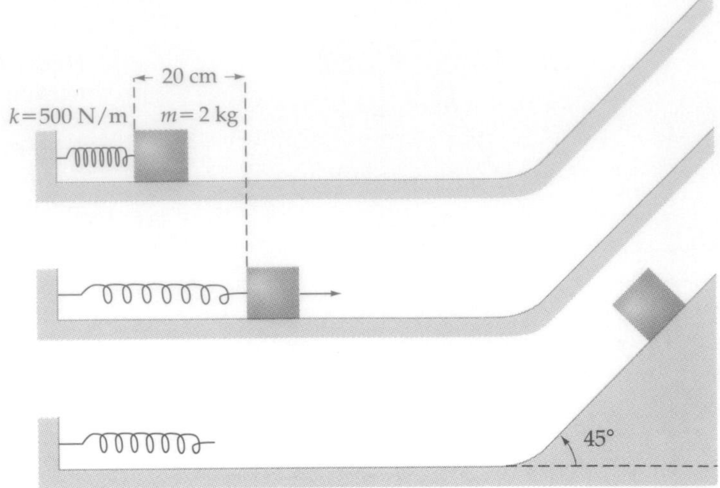

Figure 6-20 Block pushed against a spring for Example 6-11. In this example, the initial potential energy in the spring is converted first into kinetic energy and then into gravitational potential energy.

Example 6-11

A 2-kg block is pushed against a spring that has a force constant of 500 N/m, compressing it 20 cm. The block is then released, and the spring projects it along a frictionless horizontal surface and then up a frictionless incline of angle 45° as shown in Figure 6-20. How far up the incline does the block travel?

After the spring is released, the only forces that do work are the force exerted by the spring and the force of gravity. Since both these forces are conservative, the total mechanical energy of the block–spring–earth system is conserved. In this case, the total mechanical energy consists of kinetic energy of the block $\frac{1}{2}mv^2$, the potential energy of the spring $\frac{1}{2}kx^2$, and the gravitational potential energy mgh. We choose $h = 0$ at the horizontal surface. The initial energy is then just the potential energy of the spring:

$$E_i = \tfrac{1}{2}kx^2 = \tfrac{1}{2}(500\text{ N/m})(0.20\text{ m})^2 = 10\text{ J}$$

When the block leaves the spring, its kinetic energy is 10 J, the initial potential energy of the compressed spring. As the block slides up the incline, its speed decreases until it is zero at its maximum height h. At this height, the block is at rest so its energy is entirely gravitational potential energy:

$$E_f = mgh$$

Setting the final energy equal to the initial energy, we obtain

$$mgh = 10\text{ J}$$

$$h = \frac{10\text{ J}}{mg} = \frac{10\text{ J}}{(2\text{ kg})(9.81\text{ N/kg})} = 0.51\text{ m}$$

The distance s the block travels up the incline is found from

$$\frac{h}{s} = \sin 45° = 0.707$$

$$s = \frac{h}{\sin 45°} = \frac{0.51\text{ m}}{0.707} = 0.721\text{ m}$$

Exercise

Find the speed of the block in Example 6-11 after it leaves the spring. (Answer: 3.16 m/s)

Example 6-12

A spring with a force constant of k hangs vertically. A block of mass m is attached to the unstretched spring and allowed to fall from rest. Find the maximum distance the block falls before it begins moving upward.

Again, we have two kinds of potential energy, gravitational potential energy and the potential energy of the stretched spring. The initial and final positions of the block are shown in Figure 6-21. We let y be the distance the block falls, and we choose the gravitational potential energy of the block to be zero at the original position $y = 0$. The initial potential energy of the spring is zero because the spring is unstretched at this position. Since the block is initially at rest, its kinetic energy is initially zero. The total initial energy is thus zero:

$$E_i = K_i + U_i = 0$$

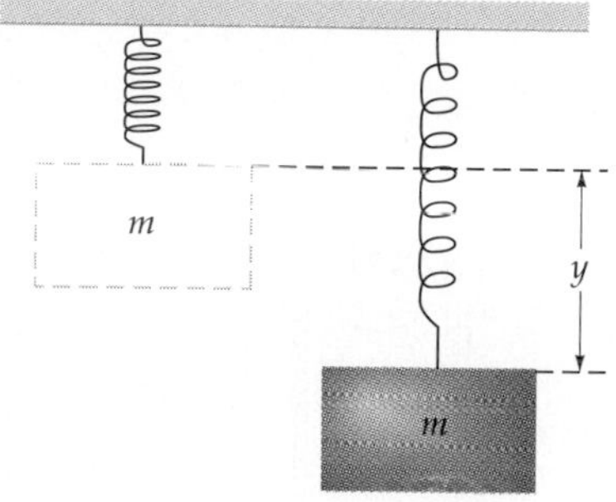

Figure 6-21 Block attached to an unstretched spring and dropped for Example 6-12. Gravitational potential energy is converted into the kinetic energy of the block plus the potential energy of the spring. At the lowest point, where the block is momentarily at rest, the loss in gravitational potential energy of the block equals the gain in potential energy of the spring.

After the block has fallen some distance y, it has kinetic energy $\frac{1}{2}mv^2$. The gravitational potential energy is $-mgy$, and the potential energy of the spring is $\frac{1}{2}ky^2$. The potential energy of the system is then

$$U = -mgy + \tfrac{1}{2}ky^2$$

and the total energy is

$$E = K + U = \tfrac{1}{2}mv^2 - mgy + \tfrac{1}{2}ky^2$$

Since mechanical energy is conserved, the total energy always equals the initial energy, which is zero. Setting $E = E_i = 0$, we obtain

$$\tfrac{1}{2}mv^2 - mgy + \tfrac{1}{2}ky^2 = 0 \qquad 6\text{-}27$$

Equation 6-27 relates the speed v to the distance dropped y. As the block drops, its speed first increases, then reaches some maximum value, and then decreases until it is again zero when the block is at its lowest point, which we will call y_m. We can find y_m by setting $v = 0$ in Equation 6-27:

$$-mgy_m + \tfrac{1}{2}ky_m^2 = 0$$

The two solutions to this equation are $y_m = 0$, which corresponds to the initial position, and

$$y_m = \frac{2mg}{k}$$

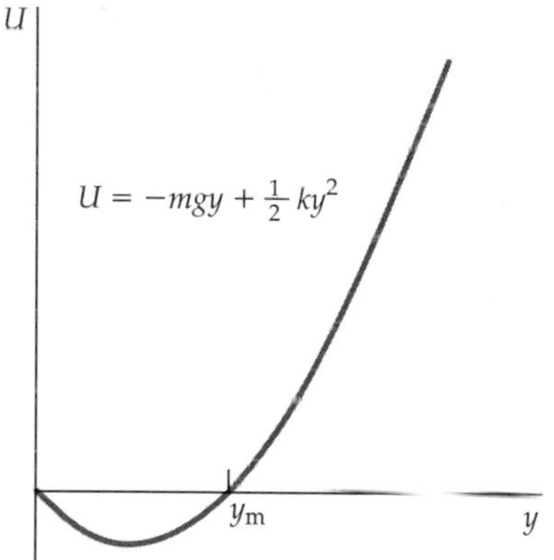

Figure 6-22 Potential-energy function $U = -mgy + \frac{1}{2}ky^2$ versus y for Example 6-12. The initial potential energy is chosen to be zero, and since the block is initially at rest, the total energy is zero. At the minimum in the potential-energy curve, the kinetic energy of the block is maximum. The value of y at this minimum point is $y = mg/k$, which is found by setting $dU/dy = 0$.

Figure 6-22 shows the plot of the potential energy $U = -mgy + \frac{1}{2}ky^2$ versus the displacement y for Example 6-12. The potential energy is initially zero, then decreases to some minimum value, at which time the kinetic energy of the block is maximum, and then increases back to zero, at which time the block is at rest at its lowest point. It is instructive to find the value of y for which the potential energy is minimum. At this point, the slope dU/dy is zero, and the net force acting on the block $F_y = -dU/dy$ is zero. We can find the value of y for which the potential energy is minimum by computing the derivative dU/dy and setting it equal to zero. Computing dU/dy gives

$$\frac{dU}{dy} = \frac{d}{dy}(-mgy + \tfrac{1}{2}ky^2) = -mg + ky$$

The net force acting on the block is thus

$$F_y = -\frac{dU}{dy} = mg - ky$$

This force is zero when $y = mg/k$, which is the value of y for which the upward force exerted by the spring on the block equals the weight of the block. This is the equilibrium position of the block hanging on the spring.

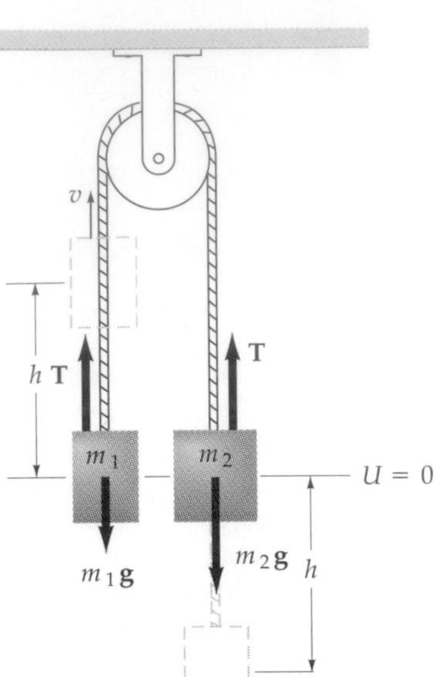

Figure 6-23 Atwood's machine for Example 6-13.

Example 6-13

Two blocks of mass m_1 and m_2 are initially at rest attached to a light string that passes over a frictionless peg as shown in Figure 6-23. Find the speed of either block when the heavier one falls a distance h.

This device, called *Atwood's machine,* was developed in the eighteenth century to measure the acceleration of gravity g. As we will see, if the values of m_1 and m_2 are not too different, the acceleration of either object is a small fraction of g. It could be easily measured with the rather crude timing devices available in the eighteenth century, whereas a direct measurement of g was difficult if not impossible. (In his experiment, Atwood used a pulley rather than a peg. If the mass of the pulley can be neglected, the analysis is the same as for a frictionless peg.)

Since the peg is frictionless, the tension in the string is uniform. Provided that the string does not stretch, the total work done by the tension is zero because when the lighter object moves upward in the direction of the tension, the heavier object moves downward the same distance in the opposite direction. The only force that does work on the two-object system is the force of gravity. Therefore, the total mechanical energy of the system is conserved.

Let us assume that m_2 is greater than m_1 and choose the potential energy of the system to be zero at the initial position of the blocks. Since they are at rest at this position, the total energy is zero. Calling this initial energy E_i, we have

$$E_i = 0$$

Let v be the speed of m_1 after it has moved up a distance h. Then its kinetic energy is $\frac{1}{2}m_1v^2$. Since the connecting string does not stretch, m_2 must move down the same distance h and acquire the same speed v. Its kinetic energy is $\frac{1}{2}m_2v^2$. Since m_1 moves up a distance h and m_2 moves down a distance h, the potential energy of the system is $m_1gh - m_2gh$. The energy at this time E_f is thus

$$E_f = \tfrac{1}{2}m_1v^2 + \tfrac{1}{2}m_2v^2 + m_1gh - m_2gh$$

Setting the final energy equal to the initial energy, we obtain

$$\tfrac{1}{2}m_1v^2 + \tfrac{1}{2}m_2v^2 + m_1gh - m_2gh = 0$$

or

$$\tfrac{1}{2}(m_1 + m_2)v^2 = (m_2 - m_1)gh \qquad 6\text{-}28$$

Note that the left side of Equation 6-28 is the kinetic energy gained by the system and the right side is the net potential energy lost. Solving for v^2 gives

$$v^2 = \frac{2(m_2 - m_1)}{(m_1 + m_2)}gh \qquad 6\text{-}29$$

which relates the speed of either mass to the distance it moves. Comparing Equation 6-29 with the constant-acceleration equation $v^2 = 2ah$, we see that the acceleration is given by

$$a = \frac{(m_2 - m_1)}{(m_1 + m_2)}g$$

In the laboratory experiment, the acceleration a is measured, and the unknown value of g is then obtained from

$$g = \frac{(m_1 + m_2)}{(m_2 - m_1)} a$$

This problem could also be solved by applying Newton's second law to each of the bodies and eliminating the tension T from the two equations obtained.

Exercise
What is the magnitude of the acceleration of either block in Example 6-13 if the masses are $m_1 = 3$ kg and $m_2 = 5$ kg? (Answer: $a = 0.25\ g = 2.45\ \text{m/s}^2$)

We have seen that the law of conservation of mechanical energy as expressed in Equation 6-25 can be used as an alternative to Newton's laws for solving certain problems in mechanics. Since this equation was derived from Newton's laws, any problem that can be solved using the conservation of mechanical energy can also be solved directly from Newton's laws. In spite of this, energy conservation is an extremely useful tool for the analysis of problems.

6-7 The Generalized Work–Energy Theorem

When both conservative and nonconservative forces do work, the total mechanical energy of a system does not remain constant. Let us again consider a system in which work is done on only one of the particles. Consider the particle to be acted on by a nonconservative force $\mathbf{F}_{nc}$ and two conservative forces $\mathbf{F}_1$ and $\mathbf{F}_2$ so that the net force is

$$\mathbf{F}_{net} = \mathbf{F}_{nc} + \mathbf{F}_1 + \mathbf{F}_2$$

(Our discussion can be easily extended to include any number of forces.) According to the work–energy theorem, the total work done by these forces equals the change in the kinetic energy of the particle (and therefore of the system):

$$W_{total} = \int \mathbf{F}_{nc} \cdot d\mathbf{s} + \int \mathbf{F}_1 \cdot d\mathbf{s} + \int \mathbf{F}_2 \cdot d\mathbf{s} = W_{nc} + W_1 + W_2 = \Delta K \qquad 6\text{-}30$$

where W_{nc} is the work done by the nonconservative force, W_1 is that done by force $\mathbf{F}_1$, and W_2 is that done by force $\mathbf{F}_2$. For each conservative force we define a potential-energy function U_i in the usual way (Equation 6-17):

$$W_i = -\Delta U_i$$

Equation 6-30 can then be written

$$W_{nc} - \Delta U_1 - \Delta U_2 = \Delta K$$

or

$$W_{nc} = \Delta U_1 + \Delta U_2 + \Delta K = \Delta E \qquad 6\text{-}31$$

Generalized work–energy theorem

where

$$E = U_1 + U_2 + K$$

is the total mechanical energy of the system. Equation 6-31 is the **generalized work–energy theorem:**

> The work done by a nonconservative force acting on a particle equals the change in the total mechanical energy of the system.

This modified form of the work–energy theorem is the most convenient in many applications since explicit calculation of the work done is required only for the nonconservative forces. When there is no work done by nonconservative forces, the total mechanical energy is conserved, that is, it remains constant.

Work done by kinetic friction

We have derived the generalized work–energy theorem for a particle, but the most common nonconservative force is that of kinetic friction, which is a force exerted by one surface on another. As we have stated, we must be careful when we calculate the work done by a force acting on an extended object because the displacement of the point of application of the force is not in general equal to the displacement of the object. Consider a block on a horizontal table for which the coefficient of kinetic friction is μ_k. Suppose the block has an initial horizontal velocity v_i, and it slides a distance Δx before coming to rest. The initial mechanical energy of the block is its initial kinetic energy

$$K_i = \tfrac{1}{2}mv_i^2$$

The final mechanical energy of the block is zero. The work done *on the block* by the force of friction is not $-f\,\Delta x$, where f is the force of friction, because the displacement of the point of application of the force is not the same as the displacement of the block Δx. However, the total mechanical energy change *is* equal to $-f\,\Delta x$. We can see this by applying Newton's second law to the block. (Since the displacement of all parts of the *block* is the same, we may treat the block as a particle for the application of Newton's second law.) We have

$$F_{\text{net}} = ma$$

$$-f = ma$$

Multiplying both sides by Δx, we obtain

$$-f\,\Delta x = ma\,\Delta x = m(\tfrac{1}{2}v_f^2 - \tfrac{1}{2}v_i^2) = -\tfrac{1}{2}mv_i^2$$

where we have used the constant-acceleration formula $2a\,\Delta x = v_f^2 - v_i^2$, and $v_f = 0$. It can be shown that $-f\,\Delta x$ does equal the work done by kinetic friction on the block *plus* the work done by kinetic friction on the table.* We may thus use the generalized work–energy equation in problems involving kinetic friction if we use $-f\,\Delta x$ for the work done by the nonconservative force and recognize that this is the total work done on the two objects that are sliding relative to each other.

Example 6-14

A 5-kg sled is sliding with an initial speed of 4 m/s. If the coefficient of friction between the sled and the snow is 0.14, how far will the sled slide before coming to rest?

The initial mechanical energy of the system is the original kinetic energy of the sled:

$$E_i = K_i = \tfrac{1}{2}mv^2 = \tfrac{1}{2}(5\text{ kg})(4\text{ m/s})^2 = 40\text{ J}$$

*The work done by kinetic friction is examined in detail in "Work and Heat Transfer in the Presence of Sliding Friction" by B. A. Shewood and W. H. Bernard, *American Journal of Physics,* **52,** 1001 (1984)

The final mechanical energy is zero. The change in the mechanical energy of the sled–snow system is thus

$$\Delta E = E_f - E_i = -E_i = -40 \text{ J}$$

The only force that does work is the nonconservative force of kinetic friction. The normal force exerted by the snow on the sled equals the weight of the sled,

$$F_n = mg = (5 \text{ kg})(9.81 \text{ N/kg}) = 49.0 \text{ N}$$

The magnitude of the frictional force on the sled is then

$$f = \mu_k F_n = (0.14)(49.0 \text{ N}) = 6.86 \text{ N}$$

When the sled moves a distance Δx, the total work done by friction on the sled and on the snow is

$$W_{nc} = f_x \, \Delta x = (-6.86 \text{ N}) \, \Delta x$$

Setting the total work done by the nonconservative force equal to the change in the mechanical energy of the system we obtain

$$W_{nc} = \Delta E = -40 \text{ J}$$

$$(-6.86 \text{ N}) \, \Delta x = -40 \text{ J}$$

$$\Delta x = \frac{-40 \text{ J}}{-6.86 \text{ N}} = 5.83 \text{ m}$$

Example 6-15

A horizontal force of 25 N is applied to a 4-kg box, which is initially at rest on a rough horizontal table. The coefficient of kinetic friction between the box and table is 0.35. Find the speed of the box after it has been pushed 3 m.

In this problem, we have two nonconservative forces acting—the applied force and the force of kinetic friction. The applied force does work on the box given by

$$W_{app} = F_{app} \, \Delta x = (25 \text{ N})(3 \text{ m}) = 75 \text{ J}$$

Since the normal force exerted by the table on the box equals the weight of the box, the force of kinetic friction exerted on the box is

$$f = \mu_k F_n = \mu_k mg = (0.35)(4 \text{ kg})(9.81 \text{ N/kg}) = 13.7 \text{ N}$$

The total work done by the force of kinetic friction on the box and on the table is given by

$$W_f = -f \, \Delta x = -(13.7 \text{ N})(3 \text{ m}) = -41.1 \text{ J}$$

The total work done by nonconservative forces is thus

$$W_{nc} = W_{app} + W_f = 75 \text{ J} - 41.1 \text{ J} = 33.9 \text{ J}$$

If we take the potential energy of the box to be zero on the table, the total mechanical energy of this system is just the kinetic energy of the box. The generalized work–energy theorem then gives

$$W_{nc} = \Delta E = \Delta K = \tfrac{1}{2}mv^2 = 33.9 \text{ J}$$

$$v = \sqrt{\frac{2(33.9 \text{ J})}{m}} = \sqrt{\frac{2(33.9 \text{ J})}{4 \text{ kg}}} = 4.12 \text{ m/s}$$

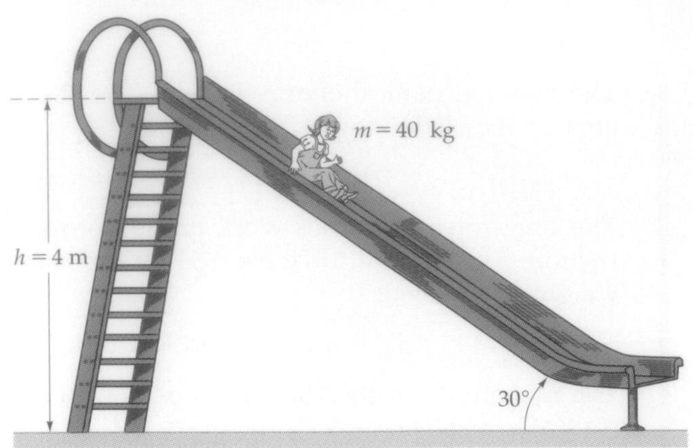

Figure 6-24 Child on a slide for Example 6-16. In this example, mechanical energy is not conserved because of friction.

Example 6-16

A child of mass 40 kg slides down a rough slide inclined at 30° (Figure 6-24). The coefficient of kinetic friction between the child and the slide is $\mu_k = 0.2$. If the child starts from rest at the top of the slide, a height 4 m above the bottom, how fast is she traveling when she reaches the bottom?

The forces acting on the child are gravity, the normal force F_n, and the frictional force $f = \mu_k F_n$ exerted by the slide. The normal force exerted by the slide is $F_n = mg \cos 30°$. Since the normal force is perpendicular to the motion, it does no work. The work done by the nonconservative force of friction is the negative of the force f times the distance s traveled. Since the original height h is 4 m and $\sin 30° = h/s = 0.5$, the distance s is 8 m. The work done by friction is thus

$$W_f = -fs = -(\mu_k mg \cos 30°)s$$
$$= -(0.2)(40 \text{ kg})(9.81 \text{ N/kg})(0.866)(8 \text{ m}) = -544 \text{ J}$$

According to the work–energy theorem, the −544 J of work done by friction equals the change in the total energy of the child–slide–earth system. Since the kinetic energy K_i of the child at the top of the slide is zero, the original energy of the system is its potential energy U_i. If we choose the potential energy to be zero at the bottom of the slide, the original energy is

$$E_i = U_i = mgh = (40 \text{ kg})(9.81 \text{ N/kg})(4 \text{ m}) = 1570 \text{ J}$$

The work–energy theorem then gives

$$W_{nc} = \Delta E = E_f - E_i = E_f - 1570 \text{ J} = -544 \text{ J}$$

The final energy is therefore

$$E_f = 1570 \text{ J} - 544 \text{ J} = 1026 \text{ J}$$

When the child reaches the bottom of the slide the final energy is the kinetic energy of the child $\frac{1}{2}mv^2$. Then

$$E_f = K_f = \tfrac{1}{2}mv^2 = 1026 \text{ J}$$

Solving for v, we obtain

$$v = \sqrt{\frac{2K_f}{m}} = \sqrt{\frac{2(1026 \text{ J})}{40 \text{ kg}}} = 7.16 \text{ m/s}$$

The total mechanical energy of the child–slide–earth system is thus decreased from its original value of 1570 J to its final value of 1570 J − 544 J = 1026 J, which equals the kinetic energy of the child at the bottom, $\frac{1}{2}mv^2$.

Example 6-17

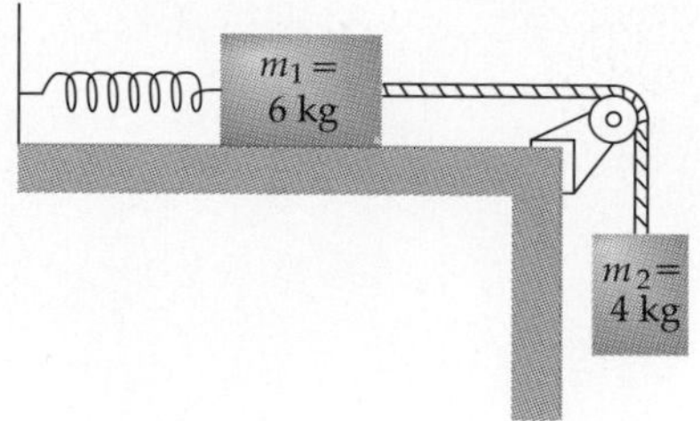

Figure 6-25 Two blocks connected by a light string for Example 6-17. The 6-kg block is pushed against the spring and released. As the two blocks move, the potential energy of the spring decreases, the gravitational potential energy of the 4-kg block decreases, and mechanical energy is lost because of friction.

A 4-kg block hangs by a light string that passes over a smooth peg and is connected to a 6-kg block that rests on a rough table (Figure 6-25). The coefficient of kinetic friction is $\mu_k = 0.2$. The 6-kg block is pushed against a spring, which has a force constant of 600 N/m, compressing it 30 cm. It is then released. Find the speed of the blocks when the 4-kg block has fallen a distance of 40 cm.

We consider the system as consisting of the earth, the table, the spring, and the two blocks $m_1 = 6$ kg and $m_2 = 4$ kg, and we choose the initial gravitational potential energy to be zero. The initial energy is then just the potential energy of the spring:

$$E_i = \tfrac{1}{2}kx^2 = \tfrac{1}{2}(600\ \text{N/m})(0.30\ \text{m})^2 = 27\ \text{J}$$

The work done on this system by friction as the 6-kg block moves a distance $x_1 = 0.40$ m is

$$\begin{aligned} W_{nc} &= -\mu_k m_1 g x_1 \\ &= -(0.2)(6\ \text{kg})(9.81\ \text{N/kg})(0.40\ \text{m}) = -4.7\ \text{J} \end{aligned}$$

The work–energy theorem then gives

$$W_{nc} = E_f - E_i = -4.7\ \text{J}$$

$$E_f = E_i - 4.7\ \text{J} = 27\ \text{J} - 4.7\ \text{J} = 22.3\ \text{J}$$

After the 4-kg block has fallen a distance y, the two blocks are moving with the same speed v, and the total energy of the system is

$$E_f = \tfrac{1}{2}(m_1 + m_2)v^2 - m_2 g y$$

For $y = 40$ cm $= 0.4$ m, we have

$$\begin{aligned} E_f &= \tfrac{1}{2}(6\ \text{kg} + 4\ \text{kg})v^2 - (4\ \text{kg})(9.81\ \text{N/kg})(0.4\ \text{m}) \\ &= (5\ \text{kg})v^2 - 15.7\ \text{J} \end{aligned}$$

Setting this equal to 22.3 J and solving for v, we obtain

$$\begin{aligned} E_f &= (5\ \text{kg})v^2 - 15.7\ \text{J} \\ &= 22.3\ \text{J} \\ v &= 2.76\ \text{m/s} \end{aligned}$$

Like the law of conservation of mechanical energy, the work–energy theorem of Equation 6-31 can be used as an alternative to Newton's laws for solving certain problems in mechanics. Since Equation 6-31 was derived from Newton's laws, any problem that can be solved using the work–energy theorem can also be solved directly from Newton's laws. Like energy conservation, the work–energy theorem is a useful tool for the analysis of many problems.

Questions

10. If a pulley with nonnegligible mass is used rather than a frictionless peg in Atwood's machine, the pulley has kinetic energy of rotation (which will be studied in Chapter 8). Would you expect the inclusion of this kinetic energy to increase or decrease the speed v for a given distance h?

11. What are the advantages and disadvantages of solving mechanics problems using energy methods compared with using Newton's laws?

6-8 The Conservation of Energy

In the macroscopic world, nonconservative forces are always present to some extent, the most common being frictional forces. Another type of nonconservative force is that involved in the deformations of objects. For example, if a spring is stretched beyond its elastic limit, it becomes permanently deformed, and the work done in stretching the spring is not recovered when the spring is released. Some of the work done in deforming the spring is dissipated into thermal energy, so the spring becomes warmer. You may have observed that when you bend a coat hanger back and forth until it breaks, the coat hanger becomes warm.

Since mechanical energy is often not conserved, the importance of energy was not realized until the nineteenth century, when it was discovered that the disappearance of macroscopic mechanical energy is accompanied by the appearance of thermal energy, usually indicated by an increase in temperature. We now know that, on the microscopic scale, this thermal energy consists of the kinetic and potential energies of the molecules in the system.

When we carefully define a system—for example, several objects and their local surroundings—we find that even when thermal energy and other types of energy, such as chemical energy, are included, the total energy of the system does not always remain constant. The energy of a system often decreases because of some form of radiation, such as sound waves from a collision between two objects, water waves produced by a ship, or electromagnetic waves produced by accelerated charges in a radio antenna. The energy of a system can also increase through the absorption of radiant energy. For example, the earth absorbs radiant energy from the sun. However, *the increase or decrease in the energy of a system can always be accounted for by the appearance or disappearance of some kind of energy somewhere else.* This experimental result is known as the **law of conservation of energy.** It is one of the most important laws in all science. Let E_{sys} be the total energy of a given system, E_{in} be the energy put into the system, and E_{out} be the energy that leaves the system. The law of conservation of energy then states:

Law of conservation of energy

$$E_{in} - E_{out} = \Delta E_{sys} \qquad 6\text{-}32$$

A common method of putting energy into a system is to do work on it. When work is done on a system, energy is transferred to the system from the agent or system doing the work. Energy put into a system may appear as an increase in the mechanical energy of the system, such as when a nonconservative force acts on a single particle within the system (the work–energy theorem), or it may appear as internal energy of the system. Energy can also be transferred into or out of a system in the form of heat. We will discuss internal energy and heat in more detail in Chapter 16 when we study the first law of thermodynamics, which is essentially a statement of the law of conservation of energy.

Example 6-18

A ball of putty of mass m is released from rest from a height h above the floor. It falls to the floor. (Plop!) Find the work done on the ball and discuss the application of the law of conservation of energy to this situation.

The two forces acting on the ball are the force of gravity and the contact force exerted by the floor. The work done by gravity is $+mgh$.

Geothermal energy is released dramatically in the eruption of Cerro Negro in Nicaragua.

Niagara Falls. The potential energy of the water at the top of the falls is used to produce electrical energy.

The work done by the floor is zero because the point of application of the force does not move. Thus the total work done on the ball is mgh. Energy is transferred to the ball by the work done on it by gravity. This energy appears as kinetic energy of the ball before it hits the floor and as thermal energy within the ball after. The ball warms slightly and the energy is eventually transferred to the ball's surroundings as heat.

If we consider a system consisting of the ball and earth, there is no work done on the system because the work done by gravity is internal to the system. The original potential energy of the system is converted into internal thermal energy of the ball. Note that we *cannot* treat the ball of putty as a particle and apply the work–energy theorem. Although the only work done is by a conservative force, the total mechanical energy of the ball is not conserved.

There are situations in which some of a system's internal energy is converted into mechanical energy with no work being done by any outside agent. Consider a car that starts from rest and accelerates along a horizontal road such that the wheels do not slip. The net force on the car that accelerates it is the force of static friction exerted by the road. This force does no work. According to our definition of work, the point of application of the force must move through a distance. Since the road and the tire are always instantaneously at rest, no energy is transferred to the car by the road. The increase in the kinetic energy of the car comes from the chemical energy in the fuel that is being burned in the car engine. According to the work–energy theorem, the total energy of the car does not change. Its mechanical kinetic energy increases, so its internal chemical energy must decrease by

(*a*)

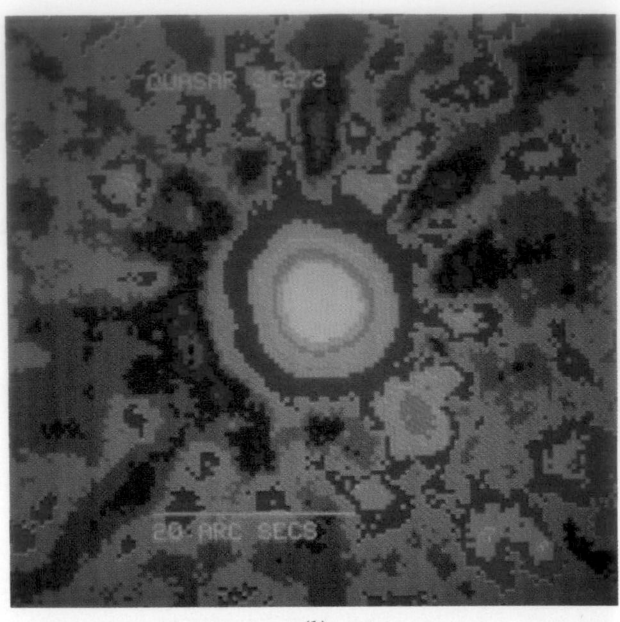

(*b*)

(*a*) This entire pizza loaded with meat and cheese contains only about 16 megajoules of energy. (*b*) The Quasar 3C 273 is shown imaged via x-ray energy. The x-ray energy emitted by this quasar is more than a million times that emitted by the entire Milky Way.

the same amount. Walking provides a similar example. To walk forward, you push back on the floor and the floor pushes forward on you with the force of static friction. This force accelerates you forward, but it does no work. There is no displacement of the point of application of the force and no energy is transferred from the floor to your body. The kinetic energy of your body comes from the conversion of chemical energy in your body derived from the food you eat.

Example 6-19

A man of mass m walks with a small constant speed up a flight of stairs to a height h. Find the work done on the man, and discuss the application of energy conservation to this situation.

There are two forces exerted on the man, the force of gravity exerted by the earth, and the contact force of the stairs on the man's feet. The work done by gravity is $-mgh$. This work is negative because the force is in the opposite direction of the displacement. The work done by the stairs is zero because the point of application of this force does not undergo any displacement. Thus the total work done on the man is $-mgh$. Since this is less than zero, we conclude from the law of conservation of energy that (neglecting any heat losses) the internal energy of the man decreases by mgh. (Actually, heat losses are not negligible for this problem. Because the body is relatively inefficient, the amount of chemical energy converted in the man's body will be considerably greater than mgh. The excess energy is eventually transferred from the man to his surroundings as heat.)

If we consider a system consisting of the man and the earth, there is no work done on the system because the work done by gravity is internal to this system. The mechanical energy of this system is increased by the increase in potential energy mgh. This increase in mechanical potential energy comes from a decrease in the internal chemical energy of the man. Again we note that we cannot treat the man as a particle. In this case the total mechanical energy of the man–earth system is increased even though there is no work done by any nonconservative force acting on the system.

6-9 Power

Power is the rate at which energy is transferred from one system to another. Consider a particle with instantaneous velocity $\mathbf{v}$. In a short time interval dt, the particle has a displacement $d\mathbf{s} = \mathbf{v}\, dt$. The work done by a force $\mathbf{F}$ acting on the particle during this time interval is

$$dW = \mathbf{F}\cdot d\mathbf{s} = \mathbf{F}\cdot\mathbf{v}\, dt$$

The rate at which the force does work is the power input P of the force:

$$P = \frac{dW}{dt} = \mathbf{F}\cdot\mathbf{v} \qquad 6\text{-}33$$

Power defined

The SI unit of power, one joule per second, is called a **watt** (W):

$$1\ \text{J/s} = 1\ \text{W} \qquad 6\text{-}34$$

Power should not be confused with work or energy. A car is said to be powerful if it can convert the chemical energy in its fuel into kinetic energy (or potential energy if the car is climbing a hill) in a short period of time. You can increase the chemical energy of a car by increasing the amount of fuel it carries. To increase its power, you must increase the *rate* at which the fuel is burned by increasing the number or size of the cylinders. When you pay your electric or gas bill to the power company you are paying for energy, not power. You are usually charged by the kilowatt-hour (kW·h). A **kilowatt-hour** of energy is

$$1\ \text{kW}\cdot\text{h} = (10^3\ \text{W}) \times (3600\ \text{s}) = 3.6 \times 10^6\ \text{W}\cdot\text{s} = 3.6\ \text{MJ} \qquad 6\text{-}35$$

In the U.S. customary system, the unit of energy is the foot-pound and the unit of power is the foot-pound per second. A common multiple of this unit is called a **horsepower** (hp):

$$1\ \text{hp} = 550\ \text{ft}\cdot\text{lb/s} = 746\ \text{W}$$

In this power plant in Kansas, energy stored in the fossil fuel coal (black area at the right) is released by burning the coal to produce steam; the steam is then used to drive turbines to produce electricity. The excess heat is dissipated by the cooling towers.

Example 6-20

A small motor is used to power a lift that raises a load of bricks weighing 800 N to a height of 10 m in 20 s. What is the minimum power the motor must provide?

If the bricks are lifted without acceleration, the upward force equals the force of gravity, which is 800 N. The speed of the bricks is (10 m)/(20 s) = 0.5 m/s. Since the applied force is in the direction of motion, the power input of the force is

$$P = Fv = (800\ \text{N})(0.5\ \text{m/s}) = 400\ \text{N·m/s} = 400\ \text{J/s} = 400\ \text{W}$$

If there are no mechanical energy losses—to frictional forces, for example—the motor must have a power output of 400 W, which is a little more than $\frac{1}{2}$ horsepower.

Example 6-21

A 1000-kg car travels at a constant speed of 100 km/h = 28 m/s = 62 mi/h up a 10 percent grade. (A 10 percent grade means that the road is inclined at an angle θ given by $\tan\theta = 0.1$; that is, for each 10 m traveled horizontally, the gain in elevation is 1 m.) The total frictional force (rolling friction plus air drag) acting on the car is 700 N. Neglecting internal energy losses, what is the minimum power that must be delivered by the car's engine?

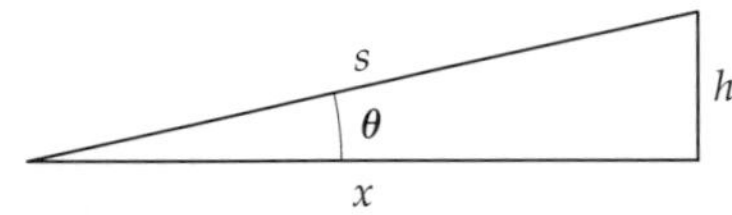

$\tan\theta = h/x \sim \sin\theta = h/s$

Figure 6-26 Relation between height climbed and distance traveled for car driving up a 10 percent grade. Here $\tan\theta = 0.10$ is given, and $\sin\theta \approx \tan\theta$ because the angle is small.

Some of the power delivered by the car's engine goes into increasing the potential energy of the car as it climbs the hill, and some goes into work against friction. From Figure 6-26 we can see that when the car travels a distance s along the road, it climbs a height h, which is related to s by $h = s\sin\theta \approx s\tan\theta = s/10$. We can use the approximation $\tan\theta \approx \sin\theta$ because the angle is small. The potential energy of the car is thus

$$U = mgh = (0.1)mgs$$

The rate of change of the potential energy is

$$\frac{dU}{dt} = mg\frac{dh}{dt} = (0.1)mg\frac{ds}{dt} = (0.1)mgv$$

where $v = ds/dt$ is the speed of the car. Substituting the numerical values for m and g, we obtain

$$\frac{dU}{dt} = (0.1)(1000\ \text{kg})(9.81\ \text{N/kg})(28\ \text{m/s}) = 27.5\ \text{kW}$$

The power input of the force of friction is

$$P_{\text{f}} = \mathbf{f}\cdot\mathbf{v} = -fv = -(700\ \text{N})(28\ \text{m/s}) = -19.6\ \text{kW}$$

The total power that the engine must deliver is thus 27.5 kW + 19.6 kW = 47.1 kW = 63.1 hp. Actually, the power needed is considerably greater than this because a car is only about 15 percent efficient; that is, only about 15 percent of the power produced by the engine is delivered to power the car. The other 85 percent is lost to the cooling system, heat exhaust, and internal friction.

Summary

1. The work done by a constant force is the product of the component of the force in the direction of motion and the displacement of the point of application of the force:

$$W = F \cos \theta \, \Delta x = F_x \, \Delta x$$

2. The work done by a variable force equals the area under the force-versus-distance curve:

$$W = \int_{x_1}^{x_2} F_x \, dx = \text{area under the } F_x\text{-versus-}x \text{ curve}$$

3. Kinetic energy is energy associated with the motion of an object and is related to its mass and speed by

$$K = \tfrac{1}{2}mv^2$$

4. The total work done on a particle equals the change in the kinetic energy of the particle. This is the work–energy theorem:

$$W_{\text{total}} = \Delta K = \tfrac{1}{2}mv_f^2 - \tfrac{1}{2}mv_i^2$$

5. The SI unit of work and energy is the joule (J):

$$1 \text{ J} = 1 \text{ N}\cdot\text{m}$$

6. The dot product or scalar product of two vectors is defined by

$$\mathbf{A}\cdot\mathbf{B} = AB \cos \phi$$

where ϕ is the angle between the vectors. In terms of the components of the vectors, the dot product is

$$\mathbf{A}\cdot\mathbf{B} = A_xB_x + A_yB_y + A_zB_z$$

The work done on a particle by a force **F** for a small displacement $d\mathbf{s}$ is written

$$dW = \mathbf{F}\cdot d\mathbf{s}$$

and the work done on a particle that moves from point 1 to point 2 is

$$W = \int_{s_1}^{s_2} \mathbf{F}\cdot d\mathbf{s}$$

7. A force is conservative if the total work it does on a particle is zero when the particle moves along any closed path that returns it to its initial position. The work done by a conservative force on a particle is independent of how the particle moves from one point to another.

8. The potential energy of a system is the energy associated with the configuration of the system. The change in the potential energy of a system is defined as the negative of the work done by a conservative force acting on the system:

$$dU = -\mathbf{F}\cdot d\mathbf{s}$$

$$\Delta U = U_2 - U_1 = -W = -\int_{s_1}^{s_2} \mathbf{F}\cdot d\mathbf{s}$$

The work done on a system by a conservative force equals the decrease in the potential energy of the system. The absolute value of potential energy is unimportant. Only changes in potential energy are important.

9. The gravitational potential energy of an object of mass m at a height y above some reference point is

$$U = mgy$$

The potential energy of a spring with a force constant k when it is stretched or compressed a distance x from equilibrium is given by

$$U = \tfrac{1}{2}kx^2$$

10. In one dimension, a conservative force equals the negative derivative of the potential-energy function associated with it:

$$F_x = -\frac{dU}{dx}$$

At a minimum on the curve of the potential-energy function versus the displacement, the force is zero and the system is in stable equilibrium. At a maximum, the force is zero and the system is in unstable equilibrium. A conservative force always tends to accelerate a particle toward a position of lower potential energy.

11. If only conservative forces do work on an object, the sum of the kinetic and potential energies of the object remains constant:

$$E = K + U = \tfrac{1}{2}mv^2 + U = \text{constant}$$

This is the law of conservation of mechanical energy.

12. The work done by a nonconservative force acting on a particle equals the change in the total mechanical energy of the system:

$$W_{\text{nc}} = \Delta(U + K) = \Delta E$$

This is the generalized work–energy theorem.

The conservation of mechanical energy and the generalized work–energy theorem can be used as an alternative to Newton's laws to solve mechanics problems that require the determination of the speed of a particle as a function of its position.

13. The total energy of a system can include other types of energy such as thermal energy or internal chemical energy as well as mechanical energy. The energy of a system can be changed by various means such as emission or absorption of radiation, work done on the system, or heat transfer. The increase or decrease in the energy of the system can always be accounted for by the appearance or disappearance of some kind of energy somewhere else, an experimental result known as the law of conservation of energy:

$$E_{\text{in}} - E_{\text{out}} = \Delta E_{\text{sys}}$$

14. Power is the rate at which energy is transferred from one system to another. If a force **F** acts on a particle that moves with velocity **v** the power input of the force is

$$P = \frac{dW}{dt} = \mathbf{F}\cdot\mathbf{v}$$

The SI unit of power is the watt (W), which equals one joule per second. A common energy unit is the kilowatt-hour, which equals 3.6 megajoules.

Suggestions for Further Reading

"Energy", Chapter 7 of *World Resources 1988–89: A Report by the World Resources Institute and the International Institute for Environment and Development in Collaboration with the United Nations Environment Programme,* Basic Books, Inc., New York, 1988.

Charts, tables, and text present trends in energy use and conservation around the world. This book also contains similar information on population, food, forests, wildlife, the atmosphere, and the oceans.

Dyson, Freeman J.: "Energy in the Universe," *Scientific American,* September 1971, p. 50.

This article introduces a Scientific American *special issue on energy by examining the different forms in which energy manifests itself and how it is converted from one form to another.*

Gosz, James R., Richard T. Holms, Gene E. Likens, and F. Herbert Bermann: "The Flow of Energy in a Forest Ecosystem," *Scientific American,* March 1978, p. 92.

The various ways in which energy from the sun is transferred among living things in a forest and how efficiently they use it are examined.

Heinrich, Bernd: "The Energetics of the Bumblebee," *Scientific American,* April 1973, p. 96.

Kingsolver, Joel G.: "Butterfly Engineering," *Scientific American,* August 1985, p. 106.

These two articles describe how bumblebees and butterflies metabolize nectar and absorb thermal energy from the sun to achieve body temperatures high enough to enable them to fly.

Walker, Jearl: "The Amateur Scientist: Fly Casting Illuminates the Physics of Fishing," *Scientific American,* July 1985, p. 122.

The conservation of kinetic energy helps to explain how it is possible to cast a fly so far; a consideration of drag makes the model more realistic; and the forces involved in fighting a fish once it is on the line are analyzed.

Review

A. Objectives: After studying this chapter, you should:

1. Know the definitions of work, kinetic energy, potential energy, and power.

2. Be able to distinguish between conservative and nonconservative forces, and know the criterion for a force to be conservative.

3. Be able to find the potential-energy function from a given conservative force as a function of position.

4. Be able to state the law of conservation of mechanical energy and use it in solving problems.

5. Be able to state the work–energy theorem and use it in solving problems.

B. Define, explain, or otherwise identify:

Work
Energy
Joule
Foot-pound
Kinetic energy
Work–energy theorem
Scalar product
Dot product
Potential energy
Conservative force
Gravitational potential energy
Potential energy of a spring
Stable equilibrium
Unstable equilibrium
Total mechanical energy
Law of conservation of mechanical energy
Generalized work–energy theorem
Law of conservation of energy
Power
Watt
Kilowatt-hour
Horsepower

C. True or false: If the statement is true, explain why it is true. If it is false, give a counterexample.

1. Only the net force acting on an object can do work.

2. No work is done on a particle that remains at rest.

3. Work is the area under the force-versus-time curve.

4. A force that is always perpendicular to the velocity of a particle does no work on the particle.

5. A kilowatt-hour is a unit of power.

6. Only conservative forces can do work.

7. If only conservative forces act, the kinetic energy of a particle does not change.

8. The work done by a conservative force decreases the potential energy associated with that force.

Problems

Level I

6-1 Work and Kinetic Energy: Motion in One Dimension with Constant Forces

1. A 10-g bullet has a speed of 1.2 km/s. (*a*) What is its kinetic energy in joules? (*b*) What is the bullet's kinetic energy if its speed is halved? (*c*) If its speed is doubled?

2. Find the kinetic energy in joules of (*a*) a 0.145-kg baseball moving with a speed of 40 m/s and (*b*) a 60-kg jogger running at a steady pace of 9 minutes per mile.

3. From the conversion factors for converting newtons to pounds and meters to feet, derive the relation between the joule and the foot-pound given in Equation 6-3.

4. A 5-kg box is raised from rest a distance of 4 m by a vertical force of 80 N. Find (*a*) the work done by the force, (*b*) the work done by gravity, and (*c*) the final kinetic energy of the box.

6-2 Work Done by a Variable Force

5. A 2-kg particle is moving with a speed of 3 m/s when it is at $x = 0$. It is subjected to a single force F_x that varies with position as shown in Figure 6-27. (*a*) What is the kinetic energy of the particle when it is at $x = 0$? (*b*) How much work is done by the force as the particle moves from $x = 0$ to $x = 4$ m? (*c*) What is the speed of the particle when it is at $x = 4$ m?

Figure 6-27 Problem 5.

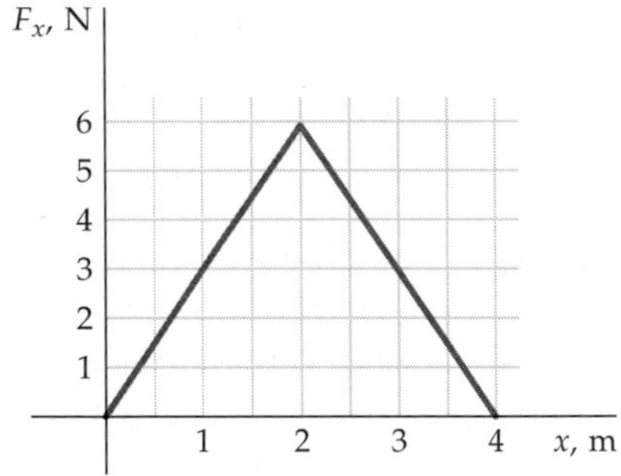

6. A 4-kg particle is initially at rest at $x = 0$. It is subjected to a single force F_x that varies with position as shown in Figure 6-28. Find the work done by the force as the particle moves (*a*) from $x = 0$ to $x = 3$ m and (*b*) from $x = 3$ m to $x = 6$ m. Find the kinetic energy of the particle when it is at (*c*) $x = 3$ m and (*d*) $x = 6$ m.

7. A force F_x is related to the position of a particle by the formula $F_x = Cx^3$, where C is a constant. Find the work done by this force on the particle when the particle moves from $x = 0$ to $x = 3$ m.

6-3 Work and Energy in Three Dimensions and the Dot Product

8. A block of mass 6 kg slides down a frictionless incline. The angle of the incline is 60°. (*a*) List all the forces acting on the block, and find the work done by each force when the block slides 2 m (measured along the incline). (*b*) What is the total work done on the block? What is the speed of the block after it has slid 2 m if (*c*) it starts from rest and (*d*) it starts with an initial speed of 3 m/s?

9. A 100-kg cart is raised up a 1-m step by being rolled up an incline formed by a plank of length L that has been laid from the lower level to the top of the step. (Assume the rolling is equivalent to sliding without friction.) (*a*) Find the force parallel to the incline needed to push the cart up without acceleration for $L = 3$, 4, and 5 m. (*b*) Calculate directly from Equation 6-15 the work needed to push the cart up the incline for each value of L. (*c*) Since the work found in (*b*) is the same for each value of L, what advantage, if any, is there in choosing one length over another?

10. A 2-kg object attached to a string moves with a speed of 1.5 m/s in a circle of radius 3 m on a frictionless horizontal surface. (*a*) Find the tension in the string. (*b*) List the forces acting on the object, and find the work done by each force during 1 revolution.

11. Two vectors **A** and **B** have magnitudes of 6 m and make an angle of 60° with each other. Find **A·B**.

12. Find **A·B** for the following vectors: (*a*) $\mathbf{A} = 2\mathbf{i} - 6\mathbf{j}$, $\mathbf{B} = -3\mathbf{i} + 2\mathbf{j}$; (*b*) $\mathbf{A} = 4\mathbf{i} + 4\mathbf{j}$, $\mathbf{B} = 2\mathbf{i} - 3\mathbf{j}$; and (*c*) $\mathbf{A} = 3\mathbf{i} + 4\mathbf{j}$, $\mathbf{B} = 4\mathbf{i} - 3\mathbf{j}$.

13. Find the angles between the vectors **A** and **B** in Problem 12.

14. A 2-kg object is given a displacement $\Delta\mathbf{s} = 3\text{ m}\mathbf{i} + 3\text{ m}\mathbf{j} - 2\text{ m}\mathbf{k}$ along a straight line. During the displacement, a constant force $\mathbf{F} = 2\text{ N}\mathbf{i} - 1\text{ N}\mathbf{j} + 1\text{ N}\mathbf{k}$ acts on the object. (*a*) Find the work done by **F** for this displacement. (*b*) Find the component of **F** in the direction of the displacement.

15. (*a*) Given any vector $\mathbf{A} = A_x\mathbf{i} + A_y\mathbf{j} + A_z\mathbf{k}$, prove that the x component of **A** is given by **A·i**. (*b*) Find the unit vector that is parallel to **A**. (*c*) Find the component of the vector $2\mathbf{i} + \mathbf{j} + \mathbf{k}$ in the direction of the vector $3\mathbf{i} + 4\mathbf{j}$.

Figure 6-28 Problem 6.

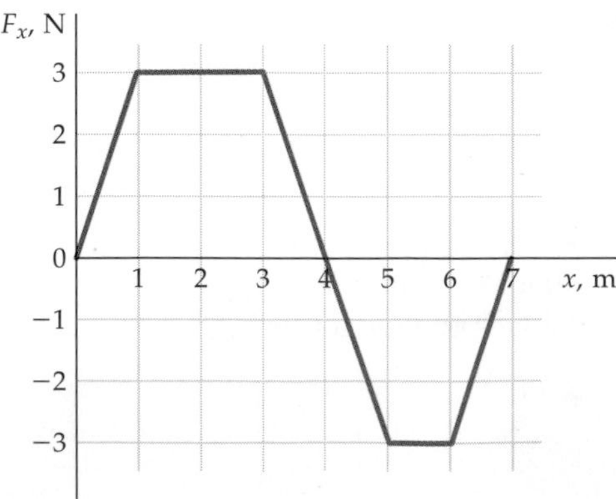

6-4 Work and Energy for Particle Systems: Potential Energy

16. A man of mass 80 kg climbs up stairs to a height of 5 m. What is the increase in the potential energy of the man?

17. State which of the following forces are conservative and which are nonconservative: (*a*) the frictional force exerted on a sliding box, (*b*) the force exerted by a linear spring that obeys Hooke's law, (*c*) the force of gravity, and (*d*) the wind resistance on a moving car.

18. A 2-kg book is held at a height of 20 m above the ground and is released at $t = 0$. (*a*) What is the original potential energy of the book relative to the ground? (*b*) From Newton's laws, find the distance the book falls in 1 s and its speed at $t = 1$ s. (*c*) Find the potential energy and the kinetic energy of the book at $t = 1$ s. (*d*) Find the kinetic energy and the speed of the book just before it hits the ground.

19. A 2-kg box slides down a long, frictionless incline of angle 30°. It starts from rest at time $t = 0$ at the top of the incline at a height of 20 m above the ground. (*a*) What is the original potential energy of the box relative to the ground? (*b*) From Newton's laws, find the distance the box travels in 1 s and its speed at $t = 1$ s. (*c*) Find the potential energy and the kinetic energy of the box at $t = 1$ s. (*d*) Find the kinetic energy and the speed of the box just as it reaches the bottom of the incline.

20. How high must an object be lifted to gain an amount of potential energy equal to the kinetic energy it has when it is moving at a speed of 20 m/s?

21. A force $F_x = 4$ N is constant. (*a*) Find the potential-energy function U associated with this force for an arbitrary choice of where $U = 0$. (*b*) Find U such that $U = 0$ at $x = 6$ m. (*c*) Find U such that $U = 12$ J at $x = 6$ m.

22. A spring has a force constant of $k = 10^4$ N/m. How far must it be stretched for its potential energy to be (*a*) 50 J and (*b*) 100 J?

6-5 Potential Energy and Equilibrium in One Dimension

23. Figure 6-29 shows the plot of a potential-energy function U versus x. (*a*) At each point indicated, state whether the force F_x is positive, negative, or zero. (*b*) At which point does the force have the greatest magnitude? (*c*) Identify any equilibrium points, and state whether the equilibrium is stable, unstable, or neutral.

Figure 6-29 Problem 23.

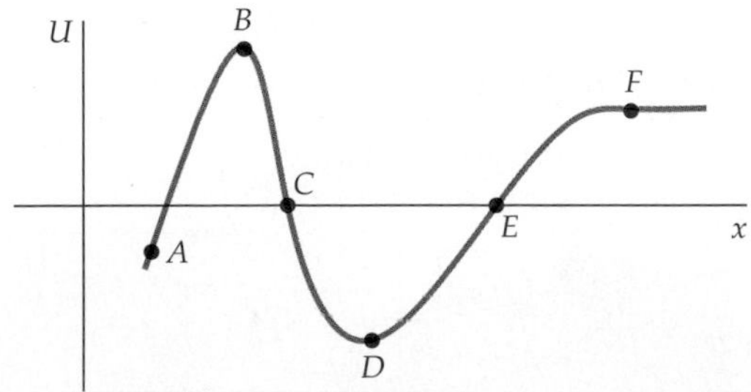

24. (*a*) Find the force F_x associated with the potential-energy function $U = Ax^4$, where A is a constant. (*b*) At what point or points is the force zero?

25. A potential-energy function is given by $U = C/x$, where C is a positive constant. (*a*) Find the force F_x as a function of x. (*b*) Is this force directed toward the origin or away from it? (*c*) Does the potential energy increase or decrease as x increases? (*d*) Answer parts (*b*) and (*c*) if C is a negative constant.

26. On the potential-energy curve for U versus y shown in Figure 6-30, the segments AB and CD are straight lines. Sketch a plot of the force F_y versus y.

Figure 6-30 Problem 26.

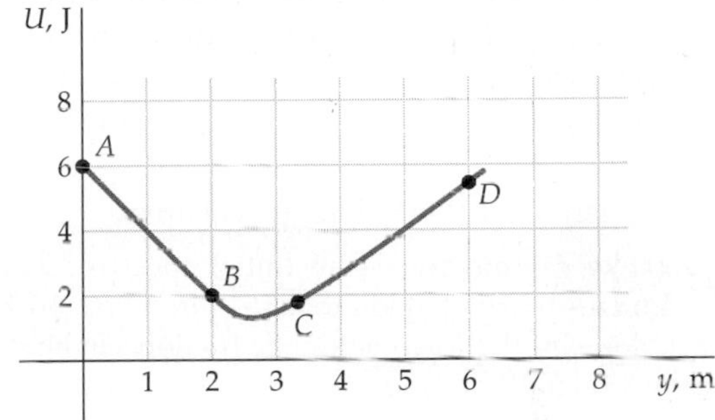

6-6 The Conservation of Mechanical Energy

27. A woman places a 2-kg block against a horizontal spring of force constant $k = 300$ N/m and compresses it 9 cm. (*a*) Find the work done by the woman and the work done by the spring. (*b*) The block is released and leaves the spring when the spring is at its original position. Find the speed of the block when it leaves the spring.

28. A 3-kg block slides along a frictionless horizontal surface with speed of 7 m/s. After sliding a distance of 2 m, it encounters a frictionless ramp inclined at an angle of 40° to the horizontal. How far up the ramp does the block slide before coming momentarily to rest?

29. The 3-kg object in Figure 6-31 is released from rest at a height of 5 m on a curved frictionless ramp. At the foot of the ramp is a spring of force constant $k = 400$ N/m. The object slides down the ramp and into the spring, compressing it a distance x before coming momentarily to rest. (*a*) Find x. (*b*) What happens to the object after it comes to rest?

Figure 6-31 Problem 29.

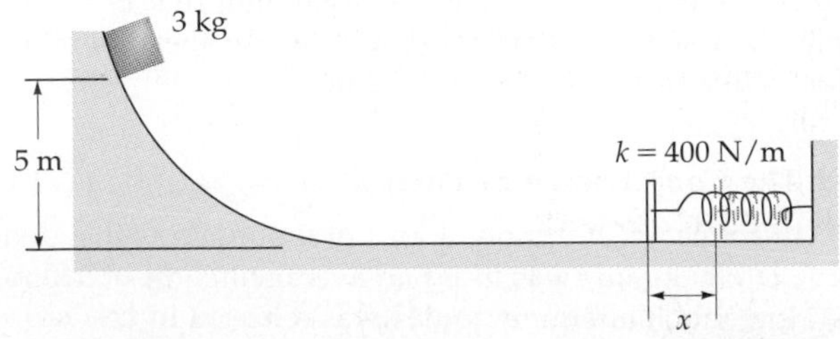

30. The system shown in Figure 6-32 is at rest when the lower string is cut. Find the speed of the objects when they are at the same height.

31. An object slides along the frictionless track shown in Figure 6-33. Initially, it is at point P, headed downhill with speed v_0. Describe the motion of the object in as much detail as you can for (*a*) $v_0 = 7$ m/s and (*b*) $v_0 = 12$ m/s. (*c*) What is the least speed needed by the object to get past point Q?

32. In Figure 6-34, the blocks are initially at rest. Choose $U = 0$ at this initial position. (*a*) Write an expression for the total mechanical energy of the system after the 2-kg block has fallen a distance y. (*b*) Find the speed of the 2-kg mass after it has fallen from rest a distance of 2 m, assuming no friction.

33. A 15-g ball is shot from a spring gun whose spring has a force constant of 600 N/m. The spring can be compressed 5 cm. How high can the ball be shot if the gun is aimed vertically?

6-7 The Generalized Work–Energy Theorem

34. A 2000-kg car moving at an initial speed of 25 m/s along a horizontal road skids to a stop in 60 m. (*a*) Find the work done by the force of kinetic friction. (*b*) Find the coefficient of kinetic friction between the tires and the road.

35. An 8-kg sled is initially at rest on a horizontal road. The coefficient of kinetic friction between the sled and the road is 0.4. The sled is pulled a distance of 3 m by a force of 40 N applied to the sled at an angle of 30° to the horizontal. (*a*) Find the work done by the applied force. (*b*) Find the work done by friction. (*c*) Find the change in the kinetic energy of the sled. (*d*) Find the speed of the sled after it has traveled 3 m.

36. Suppose that the surfaces in Problem 28 are not frictionless and that the coefficient of kinetic friction between the block and the surfaces is 0.30. Find (*a*) the speed of the block when it reaches the ramp and (*b*) the distance up the ramp the block slides before coming momentarily to rest.

37. The 2-kg block in Figure 6-35 slides down a frictionless curved ramp from rest at a height of 3 m. It then slides 9 m on a rough horizontal surface before coming to rest. (*a*) What is the speed of the block at the bottom of the ramp? (*b*) How much work is done by friction? (*c*) What is the coefficient of friction between the block and the horizontal surface?

38. The coefficient of kinetic friction between the 4-kg block and the table in Figure 6-34 is 0.35. (*a*) Find the work done by friction when the 2-kg block falls a distance y. (*b*) Find the total mechanical energy E of the system after the 2-kg block falls a distance y, assuming that $E = 0$ initially. (*c*) Use your result for (*b*) to find the speed of either block after the 2-kg block falls 2 m.

6-8 The Conservation of Energy

39. In a volcanic eruption, 4 km^3 of mountain with a density of 1600 kg/m^3 was lifted an average height of 500 m. (*a*) How much energy in joules was released in this eruption? (*b*) The energy released by bombs is measured in megatons of TNT, where 1 megaton of TNT = 4.2×10^{15} J. Convert your answer for (*a*) to megatons of TNT.

Figure 6-32 Problem 30.

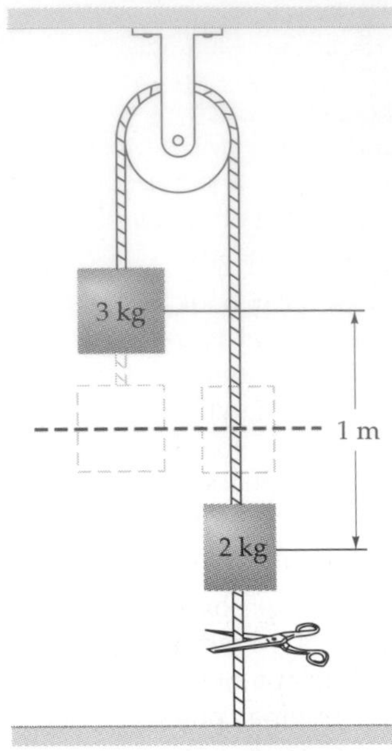

Figure 6-33 Problem 31.

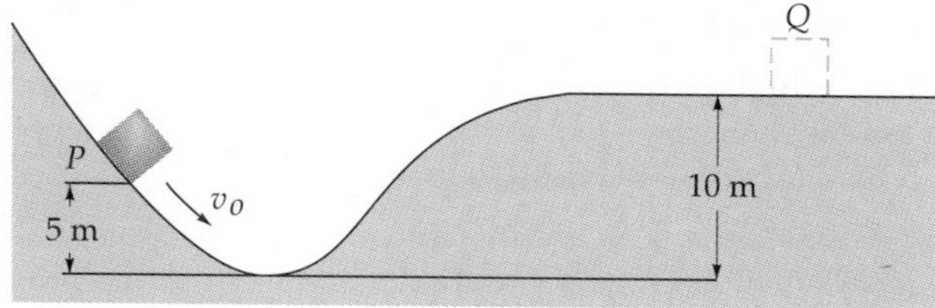

Figure 6-34 Problems 32 and 38.

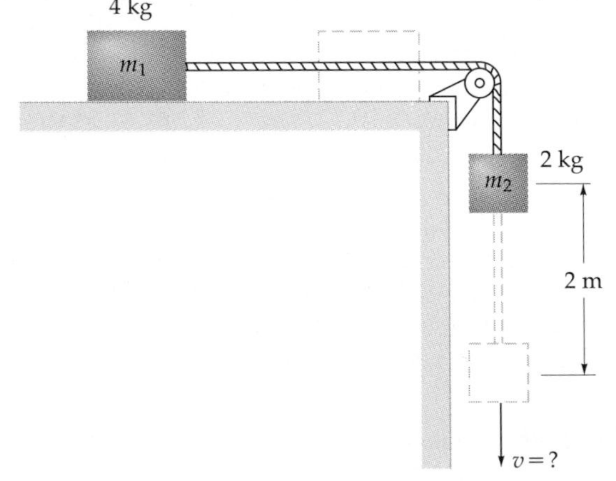

Figure 6-35 Problem 37.

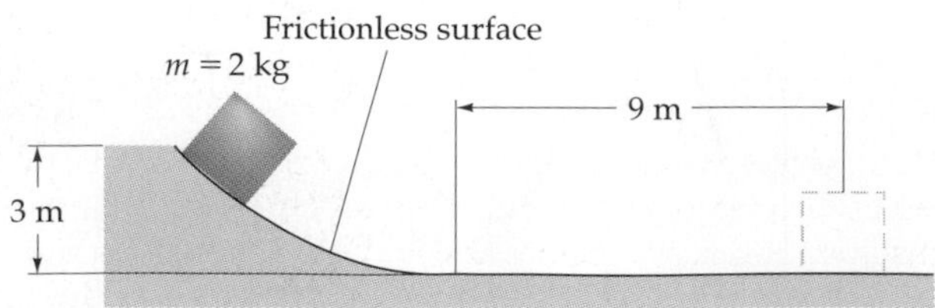

40. An 80-kg physics student climbs a 120-m hill. (*a*) What is the increase in the gravitational potential energy of the student? (*b*) Where does this energy come from? (*c*) The student's body is 20 percent efficient; that is, for every 20 J that are converted to mechanical energy, 100 J of internal energy are expended, with 80 J going into heat. How much internal energy is expended by the student during the climb?

41. A 70-kg skater pushes off the wall of a skating rink, acquiring a speed of 4 m/s. (*a*) How much work is done on the skater? (*b*) What is the change in the mechanical energy of the skater? (*c*) Discuss the conservation of energy as applied to the skater.

6-9 Power

42. A 4-kg box is lifted by a force equal to the weight of the box. The box moves upward at a constant velocity of 2 m/s. (*a*) What is the power input of the force? (*b*) How much work is done by the force in 3 s?

43. A constant horizontal force $F = 3$ N drags a box along a rough horizontal surface at a constant speed v. The force does work at the rate of 5 W. (*a*) What is the speed v? (*b*) How much work is done by F in 3 s?

44. A single force of 5 N acts in the x direction on a 12-kg object. (*a*) If the object starts from rest at $x = 0$ at time $t = 0$, find its velocity v as a function of time. (*b*) Write an expression for the power input as a function of time. (*c*) What is the power input of the force at time $t = 3$ s?

45. Water flows over Victoria Falls, which is 100 m high, at an average rate of 1.4×10^6 kg/s. If all the potential energy of the water were converted into electric energy, how much power could be produced by these falls?

46. On a clear day, solar energy falls on a certain house at an average rate of 400 W/m^2 for 8 h. How much energy is collected by a large glass window of area 40 m^2?

47. Our bodies convert internal chemical energy into work and heat at the rate of about 100 W, which is called our metabolic rate. (*a*) How much internal chemical energy do we use in 24 h? (*b*) The energy comes from the food that we eat and is usually measured in kilocalories, where 1 kcal = 4.184 kJ. How many kilocalories of food energy must we ingest per day if our metabolic rate is 100 W?

48. Find the power input of a force **F** acting on a particle that moves with a velocity **v** for (*a*) **F** = 3 N**i** + 4 N**k**, **v** = 2 m/s **i**; (*b*) **F** = 5 N**i** − 6 N**j**, **v** = −5 m/s **i** + 4 m/s **j**; and (*c*) **F** = 2 N**i** + 4 N**j**, **v** = 6 m/s **i** + 3 m/s **j**.

Level II

49. A child is swinging from a suspended rope 4.0 m long that will break if the tension equals twice the weight of the child. (*a*) What is the greatest angle θ_0 the rope can make with the vertical during a swing if the rope is not to break? (*b*) If the angle slightly exceeds that found in (*a*), what is the speed of the child when the rope breaks?

50. A pendulum of length L has a bob of mass m. It is released from some angle θ_1. The string hits a peg at a distance x directly below the pivot as shown in Figure 6-36, effectively shortening the length of the pendulum.

Figure 6-36 Problem 50.

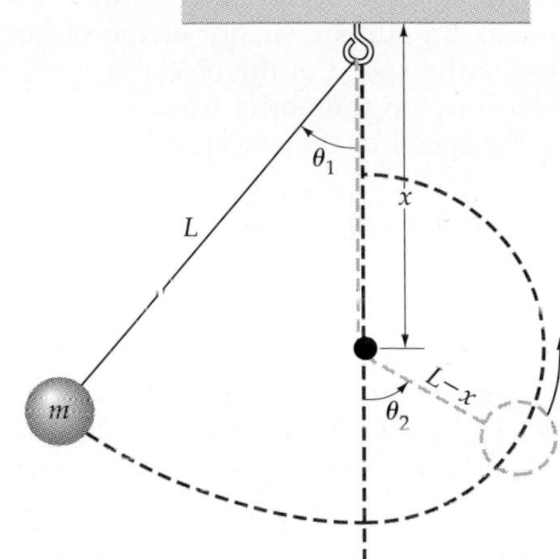

Find the maximum angle θ_2 between the string and the vertical when the bob is to the right of the peg.

51. A force F_x acting on a particle is shown as a function of x in Figure 6-37. (*a*) From the graph, calculate the work done by the force when the particle moves from $x = 0$ to the following values of x: −4, −3, −2, −1, 0, 1, 2, 3, and 4 m. (*b*) Plot the potential energy U versus x for the range of values of x from −4 m to +4 m, assuming that $U = 0$ at $x = 0$.

Figure 6-37 Problem 51.

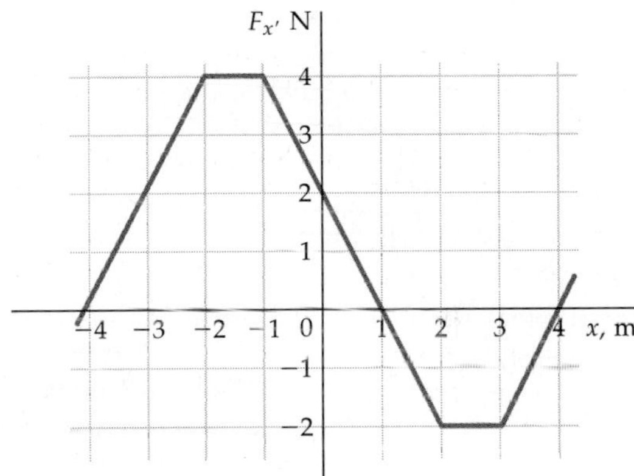

52. Repeat Problem 51 for the force F_x shown in Figure 6-38.

Figure 6-38 Problem 52.

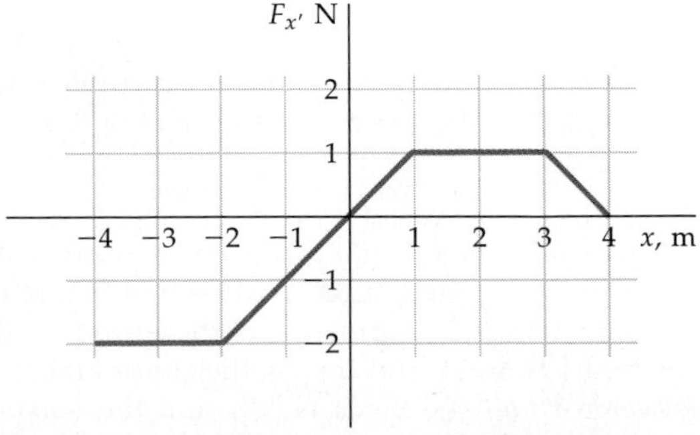

53. A 3-kg object is moving with a speed of 1.50 m/s in the x direction. When it passes the origin, it is acted on by

a single force F_x, which varies with x as shown in Figure 6-39. (*a*) Find the work done by the force from $x = 0$ to $x = 2$ m. (*b*) What is the kinetic energy of the object at $x = 2$ m? (*c*) What is the speed of the object at $x = 2$ m? (*d*) Find the work done on the object from $x = 0$ to $x = 4$ m. (*e*) What is the speed of the object at $x = 4$ m?

Figure 6-39 Problem 53.

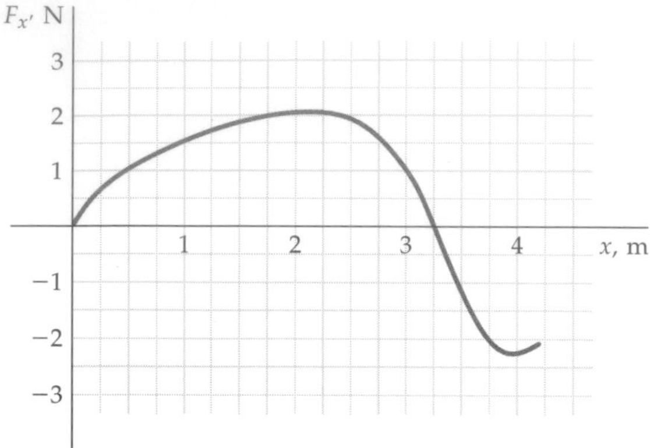

54. A T-bar tow is required to pull 80 skiers up a 600-m slope inclined at 15° to the horizontal at a speed of 2.5 m/s. The coefficient of kinetic friction is 0.06. Find the motor power required if the mass of the average skier is 75 kg.

55. A 0.2-kg ball is thrown upward with an initial speed of 24 m/s at an angle of 30° with the horizontal from the edge of a 60-m cliff. Neglecting air resistance, find (*a*) the initial kinetic energy of the ball, (*b*) the potential energy and kinetic energy of the ball when it is at its highest point, (*c*) the kinetic energy of the ball when it reaches the ground 60 m below the cliff, and (*d*) the speed of the ball just before it strikes the ground. Assume that $U = 0$ at the top of the cliff.

56. A small block is attached to a rubberlike material that exerts a force $F_x = -kx - ax^2$ when it is stretched a distance x ($x > 0$), where k and a are constants. Find the work done on the block by the material when it is stretched from $x = 0$ to $x = A$.

57. The force acting on a particle moving along the x axis is given by $F_x = -ax^2$, where a is a constant. Calculate the potential-energy function U relative to $U = 0$ at $x = 0$, and sketch a graph of U versus x.

58. A ski jumper of mass 70 kg starts from rest at A in Figure 6-40. His speed is 30.0 m/s at B and 23.0 m/s at C, where the distance BC is 30 m. (*a*) How much work is done by friction on the skier as he moves from B to C? (*b*) Find the greatest height above the level of C attained by the jumper.

59. A 2-kg block is released 4 m from a massless spring with a force constant $k = 100$ N/m that is fixed along a frictionless plane inclined at 30° as shown in Figure 6-41. (*a*) Find the maximum compression of the spring. (*b*) If the plane is not frictionless and the coefficient of kinetic friction between it and the block is 0.2, find the maximum compression. (*c*) For the rough incline, how far up the incline will the block travel after leaving the spring?

60. A train with a total mass of 2×10^9 g rises 707 m in a travel distance of 62 km at an average speed of 15.0 km/h. If the frictional force is 0.8 percent of the weight, find (*a*) the kinetic energy of the train, (*b*) the total change in its potential energy, (*c*) the work done on the train by friction, and (*d*) the power output of the train's engines.

61. A woman carries a large bucket of water to the top of a 40-m tower at a constant upward speed. The bucket has a mass of 10 kg and initially contains 30 kg of water, but it leaks at a constant rate and only 10 kg of water are in the bucket at the top of the tower. (*a*) Write an expression for the mass of the bucket plus water as a function of the height y climbed. (*b*) Find the work done by the woman on the bucket.

62. Two men stand at the edge of a 10-m cliff. Simultaneously, they throw balls at an initial speed of 10 m/s, one straight up and the other straight down. (*a*) Neglecting air resistance, find the speed of each ball when it reaches the bottom of the cliff. (*b*) Discuss qualitatively how the speeds would compare if air resistance were not neglected.

63. A 2-kg box is projected with an initial speed of 3 m/s up a rough plane inclined at 60° to the horizontal. The coefficient of kinetic friction is 0.3. (*a*) List all the forces acting on the box, and find the work done by each as the box slides up the plane. (*b*) How far up the plane does the box slide before it stops momentarily? (*c*) Find the work done by each force acting on the box as it slides back down the plane. (*d*) Find its speed when it reaches its initial position.

64. One expects accelerating to consume more energy than driving at a constant speed. (*a*) Calculate the energy required to give a 1200-kg car a speed of 50 km/h neglect-

Figure 6-40 Problem 58.

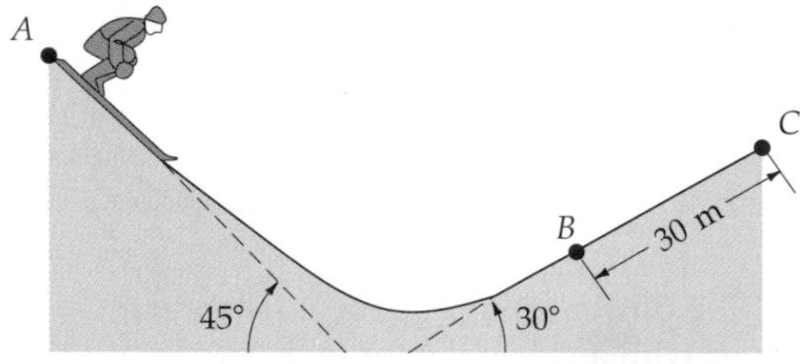

Figure 6-41 Problem 59.

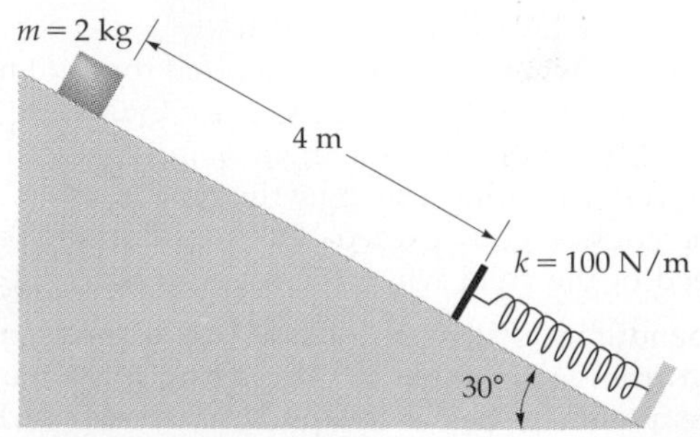

ing friction. (*b*) If friction of all types results in a retarding force of 300 N at a speed of 50 km/h, what is the energy needed to move the car a distance of 300 m at a constant speed of 50 km/h? (*c*) Assuming the energy losses due to friction in part (*a*) are 75 percent of those found in part (*b*), estimate the ratio of the energy consumption for the two cases considered.

65. A particle of mass m moves in a horizontal circle of radius r on a rough table. It is attached to a string fixed at the center of the circle. The speed of the particle is initially v_0. After completing one full trip around the circle, the speed of the particle is $\frac{1}{2}v_0$. (*a*) Find the work done by friction during that one revolution in terms of m, v_0, and r. (*b*) What is the coefficient of kinetic friction? (*c*) How many more revolutions will the particle make before coming to rest?

66. Water from behind a dam flows through a large turbine at a rate of 1.5×10^6 kg/min. The turbine is located 50 m below the surface of the reservoir, and the water leaves the turbine with a speed of 5 m/s. (*a*) Neglecting any energy dissipation, what is the power output of the turbine? (*b*) How many U.S. citizens would be supplied with energy by this dam if each citizen uses 3×10^{11} J of energy per year?

67. An elastic string has a natural length L_0 and a spring constant k. When an object of mass m is hanging from it vertically, the string stretches by y_0. One end of the string is attached to the top of a frictionless plane inclined at 30° to the horizontal. With the string lying along the incline, the object of mass m is attached to it and is released with the string in its unstretched condition. How far does the object slide down the incline before coming to rest for the first time?

68. Vectors **A**, **B**, and **C** form a triangle as shown in Figure 6-42. The angle between **A** and **B** is θ, and the vectors are related by $\mathbf{C} = \mathbf{A} - \mathbf{B}$. Compute $\mathbf{C}\cdot\mathbf{C}$ in terms of A, B, and θ, and derive the law of cosines, $C^2 = A^2 + B^2 - 2AB\cos\theta$.

Figure 6-42 Problem 68.

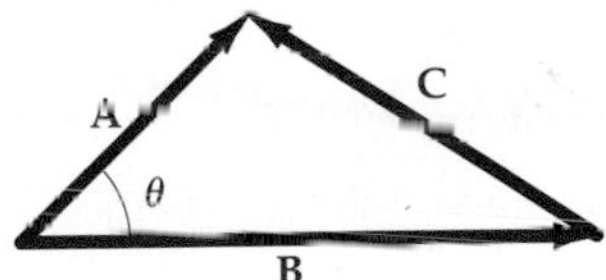

69. A car of mass 1500 kg traveling at 24 m/s is at the foot of a hill that rises 120 m in 2.0 km. At the top of the hill, the speed of the car is 10 m/s. Find the average power delivered by the car's engine, neglecting any frictional losses.

70. A straight rod of negligible mass is mounted on a frictionless pivot as shown in Figure 6-43. Masses m_1 and m_2 are suspended at distances l_1 and l_2 as shown. (*a*) Write an expression for the gravitational potential energy of the masses as a function of the angle θ made by the rod and the horizontal. (*b*) For what angle θ is the potential energy a minimum? Is the statement "systems tend to move toward minimum potential energy" consistent with your result? (*c*) Show that if $m_1 l_1 = m_2 l_2$, the potential energy is the same for all values of θ. (When this holds, the system will balance at any angle θ. This result is known as *Archimedes' law of the lever*.)

Figure 6-43 Problem 70.

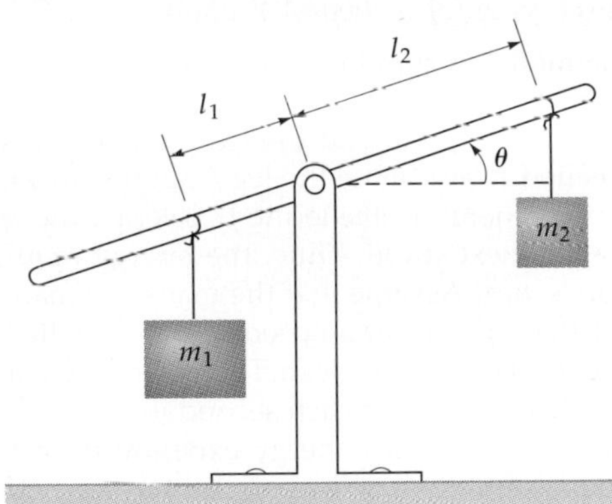

71. A force $\mathbf{F} = (2x^2\ \text{N/m}^2)\mathbf{i}$ is applied to a particle. Find the work done on the particle as it moves a total distance of 5 m (*a*) parallel to the y axis from point $x = 2$ m, $y = 2$ m to point $x = 2$ m, $y = 7$ m and (*b*) in a straight line from $x = 2$ m, $y = 2$ m to $x = 5$ m, $y = 6$ m.

72. In a volcanic eruption, a 200-g piece of porous volcanic rock is thrown vertically upward with an initial speed of 40 m/s. It travels upward a distance of 50 m before it begins to fall back to the earth. (*a*) What is the initial kinetic energy of the rock? (*b*) How much work does air friction do on the rock? (*c*) If air friction does seven-tenths as much work on the rock on the way down as on the way up, what is the velocity of the rock when it returns to its initial position?

73. Two blocks of equal mass M are attached to the ends of a very light string hung over two frictionless pegs as shown in Figure 6-44. A third block of mass m is attached to the middle of the string between the pegs. (*a*) Find the potential energy of the system as a function of the distance y shown in the figure. (*b*) Find the equilibrium distance y_0 using the potential-energy function. Check your answer by analyzing the forces.

Figure 6-44 Problem 73.

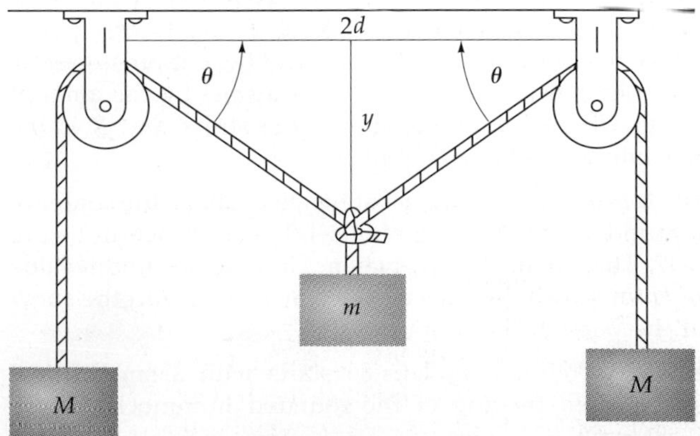

74. A large nuclear power plant produces 3000 MW of power by nuclear fission, which converts mass m into energy according to Einstein's equation $E = mc^2$, where $c = 3 \times 10^8$ m/s is the speed of light. (*a*) How much mass is converted into energy in one year? (*b*) In a coal-burning power plant, each kilogram of coal releases 31 MJ of energy when burned. How many kilograms of coal are needed each year for a 3000-MW plant?

75. In one model of running, the energy expended is assumed to go into accelerating and decelerating the legs. If the mass of the leg is m and the running speed is v, the energy needed to accelerate the leg from rest to v is $\frac{1}{2}mv^2$, and the same energy is needed to decelerate the leg back to rest for the next stride. Thus, the energy required for each stride is mv^2. Assume that the mass of a man's leg is 10 kg and that he runs at a speed of 3 m/s with 1 m between one footfall and the next. Therefore, the energy he must provide to his legs in each second is $3 \times mv^2$. Calculate the rate of the man's energy expenditure using this model and assuming that his muscles have an efficiency of 25 percent.

76. A rope with a mass per unit length of u lies coiled on the floor. (*a*) If a length y of the rope is picked up, what force F is required to hold it above the floor as shown in Figure 6-45? (*b*) Integrate $F\,dy$ from $y = 0$ to $y = L$ to find the total work needed to pick up a length L of the rope. (*c*) What is the potential energy of the rope when a length L has been picked up?

Level III

77. A small block of mass m slides without friction along the loop-the-loop track shown in Figure 6-46. The block starts from point P a distance h above the bottom of the loop. (*a*) What is the kinetic energy of the block when it reaches the top of the loop? (*b*) What is its acceleration at the top of the loop, assuming that it stays on the track? (*c*) What is the least value of h for which the block will reach the top of the loop without leaving the track?

78. A pendulum consists of a bob of mass m attached to a light rod of length L and negligible mass. The other end of the rod is fixed to a frictionless pivot. The pendulum is released when the bob is directly over the pivot. Find the force exerted by the rod on the bob when (*a*) the bob is at its lowest point, (*b*) the rod is at an angle of 30° below the horizontal, and (*c*) the rod is at an angle of 30° above the horizontal.

79. A pendulum bob is released from rest at an initial angle of θ_0 measured from $\theta = 0$ at the lowest point. Prove that the tension in the string at the bottom is greater than the tension when the bob is at rest by the amount $3K/L$, where L is the length of the string and K is the maximum kinetic energy of the bob.

80. A skier starts from rest at height H above the center of a rounded hummock of radius 4.0 m as shown in Figure 6-47. There is negligible friction. Find the maximum value of H for which the skier remains in contact with the snow at the peak of the hummock.

81. The skier of Problem 80 starts with a small initial speed from the top of the rounded hummock (Figure 6-47). If friction can be neglected, find (*a*) his speed v as a function of the angle θ, and (*b*) the angle θ at which he loses contact with the slope.

82. A 5-kg block is held against a spring of force constant 20 N/cm, compressing it 3 cm. The block is released and the spring extends, pushing the block along a rough horizontal surface. The coefficient of friction between the surface and the block is 0.2. (*a*) Find the work done on the block by the spring as it extends from its compressed position to its equilibrium position. (*b*) Find the work done by friction on the block while it moves the 3 cm to the equilibrium position of the spring. (*c*) What is the speed of the block when the spring is at its equilibrium position? (*d*) If the block is not attached to the spring, how far will it slide along the rough surface before coming to rest?

Figure 6-45 Problem 76.

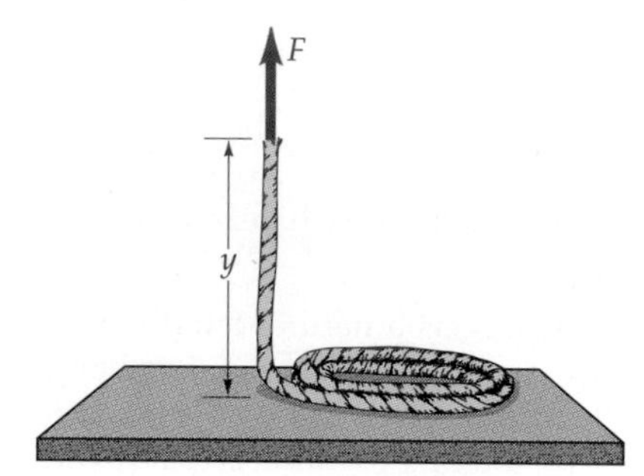

Figure 6-46 Problem 77.

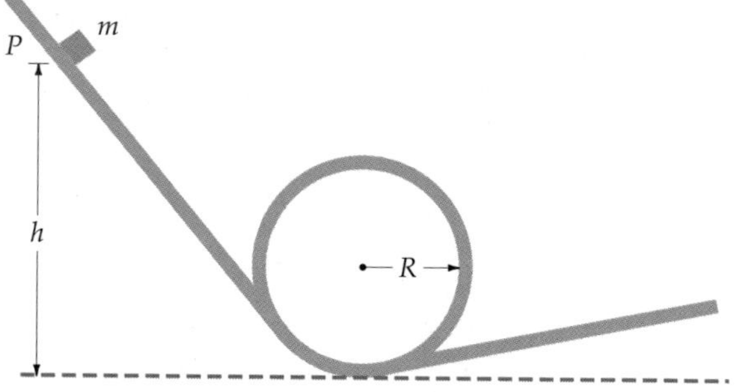

Figure 6-47 Problems 80 and 81.

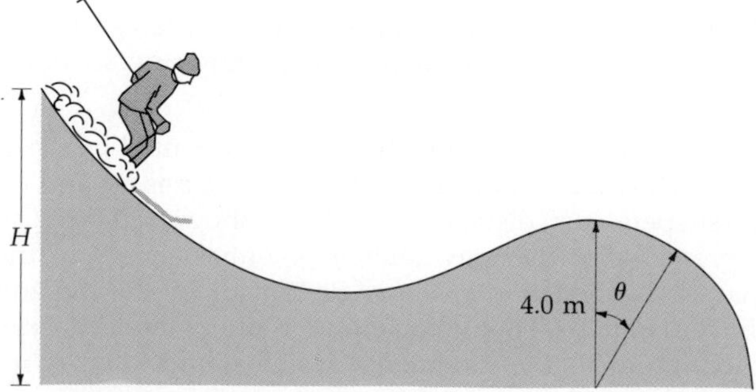

83. The potential-energy function of a particle of mass 4 kg in a force field is described by

$$U = 3x^2 - x^3 \qquad x \le 3 \text{ m}$$

$$U = 0 \qquad x \ge 3 \text{ m}$$

where U is in joules and x is in meters. (*a*) For what values of x is the force F_x zero? (*b*) Sketch a plot of U versus x. (*c*) Discuss the stability of the equilibrium for the values of x found in (*a*). (*d*) If the total energy of the particle is 12 J, what is its speed at $x = 2$ m?

84. A force is given by $F_x = Ax^{-3}$, where $A = 8 \text{ N·m}^3$. (*a*) For positive values of x, does the potential energy associated with this force increase or decrease with increasing x? You can determine the answer to this question by imagining what happens to a particle that is placed at rest at some point x and is then released. (*b*) Find the potential-energy function U associated with this force such that U approaches zero as x approaches infinity. (*c*) Sketch U versus x.

85. A force in the xy plane is given by $\mathbf{F} = A(10a\mathbf{i} + 3x\mathbf{j})$, where A and a are constants, F is in newtons, and x is in meters. Suppose that the force acts on a particle as it moves from an initial position $x = 4$ m, $y = 1$ m to a final position $x = 4$ m, $y = 4$ m. Show that this force is not conservative by computing the work done by the force for at least two different paths.

86. A force in the xy plane is given by $\mathbf{F} = (F_0/r)(y\mathbf{i} - x\mathbf{j})$, where F_0 is a constant and $r = \sqrt{x^2 + y^2}$. (*a*) Show that the magnitude of this force is F_0 and that its direction is perpendicular to $\mathbf{r} = x\mathbf{i} + y\mathbf{j}$. (*b*) Find the work done by this force on a particle that moves in a circle of radius 5 m centered at the origin.

87. A theoretical formula for the potential energy associated with the nuclear force between two protons, two neutrons, or a neutron and a proton is the *Yukawa potential:*

$$U = -U_0\left(\frac{a}{x}\right)e^{-x/a}$$

where U_0 and a are constants. (*a*) Sketch U versus x using $U_0 = 4$ pJ and $a = 2.5$ fm. (*b*) Find the force F_x. (*c*) Compare the magnitude of the force at the separation $x = 2a$ to that at $x = a$. (*d*) Compare the magnitude of the force at the separation $x = 5a$ to that at $x = a$.

88. The bob of a pendulum of length L is pulled aside so the string makes an angle θ_0 with the vertical and released. In Example 6-10, the conservation of energy was used to obtain the speed of the bob at the bottom of its swing. In this problem, you are to obtain the same result using Newton's second law. (*a*) Show that the tangential component of Newton's second law gives $dv/dt = -g \sin \theta$, where v is the speed and θ is the angle made by the string and the vertical. (*b*) Show that v can be written

$$v = L\frac{d\theta}{dt}$$

(*c*) Use this result and the chain rule for derivatives to obtain

$$\frac{dv}{dt} = \frac{dv}{d\theta}\frac{d\theta}{dt} = \frac{dv}{d\theta}\frac{v}{L}$$

(*d*) Combine the results of (*a*) and (*c*) to obtain

$$v\,dv = -gL \sin \theta \, d\theta$$

(*e*) Integrate the left side of the equation in part (*d*) from $v = 0$ to the final speed v and the right side from $\theta = \theta_0$ to $\theta = 0$, and show that the result is equivalent to $v = \sqrt{2gh}$, where h is the original height of the bob above the bottom.

89. (*a*) Using $v = dx/dt$, show that Equation 6-26*b* can be written

$$\frac{dx}{\sqrt{E - U}} = \sqrt{\frac{2}{m}}\,dt$$

For one-dimensional motion, U is a function of x, so the left side of this equation depends only on x (and not on t) and the right side depends only on t. (*b*) Apply this result to a particle of mass m on a spring with a force constant k oscillating with an amplitude A so that $U = \frac{1}{2}kx^2$ and $E = \frac{1}{2}kA^2$ to obtain the equation

$$\frac{dx}{\sqrt{A^2 - x^2}} = \sqrt{\frac{k}{m}}\,dt$$

(*c*) Integrate both sides of this equation to obtain an expression relating the position of the mass x to the time t, assuming that $x = A$ at time $t = 0$.

Chapter 7

Systems of Particles and the Conservation of Momentum

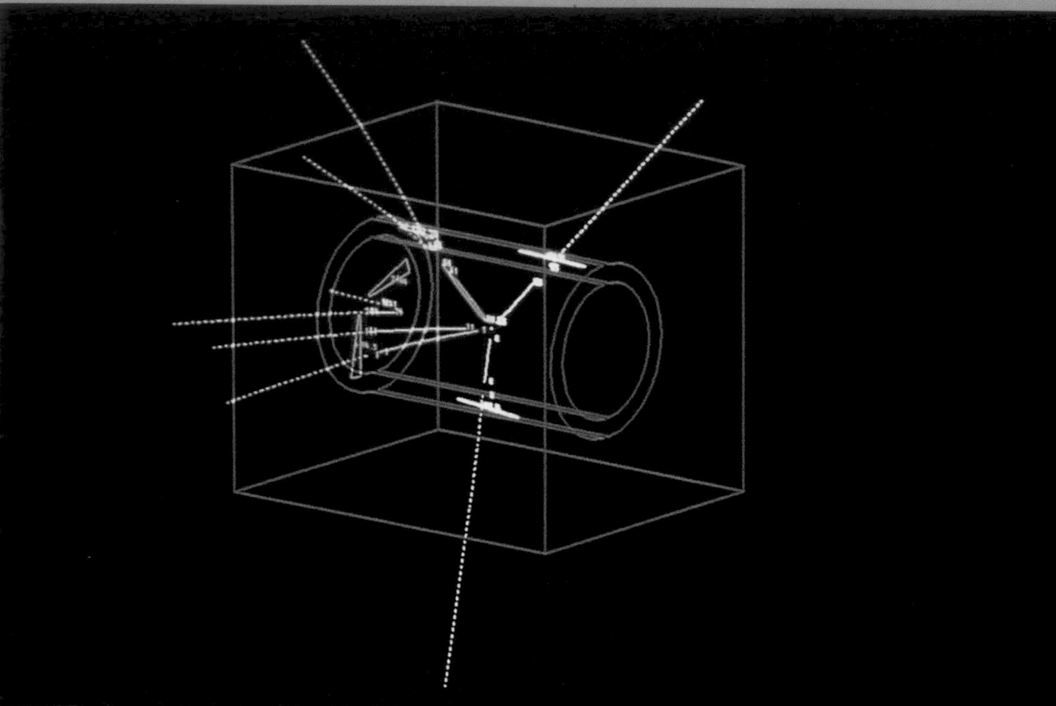

The collisions of elementary particles provide information about the basic interactions and the structure of matter. In this photograph, a proton and an antiproton collide producing a shower of other particles including the rare Z particle. The Z particle, which is electrically neutral, leaves no track, but it quickly decays into other particles, which are charged and do leave tracks. In this case it decays into an electron (pink track) and a positron (white track). The energy of the electron and positron can be measured by the curvature of their tracks in a magnetic field. The total energy of these decay particles equals $m_Z c^2$, where m_Z is the predicted mass of the Z particle.

Although our discussion of Newton's laws has been in terms of the motion of point particles, many of our applications have been to extended objects, such as blocks, balls, and even automobiles. We will justify these applications in this chapter, where we consider an extended object to be a system of point particles and assume that Newton's laws apply to each particle. We will show that there is one point, called the **center of mass,** of a system that moves as if the mass of the system were concentrated at that point and the external forces acting on the system were acting exclusively on that point. The motion of any object or system of particles, no matter how complex, can be described in terms of the motion of the center of mass (which may be thought of as the bulk motion of the system) plus the motion of the individual particles in the system relative to the center of mass.

We will also introduce an important new quantity, momentum, which is the product of the mass of a particle and its velocity. Momentum is important because the total momentum of a system of particles remains constant if the net external force acting on the system is zero. Like energy, momentum is a conserved quantity in an isolated system. Momentum conservation is very useful in analyzing collisions between billiard balls, automobiles, and subatomic particles in a nuclear reaction. It is also useful in analyzing the motion of a jet plane or rocket, and the recoil of a rifle.

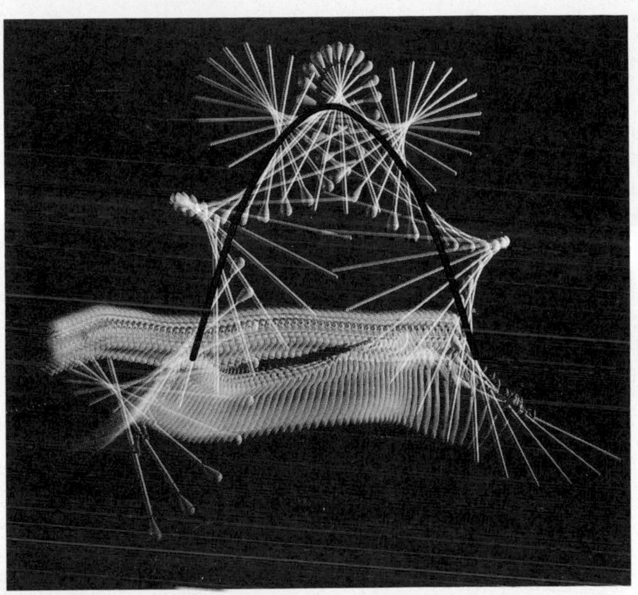

Figure 7-1 Multiflash photograph of a baton thrown into the air. The center of mass follows the same simple parabolic path it would if it were a single point particle.

Figure 7-1 is a multiflash photograph of a baton thrown into the air. Although the motion of the baton is complicated, the motion of one point, the center of mass, is simple. While the baton is in the air, the center of mass follows a parabolic path, the same path that would be followed by a point particle. We will show in general that the acceleration of the center of mass of a system of particles equals the net external force acting on the system divided by the total mass of the system. For the baton thrown into the air, the acceleration of the center of mass is **g** downward.

7-1 The Center of Mass

Let us first consider a simple system of two particles in one dimension. Let x_1 and x_2 be the coordinates of the particles relative to some arbitrarily chosen origin. The center-of-mass coordinate X_{cm} is then defined by

$$MX_{cm} = m_1x_1 + m_2x_2 \qquad 7\text{-}1$$

where $M = m_1 + m_2$ is the total mass of the system. For this case of just two particles, the center of mass lies at some point on the line joining the particles (Figure 7-2*a*). This is easily seen if we choose our origin to coincide with one of the particles, say m_1 (Figure 7-2*b*). Then x_2 is the distance d between the particles. The center-of-mass coordinate for this choice of origin is then obtained from Equation 7-1:

$$MX_{cm} = m_1x_1 + m_2x_2 = m_1(0) + m_2d$$

$$X_{cm} = \frac{m_2}{M}d = \frac{m_2}{m_1 + m_2}d \qquad 7\text{-}2$$

For particles of equal mass, the center of mass is halfway between the particles. Otherwise, it is closer to the particle of greater mass (Figure 7-2*c*).

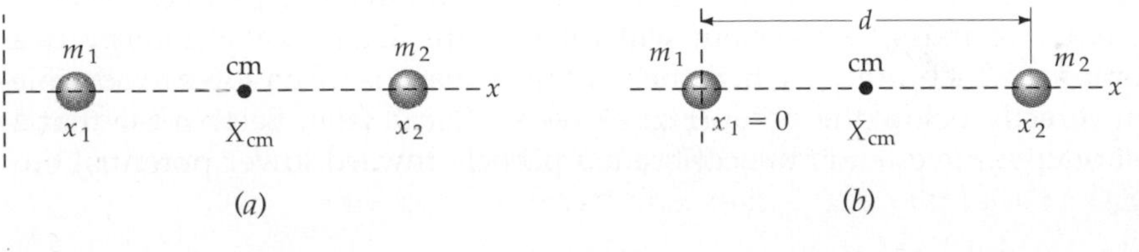

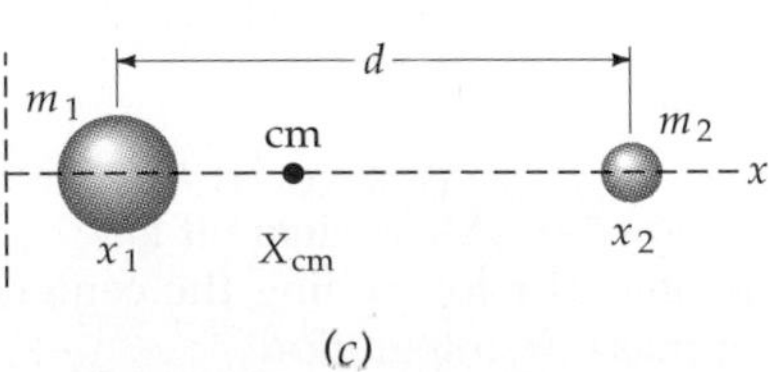

Figure 7-2 The center of mass of a two-particle system. (*a*) If the particles have equal masses, the center of mass is midway between them. (*b*) If the origin is chosen at m_1, the position of the center of mass is given by $X_{cm} = m_2d/(m_1 + m_2)$. (*c*) If the particles have unequal masses, the center of mass is closer to the more massive particle.

Exercise

A 4-kg mass is at the origin and a 2-kg mass is at $x = 6$ cm. Find X_{cm}. (Answer: $X_{cm} = 2$ cm)

We can generalize from our special case of two particles in one dimension to many particles in three dimensions. If we have N particles, the x coordinate of the center of mass X_{cm} is defined by

$$MX_{cm} = m_1x_1 + m_2x_2 + m_3x_3 + \cdots + m_Nx_N = \sum_i m_ix_i \qquad \text{7-3a}$$

where again $M = \Sigma\, m_i$ is the total mass of the system. Similar equations define the y and z coordinates of the center of mass:

$$MY_{cm} = \sum_i m_iy_i \qquad \text{7-3b}$$

$$MZ_{cm} = \sum_i m_iz_i \qquad \text{7-3c}$$

In vector notation, if $\mathbf{r}_i = x_i\mathbf{i} + y_i\mathbf{j} + z_i\mathbf{k}$ is the position vector of the ith particle, the position vector of the center of mass $\mathbf{R}_{cm}$ is given by

Center of mass defined, discrete systems

$$M\mathbf{R}_{cm} = \sum_i m_i\mathbf{r}_i \qquad \text{7-4}$$

where $\mathbf{R}_{cm} = X_{cm}\mathbf{i} + Y_{cm}\mathbf{j} + Z_{cm}\mathbf{k}$.

For a continuous body, the sum in Equation 7-4 is replaced by an integral:

Center of mass defined, continuous systems

$$M\mathbf{R}_{cm} = \int \mathbf{r}\, dm \qquad \text{7-5}$$

where dm is an element of mass that is located at position $\mathbf{r}$ as shown in Figure 7-3.

If two point masses are connected by a light rod of negligible mass, the system will balance on a pivot at the center of mass (Figure 7-4*a*). This follows from the fact that the gravitational potential energy of a system of particles is the same as it would be if the total mass were concentrated at the center of mass as we will now show. Let y_i be the height above the earth of the ith particle in a general system. The potential energy of the system is

$$U = \sum_i m_igy_i = g\sum_i m_iy_i$$

But

$$\sum_i m_iy_i = MY_{cm}$$

so

$$U = MgY_{cm} \qquad \text{7-6}$$

If we try to balance our two particles on a light rod at some point other than the center of mass, the system will rotate until the potential energy is a minimum, which occurs when the center of mass is at its lowest possible point directly below the pivot (Figure 7-4*b*). (Recall from Section 6-5 that a conservative force tends to accelerate a particle toward lower potential energy.)

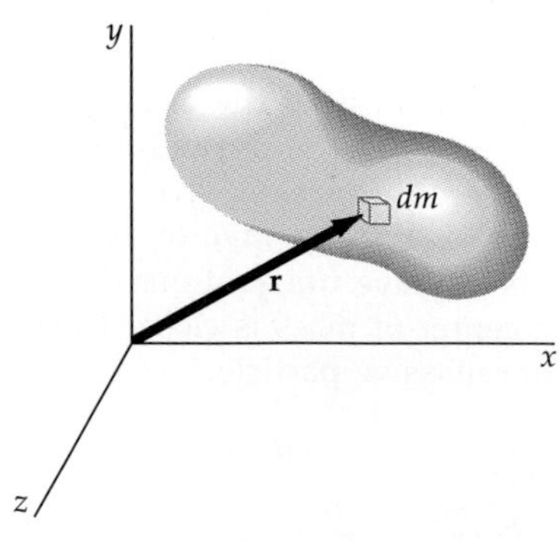

Figure 7-3 Mass element dm located at $\mathbf{r}$ for finding the center of mass by integration.

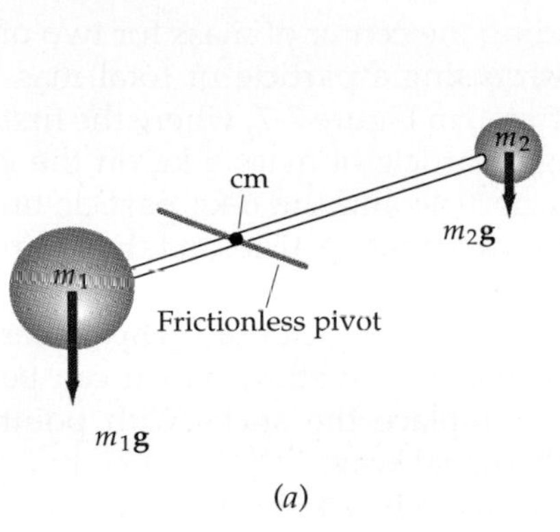

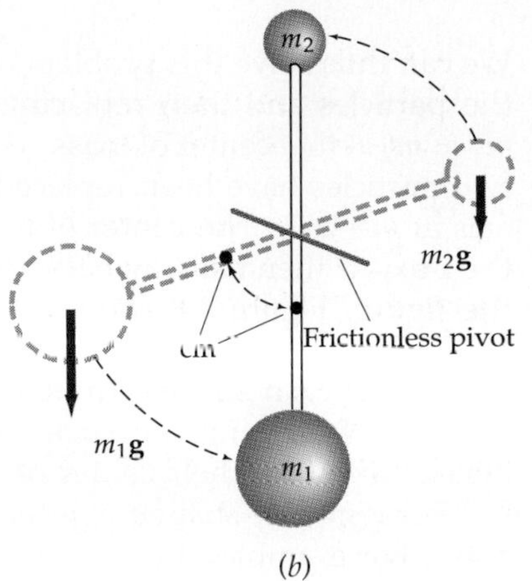

Figure 7-4 (a) Two masses connected by a light rod balance if a pivot is placed at the center of mass. (b) If the pivot is not placed at the center of mass, the potential energy of the system is least when the center of mass is directly below the pivot.

Figure 7-5 illustrates a simple experimental method for finding the center of mass of any plane object, such as a board or metal sheet. We suspend the object from a frictionless pivot at any point on the object. It will hang in equilibrium with its center of mass directly below the pivot. Thus, if we draw a vertical line downward from the pivot, the center of mass must lie somewhere on the line. We can then suspend the object from another point and draw a second vertical line. The center of mass will lie at the intersection of the two lines.

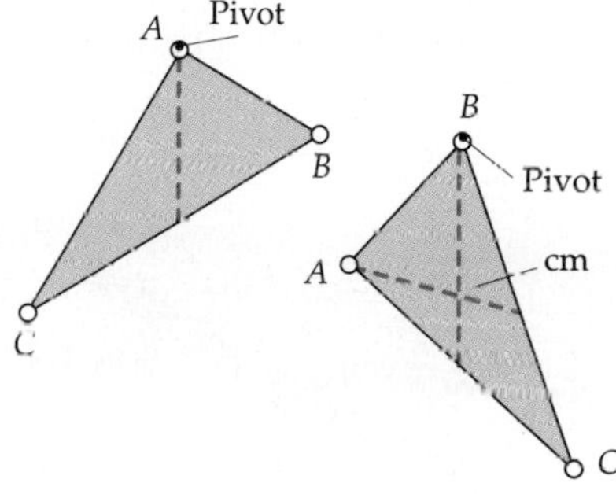

Figure 7-5 The center of mass of an irregularly shaped plane object lies directly below any point about which the object is pivoted. This property can be used to locate the center of mass of the object.

Example 7-1

Find the center of mass of a system consisting of three particles: $m_1 = 2$ kg at the origin, $m_2 = 4$ kg on the y axis at $y = 3$ m, and $m_3 = 6$ kg on the x axis at $x = 4$ m (Figure 7-6).

Using Equation 7-3a, we have

$$MX_{\text{cm}} = \sum_i m_i x_i = m_1x_1 + m_2x_2 + m_3x_3$$

where $M = m_1 + m_2 + m_3 = 12$ kg. Then

$$X_{\text{cm}} = \frac{\sum_i m_i x_i}{M} = \frac{(2\text{ kg})(0) + (4\text{ kg})(0) + (6\text{ kg})(4\text{ m})}{12\text{ kg}} = \frac{24\text{ kg}\cdot\text{m}}{12\text{ kg}} = 2\text{ m}$$

Similarly, from Equation 7-3b we have

$$Y_{\text{cm}} = \frac{\sum_i m_i y_i}{M} = \frac{(2\text{ kg})(0) + (4\text{ kg})(3\text{ m}) + (6\text{ kg})(0)}{12\text{ kg}} = 1\text{ m}$$

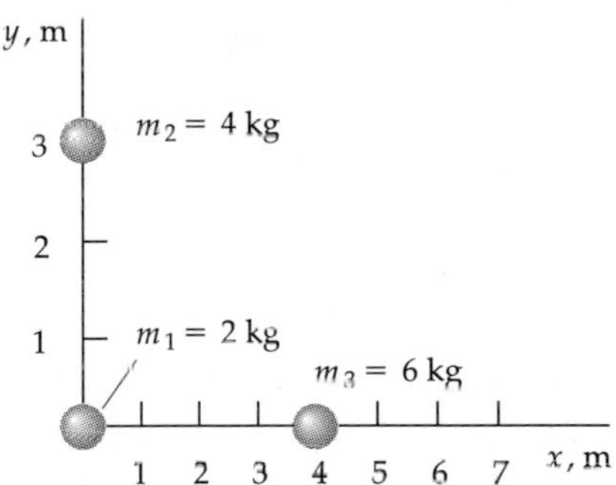

Figure 7-6 Three-particle system for Example 7-1.

There is a useful technique for solving Example 7-1 and other more complex problems. For instance, Equation 7-4 for this problem is

$$M\mathbf{R}_{\text{cm}} = m_1\mathbf{r}_1 + m_2\mathbf{r}_2 + m_3\mathbf{r}_3$$

The first two terms on the right side of this equation are related to the center of mass of the first two particles:

$$m_1\mathbf{r}_1 + m_2\mathbf{r}_2 = (m_1 + m_2)\mathbf{r}'_{\text{cm}}$$

where $\mathbf{r}'_{\text{cm}}$ is the center of mass of the first two particles alone. The center of mass of the three-particle system can thus be written

$$M\mathbf{R}_{\text{cm}} = (m_1 + m_2)\mathbf{r}'_{\text{cm}} + m_3\mathbf{r}_3$$

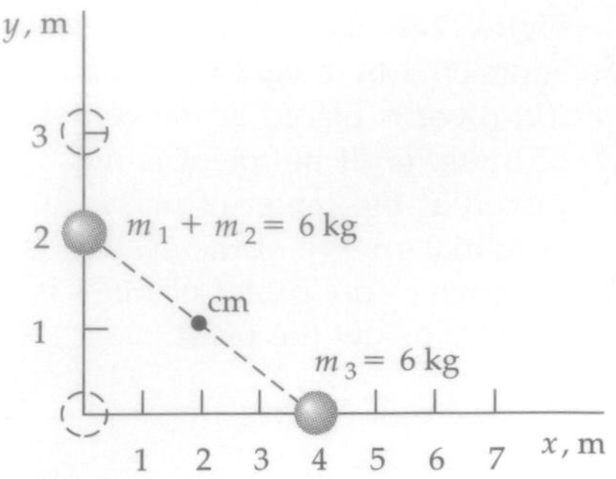

Figure 7-7 Figure 7-6 with the particles m_1 and m_2 replaced by a single particle of mass $m_1 + m_2$ located at the center of mass of the original two particles. The three-particle system has now been reduced to a two-particle system.

We can thus solve this problem by first finding the center of mass for two of the particles and then replacing them with a single particle of total mass $m_1 + m_2$ at that center of mass. This is illustrated in Figure 7-7, where the first two particles have been replaced by a single particle of mass 6 kg on the y axis at $y = 2$ m. The center of mass of this particle and the 6-kg particle on the x axis is then halfway between the two equal mass particles, as shown in the figure. Figure 7-8 shows how this technique can be used to find the center of mass of two sticks whose mass is uniformly distributed. The center of mass of each stick separately is at the center of the stick, which can be found by balancing the stick. We can then replace the sticks with point masses and find their center of mass in the usual way.

For regularly shaped objects, we can use symmetry to find the center of mass. For example, the center of mass of a uniform cylinder or sphere is at the geometric center of the object.

We now illustrate the use of integration to find the center of mass.

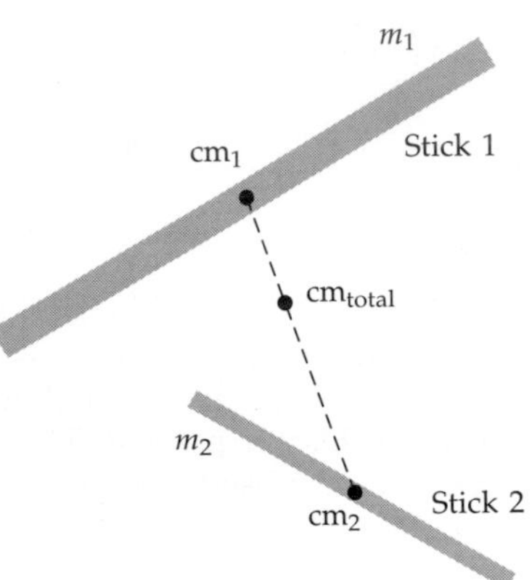

Figure 7-8 The center of mass of a system consisting of two uniform sticks can be found by treating each stick as a point particle at its individual center of mass.

Example 7-2

Find the center of mass of a uniform stick of mass M and length L.

This simple example, in which we already know the result from symmetry, illustrates the technique for setting up the integration indicated in Equation 7-5. We first set up a coordinate system with the x axis along the stick and one end of the stick at the origin as in Figure 7-9. Let us call the mass per unit length of the stick λ. Since the stick is uniform, $\lambda = M/L$. In the figure, we have indicated a mass element dm of length dx at a distance x from the origin. Since M is the total mass of the stick, the mass of an element of length dx is

$$dm = M\frac{dx}{L} = \frac{M}{L}\,dx = \lambda\,dx$$

Equation 7-5 thus gives

$$MX_{\text{cm}} = \int x\,dm = \int_0^L x\lambda\,dx = \frac{\lambda x^2}{2}\bigg|_0^L = \frac{\lambda L^2}{2}$$

Using $\lambda = M/L$, we obtain the expected result

$$X_{\text{cm}} = \frac{\lambda L^2}{2M} = \frac{M}{L}\left(\frac{L^2}{2M}\right) = \frac{1}{2}L$$

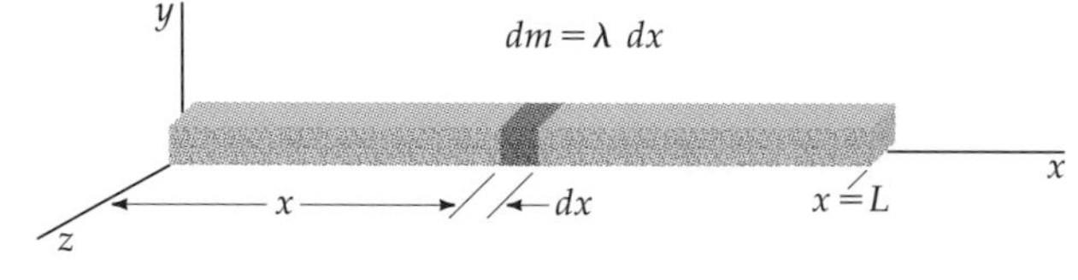

Figure 7-9 Calculation of the center of mass of a uniform stick by integration. The mass element dm at a distance x from the origin is treated as a point particle.

Example 7-3

Find the center of mass of a semicircular hoop.

This calculation is easiest if we choose the origin to be on the line of symmetry of the hoop (the y axis) at the center of curvature as shown in Figure 7-10. Then $X_{\text{cm}} = 0$ because of symmetry. (For every mass element at $+x$, there is an equal mass element at $-x$.) However, Y_{cm} is not zero because all of the mass is at positive values of y. In the figure, we have indicated a mass element of length $ds = R\,d\theta$. Since the total length

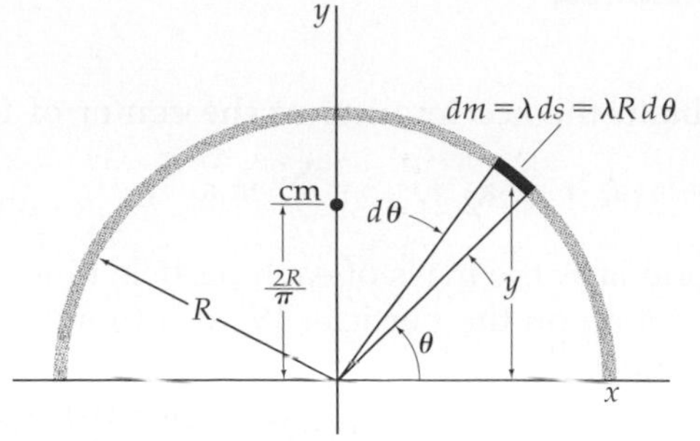

Figure 7-10 Geometry for the calculation of the center of mass of a semicircular hoop by integration. The center of mass lies on the y axis.

of the hoop is πR, the mass per unit length is $\lambda = M/\pi R$, where M is the total mass. The mass of the element chosen is thus

$$dm = \lambda\ ds = \lambda R\ d\theta$$

The y coordinate of the mass element is related to the angle θ by $y = R \sin \theta$. The angle θ varies from 0 to π. We thus have

$$MY_{\text{cm}} = \int y\ dm = \int y\lambda\ ds = \int y\lambda R\ d\theta$$

$$= \int_0^{\pi} (R \sin \theta)\lambda R\ d\theta = R^2\lambda \int_0^{\pi} \sin \theta\ d\theta = 2R^2\lambda$$

since

$$\int_0^{\pi} \sin \theta\ d\theta = -\cos \theta \Big|_0^{\pi} = 2$$

Using $\lambda = M/\pi R$, we have

$$MY_{\text{cm}} = 2R^2 \frac{M}{\pi R}$$

$$Y_{\text{cm}} = \frac{2R}{\pi}$$

The center of mass is not within the body in this case.

Figure 7-11 shows how the location of the center of mass can be found by suspending a semicircular hoop from one end and then from some other point.

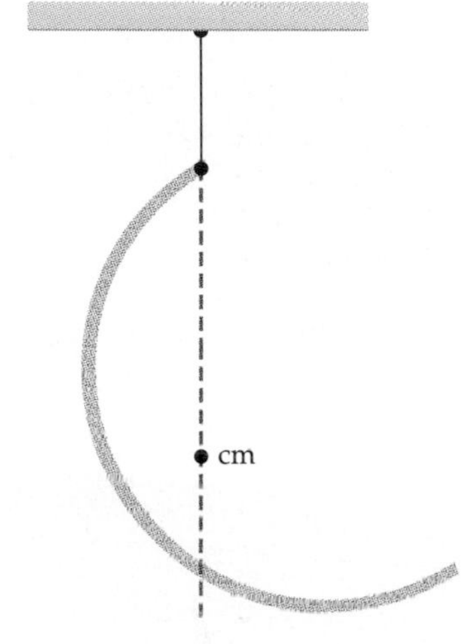

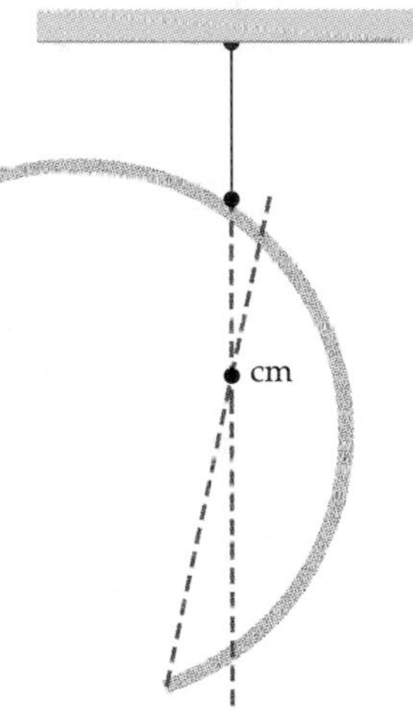

Figure 7-11 The center of mass of a semicircular hoop can be found by suspending it from two points.

Question

1. Give an example of a three-dimensional object that has no mass at its center of mass.

7-2 Motion of the Center of Mass of a System

The motion of an extended object such as the baton in Figure 7-1 is complicated. To describe it, we first describe the motion of the center of mass. We find the velocity of the center of mass by differentiating Equation 7-4 with respect to time. We have

$$M\mathbf{R}_{\text{cm}} = \sum_i m_i \mathbf{r}_i$$

$$M\frac{d\mathbf{R}_{\text{cm}}}{dt} = m_1 \frac{d\mathbf{r}_1}{dt} + m_2 \frac{d\mathbf{r}_2}{dt} + \cdots = \sum_i m_i \frac{d\mathbf{r}_i}{dt}$$

or

$$M\mathbf{V}_{\text{cm}} = m_1\mathbf{v}_1 + m_2\mathbf{v}_2 + \cdots = \sum_i m_i\mathbf{v}_i \qquad 7\text{-}7$$

We differentiate again to obtain the acceleration of the center of mass:

$$M\mathbf{A}_{\text{cm}} = m_1\mathbf{a}_1 + m_2\mathbf{a}_2 + \cdots = \sum_i m_i\mathbf{a}_i \qquad 7\text{-}8$$

According to Newton's second law, the mass of each particle times its acceleration equals the net force acting on the particle. We can therefore replace the quantity $m_i\mathbf{a}_i$ with $\mathbf{F}_i$, the net force acting on the ith particle.

The forces acting on a particle can be separated into two categories: internal forces due to interactions with other particles within the system and external forces due to agents outside the system:

$$\mathbf{F}_i = m_i\mathbf{a}_i = \mathbf{F}_{i,\text{int}} + \mathbf{F}_{i,\text{ext}}$$

Substituting this into Equation 7-8, we obtain

$$M\mathbf{A}_{\text{cm}} = \sum_i \mathbf{F}_{i,\text{int}} + \sum_i \mathbf{F}_{i,\text{ext}} \qquad 7\text{-}9$$

According to Newton's third law, for each internal force acting on one particle, there is an equal but opposite force acting on another particle. For example, if particle m_1 exerts a force on particle m_2, particle m_2 exerts an equal and opposite force on m_1. The internal forces thus occur in pairs of equal and opposite forces. When we sum over all the particles in the system, the internal forces cancel out and we are left with only the external forces. Equation 7-9 then becomes

Newton's second law for a system

$$\mathbf{F}_{\text{net,ext}} = \sum_i \mathbf{F}_{i,\text{ext}} = M\mathbf{A}_{\text{cm}} \qquad 7\text{-}10$$

This equation states that the net external force acting on the system equals the total mass M of the system times the acceleration of the center of mass $\mathbf{A}_{\text{cm}}$. This has the same form as Newton's second law for a single particle of mass M located at the center of mass and under the influence of the net external force. So

> The center of mass of a system moves like a particle of mass $M = \Sigma\, m_i$ under the influence of the net external force acting on the system.

This theorem is important because it shows us how to describe the motion of one point, the center of mass, for any system of particles no matter how extensive the system may be. The center of mass of the system behaves just like a single particle that is subject only to the external forces. The individual motions of the particles of the system are usually much more complex. For example, the motion of a pair of masses connected by a spring and thrown into the air is quite involved. The masses tumble and turn as they move and oscillate along the line joining them. But the center of mass moves as if it were a single particle—it follows a simple parabolic trajectory.

This theorem for the motion of the center of mass is the justification for our earlier treatment of large objects as point particles. All large objects can be considered to be made up of many small particles whose motions are governed by Newton's laws. No matter how complicated the motion of the object, the center of mass moves like a simple particle. The baton thrown into the air in Figure 7-1 is an example. The only external force acting is that of gravity so the center of mass of the baton moves in a simple parabolic path. We note that Equation 7-10 does not give a complete description of the complicated motion of the baton. It applies only to the motion of one point, the center of mass.

Figure 7-12 shows two particles of equal mass m connected by a massless spring on a frictionless table. An external force $\mathbf{F}_{\text{app}}$ is applied to one of the particles. The vertical force of gravity acting on each particle is balanced by the normal force exerted by the table. The net external force acting on this system is thus the applied force $\mathbf{F}_{\text{app}}$. Since the total mass of the system is $2m$, the center of mass of this system, which lies midway between the particles, has an acceleration given by

$$\mathbf{A}_{\text{cm}} = \frac{\mathbf{F}_{\text{app}}}{2m}$$

If we have the same two masses but no spring, that is, if we have two unconnected masses on a table, and an external force $\mathbf{F}_{\text{app}}$ is applied to one of them (Figure 7-13), the acceleration of the center of mass will be the same as if the masses were connected by the spring. However, the internal motions of the masses of the two systems will be quite different. Thus, Equation 7-10 applies to any system even if the parts of the system are not interacting, and even if the external forces are applied to only parts of the system.

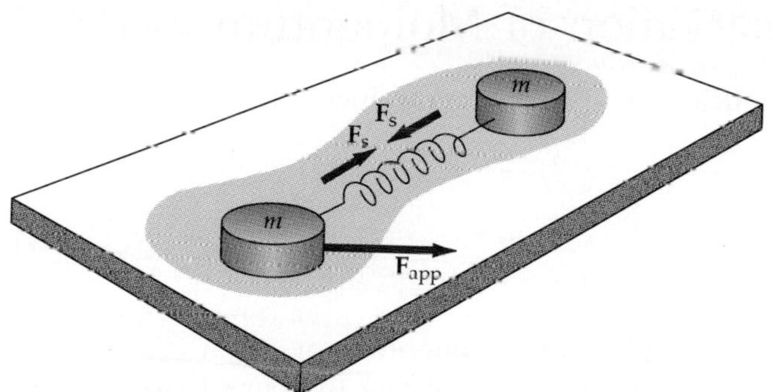

Figure 7-12 Two particles of equal mass m connected by a spring. The forces exerted by the spring are internal forces. These forces cancel when all the forces acting on the system are added up. The net force acting on the two-particle system is therefore the applied force $\mathbf{F}_{\text{app}}$.

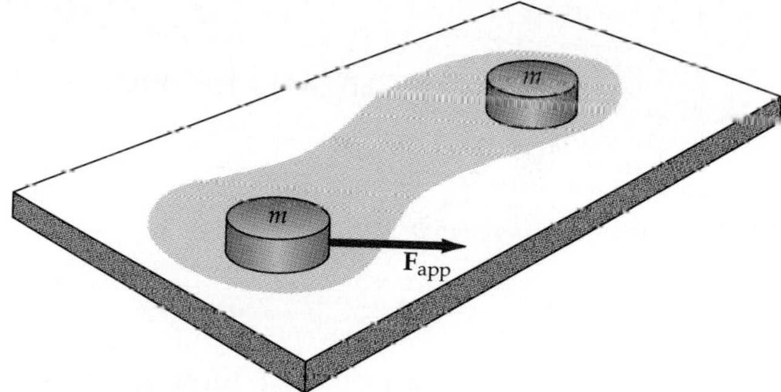

Figure 7-13 The same two particles as Figure 7-12 without the spring. The net force acting on the two-particle system is the applied force $\mathbf{F}_{\text{app}}$, even though this force acts on only one of the particles. The acceleration of the center of mass of the system is $\mathbf{F}_{\text{app}}/2m$.

Figure 7-14 shows a projectile that explodes into two equal parts at the top of its flight so that one piece falls straight down. Since the only external force acting on the projectile whether it is in one piece or two is that of gravity, the center of mass continues on its parabolic path just as if there had been no explosion.

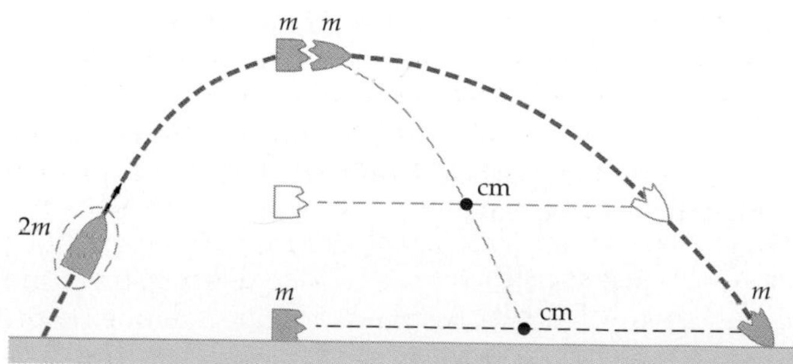

Figure 7-14 A projectile that explodes into two equal pieces at the top of its trajectory. The center of mass, which is midway between the fragments, continues on its original parabolic path.

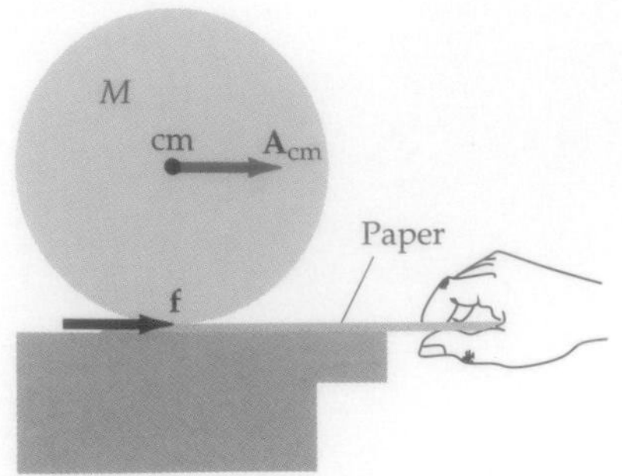

Figure 7-15 A cylinder resting on a piece of paper. When the paper is pulled to the right, the net external force acting on the cylinder is that of friction to the right, so the center of mass of the cylinder must accelerate toward the right. The cylinder rolls backward relative to the paper because the acceleration of the paper is greater than that of the cylinder.

Figure 7-15 shows a cylinder resting on a piece of paper on a table. When the paper is pulled to the right, the cylinder rolls backward on the paper. However, if you mark the original position of the cylinder on the table, you will notice that the cylinder moves to the right relative to the table. (That is, the paper moves further to the right than the cylinder does. So, relative to the paper, the cylinder moves to the left.) This must be because the net external force acting on the cylinder is that of friction between it and the paper, and this force is to the right. Thus, the center of mass of the cylinder must accelerate to the right in the direction of the net force.

Questions

2. A self-made man is said to have pulled himself up by his bootstraps. Discuss this act from the point of view of internal and external forces and the motion of the center of mass of the system.
3. If only external forces can cause the center of mass of a system to accelerate, how can an automobile be accelerated by its motor?

7-3 The Conservation of Momentum

The **momentum** of a particle is defined as the product of its mass and velocity:

Momentum of a particle defined

$$\mathbf{p} = m\mathbf{v} \tag{7-11}$$

Momentum is a vector quantity. The momentum of a particle may be thought of as measuring the difficulty of bringing the particle to rest. For example, a heavy truck has more momentum than a light car traveling at the same speed. It takes a greater force to stop the truck than it does the car in a given time. (The quantity $m\mathbf{v}$ is sometimes referred to as the *linear* momentum of a particle to distinguish it from the *angular* momentum, which will be discussed in Chapter 8.) Newton's second law can be written in terms of the momentum of a particle. Differentiating Equation 7-11, we obtain

$$\frac{d\mathbf{p}}{dt} = \frac{d(m\mathbf{v})}{dt} = m\frac{d\mathbf{v}}{dt} = m\mathbf{a}$$

Substituting the force $\mathbf{F}_{\text{net}}$ for $m\mathbf{a}$, we obtain

$$\mathbf{F}_{\text{net}} = \frac{d\mathbf{p}}{dt} \tag{7-12}$$

Thus, the net force acting on a particle equals the time rate of change of the particle's linear momentum. Newton's original statement of his second law was in fact in this form.

The concept of momentum is important because, if no net external force acts on a system of particles, the total momentum of the system is conserved; that is, it remains constant over time. Consider two particles that exert equal and opposite forces on each other but have no other forces acting on them. If $\mathbf{F}_{12}$ is the force exerted by particle 1 on particle 2 and $\mathbf{F}_{21}$ is that exerted by particle 2 on particle 1, we have

$$\mathbf{F}_{21} = \frac{d\mathbf{p}_1}{dt}$$

and

$$\mathbf{F}_{12} = \frac{d\mathbf{p}_2}{dt}$$

Adding these equations and using the fact that by Newton's third law $\mathbf{F}_{12} = -\mathbf{F}_{21}$, we obtain

$$0 = \frac{d\mathbf{p}_1}{dt} + \frac{d\mathbf{p}_2}{dt}$$

$$\frac{d(\mathbf{p}_1 + \mathbf{p}_2)}{dt} = 0$$

or

$$\mathbf{p}_1 + \mathbf{p}_2 = \text{constant}$$

This result can be extended to systems of many particles. The total momentum **P** of the system is the sum of the momenta of the individual particles:

$$\mathbf{P} = \sum_i m_i\mathbf{v}_i = \sum_i \mathbf{p}_i$$

According to Equation 7-7, the total momentum of the system equals the total mass M times the velocity of the center of mass:

$$\mathbf{P} = \sum_i m_i\mathbf{v}_i = M\mathbf{V}_{\text{cm}} \qquad 7\text{-}13$$

Total momentum of a system

Differentiating this equation, we obtain

$$\frac{d\mathbf{P}}{dt} = M\frac{d\mathbf{V}_{\text{cm}}}{dt} = M\mathbf{A}_{\text{cm}}$$

But according to Equation 7-10, the mass times the acceleration of the center of mass equals the net external force acting on the system. Thus,

$$\sum_i \mathbf{F}_{i,\text{ext}} = \mathbf{F}_{\text{net,ext}} = \frac{d\mathbf{P}}{dt} \qquad 7\text{-}14$$

When the net external force acting on a system of particles is zero, the rate of change of the total momentum is zero, and the total momentum of the system remains constant:

$$\mathbf{P} = M\mathbf{V}_{\text{cm}} = \sum_i m_i\mathbf{v}_i = \text{constant} \qquad 7\text{-}15$$

Conservation of momentum

This result is known as the **law of conservation of momentum:**

> If the net external force on a system is zero, the velocity of the center of mass of the system is constant and the total momentum of the system is conserved; that is, it remains constant.

This law is one of the most important in physics. It applies, for example, to any system isolated from its surroundings so that there are no external forces acting on it. It is more widely applicable than the law of conservation of mechanical energy because internal forces exerted by one particle in the system on another are often not conservative. Thus, these internal forces can change the total mechanical energy of the system, but since they always occur in pairs, they cannot change the total momentum of the system. The conservation of momentum is particularly useful when considering collisions, which we will do in Section 7-6.

Example 7-4

A man of mass 70 kg and a boy of mass 35 kg are standing together on a smooth ice surface for which friction is negligible. If they push each other apart and the man moves away with speed of 0.3 m/s relative to the ice, how far apart are they after 5 s (Figure 7-16)?

We take the man and boy together to be the system. The force exerted by the man on the boy is equal and opposite to that exerted by the boy on the man. The force of gravity on each is balanced by a corresponding normal force exerted by the ice. Since there is no friction, the net external force on the system is zero. The acceleration of the center of mass is therefore zero, and since it is initially at rest, it remains at rest at its initial position. Since the man and boy are standing at rest initially, the total momentum of the system is zero, and since the net external force is zero, the total momentum remains zero. Therefore, after they push each other, they must have equal and opposite momenta. Since the motion takes place in one dimension, we can take care of the vector nature of the momenta by choosing the momentum to the right to be positive and that to the left to be negative. If the man is moving to the right, his momentum is

$$p_{\mathrm{m}} = m_{\mathrm{m}}v_{\mathrm{m}} = (70\ \mathrm{kg})(0.3\ \mathrm{m/s}) = 21\ \mathrm{kg{\cdot}m/s}$$

The momentum of the boy is

$$p_{\mathrm{b}} = m_{\mathrm{b}}v_{\mathrm{b}} = (35\ \mathrm{kg})v_{\mathrm{b}}$$

Setting the total momentum equal to zero, we obtain

$$p_{\mathrm{m}} + p_{\mathrm{b}} = 21\ \mathrm{kg{\cdot}m/s} + (35\ \mathrm{kg})v_{\mathrm{b}} = 0$$

$$v_{\mathrm{b}} = -\frac{21\ \mathrm{kg{\cdot}m/s}}{35\ \mathrm{kg}} = -0.6\ \mathrm{m/s}$$

Since the man has twice the mass of the boy and moves in one direction with a speed of 0.3 m/s, the boy moves in the opposite direction with a speed of 0.6 m/s. After 5 s, the man has moved 1.5 m and the boy has moved 3 m, so they are 4.5 m apart.

Note that the mechanical energy of this system is not conserved. The forces the man and boy exert on each other are not conservative. In this case, the mechanical energy of the system is increased as they push each other apart since the kinetic energy was initially zero and the potential energy does not change. This energy comes from a decrease in internal chemical energy of the man and boy.

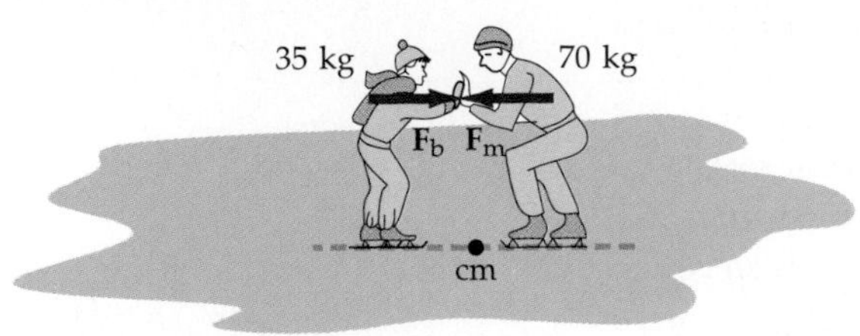

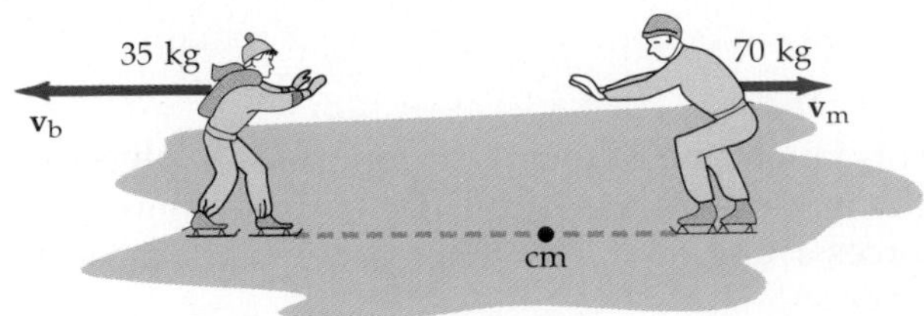

Figure 7-16 The net external force acting on the man–boy system is zero, so the total momentum of the system is conserved and remains zero. The momentum of the man to the right is equal in magnitude to that of the boy to the left. Since the mass of the man is twice that of the boy, his speed must be half that of the boy.

Exercise

Find the final kinetic energy of the man–boy system in Example 7-4. (Answer: 9.45 J)

Example 7-5

A bullet of mass 0.01 kg moves horizontally with a speed of 400 m/s and embeds itself in a block of mass 0.39 kg that is initially at rest on a frictionless table (Figure 7-17). Find (*a*) the final velocity of the bullet and block and (*b*) the initial and final mechanical energy of the bullet–block system.

(*a*) Since there are no external horizontal forces on the bullet–block system, the horizontal component of the momentum of the system is conserved. The total initial horizontal momentum P_i before the bullet strikes the block is just that of the bullet

$$P_i = m_1 v_1 = (0.01\text{ kg})(400\text{ m/s}) = 4\text{ kg·m/s}$$

Afterward, the bullet and block move together with a common velocity V_f. The total final momentum P_f is

$$P_f = (m_1 + m_2)V_f = (0.4\text{ kg})V_f$$

Since the total momentum is conserved, the final momentum equals the initial momentum:

$$(0.4\text{ kg})V_f = 4\text{ kg·m/s}$$

$$V_f = 10\text{ m/s}$$

Since the bullet and block both move together with this final velocity, the center of mass must also move with this velocity.

We could have found the velocity of the center of mass from Equation 7-13:

$$P = MV_{cm} = m_1 v_1 + m_2 v_2$$

$$= (0.01\text{ kg})(400\text{ m/s}) + (0.39\text{ kg})(0)$$

$$= 4\text{ kg·m/s}$$

$$V_{cm} = \frac{4\text{ kg·m/s}}{0.01\text{ kg} + 0.39\text{ kg}} = 10\text{ m/s}$$

(*b*) The initial mechanical energy of the bullet–block system is the kinetic energy of the bullet:

$$E_i = K_i = \tfrac{1}{2}m_1 v_1^2 = \tfrac{1}{2}(0.01\text{ kg})(400\text{ m/s})^2 = 800\text{ J}$$

Since the potential energy of the system does not change, the final mechanical energy is the kinetic energy of the bullet and block moving together:

$$E_f = K_f = \tfrac{1}{2}(0.4\text{ kg})(10\text{ m/s})^2 = 20\text{ J}$$

In this case, most of the initial mechanical energy (780 J out of 800 J) is lost because of the large nonconservative forces between the bullet and the block. Some of the energy goes into thermal energy, some goes into producing permanent deformations of the block and bullet, and a small amount goes into sound waves produced when the bullet strikes the block. A bullet embedding itself in a block is an example of an inelastic collision, which we will study in Section 7-6.

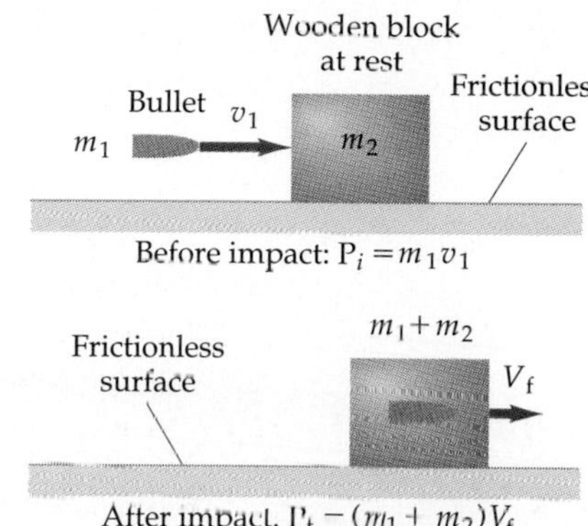

Figure 7-17 Bullet striking a block for Example 7-5. Since there is no net external force acting on the system, the momentum of the system and the velocity of the center of mass are constant.

Questions

4. As a pendulum bob swings back and forth, is the momentum of the bob conserved? Explain why or why not.
5. How is the recoil of a rifle or a cannon related to momentum conservation?
6. A man is stranded in the middle of an ice rink that is perfectly frictionless. How can he get to the edge?
7. A girl jumps from a boat to a dock. Why does she have to jump with more energy than she would need if she were jumping the same distance from one dock to another?

7-4 The Center-of-Mass Reference Frame

As we have seen, when the net external force on a system is zero, the velocity of the center of mass is constant. It is often convenient to choose a coordinate system with the origin at the center of mass. Then, relative to the original coordinate system, this coordinate system moves with a constant velocity V_{cm}. The reference frame attached to the center of mass is called the **center-of-mass reference frame.** Relative to this frame, the velocity of the center of mass is zero. Since the total momentum of a system equals the total mass times the velocity of the center of mass, the total momentum is also zero in the center-of-mass frame. This frame is also called the **zero-momentum reference frame.**

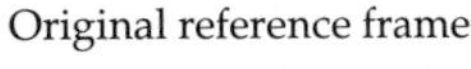

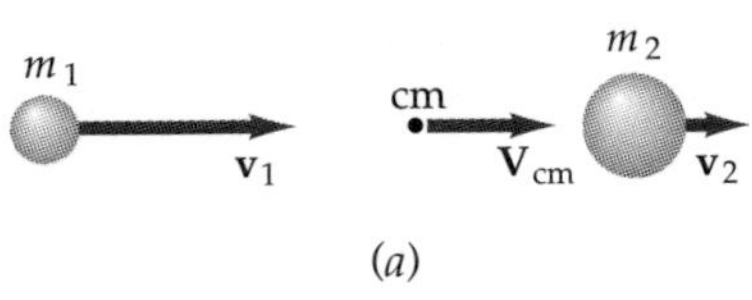

(a)

Center-of-mass reference frame

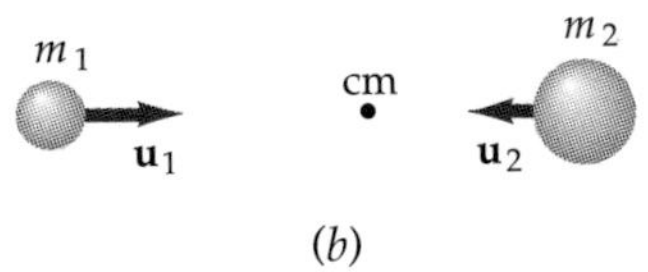

(b)

Figure 7-18 (*a*) Two particles moving in a general reference frame in which the center of mass has a velocity $\mathbf{V}_{cm}$. (*b*) In the center-of-mass reference frame, the center of mass is at rest and the particles have equal and opposite momenta. The velocities in the two frames are related by $\mathbf{u}_1 = \mathbf{v}_1 - \mathbf{V}_{cm}$ and $\mathbf{u}_2 = \mathbf{v}_2 - \mathbf{V}_{cm}$.

In the analysis of problems, it is sometimes convenient to transform from a given frame to the center-of-mass frame and back. This is straightforward. We transform from some original frame to the center-of-mass frame by subtracting the velocity of the center of mass $\mathbf{V}_{cm}$ in the original frame from the velocity of each particle in that frame. Consider, for example, a simple two-particle system in a reference frame in which one particle of mass m_1 is moving with a velocity $\mathbf{v}_1$ and a second particle of mass m_2 is moving with a velocity $\mathbf{v}_2$ (Figure 7-18). The velocities in the center-of-mass frame are $\mathbf{u}_1$ and $\mathbf{u}_2$, given by

$$\mathbf{u}_1 = \mathbf{v}_1 - \mathbf{V}_{cm} \qquad \text{7-16}a$$

and

$$\mathbf{u}_2 = \mathbf{v}_2 - \mathbf{V}_{cm} \qquad \text{7-16}b$$

where, as usual, the velocity of the center of mass in the original frame is

$$\mathbf{V}_{cm} = \frac{m_1\mathbf{v}_1 + m_2\mathbf{v}_2}{m_1 + m_2}$$

Since the total momentum is zero in the center-of-mass frame, the particles have equal and opposite momenta in this frame.

Example 7-6

A 2.5-kg box moves with velocity $\mathbf{v}_1 = 10$ m/s **i** and a 3.5-kg box moves with velocity $\mathbf{v}_2 = -2$ m/s **i**. Find (*a*) the total momentum, (*b*) the velocity of the center of mass, and (*c*) the velocity of each box in the center-of-mass reference frame.

(*a*) The total momentum is

$$\mathbf{P} = m_1\mathbf{v}_1 + m_2\mathbf{v}_2 = (2.5\text{ kg})(10\text{ m/s }\mathbf{i}) + (3.5\text{ kg})(-2\text{ m/s }\mathbf{i})$$
$$= (25\text{ kg·m/s} - 7\text{ kg·m/s})\,\mathbf{i} = 18\text{ kg·m/s }\mathbf{i}$$

(*b*) The velocity of the center of mass is then

$$\mathbf{V}_{cm} = \frac{\mathbf{P}}{M} = \frac{m_1\mathbf{v}_1 + m_2\mathbf{v}_2}{m_1 + m_2} = \frac{18\ \text{kg·m/s}\ \mathbf{i}}{(2.5\ \text{kg} + 3.5\ \text{kg})} = 3\ \text{m/s}\ \mathbf{i}$$

(*c*) The velocities of the boxes in the center-of-mass reference frame are

$$\mathbf{u}_1 = \mathbf{v}_1 - \mathbf{V}_{cm} = 10\ \text{m/s}\ \mathbf{i} - 3\ \text{m/s}\ \mathbf{i} = 7\ \text{m/s}\ \mathbf{i}$$

and

$$\mathbf{u}_2 = \mathbf{v}_2 - \mathbf{V}_{cm} = -2\ \text{m/s}\ \mathbf{i} - 3\ \text{m/s}\ \mathbf{i} = -5\ \text{m/s}\ \mathbf{i}$$

As a check, we compute the total momentum in the center-of-mass frame

$$\mathbf{P}_{cm} = m_1\mathbf{u}_1 + m_2\mathbf{u}_2 = (2.5\ \text{kg})(7\ \text{m/s}\ \mathbf{i}) + (3.5\ \text{kg})(-5\ \text{m/s}\ \mathbf{i})$$
$$= 17.5\ \text{kg·m/s}\ \mathbf{i} - 17.5\ \text{kg·m/s}\ \mathbf{i} = 0$$

In the center-of-mass reference frame, the total momentum is zero.

Example 7-7

Analyze the collision in Example 7-5 in the center-of-mass frame.

The velocity of the center of mass in Example 7-5 was found to be $V_{cm} = 10$ m/s. The velocity of the bullet in the center-of-mass frame is thus

$$u_1 = v_1 - V_{cm} = 400\ \text{m/s} - 10\ \text{m/s} = 390\ \text{m/s}$$

and that of the block is

$$u_2 = v_2 - V_{cm} = 0 - 10\ \text{m/s} = -10\ \text{m/s}$$

In the center-of-mass frame, the initial momentum is zero:

$$P_i = m_1u_1 + m_2u_2 = (0.01\ \text{kg})(390\ \text{m/s}) + (0.39\ \text{kg})(-10\ \text{m/s})$$
$$= 3.90\ \text{kg·m/s} - 3.90\ \text{kg·m/s} = 0$$

After the collision, the two particles stick together and the total momentum is zero, so the velocity of both particles is zero.

The initial kinetic energy of the system in the center-of-mass frame is

$$K_i = \tfrac{1}{2}m_1u_1^2 + \tfrac{1}{2}m_2u_2^2$$
$$= \tfrac{1}{2}(0.01\ \text{kg})(390\ \text{m/s})^2 + \tfrac{1}{2}(0.39\ \text{kg})(-10\ \text{m/s})^2$$
$$= 760.5\ \text{J} + 19.5\ \text{J} = 780\ \text{J}$$

The final kinetic energy in the center-of-mass frame is zero because both particles are at rest in this frame. Thus, all the initial mechanical energy is lost in this frame. Note that the mechanical energy lost is 780 J, the same as was calculated in the original reference frame in which the block was initially at rest.

7-5 Kinetic Energy of a System of Particles

Although the total linear momentum of a system of particles must be constant if the net external force on the system is zero, the total mechanical energy of the system may change. As we saw in Examples 7-4 and 7-5, internal forces that cannot change the total momentum may be nonconservative and thus change the total mechanical energy of the system. In Example

7-4 the mechanical energy increased, whereas in Example 7-5 the mechanical energy decreased. There is an important theorem concerning the kinetic energy of a system of particles that allows us to treat the energy of complex systems more easily and gives us insight into energy changes within a system:

The kinetic energy of a system of particles can be written as the sum of two terms: (1) the kinetic energy associated with the motion of the center of mass, $\frac{1}{2}MV_{\rm cm}^2$, where M is the total mass of the system; and (2) the kinetic energy associated with the motion of the particles of the system relative to the center of mass, $\Sigma\, \frac{1}{2}m_iu_i^2$, where u_i is the velocity of the ith particle relative to the center of mass.

The kinetic energy of motion relative to the center of mass is the kinetic energy as observed in the center-of-mass reference frame.

We will now prove this theorem. The kinetic energy of a system of particles is the sum of the kinetic energies of the individual particles:

$$K = \sum_i \tfrac{1}{2}m_i\mathbf{v}_i^2 = \sum_i \tfrac{1}{2}m_i(\mathbf{v}_i\cdot\mathbf{v}_i)$$

where $\mathbf{v}_i$ is the velocity of the ith particle. From Equations 7-16*a*, and *b*, we know that the velocity of the ith particle can be written as the sum of the velocity of the center of mass $\mathbf{V}_{\rm cm}$ and the velocity of the particle relative to the center of mass $\mathbf{u}_i$, which is the velocity of the particle in the center-of-mass frame:

$$\mathbf{v}_i = \mathbf{V}_{\rm cm} + \mathbf{u}_i$$

The kinetic energy of the system is then

$$K = \sum_i \tfrac{1}{2}m_i(\mathbf{v}_i\cdot\mathbf{v}_i) = \sum_i \tfrac{1}{2}m_i(\mathbf{V}_{\rm cm} + \mathbf{u}_i)\cdot(\mathbf{V}_{\rm cm} + \mathbf{u}_i)$$

$$= \sum_i \tfrac{1}{2}m_i\mathbf{V}_{\rm cm}^2 + \sum_i \tfrac{1}{2}m_i\mathbf{u}_i^2 + \mathbf{V}_{\rm cm}\cdot\sum_i m_i\mathbf{u}_i$$

where in the last term we have removed $\mathbf{V}_{\rm cm}$ from the sum because it is the same for each particle, that is, it refers to the system and not any particular particle. The quantity $\Sigma m_i\mathbf{u}_i$ is the momentum of the system as seen in the center-of-mass reference frame, which is zero. Setting this term equal to zero, we obtain our desired result:

Kinetic energy of a system

$$K = \tfrac{1}{2}MV_{\rm cm}^2 + K_{\rm rel} \qquad 7\text{-}17$$

where $M = \Sigma\, m_i$ is the total mass of the system and

$$K_{\rm rel} = \sum_i \tfrac{1}{2}m_i\mathbf{u}_i^2 \qquad 7\text{-}18$$

is the kinetic energy of motion relative to the center of mass; that is, it is the kinetic energy as seen in the center-of-mass reference frame. This energy is the same in any reference frame because it depends only on the speeds of the particles relative to the center of mass. However, the energy associated with the motion of the center of mass depends on the reference frame. It is $\frac{1}{2}M\mathbf{V}_{\rm cm}^2$ in any frame in which the center of mass is moving with a velocity $\mathbf{V}_{\rm cm}$. In the center-of-mass frame, it is zero. In that case, all the kinetic energy is the energy of relative motion.

We will have many occasions to use Equation 7-17 in the following chapters. For example, the kinetic energy of a rolling ball will be written as the sum of $\frac{1}{2}MV_{cm}^2$ and the energy of relative motion, which in this case is the energy of rotation.

If there are no external forces acting on the system, the velocity of the center of mass is constant and the first term in Equation 7-17 does not change. Only the relative energy can increase or decrease in an isolated system. We will have occasion to use this result when we study collisions in Section 7-6. In collisions, the external forces are negligible, so the kinetic energy of the center of mass remains constant. We saw an example of this in Examples 7-5 and 7-7, in which a bullet embeds itself in a block. In the frame of Example 7-5, in which the block is originally at rest, the initial energy was found to be 800 J. In the center-of-mass frame of Example 7-7, the initial energy was found to be 780 J. All of this mechanical energy was lost in the collision. The energy of the center of mass (in the original frame), $\frac{1}{2}MV_{cm}^2 = \frac{1}{2}(0.4\text{ kg})(10\text{ m/s})^2 = 20\text{ J}$, was not changed by the collision.

Pseudowork

The kinetic energy associated with motion of the center of mass can be related to the net force acting on a system and the displacement of the center of mass. From Newton's second law for a system of particles, we have (Equation 7-10)

$$\mathbf{F}_{\text{net}} = M\mathbf{A}_{\text{cm}} = M\frac{d\mathbf{V}_{\text{cm}}}{dt}$$

If we take the dot product of the net force and the displacement of the center of mass $d\mathbf{R}_{\text{cm}}$, we have

$$\mathbf{F}_{\text{net}}\cdot d\mathbf{R}_{\text{cm}} = M\frac{d\mathbf{V}_{\text{cm}}}{dt}\cdot d\mathbf{R}_{\text{cm}} = M\frac{d\mathbf{R}_{\text{cm}}}{dt}\cdot d\mathbf{V}_{\text{cm}} = M\mathbf{V}_{\text{cm}}\cdot d\mathbf{V}_{\text{cm}}$$

Integrating over some finite displacement, we obtain

$$\int \mathbf{F}_{\text{net}}\cdot d\mathbf{R}_{\text{cm}} = \int_{V_{\text{cm},1}}^{V_{\text{cm},2}} MV_{\text{cm}}\,dV_{\text{cm}} = \tfrac{1}{2}MV_{\text{cm},2}^2 - \tfrac{1}{2}MV_{\text{cm},1}^2 \qquad 7\text{-}19$$

Equation 7-19 looks like the work–energy theorem, but the left side does *not* represent work done on the system because the displacement $d\mathbf{R}_{\text{cm}}$ is the displacement of the center of mass, which is not generally equal to the displacement of the point of application of any of the forces acting on the system. In addition, the right side of Equation 7-19 is not the change in the total energy of the system, but merely the change in the energy associated with the motion of the center of mass. The quantity $\int \mathbf{F}_{\text{net}}\cdot d\mathbf{R}_{\text{cm}}$ is called **pseudowork.** We can understand Equation 7-19 from a simple example of a car accelerating from rest along a road, such that its wheels do not slip. The net force acting on the car is the force of static friction **f** acting in the direction of motion of the car. If V_{cm} is the speed of the center of mass of the car after it has traveled a distance Δx, Equation 7-19 gives

$$f\,\Delta x = \Delta K_{\text{cm}} = \tfrac{1}{2}MV_{\text{cm}}^2$$

This is the same result as is obtained using Newton's second law for a system of particles (Equation 7-10). The pseudowork done by the force of static friction exerted by the road on the car has nothing to do with any energy transfer between the road and the car. In fact, the road does not do any work on the car. The distance Δx is the displacement of the center of mass of the car and not of the point of application of the frictional force which is static. The increase in the mechanical energy of the car results from the conversion of internal chemical energy of the fuel in the car.

7-6 Collisions in One Dimension

In a collision, two objects approach each other, interact strongly, and move apart. Before the collision, when they are far apart, the objects move with constant velocities. After the collision, they move with constant but altered velocities. Usually, we wish to find the final velocities of the objects when their initial velocities and the characteristics of the collision are given.

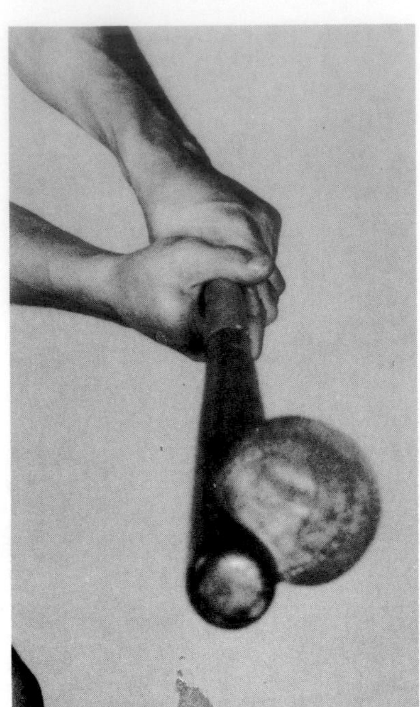

Collision of baseball with bat. Note the deformation of the ball due to the large force exerted by the bat during the brief time of contact. As the ball leaves the bat, it springs back to its original shape, converting the elastic potential energy of deformation into kinetic energy.

A collision can be brief, like the click of two billiard balls, or it can take centuries, as when two stars collide in space. However, in all collisions, the objects interact strongly with each other only during this collision time. When there are external forces present, these forces are much smaller than the forces of interaction during the collision and can be neglected. For example, billiard balls are gradually slowed down by friction with the table, but the forces of friction are negligible compared to the forces of interaction during their brief collision.

When the total kinetic energy of the two objects is the same after the collision as before, the collision is said to be an **elastic collision.** When the total kinetic energy is not the same after the collision, the collision is said to be an **inelastic collision.** Inelastic collisions occur between macroscopic systems when nonconservative forces act that change the mechanical energy of the system. An example is a ball of putty dropped onto the floor. Inelastic collisions occur between microscopic systems when one of the systems is left in a different state of internal energy. An example is the scattering of an electron by an atom. If the internal energy of the atom is not changed, the total kinetic energy of the atom and the electron is the same after the collision as before and the collision is elastic. However, sometimes the internal energy of the atom is changed (that is, the atom is excited to a higher internal energy state), and the total kinetic energy of the atom and the electron is less after the collision, in which case the collision is inelastic.

In an inelastic collision the kinetic energy relative to the center of mass is changed, but the kinetic energy of the center of mass $\frac{1}{2}MV_{\text{cm}}^2$ remains constant because the velocity of the center of mass is unchanged when the external forces acting on the system are negligible. In one special case, all the relative energy is lost and the objects move together with the velocity of the center of mass. Such a collision is called a **perfectly inelastic collision.** The bullet embedding itself in a block in Example 7-5 is a typical example of a perfectly inelastic collision. In most collisions, only part of K_{rel} is lost; these collisions are neither elastic nor perfectly inelastic.

In this section, we consider collisions in one dimension only. Figure 7-19 shows an object of mass m_1 moving with an initial velocity $v_{1\text{i}}$ toward a second object of mass m_2 moving with an initial velocity $v_{2\text{i}}$ (which we assume to be less than $v_{1\text{i}}$ so that the bodies collide). Let $v_{1\text{f}}$ and $v_{2\text{f}}$ be the final velocities of the objects after the collision. The velocities can be positive or negative, depending on whether the objects are moving to the right or to the left. Conservation of momentum gives

$$m_1v_{1\text{f}} + m_2v_{2\text{f}} = m_1v_{1\text{i}} + m_2v_{2\text{i}} \qquad 7\text{-}20$$

Equation 7-20 provides one relation between the two unknown velocities $v_{1\text{f}}$ and $v_{2\text{f}}$ (if the initial velocities $v_{1\text{i}}$ and $v_{2\text{i}}$ are given). To find these unknowns, we must have a second relation. The second relation comes from energy considerations.

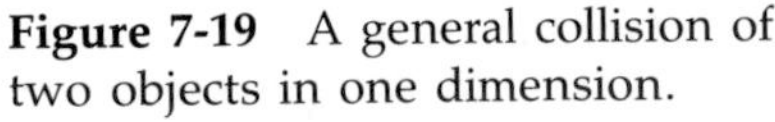

Figure 7-19 A general collision of two objects in one dimension.

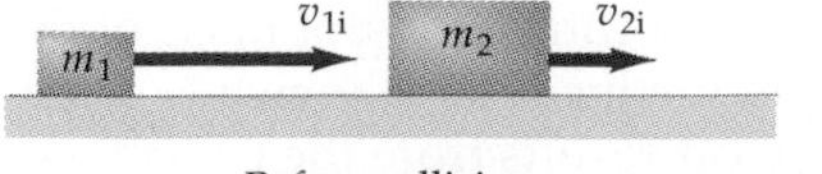

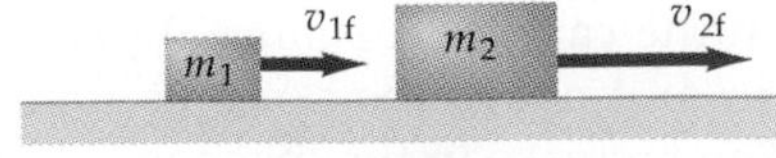

Elastic Collisions

For elastic collisions, the initial and final energies are equal. If there is no change in the internal potential energy of the system, the final kinetic energy equals the initial kinetic energy:

$$\tfrac{1}{2}m_1v_{1f}^2 + \tfrac{1}{2}m_2v_{2f}^2 = \tfrac{1}{2}m_1v_{1i}^2 + \tfrac{1}{2}m_2v_{2i}^2 \qquad 7\text{-}21$$

Equations 7-20 and 7-21 are sufficient to determine the final velocities of the two objects, from their initial velocities. However, the quadratic nature of Equation 7-21 presents algebraic difficulties in solving elastic collision problems. These difficulties can be avoided by combining the equations to obtain a third equation that is linear in the velocities. Rearranging Equation 7-21, we have

$$m_2(v_{2f}^2 - v_{2i}^2) = m_1(v_{1i}^2 - v_{1f}^2)$$

or

$$m_2(v_{2f} - v_{2i})(v_{2f} + v_{2i}) = m_1(v_{1i} - v_{1f})(v_{1i} + v_{1f}) \qquad 7\text{-}22$$

Similarly, Equation 7-20 can be rearranged to yield

$$m_2(v_{2f} - v_{2i}) = m_1(v_{1i} - v_{1f}) \qquad 7\text{-}23$$

Dividing Equation 7-22 by Equation 7-23, we obtain

$$v_{2f} + v_{2i} = v_{1i} + v_{1f}$$

which can be written

$$v_{2f} - v_{1f} = -(v_{2i} - v_{1i}) \qquad 7\text{-}24$$

The relative velocity $v_2 - v_1$ is the velocity of object 2 as seen from object 1. If the objects are to collide, $v_{2i} - v_{1i}$ must be negative (provided that object 1 is on the left). Then $-(v_{2i} - v_{1i})$ is the speed of approach of the two objects. After the collision, the objects move away from each other, so $v_{2f} - v_{1f}$ must be positive. It is the speed of recession after the collision. Equation 7-24 states the following important result:

> For elastic collisions, the relative speed of recession after the collision equals the relative speed of approach before the collision.

In solving problems involving elastic collisions, it is nearly always easiest to use Equations 7-20 and 7-24 to find the final velocities, thereby avoiding the quadratic terms in the conservation-of-energy equation 7-21. It should be remembered that Equation 7-24 was derived using the conservation of mechanical energy, so it applies only to elastic collisions.

Example 7-8

A 4-kg block moving to the right at 6 m/s makes an elastic collision with a 2-kg block moving to the right at 3 m/s (Figure 7-20). Find the velocities v_{1f} and v_{2f}.

The conservation-of-momentum Equation 7-20 gives

$$(4\text{ kg})v_{1f} + (2\text{ kg})v_{2f} = (4\text{ kg})(6\text{ m/s}) + (2\text{ kg})(3\text{ m/s}) = 30\text{ kg·m/s}$$

or

$$4v_{1f} + 2v_{2f} = 30\text{ m/s}$$

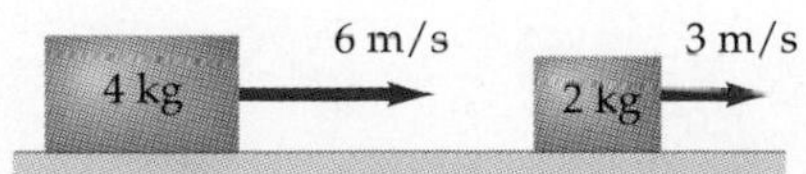

Figure 7-20 Collision of two blocks for Example 7-8.

The velocity of block 2 relative to block 1 before the collision is

$$v_{2i} - v_{1i} = 3 \text{ m/s} - 6 \text{ m/s} = -3 \text{ m/s}$$

The speed of approach is thus 3 m/s, which must equal the speed of recession. Equation 7-24 then gives

$$v_{2f} - v_{1f} = -(-3 \text{ m/s}) = +3 \text{ m/s}$$

Combining this equation with $4v_{1f} + 2v_{2f} = 30$ m/s and solving for the two final velocities, we obtain

$$v_{2f} = 7 \text{ m/s} \qquad \text{and} \qquad v_{1f} = 4 \text{ m/s}$$

To check our results, we can compute the initial and final kinetic energies of the blocks:

$$K_i = \tfrac{1}{2}(4 \text{ kg})(6 \text{ m/s})^2 + \tfrac{1}{2}(2 \text{ kg})(3 \text{ m/s})^2 = 72 \text{ J} + 9 \text{ J} = 81 \text{ J}$$

$$K_f = \tfrac{1}{2}(4 \text{ kg})(4 \text{ m/s})^2 + \tfrac{1}{2}(2 \text{ kg})(7 \text{ m/s})^2 = 32 \text{ J} + 49 \text{ J} = 81 \text{ J}$$

Our results are consistent with conservation of energy.

In the center-of-mass reference frame, an elastic collision is particularly simple. The velocity of each object is merely reversed by the collision. We can prove this by writing the initial and final kinetic energies in terms of the momenta of the bodies. Let u_{1i} and u_{2i} be the initial velocities relative to the center of mass. The initial momentum of the first object is

$$p_{1i} = m_1 u_{1i}$$

and that of the second object is

$$p_{2i} = m_2 u_{2i}$$

Since the total momentum in the center-of-mass reference frame is zero, we have

$$p_{2i} = -p_{1i} \tag{7-25}$$

The kinetic energy of an object can be written in terms of its momentum:

$$K = \tfrac{1}{2}mu^2 = \frac{(mu)^2}{2m}$$

or

$$K = \frac{p^2}{2m} \tag{7-26}$$

The initial kinetic energy of the two bodies is thus

$$K_i = \frac{p_{1i}^2}{2m_1} + \frac{p_{2i}^2}{2m_2} = p_{1i}^2\left(\frac{1}{2m_1} + \frac{1}{2m_2}\right)$$

where we have used $p_{2i}^2 = p_{1i}^2$. Similarly, the final kinetic energy written in terms of the final momenta of the bodies is

$$K_f = \frac{p_{1f}^2}{2m_1} + \frac{p_{2f}^2}{2m_2} = p_{1f}^2\left(\frac{1}{2m_1} + \frac{1}{2m_2}\right)$$

where again $p_{2f}^2 = p_{1f}^2$. Setting the final kinetic energy equal to the initial kinetic energy gives

$$p_{1f}^2 = p_{1i}^2$$

or

$$p_{1f} = \pm p_{1i} \qquad \text{7-27}a$$

This also implies that

$$p_{2f} = \pm p_{2i} \qquad \text{7-27}b$$

The plus signs in Equations 7-27*a* and *b* correspond to no collision at all, which would be the case if the bodies were initially heading away from each other. Since only conservation of momentum and conservation of energy were used to derive these equations, it is not surprising that the results include the case of no collision. If there is a collision, we have

$$p_{1f} = -p_{1i} \quad \text{and} \quad p_{2f} = -p_{2i} \qquad \text{7-28}$$

Since the velocity of each object is its momentum divided by its mass, Equation 7-28 implies

$$u_{1f} = -u_{1i} \quad \text{and} \quad u_{2f} = -u_{2i} \qquad \text{7-29}$$

which is the result we desired to prove. Thus, in an elastic collision in one dimension, each object is merely turned around in the center-of-mass reference frame and leaves the collision with the same speed and energy it had before the collision.

(*a*) Initial conditions

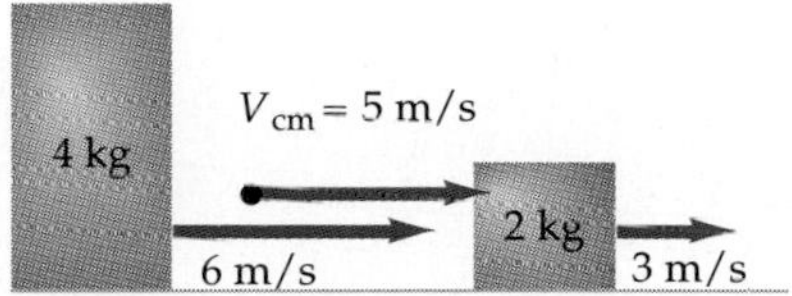

(*b*) Step 1; Transform to the center-of-mass frame by subtracting V_{cm}

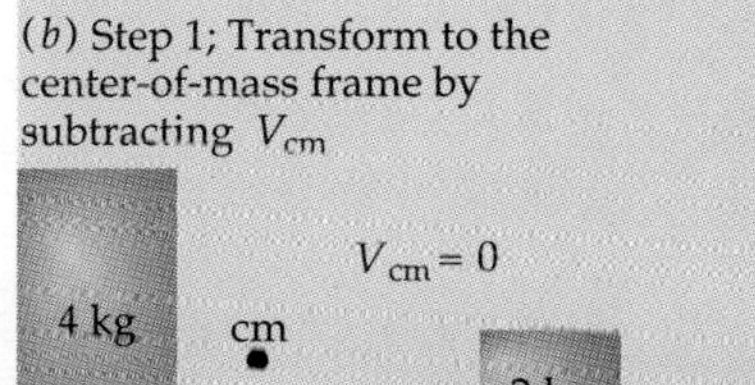

(*c*) Step 2; Solve collision

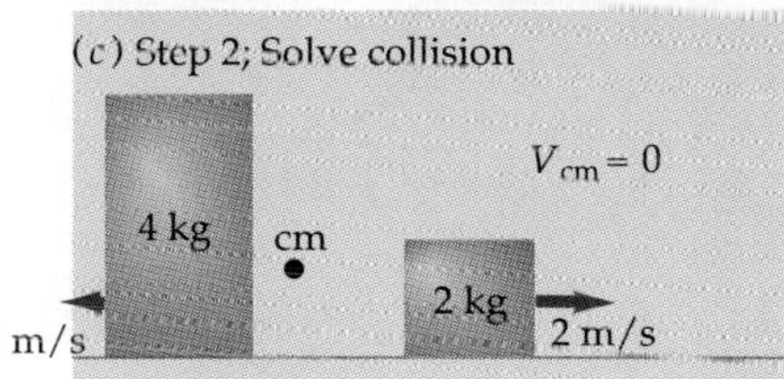

(*d*) Step 3; Transform back to the original frame by adding V_{cm}

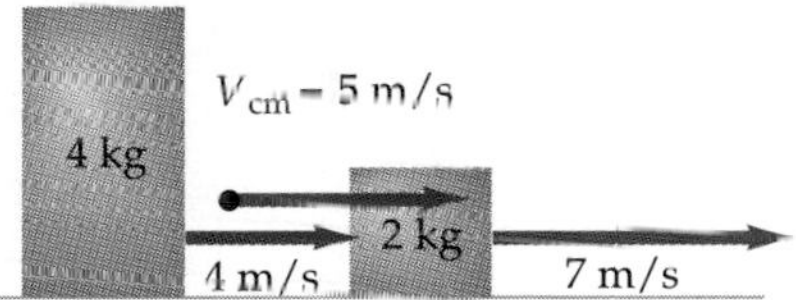

Figure 7-21 Finding the final velocities after the collision in Example 7-8 by transferring to the center-of-mass reference frame. (*a*) In the original frame, the center of mass is moving to the right at 5 m/s. (*b*) We transfer to the center-of-mass frame by subtracting 5 m/s from the initial velocity of each block. The center of mass is at rest in this frame, and the blocks move toward each other with equal and opposite momenta. (*c*) Each block has its velocity reversed in an elastic collision. (*d*) We transfer back to the original frame by adding 5 m/s to the final velocity of each block.

Example 7-9

Find the final velocities for the elastic collision in Example 7-8 by transforming to the center-of-mass reference frame.

We first find the velocity of the center of mass in the original reference frame (Figure 7-21*a*):

$$MV_{cm} = m_1 v_{1i} + m_2 v_{2i}$$

$$(6\text{ kg})V_{cm} = (4\text{ kg})(6\text{ m/s}) + (2\text{ kg})(3\text{ m/s}) = 30\text{ kg}\cdot\text{m/s}$$

$$V_{cm} = 5\text{ m/s}$$

We now transform to the center-of-mass reference frame (Figure 7-21*b*) by subtracting the velocity of the center of mass from the initial velocities of the blocks in the original frame. We have

$$u_{1i} = v_{1i} - V_{cm} = 6\text{ m/s} - 5\text{ m/s} = 1\text{ m/s}$$

and

$$u_{2i} = v_{2i} - V_{cm} = 3\text{ m/s} - 5\text{ m/s} = -2\text{ m/s}$$

Figures 7-21*b* and *c* show the collision in the center-of-mass frame. Since each object is reversed by the collision, the final velocities in this frame are

$$u_{1f} = -u_{1i} = -(1\text{ m/s}) = -1\text{ m/s}$$

and

$$u_{2f} = -u_{2i} = -(-2\text{ m/s}) = +2\text{ m/s}$$

We now transform back into the original frame by adding $V_{cm} = 5$ m/s to each velocity (Figure 7-21*d*):

$$v_{1f} = u_{1f} + V_{cm} = -1\text{ m/s} + 5\text{ m/s} = 4\text{ m/s}$$

and

$$v_{2f} = u_{2f} + V_{cm} = 2\text{ m/s} + 5\text{ m/s} = 7\text{ m/s}$$

which are the same as the results we found in Example 7-8.

Example 7-10

An object of mass m_1 moving with speed v_{1i} makes an elastic collision with a second object of mass m_2 that is initially at rest (Figure 7-22). What is the velocity of the second object after the collision?

This problem is most easily worked in the center-of-mass frame. We first find the velocity of the center of mass in the original frame by setting the total momentum equal to $m_1 v_{1i}$:

$$(m_1 + m_2)V_{cm} = m_1 v_{1i}$$

$$V_{cm} = \frac{m_1 v_{1i}}{m_1 + m_2}$$

To transform to the center-of-mass frame, we subtract V_{cm} from the velocity of each object in the original frame. Since the second object was originally at rest, its velocity in the center-of-mass frame is

$$u_{2i} = -V_{cm}$$

After the collision, the velocity of the second object is

$$u_{2f} = +V_{cm}$$

We transform back to the original frame by adding V_{cm} to each velocity in the center-of-mass frame. The final velocity of the object that was originally stationary is therefore

$$v_{2f} = +2V_{cm} = \frac{2m_1 v_{1i}}{m_1 + m_2} \qquad \text{7-30}$$

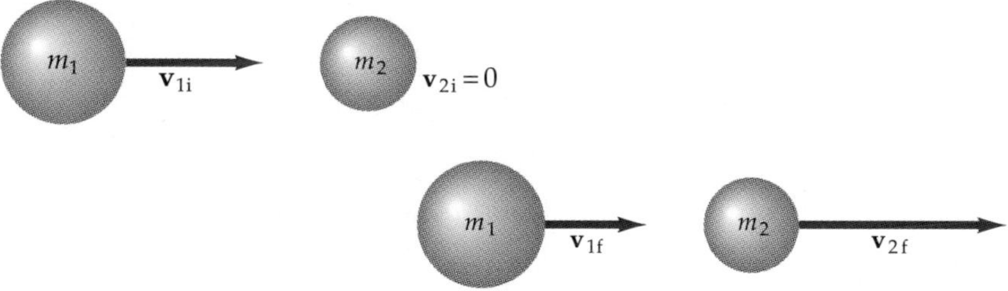

Figure 7-22 Elastic collision of an object of mass m_1 moving with a velocity $\mathbf{v}_{1i}$ and a second object that is initially at rest for Example 7-10.

If the first object in Example 7-10 is much more massive than the second one that is originally at rest, we can neglect m_2 compared to m_1 in the denominator of Equation 7-30. Then the final velocity of the second object is approximately

$$v_{2f} \approx 2v_{1i}$$

We can understand this as follows. The massive object will not be affected very much by a collision with a much less massive stationary object. A cannonball, for example, will hardly be slowed down if it collides with a stationary beach ball. Before the collision, the relative velocity of approach is v_{1i}. Then after the collision, the relative velocity of recession must be v_{1i}. For a first approximation, we can neglect any change in the velocity of the first object. Since it continues to move with a velocity v_{1i}, the velocity of the smaller object must be $2v_{1i}$.

An example of such a collision is that of a golf club with a golf ball. Careful measurement of the distances the club and the ball travel between successive flashes in Figure 7-23 shows that the speed of the ball is indeed twice that of the club.

Although energy is conserved in an elastic collision, energy is transferred from one object to the other. In cases like Example 7-10, in which one object is initially stationary, the energy transferred to the initially stationary

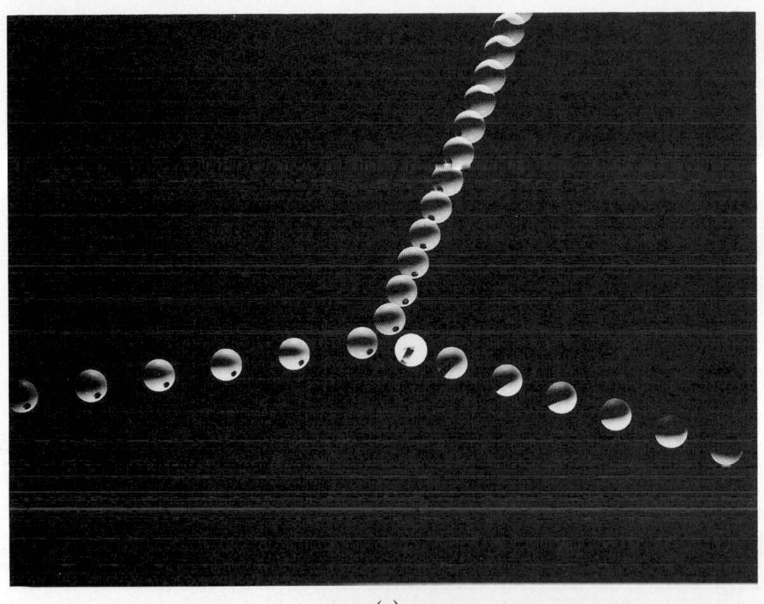

(*a*)

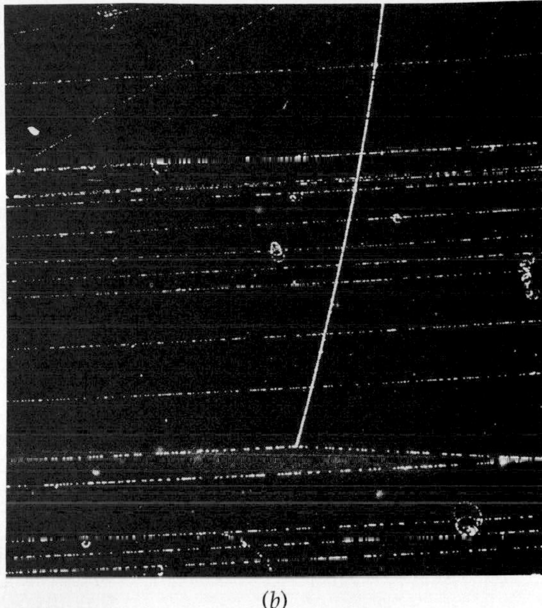

(*b*)

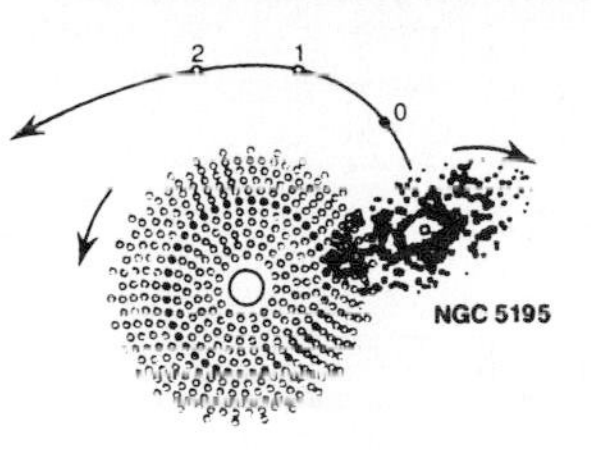

(*c*)

(*d*)

(*a*) Multiflash photograph of an elastic, off-center collision of two balls of equal mass. The dotted ball is incident from the left and strikes the striped ball, which is initially at rest. The final velocities of the two balls are perpendicular to each other. (*b*) Proton-proton collision in a liquid hydrogen bubble chamber. The incident proton enters from the left and interacts with a stationary proton in the chamber. The two particles then move off at right angles after the collision. The slight curvature of the tracks is due to a magnetic field. (*c*) Computer simulation of the collision of two galaxies [shown in (*d*)] at four different times. The time scale is in units of 100 million years. (*d*) Photograph from the Kitt Peak National Observatory of the collision of the large spiral galaxy NGC 5194 (bottom), which has a mass of about 10^{11} solar masses, and a much younger galaxy NGC 5195 (top), which is about three times less massive.

7-8 Impulse and Time Average of a Force

In our study of collisions, we have said little about the forces of interaction between the bodies except that they are usually very large and act for a very short time. Figure 7-30 shows the variation over time in the magnitude of a typical force exerted by one body on another during a collision. Before time t_i the bodies are apart and the force is zero. When the bodies come into contact, the force rises steeply and then falls back to zero at time t_f, when the bodies separate. The time of contact $\Delta t = t_f - t_i$ is usually very small, perhaps only about a millisecond. The **impulse I** of the force is a vector defined by

Impulse defined

$$\mathbf{I} = \int_{t_i}^{t_f} \mathbf{F}\, dt \qquad 7\text{-}44$$

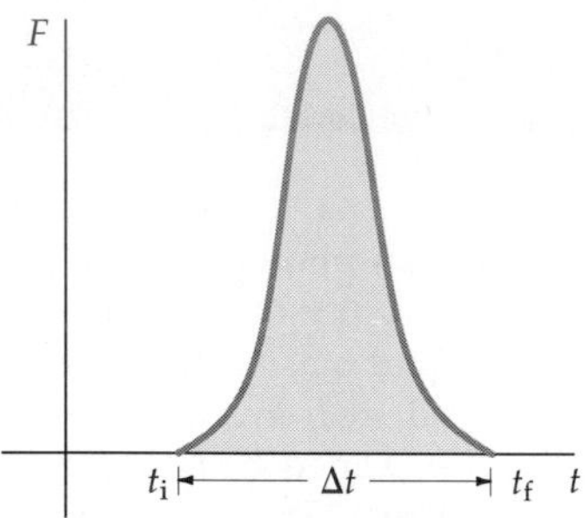

Figure 7-30 Typical time variation of the force exerted by one object on another during a collision. The force becomes very large, but it acts for only a very short time. The area under the *F*-versus-*t* curve is the magnitude of the impulse of the force. The impulse exerted on an object equals the change in its momentum.

The area under the *F*-versus-*t* curve is the magnitude of the impulse of the force. Assuming that **F** is the net force and using Newton's second law $\mathbf{F} = d\mathbf{p}/dt$, we can see that the impulse equals the total change in momentum during the time interval:

$$\mathbf{I} = \int_{t_i}^{t_f} \mathbf{F}\, dt = \int_{t_i}^{t_f} \frac{d\mathbf{p}}{dt}\, dt = \mathbf{p}_f - \mathbf{p}_i = \Delta\mathbf{p} \qquad 7\text{-}45$$

From Equation 7-45, we can see that the units of impulse are newton-seconds or kilogram-meters per second.

For a general force **F**, the impulse depends on the times t_i and t_f, but the forces that occur in collisions are zero except during a very small time interval as shown in Figure 7-30. For these forces, the impulse does not depend on the time interval as long as t_i is any time before the collision and t_f is any time afterward. It is for this type of force that the concept of impulse is most useful.

The **time average of a force** for the interval $\Delta t = t_f - t_i$ is defined as

Time average of a force defined

$$\mathbf{F}_{av} = \frac{1}{\Delta t} \int_{t_i}^{t_f} \mathbf{F}\, dt = \frac{\mathbf{I}}{\Delta t} \qquad 7\text{-}46$$

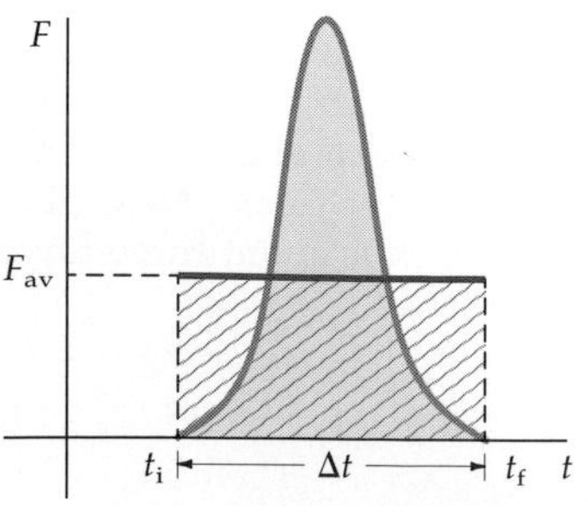

Figure 7-31 The average force F_{av} for some time interval is the constant force that gives the same impulse as the actual force during that interval. The rectangular area $F_{av}\,\Delta t$ is the same as the area under the *F*-versus-*t* curve.

The average force is the constant force that gives the same impulse as the actual force in the time interval Δt. F_{av} is shown in Figure 7-31. It is often useful to compute the average force for a collision in order to compare it with other forces, such as frictional forces or the force of gravity. The average force can often be estimated by finding the time of the collision from a reasonable estimate of the distance traveled by one of bodies during the collision.

Example 7-14

An egg of mass 50 g rolls off a table 1 m high and splatters on the floor. (*a*) Find the impulse exerted on the egg by the floor. (*b*) Assuming that the egg travels 2 cm (about half the smaller diameter of a typical egg) after it first comes into contact with the floor, estimate the collision time and the average force exerted by the floor on the egg.

(*a*) We can find the speed of the egg just before it hits the floor from the constant acceleration formula $v^2 = 2gy$, where g is the acceleration of gravity, and y is the distance the egg falls. For $y = 1$ m,

$$v^2 = 2gy = 2(9.81 \text{ m/s}^2)(1 \text{ m}) = 19.6 \text{ m}^2/\text{s}^2$$

The speed is then

$$v = \sqrt{19.6} \text{ m/s} \approx 4.4 \text{ m/s}$$

The momentum of the egg just before it hits the floor is downward and has the magnitude

$$p_i = mv = (0.05 \text{ kg})(4.4 \text{ m/s}) = 0.22 \text{ kg·m/s}$$

Since the final momentum is zero, the change in momentum is 0.22 kg·m/s in the upward direction. Thus, the impulse exerted by the floor on the egg is upward and has a magnitude of 0.22 kg·m/s = 0.22 N·s.

(*b*) Taking $\frac{1}{2}(4.4 \text{ m/s}) = 2.2$ m/s for the average speed of the egg during the collision, the time of the collision is estimated to be

$$\Delta t = \frac{\Delta y}{v_{av}} = \frac{0.02 \text{ m}}{2.2 \text{ m/s}} = 0.009 \text{ s} = 9 \text{ ms}$$

The average force is then

$$F_{av} = \frac{\Delta(mv)}{\Delta t} = \frac{(0.05 \text{ kg})(4.4 \text{ m/s})}{0.009 \text{ s}} \approx 24 \text{ N}$$

This is about 50 times the weight of the egg.

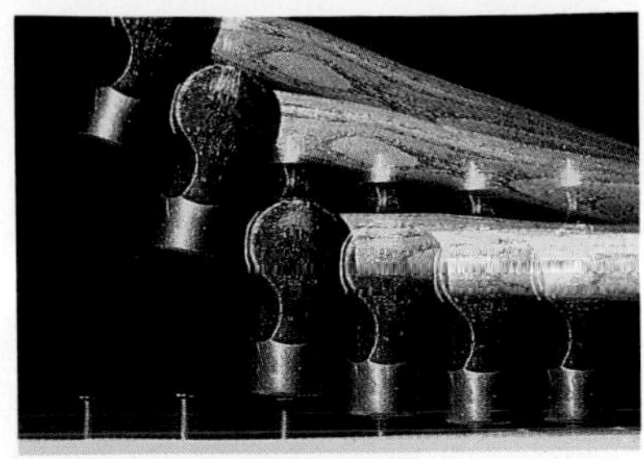

The hammer delivers a large impulsive force to drive the nail into the board.

Example 7-15

Estimate the force exerted by the seat belt on an 80-kg car driver when the car, originally moving at 25 m/s (about 56 mi/h), crashes into a fixed object.

We assume that the car travels about 1 m as the front end of the car crumples during the collision. This is also the distance traveled by the driver during the collision if he is wearing a seat belt. We also assume that the acceleration of the car is uniform as it crashes. Therefore, the average speed of the car during the collision is one-half of the initial speed, or 12.5 m/s. The time of the collision is then

$$\Delta t = \frac{1 \text{ m}}{12.5 \text{ m/s}} = 0.08 \text{ s}$$

The average acceleration is

$$a_{av} = \frac{\Delta v}{\Delta t} = \frac{25 \text{ m/s}}{0.08 \text{ s}} = 312 \text{ m/s}^2$$

This acceleration is about $32g$, that is, about 32 times the free-fall acceleration of gravity. The average force exerted by the seat belt on the driver is then

$$F_{av} = ma_{av} = (80 \text{ kg})(312 \text{ m/s}^2) = 25{,}000 \text{ N}$$

This force is great enough to break the driver's ribs and cause other chest injuries, but he may survive the crash. Were he not wearing a seat belt, he would continue to move at 25 m/s until he hit the dashboard or windshield. His stopping distance would then be considerably less than 1 m, with the acceleration and force being correspondingly greater.

Example 7-16

What are reasonable magnitudes for the impulse I, average force F_{av}, and collision time Δt for a golf club hitting a golf ball? A typical golf ball has a mass $m = 45$ g and a radius $r = 2$ cm. For a typical drive, a reasonable range is $R = 160$ m (about 175 yd).

If air resistance is neglected, the speed v_0 of the ball when it leaves the club is related to the range R by

$$R = \frac{v_0^2}{g} \sin 2\theta_0$$

where θ_0 is the angle of the projection. Taking $\theta_0 = 45°$ (which corresponds to the maximum range), we find

$$v_0^2 = \frac{Rg}{\sin 2\theta_0} = \frac{(160\text{ m})(9.81\text{ m/s}^2)}{1} \approx 1600\text{ m}^2/\text{s}^2$$

$$v_0 = 40\text{ m/s}$$

The magnitude of the impulse is thus

$$I = \int F\,dt = \Delta p = mv_0 = (0.045\text{ kg})(40\text{ m/s})$$
$$= 1.8\text{ kg·m/s} = 1.8\text{ N·s}$$

A reasonable estimate for the distance traveled by the ball while it is in contact with the golf club is the radius of the golf ball, so $\Delta x = 2$ cm. Since the ball starts from rest and its final speed is 40 m/s, its average speed is 20 m/s (assuming constant acceleration). The collision time, that is, the time needed for the ball to travel 2 cm at an average speed of 20 m/s, is

$$\Delta t = \frac{\Delta x}{v_{av}} = \frac{0.02\text{ m}}{20\text{ m/s}} = 0.001\text{ s}$$

The magnitude of the average force is therefore

$$F_{av} = \frac{I}{\Delta t} = \frac{1.8\text{ N·s}}{0.001\text{ s}} = 1800\text{ N}$$

We can see that the average force exerted by the club, 1800 N, is much larger than any of the other forces acting on the ball. For example, the weight of the ball is only about 0.44 N, and the frictional forces exerted on the ball by the grass or tee are even less, assuming the coefficient of friction is less than 1.

Questions

10. Why can friction and the force of gravity usually be neglected in collision problems?
11. Explain why a safety net can save the life of a circus performer.
12. How might you estimate the collision time of a baseball and bat?
13. Why might a wine glass survive a fall onto a carpet but not onto a concrete floor?

7-9 Jet Propulsion

Jet propulsion is an interesting application of Newton's third law and the conservation of momentum. It is, for instance, the means by which the squid and octopus propel themselves. They squirt water from their bodies with great force, and the expelled water exerts an equal and opposite force on the squid or octopus, propelling it forward. A rocket gets its thrust by burning fuel and exhausting the gas created out the back. The rocket exerts a force on the exhaust gas, and by Newton's third law, the gas exerts an equal and opposite force on the rocket, propelling it forward. The momentum lost by the expelled gas is equal to the momentum gained by the rocket. A misconception prevalent before rockets in space were commonplace was that a rocket needs air to push against. This is not true. The rocket pushes against its own exhaust gas, which pushes back against the rocket. Jet propulsion is in fact more efficient in empty space where there is no air resistance.

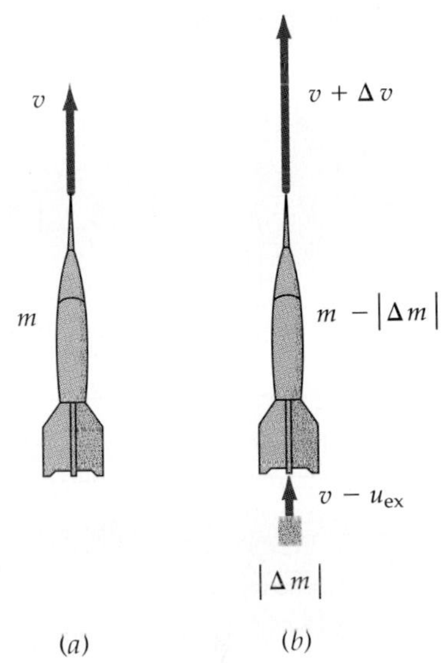

Figure 7-32 (*a*) Rocket moving with an initial speed v. (*b*) After a time interval Δt, the rocket has a mass $m - |\Delta m|$ and moves with a speed $v + \Delta v$. Gas exhausted from the rocket at a speed u_{ex} relative to the rocket is moving with speed $v - u_{ex}$. The change in the momentum of the rocket–exhaust system equals the impulse $F_{ext}\,\Delta t$.

We will develop an equation that describes the motion of a rocket. The description is complicated because the mass of the rocket changes continuously as it burns fuel and expels the exhaust gas. The easiest approach is to compute the change in the momentum of the total system (including the exhaust gas) for some time interval and set this change equal to the impulse exerted on the system by the external forces acting on it. Let F_{ext} be the net external force acting on the rocket, m be the mass of the rocket (plus the unburned fuel inside), and v be the speed of the rocket relative to the earth at time t (Figure 7-32*a*). At a later time $t + \Delta t$, the rocket has expelled gas of mass $|\Delta m|$ (Figure 7-32*b*). We use the absolute-value sign because the mass of the expelled gas is equal in magnitude to the change in the mass of the rocket Δm, which is negative. Thus, at time $t + \Delta t$, the rocket has a mass $m - |\Delta m|$ and is moving at a speed $v + \Delta v$. If the gas is exhausted at a speed u_{ex} *relative to the rocket*, its velocity at time $t + \Delta t$ relative to the earth is $v - u_{ex}$. The initial momentum of the system at time t is

$$P_i = mv$$

The momentum of the system at time $t + \Delta t$ is

$$\begin{aligned}P_f &= (m - |\Delta m|)(v + \Delta v) + |\Delta m|\,(v - u_{ex})\\ &= mv + m\,\Delta v - v\,|\Delta m| - |\Delta m|\,\Delta v + v\,|\Delta m| - u_{ex}\,|\Delta m|\\ &\approx mv + m\,\Delta v - u_{ex}\,|\Delta m|\end{aligned}$$

where we have dropped the term $|\Delta m|\,\Delta v$, which is the product of two very small quantities and therefore negligible compared with the others if the time interval Δt is very small. Computing the change in momentum and setting it equal to the impulse, we have

$$\Delta P = P_f - P_i = m\,\Delta v - u_{ex}\,|\Delta m| = F_{ext}\,\Delta t$$

We now divide by the time interval and take the limit as Δt approaches zero. The term $\Delta v/\Delta t$ approaches the derivative dv/dt, which is the acceleration, and the term $|\Delta m|/\Delta t$ approaches $|dm/dt|$, the absolute value of the rate of change of the mass of the rocket. This gives us the **rocket equation:**

$$m\,\frac{dv}{dt} = u_{ex}\left|\frac{dm}{dt}\right| + F_{ext} \qquad 7\text{-}47$$

Rocket equation

The quantity $u_{ex}\,|dm/dt|$ is called the **thrust** of the rocket:

$$F_{th} = u_{ex}\left|\frac{dm}{dt}\right| \qquad 7\text{-}48$$

When the rocket moves near the surface of the earth, the external force F_{ext} is the weight of the rocket. It is negative in Equation 7-47 because it is directly opposite the direction of the velocity, provided that the rocket is moving upward. Thus, the thrust must be greater than the weight of the rocket if the rocket is to accelerate upward. After we substitute $F_{ext} = -mg$ and divide by m, Equation 7-47 becomes

$$\frac{dv}{dt} = -g + \frac{u_{ex}}{m}\left|\frac{dm}{dt}\right| \qquad 7\text{-}49$$

To solve Equation 7-49 for the velocity v, we must know the exhaust speed relative to the rocket u_{ex} and the rate the rocket burns fuel $|dm/dt|$. The solution of this equation is complicated by the fact that m is not constant but a function of time. For example, if the rocket burns fuel at a constant rate R, the mass of the rocket at any time is $m = m_i - Rt$, where m_i is the initial mass. Since dm/dt is negative, we write $|dm/dt| = -dm/dt$. Equation 7-49 is then

$$\frac{dv}{dt} = -g - \frac{u_{ex}}{m}\frac{dm}{dt}$$

or

$$dv = -g\, dt - u_{ex}\frac{dm}{m}$$

Assuming that g is constant and integrating from $t = 0$ to $t = t_b$ when the fuel is completely burned, we obtain

$$\int_{v_i}^{v_f} dv = -\int_0^{t_b} g\, dt - u_{ex}\int_{m_i}^{m_f}\frac{dm}{m}$$

$$v_f - v_i = -gt_b - u_{ex}\ln\frac{m_f}{m_i}$$

where we have used $\int(dm/m) = \ln m$ (Appendix A). Using $-\ln m_f/m_i = \ln(m_i/m_f)$, we obtain

$$v_f - v_i = +u_{ex}\ln\frac{m_i}{m_f} - gt_b \qquad 7\text{-}50$$

Equation 7-50 gives the change in velocity for a rocket moving in a constant gravitational field in terms of the exhaust speed, the time for the fuel to burn t_b, and the ratio of the initial mass to final mass. For a rocket moving in free space with no external force, the change in velocity is given by

$$v_f - v_i = +u_{ex}\ln\frac{m_i}{m_f} \qquad \text{(no external force)} \qquad 7\text{-}51$$

The mass of the rocket without any fuel is called the **payload.** If the payload is just 10 percent of the total initial mass, that is, if 90 percent of the initial mass is fuel, the ratio m_i/m_f when the fuel is spent will be 10. For a rocket moving with $v_i = 0$ and no external force, the final speed will then be

$$v_f = u_{ex}\ln 10 = 2.3u_{ex}$$

The logarithmic nature of Equations 7-50 and 7-51 severely restricts the final speeds attainable. For example, if we increase the fuel so that the payload is only 1 percent of the initial mass, the final speed with no external force is $4.6u_{ex}$, just twice that for a 10-percent payload.

(*a*)

(*b*)

(*a*) Launch of a satellite, using a Thor booster rocket. (*b*) The octopus also moves by jet propulsion. However, instead of burning fuel like a rocket, the octopus takes in water and expels it at a high velocity.

Example 7-17

The Saturn V rocket used in the Apollo moon-landing program had an initial mass m_i of 2.85×10^6 kg, a payload of 27 percent, a burn rate $|dm/dt|$ of 13.84×10^3 kg/s, and a thrust F_{th} of 34×10^6 N. Find (*a*) the exhaust speed, (*b*) the burn time t_b, (*c*) the initial acceleration at lift-off, (*d*) the acceleration at burnout t_b, and (*e*) the final speed of the rocket.

(*a*) The exhaust speed is obtained from Equation 7-48:

$$u_{ex} = \frac{F_{th}}{|dm/dt|} = \frac{34 \times 10^6 \text{ N}}{13.84 \times 10^3 \text{ kg/s}} = 2.46 \text{ km/s}$$

(*b*) Since the payload is 27 percent, the mass of fuel burned is 73 percent of the initial mass or $m_{fuel} = (0.73)(2.85 \times 10^6 \text{ kg}) = 2.08 \times 10^6$ kg. The time to burn this much fuel at a rate of 13.84×10^3 kg/s is

$$t_b = \frac{2.08 \times 10^6 \text{ kg}}{13.84 \times 10^3 \text{ kg/s}} = 150 \text{ s}$$

(*c*) The initial acceleration is

$$\frac{dv}{dt} = -g + \frac{u_{ex}}{m_i}\left|\frac{dm}{dt}\right|$$

$$= -9.81 \text{ m/s}^2 + \frac{2.46 \text{ km/s}}{2.85 \times 10^6 \text{ kg}}(13.84 \times 10^3 \text{ kg/s})$$

$$= -9.81 \text{ m/s}^2 + 11.95 \text{ m/s}^2 = 2.14 \text{ m/s}^2 \approx 0.21g$$

(*d*) At burnout, the final mass is $m_f = (0.27)(2.85 \times 10^6\ \text{kg}) = 7.70 \times 10^5$ kg, so the acceleration is

$$\frac{dv}{dt} = -g + \frac{u_{ex}}{m_f}\left|\frac{dm}{dt}\right|$$

$$= -9.81\ \text{m/s}^2 + \frac{2.46\ \text{km/s}}{7.70 \times 10^5\ \text{kg}}(13.84 \times 10^3\ \text{kg})$$

$$= -9.81\ \text{m/s}^2 + 44.26\ \text{m/s}^2$$

$$= 34.4\ \text{m/s}^2 \approx 3.5\ \text{g}$$

(*e*) The final speed of the rocket is

$$v_f = +u_{ex} \ln \frac{m_i}{m_f} - gt_b$$

$$= (2.46\ \text{km/s}) \ln \frac{m_i}{0.27m_i} - (9.81\ \text{m/s}^2)(150\ \text{s})$$

$$= 3.22\ \text{km/s} - 1.47\ \text{km/s}$$

$$= 1.75\ \text{km/s}$$

Summary

1. The position of the center of mass of a system of particles is defined by

$$M\mathbf{R}_{cm} = m_1\mathbf{r}_1 + m_2\mathbf{r}_2 + \cdots = \sum_i m_i\mathbf{r}_i$$

where M is the total mass of the system and $\mathbf{R}_{cm}$ is the position vector from the chosen origin to the center of mass.

2. The center of mass of a system moves like a particle having a mass equal to the total mass of the system under the influence of the net external force acting on the system:

$$\mathbf{F}_{net,ext} = M\mathbf{A}_{cm}$$

3. The momentum of a particle is defined as its mass times its velocity:

$$\mathbf{p} = m\mathbf{v}$$

Newton's second law states that the net force acting on a particle equals the time rate of change of its momentum:

$$\mathbf{F}_{net} = \frac{d\mathbf{p}}{dt}$$

The kinetic energy of a particle can be written in terms of its momentum:

$$K = \frac{p^2}{2m}$$

4. The total mass of a system times the velocity of the center of mass equals the total momentum of the system:

$$\mathbf{P} = \sum_i m_i\mathbf{v}_i = M\mathbf{V}_{cm}$$

5. If the net external force acting on a system is zero, the total momentum of the system is conserved. This is the law of conservation of momentum.

6. The center-of-mass reference frame is the frame that moves with the center of mass. In this frame, the total momentum is zero.

7. The kinetic energy of a system of particles can be written as the sum of the kinetic energy associated with the motion of the center of mass, $\frac{1}{2}MV_{cm}^2$, and the energy of motion relative to the center of mass, $K_{rel} = \Sigma\, \frac{1}{2}m_i u_i^2$, which is the kinetic energy as observed in the center-of-mass frame:

$$K = \tfrac{1}{2}MV_{cm}^2 + K_{rel}$$

8. An elastic collision is one in which the total kinetic energy of the two objects is the same before and after the collision. An inelastic collision is one in which the total kinetic energy is changed by the collision.

9. In a perfectly inelastic collision, the bodies stick together and move with the velocity of the center of mass.

10. For an elastic collision, the relative speed of recession of the bodies after the collision equals the relative speed of approach before the collision.

11. The coefficient of restitution e is a measure of the elasticity of a collision and is defined as the ratio of the relative speed of recession to the relative speed of approach. For an elastic collision, $e = 1$; for a perfectly inelastic collision, $e = 0$.

12. The impulse of a force is defined as the integral of the force over the time interval during which the force acts. The impulse of the net force equals the total change in the momentum of a particle:

$$\mathbf{I} = \int_{t_i}^{t_f} \mathbf{F}\, dt = \Delta\mathbf{p}$$

The time average of a force for the interval $\Delta t = t_f - t_i$ is

$$\mathbf{F}_{av} = \frac{1}{\Delta t}\int_{t_i}^{t_f} \mathbf{F}\, dt = \frac{\mathbf{I}}{\Delta t}$$

13. A rocket gets its thrust by burning fuel and exhausting the resulting gases. The force exerted by the exhaust gases on the rocket propels the rocket forward. The equation for motion of a rocket obtained from Newton's laws is

$$m\frac{dv}{dt} = u_{ex}\left|\frac{dm}{dt}\right| + F_{ext}$$

In this equation, u_{ex} is the speed of the exhaust gas relative to the rocket. If the external force of gravity is constant, the final velocity of the rocket is related to the exhaust speed of the gas and the initial and final mass of the rocket:

$$v_f = +u_{ex} \ln \frac{m_i}{m_f} - gt_b$$

where t_b is the burn time.

Discovery of the Neutrino

Ralph A. Llewellyn
University of Central Florida

The concept of energy conservation was first set forth by the great biologist and physicist Herman von Helmholtz in 1847 when he generalized the results of a series of painstaking experiments that had been conducted some years previously by James Joule. The latter had shown that when energy is converted from one form into another no energy was either lost or created. Joule demonstrated, for example, that the kinetic energy of a paddle wheel turning in water is converted into thermal energy of the water. Similarly, the mechanical energy of a piston is converted into potential energy of a gas when the gas is compressed by the piston. That the total amount of energy of a closed system remains unchanged was tested and confirmed by innumerable experiments in the years following. Even Albert Einstein's discovery of the equivalence of mass and energy, embodied in the famous equation $E = mc^2$, did not imply a violation of the law of conservation of energy. It simply added a new kind of conversion, that of energy to mass and mass to energy, to those already known.

The law of conservation of energy was based entirely on experimental observation. There existed no fundamental physical theory that *predicted* the conservation of total energy. Nor, in fact, does such a theory or equation exist now. It was in this context that, in 1896, Henri Becquerel discovered radioactivity. What Becquerel observed was the emission of previously unknown "rays" from certain materials (initially, salts containing uranium). This strange radiation was subsequently shown—by Becquerel, Ernest Rutherford, and others—to consist of three distinct types, called alpha, beta, and gamma rays, which differed in their ability to penetrate matter. They had other significant differences, too, but shared the common characteristic that the energy of each individual ray was very large.

Following Rutherford's discovery of the atomic nucleus and Niels Bohr's use of the nuclear model in his explanation of the structure of atoms (both in 1913), it became clear that the large energies of the radioactive rays meant that they were associated with events occurring inside the nucleus itself. The study of radioactivity therefore afforded scientists the exciting opportunity to probe the internal structure of the nucleus. Precise measurements of the energy balances in the nuclear events involving alpha, beta, and gamma rays* were high priority experiments, since data from corresponding experiments on the energy of electromagnetic radiation (light) emitted by atoms had formed the foundation of Bohr's successful theory of atomic structure.

For any given nuclear decay involving only the emission of a gamma ray, the gamma rays were observed to all have the same energy (that is, to be "monoenergetic"). In particular, the energy of the gamma ray (E_γ) was exactly equal to the energy of the radioactive nucleus before emission (E_i) minus the nucleus after emission (E_f):

$$E_\gamma = E_i - E_f$$

Thus, the emission of gamma rays by nuclei conserved energy.

The same result held true for any given nuclear decay involving only the emission of an alpha ray. For a given nuclear decay, the alpha rays were monoenergetic, with energy E_α given by

$$E_\alpha = E_i - E_f$$

This again indicated that energy was conserved in the nuclear decay.

However, those nuclear events resulting in the emission of a beta ray appeared different: When a nucleus decayed from some specific initial state of energy E_i to a specific final state of energy E_f, the

*It is now known that alpha rays are helium nuclei, beta rays are electrons or positrons (antielectrons), and gamma rays are high-energy electromagnetic radiation.

Ralph A. Llewellyn has been Professor of Physics at the University of Central Florida since 1980. After earning a Ph.D. in experimental nuclear physics at Purdue University in 1962, he taught and chaired physics departments at Rose–Hulman Institute of Technology and Indiana State University. He has also been a member of the scientific staff of the National Academy of Sciences/National Research Council, where he directed studies on alternative energy sources, energy conservation, and world energy resources. When not doing physics, his recreational interests include photography and bicycle racing.

beta rays were not monoenergetic but were emitted in a continuous spectrum of energies from 0 to $E_{max} = E_i - E_f$. (Also, the beta rays were subsequently discovered to be electrons.) The energy balance equation for beta decay was, therefore,

$$E_\beta \leq E_{max} = E_i - E_f$$

The difference between the initial and final states of the nucleus ($E_i - E_f$) was generally *not* the energy present in the beta ray (E_β), which varied. This seemed to indicate that energy was not being conserved during beta decay.

This presented physics with a monumental problem, one capable of severely damaging the theoretical basis of the science. After all of these years it is difficult for us to realize how serious this matter was. Many physicists were prepared to abandon the law of conservation of energy, at least for nuclear events, on the logical grounds that a general law founded on experimental results must be discarded if further experiment fails to confirm it. Notable among these physicists was Neils Bohr himself. The proposal was made that the energy principle be rewritten as

$$E_{\substack{\text{final} \\ \text{system}}} \leq E_{\substack{\text{initial} \\ \text{system}}}$$

This form would preserve energy conservation in all experiments, including the disappearance of energy in beta decay (with the "less than" inequality), and still disallow the *creation* of energy.

That this was not a good solution to the problem was pointed out by Wolfgang Pauli. In a letter to Lise Meitner and Hans Geiger in December 1930, Pauli noted that beta decay seemed to imply not only a failure of the conservation of energy but also the failure of the conservation of linear momentum (Equation 7-15) and angular momentum (to be discussed in Chapter 8). These failures were no less serious: Furthermore, inserting the inequality sign into the energy conservation law did not save the others.

In the same letter, Pauli went on to suggest a way out of the dilemma. He made the proposal that a new fundamental particle existed and that one (or more) of these new particles was emitted along with the electron in beta decay. Regarding the new particle, he made several predictions. It would have no electric charge, since electric charge was already conserved in beta decay without it. It would have an extremely high ability to penetrate matter, since no one had ever seen one. (This high penetrability would imply a weak interaction with matter, which in turn would imply that the particle would be hard to detect.) Its mass would be zero, or nearly so, since beta rays with energy nearly equal to the maximum possible value (E_{max}) were occasionally seen in beta decays.

The proposed new particle would carry energy equal to the difference $E_{max} - E_\beta$ in each individual beta decay, thus preserving the law of conservation of energy. Its physical behavior would also be consistent with the preservation of linear momentum and angular momentum.

To appreciate how innovative Pauli's proposal was, one must remember that—at the time he made it—only two fundamental particles were known, the electron and the proton. The existence of both had been deduced from their appearance and behavior in many, many experiments. No one had ever before *invented* a new particle. Pauli acknowledged that his solution might appear an unlikely one, but he went on to say, "Nothing ventured, nothing gained" and to urge the "dear radioactive folks [to] put it to the test and judge."

Pauli went public with his proposal for this strange new particle at a meeting of the American Physical Society in June 1931. Despite his substantial reputation as a theoretical physicist, the response was decidedly underwhelming. Most physicists were highly skeptical of the proposal for a new, apparently undetectable particle. A consensus emerged that Bohr was correct. For some reason, energy was not conserved in nuclear processes.

In 1933, Pauli was encouraged by James Chadwick's discovery a few months earlier of the neutron (a new uncharged particle with mass nearly the same as the proton). Pauli then argued forcefully against accepting a breakdown of the conservation laws in beta decay:

> The electric charge is conserved in the process, and I don't see why conservation of charge should be more fundamental than conservation of energy and momentum.—If the conservation laws were not valid, one would have to conclude from these relations that a beta disintegration occurs always with a loss of energy and never a gain; this conclusion implies an irreversibility of these processes with respect to time, which doesn't seem to me at all acceptable.*

Enrico Fermi was much attracted by Pauli's proposed new particle, which he called the *neutrino* (meaning "little neutral one" in Italian) in order to distinguish it from Chadwick's massive neutron. Fermi developed an entirely new theory of beta decay incorporating the neutrino. Fermi's theory was enormously successful in accounting accurately for experimental observations involving beta decay.

*Wolfgang Pauli, "Structures et Propriétés des Noyaux Atomiques," *Proceedings of the Solvay Congress 1933*, Paris, Gauthier-Villars, 1934, p. 324.

Continued

This theory all but guaranteed the existence of the neutrino with precisely the properties that Pauli had suggested, despite—as was claimed at the time—the utter impossibility of its observation in any experiment.

Time passed. Our knowledge of the fundamental particles that make up the atom, the forces that act between them, and the conservation laws that govern their interactions, all expanded and improved. Physicists began to realize that the neutrino played a far more central role in particle physics than even Pauli had at first suspected. As a consequence, it became a matter of high priority to verify that the neutrino existed and that it behaved according to the assumptions that were rapidly being incorporated into nuclear and particle theories. In 1949, Chalmers Sherwin measured the linear momentum of beta rays and simultaneously measured the linear momentum of the recoiling nuclei. He showed that a single massless particle, corresponding to Pauli's neutrino, emitted along with the beta ray could precisely satisfy both the conservation of linear momentum and the conservation of energy.

That neutrinos interact exceedingly weakly with matter means, for example, that the average distance a typical neutrino will travel from where it is created to where it first interacts with matter is of the order of 1000 light-years. This makes neutrinos very hard to detect in *any* experiment. In order to have a reasonable chance of detecting even a few neutrinos, an enormous number of them must pass through whatever experimental apparatus is set up to detect them.

It was not until 25 years after Pauli first proposed the neutrino that the advent of large nuclear reactors made available sources of neutrinos in the numbers necessary to make detection experiments conceivable. Finally, in 1956, Clyde Cowan and Frederick Reines, using the very high neutrino flux (about 10^{15} neutrinos per square centimeter per second) from a reactor at Savannah River, were able to detect the unmistakable "signature" of a nuclear event that could only have been initiated by a neutrino (actually, by an antineutrino, but that's another story!). Thus was the last element of doubt regarding Pauli's proposal effectively removed.

Since that time the complex, fundamental properties of the neutrino have continued to be the focus of active theoretical and experimental study. We believe that the neutrinos (we're now sure there are at least six different kinds) are among the very few, truly fundamental particles. By "fundamental," we mean that they do not themselves decay. Neutrino experiments, once considered impossible, are now performed at all the world's major particle accelerators and are enabling important advances in our search for deeper understanding of the physical universe. Neutrino astronomy is an area of active research because neutrinos, due to the weakness of their interaction with matter, are the only particles that can easily escape the interior of the densest stars. They bring information about the genesis of stars and the earliest moments of the universe.

And it all began with a letter written by Wolfgang Pauli. "Nothing ventured nothing gained" indeed!

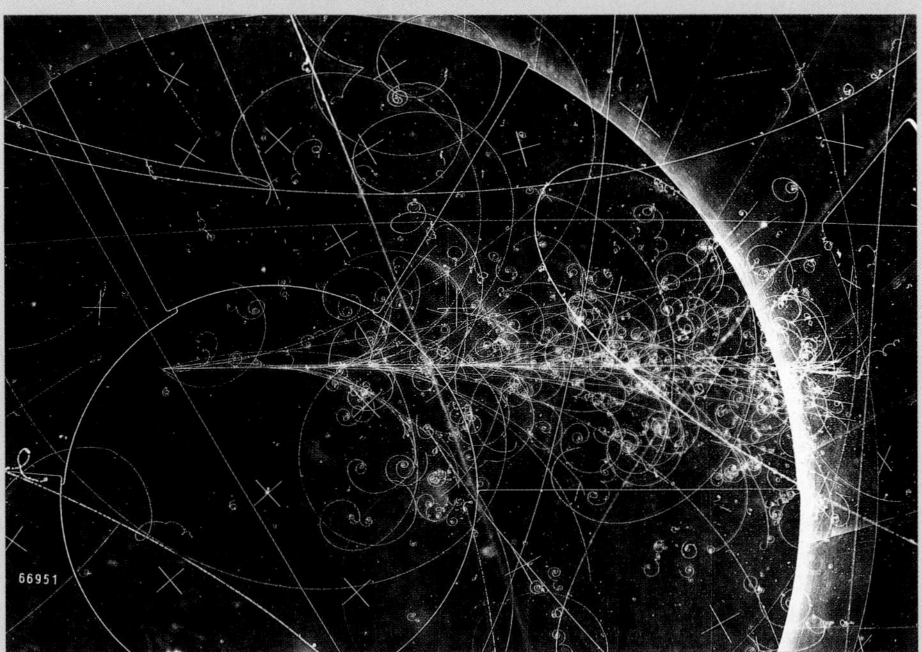

This photograph shows the tracks of a complicated spray of particles that are emitted when a neutrino (entering from the bottom) collides with a proton in the Big European Bubble Chamber at CERN.

Suggestions for Further Reading

Damask, Arthur C. "Forensic Physics of Vehicle Accidents," *Physics Today*, March 1987, p. 36.

The law of conservation of momentum and the concept of the center of mass are extremely useful in reconstructing collisions between vehicles or between a vehicle and a pedestrian.

Walker, Jearl: "The Amateur Scientist: Success in Racquetball Is Enhanced by Knowing the Physics of the Collision of Ball with Wall," *Scientific American*, September 1984, p. 215.

The laws of conservation of momentum and energy are applied to the collisions of a racquetball.

Review

A. Objectives: After studying this chapter, you should:

1. Be able to find the center of mass of a system of discrete particles and of a continuous object.

2. Be able to state Newton's second law for a system of particles.

3. Be able to use the conservation of momentum to solve problems.

4. Know the definition of the center-of-mass reference frame and be able to transform into it and out of it to work collision problems.

5. Know that the kinetic energy of a system can be written as the sum of the kinetic energy of the motion of the center of mass and the kinetic energy of the motion of the system relative to the center of mass.

6. Be able to solve elastic and inelastic collision problems in one dimension.

7. Know the relation between impulse and momentum, and be able to use it to estimate the magnitude of the large forces that occur in collisions.

8. Be able to apply Newton's laws to rocket motion and to derive the equation for the motion of a rocket.

B. Define, explain, or otherwise identity:

Center of mass
Newton's second law for a system
Momentum
Law of conservation of momentum
Center-of-mass reference frame
Zero-momentum reference frame
Elastic collision
Inelastic collision
Perfectly inelastic collision
Ballistic pendulum
Coefficient of restitution
Impulse
Time-average force
Rocket equation
Thrust
Payload

C. True or false: If the statement is true, explain why. If it is false, give a counterexample.

1. The momentum of a heavy object is greater than that of a light object moving at the same speed.

2. In a perfectly inelastic collision, all the kinetic energy of the bodies is lost.

3. The momentum of a system may be conserved even when mechanical energy is not.

4. Kinetic energy is conserved in an elastic collision.

5. In an elastic collision, the relative speed of recession after the collision equals the relative speed of approach before the collision.

6. The velocity of the center of mass of a system equals the total momentum of the system divided by its total mass.

Problems

Level I

7-1 The Center of Mass

1. Three point masses of 2 kg each are located on the x axis at the origin, $x = 0.20$ m and $x = 0.50$ m. Find the center of mass of the system.

2. A 24-kg child is 20 m from an 86-kg man. Where is the center of mass of this system?

3. Three objects of 2 kg each are located as follows: object 1 is at $x = 10$ cm, $y = 0$; object 2 is at $x = 0$, $y = 10$ cm; and object 3 is at $x = 10$ cm, $y = 10$ cm. Find the location of the center of mass.

4. Find the center of mass x_{cm} of the three masses shown in Figure 7-33.

5. The uniform sheet of plywood in Figure 7-34 has a mass of 20 kg. Find the x and y coordinates of the center of mass.

6. Alley Oop's club-axe consists of a symmetrical 8-kg stone attached to the end of a uniform 2.5-kg stick. The dimensions of the club-axe are shown in Figure 7-35. How far is the center of mass from the handle end of the club-axe?

7. Three small balls A, B, and C, with masses of 300, 100, and 100 g, respectively, are connected by massless rods. The balls are located as shown in Figure 7-36. What are the coordinates of the center of mass?

7-2 Motion of the Center of Mass of a System

8. Two 3-kg masses have velocities $\mathbf{v}_1 = 2$ m/s $\mathbf{i}$ + 3 m/s $\mathbf{j}$ and $\mathbf{v}_2 = 4$ m/s $\mathbf{i}$ − 6 m/s $\mathbf{j}$. Find the velocity of the center of mass.

9. A 1500-kg car is moving to the west with a speed of 20 m/s, and a 3000-kg truck is moving east with a speed of 16 m/s. Find the velocity of the center of mass of the system.

10. A force $\mathbf{F} = 12$ N$\mathbf{i}$ is applied to the 300-g ball in Figure 7-36. What is the acceleration of the center of mass?

11. A constant force $\mathbf{F} = 26$ N$\mathbf{i}$ is applied to the 8-kg mass of Problem 4 at $t = 0$. (*a*) Find the velocity of the center of mass at $t = 5$ s. (*b*) What is the location of the center of mass at $t = 5$ s?

7-3 The Conservation of Momentum

12. An open railroad car of mass 20,000 kg is rolling without friction along a track with a speed of 5 m/s when it starts to rain. After the car has collected 2000 kg of water, what is its new speed?

13. A girl of mass 50 kg jumps out of the bow of a 250-kg canoe that is initially at rest. If her velocity is 7.5 m/s to the right, what is the velocity of the canoe after she jumps?

14. Two masses of 5 kg and 10 kg resting on a frictionless table are connected by a compressed spring. When the spring is released, the smaller mass has a velocity of 8 m/s to the left. What is the velocity of the larger mass?

15. A model-train car of mass 250 g traveling with a speed of 0.50 m/s couples to another car of mass 400 g that is initially at rest. What is the speed of the cars after they have coupled together?

7-4 The Center-of-Mass Reference Frame and

7-5 Kinetic Energy of a System of Particles

16. (*a*) Find the total kinetic energy of the two model-train cars of Problem 15 before they couple. (*b*) Find the initial velocities of the two cars relative to the center of mass of the system, and use them to calculate the initial kinetic energy of the system relative to the center of mass. (*c*) Find the kinetic energy of the center of mass. (*d*) Compare your answers for parts (*b*) and (*c*) with that for part (*a*).

17. Describe how a solid ball can move such that (*a*) its total kinetic energy is just the energy of motion of its center of mass and (*b*) its total kinetic energy is the energy of its motion relative to its center of mass.

Figure 7-33 Problems 4 and 11.

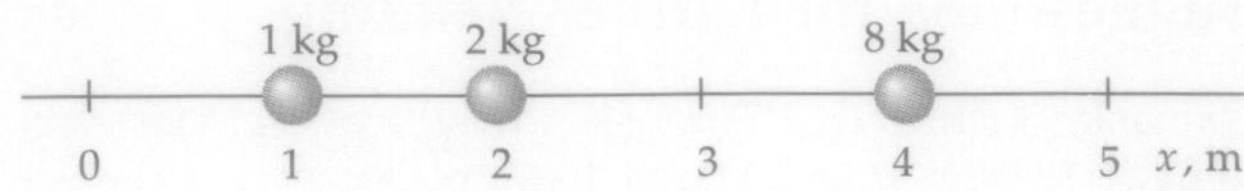

Figure 7-34 Problem 5.

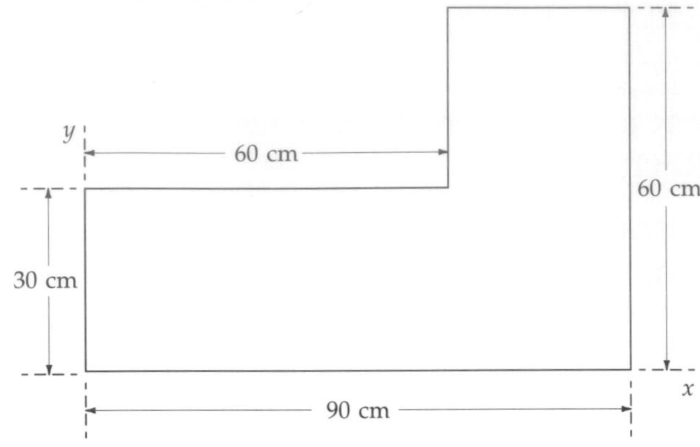

Figure 7-35 Problem 6.

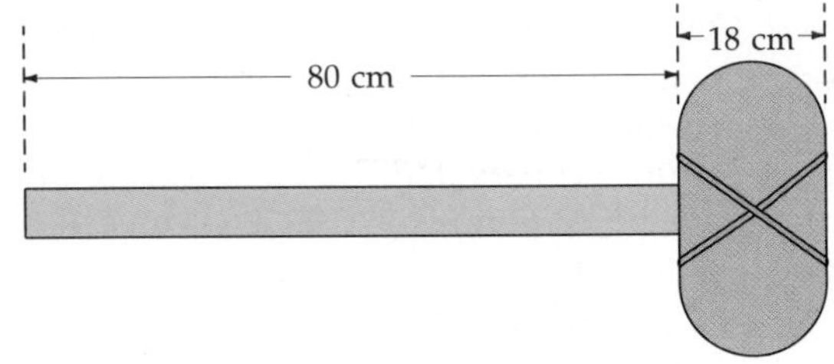

Figure 7-36 Problem 7 and 10.

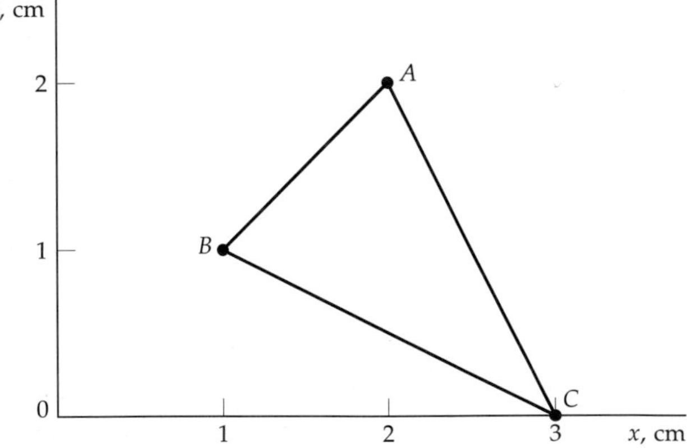

18. Two bowling balls are moving with the same velocity, but one just slides down the alley, whereas the other rolls down the alley. Which ball has more energy?

19. A 3-kg block is traveling to the right at 5 m/s, and a second 3-kg block is traveling to the left at 2 m/s. (*a*) Find the total kinetic energy of the two blocks in this reference frame. (*b*) Find the velocity of the center of mass of the two-body system. (*c*) Find the velocities of the two blocks relative to the center of mass. (*d*) Find the kinetic energy of the motion of the blocks relative to the center of mass. (*e*) Show that your answer for part (*a*) is greater than your answer for part (*d*) by the kinetic energy of the center of mass.

20. Repeat Problem 19 for a second block having a mass of 5 kg and moving to the right at 3 m/s.

7-6 Collisions in One Dimension

21. A 150-g glob of clay is thrown horizontally with a speed of 5 m/s at a 1-kg block that is initially at rest on a frictionless surface. If the clay sticks to the block, what is the speed of the combined system?

22. A 2000 kg car traveling to the right at 30 m/s is chasing a second car of the same mass that is traveling to the right at 10 m/s. (*a*) If the two cars collide and stick together, what is their speed just after the collision? (*b*) What fraction of the initial kinetic energy of the cars is lost during this collision? Where does it go?

23. An 85-kg running back moving at 7 m/s makes a perfectly inelastic collision with a 105-kg linebacker who is initially at rest. What is the speed of the players just after their collision?

24. Find the final velocity of each block in Problem 19 if they make (*a*) a perfectly inelastic collision and (*b*) an elastic collision.

25. Repeat Problem 24 using the blocks in Problem 20.

26. A 5.0-kg object with a speed of 4.0 m/s collides head-on with a 10-kg object moving toward it with a speed of 3.0 m/s. (*a*) If the 10-kg object stops dead after the collision, what is the final speed of the 5-kg object? (*b*) Is the collision elastic?

27. A heavy ball of mass 5 kg hits an 85-kg man in the chest and bounces directly back at a speed of 2 m/s. (*a*) If the initial speed of the ball is 8 m/s and the man is initially at rest, find the speed imparted to the man by the collision. (*b*) Is this an elastic or inelastic collision?

28. According to the official rules of racquetball, a ball acceptable for tournament play must bounce to a height of between 173 and 183 cm when dropped from a height of 254 cm at room temperature. What is the acceptable range of values for the coefficient of restitution for the racquetball–floor system?

29. A ball bounces to 80 percent of its original height. (*a*) What fraction of its mechanical energy is lost each time it bounces? (*b*) What is the coefficient of restitution of the ball–floor system?

7-7 Collisions in Three Dimensions

30. A 1500-kg car traveling north at 70 km/h collides at an intersection with a 2000-kg car traveling west at 55 km/h. The two cars stick together. (*a*) What is the total momentum of the system before the collision? (*b*) Find the magnitude and direction of the velocity of the wreckage just after the collision.

31. In a pool game, the cue ball, which has an initial speed of 5 m/s, makes an elastic collision with the 8 ball, which is initially at rest. After the collision, the 8 ball moves at an angle of 30° with the original direction of the cue ball. (*a*) Find the direction of motion of the cue ball after the collision. (*b*) Find the speed of each ball. Assume that the balls have equal mass.

32. Figure 7-37 shows the results of a collision of two objects of unequal mass. (*a*) Find the speed v_2 of the larger mass after the collision and the angle θ_2. (*b*) Show that the collision is elastic.

Figure 7-37 Problem 32.

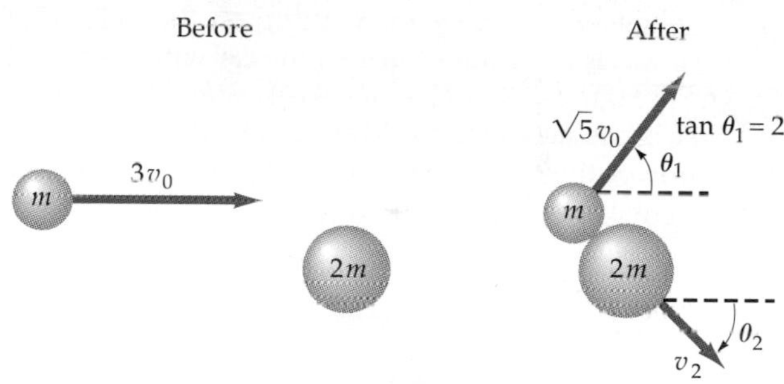

7-8 Impulse and the Time Average of a Force

33. A soccer ball of mass 0.43 kg leaves the foot of the kicker with an initial speed of 25 m/s. (*a*) What is the impulse imparted to the ball by the kicker? (*b*) If the foot of the kicker is in contact with the ball for 0.008 s, what is the average force exerted by the foot on the ball?

34. A 0.3-kg brick is dropped from a height of 8 m. It hits the ground and comes to rest. (*a*) What is the impulse exerted by the ground on the brick? (*b*) If it takes 1.3 ms from the time the brick first touches the ground until it comes to rest, what is the average force exerted by the ground on the brick?

35. When a 0.15-kg baseball is hit, its velocity changes from +20 m/s to −20 m/s. (*a*) What is the magnitude of the impulse delivered by the bat to the ball? (*b*) If the baseball is in contact with the bat for 1.3 ms, what is the average force exerted by the bat on the ball?

36. A 300-g handball moving with a speed of 5.0 m/s strikes the wall at an angle of 40° and then bounces off with the same speed at the same angle. It is in contact with the wall for 2 ms. What is the average force exerted by the ball on the wall?

37. In Problem 23, if the time of the collision is 0.1 s, what is the average force each player exerts on the other?

38. In Problem 27, if the time of the collision is 0.05 s, what is the average force exerted by the ball on the man?

7-9 Jet Propulsion

39. A rocket burns fuel at a rate of 200 kg/s and exhausts the gas at a relative speed of 6 km/s. Find the thrust of the rocket.

40. The payload of a rocket is 5 percent of its total mass, the rest being fuel. If the rocket starts from rest and moves with no external forces acting on it, what is its final velocity if the exhaust velocity of its gas is 5 km/s?

41. A rocket moves in free space with no external forces acting on it. It starts from rest and has an exhaust speed of 3 km/s. Find the final velocity if the payload is (*a*) 20 percent, (*b*) 10 percent, (*c*) 1 percent.

Level II

42. A small car of mass 800 kg is parked behind a small truck of mass 1600 kg on a level road. The brakes of both the car and the truck are off so that they are free to roll with negligible friction. A man sitting on the tailgate of the truck exerts a constant force on the car with his feet as shown in Figure 7-38. The car accelerates at 1.2 m/s^2. (*a*) What is the acceleration of the truck? (*b*) What is the magnitude of the force exerted on either the truck or the car?

Figure 7-38 Problem 42.

43. A 16-g bullet is fired into the bob of a ballistic pendulum of mass 1.5 kg. When the bob is at its maximum height, the strings make an angle of 60° with the vertical. The length of the pendulum is 2.3 m. Find the speed of the bullet.

44. A 3-kg body moving at 4 m/s makes an elastic collision with a stationary body of mass 2 kg. Use conservation of momentum and the fact that the relative velocity of recession equals the relative velocity of approach to find the velocity of each body after the collision. Check your answer by calculating the initial and final kinetic energies of each body.

45. A 6-kg projectile is launched at an angle of 30° with the horizontal and an initial speed of 40 m/s. At the top of its flight, it explodes into two parts with masses of 2 and 4 kg. The fragments move horizontally just after the explosion, and the 2-kg piece lands back at the launch site. (*a*) Where does the 4-kg piece land? (*b*) Compute the kinetic energy of the projectile just before the explosion and the total kinetic energy of the fragments just after the explosion to find the energy of the explosion.

46. A 3-kg bomb slides along a frictionless horizontal plane in the x direction at 6 m/s. It explodes into two pieces, one of mass 2 kg and the other of mass 1 kg. The 1-kg piece moves along the horizontal plane in the y direction at 4 m/s. (*a*) Find the velocity of the 2-kg piece. (*b*) What is the velocity of the center of mass after the explosion?

47. A 2-kg body moving at 3 m/s to the right collides with a 3-kg body moving at 2 m/s to the left. The coefficient of restitution is 0.4. Find the velocity of each body after the collision.

48. A 2-kg body moving at 6 m/s collides with a 4-kg body that is initially at rest. After the collision, the 2-kg body moves backward at 1 m/s. (*a*) Find the velocity of the 4-kg body after the collision. (*b*) Find the energy lost in the collision. (*c*) What is the coefficient of restitution for this collision?

49. The ratio of the mass of the earth to the mass of the moon is $M_E/m_m = 81.3$. The radius of the earth is about 6370 km, and the distance from the earth to the moon is about 384,000 km. (*a*) Locate the center of mass of the earth–moon system relative to the surface of the earth. (*b*) What external forces act on the earth–moon system? (*c*) In what direction is the acceleration of the center of mass of this system? (*d*) Assume that the center of mass of this system moves in a circular orbit around the sun. How far must the center of the earth move in the radial direction (toward or away from the sun) during the 14 days between the time the moon is farthest from the sun (full moon) and the time it is closest to the sun (new moon)?

50. A circular plate of radius r has a circular hole cut out of radius $r/2$ (Figure 7-39). Find the center of mass of the plate. *Hint:* The hole can be represented by two disks superimposed, one of mass m and the other of mass $-m$.

51. Using the hint from Problem 50, find the center of mass of a solid sphere of radius r that has a spherical cavity of radius $r/2$ as shown in Figure 7-40.

52. A peculiar baseball bat of length L has a linear density (mass per unit length) given by $\lambda = \lambda_0(1 + x^2/L^2)$. Find the x coordinate of the center of mass in terms of L. See Figure 7-41.

53. A 40-kg girl gets on her 10-kg wagon on level ground with two 5-kg bricks. She throws the bricks horizontally off the back of the wagon one at a time at a speed of 7 m/s relative to herself. (*a*) How fast does she go after throwing the second brick? (*b*) How fast would she go if she threw both bricks at the same time at 7 m/s relative to herself?

54. A hammer of mass 0.8 kg is used to drive nails of mass 30 g into wood. When the hammer has an impact speed of 0.5 m/s, the nail penetrates 2.0 cm at a single blow. Find (*a*) the common speed of the hammer and nail immediately after impact, assuming the collision is perfectly inelastic; (*b*) the time the nail is in motion, assuming that its initial speed is acquired in a negligible time, after

Figure 7-39 Problem 50.

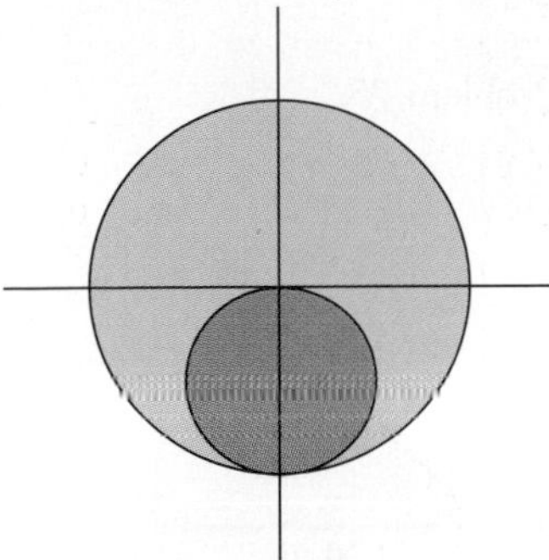

Figure 7-40 Problem 51.

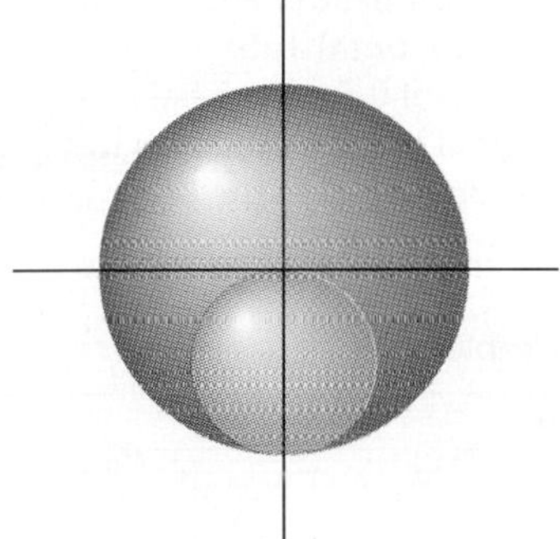

Figure 7-41 Problem 52.

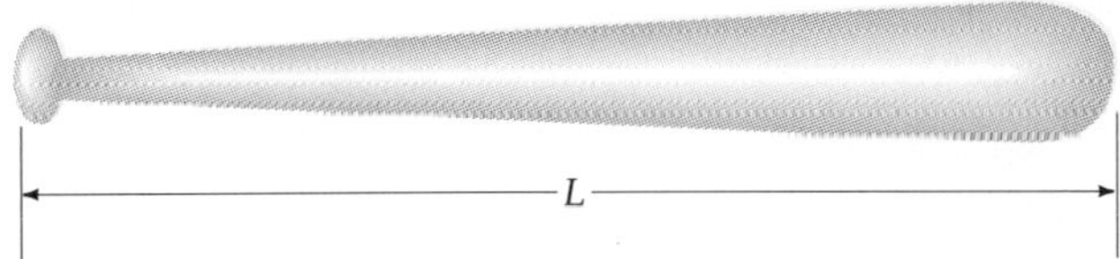

which there is uniform deceleration; and (*c*) the average resisting force of the wood against the nail as it penetrates.

55. You throw a 150 g ball to a height of 10 m. (*a*) Use a reasonable value for the distance the ball moves while it is in your hand to calculate the average force exerted by your hand and the time the ball is in your hand while you throw it. (*b*) Is it reasonable to neglect the weight of the ball while it is being thrown?

56. A 2000-kg car traveling at 90 km/h crashes into a concrete wall that does not give at all. (*a*) Estimate the time of the collision, assuming that the center of the car travels halfway to the wall with constant deceleration. (Use any reasonable length for the car.) (*b*) Estimate the average force exerted by the wall on the car.

57. A ball moving at 10 m/s makes an off-center elastic collision with another ball of equal mass that is initially at rest. The incoming ball is deflected at an angle of 30° from its original direction of motion. Find the velocity of each ball after the collision.

58. A proton of mass m makes a head-on elastic collision with a stationary carbon nucleus of mass $12m$. The speed of the proton is 300 m/s. (*a*) Find the velocity of the center of mass of the system. (*b*) Find the velocity of the proton after the collision in the center-of-mass reference frame. (*c*) Find the velocity of the proton after the collision in the lab reference frame.

59. A handball of mass 300 g is thrown straight against a wall with a speed of 8 m/s. It rebounds with the same speed. (*a*) What impulse is delivered to the wall? (*b*) If the ball is in contact with the wall for 0.003 s, what average force is exerted on the wall by the ball? (*c*) The ball is caught by a player who brings it to rest. In the process, her hand moves back 0.5 m. What is the impulse received by the player? (*d*) What average force was exerted on the player by the ball?

60. Wistful Willie is skating along on very smooth ice at 3 m/s, daydreaming of nothing in particular, when he collides head-on with Dennis the Menace, who is moving in the opposite direction at 5 m/s. Willie's mass is 80 kg and Dennis's mass is 30 kg. After the collision, Willie and Dennis stick together. (*a*) Find the common speed of the boys after the collision. (*b*) Find the velocity of their center of mass before the collision. (*c*) Find the velocity of their center of mass after the collision. (*d*) If the time of impact is 0.1 s, what is the average force exerted on Willie by Dennis? *Note:* After 0.1 s, the boys move along together but they no longer exert forces on one another.

61. A 13-kg block is at rest on a level floor. A 400-g glob of putty is thrown at the block such that it travels horizontally, hits the block, and sticks to it. The block and putty slide 15 cm along the floor. If the coefficient of sliding friction is 0.4, what is the initial speed of the putty?

62. A bullet of mass m_1 is fired with a speed v into the bob of a ballistic pendulum of mass m_2. The bob is attached to a very light rod of length L that is pivoted at the other end. The bullet is stopped in the bob. Find the minimum v such that the bob will swing through a complete circle.

63. A bullet of mass m_1 is fired with a speed v into the bob of a ballistic pendulum of mass m_2. Find the maximum height h attained by the bob if the bullet passes through the bob and emerges with a speed $v/2$.

64. The force acting on a particle of mass 5.0 kg is given by the function $F(t) = 3t^2$ N. If the particle starts from rest, what will its velocity be after 5 s?

65. A rocket has initial mass of 30,000 kg of which 20 percent is the payload. It burns fuel at a rate of 200 kg/s and exhausts its gas at a relative speed of 1.8 km/s. Find (*a*) the thrust of the rocket, (*b*) the time until burnout, and (*c*) its final speed assuming it moves upward near the surface of the earth where the gravitational field g is constant.

66. Find the initial acceleration and the acceleration at burnout for the rocket in Problem 65.

Level III

67. The great limestone caverns were formed by dripping water. (*a*) If water droplets of 10 mL fall from a height of 5 m at a rate of 10 per minute, what is the average force exerted on the limestone floor by the droplets of water? (*b*) Compare this force to the weight of a water droplet.

68. A favorite game at picnics is the egg toss. Two people toss a raw egg back and forth as they move further apart. If the force required to break the egg's shell is about 5 N and the mass of the egg is 50 g, estimate the maximum separation distance for the egg throwers. Make whatever assumptions seem reasonable.

69. A rocket for mining the asteroid belt between Mars and Jupiter is designed like a large scoop. Its mass m is 10^5 kg, and it is traveling at an initial velocity of 10^4 m/s. It overtakes asteroids traveling at 100 m/s. If it scoops up asteroids at the rate of 100 kg/s, what will its speed be after 1 h of mining?

70. What thrust must the mining rocket in Problem 71 have for it to continue travel at a constant speed? Ignore any decrease in the mass of the rocket due to the exhaust of gases.

71. In the billiards shot shown in Figure 7-42, the initial direction of the cue ball is perpendicular to the line joining the centers of the other two balls, which are touching. The cue ball strikes both balls simultaneously. Use the symmetry of the situation along with the conservation of energy and momentum to find the final velocities of all three balls.

Figure 7-42 Problem 71.

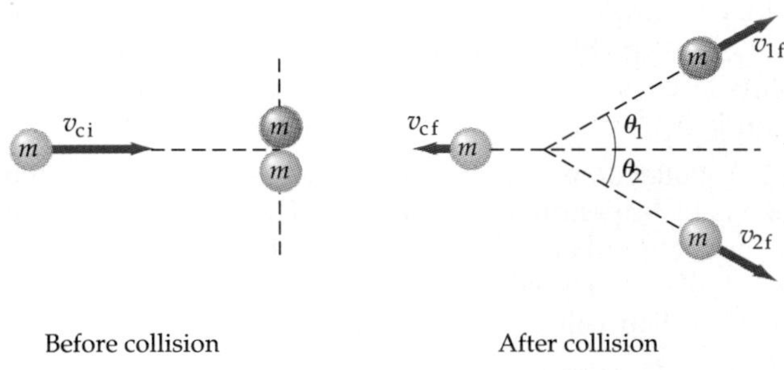

72. Show that in a one-dimensional collision between two particles, the fractional relative-energy loss is related to the coefficient of restitution by

$$-\frac{\Delta K_{rel}}{K_{rel}} = 1 - e^2$$

73. Show that for an elastic collision between two particles of equal mass when one of the particles is initially at rest, the energy transferred to the originally stationary particle is $(\sin^2 \theta)E_0$, where E_0 is the initial energy of the incoming particle, and θ is the angle of its deflection.

74. Use integration to find the center of mass of a uniform plate cut in the shape of a half circle. Choose the origin to be at the center of curvature of the plate, with the y axis on the bisector.

75. Use integration to find the center of mass of the right isosceles triangle shown in Figure 7-43.

Figure 7-43 Problem 75.

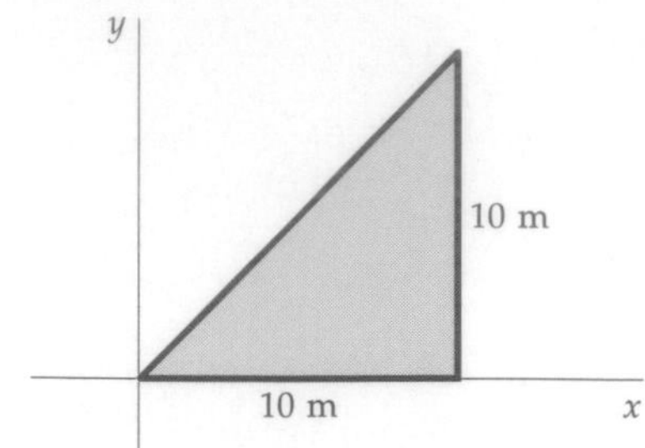

76. A stream of glass beads, each with a mass of 0.5 g, comes out of a horizontal tube at 100 per second. The beads fall a distance of 0.5 m to a balance pan and bounce back to their original height as shown in Figure 7-44. How much mass must be placed in the other pan of the balance to keep the pointer at zero?

Figure 7-44 Problem 76.

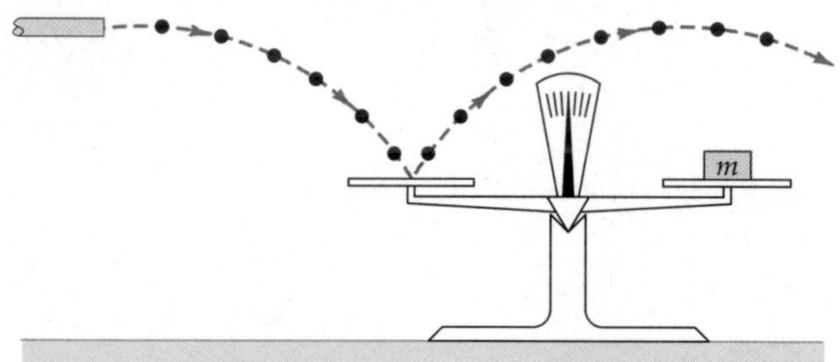

77. A rocket has a payload of 5000 kg and a fuel supply of 20,000 kg. It starts from rest, and it burns fuel at a rate of 200 kg/s and exhausts the gases at $u_{ex} = 6$ km/s. (*a*) Find the final velocity of the rocket if it is in free space where there is no gravity. (*b*) Find the velocity after the fuel burns up if the rocket moves against a uniform gravitational field g. (*c*) In part (*b*), if the rocket starts from the earth's surface, is it reasonable to neglect the variation in g with height above the earth's surface?

78. A particle has an initial speed v_0. It collides with a second particle that is at rest and is deflected through an angle ϕ. Its speed after the collision is v. The second particle recoils. Its velocity makes an angle θ with the initial direction of the first particle. (*a*) Show that

$$\tan \theta = \frac{v \sin \phi}{v_0 - v \cos \phi}$$

(*b*) Do you have to assume that the collision is either elastic or inelastic to get the result in part (*a*)?

Chapter 8

Rotation

Star tracks in a time exposure of the night sky.

There are many examples of rotational motion in the world around us. Merry-go-rounds rotate, as do high divers as they somersault and ice skaters as they pirouette. The earth rotates about its axis. Balls and cylinders rotate as they roll. Molecules, atoms, and even electrons rotate.

In this chapter, we will learn how to describe rotational motion. First we will define the kinematic quantities angular displacement, angular velocity, and angular acceleration, quantities that are the rotational analogs of displacement, velocity, and acceleration, which we use to describe linear motion. We will see that the rotation of a rigid body with constant angular acceleration is described by equations that are the rotational analogs of the equations we have already dealt with for linear motion with constant acceleration. We will then study the dynamics of rotational motion and the rotational analog of Newton's second law of motion. Again, we will find that dynamic quantities we encountered in our study of linear motion—force, mass, and momentum—have corresponding rotational analogs—torque, moment of inertia, and angular momentum.

8-1 Angular Velocity and Angular Acceleration

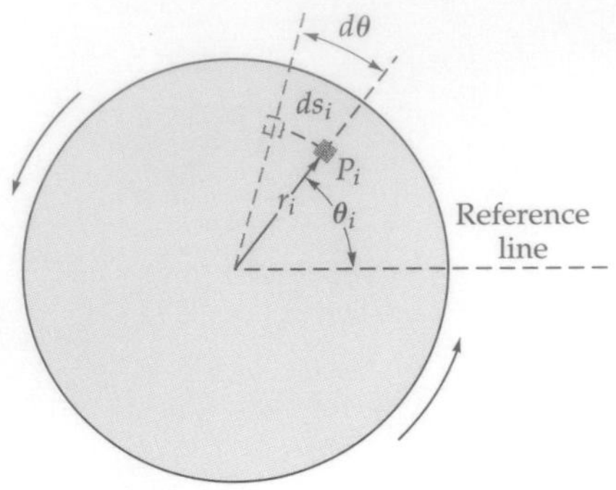

Figure 8-1 A disk rotating about a fixed axis through its center. The distance ds_i moved by the ith particle in some time interval depends on r_i, but the angular displacement $d\theta$ is the same for all the particles on the disk.

Figure 8-1 shows a disk that is free to rotate about its axis, which is through its center and perpendicular to the disk. When the disk rotates, different parts of it move at different speeds. For example, a point near the rim moves faster than a point near the axis. It therefore makes no sense to speak of the speed or velocity of the disk as it rotates. However, when a point near the rim moves through a complete circle, so does any other point on the disk. In a given time, the angle swept out by a line from the axis of rotation to any point on the disk is the same whether the point is on the rim, near the axis, or at any other place on the disk. The angle swept out by a radial line on the disk is thus a characteristic of the disk as a whole, as is the rate at which the angle changes.

To develop the concepts of angular velocity and angular acceleration, we consider the disk to be made up of many small point particles. As the disk turns, the distance between any two particles remains fixed. Such a system is called a **rigid body.**

Let us focus our attention on a typical particle of mass m_i on the disk. We can specify the position of the particle P_i by its distance r_i from the center of the disk and the angle θ_i between a line from the center to the particle and a reference line that is fixed in space, as shown in Figure 8-1. During a small time interval dt, the particle moves along the arc of a circle a distance ds_i given by

$$ds_i = v_i\, dt \tag{8-1}$$

where v_i is the speed of the particle. During this time, the radial line to the particle sweeps out an angle $d\theta$ relative to a line fixed in space. The measure of this angle in radians is the arc length ds_i divided by r_i:

$$d\theta = \frac{ds_i}{r_i} \tag{8-2}$$

(If you are unfamiliar with the radian measure of angles, you should read the trigonometry review section in Appendix A.) Although the distance ds_i varies from particle to particle, the angle $d\theta$ swept out in a given time is the same for all the particles of the disk. It is called the **angular displacement** of the disk. When the disk makes one complete revolution, the arc length Δs_i is $2\pi r_i$ and the angular displacement is

$$\Delta\theta = \frac{2\pi r_i}{r_i} = 2\pi \text{ rad} = 360° = 1 \text{ rev}$$

The rate of change of the angle with respect to time, $d\theta/dt$, is the same for all the particles of the disk. It is called the **angular velocity** ω of the disk:

Angular velocity defined

$$\omega = \frac{d\theta}{dt} \tag{8-3}$$

The angular velocity ω is positive for counterclockwise rotation, for which θ is increasing, and negative for clockwise rotation, for which θ is decreasing. (Later in this chapter, we will see that, for general rotations, the angular velocity is a vector quantity that points along the axis of rotation. For rotation about a fixed axis, there are but two possible directions for the angular velocity, corresponding to clockwise or counterclockwise rotation. This is analogous to linear motion in one dimension in which the direction of the velocity is either positive or negative.) The units of angular velocity are radians per second. Since a radian is a dimensionless unit, the dimensions of angular velocity are those of reciprocal time (T^{-1}). The magnitude of the angular velocity is called the **angular speed.** Although the angular motion of a disk is often described using other units, such as revolutions per minute, it is important to remember that many of the equations we will use involving the angular velocity ω are valid only when the angles are expressed in radians and the angular velocity, in radians per second. To convert from revolutions to radians, we use

$$1 \text{ rev} = 2\pi \text{ rad}$$

Exercise

A record rotates at 33.3 revolutions per minute. What is its angular speed? (Answer: 3.49 rad/s)

The rate of change of angular velocity with respect to time is called the **angular acceleration** α:

$$\alpha = \frac{d\omega}{dt} = \frac{d^2\theta}{dt^2} \qquad 8\text{-}4$$

Angular acceleration defined

The units of angular acceleration are radians per second per second (rad/s^2). The angular acceleration is positive if the angular velocity ω is increasing and negative if ω is decreasing.

We can relate the tangential velocity of a particle on the disk to the angular velocity of the disk using Equations 8-2 and 8-3:

$$v_{it} = \frac{ds_i}{dt} = \frac{r_i\,d\theta}{dt} = r_i\omega \qquad 8\text{-}5$$

A tiny motor called a "wobble motor." This motor, which has a diameter of the order of a milli meter, has achieved angular speeds in excess of 120,000 rev/min. The edge of a dime is visible in the background

Similarly, the tangential acceleration of a particle on the disk is related to the angular acceleration of the disk:

$$a_{it} = \frac{dv_i}{dt} = r_i \frac{d\omega}{dt}$$

Thus

$$a_{it} = r_i\alpha \qquad 8\text{-}6$$

Each particle of the disk also has a radial acceleration, the centripetal acceleration, which points inward along the radial line and has the magnitude

$$a_{ic} = \frac{v_i^2}{r_i} = \frac{(r_i\,\omega)^2}{r_i} = r_i\,\omega^2 \qquad 8\text{-}7$$

Exercise

A point on the rim of a record is 15 cm from the axis of rotation. Find the tangential velocity v_t, the tangential acceleration a_t, and the centripetal acceleration a_c of the point if the record is rotating at a constant angular velocity of 3.49 rad/s. (Answers: $v_t = 52.4$ cm/s, $a_t = 0$, $a_c = 183$ cm/s^2)

The angular displacement θ, angular velocity ω, and angular acceleration α are analogous to the linear displacement x, linear velocity v, and linear acceleration a in one-dimensional motion. Because of the similarity of the definitions of the rotational and linear quantities, much of what we learned in Chapter 2 will be useful in treating problems involving the rotation of a rigid body. For example, the equations for constant angular acceleration are the same as Equations 2-9 through 2-13, with θ replacing x, ω replacing v, and α replacing a. Thus,

$$\omega = \omega_0 + \alpha t \qquad 8\text{-}8$$

is the rotational analog of

$$v = v_0 + at$$

and

$$\theta = \theta_0 + \omega_0 t + \tfrac{1}{2}\alpha t^2 \qquad 8\text{-}9$$

is the rotational analog of

$$x = x_0 + v_0 t + \tfrac{1}{2}at^2$$

where ω_0 and θ_0 are the initial values of the angular velocity and angular position, respectively. As with the constant-linear-acceleration equations, we can eliminate time from these equations to obtain an equation relating the angular displacement, angular velocity, and angular acceleration:

$$\omega^2 = \omega_0^2 + 2\alpha(\theta - \theta_0) \qquad 8\text{-}10$$

which is the rotational analog of

$$v^2 = v_0^2 + 2a(x - x_0)$$

Example 8-1

A disk rotates with constant angular acceleration $\alpha = 2\ \text{rad/s}^2$. If the disk starts from rest, how many revolutions does it make in 10 s?

This is analogous to the linear problem of finding the distance traveled by a particle in a given time if it starts from rest with constant acceleration. The number of revolutions is related to the angular displacement by the fact that each revolution is an angular displacement of 2π rad. Thus, we need to find the angular displacement $\theta - \theta_0$ in radians for a time of 10 s and multiply by the conversion factor (1 rev)/(2π rad).

Equation 8-9 relates the angular displacement to the time. We are given $\omega_0 = 0$ (the disk starts from rest). Thus,

$$\theta - \theta_0 = \omega_0 t + \tfrac{1}{2}\alpha t^2 = 0 + \tfrac{1}{2}(2\ \text{rad/s}^2)(10\ \text{s})^2 = 100\ \text{rad}$$

The number of revolutions is therefore

$$100\ \text{rad} \times \frac{1\ \text{rev}}{2\pi\ \text{rad}} = 15.9\ \text{rev}$$

Example 8-2

Find the angular speed of the disk in Example 8-1 after 10 s.

Using Equation 8-8 we have

$$\omega = \omega_0 + \alpha t = 0 + (2\ \text{rad/s}^2)(10\ \text{s}) = 20\ \text{rad/s}$$

As a check of this result and that of the previous example, we can also find the angular speed from Equation 8-10:

$$\omega^2 = 2\alpha(\theta - \theta_0) = 2(2\ \text{rad/s}^2)(100\ \text{rad}) = 400\ \text{rad}^2/\text{s}^2$$

or

$$\omega = \sqrt{400\ \text{rad}^2/\text{s}^2} = 20\ \text{rad/s}$$

Question

1. Two points are on a disk turning at constant angular velocity, one point on the rim and the other halfway between the rim and the axis. Which point moves the greater distance in a given time? Which turns through the greater angle? Which has the greater speed? The greater angular velocity? The greater tangential acceleration? The greater angular acceleration? The greater centripetal acceleration?

8-2 Torque and Moment of Inertia

When you wish to set a top spinning you twist it. In Figure 8-2*a*, a disk resting flat on a horizontal surface is set spinning by the forces $\mathbf{F}_1$ and $\mathbf{F}_2$ exerted at the edge of the disk. Note that the locations of the points of application of these forces are important. The same two forces applied so that their lines of action pass through the center of the disk, as in Figure 8-2*b*, will not set the disk spinning. (The line of action of a force is the line along which the force acts.) The perpendicular distance between the line of action of a force and the axis of rotation is called the **lever arm** ℓ of the force. The product of a force times its lever arm is called a **torque** τ. The torque exerted on an object by a force is the quantity that affects the angular velocity of the object.

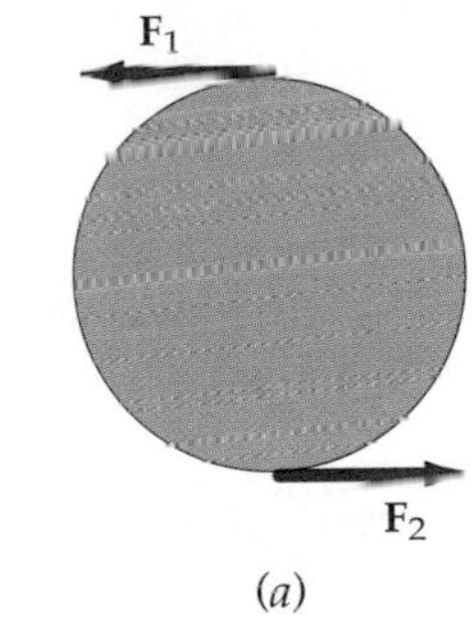

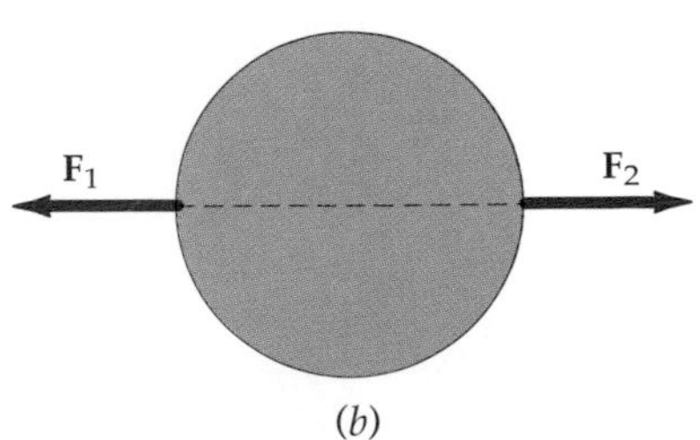

Figure 8-2 (*a*) The forces $\mathbf{F}_1$ and $\mathbf{F}_2$ will set the disk spinning about its axis. (*b*) If the same two forces are applied such that their lines of action pass through the center of the disk, they will not set the disk spinning.

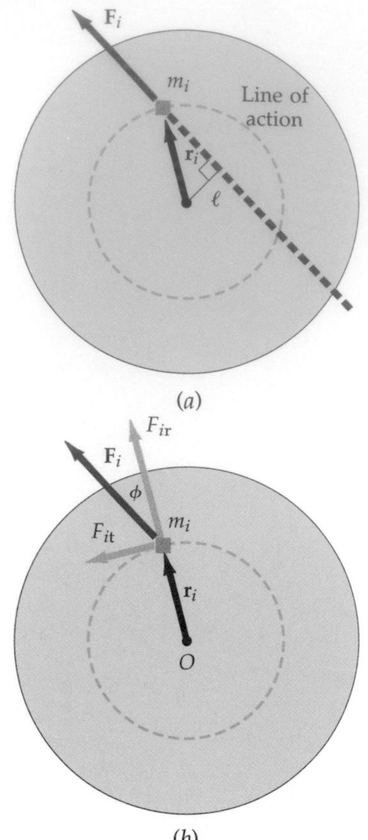

Figure 8-3 A force $\mathbf{F}_i$ acting on the ith particle of a disk pivoted about its center. (*a*) The lever arm ℓ is the perpendicular distance from the line of action of the force to the axis of rotation. The axis of rotation here is directly into the page. (*b*) The force is resolved into a radial component $F_{ir} = F_i \cos \phi$ and a perpendicular component $F_{it} = F_i \sin \phi$. The radial component does not affect the rotation of the disk.

Figure 8-3*a* shows a single force $\mathbf{F}_i$ acting on the ith particle of a disk. The lever arm of this force is $\ell = r_i \sin \phi$, where ϕ is the angle between the force $\mathbf{F}_i$ and the position vector $\mathbf{r}_i$ to the point of application of the force. The torque exerted by this force is thus

$$\tau_i = F_i\ell = F_i r_i \sin \phi \qquad 8\text{-}11$$

In Figure 8-3*b*, we have resolved the force $\mathbf{F}_i$ into two components, $F_{ir} = F_i \cos \phi$ along the radial line $\mathbf{r}_i$ and $F_{it} = F_i \sin \phi$ perpendicular to the radial line. The radial component F_{ir} has no effect on the rotation of the disk. The torque exerted by the force $\mathbf{F}_i$ can be written in terms of F_{it}. From Equation 8-11, we have

$$\tau_i = F_i\ell = F_i r_i \sin \phi = F_{it} r_i \qquad 8\text{-}12$$

We will now show that the angular acceleration of a rigid body is proportional to the net torque acting on it. Let us assume that $\mathbf{F}_i$ is the net external force acting on the ith particle. The tangential acceleration of the ith particle is, by Newton's second law,

$$F_{it} = m_i a_{it} = m_i r_i \alpha \qquad 8\text{-}13$$

where we have used the relation $a_{it} = r_i\alpha$ between the tangential acceleration of the ith particle and the angular acceleration of the object (Equation 8-6). If we multiply each side by r_i, we obtain

$$r_i F_{it} = m_i r_i^2 \alpha \qquad 8\text{-}14$$

The left side of Equation 8-15 is the torque $\tau_i = r_i F_{it}$ exerted by the force $\mathbf{F}_i$ about the pivot O. We thus have

$$\tau_i = m_i r_i^2 \alpha \qquad 8\text{-}15$$

If we now sum over all the particles in the object, we obtain

$$\sum_i \tau_i = \sum_i m_i r_i^2 \alpha \qquad 8\text{-}16$$

The quantity $\Sigma\tau_i$ is the net torque acting on the object, which we will denote by τ_{net}. For a rigid body, the angular acceleration is the same for all the particles of the object and can therefore be taken out of the sum. The quantity $\Sigma m_i r_i^2$ is a property of the object and the axis of rotation called the **moment of inertia** I:

Moment of inertia defined

$$I = \sum_i m_i r_i^2 \qquad 8\text{-}17$$

The torque exerted by a wrench on a nut is proportional to the force and to the lever arm. Charlie Chaplin could exert a greater torque with the same force if he held the wrenches nearer their ends.

In this equation, the distance r_i is the distance from the ith particle *to the axis of rotation*. Usually, this distance is not the same as the distance from the ith particle to the origin, though for a disk with the origin at the center on the axis, these distances are the same. The moment of inertia is a measure of the resistance of an object to changes in its rotational motion. It depends on the distribution of the mass of the object relative to the axis of rotation of the object. It is a property of the object (and the axis of rotation), just as the mass m is a property of the object that measures its resistance to changes in translational motion. For systems consisting of a small number of discrete particles, we can compute the moment of inertia about a given axis directly from Equation 8-17. For the more common case of a continuous object, such as a wheel, the moment of inertia about a given axis is computed using calculus. We will illustrate such calculations in Section 8-4. Table 8-1 lists the moments of inertia of various uniform objects.

In terms of the moment of inertia, Equation 8-16 becomes

$$\tau_{\text{net}} = I\alpha \tag{8-18}$$

Newton's second law for rotation

Equation 8-18 is the rotational analog of Newton's second law for linear motion:

$$\mathbf{F}_{\text{net}} = m\mathbf{a}$$

The net torque is analogous to the net force, the moment of inertia is analogous to the mass, and the angular acceleration is analogous to the linear acceleration.

In Chapter 7, we saw that the net force acting on a system of particles is equal to the net *external* force acting on the system because, by Newton's

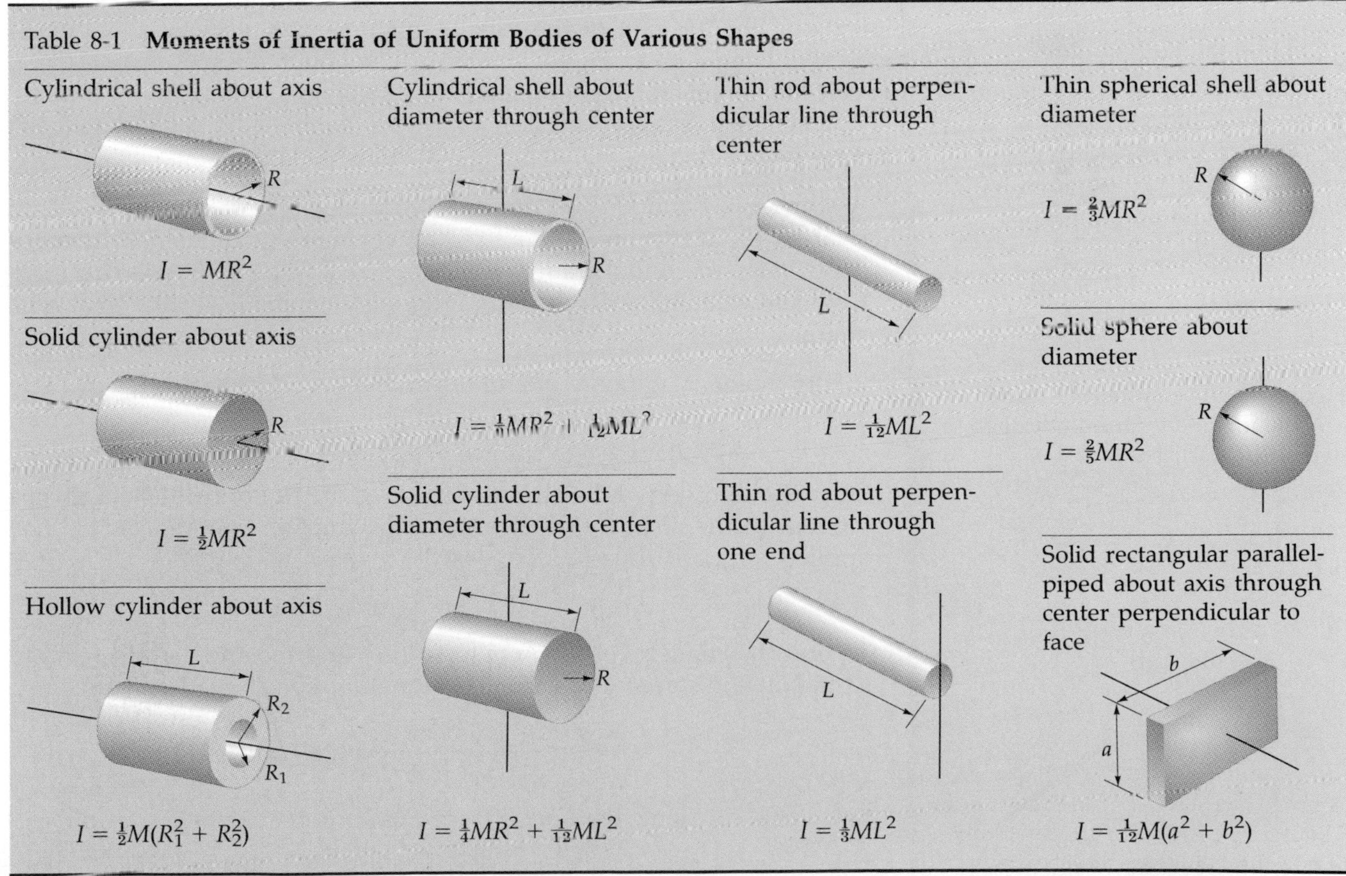

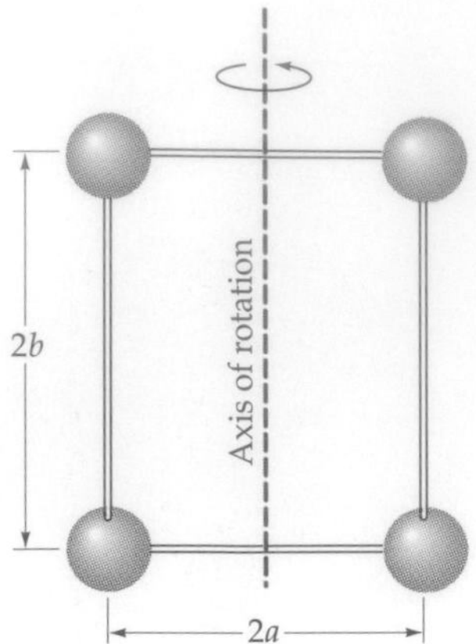

Figure 8-4 Four particles of equal mass connected by massless rods and rotating about an axis through the plane of the particles and through their center of mass for Example 8-3.

third law, the internal forces exerted by the particles within the system on one another cancel in pairs. The treatment of internal torques exerted by the particles within a system on one another is somewhat complicated and will be deferred until Section 8-7. There we will show that the internal torques sum to zero if the force exerted by one particle on another acts along the line joining the particles. Assuming that this holds for all the internal forces, we find that the net torque acting on a system is also equal to the net *external* torque acting on the system.

Example 8-3

Four particles of mass m are connected by massless rods to form a rectangle of sides $2a$ and $2b$ as shown in Figure 8-4. The system rotates about an axis in the plane of the figure through the center. Find the moment of inertia about this axis.

From the figure, we can see that the distance from each particle to the axis of rotation is a. The moment of inertia of each particle about this axis is therefore ma^2, and since there are four particles, the total moment of inertia of the body is

$$I = 4ma^2$$

The distance b plays no role at all because it is not related to the distance from any mass to the axis of rotation.

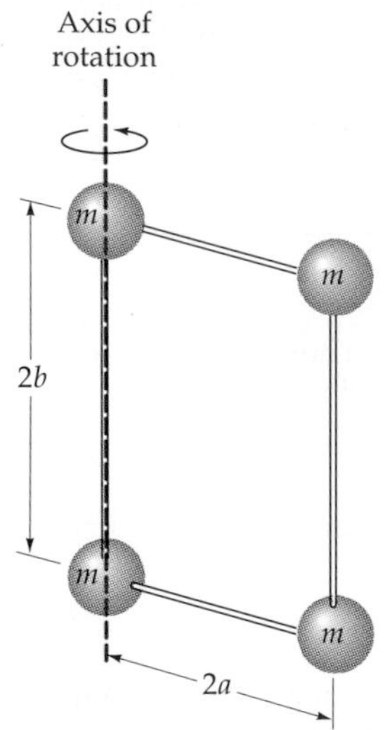

Figure 8-5 The system of Figure 8-4 rotating about an axis through two of the particles.

Exercise

Find the moment of inertia of the system of Example 8-3 for rotation about an axis parallel to the first axis but passing through two of the masses as shown in Figure 8-5. (Answer: $I = 8ma^2$)

Example 8-3 and the exercise illustrate the fact that the moment of inertia depends on the location of the axis of rotation. The moment of inertia is larger about the axis in Figure 8-5 than it is about the axis parallel to it through the center of mass in Figure 8-4.

Example 8-4

A string is wound around the rim of a uniform disk that is pivoted to rotate without friction about a fixed axis through its center. The mass of the disk is 3 kg, and its radius is 25 cm. The string is pulled with a force **F** of magnitude 10 N (Figure 8-6). If the disk is initially at rest, what is its angular velocity after 5 s?

The moment of inertia of a uniform disk about its axis is (from Table 8-1)

$$I = \tfrac{1}{2}MR^2 = \tfrac{1}{2}(3\text{ kg})(0.25\text{ m})^2 = 9.38 \times 10^{-2}\text{ kg·m}^2$$

Since the direction of the string as it leaves the rim of the disk is always tangent to the disk, the lever arm of the force it exerts is just R. The applied torque is thus

$$\tau = FR = (10\text{ N})(0.25\text{ m}) = 2.5\text{ N·m}$$

To find the angular velocity, we must first find the angular acceleration from Newton's second law for rotational motion (Equation 8-18):

$$\alpha = \frac{\tau_{\text{net}}}{I} = \frac{2.5\text{ N·m}}{0.0938\text{ kg·m}^2} = 26.7\text{ rad/s}^2$$

Since α is constant, we find ω from Equation 8-8, setting $\omega_0 = 0$:

$$\omega = \omega_0 + \alpha t = 0 + (26.7\text{ rad/s}^2)(5\text{ s}) = 133\text{ rad/s}$$

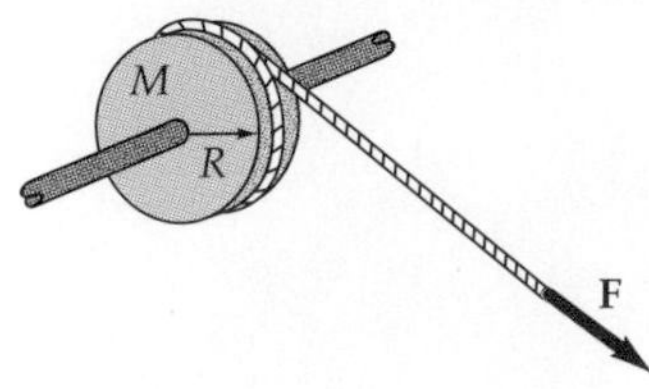

Figure 8-6 A string wrapped around a disk for Example 8-4.

Example 8-5

An object of mass m is tied to a light string wound around a wheel of moment of inertia I and radius R (Figure 8-7). The wheel bearing is frictionless, and the string does not slip on the rim. Find the tension in the string and the acceleration of the object.

The only force exerting a torque on the wheel is the tension **T** in the string, which has a lever arm R and produces a clockwise rotation. Taking the clockwise direction to be positive, we have

$$TR = I\alpha \qquad 8\text{-}19$$

Two forces act on the suspended object, the upward tension **T**, and the downward force of gravity $m\mathbf{g}$. Taking the downward direction to be positive, so that a and α both have the same sign, we have from Newton's second law

$$mg - T = ma \qquad 8\text{-}20$$

There are three unknowns, T, a, and α, in these two equations. The string provides a constraint by which we can relate a and α. Since the string does not slip, its speed is equal to the speed of a point on the rim, and its acceleration is the same as the tangential acceleration of a point on the rim. Its acceleration is thus

$$a = R\alpha \qquad 8\text{-}21$$

Substituting a/R for α in Equation 8-19 gives

$$TR = I\frac{a}{R}$$

or

$$a = \frac{TR^2}{I} \qquad 8\text{-}22$$

Substituting this result for a into Equation 8-20 gives

$$mg - T = m\frac{TR^2}{I}$$

or

$$T\left(1 + \frac{mR^2}{I}\right) = mg$$

$$T = \frac{I}{I + mR^2}mg$$

We can use this value for T in Equation 8-22 to find a:

$$a = \frac{mR^2}{I + mR^2}g$$

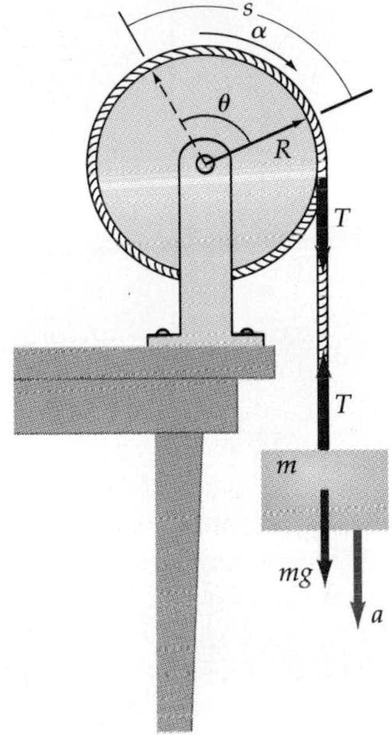

Figure 8-7 Object of mass m tied to a string wrapped around a rotating wheel for Example 8-5. Since the string does not slip, the acceleration of the object equals the tangential acceleration of a point on the rim, $a = R\alpha$.

Questions

2. Can an object rotate if there is no torque acting?
3. Can a given rigid body have more than one moment of inertia?
4. Does an applied net torque always increase the angular speed of an object?
5. If the angular velocity of an object is zero at some instant, does this mean that the net torque on the object must be zero?

8-3 Rotational Kinetic Energy

When a rotating object turns through a small angular displacement $d\theta$, the ith particle moves through a distance $ds_i = r_i\,d\theta$. If a force $\mathbf{F}_i$ acts on the ith particle, it does work

$$dW_i = F_{it}\,ds_i = F_{it}r_i\,d\theta = \tau_i\,d\theta$$

In general, the work done by a torque τ when an object turns through a small angle $d\theta$ is

$$dW = \tau\,d\theta \qquad 8\text{-}23$$

Equation 8-23 is analogous to the similar result for linear motion in one dimension, $dW = F_s\,ds$. The rate at which the torque does work is the power input of the torque:

$$P = \frac{dW}{dt} = \tau\frac{d\theta}{dt}$$

or

Power

$$P = \tau\omega \qquad 8\text{-}24$$

Equation 8-24 is the rotational analog of $P = F_s v_s$.

The total work done on a system equals the change in the kinetic energy of the system (if the system has no change in potential energy and there are no energy losses). For an object rotating about an axis through its center of mass, the kinetic energy of the object is the kinetic energy relative to the center of mass, K_{rel}, as discussed in Chapter 7. This kinetic energy is the sum of the kinetic energies of the individual particles in the body:

$$K = \sum_i \tfrac{1}{2}m_i v_i^2 = \sum_i \tfrac{1}{2}m_i(r_i\omega)^2 = \tfrac{1}{2}\sum_i m_i r_i^2\omega^2$$

or

Kinetic energy of rotation

$$K = \tfrac{1}{2}I\omega^2 \qquad 8\text{-}25$$

Equation 8-25 is the rotational analog of $K = \frac{1}{2}mv^2$ for linear motion.

The Archimedes screw is a device for lifting water. The rotational work done by the torque exerted at the handle is converted into increased potential energy of the water.

Example 8-6

A uniform disk of mass 3 kg and radius 12 cm rotates at 480 rev/min. Find its kinetic energy.

From Table 8-1, the moment of inertia of a uniform disk is given by

$$I = \tfrac{1}{2}mR^2 = \tfrac{1}{2}(3\text{ kg})(0.12\text{ m})^2 = 0.0216\text{ kg}\cdot\text{m}^2$$

The angular velocity is

$$\omega = \left(\frac{480\text{ rev}}{60\text{ s}}\right)\left(\frac{2\pi\text{ rad}}{1\text{ rev}}\right) = 50.3\text{ rad/s}$$

The kinetic energy is then

$$K = \tfrac{1}{2}I\omega^2 = \tfrac{1}{2}(0.0216\text{ kg}\cdot\text{m}^2)(50.3\text{ rad/s})^2 = 27.3\text{ J}$$

Note that we drop the dimensionless unit radian and use 1 kg·m^2/s^2 = 1 J.

The Crab Pulsar is one of the fastest rotating neutron stars known, but it is slowing down. It appears to blink on (left) and off (right) like a rotating spotlight about 30 times per second, but the period is increasing by about 10^{-5} seconds per year. The loss in rotational energy, which is equivalent to the power output of 100,000 suns, appears as light emitted by electrons accelerated in the magnetic field of the pulsar.

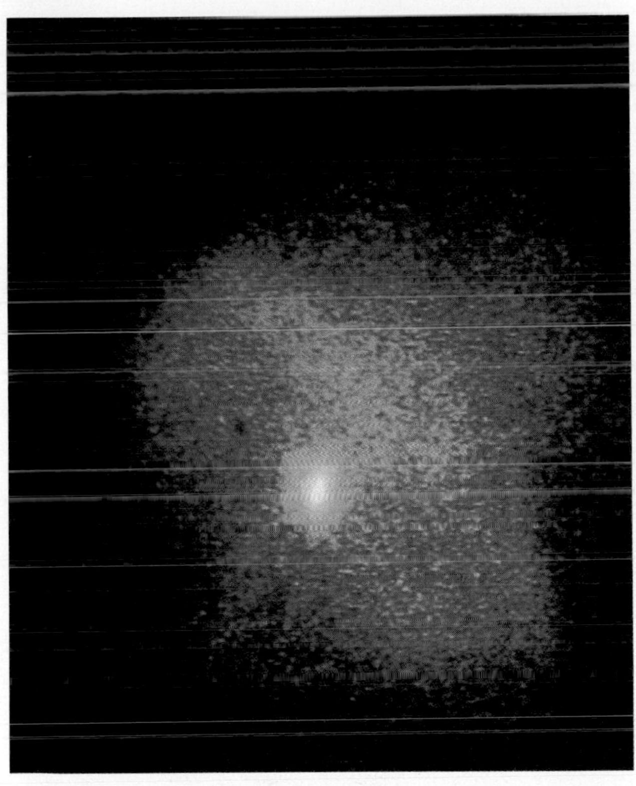

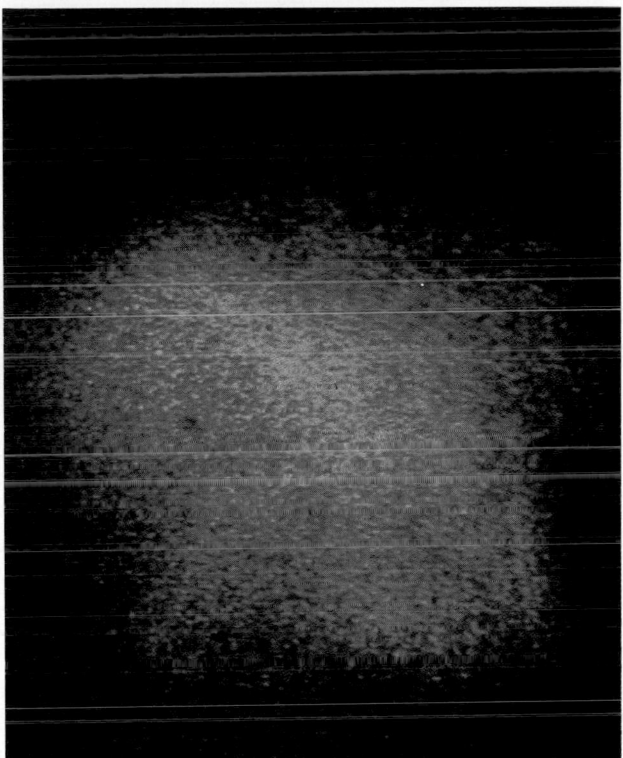

Example 8-7

A car engine delivers 380 N·m of torque at 3200 rev/min. Find the power output of the engine.

The angular speed corresponding to 3200 rev/min is

$$\omega = \left(\frac{3200\text{ rev}}{\text{min}}\right)\left(\frac{2\pi\text{ rad}}{1\text{ rev}}\right)\left(\frac{1\text{ min}}{60\text{ s}}\right) = 335\text{ rad/s}$$

The power output of the engine is given by Equation 8-25:

$$P = \tau\omega = (380\text{ N}\cdot\text{m})(335\text{ rad/s}) = 127\text{ kW}$$

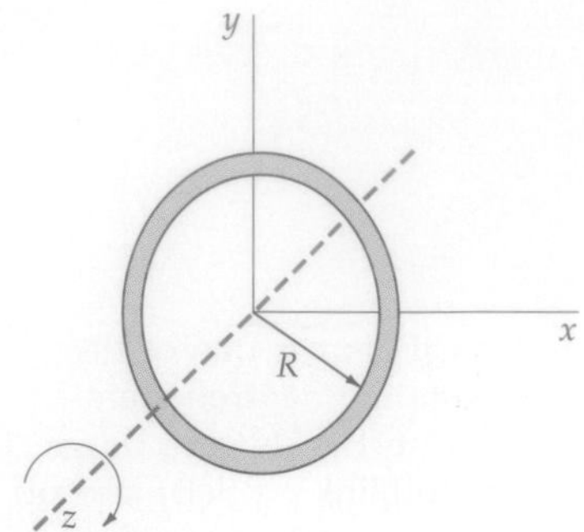

Figure 8-8 A hoop rotating about an axis perpendicular to the plane of the hoop and through its center. Since all the mass of the hoop is at a distance R from the axis, the moment of inertia is MR^2.

8-4 Calculating the Moment of Inertia

For continuous bodies, the sum in Equation 8-17 for the moment of inertia is replaced by an integral:

$$I = \int r^2\,dm \tag{8-26}$$

where r is the distance of the mass element dm from the axis of rotation.

Example 8-8

Find the moment of inertia of a hoop of mass M and radius R about an axis through the center and perpendicular to the plane of the hoop (Figure 8-8).

In this case, all the mass is at a distance $r = R$, so the moment of inertia is simply

$$I = \int r^2\,dm = R^2 \int dm = MR^2$$

Example 8-9

Find the moment of inertia of a stick of uniform density about an axis perpendicular to the stick through one end.

The mass element dm is shown in Figure 8-9. It is at a distance x from the axis of rotation. Since the total mass M is uniformly distributed along the length L, the linear mass density is $\lambda = M/L$. Thus, $dm = \lambda\,dx = (M/L)\,dx$. The moment of inertia about the y axis is

$$I_y = \int_0^L x^2\,dm = \int_0^L x^2 \frac{M}{L}\,dx = \frac{M}{L}\int_0^L x^2\,dx$$

$$= \frac{M}{L}\frac{1}{3}x^3\Big|_0^L = \frac{ML^3}{3L} = \frac{1}{3}ML^2$$

The moment of inertia about the z axis is also $\frac{1}{3}ML^2$, and that about the x axis is zero, if all the mass is right on the x axis.

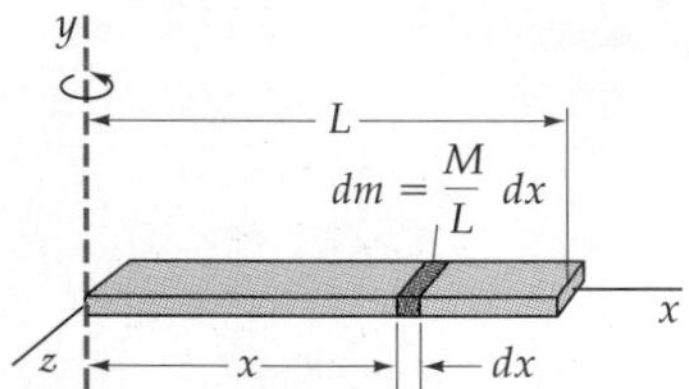

Figure 8-9 Geometry for setting up the integral to calculate the moment of inertia of a uniform stick rotating about an axis perpendicular to the stick and through one end.

Example 8-10

Find the moment of inertia of a uniform disk about the axis through its center and perpendicular to the plane of the disk.

We expect that I will be smaller than MR^2 since the mass of a disk is not concentrated at $r = R$ like that of a hoop but, rather, is uniformly distributed from $r = 0$ to $r = R$. We calculate I by taking mass elements dm as shown in Figure 8-10. Each mass element is a hoop of radius r and thickness dr. The moment of inertia of each element is $r^2\,dm$. Since the area of each element is $dA = 2\pi r\,dr$, the mass of each element is

$$dm = \frac{M}{A}\,dA = \frac{M}{A}2\pi r\,dr$$

where $A = \pi R^2$ is the area of the disk. We thus have

$$I = \int r^2\,dm = \int_0^R r^2 \frac{M}{A} 2\pi r\,dr = \frac{2\pi M}{\pi R^2}\int_0^R r^3\,dr = \frac{2M}{R^2}\frac{R^4}{4} = \frac{1}{2}MR^2$$

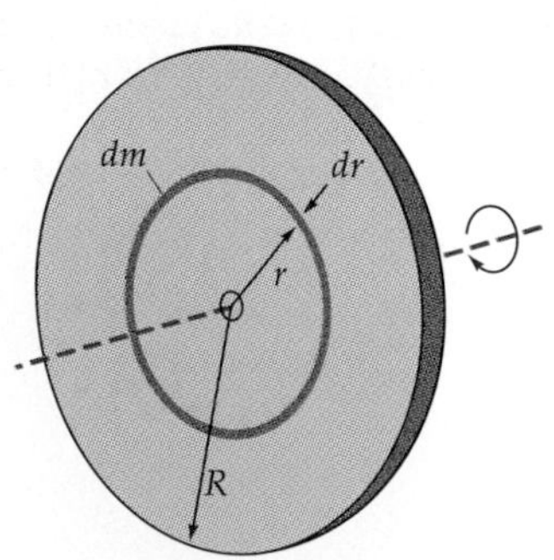

Figure 8-10 Geometry for setting up the integral to calculate the moment of inertia of a uniform disk rotating about an axis through its center and perpendicular to the plane of the disk.

Example 8-11

Find the moment of inertia of a cylinder of uniform density about its axis (Figure 8-11).

We can consider the cylinder to be a set of disks, each with mass m_i and moment of inertia $\frac{1}{2}m_iR^2$. Then the moment of inertia of the complete cylinder is

$$I = \sum_i \tfrac{1}{2}m_iR^2 = \tfrac{1}{2}R^2 \sum_i m_i = \tfrac{1}{2}MR^2$$

where $M = \Sigma\, m_i$ is the total mass of the cylinder.

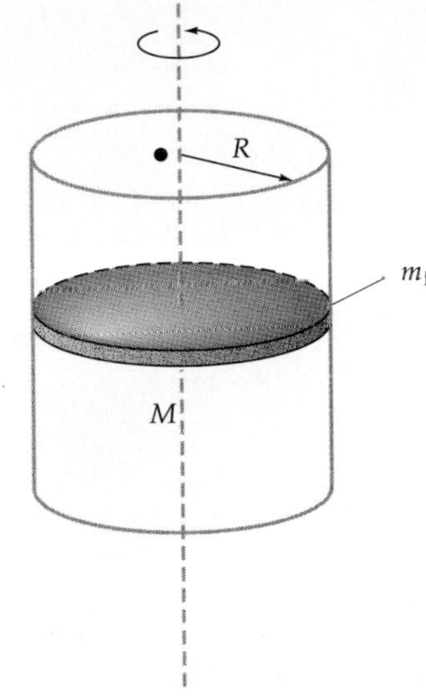

Figure 8-11 A cylinder rotating about its axis can be considered to be a stack of disks of mass m_i. Since each disk has a moment of inertia $\frac{1}{2}m_iR^2$, the moment of inertia of the cylinder is $\frac{1}{2}MR^2$.

Example 8-12

Find the moment of inertia of a sphere of uniform density about an axis through its center.

We calculate this moment of inertia by treating the sphere as a set of disks as shown in Figure 8-12. We consider a disk element at height x above the center. The radius of the disk element is

$$r = \sqrt{R^2 - x^2}$$

The volume of the disk is the area times the height, $dV = \pi r^2\, dx$. If M is the total mass of the sphere and V is its volume, the mass per unit volume is M/V. The mass of each disk is then

$$dm = \frac{M}{V}\, dV = \frac{M}{V}\pi r^2\, dx = \frac{M}{V}\pi(R^2 - x^2)\, dx$$

The moment of inertia of each disk element is

$$dI = \frac{1}{2}r^2\, dm = \frac{1}{2}(R^2 - x^2)\left[\frac{M}{V}\pi(R^2 - x^2)\, dx\right]$$

$$= \frac{1}{2}\frac{M}{V}\pi(R^2 - x^2)^2\, dx$$

When x varies from 0 to R, the top half of the sphere is covered. The total moment of inertia of the sphere is therefore twice the integral of dI from $x = 0$ to $x = R$:

$$I = 2\int_0^R \frac{1}{2}\frac{M}{V}\pi(R^2 - x^2)^2\, dx$$

The integral in this equation can be evaluated by writing $(R^2 - x^2)^2 = R^4 - 2R^2x^2 + x^4$ and integrating term by term. The result is

$$\int_0^R (R^2 - x^2)^2\, dx = \frac{8R^5}{15} \qquad \text{8-27}$$

The moment of inertia of the sphere is thus

$$I = \frac{\pi M}{V}\frac{8R^5}{15} = \frac{2}{5}MR^2$$

where we have used $V = \frac{4}{3}\pi R^3$.

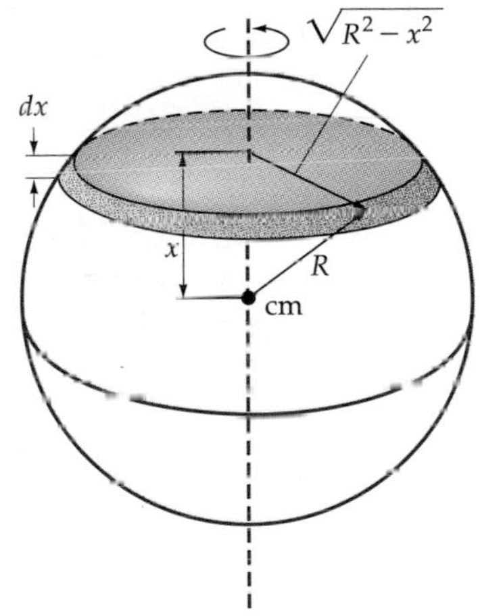

Figure 8-12 Geometry for the calculation of the moment of inertia of a uniform sphere rotating about a diameter (Example 8-12). The sphere can be thought of as a set of disks of varying radii.

Exercise

Derive Equation 8-27 by expanding the integrand and integrating term by term.

The Parallel-Axis Theorem

We can often simplify the calculation of the moments of inertia for various bodies by using general theorems that relate the moment of inertia about one axis of the body to that about another axis. The parallel-axis theorem relates the moment of inertia about an axis through the center of mass of a body to that about a second, parallel axis. Let I_{cm} be the moment of inertia about an axis through the center of mass of a body and I be that about a parallel axis a distance h away. The parallel-axis theorem states that

$$I = I_{cm} + Mh^2$$

where M is the total mass of the body. Example 8-3 and the exercise following it illustrated a special case of this theorem with $h = a$, $M = 4m$, and $I_{cm} = 4ma^2$.

We can prove this theorem using the result developed in Chapter 7 that the kinetic energy of a system of particles is the sum of the kinetic energy of the motion of the center of mass plus the kinetic energy of the motion relative to the center of mass:

$$K = \tfrac{1}{2}MV_{cm}^2 + K_{rel} \qquad \text{8-28}$$

Consider a rigid object rotating with an angular velocity ω about an axis a distance h away from a parallel axis through the center of mass as shown in Figure 8-13. When the body rotates through an angle $\Delta\theta$ measured about the axis of rotation, it rotates through the same angle $\Delta\theta$ measured about any other parallel axis. The motion of the object relative to the center of mass is thus a rotation about the center-of-mass axis with the same angular velocity ω. The kinetic energy of the motion relative to the center of mass is thus

$$K_{rel} = \tfrac{1}{2}I_{cm}\omega^2$$

The velocity of the center of mass relative to any point on the axis of rotation is

$$V_{cm} = h\omega$$

The kinetic energy of the motion of the center of mass is thus

$$\tfrac{1}{2}MV_{cm}^2 = \tfrac{1}{2}M(h\omega)^2 = \tfrac{1}{2}M\omega^2h^2$$

When the total kinetic energy of the body is written as $\tfrac{1}{2}I\omega^2$, Equation 8-28 becomes

$$K = \tfrac{1}{2}I\omega^2 = \tfrac{1}{2}M\omega^2h^2 + \tfrac{1}{2}I_{cm}\omega^2 = \tfrac{1}{2}(Mh^2 + I_{cm})\omega^2$$

Thus

Parallel-axis theorem

$$I = Mh^2 + I_{cm} \qquad \text{8-29}$$

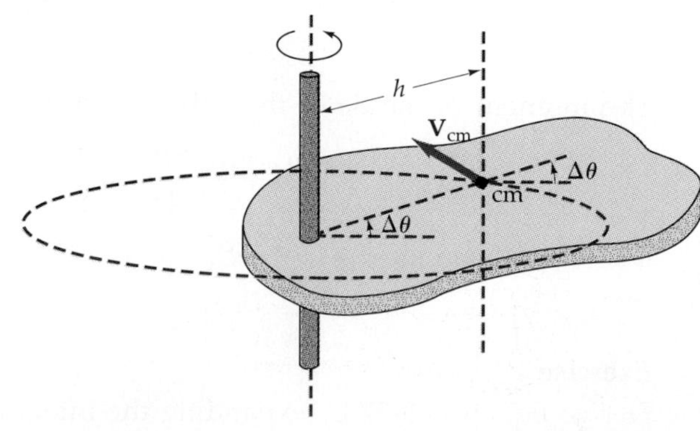

Figure 8-13 An object rotating about an axis parallel to an axis through the center of mass and a distance h from it.

Example 8-13

Find the moment of inertia of the hoop of Example 8-8 about an axis perpendicular to the hoop but through its edge as shown in Figure 8-14.

This calculation is easily done using the parallel-axis theorem with $h = R$ and the result from Example 8-8 that $I_{cm} = MR^2$:

$$I = I_{cm} + Mh^2 = MR^2 + MR^2 = 2MR^2$$

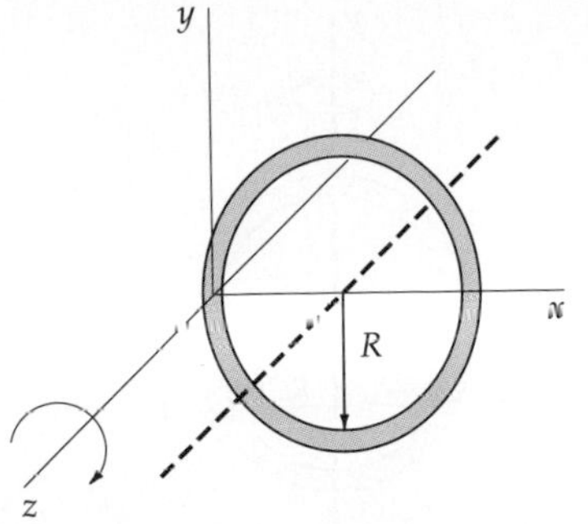

Figure 8-14 A hoop rotating about an axis that is perpendicular to the plane of the hoop and passes through the edge of the hoop for Example 8-13. The moment of inertia is found to be $2MR^2$ using the parallel-axis theorem.

Example 8-14

Find the moment of inertia of a stick of uniform density about the y' axis through the center of mass (Figure 8-15).

In Example 8-9, we found the moment of inertia about an axis through the end of a uniform stick to be $\frac{1}{3}ML^2$. Since this axis is at a distance $h = \frac{1}{2}L$ from the center of mass of the stick, the parallel-axis theorem gives

$$I = I_{cm} + M(\tfrac{1}{2}L)^2 = \tfrac{1}{3}ML^2$$

or

$$I_{cm} = \tfrac{1}{3}ML^2 - \tfrac{1}{4}ML^2 = \tfrac{1}{12}ML^2$$

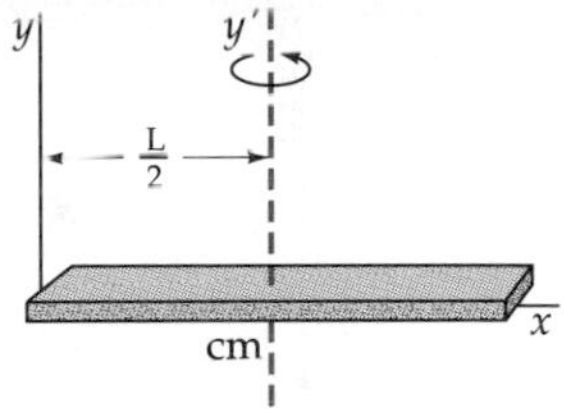

Figure 8-15 A uniform stick rotating about an axis perpendicular to the stick and passing through its center of mass for Example 8-14. The moment of inertia is found to be $\frac{1}{12}ML^2$ using the parallel-axis theorem.

The Perpendicular-Axis Theorem

The perpendicular-axis theorem applies only to plane figures. It relates the moments of inertia about two perpendicular axes in the plane of the figure to the moment of inertia about a third axis that is perpendicular to the plane of the figure. If x, y, and z are perpendicular axes for a figure that lies in the xy plane, the perpendicular-axis theorem states that the moment of inertia about the z axis equals the sum of the moments of inertia about the x and y axes.

Figure 8-16 shows a figure in the xy plane. The distance of some mass element dm from the x axis is y. The moment of inertia about the x axis is thus $I_x = \int y^2\,dm$. Similarly, the moment of inertia about the y axis is $I_y = \int x^2\,dm$. The moment of inertia about the z axis perpendicular to the figure is

$$I_z = \int r^2\,dm$$

But for each element, $r^2 = x^2 + y^2$. Thus,

$$I_z = \int r^2\,dm = \int (x^2 + y^2)\,dm = \int x^2\,dm + \int y^2\,dm$$

or

$$I_z = I_y + I_x \qquad \text{8-30}$$

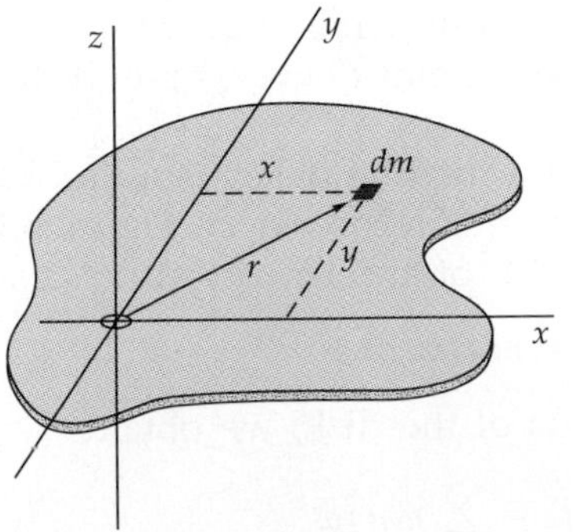

Figure 8-16 A plane figure with the z axis perpendicular to the plane. The moment of inertia about the z axis equals the sum of the moments of inertia about the x and y axes.

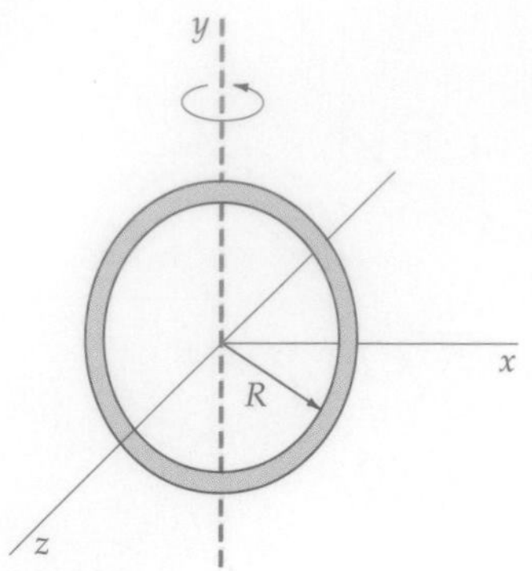

Figure 8-17 A hoop rotating about an axis through its center and in the plane of the hoop for Example 8-15. The moment of inertia is $\frac{1}{2}MR^2$, which is found using the perpendicular-axis theorem.

Example 8-15

Find the moment of inertia of the hoop in Example 8-8 about an axis that is a diameter of the hoop (Figure 8-17).

We take the hoop to be in the xy plane with the origin at the center of the hoop. By symmetry $I_x = I_y$. Since we have already found I_z in Example 8-8 to be MR^2, we have

$$I_z = I_y + I_x$$
$$= 2I_x = MR^2$$

Therefore,

$$I_x = I_y = \tfrac{1}{2}MR^2$$

Question

6. Through what point in a body must the axis of rotation pass if the moment of inertia is to be a minimum?

8-5 Angular Momentum

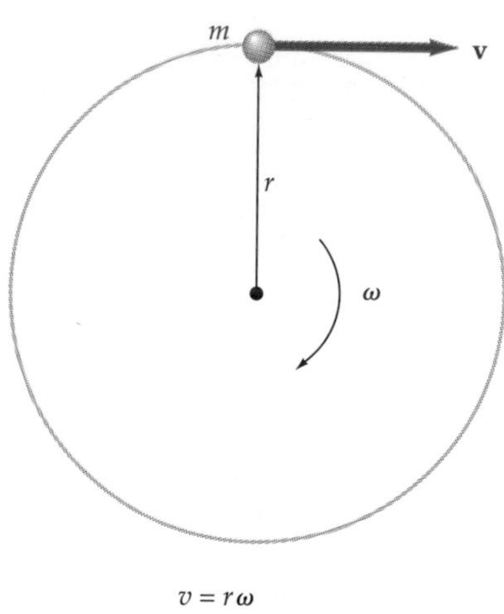

Figure 8-18 The angular momentum of a particle moving in a circle is $L = mvr = mr^2\omega$.

In Chapter 7 we defined the linear momentum $\mathbf{p}$ of a particle and showed that Newton's second law can be written

$$\mathbf{F}_{\text{net}} = \frac{d\mathbf{p}}{dt} \qquad 8\text{-}31$$

The **angular momentum** of a particle is defined as follows. For a particle moving in a circle of radius r with an angular velocity ω (Figure 8-18), the angular momentum L relative to the center of the circle is defined as the product of the magnitude of the linear momentum mv and the radius r:

$$L = mvr = m(r\omega)r$$
$$= mr^2\omega = I\omega \qquad 8\text{-}32$$

where $I = mr^2$ is the moment of inertia of the particle about an axis through the center of the circle and perpendicular to the plane of motion. The direction of L is the same as that of ω. For counterclockwise motion, ω and L are usually taken to be positive; then for clockwise motion, they are negative.

For general motion, the angular momentum of a particle relative to the origin O is defined as

Angular momentum defined

$$L = mvr_\perp = mvr \sin\theta \qquad 8\text{-}33$$

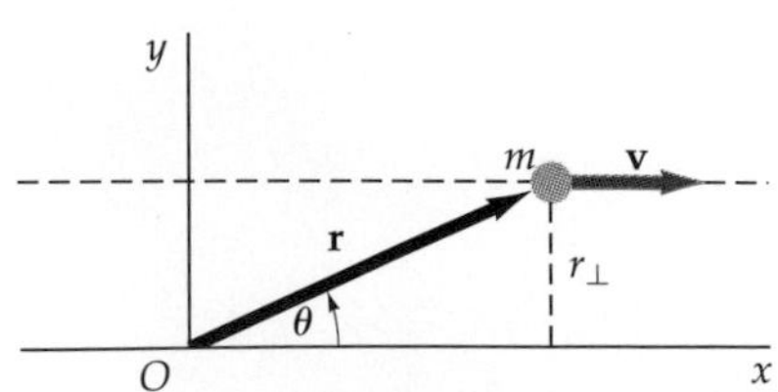

Figure 8-19 A particle of mass m moving with a velocity $\mathbf{v}$ along a line a distance $r_\perp$ from the origin O. The angular momentum of the particle about the point O is $L = mvr_\perp$.

where v is the velocity of the particle and $r_\perp = r \sin\theta$ is the perpendicular distance from the line of motion to O as shown in Figure 8-19. Note that the particle has angular momentum about point O even though the particle is not moving in a circle.

The total angular momentum of a rotating object is found by summing the angular momenta of all the elements in the object. Figure 8-20 shows a rotating disk. The angular momentum of an element of mass m_i is

$$L_i = m_i r_i^2 \omega$$

When we sum over all the elements of the disk, we obtain

$$L = \sum_i L_i = \sum_i m_i r_i^2 \omega$$

or

$$L = I\omega \qquad 8\text{-}34$$

Angular momentum of a rotating object

Equation 8-34 is the rotational analog of the equation for linear momentum, $p = mv$. It holds for objects rotating about a fixed axis and also for objects rotating about an axis that is moving such that it remains parallel to itself, such as when a ball or cylinder rolls along a line.

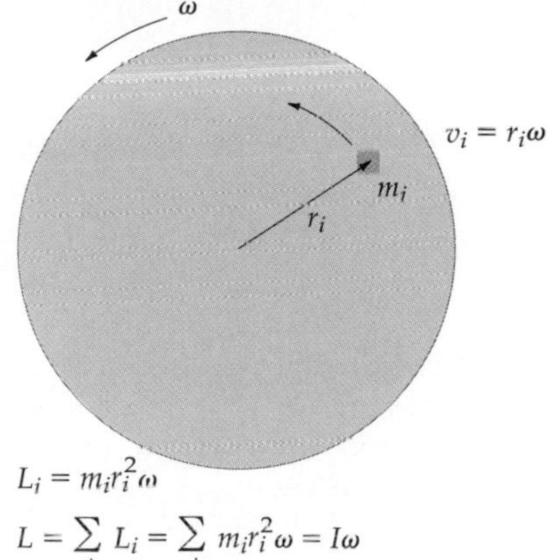

Figure 8-20 The angular momentum of a disk is calculated by summing the angular momenta of all the mass elements m_i: $L = \Sigma\, L_i = \Sigma\, m_i r_i^2 \omega = I\omega$.

Example 8-16

A 2.4-kg particle moves in a circle of radius 1.5 m with a speed of 3 m/s. (*a*) Find the angular momentum of the particle relative to the center of the circle. (*b*) Find the angular momentum relative to the origin of the same particle moving at 3 m/s along the line $y = 1.5$ m.

(*a*) From Equation 8-32, we obtain

$$L = mvr = (2.4\text{ kg})(3\text{ m/s})(1.5\text{ m}) = 10.8\text{ kg·m}^2\text{/s}$$

We could also compute the angular momentum using Equation 8-34. The moment of inertia of the particle relative to an axis through the center of the circle and perpendicular to the plane of the circle is $I = mr^2 = (2.4\text{ kg})(1.5\text{ m})^2 = 5.40\text{ kg·m}^2$, and the angular speed is $\omega = v/r = (3\text{ m/s})/(1.5\text{ m}) = 2$ rad/s. The angular momentum is then

$$L = I\omega = (5.40\text{ kg·m}^2)(2\text{ rad/s}) = 10.8\text{ kg·m}^2\text{/s}$$

(*b*) For the same particle moving along the line $y = 1.5$ m, $r_\perp = 1.5$ m. The angular momentum of the particle is then (Equation 8-33)

$$L = mvr_\perp = (2.4\text{ kg})(3\text{ m/s})(1.5\text{ m}) = 10.8\text{ kg·m}^2\text{/s}$$

The angular momentum is the same whether the particle is moving in a circle of radius 1.5 m or is moving in a straight line that is a perpendicular distance of 1.5 m from the origin.

Newton's second law for rotation can be stated in a form analogous to Equation 8-31:

$$\tau_{\text{net}} = \frac{dL}{dt} = \frac{d(I\omega)}{dt} \qquad 8\text{-}35$$

where τ_{net} is the net external torque exerted on the system.*

The net external torque acting on a system equals the rate of change of the angular momentum of the system.

For a rigid body, the moment of inertia is constant and Equation 8-35 becomes

$$\tau_{\text{net}} = \frac{d(I\omega)}{dt} = I\frac{d\omega}{dt} = I\alpha$$

which is the same as Equation 8-18. However, for a general system of particles, *the moment of inertia need not remain constant.* Equation 8-35 holds whether or not the moment of inertia is constant.

*The net torque is also the net external torque because the internal torques sum to zero, as will be discussed in Section 8-7.

Example 8-17

An object of mass m hangs from a string wrapped around a wheel of radius R and moment of inertia I as in Example 8-5 (Figure 8-7, page 235). Use Equation 8-35 to find the angular acceleration of the wheel.

Let v be the speed of the object and ω be the angular speed of the wheel at some time. Since the string does not slip, these quantities are related by $v = R\omega$. Relative to the center of the wheel, the angular momentum of the object is

$$L_o = mvR = mR^2\omega$$

and the angular momentum of the wheel is

$$L_w = I\omega$$

The total angular momentum of the system is thus

$$L = L_o + L_w = (I + mR^2)\omega$$

The only torque acting on the system relative to the center of the wheel is that due to the force of gravity mg acting on the object. Since the lever arm of this force relative to the center of the wheel is R, the torque is

$$\tau = mgR$$

Setting the torque equal to the rate of change of the angular momentum, we obtain

$$mgR = \frac{dL}{dt} = \frac{d}{dt}[(I + mR^2)\omega]$$

$$= (I + mR^2)\frac{d\omega}{dt}$$

$$= (I + mR^2)\alpha$$

The angular acceleration of the wheel is therefore

$$\alpha = \frac{mgR}{I + mR^2}$$

This agrees with the result found in Example 8-5 for the linear acceleration of the object $a = R\alpha$.

When the net external torque acting on a system is zero, we have

$$\frac{dL}{dt} = 0$$

or

$$L = \text{constant} \qquad 8\text{-}36$$

Equation 8-36 is a statement of the **law of conservation of angular momentum.**

Conservation of angular momentum

If the net external torque acting on a system is zero, the total angular momentum of the system is constant.

This is analogous to the law of conservation of linear momentum, which states that when the net external *force* acting on a system is zero, the total *linear* momentum is constant. Table 8-2 lists the rotational equations developed in this chapter along with the analogous equations for linear motion.

Table 8-2 **Comparison of Linear Motion and Rotational Motion**

Linear motion		Rotational motion	
Displacement	Δx	Angular displacement	$\Delta\theta$
Velocity	$v = \frac{dx}{dt}$	Angular velocity	$\omega = \frac{d\theta}{dt}$
Acceleration	$a = \frac{dv}{dt} = \frac{d^2x}{dt^2}$	Angular acceleration	$\alpha = \frac{d\omega}{dt} = \frac{d^2\theta}{dt^2}$
Constant-acceleration equations	$v = v_0 + at$ $\Delta x = v_{av}\,\Delta t$ $v_{av} = \frac{1}{2}(v_0 + v)$ $x = x_0 + v_0t + \frac{1}{2}at^2$ $v^2 = v_0^2 + 2a\,\Delta x$	Constant–angular-acceleration equations	$\omega = \omega_0 + \alpha t$ $\Delta\theta = \omega_{av}\,\Delta t$ $\omega_{av} = \frac{1}{2}(\omega_0 + \omega)$ $\theta = \theta_0 + \omega_0 t + \frac{1}{2}\alpha t^2$ $\omega^2 = \omega_0^2 + 2\alpha\,\Delta\theta$
Mass	m	Moment of inertia	I
Momentum	$p = mv$	Angular momentum	$L = I\omega$
Force	F	Torque	τ
Kinetic energy	$K = \frac{1}{2}mv^2$	Kinetic energy	$K = \frac{1}{2}I\omega^2$
Power	$P = Fv$	Power	$P = \tau\omega$
Newton's second law	$F_{net} = \frac{dp}{dt} = ma$	Newton's second law	$\tau_{net} = \frac{dL}{dt} = I\alpha$

Figure 8-21 Because the torque exerted by the ice is small, the angular momentum of the skater is approximately constant. When she reduces her moment of inertia by drawing in her arms, her angular velocity increases.

Figure 8-22 Multiflash photograph of a diver. The diver's center of mass moves along a parabolic path after he leaves the board. The angular momentum is provided by the initial torque due to the force of the board. This force does not pass through the diver's center of mass if he leans forward as he jumps. If the diver wanted to undergo $1\frac{1}{2}$ revolutions in the air, he would draw in his arms and legs, decreasing his moment of inertia to increase his angular velocity.

If a system is isolated from its surroundings, so that there are no external forces or torques acting on it, three quantities are conserved: energy, linear momentum, and angular momentum. The law of conservation of angular momentum is a fundamental law of nature. Even on the microscopic scale of atomic and nuclear physics, where newtonian mechanics does not hold, the angular momentum of an isolated system is constant over time. There are many examples of the conservation of angular momentum in everyday life. Consider, for example, the ice skater spinning on the tips of her skates in Figure 8-21. Because the torque exerted by the ice is small, the angular momentum of the skater is approximately constant. When she draws her arms in toward her body, the moment of inertia of her body about a vertical axis through her body is reduced. Since her angular momentum $L = I\omega$ must remain constant, as I decreases her angular velocity ω increases; that is, she spins at a faster rate. Figure 8-22 is a multiflash photograph of a diver. His center of mass moves along a parabolic path after he leaves the board. The angular momentum is provided by the initial torque due to the force of the board. (This force does not pass through the diver's center of mass if he leans forward as he jumps.) If the diver wants to somersault in the air, he draws in his arms and legs, thereby decreasing his moment of inertia and increasing his angular velocity.

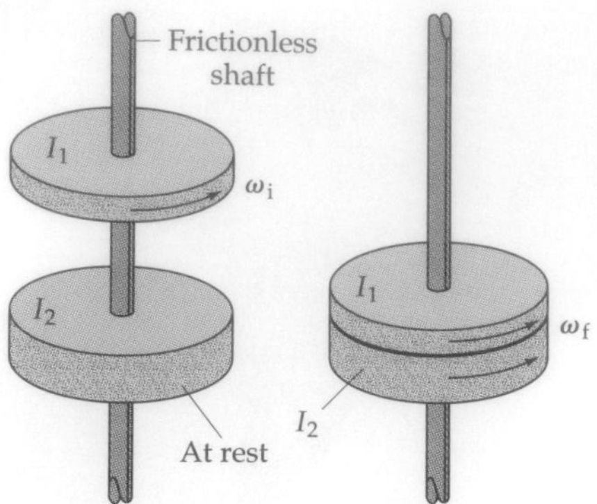

Figure 8-23 Inelastic rotational collision for Example 8-18. Since the only torques acting are internal to the system, angular momentum is conserved.

Example 8-18

A disk with a moment of inertia I_1 is rotating with an initial angular velocity ω_i about a frictionless shaft. It drops onto another disk with a moment of inertia I_2 that is initially at rest on the same shaft (see Figure 8-23). Because of surface friction, the two disks eventually attain a common angular velocity ω_f. Find this final common angular velocity.

Each disk exerts a torque on the other, but there is no torque external to the two-disk system, and so angular momentum of the system is conserved. Initially, the total angular momentum of the system L_i is the angular momentum of the first disk

$$L_i = I_1\omega_i$$

When both disks are rotating together, the total angular momentum is

$$L_f = I_1\omega_f + I_2\omega_f$$
$$= (I_1 + I_2)\omega_f$$

Setting the final angular momentum equal to the initial angular momentum, we obtain

$$(I_1 + I_2)\omega_f = I_1\omega_i$$

The final angular velocity is thus

$$\omega_f = \frac{I_1}{I_1 + I_2}\omega_i$$

In the collision of the two disks in Example 8-18, mechanical energy is not conserved. We can see this by writing the energy in terms of the angular momentum. In general, if a system with a moment of inertia I is rotating with an angular velocity ω, its kinetic energy is

$$K = \frac{1}{2}I\omega^2 = \frac{(I\omega)^2}{2I}$$

But $I\omega$ is the angular momentum L. The kinetic energy is therefore related to the angular momentum by

$$K = \frac{L^2}{2I} \qquad \text{8-37}$$

The initial kinetic energy in Example 8-18 is

$$K_i = \frac{L_i^2}{2I_1}$$

and the final kinetic energy is

$$K_f = \frac{L_f^2}{2(I_1 + I_2)}$$

Since the final angular momentum L_f equals the initial angular momentum L_i, the final kinetic energy is less than the initial kinetic energy by the factor $I_1/(I_1 + I_2)$. This interaction of the disks is analogous to a one-dimensional inelastic collision of two objects.

The rotating plates in the transmission of a truck make inelastic collisions when engaged.

Example 8-19

A merry-go-round of radius 2 m and moment of inertia 500 kg·m^2 is rotating about a frictionless pivot. It makes one revolution every 5 s. A child of mass 25 kg originally standing at the center walks out to the rim. Find the new angular speed of the merry-go-round.

Since there is no external torque acting on the child–merry-go-round system, the angular momentum of the system remains constant. Originally, the child is at the center, where he has no appreciable moment of inertia about the axis and therefore no angular momentum. When he is at the rim, he has an angular momentum $I_c\omega_f$, where $I_c = mr^2$ is the moment of inertia of the child relative to the axis of the merry-go round, and ω_f is the final angular velocity of the child and the merry-go-round. Since the mass of the child is 25 kg and r is 2 m, the moment of inertia of the child when he is at the rim is

$$I_c = (25\text{ kg})(2\text{ m})^2 = 100\text{ kg}\cdot\text{m}^2$$

If I_m is the moment of inertia of the merry-go-round and ω_i is its initial angular velocity, we have from the conservation of angular momentum

$$L_i = L_f$$

$$I_m\omega_i = I_m\omega_f + I_c\omega_f = (I_m + I_c)\omega_f$$

$$\omega_f = \frac{I_m}{I_m + I_c}\omega_i = \frac{500\text{ kg}\cdot\text{m}^2}{(500 + 100)\text{ kg}\cdot\text{m}^2}\omega_i = \frac{5}{6}\omega_i$$

Since the merry-go-round originally makes one revolution every 5 s, its initial angular velocity is $\frac{1}{5}$ rev/s or 0.4π rad/s. The final angular velocity is therefore

$$\omega_f = (\tfrac{5}{6})\tfrac{1}{5}\text{ rev/s} = \tfrac{1}{6}\text{ rev/s}$$

(There is no particular need here to convert the given units of angular velocity to radians per second.) Thus, after the child reaches the rim, the merry-go-round rotates at one revolution every 6 s.

When the child is at the center of the merry-go-round, he is at rest. As he walks out, he begins moving in a circle. The force that accelerates the child is the force of friction exerted on him by the merry-go-round. This force has a component that is tangential to the circle that produces a torque that increases the angular momentum of the child. The child exerts an equal and opposite frictional force on the merry-go round. The torque associated with this force decreases the angular momentum of the merry-go-round.

(a)

(b)

(a) The conservation of angular momentum has a dramatic impact on astronauts in space. Here, astronaut Dale Gardner prepares to capture a spinning Westar VI communications satellite. (b) Gardner used a large tool called the apogee kick motor, a capture device, to stabilize the satellite sufficiently to capture it for return to earth in the cargo bay of the Discovery.

As in Example 8-18, we can calculate the initial and final mechanical energy of the child and the merry-go-round. Since the angular momentum does not change and the final moment of inertia of the system is greater than the initial momentum of inertia, we can see from Equation 8-37 that the total kinetic energy decreases. The detailed analysis of the energy transfers in this system is complicated. As the child moves outward, his tangential velocity increases. With each step, he makes an inelastic collision with a part of the merry-go-round that is further out and therefore moving faster than he is. Mechanical energy is lost in each of these inelastic collisions. Conversely, if the child walks inward, the moment of inertia of the child–merry-go-round system decreases, so from Equation 8-37, the total kinetic energy of the system must increase. This energy comes from the internal chemical energy of the child.

Example 8-20

The same child as in Example 8-19 runs along a path tangential to the rim of the merry-go-round, which is initially at rest, with an initial speed v = 2.5 m/s and jumps on the merry-go-round (Figure 8-24). What is the final angular velocity of the child and the merry-go-round together?

Figure 8-24 A child runs tangentially to the rim of a merry-go-round and jumps on for Example 8-20. If the pivot is frictionless, angular momentum is conserved.

We do not expect mechanical energy to be conserved because the child makes an inelastic collision with the rim of the merry-go-round. Linear momentum is not conserved either. The pivot of the merry-go-round exerts an impulse during the collision, but because it is frictionless, it cannot exert any torque. Thus, angular momentum about the pivot is conserved. (This example demonstrates that we need not have circular motion to have angular momentum.) The initial angular momentum of the child relative to the pivot of the merry-go-round is

$$L_i = mvR = (25 \text{ kg})(2.5 \text{ m/s})(2 \text{ m})$$
$$= 125 \text{ kg}\cdot\text{m}^2/\text{s}$$

When the child is on the merry-go-round, the angular momentum of the child plus the merry-go-round is

$$L_f = (I_c + I_m)\omega_f$$
$$= (100\ \text{kg}\cdot\text{m}^2 + 500\ \text{kg}\cdot\text{m}^2)\omega_f$$

where we have used $I_c = 100\ \text{kg}\cdot\text{m}^2$ and $I_m = 500\ \text{kg}\cdot\text{m}^2$ from Example 8-19. Setting the final angular momentum equal to the initial angular momentum, we have

$$(600\ \text{kg}\cdot\text{m}^2)\omega_f = 125\ \text{kg}\cdot\text{m}^2/\text{s}$$
$$\omega_f = \frac{125\ \text{kg}\cdot\text{m}^2/\text{s}}{600\ \text{kg}\cdot\text{m}^2}$$
$$= 0.208\ \text{rad/s}$$

Note that the unit of ω_f comes out to be s^{-1}, which we have written as rad/s to conform with the usual unit for angular velocity.

Exercise
Calculate the initial and final kinetic energies of the child–merry-go-round system in Example 8-20. (Answer: $K_i = 78.2$ J, $K_f = 13.0$ J)

Questions

7. A woman sits on a spinning piano stool with her arms folded. If she extends her arms, what happens to her angular velocity?
8. In Example 8-19, does the merry-go-round do work on the child?
9. Is it easier to crawl radially outward or radially inward on a rotating merry-go-round? Why?
10. Folk wisdom says that a cat always lands on its feet. If a cat starts falling with its feet up, how can it land on its feet without violating the law of conservation of angular momentum?

8-6 Rolling Objects

Figure 8-25 shows a ball of radius R rolling along a plane surface. If the ball rolls without slipping, its rotational and translational motions are simply related. As the ball turns through the angle ϕ shown in the figure, the point of contact between the ball and the plane moves a distance s that is related to ϕ by

$$s = R\phi \qquad 8\text{-}38$$

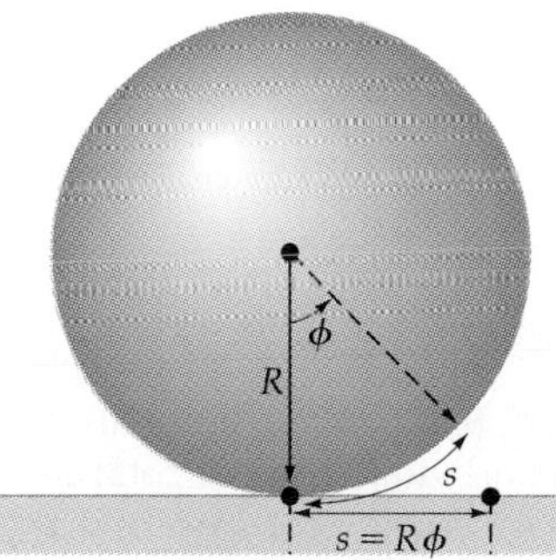

Figure 8-25 When a ball rolls without slipping, the center of mass moves a distance $s = R\phi$ as the ball rotates through an angle ϕ.

Since the center of the ball lies directly over the point of contact, the center of mass of the ball also moves through the distance s. The velocity of the center of mass and the angular velocity of rotation are therefore related by

$$V_{cm} = \frac{ds}{dt} = R\frac{d\phi}{dt}$$

or

$$V_{cm} = R\omega \qquad 8\text{-}39$$

Rolling condition

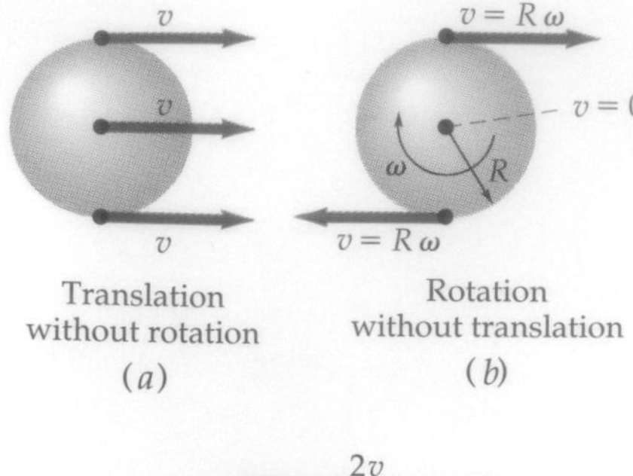

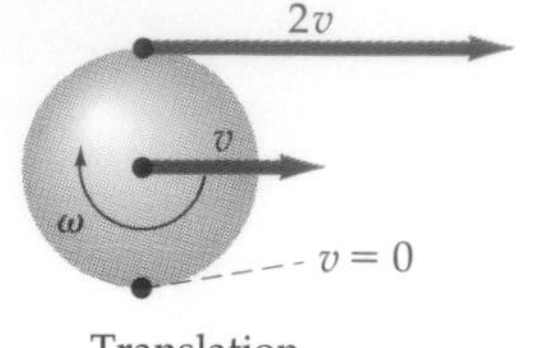

Figure 8-26 (*a*) Translation without rotation. The top and bottom of the ball move with the same velocity. (*b*) Rotation without translation. The top of the ball moves to the right with a speed $v = R\omega$ relative to the center, which is at rest. The bottom moves to the left with the same speed relative to the center. (*c*) Rolling without slipping is a combination of translation and rotation. If the center moves with speed v, the top moves with speed $2v$ and the bottom of the ball is momentarily at rest.

Equation 8-39 (or the equivalent Equation 8-38) is called the **rolling condition.** It holds whenever a ball or cylinder rolls without slipping. We can also write the rolling condition in terms of the linear and angular accelerations:

$$\frac{dV_{cm}}{dt} = R\frac{d\omega}{dt}$$

or

$$A_{cm} = R\alpha \qquad 8\text{-}40$$

When a ball rotates with an angular speed ω, a point on the edge of the ball has a speed $R\omega$ relative to the center of the ball. Since the center of the ball is moving with a speed $R\omega$ relative to the surface, and since the point of the ball in contact with the surface is moving backward with this same speed relative to the center of the ball, the point in contact is instantaneously at rest relative to the surface (see Figure 8-26). If a frictional force is exerted by the surface on the ball, it is static friction and no energy is dissipated.

Example 8-21

A homogeneous sphere of radius 12 cm and mass 30 kg is rolling, without slipping, on a horizontal floor at 2 m/s. How much work is required to stop the sphere?

The work needed to stop the sphere equals the initial kinetic energy of the sphere, which is the kinetic energy of translation of the center of mass, $\frac{1}{2}mv^2$, plus the kinetic energy of rotation relative to the center of mass, $\frac{1}{2}I_{cm}\omega^2$. We use $I_{cm} = \frac{2}{5}mR^2$ for the moment of inertia of the sphere. The rolling condition is $v = R\omega$. Thus, we obtain for the total kinetic energy

$$K = \tfrac{1}{2}mv^2 + \tfrac{1}{2}I_{cm}\omega^2 = \tfrac{1}{2}mv^2 + \tfrac{1}{2}(\tfrac{2}{5}mR^2)\left(\frac{v}{R}\right)^2$$

$$= \tfrac{1}{2}mv^2 + \tfrac{1}{5}mv^2 = \tfrac{7}{10}mv^2 = \tfrac{7}{10}(30\text{ kg})(2\text{ m/s})^2 = 84\text{ J}$$

We can apply these ideas to several interesting problems. We will first consider what happens when a bowling ball is thrown such that initially it has no rotation. As the ball slides along the floor, a force of sliding friction opposes its motion (Figure 8-27). This frictional force causes the ball to rotate, and it also reduces the linear velocity of the center of mass of the ball. The angular velocity increases and the linear velocity decreases until the rolling condition $V_{cm} = R\omega$ is met. Then the ball rolls without slipping, and there is no sliding friction between the surfaces.

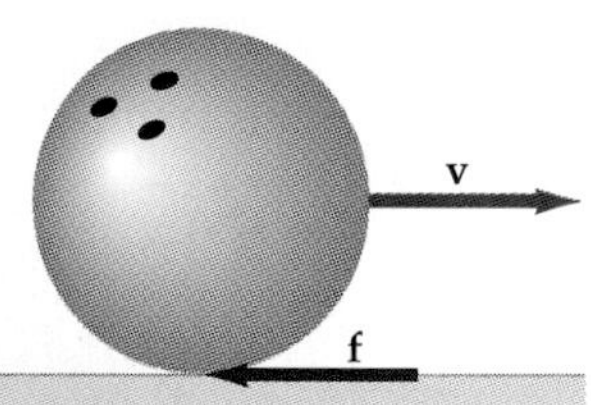

Figure 8-27 A bowling ball moving with no initial rotation. The frictional force **f** exerted by the floor reduces the speed of the center of mass and increases the angular speed until the rolling condition $v = R\omega$ is reached.

Example 8-22

A bowling ball of mass M and radius R is thrown such that the instant it touches the floor it is moving horizontally with speed $v_0 = 5$ m/s and is not rotating. The coefficient of kinetic friction between the ball and the floor is $\mu_k = 0.3$. Find (*a*) the time the ball slides before the rolling condition is met and (*b*) the distance the ball slides before it rolls without slipping.

(*a*) While the ball is sliding, the net force on the ball is the force of kinetic friction $\mu_k Mg$ in the direction opposite its velocity. The acceleration of the center of mass of the ball has the magnitude $a = F_{net}/M = \mu_k g$. At any time t while the ball is sliding, the velocity of the ball is given by

$$v = v_0 - at = v_0 - \mu_k gt$$

The net torque about the center of mass of the ball, while the ball is sliding, is that due to kinetic friction. Its magnitude is

$$\tau = \mu_k MgR$$

From Table 8-1, the moment of inertia of the ball is $I = \frac{2}{5}MR^2$. The angular acceleration of the ball is then

$$\alpha = \frac{\tau}{I} = \frac{\mu_k MgR}{\frac{2}{5}MR^2} = \frac{5}{2}\left(\frac{\mu_k g}{R}\right)$$

While the ball is sliding, the angular velocity of the ball at any time t is

$$\omega = \alpha t = \frac{5}{2}\left(\frac{\mu_k g}{R}\right)t$$

At some time $t = t_1$, the rolling condition $v = R\omega$ is met and the ball stops sliding. Setting v equal to $R\omega$ at time $t = t_1$ gives

$$v = v_0 - \mu_k g t_1 = R\omega = \tfrac{5}{2}\mu_k g t_1$$

$$v_0 = \tfrac{5}{2}\mu_k g t_1 + \mu_k g t_1 = \tfrac{7}{2}\mu_k g t_1$$

Solving for t_1, we obtain

$$t_1 = \frac{2v_0}{7\mu_k g} = \frac{2(5\ m/s)}{7(0.3)(9.81\ m/s^2)} = 0.485\ \text{s}$$

(*b*) Since the acceleration of the center of mass of the ball is constant, the average velocity while it is sliding is the average of the initial and final velocities. From the equations above, we see that, at $t = t_1$, $v = \frac{5}{7}v_0$. The distance traveled during this time is thus

$$s = v_{\text{av}}t_1 = \tfrac{1}{2}(v_0 + v)t_1 = \tfrac{1}{2}(v_0 + \tfrac{5}{7}v_0)$$

$$= \tfrac{1}{2}(\tfrac{12}{7})(5\ \text{m/s})(0.485\ \text{s}) = 2.08\ \text{m}$$

We will next consider the problem of where to strike the cue ball when playing pool or billiards. If the cue ball is struck with a horizontal force **F** from the cue stick such that the line of action of **F** passes through the ball's center of mass, the ball will begin to move initially with no rotation like the bowling ball we just considered. If the ball is struck such that the line of action is below the center of mass, it will start moving with back spin. The force of sliding friction will reduce the back spin and will eventually produce forward spin that increases until the rolling condition of Equation 8-39 is met. Where should the ball be struck so that it will roll without slipping from the beginning? We can find this by requiring the initial linear acceleration and angular acceleration of the ball produced by the cue stick to obey the rolling condition. Let **F** be the force exerted by the cue stick at a height x above the center of the ball (Figure 8-28). Since the cue stick imparts a very large impulsive force for a very short time, we can neglect friction during this time. Newton's second law for linear motion gives

$$F = ma = mR\alpha \tag{8-41}$$

where we have substituted the rolling condition $a = R\alpha$. Newton's second law for rotation gives

$$\tau = Fx = I\alpha \tag{8-42}$$

Dividing Equation 8-42 by Equation 8-41 to eliminate F and α gives

$$x = \frac{I}{mR}$$

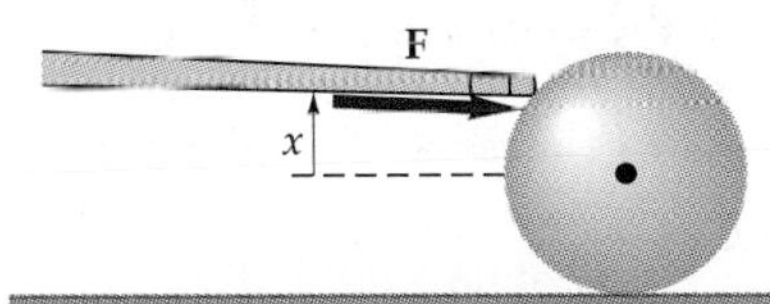

Figure 8-28 Cue stick hitting the cue ball such that the line of action is a distance x above the center. If x is chosen properly, the cue ball will roll without slipping from the beginning.

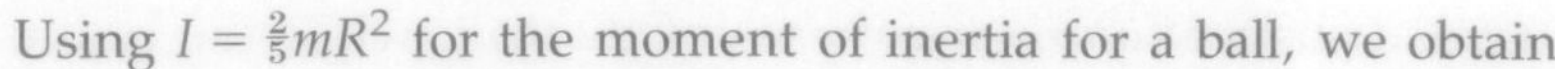

Using $I = \frac{2}{5}mR^2$ for the moment of inertia for a ball, we obtain

$$x = \tfrac{2}{5}R$$

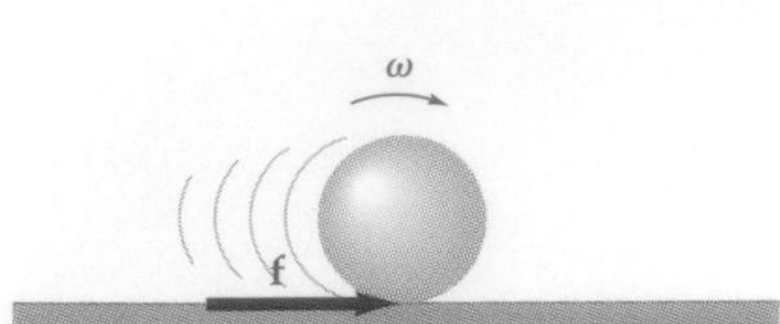

Figure 8-29 Cue ball with top spin. The force of friction increases the speed of the center of mass and reduces the angular speed until the rolling condition is met.

Thus, if the ball is struck such that the line of action of the force is $\frac{2}{5}R$ above the center, it will begin to roll without slipping. If the line of action is higher than this, the ball will have top spin. There will then be a force of friction in the direction of motion (Figure 8-29) that increases the linear velocity of the ball and decreases its rotational velocity until the rolling condition is met.

The final problem we will consider is that of spheres, hoops, or cylinders rolling down an incline. Our analysis of this type of problem is simplified by an important theorem concerning the angular momentum of a system relative to the center of mass:

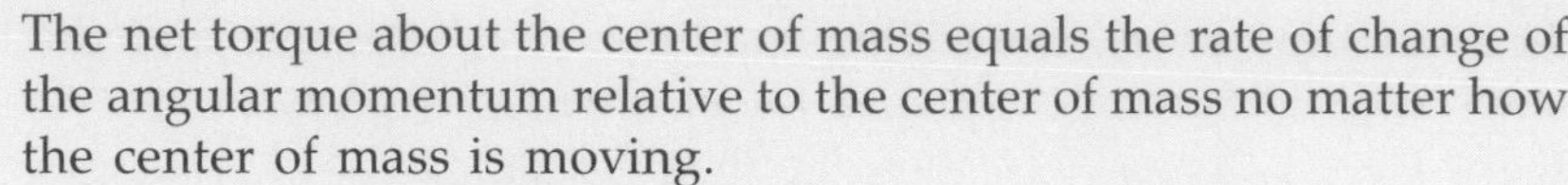

The net torque about the center of mass equals the rate of change of the angular momentum relative to the center of mass no matter how the center of mass is moving.

$$\tau_{\text{net,cm}} = \frac{dL_{\text{cm}}}{dt} \tag{8-43}$$

This is the same as Equation 8-35 except that the torque and the angular momentum are computed relative to the center of mass rather than relative to some fixed point. This theorem is important because, for example, when a ball rolls down an incline, the center of mass is accelerating, and we would not necessarily expect Newton's second law for rotation to hold in a noninertial reference frame. Equation 8-43 is true because of the special properties of the center of mass of a system. A proof of this theorem is given in most intermediate-level mechanics books.*

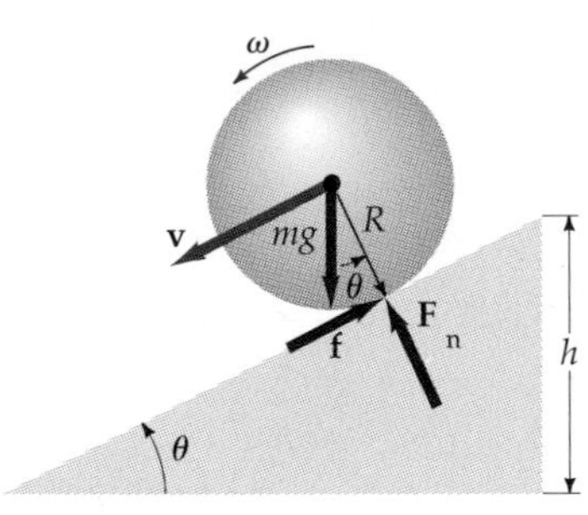

Figure 8-30 Forces acting on a sphere rolling down an incline. The frictional force acting up the incline exerts a torque about the center of mass of the sphere that increases its angular velocity so that the rolling condition is maintained as the sphere accelerates down the incline.

Figure 8-30 shows a sphere of mass m and radius R rolling down an incline. (Our analysis applies equally to a cylinder or hoop.) The forces acting are the weight $m\mathbf{g}$ downward, the normal force $\mathbf{F}_n$ that balances the normal component of the weight, and the force of friction $\mathbf{f}$ acting up the incline. If the object rolls without slipping, the friction is static, as previously discussed. As the object accelerates down the incline, the angular velocity of rotation must increase if it is to roll without slipping. Thus, the angular momentum relative to the center of mass must increase. This increase is due to the torque exerted by the frictional force. (The weight and normal force both act through the center of mass and therefore exert no torque about it.) The angular momentum about the center of mass is

$$L_{\text{cm}} = I_{\text{cm}}\omega$$

where I_{cm} is the moment of inertia about the center of mass, which is constant. The torque exerted by friction is fR. Setting the torque equal to the rate of change of the angular momentum, we have

$$\tau_{\text{net}} = fR = \frac{dL_{\text{cm}}}{dt} = I_{\text{cm}}\frac{d\omega}{dt}$$

or

$$fR = I_{\text{cm}}\alpha \tag{8-44}$$

The linear acceleration of the center of mass of the object is down the incline. The net force down the incline is $mg \sin\theta - f$. Newton's second law therefore gives

$$mg \sin\theta - f = mA_{\text{cm}} \tag{8-45}$$

*See, for example, G. R. Fowles, *Analytical Mechanics*, pp. 201–203, Holt, New York, 1977.

We can solve Equations 8-44 and 8-45 for the unknowns α, A_{cm}, and f using Equation 8-40 for the rolling condition. Substituting A_{cm}/R for α in Equation 8-44, we have

$$fR = I_{cm}\frac{A_{cm}}{R}$$

$$f = \frac{I_{cm}}{R^2}A_{cm} \qquad 8\text{-}46$$

Substituting this result into Equation 8-45 gives

$$mg \sin \theta - \frac{I_{cm}}{R^2}A_{cm} = mA_{cm}$$

$$A_{cm} = \frac{mg \sin \theta}{m + I_{cm}/R^2} \qquad 8\text{-}47$$

For a sphere, $I_{cm} = \frac{2}{5}mR^2$, so the acceleration is

$$A_{cm} = \frac{mg \sin \theta}{m + \frac{2}{5}m} = \frac{5}{7}g \sin \theta \qquad \text{(sphere)} \qquad 8\text{-}48$$

For a cylinder, $I_{cm} = \frac{1}{2}mR^2$, so the acceleration is

$$A_{cm} = \frac{mg \sin \theta}{m + \frac{1}{2}m} = \frac{2}{3}g \sin \theta \qquad \text{(cylinder)} \qquad 8\text{-}49$$

For a hoop, $I_{cm} = mR^2$, so the acceleration is

$$A_{cm} = \frac{1}{2}g \sin \theta \qquad \text{(hoop)} \qquad 8\text{-}50$$

Note that the accelerations given by Equations 8-48, 8-49, and 8-50 are independent of the radius of the object. If we release a sphere, a cylinder, and a hoop at the top of an incline and if they all roll without slipping, the sphere will reach the bottom first because it has the greatest acceleration. The cylinder will be second and the hoop third (Figure 8-31). If an object could slide down the incline without friction, it would reach the bottom before any of the rolling objects.

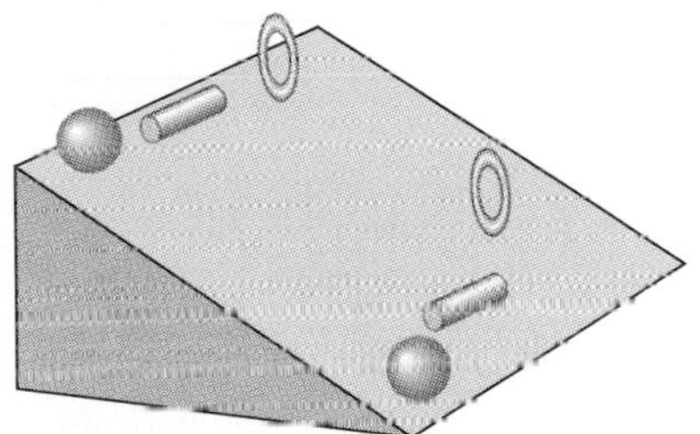

Figure 8-31 A sphere, a cylinder, and a hoop are started together from rest at the top of an incline. The sphere reaches the bottom first, followed by the cylinder and then the hoop.

We can use these results and Equation 8-46 to find the force of friction. For a cylinder with $I_{cm} = \frac{1}{2}mR^2$, for example, we have

$$f = \frac{I_{cm}}{R^2}A_{cm} = \frac{\frac{1}{2}mR^2}{R^2}\left(\frac{2}{3}g \sin \theta\right)$$

or

$$f = \frac{1}{3}mg \sin \theta \qquad 8\text{-}51$$

Note that we have found the force of friction without considering the coefficient of friction. Since the cylinder is rolling without slipping, the friction is static friction, which is usually not equal to its maximum value $\mu_s F_n$, where μ_s is the coefficient of static friction. For the general case,

$$f \leq \mu_s F_n = \mu_s mg \cos \theta$$

Using Equation 8-51 for f for a cylinder that is rolling without slipping, we have

$$f = \frac{1}{3}mg \sin \theta \leq \mu_s mg \cos \theta$$

or

$$\tan \theta \leq 3\mu_s \qquad 8\text{-}52$$

Thus, if the tangent of the angle of incline is greater than $3\mu_s$, the cylinder will slip as it moves down the incline.

Exercise

A cylinder rolls down a plane inclined at $\theta = 50°$. What is the minimum value of the coefficient of static friction for which the cylinder will roll without slipping? (Answer: 0.40)

Exercise

Find (*a*) the force of friction for a hoop rolling down an incline and (*b*) the maximum value of tan θ for which the hoop will roll without slipping. [Answer: (*a*) $f = \frac{1}{2}mg \sin \theta$, (*b*) $\tan \theta \leq 2\mu_s$]

The linear acceleration of any object rolling down an incline is less than $g \sin \theta$ because of the frictional force directed up the incline. We can find the speed of the object at the bottom of the incline by using the constant-acceleration formulas or from the conservation of mechanical energy. (Since the friction is static, there is no dissipation of mechanical energy.) We will use the conservation of mechanical energy here. At the top of the incline, the total energy is potential energy mgh. At the bottom, the total energy is kinetic energy, which is the translational kinetic energy of the center of mass, $\frac{1}{2}mv^2$, plus the kinetic energy of rotation relative to the center of mass, $\frac{1}{2}I_{cm}\omega^2$. The conservation of energy therefore gives

$$\frac{1}{2}mv^2 + \frac{1}{2}I_{cm}\omega^2 = mgh$$

We can use the rolling condition to eliminate either v or ω. Substituting $\omega = v/R$, we obtain

$$\frac{1}{2}mv^2 + \frac{1}{2}I_{cm}\left(\frac{v}{R}\right)^2 = mgh$$

$$v^2 = \frac{2mgh}{m + I_{cm}/R^2} \qquad \text{8-53}$$

For a cylinder with $I_{cm} = \frac{1}{2}mR^2$, for example, this gives us

$$v^2 = \frac{2mgh}{m + \frac{1}{2}m} = \frac{4}{3}gh$$

Note that this result is independent of the radius of the cylinder and that it is less than $2gh$, which would be the result for an object sliding down the incline with no friction.

Water spouts off the Grand Bahama Islands offer a stunning visualization of rotational motion.

8-7 The Vector Nature of Rotation and the Cross Product

For general rotational motion, torque, angular velocity, and angular momentum are vector quantities. We have not had to worry about the vector nature of these quantities in our discussion so far because we have considered only rotation about a fixed axis (or in the case of rolling objects, about an axis that remains parallel to a fixed direction in space). This rotation was therefore analogous to linear motion in one dimension, where we could use a plus or minus sign to indicate the direction of the velocity or acceleration. When the axis of rotation is not fixed in space, the vector nature of rotational quantities becomes important.

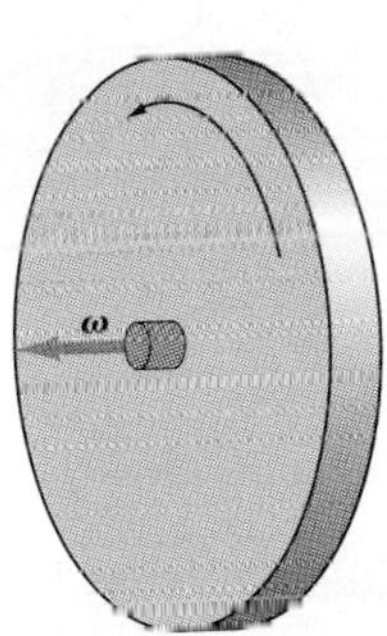

Figure 8-32 A disk rotating about an axis through its center and perpendicular to its plane. All directions in the plane of the disk are equivalent. The direction of the angular velocity $\boldsymbol{\omega}$ is chosen to be along the axis of rotation according to the right-hand rule.

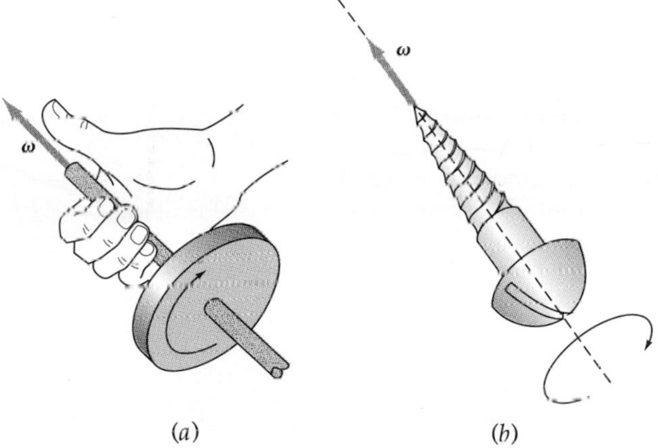

Figure 8-33 (*a*) The right-hand rule for determining the direction of the angular velocity $\boldsymbol{\omega}$. When the fingers of the right hand curl in the direction of rotation, the thumb points in the direction of $\boldsymbol{\omega}$. (*b*) The direction of $\boldsymbol{\omega}$ is that of the advance of a rotating right-hand screw.

Consider the rotating disk in Figure 8-32. Since all directions in the plane of the disk are equivalent by symmetry, we describe the "direction" of rotation by giving the direction of the axis of rotation. We therefore choose the angular velocity vector $\boldsymbol{\omega}$ to be along the axis of rotation. We arbitrarily choose the direction of $\boldsymbol{\omega}$ to be given by the **right-hand rule,** which is illustrated in Figure 8-33*a*. If the axis of rotation is grasped by the right hand with the fingers following the rotation, the extended thumb points in the direction of $\boldsymbol{\omega}$. The direction of $\boldsymbol{\omega}$ is also that of the advance of a rotating right-hand screw, as shown in Figure 8-33*b*. Thus, if the rotation is counterclockwise, as in Figure 8-32, $\boldsymbol{\omega}$ is outward; if it is clockwise, $\boldsymbol{\omega}$ is inward.

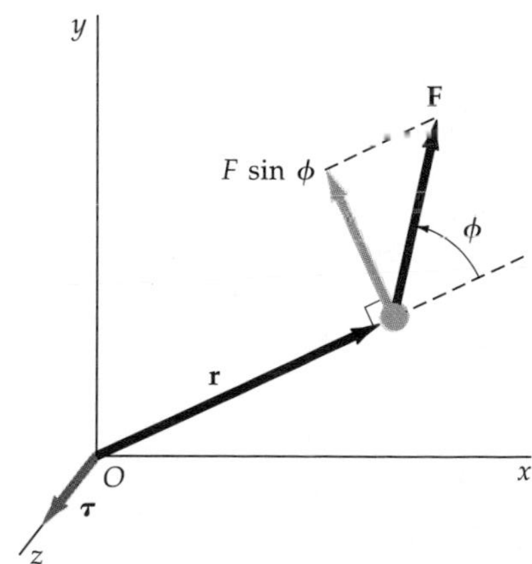

Figure 8-34 The torque $\boldsymbol{\tau}$ exerted by a force **F** acting on a particle at position **r** is perpendicular to both **F** and **r** and has a magnitude $Fr \sin \phi$.

Figure 8-34 shows a force **F** acting on a particle at some position **r** relative to the origin O. The torque exerted by this force relative to the origin O is defined as a vector of magnitude $Fr \sin \phi$, where ϕ is the angle between **F** and **r**, that has a direction perpendicular to the plane formed by **F** and **r**. This is the same definition of torque we have been using, except that we have now assigned the torque a direction. If **F** and **r** are in the xy plane, as in the figure, the torque is along the z axis. The torque is written as the **vector product** or **cross product** of **r** and **F**:

$$\boldsymbol{\tau} = \mathbf{r} \times \mathbf{F} \qquad 8\text{-}54$$

The cross product of two vectors **A** and **B** is defined to be a vector **C** whose magnitude equals the area of the parallelogram formed by the two vectors (Figure 8-35*a*). The direction of **C** is perpendicular to the plane containing **A** and **B** in the sense given by the right-hand rule as **A** is rotated into **B** through the smaller angle between these vectors (Figure 8-35*b*). If ϕ is the angle between the two vectors and $\hat{\mathbf{n}}$ is a unit vector that is perpendicular to each in the sense described, the cross product of **A** and **B** is

Cross product defined

$$\mathbf{A} \times \mathbf{B} = (AB \sin \phi)\hat{\mathbf{n}} \tag{8-55}$$

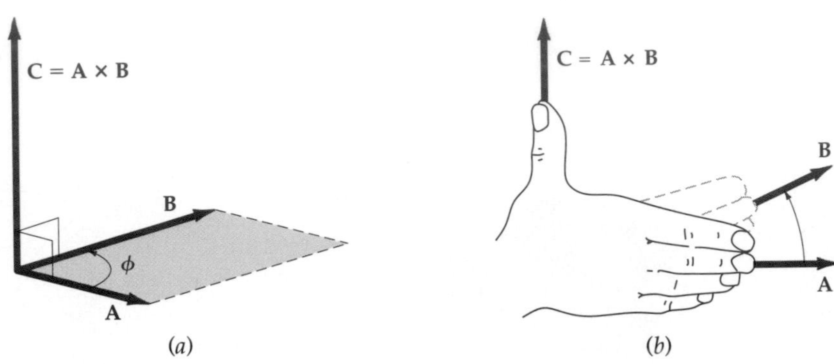

Figure 8-35 (*a*) The vector product **A × B** is a vector **C** that is perpendicular to both **A** and **B** and has a magnitude $AB \sin \phi$, which equals the area of the parallelogram shown. (*b*) The direction of **A × B** is given by the right-hand rule when **A** is rotated into **B** through the angle ϕ.

If **A** and **B** are parallel, **A × B** is zero. It follows from the definition of the cross product (Equation 8-55) that

$$\mathbf{A} \times \mathbf{A} = 0 \tag{8-56}$$

and

$$\mathbf{A} \times \mathbf{B} = -\mathbf{B} \times \mathbf{A} \tag{8-57}$$

It should be noted that the order in which two vectors are multiplied when determining the cross product is significant. Unlike the multiplication of ordinary numbers, changing the order of multiplication of the two vectors in a cross product changes the sign of the result. Some properties of the cross product of two vectors are

1. The cross product obeys a distributive law under addition:

$$\begin{aligned}&\mathbf{A} \times (\mathbf{B} + \mathbf{C}) \\ &= \mathbf{A} \times \mathbf{B} + \mathbf{A} \times \mathbf{C}\end{aligned} \tag{8-58}$$

2. If **A** and **B** are functions of some variable such as t, the derivative of **A × B** follows the usual product rule for derivatives:

$$\frac{d}{dt}(\mathbf{A} \times \mathbf{B}) = \mathbf{A} \times \frac{d\mathbf{B}}{dt} + \frac{d\mathbf{A}}{dt} \times \mathbf{B} \tag{8-59}$$

   Note that we must keep the order straight here since, for example, $\mathbf{B} \times d\mathbf{A}/dt = -(d\mathbf{A}/dt) \times \mathbf{B}$.

3. The unit vectors **i**, **j**, and **k**, which are mutually perpendicular, have cross products given by:

$$\mathbf{i} \times \mathbf{j} = \mathbf{k} \qquad \mathbf{j} \times \mathbf{k} = \mathbf{i} \qquad \mathbf{k} \times \mathbf{i} = \mathbf{j}$$

   (See Figure 8-36.)

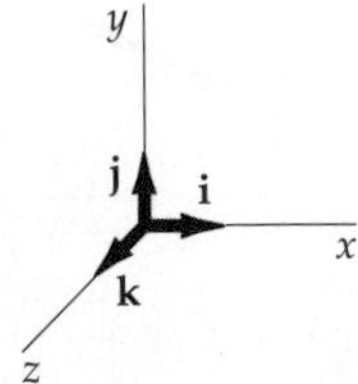

Figure 8-36 The unit vectors **i**, **j**, and **k** are mutually perpendicular and have magnitude 1. The cross product of **i** and **j** is $\mathbf{i} \times \mathbf{j} = (1)(1) \sin 90°\mathbf{k} = \mathbf{k}$. Similarly $\mathbf{j} \times \mathbf{k} = \mathbf{i}$, and $\mathbf{k} \times \mathbf{i} = \mathbf{j}$.

And,

$$\mathbf{i} \times \mathbf{i} = \mathbf{j} \times \mathbf{j} = \mathbf{k} \times \mathbf{k} = 0$$

We can use these results to express the cross product of two vectors in terms of the vectors' rectangular components:

$$\begin{aligned}\mathbf{A} \times \mathbf{B} &= (A_x\mathbf{i} + A_y\mathbf{j} + A_z\mathbf{k}) \times (B_x\mathbf{i} + B_y\mathbf{j} + B_z\mathbf{k}) \\ &= A_xB_x\mathbf{i} \times \mathbf{i} + A_xB_y\mathbf{i} \times \mathbf{j} + A_xB_z\mathbf{i} \times \mathbf{k} + A_yB_x\mathbf{j} \times \mathbf{i} + A_yB_y\mathbf{j} \times \mathbf{j} \\ &\qquad + A_yB_z\mathbf{j} \times \mathbf{k} + A_zB_x\mathbf{k} \times \mathbf{i} + A_zB_y\mathbf{k} \times \mathbf{j} + A_zB_z\mathbf{k} \times \mathbf{k} \\ &= 0 + A_xB_y\mathbf{k} + A_xB_z(-\mathbf{j}) + A_yB_x(-\mathbf{k}) + 0 + A_yB_z\mathbf{i} \\ &\qquad + A_zB_x\mathbf{j} + A_zB_y(-\mathbf{i}) + 0\end{aligned}$$

or,

$$\mathbf{A} \times \mathbf{B} = (A_yB_z - A_zB_y)\mathbf{i} + (A_zB_x - A_xB_z)\mathbf{j} + (A_xB_y - A_yB_x)\mathbf{k} \qquad 8\text{-}60a$$

This result can also be written in terms of a determinant:

$$\mathbf{A} \times \mathbf{B} = \begin{vmatrix} \mathbf{i} & \mathbf{j} & \mathbf{k} \\ A_x & A_y & A_z \\ B_x & B_y & B_z \end{vmatrix} \qquad 8\text{-}60b$$

The direction of rotation is changed by this bevel gear in a diesel engine.

4. For the special case in which **A** and **B** are in the xy plane, $A_z = B_z = 0$ and $\mathbf{A} \times \mathbf{B}$ has only a z component:

$$\mathbf{A} \times \mathbf{B} = (A_xB_y - A_yB_x)\mathbf{k} \quad \text{(for } \mathbf{A} \text{ and } \mathbf{B} \text{ in the } xy \text{ plane)} \qquad 8\text{-}61$$

Example 8-23

Find $\mathbf{A} \times \mathbf{B}$ for $\mathbf{A} = 5\mathbf{i} + 6\mathbf{j}$ and $\mathbf{B} = 3\mathbf{i} - 2\mathbf{j}$.

Computing the cross product directly, we have

$$\begin{aligned}\mathbf{A} \times \mathbf{B} &= (5\mathbf{i} + 6\mathbf{j}) \times (3\mathbf{i} - 2\mathbf{j}) \\ &= 15(\mathbf{i} \times \mathbf{i}) - 10(\mathbf{i} \times \mathbf{j}) + 18(\mathbf{j} \times \mathbf{i}) - 12(\mathbf{j} \times \mathbf{j}) \\ &= 15(0) - 10(\mathbf{k}) + 18(-\mathbf{k}) - 12(0) \\ &= -10\mathbf{k} - 18\mathbf{k} \\ &= -28\mathbf{k}\end{aligned}$$

Alternatively, since both **A** and **B** are in the xy plane, we may use Equation 8-61

$$\begin{aligned}\mathbf{A} \times \mathbf{B} &= (A_xB_y - A_yB_x)\mathbf{k} \\ &= [(5)(-2) - (6)(3)]\mathbf{k} \\ &= -28\mathbf{k}\end{aligned}$$

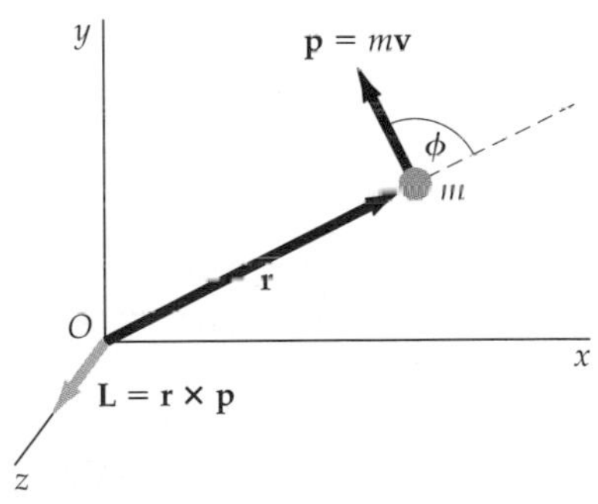

Figure 8-37 A particle with a momentum **p** at position **r** has angular momentum relative to the origin O given by $\mathbf{L} = \mathbf{r} \times \mathbf{p}$. For **r** and **p** in the xy plane, as shown here, **L** is along the z axis.

The angular momentum of a particle can be written as a vector cross product. Figure 8-37 shows a particle of mass m moving with a velocity **v** at a position **r** relative to the origin O. The linear momentum of the particle is $\mathbf{p} = m\mathbf{v}$. The angular momentum **L** of the particle relative to the origin O is defined to be the cross product of **r** and **p**:

$$\mathbf{L} = \mathbf{r} \times \mathbf{p} \qquad 8\text{-}62$$

Angular momentum of a particle defined

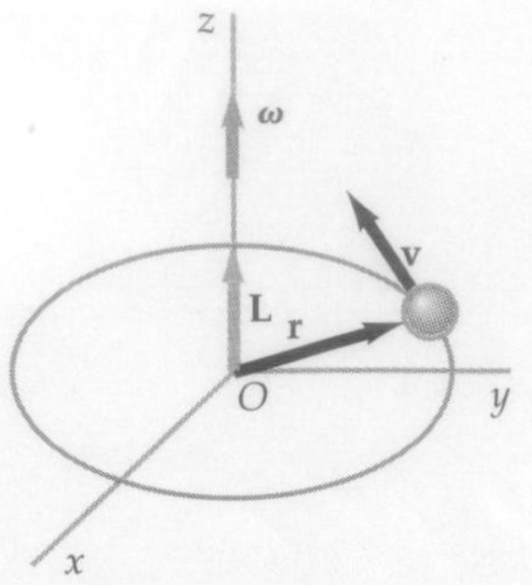

Like torque, angular momentum is defined relative to a point in space. Figure 8-38 shows a particle moving in a circle in the xy plane with the center of the circle at the origin. The speed v of the particle and the magnitude of its angular velocity ω are related by $v = r\omega$. The angular momentum of the particle relative to the center of the circle is

$$\mathbf{L} = \mathbf{r} \times \mathbf{p} = \mathbf{r} \times m\mathbf{v} = (rmv \sin 90°)\mathbf{k} = rmv\mathbf{k} = mr^2\omega\mathbf{k}$$

The angular momentum is in the same direction as the angular velocity. Since mr^2 is the moment of inertia for a single particle about the z axis, we have

$$\mathbf{L} = mr^2\boldsymbol{\omega} = I\boldsymbol{\omega}$$

Figure 8-38 A particle moving in a circle has an angular momentum relative to the center of the circle that is parallel to the angular velocity and is given by $\mathbf{L} = I\boldsymbol{\omega}$.

Figure 8-39 shows the angular-momentum vector $\mathbf{L}'$ for the same particle moving in the same circle but with $\mathbf{L}'$ computed relative to a point on the z axis that is not at the center of the circle. In this case, the angular momentum is not parallel to the angular velocity $\boldsymbol{\omega}$, which is along the z axis. In Figure 8-40, we have added a second particle of equal mass moving in the same circle. The angular-momentum vectors $\mathbf{L}_1'$ and $\mathbf{L}_2'$ are shown relative to the same point as in Figure 8-39. The total angular momentum $\mathbf{L}_1'$ and $\mathbf{L}_2'$ of the two-particle system is again parallel to the angular velocity $\boldsymbol{\omega}$. In this case the axis of rotation, the z axis, passes through the center of mass of the two-particle system, and the mass distribution is symmetric about this axis. Such an axis is called a **symmetry axis.** For any system of particles that rotates about a symmetry axis, the total angular momentum (which is the sum of the angular momenta of the individual particles) is parallel to the angular velocity and is given by

Angular momentum of a system

$$\mathbf{L} = I\boldsymbol{\omega} \qquad 8\text{-}63$$

We will now show that Newton's second law implies that the rate of change of the angular momentum of a particle equals the net torque acting on the particle. If we have a number of forces acting on a particle, the net torque relative to the origin O is the sum of the torques due to each force:

$$\boldsymbol{\tau}_{\text{net}} = (\mathbf{r} \times \mathbf{F}_1) + (\mathbf{r} \times \mathbf{F}_2) + \cdots = \mathbf{r} \times \mathbf{F}_{\text{net}}$$

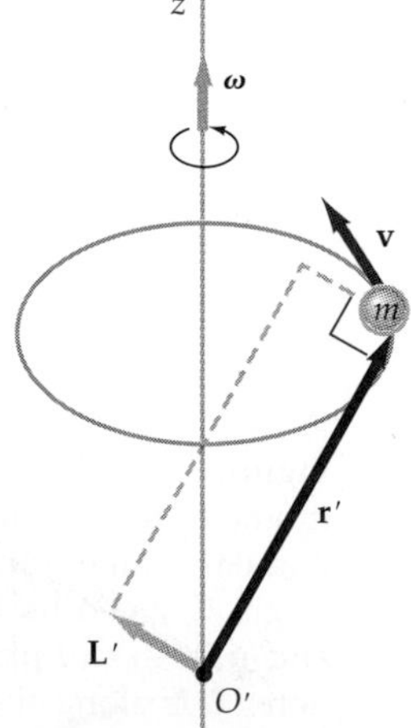

Figure 8-39 The angular momentum $\mathbf{L}'$ of a particle moving in a circle calculated relative to a point not at the center of the circle. In this case, $\mathbf{L}'$ is not parallel to the angular velocity $\boldsymbol{\omega}$.

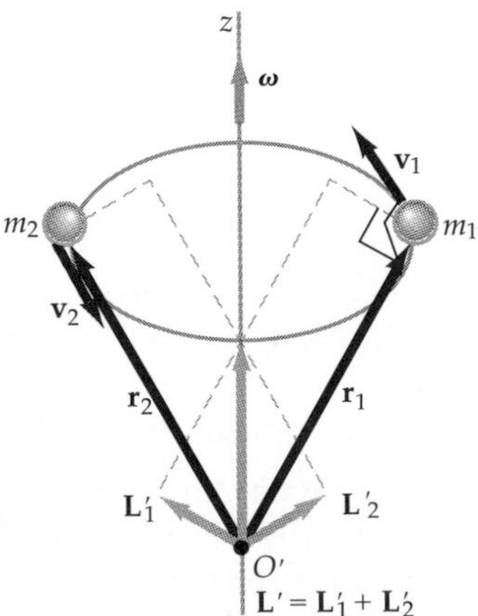

Figure 8-40 Angular-momentum vectors for two particles moving in a circle calculated relative to a point on an axis about which the mass distribution is symmetrical. The total angular momentum $\mathbf{L}_1' + \mathbf{L}_2'$ is parallel to the angular velocity $\boldsymbol{\omega}$.

According to Newton's second law, the net force equals the rate of change of the linear momentum $d\mathbf{p}/dt$. Thus,

$$\tau_{net} = \mathbf{r} \times \mathbf{F}_{net} = \mathbf{r} \times \frac{d\mathbf{p}}{dt} \qquad \text{8-64}$$

We can compute the rate of change of the angular momentum using the product rule for derivatives:

$$\frac{d\mathbf{L}}{dt} = \frac{d}{dt}(\mathbf{r} \times \mathbf{p}) = \left(\frac{d\mathbf{r}}{dt} \times \mathbf{p}\right) + \mathbf{r} \times \frac{d\mathbf{p}}{dt}$$

The first term on the right of this equation is zero because

$$\frac{d\mathbf{r}}{dt} \times \mathbf{p} = \mathbf{v} \times m\mathbf{v} = 0$$

Thus,

$$\frac{d\mathbf{L}}{dt} = \mathbf{r} \times \frac{d\mathbf{p}}{dt} \qquad \text{8-65}$$

Comparing Equations 8-64 and 8-65 we have

$$\tau_{net} = \frac{d\mathbf{L}}{dt} \qquad \text{8-66}$$

The net torque acting on a system of particles is the sum of the individual torques acting on the system. The generalization of Equation 8-66 to a system of particles is thus

$$\sum_i \tau_i = \sum_i \frac{d\mathbf{L}_i}{dt} = \frac{d}{dt}\sum_i \mathbf{L}_i = \frac{d\mathbf{L}}{dt}$$

In this equation, the sum of the torques may include internal torques as well as those due to forces external to the system. However, it is found experimentally that, if there is no net external torque, the angular momentum of a system is conserved. Thus, the sum of the internal torques must be zero. Therefore,

$$\tau_{net,ext} = \frac{d\mathbf{L}}{dt} \qquad \text{8-67}$$

Newton's second law for rotation

The net external torque acting on a system equals the rate of change of the angular momentum of a system.

When the net external torque is zero, the angular momentum of the system is conserved. Equation 8-67 is the rotational analog of $\mathbf{F}_{net,ext} = d\mathbf{p}/dt$ for linear motion.

The law of conservation of angular momentum is an experimental law that is independent of Newton's laws of motion. However, the fact that the internal torques of a system cancel is suggested by Newton's third law. Consider the two particles shown in Figure 8-41. Let $\mathbf{F}_1$ be the force exerted by particle 2 on particle 1 and $\mathbf{F}_2$ be that exerted by particle 1 on particle 2. By Newton's third law, $\mathbf{F}_2 = -\mathbf{F}_1$. The sum of the torques exerted by these forces about the origin O is

$$\tau_1 + \tau_2 = (\mathbf{r}_1 \times \mathbf{F}_1) + (\mathbf{r}_2 \times \mathbf{F}_2) = (\mathbf{r}_1 \times \mathbf{F}_1) + (\mathbf{r}_2 \times -\mathbf{F}_1)$$
$$= (\mathbf{r}_1 - \mathbf{r}_2) \times \mathbf{F}_1$$

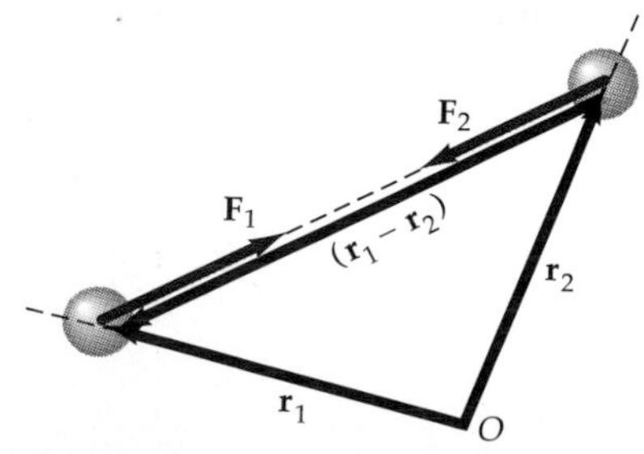

Figure 8-41 The internal forces $\mathbf{F}_1$ and $\mathbf{F}_2$ produce no torque about O if they act along the line joining the particles.

The vector $\mathbf{r}_1 - \mathbf{r}_2$ is along the line joining the two particles. If $\mathbf{F}_1$ acts parallel to the line joining m_1 and m_2, $\mathbf{F}_1$ and $\mathbf{r}_1 - \mathbf{r}_2$ are either parallel or antiparallel, and

$$(\mathbf{r}_1 - \mathbf{r}_2) \times \mathbf{F}_1 = 0$$

If this is true for all the internal forces, the internal torques cancel in pairs.

Questions

11. What is the angle between a particle's linear momentum $\mathbf{p}$ and its angular momentum $\mathbf{L}$?
12. A particle moves along a straight line at constant speed. How does its angular momentum about any point vary over time?
13. A particle moving at constant velocity has zero angular momentum about a particular point. Show that the particle either has passed through that point or will pass through it.

8-8 Motion of a Gyroscope

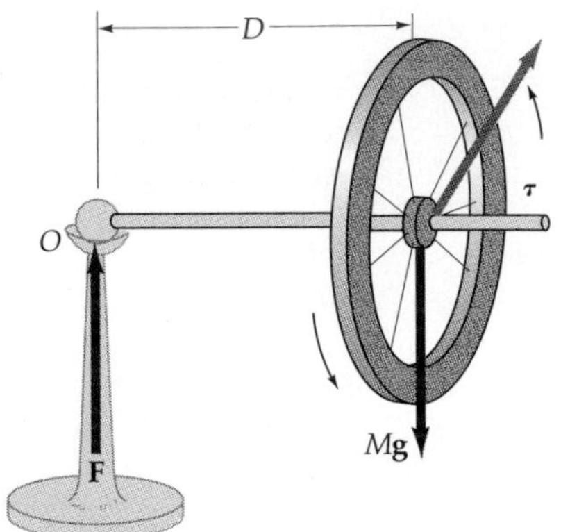

Figure 8-42 A bicycle-wheel gyroscope. The force of gravity $M\mathbf{g}$ produces a torque about the pivot that is perpendicular to the axle. This torque causes a change in the angular momentum in the direction of the torque. If the wheel is not spinning and is released, it just falls down. In this case, the change in angular momentum produces angular momentum due to the motion of the center of mass of the wheel.

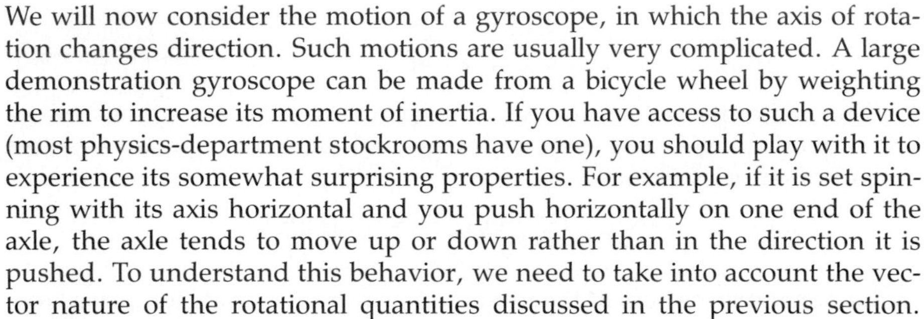

We will now consider the motion of a gyroscope, in which the axis of rotation changes direction. Such motions are usually very complicated. A large demonstration gyroscope can be made from a bicycle wheel by weighting the rim to increase its moment of inertia. If you have access to such a device (most physics-department stockrooms have one), you should play with it to experience its somewhat surprising properties. For example, if it is set spinning with its axis horizontal and you push horizontally on one end of the axle, the axle tends to move up or down rather than in the direction it is pushed. To understand this behavior, we need to take into account the vector nature of the rotational quantities discussed in the previous section.

Figure 8-42 shows a gyroscope consisting of a bicycle wheel that is free to turn on an axle. The axle is pivoted at a point a distance D from the center of the wheel, but it is free to turn in any direction. We will try to achieve a qualitative understanding of the complicated motion of such a system by using Newton's second law for rotation as expressed in Equation 8-67:

$$\boldsymbol{\tau}_{\text{net,ext}} = \frac{d\mathbf{L}}{dt}$$

All that we really need to remember is that the *change* in the angular momentum of the wheel must be in the direction of the net torque acting on the wheel.

When the axle is held horizontal and is then released, the wheel simply falls if it is not initially spinning. The torque about point O is MgD in the horizontal direction, perpendicular to the axle as shown in Figure 8-42. As the wheel falls, it has an angular momentum in the direction of the torque. This angular momentum is due to motion of the wheel itself; that is, it is associated with the motion of the center of mass of the wheel. (Since the center of mass accelerates downward, the upward force F exerted by the support at O is evidently less than Mg.)

We will now consider what happens when the wheel is initially set spinning, and the axle is held horizontal and is then released. The wheel now has a large angular momentum along its axle (to the right in the figure) due to its spin. It is quite easy in practice to make this spin angular momentum very large so that, to a first approximation, we can neglect the contribution due to the motion of the center of mass of the wheel. If the wheel were to fall as before, the axle would point downward, resulting in a large component of angular momentum in the downward direction. But there is no torque in the downward direction. The torque is horizontal. In Figure 8-43, a large component of the angular momentum is shown along the axis of the wheel. The change in the angular momentum $d\mathbf{L}$ in the direction of the torque is also indicated. What actually happens when the axle of the spinning wheel is held horizontal and is then released is that the axle moves in a horizontal plane in the direction into the paper. The wheel must move this way when it is spinning so that the *change* in the angular momentum is in the direction of the net torque. This motion, which is often quite surprising when first encountered, is called **precession.**

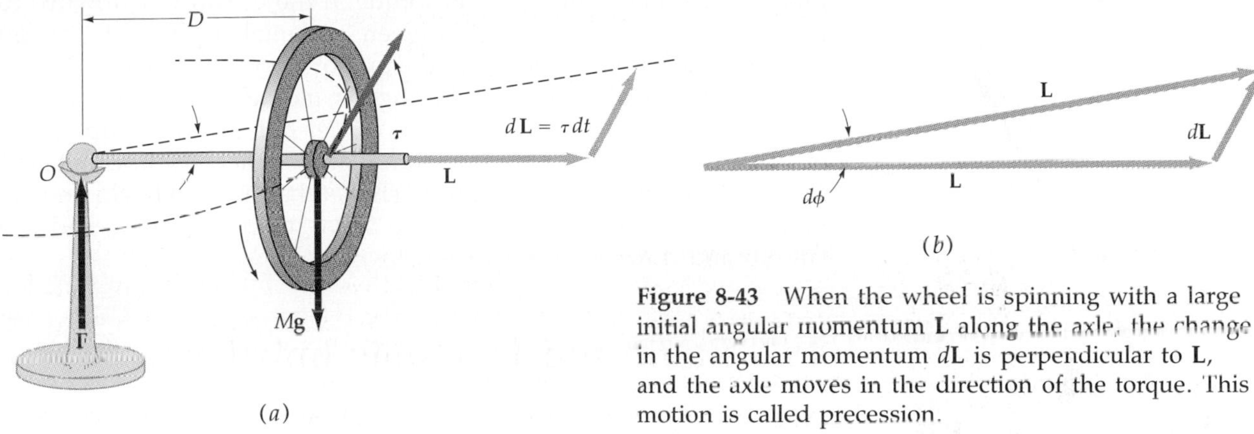

Figure 8-43 When the wheel is spinning with a large initial angular momentum **L** along the axle, the change in the angular momentum $d\mathbf{L}$ is perpendicular to **L**, and the axle moves in the direction of the torque. This motion is called precession.

We can calculate the angular velocity of precession as follows. In a small time interval dt, the change in the angular momentum has a magnitude dL:

$$dL = \tau\, dt - MgD\, dt$$

From Figure 8-43, the angle $d\phi$ through which the axle moves is

$$d\phi = \frac{dL}{L} = \frac{MgD\, dt}{L}$$

The angular velocity of the precession is thus

$$\omega_p = \frac{d\phi}{dt} = \frac{MgD}{L} \tag{8-68}$$

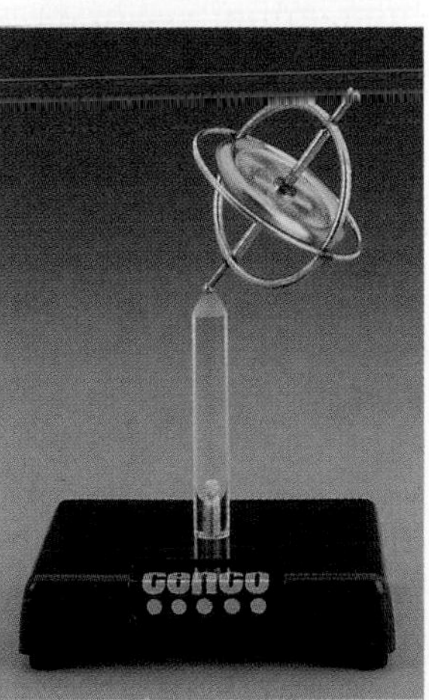

(*Left*) A rotating bicycle wheel supported by a string is used by Professor Minnix to demonstrate the precession of a gyroscope. (*Right*) A simple demonstration gyroscope.

If the angular momentum due to the spin of the wheel is very large, the precession will be very slow; that is, the angular velocity of the precession associated with the axle of the wheel moving in a horizontal plane will be very small.

If you perform this experiment, you will note a small up and down oscillation of the axle. This motion is called **nutation.** It can be eliminated by giving the axle a slight horizontal push when it is released. We can understand nutation qualitatively from a consideration of the directions of the angular momentum of the wheel and the torques exerted on it. Even when the precession is slow, there is a small amount of angular momentum associated with the motion of the center of mass of the wheel. The direction of this small angular momentum is upward. If the axle is simply released with no push, there is no component of torque in the upward or downward direction. Thus, unless the wheel is given an initial push (and thus an initial torque in the upward direction), the axle of the wheel will dip down slightly to provide a small component of angular momentum in the downward direction associated with the spin to cancel out the small upward angular momentum associated with the motion of the wheel itself. A detailed analysis of the motion of the gyroscope shows that if the wheel is not given an initial push, the axle will initially dip down, then overshoot, and then oscillate up and down.

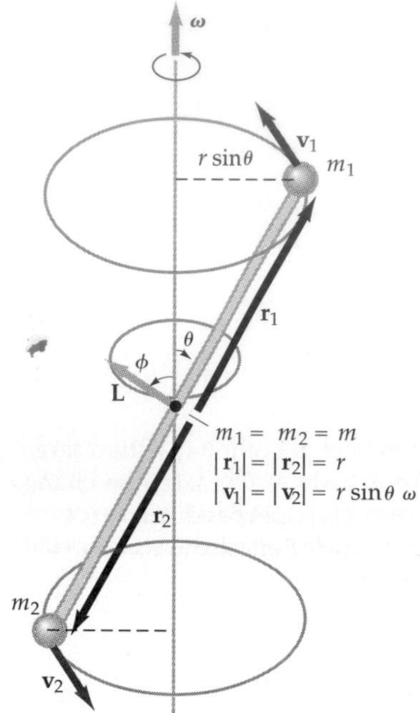

Figure 8-44 Particles of equal mass rotating about an axis through their center of mass that is not a symmetry axis. The angular momentum **L** is perpendicular to the line joining the particles and makes an angle ϕ with the angular velocity $\boldsymbol{\omega}$. Since **L** changes direction as the system rotates, a torque must be exerted by the bearings.

8-9 Static and Dynamic Imbalance

When the axis of rotation is not a symmetry axis, the angular momentum is not necessarily parallel to the angular velocity. Figure 8-44 shows a simple dumbbell consisting of equal masses connected by a light rod that is rotating about an axis which passes through the center of mass but is not perpendicular to the rod. Each mass is moving in a circle of radius $r \sin \theta$ with a speed $(r \sin \theta)\omega$, where θ is the angle between the rod and the axis of rotation and ω is the angular velocity. The angular momentum of mass 1 is $\mathbf{L}_1 = \mathbf{r}_1 \times m\mathbf{v}_1$. Its direction is in the plane of the paper, perpendicular to the rod, and its magnitude is $L_1 = (mr^2 \sin \theta)\omega$. The angular momentum of mass 2 is $\mathbf{L}_2 = \mathbf{r}_2 \times m\mathbf{v}_2$. Since $\mathbf{r}_2 = -\mathbf{r}_1$ and $\mathbf{v}_2 = -\mathbf{v}_1$, $\mathbf{L}_2 = \mathbf{L}_1$. The total angular momentum of the system is thus $(2mr^2 \sin \theta)\omega$ in the direction perpendicular to the rod as shown in the figure. It is not parallel to $\boldsymbol{\omega}$, which is along the axis of rotation. As the dumbbell rotates, the angular-momentum vector rotates with its tip moving in a circle as shown in the figure. If the angular velocity is constant, the angular momentum has a constant magnitude, but its direction changes. Thus, even if the angular velocity is constant, the angular-momentum vector changes, so there must be a net torque on the system. This torque is exerted by the bearings supporting the system. The rotating system exerts an equal and opposite torque on the support bearings, causing wear. Such a system is said to be **dynamically imbalanced.** A calculation of the rate of change of the angular momentum shows that it is proportional to ω^2. Thus, the greater the angular velocity, the greater the wear on the bearings. Note that there is no static imbalance. If we pivot the dumbbell at its center of mass, it will be in static equilibrium at any orientation. If we have a system of unknown mass distribution (such as an automobile wheel), we cannot detect or correct dynamic imbalance by static-imbalance methods.

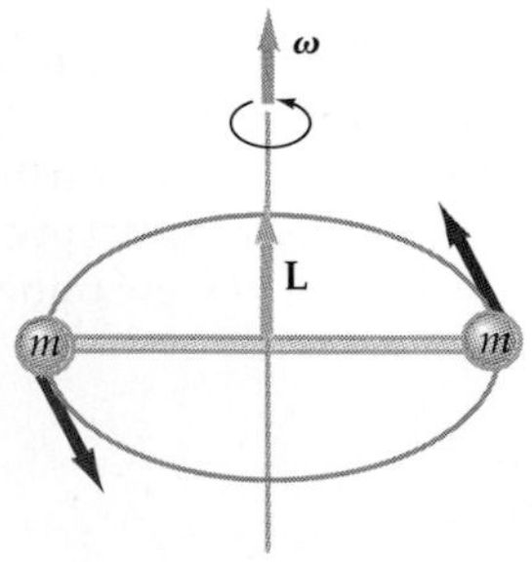

Figure 8-45 This arrangement of two rotating particles of equal mass is dynamically balanced. The axis of rotation is now a symmetry axis and **L** is parallel to $\boldsymbol{\omega}$. As the system rotates, **L** does not change.

We can correct the dynamic imbalance of our two-mass system by making the angle θ equal to 90°, as in Figure 8-45, or by adding masses, as in Figure 8-46. Then **L** and $\boldsymbol{\omega}$ will be parallel, and no torque will be required to maintain a constant angular velocity.

Figure 8-47 shows a wheel rotating about an axis that is off center but parallel to the symmetry axis through the center of mass and a distance h

from it. The angular momentum about a point on the axis of rotation in the plane of the wheel is the sum of the angular momentum about the center of mass and the angular momentum associated with the motion of the center of mass, both of which are parallel to the axis of rotation, which is the direction of $\boldsymbol{\omega}$. If the wheel rotates with constant angular velocity, the angular momentum is constant and no torque is required. However, since the center of mass moves in a circle of radius h, there must be a net force on the wheel of magnitude $MV_{cm}^2/h = Mh\omega^2$ directed toward the axis of rotation from the center of mass. This force is exerted by the bearings. Again, the equal but opposite force exerted on the bearings causes wear on them, particularly if ω is large. This **static imbalance** can be detected and corrected by pivoting the wheel on a horizontal axle and balancing it so that the axis of rotation passes through the center of mass.

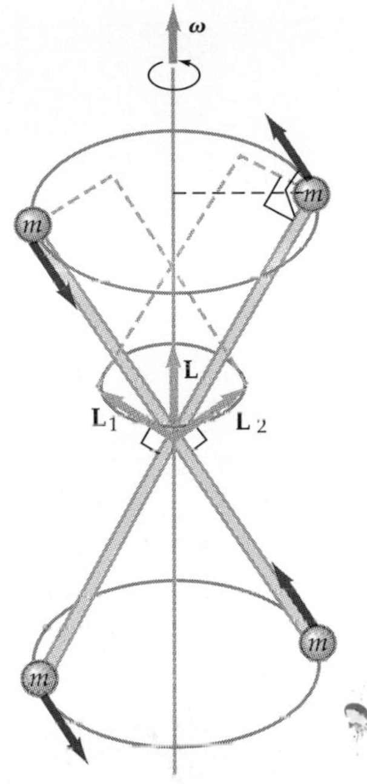

Figure 8-46 This arrangement is also dynamically balanced. Here, too, **L** is parallel to $\boldsymbol{\omega}$ and does not change as the system rotates.

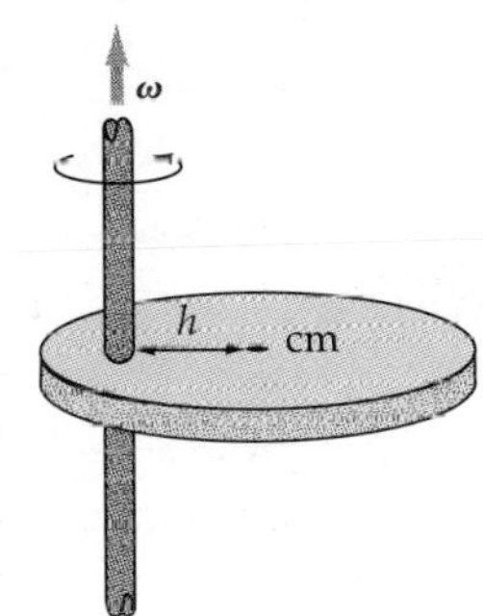

Figure 8-47 This system is dynamically balanced because **L** is parallel to $\boldsymbol{\omega}$, but it is statically imbalanced. As the system rotates, the bearings exert no torque, but they must exert a force to accelerate the center of mass.

Summary

1. The angular velocity of an object is the rate of change of the angle between a line fixed on the object and one fixed in space:

$$\omega = \frac{d\theta}{dt}$$

The angular acceleration is the rate of change of the angular velocity:

$$\alpha = \frac{d\omega}{dt} = \frac{d^2\theta}{dt^2}$$

2. The linear speed of a particle a distance r from the axis of rotation is related to the angular velocity of the particle by

$$v = r\omega$$

Similarly, the tangential acceleration of such a particle is related to the angular acceleration of the particle by

$$a_t = r\alpha$$

The centripetal acceleration of such a particle is given by

$$a_c = \frac{v^2}{r} = r\omega^2$$

3. The equations for rotation of a rigid body with constant angular acceleration are

$$\omega = \omega_0 + \alpha t$$

$$\theta = \theta_0 + \omega_0 t + \tfrac{1}{2}\alpha t^2$$

and

$$\omega^2 = \omega_0^2 + 2\alpha(\theta - \theta_0)$$

These equations are analogous to those for motion with constant acceleration in one dimension.

4. The torque exerted by a force on a particle is defined as the product of the force and the lever arm. Newton's second law for the rotation of a rigid body about a fixed axis is

$$\tau_{net} = I\alpha$$

where I is the moment of inertia of a system of particles, which is defined as

$$I = \sum_i m_i r_i^2$$

The moment of inertia for rotational motion is analogous to the mass in translational motion.

5. The kinetic energy of a rotating body is given by

$$K = \tfrac{1}{2} I\omega^2$$

and the power input of a torque is given by

$$P = \tau\omega$$

6. The angular momentum of a system of particles with a common angular speed ω is

$$L = I\omega$$

7. The generalized form of Newton's second law for rotational motion is

$$\tau_{\text{net}} = \frac{dL}{dt} = \frac{d(I\omega)}{dt}$$

8. In an isolated system, the net torque is zero and the angular momentum of the system is conserved.

9. When a ball or cylinder of radius R rolls without slipping, the velocity of the center of mass of the body is related to the angular velocity by the rolling condition:

$$V_{\text{cm}} = R\omega$$

The acceleration is similarly related to the angular acceleration by

$$A_{\text{cm}} = R\alpha$$

10. The vector product or cross product of two vectors is defined by

$$\mathbf{A} \times \mathbf{B} = (AB \sin \phi)\hat{\mathbf{n}}$$

where ϕ is the angle between the vectors and $\hat{\mathbf{n}}$ is a unit vector that is perpendicular to the plane of $\mathbf{A}$ and $\mathbf{B}$ in the sense given by the right-hand rule as $\mathbf{A}$ is rotated into $\mathbf{B}$.

11. When the axis of rotation changes direction in space, the vector nature of rotational quantities is important. The direction of the angular velocity $\boldsymbol{\omega}$ is along the axis of rotation in the sense given by the right-hand rule. The torque exerted by a force $\mathbf{F}$ about some point P is

$$\boldsymbol{\tau} = \mathbf{r} \times \mathbf{F}$$

If a particle has a linear momentum $\mathbf{p} = m\mathbf{v}$, its angular momentum relative to some point P is

$$\mathbf{L} = \mathbf{r} \times \mathbf{p}$$

The angular momentum of a system of particles rotating about a symmetry axis through the center of mass is

$$\mathbf{L} = I\boldsymbol{\omega}$$

12. The complicated motion of a gyroscope can be understood in terms of the directional properties of torque and angular momentum. When a system such as the bicycle-wheel gyroscope has a large initial angular momentum and the net torque acting is perpendicular to the angular momentum, the system moves such that the change in the angular momentum is in the direction of the torque. This motion is called precession.

13. When a body rotates about an axis that is not a symmetry axis, the angular momentum may not be parallel to the angular velocity, in which case the body is dynamically imbalanced and the direction of the angular momentum changes even when the body rotates at constant angular velocity.

Suggestions for Further Reading

Frohlich, Cliff: "The Physics of Somersaulting and Twisting," *Scientific American,* March 1980, p. 154.

Though they may appear to, the midair rotations of divers and cats do not violate the law of conservation of angular momentum. With the aid of excellent illustrations, this article explains why.

Jones, David E. H.: "The Stability of the Bicycle," *Physics Today,* vol. 23, no. 4, 1970.

An amusing account of the author's attempt to test experimentally the idea, among others, that the gyroscopic effect of the spinning front wheel of a bicycle is responsible for its stability.

Laws, Kenneth: "The Physics of Dance," *Physics Today,* February 1985, p. 25.

Walker, Jearl: "The Amateur Scientist: The Essence of Ballet Maneuvers Is Physics," *Scientific American,* June 1982, p. 146.

Both articles present analyses of various dance movements from a physical point of view, showing that dance may be appreciated on more than just a purely esthetic level. Various jumps and turns are discussed, as well as the dancer's problem of maintaining balance.

Walker, Jearl: "The Amateur Scientist: The Physics of Spinning Tops, Including Some Far-Out Ones," *Scientific American,* March 1981, p. 182.

Precession, nutation, and other, more unusual aspects of top mechanics are explained in a nonmathematical fashion.

Walker, Jearl: "The Amateur Scientist: Thinking About Physics While Scared to Death (on a Falling Roller Coaster)," *Scientific American,* October 1983, p. 162.

Momentum and its contribution to a thrilling roller-coaster ride, the interconversion of potential and kinetic energy, and the rotational dynamics of several amusement-park rides are discussed.

Walker, Jearl: "The Amateur Scientist: Why Sidespin Helps the Bowler—and How to Keep Scoring Strikes," *Scientific American,* March 1988, p. 110.

A look at the theory behind good bowling technique.

Review

A. Objectives: After studying this chapter, you should:

1. Be able to state the definitions of angular velocity, angular acceleration, and moment of inertia.

2. Be able to state the equations for rotation with constant angular acceleration and use these equations to work problems.

3. Be able to state Newton's second law for rotational motion and apply it to problems.

4. Be able to calculate the moment of inertia for discrete or continuous mass distributions.

5. Be able to apply the conservation of angular momentum to problems in which the moment of inertia changes.

6. Be able to state the condition for rolling without slipping and apply it to problems involving translation and rotation.

7. Be able to describe qualitatively the motion of a gyroscope.

B. Define, explain, or otherwise identify:

Rigid body
Angular displacement
Angular velocity
Angular speed
Angular acceleration
Lever arm
Torque
Moment of inertia
Parallel-axis theorem
Perpendicular-axis theorem
Angular momentum
Law of conservation of angular momentum
Rolling condition
Right-hand rule
Vector product
Cross product
Symmetry axis
Precession
Nutation
Dynamic imbalance
Static imbalance

C. True or false: If the statement is true, explain why it is true. If it is false, give a counterexample.

1. Angular velocity and linear velocity have the same dimensions.

2. All parts of a rotating wheel have the same angular velocity.

3. All parts of a rotating wheel have the same angular acceleration.

4. The moment of inertia of a body depends on the location of the axis of rotation.

5. The moment of inertia of a body depends on the angular velocity of the body.

6. If the net torque on a body is zero, the angular momentum must be zero.

7. If the net torque on a rotating system is zero, the angular velocity of the system cannot change.

Problems

Level I

8-1 Angular Velocity and Angular Acceleration

1. A particle moves in a circle with a radius of 100 m with a constant speed of 20 m/s. (*a*) What is its angular velocity in radians per second about the center of the circle? (*b*) How many revolutions does it make in 30 s?

2. A wheel starts from rest with a constant angular acceleration of 2 rad/s^2. (*a*) What is its angular velocity after 5 s? (*b*) Through what angle has the wheel turned after 5 s? (*c*) How many revolutions has it made in 5 s? (*d*) After 5 s, what is the speed and acceleration of a point 0.3 m from the axis of rotation?

3. A turntable rotating at $33\frac{1}{3}$ rev/min is shut off. It brakes with constant angular acceleration and comes to rest in 2 min. (*a*) Find the angular acceleration. (*b*) What is the average angular velocity of the turntable? (*c*) How many revolutions does it make before stopping?

4. A disk of radius 10 cm begins rotating about its axis from rest with a constant angular acceleration of 10 rad/s^2. At $t = 5$ s, what are (*a*) the angular velocity of the disk and (*b*) the tangential acceleration a_t and the centripetal acceleration a_c of a point on the edge of the disk?

5. A spot on a record is on the radial line $\theta = 0°$ relative to a point fixed in space. At $t = \frac{1}{4}$ s after the turntable is turned on, the spot has advanced to $\theta = 10°$. Assuming constant angular acceleration, how long will it be before the record is rotating at 33.3 rev/min?

6. A Ferris wheel of radius 12 m rotates once in 25 s. (*a*) What is its angular velocity in radians per second? (*b*) What is the linear speed of a passenger? (*c*) What is the centripetal acceleration of a passenger?

7. A cyclist starts from rest and pedals such that the wheels of his bike have a constant angular acceleration. After 10 s, the wheels have made 5 rev. (*a*) What is the angular acceleration of the wheels? (*b*) What is the angular velocity of the wheels after 10 s? (*c*) If the radius of the wheel is 36 cm, and the wheel rolls without slipping, how far has the cyclist traveled in 10 s?

8-2 Torque and Moment of Inertia

8. A 0.5-kg particle falls freely under the influence of gravity. (*a*) It is at $y = 10$ m and $x = 2$ m at time t_1. What is the torque about the origin exerted by gravity on the particle at this time? (*b*) At some later time, the particle is at $y = 0$ and $x = 2$ m. What is the torque about the origin at this time?

9. A disk-shaped grindstone of mass 2 kg and radius 7 cm is spinning at 700 rev/min. After the power is shut off, a woman continues to sharpen her axe by holding it against the grindstone for 10 s until the grindstone stops rotating. (*a*) Find the angular acceleration of the grindstone, assuming it to be constant. (*b*) What is the torque exerted by the axe on the grindstone? (Assume there are no other frictional torques.)

10. To start a playground merry-go-round rotating, a rope is wrapped around it and pulled. A force of 200 N is exerted on the rope for 10 s. During this time, the merry-go-round, which has a radius of 2 m, makes one complete rotation. (*a*) Find the angular acceleration of the merry-go-round, assuming it to be constant. (*b*) What torque is exerted by the rope on the merry-go-round? (*c*) What is the moment of inertia of the merry-go-round?

11. A rope is wrapped around a 3-kg cylinder of radius 10 cm that is free to turn about its axis. The rope is pulled with a force of 15 N. The cylinder is initially at rest at $t = 0$. (*a*) Find the torque exerted by the rope, and the angular acceleration of the cylinder. (*b*) Find the angular velocity of the cylinder at time $t = 4$ s.

8-3 Rotational Kinetic Energy

12. The particles in Figure 8-48 are connected by a very light rod whose moment of inertia may be neglected. They rotate about the y axis with an angular velocity of 2 rad/s. (*a*) Find the speed of each particle, and use it to calculate the kinetic energy of this system directly from $\Sigma\frac{1}{2}m_i v_i^2$. (*b*) Find the moment of inertia about the y axis, and calculate the kinetic energy from $E_K = \frac{1}{2}I\omega^2$.

Figure 8-48 Problem 12.

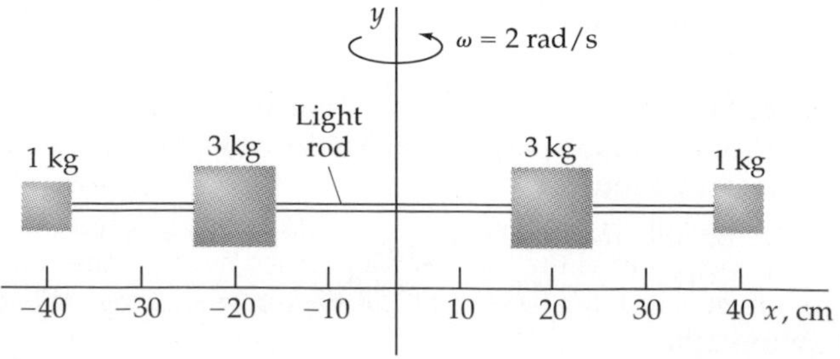

13. The four particles at the corners of the square in Figure 8-49 are connected by massless rods. The masses of the particles are $m_1 = m_3 = 3$ kg and $m_2 = m_4 = 4$ kg. The length of a side of the square is $L = 2$ m. (*a*) Find the moment of inertia of the system about an axis that is perpendicular to the plane of the particles and passes through m_4. (*b*) How much work is required to produce a rotation of 2 rad/s about this axis?

Figure 8-49 Problems 13, 17, and 19.

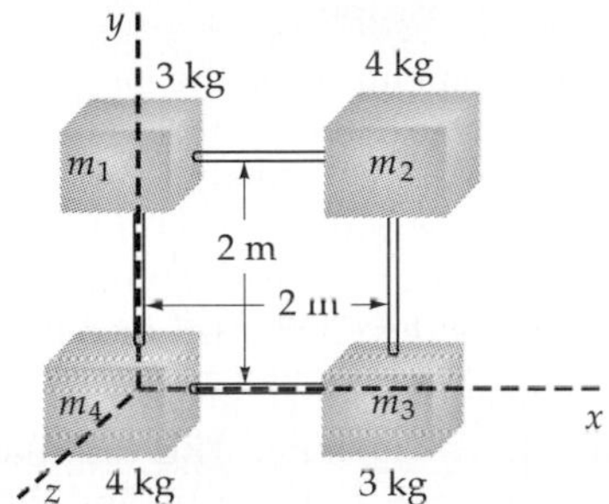

14. Four 2-kg particles are located at the corners of a rectangle of sides 3 and 2 m (Figure 8-50). (*a*) Find the moment of inertia of this system about an axis that is perpendicular to the plane of the masses and passes through one of the masses. (*b*) The system is set rotating about this axis with a kinetic energy of 184 J. Find the number of revolutions the system makes per minute.

Figure 8-50 Problems 14 and 20.

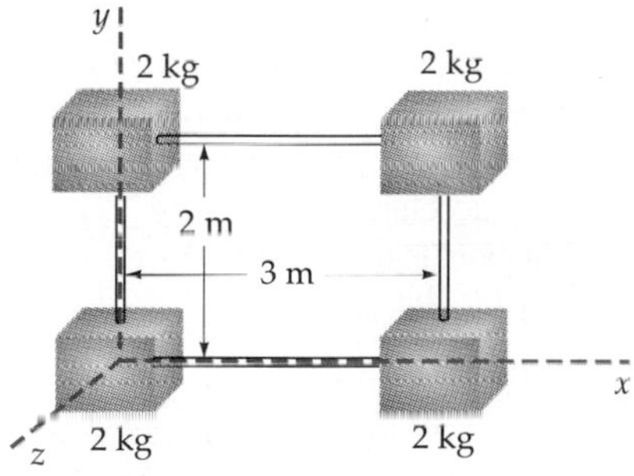

15. A solid ball of mass 1.2 kg and diameter 16 cm is rotating about its diameter at 90 rev/min. (*a*) What is the kinetic energy of rotation? (*b*) If an additional 2 J of energy are supplied to the rotational energy, what is the new angular speed of the ball?

16. An engine develops 500 N·m of torque at 3500 rev/min. Find the power developed by the engine.

8-4 Calculating the Moment of Inertia

17. Use the parallel-axis theorem and your results for Problem 13 to find the moment of inertia of the four-particle system in Figure 8-49 about an axis that is perpendicular to the plane of the masses and passes through the center of mass of the system. Check your result by direct computation.

18. A tennis ball has a mass of 57 g and a diameter of 7 cm. Find the moment of inertia about its diameter. Assume the ball is a thin spherical shell.

19. (*a*) Find the moment of inertia I_x for the four-particle system of Figure 8-49 about the x axis, which passes through m_3 and m_4. (*b*) Find I_y for the system about the y axis, which passes through m_1 and m_4. (*c*) Use Equation 8-30 to calculate the moment of inertia I_z about the z axis, which passes through m_4 and is perpendicular to the plane of the figure.

20. (*a*) Use the parallel-axis theorem to find the moment of inertia about an axis that is parallel to the z axis and passes through the center of mass of the system in Figure 8-50. (*b*) Let x' and y' be axes in the plane of the figure that pass through the center of mass and are parallel to the sides of the rectangle. Compute $I_{x'}$ and $I_{y'}$, and use your results and Equation 8-30 to check your result for part (*a*).

21. Use Equation 8-30 to find the moment of inertia of a disk of radius R and mass M about an axis in the plane of the disk that passes through its center (Figure 8-51).

22. Use the parallel-axis theorem to find the moment of inertia of a solid sphere of mass M and radius R about an axis that is tangent to the sphere (Figure 8-52).

Figure 8-51 Problem 21. **Figure 8-52** Problem 22.

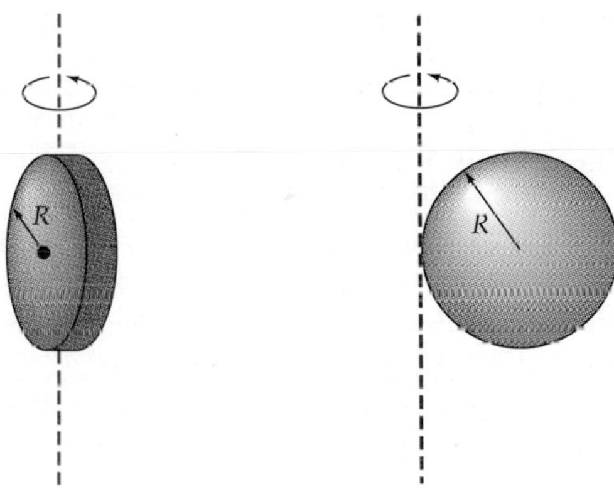

8-5 Angular Momentum

23. A 3-kg particle moves at a constant speed of 4 m/s around a circle of radius 5 m. (*a*) What is its angular momentum about the center of the circle? (*b*) What is its moment of inertia about an axis through the center of the circle and perpendicular to the plane of the motion? (*c*) What is the angular velocity of the particle?

24. A 3-kg particle moves at constant speed of 4 m/s along a straight line. (*a*) What is its angular momentum about a point 5 m from the line? (*b*) Describe qualitatively how its angular velocity about that point varies with time.

25. A particle travels in a circular path. (*a*) If its linear momentum p is doubled, how is its angular momentum affected? (*b*) If the radius of the circle is doubled but the speed is unchanged, how is the angular momentum of the particle affected?

26. A 2-kg particle moves in a circle of radius 3 m. Its angular momentum relative to the center of the circle depends on time according to $L = (4\ \text{N}\cdot\text{m})t$. (*a*) Find the torque acting on the particle. (*b*) Find the angular velocity of the particle as a function of time.

27. A planet moves in an elliptical orbit about the sun with the sun at one focus of the ellipse (Figure 8-53). (*a*) What is the torque produced by the gravitational force of attraction of the sun for the planet? (*b*) At position *A*, the planet is a distance r_1 from the sun and is moving with a speed v_1 perpendicular to the line from the sun to the planet. At position *B*, it is at distance r_2 and is moving with speed v_2, again perpendicular to the line from the sun to the planet. What is the ratio of v_1 to v_2 in terms of r_1 and r_2?

Figure 8-53 Problem 27.

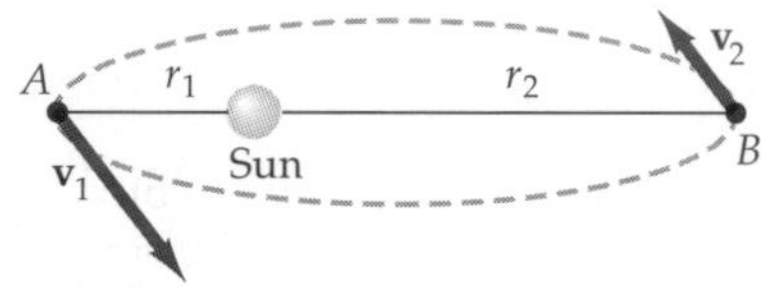

8-6 Rolling Objects

28. A homogeneous cylinder of radius 15 cm and mass 50 kg is rolling without slipping along a horizontal floor at 6 m/s. How much work is needed to stop the cylinder?

29. Find the percentages of the total kinetic energy associated with rotation and with translation for an object that is rolling without slipping if the object is (*a*) a uniform sphere, (*b*) a uniform cylinder, and (*c*) a hoop.

30. A hoop of radius 0.50 m and mass 0.8 kg is rolling without slipping at a speed of 20 m/s toward an incline of slope 30°. How far up the incline will the hoop roll, assuming it rolls without slipping?

31. A ball rolls without slipping down an incline of angle θ. Find, in terms of the coefficient of friction μ_s, (*a*) the acceleration of the ball, (*b*) the force of friction, and (*c*) the maximum angle of the incline for which the ball will roll without slipping.

32. A ball rolls without slipping along a horizontal plane. Show that the frictional force acting on the ball must be zero. *Hint:* Consider a possible direction for the action of the frictional force and what effects such a force would have on the velocity of the center of mass and on the angular velocity.

8-7 The Vector Nature of Rotation and the Cross Product

33. A force of magnitude *F* is applied horizontally in the negative *x* direction to the rim of a disk of radius *R* as shown in Figure 8-54. Write **F** and **r** in terms of the unit vectors **i**, **j**, and **k**, and compute the torque produced by the force about the origin at the center of the disk.

34. Compute the torque about the origin for the force $\mathbf{F} = -mg\mathbf{j}$ acting on a particle at $\mathbf{r} = x\mathbf{i} + y\mathbf{j}$, and show that this torque is independent of the *y* coordinate.

35. Find $\mathbf{A} \times \mathbf{B}$ for (*a*) $\mathbf{A} = 6\mathbf{i}$ and $\mathbf{B} = 6\mathbf{i} + 6\mathbf{j}$, (*b*) $\mathbf{A} = 6\mathbf{i}$ and $\mathbf{B} = 6\mathbf{i} + 6\mathbf{k}$, and (*c*) $\mathbf{A} = 3\mathbf{i} + 3\mathbf{j}$ and $\mathbf{B} = -2\mathbf{i} + 2\mathbf{j}$.

36. Under what conditions is the magnitude of $\mathbf{A} \times \mathbf{B}$ equal to $\mathbf{A} \cdot \mathbf{B}$?

37. A particle of mass 3 kg moves with velocity $\mathbf{v} = 2\ \text{m/s}\ \mathbf{i}$ along the line $z = 0$, $y = 4.3$ m. (*a*) Find the angular momentum **L** relative to the origin when the particle is at $x = 12$ m, $y = 4.3$ m. (*b*) A force $\mathbf{F} = -3\ \text{N}\ \mathbf{i}$ is applied to the particle. Find the torque relative to the origin due to this force.

8-8 Motion of a Gyroscope

38. The angular momentum of the propeller of a small airplane points forward. (*a*) As the plane takes off, the nose lifts up and the airplane tends to veer to one side. To which side does it veer and why? (*b*) If the plane is flying horizontally and suddenly turns to the right, does the nose of the plane tend to move up or down? Why?

39. A bicycle wheel of radius 30 cm is mounted at the middle of an axle 60 cm long. The tire and rim weigh 36 N. The wheel is spun at 10 rev/s, and the axle is then placed in a horizontal position with one end resting on a pivot. (*a*) What is the angular momentum due to the spinning of the wheel? (Treat the wheel as a hoop.) (*b*) What is the angular velocity of precession? (*c*) How long does it take for the axle to swing through 360° around the pivot? (*d*) What is the angular momentum associated with the motion of the center of mass, that is, due to the precession? In what direction is this angular momentum?

40. A uniform disk of mass 2 kg and radius 6 cm is mounted in the center of a 10-cm axle and spun at 900 rev/min. The axle is then placed in a horizontal position with one end resting on a pivot. The other end is given an initial horizontal velocity such that the precession is smooth with no nutation. (*a*) Find the angular velocity of precession. (*b*) What is the speed of the center of mass during the precession? (*c*) What are the magnitude and direction of the acceleration of the center of mass? (*d*) What are the vertical and horizontal components of the force exerted by the pivot?

8-9 Static and Dynamic Imbalance

41. A uniform disk of radius 30 cm, thickness 3 cm, and mass 5 kg rotates at $\omega = 10$ rad/s about an axis parallel to the symmetry axis but 0.5 cm from that axis. (*a*) Find the net force on the bearings due to this imbalance. (*b*) Where should a 100-g mass be placed on the disk to correct this problem?

Figure 8-54 Problem 33.

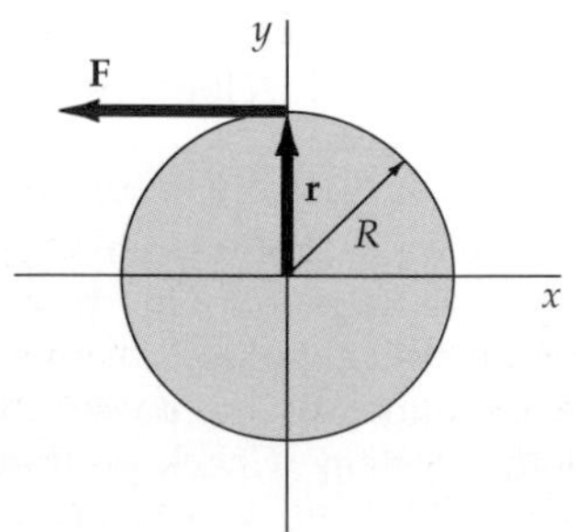

Level II

42. A flywheel is a uniform disk of mass 100 kg and radius of 1.2 m. It rotates with an angular velocity of 1200 rev/min. (*a*) A constant tangential force is applied at a radial distance of 0.5 m. What work must this force do to stop the wheel? (*b*) If the wheel is brought to rest in 2 min, what torque does the force produce? What is the magnitude of the force? (*c*) How many revolutions does the wheel make in these 2 min?

43. A uniform disk of radius 0.12 m and mass 5 kg is pivoted such that it rotates freely about its axis. A string wrapped around the disk is pulled with a force of 20 N (Figure 8-55). (*a*) What is the torque exerted on the disk? (*b*) What is the angular acceleration of the disk? (*c*) If the disk starts from rest, what is its angular velocity after 3 s? (*d*) What is its kinetic energy after 3 s? (*e*) What is its angular momentum after 3 s? (*f*) Find the total angle θ the disk turns through in 3 s, and (*g*) show that the work done by the torque $\tau\theta$ equals the kinetic energy.

Figure 8-55 Problem 43.

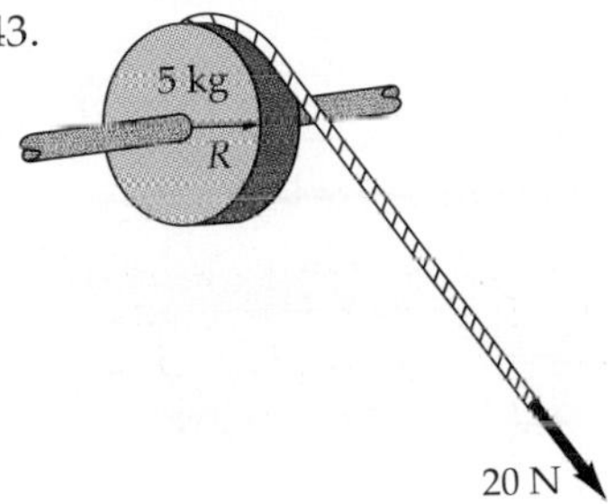

44. A 2000-kg block is lifted by a steel cable that passes over a pulley to a motor-driven winch (Figure 8-56). The radius of the winch drum is 30 cm, and the moment of inertia of the pulley is negligible. (*a*) What force must be exerted by the cable to lift the block at a constant velocity of 8 cm/s? (*b*) What torque does the cable exert on the winch drum? (*c*) What is the angular velocity of the winch drum? (*d*) What power must be developed by the motor to drive the winch drum?

Figure 8-56 Problem 44.

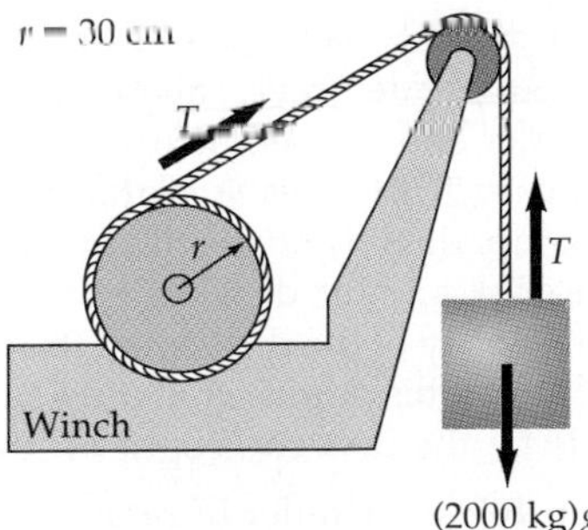

45. A homogeneous cylinder of mass 100 kg and radius 0.3 m is mounted such that it turns without friction on its fixed symmetry axis. It is rotated by a drive belt that wraps around its perimeter and exerts a constant torque. At time $t = 0$, its angular velocity is zero. At time $t = 30$ s, its angular velocity is 600 rev/min. (*a*) What is its angular momentum at $t = 30$ s? (*b*) At what rate is the angular momentum increasing? (*c*) What is the torque acting on the cylinder? (*d*) What is the magnitude of the force acting on the rim of the cylinder?

46. A particle is traveling with a constant velocity **v** along a line that is a distance b from the origin O. Let dA be the area swept out by the position vector from O to the particle in time dt (Figure 8-57). Show that dA/dt is constant in time and equal to $\frac{1}{2}L/m$, where L is the angular momentum of the particle about the origin.

Figure 8-57 Problem 46.

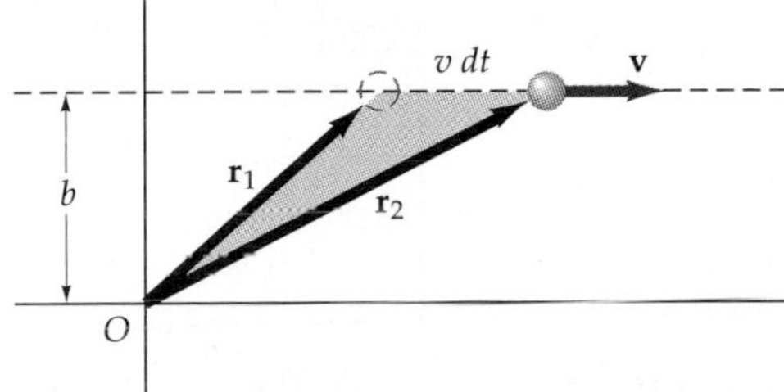

47. A 15-g coin of diameter 1.5 cm is spinning at 10 rev/s about a vertical diameter at a fixed point on a tabletop. (*a*) What is the angular momentum of the coin about its center of mass? (*b*) What is its angular momentum about a point on the table 10 cm from the coin? If the coin spins about a vertical diameter at 10 rev/s, but it also travels in a straight line across the tabletop at 5 cm/s, (*c*) what is the angular momentum of the coin about a point on the line of motion? (*d*) What is the angular momentum of the coin about a point 10 cm from the line of motion? (There are two answers to this question. Explain why and give both.)

48. (*a*) Assuming that the incline in Figure 8-58 is frictionless and that the string passes through the center of mass of m_2, find the net torque acting on the system (the two masses and the pulley) about the center of the pulley. (*b*) Write an expression for the total angular momentum of the system about the center of the pulley when the masses are moving with a speed v. Assume the pulley has a moment of inertia I and a radius r. (*c*) Find the acceleration of the masses from your results for parts (*a*) and (*b*) by setting the net torque equal to the rate of change of the angular momentum of the system.

Figure 8-58 Problem 48.

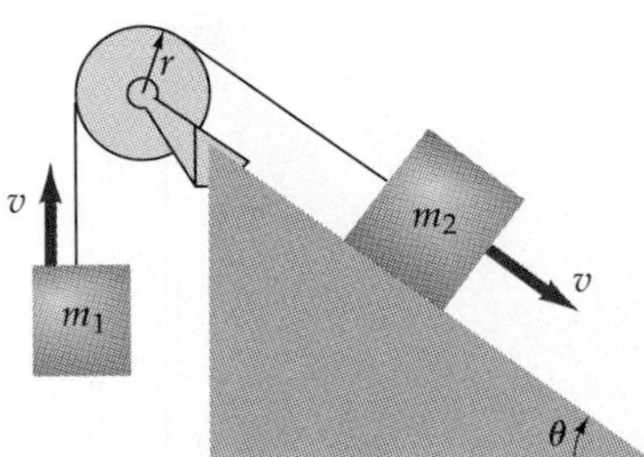

49. Calculate the kinetic energy of rotation of the earth, and compare it with the kinetic energy of motion of the earth's center of mass. Assume the earth to be a homogeneous sphere of mass 6.0×10^{24} kg and radius 6.4×10^{6} m. The radius of the earth's orbit is 1.5×10^{11} m.

50. A 2-kg mass attached to a string of length 1.5 m moves in a horizontal circle as a conical pendulum (Figure 8-59). The string makes an angle $\theta = 30°$ with the vertical. (*a*) Show that the angular momentum of the mass about the point of support *P* has a horizontal component toward the center of the circle as well as a vertical component, and find these components. (*b*) Find the magnitude of $d\mathbf{L}/dt$, and show that it equals the magnitude of the torque exerted by gravity about the point of support.

Figure 8-59 Problem 50.

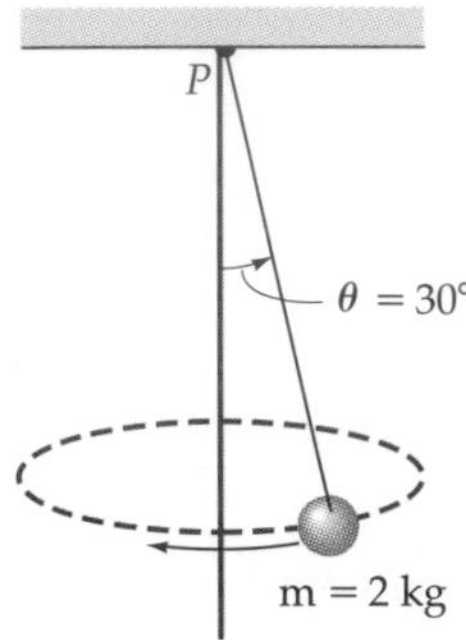

51. A wheel mounted on an axis that is not frictionless is initially at rest. A constant external torque of 50 N·m is applied to the wheel for 20 s. At the end of the 20 s, the wheel has an angular velocity of 600 rev/min. The external torque is then removed, and the wheel comes to rest after 120 s more. (*a*) What is the moment of inertia of the wheel? (*b*) What is the frictional torque, which is assumed to be constant?

52. A 1-Mg car is being unloaded by a winch. At the moment shown in Figure 8-60, the gearbox shaft of the winch breaks, and the car falls from rest. During the car's fall there is no slipping between the rope, the pulley, and the winch drum. The moment of inertia of the winch drum is 320 kg·m^2 and that of the pulley is 4 kg·m^2. The radius of the winch drum is 0.80 m and that of the pulley is 0.30 m. Find the speed of the car as it hits the water.

Figure 8-60 Problem 52.

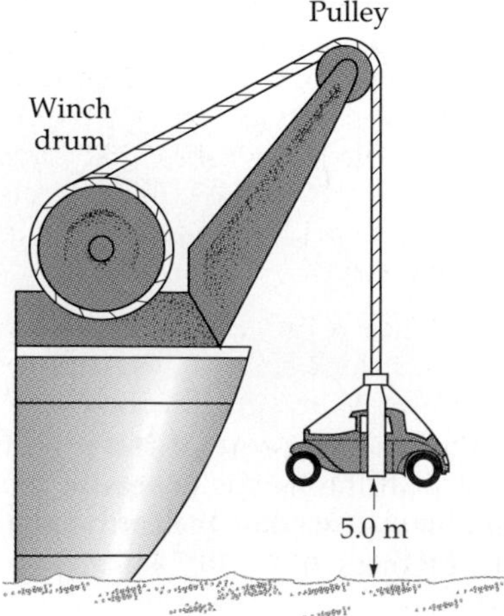

53. A typical car engine delivers about 2 MJ of mechanical energy per kilometer on the average. A car is designed to use the energy stored in a flywheel in a vacuum container. If the mass of the flywheel is not to exceed 100 kg and its angular velocity is not to exceed 400 rev/s, find the smallest radius of the flywheel for which the car can travel 300 km without the flywheel having to be recharged. (Assume the flywheel to be a uniform cylinder.)

54. The system in Figure 8-61 is released from rest. The 30-kg block is 2 m above the floor. The pulley is a uniform disk with a radius of 10 cm and mass of 5 kg. Find (*a*) the speed of the 30-kg block just before it hits the floor, (*b*) the angular speed of the pulley at that time, (*c*) the tensions in the strings, and (*d*) the time it takes for the 30-kg block to reach the floor. Assume the string does not slip on the pulley.

Figure 8-61 Problem 54.

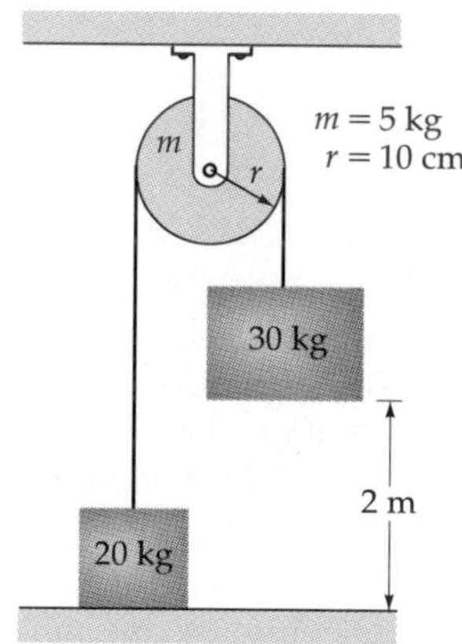

55. A meterstick is pivoted at one end such that it can swing freely in a vertical plane. It is released from rest in a horizontal position. (*a*) What is the angular velocity of the stick when it is vertical? (*b*) What force is exerted by the pivot when the stick is vertical? (Answer in terms of the mass *m* of the stick.)

56. A uniform disk of mass *M* and radius *R* is pivoted such that it can rotate freely about a horizontal axis through its center and perpendicular to the plane of the disk. A small particle of mass *m* is attached to the rim of the disk at the top, directly above the pivot. The system is given a gentle start, and the disk begins to rotate. (*a*) What is the angular velocity of the disk when the particle is at its lowest point? (*b*) At this point, what force must be exerted on the particle by the disk to keep it on the disk?

57. A uniform, 100-kg cylinder of radius 0.60 m is placed flat on some smooth ice. Two skaters wind ropes around the cylinder in the same sense. The skaters then pull on their ropes as they skate away, exerting constant forces of 40 N and 60 N, respectively, for 5 s (Figure 8-62). Describe the motion of the cylinder. That is, what are its acceleration and velocity and the position of its center of mass as functions of time, and what are its angular acceleration and angular velocity as functions of time?

Figure 8-62 Problem 57.

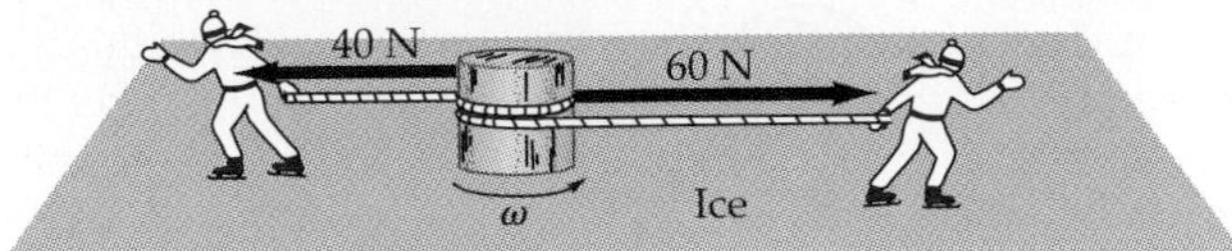

58. A vertical grinding wheel is a uniform disk of mass 50 kg and radius 40 cm. It has a handle of radius 60 cm of negligible mass. A 20-kg load is attached to the handle when it is in the horizontal position. Find (*a*) the initial angular acceleration of the wheel and (*b*) the maximum angular velocity of the wheel.

59. A uniform sphere of mass M and radius R is free to rotate about a horizontal axis through its center. A string is wrapped around the sphere and is attached to a body of mass m as shown in Figure 8-63. Find (*a*) the acceleration of the body and (*b*) the tension in the string.

Figure 8-63 Problem 59.

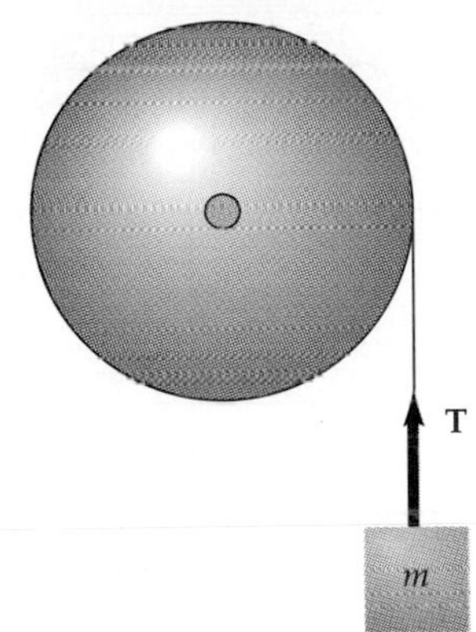

60. A uniform rectangular plate has mass m and sides a and b. (*a*) Show by integration that the moment of inertia of the plate about an axis that is perpendicular to the plate and passes through one corner is $\frac{1}{3}m(a^2 + b^2)$. (*b*) What is the moment of inertia about an axis that is perpendicular to the plate and passes through its center of mass?

61. A uniform ball of radius r rolls without slipping along the loop the loop track in Figure 8-64. The ball starts at rest at a height h above the bottom of the loop. (*a*) If it is not to leave the track at the top of the loop, what is the least value h can have (in terms of the radius R of the loop)? (*b*) What would h have to be if, instead of rolling, the ball slides without friction?

Figure 8-64 Problem 61.

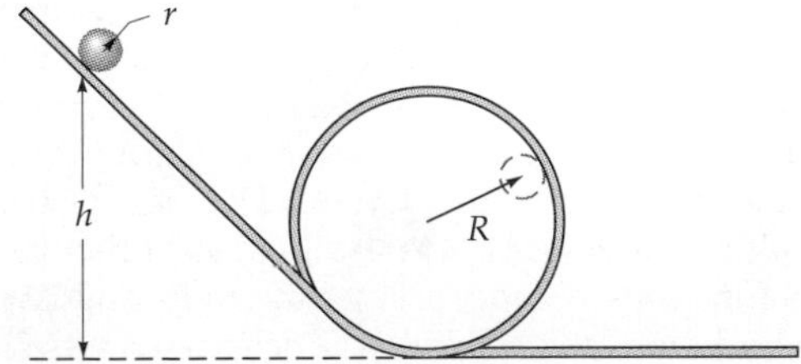

62. A particle moves in a circle of radius r with an angular velocity $\boldsymbol{\omega}$. (*a*) Show that its velocity is $\mathbf{v} = \boldsymbol{\omega} \times \mathbf{r}$. (*b*) Show that its centripetal acceleration is $\mathbf{a}_c = \boldsymbol{\omega} \times \mathbf{v} = \boldsymbol{\omega} \times (\boldsymbol{\omega} \times \mathbf{r})$.

63. A uniform cylinder of mass M and radius R has a string wrapped around it. The string is held fixed, and the cylinder falls vertically as shown in Figure 8-65. (*a*) Show that the acceleration of the cylinder is downward with a magnitude $a = 2g/3$. (*b*) Find the tension in the string.

Figure 8-65 Problem 63.

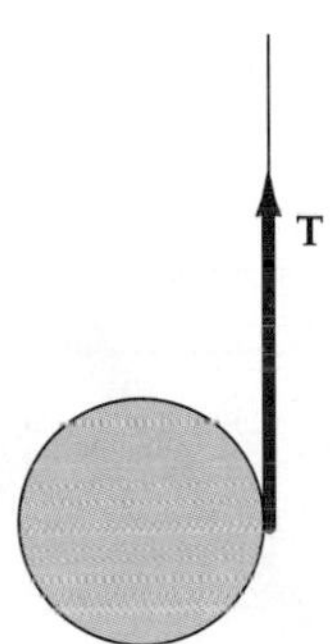

64. A car is powered by the energy stored in a single flywheel with an angular momentum **L**. Discuss the problems that would arise for various orientations of **L** and various maneuvers of the car. For example, what would happen if **L** points vertically upward and the car travels over a hilltop or through a valley, or what would happen if **L** points forward or to one side and the car attempts to turn to the left or right. In each case, consider the direction of the torque exerted on the car by the road.

65. A man stands on a frictionless platform that is rotating with an angular speed of 2.0 rev/s. His arms are outstretched, and he holds a heavy weight in each hand. The moment of inertia of the man, the extended weights, and the platform is 5 kg·m^2. When the man pulls the weights inward toward his body the moment of inertia decreases to 2 kg·m^2. (*a*) What is the resulting angular speed of the platform? (*b*) What is the change in kinetic energy of the system? (*c*) Where did this increase in energy come from?

66. A man of mass m stands at the edge of a rotating circular turntable. The turntable has a radius R and a moment of inertia I, and it rotates without friction. The angular velocity about the vertical axis through the center of the turntable is 8 rad/s. The man walks radially inward. What is the final angular velocity of the system? Assume that $mR^2 = 3I$ and that the man has a moment of inertia of $I/10$ when he is standing at the center.

67. A professor stands at the center of a turntable that can rotate without friction. She begins to rotate a heavy ball on the end of a 0.8-m chain about her head. The ball has a mass of 2 kg, and it makes one revolution every 3 s. The professor and platform have a moment of inertia of 0.5 kg·m^2. (*a*) What is the angular speed of the professor? (*b*) What is the total kinetic energy of the ball, professor, and platform?

68. A 0.1-kg yo-yo consists of two solid disks of radius 10 cm joined together by a massless rod of radius 1 cm and a string wrapped around the rod. One end of the string is held fixed and is under constant tension T as the yo-yo is released. Find the acceleration of the yo-yo and the tension T.

69. A uniform boom 5.0 m long and having a total mass of 150 kg is connected to the ground by a hinge at the bottom and is supported by a horizontal cable as shown in Figure 8-66. (*a*) What is the tension in the cable? (*b*) If the cable is cut, what is the angular acceleration of the boom the instant the cable is cut? (*c*) What is the angular velocity of the boom when it is horizontal?

Figure 8-66 Problem 69.

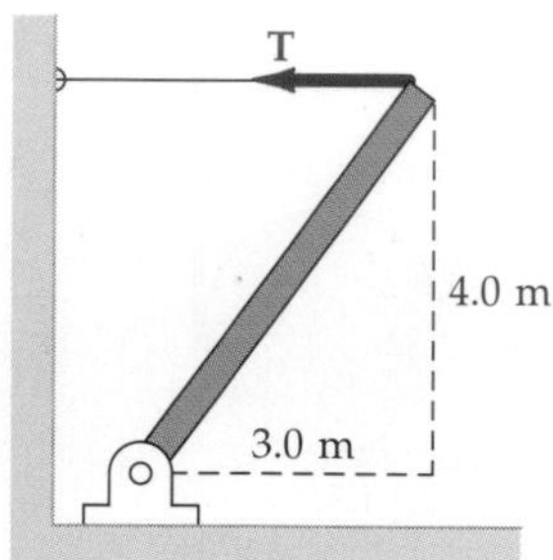

70. A cue ball of radius r is initially at rest on a horizontal pool table. It is struck by a horizontal cue stick that delivers an impulse of magnitude $P_0 = F_{av}\,\Delta t$. (We use P_0 for the impulse rather than I_0 to avoid confusion with the moment of inertia I.) The stick strikes the ball at a point h above the ball's point of contact with the table (Figure 8-67). Show that the ball's initial angular velocity ω_0 is related to the initial linear velocity of its center of mass v_0 by $\omega_0 = 5v_0(h - r)/2r^2$.

Figure 8-67 Problem 70.

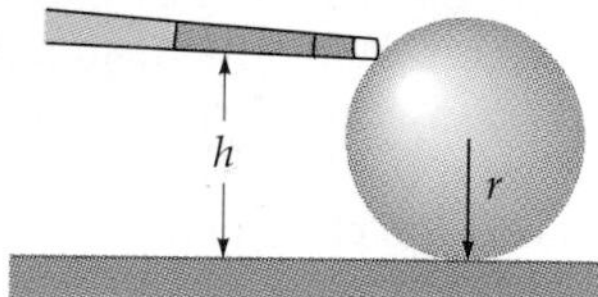

71. The sun's radius is 6.96×10^8 m and it rotates with a period of 25.3 d. Estimate its new period of rotation if it collapses with no loss of mass to become a neutron star of radius 5 km.

72. A uniform cylinder of mass m_1 and radius R is pivoted on frictionless bearings. A string wrapped around the cylinder connects to a mass m_2, which is on a frictionless incline of angle θ as shown in Figure 8-68. The system is released from rest with m_2 a height h above the bottom of the incline. (*a*) What is the acceleration of m_2? (*b*) What is the tension in the string? (*c*) What is the total energy of the system when m_2 is at height h? (*d*) What is the total energy when m_2 is at the bottom of the incline and has a speed v? (*e*) What is the speed v? (*f*) Evaluate your answers for the extreme cases of $\theta = 0°$, $\theta = 90°$, and $m_1 = 0$.

Figure 8-68 Problem 72.

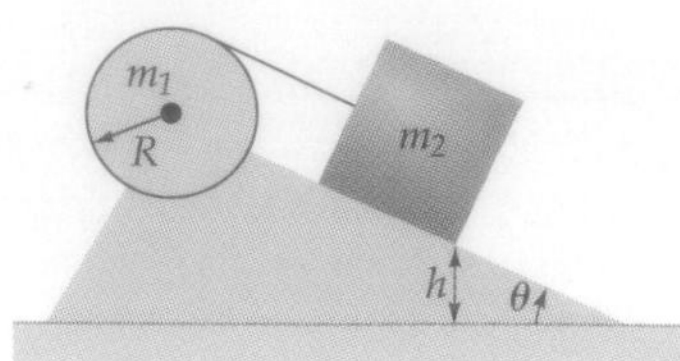

73. An Atwood's machine has two objects of mass $m_1 =$ 500 g and $m_2 = 510$ g, connected by a string of negligible mass that passes over a frictionless pulley (Figure 8-69). The pulley is a uniform disk with a mass of 50 g and a radius of 4 cm. The string does not slip on the pulley. (*a*) Find the acceleration of the objects. (*b*) What is the tension in the string supporting m_1? In the string supporting m_2? By how much do they differ? (*c*) What would your answers have been if you had neglected the motion of the pulley?

Figure 8-69 Problem 73.

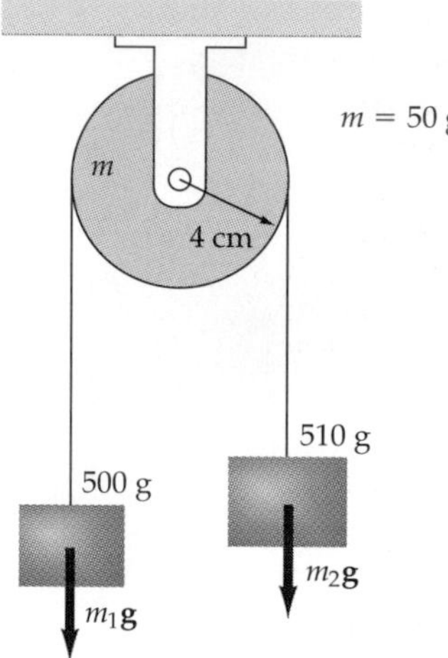

74. Two objects are attached to ropes that are attached to wheels on a common axle as shown in Figure 8-70. The total moment of inertia of the two wheels is 40 kg·m^2. The radii of the wheels are $R_1 = 1.2$ m and $R_2 = 0.4$ m. (*a*) If $m_1 = 24$ kg, find m_2 such that the system is in equilibrium. (*b*) If 12 kg is gently added to the top of m_1, find the angular acceleration of the wheels and the tensions in the ropes.

Figure 8-70 Problem 74.

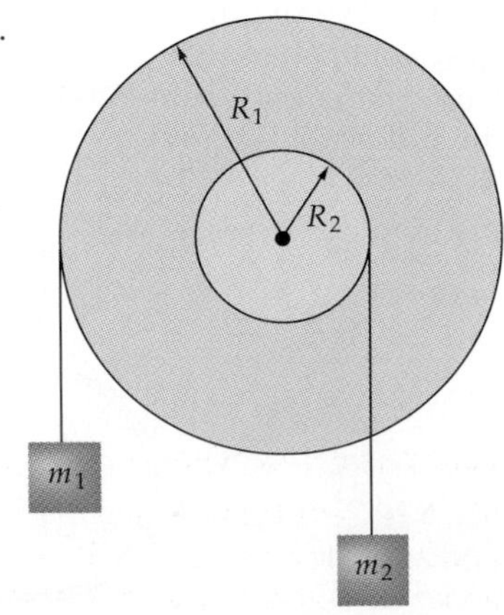

Level III

75. A mass m is attached to a light string that passes through a small hole in a frictionless tabletop. Initially, the mass is sliding with a speed v_0 in a circle of radius r_0 about the hole. A woman under the table now begins to pull the string slowly in. (*a*) Show that when the mass is moving in a circle of radius r, the tension in the string is $T = L_0^2/mr^3$, where $L_0 = mv_0r_0$ is the initial angular momentum. (*b*) The string is pulled in until the radius of the circular orbit is r_f. Using the result of part (*a*), calculate the work done by integrating $\mathbf{T}\cdot d\mathbf{r}$, and show that the work done is $(L_0^2/2m)(r_f^{-2} - r_0^{-2})$. (*c*) What is the velocity v_f when the radius of the circle is r_f? Show that the work done as calculated in part (*b*) equals the change in kinetic energy $\frac{1}{2}mv_f^2 - \frac{1}{2}mv_0^2$. (*d*) If $m = 0.5$ kg, $r_0 = 1$ m, $v_0 = 3$ m/s, and the maximum tension the string can withstand without breaking is 200 N, find the minimum of r_f before the string breaks.

76. (*a*) Assuming the earth to be a homogeneous sphere of radius r and mass m, show that the period T of the earth's rotation about its axis is related to its radius by $T = (4\pi m/5L)r^2$, where L is the angular momentum of the earth due to its rotation. (*b*) Suppose that the radius r changes by a very small amount Δr due to some internal effect such as thermal expansion. Show that the fractional change in the period ΔT is given approximately by $\Delta T/T = 2\,\Delta r/r$. *Hint*: Use the differentials dr and dT to approximate the changes in these quantities. (*c*) By how many kilometers would the earth need to expand for the period to change by $\frac{1}{4}$ d/y so that leap years would no longer be necessary?

77. The polar ice caps contain about 2.3×10^{19} kg of ice. This mass contributes essentially nothing to the moment of inertia of the earth because it is located at the poles, close to the axis of rotation. Estimate the change in the length of the day that would be expected if the polar ice caps were to melt and the water were distributed uniformly over the surface of the earth. (The moment of inertia of a spherical shell of mass m and radius r is $\frac{2}{3}mr^2$.)

78. Show that the moment of inertia of a spherical shell of radius R and mass m is $\frac{2}{3}mR^2$. This can be done by direct integration or, more easily, by finding the increase in the moment of inertia of a solid sphere when its radius changes. To do this, first show that the moment of inertia of a solid sphere of density ρ is $I = \frac{8}{15}\pi\rho R^5$. Then compute the change dI in I for a change dR, and use the fact that the mass of the shell is $m = 4\pi R^2\rho\, dR$.

79. Figure 8-71 shows a pair of uniform spheres, each of mass 500 g and radius 5 cm. They are mounted on a uniform rod that has a length $L = 30$ cm and a mass $m = 60$ g. (*a*) Calculate the moment of inertia of this system about an axis perpendicular to the rod through the center of the rod using the approximation that the two spheres can be treated as point particles that are 20 cm from the axis of rotation and that the mass of the rod is negligible. (*b*) Calculate the moment of inertia exactly and compare your result with your answer for part (*a*).

80. A hollow cylinder has mass m, an outside radius R_2, and an inside radius R_1. Show that its moment of inertia about its symmetry axis is given by $I = \frac{1}{2}m(R_2^2 + R_1^2)$.

81. A bowling ball of mass M and radius R is thrown such that at the instant it touches the floor it is moving horizontally with a speed v_0 and is not rotating. It slides for a time t_1 a distance s_1 before it begins to roll without slipping. (*a*) If μ_k is the coefficient of sliding friction between the ball and the floor, find s_1, t_1, and the final speed v_1 of the ball. (*b*) Evaluate these quantities for $v_0 = 8$ m/s and $\mu_k = 0.4$.

82. A uniform spherical ball is set rotating about a horizontal axis with an angular speed ω_0 and is placed on the floor. If the coefficient of sliding friction between the ball and the floor is μ_k, find the speed of the center of mass of the ball when it begins to roll without slipping.

83. A uniform rod of mass M and length L is pivoted at one end and hangs as shown in Figure 8-72 such that it is free to rotate about its pivot without friction. It is struck by a horizontal force that delivers an impulse $P_0 = F_{av}\,\Delta t$ at a distance x below the pivot as shown. (*a*) Show that the initial speed of the center of mass of the rod is given by $V_0 = 3P_0x/2ML$. (*b*) Find the impulse delivered by the pivot, and show that this impulse is zero if $x = 2L/3$. This point is called the *center of percussion* of the rod.

Figure 8-71 Problem 79.

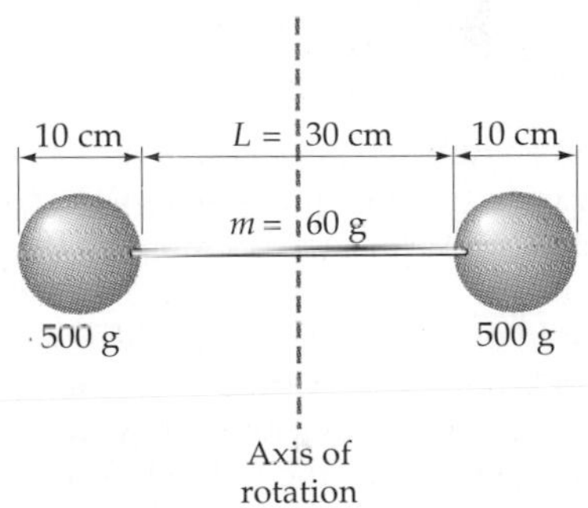

Figure 8-72 Problem 83.

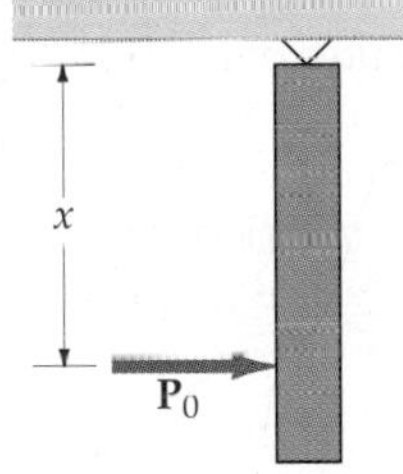

84. The rod in Problem 83 is struck by a piece of putty of mass $m = M/6$ that is moving horizontally with an initial speed v_0. The putty strikes the end of the rod and sticks to it. (*a*) Find the initial angular velocity of the rod with the putty attached just after the collision. (*b*) What fraction of the initial energy of the putty is lost in this inelastic collision?

85. Two heavy, uniform disks are connected by a short rod. The system is placed on an inclined plane such that the disks hang over the sides. Each disk has a mass of 20 kg and a radius of 30 cm, and the solid connecting rod has a radius of 2 cm and a mass of 1 kg. The plane is inclined at 30° and the rod rolls on the plane without slipping. (*a*) What is the acceleration of the system down the plane? (*b*) What is the angular acceleration of the system? (*c*) If the system starts from rest, what is its kinetic energy of translation when it has rolled 2 m down the incline? (*d*) What is the kinetic energy of rotation for the 2-m displacement?

86. A heavy, homogeneous cylinder has a mass m and a radius R. It is accelerated by a force **T**, which is applied through a rope wound around a light drum of radius r that is attached to the cylinder (Figure 8-73). The coefficient of static friction is sufficient for the cylinder to roll without slipping. (*a*) Find the frictional force. (*b*) Find the acceleration a of the center of the cylinder. (*c*) Is it possible to choose r so that a is greater than T/m? How? (*d*) What is the direction of the frictional force in the circumstances of part (*c*)?

Figure 8-73 Problem 86.

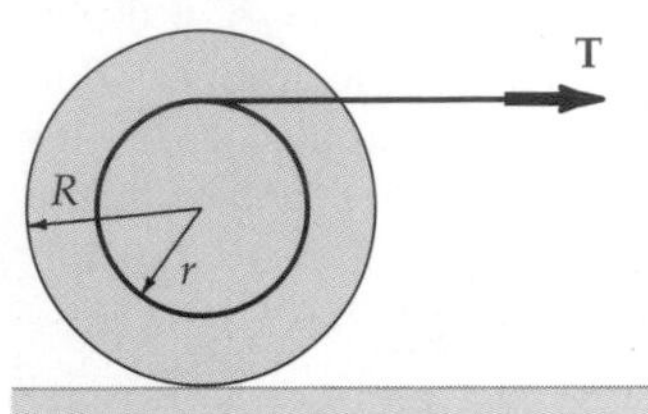

87. A uniform solid ball has a mass of 20 g and a radius of 5 cm. It rests on a horizontal surface. A sharp force is applied to the ball in the horizontal direction 9 cm above the horizontal surface. The force rises linearly from 0 to a peak value of 40,000 N in 10^{-4} s and then decreases linearly to 0 in 10^{-4} s. (*a*) What is the velocity of the ball after impact? (*b*) What is the angular velocity of the ball after impact? (*c*) What is the velocity of the ball when it begins to roll without sliding? (*d*) For how long does the ball slide on the surface? Assume $\mu_k = 0.5$.

88. A 0.3-kg billiard ball of radius 3 cm is given a sharp blow by a cue stick. The applied impulse is horizontal and passes through the center of the ball. The initial velocity of the ball is 4 m/s. The coefficient of kinetic friction is 0.6. (*a*) For how many seconds does the ball slide before it begins to roll without slipping? (*b*) How far does it slide? (*c*) What is its velocity once it begins rolling without slipping?

89. A uniform stick of length L and mass M is hinged at one end. It is released from rest at an angle θ_0 with the vertical. Show that when the angle with the vertical is θ, the hinge exerts a force F_r along the stick and a force F_t perpendicular to the stick given by

$$F_r = \tfrac{1}{2}Mg(5 \cos \theta - 3 \cos \theta_0)$$

and

$$F_t = \tfrac{1}{4}Mg \sin \theta$$

90. A wheel of radius R rolls without slipping at a speed V. (*a*) Show that the x and y coordinates of point P in Figure 8-74 are $r_0 \cos \theta$ and $R + r_0 \sin \theta$, respectively. (*b*) Show that the total velocity **v** of point P has the components $v_x = V + (r_0 V \sin \theta)/R$ and $v_y = -(r_0 V \cos \theta)/R$. (*c*) Show that **v** and **r** are perpendicular to each other by calculating their scalar product. (*d*) Show that $v = r\omega$, where $\omega = V/R$ is the angular velocity of the wheel. (*e*) These results demonstrate that, in the case of rolling without slipping, the motion is the same as if the rolling object were instantaneously rotating about the point of contact with an angular speed $\omega = V/R$. Calculate the kinetic energy of the wheel, assuming that it is in pure rotation about point O, and show that the result is the same as that calculated by summing the translational kinetic energy of the center of mass and the rotational kinetic energy of the wheel about its center of mass.

Figure 8-74 Problem 90.

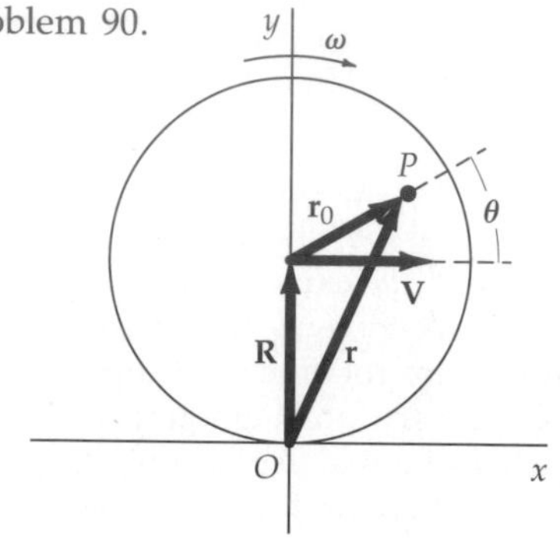

91. A bowling ball of radius R is given an initial velocity v_0 down the lane and a forward spin $\omega_0 = 3v_0/R$. The coefficient of kinetic friction is μ_k. (*a*) What is the speed of the ball when it begins to roll without slipping? (*b*) For how long does the ball slide before it begins to roll without slipping? (*c*) What distance does the ball slide down the lane before it begins rolling without slipping?

92. A billiard ball initially at rest is given a sharp impulse by a cue stick. The impulse is horizontal and is applied at a distance $2R/3$ below the centerline (Figure 8-75). The initial speed of the ball is v_0. (*a*) What is the initial angular speed ω_0? (*b*) What is the speed of the ball once it begins to roll without slipping? (*c*) What is the initial kinetic energy of the ball? (*d*) What is the frictional work done on the ball as it slides on the table?

Figure 8-75 Problem 92.

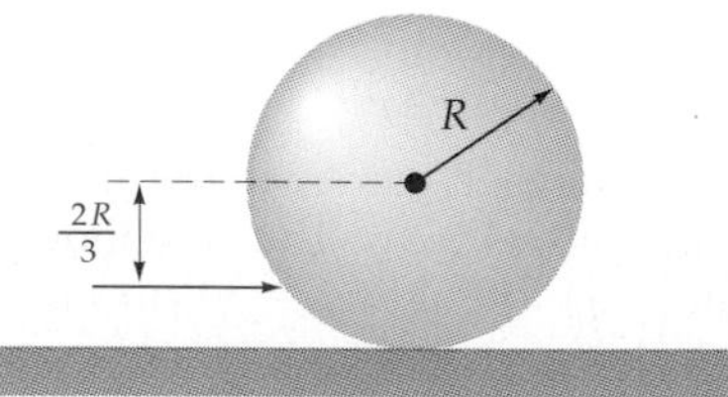

93. A marble rolls from rest without sliding from the top of a large sphere, which is held fixed. Find the angle from the top of the sphere of the point where the marble breaks contact with the sphere. The radius of the marble is 1 cm and that of the sphere is 80 cm.

Chapter 9

Static Equilibrium of a Rigid Body

The Oakland–San Francisco Bay Bridge. The part shown here is a suspension bridge supported by steel cables from one anchor post to another.

If an object is stationary and remains stationary, it is said to be in static equilibrium. Being able to determine the forces acting on an object in static equilibrium has many important applications, particularly in engineering. For example, the forces exerted by the cables of a suspension bridge must be known so that the cables can be designed to be strong enough to support the bridge. Similarly, cranes must be designed to not topple over when lifting a weight.

In this chapter, we will first examine the conditions necessary for static equilibrium to exist. We will then define the center of gravity and look at some examples of how to calculate the forces necessary for a body to be in static equilibrium. Finally, we will discuss the stability of equilibrium.

9-1 Conditions for Equilibrium

In Chapter 4, we found that a necessary condition for a particle to remain stationary is that the net force acting on the particle be zero. Under this condition, the particle does not accelerate, and if its velocity is initially zero, the particle remains at rest. Since the acceleration of the center of mass of an object equals the net force acting on the object divided by the total mass of the object, this condition is also necessary for a rigid body to be in equilibrium. However, even when the center of mass of an object is at rest, the object may rotate. Thus, it is also necessary that the net torque about the center of mass be zero. If the center of mass of an object is at rest and there is no rotation about it, there can be no rotation about any point. Thus, for static equilibrium to exist, the net torque acting on an object must be zero about any point. This condition is often useful in solving problems because it allows us to choose any convenient point for computing torques.

Conditions for equilibrium

The two necessary conditions for a rigid body to be in static equilibrium are therefore

1. The net external force acting on the body must be zero:

$$\mathbf{F}_{\text{net}} = 0 \quad \text{9-1}$$

2. The net external torque about any point must be zero:

$$\tau_{\text{net}} = 0 \quad \text{9-2}$$

As we have seen, we can describe the vector nature of rotation about a fixed axis as being positive or negative. We choose either clockwise rotation or counterclockwise rotation to be positive and the other to be negative. An alternative statement of our second condition is then

For static equilibrium to exist, the sum of the torques that tend to produce clockwise rotation about any point must equal the sum of the torques that tend to produce counterclockwise rotation about that point.

We will refer to a torque that tends to produce clockwise rotation as a clockwise torque and one that tends to produce counterclockwise rotation as a counterclockwise torque.

Two massive, counter-weighted cranes lift a refinery tower to its vertical position.

This bridge in Oregon uses semicircular steel arches for support on the left and a cantilevered span at the right.

Example 9-1

A 3-m board of negligible weight rests with its ends on scales as shown in Figure 9-1*a*. A small, 60-N weight rests on the board 2.5 m from the left end and 0.5 m from the right end. Find the readings on the scales.

Figure 9-1*b* shows the free-body diagram for the board. The force $\mathbf{F}_L$ is that exerted by the scale at the left end of the board. Since the board exerts an equal but opposite force on the scale, the magnitude of $\mathbf{F}_L$ is the reading on the left scale. Similarly, the magnitude of $\mathbf{F}_R$ is the reading on the right scale. Taking up to be positive, we have from our first condition for equilibrium (the net force must be zero)

$$F_L + F_R - 60\text{ N} = 0$$

We can get a second relation between F_L and F_R by considering torques. If we consider the point at the weight to be the "pivot" point, we have two torques, a clockwise torque of magnitude $F_L(2.5\text{ m})$ and a counterclockwise torque of magnitude $F_R(0.5\text{ m})$. Choosing counterclockwise to be positive, we have

$$0.5F_R - 2.5F_L = 0$$

or

$$F_R = 5F_L$$

Substituting this result into the previous equation, we obtain

$$F_L + 5F_L = 60\text{ N}$$

or

$$F_L = 10\text{ N}$$

and

$$F_R = 60\text{ N} - F_L = 50\text{ N}$$

The scale readings are thus 10 N for the left scale and 50 N for the right one. The scale on the right supports the greater weight, as expected.

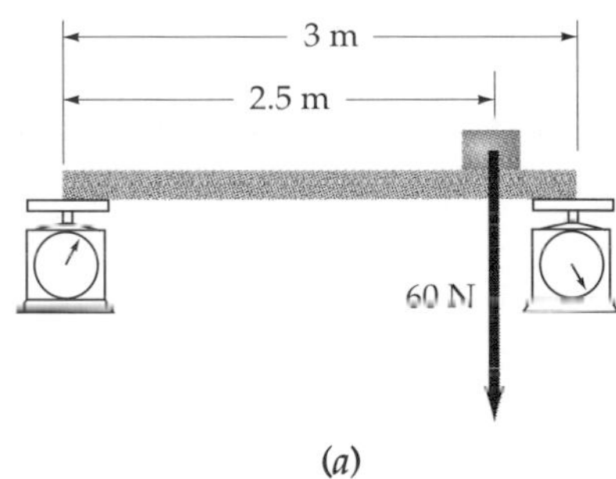

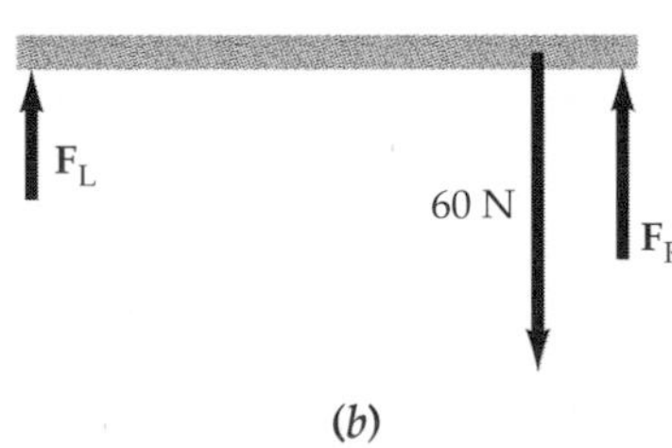

Figure 9-1 (*a*) Board of negligible weight on scales for Example 9-1. (*b*) Free-body diagram for the board.

This nearly completed globe sports complex in Stockholm achieves stability and support from its spherical shape.

Although there is nothing incorrect in this solution, there is an easier way to solve the problem without having to solve two equations for two unknowns. If we compute the torques about a point on the line of action of one of the unknown forces, that force will not enter into the equation because its lever arm will be zero. For instance, let us first consider the torques about the left scale. The weight produces a clockwise torque of magnitude (60 N)(2.5 m) = 150 N·m, and $\mathbf{F}_R$ produces a counterclockwise torque F_R(3 m). Setting the net torque equal to zero, we get

$$F_R(3\text{ m}) - 150\text{ N}\cdot\text{m} = 0$$

or

$$F_R = 50\text{ N}$$

We can then find F_L immediately from $F_L = 60\text{ N} - 50\text{ N} = 10\text{ N}$. Alternatively, we can consider the torques about the right scale. We then have $(60\text{ N})(0.5\text{ m}) - F_L(3\text{ m}) = 0$ or $F_L = 10\text{ N}$. Whenever there are two ways of solving a problem, it is a good idea to use one method to check the results of the other.

In general, if we have several unknown forces, we can reduce the work involved in solving the problem by computing the torques about a point on the line of action of one of the unknown forces so that that force does not enter into the equation.

Question

1. If the net torque about some point is zero, must it be zero about any other point? Explain.

9-2 The Center of Gravity

When two or more parallel forces act on an object, they can be replaced by a single equivalent force that is equal to the sum of the forces and is applied at a point such that the torque produced by the single equivalent force equals

Example 9-4

A sign of mass 20 kg hangs from the end of a rod of length 2 m and mass 4 kg (Figure 9-6*a*). A wire is attached to the end of the rod and to a point 1 m above point *O*. Find the tension in the wire and the force exerted by the wall on the rod at point *O*.

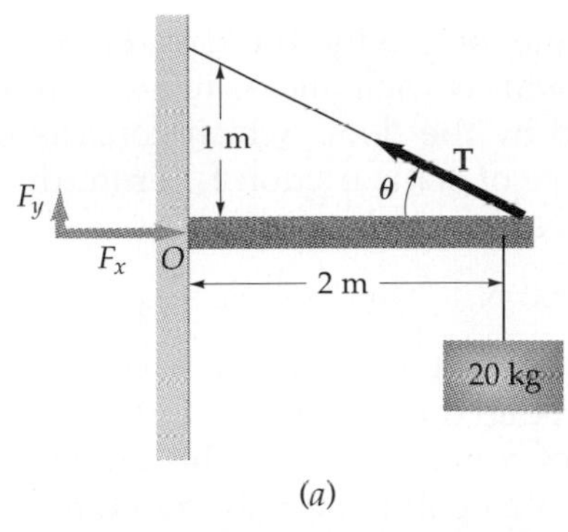

(*a*)

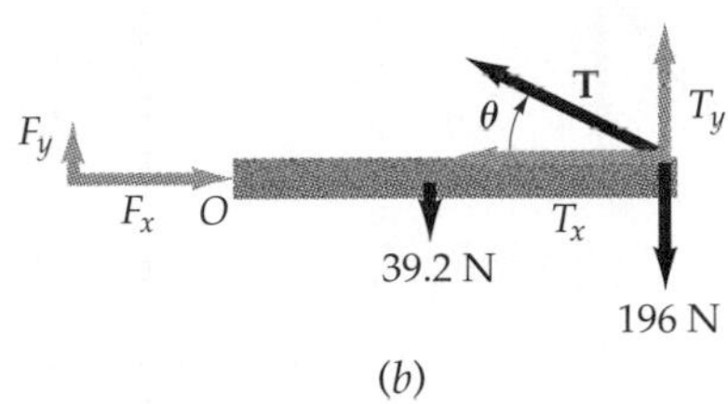

(*b*)

Figure 9-6 (*a*) A sign hanging from a rod for Example 9-4. (*b*) Free-body diagram showing the forces acting on the rod.

Figure 9-6*b* shows the forces acting on the rod. Since $g = 9.81$ N/kg, the weight of the sign is 196 N and that of the rod is 39.2 N. The force exerted by the wall has the components F_x and F_y. The tension **T** has been resolved into its *x* and *y* components. Since we do not know the force exerted by the wall, we choose point *O* for computing torques. The weight of the sign and that of the rod produce clockwise torques about *O*, and the *y* component of the tension in the wire produces a counterclockwise torque about *O*. Setting these torques equal, we have

$$T_y(2\text{ m}) = (196\text{ N})(2\text{ m}) + (39.2\text{ N})(1\text{ m}) = 431\text{ N}\cdot\text{m}$$

$$T_y = 215.5\text{ N}$$

The *x* component of the tension can be related to T_y and the angle θ. From Figure 9-6*b*, we have

$$\frac{T_y}{-T_x} = \tan\theta$$

And, from Figure 9-6*a*, we have

$$\tan\theta = \frac{1\text{ m}}{2\text{ m}} = \frac{1}{2}$$

Thus,

$$T_x = -2T_y = -2(215.5\text{ N}) = -431\text{ N}$$

(T_x is negative because it points to the left.) The magnitude of the tension is

$$T = \sqrt{T_x^2 + T_y^2} = \sqrt{(-431\text{ N})^2 + (215.5\text{ N})^2} = 482\text{ N}$$

The force exerted by the wall on the rod at *O* is found from the first condition for equilibrium. That is, the horizontal component F_x must equal 431 N to balance the horizontal component of the tension. Setting the upward forces equal to the downward forces, we have

$$F_y + T_y = 196\text{ N} + 39.2\text{ N} = 235\text{ N}$$

$$F_y = 235\text{ N} - T_y = 235\text{ N} - 215.5\text{ N} = 19.5\text{ N}$$

This mobile by Alexander Calder hangs suspended over the courtyard of the East Wing of the National Gallery in Washington, D.C.

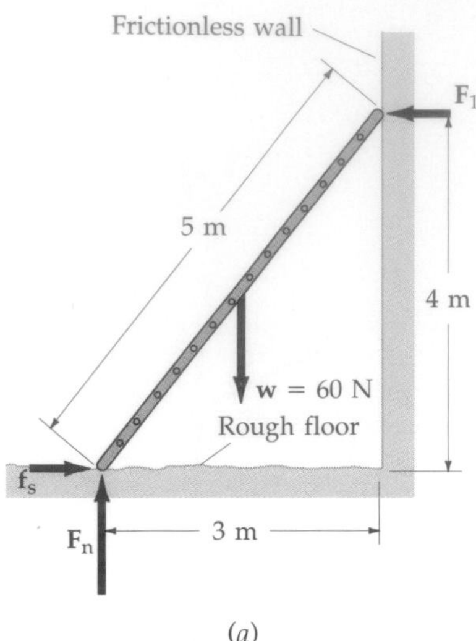

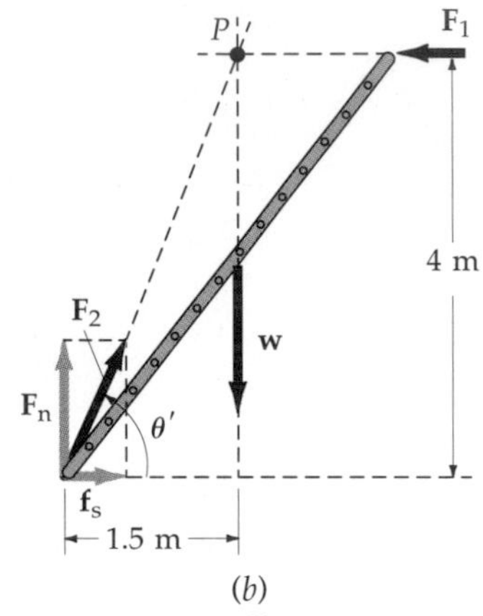

Figure 9-7 (*a*) A ladder on a rough floor leaning against a frictionless wall for Example 9-5. (*b*) Free-body diagram for the ladder. Since the ladder is in static equilibrium under the influence of three nonparallel forces, $\mathbf{F}_2$, $\mathbf{w}$, and $\mathbf{F}_1$, the lines of action of the three forces must all intersect at a common point P.

Example 9-5

A uniform, 5-m ladder weighing 60 N leans against a frictionless vertical wall (Figure 9-7*a*). The foot of the ladder is 3 m from the wall. What is the minimum coefficient of static friction necessary between the ladder and the floor if the ladder is not to slip?

The forces acting on the ladder are the force due to gravity $\mathbf{w}$ acting downward at the ladder's center of gravity, the force $\mathbf{F}_1$ exerted horizontally by the wall (since the wall is frictionless, it exerts only a normal force), and the force exerted by the floor, which consists of a normal force $\mathbf{F}_n$ and a horizontal force of static friction $\mathbf{f}_s$. From the first condition for static equilibrium, we have

$$F_n = w = 60 \text{ N} \qquad \text{and} \qquad F_1 = f_s$$

Since we know neither f_s nor F_1, we must use the second condition for static equilibrium and compute the torques about some convenient point. We choose the point of contact between the ladder and the floor because both $\mathbf{F}_n$ and $\mathbf{f}_s$ act at this point and will therefore not appear in our torque equation. The torque exerted by the force of gravity about this point is clockwise, with a magnitude of 60 N times the lever arm 1.5 m. The torque exerted by $\mathbf{F}_1$ about this point is counterclockwise, with a magnitude of F_1 times the lever arm 4 m. The second condition for static equilibrium thus gives

$$F_1(4 \text{ m}) - (60 \text{ N})(1.5 \text{ m}) = 0$$

$$F_1 = 22.5 \text{ N}$$

This equals the magnitude of the frictional force. Since the frictional force $\mathbf{f}_s$ is related to the normal force $\mathbf{F}_n$ by

$$f_s \leq \mu_s F_n$$

we have

$$\mu_s \geq \frac{f_s}{F_n} = \frac{22.5 \text{ N}}{60 \text{ N}} = 0.375$$

where μ_s is the coefficient of static friction.

There is another way to solve the problem in Example 9-5. Let $\mathbf{F}_2 = \mathbf{f}_s + \mathbf{F}_n$ be the force exerted on the ladder by the floor. The ratio f_s/F_n is the cotangent of the angle θ' between the force $\mathbf{F}_2$ and the horizontal. We can find this angle as follows. If we extend the lines of action of the forces $\mathbf{w}$ and $\mathbf{F}_1$, they meet at point P in Figure 9-7*b*. The torques exerted by these two forces about this point are zero. Since $\mathbf{F}_2$ is the only other force acting, it must also exert zero torque about point P, so its line of action must also pass through this point. Note that the force $\mathbf{F}_2$ is *not* along the line of the ladder. The force exerted by the ladder on the floor, which equals $-\mathbf{F}_2$, is not generally along the line of the ladder. This differs from a tension force in a string, which is always along the string. From the figure, we can see that if the line of action of $\mathbf{F}_2$ passes through point P, the cotangent of the angle made by $\mathbf{F}_2$ with the horizontal is

$$\cot\theta' = \frac{1.5 \text{ m}}{4 \text{ m}} = 0.375 = \frac{f_s}{F_n} \leq \mu_s$$

We note that

> If an object is in static equilibrium under the influence of three nonparallel forces, the lines of action of the forces must intersect at one point.

Example 9-6

A ladder of length L and negligible weight leans against a frictionless vertical wall making an angle θ with the horizontal. The coefficient of friction between the ladder and the floor is μ_s. A man of weight W climbs up the ladder. Show that the maximum distance s he can climb before the ladder slips is given by $s = \mu_s L \tan \theta$.

Figure 9-8 shows the man on the ladder and the forces involved. When the man is on the ladder, the clockwise torque exerted by his weight about the support point at the floor is balanced by the counterclockwise torque due to the force $\mathbf{F}_1$ exerted by the wall. As he climbs the ladder, both these torques increase, so the force $\mathbf{F}_1$ must increase. But the horizontal force $\mathbf{F}_1$ is balanced by the force of friction $\mathbf{f}_s$, which is limited to the value $\mu_s F_n$. Thus, there is a maximum distance s up the ladder for which equilibrium can be maintained. In the figure, the lines of action of the weight and the force $\mathbf{F}_1$ have been extended so that they meet at point P. The line of action of the force exerted by the floor on the ladder $\mathbf{F}_2 = \mathbf{F}_n + \mathbf{f}_s$ must also pass through point P. In the figure, θ is the angle made by the ladder with the horizontal, and θ' is that made by $\mathbf{F}_2$ with the horizontal. Since the friction is limiting, we have

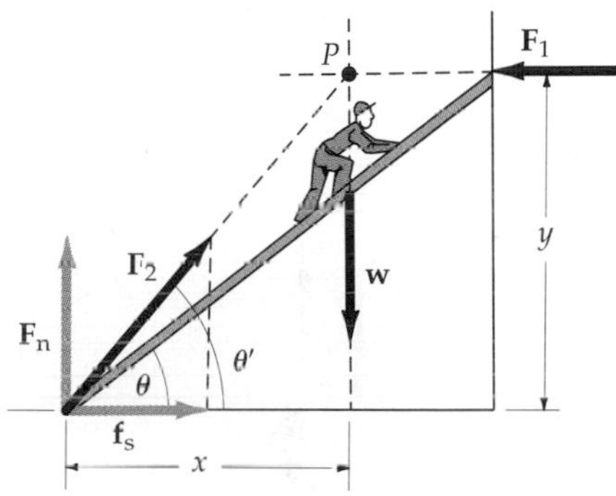

Figure 9-8 A man climbing on a ladder of negligible weight. The lines of action of the three forces acting on the ladder must intersect at a common point P.

$$f_s = \mu_s F_n$$

or

$$\mu_s = \frac{f_s}{F_n} = \cot \theta' \qquad 9\text{-}6$$

If y is the height of the top of the ladder above the floor and x is the horizontal distance from the point of contact between the floor and the ladder to the man, we have from the geometry of the figure

$$\tan \theta' = \frac{y}{x} \qquad 9\text{-}7$$

$$\sin \theta = \frac{y}{L} \qquad 9\text{-}8$$

and

$$\cos \theta = \frac{x}{s} \qquad 9\text{-}9$$

Combining Equations 9-6 and 9-7, we have

$$x = \frac{y}{\tan \theta'} = y \cot \theta' = \mu_s y$$

Using this result in Equation 9-9, we have for the distance s

$$s = \frac{x}{\cos \theta} = \frac{\mu_s y}{\cos \theta} = \frac{\mu_s L \sin \theta}{\cos \theta} = \mu_s L \tan \theta$$

where we have used $y = L \sin \theta$ from Equation 9-8.

Questions

4. Can a ladder be in equilibrium standing on a frictionless floor and leaning against a wall?
5. If an object is in equilibrium under three forces, must the forces all be in the same plane?

9-4 Couples

We have seen that we can replace a set of parallel forces with a single force, equal to the sum of the parallel forces, that acts at such a point so as to produce the same rotational torque about any point as the parallel forces. This idea was illustrated in Figure 9-2 for two parallel forces. Further, we used this idea when we replaced the forces of gravity acting on the various parts of an object with a single force, the weight of the object, acting at the center of gravity. However, two forces that are equal in magnitude but opposite in direction and have different lines of action cannot be replaced by a single force. Such a pair of forces, called a **couple,** tends to produce rotation, but their net force is zero. Consider the couple shown in Figure 9-9. The torque produced by these forces about point O is

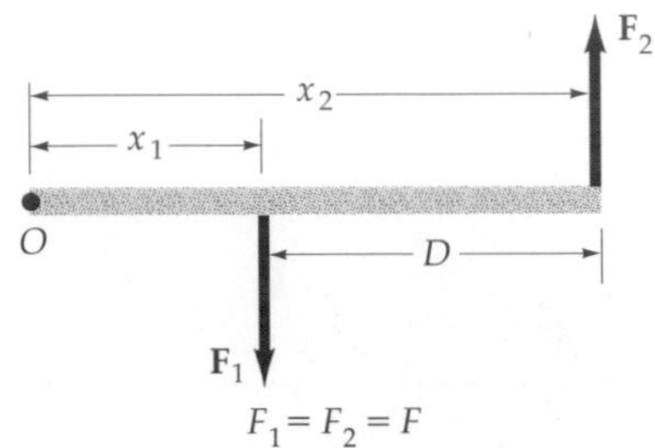

Figure 9-9 Two equal and opposite forces constitute a couple. The torque exerted by a couple has the same value FD about any point in space.

$$\tau = Fx_2 - Fx_1 = F(x_2 - x_1) = FD \qquad 9\text{-}10$$

where F is the magnitude of either force and $D = x_2 - x_1$ is the distance between them. This result does not depend on the choice of the point O.

> The torque produced by a couple is the same about all points in space.

Two unequal, antiparallel forces, such as those shown in Figure 9-10, can be replaced by a single force that is equal to the net force acting at the center of mass plus a couple whose torque equals the torque about the center of mass of the original forces. In general, any number of forces can be replaced by a single force equal to the net force and a couple. Since the net force exerted by a couple is zero, it can be balanced only by a second couple that exerts an equal but opposite torque. For example, the forces $\mathbf{F}_n$ and $\mathbf{w}$ in Figure 9-7*b* for Example 9-5 form a couple with a torque of (60 N)(1.5 m) = 90 N·m. It is balanced by the forces $\mathbf{f}_s$ and $\mathbf{F}_1$, which also form a couple. Since the separation of these forces is $D = 4$ m, their magnitude must be (90 N·m)/(4 m) = 22.5 N, as found in that example.

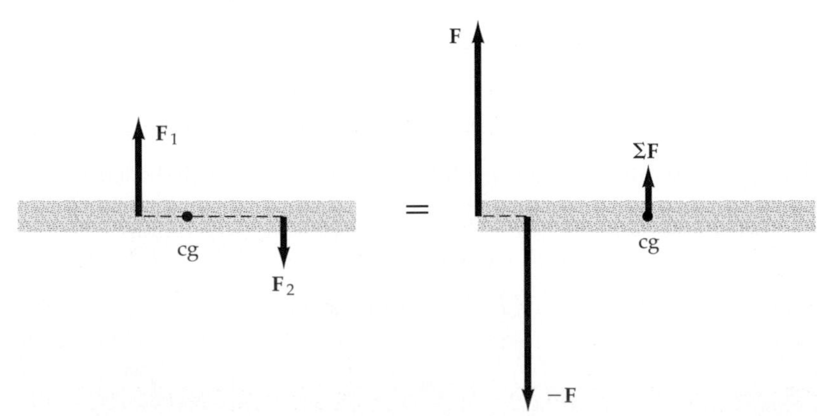

Figure 9-10 Two antiparallel forces of unequal magnitude can be replaced by a single force $\Sigma\mathbf{F}$ acting at the center of gravity plus a couple whose torque is equal to the torque exerted by the original forces about the center of gravity.

9-5 Stability of Equilibrium

The equilibrium of an object can be classified according to three categories: stable, unstable, or neutral. **Stable equilibrium** occurs when the torques or forces that arise from a small displacement of the object urge the object back toward its equilibrium position. Stable equilibrium is illustrated in Figures 9-11 and 9-12. When the box in Figure 9-11 is rotated slightly about one end, the resulting torque about the pivot point tends to restore the box to its original position. Similarly, if the marble resting at the bottom of the hemispherical bowl in Figure 9-12 is displaced slightly, the tangential component of its weight urges it back toward its initial position.

Unstable equilibrium, illustrated in Figures 9-13 and 9-14, occurs when the forces or torques that arise from a small displacement of the object urge the object away from its equilibrium position. A slight rotation of the narrow stick in Figure 9-13 causes it to fall over because the torque due to its weight urges it away from its original position. Similarly, if the marble on the overturned bowl in Figure 9-14 is displaced slightly, the tangential component of the weight urges the marble away from its original position.

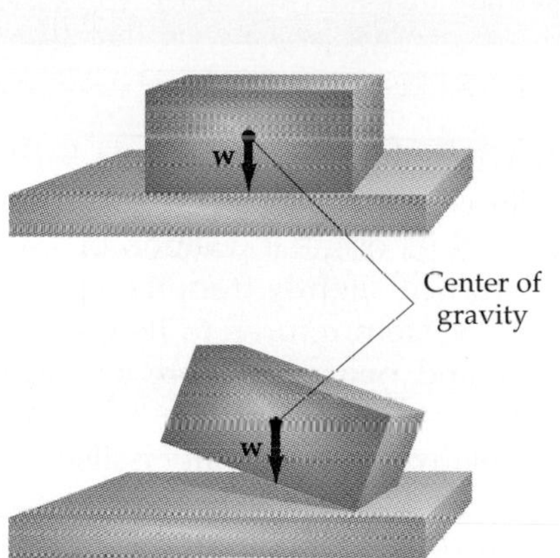

Figure 9-11 An example of stable equilibrium. When the box is rotated slightly, the torque exerted by the weight about the pivot point tends to restore the box to its original position.

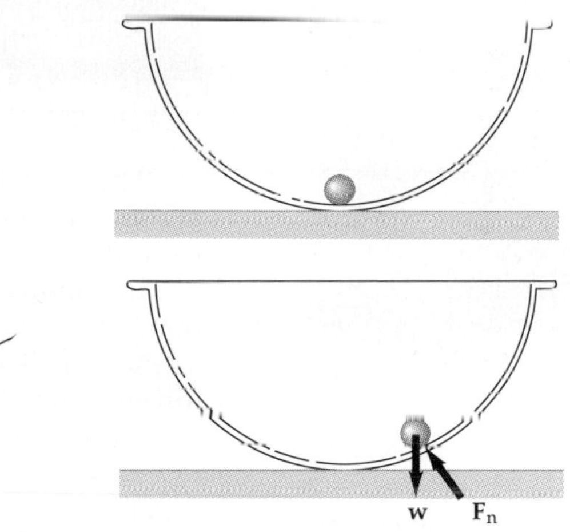

Figure 9-12 Another example of stable equilibrium. If the marble is displaced slightly in any direction, there is a net force that pushes the marble back toward its original position.

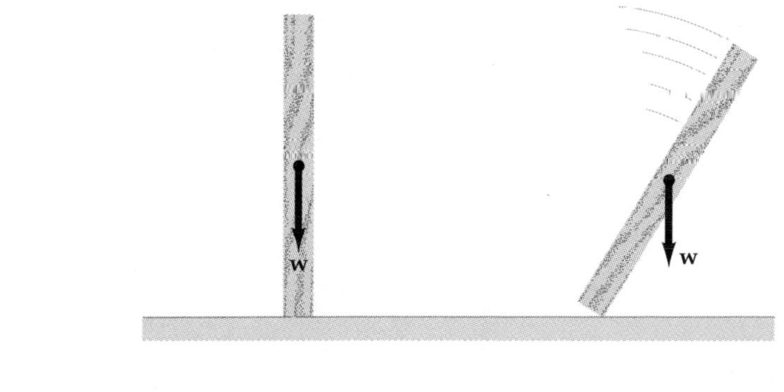

Figure 9-13 An example of unstable equilibrium. If the stick is rotated slightly, the torque exerted by the weight about the pivot point tends to rotate the stick away from its original position.

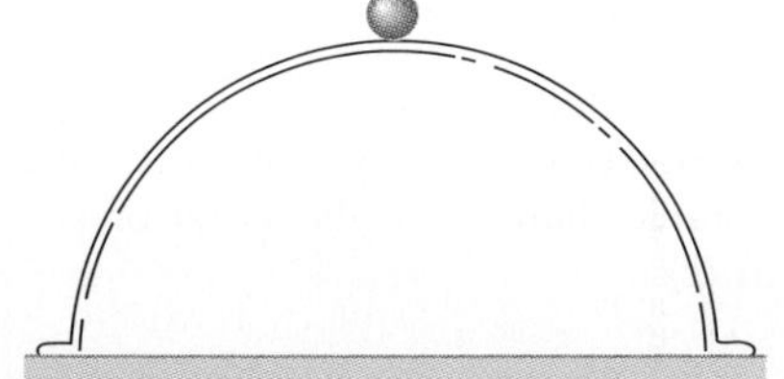

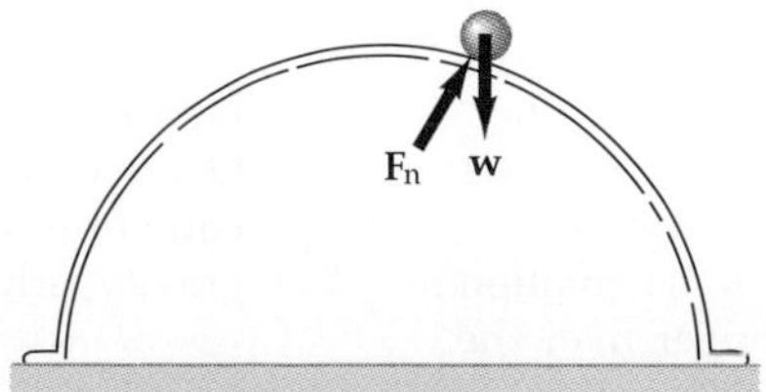

Figure 9-14 Another example of unstable equilibrium. If the marble is displaced slightly, there is net force that pushes the marble further from its original position.

The gymnast improves her balance by bringing her center of gravity as close to the base of support as possible.

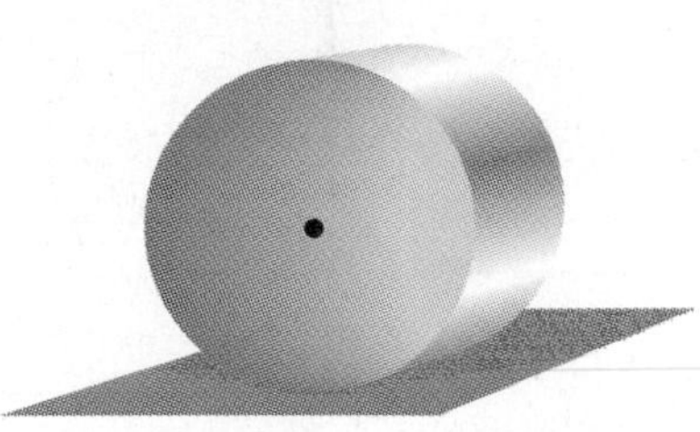

Figure 9-15 An example of neutral equilibrium. If the cylinder is rotated slightly, it is again in equilibrium. There are no torques or forces that tend to rotate the cylinder either toward or away from its original position.

The cylinder resting on a horizontal surface in Figure 9-15 illustrates **neutral equilibrium.** If the cylinder is rotated slightly, there is no torque or force that urges it either back toward its original position or away from it.

In summary, if a system is disturbed slightly from its equilibrium position, the equilibrium is stable if the system returns to its original position, unstable if it moves farther away, and neutral if there are no torques or forces urging it in either direction.

Since "disturbed slightly" is a relative term, stability is also relative. One example of equilibrium may be more or less stable than another. Figure 9-16*a* shows a stick that is not as narrow as that in Figure 9-13 balanced on end. Here, if the disturbance is very small (Figure 9-16*b*), the stick will move back toward its original position, but if the disturbance is great enough that the center of gravity no longer lies over the base of support (Figure 9-16*c*), the stick will fall over. We can improve the stability of a system by either lowering its center of gravity or widening its base of support. Figure 9-17 shows a nonuniform stick that has its center of gravity near one end. If the

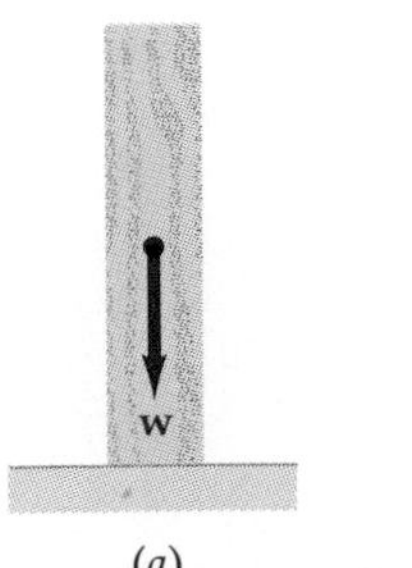

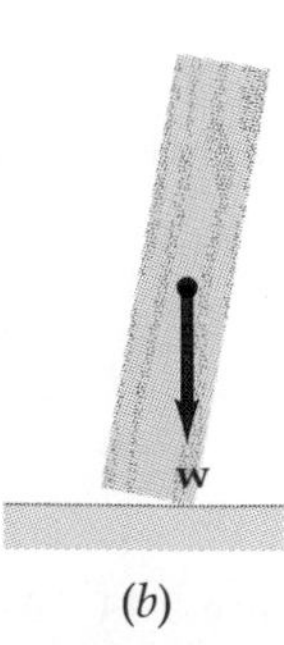

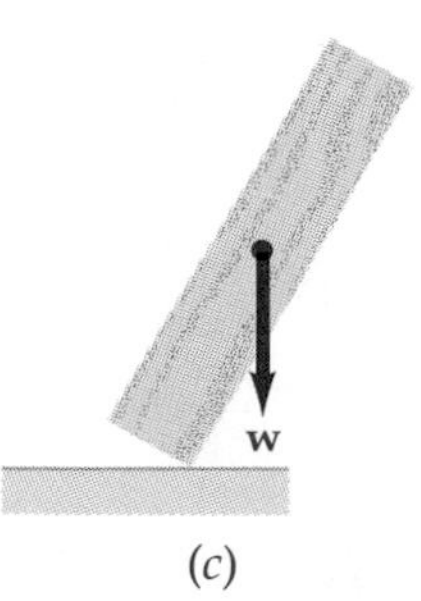

Figure 9-16 Stability of equilibrium is relative. If the stick in (*a*) is rotated slightly, as in (*b*), it returns to its original equilibrium position as long as the center of gravity lies over the base of support. (*c*) If the rotation is too great, the center of gravity is no longer over the base of support, and the stick falls over.

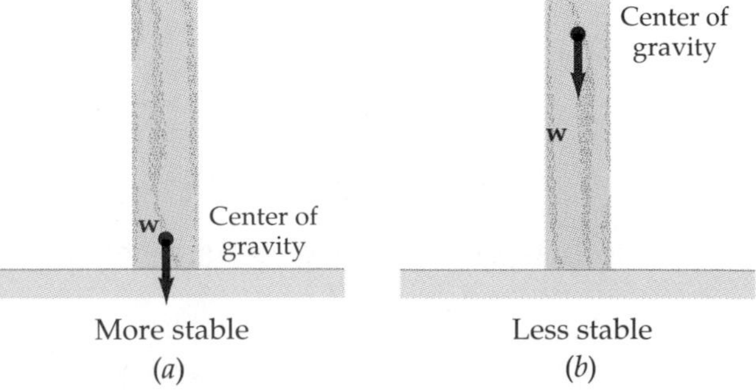

Figure 9-17 When a nonuniform stick rests on its heavy end with its center of gravity low, as in (*a*), the equilibrium is more stable than when its center of gravity is high, as in (*b*).

stick is stood on its heavy end so that its center of gravity is low (Figure 9-17*a*), it is much more stable than if it is stood on the other end so that its center of gravity is high (Figure 9-17*b*). In Figure 9-18, the center of gravity lies below the point of support of the system. This system is stable for any displacement because the resulting torque always rotates the system back toward its equilibrium position.

A human being standing or walking upright has difficulty maintaining balance because the center of gravity must be kept over the base of support, the feet, which is relatively small. An animal walking on four legs has a much easier time because its base of support is larger and its center of gravity is lower.

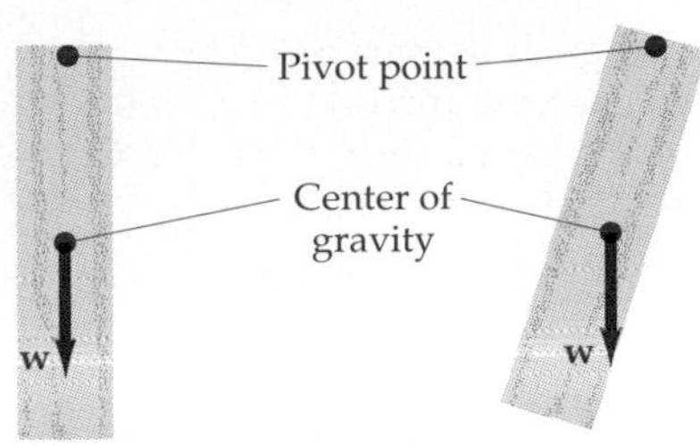

Figure 9-18 When a stick is pivoted such that its center of gravity is below the pivot point, the equilibrium is stable no matter how far the stick is displaced from equilibrium.

Questions

6. How should a hiker with a heavy backpack stand so as to maintain balance?
7. How do linemen in football improve their balance?
8. Why is it so much more difficult to stand on one foot than on two?

(*a*)

(*b*)

(*a*) The tightrope walker uses a long, heavy pole to help maintain balance. (*b*) In this fascinating demonstration of stable equilibrium, the center of gravity of the two-fork system lies under the point of support of the toothpick.

Summary

1. The two conditions that are necessary for a rigid body to be in static equilibrium are
   1. The net external force acting on the body must be zero:
   $$\mathbf{F}_{\text{net}} = 0$$
   2. The net external torque about any point must be zero:
   $$\tau_{\text{net}} = 0$$

   An alternative statement of the second condition is that the sum of the torques that tend to produce clockwise rotation about any point must equal the sum of the torques that tend to produce counterclockwise rotation about that point.

2. When two or more parallel forces act on an object, they can be replaced by a single equivalent force that is equal to the sum of the forces and is applied at a point such that the torque produced by the single equivalent force equals the net torque produced by the original forces.

3. The force of gravity exerted on the various parts of an object can be replaced by a single force, the total weight of the object W, acting at the center of gravity. The x coordinate of the center of gravity X_{cg} relative to some origin is given by

$$X_{cg}W = \sum_i w_i x_i$$

When the acceleration of gravity is the same at all points of an object, the center of gravity coincides with the center of mass.

4. If an object is in static equilibrium under the influence of three nonparallel forces, the lines of action of the forces must intersect at one point.

5. Two equal and opposite forces that have different lines of action constitute a couple. The torque produced by a couple is the same about all points in space. Its magnitude is the product of the magnitude of either force times the distance between their lines of action.

6. The equilibrium of an object can be classified according to three categories: stable, unstable, or neutral. An object resting on some surface will be in equilibrium if its center of gravity lies over its base of support. The stability of the equilibrium of an object can be improved by lowering its center of gravity or by increasing the size of its base of support.

Suggestions for Further Reading

Mark, Robert: "The Structural Analysis of Gothic Cathedrals," *Scientific American*, November 1972, p. 90.

This article discusses the use of the technique of optical stress analysis of scale models in an analysis of the static and wind-loaded behaviors of Gothic cathedrals.

Steinman, David B.: "Bridges," *Scientific American*, November 1954, p. 60.

This article presents a history of bridge building, an empirical discipline until the eighteenth century, when physical principles began to be studied and applied to it. The aerodynamics of the Tacoma Narrows Bridge and the reasons for its collapse are also considered.

Review

A. Objectives: After studying this chapter, you should:

1. Be able to state the conditions necessary for static equilibrium to exist and apply them to problems.

2. Be able to discuss stable, unstable, and neutral equilibrium.

B. Define, explain, or otherwise identify:

Center of gravity
Couple
Stable equilibrium
Unstable equilibrium
Neutral equilibrium

C. True or false: If the statement is true, explain why it is true. If it is false, give a counterexample.

1. $\Sigma \mathbf{F} = 0$ is sufficient for static equilibrium to exist.

2. $\Sigma \mathbf{F} = 0$ is necessary for static equilibrium to exist.

3. In static equilibrium, the net torque about any point is zero.

4. The center of gravity is always at the geometric center of a body.

5. An object is in equilibrium only when there are no forces acting on it.

Problems

Level I

9-1 Conditions for Equilibrium

1. A seesaw consists of a 4-m board pivoted at the center. A 28-kg child sits on one end of the board. Where should a 40-kg child sit to balance the seesaw?

2. In Figure 9-19, Misako is about to do a push-up. Her center of gravity lies directly above point P on the floor, which is 0.9 m from her feet and 0.6 m from her hands. If her mass is 54 kg, what is the force exerted by the floor on her hands?

Figure 9-19 Problem 2.

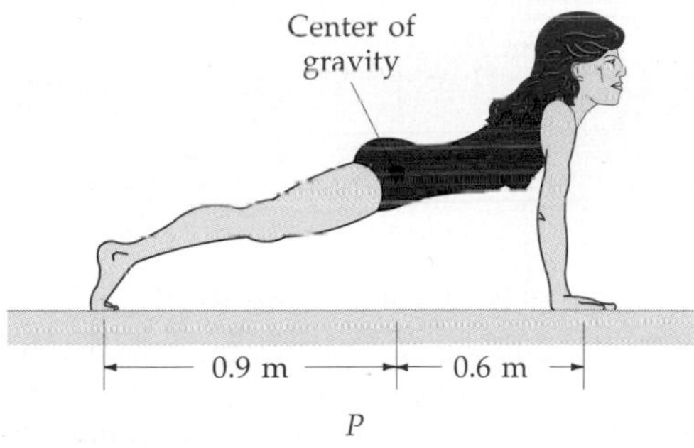

3. Juan and Betina are carrying a 60-kg weight on a 4-m board as shown in Figure 9-20. The mass of the board is 10 kg. Since Juan spends most of his time reading cookbooks, whereas Betina regularly does push-ups, they place the weight 2.5 m from Juan and 1.5 m from Betina. Find the force in newtons exerted by each to carry the weight.

Figure 9-20 Problem 3.

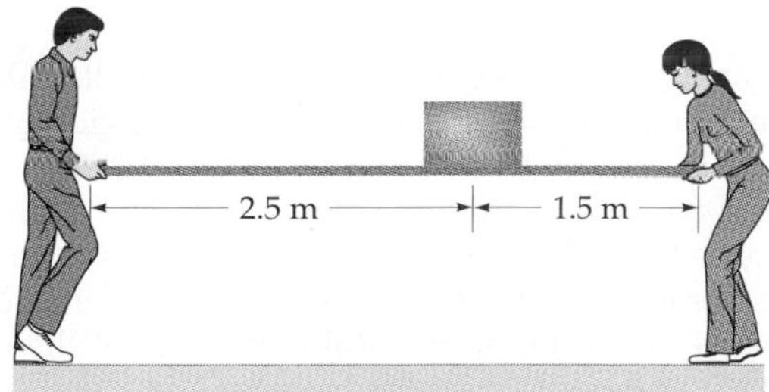

4. Misako wishes to measure the strength of her biceps muscle by exerting a force on a test strap as shown in Figure 9-21. The strap is 28 cm from the pivot point at the elbow, and her biceps muscle is attached at a point 5 cm from the pivot point. If the scale reads 18 N when she exerts her maximum force, what force is exerted by the biceps muscle?

Figure 9-21 Problem 4.

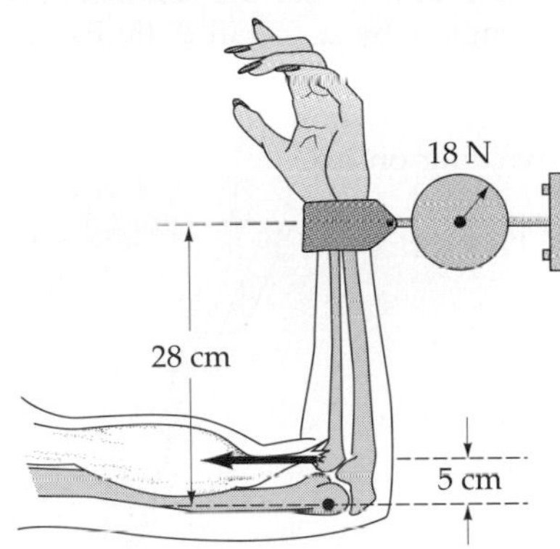

5. A 90-N board 12 m long rests on two supports, each 1 m from the end of the board. A 360-N block is placed on the board 3 m from one end as shown in Figure 9-22. Find the force exerted by each support on the board.

Figure 9-22 Problem 5.

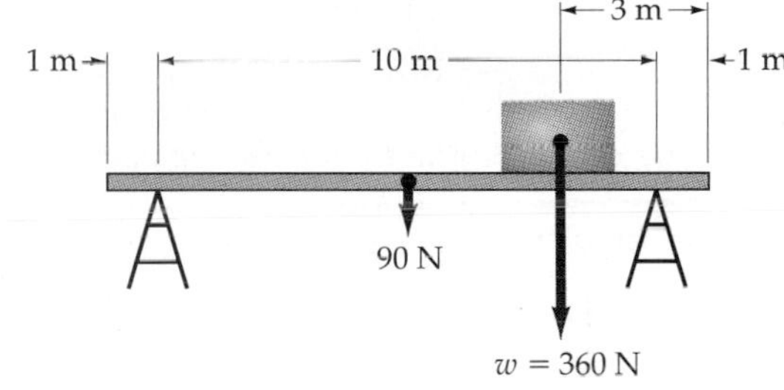

6. Figure 9-23 shows a mobile consisting of four weights hanging on three rods of negligible mass. Find the value of each of the unknown weights if the mobile is to balance. *Hint:* Find weight w_1 first.

Figure 9-23 Problem 6.

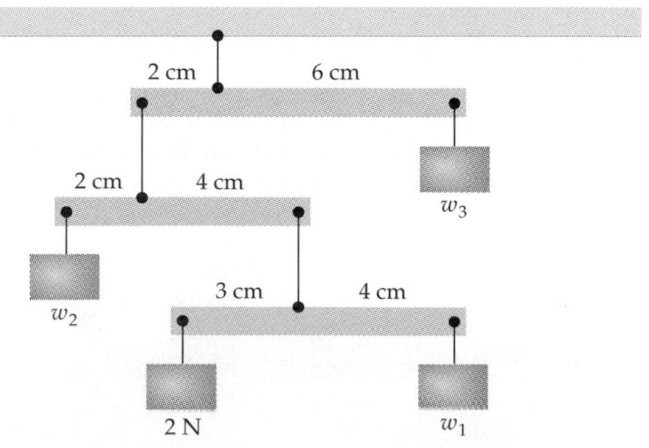

7. A crutch is pressed against the sidewalk with a force $\mathbf{F}_c$ along its own direction as shown in Figure 9-24. This force is balanced by a normal force $\mathbf{F}_n$ and a frictional force $\mathbf{f}_s$, as shown in the figure. (*a*) Show that when the force of friction is at its maximum value, the coefficient of friction is related to the angle θ by $\mu_s = \tan\theta$. (*b*) Explain how this result applies to the forces on your foot when you are not using a crutch. (*c*) Why is it advantageous to take short steps when walking on ice?

Figure 9-24 Problem 7.

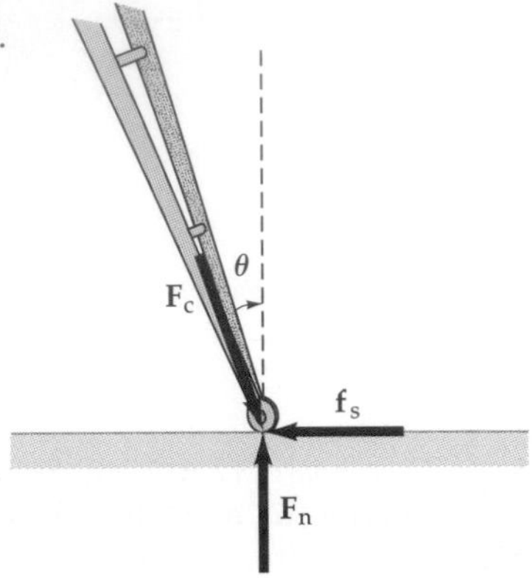

9-2 The Center of Gravity

8. Two spheres of radius R rest on a horizontal table with their centers a distance $4R$ apart. One sphere has twice the weight of the other sphere. Where is the center of gravity of this system?

9. Uniform wire is bent into the shapes shown in Figure 9-25. Find the center of gravity for each shape.

Figure 9-25 Problem 9.

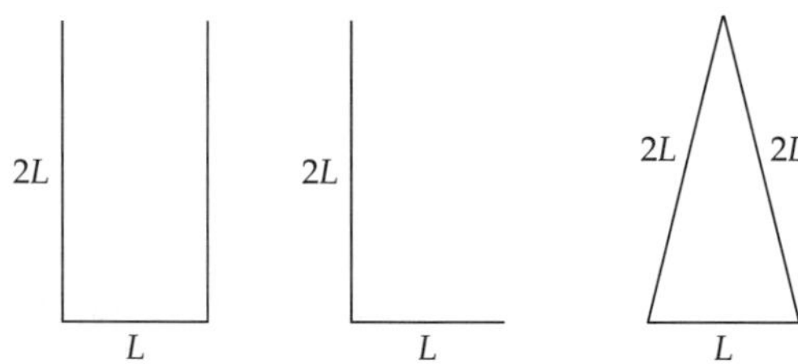

10. An automobile has 58 percent of its weight on the front wheels. The front and back wheels are separated by 8 m. Where is the center of gravity located with respect to the front wheels?

11. The height of the center of gravity of a man standing erect is determined by weighing the man as he lies on a board of negligible weight supported by two scales as shown in Figure 9-26. If the man's height is 188 cm and the left scale reads 445 N and the right scale reads 400 N, where is his center of gravity relative to his feet?

Figure 9-26 Problem 11.

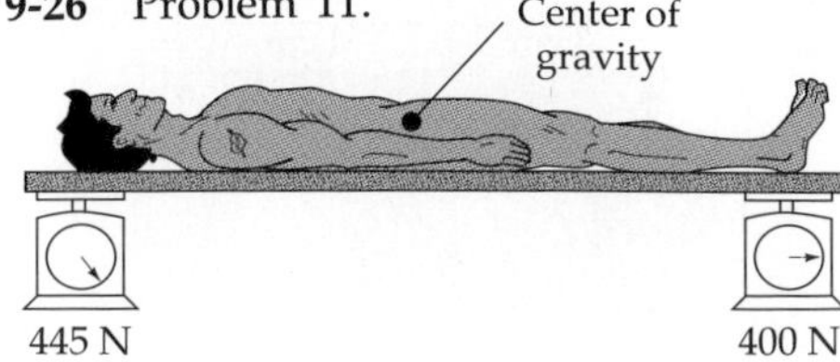

12. Each of the objects shown in Figure 9-27 is suspended from the ceiling by a thread attached to the point marked × on the object. Describe the orientation of each suspended object with a diagram.

Figure 9-27 Problem 12.

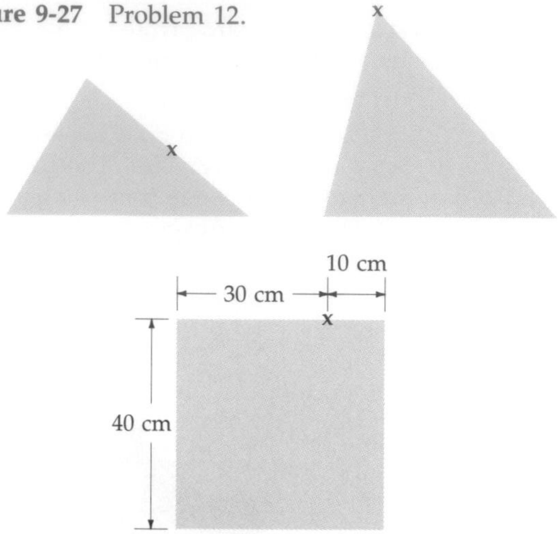

9-3 Some Examples of Static Equilibrium

13. A uniform 10-m log of mass 100 kg rests in a horizontal position on two supports. One support is 2 m from the left end of the log and the other is 4 m from the right end. Find the forces exerted on the log by the supports.

14. A 10-m beam of mass of 300 kg extends over a ledge as shown in Figure 9-28. The beam is not held down but simply rests on the surface. A 60-kg student has positioned the beam so that he can walk to the end of it. How far from the edge of the ledge can the beam extend?

Figure 9-28 Problem 14.

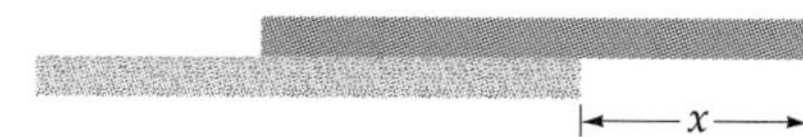

15. A man uses a crowbar that is 1 m long to lift a heavy crate off the ground. The crowbar has a rigid fulcrum 10 cm from one end as shown in Figure 9-29. (*a*) If the man exerts a downward force of 600 N on one end of the crowbar, what is the upward force exerted on the crate by the other end? (*b*) The ratio of the forces at the ends of the crowbar is called the mechanical advantage of the crowbar. What is the mechanical advantage?

Figure 9-29 Problem 15.

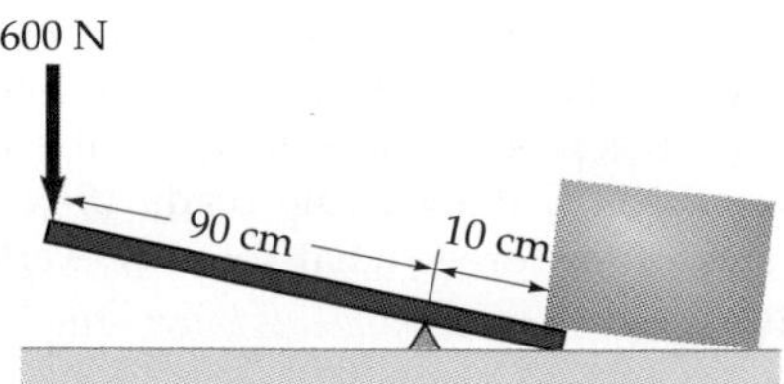

16. A gravity board for locating the center of gravity of a person consists of a horizontal board supported by a fulcrum at one end and by a scale at the other end. A physics student lies horizontally on the board with the top of his head above the fulcrum point as shown in Figure 9-30. The scale is 2 m from the fulcrum. The student has a mass of 70 kg, and when he is on the gravity board, the scale advances 250 N. Where is the center of gravity of the student?

Figure 9-30 Problem 16.

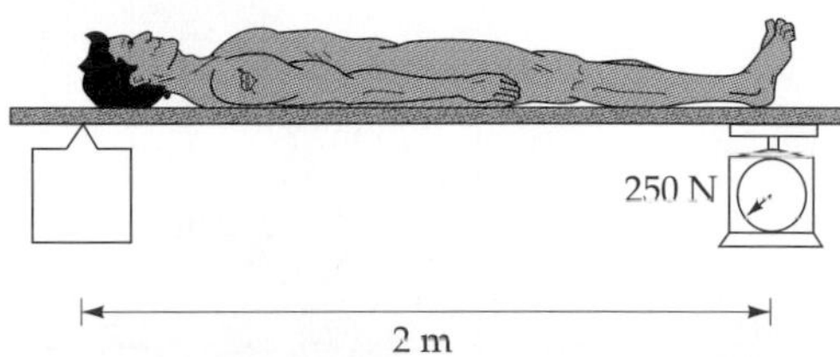

9-4 Couples

17. Two 80-N forces are applied to opposite corners of a rectangular plate as shown in Figure 9-31. Find the torque produced by this couple.

18. Resolve each force in Problem 17 into its horizontal and vertical components, producing two couples. The algebraic sum of the two component couples equals the resultant couple. Use this result to find the perpendicular distance between the lines of action of the two forces.

Figure 9-31 Problems 17 and 18.

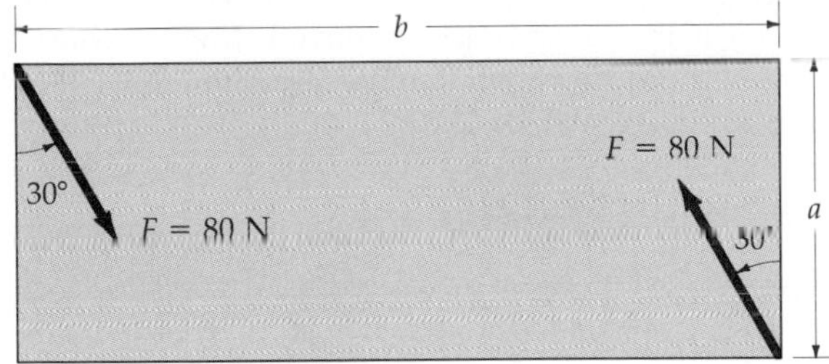

19. A uniform cube of side a and mass M rests on a horizontal surface. A horizontal force **F** is applied to the top of the cube as shown in Figure 9-32. This force is not sufficient to move or tip the cube. (*a*) Show that the force of static friction exerted by the surface and the applied force constitute a couple, and find the torque exerted by the couple. (*b*) This couple is balanced by the couple consisting of the normal force exerted by the surface and the weight of the cube. Use this fact to find the effective point of application of the normal force when $F = Mg/3$. (*c*) What is the greatest magnitude of **F** for which the cube will not tip?

Figure 9-32 Problem 19.

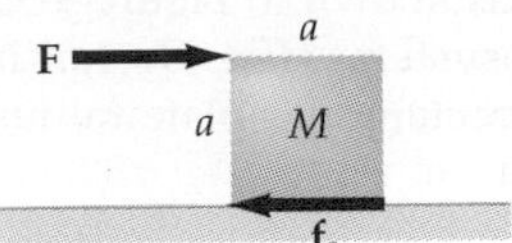

9-5 Stability of Equilibrium

There are no problems for this section.

Level II

20. A 3-m board of mass 5 kg is hinged at one end. A force **F** is applied vertically at the other end to lift a 60-kg box, which rests on the board 80 cm from the hinge as shown in Figure 9-33. (*a*) Find the magnitude of the force needed to hold the board stationary at $\theta = 30°$. (*b*) Find the force exerted by the hinge at this angle. (*c*) Find the force **F** and the force exerted by the hinge if **F** is exerted perpendicular to the board when $\theta = 30°$.

Figure 9-33 Problem 20.

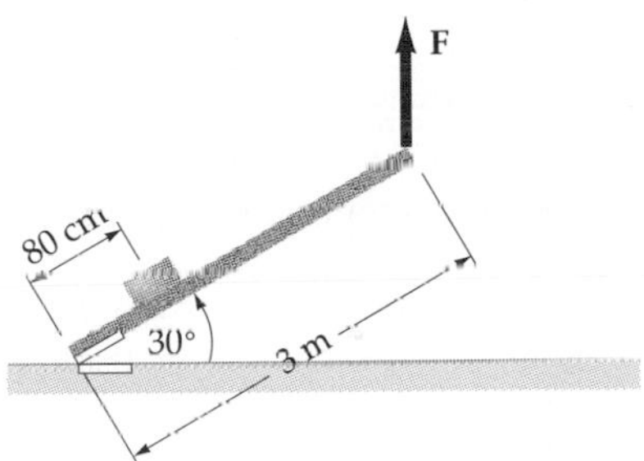

21. A square plate is produced by welding together four smaller square plates, each of side a as shown in Figure 9-34. Plate 1 weighs 40 N; plate 2, 60 N; plate 3, 30 N; and plate 4, 50 N. Find the center of gravity (x_{cg}, y_{cg}).

Figure 9-34 Problem 21.

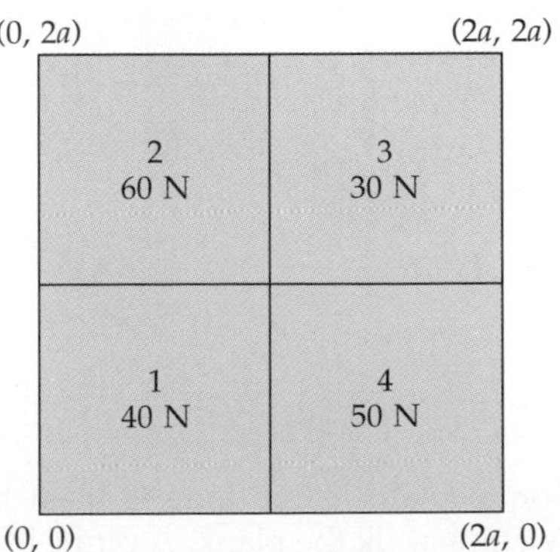

22. A uniform rectangular plate has a circular section of radius R cut out as shown in Figure 9-35. Find the center of gravity of the system. *Hint:* Do not integrate. Use superposition of a rectangular plate minus a circular plate.

Figure 9-35 Problem 22.

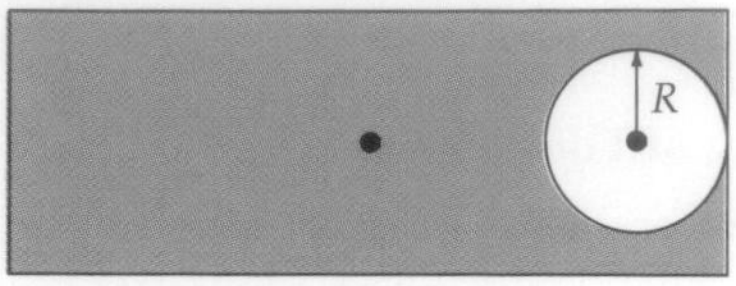

23. A cylinder of weight **W** is supported by a frictionless trough formed by a plane inclined at 30° to the horizontal on the left and one inclined at 60° on the right as shown in Figure 9-36. Find the force exerted by each plane on the cylinder.

Figure 9-36 Problem 23.

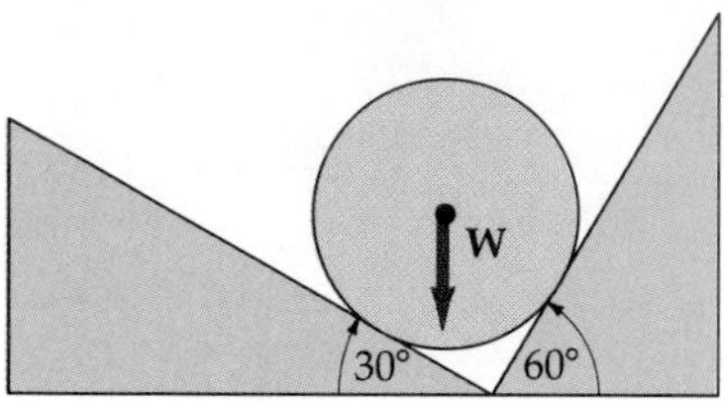

24. An 80-N weight is supported by a cable attached to a strut hinged at point A (Figure 9-37). The strut is supported by a second cable under tension $\mathbf{T}_2$. The mass of the strut is negligible. (*a*) What are the three forces acting on the strut? (*b*) Show that the vertical component of the tension $\mathbf{T}_2$ must equal 80 N. (*c*) Find the force exerted on the strut by the hinge.

25. Find the force exerted on the strut by the hinge at A for the arrangement in Figure 9-38 if (*a*) the strut is weightless and (*b*) the strut weighs 20 N.

Figure 9-37 Problem 24. **Figure 9-38** Problem 25.

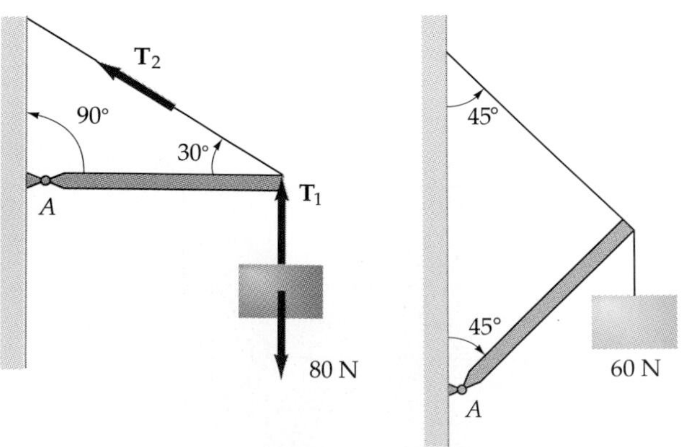

26. A horizontal plank 8.0 m long is used by pirates to make their victims walk the plank. A pirate of mass 105 kg stands on the shipboard end of the plank to prevent it from tipping. Find the maximum distance the plank can overhang for a 63-kg victim to be able to walk to the end if (*a*) the mass of the plank is negligible and (*b*) the mass of the plank is 25 kg.

27. A box that is 2 by 1 by 1 m of uniform mass is placed on end on a hinged rough plank as shown in Figure 9-39. The plank is inclined at an angle θ, which is slowly increased. The coefficient of friction is large enough to prevent the box from sliding before it tips over. Find the greatest angle that can be attained without tipping the box over.

Figure 9-39 Problem 27.

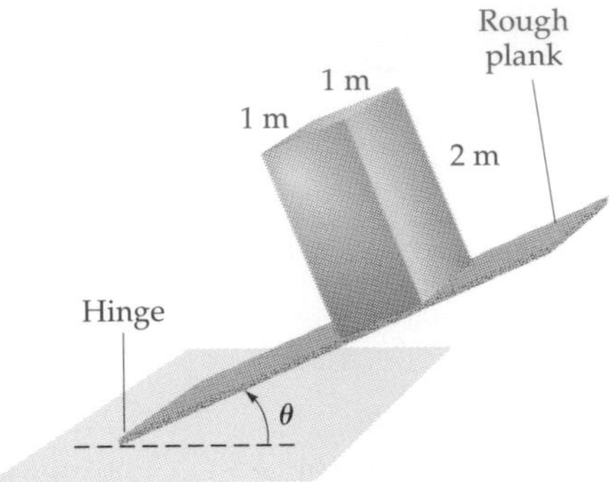

28. A uniform 18-kg door, 2.0 m high by 0.8 m wide, is hung from two hinges that are 20 cm from the top and 20 cm from the bottom. If each hinge supports half the weight of the door, find the magnitude and direction of the horizontal components of the forces exerted by the two hinges on the door.

29. A 900-N boy sits on top of a ladder of negligible weight that rests on a frictionless floor (Figure 9-40). There is a cross brace halfway up the ladder. The angle at the apex is $\theta = 30°$. (*a*) What is the force exerted by the floor on each leg of the ladder? (*b*) Find the tension in the cross brace. (*c*) If the cross brace is moved down toward the bottom of the ladder (maintaining the same angle θ), will its tension be greater or less?

Figure 9-40 Problem 29.

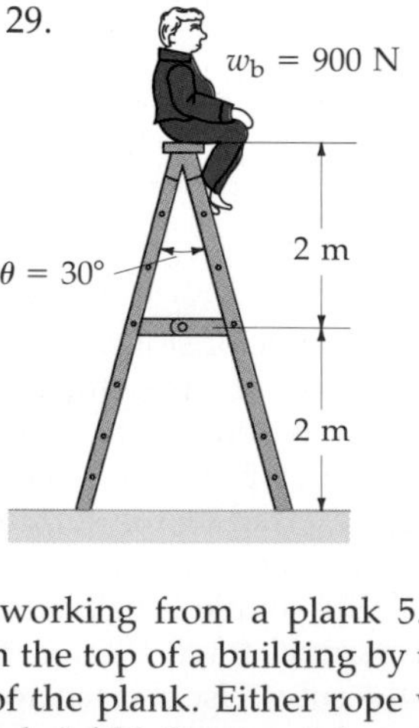

$w_b = 900$ N

$\theta = 30°$

2 m

2 m

30. Two painters are working from a plank 5.0 m long that is suspended from the top of a building by two ropes attached to the ends of the plank. Either rope will break when its tension exceeds 1 kN. Painter A (mass 80 kg) is working at a distance of 1.0 m from one end. Find the range of positions available to painter B if her mass is 60 kg and the plank has mass of 20 kg.

31. A wheel of mass M and radius R rests on a horizontal surface against a step of height h ($h < R$). The wheel is to be raised over the step by a horizontal force **F** applied to the axle of the wheel (Figure 9-41). Find the force **F** necessary to raise the wheel over the step.

Figure 9-41 Problem 31.

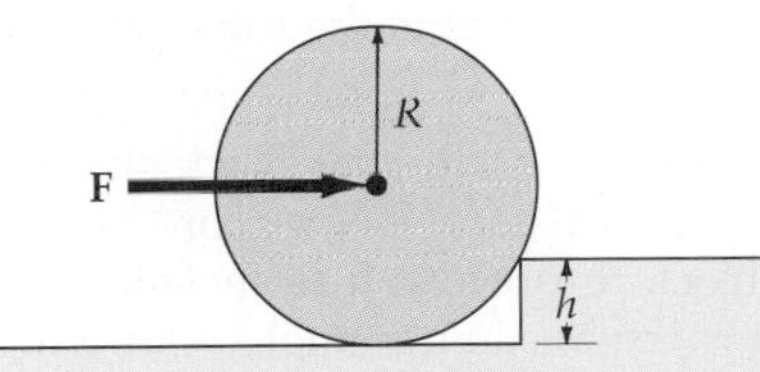

32. One end of a uniform, 100-kg beam 10 m long is hinged at a vertical wall. It is held horizontal by a cable that connects to the beam 6 m from the wall as shown in Figure 9-42. A 400-kg weight is suspended from the free end of the beam. (*a*) What is the tension in the cable? (*b*) What is the horizontal force on the hinge? (*c*) What is the vertical force on the beam at the hinge?

Figure 9-42 Problems 32 and 33.

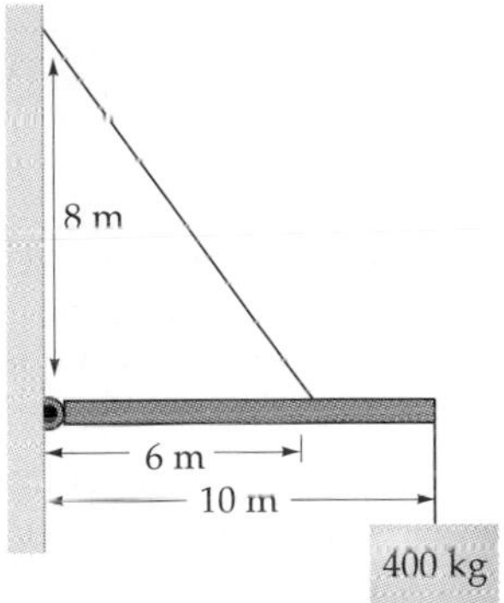

33. The cable in Figure 9-42 must remain attached to the wall 8 m above the hinge, but its length can vary so that it can be connected to the beam at various distances x from the wall. How far from the wall should it be attached so that the force on the hinge has no vertical component?

34. A tall, uniform, rectangular block sits on an inclined plane as shown in Figure 9-43. If $\mu_s = 0.4$, does the block slide or fall over as the angle θ is slowly increased?

Figure 9-43 Problem 34.

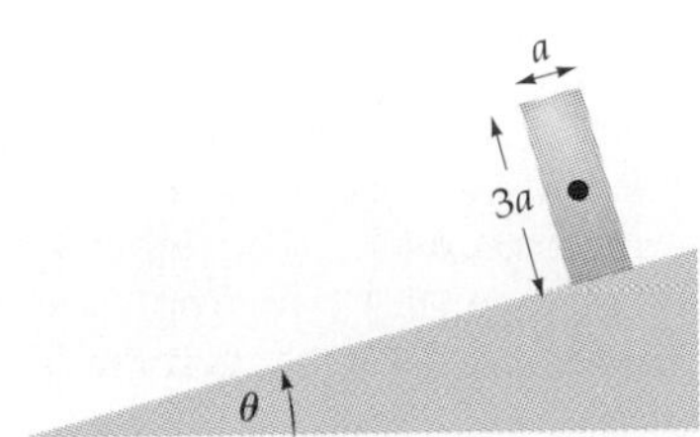

35. A cylinder of mass M and radius R rolls against a step of height h as shown in Figure 9-44. When a horizontal force **F** is applied to the top of cylinder, the cylinder remains at rest. (*a*) What is the normal force exerted by the floor on the cylinder? (*b*) What is the horizontal force exerted by the corner of the step on the cylinder? (*c*) What is the vertical component of the force exerted by the corner on the cylinder?

Figure 9-44 Problems 35 and 36.

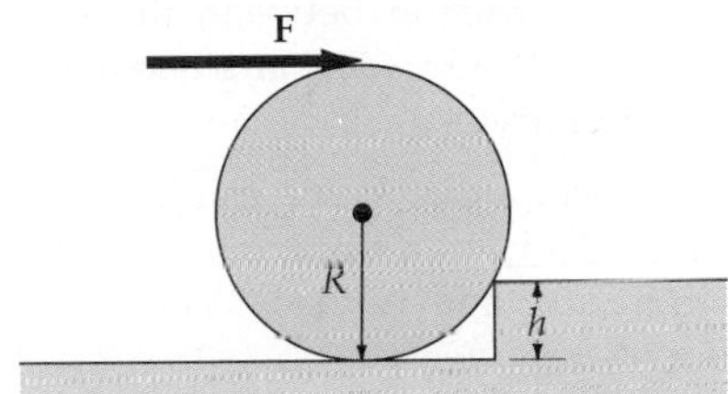

36. For the cylinder in Problem 35, find the minimum horizontal force **F** that will roll the cylinder over the step if the cylinder does not slide on the corner.

37. A strong man holds one end of a 3-m rod of mass 5 kg at rest in a horizontal position. (*a*) What total force does the man exert on the rod? (*b*) What total torque does the man exert on the rod? (*c*) If you approximate the effort of the man with two forces that act in opposite directions and are separated by the width of the man's hand, which is taken to be 10 cm, what are the magnitudes and directions of the two forces?

38. A large gate weighing 200 N is supported by hinges at the top and bottom and is further supported by a wire as shown in Figure 9-45. (*a*) What must the tension in the wire be for the force on the upper hinge to have no horizontal component? (*b*) What is the horizontal force on the lower hinge? (*c*) What are the vertical forces on the hinges?

Figure 9-45 Problem 38.

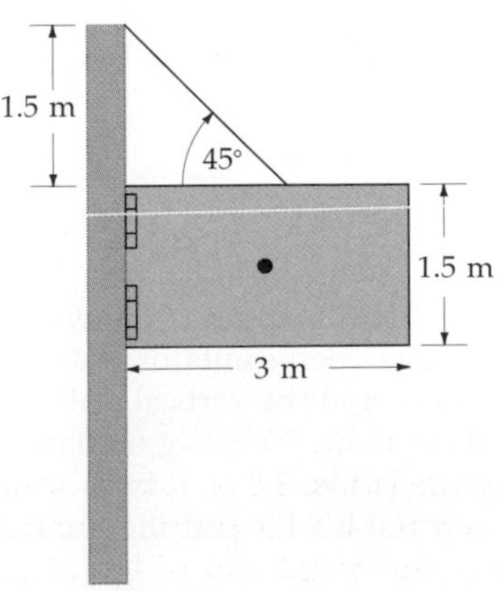

39. The diving board shown in Figure 9-46 has a mass of 30 kg. Find the force on the supports when a 70-kg diver stands at the end of the diving board. Give the direction of each support force as a tension or a compression.

Figure 9-46 Problem 39.

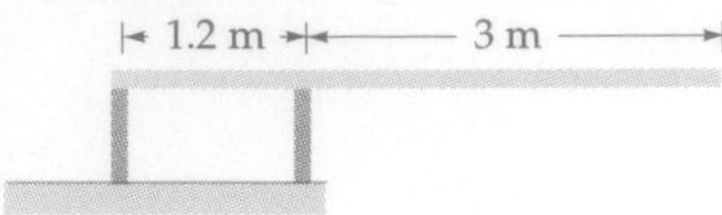

40. A ladder rests against a frictionless vertical wall. The coefficient of static friction between the ladder and the floor is 0.3. What is the smallest angle at which the ladder will remain stationary?

41. A uniform rectangular box 3 m high sits on its base, which is 1 m by 1 m, on a horizontal platform in a truck that is moving along a level road. What deceleration will cause the box to tip over?

Level III

42. A uniform log with a mass of 100 kg, a length of 4 m, and a radius of 12 cm is held in an inclined position as shown in Figure 9-47. The coefficient of static friction between the log and the horizontal surface is 0.6. The log is on the verge of slipping to the right. Find the tension in the support wire and the angle the wire makes with the vertical wall.

Figure 9-47 Problem 42.

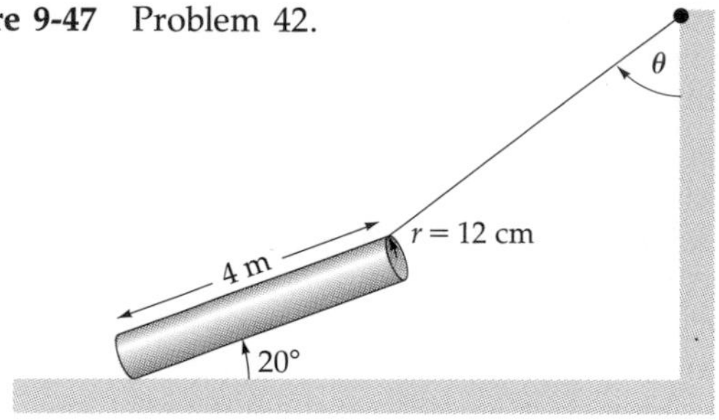

43. Find the center of gravity of a uniform wire that has been bent into a semicircle as shown in Figure 9-48. Use the coordinate system given in the figure.

Figure 9-48 Problem 43.

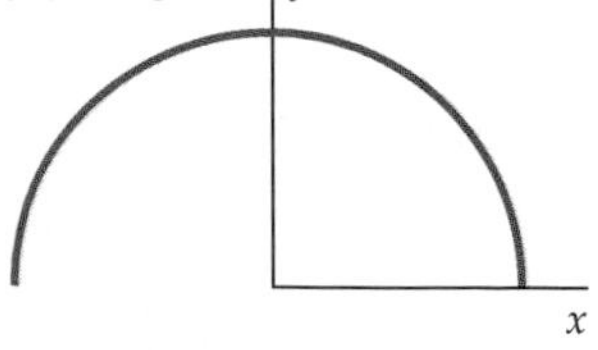

44. A uniform ladder of length L and weight 200 N leans against a wall. The coefficients of static friction are 0.4 between the ladder and the vertical wall and 0.7 between the ladder and the floor. An 80-kg firefighter is four-fifths of the way up the ladder when it begins to slip. Find the angle θ between the ladder and the ground.

45. The legs of a tripod make equal angles of 90° with each other at the apex, where they join together. A 100-kg block hangs from the apex. What are the compressional forces in the three legs?

46. Six identical bricks are stacked one on top of the other lengthwise to produce a leaning tower with the maximum offset that will still allow the tower to stand. (*a*) Starting from the top, give the maximum possible offset for each successive brick. (*b*) What is the total protrusion or offset of the six bricks?

47. A tall, uniform, rectangular block sits on an inclined plane as shown in Figure 9-49. A cord is attached to the top of the block to prevent it from falling down the incline. What is the maximum angle θ for which the block will not slide on the incline? Let b/a be 4 and $\mu_s = 0.8$.

Figure 9-49 Problem 47.

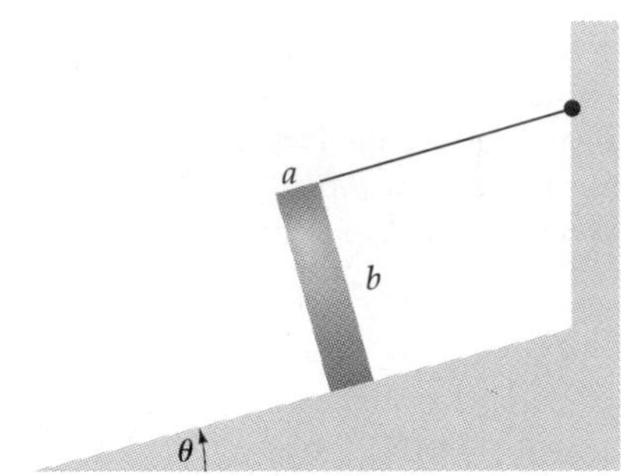

48. A thin rail of length 10 m and mass 20 kg is supported at a 30° incline. One support is 2 m and the other is 6 m from the lower end of the rail. Friction prevents the rail from sliding off the supports. Find the force (magnitude and direction) exerted on the rail by each support.

49. A ladder leans against a large smooth sphere of radius R that is fixed in place on a horizontal surface. The ladder makes an angle of 60° with the horizontal surface and has a length $5R/2$. (*a*) What is the force that the sphere exerts on the ladder? (*b*) What is the frictional force that prevents the ladder from slipping? (*c*) What is the normal force that the horizontal surface exerts on the ladder?

50. A uniform sphere of radius R and mass M is held at rest on an inclined plane of angle θ by a horizontal string as shown in Figure 9-50. Let $R = 20$ cm, $M = 3$ kg, and $\theta = 30°$. (*a*) Find the tension in the string. (*b*) What is the normal force exerted on the sphere by the inclined plane? (*c*) What is the frictional force acting on the sphere?

Figure 9-50 Problem 50.

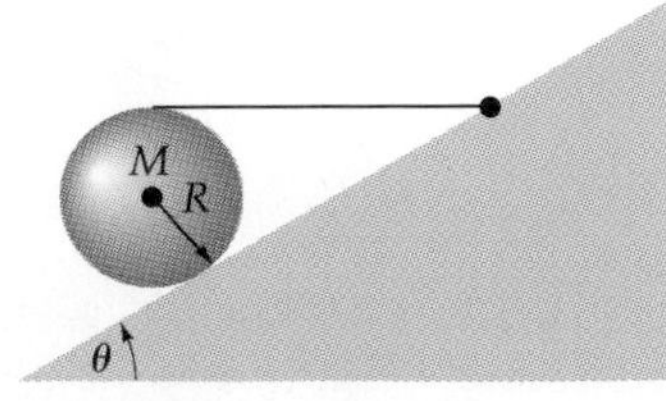

Chapter 10

Gravity

A mechanical model of the solar system, called an orrery, in the collection of Historical Scientific Instruments at Harvard University.

The gravitational force is the weakest of the four basic interactions occurring among elementary particles discussed in Chapter 4. It is negligibly small in the interactions of elementary particles. It is also difficult to observe the gravitational force between objects of everyday life even though their masses may be thousands of kilograms. Yet gravity is of primary importance when we consider interactions that involve very large objects such as planets, moons, and stars. It is gravity that binds us to the earth and keeps the earth and the other planets within the solar system. The gravitational force plays an important role in the evolution of stars and in the behavior of galaxies. In a sense, it is gravity that holds the universe together. In this chapter, we will study the force of gravity in some detail. We begin by stating Kepler's empirical laws of planetary motion and then discuss how these laws are related to Newton's law of gravity.

10-1 Kepler's Laws

The nighttime sky with its myriad stars and shining planets has always fascinated humankind. The apparent movements of the stars and planets relative to the earth have been observed and charted by astronomers for many centuries.

The apparent motions of the planets as seen from the earth. This photograph was taken in the Munich planetarium, where the motions of the planets over a period of seventeen years are being simulated.

A model of the universe, published by Ptolemy in about 140 A.D., had the earth at the center, with the sun and stars moving around the earth in simple circles and the planets moving around the earth in more complicated paths consisting of small circles, called epicycles, superimposed on larger circles (Figure 10-1). This somewhat complicated model was in basic agreement with the naked-eye observations of that time and was universally accepted for over 14 centuries. It was replaced in 1543 by the simpler but highly controversial model of Copernicus in which the sun and other stars were fixed and the planets, including the earth, revolved in circles around the sun. (This model was controversial because it did not place the earth at the center of the universe, contrary to the religious doctrine of the day.)

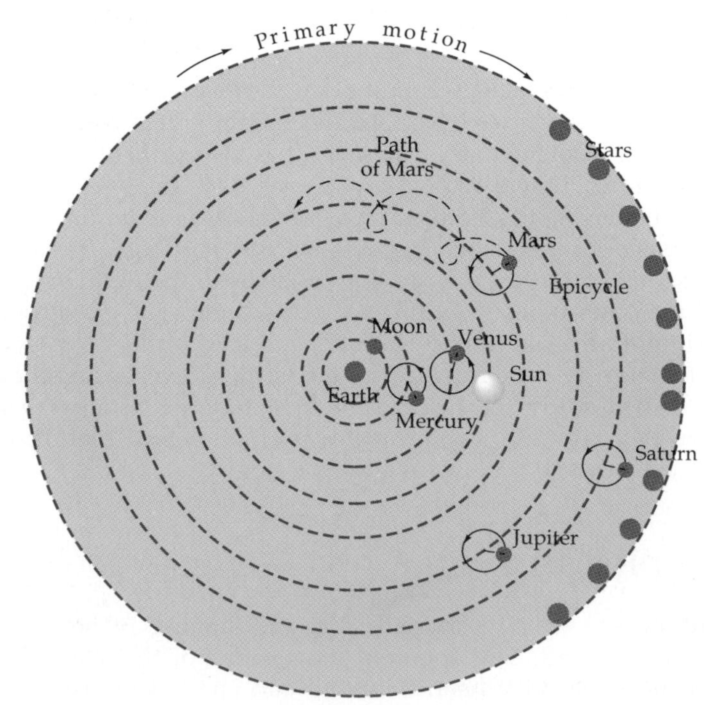

Figure 10-1 Ptolemy's model of the universe. The primary motion of the planets and stars is clockwise with a period of 1 day. To account for the motion of the planets relative to the stars, the planets are given a smaller counterclockwise angular velocity. This model is called a geocentric model because it places the earth at the center of the universe.

Toward the end of the sixteenth century, the astronomer Tycho Brahe studied the motions of the planets and made observations that were considerably more accurate than those previously available. Using Brahe's data, Johannes Kepler, after much trial and error, discovered that the actual paths of the planets about the sun were ellipses (Figure 10-2). He also showed that the planets did not move with constant speed but moved faster when close to the sun than when farther away. Finally, Kepler developed a precise mathematical relation between the period of a planet and its average distance from the sun. Kepler stated his results in three empirical laws of planetary motion. Ultimately, these laws provided the basis for Newton's discovery of the law of gravity.

Kepler's three laws are

Kepler's laws

Law 1. All planets move in elliptical orbits with the sun at one focus.

Law 2. A line joining any planet to the sun sweeps out equal areas in equal times.

Law 3. The square of the period of any planet is proportional to the cube of the planet's mean distance from the sun.

Figure 10-3 shows an **ellipse.** The points labeled F are called the foci. The distance a is called the semimajor axis, and b is called the semiminor axis. You can draw an ellipse by taking a piece of string, fixing each end at a focus (for example, with thumb tacks), and using it to guide your pencil as is shown in the figure. As the foci are moved closer together, the ellipse begins to resemble a circle. Indeed, a circle is a special case of an ellipse in which the

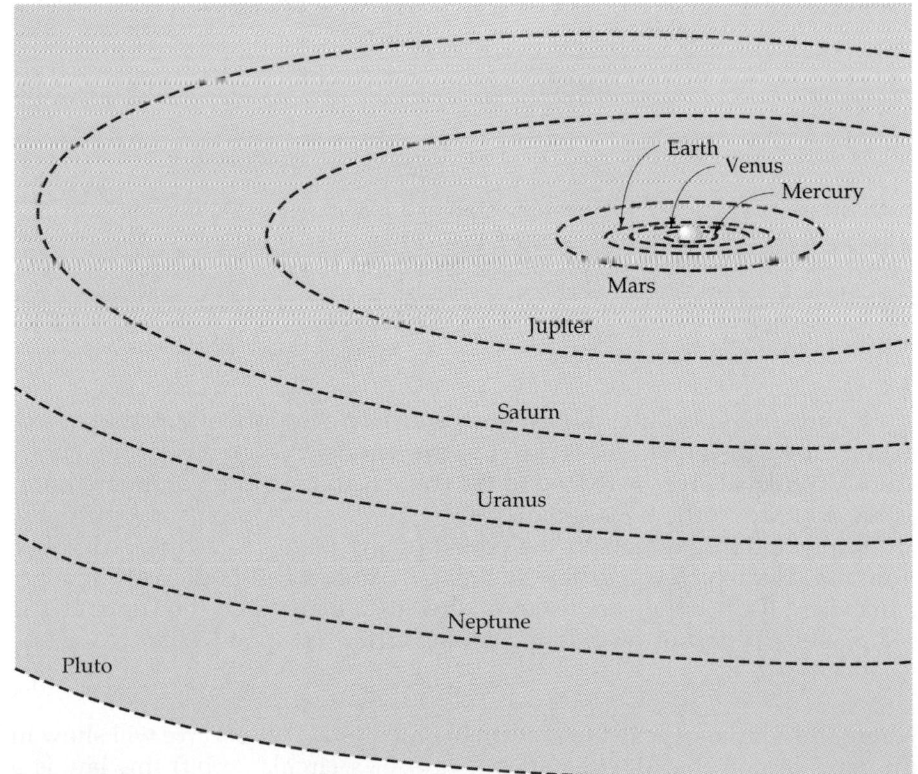

Figure 10-2 Orbits of the planets around the sun.

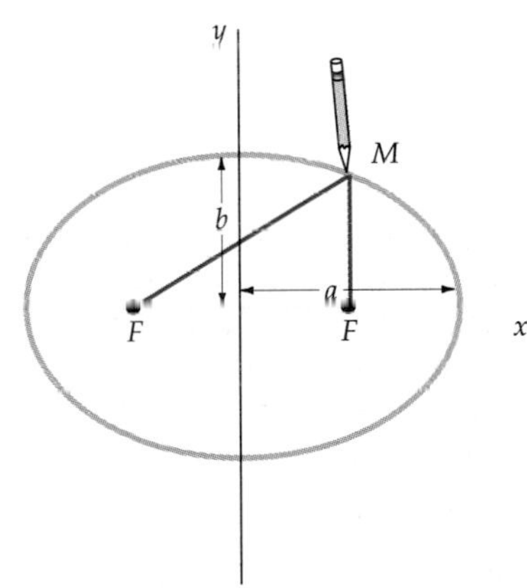

Figure 10-3 An ellipse is the locus of points for which the sum of the distances from two foci F is constant. Such a figure can be drawn by fastening a string at the foci and using it to guide a pencil (shown at M) to trace out the ellipse. The distance a is called the semimajor axis, and b is called the semiminor axis of the ellipse. If the foci coincide, a and b are equal and the ellipse is a circle.

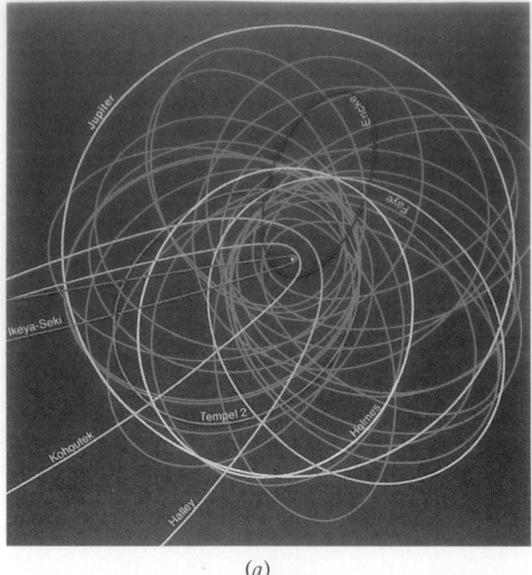

(a)

(b)

(a) The tightly woven orbits about the sun of many of the known comets, including Halley's comet, are depicted in this drawing. (b) Halley's comet as photographed over Duxbury Beach, Massachusetts, in March 1987.

two foci coincide. Figure 10-4 shows an elliptical path of a planet with the sun at one focus. The point P, at which the planet is closest to the sun, is called the perihelion, while point A, the farthest distance, is called the aphelion. The earth's orbit is nearly circular, with the distance to the sun at perihelion being 91.5 million miles and that at aphelion being 94.5 million miles. The semimajor axis equals half the sum of these distances, which is 93 million miles for the earth's orbit. This turns out to be the mean distance from the earth to the sun during its orbit.

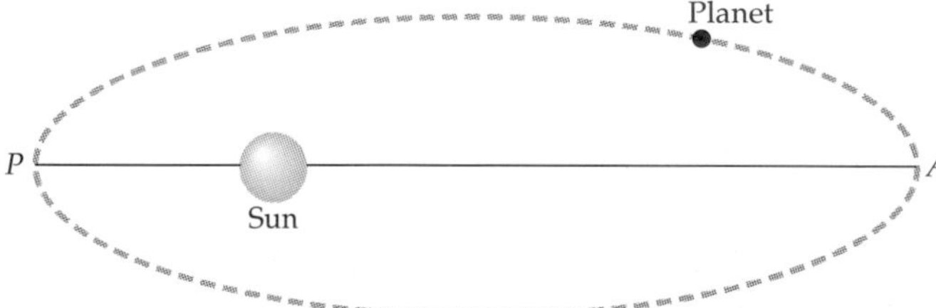

Figure 10-4 Elliptical path of a planet with the sun at one focus. Point P, where the planet is closest to the sun, is called the perihelion, and point A, where it is farthest, is called the aphelion. The average distance between the planet and the sun equals the semimajor axis.

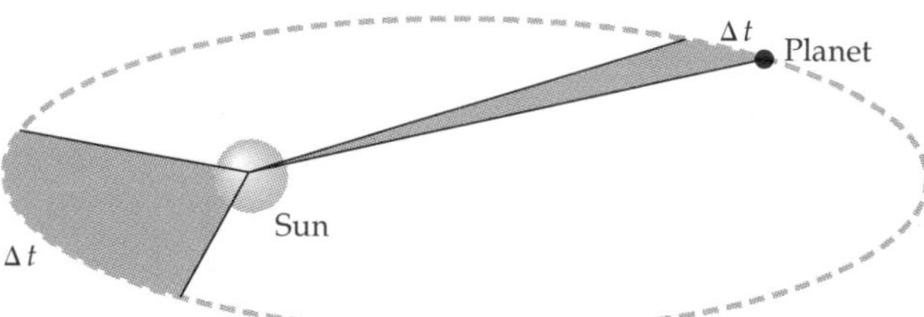

Figure 10-5 When a planet is close to the sun, it moves faster than when it is farther away. The areas swept out in a given time interval are equal.

Figure 10-5 illustrates Kepler's second law, the law of equal areas. A planet moves faster when it is closer to the sun than when it is farther away. The law of equal areas is related to the conservation of angular momentum, as we will see in the next section.

Kepler's third law relates the period of any planet to its mean distance from the sun, which equals the semimajor axis of its elliptical path. In algebraic form, if r is the mean distance between a planet and the sun and T is the planet's period of revolution, Kepler's third law states that

$$T^2 = Cr^3 \qquad \text{10-1}$$

where the constant C has the same value for all the planets. We will show in the next section that (for the special case of a circular orbit) this law is a simple consequence of the fact that the force exerted by the sun on a planet varies inversely with the square of the distance from the sun to the planet.

Example 10-1

The mean distance of Jupiter from the sun is 5.20 astronomical units (AU), where

$$1 \text{ AU} = 1.50 \times 10^{11} \text{ m} = 93.0 \times 10^{6} \text{ mi}$$

is the mean distance between the earth and the sun. What is the period of Jupiter?

From Kepler's third law, the square of the period is proportional to the cube of the mean distance from the sun. Taking the square root of both sides of Equation 10-1, we have

$$T = \sqrt{C}r^{3/2}$$

If T_E and r_E are the period and mean distance for the earth, and T_J and r_J are the period and mean distance for Jupiter, we have

$$\frac{T_J}{T_E} = \frac{\sqrt{C}\, r_J^{3/2}}{\sqrt{C}\, r_E^{3/2}}$$

$$= \left(\frac{r_J}{r_E}\right)^{3/2} = \left(\frac{5.20 \text{ AU}}{1 \text{ AU}}\right)^{3/2}$$

$$T_J = (5.20)^{3/2} T_E = 11.9(1 \text{ y}) = 11.9 \text{ y}$$

Exercise

The period of Neptune is 164.8 y. What is its mean distance from the sun? (Answer: 30.1 AU)

10-2 Newton's Law of Gravity

Although Kepler's laws were an important step in understanding the motion of the planets, they were still merely empirical rules obtained from the astronomical observations of Brahe. It remained for Newton to take the giant step forward and attribute the acceleration of a planet in its orbit to a force exerted by the sun on the planet that varied inversely with the square of the distance between the sun and the planet. Others besides Newton had proposed that such a force existed, but Newton was able to prove that a force that varied inversely with the square of the separation distance would result in the elliptical orbits observed by Kepler. He then made the bold assumption that such a force existed between any two objects in the universe. (Before Newton, it was not even generally accepted that the laws of physics observed on earth were applicable to the heavenly bodies.)

Newton's law of gravity postulates that every object exerts an attractive force on every other object that is proportional to the masses of the two objects and inversely proportional to the square of the distance separating them. Newton's law of gravity can be written as a simple vector equation. Let m_1 and m_2 be two point masses separated by a distance r_{12}, which is the magnitude of the vector $\mathbf{r}_{12}$ pointing from mass m_1 to m_2 (Figure 10-6). The force $\mathbf{F}_{12}$ exerted by mass m_1 on m_2 is then

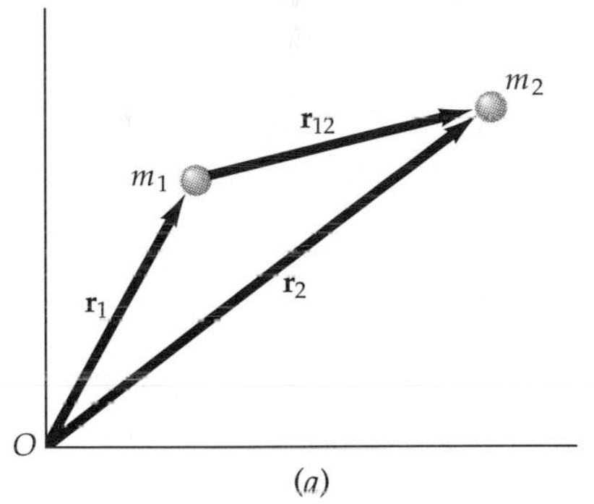

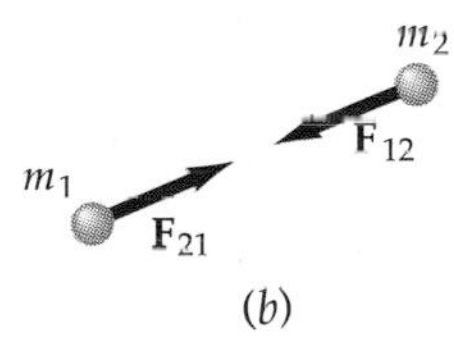

Figure 10-6 (a) A particle of mass m_1 at position $\mathbf{r}_1$ and a second particle of mass m_2 at $\mathbf{r}_2$. The vector $\mathbf{r}_{12}$ points from m_1 to m_2. (b) The force $\mathbf{F}_{12}$ exerted by m_1 on m_2 is toward m_1 and varies inversely as the square of the distance between the two particles.

$$\mathbf{F}_{12} = -\frac{Gm_1m_2}{r_{12}^2}\hat{\mathbf{r}}_{12} \qquad 10\text{-}2$$

Newton's law of gravity

where $\hat{\mathbf{r}}_{12} = \mathbf{r}_{12}/r_{12}$ is a unit vector pointing from m_1 to m_2 and G is the **universal gravitational constant** and has the value

$$G = 6.67 \times 10^{-11} \text{ N}\cdot\text{m}^2/\text{kg}^2 \qquad 10\text{-}3$$

The force $\mathbf{F}_{21}$ exerted by m_2 on m_1 is the negative of $\mathbf{F}_{12}$, by Newton's third law. That is, $\mathbf{F}_{21}$ is equal in magnitude to $\mathbf{F}_{12}$ but opposite in direction. The magnitude of the gravitational force exerted by a particle of mass m_1 on another particle of mass m_2 a distance r away is thus given by

$$F = \frac{Gm_1m_2}{r^2} \tag{10-4}$$

Newton published his theory of gravitation in 1686, but it was not until a century later that an accurate experimental determination of G was made by Cavendish, as will be discussed in Section 10-3.

We can use the known value of G to compute the gravitational attraction between two ordinary objects.

Example 10-2

Find the force of attraction between two balls, each of mass 1 kg, when their centers are 10 cm apart.

We can treat each ball as a point mass. The magnitude of the force on either ball exerted by the other is then

$$F = \frac{(6.67 \times 10^{-11}\ \text{N}\cdot\text{m}^2/\text{kg}^2)(1\ \text{kg})(1\ \text{kg})}{(0.1\ \text{m})^2}$$

$$= 6.67 \times 10^{-9}\ \text{N}$$

This example demonstrates that the gravitational force exerted by an object of ordinary size on another such object is extremely small. For example, the weight of a 1-kg object is 9.81 N, more than a billion times the force calculated in Example 10-2. We can usually neglect the gravitational force between objects compared with the other forces acting on them. The gravitational attraction can be noticed only if at least one of the objects is extremely massive, as with an ordinary object and the earth, or if great care is taken to eliminate other forces on the objects, as Cavendish did in determining G.

Exercise

Find the gravitational force that attracts a 65-kg boy to a 50-kg girl when they are 0.5 m apart, assuming that they are point masses. (Answer: 8.67×10^{-7} N)

Newton showed that, in general, when an object (such as a planet or comet) moves around a force center (such as the sun) to which it is attracted by a force that varies as $1/r^2$, the path of the object is either an ellipse, a parabola, or a hyperbola. The parabolic and hyperbolic paths apply to objects (if there are any) that make one pass by the sun and never return. Such orbits are not closed orbits. The only possible closed orbits in an inverse-square force field are ellipses. (A circle is a special case of an ellipse.) Thus, Kepler's first law is a direct consequence of Newton's law of gravity.

Kepler's second law, the law of equal areas, follows from the fact that the force exerted by the sun on a planet is directed toward the sun. Such a force is called a **central force.** Since the force on a planet is along the line from the planet to the sun, it has no torque about the sun. We know from our study of angular momentum that, if the net torque on an object is zero, the angular momentum of the object is conserved. Figure 10-7 shows a planet moving in an elliptical orbit around the sun. In time dt, the planet moves a distance $v\,dt$ and sweeps out the area indicated in the figure. This is half the area of the parallelogram formed by the vectors $\mathbf{r}$ and $\mathbf{v}\,dt$, which is

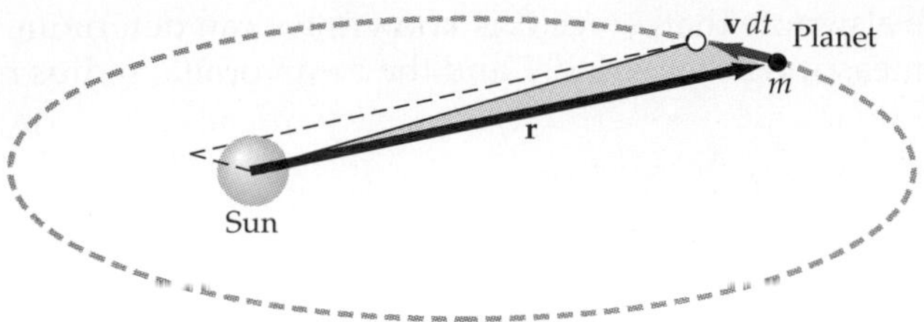

Figure 10-7 The area swept out by a planet in time dt is half the area of the parallelogram indicated, which is the magnitude of the cross product $\mathbf{r} \times \mathbf{v}\, dt$. The area is thus proportional to the angular momentum of the planet about the sun.

$\mathbf{r} \times \mathbf{v}\, dt$. Thus, the area dA swept out in time dt by the radius vector $\mathbf{r}$ is

$$dA = \frac{1}{2}|\mathbf{r} \times \mathbf{v}\, dt| = \frac{1}{2m}|\mathbf{r} \times m\mathbf{v}\, dt|$$

or

$$dA = \frac{1}{2m} L\, dt \qquad \text{10-5}$$

where $\mathbf{L} = \mathbf{r} \times m\mathbf{v}$ is the angular momentum of the planet relative to the sun. The area swept out in a given time interval dt is therefore proportional to the angular momentum L. Since L is constant during the motion of the planet, the area swept out in a given time interval dt is the same for all parts of the orbit, which is Kepler's second law.

We will now show that Newton's law of gravity implies Kepler's third law for the special case of a circular orbit. We consider a planet moving about the sun with speed v in a circular orbit of radius r. Since the planet is moving in a circle, we know it has centripetal acceleration v^2/r. This acceleration is provided by the force of attraction between the sun and the planet, which is given by Newton's law of gravity. From Newton's second law of motion, we have

$$F = m_p a$$

$$\frac{GM_s m_p}{r^2} = m_p \frac{v^2}{r} \qquad \text{10-6}$$

where M_s is the mass of the sun and m_p is that of the planet. Solving for v^2, we obtain

$$v^2 = \frac{GM_s}{r} \qquad \text{10-7}$$

We can now relate the speed v of the planet to its period T. Since the planet moves a distance $2\pi r$ in time T, its speed is

$$v = \frac{2\pi r}{T} \qquad \text{10-8}$$

Substituting this expression for v in Equation 10-7, we obtain

$$v^2 = \frac{4\pi^2 r^2}{T^2} = \frac{GM_s}{r}$$

or

$$T^2 = \frac{4\pi^2}{GM_s} r^3 \qquad \text{10-9}$$

Kepler's third law

Equation 10-9 is Kepler's third law. For a general elliptical path, the radius r is replaced by the mean distance between the planet and the sun, which equals the semimajor axis of the ellipse. This equation also applies to the orbits of the satellites of any planet if we replace the mass of the sun M_s with the mass of the planet. For example, it applies to the earth's moon and to all the artificial satellites orbiting the earth if the sun's mass M_s is replaced by

Johannes Kepler (1571–1630).

the earth's mass M_E. We also note that, since G is known, we can determine the mass of a planet by measuring the period T and the mean orbital radius r of a moon orbiting it.

Example 10-3

Mars has a moon with a period of 460 min and a mean orbital radius of 9.4 Mm. What is the mass of Mars?

Replacing M_s in Equation 10-9 with the mass of Mars M and using $r = 9.4 \times 10^6$ m, $T = 460(60)$ s, and $G = 6.67 \times 10^{-11}\ \text{N}\cdot\text{m}^2/\text{kg}^2$, we obtain

$$M = \frac{4\pi^2 r^3}{GT^2} = \frac{4\pi^2(9.4 \times 10^6\ \text{m})^3}{(6.67 \times 10^{-11}\ \text{N}\cdot\text{m}^2/\text{kg}^2)[460(60)\ \text{s}]^2} = 6.45 \times 10^{23}\ \text{kg}$$

As a further check on the validity of the inverse-square nature of the gravitational force, Newton compared the acceleration of the moon in its orbit with the acceleration of objects near the surface of the earth (such as the legendary apple). He made the bold assumption that the force that causes the moon to circle the earth has the same origin as the force that causes objects near the earth's surface to fall toward the earth, namely the gravitational attraction due to the earth. He first assumed that the earth and moon could be treated as point particles with their total masses concentrated at their centers. Since the distance to the moon is about 60 times the radius of the earth, the acceleration of objects near the surface of the earth (g = 9.81 m/s^2) should be $60^2 = 3600$ times the acceleration of the moon. The moon's centripetal acceleration can be calculated from its known distance from the center of the earth and its period:

$$a_m = \frac{v^2}{r} = \frac{(2\pi r/T)^2}{r} = \frac{4\pi^2 r}{T^2}$$

For $r = 3.84 \times 10^8$ m and $T = 27.3$ d, the result is $a_m = 2.72 \times 10^{-3}$ m/s^2. Comparing this result with g, we have

$$\frac{g}{a_m} = \frac{9.81\ \text{m/s}^2}{2.72 \times 10^{-3}\ \text{m/s}^2} = 3607 \approx 60^2$$

The calculations agreed "pretty nearly," according to Newton: "I thereby compared the force requisite to keep the Moon in her orb with the force of gravity at the surface of the Earth, and found them answer pretty nearly."

The assumption that the earth and moon can be treated as point particles in the calculation of the force on the moon is reasonable because the moon is far from the earth compared with the radius of either the earth or the moon, but such an assumption is certainly questionable when applied to the force exerted by the earth on an object near its surface. After considerable effort, Newton was able to prove that the force exerted by any spherically symmetric object on a point mass either on or outside its surface is the same as if all the mass of the object were concentrated at its center. The proof involves integral calculus, which Newton developed to solve this problem. This proof is given in Section 10-7.

The force exerted by the earth on any mass m at a distance r from the center of the earth is directed toward the earth and has the magnitude given by Equation 10-4 with m_1 equal to the mass of the earth M_E and m_2 replaced by m:

$$F = \frac{GM_E m}{r^2} \tag{10-10}$$

As mentioned in Chapter 4, the gravitational force on a mass divided by the mass is called the gravitational field. The gravitational field of the earth at a distance r (with r greater than the radius of the earth) points toward the earth and has the magnitude $g(r)$ given by

$$g(r) = \frac{F}{m} = \frac{GM_E}{r^2} \tag{10-11}$$

Gravitational field of the earth

The force exerted by the earth on an object at some height h above the surface of the earth is given by Equation 10-10 with $r = R_E + h$. The force exerted by the earth on an object of mass m near the earth's surface is also given by Equation 10-10 but with $r = R_E$:

$$F = \frac{GM_E m}{R_E^2}$$

When this is the only force acting on the object, the object is in free fall and has an acceleration

$$a = \frac{F}{m} = \frac{GM_E}{R_E^2} = g = 9.81 \text{ m/s}^2 \tag{10-12}$$

Since $g = 9.81 \text{ m/s}^2$ is easily measured and the radius of the earth is known, Equation 10-12 can be used to determine either the constant G or the mass of the earth M_E if one of these quantities is known. Newton estimated the value of G from an estimate of the mass of the earth. When Cavendish determined G some 100 years later by measuring the force between small spheres of known mass and separation, he called his experiment "weighing the earth."

Example 10-4

What is the free-fall acceleration of an object 200 km above the earth's surface?

The force on the object is given by Equation 10-10 with $r = R_E +$ 200 km. The acceleration is then

$$a = \frac{F}{m} = \frac{GM_E}{r^2}$$

Rather than substituting the values of G and M_E into this equation, we can use Equation 10-12 to write the product GM_E in terms of the radius of the earth and the free-fall acceleration of gravity g near the surface of the earth:

$$GM_E = gR_E^2 \tag{10-13}$$

The acceleration at distance r is then

$$a = \frac{GM_E}{r^2} = \frac{gR_E^2}{r^2} = g\frac{R_E^2}{r^2} \tag{10-14}$$

Using $R_E = 6.37$ Mm $= 6370$ km and $r = R_E + 200$ km $= 6570$ km, we obtain

$$a = (9.81 \text{ m/s}^2)\left(\frac{6370 \text{ km}}{6570 \text{ km}}\right)^2$$

$$= 9.22 \text{ m/s}^2$$

Exercise

At what distance h above the surface of the earth is the acceleration of gravity half its value at sea level? (Answer: $h = 2640$ km)

Example 10-5

A satellite orbits the earth in a circular orbit. Find its period (*a*) if the satellite is just above the surface of the earth and (*b*) if the satellite is at an altitude of 300 km. (Assume that air resistance can be neglected.)

(*a*) We can apply Equation 10-9 (Kepler's third law) to satellites orbiting the earth if we replace the mass of the sun M_s with the mass of the earth M_E. We then have

$$T^2 = \frac{4\pi^2}{GM_E} r^3$$

Again, it is convenient to replace GM_E with gR_E^2:

$$T^2 = \frac{4\pi^2}{gR_E^2} r^3$$

If the satellite is just above the surface of the earth, $r = R_E$ and

$$T^2 = \frac{4\pi^2}{gR_E^2} R_E^3 = \frac{4\pi^2 R_E}{g}$$

Therefore

$$T = 2\pi\sqrt{\frac{R_E}{g}} = 2\pi\sqrt{\frac{6.37 \times 10^6 \text{ m}}{9.81 \text{ m/s}^2}} = 5.06 \times 10^3 \text{ s} = 84.4 \text{ min}$$

(*b*) At an altitude of 300 km above the earth's surface, $r = 6370$ km + 300 km = 6670 km = 6.67×10^6 m. Since T is proportional to $r^{3/2}$, we can find T at this distance from

$$T = (84.4 \text{ min})\left(\frac{r}{R_E}\right)^{3/2}$$

$$= (84.4 \text{ min})\left(\frac{6.67 \times 10^6 \text{ m}}{6.37 \times 10^6 \text{ m}}\right)^{3/2}$$

$$= 90.4 \text{ min}$$

Exercise

Find the radius of a circular orbit of a satellite that orbits the earth with a period of one day. (If such a satellite is in orbit over the equator and moves in the same direction as the rotation of the earth, it appears stationary relative to the earth. Most satellites are parked in such an orbit, which is called a geosynchronous orbit.) (Answer: $r = 6.63R_E = 4.22 \times 10^7$ m)

Questions

1. Why don't you feel the gravitational attraction of a large building when you walk near it?
2. Astronauts orbiting the earth in a satellite 300 km above the surface of the earth feel weightless. Why? Is the force of gravity exerted by the earth on them negligible at this height?

10-3 Measurement of the Universal Gravitational Constant

A knowledge of the universal gravitational constant G is not only of intrinsic interest, but it also has practical applications in the determination of the density distributions of the interiors of the earth, the moon, other planets, and distant stars. The first measurement of G was made by Henry Cavendish in 1798. Figure 10-8*a* is a schematic sketch of the apparatus Cavendish used to measure the gravitational force between two small objects. Two small spheres, each of mass m_2, are at the ends of a light rod suspended by a fine fiber. A torque is required to turn the two spheres through the angle θ from their equilibrium position because the fiber must be twisted. Careful measurement shows that the torque required to turn the fiber through a given angle is proportional to the angle. The constant of proportionality, called the torsion constant, can be determined, and the fiber and the suspended masses can be used to measure very small torques. This arrangement, called a torsion balance, was invented in the eighteenth century by John Michell. Charles Augustin de Coulomb used a similar device in 1785 to determine the electric force between charged particles, which is now referred to as Coulomb's law (Chapter 18). Cavendish used a refined and especially sensitive torsion balance in his determination of G.

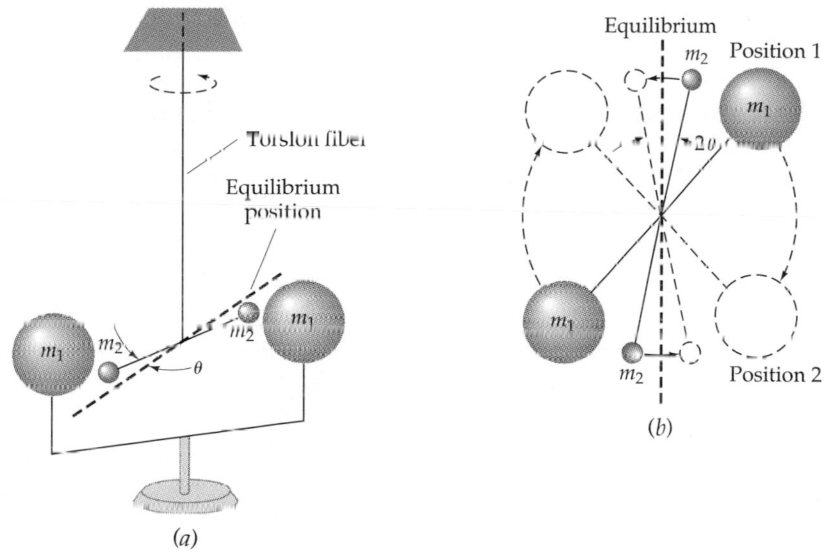

Figure 10-8 (*a*) Schematic drawing of Cavendish's apparatus for measuring G. Because of the gravitational attraction of the large spheres of mass m_1 for the nearby small spheres of mass m_2, the torsion fiber is turned through a very small angle θ from its equilibrium position. (*b*) The apparatus as seen from above. The large spheres are reversed so that they are at the same distance from the equilibrium position of the balance but on the other side. The fiber then turns through the angle 2θ. Once the torsion constant of the fiber has been determined, measurement of this angle makes it possible to determine the force exerted by m_1 on m_2, which in turn allows G to be determined.

In Cavendish's experiment two large spheres, each of mass m_1, are placed near the small spheres as shown in Figure 10-8*a*. The apparatus is allowed to come to equilibrium. Since the apparatus is so sensitive and the gravitational force is so small, this takes hours. Instead of measuring the deflection angle directly, Cavendish reversed the positions of the large spheres as shown by the dashed lines in Figure 10-8*b*. If the torsion balance is again allowed to come to equilibrium, it will turn through angle 2θ in response to the reversal of the torque. Once the torsion constant has been determined, the forces between the masses m_1 and m_2 can be determined from the measurement of this angle. When the masses and their separations are known, G can be calculated. Cavendish obtained a value for G within about 1 percent of the presently accepted value given by Equation 10-3.

Cavendish's measurement of G has been repeated by other experimenters with various improvements and refinements. In one variation, the torsion balance is used as an oscillator, and the period of oscillation, which

Original drawing of the Cavendish apparatus.

depends on the attraction of the nearby masses, is measured. (We will study oscillations in Chapter 12.) In another, the attracting masses are moved back and forth around the torsion balance in resonance with the oscillations and the amplitude of the vibration is measured. In a different kind of procedure, G is determined by measuring the effect of a large attracting mass on one of the masses of a very sensitive, equal-arm chemical balance. All measurements of G are difficult because of the extreme weakness of the gravitational attraction. Consequently, the value of G is known today only to about 1 part in 10,000. Although G was one of the first physical constants ever measured, it remains one of the least accurately known. Table 10-1 lists some of the results of the various measurements of G.

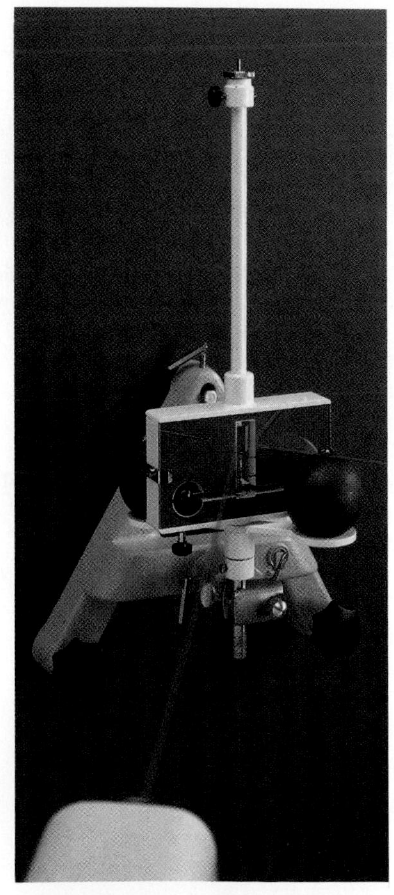

Gravitational Torsion Balance used in student laboratories for the measurement of G. A tiny angular deflection of the balance results in a large angular deflection of the laser beam that reflects from a mirror on the balance.

Table 10-1 **Measurements of *G***

Experimenter	Year	Method	G (10^{-11} N·m^2/kg^2)
Cavendish	1798	Torsion balance, deflection	6.754
Poynting	1891	Common balance	6.698
Boys	1895	Torsion balance, deflection	6.658
Von Eötvös	1896	Torsion balance, deflection	6.65
Heyl	1930	Torsion balance, period	
		Gold	6.678
		Platinum	6.664
		Glass	6.674
Zahradniček	1933	Torsion balance, resonance	6.659
Heyl and Chrzanowski	1942	Torsion balance, period	6.673
Luther and Towler	1982	Torsion balance, period	6.6726

The very small magnitude of G means that the gravitational force exerted by one object of ordinary size on another such object is extremely small, as we saw in Example 10-2. Such forces can be observed only if extreme care is taken to balance all other forces acting on the objects, as must be done to measure G.

10-4 Gravitational and Inertial Mass

The property of an object responsible for the gravitational force it exerts on another object is called its **gravitational mass.** On the other hand, the property of an object that measures its resistance to acceleration is called its **inertial mass.** We have used the same symbol m for these two properties because, experimentally, the gravitational and inertial masses of an object are equal. The fact that the gravitational force exerted by an object is proportional to its inertial mass is a characteristic unique to the force of gravity among all the forces we know, and it is a matter of considerable interest. One consequence is that all objects near the surface of the earth fall with the same acceleration if air resistance is neglected. This fact has seemed surprising to all since it was discovered. The well-known story of how Galileo demonstrated it by dropping objects from the Leaning Tower of Pisa is just one example of the excitement this discovery aroused in the sixteenth century.

We could easily imagine that the gravitational and inertial masses of an object were not the same. Suppose we write m_G for the gravitational mass and m for the inertial mass. The force exerted by the earth on an object near its surface would then be

$$F = \frac{GM_E m_G}{R_E^2} \tag{10-15}$$

where M_E is the gravitational mass of the earth. The free-fall acceleration of the object near the earth's surface would then be

$$a = \frac{F}{m} = \left(\frac{GM_E}{R_E^2}\right)\frac{m_G}{m} \tag{10-16}$$

If gravity were just another property of matter like color or hardness, it might be reasonable to expect that the ratio m_G/m would depend on such things as the chemical composition of the object, its temperature, or some other physical characteristics of the object. The free-fall acceleration would then be different for different objects. The experimental fact, however, is that a is the same for all bodies. This means that the ratio m_G/m is the same for every object. Since this is the case, we need not maintain the distinction between m_G and m and can set $m_G = m$. (This amounts to choosing the ratio to be 1, which in turn determines the magnitude and units of G in the law of gravity.) We must keep in mind, however, that the equivalence of gravitational and inertial mass is an experimental law, that is limited by the accuracy of experiment. Experiments testing this equivalence were carried out by Simon Stevin in the 1580s. Galileo publicized this law widely, and his contemporaries made considerable improvements in the experimental accuracy with which the law was established.

The most precise early comparisons of gravitational and inertial mass were made by Newton. Through experiments using simple pendulums rather than falling bodies, Newton was able to establish the equivalence between gravitational and inertial mass to an accuracy of about 1 part in 1000. Experiments comparing gravitational and inertial mass have improved steadily over the years. Their equivalence is now established to about 1 part in 10^{12}. Thus, the equivalence of gravitational and inertial mass is one of the best established of all physical laws.

The equivalence of gravitational and inertial mass is the basis for the principle of equivalence, which is the foundation of Einstein's general theory of relativity. We will study this theory in Chapter 34.

10-5 Escaping the Earth

In the last three decades, the idea of escaping from the earth's gravitational field has changed from fantasy to reality. Space probes have been sent out to the far reaches of the solar system. Some of these probes are expected to orbit the sun, whereas others will leave the solar system and drift on into outer space. In this section, we will look at the problem of escaping from the gravitational field of the earth (or of the sun). We will see that there is a minimum initial speed, called the escape speed, that is required of an object to escape from the earth, and we will calculate this escape speed using Newton's law of gravity.

Near the surface of the earth, the force of attraction between the earth and some object of mass m is a constant mg, which is independent of the height of the object above the earth's surface. The gravitational field near the surface of the earth is said to be uniform. If we project an object vertically upward with initial speed v in a uniform gravitational field, it will rise to a maximum height h that we can easily calculate from the conservation of energy. (We will neglect air resistance, which is important for practical calculations but does not affect the essential ideas of this discussion.) If the gravitational potential energy is zero at the earth's surface, the initial potential energy of the object is zero and the initial kinetic energy is $\frac{1}{2}mv^2$. At the maximum height h, the kinetic energy is zero and the potential energy is mgh. The conservation of energy gives

$$\tfrac{1}{2}mv^2 = mgh \qquad 10\text{-}17$$

or

$$v^2 = 2gh \qquad 10\text{-}18$$

According to Equation 10-18, if we increase the initial speed of the object, the maximum height attained by the object increases. But for any initial speed, no matter how great, there is some maximum height h. That is, there is no initial speed v great enough to allow an object to escape from a uniform gravitational field.

However, we know from Newton's law of gravity that the gravitational field of the earth is not uniform but decreases as $1/r^2$, where r is the distance from the center of the earth. If we project an object upward with a very large initial speed so that the object moves a distance comparable to the radius of the earth, we must take into account the decrease in the gravitational force on the object to calculate correctly the maximum height the object attains.

(*a*) Proposed orbit of the Galileo satellite launched in 1989 and due to arrive at Jupiter in 1995. (*b*) The Galileo satellite, which will orbit Jupiter and send back high quality images and instrument data about the planet and its natural moons.

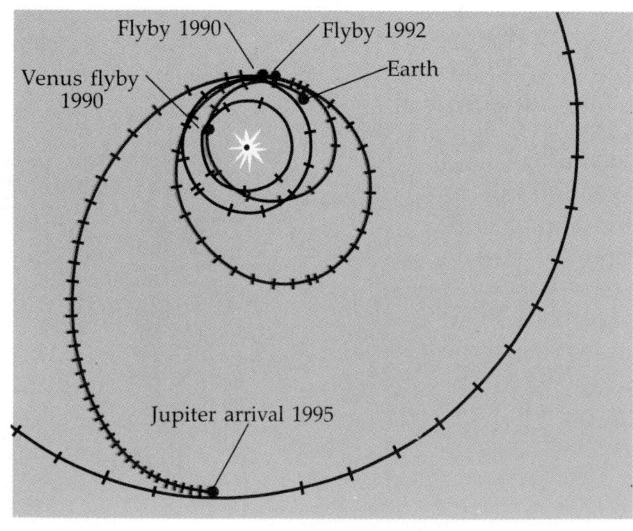

(*a*)

(*b*)

Thus, in using conservation of energy, we need to find the correct expression for the potential energy of an object in the gravitational field of the earth, which varies as $1/r^2$.

We first show that any central force whose magnitude depends only on the radial coordinate r is conservative. We do this by showing that the work done by such a force on a particle that moves from some initial position P_1 to some final position P_2 is independent of the path followed and therefore depends only on the initial and final positions. We consider the general central force

$$\mathbf{F} = F_r\hat{\mathbf{r}}$$

where $\hat{\mathbf{r}}$ is a unit vector in the radial direction and F_r depends only on r. The work done by this force on a particle whose displacement is $d\mathbf{s}$ is

$$dW = \mathbf{F}\cdot d\mathbf{s} = (F_r\hat{\mathbf{r}})\cdot d\mathbf{s} = F_r\ dr$$

(In general, the displacement $d\mathbf{s}$ has both a radial component $dr\ \hat{\mathbf{r}}$ and a component perpendicular to $\hat{\mathbf{r}}$. However, since $\mathbf{F}$ has only a radial component, the dot product $\mathbf{F}\cdot d\mathbf{s} = F_r\ dr$.) Figure 10-9 shows two paths connecting the initial and final positions of the particle. The work done by $\mathbf{F}$ along path 1, which is a simple radial path, is the integral of $F_r\ dr$ from r_1 to r_2. Path 2 is made up of radial segments and arc segments that are perpendicular to the radial lines. No work is done along these perpendicular segments because $\mathbf{F}$ and $d\mathbf{s}$ are perpendicular and therefore $\mathbf{F}\cdot d\mathbf{s} = 0$. The total work done along the radial segments of path 2 is the same as that done along path 1 because F_r depends only on r. Thus, the work done is independent of the path, so the force $\mathbf{F}$ is conservative.

To find the potential energy function $U(r)$ for the system consisting of a particle of mass m and the earth, we use the general definition of potential energy (Equation 6-17):

$$dU = -\mathbf{F}\cdot d\mathbf{s} = -\left(-\frac{GM_{\mathrm{E}}m}{r^2}\ \hat{\mathbf{r}}\right)\cdot d\mathbf{s}$$

where $\mathbf{F}$ is the force exerted by the earth on the particle and $d\mathbf{s}$ is a general displacement of the particle. The change in the potential energy of the object–earth system for a displacement $d\mathbf{s}$ is

$$dU = -\mathbf{F}\cdot d\mathbf{s} = -F_r\ dr = -\left(-\frac{GM_{\mathrm{E}}m}{r^2}\right)dr = +\frac{GM_{\mathrm{E}}m}{r^2}\ dr \qquad \text{10-19}$$

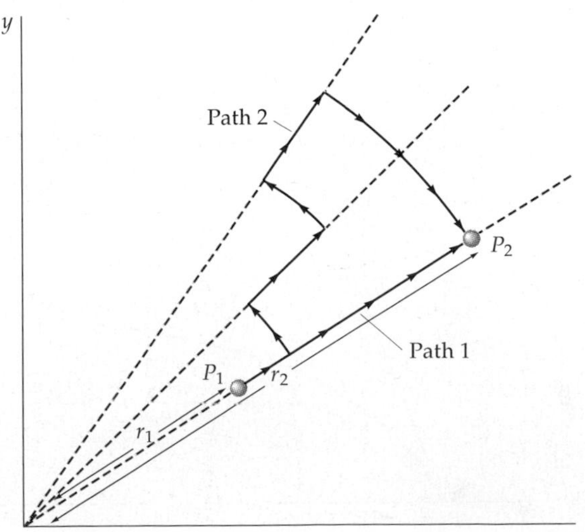

Figure 10-9 As a particle moves from point P_1 to point P_2, the work done on it by a central force $\mathbf{F}$ that depends only on r is the same along the radial path 1 as it is along path 2, which is made up of radial segments and arc segments. The work done along the arc segments is zero because $\mathbf{F}$ is perpendicular to $d\mathbf{s}$.

When the particle is displaced from some initial position r_1 to some final position r_2, the change in the potential energy is

$$U(r_2) - U(r_1) = \int_{r_1}^{r_2} dU = \int_{r_1}^{r_2} \frac{GM_E m}{r^2}\,dr$$

$$= -\frac{GM_E m}{r}\bigg|_{r_1}^{r_2} = \frac{GM_E m}{r_1} - \frac{GM_E m}{r_2} \qquad 10\text{-}20$$

If we choose the potential energy to be zero when the particle is at the earth's surface ($r_1 = R_E$), we have

$$U(r_2) - 0 = \frac{GM_E m}{R_E} - \frac{GM_E m}{r_2}$$

At some general position r ($r > R_E$), the potential-energy function $U(r)$ is thus

Gravitational potential energy with U = 0 at earth's surface

$$U(r) = \frac{GM_E m}{R_E} - \frac{GM_E m}{r} \qquad U = 0 \text{ at } r = R_E \qquad 10\text{-}21$$

If we write this potential-energy function in terms of the distance above the surface of the earth, $y = r - R_E$, we can compare this function with mgy, which we obtained previously for an object near the surface of the earth at a height y above the surface. Putting the two terms in Equation 10-21 over a common denominator $R_E r$, we obtain

$$U(r) = \frac{GM_E m}{R_E} - \frac{GM_E m}{r} = \frac{GM_E m}{R_E r}(r - R_E)$$

Using $y = r - R_E$, and writing $GM_E = gR_E^2$ from Equation 10-13, we have

$$U(r) = \frac{GM_E m}{R_E r} y = m\left(\frac{GM_E}{R_E^2}\right) y \frac{R_E}{r} = mgy\frac{R_E}{r} \qquad 10\text{-}22$$

The potential energy is thus mgy times the quantity R_E/r. Near the surface of the earth, r and R_E are approximately equal, and the potential energy is approximately mgy, as obtained by assuming the gravitational force to be constant. Figure 10-10 shows this potential energy as a function of r. The blue solid line is the potential-energy function $mgy = mg(r - R_E)$ for a constant gravitational force. We note that the actual potential-energy function for the varying gravitational force does not become infinitely large as r in-

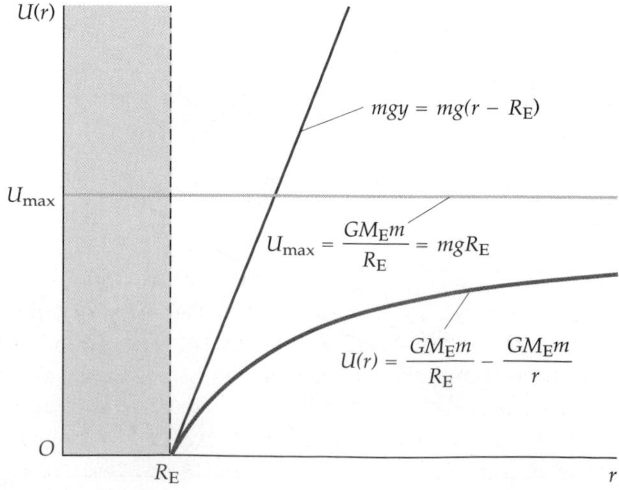

Figure 10-10 Graph of the gravitational potential energy $U(r)$ versus r. For r just slightly greater than R_E, $U(r)$ is approximately equal to $mgy = mg(r - R_E)$, which is indicated by the blue solid line. At large values of r, $U(r)$ approaches U_{max}.

creases. Instead, as r becomes larger and larger, the second term in Equation 10-21 gets smaller and smaller, and the potential energy approaches a maximum value U_{max} given by

$$U_{max} = \frac{GM_E m}{R_E} = mgR_E \qquad 10\text{-}23$$

where we have again used $GM_E = gR_E^2$. Note that this result can also be obtained from Equation 10-22 using $y \approx r$ for very large r.

Example 10-6

A projectile is fired straight up from the surface of the earth with an initial speed $v_1 = 8$ km/s. Find the maximum height it reaches, neglecting air resistance.

We use conservation of energy. Initially, at the surface of the earth, the potential energy of the projectile is zero and its kinetic energy is $\frac{1}{2}mv_1^2$. When it is at a height $h = r - R_E$, its potential energy is given by Equation 10-21. We wish to find the height for which its final speed $v_2 = 0$. Conservation of energy gives

$$\tfrac{1}{2}mv_1^2 + U_1 = \tfrac{1}{2}mv_2^2 + U_2$$

$$\tfrac{1}{2}mv_1^2 + 0 = 0 + \left(\frac{GM_E m}{R_E} - \frac{GM_E m}{r}\right)$$

Simplifying this equation, we obtain

$$v_1^2 = \frac{2GM_E}{R_E} - \frac{2GM_E}{r} = \frac{2GM_E}{R_E}\left(1 - \frac{R_E}{r}\right)$$

We can simplify this further by using $GM_E = gR_E^2$. Then

$$v_1^2 = 2gR_E\left(1 - \frac{R_E}{r}\right)$$

It is sufficient to solve for R_E/r:

$$1 - \frac{R_E}{r} = \frac{v_1^2}{2gR_E}$$

$$\frac{R_E}{r} = 1 - \frac{v_1^2}{2gR_E}$$

$$= 1 - \frac{(8000 \text{ m/s})^2}{2(9.81 \text{ m/s}^2)(6.37 \times 10^6 \text{ m})}$$

$$= 1 - 0.512 = 0.488$$

Then

$$r = \frac{R_E}{0.488} = 2.05R_E$$

The maximum height reached is therefore $h = r - R_E = 1.05R_E$.

We are now ready to find the initial speed an object must have to escape from the earth's gravitational field. If we project an object upward from the earth with some initial kinetic energy, the kinetic energy decreases and the potential energy increases as the object rises. But the potential energy cannot increase by more than U_{max}, as given by Equation 10-23. Therefore, this is the most that the kinetic energy can decrease. If the initial kinetic energy is greater than U_{max}, the object will still have some kinetic energy when r is very great (or even when r is infinite). Thus, the object will escape from the earth if the initial kinetic energy is greater than U_{max}. The critical speed v_e associated with this critical energy is called the **escape speed.** It is found from

$$\tfrac{1}{2}mv_e^2 = U_{max} = \frac{GM_E m}{R_E} = mgR_E$$

or

Escape speed

$$v_e = \sqrt{\frac{2GM_E}{R_E}} = \sqrt{2gR_E} \qquad 10\text{-}24$$

Using $g = 9.81\ \text{m/s}^2$ and $R_E = 6.37$ Mm, we obtain

$$v_e = \sqrt{2(9.81\ \text{m/s}^2)(6.37 \times 10^6\ \text{m})} = 11.2\ \text{km/s}$$

This is about 6.95 mi/s or about 25,000 mi/h.

The energy $U_{max} = mgR_E$ is called the **binding energy.** If the kinetic energy of an object at the earth's surface is less than the binding energy, the object will not leave the earth but will rise to some maximum height h and then fall back to earth. If its initial kinetic energy is greater than the binding energy, the object will continue moving away from the earth forever. Thus, the escape speed is just that speed corresponding to a kinetic energy that is equal to the binding energy. An earth–mass system is said to be bound or unbound depending on whether the kinetic energy of the mass at the earth's surface is less or greater than the binding energy.

The magnitude of the escape speed for a planet or moon relative to the thermal speeds of gas molecules determines the kind of atmosphere a planet or moon can have. The average kinetic energy of gas molecules, $(\frac{1}{2}mv^2)_{av}$, is proportional to the absolute temperature T (Chapter 15). The average speed of a gas molecule then depends on the temperature and varies inversely with the mass of the molecule. At the surface of the earth, the speeds of oxygen and nitrogen molecules are much lower than the escape speed, so these gases can exist in our atmosphere. For the lighter molecules hydrogen and helium, however, a considerable fraction of the molecules have speeds greater than the escape speed. Hydrogen and helium gases are therefore not found in our atmosphere. The escape speed at the surface of the moon can be calculated from Equation 10-24, with the acceleration of gravity on the moon and the radius of the moon replacing g and R_E. The escape speed for the moon is 2.3 km/s, considerably smaller than that for the earth and too small for any atmosphere to exist.

Example 10-7

Find the escape speed at the surface of Mercury, which has a mass $M = 3.31 \times 10^{23}$ kg and a radius $R = 2.44$ Mm.

Equation 10-24 can be used for the escape speed at the surface of Mercury if we replace the mass of the earth M_E with the mass of Mercury, and replace the radius of the earth R_E with the radius of Mercury. Then

$$v_e = \sqrt{\frac{2GM}{R}} = \sqrt{\frac{2(6.67 \times 10^{-11}\ \text{N}\cdot\text{m}^2/\text{kg}^2)(3.31 \times 10^{23}\ \text{kg})}{2.44 \times 10^6\ \text{m}}}$$

$$= 4.25 \times 10^3\ \text{m/s} = 4.25\ \text{km/s}$$

Questions

3. What is the effect of air resistance on the escape speed near the earth's surface?
4. Would it be possible in principle for the earth to escape from the solar system?

10-6 Potential Energy, Total Energy, and Orbits

Earth as seen from Apollo 11 orbiting the moon on July 31, 1969.

For the gravitational potential-energy function of Equation 10-21, we chose the potential energy to be zero at the surface of the earth. Since only changes in potential energy are important, we could have chosen the potential energy to be zero at any arbitrary position. Although it seems natural to choose the potential energy to be zero at the earth's surface when we are working problems involving objects near the earth's surface, there are many situations in which this is not at all the most convenient choice. For example, when considering the potential energy associated with a planet and the sun, there is no reason to want the potential energy to be zero at the surface of the sun. In fact, it is nearly always more convenient to choose the gravitational potential energy of a two-object system to be zero when the separation of the two objects is infinite. For a general two-object system consisting of one object of mass m and another of mass M, the change in the potential energy when the system moves from a separation of r_1 to a separation of r_2 is given by Equation 10-20:

$$U(r_2) - U(r_1) = \frac{GMm}{r_1} - \frac{GMm}{r_2}$$

Setting r_2 equal to some general separation r and $r_1 = \infty$, we have

$$U(r) - U(\infty) = -\frac{GMm}{r}$$

Choosing U to be zero at $r = \infty$ then gives

$$U(r) = -\frac{GMm}{r} \qquad U = 0 \text{ at } r = \infty \qquad \text{10-25}$$

Gravitational potential energy with $U = 0$ at infinite separation

(This choice for zero potential energy seems to have the disadvantage of making the potential energy always negative, but this is not really a disadvantage. It is like choosing the potential energy to be zero at the ceiling of a room rather than at the floor when you are doing a problem involving an object in a room.)

Figure 10-11 is a plot of $U(r)$ versus r for this choice of $U = 0$ at $r = \infty$ for an object of mass m and the earth of mass M_E. This function begins at the negative value $U = -(GM_Em)/R_E = -mgR_E$ at the earth's surface and increases as r increases, approaching zero at infinite r. The maximum increase in potential energy is still mgR_E, as it was when U was chosen to be zero at the earth's surface. The criterion for an object to escape is also the same: the kinetic energy at the earth's surface must be greater than or equal to mgR_E. Since the potential energy at the earth's surface is $-mgR_E$, the total energy $E = K + U$ must be greater than or equal to zero for the object to escape.

In Figure 10-11, two possible values for the total energy E are shown on the graph: E_1, which is negative, and E_2, which is positive. A negative total

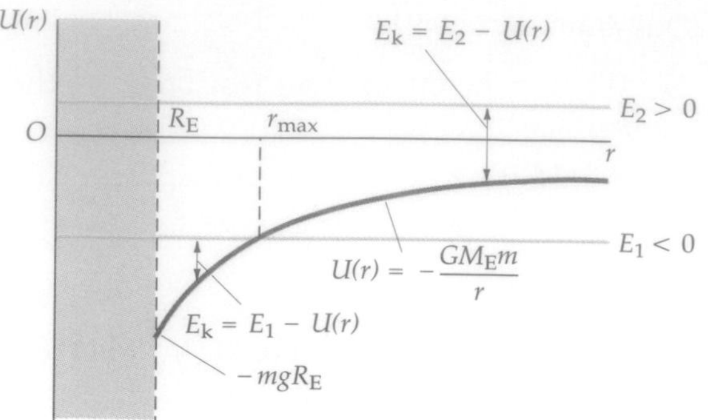

Figure 10-11 The same potential-energy curve as in Figure 10-10, except that $U(r)$ is chosen to be zero at $r = \infty$. For this choice, a particle is bound if its total energy E is less than zero and unbound if E is greater than zero.

energy merely means that the kinetic energy at the earth's surface is less than mgR_E, so the magnitude of the kinetic energy is never greater than the magnitude of the potential energy. From the figure, we can see that when the total energy is negative, the total-energy line intersects the potential-energy curve at some maximum separation r_{max} and the system is bound. On the other hand, when the total energy is positive, there is no intersection and the system is unbound. The criteria for bound and unbound systems can be simply stated:

If $E < 0$, the system is bound.

If $E > 0$, the system is unbound.

The binding energy is, of course, the same regardless of the choice of where the potential energy is zero.

When we choose the gravitational potential energy to be zero at infinite separation, we can discuss the kind of orbit taken by an object that moves in the gravitational field of the sun. The potential energy of an object of mass m at a distance r from the sun is

$$U(r) = -\frac{GM_sm}{r} \qquad 10\text{-}26$$

where M_s is the mass of the sun. The kinetic energy of the object is $\frac{1}{2}mv^2$. If the total energy, kinetic plus potential, is less than zero, the orbit will be an ellipse (or a circle), and the object will be bound to the sun. That is, it cannot escape from the gravitational field of the sun. On the other hand, if the total energy is positive, the orbit will be a hyperbola, which means that the object will make one swing around the sun and leave, never to return again. If the total energy is exactly zero, the orbit will be a parabola, and again the object will escape. That is, when the total energy is zero or positive, the object is not bound to the gravitational field of the sun.

Example 10-8

Show that the total energy of a satellite in a circular orbit equals half its potential energy.

The total energy of an orbiting satellite is the sum of its potential energy and its kinetic energy. If it is a distance r from the center of the earth, its potential energy is

$$U = -\frac{GM_Em}{r}$$

Since it is moving in a circle of radius r, it has a centripetal acceleration of magnitude v^2/r. From Newton's second law, $\mathbf{F} = m\mathbf{a}$, we have

$$\frac{GM_E m}{r^2} = \frac{mv^2}{r}$$

The kinetic energy of the satellite is thus

$$K = \frac{1}{2}mv^2 = \frac{1}{2}\frac{GM_E m}{r}$$

Thus, the kinetic energy has a magnitude half that of the potential energy. This result holds for any circular orbit in an inverse-square force field. The total energy of the satellite is thus

$$E = K + U = \frac{1}{2}\frac{GM_E m}{r} - \frac{GM_E m}{r} = -\frac{1}{2}\frac{GM_E m}{r} = \frac{1}{2}U(r)$$

On the energy diagram in Figure 10-11, the total energy of a satellite orbiting in a circle would be halfway between the potential energy $U(r)$ and zero. The minimum additional energy such a satellite must be given to escape from the earth is thus $(\frac{1}{2}GM_E m)/r$. That is, the kinetic energy of the orbiting satellite must be doubled.

Example 10-9

A satellite of mass 450 kg orbits the earth in a circular orbit at 6.83 Mm above the earth's surface. Find (*a*) the potential energy, (*b*) the kinetic energy, and (*c*) total energy of the satellite.

(*a*) The distance between the satellite and the center of the earth is

$$r = R_E + h = 6.37 \text{ Mm} + 6.83 \text{ Mm} = 13.2 \text{ Mm}$$

From Equation 10-26, the potential energy of the satellite is

$$U = -\frac{GM_E m}{r}$$
$$= -\frac{(6.67 \times 10^{-11} \text{ N·m}^2/\text{kg}^2)(5.98 \times 10^{24} \text{ kg})(450 \text{ kg})}{1.32 \times 10^7 \text{ m}} = -13.6 \text{ GJ}$$

(*b*) In the previous example, we found that in a circular orbit the kinetic energy is half the magnitude of the potential energy. The kinetic energy is therefore

$$K = \frac{1}{2}\frac{GM_E m}{r} = -\frac{1}{2}U = -\frac{1}{2}(-13.6 \text{ GJ}) = 6.80 \text{ GJ}$$

(*c*) The total energy is the sum of the potential and kinetic energies.

$$E = K + U = 6.80 \text{ GJ} + (-13.6 \text{ GJ}) = -6.80 \text{ GJ}$$

The total energy equals the negative of the kinetic energy.

Question

5. An object (say, a newly discovered comet) enters the solar system and makes a pass around the sun. How can we tell if the object will return many years later, or if it will never return?

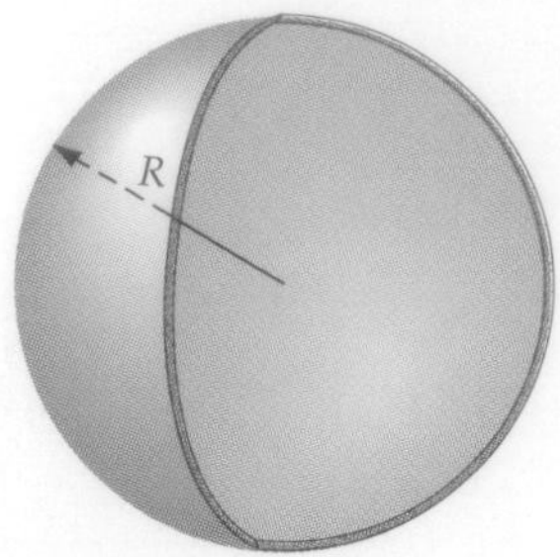

Figure 10-12 A uniform spherical shell of mass M and radius R. The force exerted by the shell on a mass outside the shell is the same as if the shell were a point mass at the center. The force exerted by the shell on a mass inside the shell is zero.

10-7 The Gravitational Field of a Spherical Shell and a Solid Sphere

One of Newton's motivations for developing calculus was to be able to prove that the gravitational field of a solid sphere at a point outside the sphere is the same as if all the mass of the sphere were concentrated at its center. This proof is not difficult, but it appears so because the geometry is somewhat involved, and the use of spherical coordinates may be unfamiliar. Before we give this proof, we will discuss some of the important results derived from it.

Figure 10-12 shows a uniform spherical shell of mass M and radius R. We will show below that the force exerted by this shell on a point mass m_0 located a distance r from the center of the shell is given by

$$\mathbf{F} = -\frac{GMm_0}{r^2}\,\hat{\mathbf{r}} \qquad r > R$$

and

$$\mathbf{F} = 0 \qquad r < R$$

That is, if m_0 is outside the shell, the force exerted on it by the shell is the same as if all the mass of the shell were concentrated at its center, whereas if m_0 is inside the shell, the shell exerts no force on it. This surprising result can be understood from Figure 10-13, which shows a point mass m_0 inside a spherical shell. In this figure, the masses of the shell segments m_1 and m_2 are related by

$$m_2 = m_1\left(\frac{r_2^2}{r_1^2}\right)$$

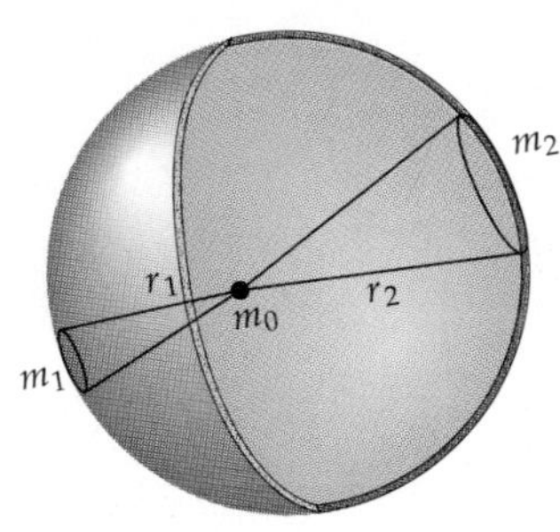

Figure 10-13 A point mass m_0 inside a uniform spherical shell feels no net force. The attraction of the part of the shell of mass m_1 is balanced by that of mass m_2, which is farther away but larger.

Since the force due to each mass is proportional to $1/r^2$, the force due to the smaller mass on the left is exactly balanced by that due to the larger mass on the right.

The gravitational field $\mathbf{g}$ of the shell is the force exerted on m_0 divided by m_0:

$$\mathbf{g} = \frac{\mathbf{F}}{m_0}$$

The gravitational field of a spherical shell of mass M and radius R is thus

Gravitational field of a spherical shell

$$\mathbf{g} = -\frac{GM}{r^2}\,\hat{\mathbf{r}} \qquad r > R \qquad \text{10-27}a$$

$$\mathbf{g} = 0 \qquad r < R \qquad \text{10-27}b$$

The gravitational field of a uniform spherical shell of mass M at a point outside the shell is the same as that due to a point mass M at the center of the shell. Inside a uniform spherical shell, the gravitational field is zero.

We will now use this result to find the gravitational field of a solid sphere. For a point some distance r, outside the sphere, the problem is very simple. We merely consider the solid sphere to consist of a continuous set of spherical shells. Since the field due to each shell is the same as if its mass were concentrated at the center of the shell, the field due to the entire sphere is the same as if the entire mass of the sphere were concentrated at its center:

$$g_r = -\frac{GM}{r^2} \qquad r > R \tag{10-28}$$

This result holds whether or not the sphere has a constant density as long as the density depends only on r so that spherical symmetry is maintained.

We can also find the gravitational field of a solid sphere of constant density at a point inside the sphere at a distance r from the center, where r is less than the radius R of the sphere. This would apply, for example, to finding the weight of an object at the bottom of a deep mine shaft. As we have seen, the field inside a spherical shell is zero. Thus, in Figure 10-14, the mass of the sphere outside r exerts no force at or inside r. Therefore, only the mass M' within the radius r contributes to the gravitational field at r. This mass produces a field equal to that of a point mass M' at the center of the sphere. The fraction of the total mass of the sphere within r is equal to the ratio of the volume of a sphere of radius r to that of a sphere of radius R. Thus, for a uniform mass distribution, if M is the total mass of the sphere, M' is given by

$$M' = \frac{\frac{4}{3}\pi r^3}{\frac{4}{3}\pi R^3}M = \frac{r^3}{R^3}M \tag{10-29}$$

The gravitational field at the distance r is thus

$$g_r = -\frac{GM'}{r^2} = -\frac{(GMr^3)/R^3}{r^2}$$

or

$$g_r = -\frac{GM}{R^3}r \qquad r < R \tag{10-30}$$

The magnitude of the field increases with distance r inside the sphere. Figure 10-15 shows a plot of the field g_r as a function of r for a solid sphere of constant mass density.

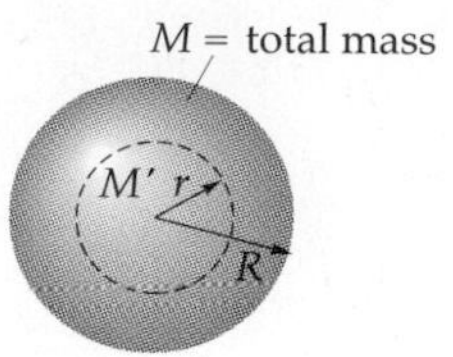

Figure 10-14 A uniform solid sphere of mass M. The gravitational field outside the sphere is the same as if the sphere were a point mass at the origin. Inside the sphere at some distance $r < R$, only the mass M' that is within r contributes to the gravitational field.

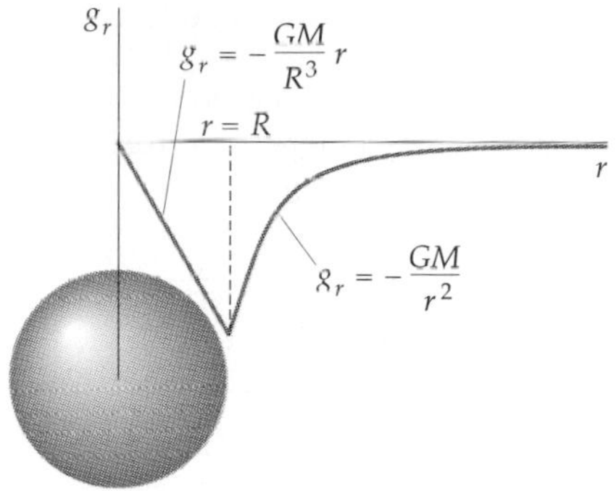

Figure 10-15 Plot of g_r versus r for a uniform solid sphere of mass M. The magnitude of the field increases with r inside the sphere and decreases as $1/r^2$ outside.

The rings and three moons of Saturn. Because of the $1/r^2$ dependence, the gravitational force on the near side of a satellite is much greater than that on the far side if the orbit is very near the planet. This force difference, called a **tidal force,** is great enough to break up large objects that are close to the planet, hence the rings of small particles orbiting close to Saturn. The two moons shown orbit at much greater distances where the tidal forces are much smaller.

Derivation of the Equation for the Gravitational Field of a Spherical Shell (Optional)

We will derive the equation for the gravitational field of a spherical shell in two steps. We will first find the gravitational field on the axis of a ring of uniform mass. We then apply our result to a spherical shell, which we can consider to be a set of coaxial rings.

Figure 10-16 shows a ring of total mass m and radius a and a point mass m_0 on the axis of the ring a distance x from its center. We have chosen an element of mass dm on the ring that is small enough to be considered a point particle. The element is at a distance s from the point mass on the axis, and the line joining the element and the point mass makes an angle α with the axis of the ring.

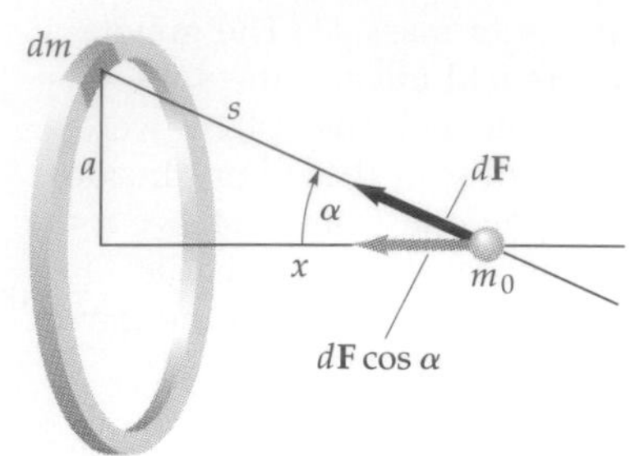

Figure 10-16 A uniform ring of total mass m and a point mass m_0 on its axis. The force on m_0 exerted by the element dm is toward the element. The total force exerted on m_0 by the ring is along the axis of the ring.

The force exerted by the element on m_0 is toward the element and has the magnitude dF given by

$$dF = \frac{G(dm)m_0}{s^2}$$

From the symmetry of the figure, we can see that when we sum over all the elements of the ring, the net force will be along the axis of the ring; that is, the perpendicular components will sum to zero. For example, the perpendicular component of the force on m_0 shown in the figure will be canceled by that due to another element of the ring directly opposite the one shown. The net force on m_0 will therefore be in the negative x direction. The x component of the force due to the element dm is

$$dF_x = -dF \cos\alpha = -\frac{G\,(dm)m_0}{s^2}\cos\alpha$$

The gravitational field dg_x due to the ring at the position of the point mass is this force divided by m_0:

$$dg_x = \frac{dF_x}{m_0} = -\frac{G\,dm}{s^2}\cos\alpha$$

We obtain the total field by summing over all the elements of the ring:

$$g_x = -\int \frac{G\,dm}{s^2}\cos\alpha$$

Since s and α are the same for all points on the ring, they are constants as far as the integration is concerned. Thus,

$$g_x = -\frac{G}{s^2}\cos\alpha \int dm = -\frac{Gm}{s^2}\cos\alpha \qquad \text{10-31}$$

where $m = \int dm$ is the total mass of the ring. We can write this result in terms of the distance x by using

$$s^2 = x^2 + a^2$$

and

$$\cos\alpha = \frac{x}{s} = \frac{x}{\sqrt{x^2 + a^2}}$$

Then,

$$g_x = -\frac{Gmx}{(x^2 + a^2)^{3/2}} \qquad \text{10-32}$$

We will now use this result to calculate the gravitational field of a spherical shell of mass M and radius R at a point a distance r from the center of the shell. We first consider the case in which the point is outside the shell. By

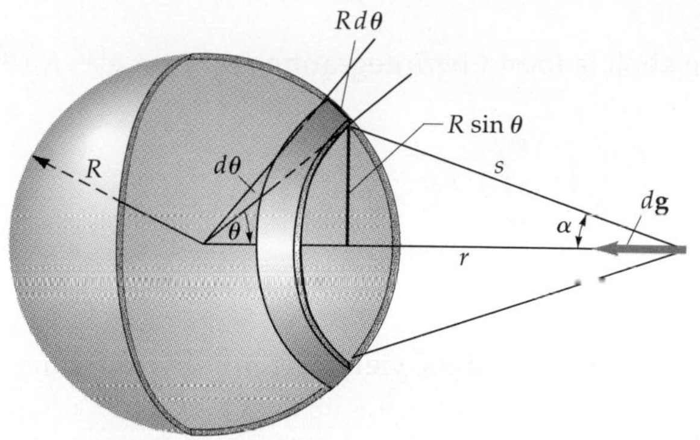

Figure 10-17 A uniform spherical shell of radius R and total mass M. The strip shown can be considered to be a ring of width $R\,d\theta$ and circumference $2\pi\,R \sin\theta\,d\theta$. The gravitational field of the shell is found by summing over all such strips on the shell, which is done by integrating from $\theta = 0$ to $\theta = 180°$.

symmetry, the field must be radial. The geometry is shown in Figure 10-17. We choose for our element of mass the strip shown, which can be considered to be a ring. The radius of the strip is $R \sin\theta$ and the width is $R\,d\theta$. If the total area of the shell is A, the area dA of the strip is the circumference, $2\pi R \sin\theta$, times the width, $R\,d\theta$. If M is the total mass of the shell, the mass of the strip of area dA is

$$dM = M\,\frac{dA}{A} = M\frac{2\pi R^2 \sin\theta\,d\theta}{4\pi R^2}$$

or

$$dM = \frac{M}{2}\sin\theta\,d\theta \qquad 10\text{-}33$$

where we have used $A = 4\pi R^2$ for the total area of the shell. Using Equation 10-31 for the field of a ring, we obtain

$$dg_r = -\frac{G\,dM}{s^2}\cos\alpha$$

Using Equation 10-33 for dM, we have

$$dg_r = -\frac{GM\sin\theta\,d\theta}{2s^2}\cos\alpha \qquad 10\text{-}34$$

Before integrating over the entire shell, we must eliminate two of the three related variables s, θ, and α. It turns out to be easiest to write everything in terms of s, which varies from $s = r - R$ at $\theta = 0$ to $s = r + R$ at $\theta = 180°$. By the law of cosines, we have

$$s^2 = r^2 + R^2 - 2rR\cos\theta$$

Differentiating gives

$$2s\,ds = +2rR\sin\theta\,d\theta$$

or

$$\sin\theta\,d\theta = \frac{s\,ds}{rR}$$

An expression for $\cos\alpha$ can be obtained by applying the law of cosines to the same triangle. We have

$$R^2 = s^2 + r^2 - 2sr\cos\alpha$$

or

$$\cos\alpha = \frac{s^2 + r^2 - R^2}{2sr}$$

Substituting these results into Equation 10-34 gives

$$dg_r = -\frac{GM}{2s^2}\,\frac{s\,ds}{rR}\,\frac{s^2 + r^2 - R^2}{2sr} = -\frac{GM}{4r^2R}\left(1 + \frac{r^2 - R^2}{s^2}\right)ds \qquad 10\text{-}35$$

The field due to the entire shell is found by integrating from $s = r - R$ ($\theta = 0$) to $s = r + R$ ($\theta = 180°$).

$$g_r = -\frac{GM}{4r^2R}\int_{r-R}^{r+R}\left(1 + \frac{r^2 - R^2}{s^2}\right)ds$$

$$= -\frac{GM}{4r^2R}\left[s - \frac{r^2 - R^2}{s}\right]_{r-R}^{r+R}$$

Substitution of the upper and lower limits yields $4R$ for the quantity in brackets. Thus,

$$g_r = -\frac{GM}{r^2}$$

which is Equation 10-27*a*.

We will now calculate the gravitational field of the shell at a point located inside the shell. This calculation is identical to that for a point outside the shell except that s now varies from $R - r$ to $R + r$. Thus,

$$g_r = -\frac{GM}{4r^2R}\left[s - \frac{r^2 - R^2}{s}\right]_{R-r}^{R+r}$$

Substitution of these upper and lower limits yields 0. Therefore,

$$g_r = 0$$

which is Equation 10-27*b*.

Question

6. Explain why the gravitational field increases with r rather than decreasing as $1/r^2$ as one moves out from the center inside a solid sphere of uniform mass.

Summary

1. Kepler's three laws are

Law 1. All planets move in elliptical orbits with the sun at one focus.
Law 2. A line joining any planet to the sun sweeps out equal areas in equal times.
Law 3. The square of the period of any planet is proportional to the cube of the planet's mean distance from the sun.

Kepler's laws can be derived from Newton's law of gravity. The first and third laws follow from the fact that the force exerted by the sun on the planets varies inversely as the square of the separation distance. The second law follows from the fact that the force exerted by the sun on a planet is along the line joining them and so the angular momentum of the planet is conserved. Kepler's laws also hold for any object orbiting another in an inverse-square field, such as a satellite orbiting a planet.

2. Newton's law of gravity postulates that every object exerts an attractive force on every other object that is proportional to the masses of the two objects and inversely proportional to the square of the distance separating them. The gravitational force exerted by a particle of mass m_1 on another particle of mass m_2 a distance r away is given by

$$\mathbf{F}_{12} = -\frac{Gm_1m_2}{r_{12}^2}\hat{\mathbf{r}}_{12}$$

where G, the universal gravitational constant, has the value

$$G = 6.67 \times 10^{-11}\ \mathrm{N{\cdot}m^2/kg^2}$$

3. The gravitational force exerted by the earth on a mass m a distance r from the center of the earth is directed toward the earth and has magnitude

$$F = \frac{GM_Em}{r^2}$$

The gravitational force on a mass divided by the mass is called the gravitational field. The gravitational field of the earth points toward the earth and has magnitude

$$g(r) = \frac{GM_E}{r^2}$$

4. The fact that all bodies have the same free-fall acceleration near the surface of the earth implies that the gravitational mass and the inertial mass of an object are equal. This equivalence of gravitational and inertial mass has been established experimentally to a very high degree of accuracy.

5. The gravitational potential energy of an object of mass m at a distance r from the center of the earth is given by

$$U(r) = U_{max} - \frac{GM_Em}{r} \qquad U = 0 \text{ at } r = R_E$$

where $U_{max} = GM_Em/R_E = mgR_E$ and the potential energy is chosen to be zero at the earth's surface. If an object is given an initial kinetic energy at the earth's surface greater than or equal to U_{max}, the object will escape from the earth. In the absence of air resistance, the speed needed for escape is approximately 11.2 km/s.

6. When the gravitational potential energy is chosen to be zero for an infinite separation of the bodies, $U(r)$ is given by

$$U(r) = -\frac{GMm}{r} \qquad U = 0 \text{ at } r = \infty$$

and the orbits of objects in a gravitational field can be simply classified. If the total energy of an orbiting object is less than zero, the object is bound and the orbit is an ellipse. If the total energy is greater than or equal to zero, the object is not bound and the orbit is a parabola (for zero energy) or a hyperbola (for positive energy).

7. The gravitational field of a uniform spherical shell outside the shell is the same as if all the mass of the shell were concentrated at the center. The field inside the shell is zero. These results are obtained directly from Newton's law of gravity by integration. The gravitational field of a uniform solid sphere of mass M and radius R at a point a distance r from the center of the sphere is given by

$$g_r = -\frac{GM}{r^2} \qquad r > R$$

$$g_r = -\frac{GM}{R^3}r \qquad r < R$$

Isaac Newton (1642–1727)

A.P. French
Massachusetts Institute of Technology

When one considers all that Newton achieved, and the cultural and scientific environment in which he achieved it, there is reason to regard him as the greatest scientist—and perhaps the greatest genius—who has ever lived.

Despite the godlike status that some have tried to ascribe to him, he was of course a fallible human being. He groped and fumbled, as all scientists do, in approaching a new problem; he even fudged things a bit sometimes in trying to reconcile theory with observation. But the extent and profundity of his discoveries remain without parallel.

This would not have been predictable from his beginnings as a somewhat unloved child growing up in the English countryside, though he did demonstrate practical ingenuity and a probing curiosity, and was a good enough student to be admitted to Cambridge University. It was there that the real Newton emerged, with astonishing suddenness, as he approached the end of his undergraduate career. This young man, still an undergraduate, devoured most of what was then known in mathematics and began making his own original contributions. His physics began at about the same time with investigations into optics, and through a succession of controlled experiments he built up a picture of what light is and how it behaves. Here his powers as an experimentalist revealed themselves—the brilliantly analytical mind behind the experiments that it devised.

His first concern seems to have been with the way in which the eye perceives color. The eye as an image-forming device was already well understood, but color perception was much more mysterious. (It still is.) One of the 21-year-old Newton's first experiments, in 1664, took the form of pressing one side of his eyeball with a fingertip and observing the colored rings that appeared to be formed around the pressure point. A little later he did the truly horrifying experiment of thrusting a bodkin into his eye socket, between the bone and the eyeball, to study the same phenomenon closer to the center of the retina (Figure 1). He also stared directly at the Sun and studied the subsequent color impressions he obtained by looking at light or dark objects. It is a miracle that he did not blind himself or do himself serious injury.

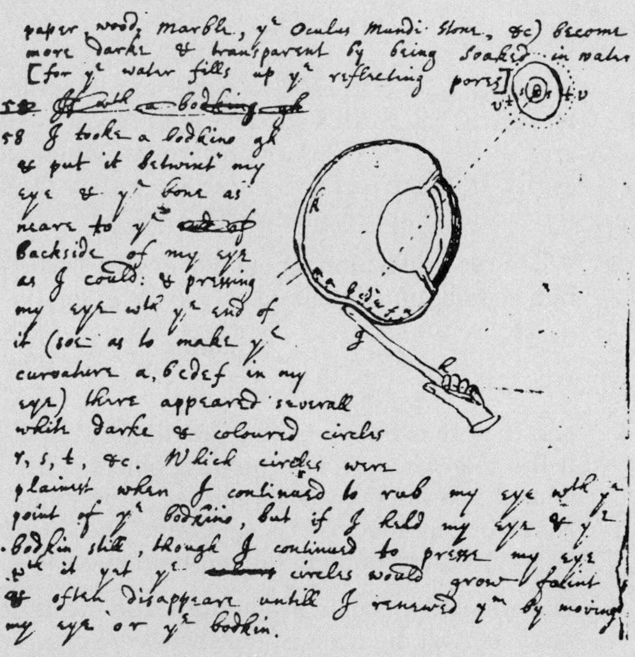
paper, wood, marble, y^e^ Oculus Mundi stone, &c) become more darke & transparent by being soaked in water [for y^e^ water fills up y^e^ reflecting pores]
58 I tooke a bodkine gh
& put it betwixt my
eye & ye bone as
neare to ye backside of my eye
as I could: & pressing
my eye wth ye end of
it (soe as to make y^e^
curvature a, bcdef in my
eye) there appeared severall
white darke & coloured circles
r, s, t, &c. Which circles were
plainest when I continued to rub my eye wth y^e^
point of ye bodkine, but if I held my eye & y^e^
bodkin still, though I continued to presse my eye
wth it yet ye circles would grow faint
& often disappeare untill I renewed y^m^ by moving
my eye or ye bodkin.

Figure 1 Newton's experiment on colors produced by pressing his eyeball.

He then concentrated on investigating the refraction of light through prisms—in particular the white light from the Sun (Figure 2). It had been the belief that colors were the result of the modification of pure white light by materials. What Newton con-

A. P. French was educated at Cambridge University. He was a member of the British Mission at Los Alamos in the Manhattan Project during World War II. From 1948 to 1955 he taught physics and did nuclear research at the Cavendish Laboratory, Cambridge. He then emigrated to the U.S.A. and joined the faculty of the University of South Carolina. In 1962 he moved to M.I.T., where he has been a member of the permanent faculty since 1964. His chief interests for many years have been in curriculum development, teaching, and the history of physics. He was President of the American Association of Physics Teachers in 1985–1986.

Figure 2 One of Newton's first experiments in forming an extended spectrum (from Voltaire, *The Elements of Sir Isaac Newton's Philosophy* 97).

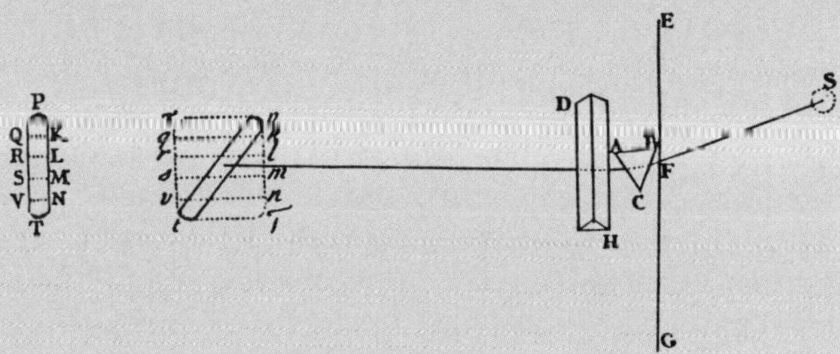

Figure 3 Newton's crossed-prism experiment (from Newton, *Opticks* 36).

cluded was that the various colors were already present in the white light, that this was itself complex, and that the prism merely resolved it into its components. In all kinds of ways he sought to demonstrate the correctness of his analysis. One very beautiful experiment was to pass the light emerging from a first prism through a second prism oriented at right angles to it (Figure 3). This showed that no further modification of the light occurred; once the separation of the white light into its various components had been achieved with the first prism, the second prism merely refracted the individual colors by differing amounts and produced a final spectrum along a diagonal line, rather than a pattern filling a rectangle.

Newton's sustained and many-faceted attack on a single problem was something for which there was essentially no precedent or parallel in the scientific literature of his time. It was something qualitatively different from the mere collection of precise observations. Even in our own day, his approach could be taken as a model for the exploration of a new phenomenon.

In 1669, at the age of twenty-six, Newton was appointed Professor of Mathematics at Cambridge University. He chose to give his inaugural course of lectures on the subject of optics, and their final section—another Newtonian *tour de force*—was a quantitative explanation of the rainbow. A year or two later he designed and constructed the first reflecting telescope, the metal mirror of which was not only ground and polished by him, but was made of an alloy evolved during the chemical (and alchemical) researches that he had been conducting at the same time. All of this work was described in his *Opticks*, which was completed in the 1670s but not published until 1704. It is essentially a blow-by-blow account of his work in this field, written in an easy and conversational style, and makes excellent reading.

His most glorious achievements, however, were in mechanics, and were published in detail in his masterpiece, *Mathematical Principles of Natural Philosophy*, universally known as the *Principia*. Its style is in total contrast to the *Opticks*. Quite apart from the fact that its first editions were in classical Latin, the whole presentation is forbidding. Using pure geometry with a virtuosity that few mathematicians today could emulate, Newton embodied his arguments in a Euclidean framework of axioms, theorems, lemmas, and corollaries. At least, this is the general character of the first two parts. In the third part, "The System of the World," Newton proceeded to apply his theory to all the celestial phenomena that were known or knowable in his time. It was a dazzling performance.

Although the *Principia* was not published until 1687, Newton's conjectures about the motions of the Moon and the planets had begun about twenty years earlier, soon after his graduation. There seems to be no reason to doubt the basic truth of the story of Newton and the apple: that in 1666, having left Cambridge for a while on account of the Great Plague, he was moved by the fall of an apple to speculate if the Moon itself was falling toward the earth in a similar way. He was at that time already aware of Kepler's discovery (1619) that the squares of the periodic times of the planets are in proportion to the cubes of their mean distances from the Sun. Newton deduced, to quote his own words (written before 1669), that "the endeavour of receding from the Sun will be reciprocally as the squares of the distances from the Sun"—that is, an inverse-square law of "centrifugal force." The germ of Newton's theory of universal gravitation was undoubtedly present here, but his full understanding and exploitation of the theory did not take place until, in the space of about two years (starting in 1684), he wrote it all out in the *Principia* (chiefly at the urging of his devoted admirer Edmond Halley, who financed the publication of the work out of his own pocket).

Continued

Newton had special reason for dividing the *Principia* into three separate parts ("Books"). Book I, "The Motion of Bodies," is a development of the basic dynamics of particles, and embodies the theoretical substructure for the celestial mechanics of Book III. Book II, "The Motion of Bodies in Resisting Mediums," had as its main result, and its main goal, a demonstration of the invalidity of Descartes' hypothesis that the circular motions of heavenly bodies resulted from their being swept around in huge fluid vortices. So Book I is concerned with the general laws of motion, and Book II with fluid mechanics (in which Newton solved a number of important problems, such as aerodynamic drag and a calculated value for the speed of sound).

In Book I Newton began with his ideas about space and time. Although he recognized that the only measurements we can make of positions and velocities are relative ones, he believed that space itself—and acceleration, the central quantity in his laws of motion—must be absolute. The proof? Take a bucket of water hanging at the end of a twisted rope, and let it spin (he did the experiment). Initially, the bucket turns, but the water remains at rest and its surface is flat. Later, the water picks up the rotation and its surface becomes concave. If the bucket is then suddenly stopped, the water continues to rotate and its surface remains concave. Clearly, the important feature is not the relative rotation of bucket and water, and Newton concluded that what matters is the rotation of the water (and the associated radial acceleration) within a fundamental reference frame, closely related to the space defined by the fixed stars. (The bulging of the Earth in its own rotation points to the same conclusion.) The deep implications of this have continued to be debated into our own day.

Newton's laws of motion themselves (Chapter 4) were based on some very simple experiments: observations of collisions between different kinds and sizes of objects. From them, Newton extracted his quantitative concept of momentum and the role of force in changing the momentum of a given body. He also demonstrated one of the most important and fundamental relationships in physics, the proportionality of weight to mass. He did this by showing that pendulums of the same length all had the same period of swing, regardless of the mass or composition of the pendulum bob. This result is, in effect, a more refined and precise test of the fact that different objects have equal accelerations under gravity down any given slope (including free vertical fall). Coupled with Newton's second law, this result implied that gravitational force is proportional to inertial mass—to an accuracy, Newton claimed, of about one part in a thousand.

His work in optics and in basic mechanics (not to mention pure mathematics and the invention of calculus) would already have been enough to mark Newton as one of the greatest scientists of all time. But the crowning glory of his work was surely the third book of the *Principia,* in which he analyzed the workings of the universe as a magnificent machine.

Newton began, perhaps surprisingly, with the satellites of Jupiter. The discovery of the four principal satellites by Galileo in 1610, using one of the earliest telescopes, was one of the great events in astronomical history—the first "additions" to the solar system since antiquity. Galileo himself had recorded their positions, night after night, for many months. As a result he was able to state their orbital periods with considerable accuracy. The data showed clearly that Jupiter's satellites, like the planets going around the sun, conformed to Kepler's third law (Section 10-1). For Newton, this was powerful evidence that a truly universal inverse-square law of gravitation was at work. He asked John Flamsteed, the first Astronomer Royal (and a man whom Newton treated very shabbily) to provide up-to-date observations. In fact, Flamsteed's results were not greatly different from those of Galileo, who must have been a magnificent observer. In 1687 Giovanni Cassini provided similar data for the five recently discovered satellites of Saturn.

From all this information, plus his fundamental law of gravitation, Newton was able to compare the masses of Earth, Sun, Jupiter, and Saturn, because the motion of any satellite involves a force proportional to the central mass that attracts it, and this fact is reflected in the orbital period. Newton, armed with his theories, had thus gained access to hitherto unknowable facts about the universe. This was exploration on a grand scale.

Another of Newton's great achievements was his application of universal gravitation to the motions of comets. Newton had first undertaken detailed observations of a comet in 1664, while he was still an undergraduate, but his interest in such matters subsided until December 1680, when a spectacular comet became visible, moving away from the Sun toward the outer reaches of the solar system. Newton himself charted its course. At the same time John Flamsteed was following it from the Royal Observatory at Greenwich. Flamsteed was also familiar with a less brilliant comet that had been seen approaching the sun only a month earlier. He proposed to Newton that these were one and the same object. Newton at first strongly resisted this suggestion; such a complete reversal of direction seemed inexplicable to him at the time. But the idea grew on him, and when, in the process of writing the *Principia* in 1685, he made a full assault on the problem of comets, he took the comet of 1680/81 as his chief case. Meanwhile, in 1682, his friend Halley had observed the comet that came to bear his name.

Comets, in general, are such small objects that they are invisible until they enter the inner regions of the solar system, where their trajectories can be excellently fitted as parabolas. In fact, as Newton recognized, the orbit of a comet is either hyperbolic, in which case it makes only one close encounter with the sun, or a very elongated ellipse, in which case it returns periodically (like Halley's comet, with its period of 76 years).

Another of Newton's great calculations involved the Earth itself. Using once again the inverse-square law of gravitational attraction, he showed how the Moon and the Sun would combine to create a double bulge in the ocean, with the maxima diametrically opposite one other. A given place on the spinning Earth would then encounter two maxima and two minima of ocean level every day. Newton knew that the actual behavior of the tides is much more complicated than this, but he was able to explain the way in which the effects of Sun and Moon combine differently at different times to produce the monthly cycle of tides of different heights. A much simpler problem that Newton solved was that of the Earth's shape—an oblate spheroid whose equatorial radius, because of "centrifugal force," is about one part in 300 bigger than the polar radius.

In all of the problems described up to this point, Newton felt himself to be on pretty solid ground; but no account of his role as an explorer of the universe would be complete without some mention of his more tentative speculations. They reveal a mind reaching out in all directions to understand how the universe is put together and how it works. Newton's curiosity in these matters knew no limits, and his brilliant scientific imagination is always in evidence.

The *Principia* itself contains some fascinating examples. Several of them have a bearing on the ages of the Earth and the solar system. In his discussion of the heating of comets by the sun (which creates their tails through vaporization of material, combined with the solar wind) Newton offers an estimate of how long it would have taken for the Earth to cool down if it had started out as a red-hot ball. His result was at least 50,000 years—far shorter than Baron William Kelvin calculated on the same basis two centuries later, but disturbingly long for any of Newton's contemporaries who might have been tempted to believe Bishop Ussher's date of 4004 B.C. for the Creation.

A more definite conjecture concerned the persistence of planetary motions. Astronomical records made it certain that these motions had continued undiminished for thousands of years; clearly they experienced no detectable resistance. Was space simply a void? Newton was not sure, but certainly there could be no "gross matter." This fitted in well with what Newton had calculated about the Earth's atmosphere. He had discovered the theoretical law of an exponential decrease of atmospheric pressure and density with height. Both would fall by a factor of a million for each 75 miles or so of altitude, and hence the space between the planets must be an essentially perfect vacuum. And yet . . . If there was no intervening medium, how was the force of gravity transmitted? He expressed his thoughts on this in a famous letter to Richard Bentley (see p. 89).

At the end of his *Opticks*, in a series of what he called Queries, Newton set down a number of his final speculations about the physical world. Many, but not all, are indeed concerned with optics. The next to last Query seems to foreshadow Einstein: "Are not gross Bodies and Light convertible into one another, and may not Bodies receive much of their Activity from the Particles of Light which enter their Composition?" It sounds just like a qualitative statement of $E = mc^2$! Of course it was not, and could not have been; Newton's concern here was to compare the absorption and emission of light with various other chemical and biological transformations. But it is yet one more example of his boundless capacity for speculation.

The last Query introduced a lengthy discussion in which Newton surveyed the whole range of the forces of nature. He began: "Have not the small Particles of Bodies certain Powers, Virtues, or Forces, by which they act at a distance, not only upon the Rays of Light for reflecting, refracting, and inflecting them, but also upon one another for producing a great Part of the Phenomena of Nature?" He then went on to discuss chemical reactions, surface tension phenomena, and the cohesion of solids, and speculated on the possible role of electric and magnetic forces in producing them. He expressed a belief in a fundamental uniformity of physical phenomena, saying: "And thus Nature will be very comformable to her self and very simple, performing all the great Motions of the Heavenly Bodies by the Attraction of Gravity which intercedes those Bodies, and almost all the small ones of their Particles by some other attractive and repelling Powers which intercede the Particles." It was a great vision, which he had done more than any other person to make possible. It is still not complete, more than three centuries later, but this unification of the forces of nature remains the physicist's ultimate dream.

Suggestions for Further Reading

Cohen, I. Bernard: "Newton's Discovery of Gravity," *Scientific American*, March 1981, p. 166.

A historical article in which the development of Newton's ideas leading to his formulation of the law of gravity is put into the context of the ideas of other natural philosophers of the time.

Shamos, Morris H.: "The Law of Gravitation—Henry Cavendish," in *Great Experiments in Physics*, Henry Holt and Company, New York, 1959. Reprinted by Dover, 1987.

Cavendish's description of his experiments, with editorial annotations for clarity and a biographical sketch.

Whipple, Fred L., and J. Allen Hynek: "Observations of Satellite I," *Scientific American*, December 1957, p. 37.

An account of the American effort to determine precisely the orbit of the first artificial earth satellite, launched by the U.S.S.R. in 1957.

Review

A. Objectives: After studying this chapter, you should:

1. Be able to state Kepler's three empirical laws of planetary motion.

2. Be able to derive Kepler's third law of planetary motion, $T^2 \propto r^3$, for circular orbits.

3. Be able to discuss gravitational and inertial mass.

4. Be able to give the value for the speed needed to escape from the earth.

5. Be able to sketch a graph of gravitational potential energy as a function of separation distance and to discuss the calculation of the escape speed.

6. Be able to discuss the relationship between the total energy of an orbiting body and the type of orbit.

7. Be able to describe the gravitational field of a spherical shell and of a solid sphere.

B. Define, explain, or otherwise identify:

Kepler's laws
Ellipse
Newton's law of gravity
Universal gravitational constant
Central force
Gravitational mass
Inertial mass
Escape speed
Binding energy

C. True or false: If the statement is true, explain why it is true. If it is false, give a counterexample.

1. Kepler's law of equal areas implies that gravity varies inversely with the square of the distance.

2. The planet closest to the sun, on the average, has the shortest period of revolution about the sun.

3. The force that causes an apple to fall has the same origin as the force that causes the moon to move in a circle about the earth.

Problems

Level I

10-1 Kepler's Laws

1. Suppose a small planet with a period of 5 y were discovered. What would be its mean distance from the sun?

2. Halley's comet has a period of about 76 y. What is its mean distance from the sun?

3. The comet Kohoutek has a period estimated to be at least 10^6 y. What is its mean distance from the sun?

4. The radius of the earth's orbit is 1.496×10^{11} m and that of Uranus is 2.87×10^{12} m. What is the period of Uranus?

5. The mean distance of Pluto from the sun is 39.5 AU. Find the period of Pluto.

6. A planet moves around a massive sun with constant angular momentum. When the planet is at perihelion, it has a speed of 5×10^4 m/s and is 1.0×10^{15} m from the sun. The orbital radius increases to 2.2×10^{15} m at aphelion. What is the planet's speed at aphelion?

7. A comet orbits the sun with constant angular momentum. It has a maximum radius of 150 AU, where its speed is 7×10^3 m/s. The comet's closest approach to the sun is 0.4 AU. What is its speed at perihelion?

10-2 Newton's Law of Gravity

8. One of Jupiter's moons, Io, has a mean orbital radius of 422 Mm and a period of 1.53×10^5 s. (*a*) Find the mean orbital radius of another of Jupiter's moons, Callisto, whose period is 1.44×10^6 s. (*b*) Use the known value of G to compute the mass of Jupiter.

9. Uranus has a moon, Umbriel, whose mean orbital radius is 267 Mm and whose period is 3.58×10^5 s. (*a*) Find the period of another of Uranus's moons, Oberon, whose mean orbital radius is 586 Mm. (*b*) Use the known value of G to find the mass of Uranus.

10. The mass of Saturn is 5.69×10^{26} kg. (*a*) Find the period of its moon Mimas, whose mean orbital radius is 186 Mm. (*b*) Find the mean orbital radius of its moon Titan, whose period is 1.38×10^6 s.

11. Calculate the mass of the earth from the period of the moon $T = 27.3$ d, its mean orbital radius $r_m = 384$ Mm, and the known value of G.

12. Use the period of the earth (1 y), its mean orbital radius (1.496×10^{11} m), and the value of G to calculate the mass of the sun.

13. Calculate the mass of the earth using the known values of G, g, and R_E.

14. A body is dropped from a height of 6.37 Mm above the surface of the earth. What is its initial acceleration?

15. Suppose you land on a planet of another solar system that has the same mass per unit volume as the earth but has 10 times the earth's radius. What would you weigh on this planet compared with what you weigh on earth?

10-3 Measurement of the Universal Gravitational Constant

16. The masses in a Cavendish apparatus are $m_1 = 10$ kg and $m_2 = 10$ g, the separation of their centers is 5 cm, and the rod separating the two small masses is 20 cm long. (*a*) What is the force of attraction between the large and small masses? (*b*) What torque must be exerted by the suspension to balance these forces?

17. The masses in a Cavendish apparatus are $m_1 = 12$ kg and $m_2 = 15$ g, and the separation of their centers is 5 cm. (*a*) What is the force of attraction between these two masses? (*b*) If the rod separating the two small masses is 18 cm long, what torque must be exerted by the suspension to balance the torque exerted by gravity?

10-4 Gravitational and Inertial Mass

18. A standard object, which is defined as having a mass of exactly 1 kg, is given an acceleration of 2.6587 m/s^2 when a certain force is applied to it. A second object of unknown mass acquires an acceleration of 1.1705 m/s^2 when the same force is applied to it. (*a*) What is the mass of the second object? (*b*) Is the mass that you determined in part (*a*) gravitational or inertial mass?

19. The weight of a standard object, which is defined as having a mass of exactly 1 kg, is measured to be 9.81 N. In the same laboratory, a second object weighs 56.6 N. (*a*) What is the mass of the second object? (*b*) Is the mass you determined in part (*a*) gravitational or inertial mass?

10-5 Escaping the Earth

20. The planet Saturn has a mass 95.2 times that of the earth and a radius 9.47 times that of the earth. Find the escape speed for objects near the surface of Saturn.

21. Find the escape speed for a rocket leaving the moon. The acceleration of gravity on the moon is 0.166 times that on earth, and the moon's radius is $0.273R_E$.

22. Suppose that in space there is a duplicate earth, except that it has no atmosphere, is not rotating, and is not in motion around any sun. What initial velocity must a spacecraft have to travel vertically upward a distance above the surface of the planet equal to one earth radius?

10-6 Potential Energy, Total Energy, and Orbits

23. (*a*) Taking the potential energy to be zero at infinite separation, find the potential energy of a 100-kg object at the surface of the earth. (Use 6.37 Mm for the earth's radius.) (*b*) Find the potential energy of the same object at a height above the earth's surface equal to the earth's radius. (*c*) What would be the escape speed for a body projected from this height?

24. A satellite with a mass of 300 kg moves in a circular orbit 5×10^7 m above the earth's surface. (*a*) What is the gravitational force on the satellite? (*b*) What is the speed of the satellite? (*c*) What is the period of the satellite?

"Castle in the Pyrenees" by René Magritte.

10-7 The Gravitational Field of a Spherical Shell and a Solid Sphere

25. A spherical shell has a radius of 2 m and a mass of 300 kg. What is the gravitational field (*a*) 0.5 m, (*b*) 1.9 m, and (*c*) 2.5 m from the center of the shell?

26. A spherical shell has a radius of 2 m and a mass of 300 kg, and its center is located at the origin of a coordinate system. Another spherical shell with a radius of 1 m and mass 150 kg is inside the larger shell with its center at 0.6 m on the x axis. What is the gravitational force of attraction between the two shells?

Level II

27. A point mass m_0 is initially at the surface of a large sphere of mass M and radius R. How much work is needed to remove it to a very large distance away from the large sphere?

28. Two concentric uniform spherical shells have masses M_1 and M_2 and radii a and $2a$ as shown in Figure 10-18. (*a*) What is the gravitational force on a point mass m located a distance $3a$ from the center of the shells? (*b*) What is the force on the point mass if it is located a distance $1.9a$ from the center of the shells? (*c*) What is the force on the point mass if it is located a distance $0.9a$ from the center of the shells?

Figure 10-18 Problem 28.

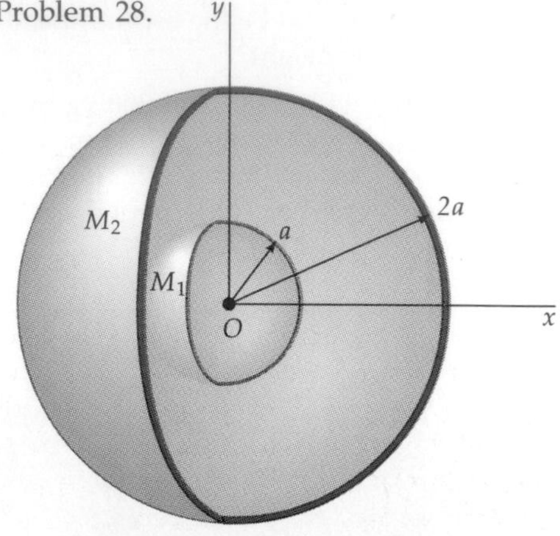

29. The inner spherical shell in Problem 28 is shifted such that its center is now at $x = 0.8a$. The points $3a$, $1.9a$, and $0.9a$ lie along the same radial line from the center of the larger spherical shell. (*a*) What is the force on m at $3a$? (*b*) What is the force on m at $1.9a$? (*c*) What is the force on m at $0.9a$?

30. A uniform sphere of radius 100 m and density 2000 kg/m^3 is in free space far from other massive objects. (*a*) Find the gravitational field outside of the sphere as a function of r. (*b*) Find the gravitational field inside the sphere as a function of r.

31. A particle is projected from the surface of the earth with a speed equal to twice the escape speed. When it is very far from the earth, what is its speed? (Neglect air resistance.)

32. What initial speed should a particle be given if it is to have a final speed when it is very far from the earth equal to the escape speed?

33. A space probe sent from the earth is to have a speed of 50 km/s when it is very far from the earth. What speed is needed for the probe at the surface of the earth? (Neglect air resistance.)

34. An object is dropped from rest from a height of 4×10^6 m above the surface of the earth. If there were no air resistance, what would its speed be when it strikes the earth?

35. An object is projected upward from the surface of the earth with an initial speed of 4 km/s. Find the maximum height it reaches. (Neglect air resistance.)

36. (*a*) Calculate the energy in joules necessary to launch a 1-kg mass from the earth with the escape speed. (*b*) Convert this energy to kilowatt-hours. (*c*) If energy can be obtained at 10 cents per kilowatt-hour, what is the minimum cost of giving an 80-kg astronaut enough energy to escape the earth's gravitational field?

37. Two planets of equal mass orbit a much more massive star (Figure 10-19). Planet m_1 moves in a circular orbit of radius 100 Gm with a period of 2 y. Planet m_2 moves in an elliptical orbit, with its closest distance $r_1 = 100$ Gm and its farthest distance $r_2 = 180$ Gm. (*a*) Using the fact that the mean radius of an elliptical orbit is the length of the semimajor axis, find the period of m_2's orbit. (*b*) What is the mass of the star? (*c*) Which planet has the greater speed at point P? Which has the greater total energy? (*d*) How does the speed of m_2 at point P compare with its speed at point A?

Figure 10-19 Problem 37.

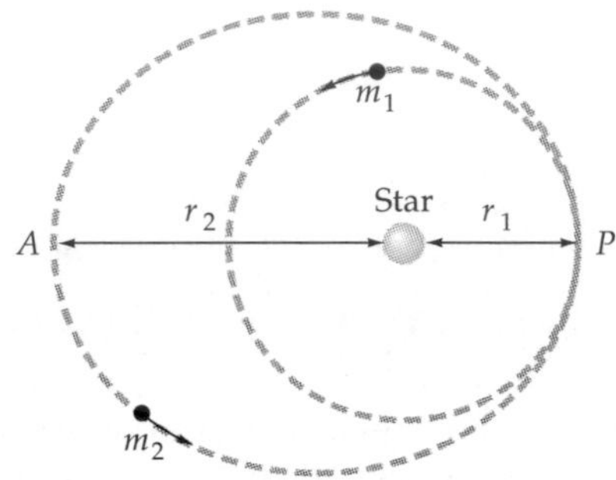

38. A thick spherical shell of mass M and uniform density has an inner radius R_1 and an outer radius R_2. Find the gravitational field g_r as a function of r for all possible values of r.

39. Two point particles of mass m are on the y axis at $y = +a$ and $y = -a$ as shown in Figure 10-20. (*a*) Show that the force exerted by these two particles on a third particle of mass m_0 located on the x axis at a distance x from the origin is given by

$$\mathbf{F} = -\frac{2Gmm_0x}{(x^2 + a^2)^{3/2}}\,\mathbf{i}$$

(*b*) What is the gravitational field $\mathbf{g}$ on the x axis due to the two particles on the y axis? (*c*) Show that g_x due to the two particles on the y axis is approximately $-2Gm/x^2$ when x is much greater than a. (*d*) Show that the maximum value of $|g_x|$ occurs at the points $x = \pm a/\sqrt{2}$.

Figure 10-20 Problem 39.

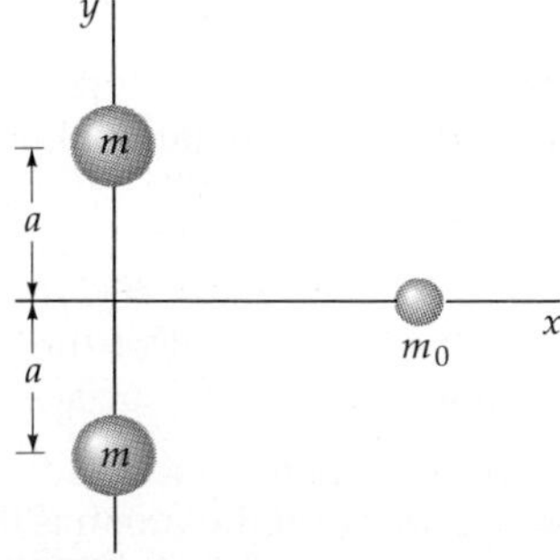

40. (*a*) Sketch a plot of the gravitational field g_x versus x due to a uniform ring of mass M and radius R whose axis is the x axis. (*b*) At what points is the magnitude of g_x maximum?

41. Five equal masses M are equally spaced on the arc of a semicircle of radius R as shown in Figure 10-21. A mass m is located at the center of curvature of the arc. If M is 3 kg, m is 2 kg, and R is 10 cm, what is the force on m due to the five masses?

Figure 10-21 Problem 41.

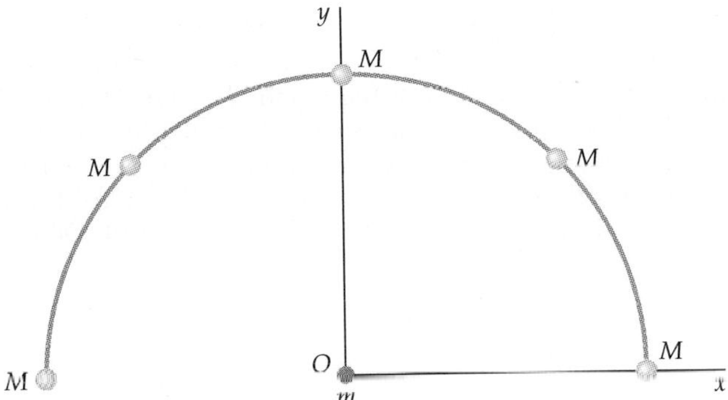

42. (*a*) Show that the gravitational field of a ring of uniform mass is zero at the center of the ring. (*b*) Figure 10-22 shows a point P in the plane of the ring but not at its center. Consider two elements of the ring of length s_1 and s_2 at distances of r_1 and r_2, respectively. What is the ratio of the masses of these elements? Which produces the greater gravitational field at point P? What is the direction of the field at point P due to these elements? (*c*) What is the direction of the gravitational field at point P due to the entire ring? (*d*) Suppose that the gravitational field varied as $1/r$ rather than $1/r^2$. What would be the net gravitational field at point P due to the two elements? (*e*) How would your answers to parts (*b*) and (*c*) differ if point P were inside a spherical shell of uniform mass rather than inside a plane circular ring?

Figure 10-22 Problem 42.

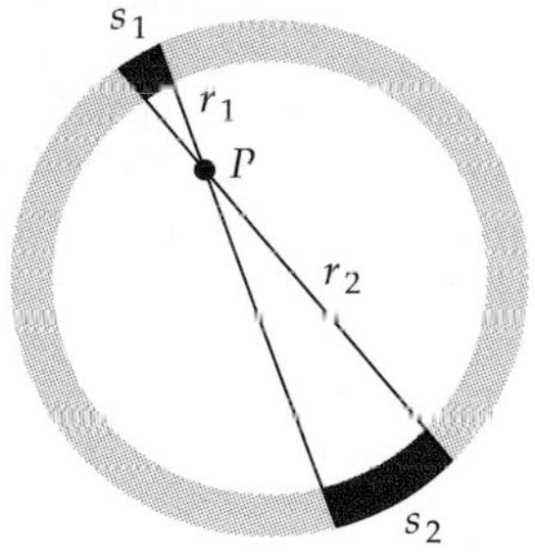

43. A superconducting gravity meter can measure changes in gravity of the order $\Delta g/g = 10^{-11}$. (*a*) Estimate the maximum range at which an 80-kg person can be detected by this gravity meter. Assume that the gravity meter is stationary, and that the person's mass can be considered to be concentrated at his or her center of gravity. (*b*) What vertical change in the position of the gravity meter in the earth's gravitational field is detectable?

44. A spaceship traveling on a line from the earth to the moon passes through a point where the force of gravity of the earth and that of the moon exactly cancel. How far from the center of the earth is this special point?

45. During a solar eclipse, when the moon is between the earth and the sun, a student's weight is affected by the gravitational pull of the moon and the sun. (*a*) If the pull of the earth on the student is 800 N, what is the force of the moon on the student? (*b*) What is the force of the sun on the student? (*c*) What percentage correction due to the pull of the sun and moon should be applied to the reading of a very accurate scale to obtain the student's weight?

46. At the airport, a physics student weighs 800 N. The student boards a jet plane that rises to an altitude of 15,000 m. What is the student's loss in weight?

47. Suppose the earth were a sphere of uniform mass. If there were a deep elevator shaft going 15,000 m into the earth, what would be the loss in weight at the bottom of this deep shaft for a student who weighs 800 N at the surface of the earth?

Level III

48. Assume that the earth is a sphere of uniform mass and that a small-diameter hole has been drilled from the surface to the center. (*a*) How much work would be needed to bring a small object of mass m from the center of the earth to the surface? (*b*) If the object were dropped into the hole at the surface of the earth, what would its speed be when it reaches the earth's center?

49. A uniform sphere of mass M is located near a thin, uniform rod of mass m and length L as shown in Figure 10-23. Find the gravitational force of attraction exerted by the sphere on the rod.

Figure 10-23 Problem 49.

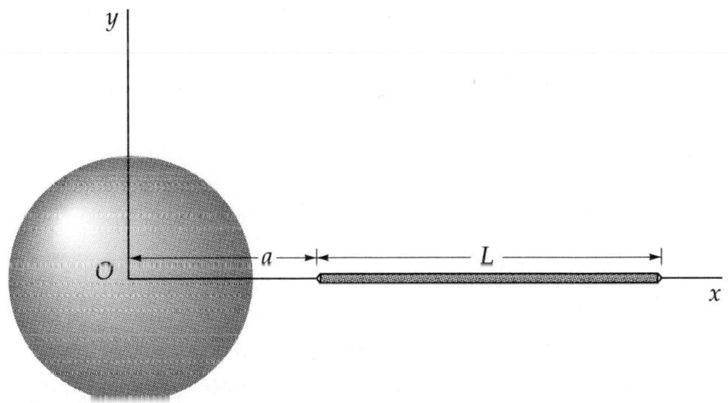

50. A uniform rod of mass $M = 20$ kg and length $L = 5$ m is bent into a semicircle. What is the gravitational force exerted by the rod on a point mass $m = 0.1$ kg located at the center of the circular arc?

51. A spherical shell has a radius R and a mass M. (*a*) Write expressions for the force exerted by the shell on a point mass m_0 when m_0 is outside the shell and when it is inside the shell. (*b*) What is the potential-energy function $U(r)$ for this system when the mass m_0 is at a distance r ($r \geq R$) if $U = 0$ at $r = \infty$? Evaluate this function at $r = R$. (*c*) Using the general relation for $dU = -\mathbf{F}\cdot d\mathbf{r} = -F_r\, dr$, show that U is constant everywhere inside the shell. Using the fact that U is continuous everywhere, including at $r = R$, find the value of the constant U inside the shell. (*d*) Sketch $U(r)$ versus r for all possible values of r.

52. A uniform rod of mass M and length L lies along the x axis with its center at the origin. Consider an element of

length dx at a distance x from the origin. (*a*) Show that this element produces a gravitational field at a point x_0 on the x axis ($x_0 > \frac{1}{2}L$) given by

$$dg_x = -\frac{GM}{L(x_0 - x)^2}\,dx$$

(*b*) Integrate this result over the length of the rod to find the total gravitational field at the point x_0 due to the rod. (*c*) What is the force on an object of mass m_0 at x_0?

53. In this problem, you are to find the gravitational potential energy of the rod in Problem 52 and a point mass m_0 that is on the x axis at x_0. (*a*) Show that the potential energy of an element of the rod dm and m_0 is given by

$$dU = -\frac{GM_0\,dm}{x_0 - x} = -\frac{GMm_0}{L(x_0 - x)}\,dx$$

where $U = 0$ at $x_0 = \infty$. (*b*) Integrate your result for part (*a*) over the length of the rod to find the total potential energy for the system. Write your result as a general function $U(x)$ by setting x_0 equal to a general point x. (*c*) Compute the force on m_0 at a general point x from $F_x = -dU/dx$, and compare your result with that for part (*c*) of Problem 52.

54. Our galaxy can be considered to be a large disk of radius R and mass M of approximately uniform mass density. (*a*) Find the gravitational potential energy of a 1-kg mass on the axis of the disk at a distance x from it due to a ring element of radius r and thickness dr. (*b*) Integrate your result for part (*a*) to find the total gravitational potential energy of a 1-kg mass at a distance x due to the disk. (*c*) From $F_x = -dU/dx$ and your result for part (*b*), find the gravitational field g_x on the axis of the disk.

55. A sphere of radius R has its center at the origin. It has a uniform mass density ρ_0, except that there is a spherical cavity in it of radius $r = \frac{1}{2}R$ centered at $x = \frac{1}{2}R$ as shown in Figure 10-24. Find the gravitational field at points on the x axis for $|x| > R$. (*Hint*: The cavity may be thought of as a sphere of mass $m = \frac{4}{3}\pi r^3 \rho_0$ plus a sphere of mass $-m$.)

Figure 10-24 Problem 55.

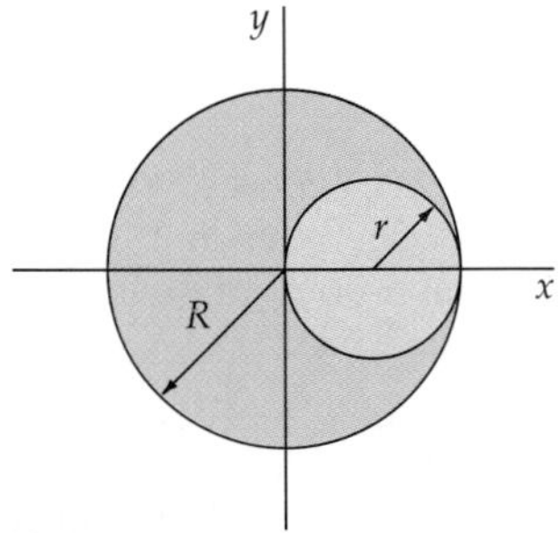

56. For the sphere with the cavity in it in Problem 55, show that the gravitational field inside the cavity is uniform, and find its magnitude and direction.

57. A straight, smooth tunnel is dug through a spherical planet whose mass density ρ_0 is constant. The tunnel passes through the center of the planet and is perpendicular to the planet's axis of rotation, which is fixed in space. The planet rotates with an angular velocity ω such that objects in the tunnel have no acceleration relative to the tunnel. Find ω.

58. A plumb bob near a large mountain is slightly deflected from the vertical by the gravitational attraction of the mountain. Estimate the order of magnitude of the angle of deflection using any assumptions you like.

59. Both the sun and the moon exert gravitational forces on the oceans of the earth, causing tides. (*a*) Show that the ratio of the force exerted by the sun to that exerted by the moon is $M_s r_m^2/M_m r_s^2$, where M_s and M_m are the masses of the sun and moon and r_s and r_m are the distances from the earth to the sun and to the moon. Evaluate this ratio. (*b*) Even though the sun exerts a much greater force on the oceans than the moon does, the moon has a greater effect on the tides because it is the difference in the force from one side of the earth to the other that is important. Differentiate the expression

$$F = \frac{Gm_1m_2}{r^2}$$

to calculate the change in F due to a small change in r. Show that $dF/F = (-2\,dr)/r$. (*c*) During one full day, the distance from the sun or moon to an ocean changes (because of the earth's rotation) by at most the diameter of the earth. Show that for a small change in distance, the change in the force exerted by the sun is related to the change in the force exerted by the moon by

$$\frac{\Delta F_s}{\Delta F_m} \approx \frac{M_s r_m^3}{M_m r_s^3}$$

and calculate this ratio.

60. The solid surface of the earth has a density of about 3000 kg/m^3. A spherical deposit of heavy metals with a density of 8000 kg/m^3 and a radius of 1000 m is centered 2000 m below the surface. Find $\Delta g/g$ directly above this deposit, where Δg is the increase in the gravitational field due to the deposit.

61. Two identical spherical hollows are made in a lead sphere of radius R. The hollows have a radius $R/2$. They touch the outside surface of the sphere and its center as shown in Figure 10-25. The mass of the lead sphere before hollowing was M. (*a*) Find the force of attraction of a small sphere of mass m to the lead sphere at the position shown in the figure. (*b*) What is the attractive force if m is located right at the surface of the lead sphere?

Figure 10-25 Problem 61.

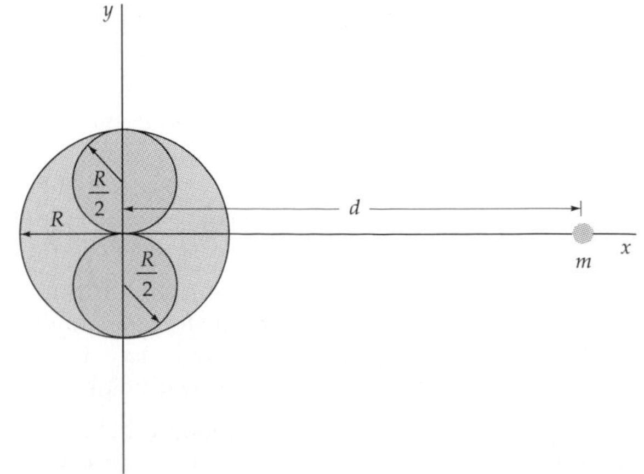

Chapter 11

Mechanics of Solids and Fluids

Scanning-electron micrograph of a fiber-reinforced ceramic composite that was loaded to near the point of complete failure of the ceramic material. The fibers bridge the cracks allowing the composite to carry a high load despite localized damage. Such reinforced materials are used in turbine motors that rotate at over 100,000 rev/min.

The states of matter in bulk can be conveniently divided into solids and fluids. Solids tend to be rigid and to maintain their shape, whereas fluids do not maintain their shape but flow. Fluids include both liquids, which flow under gravity until they occupy the lowest possible regions of their containers, and gases, which expand to fill their containers regardless of their shapes. The distinction between solids and liquids is not sharp. Although ice is considered a solid, the flow of glaciers is well known. Similarly, glasses, and even rocks under great pressure, tend to flow slightly over long periods of time.

In this chapter, we will look at some of the mechanical properties of solids and of fluids at rest and in motion.

11-1 Density

An important property of a substance is the ratio of its mass to its volume, which is called its **density:**

$$\text{Density} = \frac{\text{mass}}{\text{volume}}$$

The Greek letter ρ (rho) is usually used to denote density:

Density defined

$$\rho = \frac{m}{V} \tag{11-1}$$

Because the gram was originally defined as the mass of one cubic centimeter of water, the density of water in cgs units is 1 g/cm^3. Converting to SI units of kilograms per cubic meter, we obtain for the density of water

$$\rho = \frac{1\ \text{g}}{\text{cm}^3} \times \frac{\text{kg}}{10^3\ \text{g}} \times \left(\frac{100\ \text{cm}}{\text{m}}\right)^3 = 10^3\ \text{kg/m}^3 \tag{11-2}$$

The density of water varies with temperature. Equation 11-2 gives its maximum value, which occurs at 4°C. A convenient unit of volume is the **liter** (L):

$$1\ \text{L} = 10^3\ \text{cm}^3 = 10^{-3}\ \text{m}^3$$

In terms of this unit, the density of water is 1.00 kg/L.

When an object's density is greater than that of water, it will sink in water. When its density is less, it will float. In fact, we will show in Section 11-4 that, for floating objects, the fraction of the volume of an object that is submerged in any liquid equals the ratio of the density of the object to that of the liquid. For example, ice has a density of approximately 0.92 g/cm^3, and it floats in water with about 92 percent of its volume submerged. Table 11-1 lists the densities of some common materials. Since the density of water in cgs units is 1 g/cm^3, these units are slightly more convenient to use than SI units. The ratio of the density of a substance to that of water is called the **specific gravity** of the substance. The specific gravity is a dimensionless number that equals the magnitude of the density when it is expressed in grams per cubic centimeter (or in kilograms per liter). The specific gravity of a substance can be found by dividing its density by 10^3 kg/m^3. For example, the specific gravity of aluminum is 2.7, and that of ice is 0.92. The specific gravities of objects that sink in water range from 1 to about 22.5 (for the densest element, osmium).

A glacier flows into Glacier Bay, Alaska.

Although most solids and liquids expand slightly when heated and contract slightly when subjected to an increase in external pressure, these changes in volume are relatively small, so we can say that the densities of most solids and liquids are approximately independent of temperature and pressure. On the other hand, the density of a gas depends strongly on the pressure and temperature, so the temperature and pressure must be specified when giving the densities of gases. In Table 11-1, the densities are given at standard conditions (atmospheric pressure at sea level and a temperature of 0°C). Note that the densities of gases are considerably less than those of liquids or solids. For example, the density of water is about 800 times that of air under standard conditions.

Exercise

A solid metal cube 8 cm on a side has a mass of 4.08 kg. (*a*) What is the density of the cube? (*b*) If the cube is made from a single element listed in Table 11-1, what is the element? [Answers: (*a*) 7.97 kg/L, (*b*) iron]

In the U.S. customary system, the term **weight density,** which is defined as the ratio of the weight of an object to its volume, is often used. The weight density is the product of the density ρ and the acceleration due to gravity g:

$$\rho g = \frac{w}{V} = \frac{mg}{V} \tag{11-3}$$

The weight density of water is

$$\rho_w g = 62.4 \text{ lb/ft}^3 \qquad 11\text{-}4$$

The weight density of any other material can be found by multiplying its specific gravity by 62.4 lb/ft³.

Exercise

Find the weight density of copper. (Answer: 559 lb/ft³)

Example 11-1

A lead brick is 5 by 10 by 20 cm. How much does it weigh?

The volume of the brick is

$$V = (5 \text{ cm})(10 \text{ cm})(20 \text{ cm}) = 1000 \text{ cm}^3 = 10^{-3} \text{ m}^3$$

From Table 11-1, the density of lead is 11.3×10^3 kg/m³. Its mass is therefore

$$m = \rho V = (11.3 \times 10^3 \text{ kg/m}^3)(10^{-3} \text{ m}^3) = 11.3 \text{ kg}$$

and its weight is

$$w = mg = (11.3 \text{ kg})(9.81 \text{ N/kg}) = 111 \text{ N} \approx 25 \text{ lb}$$

Example 11-2

A 200-mL flask is filled with water at 4°C. When the flask is heated to 80°C, 6 g of water spill out. What is the density of water at 80°C? (Assume that the expansion of the flask is negligible.)

Since the density of water at 4°C is 1 g/cm³ and 200 mL = 200 cm³, the mass of the water originally in the flask is

$$m = \rho V = (1 \text{ g/cm}^3)(200 \text{ cm}^3) = 200 \text{ g}$$

Since 6 g spill out of the flask at 80°C, the mass of the remaining water is 200 g − 6 g = 194 g. This mass of water now occupies 200 mL, so the density of water at 80°C is

$$\rho = \frac{m}{V} = \frac{194 \text{ g}}{200 \text{ cm}^3} = 0.97 \text{ g/cm}^3$$

Table 11-1 **Densities of Selected Substances***

Substance	Density, kg/m³
Aluminum	2.70×10^3
Bone	$1.7–2.0 \times 10^3$
Brick	$1.4–2.2 \times 10^3$
Cement	$2.7–3.0 \times 10^3$
Copper	8.93×10^3
Earth (average)	5.52×10^3
Glass (common)	$2.4–2.8 \times 10^3$
Gold	19.3×10^3
Ice	0.92×10^3
Iron	7.96×10^3
Lead	11.3×10^3
Wood (oak)	$0.6–0.9 \times 10^3$
Alcohol (ethanol)	0.806×10^3
Gasoline	0.68×10^3
Mercury	13.6×10^3
Seawater	1.025×10^3
Water	1.00×10^3
Air	1.293
Helium	0.1786
Hydrogen	0.08994
Steam (100°C)	0.6

*t = 0°C and P = 1 atm unless otherwise indicated.

Question

1. What is the approximate specific gravity of your body?

11-2 Stress and Strain

If a solid object is in equilibrium but is subjected to forces that tend to stretch, shear, or compress it, the shape of the object changes. If the object returns to its original shape when the forces are removed, it is said to be elastic. Most objects are elastic for forces up to a certain limit called the **elastic limit.** If the forces are too great and the elastic limit is exceeded, the object does not return to its original shape but is permanently deformed.

Figure 11-1*a* shows a solid bar subjected to a stretching or tensile force **F** to the right and an equal but opposite force to the left. In Figure 11-1*b*, we concentrate on a small element of the bar of length L. Since this element is in equilibrium, the forces exerted on it by adjacent elements to the right must

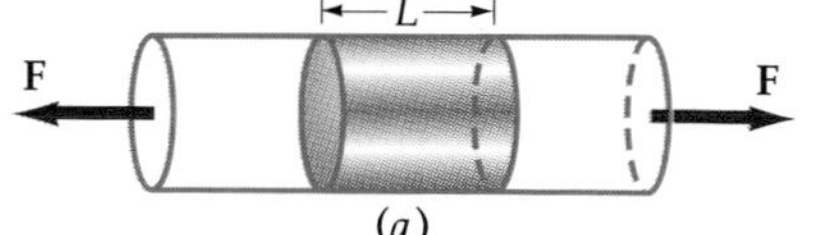

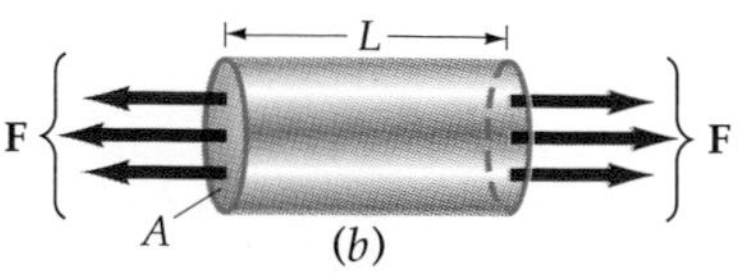

Figure 11-1 (*a*) Solid bar subjected to a stretching force **F**. (*b*) A small length of the bar. The force per unit area is the stress S.

equal those exerted by adjacent elements to the left. If the element is not too near the end of the bar, these forces will be distributed uniformly over the cross-sectional area of the bar. The ratio of the force F to the cross-sectional area A is called the **tensile stress:**

$$\text{Stress} = \frac{F}{A} \qquad 11\text{-}5$$

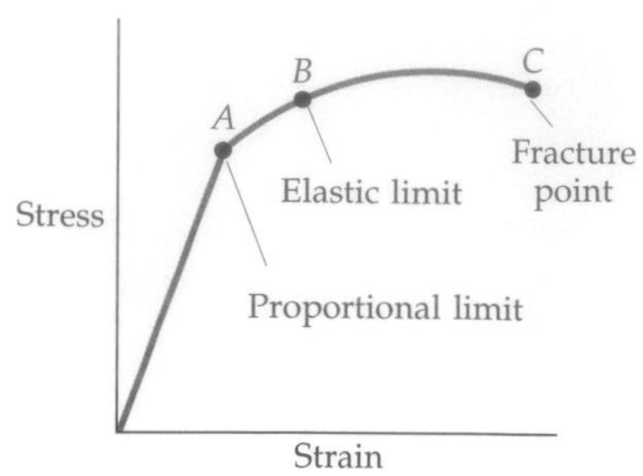

Figure 11-2 A graph of stress versus strain. Up to point A, the strain is proportional to the stress. Beyond the elastic limit at point B, the bar will not return to its original length when the stress is removed. At point C, the bar fractures.

The forces exerted on the bar tend to stretch the bar. The fractional change in the length of the bar $\Delta L/L$ is called the **strain:**

$$\text{Strain} = \frac{\Delta L}{L} \qquad 11\text{-}6$$

Figure 11-2 shows a graph of the strain versus the stress for a typical solid bar. The graph is linear up to point A. The result that strain varies linearly with stress is known as Hooke's law. (This is the same behavior as that of a coiled spring for small stretching.) Point B in Figure 11-2 is the elastic limit of the material. If the bar is stretched beyond this point, it will not return to its original length but is permanently deformed. If an even greater stress is applied, the material eventually breaks, as indicated by point C. The ratio of the stress to the strain in the linear region of the graph is a constant called **Young's modulus** Y:

Young's modulus defined

$$Y = \frac{\text{stress}}{\text{strain}} = \frac{F/A}{\Delta L/L} \qquad 11\text{-}7$$

The units of Young's modulus are newtons per square meter (or pounds per square inch). Approximate values of Young's modulus for various materials are listed in Table 11-2.

Table 11-2 **Young's Modulus Y and Strengths of Various Materials***

Material	Y, GN/m^2†	Tensile strength, MN/m^2	Compressive strength, MN/m^2
Aluminum	70	90	
Bone			
Tensile	16	200	
Compressive	9	—	270
Brass	90	370	
Concrete	23	2	17
Copper	110	230	
Iron (wrought)	190	390	
Lead	16	12	
Steel	200	520	520

*These values are representative. Actual values for particular samples may differ.
†1 GN = 10^3 MN = 10^9 N.

Example 11-3

A certain man's biceps muscle has a maximum cross-sectional area of 12 cm^2 = 1.2×10^{-3} m^2. What is the stress in the muscle if it exerts a force of 300 N?

From the definition of tensile stress, we have

$$\text{Stress} = \frac{F}{A} = \frac{300\ \text{N}}{1.2 \times 10^{-3}\ \text{m}^2} = 2.5 \times 10^5\ \text{N/m}^2$$

The maximum stress that can be exerted is approximately the same for all muscles. Greater forces can be exerted by muscles with greater cross-sectional areas.

Example 11-4

A 500-kg mass is hung from a 3-m steel wire with a cross-sectional area of 0.15 cm^2. By how much does the wire stretch?

The weight of a 500-kg mass is

$$mg = (500 \text{ kg})(9.81 \text{ N/kg}) = 4.90 \times 10^3 \text{ N}$$

The stress of the wire is

$$S = \frac{F}{A} = \frac{4.9 \times 10^3 \text{ N}}{0.15 \text{ cm}^2} = 3.27 \times 10^4 \text{ N/cm}^2 = 3.27 \times 10^8 \text{ N/m}^2$$

From Table 11-2, we find Young's modulus for steel to be about 2.0×10^{11} N/m^2. The strain is therefore

$$\frac{\Delta L}{L} = \frac{S}{Y} = \frac{3.27 \times 10^8 \text{ N/m}^2}{2.0 \times 10^{11} \text{ N/m}^2} = 1.63 \times 10^{-3}$$

Since the wire is 300 cm long, the amount it stretches is

$$\Delta L = (1.63 \times 10^{-3})L = (1.63 \times 10^{-3})(300 \text{ cm}) = 0.49 \text{ cm}$$

Exercise

A wire 1.5 m long has a cross-sectional area of 2.4 mm^2. It is hung vertically and stretches 0.32 mm when a 10-kg block is attached to it. Find (*a*) the stress, (*b*) the strain, and (*c*) Young's modulus for the wire. [Answers: (*a*) 4.09×10^7 N/m^2, (*b*) 2.13×10^{-4}, (*c*) 192 GN/m^2]

If a bar is subject to forces that tend to compress it rather than stretch it, the stress is called **compressive stress.** For many (but not all) materials, Young's modulus for compressive stress is the same as that for tensile stress if ΔL in Equation 11-7 is taken to be the decrease in the length of the bar.

(*a*) These flying buttresses make use of the compressive strength and weight of stone to support Notre Dame Cathedral in Paris. (*b*) A sharp karate chop to the cinderblocks is enough to shatter them indicating that their resistance to a "bending load" is much less than their compressive strength.

(*a*)

(*b*)

(Bone is an important exception that has different Young's moduli for compression and extension.) If the tensile or compressive stress is too great, the bar breaks. The stress at which breakage occurs is called the **tensile strength,** or in the case of compression, the **compressive strength.** Approximate values of the tensile and compressive strengths of various materials are also listed in Table 11-2.

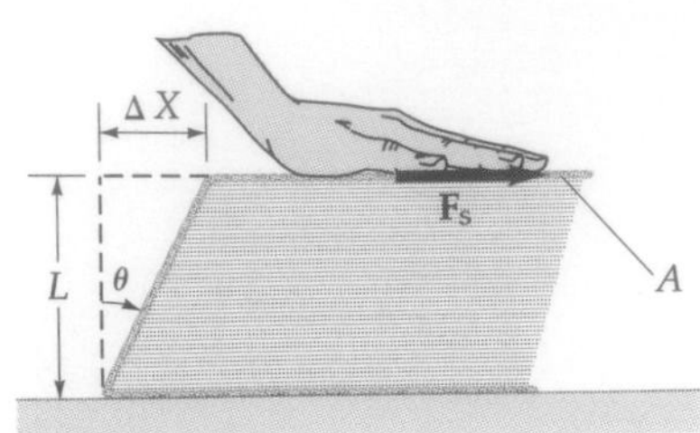

Figure 11-3 The application of the horizontal force $\mathbf{F}_s$ to the book causes a shearing stress defined as the force per unit area. The ratio $\Delta X/L = \tan\theta$ is the shear strain.

In Figure 11-3, a force $\mathbf{F}_s$ is applied tangentially to the top of a book. Such a force is called a **shear force.** The ratio of the shear force F_s to the area A is called the **shear stress:**

$$\text{Shear stress} = \frac{F_s}{A}$$

A shear stress tends to deform the book as shown in Figure 11-3. The ratio $\Delta X/L$ is called the **shear strain:**

$$\text{Shear strain} = \frac{\Delta X}{L} = \tan\theta$$

where θ is the shear angle shown in the figure. The ratio of the shear stress to the shear strain is called the **shear modulus** M_s:

Shear modulus defined

$$M_s = \frac{\text{shear stress}}{\text{shear strain}} = \frac{F_s/A}{\Delta X/L} = \frac{F_s/A}{\tan\theta} \qquad \text{11-8}$$

Table 11-3 **Approximate Values of the Shear Modulus M_s of Various Materials**

Material	M_s, GN/m^2
Aluminum	30
Brass	36
Copper	42
Iron	70
Lead	5.6
Steel	84
Tungsten	150

The shear modulus is also known as the **torsion modulus.** The fact that this modulus is approximately constant for small stresses, which implies that the shear strain varies linearly with the shear stress, is Hooke's law for torsional stress. In a torsion balance, such as that used in Cavendish's apparatus for measuring the universal gravitational constant G, the torque (which is related to the stress) is proportional to the angle of twist (which equals the strain for small angles). Approximate values of the shear modulus for various materials are listed in Table 11-3.

11-3 Pressure in a Fluid

Fluids differ from solids in that they are unable to support a shear stress. Thus, they deform to fill a container of any shape. When a body is submerged in a fluid such as water, the fluid exerts a force perpendicular to the surface of the body at each point on the surface. If the body is small enough so that we can neglect any differences in the depth of the fluid, the force per unit area exerted by the fluid is the same at each point on the surface of the body. This force per unit area is called the **pressure** P of the fluid:

Pressure defined

$$P = \frac{F}{A} \qquad \text{11-9}$$

The SI unit of pressure is the newton per square meter (N/m^2), which is called the **pascal** (Pa):

$$1\text{ Pa} = 1\text{ N/m}^2 \qquad \text{11-10}$$

In the U.S. customary system, pressure is usually given in pounds per square inch (lb/in^2). Another common unit of pressure is the atmosphere (atm), which is approximately the air pressure at sea level. The atmosphere is now defined to be 101.325 kilopascals, which is approximately 14.70 lb/in^2:

$$1 \text{ atm} = 101.325 \text{ kPa} = 14.70 \text{ lb/in}^2 \qquad 11\text{-}11$$

Other units of pressure in common use will be discussed later.

The pressure due to a fluid pressing against a body tends to compress the body. The ratio of the pressure to the fractional decrease in volume ($-\Delta V/V$) is called the **bulk modulus** B:

$$B = -\frac{P}{\Delta V/V} \qquad 11\text{-}12$$

Bulk modulus defined

Since all materials decrease in volume when they are subjected to external pressure, a minus sign is introduced in Equation 11-12 to make B positive. The pressure exerted by a fluid is equivalent to a compressive stress, and the fractional decrease in the volume ($-\Delta V/V$) is the compressive strain. The inverse of the bulk modulus is called **compressibility** k:

$$k = \frac{1}{B} = -\frac{\Delta V/V}{P} \qquad 11\text{-}13$$

The more difficult material is to compress, the smaller the fractional change $\Delta V/V$ for a given pressure and, hence, the smaller the compressibility k. The concepts of the bulk modulus and compressibility apply to liquids and gases as well as to solids. Solids and liquids are relatively incompressible; that is, they have small values of compressibility and large values for the bulk modulus. These values are also relatively independent of temperature and pressure. Gases, on the other hand, are easily compressed, and the values of B and k depend strongly on the pressure and temperature. Table 11-4 lists values of the bulk modulus for various materials.

Table 11-4 **Approximate Values of the Bulk Modulus B of Various Materials**

Material	B, GN/m^2
Aluminum	70
Brass	61
Copper	140
Iron	100
Lead	7.7
Mercury	27
Steel	160
Tungsten	200
Water	2.0

As any scuba diver knows, the pressure in a lake or ocean increases as depth increases. Similarly, the pressure of the atmosphere decreases as altitude increases. (This is why aircraft cabins must be pressurized.) For a liquid such as water whose density is constant throughout, the pressure increases linearly with depth. We can see this easily by considering the column of liquid of height h and cross-sectional area A shown in Figure 11-4. The pressure at the bottom of the column must be greater than the pressure at the top of the column to support the weight of the column. The mass of this liquid column is

$$m = \rho V = \rho A h$$

and its weight is

$$w = mg = \rho A h g$$

If P_0 is the pressure at the top and P is the pressure at the bottom, the net upward force exerted by this pressure difference is $PA - P_0 A$. Setting this net upward force equal to the weight of the column, we obtain

$$PA - P_0 A = \rho A h g$$

or

$$P = P_0 + \rho g h \qquad (\rho \text{ constant}) \qquad 11\text{-}14$$

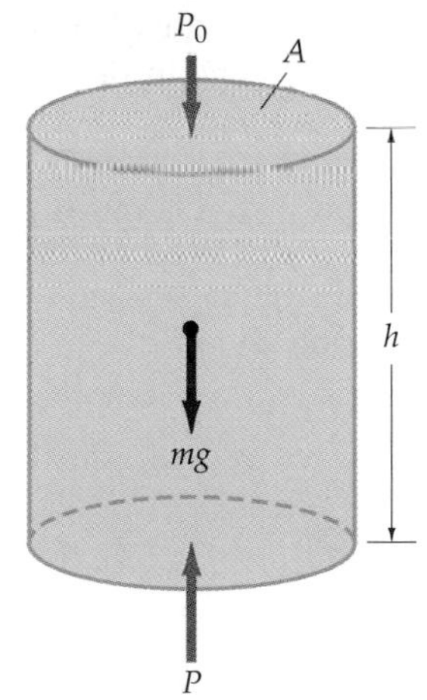

Figure 11-4 Column of water of height h and cross-sectional area A. The pressure P at the bottom must be greater than the pressure P_0 at the top to balance the weight of the water.

Example 11-5

Find the pressure at a depth of 10 m below the surface of a lake if the pressure at the surface is 1 atm.

Using Equation 11-14 with $P_0 = 1$ atm $= 101$ kPa, $\rho = 10^3$ kg/m^3, and $g = 9.81$ N/kg, we have

$$\begin{aligned} P &= 101 \text{ kPa} + (10^3 \text{ kg/m}^3)(9.81 \text{ N/kg})(10 \text{ m}) \\ &= 101 \text{ kPa} + (9.81 \times 10^4 \text{ N/m}^2) \\ &= 101 \text{ kPa} + 98.1 \text{ kPa} = 199 \text{ kPa} = 1.97 \text{ atm} \end{aligned}$$

The pressure at a depth of 10 m is nearly twice that at the surface.

The result that the pressure at a depth h is greater than that at the top by the amount ρgh holds for a liquid in any container, independent of the shape of the container. Furthermore, the pressure is the same at all points at the same depth. Thus, if we increase P_0, say, by pressing down on the top surface with a piston, the increase in pressure is the same throughout the liquid. This is known as **Pascal's principle,** named after Blaise Pascal (1623–1662):

Pascal's principle

> Pressure applied to an enclosed liquid is transmitted undiminished to every point in the fluid and to the walls of the container.

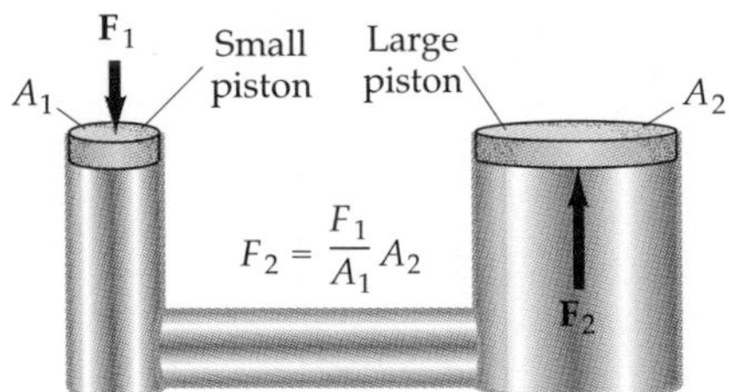

Figure 11-5 Hydraulic lift. A small force $\mathbf{F}_1$ on the small piston produces a change in pressure that is transmitted by the liquid to the large piston. Since the area of the large piston is much greater than that of the small piston, the force $\mathbf{F}_2$ on the large piston is much greater than $\mathbf{F}_1$.

A common application of Pascal's principle is the hydraulic lift shown in Figure 11-5. When force $\mathbf{F}_1$ is applied to the smaller piston, the pressure in the liquid is increased by F_1/A_1. The upward force exerted by the liquid on the larger piston is this increase in pressure times the area A_2. Calling this force $\mathbf{F}_2$, we have

$$F_2 = \frac{F_1}{A_1} A_2 = \frac{A_2}{A_1} F_1$$

If A_2 is much greater than A_1, a small force $\mathbf{F}_1$ can be used to exert a much larger force $\mathbf{F}_2$ to lift a weight placed on the larger piston.

Example 11-6

The large piston in a hydraulic lift has a radius of 20 cm. What force must be applied to the small piston of radius 2 cm to raise a car of mass 1500 kg?

The weight of the car is

$$mg = (1500 \text{ kg})(9.81 \text{ N/kg}) = 1.47 \times 10^4 \text{ N}$$

The force that must be applied is therefore

$$\begin{aligned} F_1 &= \frac{A_1}{A_2} F_2 = \frac{\pi r_1^2}{\pi r_2^2} mg \\ &= \frac{(2 \text{ cm})^2}{(20 \text{ cm})^2} (1.47 \times 10^4 \text{ N}) = 147 \text{ N} \approx 33 \text{ lb} \end{aligned}$$

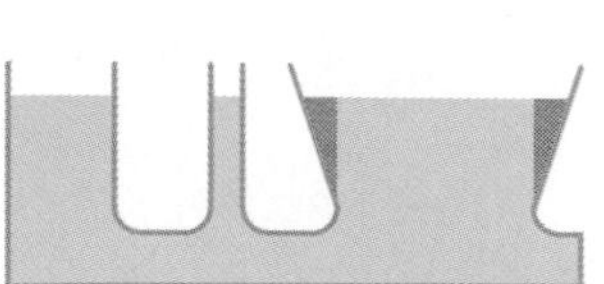

Figure 11-6 The hydrostatic paradox. The water level is the same regardless of the shape of the vessel. The shaded portion of the water is supported by the sides of the container.

Figure 11-6 shows water in a container with sections of different shapes. At first glance, it might seem that the pressure in the largest section of the container would be greatest and that the water would therefore be forced into the smallest section of the container to a greater height. That this does not happen is known as the hydrostatic paradox. The pressure depends only on the depth of the water, not on the shape of the container, so at the same height the pressure is the same in all parts of the container, as can be shown experimentally. Although the water in the largest section of the container weighs more than that in the smaller sections, some of this weight is supported by the normal force exerted by the sides of the largest section of the container, which in this case has a component upward. In fact, the shaded portion of the water is completely supported by the sides of the container.

We can use the result that the pressure difference is proportional to the depth of the fluid to measure unknown pressures. Figure 11-7 shows the simplest pressure gauge, the open-tube manometer. The top of the tube is open to the atmosphere at pressure P_{at}. The other end of the tube is at pressure P, which is to be measured. The difference $P - P_{at}$ is equal to ρgh, where ρ is the density of the liquid in the tube. The difference between the "absolute" pressure P and atmospheric pressure P_{at} is called the **gauge pressure.** The pressure you measure in your automobile tire is gauge pressure. When the tire is entirely flat, the gauge pressure is zero, and the absolute pressure in the tire is atmospheric pressure. The absolute pressure is obtained from gauge pressure by adding atmospheric pressure to it:

$$P = P_{gauge} + P_{at} \qquad 11\text{-}15$$

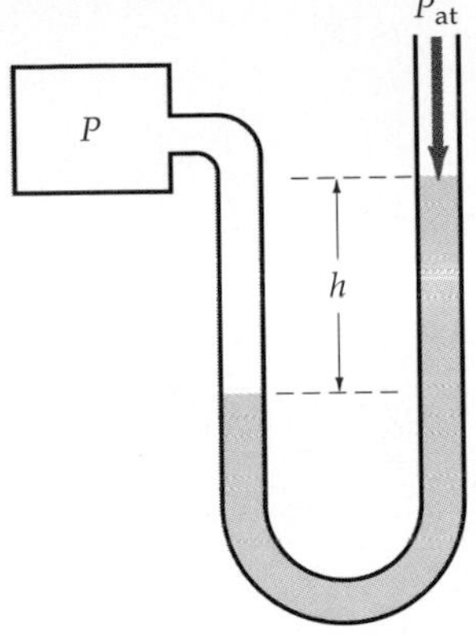

Figure 11-7 Open-tube manometer for measuring an unknown pressure P. The difference $P - P_{at}$ equals ρgh.

Figure 11-8 shows a mercury U-tube barometer, which is used to measure atmospheric pressure. The top end of the tube has been closed off and evacuated so that the pressure there is zero. The other end is open to the atmosphere at pressure P_{at}. The pressure P_{at} is given by $P_{at} = \rho gh$, where ρ is the density of mercury.

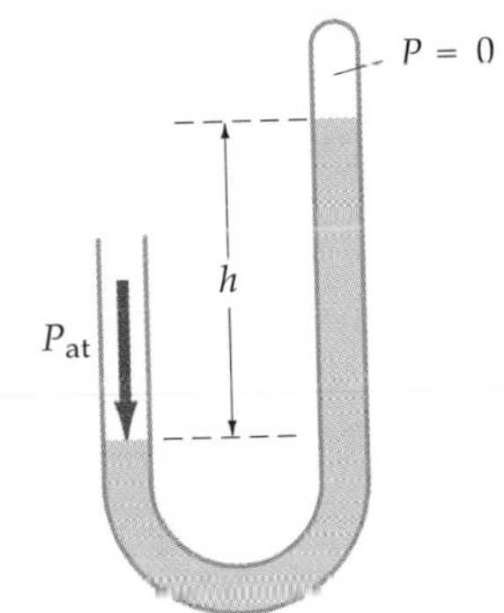

Figure 11-8 U-tube barometer for measuring atmospheric pressure P_{at}.

Example 11-7

At 0°C the density of mercury is 13.595×10^3 kg/m^3. What is the height of the mercury column in a U-tube barometer if the pressure is 1 atm = 101.325 kPa?

We have

$$h = \frac{P}{\rho g} = \frac{1.01325 \times 10^5 \text{ N/m}^2}{(13.595 \times 10^3 \text{ kg/m}^3)(9.81 \text{ N/kg})}$$

$$= 0.7597 \text{ m} \approx 760 \text{ mm}$$

In practice, pressure is often measured in millimeters of mercury (commonly called **torr,** after the Italian physicist Torricelli), in inches of mercury (written inHg), and in inches or feet of water (written inH_2O or ftH_2O). The various units of pressure are related as follows:

$$1 \text{ atm} = 760 \text{ mmHg} = 760 \text{ torr} = 29.9 \text{ inHg}$$

$$= 33.9 \text{ ftH}_2\text{O} = 101.325 \text{ kPa} = 14.7 \text{ lb/in}^2 \qquad 11\text{-}16$$

$$1 \text{ mmHg} = 1 \text{ torr} = 1.316 \times 10^{-3} \text{ atm} = 133.3 \text{ Pa} \qquad 11\text{-}17$$

Other units commonly used on weather maps are the **bar** and the **millibar,** which are defined as

$$1 \text{ bar} = 10^3 \text{ millibars} = 100 \text{ kPa} \qquad 11\text{-}18$$

A pressure of 1 bar is just slightly less than 1 atm.

Example 11-8

The average (gauge) pressure in the aorta is about 100 mmHg. Convert this average blood pressure to pascals and pounds per square inch.

We can use the conversion factors implied in Equation 11-16. For example, we have 760 mmHg = 101.325 kPa. Thus,

$$P = 100 \text{ mmHg} \left(\frac{101.325 \text{ kPa}}{760 \text{ mmHg}}\right) = 13.3 \text{ kPa}$$

and

$$P = 100 \text{ mmHg} \left(\frac{14.7 \text{ lb/in}^2}{760 \text{ mmHg}}\right) = 1.93 \text{ lb/in}^2$$

Exercise

Convert a pressure of 45 kPa to (*a*) mmHg and (*b*) atmospheres. [Answers: (*a*) 338 mmHg, (*b*) 0.444 atm]

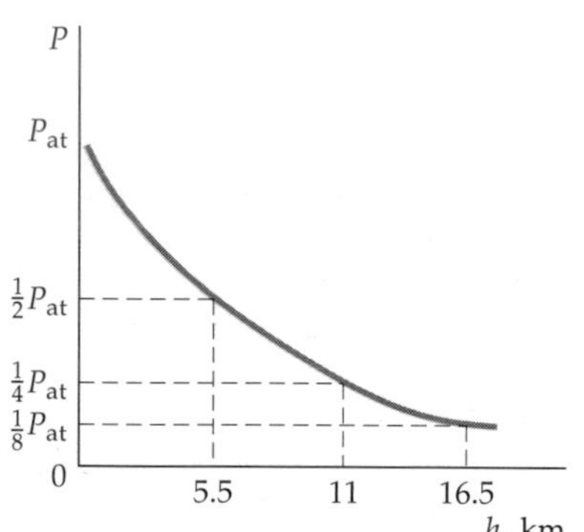

Figure 11-9 Variation in pressure with height above the earth's surface. For each increase in height of 5.5 km, the pressure decreases by half. This is an example of an exponential decrease.

The relation between pressure and altitude for a gas such as air is more complicated than the relation between pressure and depth for a liquid because the density of a gas is not constant but depends on the pressure. In fact, to a good approximation, the density of a gas is proportional to the pressure. The pressure in a column of air decreases with height as you go up from the surface of the earth, like the pressure in a water column decreases as you go up from the bottom, but unlike the case with water pressure, the decrease in air pressure is not linear with distance. Instead, the air pressure decreases by a constant fraction for a given increase in height, as is shown in Figure 11-9. This type of decrease is called an exponential decrease. At a height of about 5.5 km (18,000 ft), the air pressure is half its value at sea level. If we go up another 5.5 km to an altitude of 11 km (a typical altitude for airliners), the pressure is again halved so that it is one-fourth its value at sea level, and so on. Since the density of air is proportional to the pressure, the density of air also decreases with altitude. Thus, for example, there is less oxygen available on a mountain than at normal elevations, which makes exercising difficult. At very high altitudes, such as those at which jet planes fly, the cabins must be pressurized.

11-4 Buoyancy and Archimedes' Principle

If a heavy object submerged in water is "weighed" by suspending it from a spring scale (Figure 11-10*a*), the scale reads less than when the object is weighed in air. This is because the water exerts an upward force that partially balances the force of gravity. This force is even more evident when we submerge a piece of cork. When completely submerged, the cork experiences an upward force from the water pressure that is greater than the force of gravity, so it accelerates up toward the surface, where it floats partially submerged. The force exerted by a fluid on a body submerged in it is called the **buoyant force.** It depends on the density of the fluid and the volume of the body but not on the composition or shape of the body and is equal in magnitude to the weight of the fluid displaced by the body.

Archimedes' principle

A body wholly or partially submerged in a fluid is buoyed up by a force equal to the weight of the displaced fluid.

This result is known as **Archimedes' principle.**

Archimedes (287–212 B.C.) had been given the task of determining whether a crown made for King Hieron II was of pure gold or whether it contained some cheaper metal, such as silver. The problem was to determine the density of the irregularly shaped crown without destroying it. As the story goes, Archimedes came upon the solution while bathing and immediately rushed naked through the streets of Syracuse shouting "Eureka" ("I have found it"). This flash of insight preceded Newton's laws, from which Archimedes' principle can be derived, by some 1900 years. What Archimedes found was a simple and accurate way to determine the specific gravity of the crown, which he could then compare with the specific gravity of gold. The specific gravity of an object is the weight of the object in air divided by the weight of an equal volume of water:

$$\text{Specific gravity} = \frac{\text{weight of object in air}}{\text{weight of equal volume of water}}$$

But, according to Archimedes' principle, the weight of an equal volume of water equals the buoyant force on the object when it is submerged. It is therefore equal to the *loss* in weight of the object when it is weighed while submerged in water. Thus,

$$\text{Specific gravity} = \frac{\text{weight of object in air}}{\text{weight loss when submerged in water}} \qquad 11\text{-}19$$

The specific gravity of the crown could be determined by weighing the crown in air and then again when it was submerged in water.

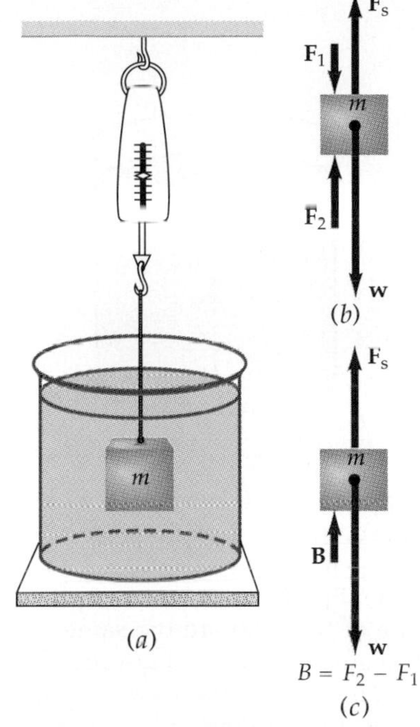

Figure 11-10 (*a*) Weighing an object submerged in a fluid. (*b*) Free-body diagram showing the weight, the force of the spring $\mathbf{F}_s$, and the forces $\mathbf{F}_1$ and $\mathbf{F}_2$ exerted by the surrounding fluid. (*c*) The buoyant force $B = F_2 - F_1$ is the net force exerted on the object by the fluid. It is upward because the pressure at the bottom of the object is greater than that at the top.

Example 11-9

The specific gravity of gold is 19.3. If a crown made of pure gold weighs 8 N in air, what should its weight be when it is submerged in water?

From Equation 11-19, we have for the weight loss in water

$$\text{Weight loss} = \frac{\text{weight in air}}{\text{specific gravity}} = \frac{8\text{ N}}{19.3} = 0.415\text{ N}$$

The crown should therefore weigh 8 N − 0.415 N = 7.59 N.

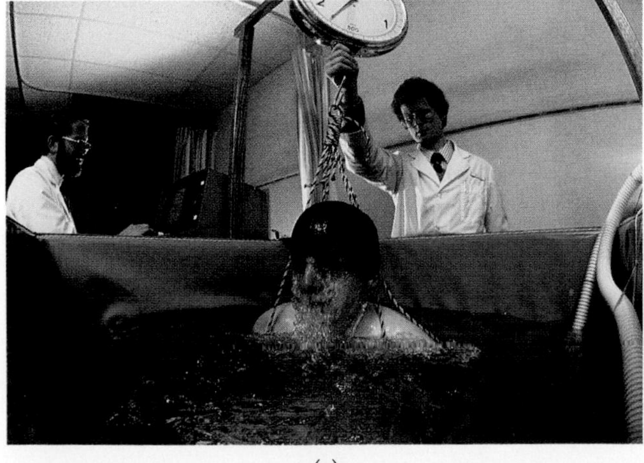

(*a*)

(*b*)

(*a*) To determine the percentage of fat in this man's body, the density of his body is measured by weighing him while he is submerged under water. (*b*) Archimedes in his bath, at the legendary moment when he discovered how to measure the density of an object.

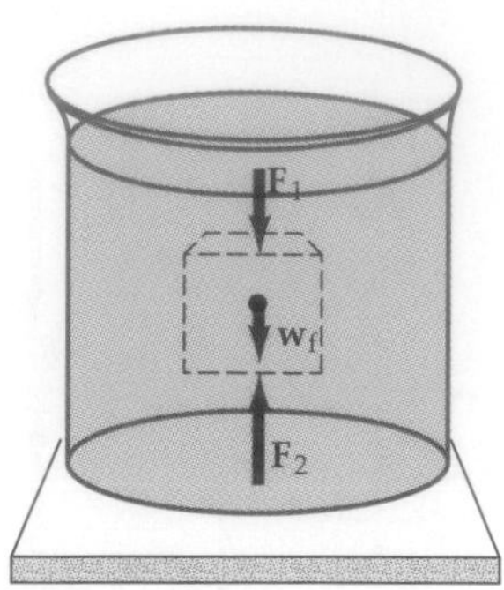

Figure 11-11 Figure 11-10 with the submerged body replaced by an equal volume of fluid. The forces $\mathbf{F}_1$ and $\mathbf{F}_2$ due to the pressure of the fluid are the same as in Figure 11-10. The buoyant force is thus equal to the weight $\mathbf{w}_f$ of the displaced fluid.

Hot-air balloons rising in the night sky over Albuquerque during a balloon festival.

We can derive Archimedes' principle from Newton's laws by considering the forces acting on a portion of a fluid and noting that in static equilibrium the net force must be zero. Figure 11-10*b* shows the vertical forces acting on an object being weighed while submerged. These are the force of gravity **w** down, the force of the spring scale $\mathbf{F}_s$ acting up, a force $\mathbf{F}_1$ acting down because of the fluid pressing on the top surface of the object, and a force $\mathbf{F}_2$ acting up because of the fluid pressing on the bottom surface of the object. Since the spring scale reads a force less than the weight, the force $\mathbf{F}_2$ must be greater in magnitude than the force $\mathbf{F}_1$. The difference in magnitude of these two forces is the buoyant force $B = F_2 - F_1$. The buoyant force occurs because the pressure of the fluid at the bottom of the object is greater than that at the top.

In Figure 11-11 the spring scale has been eliminated and the submerged object has been replaced by an equal volume of fluid (indicated by the dotted lines). The buoyant force $B = F_2 - F_1$ acting on this volume of fluid is the same as the buoyant force that acted on our original object since the fluid surrounding the space is the same. Because this volume of fluid is in equilibrium, the net force acting on it must be zero. The upward buoyant force thus equals the downward weight of this volume of fluid:

$$B = w_f \qquad 11\text{-}20$$

Note that this result does not depend on the shape of the submerged object. If we consider any irregularly shaped portion of fluid, there must be a buoyant force acting on it due to the surrounding fluid that is equal to the weight of that portion. Thus, we have derived Archimedes' principle.

We can see from Archimedes' principle that an object will float in fluid if the density of the object is less than that of the fluid. If ρ_f is the density of a fluid, a volume V of the fluid then has a mass $\rho_f V$ and a weight

$$w_f = \rho_f g V = B$$

The weight of the object can be written

$$w_o = \rho g V$$

where ρ is the density of the object. If the density of the object is greater than that of the fluid, the weight of the object will be greater than the buoyant force, and the object will sink unless supported. If ρ is less than ρ_f, the buoyant force will be greater than the weight of the object, and the object will accelerate up to the top of the fluid unless held down. At the top, it will float in equilibrium with a fraction of its volume submerged such that the weight of the fluid it displaces equals the weight of the object.

Example 11-10

A cork has a density of 200 kg/m^3. Find the fraction of the volume of the cork that is submerged when the cork floats in water.

Let V be the volume of the cork and V' be the volume that is submerged when it floats in water. The weight of the cork is $\rho g V$, and the buoyant force is $\rho_w g V'$. Since the cork is in equilibrium, the buoyant force equals the weight:

$$B = w$$

$$\rho_w g V' = \rho g V$$

The fraction of the cork submerged is thus

$$\frac{V'}{V} = \frac{\rho}{\rho_w} = \frac{200\ \text{kg/m}^3}{1000\ \text{kg/m}^3} = \frac{1}{5} \qquad 11\text{-}21$$

Thus, one-fifth of the cork is submerged.

For any object floating in a fluid, Equation 11-21 gives the fraction of the object that is submerged if we replace ρ_w with ρ_f, the density of the fluid. Since the density of ice is 920 kg/m^3 and that of seawater is 1025 kg/m^3, the fraction of an iceberg that is submerged in seawater is

$$\frac{V'}{V} = \frac{\rho}{\rho_f} = \frac{920 \text{ kg/m}^3}{1025 \text{ kg/m}^3} = 0.898$$

The great danger of icebergs to ships is related to the fact that only about 10 percent of an iceberg is visible above the water.

When an object submerged in a fluid is weighed, as in Figure 11-10*a*, the measured weight F_s, the reading on the spring scale, is less than the true weight w_o because of the upward buoyant force B:

$$F_s = w_o - B$$

Writing these forces in terms of the volume V of the object and the densities ρ of the object and ρ_f of the fluid, we have

$$F_s = \rho g V - \rho_f g V = \rho g V\left(1 - \frac{\rho_f}{\rho}\right)$$

or

$$F_s = w_o\left(1 - \frac{\rho_f}{\rho}\right) \qquad \text{11-22}$$

Thus, the measured weight F_s of the submerged object is less than the true weight w_o by the factor $(1 - \rho_f/\rho)$.

Example 11-11

A block of an unknown material weighs 3 N in air and 1.89 N when submerged in water, as in Figure 11-10*a*. (*a*) What is the material? (*b*) What correction must be made for the buoyancy of air when the block is weighed in air? (The density of air is approximately 1.3 kg/m^3.)

(*a*) We determine the material of the block by finding its density. According to Equation 11-19, the specific gravity of an object equals the weight of the object in air divided by the loss in weight when the object is submerged in water. Since for this example the loss in weight is 3 N − 1.89 N = 1.11 N, we have

$$\text{Specific gravity} = \frac{3 \text{ N}}{1.11 \text{ N}} = 2.70$$

The density of this material is thus 2.70 times that of water or 2.70×10^3 kg/m^3. Comparing this with the densities listed in Table 11-1, we see that the block is probably made of aluminum.

(*b*) The correction that must be made when an aluminum block is weighed in air is obtained from Equation 11-22. For aluminum in air, which has a density $\rho_f = 1.3$ kg/m^3, we have

$$\frac{\rho_f}{\rho} = \frac{1.3 \text{ kg/m}^3}{2.7 \times 10^3 \text{ kg/m}^3} = 4.8 \times 10^{-4}$$

and

$$1 - \frac{\rho_f}{\rho} = 1 - 0.00048$$

This differs from 1 by only 0.048 percent, showing that we can usually neglect the buoyant force of air.

Exercise

A piece of lead (specific gravity = 11.3) weighs 80 N in air. What does it weigh when submerged in water? (Answer: 72.9 N)

Questions

2. How can you estimate your average density at a swimming pool?
3. Why can you see "only the tip of the iceberg"?
4. Smoke usually rises from a smokestack, but it may sink on a very humid day. What can be concluded about the relative densities of humid air and dry air?
5. Why is it easier to float in salt water than in fresh water?
6. Fish can adjust their volume by varying the amount of oxygen and nitrogen gas (obtained from the blood) in a thin-walled sac under their spinal column called a swim bladder. Explain how this helps them swim.
7. A certain object has a density just slightly less than that of water so that it floats almost completely submerged. However, the object is more compressible than water. What happens if the floating object is given a slight push to submerge it?

11-5 Surface Tension and Capillarity

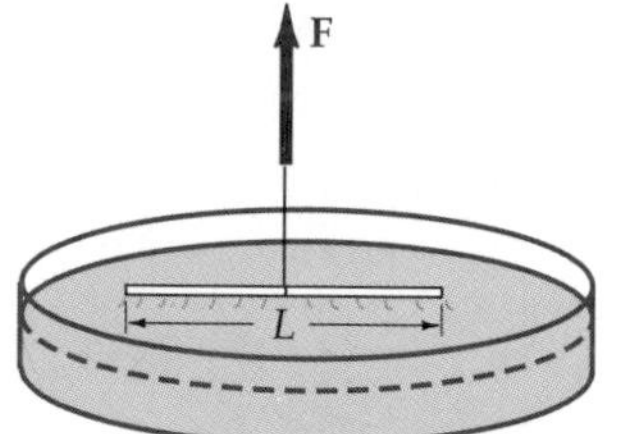

Figure 11-12 A needle of length L being lifted off the surface of a liquid. Surface tension exerts a force on the needle toward the surface.

A needle can be made to "float" on a water surface if it is placed there carefully. The forces that support the needle are not buoyant forces but are instead due to **surface tension.** In the interior of a liquid, a molecule is surrounded on all sides by other molecules, but at the surface, there are no molecules above the surface molecules. If a surface molecule is raised slightly, the molecular bonds between it and the adjacent molecules are stretched, and there is a restoring force that tends to pull the molecule back toward the surface. Similarly, when a needle is placed carefully on the surface, the surface molecules are depressed slightly and the adjacent molecules exert an upward restoring force on them, supporting the needle. Thus, the surface of a liquid is like a stretched elastic membrane. The force necessary to break the surface can be measured by lifting a needle off of the surface as shown in Figure 11-12. This force is found to be proportional to the length of the surface broken, which is twice the length of the needle since there is a surface film on both sides of the needle. If the needle has a mass m and length L, the force F needed to lift it off the surface is

$$F = \gamma 2L + mg \tag{11-23}$$

where γ is the **coefficient of surface tension,** which is the force per unit length exerted by the film. The value of γ for water is about 0.073 N/m. It is because of surface tension that small droplets of a liquid tend to be spherical.

(*a*)

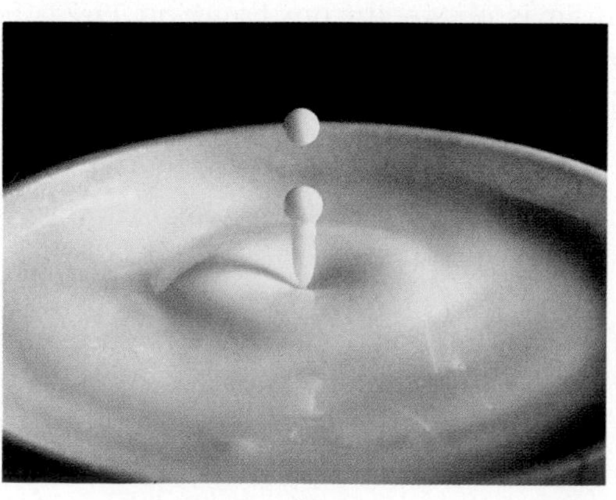

(*b*)

(*a*) The formation of a water drop. The nearly spherical shape of the drop is the result of surface tension. (*b*) A spherical drop rises from a liquid as a result of another drop falling into it.

As the drop is formed, surface tension pulls the surface together, minimizing the surface area and making the drop spherical.

Exercise

A wire of length 12.0 cm is pulled off a water surface with its length parallel to the surface. What force in addition to the weight of the wire is needed? (Answer: 0.0175 N)

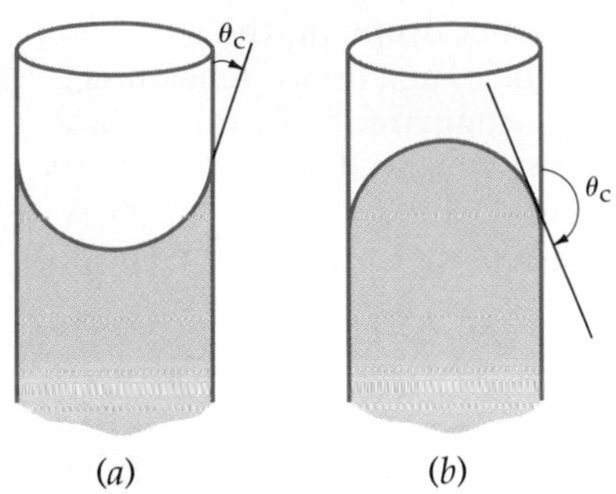

Figure 11-13 (*a*) Surface of a liquid in a narrow tube for the case in which the adhesive forces are greater than the cohesive forces. The surface is concave upwards, and the contact angle θ_c is less than 90°. (*b*) When the cohesive forces are greater than the adhesive forces, the surface of the liquid is convex, and the contact angle θ_c is greater than 90°.

The attractive forces between a molecule in a liquid and other molecules in the liquid are called **cohesive forces.** The force between a molecule of liquid and another substance, such as the wall of a thin tube, is called an **adhesive force.** When the adhesive forces are large relative to the cohesive forces, as in the case of water and a glass surface, the liquid is said to wet the surface of the other substance. In this case, the surface of a column of liquid in a tube is concave upward as shown in Figure 11-13*a*. The contact angle θ_c between the wall and the surface indicates the relative strength of the cohesive and adhesive forces. For a liquid that wets the surface, the contact angle is less than 90°, as in Figure 11-13*a*. When the adhesive forces are small relative to the cohesive forces, as is the case for mercury and glass, the liquid does not wet the surface, and the surface is convex as shown in Figure 11-13*b*. In this case, the contact angle is greater than 90°. Cohesive and adhesive forces are difficult to calculate theoretically, but the angles of contact θ_c in Figures 11-13*a* and *b* can be measured. For water and glass, the contact angle is approximately 0°. For mercury and glass, it is about 140°.

When the surface of a liquid is concave upward, the surface tension at the wall of the tube has a component upward as shown in Figure 11-14. The liquid will rise in the tube until the net upward force on it due to the surface tension is balanced by the weight of the liquid. This rise is called **capillary action,** or just **capillarity,** and the tube is called a capillary tube. (The smallest blood vessels are also called capillaries.) In Figure 11-14, the liquid has risen to a height h in a thin capillary tube of radius r. The tube is open to atmospheric pressure at the top. The force holding the liquid up is the vertical component of the surface tension, $F \cos \theta_c$. Since the length of the contact surface is $2\pi r$, this vertical force is $\gamma 2\pi r \cos \theta_c$. If the slight curvature of the surface is neglected, the volume of the liquid in the tube is $\pi r^2 h$. Setting the net upward force equal to the weight, we get

$$\gamma 2\pi r \cos \theta_c = \rho(\pi r^2 h)g$$

or

$$h = \frac{2\gamma \cos \theta_c}{\rho r g} \qquad \text{11-24}$$

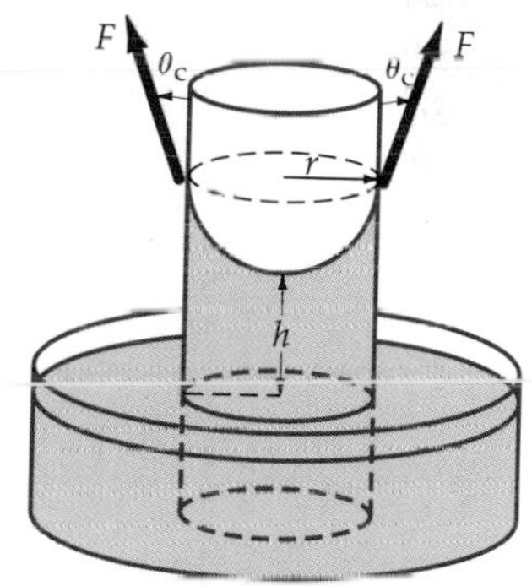

Figure 11-14 Liquid rising in a capillary tube. The upward force due to surface tension supports the weight of the column of liquid.

Capillarity is responsible for the rise of a liquid in blotting paper and for the rise of oil in the wick of a lamp. An important effect of capillarity is the holding of water in soil in the small spaces between soil particles. Were it not for capillarity, all the rain water would trickle down to the water table, leaving the upper soil dry. Farming could then only be done in swampy areas, as is the case for rice farming.

Example 11-12

How high will water rise in a tube of radius 0.1 mm if the contact angle is zero?

Using $\gamma = 0.073$ N/m for water, we have from Equation 11-24

$$h = \frac{2(0.073\ \text{N/m}) \cos 0°}{(1000\ \text{kg/m}^3)(0.0001\ \text{m})(9.81\ \text{N/kg})} = 0.149\ \text{m} = 14.9\ \text{cm}$$

(*a*) Water drops on the surface of a plant. The surface tension is large compared with the weight of the smaller drops, so they are nearly spherical in shape. (*b*) A water strider rests on the surface of a lake.

Example 11-13

The surface tension for mercury is 0.465 N/m and $\theta_c = 140°$. A glass capillary tube of radius 3 mm has been placed in a bowl of mercury. What is the height of the mercury in the tube relative to that in the bowl?

From Equation 11-24, we obtain

$$h = \frac{2(0.465 \text{ N/m}) \cos 140°}{(13.6 \times 10^3 \text{ kg/m}^3)(3 \times 10^{-3} \text{ m})(9.81 \text{ N/kg})}$$

$$= -1.78 \times 10^{-3} \text{ m} = -1.78 \text{ mm}$$

The mercury in the capillary tube is depressed 1.78 mm below the surface of the mercury in the bowl.

Question

8. A water bug walks on the surface of the lake. What keeps the bug from sinking?

11-6 Fluids in Motion and Bernoulli's Equation

The general flow of a fluid can be very complicated. Consider, for example, the rise of smoke from a burning cigarette (Figure 11-16 on facing page). At first the smoke rises in a regular stream, but soon turbulence sets in and the smoke begins to swirl irregularly. Turbulent flow is very difficult to treat even qualitatively. We will therefore consider only nonturbulent, steady-state flow as, for example, the smooth rising of the smoke before it becomes turbulent.

We will first consider a fluid that is flowing with no dissipation of mechanical energy. Such a fluid is called nonviscous. (We will treat viscosity in the next section.) We will also assume that the fluid is incompressible, which is a good approximation for most liquids. In an incompressible fluid, the density is constant throughout the fluid.

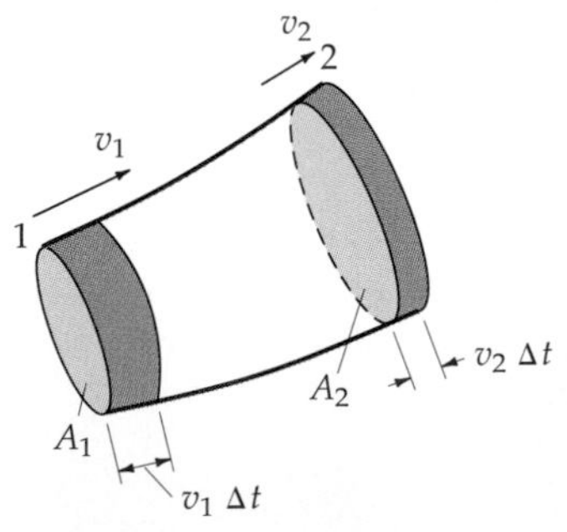

Figure 11-15 An incompressible fluid flowing in a tube of varying cross section. The shaded volumes are equal.

Figure 11-15 shows a fluid flowing in a tube of varying cross-sectional area. The shading on the left indicates the volume of fluid flowing into the tube at point 1 in some time Δt. If the speed of the fluid at this point is v_1 and the cross-sectional area of the tube is A_1, the volume flowing into the tube in time Δt is

$$\Delta V = A_1 v_1 \, \Delta t$$

Since we are assuming the fluid is incompressible, an equal volume of fluid must flow out of the tube at point 2 as indicated by the shading on the right.

If the speed of the fluid at this point is v_2 and the cross-sectional area is A_2, the volume is $\Delta V = A_2 v_2 \, \Delta t$. Since these volumes are equal, we have

$$A_1 v_1 \, \Delta t = A_2 v_2 \, \Delta t$$

$$A_1 v_1 = A_2 v_2 \qquad \text{11-25}$$

The quantity Av is called the **volume flow rate** I_V. The dimensions of I_V are those of volume per time. In the steady flow of an incompressible fluid, the volume flow rate is the same at any point in the fluid:

$$I_V = vA = \text{constant} \qquad \text{11-26}$$

Continuity equation

Equation 11-26 is called the **continuity equation.**

Example 11-14

Blood flows in an aorta of radius 1.0 cm at 30 cm/s. What is the volume flow rate?

From Equation 11-26,

$$I_V = vA = (0.30 \text{ m/s})\pi(0.01 \text{ m})^2$$
$$= 9.42 \times 10^{-5} \text{ m}^3/\text{s}$$

It is customary to give the pumping rate of the heart in liters per minute. Using 1 L = 10^{-3} m^3 and 1 min = 60 s, we have

$$I_V = (9.42 \times 10^{-5} \text{ m}^3/\text{s}) \times \frac{1 \text{ L}}{10^{-3} \text{ m}^3} \times \frac{60 \text{ s}}{1 \text{ min}}$$
$$= 5.65 \text{ L/min}$$

(*a*)

(*b*)

Figure 11-16 (*a*) Smoke rising from a cigarette. The simple streamlined flow quickly becomes turbulent. (*b*) Vortex of turbulent flow formed at the tip of a fan blade by heated air above an alcohol lamp.

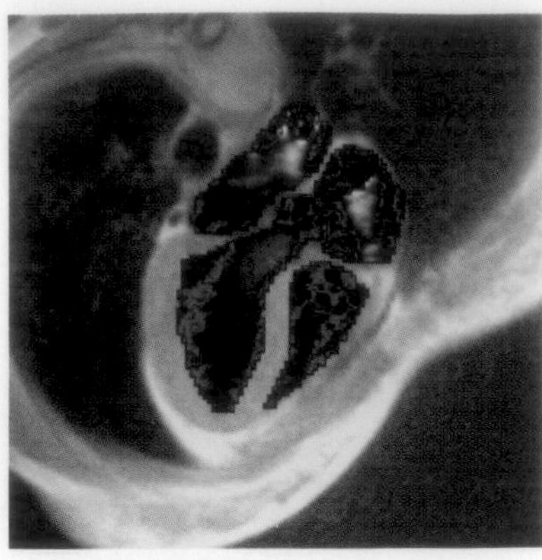

False-color view of turbulence of blood flowing into and out of the heart as seen by Magnetic Resonance Imaging. Systolic ejection from the left ventricle into the aorta is seen in red, and diastolic filling of the ventricles is in blue.

Example 11-15

Blood flows from a large artery of radius 0.3 cm, where its speed is 10 cm/s, into a region where the radius has been reduced to 0.2 cm because of thickening of the walls (arteriosclerosis). What is the speed of the blood in the narrower region?

If v_1 and v_2 are the initial and final speeds and A_1 and A_2 are the initial and final areas, Equation 11-25 gives

$$v_2 = \frac{A_1}{A_2} v_1 = \frac{\pi(0.3\text{ cm})^2}{\pi(0.2\text{ cm})^2}(10\text{ cm/s}) = 22.5\text{ cm/s}$$

Exercise

Water is flowing at 12 m/s in a horizontal pipe. If the pipe widens to twice its original diameter, what is the flow speed in the wide section? (Answer: 3 m/s)

We will now consider a fluid flowing in a tube that varies in elevation as well as in cross-sectional area as shown in Figure 11-17. We apply the work-energy theorem to the fluid that is contained initially between points 1 and 2 in Figure 11-17*a*. After some time Δt, this fluid will have moved along the tube and will be contained in the region between points 1′ and 2′ in Figure 11-17*b*. The only change between Figures 11-17*a* and *b* is for the dark green portions of the fluid mass. Let $\Delta m = \rho\,\Delta V$ be the mass of this fluid. The net effect on the fluid in time Δt is that the mass of fluid Δm is lifted from height y_1 to height y_2 and its speed is increased from v_1 to v_2. The change in the potential energy of this mass is

$$\Delta U = \Delta m\, gy_2 - \Delta m\, gy_1 = \rho\,\Delta V\, g(y_2 - y_1)$$

and the change in its kinetic energy is

$$\Delta K = \tfrac{1}{2}(\Delta m)v_2^2 - \tfrac{1}{2}(\Delta m)v_1^2 = \tfrac{1}{2}\rho\,\Delta V\,(v_2^2 - v_1^2)$$

The fluid following the mass of fluid Δm in the pipe (to the left of it) exerts a force on it to the right in the figure of magnitude $F_1 = P_1A_1$, where P_1 is the pressure at point 1. This force does work

$$W_1 = F_1\,\Delta x_1 = P_1A_1\,\Delta x_1 = P_1\,\Delta V$$

At the same time, the fluid preceding it (to the right) exerts a force $F_2 = P_2A_2$ on it to the left in the figure. This force does negative work because it opposes the motion:

$$W_2 = -F_2\,\Delta x_2 = -P_2A_2\,\Delta x_2 = -P_2\,\Delta V$$

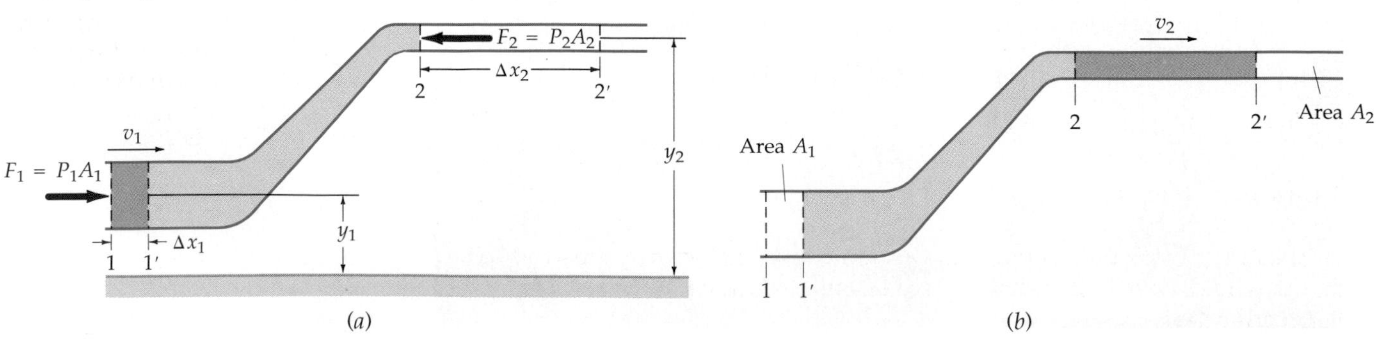

Figure 11-17 Fluid moving in a pipe that varies in both height and cross-sectional area. The total work done by the forces $F_1 = P_1A_1$ and $F_2 = P_2A_2$ has the effect of raising the portion of the fluid shaded in dark green from height y_1 to y_2 and changing its speed from v_1 to v_2.

The total work done by these forces is

$$W_{\text{total}} = P_1\,\Delta V - P_2\,\Delta V = (P_1 - P_2)\,\Delta V$$

The work-energy theorem gives

$$W_{\text{total}} = \Delta U + \Delta K$$

so

$$(P_1 - P_2)\,\Delta V = \rho\,\Delta V\,g(y_2 - y_1) + \tfrac{1}{2}\rho\,\Delta V\,(v_2^2 - v_1^2)$$

If we divide by ΔV, we obtain

$$P_1 - P_2 = \rho g y_2 - \rho g y_1 + \tfrac{1}{2}\rho v_2^2 - \tfrac{1}{2}\rho v_1^2$$

When we collect all the quantities having a subscript 1 on one side and those having a subscript 2 on the other, this equation becomes

$$P_1 + \rho g y_1 + \tfrac{1}{2}\rho v_1^2 = P_2 + \rho g y_2 + \tfrac{1}{2}\rho v_2^2 \qquad \text{11-27}a$$

This result can be restated as

$$P + \rho g y + \tfrac{1}{2}\rho v^2 = \text{constant} \qquad \text{11-27}b$$

Bernoulli's equation

which means that this combination of quantities has the same value at any point along the pipe. Equation 11-27 is known as **Bernoulli's equation** for the steady, nonviscous flow of an incompressible fluid. To some extent, we can also apply Bernoulli's equation to compressible fluids such as gases.

A special application of Bernoulli's equation is for a fluid at rest. Then $v_1 = v_2 = 0$, and we obtain

$$P_1 - P_2 = \rho g(y_2 - y_1) = \rho g h$$

where $h = y_2 - y_1$ is the difference in height between points 2 and 1. This is the same as Equation 11-14.

We now give some examples using Bernoulli's equation in nonstatic situations.

Example 11-16

A large tank of water has a small hole a distance h below the water surface as shown in Figure 11-18. Find the speed of the water as it flows out the hole.

We apply Bernoulli's equation to points a and b in the figure. Since the diameter of the hole is much smaller than the diameter of the tank, we can neglect the velocity of the water at the top (point a). We then have

$$P_a + \rho g y_a = P_b + \tfrac{1}{2}\rho v_b^2 + \rho g y_b$$

Since both points a and b are open to the atmosphere, the pressures P_a and P_b are both equal to atmospheric pressure. Therefore,

$$v_b^2 = 2g(y_a - y_b) = 2gh$$

and

$$v_b = \sqrt{2gh}$$

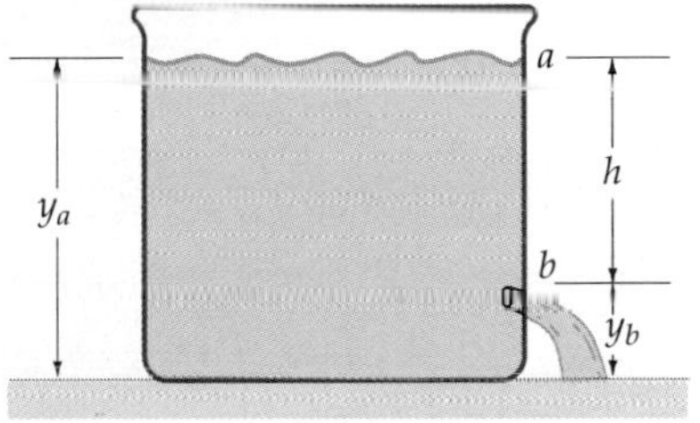

Figure 11-18 Water tank with small hole near the bottom. The speed of the water emerging from the hole is the same as if the water had dropped in free fall a distance $h = y_a - y_b$, a result known as Torricelli's law.

In Example 11-16, the water emerges from the hole with a speed equal to that it would have if it dropped in free fall a distance h. This is known as **Torricelli's law.**

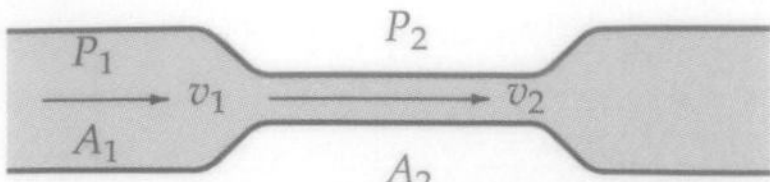

Figure 11-19 Constriction in a pipe carrying a moving fluid. The pressure is smaller in the narrow section of the pipe where the fluid is moving faster.

In Figure 11-19, water is flowing through a horizontal pipe that has a constricted section. Since both sections of the pipe are at the same elevation, $y_1 = y_2$ in Equation 11-27*a*. Then Bernoulli's equation becomes

$$P + \tfrac{1}{2}\rho v^2 = \text{constant} \qquad 11\text{-}28$$

When the fluid moves into the constriction, the area A gets smaller, so the speed v must get larger since Av remains constant. But according to Equation 11-28, if the speed gets larger, the pressure must get smaller if $P + \frac{1}{2}\rho v^2$ is to remain constant. Thus, the pressure in the constriction is reduced. Equation 11-28 is an important result that applies to many situations in which we can ignore changes in height.

Venturi effect

When the speed of a fluid increases, the pressure drops.

This result is often referred to as the **Venturi effect.**

Example 11-17

Water moves through a pipe at 4 m/s under a pressure of 200 kPa. The pipe narrows to half its original diameter. Find (*a*) the speed and (*b*) the pressure of the water in the narrower part of the pipe.

(*a*) Since the area of the pipe is proportional to the square of the diameter, the area of the narrow part of the pipe is one-fourth that of the original. Then, from the continuity equation $I_V = vA =$ constant, the speed in the narrow part must be 4 times that in the wide part or 16 m/s.

(*b*) To find the pressure in the narrow part of the pipe, we use Equation 11-28:

$$P_1 + \tfrac{1}{2}\rho v_1^2 = P_2 + \tfrac{1}{2}\rho v_2^2$$

$$200 \text{ kPa} + \tfrac{1}{2}(1000 \text{ kg/m}^3)(4 \text{ m/s})^2 = P_2 + \tfrac{1}{2}(1000 \text{ kg/m}^3)(16 \text{ m/s})^2$$

$$200 \text{ kPa} + 8000 \text{ Pa} = P_2 + 128{,}000 \text{ Pa}$$

$$P_2 = 200 \text{ kPa} + 8 \text{ kPa} - 128 \text{ kPa} = 80 \text{ kPa}$$

Note that when we put ρ and v in their proper SI units, the units of $\frac{1}{2}\rho v^2$ are pascals. We can check this by writing out the units of ρv^2:

$$\frac{\text{kg}}{\text{m}^3} \times \frac{\text{m}^2}{\text{s}^2} = \frac{\text{kg·m/s}^2}{\text{m}^2} = \frac{\text{N}}{\text{m}^2} = \text{Pa}$$

where we have used 1 kg·m/s^2 = 1 N.

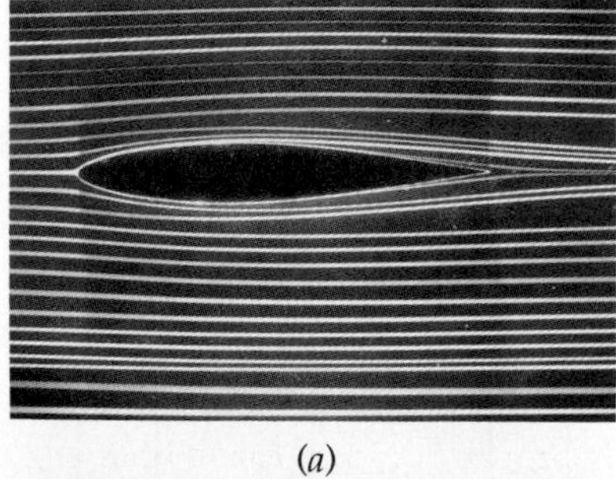

(*a*)

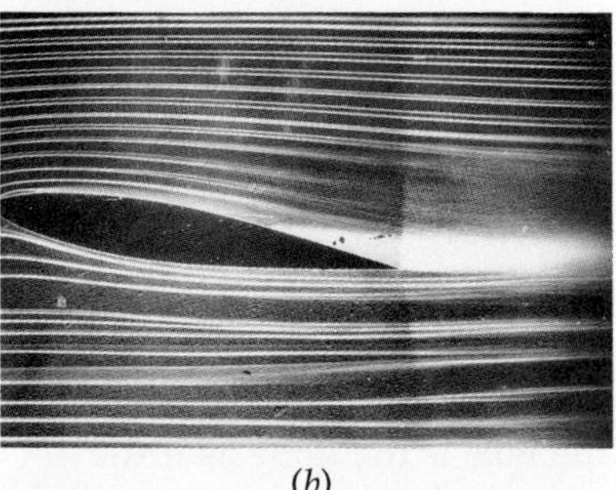

(*b*)

Air flow around a wing foil. (*a*) When the wing is horizontal, the flow is uniform and the pressure above the wing is the same as below it. (*b*) When the wing is tilted, the pressure is greater at the bottom of the wing than at the top of the wing, giving rise to the lift.

The Venturi effect can be used qualitatively to understand the lift of an airplane wing and the curve of the path of a baseball. An airplane wing is designed so that the air moves faster over the top of the wing than it does under the bottom of the wing, thus making the air pressure less on top of the wing than underneath. This difference in pressure results in a net force upward on the wing. Figure 11-20*a* shows a top view of the motion of a baseball thrown by a right-handed pitcher throwing a curve ball. As the ball spins, it tends to drag the air around with it. Figure 11-20*b* is drawn from the point of view of the ball being stationary (but spinning) with the air rushing past it. The air movement caused by the drag of the spinning ball adds to the velocity of the air rushing by on the left side of the ball and subtracts from it

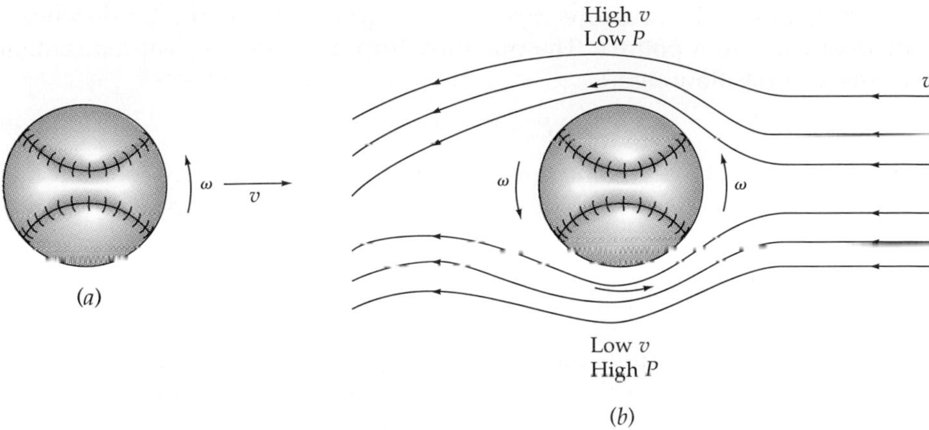

Figure 11-20 (*a*) Top view of baseball thrown with a counterclockwise spin, as would be the case for a curve ball thrown by a right-handed pitcher. (*b*) In the frame of the ball, the ball is stationary (but spinning) and the air rushes past it. Because of its rough cover, the spinning ball drags the air around with it, making the air speed greater on the left side and less on the right. The pressure is therefore less on the left side, where the air speed is greater, so the ball curves to the left.

on the right. Thus, the speed of the air is greater on the left side of the ball than on the right, so according to Equation 11-28, the pressure on the left is less than that on the right. The ball therefore curves to the left.

Figure 11-21 shows an atomizer. When the bulb is squeezed, air is forced through the constriction in the horizontal tube, which reduces the pressure there below atmospheric pressure. Because of the resulting pressure difference, the liquid in the jar is pumped up through the vertical tube, enters the airstream, and emerges from the nozzle as a fine spray. A similar effect occurs in the carburetor of a gasoline engine.

Although Bernoulli's equation is very useful for qualitative descriptions of many features of fluid flow, such descriptions are often grossly inaccurate when compared with the quantitative results of experiments. Of course, gases like air are hardly incompressible, and liquids like water have viscosity, which invalidates the assumption of the conservation of mechanical energy. In addition, it is often difficult to maintain steady-state, streamlined flow without turbulence.

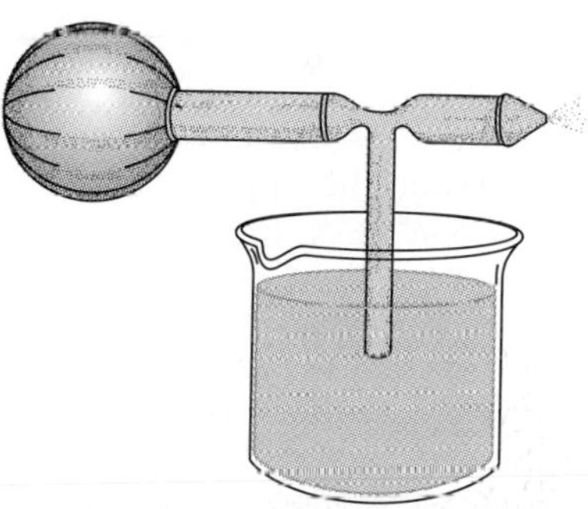

Figure 11-21 Atomizer. Squeezing the bulb at the left forces air through the narrow constriction. Because the pressure in the constriction is below atmospheric pressure, the liquid in the container below is forced into the airstream, resulting in a fine spray of droplets.

Questions

9. In Figure 11-17, the fluid is accelerated to a greater speed as it enters the narrow part of the pipe. What forces act on the fluid to produce this acceleration?

10. In a department store, a beach ball is supported by the airstream from a hose connected to the exhaust of a vacuum cleaner. Does the air blow under or over the ball to support it? Why?

11-7 Viscous Flow

According to Bernoulli's equation, when a fluid flows steadily through a long, narrow, horizontal pipe of constant cross section, the pressure along the pipe will be constant. In practice, however, we observe a pressure drop as we move along the direction of the flow. Looking at this in another way, a pressure difference is required to push a fluid through a horizontal pipe. This pressure difference is needed because of the drag force that is exerted by the pipe on the layer of fluid next to it and because of the drag force exerted by each layer of the fluid on an adjacent layer that is moving with a different velocity. These drag forces are called **viscous forces.** As a result of these viscous forces, the velocity of the fluid is not constant across the diameter of the pipe. Instead, it is greatest near the center of the pipe and least near the edge, where the fluid is in contact with the walls of the pipe (Figure

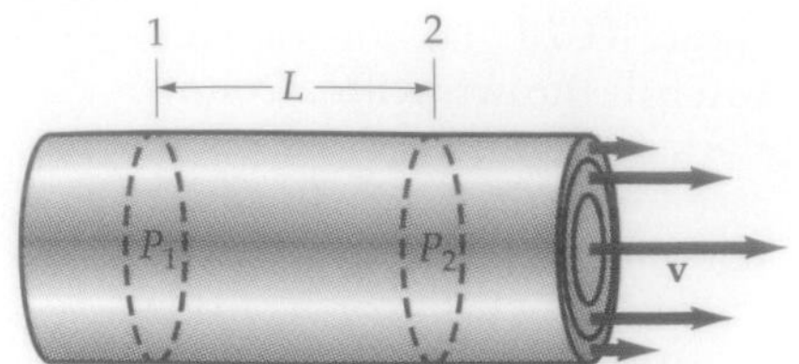

Figure 11-22 When a viscous fluid flows through a pipe, the speed is greatest at the center of the pipe. Near the walls of the pipe, the fluid tends to remain stationary.

11-22). Let P_1 be the pressure at point 1 and P_2 be that at point 2, a distance L downstream from point 1. The pressure drop $\Delta P = P_1 - P_2$ is proportional to the volume flow rate:

$$\Delta P = P_1 - P_2 = I_V R \qquad 11\text{-}29$$

where $I_V = vA$ is the volume flow rate and the proportionality constant R is the resistance to flow, which depends on the length of the pipe L, the radius r, and the viscosity of the fluid.

Example 11-18

As blood flows from the aorta through the major arteries, the small arteries, the capillaries, and the veins to the right atrium, the (gauge) pressure drops from about 100 torr to zero. If the volume flow rate is 0.8 L/s, find the total resistance of the circulatory system.

In Example 11-8, we found that 100 torr = 13.3 kPa = 1.33×10^4 N/m^2. Using 1 L = 1000 cm^3 = 10^{-3} m^3, we have from Equation 11-29

$$R = \frac{\Delta P}{I_V} = \frac{1.33 \times 10^4 \text{ N/m}^2}{8 \times 10^{-4} \text{ m}^3\text{/s}} = 1.66 \times 10^7 \text{ N·s/m}^5$$

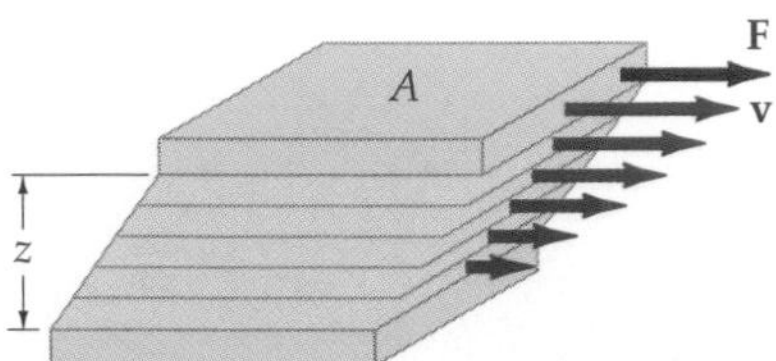

Figure 11-23 Two plates of equal area with a viscous fluid between them. When the upper plate is moved relative to the lower one, each layer of fluid exerts a drag force on adjacent layers. The force needed to pull the upper plate is directly proportional to v and the area A and inversely proportional to z, the separation between the plates.

To define the coefficient of viscosity of a fluid, we consider a fluid that is confined between two parallel plates, each of area A, separated by a distance z as shown in Figure 11-23. The upper plate is pulled at a constant speed v by a force **F** while the bottom plate is held at rest. A force is needed to pull the upper plate because the fluid next to the plate exerts a viscous drag force opposing its motion. The speed of the fluid between the plates is essentially v near the upper plate and zero near the lower plate, and it varies linearly with separation between the plates. The force **F** is found to be directly proportional to v and A and inversely proportional to the plate separation z. The proportionality constant is the **coefficient of viscosity** η:

$$F = \eta \frac{vA}{z} \qquad 11\text{-}30$$

The SI unit of viscosity is the N·s/m^2 = Pa·s. An older cgs unit still in common use is the **poise** after the French physicist Poiseuille. These units are related by

$$1 \text{ Pa·s} = 10 \text{ poise} \qquad 11\text{-}31$$

Table 11-5 **Coefficients of Viscosity of Various Fluids**

Fluid	t, °C	η, mPa·s
Water	0	1.8
	20	1.00
	60	0.65
Blood (whole)	37	4.0
Engine oil (SAE 10)	30	200
Glycerin	0	10,000
	20	1,410
	60	81
Air	20	0.018

Table 11-5 lists the coefficients of viscosity for various fluids at various temperatures. Generally, the viscosity of a liquid increases as the temperature decreases. Thus, in cold climates, a thinner grade of oil is used to lubricate automobile engines in the winter than in summer.

The resistance to flow R in Equation 11-29 for steady flow through a circular tube of radius r can be shown to be

$$R = \frac{8\eta L}{\pi r^4} \qquad 11\text{-}32$$

Equations 11-29 and 11-32 can be combined to give the pressure drop over a length L of a circular tube of radius r:

Poiseuille's law

$$\Delta P = \frac{8\eta L}{\pi r^4} I_V \qquad 11\text{-}33$$

Equation 11-33 is known as **Poiseuille's law.** Note the inverse r^4 dependence of the pressure drop. If the radius of the tube is halved, the pressure drop for a given volume flow rate is increased by a factor of 16; or a pressure 16 times as great is needed to pump the fluid through the tube at the original volume flow rate. Thus, if the diameter of the blood vessels or arteries is reduced for some reason, either the volume flow rate of the blood is greatly reduced, or the heart must work much harder to maintain the volume flow rate. For water flowing through a long garden hose, the pressure drop is fixed. It equals the difference in pressure between that at the water source and atmospheric pressure at the open end. The volume flow rate is then proportional to the fourth power of the radius. Thus, if the radius is halved, the volume flow rate drops by a factor of 16.

Poiseuille's law applies only to the laminar (nonturbulent) flow of a fluid of constant viscosity that is independent of the velocity of the fluid. Blood is a complex fluid consisting of solid particles of various shapes suspended in a liquid. The red blood cells, for example, are disk-shaped objects that are randomly oriented at low velocities but tend to become oriented at high velocities to facilitate the flow. Thus, the viscosity of blood decreases as the flow velocity increases, so Poiseuille's law is not strictly valid. Nevertheless, Poiseuille's law is a good approximation that is very useful for obtaining a qualitative understanding of blood flow.

When the flow velocity of a fluid becomes sufficiently great, laminar flow breaks down and turbulence sets in. The critical velocity above which the flow through a tube is turbulent depends on the density and viscosity of the fluid and on the radius of the tube. The flow of a fluid can be characterized by a dimensionless number called the **Reynolds number** N_R, which is defined by

$$N_R = \frac{2r\rho v}{\eta} \qquad 11\text{-}34$$

where v is the average velocity of the fluid. Experiments have shown that the flow will be laminar if the Reynolds number is less than about 2000 and turbulent if it is greater than 3000. Between these values, the flow is unstable and may change from one type to the other.

This turbulent motion of air around a vibrating cylinder in a wind tunnel is called **vortex shedding.** The cylinder was oscillating normal to the oncoming flow of air and the vortices were shed periodically in unison with the oscillations. The Reynolds number was 200. A sheet of minute aerosol particles was introduced into the wind tunnel upstream of the cylinder in order to make the flow pattern visible.

Example 11-19

Calculate the Reynolds number for blood flowing at 30 cm/s through an aorta of radius 1.0 cm. Assume that blood has a viscosity of 4 mPa·s and a density of 1060 kg/m^3.

Since the Reynolds number is dimensionless, we can use any set of units as long as we are consistent. Putting each quantity in Equation 11-34 in SI units, we get

$$N_R = \frac{2r\rho v}{\eta} = \frac{2(0.01\ \text{m})(1060\ \text{kg/m}^3)(0.3\ \text{m/s})}{4 \times 10^{-3}\ \text{Pa}\cdot\text{s}} = 1590$$

Since the Reynolds number is less than 2000, this flow will be laminar rather than turbulent.

Summary

1. The density of a substance is the ratio of its mass to its volume:

$$\text{Density} = \frac{\text{mass}}{\text{volume}}$$

$$\rho = \frac{m}{V}$$

The specific gravity of a substance is the ratio of its density to that of water. An object sinks or floats in a given fluid depending on whether its density is greater than or less than that of the fluid. The densities of most solids and liquids are approximately independent of temperature and pressure, whereas those of gases depend strongly on these quantities. Weight density is the density times g. The weight density of water is 62.4 lb/ft^3.

2. Tensile stress is the force per unit area applied to a body:

$$\text{Stress} = \frac{F}{A}$$

Strain is the fractional change in the length of the body:

$$\text{Strain} = \frac{\Delta L}{L}$$

Young's modulus is the ratio of stress to strain:

$$Y = \frac{\text{stress}}{\text{strain}} = \frac{F/A}{\Delta L/L}$$

The shear modulus is the ratio of the shear stress to the shear strain:

$$M_s = \frac{\text{shear stress}}{\text{shear strain}} = \frac{F_s/A}{\Delta X/L}$$

The (negative) ratio of the pressure to the fractional change in volume of an object is called its bulk modulus:

$$B = -\frac{P}{\Delta V/V}$$

The inverse of this ratio is the compressibility k.

3. Pressure of a fluid is the force per unit area exerted by the fluid:

$$P = \frac{F}{A}$$

The SI unit of pressure is the pascal (Pa), which is a newton per square meter:

$$1 \text{ Pa} = 1 \text{ N/m}^2$$

Many other units of pressure, such as atmosphere, bar, torr, pound per square inch, or millimeter of mercury, are often used. These units are related by

$$1 \text{ atm} = 101.325 \text{ kPa} = 760 \text{ mmHg} = 760 \text{ torr}$$
$$= 29.9 \text{ inHg} = 33.9 \text{ ftH}_2\text{O} = 14.71 \text{ lb/in}^2$$

Gauge pressure is the difference between the absolute pressure and atmospheric pressure.

4. Pascal's principle states that pressure applied to an enclosed liquid is transmitted undiminished to every point in the fluid and to the walls of the container.

5. In a liquid, such as water, the pressure increases linearly with depth:

$$P = P_0 + \rho g h$$

In a gas such as air, pressure decreases exponentially with altitude.

6. Archimedes' principle states that a body wholly or partially submerged in a fluid is buoyed up by a force equal to the weight of the displaced fluid.

7. Objects can be supported on the surface of a less dense fluid because of surface tension, which is the result of molecular forces at the surface of the fluid. These molecular forces are also responsible for the rise of a liquid in a thin tube, which is known as capillarity.

8. For the steady-state flow of an incompressible fluid, the volume flow rate is the same throughout the fluid:

$$I_V = vA = \text{constant}$$

This is called the continuity equation.

9. Bernoulli's equation

$$P + \rho g y + \tfrac{1}{2}\rho v^2 = \text{constant}$$

applies to steady-state, nonviscous flow without turbulence in which mechanical energy is conserved. For situations in which we can ignore changes in height, we have the important result that when the speed of the fluid increases, the pressure drops. This result, known as the Venturi effect, can be used to explain qualitatively the lift on an airplane wing and the curve of a baseball's path.

10. In viscous flow through a tube, the pressure drop is proportional to the volume flow rate and to the resistance, which in turn is inversely proportional to the fourth power of the radius of the tube:

$$\Delta P = I_V R = \frac{8\eta L}{\pi r^4} I_V$$

This is Poiseuille's law.

Aerodynamics of Bicycling

Robert G. Hunt
Johnson County Community College

Aerodynamics is the systematic study of forces exerted by air or other gases. In this chapter you have been introduced to some of the basic concepts of viscous flow of fluids; this essay will extend those ideas to illustrate some important technical aspects of bicycling.

One of the attractions of bicycling is the aerodynamic interaction, the feel of the "wind" as we ride. Another satisfying aspect is the ease with which a bicycle is set into motion and the effortless coasting of long distances because of the low levels of mechanical friction. These two factors oppose one another, with aerodynamic drag limiting speeds obtained.

As we start riding, we exert forces on the pedals that result in a forward-directed force that causes acceleration. This motion can be described by using the kinematics and dynamics developed in Chapters 2 through 6, if the net force can be determined. This, however, becomes a complicated problem. So far in this course, we have primarily considered constant or uniformly changing forces, but the forces exerted in this case are more complicated.

Robert G. Hunt is currently a physics instructor at Johnson County Community College, Overland Park, Kansas. He has taught 14 years at the high school and community college level. His undergraduate work was at the University of Kansas and his two Master's degrees are from Rensselaer Polytechnic Institute. He has worked as a chemical physicist at Midwest Research Institute and more recently as an environmental policy consultant and as a developer of environmental curricula at Franklin Associates. He is a lifelong bicycling enthusiast, and has developed a number of demonstrations and lab experiments using bicycles.

When we start riding on a flat terrain, we exert relatively large forces on the pedals, perhaps even standing on them to get started. The bicycle accelerates forward, described by Newton's second law $\mathbf{F} = m\mathbf{a}$. The bicycle continues to accelerate as we exert forces on the pedals, but eventually a level of physiological riding comfort is reached; for most people this is near one revolution of the pedals per second. As this point is approached, the force on the pedals is reduced gradually, until a constant speed is achieved. This terminal speed is typically 4 to 7 m/s (9 to 16 mi/h) for recreational riding on a ten- or three-speed bicycle.

Constant speed means that the acceleration is zero, so the net force is zero. Thus, the propelling force is balanced by the retarding forces. The two major retarding forces are mechanical friction and the aerodynamic drag on the bicycle and the rider. Mechanical friction includes that in the bicycle bearings, chains, and rolling tires. This typically accounts for about one-third or less of the retarding forces. The dominant forces arise from aerodynamic drag.

Earlier in this chapter we learned about fluid flow confined to tubes or pipes. For bicycles, the viscous fluid (air) is unconfined. It impinges directly onto the rider and the bicycle, and an analysis of the aerodynamic drag requires a different approach.

Aerodynamic drag occurs for two reasons. One is pressure that results from air molecules striking a surface and bouncing off, undergoing a momentum change and exerting a normal force on the surface. The other type of force arises from the sliding motion of air molecules along the surface. This is called air friction; it arises because air molecules collide with rough surfaces, and also because they have a weak chemical attraction for the surface. This causes drag forces to be exerted along the surface as the air rushes past.

Fluid flow can be either laminar or turbulent (see Section 11-7), depending on many factors such as speed, surface roughness, and the type of surface material. At sufficiently slow speeds, the flow will be laminar; this results in quite low drag forces. As the relative speed of the air and the surface increases, the laminar flow becomes unstable and layers of air begin to separate. The flow then becomes turbulent, characterized by whirling eddies of air (Figure 1). Turbulent boundary layers have much higher drag than laminar layers, but the highest drag is caused by instability at air velocities in the transition region between laminar and turbulent flow. In fact, aerodynamic drag can be 5 times greater in the transition speed ranges than aerodynamic drag for purely turbulent flow.

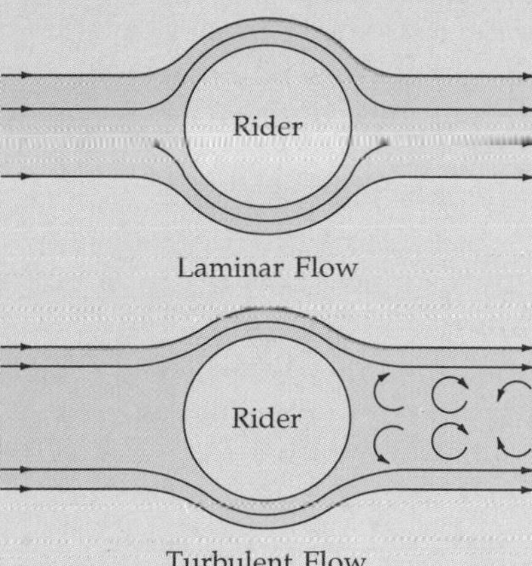

Figure 1 Air flow around a bicycle rider. The rider, moving left, is simulated here by a circular cylinder.

It is this transition region that must be avoided to achieve low drag forces. The limits of the transition region are defined by the Reynolds number (Section 11-7). For analytical purposes, a bicycle rider can be simulated by a series of stacked cylinders of circular cross section. For this geometry, the transition region occurs at a Reynolds number of about 4×10^5. Unfortunately for bicycle riders, this occurs on very smooth surfaces at speeds of about 4 to 6 m/s, which is where most riding takes place. This means that very smooth surfaces, which encourage laminar flow, should be avoided except in some special cases, to be discussed later. This may seem surprising at first, but experienced riders report noticeable increases in drag when they don very smooth plastic rain gear. An analogy can be seen in the design of golf balls. They are intentionally dimpled to make them rough because a roughened surface ensures turbulent air flow; this allows the golf ball to travel longer distances than it would with a smooth surface.

Let's now look at some quantitative aspects of aerodynamic drag for recreational bicycling. The drag forces retarding forward motion can be characterized by

$$F_D = \tfrac{1}{2} C_D \rho A v^2 \qquad (1)$$

where F_D is the drag force, C_D is a dimensionless number known as the drag coefficient, ρ is the air density, A is the projected frontal area (usually the cross-sectional area perpendicular to the general direction of travel), and v is the relative velocity of the air and the surface over which the air is flowing. This equation includes both pressure and air friction effects.

Like many practical applications of physics, the equation for determining the drag force provides an estimate, not "the" answer. The primary reason for this is that aerodynamic drag is a phenomenon for which no simple model exists. For example, C_D cannot be calculated for most real objects, but is usually inferred from experimental data, often from measurements made in wind tunnels. The value obtained for C_D depends in a complicated way on shape, speed, type of material, temperature, and so on. However, for some applications, such as bicycle riding, C_D is relatively constant over the range of speeds of interest. This simplifies our approximation.

In Equation 1, ρ, as well as C_D, can be considered constant for a given bicycle and rider. The other two terms—A and v—are important variables. Speed is usually continuously changing while riding, and the square dependence makes it the dominant variable. The frontal area term includes both bicycle and rider. While the area of the bicycle will be constant during a ride, the rider may frequently change position. The rider's frontal area is several times greater than that of the bicycle and therefore is the most important component determining frontal area. The rider's clothes are also a factor; tight fitting clothes present a smaller frontal area than those that are loose fitting.

For a casual ride on a warm spring day, the following are typical:

$$C_D = 0.90 \qquad \rho = 1.3\ \text{kg/m}^3$$
$$A = 0.45\ \text{m}^2 \qquad v = 4.0 \text{ to } 7.0\ \text{m/s}$$

Figure 2 plots data points derived using Equation 1 and the above values for C_D, A, ρ, and v. Aerodynamic drag force increases from 4 N at 4 m/s to over 12 N at 7 m/s. This figure also shows the power necessary to sustain a given speed considering only

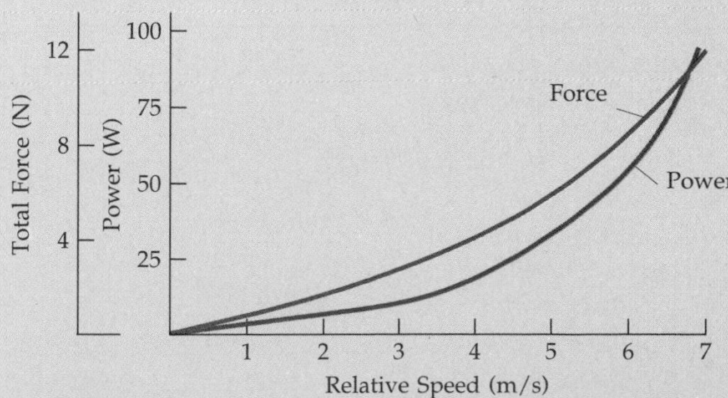

Figure 2 Force and power required to overcome aerodynamic drag at various speeds.

Continued

the drag force. Power can be found from $P = \mathbf{F} \cdot \mathbf{v}$ (Equation 6-33). For the situation we are considering here, the force and velocity are in the same direction, so Equation 6-33 reduces to $P = Fv$. From Equation 1 we have

$$P = \tfrac{1}{2}C_D \rho A v^3 \quad (2)$$

That is, the power required to sustain motion depends on the cube of the speed and so increases rapidly as the speed increases.

Although aerodynamic drag is a dominant factor for bicycle riding, it is interesting to put it into perspective and to relate it to our own senses. We will now translate the results shown in Figure 2 into three physiological parameters to which we can all relate. These are (1) the average force exerted on the pedals, (2) the average rate at which we expend energy (i.e., our power output), and (3) food energy requirements.

From experimental data obtained with inexpensive mass-marketed ten- and three-speed bicycles (of the kind typically used by students), Table 1 was prepared. The mechanical bearing, chain, and rolling tire friction was added to the aerodynamic drag forces to find a total retarding force (second column). The drag forces (which can be obtained from Figure 2) range from $\frac{2}{3}$ of the total force at 4 m/s to $\frac{3}{4}$ of the total at 7 m/s. Provided that the bicycle is in its highest gear, measurements show that the average downward force on the pedals (third column) is 14.2 times the total retarding force. The average downward force is also reported as a percent of the rider's weight for a 68-kg rider (typical for cross-country riders). This translates to a body weight of 668 N (150 lb). To maintain the selected speeds requires 13 to 34 percent of body weight be exerted downward on the pedals. Compare this to the 20 percent body weight exertion that many people find comfortable for sustained riding.

Finally, the mechanical power output (calculated by $P = Fv$) is shown in the fifth column. We may estimate food requirements, assuming that the rider's body is 25 percent efficient at converting food energy into useful work. This illustrates the efficiency of locomotion by bicycle. Traveling at a sustained leisurely rate of 5 m/s, a rider could travel 18 km in 1 h on only 150 food calories, about that contained in two slices of plain bread.

Let's now examine how riders can (intentionally or unintentionally) affect aerodynamic drag. An easily altered factor is the frontal area of the rider. As a rider shifts position and moves about, the area changes. The values range from 0.3 to 0.6 m^2. The smaller value corresponds to a racing posture, crouching low and wearing tightly fitting clothes (Figure 3). A more comfortable posture in street clothes provides a frontal area of 0.45 m^2 (the value used in our sample calculation above). For an upright, touring, or standing posture the area becomes 0.6 m^2. Thus, the range of possible areas leads to a variation of a factor of two in the aerodynamic drag force.

Another important variable is the relative speed of the air and the surface. In our sample calculation, we assumed that the air was still (something it rarely is). If the wind blows from behind, then the drag will provide a forward force. In fact, if the tail wind equals the bicycle speed, the relative speed is zero and there is no aerodynamic drag on the rider. On the other hand, a head wind increases the drag force. Suppose the rider tries to maintain a ground speed of 7 m/s against a head wind of 7 m/s. The relative speed between the air and the surface will then be 14 m/s, resulting in a drag force of 49 N (from Equation 1). At a bicycle speed of 7 m/s the mechanical friction is about 4 N for a total retarding force of 53 N. Table 2 is a summary of the forces and other consequences to be expected. Much greater

Table 1 **Estimated Force and Power for Casual Recreational Bicycling**

Speed, m/s	Total force to sustain motion, N	Average force on pedal (bike in high gear): In newtons	Average force on pedal (bike in high gear): As percentage of weight	Power output of rider, W
4	6	85	13	24
5	9	126	19	45
6	12	170	25	72
7	16	227	34	112

Table 2 **Consequences of Riding into a 7-m/s Head Wind at Two Bicycle Speeds**

	Bicycle speed: 7 m/s (high gear)	Bicycle speed: 3 m/s (medium gear)
Relative speed, m/s	14	10
Drag force, N	49	25
Total force, N	53	27
Force on pedal, N	753	189
Power output, W	371	81

Figure 3 A rider can intentionally lower aerodynamic drag by acquiring a racing posture (crouching low and wearing tightly fitting clothes).

forces are needed on the pedal than before—753 N are needed, compared to 227 N in still air. It is now necessary to stand with full weight on the pedals, and pull upward on the handles to achieve an even greater downward force on the pedals. The power requirement is now 371 W, well in excess of sustainable limits (of no more than 100 W).

In the above case, riders will normally decrease speed, and shift to a lower gear to reduce the force necessary to continue in the forward direction. Generally bicycle speed is reduced by about one-half of the wind speed by changing to a lower gear. The second column of Table 2 shows the results of shifting and slowing to 3 m/s. The forces on the pedal are reduced to a much more comfortable 189 N (28% of body weight), and to a sustainable power level of 81 W. Riders tend to adjust to changing aerodynamic conditions by changing speed and gears.

Still another factor that influences drag is the air density. While this will be relatively constant at a given location, it does change with altitude (Section 11-3) and temperature. For example, in the "mile-high" city of Denver, air density is 10 percent lower than at sea level, leading to a 10 percent reduction in aerodynamic drag. At even higher altitudes, drag may be reduced by another 10 percent. Temperature is less variable, but warm air has a lower density than cool air. A 5°C increase in air temperature produces a 1.5 percent reduction in drag.

Aerodynamic drag can also be reduced by lowering the drag coefficient. Some specially designed bicycles have a C_D of 0.1 or lower. This is accomplished by streamlining, carefully designing the shape of the bicycle so that laminar flow is preserved during the ride. This requires surrounding the bicycle and rider with a shaped surface, as shown in Figure 4. Frontal area is minimized by having the rider lie flat. However, these bicycles are quite expensive to build, are not comfortable to ride, and have never gained wide popularity.

There are many other interesting aspects of bicycling aerodynamics, which you are encouraged to pursue. Two good references for additional reading are listed with the Further Readings at the end of the chapter.

Figure 4 Jim Glover is shown here riding a bicycle designed by Alex Moulton, with which he set the world speed record of 82.53 km/h at Vancouver in 1986.

Suggestions for Further Reading

Hazen, David C., and Rudolf F. Lehnert: "Low-Speed Flight," *Scientific American*, April 1956, p. 46.

How an airplane wing generates lift and how viscosity of air helps to determine the minimum possible flying speed of an airplane.

Kyle, Chester R.: "How Wind Affects Cycling," *Bicycling*, May 1988, p. 194.

The effects of wind on the speed of bicycle riding.

Walker, Jearl: "The Amateur Scientist: Looking into the Ways of Water Striders, the Insects that Walk (and Run) on Water," *Scientific American*, November 1983, p. 188.

Walker explains how surface tension enables water striders to stay dry and how they are able to propel themselves.

Walker, Jearl: "The Amateur Scientist: What Forces Shape the Behavior of Water as a Drop Meanders Down a Windowpane?" *Scientific American*, September 1985, p. 138.

The effects of surface tension on both resting drops and small streams on glass or plexiglass are investigated.

Whitt, Frank Rowland, and David Gordon Wilson: *Bicycling Science*, second edition, MIT Press, Cambridge, 1982.

A comprehensive technical treatment of many aspects of bicycling.

Review

A. Objectives: After studying this chapter, you should:

1. Be able to state the definitions of density, specific gravity, weight density, Young's modulus, and the bulk modulus.

2. Be able to explain the buoyancy of objects, such as boats and hot-air balloons, using Archimedes' principle.

3. Be able to work problems involving buoyant forces on submerged or floating objects.

4. Be able to discuss the rise of a liquid in a capillary tube in terms of the surface tension of the liquid.

5. Be able to explain qualitatively the lift of an airplane wing and the curving of the path of a baseball's flight using the Venturi effect.

B. Define, explain, or otherwise identify:

Density
Liter
Specific gravity
Weight density
Elastic limit
Tensile stress
Strain
Young's modulus
Compressive stress
Tensile strength
Compressive strength
Shear force
Shear stress
Shear strain
Shear modulus
Torsion modulus
Pressure
Pascal
Bulk modulus
Compressibility
Pascal's principle
Gauge pressure
Torr
Bar
Millibar
Buoyant force
Archimedes' principle
Surface tension
Coefficient of surface tension
Cohesive forces
Adhesive forces
Capillarity
Volume flow rate
Continuity equation
Bernoulli's equation
Torricelli's law
Venturi effect
Viscous forces
Coefficient of viscosity
Poise
Poiseuille's law
Reynolds number

C. True or false: If the statement is true, explain why it is true. If it is false, give a counterexample.

1. The buoyant force on a submerged object depends on the shape of the object.

2. Young's modulus has the same dimensions as pressure.

3. If the density of an object is greater than that of water, it cannot float on the surface of the water.

4. In viscous flow, the pressure drop along a pipe is proportional to the flow rate.

Problems

Level I

11-1 Density

1. A copper cylinder is 6 cm long and has a radius of 2 cm. Find its mass.

2. Find the mass of a lead sphere of radius 2 cm.

3. Find the mass of air in a room 4 m by 5 m by 4 m.

4. A small flask used for measuring densities of liquids (called a pyconmeter) has a mass of 22.71 g. When it is filled with water, the total mass is 153.38 g, and when it is filled with milk, the total mass is 157.67 g. Find the density of milk.

5. A solid oak door is 200 cm high, 75 cm wide, and 4 cm thick. How much does it weigh?

6. A 60-mL flask is filled with mercury at 0°C. When the temperature rises to 80°C, 1.47 g of mercury spills out of the flask. Assuming the volume of the flask is constant, find the density of mercury at 80°C if its density at 0°C is 13,645 kg/m^3.

11-2 Stress and Strain

7. A 50-kg ball is suspended from a steel wire of length 5 m and radius 2 mm. By how much does the wire stretch?

8. Copper has a breaking stress of about 3×10^8 N/m^2. (*a*) What is the maximum load that can be hung from a copper wire of diameter 0.42 mm? (*b*) If half this maximum load is hung from the copper wire, by what percentage of its length will it stretch?

9. As a runner's foot touches the ground, the shearing force acting on an 8-mm-thick sole is as shown in Figure 11-24. If the force of 25 N is distributed over an area of 15 cm^2, find the angle of shear θ shown, given that the shear modulus of the sole is 1.9×10^5 N/m^2.

11-3 Pressure in a Fluid

10. Barometer readings are commonly given in inches of mercury. Find the pressure in inHg equal to 101 kPa.

11. The pressure on the surface of a lake is atmospheric pressure P_{at} = 101 kPa. (*a*) At what depth is the pressure twice atmospheric pressure? (*b*) If the pressure at the top of a deep pool of mercury is P_{at}, at what depth is the pressure $2P_{at}$?

12. Find the (*a*) absolute pressure and (*b*) the gauge pressure at the bottom of a diving pool of depth 5.0 m.

13. The top of a card table is 80 cm by 80 cm. What is the force exerted on it by the atmosphere? Why doesn't the table collapse?

14. When a woman in high heels takes a step, she momentarily places her entire weigh on one heel of her shoe, which has a radius of 0.4 cm. If her mass is 56 kg, what is the pressure exerted on the floor by her heel?

15. A hydraulic lift is used to raise an automobile of mass 1500 kg. The radius of the shaft of the lift is 8 cm and that of the piston is 1 cm. How much force must be applied to the piston to raise the automobile?

16. Blood flows into the aorta through a circular opening of radius 0.9 cm. If the blood pressure is 120 torr, how much force must be exerted by the heart?

17. A car misses a turn and sinks into a shallow lake to a depth of 8 m. (*a*) If the area of the car door is 0.5 m^2, what is the force exerted on the outside of the door by the water? (*b*) What is the force exerted on the inside of the door by the air, assuming the inside of the car is at atmospheric pressure? (*c*) What should the occupant do to get the door open?

18. What pressure is required to reduce the volume of 1 kg of water from 1.00 L to 0.99 L?

19. Seawater has a bulk modulus of 2.3×10^9 N/m^2. Find the density of seawater at a depth where the pressure is 800 atm if the density at the surface is 1024 kg/m^3.

20. A 1500-kg car rests on four tires, each of which is inflated to a gauge pressure of 200 kPa. What is the area of contact of each tire with the road, assuming the four tires support the weight equally?

21. Blood plasma flows from a bag through a tube into a patient's vein, where the blood pressure is 12 mmHg. The specific gravity of blood plasma at 37°C is 1.03. What is the minimum elevation the bag must have so that the pressure of the plasma as it flows into the vein is at least 12 mmHg?

22. In the seventeenth century, Pascal performed the experiment shown in Figure 11-25. A wine barrel was filled with water and was then connected to a long tube. Water was added to the tube until the barrel burst. (*a*) If the radius of the lid was 20 cm and the height of the water in the tube was 12 m, calculate the force exerted on the lid. (*b*) If the tube had an inner radius of 3 mm, what mass of water in the tube caused the pressure that burst the barrel?

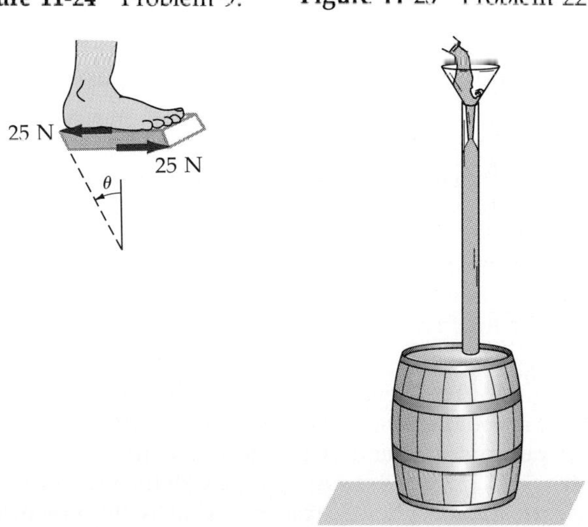

Figure 11-24 Problem 9. **Figure 11-25** Problem 22.

11-4 Buoyancy and Archimedes' Principle

23. A 500-g piece of copper (specific gravity 9.0) is suspended from a spring scale and is submerged in water (Figure 11-26). What force does the spring scale read?

Figure 11-26 Problem 23.

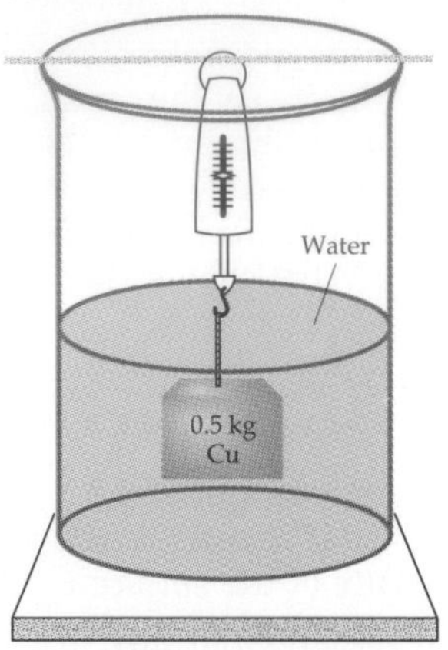

24. When a 60-N stone is attached to a spring scale and is submerged in water, the scale reads 40 N. Find the specific gravity of the stone.

25. A block of an unknown material weighs 5 N in air and 4.55 N when submerged in water. (*a*) What is the density of the material? (*b*) Of what material is the block made?

26. A 5-kg iron block is attached to a spring scale and is submerged in a fluid of unknown density. The spring scale reads 6.16 N. What is the density of the fluid?

11-5 Surface Tension and Capillarity

27. When a capillary tube with a diameter of 0.8 mm is dipped into methanol, the methanol rises to a height of 15.0 mm. If the angle of contact is zero, find the surface tension of methanol (specific gravity 0.79).

28. Tiny tubes called xylem carry nutrients upward in a plant. Each tube has a radius of about 0.01 mm. How high will water rise by capillary action in such a tube, assuming the contact angle to be zero?

11-6 Fluids in Motion and Bernoulli's Equation

29. Blood flows in an aorta of radius 9 mm at 30 cm/s. (*a*) Calculate the volume flow rate in liters per minute. (*b*) Although the cross-sectional area of a capillary is much smaller than that of the aorta, there are many capillaries, so their total cross-sectional area is much larger. If all the blood from the aorta flows into the capillaries and the speed of flow through the capillaries is 1.0 mm/s, calculate the total cross-sectional area of the capillaries.

30. Water flows through a 3-cm-diameter hose at 0.65 m/s. The diameter of the nozzle is 0.30 cm. (*a*) At what speed does the water pass through the nozzle? (*b*) If the pump at one end of the hose and the nozzle at the other end are at the same height, and if the pressure at the nozzle is atmospheric pressure, what is the pressure at the pump?

31. Water is flowing at 3 m/s in a horizontal pipe under a pressure of 200 kPa. The pipe narrows to half its original diameter. (*a*) What is the speed of flow in the narrow section? (*b*) What is the pressure in the narrow section? (*c*) How do the volume flow rates in the two sections compare?

32. The pressure in a section of horizontal pipe with a diameter of 2 cm is 142 kPa. Water flows through the pipe at 2.80 L/s. What should the diameter of a constricted section of the pipe be for the pressure there to be 101 kPa?

33. During very high winds, the atmospheric pressure inside a house may blow off the roof because of the reduced pressure outside. Calculate the force on a square roof 15 m on a side if the wind speed over the roof is 30 m/s.

11-7 Viscous Flow

34. A horizontal tube with an inside diameter of 1.2 mm and a length of 25 cm has water flowing through it at 0.30 mL/s. Find the pressure difference required to drive this flow if the viscosity of water is 1.00 mPa·s.

35. Find the diameter of a tube that would give double the flow rate for the pressure difference in Problem 34.

36. Blood takes about 1.0 s to pass through a 1-mm-long capillary of the human circulatory system. If the diameter of the capillary is 7 μm and the pressure drop is 2.60 kPa, find the viscosity of blood.

Level II

37. Many people believe that if they float the top of a flexible snorkel tube out of the water (Figure 11-27), they will be able to breathe with it while walking under water. However, the water pressure opposes the expansion of the chest and the inflation of the lungs. Suppose you can just breath while lying on the floor with a 400-N weight on your chest. How far below the surface of the water could your chest be for you still to be able to breathe, assuming your chest has a frontal area of 0.09 m^2?

Figure 11-27 Problem 37.

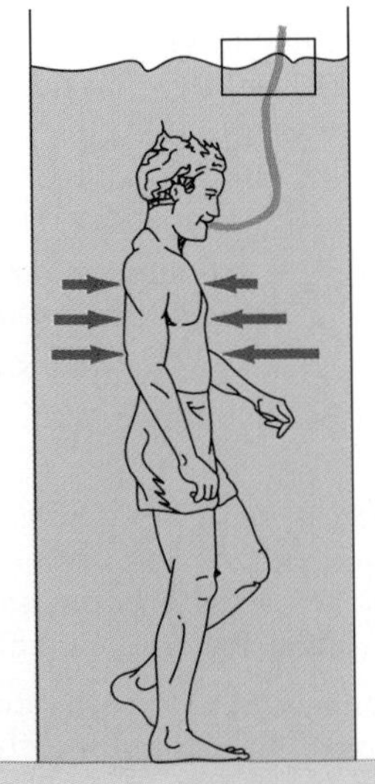

38. When the ground becomes saturated with water during floods, pressure develops in the ground similar to that in water. This pressure forces water through the joints in concrete-block cellar walls. If this happens quickly enough to fill up the cellar with water, there may be no further damage. Otherwise, the upward pressure on the cellar floor may float the house like a ship. What upward force would be exerted on a 10 m by 10 m basement floor if the floor were 2 m below the surface of the water?

39. Under standard conditions, the density of air is 1.29 kg/m^3 and that of helium is 0.178 kg/m^3. A helium balloon lifts a basket and cargo of total weight 2000 N. What must be the volume of the balloon?

40. Figure 11-28 shows a small sphere that is supported on the surface of water by surface tension. The upward force due to surface tension is $2\pi r\gamma \cos \theta_c$. An insect of mass 0.002 g is supported on the surface of a lake by its six legs, each of which has an approximately spherical base of radius 0.02 mm. (*a*) Using

$$\tfrac{1}{6}w = \tfrac{1}{6}mg = 2\pi r\gamma \cos \theta_c$$

where w is the weight of the insect and the factor $\frac{1}{6}$ is used because the six legs support the insect equally, calculate the angle θ_c. (*b*) If the mass of the insect increases with no increase in r, the angle θ_c decreases until the critical value $\theta_c = 0$ is reached. If the mass increases further, the insect cannot be supported by surface tension. Find the critical mass of the insect that can be supported by legs of this size.

Figure 11-28 Problems 40 and 59.

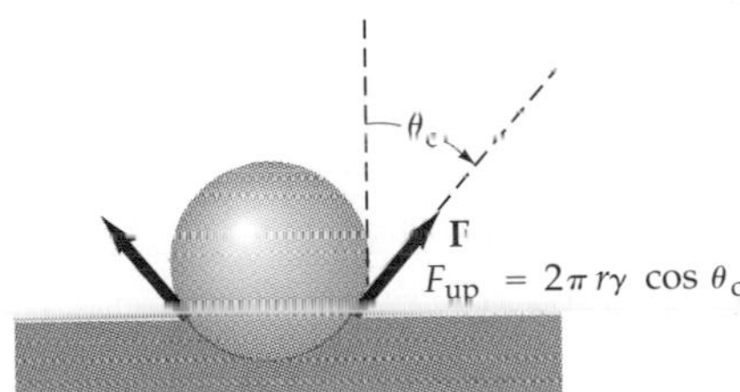

41. A large tank of water is tapped a distance h below the water surface by a small pipe as shown in Figure 11-29. Find the distance x reached by the water flowing out the pipe.

Figure 11-29 Problems 41 and 64.

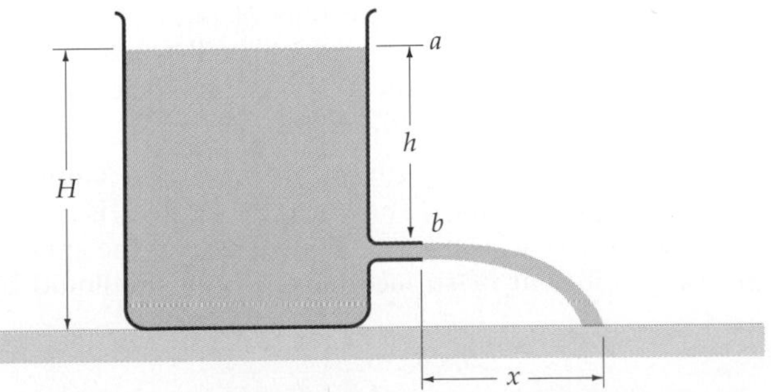

42. A force F is applied to a long wire of length L and cross-sectional area A. Show that if the wire is considered to be a spring, the force constant k is given by $k = AY/L$ and that the energy stored in the wire is $U = \frac{1}{2}F\,\Delta L$, where Y is Young's modulus and ΔL is the amount the wire has stretched.

43. A manometer is a U-tube that contains a liquid and is designed to measure small pressure differences between its two arms. If an oil ($\rho = 900$ kg/m^3) manometer can be read to ± 0.05 mm, what is the smallest pressure change that can be detected?

44. A 3-m by 3-m raft is 11 cm thick and is made of wood with a specific gravity of 0.6. How many 70-kg people can stand on the raft without getting their feet wet when the water is calm?

45. An object has neutral buoyancy when its density equals that of the liquid in which it is submerged, which means that it neither floats nor sinks. If the average density of an 85-kg diver is 0.96 kg/L, what mass of lead should be added to give him neutral buoyancy?

46. When you weigh yourself in air, the scale reading is less than the force exerted on you by gravity because of the buoyant force of air. Estimate the correction to the scale reading you should make to get your true weight.

47. A beaker of mass 1 kg containing 2 kg of water rests on a scale. A 2-kg block of aluminum (specific gravity 2.70) suspended from a spring scale is submerged in the water as shown in Figure 11-30. Find the readings of both scales.

Figure 11-30 Problem 47.

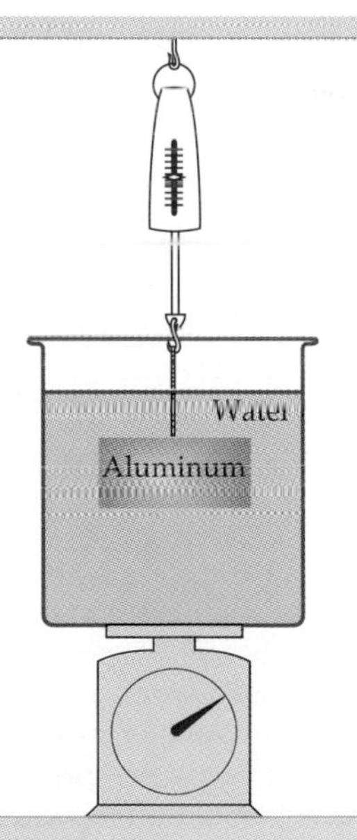

48. In Example 11-6, a force of 147 N is applied to a small piston to lift a car that weighs 14,700 N. Show that this does not violate the law of conservation of mechanical energy by showing that, when the car is lifted some distance h, the work done by the force acting on the small piston equals the work done by the large piston on the car.

49. A large piece of cork weighs 0.285 N in air. When held submerged under water by a spring scale as in Figure 11-31, the spring scale reads 0.855 N. Find the density of the cork.

Figure 11-31 Problem 49.

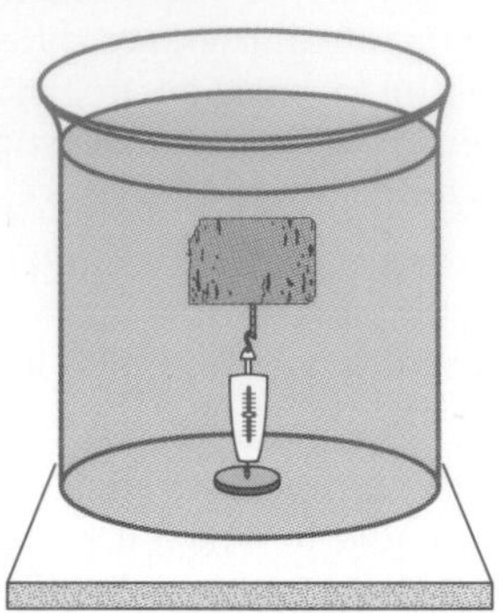

50. A building is to be demolished by a 400-kg steel ball swinging on the end of a 30-m steel wire of diameter 5.0 mm hanging from a tall crane. As the ball is swung through an arc from side to side, the wire makes an angle of 50° with the vertical at the top of the swing. Find the amount by which the wire is stretched at the bottom of the swing.

51. A firefighter holds a hose with a bend in it as shown in Figure 11-32. Water is expelled from the hose in a stream of radius 1.5 cm at a speed of 30 m/s. (*a*) What mass of water emerges from the hose in 1 s? (*b*) What is the horizontal momentum of this water? (*c*) Before reaching the bend, the water has momentum upward, whereas afterward, its momentum is horizontal. Draw a vector diagram of the initial and final momentum vectors, and find the change in the momentum of the water at the bend in 1 s. From this, find the force exerted on the water by the hose.

Figure 11-32 Problem 51.

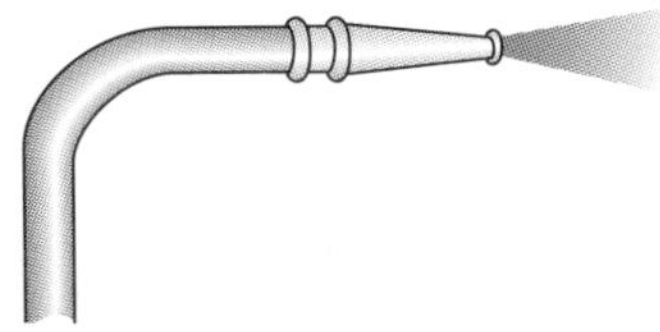

52. A fountain designed to spray a column of water 12 m into the air has a 1-cm-diameter nozzle at ground level. The water pump is 3 m below the ground. The pipe to the nozzle has a diameter of 2 cm. Find the necessary pump pressure (neglecting the viscosity of the water).

53. A rectangular dam 30 m wide supports a body of water to a height of 25 m. (*a*) Neglecting atmospheric pressure, find the total force due to water pressure acting on a thin strip of the dam of height dy located at a depth y. (*b*) Integrate your result in part (*a*) to find the total horizontal force exerted by the water on the dam. (*c*) Why is it reasonable to neglect atmospheric pressure?

54. When a rubber strip with a 3 by 1.5 mm cross section is suspended vertically and various masses are attached to it, a student obtains the following data for length versus load:

Load, g	0	100	200	300	400	500
Length, cm	5.0	5.6	6.2	6.9	7.8	10.0

(*a*) Find Young's modulus for the rubber strip for small loads. (*b*) Find the energy stored in the strip when the load is 150 g. (See Problem 42.)

55. The steel E string of a violin is under a tension of 53 N. The diameter of the string is 0.20 mm, and its length under tension is 35.0 cm. Find (*a*) the unstretched length of this string, and (*b*) the work needed to stretch the string. (See Problem 42.)

Level III

56. A ship sails from seawater (specific gravity 1.03) into fresh water and therefore sinks slightly. When its load of 600 Mg is removed, it returns to its original level. Assuming the sides of the ship are vertical at the water line, find the mass of the ship before it was unloaded.

57. The hydrometer shown in Figure 11-33 is a device for measuring the density of liquids. The bulb contains lead shot, and the density can be read directly from the liquid level on the stem after the hydrometer has been calibrated. The bulb's volume is 20 mL, the stem is 15 cm long and has a diameter of 5.00 mm, and the mass of the glass is 6.0 g. (*a*) What mass of lead shot must be added so that the least density of liquid that can be measured is 0.9 kg/L? (*b*) What is the maximum density of liquid that can be measured?

Figure 11-33 Problem 57.

58. A hollow can with a small hole of radius 0.1 mm is pushed under water. At what depth will water start to flow into the can through the hole if the coefficient of surface tension of water is 0.073 N/m?

59. A small sphere rests on the surface of a liquid as shown in Figure 11-28. (*a*) Show that the angle θ_c is related to the radius r of the sphere, the density ρ of the sphere, and the coefficient of surface tension γ of the liquid by

$$\cos \theta_c = \frac{2}{3} \frac{r^2 \rho g}{\gamma}$$

(*b*) Find the radius of the largest copper sphere that can rest on water without sinking.

60. The volume of a cone of height h and base radius r is $V = \frac{1}{3}\pi r^2 h$. A conical vessel of height 25 cm resting on its base of radius 15 cm is filled with water. (*a*) Find the volume and weight of the water in the vessel. (*b*) Find the force exerted by the water on the base of the vessel. Explain how this force can be greater than the weight of the water.

61. The water in the cylindrical container in Figure 11-34 drains through a horizontal capillary of diameter 0.50 mm. Calculate the length of time needed for the water in the cylinder to drop from 10.0 cm to 5.0 cm if the viscosity of water is 1.00 mPa·s.

Figure 11-34 Problem 61.

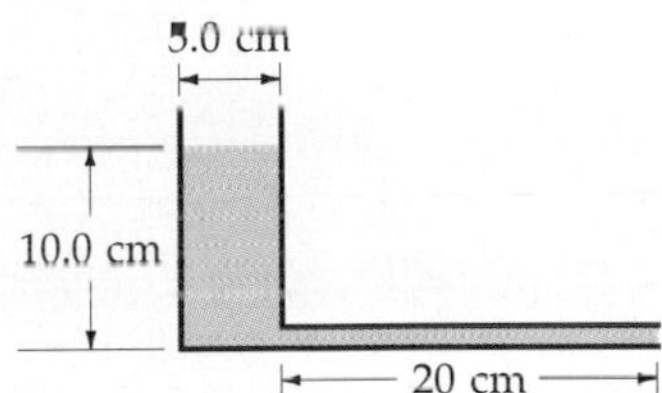

62. In this problem you are to prove that, for a spherical balloon of radius R with a surface tension γ, the pressure inside the balloon is given by $P = P_{at} + 2\gamma R$. First draw a line around the circumference of the balloon and consider the forces exerted on the right hemisphere of the balloon (Figure 11-35). The force to the right is due to the excess pressure against the hemispherical surface of the balloon, whereas the force to the left is due to the surface tension acting on the line.

63. A large beer keg of height H and cross-sectional area a_1 is filled with beer. The top is open to atmospheric pressure. At the bottom is a spigot opening of area A_2, which is much smaller than A_1. (*a*) Show that when the height of the beer is h, the speed of the beer leaving the spigot is approximately $\sqrt{2gh}$. (*b*) Show that for the approximation $A_2 \ll A_1$, the rate of change of the height h of the beer is given by

$$\frac{dh}{dt} = -\frac{A_2}{A_1}(2gh)^{1/2}$$

(*c*) Find h as a function of time if $h = H$ at $t = 0$. (*d*) Find the total time needed to drain the keg if $H = 2$ m, $A_1 = 0.8$ m^2, and $A_2 = (10^{-4})A_1$.

64. In Figure 11-29, (*a*) find the distance x at which the water strikes the ground as a function of h and H. (*b*) Show that there are two values of h that are equidistant from the point $h = \frac{1}{2}H$ that give the same distance x. (*c*) Show that x is a maximum when $h = \frac{1}{2}H$. What is the value of this maximum distance x?

Figure 11-35 Problem 62.

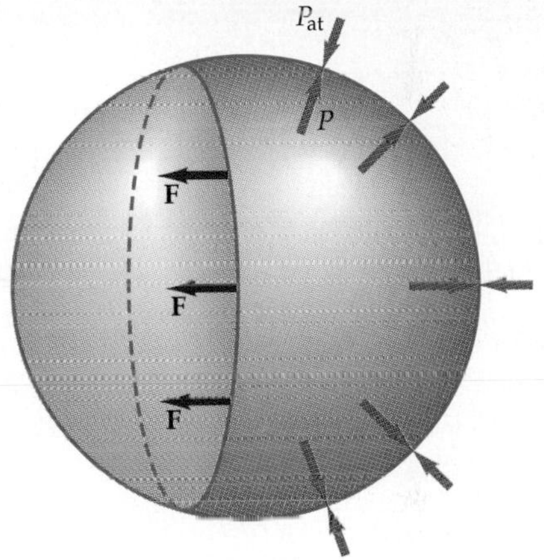

Part 2 Oscillations and Waves

Vibration patterns in a violin. The vibrations of the violin strings are transmitted via the bridge to the body of the violin. The vibrations of the body, made visible here by interference effects, resonate with the vibrations of the strings, thus giving the violin its tone.

Chapter 12

Oscillations

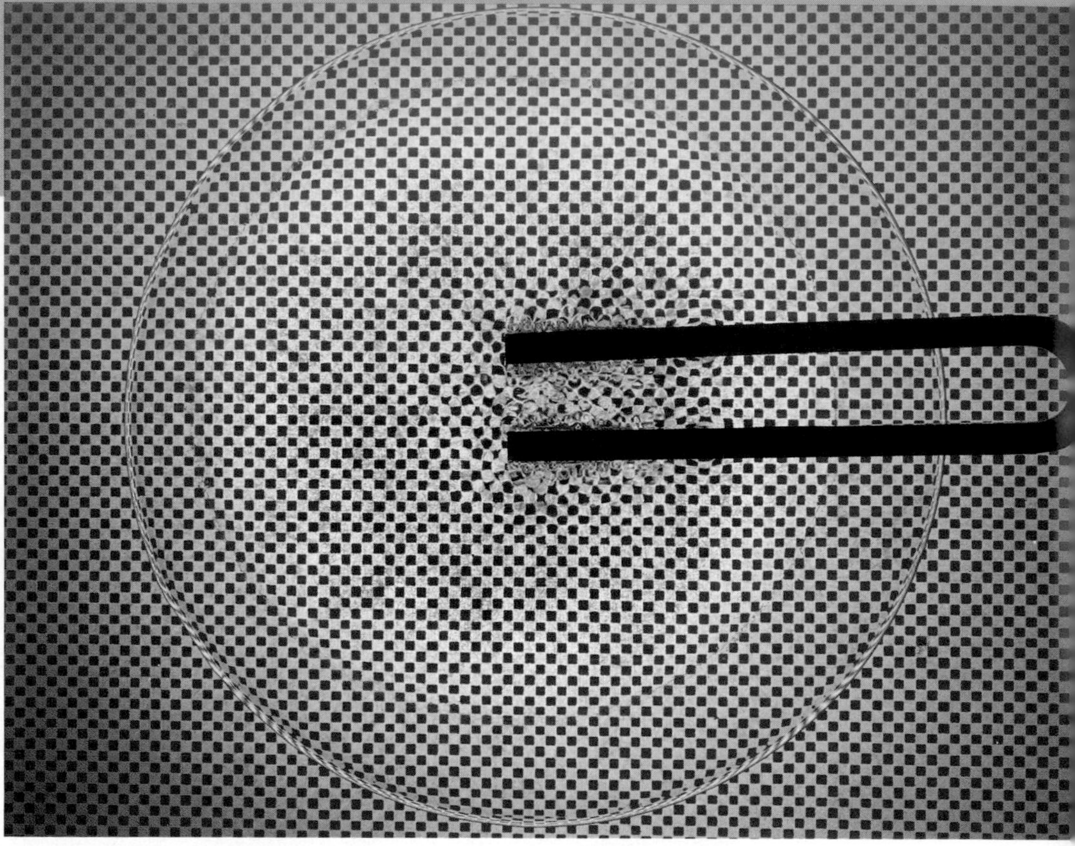

A vibrating tuning fork touching a water surface.

Oscillation occurs when a system is disturbed from a position of stable equilibrium. The most recognizable characteristic of oscillatory motion is that the motion is **periodic,** that is, it repeats itself. There are many familiar examples of oscillations: small boats bob up and down, clock pendulums swing back and forth, and the strings and reeds of musical instruments vibrate. Other, less familiar examples are the oscillations of air molecules in a sound wave and the oscillations of electric currents in radios or television sets.

Wave motion is closely related to oscillatory motion. Sound waves, for example, are produced by a vibrating string (such as a violin string), a vibrating oboe reed, a vibrating drum head, or the vibrations of your vocal cords when you speak. In each case, the vibrating system sets up oscillations in the nearby air molecules, and these oscillations are propagated through the air (or some other medium, such as water or a solid).

In this chapter, we will study some of the fundamentals of oscillating systems. In Chapters 13 and 14, we will study mechanical waves—waves on strings and sound waves.

12-1 Simple Harmonic Motion: A Mass on a Spring

A common and very important kind of oscillatory motion is simple harmonic motion. When an object is displaced from its equilibrium position, simple harmonic motion occurs if there is a restoring force that is proportional to the displacement. This condition is nearly always met, at least approximately, when the displacement from equilibrium is small.

A typical system that exhibits simple harmonic motion is an object attached to a spring, as illustrated in Figure 12-1. In equilibrium, the spring exerts no force on the object. When the object is displaced an amount x from its equilibrium position, the spring exerts a force $-kx$, as given by Hooke's law (Section 4-5):

$$F_x = -kx \qquad 12\text{-}1$$

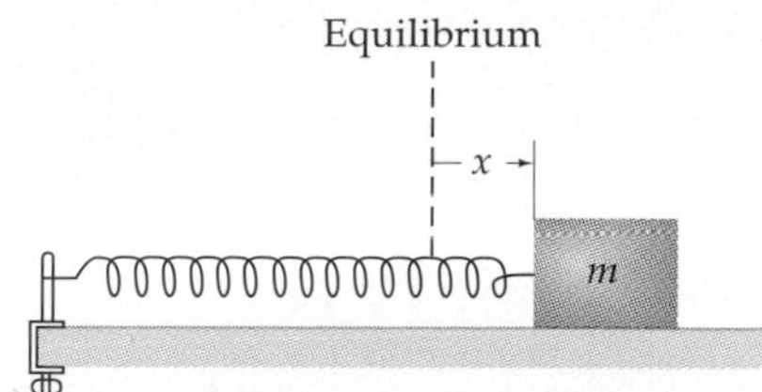

Figure 12-1 An object attached to a spring resting on a frictionless table. The displacement x is measured from the equilibrium position. The displacement can be positive or negative depending on whether the spring is stretched or compressed from its natural length.

The minus sign in Hooke's law arises because the force is opposite the direction of the displacement. If we choose x to be positive for displacements to the right, the force is negative (to the left) when x is positive and positive (to the right) when x is negative. Combining Equation 12-1 with Newton's second law, we have

$$F_x = -kx = ma = m\frac{d^2x}{dt^2}$$

or

$$a = \frac{d^2x}{dt^2} = -\left(\frac{k}{m}\right)x \qquad 12\text{-}2$$

The acceleration is proportional to the displacement and is oppositely directed. This is a general characteristic of simple harmonic motion and can, in fact, be used to identify systems that will exhibit simple harmonic motion:

Conditions for simple harmonic motion

Whenever the acceleration of an object is proportional to its displacement and is oppositely directed, the object will move with simple harmonic motion.

If we displace an object from equilibrium and release it, it oscillates back and forth about its equilibrium position. The time for the object to make a complete oscillation is called the **period** T. The reciprocal of the period is the **frequency** f, which is the number of oscillations per second:

$$f = \frac{1}{T} \qquad 12\text{-}3$$

The unit of frequency is the reciprocal second (s^{-1}), which is called a **hertz** (Hz). For example, if the time for one complete oscillation is 0.25 s, the frequency is 4 Hz.

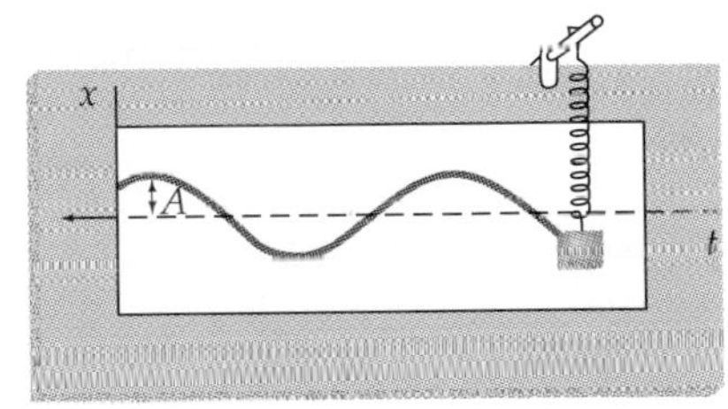

Figure 12-2 An object oscillating on a vertical spring. As the paper is moved with constant speed to the left, the marking pen attached to the object on the spring traces out the displacement x as a function of time t. (For this case, we have chosen x to be positive when the spring is compressed.)

For an oscillating object, the displacement x as a function of time t can be obtained experimentally. For example, let us attach a marking pen to an object that is attached to a vertical spring and place the pen so it can write on a strip of paper that can be moved perpendicular to the direction of oscillation as shown in Figure 12-2. We then displace the object a distance A and begin to pull the paper to the left with constant speed when we release the object. The pen traces out the sinusoidal curve shown in the figure. The equation for this curve is

Simple harmonic motion defined

$$x = A\cos(\omega t + \delta) \qquad 12\text{-}4$$

where A, ω, and δ are constants. Motion in which the position varies with time according to Equation 12-4 is by definition **simple harmonic motion.** Note that $\cos(\omega t + \delta) = \sin(\omega t + \delta + \pi/2)$. Whether the equation is expressed as a cosine function or a sine function depends merely on when we choose $t = 0$. The maximum displacement from equilibrium is called the **amplitude** A. The argument of the cosine function, $\omega t + \delta$, is called the **phase** of the motion, and the constant δ is called the **phase constant.** During one complete cycle of the motion, the phase increases by 2π. At the end of the cycle, the object again has the same position and velocity as it had at the start of the cycle since $\cos(\omega t + \delta + 2\pi) = \cos(\omega t + \delta)$. We can determine the period T from the fact that the phase at time $t + T$ is just 2π plus the phase at time t:

$$\omega(t + T) + \delta = 2\pi + \omega t + \delta$$

or

$$\omega T = 2\pi$$

so

$$T = \frac{2\pi}{\omega} \qquad 12\text{-}5$$

From Equation 12-3, we obtain for the frequency

Frequency, period, and angular frequency

$$f = \frac{1}{T} = \frac{\omega}{2\pi} \qquad 12\text{-}6$$

The constant $\omega = 2\pi f$ is called the **angular frequency.** It has units of radians per second and dimensions of inverse time, the same as angular velocity, which is also designated by ω. In terms of the frequency or period, Equation 12-4 may be rewritten as

$$x = A\cos(2\pi f t + \delta) = A\cos\left(\frac{2\pi t}{T} + \delta\right) \qquad 12\text{-}7$$

The phase constant δ depends on when we choose $t = 0$. If we choose $t = 0$ when $x = A$, as in Figure 12-3a, the phase constant is zero and $x = A\cos\omega t$. On the other hand, if we choose $t = 0$ when $x = 0$, δ is either $\pi/2$ or $3\pi/2$, depending on whether x is increasing or decreasing at $t = 0$. For example, if x is increasing at $t = 0$ as in Figure 12-3b, $\delta = 3\pi/2$ and

$$x = A\cos\left(\omega t + \frac{3\pi}{2}\right) = A\sin\omega t$$

The general relation between the initial position x_0 and the constants A and δ is found by setting $t = 0$ in Equation 12-4. Then

$$x_0 = A\cos\delta \qquad 12\text{-}8$$

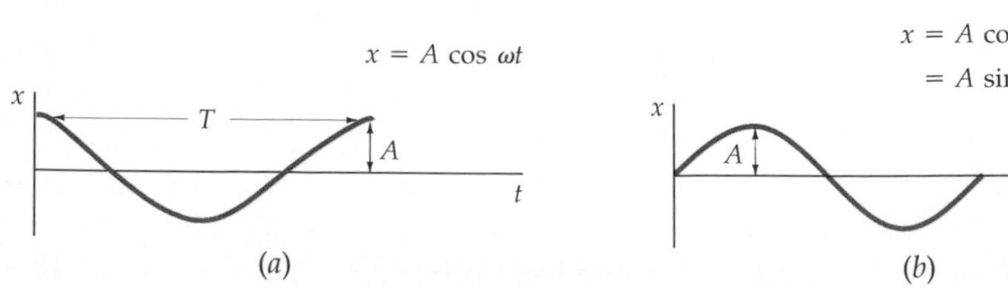

Figure 12-3 (a) Graph of x versus t for $\delta = 0$. At $t = 0$, the displacement has its maximum value A, and the object is at rest, which is indicated by the fact that the slope is zero. The time for one complete oscillation is the period T. (b) Graph of x versus t for $\delta = 3\pi/2$. At $t = 0$, the particle is at $x = 0$ and is moving toward positive x, which is indicated by the fact that the slope is positive.

We can show that x as given by Equation 12-4 is a solution to Equation 12-2 by differentiating x twice. The first derivative of x with respect to time gives the velocity v:

$$v = \frac{dx}{dt} = -A\omega \sin(\omega t + \delta) = A\omega \cos\left(\omega t + \delta + \frac{\pi}{2}\right) \qquad 12\text{-}9$$

The phase of the velocity differs from that of the position by $\pi/2$ rad = 90°. When $\cos(\omega t + \delta)$ has the value +1 or −1, $\sin(\omega t + \delta) = 0$. Thus, when x is at its maximum or minimum value, the velocity is zero. Similarly, when $\sin(\omega t + \delta)$ has the value +1 or −1, $\cos(\omega t + \delta) = 0$. The speed is maximum when the object passes through its equilibrium position $x = 0$. We can relate the initial velocity v_0 to the constants A and δ by setting $t = 0$ in Equation 12-9. We obtain

$$v_0 = -A\omega \sin \delta \qquad 12\text{-}10$$

Differentiating the velocity (Equation 12-9) with respect to time gives the acceleration of the object.

$$a = \frac{dv}{dt} = \frac{d^2x}{dt^2} = -\omega^2 A \cos(\omega t + \delta) \qquad 12\text{-}11$$

or

$$a = -\omega^2 x \qquad 12\text{-}12$$

Comparing this with $a = -(k/m)x$ for a mass on a spring (Equation 12-2), we see that $x = A\cos(\omega t + \delta)$ is a solution of $d^2x/dt^2 = -(k/m)x$ if the angular frequency ω is related to the force constant k and the mass m by

$$\omega^2 = \frac{k}{m} \qquad 12\text{-}13$$

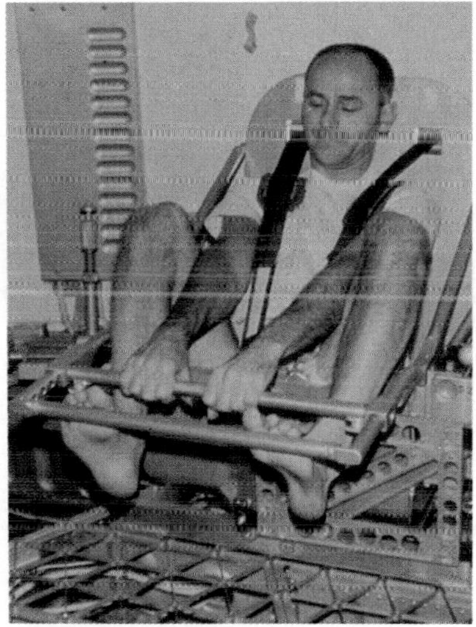

Astronaut Alan L. Bean measures his body mass during the second Skylab mission. The total mass of the astronaut plus the apparatus is related to his frequency of vibration by Equation 12-14.

The frequency and period of a mass on a spring are thus related to the force constant k and the mass m by

$$f = \frac{\omega}{2\pi} = \frac{1}{2\pi}\sqrt{\frac{k}{m}} \qquad 12\text{-}14$$

$$T = \frac{1}{f} = 2\pi\sqrt{\frac{m}{k}} \qquad 12\text{-}15$$

Period and frequency for an object on a spring

From this result, we can see that when k is large, as in the case of a stiff spring, the frequency is large. Similarly, if the mass is large, the frequency is small.

We have demonstrated by direct substitution that $x = A\cos(\omega t + \delta)$ is a solution of the differential equation $d^2x/dt^2 = -\omega^2 x$. Rather than solving the differential equation directly, we guessed the solution by using a physical picture of a pen attached to a mass on a spring so that it traces on paper. Although there are mathematical techniques for solving differential equations directly, they work only for some kinds of differential equations. There is no general method for solving an arbitrary differential equation. Solutions to differential equations are often based on guesswork aided by physical insight. A direct solution of the equation $d^2x/dt^2 = -\omega^2 x$ using the conservation of mechanical energy is outlined in Problem 89 of Chapter 6.

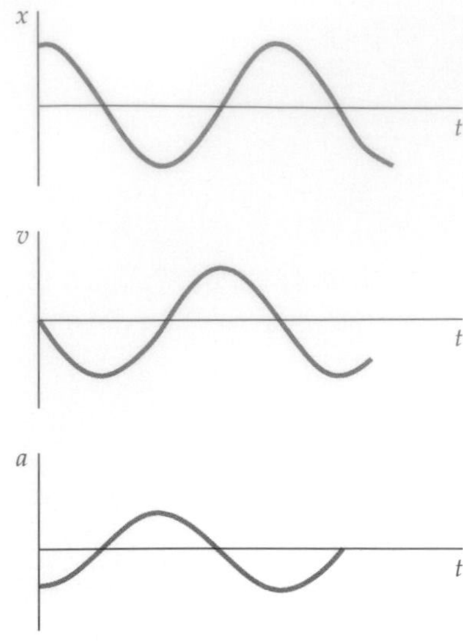

Figure 12-4 Plots of the displacement x, the velocity v, and the acceleration a as functions of time t for $\delta = 0$.

The variations with time of the displacement x, the velocity v, and the acceleration a for the special case of $\delta = 0$ are plotted in Figure 12-4. For $\delta = 0$, Equations 12-4, 12-9, and 12-11 become

$$x = A \cos \omega t \tag{12-16a}$$

$$v = -\omega A \sin \omega t \tag{12-16b}$$

and

$$a = -\omega^2 A \cos \omega t \tag{12-16c}$$

Initially, at time $t = 0$, the displacement is maximum, the velocity is zero, and the acceleration is negative and equal to $-\omega^2 A$. The velocity then becomes negative as the object moves back toward its equilibrium position. The object is at its equilibrium position $x = 0$ when $\cos \omega t = 0$. The acceleration at these times is also zero, and the speed has its maximum value of ωA because, when $\cos \omega t$ is zero, $\sin \omega t$ is either 1 or -1.

Example 12-1

A particle has displacement x given by

$$x = 0.3 \cos \left(2t + \frac{\pi}{6}\right)$$

where x is in meters and t is in seconds. (*a*) What are the frequency, period, amplitude, angular frequency, and phase constant of the motion? (*b*) Where is the particle at $t = 1$ s? (*c*) Find the velocity and acceleration at any time t. (*d*) Find the initial position and velocity of the particle.

(*a*) Comparing this equation with Equation 12-4, we see that the angular frequency $\omega = 2$ rad/s, the amplitude $A = 0.3$ m, and the phase constant $\delta = \pi/6$ rad. The frequency is thus $f = \omega/2\pi = 0.318$ Hz, and the period is $T = 1/f = 3.14$ s.

(*b*) At $t = 1$ s, the position of the particle is

$$x = 0.3 \cos \left[2(1) + \frac{\pi}{6}\right] = -0.245 \text{ m}$$

(*c*) The velocity is obtained from

$$v = \frac{dx}{dt} = -0.3 \sin \left(2t + \frac{\pi}{6}\right) \frac{d(2t)}{dt}$$

$$= -0.6 \sin \left(2t + \frac{\pi}{6}\right)$$

We differentiate again to obtain the acceleration

$$a = \frac{dv}{dt} = \frac{d}{dt}\left[-0.6 \sin \left(2t + \frac{\pi}{6}\right)\right]$$

$$= -0.6 \cos \left(2t + \frac{\pi}{6}\right) \frac{d(2t)}{dt} = -1.2 \cos \left(2t + \frac{\pi}{6}\right)$$

(*d*) We find the initial position and the initial velocity by substituting $t = 0$ into the expressions for x and v. We obtain

$$x_0 = 0.3 \cos \frac{\pi}{6} = 0.260 \text{ m}$$

and

$$v_0 = -0.6 \sin \frac{\pi}{6} = -0.300 \text{ m/s}$$

Exercise

A 0.8-kg object is attached to a spring of force constant $k = 400$ N/m. Find the frequency and period of the motion of the object when it is displaced from equilibrium. (Answer: $f = 3.56$ Hz, $T = 0.281$ s)

Example 12-2

A 2-kg object stretches a spring 10 cm when it hangs vertically in equilibrium (Figure 12-5*a*). The object is then attached to the same spring while the object is resting on a frictionless table and the spring is fixed at one end as in Figure 12-1. The object is held a distance 5 cm from the equilibrium position and is released at $t = 0$. Find the amplitude A, the angular frequency ω, the frequency f, and the period T.

The force constant of the spring is determined from the first measurement. In equilibrium, the downward force mg must equal the upward force kx_0, where x_0 is the amount the spring is stretched from its natural length (Figure 12-5). Using $g = 9.81$ N/kg, $m = 2$ kg, and $x_0 = 10$ cm $=$ 0.1 m, we have

$$k = \frac{mg}{x_0} = \frac{(2\text{ kg})(9.81\text{ N/kg})}{0.1\text{ m}} = 196\text{ N/m}$$

In the horizontal position, the spring is initially stretched 5 cm from equilibrium and released. Since the initial velocity v_0 is zero, the phase constant $\delta = 0$ (Equation 12-10), and the displacement x is given by Equation 12-16*a*: $x = A\cos\omega t$. The initial displacement is 5 cm so $A =$ 5 cm.

The angular frequency is found from Equation 12-13:

$$\omega = \sqrt{\frac{k}{m}}$$

$$= \sqrt{\frac{196\text{ N/m}}{}} = 9.90\text{ rad/s}$$

The frequency is then

$$f = \frac{\omega}{2\pi} = \frac{9.90\text{ rad/s}}{2\pi} = 1.58\text{ Hz}$$

The period is the reciprocal of the frequency:

$$T = \frac{1}{f} = \frac{1}{1.58\text{ Hz}} = 0.63\text{ s}$$

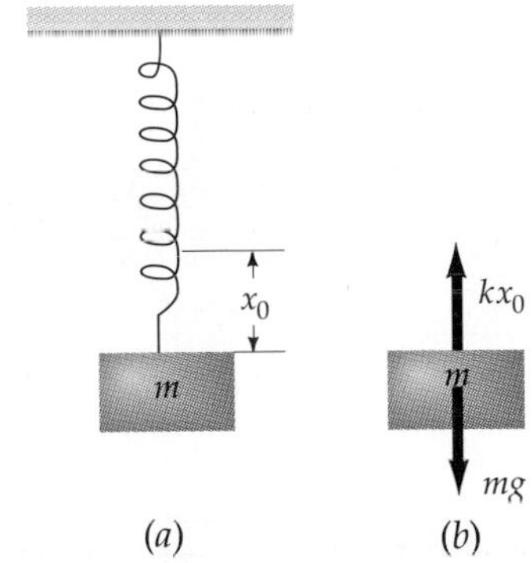

Figure 12-5 (*a*) An object hanging in equilibrium from a spring for Example 12-2. (*b*) Free-body diagram of the object. In equilibrium, the weight of the object mg is balanced by the upward force of magnitude kx_0 exerted by the spring.

Example 12-3

What is the maximum speed of the object on the spring in Example 12-2, and when does this maximum speed first occur?

The velocity of the object at any time is given by Equation 12-16*b*. The maximum value of the velocity occurs when $\sin\omega t = 1$ (or -1). The magnitude of the velocity at this time is

$$v_{max} = \omega A = (9.90\text{ rad/s})(5\text{ cm}) = 49.5\text{ cm/s}$$

This maximum speed first occurs after one-fourth of a period, when the object first reaches equilibrium. This is $(0.63\text{ s})/4 = 0.16$ s after the object is released. At this time, $\sin\omega t = +1$, and the velocity is negative, which means that the object is moving to the left.

Example 12-4

A second spring, identical to that in Example 12-2, is attached to a second object, also of mass 2 kg. This spring is stretched a distance 10 cm from equilibrium and is released at the same time as the first, which is again stretched to 5 cm. Which object reaches the equilibrium position first?

Figure 12-6 shows the initial positions of the objects, and Figure 12-7 shows a sketch of the position functions for the two objects. Both the springs have the same force constant and the masses of the two objects are equal. Only the amplitudes differ. But, according to Equation 12-15, the frequency and period depend only on k and m and not on the amplitude. Thus, the frequencies and periods of the motions are equal and both objects reach their equilibrium positions at the same time. The second object has twice as far to go to reach equilibrium as the first object, but it also begins with twice the initial acceleration.

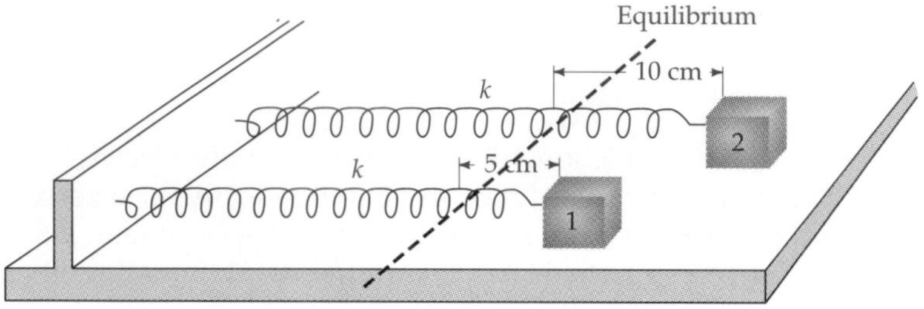

Figure 12-6 Two objects on identical springs released simultaneously. They reach their equilibrium positions at the same time because the period does not depend on the amplitude. It depends only on the mass of the object and force constant of the spring, which are the same for both systems.

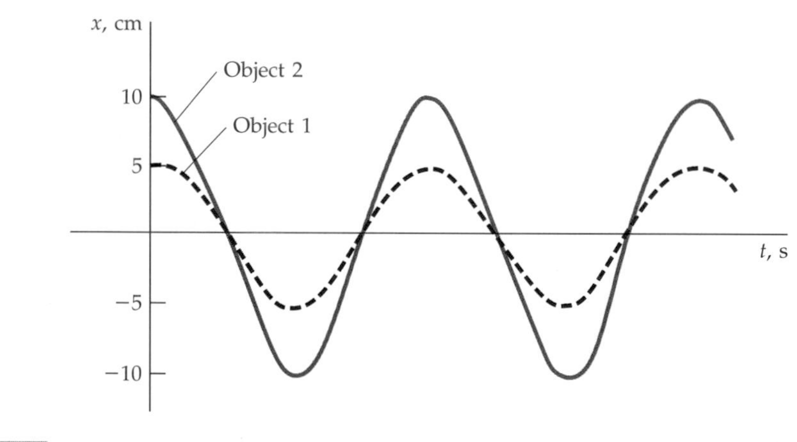

Figure 12-7 Plots of position versus time for the objects in Figure 12-6.

Example 12-4 illustrates a very important general property of simple harmonic motion:

> In simple harmonic motion, the frequency and period are independent of the amplitude.

The fact that the frequency in simple harmonic motion is independent of the amplitude has important consequences in many fields. In music, for example, it means that the pitch (which corresponds to the frequency) of a note struck on the piano does not depend on how loudly the note is played (which corresponds to the amplitude). The pitch does not change as the amplitude of the note decreases. For many musical instruments, there is a slight dependence of the frequency on amplitude. The pitch of an oboe reed, for example, depends slightly on how hard it is blown because the vibration is not exactly simple harmonic. However, this effect can be corrected for by a skilled musician. If changes in amplitude had a large effect on the frequency, musical instruments would be unplayable.

Example 12-5

If the object in Example 12-2 is initially at $x_0 = 3$ cm and has an initial velocity $v_0 = -25$ cm/s, find the amplitude and phase constant for the motion.

Equations 12-8 and 12-10 relate the initial position and velocity to the amplitude and phase constant. We have

$$x_0 = A \cos \delta$$

and

$$v_0 = -\omega A \sin \delta$$

Then

$$\frac{v_0}{x_0} = \frac{-\omega A \sin \delta}{A \cos \delta} = -\omega \tan \delta$$

Using $\omega = 9.90$ rad/s from Example 12-2 and $x_0 = 3$ cm and $v_0 = -25$ cm/s, we obtain

$$\tan \delta = -\frac{v_0}{\omega x_0} = -\frac{-25 \text{ cm/s}}{(9.90 \text{ rad/s})(3 \text{ cm})} = 0.842$$

$$\delta = 0.70 \text{ rad}$$

The amplitude can be found from

$$A = \frac{x_0}{\cos \delta} = \frac{3 \text{ cm}}{\cos 0.70} = 3.9 \text{ cm}$$

Questions

1. How far does a particle oscillating with an amplitude A move in one full period? What is its displacement after one full period?
2. If you know that the speed of an oscillator of amplitude A is zero at certain times, can you say exactly what its displacement is at those times?
3. What is the magnitude of the acceleration of an oscillator of amplitude A and frequency f when its speed is maximum? When its displacement is maximum?
4. Can the acceleration and the displacement of a simple harmonic oscillator ever be in the same direction? The acceleration and the velocity? The velocity and the displacement? Explain.
5. The effect of the mass of a spring on the motion of an object attached to it is usually neglected. Describe qualitatively its effect when it is not neglected.

12-2 Simple Harmonic Motion and Circular Motion

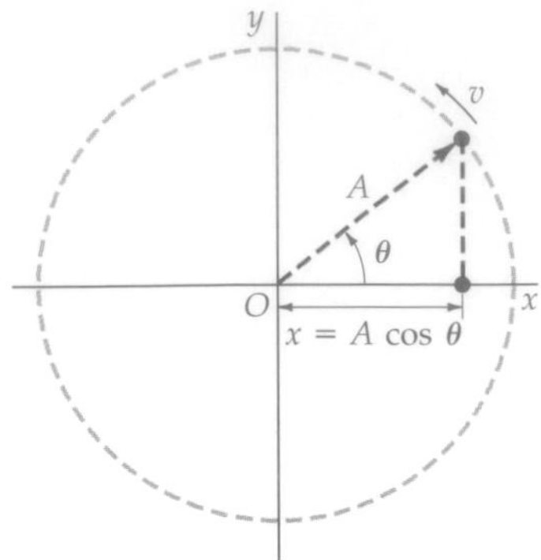

Figure 12-8 A particle moving with constant speed v in a circle of radius A. The angle θ increases with time, $\theta = \omega t + \delta$, where ω is the angular velocity of the circular motion. The x component of the motion is simple harmonic.

There is a simple but important mathematical relation between simple harmonic motion and circular motion with constant speed. Consider a particle moving with constant speed v in a circle of radius A as shown in Figure 12-8. Its angular velocity ω is constant and is related to its speed by $\omega = v/A$. The angular displacement of the particle relative to the x axis is given by

$$\theta = \omega t + \delta$$

where δ is the angular displacement at time $t = 0$. From the figure we see that the x component of the position of the particle is given by

$$x = A \cos \theta = A \cos (\omega t + \delta)$$

which is the same as Equation 12-4. Thus,

> The projection on a straight line of a particle moving with uniform circular motion is simple harmonic motion.

The frequency and period of the circular motion are the same as the corresponding frequency and period of the projected simple harmonic motion. The relation between circular motion and simple harmonic motion can be demonstrated with a turntable and an object hanging on a spring as shown in Figure 12-9. The shadow of the peg and that of the object are projected on a screen. If the period of the rotating turntable is adjusted to be equal to the period of the oscillating object, and the amplitude of the spring system equals the radius of the turntable, the shadows move together.

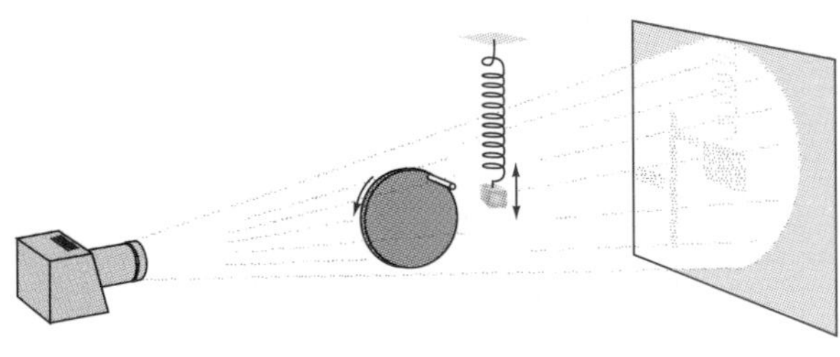

Figure 12-9 The shadow of a peg on a turntable and the shadow of an object on a spring are projected onto a screen. When the period of rotation of the turntable equals the period of oscillation of the object on the spring, the shadows move together.

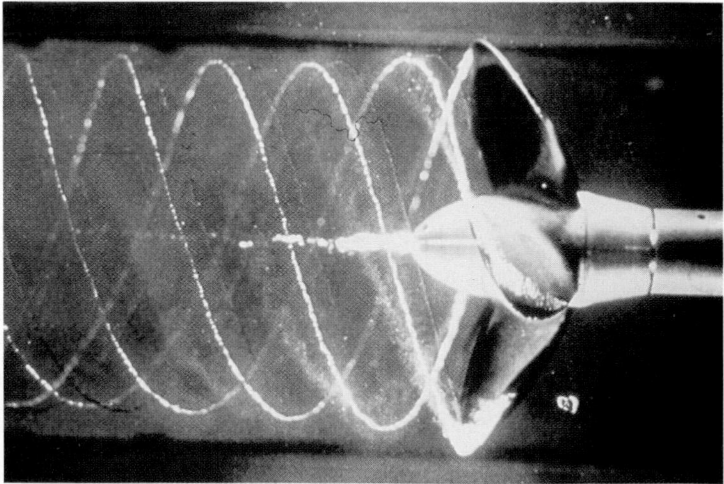

Sinusoidal pattern produced by a rotating turbine blade as it moves through water.

The projection of the circular motion on the y axis is $y = A \sin \theta = A \sin (\omega t + \delta) = A \cos (\omega t + \delta - \pi/2)$. We can therefore consider simple harmonic motion to be one component of circular motion. Alternatively, the circular motion of a particle can be considered to be the combination of perpendicular simple harmonic motions having the same amplitude and frequency but a relative phase difference of $\pi/2$.

12-3 Energy in Simple Harmonic Motion

When an object oscillates on a spring, the kinetic energy of the object and potential energy of the object–spring system both vary with time, while their sum, the total energy, is constant (assuming no friction). The potential energy for a spring of force constant k stretched a distance x from equilibrium is given by Equation 6-19:

$$U = \tfrac{1}{2}kx^2 \qquad 12\text{-}17$$

The kinetic energy of an object of mass m moving with speed v is

$$K = \tfrac{1}{2}mv^2 \qquad 12\text{-}18$$

The total energy is the sum of the potential and kinetic energies:

$$E_{\text{total}} = U + K = \tfrac{1}{2}kx^2 + \tfrac{1}{2}mv^2 \qquad 12\text{-}19$$

When the displacement is maximum, $x = A$, the velocity is zero, and the total energy is

$$E_{\text{total}} = \tfrac{1}{2}kA^2 \qquad 12\text{-}20$$

Total energy in simple harmonic motion

This equation gives an important general property of simple harmonic motion:

> The total energy in simple harmonic motion is proportional to the square of the amplitude.

If we begin with the object at its maximum displacement, the total energy is originally just potential energy. As the object moves toward equilibrium, its kinetic energy increases and the potential energy of the system decreases. At the equilibrium point, the speed of the object is maximum, the potential energy of the system is zero, and the total energy is kinetic. As the object moves past the equilibrium point, its kinetic energy begins to decrease and the potential energy of the system increases until the object is again at its maximum displacement (in the other direction), at which time it stops momentarily, so its kinetic energy is again zero and the potential energy of the system is again maximum. At all times, the sum of the potential and kinetic energies is constant.

In Figure 12-10, the potential energy U is graphed as a function of x. This curve is a parabola with its minimum at the equilibrium position. (This is the same plot as Figure 6-14.) The total energy E_{total} is constant and is therefore

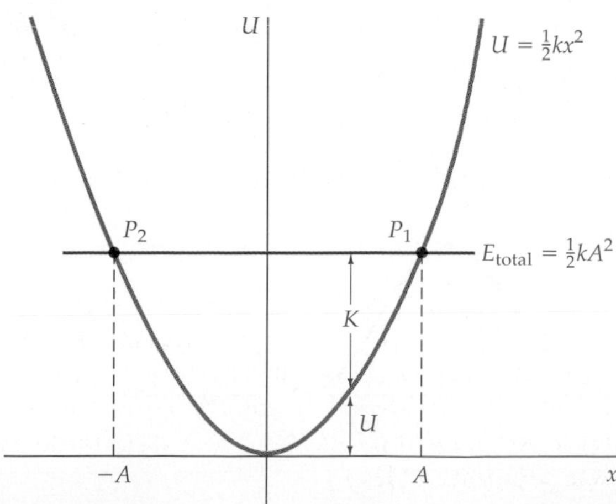

Figure 12-10 The potential-energy function $U = \frac{1}{2}kx^2$ for an object of mass m on a spring of force constant k. The horizontal line represents the total energy for an amplitude of A, $E_{\text{total}} = \frac{1}{2}kA^2$. The kinetic energy K is represented by the vertical distance $K = E_{\text{total}} - U$.

plotted as a straight line. This line intersects the potential-energy curve at $x = A$ and $x = -A$. The motion is restricted to $-A \leq x \leq +A$ since the total energy must be greater than or equal to the potential energy. The kinetic energy at any value of x is $K = E_{total} - U$.

We can demonstrate that the total energy for an object oscillating on a spring is constant by substituting the expressions for x and v in Equations 12-4 and 12-9 into Equations 12-17 and 12-18 for the potential and kinetic energies. We have

$$U = \tfrac{1}{2}k[A \cos(\omega t + \delta)]^2$$

or

$$U = \tfrac{1}{2}kA^2 \cos^2(\omega t + \delta) \qquad 12\text{-}21$$

and

$$K = \tfrac{1}{2}m[-A\omega \sin(\omega t + \delta)]^2$$

Using $\omega^2 = k/m$ from Equation 12-13, the kinetic energy can be written

$$K = \tfrac{1}{2}kA^2 \sin^2(\omega t + \delta) \qquad 12\text{-}22$$

Adding the potential and kinetic energies to get the total energy, we obtain

$$\begin{aligned} E_{total} &= \tfrac{1}{2}kA^2 \cos^2(\omega t + \delta) + \tfrac{1}{2}kA^2 \sin^2(\omega t + \delta) \\ &= \tfrac{1}{2}kA^2[\cos^2(\omega t + \delta) + \sin^2(\omega t + \delta)] \\ &= \tfrac{1}{2}kA^2 \end{aligned}$$

where we have used the trigonometric identity $\cos^2\theta + \sin^2\theta = 1$.

In terms of the total energy, the potential and kinetic energies in simple harmonic motion are

$$U = E_{total} \cos^2\theta$$

and

$$K = E_{total} \sin^2\theta$$

where $\theta = \omega t + \delta$. Figure 12-11*a* and *b* show plots of U and K as functions of t for $\delta = 0$. These curves have the same form, except that U is maximum when K is zero and U is zero when K is maximum. Their average values over one or more cycles are equal, and since $U + K = E_{total}$, these average values are given by

$$U_{av} = K_{av} = \tfrac{1}{2}E_{total} \qquad 12\text{-}23$$

Figure 12-11 Plots of (*a*) the potential energy U and (*b*) the kinetic energy K as functions of t for $\delta = 0$. Both functions have the same shape and the same maximum value E_{total}. For each, the average value, indicated by the dashed horizontal line, is half the maximum value.

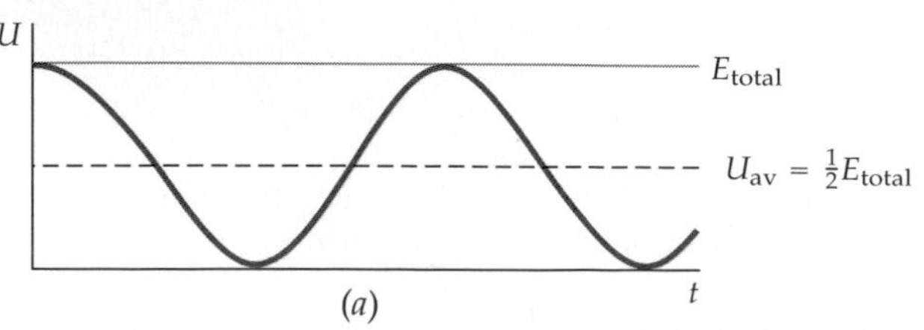

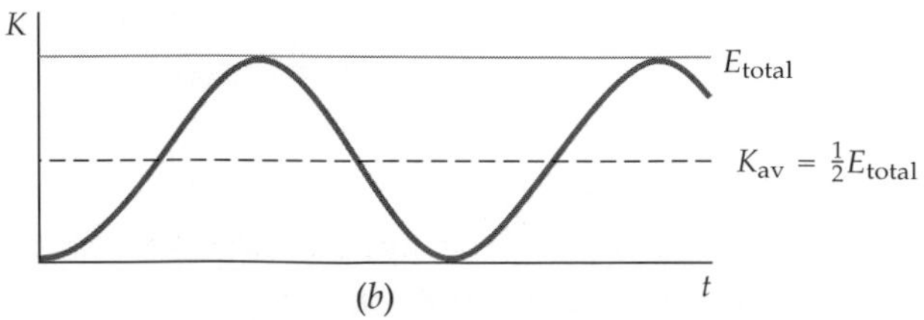

Example 12-6

A 3-kg object attached to a spring oscillates with an amplitude of 4 cm and a period of 2 s. (*a*) What is the total energy? (*b*) What is the maximum speed of the object?

(*a*) The total energy is $\frac{1}{2}kA^2$. The force constant of the spring, k, is related to the period by

$$T = 2\pi\sqrt{\frac{m}{k}}$$

so that

$$k = \frac{(2\pi)^2 m}{T^2} = \frac{(4\pi^2)(3\text{ kg})}{4\text{ s}^2} = 29.6\text{ N/m}$$

The total energy is therefore

$$E_{total} = \tfrac{1}{2}kA^2 = \tfrac{1}{2}(29.6\text{ N/m})(0.04\text{ m})^2 = 2.37 \times 10^{-2}\text{ J}$$

(*b*) We use the total energy to find the maximum speed. When the speed is maximum, the potential energy is zero and the total energy is the kinetic energy:

$$E_{total} = \tfrac{1}{2}mv_{max}^2 = 2.37 \times 10^{-2}\text{ J}$$

The maximum speed is therefore

$$v_{max} = \sqrt{\frac{2E_{total}}{m}} = \sqrt{\frac{2(2.37 \times 10^{-2}\text{ J})}{3\text{ kg}}} = 0.126\text{ m/s}$$

We could have found the maximum speed from Equation 12-16*b* as we did in Example 12-3:

$$v_{max} = \omega A(\sin \omega t)_{max} = \omega A$$

Using $\omega = 2\pi/T$ (Equation 12-6), we have

$$v_{max} = \frac{2\pi A}{T} = \frac{2(3.14)(4\text{ cm})}{2\text{ s}} = 12.6\text{ cm/s} = 0.126\text{ m/s}$$

Exercise

An object of mass 2 kg is attached to a spring of force constant 40 N/m. It is moving at 25 cm/s when it is at its equilibrium position. (*a*) What is the total energy of the object? (*b*) What is the amplitude of the motion? [Answers: (*a*) $E = 0.0625$ J, (*b*) $A = 5.59$ cm]

Question

6. If the amplitude of a simple harmonic oscillator is tripled, by what factor is the energy changed?

12-4 An Object on a Vertical Spring

When an object hangs from a vertical spring, as in Figure 12-12, there is a force mg downward in addition to the force of the spring $F_s = -ky$, where we are assuming that y is measured downward from the unstretched position of the spring. Newton's second law gives

$$m\frac{d^2y}{dt^2} = -ky + mg \qquad \text{12-24}$$

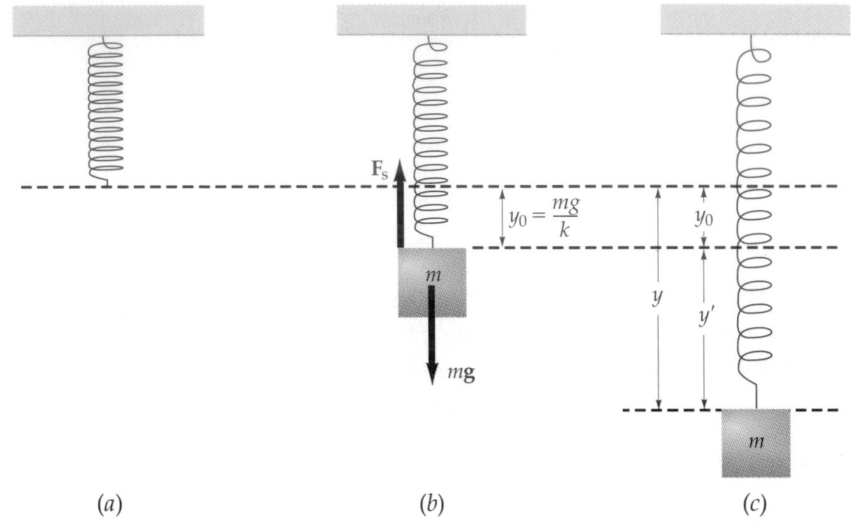

Figure 12-12 (*a*) An unstretched vertical spring. (*b*) The spring stretches an amount $y_0 = mg/k$ when an object of mass m hangs from it in equilibrium. (*c*) The object oscillates about its equilibrium position $y = y_0$ with a displacement $y' = y - y_0$.

This differs from Equation 12-2 because of the constant term mg. We handle this extra term by changing to a new variable $y' = y - y_0$, where y_0 is the amount the spring is stretched when the object is in equilibrium. When the object is in equilibrium at $y = y_0$, Equation 12-24 becomes

$$0 = -ky_0 + mg$$

or

$$y_0 = \frac{mg}{k} \qquad \text{12-25}$$

Since y' differs from y by only a constant, we have

$$\frac{dy'}{dt} = \frac{dy}{dt}$$

and

$$\frac{d^2y'}{dt^2} = \frac{d^2y}{dt^2}$$

Substituting d^2y'/dt^2 for d^2y/dt^2 and $y' + y_0$ for y in Equation 12-24, we obtain

$$m\frac{d^2y'}{dt^2} = -k(y' + y_0) + mg$$

$$= -ky' - ky_0 + mg$$

But $ky_0 = mg$, so

$$m \frac{d^2y'}{dt^2} = -ky' \qquad \text{12-26}$$

which has the familiar solution

$$y' = A \cos(\omega t + \delta)$$

Thus, the effect of the gravitational force mg is merely to shift the equilibrium position from $y = 0$ to $y' = 0$. When the object is displaced from this equilibrium position by the amount y', the unbalanced force is $-ky'$. The object oscillates about this equilibrium position with an angular frequency $\omega = \sqrt{k/m}$, the same as that for an object on a horizontal spring.

The potential energy of the spring relative to $U_s = 0$ at $y = 0$ is $\frac{1}{2}ky^2 = \frac{1}{2}k(y' + y_0)^2$. At the equilibrium point, it is $\frac{1}{2}ky_0^2$. If, instead, we choose the potential energy of the spring to be zero at the equilibrium position ($y = y_0$, $y' = 0$), the potential energy of the spring is

$$\begin{aligned} U_s &= \tfrac{1}{2}k(y' + y_0)^2 - \tfrac{1}{2}ky_0^2 \\ &= \tfrac{1}{2}ky'^2 + ky_0y' + \tfrac{1}{2}ky_0^2 - \tfrac{1}{2}ky_0^2 = \tfrac{1}{2}ky'^2 + ky_0y' \end{aligned}$$

But $ky_0 = mg$. Thus the potential energy *of the spring* relative to $U_s = 0$ at $y' = 0$ is

$$U_s = \tfrac{1}{2}ky'^2 + mgy' \qquad (U_s = 0 \text{ at } y' = 0)$$

When the spring is stretched by an amount y', the object is lowered by $\Delta h = -y'$, and the gravitational potential energy is changed by $mg(\Delta h) = -mgy'$. If the gravitational potential energy is also chosen to be zero at equilibrium ($y = y_0$, $y' = 0$), the gravitational potential energy at y' is

$$U_g = -mgy' \qquad (U_g = 0 \text{ at } y' = 0)$$

The total potential energy of the system relative to $U = 0$ at $y' = 0$, including both the potential energy of the spring and the gravitational potential energy, is therefore

$$\begin{aligned} U &= U_s + U_g \\ &= (\tfrac{1}{2}ky'^2 + mgy') - mgy' \end{aligned}$$

or

$$U = \tfrac{1}{2}ky'^2 \qquad (U = 0 \text{ at } y' = 0) \qquad \text{12-27}$$

Total potential energy of object on a vertical spring

Thus, if we measure the displacement from the equilibrium position, we can forget about the effect of gravity.

Exercise

A 4-kg object is suspended from a spring of force constant $k = 200$ N/m. (*a*) Find the amount y_0 the spring is stretched when in equilibrium. (*b*) Find the total potential energy, including gravitational potential energy, when the spring is stretched an additional 12 cm from equilibrium. (Assume $U = 0$ at $y = y_0$.) (*c*) Find the period of oscillation. [Answers: (*a*) 19.6 cm (*b*) 1.44 J, (*c*) 0.889 s]

12-5 Pendulums

The Simple Pendulum

A familiar example of oscillatory motion is that of a pendulum. The motion of a pendulum is simple harmonic motion only if the amplitude of the motion is small. Figure 12-13 shows a simple pendulum consisting of a string of length L and a bob of mass m. The forces acting on the bob are its weight $m\mathbf{g}$ and the tension $\mathbf{T}$ in the string. When the string makes an angle ϕ with the vertical, the weight has the components $mg \cos \phi$ along the string and $mg \sin \phi$ perpendicular to the string in the direction of decreasing ϕ. Let s be the arc length measured from the bottom of the circle. The arc length is related to the angle ϕ by

$$s = L\phi \tag{12-28}$$

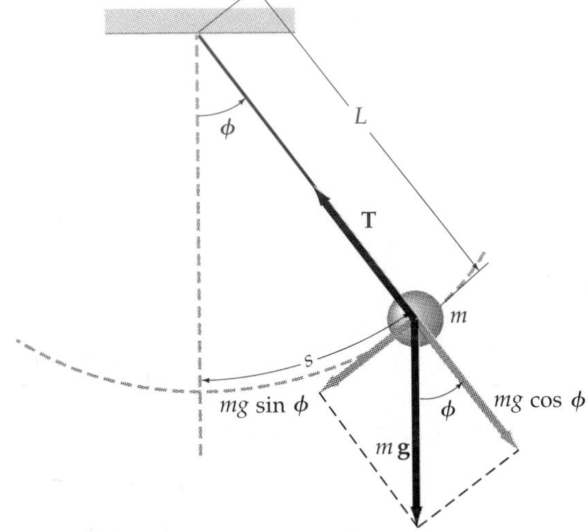

Figure 12-13 A simple pendulum. The forces on the bob are its weight $m\mathbf{g}$ and the tension $\mathbf{T}$. The tangential component of the net force is $-mg \sin \phi = -mg \sin (s/L)$. For small displacements, the motion of the pendulum is approximately simple harmonic motion.

The tangential component of the acceleration of the bob is d^2s/dt^2. The tangential component of Newton's second law is

$$\Sigma F_t = -mg \sin \phi = m \frac{d^2s}{dt^2}$$

or

$$\frac{d^2s}{dt^2} = -g \sin \phi = -g \sin \frac{s}{L} \tag{12-29}$$

If s is much less than L, the angle $\phi = s/L$ is small, and we can approximate $\sin \phi$ by the angle ϕ. Using $\sin (s/L) \approx s/L$ in Equation 12-29, we have

$$\frac{d^2s}{dt^2} = -\frac{g}{L} s \tag{12-30a}$$

We can see that for angles small enough for $\sin \phi \approx \phi$ to be valid, the acceleration is proportional to the displacement. The motion of the pendulum is thus approximately simple harmonic motion for small displacements. Equation 12-30*a* can be written

$$\frac{d^2s}{dt^2} = -\omega^2 s \tag{12-30b}$$

where

$$\omega^2 = \frac{g}{L} \tag{12-31}$$

The solution of Equation 12-30*b* is $s = s_0 \cos (\omega t + \delta)$, where s_0 is the maximum displacement measured along the arc of the circle. The period of the motion is

$$T = \frac{2\pi}{\omega} = 2\pi\sqrt{\frac{L}{g}} \qquad 12\text{-}32$$

Period of a simple pendulum

According to Equation 12-32 the greater the length of a pendulum, the greater the period, which is consistent with experimental observation. Note that the period does not depend on the mass. This is because the restoring force is proportional to the mass. The acceleration $a = F/m$ is therefore independent of the mass. Note also that the frequency and period are independent of the amplitude of oscillation, a general feature of simple harmonic motion.

It is often more convenient to describe the motion of a simple pendulum in terms of its angular displacement ϕ. Using $s = L\phi$ in Equation 12-29, we have

$$\frac{d^2(L\phi)}{dt^2} = -g \sin \phi$$

or

$$\frac{d^2\phi}{dt^2} = -\frac{g}{L} \sin \phi \qquad 12\text{-}33$$

which for small ϕ becomes

$$\frac{d^2\phi}{dt^2} = -\frac{g}{L}\phi = -\omega^2\phi \qquad 12\text{-}34$$

The solution of Equation 12-34 is

$$\phi = \phi_0 \cos (\omega t + \delta) \qquad 12\text{-}35$$

where $\phi_0 = s_0/L$ is the maximum angular displacement. The criterion for simple harmonic motion stated in terms of these angular quantities is that the angular acceleration must be proportional to the angular displacement and oppositely directed, as in Equation 12-34.

Example 12-7

Find the period of a pendulum of length 1 m.

From Equation 12-32, we have

$$T = 2\pi\sqrt{\frac{1 \text{ m}}{9.81 \text{ m/s}^2}} = 2.01 \text{ s}$$

The validity of this result can be easily demonstrated experimentally.

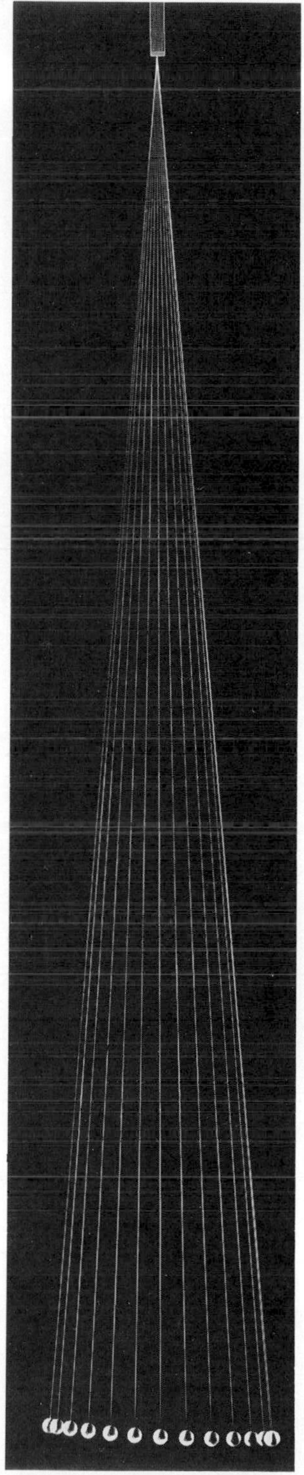

Motion of a simple pendulum. The bob is shown at equal time intervals. It moves faster at the bottom, as shown by the greater spacing of the images.

The acceleration due to gravity can be easily measured using a simple pendulum. You need only measure the length L with a meterstick and the period T by determining the time for one oscillation. (One usually measures the time for n oscillations and then divides by n to reduce the error in the time measurement.) The acceleration due to gravity is then determined by solving Equation 12-32 for g:

$$g = \frac{4\pi^2 L}{T^2} \qquad 12\text{-}36$$

When the amplitude of vibration is not small, the motion of a pendulum is periodic, but it is not simple harmonic. In particular, the period does have a slight dependence on the amplitude. This dependence is usually expressed in terms of the angular amplitude ϕ_0. For amplitudes that are not necessarily small, the period can be shown to be given by

$$T = T_0\left[1 + \frac{1}{2^2}\sin^2\frac{1}{2}\phi_0 + \frac{1}{2^2}\left(\frac{3}{4}\right)^2\sin^4\frac{1}{2}\phi_0 + \cdots\right] \qquad 12\text{-}37$$

where $T_0 = 2\pi\sqrt{L/g}$ is the period for very small amplitudes.

Example 12-8

A simple pendulum clock is calibrated to keep accurate time at an angular amplitude of $\phi_0 = 10°$. When the amplitude has decreased to the point that it is very small, how much time does the clock gain in one day?

From Equation 12-37, the original period is approximately

$$T \approx T_0(1 + \tfrac{1}{4}\sin^2 5°)$$

since for $\phi_0 = 10°$, $\sin^4 \frac{1}{2}\phi_0$ in Equation 12-37 is much smaller than $\sin^2 \frac{1}{2}\phi_0$. When the amplitude is very small, the period is T_0. Since this is less than T, the frequency is greater and the clock gains time. The difference between these two periods is

$$T - T_0 = \frac{T_0}{4}\sin^2 5°$$

$$= \frac{T_0}{4}(0.0872)^2 \approx (2 \times 10^{-3})T_0$$

The percentage change in the period is therefore

$$\frac{T - T_0}{T_0} \times 100\% = (2 \times 10^{-3}) \times 100\% = 0.2\%$$

This is a very small percentage change. However, it leads to an inaccuracy that is intolerable by today's time-keeping standards. The number of minutes in one day is

$$\frac{24 \text{ h}}{1 \text{ d}} \times \frac{60 \text{ min}}{1 \text{ h}} = 1440 \text{ min/d}$$

A gain of 0.2 percent, then, leads to an accumulated error of $(2 \times 10^{-3})(1440 \text{ min}) = 2.88$ min in 1 day. The clock thus gains nearly three minutes each day. For this reason, pendulum clocks are designed to maintain a constant amplitude.

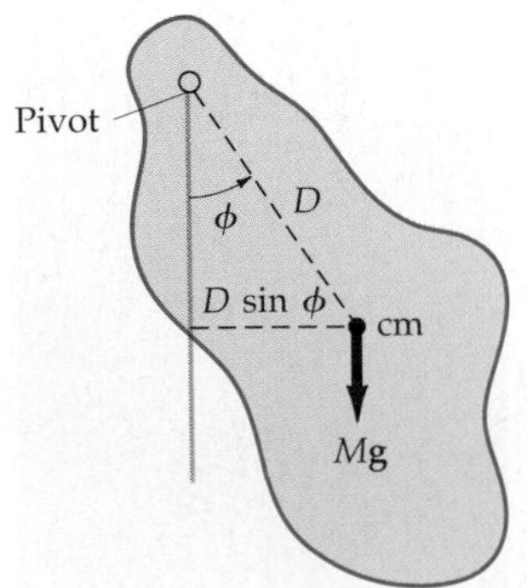

Figure 12-14 A physical pendulum. The torque about the pivot due to the force of gravity has magnitude $MgD \sin\phi$ and tends to decrease ϕ.

The Physical Pendulum

A rigid body that is suspended from some point other than its center of mass will oscillate when displaced from its equilibrium position. Such a system is called a **physical pendulum.** Consider a plane figure suspended from a point a distance D from its center of mass and displaced from equilibrium by the angle ϕ as shown in Figure 12-14. The torque about the point of suspension has a magnitude $MgD \sin\phi$ and tends to decrease ϕ. The angular acceleration α of the body is related to the torque by

$$\tau = I\alpha = I\frac{d^2\phi}{dt^2}$$

where I is the moment of inertia about the point of suspension. Substituting $-MgD \sin\phi$ for the net torque, we obtain

$$-MgD \sin\phi = I\frac{d^2\phi}{dt^2}$$

or

$$\frac{d^2\phi}{dt^2} = -\frac{MgD}{I}\sin\phi \qquad 12\text{-}38$$

For a simple pendulum, $I = ML^2$ and $D = L$, so Equation 12-38 is the same as Equation 12-34. Again, the motion is approximately simple harmonic motion if the angular displacements are small so that the approximation $\sin\phi \approx \phi$ holds. In this case, we have

$$\frac{d^2\phi}{dt^2} = -\frac{MgD}{I}\phi = -\omega^2\phi \qquad 12\text{-}39$$

where $\omega^2 = MgD/I$. The period is

$$T = \frac{2\pi}{\omega} = 2\pi\sqrt{\frac{I}{MgD}} \qquad 12\text{-}40$$

Period of a physical pendulum

For large amplitudes, the period is given by Equation 12-37, where T_0 is now given by Equation 12-40.

Equation 12-40 can be used to measure the moment of inertia of a plane figure. The center of mass can be located by suspending the body from two different points, as discussed previously. Then, to find the moment of inertia about some point, we suspend the body from that point and measure the period of oscillation. The moment of inertia is obtained from

$$I = \frac{MgDT^2}{4\pi^2} \qquad 12\text{-}41$$

Example 12-9

What is the period for small angular displacements of a uniform stick of length L that is pivoted at one end?

The moment of inertia of a uniform stick of length L about an axis at one end was shown in Chapter 8 to be $I = \frac{1}{3}ML^2$. The distance from the pivot to the center of mass is $D = \frac{1}{2}L$. Substituting these values into Equation 12-40, we obtain

$$T = 2\pi\sqrt{\frac{\frac{1}{3}ML^2}{Mg(\frac{1}{2}L)}} = 2\pi\sqrt{\frac{2L}{3g}}$$

Exercise

What is the period for small angular displacements of a meterstick that is pivoted at one end? (Answer: 1.64 s)

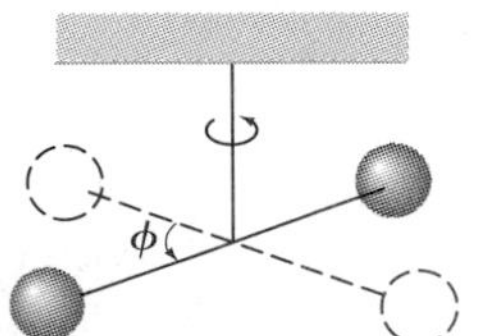

Figure 12-15 A torsional pendulum. When the wire is twisted through an angle ϕ, there is a restoring torque of magnitude $-\kappa\phi$ that tends to twist the wire back to its equilibrium position.

The Torsional Pendulum

Figure 12-15 shows a **torsional pendulum,** which consists of a body suspended by a wire attached to a fixed point. When the wire is twisted through some angle ϕ, it exerts a restoring torque that is proportional to ϕ:

$$\tau = -\kappa\phi \qquad 12\text{-}42$$

The constant of proportionality κ is called the **torsion constant.** It can be found by applying a known torque to twist the wire and then measuring the resulting angular displacement ϕ. If I is the moment of inertia of the body about the axis along the wire, Newton's second law for rotational motion gives

$$\tau = -\kappa\phi = I\,\frac{d^2\phi}{dt^2}$$

or

$$\frac{d^2\phi}{dt^2} = -\frac{\kappa}{I}\,\phi = -\omega^2\phi \qquad \text{12-43}$$

Equation 12-43 describes simple harmonic motion with an angular frequency $\omega = \sqrt{\kappa/I}$. The period of the motion is

Period of a torsional pendulum

$$T = 2\pi\sqrt{\frac{I}{\kappa}} \qquad \text{12-44}$$

Note that we have not made a small-angle approximation. The motion of a torsional pendulum is simple harmonic motion as long as the restoring torque is proportional to the twist angle. Such is the case as long as the elastic limit of the wire for a shear stress is not exceeded. The balance wheel in a watch is a torsional pendulum as is the Cavendish torsional balance.

Question

7. The length of the string or wire supporting a pendulum increases slightly when its temperature is raised. How would this affect a clock operated by a simple pendulum? By a torsional pendulum?

12-6 General Motion near Equilibrium

Whenever a particle is displaced from a position of stable equilibrium, the motion of the particle is simple harmonic if the displacements are small enough. Examples are the various pendulums just discussed.

Figure 12-16*a* is a graph of an arbitrary net force as a function of position. At positions x_1 and x_2, the force is zero, so these are positions of equilibrium. However, the equilibrium at x_2 is unstable. If the particle is displaced slightly in the positive x direction, the force is positive, whereas if it is displaced slightly in the negative x direction, the force is negative. In either case, the force accelerates the particle away from its equilibrium position. The equilibrium at x_1 is stable. If the particle is displaced slightly from equilibrium in either direction, the force accelerates it back toward equilibrium.

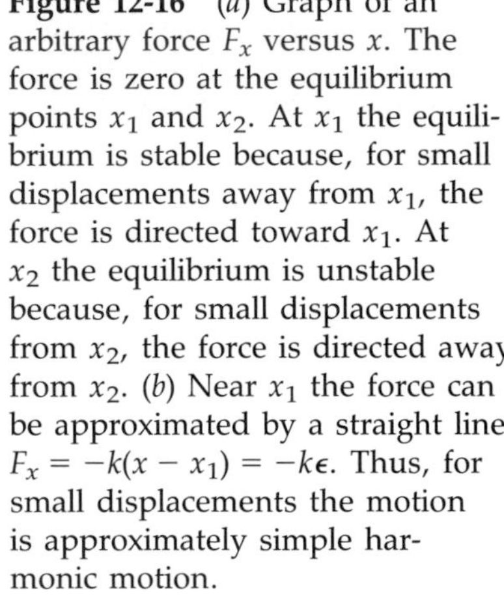

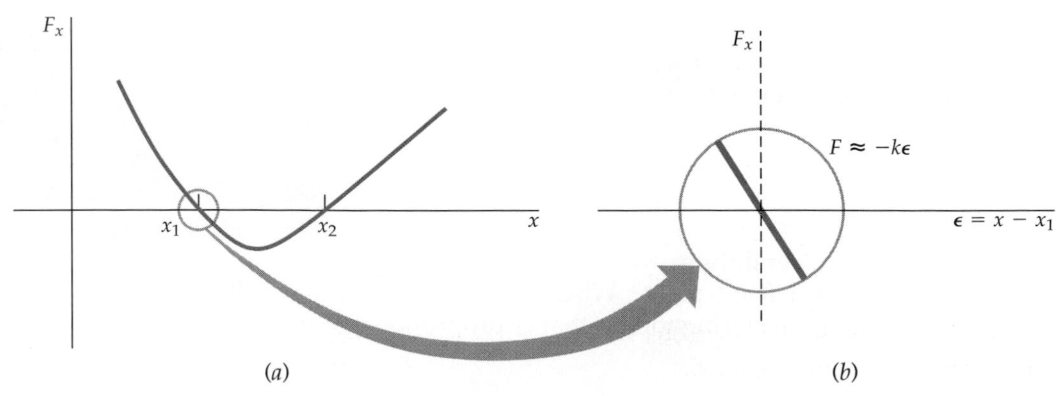

Figure 12-16 (*a*) Graph of an arbitrary force F_x versus x. The force is zero at the equilibrium points x_1 and x_2. At x_1 the equilibrium is stable because, for small displacements away from x_1, the force is directed toward x_1. At x_2 the equilibrium is unstable because, for small displacements from x_2, the force is directed away from x_2. (*b*) Near x_1 the force can be approximated by a straight line $F_x = -k(x - x_1) = -k\epsilon$. Thus, for small displacements the motion is approximately simple harmonic motion.

The particle therefore oscillates about the equilibrium point. Although the plot of force versus displacement is generally a curve, it will approximate a straight line if we consider it over a small enough interval, as shown in Figure 12-16*b*. Let $\epsilon = x - x_1$ be the displacement from equilibrium. The equation of the force for small ϵ is

$$F_x = -k\epsilon \tag{12-45}$$

where k is the magnitude of the slope of F_x versus x near x_1. Since the force is proportional to the displacement and is oppositely directed, the motion will be simple harmonic.

We can also examine the motion from the point of view of the potential-energy function U associated with the force. Figure 12-17 is a graph of U versus x for the same force as in Figure 12-16. As discussed in Chapter 6, the maximum at x_2 corresponds to unstable equilibrium, whereas the minimum at x_1 corresponds to stable equilibrium. The dashed curve in this figure is a parabolic curve that approximately fits the potential-energy curve near the stable equilibrium point. As long as the displacements from equilibrium are not too great, U is approximately parabolic. The general equation for a parabola can be written

$$U = A + B(x - x_1)^2$$

where A and B are constants. The force is related to the potential-energy curve by Equation 6-20:

$$F_x = -\frac{dU}{dx}$$

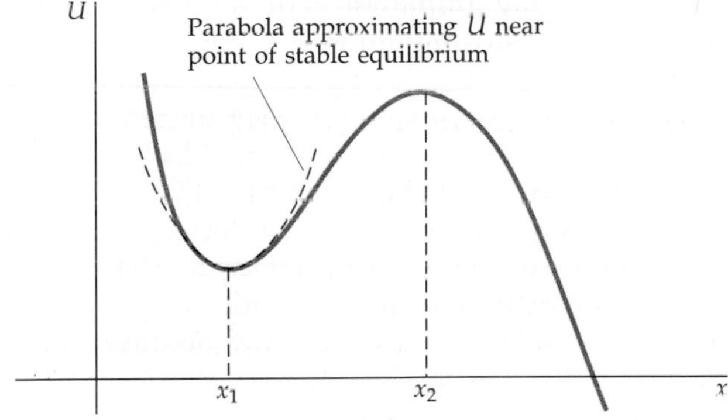

Figure 12-17 Graph of the potential-energy function U corresponding to the force in Figure 12-16. The minimum at x_1 indicates stable equilibrium, whereas the maximum at x_2 indicates unstable equilibrium. Near x_1 the curve is approximately parabolic, as it is in simple harmonic motion.

Therefore,

$$F_x = -\frac{dU}{dx} = -2B(x - x_1)$$

If we set $2B = k$, this equation reduces to

$$F_x = -\frac{dU}{dx} = -k(x - x_1) = -k\epsilon$$

which is Equation 12-45.

Question

8. Give several examples of familiar motions that are either exactly or approximately simple harmonic motions.

A ball bouncing into a water tank. The motion is damped, that is, the energy of each bounce is less than that of the previous bounce because of the resistance of the water.

12-7 Damped Oscillations

In all real oscillatory motions, mechanical energy is dissipated because of some kind of frictional force. Left to itself, a spring or a pendulum eventually stops oscillating. When the mechanical energy of oscillatory motion decreases with time, the motion is said to be **damped.** If the frictional or damping forces are small, the motion is nearly periodic, except that the amplitude decreases slowly with time as shown in Figure 12-18. If the amplitude decreases slowly with time the energy does also, since the energy of an oscillator is proportional to the square of the amplitude. For the case of small damping shown in Figure 12-18, both the amplitude and the energy of the oscillations decrease by a constant percentage in a given time interval. For example, the energy of a slightly damped pendulum may decrease by 10 percent every minute. Thus, after 1 minute the energy would be 90 percent of the original energy, after 2 minutes it would be 90 percent of 90 percent or 81 percent of the original energy, and so on. This type of decrease is called exponential decrease. (We encountered exponential decrease in Section 11-3 when we studied the decrease in air pressure with altitude.)

In Figure 12-19, the oscillation of the body is damped because of the motion of the plunger immersed in the liquid. The rate of energy loss can be varied by changing the size of the plunger or the viscosity of the liquid. Although the detailed analysis of a damping force like that for this system is usually complicated, we can often represent such a force by an empirical

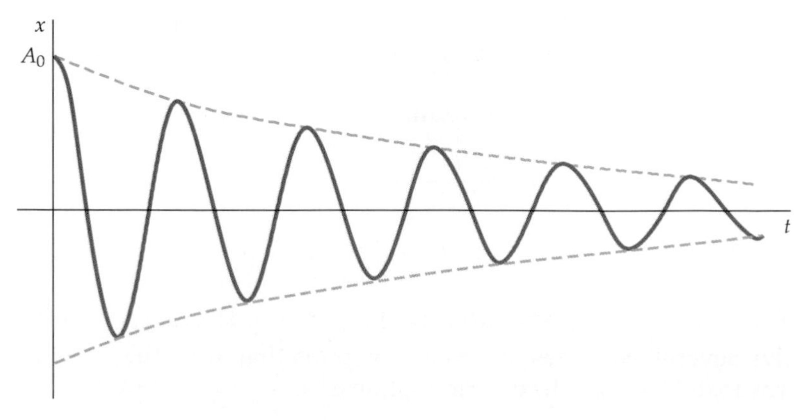

Figure 12-18 A graph of displacement versus time for a slightly damped oscillator. The motion is approximately simple harmonic motion with an amplitude that decreases slowly with time.

expression that is in reasonable accord with experimental results and for which the mathematics is comparatively simple. The simplest and most common representation of such a force is one that is proportional to the velocity of the object but in the opposite direction:

$$\mathbf{F}_{\text{d}} = -b\mathbf{v}$$

where b is a constant that describes the amount of damping. This is an example of a drag force, which we discussed in Section 5-2. Since a drag force is always opposite the direction of motion, the work done by this nonconservative force is always negative. Thus, it causes the mechanical energy of the system to decrease. Newton's second law applied to the motion of an object of mass m on a spring of force constant k when the damping force is $-bv$ is

$$F_x = ma_x$$

$$-kx - bv = m\frac{dv}{dt} \qquad \text{12-46}$$

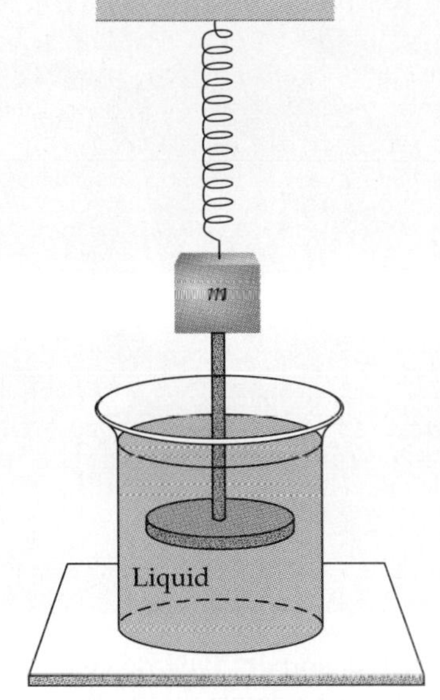

Figure 12-19 A damped oscillator. The motion is damped by the plunger immersed in the liquid.

We can obtain a qualitative understanding of the behavior of a damped oscillator without solving Equation 12-46 in detail. If the damping is small, we expect the body to oscillate with an angular frequency ω' that is approximately equal to the undamped frequency $\omega_0 = \sqrt{k/m}$ and we expect the amplitude to decrease slowly. In simple harmonic motion, the total mechanical energy oscillates between potential and kinetic energy. The average values of the potential and kinetic energies over one cycle are equal, and the total energy is twice the average value of either the potential or the kinetic energy. We can therefore write

$$E = 2(\tfrac{1}{2}mv^2)_{\text{av}} = m(v^2)_{\text{av}} \qquad \text{12-47}$$

For a slightly damped oscillator, only a very small function of the mechanical energy is lost during one cycle so that the total energy decreases slowly with time. The instantaneous rate of change of the total mechanical energy equals the power input of the damping force:

$$P = \frac{dE}{dt} = \mathbf{F}_{\text{d}} \cdot \mathbf{v} = -bv^2 \qquad \text{12-48}$$

The power input is negative, indicating that energy is leaving the system. If we replace v^2 in Equation 12-48 with its average value $(v^2)_{\text{av}} = E/m$ from Equation 12-47, we obtain

$$\frac{dE}{dt} = -\frac{b}{m}E \qquad \text{12-49}$$

Equation 12-49 describes an **exponential decrease.** The rate of decrease of the energy is proportional to the energy, so the fractional decrease, $-dE/E = (b/m)\,dt$, is the same for any given time interval.

Equation 12-49 may be solved by direct integration. We first rewrite it as

$$\frac{dE}{E} = -\frac{b}{m}\,dt \qquad \text{12-50}$$

Integrating both sides, we obtain

$$\ln E = -\frac{b}{m}t + C$$

where C is an arbitrary constant of integration. Taking the exponential of each side, we obtain

$$E = e^{-bt/m+C} = e^{C}e^{-bt/m} = E_0e^{-bt/m}$$

Shock absorbers (yellow cylinders) are used to damp the oscillations of this truck.

where $E_0 = e^C$ is another constant, which is the energy at time $t = 0$. The solution of Equation 12-49 is thus

$$E = E_0 e^{-(b/m)t} = E_0 e^{-t/\tau} \tag{12-51}$$

where the **time constant**

$$\tau = \frac{m}{b} \tag{12-52}$$

is the time it takes for the energy to decrease by the factor $1/e$. When the damping is small, b is small and τ is large, and the oscillator will lose only a very small fraction of its energy during one oscillation. In this case, we can obtain the energy loss per period by replacing the differentials dE and dt in Equation 12-50 by the differences ΔE and Δt and setting $\Delta t = T$, the period. We then have

$$\frac{\Delta E}{E} = -\frac{b}{m}T \tag{12-53}$$

The damping of a slightly damped oscillator is usually described by a dimensionless quantity Q called the **quality factor** or **Q factor.** If E is the total energy and $|\Delta E|$ is the energy loss in one period, the Q factor is defined as

Q factor defined

$$Q = 2\pi\frac{E}{|\Delta E|} \tag{12-54}$$

The Q factor is thus inversely proportional to the fractional energy loss per cycle:

$$\frac{|\Delta E|}{E} = \frac{2\pi}{Q} \tag{12-55}$$

Using Equations 12-53 and 12-54, we can relate the Q factor to the damping constant and the time constant:

$$Q = 2\pi\frac{E}{|\Delta E|} = 2\pi\frac{m}{bT} = 2\pi\frac{\tau}{T} \tag{12-56}$$

Example 12-10

A simple pendulum loses 1 percent of its energy during each oscillation. What is the Q factor?

Since the energy loss is 1 percent,

$$\frac{|\Delta E|}{E} = \frac{1}{100}$$

The Q factor is therefore

$$Q = 2\pi\frac{E}{|\Delta E|} = 2\pi(100) = 628$$

Since the energy of an oscillator is proportional to the square of its amplitude, we can use Equation 12-51 to find the time dependence of the amplitude of a slightly damped oscillator. If A is the amplitude at some time t and A_0 is the amplitude at $t = 0$, we have

$$\frac{E}{E_0} = \frac{A^2}{A_0^2}$$

Then, from Equation 12-51,

$$\frac{A^2}{A_0^2} = e^{-(b/m)t}$$

or

$$A = A_0 e^{-(b/2m)t} \qquad \text{12-57}$$

Thus, the amplitude decreases exponentially with time. The dashed curves in Figure 12-18 correspond to $x = A$ and $x = -A$ where A is given by Equation 12-57.

The exact solution of Equation 12-46 can be found using standard methods for solving differential equations. The solution for the case of small damping is

$$x = A_0 e^{-(b/2m)t} \cos(\omega' t + \delta) \qquad \text{12-58}$$

where A_0 is the maximum amplitude and the frequency ω' is related to the angular frequency $\omega_0 = \sqrt{k/m}$ by

$$\omega' = \omega_0 \sqrt{1 - \left(\frac{b}{2m\omega_0}\right)^2} = \omega_0 \sqrt{1 - \frac{1}{4Q^2}} \qquad \text{12-59}$$

Thus, we see that our qualitative observations were correct. For small damping, the frequency is very nearly equal to the undamped frequency, and the amplitude decreases exponentially with time. Equation 12-59 is given only to illustrate that, for all practical purposes, the frequency of a slightly damped oscillator is the same as that of the undamped oscillator. For example, for a moderately damped oscillator with $Q = 10$, ω and ω_0 differ by only about 0.1 percent.

Equation 12-58 is the solution of Equation 12-46 only if the damping is small. If the damping is gradually increased, it eventually reaches a critical value such that no oscillation occurs. The critical value of the damping constant is

$$b_c = 2m\omega_0 \qquad \text{12-60}$$

If b is equal to or greater than this critical value, the system does not oscillate at all but merely returns to its equilibrium position. The greater the damping, the longer it takes the system to return to equilibrium. When $b = b_c$, the system is said to be **critically damped,** and it returns to equilibrium in the shortest time with no oscillation. When b is greater than b_c the system is overdamped. Figure 12-20 shows plots of the displacement versus time for a critically damped and an overdamped oscillator.

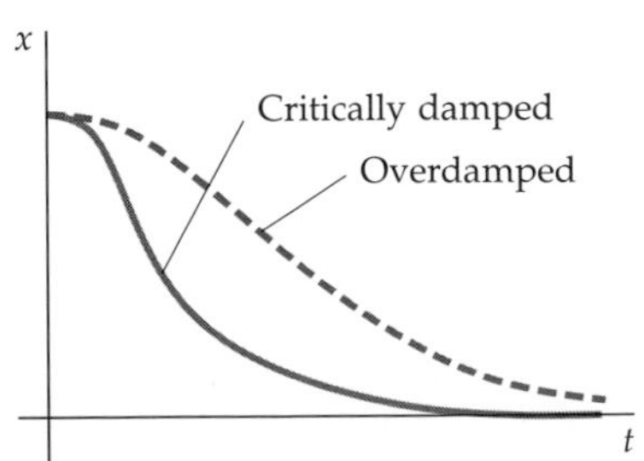

Figure 12-20 Plots of displacement versus time for a critically damped and an overdamped oscillator.

In many practical applications, critical damping or nearly critical damping is used to avoid oscillations and yet have the system return to equilibrium quickly. One example is the use of shock absorbers to damp the oscillations of an automobile on its springs. You can test the damping on your car by pushing down on the front or back of the car and releasing it. If the car moves back up to equilibrium with no oscillation, the system is critically damped or overdamped. Usually, you will note one or two oscillations, indicating that the damping is just under the critical value.

12-8 Driven Oscillations and Resonance

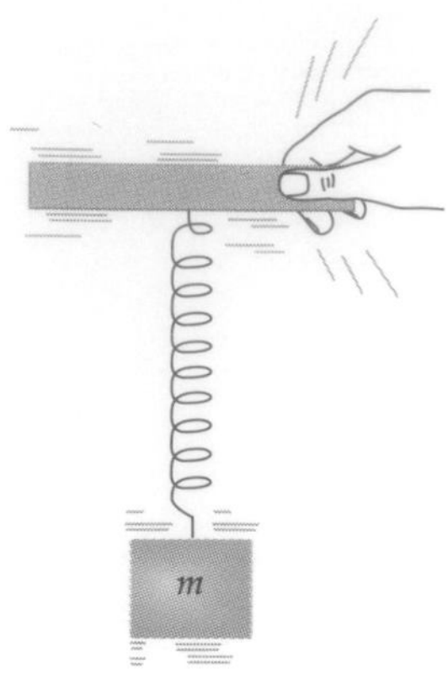

Figure 12-21 An object on a vertical spring can be driven by moving the point of support up and down.

We have seen that in damped oscillations, energy is continually dissipated and the amplitude of the oscillations decreases. To keep a damped system going, energy must be put into the system. When this is done, the oscillator is said to be driven or forced. For example, when you sit on a swing and keep it going by "pumping," that is, by moving your body and legs, you are driving an oscillator. If you put energy into the system at a greater rate than it is being dissipated by damping, the energy increases with time, which is indicated by an increase in the amplitude. If you put energy in at the same rate it is being dissipated, the amplitude remains constant in time.

Figure 12-21 shows a system consisting of an object on a spring that is being driven by moving the point of support up and down. Similarly, a simple pendulum can be driven by moving the support back and forth. (You should try some simple experiments with one of these systems to familiarize yourself with the behavior of driven oscillators.) If the point of support of an object on a spring or a simple pendulum is moved with simple harmonic motion of small amplitude and angular frequency ω, the system will begin oscillating. At first the motion is complicated, but eventually a steady state is reached in which the system oscillates with the same frequency as that of the driver and with a constant amplitude and, therefore, with constant energy. In the steady state, the energy put into the system by the driving force during one cycle equals the energy dissipated per cycle because of the damping.

The amplitude, and therefore the energy, of a system in the steady state depends not only on the amplitude of the driver, but also on its frequency. The **natural frequency** of an oscillator is defined as the frequency of that oscillator when no driving or damping forces are present. (The natural angular frequency of a spring, for example, is $\omega_0 = \sqrt{k/m}$.) If the driving frequency is equal (or approximately equal) to the natural frequency of the system, the system will oscillate with an amplitude much larger than the

The swaying of the Citicorp Building in New York during high winds is reduced by this tuned-mass damper, which oscillates at the same frequency as the building but 180° out of phase.

amplitude of the driving force. This phenomenon is called **resonance.** When the driving frequency equals that natural frequency of the oscillator, the energy absorbed by the oscillator is maximum. The natural frequency of the system is thus called the **resonance frequency** of the system.*

The average rate at which energy is absorbed over a cycle equals the average power delivered by the driving force. Figure 12-22 shows plots of the average power delivered to an oscillator as a function of the driving frequency for two different values of damping. These curves are called **resonance curves.** When the damping is small (large Q), the oscillator absorbs much more energy from the driving force at or near the resonance frequency than it does at other frequencies. The width of the peak of the resonance curve is correspondingly narrow, and we speak of the resonance as being sharp. When the damping is large (small Q) the oscillator still absorbs more energy at or near resonance than at other frequencies, but the difference is not as great and the resonance curve is broad. The width of each resonance curve $\Delta\omega$ is indicated in the figure. For relatively small damping, the ratio of the resonance frequency ω_0 to the width of the resonance can be shown to be equal to the Q factor (see Problem 74):

$$Q = \frac{\omega_0}{\Delta\omega} = \frac{f_0}{\Delta f} \qquad 12\text{-}61$$

Q factor for small damping

Thus, the Q factor is a direct measure of the sharpness of resonance.

We can treat a driven oscillator mathematically by assuming that, in addition to the restoring force and a damping force, the oscillator is subject to an external force, the driving force, that varies harmonically with time:

$$F_{ext} = F_0 \cos \omega t \qquad 12\text{-}62$$

where ω is the angular frequency of the driving force, which is generally not related to the natural angular frequency of the system ω_0.

An object of mass m attached to a spring of force constant k and subject to a damping force $-bv$ and an external force $F_0 \cos \omega t$ then obeys the equation of motion given by

$$\Sigma F = -kx - bv + F_0 \cos \omega t = m\frac{dv}{dt}$$

or

$$m\frac{d^2x}{dt^2} + b\frac{dx}{dt} + m\omega_0^2 x = F_0 \cos \omega t \qquad 12\text{-}63$$

where we have used $k = m\omega_0^2$ and $dv/dt = d^2x/dt^2$.

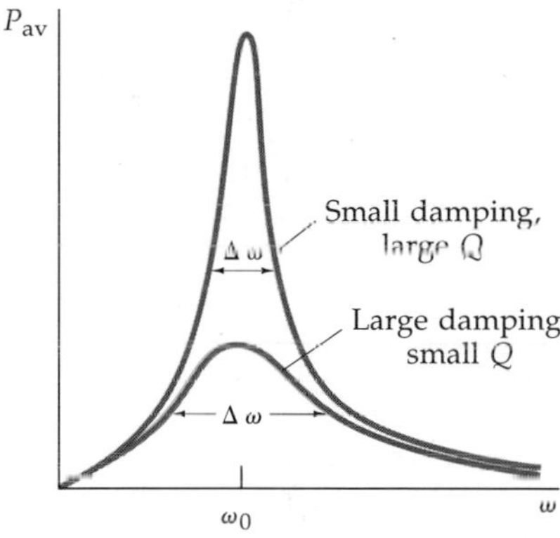

Figure 12-22 Plots of the average power delivered to an oscillator by a sinusoidal driving force versus the driving angular frequency ω for different values of damping. Resonance occurs when the (angular) frequency of the driving force equals the natural (angular) frequency of the system, ω_0. The resonance is sharp when the damping is small.

We will not attempt to solve Equation 12-63. Instead, we will discuss its general solution qualitatively. The solution of Equation 12-63 consists of two parts, the **transient solution** and the **steady-state solution.** The transient part of the solution is identical to that for a damped oscillator given by Equation 12-58. The constants in this part of the solution depend on the initial conditions. Over time, this part of the solution becomes negligible because of the exponential decrease of the amplitude. We are then left with

*When describing an oscillator mathematically, the angular frequency ω is more convenient than the frequency $f = \omega/2\pi$. Since ω and f are proportional, most statements concerning angular frequency also hold for frequency. For example, when the driving (angular) frequency ω equals the natural (angular) frequency ω_0, the driving frequency $f = \omega/2\pi$ equals the natural frequency $f_0 = \omega_0/2\pi$. In verbal descriptions we usually omit the word "angular" when such omission will not cause confusion.

the steady-state solution, which does not depend on the initial conditions. It can be written as

$$x = A \cos(\omega t - \delta) \qquad \text{12-64}$$

where the angular frequency ω is the same as that of the driving force and the amplitude A and phase constant δ are given by

$$A = \frac{F_0}{\sqrt{m^2(\omega_0^2 - \omega^2)^2 + b^2\omega^2}} \qquad \text{12-65}$$

and

$$\tan \delta = \frac{b\omega}{m(\omega_0^2 - \omega^2)} \qquad \text{12-66}$$

Comparing Equations 12-64 and 12-62, we can see that the displacement and the driving force oscillate with the same frequency but differ in phase by δ. The negative sign in the phase of Equation 12-64 is introduced so that the phase constant δ is positive. When the driving frequency ω is much less than the natural frequency ω_0, $\delta \approx 0$ as can be seen from Equation 12-66. At resonance, $\delta = \pi/2$, and when ω is much greater than ω_0, $\delta \approx \pi$. At resonance, $\omega = \omega_0$, and the amplitude A is maximum. (The difference in phase between the displacement and the driving force can be demonstrated by holding a meterstick at one end as a pendulum and moving the end back and forth to drive it.)

The velocity of the object in the steady state is obtained by differentiating Equation 12-64 with respect to time:

$$v = \frac{dx}{dt} = -A\omega \sin(\omega t - \delta)$$

At resonance, the velocity is in phase with the driving force:

$$v = -A\omega \sin\left(\omega t - \frac{\pi}{2}\right) = +A\omega \cos \omega t$$

Thus, at resonance, the object is always moving in the direction of the driving force, as would be expected for maximum power input.

Extended objects have more than one resonance frequency. The resonance vibration patterns of the top plate of a classical guitar are shown by these holographic interferograms. The frequency and Q value for each resonance are given.

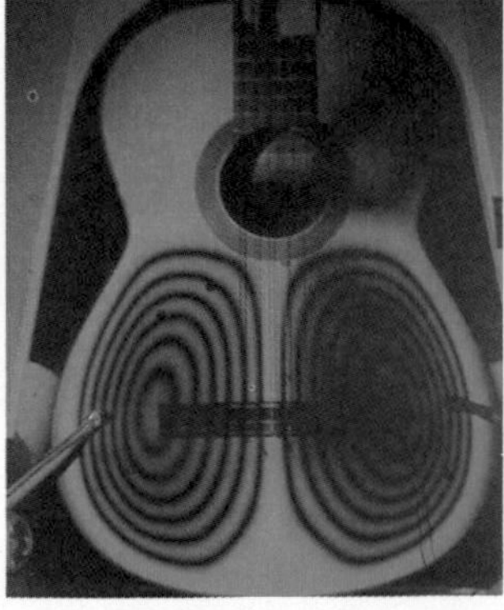

268 Hz ($Q = 52$)

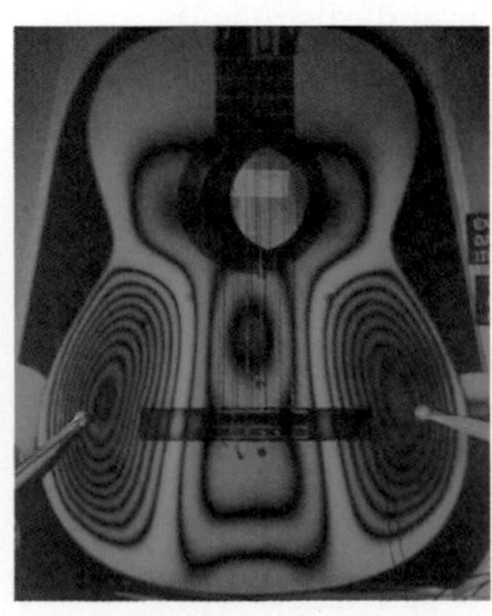

553 Hz ($Q = 66$)

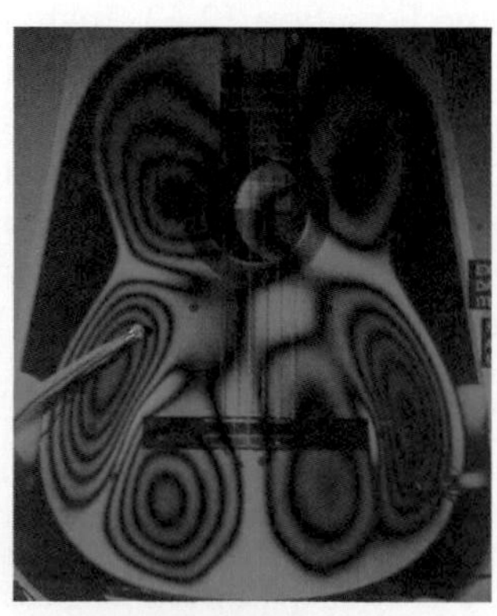

672 Hz ($Q = 61$)

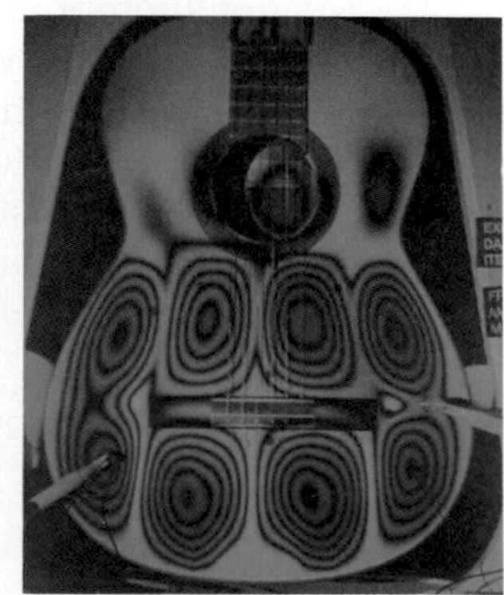

1010 Hz ($Q = 80$)

There are many familiar examples of resonance. When you sit on a swing and pump, the driving force is not simple harmonic. However, it is periodic, and you learn intuitively to pump with the same frequency as the natural frequency of the swing. When soldiers march across a small bridge, they often break step because it is possible that the frequency of their step might be near one of the resonance frequencies of the bridge. Many machines vibrate because they have rotating parts that are not in perfect balance. (Observe a washing machine in the spin cycle for an example.) If such a machine is attached to a structure that can vibrate, the structure becomes a driven oscillatory system that is set in motion by the machine. If the structure resonates with the applied vibration, considerable damage can occur, or it may produce unwanted sound throughout the building. Great engineering efforts are put into balancing the rotary parts of such machines, damping their vibrations, and isolating them from building supports.

A glass with low damping can be broken by an intense sound wave at a frequency equal to or very nearly equal to the natural frequency of vibration of the glass. This is often done in physics demonstrations using an audio oscillator and an amplifier. You can compare the relative damping of good crystal glasses by tapping them and listening to how long they ring. The longer the glass rings, the lower the damping. There are many stories about singers breaking a glass with a loud high note. Not one of these stories has ever been documented, however. The problem with a singer trying to break a glass is that it is nearly impossible for a singer to concentrate enough energy in the narrow frequency range near resonance. When we study the vibration of piano and violin strings in Chapter 13, we will see that such systems have a series of natural frequencies rather than just a single natural frequency and can resonate at any one of these natural frequencies.

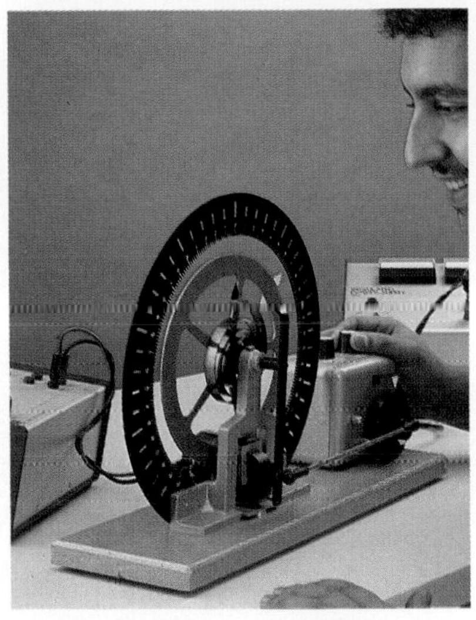

A student laboratory apparatus for studying resonance. The inner copper rim attached to a coiled spring constitutes a torsion pendulum, which can be driven by the motor at the right. The amplitude of the oscillation is measured by the protractor which is the outer rim of the apparatus. The electromagnet at the bottom provides damping that can be varied.

Question

9. Give some examples of common systems that can be considered to be driven oscillators.

Summary

1. In simple harmonic motion, the acceleration is proportional to the displacement and is oppositely directed. If x is the displacement, the acceleration is

$$a_x = -\omega^2 x$$

where ω is the angular frequency of oscillation, which is related to the frequency f by

$$f = \frac{\omega}{2\pi}$$

2. The period of oscillation is the reciprocal of the frequency:

$$T = \frac{1}{f}$$

The period and frequency in simple harmonic motion are independent of the amplitude. For the motion of an object of mass m on a spring of force constant k, the period is given by

$$T = 2\pi\sqrt{\frac{m}{k}}$$

The period of the motion of a simple pendulum of length L is

$$T = 2\pi\sqrt{\frac{L}{g}}$$

3. The position function x for simple harmonic motion of amplitude A and angular frequency ω is given by

$$x = A \cos (\omega t + \delta)$$

where δ is the phase constant, which depends on the choice of time $t = 0$. The velocity of the particle is given by

$$v = -\omega A \sin (\omega t + \delta)$$

4. When a particle moves in a circle with constant speed, the x and y components of its position vary with simple harmonic motion.

5. The total energy in simple harmonic motion is proportional to the square of the amplitude. For a mass on a spring of force constant k, it is given by

$$E_{\text{total}} = \tfrac{1}{2}kA^2$$

The potential and kinetic energies for a mass oscillating on a spring are given by

$$U = E_{\text{total}} \cos^2 (\omega t + \delta)$$

and

$$K = E_{\text{total}} \sin^2 (\omega t + \delta)$$

The average value of either the potential or kinetic energy is half the total energy.

6. In the oscillations of real systems, the motion is damped because of frictional forces or other forces that dissipate energy. If the damping is greater than some critical value, the system does not oscillate but merely returns to its equilibrium position if disturbed. The motion of a slightly damped system is nearly simple harmonic with an amplitude that decreases exponentially with time. The energy of such a system also decreases exponentially with time. For a slightly damped oscillator, the damping is measured by the Q factor:

$$Q = 2\pi\frac{E}{|\Delta E|}$$

where E is the total energy and ΔE is the energy loss per period.

7. When a slightly damped system is driven by an external force that varies sinusoidally with time, the system oscillates with frequency equal to the driving frequency and an amplitude that depends on the driving frequency. If the driving frequency is equal to or nearly equal to the natural frequency of the system, the system oscillates with a large amplitude. This is called resonance. The Q factor is a measure of the sharpness of the resonance. Systems with low damping and therefore high Q factors show a sharply peaked resonance curve. The ratio of the resonance frequency to the width of the resonance curve equals the Q factor:

$$Q = \frac{\omega_0}{\Delta\omega} = \frac{f_0}{\Delta f}$$

Chaos: An Orderly Disorder

James S. Walker
Washington State University,
Pullman, Washington

The ability to predict the future of a physical system is of central importance in physics. Given the appropriate initial conditions one can calculate the trajectory of a baseball, the orbit of a satellite, or the motion of a clock pendulum. So pervasive is this type of analysis in physics that it is sometimes tempting to reiterate the claim of the 18th century French mathematician Pierre Simon de Laplace that, given the initial conditions of the universe, one could predict its future for the rest of time. Leaving aside, for the moment, the problems associated with gathering and storing the initial conditions of every particle in the universe, and leaving aside also uncertainties that are inherent in quantum mechanics, there still remains a fundamental obstacle to predicting the future—chaos.

Chaos, according to many dictionaries, is a condition of total disorder or confusion—certainly the sort of situation to cloud the future. The meaning ascribed to chaos in physics and mathematics is similar in spirit, though more precise. For example, a system undergoing chaotic motion never repeats itself, but rather, continually does something different; thus the motion may appear quite random and disordered. Nonetheless, chaotic motion is far from totally disordered and instead displays a definite structure that is readily apparent.

James S. Walker, an associate professor of physics at Washington State University, is a theorist who studied critical phenomena and phase transitions for many years before his research turned, appropriately enough, to chaos. He received his bachelor's, master's, and doctorate degrees in physics all from the University of Washington, obtaining the Ph.D. in 1978. Before taking his current position, he did postdoctoral work at the University of Pennsylvania, the Massachusetts Institute of Technology, and the University of California at San Diego. When not rafting in Grand Canyon, or looking through his telescope on a starry night, Walker enjoys hiking, body surfing, kayaking, and juggling.

Another aspect of chaos, as it arises in physics, is an extreme sensitivity to initial conditions. Imagine, for example, trying to stand a straight pin on end. If placed exactly vertically the pin is in equilibrium. The equilibrium is unstable, however, and even the slightest disturbance—a breath of air, the sound of a pin dropping—will upset it and cause the pin to fall in one direction or another. The vertical position of the pin is an example of a state exhibiting extreme sensitivity to initial conditions, where, if one ignores even the minute puff of wind produced by a butterfly in the room, it is no longer possible to predict with certainty in which direction the pin will fall. Chaotic motion is like this, except that in chaos every state of the system, that is, every part of its motion, is just as sensitive to disturbance as is the vertical pin. As a result, even the smallest errors in measuring the initial conditions of a chaotic system lead to enormous errors later on.

The field of chaos is one of the youngest and most rapidly developing in physics today. There are many new aspects of chaos waiting to be discovered, but already much progress has been made. The inherent unpredictability of the weather, for example, is recognized as an example of chaotic motion associated with the transfer of heat in the atmosphere. While specific long-range forecasts are impossible, the general characteristics of weather—the climate—are well represented by an underlying chaotic structure. Many mechanical systems are also known to be chaotic, ranging from such mundane examples as the dripping of water from a leaky faucet to more exotic cases, like the chaotic tumbling motion of Hyperion, a moon of Saturn. In addition, chaos occurs in fluctuations of animal populations from generation to generation, in electric circuits, and even in our own bodies; systems such as the heart and even the brain show indications of chaotic behavior.

The Bender Bouncer

Fortunately, the basic properties of chaos—order within the disorder and sensitivity to initial conditions—can be seen quite clearly even in very simple physical systems. Consider, for example, a driven, damped harmonic oscillator, as shown in Figure 1. Here a spring of force constant k is attached to a mass m and driven vertically with angular frequency ω. A strip chart is arranged so that it records the periodic driving force applied to the spring, as well as the resulting displacement x of the mass. The

Continued

equation of motion is essentially that given in Equation 12-63:

$$m\frac{d^2x}{dt^2} + b\frac{dx}{dt} + kx = F \cos(\omega t)$$

where b is the damping coefficient and F is the magnitude of the driving force. (All results presented in this essay use $m = 4$ kg, $b = 1$ N · s/m, $k = 1$ N/m, and $F = 1$ N.) In Section 12-8, this system was studied for the usual situation, in which the mass is allowed to oscillate freely and for which the displacement as a function of time is simply

$$x(t) = A \cos(\omega t - \delta)$$

where A is the amplitude of motion and δ is the phase shift. This result for $x(t)$ clearly leads to periodic, predictable motion.

The simple behavior of this system is directly related to the fact that the restoring force kx is linear in the displacement; in fact, chaotic motion is found only in systems that are nonlinear. For example, if the restoring force of the spring included a nonlinear term like $k'x^2$, chaotic behavior would be possible. Of course, any real spring deviates from linearity for large enough displacements, thus nonlinear behavior is virtually unavoidable in the real world. Yet in physics one often studies linear systems almost to the exclusion of others, since their solutions and behaviors are relatively straightforward. This is certainly a reasonable approach, and in many situations the results for a linear system give a good indication of what may be expected in a system that is slightly nonlinear. But this is not always the case, and as pointed out by Robert May, a pioneer in chaos research, "Not only in research, but also in the everyday world of politics and economics, we would all be better off if more people realized that simple nonlinear systems do not necessarily possess simple dynamical properties."

An interesting way to make this mass-spring system nonlinear is to introduce a barrier which blocks the motion of the mass, as indicated in Figure 1. This system I refer to as the "Bender Bouncer," in honor of my friend and colleague Paul A. Bender, who first brought it to my attention. The barrier is considered to be infinitely massive and to undergo perfectly elastic collisions with the mass m; thus, it simply sends the mass back in the direction from which it came, traveling with the same speed. Clearly, the force acting on the mass is no longer a linear function of the displacement, since the barrier acts on the mass by giving it a sharp impulse.

With such simple characteristics for the mass–barrier collisions, it might seem that the barrier plays an essentially trivial role in the system; indeed, for many values of the driving frequency, the resulting motion is simple and periodic, as in the examples shown on the strip chart in Figure 1. However, for certain ranges of ω the motion is no longer periodic, but instead, never repeats. In short, the effect of the barrier is to produce chaos in this otherwise simple system.

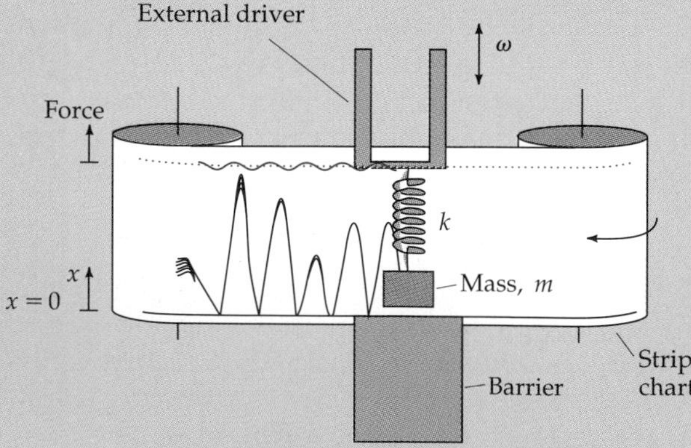

Figure 1 A driven, damped harmonic oscillator—set up so that its motion is recorded on a strip chart. Notice that the mass is not free to oscillate, but instead bounces elastically off an infinitely massive barrier. The tracings shown on the strip chart are obtained for $\omega = 1$ rad/s.

Chaotic Bouncing

A clear indication of chaos in the system is shown in Figure 2*a*, where the amplitude of oscillation is plotted as a function of the driving frequency ω. The tan-colored curve shows the result for the freely swinging oscillator, as discussed in Chapter 12. Also shown are the results for the bouncing oscillator with the barrier placed at $x = 0$, which are obtained as follows: Let the system run for many cycles so that the motion settles down, as in the examples in Figure 1; then for the next 100 bounces, plot the largest value of x that occurs between successive bounces. The result is a series of peaks separated by regions of scattered points. The smooth lines tracing the peaks result from periodic motion—each of the 100 bounces is of the same height; thus, for such frequencies, a single value of x is plotted 100 times over. Between the peaks are chaotic regions where each bounce is a different height, scattering the 100 plotted points like pepper grains over a range of amplitudes.

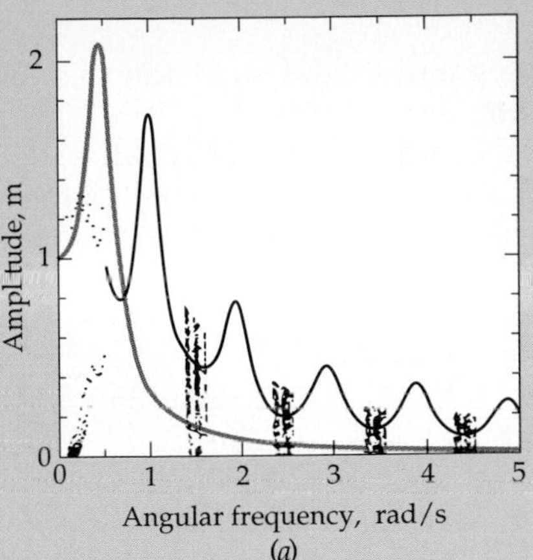

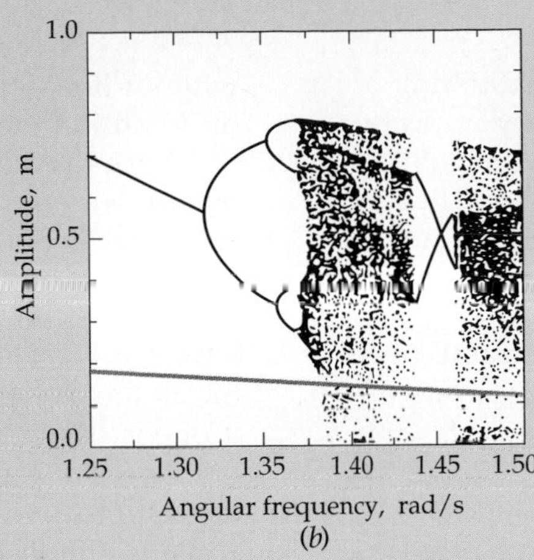

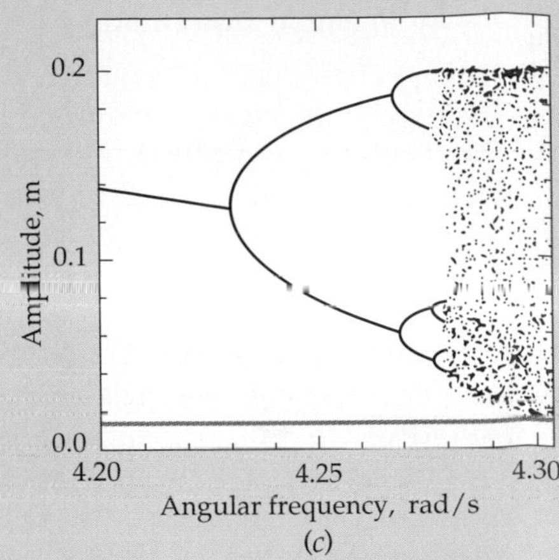

Figure 2 Amplitude versus angular frequency ω for the freely oscillating mass–spring system (tan-colored line) and the Bender bouncer (black line and scattered points). (*a*) Over a large range of angular frequency, the bouncing oscillator responds with a series of peaks (corresponding to periodic motion) separated by regions of scattered points (indicating chaotic motion). Parts (*b*) and (*c*) show details of the behavior near ω = 1.5 rad/s and ω = 4.3 rad/s, respectively, showing a progression of ever more closely spaced bifurcations leading to chaotic motion.

To show the chaotic behavior more clearly, the first chaotic region near ω = 1.5 rad/s is shown in detail in Figure 2*b*; in Figure 2*c* the chaotic region near ω = 4.3 rad/s is illustrated. As complex as this behavior is, and it is simply the response of a bouncing mass on a spring driven at different frequencies, note that the chaotic region is not simply random, but shows structure in the form of shading, as well as occasional clear "windows" where periodic motion returns. In particular, looking at Figure 2*b* one sees that simple periodic motion is observed for frequencies near 1.25 rad/s, with each bounce the same height. At roughly 1.325 rad/s the single line bifurcates, that is, it splits in two as the mass now bounces high, then low, then high, and so on. An example of such motion, for ω = 1.35 rad/s, is shown in Figure 3*a*. Notice that it takes two bounces for the motion to repeat (i.e., high, then low), thus the motion is still periodic, though the period has doubled.

Another period doubling is evident at about 1.3625 rad/s, and the mass now bounces to four different heights. With each doubling of the period it takes longer to see the "repeating pattern" of the motion, which thus appears more complex. Greater resolution in this frequency range shows, in fact, that an *infinite* progression of period doublings occurs ever more closely spaced in frequency—it's like walking forward with steps that are each a fraction the size of the one before, in which case an infinite number of steps cover only a finite distance. At a frequency of roughly 1.37 rad/s, an infinite number of "frequency steps" (period doublings) has been taken, and the period is now infinite; or to put it another way, the mass now bounces in a chaotic fashion, never repeating its motion. An example of the chaotic motion is shown in Figure 3*b*, for ω = 1.5 rad/s.

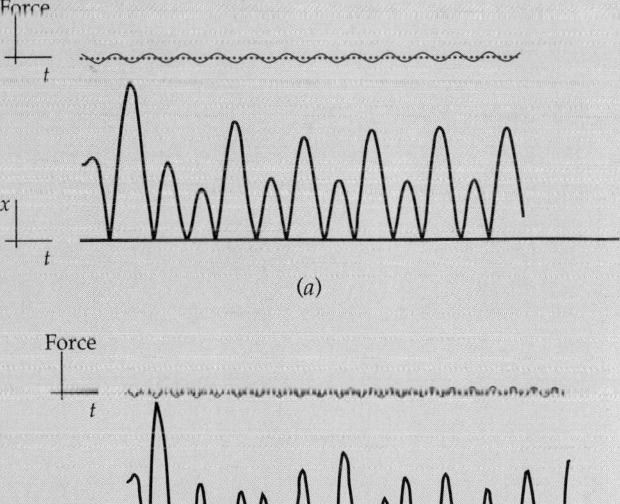

Figure 3 Position versus time plots for Bender bouncer. (*a*) ω = 1.35 rad/s. In this case, the system's motion repeats after every two bounces, which takes two cycles of the driving force. Compare with Figure 1, where the motion repeats with every cycle of the driving force. (*b*) ω = 1.5 rad/s. At this frequency the system's motion is chaotic—it never repeats.

Continued

Sensitivity to Initial Conditions

Once the system becomes chaotic, the scattering of bounce heights in Figure 2*b* and *c* is very intricate, and the dictionary definition of chaos as a condition of total disorder seems fairly appropriate. But in addition to the intricate structure, chaotic motion also displays an extreme sensitivity to initial conditions, in contrast to simple periodic motion in which all initial conditions evolve toward the same final motion. If several periodic oscillators are started off, each slightly differently, their motions quickly converge, and they spend the rest of time moving in perfect unison (as in Figure 1). Different starting conditions for a chaotic system, however, such as the bouncing oscillator at $\omega = 1.5$ rad/s, diverge to very different motions. Recall the situation with the vertical straight pin and the fact that different starting conditions arbitrarily close may still result in the pin falling in different directions. Each point in chaotic motion is just as sensitive to initial conditions, thus a set of starting points clustered together will, within a few bounces, be spread out over a range of amplitudes.

An example of this sensitivity can be seen in Figure 4, which shows five different initial conditions chosen so close together that, to begin with, they cannot be distinguished at this scale. Imagine these different starting conditions to represent a system of five different oscillators that start off moving in almost the same way. Even so, with each bounce the different motions diverge further from one another.

If each of these five starting conditions were also composed of five even more closely spaced starting conditions, it is clear that with more bounces these additional starting conditions would split off in the same way. It is like shredding a rope, splitting it apart into individual strands, then separating the strands into threads, then the threads into fibers, and so on. Chaotic motion does just this, but without limit—the separating and separating goes on to an infinitely fine level.

Since the starting conditions of an oscillator are not known exactly in any physical experiment, it follows that in a very short time the initial uncertainties will be magnified, and magnified again and again, until the mass could be located anywhere. In a practical sense, then, it is simply not meaningful to ask the question, "Where is the mass at time t?" if one is dealing with a chaotic oscillator.

Another aspect of sensitivity to initial conditions appears when the barrier is placed at a value of x that is different from zero. Consider, for example, the fantastically complex behavior in Figure 5*a*, which shows amplitude versus angular frequency for the barrier at $x = 0.5$. The blue line and points, taken by themselves, would seem to suggest a standard bifurcation route to chaos; but over the range of angular frequencies from about 2.22 rad/s to 2.55 rad/s, another type of motion with three amplitudes (shown in black) also exists. Beyond 2.55 rad/s, each of these three amplitudes undergoes its own bifurcations, until the system finally becomes chaotic.

Of particular interest is the range of angular frequencies where both types of motions are found. For any given frequency in this range, the final motion depends on exactly how the system is started. If the mass is always dropped from rest, then the final motion is found to depend on the height from which the mass is dropped and on the precise time at which it is dropped. In principle, the mass may be dropped from any height, and the time of the drop may occur at any point during a given cycle of the driving force; that is, one need only consider t in the interval $0 \le t \le T$, where T is the period of the driving force. What one finds is that, even though both of these motions are periodic rather than chaotic, the system is still extremely sensitive to initial condi-

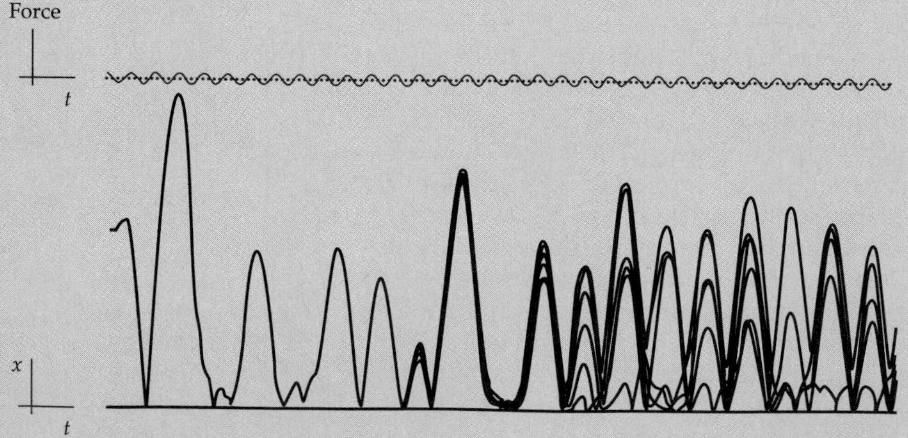

Figure 4 Position versus time plot for $\omega = 1.5$ rad/s, showing the motion for a system of five bouncing oscillators with closely spaced initial conditions. For the first few bounces, the oscillators stay in step, but with time the motions become very different.

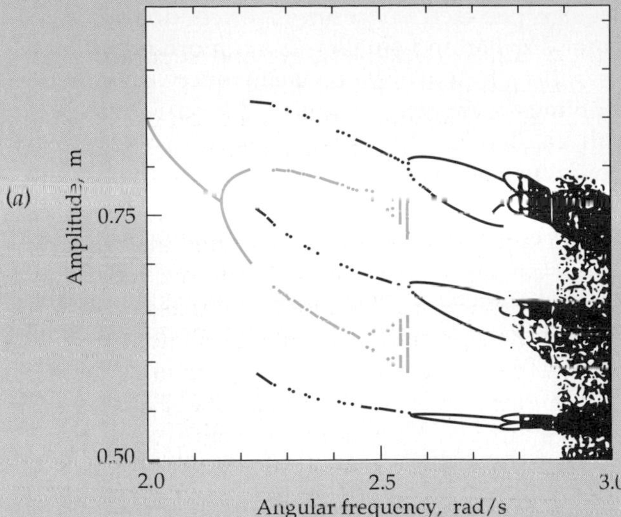

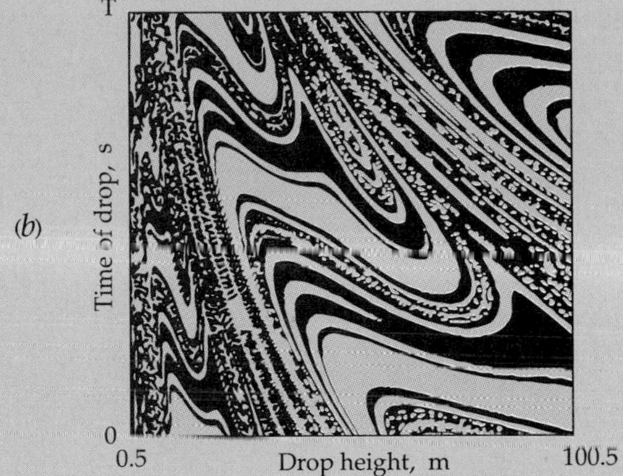

Figure 5 A different type of sensitivity to initial conditions occurs when the barrier is placed at $x = 0.5$. (*a*) The amplitude versus angular frequency plot shows two different types of motion, indicated by the blue and black points, existing in the same angular frequency range. (*b*) Initial conditions determine which motion the system displays. This plot, with $\omega = 2.4$ rad/s, shows the initial conditions that correspond to the motion indicated by the blue or the black points of part (*a*).

tions in determining which will be its final motion. This is shown clearly in Figure 5*b*, corresponding to $\omega = 2.4$ rad/s. The blue and black regions show the initial conditions that result in the blue or black motions, respectively. Again, in contrast to the apparent simplicity of this system, one finds an incredibly rich and complex structure describing its behavior.

Calculator Chaos

The Bender Bouncer can be set up experimentally very easily, and it can also be simulated on a desktop microcomputer. But if you have a pocket calculator handy, you can explore chaos for yourself, right now, using a very simple mathematical relation. Consider the following nonlinear expression, $x' = wx(1 - x)$, known as the logistic equation, where $0 \leq x \leq 1$ and w is an adjustable parameter. As an example, let $w = 2.9$, then pick any allowed value of x, say $x = 0.4$. Calculate x' as defined above to find $x' = 0.696$. Next, take x' as your starting value, and calculate x'' in the same manner as above to obtain $x'' = 0.614$ as the next iteration. Continue this procedure for a few more iterations, and soon you notice the values of x settle down to approximately 0.655. Try different starting values for x, and again the same limiting value of 0.655 is found.

Do you see the analogy with the mass-spring system discussed above? The value of x is like the maximum displacement between successive bounces, as plotted in Figure 2*a*, *b*, and *c*. What you have just observed is analogous to Figure 1, where different starting conditions settle down within a few cycles to a fixed and repeating bounce height.

Now try $w = 3.3$. In a few iterations, it becomes clear that x alternates between a high value, 0.824, and a low value, 0.480; that is, the system has undergone a period doubling similar to the period doublings in Figure 2*b* and 2*c*. This is analogous to the motion shown in Figure 3*a*. If one continues this procedure for different values of w and plots the stable values of x that result after several iterations, the result is as shown in Figure 6. Notice the remarkable similarity with Figure 2*c*, which emphasizes the fact that w and the angular frequency play analogous roles. As in the mass–spring system, the

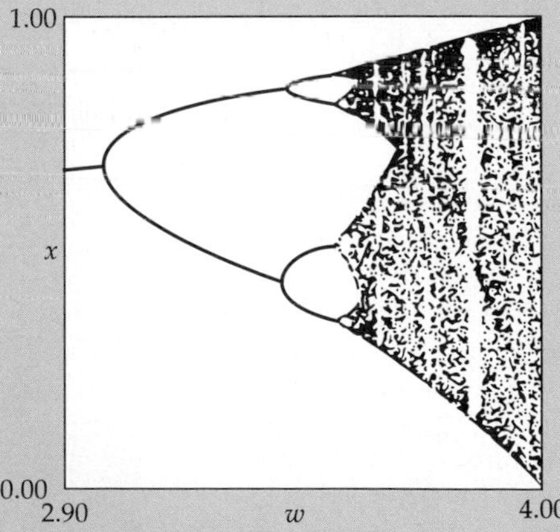

Figure 6 Results for the logistic equation $x' = wx(1 - x)$. This nonlinear equation is iterated 200 times before plotting begins, then the next 100 values of x are plotted. The bifurcations and period doublings shown here are very similar to the behavior displayed by the Bender bouncer in Figure 2*c*.

Continued

This striking image shows a small portion of a mathematical object known as the Mandelbrot set. The set occupies those regions shown in black, whereas the various colors represent regions just outside the set. Notice that the Mandelbrot set consists primarily of circular structures with the circles coming in all different sizes; from the large circle, to the smaller circles budding from it, to the even smaller circles on the buds, and so on down to circles of arbitrarily small size. This repetition of a basic shape to ever smaller sizes is one of the hallmarks of objects known as fractals. Fractals occur frequently in the study of chaos; for example, the size of the circles in the Mandelbrot set correspond exactly to the distance between successive period doublings in the logistic equation. Not surprisingly, the structure shown in Figure 5b is also a fractal.

distance between successive period doublings becomes smaller and smaller. Even more remarkably, the ratio of distances between successive period doublings approaches *exactly* the same value for both the Bender Bouncer and the logistic equation, providing just one example of the many common features shared among chaotic systems. Though the logistic equation is a simple mathematical rule and may be studied without reference to a particular physical system, it nonetheless serves as a useful model for the complex situations that can actually arise in nature.

Try now a value of w slightly greater than 3.56994571 . . . , which is where chaos sets in. Notice that the sequence of x values never repeats—you are now observing chaotic behavior and are on the road to further explorations.

Though this has been a very brief introduction to some of the unusual and extraordinary features of the orderly disorder known as chaos, it is clear that the ability to predict the future of a physical system is something that should not be taken for granted. And not only does chaos provide scientists with a new way of looking at the world, describing behavior in a wide variety of systems, but, as we have seen, there is great esthetic appeal in the fantastically complex geometrical forms associated with these systems. As the French chaotician D. Ruelle has said, "These systems of curves, these clouds of points, sometimes evoke galaxies or fireworks, other times quite weird and disturbing blossomings. There is a whole world of forms still to be explored, and harmonies still to be discovered." And, I might add, this exploration is open to us all.

When fluids are mixed the results are often chaotic, as in the experiment shown here. A small blob of yellow fluorescent tracer is placed in a container filled with glycerine, then the glycerine is stirred in a simple, repetitive fashion. As a result the blob is repeatedly stretched and folded, stretched and folded—much like a strand of salt-water taffy—giving rise to an intricately convoluted structure. An analogous "stretching and folding" occurs with the initial conditions of the Bender bouncer, as is evident in Figure 5b.

Suggestion for Further Reading

Pierce, John R.: "Sine Waves and Resonance," in *The Science of Musical Sound,* Scientific American Books, Inc., distributed by W.H. Freeman and Company, New York, 1983.

See description of the book under Suggestions for Further Reading, Chapter 13.

Review

A. Objectives: After studying this chapter, you should:

1. Be able to describe the general characteristics of simple harmonic motion.

2. Be able to describe the relationship between simple harmonic motion and circular motion.

3. Be able to work problems involving springs and pendulums.

4. Be able to discuss the general motion of any system near equilibrium.

5. Be able to describe the motion of a damped oscillator.

6. Be able to discuss the motion of a driven oscillator and sketch a typical resonance curve for various values of the Q factor.

B. Define, explain, or otherwise identify:

Periodic motion	Damped oscillations
Simple harmonic motion	Exponential decrease
Period	Q factor
Frequency	Critical damping
Hertz	Resonance
Amplitude	Resonance frequency
Phase	Resonance curve
Phase constant	Transient solution
Angular frequency	Steady-state solution
Physical pendulum	
Torsional pendulum	
Torsion constant	

C. True or false: If the statement is true, explain why it is true. If it is false, give a counterexample.

1. In simple harmonic motion, the period is proportional to the square of the amplitude.

2. In simple harmonic motion, the frequency does not depend on the amplitude.

3. In simple harmonic motion, the total energy is proportional to the square of the amplitude.

4. The motion of a simple pendulum is simple harmonic for any initial angular displacement.

5. The motion of a simple pendulum is periodic for any initial angular displacement.

6. If the acceleration of a particle is proportional to the displacement and oppositely directed, the motion is simple harmonic.

7. The energy of a damped, undriven oscillator decreases exponentially with time.

8. Resonance occurs when the driving frequency equals the natural frequency.

9. If the Q value is high, the resonance is sharp.

Problems

Level I

12-1 Simple Harmonic Motion: A Mass on a Spring

1. A 2-kg object is attached to a horizontal spring of force constant $k = 5$ kN/m. The spring is stretched 10 cm from equilibrium and released. Find (*a*) the frequency, (*b*) the period, and (*c*) the amplitude of the motion. (*d*) What is the maximum speed? (*e*) What is the maximum acceleration? (*f*) When does the object first reach its equilibrium position? What is its acceleration at this time?

2. Answer the questions in Problem 1 for a 3-kg object attached to a spring of force constant $k = 600$ N/m when the spring is initially stretched 8 cm from equilibrium.

3. A 3-kg object attached to a horizontal spring oscillates with an amplitude $A = 10$ cm and a frequency $f = 2$ Hz. (*a*) What is the force constant of the spring? (*b*) What is the period of the motion? (*c*) What is the maximum speed of the object? (*d*) What is its maximum acceleration?

4. An object on a horizontal spring oscillates with a period of 4 s. If the object is suspended from the spring vertically, by how much is the spring stretched from its natural length when the object is in equilibrium?

5. An 80-kg person steps into a car of mass 2400 kg, causing it to sink 2.50 cm on its springs. Assuming no damping, with what frequency will the car and passenger vibrate on the springs?

6. A 5-kg object oscillates on a horizontal spring with an amplitude of 4 cm. Its maximum acceleration is 24 m/s^2. Find (*a*) the force constant *k*, (*b*) the frequency, and (*c*) the period of the motion.

7. An object oscillates with an amplitude of 6 cm on a horizontal spring of force constant 2 kN/m. Its maximum speed is 2.20 m/s. Find (*a*) the mass of the object, (*b*) the frequency of the motion, and (*c*) the period of the motion.

8. The position of a particle is given by

$$x = (5 \text{ cm}) \cos 4\pi t$$

where *t* is in seconds. What is (*a*) the frequency, (*b*) the period, and (*c*) the amplitude of the particle's motion? (*d*) What is the first time after $t = 0$ that the particle is at its equilibrium position? In what direction is it moving at that time?

9. (*a*) What is the maximum speed of the particle in Problem 8? (*b*) What is its maximum acceleration?

10. What is the phase constant δ in Equation 12-4 if the position of the oscillating particle at time $t = 0$ is (*a*) 0, (*b*) $-A$, (*c*) A, (*d*) $A/2$?

11. A particle has a displacement of $x = 0.4 \cos (3t + \pi/4)$ where *x* is in meters and *t* is in seconds. (*a*) Find the frequency *f* and period *T* of the motion. Where is the particle at (*b*) $t = 0$? (*c*) At $t = 0.5$ s?

12. (*a*) Find an expression for the velocity of the particle whose position is given in Problem 10. (*b*) What is the velocity at time $t = 0$? (*c*) What is the maximum velocity? (*d*) At what time after $t = 0$ does this maximum velocity first occur?

13. A particle of mass *m* begins at rest from $x = +25$ cm and oscillates about its equilibrium position at $x = 0$ with a period of 1.5 s. Write equations for (*a*) the position *x* versus the time *t*, (*b*) the velocity *v* versus *t*, and (*c*) the acceleration *a* versus *t*.

14. Work Problem 13 for the case in which the particle is initially at $x = 25$ cm and is moving with velocity $v_0 = +50$ cm/s.

15. Find (*a*) the maximum speed and (*b*) the maximum acceleration of the particle in Problem 13. (*c*) What is the first time that the particle is at $x = 0$ and moving to the right?

12-2 Simple Harmonic Motion and Circular Motion

16. A particle moves in a circle of radius 40 cm with a constant speed of 80 cm/s. Find (*a*) the frequency and (*b*) the period of the motion. (*c*) Write an equation for the *x* component of the position of the particle as a function of time *t*, assuming that the particle is on the *x* axis at time $t = 0$.

17. A particle moves in a circle of radius 15 cm, making 1 rev every 3 s. (*a*) What is the speed of the particle? (*b*) What is its angular velocity ω? (*c*) Write an equation for the *x* component of the position of the particle as a function of time *t*, assuming that the particle is on the *x* axis at time $t = 0$.

12-3 Energy in Simple Harmonic Motion

18. Find the total energy of the object oscillating on the spring in Problem 1.

19. Find the total energy of the object oscillating on the spring in Problem 2.

20. A 1.5-kg object oscillates with simple harmonic motion on a spring of force constant $k = 500$ N/m. Its maximum speed is 70 cm/s. (*a*) What is the total energy? (*b*) What is the amplitude of the oscillation?

21. A 3-kg object oscillating on a spring of force constant 2 kN/m has a total energy of 0.9 J. (*a*) What is the amplitude of the motion? (*b*) What is the maximum speed?

22. An object oscillates on a spring with an amplitude of 4.5 cm. Its total energy is 1.4 J. What is the force constant of the spring?

12-4 An Object on a Vertical Spring

23. A 2.5-kg object hangs from a vertical spring of force constant 600 N/m. It oscillates with an amplitude of 3 cm. When the object is at its maximum downward displacement, find (*a*) the total energy of the system, (*b*) the gravitational potential energy, and (*c*) the potential energy in the spring. (*d*) What is the maximum kinetic energy of the object? (Choose $U = 0$ when the object is in equilibrium.)

24. A 1.5-kg object that stretches a spring 2.8 cm from its natural length when hanging at rest oscillates with an amplitude of 2.2 cm. Find (*a*) the total energy of the system. At maximum downward displacement find (*b*) the gravitational potential energy and (*c*) the potential energy in the spring. (*d*) What is the maximum kinetic energy of the object? (Choose $U = 0$ when the object is in equilibrium.)

25. A 1.2-kg object hanging from a spring of force constant 300 N/m oscillates with a maximum speed of 30 cm/s. (*a*) What is its maximum displacement? When the object is at its maximum displacement, find (*b*) the total energy of the system, (*c*) the gravitational potential energy, and (*d*) the potential energy in the spring.

12-5 Pendulums

26. Find the length of a simple pendulum if the period is 5 s at a point where $g = 9.81$ m/s^2.

27. What would be the period of the pendulum in Problem 26 on the moon, where the acceleration due to gravity is one-sixth that on earth?

28. If the period of a pendulum 70 cm long is 1.68 s, what is the value of *g* at the location of the pendulum.

29. A pendulum set up in the stairwell of a 10-story building consists of a heavy weight suspended on a 34.0-m wire. If $g = 9.81$ m/s^2, what is the period of oscillation?

30. A small particle of mass *m* slides without friction in a spherical bowl of radius *r*. (*a*) Show that the motion of the particle is the same as if it were attached to a string of length *r*. (*b*) Figure 12-23 shows a particle of mass m_1 that is displaced a small distance s_1 from the bottom of the bowl, where s_1 is much smaller than *r*. A second particle

Figure 12-23 Problem 30.

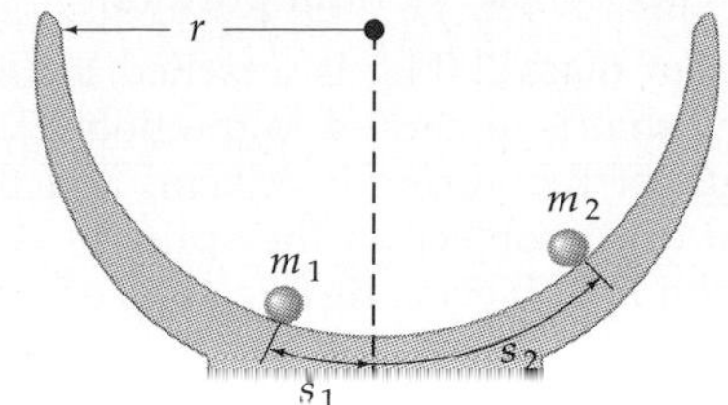

of mass m_2 is displaced in the opposite direction a distance $s_2 = 3s_1$, where s_2 is also much smaller than r. If the particles are released at the same time, where do they meet? Explain.

31. A thin disk of mass 5 kg and radius 20 cm is suspended by a horizontal axis perpendicular to the disk through its rim. The disk is displaced slightly from equilibrium and released. Find the period of the subsequent simple harmonic motion.

32. A circular hoop of radius 50 cm is hung on a narrow horizontal rod and allowed to swing in the plane of the hoop. What is the period of its oscillation, assuming that the amplitude is small?

33. A 3-kg plane figure is suspended at a point 10 cm from its center of mass. When it is oscillating with small amplitude, the period of oscillation is 2.6 s. Find the moment of inertia I about an axis perpendicular to the plane of the figure through the pivot point.

12-6 General Motion near Equilibrium

There are no problems for this section.

12-7 Damped Oscillations

34. An oscillator has a Q factor of 200. By what percentage does its energy decrease during one period?

35. A 2-kg object oscillates with an initial amplitude of 3 cm on a spring of force constant $k = 400$ N/m. Find (*a*) the period and (*b*) the total initial energy. (*c*) If the energy decreases by 1 percent per period, find the damping constant b and the Q factor.

36. Show that the ratio of the amplitudes for two successive oscillations is constant for a damped oscillator.

37. An oscillator has a period of 3 s. Its amplitude decreases by 5 percent during each cycle. (*a*) By how much does its energy decrease during each cycle? (*b*) What is the time constant τ? (*c*) What is the Q factor?

12-8 Driven Oscillations and Resonance

38. Find the resonance frequency for each of the three systems shown in Figure 12-24.

39. A damped oscillator loses 2 percent of its energy during each cycle. (*a*) What is its Q factor? (*b*) If its resonance frequency is 300 Hz, what is the width of the resonance curve $\Delta\omega$ when the oscillator is driven?

40. A 2-kg object oscillates on a spring of force constant $k = 400$ N/m. The damping constant $b = 2.00$ kg/s. The system is driven by a sinusoidal force of maximum value 10 N and angular frequency $\omega = 10$ rad/s. (*a*) What is the amplitude of the oscillations? (*b*) If the driving frequency is varied, at what frequency will resonance occur? (*c*) Find the amplitude of oscillation at resonance. (*d*) What is the width of the resonance curve $\Delta\omega$?

Level II

41. A 3-kg object oscillates on a spring with an amplitude of 8 cm. Its maximum acceleration is 3.50 m/s^2. Find the total energy.

42. The period of an oscillating particle is 8 s, and its amplitude is 12 cm. At $t = 0$, it is at its equilibrium position. Find the distance traveled during the interval (*a*) $t = 0$ to $t = 2$ s, (*b*) $t = 2$ s to $t = 4$ s, (*c*) $t = 0$ to $t = 1$ s, and (*d*) $t = 1$ s to $t = 2$ s.

43. The period of an oscillating particle is 8 s. At $t = 0$, the particle is at $x = A = 10$ cm. (*a*) Sketch x as a function of t. (*b*) Find the distance traveled in the first second, the second second, the third second, and the fourth second after $t = 0$.

44. An object of mass m is supported by a vertical spring of force constant 1800 N/m. When pulled down 2.5 cm from equilibrium and released from rest, the object oscillates at 5.5 Hz. (*a*) Find m. (*b*) Find the amount the spring is stretched from its natural length when the object is in equilibrium. (*c*) Write expressions for the displacement x, the velocity v, and the acceleration a as functions of time t.

45. A spring is hanging vertically. An object of unknown mass is hung on the end of the unstretched spring and is released from rest. If it falls 3.42 cm before first coming to rest, find the period of the motion. (Assume $g = 9.81$ m/s^2.)

46. In heavy seas, the bow of a destroyer undergoes a simple harmonic vertical pitching motion with a period of 8.0 s and an amplitude of 2.0 m. (*a*) What is the maximum vertical velocity of the destroyer's bow? (*b*) What is its maximum acceleration? (*c*) An 80-kg sailor is standing on a scale in the bunkroom in the bow. What are the maximum and minimum readings on the scale in newtons?

47. Military specifications often call for electronic devices to be able to withstand accelerations of $10g = 98.1$ m/s^2. To make sure their products meet this specification, manufacturers test them using a shaking table that can vibrate a device at various specified frequencies and amplitudes. If a device is given a vibration of amplitude 1.5 cm, what should its frequency be?

Figure 12-24 Problem 38.

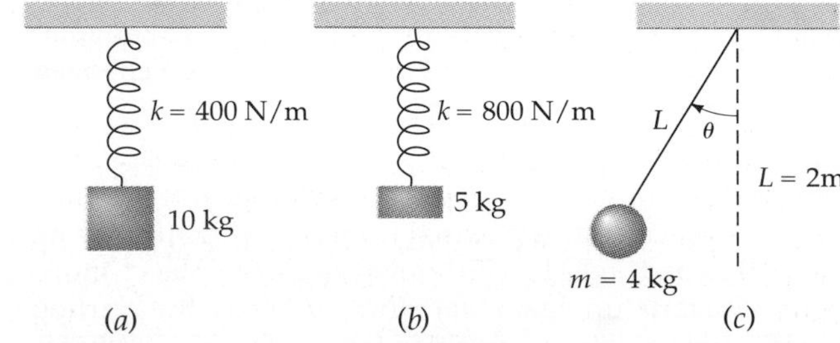

48. A block of wood slides on a frictionless horizontal surface. It is attached to a spring and oscillates with a period of 0.8 s. A second block rests on top of the first. The coefficient of static friction between the two blocks is 0.25. (*a*) If the amplitude of oscillation is 1 cm, will the block on top slip? (*b*) What is the greatest amplitude of oscillation for which the top block will not slip?

49. A child on a swing oscillates with a period of 3 s. The mass of the child and swing is 35 kg. The swing is pushed by a patient father such that the amplitude remains constant. At the bottom of the swing, the speed of the child and swing is 2.0 m/s. (*a*) What is the total energy of the child and swing? (*b*) If $Q = 20$, how much energy is dissipated during each oscillation of the swing? (*c*) What is the power input of the father? *Note:* Pushing a swing is usually not done sinusoidally. However, to maintain a steady amplitude, the energy lost per cycle due to damping must be replaced by an external energy source. This energy may come from the child's "pumping" or from someone else's pushing the swing.

50. A damped oscillator loses 2 percent of its energy during each cycle. (*a*) How many cycles elapse before half its original energy is dissipated? (*b*) What is the Q factor? (*c*) If the natural frequency is 100 Hz, what is the width of the resonance curve when the oscillator is driven?

51. An oscillator has a Q factor of 20. (*a*) By what fraction does the energy decrease during each cycle? (*b*) Use Equation 12-59 to find the percentage difference between ω' and ω_0. *Hint:* Use the approximation $(1 + x)^{1/2} \approx 1 + \frac{1}{2}x$ for small x.

52. For a child on a swing, the amplitude drops by a factor of $1/e$ in about eight periods if no energy is fed in. Estimate the Q factor for this system.

53. It has been stated that the vibrating earth has a resonance period of 54 min and a Q factor of about 400 and that after a large earthquake, the earth "rings" (continues to vibrate) for about 2 months. (*a*) Find the percentage of the energy of vibration lost to damping forces during each cycle. (*b*) Show that after n periods, the energy is $E_n = (0.984)^n E_0$, where E_0 is the original energy. (*c*) If the original energy of vibration of an earthquake is E_0, what is the energy after 2 days?

54. An object of mass m_1 sliding on a frictionless horizontal surface is attached to a spring of force constant k. It oscillates with an amplitude A. When the spring is at its greatest extension and the mass is instantaneously at rest, a second object of mass m_2 is placed on top of it. (*a*) What is the smallest value the coefficient of static friction μ_s can have if the second object is not to slip on the first? (*b*) Explain how the total energy E, the amplitude A, the angular frequency ω, and the period T of the system are changed by placing m_2 on m_1.

55. A spring of force constant $k = 250$ N/m is suspended from a rigid support. An object of mass 1 kg is attached at the bottom end and is released from rest when the spring is unstretched. (*a*) How far down does the object move before it starts up again? (*b*) How far below the starting point is the equilibrium position for the object? (*c*) What is the period of oscillation? (*d*) What is the speed of the object when it first reaches its equilibrium position? When does it first reach its equilibrium position?

56. An object of mass 2.0 kg is attached to the top of a vertical spring that is anchored to the floor. The uncompressed length of the spring is 8.0 cm, and the equilibrium position of the object on the spring is 5.0 cm from the floor. When the object is resting at its equilibrium position, it is given a downward impulse with a hammer such that its initial speed is 0.3 m/s. (*a*) To what maximum height above the floor does the object eventually rise? (*b*) How long does it take for the object to reach its maximum height the first time? (*c*) Does the spring ever become uncompressed? What minimum initial velocity must be given to the object for the spring to be uncompressed at some time?

57. A spider of mass 0.36 g sits in the middle of its horizontal web, which sags 3.00 mm under its weight. Estimate the frequency of vertical vibration for this system.

58. The acceleration due to gravity g varies with location on the earth because of the earth's rotation and because the earth is not exactly spherical. This was first discovered in the seventeenth century, when it was noted that a pendulum clock carefully adjusted to keep correct time in Paris lost about 90 s/d near the equator. (*a*) Show that a small change in the acceleration of gravity Δg produces a small change in the period ΔT of a pendulum given by

$$\frac{\Delta T}{T} \approx -\frac{1}{2}\frac{\Delta g}{g}$$

(Use differentials to approximate ΔT and Δg.) (*b*) How great a change in g is needed to account for a change in the period of 90 s/d?

Level III

59. A 20-kg ball is hung on a steel wire of 1.00 mm diameter and 3.00 m length. When caused to vibrate vertically, it is found to have a frequency of 8.14 Hz. (*a*) Find the force constant k of the wire. (*b*) Show that Young's modulus for the wire is related to the force constant by $Y = kL/A$. (*c*) Find Young's modulus for this wire.

60. A pendulum clock that has run down to a very small amplitude gains 5 min each day. What angular amplitude should the pendulum have to keep the correct time?

61. An object of mass 2 kg resting on a frictionless horizontal surface is attached to a spring of force constant 600 N/m. A second object of mass 1 kg slides along the surface toward the first object at 6 m/s. (*a*) Find the amplitude of oscillation if the objects make a perfectly inelastic collision and remain together on the spring. What is the period of oscillation? (*b*) Find the amplitude and period of oscillation if the collision is elastic. (*c*) For each type of collision, write an expression for the position x as a function of time t for the object attached to the spring, assuming that the collision occurs at time $t = 0$. What is the impulse delivered to the 2-kg object in each case?

62. A small block of mass m_1 rests on a piston that is vibrating vertically with simple harmonic motion given by $y = A \sin \omega t$. (*a*) Show that the block will leave the piston

if $\omega^2 A > g$. (*b*) If $\omega^2 A = 3g$ and $A = 15$ cm, at what time will the block leave the piston?

63. A block suspended from a spring is oscillating vertically with a frequency of 4 Hz and an amplitude of 7 cm. A very small bit of rock is placed on top of the oscillating block just as it reaches its lowest point. Assume that the rock has no effect on the oscillation. (*a*) At what distance above the block's equilibrium position does the rock lose contact with the block? (*b*) What is the speed of the rock when it leaves the block? (*c*) What is the greatest distance above the block's equilibrium position reached by the rock?

64. Show that for the situations in Figure 12-25*a* and *b*, the object oscillates with a frequency $f = (1/2\pi)\sqrt{k_{eff}/m}$, where k_{eff} is given by (*a*) $k_{eff} = k_1 + k_2$ and (*b*) $1/k_{eff} = 1/k_1 + 1/k_2$. *Hint:* Find the net force F on the object for a small displacement x and write $F = -k_{eff}x$. Note that in (*b*) the springs stretch by different amounts, the sum of which is x.

Figure 12-25 Problem 64.

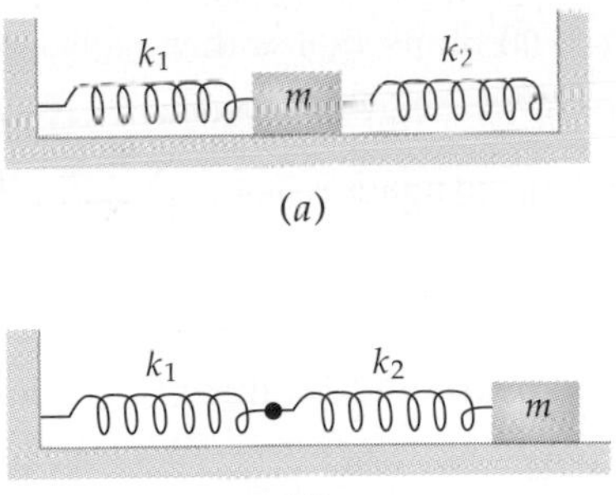

(*a*)

(*b*)

65. A straight tunnel is dug through the earth as shown in Figure 12-26. Assume the walls of the tunnel are frictionless. (*a*) The gravitational force exerted by the earth on a particle of mass m at a distance r from the center of the earth when $r < R_E$ is $F_r = -(GmM_E/R_E^3)r$, where M_E is the mass of the earth and R_E is its radius. Show that the net force on a particle of mass m at a distance x from the middle of the tunnel is given by $F_x = -(GmM_E/R_E^3)x$, and that the motion of the particle is therefore simple harmonic motion. (*b*) Show that the period of the motion is given by $T = 2\pi\sqrt{R_E/g}$ and find its value in minutes. (This is the same period as that of a satellite orbiting near the surface of the earth, and it is independent of the length of the tunnel.)

Figure 12-26 Problem 65.

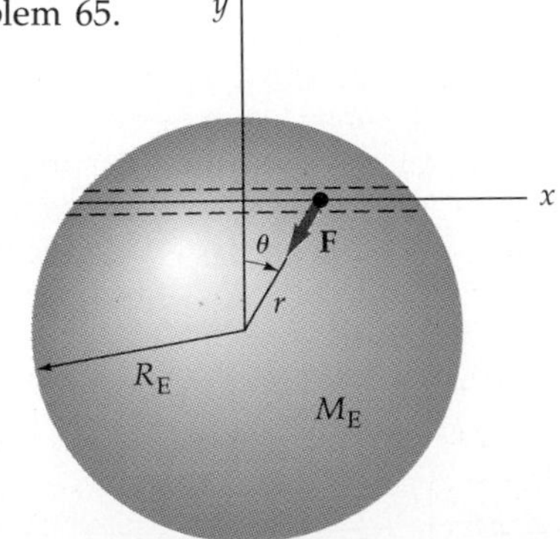

66. A simple pendulum of length L is released from rest from an angle θ_0. (*a*) Assuming the motion of the pendulum to be simple harmonic motion, find its speed as it passes through $\theta = 0$. (*b*) Using the conservation of energy, find this speed exactly. (*c*) Show that your results for parts (*a*) and (*b*) are the same when θ_0 is small. (*d*) Find the difference in your results for $\theta_0 = 0.20$ rad and $L = 1$ m.

67. A 3-kg sphere dropped in the air has a terminal speed of 25 m/s. (Assume the drag force is $-bv$.) The sphere is attached to a spring of force constant $k = 400$ N/m, and it oscillates with an initial amplitude of 20 cm. (*a*) What is Q? (*b*) When will the amplitude be 10 cm? (*c*) How much energy will have been lost when the amplitude is 10 cm?

68. A plane object has moment of inertia I about its center of mass. When pivoted at point P_1, as shown in Figure 12-27, it oscillates about the pivot with a period T. There is a second point P_2 on the opposite side of the center of mass about which the object can be pivoted so that the period of oscillation is also T. Show that $h_1 + h_2 = gT^2/4\pi^2$.

Figure 12-27 Problem 68.

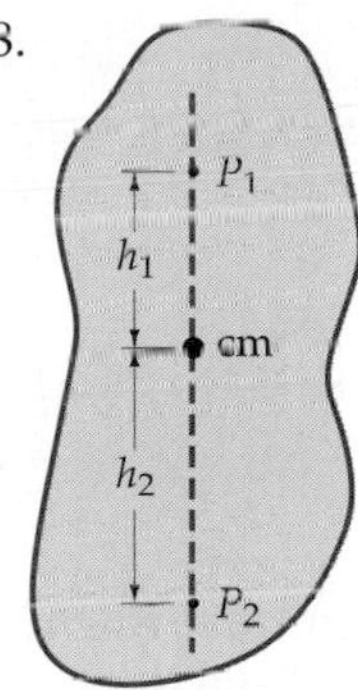

69. A physical pendulum consists of a spherical bob of radius r and mass m suspended from a string (Figure 12-28). The distance from the center of the sphere to the point of support is L. When r is much less than L, such a pendulum is often treated as a simple pendulum of length L. (*a*) Show that the period for small oscillations is given by

$$T = T_0\sqrt{1 + \frac{2r^2}{5L^2}}$$

where $T_0 = 2\pi\sqrt{L/g}$ is the period of a simple pendulum of length L. (*b*) Show that when r is much smaller than L, the period is approximately $T \approx T_0(1 + r^2/5L^2)$. (*c*) If $L = 1$ m and $r = 2$ cm, find the error when the approximation $T = T_0$ is used for this pendulum. How large must the radius of the bob be for the error to be 1 percent?

Figure 12-28 Problem 69.

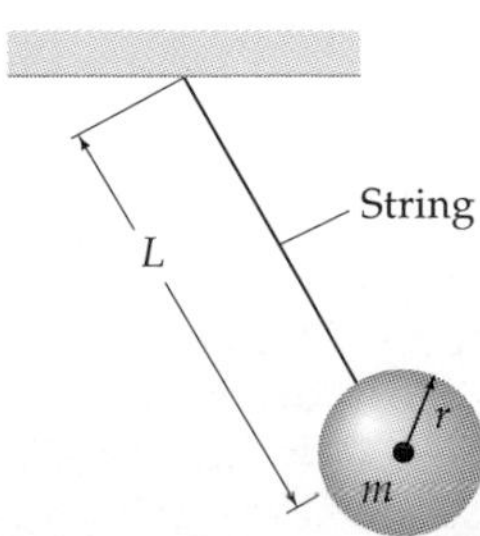

70. A damped oscillator has a frequency ω' that is 10 percent less than its undamped frequency. (*a*) By what factor is its amplitude decreased during each oscillation? (*b*) By what factor is its energy reduced during each oscillation?

71. Show by direct substitution that Equation 12-58 is a solution of Equation 12-46.

72. A block of mass m on a horizontal table is attached to a spring of force constant k as shown in Figure 12-29. The coefficient of kinetic friction between the block and the table is μ_k. The spring is stretched a distance A and released. (*a*) Apply Newton's second law to the block to obtain an equation for its acceleration d^2x/dt^2 for the first half cycle, during which the block is moving to the left. Show that the resulting equation can be written $d^2x'/dt^2 = -\omega^2 x'$, where $x = 0$ at the equilibrium position of the spring, and $x' = x - x_0$, with $x_0 = \mu_k mg/k = \mu_k g/\omega^2$. (*b*) Repeat part (*a*) for the second half cycle as the block moves to the right, and show that $d^2x''/dt^2 = -\omega^2 x''$, where $x'' = x + x_0$ and x_0 has the same value. (*c*) Sketch the first few cycles for $A = 10x_0$.

Figure 12-29 Problem 72.

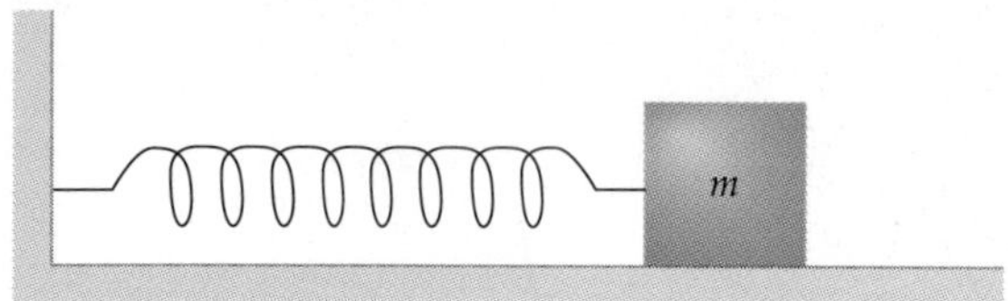

73. In this problem, you will derive the expression for the average power delivered by a driving force to a driven oscillator (Figure 12-22). (*a*) Show that the instantaneous power input of the driving force is given by

$$P = Fv = -A\omega F_0 \cos \omega t \sin (\omega t - \delta)$$

(*b*) Use the trigonometric identity $\sin(\theta_1 - \theta_2) = \sin\theta_1 \cos\theta_2 - \cos\theta_1 \sin\theta_2$ to show that this can be written

$$P = A\omega F_0 \sin\delta \cos^2 \omega t - A\omega F_0 \cos\delta \cos\omega t \sin\omega t$$

(*c*) Show that the average value of the second term in your result for part (*b*) over one or more periods is zero and that therefore

$$P_{av} = \tfrac{1}{2}A\omega F_0 \sin\delta$$

(*d*) From Equation 12-66 for tan δ, construct a right triangle in which the side opposite the angle δ is $b\omega$ and the side adjacent is $m(\omega_0^2 - \omega^2)$, and use this triangle to show that

$$\sin\delta = \frac{b\omega}{\sqrt{m^2(\omega_0^2 - \omega^2)^2 + b^2\omega^2}} = \frac{b\omega A}{F_0}$$

(*e*) Use your result for part (*d*) to eliminate ωA so that the average power input can be written

$$P_{av} = \frac{1}{2}\frac{F_0^2}{b}\sin^2\delta = \frac{1}{2}\frac{b\omega^2 F_0^2}{m^2(\omega_0^2 - \omega^2)^2 + b^2\omega^2} \qquad \text{12-67}$$

74. In this problem you are to use the result of Problem 73 to derive Equation 12-61, which relates the width of the resonance curve to the Q value when the resonance is sharp. At resonance, the denominator in Equation 12-67 is $b^2\omega_0^2$ and P_{av} has its maximum value. For a sharp resonance, the variation in ω in the numerator in Equation 12-67 can be neglected. Then the power input will be half its maximum value at the values of ω, for which the denominator is $2b^2\omega_0^2$. (*a*) Show that ω then satisfies

$$m^2(\omega - \omega_0)^2(\omega + \omega_0)^2 = b^2\omega_0^2$$

(*b*) Using the approximation $\omega + \omega_0 \approx 2\omega_0$, show that

$$\omega - \omega_0 \approx \pm\frac{b}{2m}$$

(*c*) Show from Equation 12-56, that Q is related to b and m by

$$Q = \frac{m\omega_0}{b}$$

(*d*) Combine the results of (*b*) and (*c*) to show that there are two values of ω for which the power input is half that at resonance, and they are given by

$$\omega_1 = \omega_0 - \frac{\omega_0}{2Q} \quad \text{and} \quad \omega_2 = \omega_0 + \frac{\omega_0}{2Q}$$

Therefore, $\omega_2 - \omega_1 = \Delta\omega = \omega_0/Q$, which is equivalent to Equation 12-61.

Chapter 13

Waves on a String

A bass viol made by Henry Jaye in London, 1619.

Wave motion can be thought of as the transport of energy and momentum from one point in space to another without the transport of matter. In mechanical waves, such as waves on a string or sound waves in air, the energy and momentum are transported by means of a disturbance in the medium. A violin string is plucked or bowed, and the disturbance in the shape of the string is propagated along the string. At the same time, the vibrating string produces a slight change in the pressure of the adjacent air, and this pressure change is propagated as a sound wave through the air. In both cases, the disturbance is propagated because of the elastic properties of the medium. On the other hand, in electromagnetic waves (such as light, radio, television, or x rays) the energy and momentum are carried by electric and magnetic fields that can propagate through a vacuum. Light waves and other electromagnetic waves, which will be studied in Chapters 29 to 33, are produced by oscillating electric charges in atoms and molecules, or perhaps a transmitting radio antenna.

Although the variety of wave phenomena observed in nature is immense, many features are common to all kinds of waves, and others are shared by a wide range of wave phenomena. In this chapter, we will study waves on strings, which are common in our everyday experience and are easily visualized. In Chapter 14, we will study sound waves, which are also common in our everyday experience but are not as easily visualized. Many of the ideas and results discussed here will be applied later when we study more abstract kinds of waves such as light and other electromagnetic waves.

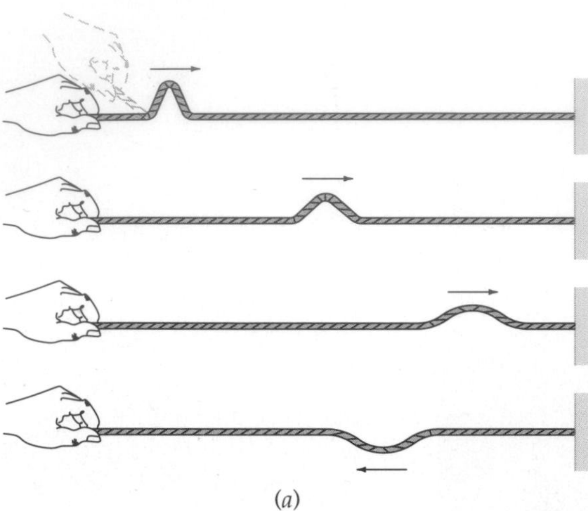

(a)

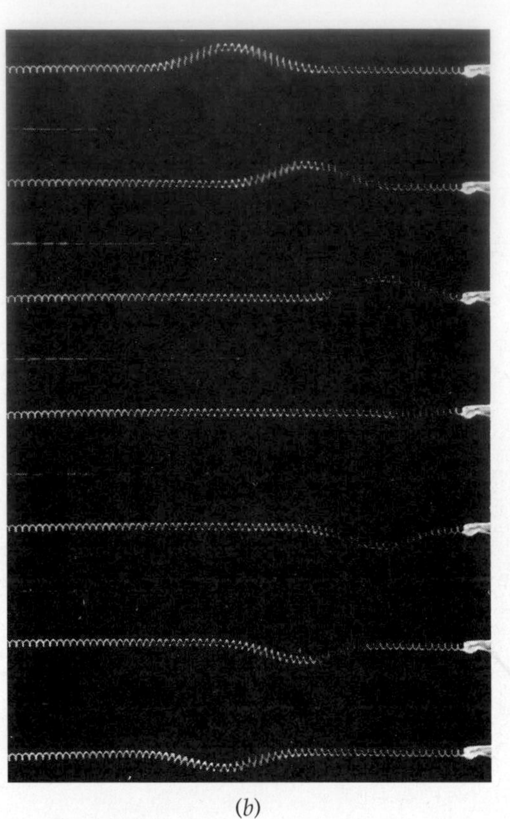

(b)

Figure 13-1 (*a*) A wave pulse moving to the right on a stretched string. When the pulse arrives at the rigid support, it is reflected and inverted. (*b*) Photograph of a similar pulse on a spring.

13-1 Wave Pulses

When a string (or spring) stretched under tension is given a flip, as in Figure 13-1, its shape changes over time in a regular way. The bump produced by the flip of the string travels down the string as a wave pulse. The disturbance in the medium, in this case, is the change in the shape of the string from its equilibrium shape, which is that of a taut string. The wave pulse travels down the string at a definite speed that depends on the tension in the string and on its mass density (mass per unit length). As it moves, the pulse may change shape. For example, it may gradually spread out. This effect is called **dispersion.** We will usually neglect dispersion in this chapter and assume that the wave pulse travels with an approximately constant shape. The fate of the pulse at the other end of the string depends on how the string is fastened there. If it is tied to a rigid support, as in Figure 13-1, the pulse will be reflected and will return inverted. When the pulse arrives at the fixed support, it exerts an upward force on the support. The fixed support exerts an equal and opposite downward force on the string, causing the pulse to invert upon reflection.

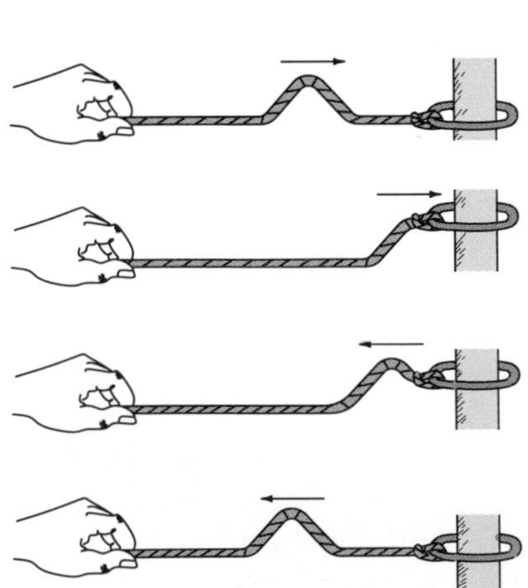

Figure 13-2 A free end for a string can be approximated by tying the string to a ring that moves on a frictionless post. A wave pulse traveling on the string is reflected at the free end without inversion.

In Figure 13-2, the string is tied to a smooth ring of negligible mass that is free to move vertically on a post. This arrangement approximates a free end for the string. When the pulse arrives, it exerts an upward force on the ring and the ring accelerates upward. The ring overshoots the height of the pulse, causing a reflected pulse that is not inverted.

If the string is tied to another string of different mass density, part of the pulse will be transmitted and part will be reflected. If the second string is

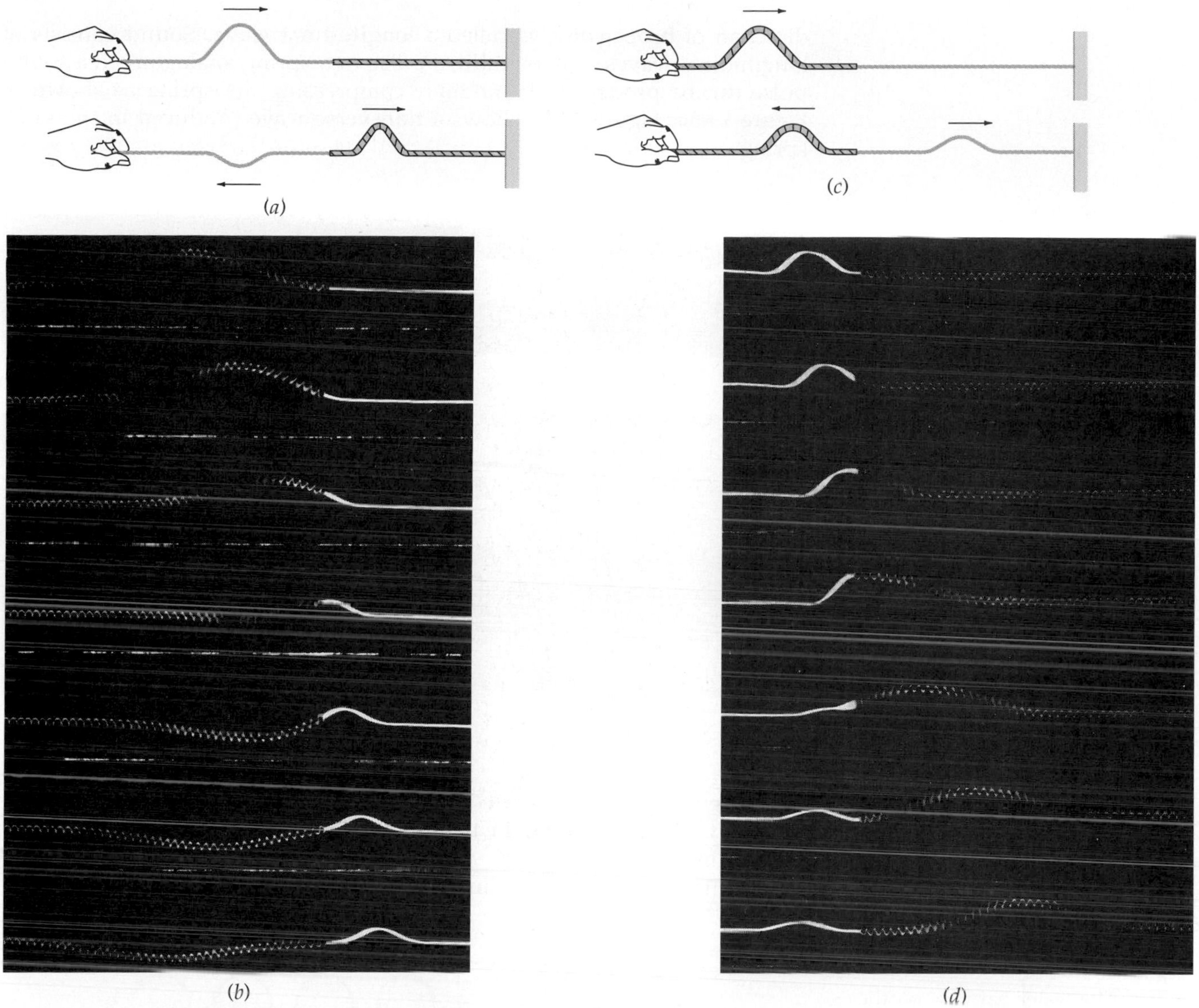

Figure 13-3 (*a*) A wave pulse traveling on a light string attached to a heavier string in which the wave speed is smaller. The reflected pulse is inverted, whereas the transmitted pulse is not. (*b*) Photograph of a similar pulse on a light spring attached to a heavier spring. (*c*) A wave pulse traveling on a heavy string attached to a light string in which the wave speed is greater. In this case, the reflected pulse is not inverted. (*d*) Photograph of a similar pulse on a heavy spring attached to a lighter spring.

heavier than the first, the reflected part of the pulse will be inverted (Figure 13-3*a*). If the second string is lighter, the reflected pulse will not be inverted (Figure 13-3*c*). In either case, the transmitted pulse is not inverted.

It is important to understand that it is not the mass elements of the string that are transported in wave motion but rather the disturbance in the shape of the string caused by flipping one end. The mass elements of the string, in fact, move in a direction perpendicular to the string and thus perpendicular to the direction of motion of the pulse (Figure 13-4). A wave in which the disturbance is perpendicular to the direction of propagation is called a **transverse wave.** A wave in which the disturbance is parallel to the

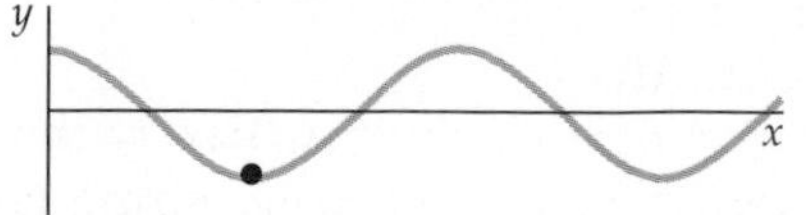

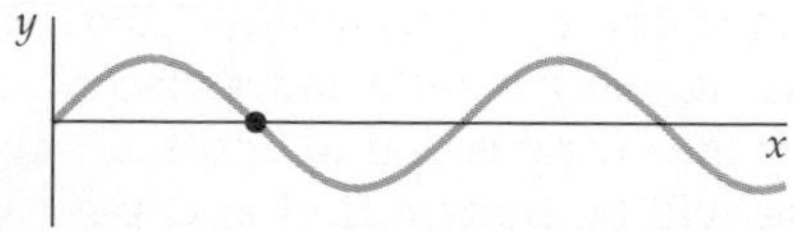

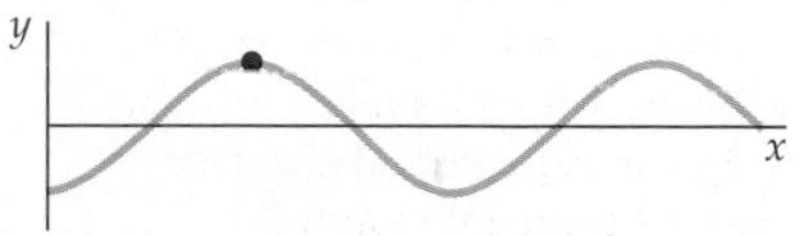

Figure 13-4 Three successive drawings of a wave on a string traveling to the right. An element of the string moves up and down in simple harmonic motion.

direction of propagation is called a **longitudinal wave.** Sound consists of longitudinal waves. A longitudinal pulse in a spring analogous to a sound pulse can be produced by suddenly compressing the spring as shown in Figure 13-5*a*. Figure 13-5*b* shows a transverse wave produced in the same spring.

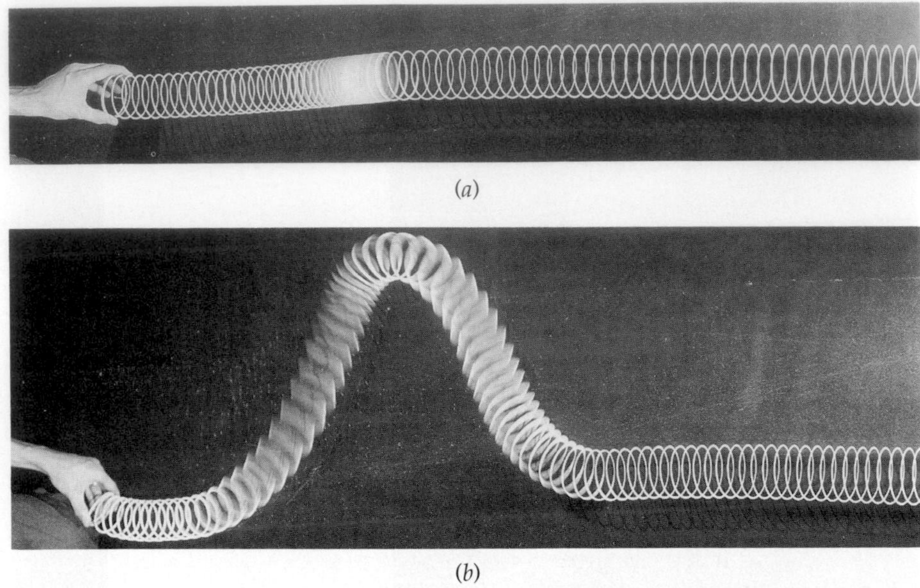

Figure 13-5 (*a*) Longitudinal wave pulse in a spring. The disturbance is in the direction of the motion of the wave. (*b*) Transverse wave pulse on the same spring. The disturbance is perpendicular to the direction of the motion of the wave.

Water waves are neither completely transverse nor completely longitudinal but a combination of the two. Figure 13-6 shows the motion of a water particle in a water wave. The water particle moves in a circular path with both transverse and longitudinal components.

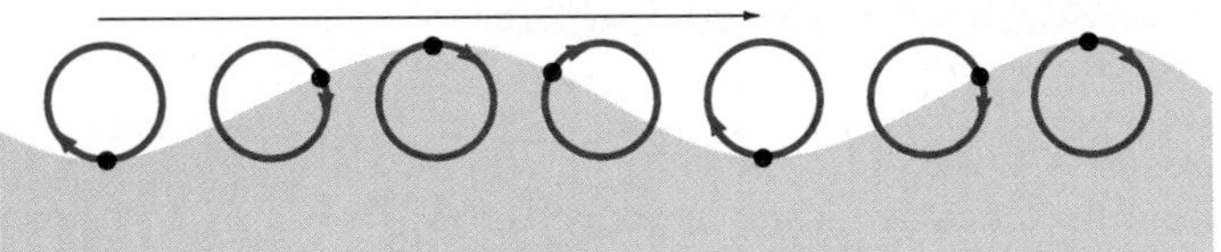

Figure 13-6 Surface waves on water. The water particles on the surface move in circular paths having both longitudinal and transverse components.

Figure 13-7 shows a pulse on a string at time $t = 0$. The shape of the string at this time can be represented by some function $y = y(x)$. At some later time, the pulse is farther down the string, so its shape then is some other function of x. We will neglect dispersion and assume that the pulse does not vary in shape. Let us introduce a new coordinate system with an origin O' that moves with the speed of the pulse v. In this reference frame,

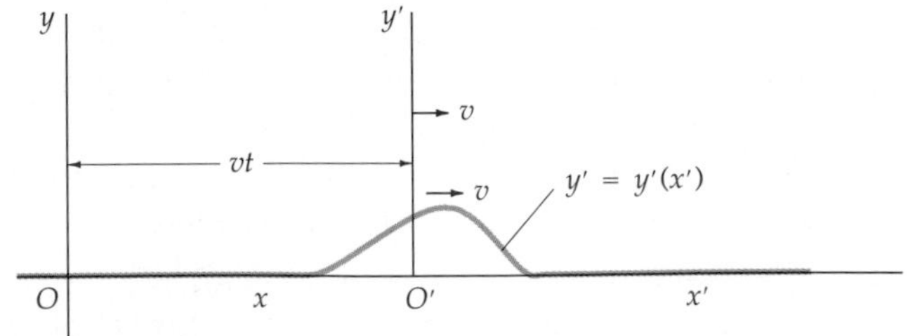

Figure 13-7 A wave pulse moving without change in shape in the positive x direction with a speed v relative to the origin O. In the primed coordinate system moving with the pulse, the wave function is $y' = y'(x')$ at all times. In the original, unprimed system, the wave function is $y = y(x - vt)$.

the pulse is stationary. The shape of the string is $y' = y'(x')$ for all times. The coordinates of the two reference frames are related by

$$y = y'$$

and

$$x = x' + vt$$

Thus, the shape of the string in the original frame O can be written

$$y = y(x - vt) \qquad \text{wave moving right} \qquad 13\text{-}1$$

This same line of reasoning for a pulse moving to the left leads to

$$y = y(x + vt) \qquad \text{wave moving left} \qquad 13\text{-}2$$

In both of these expressions, v is the speed of propagation of the wave. The function $y = y(x - vt)$ is called the **wave function.** For waves on a string, the wave function represents the transverse displacement of the string at point x and time t. Hence, the wave function is a function of the two variables x and t.

Figure 13-8 shows wave pulses moving in opposite directions on a string. The shape of the string when they meet can be found by adding the displacements produced by each pulse separately as shown in the figure. The property of wave motion that the resultant wave is the sum of two or more individual waves is called the **principle of superposition.** Mathematically, if $y_1(x - vt)$ is the wave function for the pulse moving to the right and $y_2(x + vt)$ is that for the pulse moving to the left, the total wave function when there are two pulses on the string is just the algebraic sum of the individual wave functions:

$$y(x, t) = y_1(x - vt) + y_2(x + vt) \qquad 13\text{-}3$$

The principle of superposition holds only for small wave pulses for which the height of the pulse is small compared with its length. In Section 13-8, we will see that the principle of superposition follows from the fact that the wave equation is linear for small transverse displacements.

In the special case where two pulses are identical except that one is inverted relative to the other, as in Figure 13-8b, there will be one time when the pulses exactly overlap and add to zero. At this time the string is horizontal, but it is not at rest. A short time later the pulses emerge, each continuing in its original direction. Superposition is a characteristic property of wave motion. There is no analogous situation in particle motion; that is, two particles never overlap or add together in this way. Superposition is unique to wave motion.

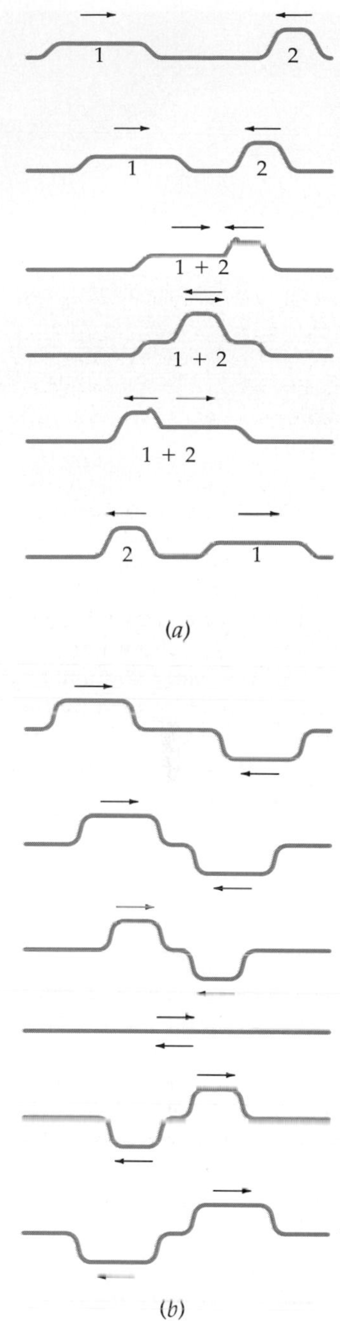

Figure 13-8 Wave pulses moving in opposite directions on a string. The shape of the string when the pulses meet is found by adding the displacements of each separate pulse. This property of wave motion is called superposition. (a) Superposition of pulses having displacements in the same direction. (b) Superposition of pulses having opposite displacements. Here the algebraic addition of the displacements amounts to the subtraction of the magnitudes.

Questions

1. Give some examples of wave pulses that occur in nature in addition to those mentioned in the text. In each case, explain what kind of disturbance from equilibrium occurs.
2. Consider a long line of cars equally spaced by one car length and moving slowly with the same speed. One car suddenly slows to avoid a dog and then speeds up until it is again one car length behind the car ahead. Discuss how the space between cars propagates back along the line. How is this like a wave pulse? Is there any transport of energy? What does the speed of propagation depend on?
3. When two waves moving in opposite directions superimpose as in Figure 13-8, does either impede the progress of the other?

Ocean waves breaking as they approach shallow water. The speed of a water wave depends on the depth of the water. As the depth of the water suddenly decreases, as evident here by the change in color, the back of the wave overtakes the front and the wave "breaks."

13-2 Speed of Waves

A general property of waves is that their speed depends on the properties of the medium but is independent of the motion of the source of the waves relative to the medium. For example, the speed of a wave on a string depends only on the properties of the string. Similarly, the speed of a sound wave produced by a train whistle depends only on the properties of the air and not on the motion of the train.

If we send wave pulses down a long rope, we can easily demonstrate that the speed of propagation of the wave pulses increases as we increase the tension in the rope. Futhermore, if we have two ropes, a light rope and a heavy rope, under the same tension, wave pulses will propagate more slowly in the heavy rope. Thus, the speed of propagation v of waves on a string or rope is related to the tension F and the mass per unit length μ. (We use F for tension here rather than T so as not to confuse the tension with the period T used in the next section.) We can obtain this relation using Newton's laws.

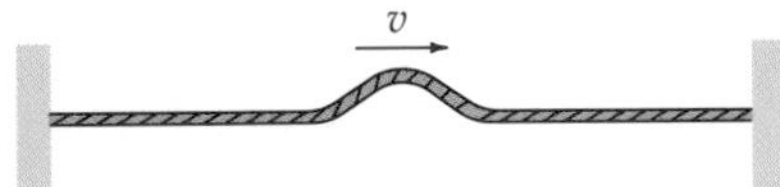

Figure 13-9 A wave pulse moving with a speed v along a string. If the pulse is small enough, the tension in the string does not change.

Figure 13-9 shows a pulse traveling along a string with a speed v to the right. We assume that the pulse is small compared to the length of the string so that, to a good approximation, the tension is constant along the string and is the same as when there is no pulse. It is convenient to consider the pulse in a reference frame moving with speed v to the right. In this frame, the pulse is stationary and the string moves with a speed v to the left. Figure 13-10 shows a small segment of the string of length Δs. If the segment is small enough, we can consider it to be part of a circular arc of radius R. The segment is thus moving in a circle of radius R with a speed v, and it has a centripetal acceleration v^2/R. Let the angle subtended by the string be θ:

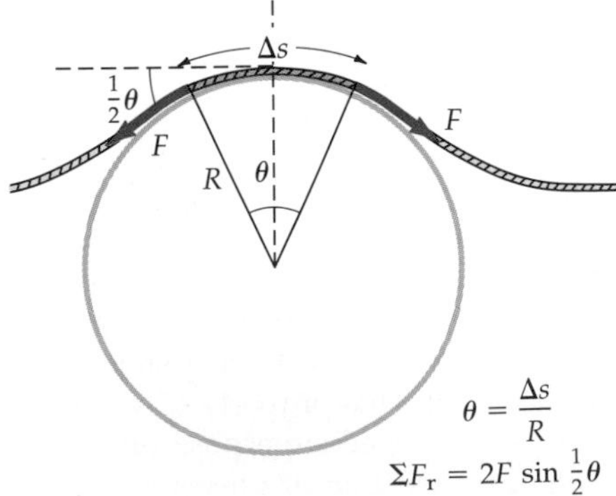

Figure 13-10 In a frame in which the wave pulse of Figure 13-9 is at rest, the string is moving with a speed v to the left. A small segment of the string of length Δs is moving in a circular arc of radius R. The centripetal acceleration of the segment is provided by the radial components of the tension.

$$\theta = \frac{\Delta s}{R}$$

The forces acting on the segment are the tension F at each end. The horizontal components of these forces are equal and opposite and thus cancel. The vertical components of these forces point toward the center of the circular arc. The sum of these radial forces provides the centripetal acceleration. The net radial force acting on the segment is

$$\Sigma F_r = 2F \sin \tfrac{1}{2}\theta \approx 2F(\tfrac{1}{2}\theta) = F\theta$$

where we have assumed that θ is small enough to permit the approximation $\sin \frac{1}{2}\theta \approx \frac{1}{2}\theta$. If μ is the mass per unit length of the string, the mass of a segment of length $\Delta s = R\theta$ is

$$m = \mu \, \Delta s = \mu R \theta$$

Setting the net radial force equal to the mass times the centripetal acceleration gives

$$F\theta = \mu R\theta \frac{v^2}{R}$$

Canceling the common factor θ and solving for v, we obtain

$$v = \sqrt{\frac{F}{\mu}}$$

Since this speed is independent of R and θ, this result holds for all segments of the string. However, the derivation depends on angle θ being small, which will be true if the height of the pulse is small compared to its length or, alternatively, if the slope of the string at any point is small.

In the original frame, the string is fixed, and the pulse moves with speed

$$v = \sqrt{\frac{F}{\mu}} \qquad 13\text{-}4$$

Speed of waves on a string

The dependence of the speed on F/μ is not surprising when we realize that it is essentially the tension that accelerates the mass elements in a rope to provide the wave pulse.

Example 13-1

The tension in a string is provided by hanging an object of mass 3 kg at one end as shown in Figure 13-11. The length of the string is 2.5 m and its mass is 50 g. What is the speed of waves on the string?

The tension in the string is

$$F = mg = (3 \text{ kg})(9.81 \text{ N/kg}) = 29.4 \text{ N}$$

The mass per unit length is

$$\mu = \frac{m}{L} = \frac{0.05 \text{ kg}}{2.5 \text{ m}} = 0.02 \text{ kg/m}$$

The speed is therefore

$$v = \sqrt{\frac{F}{\mu}} = \sqrt{\frac{29.4 \text{ N}}{0.02 \text{ kg/m}}} = 38.3 \text{ m/s}$$

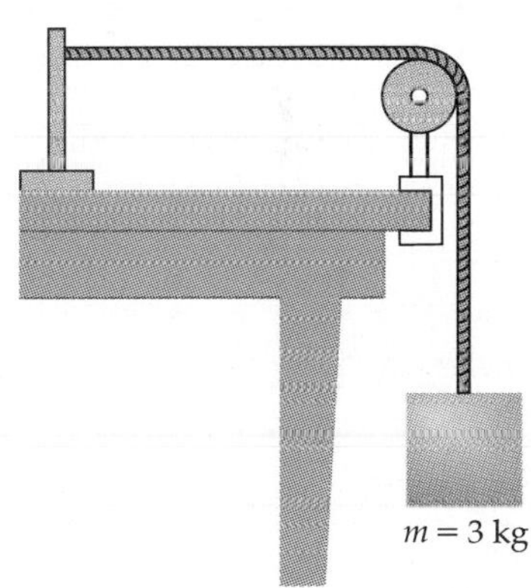

Figure 13-11 Example 13-1. The tension in the string is provided by the weight of the block.

Exercise

If the 3-kg mass in Example 13-1 is replaced with a 6-kg mass, what is the speed of the waves on the string? (Answer: 54.2 m/s)

Question

4. Two strings are stretched between the same two posts. One weighs twice as much as the other. How should their tensions be adjusted so that waves travel along each at the same speed?

13-3 Harmonic Waves

If we move one end of a string up and down in simple harmonic motion (for example, by attaching it to a vibrating tuning fork), a sinusoidal wave train propagates along the string. Such a wave is called a **harmonic wave.** The shape of the string at some instant in time is that of a sine function, as shown in Figure 13-12. (Again, whether this is a sine function or a cosine function depends merely on where the origin is chosen on the x axis.) Such a figure can be obtained by taking a snapshot of the string. The distance between two successive wave crests is called the **wavelength** λ. The wavelength is the distance in space after which the wave repeats itself. As the wave propagates down the string, each point on the string moves up and down, that is, perpendicular to the direction of propagation, in simple harmonic motion with the frequency f of the tuning fork or whatever agent is driving the end of the string. There is a simple relation among the frequency f, the wavelength λ, and the speed v of the harmonic wave. During one period $T = 1/f$, the wave moves a distance of one wavelength, so the speed is given by

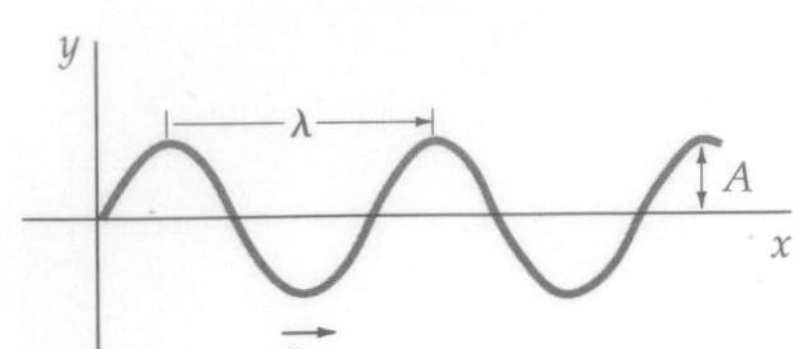

Figure 13-12 Harmonic wave at some instant in time. A is the amplitude and λ is the wavelength. For waves on a string, this figure can be obtained by taking a snapshot of the string.

$$v = \frac{\lambda}{T} = f\lambda \qquad 13\text{-}5$$

Since Equation 13-5 arises simply from the definitions of wavelength and frequency, it applies to all types of harmonic waves. Because the speed of wave propagation is determined by the properties of the medium, the wavelength is determined by the frequency of the source through $\lambda = v/f$. The greater the frequency, the smaller the wavelength.

The sine function that describes the displacement shown in Figure 13-12 is

$$y(x) = A \sin kx \qquad 13\text{-}6$$

where A is the **amplitude** and k is a constant called the **wave number.** The wave number is related to the wavelength. If we move from some point x_1 to another point a distance one wavelength away, $x_2 = x_1 + \lambda$, the argument of the sine function changes by 2π. We thus have

$$k(x_1 + \lambda) = kx_1 + 2\pi$$

$$k\lambda = 2\pi$$

or

$$k = \frac{2\pi}{\lambda} \qquad 13\text{-}7$$

To describe a wave that is traveling to the right with a speed v, we replace x in Equation 13-6 with $x - vt$ as we did with wave pulses in Section 13-1. Thus, the wave function for a wave traveling with a speed v to the right is written

$$y(x, t) = A \sin k(x - vt) = A \sin (kx - kvt)$$

or

Harmonic wave function

$$y(x, t) = A \sin (kx - \omega t) \qquad 13\text{-}8$$

where

$$\omega = kv \qquad 13\text{-}9$$

is the angular frequency, which is related to the frequency f and the period T, as in Chapter 12, by

$$\omega = 2\pi f = \frac{2\pi}{T} \qquad 13\text{-}10$$

Substituting $\omega = 2\pi f$ into Equation 13-9 and using $k = 2\pi/\lambda$, we obtain

$$2\pi f = kv = \frac{2\pi}{\lambda} v$$

or

$$v = f\lambda$$

which is Equation 13-5. Equation 13-8 can also be written in terms of the period T and the wavelength λ. Substituting in $k = 2\pi/\lambda$ and $\omega = 2\pi/T$, we obtain

$$y(x, t) = A \sin \left[2\pi\left(\frac{x}{\lambda} - \frac{t}{T}\right)\right] \qquad 13\text{-}11$$

Harmonic wave function in terms λ and T

From this form of the equation we can see that if t advances by one period T or if x advances by one wavelength λ, the argument of the sine function changes by 2π and $y(x, t)$ has the same value as before.

Note that a string segment at a fixed point x moves up and down with simple harmonic motion with angular frequency ω.

Example 13-2

The wave function for a harmonic wave on a string is

$$y(x, t) = 0.03 \sin (2.2x - 3.5t)$$

where y and x are in meters and t is in seconds. Find the amplitude, the wavelength, the frequency, the period, and the speed of the wave.

Comparing this wave function with Equation 13-8, we can see that the amplitude $A = 0.03$ m, the wave number $k = 2.2\ \text{m}^{-1}$, and the angular frequency $\omega = 3.5\ \text{s}^{-1}$. Then the wavelength $\lambda = 2\pi/k = 2.86$ m, and the period $T = 2\pi/\omega = 1.80$ s. The speed of the wave is therefore

$$v = f\lambda = \frac{\lambda}{T} = \frac{2.86\ \text{m}}{1.80\ \text{s}} = 1.59\ \text{m/s}$$

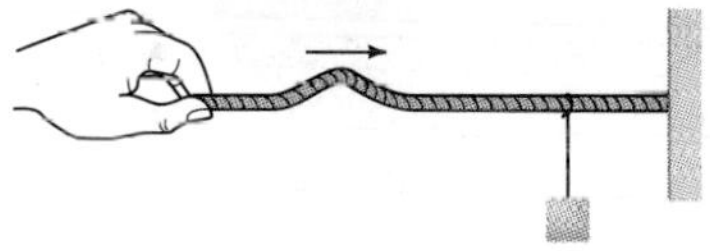

Figure 13-13 That waves transmit energy is indicated by the upward motion of the weight when the wave pulse arrives.

13-4 Energy Transmitted by Waves

As a wave moves along a string, energy is transmitted. We can demonstrate that energy is transported by a wave pulse by hanging a weight on a string under tension and giving the string a flip at one end (Figure 13-13). When the pulse arrives at the weight, the weight is momentarily lifted. Thus, the energy introduced by the flip at the end of the string is transmitted along the string and is received by the weight.

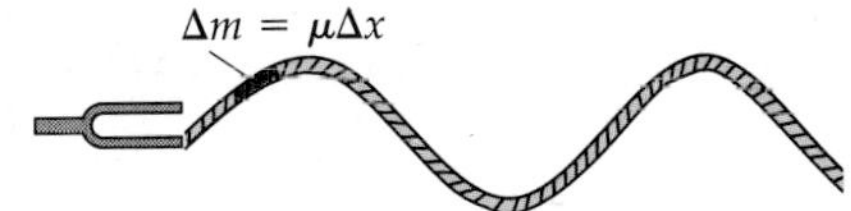

Figure 13-14 The vibrating tuning fork at the left generates a harmonic wave that travels to the right along the string. A segment of the string of mass $\Delta m = \mu\ \Delta x$ is oscillating perpendicular to the string with simple harmonic motion. The energy of this segment is the energy of simple harmonic motion, which is $\frac{1}{2}kA^2 = \frac{1}{2}(\Delta m)\ \omega^2 yA^2$, where A is the amplitude and $k = \Delta m\ \omega^2$ is the effective force constant.

Figure 13-14 shows a harmonic wave of amplitude A and angular frequency ω generated at the left by a vibrating tuning fork. Each element of the string is undergoing simple harmonic motion of amplitude A and angular frequency ω. In Chapter 12, we saw that the total energy of a mass oscillating on a spring is $\frac{1}{2}kA^2$, where $k = m\omega^2$ is the force constant and A is the amplitude. (Do not confuse the force constant k with the wave number k.) Applying this to a segment of the string of mass $\Delta m = \mu\ \Delta x$ oscillating perpendicular to the string with an amplitude A, we have

$$\Delta E = \tfrac{1}{2}(\Delta m)\omega^2 A^2 = \tfrac{1}{2}\mu\omega^2 A^2\ \Delta x \qquad 13\text{-}12$$

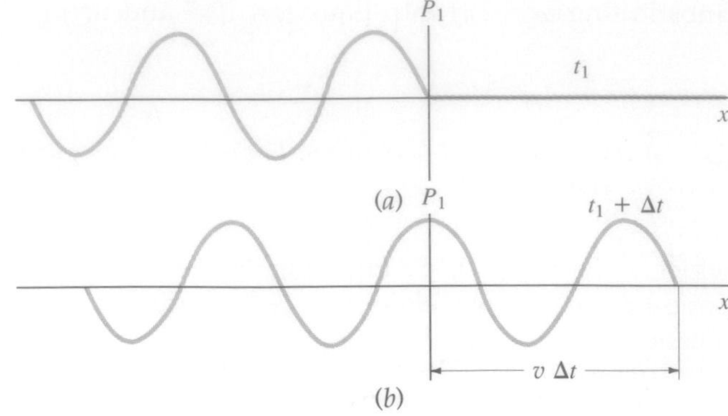

Figure 13-15 Energy transmitted by a wave on a string. (*a*) At time t_1, the wave has just reached point P_1. The string to the left of this point has energy due to the simple harmonic motion of its elements, whereas the string to the right of P_1 has no energy. (*b*) In an additional time Δt, the wave travels a distance $v\ \Delta t$. The energy transmitted past point P_1 in time Δt is the energy in the length $v\ \Delta t$.

In Figure 13-15*a*, a harmonic wave traveling to the right has just reached point P_1 at time t_1. Each segment of string Δx to the left of P_1 has an energy ΔE given by Equation 13-12. After a time Δt, the wave travels a distance $\Delta x = v\ \Delta t$ as shown in Figure 13-15*b*. The energy transmitted past point P_1 during this time is thus

$$\Delta E = \tfrac{1}{2}\mu\omega^2 A^2\ \Delta x = \tfrac{1}{2}\mu\omega^2 A^2 v\ \Delta t \qquad 13\text{-}13$$

The rate at which energy is transmitted is the power transmitted:

Power transmitted by a harmonic wave

$$P = \frac{dE}{dt} = \tfrac{1}{2}\mu\omega^2 A^2 v \qquad 13\text{-}14$$

Thus, the power transmitted is proportional to the square of the amplitude, to the square of the frequency, and to the wave speed.

Example 13-3

Waves of wavelength 35 cm and amplitude 1.2 cm move along a 15-m string that has a mass of 80 g and a tension of 12 N. (*a*) What is the total energy of the waves in the string? (*b*) Find the power transmitted past a given point on the string.

(*a*) The linear mass density of the string is $\mu = m/L = (0.08\ \text{kg})/(15\ \text{m}) = 5.33 \times 10^{-3}\ \text{kg/m}$. The speed of the waves on the string is then

$$v = \sqrt{\frac{F}{\mu}} = \sqrt{\frac{12\ \text{N}}{5.33 \times 10^{-3}\ \text{kg/m}}} = 47.4\ \text{m/s}$$

The angular frequency of the waves is

$$\omega = 2\pi f = 2\pi\frac{v}{\lambda} = 2\pi\frac{47.4\ \text{m/s}}{0.35\ \text{m}} = 851\ \text{rad/s}$$

The total energy of the waves on the string is then given by Equation 13-13:

$$\Delta E = \tfrac{1}{2}\mu\omega^2 A^2\ \Delta x$$
$$= \tfrac{1}{2}(5.33 \times 10^{-3}\ \text{kg/m})(851\ \text{rad/s})^2(0.012\ \text{m})^2(15\ \text{m}) = 4.17\ \text{J}$$

(*b*) The power transmitted past a point on the string is

$$P = \tfrac{1}{2}\mu\omega^2 A^2 v$$
$$= \tfrac{1}{2}(5.33 \times 10^{-3}\ \text{kg/m})(851\ \text{rad/s})^2(0.012\ \text{m})^2(47.4\ \text{m/s}) = 13.2\ \text{W}$$

13-5 Superposition and Interference of Harmonic Waves

The superposition of harmonic waves is called **interference.** The interference of harmonic waves depends on the phase difference between the waves. Let $y_1(x, t)$ be the wave function for a harmonic wave traveling to the right with amplitude A, angular frequency ω, and wave number k:

$$y_1 = A \sin (kx - \omega t) \tag{13-15}$$

For this wave function, we have chosen $t = 0$ when the displacement is zero at $x = 0$. [This choice is convenient, but it is not mandatory. If, for example, we chose $t = 0$ when the displacement was maximum at $x = 0$, we would write $y_1 = A \cos (kx - \omega t) = A \sin (kx - \omega t - \pi/2)$.] If we have a second harmonic wave that is also traveling to the right with the same amplitude, frequency, and wave number and we have already chosen the zero time, the general equation for its wave function can be written

$$y_2 = A \sin (kx - \omega t + \delta) \tag{13-16}$$

where δ is the phase constant. The two waves described by Equations 13-15 and 13-16 differ in phase by δ. Figure 13-16 shows a plot of two wave functions versus position for a fixed time $t = 0$. The resultant wave is the sum

$$y_1 + y_2 = A \sin (kx - \omega t) + A \sin (kx - \omega t + \delta) \tag{13-17}$$

If the phase difference is zero, that is, if the waves are in phase, the resultant wave has an amplitude twice that of either wave:

$$y_1 + y_2 = 2A \sin (kx - \omega t) \qquad \delta = 0$$

This type of superposition is called **constructive interference** (Figure 13-17). On the other hand, if the phase difference is π rad $= 180°$, the waves are out of phase. Then

$$y_1 + y_2 = A \sin (kx - \omega t) + A \sin (kx - \omega t + \pi)$$
$$= A \sin (kx - \omega t) - A \sin (kx - \omega t)$$

or

$$y_1 + y_2 = 0 \qquad \delta = \pi \text{ rad} = 180°$$

where we have used

$$A \sin (kx - \omega t + \pi) = -A \sin (kx - \omega t)$$

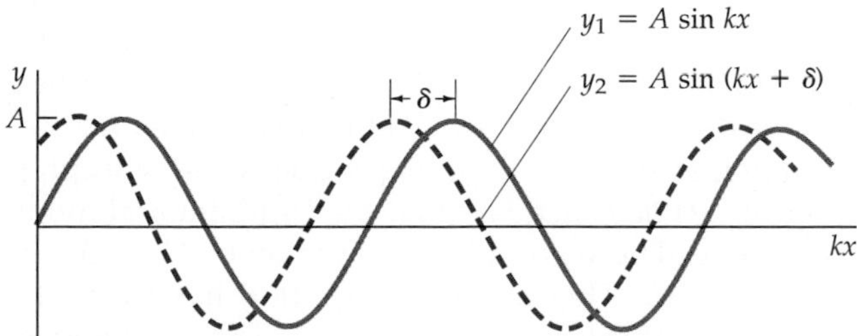

Figure 13-16 Displacement versus position for two harmonic waves having the same amplitude, frequency, and wavelength but differing in phase by δ.

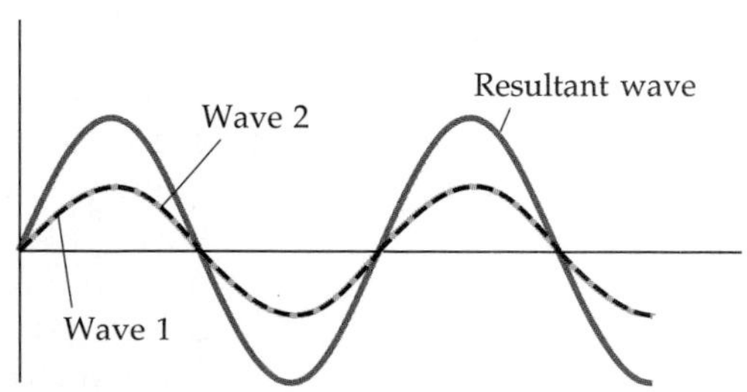

Figure 13-17 Constructive interference. When two waves are in phase, they add. The amplitude of the resultant wave is the sum of the amplitudes of the individual waves. For waves of equal amplitude, as shown here, the resultant amplitude is twice that of either wave.

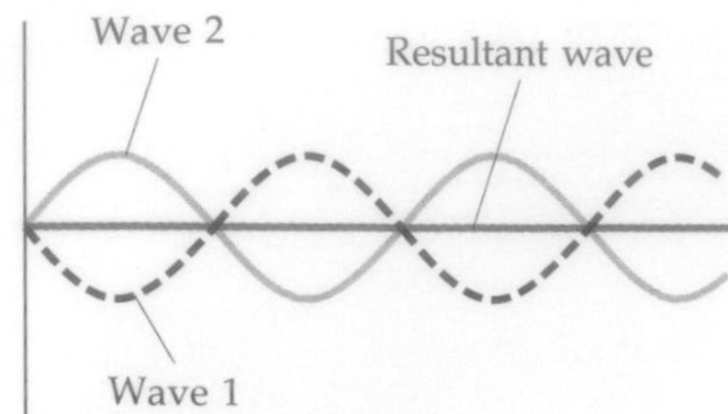

Figure 13-18 Destructive interference. When two waves have a phase difference of π, the amplitude of the resultant wave is the difference of the amplitudes of the individual waves. If the original waves have equal amplitudes, they cancel completely, giving zero for the amplitude of the resultant wave.

When the waves differ in phase by 180° (π rad) or any odd integer times 180°, the waves cancel if their amplitudes are equal. This type of superposition is called **destructive interference** (Figure 13-18).

For a general phase difference δ, we can simplify Equation 13-17 by using the trigonometric identity

$$\sin \theta_1 + \sin \theta_2 = 2 \cos \tfrac{1}{2}(\theta_1 - \theta_2) \sin \tfrac{1}{2}(\theta_1 + \theta_2) \qquad 13\text{-}18$$

For this case, $\theta_1 = kx - \omega t$ and $\theta_2 = kx - \omega t + \delta$ so that

$$\tfrac{1}{2}(\theta_1 - \theta_2) = -\tfrac{1}{2}\delta$$

and

$$\tfrac{1}{2}(\theta_1 + \theta_2) = kx - \omega t + \tfrac{1}{2}\delta$$

Thus, Equation 13-17 becomes

$$y_1 + y_2 = 2A \cos \tfrac{1}{2}\delta \sin (kx - \omega t + \tfrac{1}{2}\delta) \qquad 13\text{-}19$$

where we have used $\cos(-\tfrac{1}{2}\delta) = \cos \tfrac{1}{2}\delta$. From Equation 13-19, we see that, in general, the superposition of two harmonic waves results in another harmonic wave having the same wave number and frequency. It differs in phase from both the original waves, and its amplitude is $2A \cos \tfrac{1}{2}\delta$. If the two waves are in phase, the phase difference $\delta = 0$, $\cos 0 = 1$, and the amplitude of the resultant wave is $2A$. If they are 180° out of phase, $\delta = \pi$ rad, $\cos(\pi/2) = 0$, and the amplitude of the resultant wave is zero.

Exercise

Two waves with the same frequency, wavelength, and amplitude are traveling in the same direction. If they differ in phase by $\pi/2$ and each has amplitude of 4.0 cm, what is the amplitude of the resultant wave? (Answer: 5.66 cm)

13-6 Standing Waves

When waves are confined in space, like waves on a piano string, there are reflections at both ends, and there are therefore waves traveling in both directions. These waves combine according to the principle of superposition. For a given string, there are certain frequencies for which superposition results in a stationary vibration pattern called a **standing wave.**

String Fixed at Both Ends

If we fix both ends of a long string and move a portion of the string up and down with simple harmonic motion of small amplitude, we find that at certain frequencies standing-wave patterns such as those shown in Figure 13-19 are produced. The frequencies that produce these patterns are called the **resonance frequencies** of the string system. The lowest resonance frequency is called the fundamental frequency f_1. It produces the standing-wave pattern shown in Figure 13-19*a* which is called the **fundamental** mode of vibration or the **first harmonic.** The second lowest frequency f_2 produces the pattern shown in Figure 13-19*b*. This mode of vibration has twice the fundamental frequency and is called the second harmonic. The third lowest frequency f_3 is three times the fundamental frequency, and it produces the third harmonic pattern shown in Figure 13-19*c*. We note from Figure 13-19 that, for each harmonic, there are certain points on the string that do not move. For example, the midpoint in Figure 13-19*b* does not move. Such points are called **nodes.** Midway between each pair of nodes is a point of maximum amplitude of vibration called an **antinode.** Both fixed ends of the string are, of course, nodes. If one end is attached to a tuning fork rather

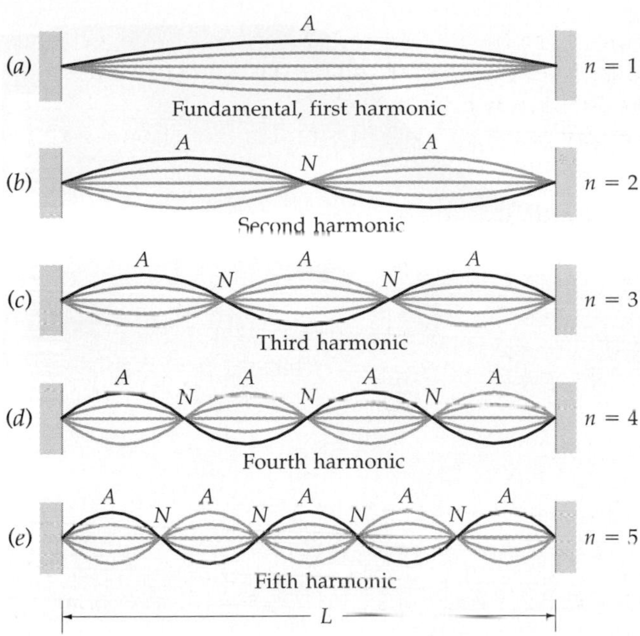

Figure 13-19 Standing waves on a string that is fixed at both ends. The points labeled A are antinodes and those labeled N are nodes. In general, the nth harmonic has n antinodes.

than being fixed, it will still be approximately a node because the amplitude of the vibration at that end is much smaller than the amplitude at the antinodes. We note that the fundamental or first harmonic has one antinode, the second harmonic has two antinodes, and so on.

We can relate the resonance frequencies to the wave speed in the string and the length of the string. We can see from Figure 13-19 that the length of the string L equals one-half wavelength for the first harmonic, two half wavelengths for the second harmonic, $\frac{3}{2}\lambda$ for the third harmonic, and so forth. In general, for the nth harmonic, we have

$$L = n\frac{\lambda_n}{2} \qquad n = 1, 2, 3, \ldots \qquad \text{13-20}$$

Standing-wave condition, both ends fixed

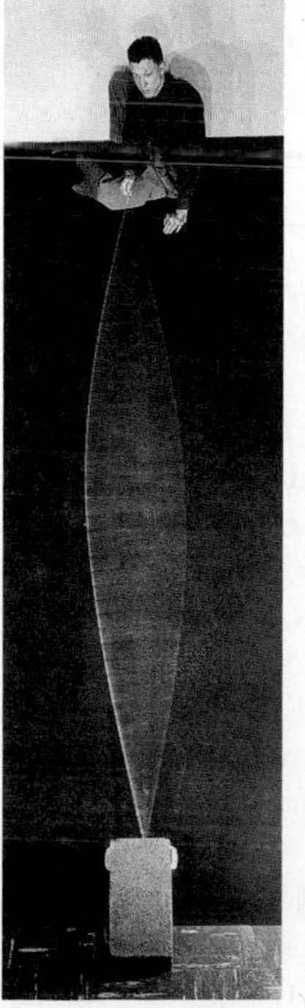
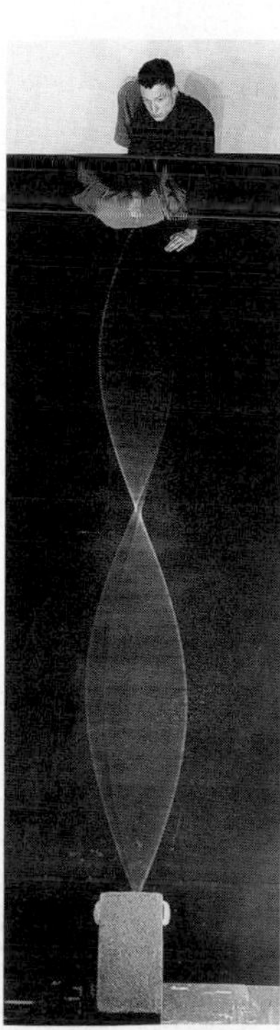
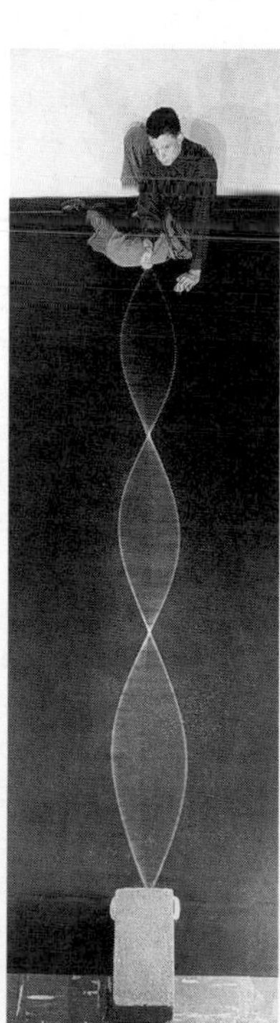
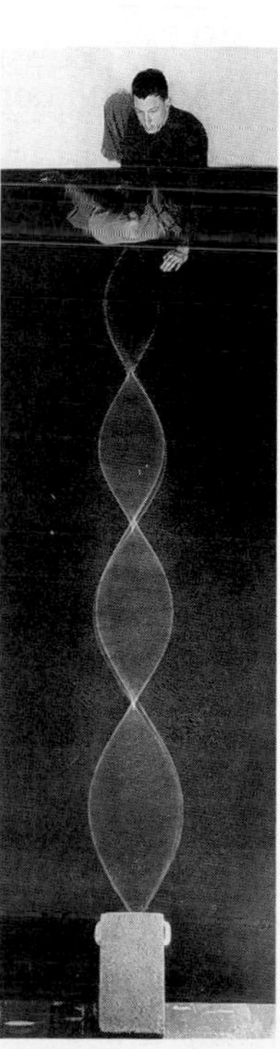

The first four harmonics of standing waves on a string fixed at both ends.

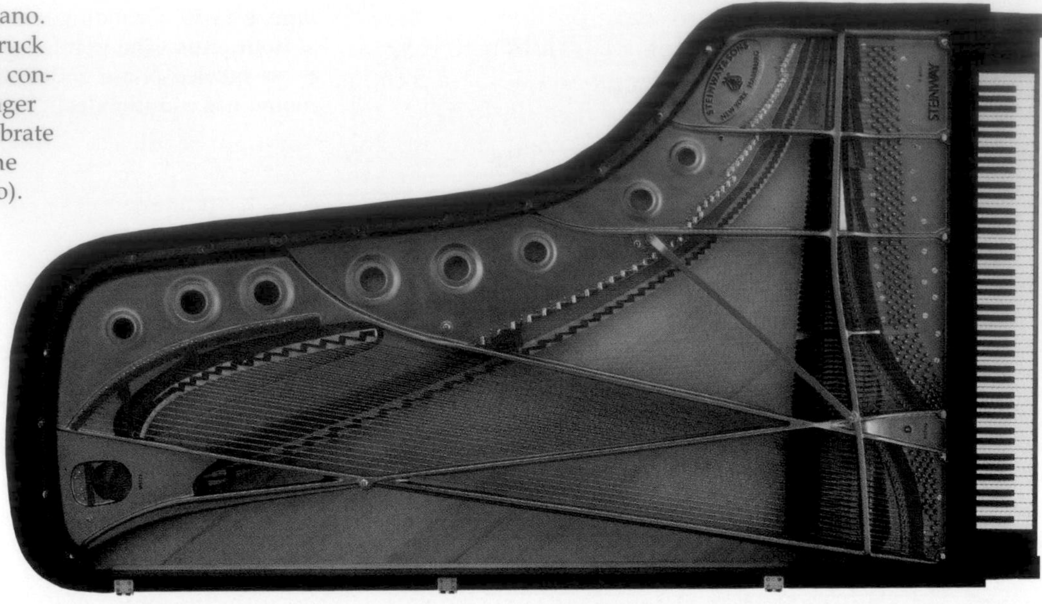

A classic Steinway grand piano. The strings vibrate when struck by the hammers, which are controlled by the keys. The longer strings (bottom of photo) vibrate at lower frequencies than the shorter strings (top of photo).

This result is known as the **standing-wave condition.** We can find the frequency of the nth harmonic from the fact that the wave speed v equals the frequency f times the wavelength. Thus,

$$f_n = \frac{v}{\lambda_n} = \frac{v}{(2L/n)}$$

or

Resonance frequencies, both ends fixed

$$f_n = n\frac{v}{2L} = nf_1 \qquad n = 1, 2, 3, \ldots \qquad 13\text{-}21$$

where

$$f_1 = \frac{v}{2L}$$

is the fundamental frequency. Since the wave speed v is $\sqrt{F/\mu}$, where F is the tension in the string and μ is its mass per unit length,

$$f_1 = \frac{1}{2L}\sqrt{\frac{F}{\mu}} \qquad 13\text{-}22$$

An easy way to remember the resonance frequencies given by Equation 13-21 is to sketch Figure 13-19 to remind yourself of the standing-wave condition $L = n\lambda_n/2$ and then use $f = v/\lambda$.

We can understand the production of standing waves in terms of resonance. Let us consider a string of length L that is attached at one end to a vibrating tuning fork, or some other vibrating body, and is fixed at the other end. The first wave sent out by the vibrating tuning fork travels down the string at a distance L to the fixed end, where it is reflected and inverted. It then travels back to the tuning fork, where it is again reflected at the tuning fork. It is again inverted and starts back down the string. If the time it takes for the wave to travel the complete distance $2L$ is exactly equal to the period of the vibrating fork, the twice-reflected wave will exactly overlap the second wave produced by the fork. The two waves will interfere constructively, which means that they will add together to produce a wave having twice the amplitude of either wave. The combined wave will travel down the string and back and add to the third wave produced by the fork, and so on. The amplitude of the combined wave will continue to grow as the string absorbs energy from the tuning fork. Various damping effects, such as the loss of energy during reflection or the imperfect flexibility of the string, put a limit on the maximum amplitude that can be reached. Still, this maximum amplitude is much larger than that of the tuning fork. Thus the tuning fork is in resonance with the string. Resonance will also occur if the time it takes for the first wave to travel the distance $2L$ is twice the period of the vibrating tuning fork or any integer n times the period. Since the time for a wave to travel the distance $2L$ is $2L/v$, where v is the wave speed, we can write the resonance condition as

$$\frac{2L}{v} = nT = \frac{n}{f}$$

or

$$f = n\frac{v}{2L}$$

where $T = 1/f$ is the period. This is the same result we found by fitting an integral number of half-wavelengths into the distance L.

The resonance frequencies given by Equation 13-21 are also called the **natural frequencies** of the string. When the frequency of the tuning fork is not one of the natural frequencies of the vibrating string, standing waves are not produced. After the first wave has traveled the distance $2L$ and is reflected from the fork, it differs in phase from the wave being generated at the fork (Figure 13-20). The reflected wave combines with this new wave to produce a resultant wave whose amplitude may be greater or smaller than that of the original wave, depending on the phase difference. When this resultant wave has traveled the distance $2L$ and is again reflected at the fork, it will differ in phase from the next wave generated. In some cases, the new resultant wave will have a greater amplitude than the first, whereas in other cases the new amplitude will be less. On the average, however, the amplitude will not increase but will remain on the order of the amplitude of the first wave generated, which is the amplitude of the tuning fork and is very small compared with the amplitudes attained at resonance frequencies.

Figure 13-20 Waves on a string produced by a tuning fork whose frequency is not in resonance with the natural frequencies of the string. The wave leaving the tuning fork for the first time (dashed line) and that leaving for the second time after being reflected twice (grey line) are not in phase and so do not interfere constructively. The resultant wave (black line) has about the same amplitude as the individual waves, which is about the amplitude of the fork.

Only when the frequency of the tuning fork equals one of the natural frequencies of the string given by Equation 13-21 will the waves add in phase and the amplitude build up. This resonance phenomenon is analogous to the resonance of a simple harmonic oscillator with a harmonic driving force we discussed in Section 12-8. There we saw that if the frequency of the driving force equals the natural frequency of a simple harmonic oscillator, the oscillator absorbs the maximum amount of energy from the driving force. Note, however, that a vibrating string has not just one natural frequency but a sequence of natural frequencies that are integral multiples of the fundamental frequency. This sequence is called a **harmonic series.** The first frequency f_1 is termed the first harmonic, the second frequency $f_2 = 2f_1$ the second harmonic, and so on. Each vibration pattern is called a **mode** of vibration. In the terminology often used in music, the second harmonic is called the first **overtone,** the third harmonic the second overtone, and so on.

Turbulent winds set up standing waves in the Tacoma Narrows suspension bridge, leading to its collapse on November 7, 1940, just four months after it had been opened for traffic.

Example 13-4

A string is stretched between two fixed supports 1 m apart and the tension is adjusted until the fundamental frequency of the string is 440 Hz. What is the speed of transverse waves on the string?

From the standing-wave condition (Equation 13-20), the wavelength for the first harmonic is $\lambda = 2L = 2$ m. Hence, the wave speed is

$$v = \lambda f = (2\text{ m})(440\text{ Hz}) = 880\text{ m/s}$$

Exercise

The speed of transverse waves on a stretched string is 200 m/s. If the string is 5 m long, find the frequencies of the fundamental and the second and third harmonics. (Answer: $f_1 = 20$ Hz, $f_2 = 40$ Hz, $f_3 = 60$ Hz)

Example 13-5

A string of length 3 m and mass density 0.0025 kg/m is fixed at both ends. One of its resonance frequencies is 252 Hz. The next higher resonance frequency is 336 Hz. Find (*a*) the fundamental frequency and (*b*) the tension in the string.

(*a*) According to Equation 13-21, each resonance frequency is an integer times the fundamental frequency. The ratio of two successive resonance frequencies must then be the ratio of two successive integers. The ratio of the frequencies given is $\frac{336}{252} = 1.33$, which is the ratio of 4 to 3. Thus 336 Hz must be the fourth harmonic and 252 Hz must be the third harmonic. The fundamental frequency is thus

$$f_1 = \frac{1}{3}f_3 = \frac{252\text{ Hz}}{3} = 84\text{ Hz}$$

(*b*) We find the tension in the string by first finding the wave speed. Since the wavelength of the fundamental is $\lambda = 2L = 6$ m, the wave speed is $v = f\lambda = (84\text{ Hz})(6\text{ m}) = 504$ m/s. The tension is then found from

$$v = \sqrt{\frac{F}{\mu}}$$

$$F = \mu v^2 = (0.0025\text{ kg/m})(504\text{ m/s})^2 = 635\text{ N}$$

String Fixed at One End

Standing waves can also be produced on a string that has one end free instead of both ends fixed (Figure 13-21). The standing-wave patterns for such a string are shown in Figure 13-22. Note that the free end is an antinode. In the fundamental mode of vibration, the length of the string L equals $\lambda/4$. In the next highest mode, $L = 3\lambda/4$. The standing-wave condition can thus be written

$$L = n\frac{\lambda}{4} \qquad n = 1, 3, 5, \ldots \tag{13-23}$$

Standing-wave condition, one end free

The resonance frequencies are therefore given by

$$f_n = n\frac{v}{4L} = nf_1 \qquad n = 1, 3, 5, \ldots \tag{13-24}$$

Resonance frequencies, one end free

where

$$f_1 = \frac{v}{4L} \tag{13-25}$$

is the fundamental frequency. The natural frequencies of this system occur in the ratios 1:3:5:7: . . . , which means that the even harmonics are missing. Again, an easy way to remember the resonance frequencies given by Equation 13-24 is to sketch Figure 13-22 to remind yourself of the standing-wave condition and use $f = v/\lambda$.

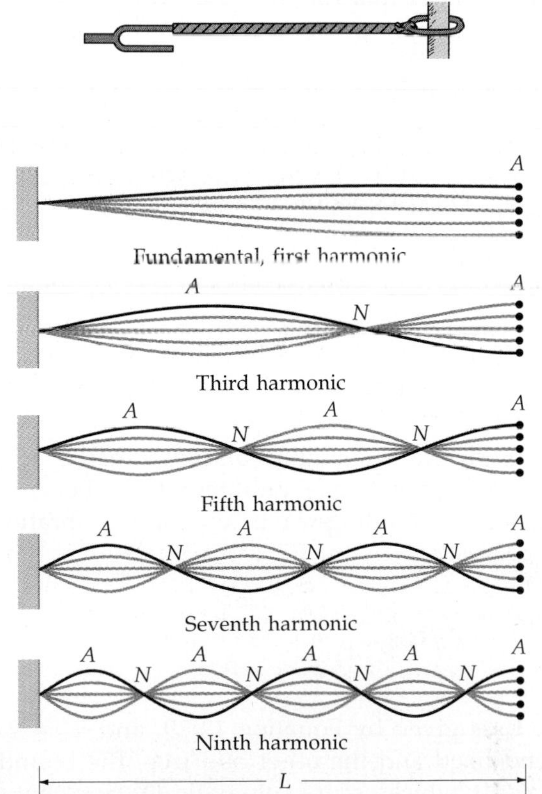

Figure 13-21 An approximation of a string fixed at one end and free at the other end can be produced by connecting the "free" end of the string to a ring that is free to move on a post. Since the amplitude of the tuning fork is very small, the end attached to the fork is approximately fixed.

Figure 13-22 Standing waves on a string fixed at only one end. The free end is an antinode.

Standing-Wave Functions

We can find the equations for the wave functions illustrated in Figures 13-19 and 13-22 and also rederive the standing-wave conditions by considering the resultant wave formed by the superposition of a wave traveling to the right and a wave traveling to the left. Let the displacement of the wave traveling to the right be y_R and that of the wave traveling to the left be y_L. Assume that the amplitudes are equal. Then

$$y_R = A \sin (kx - \omega t)$$

and

$$y_L = A \sin (kx + \omega t)$$

where $k = 2\pi/\lambda$ is the wave number and $\omega = 2\pi f$ is the angular frequency. The sum of these two waves is

$$y(x, t) = y_R + y_L = A \sin (kx - \omega t) + A \sin (kx + \omega t)$$

Again, we use the identity $\sin \theta_1 + \sin \theta_2 = 2 \cos \frac{1}{2}(\theta_1 - \theta_2) \sin \frac{1}{2}(\theta_1 + \theta_2)$. If we let $\theta_1 = kx + \omega t$ and $\theta_2 = kx - \omega t$, we have

$$\tfrac{1}{2}(\theta_1 + \theta_2) = kx$$

and

$$\tfrac{1}{2}(\theta_1 - \theta_2) = \omega t$$

Then

$$y(x, t) = 2A \cos \omega t \sin kx \qquad 13\text{-}26$$

If the string is fixed at $x = 0$ and $x = L$, we have the following **boundary conditions** on the wave function:

$$y(x, t) = 0 \qquad \text{at } x = 0 \text{ and } x = L \qquad 13\text{-}27$$

for all times t. The boundary condition at $x = 0$ is automatically met because $\sin kx = 0$ at $x = 0$. The boundary condition at $x = L$ is met only for those particular values of the wave number that satisfy

$$\sin kL = 0 \qquad 13\text{-}28$$

The values k_n that satisfy this equation are given by

$$k_n L = n\pi \qquad n = 1, 2, 3, \ldots \qquad 13\text{-}29$$

In terms of the wavelength $\lambda = 2\pi/k$, Equation 13-29 is

$$\frac{2\pi}{\lambda_n} L = n\pi$$

or

$$n \frac{\lambda_n}{2} = L$$

which is the same as the standing-wave condition discussed earlier (Equation 13-20). The length of the string must equal an integral number of half-wavelengths. Equation 13-26 for the standing-wave function for vibration in the nth harmonic can thus be written

Standing-wave functions

$$y_n(x, t) = A_n \cos \omega_n t \sin k_n x \qquad 13\text{-}30$$

where A_n is the amplitude, k_n is given by Equation 13-29, and $\omega_n = 2\pi f_n$.

For a string with one end fixed and the other end free, the boundary conditions are that $y = 0$ at $x = 0$, which occurs automatically, and that y be a maximum or minimum at $x = L$. That is, the point $x = L$ must be an antinode. This occurs if

$$\sin k_nL = \pm 1$$

or

$$k_nL = n\frac{\pi}{2} \qquad n = 1, 3, 5, \ldots$$

In terms of the wavelength, this condition is $2\pi L/\lambda_n = n\pi/2$ or

$$L = n\frac{\lambda_n}{4} \qquad n = 1, 3, 5, \ldots \qquad \text{13-31}$$

which is the same as Equation 13-23.

Example 13-6

A string fixed at one end only is vibrating in its third harmonic. Its wave function is given by $y(x, t) = 0.015 \cos 189t \sin 0.262x$ where y and x are in meters and t is in seconds. (*a*) What is the frequency of vibration? (*b*) What is the length of the string?

(*a*) Since the angular frequency is $\omega = 189$ rad/s, the frequency is $f = \omega/2\pi = (189 \text{ rad/s})/(6.28) = 30.1$ Hz. The length of the string is related to the wavelength of the third harmonic by Equation 13-31,

$$L = n\frac{\lambda_n}{4} = 3\left(\frac{\lambda_3}{4}\right)$$

(*b*) Since the wave number is $k = 0.262 \text{ m}^{-1}$, the wavelength is $\lambda_3 = 2\pi/k = 6.28/(0.262 \text{ m}^{-1}) = 24.0$ m. The length of the string is therefore

$$L = 3\left(\frac{\lambda_3}{4}\right) = 3\left(\frac{24 \text{ m}}{4}\right) = 18.0 \text{ m}$$

13-7 The Superposition of Standing Waves

In general, a vibrating system such as a string fixed at both ends does not vibrate in a single harmonic mode. Instead, the motion consists of a mixture of the allowed harmonics. For example, the general motion of a string fixed at both ends can be described by the wave function

$$y(x, t) = \sum_n A_n \cos(\omega_n t + \delta_n) \sin k_n x \qquad \text{13-32}$$

where k_n satisfies Equation 13-29, $\omega_n = k_n v$, and A_n and δ_n are constants. The wave function in Equation 13-32 is a linear combination of the harmonic wave functions as described by Equation 13-30. Since the energy in a wave is proportional to the square of the amplitude, the quantity A_n^2 describes the fraction of the energy associated with the nth harmonic. The constants A_n and δ_n depend on the initial position and velocity of the string. A piano string, for example, is set into motion by striking it with a hammer. The location of the point of impact of the hammer on the string determines the relative values of A_n, which in turn describe the relative amounts of energy in each harmonic component. A harp string, on the other hand, is plucked. If a string is plucked at the center and released as in Figure 13-23, the initial shape of the string is symmetric about the point $x = \frac{1}{2}L$. The motion of the string after it has been released will remain symmetric about this point. Only the odd harmonics, which are also symmetric about $x = \frac{1}{2}L$, will be excited. The even harmonics, which are antisymmetric about $x = \frac{1}{2}L$, are not excited.

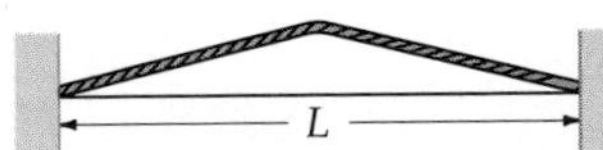

Figure 13-23 A string plucked at the center. When it is released, its vibration is a linear combination of standing waves.

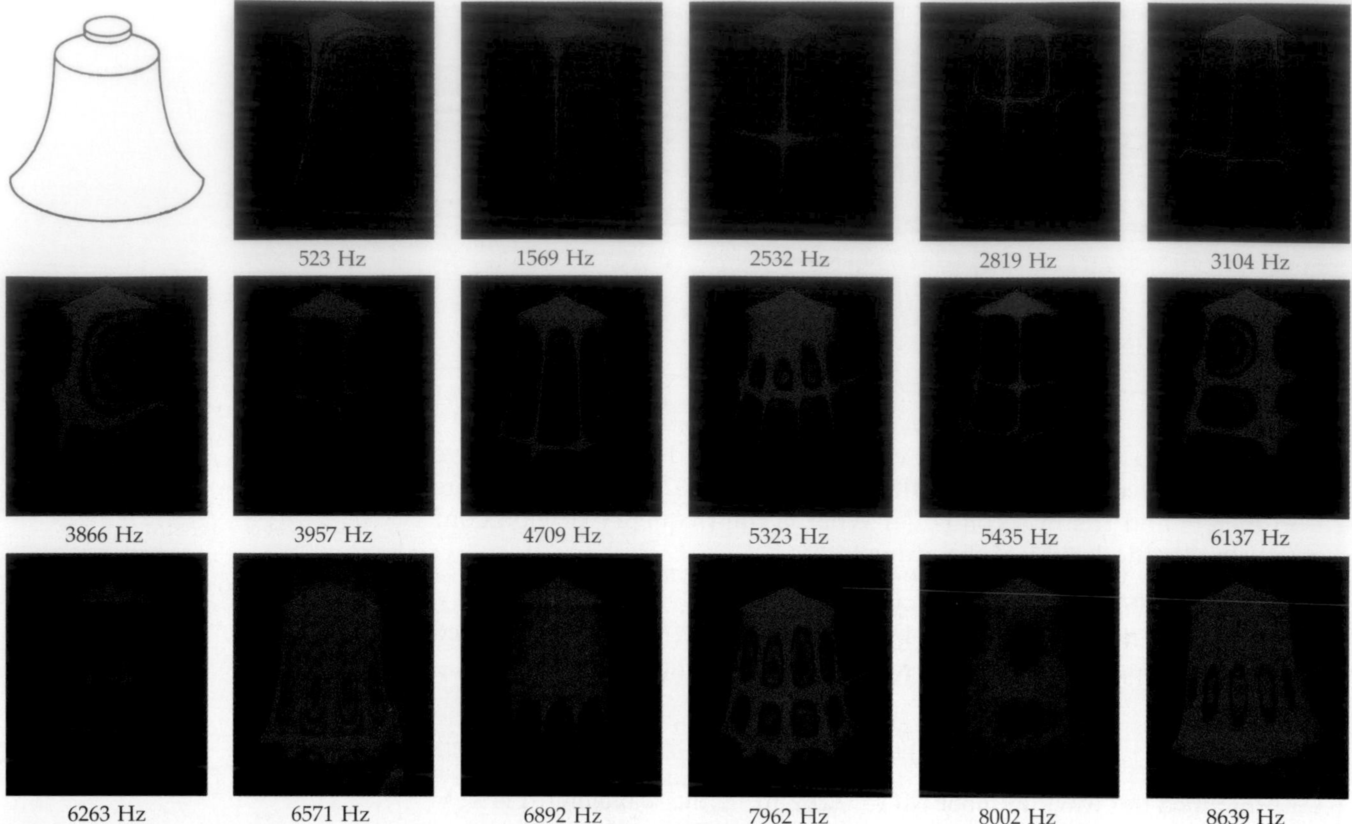

Hologram interferograms showing standing waves in a handbell. The "bull's eyes" locate the antinodes.

The constant A_n is zero for every even n. (The shapes of the first four harmonics are shown in Figure 13-24.) Most of the energy of the plucked string is associated with the fundamental, but small amounts of energy are associated with the third and higher odd harmonic modes. Figure 13-25 shows an approximation to the initial shape of the string using the superpositions of only the first three odd harmonics.

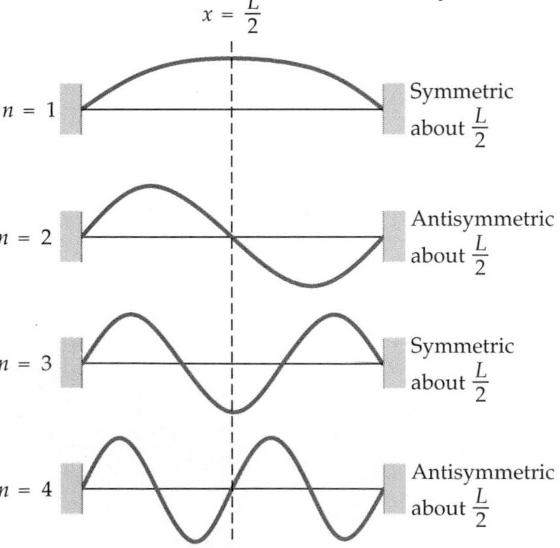

Figure 13-24 The first four harmonics for a string fixed at both ends. The odd harmonics are symmetrical about the center of the string, whereas the even harmonics are not. When a string is plucked at the center, it vibrates only in its odd harmonics.

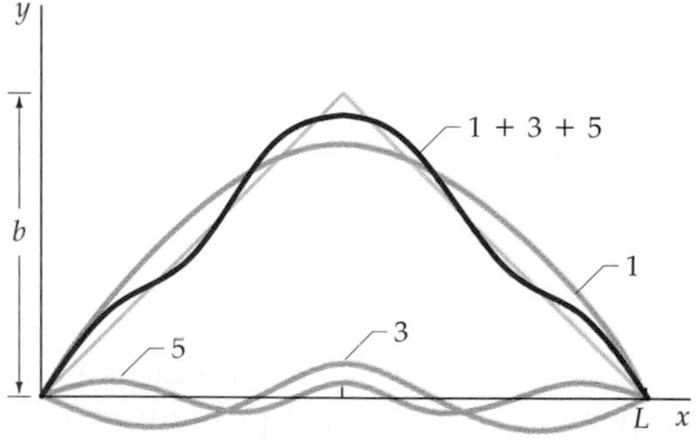

Figure 13-25 Synthesis of a string plucked at the center, as in Figure 13-23, using only the first three odd harmonics. The heavy colored line is an approximation of the original shape of the string based on the first three odd harmonics. The height of the string is exaggerated in this drawing to show the relative amplitudes of the harmonics. Most of the energy is associated with the fundamental, but there is some energy in the third, fifth, and other odd harmonics.

13-8 The Wave Equation (Optional)

A general wave function $y(x, t)$ is a solution of a differential equation called the wave equation. The wave equation follows directly from Newton's laws.

In Figure 13-26, one segment of a string has been isolated. Our derivation will apply only if the wave is small enough in amplitude for the angle between the string and the horizontal (the original direction of the string with no wave) to be small. In this case, the length of the segment is approximately Δx and its mass is $\mu\ \Delta x$. The string segment moves vertically. Its acceleration is the second derivative of $y(x, t)$ with respect to t for constant x.

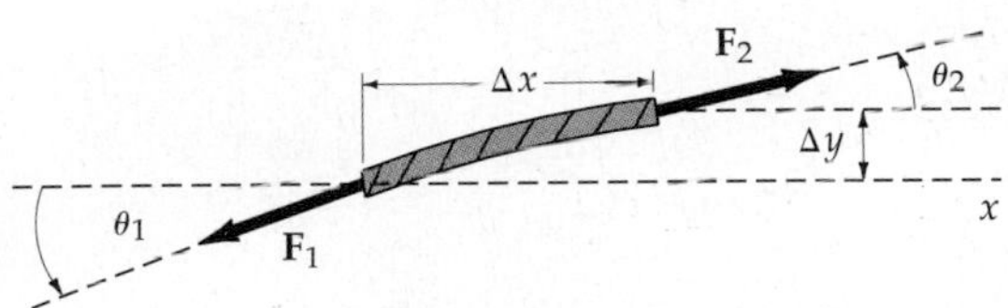

Figure 13-26 Segment of a stretched string used for the derivation of the wave equation. The net vertical force on the segment is $F \sin \theta_2 - F \sin \theta_1$, where F is the tension in the string. The wave equation is derived by applying Newton's second law to the segment.

A derivative of a function of two variables with respect to one of the variables with the other held constant is called a **partial derivative.** The partial derivative of y with respect to t is written $\partial y/\partial t$. The second partial derivative of y with respect to t is written $\partial^2 y/\partial t^2$. The net vertical force is

$$\sum F = F \sin \theta_2 - F \sin \theta_1$$

where θ_2 and θ_1 are the angles shown in Figure 13-26, and F is the tension in the string. Since the angles are assumed to be small, we may approximate $\sin \theta$ by $\tan \theta$. Thus, the net vertical force on the string segment can be written

$$\sum F = F(\sin \theta_2 - \sin \theta_1) \approx F(\tan \theta_2 - \tan \theta_1)$$

The tangent of the angle made by the string with the horizontal is the slope of the curve formed by the string. If we call this slope S, we have

$$S = \tan \theta = \frac{\partial y}{\partial x}$$

Then

$$\sum F = F(S_2 - S_1) = F\ \Delta S$$

where S_1 and S_2 are the slopes of either end of the string segment, and ΔS is the change in the slope. Setting this net force equal to the mass $\mu\ \Delta x$ times the acceleration $\partial^2 y/\partial t^2$ gives

$$F\ \Delta S = \mu\ \Delta x \frac{\partial^2 y}{\partial t^2}$$

or

$$F \frac{\Delta S}{\Delta x} = \mu \frac{\partial^2 y}{\partial t^2} \qquad \text{13-33}$$

In the limit $\Delta x \to 0$, we have

$$\lim_{\Delta x \to 0} \frac{\Delta S}{\Delta x} = \frac{\partial S}{\partial x} = \frac{\partial}{\partial x}\frac{\partial y}{\partial x} = \frac{\partial^2 y}{\partial x^2}$$

Thus, Equation 13-33 becomes

$$\frac{\partial^2 y}{\partial x^2} = \frac{\mu}{F}\frac{\partial^2 y}{\partial t^2} \qquad \text{13-34a}$$

Wave equation

Equation 13-34 is the **wave equation** for a stretched string. It is important to remember that this equation holds only for small angles and thus for small displacements $y(x, t)$.

We now show that the wave equation is satisfied by any function of either $x - vt$ or $x + vt$. Let $\alpha = x - vt$ and consider any wave function

$$y = y(x - vt) = y(\alpha)$$

We will use y' for the derivative of y with respect to α. Then, by the chain rule for derivatives, we have

$$\frac{\partial y}{\partial x} = \frac{\partial y}{\partial \alpha}\frac{\partial \alpha}{\partial x} = y'\frac{\partial \alpha}{\partial x}$$

and

$$\frac{\partial y}{\partial t} = \frac{\partial y}{\partial \alpha}\frac{\partial \alpha}{\partial t} = y'\frac{\partial \alpha}{\partial t}$$

Since $\partial\alpha/\partial x = 1$ and $\partial\alpha/\partial t = -v$, we have

$$\frac{\partial y}{\partial x} = y'$$

and

$$\frac{\partial y}{\partial t} = -vy'$$

Taking the second derivatives, we obtain

$$\frac{\partial^2 y}{\partial x^2} = y''$$

$$\frac{\partial^2 y}{\partial t^2} = -v\frac{\partial y'}{\partial t} = -v\frac{\partial y'}{\partial \alpha}\frac{\partial \alpha}{\partial t} = +v^2 y''$$

Thus,

Wave equation

$$\frac{\partial^2 y}{\partial x^2} = \frac{1}{v^2}\frac{\partial^2 y}{\partial t^2} \qquad \text{13-34}b$$

Comparing Equations 13-34*a* and *b*, we see that the speed of propagation of the wave is

$$v = \sqrt{\frac{F}{\mu}} \qquad \text{13-35}$$

Example 13-7

Show by explicitly calculating the derivatives that $y = A \sin(kx - \omega t)$ satisfies Equation 13-34*b*.

Taking the partial derivative of y with respect to x, we obtain

$$\frac{\partial y}{\partial x} = \frac{\partial}{\partial x}[A \sin(kx - \omega t)] = A \cos(kx - \omega t)\frac{\partial(kx)}{\partial x}$$

$$= kA \cos(kx - \omega t)$$

Taking the second partial derivative with respect to x, we obtain

$$\frac{\partial^2 y}{\partial x^2} = -k^2 A \sin(kx - \omega t)$$

Similarly, the derivatives with respect to t are

$$\frac{\partial y}{\partial t} = \frac{\partial}{\partial t}[A \sin(kx - \omega t)] = A \cos(kx - \omega t)\frac{\partial(-\omega t)}{\partial t}$$

$$= -\omega A \cos(kx - \omega t)$$

and

$$\frac{\partial^2 y}{\partial t^2} = -\omega^2 A \sin(kx - \omega t)$$

Equation 13-34b then gives

$$-k^2 A \sin(kx - \omega t) = \frac{1}{v^2}[-\omega^2 A \sin(kx - \omega t)]$$

which holds if $v = \omega/k$.

An important property of the wave equation is that it is linear; that is, the function $y(x, t)$ and its derivatives occur only to the first power. An important property of linear equations is that if y_1 and y_2 are two solutions of the equation, the linear combination

$$y_3 = C_1 y_1 + C_2 y_2 \qquad \text{13-36}$$

where C_1 and C_2 are any constants, is also a solution. This can be shown by the direct substitution of y_3 into the wave equation. This result is the mathematical statement of the principle of superposition. If any two waves satisfy a wave equation, their sum also satisfies the same wave equation. Again, the principle of superposition holds for waves on a string only if the amplitudes of the waves are small, so that the approximation used in deriving the wave equation, $\sin \theta \approx \tan \theta$, holds.

Exercise

Show that any function $y(x + vt)$ satisfies Equation 13-34b.

Exercise

Show that the function y_3 given by Equation 13-36 satisfies Equation 13-34.

Summary

1. Wave motion is the propagation of a disturbance in a medium. In transverse waves, such as waves on a string, the disturbance is perpendicular to the direction of propagation. In longitudinal waves, such as sound waves, the disturbance is along the direction of propagation. Both energy and momentum are transported by a wave.

2. The speed of a wave depends on the density and elastic properties of the medium. It is independent of the motion of the wave source. The speed of waves on a string is related to the tension F in the string and its mass per unit length μ by

$$v = \sqrt{\frac{F}{\mu}}$$

3. When two or more waves meet at the same place they superimpose; that is, the disturbances add algebraically. The principle of superposition holds for waves on a string if the transverse displacements are not too large.

4. In harmonic waves, the disturbance varies sinusoidally in time and space. In a harmonic wave on a string, a segment of the string oscillates with simple harmonic motion in a direction perpendicular to the direction of the wave. The distance between successive wave crests is the wavelength λ. The wave function $y(x, t)$ for a harmonic wave is

$$y(x, t) = A \sin (kx - \omega t)$$

where A is the amplitude, k is the wave number, which is related to the wavelength by

$$k = \frac{2\pi}{\lambda}$$

and ω is the angular frequency, which is related to the frequency f by

$$\omega = 2\pi f$$

The speed of a harmonic wave equals the frequency times the wavelength:

$$v = f\lambda = \frac{\omega}{k}$$

5. The power transmitted by a harmonic wave is proportional to the square of the amplitude of the wave and is given by

$$P = \tfrac{1}{2}\mu\omega^2 A^2 v$$

6. The superposition of harmonic waves is called interference. If the waves are in phase or differ in phase by an integer times 2π, the amplitudes of the waves add and the interference is constructive. If the waves differ in phase by π or by an odd integer times π, the amplitudes subtract and the interference is destructive.

7. When waves are confined in space, standing waves occur. For a string fixed at both ends, the standing-wave condition can be found by fitting waves into the string with a node at each end. The result is that an integral number of half-wavelengths must fit into the length of the string. The standing-wave condition in this case is

$$L = n\frac{\lambda_n}{2} \qquad n = 1, 2, 3, \ldots$$

The allowed waves form a harmonic series, with the frequencies given by

$$f_n = nf_1 \qquad n = 1, 2, 3, \ldots$$

where $f_1 = v/2L$ is the lowest frequency, which is called the fundamental. The wave functions for these standing waves are of the form

$$y_n(x, t) = A_n \cos \omega_n t \sin k_n x$$

where $k_n = 2\pi/\lambda_n$ and $\omega_n = 2\pi f_n$.

If the string has one end fixed and one end free, there is a node at one end and an antinode at the other. The standing-wave condition in this case is

$$L = n\frac{\lambda_n}{4} \qquad n = 1, 3, 5, \ldots$$

Only the odd harmonics are present. Their frequencies are given by

$$f_n = nf_1 \qquad n = 1, 3, 5, \ldots$$

where $f_1 = v/4L$.

8. In general, a vibrating system, such as a string fixed at both ends, does not vibrate in a single harmonic mode but in a mixture of the allowed harmonics.

9. The wave function for waves on a string obeys the wave equation, which relates the spatial derivatives of the wave function to the time derivatives:

$$\frac{\partial^2 y}{\partial x^2} = \frac{1}{v^2}\frac{\partial^2 y}{\partial t^2}$$

The wave equation follows from Newton's second law applied to a segment of a vibrating string.

Suggestions for Further Reading

Pierce, John R.: "Periodicity, Pitch and Waves," in *The Science of Musical Sound*, Scientific American Books Inc., New York, 1983.

Traveling and standing waves on a stretched string form the basis of this discussion of the pitch of sound.

Schelleng, John C.: "The Bowed String," *Scientific American*, January 1974. Also in *The Physics of Music*, Scientific American Inc., New York, 1978.

Measurements of the position and velocity of points on a violin string, as the bow is drawn across it, result in new understanding of the string's peculiar standing-wave motion.

Review

A. Objectives. After studying this chapter, you should:

1. Be able to state the quantities upon which the speed of a wave on a string depend.

2. Be able to state the relationship between the speed, frequency, and wavelength of a harmonic wave.

3. Be able to sketch standing-wave patterns for a string fixed at both ends and for a string with one end free and from these sketches obtain the possible frequencies for standing waves.

4. Know that the energy in a harmonic wave is proportional to the square of the amplitude of the wave.

B. Define, explain, or otherwise identify:

Dispersion
Transverse wave
Longitudinal wave
Wave function
Principle of superposition
Harmonic wave
Wavelength
Amplitude
Wave number
Phase difference
Interference
Constructive interference
Destructive interference
Standing wave
Resonance frequencies
Fundamental
First harmonic
Node
Antinode
Standing-wave condition
Natural frequencies
Harmonic series
Mode
Overtone
Boundary condition
Partial derivative
Wave equation

C. True or false: If the statement is true, explain why it is true. If it is false, give a counterexample.

1. Wave pulses on strings are transverse waves.

2. When a wave pulse is reflected, it is always inverted.

3. The speed of a harmonic wave on a string is proportional to its wavelength.

4. The power transmitted by a wave is proportional to the square of the amplitude of the wave.

5. The frequency of the third harmonic is three times that of the first harmonic.

Problems

Level I

13-1 Wave Pulses

1. Figure 13-27 shows a wave pulse at time $t = 0$. The pulse is moving to the right without dispersion at 2 cm/s. Sketch the shape of the string at times $t = 1$, 2, and 4 s.

Figure 13-27 Problems 1, 2, and 3.

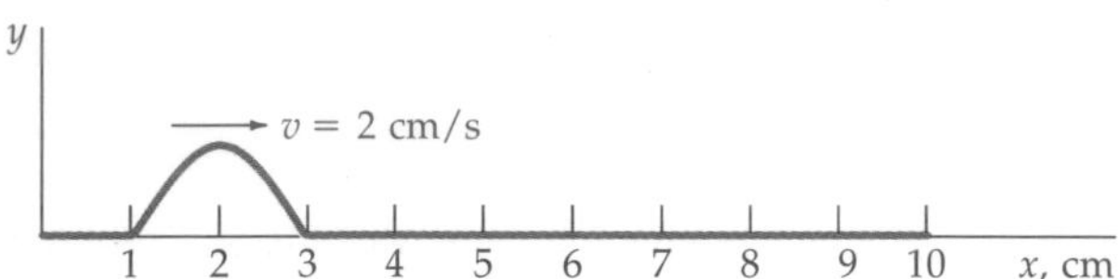

2. Figure 13-27 shows a wave pulse at time $t = 0$ moving to the right. At this particular time, which segments of the string are moving up? Which are moving down? Is there any segment of the string at the pulse that is instantaneously at rest? Answer these questions by sketching the pulse at a slightly later time and a slightly earlier time to see how the segments of the string are moving.

3. Make a sketch of the velocity of each string segment versus position for the pulse shown in Figure 13-27.

4. Two rectangular wave pulses are traveling in opposite directions along a string. At $t = 0$, the two pulses are as shown in Figure 13-28. Sketch the wave functions for $t =$ 1, 2, and 3 s.

Figure 13-28 Problems 4 and 5.

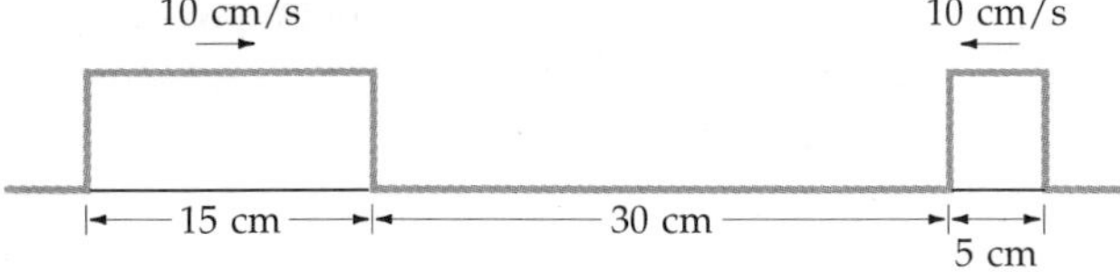

5. Repeat Problem 4 for the case in which the pulse on the right in Figure 13-28 is inverted.

6. The following wave functions represent traveling waves:

(*a*) $y_2(x, t) = A \cos k(x + 34t)$

(*b*) $y_3(x, t) = Ae^{-k(x-20t)}$

(*c*) $y_1(x, t) = \dfrac{B}{C + (x - 10t)^2}$

where x is in meters, t is in seconds, and A, k, B, and C are constants that have the proper units for y to be in meters. Give the direction of propagation and the speed of the wave for each wave function.

7. At time $t = 0$, the shape of a wave pulse on a string is given by the function

$$y(x, 0) = \frac{0.12 \text{ m}^3}{(2.00 \text{ m})^2 + x^2}$$

where x is in meters. (*a*) Sketch $y(x, 0)$ versus x. Give the wave function $y(x, t)$ at a general time t if (*b*) the pulse is moving in the positive x direction with a speed of 10 m/s and (*c*) the pulse is moving in the negative x direction with a speed of 10 m/s.

13-2 Speed of Waves

8. Show that if F is in newtons and μ is in kilograms per meter, the units of $\sqrt{F/\mu}$ are meters per second.

9. A steel wire 7 m long has a mass of 100 g. It is under a tension of 900 N. What is the speed of a transverse wave pulse on this wire?

10. Transverse waves travel at 150 m/s on a wire of length 80 cm that is under a tension of 550 N. What is the mass of the wire?

11. A wave pulse propagates along a wire in the positive x direction at 20 m/s. What will the pulse velocity be (*a*) if we double the length of the wire but keep the tension and mass per unit length constant? (*b*) If we double the tension while holding the length and mass per unit length constant? (*c*) If we double the mass per unit length while holding the other variables constant?

12. A steel piano wire is 0.7 m long and has a mass of 5 g. It is stretched with a tension of 500 N. (*a*) What is the speed of transverse waves on the wire? (*b*) To reduce the wave speed by a factor of 2 without changing the tension, what mass of copper wire would have to be wrapped around the steel wire?

13. In a common lecture demonstration of wave pulses, a piece of rubber tubing is tied at one end to a fixed post and is passed over a pulley to a weight hanging at the other end. Suppose that the distance from the fixed support to the pulley is 10 m, the mass of this length of tubing is 0.7 kg, and the suspended weight is 110 N. If the tubing is given a transverse blow at one end, how long will it take the resulting pulse to reach the other end?

14. A long glass tube has a semicircular bend of radius 8 cm. A string of mass density 0.04 kg/m under a tension of 20 N is pulled through the tube. (*a*) At what speed should the string be pulled for it to pass through the tube without touching the sides? (*b*) What is the magnitude of the acceleration of the string segments as they pass through the semicircular bend? (*c*) Draw a diagram indicating the forces acting on a string segment in the bend that provide this acceleration.

15. The cable of a ski lift runs 400 m up a mountain and has a mass of 80 kg. When the cable is struck with a transverse blow at one end, the return pulse is detected 12 s later. (*a*) What is the speed of the wave? (*b*) What is the tension in the cable?

13-3 Harmonic Waves

16. One end of a string 6 m long is moved up and down with simple harmonic motion at a frequency of 60 Hz. The waves reach the other end of the string in 0.5 s. Find the wavelength of the waves on the string.

17. Equation 13-8 expresses the displacement of a harmonic wave as a function of x and t in terms of the wave parameters k and ω. Write the equivalent expressions that contain the following pairs of parameters instead of k and ω: (*a*) k and v, (*b*) λ and f, (*c*) λ and T, (*d*) λ and v, and (*e*) f and v.

18. Equation 13-5 applies to all types of periodic waves, including electromagnetic waves such as light waves and microwaves, which travel at 3×10^8 m/s in a vacuum. (*a*) The range of wavelengths of light to which the eye is sensitive is about 4×10^{-7} to 7×10^{-7} m. What are the frequencies that correspond to these wavelengths? (*b*) Find the frequency of a microwave that has a wavelength of 3 cm.

19. A harmonic wave with a frequency of 60 Hz and an amplitude of 0.02 m travels along a string to the right with a speed of 10 m/s. Write a suitable wave function for this wave.

20. The wave function for a harmonic wave on a string is $y(x, t) = 0.001 \sin (62.8x + 314t)$, where y and x are in meters and t is in seconds. (*a*) In what direction does this wave travel, and what is its speed? (*b*) Find the wavelength, frequency, and period of this wave. (*c*) What is the maximum displacement of any string segment?

21. A harmonic wave travels along an infinite uniform rope under constant tension. The rope is marked off in meter intervals. At the 0-meter mark, the rope is observed to reach its maximum transverse displacement of 50 cm every 5 s. The distance between the maxima at any instant in time is 50 m. Give the wave function of this wave, assuming that it is harmonic, has a maximum displacement at $x = 0$ when $t = 0$, and is moving along the rope from left to right.

13-4 Energy Transmitted by Waves

22. Waves of frequency 200 Hz and amplitude 1 cm move along a 20-m string that has a mass of 0.06 kg and a tension of 50 N. (*a*) What is the total energy of the waves on the string? (*b*) Find the power transmitted past a given point on the string.

23. A harmonic wave on a string with a mass per unit length of 0.05 kg/m and a tension of 80 N has an amplitude of 5 cm. Each section of the string moves with simple harmonic motion at a frequency of 10 Hz. Find the power propagated along the string.

24. A rope 2 m long has a mass of 0.1 kg. The tension is 60 N. A power source at one end sends a harmonic wave with an amplitude of 1 cm down the rope. The wave is extracted at the other end without any reflection. What is the frequency of the power source if the power transmitted is 100 W?

13-5 Superposition and Interference of Harmonic Waves

25. Two waves traveling on a string in the same direction both have a frequency of 100 Hz, a wavelength of 2 cm, and an amplitude of 0.02 m. What is the amplitude of the resultant wave if the original waves differ in phase (*a*) by $\pi/6$? (*b*) By $\pi/3$?

26. What is the phase difference between the two waves of Problem 25 if the amplitude of the resultant wave is 0.02 m, the same as the amplitude of each original wave?

27. Two waves having the same frequency, wavelength, and amplitude are traveling in the same direction. If they differ in phase by $\pi/2$ and each has an amplitude of 0.05 m, find the amplitude of the resultant wave.

13-6 Standing Waves

28. A string fixed at both ends is 3 m long. It resonates in its second harmonic at a frequency of 60 Hz. What is the speed of transverse waves on the string?

29. A string 3 m long and fixed at both ends is vibrating in its third harmonic. The maximum displacement of any point on the string is 4 mm. The speed of transverse waves on this string is 50 m/s. (*a*) What are the wavelength and frequency of this wave? (*b*) Write the wave function for this wave.

30. A 5-g steel wire 1.4 m long is fixed at both ends and has a tension of 968 N. (*a*) Find the speed of transverse waves on the wire. (*b*) Find the wavelength and frequency of the fundamental. (*c*) Find the frequencies of the second and third harmonics.

31. A rope 4 m long is fixed at one end and is attached to a light string at the other end so that it is free to move there. The speed of waves on the rope is 20 m/s. Find the frequency of (*a*) the fundamental, (*b*) the first overtone, and (*c*) the second overtone.

32. Middle C on the equal-temperament scale used by modern instrument makers has a frequency of 261.63 Hz. If this is the fundamental frequency of a 7 g piano wire that is 80 cm long, what should the tension in the wire be?

33. A piano wire without windings has a fundamental frequency of 200 Hz. When it is wound with wire, its linear mass density is doubled. What is its new fundamental frequency assuming that the tension is unchanged?

34. The length of the B string on a certain guitar is 60 cm. It vibrates at 247 Hz. (*a*) What is the speed of transverse waves on the string? (*b*) If the linear mass density of the guitar string is 0.01 g/cm, what should its tension be when it is in tune?

35. The wave function $y(x, t)$ for a certain standing wave on a string fixed at both ends is given by $y(x, t) = 0.30 \sin 0.20x \cos 300t$, where y and x are in centimeters and t is in seconds. (*a*) What are the wavelength and frequency of these waves? (*b*) What is the speed of transverse waves on this string? (*c*) If the string is vibrating in its fourth harmonic, how long is it?

36. The wave function $y(x, t)$ for a certain standing wave on a string fixed at both ends is given by $y(x, t) = 0.5 \sin 0.025x \cos 500t$, where y and x are in centimeters and t is in seconds. (*a*) Find the speed and amplitude of the two traveling waves that result in this standing wave. (*b*) What is the distance between successive nodes on the string? (c) What is the shortest possible length of the string?

37. A string 2.51 m long has the wave function given in Problem 36. (a) Sketch the position of the string at the times $t = 0$, $t = \frac{1}{4}T$, $t = \frac{1}{2}T$, and $t = \frac{3}{4}T$, where $T = 1/f$ is the period of the vibration. (b) Find T in seconds. (c) At a time t when the string is horizontal, that is, $y(x) = 0$ for all x, what has become of the energy in the wave?

38. A 160-g rope 4 m long is fixed at one end and is tied to a light string at the other end. Its tension is 400 N. (a) What are the wavelengths of the fundamental and the next two harmonics? (b) What are the frequencies of these standing waves?

39. A string fixed at one end only is vibrating in its fundamental mode. The wave function is $y(x, t) = 0.02 \sin 2.36x \cos 377t$, where y and x are in meters and t is in seconds. (a) What is the wavelength of the wave? (b) What is the length of the string? (c) What is the speed of transverse waves on the string?

40. A string 5 m long that is fixed at one end only is vibrating in its fifth harmonic with a frequency of 400 Hz. The maximum displacement of any segment of the string is 3 cm. (a) What is the wavelength of this wave? (b) What is the wave number k? (c) What is the angular frequency? (d) Write the wave function for this standing wave.

41. Three successive resonance frequencies for a certain string are 75, 125, and 175 Hz. (a) Find the ratios of each pair of successive resonance frequencies. (b) How can you tell that these frequencies are for a string fixed at one end only rather than for a string fixed at both ends? (c) What is the fundamental frequency? (d) Which harmonics are these resonance frequencies? (e) If the speed of transverse waves on this string is 400 m/s, find the length of the string.

13-7 The Superposition of Standing Waves

There are no problems for this section.

13-8 The Wave Equation (Optional)

42. Show explicitly that the following functions satisfy the wave equation: (a) $y(x, t) = (x + vt)^3$; (b) $y(x, t) = Ae^{ik(x-vt)}$, where A and k are constants and $i = \sqrt{-1}$; and (c) $y(x, t) = \ln(x - vt)$.

43. Show that the function $y = A \sin kx \cos \omega t$ satisfies the wave equation.

Level II

44. The G string on a violin is 30 cm long. When played without fingering, it vibrates at a frequency of 196 Hz. The next higher notes on the scale are A (220 Hz), B (247 Hz), C (262 Hz), and D (294 Hz). How far from the end of the string must a finger be placed to play each of these notes?

45. A string with a mass density of 4×10^{-3} kg/m is under a tension of 360 N and is fixed at both ends. One of its resonance frequencies is 375 Hz. The next higher resonance frequency is 450 Hz. (a) What is the fundamental frequency of this string? (b) Which harmonics are the ones given? (c) What is the length of the string?

46. A string fastened at both ends has successive resonances with wavelengths of 0.54 m for the nth harmonic and 0.48 m for the $(n + 1)$th harmonic. (a) Which harmonics are these? (b) What is the length of the string?

47. A tuning fork attached to a stretched wire generates transverse waves. The vibration of the fork is perpendicular to the wire. Its frequency is 400 Hz, and the amplitude of its oscillation is 0.50 mm. The wire has linear mass density of 0.01 kg/m and is under a tension of 1 kN. Assume that there are no reflected waves. (a) Find the period and frequency of waves on the wire (b) What is the speed of the waves? (c) What are the wavelength and wave number? (d) Write a suitable wave function for the waves on the wire. (e) Calculate the maximum speed and acceleration of a point on the wire. (f) At what average rate must energy be supplied to the fork to keep it oscillating at a steady amplitude?

48. In a lecture demonstration of standing waves, a string is attached to a tuning fork that vibrates at 60 Hz and sets up transverse waves of that frequency on the string. The other end of the string passes over a pulley, and the tension is varied by attaching weights to that end. The string has approximate nodes at the tuning fork and at the pulley. (a) If the string has linear mass density of 8 g/m and is 2.5 m long (from the tuning fork to the pulley), what must the tension be for the string to vibrate in its fundamental mode? (b) Find the tensions necessary for the string to vibrate in each of its first three overtones.

49. (a) For the wave function given in Problem 39, find the velocity of a string segment at some point x as a function of time. (b) Which point has the greatest speed at any time? What is the maximum speed of this point? (c) Find the acceleration of a string segment at some point x as a function of time. (d) Which point has the greatest acceleration? What is the maximum acceleration of this point?

50. Two wires of different linear mass densities are soldered together end to end and are then stretched under a tension F (the same in both wires). The wave speed in the first wire is twice that in the second wire. When a harmonic wave traveling in the first wire is reflected at the junction of the wires, the reflected wave has half the amplitude of the transmitted wave. (a) Assuming no loss in the wire, what fraction of the incident power is reflected at the junction and what fraction is transmitted? (b) If the amplitude of the incident wave is A, what are the amplitudes of the reflected and transmitted waves?

51. Power is to be transmitted along a stretched wire by means of transverse harmonic waves. The wave speed is 10 m/s, and the linear mass density of the wire is 0.01 kg/m. The power source oscillates with an amplitude of 0.50 mm. (a) What average power is transmitted along the wire if the frequency is 400 Hz? (b) The power transmitted can be increased by increasing the tension in the wire, the frequency of the source, or the amplitude of the waves. How would each of these quantities have to be changed to effect an increase in power by a factor of 100 if it is the only quantity changed? (c) Which of the quantities would probably be the easiest to change?

52. A 2-m string is fixed at one end and is vibrating in its third harmonic. The greatest displacement of any segment of the string is 3 cm. The frequency of vibration is 100 Hz. (*a*) Write the wave function for this vibration. (*b*) Write an expression for the kinetic energy of a segment of the string of length dx at a point x at some time t. At what time is this kinetic energy maximum? What is the shape of the string at this time? (*c*) Find the maximum kinetic energy of the string by integrating your expression for part (*b*) over the total length of the string. (*d*) Find the potential energy of a segment of the string, and compute the maximum potential energy of the string by integration. *Hint:* Remember that the potential energy of a mass m in simple harmonic motion of angular frequency ω is $\frac{1}{2}m\omega^2y^2$, where y is the displacement.

53. The kinetic energy of a segment Δm of a vibrating string is given by

$$\Delta K = \tfrac{1}{2}\Delta m\left(\frac{\partial y}{\partial t}\right)^2 = \tfrac{1}{2}\mu\left(\frac{\partial y}{\partial t}\right)^2 \Delta x$$

(*a*) Find the total kinetic energy of the nth mode of vibration of a string of length L fixed at both ends. (*b*) Give the maximum kinetic energy of the string. (*c*) What is the wave function when the kinetic energy has its maximum value? (*d*) Show that the maximum kinetic energy in the nth mode is proportional to $n^2A_n^2$.

54. (*a*) Compute the derivative of the speed of a wave on a string with respect to the tension dv/dF, and show that the differentials dv and dF obey $dv/v = \frac{1}{2}\,dF/F$. (*b*) A wave moves with a speed of 300 m/s on a wire that is under a tension of 500 N. Using dF to approximate a change in tension, find how much the tension must be changed to increase the speed to 312 m/s.

55. (*a*) Show that, if the tension in a string fixed at both ends is changed by a small amount dF, the frequency of the fundamental is changed by approximately df, where $df/f = \frac{1}{2}\,dF/F$. (See Problem 54.) Does this result apply to all harmonics? (*b*) Use this result to find the percentage change in the tension needed to increase the frequency of the fundamental of a piano wire from 260 to 262 Hz.

56. Three waves with the same frequency, wavelength, and amplitude are traveling in the same direction. The three waves are given by

$$y_1(x,\,t) = 0.05\sin\left(kx - \omega t - \frac{\pi}{3}\right)$$

$$y_2(x,\,t) = 0.05\sin\,(kx - \omega t)$$

$$y_3(x,\,t) = 0.05\sin\left(kx - \omega t + \frac{\pi}{3}\right)$$

Find the resultant wave.

57. A very long wire can be vibrated up and down with a mechanical motor to produce waves traveling down the wire. At the far end of the wire, the traveling waves are cleverly absorbed so that no reflected waves are returned to the motor. The wave speed is observed to be 240 m/s, the maximum transverse displacement of the wire is 1 cm, and the distance between maxima is 3.0 m. (*a*) Write a wave function to represent the wave propagating down this wire. (*b*) What is the frequency of vibration of the motor? (*c*) What is the period of the transverse oscillations of the wire? (*d*) What is the maximum transverse velocity of a small insect clinging to the wire?

58. Two connected wires with linear mass densities that are related by $\mu_1 = 3\mu_2$ are under the same tension. When the wires oscillate at a frequency of 120 Hz, waves of wavelength 10 cm travel down the first wire with the linear density of μ_1. (*a*) What is the wave speed in the first wire? (*b*) What is the wave speed in the second wire? (*c*) What is the wavelength in the second wire?

59. A violin string of length 40 cm and mass 1.2 g has a frequency of 500 Hz when it is vibrating in its fundamental mode. (*a*) What is the wavelength of the standing wave on the string? (*b*) What is the tension in the string? (*c*) Where should you place your finger to increase the frequency to 650 Hz?

60. Two traveling wave pulses on a string are represented by the wave functions

$$y_1(x,\,t) = \frac{0.02\text{ m}^3}{2\text{ m}^2 + (x - 2t)^2}$$

and

$$y_2(x,\,t) = \frac{-0.02\text{ m}^3}{2\text{ m}^2 + (x + 2t)^2}$$

where x is in meters and t is in seconds. (*a*) Sketch each wave function separately as a function of x at $t = 0$, and describe the behavior of each as time increases. (*b*) Find the resultant wave function at $t = 0$. (*c*) Find the resultant wave function at $t = 1$ s. (*d*) Sketch the resultant wave function at $t = 1$ s.

61. A transverse wave of frequency 40 Hz propagates down a string. Two points 5 cm apart are out of phase by $\pi/6$. (*a*) What is the wavelength of the wave? (*b*) What is the phase difference between two displacements at a given point for times 5 ms apart? (*c*) What is the wave velocity?

62. A standing wave on a rope is represented by the following wave function:

$$y(x,\,t) = 0.02\sin\frac{\pi x}{2}\cos 40\pi t$$

where x and y are in meters and t is in seconds. (*a*) Write wave functions for two traveling waves that when superimposed will produce the resultant standing-wave pattern. (*b*) What is the distance between the nodes of the standing wave? (*c*) What is the velocity of a segment of the rope $x = 1$ m? (*d*) What is the acceleration of a segment of the rope at $x = 1$ m?

Level III

63. The wave functions of two standing waves on a string of length L are

$$y_1(x,\,t) = A_1\cos\omega_1 t\sin k_1 x$$

$$y_2(x,\,t) = A_2\cos\omega_2 t\sin k_2 x$$

where $k_n = n\pi/L$ and $\omega_n = n\omega_1$. The wave function of the resultant wave is

$$y_r(x, t) = y_1(x, t) + y_2(x, t)$$

(*a*) Find the velocity of a segment dx of the string. (*b*) Find the kinetic energy of this segment. (*c*) By integration, find the total kinetic energy of the resultant wave. Notice the disappearance of the cross terms so that the total kinetic energy is proportional to $(n_1A_1)^2 + (n_2A_2)^2$.

64. A 2-m wire fixed at both ends is vibrating in its fundamental mode. The tension in the wire is 40 N and the mass of the wire is 0.1 kg. The midpoint of the wire has an amplitude of 2 cm. (*a*) Find the maximum kinetic energy of the wire. (*b*) At the instant the transverse displacement is given by (0.02 m) sin $(\pi x/2)$, what is the kinetic energy of the wire? (*c*) At what position on the wire does the kinetic energy per unit length have its largest value? (*d*) Where does the potential energy per unit length have its maximum value?

65. If a loop of chain is spun at high speed, it will roll like a hoop without collapsing. Consider a chain of linear mass density μ that is rolling without slipping at a high speed v_0. (*a*) Show that the tension in the chain is $F = \mu v_0^2$. (*b*) If the chain rolls over a small bump, a transverse wave pulse will be generated in the chain. At what speed will it travel along the chain? (*c*) How far around the loop (in degrees) will a transverse wave pulse travel in the time the hoop rolls through one complete revolution?

66. A long rope with a mass per unit length of 0.1 kg/m is under a constant tension of 10 N. A motor at the point $x = 0$ drives one end of the rope with harmonic motion at 5 oscillations per second and an amplitude of 4 cm. (*a*) What is the wave speed? (*b*) What is the wavelength? (*c*) What is the maximum transverse linear momentum of a 1-mm segment of the rope? (*d*) What is the maximum net force on a 1-mm segment of the rope?

67. A heavy rope 3 m long is attached to the ceiling and is allowed to hang freely. (*a*) Show that the speed of transverse waves on the rope is independent of its mass and length but does depend on the distance y from the bottom according to the formula $v = \sqrt{gy}$. (*b*) If the bottom end of the rope is given a sudden sideways displacement, how long does it take the resulting wave pulse to go to the ceiling, reflect, and return to the bottom of the rope?

68. Two wires of different linear mass densities are joined at $x = 0$ and are under constant tension. A wave y_{in} is incident on the junction from the left at $x = 0$, where

$$y_{in} = A_{in} \cos (k_1x - \omega t)$$

This wave is partly reflected and partly transmitted at $x = 0$. The transmitted wave

$$y_t = A_t \cos (k_2x - \omega t)$$

moves to the right at $x > 0$. The reflected wave

$$y_r = A_r \cos (k_1x + \omega t)$$

moves to the left at $x < 0$. The boundary conditions at $x = 0$ are that the resultant displacement and the resultant slope $\partial y/\partial x$ are both continuous. (*a*) Find the amplitude of the reflected and transmitted waves. (*b*) Verify that mechanical energy is conserved. (*c*) If the wire at $x > 0$ is denser than the wire at $x < 0$, find the phase change between the incident wave and the reflected wave.

69. The linear mass density of a nonuniform wire under constant tension decreases gradually along the wire so that an incident wave is transmitted without reflection. The wire is uniform for $-\infty \leq x \leq 0$. In this region, a transverse wave has the form $y(x, t) = 0.003 \cos (25x - 50t)$ where y and x are in meters and t is in seconds. From $x = 0$ to $x = 20$ m, the linear mass decreases gradually from μ_1 to $\mu_1/4$. For $20 \leq x \leq \infty$, the linear mass density is constant at $\mu_1/4$. (*a*) What is the wave velocity for large values of x? (*b*) What is the amplitude of the wave for large values of x? (*c*) Give $y(x, t)$ for $20 \leq x \leq \infty$.

Chapter 14

Sound

Sound waves from a telephone receiver spreading out in air. The waves have been made visible by sweeping out the space in front of the receiver with a light bulb that has a brightness controlled by a microphone.

Sound waves are *longitudinal* waves of compression and rarefaction in a gas, liquid, or solid medium. They are produced when a body, such as a tuning fork or violin string, vibrates and causes a disturbance in the density of a medium. The disturbance is propagated through the medium by the interactions of its molecules. The vibration of the molecules is along the direction of propagation of the wave. As in the case of waves on a string, only the *disturbance* is propagated; the molecules themselves merely vibrate back and forth about their equilibrium positions.

In a gas, density and pressure are closely related. Therefore, sound waves in a gas, such as air, can be thought of as either density waves or pressure waves. In waves on a string, the transverse displacement of the string is described by a wave function $y(x \pm vt)$. The wave functions for sound waves analogous to the transverse displacement of a string are the longitudinal displacement of gas molecules from their equilibrium position $s(x \pm vt)$ or a related function—the change in the gas pressure $p(x \pm vt)$.

Some of the properties of sound waves discussed in this chapter, such as interference, diffraction, reflection, and refraction, are applicable to all types of waves and will be encountered again when we study optics.

14-1 Speed of Sound Waves

The speed of sound waves, like the speed of waves on strings, depends on the properties of the medium. For sound waves in a fluid such as air or water, the speed v is given by

$$v = \sqrt{\frac{B}{\rho}} \qquad 14\text{-}1$$

where ρ is the equilibrium density of the medium and B is the bulk modulus (Section 11-3). The derivation of this result from Newton's laws is given at the end of this section. For sound waves in a long, solid rod, the bulk modulus is replaced by Young's modulus Y (Section 11-2):

$$v = \sqrt{\frac{Y}{\rho}} \qquad 14\text{-}2$$

Comparing Equations 14-1 and 14-2 for the speed of sound waves with Equation 13-4 for the speed of waves on a string, we can see that, in general, the speed of waves depends on (1) an elastic property of the medium—the tension for string waves and the bulk modulus or Young's modulus for sound waves—and (2) an inertial property of the medium—the linear mass density or the volume mass density.

For sound waves in a gas such as air, the bulk modulus is proportional to the pressure, which is in turn proportional to the density ρ and to the absolute temperature T of the gas. The ratio B/ρ is thus independent of both the volume and pressure and is merely proportional to the absolute temperature T. In Chapter 16, we will show that, for sound waves in a gas, Equation 14-1 is equivalent to

Speed of sound in a gas

$$v = \sqrt{\frac{\gamma RT}{M}} \qquad 14\text{-}3$$

In this equation, T is the absolute temperature measured in kelvins (K), which is related to the Celsius temperature t_C by

$$T = t_C + 273 \qquad 14\text{-}4$$

The constant R is the universal gas constant, which has the value

$$R = 8.314 \text{ J/mol·K} \qquad 14\text{-}5$$

The constant M is the molar mass of the gas (that is, the mass of 1 mol of the gas), which for air is

$$M = 29 \times 10^{-3} \text{ kg/mol}$$

and γ is a constant that depends on the kind of gas and has the value 1.4 for air.

Example 14-1

Calculate the speed of sound in air at (*a*) 0°C and (*b*) 20°C.

(*a*) The absolute temperature corresponding to the Celsius temperature 0°C is

$$T = t_C + 273 = 0 + 273 = 273 \text{ K}$$

The speed of sound at 0°C is therefore

$$v = \sqrt{\frac{\gamma RT}{M}} = \sqrt{\frac{(1.4)(8.31 \text{ J/mol·K})(273 \text{ K})}{29.0 \times 10^{-3} \text{ kg/mol}}} = 331 \text{ m/s}$$

(*b*) To find the speed at 20°C = 293 K, we note that the speed of sound is proportional to the square root of the absolute temperature. Its value at 293 K, v_{293}, is therefore related to its value 273 K, v_{273}, by

$$\frac{v_{293}}{v_{273}} = \frac{\sqrt{293}}{\sqrt{273}}$$

or

$$v = \sqrt{\tfrac{293}{273}}\,(331 \text{ m/s}) = 343 \text{ m/s}$$

Exercise

For helium, $M = 4 \times 10^{-3}$ kg/mol and $\gamma = 1.67$. What is the speed of sound waves in helium at 20°C? (Answer: 1.01 km/s)

Question

1. Although the densities of most solids are more than 1000 times that of air, the speed of sound in solids is usually greater than in air. Why?

Derivation of Equation for the Speed of Sound (Optional)

Consider a fluid of density ρ and pressure P in a long tube as shown in Figure 14-1. As a source of a longitudinal wave in the fluid, we imagine a piston of area A. We suddenly compress the fluid by moving the piston to the right for a short time Δt, which increases the pressure to the left of the fluid by an amount ΔP. The piston collides with molecules in the fluid, which in turn collide with other molecules, and a disturbance in the form of a wave pulse is propagated down the tube. We make the simplifying assumption that the piston moves with a constant speed u for the time Δt. We further assume that the piston speed u is much less than the wave speed v. (It is important not to confuse these two speeds.) In time Δt the piston moves a distance $u\,\Delta t$, and the wave pulse moves a distance $v\,\Delta t$. We assume in our model that the effect of the motion of the piston is to give a speed u to all the fluid from the piston to the leading edge of the pulse, a distance $v\,\Delta t$ from the original position of the piston. This distance is based on the fact that the wave speed v is the speed at which the disturbance is propagated through the fluid. Thus, in time Δt, the greatest distance the disturbance can penetrate into the fluid ahead of the piston is $v\,\Delta t$. The assumption that all the fluid in this region moves with the same speed u amounts to assuming a rectangular shape for the wave pulse. We can calculate the speed of the pulse by setting the change in the momentum of the fluid equal to the impulse acting on the fluid due to the increased pressure for the time Δt. If A is the area of the piston, the impulse is

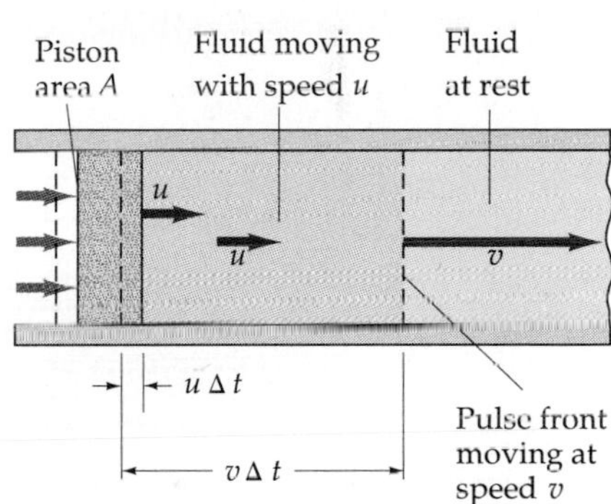

Figure 14-1 A longitudinal wave pulse is produced by suddenly moving the piston to the right with speed u. After a short time Δt, the piston has moved a distance $u\,\Delta t$ and the pulse has moved a distance $v\,\Delta t$. If a rectangular pulse is assumed, the fluid in the shaded region of length $v\,\Delta t$ is moving with speed u.

$$\text{Impulse} = F\,\Delta t = A\,\Delta P\,\Delta t$$

The mass of fluid set into motion is the density ρ times the volume $Av\,\Delta t$. The change in momentum is this mass times the velocity u:

$$\text{Momentum change} = \rho(Av\,\Delta t)u$$

Equating the momentum change to the impulse gives

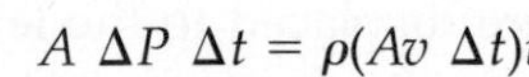

$$A\,\Delta P\,\Delta t = \rho(Av\,\Delta t)u$$

or

$$\Delta P = \rho v u \qquad \text{14-6}$$

The change in pressure is related to the decrease in volume of the fluid by the bulk modulus (Equation 11-12):

$$\Delta P = B\,\frac{-\Delta V}{V}$$

The original volume of the fluid under consideration is $V = Av\,\Delta t$, and the change in the volume swept out by the piston is $\Delta V = -Au\,\Delta t$. Thus,

$$\frac{-\Delta V}{V} = \frac{Au\,\Delta t}{Av\,\Delta t} = \frac{u}{v}$$

and

$$\Delta P = \frac{Bu}{v} \qquad 14\text{-}7$$

Using this result for ΔP in Equation 14-6, we obtain

$$\frac{Bu}{v} = \rho v u$$

or

$$v = \sqrt{\frac{B}{\rho}}$$

which is Equation 14-1.

14-2 Harmonic Sound Waves

Harmonic sound waves can be generated by a source that is vibrating with simple harmonic motion, such as a tuning fork, or a speaker driven by an audio oscillator. The vibrating source causes the air molecules next to it to oscillate with simple harmonic motion about their equilibrium position. These molecules collide with neighboring molecules, causing them to oscillate, thereby propagating the sound wave. The displacement of the molecules $s(x,t)$ for a harmonic wave can be written

$$s(x,t) = s_0 \sin\,(kx - \omega t) \qquad 14\text{-}8$$

where s_0 is the maximum displacement of a gas molecule from its equilibrium position, k is the wave number

$$k = \frac{2\pi}{\lambda} \qquad 14\text{-}9$$

and ω is the angular frequency

$$\omega = 2\pi f = \frac{2\pi}{T} \qquad 14\text{-}10$$

As with all harmonic waves, the speed of the wave equals the frequency times the wavelength:

$$v = f\lambda = \frac{\omega}{k} \qquad 14\text{-}11$$

The displacement from equilibrium given by Equation 14-8 is along the direction of the motion of the wave; that is, sound is a longitudinal wave. These displacements lead to variations in density of the air and in the air pressure. Figure 14-2*a* shows a graph of the displacement as a function of position at some fixed time. Points x_1 and x_3 are points of zero displacement at this time. Just to the left of point x_1, the displacement is negative, indicating that the gas molecules are displaced to the left, away from point x_1, at this time. Just to the right of x_1, the displacement is positive, indicating that the molecules are displaced to the right, which is again away from point x_1. Figure 14-2*b* shows some representative molecules near points x_1, x_2, and x_3

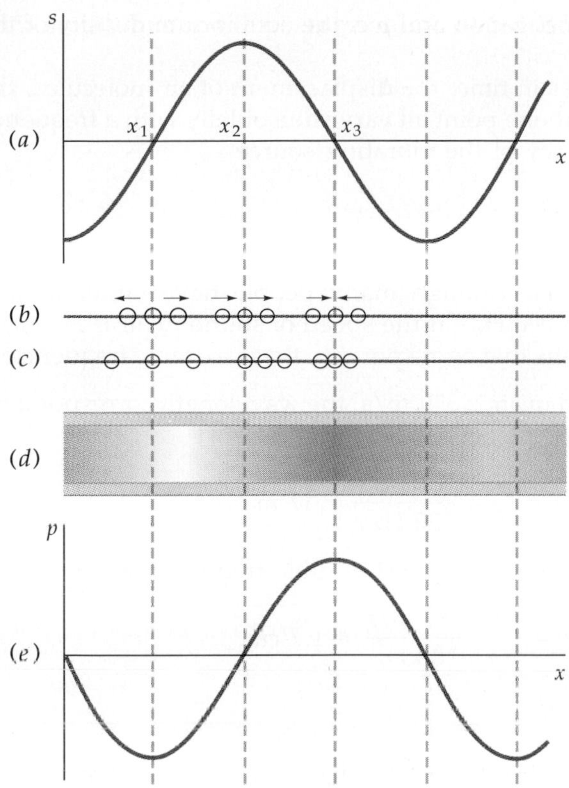

Figure 14-2 (*a*) Displacement from equilibrium of air molecules in a harmonic sound wave versus position at some instant. The molecules are at their equilibrium positions at points x_1 and x_3 and have maximum displacements at point x_2. (*b*) Some representative molecules equally spaced at their equilibrium positions before the sound wave arrives. The arrows indicate the direction of the displacement that will be caused by the sound wave. (*c*) Molecules near points x_1, x_2, and x_3 after the sound wave arrives. (*d*) Density of the air at this time. The density is maximum at x_3 and minimum at x_1, which are both points of zero displacement. (*e*) Pressure change versus position. The pressure change and displacement are 90° out of phase.

when there is no sound wave. These molecules are equally spaced. The arrows in this figure show the direction of displacements that will be caused by the sound wave. In Figure 14-2*c*, these displacements have occurred. At point x_1, the density is a minimum because the gas molecules on both sides are displaced away from that point. At point x_3 the density is a maximum because the molecules on both sides of that point are displaced toward point x_3. At point x_2 the displacement function is a maximum. The displacements to the left and to the right of this point are both positive and of equal magnitude. Thus, the density does not change near this point. Since the pressure in a gas is proportional to the density, the change in pressure is maximum where the change in density is maximum. Figure 14-2*e* indicates the change in pressure for this displacement wave. It could also represent the change in density for this wave. We can see from these figures that the pressure or density wave is 90° out of phase with the displacement wave. When the displacement is zero, the pressure and density changes are either maximum or minimum. When the displacement is a maximum or minimum, the pressure and density changes are zero. At the end of this section, we will show that a displacement wave given by Equation 14-8 implies a pressure wave given by*

$$p = p_0 \sin (kx - \omega x - \pi/2) \qquad 14\text{-}12$$

where p stands for the *change* in pressure from the equilibrium pressure and p_0 is the maximum value of this change. We will also see that the maximum pressure amplitude p_0 is related to the maximum displacement amplitude s_0 by

$$p_0 = \rho \omega v s_0 \qquad 14\text{-}13$$

*In the arguments of sine or cosine functions, we will always express phase angles in radians, which are dimensionless. However, in verbal descriptions it is often more convenient to describe phase differences in degrees. Thus we say that "two waves are 90° out of phase" rather than "two waves are out of phase by $\pi/2$ rad."

where v is the speed of propagation and ρ is the equilibrium density of the gas.

As a sound wave moves in time, the displacement of air molecules, the density, and the pressure at one point all vary sinusoidally with a frequency f, which equals the frequency of the vibrating source.

Example 14-2

The human ear can hear sound of frequencies that range from about 20 Hz to about 20,000 Hz (though many people have rather limited hearing above about 15,000 Hz). If the speed of sound in air is 340 m/s, what are the wavelengths that correspond to these extreme frequencies?

If the speed of sound in air is 340 m/s, the wavelength corresponding to the lowest audible frequency is

$$\lambda = \frac{v}{f} = \frac{340\ \text{m/s}}{20\ \text{Hz}} = 17\ \text{m}$$

and that corresponding to the highest audible frequency is

$$\lambda = \frac{v}{f} = \frac{340\ \text{m/s}}{20\ \text{kHz}} = 1.7\ \text{cm}$$

Exercise

If the speed of sound in air is 340 m/s, what is the frequency of a sound wave whose wavelength is 2 m? (Answer: 170 Hz)

Derivation of Equations 14-12 and 14-13 (Optional)

For a sound wave in a gas, we can relate the change in pressure to the displacement by noting that a displacement changes the volume of a given mass of gas, and that changes in pressure and volume are related by the bulk modulus. Since we are using p for a change in pressure here, we have from the definition of the bulk modulus

$$p = -B\,\frac{\Delta V}{V}$$

Using $B = \rho v^2$ from Equation 14-1, we have

$$p = -\rho v^2\,\frac{\Delta V}{V} \tag{14-14}$$

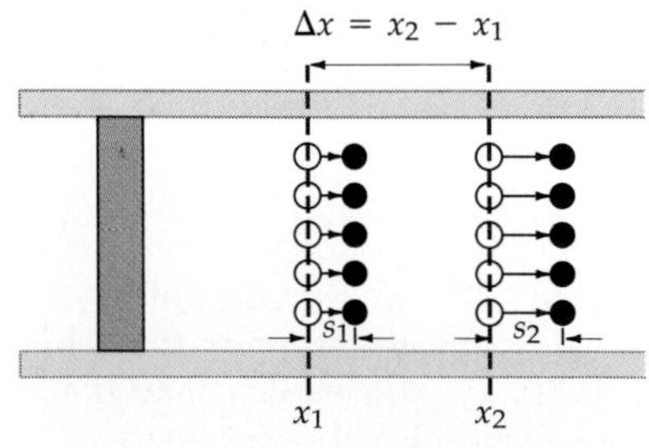

Figure 14-3 Change in the volume of a given mass of gas due to a variation in the displacement with position. The original volume is $V = A\,\Delta x$, and the change in volume is $\Delta V = A\,\Delta s$, where A is the cross-sectional area of the tube.

Let us consider the mass of gas in the tube shown in Figure 14-3 that is initially between points x_1 and x_2. It occupies a volume $V = A(x_2 - x_1) = A\,\Delta x$, where A is the cross-sectional area of the tube and $\Delta x = x_2 - x_1$. In general, a displacement of a given mass of gas causes a change in its volume. If the displacement is the same at points x_1 and x_2, there will be no change in the volume, but if the displacement is different at these two points, the volume will change. For example, if the displacement is greater at x_2 than at x_1, as is the case in Figure 14-3, the volume will increase. If $\Delta s = s(x_2, t_0) - s(x_1, t_0)$ is the difference in the displacement at these two points, the change in the volume is

$$\Delta V = A\,\Delta s$$

Substituting this and $V = A\,\Delta x$ into Equation 14-14, we obtain

$$p = -\rho v^2\,\frac{A\,\Delta s}{A\,\Delta x} = -\rho v^2\,\frac{\Delta s}{\Delta x} \tag{14-15}$$

In the limit as Δx goes to zero, the ratio $\Delta s/\Delta x$ becomes the derivative of s with respect to x. This is a partial derivative because x is also a function of t, which we are holding constant. Replacing $\Delta s/\Delta x$ with the notation for the partial derivative $\partial s/\partial x$, we have

$$p = -\rho v^2 \frac{\partial s}{\partial x}$$

We compute $\partial s/\partial x$ from Equation 14-8:

$$s(x, t) = s_0 \sin (kx - \omega t)$$

$$\frac{\partial s}{\partial x} = s_0 k \cos (kx - \omega t)$$

Hence,

$$p = -\rho v^2 s_0 k \cos (kx - \omega t)$$

$$= +k\rho v^2 s_0 \sin \left(kx - \omega t - \frac{\pi}{2}\right)$$

$$= p_0 \sin \left(kx - \omega t - \frac{\pi}{2}\right) \qquad \text{14-16}$$

where

$$p_0 = k\rho v^2 s_0$$

But according to Equation 14-11, $kv = \omega$. Therefore,

$$p_0 = \rho\omega v s_0 \qquad \text{14-17}$$

and Equations 14-16 and 14-17 are identical to Equations 14-12 and 14-13.

14-3 Waves in Three Dimensions: Intensity

So far, we have discussed one-dimensional waves, that is, waves that propagate in a straight line. Figure 14-4 shows two-dimensional circular waves on the surface of water in a ripple tank. These waves are generated by a point source moving up and down with simple harmonic motion. The wavelength, in this case, is the distance between successive wave crests, which are concentric circles. These circles are called **wavefronts.** For a point source of sound, the waves are emitted in three dimensions. They move out in all directions, and the wavefronts are concentric spherical surfaces.

Figure 14-4 Circular wavefronts diverging from a point source in a ripple tank.

The motion of any set of wavefronts can be indicated by **rays,** which are directed lines perpendicular to the wavefronts (Figure 14-5). For circular or spherical waves, the rays are radial lines.

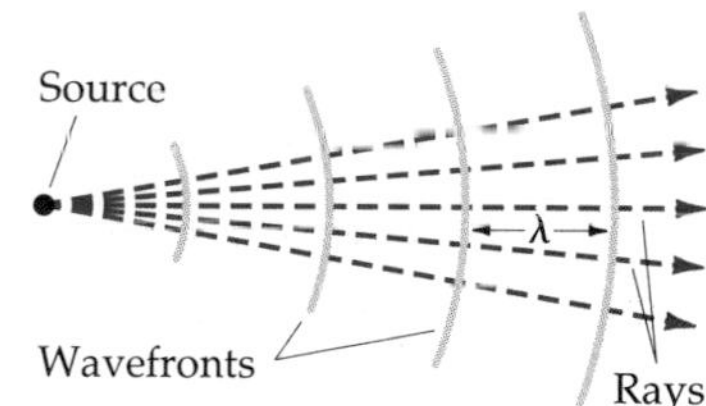

Figure 14-5 The motion of wavefronts can be represented by rays drawn perpendicular to the wavefronts. For a point source, the rays are radial lines diverging from the source.

If a point source emits waves uniformly in all directions, the energy at a distance r from the source will be distributed uniformly on a spherical shell of radius r and area $4\pi r^2$. If P is the power emitted by the source, the power per unit area at a distance r from the source will be $P/4\pi r^2$. The average power per unit area that is incident perpendicular to the direction of propagation is called the **intensity:**

$$I = \frac{P_{av}}{A} \qquad \text{14-18}$$

The units of intensity are watts per square meter. At a distance r from a point source, the intensity is

$$I = \frac{P_{av}}{4\pi r^2} \qquad \text{14-19}$$

Intensity due to a point source

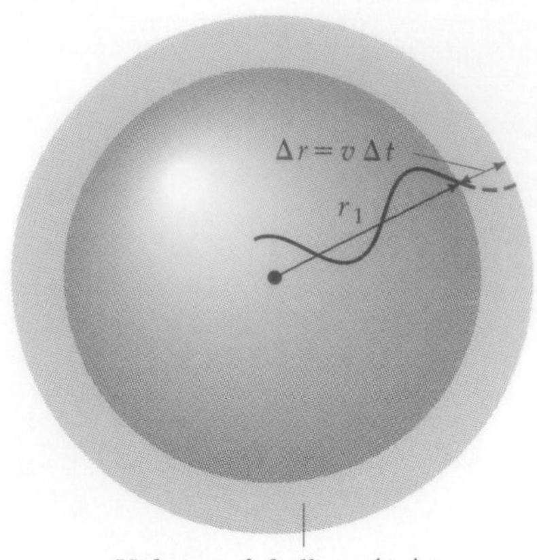

Figure 14-6 A spherical wave that has just reached the distance r_1 at this time. All the energy is contained in the sphere of this radius. In some time Δt, the wave moves out a distance $v\,\Delta t$ and occupies the additional volume $Av\,\Delta t$ of the spherical shell. The energy transmitted past r_1 is $\eta Av\,\Delta t$, where η is the energy per unit volume and A is the average surface area of the spherical shell.

The intensity of a three-dimensional wave varies inversely with the square of the distance from a point source.

There is a simple relation between the intensity of a wave and the energy per unit volume in the medium carrying the wave. Consider the spherical wave that has just reached the radius r_1 in Figure 14-6. The volume inside the radius r_1 contains energy because the particles in that region are oscillating with simple harmonic motion. The region outside r_1 contains no energy because the wave has not yet reached that region. After a short time Δt, the wave moves out past r_1 a short distance $\Delta r = v\,\Delta t$. The total energy in the medium is increased by the energy in the spherical shell of surface area A, thickness $v\,\Delta t$, and volume $\Delta V = Av\,\Delta t$. The additional energy in the spherical shell is

$$\Delta E = \eta\,\Delta V = \eta Av\,\Delta t$$

where η is the average energy per unit volume in the shell that now contains energy. The rate of increase of energy is the power passing into the shell. The source of this energy is at the center of the sphere from which the wave is radiating. Thus, the average incident power is

$$P_{\text{av}} = \frac{\Delta E}{\Delta t} = \eta Av$$

and the intensity of the wave is

$$I = \frac{P_{\text{av}}}{A} = \eta v \qquad \text{14-20}$$

Thus, the intensity equals the product of the wave speed v and the average energy per unit volume η. This result applies to all waves.

The energy of a sound wave in air is the energy of oscillation of air molecules that vibrate in simple harmonic motion along the direction of wave propagation. As found in Chapter 12, the total energy of a mass m oscillating in simple harmonic motion with angular frequency ω and amplitude A is $E = \frac{1}{2}m\omega^2A^2$. Writing s_0 for the amplitude and $\Delta m = \rho\,\Delta V$ for the mass, where ρ is the density, we have for the energy of a sound wave in volume element ΔV

$$\Delta E = \tfrac{1}{2}\rho\omega^2 s_0^2\,\Delta V$$

The energy density η is thus

$$\eta = \frac{\Delta E}{\Delta V} = \tfrac{1}{2}\rho\omega^2 s_0^2 \qquad \text{14-21}$$

and the intensity is

$$I = \eta v = \tfrac{1}{2}\rho\omega^2 s_0^2 v = \frac{1}{2}\frac{p_0^2}{\rho v} \qquad \text{14-22}$$

where we have used $s_0 = p_0/\rho\omega v$ from Equation 14-13 to relate the displacement amplitude s_0 to the pressure amplitude p_0. This result—that the intensity of a sound wave is proportional to the square of the amplitude—is a general property of harmonic waves.

The human ear can accommodate a rather large range of sound-wave intensities, from about 10^{-12} W/m^2 (which is usually taken to be the threshold of hearing) to about 1 W/m^2 (which produces a sensation of pain in most people). The pressure variations that correspond to these extreme intensities are about 3×10^{-5} Pa for the hearing threshold and 30 Pa for the pain threshold. (Recall that a pascal is a newton per square meter.) These small pressure variations are superimposed on the normal atmospheric pressure of about 101 kPa.

Loudness and Intensity Level

Because of the enormous range of intensities to which the ear is sensitive and because the psychological sensation of loudness does not vary directly with intensity but more nearly logarithmically, a logarithmic scale is used to describe the intensity level of a sound wave. The **intensity level** β measured in **decibels** (dB) is defined by

$$\beta = 10 \log \frac{I}{I_0} \qquad 14\text{-}23$$

Intensity level in dB

where I is the intensity of the sound and I_0 is a reference level, which we will take to be the threshold of hearing:

$$I_0 = 10^{-12} \text{ W/m}^2 \qquad 14\text{-}24$$

On this scale, the threshold of hearing is

$$\beta = 10 \log \frac{I_0}{I_0} = 0 \text{ dB}$$

and the pain threshold is

$$\beta = 10 \log \frac{1}{10^{-12}} = 10 \log 10^{12} = 120 \text{ dB}$$

Thus, the range of sound intensities from 10^{-12} W/m^2 to 1 W/m^2 corresponds to a range of intensity levels from 0 dB to 120 dB. Table 14-1 lists the intensity levels of some common sounds.

Exercise

What is the ratio of the intensity of a 90-dB sound to that of a 60-dB sound? (Answer: 1000)

Table 14-1 **Intensity and Intensity Level of Some Common Sounds** ($I_0 = 10^{-12}$ W/m^2)

Source	I/I_0	dB	Description
	10^0	0	Hearing threshold
Normal breathing	10^1	10	Barely audible
Rustling leaves	10^2	20	
Soft whisper (at 5 m)	10^3	30	Very quiet
Library	10^4	40	
Quiet office	10^5	50	Quiet
Normal conversation (at 1 m)	10^6	60	
Busy traffic	10^7	70	
Noisy office with machines; average factory	10^8	80	
Heavy truck (at 15 m); Niagara Falls	10^9	90	Constant exposure endangers hearing
Old subway train	10^{10}	100	
Construction noise (at 3 m)	10^{11}	110	
Rock concert with amplifiers (at 2 m); jet takeoff (at 60 m)	10^{12}	120	Pain threshold
Pneumatic riveter; machine gun	10^{13}	130	
Jet takeoff (nearby)	10^{15}	150	
Large rocket engine (nearby)	10^{18}	180	

Example 14-3

A barking dog delivers about 1 mW of power. (*a*) If this power is uniformly distributed in all directions, what is the sound intensity level at a distance of 5 m? (*b*) What would the intensity level be of two dogs barking at the same time if each delivered 1 mW of power?

(*a*) The intensity at a distance of 5 m is the power divided by the area (Equation 14-19):

$$I = \frac{P}{4\pi r^2} = \frac{10^{-3}\ \text{W}}{4\pi(5\ \text{m})^2} = 3.18 \times 10^{-6}\ \text{W/m}^2$$

The intensity level at this distance is

$$\beta = 10 \log \frac{I}{I_0} = 10 \log \frac{3.18 \times 10^{-6}}{10^{-12}}$$

$$= 10 \log (3.18 \times 10^6)$$

$$= 10 (\log 3.18 + \log 10^6)$$

$$= 10(0.50 + 6) = 65.0\ \text{dB}$$

(*b*) If there are two dogs barking at the same time, the intensity will be 2 times as great, or

$$I = 2 \times (3.18 \times 10^{-6}\ \text{W/m}^2) = 6.36 \times 10^{-6}\ \text{W/m}^2$$

Then I/I_0 will be

$$\frac{I}{I_0} = 6.36 \times 10^6$$

and the intensity level will be

$$\beta = 10 \log (6.36 \times 10^6) = 68.0\ \text{dB}$$

We can see from this example that if the intensity is doubled, the intensity level increases by 3 dB.

Example 14-4

A sound absorber attenuates the sound level by 30 dB. By what factor is the intensity decreased?

From Table 14-1, we can see that for every 10-dB drop in the intensity level, the intensity decreases by a factor of 10. Thus, a drop of 30 dB corresponds to a decrease in intensity by a factor of $10^3 = 1000$.

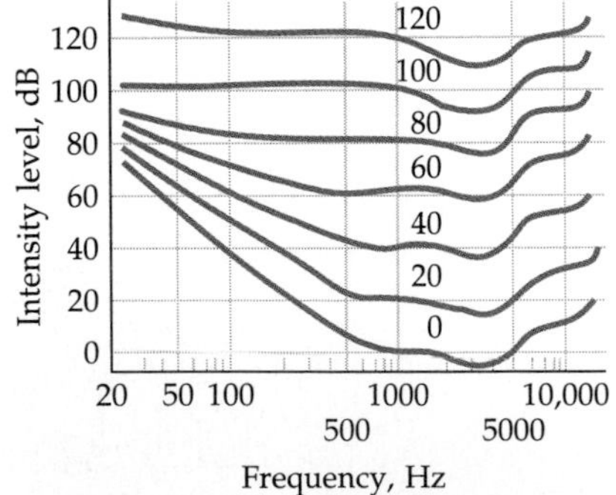

Figure 14-7 Intensity level versus frequency for sounds of equal loudness. The lowest curve is below the threshold for hearing of all but about 1 percent of the population. The second lowest curve is approximately the hearing threshold for about 50 percent of the population.

The sensation of loudness depends on the frequency as well as the intensity of a sound. Figure 14-7 is a plot of intensity level versus frequency for sounds of equal loudness to the human ear. (In this figure, the frequency is plotted on a logarithmic scale to display the wide range of frequencies from 20 Hz to 10 kHz.) The lowest curve corresponds to the threshold of hearing for someone with a very good ear. We can see from this curve that the hearing threshold is 0 dB at 1 kHz, but it is about 50 dB at 60 Hz. About 1 percent of the population have a hearing threshold this low. The second lowest curve is a more typical hearing threshold curve that applies to about 50 percent of the population. The upper curve represents the pain threshold. Note that it does not vary as much with frequency as do the lower-level curves. We note from this figure that the human ear is most sensitive at about 4 kHz for all intensity levels.

14-4 Interference: Beats

In Section 13-5, where we discussed the interference of harmonic waves, we saw that the superposition of two harmonic waves of equal amplitude results in a third harmonic wave whose amplitude depends on the phase difference between the original waves. If the two waves are in phase, the interference is constructive and the amplitude of the resultant wave is twice that of either of the original waves. If the two waves differ in phase by 180°, the interference is destructive and the waves cancel. For a phase difference δ, the pressure of the resultant wave is (Equation 13-19)

$$p_1 + p_2 = 2p_0 \cos\left(\tfrac{1}{2}\delta\right) \sin\left(kx - \omega t + \tfrac{1}{2}\delta\right) \qquad 14\text{-}25$$

A common cause of a phase difference between two sound waves is different path lengths between the sources of waves and the point of interference. Suppose we have two sources that are emitting harmonic waves of the same frequency and wavelength and are oscillating in phase; that is, when a positive crest leaves one source, a positive crest leaves the other source at the same time. If the path difference to some point is one wavelength, as is the case in Figure 14-8, the interference is constructive. The two waves fall on top of one another with their maxima occurring at the same time and their minima also occurring at the same time. If the waves have equal amplitudes, the amplitude of the resulting wave will be twice that of either original wave. We can see from the figure that a path difference of one wavelength or any integral number of wavelengths is equivalent to no path difference at all. If the path difference is one-half wavelength or an odd number of half wavelengths, as in Figure 14-9, the maximum of one wave falls at the minimum of the other and the interference is destructive. For the general case, we can write the wave functions

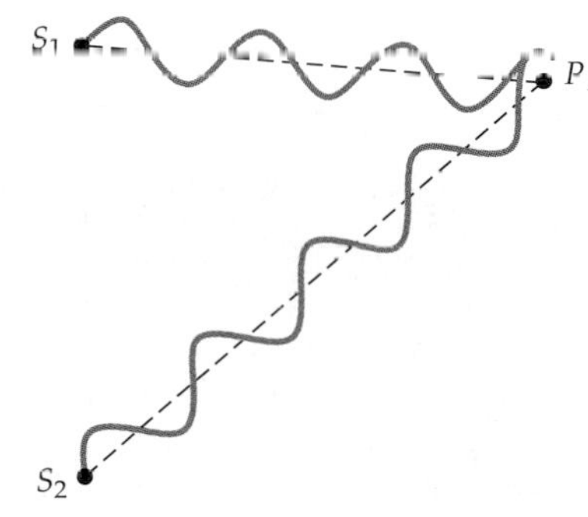

Figure 14-8 Waves from two sources S_1 and S_2 meeting at a point P_1 that is farther from one source than the other by one wavelength λ. The waves are in phase at P_1 and interfere constructively. They will also interfere constructively at any other point for which the path difference is zero or an integral number of wavelengths.

$$p_1 = p_0 \sin(kx_1 - \omega t)$$

and

$$p_2 = p_0 \sin(kx_2 - \omega t)$$

The phase difference for these two wave functions is

$$\delta = (kx_2 - \omega t) - (kx_1 - \omega t) = k(x_2 - x_1) = k\,\Delta x$$

Using $k = 2\pi/\lambda$, we have

$$\delta = 2\pi\frac{\Delta x}{\lambda} = (360°)\frac{\Delta x}{\lambda} \qquad 14\text{-}26$$

Phase difference due to path difference

Figure 14-9 Waves from two sources S_1 and S_2 meeting at a point P_2 that is farther from one source than the other by $\frac{1}{2}\lambda$. The sources are in phase, but the waves at P_2 are out of phase by 180° and therefore interfere destructively. If the waves are of equal amplitude at P_2, they will cancel completely at this point. They will also interfere destructively at any other point for which the path difference is $\frac{1}{2}\lambda$ or any odd number of half wavelengths.

Example 14-5

Two sound sources oscillate in phase. At a point 5.00 m from one source and 5.17 m from the other, the amplitude of the sound from each source separately is p_0. Find the amplitude of the resultant wave if the frequency of the sound waves is (*a*) 1000 Hz, (*b*) 2000 Hz, and (*c*) 500 Hz. (Use 340 m/s for the speed of sound.)

In all the cases, the path difference is $\Delta x = 5.17\text{ m} - 5.00\text{ m} = 0.17\text{ m} = 17\text{ cm}$.

(*a*) If the frequency is 1000 Hz, the wavelength is

$$\lambda = \frac{v}{f} = \frac{340\text{ m/s}}{1000\text{ Hz}} = 0.34\text{ m} = 34\text{ cm} = 2\,\Delta x$$

Since the path difference is one-half the wavelength, the phase difference is π and the resultant amplitude will be 0.

(*b*) When the frequency is 2000 Hz, the wavelength is

$$\lambda = \frac{340\text{ m/s}}{2000\text{ Hz}} = 17\text{ cm} = \Delta x$$

In this case, the path difference is one wavelength and the resultant amplitude is $2p_0$.

(*c*) For $f = 500$ Hz, the wavelength is

$$\lambda = \frac{340\text{ m/s}}{500\text{ Hz}} = 68\text{ cm} = 4\,\Delta x$$

The phase difference is then

$$\delta = 2\pi\frac{\Delta x}{\lambda} = 2\pi\,\frac{\Delta x}{4\,\Delta x} = \frac{\pi}{2}$$

The factor $2p_0 \cos \frac{1}{2}\delta$ in Equation 14-25 is the amplitude of the resulting wave. For $\delta = \pi/2$, the amplitude is

$$2p_0 \cos \tfrac{1}{2}\left(\frac{\pi}{2}\right) = 2p_0 \cos\left(\frac{\pi}{4}\right) = 1.41\,p_0$$

Figure 14-10*a* shows the wave pattern produced by two point sources separated by a small distance that are oscillating in phase, each producing circular waves of wavelength λ. We can construct a similar pattern with a compass by drawing circular arcs representing the wave crests from each source at some particular time (Figure 14-10*b*). At the points where the crests from each source overlap, the waves interfere constructively. At these points, the paths for the waves from the two sources are either equal in length or differ by an integral number of wavelengths. The dashed lines

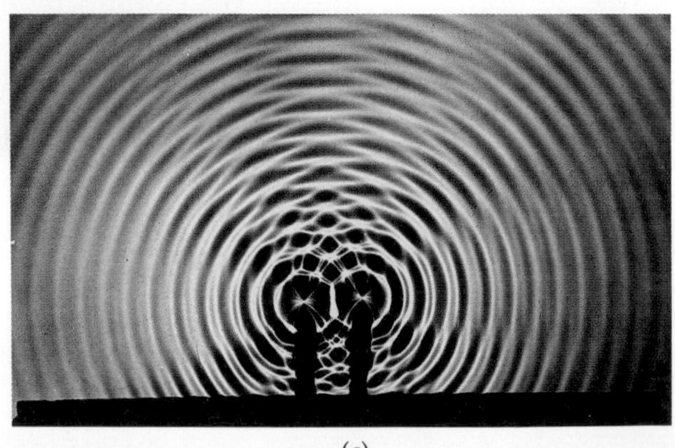

(*a*)

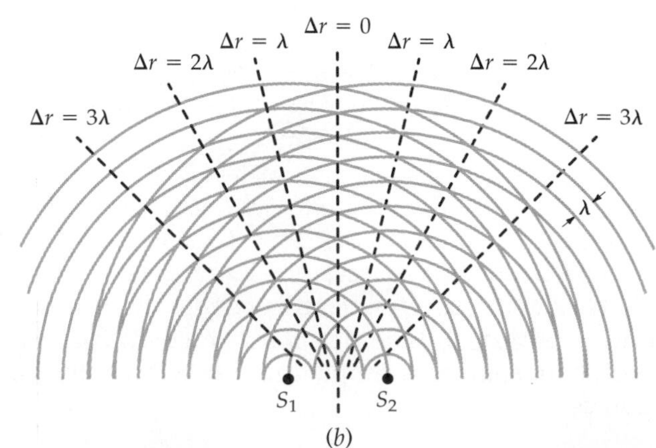

(*b*)

Figure 14-10 (*a*) Interference pattern of two point sources oscillating in phase in a ripple tank. (*b*) Geometric construction of the interference pattern in (*a*). The waves interfere constructively at the points of intersection. These points occur wherever the path lengths from the two sources differ by an integral number of wavelengths.

indicate the points that are equidistant from the sources or whose path differences are one wavelength, two wavelengths, or three wavelengths. If we assume that the two sources generate waves of equal amplitude, the resulting amplitude at points where the waves interfere constructively will be twice the amplitude of either wave separately. Since the energy of a wave is proportional to the square of the amplitude, the energy of the resultant wave where the waves interfere constructively will be 4 times that of either wave separately. Between each pair of interference maxima is an interference minimum for which the path difference is an odd number of half wavelengths. The lines along which the waves completely cancel are called nodes or nodal lines. At any point between the maxima and minima, the amplitude of the resultant wave is given by $A = 2p_0 \cos \frac{1}{2}\delta$, where p_0 is the amplitude of each wave separately, and δ is related to the path difference by Equation 14-26.

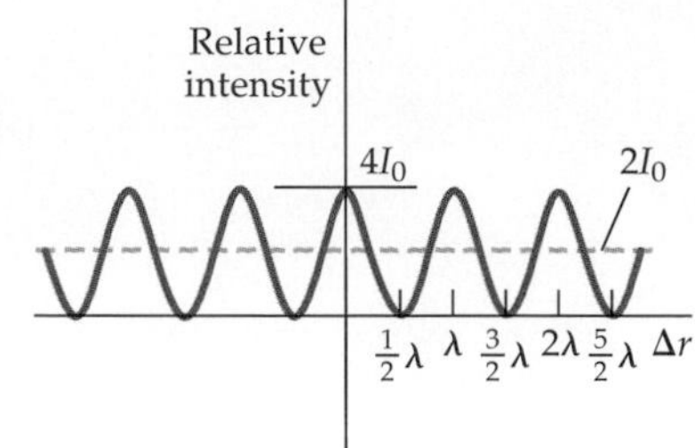

Figure 14-11 Plot of relative intensity versus path difference for two sources that are in phase. When the path difference is an integral number of wavelengths, the intensity is $4I_0$, where I_0 is the intensity due to each source individually. When the path difference is an odd number of half wavelengths, the intensity is zero. The dashed line indicates the average intensity, which is twice that due to either source alone.

Figure 14-11 shows the intensity of the resultant wave from two sources as a function of path difference. At points where the interference is constructive, the intensity is 4 times that due to either source alone because the amplitude of the resultant wave is twice that of either wave alone. At points of destructive interference, the intensity is zero. The average intensity, shown by the dashed line in the figure, is twice the intensity due to either source alone, a result required by the conservation of energy. We can see that, as a result of the interference of the waves from the two sources, the energy is redistributed in space.

The interference of two sound sources can be demonstrated by driving two separated speakers with the same amplifier (so that they are always in phase) fed by an audio-signal generator. Moving about the room, one can detect by ear the positions of constructive and destructive interference. (In this demonstration, the sound intensity will not be quite zero at the points of destructive interference of the direct sound waves because of sound reflections from the walls or objects in the room.)

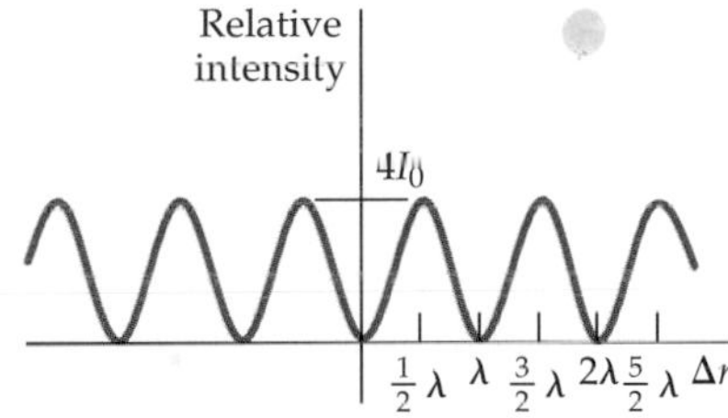

Figure 14-12 Plot of relative intensity versus path difference for two sources that are 180° out of phase. The pattern is the same as that in Figure 14-11 except that it is shifted by one-half wavelength.

Two sources need not be in phase to produce an interference pattern. Figure 14-12 shows a plot of the intensity versus path difference produced by two sources that are 180° out of phase. (Two sound sources that are in phase can be made to be out of phase by 180° merely by switching the leads to one of the speakers.) The pattern is the same as that in Figure 14-11 except that the maxima and minima are interchanged. At points equidistant from the sources or those for which the distance differs by an integral number of wavelengths the interference is destructive because the waves are 180° out of phase and cancel (Figure 14-13). At points where the path difference is an odd number of half wavelengths, the waves are now in phase because the 180° phase difference of the sources is offset by the 180° phase difference due to the path difference.

No matter what the phase difference between the sources might be, similar interference patterns will be produced as long as the phase difference is constant over time. Two sources that are in phase or have a constant phase difference are said to be **coherent sources.** Coherent sources of water waves in a ripple tank are easy to produce by driving both sources with the same motor. Coherent sound sources are obtained by driving two speakers with the same signal source and amplifier. Wave sources whose phase difference is not constant over time but varies randomly are said to be incoherent sources. There are many examples of incoherent sources, such as two speakers driven by different amplifiers or two violins bowed by different players. In general, any two separate light sources, such as two candles, are incoherent. For incoherent sources, the interference at a particular point varies rapidly back and forth from constructive to destructive and no infererence pattern is observed. The resultant intensity of waves from two or more incoherent sources is simply the sum of the intensities due to the individual sources.

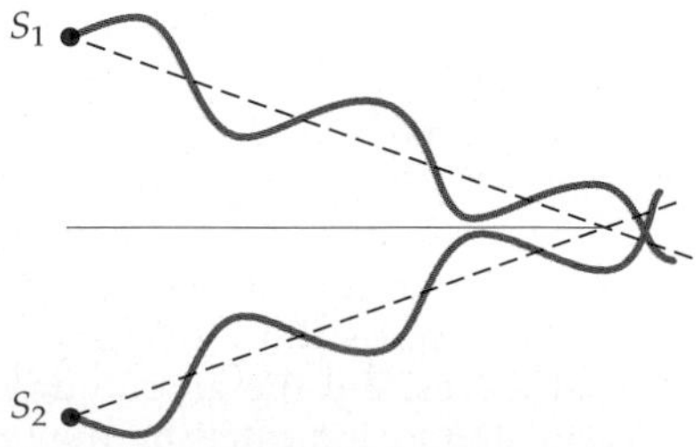

Figure 14-13 At a point equidistant from two sources that are out of phase by 180°, the waves cancel.

Beats

The interference of two waves of different but nearly equal frequencies produces the interesting phenomenon known as **beats.** A familiar example is the beats produced by the sound waves from two tuning forks or two guitar strings of nearly equal but not identical frequencies. What is heard is a tone whose intensity varies alternately between loud and soft. The frequency of this variation in intensity is called the **beat frequency.**

Let us consider two sound waves with angular frequencies of ω_1 and ω_2 having the same pressure amplitude p_0. We consider how the resultant wave behaves over time at a single point in space. We therefore neglect the spatial dependence of the wave, which merely contributes a phase constant. The pressure at the ear due to either wave acting alone will be a simple harmonic function of the type

$$p_1 = p_0 \sin \omega_1 t$$

or

$$p_2 = p_0 \sin \omega_2 t$$

where we have chosen sine functions for convenience and have assumed that the waves are in phase at time $t = 0$. Using

$$\sin \theta_1 + \sin \theta_2 = 2 \cos \tfrac{1}{2}(\theta_1 - \theta_2) \sin \tfrac{1}{2}(\theta_1 + \theta_2)$$

for the sum of two sine functions, we obtain for the resultant wave

$$\begin{aligned} p &= p_0 \sin \omega_1 t + p_0 \sin \omega_2 t \\ &= 2p_0 \cos \tfrac{1}{2}(\omega_1 - \omega_2)t \sin \tfrac{1}{2}(\omega_1 + \omega_2)t \end{aligned}$$

If we write $\omega_{av} = \frac{1}{2}(\omega_1 + \omega_2)$ for the average angular frequency and $\Delta\omega = \omega_1 - \omega_2$ for the difference in angular frequencies, the resultant wave function is

$$\begin{aligned} p &= 2p_0 \cos (\tfrac{1}{2}\,\Delta\omega\, t) \sin \omega_{av} t \\ &= 2p_0 \cos (2\pi\tfrac{1}{2}\,\Delta f\, t) \sin 2\pi f_{av} t \end{aligned} \qquad \text{14-27}$$

where $\Delta f = \Delta\omega/2\pi$ and $f_{av} = \omega_{av}/2\pi$.

Figure 14-14 shows a plot of the pressure variation as a function of time. The waves are originally in phase and add constructively at time $t = 0$. Because their frequencies differ, the waves gradually become out of phase, and at time t_1 they are 180° out of phase and interfere destructively. (Complete cancellation occurs only when the pressure amplitudes of the two waves are equal.) An equal time interval later (time t_2 in the figure), the two waves will again be in phase and will interfere constructively. The greater the difference in frequencies of the two waves, the sooner they become out of phase and then back in phase again.

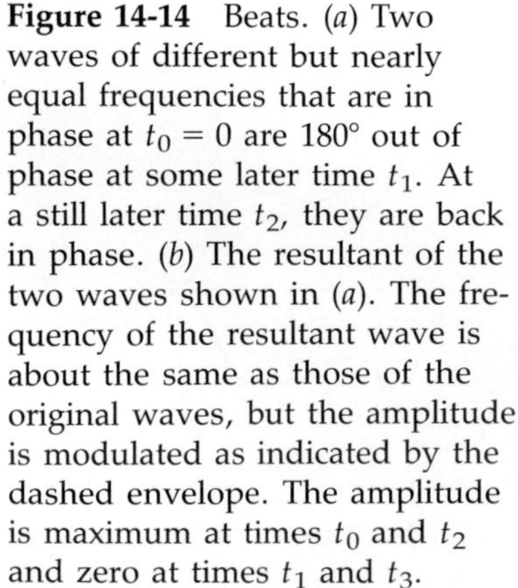

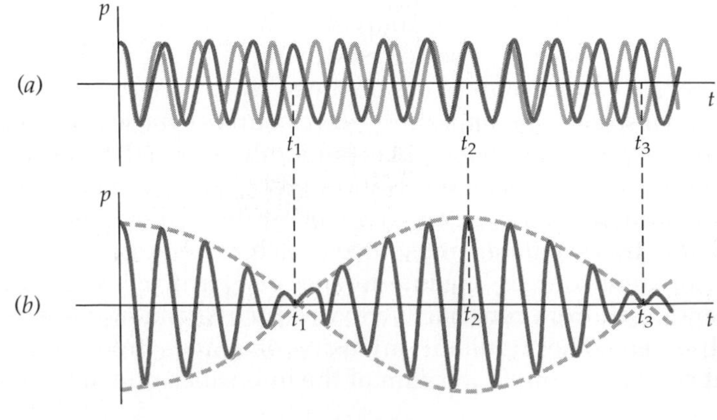

Figure 14-14 Beats. (*a*) Two waves of different but nearly equal frequencies that are in phase at $t_0 = 0$ are 180° out of phase at some later time t_1. At a still later time t_2, they are back in phase. (*b*) The resultant of the two waves shown in (*a*). The frequency of the resultant wave is about the same as those of the original waves, but the amplitude is modulated as indicated by the dashed envelope. The amplitude is maximum at times t_0 and t_2 and zero at times t_1 and t_3.

The ear hears the average frequency $f_{av} = \frac{1}{2}(f_1 + f_2)$ with an amplitude of $2p_0 \cos(2\pi\frac{1}{2}\,\Delta f\,t)$. The amplitude thus oscillates with the frequency $\frac{1}{2}\,\Delta f$. Since the intensity is proportional to the square of the amplitude, the sound is loud whenever the amplitude is either maximum or minimum. Since the maximum amplitude occurs with frequency $\frac{1}{2}\,\Delta f$, the frequency of maximum and minimum amplitudes is just twice this or Δf. Thus, the **beat frequency** equals the difference in the individual frequencies of the two waves:

$$f_{\text{beat}} = \Delta f \qquad \text{14-28}$$

The closer the two waves are in frequency, the longer it takes for them to move out of phase and back in phase, so the closer the waves are in frequency, the smaller the beat frequency. For example, if we simultaneously strike two tuning forks having the frequencies 241 Hz and 243 Hz, we will hear a pulsating tone at the average frequency of 242 Hz that has a maximum intensity 2 times per second; that is, the beat frequency will be 2 Hz.

The phenomenon of beats is often used to compare an unknown frequency with a known frequency, as when a tuning fork is used to tune a piano string. The ear can detect up to about 15 to 20 beats per second. Above this frequency, the fluctuations in loudness are too rapid to be heard. Beats are also used to detect small frequency changes such as those that are produced when a radar beam is reflected from a moving car. The shift in frequency of the reflected beam is due to the doppler effect (Section 14-9), which depends on the speed of the car. The speed can be determined by measuring the beats produced when the reflected radar beam is combined with the original radar beam.

An interesting phenomenon somewhat related to beats is the moiré pattern that is produced when two sets of parallel lines of slightly different spacing overlap (Figure 14-15). Moiré patterns can often be observed when one looks through two parallel fences along a road or through two window screens.

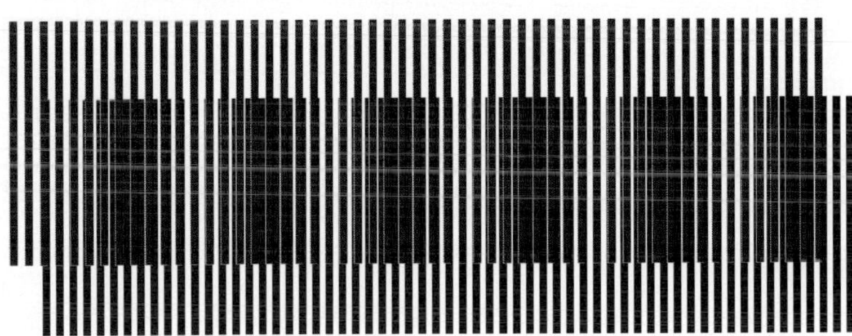

Figure 14-15 Moiré pattern produced by the overlap of two sets of parallel lines where the spacing of one set differs slightly from that of the other.

Example 14-6

When a 440-Hz tuning fork is struck simultaneously with the playing of the note A on a guitar, 3 beats per second are heard. After the guitar string is tightened slightly to increase its frequency, the beat frequency increases to 6 beats per second. What is the frequency of the guitar string after it is tightened?

Since 3 beats per second were heard initially, the original frequency of the guitar string was either 443 Hz or 437 Hz. Had it been 437 Hz, increasing the string's frequency by slightly tightening it would decrease the beat frequency because the frequency of the string then would be closer to that of the tuning fork. Since the beat frequency increases to 6 beats per second, the original frequency has to have been 443 Hz and the new frequency is 446 Hz. To tune the string to 440 Hz, the string should be loosened slightly to lower its frequency.

Questions

2. When musical notes are sounded together to make chords, beats are not heard. Why not?
3. About how accurately do you think you can tune a piano string to a tuning fork?

14-5 Standing Sound Waves

Much of what we learned about standing waves on strings in Chapter 13 can be applied to standing sound waves. Figure 14-16 shows a tube of air closed at the right end and fitted with a movable piston at the left end. Because the air cannot vibrate past the closed end, this point must be a displacement node. If the piston at the left end vibrates with an amplitude that is small relative to the amplitude of the waves in the air, then that end is also approximately a displacement node. (This approximation is similar to that for the driven end of a vibrating string.)

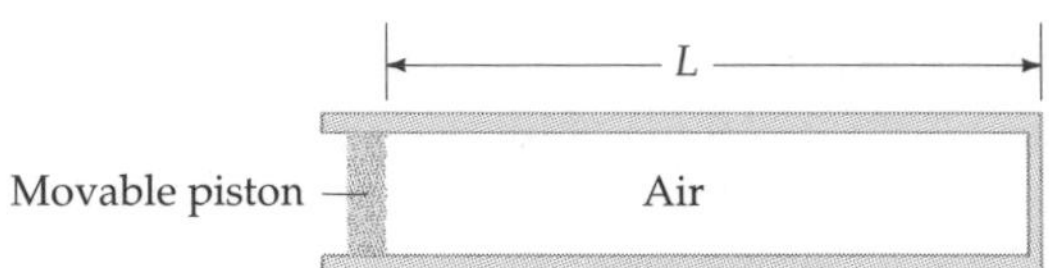

Figure 14-16 Air confined to a tube closed at both ends. There is a displacement node at each end if the amplitude of oscillation of the piston is small. The condition for standing sound waves is the same as that for standing waves on a string fixed at both ends.

The standing-wave condition for this system is the same as that for a string fixed at both ends, and all the same equations apply. The length of the tube L must contain an integral number of half wavelengths. That is, the allowed wavelengths are those that can be fitted into the length of the tube such that there are displacement nodes at each end. Thus,

$$L = n\frac{\lambda_n}{2} \qquad n = 1, 2, 3, \ldots \tag{14-29}$$

The allowed frequencies are then

$$f_n = \frac{v}{\lambda_n} = \frac{v}{(2L/n)}$$

or

Standing-wave frequencies, both ends closed

$$f_n = n\frac{v}{2L} = nf_1 \qquad n = 1, 2, 3, \ldots \tag{14-30}$$

where

$$f_1 = \frac{v}{2L}$$

is the fundamental frequency.

As we have seen, a sound wave can be thought of as a pressure wave; that is, the longitudinal vibrations of the air molecules back and forth cause variations in the air pressure. These pressure variations are sinusoidal if the displacement is sinusoidal, as in a harmonic wave. However, the pressure and displacement variations in a sound wave are 90° out of phase. Thus, in a standing sound wave, the displacement nodes are pressure antinodes and vice versa.

There are no musical instruments based on tubes that are closed at both ends. However, there are electrical devices—antennas and cavity resonators, for example—that are analogous to this case.

Example 14-7

If the speed of sound is 340 m/s, what are the allowed frequencies and wavelengths for standing sound waves in a closed tube 1 m long?

The fundamental frequency is the lowest allowed frequency, which corresponds to the longest wavelength

$$\lambda_1 = 2L = 2 \text{ m}$$

Thus, the fundamental frequency is

$$f_1 = \frac{v}{\lambda_1} = \frac{340 \text{ m/s}}{2 \text{ m}} = 170 \text{ Hz}$$

The frequencies of the other harmonics are

$$f_2 = 2f_1 = 2(170 \text{ Hz}) = 340 \text{ Hz}$$

$$f_3 = 3f_1 = 3(170 \text{ Hz}) = 510 \text{ Hz}$$

$$f_n = nf_1 = 170n \text{ Hz}$$

The other wavelengths are

$$\lambda_2 = \frac{2L}{2} = 1 \text{ m}$$

$$\lambda_n = \frac{2L}{n} = \frac{2 \text{ m}}{n}$$

If the end of the tube at the right in Figure 14-16 is not closed but is open to the atmosphere, that open end is a displacement antinode. (It is also a pressure node since the pressure is fixed at atmospheric pressure.) The standing-wave condition for this system is the same as that for a string with one end fixed and one end free. The length of the tube must equal an odd integer times $\lambda/4$. That is, the wavelength of the fundamental mode is 4 times the length of the tube, and only the odd harmonics are present. Thus,

$$L = n\frac{\lambda_n}{4} \qquad n = 1, 3, 5, \ldots \tag{14-31}$$

The allowed frequencies are therefore

$$f_n = n\frac{v}{4L} = nf_1 \qquad n = 1, 3, 5, \ldots \tag{14-32}$$

Standing-wave frequencies, one end open

where

$$f_1 = \frac{v}{4L} \tag{14-33}$$

is the fundamental frequency.

Exercise

Find the three lowest frequencies for a tube of length 1 m that is closed at one end and open at the other.
(Answer: $f_1 = 85$ Hz, $f_3 = 3f_1 = 255$ Hz, $f_5 = 5f_1 = 425$ Hz)

The result that the open end of the tube is a displacement antinode (and pressure node) is based on the assumption that the sound wave in the tube is a one-dimensional wave, which is approximately true if the diameter of the tube is very small compared with the wavelength of the sound wave. In practice, the displacement antinode and pressure node lie slightly beyond the open end of the tube. The effective length of the tube is therefore somewhat longer than the true length. If L is the true length of the tube and ΔL is

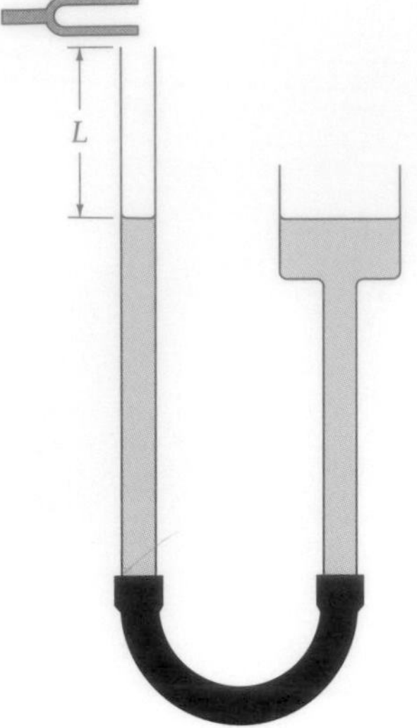

Figure 14-17 Apparatus for determining the speed of sound in air. Sound waves of the frequency of the tuning fork are excited in the tube on the left, whose length L can be adjusted by adjusting the level of the water. Resonance occurs when the effective length of the tube equals $\frac{1}{4}\lambda$, $\frac{3}{4}\lambda$, $\frac{5}{4}\lambda$, and so on, where λ is the wavelength of the sound.

the end correction for the actual position of the displacement antinode or pressure node, the effective length of the tube is

$$L_{\text{eff}} = L + \Delta L$$

The end correction ΔL is of the order of the radius of the tube. The distance between two successive nodes or antinodes is still $\frac{1}{2}\lambda$, but the distance from the open end of the tube to the first displacement node is slightly less than $\frac{1}{4}\lambda$ because of the end correction.

Figure 14-17 shows an apparatus for measuring the speed of sound in air. A narrow vertical tube on the left is partially filled with water. Since the level of the water can be adjusted, the length of the air column can be adjusted. A tuning fork of frequency f_0 vibrating at the open end of the tube sends sound waves of wavelength $\lambda_0 = v/f_0$ down the tube. They are reflected at the adjustable bottom, the water level. When the water level is such that the standing-wave condition is met, resonance occurs and the increased energy of the sound wave can be detected by ear. If ΔL is the end correction for the tube, the tube will resonate when its length is L_1, which is given by

$$L_1 + \Delta L = \tfrac{1}{4}\lambda_0$$

It will also resonate at lengths L_3, L_5, and so on, which are given by

$$L_3 + \Delta L = \tfrac{3}{4}\lambda_0$$

$$L_5 + \Delta L = \tfrac{5}{4}\lambda_0$$

and so on. The wavelength can be measured by measuring the distance between two consecutive resonance lengths. For example,

$$L_3 - L_1 = \tfrac{1}{2}\lambda_0$$

Then the speed can be found from

$$v = f_0\lambda_0 = 2f_0(L_3 - L_1)$$

Example 14-8

When a tuning fork of frequency 500 Hz is held above the tube in Figure 14-17, resonances are found when the water level is at distances 16.0, 50.5, 85.0, and 119.5 cm from the top of the tube. (*a*) What is the speed of sound in air? (*b*) How far from the end of the tube is the pressure node?

(*a*) The distance between the water levels for two successive resonances is 34.5 cm. The wavelength is thus twice this distance or 69.0 cm = 0.690 m. From these data, the speed of sound in air is

$$v = f\lambda = (500\text{ Hz})(0.690\text{ m}) = 345\text{ m/s}$$

(*b*) Since the distance to the first position of resonance is 16.0 cm and $\lambda/4 = (69.0\text{ cm})/4 = 17.25$ cm, the distance from the end of the tube to the pressure node or displacement antinode is

$$\Delta L = \tfrac{1}{4}\lambda - L = 17.25\text{ cm} - 16.0\text{ cm} = 1.25\text{ cm}$$

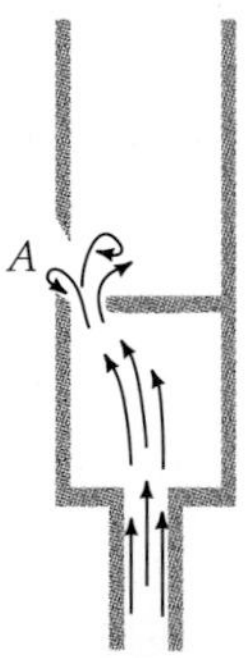

Figure 14-18 Flue-type organ pipe. A stream of air is blown against the edge causing a swirling motion of the air near A that excites standing waves in the pipe. There is a pressure node near point A, which is open to the atmosphere. The resonance frequencies of the pipe depend on the length of the pipe and on whether the top of the pipe is open or closed.

An organ pipe is a familiar example of the use of standing waves in air columns. In the flue-type organ pipe (Figure 14-18), a stream of air is directed against the sharp edge of an opening (point A in the figure). The complicated swirling motion of the air near the edge sets up vibrations in the air column. The resonance frequencies of the pipe depend on the length of the pipe and on whether the top is closed or open.

In a closed organ pipe, there is a displacement node at the closed end and an antinode near the opening (point A in Figure 14-18). The resonance frequencies for such a pipe are therefore those for a tube open at one end and closed at the other. The wavelength of the fundamental is approximately 4 times the length of the pipe, and only the odd harmonics are present. The musical instruments in the clarinet family are effectively open–closed cylinders with odd harmonics for the lowest notes. Figure 14-19 shows the standing-wave patterns in a pipe open at one end and closed at the other.

In an open organ pipe, there is a displacement antinode (and pressure node) near both ends of the pipe. The resonance frequencies for a pipe open at both ends are the same as for a pipe closed at both ends except that there is an end correction at each end. The wavelength of the fundamental is 2 times the effective length of the pipe, and all harmonics are present. Flutes and recorders behave as open–open pipes with all the harmonics present. The standing-wave patterns for a pipe open at both ends are shown in Figure 14-20.

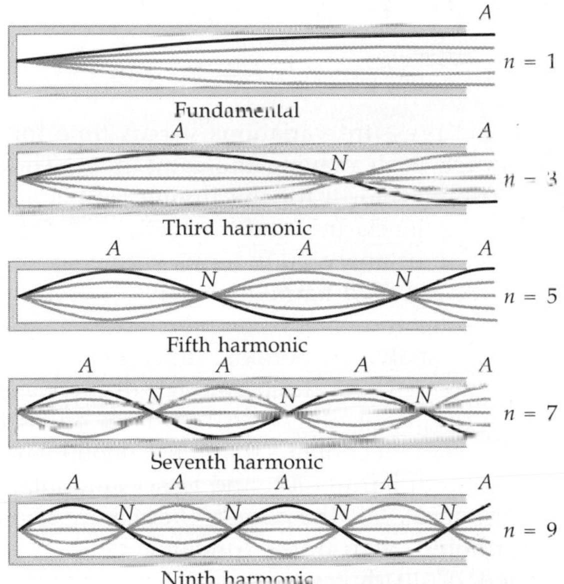

Figure 14-19 Standing-wave patterns in pipe that is open at one end and closed at the other. These plots are for the displacement, which has a node at the closed end and an antinode at the open end. As is the case for a string fixed at one end, only the odd harmonics are present.

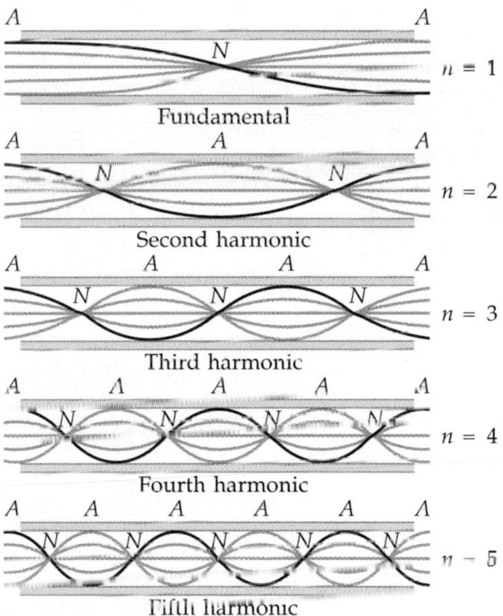

Figure 14-20 Standing-wave patterns in a pipe that is open at both ends. The displacement has an antinode at each end. As is the case for a string fixed at both ends, all harmonics are present.

As with vibrating strings, the general motion of the air in an organ pipe contains a mixture of harmonics. In addition, most musical instruments are much more complicated than simple cylindrical tubes. The conical tube, which is the basis for the oboe, bassoon, English horn, and saxophone, has a complete harmonic series with its fundamental wavelength equal to twice the length of the cone. Brass instruments are combinations of cones and cylinders. The fact that they have nearly harmonic series is a triumph of educated trial and error rather than mathematical calculation. The analysis of these instruments is extremely complex.

Questions

4. How do the resonance frequencies of an organ pipe change when the air temperature increases?

14-6 Harmonic Analysis and Synthesis

When an oboe and a violin play the same note, say, concert A, they sound quite different. Both notes have the same pitch, a physiological sensation of the highness or lowness of the note that is strongly correlated with frequency—the higher the frequency, the higher the pitch. However, the notes differ in what is called **tone quality.** The principal reason for the difference in tone quality is that, although both the violin and oboe are producing vibrations at the same fundamental frequency, 440 Hz for concert A, each instrument is also producing harmonics whose relative intensities depend on the instrument and how it is played. If each instrument produced only the fundamental frequency, the sound would be the same for each.

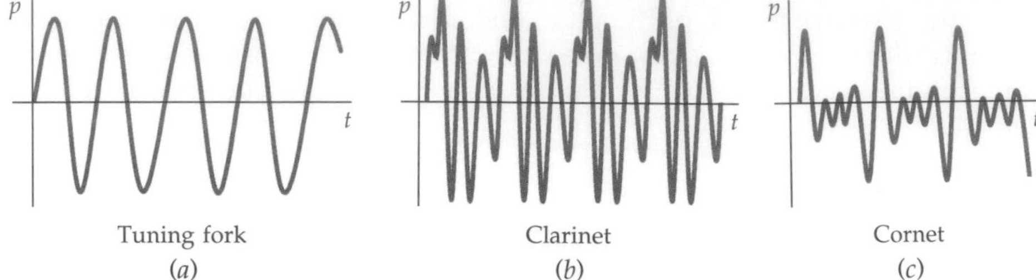

Figure 14-21 Waveforms of (*a*) a tuning fork, (*b*) a clarinet, and (*c*) a cornet, each at a fundamental frequency of 440 Hz and all at approximately the same intensity.

Figure 14-21 shows plots of the pressure variations versus time for a tuning fork, a clarinet, and a cornet, each playing the same note. These patterns are called **waveforms.** The waveform for the tuning fork is very nearly a pure sine wave, but those for the clarinet and the cornet are clearly not simple sine waves. Waveforms can be analyzed in terms of the harmonics that comprise them. Such analysis is called **harmonic analysis.** (It is also sometimes called **Fourier analysis** after the French mathematician Fourier who developed the mathematics for analyzing periodic functions.) Figure 14-22 shows a plot of the relative intensities of the harmonics of the waveforms in Figure 14-21. The waveform of the tuning fork contains only the fundamental frequency. That of the clarinet contains the fundamental, large amounts of the third, fifth, and seventh harmonics, and lesser amounts of the second, fourth, and sixth harmonics. For the cornet, there is more energy in the second and third harmonics than in the fundamental.

Although the harmonic content of steady tones is important in identifying the instrument, other factors are also important. These include

1. The attack (the way the tone begins)
2. The presence or absence of vibrato and tremolo (variations in pitch and loudness)
3. The rate of buildup of the harmonics
4. The decay (the rate at which the tone decreases from its peak level)
5. The release (the rate at which the sound level drops off at the end of the note)

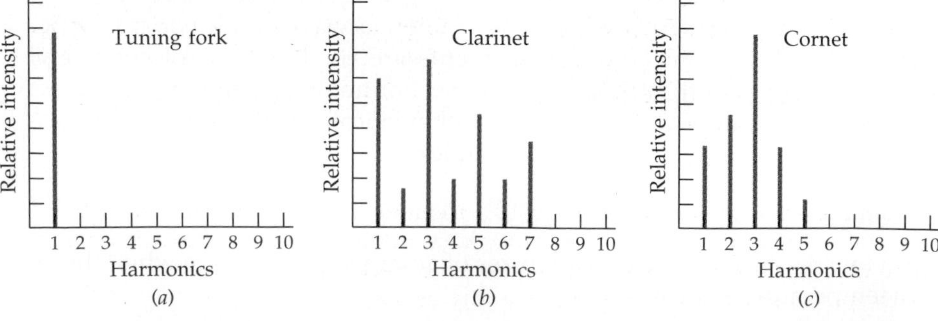

Figure 14-22 Relative intensities of the harmonics in the waveforms shown in Figure 14-21 for (*a*) the tuning fork, (*b*) the clarinet, and (*c*) the cornet.

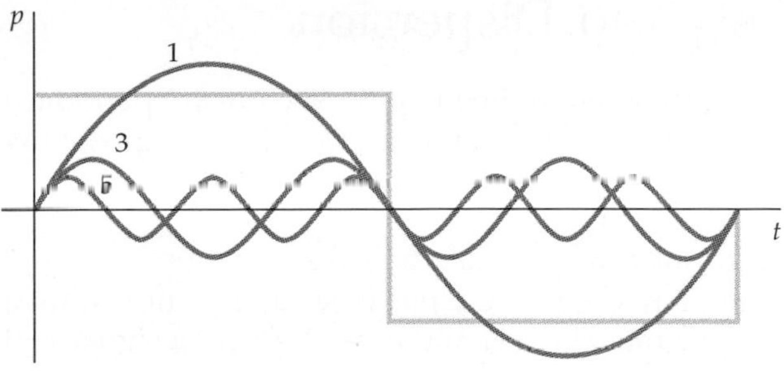

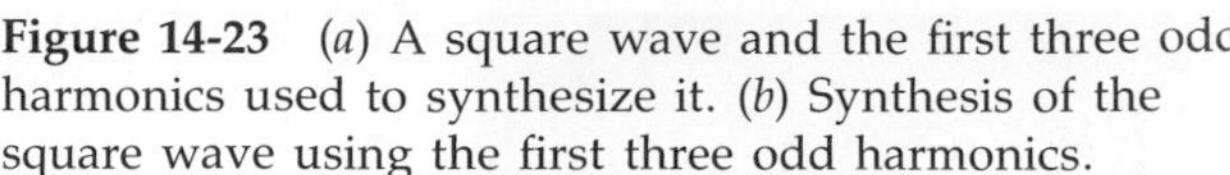

Figure 14-23 (*a*) A square wave and the first three odd harmonics used to synthesize it. (*b*) Synthesis of the square wave using the first three odd harmonics.

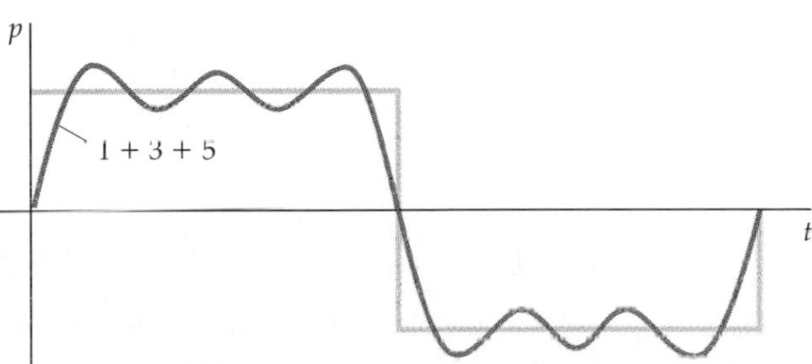

The inverse of harmonic analysis is **harmonic synthesis,** which is the construction of a periodic wave from its harmonic components. Figure 14-23 shows a square wave that can be produced by an electronic square-wave generator and the approximate synthesis of the square wave using only three harmonics. The more harmonics used in a synthesis, the closer the approximation is to the actual waveform. The relative amplitudes of the harmonics needed to synthesize the square wave are shown in Figure 14-24.

An electronic music synthesizer is a device that produces a series of harmonics whose relative amplitudes can be adjusted so that the combination produces the desired waveform. In addition, each note can be shaped by adjusting such parameters as the attack, decay, vibrato, tremolo, and release. Sophisticated modern synthesizers can produce tones that sound like any instrument in an orchestra.

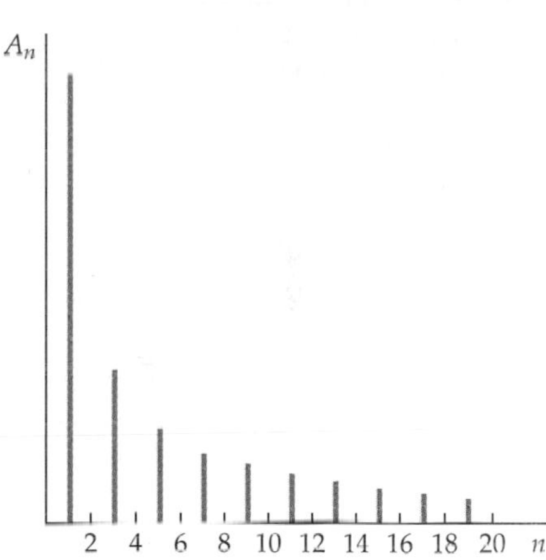

Figure 14-24 Relative amplitudes A_n of the harmonics needed to synthesize the square wave shown in Figure 14-23. The more harmonics that are used, the closer the approximation is to the square wave.

(*a*)

(*b*)

(a) Synthesizer designed by Heinrich von Helmholtz and constructed by Rudolph Koenig. Each electrically driven tuning fork has a resonator that can be mechanically opened or closed to vary the intensity of that harmonic. (b) In the more modern SY 77 Yamaha synthesizer, the harmonics are produced and mixed electronically.

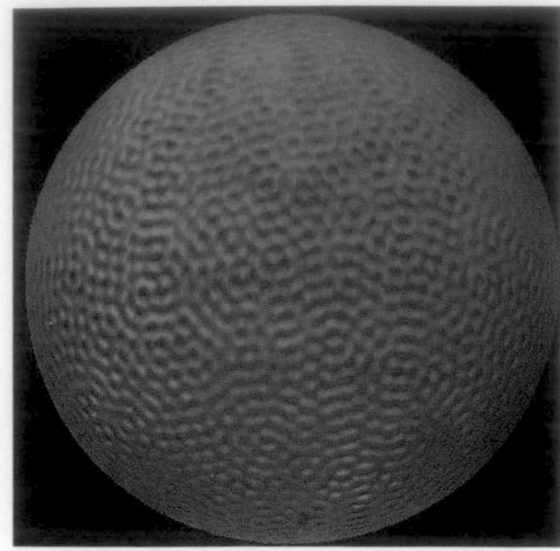

Standing sound waves on the surface of the sun. The surface of the sun is observed to oscillate with a period of about 5 min. Out of the ten million modes of oscillation known to exist, a combination of approximately 100 modes is illustrated here. The displacements of the surface are exaggerated by a factor of 1000. The period of oscillation for each mode contains information about the structure and dynamics of the solar interior.

14-7 Wave Packets and Dispersion

The complicated waveforms discussed in the previous section are periodic in time. Functions that are not periodic, such as pulses, can also be represented by harmonic (sinusoidal) wave functions, but a continuous distribution of frequencies is needed.

The characteristic feature of a wave pulse that distinguishes it from a periodic wave of a single frequency is that the pulse has a beginning and an end, whereas a harmonic wave repeats over and over. To send a signal with a wave, a pulse consisting of a group of waves of different frequencies is needed rather than a harmonic wave of a single frequency. Such a group of waves is called a **wave packet.** There is an important relation between the distribution of frequencies of the harmonic functions that make up a wave packet describing a pulse, and the time duration of the pulse. If the duration Δt is very short, the range of frequencies $\Delta\omega$ is very large. The general relation between Δt and $\Delta\omega$ is

$$\Delta\omega\,\Delta t \sim 1 \qquad 14\text{-}34$$

The exact value of this product depends on just how the quantities $\Delta\omega$ and Δt are defined. For any reasonable definitions, $\Delta\omega$ is of the order of $1/\Delta t$. We can understand this relation qualitatively from our discussion of beats (Section 14-4). The envelope in Figure 14-14*b* resembles a wave pulse, in that it has a beginning and an end; however, it repeats over and over in time. If the two waves that produce the envelope are very close in frequency, that is, if $\Delta\omega$ is very small, the time interval $\Delta t = t_3 - t_1$ (between the first time t_1 when the waves are 180° out of phase and the next time t_3) will be large. On the other hand, if the frequency difference between the two waves is large, they will quickly be out of phase again, and Δt will be small.

A wave pulse produced by a source of short duration Δt has a narrow width in space $\Delta x = v\,\Delta t$, where v is the wave speed. Each harmonic wave of frequency ω has a wave number $k = \omega/v$. A range of frequencies $\Delta\omega$ implies a range of wave numbers $\Delta k = \Delta\omega/v$. Substituting $v\,\Delta k$ for $\Delta\omega$ in Equation 14-34 gives

$$v\,\Delta k\,\Delta t \sim 1$$

or

$$\Delta k\,\Delta x \sim 1 \qquad 14\text{-}35$$

where again the exact value for the product depends on the precise definitions of Δk and Δx.

The relations expressed by Equations 14-34 and 14-35 are important characteristics of wave packets that apply to all types of waves. Since information cannot be transported by a single harmonic wave, which has no beginning or end in time, the transmission of short pulses depends on the ability to transmit a wide range of frequencies.

If a wave packet is to maintain its shape as it travels, all the component harmonic waves that make up the packet must travel the same speed. This occurs if the speed of harmonic waves does not depend on wavelength or frequency. A medium for which wave speed is independent of wavelength or frequency is called a **nondispersive medium.** Air is a nondispersive medium for sound waves, but solids and liquids generally are not.

Sometimes the wave speed in a dispersive medium depends only slightly on the frequency and wavelength. Under such conditions a wave packet will change shape as it travels, but it will travel a considerable distance as a recognizable packet. However, the speed of the packet, for example, the speed of the center of the packet, is not the same as the (average) speed of the individual component harmonic waves. The average speed of the harmonic waves is called the **phase velocity,** and the speed of the packet is called the **group velocity.**

14-8 Reflection, Refraction, and Diffraction

In a homogeneous medium, such as air at constant density, a wave travels in straight lines in the direction of the rays. At a great distance from a point source, a small part of the wavefront can be approximated by a plane, and the rays are approximately parallel lines; such a wave is called a **plane wave** (Figure 14-25). The two-dimensional analog of a plane wave is a line wave, which is a small part of a circular wavefront at a great distance from the source. Such waves can also be produced in a ripple tank by a line source as in Figure 14-26. A plane wave propagates in a straight line in the direction of the rays—much like a beam of particles.

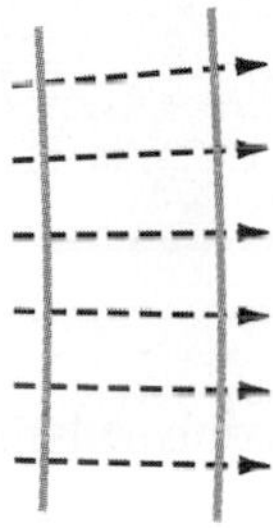

Figure 14-25 Plane waves. At great distances from a point source, the wavefronts are approximately parallel planes, and the rays are parallel lines perpendicular to the wavefronts.

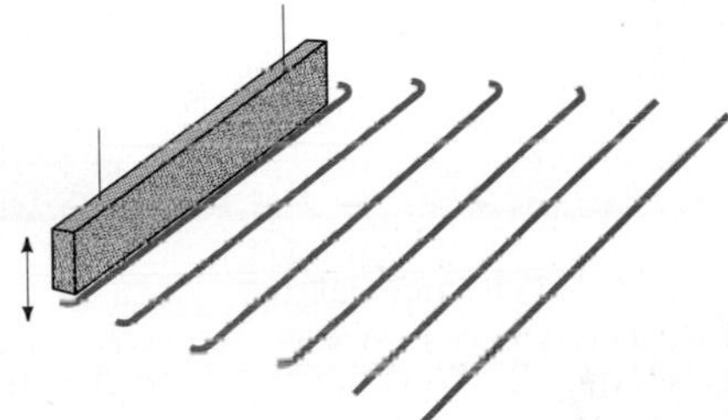

Figure 14-26 A two-dimensional analog of a plane wave can be generated in a ripple tank by a flat board that oscillates up and down in the water to produce wavefronts that are straight lines.

When a wave is incident on a boundary surface that separates two regions of differing wave speed, part of the wave is reflected and part is transmitted. This occurs, for example, when a sound wave in air strikes a solid or liquid surface. The reflected ray makes an angle with the normal to the surface equal to that of the incident ray, whereas the transmitted ray is bent toward or away from the normal—depending on whether the wave speed in the second medium is less or greater than that in the incident medium (Figure 14-27). The bending of the transmitted ray is called **refraction.** We will study refraction when we study light in Chapter 30.

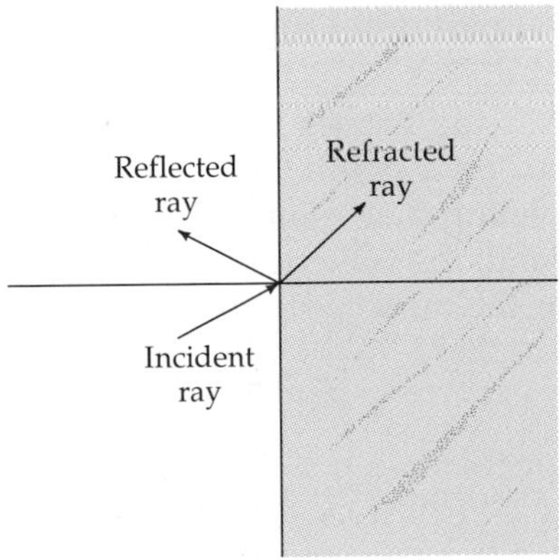

Figure 14-27 A wave striking a boundary surface between two media in which the wave speed differs. Part of the wave is reflected and part is transmitted. The change in direction of the transmitted ray is called refraction.

The amount of sound energy reflected from a surface depends on the surface. Flat walls, floors, and ceilings make good reflectors; whereas less rigid and porous materials, such as cloth in draperies and furniture coverings, absorb much of the incident sound. The reflection of sound waves plays an important role in the design of a lecture hall, a library, or a music auditorium. In a lecture hall with many flat reflecting surfaces, speech is difficult to understand because of the many echoes that arrive at different times at the listener's ear. Absorbent material is often placed on the walls

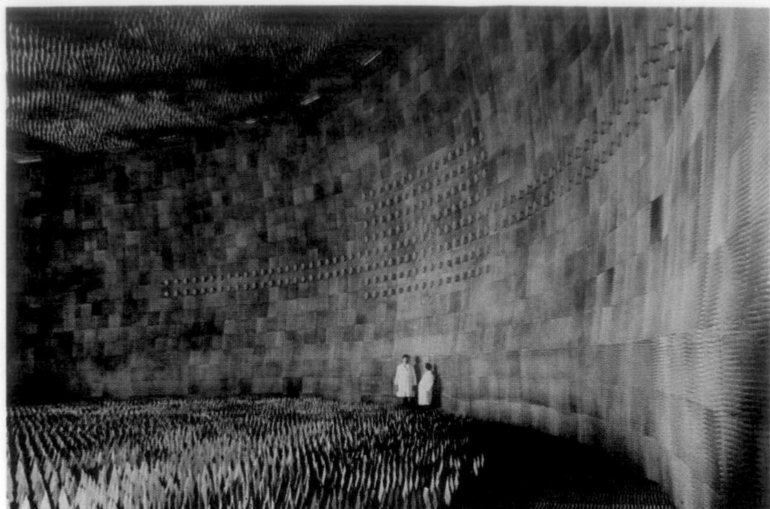

(*Left*) This anechoic chamber at the Naval Research Laboratory is used in the testing of electronic equipment. (*Right*) Davis Symphony Hall in San Francisco. The plastic reflectors above the orchestra reflect the sound waves out toward the audience.

and ceiling to reduce such reflections. In a concert hall, a reflecting shell is placed behind the orchestra, and reflecting panels are hung from the ceiling to reflect and direct the sound back toward the listeners.

When a portion of a wave is cut off by an obstruction, the propagation of the wave is complicated. The portion of the wavefront that is not obstructed does not simply propagate in the direction of the straight rays as might be expected. Figure 14-28 shows plane waves in a ripple tank meeting a barrier with a small opening. The waves to the right of the barrier are not confined to the narrow angle of the rays from the source that can pass through the opening; instead, they are circular waves, just as if there were a point source at the opening.

The propagation of a wave is thus quite different from the propagation of a stream of particles. In Figure 14-29*a*, the arrows indicate streams of particles that hit a barrier with a small opening, or aperture. Those particles that get through the opening will be confined to a small angle. In Figure 14-29*b*, the arrows indicate the rays that describe the propagation of circular waves toward the barrier. In passing beyond the barrier, the rays bend around the edges of the small opening. This bending of the rays, which always occurs to some extent when part of a wavefront is limited, is called **diffraction.**

In Figure 14-28, the opening in the barrier is smaller than the wavelength and could be considered to be a point. Figure 14-30 shows plane waves hitting a barrier with an opening that is much larger than the wavelength. Near the edges of the opening, the wavefronts are distorted, and the waves appear to bend slightly. For the most part, however, the wavefronts are not affected.

Figure 14-28 Plane waves in a ripple tank meeting a barrier with a small opening. The waves to the right of the barrier are circular waves that are concentric about the opening just as if there were a point source at the opening.

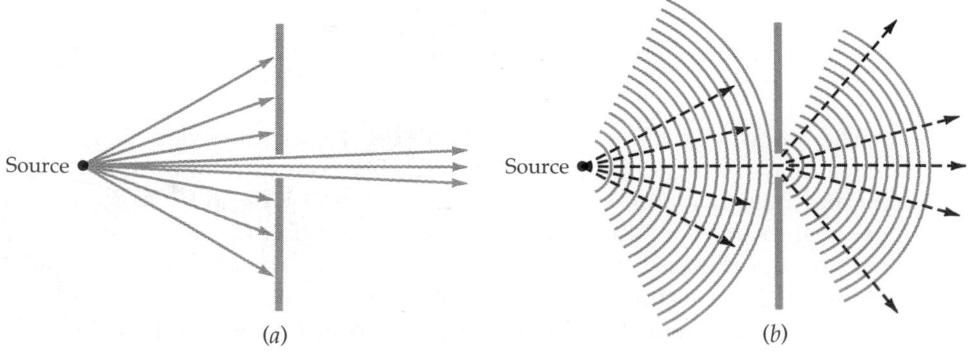

Figure 14-29 Comparison of the transmission through a narrow opening in a barrier of (*a*) a beam of particles and (*b*) a wave. In (*a*), the transmitted particles are confined to a narrow angle. In (*b*), the opening acts as a point source of circular waves that are radiated to the right in a much wider angle than are the particles in (*a*).

If an aperture or obstacle is large compared with the wavelength, the bending of the wavefront is not noticeable and the wave propagates in straight lines or rays, much as a beam of particles does.

Ray approximation

The result is known as the **ray approximation.** Because the wavelengths of audible sound range from a few centimeters to several meters (Example 14-2) and are often large compared with apertures and obstacles (doors or windows, for example), the bending of sound waves around corners is a common phenomenon. On the other hand, the wavelengths of visible light range from about 4×10^{-7} to 7×10^{-7} m. Because these wavelengths are so small compared with the size of ordinary objects and apertures, the diffraction of light is not easily noticed, so light appears to travel in straight lines.

The effects of diffraction place a limitation on the ability to locate small objects or identify their details by reflecting waves off of them. No appreciable reflection of waves occurs unless an object is at least as large as the wavelength. Thus, no detail can be observed on a scale smaller than the wavelength used. For example, when an object is located using waves of any kind, the position of the object is uncertain by an amount that is of the order of the wavelength used. Since the smallest wavelength of audible sound is about 2 cm (corresponding to a frequency of 17 kHz), an object cannot be located to better than ± 2 cm using audible sound. Sound waves with frequencies above 20,000 Hz are called **ultrasonic waves.** Ships use ultrasonic waves to detect submarines and other submerged objects with a device called sonar (from *so*und *n*avigation *a*nd *r*anging). Ultrasonic waves are used in medicine for diagnostic purposes. For example, a "sonogram," a picture constructed from ultrasonic waves, is taken of a pregnant woman to measure the size of the fetus or to detect any abnormalities. Because of their very small wavelengths, narrow beams of ultrasonic waves can be sent out and reflected from small objects. Bats emit ultrasonic waves and detect

Figure 14-30 Plane waves in a ripple tank meeting a barrier with an opening that is much larger than the wavelength. The barrier has a noticeable effect only near the edges of the opening.

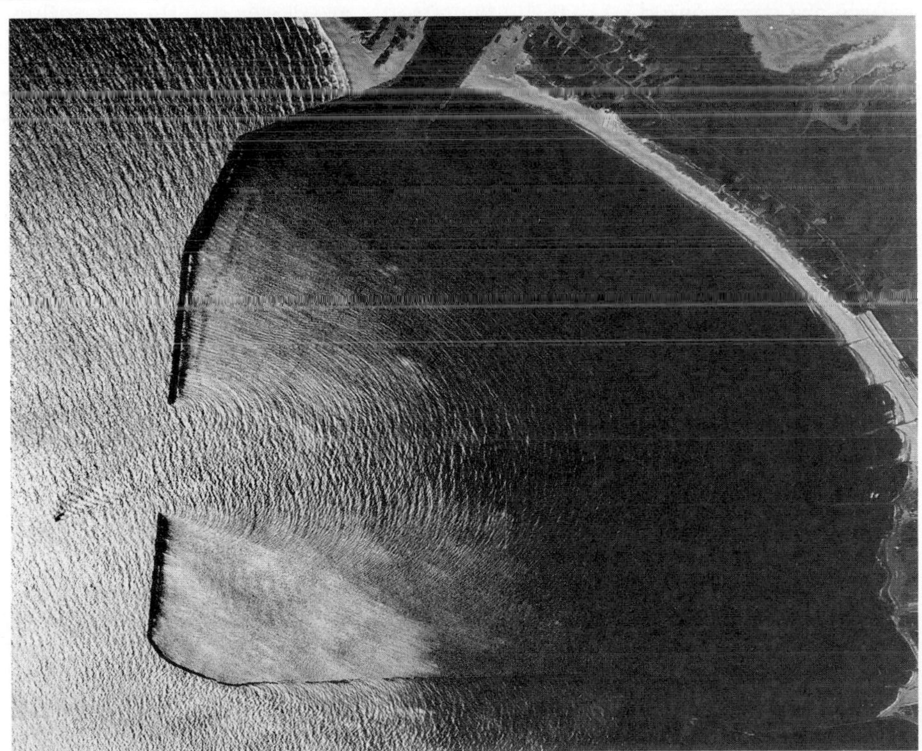

Water waves on the surface of the ocean diffract as they encounter an aperture formed by the breakwater.

their reflection to locate prey such as moths. They can hear frequencies up to about 120,000 Hz, which corresponds to a wavelength of (340 m/s)/(120 kHz) = 2.8 mm. The shortest wavelength of visible light is about 4×10^{-7} m, which is much smaller than most objects but is about 4000 times larger than the diameter of a typical atom. Thus, atoms can never be seen with visible light.

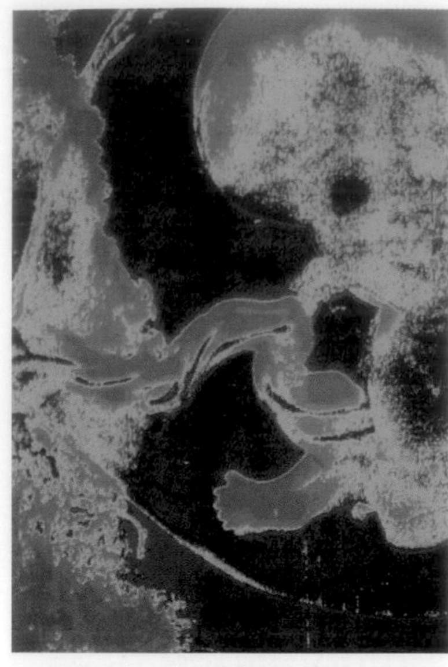

(*a*)

(*b*)

(*a*) Sonogram of a pregnant woman showing the fetus in the womb. (*b*) Remains of the American schooner Hamilton sunk in 300 feet of water in Lake Ontario in the War of 1812. The black outline of the ship results from processing of low-intensity signals. The red, yellow, and green are false colors resulting from the processing of moderate to high intensity signals.

Question

5. You may have noticed that when you hear music coming from another room around a corner, there is a loss in the high frequencies compared to the rest of the sound. Why does this occur?

14-9 The Doppler Effect

When a wave source and a receiver are moving relative to each other, the frequency observed by the receiver is not the same as that of the source. When they are moving toward each other, the observed frequency is greater than the source frequency; when they are moving away from each other, the observed frequency is less than the source frequency. This is called the **doppler effect.** A familiar example is the change in the pitch of a car horn as the car approaches or recedes.

The change in frequency of a sound wave is slightly different depending on whether the source or the receiver is moving relative to the medium. When the source moves, the wavelength changes, and the new frequency f' is found by first finding the new wavelength λ and then computing $f' = v/\lambda$. On the other hand, when the source is stationary and the receiver moves, the frequency is different simply because the receiver moves past more or fewer waves in a given time.

We will first consider the case of a moving source. Figure 14-31*a* shows waves in a ripple tank produced by a source moving to the right with a speed less than the wave speed. We can see that the waves in front of the source are compressed so that the wavefronts are closer together than they would be for a stationary source, whereas behind the source, they are farther apart. We can calculate the wavelength in front of the source λ_f and that behind the source λ_b as follows. Let the frequency of the source be f_0. In some time Δt, the source produces a number of waves N given by $N = f_0\,\Delta t$. The first wavefront moves a distance $v\,\Delta t$ while the source moves a distance $u_s\,\Delta t$, where u_s is the speed of the source relative to the medium (Figure 14-31*b*). Since these N waves are contained in the distance $(v - u_s)\,\Delta t$, the wavelength in front of the source is obtained by dividing this distance by N:

$$\lambda_f = \frac{(v - u_s)\,\Delta t}{N} = \frac{(v - u_s)\,\Delta t}{f_0\,\Delta t}$$

or

$$\lambda_f = \frac{v - u_s}{f_0} = \frac{v}{f_0}\left(1 - \frac{u_s}{v}\right) \qquad \text{14-36}$$

Behind the source, the N waves are contained in a distance $(v + u_s)\,\Delta t$, so the wavelength behind the source is

$$\lambda_b = \frac{(v + u_s)\,\Delta t}{f_0\,\Delta t}$$

or

$$\lambda_b = \frac{v + u_s}{f_0} = \frac{v}{f_0}\left(1 + \frac{u_s}{v}\right) \qquad 14\text{-}37$$

The speed v of the waves depends only on the properties of the medium and not on the motion of the source. For a source approaching the receiver, the frequency f' at which the waves pass a point at rest relative to the medium is thus

$$f' = \frac{v}{\lambda_f} = \frac{f_0}{1 - u_s/v} \quad \text{source approaching} \qquad 14\text{-}38a$$

For a source receding from the receiver, the frequency is

$$f' = \frac{v}{\lambda_b} = \frac{f_0}{1 + u_s/v} \quad \text{source receding} \qquad 14\text{-}38b$$

When the source is at rest and the receiver is moving relative to the medium, there is no change in the wavelength, but the frequency with which waves pass the receiver is increased when the receiver moves toward the source and decreased when the receiver moves away from the source. The number of waves passing a stationary receiver in time Δt is the number in the distance $v\,\Delta t$, which is $v\,\Delta t/\lambda$. When the receiver moves toward the source with speed u_r, it passes an additional number of waves $u_r\,\Delta t/\lambda$ (Figure 14-32). The total number of waves passing the receiver in time Δt is then

$$N = \frac{v\,\Delta t + u_r\,\Delta t}{\lambda} = \frac{v + u_r}{\lambda}\,\Delta t$$

The frequency observed is this number of waves divided by the time interval:

$$f' = \frac{N}{\Delta t} = \frac{v + u_r}{\lambda}$$

or

$$f' = f_0\left(1 + \frac{u_r}{v}\right) \quad \text{receiver approaching} \qquad 14\text{-}39a$$

If the receiver moves away from the source with speed u_r, similar reasoning leads to the observed frequency

$$f' = f_0\left(1 - \frac{u_r}{v}\right) \quad \text{receiver receding} \qquad 14\text{-}39b$$

When both the source and the receiver are moving relative to the medium, the results expressed in Equations 14-38 and 14-39 can be combined:

$$f' = \frac{(1 \pm u_r/v)}{(1 \pm u_s/v)} f_0 \qquad 14\text{-}40$$

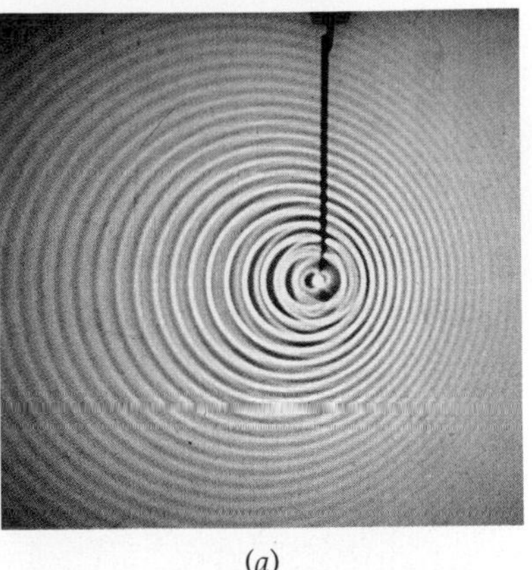

(a)

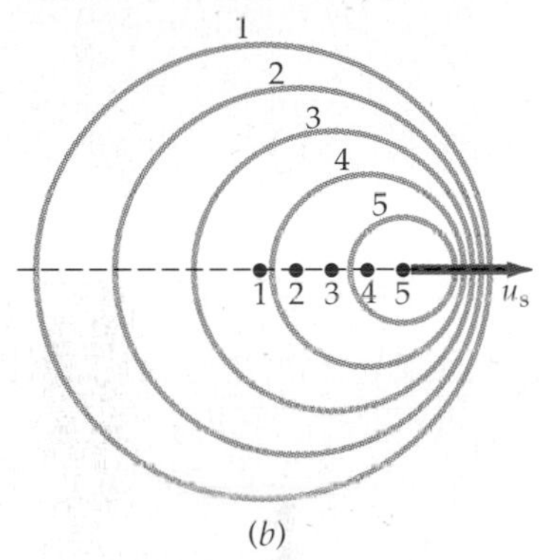

(b)

Figure 14-31 (a) Waves in a ripple tank produced by a point source moving to the right with a speed that is less than the wave speed. The wavefronts are closer together in front of the source and farther apart behind the source than they would be for a stationary source. (b) Successive wavefronts emitted by a point source moving with speed u_s to the right. Each numbered wavefront was emitted when the source was at the correspondingly numbered position.

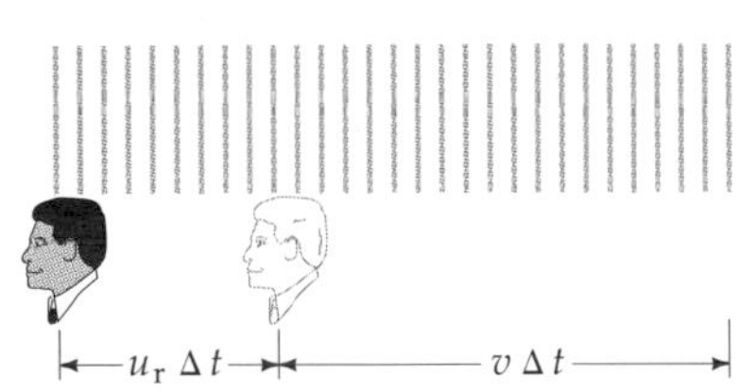

Figure 14-32 The number of waves passing a stationary receiver in time Δt is the number in the distance $v\,\Delta t$, where v is the wave speed. If the receiver moves toward the source with speed u_r, he passes the additional number of waves in the distance $u_r\,\Delta t$. The frequency of the waves received is therefore increased.

The correct choices for the plus or minus signs are most easily obtained by remembering that the frequency increases when the source and receiver are moving toward the other, whereas it decreases when they are moving away from each other. Thus, for example, if the source is moving toward the receiver and the receiver is moving toward the source, the plus sign is used in the numerator and the minus sign is used in the denominator.

In general, the observed frequency f' can always be written

$$f' = \frac{v'}{\lambda} \tag{14-41}$$

where v' is the speed of the waves relative to the observer, and λ is the wavelength in the medium, that is, the distance between successive wave crests as measured by an observer at rest relative to the medium.

If a medium is moving (for example, if there is a wind blowing), the wave speed v is replaced by $v' = v \pm u_w$, where u_w is the speed of the medium.

Example 14-9

The frequency of a car horn is 400 Hz. What frequency is observed if the car moves through still air toward a stationary receiver with a speed $u_s = 34$ m/s (about 76 mi/h)? Take the speed of sound in air to be 340 m/s.

According to Equation 14-36, the wavelength in front of the car is

$$\lambda_f = \frac{v - u_s}{f_0} = \frac{340 - 34\ \text{m/s}}{400\ \text{s}^{-1}} = \frac{306}{400}\ \text{m} = 0.765\ \text{m}$$

The frequency observed is thus

$$f' = \frac{v}{\lambda_f} = \frac{340\ \text{m/s}}{0.765\ \text{m}} = 444\ \text{Hz}$$

Note that the observed frequency is 11 percent higher than the original frequency of the car horn.

Example 14-10

The horn of a stationary car has a frequency of 400 Hz. What frequency is observed by a receiver moving through still air toward the car at 34 m/s?

For a moving receiver, the wavelength does not change; the receiver merely passes more waves in a given time. The observed frequency is (Equation 14-39*a*):

$$f' = f_0\left(1 + \frac{u_r}{v}\right) = (400\ \text{Hz})\left(1 + \frac{34}{340}\right) = (400\ \text{Hz})(1.10) = 440\ \text{Hz}$$

In this case, the increase in the observed frequency is 10 percent.

Exercise

As a train moving at 90 km/h is approaching a stationary listener, it blows its horn, which has a frequency of 630 Hz. (*a*) What is the wavelength of the sound waves in front of the train? (*b*) What frequency is heard by the listener? (Use 340 m/s for the speed of sound.) [Answers: (*a*) $\lambda_f = 0.5$ m, (*b*) $f' = 680$ Hz]

A familiar example of the use of the doppler effect is the use of radar by the police to measure the speed of a car. Electromagnetic waves emitted by the radar transmitter strike the moving car. The car acts as both a moving receiver and a moving source as the waves reflect off it back to the radar receiver. Another application involves the famous red shift in the light from distant galaxies. Because the galaxies are moving away from us, the light they emit is shifted toward the longer, red wavelengths. The speed of the galaxies relative to us can be determined by measuring this shift.

In Examples 14-9 and 14-10, we saw that the magnitude of the doppler shift in frequency depends on whether it is the source or the receiver that is moving relative to the medium. In Example 14-9, the source moved with a speed of 34 m/s relative to still air, and the frequency shifted from 400 to 444 Hz, a difference of 11 percent. In Example 14-10, the receiver moved with a speed of 34 m/s relative to still air, and the frequency shifted from 400 to 440 Hz, a difference of 10 percent. Though these shifts are approximately the same, they are not exactly equal.

The frequency f' observed for a source moving with a speed u toward a receiver is, from Equation 14-38*a*,

$$f' = \frac{f_0}{1 - u/v} = f_0\left(1 - \frac{u}{v}\right)^{-1} \qquad \text{14-42}a$$

where we have dropped the subscripts on the u for simplicity. We can compare this with Equation 14-39*a* for an approaching receiver by using the binomial expansion:

$$(1 + x)^n = 1 + nx + \frac{n(n-1)}{2}x^2 + \cdots$$

For values of x much less than 1, we can approximate using only the first few terms. For $x = -u/v$ and $n = -1$, we have

$$\left(1 - \frac{u}{v}\right)^{-1} = 1 + (-1)\left(-\frac{u}{v}\right) + \frac{(-1)(-2)}{2}\left(-\frac{u}{v}\right)^2 + \cdots$$

$$\approx 1 + \frac{u}{v} + \frac{u^2}{v^2}$$

Thus, Equation 14-42*a* is approximately

$$f' \approx f_0\left(1 + \frac{u}{v} + \frac{u^2}{v^2}\right) \qquad \text{14-42}b$$

On the other hand, if the receiver moves toward a stationary source with a speed u, the observed frequency from Equation 14-39*a* is

$$f' = f_0\left(1 + \frac{u}{v}\right) \qquad \text{14-43}$$

Comparing Equations 14-42*b* and 14-43, we can see that the *difference* in the frequency shift between the source moving with a speed u and the receiver moving with a speed u is of the order of $(u/v)^2$. Often this difference can be neglected because u is usually much smaller than v. Then the doppler shift in frequency can be written

$$\frac{\Delta f}{f_0} \approx \pm\frac{u}{v} \qquad \text{14-44}$$

where $\Delta f = f' - f_0$. The plus sign is used when the source and receiver are approaching each other with a relative speed u and the minus sign is used when they are receding.

The small difference in the doppler shift for a moving source or moving receiver is real, and it is of theoretical importance because it shows that these two situations are really different. Not only is the motion of the source and receiver relative to each other important, but so is the "absolute" motion of both relative to the medium. Thus, if we can measure the doppler shift in frequency to the order of $(u/v)^2$, we can tell whether it is the source or receiver that is moving relative to the medium. If you move relative to still air, you feel air rushing past you. In your reference frame, there is a wind. For sound waves in air, therefore, we can tell whether the source or receiver is moving by noting if there is a wind in the reference frame of the source or the receiver.

A problem arises when we apply the doppler effect to light and other electromagnetic waves that propagate through a vacuum. Our equations for the doppler effect for sound seem to imply that we could detect absolute motion relative to a vacuum if we could measure the doppler shift accurately enough. This contradicts the principle of relativity enunciated by Albert Einstein that states that it is impossible to detect absolute motion. It turns out that for light and other electromagnetic waves, our equations for the doppler effect are only approximately correct. When the proper relativistic corrections are made, the doppler shift for electromagnetic waves is the same for a moving source as for a moving receiver; that is, it depends only on the relative motion of the source and receiver.

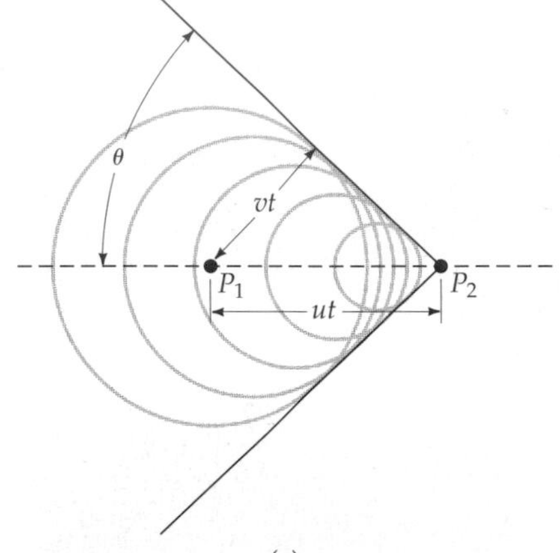

(a)

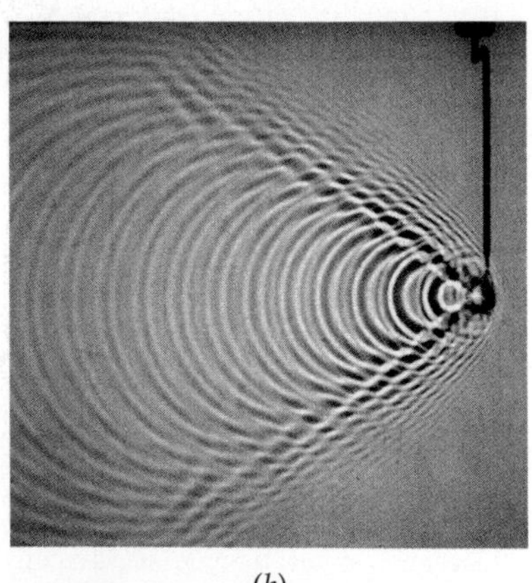
(b)

Figure 14-33 (*a*) A source moving from P_1 to P_2 with a speed u that is greater than the wave speed v. The envelope of the wavefronts forms a cone with the source at the apex. The angle θ of this cone is given by $\sin\theta = v/u$. (*b*) Waves in a ripple tank produced by a source moving with a speed greater than the wave speed.

In our derivations of the doppler-shift expressions, we have assumed that the speed u of the source or receiver is less than the wave speed v. If the receiver moves toward the source faster than the wave speed, there is no problem. Equation 14-39*a* for the observed frequency still holds. If the receiver moves away from the source faster than the wave speed, the waves never reach the receiver. (According to the special theory of relativity, a receiver cannot move faster than the speed of light c in a vacuum.) If a sound source moves with speed greater than the wave speed, there will be no waves in front of the source. Behind the source, the waves pile up on top of one another to form a shock wave that is heard as a sonic boom when it arrives at the receiver. The shock wave is confined to a cone that narrows as u increases. We can calculate the angle of this cone from Figure 14-33*a*, which shows a source originally at point P_1 moving to the right with speed u. After some time t, the wave emitted from point P_1 will have traveled a distance vt. The source will have traveled a distance ut and will be at point P_2. The line from this new position of the source to the wavefront emitted when the source was at P_1 makes an angle θ with the path of the source, given by

$$\sin\theta = \frac{vt}{ut} = \frac{v}{u} \qquad 14\text{-}45$$

The ratio of the source speed u to the wave speed v is called the **Mach number:**

$$\text{Mach number} = \frac{u}{v} \qquad 14\text{-}46$$

Equation 14-45 also applies to the electromagnetic radiation given off when a charged particle moves in a medium with speed u that is greater than the speed of light v in that medium.* This radiation, called Čerenkov radiation, is confined to a cone with an angle given by Equation 14-45.

*It is impossible for a particle to move faster than c, the speed of light in a vacuum. In a medium such as glass, however, electrons and other particles can move faster than the speed of light in that medium.

(*a*)

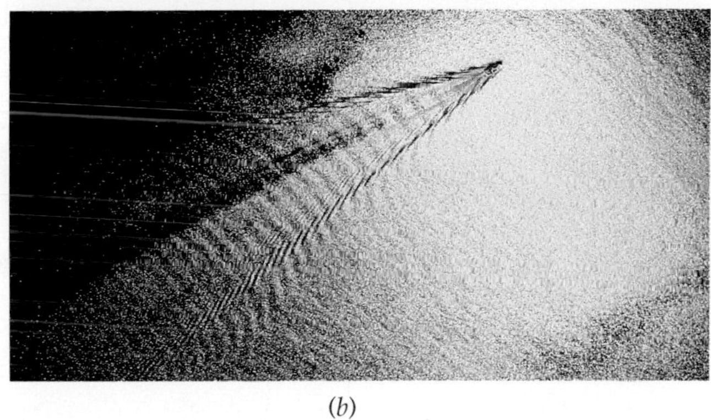

(*b*)

(*c*)

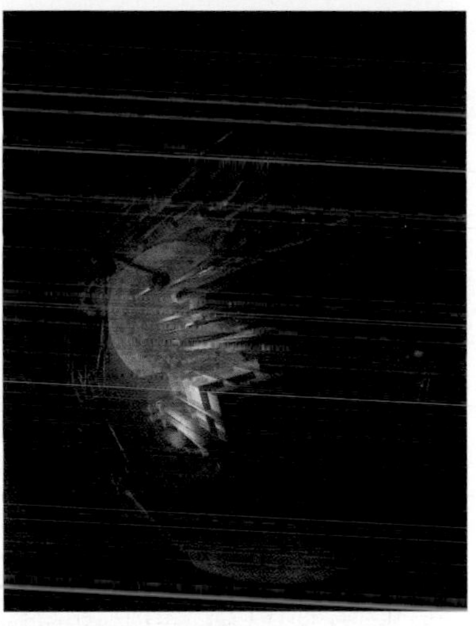

(*d*)

(*a*) Shock waves from a supersonic airplane. (*b*) Bow waves from a boat. (*c*) Shock waves produced by a bullet traversing a helium balloon. (*d*) Glow of a small reactor produced by Čerenkov radiation.

Question

6. If the source and receiver are at rest relative to each other but the wave medium is moving relative to them, will there be any doppler shift in frequency?

Summary

1. Sound waves are longitudinal waves of compression and rarefaction. In a fluid, they travel with speed

$$v = \sqrt{\frac{B}{\rho}}$$

where B is the bulk modulus and ρ is the equilibrium density of the fluid. The speed of sound in a gas is related to the absolute temperature by

$$v = \sqrt{\frac{\gamma RT}{M}}$$

where the absolute temperature T is related to the Celsius temperature t_C by

$$T = t_C + 273$$

and $R = 8.314$ J/mol·K is the universal gas constant, M is the molar mass (mass per mole), and γ is a constant that depends on the kind of gas and has the value 1.4 for air. In a solid, the speed of sound is related to Young's modulus Y and the density ρ by

$$v = \sqrt{\frac{Y}{\rho}}$$

2. Sound waves can be considered to be either displacement waves or pressure waves. In a harmonic sound wave, the pressure amplitude p_0 is related to the displacement amplitude s_0 by

$$p_0 = \rho\omega v s_0$$

where ω is the angular frequency, ρ is the density of the medium, and v is the speed of the wave. The human ear is sensitive to sound waves of frequencies in the range of about 20 Hz to 20 kHz.

3. The intensity of a wave is the power divided by the area. The intensity of spherical waves from a point source varies inversely as the square of the distance from the source:

$$I = \frac{P_{\text{av}}}{4\pi r^2}$$

The intensity of a harmonic wave is proportional to the square of the amplitude of the wave.

4. Sound intensity levels are measured on a logarithmic scale. The sound intensity level β in decibels (dB) is related to the intensity I by

$$\beta = 10 \log \frac{I}{I_0}$$

where $I_0 = 10^{-12}\ \text{W/m}^2$, which is approximately the threshold of hearing. On this scale, the threshold of hearing is 0 dB and the pain threshold is 120 dB.

5. Two sources that are in phase or have a constant phase difference are coherent. Interference is observed only for waves from coherent sources. Waves from incoherent sources have a phase difference that varies randomly in time, so the interference at a particular point varies back and forth from constructive to destructive and no interference pattern is observed. A common cause of a phase difference between two waves is a difference in the path length traveled by the waves. A path difference Δx produces a phase difference δ given by

$$\delta = 2\pi \frac{\Delta x}{\lambda}$$

6. Beats are the result of the interference of two waves of slightly different frequencies. The beat frequency equals the difference in the frequencies of the two waves:

$$f_{\text{beat}} = \Delta f$$

7. When sound waves are confined in space, as in an organ pipe, standing waves occur. For a pipe that is either open or closed at both ends, the standing-wave condition can be found by fitting waves into the pipe with a node at each end (closed at both ends) or an antinode at each end (open at both ends). An integral number of half wavelengths must fit into the length of the pipe. The standing-wave condition is thus

$$n \frac{\lambda_n}{2} = L \qquad n = 1, 2, 3, \ldots$$

The allowed frequencies are

$$f_n = nf_1 \qquad n = 1, 2, 3, \ldots$$

where $f_1 = v/2L$ is the fundamental frequency. If the pipe is closed at one end and open at the other end, there is a node at one end and an antinode at the other. The standing-wave condition in this case is

$$n \frac{\lambda_n}{4} = L \qquad n = 1, 3, 5, \ldots$$

Only the odd harmonics are present. The allowed frequencies are

$$f_n = nf_1 \qquad n = 1, 3, 5, \ldots$$

where $f_1 = v/4L$.

8. Sounds of different tone quality contain different mixtures of harmonics. The analysis of a particular tone in terms of its harmonic content is called harmonic analysis. Harmonic synthesis is the construction of a tone by adding the proper mixture of harmonics.

9. A wave pulse can be represented by a continuous distribution of harmonic waves. If the duration of the pulse in time is small, a wide range of frequencies is needed. The range of frequencies $\Delta\omega$ is related to the width in time Δt by

$$\Delta\omega\ \Delta t \sim 1$$

Similarly, the range of wave numbers Δk is related to the width in space Δx by

$$\Delta k\ \Delta x \sim 1$$

10. In a nondispersive medium, the wave speed does not depend on the frequency or wavelength, and a pulse travels without change in shape. In a dispersive medium, the wave speed does depend on the wavelength and frequency, and a pulse changes shape as it moves. In a dispersive medium, the velocity of the pulse, called the group velocity, is not the same as the phase velocity, which is the average velocity of the harmonic components of the pulse.

11. Waves can be reflected, refracted, and diffracted. Refraction is the change in direction of a wave that occurs when the wave speed changes because the medium changes. Diffraction is the bending of a wave around an obstacle or an edge of an aperture. It occurs whenever the wavefront is limited. When the obstacles or apertures are large compared with wavelength, diffraction is negligible and the wave propagates in straight lines like a beam of particles. This is known as the ray approximation. Because of diffraction, waves can be used to locate an object only to within a wavelength or so.

12. When a sound source and receiver are in relative motion, the observed frequency is increased if they move toward each other and decreased if they move away from each other. This is known as the doppler effect. The observed frequency f' is related to the source frequency f_0 by

$$f' = \frac{1 \pm u_r/v}{1 \pm u_s/v} f_0$$

When the relative speed of the source or receiver u is much less than the wave speed v, the doppler shift is approximately the same whether it is the source or the receiver that is moving and is given by

$$\frac{\Delta f}{f_0} \approx \pm\frac{u}{v}$$

Seismic Waves

Jack L. Flinner
Mankato State University

A bat swoops on a moth in the dark of night. A minesweeper hunts for submerged mines. A doctor carefully examines a fetus being "scanned" in its mother's womb. In each of these cases sound waves are emitted, reflected, analyzed, and interpreted. In a way, they resemble the method used to estimate the distance to a large reflecting wall, such as a cliff across a valley, by timing the echo of a loud yell "Hello-o-o-o." To find the distance to the cliff, you simply multiply the speed of sound in air by one-half the time that elapses between the shout and the arrival of its echo. Reflection of a sound wave by an obstacle creates a pattern rich in information. Along with its distance, the shape or the speed of the object relative to the observer can be determined from a careful analysis of this pattern. Correct interpretation of the information is to the clear advantage of both bat and minesweeper bent on survival.

The center of the earth is approximately 6370 km below where you are sitting. The deepest hole ever drilled managed to get only 15 km of the way there. Our information about the structure and composition of the material along the remaining 6355 km has been derived primarily from the analysis of the behavior of seismic waves. Seismic waves are waves of very low frequency that travel through the earth. Sudden displacements of crustal segments that are bounded by fault zones can produce seismic waves; these are often of catastrophic dimensions affecting both people and structures. Seismic waves can also be generated in the earth's upper crust by explosives and other artificial devices. These enable geologists and geophysicists to gain knowledge of the earth's interior and assist in the search for new fossil fuel resources.

The hundreds of thousands of earthquakes that occur each year (most, thankfully, too minor for us to notice—see Table 1) provide ample opportunity for the production and study of seismic waves. The device used to detect these waves is called a seismograph. As you might guess, the calculations needed to extract information from these data are not as simple as those done when you holler across a valley. Like all waves, the speed of seismic waves depends on the properties of the medium through

Jack L. Flinner is Professor of Physics at Mankato State University in Mankato, Minnesota. He earned his Bachelor's degree from Wittenberg University, Springfield, Ohio in 1953; his M.S. from the University of Illinois, Urbana–Champaign, in 1955; and his Ph.D. from Ohio State University in 1965. His research was in low-energy nuclear physics. He taught at Wittenberg University for twelve years and has been at Mankato State University since 1969. He served as chairperson of the department from 1975 to 1981 and then returned to teaching and to research with the Van de Graaff accelerator using the PIXE method to investigate the L-shell spectrum of high-Z elements. He has served on such projects as the establishments of an East Asia studies program at Wittenberg, as the Minnesota State University representative on the committee to propose a state school of optometry, and then as a representative on the Minnesota State Legislature ad hoc committee on science and technology.

Table 1 **Earthquake Magnitudes and Frequencies for the Entire Earth and their Damaging Effects**

Richter magnitude	Number per year	Characteristic effects of shocks in populated areas
<3.4	800,000	Recorded only by seismographs
3.5–4.2	30,000	Felt by some people
4.3–4.8	4800	Felt by many people
4.9–5.4	1400	Felt by everyone
5.5–6.1	500	Slight building damage
6.2–6.9	100	Much building damage
7.0–7.3	15	Serious damage, bridges twisted, walls fractured
7.4–7.9	4	Great damage, buildings collapse
>8.0	One every 5–10 yr	Total damage, waves seen on ground surface, objects thrown in the air

Source: From Skinner and Porter (1987) *Physical Geology*, p. 456, John Wiley & Sons, Inc.

which they propagate—in particular, on the rigidity (for example, the bulk modulus) and on the density of the medium. Also, at a boundary surface between two media in which the wave speed differs, a seismic wave is partially reflected and partially refracted. Further complications arise due to the fact that the media through which seismic waves travel can be solid, liquid, or gaseous. The physical properties of the medium have profound effects on the transmission pattern of seismic waves. The rigidity and density of the materials that make up the earth vary with depth, and their values are of particular importance for an accurate analysis. Furthermore, the intensity of the reflected (or deflected) waves can be extremely small, so the sensitivity of the detectors used must be correspondingly large.

There are four types of seismic waves that fall into two categories: body waves (P for primary, and S for secondary) and surface waves (Love and Rayleigh). All four types of waves are produced by an earthquake. Body waves travel through the earth outward from the focus of an earthquake (Figure 1). P waves are longitudinal waves—like sound waves (refer to Figure 13-5 in the main text). They travel at speeds up to 14 km/s and can pass through solids, liquids, and gases. Since they travel faster than S waves they are the first to arrive at an earthquake detector (hence "primary"). S waves are transverse shear waves that travel at speeds up to 3.5 km/s (refer to Figure 13-5 in the main text). They can travel only through solids because liquids and gases cannot support shear stresses. Love and Rayleigh waves remain at the earth's surface. Love waves are torsional waves. They produce momentary distortions of the earth's surface like those you could produce on a spherical balloon globe if you grasped it at the poles and gave a twist with each hand in opposite directions. Rayleigh waves are similar to the rise and swell of ocean waves. They produce momentary distortions like those you could produce on a spherical balloon globe by pushing the poles in or pulling them out.

P and S waves propagate through the earth at substantially different speeds, so their separate arrival times at seismograph stations can be used to locate earthquake epicenters with great precision. Travel-time graphs for P and S waves have been constructed (Figure 2), first using earthquake epicenters whose locations were relatively well established from physical evidence, and later by the use of data from nuclear explosions whose detonation times and locations were precisely known. When P and S seismic waves are recorded at a seismograph station, the time interval between the early arrival of the P wave and the later arrival of the S wave is

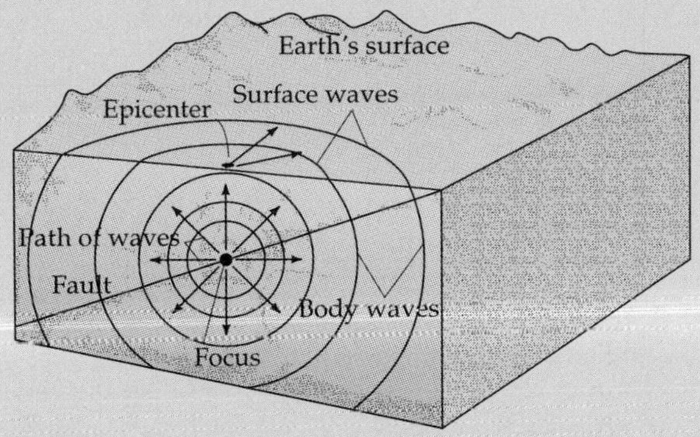

Figure 1 The body waves move radially outward from the focus of an earthquake submerged within the earth. The epicenter, which lies at the earth's crust directly above the focus, is the center of the surface waves that radiate from that point.

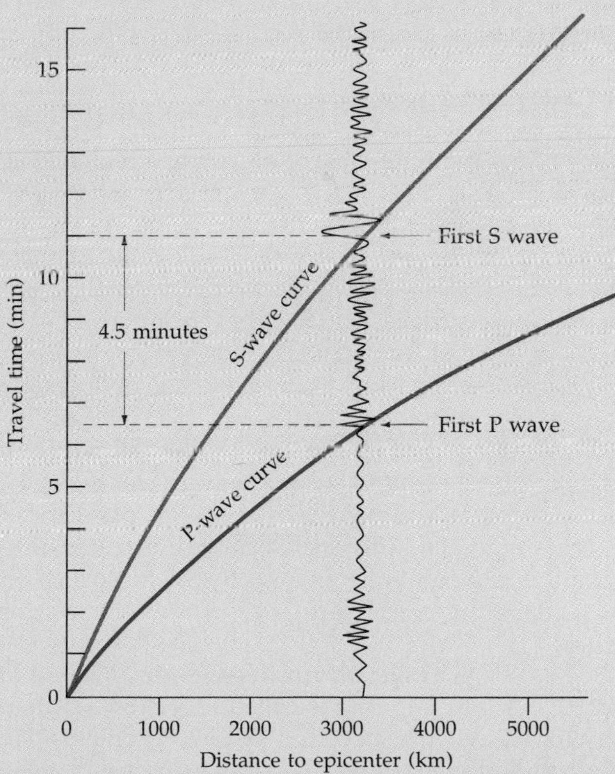

Figure 2 A travel-time graph can be used to determine the distance of a seismograph station from an earthquake's epicenter.

Continued

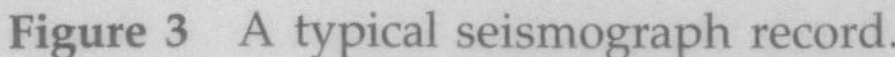

Figure 3 A typical seismograph record.

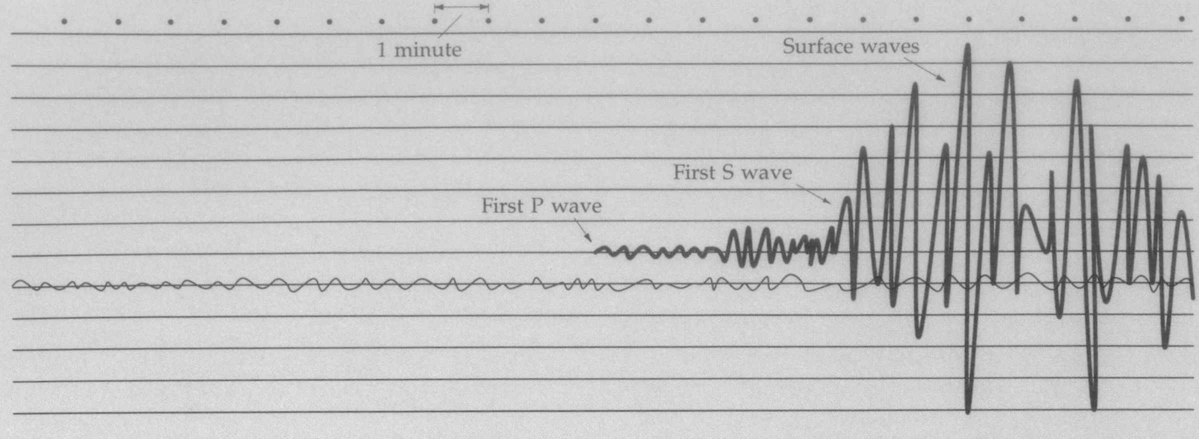

easily read from the seismograph (Figure 3). This time interval can be entered on the travel-time graph, allowing the distance to the recording site to be read directly from the graph. The precise location of the epicenter is determined from the intersection point provided by distance data from three or more stations. Clustered seismograph arrays and high-speed computers, used to discriminate between nuclear explosions and natural earthquakes, have proved very useful in the accurate determination of epicenters.

The earth's average density is 5.5 g/cm^3. At the crust, the rock density is about 3 g/cm^3, so some parts of the earth must have much greater densities. Seismographic readings from all over the world have enabled geologists to build a model of the structure of the earth. It shows not only that the most dense rocks are at the center with the least dense at the surface, but also that the interior of the earth is divided into regions (Figure 4).

Seismic waves can also be produced deliberately to map the subsurface rock strata of geological basins in order to find likely sites for oil or gas accumulations. These induced seismic waves (and their correct interpretation) have a vital role to play in our energy-intensive industrial society; they often provide the only cost-effective means of exploring for oil and gas reserves concealed within the earth's crust.

The use of artificially produced seismic waves to search for fuel deposits is called reflection seismology, or seismic prospecting. It has been practiced in the United States since 1923. Reflection seismology simply stated is the generation of minor, artificial earthquakes to produce seismic waves so that their reflection and refraction patterns can be studied. The method used to produce the waves depends on whether the survey is taking place on land or at sea. Options include using dynamite blasts in shallow boreholes; deploying large, low-intensity hydraulic vibrators (particularly useful in populated or ecologically sensitive areas); dropping heavy weights;

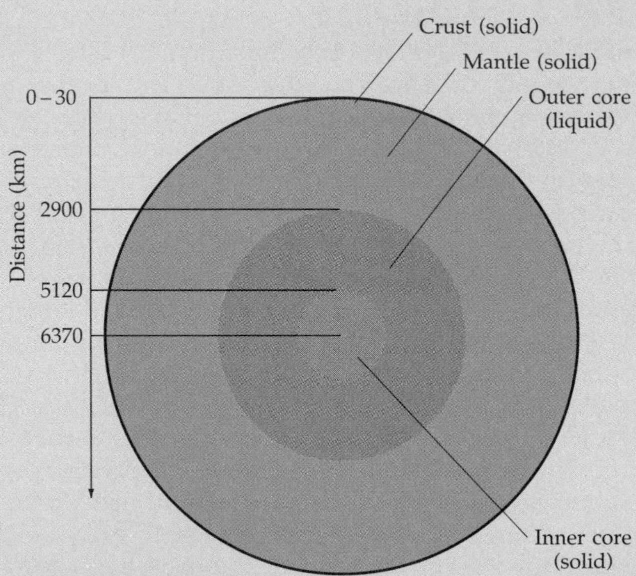

Figure 4 Geologists have constructed this model of the interior of the earth on the basis of the refraction and transmission patterns of seismic body waves (S and P). The inability of S waves to penetrate the large interior of the earth indicates that the outer core is liquid. The transmission patterns of those P waves that are not too strongly refracted by the outer core indicate that the inner core is solid.

striking the ground with an air hammer; or using an array of air guns. Similar effects are produced by each method. The waves that are produced travel through the earth's crust and are reflected back toward the surface at the boundaries of various strata. The arrival times of seismic waves at the earth's surface is recorded by detectors (called geophones). The arrival times, amplitudes, and frequencies of the seismic waves are then stored on computers (Figure 5).

The very fast, large-memory computers, which have become available in recent years, not only make it possible to store a seismic reflection profile (Figure 6) that depicts a two-dimensional slice through the earth's crust, but also make it possible to move the location of the seismic energy sources—thus producing a series of profiles that can be used to generate a three-dimensional representation. The two-dimensional profiles indicate the discontinuities encountered in the density and rigidity in the different layers of the earth's crust and can be interpreted to reveal the nature of geologic configurations, some of which are known to be particularly

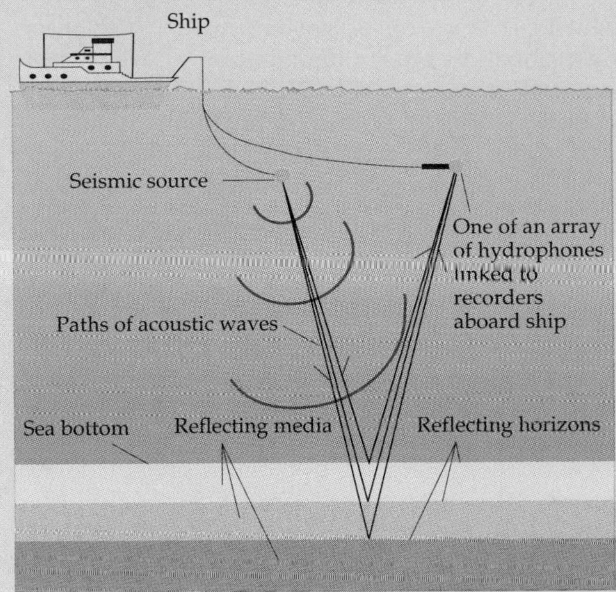

Figure 5 Seismic reflection studies at sea involve a survey ship towing an energy source nearby, which is used to generate seismic waves. It also tows cables up to 5 km long that carry hydrophones that detect the reflected seismic waves returning from the crust.

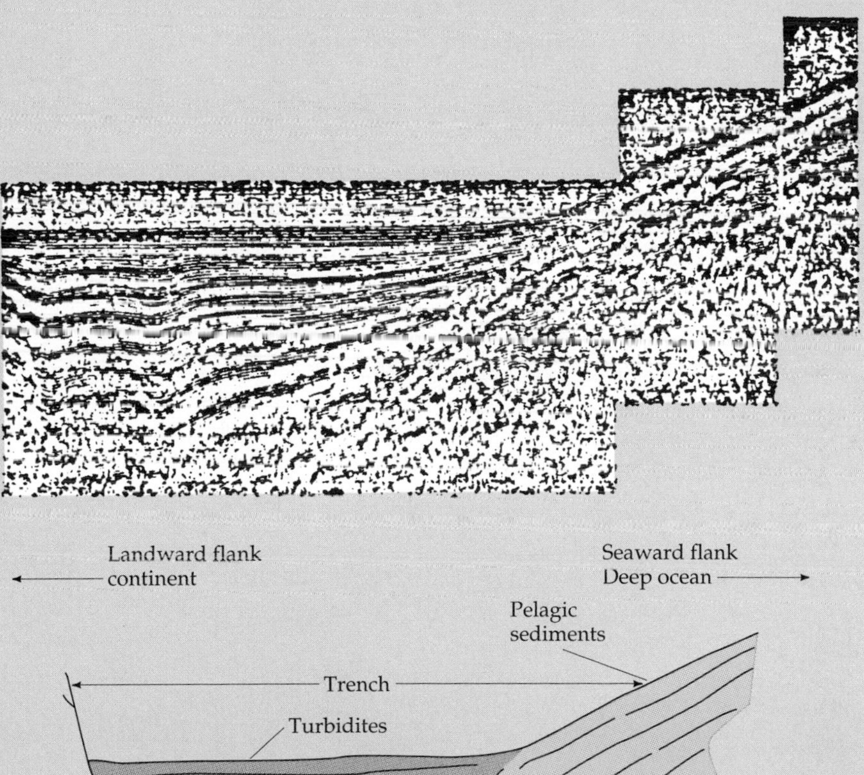

Figure 6 A seismic reflection profile together with an interpretation of the various layers revealed.

Continued

likely to contain oil or gas (Figure 7). The three-dimensional representation shows even more clearly the various rock layers and makes it possible to study carefully a volume of crust of particular interest. Once a pocket of gas has been found, for instance, its size can be determined by rotating its image and "viewing" it from different angles. The computers and software used today by geologists have become very sophisticated and make such rotations possible. It is also possible, by digital processing, to reduce the effect of the scattering of the wavefronts that emerge from the features being analyzed. This process, called migration, is roughly equivalent to focusing the lens of a camera or microscope, thus enabling one to examine finer geological features. This "focusing" becomes more important as the larger oil fields are depleted, and interest is shifted to smaller reserves.

Computers are also used to store data about local geology, such as rock porosity for water and oil, that are important in determining the ease with which oil can be extracted and the likelihood of oil or gas in a new find being economically retrievable. The geologist is thereby able to process and utilize vast amounts of data rapidly and effectively and to search for those features that are characteristic of a promising new fuel source before any investment is made in an expensive exploratory drilling. (The present state of the art in reflection seismology is, indeed, a beautiful example of equipment and technology being available soon enough to help solve an increasingly difficult problem.)

Reflection seismology is one of many applications of the reflection and refraction of waves as they travel through matter. As the power and speed of computers grow, all aspects of seismology are expected to play increasingly important and sophisticated roles in studies of the earth's crust—its features, its often unpredictable behavior, and its hidden resources.

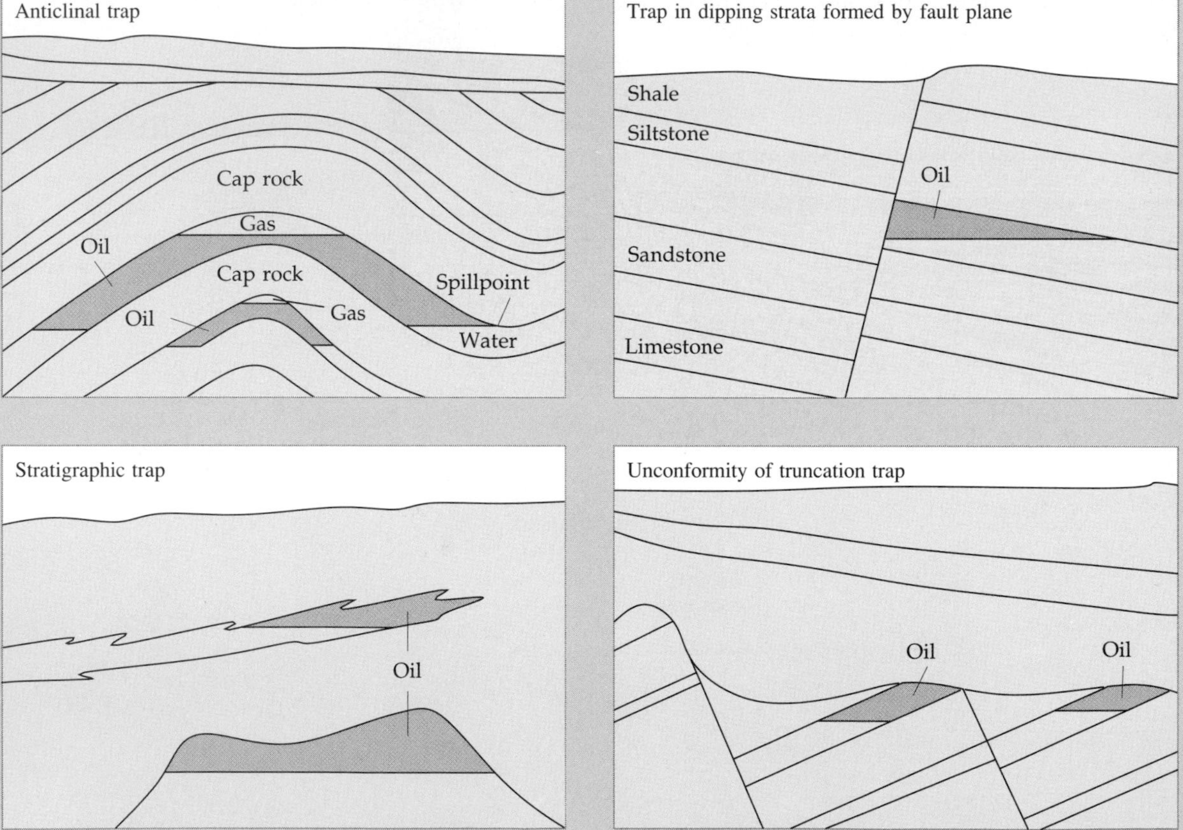

Figure 7 Some of the rock structures that can trap hydrocarbons, and schematic versions of their appearance in profiles.

Suggestions for Further Reading

Boore, David M.: "The Motion of the Ground in Earthquakes," *Scientific American*, December 1977, p. 68.

Sudden slippage of a continental plate relative to its neighbor along a fault line excites mechanical waves in the earth. The study of this phenomenon may lead to improved building designs for earthquake zones.

Devey, Gilbert B., and Peter N. T. Wells: "Ultrasound in Medical Diagnosis," *Scientific American*, May 1978, p. 98.

The physics and technology of imaging using ultrahigh-frequency sound waves and the interpretation of the medical images produced are discussed.

Fletcher, Neville H., and Suzanne Thwaites: "The Physics of Organ Pipes," *Scientific American*, January 1983, p. 94.

Modern measurements show that the competing theories of nineteenth-century physicists Helmholtz and Rayleigh to explain the sound production of organ pipes are both correct but apply to different pressure ranges.

The Physics of Music, readings from *Scientific American* with an introduction by Carleen M. Hutchins, Scientific American, Inc., distributed by W. H. Freeman and Co., New York, 1978.

This collection of articles includes analyses of the acoustics of the singing voice, the piano, woodwinds, brasses, violins, and the bowed string as well as architectural acoustics.

Jeans, Sir James: *Science and Music*, The University Press, Cambridge, 1953.

The basic features of sound production by vibrating strings and columns of air as well as harmony are well treated for the general reader in this book originally published in 1937.

Matthews, Max V. and John R. Pierce: "The Computer as a Musical Instrument," *Scientific American*, February 1981, p. 126.

Analysis and synthesis of waveforms has helped to determine what constitutes a "pleasing" musical sound. This article describes some of the history of computers in music generation.

Pierce, John R.: *The Science of Musical Sound*, Scientific American Books, Inc., distributed by W. H. Freeman and Co., New York, 1983.

This excellent book discusses the physical and mathematical aspects of sound waves that underlie our experience of music as well as the quantifiable aspects of sound perception.

Quate, Calvin F.: "The Acoustic Microscope," *Scientific American*, October 1979, p. 62.

The development of an acoustic microscope employing sound with a frequency 1000 times higher than that used in medical imaging technology is described. The images are comparable in quality to light micrographs.

Review

A. Objectives: After studying this chapter, you should:

1. Know how the speed of sound in air depends on the absolute temperature, and know the approximate value in meters per second of the speed of sound in air at ordinary temperatures.

2. Be able to calculate the intensity level in decibels from the intensity in watts per square meter and vice versa.

3. Be able to sketch the interference pattern produced by two coherent sources.

4. Know that the beat frequency equals the difference between the two frequencies producing the beats.

5. Be able to explain what is meant by the terms coherent sources and diffraction.

6. Be able to sketch standing-wave patterns for sound waves in a tube open at both ends, closed at both ends, and closed at one end and from these sketches determine the possible frequencies for standing waves.

7. Be able to derive expressions for the doppler frequency shift for a moving source or a moving receiver and use these expressions in working problems.

B. Define, explain, or otherwise identify:

Wavefront	Wave packet
Ray	Nondispersive medium
Intensity	Dispersive medium
Intensity level	Phase velocity
Decibel	Group velocity
Coherent sources	Plane wave
Beats	Reflection
Beat frequency	Refraction
Tone quality	Diffraction
Waveform	Ray approximation
Harmonic analysis	Ultrasonic waves
Fourier analysis	Doppler effect
Harmonic synthesis	Mach number

C. True or false: If the statement is true, explain why it is true. If it is false, give a counterexample.

1. Sound waves in air are transverse waves of compression and rarefaction.

2. The speed of sound at 20°C is twice that at 5°C.

3. A 60-dB sound has twice the intensity of a 30-dB sound.

4. The waves from two coherent sources that are in phase interfere constructively everywhere in space.

5. Two wave sources that are out of phase by 180° are incoherent.

6. Interference patterns are observed only for coherent sources.

7. Diffraction occurs only for transverse waves.

8. The beat frequency between two sound waves of nearly equal frequencies equals the difference in the frequencies of the individual sound waves.

9. The frequency of the fifth harmonic is 5 times that of the fundamental.

10. In a pipe that is open at one end and closed at the other, the even harmonics are not excited.

11. The doppler shift in sound waves depends only on the relative motion of the source and receiver.

Problems

Unless a problem specifies otherwise, take the speed of sound to be 340 m/s *in air and* 1500 m/s *in water.*

Level I

14-1 Speed of Sound Waves

1. Show that $\sqrt{B/\rho}$ has the units of meters per second if B has the correct SI units and ρ is in kilograms per cubic meter.

2. The bulk modulus for water is 2.0×10^9 N/m^2. Use it to find the speed of sound in water.

3. Young's modulus for aluminum is 7.0×10^{10} N/m^2. The density of aluminum is 2.7×10^3 kg/m^3. Find the speed of sound in an aluminum rod.

4. Calculate the speed of sound waves in hydrogen gas at $T = 300$ K. (Take $M = 2$ g/mol and $\gamma = 1.4$.)

5. The speed of sound in mercury is 1410 m/s. What is the bulk modulus for mercury ($\rho = 13.6 \times 10^3$ kg/m^3)?

14-2 Harmonic Sound Waves

6. A sound wave in air produces a pressure variation given by

$$p(x, t) = 0.75 \cos \frac{\pi}{2} (x - 340t)$$

where p is in pascals, x is in meters, and t is in seconds. (*a*) What is the pressure amplitude, (*b*) the wavelength, (*c*) the frequency and (*d*) the speed of the sound wave?

7. (*a*) Middle C on the musical scale has a frequency of 262 Hz. What is the wavelength of this note in air? (*b*) The frequency of the C an octave above middle C is twice that of middle C. What is the wavelength of this note in air?

8. What is the displacement amplitude for a sound wave having a frequency of 100 Hz and a pressure amplitude of 10^{-4} atm?

9. The displacement amplitude of a sound wave of frequency 300 Hz is 10^{-7} m. What is the pressure amplitude of this wave?

10. (*a*) Find the displacement amplitude of a sound wave of frequency 500 Hz at the pain-threshold pressure amplitude of 29 Pa. (*b*) Find the displacement amplitude of a sound wave with the same pressure amplitude but a frequency of 1 kHz.

11. A typical loud sound wave with a frequency of 1 kHz has a pressure amplitude of about 10^{-4} atm. (*a*) At $t = 0$, the pressure is a maximum at some point x_1. What is the displacement at that point at $t = 0$? (*b*) What is the maximum value of the displacement at any time and place? (Take the density of air to be 1.29 kg/m^3.)

12. (*a*) Find the displacement amplitude of a sound wave of frequency 500 Hz at the threshold-of-hearing pressure amplitude of 2.9×10^{-5} Pa. (*b*) Find the displacement amplitude of a wave of the same pressure amplitude but a frequency of 1 kHz.

14-3 Waves in Three Dimensions: Intensity

13. A piston at one end of a long tube filled with air at room temperature and normal pressure oscillates with a frequency of 500 Hz and an amplitude of 0.1 mm. The area of the piston is 100 cm^2. (*a*) What is the pressure amplitude of the sound waves generated in the tube? (*b*) What is the intensity of the waves? (*c*) What average power is required to keep the piston oscillating (neglecting friction)?

14. What is the intensity level in decibels of a sound wave of intensity (*a*) 10^{-10} W/m^2 and (*b*) 10^{-2} W/m^2?

15. Find the intensity of a sound wave if (*a*) $\beta = 10$ dB and (*b*) $\beta = 3$ dB. (*c*) Find the pressure amplitudes of sound waves in air for each of these intensities.

16. Show that if the intensity is doubled, the intensity level increases by 3.0 dB.

17. What fraction of the acoustic power of a noise would have to be eliminated to lower its sound intensity level from 90 to 70 dB?

14-4 Interference: Beats

18. Two sound sources oscillate in phase with the same amplitude A. They are separated in space by $\frac{1}{3}\lambda$. What is the amplitude of the resultant wave from the two sources at a point that is on the line that passes through the sources but is not between the sources?

19. Two sound sources oscillate in phase with a frequency of 100 Hz. At a point 5.00 m from one source and 5.85 m from the other, the amplitude of the sound from each source separately is A. (*a*) What is the phase difference in the sound waves from the two sources at that point? (*b*) What is the amplitude of the resultant wave at that point?

20. With a compass, draw circular arcs representing wave crests originating from each of two point sources a distance $d = 6$ cm apart for $\lambda = 1$ cm. Connect the intersections corresponding to points of constant path difference, and label the path difference for each line. (See Figure 14-10.)

21. Two loudspeakers are separated by a distance of 6 m. A listener sits directly in front of one speaker at a distance of 8 m from it such that the two speakers and the listener form a right triangle. (*a*) Find the two lowest frequencies for which the path difference from the speakers to the listener is an odd number of half wavelengths. (*b*) Why can these frequencies be heard even if the speakers are driven in phase by the same amplifier?

22. Two violinists standing a few feet apart are playing the same notes. Are there places in the room where certain notes are not heard because of destructive interference? Explain.

23. Two speakers separated by some distance emit sound waves of the same frequency. At some point P, the intensity due to each speaker separately is I_0. The path distance from P to one of the speakers is $\frac{1}{2}\lambda$ greater than that from P to the other speaker. What is the intensity at P if (*a*) the speakers are coherent and in phase, (*b*) the speakers are incoherent, and (*c*) the speakers are coherent but have a phase difference of π rad?

24. Answer the questions of Problem 23 for a point P' for which the distance to the far speaker is 1λ greater than the distance to the near speaker. Assume that the intensity at point P' due to each speaker separately is again I_0.

25. Two speakers separated by some distance emit sound waves of the same frequency, but the waves from speaker 1 lead those from speaker 2 in phase by 90°. Let r_1 be the distance from some point to speaker 1 and r_2 be the distance from that point to speaker 2. Find the smallest value of $r_2 - r_1$ such that the sound at the point will be (*a*) maximum and (*b*) minimum. (Express your answers in terms of the wavelength.)

26. Two tuning forks have frequencies of 256 and 260 Hz. If the forks are vibrating at the same time, what is the beat frequency?

27. When a violin string is played (without fingering) simultaneously with a tuning fork of frequency 440 Hz, beats are heard at the rate of 3 per second. When the tension in the string is increased slightly, the beat frequency decreases. What was the initial frequency of the violin string?

28. Two tuning forks are struck simultaneously, and 4 beats per second are heard. The frequency of one fork is 500 Hz. (*a*) What are the possible values for the frequency of the other fork? (*b*) A piece of wax is placed on the 500-Hz fork to lower its frequency slightly. Explain how the measurement of the new beat frequency can be used to determine which of your answers to part (*a*) is the correct frequency of the second fork.

14-5 Standing Sound Waves

29. Calculate the fundamental frequency for a 10-m organ pipe that is (*a*) open at both ends and (*b*) closed at one end.

30. The normal range of hearing is about 20 to 20,000 Hz. What is the greatest length of an organ pipe that would have its fundamental note in this range if (*a*) it is closed at one end and (*b*) it is open at both ends?

31. The shortest pipes used in organs are about 7.5 cm long. (*a*) What is the fundamental frequency of a pipe this long that is open at both ends? (*b*) For such a pipe, what is the highest harmonic that is within the audible range (see Problem 30)?

32. The space above the water in a tube like that shown in Figure 14-17 is 120 cm long. Near the open end, there is a loudspeaker that is driven by an audio oscillator whose frequency can be varied from 10 to 5000 Hz. (*a*) What is the lowest frequency of the oscillator that will produce resonance within the tube? (*b*) What is the highest frequency that will produce resonance? (*c*) How many different frequencies of the oscillator will produce resonance? (Neglect the end correction.)

33. The ear canal, which is about 2.5 cm long, roughly approximates a pipe that is open at one end and closed at the other. (*a*) What are the resonance frequencies of the ear canal? (*b*) Describe the possible effect of the resonance modes of the ear canal on the threshold of hearing.

14-6 Harmonic Analysis and Synthesis

There are no problems for this section.

14-7 Wave Packets and Dispersion

34. Information for use by computers is transmitted along a cable in the form of short electric pulses at the rate of 10^5 pulses per second. (*a*) What is the maximum duration of each pulse if no two pulses overlap? (*b*) What is the range of frequencies to which the receiving equipment must respond?

35. A tuning fork of frequency f_0 begins vibrating at time $t = 0$ and is stopped after a time interval Δt. The waveform of the sound at some later time is shown as a function of x in Figure 14-34. Let N be the (approximate) number of cycles in this waveform. (*a*) How are N, f_0, and Δt related? (*b*) If Δx is the length in space of this wave packet, what is the wavelength in terms of Δx and N? (*c*) What is the wave number k in terms of N and Δx? (*d*) The number N is uncertain by about ± 1 cycle. Use Figure 14-34 to explain why. (*e*) Show that the uncertainty in the wave number due to the uncertainty in N is $2\pi/\Delta x$.

Figure 14-34 Problem 35

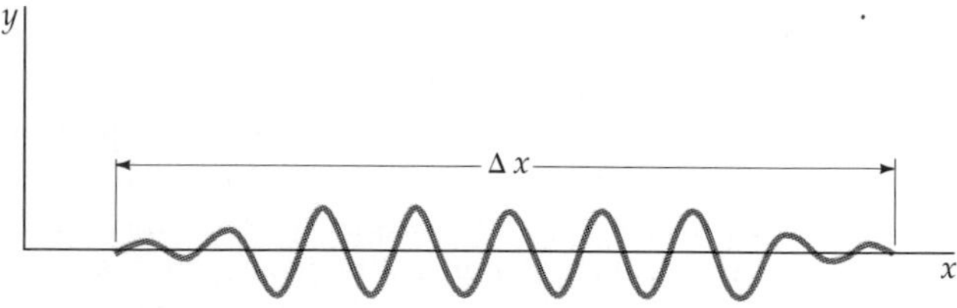

14-8 Reflection, Refraction, and Diffraction

36. If a wavelength is much larger than the diameter of a loudspeaker, the speaker radiates in all directions, much like a point source. On the other hand, if the wavelength is much smaller than the diameter, the sound travels in approximately a straight line in front of the speaker. Find the frequency of a sound wave that has a wavelength (*a*) 10 times and (*b*) one-tenth the diameter of a 30-cm speaker. (*c*) Do the same for a 6-cm speaker.

14-9 The Doppler Effect

37. This problem is a doppler-effect analogy. A conveyor belt moves to the right with a speed $v = 300$ m/min. A very fast piemaker puts pies on the belt at a rate of 20 per minute, and they are received at the other end by a pie eater. (*a*) If the piemaker is stationary, find the spacing λ between the pies and the frequency f with which they are received by the stationary pie eater. (*b*) The piemaker now walks with a speed of 30 m/min toward the receiver while continuing to put pies on the belt at 20 per minute. Find the spacing of the pies and the frequency with which they are received by the stationary pie eater. (*c*) Repeat your calculations for a stationary piemaker and pie eater who moves toward the piemaker at 30 m/min.

38. For the situation described in Problem 37, derive general expressions for the spacing of the pies λ and the frequency f with which they are received by the pie eater in terms of the speed of the belt v, the speed of the sender u_s, the speed of the receiver u_r, and the frequency f_0 with which the piemaker places pies on the belt.

In Problems 39 through 44, a source emits sounds of frequency 200 Hz *that travel through still air at* 340 m/s.

39. The source moves with a speed of 80 m/s relative to still air toward a stationary listener. (*a*) Find the wavelength of the sound between the source and the listener. (*b*) Find the frequency heard by the listener.

40. Consider the situation in Problem 39 from the reference frame in which the source is at rest. In this frame, the listener moves toward the source with a speed of 80 m/s, and there is a wind blowing at 80 m/s from the listener to the source. (*a*) What is the speed of the sound from the source to the listener in this frame? (*b*) Find the wavelength of the sound between the source and the listener. (*c*) Find the frequency heard by the listener.

41. The source moves away from the stationary listener at 80 m/s. (*a*) Find the wavelength of the sound waves between the source and the listener. (*b*) Find the frequency heard by the listener.

42. The listener moves at 80 m/s relative to still air toward the stationary source. (*a*) What is the wavelength of the sound between the source and the listener? (*b*) What is the frequency heard by the listener?

43. Consider the situation in Problem 42 in a reference frame in which the listener is at rest. (*a*) What is the wind velocity in this frame? (*b*) What is the speed of the sound from the source to the listener in this frame, that is, relative to the listener? (*c*) Find the wavelength of the sound between the source and the listener in this frame. (*d*) Find the frequency heard by the listener.

44. The listener moves at 80 m/s relative to the still air away from the stationary source. Find the frequency heard by the listener.

45. A whistle of frequency 500 Hz moves in a circle of radius 1 m at 3 rev/s. What are the maximum and minimum frequencies heard by a stationary listener in the plane of the circle and 5 m away from its center?

46. Ocean waves move toward the beach with a speed of 8.9 m/s and a crest-to-crest separation of 15.0 m. You are in a small boat anchored off shore. (*a*) What is the frequency of the ocean waves? (*b*) You now lift anchor and head out to sea at a speed of 15 m/s. What frequency of the waves do you observe?

47. A boat traveling at 10 m/s on a still lake makes a bow wave at an angle of 20° with its direction of motion. What is the speed of the bow wave?

48. A jet is traveling at Mach 2.5 at an altitude of 5000 m. (*a*) What is the angle that the shock wave makes with the track of the jet? (Assume the speed of sound at this altitude is still 340 m/s.) (*b*) Where is the jet when a person on the ground hears the shock wave?

Level II

49. Two parallel tubes with the same diameter are filled with gas at the same pressure and temperature. One tube contains H_2, and the other contains O_2. (*a*) If sound waves traveling in the tubes have the same displacement amplitude and frequency, how do their intensities compare? (*b*) If the waves have the same frequency and pressure amplitude, how do their intensities compare? (*c*) If the waves have the same frequency and intensity, how do their pressure amplitudes and displacement amplitudes compare?

50. A common rule of thumb for finding the distance to a lightning flash is to begin counting when the flash is observed and stop when the thunder clap is heard. The number of seconds counted is then divided by 3 to get the distance in kilometers. Why is this justified? How accurate is this procedure? What is the velocity of sound in kilometers per second? Is a correction for the time it takes for the light to reach you important? (The speed of light is about 3×10^8 m/s.)

51. A method for measuring the speed of sound using an ordinary watch with a second hand is to stand some distance from a large flat wall and clap your hands rhythmically in such a way that the echo from the wall is heard halfway between every two claps. Show that the speed of sound is given by $v = 4LN$, where L is the distance to the wall and N is the number of claps per second. What is a reasonable value for L for this experiment to be feasible? (If you have access to a flat wall outdoors somewhere, try this method and compare your result with the standard value for the speed of sound.)

52. A man drops a stone from a high bridge and hears it strike the water below exactly 4 s later. (*a*) Estimate the distance to the water based on the assumption that the travel time for the sound to reach the man is negligible. (*b*) Improve your estimate by using your result from part (*a*) for the distance to the water to estimate the time it takes for sound to travel this distance and then calculate the distance the rock falls in 4 s minus this time. (*c*) Calculate the exact distance and compare your result with your previous estimates.

53. Kundt's method was an early procedure for determining the speed of sound in different gases. A cylindrical glass tube is placed in a horizontal position, and a quantity of light powder is spread along the bottom of the tube. One end of the tube is closed by a piston that is attached to an oscillator of a known frequency f such as a tuning fork. The other end is closed by a piston whose position can be varied. While the first piston is made to oscillate at the frequency f, the movable piston's position is adjusted until resonance occurs. When this happens, the powder collects in equally spaced piles along the bottom of the tube. (*a*) Explain why the powder collects in this way. (*b*) Derive a formula that gives the speed of sound in the gas in terms of f and the distance between the piles of powder. (*c*) Give suitable values for the frequency f and the distance between the piles of powder. (*d*) Give suitable values of the frequency f and the length L of the tube for which the speed of sound could be measured in either air or helium.

54. Three successive resonance frequencies in an organ pipe are 1310, 1834, and 2358 Hz. (*a*) Is the pipe closed at one end or open at both ends? (*b*) What is the fundamental frequency? (*c*) What is the length of the pipe?

55. A wire of mass 1 g and length 50 cm is stretched with a tension of 440 N. It is placed near the open end of the tube in Figure 14-17 and is stroked with a violin bow so that it oscillates at its fundamental frequency. The water level in the tube is lowered until a resonance is first obtained at 18 cm below the top of the tube. Use the data given to determine the speed of sound in air. Why is this method not very accurate?

56. A 460-Hz tuning fork causes resonance in the tube in Figure 14-17 when the top of the tube is 18.3 cm and 55.8 cm above the water surface. (*a*) Find the speed of sound in air. (*b*) What is the end correction to adjust for the fact that the antinode does not occur exactly at the end of the open tube?

57. (*a*) Compute the derivative of the velocity of sound with respect to the absolute temperature, and show that the differentials dv and dT obey $dv/v = \frac{1}{2}\,dT/T$. (*b*) Use this result to compute the percentage change in the velocity of sound when the temperature changes from 0 to 27°C. (*c*) If the speed of sound is 331 m/s at 0°C, what is it (approximately) at 27°C? How does this approximation compare with the result of an exact calculation?

58. (*a*) Show that if the temperature changes by a small amount ΔT, the fundamental frequency of an organ pipe changes by approximately Δf, where $\Delta f/f = \frac{1}{2}\,\Delta T/T$. (*b*) Suppose an organ pipe that is closed at one end has a fundamental frequency of 200 Hz when the temperature is 20°C. What will its fundamental frequency be when the temperature is 30°C? (Ignore any change in the length of the pipe due to thermal expansion.)

59. Two loudspeakers emit sound waves in phase at 170 Hz. An observer sits at 8 m from one speaker and 11 m from the other. The intensity level from either speaker acting alone is 60 dB. (*a*) Find the observed intensity level when both speakers are on together. (*b*) Find the observed intensity level when both speakers are on together but the leads to one have been reversed so that the speakers are 180° out of phase. (*c*) Find the observed intensity level when both speakers are on together and in phase but the frequency is 85 Hz.

60. Instructions for connecting stereo speakers to an amplifier correctly so that they are in phase are as follows: "After both speakers are connected, play a monophonic record or program with the bass control turned up and the treble control turned down. While listening to the speakers, turn the balance control so that first one speaker is heard separately, then the two together, and then the other separately. If the bass is stronger when both speakers play together, they are connected properly. If the bass is weaker when both play together than when each plays separately, reverse the connections on one speaker." Explain why this method works. In particular, explain why a stereo source is not used and why only the bass is compared.

61. Two identical speakers emit sound waves of frequency 680 Hz uniformly in all directions. The total audio output of each speaker is 1 mW. A point P is 2.00 m from one speaker and 3.00 m from the other. (*a*) Find the intensities I_1 and I_2 from each speaker separately at point P. (*b*) If the speakers are driven coherently and are in phase, what is the intensity at point P? (*c*) If they are driven coherently but are 180° out of phase, what is the intensity at point P? (*d*) If the speakers are incoherent, what is the intensity at point P?

62. A radar device emits microwaves with a frequency of 2.00 GHz. When the waves are reflected from a car moving directly away from the emitter, the beat frequency is 293 beats per second. Find the speed of the car.

63. Hovering over the pit of hell, the devil observes that as a student falls past (with the terminal velocity), the frequency of his scream decreases from 842 to 820 Hz. (*a*) Find the speed of descent of the student. (*b*) The student's scream generates beats when it is mixed with its echo from the bottom of the pit. Find the number of beats per second heard by the student. (*c*) Find the number of beats per second heard by the devil after the student has passed by.

64. The ratio of the frequencies of a note to those of the semitone above it on the diatonic scale is about 15:16. Find the speed of a car for which the tone of its horn will drop a semitone as it passes you.

65. A loudspeaker diaphragm 30 cm in diameter is vibrating at 1 kHz with an amplitude of 0.020 mm. Assuming that the air molecules in the vicinity have this same amplitude of vibration, find (*a*) the pressure amplitude immediately in front of the diaphragm, (*b*) the sound intensity, and (*c*) the acoustic power being radiated.

66. A stationary destroyer is equipped with sonar that sends out pulses of sound at 40 MHz. Reflected pulses are received from a submarine directly below with a time delay of 80 ms at a frequency of 39.958 MHz. If the speed of sound in seawater is 1.54 km/s, find (*a*) the depth of the submarine and (*b*) its vertical speed.

67. Two students with vibrating 440-Hz tuning forks walk away from each other with equal speeds. How fast must they walk to hear a beat frequency of 2 Hz?

68. A physics student walks down a long hall carrying a vibrating 512-Hz tuning fork. The end of the hall is closed so that sound reflects from it. The student hears 4 beats per second. How fast is the student walking?

69. A spherical source radiates sound uniformly in all directions. At a distance of 10 m, the sound intensity level is 80 dB, with 0 dB corresponding to 10^{-12} W/m^2. (*a*) At what distance from the source is the intensity level 60 dB? (*b*) What power is radiated by this source?

70. A plane, harmonic, acoustical wave that oscillates in air with an amplitude of 10^{-6} m has an intensity of 100 dB. What is the frequency of the sound wave?

71. A loudspeaker at a rock concert generates 10^{-2} W/m^2 at 20 m at a frequency of 1 kHz. Assume that the speaker spreads its energy uniformly in the forward hemisphere and that no energy is reflected from the ground or elsewhere so that the intensity is given by $I = P/2\pi r^2$. (*a*) What is the intensity level at 20 m? (*b*) What is the total acoustic power output of the speaker? (*c*) At what distance will the intensity level be at the pain threshold of 120 dB? (*d*) What is the intensity level at 30 m?

72. When a pin of mass 0.1 g is dropped from a height of 1 m, 0.05 percent of its energy is converted into a sound pulse with a duration of 0.1 s. (*a*) Estimate the range at which the dropped pin can be heard if the minimum audible intensity is 10 pW/m^2. (*b*) Your result in (*a*) is much too large in practice because of background noise. If you assume that the intensity level must be at least 40 dB for the sound to be heard, estimate the range at which the dropped pin can be heard. (In both parts, assume that the intensity is $P/4\pi r^2$.)

73. Normal human speech has a sound intensity level of about 65 dB at 1 m. Estimate the power of human speech.

74. Three noise sources produce intensity levels of 70, 73, and 80 dB when acting separately. When the sources act together, their intensities add. (There is no interference between the sound waves from the different sources because the relative phase differences vary randomly.) (*a*) Find the sound intensity level in decibels when the three sources act at the same time. (*b*) Discuss the effectiveness of eliminating the two least intense sources in reducing the intensity level of the noise.

75. An article on noise pollution claims that sound intensity levels in large cities have been increasing by about 1 dB annually. (*a*) To what percentage increase in intensity does this correspond? Does this increase seem reasonable? (*b*) In about how many years will the intensity of sound double if it increases at 1 dB annually?

76. A column of precision marchers keeps in step by listening to the band positioned at the head of the column. The beat of the music is for 100 paces per minute. A television camera shows that only the marchers at the front and the rear of the column are actually in step. The marchers in the middle section are striding forward with the left foot when those at the front and rear are striding forward with the right foot. The marchers are so well trained, however, that they are all certain they are in proper step with the music. Explain the source of the problem, and calculate the length of the column.

Level III

77. Everyone at a party is talking equally loudly. If only one person were talking, the sound level would be 72 dB. Find the sound level when all 38 people are talking.

78. The noise level in an empty examination hall is 40 dB. When 100 students are writing an exam, the sounds of heavy breathing and pens traveling rapidly over paper cause the noise level to rise to 60 dB (not counting the occasional groans). Assuming that each student contributes an equal amount of noise power, find the noise level to the nearest decibel when 50 students have left.

79. Two point sources that are in phase are separated by a distance d. An interference pattern is detected along a line parallel to that through the sources and a large distance D from the sources as shown in Figure 14-35. (*a*) Show that the path difference from the two sources to some point on the line at an angle θ is given approximately by

$$\Delta x = d \sin \theta$$

(*b*) Show that the distance y_m from the central maximum point to the mth interference maximum is given approximately by

$$y_m = m\,\frac{D\lambda}{d}$$

Figure 14-35 Problems 79–81.

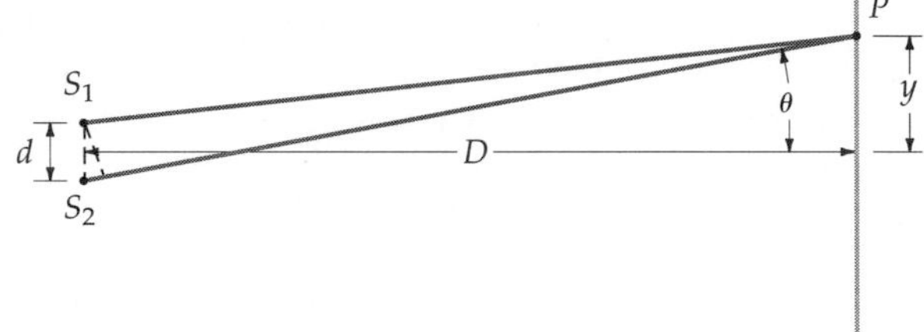

80. Two loudspeakers are driven in phase by an audio amplifier at a frequency of 600 Hz. The speakers are on the y axis, one at $y = +1.00$ m and the other at $y = -1.00$ m. A listener begins at $y = 0$ and walks along a line parallel to the y axis at a very large distance D away. (See Problem 79.) (*a*) At what angle θ will she first hear a minimum in the sound intensity? (*b*) At what angle will she first hear a maximum (after $\theta = 0$)? (*c*) How many maxima can she possibly hear if she keeps walking in the same direction?

81. Two sound sources, driven in phase by the same amplifier, are 2 m apart on the y axis. At a point a very large distance from the y axis, constructive interference is first heard at an angle $\theta_1 = 0.140$ rad with the x axis and is next heard at $\theta_2 = 0.283$ rad. (See Figure 14-35.) (*a*) What is the wavelength of the sound waves from the sources? (*b*) What is the frequency of the sources? (*c*) At what other angles is constructive interference heard? (*d*) What is the smallest angle for which the sound waves completely cancel?

82. Two sources have a phase difference δ_0 that is proportional to time: $\delta_0 = Ct$, where C is a constant. The amplitude of the wave from each source at some point P is A_0. (*a*) Write the wave functions for each of the two waves at point P, assuming this point to be a distance x_1 from one source and $x_1 + \Delta x$ from the other. (*b*) Find the resultant wave function, and show that its amplitude is $2A_0 \cos \frac{1}{2}(\delta + \delta_0)$, where δ is the phase difference at P due to the path difference. (*c*) Sketch the intensity at point P versus time for a zero path difference. (Let I_0 be the intensity due to each wave separately.) What is the time average of the intensity? (*d*) Make the same sketch for the intensity at a point for which the path difference is $\frac{1}{2}\lambda$.

83. In this problem, you will derive a convenient formula for the speed of sound in air at temperature t in Celsius degrees. Begin by writing the temperature as $T = T_0 + \Delta T$, where $T_0 = 273$ K corresponds to 0°C and $\Delta T = t$, the Celsius temperature. The speed of sound is a function of T, $v(T)$. To a first-order approximation, you can write

$$v(T) \approx v(T_0) + \left(\frac{dv}{dT}\right)_{T_0} \Delta T$$

where $(dv/dT)_{T_0}$ is the derivative evaluated at $T = T_0$. Compute this derivative, and show that the result leads to

$$v \approx (331 \text{ m/s})\left(1 + \frac{t}{2T_0}\right) = (331 + 0.606t) \text{ m/s}$$

84. While studying physics in her dorm room, a student is listening to the live radio broadcast of a baseball game. She is 1.6 km due south of the baseball field. Over her radio, the student hears a noise generated by the electromagnetic pulse of a lightning bolt. Two seconds later, she hears over the radio the thunder picked up by the microphone at the baseball field. Four seconds after she hears the noise of the electromagnetic pulse over the radio, thunder rattles her windows. Where, relative to the ballpark, did the lightning bolt occur?

85. Two identical sound sources have a frequency of 500 Hz. The coordinates of the sources are (0, 1) m and (0, −1) m. A detector 80 m from the origin is free to revolve in the xy plane with a radius of 80 m. The first maxima in intensity is detected when the detector is at (80, 0) m. (*a*) Find the coordinates of the detector for the first five maxima in order of decreasing positive values of x. (*b*) Find the coordinates for the first four minima in order of decreasing positive values of x subject to the condition $\sqrt{x^2 + y^2} = 80$ m.

86. A radio telescope consists of two antennas separated by a distance of 200 m. Both antennas are tuned to a particular frequency, such as 20 MHz. The signals from each antenna are fed into a common amplifier, but one signal first passes through a phase adjuster that delays its phase by a chosen amount so that the telescope can "look" in different directions. When the phase delay is zero, plane radio waves that are incident vertically on the antennas produce signals that add constructively at the amplifier. What should the phase delay be so that signals coming from an angle $\theta = 10°$ with the vertical (in the plane formed by the vertical and the line joining the antennas) will add constructively at the amplifier?

87. A coiled spring, such as a Slinky, is stretched to a length L. It has a force constant k and a mass m. (*a*) Show that the velocity of longitudinal compression waves along the spring is given by $v = L\sqrt{k/m}$. (*b*) Show that this is also the velocity of transverse waves along the spring if the natural length of the spring is much less than L.

88. A physics student drops a vibrating 440-Hz tuning fork down an elevator shaft of a tall building. When the student hears a frequency of 400 Hz, how far has the tuning fork fallen?

Part 3 Thermodynamics

Two red intersecting laser beams are used here to study the combustion of coal–water slurries in a conventional power conversion device. The test material is injected into the combustion reactor (blue flame), giving off a yellow-orange emission as it ignites and burns. The laser light is used to measure the particle size of combustible material.

Chapter 15

Temperature

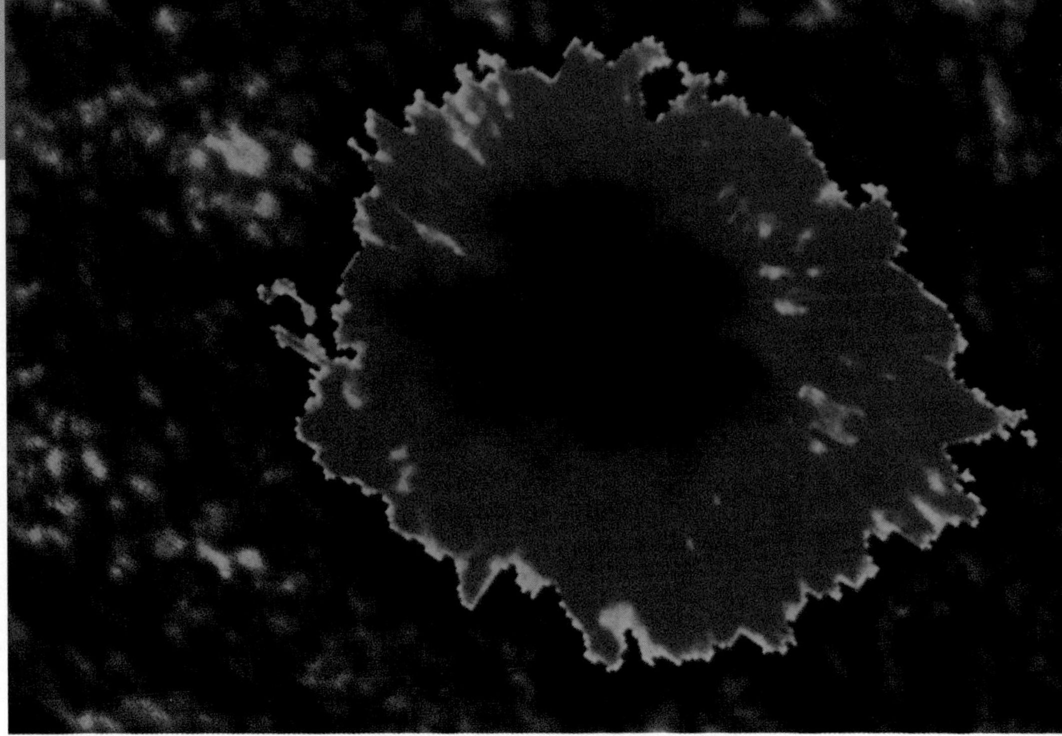

Sunspots appear on the surface of the sun when streams of gases slowly erupt from deep within the star. The solar "flower" is 10,000 miles in diameter. The temperature variation, indicated by color changes made by computer color enhancement, is not fully understood. The central portion of the sunspot is cooler than the outer regions as indicated by the dark area. The temperature at the sun's core is of the order of 10^6 K, whereas at the surface the temperature is only about 6000 K.

Thermodynamics is the study of temperature, heat, and the exchange of energy. It has practical applications in all branches of science and engineering as well as in many aspects of daily life, from dealing with the weather to cooking. In this chapter, we will study temperature and some of the thermal properties of matter.

Temperature is familiar to us all as the measure of the hotness or coldness of objects. More precisely, it is a measure of the average internal molecular kinetic energy of an object. The definition and determination of temperature are subtle. For example, it is quite difficult to define temperature so that different thermometers will agree with each other in the measurement of the temperature of a substance. However, the properties of gases at low densities do allow us to define a temperature scale and to construct gas thermometers that do agree. In Chapter 17, we will see that the second law of thermodynamics can be used to define a temperature scale that is independent of the properties of any substance and can be chosen to coincide with the temperature scale defined using the properties of gases.

15-1 The Celsius and Fahrenheit Temperature Scales

We are born with built-in thermometers. Our sense of touch can usually tell us if an object is hot or cold. Early in childhood we learn that to make a cold object warmer we place it in contact with a hot object. To warm water, we put the pot on a flame. To make a warm object cool, we place it in contact with a cold object. To chill a drink, we place the bottle in an ice bucket.

When an object is heated or cooled, some of its physical properties change. For example, most solids and liquids expand when they are heated. A gas, if it is allowed to, will also expand when it is heated, or, if its volume is kept constant, its pressure will rise. If an electrical conductor is heated, its electrical resistance changes. A physical property that changes with temperature is called a **thermometric property.** A change in a thermometric property indicates a change in the temperature of the object.

Suppose we place a warm copper bar in close contact with a cold iron bar. The copper bar contracts slightly, indicating that it is cooling, and the iron bar expands slightly, indicating that it is warming. The two bars are said to be in **thermal contact.** Eventually this process stops; that is, neither bar changes anymore in length. When this happens the two bars are said to be in **thermal equilibrium** with each other.

Now, suppose that instead of placing the bars in contact with each other, we first place the warm copper bar in a cool lake. The bar cools down and the water in the lake warms slightly, but the lake is large enough that the change in its temperature is negligible. Again, this process eventually stops and the bar and water are in thermal equilibrium. Now suppose we place the cold iron bar in the lake far away from the copper bar so that the two bars are not in thermal contact. The iron bar warms until it and the lake water are also in thermal equilibrium. If we remove the iron and copper bars and place them in contact with each other, we find that their lengths do not change. They are in thermal equilibrium with each other. This is such a common experience that you are probably not surprised. However, there is no logical way to deduce this result. It is an important statement of the thermal behavior of all objects.

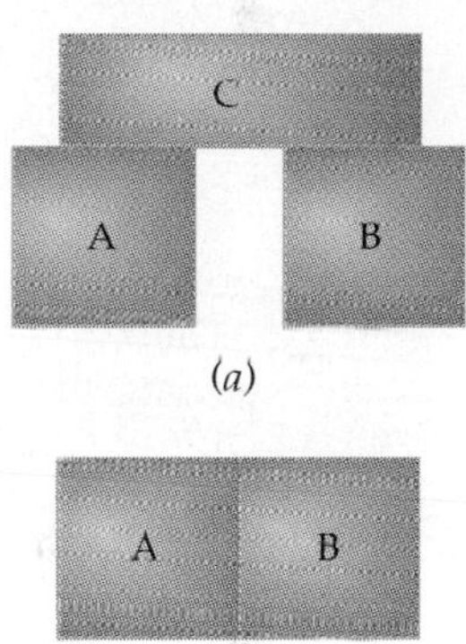

Figure 15-1 The zeroth law of thermodynamics. (*a*) Systems *A* and *B* are in thermal contact with system *C* but not with each other. When *A* and *B* are each in thermal equilibrium with *C*, they are in thermal equilibrium with each other, which can be checked by placing them in contact with each other as in (*b*).

If two objects are in thermal equilibrium with a third, then they are in thermal equilibrium with each other.

This is often called the **zeroth law of thermodynamics.** It is illustrated in Figure 15-1. The zeroth law of thermodynamics allows us to define a temperature scale. Two objects in thermal equilibrium with each other are said to have the same temperature.

Any thermometric property can be used to establish a temperature scale and construct a thermometer. Figure 15-2 shows a common mercury thermometer, which consists of a glass bulb and tube containing a fixed amount of mercury. When the mercury is heated by placing the thermometer in contact with a warmer body, the mercury expands more than the glass, and the length of the mercury column increases. Temperatures are measured by comparing the end of the mercury column with markings on the glass tube. These markings are determined as follows. The thermometer is first placed in ice and water in equilibrium at a pressure of 1 atm.* If the thermometer is

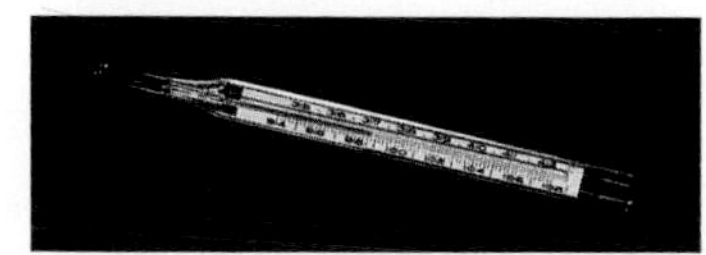

Figure 15-2 A simple mercury thermometer. When in thermal contact with a warmer body, the mercury expands and the column rises.

*Water and ice in equilibrium provide a constant-temperature bath. When ice is placed in warm water, the water cools as some of the ice melts. Eventually, thermal equilibrium is reached and no more ice melts. If the system is heated slightly, some more of the ice melts, but the temperature does not change as long as some ice remains.

(a)

(b)

(c)

(a) The temperature in the interior of a furnace is measured by a radiation thermometer. (b) Various temperature probes for use with a laboratory thermocouple digital thermometer. (c) Thermometers built around 1660 in Florence, Italy.

initially warmer than the ice water, the length of the mercury column will decrease, but it will eventually cease changing. The thermometer is now in thermal equilibrium with the ice water. The position of the mercury column is marked on the glass tube. This is the **ice-point temperature** (also called the **normal freezing point** of water). Next, the thermometer is placed in boiling water at a pressure of 1 atm, and the length of the mercury column increases until the thermometer is in thermal equilibrium with the boiling water. The new position of the column is marked. This is the **steam-point temperature** (also called the **normal boiling point** of water).

The **Celsius temperature scale** (formerly called the centigrade scale) is constructed by defining the ice-point temperature to be zero degrees Celsius (0°C) and the steam-point temperature to be 100°C. The space on the glass tube between the ice-point mark and the steam-point mark is then divided into 100 equal intervals or degrees, and the degree markings are extended below the ice-point mark and above the steam-point mark. The temperature of another system can now be measured by placing the mercury thermometer in thermal contact with it, waiting for thermal equilibrium to be established, and noting the position of the mercury column. If L_t is the length of the mercury column, the Celsius temperature t_C is given by

$$t_C = \frac{L_t - L_0}{L_{100} - L_0} \times 100^\circ \qquad 15\text{-}1$$

where L_0 is the length of the mercury column when the thermometer is in an ice bath and L_{100} is its length when the thermometer is in a steam bath. The normal temperature of the human body measured on the Celsius scale is about 37°C.

The **Fahrenheit temperature scale** is constructed by defining the ice-point temperature to be 32°F and the steam-point temperature to be 212°F.* Because the Fahrenheit scale is in common use in the United States and the Celsius scale is used in scientific work and throughout the rest of the world, we often need to convert temperatures between these two scales. We note that there are 100 Celsius degrees and 180 Fahrenheit degrees between the ice point and the steam point. A temperature change of one Fahrenheit degree is therefore smaller than a change of one Celsius degree. A temperature change of one Celsius degree (written 1 C° to distinguish it from a temperature of 1°C) equals a change of $\frac{9}{5}$ Fahrenheit degrees. To convert a temperature given on one scale to that of the other scale, we must also take

*Fahrenheit wanted all measurable temperatures to be positive. Originally, he chose 0°F for the coldest temperature he could obtain with a mixture of ice and saltwater, and 96°F (a convenient number with many factors for subdivision) for the temperature of the human body. He then modified his scale slightly to make the ice-point and steam-point temperatures whole numbers. This resulted in the temperature of the human body being between 98° and 99°F.

into account the fact that the zero temperatures of the two scales are not the same. The general relation between a Fahrenheit temperature t_F and a Celsius temperature t_C is

$$t_C = \tfrac{5}{9}(t_F - 32°) \qquad 15\text{-}2$$

Fahrenheit–Celsius conversion

Example 15-1

Find the temperature on the Celsius scale equivalent to 41°F.

From Equation 15-2, we obtain

$$t_C = \tfrac{5}{9}(41° - 32°) = \tfrac{5}{9}(9°) = 5°C$$

We can understand this calculation as follows. A temperature of 41°F is 9 Fahrenheit degrees above the ice point (41 − 32 = 9). But 9 Fahrenheit degrees above the ice point is equivalent to 5 Celsius degrees above the ice point. Since the ice point on the Celsius scale is 0°C, a temperature of 41°F corresponds to 5°C.

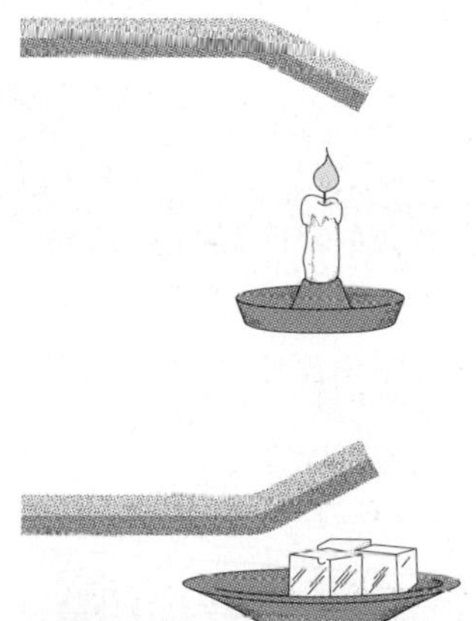

Figure 15-3 A bimetallic strip. When heated or cooled the two metals expand or contract by different amounts, causing the strip to bend.

Example 15-2

Find the temperature on the Fahrenheit scale equivalent to −10°C.

The temperature −10°C is 10 C° below the ice point. This is equivalent to 18 F° below the ice point on the Fahrenheit scale. Since the Fahrenheit ice point is 32°F, −10°C is equivalent to 14°F. This result can also be obtained directly from Equation 15-2 by substituting $t_C = -10°C$.

Exercise

(*a*) Find the Celsius temperature equivalent to 68°F. (*b*) Find the Fahrenheit temperature equivalent to −40°C. [Answers: (*a*) 20°C, (*b*) −40°F]

Other thermometric properties can be used to set up thermometers and construct temperature scales. Figure 15-3 shows a bimetallic strip consisting of two different metals bonded together. When the strip is heated or cooled, it bends to accommodate the difference in the thermal expansion of the two metals. One application of a bimetallic strip is in a thermostat to regulate a furnace. The bending of the strip is used to make or break an electrical contact to turn the furnace on or off (Figure 15-4). Figure 15-5 shows a thermometer consisting of a bimetallic coil with a pointer attached to indicate the temperature. When the thermometer is heated, the coil bends and the pointer moves to the right. The thermometer is calibrated in the same way as a mercury thermometer, that is, by determining the ice point and the steam point and dividing the interval between them into 100 Celsius degrees (or 180 Fahrenheit degrees).

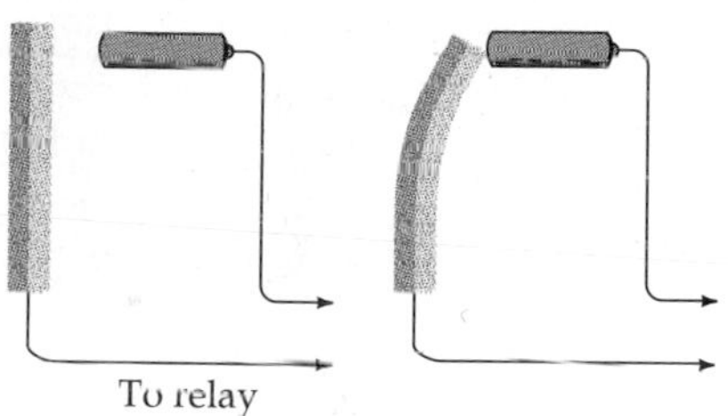

Figure 15-4 A bimetallic strip can be used in a thermostat to make or break an electrical contact.

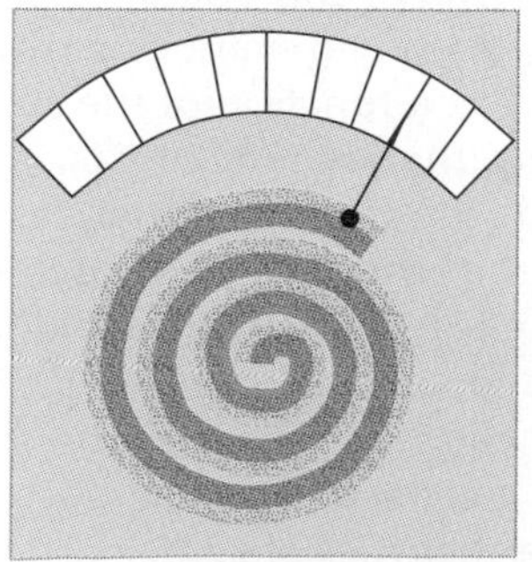

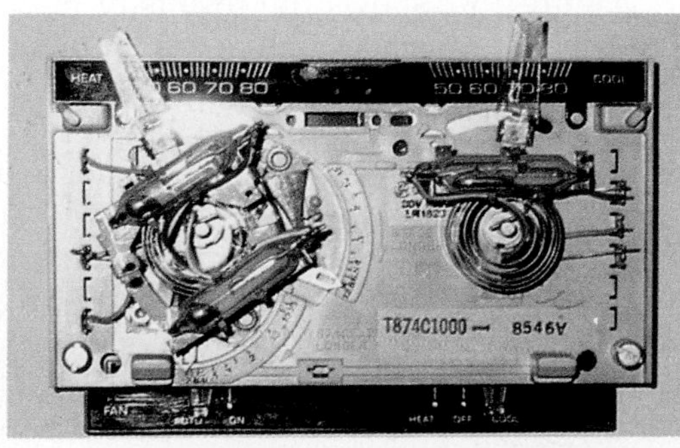

Figure 15-5 (*Left*) A thermometer using a bimetallic coil. (*Right*) An office thermostat. The coil on the right controls the heater. When the coil expands or contracts, the tube mounted on it tilts, and mercury in the tube slides to open or close an electrical switch. A slide lever (at the top) mounted on the coil is used to set the desired temperature. The coil on the left (partially obscured by two mercury switches) is used in a similar way to control the air conditioner.

Questions

1. How can you determine if two bodies are in thermal equilibrium with each other if it is impossible to put them into thermal contact with each other?
2. Which is greater, an increase in temperature of 1 C° or one of 1 F°?
3. One body is at −2°C and another is at +20°F. Which body is colder?
4. Distinguish between 1°C and 1 C°.

15-2 Gas Thermometers and the Absolute Temperature Scale

Mercury thermometers and bimetallic thermometers are commonly used. If they are calibrated in ice water and steam as described in Section 15-1, they must agree at 0°C and 100°C. But there is no reason to expect them to agree at any other temperature such as 50°C. Indeed, if they did it would mean that the thermal expansion of mercury, glass, and the metals used in the bimetallic thermometer varied with temperature in precisely the same way. In fact, at temperatures between 0 and 100°C different thermometers give slightly different readings, though the agreement is good enough for most purposes. But at temperatures well above the steam point or below the ice point, the agreement becomes very poor. Furthermore, mercury freezes at very low temperatures and vaporizes at very high temperatures. So the question of the measurement of temperatures other than 0°C or 100°C is more than just one of accuracy. How are such temperatures defined? Which thermometer is right?

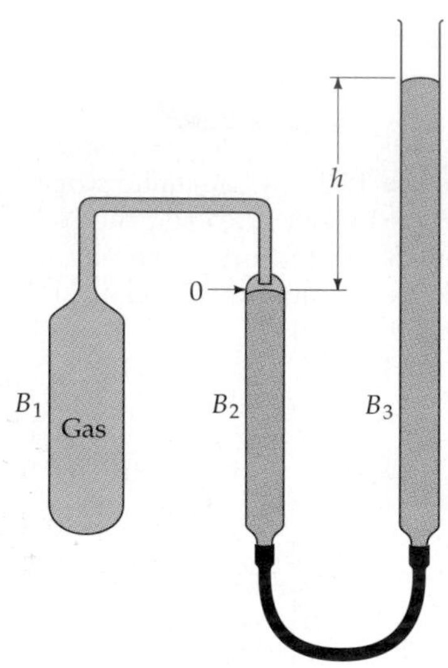

Figure 15-6 A constant-volume gas thermometer. The volume is kept constant by raising or lowering tube B_3 such that the mercury in tube B_2 remains at the zero mark. The temperature is chosen to be proportional to the pressure of the gas in tube B_1, which is indicated by the height h of the mercury column in tube B_3.

There is one group of thermometers for which measured temperatures are in very close agreement even far from the calibration points. These are the gas thermometers. In one version, the **constant-volume gas thermometer,** the volume of the gas is kept constant, and the pressure of the gas is used as the thermometric property. Figure 15-6 shows a constant-volume gas thermometer. The volume of the gas in tube B_1 is kept constant by raising or lowering tube B_3 so that the mercury in tube B_2 remains at the zero mark. The pressure of the gas is found by reading the height h of the mercury column in tube B_3. As we learned in Chapter 11, a column of mercury 760 mm high is equivalent to a pressure of 1 atm. The ice-point pressure and the steam-point pressure are determined in the usual way, and the interval between them is divided into 100 equal degrees (for the Celsius scale).

If the pressure in a constant-volume gas thermometer is P_0 in an ice bath, P_{100} in a steam bath, and P_t in a bath whose temperature is to be determined, that temperature in degrees Celsius is defined to be

$$t_C = \frac{P_t - P_0}{P_{100} - P_0} \times 100^{\circ} \qquad 15\text{-}3$$

Figure 15-7 shows the results of measurements of the boiling point of sulfur using constant-volume gas thermometers filled with various gases. In this figure, the measured temperature is plotted as a function of the steam-point pressure, P_{100}, which is varied by varying the amount of gas in the thermometer. As the amount of gas is reduced, the density of the gas decreases and steam-point pressure decreases. From this figure, we see that the agreement among the thermometers is very close no matter what kind of gas is used, as long as the density of the gas is low. Indeed, in the limit as the gas density goes to zero, all gas thermometers give exactly the same value for the temperature, independent of the gas used. Because this temperature measurement does not depend on the properties of some particular

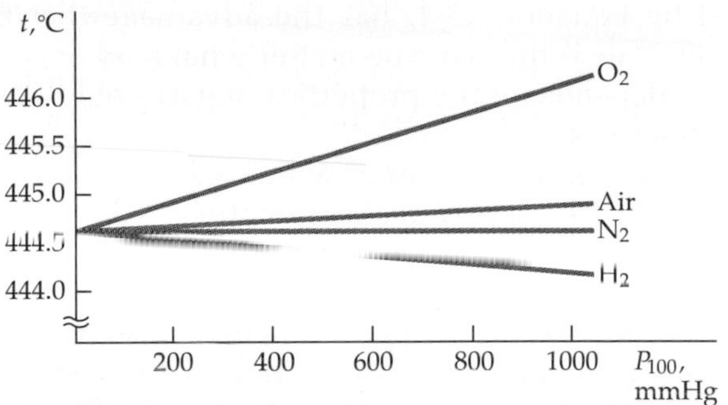

Figure 15-7 Temperature of the boiling point of sulfur measured with constant-volume gas thermometers filled with various gases. The pressure at the steam point of water P_{100} is varied by varying the amount of gas in the thermometers. As the amount of gas is reduced, the temperatures measured by all the thermometers approach the value 444.60°C.

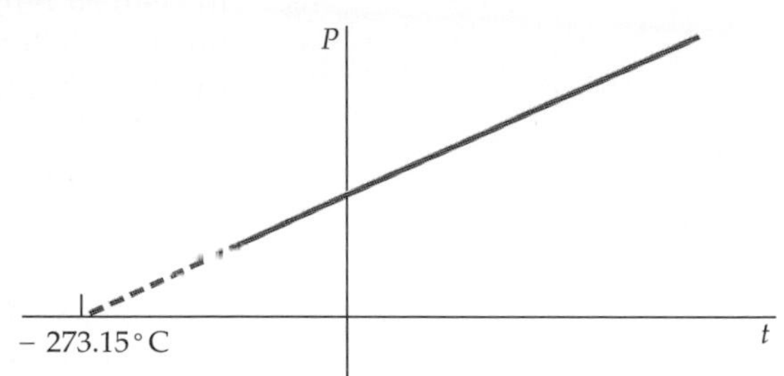

Figure 15-8 Plot of pressure versus temperature as measured by a constant-volume gas thermometer. When extrapolated to zero pressure, the plot intersects the temperature axis at the value -273.15°C.

substance, like mercury, low-density gas thermometers are used to define temperature. This choice is reinforced by theoretical considerations, which we will discuss later.

Let us now consider a series of temperature measurements with a constant-volume gas thermometer that has a very small but fixed amount of gas. According to Equation 15-3, the pressure in the thermometer P_t varies linearly with the measured temperature t_C. Figure 15-8 shows a plot of measured temperature versus pressure in a constant-volume gas thermometer. When we extrapolate this straight line to zero pressure, the temperature approaches -273.15°C. This limit is the same no matter what kind of gas is used, but there is often considerable experimental uncertainty in the measurement of this number because of the difficulty in duplicating the ice-point and steam-point states with high precision in different laboratories. Because of these difficulties, a temperature scale based on a single fixed point was adopted in 1954 by the International Committee on Weights and Measures. A reference state that is much more precisely reproducible than either the ice point or the steam point is the triple point of water—the single temperature and pressure at which water, water vapor, and ice coexist in equilibrium. If we put water, ice, and water vapor in a sealed container with no air, the system will eventually reach a state of equilibrium where no ice will melt or evaporate, no water will freeze or evaporate, and no vapor will condense or freeze. This equilibrium state, called the **triple point of water,** occurs at a specific pressure and temperature. The triple-point pressure is 4.58 mmHg, and the temperature is 0.01°C. The **ideal-gas temperature scale** is defined so that the temperature of the triple point is 273.16 kelvins (K) (the **kelvin,** formerly called the degree kelvin, is a degree unit that is the same size as the Celsius degree). The temperature T of any other state is defined to be proportional to the pressure in a constant-volume gas thermometer:

H_2O at its triple point. The container is a hollow cylindrical shell that is sealed and evacuated. It contains water, ice, and water vapor in equilibrium. The cylindrical well in the center is filled with water and contains an aluminum bushing for inserting a thermometer. The container, stored at the National Bureau of Standards, is in an ice-water bath that is just slightly below the triple-point temperature so that the water freezes very slowly.

$$T = \frac{273.16\text{ K}}{P_3} P \qquad 15\text{-}4$$

where P is the pressure of the gas in the thermometer when the thermometer is in thermal equilibrium with the system whose temperature is being measured and P_3 is the pressure of the gas in the thermometer when the thermometer is immersed in a water–ice–vapor bath at its triple point. The value of P_3 depends on the amount of gas in the thermometer. The ideal-gas temperature scale, defined by Equation 15-4, has the advantage that the measured temperature of any state is the same no matter what kind of gas is used. This temperature scale depends on the properties of gases but not on the properties of any particular gas.

The lowest temperature that can be measured with a gas thermometer using helium for the gas is about 1 K. Below this temperature helium liquefies. (All other gases liquefy at higher temperatures.) In Chapter 17, we will see that the second law of thermodynamics can be used to define the **absolute temperature scale** independent of the properties of any substance, and with no limitations on the range of temperatures that can be measured. As a result, it has been possible to measure temperatures as low as a millionth of a kelvin. The absolute scale so defined is identical to that defined by Equation 15-4 for the range of temperatures for which gas thermometers can be used. The symbol T is used when referring to the absolute temperature.

Because the Celsius degree and the kelvin are the same size, temperature *differences* are the same on both the Celsius scale and the absolute temperature scale (also called the **Kelvin scale**). That is, a temperature *change* of 1 K is identical to a temperature *change* of 1 C°. As a result, the only difference between the two scales lies in the choice of the zero temperature. To convert from degrees Celsius to kelvins, we merely add 273.15:

Celsius–absolute conversion

$$T = t_C + 273.15 \tag{15-5}$$

(For most purposes, we can round off the temperature of absolute zero to −273°C, so we can simply add 273 to the Celsius temperature to get the absolute temperature.) To convert from kelvins to degrees Fahrenheit, it is usually easiest to convert first to degrees Celsius by subtracting 273 and then use Equation 15-2. Although the Celsius and Fahrenheit scales are convenient for everyday use, the absolute scale is much more convenient for scientific purposes, partly because many formulas are more simply expressed when the absolute scale is used, and partly because the absolute temperature can be given a more fundamental interpretation.

Example 15-3

What is the Kelvin temperature corresponding to 70°F?

From Equation 15-2, the Celsius temperature equal to 70°F is

$$t_C = \tfrac{5}{9}(70° - 32°) = \tfrac{5}{9}(38°) = 21°\text{C}$$

The Kelvin temperature is then obtained from Equation 15-5:

$$T = t_C + 273 = 21 + 273 = 294 \text{ K}$$

Exercise

Find the Fahrenheit temperature corresponding to 300 K. (Answer: 80.6°F)

Questions

5. Why might the Celsius and Fahrenheit scales be more convenient than the absolute scale for ordinary, nonscientific purposes?
6. The temperature of the interior of the sun is said to be about 10^7 degrees. Do you think that this is degrees Celsius or kelvins, or does it matter?

15-3 Thermal Expansion

When the temperature of an object increases, the object usually expands. Consider a long rod of length L at a temperature T. When the temperature changes by ΔT, the change in length ΔL is proportional to ΔT and to the original length L:

$$\Delta L = \alpha L \, \Delta T \qquad 15\text{-}6$$

where α is called the **coefficient of linear expansion.** It is the ratio of the fractional change in length to the change in temperature:

$$\alpha = \frac{\Delta L/L}{\Delta T} \qquad 15\text{-}7$$

Its units are reciprocal Celsius degrees (1/C°) or reciprocal kelvins (1/K). The coefficient of linear expansion for a solid or liquid usually does not vary much with pressure, but it may vary with temperature. Equation 15-7 gives the average value of α over the temperature interval ΔT. The coefficient of linear expansion at a particular temperature T is found by taking the limit as ΔT approaches zero:

$$\alpha = \lim_{\Delta T \to 0} \frac{\Delta L/L}{\Delta T} = \frac{1}{L}\frac{dL}{dT} \qquad 15\text{-}8$$

In most cases, sufficient accuracy is obtained by using the average value of α over a wide temperature range.

The **coefficient of volume expansion** β is defined similarly as the ratio of the fractional change in volume to the change in temperature (at constant pressure):

$$\beta = \lim_{\Delta T \to 0} \frac{\Delta V/V}{\Delta T} = \frac{1}{V}\frac{dV}{dT} \qquad 15\text{-}9$$

Like α, β does not usually vary with pressure for solids and liquids but may vary with temperature. Average values for α and β for various substances are given in Table 15-1.

Table 15-1 **Approximate Values of the Coefficients of Thermal Expansion for Various Substances**

Material	α, K^{-1}
Aluminum	24×10^{-6}
Brass	19×10^{-6}
Carbon	
Diamond	1.2×10^{-6}
Graphite	7.9×10^{-6}
Copper	17×10^{-6}
Glass	
Ordinary	9×10^{-6}
Pyrex	3.2×10^{-6}
Ice	51×10^{-6}
Invar	1×10^{-6}
Steel	11×10^{-6}
Material	**β, K^{-1}**
Acetone	1.5×10^{-3}
Air	3.67×10^{-3}
Alcohol	1.1×10^{-3}
Mercury	0.18×10^{-3}
Water (20°C)	0.207×10^{-3}

We can show that for a given material, the coefficient of volume expansion is 3 times the coefficient of linear expansion. Consider a box of dimensions L_1, L_2, and L_3. Its volume at a temperature T is

$$V = L_1 L_2 L_3$$

The rate of change of the volume with respect to temperature is

$$\frac{dV}{dT} = L_1 L_2 \frac{dL_3}{dT} + L_1 L_3 \frac{dL_2}{dT} + L_2 L_3 \frac{dL_1}{dT}$$

Dividing each side of the equation by the volume, we obtain

$$\beta = \frac{1}{V}\frac{dV}{dT} = \frac{1}{L_3}\frac{dL_3}{dT} + \frac{1}{L_2}\frac{dL_2}{dT} + \frac{1}{L_1}\frac{dL_1}{dT}$$

Since each term on the right side of the above equation equals α, we have

$$\beta = 3\alpha \qquad 15\text{-}10$$

The increase in size of any part of a body for a given temperature change is proportional to the original size of that part of the body. Thus, if we increase the temperature of a steel ruler, for example, the effect will be similar to that of a (very slight) photographic enlargement. Lines that were previously equally spaced will still be equally spaced, but the spaces will be larger. Likewise, the width of the ruler will be slightly larger. If the ruler has a hole in it, the hole will get larger, just as it would in a photographic enlargement.

Expansion joint allows the bridge to expand or contract without stress as the temperature changes.

Though most materials expand when heated, water between 0 and 4°C is an important exception. Figure 15-9 shows the volume occupied by 1 g of water as a function of temperature. The volume is minimum, and therefore the density is maximum, at 4°C. Thus, when water is heated at temperatures below 4°C, it contracts rather than expands. This property has important consequences for the ecology of lakes. At temperatures above 4°C, the water in a lake becomes denser as it cools and sinks to the bottom. But at temperatures below 4°C, the water becomes less dense as it cools, so it stays at the surface. Ice therefore forms first on the surface of a lake and, being less dense than water, it remains there and acts as a thermal insulator for the water below. If water contracted when it freezes, as most substances do, ice would sink and new water would be exposed at the surface to freeze. Lakes would fill with ice from the bottom up and would be much more likely to become completely frozen in the winter, killing the fish and other aquatic life.

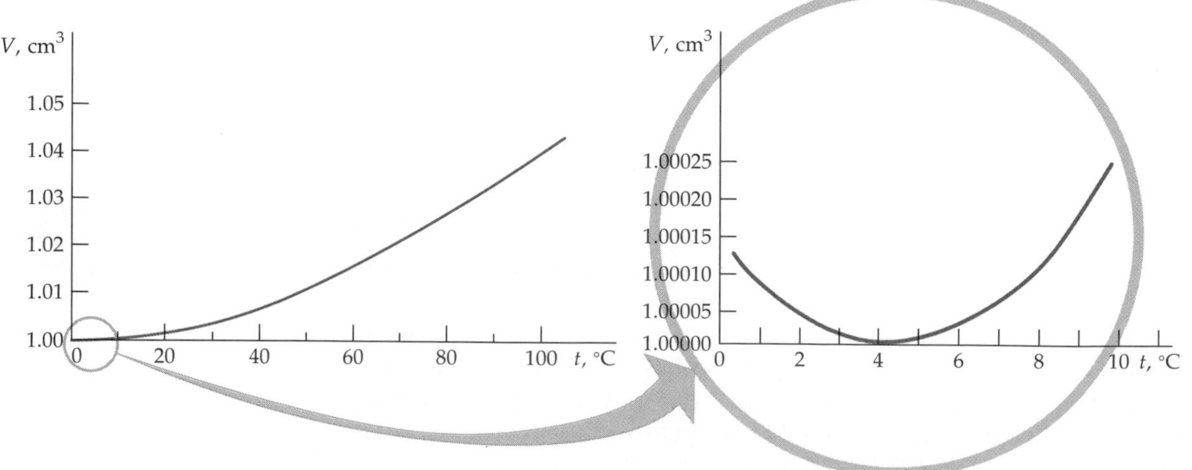

Figure 15-9 Volume of 1 g of water at atmospheric pressure versus temperature. The minimum volume, which corresponds to the maximum density, occurs at 4°C.

Example 15-4

A steel bridge is 1000 m long. By how much does it expand when the temperature rises from 0 to 30°C?

From Table 15-1, the coefficient of linear expansion for steel is $11 \times 10^{-6}\ \mathrm{K}^{-1}$. The change in length for a 30-C° = 30-K rise in temperature is thus

$$\Delta L = \alpha L\ \Delta T = (11 \times 10^{-6}\ \mathrm{K}^{-1})(1000\ \mathrm{m})(30\ \mathrm{K})$$
$$= 0.33\ \mathrm{m} = 33\ \mathrm{cm}$$

To relieve the stresses that would occur if such expansions were not allowed for, expansion joints must be included in the bridge. We can calculate the stress that would result if the bridge could not expand by using Equation 11-7 for Young's modulus:

$$Y = \frac{\text{stress}}{\text{strain}} = \frac{F/A}{\Delta L/L}$$

$$\frac{F}{A} = Y\frac{\Delta L}{L}$$
$$= (2 \times 10^{11}\ \mathrm{N/m^2})\frac{0.33\ \mathrm{m}}{1000\ \mathrm{m}} = 6.6 \times 10^{7}\ \mathrm{N/m^2}$$

where we have used the value of Young's modulus for steel from Table 11-2. This stress is about one-third of the breaking stress for steel under compression. A compression stress of this magnitude would cause a steel bridge to buckle and become permanently deformed.

Example 15-5

A 1-L glass flask is filled to the brim with alcohol at 10°C. If the temperature warms to 30°C, how much alcohol spills out of the flask?

Using $\Delta T = 20$ C° = 20 K and the expansion coefficients from Table 15-1, we obtain for the change in the volume of the glass flask

$$\Delta V_{\mathrm{g}} = \beta V\ \Delta T = 3\alpha V\ \Delta T$$
$$= 3(9 \times 10^{-6}\ \mathrm{K}^{-1})(1\ \mathrm{L})(20\ \mathrm{K})$$
$$= 5.4 \times 10^{-4}\ \mathrm{L} = 0.54\ \mathrm{mL}$$

and for the change in the volume of the alcohol

$$\Delta V_{\mathrm{a}} = \beta V\ \Delta T = (1.1 \times 10^{-3}\ \mathrm{K}^{-1})(1\ \mathrm{L})(20\ \mathrm{K})$$
$$= 2.2 \times 10^{-2}\ \mathrm{L} = 22.0\ \mathrm{mL}$$

The amount of alcohol that spills out is thus 22.0 mL − 0.54 mL ≈ 21.5 mL.

Example 15-6

A copper bar is heated to 300°C and is then clamped rigidly between two fixed points so that it can neither expand nor contract. If the breaking stress of copper is 230 $\mathrm{MN/m^2}$, at what temperature will the bar break as it cools?

In this example, the change in length ΔL that would occur if the bar contracted as it cooled is offset by an equal stretching due to tensile stress in the bar. From Equation 11-7, for Young's modulus, we have for the stretching ΔL caused by a tensile stress F/A

$$\Delta L = L\frac{F/A}{Y}$$

Setting this equal to the change in length that would occur if the bar could contract, we have

$$\Delta L = L\alpha\,\Delta T = L\frac{F/A}{Y}$$

Solving for ΔT and using $Y = 110\ \mathrm{GN/m^2}$ from Table 11-2, $\alpha = 17 \times 10^{-6}\ \mathrm{K^{-1}}$ from Table 15-1, and $F/A = 230\ \mathrm{MN/m^2}$ for the breaking stress of copper, we obtain

$$\Delta T = \frac{F/A}{\alpha Y} = \frac{230 \times 10^6\ \mathrm{N/m^2}}{(17 \times 10^{-6}\ \mathrm{K^{-1}})(110 \times 10^9\ \mathrm{N/m^2})} = 123\ \mathrm{K} = 123\ \mathrm{C°}$$

Since the original temperature is 300°C, the temperature at which the bar will break is

$$t = 300°\mathrm{C} - 123\ \mathrm{C°} = 177°\mathrm{C}$$

Question

7. If mercury and glass had the same coefficient of thermal expansion, could a mercury thermometer work?

15-4 The Ideal-Gas Law

If we compress a gas while keeping its temperature constant, we find that the pressure increases as the volume decreases. Similarly, if we cause a gas to expand at constant temperature, its pressure decreases as its volume increases. To a good approximation, the pressure of a gas varies inversely with the volume. This implies that, at a constant temperature, the product of the pressure and the volume of a gas is constant. This result was discovered experimentally by Robert Boyle (1627–1691), a younger contemporary of Galileo and an older one of Newton, and is known as **Boyle's law:**

$$PV = \text{constant} \qquad \text{constant temperature}$$

This law holds approximately for all gases at low densities. But according to Equation 15-4, the absolute temperature of a gas at low densities is proportional to the pressure at constant volume. Similarly, the absolute temperature is proportional to the volume of a gas if the pressure is kept constant, a result discovered experimentally by Jacques Charles (1746–1823) and Gay-Lussac (1778–1850). Thus, at low densities, the product PV is very nearly proportional to the temperature T:

$$PV = CT \qquad 15\text{-}11$$

where C is a constant of proportionality appropriate to a particular body of gas.

Now, suppose we have two identical containers, each holding the same amount of the same kind of gas at the same temperature. Each has a volume V described by Equation 15-11. If we combine the two containers, we end up with twice the volume of gas at the same pressure P and temperature T. It follows from Equation 15-11 that C must have increased by a factor of 2. In other words, C is proportional to the amount of gas. We can therefore write

$$C = kN$$

where N is the number of molecules of the gas and k is a constant. Equation 15-11 then becomes

$$PV = NkT \tag{15-12}$$

The constant k is called **Boltzmann's constant.** It is found experimentally to have the same value for any kind or amount of gas. Its value in SI units is

$$k = 1.381 \times 10^{-23} \text{ J/K} \tag{15-13}$$

It is often convenient to write the amount of a gas in terms of the number of moles. A **mole** (mol) of any substance is the amount of that substance that contains Avogadro's number of atoms or molecules. **Avogadro's number** N_A is defined as the number of carbon atoms in 12 grams of ^{12}C. The value of Avogadro's number is

$$N_A = 6.022 \times 10^{23} \text{ molecules/mol} \tag{15-14}$$

If we have n moles of a substance, the number of molecules is

$$N = nN_A \tag{15-15}$$

Equation 15-12 is then

$$PV = nN_AkT = nRT \tag{15-16}$$

where

$$R = kN_A \tag{15-17}$$

is called the **universal gas constant.** Its value, which is the same for all gases, is

$$R = 8.314 \text{ J/mol·K} = 0.08206 \text{ L·atm/mol·K} \tag{15-18}$$

Figure 15-10 shows plots of PV/nT versus the pressure P for several gases. (In these plots, the pressure is varied by varying the amount of gas.) We can see that for real gases, PV/nT is very nearly constant over a rather large range of pressures. (Even for oxygen, which has the largest variation in this graph, the variation is only about 1 percent between 0 and 5 atm.) We define an **ideal gas** as one for which PV/nT is constant for all pressures. For an ideal gas, the pressure, volume, and temperature are related by

$$PV = nRT \tag{15-19}$$

Ideal-gas law

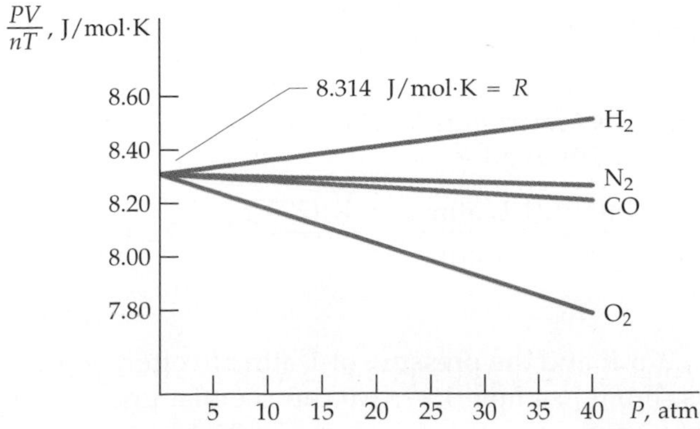

Figure 15-10 Plot of PV/nT versus P for real gases. As the density of the gas is reduced, thereby reducing the pressure, the ratio PV/nT approaches the same value, 8.314 J/mol·K, for all gases. This value is the universal gas constant R. The ideal-gas equation, $PV = nRT$, is a good approximation for all real gases at low pressures, that is, at pressures up to a few atmospheres.

The mass of 1 mol is called the **molar mass** M. (The terms *molecular weight* or *molecular mass* are also sometimes used.) The molar mass of ^{12}C is, by definition, 12 g/mol or 12×10^{-3} kg/mol. Molar masses of the elements are given in the periodic table in Appendix E. The molar mass of a molecule, such as CO_2, is the sum of the molar masses of the elements in the molecule. Since the molar mass of oxygen is 16 g/mol (actually 15.999 g/mol), the molar mass of O_2 is 32 g/mol and that of CO_2 is 12 + 32 = 44 g/mol.

The mass of n moles of gas is given by

$$m = nM$$

The density ρ of an ideal gas is

$$\rho = \frac{m}{V} = \frac{nM}{V}$$

Using $n/V = P/RT$ from Equation 15-19, we can write

$$\rho = \frac{M}{RT}P \qquad 15\text{-}20$$

At a given temperature, the density of an ideal gas is proportional to the pressure.

Equation 15-19, which relates P, V, and T for a given amount of gas, is called an **equation of state.** The state of a gas of constant mass is determined by any two of the three variables P, V, and T. For example, if P and V are known, then the temperature T is determined by some function $T(P, V)$. The particular function, the equation of state, for an ideal gas is given by Equation 15-19. The concept of an ideal gas is an extrapolation of the behavior of real gases at low densities and pressures to ideal behavior. As can be seen from Figure 15-10, at reasonably low pressures real gases differ little from an ideal gas. Thus, Equation 15-19 is quite useful for describing the properties of real gases. At higher gas densities and pressures, corrections must be made to Equation 15-19 if it is to apply to real gases. One set of these corrections is discussed in Section 15-6.

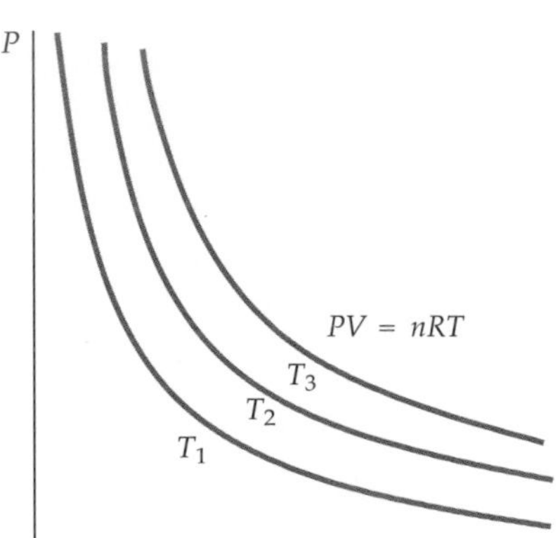

Figure 15-11 Isotherms on the *PV* diagram for a gas. For an ideal gas, these curves are hyperbolas given by $PV = nRT$.

Figure 15-11 shows plots of P versus V for various temperatures T. These curves are called **isotherms.** The isotherms for an ideal gas are hyperbolas.

For a fixed amount of gas, we can see from Equation 15-19 that the quantity PV/T is constant. If we use the subscripts 1 for the initial values and 2 for the final values, we have

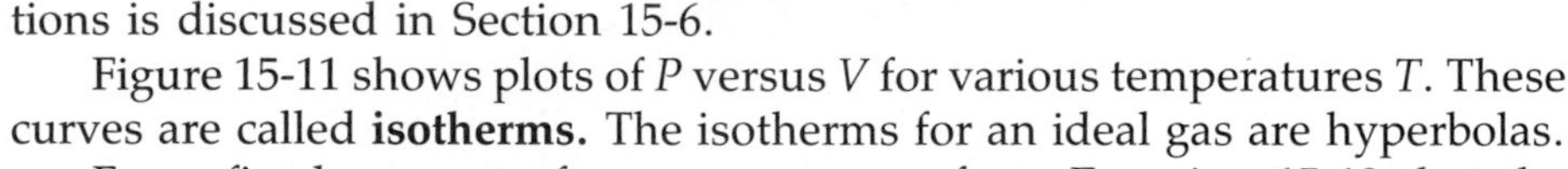

$$\frac{P_2V_2}{T_2} = \frac{P_1V_1}{T_1} \qquad 15\text{-}21$$

Example 15-7

What volume is occupied by 1 mol of gas at a temperature of 0°C and a pressure of 1 atm?

The absolute temperature corresponding to 0°C is 273 K. From the ideal-gas law (Equation 15-19), we have

$$V = \frac{nRT}{P} = \frac{(1 \text{ mol})(0.0821 \text{ L·atm/mol·K})(273 \text{ K})}{1 \text{ atm}} = 22.4 \text{ L}$$

The temperature of 0°C = 273 K and the pressure of 1 atm are often referred to as **standard conditions.** From Example 15-7, we can see that under standard conditions, 1 mol of any gas occupies a volume of 22.4 L.

Exercise

Find (*a*) the number of moles n and (*b*) the number of molecules N in 1 cm^3 of a gas under standard conditions.
[Answers: (*a*) $n = 4.46 \times 10^{-5}$ mol, (*b*) $N = 2.68 \times 10^{19}$ molecules]

Example 15-8

The molar mass of hydrogen is 1.008 g/mol. What is the mass of one hydrogen atom?

Since there are N_A hydrogen atoms in 1 mol, the mass m of one atom is

$$m = \frac{1.008 \text{ g/mol}}{6.022 \times 10^{23} \text{ atoms/mol}} = 1.67 \times 10^{-24} \text{ g/atom}$$

We can see from this example that Avogadro's number is approximately the reciprocal of the mass of the hydrogen atom measured in grams.

Example 15-9

A gas has a volume of 2 L, a temperature of 30°C, and a pressure of 1 atm. It is heated to 60°C and compressed to a volume of 1.5 L. Find its new pressure.

From Equation 15-21, we have

$$P_2 = \frac{T_2 V_1}{T_1 V_2} P_1$$

Since we are dealing with ratios, we can express the pressure and volume in any units, but we must remember that the ideal-gas equation holds only if the temperatures are absolute temperatures. Using $T_1 = 273 + 30 = 303$ K and $T_2 = 273 + 60 = 333$ K, we have

$$P_2 = \frac{(333 \text{ K})(2 \text{ L})}{(303 \text{ K})(1.5 \text{ L})}(1 \text{ atm}) = 1.47 \text{ atm}$$

Example 15-10

One hundred grams of CO_2 occupies a volume of 55 L at a pressure of 1 atm. (*a*) Find the temperature. (*b*) If the volume is increased to 80 L and the temperature is kept constant, what is the new pressure?

(*a*) We can find the temperature from the ideal-gas equation (Equation 15-19) if we first find the number of moles. Since the molar mass of CO_2 is 44 g/mol, the number of moles is

$$n = \frac{m}{M} = \frac{100 \text{ g}}{44 \text{ g/mol}} = 2.27 \text{ mol}$$

The absolute temperature is then

$$T = \frac{PV}{nR} = \frac{(1 \text{ atm})(55 \text{ L})}{(2.27 \text{ mol})(0.0821 \text{ L} \cdot \text{atm/mol} \cdot \text{K})} = 295 \text{ K}$$

(*b*) Using Equation 15-21 with $T_2 = T_1$, we have

$$P_2 V_2 = P_1 V_1$$

$$P_2 = \frac{V_1}{V_2} P_1 = \frac{55 \text{ L}}{80 \text{ L}}(1 \text{ atm}) = 0.688 \text{ atm}$$

15-5 The Molecular Interpretation of Temperature: The Kinetic Theory of Gases

In previous sections of this chapter, the behavior of gases has been described in terms of P, V, and T, quantities that are called **macroscopic variables.** Such variables describe the macroscopic state of a substance. To describe the *microscopic* state of a gas would require giving the coordinates and velocities of all its molecules—a truly impossible task. In this section, we will show how the macroscopic description can be related to simple averages of microscopic quantities. In particular we will show that the absolute temperature of a gas is a measure of the average kinetic energy of the gas molecules. We do this by using a simple model to calculate the pressure exerted by the gas on the walls of its container.

In this microscopic point of view, called the **kinetic theory of gases,** the pressure of a gas is the result of the collisions between the gas molecules and the walls of their container. We can calculate this pressure by calculating the rate of change of momentum of the gas molecules due to collisions with a wall of the container. By Newton's second law, the rate of change of momentum equals the force exerted by the wall on the gas molecules:

$$\mathbf{F} = \frac{d\mathbf{p}}{dt}$$

By Newton's third law, this force equals the force exerted by the molecules on the wall. The force per unit area equals the pressure.

We begin by making the following assumptions:

1. The gas consists of a large number of molecules that make elastic collisions with each other and with the walls of the container.
2. The molecules are separated, on the average, by distances that are large compared with their diameters, and they exert no forces on each other except when they collide.
3. In the absence of external forces (the molecules are moving fast enough that we can neglect gravity), there is no preferred position for a molecule in the container, and there is no preferred direction for the velocity vector.

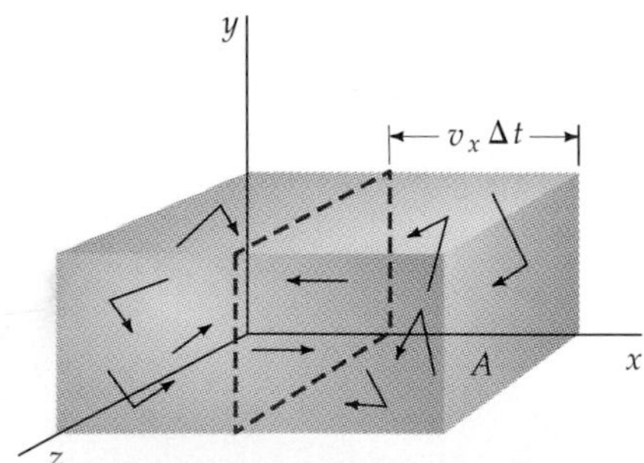

Figure 15-12 Gas molecules in a rectangular container. In a time interval Δt, the molecules at a distance $v_x\,\Delta t$ from the right wall will hit the right wall if they are moving to the right. The number of molecules within this distance is proportional to v_x and to the number of molecules per unit volume.

The second assumption, that the gas molecules are, on the average, far apart, is equivalent to assuming a very low gas density, which as we saw in the last section is the same as assuming that the gas is an ideal gas. Since momentum is conserved, the collisions the molecules make with each other have no effect on the total momentum in any direction, so we can neglect these collisions.

Let us assume that we have a rectangular container of volume V containing N molecules, each of mass m and moving with a speed v. We wish to calculate the force exerted by these molecules on the right wall, which is perpendicular to the x axis. The x component of the momentum of a molecule before it hits the wall is $+mv_x$, and after it makes an elastic collision with the wall it is $-mv_x$. The magnitude of the change in momentum during the collision of one molecule with the wall is thus $2mv_x$. The total change in the momentum of all the molecules during some time interval Δt is $2mv_x$ times the number of molecules that hit the wall during the interval. Figure 15-12 depicts gas molecules in a rectangular container. The number of molecules that hit the right wall of area A is the number that are within the distance $v_x\,\Delta t$ and are moving to the right. This number is the number of molecules per unit volume N/V times the volume $v_x \, \Delta t \, A$ times $\frac{1}{2}$, because on the average, half the molecules will be moving to the right and half to the left. The total change in the momentum Δp of the gas molecules in the time interval Δt is this number times $2mv_x$, the change in momentum per molecule:

$$\Delta p = \frac{1}{2}\frac{N}{V}(v_x \, \Delta t \, A)2mv_x = \frac{N}{V}mv_x^2 \, A \, \Delta t$$

The force exerted by the wall on the molecules and by the molecules on the wall is this change in momentum divided by the time interval Δt. The pressure is this force divided by the area:

$$P = \frac{F}{A} = \frac{1}{A}\frac{\Delta p}{\Delta t}$$

Dividing the change in momentum by the time Δt and by the area A, we obtain

$$P = \frac{N}{V}mv_x^2 \qquad \text{15-22}$$

and

$$PV = Nmv_x^2 \qquad \text{15-23}$$

To allow for the fact that all the molecules in a container do not have the same speed, we merely replace v_x^2 with the average value $(v_x^2)_{av}$. Then, if we write Equation 15-23 in terms of the kinetic energy $\frac{1}{2}mv_x^2$ associated with motion along the x axis, we have

$$PV = 2N(\tfrac{1}{2}mv_x^2)_{av} \qquad \text{15-24}$$

Comparing this with Equation 15-12, which was obtained experimentally for any gas at very low densities, we can see that

$$PV = NkT = 2N(\tfrac{1}{2}mv_x^2)_{av}$$

or

$$(\tfrac{1}{2}mv_x^2)_{av} = \tfrac{1}{2}kT \qquad \text{15-25}$$

Thus, the average kinetic energy associated with motion along the x axis is $\frac{1}{2}kT$. But there is nothing special about the x direction. On the average

$$(v_x^2)_{av} = (v_y^2)_{av} = (v_z^2)_{av}$$

and

$$(v^2)_{av} = (v_x^2)_{av} + (v_y^2)_{av} + (v_z^2)_{av} = 3(v_x^2)_{av}$$

Writing $(v_x^2)_{av} = \frac{1}{3}(v^2)_{av}$ and K_{av} for the average kinetic energy of the molecules, Equation 15-25 becomes

$$K_{av} = (\tfrac{1}{2}mv^2)_{av} = \tfrac{3}{2}kT \qquad \text{15-26}$$

The average kinetic energy of the molecules is $\frac{3}{2}kT$. The absolute temperature is thus a measure of the average translational kinetic energy of the molecules. We include the word "translational" because the molecules may also have rotational or vibrational kinetic energy. Only the translational kinetic energy is relevant to the calculation of the pressure exerted by a gas on the walls of its container. The total translational kinetic energy of n moles of a gas containing N molecules is

$$K = N(\tfrac{1}{2}mv^2)_{av} = \tfrac{3}{2}NkT = \tfrac{3}{2}nRT \qquad \text{15-27}$$

Thus, the translational kinetic energy is $\frac{3}{2}kT$ per molecule and $\frac{3}{2}RT$ per mole.

We can use these results to estimate the order of magnitude of the speeds of the molecules in a gas. The average value of v^2 is, by Equation 15-26,

$$(v^2)_{av} = \frac{3kT}{m} = \frac{3N_A kT}{N_A m} = \frac{3RT}{M}$$

where $M = N_A m$ is the molar mass. The square root of $(v^2)_{av}$ is the **root mean square** (rms) speed:

$$v_{rms} = \sqrt{(v^2)_{av}} = \sqrt{\frac{3kT}{m}} = \sqrt{\frac{3RT}{M}} \qquad 15\text{-}28$$

Note that Equation 15-28 is similar to Equation 14-3 for the speed of sound in a gas:

$$v_{sound} = \sqrt{\frac{\gamma RT}{M}} \qquad 15\text{-}29$$

where $\gamma = 1.4$ for air. This should not be surprising since a sound wave in air is a pressure disturbance that is propagated by collisions between air molecules.

Example 15-11

Oxygen gas (O_2) has a molar mass of about 32 g/mol and hydrogen gas (H_2) has a molar mass of about 2 g/mol. Calculate (*a*) the rms speed of an oxygen molecule and (*b*) the rms speed of a hydrogen molecule when the temperature is 300 K.

(*a*) To make the units come out right, we must express the molecular mass of O_2 in Equation 15-28 in kilograms per mole. We then have

$$v_{rms} = \sqrt{\frac{3RT}{M}} = \sqrt{\frac{3(8.31\text{ J/mol·K})(300\text{ K})}{32 \times 10^{-3}\text{ kg/mol}}} = 483\text{ m/s}$$

(*b*) Since the molar mass of hydrogen is one-sixteenth that of oxygen and v_{rms} is proportional to $1/\sqrt{M}$, the rms speed of hydrogen is 4 times that of oxygen or about 1.93 km/s.

We can see from Example 15-11 that the rms speed of oxygen molecules is somewhat greater than but of the same order of magnitude as the speed of sound in air, which at 300 K is about 347 m/s.

The Distribution of Molecular Speeds

As we have noted, all the molecules in a gas do not have the same speed. The distribution of the molecular speeds of a gas can be measured directly using the apparatus illustrated in Figure 15-13. A substance is vaporized in an oven, and the vapor molecules are allowed to escape through a hole in the oven wall into a vacuum chamber. The molecules are collimated into a narrow beam by a series of slits (not shown in the figure). The beam is aimed at a detector that counts the number of molecules that are incident on it in a given period of time. Most of the beam is stopped by a rotating cylinder. Small helical slits in the cylinder (only one of which is depicted in the figure) allow the passage of molecules that have a narrow range of speeds that is determined by the angular velocity of rotation of the cylinder. The number of molecules in each range of speeds is measured by varying the angular velocity of the cylinder and counting the number of molecules that reach the detector for each angular velocity.

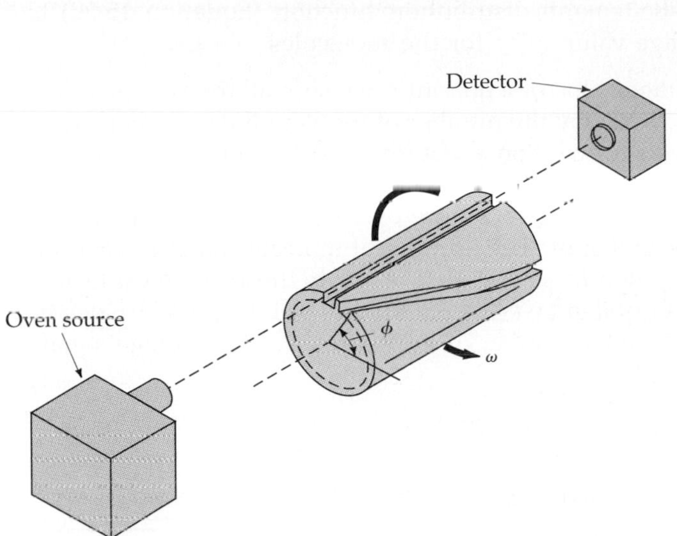

Figure 15-13 Schematic diagram of the apparatus for determining the speed distribution of the molecules of a gas. The molecules emerge from the oven and are collimated into a narrow beam by a series of slits (not shown). If the cylinder is not rotating, the beam passes through the straight slit and hits the detector. When the cylinder is rotating, only those molecules in a narrow speed range determined by the angular speed of rotation ω pass through the helical slit and reach the detector.

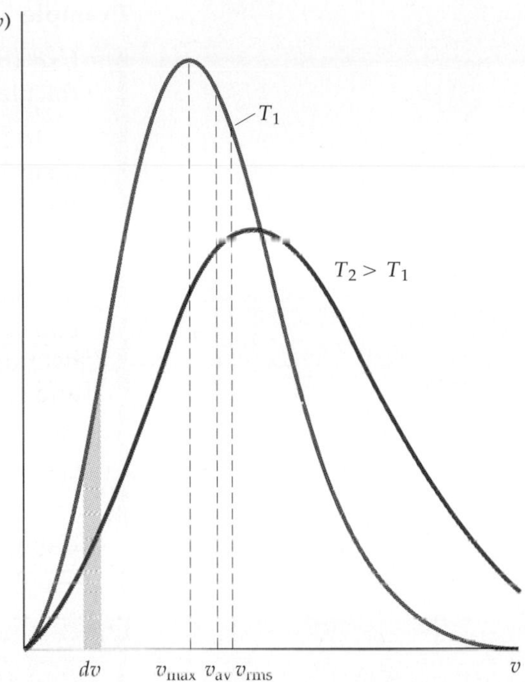

Figure 15-14 Distributions of molecular speeds in a gas at two temperatures, T_1 and $T_2 > T_1$. The shaded area $f(v)\,dv$ equals the fraction of the number of molecules having a particular speed in a narrow range of speeds dv. The mean speed v_{av} and the rms speed v_{rms} are both slightly greater than the most probable speed v_{max}.

Figure 15-14 shows the distributions of molecular speeds in a gas for two different temperatures. In this figure the quantity $f(v)$, called the **Maxwell–Boltzmann distribution function,** is defined as follows: If N is the total number of molecules, the number that have speeds in the range between v and $v + dv$ is dN, given by

$$dN = N\,f(v)\,dv \tag{15-30}$$

The fraction $dN/N = f(v)\,dv$ in a particular range dv is illustrated by the shaded region in the figure. The speed v_{max} at which the distribution is maximum is the most probable speed. The mean speed v_{av} and the rms speed v_{rms} are somewhat greater than the most probable speed.

The Maxwell–Boltzmann distribution function can be derived using statistical mechanics. The result is

$$f(v) = \frac{4}{\sqrt{\pi}}\left(\frac{m}{2kT}\right)^{3/2} v^2 e^{-mv^2/2kT} \tag{15-31}$$

The most probable speed v_{max} is that speed for which $f(v)$ is maximum. It is left as a problem to show that

$$v_{max} = \sqrt{\frac{2kT}{m}} \tag{15-32}$$

Comparing Equation 15-32 with Equation 15-28, we see that the most probable speed is slightly less than the rms speed.

Example 15-12

Use the Maxwell-Boltzmann distribution function (Equation 15-30) to calculate the average value of v^2 for the molecules in a gas.

To find the average value of a quantity, we sum all the values of that quantity and then divide by the number of terms in the sum. For example, to find the average score on a test given to N students, we add all the scores and divide by N. An alternative to summing each score obtained is to multiply each possible score by the number of students that received that score and sum. The number of molecules that have a certain speed v in the range dv is $Nf(v)\,dv$. We obtain the average value of v^2 by multiplying this number by v^2, summing over all the possible speeds, and dividing by the total number of molecules N. The average square speed is thus given by

$$(v^2)_{\text{av}} = \frac{1}{N}\int_0^\infty v^2\,Nf(v)\,dv = \int_0^\infty v^2 f(v)\,dv$$

Using Equation 15-31 for $f(v)$, we obtain

$$(v^2)_{\text{av}} = \int_0^\infty v^2\,\frac{4}{\sqrt{\pi}}\left(\frac{m}{2kT}\right)^{3/2} v^2 e^{-mv^2/2kT}\,dv$$

$$= \frac{4}{\sqrt{\pi}}\left(\frac{m}{2kT}\right)^{3/2}\int_0^\infty v^4 e^{-mv^2/2kT}\,dv$$

This integral can be found in tables. Its value is

$$\int_0^\infty v^4 e^{-mv^2/2kT}\,dv = \frac{3}{8}\sqrt{\pi}\left(\frac{2kT}{m}\right)^{5/2}$$

Combining these results, we obtain

$$(v^2)_{\text{av}} = \frac{4}{\sqrt{\pi}}\left(\frac{m}{2kT}\right)^{3/2}\frac{3}{8}\sqrt{\pi}\left(\frac{2kT}{m}\right)^{5/2} = \frac{3kT}{m}$$

which agrees with $v_{\text{rms}} = \sqrt{3kT/m}$ from Equation 15-28.

In Example 15-11, we found that the rms speed of hydrogen molecules is about 1.93 km/s. This is about one-sixth of the escape speed at the surface of the earth, which we found to be 11.2 km/s in Section 10-5. So why is there no free hydrogen in the earth's atmosphere? As we can see from Figure 15-14, a considerable fraction of the molecules of a gas in equilibrium at some temperature has speeds greater than the rms speed. When the rms speed of the molecules of a particular gas is only about one-sixth of the escape speed for a planet, enough of the molecules have speeds greater than the escape speed so that the gas cannot exist in the atmosphere of that planet. Thus, no hydrogen. The rms speed of oxygen molecules, on the other hand, is about one-fourth that of hydrogen molecules, which makes it much less than the escape speed at the surface of the earth. Therefore, very few oxygen molecules have speeds greater than the escape speed, so oxygen is found in the earth's atmosphere.

The Maxwell–Boltzmann distribution of speed as given by Equation 15-31 can also be written as an energy distribution. We write the number of molecules with energy E in the range between E and $E + dE$ as

$$dN = NF(E)\,dE$$

where $F(E)$ is the energy distribution function. This will be the same number as given by Equation 15-30 if the energy E is related to the speed v by $E = \frac{1}{2}mv^2$. Then

$$dE = mv\,dv$$

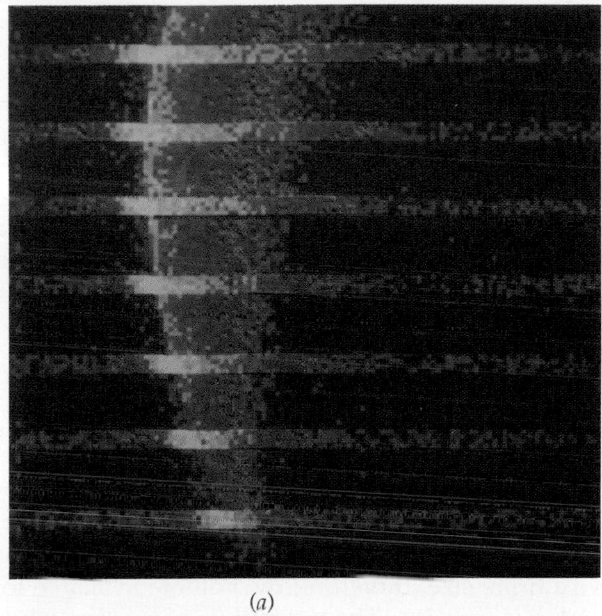
(a)

(b)

(a) The atmosphere of Venus is almost entirely CO_2. However, measurements by the Pioneer Venus Orbiter show an atomic hydrogen cloud surrounding Venus. The crescent-shaped image shows atomic oxygen, and the bars show hydrogen extending far above the atmosphere. Since the escape speed on Venus is 10.3 km/s, slightly smaller than that on earth, and the temperature of its atmosphere is considerably warmer, all of the hydrogen in the atmosphere at the time of the formation of Venus should have escaped by now. (b) Jupiter as seen from about twelve million miles. Because the escape speed at the surface of Jupiter is about 60 km/s, Jupiter easily retains hydrogen in its atmosphere.

and

$$Nf(v)\,dv = NF(E)\,dE$$

We can write

$$f(v)\,dv = Cv^2e^{-mv^2/2kT}\,dv = Cve^{-E/kT}v\,dv = C\left(\frac{2E}{m}\right)^{1/2}e^{-E/kT}\,\frac{dE}{m}$$

where $C = (4/\sqrt{\pi})(m/2kT)^{3/2}$ from Equation 15-31. The energy distribution function $F(E)$ is thus given by

$$F(E) = \frac{4}{\sqrt{\pi}}\left(\frac{m}{2kT}\right)^{3/2}\left(\frac{2}{m}\right)^{1/2}\frac{1}{m}\,E^{1/2}e^{-E/kT}$$

Simplifying we obtain **Maxwell–Boltzmann energy distribution function:**

$$F(E) = \frac{2}{\sqrt{\pi}}\left(\frac{1}{kT}\right)^{3/2}E^{1/2}e^{-E/kT} \qquad 15\text{-}33$$

The energy distribution function is thus proportional to $E^{1/2}e^{-E/kT}$

Questions

8. By what factor must the absolute temperature of a gas be increased to double the rms speed of its molecules?
9. How does the average translational kinetic energy of a molecule of a gas change if the pressure is doubled while the volume is kept constant? If the volume is doubled while the pressure is kept constant?
10. Why wouldn't you expect all molecules in a gas to have the same speed?
11. Two different gases are at the same temperature. What can you say about the rms speeds of the gas molecules? What can you say about the average kinetic energies of the molecules?
12. If a gas is heated at constant volume, explain in terms of molecular motion why its pressure on the walls of its container increases.
13. If the volume of a gas is reduced at constant temperature, explain in terms of molecular motion why its pressure on the walls of its container increases.

15-6 The van der Waals Equation and Liquid–Vapor Isotherms

Although most gases behave like an ideal gas at ordinary pressures, this ideal behavior breaks down at sufficiently high pressures or low temperatures when the density of the gas is high and the molecules are not, on the average, far apart. An equation of state called the **van der Waals equation** describes the behavior of many real gases over a wide range of pressures more accurately than does the ideal-gas equation of state ($PV = nRT$). The van der Waals equation for n moles of gas is

The van der Waals equation of state

$$\left(P + \frac{an^2}{V^2}\right)(V - bn) = nRT \qquad 15\text{-}34$$

The constant b in this equation arises because the gas molecules are not point particles but have a finite size; therefore, the volume available for the molecules to move about in is reduced. The magnitude of b is the volume of one mole of gas molecules. The term an^2/V^2 arises from the attraction of the gas molecules for each other. As a molecule approaches the wall of the container, it is pulled back by the molecules surrounding it with a force that is proportional to the density of those molecules n/V. Since the number of molecules that hit the wall in a given time is also proportional to the density of the molecules, the decrease in pressure due to the attraction of the molecules for each other is proportional to the square of the density and therefore to n^2/V^2. The constant a depends on the gas and is small for inert gases that have small chemical interactions. The terms bn and an^2/V^2 are both negligible when the volume V is large, that is, at low densities. Thus, at low densities the van der Waals equation approaches the ideal-gas law, whereas at high densities it provides a much better description of the behavior of real gases.

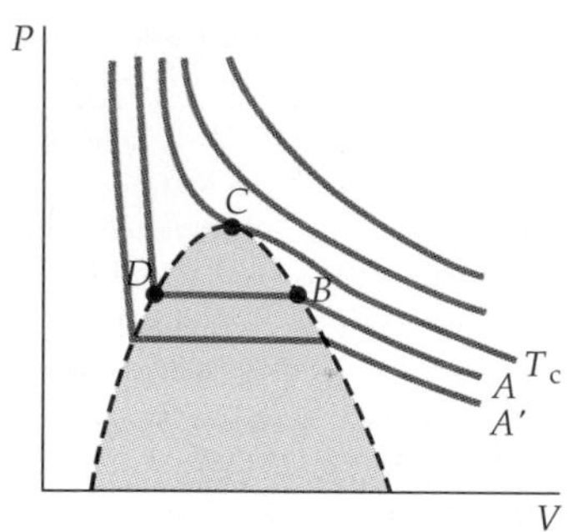

Figure 15-15 Isotherms on the PV diagram for a real substance. For temperatures above the critical temperature T_c, the substance remains a gas at all pressures and is described by the van der Waals equation. The pressure for the horizontal portions of the curves in the shaded region is the vapor pressure, which is the pressure at which the vapor and liquid are in equilibrium. To the left of the shaded region for temperatures below the critical temperature, the substance is a liquid and is nearly incompressible.

Figure 15-15 shows P-versus-V isothermal curves for a real substance at various temperatures. For temperatures above some critical temperature T_c, these curves are described quite accurately by the van der Waals equation and can be used to determine the constants a and b. For example, the values of these constants that give the best fit to the experimental curves for nitrogen are $a = 0.14\ \text{Pa}\cdot\text{m}^6/\text{mol}^2$ and $b = 39.1\ \text{cm}^3/\text{mol}$. This volume of $39.1\ \text{cm}^3$ per mole is about 0.2 percent of the volume of $22{,}400\ \text{cm}^3$ occupied by 1 mol of nitrogen under standard conditions. Since the molar mass of nitrogen is 28 g/mol, if 1 mol of nitrogen molecules were packed into a volume of $39.1\ \text{cm}^3$, the density would be

$$\rho = \frac{M}{V} = \frac{28\ \text{g}}{39.1\ \text{cm}^3} = 0.72\ \text{g/cm}^3$$

which compares favorably with the density of liquid nitrogen, $0.80\ \text{g/cm}^3$.

Since the constant b is the actual volume of 1 mol of gas molecules, its value can be used to estimate the size of a molecule. For instance, if 1 mol = N_A molecules of nitrogen has a volume of $39.1\ \text{cm}^3$, the volume of one nitrogen molecule is

$$V = \frac{b}{N_A} = \frac{39.1\ \text{cm}^3/\text{mol}}{6.02 \times 10^{23}\ \text{molecules/mol}}$$

$$= 6.50 \times 10^{-23}\ \text{cm}^3/\text{molecule}$$

If we assume that each molecule is a sphere of diameter d occupying a cubic volume of side d, we obtain

$$d^3 = 6.50 \times 10^{-23}\ \text{cm}^3$$

or

$$d = 4.0 \times 10^{-8}\ \text{cm}$$

which is a reasonable estimate for the diameter of a molecule.

At temperatures below T_c, the van der Waals equation describes those portions of the isotherms outside the shaded region in Figure 15-15 but not those portions inside the shaded region. Suppose we have a gas at a temperature below T_c that initially has a low pressure and a large volume. We begin to compress the gas while holding the temperature constant (isotherm A in the figure). At first the pressure rises, but when we reach point B on the dashed curve, the pressure ceases to rise and the gas begins to liquefy at constant pressure. Along the horizontal line BD in the figure, the gas and liquid are in equilibrium. As we continue to compress the gas, more and more gas liquefies until at point D on the dashed curve we have only liquid. Then, if we try to compress the substance further, the pressure rises sharply because a liquid is nearly incompressible.

The constant pressure at which a liquid is in equilibrium with its own vapor at a given temperature is called the **vapor pressure.** When this equilibrium exists and the liquid is heated slightly (or the pressure is reduced slightly), the liquid boils. We can see from the figure that the vapor pressure for a gas depends on its temperature. If we had started compressing the gas at a lower temperature, as with isotherm A' in Figure 15-15, the vapor pressure would be lower, as is indicated by the horizontal constant-pressure line for A' at a lower value of pressure. The temperature for which the vapor pressure for a substance equals 1 atm is the **normal boiling point** of that substance. For example, the temperature at which the vapor pressure of water is 1 atm is 373 K = 100°C, so this temperature is the normal boiling point of water. At high altitudes, such as on the top of a mountain, the pressure is less than 1 atm, and water then boils at a temperature lower than 373 K. Table 15-2 gives the vapor pressures of water at various temperatures.

At temperatures greater than the critical temperature T_c, a gas will not liquefy at any pressure. The critical temperature for water vapor is 647 K = 374°C. The point at which the critical isotherm intersects the dashed curve in Figure 15-15 (point C) is called the critical point.

Table 15-2 **Vapor Pressure of Water versus Temperature**

t, °C	P, mmHg	P, kPa
0	4.581	0.611
10	9.209	1.23
15	12.653	1.69
20	17.535	2.34
30	31.827	4.24
40	55.335	7.38
50	92.55	12.3
60	149	19.9
70	233.8	31.2
80	355	47.4
90	526	70.1
100	760	101.3
110	1074	143.3
120	1489	198.5
130	2026	270.1

Critical temperature demonstration. (*a*) Below the critical temperature, the liquid is in equilibrium with its saturated vapor, and a clear meniscus separating the two is visible. (*b*) At the critical point, there is no distinction between the liquid and gas phases. (*c*) As the gas cools, clusters of molecules of increasing size form and redissolve. Eventually, the liquid–gas interface with a meniscus returns as in the first picture.

(*a*)

(*b*)

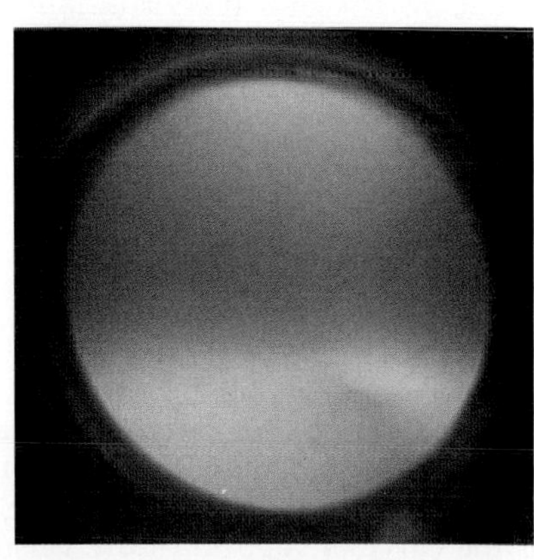

(*c*)

15-7 Phase Diagrams

Consider what happens if we put a liquid such as water in an evacuated container that is sealed so that its volume remains constant. At first, some of the water will evaporate, and water-vapor molecules will fill the previously empty space in the container. Some of the water-vapor molecules will hit the liquid surface and condense again into liquid water. Initially, the rate of evaporation will be greater than the rate of condensation, and the density of the water-vapor molecules will increase. But as the number of water-vapor molecules increases, the condensation rate will increase until it equals the rate of evaporation and equilibrium is established. The pressure of the water vapor at equilibrium is the vapor pressure for that temperature. If we now heat the container to a greater temperature, more liquid will evaporate and a new equilibrium will be established at a higher vapor pressure.

Figure 15-16 is a plot of pressure versus temperature at a constant volume for water. Such a plot is called a **phase diagram.** The portion of the diagram between points *O* and *C* shows the vapor pressure versus the temperature. As we continue to heat the container, the density of the liquid decreases and the density of the vapor increases. At point *C* on the diagram, these densities are equal. Point *C* is the critical point, which was also shown in Figure 15-15 on the isotherm labeled T_c. At this point and above it, there is no distinction between the liquid and the gas.* Critical temperatures for various substances are listed in Table 15-3.

Table 15-3 **Critical Temperatures T_c for Various Substances**

Substance	T_c, K
Argon	150.8
Carbon dioxide	304.2
Chlorine	417.2
Helium	5.3
Hydrogn	33.3
Neon	44.4
Nitric oxide	180.2
Oxygen	154.8
Sulfur dioxide	430.9
Water	647.4

If we now cool our container, some of the vapor condenses into a liquid as we move back down the curve *OC* until the substance reaches point *O* in Figure 15-16. At this point, the liquid begins to solidify. Point *O* is the triple point, that one point at which the vapor, liquid, and solid phases of a substance can coexist in equilibrium. Every substance has a unique triple point, with a specific temperature and pressure. As we saw in Section 15-2, the triple-point temperature for water is 273.16 K = 0.01°C and the triple-point pressure is 4.58 mmHg.

At temperatures and pressures below the triple point, the liquid cannot exist. The curve *OA* in the phase diagram of Figure 15-16 is the locus of pressures and temperatures for which the solid and vapor coexist in equilibrium. The direct change from a solid to a vapor is called **sublimation.** You can observe sublimation if you put ice cubes in the freezer compartment of a refrigerator (especially a self-defrosting refrigerator). The ice cubes eventually disappear due to sublimation. Because atmospheric pressure is well above the triple-point pressure of water, equilibrium is never established

*Often the word "vapor" is used if the temperature is below the critical temperature, and the word "gas" is used if the temperature is above the critical temperature, though there is no need for such a distinction.

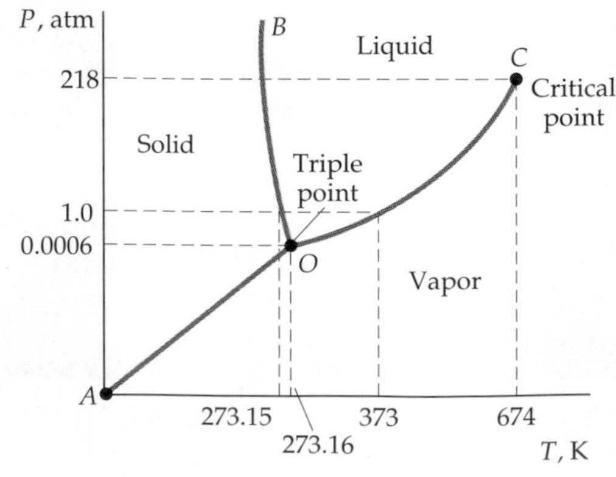

Figure 15-16 Phase diagram for water. The pressure and temperature scales are not linear but are compressed to show the interesting points. Curve *OC* is the curve of vapor pressure versus temperature. Curve *OB* is the melting curve, and curve *OA* is the sublimation curve.

between the ice and water vapor. The triple-point temperature and pressure of carbon dioxide (CO_2) are 216.55 K and 3880 mmHg, which means that liquid CO_2 can exist only at pressures above 3880 mmHg = 5.1 atm. Thus, at ordinary atmospheric pressures, liquid carbon dioxide cannot exist at any temperature. Solid carbon dioxide sublimates directly into gaseous CO_2 without going through the liquid phase; hence, the name "dry ice."

The curve *OB* in Figure 15-16 is the melting curve separating the liquid and solid phases. For a substance such as water for which the melting temperature decreases as the pressure increases, curve *OB* slopes upward to the left from the triple point, as in this figure. For most other substances, the melting temperature increases as the pressure increases. For such a substance, curve *OB* slopes upward to the right from the triple point.

Boiling is a rapid form of vaporization in which bubbles of vapor form inside the liquid. To form a bubble, the vapor must have enough pressure to push the liquid back against the pressure applied to the top of the liquid (normally atmospheric pressure) plus the pressure due to the liquid above the bubble. If the applied pressure is increased, as in a pressure cooker, the vapor pressure required for boiling is increased. Therefore, the temperature of the liquid must be raised further before boiling can occur.

For a molecule to escape from the surface of a liquid, energy is required to break the molecular bonds responsible for surface tension, which we discussed in Chapter 11. Vaporization is therefore a cooling process for the liquid left behind. If water is boiled in the usual way through the input of heat, the cooling effect keeps the temperature of the liquid constant at the boiling point. This is the reason that the boiling point of a substance can be used to calibrate thermometers. However, water can also be caused to boil without adding heat by evacuating the air above it and thereby lowering the applied pressure. The energy needed for vaporization is then taken from the liquid left behind. As a result, the liquid will cool down, even to the point that ice forms on the top of the boiling water.

Humidity

Air is 78 percent nitrogen and 21 percent oxygen plus small amounts of other gases, such as argon, carbon dioxide, and water vapor. The pressure exerted by air is the sum of the partial pressures exerted by each of the various gases that comprise it. In a given volume of air, the partial pressure of any particular gas—nitrogen, oxygen, water vapor, or whatever—is the same as if that gas occupied that volume alone. That is, the presence of other gases does not alter the partial pressure of any given gas. This result is known as **Dalton's law.** Dalton's law can be understood in terms of the kinetic-theory model discussed in Section 15-5. The collisions of one type of gas molecule with the walls of a container are not affected by the presence of other types of gas molecules.

If more water vapor is added to a given volume of air at a given temperature, the partial pressure of the water vapor is increased. When this partial pressure equals the vapor pressure for that temperature, the air is said to be saturated. The water vapor then begins to condense into liquid water if the temperature is above the melting point or into ice crystals (snow or frost) if the temperature is below the melting point. The ratio of the partial pressure of water vapor to the vapor pressure for a particular temperature is called the **relative humidity,** which is usually expressed as a percentage:

$$\text{Relative humidity} = \frac{\text{partial pressure}}{\text{vapor pressure}} \times 100\% \qquad 15\text{-}35$$

Relative humidity defined

Dew drops form on these spiderwebs when the dew point is above 0°C.

Frost forms on a window when the dew point is below 0°C.

The relative humidity can be increased either by increasing the amount of water vapor in the air at a given temperature or by lowering the temperature and thereby lowering the vapor pressure. The temperature at which air becomes saturated with water vapor, that is, the temperature at which relative humidity equals 100 percent, is called the **dew point.** When the surface of the earth cools below the dew point at night due to radiation (which will be discussed in the next chapter), dew forms if the dew point is above 0°C and frost forms if it is below 0°C.

Example 15-13

On a humid, 20°C day, the dew point is measured by cooling a metal container until moisture forms on its surface. This happens when the temperature of the container is 15°C. What is the relative humidity?

At the dew point of 15°C, the partial pressure of the water vapor in the air equals the vapor pressure of water, which according to Table 15-2 is 1.69 kPa. This, then, is the original partial pressure of the water vapor at the original temperature of 20°C. Since the vapor pressure at 20°C is 2.34 kPa (Table 15-2, again), the relative humidity by Equation 15-35 is

$$\text{Relative humidity} = \frac{1.69\text{ kPa}}{2.34\text{ kPa}} \times 100\% = 72.2\%$$

Questions

14. At high altitudes, as in the mountains, cooking things in boiling water takes longer than at sea level. Why?
15. What is the advantage of a pressure cooker?
16. When a pan of water is heated, just before the water really boils small bubbles form on the bottom but disappear as they rise. Later, when the water is really boiling, the bubbles grow in size as they rise. Explain.

Summary

1. A temperature scale can be constructed by choosing some thermometric property and defining the temperature to vary linearly with that property using two fixed points, such as the ice point and steam point of water. On the Celsius scale, the ice point is defined to be 0°C and the steam point 100°C. On the Fahrenheit scale, the ice point is 32°F and the steam point 212°F. Temperatures on the Fahrenheit and Celsius scales are related by

$$t_C = \tfrac{5}{9}(t_F - 32°)$$

2. Different thermometers do not always agree with each other on the measurement of a temperature except at the fixed points. Gas thermometers have the property that they all agree with each other in the measurement of any temperature as long as the density of the gas in the thermometers is very low. The ideal-gas temperature T is defined by

$$T = \frac{273.16\ \text{K}}{P_3}P$$

where P is the pressure of the gas in the thermometer when the thermometer is in thermal equilibrium with the system whose temperature is being measured, and P_3 is the pressure when the thermometer is immersed in a water–ice–vapor bath at its triple point. The absolute or kelvin temperature scale is the same as the ideal-gas scale in the range of temperatures for which gas thermometers can be used. The absolute temperature is related to the Celsius temperature by

$$T = t_C + 273.15\ \text{K}$$

3. The coefficient of linear expansion is the ratio of the fractional change in length to the change in temperature:

$$\alpha = \frac{\Delta L/L}{\Delta T}$$

The coefficient of volume expansion, which is the ratio of the fractional change in volume to the change in temperature, is three times the coefficient of linear expansion:

$$\beta = \frac{\Delta V/V}{\Delta T} = 3\alpha$$

4. At low densities, all gases obey the ideal-gas law:

$$PV = nRT$$

where

$$R = 8.314\ \text{J/mol}\cdot\text{K}$$

is the universal gas constant, which is related to Avogadro's number N_A and Boltzmann's constant k by

$$R = kN_A$$

Avogadro's number is

$$N_A = 6.022 \times 10^{23} \text{ molecules/mol}$$

and Boltzmann's constant is

$$k = 1.381 \times 10^{-23} \text{ J/K}$$

A form of the ideal-gas law that is useful for solving problems involving a fixed amount of gas is

$$\frac{P_2V_2}{T_2} = \frac{P_1V_1}{T_1}$$

5. The absolute temperature T is a measure of the average molecular energy. For an ideal gas, the average translational kinetic energy of the molecules is

$$K_{av} = (\tfrac{1}{2}mv^2)_{av} = \tfrac{3}{2}kT$$

The total translational kinetic energy of n moles of a gas containing N molecules is given by

$$K = N(\tfrac{1}{2}mv^2)_{av} = \tfrac{3}{2}NkT = \tfrac{3}{2}nRT$$

The rms speed of a molecule of a gas is related to the absolute temperature by

$$v_{rms} = \sqrt{(v^2)_{av}} = \sqrt{\frac{3kT}{m}} = \sqrt{\frac{3RT}{M}}$$

where m is the mass of the molecule and M is the molar mass.

6. The van der Waals equation of state describes the behavior of real gases over a wide range of temperatures and pressures:

$$\left(P + \frac{an^2}{V^2}\right)(V - bn) = nRT$$

It takes into account the space occupied by the gas molecules themselves and the attraction of the molecules for one another.

7. Vapor pressure is the pressure at which the liquid and gas phases of a substance are in equilibrium at a given temperature. The liquid boils at that temperature for which the external pressure equals the vapor pressure.

8. The triple point is the particular temperature and pressure at which the gas, liquid, and solid phases of a substance can coexist. At temperatures and pressures below the triple point, the liquid phase of a substance cannot exist.

9. Relative humidity is the ratio of the partial pressure of water vapor in air to its vapor pressure at a particular temperature.

Suggestions for Further Reading

Hall, Marie Boas: "Robert Boyle," *Scientific American*, August 1967, p. 96.

This English investigator of the seventeenth century performed careful experiments that led to his discovery of the inverse relationship between pressure and volume in a gas, and he obtained numerous results in chemistry, guided by his views on the "mechanical" nature of matter.

Romer, Robert H: "Temperature Scales: Celsius, Fahrenheit, Kelvin, Reamur, and Romer," *The Physics Teacher*, vol. 20, no. 7, 1982, p. 450.

A compact and well-researched history of the development of temperature scales.

Review

A. Objectives: After studying this chapter, you should:

1. Be able to state the zeroth law of thermodynamics.

2. Be able to discuss how a mercury thermometer is calibrated using two fixed points for either the Celsius or Fahrenheit temperature scales.

3. Be able to convert temperatures from Celsius to Fahrenheit and from Fahrenheit to Celsius.

4. Be able to define the ideal-gas temperature scale.

5. Be able to calculate the linear expansion and volume expansion of a substance given the change in its temperature.

6. Be able to work problems using the ideal-gas equation, $PV = nRT$.

7. Be able to discuss the molecular interpretation of temperature.

8. Be able to discuss the van der Waals equation.

9. Be able to sketch the liquid–vapor isotherms for a real substance and indicate the critical point on a PV diagram.

10. Be able to sketch the phase diagram for a real substance.

11. Be able to interpret the parts of a phase diagram.

B. Define, explain, or otherwise identify:

Thermodynamics
Temperature
Thermal contact
Thermometric property
Thermal equilibrium
Zeroth law of thermodynamics
Ice-point temperature
Normal freezing point
Steam-point temperature
Normal boiling point
Celsius temperature scale
Fahrenheit temperature scale
Constant-volume gas thermometer
Triple point of water
Ideal-gas temperature scale
Kelvin
Absolute temperature scale
Coefficient of linear expansion
Coefficient of volume expansion
Boyle's law
Boltzmann's constant
Mole
Avogadro's number
Universal gas constant
Ideal gas
Molar mass
Equation of state
Isotherm
Standard conditions
Macroscopic variables
Kinetic theory of gases
RMS speed
Maxwell–Boltzmann distribution function
Maxwell–Boltzmann energy distribution function
Van der Waals equation
Vapor pressure
Normal boiling point
Phase diagram
Sublimation
Dalton's law
Relative humidity
Dew point

C. True or false: If the statement is true, explain why it is true. If it is false, give a counterexample.

1. Two objects in thermal equilibrium with each other must be in thermal equilibrium with a third object.

2. All materials expand when they are heated.

3. The Farhrenheit and Celsius temperature scales differ only in the choice of the zero temperature.

4. The kelvin is the same size as the Celsius degree.

5. All thermometers give the same result when measuring the temperature of a particular system.

6. If the pressure of a gas increases, the temperature must increase.

7. The absolute temperature of a gas is a measure of the average translational kinetic energy of the gas molecules.

8. The temperature at which water boils depends on the pressure.

9. The vapor pressure of a gas depends on the temperature.

Problems

Level I

15-1 The Celsius and Fahrenheit Temperature Scales

1. A certain ski wax is rated for use between −12 and −7°C. What is this temperature range on the Fahrenheit scale?

2. The melting point of gold (Au) is 1945.4°F. Express this temperature in degrees Celsius.

3. The highest and lowest temperatures ever recorded in the United States are 134°F (in California in 1913) and −80°F (in Alaska in 1971). Express these temperatures on the Celsius scale.

4. What is the Celsius temperature corresponding to the normal temperature of the human body of 98.6°F?

5. The length of the column of a mercury thermometer is 4.0 cm when the thermometer is immersed in ice water and 24.0 cm when the thermometer is immersed in boiling water. (*a*) What should the length be at room temperature, 22.0°C? (*b*) The mercury column is 25.4 cm long when the thermometer is immersed in a chemical solution. What is the temperature of the solution?

15-2 Gas Thermometers and the Absolute Temperature Scale

6. The temperature of the interior of the sun is about 10^7 K. What is this temperature on (*a*) the Celsius scale and (*b*) the Fahrenheit scale?

7. The boiling point of nitrogen N_2 is 77.35 K. Express this temperature in degrees Fahrenheit.

8. The pressure of a constant-volume gas thermometer is 0.400 atm at the ice point and 0.546 atm at the steam point. (*a*) When the pressure is 0.100 atm, what is the temperature? (*b*) What is the pressure at 444.6°C, the boiling point of sulfur?

9. A constant-volume gas thermometer reads 50 torr at the triple point of water. (*a*) What will the pressure be when the thermometer measures a temperature of 300 K? (*b*) What ideal-gas temperature corresponds to a pressure of 678 torr?

10. A constant-volume gas thermometer has a pressure of 30 torr when it reads a temperature of 373 K. (*a*) What is its triple-point pressure P_3? (*b*) What temperature corresponds to a pressure of 0.175 torr?

15-3 Thermal Expansion

11. A steel ruler has a length of 30 cm at 20°C. What is its length at 100°C?

12. A bridge 100 m long is built of steel. If it is built as a single, continuous structure, how much will its length change from the coldest winter days (−30°C) to the hottest summer days (40°C)?

13. A steel tape is placed around the earth at the equator when the temperature is 0°C. What will the clearance between the tape and the ground (assumed to be uniform) be if the temperature of the tape rises to 30°C? Neglect the expansion of the earth.

15-4 The Ideal-Gas Law

14. (*a*) If 1 mol of a gas in a container occupies a volume of 10 L at a pressure of 1 atm, what is the temperature of the gas in kelvins? (*b*) The container is fitted with a piston so that the volume can change. When the gas is heated at constant pressure, it expands to a volume of 20 L. What is the temperature of the gas in kelvins? (*c*) The volume is fixed at 20 L, and the gas is heated at constant volume until its temperature is 350 K. What is the pressure of the gas?

15. A gas is kept at constant pressure. If its temperature is changed from 50 to 100°C, by what factor does the volume change?

16. A pressure as low as 1×10^{-8} torr can be achieved using an oil diffusion pump. How many molecules are there in 1 cm^3 of a gas at this pressure if its temperature is 300 K?

17. A 10-L vessel contains gas at 0°C and a pressure of 4 atm. How many moles of gas are there in the vessel? How many molecules?

18. How many moles of lead ($M = 207.2$ g/mol) are there in a lead brick 5 cm by 10 cm by 20 cm?

19. A room is 6 m by 5 m by 3 m. (*a*) If the air pressure in the room is 1 atm and the temperature is 300 K, find the number of moles of air in the room. (*b*) If the temperature rises by 5 K and the pressure remains constant, how many moles of air leave the room?

15-5 The Molecular Interpretation of Temperature: The Kinetic Theory of Gases

20. (*a*) Find v_{rms} for an argon atom if 1 mol of the gas is confined to a 1-L container at a pressure of 10 atm. (For argon, $M = 40 \times 10^{-3}$ kg/mol.) (*b*) Compare this with v_{rms} for a helium atom under the same conditions. (For helium, $M = 4 \times 10^{-3}$ kg/mol.)

21. Find the total translational kinetic energy of 1 L of oxygen gas held at a temperature of 0°C and a pressure of 1 atm.

22. At what temperature will the rms speed of an H_2 molecule equal 332 m/s?

23. Find the rms speed and the average kinetic energy of a hydrogen atom at a temperature of 10^7 K. (At this temperature, which is of the order of the temperature in the interior of stars, the hydrogen is ionized and consists of a single proton.)

15-6 The van der Waals Equation and Liquid–Vapor Isotherms

24. (*a*) Calculate the volume of 1 mol of steam at 100°C and a pressure of 1 atm, assuming that it is an ideal gas. (*b*) Find the temperature at which the steam will occupy the volume found in part (*a*) if it obeys the van der Waals equation with $a = 0.55$ Pa·m^6/mol^2 and $b = 30$ cm^3.

25. Use the values in Table 15-2 to draw a graph of the vapor pressure of water versus temperature. From your graph, find (*a*) the temperature at which water boils on a mountain where the atmospheric pressure is 70 kPa, (*b*) the temperature at which water will boil in a container in which the pressure has been reduced to 0.5 atm, and (*c*) the pressure at which water will boil at 115°C.

15-7 Phase Diagrams

26. Which gases in Table 15-3 can not be liquefied by applying pressure at 20°C?

27. On a certain day, the temperature is 30°C and the relative humidity is 80 percent. What is the partial pressure of the water vapor in the air?

28. If the temperature is 20°C and the dew point is 10°C, what is the relative humidity?

29. If the partial pressure of water vapor is 3.00 kPa when the temperature is 30°C, what is the relative humidity?

Level II

30. A cubic metal box of sides 20 cm contains air at a pressure of 1 atm and a temperature of 300 K. The box is sealed so that the volume is constant, and it is heated to a temperature of 400 K. Find the net force on each wall of the box.

31. The escape velocity on Mars is 5.0 km/s, and the surface temperature is typically 0°C. Calculate the rms speeds for (*a*) H_2, (*b*) O_2, and (*c*) CO_2 at this temperature. (*d*) If the rms speed of a gas is greater than about one-sixth of the escape velocity of a planet, virtually all of the molecules of that gas will have escaped the atmosphere of the planet by the present time. Based on this criterion, are H_2, O_2, and CO_2 likely to be found in Mars's atmosphere?

32. Repeat Problem 31 for Jupiter, whose escape velocity is 60 km/s and whose temperature is typically −150°C.

33. At what temperature do the Fahrenheit and Celsius temperature scales give the same reading?

34. A solid-state temperature transducer is essentially a linear amplifier whose gain is linearly temperature dependent. If the gain is 25 at 20°C and 60 at 70°C, what is the temperature when the gain is 45?

35. An aluminum rod is measured with a steel tape measure at 25°C. Its length is found to be 75 cm. What will the tape indicate the length of the rod to be when both the tape and the rod are at (*a*) 0°C and (*b*) 50°C?

36. (*a*) From the definition of β, the coefficient of volume expansion (at constant pressure), show that $\beta = 1/T$ for an ideal gas. (*b*) The experimentally determined value of β for N_2 gas at 0°C is 0.003673 K^{-1}. Compare this value with the theoretical value $\beta = 1/T$, assuming that N_2 is an ideal gas.

37. One way to construct a device with two points whose separation remains the same in spite of temperature changes is to bolt together one end of two rods having different coefficients of linear expansion as in the arrangement shown in Figure 15-17. (*a*) Show that the distance L will not change with temperature if the lengths L_A and L_B are chosen such that $L_A/L_B = \alpha_B/\alpha_A$. (*b*) If material B is steel, material A is brass, and $L_A = 250$ cm at 0°C, what is the value of L?

Figure 15-17 Problem 37.

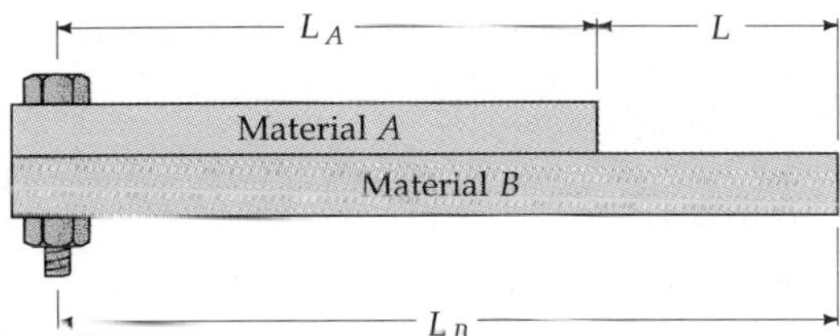

38. For the phase diagram given in Figure 15-18, state what changes (if any) occur for each line segment AB, BC, CD, and DE—in (*a*) volume and (*b*) phase. (*c*) For what type of substance would OH be replaced by OG? (*d*) What is the significance of point F?

Figure 15-18 Problems 38 and 40.

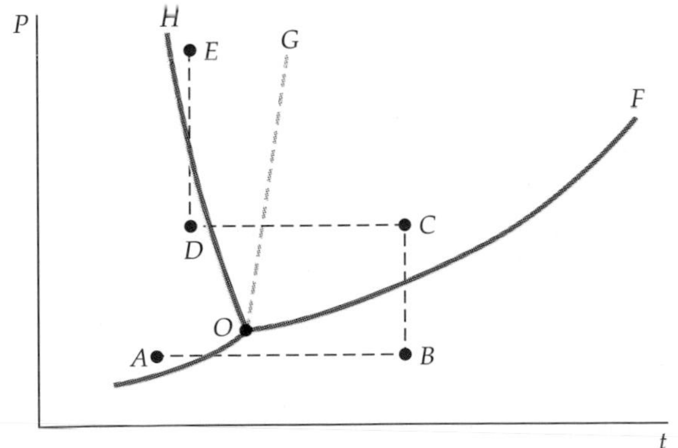

39. The van der Waals constants for helium are $a = 0.03412$ L^2·atm/mol and $b = 0.0237$ L/mol. Use these data to find the volume in cubic centimeters occupied by one helium atom and to estimate the radius of the atom.

40. The phase diagram in Figure 15-18 can be interpreted to yield information on how the boiling and melting points of water change with altitude. (*a*) Explain how this information can be obtained. (*b*) How might this information affect cooking procedures in the mountains?

41. An automobile tire is filled to a gauge pressure of 200 kPa when its temperature is 20°C. (Gauge pressure is the difference between the actual pressure and atmospheric pressure.) After the car has been driven at high speeds, the tire temperature has increased to 50°C. (*a*) Assuming that the volume of the tire has not changed, find the gauge pressure of the air in the tire (assuming air is an ideal gas). (*b*) Calculate the gauge pressure if the tire expands such that the volume increases by 10 percent.

42. A 1-km steel railroad rail is fastened securely at both ends when the temperature is 20°C. As the temperature increases, the rail begins to buckle. When the temperature is 25°C, how high is the buckle, assuming it is triangularly shaped?

43. A car has a 60 L steel gas tank filled to the top with gasoline when the temperature is 10°C. The coefficient of volume expansion of gasoline is $\beta = 0.900 \times 10^{-3}$ K^{-1}. Taking the expansion of the steel tank into account, how much gasoline spills out of the tank when the car is parked in the sun and its temperature rises to 25°C?

44. (*a*) Define a coefficient of area expansion. (*b*) Calculate it for a square and a circle, and show that it is 2 times the coefficient of linear expansion.

45. A container fitted with a piston holds 1 mol of gas. Its initial pressure and temperature are 2 atm and 300 K. The gas is allowed to expand at constant temperature until the pressure is 1 atm. The gas is then compressed and heated at the same time until it is back to its original volume, at which point the pressure is 2.5 atm. What is its final temperature?

46. A scuba diver at 40 m below the surface of a lake, where the temperature is 5°C, releases an air bubble with a volume of 15 cm^3. The bubble rises to the surface, where the temperature is 25°C. What is the volume of the bubble right before it breaks the surface? *Hint:* Remember that the pressure also changes.

47. Oxygen (O_2) is confined to a cubic container, 15 cm on a side, at a temperature of 300 K. Compare the average kinetic energy of a molecule of the gas to the change in its gravitational potential energy if it falls from the top of the container to the bottom.

48. Ultra-high-vacuum (UHV) laboratory systems can produce pressures of 10^{-12} torr. At a temperature of 20°C, how many molecules are there in 1 cm^3 at this pressure?

49. Ten objects have the following speeds:

Speed, m/s	2	5	6	8
Number of objects	3	3	3	1

Calculate (*a*) the average speed and (*b*) the rms speed of the objects.

50. Show that $f(v)$ given by Equation 15-31 is maximum when $v = \sqrt{2kT/m}$. *Hint:* Set $df/dv = 0$ and solve for v.

51. Since $f(v)\,dv$ gives the fraction of molecules that have speeds in the range dv, the integral of $f(v)\,dv$ over all the possible ranges of speeds must equal 1. Given the integral

$$\int_0^\infty v^2 e^{-av^2}\,dv = \frac{\sqrt{\pi}}{4}\,a^{-3/2}$$

show that

$$\int_0^\infty f(v)\,dv = 1$$

where $f(v)$ is given by Equation 15-31.

52. Given the integral

$$\int_0^\infty v^3\,e^{-av^2}\,dv = \frac{a^{-2}}{2}$$

calculate the average speed v_{av} of molecules in a gas using the Maxwell–Boltzmann distribution function (Equation 15-31).

Level III

53. At the critical point on the critical isotherm, dP/dV and d^2P/dV^2 are both zero. Show that for a van der Waals gas, the critical volume $V_c = 3b$ per mole.

54. (*a*) For a van der Waals gas, show that the critical temperature is $8a/27Rb$ and the critical pressure is $a/27b^2$. (*b*) Rewrite the van der Waals equation of state in terms of the reduced variable $V_r = V/V_c$, $P_r = P/P_c$, and $T_r = T/T_c$.

55. A thermistor is a solid-state device whose resistance varies greatly with temperature. Its temperature dependence is given approximately by $R = R_0 e^{B/T}$, where R is in ohms, T is in kelvins, and R_0 and B are constants that can be determined by measuring R at calibration points such as the ice point and the steam point. (*a*) If $R = 7360$ ohms at the ice point and 153 ohms at the steam point, find R_0 and B. (*b*) What is the resistance of the thermistor at $t =$ 98.6°F? (*c*) What is the rate of change of the resistance with temperature (dR/dT) at the ice point and the steam point? At which temperature is the thermistor most sensitive?

56. A grandfather's clock is calibrated at a temperature of 20°C. (*a*) On a hot day, when the temperature is 30°C, does the clock "run" fast or slow? (*b*) How much does it gain or lose in a 24-h period? Assume that the pendulum is a thin brass rod with a heavy mass attached to the end.

57. At a temperature of 20°C, a steel bar of radius 2.2 cm and length 60 cm is jammed horizontally perpendicular between two vertical concrete walls. With a blow torch, the temperature of the bar is raised to 60°C. Find the force exerted by the bar on each wall.

58. A steel tube has an outside diameter of 3.000 cm at room temperature (20°C). A brass tube has an inside diameter of 2.997 cm at the same temperature. To what temperature must the ends of the tubes be heated if the steel tube is to be inserted into the brass tube?

59. A constant-volume gas thermometer with a triple-point pressure $P_3 = 500$ torr is used to measure the boiling point of some substance. When the thermometer is placed in thermal contact with the boiling substance, its pressure is 734 torr. Some of the gas in the thermometer is then allowed to escape so that its triple-point pressure is 200 torr. When it is again placed in thermal contact with the boiling substance, its pressure is 293.4 torr. Again, some of the gas is removed from the thermometer so that its triple-point pressure is 100 torr. When it is placed in thermal contact with the boiling substance once again, its pressure is 146.65 torr. Find the ideal-gas temperature of the boiling substance.

60. At the surface of the sun, the temperature is about 6000 K, and all the substances present are gaseous. From data given by the light spectrum of the sun, it is known that most elements are present. (*a*) What is the average kinetic energy of translation of an atom at the surface of the sun? (*b*) What is the range of rms speeds at the surface of the sun if the atoms present range from hydrogen ($M =$ 1 g/mol) to uranium ($M = 238$ g/mol)?

61. In Section 10-5, we found that the escape speed at the surface of a planet of radius R is $v_e = \sqrt{2gR}$, where g is the acceleration of gravity. (*a*) At what temperature is v_{rms} for O_2 equal to the escape speed for the earth? (*b*) At what temperature is v_{rms} for H_2 equal to the escape speed for the earth? (*c*) Temperatures in the upper atmosphere reach 1000 K. How does this account for the low abundance of hydrogen in the earth's atmosphere? (*d*) Compute the temperatures for which the rms speeds of O_2 and H_2 are equal to the escape velocity at the surface of the moon, where g is about one-sixth of its value on earth and $R = 1738$ km. How does this account for the absence of an atmosphere on the moon?

Chapter 16

Heat and the First Law of Thermodynamics

Steel ingots in a twin-tube tunnel furnace. The three 53-cm diameter carbon steel ingots seen here have been heated for about 7 hours to approximately 1340°C. Each 3200-kg ingot sits on a furnace car that transports it through the 81-m furnace, which is divided into twelve separate heating zones so that the temperature of the ingot is increased gradually to prevent cracking. The ingots, glowing a yellow-whitish color, exit the furnace to be milled into large, heavy-walled pipes.

Heat is energy that is transferred from one object to another because of a difference in temperature. In the seventeenth century, Galileo, Newton, and other scientists generally supported the theory of the ancient Greek atomists, who considered heat to be a manifestation of molecular motion. In the next century, methods were developed for making quantitative measurements of the amount of heat that leaves or enters a body, and it was found that when two bodies are in thermal contact, the amount of heat that leaves one body equals the amount that enters the other. This discovery led to the development of an apparently successful theory of heat as a conserved material substance—an invisible fluid called "caloric"—that was neither created nor destroyed but merely flowed out of one body and into another.

The caloric theory of heat served quite well in the description of heat transfer but eventually was discarded when it was observed that caloric apparently could be created endlessly by friction with no corresponding disappearance of caloric somewhere else. In other words, the principle of conservation of caloric, which had been the experimental foundation of this theory of heat, proved to be false.

The first clear observations showing that caloric cannot be conserved were made at the end of the eighteenth century by Benjamin Thompson, an American who emigrated to Europe, became the director of the Bavarian arsenal, and was given the title Count Rumford. Thompson supervised the boring of cannon for Bavaria. Because of the heat generated by the boring tool, water was used for cooling. The water had to be replaced continually because it boiled away during the boring. According to the caloric theory, as the metal from the bore was cut into small chips, its ability to retain caloric was decreased. It therefore released caloric to the water, heating it and causing it to boil. Thompson noticed, however, that even when the drill was too dull to cut the metal, the water still boiled away as long as the drill was turning. Apparently, heat was being produced merely by friction and could be produced endlessly. Based on this observation, Thompson suggested that heat is not a substance that is conserved but rather is some form of motion that was communicated from the bore to the water. He showed, in fact, that the heat produced was approximately proportional to the work done by the boring tool.

The caloric theory continued to be the leading theory of heat for some 40 years after Thompson's work, but it gradually weakened as more and more examples of the nonconservation of heat were observed. The modern, mechanical theory of heat did not emerge until the 1840s. In this view, heat is energy that is transferred from one object to another because of a difference in temperature. (The internal energy of a system is often referred to as thermal energy. When a warm system is in contact with a cooler system, internal energy is transferred from the warm system to the cooler system in the form of heat.) The most varied and precise early experiments demonstrating this were performed, starting in the late 1830s, by James Joule (1818–1889), after whom the SI unit of energy is named. Joule showed that the appearance or disappearance of a given quantity of thermal energy is always accompanied by the disappearance or appearance of an equivalent quantity of mechanical energy. The experiments of Joule and others showed that neither thermal energy nor mechanical energy is conserved independently but that the mechanical energy lost always equals the thermal energy produced. What is conserved is the total mechanical energy plus thermal energy.

16-1 Heat Capacity and Specific Heat

When heat energy is added to a substance, the temperature of the substance usually rises. (An exception occurs during a change of phase, such as when water freezes or evaporates, which we will discuss in the next section.) The amount of heat energy Q needed to raise the temperature of a substance is proportional to the temperature change and to the mass of the substance:

$$Q = C\,\Delta T = mc\,\Delta T \qquad 16\text{-}1$$

where C is the **heat capacity** of the substance, which is defined as the heat energy needed to raise the temperature of a substance by one degree. The **specific heat** c is the heat capacity per unit mass:

$$c = \frac{C}{m} \qquad 16\text{-}2$$

The historical unit of heat energy, the **calorie,** was originally defined to be the amount of heat energy needed to raise the temperature of one gram of water one Celsius degree (or one kelvin since the Celsius degree and the kelvin are the same size). The kilocalorie is then the amount of heat energy needed to raise the temperature of one kilogram of water by one Celsius degree. (The "calorie" used in measuring the energy equivalent of foods is actually the kilocalorie.) Since we now recognize that heat is just another form of energy, we do not need any special units for heat that differ from other energy units. The calorie is now defined in terms of the SI unit of energy, the joule:

$$1 \text{ cal} = 4.184 \text{ J} \qquad 16\text{-}3$$

The U.S. customary unit of heat is the **Btu** (for British thermal unit), which was originally defined to be the amount of energy needed to raise the temperature of one pound of water by one Fahrenheit degree. The Btu is related to the calorie and to the joule by

$$1 \text{ Btu} = 252 \text{ cal} = 1.054 \text{ kJ} \qquad 16\text{-}4$$

From the original definition of the calorie, the specific heat of water is

$$c_{\text{water}} = 1 \text{ cal/g·C°} = 1 \text{ kcal/kg·C°}$$

$$= 1 \text{ kcal/kg·K} = 4.184 \text{ kJ/kg·K} \qquad 16\text{-}5a$$

Similarly from the definition of the Btu, the specific heat of water in U. S. customary units is

$$c_{\text{water}} = 1 \text{ Btu/lb·F°} \qquad 16\text{-}5b$$

Careful measurement shows that the specific heat of water varies slightly with temperature, but this variation is only about 1 percent over the entire temperature range from 0 to 100°C. We will usually neglect this small variation and take the specific heat of water to be 1 kcal/kg·K = 4.18 kJ/kg·K.

The heat capacity per mole is called the **molar heat capacity** C_m. The molar heat capacity equals the specific heat (heat capacity per unit mass) times the molar mass M (the mass per mole):

$$C_m = Mc \qquad 16\text{-}6$$

The heat capacity of n moles of a substance is then

$$C = nC_m \qquad 16\text{-}7$$

Table 16-1 lists the specific heats and molar heat capacities of some solids and liquids. (Note that the molar heat capacities of all the metals are about the same. We will discuss the significance of this in Section 16-7.)

Table 16-1 **Specific Heat and Molar Heat Capacities for Various Solids and Liquids at 20°C**

Substance	c, kJ/kg·K	c, kcal/kg·K or Btu/lb·F°	C_m, J/mol·K
Aluminum	0.900	0.215	24.3
Bismuth	0.123	0.0294	25.7
Copper	0.386	0.0923	24.5
Gold	0.126	0.0301	25.6
Ice (−10°C)	2.05	0.49	36.9
Lead	0.128	0.0305	26.4
Silver	0.233	0.0558	24.9
Tungsten	0.134	0.0321	24.8
Zinc	0.387	0.0925	25.2
Alcohol (ethyl)	2.4	0.58	111
Mercury	0.140	0.033	28.3
Water	4.18	1.00	75.2

Example 16-1

How much heat is needed to raise the temperature of 3 kg of copper by 20 C°?

From Table 16-1, the specific heat of copper is 0.386 kJ/kg·K. Equation 16-1 then gives

$$Q = mc\,\Delta T = (3\text{ kg})(0.386\text{ kJ/kg·K})(20\text{ K}) = 23.2\text{ kJ}$$

Note we use 20 C° = 20 K. Alternatively, we could express the specific heat as 0.386 kJ/kg·C° and write the temperature change as 20 C°.

Exercise

A 2-kg aluminum block is originally at 10°C. If 36 kJ of heat energy are added to the block, what is its final temperature? (Answer: 30°C)

We can see from Table 16-1 that the specific heat of water is considerably larger than that of other substances. For example, it is more than 10 times the specific heat of copper. Because of its very large heat capacity, water is an excellent material for storing thermal energy, as in a solar heating system. It is also an excellent coolant. A large body of water, such as a lake or an ocean, tends to moderate the variations of temperature near it because it can absorb or release large quantities of thermal energy while undergoing only a very small change in temperature.

Since the specific heat of water is practically constant over a wide range of temperatures, the specific heat of an object can be conveniently measured by heating the object to some easily measured temperature, placing it in a water bath of known mass and temperature, and measuring the final equilibrium temperature. If the whole system is isolated from its surroundings, the heat leaving the object equals the heat entering the water and its container. This procedure is called **calorimetry,** and the insulated water container is called a **calorimeter.** Let m be the mass of the object, c its specific heat, and T_{io} its initial temperature. If T_f is the final temperature of the object in its water bath, the heat leaving the object is

$$Q_{out} = mc(T_{io} - T_f)$$

Similarly, if T_{iw} is the initial temperature of the water and container, and T_f is their final temperature (the final temperature of the object and water are the same, since they eventually come to equilibrium), the heat absorbed by the water and container is

$$Q_{in} = m_w c_w (T_f - T_{iw}) + m_c c_c (T_f - T_{iw})$$

where m_w and $c_w = 4.18$ kJ/kg·K are the mass and specific heat of the water, and m_c and c_c are the mass and specific heat of the container. Note that in these equations we have chosen to write the temperature differences so that both the heat in and the heat out are positive quantities. (We know that if we place a heated object in cooler water, the final temperature T_f will be greater than the initial temperature of the water bath and less than the initial temperature of the object.) Since these amounts of heat are equal, the specific heat c of the object can be calculated by setting the heat out of the object equal to the heat into the water and its container.

$$Q_{out} = Q_{in}$$

$$mc(T_{io} - T_f) = m_w c_w (T_f - T_{iw}) + m_c c_c (T_f - T_{iw}) \qquad 16\text{-}8$$

Since only temperature differences occur in Equation 16-8 and since the Kelvin and Celsius degree are the same size, the temperatures can all be measured on either the Celsius or Kelvin scale without affecting the result.

Example 16-2

Lead shot of mass 600 g is heated to 100°C and placed in an aluminum can of mass 200 g which contains 500 g of water initially at 17.3°C. The specific heat of the aluminum can is 0.900 kJ/kg·K. The final equilibrium temperature of the mixture is 20.0°C. What is the specific heat of lead?

Since the temperature change of the water is 20.0°C − 17.3°C = 2.7 C° = 2.7 K, the heat absorbed by the water is

$$Q_w = m_w c_w \, \Delta T_w = (0.5 \text{ kg})(4.18 \text{ kJ/kg·K})(2.7 \text{ K}) = 5.64 \text{ kJ}$$

Similarly, the heat absorbed by the aluminum container is

$$Q_c = m_c c_c \, \Delta T_c = (0.2 \text{ kg})(0.900 \text{ kJ/kg·K})(2.7 \text{ K}) = 0.486 \text{ kJ}$$

The temperature change of the lead is 100°C − 20.0°C = 80.0 C° = 80.0 K, and the heat given off by the lead is

$$Q_{Pb} = m_{Pb} c_{Pb} \, \Delta T_{Pb} = (0.6 \text{ kg})(c_{Pb})(80.0 \text{ K}) = (48.0 \text{ kg·K})c_{Pb}$$

Setting the heat given off by the lead equal to that absorbed by the water and its container, we obtain

$$Q_{Pb} = Q_w + Q_c$$

$$(48.0 \text{ kg·K})c_{Pb} = 5.64 \text{ kJ} + 0.486 \text{ kJ} = 6.13 \text{ kJ}$$

Solving for c_{Pb} gives

$$c_{Pb} = \frac{6.13 \text{ kJ}}{48.0 \text{ kg·K}} = 0.128 \text{ kJ/kg·K}$$

Note that the specific heat of lead is considerably less than that of water.

The amount of heat needed to raise the temperature of a substance a given amount depends on whether the substance is allowed to expand while it is being heated. If it is kept from expanding, all the added heat will go into increasing the temperature. If it is allowed to expand, however, it will do work on its surroundings—on the air, if on nothing else. Therefore, more heat energy will be needed to do the work as well as raise the temperature. Thus, for all substances that expand when they are heated, the heat capacity or specific heat at constant volume c_v is less than the heat capacity or specific heat at constant pressure c_p. It is very difficult to prevent a solid or a liquid from expanding when it is heated. However, the expansion, and therefore the work done, is usually very small, so the difference between c_v and c_p can be ignored in most cases. Since solids and liquids are usually heated at constant (atmospheric) pressure, the values listed in Table 16-1 are those for constant pressure.

For gases, the situation is quite different. Gases expand significantly when they are heated at constant pressure, so they do a significant amount of work. Thus, there is a large difference between the specific heat or the heat capacity of a gas at constant pressure and that at constant volume. (It is quite easy to keep the volume of a gas constant while it is being heated, as when a gas is heated in a closed container in a laboratory.) We will calculate the difference in the heat capacities of an ideal gas in Section 16-7.

Questions

1. A potato wrapped in aluminum foil is baked in an oven; it is taken out and its foil removed. The foil cools much faster than the potato. Why?
2. Body *A* has twice the mass and twice the specific heat of body *B*. If they are supplied with equal amounts of heat, how do the changes in their temperatures compare?

(a)

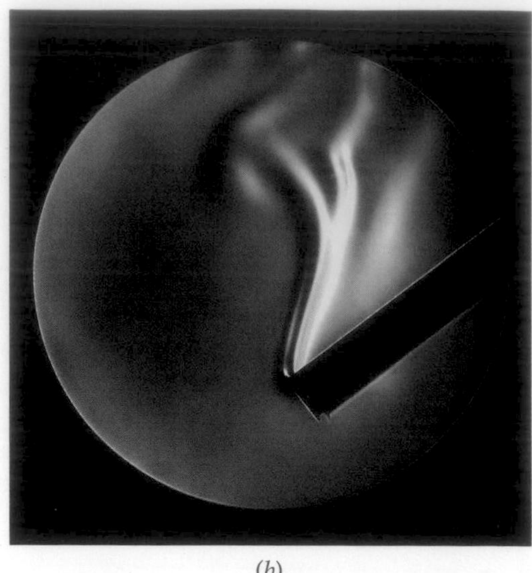

(b)

(*a*) Crystal spires of KCl condensed from gas at 550°C. (*b*) Gasoline vapor made visible by Schlieren photography, which is sensitive to very slight changes in density. (*c*) Color changes produced by changes in temperature are characteristic of liquid crystals such as those shown here. A liquid crystal is a phase between a solid and a liquid—having fluid properties of a liquid and optical properties of a solid. They are observed in organic compounds and occur widely in living matter. (*d*) Ice floating on a thawing lake is an example of a liquid and a solid in thermal equilibrium. (*e*) Carbon dioxide subliming from solid to gas.

16-2 Change of Phase and Latent Heat

When heat is supplied to a substance at constant pressure, the result is usually an increase in the temperature of the substance. Sometimes, though, a substance can absorb large amounts of heat without undergoing any change in its temperature. This happens during a **phase change,** that is, when the physical condition of the substance is changing from one form to another. Types of phase changes are **fusion,** the change of a liquid to a solid (as in the freezing of water into ice); **vaporization,** the change of a liquid to a vapor or gas (as in the evaporation of water); and **sublimation,** the change of a solid directly into a gas (as in the vaporization of moth balls or solid carbon dioxide, frequently called dry ice). There are also other types of phase changes, as when a solid changes from one crystalline form to another.

Phase changes can be understood in terms of molecular theory. An increase in the temperature of a substance reflects an increase in the kinetic energy of motion of its molecules. When a substance changes from liquid to gaseous form, its molecules, which are close together in the liquid form, are moved far apart from each other. This requires that work be done against the attractive forces that hold the molecules close together; that is, energy must be supplied to the molecules to separate them. This energy goes into an increase in the potential energy of the molecules rather than an increase in their kinetic energy. Therefore, the temperature of the substance, which is a measure of the average kinetic energy of its molecules, does not change.

For a pure substance, a change in phase at a given pressure occurs only at a particular temperature. For example, pure water at a pressure of 1 atm changes from solid to liquid at 0°C and from liquid to gas at 100°C. The first temperature is the normal melting point of water and the second is the normal boiling point.

A specific quantity of heat energy is required to change the phase of a given amount of substance. The heat required is proportional to the mass of the substance. The heat required to melt a substance of mass m with no change in its temperature is

$$Q = mL_f \tag{16-9}$$

(c)

(d)

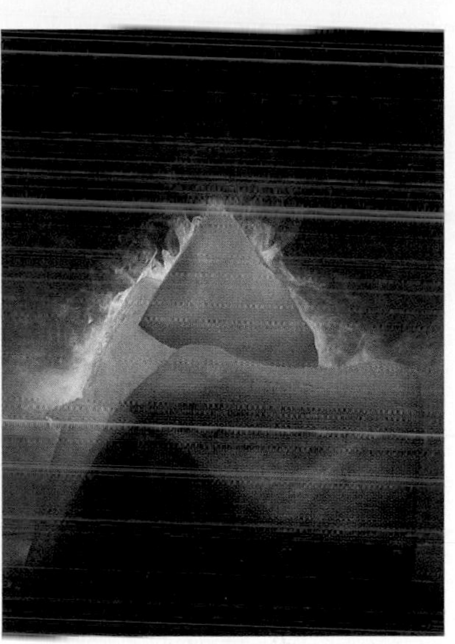

(e)

where L_f is called the **latent heat of fusion** of the substance. For the melting of ice to water at a pressure of 1 atm, the latent heat of fusion is 333.5 kJ/kg = 79.7 kcal/kg. When the phase change is from liquid to gas, the heat required is

$$Q = mL_v \tag{16-10}$$

where L_v is the **latent heat of vaporization.** For water at a pressure of 1 atm, the latent heat of vaporization is 2.26 MJ/kg = 540 kcal/kg. Table 16-2 gives the normal melting and boiling points and the latent heats of fusion and vaporization at 1 atm for various substances.

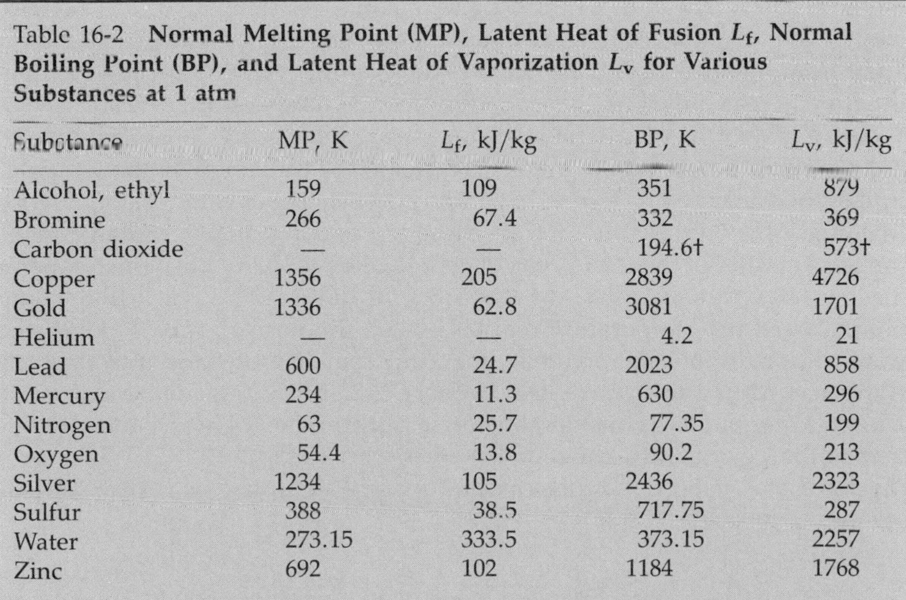

Table 16-2 **Normal Melting Point (MP), Latent Heat of Fusion L_f, Normal Boiling Point (BP), and Latent Heat of Vaporization L_v for Various Substances at 1 atm**

Substance	MP, K	L_f, kJ/kg	BP, K	L_v, kJ/kg
Alcohol, ethyl	159	109	351	879
Bromine	266	67.4	332	369
Carbon dioxide	—	—	194.6†	573†
Copper	1356	205	2839	4726
Gold	1336	62.8	3081	1701
Helium	—	—	4.2	21
Lead	600	24.7	2023	858
Mercury	234	11.3	630	296
Nitrogen	63	25.7	77.35	199
Oxygen	54.4	13.8	90.2	213
Silver	1234	105	2436	2323
Sulfur	388	38.5	717.75	287
Water	273.15	333.5	373.15	2257
Zinc	692	102	1184	1768

†These values are for sublimation. Carbon dioxide does not have a liquid state at 1 atm.

Example 16-3

If 1 kg of ice at −20°C is heated at a pressure of 1 atm until all the ice has been changed into steam, how much heat is required?

Provided that the heat capacity of ice is constant and is equal to 2.05 kJ/kg·K (Table 16-1), the heat energy needed to raise the temperature of the ice from −20 to 0°C is

$$Q_1 = mc\ \Delta T = (1\ \text{kg})(2.05\ \text{kJ/kg·K})(20\ \text{K}) = 41\ \text{kJ}$$

The latent heat of fusion for ice is 334 kJ/kg (Table 16-2), so the heat needed to melt 1 kg of ice is

$$Q_2 = mL_f = (1\ \text{kg})(334\ \text{kJ/kg}) = 334\ \text{kJ}$$

The heat needed to raise the temperature of the resulting 1 kg of water from 0 to 100°C is

$$Q_3 = mc\ \Delta T = (1\ \text{kg})(4.18\ \text{kJ/kg·K})(100\ \text{K}) = 418\ \text{kJ}$$

where we have neglected any variation in the heat capacity of water over this range of temperatures. Finally, the heat needed to vaporize 1 kg of water at 100°C is

$$Q_4 = mL_v = (1\ \text{kg})(2.26 \times 10^3\ \text{kJ/kg}) = 2.26\ \text{MJ}$$

The total amount of heat required is

$$\begin{aligned} Q &= Q_1 + Q_2 + Q_3 + Q_4 \\ &= 0.041\ \text{MJ} + 0.334\ \text{MJ} + 0.418\ \text{MJ} + 2.26\ \text{MJ} \\ &= 3.05\ \text{MJ} \end{aligned}$$

Note that most of the heat supplied in this process was needed to change the phases of the water, not to raise its temperature.

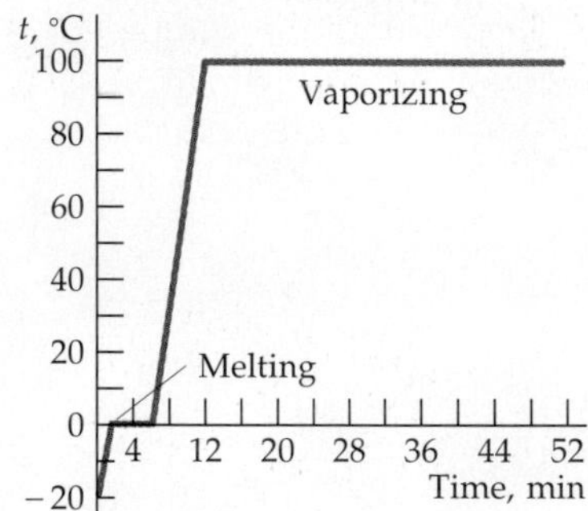

Figure 16-1 Plot of temperature versus time when heat is added to 1 kg of ice originally at −20°C at a constant rate of 1 kJ/s. The temperature first rises to 0°C, at which point the ice melts at constant temperature. When the ice has melted, the temperature of the water rises to 100°C. Then the water vaporizes at constant temperature. If heat were added after all the water has turned to steam, the temperature of the steam would rise.

Figure 16-1 shows a graph of the temperature versus time for Example 16-3, provided that the heat is added at a constant rate of 1 kJ/s. The temperature of the ice increases at a steady rate until it reaches 0°C, the normal melting point. The temperature then remains constant at 0°C as the ice melts. Note that it takes longer to melt the ice than it does to raise its temperature from −20 to 0°C. As calculated in the example, 334 kJ is needed to melt the ice whereas only 41 kJ is needed to raise its temperature by 20 C°. So at 1 kJ/s, it takes only 41 s to raise the temperature of the ice 20 C°, but it takes 334 s to melt the ice. When all the ice is melted to water at 0°C, the temperature of the water begins to rise. Since the specific heat of water is about twice that of ice, the temperature of the water increases at a slower rate than does the temperature of the ice. As calculated, it takes 418 kJ to raise the temperature by 100 C°, so at 1 kJ/s, it takes 418 s for the water to reach the boiling point. When the temperature reaches 100°C, the normal boiling point, the water vaporizes (boils), and the temperature remains constant until all of the water has turned to vapor (steam). Since 2.26 MJ are needed to vaporize 1 kg of water, it takes considerably longer to vaporize the water than it does to melt the ice or to raise the temperature of the water by 100 C°. When all of the water has vaporized, the temperature will again rise as heat is added.

Exercise

An 830-g piece of lead is heated to its melting point of 600 K. How much additional heat energy must be added to melt the lead?
(Answer: 20.5 kJ)

16-3 The Transfer of Thermal Energy

Thermal energy is transferred from one place to another by three processes: conduction, convection, and radiation. In **conduction,** thermal energy is transferred by interactions among atoms or molecules, though there is no transport of the atoms or molecules themselves. For example, if one end of a solid bar is heated, the atoms in the heated end vibrate with greater energy than those at the cooler end. Because of the interaction of the more energetic atoms with their neighbors, this energy is transported along the bar. If the solid is a metal, the transport of thermal energy is helped by free electrons, which move throughout the metal, receiving and giving off thermal energy as they collide with the atoms of the metal. In a gas, heat is conducted by the direct collisions of the gas molecules. Molecules in the warmer part of the gas have a higher average energy than the molecules in the cooler part of the gas. When the high-energy molecules collide with the low-energy molecules, some of the energy of the high-energy molecules is transferred to the low-energy molecules.

In **convection,** heat is transported by direct mass transport. For example, when air near the floor is warmed, it expands and rises because of its lower density. The thermal energy in this warm air is thus transported from the floor to the ceiling along with the mass of warm air.

In **radiation,** energy is emitted and absorbed by bodies in the form of electromagnetic radiation. This radiation moves through space at the speed of light. Thermal radiation, light waves, radio waves, television waves, and x rays are all forms of electromagnetic radiation that differ from one another only in their wavelengths and frequencies. All bodies emit and absorb electromagnetic radiation. When a body is in thermal equilibrium with its surroundings, it emits and absorbs energy at the same rate. However, if it is warmed to a higher temperature than its surroundings, it radiates away more energy than it absorbs, thereby cooling down as its surroundings warm up.

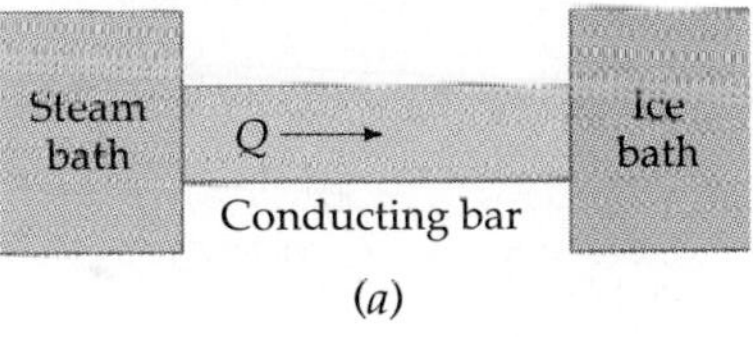

(a)

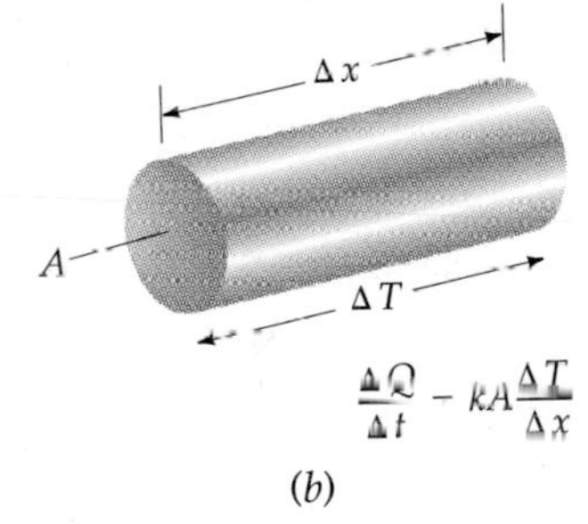

(b)

Figure 16-2 (a) A conducting bar with its ends at two different temperatures. (b) A section of the bar of length Δx. The rate at which thermal energy is conducted across the section is proportional to the cross-sectional area and the temperature difference and is inversely proportional to the thickness of the section.

Conduction and Convection

Figure 16-2 shows a solid bar of cross-sectional area A. If we keep one end of the bar at a high temperature (for example, in a steam bath) and the other end at a low temperature (for example, in an ice bath), thermal energy is continually conducted down the bar from the hot end to the cold end. In the steady state, the temperature varies uniformly (if the bar is uniform) from the hot end to the cold end. The rate of change of the temperature along the bar $\Delta T/\Delta x$ is called a **temperature gradient**. Let us consider a small portion of the bar, a section of thickness Δx, and let ΔT be the temperature difference across the section (Figure 16-2b). If ΔQ is the amount of thermal energy conducted through the section in some time Δt, the rate of conduction of thermal energy $\Delta Q/\Delta t$ is called the **thermal current** I. Experimentally, it is found that the thermal current is proportional to the temperature gradient and to the cross-sectional area A:

$$I = \frac{\Delta Q}{\Delta t} = kA\frac{\Delta T}{\Delta x} \qquad 16\text{-}11$$

Thermal conduction

The proportionality constant k, called the **coefficient of thermal conductivity** or just the **thermal conductivity,** depends on the composition of the bar. In SI units, thermal current is expressed in watts (joules per second), and the thermal conductivity has units of watts per meter-kelvin (though in some tables, the energy may be given in calories or kilocalories and the

Table 16-3 **Thermal Conductivities k for Various Materials**

Material	k, W/m·K	k, Btu·in/h·ft^2·F°
Air (27°C)	0.026	0.18
Ice	0.592	4.11
Water (27°C)	0.609	4.22
Aluminum	237	1644
Copper	401	2780
Gold	318	2200
Iron	80.4	558
Lead	353	2450
Silver	429	2980
Steel	46	319
Oak	0.15	1.02
White Pine	0.11	0.78
Concrete	0.19–1.3	6–9
Glass	0.7–0.9	5–6

thickness in centimeters). In practical calculations in the United States—for finding the heat conducted through the walls of a room, for example—the thermal current is usually expressed in Btu per hour, the area in square feet, the thickness in inches, and the temperature in degrees Fahrenheit. The thermal conductivity is then given in Btu·in/h·ft^2·F°. Table 16-3 lists thermal conductivities for various materials in both SI and U.S. customary units.

If we solve Equation 16-11 for the temperature difference, we obtain

$$\Delta T = \frac{\Delta x}{kA} I \qquad 16\text{-}12$$

or

$$\Delta T = IR \qquad 16\text{-}13$$

where $\Delta x/kA$ is called the **thermal resistance** R:

Thermal resistance

$$R = \frac{\Delta x}{kA} \qquad 16\text{-}14$$

Equation 16-13 is of the same form as Equation 11-29 for the viscous flow of a fluid through a pipe, except that I now stands for the flow of thermal energy rather than the flow of a fluid, R is now the thermal resistance, and the pressure difference ΔP is replaced by the temperature difference ΔT.

Exercise

Calculate the thermal resistance of an aluminum slab of cross-sectional area 15 cm^2 and thickness 2 cm. (Answer: 0.0563 K/W)

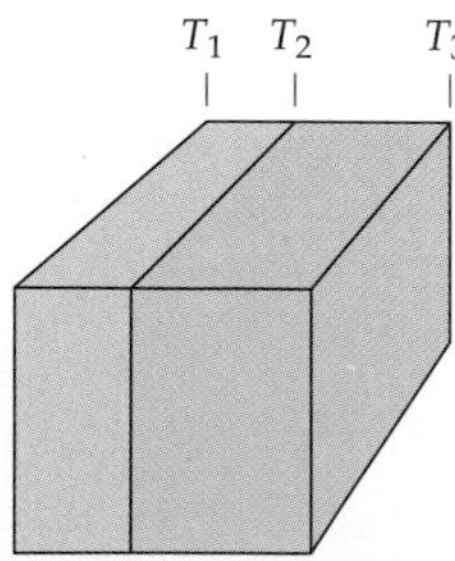

Figure 16-3 Two thermally conducting slabs of different materials in series. The equivalent thermal resistance of the slabs in series is the sum of their individual thermal resistances. The thermal current is the same through both slabs.

In many practical problems, we are interested in the flow of heat through two or more conductors (or insulators) in series. For example, we may wish to know the effect of adding insulating material of a certain thickness and thermal conductivity to the space between two layers of plasterboard. Figure 16-3 shows two thermally conducting slabs of the same cross-sectional area but of different materials and different thicknesses. Let T_1 be the temperature on the warm side, T_2 be that at the interface between the slabs, and T_3 be that on the cool side. Under the conditions of steady-state heat flow, the thermal current must be the same through both slabs. This follows from energy conservation; in the steady state, the energy going in must equal that coming out. If R_1 and R_2 are the thermal resistances of the two slabs, we have from Equation 16-13 for each slab

$$T_1 - T_2 = IR_1$$

and

$$T_2 - T_3 = IR_2$$

Adding these equations gives

$$\Delta T = T_1 - T_3 = (R_1 + R_2)I = R_{eq}I \qquad 16\text{-}15$$

where R_{eq} is the equivalent resistance. Thus, for thermal resistances in series, the equivalent resistance is the sum of the individual resistances:

$$R_{eq} = R_1 + R_2 + \cdots \qquad 16\text{-}16$$

Resistances in series

This result can be applied to any number of resistances in series. It is the same result we will obtain later for electrical resistances in series.

In the building industry, the thermal resistance in U.S. customary units for a square foot of material is called the ***R* factor,** which we will designate by R_f. The R factor is merely the thickness of the material divided by its thermal conductivity:

$$R_f = \frac{\Delta x}{k} = RA \qquad 16\text{-}17$$

The reciprocal of the thermal conductivity in U.S. customary units is the R factor per inch of material. Table 16-4 lists R factors for several materials. To find the rate of heat conduction per unit area in Btu/h·ft² through several slabs of materials, we divide temperature difference in Fahrenheit degrees by the sum of the R factors. From Equation 16-15, we have

$$\Delta T = R_{eq}I = R_{eq}A\left(\frac{I}{A}\right) = R_{f,eq}\left(\frac{I}{A}\right)$$

or

$$\frac{I}{A} = \frac{\Delta T}{R_{f,eq}} \qquad 16\text{-}18$$

Table 16-4 ***R* Factors $\Delta x/k$ for Various Building Materials**

Material	Thickness, in	R_f, h·ft²·F°/Btu
Building board		
Gypsum or plasterboard	0.375	0.32
Plywood (Douglas fir)	0.5	0.62
Plywood or wood panels	0.75	0.93
Particle board, medium density	1.0	1.06
Finish flooring materials		
Carpet and fibrous pad	1.0	2.08
Tile		0.5
Wood, hardwood finish	0.75	0.68
Roof insulation	1.0	2.8
Roofing		
Asphalt roll roofing		0.15
Asphalt shingles		0.44
Windows		
Single-pane		0.9
Double-pane		1.8

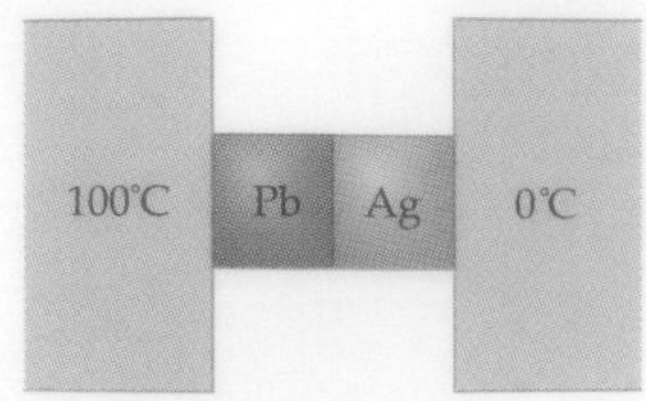

Figure 16-4 Two thermally conducting cubes in series for Example 16-4.

Example 16-4

Figure 16-4 shows two metal cubes, with 2-cm sides, between two walls, one held at 100°C and the other at 0°C. The cubes are lead and silver. Find (*a*) the total thermal current through the cubes and (*b*) the temperature at the interface.

(*a*) We first find the thermal resistance of each cube using Equation 16-14 and values for the thermal conductivity *k* from Table 16-3. For the lead cube, we have

$$R_{\text{Pb}} = \frac{\Delta x}{kA} = \frac{0.02\text{ m}}{(353\text{ W/m·K})(0.02\text{ m})^2} = 0.142\text{ K/W}$$

Similarly, the resistance of the silver cube is

$$R_{\text{Ag}} = \frac{0.02\text{ m}}{(429\text{ W/m·K})(0.02\text{ m})^2} = 0.117\text{ K/W}$$

The equivalent thermal resistance for the two cubes in series is thus

$$R_{\text{eq}} = R_{\text{Pb}} + R_{\text{Ag}} = 0.142 + 0.117 = 0.259\text{ K/W}$$

The thermal current is then given by

$$I = \frac{\Delta T}{R_{\text{eq}}} = \frac{100\text{ K}}{0.259\text{ K/W}} = 386\text{ W}$$

(*b*) We find the temperature at the interface by noting that the temperature drop between the end at 100°C and the interface must equal the current *I* times the thermal resistance of the lead cube. Calling the temperature at the interface t_{if}, we have

$$100°\text{C} - t_{\text{if}} = IR_{\text{Pb}} = (386\text{ W})(0.142\text{ K/W}) = 54.8\text{ K} = 54.8\text{ C°}$$

The temperature at the interface is thus

$$t_{\text{if}} = 100°\text{C} - 54.8\text{ C°} = 45.2°\text{C}$$

Example 16-5

A 60- by 20-ft roof is made of 1-in pine board covered with asphalt shingles. (*a*) If the overlap in the shingles is neglected, at what rate is heat conducted through the roof when the temperature inside is 70°F and that outside is 40°F? (*b*) By what factor is the heat loss reduced if 2 in of roof insulation are added?

(*a*) From Table 16-3, the thermal conductivity of pine board in U.S. customary units is 0.78 Btu·in/h·ft²·F°. The *R* factor for a board 1 in thick is therefore

$$R_{\text{f}} = \frac{\Delta x}{k} = \frac{1\text{ in}}{0.78\text{ Btu·in/h·ft}^2\text{·F°}} = 1.28\,\frac{\text{h·ft}^2\text{·F°}}{\text{Btu}}$$

From Table 16-4, the *R* factor for asphalt shingles is 0.44 h·ft²·F°/Btu. The *R* factor for the combination is therefore 1.28 + 0.44 = 1.72 h·ft²·F°/Btu. For a temperature difference of 30 F°, the heat conducted per square foot is

$$\frac{I}{A} = \frac{\Delta T}{R_{\text{f,eq}}} = \frac{30\text{ F°}}{1.72\text{ h·ft}^2\text{·F°/Btu}} = 17.4\,\frac{\text{Btu}}{\text{h·ft}^2}$$

Thus, the rate of conduction through the area of 60 by 20 ft = 1200 ft² is

$$I = (17.4\text{ Btu/h·ft}^2)(1200\text{ ft}^2) = 21{,}000\text{ Btu/h}$$

(*b*) From Table 16-4, the R factor for roof insulation is 2.8 h·ft^2·F°/Btu per inch, so for 2 in of insulation, the R factor is 5.6 h·ft^2·F°/Btu. Adding the insulation thus increases the R factor from 1.72 h·ft^2·F°/Btu to 1.72 + 5.6 = 7.32 h·ft^2·F°/Btu. Since 7.32/1.72 = 4.26, the rate at which heat is conducted through the roof is decreased by a factor of 4.26. Thus, for the temperature difference given, the rate of heat conduction through the roof with the insulation will be reduced to (21,000 Btu/h)/4.26 = 4900 Btu/h.

To calculate the amount of heat leaving a room by conduction in a given time, we need to know how much leaves through the walls, the windows, the floor, and the ceiling. Such a problem involves what are called "parallel paths" for heat flow. The walls, windows, floor, and ceiling all represent independent paths for heat loss to the outside. The temperature difference is the same for each path, but the thermal current is different. The total thermal current is the sum of the thermal currents through each of the independent or parallel paths:

$$I_{\text{total}} = I_1 + I_2 + \cdots = \frac{\Delta T}{R_1} + \frac{\Delta T}{R_2} + \cdots$$

or

$$I_{\text{total}} = \frac{\Delta T}{R_{\text{eq}}} \qquad 16\text{-}19$$

where the equivalent thermal resistance is given by

$$\frac{1}{R_{\text{eq}}} = \frac{1}{R_1} + \frac{1}{R_2} + \cdots \qquad 16\text{-}20$$

Resistances in parallel

We will encounter this equation again when we study electric conduction through parallel resistances.

The thermal conductivity of air is very small compared with that of solid materials, so air is a very good insulator. However, when there is a large air gap—say, between a storm window and the inside window—the insulating

The effect of insulation can be seen by the snow on the roof of this van. The roof over the driver's compartment is insulated but the rest of the roof is not. Three large braces across the back roof evidently act to insulate the cool metal roof from the warm air in the van.

efficiency of air is greatly reduced because of convection. Whenever there is a temperature difference between different parts of the air space, convection currents act quickly to equalize the temperature, so the effective conductivity is greatly increased. For storm windows, air gaps of about 1 to 2 cm are optimal. Wider air gaps actually reduce the thermal resistance of a double-pane window because of convection.

The insulating properties of air can be used very effectively if the air can be trapped in small pockets that are separated from each other so that convection cannot take place. Goose down, for example, is a good thermal insulator because it traps air so that it cannot circulate and transport heat by convection. Another example is Styrofoam, a cellular material in which tiny pockets of air are separated by poorly conducting cell walls that prevent convection. The thermal conductivity of such a material is essentially the same as that of air.

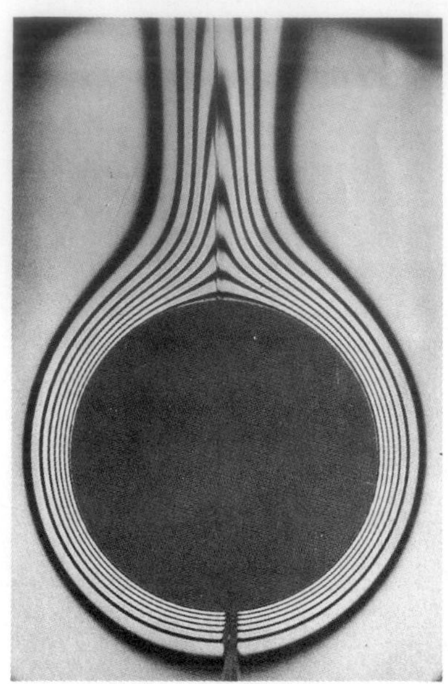

Convection from a heated cylinder. This interferometer photograph shows isotherms as the cylinder cools by convection.

If you touch the inside glass surface of a window when it is cold outside, you will observe that the surface is considerably colder than the inside air. The thermal resistance of a window is due mainly to thin insulating air films that adhere to either side of the glass surface. The thickness of the glass has little effect on the overall thermal resistance. An air film typically adds an R factor of about 0.45. Thus, the R factor of a window with N separated glass layers is approximately $0.9N$ because of the two sides of each layer. Under windy conditions, the outside air film may be greatly decreased, leading to a smaller R factor for the window.

It is possible to write an equation for the thermal energy transported by convection and to define a coefficient of convection, but analyses of practical problems involving convection are very difficult, so they will not be discussed here. The heat transferred from a body to its surroundings by convection is approximately proportional to the area of the body and to the difference in temperature between the body and the surrounding fluid.

Radiation

The third mechanism for the transfer of thermal energy is radiation in the form of electromagnetic waves. The rate at which an object radiates thermal energy is proportional to the area of the object and to the fourth power of its absolute temperature. This result, found empirically by Josef Stefan in 1879 and derived theoretically by Ludwig Boltzmann about five years later, is called the **Stefan–Boltzmann law:**

Stefan–Boltzmann law

$$I = e\sigma AT^4 \tag{16-21}$$

where I is the power radiated in watts, A is the area, e is the **emissivity** of the object, and σ is a universal constant called **Stefan's constant,** which has the value

$$\sigma = 5.6703 \times 10^{-8}\ \mathrm{W/m^2{\cdot}K^4} \tag{16-22}$$

The emissivity e is a fraction that varies from 0 to 1 and depends on the composition of the surface of the object.

When radiation falls on an opaque object, part of the radiation is reflected and part is absorbed. Light-colored objects reflect most visible radiation, whereas dark objects absorb most of it. The rate at which an object absorbs radiation is given by

$$I_a = e\sigma AT_0^4 \tag{16-23}$$

where T_0 is the temperature of the surroundings.

(*a*)

(*b*)

(*a*) Thermograph of a boy and his dog. The bright regions are those of high temperature. Note the dog's cold nose. (*b*) Thermograph of Niagara falls taken from an altitude of 762 m. The brighter regions indicate vegetation. The land mass to the right (West) of the spectacular Canadian (Horseshoe) Falls is Ontario, Canada.

If an object emits more radiation than it absorbs, it cools down while the surroundings absorb radiation from the object and warm up. If the object absorbs more than it emits, the object warms and the surroundings cool. When an object is in thermal equilibrium with its surroundings, $T = T_0$, and the object emits and absorbs radiation at the same rate. The net power radiated by an object at temperature T in an environment at temperature T_0 is

$$I_{net} = e\sigma A(T^4 - T_0^4) \qquad 16\text{-}24$$

An object that absorbs all the radiation incident upon it has an emissivity equal to 1 and is called a **blackbody**. A blackbody is also an ideal radiator. The concept of an ideal blackbody is important because the characteristics of the radiation emitted by such an object can be calculated theoretically. Materials such as black velvet come close to being ideal blackbodies. The best practical approximation of an ideal blackbody is a small hole leading into a cavity, such as a keyhole in a closet door (Figure 16-5). Radiation incident on the hole has little chance of being reflected back out the hole before it is absorbed by the walls of the cavity. The radiation emitted through the hole is thus characteristic of the temperature of the object.

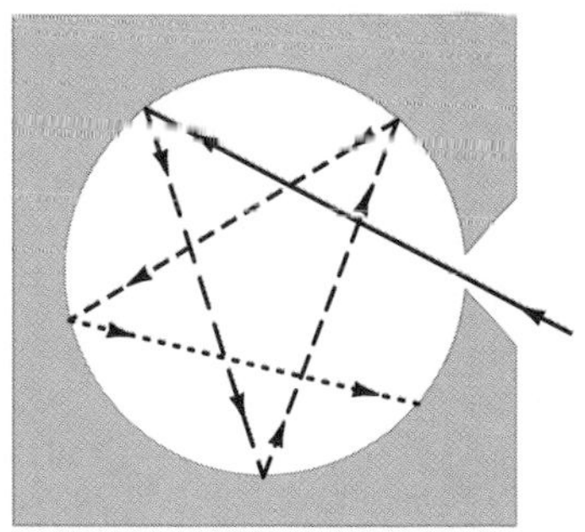

Figure 16-5 A hole in a cavity approximating an ideal blackbody. Radiation entering the cavity has little chance of leaving the cavity before it is completely absorbed. The radiation emitted through the hole is therefore characteristic of the temperature of the object.

The radiation emitted by an object at temperatures below about 600°C is not visible. Most of it is concentrated at wavelengths much longer than those of visible light. (When we study light, we will see that visible light is electromagnetic radiation with wavelengths between about 400 and 700 nm.) As an object is heated, the rate of energy emission increases (Equation 16-21), and the energy radiated extends to shorter and shorter wavelengths. Between about 600 and 700°C, enough of the radiated energy is in the visible spectrum for the body to glow a dull red. At higher temperatures, it may become bright red or even "white hot." Figure 16-6 shows the power radiated by a blackbody as a function of wavelength for several different temperatures. The wavelength at which the power is a maximum varies inversely with the temperature, a result known as **Wien's displacement law:**

Wien's displacement law

$$\lambda_{max} = \frac{2.898 \text{ mm}\cdot\text{K}}{T} \qquad 16\text{-}25$$

This law is used to determine the temperatures of stars from analyses of their radiation. It can also be used to map out the variation in temperature over different regions of the surfaces of an object. Such a map is called a **thermograph.** Thermographs can be used to detect cancer because cancerous material is at a slightly higher temperature than surrounding tissue.

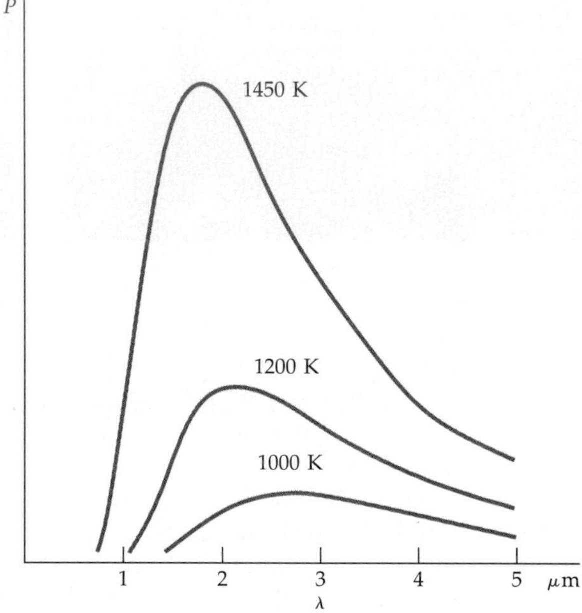

Figure 16-6 Radiated power versus wavelength for radiation emitted by a blackbody. The wavelength of the maximum power varies inversely with the absolute temperature of the blackbody.

The spectral-distribution curves shown in Figure 16-6 played an important role in the history of physics. It was the discrepancy between theoretical calculations of these curves using classical thermodynamics and the experimental measurements of them that led to Max Planck's first ideas about the quantization of energy in 1897. We will discuss this problem further in Chapter 35.

Example 16-6

The surface temperature of the sun is about 6000 K. If the sun is assumed to be a blackbody radiator, at what wavelength λ_{max} would its spectrum peak?

From Wien's displacement law (Equation 16-25), we have

$$\lambda_{max} = \frac{2.898 \text{ mm}\cdot\text{K}}{6000 \text{ K}} = 483 \times 10^{-9} \text{ m} = 483 \text{ nm}$$

This wavelength is in the visible spectrum. The blackbody radiation spectrum describes the sun's radiation fairly well, so the sun is indeed a good example of a blackbody.

Example 16-7

Calculate λ_{max} for a blackbody at room temperature, $T = 300$ K.

From Equation 16-25, we have

$$\lambda_{max} = \frac{2.898 \text{ mm}\cdot\text{K}}{300 \text{ K}} = 9.66 \times 10^{-6} \text{ m} = 9660 \text{ nm}$$

This is an infrared wavelength, far longer than the wavelengths visible to the eye. This obscures the fact that surfaces that are not black to our eyes may act as blackbodies for infrared radiation and absorption. For example, it has been found experimentally that the skin of human beings of all races is black to infrared radiation; hence, the emissivity of skin is 1.00 for its own radiation process.

Temperature regulation along the Alaska pipeline. The 1.2-m diameter pipeline carrying hot oil (up to 145°F) is rigidly attached to vertical support members that are anchored in the ground, which is typically at 30°F. A typical support member is 15 m long of which only 4 m is above ground. Because the ground unfreezes enough to lose its structural rigidity at temperatures greater than 31°F, thermal energy must be transported away from the pipe and ground. Inside the support members are heat pipes containing liquid ammonia that absorbs heat, turns to gas, and rises. The ammonia then liquefies, releasing the heat to the air via the aluminum radiator fins. It then runs back down the walls of the heat pipe.

Example 16-8

Calculate the net loss in radiated energy for a naked person in a room at 20°C, assuming the person to be a blackbody with a surface area of 1.4 m^2 and a surface temperature of 33°C = 306 K. (The surface temperature of the human body is slightly less than the internal temperature, about 37°C, because of the thermal resistance of the skin.)

Using Equation 16-24 with $T = 306$ K and $T_0 = 293$ K, we have

$$I_{net} = (1)(5.67 \times 10^{-8} \text{ W/m}^2\cdot\text{K}^4)(1.4 \text{ m}^2)[(306)^4 - (293)^4] = 111 \text{ W}$$

This is a large energy loss. It is approximately equal to the basal metabolic rate of about 120 W for a food intake of 2500 kcal per day. (See Problem 1.) We protect ourselves from such a great energy loss by wearing clothing, which, because of its low thermal conductivity, has a much lower outside temperature and therefore a much lower rate of thermal radiation.

When the absolute temperature of a body differs only slightly from that of its surroundings, the net power radiated is approximately proportional to the temperature difference. We can see this from Equation 16-24 by noting that $T^4 - T_0^4 = (T^2 + T_0^2)(T^2 - T_0^2) = (T^2 + T_0^2)(T + T_0)(T - T_0)$, and also noting that in the sums we can replace T_0 by T without changing the result much. Replacing T_0 by T in these sums, we have

$$T^4 - T_0^4 \approx (T^2 + T^2)(T + T)(T - T_0) = 4T^3 \, \Delta T$$

(This result can also be obtained by computing the differential $d(T^4) = 4T^3 \, dT$.) Thus for all mechanisms of heat transfer

Newton's law of cooling

The rate of cooling of a body is approximately proportional to the temperature difference between the body and its surroundings.

This result is known as **Newton's law of cooling.**

In many real situations, all three mechanisms for the transfer of thermal energy occur simultaneously, though one may be more effective than the others. For example, an ordinary space heater with a metal heating element heats partly by direct radiation and partly by convection. Because the emissivity of metals is fairly small, the main mechanism is convection. The heater warms the nearby air, which then rises and is replaced by cooler air. Often a fan is included in the heater to speed the convection process. (This is called "forced convection.") On the other hand, a space heater with a quartz filament heats mainly by radiation because the filament has a large emissivity.

Similarly, coffee in a ceramic coffee cup cools by radiation as well as conduction (and evaporation) because the cup has a relatively large emissivity, whereas coffee in a metal cup cools mainly by conduction (and evaporation).

Questions

3. In a cool room, a metal or marble table top feels much colder to the touch than a wood surface does even though they are at the same temperature. Why?
4. Which heat-transfer mechanisms are most important in the warming effect of a fire in a fireplace?
5. Which heat-transfer mechanism is important in the transfer of energy from the sun to the earth?
6. Do you think that it is a coincidence that the radiant energy from the sun is concentrated in the range of visible wavelengths? If not, what is the reason?

James Prescott Joule (1818–1889).

16-4 The First Law of Thermodynamics

The first law of thermodynamics is a statement of the conservation of energy. It reflects the results of many experiments relating the work done on a system, the heat added to or subtracted from the system, and the internal energy of the system. From the original definition of the calorie, we know that it takes one calorie of heat to raise the temperature of 1 g of water by 1 C°. But we can also raise the temperature of water or any other system by doing work on it without adding any heat.

Figure 16-7*a* is a schematic diagram of the apparatus Joule used in his most famous experiment in which he determined the amount of work equivalent to a given amount of heat, that is, the amount of work needed to raise the temperature of one gram of water by one Celsius degree. Once the experimental equivalence of heat and energy has been established, Joule's experiment can be described as determining the size of the calorie in the usual energy units. The water in Figure 16-7 is enclosed by insulating walls so that the temperature of the system cannot be affected by heat entering or leaving it. As the weights fall at constant speed, they turn a paddle wheel,

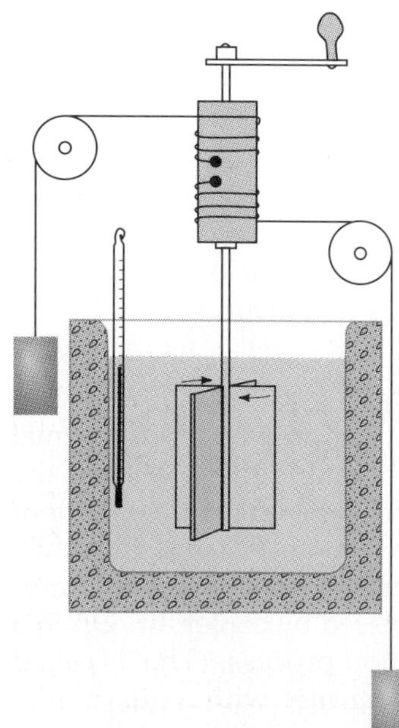

(*a*)

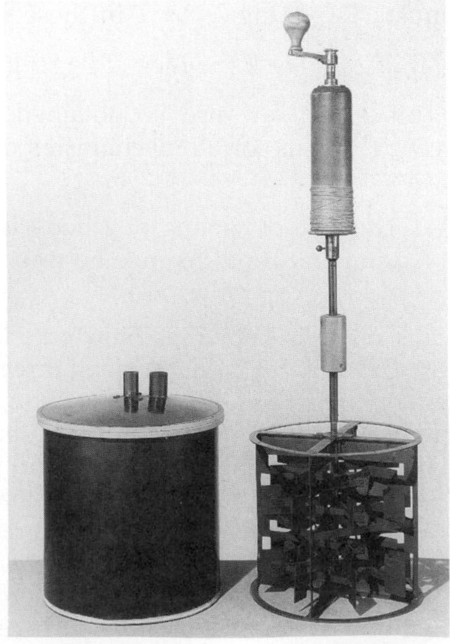

(*b*)

Figure 16-7 (*a*) Schematic diagram of the apparatus for Joule's experiment to determine the amount of work required to produce a given temperature increase in a given amount of water. The work done on the water is determined by calculating the loss in the potential energy of the weights. (*b*) Photograph of the apparatus for Joule's experiment showing the paddle wheel and its container.

which does work against the water. If the energy loss through friction in the bearings, and so forth, is negligible, the work done by the paddle wheel against the water equals the loss of mechanical energy of the falling weights. The latter is easily measured by determining the distance through which the weights fall. The result of Joule's experiment and of those of many others after his is that it takes about 4.18 units of mechanical work (joules) to raise the temperature of 1 g of water by 1 C°. The result that 4.18 J of mechanical energy is equivalent to 1 cal of heat energy was known as the **mechanical equivalence of heat**. Historically, it was customary to express heat energy in calories and then use the mechanical equivalence of heat to convert to the standard units of mechanical energy. Today, all forms of energy are usually expressed in joules.

There are other ways of doing work on this system. For example, we could merely drop the insulated container of water from some height h, letting the system make an inelastic collision with the ground, or we could do mechanical work to generate electricity and then use the electricity to heat the water (Figure 16-8). In all such experiments, the same amount of work is required to produce a given temperature change in a given system. By the conservation of energy, this work done must go into an increase in internal energy U of the system.* The change in the internal energy of the system is usually indicated by the change in the temperature of the system or by a phase change.

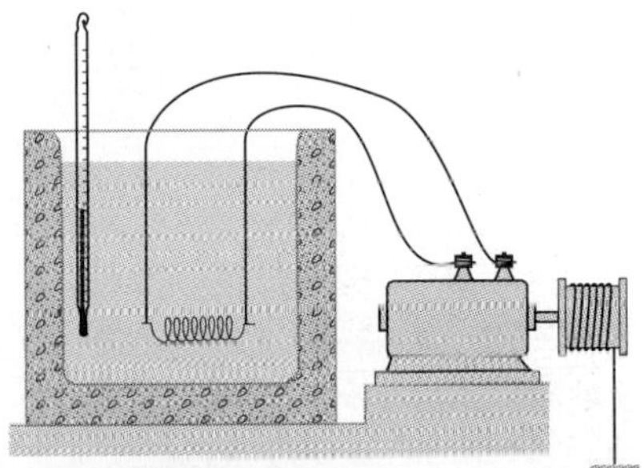

Figure 16-8 Another method of doing work on a thermally insulated container of water. Electrical work is done on the system by the generator, which is driven by the falling weight.

Example 16-9

If a thermally insulated container of water is dropped from a height h and collides inelastically with the ground, what must h be for the temperature of the water to increase by 1 C°?

For the temperature to increase by 1 C°, the internal energy of the water must increase by 4.18 J for each gram of water or 4.18 kJ for each kilogram. If the mass of the water is m, the increase in its internal energy must be m(4.18 kJ/kg). Setting this equal to the loss in potential energy mgh, we have

$$mgh = m(4.18\ \text{kJ/kg})$$

$$h = \frac{4.18\ \text{kJ/kg}}{9.81\ \text{N/kg}} = 0.426\ \text{km} = 426\ \text{m}$$

Note that this result is independent of the mass of the water because the loss in potential energy and the energy needed to raise the temperature by one degree are both proportional to the mass. Note also that the distance is rather large. This result illustrates one of the difficulties with Joule's experiment. A large amount of work must be done to produce a measurable change in temperature of the water.

Now, suppose we perform Joule's experiment but replace the insulating walls of the container with conducting walls. We find that the work needed to produce a given change in the temperature of the system is not always the same but rather depends on how much heat is added to or subtracted from the system by conduction through the walls. However, if we measure the heat added to or subtracted from the system and the work done on the system, we find that the sum of the work done on the system and the net heat added to the system is always the same for a given temperature change.

*The symbol U, which we used in previous chapters to denote potential energy, is now used for the internal energy of a system, which may include both kinetic and potential energy of the molecules in the system.

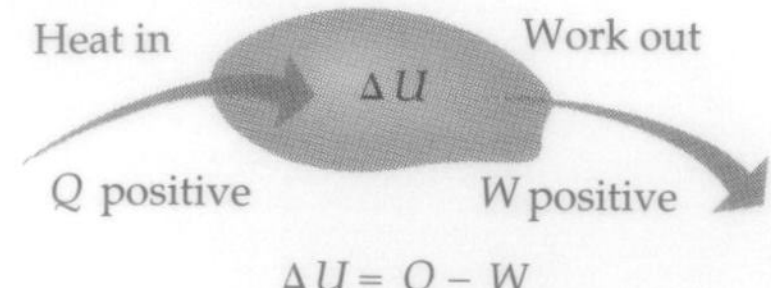

Figure 16-9 Sign conventions for quantities in the first law of thermodynamics. Heat Q is positive when it goes into the system, whereas the work W is positive when it is done by the system.

That is, the sum of the heat added and the work done on the system equals the change in the internal energy of the system. This is a statement of the first law of thermodynamics. It is customary to write W for the work done *by* the system on its surroundings. Then $-W$ is the work done *on* the system. For example, if a gas expands against a piston, doing work on the surroundings, W is positive. Conversely, if work is done to compress the gas, W is negative. The heat Q is taken to be positive if it is put *into* the system and negative if it is taken *out of* the system (Figure 16-9). Using these conventions and writing U for the internal energy of a system, the **first law of thermodynamics** is

$$Q + (-W) = \Delta U$$

or, as it is usually written:

First law of thermodynamics

$$Q = \Delta U + W \qquad 16\text{-}26$$

The net heat added to a system equals the change in the internal energy of the system plus the work done by the system.

The first law of thermodynamics is a statement of the conservation of energy. The heat energy put into a system is accounted for by the work done by the system or by the increase in internal energy of the system or by some combination of the two.

Example 16-10

A system consists of 3 kg of water at 80°C. Twenty-five kilojoules of work is done on the system by stirring it with a paddle wheel, while 15 kcal of heat is removed. (*a*) What is the change in the internal energy of the system? (*b*) What is the final temperature of the system?

(*a*) We first change the units of the heat removed from calories to joules so that both the work and heat are expressed in the same units. We then calculate the change in the internal energy of the system from the first law of thermodynamics as expressed in Equation 16-26. Since 1 kcal = 4.18 kJ, 15 kcal = 15 × 4.18 kJ = 62.7 kJ. Thus 25 kJ of work is done on the system and 62.7 kJ of heat is removed. The net heat added to the system is $Q = -62.7$ kJ, and the work done by the system is $W = -25$ kJ. Equation 16-26 gives

$$Q = \Delta U + W$$

$$-62.7 \text{ kJ} = \Delta U + (-25 \text{ kJ})$$

$$\Delta U = -62.7 \text{ kJ} + 25 \text{ kJ} = -37.7 \text{ kJ}$$

The net change in the internal energy of the system is negative because more energy is removed from the system as heat than is added to the system by doing work on it.

(*b*) Since 4.18 kJ will change the temperature of 1 kg of water 1 C°, the change in temperature of the 3 kg of water will be

$$\Delta T = \frac{-37.7 \text{ kJ}}{(4.18 \text{ kJ/kg·C°})(3 \text{ kg})} = -3.01 \text{ C°}$$

The final temperature is therefore 80°C − 3.01 C° = 76.99°C ≈ 77°C.

It is important to understand that the internal energy U of a system is a property of the system that depends on the state of the system. That is, U is a function of the state of the system, just as P, V, and T are. As we have mentioned, the state of a gas is determined by any two properties such as P and V. Suppose that a gas is in some initial state (P_1, V_1). If we compress the gas or let it expand, add or remove heat from it, or do work on it or let it do work, the gas will move through a sequence of states; that is, it will have different values of the state functions P, V, T, and U. If the gas is then returned to its original state (P_1, V_1), the temperature T and the internal energy U must equal their original values.

On the other hand, the net heat input Q and the work W done by the gas are not functions of the state of the system. It is correct to say that a system has a large amount of internal energy, but it is not correct to say that a system has a large amount of heat or a large amount of work. In Example 16-10, for instance, it is not correct to say that the system contains 62.7 kJ less heat because we removed that much heat from the system. Nor can we say that the system contains 25 kJ more work because we did that much work on the system. What we can say is that after doing that much work on the system and removing that much heat, the internal energy of the system is 37.7 kJ less than it was originally. To increase the internal energy of a system by 2 J, we may add 2 J of heat, or we may add no heat but instead do 2 J of work on the system, or we may do 3 J of work on the system and remove 1 J of heat. In other words, heat is not something that is contained in a system. Rather, it is a measure of the energy that flows from one system to another because of a difference in temperature.

For very small amounts of heat added, work done, and changes in internal energy, it is customary to write Equation 16-26 as

$$dQ = dU + dW \tag{16-27}$$

In this equation, dU is the differential of the internal-energy function. However, neither dQ nor dW is a differential of any function. Instead, dQ merely represents a small amount of heat added to the system, and dW represents a small amount of work done by the system.

Questions

7. The experiment of Joule discussed in this section involved the conversion of mechanical energy into internal energy. Give some examples in which the internal energy of a system is converted into mechanical energy.
8. Can a system absorb heat with no change in its internal energy?

16-5 The Internal Energy of an Ideal Gas

In Chapter 15, we saw that, according to a simple molecular model of a gas, the temperature T of the gas is associated with the kinetic energy K of translation of the molecules of the gas by the relation

$$K = \tfrac{3}{2}nRT$$

where n is the number of moles of gas and R is the universal gas constant. If this translational energy is taken to be the total internal energy of the gas, the internal energy will depend only on the temperature of the gas and not on its volume or pressure. Writing U for K, we have

$$U = \tfrac{3}{2}nRT \tag{16-28}$$

If the internal energy of the gas includes other kinds of energy, its value will be different from that given by Equation 16-28, and it may or may not depend on the pressure and volume of the gas. Suppose, for example, that nearby gas molecules exert attractive forces on each other. Work is then required to increase the separation of the molecules. Thus, if the average distance between the molecules is increased, the potential energy associated with the molecular attraction will increase. The internal energy of the gas will then depend on the volume of the gas as well as on its temperature.

Joule, using an apparatus like that shown in Figure 16-10*a*, performed an interesting experiment to determine whether the internal energy of a gas depends on its volume. Initially, the compartment on the left contains a gas and the compartment on the right has been evacuated. The compartments are connected by a stopcock that is initially closed. The whole system is thermally insulated from its surroundings by rigid walls so that no heat can go into or out of the system and no work can be done. When the stopcock is opened, the gas rushes into the evacuated chamber. This process is called a **free expansion.** Eventually, the gas reaches thermal equilibrium with itself. Since no work has been done and no heat has been transferred, the final internal energy of the gas must equal its initial internal energy. If the gas molecules exert attractive forces on one another, the potential energy associated with these forces will increase when the volume increases. Since energy is conserved, the kinetic energy of translation must therefore decrease, which will result in a decrease in the temperature of the gas.

When Joule did this experiment, he found the final temperature to be equal to the initial temperature. Subsequent experiments verified this result for low gas densities. That is, the result held when the amount of gas originally in the left compartment was small. This implies that for a gas at low density—that is, for an ideal gas—the temperature depends only on the internal energy, or as we usually think of it, the internal energy depends only on the temperature. However, when the experiment is done with a large amount of gas initially in the left compartment so that the density is high, the temperature after the expansion is slightly lower than that before the expansion. This indicates that there is a small attraction between the gas molecules of a real gas at high density.

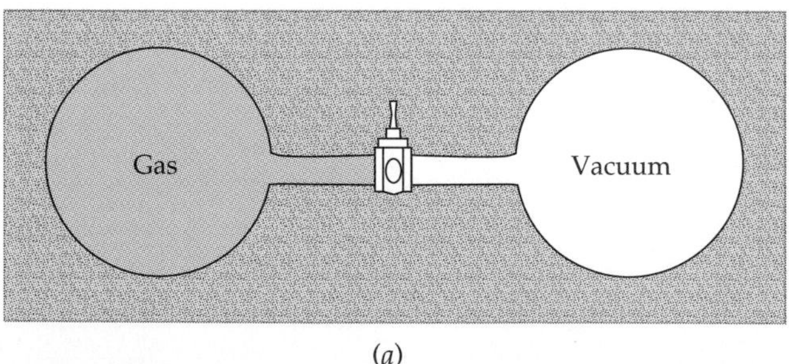

(*a*)

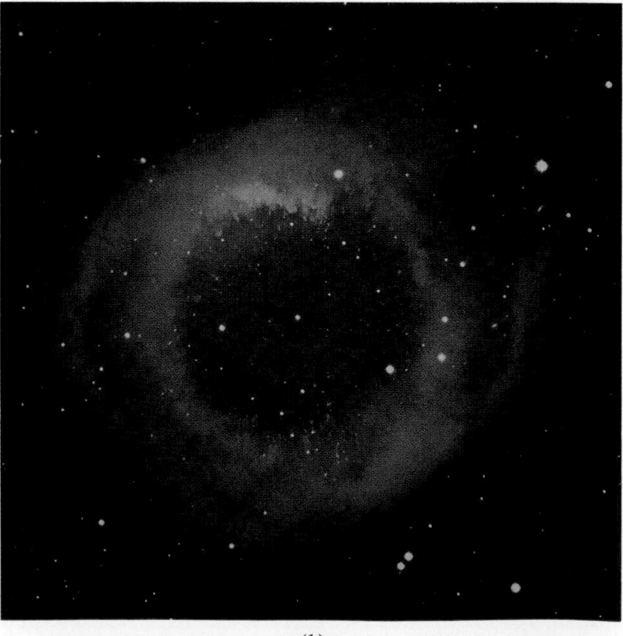

(*b*)

Figure 16-10 (*a*) Free expansion of a gas. When the stopcock is opened, the gas expands rapidly into the evacuated chamber. Since no work is done and the whole system is thermally insulated, the initial and final internal energies of the gas are equal. (*b*) The planetary nebula NGC 7293 (also known as the Helix nebula). This large-scale expansion of a hot gas occurs when an unstable star blows off its outer layers as its core shrinks to form a white dwarf star. In contrast to the free expansion in (*a*), in which the gas is insulated from its surroundings, the shell of hot gas in (*b*) is heated by ultraviolet radiation from the star.

16-6 Work and the *PV* Diagram for a Gas

In the rest of this chapter, we will look at some applications of the first law of thermodynamics. In this section, we will discuss the work done by a gas under various conditions. In our discussion, we will assume that we have an ideal gas because the equation of state for an ideal gas is simple, and it provides a fairly accurate description of real gases in many situations. Understanding the work done by a gas is important for an understanding of engines, which are used to convert thermal energy into usable work. In many engines, work is done by a gas expanding against a movable piston. For example, in a steam engine, water is heated in a boiler to produce steam. The steam then does work as it expands and drives the piston. In an automobile engine, a mixture of gasoline vapor and air is ignited, causing it to explode. The resulting high temperatures and pressures cause the gas to expand rapidly, driving a piston and doing work. Heat engines will be considered in more detail in Chapter 17 when we discuss the second law of thermodynamics.

Figure 16-11 shows a gas confined in a container with a tightly fitting piston that we will assume to be frictionless. When the piston moves, the volume of the gas changes. The temperature or pressure or both must also change since these three variables are related by the equation of state $PV = nRT$. If we suddenly push the piston in to compress the gas, the pressure will initially be greater near the piston than far from it. Until the gas settles down into thermal equilibrium with itself, we cannot determine such macroscopic variables as T, P, or U for the entire gas system. However, if we move the piston slowly in small steps and wait after each step for equilibrium to be reestablished, we can compress or expand the gas in such a way that the gas is never far from an equilibrium state. In this kind of process, called a **quasi-static process,** the gas moves through a series of equilibrium states. In practice, it is possible to approximate quasi-static processes fairly well.

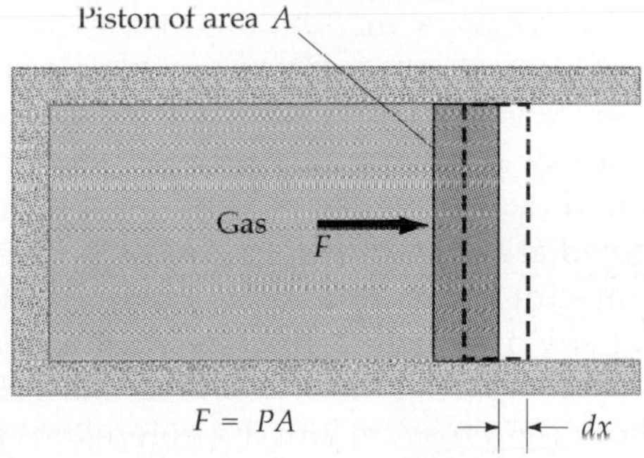

Figure 16-11 Gas confined in a thermally insulated cylinder with a movable piston. When the piston moves a distance dx, the volume of the gas changes by $dV = A\,dx$. The work done by the gas is $PA\,dx = P\,dV$, where P is the pressure.

Let us begin with a gas at a fairly high pressure, and let it expand quasi-statically. The force exerted by the gas on the piston is PA, where A is the area of the piston and P is the gas pressure, which usually changes as the gas expands. (We could keep the pressure constant by heating the gas as it expands.) We are assuming that the piston does not accelerate, so there must be an external force pushing against the piston that is also equal to PA. The piston does work on the agent providing this external force. If the piston moves a small distance dx, the work done by the gas on the piston is

$$dW = F\,dx = PA\,dx = P\,dV \qquad 16\text{-}29$$

where $dV = A\,dx$ is the increase in the volume of the gas. Equation 16-29 also holds for a compression. In this case, dV is negative, indicating that work is done on the gas. To calculate the work done by the gas during an expansion from a volume of V_1 to a volume of V_2, we need to know how the pressure varies during the expansion.

Since the state of a gas is determined by two variables such as P and V, we can represent the states of a gas on a diagram of P versus V. Each point on the PV diagram indicates a particular state of the gas. Since P and V are related to the absolute temperature T by the equation of state for an ideal gas, $PV = nRT$, there is also a temperature T associated with each point. There is also a value of U associated with each point on the PV diagram. Figure 16-12 shows a PV diagram with a horizontal line representing a series of states that all have the same value of P. This line represents an expansion at constant pressure. Such a process is called an **isobaric expansion.** For a volume change of ΔV, the work done is $P\,\Delta V$ which is equal to the shaded area under the curve in the figure. In general, the work done by the gas is equal to the area under the P-versus-V curve:

Work done by a gas

$$W = \int P\,dV = \text{area under the } P\text{-versus-}V \text{ curve} \tag{16-30}$$

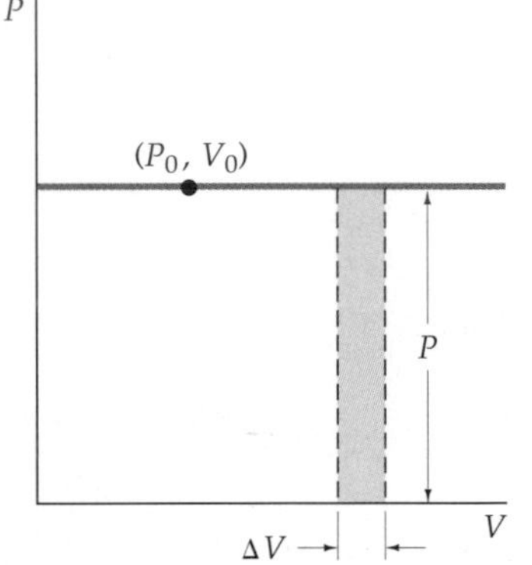

Figure 16-12 Each point on a PV diagram, such as (P_0, V_0), represents a particular state of the gas. The horizontal line represents states with a constant pressure P_0. The work done by a gas as it expands an amount ΔV is represented by the shaded area, $P_0\,\Delta V$.

Since pressures are often given in atmospheres and volumes are often given in liters, it is convenient to have a conversion factor between liter-atmospheres and joules:

$$1\ \text{L}\cdot\text{atm} = (10^{-3}\ \text{m}^3)(101.3 \times 10^3\ \text{N/m}^2) = 101.3\ \text{J} \tag{16-31}$$

Exercise

If 3 L of an ideal gas at a pressure of 2 atm is heated so that it expands at constant pressure until its volume is 5 L, what is the work done by the gas? (Answer: 405.2 J)

Figure 16-13 shows three different possible paths on a PV diagram for a gas that is originally in state (P_1, V_1) and is finally in state (P_2, V_2). We assume the gas is ideal and have chosen the original and final states to have the same temperature so that $P_1V_1 = P_2V_2 = nRT$. In Figure 16-13*a*, the gas is heated at constant pressure until its volume is V_2, after which it is cooled at constant volume until its pressure is P_2. The work done along path A is $P_1(V_2 - V_1)$ for the horizontal part of the path and zero for the constant-volume part. This work is indicated by the shaded area under the curve. In Figure 16-13*b*, the gas is first cooled at constant volume until its pressure is P_2, after which it is heated at constant pressure until its volume is V_2. The work done along path B is $P_2(V_2 - V_1)$, which is much less than that done along the path A as can be seen by comparing the shaded regions in Figures 16-13*a* and *b*. In Figure 16-13*c*, both the pressure and the volume vary along the path. Path C is one of constant temperature, PV = constant, so the process is called an **isothermal expansion.** The work done is again indicated by

Figure 16-13 Three paths on PV diagrams connecting an initial state (P_1, V_1) and a final state (P_2, V_2). The work done along each path is indicated by the shaded area.

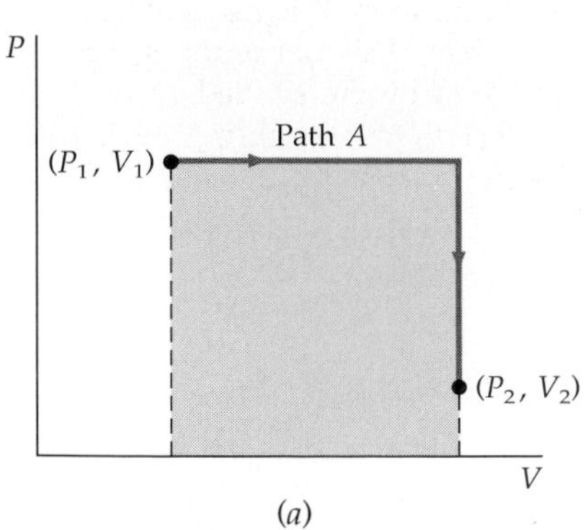

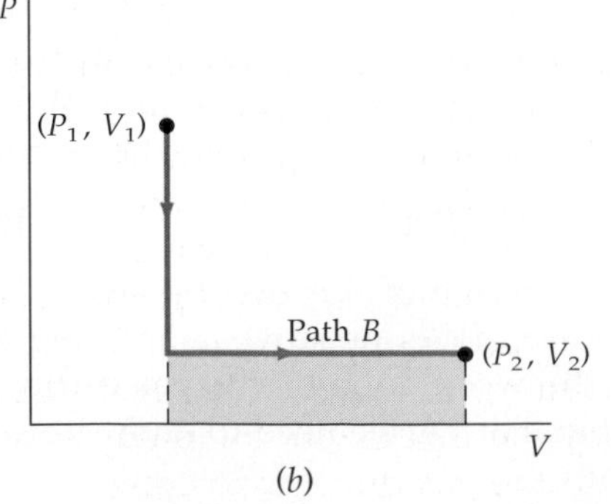

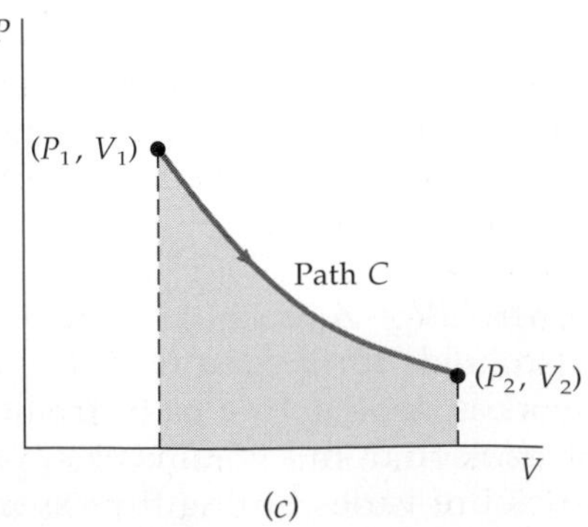

the shaded area. We can calculate this work using integral calculus. Using $P = nRT/V$, we have

$$dW = P\,dV = \frac{nRT}{V}\,dV$$

Hence, the work done by the gas as it expands isothermally from V_1 to V_2 is

$$W_{\text{isothermal}} = \int_{V_1}^{V_2} P\,dV = \int_{V_1}^{V_2} \frac{nRT}{V}\,dV$$

Since T is constant, we can remove it from the integral. We then have

$$W_{\text{isothermal}} = nRT \int_{V_1}^{V_2} \frac{dV}{V} = nRT \ln \frac{V_2}{V_1} \qquad 16\text{-}32$$

where the symbol "ln" stands for the natural logarithm, the logarithm to the base e.

In Figure 16-13, we can see that the amount of work done by the gas is different for each process illustrated. Since the change in the internal energy must be the same for each of the paths that begin at one state and end at another state, the net amount of heat added must also be different for each of the processes illustrated. In this case, the final internal energy equals the initial internal energy because we have an ideal gas and the initial and final temperatures are the same. Therefore, the net amount of heat added for each process equals the net work done by the gas. This discussion illustrates the fact that both the work done and the heat added depend on just how a system moves from one state to another, but the change in the internal energy of the system does not.

In Figure 16-14, an ideal gas undergoes a cyclic process from point A to point B to point C to point D and back to point A. (Such cyclic processes have important applications for heat engines, as we will see in the next chapter.) The gas begins at a volume of 1 L and a pressure of 2 atm and expands at constant pressure until the volume is doubled to 2 L, after which it is cooled at constant volume until its pressure is 1 atm. It is then compressed at constant pressure until its volume is again 1 L, after which it is heated at constant volume until it is back in its original state. We wish to find the total work done by the gas and the total heat added during this cycle.

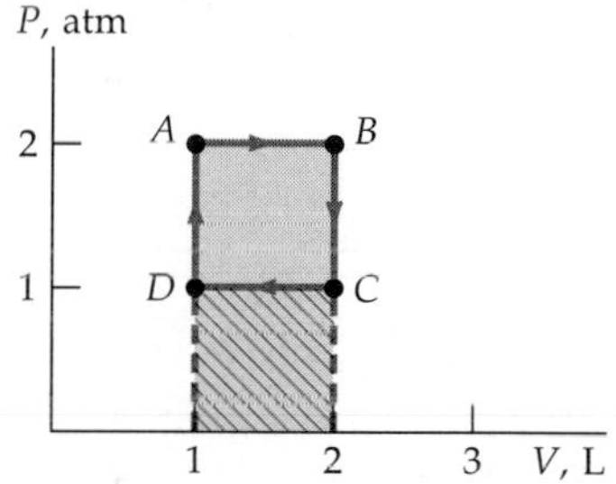

Figure 16-14 A *PV* diagram for a cyclic process in which a gas expands at constant pressure from state A to state B, is then cooled at constant volume from state B to state C, is then compressed at constant pressure from state C to state D, and is finally heated at constant volume back to state A. The work done by the gas as it expands from A to B minus that done on the gas as it is compressed from C to D is the net work done by the gas during the cycle. This net work is represented by the shaded area (without cross hatching) enclosed by the cycle.

As the gas expands from point A to point B at constant pressure, the work done by the gas is

$$W_{AB} = P(V_B - V_A) = (2\text{ atm})(2\text{ L} - 1\text{ L}) = 2\text{ L}\cdot\text{atm}$$

This work is equal to the shaded area under curve AB in Figure 16-14. Using Equation 16-31 to convert from liter-atmospheres to joules, we find this work equals 202.6 J. As the gas cools from point B to point C, the volume is constant, so no work is done. As the gas is compressed at constant pressure from point C to point D, it does negative work; that is, work must be done on the gas. The work done *by* the gas is

$$W_{CD} = P(V_D - V_C) = (1\text{ atm})(1\text{ L} - 2\text{ L}) = -1\text{ L}\cdot\text{atm} = -101.3\text{ J}$$

The magnitude of this work is equal to the area under curve CD in Figure 16-14, which is the shaded area with the cross hatching. As the gas is heated back to its original state A, the volume is again constant, so no work is done. The net work done by the gas during the entire cycle is therefore 202.6 J − 101.3 J = 101.3 J. This net work is represented in the figure by the enclosed, shaded area without cross hatching. Since the gas is back in its original state, all the state properties, P, V, T, and U, are at their original values. There has thus been no net change in the internal energy of the gas. However, 101.3 J of work has been done by the gas, so by the first law of thermodynamics, 101.3 J of heat must have been added to the gas during the cycle.

16-7 Heat Capacities and the Equipartition Theorem

The determination of the heat capacity of a substance provides information about its internal energy, which in turn provides information about its molecular structure. For all substances that expand when heated, the heat capacity at constant pressure C_p is greater than the heat capacity at constant volume C_v because of the work done by the substance as it expands at constant pressure. As was mentioned in Section 16-1, this work is usually negligible for solids and liquids, so for them C_p and C_v are approximately equal. However, for gases we must distinguish between C_p and C_v because a gas heated at constant pressure readily expands and does a significant amount of work.

When heat is added to a gas at constant volume, no work is done by or on the gas, so the heat added equals the increase in the internal energy of the gas. Writing Q_v for the heat added at constant volume, we have

$$Q_v = C_v\,\Delta T$$

Since $W = 0$, we have from the first law of thermodynamics

$$Q_v = \Delta U + W = \Delta U$$

Thus,

$$\Delta U = C_v\,\Delta T$$

Taking the limit as ΔT approaches zero, we obtain

$$dU = C_v\,dT \qquad \text{16-33}$$

and

$$C_v = \frac{dU}{dT} \qquad \text{16-34}$$

Thus, the heat capacity at constant volume is the rate of change of the internal energy with temperature.

If we add heat at constant pressure, the gas will expand and do work on its surroundings. Therefore, only part of the added heat will go into an increase in the internal energy of the gas. Since the absolute temperature of the gas T is a measure of its internal energy, more heat must be added at constant pressure than at constant volume to achieve the same increase in temperature. In other words, the heat capacity at constant pressure is greater than that at constant volume.

We will now calculate the difference $C_p - C_v$ for an ideal gas. If we use Q_p for the heat added at constant pressure, we have from the definition of C_p

$$Q_p = C_p\,\Delta T$$

From the first law of thermodynamics,

$$Q_p = \Delta U + W = \Delta U + P\,\Delta V$$

Therefore,

$$C_p\,\Delta T = \Delta U + P\,\Delta V$$

For infinitesimal changes, this becomes

$$C_p\,dT = dU + P\,dV$$

Using Equation 16-33 for dU, we obtain

$$C_p\,dT = C_v\,dT + P\,dV \qquad 16\text{-}35$$

The pressure, volume, and temperature for an ideal gas are related by

$$PV = nRT$$

Taking the differentials of both sides with $dP = 0$ for constant pressure, we obtain

$$P\,dV = nR\,dT$$

Substituting this into Equation 16-35, we obtain

$$C_p\,dT = C_v\,dT + nR\,dT$$

Therefore,

$$C_p = C_v + nR \qquad 16\text{-}36$$

For an ideal gas, the heat capacity at constant pressure is greater than that at constant volume by the amount nR.

Table 16-5 lists measured molar heat capacities C_{mp} and C_{mv} for several gases. We note from this table that the ideal gas prediction, $C_{mp} - C_{mv} = R$, holds quite well for all gases. We also note that C_{mv} is approximately $1.5R$ for all monatomic gases, $2.5R$ for all diatomic gases, and greater than $2.5R$ for gases consisting of more complex molecules. We can understand these results by considering the molecular model of a gas discussed in Section 15-5. In that section, we saw that the total translational kinetic energy of n moles of a gas is $K = \frac{3}{2}nRT$ (Equation 15-27). Thus, if the internal energy of a gas consists of translational kinetic energy only, we have

$$U = \tfrac{3}{2}nRT \qquad 16\text{-}37$$

Table 16-5 Molar Heat Capacities in J/mol·K of Various Gases at 25°C

Gas	C_{mp}	C_{mv}	$\frac{C_{mv}}{R}$	$C_{mp} - C_{mv}$	$\frac{C_{mp} - C_{mv}}{R}$
Monatomic					
He	20.79	12.52	1.51	8.27	0.99
Ne	20.79	12.68	1.52	8.11	0.98
Ar	20.79	12.45	1.50	8.34	1.00
Kr	20.79	12.45	1.50	8.34	1.00
Xe	20.79	12.52	1.51	8.27	0.99
Diatomic					
N_2	29.12	20.80	2.50	8.32	1.00
H_2	28.82	20.44	2.46	8.38	1.01
O_2	29.37	20.98	2.52	8.39	1.01
CO	29.04	20.74	2.49	8.30	1.00
Polyatomic					
CO_2	36.62	28.17	3.39	8.45	1.02
N_2O	36.90	28.39	3.41	8.51	1.02
H_2S	36.12	27.36	3.29	8.76	1.05

The heat capacities are then

$$C_v = \frac{dU}{dT} = \tfrac{3}{2}nR \qquad \text{16-38}$$

and

$$C_p = C_v + nR = \tfrac{5}{2}nR \qquad \text{16-39}$$

We can see that the experimental results in Table 16-5 agree well with these predictions for monatomic gases, but for other gases, the heat capacities are greater than those predicted by Equations 16-38 and 16-39. The internal energy for a gas consisting of diatomic or more complicated molecules is evidently greater than $\frac{3}{2}nRT$. The reason is that such molecules can have other types of energy, such as rotational or vibrational energy, in addition to translational kinetic energy.

In Section 15-5, we saw that the average kinetic energy associated with translational motion in the x direction is $\frac{1}{2}kT$ per molecule (Equation 15-25), where k is Boltzmann's constant, or, equivalently, $\frac{1}{2}RT$ per mole. Since x was an arbitrary direction, equivalent relations must also hold for the average kinetic energy of translation associated with the motion of a molecule in the y and z directions. If the energy of a molecule associated with its motion in the x direction is momentarily increased, say, by a collision between the molecule and a moving piston during a compression (Figure 16-15), collisions between that molecule and other molecules will quickly redistribute the additional energy so that, when the gas is again in equilibrium, the energy will be equally partitioned among the translational kinetic energies associated with motion in the x, y, and z directions. This sharing of the energy equally between the three terms in the translational kinetic energy is a special case of the **equipartition theorem,** a result that follows from statistical mechanics. Each coordinate, velocity component, angular velocity component, and so forth that appears in the expression for the energy of a molecule* is called a **degree of freedom.** Typical degrees of freedom are associated with the kinetic energy of translation, rotation, and vibration, and with the potential energy of vibration. The equipartition theorem states that

Equipartition theorem

When a substance is in equilibrium, there is an average energy of $\frac{1}{2}kT$ per molecule or $\frac{1}{2}RT$ per mole associated with each degree of freedom.

According to Table 16-5 nitrogen, oxygen, hydrogen, and carbon monoxide all have molar heat capacities at constant volume of about $\frac{5}{2}R$. The internal energy of n moles of any of these gases must therefore be

$$U = \tfrac{5}{2}nRT \qquad \text{16-40}$$

*Strictly speaking, the coordinate must appear as a squared term in the expression for energy.

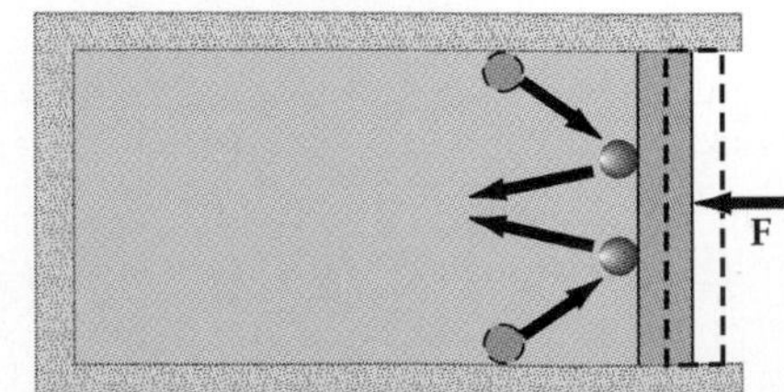

Figure 16-15 Molecules bouncing off a moving piston during a compression of the gas. The magnitude of the x component of momentum of each molecule is increased because of the collision. This results in an increase in the average kinetic energy associated with the translational motion of each molecule in the x direction, $\frac{1}{2}mv_x^2$. Collisions with other molecules quickly distribute this additional energy equally among other forms such as $\frac{1}{2}mv_y^2$ and $\frac{1}{2}mv_z^2$.

Thus, each of these gases has five degrees of freedom. About 1880, Clausius speculated that these gases must be diatomic gases that can rotate about two axes, which gives them two additional degrees of freedom. Figure 16-16 shows a rigid-dumbbell model of a diatomic molecule. The two degrees of freedom in addition to the three for translation are now known to be associated with their rotation about each of the two axes, x' and y', perpendicular to the line joining the atoms. The kinetic energy of a diatomic molecule is therefore

$$K = \tfrac{1}{2}mv_x^2 + \tfrac{1}{2}mv_y^2 + \tfrac{1}{2}mv_z^2 + \tfrac{1}{2}I_{x'}\omega_{x'}^2 + \tfrac{1}{2}I_{y'}\omega_{y'}^2$$

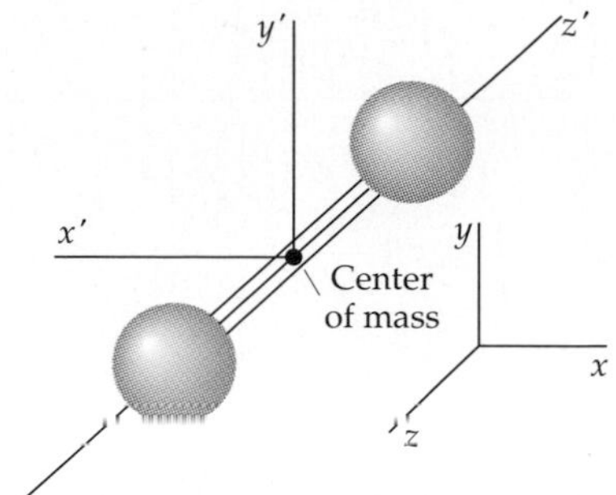

Figure 16-16 Rigid-dumbbell model of a diatomic gas molecule. If there is no rotation about the z' axis, such a molecule has five degrees of freedom, three associated with translation in the x, y, and z directions and two associated with rotation about the x' and y' axes.

According to the equipartition theorem, an average energy of $\frac{1}{2}RT$ per mole is associated with each term, which results in the total internal energy given by Equation 16-40. The heat capacity at constant volume for a diatomic gas is thus

$$C_v = \tfrac{5}{2}nR \qquad 16\text{-}41$$

In Section 16-1, we noted that all of the metals listed in Table 16-1 have approximately equal molar heat capacities. For instance, the molar heat capacity of copper is 24.5 J/mol·K, that for gold is 25.6 J/mol·K, and that for lead is 26.4 J/mol·K. Indeed, it is found experimentally that most solids have molar heat capacities approximately equal to $3R$:

$$C_m = 3R = 24.9 \text{ J/mol·K} \qquad 16\text{-}42$$

This result is known as the **Dulong–Petit law.** We can understand it by applying the equipartition theorem to the simple model for a solid shown in Figure 16-17. According to this model, a solid consists of a regular array of atoms in which each of the atoms has a fixed equilibrium position and is connected by springs to its neighbors. Each atom can vibrate in the x, y, and z directions. The total energy of an atom in a solid is thus

$$E = \tfrac{1}{2}mv_x^2 + \tfrac{1}{2}mv_y^2 + \tfrac{1}{2}mv_z^2 + \tfrac{1}{2}Kx^2 + \tfrac{1}{2}Ky^2 + \tfrac{1}{2}Kz^2$$

where K is the effective force constant of the hypothetical springs. Each atom thus has six degrees of freedom. The equipartition theorem states that a substance in equilibrium has an average energy of $\frac{1}{2}RT$ per mole for each degree of freedom. Thus, the internal energy of n moles of a solid is

$$U = 6 \times \tfrac{1}{2}nRT = 3nRT \qquad 16\text{-}43$$

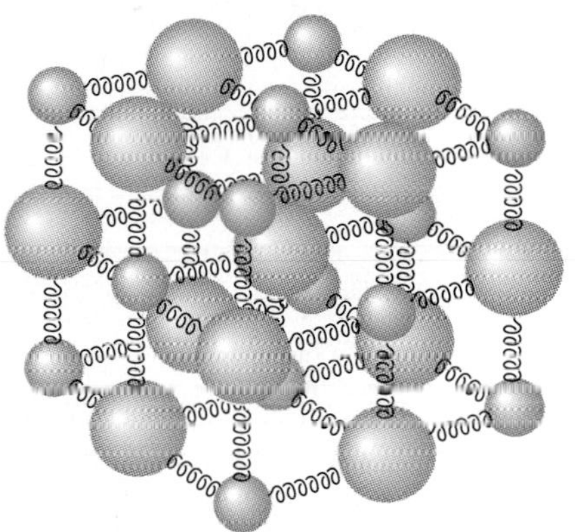
Figure 16-17 Model of a solid in which the atoms are connected to each other by springs. The internal energy of the solid consists of the kinetic and potential energies of vibration.

which means that the molar heat capacity is $3R$.

Although the equipartition theorem has had spectacular successes in explaining the measured heat capacities of gases and solids, it has had equally spectacular failures. If a diatomic gas molecule rotates about the line joining the atoms, there should be an additional degree of freedom. Similarly, if a diatomic molecule is not rigid, the two atoms should vibrate along the line joining them. We would then have two more degrees of freedom corresponding to kinetic and potential energies of vibration. According to the measured values of the molar heat capacities in Table 16-5, however, diatomic gases apparently do not rotate about the line joining them nor do they vibrate. The equipartition theorem gives no explanation for this, nor for the fact that monatomic atoms apparently do not rotate about any of the three possible perpendicular axes in space. Furthermore, the equipartition theorem predicts constant values for the heat capacities of gases, but careful measurement shows that these quantities are somewhat dependent on temperature. Finally, the equipartition theorem predicts a constant value of $3R$ for the heat capacity of solids. This result holds for many, though not all, solids at high temperatures, but it does not hold at very low temperatures.

The equipartition theorem fails because classical mechanics itself breaks down when applied to atomic and molecular systems and must be replaced by quantum mechanics. We will discuss this in more detail in Chapter 35. It is interesting to note that the successes of the equipartition theorem in explaining the measured heat capacities of gases and solids led to the first real understanding of molecular structure in the nineteenth century, whereas its failures played an important role in the development of quantum mechanics in the twentieth century.

Example 16-11

The molar mass of copper is 63.5 g/mol. Use the Dulong–Petit law to calculate the specific heat of copper.

According to the Dulong–Petit law, the molar heat capacity of a solid is

$$C_m = 3R = 3(8.31\ \text{J/mol·K}) = 24.9\ \text{J/mol·K}$$

The specific heat of copper should therefore be

$$c = \frac{C_m}{M} = \frac{24.9\ \text{J/mol·K}}{63.5\ \text{g/mol}} = 0.392\ \text{J/g·K} = 0.392\ \text{kJ/kg·K}$$

This agrees fairly closely with the measured value of 0.386 kJ/kg·K given in Table 16-1.

Exercise

The specific heat of a certain metal is measured to be 1.02 kJ/kg·K. (*a*) Calculate the molar mass of this metal, assuming that the metal obeys the Dulong–Petit law. (*b*) What is the metal? [Answers: (*a*) M = 24.4 g/mol; (*b*) the metal must be magnesium, which has a molar mass of 24.31 g/mol.]

Example 16-12

One mole of oxygen gas is heated from a temperature of 20°C and a pressure of 1 atm to a temperature of 100°C. Assume that oxygen is an ideal gas. (*a*) How much heat must be supplied if the volume is kept constant during the heating? (*b*) How much heat must be supplied if the pressure is kept constant? (*c*) How much work is done by the gas in part (*b*)?

(*a*) Using $n = 1$ mol and $R = 8.31$ J/K, the heat capacity of oxygen at constant volume is

$$C_v = \tfrac{5}{2}nR = \tfrac{5}{2}(1)(8.31\ \text{J/K}) = 20.8\ \text{J/K}$$

The heat that must be added to raise the temperature by 80 C° = 80 K is therefore

$$Q_v = C_v\,\Delta T = (20.8\ \text{J/K})(80\ \text{K}) = 1.66\ \text{kJ}$$

No work is done when the volume is kept constant, so the internal energy of the gas must increase by 1.66 kJ. Since the internal energy depends only on the temperature, this is the increase in the internal energy when the temperature of 1 mol of oxygen changes from 20°C to 100°C by any process.

(*b*) If we keep the pressure constant, the heat that must be added to raise the temperature by 80 K is

$$Q_p = C_p\,\Delta T = \tfrac{7}{2}(8.31\ \text{J/K})(80\ \text{K}) = 2.33\ \text{kJ}$$

where we have used $C_p = C_v + R = \tfrac{7}{2}R$ for 1 mol.

(*c*) Since the increase in internal energy is again 1.66 kJ, the work done by the gas is

$$W = Q - \Delta U$$
$$= 2.33 \text{ kJ} - 1.66 \text{ kJ}$$
$$= 0.67 \text{ kJ}$$

We can check this result by calculating directly the work done by the gas when the heat is added at constant pressure. Since the volume of 1 mol of any gas is 22.4 L at the standard conditions of 1 atm and 0°C = 273 K, and since the volume is proportional to the absolute temperature if the pressure is constant, the initial volume V_1 at 20°C = 293 K is

$$V_1 = (22.4 \text{ L})\frac{293 \text{ K}}{273 \text{ K}} = 24.0 \text{ L}$$

When the temperature is raised by 80 K to 373 K at constant pressure, the volume increases to V_2, which is given by

$$V_2 = (22.4 \text{ L})\frac{373 \text{ K}}{273 \text{ K}} = 30.6 \text{ L}$$

The work done is thus

$$W = P\,\Delta V = (1 \text{ atm})(30.6 \text{ L} - 24.0 \text{ L})$$
$$= 6.6 \text{ L·atm}$$

Converting this to joules, we obtain

$$W = (6.6 \text{ L·atm})\frac{101.3 \text{ J}}{1 \text{ L·atm}} = 0.67 \text{ kJ}$$

which agrees with our previous result.

16-8 The Quasi-Static Adiabatic Expansion of a Gas

A process in which no heat flows into or out of a system is called an **adiabatic process.** In this section, we will consider the quasi-static adiabatic expansion of a gas in which the gas in a thermally insulated container expands slowly against a piston, doing work on it. Since no heat enters or leaves the gas, the work done by the gas equals the decrease in the internal energy of the gas, and the temperature of the gas decreases. The curve representing this process on a *PV* diagram is shown in Figure 16-18. This curve is steeper than that for an isothermal expansion, as shown. That is, when the volume increases by some amount, the pressure decreases by more in an adiabatic expansion than it does in an isothermal expansion. This is to be expected since the temperature decreases during an adiabatic expansion.

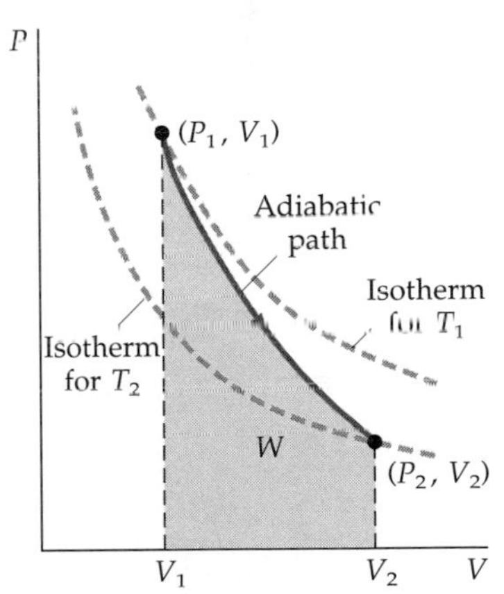

Figure 16-18 Quasi-static adiabatic expansion of an ideal gas. The dashed lines are the isotherms for the initial and final temperatures. The curve connecting the initial and final states of the adiabatic expansion is steeper than the isotherms because the temperature drops during the expansion.

We can find the equation for the adiabatic curve for an ideal gas using the equation of state and the first law of thermodynamics. We have

$$dQ = dU + dW = C_V\,dT + P\,dV = 0 \qquad 16\text{-}44$$

where we have used $dU = C_V\,dT$ from Equation 16-34. Then using $P = nRT/V$, we obtain,

$$C_V\,dT + nRT\frac{dV}{V} = 0$$

Rearranging slightly, we obtain

$$\frac{dT}{T} + \frac{nR}{C_V}\frac{dV}{V} = 0 \qquad 16\text{-}45$$

Equation 16-45 can be further simplified by noting that $C_P - C_V = nR$, so

$$\frac{nR}{C_V} = \frac{C_P - C_V}{C_V} = \frac{C_P}{C_V} - 1 = \gamma - 1$$

where γ is the ratio of the heat capacities:

$$\gamma = \frac{C_P}{C_V} \qquad 16\text{-}46$$

We then have

$$\frac{dT}{T} + (\gamma - 1)\frac{dV}{V} = 0$$

Integrating, we obtain

$$\ln T + (\gamma - 1) \ln V = \text{constant}$$

Using the properties of logarithms (Appendix A), we obtain

$$\ln (TV^{\gamma-1}) = \text{constant}$$

Then

$$TV^{\gamma-1} = \text{constant} \qquad 16\text{-}47$$

(The constants in the two preceding equations are not the same.) We can eliminate T from Equation 16-47 using $PV = nRT$. We then have

$$\frac{PV}{nR}V^{\gamma-1} = \text{constant}$$

or

Quasi-static adiabatic expansion

$$PV^{\gamma} = \text{constant} \qquad 16\text{-}48$$

Equation 16-48 relates P and V for a quasi-static adiabatic expansion. It also applies to a quasi-static adiabatic compression, in which the piston does work on the gas.

We can use Equation 16-48 to calculate the adiabatic bulk modulus for an ideal gas, which is related to the speed of sound waves in air. Differentiating Equation 16-48, we obtain

$$P\gamma V^{\gamma-1}\, dV + V^{\gamma}\, dP = 0$$

$$dP = -\frac{\gamma P\, dV}{V}$$

Recalling that the bulk modulus is the ratio of the change in pressure to the fractional change in volume, we have for the adiabatic bulk modulus

$$B_{\text{adiabatic}} = -\frac{dP}{dV/V} = \gamma P \qquad 16\text{-}49$$

The work done by the gas in the quasi-static adiabatic expansion in the PV diagram of Figure 16-18 is equal to the area under the curve. It is simply related to the change in the temperature of the gas. Setting $dQ = 0$ in Equation 16-44, we have

$$P\, dV = -C_V\, dT$$

Then

$$W_{\text{adiabatic}} = \int P\,dV = -\int C_v\,dT$$

or

$$W_{\text{adiabatic}} = -C_v\,\Delta T \qquad 16\text{-}50$$

Adiabatic work

where we have assumed that C_v is constant. (For an ideal gas, U is proportional to the absolute temperature, and therefore $C_v = dU/dT$ is a constant.) We note from Equation 16-50 that the work done by the gas depends only on the change in the absolute temperature of the gas. In a quasi-static adiabatic expansion, the gas does work, and its internal energy and temperature decrease. In a quasi-static adiabatic *compression*, work is done *on* the gas, and the internal energy and temperature increase.

We can use the ideal gas law to write Equation 16-50 in terms of the initial and final values of the pressure and volume. If T_1 is the initial temperature and T_2 is the final temperature, we have for the work done

$$W_{\text{adiabatic}} = -C_v\,\Delta T = -C_v(T_2 - T_1) = C_v(T_1 - T_2)$$

Using $PV = nRT$, we obtain

$$W_{\text{adiabatic}} = C_v\left(\frac{P_1V_1}{nR} - \frac{P_2V_2}{nR}\right) = \frac{C_v}{C_p - C_v}(P_1V_1 - P_2V_2)$$

where we have used $nR = C_p - C_v$. Dividing the numerator and denominator by C_v and writing γ for C_p/C_v, we obtain

$$W_{\text{adiabatic}} = \frac{P_1V_1 - P_2V_2}{\gamma - 1} \qquad 16\text{-}51$$

Clouds are formed when rising moist air cools below the dew point due to adiabatic expansion of the air.

Example 16-13

A quantity of air ($\gamma = 1.4$) expands adiabatically and quasi-statically from an initial pressure of 2 atm and volume of 2 L at temperature 20°C to twice its original volume. Find (*a*) the final pressure, (*b*) the final temperature, and (*c*) the work done by the gas.

(*a*) According to Equation 16-48, the quantity PV^γ remains unchanged during a quasi-static adiabatic expansion. Thus, if P_1 and V_1 are the initial pressure and volume and P_2 and V_2 are the final pressure and volume, we have

$$P_1V_1^\gamma = P_2V_2^\gamma \qquad \text{or} \qquad P_2 = P_1\left(\frac{V_1}{V_2}\right)^\gamma$$

For the given values we have

$$P_2 = (2\text{ atm})\left(\frac{2\text{ L}}{4\text{ L}}\right)^{1.4} = 0.758\text{ atm}$$

(*b*) The change in temperature is found most easily from Equation 16-47, which implies that

$$T_1V_1^{\gamma-1} = T_2V_2^{\gamma-1}$$

or

$$T_2 = T_1\left(\frac{V_1}{V_2}\right)^{\gamma-1} = (293\text{ K})\left(\frac{2\text{ L}}{4\text{ L}}\right)^{0.4} = 222\text{ K} = -51°\text{C}$$

(*c*) The work done by the gas is most easily found using Equation 16-51:

$$W_{\text{adiabatic}} = \frac{P_1V_1 - P_2V_2}{\gamma - 1}$$

$$= \frac{(2\text{ atm})(2\text{ L}) - (0.758\text{ atm})(4\text{ L})}{1.4 - 1}$$

$$= 2.42\text{ L}\cdot\text{atm}$$

Converting to joules, we obtain

$$W_{\text{adiabatic}} = 2.42\text{ L}\cdot\text{atm} \times \frac{101.3\text{ J}}{1\text{ L}\cdot\text{atm}}$$

$$= 245\text{ J}$$

Summary

1. Heat is energy that is transferred from one object to another because of a temperature difference. The heat capacity of a substance is the heat energy needed to raise the temperature of the substance by one degree. The specific heat is the heat capacity per unit mass. The calorie, originally defined as the heat necessary to raise the temperature of one gram of water by one Celsius degree, is now defined to be 4.184 joules. The specific heat of water is 4.184 kJ/kg·K.

2. The heat needed to melt a substance is the product of the mass of the substance and its latent heat of fusion L_f:

   $$Q = mL_f$$

   The heat needed to vaporize a liquid is the product of the mass of the liquid and its latent heat of vaporization L_v:

   $$Q = mL_v$$

   Both melting and vaporization occur at a constant temperature. For water, $L_f = 333.5$ kJ/kg and $L_v = 2257$ kJ/kg. The heat required to melt 1 g of ice or to vaporize 1 g of water is large compared with the heat required to raise the temperature of 1 g of water by one degree.

3. The three mechanisms by which thermal energy is transferred are conduction, convection, and radiation.

4. The rate of conduction of thermal energy is given by

   $$I = \frac{\Delta Q}{\Delta t} = kA\,\frac{\Delta T}{\Delta x}$$

   where I is the thermal current and k is the coefficient of thermal conductivity. This equation can be written

   $$\Delta T = IR$$

   where R is the thermal resistance:

   $$R = \frac{\Delta x}{kA}$$

The thermal resistance for a unit area of a slab of material is called the R factor R_f:

$$R_f = RA = \frac{\Delta x}{k}$$

The equivalent thermal resistance of a series of thermal resistances equals the sum of the individual resistances:

$$R_{eq} = R_1 + R_2 + \cdots \qquad \text{resistances in series}$$

The equivalent thermal resistance for thermal resistances in parallel is given by

$$\frac{1}{R_{eq}} = \frac{1}{R_1} + \frac{1}{R_2} + \cdots \qquad \text{resistances in parallel}$$

5. The rate at which an object radiates thermal energy is given by

$$I = e\sigma AT^4$$

where $\sigma = 5.6703 \times 10^{-8}\ \mathrm{W/m^2{\cdot}K^4}$ is Stefan's constant, and e is the emissivity, which varies between 0 and 1 depending on the composition of the surface of the object. Materials that are good heat absorbers are good heat radiators. A blackbody has an emissivity of 1. It is a perfect radiator, and it absorbs all the radiation incident upon it. The net thermal power radiated by an object at a temperature T in an environment at a temperature T_0 is given by

$$I_{net} = e\sigma A(T^4 - T_0^4)$$

The spectrum of electromagnetic energy radiated by a blackbody has a maximum at a wavelength λ_{max}, which varies inversely with the absolute temperature of the body:

$$\lambda_{max} = \frac{2.898\ \mathrm{mm{\cdot}K}}{T}$$

This is known as Wien's displacement law.

6. For all mechanisms of heat transfer, if the temperature difference between the body and its surroundings is small, the rate of cooling of a body is approximately proportional to the temperature difference. This result is known as Newton's law of cooling.

7. The first law of thermodynamics is a statement of the conservation of energy. It says that the net heat added to a system equals the change in the internal energy of the system plus the work done by the system:

$$Q = \Delta U + W$$

The internal energy of a system is a property of the state of the system, as are the pressure, volume, and temperature, but heat and work are not.

8. The internal energy of an ideal gas depends only on the absolute temperature T.

9. A quasi-static process is one that occurs slowly so that the system moves through a series of equilibrium states. A process is isobaric if the pressure remains constant, isothermal if the temperature remains constant, and adiabatic if no heat is transferred. For a quasi-static adiabatic expansion of an ideal gas, the pressure and volume are related by

$$PV^{\gamma} = \text{constant}$$

where γ is the ratio of the heat capacity at constant pressure to that at constant volume:

$$\gamma = \frac{C_p}{C_v}$$

10. When a system expands quasi-statically, the work done by the system is given by

$$W = \int P\,dV$$

The work done by a gas can be represented graphically as the area under the P-versus-V curve. It can be calculated if P is known as a function of V for the expansion. For an isothermal expansion of an ideal gas, the work done by the gas is

$$W_{\text{isothermal}} = nRT \ln \frac{V_2}{V_1}$$

For an adiabatic expansion of an ideal gas, the work done by the gas is

$$W_{\text{adiabatic}} = -C_v\,\Delta T = \frac{P_1V_1 - P_2V_2}{\gamma - 1}$$

11. The heat capacity at constant volume is related to the change in internal energy by

$$C_v = \frac{dU}{dT}$$

For an ideal gas, the heat capacity at constant pressure is greater than that at constant volume by the amount nR:

$$C_p = C_v + nR$$

The heat capacity at constant pressure is greater because a gas that is heated at constant pressure expands and does work, so it takes a greater amount of heat to achieve the same change in temperature. The heat capacity at constant volume for a monatomic gas is

$$C_v = \tfrac{3}{2}nR \qquad \text{monatomic gas}$$

For a diatomic gas, it is

$$C_v = \tfrac{5}{2}nR \qquad \text{diatomic gas}$$

12. The equipartition theorem states that when a system is in equilibrium, there is an average energy of $\frac{1}{2}kT$ per molecule or $\frac{1}{2}RT$ per mole associated with each degree of freedom. Monatomic gases have three degrees of freedom, which are associated with the kinetic energy of translation in three dimensions. Diatomic gases have two additional degrees of freedom, which are associated with their rotation about axes perpendicular to the line joining the atoms.

13. The molar heat capacity of most solids is $3R$, a result known as the Dulong–Petit law. This result can be understood by applying the equipartition theorem to a model of a solid in which each atom in the solid can vibrate in three dimensions and therefore has a total of six degrees of freedom, three associated with the kinetic energy of vibration and three with the potential energy of vibration.

The Earth's Energy Balance and Global Warming

Jerrold H. Krenz
University of Colorado

Humans have inhabited this planet for millions of years, but it is only during the past hundred years or so that—as a consequence of both population growth and industrialization—we have begun to affect our climate. The global temperature has risen by 0.5 K since 1900, the ocean levels have risen, inland glaciers are receding, polar ice caps are shrinking, and the depth to Alaskan permafrost has decreased. The combustion of fossil fuels and deforestation combine to produce an increase in atmospheric levels of carbon dioxide (CO_2); combined with other gaseous products of industrialization (for example, nitrous oxide, chlorofluorocarbons, and ozone) this promotes global warming, due to a process commonly called the "greenhouse effect."

There is still debate as to whether current climatological models prove conclusively that increased global warming has already occurred. There is, however, a consensus in the scientific community that a continuation of present human activities will result in a significant additional warming of the earth (1.5 to 5.5 K) over the next 50–100 years. Global warming of this magnitude would have profound and long-lasting effects on agriculture, wildlife, and human society.

Dr. Krenz did his undergraduate work at the State University of New York at Buffalo and his graduate work in electrical engineering (MS and PhD) at Stanford University. Before joining the faculty of the University of Colorado–Boulder, he worked several years at the General Electric Microwave Research Laboratory in Palo Alto, California.

At the University of Colorado his teaching interests have included the interaction of technology and society. During the 1970s, in response to the "energy crisis," he offered a senior-level engineering course on energy which dealt with both the potentialities and limitations of various existing and proposed energy systems. He has published several books, including two that address energy issues and include discussions of the "greenhouse effect."

In addition to trying to understand how societies use energy, Dr. Krenz has been deeply concerned with the interaction of technological developments and the nuclear arms race.

The Earth's Temperature

If the earth merely absorbed solar radiation, its temperature would continually increase. However, the earth also radiates energy into space. Through this process of absorption and radiation, an equilibrium condition is maintained that is generally referred to as the earth's energy balance. It is the *rate* at which energy arrives and leaves the earth that is balanced so, to be precise, it is actually a power balance.

To appreciate the process by which the earth is being warmed, it is necessary to understand the mechanism by which the temperature of the earth is maintained. At the earth's average distance from the sun, radiant energy from the sun arrives at a rate of 1353 W/m^2 (the solar constant, S). Since the surface temperature of the sun is approximately 6000 K, its radiation spectrum consists of very short wavelengths centered around approximately 0.5 μm. This is observed experimentally; and it also follows from Wien's displacement law if the sun is approximated as a blackbody (Section 16-3).

The solar radiation absorbed by the earth depends on the cross-sectional area of the earth as viewed from the sun, πR_E^2 (where R_E is the radius of the earth, 6.4×10^6 m). Not all the solar radiation intercepted by the earth is absorbed; a portion ($r \approx 0.3$) is reflected (this is the "earthlight" that astronauts see from space). We may therefore write the following equation:

$$\text{Power absorbed by the earth} = (1 - r)\pi R_E^2 S \text{ W} \quad (1)$$

The earth reradiates some of this absorbed power back into space. Although the average surface temperature of the earth is about 13°C (286 K), the average radiative temperature of the earth's atmosphere is only about −22°C (251 K). A relative emissivity constant e is used to account for this difference in the blackbody radiation expression ($e \approx 0.6$). The power radiated by each square meter of the earth's surface is described by the following equation:

$$\text{Radiated power density} = e\sigma T_E^4 \quad \text{W/m}^2 \quad (2)$$

where σ is the Stefan radiation constant (Equation 16-22), and T_E is the surface temperature of the earth. The earth's surface is largely covered with water with a relatively stable temperature; this means that T_E varies little from day to night. Because of this, and due to the fact that solar energy is *absorbed* only by the portion of the earth's surface

Continued

that is facing the sun but is *radiated* from the entire surface of the globe, the radiative behavior of the earth may be approximated by a spherical radiator with a uniform temperature:

$$\text{Power radiated by the earth} = 4\pi R_E^2 e\sigma T_E^4 \quad \text{W} \quad (3)$$

For an equilibrium condition, the power absorbed and radiated by the earth must be equal:

$$(1 - r)\pi R_E^2 S = 4\pi R_E^2 e\sigma T_E^4 \quad \text{W} \quad (4)$$

This equation, for the approximations introduced, is the energy or power balance of the earth. Dividing by the surface area of the earth, $4\pi R_E^2$ we have

$$\frac{(1 - r)S}{4} = e\sigma T_E^4 \quad \text{W/m}^2 \quad (5)$$

which is the average rate at which energy is absorbed and subsequently radiated by each square meter of the earth's surface ($\approx$237 W/m^2, about the strength of four 60-W light bulbs for every square meter of the earth's surface).

Other energy inputs (see Table 1) have only a slight effect on the earth's energy/power balance. Ignoring these, we may rewrite Equation 5 to solve for temperature of the earth, T_E:

$$T_E = \left[\frac{(1 - r)S}{4e\sigma}\right]^{1/4}$$

A decrease in either the reflectivity of the earth, r, or its relative emissivity, e, will result in an increase in its average temperature. It is *primarily* the properties of the gases that make up the earth's atmosphere that affect these quantities.

Some Atmospheric Gases Absorb Thermal Radiation

The atmosphere is fairly transparent to *incoming* solar radiation from the sun. Although the principal gases of the atmosphere, oxygen (O_2) and nitrogen (N_2), are transparent to the long-wavelength thermal radiation from the earth, this is *not* the case for all gases in the atmosphere. Solar radiation can therefore penetrate the earth's atmosphere, but a portion is trapped when it is reradiated as longer wavelength thermal radiation, resulting in a warming of the earth (Figure 1).

You are probably familiar with the terms "greenhouse effect" to describe global warming and "greenhouse gases" to characterize the gases that contribute to global warming by absorbing thermal radiation. It is interesting to note that these names are somewhat inappropriate as greenhouses operate by another mechanism altogether (by the reduction of convective cooling; see Section 16-3).

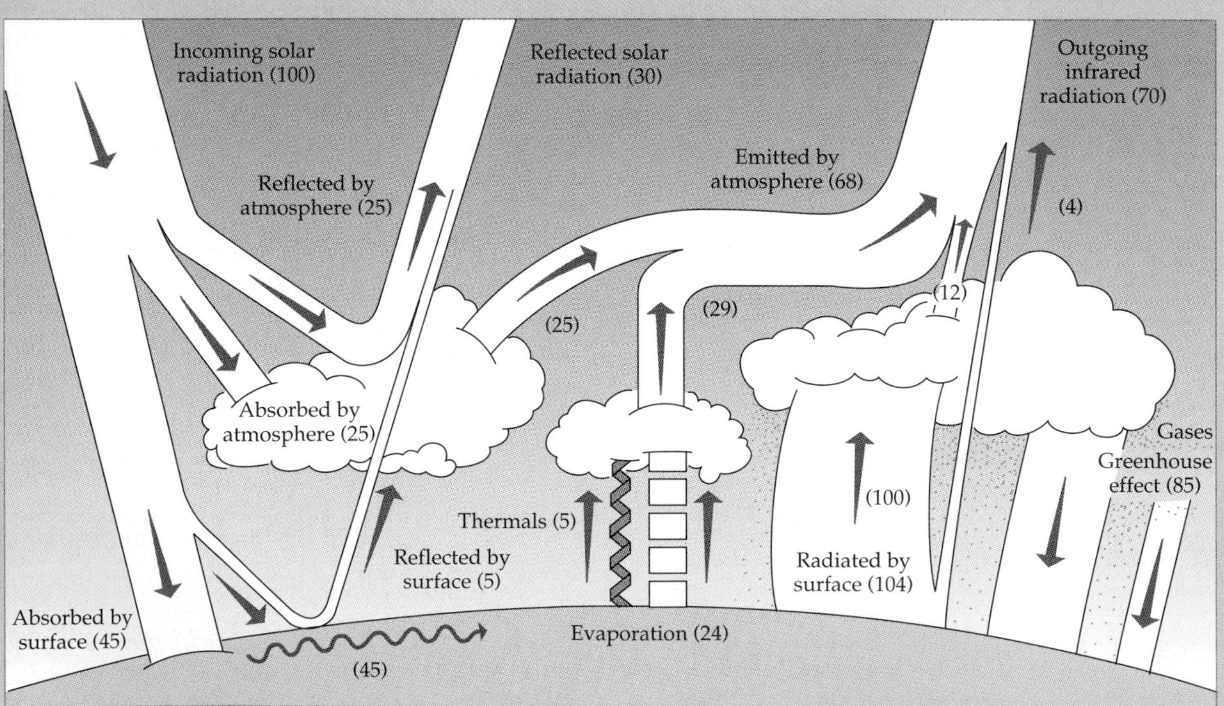

Figure 1 The earth's energy/power balance, including the effect of the atmospheric system. The quantities indicated are relative to the earth-averaged solar constant $S/4$. From Stephen H. Schneider, "The Greenhouse Effect: Science and Policy," *Science*, vol. 243, pp. 771–781. (Reprinted with the permission of the American Association for the Advancement of Science.)

The most prevalent gases that absorb thermal radiation are water vapor and carbon dioxide. Without the *benefit* of the thermal effects of water vapor and CO_2, the earth's temperature would actually be insufficient for most existing life forms. The relationship between the "greenhouse gases" is complex. As their total levels build up and the global temperature increases, the rate of water evaporation from the oceans will be enhanced. While this would appear to create a positive feedback loop by boosting the atmospheric concentrations of water vapor, it is also likely to influence the earth's cloud cover. Clouds have an as-yet poorly understood role to play in the overall equation: They increase the earth's reflectivity, thereby reducing global warming; at the same time, they reduce the rate at which thermal energy can be radiated into space thereby enhancing global warming. The effect of clouds may be further complicated by changing seasonal conditions.

Table 1 **Average Surface Contribution to the Earth's Energy Balance from Several Minor Sources of Thermal Energy**

Source	Contribution (W/m^2)
Radioisotope decay	0.06
Fuel consumption	0.010
Tidal friction	0.005

Carbon Dioxide

Contemporary levels of atmospheric CO_2 are being influenced by the combustion of fossil fuels and by the net result of global biomass accumulation or destruction. Since 1957, accurate measurements of atmospheric CO_2 have been made at Mauna Loa Observatory, Hawaii (Figure 2). Overshadowing the seasonal fluctuations, which are due primarily to seasonal variations in the total levels of plants' photosynthetic activity, there can be seen a steady and continuing increase in the concentration of atmospheric CO_2. This now stands at 350 parts per million by volume (ppmv). Based on analyses of air trapped long ago in glacial ice, the concentration of atmospheric CO_2 in 1750 was about 280 ppmv. There has therefore been a 25 percent increase since the beginning of modern industrialization.

Carbon dioxide emissions are a direct consequence of the energy conversion process, the oxidation of the carbon of the fuel. Consider the case of coal, which is primarily carbon:

$$C + O_2 \rightarrow CO_2$$

Combustion of hydrocarbon fuels, petroleum, and natural gas produces a similar effect, also creating water vapor (another "greenhouse gas") by the oxidation of the hydrogen of the fuel.

Every year approximately 5×10^{12} kg of carbon in the form of CO_2 is released to the atmosphere as a result of global consumption of fossil fuels. In 1988, on average each of us in the United States consumed about 3.6 tons of coal, 73,000 cubic feet of natural gas, and 1070 gallons of petroleum. With less than 5 percent of the world's population, the United States accounts for approximately 33 percent of the world's fossil fuel consumption. The majority of the world's inhabitants have low levels of economic activity and concurrent low energy-usage rates but have aspirations for higher standards of living. An increase in global fossil fuel usage (and hence CO_2 emissions) is therefore considered inevitable. Although increased energy efficiencies and the use of alternative energy sources may moderate the growth in fossil fuel usage, atmospheric concentrations of CO_2 of 440 to 550 ppmv are nevertheless anticipated by the year 2100. As the level of CO_2 dissolved in the oceans increases and as deforestation and acid rain take their toll, the capacity of the earth to absorb increased atmospheric CO_2 levels diminishes.

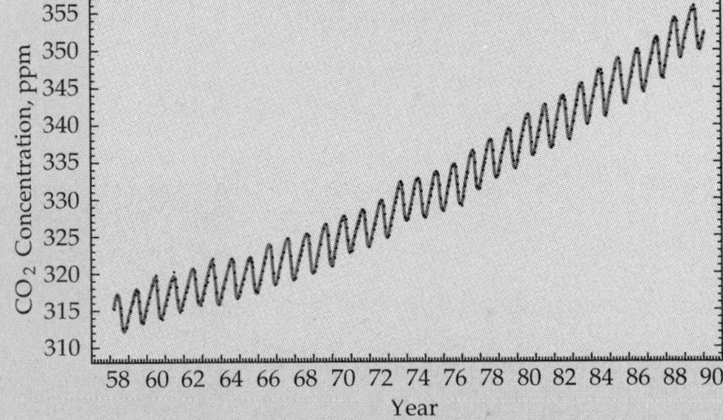

Figure 2 Concentration of atmospheric carbon dioxide at Mauna Loa Observatory, Hawaii. (Courtesy John Chin.)

Continued

Other Greenhouse Gases

Although present in trace quantities only, other gases (methane, ozone, nitrous oxide, and chlorofluorocarbons) also absorb long-wavelength thermal radiation. The concentration of methane (CH_4) in the atmosphere is less than 1 percent that of CO_2, however, each molecule of CH_4 is about twenty times as effective as a molecule of CO_2 at absorbing radiant heat. Moreover, its concentration has been increasing at a rate of 1 percent per year and has doubled over the past 250 years. A positive feedback loop could develop because increases in global temperature enhance net rates of organic decay (for example, in swamps), thereby accelerating the overall increase in atmospheric concentrations of CH_4.

Ozone (O_3) is formed as a result of photochemical processes primarily involving sunlight, methane, carbon monoxide, and nitrogen oxides (typical pollutants). The concentration of O_3 in the troposphere has increased by 10 percent (notwithstanding its *decreased* concentration in the stratosphere over the poles). A similar increase (5–10 percent) has occurred for nitrous oxide; this has arisen primarily due to the use of nitrogen-based fertilizers, deforestation, and the burning of biomass—all consequences of cultivation of new land areas.

Chlorofluorocarbons (CFCs) are not naturally present in the atmosphere. These compounds are manufactured for use as solvents, refrigerants, and spray-can propellants. They act as CO_2 does in trapping radiant energy, in addition to destroying ozone in the stratosphere. While the manufacture of CFCs is being restricted as a result of an international agreement, their combined atmospheric concentration is nevertheless expected to reach levels of 1 ppbv, owing to their long lifetimes.

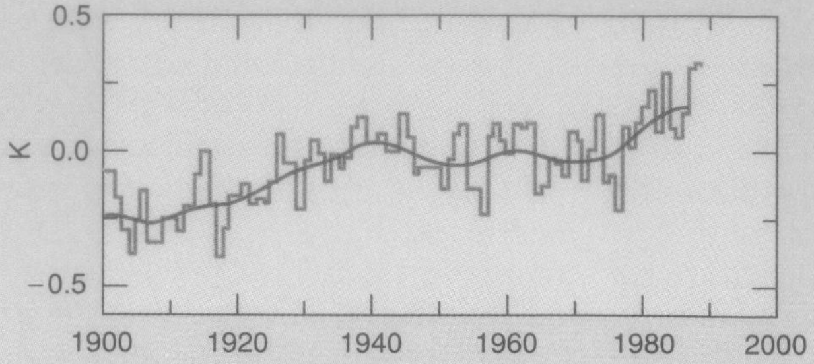

Figure 3 Global surface air temperature relative to 1950–1979 average. The data from the University of East Anglia and the UK Meteorological Office indicate a global warming of about 0.5 K this century. From Richard A. Kerr, "1988 Ties for Warmest Year," *Science*, vol. 243, p. 891. (Reprinted with the permission of the American Association for the Advancement of Science.)

Global Warming

The average global temperature has already increased by about 0.5 K since 1900 (Figure 3). The six warmest recorded years to date this century have been since 1980. More data are necessary before it will be possible to state without equivocation that a climatic change is occurring and that increases in concentrations of "greenhouse gases" are indeed the cause of the change. However, there is no doubt that their continued increase would certainly lead at some point to this effect. Deferring our response to this until analyses are conclusive will mean that the damage is more pronounced and more difficult to control or undo.

The net effect that global warming will have on a particular region is uncertain. Global precipitation patterns are expected to change dramatically, with consequent disruption of agriculture, trade, and the global economy. Thermal expansion of the ocean due to overall global warming and partial melting of land-based ice are expected to increase sea levels. Low-lying coastal and inland areas, where a large portion of the world's population lives, would be threatened with severe social, economic, and environmental repercussions.

The difference between this and previous climatic shifts is that these changes are a direct result of our own activities, not the effects of natural occurrences over which we have no control. By changing our activities we can moderate and perhaps even prevent the anticipated climatic changes. Given the economic costs, we must decide what should be done, taking into account both the present levels of uncertainty and the scale of the possible consequences of our choices. Will we have the wisdom to deal with this challenge effectively?

Sunset over the Amazon Basin as seen from the space shuttle Challenger, showing the thinness of the atmosphere.

Suggestions for Further Reading

Allen, Philip B.: "Conduction of Heat," *The Physics Teacher*, vol. 21, no. 9, 1983, p. 582.

This is an advanced article on our present understanding of heat conduction by solids and gases.

"Heat," *Scientific American* (Special Issue), September 1954.

This issue discusses many aspects of heat. It includes articles on the use of fire by early human beings, the temperature range in which life is possible, the temperatures of flames, materials that are resistant to high temperatures, chemistry at high temperatures, and the ultrahigh temperatures found in the interiors of stars.

Velarde, Manuel G., and Christiane Normand: "Convection," *Scientific American*, July 1980, p. 92.

Interesting illustrations and a discussion of Lord Rayleigh's theory of convection make this article useful, though the modern theories are quite mathematical.

Walker, Jearl: "The Amateur Scientist: Icicles Ensheathe a Number of Puzzles: Just How Does the Water Freeze?" *Scientific American*, May 1988, p. 114.

A discussion of research into a complex and beautiful change of phase.

Weaver, Jefferson Hane: "Energy," in *The World of Physics: A Small Library of the Literature of Physics from Antiquity to the Present*, Simon and Schuster, New York, 1987.

Includes "The Source of the Heat Which Is Excited by Friction" by Benjamin Thompson, Count Rumford, and writings on the mechanical equivalent of heat by J. R. Mayer and J. P. Joule.

Wilson, Mitchell: "Count Rumford," *Scientific American*, October 1960, p. 158.

This is an article about Benjamin Thompson, the highly unusual man whose experiments dealt heavy blows to the caloric theory of heat.

Review

A. Objectives: After studying this chapter, you should:

1. Be able to work calorimetry problems.

2. Be able to calculate the rate of conduction of thermal energy for various thermal resistances in series or in parallel.

3. Be able to state the Stefan–Boltzmann law of radiation and use it to calculate the power radiated by an object at a given temperature.

4. Be able to state Wien's displacement law.

5. Be able to state the first law of thermodynamics and use it in solving problems.

6. Be able to calculate the work done by a gas during various quasi-static processes and sketch the processes on a *PV* diagram.

7. Be able to state the equipartition theorem and use it to explain the molar heat capacities of gases and solids.

8. Be able to state the Dulong–Petit law and use it to estimate the heat capacity of a given solid or to calculate the molar mass of a solid from its specific heat.

B. Define, explain, or otherwise identify:

Heat
Heat capacity
Specific heat
Calorie
Btu
Molar heat capacity
Calorimetry
Calorimeter
Phase change
Fusion
Vaporization
Sublimation
Latent heat of fusion
Latent heat of vaporization
Conduction
Convection
Radiation
Temperature gradient
Thermal current
Coefficient of thermal conductivity
Thermal resistance
R factor
Stefan–Boltzmann law
Emissivity
Stefan's constant
Blackbody
Wien's displacement law
Thermograph
Newton's law of cooling
First law of thermodynamics
Free expansion
Quasi-static process
Isobaric expansion
Isothermal expansion
Equipartition theorem
Degree of freedom
Dulong–Petit law
Adiabatic process

C. True or false: If the statement is true, explain why it is true. If it is false, give a counterexample.

1. During a phase change, the temperature of a substance remains constant.

2. The heat capacity of a body is the amount of heat it can store at a given temperature.

3. The rate of conduction of thermal energy is proportional to the temperature gradient.

4. The rate at which an object radiates energy is proportional to the square of its absolute temperature.

5. When a system goes from state 1 to state 2, the amount of heat added to the system is the same for all processes.

6. When a system goes from state 1 to state 2, the work done on the system is the same for all processes.

7. When a system goes from state 1 to state 2, the change in the internal energy of the system is the same for all processes.

8. The internal energy of a given amount of an ideal gas depends only on its absolute temperature.

9. A quasi-static process is one in which there is no motion.

10. For any material that expands when heated, C_P is greater than C_V.

Problems

Level I

16-1 Heat Capacity and Specific Heat

1. A man typically consumes food having a total energy value of 2500 kcal each day. (*a*) How many joules of energy is this? (*b*) Assuming that this energy is dissipated at a steady rate during 24 h, calculate the person's power output in watts.

2. A 200-g piece of lead is heated to 90°C and is then dropped into a calorimeter containing 500 g of water that is initially at 20°C. Neglecting the heat capacity of the container, find the final temperature of the lead and water.

3. The specific heat of a certain metal can be determined by measuring the temperature change that occurs when a heated piece of the metal is placed in an insulated container made of the same material and containing water. The piece of metal has a mass of 100 g and is initially at 100°C. The container has a mass of 200 g and contains 500 g of water at an initial temperature of 20.0°C. The final temperature is 21.4°C. What is the specific heat of the metal?

4. A solar home contains 10^5 kg of concrete (specific heat = 1.00 kJ/kg·K). How much heat is given off by the concrete when it cools from 25 to 20°C?

16-2 Change of Phase and Latent Heat

5. A 200-g piece of ice at 0°C is placed in 500 g of water at 20°C. The system is in a container of negligible heat capacity and insulated from its surroundings. (*a*) What is the final equilibrium temperature of the system? (*b*) How much of the ice melts?

6. How much heat must be removed when 100 g of steam at 150°C is cooled and frozen into 100 g of ice at 0°C. (Take the specific heat of steam to be 2.01 kJ/kg·K.)

7. If 500 g of molten lead at 327°C is poured into a cavity in a large block of ice at 0°C, how much of the ice melts?

16-3 The Transfer of Thermal Energy

8. A copper bar 2 m long has a circular cross section of radius 1 cm. One end is kept at 100°C and the other end is kept at 0°C. The surface of the bar is insulated so that there is negligible heat loss through it. Find (*a*) the thermal resistance of the bar, (*b*) the thermal current I, (*c*) the temperature gradient $\Delta T/\Delta x$, and (*d*) the temperature of the bar 25 cm from the hot end.

9. A 20- by 30-ft slab of insulation has an R factor of 11. How much heat (in Btu per hour) is conducted through the slab if the temperature on one side is 68°F and that on the other side is 30°F?

10. Two metal cubes, one copper (Cu) and one aluminum (Al), with 3-cm edges are arranged as shown in Figure 16-19. Find (*a*) the thermal resistance of each cube, (*b*) the thermal resistance of the two-cube system, (*c*) the thermal current I, and (*d*) the temperature at the interface of the two cubes.

Figure 16-19 Two metal cubes in series for Problem 10.

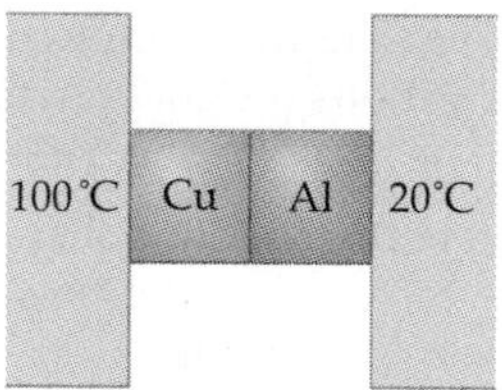

11. The same cubes as in Problem 10 are arranged as shown in Figure 16-20. Find (*a*) the thermal current carried by each cube from one side to the other, (*b*) the total thermal current, and (*c*) the equivalent thermal resistance of the two-cube system.

Figure 16-20 Two metal cubes in parallel for Problem 11.

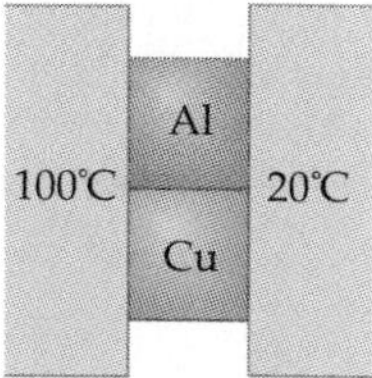

12. The heating wires of a 1-kW electric heater are "red hot" at a temperature of 900°C. Assuming that 100 percent of the heat output is due to radiation and that the wires act as blackbody radiators, what is the effective area of the radiating surface? (Assume a room temperature of 20°C.)

13. Calculate λ_{max} for a human blackbody radiator, assuming the surface temperature of the skin to be 33°C.

16-4 The First Law of Thermodynamics

14. A lead bullet initially at 30°C just melts upon striking a target. Assuming that all of the initial kinetic energy of the bullet goes into the internal energy of the bullet to raise its temperature and melt it, calculate the speed of the bullet upon impact.

15. If 400 kcal is added to a gas that expands and does 800 kJ of work, what is the change in the internal energy of the gas?

16. A lead bullet moving at 200 m/s is stopped in a block of wood. Assuming that all of the energy change goes into heating the bullet, find the final temperature of the bullet if its initial temperature is 20°C.

17. (*a*) At Niagara Falls, the water drops 50 m. If the change in potential energy goes into the internal energy of the water, compute the increase in its temperature. (*b*) Do the same for Yosemite Falls, where the water drops 740 m. (These temperature rises are not observed because the water cools by evaporation as it falls.)

16-5 The Internal Energy of an Ideal Gas

18. A certain gas consists of ions that repel each other. The gas undergoes a free expansion with no heat exchange and no work done. How does the temperature of the gas change? Why?

16-6 Work and the *PV* Diagram for a Gas

In Problems 19 through 22, the initial state of 1 mol *of an ideal gas is* $P_1 = 3$ atm, $V_1 = 1$ L, *and* $U_1 = 456$ J. *Its final state is* $P_2 = 2$ atm, $V_2 = 3$ L, *and* $U_2 = 912$ J. *All the processes are quasi-static.*

19. The gas is allowed to expand at constant pressure to a volume of 3 L. It is then cooled at constant volume until its pressure is 2 atm. (*a*) Show this process on a *PV* diagram, and calculate the work done by the gas. (*b*) Find the heat added during this process.

20. The gas is first cooled at constant volume until its pressure is 2 atm. It is then allowed to expand at constant pressure until its volume is 3 L. (*a*) Show this process on a *PV* diagram, and calculate the work done by the gas. (*b*) Find the heat added during this process.

21. The gas is allowed to expand isothermally until its volume is 3 L and its pressure is 1 atm. It is then heated at constant volume until its pressure is 2 atm. (*a*) Show this process on a *PV* diagram, and calculate the work done by the gas. (*b*) Find the heat added during this process.

22. The gas is heated and is allowed to expand such that it follows a straight-line path on a *PV* diagram from its initial state to its final state. (*a*) Show this process on a *PV* diagram, and calculate the work done by the gas. (*b*) Find the heat added during this process.

23. One mole of an ideal gas initially at a pressure of 1 atm and a temperature of 0°C is compressed isothermally and quasi-statically until its pressure is 2 atm. Find (*a*) the work needed to compress the gas and (*b*) the heat removed from the gas during the compression.

24. An ideal gas initially at 20°C and 200 kPa has a volume of 4 L. It undergoes a quasi-static, isothermal expansion until its pressure is reduced to 100 kPa. Find (*a*) the work done by the gas and (*b*) the heat added to the gas during the expansion.

16-7 Heat Capacities and the Equipartition Theorem

25. The specific heat of air at 0°C is listed in a handbook as having the value of 1.00 J/g·K measured at constant pressure. (*a*) Assuming air is an ideal gas with a molar mass $M = 29.0$ g/mol, what is its specific heat at 0°C at constant volume? (*b*) How much internal energy is there in 1 L of air at 0°C?

26. The heat capacity at constant volume of a certain amount of a monatomic gas is 49.8 J/K. (*a*) Find the number of moles of the gas. (*b*) What is the internal energy of this gas at $T = 300$ K? (*c*) What is its heat capacity at constant pressure?

27. One mole of an ideal diatomic gas is heated at constant volume from 300 to 600 K. (*a*) Find the increase in internal energy, the work done, and the heat added. (*b*) Find the same quantities if this gas is heated from 300 to 600 K at constant pressure. Use the first law of thermodynamics and your results for part (*a*) to calculate the work done. (*c*) Calculate the work done in part (*b*) directly from $dW = P\,dV$.

28. The Dulong–Petit law was originally used to determine the molecular mass of a substance from its measured heat capacity. The specific heat of a certain solid is measured to be 0.447 kJ/kg·K. (*a*) Find the molecular mass of the substance. (*b*) What element is this?

16-8 The Quasi-Static Adiabatic Expansion of a Gas

29. One mole of an ideal gas ($\gamma = \frac{5}{3}$) expands adiabatically and quasi-statically from a pressure of 10 atm and a temperature of 0°C to a pressure of 2 atm. Find (*a*) the initial and final volumes, (*b*) the final temperature, and (*c*) the work done by the gas.

30. An ideal gas at a temperature of 20°C is compressed quasi-statically and adiabatically to half its original volume. Find its final temperature if (*a*) $C_v = \frac{3}{2}nR$ and (*b*) $C_v = \frac{5}{2}nR$.

Level II

31. A 200-g aluminum calorimeter can contains 500 g of water at 20°C. Aluminum shot of mass 300 g is heated to 100°C and is then placed in the calorimeter. (*a*) Using the value of the specific heat of aluminum given in Table 16-1, find the final temperature of the system, assuming that no heat is lost to the surroundings. (*b*) The error due to heat transfer to or from the surroundings can be minimized if the initial temperature of the water and calorimeter is chosen to be $\frac{1}{2}\Delta t_w$ below room temperature, where Δt_w is the temperature change of the calorimeter and water during the measurement. Then the final temperature is $\frac{1}{2}\Delta t_w$ above room temperature. What should the initial temperature of the water and container be if the room temperature is 20°C?

32. A thermally insulated system consists of 1 mol of a diatomic ideal gas at 100 K and 2 mol of a solid at 200 K

that are separated by a rigid insulating wall. Find the equilibrium temperature of the system after the insulating wall is removed, assuming that the solid obeys the Dulong–Petit law.

33. At very low temperatures, the specific heat of a metal is given by $c = aT + bT^3$. For copper, $a = 0.0108$ J/kg·K^2 and $b = 7.62 \times 10^{-4}$ J/kg·K^4. (*a*) What is the specific heat of copper at 4 K? (*b*) How much heat is required to heat copper from 1 to 3 K?

34. A 50-g piece of aluminum at 20°C is cooled to −196°C by placing it in a large container of liquid nitrogen at that temperature. How much nitrogen is vaporized? (Assume that the specific heat of aluminum is constant and is equal to 0.90 kJ/kg·K.)

35. A piece of ice is dropped from a height H. (*a*) Find the minimum value of H such that the ice melts when it makes an inelastic collision with the ground. Assume that all the mechanical energy lost goes into melting the ice. (*b*) Is it reasonable to neglect the variation in the acceleration of gravity in doing this problem? (*c*) Comment on the reasonableness of neglecting air resistance. What effect would air resistance have on your answer?

36. An insulated cylinder fitted with a movable piston to maintain constant pressure initially contains 100 g of ice at −10°C. Heat is supplied to the contents at a constant rate by a 100-W heater. Make a graph showing the temperature of the cylinder contents as a function of time starting at $t = 0$, when the temperature is −10°C, and ending when the temperature is 110°C. (Use $c = 2.0$ kJ/kg·K for the average specific heat of ice from −10 to 0°C and of steam from 100 to 110°C.)

37. A 200-g aluminum calorimeter can contains 500 g of water at 20°C. A 100-g piece of ice cooled to −20°C is placed in the calorimeter. (*a*) Find the final temperature of the system, assuming no heat losses. (Assume that the specific heat of ice is 2.0 kJ/kg·K.) (*b*) A second 200-g piece of ice at −20°C is added. How much ice remains in the system after it reaches equilibrium? (*c*) Would your answer to part (*b*) be different if both pieces of ice were added at the same time?

38. A well-insulated bucket contains 150 g of ice at 0°C. (*a*) If 20 g of steam at 100°C is injected into the bucket; what is the final equilibrium temperature of the system? (*b*) Is any ice left?

39. A six-pack of 12-oz aluminum cans of diet cola originally at 80°F is placed in a well-insulated Styrofoam container. How many 30-g ice cubes must you add to the container so that the final temperature of the cola is 40°F? (Neglect heat losses through the container and the heat removed from the aluminum. The cola is mostly water.)

40. The surface temperature of the filament of an incandescent lamp is 1300°C. If the electric power input is doubled, what will the temperature become? *Hint:* Show that you can neglect the temperature of the surroundings.

41. A copper-bottomed saucepan containing 0.8 L of boiling water boils dry in 10 min. Assuming that all the heat flows through the flat copper bottom, which has a diameter of 15 cm and a thickness of 3.0 mm, calculate the temperature of the outside of the copper bottom while some water is still in the pan.

42. For a boiler at a power station, heat must be transferred to boiling water at the rate of 3 GW. The boiling water passes through copper pipes having a wall thickness of 4.0 mm and a surface area of 0.12 m^2 per meter length of pipe. Find the total length of pipe (actually there are many pipes in parallel) that must pass through the furnace if the steam temperature is 225°C and the external temperature of the pipes is 600°C.

43. A hemispherical igloo made of compacted snow has an inside radius of 2 m. It is desired to maintain an inside temperature of 20°C when the outside temperature is −20°C. The heat generated by the igloo inhabitants is 38 MJ/day. How thick must the igloo walls be if the thermal conductivity of the compacted snow is 0.209 W/m·K. As an approximation, assume that the inner surface area of the igloo is equal to the outer surface area.

44. A blackened, solid copper sphere of radius 4.0 cm hangs in a vacuum in an enclosure whose walls have a temperature of 20°C. If the sphere is initially at 0°C, find the rate at which its temperature changes, assuming that heat is transferred by radiation only.

45. Liquid helium is stored at its boiling point (4.2 K) in a spherical can that is separated by a vacuum space from a surrounding shield that is maintained at the temperature of liquid nitrogen (77 K). If the can is 30 cm in diameter and is blackened on the outside so it acts as a blackbody, how much helium boils away per hour?

46. The solar constant is the power received from the sun per unit area perpendicular to the sun's rays at the mean distance of the earth from the sun. Its value at the upper atmosphere of the earth is about 1.35 kW/m^2. Calculate the effective temperature of the sun if it radiates like a blackbody. (The radius of the sun is 6.96×10^8 m.)

47. A 20-kg block of ice at 0°C slides down a 5-m plane inclined at 30°. The coefficient of sliding friction between the ice and the plane is 0.05. Calculate the amount of ice that melts due to friction. (Assume that all the mechanical energy lost goes into melting the ice.)

48. It is possible to warm your hands on a cold day by rubbing them together. (*a*) Assume that the coefficient of friction between your hands is 0.5, that the normal force between your hands is 35 N, and that you rub them together at an average speed of 35 cm/s. What is the rate at which heat is generated? (*b*) Assume further that the mass of each of your hands is approximately 350 g, that the specific heat of your hands is about 4 kJ/kg·K, and that all the heat generated goes into raising the temperature of your hands. How long must you rub your hands together to produce a 5-C° increase in their temperature?

49. When an ideal gas undergoes a temperature change at constant volume, its energy changes by $\Delta U = C_v \, \Delta T$. (*a*) Explain why the result $\Delta U = C_v \, \Delta T$ holds for an ideal gas for any temperature change independent of the process. (*b*) Show explicitly that this result holds for the expansion of an ideal gas at constant pressure by first calculating the work done and showing that it can be written as $W = nR \, \Delta T$, and then by using $\Delta U = Q - W$ where $Q = C_p \, \Delta T$.

50. The heat capacity of a certain amount of a particular gas at constant pressure is greater than that at constant volume by 29.1 J/K. (*a*) How many moles of the gas are there? (*b*) If the gas is monatomic, what are C_v and C_p? (*c*) If the gas consists of diatomic molecules that rotate but do not vibrate, what are C_v and C_p?

51. One mole of a monatomic ideal gas is initially at 273 K and 1 atm. (*a*) What is its initial internal energy? Find its final internal energy and the work done by the gas when 500 J of heat are added (*b*) at constant pressure and (*c*) at constant volume.

52. One mole of an ideal monatomic gas is heated at constant volume from 300 to 600 K. (*a*) Find the heat added, the work done by the gas, and the change in its internal energy. (*b*) Find these same quantities if the gas is heated from 300 to 600 K at constant pressure.

53. One-half mole of helium is expanded adiabatically and quasi-statically from an initial pressure of 5 atm and temperature of 500 K to a final pressure of 1 atm. Find (*a*) the final temperature, (*b*) the final volume, (*c*) the work done by the gas, and (*d*) the change in the internal energy of the gas.

54. One mole of N_2 ($C_v = \frac{5}{2}R$) gas is originally at room temperature (20°C) and a pressure of 5 atm. It is allowed to expand adiabatically and quasi-statically until its pressure equals the room pressure of 1 atm. It is then heated at constant pressure until its temperature is again 20°C. During this heating, the gas expands. After it reaches room temperature, it is heated at constant volume until its pressure is 5 atm. It is then compressed at constant pressure until it is back to its original state. (*a*) Construct an accurate *PV* diagram showing each process in the cycle. (*b*) From your graph, determine the work done by the gas during the complete cycle. (*c*) How much heat is added or subtracted from the gas during the complete cycle? (*d*) Check your graphical determination of the work done by the gas in part (*b*) by calculating the work done during each part of the cycle.

55. Two moles of an ideal monatomic gas have an initial pressure $P_1 = 2$ atm and an initial volume $V_1 = 2$ L. The gas is taken through the following quasi-static cycle: It is expanded isothermally until it has a volume $V_2 = 4$ L. It is then heated at constant volume until it has a pressure $P_3 = 2$ atm. It is then cooled at constant pressure until it is back to its initial state. (*a*) Show this cycle on a *PV* diagram. (*b*) Calculate the heat added and the work done by the gas during each part of the cycle. (*c*) Find the temperatures T_1, T_2, and T_3.

56. Heat in the amount of 500 J is supplied to 2 mol of an ideal diatomic gas. (*a*) Find the change in temperature if the pressure is kept constant. (*b*) Find the work done by the gas. (*c*) Find the ratio of the final volume of the gas to the initial volume if the initial temperature is 20°C.

57. A 100-g piece of copper is heated in a furnace to a temperature *t*. The copper is then inserted into a 150-g copper calorimeter containing 200 g of water. The initial temperature of the water and calorimeter is 16°C, and the final temperature after equilibrium is established is 38°C. When the calorimeter and its contents are weighed, 1.2 g of water are found to have evaporated. What was the temperature *t*?

Level III

58. A small pond has a layer of ice 1 cm thick floating on its surface. (*a*) If the air temperature is −10°C, find the rate in centimeters per hour at which ice is added to the bottom of the layer. The density of ice is 0.917 g/cm^3. (*b*) How long does it take for a 20-cm layer to be built up?

59. A hand pump is used to inflate a bicycle tire to a gauge pressure of 70 lb/in^2. How much work must be done if each stroke of the pump is an adiabatic process? Atmospheric pressure is 1 atm, the air temperature is initially 20°C, and the volume of the air in the tire remains constant at 1 L.

60. A steam pipe of length *L* is insulated with a layer of material of thermal conductivity *k*. Find the rate of heat transfer if the temperature outside the insulation is t_1, that on the inside is t_2, the outside radius of the insulation is r_1, and the inside radius is r_2.

61. In an isothermal expansion, an ideal gas at an initial pressure P_0 expands until its volume is twice its initial volume. (*a*) Find its pressure after the expansion. (*b*) The gas is the compressed adiabatically and quasi-statically back to its original volume, at which point its pressure is $1.32P_0$. Is the gas monatomic, diatomic, or polyatomic? (*c*) How does the translational kinetic energy of the gas change in these processes?

62. Brine at −16°C circulating through copper pipes with walls 1.5 mm thick is used to keep a cold room at 0°C. The diameter of each pipe is very large compared to the thickness of its walls. By what fraction is the transfer of heat reduced when the pipes are coated with a 5-mm layer of ice?

63. A body initially at a temperature T_i cools in a room where the temperature is T_0 by convection and radiation. The body obeys Newton's law of cooling, which can be written

$$\frac{dQ}{dt} = hA(T - T_0)$$

where *A* is the area of the body and *h* is a constant called the surface coefficient of heat transfer. Show that the temperature *T* at any time *t* is given by

$$T = T_0 + (T_i - T_0)e^{-hAt/mc}$$

where *m* is the mass of the body and *c* is its specific heat.

64. Two 200-g copper containers, each holding 0.7 L of water, are connected by a 10-cm copper rod of cross-sectional area 1.5 cm^2. Initially, one container is at 60°C; the second is maintained at 0°C. (*a*) Show that the temperature t_c of the first container changes over time *t* according to

$$t_c = t_{c0}e^{-t/RC}$$

where t_{c0} is the initial temperature of the first container, *R* is the thermal resistance of the rod, and *C* is the total heat capacity of the container plus the water. (*b*) Evaluate *R*, *C*, and the "time constant" *RC*. (*c*) Show that the total amount of heat *Q* conducted in time *t* is

$$Q = Ct_{c0}(1 - e^{-t/RC})$$

(*d*) Find the time it takes for the temperature of the first container to be reduced to 30°C.

65. An ideal gas at initial volume V_1 and pressure P_1 expands quasi-statically and adiabatically to volume V_2 and pressure P_2. Calculate the work done by the gas directly by integrating $P\,dV$, and show that your result is the same as given by Equation 16-51.

66. Prove that the slope of the adiabatic curve passing through a point on the PV diagram is γ times the slope of the isothermal curve passing through the same point.

67. An ideal gas of n mol is initially at pressure P_1, volume V_1, and temperature T_h. It expands isothermally until its pressure and volume are P_2 and V_2. It then expands adiabatically until its temperature is T_c and its pressure and volume are P_3 and V_3. It is then compressed isothermally until it is at a pressure P_4 and a volume V_4, which is related to its initial volume V_1 by

$$T_c V_4^{\gamma-1} = T_h V_1^{\gamma-1}$$

The gas is then compressed adiabatically until it is back in its original state. (*a*) Assuming that each process is quasi-static, plot this cycle on a PV diagram. (This cycle is known as the Carnot cycle for an ideal gas.) (*b*) Show that the heat absorbed Q_h during the isothermal expansion at T_h is

$$Q_h = nRT_h \ln \frac{V_2}{V_1}$$

(*c*) Show that the heat rejected Q_c by the gas during the isothermal compression at T_c is

$$Q_c = nRT_c \ln \frac{V_3}{V_4}$$

(*d*) Using the result that, for an adiabatic expansion, $TV^{\gamma-1}$ is constant, show that $V_2/V_1 = V_3/V_4$. (*e*) The efficiency of a Carnot cycle is defined to be the net work done divided by the heat absorbed Q_h. Using the first law of thermodynamics, show that the efficiency is $1 - Q_c/Q_h$. (*f*) Using your results from the previous parts of this problem, show that $Q_c/Q_h = T_c/T_h$.

68. According to the Einstein model of a crystalline solid, the molar heat capacity at constant volume is given by

$$C_{mv} = 3R\left(\frac{T_E}{T}\right)^2 \frac{e^{T_E/T}}{(e^{T_E/T} - 1)^2}$$

where T_E is a characteristic temperature called the Einstein temperature and T is the temperature of the solid in kelvins. (*a*) Show that the Dulong–Petit law, $C_{mv} \approx 3R$, holds for the Einstein model when $T \gg T_E$. (*b*) For diamond, T_E is approximately 1060 K. If 1 mol of diamond is heated from 300 to 600 K, numerically integrate

$$\Delta U = \int C_{mv}\,dT$$

to find the increase in the internal energy of the diamond.

Chapter 17

The Availability of Energy

Solar energy is directed toward the solar oven at the center by this circular array of reflectors at Barstow, California.

We are continually told to conserve energy. Yet, according to the first law of thermodynamics, energy is always conserved. In a closed system, the total amount of energy cannot change. What, then, does it mean to conserve energy if the total amount of energy in the universe does not change no matter what we do? In this chapter, we will see that the first law of thermodynamics does not tell the whole story. Energy is always conserved, but some forms of energy are more useful than others. The possibility or impossibility of putting energy to use is the subject of the second law of thermodynamics. Simply stated, it is easy to convert mechanical work or the internal energy of a system completely into heat with no other changes, but it is impossible to remove heat or internal energy from a system and convert it completely into mechanical work with no other changes. This experimental fact is one statement of the second law of thermodynamics. There is thus a lack of symmetry in the roles played by heat and work that is not evident from the first law. This lack of symmetry is related to the fact that some processes are irreversible.

A common example of the conversion of mechanical energy into internal energy is work done against friction. Consider, for instance, a block with some initial kinetic energy that is sliding along a rough table. In this process, mechanical energy, the initial kinetic energy of the block, is converted into the internal energy of the block and the table. The reverse process never occurs, however. The internal energy of the block and table is never spontaneously converted into kinetic energy to send the block sliding along the table while the table and block cool. Yet, such an amazing occurrence would not violate the conservation of energy or any other physical laws we have encountered so far. It does, however, violate the second law of thermodynamics, as we will discuss shortly.

There are many irreversible processes, seemingly quite different from one another, but all related to the second law. Suppose, for example, that we pour a layer of black sand into a jar and cover it with a layer of white sand. If we now shake the jar, the sand becomes mixed. However, we cannot "unshake" the jar. That is, we cannot reverse the shaking until the black sand is again on the bottom and the white on top. Another example of an irreversible process is heat conduction. If we place a hot body in contact with a cold body, heat will flow from the hot body to the cold body until they are at the same temperature. However, the reverse does not occur. Two bodies in contact at the same temperature remain at the same temperature. Heat does not flow from one to the other, making one colder and the other warmer. The second law of thermodynamics summarizes the fact that processes of this type do not occur. There are many different ways of stating the second law. We will study several and see that they are all equivalent.

17-1 Heat Engines and the Second Law of Thermodynamics

The study of the efficiency of heat engines gave rise to the first clear statements of the second law of thermodynamics. The first practical heat engine was the steam engine, invented in the eighteenth century for pumping water out of coal mines. Today, the primary use of steam engines is in generating electrical energy. In a typical steam engine (Figure 17-1), water is heated under high pressure (typically several hundred atmospheres) until it vaporizes into steam at a high temperature (normally around 500°C). The steam expands and does work against a piston. The steam exits at a much lower temperature and is cooled further until it condenses. The water is then pumped back into the boiler to be heated again.

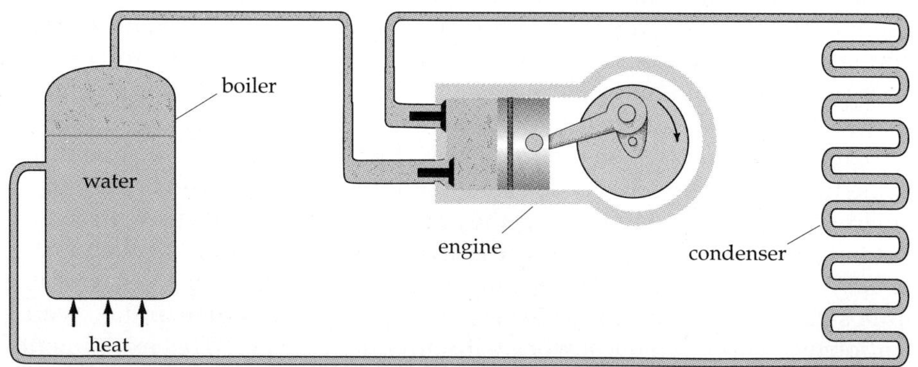

Figure 17-1 Schematic drawing of a steam engine. The high-pressure steam from the boiler does work against the piston. The steam is then exhausted at lower pressure and condenses back into water, which is recycled back to the boiler.

Figure 17-2 illustrates schematically how a simple heat engine performs work in a cyclic process. A cylinder containing gas at pressure P_1 is fitted with a piston. In step (*a*) the piston is held fixed as the gas is heated to a new pressure P_2. The pressure increases at constant volume as shown in the PV diagram in part (*e*). A weight w is then added to balance the increase in the force on the piston due to the increase in pressure. In step (*b*), more heat is added, and the gas is allowed to expand at constant pressure, lifting the

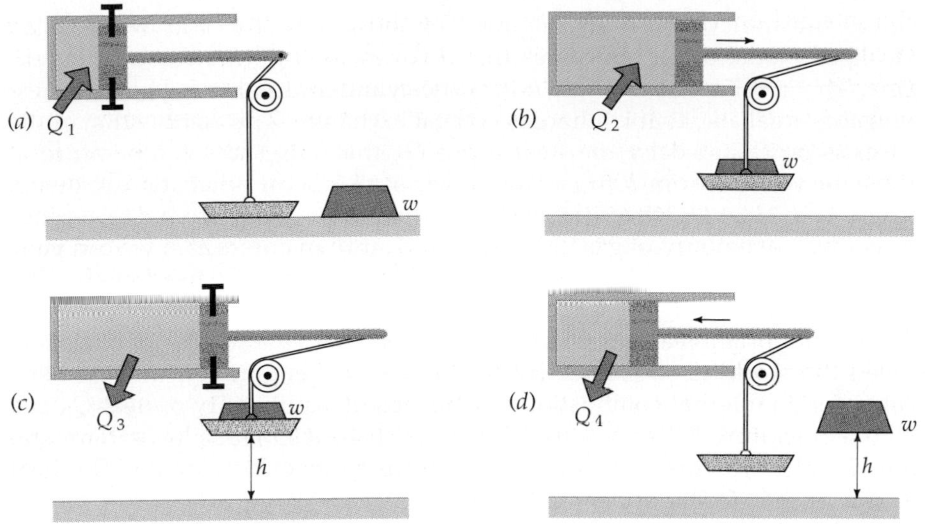

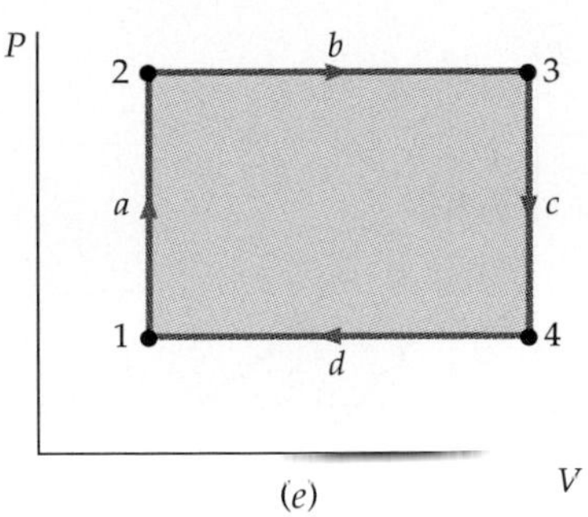

Figure 17-2 A simple heat engine that does work in a cycle. (*a*) A gas enclosed in a cylinder is heated while the piston is held fixed. The pressure increases from P_1 to P_2, as shown in the PV diagram of part (*e*). A weight w is added so that the piston is in equilibrium. (*b*) More heat is added, and the gas is allowed to expand at constant pressure, lifting the weight. (*c*) The piston is held fixed while the gas is cooled back to its original pressure. (*d*) The weight is removed, and the gas is compressed at constant pressure back to its original state. The net result of the cycle is that work has been done in lifting the weight to some height h. (*e*) The PV diagram for the simple heat engine in this figure. Work is done by the gas during step b, and on the gas in step d. The net work done by the gas is the shaded area of the diagram.

weight to some height h. The volume increases at constant pressure, as shown in part (*e*). In step (*c*), the piston is again held fixed as heat is removed, reducing the pressure back to P_1. The pressure decreases at constant volume as shown in part (*e*). The weight is then removed so that the piston is again in equilibrium. In step (*d*), the gas is compressed at constant pressure, and heat is exhausted until the gas is back to its original volume. The net result is that an amount of heat $Q_{in} = Q_1 + Q_2$ has been put into the system, some heat $Q_{out} = Q_3 + Q_4$ has been exhausted, and work $W = mgh = Q_{in} - Q_{out}$ has been performed.

Figure 17-3 is a schematic diagram of a practical engine, the internal-combustion engine used in most automobiles. With the exhaust valve closed, a mixture of gasoline vapor and air enters the combustion chamber as the piston moves down during the intake stroke. The mixture is then compressed (compression stroke), after which it is ignited by a spark from the spark plug. The hot gases then expand against the piston, doing work during the power stroke. The gases are then exhausted through the exhaust

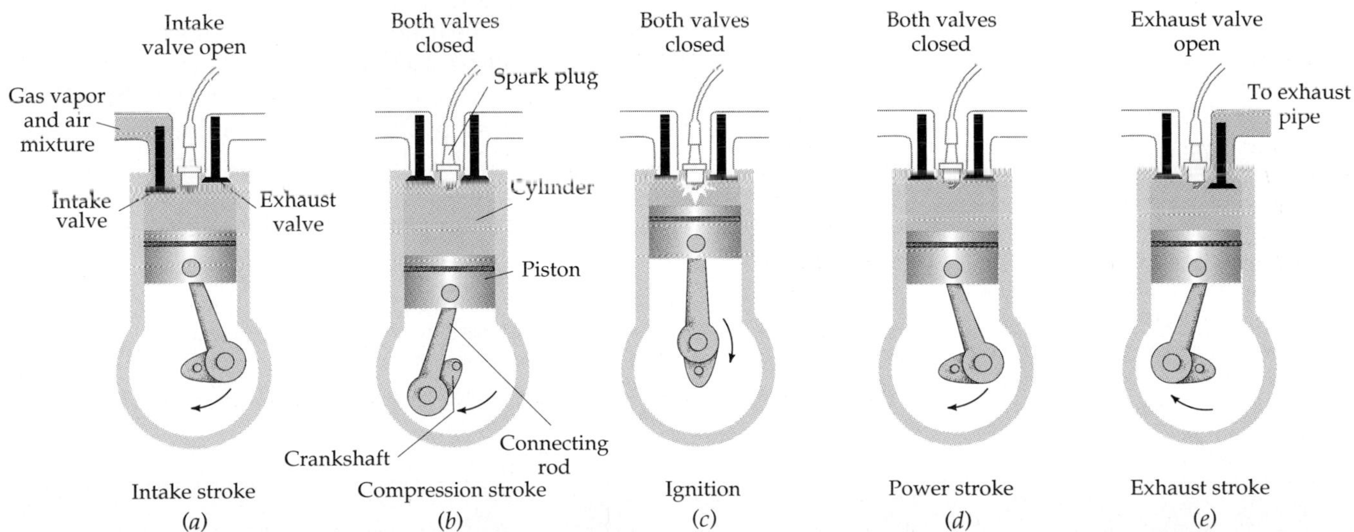

Figure 17-3 Internal-combustion engine. In (*a*), a mixture of gasoline vapor and air enters the combustion chamber as the piston moves down. The piston then moves up in (*b*), compressing the gas for ignition in (*c*). The hot gases expand, moving the piston down in (*d*), the power stroke. In (*e*), the piston moves up again to exhaust the burned gases. The cycle then repeats.

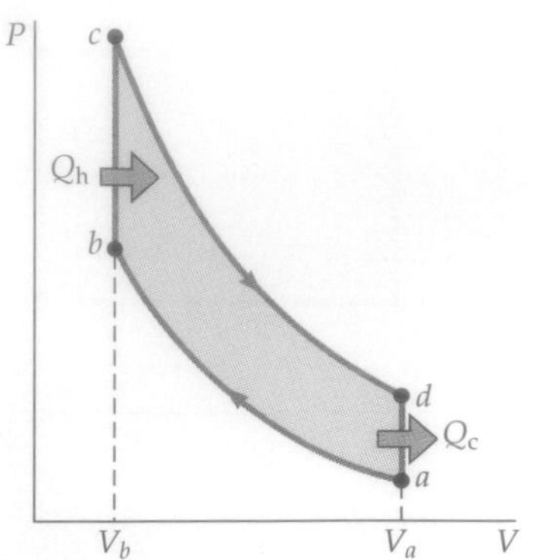

Figure 17-4 Otto cycle, representing the internal-combustion engine. The gasoline–air mixture enters at a and is adiabatically compressed to b. It is then heated (by ignition from the spark plug) at constant volume to c. The power stroke is represented by the adiabatic expansion from c to d. The cooling at constant volume from d to a represents the exhausting of the burned gases and the intake of a fresh gasoline–air mixture.

valve, and the cycle repeats. Figure 17-4 shows an *Otto cycle,* which is an idealized model of the processes that occur in the internal-combustion engine. The gasoline–air mixture enters the cylinder at point a. The compression ab is adiabatic; that is, there is no heat exchange. The combustion of the gases is represented by the heat input Q_h that raises the temperature at constant volume from b to c. The power stroke is an adiabatic expansion from c to d. Heat Q_c leaves the system during the exhaust from d to a. Since an equivalent amount of gasoline vapor and air then enters at a, we can view the process as though the same gasoline–air mixture were used again. The net work done during the cycle is indicated by the shaded area in the figure.

The important features of any **heat engine** are that a substance or system called the **working substance** (water for a steam engine, air and gasoline vapor for an internal-combustion engine) absorbs a quantity of heat Q_h at a high temperature T_h, does work W, and rejects heat $|Q_c|$* at a lower temperature T_c. The working substance then returns to its original state. The heat engine is therefore a cyclic device.

A heat engine's purpose is to convert as much heat input Q_h into work as possible. Figure 17-5 shows a heat engine in schematic form. The heat input Q_h comes from a heat reservoir at temperature T_h; the exhaust $|Q_c|$ goes into a heat reservoir at a lower temperature T_c. A **heat reservoir** is an idealized system with a heat capacity so large it can absorb or give off heat with no appreciable change in temperature. In practice, the surrounding atmosphere or lakes often act as heat reservoirs. Since the initial and final states of the engine and working substance are the same, the final internal energy of the working substance must equal its initial internal energy. Then, by the first law of thermodynamics, the work done equals the net heat absorbed:

$$W = Q_h - |Q_c| \qquad 17\text{-}1$$

The **efficiency** ϵ is defined as the ratio of the work done to the heat absorbed from the hot reservoir:

Efficiency defined

$$\epsilon = \frac{W}{Q_h} = \frac{Q_h - |Q_c|}{Q_h} = 1 - \frac{|Q_c|}{Q_h} \qquad 17\text{-}2$$

Since the heat Q_h is usually produced by burning coal, oil, or some other kind of fuel that must be paid for, heat engines are designed to have the greatest possible efficiency. The efficiencies of steam engines are typically about 40 percent, and those of internal-combustion engines may be as high as 50 percent. We can see from Equation 17-2 that we want to reject as small a fraction of the heat absorbed as possible. For perfect efficiency ($\epsilon = 1 = 100$ percent), $|Q_c| = 0$. In other words, all the heat absorbed from the hot reservoir would be converted into work and no heat would be rejected to the cold reservoir.

*According to our sign convention for the first law of thermodynamics, the heat rejected by a system is negative. Since we are interested here only in the magnitudes of the heat absorbed or rejected, we will use absolute value signs where needed.

The development and propagation of the flame in a piston engine is studied in this series of computer-enhanced schlieren photographs. The ignition begins at the left, and the flame kernel develops to the right.

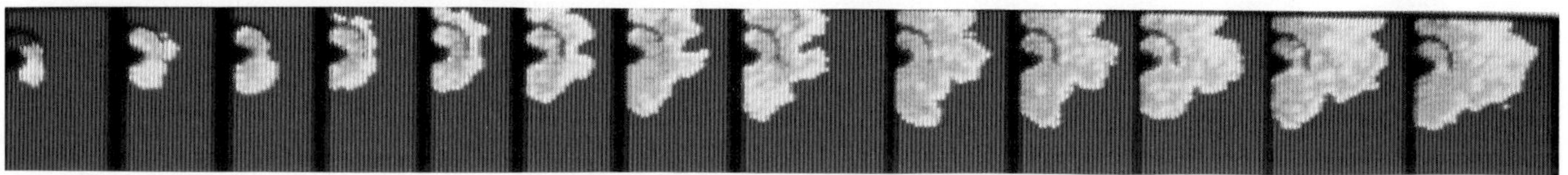

Although the efficiency of heat engines has been greatly increased since the early steam engines, it is impossible to make a perfect heat engine, that is, an engine with 100 percent efficiency.

> It is impossible for a heat engine working in a cycle to produce no other effect than that of extracting heat from a reservoir and performing an equivalent amount of work.

Second law of thermodynamics: heat-engine statement

This experimental result is the Kelvin–Planck or **heat-engine statement of the second law of thermodynamics.** It is important to include the word "cycle" in this statement because it is possible to convert heat completely into work in a noncyclic process. An ideal gas undergoing an isothermal expansion does just this. But after the expansion, the gas is not in its original state. To bring the gas back to its original state, work must be done on the gas, and some heat will be exhausted. Essentially, the second law tells us that if we want to extract energy from a heat reservoir to do work, we must have available a colder reservoir to which part of the energy can be exhausted. If this were not true, we could design a heat engine for a ship that would extract energy from the ocean (a convenient heat reservoir with an enormous supply of energy) and use it to power the ship. Unfortunately, this enormous reservoir of energy is not available for such use unless a colder reservoir is available for the exhaust.

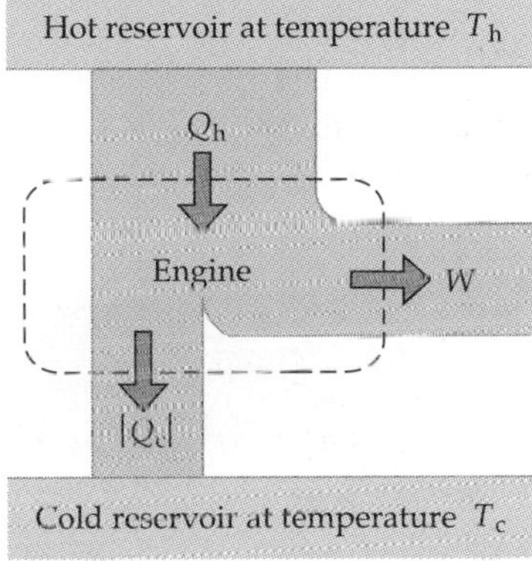

Figure 17-5 Schematic representation of a heat engine. The engine removes heat energy Q_h from a hot reservoir at a temperature T_h, does work W, and rejects heat $|Q_c|$ to a cold reservoir at a temperature T_c.

Example 17-1

A heat engine absorbs 200 J of heat from a hot reservoir, does work, and exhausts 160 J to a cold reservoir. What is its efficiency?

By the first law, the work done is

$$W = Q_h - |Q_c| = 200\text{ J} - 160\text{ J} = 40\text{ J}$$

The efficiency is thus

$$\epsilon = \frac{W}{Q_h} = \frac{40\text{ J}}{200\text{ J}} = 0.20 = 20\%$$

Exercise

A heat engine has an efficiency of 35 percent. (*a*) How much work does it perform in a cycle if it extracts 150 J of heat from a hot reservoir per cycle? (*b*) How much heat is exhausted per cycle?
[Answers: (*a*) 52.5 J, (*b*) 97.5 J]

Questions

1. Where does the energy come from in an internal-combustion engine? In a steam engine?
2. How does friction in an engine affect its efficiency?

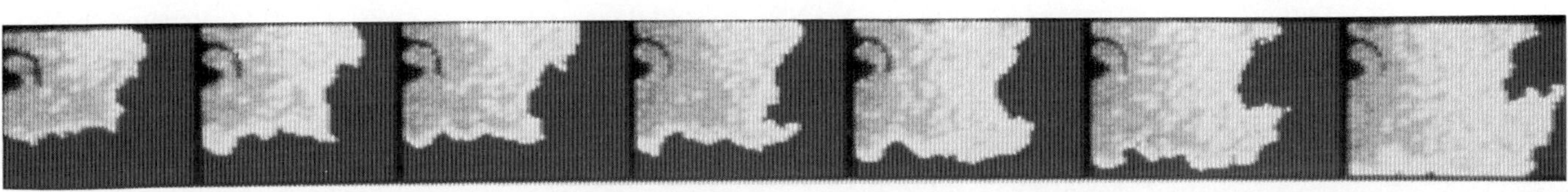

17-2 Refrigerators and the Second Law of Thermodynamics

A **refrigerator** is essentially a heat engine run backwards. Work is put into the refrigerator to extract heat from a cold reservoir and transfer it to a hot reservoir. Figure 17-6 is a schematic representation of a refrigerator. It is desirable to remove as much heat Q_c as possible while doing as little work W as possible. It is found from experience that some work must always be done. This result is the Clausius or **refrigerator statement of the second law of thermodynamics:**

Second law of thermodynamics: refrigerator statement

> It is impossible for a refrigerator working in a cycle to produce no other effect than the transfer of heat from a cold object to a hot object.

If this statement were not true, it would be possible in principle to cool our homes in the summer with a refrigerator that pumps heat to the outside without using any electricity or other energy.

A measure of the performance of a refrigerator is the ratio Q_c/W, which is called the **coefficient of performance** COP:

Coefficient of performance defined

$$\text{COP} = \frac{Q_c}{W} \qquad 17\text{-}3$$

Hot reservoir at temperature T_h

$|Q_h|$

Refrigerator

W

Q_c

Cold reservoir at temperature T_c

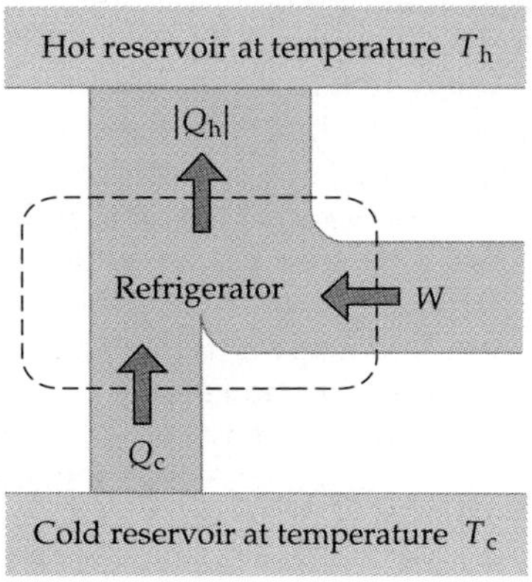

Figure 17-6 Schematic representation of a refrigerator. The refrigerator removes heat energy Q_c from a cold reservoir and rejects heat $|Q_h|$ to a hot reservoir using work W.

The greater the coefficient of performance, the better the refrigerator. Typical refrigerators have coefficients of performance of about 5 or 6. In terms of this ratio, the refrigerator statement of the second law says that the coefficient of performance of a refrigerator cannot be infinite (because W cannot be zero).

Example 17-2

A refrigerator has a coefficient of performance of 5.5. How much work is needed to make ice cubes from 1 L of water at 10°C?

The mass of 1 L of water is 1 kg. The heat energy that must be removed to reduce the temperature of the water by 10 C° = 10 K is

$$Q_1 = mc\,\Delta T = (1\text{ kg})(4.18\text{ kJ/kg·K})(10\text{ K}) = 41.8\text{ kJ}$$

The heat of fusion of water is 333.5 kJ/kg. The heat that must be removed to freeze 1 L of water into ice cubes is therefore $Q_2 = 333.5$ kJ. The total amount of heat that must be removed is thus

$$Q_c = Q_1 + Q_2 = 41.8\text{ kJ} + 333.5\text{ kJ} = 375.3\text{ kJ} \approx 375\text{ kJ}$$

From Equation 17-3, the work needed is

$$W = \frac{Q_c}{\text{COP}} = \frac{375\text{ kJ}}{5.5} = 68.2\text{ kJ}$$

Exercise

A refrigerator has a coefficient of performance of 4.0. How much heat is exhausted to the hot reservoir when 200 kJ of heat are removed from the cold reservoir? (Answer: 250 kJ)

Question

3. To cool his kitchen, a man leaves his refrigerator door open. Explain why this does not have the desired effect and, in fact, heats the kitchen.

17-3 Equivalence of the Heat-Engine and Refrigerator Statements

Although the heat-engine and refrigerator statements of the second law of thermodynamics seem quite different, they are actually equivalent. That is, if one of the statements is true, the other must also be true. This equivalence can be proven by showing that if either statement is assumed to be false, the other must also be false. We will use a numerical example to show that if the heat-engine statement is false, the refrigerator statement must also be false.

Figure 17-7*a* shows an ordinary refrigerator that uses 50 J of work to remove 100 J of energy from a cold reservoir and reject 150 J of energy to a hot reservoir. Its coefficient of performance is 2.0. If the heat-engine statement of the second law were not true, we could have a perfect heat engine that could remove energy from a single reservoir and convert it completely into work with 100 percent efficiency. We could use this perfect heat engine to remove 50 J of energy from the hot reservoir and do 50 J of work (Figure 17-7*b*). Then, if we used the perfect heat engine in conjunction with the ordinary refrigerator, we could construct a perfect refrigerator that would transfer 100 J of energy from the cold reservoir to the hot reservoir without requiring any work, as illustrated in Figure 17-7*c*. This would violate the refrigerator statement of the second law. Thus, if the heat-engine statement is false, the refrigerator statement must also be false. It can be similarly shown that if a perfect refrigerator existed, it could be used in conjunction with an ordinary heat engine to construct a perfect heat engine; see Problem 6. Thus, if the refrigerator statement is false, the heat-engine statement must also be false. It then follows that if either statement is true, the other must also be true. Therefore, the two statements are equivalent.

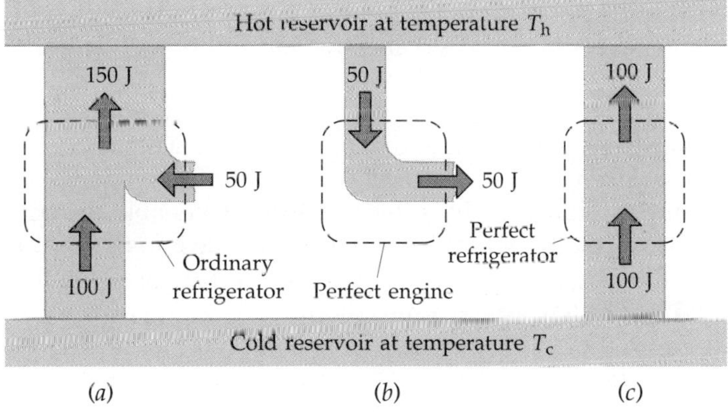

Figure 17-7 (*a*) An ordinary refrigerator that removes 100 J from a cold reservoir and rejects 150 J to the hot reservoir, requiring the input of 50 J of work. (*b*) A perfect heat engine that violates the heat-engine statement of the second law by removing 50 J from the hot reservoir and converting it completely into work. (*c*) The ordinary refrigerator in (*a*) and the perfect heat engine in (*b*) working together make a perfect refrigerator that violates the refrigerator statement of the second law by transferring 100 J of energy from the cold reservoir to the hot reservoir with no other effect.

17-4 The Carnot Engine

We have seen that, according to the second law of thermodynamics, it is impossible for a heat engine working between two heat reservoirs to be 100 percent efficient. What, then, is the maximum possible efficiency for such an engine? This question was answered in 1824 by a young French engineer, Sadi Carnot, before the first law of thermodynamics was established. Carnot found that all reversible engines working between two heat reservoirs have the same efficiency and that no engine could have a greater efficiency than that of a *reversible engine.* This result is known as the **Carnot theorem:**

Carnot theorem

> No engine working between two given heat reservoirs can be more efficient than a reversible engine working between those reservoirs.

A reversible engine working between two heat reservoirs is called a **Carnot engine.**

Let us look at what makes a process reversible or irreversible. We have already seen that the conversion of mechanical energy into heat by friction is irreversible. The reverse process of converting heat into mechanical energy with no other effects violates the heat-engine statement of the second law. Similarly, the conduction of heat from a hot object to a cold object is irreversible. The reverse process, the conduction of heat from a cold body to a hot body with no other effect, violates the refrigerator statement of the second law. A third type of irreversibility occurs when a system passes through nonequilibrium states, such as when there is turbulence in a gas or an explosion. The free expansion of a gas discussed in Section 16-5 is an example of this type of irreversibility. We never observe a free compression in which a gas in a large container moves by itself into the left or right half of the container. For a process to be reversible, we must be able to move the system back through the same equilibrium states in the reverse order.

From these considerations and our statements of the second law of thermodynamics, we can list some conditions that are necessary for a process to be reversible:

Conditions for reversibility

1. No work must be done by friction, viscous forces, or other dissipative forces that produce heat.
2. There can be no heat conduction due to a finite temperature difference.
3. The process must be quasi-static so that the system is always in an equilibrium state (or infinitesimally near an equilibrium state).

Any process that violates any of the above conditions is irreversible. Most processes in nature are irreversible. To have a reversible process, great care must be taken to eliminate frictional and other dissipative forces and to make the process quasi-static. Since this can never be done completely, a reversible process seems impossible in practice. Nevertheless, one can come very close to a reversible process, and the concept is very important in theory.

We now illustrate the Carnot theorem with a numerical example in which we show that if we assume that an engine exists that is more efficient than a reversible engine, the second law of thermodynamics is violated.

The engine in Figure 17-8*a* is defined to be reversible. It removes 100 J from the hot reservoir, does 40 J of work, and exhausts 60 J to the cold reservoir. Its efficiency is therefore 40 percent. In Figure 17-8*b*, the engine is reversed and run backwards as a refrigerator. Now, 40 J of work are done to remove 60 J from the cold reservoir and exhaust 100 J to the hot reservoir. (If

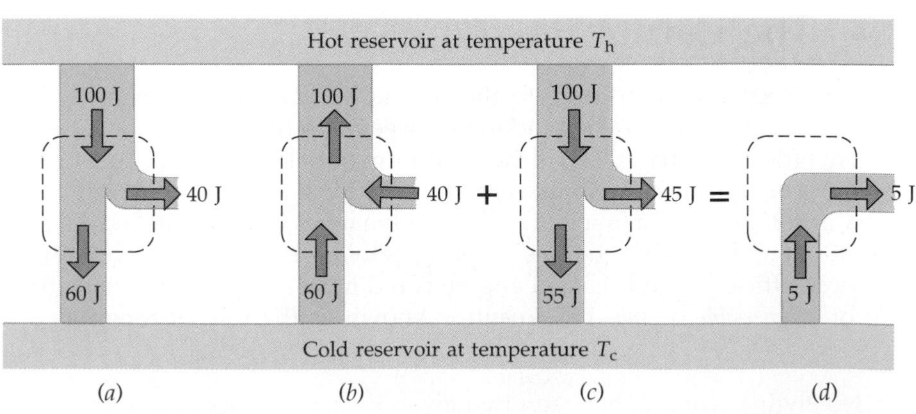

Figure 17-8 Illustration of the Carnot theorem. (*a*) A reversible heat engine with 40 percent efficiency. (*b*) The same engine runs backwards as a refrigerator. (*c*) An assumed heat engine working between the same two reservoirs with an efficiency of 45 percent, which is greater than that of the reversible engine in (*a*). (*d*) The net effect of running the engine in (*c*) in conjunction with the refrigerator in (*b*) is the same as that of a perfect heat engine that removes 5 J from the cold reservoir and converts it completely into work with no other effect. This is impossible because it violates the heat-engine statement of the second law. Thus, the reversible engine in (*a*) is the most efficient engine that can operate between these two reservoirs.

the engine were not reversible, more than 40 J of work would have to be done to remove 60 J from the cold reservoir.) Now, let us assume that there is another (nonreversible) engine working between the same two reservoirs that is more than 40 percent efficient. Figure 17-8*c* shows an engine with an efficiency of 45 percent. If we run this engine in conjunction with the refrigerator in Figure 17-8*b*, the net effect would be the same as that of a perfect engine that removes 5 J from the cold reservoir and converts it completely into work (Figure 17-8*d*). We have already seen that this is not possible because it violates the heat-engine statement of the second law of thermodynamics. Therefore, the reversible engine is the most efficient engine that can work between these two reservoirs.

If no engine can have a greater efficiency than a reversible engine, it follows that all reversible engines working between the same two reservoirs must have the same efficiency, which is called the **Carnot efficiency.** Because this efficiency is the same for all reversible engines working between these reservoirs, it must be independent of the working substances of the engines. Therefore, the Carnot efficiency depends only on the temperatures of the reservoirs.

Exhaust manifolds in the 8-cylinder Dusenberg carry heat from the engine to reduce its temperature. A typical automobile engine has an efficiency of less than 15 percent.

We can calculate the Carnot efficiency in terms of the temperatures of the reservoirs by considering a reversible engine that has an ideal gas as the working substance. A reversible cycle is called a **Carnot cycle.** Figure 17-9 shows a Carnot cycle for an ideal gas. The cycle starts at state 1 with a quasi-static isothermal absorption of heat from the hot reservoir at a temperature T_h. Since the heat is absorbed isothermally, the process could be reversed without violating the second law of thermodynamics. Work is done by the gas as it expands to state 2. From state 2 to state 3, the gas expands adiabatically, that is, with no heat exchange. If this expansion is done quasi-statically, it will be reversible. More work is done by the gas, and the temperature of the gas is reduced to T_c. The third part of the cycle is an isothermal compression at temperature T_c from state 3 to state 4. During this part of the cycle, work is done on the gas, and heat $|Q_c|$ is rejected to the cold reservoir at a temperature T_c. The last part of the cycle is an adiabatic compression from state 4 back to the original state 1. Work is again done on the gas during this compression. The net work done during the cycle is represented by the shaded area in the figure.

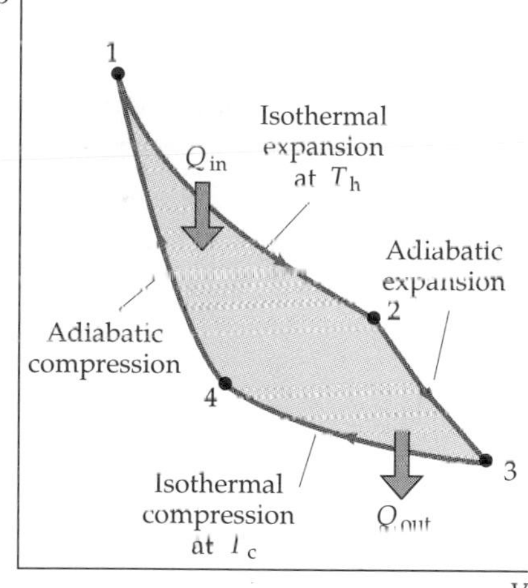

17-9 A Carnot cycle for an ideal gas. Heat is absorbed during the isothermal expansion from state 1 to state 2 at a temperature T_h. The gas expands adiabatically from state 2 to state 3 until its temperature is reduced to T_c. It then rejects heat as it is compressed isothermally to state 4, after which it is compressed adiabatically until its temperature is again T_h. All the processes are reversible.

To calculate the efficiency of this cycle, we calculate the heat absorbed from the hot reservoir and that rejected to the cold reservoir. We first consider the isothermal expansion from state 1 to state 2. Since the internal energy of an ideal gas remains constant during an isothermal expansion, the work done by the gas equals the heat absorbed from the hot reservoir. The heat absorbed is thus

$$Q_h = W = \int_1^2 P\,dV = \int_1^2 \frac{nRT_h}{V}\,dV$$

or

$$Q_h = nRT_h \ln \frac{V_2}{V_1}$$

Similarly, the heat rejected to the cold reservoir equals the work done on the gas during the isothermal compression at temperature T_c from state 3 to state 4. This work is of the same magnitude as that done by the gas if it expands from state 4 to state 3. The heat rejected is thus

$$|Q_c| = nRT_c \ln \frac{V_3}{V_4}$$

The ratio of these heats is

$$\frac{|Q_c|}{Q_h} = \frac{T_c \ln (V_3/V_4)}{T_h \ln (V_2/V_1)} \tag{17-4}$$

We can relate the volumes V_1, V_2, V_3, and V_4 using Equation 16-47 for an quasi-static adiabatic expansion:

$$TV^{\gamma-1} = \text{constant}$$

Applying this equation to the expansion from state 2 to state 3, we obtain

$$T_h V_2^{\gamma-1} = T_c V_3^{\gamma-1}$$

Similarly, for the adiabatic compression from state 4 to state 1, we have

$$T_h V_1^{\gamma-1} = T_c V_4^{\gamma-1}$$

Dividing these two equations, we obtain

$$\left(\frac{V_2}{V_1}\right)^{\gamma-1} = \left(\frac{V_3}{V_4}\right)^{\gamma-1}$$

and so $V_2/V_1 = V_3/V_4$. Then $\ln (V_2/V_1) = \ln (V_3/V_4)$, so we can cancel the logarithmic terms in Equation 17-4. We thus obtain

$$\frac{|Q_c|}{Q_h} = \frac{T_c}{T_h} \tag{17-5}$$

The Carnot efficiency ϵ_C is thus

Carnot efficiency

$$\epsilon_C = 1 - \frac{|Q_c|}{Q_h} = 1 - \frac{T_c}{T_h} \tag{17-6}$$

Equation 17-6 holds for any reversible engine working between reservoirs at the temperatures T_h and T_c. It gives the greatest possible efficiency for an engine working between these temperatures. No engine can have an efficiency greater than the Carnot efficiency. As we have seen, an engine with a greater efficiency than this would violate the second law of thermodynamics.

In Section 15-2, the ideal-gas temperature scale was defined in terms of the properties of gases at low densities. Since the Carnot efficiency depends only on the temperatures of the two heat reservoirs, it can be used to define the ratio of the temperatures of the reservoirs independent of any substance. We define the ratio of the absolute temperatures of the hot and cold reservoirs to be

$$\frac{T_c}{T_h} = 1 - \epsilon_C = \frac{|Q_c|}{Q_h} \tag{17-7}$$

To use this definition to measure the ratio of two temperatures, we have to set up a reversible engine to operate between two reservoirs at the two temperatures and carefully measure the heats absorbed from or rejected to those reservoirs during one cycle. Temperatures as low as a millionth of a kelvin have been measured in this way. The absolute temperature is completely determined by defining the ratio of two temperatures as in Equation 17-7 and choosing one fixed point, as was done for the ideal-gas temperature scale. If the absolute temperature of the triple point of water is defined to be 273.16 K, then the absolute temperature scale is identical to the ideal-gas temperature scale for the range of temperatures for which a gas thermometer can be used.

Example 17-3

A steam engine works between a hot reservoir at 100°C = 373 K and a cold reservoir at 0°C = 273 K. What is the maximum possible efficiency of this engine?

From Equation 17-6, the Carnot efficiency of the engine is

$$\epsilon_C = 1 - \frac{T_c}{T_h}$$

$$= 1 - \frac{273 \text{ K}}{373 \text{ K}}$$

$$= 0.268 = 26.8\%$$

Even though this efficiency seems to be quite low, it is the greatest efficiency possible for any engine working between these temperatures. Real engines will have lower efficiencies because of friction, heat conduction, and other irreversible processes.

We can see that it is very useful to know the Carnot efficiency. Suppose, for example, that we have a real engine that removes 100 J from a hot reservoir at 373 K, does 25 J of work, and exhausts 75 J to a cold reservoir at 273 K. Though its efficiency is only 25 percent, this is a very good engine. If we want to improve it by reducing friction and eliminating various other irreversibilities, the best efficiency we could obtain is 26.8 percent, as we calculated in Example 17-3. That is, the most work we could possibly get using an ideal engine between these reservoirs would be 26.8 J of work for every 100 J removed from the hot reservoir.

The ratio of the actual efficiency ϵ of an engine to the Carnot efficiency ϵ_C is called the **second law efficiency** ϵ_{SL}:

$$\epsilon_{SL} = \frac{\text{actual efficiency}}{\text{Carnot efficiency}} = \frac{\epsilon}{\epsilon_C} \qquad 17\text{-}8$$

The second-law efficiency of a real engine that has an efficiency of 25 percent when working between 373 K and 273 K is

$$\epsilon_{SL} = \frac{25\%}{26.8\%} = 0.93 = 93\%$$

Example 17-4

An engine removes 200 J from a hot reservoir at 373 K, does 48 J of work, and exhausts 152 J to a cold reservoir at 273 K. How much work is "lost" per cycle due to the irreversibilities of this engine?

The efficiency of this engine is ϵ = (48 J)/(200 J) = 24 percent. As we just saw in Example 17-3, the Carnot efficiency for an engine working between these two reservoirs is 26.8 percent, so this is a very good engine. Its second-law efficiency is 24/26.8 = 90 percent. A Carnot engine working between these reservoirs that removes 200 J from the hot reservoir could do 0.268 × 200 = 53.6 J of work per cycle. Since the real engine does 48 J of work per cycle, it could do only 5.6 J more work if it were an ideal engine. Thus, only 5.6 J of work is lost per cycle by this engine. This energy is, of course, not lost completely. The total energy is conserved. However, the real engine exhausts 5.6 J of energy into the cold reservoir that could have been converted into useful work if the engine were ideal, that is, reversible.

(*a*)

(*b*)

(*a*) Coal-fueled electric generating plant at Four Corners, New Mexico. (*b*) Power plant at Wairakei, New Zealand, that converts geothermal energy into electricity. (*c*) Solar energy is focused and collected individually to produce electricity by these heliostats that are being tested at Sandia National Laboratory. (*d*) Control rods are inserted into this nuclear reactor at Tihange, Belgium, (*e*) An experimental wind-powered electric generator at Sandia National Laboratory. The propeller is designed for optimum transfer of wind energy to mechanical energy.

Example 17-5

If 200 J of heat are conducted from a heat reservoir at 373 K to one at 273 K, how much work capability is "lost" in this process?

We saw in the previous example that a Carnot engine working between these two reservoirs could do 53.6 J of work if it extracted 200 J from the 373-K reservoir and exhausted to a 273-K reservoir. Thus, if 200 J is conducted directly from the hot reservoir to the cold reservoir without any work being done, 53.6 J of this energy has been "lost" in the sense that it could have been converted into useful work.

Exercise

A Carnot engine works between heat reservoirs at 500 K and 300 K. (*a*) What is its efficiency? (*b*) If it removes 200 kJ of heat from the hot reservoir, how much work does it do? [Answers: (*a*) 40%, (*b*) 80 kJ]

Exercise

A real engine works between heat reservoirs at 500 K and 300 K. It removes 500 kJ of heat from the hot reservoir and does 150 kJ of work during each cycle. (*a*) What is its efficiency? (*b*) What is its second-law efficiency? [Answers: (*a*) 30%, (*b*) 75%]

Questions

4. Why do power-plant designers try to increase the temperature of the steam fed to engines as much as possible?
5. Some think that electrical heat for the home is more efficient because all the electrical energy goes into heat in the home, whereas with a gas or oil furnace, some of the heat goes out the chimney. However, electrical energy is usually generated in a steam power plant, which is only about 40 percent efficient. In light of this fact, discuss the relative merits of electrical heat versus oil or gas heat for homes.

(c)

(d)

(e)

17-5 The Heat Pump

About 20 percent of our energy use is for the heating of our homes and office buildings. Much of this heating is done by the on-site combustion of fossil fuels (oil, coal, or natural gas). Although such combustion is relatively efficient, it is a major source of pollution in our cities. The alternative of using electric heat seems at first glance to be more efficient and less polluting, but this is actually far from the truth. Although some of our electricity is generated in hydroelectric and nuclear power plants, most comes from power plants that burn fossil fuels. This leads to just as much air pollution as the combustion of fossil fuels in homes and office buildings. In addition, the heat exhausted by power plants results in the thermal pollution of our lakes and rivers and the atmosphere. Furthermore, the Carnot efficiency of a power plant is about 50 percent, and the actual efficiency is only about 40 percent. An efficient alternative to the direct or indirect combustion of fossil fuels as a method of heating homes and office buildings in moderate climates is the heat pump.

A **heat pump** is essentially a refrigerator that is used to pump thermal energy from a cold reservoir (for example, the cold air outside a house) to a hot reservoir (for example, the hot air inside the house). If work W is done to remove heat Q_c from the cold reservoir and reject heat $|Q_h| = W + Q_c$ to the hot reservoir, the coefficient of performance (Equation 17-3) is

$$\text{COP} = \frac{Q_c}{W} \qquad 17\text{-}9$$

Using $W = |Q_h| - Q_c$, this can be written

$$\text{COP} = \frac{Q_c}{|Q_h| - Q_c} = \frac{Q_c/|Q_h|}{1 - Q_c/|Q_h|} \qquad 17\text{-}10$$

The maximum coefficient of performance is obtained using a reversible or Carnot heat pump. Then Q_c and Q_h are related by Equation 17-5:

$$\frac{Q_c}{|Q_h|} = \frac{T_c}{T_h}$$

Substituting $Q_c/|Q_h| = T_c/T_h$ into Equation 17-10, we obtain for the maximum coefficient of performance

$$\mathrm{COP}_{\mathrm{max}} = \frac{T_c/T_h}{1 - T_c/T_h} = \frac{T_c}{T_h - T_c}$$

or

$$\mathrm{COP}_{\mathrm{max}} = \frac{T_c}{\Delta T} \qquad 17\text{-}11$$

where ΔT is the difference in temperature between the hot and cold reservoirs. Real heat pumps and refrigerators have COPs less than the Carnot COP because of friction, heat conduction, and other irreversible processes.

We are usually interested in the work that must be done to exhaust a given amount of heat $|Q_h|$ into the hot reservoir. (For a house, the hot reservoir may be the hot-air supply for the heating fan.) Using $|Q_h| = Q_c + W$, we can write Equation 17-9 as

$$\mathrm{COP} = \frac{Q_c}{W} = \frac{|Q_h| - W}{W}$$

$$W(\mathrm{COP}) = |Q_h| - W$$

or

$$W = \frac{|Q_h|}{1 + \mathrm{COP}} \qquad 17\text{-}12$$

Example 17-6

An ideal heat pump is used to pump heat from the outside air at −5°C to the hot-air supply for the heating fan in a house, which is at 40°C. The pump is run by electric energy. How much work is required to pump 1 kJ of heat into the house?

From Equation 17-11, the Carnot coefficient of performance for a heat pump operating at $T_c = -5°\mathrm{C} = 268\ \mathrm{K}$ and $\Delta T = 40°\mathrm{C} - (-5°\mathrm{C}) = 45\ \mathrm{C}° = 45\ \mathrm{K}$ is

$$\mathrm{COP}_{\mathrm{max}} = \frac{T_c}{\Delta T} = \frac{268\ \mathrm{K}}{45\ \mathrm{K}} = 5.96$$

We can calculate the work needed from Equation 17-12 with $|Q_h| = 1$ kJ:

$$W = \frac{|Q_h|}{1 + \mathrm{COP}} = \frac{1\ \mathrm{kJ}}{1 + 5.96} = 0.144\ \mathrm{kJ}$$

Thus, only 0.144 kJ of work is needed to pump 1 kJ of heat into the hot-air supply in the house.

We see from this example and Equation 17-12 that the heat pump essentially multiplies the energy needed to run the pump by 1 + COP. That is, if we use 1 kJ to run a heat pump, with a COP = 6.83, we can exhaust 7.83 kJ of heat into the house.

17-6 Entropy and Disorder

We have seen that the second law of thermodynamics is related to the fact that some processes are irreversible; that is, they go in one direction only. However, there are many irreversible processes that are not easily described by the heat-engine or refrigerator statements of the second law. Examples are the free expansion of a gas or a glass falling off a table and shattering when it hits the floor.

All irreversible processes have one thing in common—the system plus its surroundings moves toward a less ordered state. This can be illustrated with a simple example. Consider a box containing a gas of mass M at a temperature T that is moving along a frictionless table with a velocity V_{cm} (Figure 17-10*a*). The total kinetic energy of the gas has two components. The first is the kinetic energy associated with the movement of the center of mass of the gas, $\frac{1}{2}MV_{cm}^2$. This is ordered mechanical energy that could be converted directly into work. For example, if a weight were attached with a string to the moving box, this energy could be used to lift the weight. The other component of the gas's kinetic energy is the energy of the motion of its molecules relative to its center of mass. This is the gas's internal thermal energy, which is related to its temperature T; it is random, nonordered energy that cannot be converted directly into work.

Now, suppose the box of gas collides inelastically with a fixed wall (Figure 17-10*b*), which is an irreversible process. Disregarding the box itself, we note that all of the ordered mechanical energy of the gas is converted into random internal energy, causing the temperature of the gas to rise. The gas still has the same total energy, but now all of it is associated with the random motion of its molecules about its center of mass, which is now at rest. Thus, the gas has become less ordered, or more disordered, and it has lost the ability to do work.

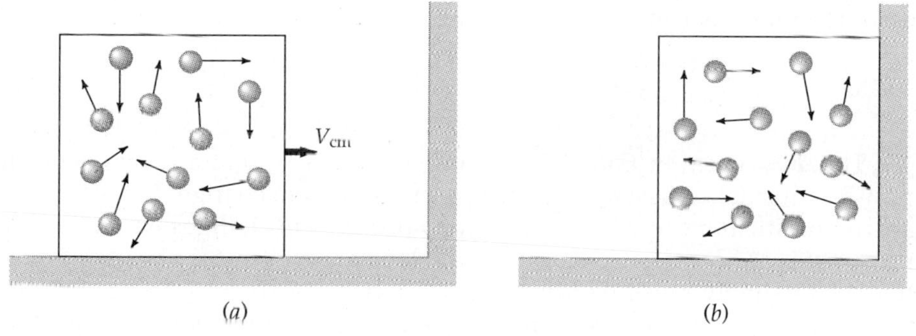

Figure 17-10 (*a*) A box of gas moving to the right with a speed V_{cm}. The kinetic energy of the molecules associated with the motion of the gas's center of mass is ordered energy. It can be converted directly into work. The kinetic energy of the molecules associated with their motion relative to the center of mass is their thermal energy, which is related to the temperature of the gas. This is disordered energy and cannot be converted directly into work. (*b*) After the box has collided inelastically with the wall, the gas molecules have the same total energy as before, but all of it is now disordered. This is indicated by the higher temperature of the gas.

There is a thermodynamic function called **entropy** S that is a measure of the disorder of a system. Like the pressure P, volume V, temperature T, and internal energy U, entropy is a function of the state of a system. As with internal energy, it is the change in entropy that is important. The change in entropy ΔS of a system when it goes from one state to another is defined as

$$\Delta S = \int \frac{dQ_{rev}}{T} \qquad 17\text{-}13$$

Entropy change defined

where dQ_{rev} is the heat that must be added to the system in a reversible process to bring it from its initial state to its final state. If heat is subtracted from the system, dQ_{rev} is negative and the entropy change of the system is negative.

We will illustrate the existence of the entropy function for an ideal gas. Let us consider an arbitrary reversible quasi-static process in which a system consisting of an ideal gas absorbs an amount of heat dQ. According to the first law, dQ is related to the change in the internal energy dU of the gas and the work done $dW = P\,dV$ by

$$dQ = dU + dW = dU + P\,dV$$

For an ideal gas, we can write dU in terms of the heat capacity, $dU = C_v\, dT$, and we can substitute nRT/V for P from the equation of state. Then

$$dQ = C_v\, dT + nRT \frac{dV}{V} \qquad \text{17-14}$$

Because W and Q are not state functions, both the work done and the heat absorbed by a system when it goes from one state to another depend on the process by which the change of state occurs. We must therefore know the path of the system on a PV diagram in order to calculate the heat absorbed by the system. This can also be seen mathematically from the fact that we cannot integrate Equation 17-14. There is no problem with the first term on the right side of the equation because $C_v = dU/dT$ is a function of temperature only. The integration of this term gives the change in the internal energy ΔU. However, we cannot integrate the second term unless we know how T depends on V. This is just another way of saying that dQ and dW are not differentials of state functions because Q and W are not state functions. But if we divide each term in Equation 17-14 by T, we obtain

$$\frac{dQ}{T} = C_v \frac{dT}{T} + nR \frac{dV}{V} \qquad \text{17-15}$$

Since C_v depends only on T, the first term on the right can still be integrated and now the second term can be integrated also. (Mathematically, the factor $1/T$ is called an integrating factor for Equation 17-14.) For simplicity, we will assume that C_v is constant. Integrating Equation 17-15, we obtain

$$\Delta S = \int \frac{dQ}{T} = C_v \ln \frac{T_2}{T_1} + nR \ln \frac{V_2}{V_1} \qquad \text{17-16}$$

The dQ_{rev} term in Equation 17-13 does not mean that a reversible heat transfer must take place for the entropy of a system to change. Indeed, there are many situations in which the entropy of a system changes when there is no transfer of heat whatsoever. (We will look at an example shortly.) Equation 17-13 simply gives us a method of calculating the entropy difference between two states of a system. Like internal energy, entropy is a state function. Therefore, the change in the entropy of a system when it moves from one state to another depends only on the system's initial and final states, not on the process by which the change occurs. However, to *calculate* the change in entropy for an irreversible process, we must first find a reversible process that connects the initial and final states, then find the heat added or subtracted during this reversible process, and use Equation 17-13. Calculating the entropy changes for some simple examples should clarify this somewhat subtle point.

We first consider a substance that is heated at constant pressure from temperature T_1 to temperature T_2. The amount of heat dQ absorbed by a substance at constant pressure, is related to its temperature change dT by

$$dQ = C_p\, dT$$

As we have seen, heat conduction between two systems that have a finite temperature difference is irreversible. We can approximate reversible heat conduction if we have a large number of heat reservoirs with temperatures ranging from T_1 to T_2 in very small steps. We then place our substance whose initial temperature is T_1 in contact with the first reservoir at a temperature just slightly greater than T_1 and let the substance absorb a small amount of heat. Since the heat transfer is approximately isothermal, the process will be approximately reversible. We then place the substance in contact with the next reservoir at a slightly higher temperature, and so on, until the final temperature T_2 is reached. When the heat is absorbed reversibly, the entropy change of the substance is

$$dS = \frac{dQ}{T} = C_P \frac{dT}{T}$$

Integrating from T_1 to T_2, we obtain the total entropy change of the substance

$$\Delta S = C_P \int_{T_1}^{T_2} \frac{dT}{T} = C_P \ln \frac{T_2}{T_1} \qquad 17\text{-}17$$

Since entropy is a state function, its change is independent of the process. If the final pressure equals the initial pressure, the entropy change of a substance is given by Equation 17-17 when it is heated by any process—whether it is reversible or irreversible. Equation 17-17 also gives the entropy change of a substance that is cooled. In this case, T_2 is less than T_1, and $\ln (T_2/T_1)$ will be negative, giving a negative entropy change.

Exercise

Find the change in entropy of 1 kg of water that is heated from 0°C to 100°C. (Answer: $\Delta S = 1.31$ kJ/K)

Next we consider a reversible isothermal expansion of an ideal gas at a temperature T from an initial volume V_1 to a final volume V_2. As we have seen before, work W is done by the gas and heat $Q = W$ is absorbed from a reservoir at temperature T. The entropy change of the gas is given by Equation 17-16 with $T_2 = T_1$:

$$\Delta S = S_2 - S_1 = nR \ln \frac{V_2}{V_1} \qquad 17\text{-}18$$

The entropy change of the gas is positive because V_2 is greater than V_1. In this process, an amount of heat $|Q|$ leaves the reservoir and enters the gas. This amount of heat equals the work done by the gas

$$|Q| = W = \int_1^2 P\,dV = nRT \int_{V_1}^{V_2} \frac{dV}{V} = nRT \ln \frac{V_2}{V_1} \qquad 17\text{-}19$$

The entropy change of the gas is $+|Q|/T$. Since the same amount of heat leaves the reservoir at temperature T, the entropy change of the reservoir is $-|Q|/T$. The net entropy change of the gas plus the reservoir is zero. This example illustrates a general result:

In a reversible process, the entropy change of the universe is zero. By "universe," we mean the system plus its surroundings.

We now consider the free expansion of an ideal gas—the irreversible process we first discussed in Section 16-5. Let the gas be at a temperature T and let it expand freely from an initial volume V_1 to a final volume V_2. As we have seen, no work is done and no heat is transferred in a free expansion. Because the process is not reversible, we cannot use $\int dQ/T$ to find the change in entropy of the gas. However, the initial and final states of the gas in the free expansion are the same as those of the gas in the isothermal expansion in the previous example. Therefore, since entropy is a state function and since the change in the entropy of a system for any process depends only on the initial and final states of the system, the entropy change for the free expansion is the same as that for the isothermal expansion. (As was noted earlier, to find the entropy change for an irreversible process, we must first find a reversible process that connects the same initial and final states.) The entropy change for the free expansion is thus given by Equation 17-18. Since the gas is a system that is isolated from its surroundings, this is also the entropy change for the universe:

$$\Delta S_u = S_2 - S_1 \qquad 17\text{-}20$$

$$= nR \ln \frac{V_2}{V_1}$$

Note that because V_2 is greater than V_1, the change in entropy of the universe for this irreversible process is positive; that is, the entropy of the universe increases. This is also a general result:

> In an irreversible process, the entropy of the universe increases.

If the final volume in the free expansion were less than the initial volume the entropy of the universe would decrease, but this does not happen. A gas does not freely contract by itself into a smaller volume. This leads us to another statement of the second law of thermodynamics:

> For any process, the entropy of the universe never decreases.

This statement is equivalent to the heat-engine and refrigerator statements.

Next, we consider an example of a block of mass m falling from a height h and making an inelastic collision with the ground. Clearly this is an irreversible process. Let the block, ground, and atmosphere all be at a temperature T, which is not significantly changed by the process. If we consider the block, ground, and atmosphere as our isolated system, there is no heat conducted into or out of the system. The state of the system has been changed because its internal energy has been increased by an amount mgh. This change is the same as if we added heat $Q - mgh$ to the system at constant temperature T. To calculate the change in entropy of the system, we thus consider a reversible process in which heat $Q_{rev} = mgh$ is added at a constant temperature T. According to Equation 17-13, the change in entropy is then

$$\Delta S = \frac{Q}{T} = \frac{mgh}{T}$$

This is also the entropy change of the universe ΔS_u.

In this process, energy is "wasted." When the block of mass m was at a height h, its potential energy mgh could have been used to do useful work, such as lifting a weight. After the inelastic collision this energy is no longer available because it has become the disordered internal energy of the block and its surroundings. (Some of it would be available if there were a reservoir at a temperature lower than that of the atmosphere T so that the energy mgh could be extracted and used to power a heat engine running between temperature T and that of the colder reservoir.) The energy that has become unavailable is equal to $T\,\Delta S_u$. This is a general result:

> In an irreversible process, energy equal to the entropy change of the universe times the temperature of the coldest available reservoir becomes unavailable for doing work.

For simplicity, we will call the energy that becomes unavailable for doing work the "work lost":

$$W_{\text{lost}} = T\,\Delta S_{\text{u}} \qquad 17\text{-}21$$

In the free expansion discussed earlier the ability to do work was also lost. In that case, the entropy change of the universe was $nR \ln (V_2/V_1)$, so the work lost was $nRT \ln (V_2/V_1)$. This is the amount of work that could have been done if the gas had expanded quasi-statically and isothermally from V_1 to V_2, as given by Equation 17-19.

Lastly, we consider heat conduction from one heat reservoir to another. Let heat $|Q|$ be conducted from a hot reservoir at a temperature T_{h} to a cold reservoir at a temperature T_{c}. The state of a heat reservoir is determined only by its temperature and its internal energy. The change in entropy of a heat reservoir due to a heat exchange is the same whether the heat exchange is reversible or not. If heat $|Q|$ is put into a reservoir at temperature T, the entropy of the reservoir increases by $|Q|/T$. If the heat is removed, the entropy of the reservoir decreases by $-|Q|/T$. In the case of heat conduction, the hot reservoir loses heat, so its entropy change is

$$\Delta S_{\text{h}} = -\frac{|Q|}{T_{\text{h}}}$$

The cold reservoir absorbs heat, so its entropy change is

$$\Delta S_{\text{c}} = +\frac{|Q|}{T_{\text{c}}}$$

The net entropy change of the universe is

$$\Delta S_{\text{u}} = \frac{|Q|}{T_{\text{c}}} - \frac{|Q|}{T_{\text{h}}}$$

According to the general result that the work lost equals the entropy change times the temperature of the coldest reservoir available, the work lost in this process is

$$W_{\text{lost}} = T_{\text{c}}\,\Delta S_{\text{u}} = |Q|\left(1 - \frac{T_{\text{c}}}{T_{\text{h}}}\right)$$

We can see that this is just the work that could have been done by a Carnot engine running between these reservoirs, removing heat $|Q|$ from the hot reservoir and doing work $W = \epsilon_{\text{C}}\,|Q|$, where $\epsilon_{\text{C}} = 1 - T_{\text{c}}/T_{\text{h}}$.

Example 17-7

During each cycle, a Carnot engine removes 100 J of energy from a reservoir at 400 K, does work, and exhausts heat to a reservoir at 300 K. Compute the entropy change of each reservoir for each cycle, and show that the entropy change of the universe is zero for this reversible process.

The efficiency of a Carnot engine working between these two reservoirs is

$$\epsilon_{\text{C}} = 1 - \frac{T_{\text{c}}}{T_{\text{h}}} = 1 - \frac{300\text{ K}}{400\text{ K}} = 0.25$$

The engine therefore exhausts 75 J to the cold reservoir during each cycle. Since the hot reservoir loses heat, its entropy change is negative:

$$\Delta S_{400} = -\frac{|Q_h|}{T_h} = -\frac{100\text{ J}}{400\text{ K}} = -0.250\text{ J/K}$$

The cold reservoir gains heat, so its entropy increases:

$$\Delta S_{300} = \frac{|Q_c|}{T_c} = \frac{75\text{ J}}{300\text{ K}} = +0.250\text{ J/K}$$

Since the engine works in a cycle, its entropy change is zero. The entropy change of the universe is therefore the sum of the entropy changes of the hot and cold reservoirs, which is zero.

Example 17-8

During each cycle, a real heat engine with a second-law efficiency of 60 percent removes 100 J from a reservoir at 400 K, does work, and exhausts heat to a reservoir at 300 K. Compute the entropy change of each reservoir for each cycle.

Since the second-law efficiency is equal to the actual efficiency divided by the Carnot efficiency, the efficiency ϵ of this engine is

$$\epsilon = \epsilon_{SL}\epsilon_C = (0.60)(0.25) = 0.15$$

where we have used the Carnot efficiency we calculated in Example 17-7. The engine therefore does $\epsilon|Q_h| = (0.15)(100\text{ J}) = 15\text{ J}$ of work and exhausts 85 J to the cold reservoir during each cycle. The entropy change of the hot reservoir is the same as in Example 17-7:

$$\Delta S_{400} = -\frac{100\text{ J}}{400\text{ K}} = -0.250\text{ J/K}$$

The entropy change of the cold reservoir is

$$\Delta S_{300} = \frac{85\text{ J}}{300\text{ K}} = +0.283\text{ J/K}$$

In this case, the entropy increase of the cold reservoir is greater than the decrease for the hot reservoir. The entropy change of the universe for one cycle is therefore

$$\begin{aligned}\Delta S_u &= \Delta S_{400} + \Delta S_{300} \\ &= -0.250\text{ J/K} + 0.283\text{ J/K} = 0.033\text{ J/K}\end{aligned}$$

17-7 Entropy and Probability

In the previous section, we saw that entropy is a measure of the disorder of a system and that, even though the entropy of a given system can decrease during an irreversible process, the net change in entropy of the universe (the system plus its surroundings) is always positive. In other words, the universe always moves toward a state of less order. In this section, we will see that the entropy of a state is related to probability. Essentially, a state of high order has a low probability, whereas a state of low order has a high probability. Thus, in an irreversible process, the universe moves from a state of low probability to one of high probability.

Let us consider a free expansion in which a gas expands from an initial volume V_1 to a final volume $V_2 = 2V_1$. The entropy change of the universe for this process is given by Equation 17-20:

$$\Delta S_u = nR \ln \frac{V_2}{V_1} = nR \ln 2 \qquad \text{17-22}$$

As we have mentioned, this process wastes energy in that the gas could have been used to absorb heat Q from a reservoir and change it completely into work W. After the free expansion, this capability is no longer there because the gas is at a greater volume and a lower pressure.

Why is this process irreversible? Why can't the gas compress by itself back into its original volume? Since there is no energy change involved, a compression would not violate the first law of thermodynamics. The reason is merely that such a compression is extremely improbable. To see this, let us begin by assuming that the gas consists of only 10 molecules, and that, initially, these molecules occupy the entire volume of their container. If we assume that each molecule moves about the total volume of the container randomly, there will be an equal chance of a given molecule being in the left or right half of the container. That is, the chance that any one particular molecule will be in the left half of the container at any given time is $\frac{1}{2}$. The chance that any two particular molecules will both be in the left half is $\frac{1}{2} \times \frac{1}{2} = \frac{1}{4}$. (This is the same as the chance that a coin flipped twice will come up heads both times.) The chance that three particular molecules will be in the left half is $\frac{1}{2} \times \frac{1}{2} \times \frac{1}{2} = (\frac{1}{2})^3 = \frac{1}{8}$. The chance that all 10 molecules will be in the left half is $(\frac{1}{2})^{10} = \frac{1}{1024}$. That is, there is 1 chance in 1024 that all 10 molecules will be in the left half of the container at any given time.

© 1987 Sidney Harris

Though the probability of all 10 molecules being on one side of the container is very small, we would not be too surprised to see it occur. For example, if we look at the gas once each second, we could expect to see it happen once in every 1024 seconds, which is about once every 17 minutes. If we started with the 10 molecules randomly distributed and then found them all in the left half of the original volume, the entropy of the universe would have decreased by $nR \ln 2$. However, this decrease is extremely small, since the number of moles n corresponding to 10 molecules is only about 10^{-23}. Still, it would violate the entropy statement of the second law of thermodynamics, which says that for any process, the entropy of the universe never decreases. Therefore, we must consider the second law to be a statement of probability.

It may be disturbing to learn that the second law of thermodynamics seems to say that irreversible processes, such as the spontaneous compression of a gas or the spontaneous conduction of heat from a cold body to a hot body, are only improbable, not impossible. But, as we have just seen, if we consider a system consisting of a very small number of molecules, there is a reasonable chance that an irreversible process will occur. However, thermodynamics itself is applicable only to macroscopic systems, that is, to systems that have a very large number of molecules. Consider, for example, trying to measure the pressure of a gas consisting of only 10 molecules. The pressure would vary wildly depending on whether no or one or two molecules were colliding with the wall of the container at the time of measurement.

As we increase the number of molecules in a system, the chance of an irreversible process occurring decreases dramatically. For example, if we have 50 molecules in a container, the chance that they will all wind up in the left half of the volume is $(\frac{1}{2})^{50} \approx 10^{-15}$. Thus, if we look at the gas once each second, we could expect to see all 50 molecules in the left half of the volume about once in every 10^{15} seconds or once in every 36 million years! For 1 mole = 6×10^{23} molecules, the chance that all will wind up in half of the volume is extremely small, so small that it is essentially zero. So, for macroscopic systems, the probability of a process resulting in a decrease in the entropy of the universe is so extremely small that the distinction between improbable and impossible becomes blurred.

We can relate the probability of a gas spontaneously compressing itself into a smaller volume to the change in its entropy. If the original volume is V_1, the probability p of finding N molecules in a smaller volume V_2 is

$$p = \left(\frac{V_2}{V_1}\right)^N$$

Taking the natural logarithm of both sides of this equation, we obtain

$$\ln p = N \ln\left(\frac{V_2}{V_1}\right) = nN_A \ln\left(\frac{V_2}{V_1}\right) \qquad 17\text{-}23$$

where n is the number of moles and N_A is Avogadro's number. The entropy change of the gas is

$$\Delta S = nR \ln\left(\frac{V_2}{V_1}\right) \qquad 17\text{-}24$$

(The entropy change is negative if V_2 is less than V_1). Comparing Equations 17-23 and 17-24, we see that

$$\Delta S = \frac{R}{N_A} \ln p = k \ln p \qquad 17\text{-}25$$

where k is Boltzmann's constant.

Summary

1. A heat engine removes heat Q_h from a hot reservoir, does work W, and exhausts heat $|Q_c|$ to a cold reservoir. The efficiency of a heat engine is the ratio of the work done to the total heat absorbed:

$$\epsilon = \frac{W}{Q_h} = 1 - \frac{|Q_c|}{Q_h}$$

2. A refrigerator uses work W to remove heat Q_c from a cold reservoir and exhausts heat $|Q_h| = Q_c + W$ to a warmer reservoir. The coefficient of performance of a refrigerator is the ratio of the heat removed to the work that must be added:

$$\text{COP} = \frac{Q_c}{W}$$

3. According to the heat-engine statement of the second law of thermodynamics, it is impossible for a heat engine working in a cycle to remove heat from a reservoir and convert it completely into work with no other effects. According to the refrigerator statement of the second law, it is impossible to transfer heat from a cold reservoir to a hot reservoir with no other effects. These two statements of the second law of thermodynamics are equivalent.

4. The Carnot theorem states that no engine working between two heat reservoirs can be more efficient than a reversible engine. A reversible engine is called a Carnot engine. The efficiency of a Carnot engine is

$$\epsilon_C = 1 - \frac{T_c}{T_h}$$

For a Carnot engine, the ratio of the heat exhausted to the cold reservoir to the heat removed from the hot reservoir is

$$\frac{|Q_c|}{Q_h} = \frac{T_c}{T_h}$$

The second-law efficiency of a heat engine is the ratio of the actual efficiency to the Carnot efficiency:

$$\epsilon_{SL} = \frac{\epsilon}{\epsilon_C}$$

5. A heat pump is a device that removes energy from a cold reservoir, such as the air outside a house, and exhausts heat into a hot reservoir, such as the air inside a house. The coefficient of performance of a Carnot refrigerator or heat pump is

$$COP_{max} = \frac{T_c}{\Delta T}$$

where $\Delta T = T_h - T_c$.

The work W needed to exhaust heat $|Q_h|$ into a hot reservoir is related to the coefficient of performance by

$$W = \frac{|Q_h|}{1 + COP}$$

6. During an irreversible process, the universe moves toward a state of less order. A measure of the disorder of a system and of the reversibility of a process is the entropy function. The change in entropy of a system is given by

$$\Delta S = \int \frac{dQ_{rev}}{T}$$

where dQ_{rev} is the heat added in a reversible process connecting the initial and final states of the system.

7. The entropy of a system can increase or decrease, but the entropy of the universe or of any other isolated system never decreases. For a reversible process, the entropy of the universe remains constant; for an irreversible process, it increases. The statement that for any process the entropy of the universe never decreases is a statement of the second law of thermodynamics that is equivalent to the heat-engine and refrigerator statements.

8. During an irreversible process, the entropy of the universe S_u increases and a certain amount of energy

$$W_{lost} = T\,\Delta S_u$$

becomes unavailable for doing work.

9. Entropy is related to probability. A highly ordered system is one of low probability and low entropy. An isolated system moves towards a state of low order, high probability, and high entropy.

Exponential Growth and Decay

Albert A. Bartlett
University of Colorado at Boulder

If a town with a population of 10,000 has a steady annual population growth of 5 percent, how large will the population be in 10, 30, and 100 y?* If I invest $100 in an account that pays 15 percent annual interest (compounded continuously), how large will the account be in 10, 30, and 100 y? These questions are real-life applications of what may be the single most important differential equation:

$$\frac{dN}{dt} = kN \tag{1}$$

This equation describes the situation in which the rate of change of a quantity dN/dt is proportional to the size N of the quantity. In our first example, N is the population of the town and $k = 0.05\ \mathrm{y}^{-1}$; in the second example N is the number of dollars in the account and $k = 0.15\ \mathrm{y}^{-1}$. In both cases t is the time in years. The meaning of k can be seen by rearranging Equation 1 to read

$$k = \frac{1}{N}\frac{dN}{dt} \tag{2}$$

Thus k is the fractional change (dN/N) in N per unit time dt. It follows that $100k$ is P, the percent growth per unit time. These two examples are cases of steady growth in which the quantity always increases by a fixed fraction (for example, 5 percent) in a fixed length of time (each year).

*These figures are for continuous compounding. Annual compounding will give growth at slightly slower rates.

Albert A. Bartlett is a professor of physics at the University of Colorado in Boulder (80309-0390) where he has been a member of the faculty since 1950. His undergraduate degree is from Colgate University and he has a Ph.D. from Harvard (nuclear physics). In 1978 he was President of the American Association of Physics Teachers. He has lectured widely on the "Forgotten Fundamentals of the Energy Crisis."

The solution of Equation 1 is the exponential function

$$N = N_0 e^{kt} \tag{3}$$

where e is the base of natural logarithms (2.718 . . .) and N_0 is the size of N at the time $t = 0$. Equation 3 describes steady growth at a fixed percent per unit time. This is sometimes called exponential growth and it occurs whenever the rate of increase in a quantity is proportional to that quantity.

In our two examples in the first paragraph, the town's population will be

$$N = 10{,}000e^{0.05t}$$

and the size of the bank account will be

$$N = 100e^{0.15t}$$

where t is the time in years.

Using Equation 3, one can show that for a given value of k, a constant time T_c is required for N to increase in size from its present value N_0 to CN_0, where C is a constant:

$$T_c = \frac{1}{k}\ln C \tag{4}$$

The time for N to double in size ($C = 2$) is

$$T_2 = \frac{1}{k}\ln 2 = \frac{0.693}{k} \tag{5}$$

which can be expressed as

$$T_2 = \frac{69.3}{P} \approx \frac{70}{P} \tag{6}$$

where $P = 100k$ is the percent growth per unit time. Table 1 shows doubling times for different steady rates of growth calculated from Equation 6.

From Table 1 we see that a quantity growing steadily at 5 percent per year will double in size in 13.9 y (steady growth at a rate of 5 percent per month will cause doubling to take place in 13.9 months). A quantity growing steadily at 15 percent per year will double in 4.62 y. We should develop the habit of doing the mental arithmetic of Equation 6 every time we see a percent growth rate in the news or elsewhere. People who do not grasp the meaning of 15 percent annual inflation will understand when you tell them that this means that prices will double every 4.6 years!

Equation 3 can be rewritten in a very useful form using T_2:

$$N = N_0 e^{kt} = N_0 2^{t/T_2} \tag{7}$$

Table 1 **Doubling Times for Various Rates of Steady Growth**

k, y^{-1}	P, %/y	T_2, y
0.01	1	69.3
0.02	2	34.7
0.03	3	23.1
0.05	5	13.9
0.07	7	9.90
0.10	10	6.93
0.15	15	4.62
0.20	20	3.47
0.30	30	2.31

In this form we can estimate the answers to the questions that opened this essay. Since the doubling time for an annual growth rate of 5 percent is roughly 14 y, it can be seen that 10 y is less than one doubling time, 30 y is approximately two doubling times, and 100 y is approximately 7 doubling times. Thus in 10 y the population will increase from 10,000 to roughly 16,000; in 30 y it will have doubled twice (to 40,000); and in 100 y it will have doubled 7 times (a factor of $2^7 = 128$), and so the population will be 1,280,000. For a 15 percent annual interest rate compounded continuously ($T_2 = 4.62$ y) the figures are shown in Table 2.

We can estimate the value of large powers of 2 by noting that

$$2^{10} = 1024 \approx 10^3$$

For example,

$$2^{22} = 2^{10} \times 2^{10} \times 2^2 \approx 10^3 \times 10^3 \times 4 \approx 4 \times 10^6 \tag{8}$$

Some very important properties of steady growth can be illustrated by the following example. Legend has it that the game of chess was invented by a mathematician who worked for a king. As a reward the mathematician asked the king for the quantity of wheat determined by placing one grain of wheat on the first square of the chessboard, on the next square doubling the one grain to make two, on the next square doubling the two grains to make four, and continuing until the doubling has been done for all the squares (see Table 3). We can use Equation 8 to get an approximate expression for this total number of grains:

$$2^{64} \approx 10^3 \times 10^3 \times 10^3 \times 10^3 \times 10^3 \times 10^3 \times 16 \approx 1.6 \times 10^{19} \text{ grains}$$

The actual value of 2^{64} is 1.84×10^{19}. This is approximately 500 times the 1980 worldwide harvest of wheat, which may be more wheat than people have harvested in the entire history of the earth. How did we get such a large number? It was simple. We started with one grain and let the number grow until it had doubled a mere 63 times. A study of Table 3 reveals another important point. The growth in any doubling time is greater than the sum of all of the preceding growth. Thus when we put 32 grains on the sixth square, the 32 is larger than the total of 31 already on the board.

In his famous speech on energy (April 18, 1977) President Carter said, "In each of those decades [the 1950s and 1960s] more oil was consumed [worldwide] than in all of mankind's previous history." This statement sounds incredible, but a look at Table 3 shows that this is a simple consequence of steady growth with a doubling time of 10 years, that is, an annual growth rate of 7 percent. This was the growth rate of world oil consumption for nearly 100 years before 1970. Oil prices increased dramatically in the 1970s and this led people to seek improved efficiency in the use of oil. As a result, the growth in consumption stopped. In the 1980s it started to rise again but at a rate less than 7 percent.

Table 2 **The Value of $100 Deposited at 15% Annual Interest (Compounded Continuously) after a Time *t***

t, y	t/T_2	$2^{t/T_2}$	Size of account
10	10/4.62 = 2.16	4.48	$448
30	30/4.62 = 6.49	90	$9000
100	100/4.62 = 21.6	3.27×10^6	$327,000,000

Table 3 **Grains of Wheat on a Chessboard**

No. of square	No. on square	Total No. on board
1	1	1
2	2	3
3	4	7
4	8	15
5	16	31
6	32	63
. . .	. . .	. . .
64	2^{63}	$2^{64} - 1$

Continued

This feature of exponential growth is extremely important. Suppose, for example, that the annual growth rate of 7 percent continued and that just as the world's supply of oil is to be exhausted, a new source of oil is discovered equal to total amount of oil consumed worldwide in the entire history of mankind. If oil consumption continued to grow at the annual rate of 7 percent, this fantastic new supply will last approximately one doubling time, or 10 years. This example makes it clear that, contrary to what prominent people often suggest, exploration and production of fossil fuels cannot meet the needs of continued growth for very long. It is interesting to calculate how long a quantity M_0 of a resource will last if it is now being consumed at a rate R_0 and if its rate of consumption is growing steadily in accord with Equation 3. It is easy to show[1,2] that the expiration time T_e is

$$T_e = \frac{1}{k} \ln \left(\frac{kM_0}{R_0} + 1 \right) \tag{9}$$

We can use Equation 9 to estimate the lifetime T_e of United States coal reserves for different steady rates of growth of coal production. The U.S. Geological Survey[3] estimates the reserve base of United States coal to be $M_0 = 379 \times 10^9$ tonnes (1 tonne = 10^3 kg). The rate of production in 1978 was $R_0 = 0.6 \times 10^9$ tonnes/y. Using these numbers in Equation 10 we find the results shown in Table 4.

If we have zero growth of coal production, United States coal would last $T_e = M_0/R_0 = 632$ y. This is the basis for news stories that say "at present rates of consumption" United States coal will last about 600 y.[4] These stories then go on to stress the national urgency of achieving rapid growth of United States coal production, but the stories don't tell how growth reduces T_e (see Table 4).

In late 1989, the U.S Geological Survey released its estimates of the amount of petroleum (discovered *and undiscovered*) that remains in the ground in the United States. If the rate of extraction of oil from the ground does not change from the 1989 level (no growth), this petroleum will last 32 years.[5] This 1989 rate of extraction was supplying only about half of our domestic consumption. The other half is imported.

If the rate of change of N is negative, the right sides of Equations 1 and 2 are negative and Equation 3 becomes

$$N = N_0 e^{-kt} \tag{10}$$

Table 4 **Lifetime of United States Coal Reserves for Different Steady Rates of Growth of Coal Production**

Annual growth rate, %	Expiration time of the reserve base
0	632
1	199
2	131
3	100
5	70
10	42
15	30

Exponential decay occurs in Newton's law of cooling (Chapter 16), where N is the temperature difference between an object and its surroundings; in the study of RC circuits (Chapter 23), where N is the charge on the capacitor; in LR circuits (Chapter 26), where N is the current in the circuit; and in radioactive decay (Chapter 40), where N is the number of radioactive nuclei that have not yet decayed.

A common example of this is the decay of radioactive substances. In this case N_0 is the number of atoms of a radioactive substance at time $t = 0$, and N is the number that have not decayed after a time t. The "half-life" is the time $T_{1/2}$ required for N to decay to $N_0/2$:

$$T_{1/2} = \frac{\ln 2}{k} = \frac{0.693}{k} \tag{11}$$

After one half-life, half the substance remains, after two half-lives, one quarter remains, and after n half-lives the fraction remaining is $(\frac{1}{2})^n$. Table 5 lists half-lives of a few radioactive species.

The exponential arithmetic that arises so naturally in physics and in many other real-life situations is incredibly important—in part because it is so poorly understood by the people who make our policies (both local and national), who follow the dictum that "growth is good," and who never ask how long growth can continue. One can see that the energy shortage was caused by growth in consumption; the energy crisis was caused by our failure to understand the arithmetic and consequences of this growth. If we try to continue to have growth in our national and global patterns of energy consumption, we can ask, "Is the next energy crisis far away?" In the words of Aldous Huxley, "Facts do not cease to exist because they are ignored."

Table 5 **Half-Lives of Some Radioactive Substances**

Substance	Half-life
Carbon-14	5735 years
Potassium-40	1.26×10^9 years
Cobalt-60	5.26 years
Strontium-90	27.7 years
Iodine-131	8.05 days
Cesium-137	30.0 years
Lead-210	22 years
Polonium-211	0.52 second
Radium-215	10^{-6} second
Radon-222	3.823 days
Radium-226	1620 years
Uranium-238	4.51×10^9 years
Plutonium-239	2.44×10^4 years

Carbon-14 is used to date archaeological remains. Potassium-40 and uranium-238 have half-lives that are comparable to the age of the earth, so they can be used in establishing the age of geological samples. Cobalt-60 is used as a source of gamma rays in medical therapy. Strontium-90 and cesium-137 are produced in nuclear reactors and in nuclear explosions. Because of their half lives they are a significant biological hazard. Radon-222 is the radioactive gas that is found in homes and other buildings. Radium-226 was discovered by Marie Curie and is used as a source of gamma rays for medical therapy. Plutonium-239 is a fissionable material that is used in nuclear weapons.

Let us summarize the points we have made about the arithmetic of steady growth (the exponential function):

1. Steady growth happens whenever the rate of increase of a quantity is proportional to the quantity (Equation 1).
2. It is characterized by a fixed doubling time (Equations 5 and 6).
3. In every doubling time, the amount of growth is greater than the total of all the preceding growth.
4. Growth for a modest number of doubling times gives astronomically large numbers.
5. If consumption of a resource is growing steadily, enormous new supplies of the resource will allow only very short extensions of the period of steady growth in the rate of consumption.

The important points of exponential decay are

1. It happens whenever the rate at which decay events take place is proportional to the quantity of material that has not yet decayed.
2. It is characterized by a constant half-life.
3. In a modest number of half-lives, the amount of material remaining undecayed will become very small and will approach zero.

It has been said that the greatest shortcoming of the human race is our inability to understand the exponential function.[6]

These topics have been developed in more detail by **1.** A. A. Bartlett, *The American Journal of Physics*, vol. 46, Sept. 1978, p. 876. **2.** A. A. Bartlett, *Journal of Geological Education*, vol. 28, Jan. 1980, p. 4. **3.** P. Averitt, "Coal Resources of the United States," U.S. Geological Survey, Bulletin 1412, Jan. 1, 1974. **4.** *Newsweek Magazine*, July 16, 1979, p. 23. **5.** *Science*, vol. 245, Sept. 22, 1989, p. 1330. **6.** A. A. Bartlett, *The Physics Teacher*, a series of articles on "The Exponential Function" was published between October 1976 and January 1979. A videotaped lecture, "The Forgotten Fundamentals of the Energy Crisis" by A. A. Bartlett is available from the Media Center, University of Colorado, Boulder 80309-0379

Suggestions for Further Reading

Angrist, Stanley W.: "Perpetual Motion Machines," *Scientific American*, January 1968, p. 114.

An amusing account of some of these machines, all of which would violate either the first or second law of thermodynamics, is presented in this article.

Atkins, P. W.: *The Second Law*, Scientific American Books, Inc., New York, 1984.

Thermodynamics as a link between the steam engine and the processes of life—a clear account with many diagrams and little mathematics.

Bennett, Charles H: "Demons, Engines and the Second Law," *Scientific American*, November 1987, p. 108.

This article gives an answer to the puzzle of Maxwell's Demon. The answer comes from the theory of computing.

Dyson, Freeman J.: "What Is Heat?" *Scientific American*, September 1954, p. 58.

This article includes a discussion of the relation of energy and entropy to heat and a short history of our evolving concepts of heat.

Ehrenberg, W.: "Maxwell's Demon," *Scientific American*, November 1967, p. 103.

Maxwell's thought experiment involving a violation of the second law of thermodynamics by a small, intelligent being has troubled physicists for over 100 years.

Penney, Terry R., and Desikan Bharathan: "Power from the Sea," *Scientific American*, January 1987, p. 86.

The developing technology of electrical power generation using heat engines that operate between warm surfaces and cooler waters below.

Shinbrot, Marvin: "Things Fall Apart," *The Sciences*, May/June 1987, p. 32.

A history of the controversies over Boltzmann's assertion that the entropy of a gas always increases—from 1872 to present-day research.

Tribus, Myron, and Edward C. McIrvine: "Energy and Information," *Scientific American*, September 1971, p. 179.

The concept of entropy can also be applied to the transmission of information, forming the basis of modern information theory.

von Baeyer, Hans Christian: "Physika: On the Cold Front," *The Sciences*, July/August 1989, p. 2.

The third law of thermodynamics states that the absolute zero of temperature is unattainable. This article describes the significance of a current experiment that tests the law.

Weaver, Jefferson Hane: "Entropy," in *The World of Physics: A Small Library of the Literature of Physics from Antiquity to the Present*, Simon and Schuster, New York, 1987.

Includes writings by Carnot, Clausius, Kelvin, Gibbs, Boltzmann, Planck, and Bridgman.

Wilson, S. S.: "Sadi Carnot," *Scientific American*, August 1981, p. 134.

A brief biography of this French investigator, who was concerned mainly with the practical improvement of steam engines and other heat engines, is presented in this article.

Review

A. Objectives: After studying this chapter, you should:

1. Be able to give the definitions of the efficiency of a heat engine and the coefficient of performance of a refrigerator.

2. Be able to give both the heat-engine and refrigerator statements of the second law of thermodynamics and illustrate their equivalence with a numerical example.

3. Be able to state the Carnot theorem and illustrate it with a numerical example.

4. Be able to give the expression for the Carnot efficiency of a heat engine.

5. Be able to discuss the concept of entropy, including its interpretation in terms of the unavailability of energy.

B. Define, explain, or otherwise identify:

Heat engine
Working substance
Heat reservoir
Efficiency
Heat-engine statement of the second law
Refrigerator
Refrigerator statement of the second law
Coefficient of performance
Carnot theorem
Carnot engine
Carnot efficiency
Carnot cycle
Second-law efficiency
Heat pump
Entropy

C. True or false: If the statement is true, explain why it is true. If it is false, give a counterexample.

1. Work can never be converted completely into heat.

2. Heat can never be converted completely into work.

3. All heat engines have the same efficiency.

4. It is impossible to transfer a given quantity of heat from a cold reservoir to a hot reservoir.

5. The coefficient of performance of a refrigerator cannot be greater than 1.

6. All Carnot engines are reversible.

7. The entropy of a system can never decrease.

8. The entropy of the universe can never decrease.

Problems

Level I

17-1 Heat Engines and the Second Law of Thermodynamics

1. An engine with 20 percent efficiency does 100 J of work in each cycle. (*a*) How much heat is absorbed in each cycle? (*b*) How much heat is rejected in each cycle?

2. An engine absorbs 400 J of heat and does 120 J of work in each cycle. (*a*) What is its efficiency? (*b*) How much heat is rejected in each cycle?

3. An engine absorbs 100 J and rejects 60 J in each cycle. (*a*) What is its efficiency? (*b*) If each cycle takes 0.5 s, find the power output of this engine in watts.

4. An engine with an output of 200 W has an efficiency of 30 percent. It works at 10 cycles/s. (*a*) How much work is done in each cycle? (*b*) How much heat is absorbed and how much is rejected in each cycle?

17-2 Refrigerators and the Second Law of Thermodynamics

5. A refrigerator absorbs 5 kJ of energy from a cold reservoir and rejects 8 kJ to a hot reservoir. (*a*) Find the coefficient of performance of the refrigerator. (*b*) The refrigerator is reversible and is run backward as a heat engine between the same two reservoirs. What is its efficiency?

17-3 Equivalence of the Heat-Engine and Refrigerator Statements

6. A certain engine running at 30 percent efficiency draws 200 J of heat from a hot reservoir. Assume that the refrigerator statement of the second law of thermodynamics is false, and show how this engine combined with a perfect refrigerator can violate the heat-engine statement of the second law.

7. A certain refrigerator takes in 500 J of heat from a cold reservoir and rejects 800 J to a hot reservoir. Assume that the heat-engine statement of the second law of thermodynamics is false, and show how a perfect engine working with this refrigerator can violate the refrigerator statement of the second law.

17-4 The Carnot Engine

8. A reversible engine working between two reservoirs at temperatures T_h and T_c has an efficiency of 30 percent. Working as a heat engine, it rejects 140 J of heat to the cold reservoir. A second engine working between the same two reservoirs also rejects 140 J to the cold reservoir. Show that if the second engine has an efficiency greater than 30 percent, the two engines working together would violate the heat-engine statement of the second law.

9. A reversible engine working between two reservoirs at temperatures T_h and T_c has an efficiency of 20 percent. Working as a heat engine, it does 100 J of work in each cycle. A second engine working between the same two reservoirs also does 100 J of work in each cycle. Show that if the efficiency of the second engine is greater than 20 percent, the two engines working together would violate the refrigerator statement of the second law.

10. A Carnot engine works between two heat reservoirs as a refrigerator. It does 50 J of work in removing 100 J from the cold reservoir and rejecting 150 J to the hot reservoir during each cycle. Its coefficient of performance $COP = Q_c/W = (100\text{ J})/(50\text{ J}) = 2$. (*a*) What is the efficiency of the Carnot engine when it works as a heat engine between the same two reservoirs? (*b*) Show that no other engine working as a refrigerator between the same two reservoirs can have a COP greater than 2.

11. A Carnot engine works between two heat reservoirs at temperatures $T_h = 300$ K and $T_c = 200$ K. (*a*) What is its efficiency? (*b*) If it absorbs 100 J from the hot reservoir during each cycle, how much work does it do? (*c*) How much heat does it reject during each cycle? (*d*) What is the COP of this engine when it works as a refrigerator between the same two reservoirs?

12. Which has a greater effect on increasing the efficiency of a Carnot engine, a 5-K increase in the temperature of the hot reservoir or a 5-K decrease in the temperature of the cold reservoir?

13. A refrigerator works between an inside temperature of 0°C and a room temperature of 20°C. (*a*) What is the largest possible coefficient of performance it can have? (*b*) If the inside of the refrigerator is to be cooled to −10°C, what is the largest possible coefficient of performance it can have, assuming the same room temperature of 20°C?

14. In each cycle, an engine removes 150 J from a reservoir at 100°C and rejects 125 J to a reservoir at 20°C. (*a*) What is the efficiency of this engine? (*b*) What is its second-law efficiency?

15. An engine has a second-law efficiency of 85 percent. In each cycle, it removes 200 kJ of heat from a hot reservoir at 500 K and exhausts heat to a cold reservoir at

200 K. (*a*) What is the efficiency of this engine? (*b*) How much work is done in each cycle? (*c*) How much heat is exhausted in each cycle?

16. An engine removes 250 J from a reservoir at 300 K and exhausts 200 J to a reservoir at 200 K. (*a*) What is its efficiency? (*b*) What is the second-law efficiency of the engine? (*c*) How much more work could be done if the engine were reversible?

17-5 The Heat Pump

17. A heat pump delivers 20 kW to heat a house. The outside temperature is $-10°C$ and the inside temperature of the hot-air supply for the heating fan is 40°C. (*a*) What is the coefficient of performance of a Carnot heat pump operating between these temperatures? (*b*) What must be the minimum power of the engine needed to run the heat pump? (*c*) If the second-law efficiency of the heat pump is 60 percent, what must the minimum power of the engine be?

18. Rework Problem 17 for an outside temperature of $-20°C$.

19. A refrigerator is rated at 370 W. (*a*) What is the maximum amount of heat it can remove in 1 min if the inside temperature of the refrigerator is 0°C and it exhausts into a room at 20°C? (*b*) If the second-law efficiency of the refrigerator is 70 percent, how much heat can it remove in 1 min?

20. Rework Problem 19 for a room temperature of 35°C.

17-6 Entropy and Disorder

21. Two moles of an ideal gas at $T = 400$ K expand quasi-statically and isothermally from an initial volume of 40 L to a final volume of 80 L. (*a*) Find the entropy change of the gas. (*b*) What is the entropy change of the universe for this process?

22. The gas in Problem 21 is taken from the same initial state ($T = 400$ K, $V_1 = 40$ L) to the same final state ($T = 400$ K, $V_2 = 80$ L) by a process that is not quasi-static. (*a*) What is the entropy change of the gas? (*b*) What can be said about the entropy change of the universe?

23. A system absorbs 200 J of heat reversibly from a reservoir at 300 K and rejects 100 J reversibly to a reservoir at 200 K as it moves from state A to state B. During this process, the system does 50 J of work. (*a*) What is the change in the internal energy of the system? (*b*) What is the change in entropy of the system? (*c*) What is the change in entropy of the universe? (*d*) If the system goes from state A to state B by a nonreversible process, how would your answers for parts (*a*), (*b*), and (*c*) differ?

24. A system absorbs 300 J from a reservoir at 300 K and 200 J from a reservoir at 400 K. It then returns to its original state, doing 100 J of work and rejecting 400 J of heat to a reservoir at a temperature T. (*a*) What is the entropy change of the system for the complete cycle? (*b*) If the cycle is reversible, what is the temperature T?

25. Two moles of an ideal gas originally at $T = 400$ K and $V = 40$ L undergo a free expansion to twice their initial volume. What is (*a*) the entropy change of the gas and (*b*) the entropy change of the universe?

26. A 5-kg block is dropped from rest at a height of 6 m above the ground. It hits the ground and comes to rest. The block, the ground, and the atmosphere are all initially at 300 K. What is the entropy change of the universe for this process?

27. If 500 J of heat is conducted from a reservoir at 400 K to one at 300 K, what is the change in entropy of the universe?

28. A 200-kg block of ice at 0°C is placed in a large lake. The temperature of the lake is just slightly higher than 0°C, and the ice melts. (*a*) What is the entropy change of the ice? (*b*) What is the entropy change of the lake? (*c*) What is the entropy change of the universe (the ice plus the lake)?

29. What is the change in entropy of 1.0 kg of water when it changes to steam at 100°C and a pressure of 1 atm?

30. How much of the 500 J of heat conducted from the 400-K reservoir to the 300-K reservoir in Problem 27 could have been converted into work using a cold reservoir at 300 K?

17-7 Entropy and Probability

There are no problems for this section.

Level II

31. (*a*) Which process is more wasteful: (1) a block moving with 500 J of kinetic energy being slowed to rest by friction when the temperature of the atmosphere is 300 K or (2) 1 kJ of heat being conducted from a reservoir at 400 K to one at 300 K? *Hint:* How much of the 1 kJ of heat could be converted into work in an ideal situation? (*b*) What is the change in entropy of the universe for each process?

32. An engine operates with 1 mol of an ideal gas for which $C_v = \frac{3}{2}R$ *and* $C_p = \frac{5}{2}R$ as its working substance. The cycle begins at $P_1 = 1$ atm and $V_1 = 24.6$ L. The gas is heated at constant volume to $P_2 = 2$ atm. It then expands at constant pressure until $V_2 = 49.2$ L. During these two steps, heat is absorbed by the gas. The gas is then cooled at constant volume until its pressure is again 1 atm. It is then compressed at constant pressure to its original state. During the last two steps, heat is rejected by the gas. All the steps are quasi-static and reversible. (*a*) Show this cycle on a PV diagram. Find the work done, the heat added, and the change in the internal energy of the gas for each step of the cycle. (*b*) Find the efficiency of the cycle.

33. An engine using 1 mol of a diatomic ideal gas performs a cycle consisting of three steps: (1) an adiabatic expansion from an initial pressure of 2.64 atm and an initial volume of 10 L to a pressure of 1 atm and a volume of 20 L, (2) a compression at constant pressure to its original volume of 10 L, and (3) heating at constant volume to its original pressure of 2.64 atm. Find the efficiency of this cycle.

34. An engine using 1 mol of an ideal gas initially at $V_1 = 24.6$ L and $T = 400$ K performs a cycle consisting of four steps: (1) an isothermal expansion at $T = 400$ K to twice its initial volume, (2) cooling at constant volume to $T = 300$ K, (3) an isothermal compression to its original volume, and (4) heating at constant volume to its original temperature of 400 K. Assume that $C_v = 21$ J/K. Sketch the cycle on a PV diagram, and find its efficiency.

35. A steam engine takes in superheated steam at 270°C and discharges condensed steam from its cylinder at 50°C. Its efficiency is 30 percent. (*a*) How does this efficiency compare with the maximum possible efficiency for these temperatures? (*b*) If the useful power output of the engine is 200 kW, how much heat does the engine discharge to its surroundings in 1 h?

36. In the cycle shown in Figure 17-11, 1 mol of an ideal gas ($\gamma = 1.4$) is initially at a pressure of 1 atm and a temperature of 0°C. The gas is heated at constant volume to $t_2 = 150$°C and is then expanded adiabatically until its pressure is again 1 atm. It is then compressed at constant pressure back to its original state. Find (*a*) the temperature t_3 after the adiabatic expansion, (*b*) the heat entering or leaving the system during each process, (*c*) the efficiency of this cycle, and (*d*) the efficiency of a Carnot cycle operating between the temperature extremes of this cycle.

Figure 17-11 Problem 36.

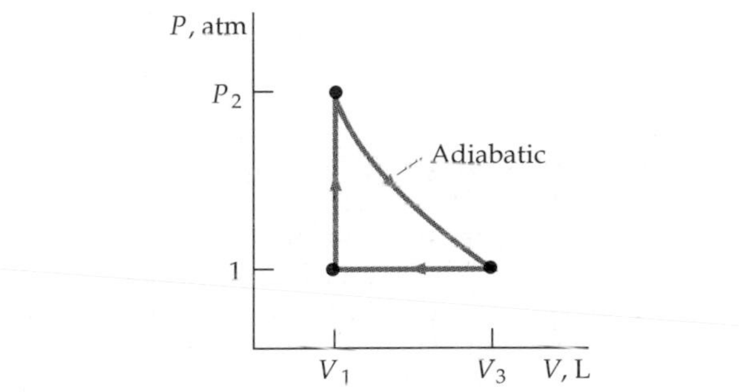

37. Helium gas ($\gamma = 1.67$) is initially at a pressure of 16 atm, a volume of 1 L, and a temperature of 600 K. It is expanded isothermally until its volume is 4 L and is then compressed at constant pressure until its volume and temperature are such that an adiabatic compression will return the gas to its original state. (*a*) Sketch this cycle on a PV diagram, and find (*b*) the volume and temperature after the isobaric compression, (*c*) the work done during each cycle, and (*d*) the efficiency of the cycle.

38. A heat engine that does the work of blowing up a balloon at a pressure of 1 atm extracts 4 kJ from a hot reservoir at 120°C. The volume of the balloon increases by 4 L, and heat is exhausted to a cold reservoir at a temperature t_c. If the second-law efficiency of the heat engine is 50 percent, find the temperature t_c.

39. A 100-g piece of ice at 0°C is placed in an insulated container with 100 g of water at 100°C. (*a*) When thermal equilibrium is established, what is the final temperature of the water? Ignore the heat capacity of the container. (*b*) Find the entropy change of the universe for this process.

40. A 1-kg block of copper at 100°C is placed in a calorimeter of negligible heat capacity containing 4 L of water at 0°C. Find the entropy change of (*a*) the copper block, (*b*) the water, and (*c*) the universe.

41. If a 2-kg piece of lead at 100°C is dropped into a lake at 10°C, find the entropy change of the universe.

42. Find the net change in entropy of the universe when 10 g of steam at 100°C and a pressure of 1 atm are introduced into a calorimeter of negligible heat capacity containing 150 g of water and 150 g of ice at 0°C.

43. A 1500-kg car traveling at 100 km/h crashes into a concrete wall. If the temperature of the air is 20°C, calculate the entropy change of the universe.

44. A Carnot engine works between two heat reservoirs at temperatures $T_h = 300$ K and $T_c = 200$ K. (*a*) What is its efficiency? (*b*) If it absorbs 100 J from the hot reservoir during each cycle, how much work does it do? (*c*) How much heat does it reject in each cycle? (*d*) What is the coefficient of performance of this engine when it works as a refrigerator between these two reservoirs?

45. A heat engine works in a cycle between reservoirs at 400 and 200 K. The engine absorbs 1000 J of heat from the hot reservoir and does 200 J of work in each cycle. (*a*) What is the efficiency of this engine? (*b*) Find the entropy change of the engine, each reservoir, and the universe for each cycle. (*c*) What would be the efficiency of a Carnot engine working between the same two reservoirs? (*d*) How much work could be done in each cycle by a Carnot engine if it absorbs 1000 J from the hot reservoir? (*e*) Show that the difference in the work done by the Carnot engine and that done by the original engine is $T_c\,\Delta S_u$, where ΔS_u is the change in entropy of the universe.

46. One mole of an ideal gas first undergoes a free expansion from $V_1 = 12.3$ L and $T_1 = 300$ K to $V_2 = 24.6$ L and $T_2 = 300$ K. It is then compressed isothermally and quasi-statically back to its original state. (*a*) What is the entropy change of the universe for the complete cycle? (*b*) How much work is wasted in this cycle? (*c*) Show that the work wasted is $T\,\Delta S_u$.

47. The cooling compartment of a refrigerator and its contents are at 5°C and have an average heat capacity of 84 kJ/K. The refrigerator exhausts heat to the room, which is at 25°C. What minimum power must the motor used to operate the refrigerator have if it is to reduce the temperature of the cooling compartment and its contents by 1 C° in 1 min?

48. (*a*) Show that the coefficient of performance of a Carnot refrigerator working between two reservoirs at temperatures T_h and T_c is related to the efficiency of a Carnot engine by COP $= T_c/(\epsilon_c T_h)$.

Level III

49. The behavior of a gasoline engine can be approximated by an *Otto cycle*. See Figure 17-4. (*a*) Compute the heat input and the heat output, and show that the efficiency can be written as

$$\epsilon = 1 - \frac{T_d - T_a}{T_c - T_b}$$

where T_a is the temperature of state a, and so forth.

(*b*) Using the relation for an adiabatic expansion or compression, $TV^{\gamma-1}$ = a constant, show that

$$\epsilon = 1 - \left(\frac{V_2}{V_1}\right)^{\gamma-1}$$

where $V_1 = V_a = V_d$ and $V_2 = V_b = V_c$. (*c*) The ratio V_1/V_2 is called the *compression ratio*. Find the efficiency of this cycle for a compression ratio of 8, using $\gamma = 1.4$. (Compression ratios much greater than this can not be achieved in real engines because of preignition problems.) (*d*) Explain why the efficiency of a real gasoline engine might be much less than that calculated in part (*c*).

50. The *diesel cycle* shown in Figure 17-12 approximates the behavior of a diesel engine. Process *ab* is an adiabatic compression, process *bc* is an expansion at constant pressure, process *cd* is an adiabatic expansion, and process *da* is cooling at constant volume. Find the efficiency of this cycle in terms of the volumes V_a, V_b, V_c, and V_d.

Figure 17-12 Diesel cycle for Problem 50.

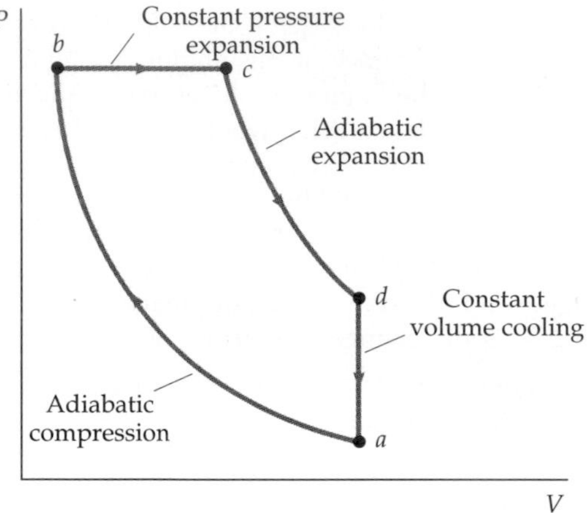

51. In the *Stirling cycle* shown in Figure 17-13, process *ab* is an isothermal compression, process *bc* is heating at constant volume, process *cd* is an isothermal expansion, and process *da* is cooling at constant volume. Find the efficiency of the Stirling cycle in terms of the temperatures T_h and T_c, and the volumes V_a and V_b.

Figure 17-13 Stirling cycle for Problem 51.

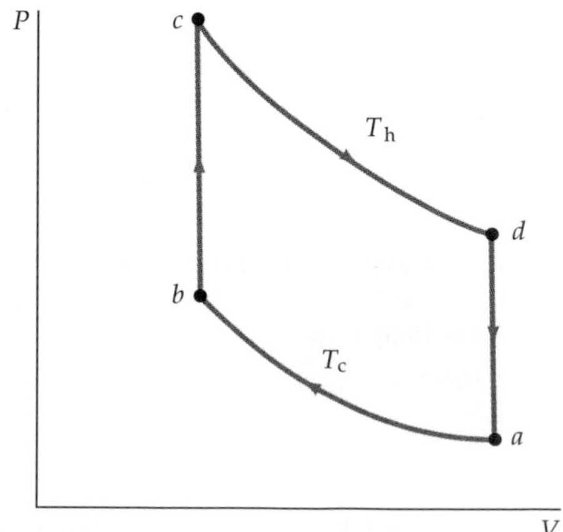

52. If two adiabatic curves intersect on a *PV* diagram, a cycle could be completed by an isothermal path between the two adiabatic curves as shown in Figure 17-14. Show that such a cycle could violate the second law of thermodynamics.

Figure 17-14 Problem 52.

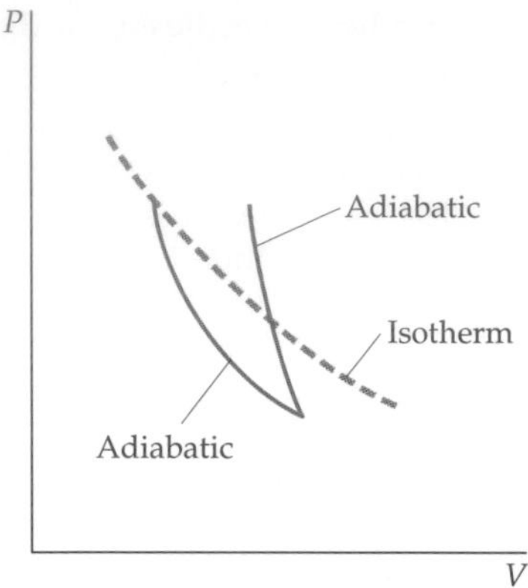

53. Using the equation for the entropy change of an ideal gas when the volume and temperature change and $TV^{\gamma-1}$ = a constant, show explicitly that the entropy change is zero for a quasi-static adiabatic expansion from state (V_1,T_1) to state (V_2,T_2).

54. (*a*) Show that if the refrigerator statement of the second law of thermodynamics were not true, the entropy of the universe could decrease. (*b*) Show that if the heat-engine statement of the second law were not true, the entropy of the universe could decrease. (*c*) An alternative statement of the second law is that the entropy of the universe can not decrease. Have you just proved that this statement is equivalent to the refrigerator and heat-engine statements?

55. One mole of an ideal monatomic gas at an initial volume V_1 = 25 L follows the cycle shown in Figure 17-15. All the processes are quasi-static. Find (*a*) the temperature of each state of the cycle, (*b*) the heat flow for each part of the cycle, and (*c*) the efficiency of the cycle.

Figure 17-15 Problem 55.

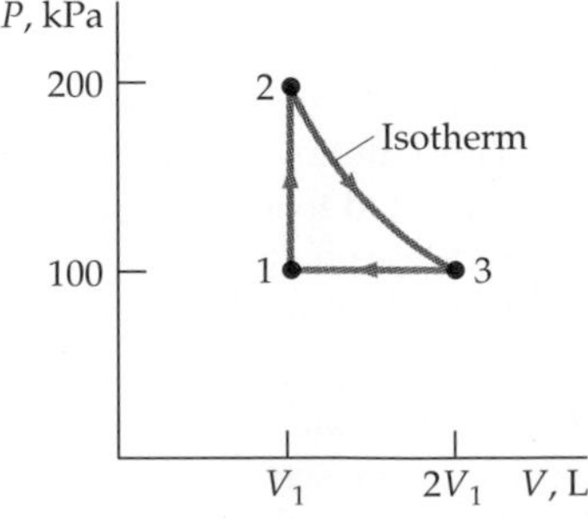

56. Suppose that two heat engines are connected in "series" such that the heat exhaust of the first engine is used as the heat input of the second engine as shown in Figure 17-16. The efficiencies of the engines are ϵ_1 and ϵ_2, respectively. Show that the net efficiency of the combination is given by

$$\epsilon_{\text{net}} = \epsilon_1 + (1 - \epsilon_1)\epsilon_2$$

57. Suppose that each engine in Figure 17-16 is an ideal reversible heat engine. Engine 1 operates between temperatures T_h and T_m and engine 2 operates between T_m and T_c, where $T_h > T_m > T_c$. Show that

$$\epsilon_{\text{net}} = 1 - \frac{T_c}{T_h}$$

This means that two reversible heat engines in "series" are equivalent to one reversible heat engine operating between the hottest and coldest reservoirs.

Figure 17-16 Problems 56 and 57.

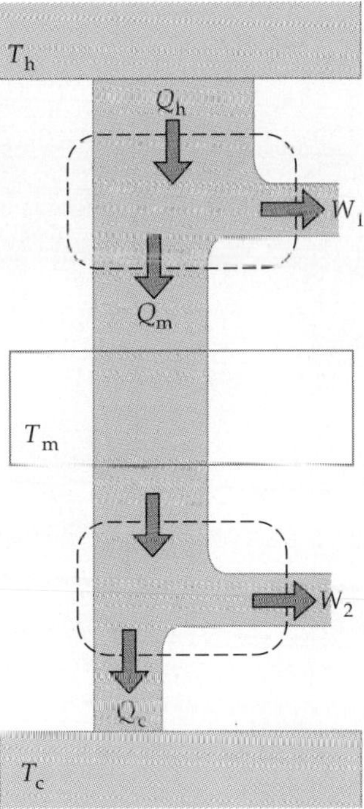

58. An ideal gas ($\gamma = 1.4$) follows the cycle shown in Figure 17-17. The temperature of state 1 is 200 K. Find (*a*) the temperatures of the other three states of the cycle and (*b*) the efficiency of the cycle.

Figure 17-17 Problem 58.

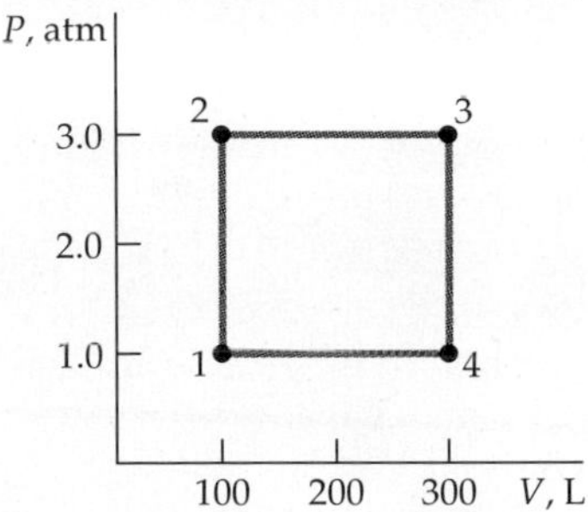

59. The Clausius equation of state is $P(V - bn) = nRT$ where b is a constant. Show that the efficiency of a Carnot cycle is the same for a gas that obeys this equation of state as it is for one that obeys the ideal-gas equation of state, $PV = nRT$.

Part 4 Electricity and Magnetism

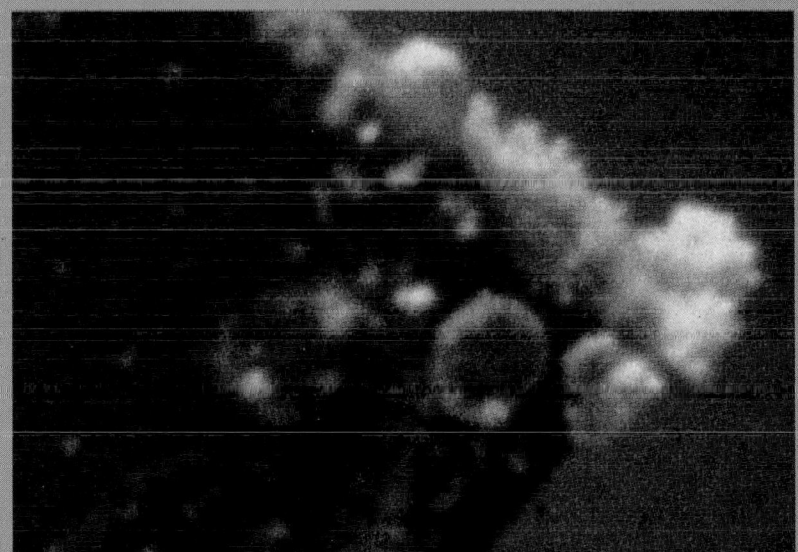

A small, cubicle permanent magnet levitates above a disk of the superconductor yttrium-barium-copper oxide, cooled by liquid nitrogen to 77 K. At temperatures below 92 K, the disk becomes superconducting. The magnetic field of the cube sets up circulating electric currents in the superconducting disk, such that the resultant magnetic field in the superconductor is zero. These currents produce a magnetic field that repels the cube.

Chapter 18

The Electric Field I: Discrete Charge Distributions

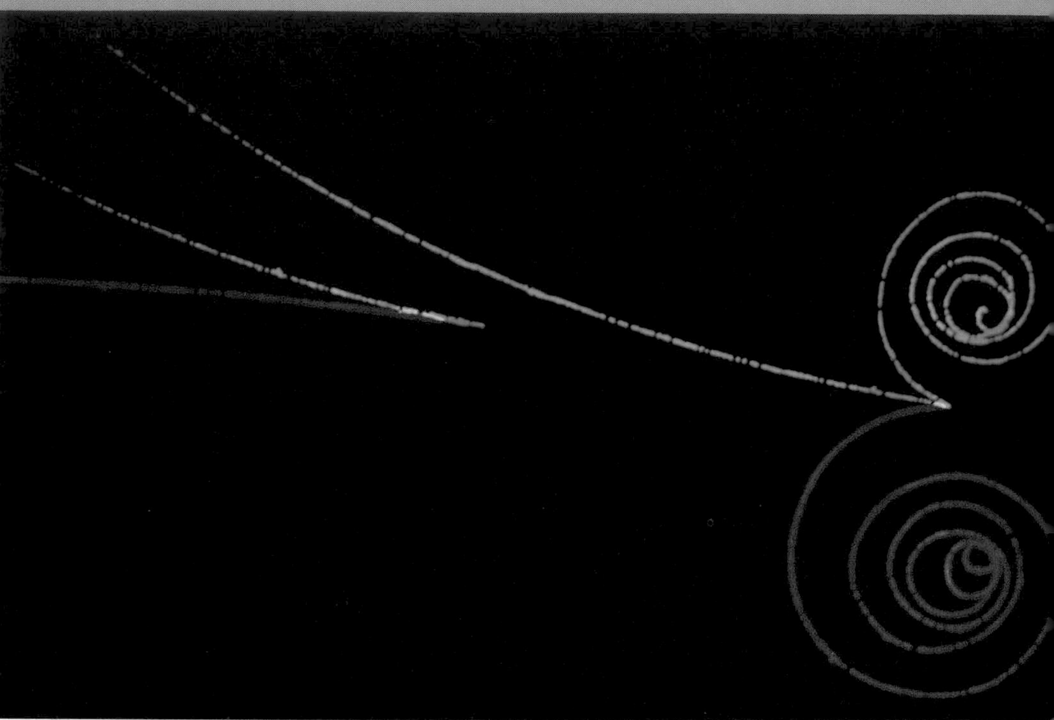

Pair production. An electron of charge $-e$ and a positron of charge $+e$ are created by the interaction of electromagnetic radiation with matter. The paths of the oppositely charged particles, made visible in a bubble chamber, are bent in opposite directions by a magnetic field.

Electricity is in such common use today that we normally give it little thought. Yet just a century ago, there were few electric lights and no electric heaters, motors, radios, or television sets. Although the practical use of electricity has been developed mostly in the twentieth century, the study of electricity has a long history. Observations of electrical attraction can be traced as far back as the time of the ancient Greeks. They noticed that after amber has been rubbed, it attracts small objects such as straw or feathers. Indeed, the word "electric" comes from the Greek word for amber, *elektron*.

In this chapter, we will begin our study of electricity with a short discussion of the concept of electric charge, followed by a brief look at conductors and insulators and how conductors can be given a charge. We will then study Coulomb's law, which describes the force exerted by one electric charge on another. Next, we will introduce the electric field and show how it can be described by electric-field lines, which indicate the magnitude and direction of the field. Last, we will discuss the behavior of point charges and electric dipoles in electric fields.

18-1 Electric Charge

Let us begin by considering a simple experiment involving electrical attraction. Suppose that we rub a plastic rod with fur and suspend the rod from a string such that it is free to rotate. If we then approach this rod with a second plastic rod that has also been rubbed with fur, we find that the rods repel each other (Figure 18-1). We can repeat the experiment and achieve the same results if we use two glass rods that have been rubbed with a piece of silk. However, if we use a plastic rod that has been rubbed with fur and a glass rod that has been rubbed with silk, we find that the rods attract each other.

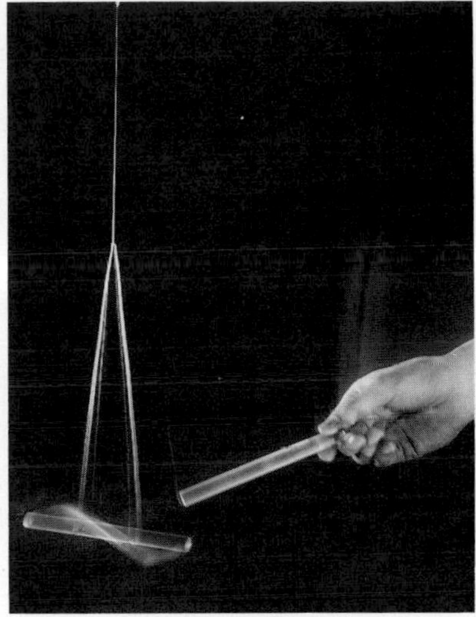

Figure 18-1 Two plastic rods that have been rubbed with fur repel each other.

When we rub the plastic rod with fur or the glass rod with silk, we cause the rod to become "electrified" or "charged." If we repeat our experiment with various types of materials, we find that we can classify all charged objects into just two groups—those that are charged like the plastic rod rubbed with fur and those that are charged like the glass rod rubbed with silk. The great American statesman and scientist Benjamin Franklin proposed a model explaining why this is so. He suggested that every object has a "normal" amount of electricity and that, when two objects are rubbed together, some of this electricity can be transferred from one of the objects to the other. This leaves one with an excess amount of electricity and the other with an equal deficiency. Franklin described the resulting charges with plus and minus signs. He chose the type of charge acquired by a glass rod when it is rubbed with a piece of silk to be positive, which meant that the piece of silk acquired a negative charge of equal magnitude. Based on Franklin's convention, then, plastic rubbed with fur acquires a negative charge and the fur acquires a positive charge of equal magnitude. As we saw in our experiment, two objects that carry the same type of charge—that is, two objects that are both positive or both negative—repel each other, and two objects that carry opposite charges attract each other (Figure 18-2).

Today, we know that when glass is rubbed with silk, electrons are transferred from the glass to the silk, leaving the silk with an excess number of electrons and the rod with a deficiency. According to Franklin's classification, which we still use, the silk is negatively charged, so the electrons are said to carry a negative charge.

We now know also that matter consists of atoms that are electrically neutral. Each atom has a tiny but massive nucleus that contains protons, each of which carries a positive charge, and neutrons, which have no charge. The number of protons in the nucleus is the atomic number Z of the element. Surrounding the nucleus are an equal number of negatively charged electrons. The electron and the proton are very different particles. For one thing, the proton is about 2000 times more massive than the electron. Yet their charges are exactly equal but opposite in sign. The charge of the proton is e and that of the electron is $-e$, where e is called the **fundamental unit of charge.** All charges occur in integral amounts of the fundamental unit of charge e. That is, **charge is quantized.** Any charge Q occurring in nature can be written $Q = \pm Ne$, where N is an integer.* The quantization of electric charge is usually not noticed because N is usually very large. For example, charging a plastic rod by rubbing it with a piece of fur typically transfers about 10^{10} electrons to the rod.

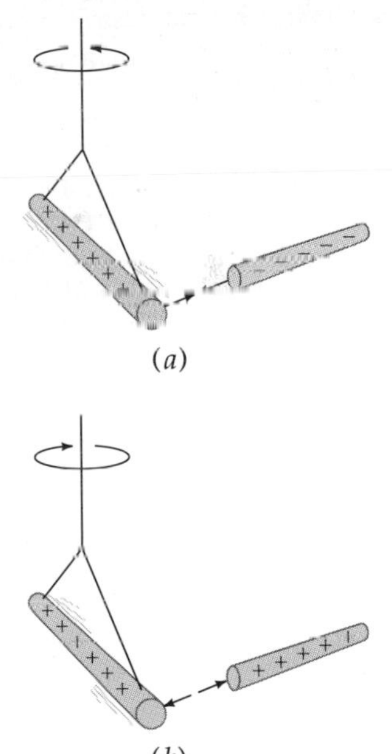

Figure 18-2 (*a*) Objects carrying charges of opposite sign attract each other. (*b*) Objects carrying charges of the same sign repel each other.

*In the quark model of elementary particles, protons, neutrons, and some other elementary particles are thought to be made up of particles called quarks that carry charges of $\pm\frac{1}{3}e$ or $\pm\frac{2}{3}e$. Apparently, quarks cannot be observed individually but only in combinations that result in a net charge of $\pm Ne$ or 0.

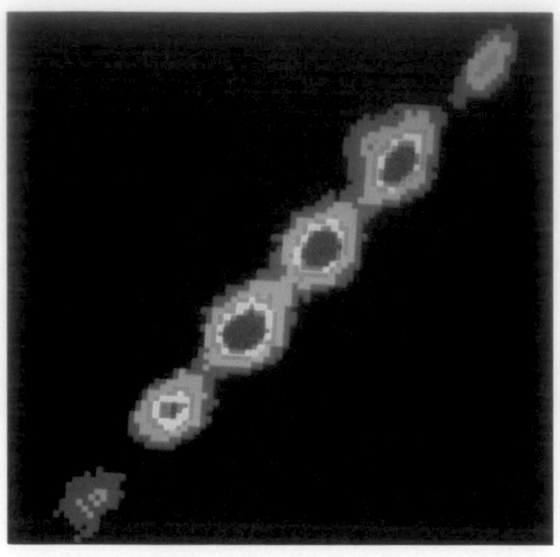

Quantized charge. These individual mercury ions are trapped in an electric field configuration called a Paul trap. In this false-color photograph, ions are preferentially located in the red areas. Neighboring ions are separated by several micrometers.

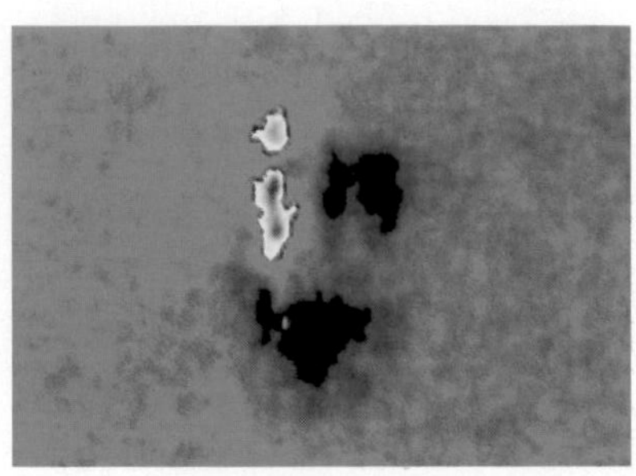

Charging by contact. A sample of plastic about 0.02 mm wide that was charged by contact with a piece of nickel. Although the plastic carries a net positive charge, regions of negative charge (dark) as well as regions of positive charge (yellow) are indicated. The photograph was taken by sweeping a charged needle of width 10^{-7} m over the sample and recording the electrostatic force on the needle.

When objects are in very close contact, as when they are rubbed against each other, electrons are transferred from one object to the other. One object is left with an excess number of electrons and is therefore negatively charged, and the other object is left lacking electrons and is therefore positively charged. In this process, charge is not created but is merely transferred. The net charge of the two objects taken together does not change. That is, *charge is conserved*. The **law of conservation of charge** is a fundamental law of nature. In certain interactions among elementary particles, charged particles such as electrons are created or annihilated. However, in all these processes, equal amounts of positive and negative charge are produced or destroyed, so the net charge of the universe is unchanged. For instance, whenever an electron of charge $-e$ is created, a particle of charge $+e$ called a *positron* is created simultaneously. (This process is called *pair production*.)

The SI unit of charge is the coulomb, which is defined in terms of the unit of electric current, the *ampere*. (The ampere is defined in terms of a magnetic-force measurement, which we will discuss in Chapter 25. It is the unit of current used in everyday electrical work.) The **coulomb** (C) is the amount of charge flowing through a cross-sectional area of a wire in one second when the current in the wire is one ampere. The fundamental unit of electric charge e is related to the coulomb by

Fundamental unit of charge

$$e = 1.60 \times 10^{-19}\ \text{C} \qquad 18\text{-}1$$

Charges from about 10 nC (1 nC = 10^{-9} C) to about 0.1 μC (1 μC = 10^{-6} C) can be produced in the laboratory by putting certain objects in intimate contact, often by simply rubbing their surfaces together. Such procedures involve the transfer of many electrons.

Example 18-1

A copper penny has a mass of 3 g. The atomic number of copper is $Z = 29$ and the atomic mass is 63.5 g/mol. What is the total charge of all the electrons in the penny?

We must first find the number of atoms in 3 g of copper. Since 1 mol of copper contains Avogadro's number of atoms and has a mass of 63.5 g, the number of atoms in 3 g of copper is

$$N = (3 \text{ g})\frac{6.02 \times 10^{23} \text{ atoms/mol}}{63.5 \text{ g/mol}} = 2.84 \times 10^{22} \text{ atoms}$$

Each atom contains $Z = 29$ electrons, so the total charge Q is

$$Q = (2.84 \times 10^{22} \text{atoms})(29 \text{electrons/atom})(-1.60 \times 10^{-19} \text{ C/electron})$$
$$= -1.32 \times 10^{5} \text{ C}$$

Question

1. After you stroke your cat, do you think the cat is positively or negatively charged?

18-2 Conductors and Insulators and Charging by Induction

In many materials, such as copper and other metals, some of the electrons are free to move about the entire material. Such materials are called **conductors.** In other materials, such as wood or glass, all the electrons are bound to nearby atoms and none can move freely. These materials are called **insulators.**

In a single atom of copper, 29 electrons are bound to the nucleus by the electrostatic attraction between the negatively charged electrons and the positively charged nucleus. The outer electrons are more weakly bound than the inner electrons because of their greater distance from the nucleus and because of the repulsion of the inner electrons. When a large number of copper atoms are combined in a piece of metallic copper, the binding of the electrons of each individual atom is changed by interactions with neighboring atoms. One or more of the outer electrons in each atom is no longer bound but is free to move throughout the whole piece of metal, much as a gas molecule is free to move about in a box. The number of free electrons depends on the particular metal, but it is typically about one per atom. The copper atom minus one of its outer electrons carries a positive charge and is called a *positive ion*. In metallic copper, the copper ions are arranged in a regular array called a *lattice*. Normally, a conductor is electrically neutral because there is a lattice ion carrying a positive charge $+e$ for each free electron carrying a negative charge $-e$. A conductor can be given a net charge by adding or removing free electrons.

Figure 18-3 An electroscope. The two gold leaves are connected to a metal rod that has a metal ball on top. When a negative charge is placed on the metal ball, the charge is conducted to the leaves and they repel each other.

Figure 18-3 shows an **electroscope,** which is a device for detecting electric charge. Two gold leaves are attached to a conducting post that has a conducting ball on top and are otherwise insulated from the container. When uncharged, the leaves hang together vertically. When the ball is touched by a negatively charged plastic rod, some of the negative charge from the rod is transferred to the ball and moves to the gold leaves, which then spread apart because of the electrical repulsion of their negative charges. Touching the ball with a positively charged glass rod also causes the leaves to spread apart. In this case, the positively charged glass rod attracts electrons from the metal ball, leaving the leaves charged positively.

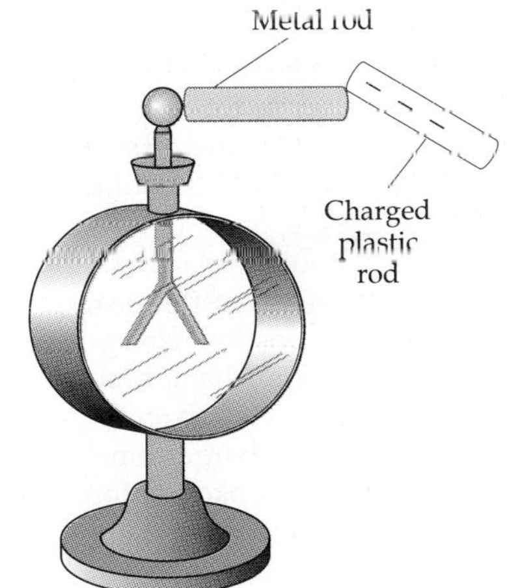

Figure 18-4 A metal rod is in contact with the metal ball on the electroscope. When the far end of the metal rod is touched by a negatively charged plastic rod, some of the charge conducts along the metal rod to the electroscope, as evidenced by the spreading of the gold leaves.

In Figure 18-4, a long metal rod is in contact with the ball of an electroscope. When the far end of the rod is touched by a charged plastic rod, the leaves of the electroscope spread apart because the electrons from the plastic rod are conducted along the metal rod to the electroscope. If the metal rod is replaced by a wooden rod and the far end of the wooden rod is touched by a charged plastic rod, nothing happens. The wooden rod is an insulator, which does not conduct electricity.

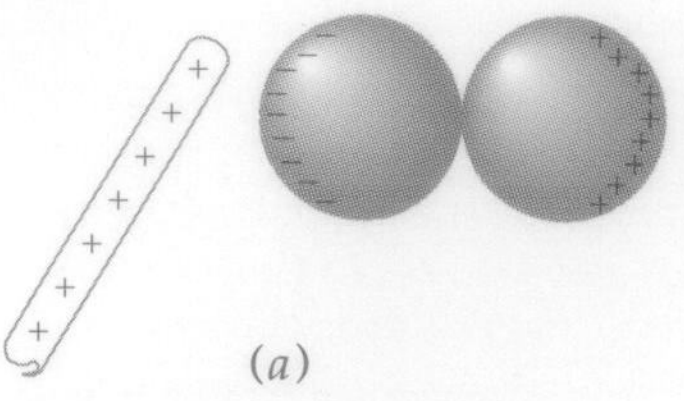

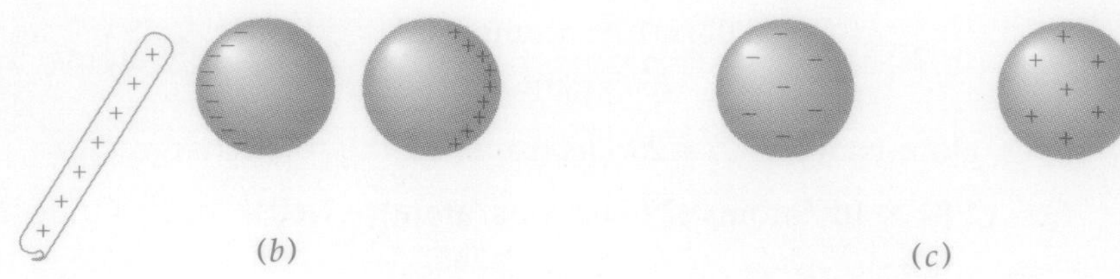

Figure 18-5 Charging by induction. (*a*) The two spherical conductors in contact become oppositely charged because the positively charged rod attracts electrons to the left sphere, leaving the right sphere positively charged. (*b*) If the spheres are now separated while the rod is in place, they retain their equal and opposite charges. (*c*) When the rod is removed and the spheres are far apart, the spheres are uniformly charged with equal and opposite charges.

A simple and practical method of charging a conductor makes use of the ready movement of free electrons in a conductor. In Figure 18-5, two uncharged metal spheres are in contact. When a charged rod is brought near one of the spheres, free electrons flow from one sphere to the other. If the rod is positively charged, it attracts the negatively charged electrons, and the sphere nearest the rod acquires electrons from the other. This leaves the near sphere with a net negative charge and the far sphere with an equal net positive charge (Figure 18-5*a*). If the spheres are separated before the rod is removed (Figure 18-5*b*), they will have equal and opposite charges (Figure 18-5*c*). A similar result is obtained with a negatively charged rod, which drives electrons from the near sphere to the far sphere. In either case, the spheres are charged without being touched by the rod, and the charge on the rod is undisturbed. This process is called **electrostatic induction** or **charging by induction.** If a charged spherical conductor is brought into contact with an identical but uncharged sphere, the charge on the first sphere will be shared equally by both conductors. If the spheres are then separated, each sphere will have half the excess charge originally on the first sphere.

A convenient large conductor is the earth itself, which for most purposes can be considered to be an infinitely large conductor. When a conductor is connected to the earth, it is said to be **grounded.** This is indicated schematically by showing a connecting wire ending in parallel horizontal lines as in Figure 18-6*b*. We can use the earth to charge a single conductor by induction. In Figure 18-6*a*, a positively charged rod is brought near an uncharged conducting sphere. Free electrons are attracted to the side near the positive rod, leaving the other side with a positive charge. If we ground the sphere while the charged rod is still present, the sphere acquires a charge opposite to that of the rod because electrons from the earth travel along the connecting wire to the sphere and neutralize the positive charge on the right side of the sphere (Figure 18-6*b*). The connection to ground is broken before the rod is removed to complete the charging by induction (Figure 18-6*c*). When the charged rod is removed, the sphere has a uniform negative charge as shown in Figure 18-6*d*.

Question

2. Can insulators as well as conductors be charged by induction?

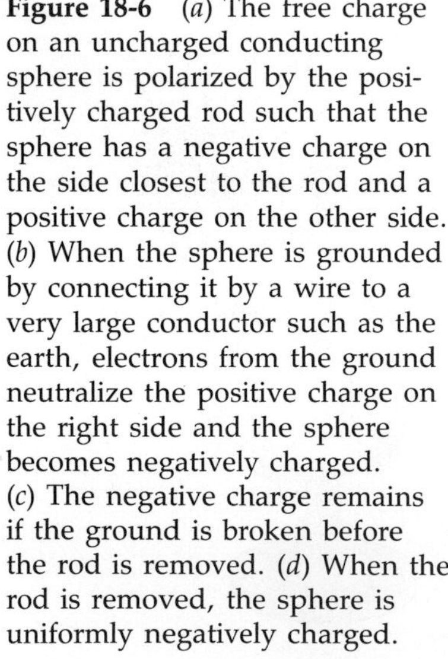

Figure 18-6 (*a*) The free charge on an uncharged conducting sphere is polarized by the positively charged rod such that the sphere has a negative charge on the side closest to the rod and a positive charge on the other side. (*b*) When the sphere is grounded by connecting it by a wire to a very large conductor such as the earth, electrons from the ground neutralize the positive charge on the right side and the sphere becomes negatively charged. (*c*) The negative charge remains if the ground is broken before the rod is removed. (*d*) When the rod is removed, the sphere is uniformly negatively charged.

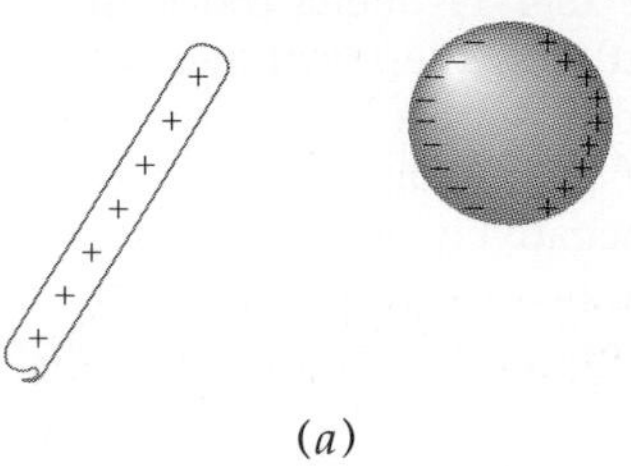

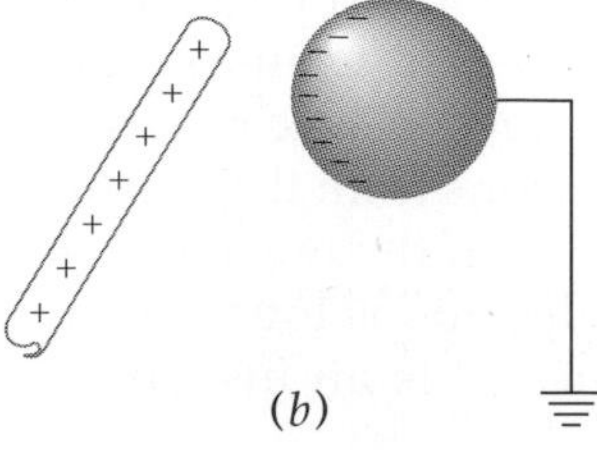

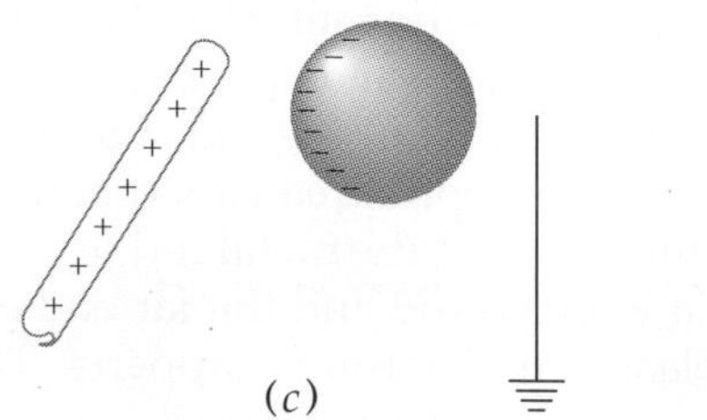

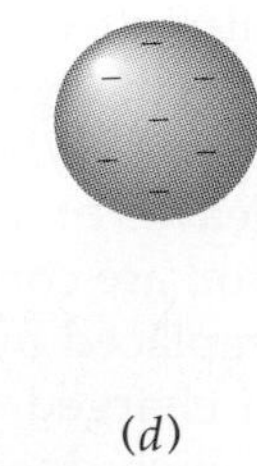

(*Left*) The lightning rod on this building is grounded to guide the electric current from a lightning discharge to the ground. (*Right*) These fashionable ladies are wearing hats with metal chains that drag along the ground supposedly to protect them from lightning.

18-3 Coulomb's Law

The force exerted by one charge on another was studied by Charles Coulomb (1736–1806) using a torsion balance of his own invention. Coulomb's experimental apparatus was essentially the same as that described for the Cavendish experiment (Chapter 10), with the masses replaced by small charged spheres. For the magnitudes of charges easily transferred by rubbing, the gravitational attraction of the spheres is completely negligible compared with their electric attraction or repulsion. In Coulomb's experiment, the charged spheres were much smaller than the distance between them so that the charges could be considered to be point charges. Coulomb used induction to produce equally charged spheres and to vary the amount of charge on the spheres. For example, beginning with charge q_0 on each sphere, he could reduce the charge to $\frac{1}{2}q_0$ by temporarily grounding one sphere to discharge it and then placing the two spheres in contact. The results of the experiments of Coulomb and others on the forces exerted by one point charge on another are summarized in **Coulomb's law:**

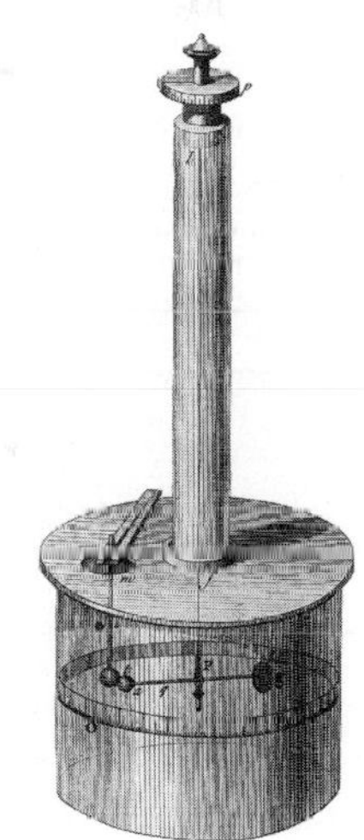

Coulomb's torsion balance.

> The force exerted by one point charge on another acts along the line joining the charges. It varies inversely as the square of the distance separating the charges and is proportional to the product of the charges. The force is repulsive if the charges have the same sign and attractive if the charges have opposite signs.

Figure 18-7 shows the forces exerted between two charges with the same sign and between two charges of opposite sign.

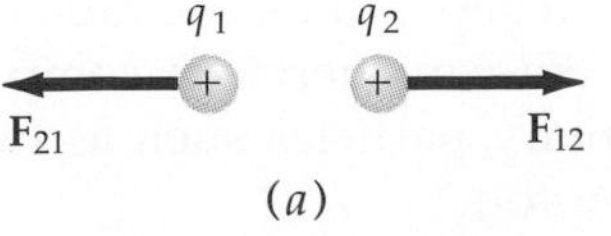

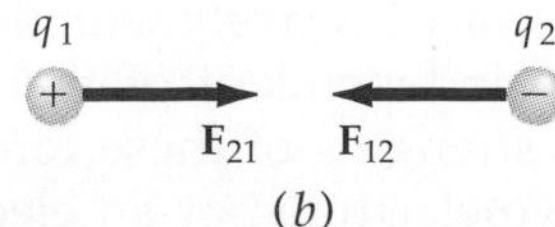

Figure 18-7 (*a*) Like charges repel, whereas (*b*) unlike charges attract.

Coulomb's law can be stated more simply using a mathematical expression. Let q_1 and q_2 be two point charges that are separated by a distance r_{12}, which is the magnitude of the vector $\mathbf{r}_{12}$ pointing from charge q_1 to q_2 (Figure 18-8). The force $\mathbf{F}_{12}$ exerted by charge q_1 on q_2 is then

Coulomb's law

$$\mathbf{F}_{12} = \frac{kq_1q_2}{r_{12}^2}\hat{\mathbf{r}}_{12} \qquad 18\text{-}2$$

where $\hat{\mathbf{r}}_{12} = \mathbf{r}_{12}/r_{12}$ is a unit vector pointing from q_1 to q_2 and k is the **Coulomb constant,** which has the value

$$k = 8.99 \times 10^9\ \text{N·m}^2/\text{C}^2 \qquad 18\text{-}3$$

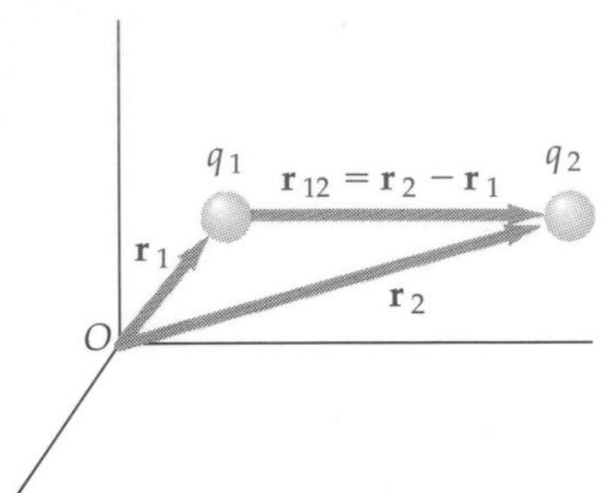

Figure 18-8 Charge q_1 at position $\mathbf{r}_1$ and charge q_2 at $\mathbf{r}_2$ relative to origin O. The force exerted by q_1 on q_2 is in the direction of the vector $\mathbf{r}_{12} = \mathbf{r}_2 - \mathbf{r}_1$ if both charges have the same sign and in the opposite direction if they have opposite signs.

By Newton's third law, the force $\mathbf{F}_{21}$ exerted by q_2 on q_1 is the negative of $\mathbf{F}_{12}$. That is, $\mathbf{F}_{21}$ is equal in magnitude to $\mathbf{F}_{12}$ but opposite in direction. The *magnitude* of the electric force exerted by a charge q_1 on another charge q_2 a distance r away is thus given by

$$F = \frac{kq_1q_2}{r^2} \qquad 18\text{-}4$$

If both charges have the same sign, that is, if both are positive or both are negative, the force is repulsive. If the two charges have opposite signs, the force is attractive. Note the similarity between Coulomb's law and Newton's law of gravity (Equation 10-2). Both are inverse-square laws. However, the gravitational force between two particles is proportional to the masses of the particles and is always attractive, whereas the electric force is proportional to the charges of the particles and may be attractive or repulsive.

Example 18-2

Two point charges of 0.05 μC each are separated by 10 cm. Find (*a*) the magnitude of the force exerted by one charge on the other and (*b*) the number of fundamental units of charge in each.

(*a*) From Coulomb's law, the magnitude of the force is

$$F = \frac{kq_1q_2}{r^2}$$

$$= \frac{(8.99 \times 10^9\ \text{N·m}^2/\text{C}^2)(0.05 \times 10^{-6}\ \text{C})(0.05 \times 10^{-6}\ \text{C})}{(0.1\ \text{m})^2}$$

$$= 2.25 \times 10^{-3}\ \text{N}$$

(*b*) The number of electrons required to produce a charge of 0.05 μC is found from

$$q = Ne$$

$$N = \frac{q}{e} = \frac{0.05 \times 10^{-6}\ \text{C}}{1.6 \times 10^{-19}\ \text{C}} = 3.12 \times 10^{11}$$

A charge of this size does not reveal that electric charge is quantized. A million electrons could be added to or subtracted from this charge without being detected by ordinary instruments.

Since both the electric force and the gravitational force between any two particles vary inversely with the square of their separation, the ratio of these forces is independent of their separation. We can therefore compare the relative strengths of these forces for elementary particles such as two protons, two electrons, or an electron and a proton.

Example 18-3

Compute the ratio of the electric force to the gravitational force exerted by one proton on another.

Since each proton has charge $+e$, the electric force is repulsive and has the magnitude

$$F_e = \frac{ke^2}{r^2}$$

The gravitational force, given by Newton's law of gravity, is attractive and has the magnitude

$$F_g = \frac{Gm_p^2}{r^2}$$

where m_p is the mass of a proton. The ratio of these forces is independent of the separation distance r:

$$\frac{F_e}{F_g} = \frac{ke^2}{Gm_p^2}$$

Substituting the values $k = 8.99 \times 10^9$ N·m²/C², $e = 1.60 \times 10^{-19}$ C, $G = 6.67 \times 10^{-11}$ N·m²/kg², and $m_p = 1.67 \times 10^{-27}$ kg, we obtain

$$\frac{F_e}{F_g} = \frac{(8.99 \times 10^9\ \text{N·m}^2/\text{C}^2)(1.60 \times 10^{-19}\ \text{C})^2}{(6.67 \times 10^{-11}\ \text{N·m}^2/\text{kg}^2)(1.67 \times 10^{-27}\ \text{kg})^2} = 1.24 \times 10^{36}$$

Exercise

In the hydrogen atom, the electron is separated from the proton by a distance of about 5.3×10^{-11} m on the average. Calculate the magnitude of the electrostatic force exerted by the proton on the electron. (Answer: 8.2×10^{-8} N)

We can see from Example 18-3 that the gravitational force between two elementary particles is so much smaller than the electric force between them (assuming they are charged) that it can always be neglected in describing interactions between them. It is only because large masses such as the earth contain almost exactly equal numbers of positive and negative charges that the gravitational force is important. If the positive and negative charges in such objects did not cancel each other, the electric forces between them would be much greater than the gravitational forces.

In a system of charges, each charge exerts a force given by Equation 18-2 on every other charge. Thus, the net force on any charge is the vector sum of the individual forces exerted on that charge by all the other charges in the system.

Example 18-4

Three point charges lie on the x axis; $q_1 = 25$ nC is at the origin, $q_2 = -10$ nC is at $x = 2$ m, and $q_0 = 20$ nC is at $x = 3.5$ m (Figure 18-9). Find the net force on q_0 due to q_1 and q_2.

The force on q_0 due to q_1, which is 3.5 m away, is given by

$$\mathbf{F}_{10} = \frac{kq_1q_0}{r_{10}^2}\,\hat{\mathbf{r}}_{10}$$

$$= \frac{(8.99 \times 10^9\ \text{N·m}^2/\text{C}^2)(25 \times 10^{-9}\ \text{C})(20 \times 10^{-9}\ \text{C})}{(3.5\ \text{m})^2}\,\mathbf{i}$$

$$= (0.367\ \mu\text{N})\mathbf{i}$$

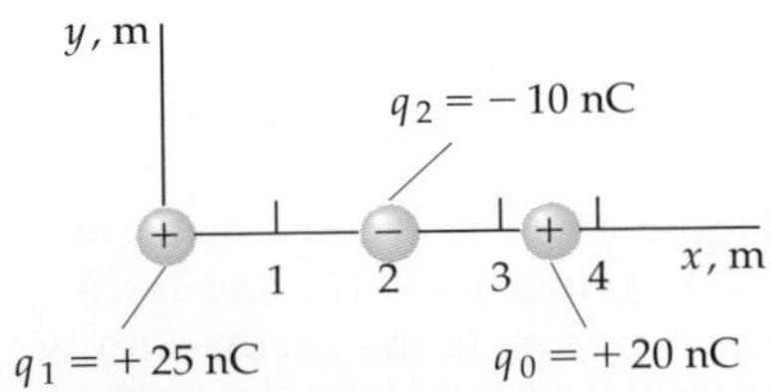

Figure 18-9 Point charges on the x axis for Example 18-4.

where we have used **i** to designate the unit vector $\hat{\mathbf{r}}_{10}$ from charge q_1 to q_0, which points in the x direction. The force on q_0 due to q_2, which is 1.5 m away, is

$$\mathbf{F}_{20} = \frac{kq_2q_0}{r_{20}^2}\,\hat{\mathbf{r}}_{20}$$

$$= \frac{(8.99 \times 10^9\ \text{N·m}^2/\text{C}^2)(-10 \times 10^{-9}\ \text{C})(20 \times 10^{-9}\ \text{C})}{(1.5\ \text{m})^2}\,\mathbf{i}$$

$$= (-0.799\ \mu\text{N})\mathbf{i}$$

The net force on the charge q_0 due to the charges q_1 and q_2 is

$$\mathbf{F}_{\text{net}} = \mathbf{F}_{10} + \mathbf{F}_{20}$$

$$= (0.367\ \mu\text{N})\mathbf{i} - (0.799\ \mu\text{N})\mathbf{i} = (-0.432\ \mu\text{N})\mathbf{i}$$

Note that in Example 18-4, the charge q_2, which is between q_1 and q_0, has no effect on the force $\mathbf{F}_{10}$ exerted by q_1 on q_0, just as the charge q_1 has no effect on the force exerted by q_2 on q_0. The net force exerted on a charge (such as q_0 in this example) by a system of charges is found by the simple superposition of the separate forces exerted by each charge in the system. This **principle of superposition of electric forces** has been verified by experiment.

If a system of charges is to remain stationary, there must be other, nonelectrical forces acting on the charges so that the net force from all sources acting on each charge is zero. In the preceding example and those that follow, we assume that there are such forces so that all the charges remain stationary.

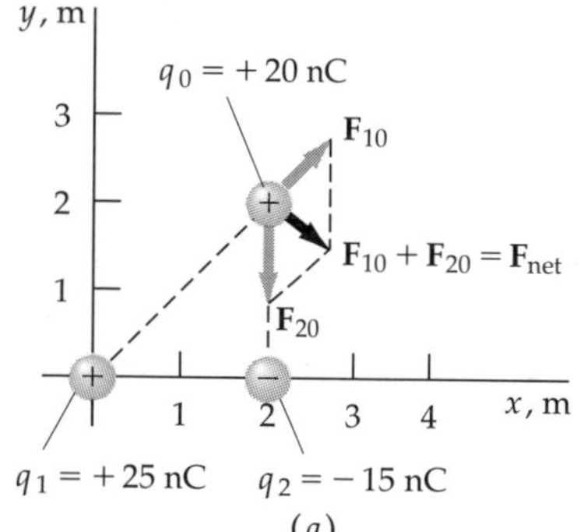

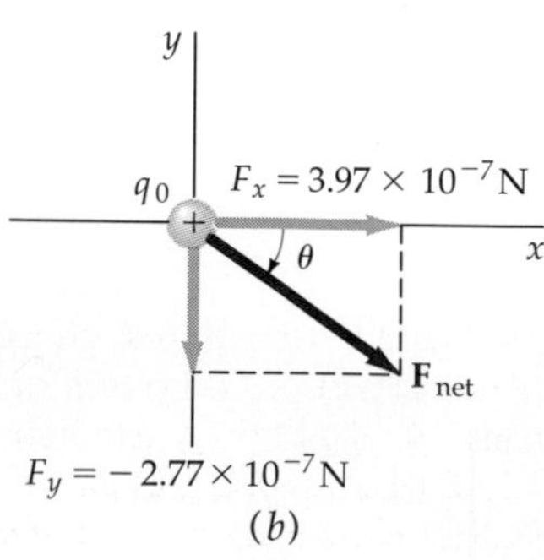

Figure 18-10 (*a*) Force diagram for Example 18-5. The net force on charge q_0 is the vector sum of the forces $\mathbf{F}_{10}$ due to q_1 and $\mathbf{F}_{20}$ due to q_2. (*b*) Diagram showing net force in (*a*) and its x and y components.

Example 18-5

Charge $q_1 = +25$ nC is at the origin, charge $q_2 = -15$ nC is on the x axis at $x = 2$ m, and charge $q_0 = +20$ nC is at the point $x = 2$ m, $y = 2$ m as shown in Figure 18-10*a*. Find the force on q_0.

Since q_2 and q_0 have opposite signs, the force exerted by q_2 on q_0 is attractive and in the negative y direction as shown in the figure. It is given by

$$\mathbf{F}_{20} = \frac{kq_2q_0}{r_{20}^2}\,\hat{\mathbf{r}}_{20}$$

$$= \frac{(8.99 \times 10^9\ \text{N·m}^2/\text{C}^2)(-15 \times 10^{-9}\ \text{C})(20 \times 10^{-9}\ \text{C})}{(2\ \text{m})^2}\,\mathbf{j}$$

$$= (-6.74 \times 10^{-7}\ \text{N})\mathbf{j}$$

The distance between q_1 and q_0 is $2\sqrt{2}$ m. The force exerted by q_1 on q_0 is

$$\mathbf{F}_{10} = \frac{kq_1q_0}{r_{10}^2}\,\hat{\mathbf{r}}_{10}$$

$$= \frac{(8.99 \times 10^9\ \text{N·m}^2/\text{C}^2)(25 \times 10^{-9}\ \text{C})(20 \times 10^{-9}\ \text{C})}{(2\sqrt{2}\ \text{m})^2}\,\hat{\mathbf{r}}_{10}$$

$$= (5.62 \times 10^{-7}\ \text{N})\,\hat{\mathbf{r}}_{10}$$

where $\hat{\mathbf{r}}_{10}$ is the unit vector that is directed along the line from q_1 to q_0.

The vector sum of these two forces is most easily found by first writing the forces in terms of their rectangular components. Since $\mathbf{F}_{10}$ makes an angle of 45° with the x and y axes, its x and y components are equal to each other and to $F_{10}/\sqrt{2}$:

$$F_{10x} = F_{10y} = \frac{5.62 \times 10^{-7}\ \text{N}}{\sqrt{2}} = 3.97 \times 10^{-7}\ \text{N}$$

The x and y components of the net force are therefore

$$F_x = F_{10x} + F_{20x} = (3.97 \times 10^{-7}\ \text{N}) + 0 = 3.97 \times 10^{-7}\ \text{N}$$

$$F_y = F_{10y} + F_{20y} = (3.97 \times 10^{-7}\ \text{N}) + (-6.74 \times 10^{-7}\ \text{N})$$

$$= -2.77 \times 10^{-7}\ \text{N}$$

The magnitude of the net force is

$$F_{\text{net}} = \sqrt{F_x^2 + F_y^2} = \sqrt{(3.97 \times 10^{-7}\ \text{N})^2 + (-2.77 \times 10^{-7}\ \text{N})^2}$$

$$= 4.84 \times 10^{-7}\ \text{N}$$

The net force points to the right and downward as shown in Figure 18-10b, making an angle θ with the x axis given by

$$\tan\theta = \frac{F_y}{F_x} = \frac{-2.77}{3.97} = -0.698$$

$$\theta = -34.9°$$

Questions

3. Discuss the similarities and differences in the properties of electric charge and gravitational mass.
4. If the sign convention for charge were changed so that the charge on the electron were positive and the charge on the proton were negative, would Coulomb's law still be written the same?

18-4 The Electric Field

The electric force exerted by one charge on another is an example of an action-at-a-distance force that is similar to the gravitational force exerted by one mass on another. To avoid the problem of action at a distance, we introduce the concept of the **electric field E.** One charge produces an electric field **E** everywhere in space, and this field exerts the force on the other charge. The force is thus exerted by the field at the position of the second charge, rather than by the first charge itself, which is some distance away.

Figure 18-11 shows a set of point charges, q_1, q_2, and q_3, arbitrarily arranged in space. If we place a charge q_0 at some point near this system of charges, there will be a force exerted on q_0 due to the other charges. The presence of the charge q_0 will generally change the original distribution of the other charges, particularly if the charges are on conductors. However, we may choose q_0 to be small enough so that its effect on the original charge distribution is negligible. We call such a small charge a **test charge** because we use it to test the field of other charges without disturbing them. The net force exerted on q_0 is the vector sum of the individual forces exerted on q_0 by each of the other charges in the system. By Coulomb's law, each of these forces is proportional to q_0, so the net force will be proportional to q_0. The electric field **E** at a point is defined as the net force on a positive test charge q_0 divided by q_0:

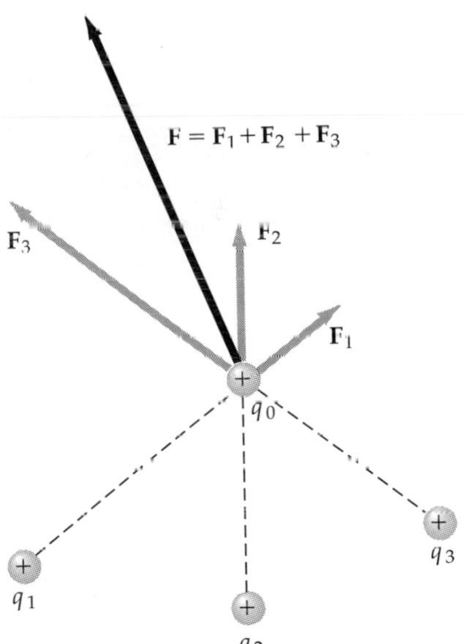

Figure 18-11 A small test charge q_0 in the vicinity of a system of charges, q_1, q_2, and q_3, experiences a force **F** that is proportional to q_0. The ratio $\mathbf{F}/q_0$ is the electric field **E** at the position of the test charge.

$$\mathbf{E} = \frac{\mathbf{F}}{q_0} \quad (q_0\ \text{small}) \qquad 18\text{-}5$$

Electric field defined

This definition is similar to that for the gravitational field of the earth, which was defined in Section 4-3 as the force per unit mass exerted by the earth on an object. The gravitational field of the earth **g** describes the property of the space around the earth such that when a mass m is placed at some point, the force exerted by the earth is $m\mathbf{g}$.

The SI unit of electric field is the newton per coulomb (N/C). Table 18-1 lists the magnitudes of some of the electric fields found in nature.

We note that the electric field is a vector and that it obeys the superposition principle. That is, the net electric field due to a system of charges can be found by computing the electric field due to each charge in the system separately and then adding these vectors to obtain the net electric field.

Table 18-1 **Some Electric Fields in Nature**

	E, N/C
In household wires	10^{-2}
In radio waves	10^{-1}
In the atmosphere	10^{2}
In sunlight	10^{3}
Under a thundercloud	10^{4}
In a lightning bolt	10^{4}
In an x-ray tube	10^{6}
At the electron in a hydrogen atom	6×10^{11}
At the surface of a uranium nucleus	2×10^{21}

The electric field **E** is a vector that describes the condition in space set up by the system of point charges. By moving a test charge q_0 from point to point, we can find **E** at all points in space (except one occupied by a charge q). The electric field **E** is thus a vector function of position.

The force exerted on a test charge q_0 at any point is related to the electric field at that point by

$$\mathbf{F} = q_0\mathbf{E} \tag{18-6}$$

Example 18-6

When a 5-nC test charge is placed at a certain point, it experiences a force of 2×10^{-4} N in the x direction. What is the electric field **E** at that point?

Since the force on the positive test charge is in the x direction, the electric field vector is also in the x direction. From its definition (Equation 18-5), the electric field is

$$\mathbf{E} = \frac{\mathbf{F}}{q_0} = \frac{(2 \times 10^{-4}\ \text{N})\mathbf{i}}{5 \times 10^{-9}\ \text{C}} = (4 \times 10^{4}\ \text{N/C})\mathbf{i}$$

Exercise

What is the force on an electron placed at the point in Example 18-6 where the electric field is $\mathbf{E} = (4 \times 10^{4}\ \text{N/C})\mathbf{i}$ [Answer: $(-6.4 \times 10^{-15}\ \text{N})\mathbf{i}$]

The electric field due to a single point charge q_i at a position $\mathbf{r}_i$ can be calculated from Coulomb's law. If we place a small, positive test charge q_0 at some point P a distance r_{i0} away, the force on it is

$$\mathbf{F}_{i0} = \frac{kq_iq_0}{r_{i0}^2}\,\hat{\mathbf{r}}_{i0}$$

where $\hat{\mathbf{r}}_{i0}$ is the unit vector that points from q_i to q_0. The electric field at point P due to charge q_i is thus

$$\mathbf{E}_i = \frac{kq_i}{r_{i0}^2}\hat{\mathbf{r}}_{i0} \qquad 18\text{-}7$$

Coulomb's law for **E** *due to a point charge*

where r_{i0} is the distance from the charge to point P called the **field point,** and $\hat{\mathbf{r}}_{i0}$ is a unit vector pointing from the charge to P. We will refer to this equation, which follows directly from Coulomb's law, as Coulomb's law for the electric field due to a single point charge. The net electric field due to a distribution of point charges is found by summing the fields due to each charge separately:

$$\mathbf{E} = \sum_i \mathbf{E}_i = \sum_i \frac{kq_i}{r_{i0}^2}\hat{\mathbf{r}}_{i0} \qquad 18\text{-}8$$

Electric field due to a system of point charges

Example 18-7

A positive charge $q_1 = +8$ nC is at the origin, and a second positive charge $q_2 = +12$ nC is on the x axis at $a = 4$ m (Figure 18-12). Find the net electric field (a) at point P_1 on the x axis at $x = 7$ m and (b) at point P_2 on the x axis at $x = 3$ m.

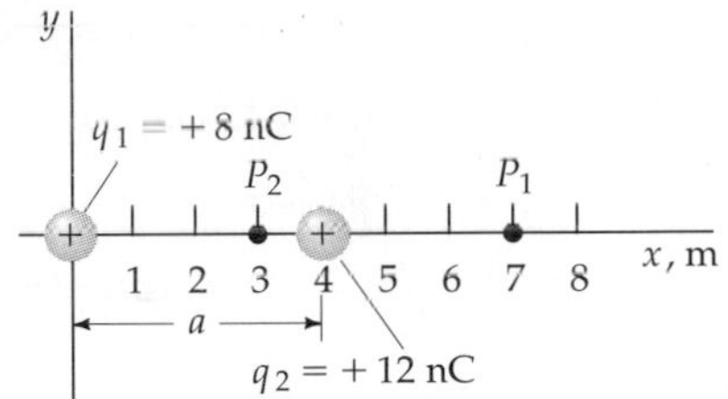

Figure 18-12 Two point charges on the x axis for Example 18-7. The net electric field is directed to the right at point P_1 and to the left at point P_2.

(a) The point P_1 at $x = 7$ m is to the right of both charges. The electric field at P_1 due to each charge is in the positive x direction. The distance from the field point to charge q_1 is $x = 7$ m and that to charge q_2 is $x - a = 7\text{ m} - 4\text{ m} = 3$ m. The net electric field at P_1 is thus

$$\mathbf{E} = \frac{kq_1}{x^2}\mathbf{i} + \frac{kq_2}{(x-a)^2}\mathbf{i}$$

$$= \frac{(8.99 \times 10^9\ \text{N}\cdot\text{m}^2/\text{C}^2)(8 \times 10^{-9}\ \text{C})}{(7\ \text{m})^2}\mathbf{i}$$

$$+ \frac{(8.99 \times 10^9\ \text{N}\cdot\text{m}^2/\text{C}^2)(12 \times 10^{-9}\ \text{C})}{(3\ \text{m})^2}\mathbf{i}$$

$$= (1.47\ \text{N/C})\mathbf{i} + (12.0\ \text{N/C})\mathbf{i} = (13.5\ \text{N/C})\mathbf{i}$$

(b) The point P_2 at $x = 3$ m is between the charges. A positive test charge placed at P_2 would experience a repulsive force to the right due to the +8-nC charge and a repulsive force to the left due to the +12-nC charge. The distance to the +8-nC charge is $x = 3$ m and that to the +12-nC charge is $a - x = 4\text{ m} - 3\text{ m} = 1$ m. The net electric field at P_2 is thus

$$\mathbf{E} = \frac{kq_1}{x^2}\mathbf{i} - \frac{kq_2}{(a-x)^2}\mathbf{i}$$

$$= \frac{(8.99 \times 10^9\ \text{N}\cdot\text{m}^2/\text{C}^2)(8 \times 10^{-9}\ \text{C})}{(3\ \text{m})^2}\mathbf{i}$$

$$- \frac{(8.99 \times 10^9\ \text{N}\cdot\text{m}^2/\text{C}^2)(12 \times 10^{-9}\ \text{C})}{(1\ \text{m})^2}\mathbf{i}$$

$$= (7.99\ \text{N/C})\mathbf{i} - (108\ \text{N/C})\mathbf{i} = (-100\ \text{N/C})\mathbf{i}$$

The electric field at point P_2 is in the negative x direction because the contribution to the field due to the +12-nC charge, which is 1 m away, is larger than that due to the +8-nC charge, which is 3 m away. As we move toward the +8-nC charge at the origin, the magnitude of the field due to the +8-nC charge increases and that due to the +12-nC charge decreases. There is one point between the charges where the net electric field due to the two charges is zero. At this point a positive test charge would experience no net force because the repulsive force to the right due to the +8-nC charge would just balance the repulsive force to the left due to the +12-nC charge. At points closer to the +8-nC charge at the origin, the electric field points in the positive x direction.

Exercise

Find the point on the x axis in Figure 18-12 where the electric field is zero. (Answer: $x = 1.80$ m)

Example 18-8

Find the electric field at a point P_3 on the y axis at $y = 3$ m for the charges in Example 18-7.

The fields due to each charge at point P_3 on the y axis are shown in Figure 18-13*a*. The field $\mathbf{E}_1$ due to the +8-nC charge is in the positive y direction and has a magnitude

$$E_1 = \frac{kq_1}{y^2} = \frac{(8.99 \times 10^9\ \text{N·m}^2/\text{C}^2)(8 \times 10^{-9}\ \text{C})}{(3\ \text{m})^2} = 7.99\ \text{N/C}$$

The field $\mathbf{E}_2$ due to the +12-nC charge is in the direction of the line from that charge to point P_3. The distance from the +12-nC charge to point P_3 is 5 m, which is found using the pythagorean theorem. The magnitude of $\mathbf{E}_2$ is

$$E_2 = \frac{(8.99 \times 10^9\ \text{N·m}^2/\text{C}^2)(12 \times 10^{-9}\ \text{C})}{(5\ \text{m})^2} = 4.32\ \text{N/C}$$

The field $\mathbf{E}_2$ has a component in the positive y direction equal to $E_2 \cos\theta$ and a component in the negative x direction equal to $-E_2 \sin\theta$. From the triangle in Figure 18-13*a*, we can see that $\cos\theta = \frac{3}{5} = 0.6$ and $\sin\theta = \frac{4}{5} = 0.8$. The x and y components of E_2 are thus

$$E_{2x} = -E_2 \sin\theta = -(4.32\ \text{N/C})(0.8) = -3.46\ \text{N/C}$$

and

$$E_{2y} = E_2 \cos\theta = (4.32\ \text{N/C})(0.6) = 2.59\ \text{N/C}$$

We obtain the x and y components of the net electric field $\mathbf{E}$ from

$$E_x = E_{1x} + E_{2x} = 0 + (-3.46\ \text{N/C}) = -3.46\ \text{N/C}$$

and

$$E_y = E_{1y} + E_{2y} = 7.99\ \text{N/C} + 2.59\ \text{N/C} = 10.6\ \text{N/C}$$

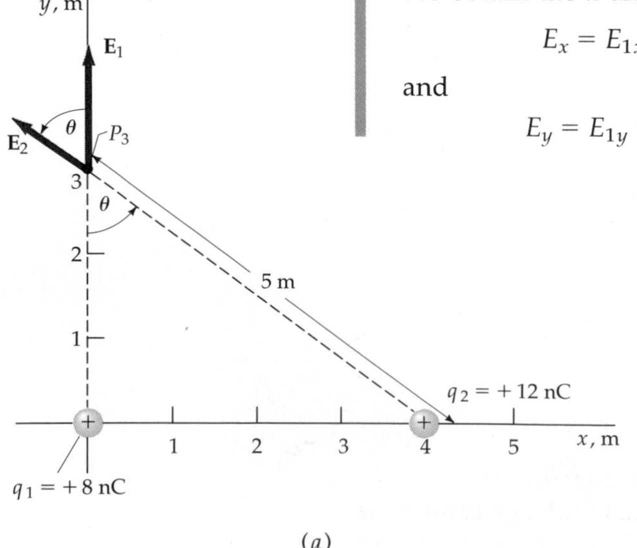

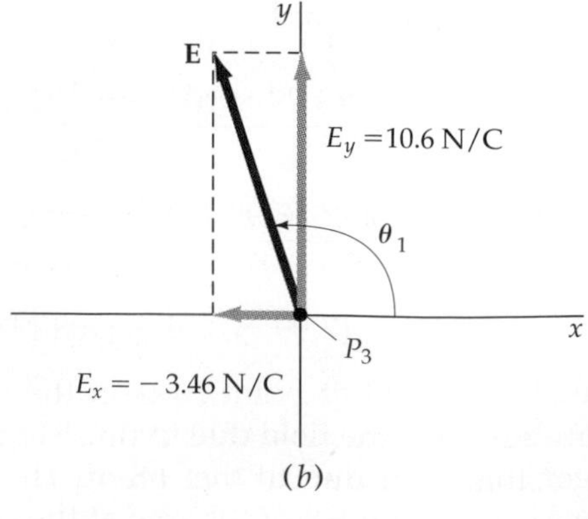

Figure 18-13 Example 18-8. (*a*) On the y axis, the electric field $\mathbf{E}_1$ due to charge q_1 is directed along the y axis, and the field $\mathbf{E}_2$ due to charge q_2 makes an angle θ with the y axis. The net electric field is the vector sum $\mathbf{E} = \mathbf{E}_1 + \mathbf{E}_2$. (*b*) The net electric field and its x and y components.

The magnitude of the net electric field is

$$E = \sqrt{E_x^2 + E_y^2} = \sqrt{(-3.46\ \text{N/C})^2 + (10.6\ \text{N/C})^2} = 11.2\ \text{N/C}$$

The net electric field **E** makes an angle θ_1 with the x axis (see Figure 18-13*b*) given by

$$\tan\theta_1 = \frac{E_y}{E_x} = \frac{10.6\ \text{N/C}}{-3.46\ \text{N/C}} = -3.06$$

$$\theta_1 = 108°$$

Example 18-9

A charge $+q$ is at $x = a$ and a second charge $-q$ is at $x = -a$ (Figure 18-14). Find the electric field on the x axis at a field point P, which is a large distance away compared to the separation of the charges.

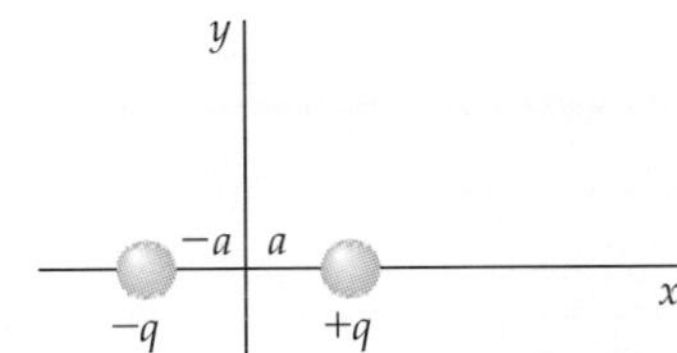

Figure 18-14 A point charge $+q$ at $x = a$ and a second point charge $-q$ at $x = -a$ for Example 18-9. This charge distribution is called an electric dipole.

The point P on the x axis is a distance $x - a$ from the positive charge and a distance $x + a$ from the negative charge. The electric field at P due to these two charges is thus

$$\mathbf{E} = \frac{kq}{(x-a)^2}\mathbf{i} + \frac{k(-q)}{(x+a)^2}\mathbf{i} = kq\mathbf{i}\left[\frac{1}{(x-a)^2} - \frac{1}{(x+a)^2}\right]$$

Putting the terms in brackets over a common denominator, we get

$$\frac{1}{(x-a)^2} - \frac{1}{(x+a)^2} = \frac{(x+a)^2 - (x-a)^2}{(x+a)^2(x-a)^2} = \frac{4ax}{(x^2-a^2)^2}$$

For $x \gg a$, we can neglect a^2 as compared with x^2 in the denominator. Therefore,

$$\frac{4ax}{(x^2-a^2)^2} \approx \frac{4ax}{x^4} = \frac{4a}{x^3}$$

Thus, the electric field at P is approximately

$$\mathbf{E} = \frac{4kqa}{x^3}\mathbf{i}$$

A system of two equal and opposite charges q separated by a small distance L is called an **electric dipole.** An electric dipole is characterized by the **electric dipole moment p,** which is a vector that points from the negative charge to the positive charge and has a magnitude that is the product of the charge q times the separation L (Figure 18-15). If **L** is the displacement vector of the positive charge from the negative charge, the dipole moment is

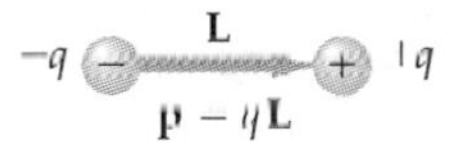

Figure 18-15 A dipole consists of two equal and opposite charges separated by some distance L. The dipole moment points from the negative charge to the positive charge and has the magnitude $p = qL$.

$$\mathbf{p} = q\mathbf{L} \qquad 18\text{-}9$$

Electric dipole moment defined

For the dipole shown in Figure 18-14, the displacement of the positive charge is $\mathbf{L} = 2a\mathbf{i}$, and the electric dipole moment is

$$\mathbf{p} = 2aq\mathbf{i}$$

In terms of the dipole moment, the electric field on the axis of the dipole at a point a great distance x away has the magnitude

$$E = \frac{2kp}{x^3} \qquad 18\text{-}10$$

Thus, the electric field far from a dipole is proportional to the dipole moment and decreases with the cube of the distance.

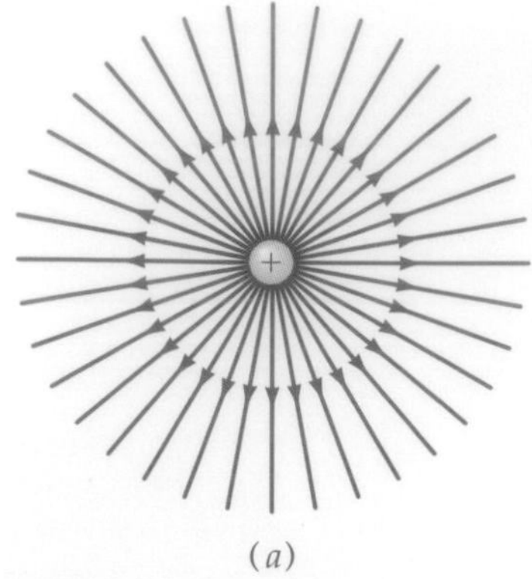

(a)

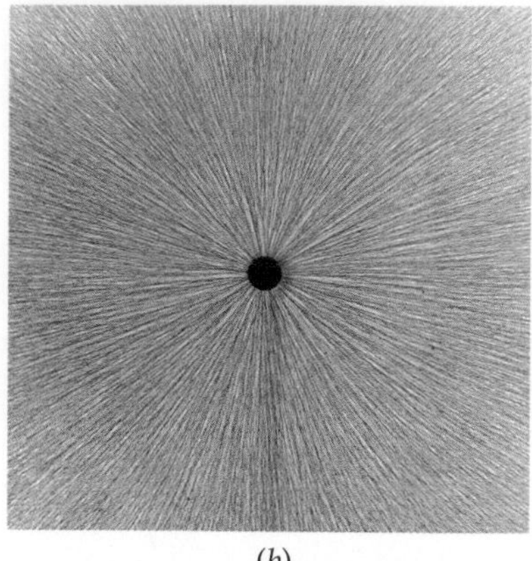

(b)

Figure 18-16 (a) Electric-field lines of a single positive point charge. If the charge were negative, the arrows would be reversed. (b) The same electric-field lines shown by bits of thread suspended in oil. The electric field of the charged object in the center induces opposite charges on the ends of each bit of thread, causing the threads to align themselves parallel to the field.

18-5 Electric-Field Lines

It is convenient to picture the electric field by drawing lines to indicate the direction of the field at any point. The field vector **E** is tangent to the line at each point and indicates the direction of the electric field at that point. Electric-field lines are also called lines of force because they show the direction of the force exerted on a positive test charge.

At any point near a positive charge, the electric field points radially away from the charge. The electric-field lines therefore diverge from a point occupied by a positive charge. Similarly, the electric field near a negative point charge points inward toward the charge, so the electric-field lines point toward a negative charge.

Figure 18-16 shows the electric-field lines of a single, positive point charge. As we move away from the charge, the electric field becomes weaker and the lines become farther apart. There is a connection between the spacing of the lines and the strength of the electric field. Consider a spherical surface of radius r with its center at the charge. We are interested in the number of lines per unit area of the sphere, which we will call the density of the lines. If we make r larger, the area of the sphere increases, but the same number of lines passes through it. The number of lines per unit area thus decreases as r increases. The area of the sphere is given by $A = 4\pi r^2$. The number of lines per unit area of the sphere thus decreases inversely with the square of the distance from the point charge. But the strength of the electric field, $E = kq/r^2$, also decreases inversely with the square of this distance. Thus, if we adopt the convention of drawing a fixed number of lines from a point charge, the number being proportional to the charge q, and if we draw the lines symmetrically about the point charge, the field strength is indicated by the density of the lines. The more closely spaced the lines, the stronger the electric field.

Figure 18-17 shows the electric-field lines for two equal positive point charges q separated by a distance a. We can sketch this pattern without calculating the field at each point. We again use the fact that the contribution to the field due to each of the charges varies as $1/r^2$, where r is the distance from that charge. At a point near one of the charges, the field is approximately due to that charge alone because the other charge is far enough away that we can ignore its contribution to the field. Thus, on a sphere of very small radius about either of the charges, the field lines are radial and equally spaced. Since the charges are equal, we draw an equal number of lines originating from each charge. At very large distances from the charges, the details of the system are not important. For example, if the two charges were 1 mm apart and we were looking at them from a point 100 km away, they

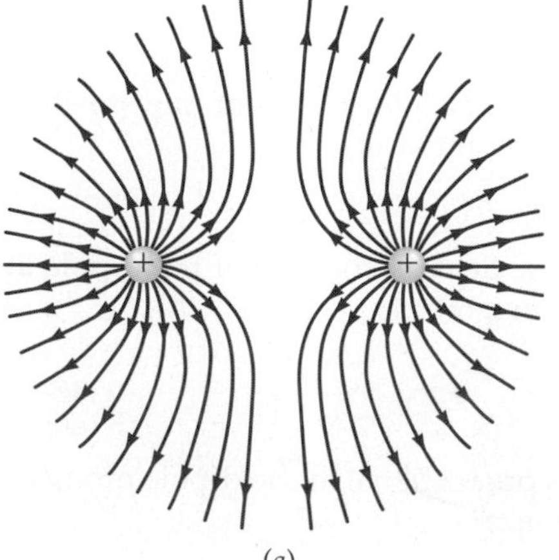

(a)

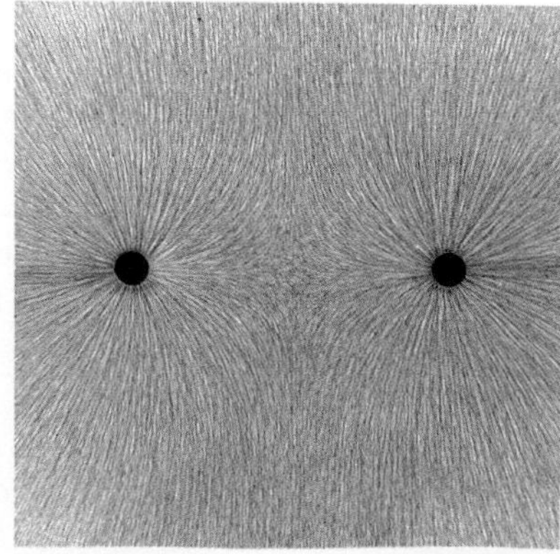

(b)

Figure 18-17 (a) Electric-field lines due to two positive point charges. The arrows would be reversed if both charges were negative. (b) The same electric-field lines shown by bits of thread in oil.

would look like a single charge. So on a sphere of radius r, where r is much greater than a, the field is approximately the same as that due to a single point charge of magnitude $2q$, and the lines are approximately equally spaced. We can see by merely looking at the figure that the electric field in the space between the charges is weak because there are few lines in this region compared with the region just to the right or left of the charges where the lines are more closely spaced. This information can, of course, also be obtained by direct calculation of the field at points in these regions.

We can apply the reasoning used in the preceding examples in drawing the electric-field lines for any system of point charges. Very near each charge, the field lines are equally spaced and leave or enter the charge radially, depending on the sign of the charge. Very far from all the charges, the detailed structure of the system is not important, so the field lines are just like those of a single point charge carrying the net charge of the system. For future reference, the rules for drawing electric-field lines can be summarized as follows:

Rules for drawing electric-field lines

1. Electric-field lines begin on positive charges and end on negative charges (or at infinity).
2. The lines are drawn symmetrically leaving or entering a charge.
3. The number of lines leaving a positive charge or entering a negative charge is proportional to the charge.
4. The density of the lines (the number of lines per unit area perpendicular to the lines) at some point is proportional to the magnitude of the field at that point.
5. At large distances from a system of charges, the field lines are equally spaced and radial as if they came from a single point charge equal to the net charge of the system.
6. No two field lines can cross.

Rule 6 follows from the fact that **E** has a unique direction at any point in space (except at a point occupied by a point charge or where $\mathbf{E} = 0$). If two lines crossed, two directions would be indicated for **E** at the point of intersection.

Figure 18-18 shows the electric-field lines due to an electric dipole. Very near the positive charge, the lines are directed radially outward. Very near the negative charge, the lines are directed radially inward. Since the charges have equal magnitudes, the number of lines that begin at the positive charge equals the number that end at the negative charge. In this case, the field is strong in the region between the charges, as indicated by the high density of field lines in this region in the figure.

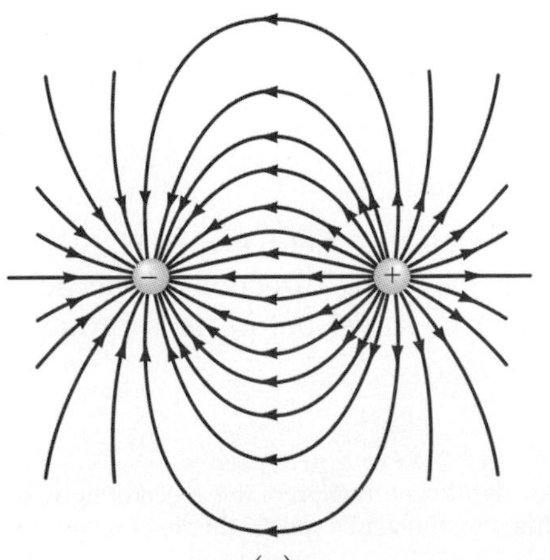

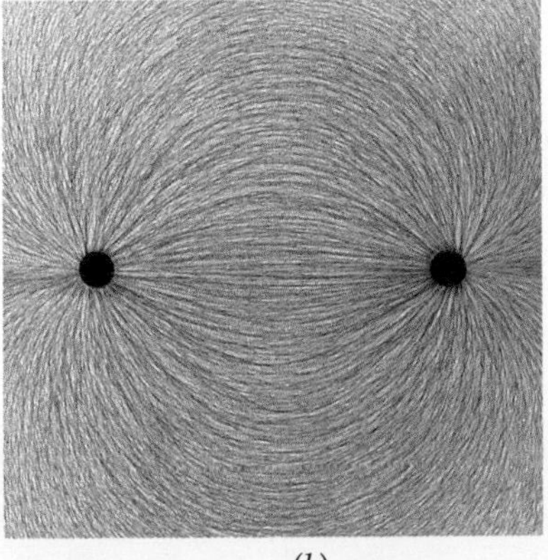

Figure 18-18 (*a*) Electric-field lines for an electric dipole. (*b*) The same field lines shown by bits of thread in oil.

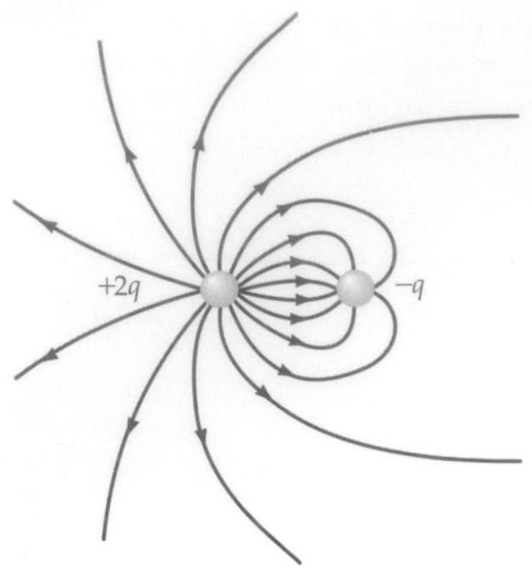

Figure 18-19 Electric-field lines for a point charge $+2q$ and a second point charge $-q$. At great distances from the charges, the lines are the same as those for a single charge $+q$.

Figure 18-19 shows the electric-field lines for a negative charge $-q$ at a distance a from a positive charge $+2q$. Since the positive charge has twice the magnitude of the negative charge, twice as many lines leave the positive charge as enter the negative charge. That is, half the lines beginning on the positive charge $+2q$ enter the negative charge $-q$, and the other half leave the system. On a sphere of radius r, where r is much larger than the separation of the charges, the lines leaving the system are approximately symmetrically spaced and point radially outward, just as they would for a single positive point charge $+q$. Thus at great distances from the charges, the system looks like a single charge $+q$. At a great distance from a system of charges, only the net charge is important. The convention indicating the electric field strength by the electric-field lines works because the electric field varies inversely as the square of the distance from a point charge.

Since the gravitational field of a point mass also varies inversely as the square of the distance, the concept of field lines is also useful for picturing the gravitational field. Near a point mass, the gravitational field lines converge toward the mass just as electric-field lines converge toward a negative charge. However, there are no points in space where gravitational field lines diverge as electric-field lines do near a positive charge because the gravitational force is always attractive, never repulsive.

18-6 Motion of Point Charges in Electric Fields

When a particle with a charge q is placed in an electric field $\mathbf{E}$, it experiences a force $q\mathbf{E}$. As we have seen, the gravitational forces acting on a particle are usually negligible in comparison with the electric forces. If the electric force is the only significant force acting on the particle, the particle has an acceleration

$$\mathbf{a} = \frac{q}{m}\mathbf{E}$$

where m is the mass of the particle.* If the electric field is known, the charge-to-mass ratio of the particle can be determined from the measured acceleration. For example, in a uniform electric field, the path of the particle is a parabola, similar to that of a projectile in a uniform gravitational field. The deflection of electrons in a uniform electric field was used by J. J. Thomson in 1897 to demonstrate the existence of electrons and to measure their charge-to-mass ratio. Familiar examples of devices that rely on the motion of electrons in electric fields are the oscilloscope and the television picture tube.

We will now look at some examples involving the motion of electrons in constant electric fields. Problems of this type can be worked using the constant-acceleration equations from Chapter 2 or the equations for projectile motion from Chapter 3.

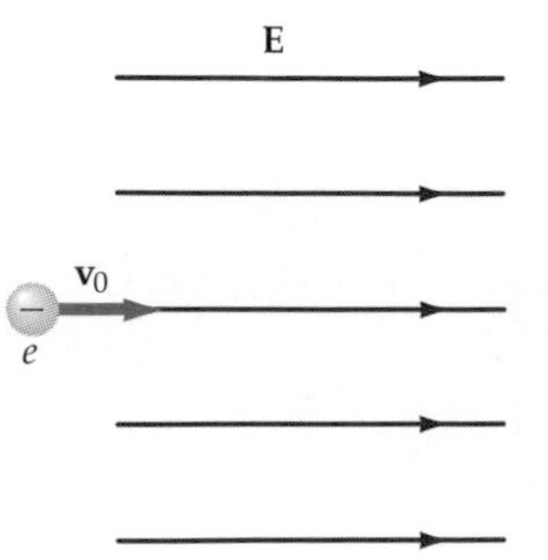

Figure 18-20 An electron projected into a uniform electric field with an initial velocity parallel to the field for Example 18-10.

Example 18-10

An electron is projected into a uniform electric field $\mathbf{E} = (1000\ \text{N/C})\mathbf{i}$ with an initial velocity $\mathbf{v}_0 = (2 \times 10^6\ \text{m/s})\mathbf{i}$ in the direction of the field (Figure 18-20). How far does the electron travel before it is brought momentarily to rest?

*Often the speed of an electron in an electric field is a significant fraction of the speed of light; in which case, Newton's laws of motion must be modified by Einstein's special theory of relativity.

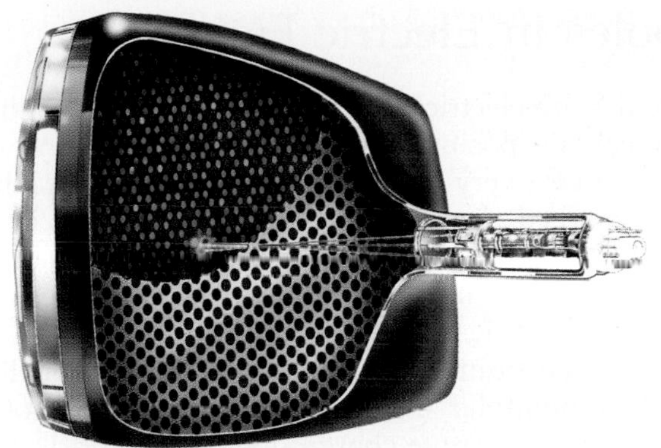

Schematic drawing of a cathode ray tube used for color television. The beams of electrons from the electron gun on the right activate phosphors on the screen at the left, giving rise to a bright spot whose color depends on the relative intensity of each beam. Electric fields between deflection plates in the gun (or magnetic fields from coils within the gun) deflect the beams. The beams sweep across the screen in a horizontal line, are then deflected downward, and sweep across again. The entire screen is covered once each 1/30 s.

Since the charge of the electron is negative, the force $-e\mathbf{E}$ acting on it is in the direction opposite that of the field. We thus have a constant-acceleration problem in which the acceleration of a particle is opposite to its initial velocity, and we are asked to find the distance the particle travels in its original direction. We can use the constant-acceleration equation relating distance to velocity:

$$v^2 = v_0^2 + 2a(x - x_0)$$

Using $x_0 = 0$, $v = 0$, $v_0 = 2 \times 10^6$ m/s, and $a = -eE/m$, we obtain

$$x = \frac{mv_0^2}{2eE} = \frac{(9.11 \times 10^{-31}\ \text{kg})(2 \times 10^6\ \text{m/s})^2}{2(1.6 \times 10^{-19}\ \text{C})(1000\ \text{N/C})} = 1.14 \times 10^{-2}\ \text{m}$$

Example 18-11

An electron is projected into a uniform electric field $\mathbf{E} = (-2000\ \text{N/C})\mathbf{j}$ with an initial velocity $\mathbf{v}_0 = (10^6\ \text{m/s})\mathbf{i}$ perpendicular to the field (Figure 18-21). (*a*) Compare the gravitational force acting on the electron to the electric force acting on it. (*b*) By how much has the electron been deflected after it has traveled 1 cm in the x direction?

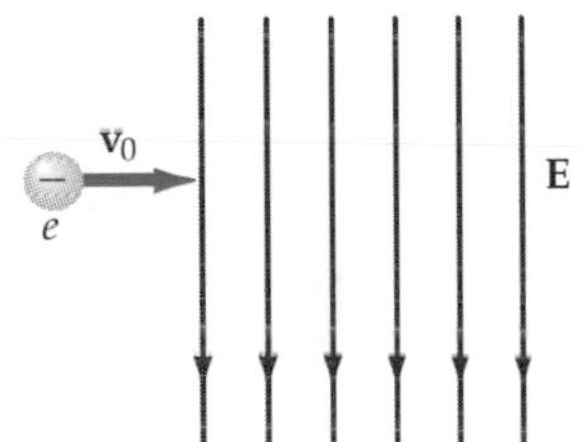

Figure 18-21 An electron projected into a uniform electric field with an initial velocity perpendicular to the field for Example 18-11.

(*a*) The electric force on the electron is $-e\mathbf{E}$, and the gravitational force on it is $m\mathbf{g}$. Since the electric field is downward, the electric force on the negative electron is upward. The gravitational force is, of course, downward. The ratio of their magnitudes is

$$\frac{F_e}{F_g} = \frac{eE}{mg} = \frac{(1.6 \times 10^{-19}\ \text{C})(2000\ \text{N/C})}{(9.1 \times 10^{-31}\ \text{kg})(9.8\ \text{N/kg})} = 3.6 \times 10^{13}$$

As is most commonly the case, the electric force is huge compared with the gravitational force, so the gravitational force is wholly negligible.

(*b*) The time it takes the electron to travel a distance of 1 cm in the x direction is

$$t = \frac{x}{v_0} = \frac{10^{-2}\ \text{m}}{10^6\ \text{m/s}} = 10^{-8}\ \text{s}$$

In this time, the electron is deflected upward, antiparallel to the field, a distance y given by

$$y = \frac{1}{2}at^2 = \frac{1}{2}\frac{eE}{m}t^2$$

Substituting in the known values of e, m, E, and t gives

$$y = 1.76 \times 10^{-2}\ \text{m} = 1.76\ \text{cm}$$

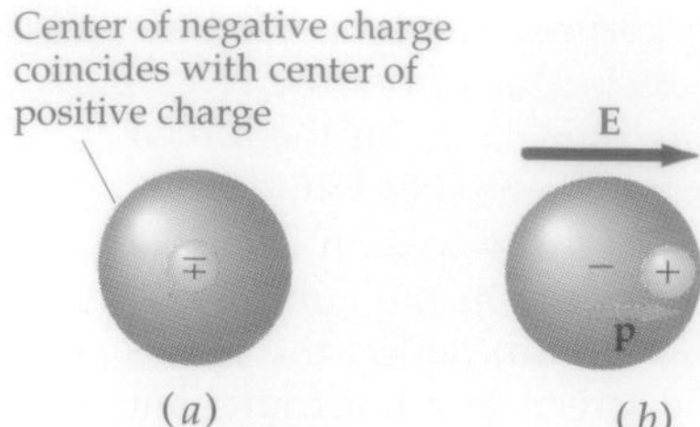

Figure 18-22 Schematic diagrams of the charge distributions of an atom or nonpolar molecule. (*a*) In the absence of an external electric field, the center of positive charge coincides with the center of negative charge. (*b*) In the presence of an external electric field, the centers of positive and negative charge are displaced, producing an induced dipole moment in the direction of the external field.

18-7 Electric Dipoles in Electric Fields

Although atoms and molecules are electrically neutral, they are affected by electric fields because they contain positive and negative charges. We can think of an atom as consisting of a very small, positively charged nucleus surrounded by a negatively charged electron cloud. Since the radius of the nucleus is about 100,000 times smaller than that of the electron cloud, we can consider it to be a point charge. In some atoms and molecules, the electron cloud is spherically symmetric, so its "center of charge" is at the center of the atom or molecule, coinciding with the positive charge. Such an atom or molecule is said to be **nonpolar.** However, in the presence of an external electric field, the center of positive charge does not coincide with the center of negative charge. The electric field exerts a force on the positively charged nucleus in the direction of the field and a force on the negatively charged electron cloud in the opposite direction. The positive and negative charges separate until the attractive force they exert on each other balances the forces due to the external electric field (Figure 18-22). Such a charge distribution behaves like an electric dipole.

The dipole moment of an atom or nonpolar molecule in an external electric field is called an **induced dipole moment.** It has the same direction as the electric field. If the electric field is uniform, there is no net force on the dipole because the forces on the positive and negative charges are equal and opposite. However, if the electric field is not uniform, there will be a net external force acting on the dipole. Figure 18-23 shows a nonpolar molecule in the electric field of a positive point charge q. The induced dipole moment is parallel to $\mathbf{E}$ in the radial direction from the point charge. The field is stronger at the center of negative charge because it is closer to the point charge, so the net force on the dipole is toward the point charge, and the dipole is attracted toward the point charge. If the point charge were negative, the induced dipole would be in the opposite direction, and the dipole would again be attracted to the point charge. The force produced by a nonuniform electric field on an electrically neutral particle is responsible for the familiar attraction of a charged comb for uncharged bits of paper. It is also responsible for the forces that hold an electrostatically charged balloon against a wall or ceiling. In this case, the charge on the balloon provides the nonuniform electric field that polarizes (that is, induces dipole moments in) the molecules of the wall or ceiling and then attracts them.

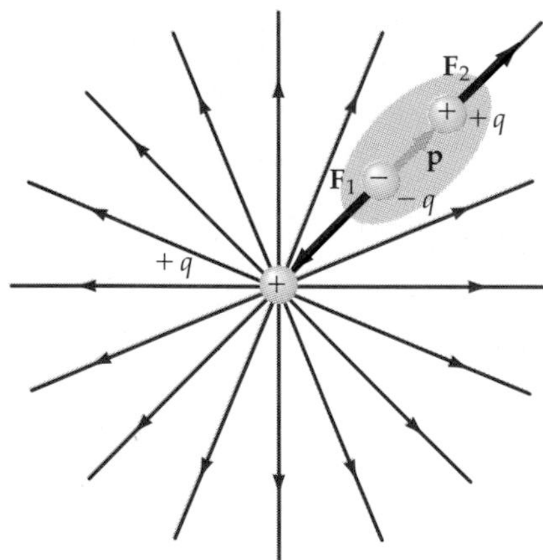

Figure 18-23 Nonpolar molecule in the nonuniform electric field of a positive point charge. The induced electric dipole moment **p** is parallel to the field of the point charge. Since the point charge is closer to the center of negative charge than to the center of positive charge, there is a net force of attraction between the dipole and the point charge.

In some molecules, the center of positive charge does not coincide with the center of negative charge even in the absence of an external electric field. These **polar molecules** have a permanent electric dipole moment. When a polar molecule is placed in a uniform electric field, there is no net force on it, but there is a torque that tends to rotate the molecule so that the dipole lines up with the field. Figure 18-24 shows the forces exerted on a dipole of moment $\mathbf{p} = q\mathbf{L}$ in a uniform electric field $\mathbf{E}$. We found in Section 9-4 that the torque produced by two equal and opposite forces, called a couple, is the same about any point in space. From the figure, we see that the torque about the negative charge has the magnitude $F_1L \sin\theta = qEL \sin\theta = pE \sin\theta$. The direction of the torque is into the paper such that it rotates the dipole moment $\mathbf{p}$ into the direction of the electric field $\mathbf{E}$. The torque can be conveniently written as the cross product of the dipole moment $\mathbf{p}$ and the electric field $\mathbf{E}$:

$$\boldsymbol{\tau} = \mathbf{p} \times \mathbf{E} \qquad 18\text{-}11$$

When the dipole rotates through an angle $d\theta$, the electric field does work

$$dW = -\tau\, d\theta = -pE \sin\theta\, d\theta$$

The minus sign arises because the torque tends to decrease θ. Setting this work equal to the decrease in potential energy, we have

$$dU = -dW = +pE \sin\theta \, d\theta$$

Integrating, we obtain

$$U = -pE \cos\theta + U_0$$

It is customary to choose the potential energy to be zero when the dipole is perpendicular to the electric field, that is, when $\theta = 90°$. Then $U_0 = 0$, and the potential energy of the dipole is

$$U = -pE \cos\theta = -\mathbf{p}\cdot\mathbf{E} \qquad 18\text{-}12$$

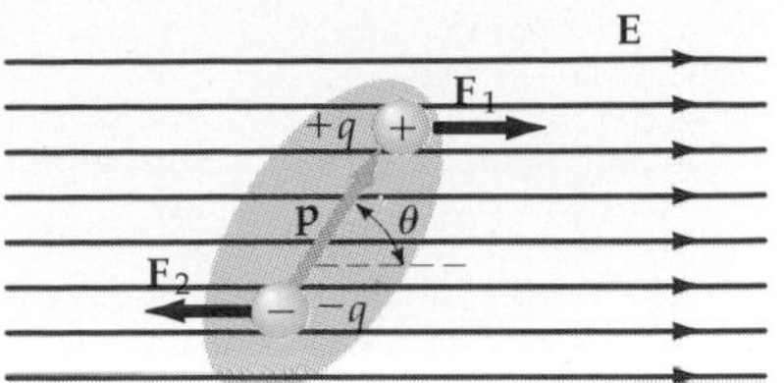

Figure 18-24 A dipole in a uniform electric field experiences equal and opposite forces that tend to rotate the dipole so that its dipole moment is aligned with the electric field.

In a nonuniform electric field, a polar molecule experiences a net force because the electric field has different magnitudes at the centers of positive and negative charge. An example of a polar molecule is HCl, which is essentially a positive hydrogen ion of charge $+e$ combined with a negative chlorine ion of charge $-e$. Another example of a polar molecule is water (Figure 18-25). The dipole moment of the water molecule is mainly responsible for the energy absorption of food in a microwave oven. Like all electromagnetic waves, microwaves have an oscillating electric field that can cause electric dipoles to vibrate. The vibration of the electric dipole moment of the water molecule in resonance with the oscillating electric field of the microwaves leads to the absorption of energy from the microwaves.

The diameter of an atom or molecule is of the order of 10^{-10} m = 0.1 nm. A convenient unit for the electric dipole moment of atoms and molecules is the fundamental electronic charge e times the distance 1 nm. For example, the dipole moment of NaCl in these units has a magnitude of about 0.2 e·nm.

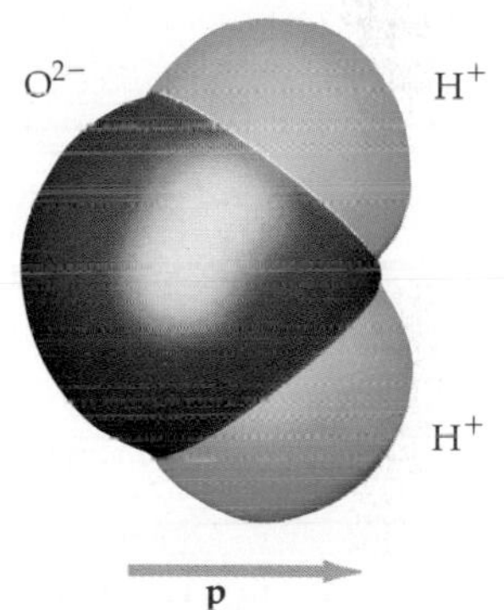

Figure 18-25 A computer-generated three-dimensional model of an H_2O molecule, consisting of oxygen ion of charge $-2e$ and two hydrogen ions of charge $+e$ each. This molecule has a permanent electric dipole moment in the direction shown.

Example 18-12

A dipole with a moment of magnitude 0.02 e·nm makes an angle of 20° with a uniform electric field of magnitude 3×10^3 N/C. Find (*a*) the magnitude of the torque on the dipole and (*b*) the potential energy of the system.

(*a*) The magnitude of the torque is

$$\tau = |\mathbf{p} \times \mathbf{E}| = pE \sin\theta$$
$$= (0.02)(1.60 \times 10^{-19}\ \text{C})(10^{-9}\ \text{m})(3 \times 10^3\ \text{N/C})(\sin 20°)$$
$$= 3.28 \times 10^{-27}\ \text{N·m}$$

(*b*) The potential energy of the system is

$$U = -\mathbf{p}\cdot\mathbf{E} = -pE\cos\theta$$
$$= -(0.02)(1.60 \times 10^{-19}\ \text{C})(10^{-9}\ \text{m})(3 \times 10^3\ \text{N/C})(\cos 20°)$$
$$= -9.02 \times 10^{-27}\ \text{J}$$

Question

5. A small, nonconducting ball with no net electric charge is suspended from a thread. When a positive charge is brought near the ball, the ball is attracted toward the charge. How does this come about? How would the situation be different if the charge brought near the ball were negative instead of positive?

Summary

1. There are two kinds of electric charge, labeled positive and negative. Electric charge always occurs in integral multiples of the fundamental unit of charge e. The charge of the electron is $-e$ and that of the proton is $+e$. Objects become charged by the transfer of electric charge from one object to another, usually in the form of electrons. Charge is conserved. It is neither created nor destroyed in the charging process, but is merely transferred.

2. The force exerted by one charge on another acts along the line joining the charges. It is proportional to the product of the charges and inversely proportional to the square of the separation. The force is repulsive if the charges have the same sign and attractive if they have opposite signs. This result is known as Coulomb's law:

$$\mathbf{F}_{12} = \frac{kq_1q_2}{r_{12}^2}\,\hat{\mathbf{r}}_{12}$$

where k is the Coulomb constant, which has the value

$$k = 8.99 \times 10^9\ \text{N·m}^2/\text{C}^2$$

3. The electric field due to a system of charges at a point is defined as the net force exerted by those charges on a positive test charge q_0 divided by q_0:

$$\mathbf{E} = \frac{\mathbf{F}}{q_0}$$

4. The electric field at some point P due to a single point charge q_i at a position $\mathbf{r}_i$ is

$$\mathbf{E}_i = \frac{kq_i}{r_{i0}^2}\,\hat{\mathbf{r}}_{i0}$$

where r_{i0} is the distance from charge q_i to the field point P and $\hat{\mathbf{r}}_{i0}$ is the unit vector pointing from q_i to P. The electric field due to several charges is the vector sum of the fields due to the individual charges:

$$\mathbf{E} = \sum_i \mathbf{E}_i = \sum_i \frac{kq_i}{r_{i0}^2}\,\hat{\mathbf{r}}_{i0}$$

5. The electric field can be represented by electric-field lines that originate on positive charges and end on negative charges. The strength of the electric field is indicated by the density of the electric-field lines.

6. An electric dipole is a system of two equal but opposite charges separated by a small distance. The dipole moment $\mathbf{p}$ is a vector with a magnitude equal to the charge times the separation that points in the direction from the negative charge to the positive charge:

$$\mathbf{p} = q\mathbf{L}$$

The electric field far from a dipole is proportional to the dipole moment and decreases with the cube of the distance.

7. In a uniform electric field, the net force on a dipole is zero, but there is a torque $\boldsymbol{\tau}$ given by

$$\boldsymbol{\tau} = \mathbf{p} \times \mathbf{E}$$

that tends to align the dipole in the direction of the field. The potential energy of a dipole in an electric field is given by

$$U = -\mathbf{p}\cdot\mathbf{E}$$

where the potential energy is taken to be zero when the dipole is perpendicular to the electric field. In a nonuniform electric field, there is a net force on a dipole.

8. Polar molecules, such as H_2O, have permanent dipole moments because their centers of positive and negative charge do not coincide. They behave like simple dipoles in an electric field. Nonpolar molecules do not have permanent dipole moments, but they acquire induced dipole moments in the presence of an electric field.

Suggestions for Further Reading

Bordeau, Sanford P.: *Volts to Hertz . . . The Rise of Electricity*, Burgess Publishing Company, Minneapolis, 1982.

This book presents a history of electrical science through discussion of the experiments and lives of the men after whom electrical and magnetic units are named. Illustrated with photographs and original engravings.

Cohn, Sherrye: "Painting the Fields of Faraday: Physics Inspired America's First Abstract Artist," *The Sciences*, November/December 1985, p. 44.

The painting "Rise of the Full Moon" by Arthur Dove, 1937, is reproduced here and discussed with reference to the field concept.

Goldhaber, Alfred Scharff, and Michael Martin Nieto: "The Mass of the Photon," *Scientific American*, May 1976, p. 86.

Strange as it may seem, tests of Coulomb's law provide an upper limit on a possible mass for the photon. This article describes the history of such tests, which began before Coulomb started his investigations and continue today.

Kevles, Daniel J.: "Robert Millikan," *Scientific American*, January 1979, p. 142.

The life and work of the second American scientist to win the Nobel prize in physics: this, in part, for his beautiful demonstration that all electrons carry the same charge, and his measurement of that charge.

Shamos, Morris H.: "The Laws of Electric and Magnetic Force—Charles Coulomb," in *Great Experiments in Physics*, Henry Holt and Co., New York, 1959. Reprinted by Dover, 1987.

Coulomb's description of his experiments in English translation, with editorial annotations for clarity and a biographical sketch.

Walker, Jearl: "The Amateur Scientist: How to Map Electronically Charged Patches with Parsley, Sage, Rosemary and Thyme," *Scientific American*, April 1988, p. 92.

Colorful experiments with charged surfaces.

Walker, Jearl: "The Amateur Scientist: The Secret of a Microwave Oven's Rapid Cooking Action Is Disclosed," *Scientific American*, February 1987, p. 134.

This article describes several mechanisms that have been suggested to explain the fact that microwaves heat water. All depend on the fact that the water molecule is polar.

Williams, Earle R.: "The Electrification of Thunderstorms," *Scientific American*, November 1988, p. 88.

The mechanisms of lightning, one of the most spectacular of natural electrical phenomena, are still in dispute.

Review

A. Objectives: After studying this chapter, you should:

1. Be able to state Coulomb's law and use it to find the force exerted by one point charge on another.

2. Know the value of the Coulomb constant in SI units.

3. Know the magnitude of the fundamental unit of electric charge e in coulombs.

4. Be able to use Coulomb's law to calculate the electric field due to a system of point charges.

5. Be able to draw the electric-field lines for simple charge distributions and to obtain information about the direction and strength of an electric field from such a diagram.

6. Be able to state the difference between a polar and a nonpolar molecule, and describe the behavior of each in uniform and nonuniform electric fields.

7. Be able to explain why bits of paper are attracted to a charged comb, and why an electrostatically charged balloon will stick to a wall.

B. Define, explain, or otherwise identify:

Charge quantization
Fundamental unit of charge
Law of conservation of charge
Coulomb
Conductors
Insulators
Electroscope

Electrostatic induction
Charging by induction
Grounded
Coulomb's law
Coulomb constant
Principle of superposition of electric forces
Electric field
Test charge
Field point
Electric dipole
Electric dipole moment
Electric-field lines
Nonpolar molecule
Induced dipole moment
Polar molecule

C. True or false: If the statement is true, explain why it is true. If it is false, give a counterexample.

1. The electric field of a point charge always points away from the charge.

2. The charge of the electron is the smallest unit of charge found.

3. Electric field lines never diverge from a point in space.

4. Electric field lines never cross at a point in space.

5. All molecules have electric dipole moments in the presence of an external electric field.

Problems

Level I

18-1 Electric Charge

1. A plastic rod is rubbed against a wool shirt, thereby acquiring a charge of $-0.8\ \mu$C. How many electrons are transferred from the wool shirt to the plastic rod?

2. A charge equal to the charge of Avogadro's number of protons ($N_A = 6.02 \times 10^{23}$) is called a *faraday*. Calculate the number of coulombs in a faraday.

3. How many coulombs of positive charge are there in 1 kg of carbon? Twelve grams of carbon contain Avogadro's number of atoms, with each atom having six protons and six electrons.

18-2 Conductors and Insulators and Charging by Induction

4. Explain, giving each step, how a positively charged insulating rod can be used to give a metal sphere (*a*) a negative charge and (*b*) a positive charge. (*c*) Can the same rod be used to simultaneously give one sphere a positive charge and another sphere a negative charge without the rod having to be recharged?

5. Two uncharged conducting spheres with their conducting surfaces in contact are supported on a large wooden table by insulated stands. A positively charged rod is brought up close to the surface of one of the spheres on the side opposite its point of contact with the other sphere. (*a*) Describe the induced charges on the two conducting spheres, and sketch the charge distributions on them. (*b*) The two spheres are separated far apart and the charged rod is removed. Sketch the charge distributions on the separated spheres.

18-3 Coulomb's Law

6. A charge $q_1 = 4.0\ \mu$C is at the origin, and charge $q_2 = 6.0\ \mu$C is on the x axis at $x = 3.0$ m. (*a*) Find the force on charge q_2. (*b*) Find the force on q_1. (*c*) How would your answers for parts (*a*) and (*b*) differ if q_2 were $-6.0\ \mu$C?

7. Three point charges are on the x axis; $q_1 = -6.0\ \mu$C is at $x = -3.0$ m, $q_2 = 4.0\ \mu$C is at the origin, and $q_3 = -6.0\ \mu$C is at $x = 3.0$ m. Find the force on q_1.

8. Two equal charges of $3.0\ \mu$C are on the y axis, one at the origin and the other at $y = 6$ m. A third charge $q_3 = 2\ \mu$C is on the x axis at $x = 8$ m. Find the force on q_3.

9. Three charges, each of magnitude 3 nC, are at the corners of a square of side 5 cm. The two charges at the opposite corners are positive, and the other is negative. Find the force exerted by these charges on a fourth charge $q = +3$ nC at the remaining corner.

10. A charge of $5\ \mu$C is on the y axis at $y = 3$ cm, and a second charge of $-5\ \mu$C is on the y axis at $y = -3$ cm. Find the force on a charge of $2\ \mu$C on the x axis at $x = 8$ cm.

18-4 The Electric Field

11. A charge of $4.0\ \mu$C is at the origin. What is the magnitude and direction of the electric field on the x axis at (*a*) $x = 6$ m and (*b*) $x = -10$ m? (*c*) Sketch the function E_x versus x for both positive and negative values of x. (Remember that E_x is negative when **E** points in the negative x direction.)

12. Two charges, each $+4\ \mu$C, are on the x axis, one at the origin and the other at $x = 8$ m. Find the electric field on the x axis at (*a*) $x = -2$ m, (*b*) $x = 2$ m, (*c*) $x = 6$ m, and (*d*) $x = 10$ m. (*e*) At what point on the x axis is the electric field zero? (*f*) Sketch E_x versus x.

13. Two equal positive charges of magnitude $q_1 = q_2 = 6.0$ nC are on the y axis at $y_1 = +3$ cm and $y_2 = -3$ cm. (*a*) What is the magnitude and direction of the electric field on the x axis at $x = 4$ cm? (*b*) What is the force exerted on a test charge $q_0 = 2$ nC placed on the x axis at $x = 4$ cm?

14. When a test charge $q_0 = 2$ nC is placed at the origin, it experiences a force of 8.0×10^{-4} N in the positive y direction. (*a*) What is the electric field at the origin? (*b*) What would be the force on a charge of -4 nC placed at the origin? (*c*) If this force is due to a charge on the y axis at $y = 3$ cm, what is the value of that charge?

15. An oil drop has a mass of 4×10^{-14} kg and a net charge of 4.8×10^{-19} C. An upward electric force just balances the downward force of gravity so that the oil drop is stationary. What is the direction and magnitude of the electric field?

16. The electric field near the surface of the earth points downward and has a magnitude of 150 N/C. (*a*) Compare the upward electric force on an electron with the downward gravitational force. (*b*) What charge should be placed on a penny of mass 3 g so that the electric force balances the weight of the penny near the earth's surface?

18-5 Electric-Field Lines

17. Figure 18-26 shows the electric-field lines for a system of two point charges. (*a*) What are the relative magnitudes of the charges? (*b*) What are the signs of the charges? (*c*) In what regions of space is the electric field strong? In what regions is it weak?

Figure 18-26 Electric-field lines for Problem 17.

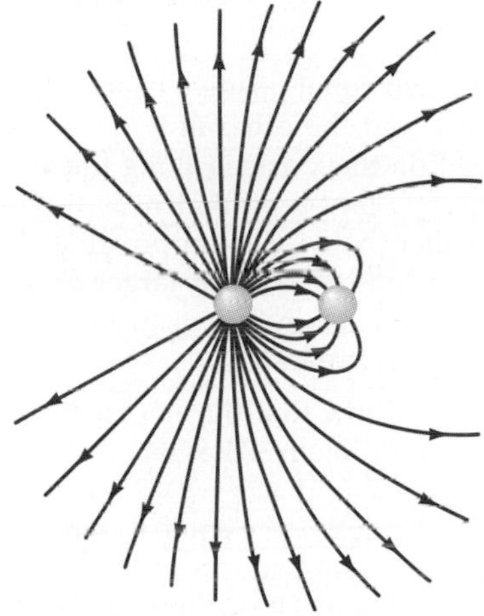

18. Two charges $+q$ and $-3q$ are separated by a small distance. Draw the electric-field lines for this system.

19. Three equal positive point charges are situated at the corners of an equilateral triangle. Sketch the electric-field lines in the plane of the triangle.

20. Two conducting spheres, each with a net positive charge, are held close together so that the electric-field lines are as shown in Figure 18-27. What is the relative charge on the small sphere compared to the large sphere?

Figure 18-27 Problem 20.

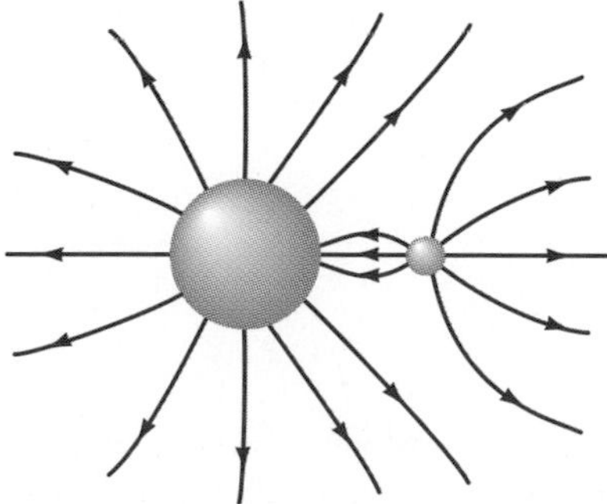

18-6 Motion of Point Charges in Electric Fields

21. In finding the acceleration of an electron or some other charged particle, the ratio of the charge to the mass of the particle is important. (*a*) Compute e/m for an electron. (*b*) What is the magnitude and direction of the acceleration of an electron in a uniform electric field with a magnitude of 100 N/C? (*c*) Nonrelativistic mechanics can be used only if the speed of the electron is significantly less than the speed of light c. Compute the time it takes for an electron placed at rest in an electric field with a magnitude of 100 N/C to reach a speed of $0.01c$. (*d*) How far does the electron travel in that time?

22. (*a*) Compute e/m for a proton, and find its acceleration in a uniform electric field with a magnitude of 100 N/C. (*b*) Find the time it takes for a proton initially at rest in such a field to reach a speed of $0.01c$ (where c is the speed of light).

23. An electron has an initial velocity of 2×10^6 m/s in the x direction. It enters a uniform electric field $\mathbf{E} = (400 \text{ N/C})\mathbf{j}$, which is in the y direction. (*a*) Find the acceleration of the electron. (*b*) How long does it take for the electron to travel 10 cm in the x direction in the field? (*c*) By how much and in what direction is the electron deflected after traveling 10 cm in the x direction in the field?

24. An electron moves in a circular orbit about a stationary proton. The centripetal force is provided by the electrostatic force of attraction between the proton and the electron. The electron has a kinetic energy of 2.18×10^{-18} J. (*a*) What is the speed of the electron? (*b*) What is the radius of the orbit of the electron?

18-7 Electric Dipoles in Electric Fields

25. Two point charges, $q_1 = 2.0$ pC and $q_2 = -2.0$ pC, are separated by 4 μm. (*a*) What is the dipole moment of this pair of charges? (*b*) Sketch the pair, and show the direction of the dipole moment.

26. A dipole of moment 0.5 e·nm is placed in a uniform electric field with a magnitude of 4.0×10^4 N/C. What is the magnitude of the torque on the dipole when (*a*) the dipole is parallel to the electric field, (*b*) the dipole is perpendicular to the electric field, and (*c*) the dipole makes an angle of 30° with the electric field? (*d*) Find the potential energy of the dipole in the electric field for each case.

Level II

27. In copper, about one electron per atom is free to move about. A copper penny has a mass of 3 g. (*a*) What percentage of the free charge would have to be removed to give the penny a charge of 15 μC? (See Example 18-1.) (*b*) What would be the force of repulsion between two pennies carrying this charge if they were 25 cm apart? Assume that the pennies are point charges.

28. A point charge of -5 μC is located at $x = 4$ m, $y = -2$ m. A second point charge of 12 μC is located at $x = 1$ m, $y = 2$ m. (*a*) Find the magnitude and direction of the electric field at $x = -1$ m, $y = 0$. (*b*) Calculate the magnitude and direction of the force on an electron at $x = -1$ m, $y = 0$.

29. A 5-μC point charge is located at $x = 1$ m, $y = 3$ m, and a -4-μC point charge is located at $x = 2$ m, $y = -2$ m. (*a*) Find the magnitude and direction of the electric field at $x = -3$ m, $y = 1$ m. (*b*) Find the magnitude and direction of the force on a proton at $x = -3$ m, $y = 1$ m.

30. A point charge of $-2.5\ \mu C$ is located at the origin. A second point charge of $6\ \mu C$ is at $x = 1$ m, $y = 0.5$ m. Find the x and y coordinates of the position at which an electron would be in equilibrium.

31. A particle leaves the origin with a speed of 3×10^6 m/s at 35° to the x axis. It moves in a constant electric field $\mathbf{E} = E_y\mathbf{j}$. Find E_y such that the particle will cross the x axis at $x = 1.5$ cm if the particle is (*a*) an electron and (*b*) a proton.

32. An electron starts at the position shown in Figure 18-28 with an initial speed $v_0 = 5 \times 10^6$ m/s at 45° to the x axis. The electric field is in the positive y direction and has a magnitude of 3.5×10^3 N/C. On which plate and at what location will the electron strike?

Figure 18-28 An electron moving in a uniform electric field for Problem 32.

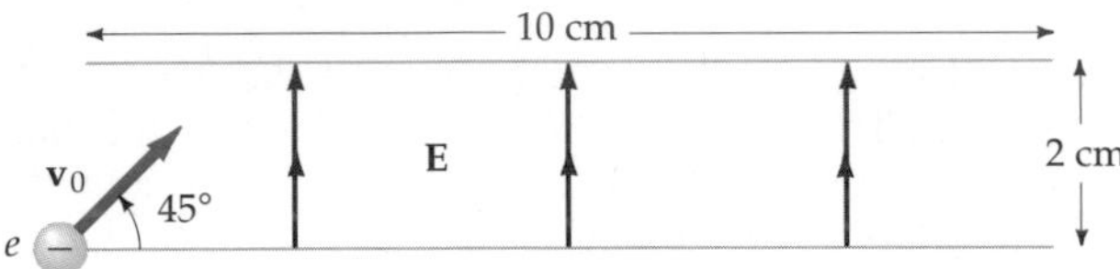

33. An electron with kinetic energy of 2×10^{-16} J is moving to the right along the axis of a cathode ray tube as shown in Figure 18-29. There is an electric field $\mathbf{E} = (2 \times 10^4\ \text{N/C})\mathbf{j}$ in the region between the deflection plates. Everywhere else, $\mathbf{E} = 0$. (*a*) How far is the electron from the axis of the tube when it reaches the end of the plates? (*b*) At what angle is the electron moving with respect to the axis? (*c*) How far from the axis will the electron be when it strikes the fluorescent screen?

Figure 18-29 An electron in a cathode ray tube for Problem 33.

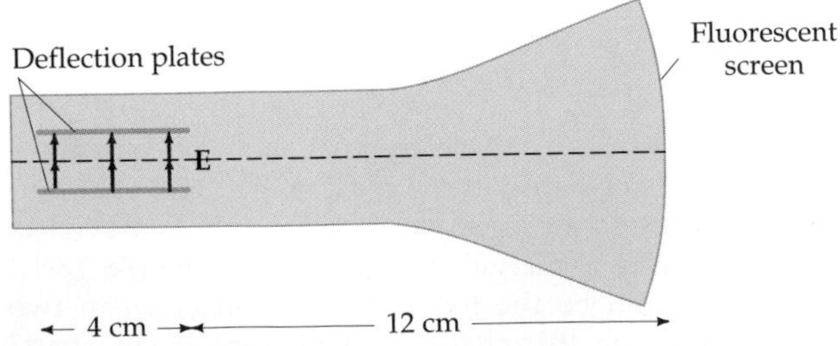

34. Four charges of equal magnitude are arranged at the corners of a square of side L as shown in Figure 18-30. (*a*) Find the magnitude and direction of the force exerted on the charge in the lower left corner by the other charges. (*b*) Show that the electric field at the midpoint of one of the sides of the square is directed along that side toward the negative charge and has a magnitude E given by

$$E = k\frac{8q}{L^2}\left(1 - \frac{\sqrt{5}}{25}\right)$$

Figure 18-30 Problem 34.

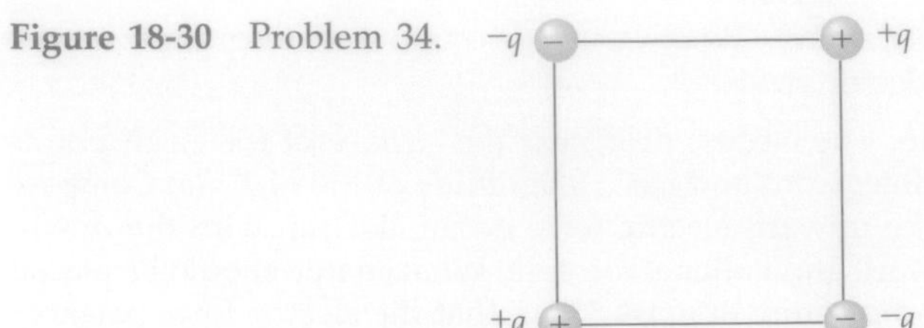

35. Two charges q_1 and q_2 have a total charge of $6\ \mu C$. When they are separated by 3 m, the force exerted by one charge on the other has a magnitude of 8 mN. Find q_1 and q_2 if (*a*) both are positive so that they repel each other and (*b*) one is positive and the other is negative so that they attract each other.

36. A positive charge Q is to be divided into two positive charges q_1 and q_2. Show that, for a given separation D, the force exerted by one charge on the other is greatest if $q_1 = q_2 = \frac{1}{2}Q$.

37. Two equal positive charges q are on the y axis, one at $y = a$ and the other at $y = -a$. (*a*) Show that the electric field on the x axis is along the x axis with $E_x = 2kqx(x^2 + a^2)^{-3/2}$. (*b*) Show that near the origin, when x is much smaller than a, E_x is approximately $2kqx/a^3$. (*c*) Show that for values of x much larger than a, E_x is approximately $2kq/x^2$. Explain why you would expect this result even before calculating it.

38. (*a*) Show that the electric field for the charge distribution in Problem 37 has its greatest magnitude at the points $x = a/\sqrt{2}$ and $x = -a\sqrt{2}$ by computing dE_x/dx and setting the derivative equal to zero. (*b*) Sketch the function E_x versus x using your results for part (*a*) of this problem and parts (*b*) and (*c*) of Problem 37.

39. An electric dipole consists of a positive charge q on the x axis at $x = a$ and a negative charge $-q$ on the x axis at $x = -a$. Find the magnitude and direction of the electric field at a point y on the y axis, and show that for $y \gg a$, the field is approximately $\mathbf{E} = -(kp/y^3)\mathbf{i}$, where p is the magnitude of the dipole moment.

40. Five equal charges Q are equally spaced on a semicircle of radius R as shown in Figure 18-31. Find the force on a charge q located at the center of the semicircle.

Figure 18-31 Problem 40.

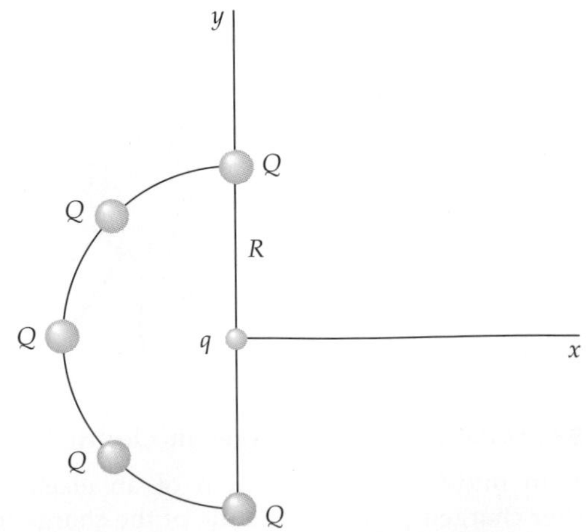

41. Two small spheres of mass m are suspended from a common point by threads of length L. When each sphere carries a charge q, each thread makes an angle θ with the vertical as shown in Figure 18-32. (a) Show that the charge q is given by

$$q = 2L \sin \theta \sqrt{\frac{mg \tan \theta}{k}}$$

where k is the Coulomb constant. (b) Find q if $m = 10$ g, $L = 50$ cm, and $\theta = 10°$.

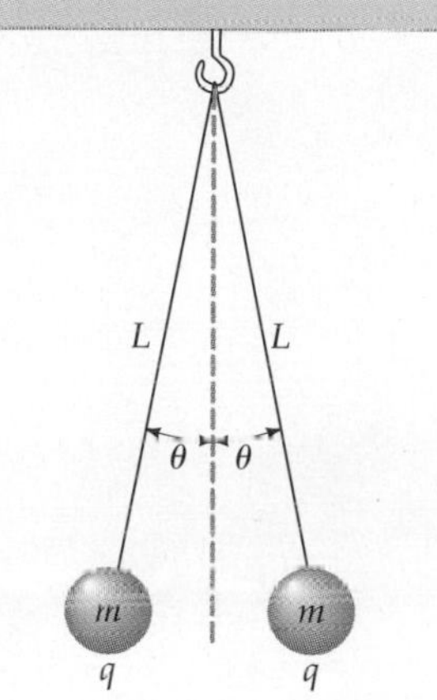

Figure 18-32 Problem 41.

42. A water molecule has its oxygen atom at the origin, one hydrogen nucleus at $x = 0.077$ nm, $y = 0.058$ nm and the other hydrogen nucleus at $x = -0.077$ nm, $y = 0.058$ nm. If the hydrogen electrons are transferred completely to the oxygen atom so that it has a charge of $-2e$, what is the dipole moment of the water molecule? This characterization of the chemical bonds of water as being totally ionic overestimates the dipole moment of a water molecule.

Level III

43. For the charge distribution in Problem 37, the electric field at the origin is zero. A test charge q_0 placed at the origin will therefore be in equilibrium. (a) Discuss the stability of the equilibrium for a positive test charge by considering small displacements from equilibrium along the x axis and small displacements along the y axis. (b) Repeat part (a) for a negative test charge. (c) Find the magnitude and sign of a charge q_0 that when placed at the origin results in a net force of zero on each of the three charges. (d) What will happen if any of the charges is displaced slightly from equilibrium?

44. Two positive point charges $+q$ are on the y axis at $y = +a$ and $y = -a$ as in Problem 37. A bead of mass m carrying a negative charge $-q$ slides along a thread that runs along the x axis. (a) Show that for small displacements of $x \ll a$, the bead experiences a restoring force that is proportional to x and therefore undergoes simple harmonic motion. (b) Find the period of the motion.

45. An electric dipole consists of two charges $+q$ and $-q$ separated by a very small distance $2a$. Its center is on the x axis at $x = x_1$, and it points along the x axis in the positive x direction. The dipole is in a nonuniform electric field, which is also in the x direction, given by $\mathbf{E} = Cx\mathbf{i}$, where C is a constant. (a) Find the force on the positive charge and that on the negative charge, and show that the net force on the dipole is $Cp\mathbf{i}$. (b) Show that, in general, if a dipole of moment $\mathbf{p}$ lies along the x axis in an electric field in the x direction, the net force on the dipole is given approximately by $(dE_x/dx)p\mathbf{i}$.

46. A positive point charge $+Q$ is at the origin, and a dipole of moment $\mathbf{p}$ is a distance r away and in the radial direction as in Figure 18-23. (a) Show that the force exerted by the electric field of the point charge on the dipole is attractive and has a magnitude of approximately $2kQp/r^3$ (see Problem 45). (b) Now assume that the dipole is centered at the origin and that a point charge Q is a distance r away along the line of the dipole. From your result for part (a) and Newton's third law, show that the magnitude of the electric field of the dipole along the line of the dipole a distance r away is approximately $2kp/r^3$.

47. A quadrupole consists of two dipoles that are close together as shown in Figure 18-33. The effective charge at the origin is $-2q$ and the other charges on the y axis at $y = a$ and $y = -a$ are each $+q$. (a) Find the electric field at a point on the x axis far away so that $x \gg a$. (b) Find the electric field on the y axis far away so that $y \gg a$.

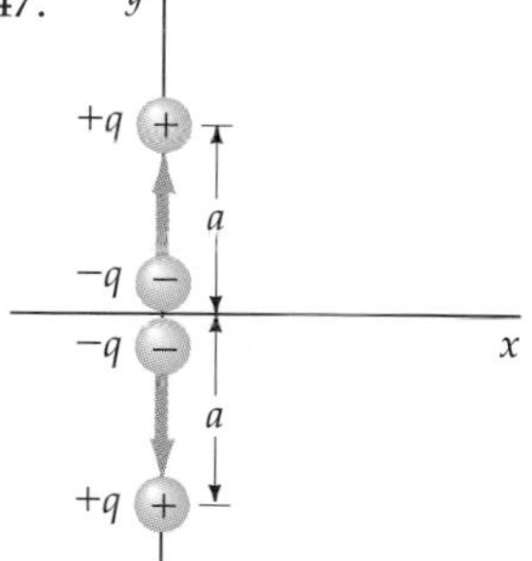

Figure 18-33 Problem 47.

Chapter 19

The Electric Field II: Continuous Charge Distributions

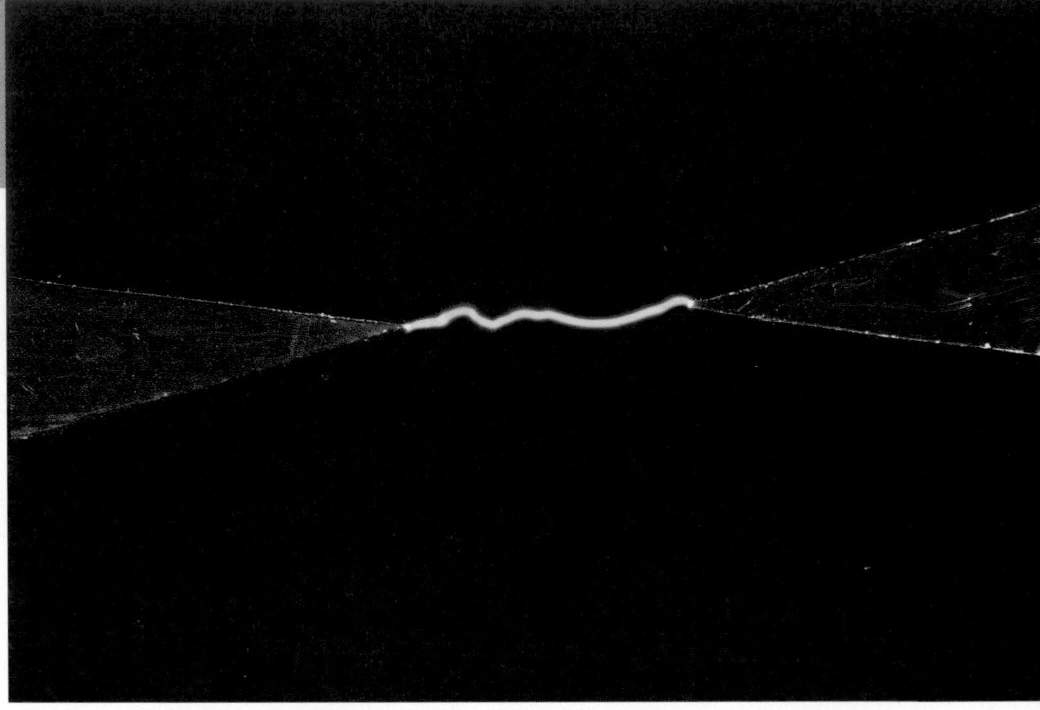

Electrical discharge between two charged conductors. The electric field near the sharp points is strong enough to strip the electrons from nearby air molecules, thus ionizing them and causing the air to conduct.

On a microscopic scale, electric charge is quantized. However, there are often situations in which many charges are so close together that the total charge can be considered to be continuously distributed in space. The use of a continuous charge density to describe a distribution of a large number of discrete charges is similar to the use of a continuous mass density to describe air, which actually consists of a large number of discrete molecules. In either case, it is usually easy to find a volume element ΔV that is large enough to contain many (billions of) individual charges or molecules and yet is small enough that replacing ΔV by a differential dV and using calculus introduces negligible error. We describe the charge per unit volume by the **volume charge density** ρ:

$$\rho = \frac{\Delta Q}{\Delta V} \qquad \text{19-1}$$

Often charge is distributed in a thin layer on the surface of an object. In such cases, we define the **surface charge density** σ as the charge per unit area:

$$\sigma = \frac{\Delta Q}{\Delta A} \qquad \text{19-2}$$

Similarly, we sometimes encounter a charge distributed along a line in space. In such cases, we define the **linear charge density** λ as the charge per unit length:

$$\lambda = \frac{\Delta Q}{\Delta L} \tag{19-3}$$

In this chapter, we will give examples of using Coulomb's law to calculate the electric field due to various types of continuous charge distributions. We will then discuss Gauss's law, which relates the electric field on a closed surface to the net charge within the surface, and use this relation to calculate the electric field for certain charge distributions that have a high degree of symmetry.

19-1 Calculation of the Electric Field from Coulomb's Law

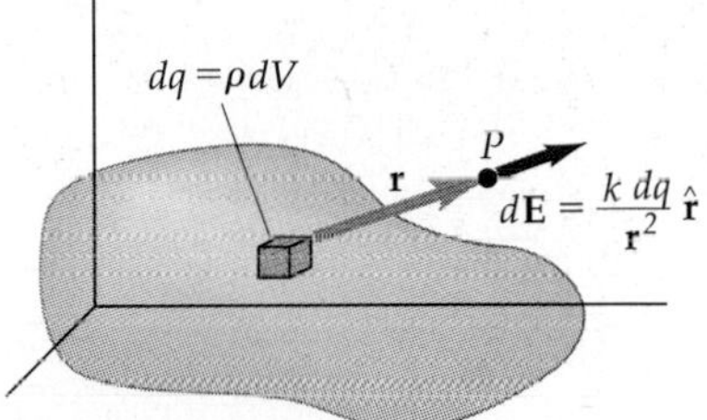

Figure 19-1 An element of charge dq produces a field $d\mathbf{E} = (k\,dq/r^2)\,\hat{\mathbf{r}}$ at point P. The field at P due to the total charge is found by integrating over the entire charge distribution.

The electric field produced by a given charge distribution can be calculated in a straightforward way using Coulomb's law. In Figure 19-1, we have chosen an element of charge $dq = \rho\, dV$ that is small enough that we may consider it to be a point charge. The electric field $d\mathbf{E}$ at a field point P due to this charge element is given by Coulomb's law:

$$d\mathbf{E} = \frac{k\,dq}{r^2}\,\hat{\mathbf{r}}$$

where r is the distance from the charge element to the field point P, and $\hat{\mathbf{r}}$ is a unit vector that points from the element to the field point. The total field at P is found by integrating this expression over the entire charge distribution, which we consider to occupy some volume V:

$$\mathbf{E} = \int_V \frac{k\,dq}{r^2}\,\hat{\mathbf{r}} \tag{19-4}$$

Electric field due to a continuous charge distribution

where $dq = \rho\, dV$. If the charge is distributed on a surface, we use $dq = \sigma\, dA$ and integrate over the surface. If the charge is along a line, we use $dq = \lambda\, dL$ and integrate over the line.

E on the Axis of a Finite Line Charge

A uniform charge Q lies along the x axis from $x = 0$ to $x = L$ as shown in Figure 19-2. The linear charge density for this charge is $\lambda = Q/L$. We wish to find the electric field produced by this line charge at some point P on the x axis at $x = x_0$, for $x_0 > L$. In the figure, we have chosen a small differential element dx at a distance x from the origin. The field point P is at a distance $r = x_0 - x$ from this charge element. The electric field due to this element of charge is directed along the x axis and has the magnitude

$$dE_x = \frac{k\,dq}{(x_0 - x)^2} = \frac{k\lambda\, dx}{(x_0 - x)^2}$$

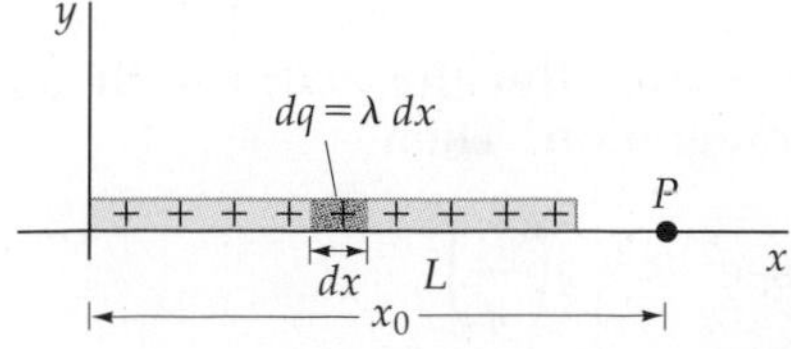

Figure 19-2 Geometry for the calculation of the electric field on the axis of a uniform line charge of charge Q, length L, and linear charge density $\lambda = Q/L$. An element $dq = \lambda\, dx$ of the line charge is treated as a point charge. The field due to this element is found from Coulomb's law, and the total field is then found by integrating from $x = 0$ to $x = L$.

We find the total field by integrating over the entire line charge from $x = 0$ to $x = L$:

$$E_x = k\lambda \int_0^L \frac{dx}{(x_0 - x)^2} = k\lambda \left[\frac{1}{x_0 - x}\right]_0^L$$

$$= k\lambda \left\{\frac{1}{x_0 - L} - \frac{1}{x_0}\right\} = k\lambda \left\{\frac{L}{x_0(x_0 - L)}\right\}$$

Using $\lambda = Q/L$, we obtain

$$E_x = \frac{kQ}{x_0(x_0 - L)} \qquad \text{19-5}$$

We can see that if L is much smaller than x_0, the electric field at x_0 is approximately kQ/x_0^2. That is, if we are sufficiently far away from the line charge, it looks like a point charge.

E on the Perpendicular Bisector of a Finite Line Charge

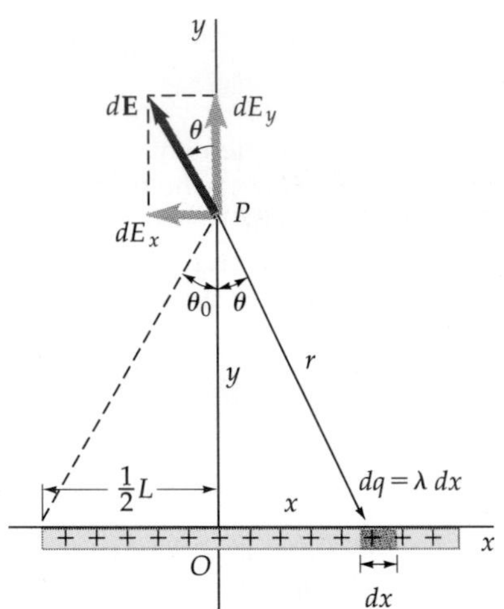

Figure 19-3 Geometry for the calculation of the electric field at a point on the perpendicular bisector of a uniform finite line charge. By symmetry, the net electric field is perpendicular to the line charge. The net field is found by integrating the expression for the perpendicular component from $\theta = 0$ to $\theta = \theta_0$ and multiplying by 2.

We will now find the electric field due to a uniform line charge of length L and total charge Q at a point P on the perpendicular bisector of the line as shown in Figure 19-3. In the figure, we have chosen a coordinate system such that the origin is at the center of the line charge, the charge is on the x axis, and the field point P is on the y axis. The charge element $dq = \lambda\, dx$ and the field $d\mathbf{E}$ that it produces are shown. The field has a component parallel to the line charge and one perpendicular to it. However, we can see from the symmetry of the charge distribution that for each charge element to the right of the origin (such as the one shown) there is one to the left that produces a parallel component of $d\mathbf{E}$ that is equal and opposite to the one shown. When we sum over all the charge elements in the line, the parallel components will sum to zero. We therefore need to calculate only the component of **E** perpendicular to the line charge.

The magnitude of the field produced by an element of charge $dq = \lambda\, dx$ is

$$|d\mathbf{E}| = \frac{k\, dq}{r^2} = \frac{k\lambda\, dx}{r^2}$$

The perpendicular component (in this case, the y component) is

$$dE_y = \frac{k\lambda\, dx}{r^2} \cos\theta \qquad \text{19-6}$$

The total field E_y is computed by integrating from $x = -\frac{1}{2}L$ to $x = +\frac{1}{2}L$. Because of the symmetry of the charge distribution, the contribution of each half of the line charge to the total field is the same, so we can get the same result by integrating from $x = 0$ to $x = \frac{1}{2}L$ and multiplying by 2. That is,

$$E_y = \int_{x=-\frac{1}{2}L}^{x=+\frac{1}{2}L} dE_y = 2\int_{x=0}^{x=\frac{1}{2}L} dE_y \qquad \text{19-7}$$

This integration can be simplified by rewriting it as an integration over the variable θ instead of the variable x. We can see from Figure 19-3 that x and θ are related by

$$x = y \tan\theta \qquad \text{19-8}$$

where y is the perpendicular distance from the line charge to the field point, which does not vary during the integration. Then

$$\frac{dx}{d\theta} = y \sec^2\theta = y\left(\frac{r}{y}\right)^2$$

The increment dx is thus related to the increment $d\theta$ by

$$dx = \frac{r^2}{y}\,d\theta$$

Substituting $r^2\,d\theta/y$ for dx in Equation 19-6, we obtain

$$dE_y = \frac{k\lambda}{y}\cos\theta\,d\theta \qquad 19\text{-}9$$

From Equation 19-8, we can see that $\theta = 0$ when $x = 0$, and that $\theta = \theta_0$ when $x = \frac{1}{2}L$, which is given by

$$\tan\theta_0 = \frac{\frac{1}{2}L}{y}$$

The total y component of the field is obtained by integrating Equation 19-9 from $\theta = 0$ to $\theta = \theta_0$ and multiplying by 2:

$$E_y = 2\int_{\theta=0}^{\theta=\theta_0} dE_y = \frac{2k\lambda}{y}\int_0^{\theta_0}\cos\theta\,d\theta$$

or

$$E_y = \frac{2k\lambda}{y}\sin\theta_0 = \frac{2k\lambda}{y}\,\frac{\frac{1}{2}L}{\sqrt{(\frac{1}{2}L)^2 + y^2}} \qquad 19\text{-}10$$

E on the perpendicular bisector of a finite line charge

where (from Figure 19-3) $\sin\theta_0$ is related to L and y by

$$\sin\theta_0 = \frac{\frac{1}{2}L}{\sqrt{(\frac{1}{2}L)^2 + y^2}}$$

When y is much greater than L, $\sin\theta_0$ is given approximately by

$$\sin\theta_0 \approx \frac{\frac{1}{2}L}{y} \qquad y \gg L$$

and E_y is approximately

$$E_y \approx \frac{k\lambda L}{y^2} = \frac{kQ}{y^2}$$

where $Q = \lambda L$ is the total charge. As expected, when we are very far away, a finite line charge looks like a point charge.

E near an Infinite Line Charge

When we are very close to a line charge or, alternatively, when the line charge is very long, so that $y \ll L$, the angle θ_0 in Figure 19-3 is approximately 90°. Substituting $\theta_0 = 90°$ into Equation 19-10 gives

$$E_y = \frac{2k\lambda}{y} \qquad 19\text{-}11$$

E at a distance y from an infinite line charge

Thus, as the distance y of a point from an infinite line charge increases, the electric field decreases as $1/y$.

Exercise

Show that Equation 19-11 has the correct units for the electric field.

Example 19-1

An infinite line charge of linear charge density $\lambda = 0.6\ \mu\text{C/m}$ lies along the z axis, and a point charge $q = 8\ \mu\text{C}$ lies on the y axis at $y = 3$ m as shown in Figure 19-4. Find the electric field at the point P on the x axis at $x = 4$ m.

The electric field at any point in space is found from the superposition of the electric field due to the line charge and that due to the point charge. At the field point P on the x axis at $x = 4$ m, the electric field due to the line charge is in the x direction and is given by

$$\mathbf{E}_\text{L} = \frac{2k\lambda}{x}\,\mathbf{i} = \frac{2(8.99 \times 10^9\ \text{N}\cdot\text{m}^2/\text{C}^2)(0.6 \times 10^{-6}\ \text{C/m})}{4\ \text{m}}\,\mathbf{i}$$
$$= (2.70\ \text{kN/C})\mathbf{i}$$

The electric field due to the point charge, which is $\sqrt{(4\ \text{m})^2 + (3\ \text{m})^2} =$ 5 m away, is

$$\mathbf{E}_\text{P} = \frac{kq}{r^2}\,\hat{\mathbf{r}} = \frac{(8.99 \times 10^9\ \text{N}\cdot\text{m}^2/\text{C}^2)(8 \times 10^{-6}\ \text{C})}{(5\ \text{m})^2} = (2.88\ \text{kN/C})\hat{\mathbf{r}}$$

where $\hat{\mathbf{r}}$ is a unit vector that points from q to the field point P. The field $\mathbf{E}_\text{P}$ makes an angle θ with the x axis as shown in the figure. The x and y components of $\mathbf{E}_\text{P}$ are

$$E_{\text{P}x} = E_\text{P} \cos\theta = (2.88\ \text{kN/C})(\tfrac{4}{5}) = 2.30\ \text{kN/C}$$

and

$$E_{\text{P}y} = -E_\text{P} \sin\theta = -(2.88\ \text{kN/C})(\tfrac{3}{5}) = -1.73\ \text{kN/C}$$

The net electric field at P has x and y components given by

$$E_x = E_{\text{L}x} + E_{\text{P}x} = 2.70\ \text{kN/C} + 2.30\ \text{kN/C} = 5.00\ \text{kN/C}$$

and

$$E_y = E_{\text{L}y} + E_{\text{P}y} = 0 + (-1.73\ \text{kN/C}) = -1.73\ \text{kN/C}$$

The net electric field therefore has a magnitude of

$$E = \sqrt{E_x^2 + E_y^2} = \sqrt{(5.00)^2 + (-1.73)^2}\ \text{kN/C} = 5.29\ \text{kN/C}$$

It is directed at an angle ϕ below the x axis given by

$$\phi = \tan^{-1}\left(-\frac{1.73}{5.00}\right) = -19.1°$$

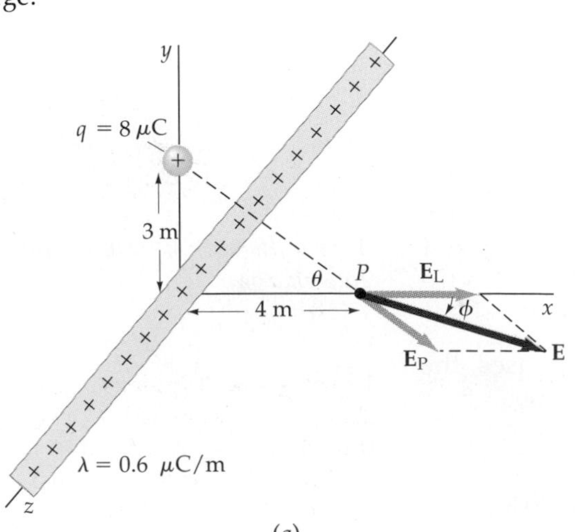

Figure 19-4 (*a*) An infinite line charge along the z axis and a point charge q on the y axis for Example 19-1. The electric field **E** at a point P on the x axis is found from the superposition of the fields due to the line charge and the point charge. The field due to the point charge is directed along the line from the point charge to the field point. (*b*) Electric-field lines near a long wire. The electric field near a high-voltage power line can be large enough to strip the electrons from air molecules, thus ionizing them and making the air a conductor. The glow resulting from the recombination of free electrons with the ions is called corona discharge.

E on the Axis of a Ring Charge

Figure 19-5 shows a uniform ring charge of radius a and total charge Q. We wish to find the electric field at a point P on the axis of the ring at a distance x from the center of the ring. The field $d\mathbf{E}$ due to the charge element dq is shown in the figure. This field has a component dE_x directed along the axis of the ring and a component $dE_\perp$ directed perpendicular to the axis. From the symmetry of the charge distribution, we can see that the net field due to the entire ring must lie along the axis of the ring; that is, the perpendicular components will sum to zero. In particular, the perpendicular component shown will be canceled by that due to the charge element on the ring directly opposite the one shown. The axial component of the field due to the charge element shown is

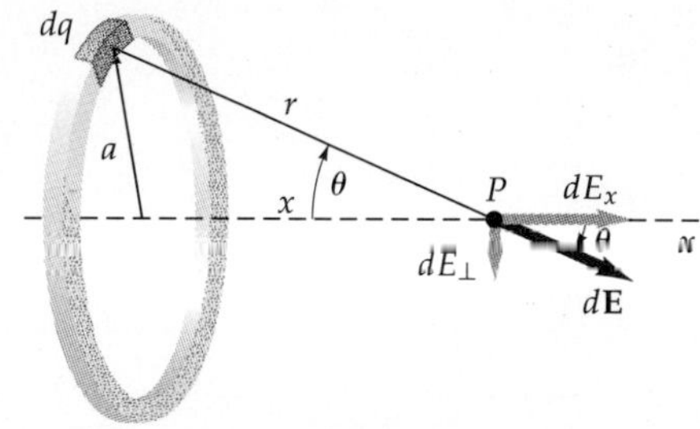

Figure 19-5 A ring charge of radius a. The electric field at point P on the x axis due to the charge element dq shown has a component along the x axis and one perpendicular to the x axis. When summed over the total ring, the perpendicular components cancel, so the net field is along the x axis.

$$dE_x = \frac{k\,dq}{r^2}\cos\theta = \frac{k\,dq}{r^2}\frac{x}{r} = \frac{k\,dq\,x}{(x^2+a^2)^{3/2}}$$

where

$$r^2 = x^2 + a^2$$

and

$$\cos\theta = \frac{x}{r} = \frac{x}{\sqrt{x^2+a^2}}$$

The field due to the entire ring of charge is

$$E_x = \int \frac{kx\,dq}{(x^2+a^2)^{3/2}}$$

Since x does not vary as we integrate over the elements of charge, we can remove it from the integral. Then

$$E_x = \frac{kx}{(x^2+a^2)^{3/2}}\int dq$$

or

$$E_x = \frac{kQx}{(x^2+a^2)^{3/2}} \qquad 19\text{-}12$$

E on the axis of a ring charge

A useful check of our result is to examine it at the extreme values of x. At $x = 0$, we obtain $E_x = 0$, which is what we should expect since for each element on the ring the field at the center is canceled by that due to the element directly opposite on the other side of the ring. When x is much larger than a, we can neglect a^2 compared with x^2 in the denominator of Equation 19-12. We then obtain $E_x \approx kQ/x^2$. That is, far from the ring, the ring looks like a point charge as expected.

E on the Axis of a Uniformly Charged Disk

Figure 19-6 shows a uniformly charged disk of radius R and total charge Q. We wish to find the electric field on the axis of the disk. Since the area of the disk is πR^2, the charge per unit area is $\sigma = Q/\pi R^2$. The electric field on the axis of the disk will be parallel to the axis. We can calculate the field by treating the disk as a set of concentric ring charges. Consider a ring of radius a and width da as shown in the figure. The area of this ring is $dA = 2\pi a\,da$, and its charge is $dq = \sigma\,dA = 2\pi\sigma a\,da$. The field produced by this ring is given by Equation 19-12 if we replace Q with $dq = 2\pi\sigma a\,da$. Thus

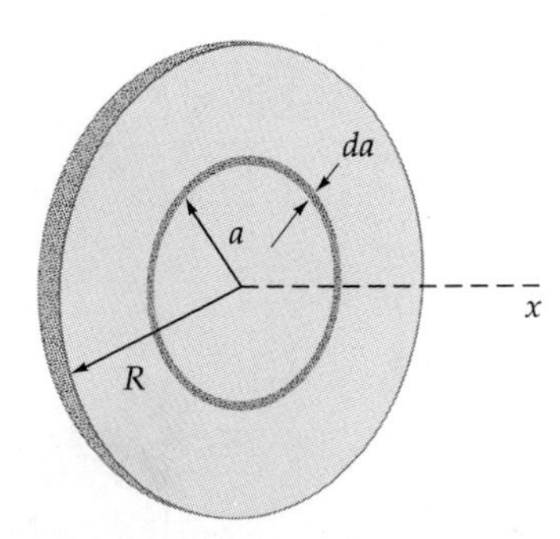

Figure 19-6 A uniform disk of charge can be treated as a set of ring charges, each of radius a and thickness da carrying a charge $dq = \sigma\,dA = (Q/\pi R^2)2\pi a\,da$.

$$dE_x = \frac{kx2\pi\sigma a\,da}{(x^2+a^2)^{3/2}}$$

The total field produced by the disk is found by integrating this expression from $a = 0$ to $a = R$:

$$E_x = \int_0^R \frac{kx2\pi\sigma a\, da}{(x^2 + a^2)^{3/2}} = kx\pi\sigma \int_0^R (x^2 + a^2)^{-3/2}\, 2a\, da$$

This integral is of the form $\int u^n\, du$, with $u = x^2 + a^2$ and $n = -\frac{3}{2}$. The integration thus gives

$$E_x = kx\pi\sigma \left[\frac{(x^2 + a^2)^{-1/2}}{-\frac{1}{2}}\right]_0^R$$

$$= -2kx\pi\sigma \left(\frac{1}{\sqrt{x^2 + R^2}} - \frac{1}{x}\right)$$

or

E on the axis of a disk charge

$$E_x = 2\pi k\sigma \left(1 - \frac{x}{\sqrt{x^2 + R^2}}\right) \qquad 19\text{-}13$$

When we are very far from the disk, we expect the disk to look like a point charge. For $x \gg R$, we can find an approximation for the second term in Equation 19-13 using the binomial expansion, $(1 + \epsilon)^n \approx 1 + n\epsilon$, for $\epsilon \ll 1$. This gives

$$\frac{x}{\sqrt{x^2 + R^2}} = \frac{x}{x(1 + R^2/x^2)^{1/2}} = \left(1 + \frac{R^2}{x^2}\right)^{-1/2} \approx 1 - \frac{R^2}{2x^2} + \cdots$$

Equation 19-13 then becomes

$$E_x \approx 2\pi k\sigma \left(1 - 1 + \frac{R^2}{2x^2} + \cdots\right) = \frac{k\pi R^2\sigma}{x^2} = \frac{kQ}{x^2}$$

where $Q = \sigma\pi R^2$ is the total charge on the disk.

E near an Infinite Plane of Charge

The interesting and important result for the field near an infinite plane of charge can be obtained from Equation 19-13 by letting either R go to infinity or x go to zero. Then

E near an infinite plane of charge

$$E_x = 2\pi k\sigma \qquad x > 0 \qquad 19\text{-}14a$$

Thus, the field due to an infinite-plane charge distribution is uniform; that is, the field does not depend on x. On the other side of the infinite plane, for negative values of x, the field points in the negative x direction, so

$$E_x = -2\pi k\sigma \qquad x < 0 \qquad 19\text{-}14b$$

As we move along the x axis, the electric field jumps from $-2\pi k\sigma\mathbf{i}$ to $+2\pi k\sigma\mathbf{i}$ when we pass through an infinite plane of charge.

Example 19-2

A disk of radius 5 cm carries a uniform surface charge density of 4 $\mu C/m^2$. Using reasonable approximations, find the electric field on the axis of the disk at distances of (*a*) 0.01 cm, (*b*) 0.03 cm, (*c*) 6 m, and (*d*) 6 cm.

(*a*) Since 0.01 cm is much less than the radius of the disk, we can approximate the disk as an infinite plane of charge and use Equation 19-14*a*. The electric field is then

$$E_x = 2\pi k\sigma$$
$$= 2\pi(8.99 \times 10^9\ \text{N·m}^2/\text{C}^2)(4 \times 10^{-6}\ \text{C/m}^2)$$
$$= 226\ \text{kN/C}$$

(*b*) Again, 0.03 cm is much less than the 5-cm radius of the disk, so the disk looks like an infinite plane from this point and the electric field is $2\pi k\sigma = 226$ kN/C.

(*c*) Since 6 m is much greater than the radius of the disk, we may treat the disk as a point charge $Q = \sigma\pi r^2 = (4\ \mu\text{C/m}^2)\pi(0.05\ \text{m})^2 = 31.4$ nC. The electric field at a distance of 6 m from such a point charge is

$$E_x = \frac{kQ}{x^2} = \frac{(8.99 \times 10^9\ \text{N·m}^2/\text{C}^2)(31.4 \times 10^{-9}\ \text{C})}{(6\ \text{m})^2}$$
$$= 7.84\ \text{N/C}$$

(*d*) Since 6 cm is neither much less than nor much greater than the radius of 5 cm, we use the exact expression given by Equation 19-13:

$$E_x = 2\pi k\sigma \left(1 - \frac{x}{\sqrt{(x^2 + R^2)}}\right)$$
$$= (226\ \text{kN/C}) \left(1 - \frac{6\ \text{cm}}{\sqrt{(6\ \text{cm})^2 + (5\ \text{cm})^2}}\right)$$
$$= (226\ \text{kN/C})(1 - 0.768) = 52.4\ \text{kN/C}$$

Note that we did not need to convert from centimeters to meters to find $x/\sqrt{x^2 + R^2}$ because the units cancel.

Exercise

Calculate the electric field to four significant figures for parts (*a*) and (*b*) of Example 19-2, and compare your results with the approximate results found in that example. [Answers: (*b*) $E_x = 225.9$ kN/C, which differs from 226 kN/C by about 0.04 percent.]

Question

1. Since electric charge is quantized, how is it possible to talk about continuous charge distributions?

19-2 Gauss's Law

The qualitative description of the electric field using electric-field lines discussed in Chapter 18 is related to a mathematical equation known as Gauss's law, which relates the electric field on a closed surface to the net charge within the surface. Electric fields arising from some symmetrical charge distributions, such as a spherical shell of charge or an infinite line of charge, can be easily calculated using Gauss's law. In this section, we give a plausible argument for Gauss's law based on the properties of electric-field lines. A rigorous derivation of Gauss's law is given in Section 19-5.

Figure 19-7 shows a surface of arbitrary shape enclosing a dipole. The number of electric-field lines coming from the positive charge, crossing the surface, and going outside of the volume enclosed by the surface depends on where the surface is drawn, but it is exactly equal to the number of lines entering the volume enclosed by the surface and ending on the negative charge. If we count the number of lines leaving as positive and the number entering as negative, the net number leaving and entering is zero. For surfaces enclosing other types of charge distributions, such as that shown in Figure 19-8, the net number of lines leaving any surface enclosing the charges is proportional to the net charge enclosed by the surface. This is a qualitative statement of Gauss's law.

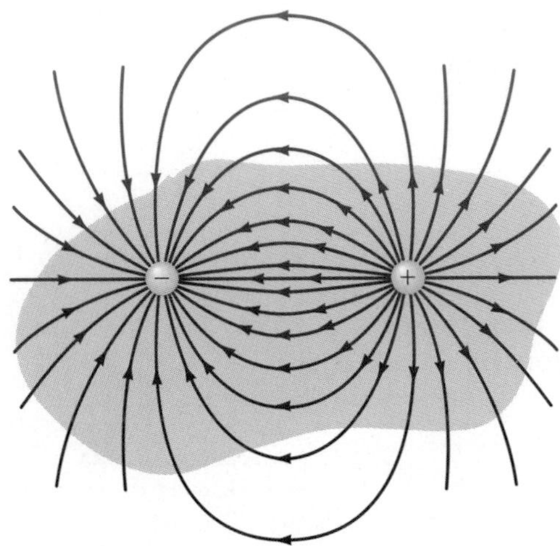

Figure 19-7 A surface of arbitrary shape enclosing an electric dipole. As long as the surface encloses both charges, the number of lines leaving the surface is exactly equal to the number of lines entering the surface no matter where the surface is drawn.

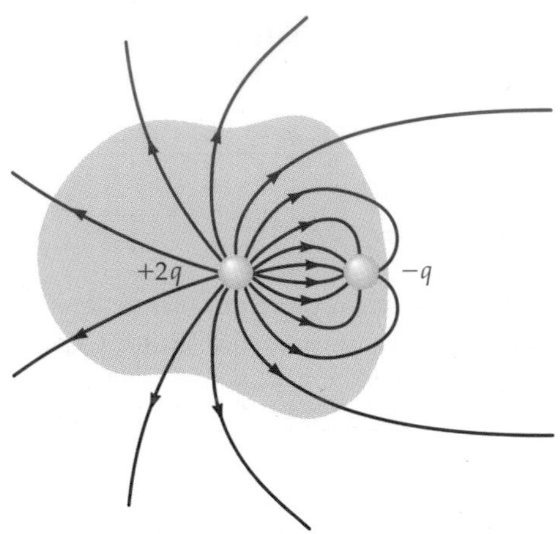

Figure 19-8 A surface of arbitrary shape enclosing the charges $+2q$ and $-q$. The field lines that end on $-q$ either do not pass through the surface or they exit once and enter once. The net number that exit is the same as that for a single charge equal to the net charge within the surface.

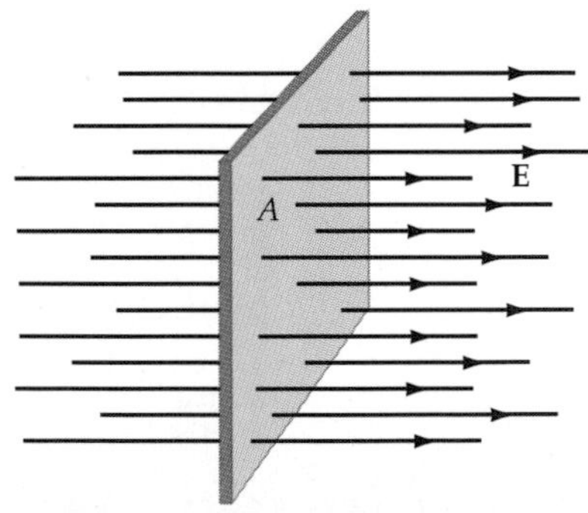

Figure 19-9 Electric-field lines of a uniform field crossing an area A that is perpendicular to the field. The product EA is the electric flux ϕ through the area.

The mathematical quantity related to the number of field lines crossing a surface is called the electric flux. Figure 19-9 shows an area A that is perpendicular to a uniform electric field. The **electric flux** ϕ through a surface of area A that is perpendicular to the field is defined as the product of the field **E** and the area A:

$$\phi = EA$$

The units of flux are newton-meters squared per coulomb ($\text{N}\cdot\text{m}^2/\text{C}$). Since the electric field is proportional to the number of lines per unit area, the flux is proportional to the number of field lines through the area.

In Figure 19-10, the surface of area A_2 is not perpendicular to the electric field **E**. The number of lines that cross area A_2 is the same as the number that cross area A_1. These areas are related by

$$A_2 \cos \theta = A_1 \qquad \text{19-15}$$

where θ is the angle between **E** and the unit vector $\hat{\mathbf{n}}$ that is perpendicular to the surface A_2 as shown in the figure. The flux through a surface that is not perpendicular to **E** is defined to be

$$\phi = \mathbf{E}\cdot\hat{\mathbf{n}}A = EA \cos \theta = E_{\text{n}}A$$

where $E_{\text{n}} = \mathbf{E}\cdot\hat{\mathbf{n}}$ is the component of the electric-field vector that is perpendicular, or normal, to the surface.

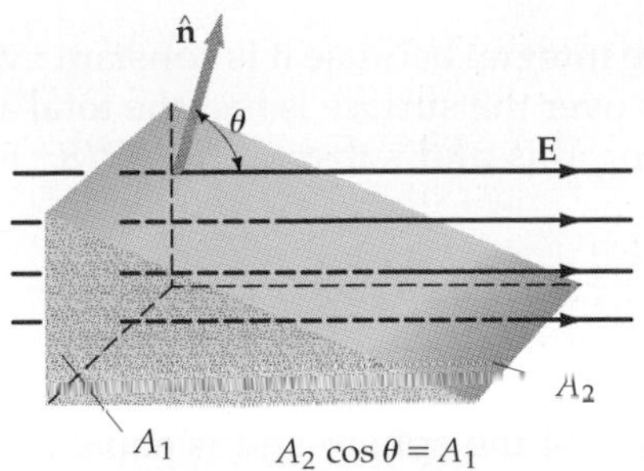

Figure 19-10 Electric-field lines of a uniform electric field that is perpendicular to the area A_1 but makes an angle θ with the unit vector $\hat{\mathbf{n}}$ that is normal to the area A_2. When $\mathbf{E}$ is not perpendicular to the area, the flux is E_nA, where $E_n = E \cos \theta$ is the component of $\mathbf{E}$ that is perpendicular to the area. The flux through A_2 is the same as that through A_1.

We can generalize our definition of electric flux to a curved surface over which the electric field may vary in magnitude or direction or both by dividing the surface into a large number of very small elements. If each element is small enough, it can be considered to be a plane, and the variation of the electric field across the element can be neglected. Let $\hat{\mathbf{n}}_i$ be the unit vector perpendicular to the ith element and ΔA_i be its area (Figure 19-11). (If the surface is curved, the unit vectors for different elements will have different directions.) The flux of the electric field through this element is

$$\Delta\phi_i = \mathbf{E}\cdot\hat{\mathbf{n}}_i\ \Delta A_i$$

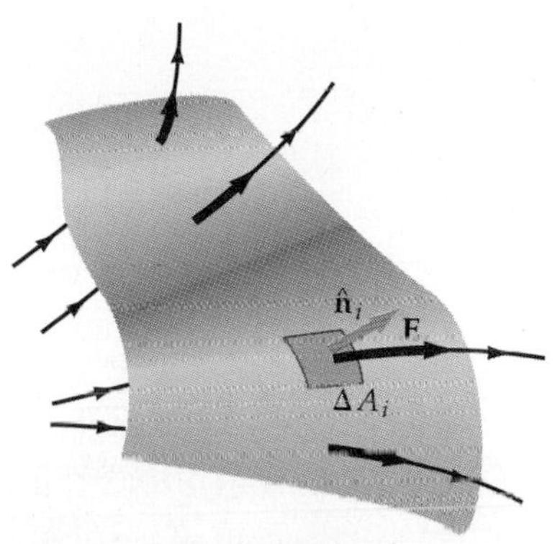

Figure 19-11 When $\mathbf{E}$ varies in either magnitude or direction, the area of the surface is divided into small elements ΔA_i. The flux through the area is computed by summing $\mathbf{E}\cdot\hat{\mathbf{n}}_i\ \Delta A_i$ over all the area elements.

The total flux through the surface is the sum of $\Delta\phi_i$ over all the elements. In the limit as the number of elements approaches infinity and the area of each element approaches zero, this sum becomes an integral. The general definition of electric flux is thus

Electric flux defined

$$\phi = \lim_{\Delta A_i \to 0} \sum_i \mathbf{E}\cdot\hat{\mathbf{n}}_i\ \Delta A_i = \int \mathbf{E}\cdot\hat{\mathbf{n}}\ dA \qquad 19\text{-}16$$

We are often interested in the flux of an electric field through a closed surface. On a closed surface, the normal unit vector $\hat{\mathbf{n}}$ is defined to be outward at each point. At a point where an electric-field line leaves the surface, $\mathbf{E}$ is directed outward and ϕ is positive; at a point where a field line enters the surface, $\mathbf{E}$ is directed inward and ϕ is negative. The total or net flux ϕ_{net} through the closed surface is positive or negative depending on whether $\mathbf{E}$ is predominantly outward or inward at the surface. Since the flux through any part of the surface is proportional to the number of lines passing through the surface, the net flux is proportional to the net number of electric-field lines leaving the surface, that is, the number going out minus the number going in. The integral over a closed surface is indicated by the symbol $\oint$. The net flux through a closed surface is therefore written

$$\phi_{net} = \oint_S \mathbf{E}\cdot\hat{\mathbf{n}}\ dA = \oint_S E_n\ dA \qquad 19\text{-}17$$

Figure 19-12 shows a spherical surface of radius R with its center at a point charge Q. The electric field everywhere on this surface is perpendicular to the surface and has the magnitude

$$E_n = \frac{kQ}{R^2}$$

The net flux through this spherical surface is

$$\phi_{net} = \oint_S E_n\ dA = E_n \oint_S dA$$

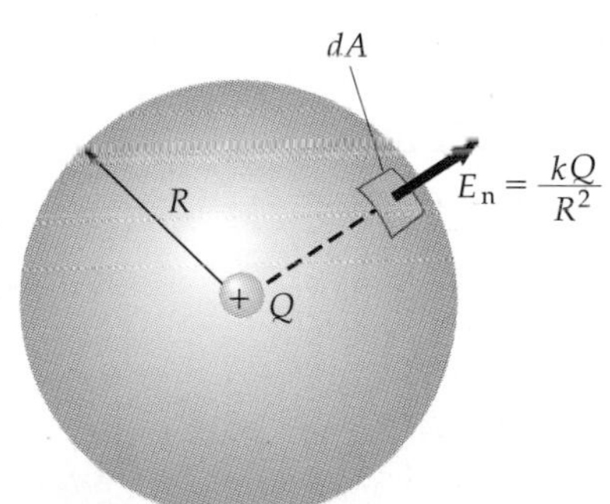

Figure 19-12 A spherical surface enclosing the point charge Q. The same number of electric-field lines that passes through this surface passes through any surface that encloses Q. The flux is easily calculated for a spherical surface. It equals E_n times the surface area $4\pi R^2$.

where we have taken E_n out of the integral because it is constant everywhere on the surface. The integral of dA over the surface is just the total area of the surface, which equals $4\pi R^2$. Using this and substituting kQ/R^2 for E_n, we obtain

$$\phi_{net} = \frac{kQ}{R^2}\, 4\pi R^2 = 4\pi kQ \tag{19-18}$$

Thus, the net flux through a spherical surface with a point charge at its center is independent of the radius of the sphere and is equal to $4\pi k$ times the magnitude of the point charge. This is consistent with our previous observation that the net number of lines going out of a surface is proportional to the net charge inside the surface. This number of lines is the same for all surfaces surrounding the charge, independent of the shape of the surface. Since the number of lines and the flux are proportional to each other, it follows that Equation 19-18 holds for the flux through any surface enclosing the point charge Q. Thus, the net flux through any surface surrounding a point charge Q equals $4\pi kQ$.

We can extend this result to systems consisting of more than one point charge. In Figure 19-13, the surface encloses two point charges q_1 and q_2, and there is a third point charge q_3 outside the surface. Since the electric field at any point on the surface is the vector sum of the electric fields produced by each of the three charges, the net flux $\phi_{net} = \oint \mathbf{E}\cdot\hat{\mathbf{n}}\, dA$ through the surface is just the sum of the fluxes due to the individual charges. The flux due to charge q_3, which is outside the surface, is zero because every field line from q_3 that enters the surface at one point leaves the surface at some other point. The net number of lines passing through a surface from a charge outside the surface is zero. The flux through the surface due to charge q_1 is $4\pi kq_1$ and that due to charge q_2 is $4\pi kq_2$. The net flux through the surface therefore equals $4\pi k(q_1 + q_2)$, which may be positive, negative, or zero depending on the signs and magnitudes of the two charges.

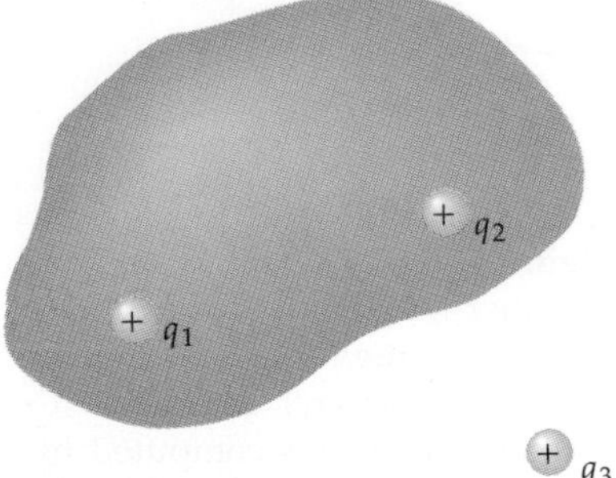

Figure 19-13 A surface enclosing point charges q_1 and q_2 but not q_3. The net flux through this surface is $4\pi k(q_1 + q_2)$.

The net flux through any surface equals $4\pi k$ times the net charge inside the surface:

Gauss's law

$$\phi_{net} = \oint_S E_n\, dA = 4\pi kQ_{inside} \tag{19-19}$$

This is **Gauss's law.** Its validity depends on the fact that the electric field due to a single point charge varies inversely with the square of the distance from the charge. It was this property of the electric field that made it possible to draw a fixed number of electric-field lines from a charge and have the density of lines be proportional to the field strength.

It is customary to write the Coulomb constant k in terms of another constant ϵ_0, which is called the **permittivity of free space:**

$$k = \frac{1}{4\pi\epsilon_0} \tag{19-20}$$

Using this notation, Coulomb's law is written

Coulomb's law in terms of ϵ_0

$$\mathbf{F}_{12} = \frac{1}{4\pi\epsilon_0}\frac{q_1 q_2}{r_{12}^2}\,\hat{\mathbf{r}}_{12} \tag{19-21}$$

and Gauss's law is written

$$\phi_{\text{net}} = \oint_S E_n \, dA = \frac{1}{\epsilon_0} Q_{\text{inside}} \qquad 19\text{-}22$$

Gauss's law in terms of ϵ_0

The value of ϵ_0 in SI units is

$$\epsilon_0 = \frac{1}{4\pi k} = \frac{1}{4\pi(8.99 \times 10^9 \text{ N}\cdot\text{m}^2/\text{C}^2)} = 8.85 \times 10^{-12} \text{ C}^2/\text{N}\cdot\text{m}^2$$

Gauss's law is valid for all surfaces and all charge distributions. For some special charge distributions that have high degrees of symmetry, it can be used to calculate the electric field as we will illustrate in the next section. The real power of Gauss's law is theoretical. For electric fields arising from static or slowly moving charges, Gauss's law and Coulomb's law are equivalent. However, Gauss's law is more general in that it also applies to electric fields arising from rapidly moving charges and accelerating charges.

Questions

2. If the electric field **E** is zero everywhere on a closed surface, is the net flux though the surface necessarily zero? What, then, is the net charge inside the surface?
3. If the net flux through a closed surface is zero, does it follow that the electric field **E** is zero everywhere on the surface? Does it follow that the net charge inside the surface is zero?
4. Is the electric field **E** in Gauss's law only that part of the electric field due to the charge inside a surface, or is it the total electric field due to all charges both inside and outside the surface?

19-3 Calculation of the Electric Field from Gauss's Law

For some highly symmetrical charge distributions, such as a uniformly charged sphere or an infinite line of charge, we can find a mathematical surface on which we know from symmetry that the electric field is constant and perpendicular to the surface. We can then easily evaluate the electric flux through this surface and use Gauss's law to relate the electric field to the charge inside the surface. A surface used to calculate the electric field from Gauss's law is called a **gaussian surface.** In this section, we will use this method to calculate the electric field due to a number of symmetrical charge distributions.

E near a Point Charge

We first use Gauss's law to find the electric field at a distance r from a point charge q. Let the point charge be at the origin. By symmetry, **E** must be radial, and its magnitude can depend only on the distance from the charge. For our gaussian surface, we choose a spherical surface of radius r centered at the charge. The normal component of **E**, $E_n = \mathbf{E}\cdot\hat{\mathbf{n}} = E_r$, has the same value everywhere on the spherical surface. The net flux through this surface is thus

$$\phi_{\text{net}} = \oint \mathbf{E}\cdot\hat{\mathbf{n}} \, dA = \oint E_r \, dA = E_r \oint dA$$

But $\oint dA = 4\pi r^2$, the total area of the spherical surface. Since the total charge inside the surface is just the point charge q, Gauss's law gives

$$E_r 4\pi r^2 = \frac{q}{\epsilon_0}$$

and

$$E_r = \frac{1}{4\pi\epsilon_0}\frac{q}{r^2}$$

We have thus derived Coulomb's law from Gauss's law. Since we originally obtained Gauss's law from Coulomb's law, we have shown that the two laws are equivalent for static charges.

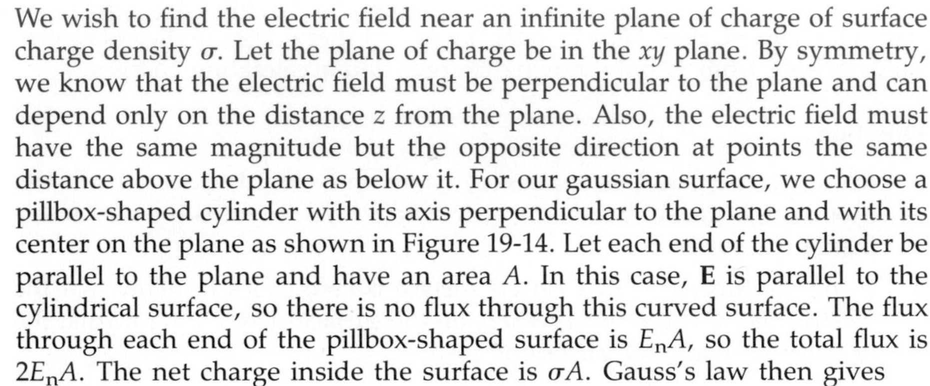

Figure 19-14 Gaussian surface for the calculation of the electric field due to an infinite plane of charge. On the upper and lower faces of this pillbox surface, **E** is perpendicular to the surface and constant in magnitude. The flux through the surface is $2E_nA$, where A is the area of each face.

E near an Infinite Plane of Charge

We wish to find the electric field near an infinite plane of charge of surface charge density σ. Let the plane of charge be in the xy plane. By symmetry, we know that the electric field must be perpendicular to the plane and can depend only on the distance z from the plane. Also, the electric field must have the same magnitude but the opposite direction at points the same distance above the plane as below it. For our gaussian surface, we choose a pillbox-shaped cylinder with its axis perpendicular to the plane and with its center on the plane as shown in Figure 19-14. Let each end of the cylinder be parallel to the plane and have an area A. In this case, **E** is parallel to the cylindrical surface, so there is no flux through this curved surface. The flux through each end of the pillbox-shaped surface is E_nA, so the total flux is $2E_nA$. The net charge inside the surface is σA. Gauss's law then gives

$$\phi_{\text{net}} = \oint E_n\, dA = \frac{1}{\epsilon_0} Q_{\text{inside}}$$

$$2E_nA = \frac{1}{\epsilon_0}\sigma A$$

or

E near an infinite plane of charge

$$E_n = \frac{\sigma}{2\epsilon_0} = 2\pi k\sigma \qquad 19\text{-}23$$

This result agrees with that we obtained by direct integration for **E** near a disk charge in the limit of infinite radius (Equation 19-14*a*).

Example 19-3

An infinite plane of surface charge density $\sigma = +4\ \text{nC/m}^2$ lies in the yz plane at the origin, and a second infinite plane of surface charge density $\sigma = -4\ \text{nC/m}^2$ lies in a plane parallel to the yz plane at $x = 2$ m. Find the electric field at (*a*) $x = 1.8$ m and (*b*) $x = 5$ m.

(*a*) The magnitude of the electric field due to either charge distribution is constant and equal to

$$E_1 = E_2 = \frac{\sigma}{2\epsilon_0} = \frac{4\times10^{-9}\ \text{C/m}^2}{2(8.85\times10^{-12}\ \text{C}^2/\text{N}\cdot\text{m}^2)} = 226\ \text{N/C}$$

The electric field due to the positive charge on the yz plane points away from the yz plane. Similarly, the electric field due to the negative charge on the plane at $x = 2$ m points toward that plane. Thus, between the planes, the magnitudes of the fields add, whereas to the right of both planes or to the left of both planes, the magnitudes subtract. The electric field at $x = 1.8$ m, which is between the planes, is therefore

$$E_x = E_1 + E_2 = 226 \text{ N/C} + 226 \text{ N/C} = 452 \text{ N/C}$$

(*b*) Since the point $x = 5$ m is to the right of both planes, the magnitudes of the fields subtract, so the net electric field is 0. The electric field lines for this charge distribution are shown in Figure 19-15.

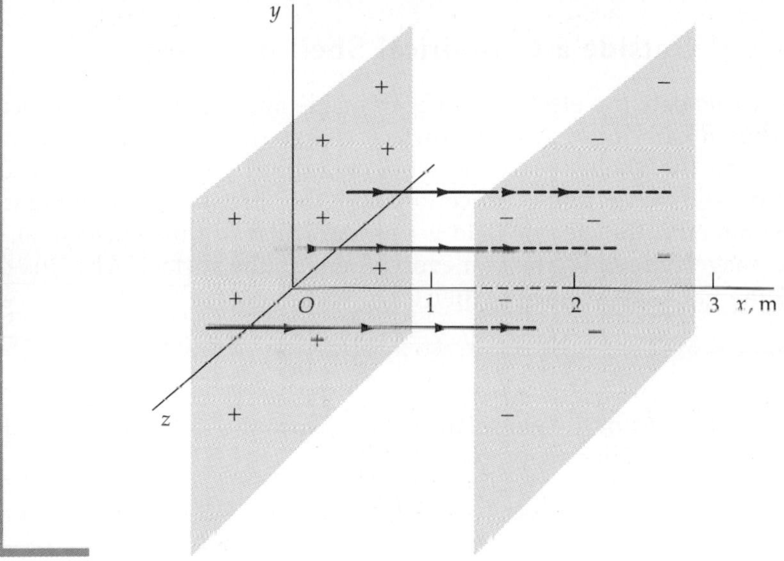

Figure 19-15 An infinite plane of charge with a positive surface charge density in the *yz* plane at the origin and a parallel infinite plane of charge with a negative surface charge density of equal magnitude at $x = 2$ m. The electric field is zero except in the region between the planes. The electric field lines begin on the positively charged plane and end on the negatively charged plane.

E near an Infinite Line Charge

We next consider the electric field at a distance r from a very long line charge of uniform linear charge density λ. In Figure 19-16, a cylindrical surface of length L and radius r has been drawn around the line. By symmetry, at points far from the ends of the line, the electric-field lines radiate out from the line uniformly (if the line charge is positive). The electric field is thus perpendicular to the cylindrical surface and has the same value E_r everywhere on the surface. The electric flux is then just the product of the electric field and the area of the cylindrical surface. There is no flux through the flat surfaces at the ends of the cylinder because $\mathbf{E}\cdot\hat{\mathbf{n}} = 0$ on these surfaces. The net charge inside this surface is the charge per unit length λ times the length L. Gauss's law then gives

$$\phi_{\text{net}} = \oint E_{\text{n}}\, dA = \frac{1}{\epsilon_0} Q_{\text{inside}}$$

$$\oint E_{\text{n}}\, dA = E_r \oint dA = \frac{\lambda L}{\epsilon_0}$$

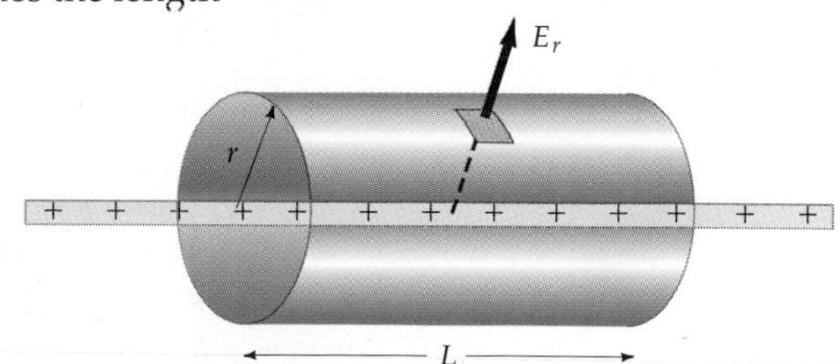

Figure 19-16 A very long, uniform line charge with a cylindrical surface enclosing part of the charge. The flux through the surface is E_r times the surface area $2\pi rL$.

Since the area of the cylindrical surface is $2\pi rL$, we have

$$E_r 2\pi rL = \frac{\lambda L}{\epsilon_0}$$

or

$$E_r = \frac{1}{2\pi\epsilon_0}\frac{\lambda}{r} = 2k\frac{\lambda}{r} \qquad \text{19-24}$$

E a distance r from an infinite line charge

This result is the same as Equation 19-11, which we obtained by direct integration over the line charge.

It is important to realize that a high degree of symmetry is needed to use Gauss's law. In the preceding calculation, we needed to assume that the field point was very far from the ends of the line charge so that E_n would be constant everywhere on the cylindrical gaussian surface. This is equivalent to assuming that, at the distance r from the line, the line charge appears to be infinitely long. If we have a line charge of finite length, we cannot assume that E_n is constant everywhere on our cylindrical surface, so we cannot use Gauss's law to calculate the electric field.

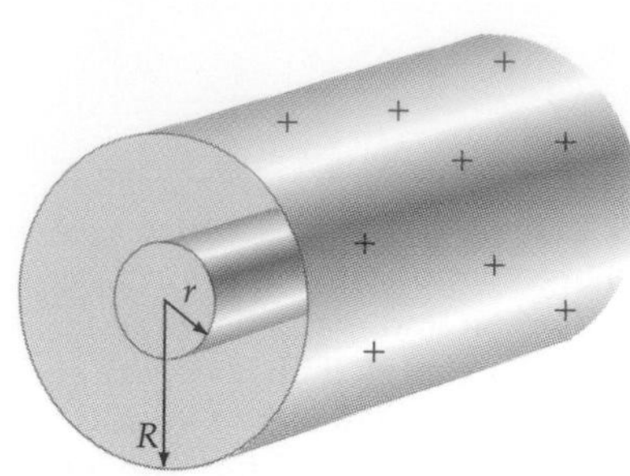

Figure 19-17 A cylindrical shell of radius R carrying a uniform surface charge density σ. To find the electric field inside the shell, we construct a cylindrical gaussian surface concentric with the shell of radius $r < R$ as shown. Since there is no charge inside the gaussian surface, the net flux through this surface is zero.

E Inside and Outside a Cylindrical Shell of Charge

We will now calculate the electric field both inside and outside a cylindrical shell of radius R carrying a uniform surface charge density σ. To calculate the field inside the shell, we construct a cylindrical gaussian surface of length L and radius $r < R$ that is concentric with the shell as shown in Figure 19-17. By symmetry, the electric field is perpendicular to this gaussian surface and its magnitude E_r is constant everywhere on the surface. The flux of **E** through the gaussian surface is then

$$\phi_{\text{net}} = \oint E_n \, dA = E_r \oint dA = E_r 2\pi r L$$

where $2\pi r L$ is the area of the gaussian surface. Since the total charge inside this surface is zero, Gauss's law gives

$$\phi_{\text{net}} = E_r 2\pi r L = 0$$

Therefore,

E inside a cylindrical shell of charge

$$E_r = 0 \qquad r < R \qquad \text{19-25}a$$

Thus, the electric field everywhere inside a cylindrical shell of charge is zero.

To find the electric field outside the cylindrical shell, we construct a cylindrical gaussian surface of radius $r > R$. Again, by symmetry, the electric field is perpendicular to this gaussian surface and its magnitude E_r is constant everywhere on the surface. The flux is again $E_r 2\pi r L$, but this time the total charge inside the surface is $\sigma 2\pi R L$. Gauss's law then gives

$$\phi_{\text{net}} = E_r 2\pi r L = \frac{\sigma 2\pi R L}{\epsilon_0}$$

Therefore,

$$E_r = \frac{\sigma R}{\epsilon_0 r}$$

Since a length L of the cylindrical shell carries a charge of $\sigma 2\pi R L$, the charge per unit length of the shell is $\lambda = \sigma 2\pi R$. Substituting $\lambda / 2\pi R$ for σ in the previous equation, we obtain

E outside a cylindrical shell of charge

$$E_r = \frac{\sigma R}{\epsilon_0 r} = \frac{1}{2\pi\epsilon_0} \frac{\lambda}{r} \qquad r > R \qquad \text{19-25}b$$

which is the same as Equation 19-24 for **E** a distance r from an infinite line charge. Thus, the field outside a cylindrical shell of charge is the same as if

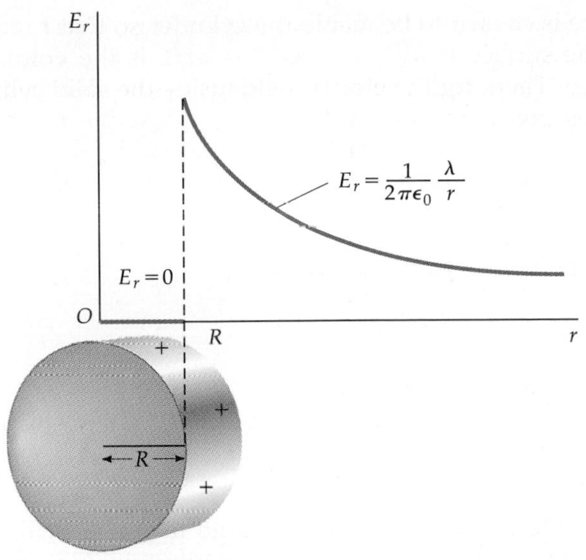

Figure 19-18 A plot of E_r versus r for a cylindrical-shell charge distribution. The electric field is discontinuous at $r = R$, where there is a surface charge of density σ. Just inside the shell the field is zero, whereas just outside the shell it has the magnitude σ/ϵ_0.

all the charge were on the axis of the cylinder. Figure 19-18 shows E_r versus r for this charge distribution. Just outside the shell at $r \approx R$, the electric field is $E_r = \sigma/\epsilon_0$. Since the field just inside the shell is zero, the electric field is discontinuous by the amount σ/ϵ_0 as we pass through the shell. This is the same result that we found for an infinite plane of charge, where the electric field is $-\sigma/2\epsilon_0$ on one side of the plane and $+\sigma/2\epsilon_0$ on the other side. It is a general result, which we will derive at the end of this section.

E Inside and Outside an Infinitely Long Solid Cylinder of Charge

Figure 19-19 shows a solid cylinder of radius R carrying a charge that is uniformly distributed throughout the volume of the cylinder with charge density ρ. As was the case with the cylindrical shell of charge, the flux through a cylindrical gaussian surface of radius r and length L is

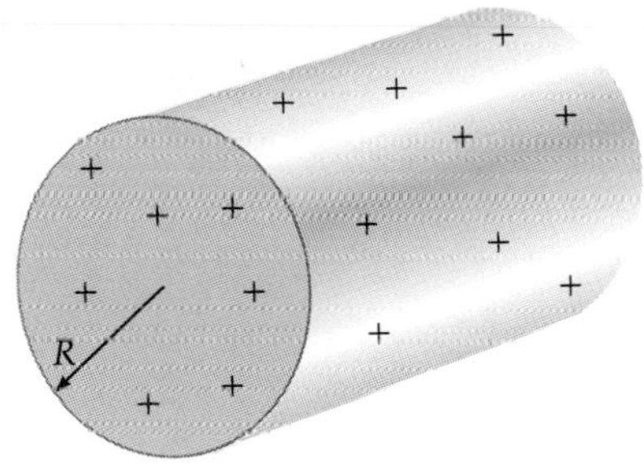

Figure 19-19 A solid cylinder carrying a uniform volume charge density ρ.

$$\phi_{\text{net}} = E_r 2\pi r L$$

If the gaussian surface is outside of the cylinder, that is, if $r > R$, the total charge inside the surface is ρ times the volume of the solid cylinder, which is $\pi R^2 L$. Gauss's law then gives

$$E_r 2\pi r L = \frac{\rho \pi R^2 L}{\epsilon_0}$$

$$E_r = \frac{\rho R^2}{2\epsilon_0 r}$$

Again, we may write this in terms of the charge per unit length along the cylinder, which is $\lambda = (\rho\pi R^2 L)/L = \rho\pi R^2$. Substituting $\lambda/\pi R^2$ for ρ in the previous equation, we obtain

$$E_r = \frac{\rho R^2}{2\epsilon_0 r} = \frac{1}{2\pi\epsilon_0}\frac{\lambda}{r} \qquad r \geq R \qquad \text{19-26}a$$

E outside a solid cylinder of charge

which is the same as Equations 19-24 and 19-25*b*. Thus, the electric field outside a solid cylinder of charge is the same as if all the charge were on the axis of the cylinder.

If the gaussian surface is chosen to be inside the cylinder so that $r < R$, the total charge inside the surface is $\rho V'$, where $V' = \pi r^2 L$ is the volume inside the gaussian surface. Then, for the electric field inside the solid cylinder of charge, Gauss's law gives

$$\phi_{\text{net}} = \frac{1}{\epsilon_0} Q_{\text{inside}}$$

$$E_r 2\pi r L = \frac{1}{\epsilon_0} \rho V' = \frac{1}{\epsilon_0} \rho \pi r^2 L$$

or

E inside a solid cylinder of charge

$$E_r = \frac{\rho}{2\epsilon_0} r = \frac{\lambda}{2\pi\epsilon_0 R^2} r \qquad r \leq R \qquad \text{19-26}b$$

Thus, the electric field inside a solid cylinder of charge increases with r. Figure 19-20 shows a plot of E_r versus r for this charge distribution. Note that E_r is continuous at $r = R$.

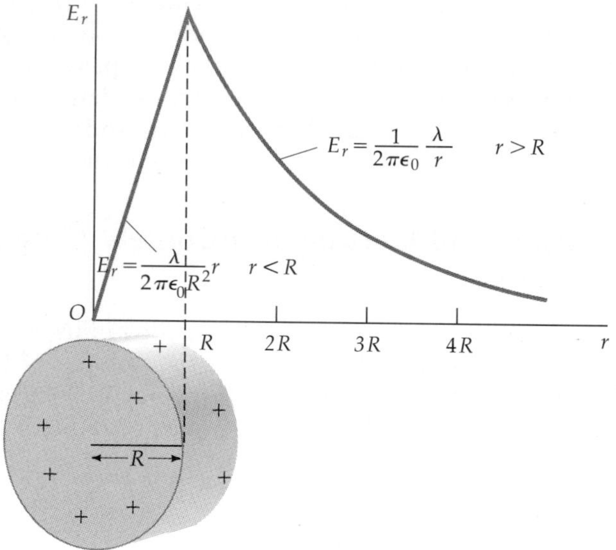

Figure 19-20 Plot of the electric field E_r due to a solid cylinder of charge of radius R versus the distance r from the axis of the cylinder. The field E_r is proportional to r for $0 < r < R$ and decreases as $1/r$ for $r > R$. The field is continuous at $r = R$.

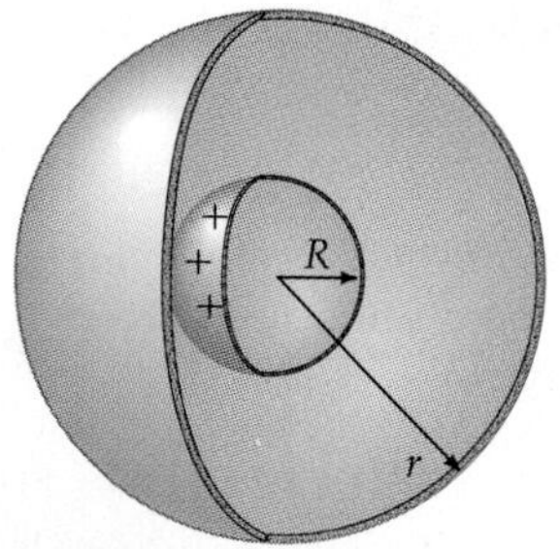

Figure 19-21 Spherical gaussian surface of radius $r > R$ for the calculation of the electric field outside a uniformly charged spherical shell of radius R. The total flux through this surface is $E_r 4\pi r^2$, and the total charge inside the surface is the total charge of the shell Q. The field is the same as if all the charge were at the center of the shell.

E Inside and Outside a Spherical Shell of Charge

We wish to find the electric field inside and outside a uniformly charged spherical shell of radius R and total charge Q. By symmetry, **E** must be radial, and its magnitude can depend only on the distance r from the center of the sphere. In Figure 19-21, we have chosen a spherical gaussian surface of radius $r > R$. Since **E** is perpendicular to this surface and constant in magnitude everywhere on it, the flux through the surface is

$$\phi_{\text{net}} = \oint E_r \, dA = E_r 4\pi r^2$$

Since the total charge inside the gaussian surface is the total charge on the shell, Q, Gauss's law gives

$$E_r 4\pi r^2 = \frac{Q}{\epsilon_0}$$

or

$$E_r = \frac{1}{4\pi\epsilon_0}\frac{Q}{r^2} \qquad r > R \qquad \text{19-27}a$$

E outside a spherical shell of charge

Thus, the electric field outside a uniformly charged spherical shell is the same as if all the charge were at the center of the shell.

If we choose a spherical gaussian surface inside the shell, where $r < R$, the net flux is again $E_r 4\pi r^2$, but the total charge inside the surface is zero. Therefore, for $r < R$, Gauss's law gives

$$\phi_{\text{net}} = E_r 4\pi r^2 = 0$$

and

$$E_r = 0 \qquad r < R \qquad \text{19-27}b$$

E inside a spherical shell of charge

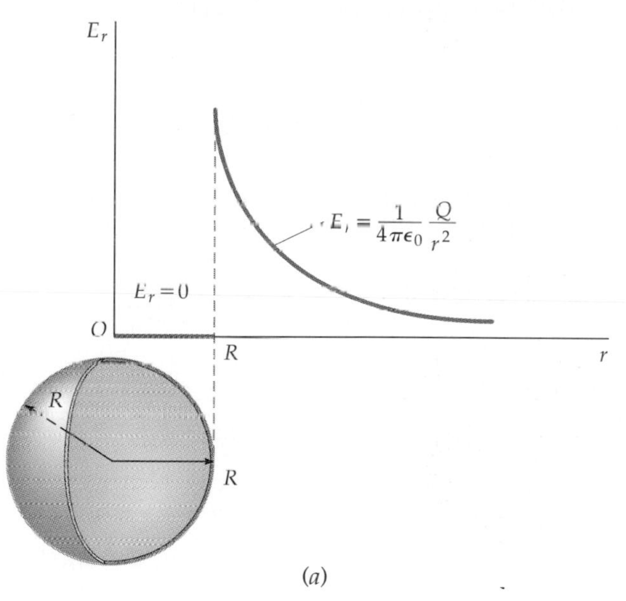

(b)

Figure 19-22 (*a*) A plot of E_r versus r for a spherical-shell charge distribution. The electric field is discontinuous at $r = R$, where there is a surface charge of density σ. Just inside the shell the field is zero, whereas just outside the shell it has the magnitude σ/ϵ_0. (*b*) The decrease in E_r over distance due to a charged spherical shell is evident by the effect of the field on the flames of these two candles. The spherical shell on the van de Graaff generator at the left carries a large negative charge that attracts the positive ions in the nearby candle flame. The flame at right, which is much farther away, is not affected.

Note the similarity of these results with those we obtained when we calculated the gravitational field due to a spherical shell of mass in Section 10-7. We could obtain these results by the direct integration of Coulomb's law, but that calculation is much more difficult. Figure 19-22 shows E_r versus r for this charge distribution. Again, note that the electric field is discontinuous at $r = R$, where the surface charge density is σ. Just inside the shell $E_r = 0$, whereas just outside the shell $E_r = Q/4\pi\epsilon_0 R^2 = \sigma/\epsilon_0$ since $\sigma = Q/4\pi R^2$.

Example 19-4

A spherical shell of radius $R = 3$ m has its center at the origin and carries a surface charge density $\sigma = 3$ nC/m^2. A point charge $q = 250$ nC is on the y axis at $y = 2$ m. Find the electric field on the x axis at (*a*) $x = 2$ m and (*b*) $x = 4$ m.

(*a*) The point on the x axis at $x = 2$ m is inside the spherical shell, so the field due to the shell is zero. The electric field at this point is due only to the point charge, which is a distance $r_1 = \sqrt{(2\text{ m})^2 + (2\text{ m})^2} = \sqrt{8}$ m away. The field makes an angle of $-45°$ with the x axis and has a magnitude of

$$E = \frac{kq}{r_1^2} = \frac{(8.99 \times 10^9\ \text{N}\cdot\text{m}^2/\text{C}^2)(250 \times 10^{-9}\ \text{C})}{(\sqrt{8}\ \text{m})^2} = 281\ \text{N/C}$$

(*b*) At the point $x = 4$ m, which is outside the spherical shell, the shell can be treated as a point charge at the origin with a magnitude $Q = \sigma 4\pi R^2 = (3\ \text{nC/m}^2)4\pi(3\ \text{m})^2 = 339$ nC. The electric field due to the shell at $x = 4$ m is in the x direction and has the magnitude

$$E_1 = E_{1x} = \frac{(8.99 \times 10^9\ \text{N}\cdot\text{m}^2/\text{C}^2)(339 \times 10^{-9}\ \text{C})}{(4\ \text{m})^2} = 190\ \text{N/C}$$

The distance from the point charge q on the y axis to the field point at $x = 4$ m is $r_2 = \sqrt{(2\text{ m})^2 + (4\text{ m})^2} = \sqrt{20}$ m. The magnitude of the electric field at $x = 4$ m due to q is

$$E_2 = \frac{(8.99 \times 10^9\ \text{N}\cdot\text{m}^2/\text{C}^2)(250 \times 10^{-9}\ \text{C})}{(\sqrt{20}\ \text{m})^2} = 112\ \text{N/C}$$

This field makes an angle θ with the x axis, where $\cos\theta = 4/\sqrt{20}$ and $\sin\theta = -2/\sqrt{20}$. The x and y components of this field are thus

$$E_{2x} = E_2 \cos\theta = (112\ \text{N/C})\left(\frac{4}{\sqrt{20}}\right) = 100\ \text{N/C}$$

and

$$E_{2y} = E_2 \sin\theta = (112\ \text{N/C})\left(-\frac{2}{\sqrt{20}}\right) = -50\ \text{N/C}$$

The x and y components of the net electric field are

$$E_x = E_{1x} + E_{2x} = 190\ \text{N/C} + 100\ \text{N/C} = 290\ \text{N/C}$$

and

$$E_y = E_{1y} + E_{2y} = 0 - 50\ \text{N/C} = -50\ \text{N/C}$$

The magnitude and direction of the net field can be found from $E = \sqrt{E_x^2 + E_y^2}$, and $\tan\theta' = E_y/E_x$.

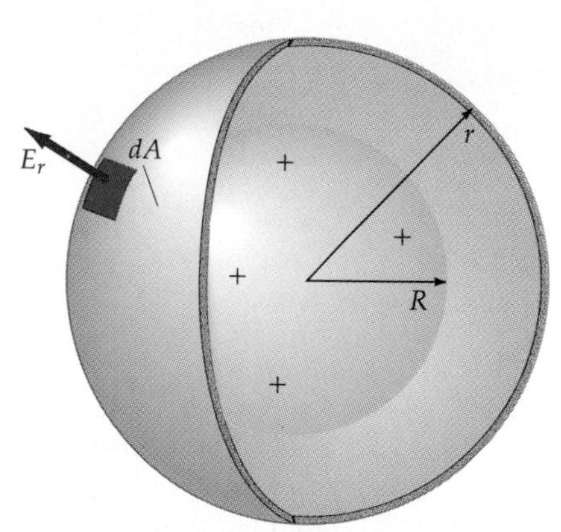

Figure 19-23 Spherical gaussian surface for the calculation of the electric field outside a uniformly charged solid sphere. The total flux through this surface is $E_r 4\pi r^2$, and the total charge inside the surface is the total charge of the sphere Q. The field is the same as if all the charge were at the center of the sphere.

E Inside and Outside a Uniformly Charged Solid Sphere

We will now calculate the electric field inside and outside a uniformly charged solid sphere of radius R carrying a total charge Q that is uniformly distributed throughout the volume of the sphere with charge density $\rho = Q/V$, where $V = \frac{4}{3}\pi R^3$ is the volume of the sphere. As was the case with the spherical shell of charge, the flux through a gaussian surface of radius r is

$$\phi_{\text{net}} = E_r 4\pi r^2$$

If the gaussian surface is outside of the sphere, as in Figure 19-23, the total charge inside the surface is Q, and Gauss's law gives

E outside a solid sphere of charge

$$E_r = \frac{1}{4\pi\epsilon_0}\frac{Q}{r^2} \qquad r \geq R \qquad \text{19-28a}$$

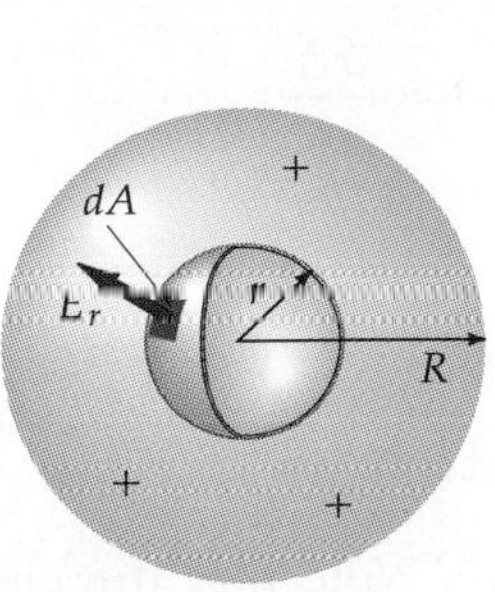

Figure 19-24 Spherical gaussian surface for the calculation of the electric field inside a uniformly charged solid sphere. The flux through the surface is again $E_r 4\pi r^2$. The total charge inside the gaussian surface is $Q(r^3/R^3)$.

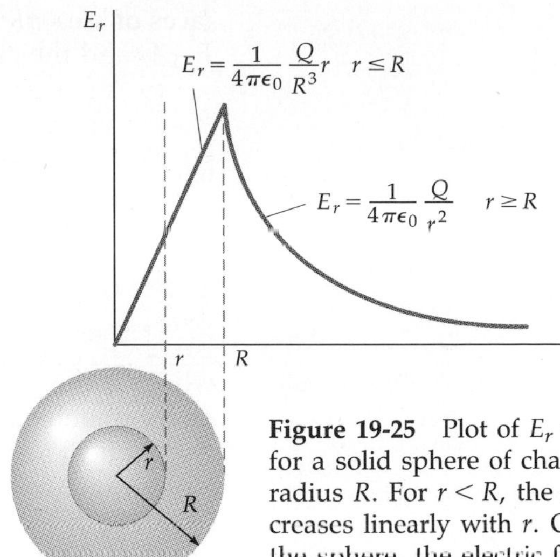

Figure 19-25 Plot of E_r versus r for a solid sphere of charge of radius R. For $r < R$, the field increases linearly with r. Outside the sphere, the electric field is the same as that due to a point charge. The field is continuous at $r = R$.

If the gaussian surface is chosen to be inside the sphere, which means that $r < R$ (Figure 19-24), the total charge inside the surface is $\rho V'$, where $V' = \frac{4}{3}\pi r^3$ is the volume inside the gaussian surface:

$$Q_{\text{inside}} = \rho V' = \frac{Q}{V}V' = \left(\frac{Q}{\frac{4}{3}\pi R^3}\right)\left(\frac{4}{3}\pi r^3\right) = Q\frac{r^3}{R^3}$$

For the electric field inside the sphere, Gauss's law then gives

$$E_r 4\pi r^2 = \frac{1}{\epsilon_0}Q\frac{r^3}{R^3}$$

or

$$E_r = \frac{1}{4\pi\epsilon_0}\frac{Q}{R^3}r \qquad r \le R \qquad \text{19-28}b$$

E inside a solid sphere of charge

Thus, the electric field inside a solid sphere of charge increases with r. Figure 19-25 shows a plot of E_r versus r for this charge distribution. Note that E_r is continuous at $r = R$. This function is sometimes used to describe the electric field of an atomic nucleus, which can be considered to be approximately a uniform solid sphere of charge.

Discontinuity of E_n

We have seen that the electric field for an infinite plane of charge, a cylindrical shell of charge, and a spherical shell of charge is discontinuous by the amount σ/ϵ_0 at a point where there is a surface charge density σ. We will now show that this is a general result for the component of the electric field that is perpendicular to a surface carrying a charge density of σ. Figure 19-26 shows a small pillbox-shaped gaussian surface with faces of area A on each side of an arbitrary surface carrying a surface charge density σ. Let the normal component of the electric field be E_{n2} on one side of the surface and E_{n1} on the other side, as shown in the figure. If we make the length of the pillbox very small compared with the radius of the faces, we can neglect the flux through the cylindrical area $2\pi RL$ compared with the flux through the

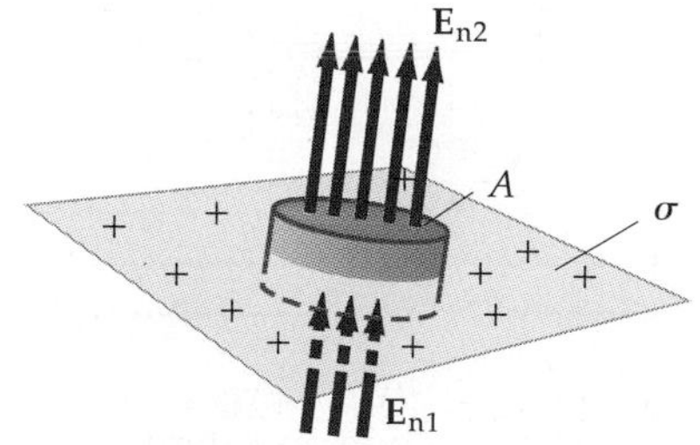

Figure 19-26 A gaussian, pillbox-shaped surface with faces of area A on each side of a surface carrying a charge density of σ. The net flux through the pillbox is $(E_{n2} - E_{n1})A$. The electric field E_{n2} on one side is greater than the electric field E_{n1} on the other side by the amount σ/ϵ_0.

faces of area πR^2. The net flux through the gaussian surface is then $E_{n2}A - E_{n1}A$, and the charge inside the surface is σA. Gauss's law gives

$$E_{n2}A - E_{n1}A = \frac{\sigma A}{\epsilon_0}$$

or

Discontinuity of E_n

$$E_{n2} - E_{n1} = \frac{\sigma}{\epsilon_0} \qquad 19\text{-}29$$

which is the result we wished to prove. Note that the electric field is *not* discontinuous at points of discontinuity of a volume charge density. The electric field just inside a solid cylinder of charge or a solid sphere of charge is the same as it is just outside such a charge distribution, as can be seen from Figures 19-20 and 19-25.

Questions

5. What information in addition to the total charge inside a surface is needed to use Gauss's law to find the electric field?
6. Explain why the electric field increases with r rather than decreasing as $1/r^2$ as one moves out from the center inside a spherical charge distribution of constant volume charge density.
7. Equation 19-10 for the electric field on the perpendicular bisector of a finite line charge is different from Equation 19-11 or 19-24 for the electric field near an infinite line charge, yet Gauss's law would seemingly give the same result for these two cases. Explain.

19-4 Charge and Field at Conductor Surfaces

The property of a conductor that is important in studying electrostatic fields is the availability of charge that is free to move about inside the conductor. In the presence of an external electric field, the free charge in a conductor moves about the conductor until it is so distributed that it creates an electric field that cancels the external field inside the conductor.* The conductor is then said to be in **electrostatic equilibrium.** Consider a charge q inside a conductor. If there is a field **E** inside the conductor, there will be a force $q\mathbf{E}$ on this charge. Therefore, if the charge is free to move—that is, if it is not bound to an atom or molecule by a stronger force—it will accelerate. Thus electrostatic equilibrium is impossible in a conductor unless the electric field is zero everywhere inside the conductor. At the surface of a conductor in equilibrium, the electric field must be perpendicular to the surface. If there were a tangential component of **E**, the free charge in the conductor would move until this component became zero.

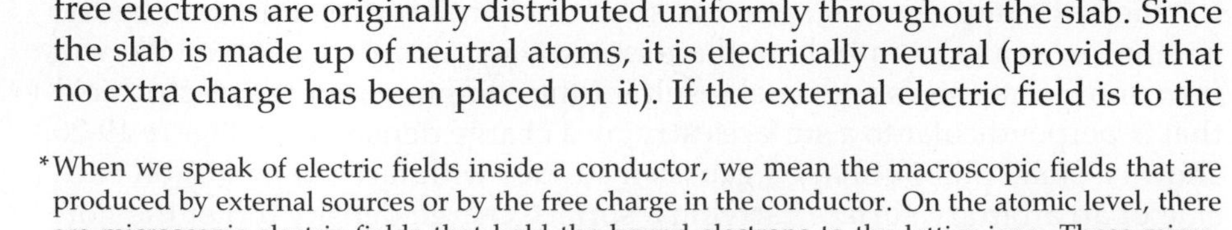

Figure 19-27 Two views of a conducting slab in an external electric field $\mathbf{E}_0$. A positive charge is induced on the right face and a negative charge is induced on the left face such that the net electric field inside the conductor is zero. The electric-field lines then end on the left face and begin again on the right face.

Figure 19-27 shows a conducting slab in an external electric field $\mathbf{E}_0$. The free electrons are originally distributed uniformly throughout the slab. Since the slab is made up of neutral atoms, it is electrically neutral (provided that no extra charge has been placed on it). If the external electric field is to the

*When we speak of electric fields inside a conductor, we mean the macroscopic fields that are produced by external sources or by the free charge in the conductor. On the atomic level, there are microscopic electric fields that hold the bound electrons to the lattice ions. These microscopic fields fluctuate wildly over time and over small distances within the atom, but they average to zero if we average over a distance that is large compared with the diameter of the atom.

right, there will be a force on each electron $\mathbf{F} = -e\mathbf{E}_0$ to the left because the electron has a negative charge, so the free electrons accordingly accelerate to the left. At the surface of the conductor, the conductor exerts forces on these electrons that keep them bound to the conductor. (If the external field is very strong, the electrons can be stripped off the surface of the conductor. In electronics this is called **field emission.** We assume here that the external field is not strong enough to overcome the forces binding the electrons to the surface.) The result is an induced negative surface charge density on the left side of the slab and an induced positive surface charge density on the right side of equal magnitude because of the movement of some of the free electrons from the right side to the left side of the slab. Together, these induced charge densities produce an electric field inside the slab that is opposite to the external field. When the induced and external fields cancel, electrostatic equilibrium is established everywhere inside the conductor so that there is no unbalanced force on the free electrons.

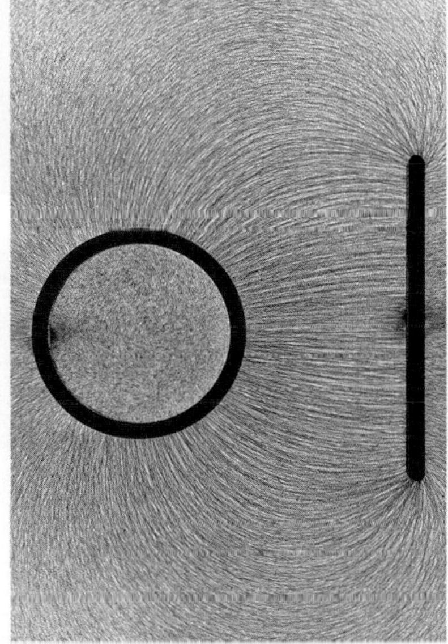

Lines of force for an oppositely charged cylinder and plate, shown by bits of fine thread suspended in oil. Note that the field lines are perpendicular to the conductors and that there are no lines inside the cylinder.

The behavior of the free charge in a conductor placed in an external electric field is the same no matter what the shape of the conductor may be. When an external field is applied, the free charge quickly moves until an equilibrium distribution is achieved such that the net electric field is zero everywhere inside the conductor. The time it takes to reach equilibrium depends on the conductor. For copper and other good conductors, it is so small that, for all practical purposes, electrostatic equilibrium is reached instantaneously.

In this section, we will use Gauss's law to show that the following results occur for conductors in electrostatic equilibrium:

1. Any net electric charge on a conductor resides on the surface of the conductor.
2. The electric field just outside the surface of a conductor is perpendicular to the surface and has the magnitude σ/ϵ_0, where σ is the surface charge density at that point on the conductor.

To obtain the first result, we consider a gaussian surface just inside the actual surface of a conductor in electrostatic equilibrium, as shown in Figure 19-28. Since the electric field is zero everywhere inside the conductor, it is zero everywhere on the gaussian surface, which is chosen to be completely within the conductor. Since $E_n = 0$ at all points on the gaussian surface, the net flux through the surface must be zero. By Gauss's law, the flux equals $1/\epsilon_0$ times the net charge inside a surface. Since the net flux is zero, there can be no net charge inside any surface lying completely within the conductor. If there is any net charge on the conductor, it must be on the conductor's surface.

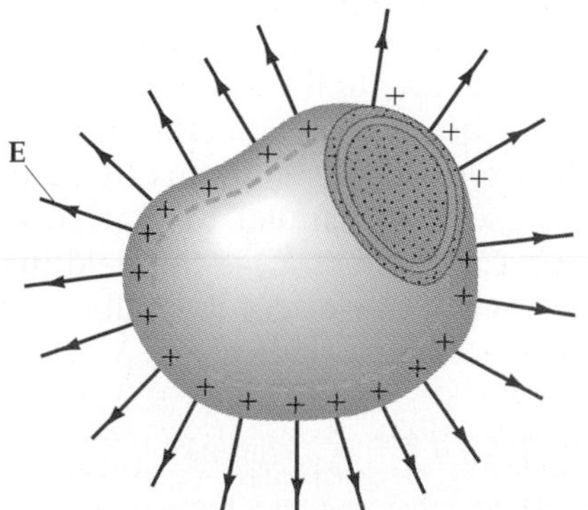

Figure 19-28 A gaussian surface (the dashed line) just inside the surface of a conductor. Since the electric field is zero inside a conductor in electrostatic equilibrium, the net flux through this surface must also be zero. Therefore, the net charge inside this surface must be zero. If there is any charge on the conductor, it must reside on the surface of the conductor. Here, the conductor carries a net positive charge.

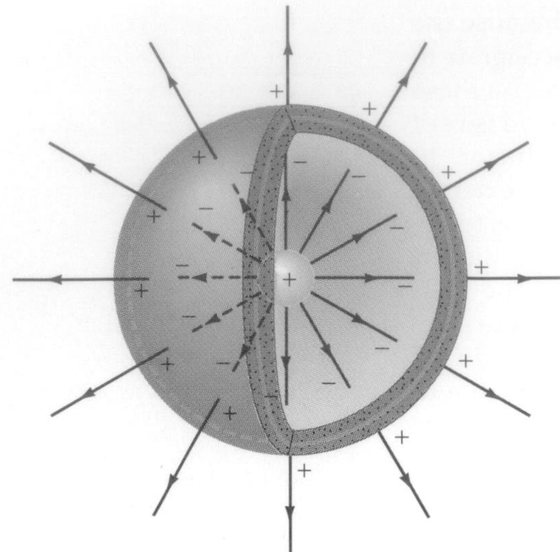

Figure 19-29 A thick spherical conducting shell with a point charge q in the cavity at the center. Since $\mathbf{E} = 0$ inside the conductor, there can be no net flux through any surface within the conductor, such as the gaussian surface indicated in blue. A surface charge $-q$ is induced on the inner surface of the shell, and the electric-field lines that begin on the point charge end on that surface. Since the conductor is neutral, an equal but opposite charge $+q$ is induced on the outer surface of the shell. The electric-field lines begin on the point charge, end on the inner surface, and begin again on the outer surface.

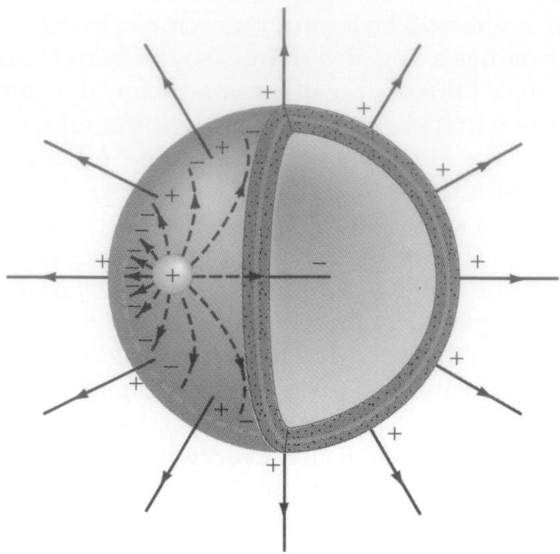

Figure 19-30 The same conductor as in Figure 19-29 with the point charge not at the center of the sphere. The charge on the outer surface and the electric-field lines outside the sphere are not affected.

Figure 19-29 shows a spherical conductor with a spherical cavity inside. At the center of the cavity is a positive point charge q. If we draw a gaussian surface within the conductor, $\mathbf{E}$ is zero everywhere on the surface, so Gauss's law tells us that there can be no net charge within the surface. A negative charge $-q$ must therefore be induced on the inside surface of the conductor. All the electric-field lines from the point charge must end on the negative charge induced on the inside surface. If the conductor has no net charge, there will be an equal induced charge $+q$ on the outside surface of the conductor. In Figure 19-30, the point charge has been moved so that it is no longer at the center of the cavity. The field lines in the cavity are altered, and the surface charge density of the induced negative charge on the inner surface is no longer uniform. However, all of the field lines still end on the induced negative charge. The positive surface charge density on the outside surface is not disturbed because it is shielded from the cavity by the conductor.

To find the electric field just outside the surface of a conductor, we consider a portion of the conductor's surface small enough to be considered flat with a surface charge density σ that is uniform over the area. We construct a pillbox-shaped gaussian surface (Figure 19-31) with one face of the cylinder just outside the conductor and parallel to its surface and the other face just inside the conductor. The electric field at the surface of the conductor in equilibrium must be perpendicular to the surface. Thus, we can take $\mathbf{E}$ to be perpendicular to the face of the pillbox. The other face of the pillbox is inside the conductor, where $\mathbf{E}$ is zero. There is no flux through the cylindrical surface of the pillbox because $\mathbf{E}$ is parallel to that surface. The flux through the pillbox is therefore $E_n A$, where E_n is the field just outside the conductor surface, and A is the area of the face of the pillbox. The net charge inside the gaussian surface is σA. Gauss's law gives

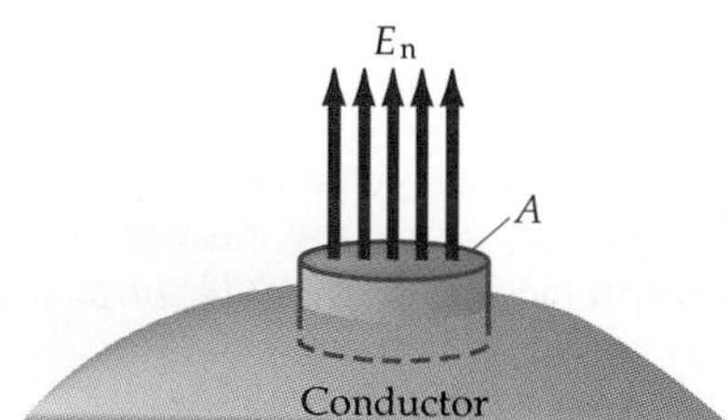

Figure 19-31 Pillbox-shaped gaussian surface for the calculation of the electric field at the surface of a conductor. There is no flux through the inside face of the pillbox because the electric field inside the conductor is zero. The flux through the outside face of the pillbox is $E_n A$.

$$\phi_{\text{net}} = \oint E_n \, dA = \frac{1}{\epsilon_0} Q_{\text{inside}}$$

$$\oint E_n \, dA = E_n \oint dA = E_n A = \frac{1}{\epsilon_0} \sigma A$$

or

$$E_n = \frac{\sigma}{\epsilon_0} \qquad 19\text{-}30$$

E_n just outside conductor surface

This result is exactly twice the field produced by an infinite plane of charge. We can understand this result from Figure 19-32, which shows a large charged conducting slab carrying a surface charge density σ on each face. Near the conducting slab, each of its faces can be considered to be an infinite plane of charge. Let E_L be the electric field due to the left face and E_R be that due to the right face. Each of these fields has the magnitude $\sigma/2\epsilon_0$. Inside the conductor, these fields are oppositely directed, so they cancel. Outside the conductor, the fields add to give a total electric field of magnitude $E = E_L + E_R = \sigma/2\epsilon_0 + \sigma/2\epsilon_0 = \sigma/\epsilon_0$. Similar but slightly more complicated reasoning can be applied to a conductor of arbitrary shape, as shown in Figure 19-33. We can consider the charge on the surface of the conductor to consist of two parts: (1) the charge in the immediate neighborhood of point P and (2) all the rest of the charge, which we shall call the distant charge. Since point P is just outside the surface, the charge in its immediate neighborhood looks like an infinite plane of charge. It produces a field of magnitude $\sigma/2\epsilon_0$ at P and a field of equal magnitude just inside the conducting surface pointing away from the surface. The rest of the charge on the conductor (or elsewhere) must produce a field $\sigma/2\epsilon_0$ inside the conductor pointing toward the surface, so that the net field inside the conductor is zero. The field due to the distant charge has the same magnitude and direction at points just inside and just outside the surface. Just inside the conductor surface, the field due to the distant charge cancels that due to the neighboring charge; but outside, the fields are in the same direction and add, giving a net field $\sigma/2\epsilon_0 + \sigma/2\epsilon_0 = \sigma/\epsilon_0$ just outside the conductor.

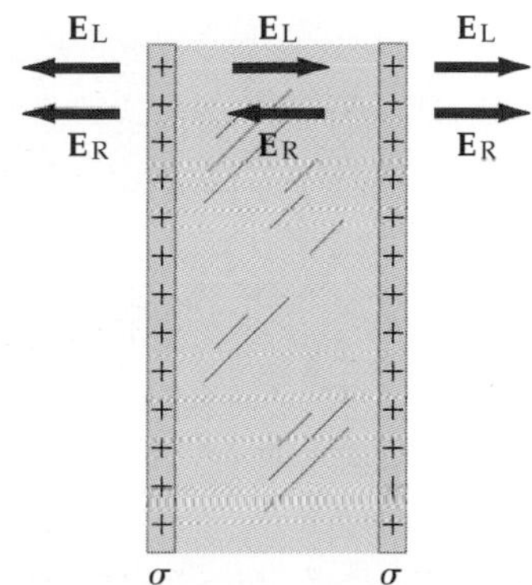

Figure 19-32 A conducting slab carrying a uniform surface charge density σ. The electric field due to the surface charge density on the left face is $\mathbf{E}_L$ and that due to the charge on the right face is $\mathbf{E}_R$. Each of these fields has a magnitude of $\sigma/2\epsilon_0$. Inside the conductor, these fields are oppositely directed and cancel. Outside the conductor, these fields are in the same direction and add to produce a field of magnitude σ/ϵ_0.

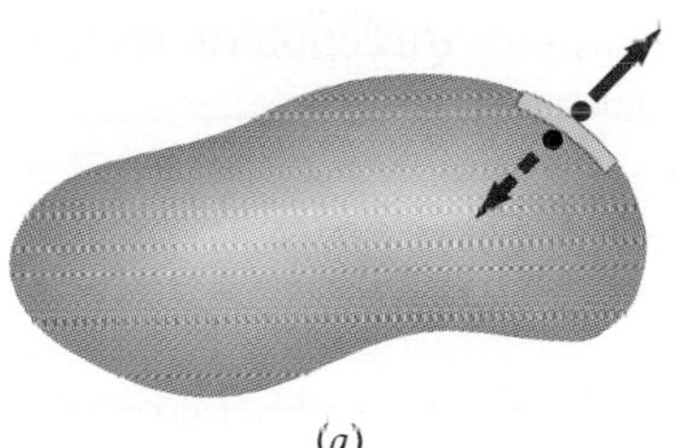

(a)

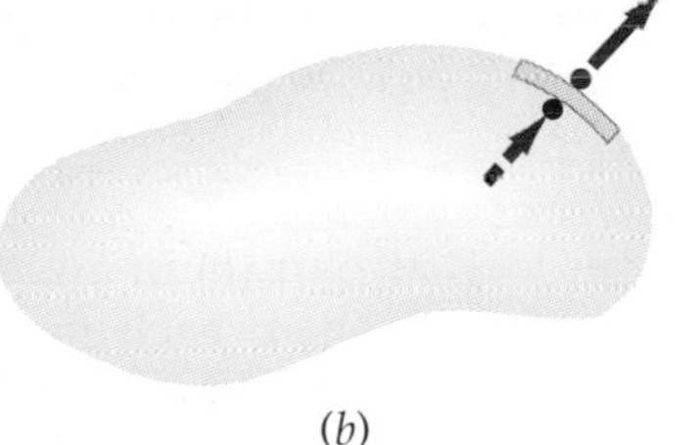

(b)

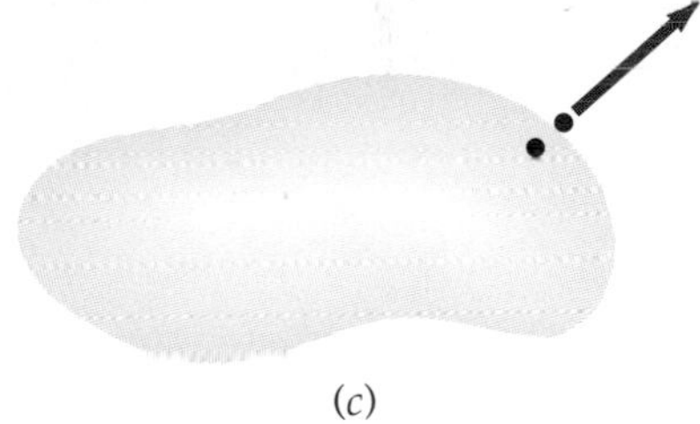

(c)

Figure 19-33 An arbitrarily shaped conductor carrying a charge on its surface. (*a*) The charge in the vicinity of point P near the surface looks like an infinite plane of charge, giving an electric field of magnitude $\sigma/2\epsilon_0$ pointing away from the surface both inside and outside the surface. (*b*) Since the net field inside the surface must be zero, the rest of the charge must produce a field of equal magnitude. (*c*) Inside the surface the fields shown in parts (*a*) and (*b*) cancel, but outside at point P they add to give $E_n = \sigma/\epsilon_0$.

Example 19-5

A square conducting slab of negligible thickness and with 4-m sides is placed in an external uniform field $\mathbf{E} = (450\ \text{kN/C})\mathbf{i}$ that is perpendicular to the faces of the slab. (*a*) Find the charge density on each face of the slab. (*b*) A net charge of 96 μC is placed on the slab. Find the new charge density on each face and the electric field near each face but far from the edges of the slab.

(*a*) Just outside the right face of the slab, the electric field is $E_n =$ 450 kN/C, so the charge density on this face is

$$\sigma_R = \epsilon_0 E_n = (8.85 \times 10^{-12}\ \text{C}^2/\text{N}\cdot\text{m}^2)(450\ \text{kN/C})$$
$$= 3.98 \times 10^{-6}\ \text{C/m}^2$$
$$= 3.98\ \mu\text{C/m}^2$$

At the left face, the electric field points toward the slab, so $E_n = -450$ kN/C. The charge density on this face is

$$\sigma_L = \epsilon_0 E_n$$
$$= (8.85 \times 10^{-12}\ \mathrm{C^2/N{\cdot}m^2})(-450\ \mathrm{kN/C})$$
$$= -3.98\ \mu\mathrm{C/m^2}$$

(*b*) The total charge of 96 μC must be distributed equally on each face of the slab so that the electric field inside the slab remains zero. Since each face has an area of 16 m^2 and carries a net charge of (96 μC)/2 = 48 μC, the additional charge density on each face is $\sigma_a = 48\ \mu\mathrm{C}/16\ \mathrm{m}^2 = 3.0\ \mu\mathrm{C/m^2}$. The net charge density on each face will therefore be

$$\sigma_R = 3.98\ \mu\mathrm{C/m^2} + 3.0\ \mu\mathrm{C/m^2} = 6.98\ \mu\mathrm{C/m^2}$$

and

$$\sigma_L = -3.98\ \mu\mathrm{C/m^2} + 3.0\ \mu\mathrm{C/m^2} = -0.98\ \mu\mathrm{C/m^2}$$

The net electric field just to the right of the slab is

$$E_{nR} = \sigma_R/\epsilon_0$$
$$= (6.98\ \mu\mathrm{C/m^2})/(8.85 \times 10^{-12}\ \mathrm{C^2/N{\cdot}m^2})$$
$$= 789\ \mathrm{kN/C}$$

Since the normal to the slab is in the positive x direction at the right face, the electric field just to the right of the slab is

$$\mathbf{E}_R = (789\ \mathrm{kN/C})\mathbf{i}$$

Just to the left of the slab the electric field is

$$E_{nL} = \sigma_R/\epsilon_0$$
$$= (-0.98\ \mu\mathrm{C/m^2})/(8.85 \times 10^{-12}\ \mathrm{C^2/N{\cdot}m^2})$$
$$= -111\ \mathrm{kN/m^2}$$

Since the normal to the slab points in the negative x direction on the left face, the electric field at this face is to the right:

$$\mathbf{E}_L = (111\ \mathrm{kN/C})\mathbf{i}$$

We can look at part (*b*) in another way. The positive charge density of 3.0 $\mu\mathrm{C/m^2}$ added to each face is equivalent to adding two planes of positive charge. Outside the slab these planes produce an electric field of magnitude

$$E = \tfrac{1}{2}\sigma/\epsilon_0 + \tfrac{1}{2}\sigma/\epsilon_0 = \sigma/\epsilon_0$$
$$= (3.0\ \mu\mathrm{C/m^2})/(8.85 \times 10^{-12}\ \mathrm{C^2/N{\cdot}m^2})$$
$$= 339\ \mathrm{kN/C}.$$

To the right of the slab, this field adds to the original field giving

$$E_R = 450\ \mathrm{kN/C} + 339\ \mathrm{kN/C} = 789\ \mathrm{kN/C}$$

To the left of the slab, this field subtracts from the original field giving

$$E_L = 450\ \mathrm{kN/C} - 339\ \mathrm{kN/C} = 111\ \mathrm{kN/C}$$

Exercise

The electric field just outside the surface of a conductor points away from the conductor and has the magnitude of 2000 N/C. What is the surface charge density on the surface of the conductor?
(Answer: 17.7 $\mathrm{nC/m^2}$)

19-5 Mathematical Derivation of Gauss's Law (Optional)

Gauss's law can be derived mathematically using the concept of the **solid angle.** Consider an area element ΔA on a spherical surface. The solid angle $\Delta\Omega$ subtended by ΔA at the center of the sphere is defined to be

$$\Delta\Omega = \frac{\Delta A}{r^2}$$

where r is the radius of the sphere. Since ΔA and r^2 both have dimensions of length squared, the solid angle is dimensionless. The unit of solid angle is the **steradian** (sr). Since the total area of a sphere is $4\pi r^2$, the total solid angle subtended by a sphere is

$$\frac{4\pi r^2}{r^2} = 4\pi \text{ steradians}$$

There is a close analogy between the solid angle and the ordinary plane angle, which is defined to be the ratio of an element of arc length of a circle Δs to the radius of the circle:

$$\Delta\theta = \frac{\Delta s}{r} \text{ radians}$$

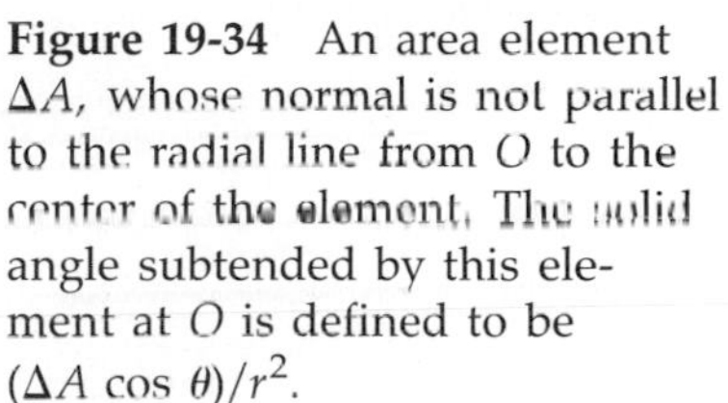

Figure 19-34 An area element ΔA, whose normal is not parallel to the radial line from O to the center of the element. The solid angle subtended by this element at O is defined to be $(\Delta A \cos\theta)/r^2$.

The total plane angle subtended by a circle is 2π radians.

In Figure 19-34, the area element ΔA is not perpendicular to the radial lines from point O. The unit vector $\hat{\mathbf{n}}$ normal to the area element makes an angle θ with the radial unit vector $\hat{\mathbf{r}}$. In this case, the solid angle subtended by ΔA at point O is defined to be

$$\Delta\Omega = \frac{\Delta A\, \hat{\mathbf{n}}\cdot\hat{\mathbf{r}}}{r^2} = \frac{\Delta A \cos\theta}{r^2} \tag{19-31}$$

Figure 19-35 shows a point charge q surrounded by a surface S of arbitrary shape. To calculate the flux through this surface, we want to find $\mathbf{E}\cdot\hat{\mathbf{n}}\,\Delta A$ for each element of area on the surface and sum over the entire surface. The flux through the area element shown is

$$\Delta\phi = \mathbf{E}\cdot\hat{\mathbf{n}}\,\Delta A = \frac{kq}{r^2}\,\hat{\mathbf{r}}\cdot\hat{\mathbf{n}}\,\Delta A = kq\,\Delta\Omega$$

The solid angle $\Delta\Omega$ is the same as that subtended by the corresponding area element of a spherical surface of any radius. The sum of the flux through the entire surface is kq times the total solid angle subtended by the closed surface, which is 4π steradians:

$$\phi_{\text{net}} = \oint \mathbf{E}\cdot\hat{\mathbf{n}}\, dA = kq \oint d\Omega = 4\pi kq = \frac{q}{\epsilon_0}$$

which is Gauss's law.

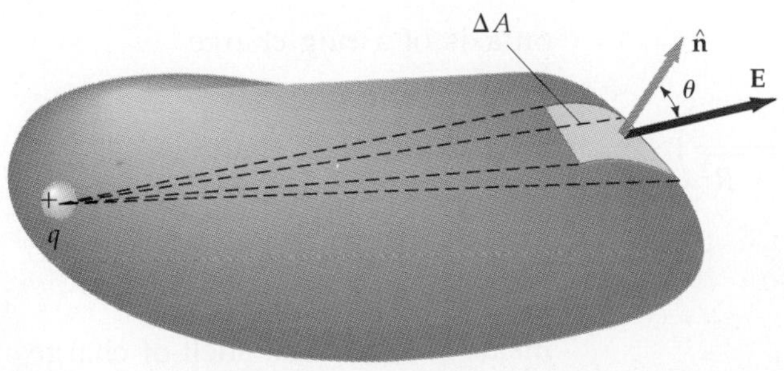

Figure 19-35 A point charge q enclosed by an arbitrary surface S. The flux through an area element ΔA is proportional to the solid angle subtended by the area element at the charge. The net flux through the surface found by summing over all the area elements is proportional to the total solid angle 4π at the charge, which is independent of the shape of the surface.

Summary

1. The electric field for continuous charge distributions can be calculated directly from Coulomb's law in the form

$$\mathbf{E} = \int_V \frac{k\,dq}{r^2}\,\hat{\mathbf{r}}$$

where $dq = \rho\,dV$ for a charge distributed throughout a volume, $dq = \sigma\,dA$ for a charge distributed on a surface, and $dq = \lambda\,dL$ for a charge distributed along a line.

2. Electric flux ϕ of a constant electric field through an area A is the product of the component of the electric field perpendicular to the area times the area:

$$\phi = \mathbf{E}\cdot\hat{\mathbf{n}}A = EA\cos\theta = E_{\mathrm{n}}A$$

For a general electric field, which may vary in space, the flux through an element of area dA is

$$d\phi = \mathbf{E}\cdot\hat{\mathbf{n}}\,dA = E\cos\theta\,dA = E_{\mathrm{n}}\,dA$$

3. The net flux through a closed surface equals $4\pi k$ times the net charge within the surface, a result known as Gauss's law:

$$\phi_{\mathrm{net}} = \oint_S E_{\mathrm{n}}\,dA = 4\pi k Q_{\mathrm{inside}}$$

Gauss's law can be used to find the electric field for highly symmetric charge distributions.

4. The Coulomb constant k is often written in terms of the permittivity of free space ϵ_0:

$$k = \frac{1}{4\pi\epsilon_0}$$

In terms of this constant, Coulomb's law is written

$$\mathbf{F}_{12} = \frac{1}{4\pi\epsilon_0}\frac{q_1q_2}{r_{12}^2}\,\hat{\mathbf{r}}_{12}$$

and Gauss's law is written

$$\phi_{\mathrm{net}} = \oint_S E_{\mathrm{n}}\,dA = \frac{1}{\epsilon_0}Q_{\mathrm{inside}}$$

5. The electric field for various charge distributions are as follows:

$$E_r = \frac{2k\lambda}{r}\sin\theta_0$$ on bisector of a finite line charge

$$E_r = \frac{1}{2\pi\epsilon_0}\frac{\lambda}{r} = 2k\frac{\lambda}{r}$$ near an infinite line charge

$$E_x = \frac{kQx}{(x^2 + a^2)^{3/2}}$$ on axis of a ring charge

$$E_x = 2\pi k\sigma\left(1 - \frac{x}{\sqrt{x^2 + R^2}}\right)$$ on axis of a disk charge

$$E_{\mathrm{n}} = \frac{\sigma}{2\epsilon_0} = 2\pi k\sigma$$ near an infinite plane of charge

$$E_r = 0 \qquad r < R$$ inside a cylindrical shell of charge

$$E_r = \frac{\sigma R}{\epsilon_0 r} = \frac{1}{2\pi\epsilon_0}\frac{\lambda}{r} \qquad r > R \qquad \text{outside a cylindrical shell of charge}$$

$$E_r = \frac{\rho R^2}{2\epsilon_0 r} = \frac{1}{2\pi\epsilon_0}\frac{\lambda}{r} \qquad r \geq R \qquad \text{outside a solid cylinder of charge}$$

$$E_r = \frac{\rho}{2\epsilon_0} r = \frac{\lambda}{2\pi\epsilon_0 R^2} r \qquad r \leq R \qquad \text{inside a solid cylinder of charge}$$

$$E_r = \frac{1}{4\pi\epsilon_0}\frac{Q}{r^2} \qquad r > R \qquad \text{outside a spherical shell of charge}$$

$$E_r = 0 \qquad r < R \qquad \text{inside a spherical shell of charge}$$

$$E_r = \frac{1}{4\pi\epsilon_0}\frac{Q}{r^2} \qquad r \geq R \qquad \text{outside a solid sphere of charge}$$

$$E_r = \frac{1}{4\pi\epsilon_0}\frac{Q}{R^3} r \qquad r \leq R \qquad \text{inside a solid sphere of charge}$$

6. At a surface carrying a surface charge density σ, the component of the electric field perpendicular to the surface is discontinuous by σ/ϵ_0:

$$E_{n2} - E_{n1} = \frac{\sigma}{\epsilon_0}$$

7. In electrostatic equilibrium, the net electric charge on a conductor resides on the surface of the conductor. The electric field just outside the surface of a conductor is perpendicular to the surface and has the magnitude σ/ϵ_0, where σ is the local surface charge density at that point on the conductor.

Suggestion for Further Reading

Reichardt, Hans: "Gauss," *The New Encylcopaedia Britannica,* 1968, vol. 19, p. 698.

A two-page biography of the German mathematician, one of the greatest mathematicians of all time, who elegantly solved problems in astronomy and geodesy as well as electromagnetism.

Review

A. Objectives: After studying this chapter, you should:

1. Be able to use Coulomb's law to calculate the electric field due to a line charge, a ring charge, and a disk charge.

2. Be able to state Gauss's law and use it to find the electric field due to an infinite plane of charge, a spherically symmetric charge distribution, and a cylindrically symmetric charge distribution.

3. Be able to show that the electric field just outside a conducting surface is σ/ϵ_0.

B. Define, explain, or otherwise identify:

Volume charge density
Surface charge density
Linear charge density
Electric flux
Gauss's law
Permittivity of free space
Gaussian surface
Electrostatic equilibrium
Field emission
Solid angle
Steradian

C. True or false: If the statement is true, explain why it is true. If it is false, give a counterexample.

1. If there is no charge in a region of space, the electric field on a surface surrounding the region must be zero everywhere.

2. Gauss's law holds only for symmetric charge distributions.

3. The electric field inside a uniformly charged spherical shell is zero.

4. In electrostatic equilibrium, the electric field inside a conductor is zero.

5. The result that $\mathbf{E} = 0$ inside a conductor can be derived from Gauss's law.

6. If the net charge on a conductor is zero, the charge density must be zero at every point on the surface of the conductor.

7. The electric field is discontinuous at all points at which the charge density is discontinuous.

8. Half of the electric field at a point just outside the surface of a conductor is due to the charge on the surface in the immediate vicinity of that point.

Problems

Level I

19-1 Calculation of the Electric Field from Coulomb's Law

1. A uniform line charge of linear charge density $\lambda = 3.5$ nC/m extends from $x = 0$ to $x = 5$ m. (*a*) What is the total charge? Find the electric field on the x axis at (*b*) $x = 6$ m, (*c*) $x = 9$ m, and (*d*) $x = 250$ m. (*e*) Find the field at $x = 250$ m, using the approximation that the charge is a point charge at the origin, and compare your result with that for the exact calculation in part (*d*).

2. Two infinite vertical planes of charge are parallel to each other and are separated by a distance $d = 4$ m. Find the electric field to the left of the planes, to the right of the planes, and between the planes when (*a*) each plane has a uniform surface charge density $\sigma = +3\ \mu\text{C/m}^2$ and (*b*) the left plane has a uniform surface charge density $\sigma = +3\ \mu\text{C/m}^2$ and that of the right plane is $\sigma = -3\ \mu\text{C/m}^2$. Draw the electric-field lines for each case.

3. A 2.75-μC charge is uniformly distributed on a ring of radius 8.5 cm. Find the electric field on the axis at (*a*) 1.2 cm, (*b*) 3.6 cm, and (*c*) 4.0 m from the center of the ring. (*d*) Find the field at 4.0 m using the approximation that the ring is a point charge at the origin, and compare your results with that for part (*c*).

4. A disk of radius 2.5 cm carries a uniform surface charge density of 3.6 $\mu\text{C/m}^2$. Using reasonable approximations, find the electric field on the axis at distances of (*a*) 0.01 cm, (*b*) 0.04 cm, (*c*) 5 m, and (*d*) 5 cm.

5. For the disk charge of Problem 4, calculate exactly the electric field on the axis at distances of (*a*) 0.04 cm and (*b*) 5 m, and compare your results with those for parts (*b*) and (*c*) of Problem 4.

6. A uniform line charge extends from $x = -2.5$ cm to $x = +2.5$ cm and has a linear charge density $\lambda = 4.5$ nC/m. (*a*) Find the total charge. Find the electric field on the y axis at (*b*) $y = 4$ cm, (*c*) $y = 12$ cm, and (*d*) $y = 4.5$ m. (*e*) Find the field at $y = 4.5$ m assuming the charge to be a point charge, and compare your result with that for part (*d*).

7. A disk of radius a lies in the yz plane with its axis along the x axis and carries a uniform surface charge density σ. Find the value of x for which $E_x = \frac{1}{2}\sigma/2\epsilon_0$.

8. A ring of radius a with its center at the origin and its axis along the x axis carries a total charge Q. Find E_x at (*a*) $x = 0.2a$, (*b*) $x = 0.5a$, (*c*) $x = 0.7a$, (*d*) $x = a$, and (*e*) $x = 2a$. (*f*) Use your results to plot E_x versus x for both positive and negative values of x.

9. Repeat Problem 8 for a disk of uniform surface charge density σ.

19-2 Gauss's Law

10. Consider a uniform electric field $\mathbf{E} = (2\ \text{kN/C})\mathbf{i}$. (*a*) What is the flux of this field through a square of side 10 cm in a plane parallel to the yz plane? (*b*) What is the flux through the same square if the normal to its plane makes a 30° angle with the x axis?

11. A single point charge $q = +2\ \mu$C is at the origin. A spherical surface of radius 3.0 m has its center on the x axis at $x = 5$ m. (*a*) Sketch electric-field lines for the point charge. Do any lines enter the spherical surface? (*b*) What is the net number of lines that leave the spherical surface, counting those that enter as negative? (*c*) What is the net flux of the electric field due to the point charge through the spherical surface?

12. An electric field is $\mathbf{E} = (200\ \text{N/C})\mathbf{i}$ for $x > 0$ and $\mathbf{E} = (-200\ \text{N/C})\mathbf{i}$ for $x < 0$. A cylinder of length 20 cm and radius 5 cm has its center at the origin and its axis along the x axis such that one end is at $x = +10$ cm and the other is at $x = -10$ cm. (*a*) What is the flux through each end? (*b*) What is the flux through the curved surface of the cylinder? (*c*) What is the net outward flux through the entire cylindrical surface? (*d*) What is the net charge inside the cylinder?

13. A positive point charge q is at the center of a cube of side L. A large number N of electric-field lines are drawn from the point charge. (*a*) How many of the field lines pass through the surface of the cube? (*b*) How many lines pass through each face, assuming that none pass through the edges or corners? (*c*) What is the net outward flux of the electric field through the cubic surface? (*d*) Use symmetry arguments to find the flux of the electric field through one face of the cube. (*e*) Which, if any, of your answers would change if the charge were inside the cube but not at its center?

14. Careful measurement of the electric field at the surface of a black box indicates that the net outward flux through the surface of the box is 6.0 kN·m^2/C. (*a*) What is the net charge inside the box? (*b*) If the net outward flux through the surface of the box were zero, could you conclude there were no charges inside the box? Why or why not?

15. A point charge $q = +2\ \mu C$ is at the center of a sphere of radius 0.5 m. (*a*) Find the surface area of the sphere. (*b*) Find the magnitude of the electric field at points on the surface of the sphere. (*c*) What is the flux of the electric field due to the point charge through the surface of the sphere? (*d*) Would your answer to part (*c*) change if the point charge were moved so that it was inside the sphere but not at the center? (*e*) What is the net flux through a cube of side 1 m that encloses the sphere?

16. Since Newton's law of gravity and Coulomb's law have the same inverse-square dependence on distance, an expression analogous in form to Gauss's law can be found for gravity. The gravitational field **g** is the force per unit mass on a test mass m_0. Then for a point mass m at the origin, the gravitational field **g** at some position **r** is

$$\mathbf{g} = -\frac{Gm}{r^2}\,\hat{\mathbf{r}}$$

Compute the flux of the gravitational field through a spherical surface of radius r centered at the origin, and show that the gravitational analog of Gauss's law is $\phi_{\text{net}} = -4\pi G m_{\text{inside}}$.

19-3 Calculation of the Electric Field from Gauss's Law

17. A spherical shell of radius 6 cm carries a uniform surface charge density $\sigma = 9\ \text{nC/m}^2$. (*a*) What is the total charge on the shell? Find the electric field at (*b*) $r = 2$ cm, (*c*) $r = 5.9$ cm, (*d*) $r = 6.1$ cm, and (*e*) $r = 10$ cm.

18. A sphere of radius 6 cm carries a uniform volume charge density $\rho = 450\ \text{nC/m}^3$. (*a*) What is the total charge of the sphere? Find the electric field at (*b*) $r = 2$ cm, (*c*) $r = 5.9$ cm, (*d*) $r = 6.1$ cm, and (*e*) $r = 10$ cm.

19. A cylindrical shell of length 12 m and radius 6 cm carries a uniform surface charge density $\sigma = 9\ \text{nC/m}^2$. (*a*) What is the total charge on the shell? Find the electric field at (*b*) $r = 2$ cm, (*c*) $r = 5.9$ cm, (*d*) $r = 6.1$ cm, and (*e*) $r = 10$ cm.

20. A cylinder of length 12 m and radius 6 cm carries a uniform volume charge density $\rho = 300\ \text{nC/m}^3$. (*a*) What is the total charge of the cylinder? Find the electric field at (*b*) $r = 2$ cm, (*c*) $r = 5.9$ cm, (*d*) $r = 6.1$ cm, and (*e*) $r = 10$ cm.

21. A spherical shell of radius R_1 carries a total charge q_1 that is uniformly distributed on its surface. A second, larger spherical shell of radius R_2 that is concentric with the first carries a charge q_2 that is uniformly distributed on its surface. (*a*) Use Gauss's law to find the electric field in the regions $r < R_1$, $R_1 < r < R_2$, and $r > R_2$. (*b*) What should the ratio of the charges q_1/q_2 and their relative signs be for the electric field to be zero for $r > R_2$? (*c*) Sketch the electric-field lines for the situation in part (*b*).

22. Consider two infinitely long, concentric cylindrical shells. The inner shell has a radius R_1 and carries a uniform surface charge density σ_1, and the outer shell has a radius R_2 and carries a uniform surface charge density σ_2. (*a*) Use Gauss's law to find the electric field in the regions $r < R_1$, $R_1 < r < R_2$, and $r > R_2$. (*b*) What should the ratio of the surface charge densities σ_2/σ_1 and their relative signs be for the electric field to be zero at $r > R_2$? What would the electric field between the shells be in this case? (*c*) Sketch the electric-field lines for the situation in part (*b*).

23. A nonuniform surface charge lies in the yz plane. At the origin, the surface charge density is $\sigma = 3.10\ \mu\text{C/m}^2$. There are various other charge distributions in space. Just to the right of the origin, the x component of the electric field is $E_x = 4.65 \times 10^5$ N/C. What is E_x just to the left of the origin?

19-4 Charge and Field at Conductor Surfaces

24. A penny is in an external electric field of magnitude 1.6 kN/C directed perpendicular to its faces. (*a*) Find the charge density on each face of the penny, assuming the faces are planes. (*b*) If the radius of the penny is 1 cm, find the total charge on one face.

25. An uncharged metal slab has square faces with 12-cm sides. It is placed in an external electric field that is perpendicular to its faces. The total charge induced on one of the faces is 1.2 nC. What is the magnitude of the electric field?

26. A charge of 6 nC is placed uniformly on a square sheet of nonconducting material of side 20 cm in the yz plane. (*a*) What is the surface charge density σ? (*b*) What is the magnitude of the electric field just to the right and just to the left of the sheet? (*c*) The same charge is placed on a square conducting slab of side 20 cm and thickness 1 mm. What is the surface charge density σ? (Assume that the charge distributes itself uniformly on the large square surfaces.) (*d*) What is the magnitude of the electric field just to the right and just to the left of each face of the slab?

27. A spherical conducting shell with zero net charge has an inner radius a and an outer radius b. A point charge q is placed at the center of the shell. (*a*) Use Gauss's law and the properties of conductors in equilibrium to find the electric field in the regions $r < a$, $a < r < b$, and $b < r$. (*b*) Draw the electric-field lines for this situation. (*c*) Find the charge density on the inner surface ($r = a$) and on the outer surface ($r = b$) of the shell.

19-5 Mathematical Derivation of Gauss's Law (Optional)

There are no problems for this section.

Level II

28. The electric field just above the surface of the earth has been measured to be 150 N/C downward. What total charge on the earth is implied by this measurement?

29. In a particular region of the earth's atmosphere, the electric field above the earth's surface has been measured to be 150 N/C downward at an altitude of 250 m and 170 N/C downward at an altitude of 400 m. Calculate the volume charge density of the atmosphere assuming it to be uniform between 250 and 400 m. (You may neglect the curvature of the earth. Why?)

30. An infinite line charge of uniform linear charge density $\lambda = -1.5\ \mu C/m$ lies parallel to the y axis at $x = -2$ m. A point charge of 1.3 μC is located at $x = 1$ m, $y = 2$ m. Find the electric field at $x = 2$ m, $y = 1.5$ m.

31. A solid sphere 1.2 m in diameter with its center on the x axis at $x = 4$ m carries a uniform volume charge of density $\rho = 5\ \mu C/m^3$. A spherical shell concentric with the sphere has a diameter of 2.4 m and a uniform surface charge density $\sigma = -1.5\ \mu C/m^2$. Calculate the magnitude and direction of the electric field at (*a*) $x = 4.5$ m, $y = 0$; (*b*) $x = 4.0$ m, $y = 1.1$ m; and (*c*) $x = 2.0$ m, $y = 3.0$ m.

32. Two infinite planes of charge lie parallel to each other and to the yz plane. One is at $x = -2$ m and has a surface charge density $\sigma = -3.5\ \mu C/m^2$. The other is at $x = 2$ m and has a surface charge density $\sigma = 6.0\ \mu C/m^2$. Find the electric field for (*a*) $x < -2$ m, (*b*) $-2\text{ m} < x < 2$ m, and (*c*) $x > 2$ m.

33. A model of an atom has a positive nuclear point charge $+Ze$ embedded in a rigid electron sphere of radius R containing a total charge of $-Ze$ uniformly distributed throughout the sphere. (*a*) For a zero external electric field, where is the equilibrium position of the nuclear point charge? (*b*) In an external electric field $\mathbf{E}_0$, where is the equilibrium position of the nuclear point charge with respect to the center of the negatively charged electron sphere? (*c*) What is the electric dipole moment induced by the field $\mathbf{E}_0$ for this atomic model?

34. Show that E_x on the axis of a ring charge of radius a has its maximum and minimum values at $x = +a/\sqrt{2}$ and $x = -a/\sqrt{2}$. Sketch E_x versus x for both positive and negative values of x.

35. A positive point charge of magnitude 2.5 μC is at the center of an uncharged spherical conducting shell of inner radius 60 cm and outer radius 90 cm. (*a*) Find the charge densities on the inner and outer surfaces of the shell and the total charge on each surface. (*b*) Find the electric field everywhere. (*c*) Repeat (*a*) and (*b*) if a net charge of +3.5 μC is placed on the shell.

36. A square conducting slab with 5-m sides carries a net charge of 80 μC. (*a*) Find the charge density on each face of the slab and the electric field just outside one face of the slab. (*b*) The slab is placed to the right of an infinite charged nonconducting plane with charge density 2.0 $\mu C/m^2$ so that the faces of the slab are parallel to the plane. Find the electric field on each side of the slab far from its edges and the charge density on each face.

37. (*a*) A finite line charge of uniform linear charge density λ lies on the x axis from $x = 0$ to $x = a$. Show that the y component of the electric field at a point on the y axis is given by

$$E_y = \frac{\lambda}{4\pi\epsilon_0 y}\sin\theta_1 = \frac{\lambda}{4\pi\epsilon_0 y}\frac{a}{\sqrt{y^2 + a^2}}$$

where θ_1 is the angle subtended by the line charge at the field point. (*b*) Show that if the line charge extends from $x = -b$ to $x = a$, the y component of the electric field at a point on the y axis is given by

$$E_y = \frac{\lambda}{4\pi\epsilon_0 y}(\sin\theta_1 + \sin\theta_2)$$

where $\sin\theta_2 = b/\sqrt{y^2 + b^2}$.

38. Imagine that a small hole has been punched through the wall of a thin, uniformly charged spherical shell whose surface charge density is σ. Find the electric field near the center of the hole.

39. An infinite plane of charge with surface charge density $\sigma_1 = 3\ \mu C/m^2$ is parallel to the xz plane at $y = -0.6$ m. A second infinite plane of charge with surface charge density $\sigma_2 = -2\ \mu C/m^2$ is parallel to the yz plane at $x = 1$ m. A sphere of radius 1 m with its center in the xy plane at the intersection of the two charged planes ($x = 1$ m, $y = -0.6$ m) has a surface charge density $\sigma_3 = -3\ \mu C/m^2$. Find the magnitude and direction of the electric field on the x axis at (*a*) $x = 0.4$ m and (*b*) $x = 2.5$ m.

40. An infinitely long cylindrical shell is coaxial with the y axis and has a radius of 15 cm. It carries a uniform surface charge density $\sigma = 6\ \mu C/m^2$. A spherical shell of radius 25 cm is centered on the x axis at $x = 50$ cm and carries a uniform surface charge density $\sigma = -12\ \mu C/m^2$. Calculate the magnitude and direction of the electric field at (*a*) the origin; (*b*) $x = 20$ cm, $y = 10$ cm; and (*c*) $x = 50$ cm, $y = 20$ cm.

41. A thick, nonconducting spherical shell of inner radius a and outer radius b has a uniform volume charge density ρ. Find the total charge and the electric field everywhere.

42. An infinite plane in the xz plane carries a uniform surface charge density $\sigma_1 = 65\ nC/m^2$. A second infinite plane carrying a uniform charge density $\sigma_2 = 45\ nC/m^2$ intersects the xz plane at the z axis and makes an angle of 30° with the xz plane as shown in Figure 19-36. Find the electric field in the xy plane at (*a*) $x = 6$ m, $y = 2$ m and (*b*) $x = 6$ m, $y = 5$ m.

Figure 19-36 Uniform surface charges on the xz plane and on a plane making an angle of 30° with the xz plane for Problem 42.

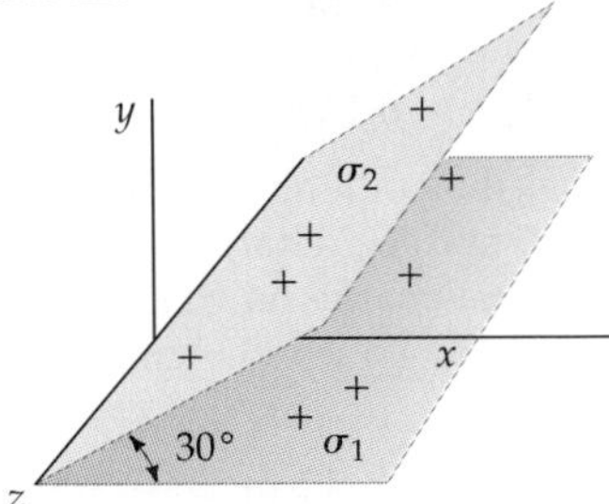

43. A ring of radius R carries a uniform, positive, linear charge density λ. Figure 19-37 shows a point P in the plane of the ring but not at the center. Consider the two elements of the ring of lengths s_1 and s_2 shown in the figure at distances r_1 and r_2, respectively, from point P. (*a*) What is the ratio of the charges of these elements? Which produces the greater field at point P? (*b*) What is the direction of the field at point P due to each element? What is the direction of the total electric field at point P? (*c*) Suppose that the electric field due to a point charge varied as $1/r$ rather than $1/r^2$. What would the electric field be at point P due to the elements shown? (*d*) How would your answers to parts (*a*), (*b*), and (*c*) differ if point P were inside a spherical shell of uniform charge and the elements were of areas s_1 and s_2?

Figure 19-37 Problem 43.

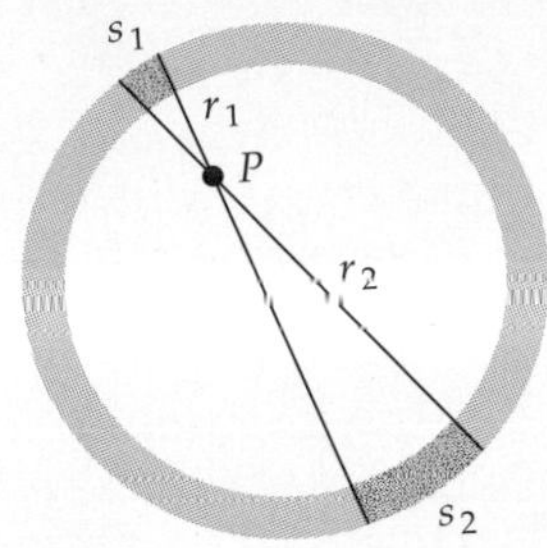

44. A disk of radius 30 cm carries a uniform charge density σ. (*a*) Compare the approximation $E = \sigma/2\epsilon_0$ with the exact expression for the electric field on the axis of the disk by computing the neglected term as a percentage of $\sigma/2\epsilon_0$ for the distances $x = 0.1$, $x = 0.2$, and $x = 3$ cm. (*b*) At what distance is the neglected term 1 percent of $\sigma/2\epsilon_0$?

45. A semi-infinite line charge of uniform linear charge density λ lies along the x axis from $x = 0$ to $x = \infty$. Find both E_x and E_y at a point on the y axis.

46. An infinite plane lies parallel to the yz plane at $x = 2$ m and carries a uniform surface charge density $\sigma = 2\ \mu C/m^2$. An infinite line charge of uniform linear charge density $\lambda = 4\ \mu C/m$ passes through the origin at 45° to the x axis in the xy plane. A sphere of volume charge density $\rho = -6\ \mu C/m^3$ and radius 0.8 m is centered on the x axis at $x = 1$ m. Calculate the magnitude and direction of the electric field in the xy plane at $x = 1.5$ m, $y = 0.5$ m.

Level III

47. An infinitely long, thick, nonconducting cylindrical shell of inner radius a and outer radius b has a uniform volume charge density ρ. Find the electric field everywhere.

48. A nonconducting solid sphere of radius R carries a volume charge density that is proportional to the distance from the center: $\rho = Ar$ for $r \leq R$, where A is a constant; $\rho = 0$ for $r > R$. (*a*) Find the total charge on the sphere by summing the charges on shells of thickness dr and volume $4\pi r^2\ dr$. (*b*) Find the electric field E_r both inside and outside the charge distribution, and sketch E_r versus r.

49. Repeat Problem 48 for a solid sphere with volume charge density $\rho = B/r$ for $r < R$ and $\rho = 0$ for $r > R$.

50. Repeat Problem 48 for a solid sphere with volume charge density $\rho = C/r^2$ for $r < R$ and $\rho = 0$ for $r > R$.

51. A uniformly charged sphere of radius R is centered at the origin with a charge of Q. Find the force on a uniformly charged line oriented radially having a total charge q with its ends at $r = R$ and $r = R + d$.

52. Two equal uniform line charges of length L lie on the x axis a distance d apart as shown in Figure 19-38. (*a*) What is the force that one line charge exerts on the other line charge? (*b*) Show that when $d \gg L$, the force tends toward the expected result of $k(\lambda L)^2/d^2$.

Figure 19-38 Problem 52.

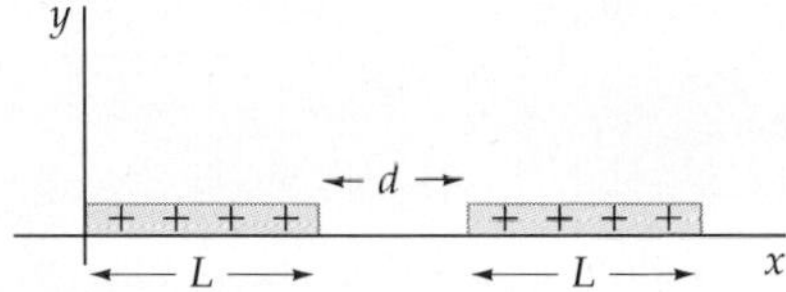

53. A line charge of linear charge density λ with the shape of a square of side L lies in the yz plane with its center at the origin. Find the electric field on the x axis at an arbitrary distance x, and compare your result to that for the field on the axis of a charged ring of radius $r = \frac{1}{2}L$ with its center at the origin and carrying the same total charge. *Hint:* Use Equation 19-10 for the field due to each segment of the square.

54. A nonconducting solid sphere of radius a with its center at the origin has a spherical cavity of radius b with its center at the point $x = b$, $y = 0$ as shown in Figure 19-39. The sphere has a uniform volume charge density ρ. Show that the electric field in the cavity is uniform and is given by $E_y = 0$, $E_x = \rho b/3\epsilon_0$. *Hint:* Replace the cavity with spheres of equal positive and negative charge densities.

55. The electrostatic force on a charge at some point is the product of the charge and the electric field due to all other charges. Consider a small charge on the surface of a conductor $\Delta q = \sigma\ \Delta A$. (*a*) Show that the electrostatic force on the charge is $\sigma^2\ \Delta A/2\epsilon_0$. (*b*) Explain why this is just half of $\Delta q\ E$, where $E = \sigma/\epsilon_0$ is the electric field just outside the conductor at that point. (*c*) The force per unit area is called the electrostatic stress. Find the electrostatic stress when a charge of 2 μC is placed on a conducting sphere of radius 10 cm.

Figure 19-39 Problem 54.

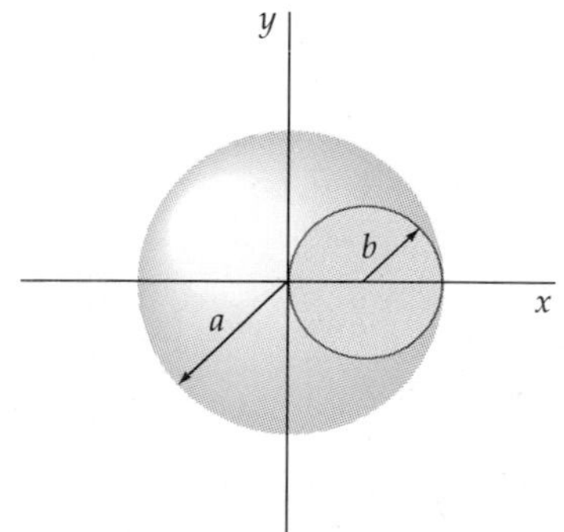

Chapter 20

Electric Potential

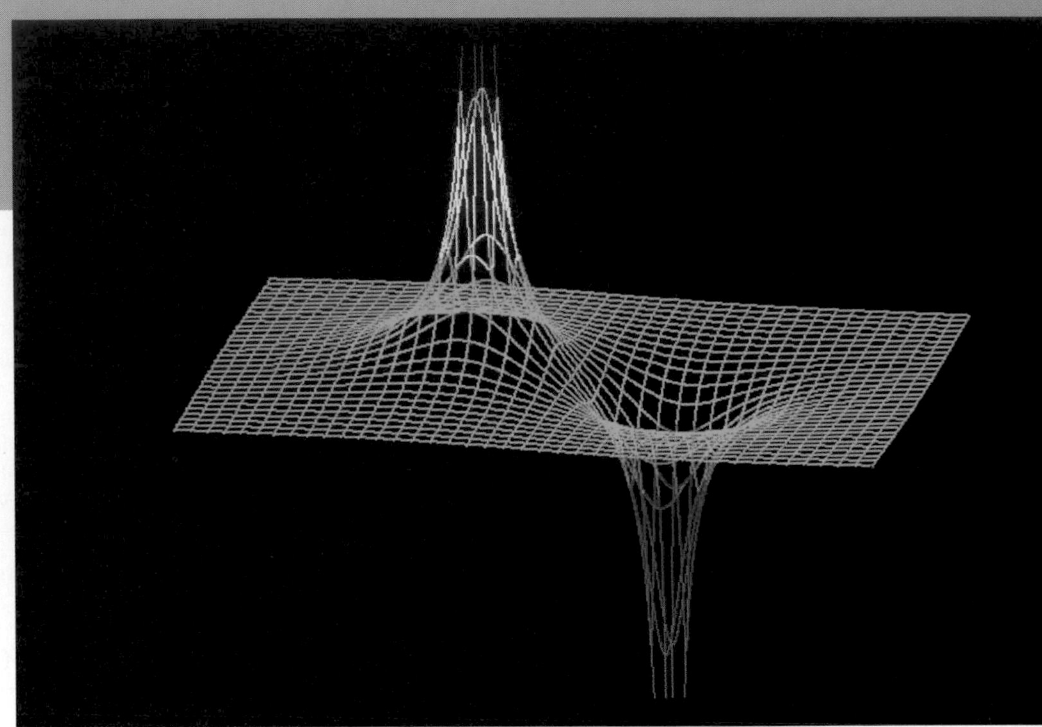

Computer plot of the electrostatic potential in the plane of an electric dipole. The potential due to each charge is proportional to the charge and inversely proportional to the distance from the charge.

In our study of mechanics, we found the concept of potential energy to be very useful. When we lift an object of mass m a vertical distance h near the earth's surface, the work we do goes into potential energy mgh of the earth-mass system. If we then drop the object, this potential energy is converted into kinetic energy. The electric force between two charges is directed along the line of the charges and depends on the inverse square of their separation, the same as the gravitational force between two masses. Like the gravitational force, the electric force is conservative. There is thus a potential-energy function associated with the electric force. As we will see, the potential energy associated with a particle in an electric field is proportional to the charge. The potential energy per unit charge is called the electric potential. Electric potential is measured in volts and is commonly called voltage. In this chapter, we will define the electric potential function V and show how to calculate the potential from a given charge distribution or from a given electric field. We will then show how the potential is related to the electric field $\mathbf{E}$ and to electrostatic potential energy. Finally, we will show that the electric potential is constant everywhere inside a conductor in an electrostatic field.

20-1 Electric Potential and Potential Difference

In general, when a conservative force **F** acts on a particle that undergoes a displacement $d\boldsymbol{\ell}$, the change in the potential-energy function dU is defined by (Equation 6-17)

$$dU = -\mathbf{F}\cdot d\boldsymbol{\ell}$$

The work done by a conservative force decreases the potential energy (Figure 20-1). The force exerted by an electric field **E** on a point charge q_0 is

$$\mathbf{F} = q_0\,\mathbf{E}$$

When the charge undergoes a displacement $d\boldsymbol{\ell}$ in an electric field **E**, the change in the electrostatic potential energy is

$$dU = -q_0\,\mathbf{E}\cdot d\boldsymbol{\ell} \qquad \text{20-1}$$

If the charge is moved from some initial point a to some final point b, the change in its electrostatic potential energy is

$$\Delta U = U_b - U_a = \int_a^b dU = -\int_a^b q_0\,\mathbf{E}\cdot d\boldsymbol{\ell} \qquad \text{20-2}$$

The potential-energy change is proportional to the test charge q_0. The potential-energy change per unit charge is called the **potential difference** dV:

$$dV = \frac{dU}{q_0} = -\mathbf{E}\cdot d\boldsymbol{\ell} \qquad \text{20-3}a$$

Potential difference defined

For a finite displacement from point a to point b, the change in potential is

$$\Delta V = V_b - V_a = \frac{\Delta U}{q_0} = -\int_a^b \mathbf{E}\cdot d\boldsymbol{\ell} \qquad \text{20-3}b$$

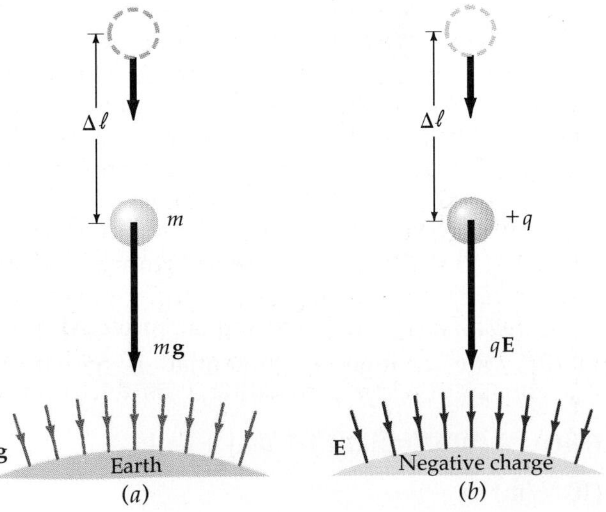

Figure 20-1 (*a*) The work done by the gravitational field on a mass decreases the gravitational potential energy. (*b*) The work done by the electric field on a positive charge $+q$ decreases the electrostatic potential energy.

The potential difference $V_b - V_a$ is the negative of the work per unit charge done by the electric field on a positive test charge when it moves from point a to point b.

Equation 20-3 defines the change in the function V, which is called the **electric potential** (or sometimes just the **potential**). As with potential energy U, only *changes* in the potential V are important. We are free to choose the potential energy or the potential to be zero at any convenient point, just as we were for mechanical potential energy. (For example, in the expression for the gravitational potential energy near the earth's surface, mgh, we can choose h to be zero at any convenient point, such as the floor or at the top of a table. For two point masses, we found that it was usually convenient to choose the gravitational potential energy to be zero when their separation was infinite.)

Since electric potential is the electrostatic potential energy per unit charge, the SI unit for potential and potential difference is the joule per coulomb, called the **volt** (V):

$$1 \text{ V} = 1 \text{ J/C} \qquad 20\text{-}4$$

Because potential difference is measured in volts, it is sometimes called **voltage.** In a 12-V car battery, the positive terminal has a potential 12 V higher than the negative terminal. If an external circuit is attached to the battery and one coulomb of charge is transferred from the positive terminal through the circuit to the negative terminal, the potential energy of the charge is decreased by $Q\,\Delta V = (1 \text{ C})(12 \text{ V}) = 12 \text{ J}$. This energy usually appears as thermal energy in the circuit.

From Equation 20-3, we can see that the dimensions of potential are also those of electric field times distance. Thus, the unit of electric field E, the newton per coulomb, is also equal to a volt per meter:

$$1 \text{ N/C} = 1 \text{ V/m} \qquad 20\text{-}5$$

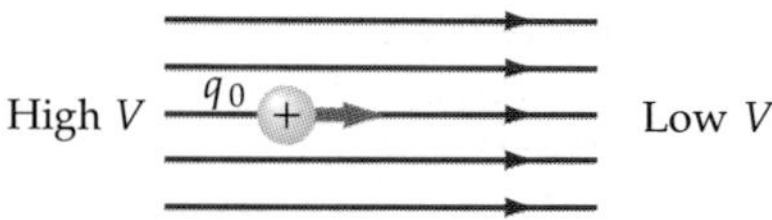

Figure 20-2 Electric-field lines point in the direction of decreasing potential. When a positive test charge q_0 is placed in an electric field, it accelerates in the direction of the field. Its kinetic energy increases and its potential energy decreases.

If we place a positive test charge q_0 in an electric field **E** and release the charge, it experiences a force in the direction of the field and will accelerate in the direction of **E** along a field line. As the kinetic energy of the charge increases, its potential energy decreases. Thus, the charge moves toward a region of lower potential energy, just as a massive body falls toward a region of lower gravitational potential energy. For a positive test charge, a region of lower potential energy is one of lower potential. Thus, as illustrated in Figure 20-2,

Electric-field lines point in the direction of decreasing electric potential.

Example 20-1

An electric field points in the positive x direction and has a constant magnitude of 10 N/C = 10 V/m. Find the potential as a function of x, assuming that $V = 0$ at $x = 0$.

The electric-field vector is given by $\mathbf{E} = (10 \text{ N/C})\mathbf{i} = (10 \text{ V/m})\mathbf{i}$. For a general displacement $d\boldsymbol{\ell}$, the change in potential is given by Equation 20-3a:

$$dV = -\mathbf{E}\cdot d\boldsymbol{\ell} = -(10 \text{ V/m})\mathbf{i}\cdot(dx\,\mathbf{i} + dy\,\mathbf{j} + dz\,\mathbf{k})$$
$$= -(10 \text{ V/m})\,dx$$

Integrating from point x_1 to x_2, we obtain the potential difference $V(x_2) - V(x_1)$:

$$V(x_2) - V(x_1) = \int_{x_1}^{x_2} dV = \int_{x_1}^{x_2} -(10\text{ V/m})\,dx$$
$$= -(10\text{ V/m})(x_2 - x_1) = (10\text{ V/m})(x_1 - x_2)$$

Since we are given that the potential is zero at $x = 0$, we have $V(x_1) = 0$ at $x_1 = 0$. Then the potential at x_2 relative to $V = 0$ at $x = 0$ is given by

$$V(x_2) - 0 = (10\text{ V/m})(0 - x_2)$$

or

$$V(x_2) = -(10\text{ V/m})x_2$$

At a general point x, the potential is

$$V(x) = -(10\text{ V/m})x$$

The potential is zero at $x = 0$ and decreases by 10 V/m in the x direction.

Example 20-2

A proton of mass 1.67×10^{-27} kg and charge 1.6×10^{-19} C is placed in a uniform electric field $\mathbf{E} = (5.0\text{ N/C})\mathbf{i} = (5.0\text{ V/m})\mathbf{i}$ and released from rest. After traveling 4 cm, how fast is it moving?

As the proton travels down the electric-field line, its potential energy decreases and its kinetic energy increases by an equal amount. According to Equation 20-3, the change in the electric potential for $\Delta x = 4$ cm $=$ 0.04 m is

$$dV = -\mathbf{E}\cdot d\boldsymbol{\ell} = -(5.0\text{ V/m }\mathbf{i})\cdot(dx\,\mathbf{i}) = -(5.0\text{ V/m})\,dx$$
$$\Delta V = -(5.0\text{ V/m})(0.04\text{ m}) = -0.20\text{ V}$$

The change in the potential energy of the proton is the product of its charge times the change in its potential (Equation 20-3):

$$\Delta U = q\,\Delta V = (1.6 \times 10^{-19}\text{ C})(-0.20\text{ V}) = -3.2 \times 10^{-20}\text{ J}$$

By conservation of energy, the loss in potential energy equals the gain in kinetic energy. Since the proton starts from rest, its gain in kinetic energy is just $\frac{1}{2}mv^2$, where v is its speed after traveling the 4 cm. We therefore have

$$\Delta K + \Delta U = 0$$
$$\Delta K = -\Delta U = -(-3.2 \times 10^{-20}\text{ J})$$
$$\tfrac{1}{2}mv^2 = 3.2 \times 10^{-20}\text{ J}$$
$$v^2 = \frac{(2)(3.2 \times 10^{-20}\text{ J})}{1.67 \times 10^{-27}\text{ kg}} = 3.83 \times 10^7\text{ J/kg}$$
$$v = \sqrt{3.83 \times 10^7\text{ J/kg}} = 6.19 \times 10^3\text{ m/s}$$

An ordinary wall outlet. The power company maintains an electric potential difference of 120 V between the slotted openings. When a heater (or other device) is plugged into the outlet, each coulomb passing from one side of the heater to the other loses 120 J of energy in the heater and connecting wires. The rounded opening is connected to the ground.

In atomic and nuclear physics, we often have elementary particles, such as electrons and protons, with charges of magnitude e moving through potential differences of several to thousands or even millions of volts. Since energy has dimensions of electric charge times electric potential, a convenient unit of energy is the product of the electron charge e times a volt. This unit is called an **electron volt** (eV). The conversion between electron volts and joules is obtained by expressing the electronic charge in coulombs:

$$1 \text{ eV} = 1.6 \times 10^{-19} \text{ C·V} = 1.6 \times 10^{-19} \text{ J} \qquad 20\text{-}6$$

In Example 20-2, the change in the potential energy of the proton after traveling 4 cm is

$$\Delta U = q\,\Delta V = e(-0.20 \text{ V}) = -0.20 \text{ eV}$$

Questions

1. Explain in your own words the distinction between electric potential and electrostatic potential energy.
2. If a test charge moves through a small distance in the direction of an electric field, does its electrostatic potential energy increase or decrease? Does your answer depend on the sign of the charge? Does the change in potential depend on the sign of the test charge?
3. In what direction can you move relative to an electric field so that the electric potential does not change?
4. A positive charge is released from rest in an electric field. Will it move toward a region of greater or smaller electric potential?

20-2 Potential Due to a System of Point Charges

The electric potential due to a point charge q at the origin can be calculated from the electric field, which is given by

$$\mathbf{E} = \frac{kq}{r^2}\,\hat{\mathbf{r}}$$

If a test charge q_0 at distance r is given a displacement $d\boldsymbol{\ell} = dr\,\hat{\mathbf{r}}$, the change in its potential energy is $dU = -q_0\,\mathbf{E}\cdot d\boldsymbol{\ell}$, and the change in the electric potential is

$$dV = -\mathbf{E}\cdot d\boldsymbol{\ell} = -\frac{kq}{r^2}\,\hat{\mathbf{r}}\cdot dr\,\hat{\mathbf{r}} = -\frac{kq}{r^2}\,dr \qquad 20\text{-}7$$

Integrating, we obtain

Potential due to a point charge

$$V = +\frac{kq}{r} + V_0 \qquad 20\text{-}8$$

where V_0 is a constant of integration.

It is customary to define the potential to be zero at an infinite distance from the point charge (that is, at $r = \infty$). Then the constant V_0 is zero, and the potential at a distance r from the point charge is

Potential due to a point charge with $V = 0$ at $r = \infty$

$$V = \frac{kq}{r} \qquad V = 0 \text{ at } r = \infty \qquad 20\text{-}9$$

The potential is positive or negative depending on the sign of the charge q.

If a test charge q_0 is released from a point a distance r from a point charge q that is held fixed at the origin, the test charge will be accelerated outward in the direction of the electric field. The work done by the electric field as the test charge moves from r to ∞ is

$$W = \int_r^\infty q_0\,\mathbf{E}\cdot d\boldsymbol{\ell} = q_0 \int_r^\infty E_r\,dr = q_0 \int_r^\infty \frac{kq}{r^2}\,dr = \frac{kqq_0}{r}$$

This work is the electrostatic potential energy of the two-charge system:

$$U = \frac{kqq_0}{r} = q_0 V$$

The potential energy is thus the work done by the electric field as the test charge moves from r to ∞. Alternatively, we may think of the potential energy as the work that must be done by an applied force $\mathbf{F}_{\text{app}} = -q_0\,\mathbf{E}$ to bring a positive test charge q_0 from an infinite distance away to a distance r from a point charge q (Figure 20-3).

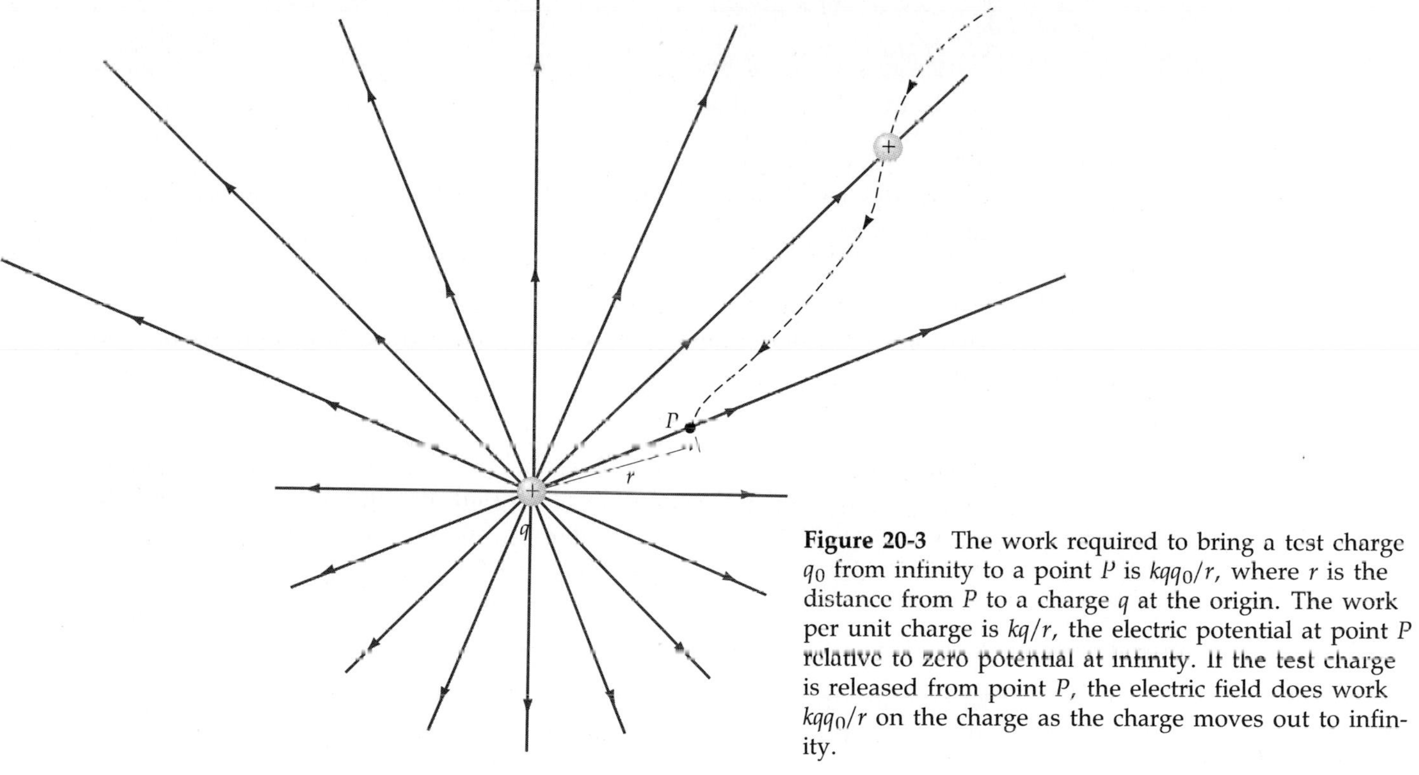

Figure 20-3 The work required to bring a test charge q_0 from infinity to a point P is kqq_0/r, where r is the distance from P to a charge q at the origin. The work per unit charge is kq/r, the electric potential at point P relative to zero potential at infinity. If the test charge is released from point P, the electric field does work kqq_0/r on the charge as the charge moves out to infinity.

The choice that the electric potential is zero at an infinite distance from a point charge is made merely for convenience. The potential energy of two charges is then zero when they are infinitely far apart. This choice is similar to that we made in our study of gravity in Chapter 10 when we chose the gravitational potential energy of two point masses to be zero when the masses were very far apart. We can also make this choice of zero potential for a system of charges as long as the system is finite, that is, as long as there are no charges an infinite distance from other charges in the system. At sufficiently great distances from any finite charge distribution, the charge distribution looks like a point charge, and the potential function V approaches that given by Equation 20-9, where q is the net charge of the distribution.

Example 20-3

(*a*) What is the electric potential at a distance $r = 0.529 \times 10^{-10}$ m from a proton? (This is the average distance between the proton and electron in the hydrogen atom.) (*b*) What is the potential energy of the electron and the proton at this separation?

(*a*) The charge of the proton is $q = 1.6 \times 10^{-19}$ C. Equation 20-9 gives

$$V = \frac{kq}{r} = \frac{(8.99 \times 10^9\ \text{N}\cdot\text{m}^2/\text{C}^2)(1.6 \times 10^{-19}\ \text{C})}{0.529 \times 10^{-10}\ \text{m}}$$

$$= 27.2\ \text{J/C} = 27.2\ \text{V}$$

(*b*) The charge of the electron is $-e = -1.6 \times 10^{-19}$ C. In electron volts, the potential energy of the electron and proton separated by a distance of 0.529×10^{-10} m is

$$U = qV = -e(27.2\ \text{V}) = -27.2\ \text{eV}$$

In SI units, the potential energy is

$$U = qV = (-1.6 \times 10^{-19}\ \text{C})(27.2\ \text{V}) = -4.35 \times 10^{-18}\ \text{J}$$

To find the potential at some point due to several point charges, we find the potential at that point due to each charge separately and sum. This follows from the superposition principle for the electric field. If $\mathbf{E}_i$ is the electric field at some point due to charge q_i, the net field at that point due to all the charges is

$$\mathbf{E} = \mathbf{E}_1 + \mathbf{E}_2 + \cdots = \sum_i \mathbf{E}_i$$

Then, from the definition of potential difference (Equation 20-3), we have for a displacement $d\boldsymbol{\ell}$

$$dV = -\mathbf{E}\cdot d\boldsymbol{\ell} = -\mathbf{E}_1\cdot d\boldsymbol{\ell} - \mathbf{E}_2\cdot d\boldsymbol{\ell} - \cdots = dV_1 + dV_2 + \cdots$$

If the charge distribution is finite, that is, if there are no charges at infinity, we may choose the potential to be zero at infinity and use Equation 20-9 for the potential due to each point charge. The potential due to a system of point charges q_i is then given by

$$V = \sum_i \frac{kq_i}{r_{i0}} \qquad 20\text{-}10$$

where the sum is taken over all the charges and r_{i0} is the distance from the ith charge to the point P at which the potential is to be found.

Example 20-4

Two equal positive point charges of magnitude +5 nC are on the x axis. One is at the origin and the other at $x = 8$ cm as shown in Figure 20-4. Find the potential at (*a*) point P_1 on the x axis at $x = 4$ cm and (*b*) point P_2 on the y axis at $y = 6$ cm.

(*a*) Point P_1 is 4 cm from each charge. Using Equation 20-10 with $q_1 = q_2 = 5$ nC and $r_{10} = r_{20} = 0.04$ m, we find for the potential at that point

$$V = \sum_i \frac{kq_i}{r_{i0}} = \frac{kq_1}{r_{10}} + \frac{kq_2}{r_{20}}$$

$$= 2 \times \frac{(8.99 \times 10^9\ \text{N}\cdot\text{m}^2/\text{C}^2)(5 \times 10^{-9}\ \text{C})}{0.04\ \text{m}} = 2250\ \text{V}$$

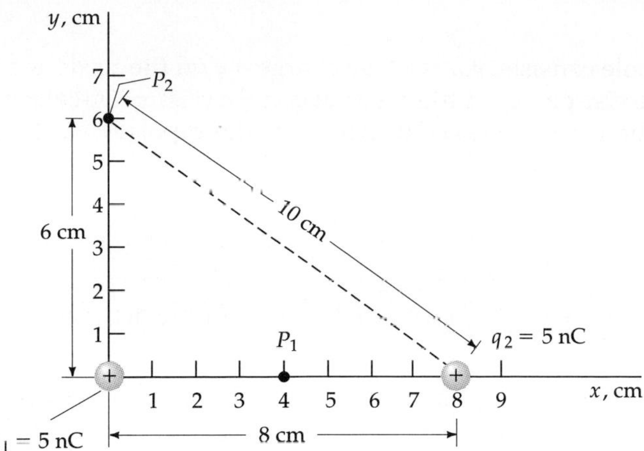

Figure 20-4 Two positive point charges on the x axis for Example 20-4. The potential is to be found at points P_1 and P_2.

Note that the electric field is zero at this point midway between the charges but the potential is not. To bring in a test charge from a long distance away to this point requires work because the electric field is zero only at the final position.

(*b*) Point P_2 is 6 cm from one charge and 10 cm from the other. The potential at this point is therefore

$$V = \frac{(8.99 \times 10^9\ \text{N}\cdot\text{m}^2/\text{C}^2)(5 \times 10^{-9}\ \text{C})}{0.06\ \text{m}} + \frac{(8.99 \times 10^9\ \text{N}\cdot\text{m}^2/\text{C}^2)(5 \times 10^{-9}\ \text{C})}{0.10\ \text{m}}$$

$$= 749\ \text{V} + 450\ \text{V} \approx 1200\ \text{V}$$

Example 20-5

A point charge q_1 is at the origin, and a second point charge q_2 is on the x axis at $x = a$ as shown in Figure 20-5. Find the potential everywhere on the x axis.

We need to divide the x axis into three regions: to the right of both charges $x > a$, between the two charges $0 < x < a$, and to the left of both charges $x < 0$. At a point P_1 to the right of both charges, the distance to q_1 is x and the distance to q_2 is $x - a$. The potential in this region is therefore

$$V = \frac{kq_1}{x} + \frac{kq_2}{x - a} \qquad x > a$$

At a point P_2 between the charges, the distance to q_1 is again x, but the distance to q_2 is $a - x$. The potential on the axis between the charges is therefore

$$V = \frac{kq_1}{x} + \frac{kq_2}{a - x} \qquad 0 < x < a$$

At a point P_3 to the left of both charges, the distance to q_1 is $-x$ (because x is negative) and the distance to q_2 is $a - x$. The potential on the axis to the left of the charges is therefore

$$V = \frac{kq_1}{-x} + \frac{kq_2}{a - x} \qquad x < 0$$

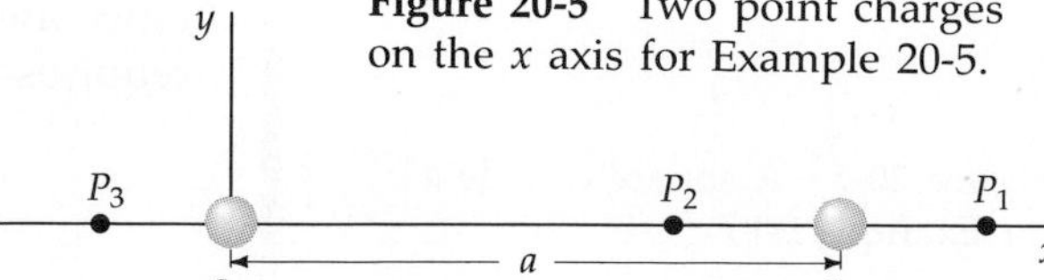

Figure 20-5 Two point charges on the x axis for Example 20-5.

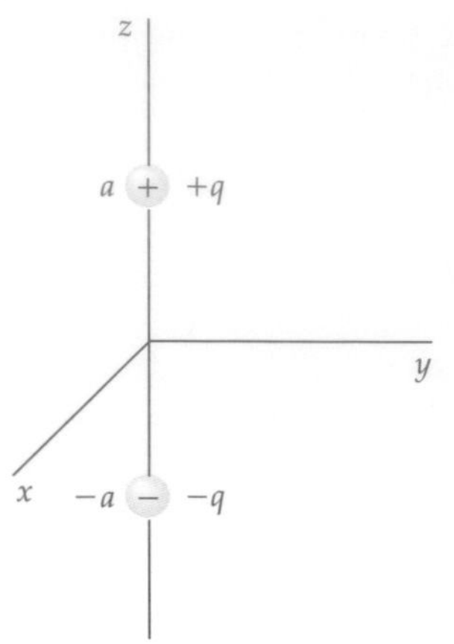

Figure 20-6 Electric dipole on the z axis for Example 20-6.

Example 20-6

An electric dipole consists of a positive charge $+q$ on the z axis at $z = +a$ and a negative charge $-q$ on the z axis at $z = -a$ (Figure 20-6). Find the potential on the z axis a great distance from the dipole.

From Equation 20-10, we have

$$V = \frac{kq}{z-a} + \frac{k(-q)}{z+a} = \frac{2kqa}{z^2 - a^2}$$

For $z \gg a$, we can neglect a^2 compared with z^2 in the denominator. We have then

$$V \approx \frac{2kqa}{z^2} = \frac{kp}{z^2} \qquad z \gg a \tag{20-11}$$

where $p = 2qa$ is the magnitude of the dipole moment.

20-3 Electrostatic Potential Energy

If we have a point charge q_1, the potential at a distance r_{12} away is given by

$$V = \frac{kq_1}{r_{12}}$$

The work required to bring up a second point charge q_2 from an infinite distance away to a distance r_{12} is $W_2 = q_2V = kq_1q_2/r_{12}$. To bring up a third charge, work must be done against the electric field produced by both q_1 and q_2. The work required to bring up a third charge q_3 to a distance r_{13} from q_1 and r_{23} from q_2 is $W_3 = kq_3q_1/r_{13} + kq_3q_2/r_{23}$. The total work required to assemble the three charges is therefore

$$W = \frac{kq_1q_2}{r_{12}} + \frac{kq_1q_3}{r_{13}} + \frac{kq_2q_3}{r_{23}}$$

This work is the **electrostatic potential energy** of the system of three point charges. It is independent of the order in which the charges are brought to their final positions. In general,

The electrostatic potential energy of a system of point charges is the work needed to bring the charges from an infinite separation to their final positions.

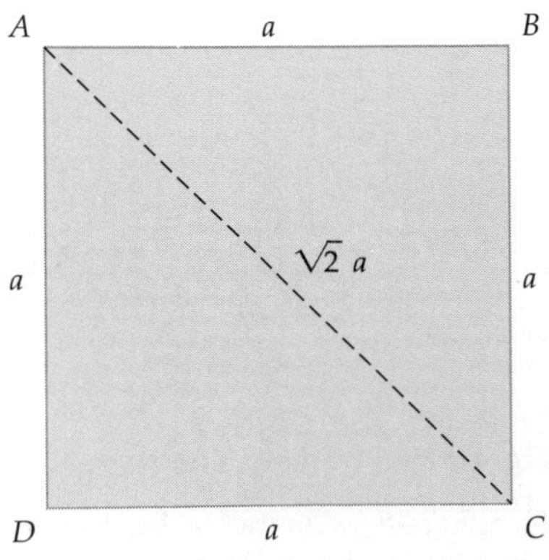

Figure 20-7 A square of side a for Example 20-7.

Example 20-7

Points A, B, C, and D are at the corners of a square of side a as shown in Figure 20-7. How much work is required to place a positive charge q at each corner of the square?

No work is needed to place the first charge at point A because the potential is zero there when the other three charges are an infinite distance away. To bring up a second charge to point B a distance a away requires work

$$W_2 = \frac{kqq}{a}$$

Point C is a distance a from point B and $\sqrt{2}a$ from point A. The potential at point C due to the charges at A and B is

$$V_C = \frac{kq}{a} + \frac{kq}{\sqrt{2}a}$$

The work required to bring a third charge q to point C is thus

$$W_3 = qV_C = \frac{kqq}{a} + \frac{kqq}{\sqrt{2}a}$$

Finally, the work needed to bring the fourth charge to point D when the other three charges are already in place is

$$W_4 = \frac{kqq}{a} + \frac{kqq}{a} + \frac{kqq}{\sqrt{2}a}$$

The total work required to assemble the four charges is

$$W_{\text{total}} = W_2 + W_3 + W_4 = \frac{4kqq}{a} + \frac{2kqq}{\sqrt{2}a} = \frac{(8 + 2\sqrt{2})kqq}{2a}$$

This work is the total electrostatic energy of the charge distribution.

20-4 Calculation of Electric Potential for Continuous Charge Distributions

In this section, we will calculate V for some important continuous charge distributions. The potential due to a continuous distribution of charge can be calculated from Equation 20-3 if the electric field is known, or it can be calculated from Equation 20-10 by choosing an element of charge dq, which we treat as a point charge, and changing the sum in Equation 20-10 to an integral:

$$V = \int \frac{k\,dq}{r} \qquad \text{20-12}$$

Potential due to a continuous charge distribution

We illustrate the application of Equation 20-12 by using it to calculate the electric potential on the axis of a uniformly charged ring and on the axis of a uniformly charged disk.

V on the Axis of a Charged Ring

Consider a uniformly charged ring of radius a and charge Q shown in Figure 20-8. In the figure, an element of charge dq is shown. The distance from this charge element to the field point P on the axis of the ring is $r = \sqrt{x^2 + a^2}$. Since this distance is the same for all elements of charge on the ring, we can remove this term from the integral in Equation 20-12. The potential at point P due to the ring is thus

$$V = \int \frac{k\,dq}{r} = \int \frac{k\,dq}{\sqrt{x^2 + a^2}}$$

$$= \frac{k}{\sqrt{x^2 + a^2}} \int dq = \frac{kQ}{\sqrt{x^2 + a^2}} \qquad \text{20-13}$$

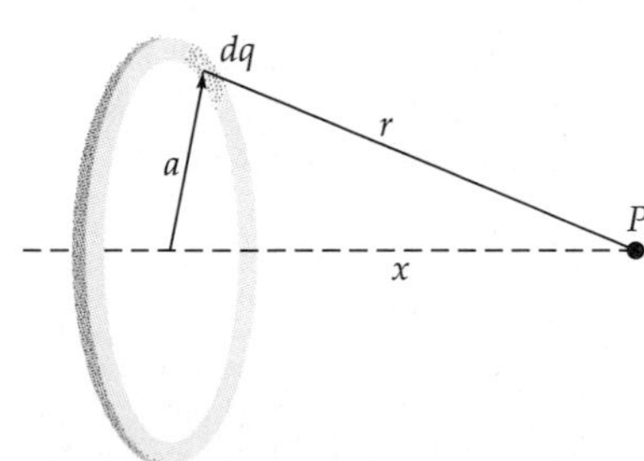

Figure 20-8 Geometry for the calculation of the electric potential at a point on the axis of a uniformly charged ring of radius a.

Example 20-8

A ring of radius 4 cm carries a uniform charge of 8 nC. A small particle of mass $m = 6$ mg $= 6 \times 10^{-6}$ kg and charge $q_0 = 5$ nC is placed at $x = 3$ cm and released. Find the speed of the charge when it is a great distance from the ring.

The potential energy of the charge q_0 at $x = 3$ cm is

$$U = q_0 V = \frac{kQq_0}{\sqrt{x^2 + a^2}}$$

$$= \frac{(8.99 \times 10^9 \text{ N·m}^2/\text{C}^2)(8 \times 10^{-9} \text{ C})(5 \times 10^{-9} \text{ C})}{\sqrt{(0.03 \text{ m})^2 + (0.04 \text{ m})^2}}$$

$$= 7.19 \times 10^{-6} \text{ J}$$

As the particle moves along the x axis away from the ring, its potential energy decreases and its kinetic energy increases. When the particle is very far from the ring, its potential energy is zero and its kinetic energy is 7.19×10^{-6} J. Its speed is then given by

$$\tfrac{1}{2}mv^2 = 7.19 \times 10^{-6} \text{ J}$$

$$v = \sqrt{\frac{2(7.19 \times 10^{-6} \text{ J})}{6 \times 10^{-6} \text{ kg}}} = 1.55 \text{ m/s}$$

Exercise

What is the potential energy of the particle in Example 20-8 when it is at $x = 9$ cm? (Answer: 3.65×10^{-6} J)

V on the Axis of a Uniformly Charged Disk

We will now use Equation 20-13 to calculate the potential on the axis of a uniformly charged disk. Let the disk have a radius R and carry a total charge Q. The surface charge density on the disk is then $\sigma = Q/\pi R^2$. We take the x axis to be the axis of the disk and treat the disk as a set of ring charges. Figure 20-9 shows a ring of radius a and width da. The area of this ring is $2\pi a\, da$, and its charge is $dq = \sigma\, dA = \sigma 2\pi a\, da$. The potential at some point P on the x axis due to this charged ring element is given by Equation 20-13:

$$dV = \frac{k\, dq}{(x^2 + a^2)^{1/2}} = \frac{k2\pi\sigma a\, da}{(x^2 + a^2)^{1/2}}$$

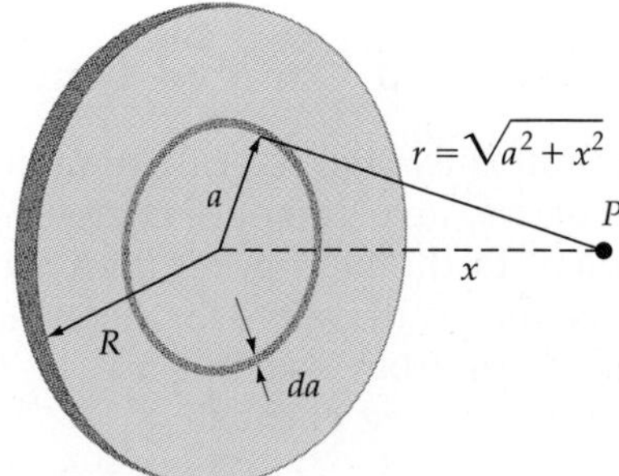

Figure 20-9 Geometry for the calculation of the electric potential at a point on the axis of a uniformly charged disk of radius R. The disk is divided into concentric rings of radius a and thickness da, each carrying charge $dq = \sigma\, dA = (Q/\pi R^2)\, 2\pi a\, da$.

The potential on the axis of the disk is found by integrating from $a = 0$ to $a = R$:

$$V = \int_0^R \frac{k2\pi\sigma a\, da}{(x^2 + a^2)^{1/2}} = k\sigma\pi \int_0^R (x^2 + a^2)^{-1/2}\, 2a\, da$$

This integral is of the form $\int u^n\,du$, with $u = x^2 + a^2$ and $n = -\frac{1}{2}$. The integration thus gives

$$V = k\sigma\pi \frac{(x^2 + a^2)^{+1/2}}{\frac{1}{2}}\Bigg|_{a=0}^{a=R} = 2\pi k\sigma[(x^2 + R^2)^{1/2} - x] \qquad \text{20-14}$$

V near an Infinite Plane of Charge: Continuity of V

If we let R become very large, our disk approaches an infinite plane. As R becomes infinite the potential function (Equation 20-14) becomes infinite. We cannot apply Equation 20-12 to charge distributions that extend to infinity, such as an infinite plane of charge or an infinite line charge, because the potential cannot be chosen to be zero at infinity. For such cases, we first find the electric field **E** (by direct integration, or from Gauss's law) and then calculate the potential from its definition, Equation 20-3. For an infinite plane of charge of density σ in the yz plane, the electric field for positive x is given by

$$\mathbf{E} = \frac{\sigma}{2\epsilon_0}\mathbf{i}$$

The potential is calculated from its definition (Equation 20-3). If the potential at the yz plane where $x = 0$ is V_0, the potential at some arbitrary positive value of x is

$$V(x) - V_0 = -\int_0^x \mathbf{E}\cdot d\boldsymbol{\ell} = -\int_0^x \frac{\sigma}{2\epsilon_0}\mathbf{i}\cdot dx\,\mathbf{i} = -\frac{\sigma}{2\epsilon_0}\int_0^x dx = -\frac{\sigma}{2\epsilon_0}x$$

or

$$V(x) = V_0 - \frac{\sigma}{2\epsilon_0}x \qquad x > 0 \qquad \text{20-15}a$$

For positive x, the potential has its maximum value V_0 at $x = 0$ and decreases linearly with distance from the plane. Since it does not approach any limiting value as x approaches infinity, we cannot choose the potential to be zero at $x = \infty$. We can, however, choose V to be zero at $x = 0$ or at any other point. For $x < 0$, the electric field is in the negative x direction and is given by

$$\mathbf{E} = -\frac{\sigma}{2\epsilon_0}\mathbf{i}$$

Repeating the calculation for the potential, using this function for the electric field, gives

$$V(x) = V_0 + \frac{\sigma}{2\epsilon_0}x \qquad x < 0 \qquad \text{20-15}b$$

Since x is negative in this equation, the potential has its maximum value of V_0 at $x = 0$ and again decreases linearly with distance from the plane. Figure 20-10 shows a plot of V versus x. Note that this function is continuous at $x = 0$, even though the electric field E_x is discontinuous there. In Chapter 19, we saw that the electric field is discontinuous by σ/ϵ_0 at a point where there is a surface charge density σ. The potential function, on the other hand, is continuous everywhere in space. We can see this from its definition. Consider two nearby points x_1 and x_2. If V_1 is the potential at x_1 and V_2 is that at x_2, the potential difference can be written

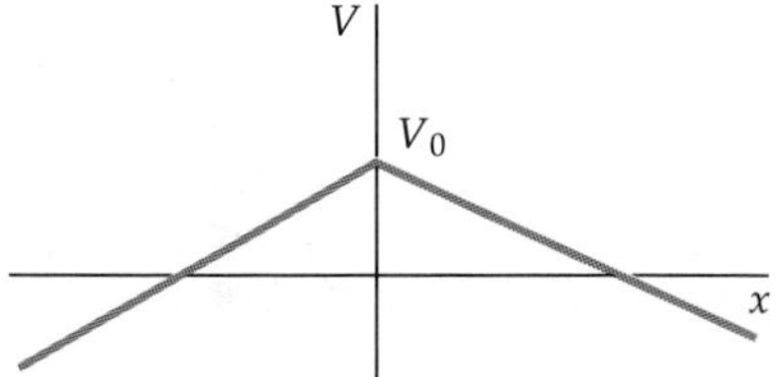

Figure 20-10 Plot of V versus x for an infinite plane of charge in the yz plane. The potential is continuous at $x = 0$ even though the electric field is not.

$$\Delta V = (E_x)_{\text{av}}\,\Delta x = (E_x)_{\text{av}}\,(x_2 - x_1)$$

where $(E_x)_{\text{av}}$ is the average value of the electric field between the points. As x_2 approaches x_1, the potential difference ΔV approaches zero as long as $(E_x)_{\text{av}}$ is not infinite. Physically, if a test charge is moved a distance Δx, the work done by the field approaches zero as Δx approaches zero, as long as the electric field is not infinite.

Example 20-9

An infinite plane of charge density σ is parallel to the yz plane at $x = -a$, and a point charge q is at the origin as shown in Figure 20-11. Find the potential at some point P a distance r from the point charge, for $x > -a$ (that is, to the right of the plane charge).

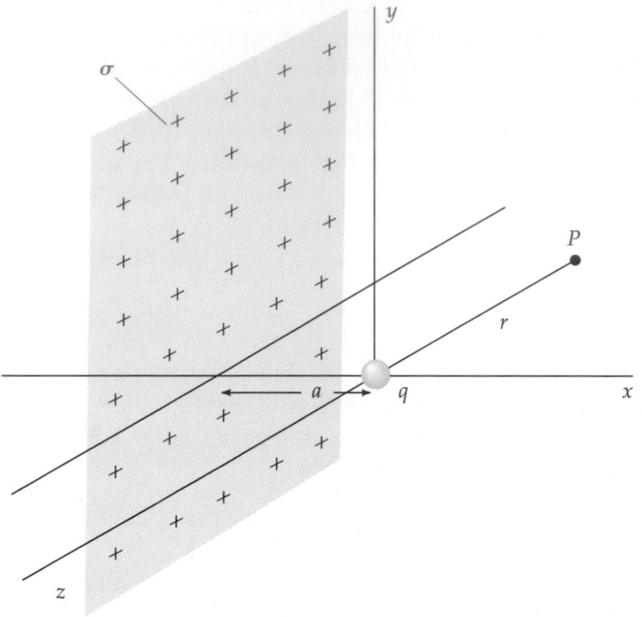

Figure 20-11 A point charge at the origin and an infinite plane of charge at $x = -a$ for Example 20-9.

The potential V_{plane} at a distance x' from the infinite plane due to the charge on the plane is given by Equation 20-15a with x' replacing x:

$$V_{\text{plane}} = A - \frac{\sigma}{2\epsilon_0}x'$$

where $x' = x + a$ and A is a constant that depends on the choice of zero potential. Because we cannot choose the potential to be zero at $r = \infty$, we use Equation 20-8 for the potential due to the point charge at the origin:

$$V_q = \frac{kq}{r} + B$$

where B is a constant that depends on the choice of zero potential. The potential due to both the infinite-plane charge and the point charge is

$$V = V_{\text{plane}} + V_q$$

$$= A - \frac{\sigma}{2\epsilon_0}x' + \frac{kq}{r} + B$$

$$= \frac{kq}{r} - \frac{\sigma}{2\epsilon_0}x' + C \qquad \text{20-16}$$

where we have combined the constants $A + B = C$. Let us choose the potential to be zero at the point at which the x axis intersects the infinite plane of charge. The coordinates of this point are $x = -a$, $y = 0$, $z = 0$. At this point $x' = 0$, and $r = a$. Equation 20-16 then gives

$$V = +\frac{kq}{a} + C = 0$$

$$C = -\frac{kq}{a}$$

which determines the constant C. At any general point, the potential is then given by

$$V = \frac{kq}{r} - \frac{kq}{a} - \frac{\sigma}{2\epsilon_0}\,x' = \frac{kq}{r} - \frac{kq}{a} - \frac{\sigma}{2\epsilon_0}\,(x + a)$$

In rectangular coordinates, $r = (x^2 + y^2 + z^2)^{1/2}$ and V is given by

$$V = \frac{kq}{(x^2 + y^2 + z^2)^{1/2}} - \frac{\sigma}{2\epsilon_0}\,(x + a) - \frac{kq}{a}$$

V Inside and Outside a Spherical Shell of Charge

We next find the potential due to a spherical shell of radius R with charge Q uniformly distributed on its surface. We are interested in the potential at all points inside and outside the shell. Since this shell is of finite extent, we could calculate the potential by direct integration of Equation 20-12, but this integration is somewhat difficult. Since the electric field for this charge distribution is easily obtained from Gauss's law, it is easiest to use Equation 20-3 to find the potential from the known electric field.*

Outside the spherical shell, the electric field is radial and is the same as if all the charge were at the origin:

$$\mathbf{E} = \frac{kQ}{r^2}\,\hat{\mathbf{r}}$$

The change in the potential for some displacement $d\boldsymbol{\ell} = dr\,\hat{\mathbf{r}}$ outside the shell is then

$$dV = -\mathbf{E}\cdot d\boldsymbol{\ell} = -\frac{kQ}{r^2}\,\hat{\mathbf{r}}\cdot dr\,\hat{\mathbf{r}} = -\frac{kQ}{r^2}\,dr$$

This is the same as Equation 20-7 for a point charge at the origin. Integrating, we obtain

$$V = \frac{kQ}{r} + V_0$$

where V_0 is the potential at $r = \infty$. Choosing the potential to be zero at $r = \infty$, we obtain

$$V = \frac{kQ}{r} \qquad r > R$$

Inside the spherical shell, the electric field is zero. The change in potential for any displacement inside the shell is therefore zero. Thus, the potential must be constant everywhere inside the shell. As r approaches R from outside the shell, the potential approaches kQ/R. Hence, the constant value of V inside must be kQ/R to make V continuous. Thus,

$$V = \begin{cases} \dfrac{kQ}{R} & r \le R \\[2ex] \dfrac{kQ}{r} & r \ge R \end{cases} \qquad \text{20-17}$$

Potential due to a spherical shell

*The calculation of the electric field $\mathbf{E}$ for a uniformly charged spherical shell by direct integration of Coulomb's law is even more difficult than the integration to find V, because V is a scalar whereas $\mathbf{E}$ is a vector. The direct calculation of $\mathbf{E}$ is similar to the calculation of the gravitational field due to a spherical shell as discussed in Section 10-7.

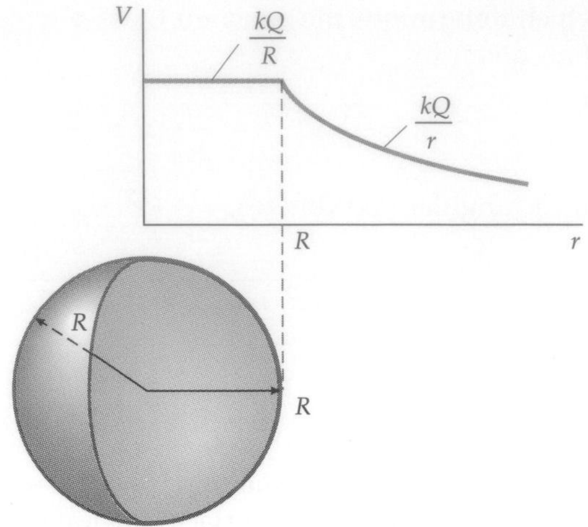

Figure 20-12 Electric potential of a uniformly charged spherical shell of radius R as a function of the distance r from the center of the shell. Inside the shell, the potential has the constant value kQ/R. Outside the shell the potential is the same as that due to a point charge at the center of the sphere.

This potential function is plotted in Figure 20-12.

A common mistake is to think that the potential must be zero inside a spherical shell because the electric field is zero. Actually, zero electric field merely implies that the potential does not change. Consider a spherical shell with a small hole so that we can move a test charge in and out of the shell. If we move the test charge from an infinite distance to the shell, the work per charge we must do is kQ/R. Inside the shell, there is no electric field, so it takes no work to move the test charge inside the shell. The total amount of work per charge it takes to bring the test charge from infinity to any point inside the shell is just the work per charge it takes to bring it up to the shell radius R, which is kQ/R. The potential is therefore kQ/R everywhere inside the shell.

Exercise

What is the potential of a spherical shell of radius 10 cm carrying a charge of 6 μC? (Answer: 5.39×10^5 V = 539 kV)

V near an Infinite Line Charge

We will now calculate the potential due to a uniform infinite line charge. Let the charge per unit length be λ. Since this charge distribution extends to infinity, we cannot use Equation 20-12 to find the potential. In Chapter 19, we found that the electric field produced by an infinite line charge points away from the line (if λ is positive) and is given by $E_r = 2k\lambda/r$. Equation 20-3 then gives for the change in potential

$$dV = -\mathbf{E}\cdot d\boldsymbol{\ell} = -E_r\,dr = -\frac{2k\lambda}{r}\,dr$$

Integrating, we obtain

$$V = V_0 - 2k\lambda \ln r \qquad \text{20-18}$$

For a positive line charge, the electric-field lines point away from the line, and the potential decreases with increasing distance from the line charge. At large values of r, the potential decreases without limit. The potential therefore cannot be chosen to be zero at $r = \infty$. (It cannot be chosen to be zero at $r = 0$ either, because $\ln r$ approaches $-\infty$ as r approaches zero.) Instead, we choose V to be zero at some distance $r = a$. Substituting $r = a$ into Equation 20-18 and setting $V = 0$, we obtain

$$V = 0 = V_0 - 2k\lambda \ln a$$

or

$$V_0 = 2k\lambda \ln a$$

Then Equation 20-18 is

$$V = 2k\lambda \ln a - 2k\lambda \ln r$$

or

$$V = -2k\lambda \ln \frac{r}{a} \qquad 20\text{-}19$$

Potential due to a line charge with $V = 0$ at $r = a$

Questions

5. In the calculation of V for a ring of charge, does it matter whether the charge Q is uniformly distributed around the ring? Would either V or E_x be different if it were not?
6. If the electric potential is constant throughout a region of space, what can you say about the electric field in that region?

20-5 Electric Field and Potential

The electric-field lines point in the direction of decreasing potential. If the potential is known, it can be used to calculate the electric field. Consider a small displacement $d\boldsymbol{\ell}$ in an arbitrary electric field **E**. The change in potential is

$$dV = -\mathbf{E}\cdot d\boldsymbol{\ell} = -E_\ell\, d\ell \qquad 20\text{-}20$$

where E_ℓ is the component of **E** parallel to the displacement. If we divide by $d\ell$, we have

$$E_\ell = -\frac{dV}{d\ell} \qquad 20\text{-}21$$

If the displacement $d\boldsymbol{\ell}$ is perpendicular to the electric field, the potential does not change. The greatest change in V occurs when the displacement $d\boldsymbol{\ell}$ is parallel or antiparallel to **E**. A vector that points in the direction of the greatest change in a scalar function and has a magnitude equal to the derivative of that function with respect to the distance in that direction is called the **gradient** of the function. The electric field **E** is the negative gradient of the potential V. The field lines point in the direction of the greatest decrease in the potential function. In vector notation, the gradient of V is written $\boldsymbol{\nabla}V$. Thus,

$$\mathbf{E} = -\boldsymbol{\nabla}V \qquad 20\text{-}22$$

Figure 20-13 shows the electric-field lines due to a point charge q at the origin. If we move a test charge perpendicular to these lines, no work is done and the potential does not change. A surface on which the electric potential is constant is called an **equipotential surface.** For the potential $V = kq/r$ produced by a point charge at the origin, the equipotential surfaces are spherical surfaces defined by $r =$ constant. Later in this chapter, we will see that the surface of any conductor in electrostatic equilibrium is an equipotential surface. The electric-field lines are always perpendicular to an equipotential surface. For a point charge at the origin, the electric-field lines are radial lines, and the equipotential surfaces are spheres. A displacement parallel to a radial electric field is written $d\boldsymbol{\ell} = dr\,\hat{\mathbf{r}}$. Equation 20-20 is then

$$dV = -\mathbf{E}\cdot d\boldsymbol{\ell} = -\mathbf{E}\cdot dr\,\hat{\mathbf{r}} = -E_r\, dr$$

and

$$E_r = -\frac{dV}{dr} \qquad 20\text{-}23$$

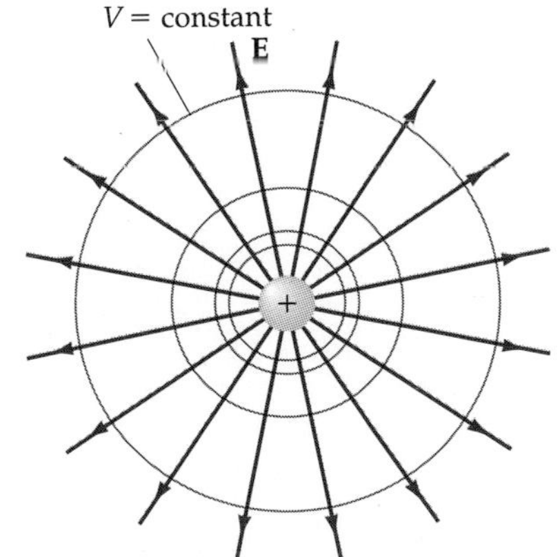

Figure 20-13 Equipotential surfaces and electric-field lines outside a point charge q. The field lines are radial and the equipotential surfaces are spherical. The electric-field lines are always perpendicular to an equipotential surface.

For any spherically symmetric charge distribution, the potential varies only with r, and the electric field is related to the potential by

$$\mathbf{E} = -\nabla V = -\frac{dV}{dr}\,\hat{\mathbf{r}} \qquad \text{20-24}$$

For a uniform electric field in the x direction, for example, that is produced by an infinite plane charge in the yz plane, the electric-field lines are parallel lines in the x direction, and the equipotential surfaces are planes parallel to the yz plane. Then, the potential function V can depend only on x. The displacement vector that is parallel to this field is given by

$$d\boldsymbol{\ell} = dx\,\mathbf{i} \qquad \text{20-25}$$

For this case, Equation 20-21 is

$$E_x = -\frac{dV}{dx}$$

and the electric field is

$$\mathbf{E} = -\frac{dV}{dx}\,\mathbf{i} \qquad \text{20-26}$$

In general, the potential function can depend on x, y, and z. The rectangular components of the electric field are related to the partial derivatives of the potential with respect to x, y, or z while the other variables are held constant. For example, the x component of the electric field is given by

$$E_x = -\frac{\partial V}{\partial x} \qquad \text{20-27}a$$

Similarly, the y and z components of the electric field are related to the potential by

$$E_y = -\frac{\partial V}{\partial y}$$

and

$$E_z = -\frac{\partial V}{\partial z} \qquad \text{20-27}b$$

Thus, Equation 20-22 in rectangular coordinates is

$$\mathbf{E} = -\nabla V = -\left(\frac{\partial V}{\partial x}\,\mathbf{i} + \frac{\partial V}{\partial y}\,\mathbf{j} + \frac{\partial V}{\partial z}\,\mathbf{k}\right) \qquad \text{20-28}$$

Example 20-10

Find the electric field for the electric potential function $V(x)$ given by $V(x) = 100\text{ V} - (25\text{ V/m})x$.

This potential function depends only on x. The electric field is found from Equation 20-26:

$$\mathbf{E} = -\frac{dV}{dx}\,\mathbf{i} = +(25\text{ V/m})\mathbf{i}$$

This electric field is uniform and in the x direction. Note that the constant 100 V in the expression for $V(x)$ has no effect on the electric field. The electric field does not depend on the choice of zero for the potential function.

Exercise

(*a*) At what point is $V = 0$ in Example 20-10? (*b*) Write the potential function corresponding to the same electric field with $V = 0$ at $x = 0$. [Answers: (*a*) $x = 4$ m, (*b*) $V = -(25\text{ V/m})\,x$]

Example 20-11

Find the electric field for the dipole charge distribution of Example 20-6.

In that example, we found the potential on the z axis at a great distance from the dipole to be

$$V = \frac{kp}{z^2}$$

where $p = 2qa$ is the magnitude of the dipole moment. The electric field at some point on the z axis is then given by

$$\mathbf{E} = -\frac{dV}{dz}\mathbf{k}$$

$$= -(-2)\frac{kp}{z^3}\mathbf{k} = \frac{2kp}{z^3}\mathbf{k}$$

which is the same as we found directly from Coulomb's law (Equation 18-10).

Example 20-12

Use the potential functions found in the previous section for the potential on the axis of a uniformly charged ring and of a uniformly charged disk to find the electric field on the axis of these charge distributions.

The potential on the axis of a uniformly charged ring of total charge Q is given by Equation 20-13:

$$V = \frac{kQ}{\sqrt{x^2 + a^2}} = kQ(x^2 + a^2)^{-1/2}$$

The electric field is then

$$\mathbf{E} = -\frac{dV}{dx}\mathbf{i} = -\left(-\frac{1}{2}\right)kQ(x^2 + a^2)^{-3/2}(2x)\mathbf{i}$$

$$= \frac{kQx}{(x^2 + a^2)^{3/2}}\mathbf{i}$$

This is the same as Equation 19-12, which we found directly from Coulomb's law.

The potential on the axis of a uniformly charged disk is given by Equation 20-14:

$$V = 2\pi k\sigma[(x^2 + R^2)^{1/2} - x]$$

Again, the electric field is found by taking the gradient of this expression:

$$\mathbf{E} = -\frac{dV}{dx}\mathbf{i} = 2\pi k\sigma\left(1 - \frac{x}{\sqrt{x^2 + R^2}}\right)\mathbf{i}$$

This is the same as Equation 19-13, which we obtained by direct calculation from Coulomb's law.

20-6 Equipotential Surfaces, Charge Sharing, and Dielectric Breakdown

We have seen that there is no electric field inside a conductor in static equilibrium. Thus, there is no force on a test charge, and no work is done on a test charge as it moves about inside a conductor. The electric potential is thus the same throughout the conductor, that is, the volume occupied by the conductor is an equipotential volume. A surface on which the potential is constant is called an equipotential surface. The surface of a conductor is an equipotential surface. If a test charge is given a displacement $d\boldsymbol{\ell}$ parallel to an equipotential surface, $dV = -\mathbf{E} \cdot d\boldsymbol{\ell} = 0$ so the electric field lines are perpendicular to an equipotential surface. Figures 20-14 and 20-15 show equipotential surfaces near a spherical conductor and near a nonspherical conductor. Note that the field lines are everywhere perpendicular to these surfaces. If we move a short distance $d\ell$ along a field line from one equipotential surface to another, the potential changes by $dV = -\mathbf{E} \cdot d\boldsymbol{\ell} = -E\, d\ell$. If E is large, equipotential surfaces with a fixed potential difference between them are closely spaced.

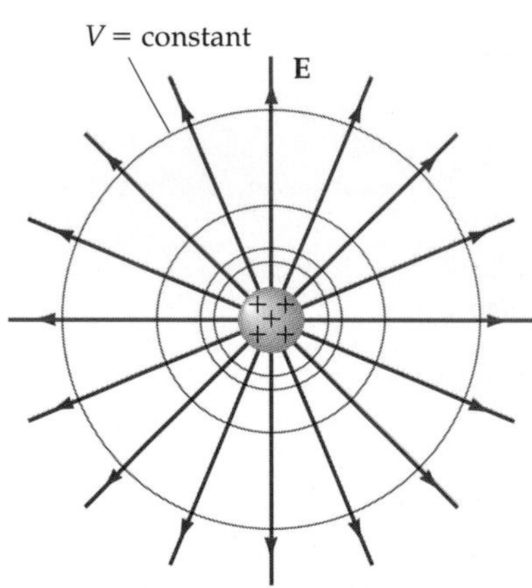

Figure 20-14 Equipotential surfaces and electric-field lines outside a uniformly charged spherical conductor. The equipotential surfaces are spherical and the field lines are radial and perpendicular to the equipotential surfaces.

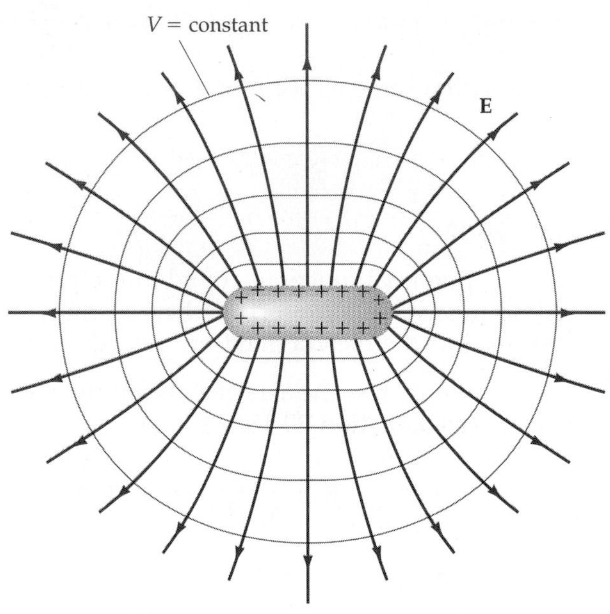

Figure 20-15 Equipotential surfaces and electric-field lines outside a nonspherical conductor. Electric-field lines are always perpendicular to equipotential surfaces.

Example 20-13

A hollow spherical conductor that is uncharged has inner radius a and outer radius b. A positive point charge $+q$ is in the cavity at the center of the sphere. Find the potential $V(r)$ everywhere, assuming that $V = 0$ at $r = \infty$.

As discussed in Chapter 19, the electric-field lines from the point charge end on the inner surface of the shell at $r = a$, where there is an induced charge $-q$ that is uniformly distributed on the inner surface. Since the conducting shell is uncharged, there is a positive charge $+q$ uniformly distributed on the outer surface at $r = b$. We thus have three charges: a point charge q at the origin, a spherical shell of total charge $-q$ and radius a, and a second spherical shell of total charge $+q$ and radius b. Outside the sphere, the electric field is the same as if the shell were not there and is given by $E_r = kq/r^2$. The electric potential outside the shell is then given by

$$V = \frac{kq}{r} \qquad r > b$$

Since the electric potential must be continuous everywhere, the potential at $r = b$ is $V = kq/b$. This is the potential everywhere inside the conducting material since the conductor is an equipotential volume. Thus,

$$V = \frac{kq}{b} \qquad a \le r \le b$$

Inside the cavity, the electric field is again $E_r = kq/r^2$. The potential for $r < a$ is therefore given by

$$V = \frac{kq}{r} + V_0$$

where V_0 is a constant. This constant is not determined by the condition $V = 0$ at $r = \infty$, because r does not go to infinity inside the cavity. Instead, we determine the constant V_0 by the condition that V must be continuous at $r = a$. Since $V = kq/b$ everywhere inside the conducting material, it must have this value at $r = a$. Thus, at $r = a$, we have

$$V = \frac{kq}{a} + V_0 = \frac{kq}{b}$$

which means that V_0 is given by

$$V_0 = \frac{kq}{b} - \frac{kq}{a}$$

Inside the cavity, the potential is thus given by

$$V = \frac{kq}{r} + \frac{kq}{b} - \frac{kq}{a} \qquad r < a$$

Figure 20-16 shows a plot of V versus r.

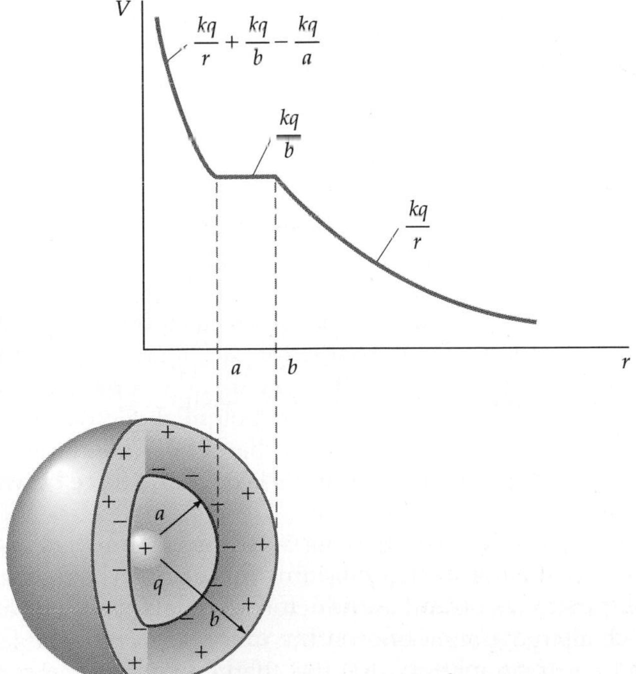

Figure 20-16 Plot of the electric potential V for a point charge at the center of an uncharged, hollow spherical conductor (Example 20-13) as a function of the distance r from the center of the cavity. Inside the conducting material, where $a \le r \le b$, the potential has the constant value kq/b. Outside the shell the potential is the same as that of a point charge.

In general, two conductors that are separated in space will not be at the same potential. The potential difference between the conductors depends on their geometrical shapes, their separation, and the net charge on each. When two conductors are brought into contact, the charge on the conductors distributes itself so that electrostatic equilibrium is established and the electric field is zero inside both conductors. While in contact, the two conductors may be considered to be a single conductor with a single equipotential surface. The transfer of charge from one conductor to another is called **charge sharing.**

Consider a spherical conductor of radius R carrying a charge $+Q$. The electric-field lines outside the conductor point radially outward, and the potential of the conductor relative to infinity is kQ/R. If we bring up a second, uncharged conductor, the potential and field lines will change. Electrons on the uncharged conductor will be attracted by the positive charge Q, leaving the near side of the uncharged conductor with a negative charge and the far side with a positive charge (Figure 20-17). This charge separation on the neutral conductor will affect the originally uniform charge distribution on the charged conductor. Although the detailed calculation of the charge distributions and potential for this case is quite complicated, we can see that some of the field lines leaving the positive conductor end on the negative charge on the near side of the neutral conductor, and an equal number of lines leave the far side of that conductor. Since electric-field lines point toward regions of lower potential, the positively charged conductor must be at a greater potential than the neutral conductor.

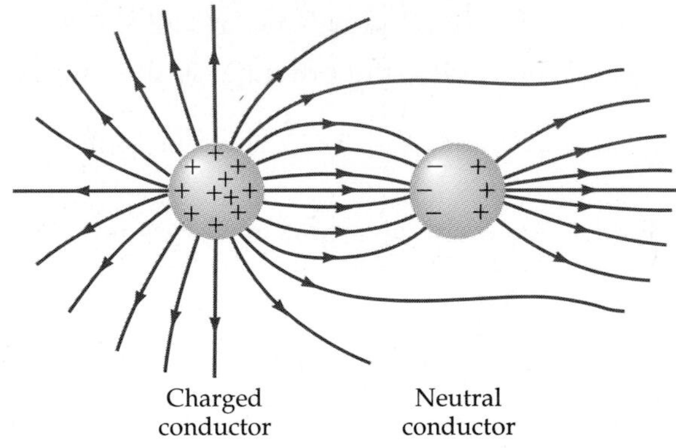

Figure 20-17 Electric-field lines for a charged spherical conductor near an uncharged spherical conductor. Some of the field lines leaving the charged conductor will end on the induced negative charge on the neutral conductor. Since electric-field lines point from regions of high potential to regions of low potential, the neutral conductor must be at a lower potential than the charged conductor.

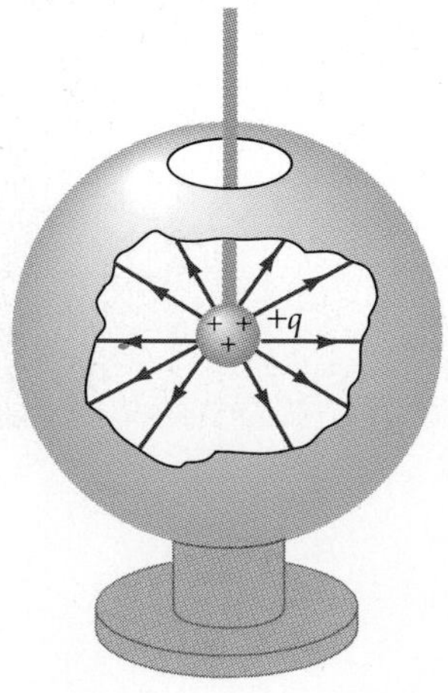

Figure 20-18 Small conductor carrying a positive charge inside a larger conductor.

If we put the two conductors in contact, positive charge will flow to the neutral conductor until both conductors are at the same potential. (Actually, negative electrons flow from the neutral conductor to the positive conductor, but it is slightly more convenient to think of this as a flow of positive charge in the opposite direction.) If the conductors are identical, they will share the original charge equally. If the conductors are now separated, each will carry charge $\frac{1}{2}Q$, and both will be at the same potential. Coulomb used this method of charge sharing to produce various charges of known ratios to some original charge in his experiment to find his law for the force between two small (point) charges.

In Figure 20-18, a small conductor carrying a positive charge q is inside the cavity of a larger conductor. In equilibrium, the electric field is zero inside the conducting material of both conductors. The electric-field lines that leave the positive charge q must end on the inner surface of the large conductor. This must occur no matter what the charge may be on the outside surface of the large conductor. Regardless of the charge on the large conductor, the small conductor in the cavity is at a greater potential because the electric-field lines go from this conductor to the larger conductor. If the conductors are now connected, say, with a fine conducting wire, *all* the charge originally on the smaller conductor will flow to the larger one. When the connection is broken, there is no charge on the small conductor in the cavity, and there are no field lines anywhere within the outer surface of the large conductor. The positive charge transferred from the smaller conductor resides completely on the outside surface of the larger conductor. If we put more positive charge on the small conductor in the cavity and again connect the conductors with a fine wire, all of the charge on the inner conductor will again flow to the outer conductor. This procedure can be repeated indefinitely. This method is used to produce large potentials in the Van de Graaff generator, in which the charge is brought to the inner surface of a larger spherical conductor by a continuous charged belt (see Figure 20-19). Work must be done by the motor driving the belt to bring the charge to the outer sphere, which is at a high potential. The greater the net charge on the outer conductor, the greater its potential.

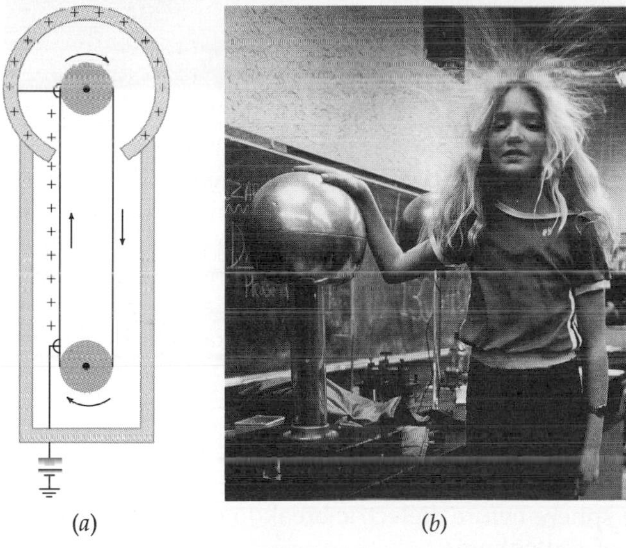

Figure 20-19 (*a*) Schematic diagram of a Van de Graaff generator. Charge leaks off the pointed conductor near the bottom onto the belt. Near the top, the charge leaks off of the belt onto the pointed conductors attached to the large spherical conductor. (*b*) This girl has been charged to a very high potential through contact with a demonstration Van de Graaff generator while standing on an insulating block. Her hair has acquired sufficient charge to show electrostatic repulsion. Some care must be taken to acquire the charge gradually and to avoid rapid discharge to prevent a painful shock. (*c*) These large demonstration Van de Graaff generators in the Boston science museum are discharging to the grounded wire cage housing the operator.

The maximum potential obtainable in this way is limited only by the fact that air molecules become ionized in very high electric fields and the air becomes a conductor. This phenomenon, which is called **dielectric breakdown,** occurs in air at electric field strengths of $E_{\max} \approx 3 \times 10^6$ V/m = 3 MV/m. The electric field strength for which dielectric breakdown occurs in a material is called the **dielectric strength** of that material. The dielectric strength of air is thus about 3 MV/m. The discharge through the conducting

(a)

(b)

(*a*) An electrical tree produced by arc discharge in a piece of plastic. The plastic is charged by an electron beam that penetrates about 0.5 cm. After the beam is turned off, the block is given a tap with a metal punch. The electrons suddenly shoot out leaving a pattern of tracks in the plastic. (*b*) A similar electrical tree produced by lightning in air.

air resulting from dielectric breakdown is called **arc discharge.** The electric shocks you receive when you touch a metal door knob after walking across a rug on a dry day is a familiar example of arc discharge. (This occurs more often on dry days because moist air will conduct away some of the charge you acquire by walking on the rug before you accumulate enough charge to reach a high potential.) Lightning is another familiar example of arc discharge.

Example 20-14

A spherical conductor has a radius of 2 m. (*a*) What maximum charge can be placed on the sphere before dielectric breakdown? (*b*) What is the maximum potential of the sphere?

(*a*) The electric field just outside the conductor carrying a surface charge σ is

$$E = \frac{\sigma}{\epsilon_0}$$

Setting this equal to the maximum electric field in air, we obtain for σ_{max}

$$E_{\text{max}} = 3 \times 10^6 \text{ N/C} = \frac{\sigma_{\text{max}}}{\epsilon_0}$$

The maximum charge on the sphere is then

$$\begin{aligned} Q &= 4\pi R^2 \sigma_{\text{max}} \\ &= 4\pi R^2(\epsilon_0 E_{\text{max}}) = 4\pi(2 \text{ m})^2(8.85 \times 10^{-12} \text{ C}^2/\text{N}\cdot\text{m}^2)(3 \times 10^6 \text{ N/C}) \\ &= 1.33 \times 10^{-3} \text{ C} \end{aligned}$$

(*b*) The maximum potential of the sphere carrying this charge is

$$\begin{aligned} V_{\text{max}} &= \frac{kQ}{R} = \frac{(8.99 \times 10^9 \text{ N}\cdot\text{m}^2/\text{C}^2)(1.33 \times 10^{-3} \text{ C})}{2 \text{ m}} \\ &= 5.98 \times 10^6 \text{ V} \end{aligned}$$

When a charge is placed on a conductor of nonspherical shape, like that in Figure 20-20*b*, the surface of the conductor will be an equipotential surface, but the surface charge density (the charge per unit area) and the electric field just outside the conductor will vary from point to point. Near a point where the radius of curvature is small, such as point *A* in the figure, the surface charge density and electric field will be large; whereas near a point where the radius of curvature is large, such as point *B* in the figure, they will be small. We can understand this qualitatively by considering the ends of the conductor to be spheres of different radii. Let σ be the surface charge density. The potential of a sphere of radius r is

$$V = \frac{kq}{r} = \frac{1}{4\pi\epsilon_0}\frac{q}{r} \tag{20-29}$$

Since the area of a sphere is $4\pi r^2$, the charge on a sphere is related to the charge density by

$$q = 4\pi r^2 \sigma$$

Substituting this expression for q into Equation 20-29, we have

$$V = \frac{1}{4\pi\epsilon_0}\frac{4\pi r^2\sigma}{r} = \frac{r\sigma}{\epsilon_0}$$

Solving for σ, we obtain

$$\sigma = \frac{\epsilon_0 V}{r} \tag{20-30}$$

Since both "spheres" are at the same potential, the one with the smaller radius must have the greater surface charge density. And since the electric field at the surface of a conductor is proportional to the surface charge density σ, the electric field is greatest at points on the conductor where the radius of curvature is least.

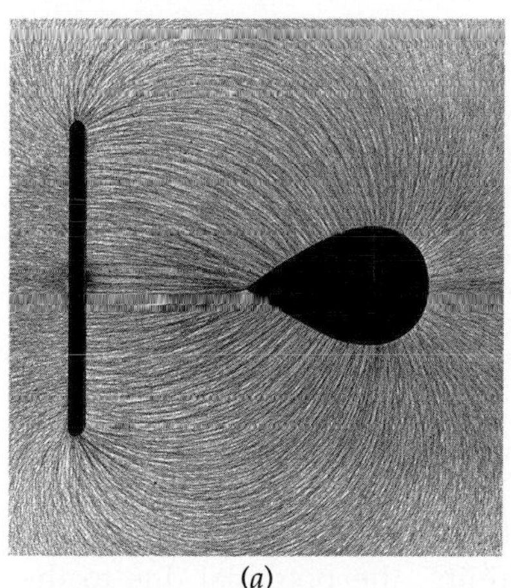

(*a*)

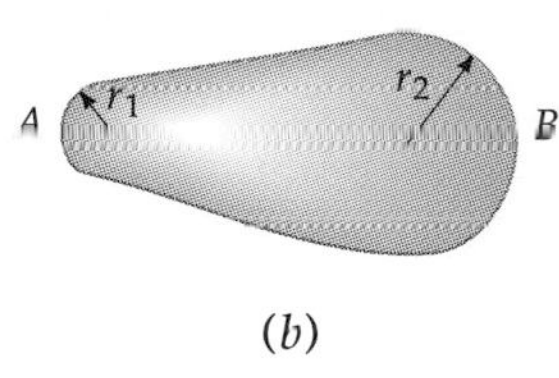

(*b*)

Figure 20-20 (*a*) Electric-field lines near a nonspherical conductor and plate carrying equal and opposite charges. The lines are shown by small bits of thread suspended in oil. The electric field is strongest near points of small radius of curvature, such as at the ends of the plate and at the pointed left side of the conductor. (*b*) A nonspherical conductor. If a charge is placed on such a conductor, it will produce an electric field that is stronger near point *A*, where the radius of curvature is small, than near point *B*, where the radius of curvature is large.

For an arbitrarily shaped conductor, the potential at which dielectric breakdown occurs depends on the smallest radius of curvature of any part of the conductor. If the conductor has sharp points of very small radius of curvature, dielectric breakdown will occur at relatively low potentials. In the Van de Graaff generator, the charge is transferred onto the belt by sharp-edged conductors near the bottom of the belt. The charge is removed from the belt by sharp-edged conductors near the top of the belt (Figure 20-19). Lightning rods at the top of a tall building draw the charge off a nearby cloud before the potential of the cloud can build up to a very large value.

Question

7. When you touch a friend after walking across a rug on a dry day, you typically draw a spark of about 2 mm. Estimate the potential difference between you and your friend before the spark.

Summary

1. Potential difference $V_b - V_a$ is defined as the negative of the work per unit charge done by the electric field when a test charge moves from point a to point b:

$$\Delta V = V_b - V_a = -\int_a^b \mathbf{E} \cdot d\boldsymbol{\ell}$$

For infinitesimal displacements this is written

$$dV = -\mathbf{E} \cdot d\boldsymbol{\ell}$$

Since only differences in electric potential are important, we can choose the potential to be zero at any convenient point. The potential at any point is the potential energy of a charge divided by the charge:

$$V = \frac{U}{q_0}$$

The SI unit of potential and potential difference is the volt (V):

$$1 \text{ V} = 1 \text{ J/C}$$

In terms of this unit, the unit for the electric field can be expressed:

$$1 \text{ N/C} = 1 \text{ V/m}$$

2. A convenient unit of energy in atomic and nuclear physics is the electron volt (eV), which is the potential energy of a particle of charge e at a point where the potential is 1 volt. The electron volt is related to the joule by

$$1 \text{ eV} = 1.6 \times 10^{-19} \text{ J}$$

3. The electric potential at a distance r from a point charge q at the origin is given by

$$V = \frac{kq}{r} + V_0$$

where V_0 is the potential at an infinite distance from the charge. When the potential is chosen to be zero at infinity, the potential due to the point charge is

$$V = \frac{kq}{r}$$

For a system of point charges, the potential is given by

$$V = \sum_i \frac{kq_i}{r_{i0}}$$

where the sum is taken over all the charges and r_{i0} is the distance from the ith charge to the point P at which the potential is to be found.

4. The electrostatic potential energy of a system of point charges is the work needed to bring the charges from an infinite separation to their final positions.

5. For a continuous distribution of charge, the potential is found by integration over the charge distribution:

$$V = \int \frac{k\,dq}{r}$$

This expression can be used only if the charge distribution is contained in a finite volume so that the potential can be chosen to be zero at infinity.

6. The electric field points in the direction of the greatest decrease in the potential. The component of **E** in the direction of a displacement $d\ell$ is related to the potential by

$$E_\ell = -\frac{dV}{d\ell}$$

A vector that points in the direction of the greatest change in a scalar function and has a magnitude equal to the derivative of that function with respect to the distance in that direction is called the gradient of the function. The electric field **E** is the negative gradient of the potential V. In vector notation, the gradient of V is written ∇V. Thus,

$$\mathbf{E} = -\nabla V$$

For any spherically symmetric charge distribution, the potential varies only with r, and the electric field is related to the potential by

$$\mathbf{E} = -\nabla V = -\frac{dV}{dr}\hat{\mathbf{r}}$$

In rectangular coordinates, the electric field is related to the potential by

$$\mathbf{E} = -\nabla V = -\left(\frac{\partial V}{\partial x}\mathbf{i} + \frac{\partial V}{\partial y}\mathbf{j} + \frac{\partial V}{\partial z}\mathbf{k}\right)$$

7. On a conductor of arbitrary shape, the surface charge density σ is greatest at points where the radius of curvature is smallest.

8. The amount of charge that can be placed on a conductor is limited by the fact that air molecules become ionized in very high electric fields, and the air becomes a conductor—a phenomenon called dielectric breakdown, which occurs in air at electric field strengths of $E_{max} \approx 3 \times 10^6$ V/m = 3 MV/m. The electric field strength for which dielectric breakdown occurs in a material is called the dielectric strength of that material. The resulting discharge through the conducting air is called arc discharge.

Electrostatics and Xerography

Richard Zallen
Virginia Polytechnic Institute and State University

There are many important and beneficial technological applications that could be included in a discussion of uses of electrostatic phenomena. For example, a powerful air-pollution preventer is the electrostatic precipitator, which years ago made life livable near cement mills and ore-processing plants and which is currently credited with extracting better than 99 percent of the ash and dust from the gases about to issue from chimneys of coal-burning power plants. The basic idea of this very effective antipollution technique is shown in Figure 1. The outer wall of a vertical metal duct is grounded, while a wire running down the center of the duct is kept at a very large negative voltage. In this concentric geometry, a very nonuniform electric field is set up, with lines of force directed radially inward toward the negative wire electrode. Close to the wire the field attains enormous values, large enough to

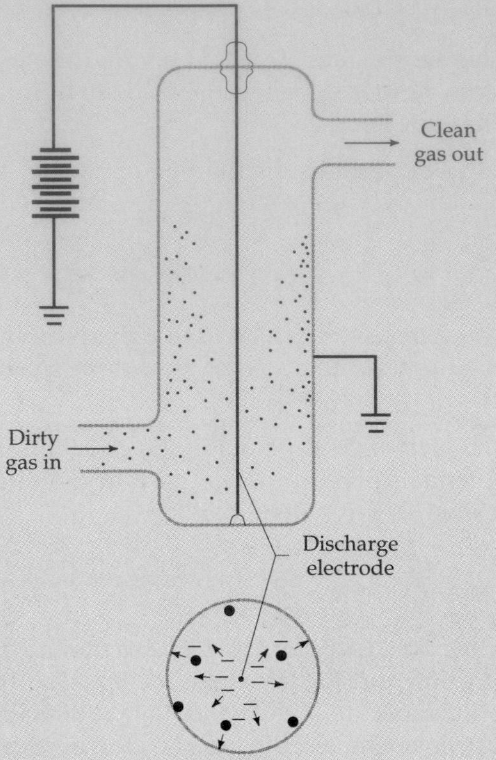

Figure 1 Schematic diagram of the use of a corona discharge in an electrostatic precipitator.

Richard Zallen received his education at Madison H.S. in Brooklyn, Rensselaer (B.S.), and Harvard (Ph.D.). He was elected a fellow of the American Physical Society in 1976. Before joining Virginia Tech in 1983, he worked for seventeen years at the Xerox Research Laboratories in Rochester, New York, which of course is where he learned about "Electrostatics and Xerography." He is married and has two children.

Throughout his career as a physicist, Professor Zallen has been involved with experimental studies of the interaction of light with solids such as semiconductors, molecular crystals, and amorphous solids. Most recently, he has been working on sol-gel systems and ion-bombarded semiconductors.

Professor Zallen is probably best known for his book *The Physics of Amorphous Solids*, Wiley, New York, 1983.

produce an electrical breakdown of air, and the normally placid mixture of neutral gas molecules is replaced by a turmoil of free electrons and positive ions. The electrons from this corona discharge are driven outward from the wire by the electric field. Most of them quickly become attached to oxygen molecules to produce negative O_2^- ions, which are also accelerated outward. As this stream of ions passes across the hot waste gas rising in the duct, small particles carried by the gas become charged by capturing ions and are pulled by the field to the outer wall. If the noxious particles are solid, they are periodically shaken down off the duct into a hopper; if they are liquid, the residue simply runs down the wall and is collected below.

Beside electrostatic precipitation, other technological examples include electrocoating with spray paints and the electrostatic separation of granular mixtures used for the removal of rock particles from minerals, garlic seeds from wheat, even rodent excreta from rice. However, the application that is the main focus of this essay is xerography, the most widely used form of electrostatic imaging, or electrophotography. This is the most familiar use of

Electrostatic precipitators are housed in the gray box-like structures at the base of these smokestacks.

electrostatics in terms of the number of people who have occasion to use plain paper copying machines in offices, libraries, and schools, and it also provides a fine example of a process utilizing a sequence of distinct electrostatic events.

The xerographic process was invented in 1937 by Chester Carlson. The term xerography, literally "dry writing," was actually adopted a bit later to emphasize the distinction from wet chemical processes. Carlson's innovative concept did not find early acceptance, and a practical realization of his idea became available only after a small company (in a famous entrepreneurial success story) risked its future in its intensive efforts to develop the process.

Four of the main steps involved in xerography are illustrated in Figure 2. In the interest of clarity the process has been oversimplified, and several subtleties (as well as gaps in our understanding) have been suppressed. Electrostatic imaging takes place on a large thin plate of a photoconducting material supported by a grounded metal backing. A photoconductor is a solid that is a good insulator *in the dark* but becomes capable of conducting electric current when exposed to light. In the dark, a uniform electrostatic charge is laid down on the surface of the photoconductor. This charging step (Figure 2*a*) is accomplished by means of a positive corona discharge surrounding a fine wire held at about +5000 V. This corona (a miniature version of, and opposite in sign to, the intense precipitator corona of Figure 1) is passed over the photoconductor surface, spraying positive ions onto it and charging it to a potential of the order of +1000 V. Since charge is free to flow within the grounded metal backing, an equal and opposite induced charge develops at the metal-photoconductor interface. In the dark the photoconductor contains no mobile charge, and the large potential difference persists across this dielectric layer, which is only 0.005 cm thick.

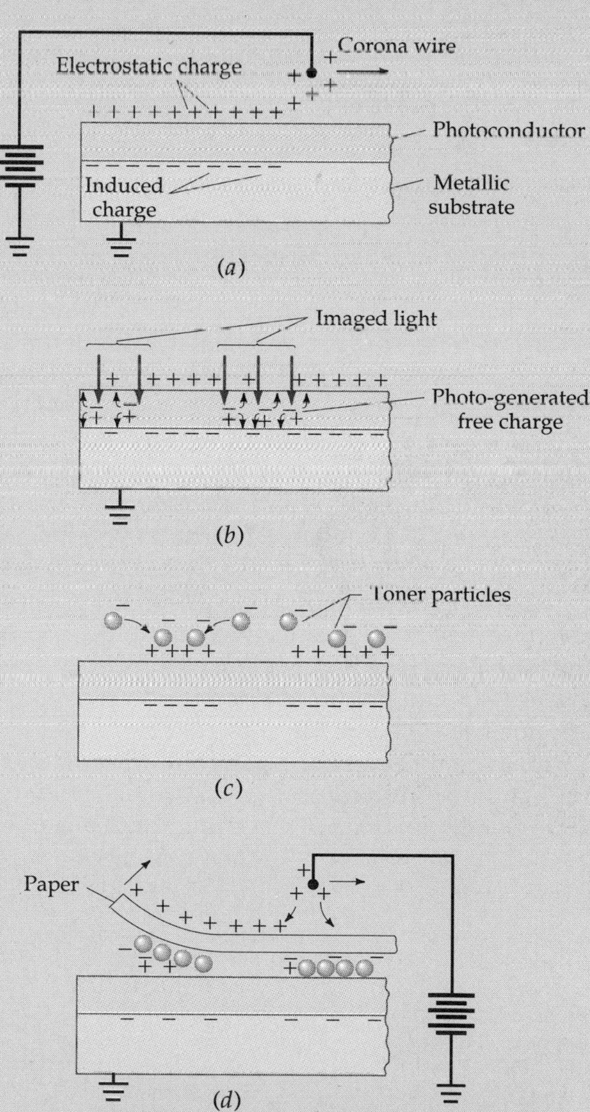

Figure 2 Steps in the xerographic process: (*a*) charging, (*b*) exposure, (*c*) development, and (*d*) transfer.

Continued

The photoconductor plate is next exposed to light in the form of an image reflected from the document being copied. What happens now is indicated in Figure 2*b*. Where light strikes the photoconductor, light quanta (photons) are absorbed, and pairs of mobile charges are created. Each photogenerated pair consists of a negative charge (an electron) and a positive charge (a hole; crudely, a missing electron). Photogeneration of this free charge depends not only on the photoconductor used and on the wavelength and intensity of the incident light but also on the electric field present. This large field (1000 V/0.005 cm = 2×10^5 V/cm = 2×10^7 V/m) helps to pull apart the mutually attracting electron-hole pairs so that they are free to move separately. The electrons then move under the influence of the field to the surface, where they neutralize positive charges, while the holes move to the photoconductor-substrate interface and neutralize negative charges there. Where intense light strikes the photoconductor, the charging step is totally undone; where weak light strikes it, the charge is partially reduced; and where no light strikes it, the original electrostatic charge remains on the surface. The critical task of converting an optical image into an electrostatic image, which is now recorded on the plate, has been completed. This latent image consists of an electrostatic potential distribution that replicates the light and dark pattern of the original document.

To develop the electrostatic image, fine negatively charged pigmented particles are brought into contact with the plate. These *toner particles* are attracted to positively charged surface regions, as shown in Figure 2*c*, and a visible image appears. The toner is then transferred (Figure 2*d*) to a sheet of paper that has been positively charged in order to attract the particles. Brief heating of the paper fuses the toner to it and produces a permanent photocopy ready for use.

Finally, to prepare the photoconductor plate for a repetition of the process, any toner particles remaining on its surface are mechanically cleaned off, and the residual electrostatic image is erased, that is, discharged, by flooding with light. The photoconductor is now ready for a new cycle, starting with the charging step. In high-speed duplicators, the photoconductor layer is often in the form of a moving continuous drum or belt, around the perimeter of which are located stations for performing the various functions of Figure 2. The speed of xerographic printing technology is presently on the order of a few copies per second.*

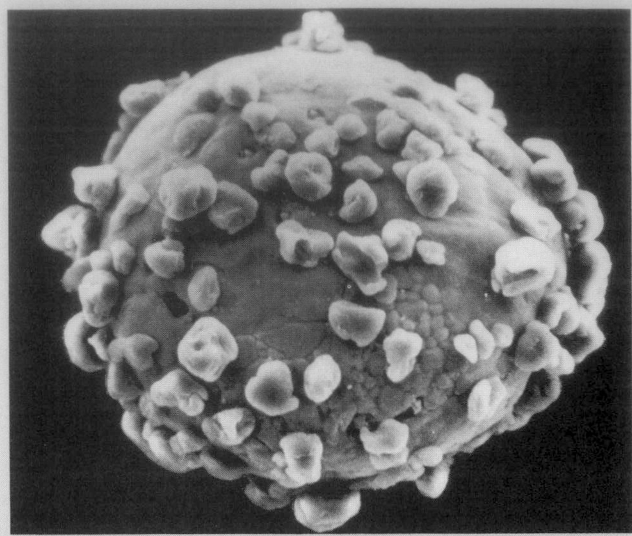

Toner particles, electrostatically attracted to a larger carrier particle.

*For further information on electrostatics in xerography, consult J. H. Dessauer and H. E. Clark (eds.), *Xerography and Related Processes*, Focal Press, New York, 1965, and R. M. Schaffert, *Electrophotography*, rev. ed., Focal Press, New York, 1973. Other applications of electrostatics are discussed in A. D. Moore, *Scientific American*, March 1972.

Suggestions for Further Reading

Moore, A. D.: "Electrostatics," *Scientific American*, March 1972, p. 46.

This article describes some modern uses of electrostatics, including precipitation of airborne industrial wastes, separation of granular solids such as minerals, efficient paint-spraying, and xerographic copying.

Rose, Peter H., and Andrew B. Wittkower: "Tandem Van de Graaff Accelerators," *Scientific American*, August 1970, p. 24.

These machines accelerate charged particles between terminals maintained at potential differences of millions of volts.

Review

A. Objectives: After studying this chapter, you should:

1. Be able to give a definition of electric potential and discuss its relation to the electric field.

2. Be able to calculate the potential difference between two points, given the electric field in the region.

3. Be able to sketch equipotential surfaces, given a pattern of electric-field lines.

4. Be able to calculate the electric potential for various charge distributions.

5. Be able to calculate the electric field from the electric-potential function.

6. Be able to discuss the phenomena of charge sharing and dielectric breakdown.

B. Define, explain, or otherwise identify:

Potential difference
Electric potential
Volt
Voltage
Electron volt
Electrostatic potential energy
Gradient
Equipotential surface
Charge sharing
Dielectric breakdown
Dielectric strength
Arc discharge

C. True or false: If the statement is true, explain why it is true. If it is false, give a counterexample.

1. If the electric field is zero in some region of space, the electric potential must also be zero in that region.

2. If the electric potential is zero in some region of space, the electric field must also be zero in that region.

3. If the electric potential is zero at a point, the electric field must also be zero at that point.

4. Electric-field lines always point toward regions of lower potential.

5. The value of the electric potential can be chosen to be zero at any convenient point.

6. In electrostatics, the surface of a conductor is an equipotential surface.

7. Dielectric breakdown occurs in air when the potential is 3×10^6 V.

Problems

Level I

20-1 Electric Potential and Potential Difference

1. A uniform electric field of 2 kN/C is in the x direction. A point charge $Q = 3\ \mu$C is released from rest at the origin. (*a*) What is the kinetic energy of the charge when it is at $x = 4$ m? (*b*) What is the change in the potential energy of the charge from $x = 0$ to $x = 4$ m? (*c*) What is the potential difference $V(4\text{ m}) - V(0)$? Find the potential $V(x)$ if $V(x)$ is chosen to be (*d*) zero at $x = 0$, (*e*) 4 kV at $x = 0$, and (*f*) zero at $x = 1$ m.

2. An infinite plane of surface charge density $\sigma = +2.5\ \mu\text{C/m}^2$ is in the yz plane. (*a*) What is the magnitude of the electric field in newtons per coulomb? In volts per meter? What is the direction of **E** for positive values of x? (*b*) What is the potential difference $V_b - V_a$ when point b is at $x = 20$ cm and point a is at $x = 50$ cm? (*c*) How much work is required by an outside agent to move a test charge $q_0 = +1.5$ nC from point a to point b?

3. A uniform electric field is in the negative x direction. Points a and b are on the x axis, a at $x = 2$ m and b at $x = 6$ m. (*a*) Is the potential difference $V_b - V_a$ positive or negative? (*b*) If the magnitude of $V_b - V_a$ is 10^5 V, what is the magnitude E of the electric field?

4. Two parallel conducting plates carry equal and opposite surface charge densities such that the electric field between them is uniform. The difference in potential between the plates is 500 V, and they are separated by 10 cm. An electron is released from rest at the negative plate. (*a*) What is the magnitude of the electric field between the plates? Is the positive or negative plate at the higher potential? (*b*) Find the work done by the electric field on the electron as the electron moves from the negative plate to the positive plate. Express your answers in both electron volts and joules. (*c*) What is the change in potential energy of the electron when it moves from the negative plate to the positive plate? What is its kinetic energy when it reaches the positive plate?

5. An electric field is given by $\mathbf{E} = ax\,\mathbf{i}$, where $\mathbf{E}$ is in newtons per coulomb, x is in meters, and a is a positive constant. (*a*) What are the SI units of a? (*b*) How much work is done by this field on a positive point charge q_0 when the charge moves from the origin to some point x? (*c*) Find the potential function $V(x)$ such that $V = 0$ at $x = 0$.

20-2 Potential Due to a System of Point Charges

6. Four 2-μC point charges are at the corners of a square of side 4 m. Find the potential at the center of the square (relative to zero potential at infinity) if (*a*) all the charges are positive, (*b*) three of the charges are positive and one is negative, and (*c*) two are positive and two are negative.

7. Three point charges are on the x axis: q_1 at the origin, q_2 at $x = 3$ m, and q_3 at $x = 6$ m. Find the potential at the point $x = 0$, $y = 3$ m if (*a*) $q_1 = q_2 = q_3 = 2\ \mu$C, (*b*) $q_1 = q_2 = 2\ \mu$C and $q_3 = -2\ \mu$C, and (*c*) $q_1 = q_3 = 2\ \mu$C and $q_2 = -2\ \mu$C.

8. Points *A*, *B*, and *C* are at the corners of an equilateral triangle of side 3 m. Equal positive charges of 2 μC are at *A* and *B*. (*a*) What is the potential at point *C*? (*b*) How much work is required to bring a positive charge of 5 μC from infinity to point *C* if the other charges are held fixed? (*c*) Answer parts (*a*) and (*b*) if the charge at *B* is replaced by a charge of $-2\ \mu$C.

9. A sphere with radius 60 cm has its center at the origin. Equal charges of 3 μC are placed at 60° intervals along the equator of the sphere. (*a*) What is the electric potential at the origin? (*b*) What is the electric potential at the north pole?

20-3 Electrostatic Potential Energy

10. A positive charge of magnitude 2 μC is at the origin. (*a*) What is the electric potential V at a point 4 m from the origin relative to $V = 0$ at infinity? (*b*) How much work must be done by an outside agent to bring a 3-μC charge from infinity to $r = 4$ m, assuming that the 2-μC charge is held fixed at the origin? (*c*) How much work must be done by an outside agent to bring the 2-μC charge from infinity to the origin if the 3-μC charge is first placed at $r = 4$ m and is then held fixed?

11. Find the electrostatic potential energy for the charge distribution described in (*a*) Problem 6*a*, (*b*) Problem 6*b*, (*c*) Problem 6*c* with like charges at opposite corners, and (*d*) Problem 6*c* with unlike charges at opposite corners.

12. Find the electrostatic potential energy for the charge distributions described in Problem 7.

13. Point charges q_1, q_2, and q_3 are at the corners of an equilateral triangle of side 2.5 m. Find the electrostatic potential energy of this charge distribution if (*a*) $q_1 = q_2 = q_3 = 4.2\ \mu$C, (*b*) $q_1 = q_2 = 4.2\ \mu$C and $q_3 = -4.2\ \mu$C, (*c*) $q_1 = q_2 = -4.2\ \mu$C and $q_3 = +4.2\ \mu$C.

20-4 Calculation of Electric Potential for Continuous Charge Distributions

14. (*a*) Sketch $V(x)$ versus x for the uniformly charged ring in the yz plane given by Equation 20-13. (*b*) At what point is $V(x)$ a maximum? (*c*) What is E_x at this point?

15. A charge of $q = +10^{-8}$ C is uniformly distributed on a spherical shell of radius 12 cm. (*a*) What is the magnitude of the electric field just outside and just inside the shell? (*b*) What is the magnitude of the electric potential just outside and just inside the shell? (*c*) What is the electric potential at the center of the shell? What is the electric field at that point?

16. A disk of radius 6.25 cm carries a uniform surface charge density $\sigma = 7.5$ nC/m^2. Find the potential on the axis of the disk at a distance (*a*) 0.5 cm, (*b*) 3.0 cm, and (*c*) 6.25 cm from the disk.

17. An infinite line charge of linear charge density $\lambda = 1.5\ \mu$C/m lies on the z axis. Find the potential at distances of (*a*) 2.0 m, (*b*) 4.0 m, and (*c*) 12 m from the line assuming that $V = 0$ at 2.5 m.

20-5 Electric Field and Potential

18. Two positive charges $+q$ are on the y axis at $y = +a$ and $y = -a$. (*a*) Find the potential V for any point on the x axis. (*b*) Use your result in part (*a*) to find the electric field at any point on the x axis.

19. A point charge $q = 3.00\ \mu$C is at the origin. (*a*) Find the potential V on the x axis at $x = 3.00$ m and at $x = 3.01$ m. (*b*) Does the potential increase or decrease as x increases? Compute $-\Delta V/\Delta x$, where ΔV is the change in potential from $x = 3.00$ m to $x = 3.01$ m and $\Delta x = 0.01$ m. (*c*) Find the electric field at $x = 3.00$ m, and compare its magnitude with $-\Delta V/\Delta x$ found in part (*b*). (*d*) Find the potential (to three significant figures) at the point $x = 3.00$ m, $y = 0.01$ m, and compare your result with the potential on the x axis at $x = 3.00$ m. Discuss the significance of this result.

20. A charge of $+3.00\ \mu$C is at the origin, and a charge of $-3.00\ \mu$C is on the x axis at $x = 6.00$ m. (*a*) Find the potential on the x axis at $x = 3.00$ m. (*b*) Find the electric field on the x axis at $x = 3.00$ m. (*c*) Find the potential on the x axis at $x = 3.01$ m, and compute $-\Delta V/\Delta x$, where ΔV is the change in potential from $x = 3.00$ m to $x = 3.01$ m and $\Delta x = 0.01$ m. Compare your result with your answer to part (*b*).

21. In the following, V is in volts and x is in meters. Find E_x when (*a*) $V(x) = 2000 + 3000x$; (*b*) $V(x) = 4000 + 3000x$; (*c*) $V(x) = 2000 - 3000x$; and (*d*) $V(x) = -2000$, independent of x.

22. The electric potential in some region of space is given by $V(x) = C_1 + C_2x^2$, where V is in volts, x is in meters, and C_1 and C_2 are positive constants. Find the electric field $\mathbf{E}$ in this region. In what direction is $\mathbf{E}$?

23. An infinite plane of charge has surface charge density 3.5 μC/m^2. How far apart are the equipotential surfaces whose potentials differ by 100 V?

24. A point charge $q = +\frac{1}{9} \times 10^{-8}$ C is at the origin. Taking the potential to be zero at $r = \infty$, locate the equipotential surfaces at 20-V intervals from 20 to 100 V, and sketch them to scale. Are these surfaces equally spaced?

20-6 Equipotential Surfaces, Charge Sharing, and Dielectric Breakdown

25. (*a*) Find the maximum net charge that can be placed on a spherical conductor of radius 16 cm before dielectric breakdown of the air occurs. (*b*) What is the potential of the sphere when it carries this maximum charge?

26. Sketch the electric-field lines and the equipotential surfaces both near to and far from the conductor shown in Figure 20-20*b*, assuming that the conductor carries some charge *Q*.

27. Find the greatest surface charge density σ_{max} that can exist on a conductor before dielectric breakdown of the air occurs.

28. If a conducting sphere is to be charged to a potential of 10,000 V, what is the smallest possible radius of the sphere such that the electric field will not exceed the dielectric strength of air?

Level II

29. A Van de Graaff generator has a potential difference of 1.25 MV between the belt and the outer shell. Charge is supplied at the rate of 200 μC/s. What minimum power is needed to drive the moving belt?

30. A uniformly charged sphere has a potential on its surface of 450 V. At a radial distance of 20 cm from this surface, the potential is 150 V. What is the radius of the sphere, and what is the charge of the sphere?

31. Four charges are at the corners of a square centered at the origin as follows: q at $(-a, +a)$; $2q$ at (a, a); $-3q$ at $(a, -a)$; and $6q$ at $(-a, -a)$. Find (*a*) the electric field at the origin and (*b*) the potential at the origin. (*c*) A fifth charge $+q$ is placed at the origin and released from rest. Find its speed when it is a great distance from the origin.

32. Two positive charges $+q$ are on the x axis at $x = +a$ and $x = -a$. (*a*) Find the potential $V(x)$ as a function of x for points on the x axis. (*b*) Sketch $V(x)$ versus x. (*c*) What is the significance of the minimum in your curve?

33. An electric field is given by $E_x = 2.0x^3$ kN/C. Find the potential difference between the points on the x axis at $x = 1$ m and $x = 2$ m.

34. Consider two infinite parallel planes of charge, one in the yz plane and the other at distance $x = a$. (*a*) Find the potential everywhere in space when $V = 0$ at $x = 0$ if the planes carry equal positive charge densities $+\sigma$. (*b*) Do the same if the charge densities are equal and opposite, with the charge in the yz plane positive.

35. In a Van de Graaff accelerator, protons are released from rest at a potential of 5 MV and travel through a vacuum to a region at zero potential. (*a*) Find the speed of the 5-MeV protons. (*b*) If the potential change occurs uniformly over a distance of 2.0 m, find the accelerating electric field.

36. Two equal positive charges are separated by a small distance. Sketch the electric-field lines and the equipotential surfaces for this system.

37. When uranium ^{235}U captures a neutron, it fissions (splits) into two nuclei (and emits several neutrons, which can cause other uranium nuclei to split). Assume that the fission products are equally charged nuclei with charge $+46e$ and that these nuclei are at rest just after fission and separated by twice their radius $2R \approx 1.3 \times 10^{-14}$ m. (*a*) Using $U = kq_1q_2/2R$, calculate the electrostatic potential energy of the fission fragments. This is approximately the energy released per fission. (*b*) About how many fissions per second are needed to produce 1 MW of power in a reactor?

38. A radioactive ^{210}Po nucleus emits an alpha particle of charge $+2e$ and energy 5.30 MeV. Assume that just after the alpha particle is formed and escapes from the nucleus, it is a distance R from the center of the daughter nucleus ^{206}Pb which has a charge $+82e$. Calculate R by setting the electrostatic potential energy of the two particles at this separation equal to 5.30 MeV.

39. An electron gun fires electrons at the screen of a television tube. The electrons start from rest and are accelerated through a potential difference of 30,000 V. What is the energy of the electrons when they hit the screen (*a*) in electron volts and (*b*) in joules. (*c*) What is the speed of impact of electrons with the screen of the picture tube?

40. Two large, parallel, nonconducting planes carry equal and opposite charge densities of magnitude σ. The planes have area A and are separated by a distance d. (*a*) Find the potential difference between the planes. (*b*) A conducting slab having thickness a and area A, the same area as the planes, is inserted between the original two planes. The slab carries no net charge. Find the potential difference between the original two planes and sketch the lines of **E** in the region between the original two planes.

41. Two concentric spherical shell conductors carry equal and opposite charges. The inner shell has radius a and charge $+q$; the outer shell has radius b and charge $-q$. Find the potential difference between the shells, $V_a - V_b$.

42. A conducting spherical shell of inner radius b and outer radius c is concentric with a small metal sphere of radius $a < b$. The metal sphere has a positive charge Q. The total charge on the conducting spherical shell is $-Q$. (*a*) What is the potential of the spherical shell? (*b*) What is the potential of the metal sphere?

43. Two very long, coaxial cylindrical shell conductors carry equal and opposite charges. The inner shell has radius a and charge $+q$; the other shell has radius b and charge $-q$. The length of each cylindrical shell is L. Find the potential difference between the shells.

44. The centers of two metal spheres of radius 10 cm are 50 cm apart on the x axis. The spheres are initially neutral, but a charge Q is transferred from one sphere to the other, creating a potential difference between the spheres of 100 V. A proton is released from rest at the surface of the positively charged sphere and travels to the negatively charged sphere. At what speed does it strike the negatively charged sphere?

45. A point charge of $+3e$ is at the origin and a second point of $-2e$ is on the x axis at $x = a$. (*a*) Sketch the potential function $V(x)$ versus x for all x. (*b*) At what point or points is $V(x)$ zero? (c) How much work is needed to bring a third charge $+e$ to the point $x = \frac{1}{2}a$ on the x axis?

46. Three equal charges lie in the xy plane. Two are on the y axis at $y = -a$ and $y = +a$, and the third is on the x axis at $x = a$. (*a*) What is the potential $V(x)$ due to these charges at a point on the x axis? (*b*) Find E_x along the x axis from the potential function $V(x)$. Evaluate your answers to (*a*) and (*b*) at the origin and at $x = \infty$ to see if they yield the expected results.

47. A charge q is at $x = 0$ and a charge $-3q$ is at $x = 1$ m. (*a*) Find $V(x)$ for a general point on the x axis. (*b*) Find the points on the x axis where the potential is zero. (c) What is the electric field at these points? (*d*) Sketch $V(x)$ versus x.

48. A rod of length L has a charge Q uniformly distributed along its length. The rod lies along the x axis with its center at the origin. (*a*) What is the electric potential as a function of position along the x axis for $x > L/2$? (*b*) Show that for $x \gg L/2$ your result reduces to that due to a point charge Q.

49. A charge of 2 nC is uniformly distributed around a ring of radius 10 cm that has its center at the origin and its axis along the x axis. A point charge of 1 nC is located at $x = 50$ cm. Find the work required to move the point charge to the origin in joules and electron volts.

50. A uniformly charged ring with a total charge of 100 μC and a radius of 0.1 m lies in the yz plane with its center at the origin. A meterstick has a point charge of 10 μC on the end marked 0 and a point charge of 20 μC on the end marked 100 cm. How much work does it take to bring the meterstick from a long distance away to a position along the x axis with the end marked 0 at $x =$ 0.2 m and the other end at $x = 1.2$ m.

51. Four equal charges Q are at the corners of a square of side L. The charges are released one at a time proceeding clockwise around the square. Each charge is allowed to reach its final speed a long distance from the square before the next charge is released. What is the final kinetic energy of (*a*) the first charge released, (*b*) the second charge released, (c) the third charge released, and (*d*) the fourth charge released?

52. Two identical uncharged metal spheres are connected by a wire as shown in Figure 20-21*a*. Two similar conducting spheres with equal and opposite charges are brought to the positions shown in Figure 20-21*b*. (*a*) Sketch the electric-field lines between spheres 1 and 3 and between spheres 2 and 4. (*b*) What can be said about the potentials V_1, V_2, V_3, and V_4 of the spheres? (*c*) If spheres 3 and 4 are connected by a wire, prove that the final charge on each must be zero.

Figure 20-21
Problem 52.

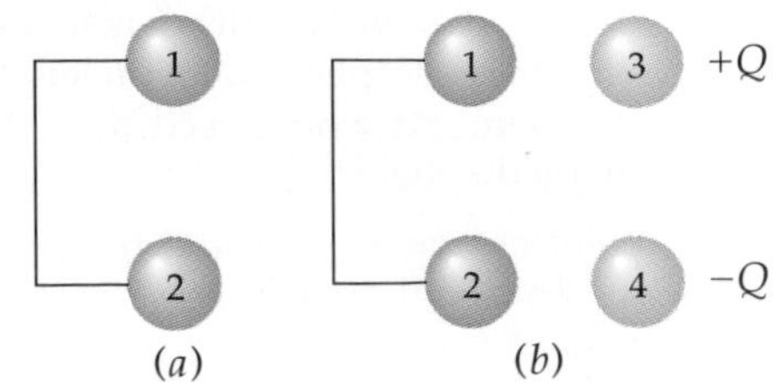

53. Three large conducting plates are parallel to one another with the outer plates connected by a wire. The inner plate is isolated and carries a charge density σ_t on the upper surface and σ_b on the lower surface, where $\sigma_t + \sigma_b = 12\ \mu\text{C/m}^2$. The inner plate is 1 mm from the top plate and 3 mm from the bottom plate. Find the surface charge densities σ_t and σ_b.

54. Show that when R is much smaller than x, the potential on the axis of a disk charge approaches kQ/x, where $Q = \sigma\pi R^2$ is the total charge on the disk. *Hint:* Write $(x^2 + R^2)^{1/2} = x(1 + R^2/x^2)^{1/2}$ and use the binomial expression.

55. A uniformly charged ring of radius a and charge Q lies in the yz plane with its axis along the x axis. A point charge Q' is placed on the x axis at $x = 2a$. (*a*) Find the potential at any point on the x axis due to the total charge $Q + Q'$. (*b*) Find the electric field for any point on the x axis.

Level III

56. A potential is given by

$$V(x, y, z) = \frac{kQ}{\sqrt{(x-a)^2 + y^2 + z^2}}$$

(*a*) Find the components E_x, E_y, and E_z of the electric field by differentiating this potential function. (*b*) What simple charge distribution might be responsible for this potential?

57. The electric potential in a region of space is given by

$$V = (2\ \text{V/m}^2)x^2 + (1\ \text{V/m}^2)yz$$

Find the electric field at the point $x = 2$ m, $y = 1$ m, $z =$ 2 m.

58. A point charge q_1 is at the origin and a second point charge q_2 is on the x axis at $x = a$ as in Example 20-5. (*a*) Calculate the electric field everywhere on the x axis from the potential function given in that example. (*b*) Find the potential at a general point on the y axis. (*c*) Use your result from (*b*) to calculate the y component of the electric field on the y axis. Compare your result with that obtained directly from Coulomb's law.

59. Consider a ball of charge of uniform volume charge density with a radius R and total charge Q. (This is a model of a proton.) The center of the ball is at the origin. Use the radial component of the electric field E_r found from Gauss's law to calculate the potential $V(r)$, assuming $V = 0$ at $r = \infty$ for (*a*) any point outside the charge $r \geq R$ and (*b*) any point inside the charge $r \leq R$. (Remember that V must be continuous at $r = R$.) (*c*) What is the potential at the origin? (*d*) Sketch V versus r.

60. In the Bohr model of the hydrogen atom, the electron moves in a circular orbit of radius r around the proton. (*a*) Find an expression for the kinetic energy of the electron as a function of r by setting the force on the electron (given by Coulomb's law) equal to ma, where a is the centripetal acceleration. Show that at any distance r the kinetic energy is half the magnitude of the potential energy. (*b*) Evaluate $\frac{1}{2}mv^2$, U, and the total energy $W = \frac{1}{2}mv^2 + U$ in electron volts for $r = 0.529 \times 10^{-10}$ m, the radius of the electron's orbit in hydrogen. The energy $|W|$ that must be supplied to the hydrogen atom to remove the electron is called the **ionization energy.**

61. (*a*) For the dipole of Example 20-6, show that the potential at a point off-axis a great distance r from the origin (Figure 20-22) is given approximately by

$$V = \frac{2kqa\cos\theta}{r^2} = \frac{kp\cos\theta}{r^2} = \frac{kpz}{r^3}$$

Hint: Show that $r_+^{-1} - r_-^{-1} \approx \Delta r/r^2$ where $\Delta r = r_+ - r_- \approx 2a\cos\theta$. (*b*) Find the x, y, and z components of the electric field at a point off-axis.

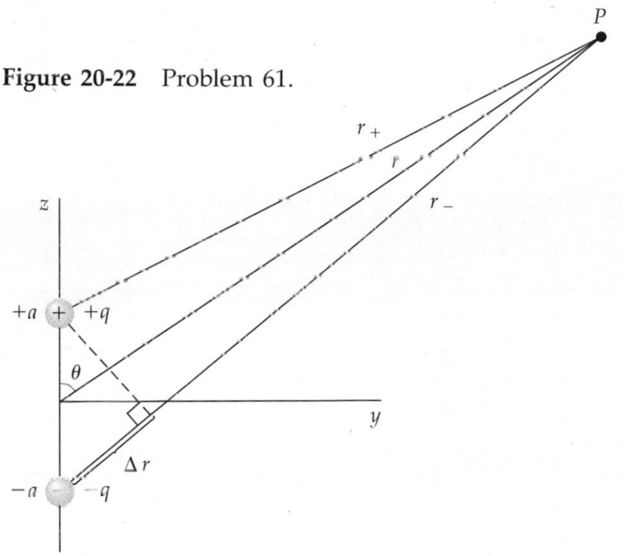

Figure 20-22 Problem 61.

62. Consider two concentric spherical metal shells of radii a and b where $b > a$. The outer shell has a charge Q, but the inner shell is grounded. This means that the inner shell is at zero potential and that electric-field lines leave the outer shell and go to infinity but other electric-field lines leave the outer shell and end on the inner shell. Find the charge on the inner shell.

63. Three concentric conducting spherical shells have radii a, b, and c such that $a < b < c$. Initially, the inner shell is uncharged, the middle shell has a positive charge Q, and the outer shell has a negative charge $-Q$. (*a*) Find the electric potential of the three shells. (*b*) If the inner and outer shells are now connected by a wire that is insulated as it passes through the middle shell, what is the electric potential of each of the three shells, and what is the final charge on each shell?

64. A nonconducting sphere of radius R has a volume charge density $\rho = \rho_0 r/R$, where ρ_0 is a constant. (*a*) Show that the total charge is $Q = \pi R^3 \rho_0$. (*b*) Show that the total charge inside a sphere of radius $r < R$ is $q = Qr^4/R^4$. (*c*) Use Gauss's law to find the electric field E_r everywhere. (*d*) Use $dV = -E_r\,dr$ to find the potential V everywhere, assuming that $V = 0$ at $r = \infty$. (Remember that V is continuous at $r = R$.)

65. A particle of mass m and charge Q is located on the x axis at $x = +a$, while a second particle of equal mass and charge $-Q$ is located on the x axis at $x = -a$. They are both released at $t = 0$. (*a*) Find the velocity of the positively charged particle as a function of its position x. (*b*) Integrate the velocity equation to find the time for the collision of the two charges.

Chapter 21

Capacitance, Dielectrics, and Electrostatic Energy

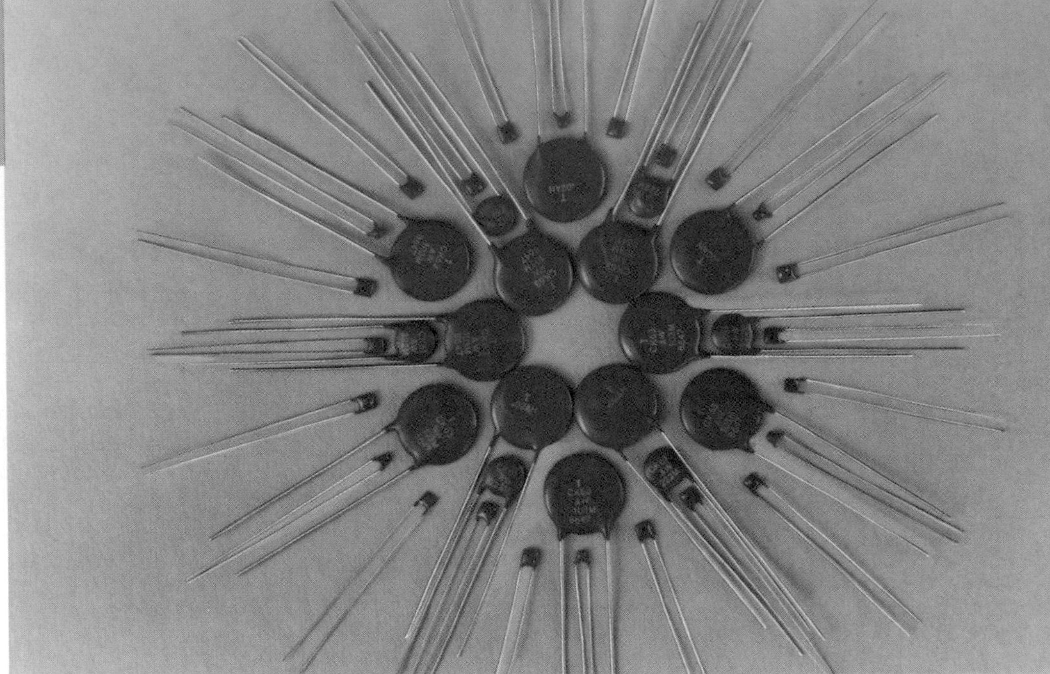

Single plate and disk ceramic capacitors for use in electronics circuits.

A **capacitor** is a useful device for storing charge and energy. It consists of two conductors, closely spaced but insulated from each other, that carry equal and opposite charges. Capacitors have many uses. The flash attachment for your camera uses a capacitor to store the energy needed to provide the sudden flash of light. Capacitors are also used to smooth out the ripples that arise when alternating current (the type of current that comes from a wall outlet) is converted into direct current for a power supply, such as that used to power your calculator or radio when the batteries are low.

The first capacitor for storing large electric charges was a jar with gold foil inside and out called a Leyden jar. It was invented in eighteenth-century Leyden (Holland) by experimenters who, while studying the effects of electric charges on people and animals, got the idea of trying to store a large amount of charge in a bottle of water. An experimenter held up a jar of water in one hand while charge was conducted to the water by a chain from a static electric generator. When he reached over to lift the chain out of the water with his other hand, he was knocked unconscious. After many experiments, it was discovered that the hand holding the jar could be replaced by metal foil on the surfaces of the jar. Benjamin Franklin realized that the device for storing charge did not have to be jar-shaped and used foil-covered window glass, called Franklin panes. With several of these connected in parallel, he stored a large charge and tried to kill a turkey with it. Instead, he knocked himself out. "I tried to kill a turkey but nearly succeeded in killing a goose," he wrote.

21-1 The Parallel-Plate Capacitor

A common capacitor is the **parallel-plate capacitor,** which utilizes two large parallel conducting plates. In practice, the plates may be thin metallic foils that are separated and insulated from one another by a thin sheet of paper. This "paper sandwich" is then rolled up to save space. When the plates are connected to a charging device, for example, a battery,* as in Figure 21-1, charge is transferred from one conductor to the other until the potential difference between the conductors due to their equal and opposite charges equals the potential difference between the battery terminals. The amount of charge on the plates depends on the potential difference and on the geometry of the capacitor, for example, on the area and separation of the plates in a parallel-plate capacitor. Let Q be the magnitude of the charge on either plate and V be the potential difference between the plates.** (When we speak of the charge on a capacitor, we mean the magnitude of the charge on either plate.) The ratio Q/V is called the **capacitance** C:

$$C = \frac{Q}{V} \tag{21-1}$$

Capacitance defined

Capacitance is a measure of the "capacity" to store charge for a given potential difference. The SI unit of capacitance is the coulomb per volt, which is called a **farad** (F) after the great English experimentalist Michael Faraday:

$$1\ \text{F} = 1\ \text{C/V} \tag{21-2}$$

Since the farad is a rather large unit, submultiples such as the microfarad ($1\ \mu\text{F} = 10^{-6}$ F) or the picofarad ($1\ \text{pF} = 10^{-12}$ F) are often used.

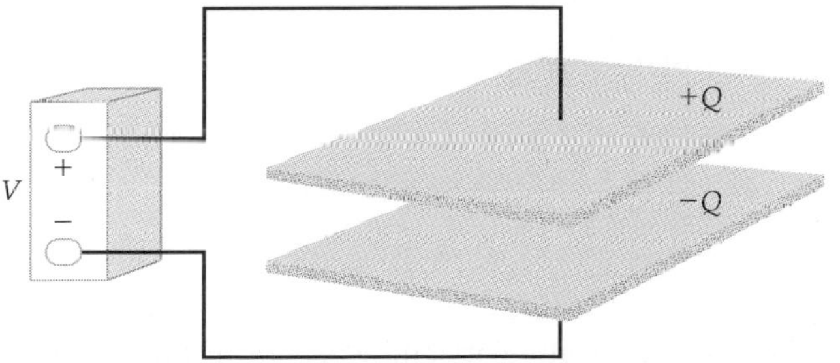

Figure 21-1 A capacitor consisting of two closely spaced parallel-plate conductors. When the conductors are connected to the terminals of a battery, the battery transfers charge from one conductor to the other until the potential difference between the conductors equals that between the battery terminals. The amount of charge transferred is proportional to the potential difference.

To calculate the capacitance of any capacitor, we first place some charge $+Q$ on one conductor and an equal and opposite charge $-Q$ on the other and then find the electric field between the conductors. We then integrate the field from one conductor to the other to find the potential difference V between the conductors. Since the potential difference is proportional to the charge Q, the capacitance $C = Q/V$ does not depend on either Q or V.

*We will discuss batteries more fully in Chapter 22. Here, all we need to know is that a battery is a device that stores and supplies electrical energy and maintains a constant potential difference V between its terminals.

**The use of V rather than ΔV for the potential difference between the plates is standard and simplifies many of the equations.

Let us consider a parallel-plate capacitor consisting of two plates of the same area A separated by a distance s, which is small compared to the length and width of the plates. We place a charge $+Q$ on one plate and $-Q$ on the other. Since the plates are close together, the electric field at any point between the plates (excluding points near the edges) is approximately equal to the field due to two equal and opposite infinite planes of charge. Each plate contributes a uniform field of magnitude $\sigma/2\epsilon_0$ (Equation 19-23), giving a total field $E = \sigma/\epsilon_0$, where $\sigma = Q/A$ is the charge per unit area on either plate. Since the field between the plates of our capacitor is uniform (Figure 21-2), the potential difference between the plates equals the field times the plate separation s:

$$V = Es = \frac{\sigma}{\epsilon_0}s = \frac{Qs}{\epsilon_0 A} \qquad \text{21-3}$$

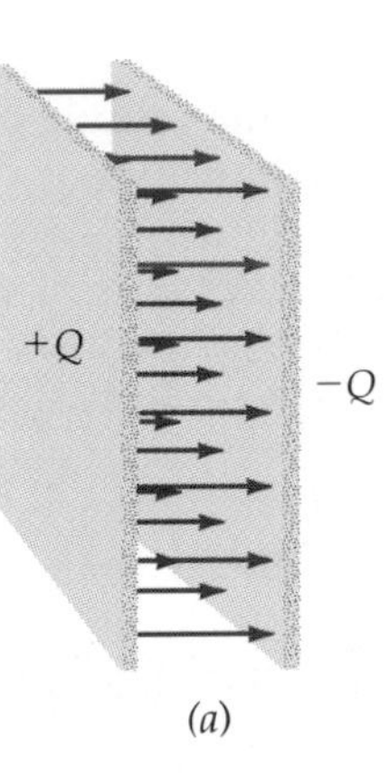

(a)

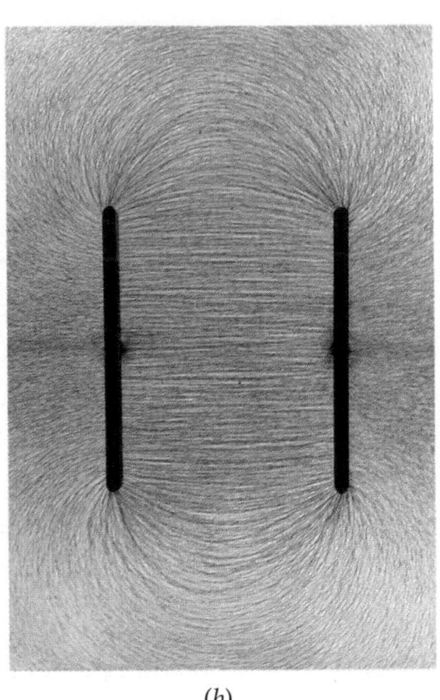

(b)

Figure 21-2 (*a*) The electric-field lines between the plates of a parallel-plate capacitor are equally spaced, indicating that the electric field there is uniform. (*b*) The electric-field lines between the plates of a parallel-plate capacitor shown by small bits of thread suspended in oil.

The capacitance of the parallel-plate capacitor is thus

Capacitance of a parallel-plate capacitor

$$C = \frac{Q}{V} = \frac{\epsilon_0 A}{s} \qquad \text{21-4}$$

Note that since V is proportional to Q, the capacitance does not depend on either the charge or the voltage of the capacitor but only on geometric factors. For a parallel-plate capacitor, the capacitance is proportional to the area of the plates and is inversely proportional to the separation distance. In general, capacitance depends on the size, shape, and geometrical arrangement of the conductors. Since capacitance is in farads and A/s is in meters, we can see from Equation 21-4 that the SI unit for the permittivity of free space ϵ_0 can also be written as a farad per meter:

$$\epsilon_0 = 8.85 \times 10^{-12} \text{ F/m} = 8.85 \text{ pF/m} \qquad \text{21-5}$$

A numerical calculation will illustrate how large a unit of capacitance the farad is.

Example 21-1

A parallel-plate capacitor has square plates of side 10 cm separated by 1 mm. (*a*) Calculate its capacitance. (*b*) If this capacitor is charged to 12 V, how much charge is transferred from one plate to another?

(*a*) Using Equation 21-4, we obtain for the capacitance

$$C = \frac{\epsilon_0 A}{s} = \frac{(8.85\ \text{pF/m})(0.1\ \text{m})^2}{0.001\ \text{m}} = 8.85 \times 10^{-11}\ \text{F}$$

$$= 88.5\ \text{pF}$$

(*b*) From the definition of capacitance (Equation 21-1), the charge transferred is

$$Q = CV = (88.5 \times 10^{-12}\ \text{F})(12\ \text{V}) = 1.06 \times 10^{-9}\ \text{C} = 1.06\ \text{nC}$$

This is the magnitude of the charge on either plate.

21-2 The Cylindrical Capacitor

A cylindrical capacitor consists of a small conducting cylinder or wire of radius a and a larger, concentric cylindrical conducting shell of radius b. A coaxial cable, such as that used for cable television, can be thought of as a cylindrical capacitor. The capacitance per unit length of a coaxial cable is important in determining the transmission characteristics of the cable. Let L be the length of the capacitor, which carries a charge $+Q$ on the inner conductor and a charge $-Q$ on the outer conductor. In Chapter 19, we found the electric field outside a long wire or cylinder of charge Q (Equation 19-24 or 19-26*a*) to be

$$E_r = \frac{1}{2\pi\epsilon_0}\frac{\lambda}{r} = \frac{Q}{2\pi\epsilon_0 L r} \qquad 21\text{-}6$$

where $\lambda = Q/L$ is the linear charge density. The field due to the charge $-Q$ on the outer cylindrical shell is zero inside the shell, as was shown using Gauss's law in Chapter 19 (Equation 19-25*a*).

The potential difference V between the conductors is found from Equation 20-3*b*. Let V_a be the potential of the inner conductor and V_b be that of the outer conductor. Then

$$V_b - V_a = -\int_a^b E_r\, dr = -\frac{Q}{2\pi\epsilon_0 L}\int_a^b \frac{dr}{r} = -\frac{Q}{2\pi\epsilon_0 L}\ln\frac{b}{a} \qquad 21\text{-}7$$

A coaxial cable is a long cylindrical capacitor with a solid wire for the inner conductor and a braided-wire shield for the outer conductor. The outer rubber coating has been pealed back here to show the conductors and the white plastic insulator that separates the conductors.

The potential is, of course, greater on the inner conductor, which carries the positive charge, since the electric-field lines point from this conductor to the outer conductor. The magnitude of this potential difference is

$$V = V_a - V_b = \frac{Q \ln (b/a)}{2\pi\epsilon_0 L}$$

and the capacitance is

$$C = \frac{Q}{V} = \frac{2\pi\epsilon_0 L}{\ln (b/a)} \qquad 21\text{-}8$$

Thus, the capacitance is proportional to the length of the conductors. The greater the length, the greater the amount of charge that can be put on the conductors for a given potential difference, because the electric field, and therefore the potential difference, depends only on the charge per unit length.

Example 21-2

A coaxial cable consists of a wire of radius 0.5 mm and an outer conducting shell of radius 1.5 mm. Find its capacitance per unit length.

Using Equation 21-8, we obtain

$$\frac{C}{L} = \frac{2\pi\epsilon_0}{\ln (b/a)} = \frac{2\pi(8.85 \text{ pF/m})}{\ln (1.5 \text{ mm}/0.5 \text{ mm})} = 50.6 \text{ pF/m}$$

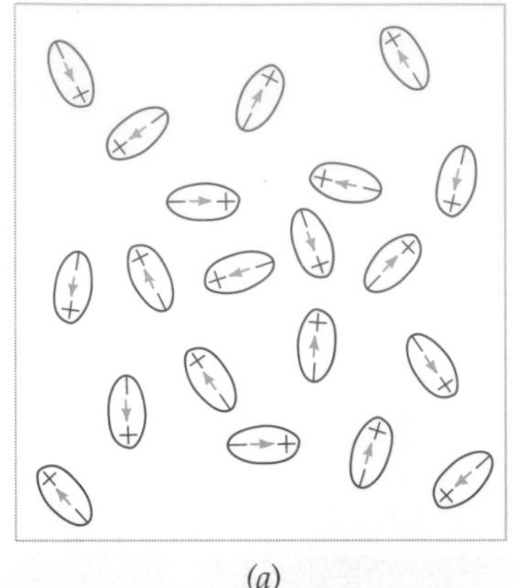

(a)

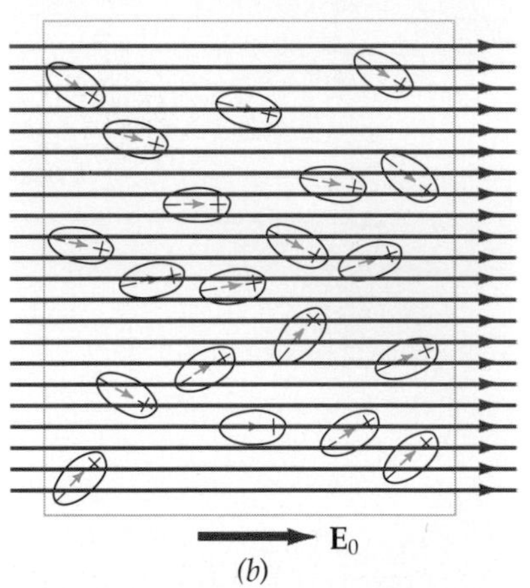

(b)

Figure 21-3 (a) The randomly oriented electric dipoles of a polar dielectric in the absence of an external electric field. (b) In the presence of an external electric field, the dipoles are partially aligned parallel to the field.

21-3 Dielectrics

A nonconducting material, such as glass, paper, or wood, is called a **dielectric.** When the space between the two conductors of a capacitor is occupied by a dielectric, the capacitance is increased by a factor κ that is characteristic of the dielectric and is called the **dielectric constant.** This was discovered experimentally by Michael Faraday. The reason for this increase is that the electric field between the plates of a capacitor is weakened by the dielectric. Thus, for a given charge on the plates, the potential difference is reduced and the ratio Q/V is increased.

A dielectric weakens the electric field between the plates of a capacitor because, in the presence of an external electric field, the molecules in the dielectric produce an additional electric field in a direction opposite to that of the external field. If the molecules of the dielectric are polar and so have permanent dipole moments, these dipole moments are normally randomly oriented (Figure 21-3*a*). In the presence of the field between the capacitor plates, however, these dipole moments experience a torque that tends to align them in the direction of the field (Figure 21-3*b*). The extent of alignment depends on the strength of the field and the temperature. At high temperatures, the random thermal motion of the molecules tends to counteract their alignment. If the molecules of the dielectric are nonpolar, they will have induced dipole moments in the presence of an external electric field in the direction of the field. A dielectric that has electric dipole moments that are predominantly in the direction of the external field is said to be polarized by the field, whether the polarization is due to the alignment of permanent dipole moments of polar molecules or to the creation of induced dipole moments in nonpolar molecules. In either case, the molecular dipoles produce an additional electric field that is in a direction opposite to that of the original field, thus weakening the original field.

The net effect of the polarization of a homogeneous dielectric is the creation of a surface charge on the dielectric faces near the plates as shown in Figure 21-4. It is this surface charge, which is bound to the dielectric, that produces an electric field opposite the direction of that due to the free charge

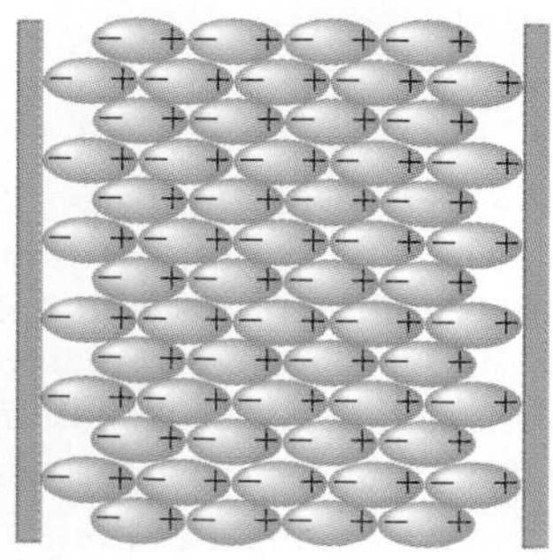

Figure 21-4 When a dielectric is placed between the plates of a capacitor, the electric field of the capacitor polarizes the molecules of the dielectric. The result is a bound charge on the surface of the dielectric that produces its own electric field that opposes the external field. The electric field between the plates is thus weakened by the dielectric.

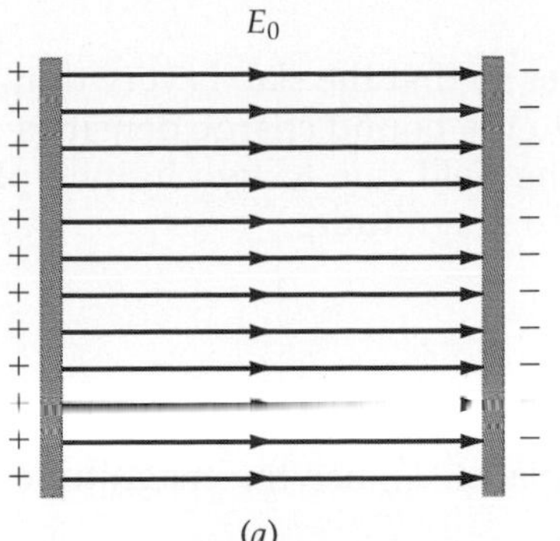

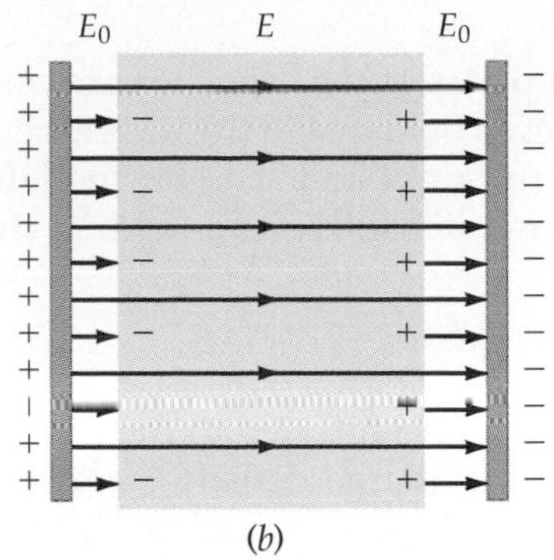

Figure 21-5 The electric field between the plates of a capacitor (*a*) with no dielectric and (*b*) with a dielectric. The surface charge on the dielectric weakens the original field between the plates.

on the conductors. Thus, the electric field between the plates is weakened as illustrated in Figure 21-5.

If the original electric field between the plates of a capacitor without a dielectric is E_0, the field in the dielectric is

$$E = \frac{E_0}{\kappa} \qquad 21\text{-}9$$

Electric field inside a dielectric

where κ is the dielectric constant. For a parallel-plate capacitor of separation s, the potential difference between the plates is

$$V = Es = \frac{E_0 s}{\kappa} = \frac{V_0}{\kappa}$$

where V is the potential difference with the dielectric and $V_0 = E_0 s$ is the original potential difference without the dielectric. The new capacitance is

$$C = \frac{Q}{V} = \frac{Q}{V_0/\kappa} = \kappa \frac{Q}{V_0}$$

or

$$C = \kappa C_0 \qquad 21\text{-}10$$

where $C_0 = Q/V_0$ is the original capacitance. The capacitance of a parallel-plate capacitor filled with a dielectric of constant κ is thus

$$C = \frac{\kappa \epsilon_0 A}{s} = \frac{\epsilon A}{s} \qquad 21\text{-}11$$

where

$$\epsilon = \kappa \epsilon_0 \qquad 21\text{-}12$$

is called the **permittivity** of the dielectric.

The charge densities on the faces of the dielectric are due to the displacement of positive and negative molecular charges near the faces caused by the external electric field of the capacitor. The charge on the dielectric is called a **bound charge** because it is bound to the molecules of the dielectric. That is, it cannot move about like the free charge on the conducting capacitor plates. Although it disappears when the external electric field disappears, it produces an electric field just like any other charge.

We will now see how the bound charge density σ_b on the surfaces of the dielectric is related to the dielectric constant κ and to the surface charge density on the plates of the capacitor, which we call the free charge density σ_f because it is free to move about the conductor. Consider a dielectric slab between the plates of a parallel-plate capacitor as shown in Figure 21-6. If

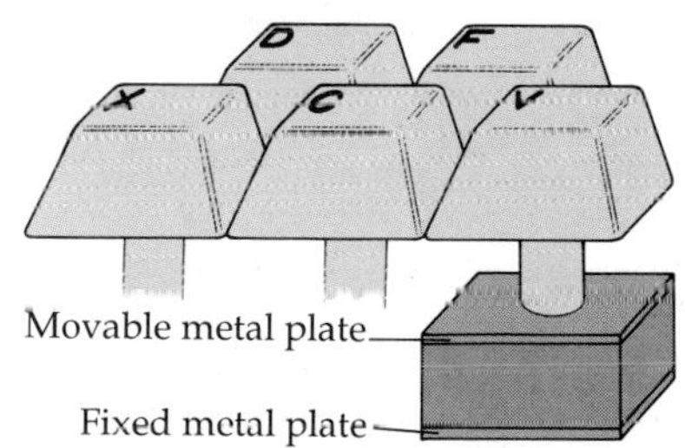

Some keyboards in computers use capacitance switching. A metal plate, mounted on a plunger attached to the key, acts as the top plate of a capacitor. When the key is depressed, the separation between the top plate and bottom plate is changed from about 5 mm to about 0.3 mm and the capacitance increases. The change in capacitance triggers the electronic circuitry to enter the information into the computer.

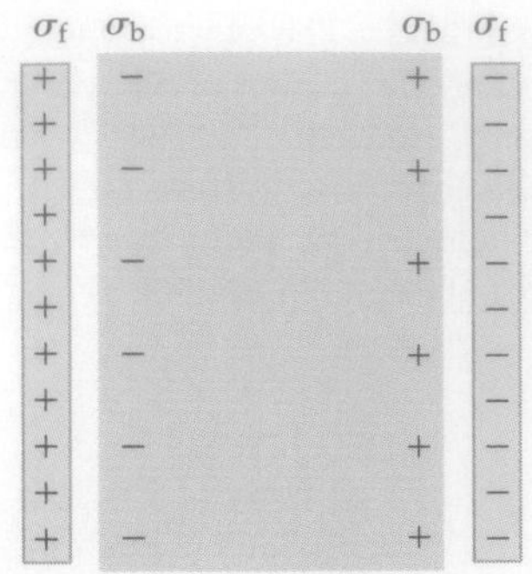

Figure 21-6 A parallel-plate capacitor with a dielectric slab between the plates. If the plates are closely spaced, each of the surface charges can be considered to be an infinite plane charge. The electric field due to the free charge on the plates is directed to the right and has a magnitude $E_0 = \sigma_f/\epsilon_0$. That due to the bound charge is directed to the left and has a magnitude $E_b = \sigma_b/\epsilon_0$.

the plates of the capacitor are close together so that the slab is very thin, the electric field inside the dielectric slab due to the bound charge densities $+\sigma_b$ on the right and $-\sigma_b$ on the left is just the field due to two infinite-plane charge densities. The field E_b thus has the magnitude

$$E_b = \frac{\sigma_b}{\epsilon_0} \qquad 21\text{-}13$$

This field is directed to the left and subtracts from the electric field E_0 due to the free charge density on the capacitor plates. E_0 has the magnitude

$$E_0 = \frac{\sigma_f}{\epsilon_0} \qquad 21\text{-}14$$

The magnitude of the net field E is the difference between these magnitudes. It also equals E_0/κ:

$$E = E_0 - E_b = \frac{E_0}{\kappa}$$

or

$$E_b = E_0\left(1 - \frac{1}{\kappa}\right) = \frac{\kappa - 1}{\kappa} E_0$$

Writing σ_b/ϵ_0 for E_b and σ_f/ϵ_0 for E_0, we obtain

$$\sigma_b = \frac{\kappa - 1}{\kappa}\sigma_f \qquad 21\text{-}15$$

The bound charge density σ_b is always less than the free charge density σ_f on the capacitor plates, and it is zero if $\kappa = 1$, which is the case when there is no dielectric.

In the preceding discussion, we assumed that the charge on the plates of the capacitor did not change when the dielectric was inserted. This would be true if the capacitor were charged and then removed from the charging source (the battery) before the insertion of the dielectric. If the dielectric is inserted while the battery is still connected, the battery will supply more charge to maintain the original potential difference. The total charge on the plates is then $Q = \kappa Q_0$. In either case, however, the capacitance is increased by the factor κ.

Exercise

The capacitor of Example 21-1 is filled with a dielectric of constant $\kappa = 2$. (*a*) Find the new capacitance. (*b*) Find the charge on the capacitor with the dielectric if the capacitor is attached to a 12-V battery. [Answers: (*a*) 177 pF, (*b*) 2.12 nC]

Exercise

The capacitor in the previous exercise is charged to 12 V without the dielectric and is then disconnected from the battery. The dielectric of constant $\kappa = 2$ is then inserted. Find the new values of the (*a*) charge Q, (*b*) the voltage V, and (*c*) the capacitance C. [Answers: (*a*) $Q = 1.06$ nC, which is unchanged; (*b*) $V = 6$ V; (*c*) $C = 177$ pF]

In addition to increasing the capacitance, a dielectric has two other functions in a capacitor. First, it provides a physical means of separating the two conductors, which must be very close together for a large capacitance since the capacitance varies inversely with the separation. Second, it increases the dielectric strength of the capacitor because the dielectric strength of a dielectric is usually greater than that of air.

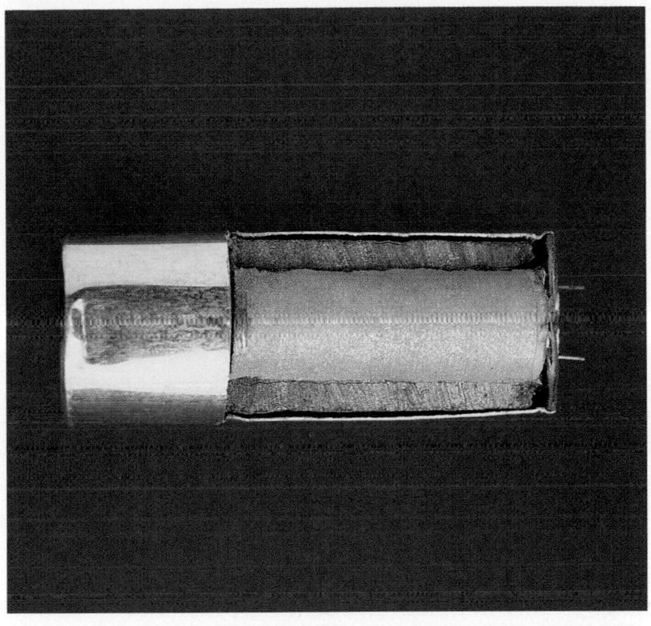

(a)

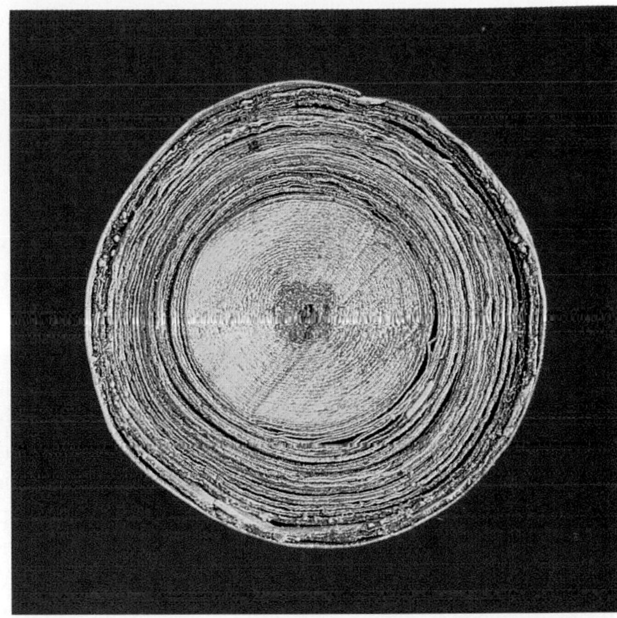

(b)

We have already seen in Chapter 20 that the dielectric strength of air is about 3 MV/m = 3 kV/mm. Fields with a magnitude greater than this cannot be maintained in air because of dielectric breakdown; that is, the air becomes ionized and conducts. Many materials have dielectric strengths greater than that of air and so allow greater potential differences between the conducting plates of a capacitor.

Examples of the three dielectric functions are provided by a parallel-plate capacitor made from two sheets of metal foil of large area (to increase the capacitance) that are separated by a thin sheet of paper. The paper increases the capacitance because of its polarization; that is, κ is greater than 1. It also provides physical separation so that the sheets can be very close together without being in electrical contact. Finally, the dielectric strength of paper is greater than that of air, so greater potential differences can be attained without dielectric breakdown. Table 21-1 lists the dielectric constants and dielectric strengths of some dielectrics. Note that for air, $\kappa \approx 1$, so for most situations we do not need to distinguish between air and a vacuum.

(c)

(a) A 200-μF capacitor used in an electronic strobe light. (b) Cross section of a foil wound capacitor. (c) A cut section of a multilayer ceramic capacitor. The bright lines are the edges of the conducting plate.

Table 21-1 **Dielectric Constants and Strengths of Various Materials**

Material	Dielectric constant κ	Dielectric strength, kV/mm
Air	1.00059	3
Bakelite	4.9	24
Glass (Pyrex)	5.6	14
Mica	5.4	10–100
Neoprene	6.9	12
Paper	3.7	16
Paraffin	2.1–2.5	10
Plexiglas	3.4	40
Polystyrene	2.55	24
Porcelain	7	5.7
Transformer oil	2.24	12
Water (20°C)	80	

Example 21-3

A parallel-plate capacitor has square plates of side 10 cm and a separation of 4 mm. A dielectric slab of constant $\kappa = 2$ has the same area as the plates but has a thickness of 3 mm. What is the capacitance (*a*) without the dielectric and (*b*) with the dielectric?

(*a*) This capacitor is the same as the one in Example 21-1 except that the plate spacing is 4 mm rather then 1 mm. Since the capacitance varies inversely with the plate separation, the capacitance without the dielectric is one-fourth the value found in Example 21-1, or $C_0 = \frac{1}{4}(88.5\text{ pF}) = 22.1\text{ pF}$.

(*b*) To find the value of the capacitance with the dielectric inserted, we place charges $+Q$ on one plate and $-Q$ on the other plate and find the electric field between the plates. We then calculate the potential difference between the plates.

In the space between the plates without the dielectric, the electric field is $E_0 = Q/\epsilon_0 A$ because the field due to the bound charges is zero. Inside the dielectric, the electric field is $E = E_0/\kappa$. The potential difference between the plates is the integral of the field over distance from one plate to the other. If s is the total separation of the plates, the thickness of the dielectric is $\frac{3}{4}s$ and that of the free space is $\frac{1}{4}s$. The potential difference between the plates is thus

$$V = E_0\left(\frac{1}{4}\right)s + \frac{E_0}{\kappa}\left(\frac{3}{4}s\right) = E_0 s\left(\frac{1}{4} + \frac{3}{4\kappa}\right) = V_0\left(\frac{\kappa + 3}{4\kappa}\right)$$

where we have used $E_0 s = V_0$, the original potential difference with no dielectric. Using $\kappa = 2$, we find the potential difference with the dielectric to be

$$V = \tfrac{5}{8}V_0$$

The new capacitance is thus

$$C = \frac{Q}{V} = \frac{Q}{\frac{5}{8}V_0} = \frac{8}{5}\frac{Q}{V_0} = \frac{8}{5}C_0$$

$$= \frac{8}{5}(22.1\text{ pF}) = 35.4\text{ pF}$$

21-4 The Storage of Electrical Energy

While a capacitor is being charged, a positive charge is transferred from the negatively charged conductor to the positively charged conductor. Since the positive conductor is at a greater potential than the negative conductor, the potential energy of the charge being transferred is increased. For example, if a small amount of charge q is transferred through a potential difference V, the potential energy of the charge is increased by the amount qV. (Remember that, by definition, potential difference is the difference in potential energy per unit charge.) Work must therefore be done to charge a capacitor. Some of this work is stored as electrostatic potential energy. At the beginning of the charging process, neither conductor is charged. There is no electric field, and both conductors are at the same potential. After the charging process, a charge Q has been transferred from one conductor to the other, and the potential difference is $V = Q/C$, where C is the capacitance.

Let q be the charge that has been transferred at some time during the process. The potential difference is then $V = q/C$. If a small amount of additional charge dq is now transferred from the negative conductor at zero potential to the positive conductor at a potential V (Figure 21-7), the potential energy of the charge is increased by

$$dU = V\,dq = \frac{q}{C}\,dq$$

The total increase in potential energy U is the sum or integral of these charges dU as q increases from zero to its final value Q (Figure 21-8):

$$U = \int dU = \int_0^Q \frac{q}{C}\,dq = \frac{1}{2}\frac{Q^2}{C}$$

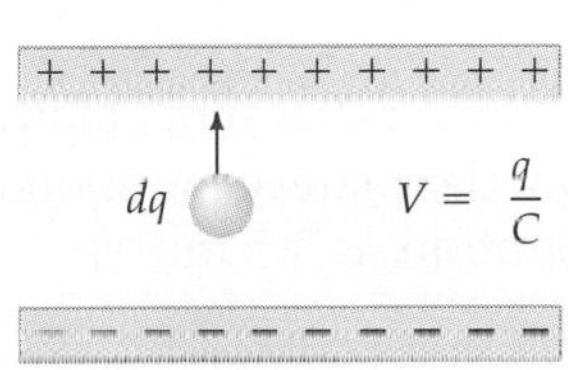

Figure 21-7 When a small amount of charge dq is moved from the negative conductor to the positive conductor, its potential energy is increased by $dU = V\,dq$, where V is the potential difference between the conductors.

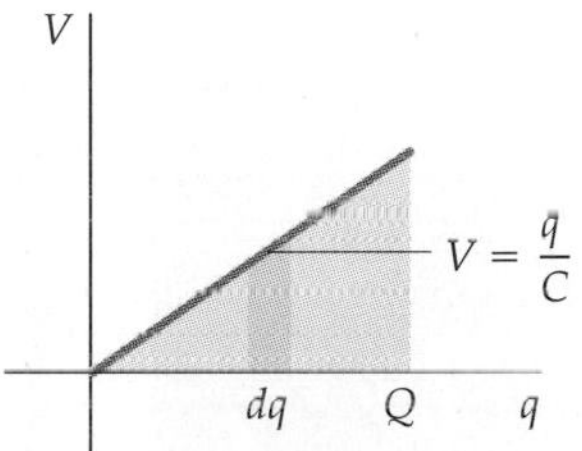

Figure 21-8 The work needed to charge a capacitor is the integral of $V\,dq$ from the original charge of $q = 0$ to the final charge of $q = Q$. This work is the area under the curve $\frac{1}{2}Q(Q/C)$.

This potential energy is the energy stored in the capacitor. Using $C = Q/V$, we can express this energy in a variety of ways:

$$U = \frac{1}{2}\frac{Q^2}{C} = \frac{1}{2}QV = \frac{1}{2}CV^2 \qquad \text{21-16}$$

Energy stored in a capacitor

Equation 21-16 is a general expression for the energy stored in a charged capacitor as electrostatic potential energy.

Exercise

A 15-μF capacitor is charged to 60 V. How much energy is stored in the capacitor? (Answer: 0.027 J)

Example 21-4

A 60-μF capacitor is charged to 12 V. The capacitor is removed from the battery and the plate separation is increased from 2.0 mm to 3.5 mm. (*a*) What is the charge on the capacitor? (*b*) How much energy was originally stored in the capacitor? (*c*) By how much is the energy increased when the plate separation is changed?

(*a*) From the definition of capacitance (Equation 21-1) the charge on the capacitor is

$$Q = CV = (60\ \mu\text{F})(12\ \text{V}) = 720\ \mu\text{C}$$

(*b*) The energy originally stored is

$$W = \frac{1}{2}QV = \frac{1}{2}(720\ \mu\text{C})(12\ \text{V}) = 4320\ \mu\text{J}$$

We could have found the energy without first finding the charge from

$$W = \frac{1}{2}CV^2 = \frac{1}{2}(60\ \mu\text{F})(12\ \text{V})^2 = 4320\ \mu\text{J}$$

(*c*) After the capacitor is removed from the battery, the charge on the plates must remain constant. When the plate separation is increased, the voltage between the plates is increased and the capacitance is decreased. We can find the increase in energy stored by either finding the increase in the voltage and using $W = \frac{1}{2}QV$ or by finding the decrease in the capacitance and using $W = \frac{1}{2}Q^2/C$. The potential difference between the plates is related to the plate separation s and the electric field E by

$$V = Es$$

The field does not change because the charge remains the same. Since the potential difference is 12 V when the plate separation is 2.0 mm, the potential difference when the separation is 3.5 mm is

$$V = (12\ \text{V})\frac{3.5\ \text{mm}}{2.0\ \text{mm}} = 21\ \text{V}$$

When the plate separation is 3.5 mm, the energy stored is

$$W = \frac{1}{2}QV = \frac{1}{2}(720\ \mu\text{C})(21\ \text{V}) = 7560\ \mu\text{J}$$

The increase in the energy stored is therefore 7560 μJ − 4320 μJ = 3240 μJ.

It is illuminating to work part *b* in another way. Because the plates of a capacitor are oppositely charged, they exert attractive forces on one another. To increase the plate separation, work must be done against these forces. Let us assume that the lower plate is held fixed and the upper plate is moved. The force on the upper plate is the charge on the plate Q times the electric field *due to the lower plate.* This field is half the total field between the plates because the charge on the upper plate also contributes equally to the field. When the potential difference is 12 V and the separation is 2 mm, the total field between the plates is

$$E = \frac{V}{s} = \frac{12\ \text{V}}{2\ \text{mm}} = 6\ \text{V/mm} = 6\ \text{kV/m}$$

The electric field due to just the charge on the bottom plate is then

$$E' = \frac{1}{2}E = 3\ \text{kV/m}$$

The force exerted on the upper plate by the bottom plate is thus

$$F = QE' = (720\ \mu\text{C})(3\ \text{kV/m}) = 2.16\ \text{N}$$

The work that must be done to move the upper plate a distance of Δs = 1.5 mm is then

$$W = F\ \Delta s = (2.16\ \text{N})(1.5\ \text{mm}) = 3.24 \times 10^{-3}\ \text{J} = 3240\ \mu\text{J}$$

This work equals the increase in the energy stored.

Capacitor bank for energy storage for the pulsed Nova laser used at Lawrence Livermore Laboratories to study fusion. Each capacitor has a capacitance of about 29 μF.

In the process of charging a capacitor, an electric field is produced between the plates. The work required to charge the capacitor can be thought of as the work required to create the electric field. That is, we can think of the energy stored in a capacitor as energy stored in the electric field, called **electrostatic field energy.** We can see this in the case of a parallel-plate capacitor that is filled with a dielectric of constant κ. Let $+Q$ be the charge on one of the plates of the capacitor. The potential difference between the plates is $V = Es$, where s is the plate separation and E is the electric field between the plates, which is related to the charge on the plates by

$$E = \frac{E_0}{\kappa} = \frac{\sigma}{\kappa\epsilon_0} = \frac{Q}{\epsilon A}$$

Substituting $Q = \epsilon AE$ and $V = Es$ into Equation 21-16, we obtain the potential energy U associated with an electric field:

$$U = \frac{1}{2}QV = \frac{1}{2}(\epsilon AE)(Es)$$
$$= \frac{1}{2}\epsilon E^2(As)$$

The quantity As is the volume of the space between the plates of the capacitor containing the electric field. The energy per unit volume is called the **energy density** η. The energy density in an electric field E is thus

$$\eta = \frac{\text{energy}}{\text{volume}} = \frac{1}{2}\epsilon E^2 \qquad 21\text{-}17$$

Energy density of an electrostatic field

Thus, the energy per unit volume of the electrostatic field is proportional to the square of the electric field. Although we obtained Equation 21-17 by considering the electric field between the plates of a parallel-plate capacitor, the result applies to any electric field. Whenever there is an electric field in space, the electrostatic energy per unit volume is given by Equation 21-17.

We can see the generality of Equation 21-17 by calculating the electrostatic field energy for a case that does not involve a capacitor, and in which the electric field is not constant. Let us consider the electrostatic potential energy of a spherical conductor of radius R that carries a charge Q. We will calculate the work needed to bring a charge from a great distance away to the conductor in the same way that we calculated the work needed to transfer a charge from one plate of a capacitor to the other. When the sphere carries a charge q, its potential relative to $V = 0$ at infinity is

$$V = \frac{1}{4\pi\epsilon_0}\frac{q}{R}$$

The work needed to bring an additional amount of charge dq from infinity to the conductor is $V\,dq$, which equals the increase in the potential energy of the conductor:

$$dU = V\,dq = \frac{1}{4\pi\epsilon_0 R}q\,dq$$

The total increase in potential energy U is the integral of dU as q increases from zero to its final value Q. Integrating from $q = 0$ to $q = Q$, we obtain

$$U = \frac{1}{4\pi\epsilon_0 R}\frac{Q^2}{2} = \frac{1}{2}QV \qquad 21\text{-}18$$

This is the electrostatic potential energy of the spherical conductor.

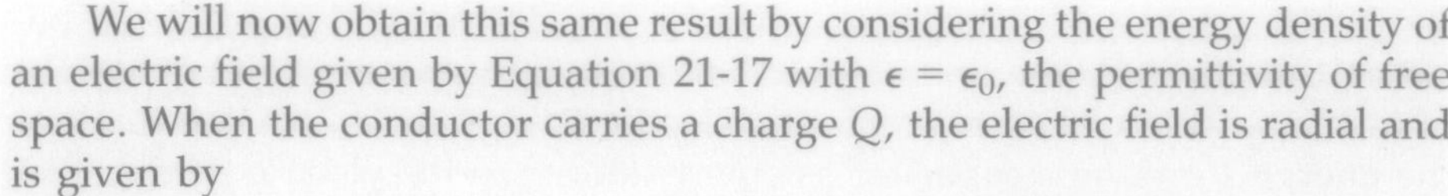

We will now obtain this same result by considering the energy density of an electric field given by Equation 21-17 with $\epsilon = \epsilon_0$, the permittivity of free space. When the conductor carries a charge Q, the electric field is radial and is given by

$$E_r = 0 \qquad r < R \text{ (inside the conductor)}$$

$$E_r = \frac{1}{4\pi\epsilon_0}\frac{Q}{r^2} \qquad r > R \text{ (outside the conductor)}$$

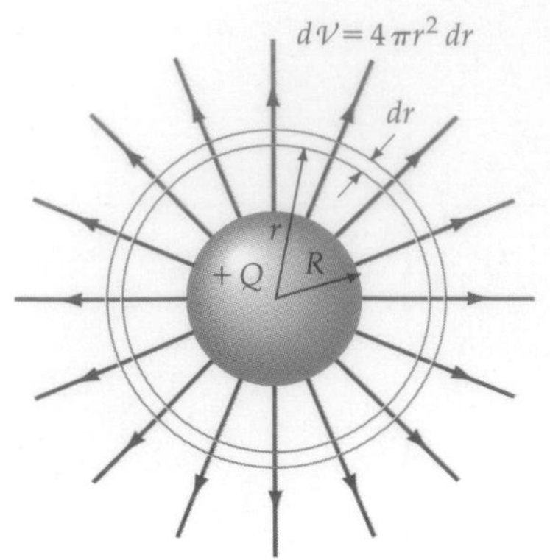

Figure 21-9 Geometry for the calculation of the electrostatic energy of a spherical conductor carrying a charge Q. The volume of the space between r and $r + dr$ is $d\mathcal{V}=4\pi r^2\,dr$. The electrostatic field energy in this volume element is $\eta\, d\mathcal{V}$, where $\eta = \frac{1}{2}\epsilon_0 E^2$ is the energy density.

Since the electric field is spherically symmetric, we choose a spherical shell for our volume element. If the radius of the shell is r and its thickness is dr, the volume is $d\mathcal{V}=4\pi r^2\,dr$ (Figure 21-9). The energy dU in this volume element is

$$dU = \eta\, d\mathcal{V} = \tfrac{1}{2}(\epsilon_0 E^2)4\pi r^2\,dr$$

$$= \frac{1}{2}\epsilon_0\left(\frac{Q}{4\pi\epsilon_0 r^2}\right)^2(4\pi r^2\,dr) = \frac{Q^2}{8\pi\epsilon_0}\frac{dr}{r^2}$$

Since the electric field is zero for $r < R$, we obtain the total energy in the electric field by integrating from $r = R$ to $r = \infty$:

$$U = \int_R^\infty \frac{Q^2}{8\pi\epsilon_0}\frac{dr}{r^2} = \frac{1}{2}\frac{Q^2}{4\pi\epsilon_0 R} = \frac{1}{2}QV \qquad 21\text{-}19$$

which is the same as Equation 21-18.

Questions

1. If the potential difference of a capacitor is doubled, by what factor does its stored electric energy change?
2. Half the charge is removed from a capacitor. What fraction of its stored energy is removed along with the charge?

21-5 Combinations of Capacitors

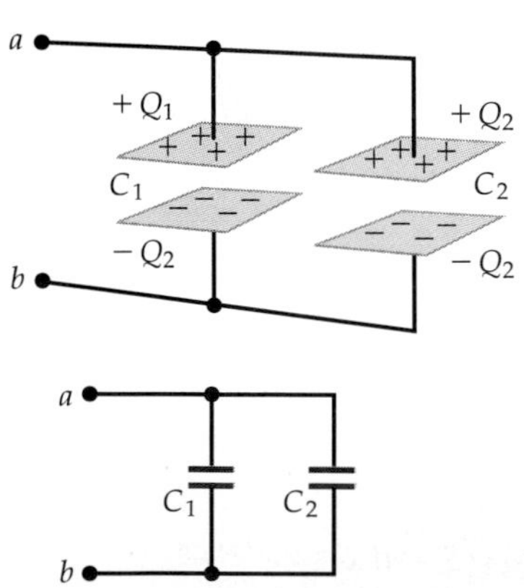

Figure 21-10 Two capacitors in parallel. The potential difference is the same across each capacitor.

Two or more capacitors are often used in combination. Figure 21-10 shows two **capacitors in parallel.** (In electric circuits, a capacitor is indicated by the symbol ⊣ ⊢.) The upper plates of the two capacitors are connected by a conducting wire and are therefore at the same potential V_a. The lower plates are also connected and are at a common potential V_b. Points a and b are connected to a battery or some other device that maintains a potential difference $V = V_a - V_b$, which is the potential difference between the plates of each capacitor. The effect of adding a second capacitor connected in this way is an increase in the capacitance. The area is essentially increased, allowing more charge to be stored for the same potential difference. If the capacitances are C_1 and C_2, the charges Q_1 and Q_2 stored on the plates are given by

$$Q_1 = C_1 V$$

and

$$Q_2 = C_2 V$$

The total charge stored is

$$Q = Q_1 + Q_2 = C_1 V + C_2 V = (C_1 + C_2)V$$

The **equivalent capacitance** is the capacitance of a single capacitor that could replace some combination of capacitors in a circuit and store the same

(a)  (b)  (c)

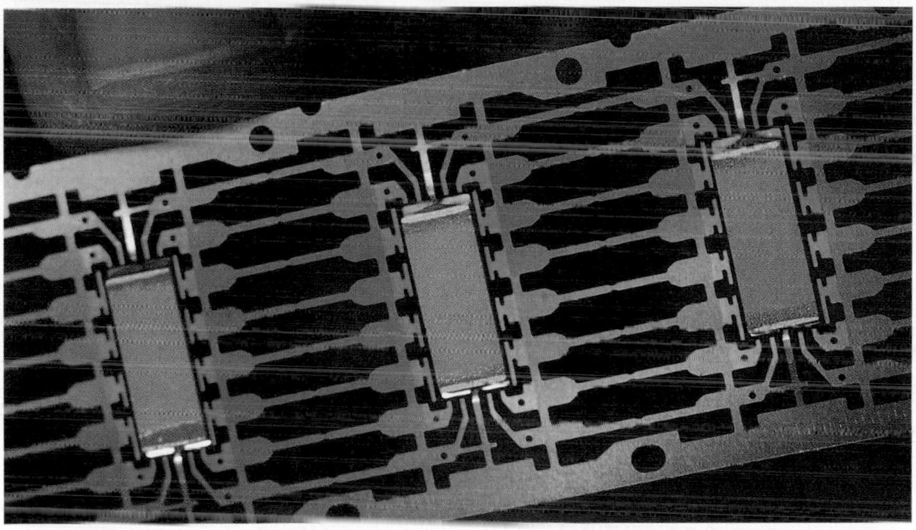

(d)

(*a*) A Leyden jar capacitor. (*b*) A variable air-gap capacitor like those that were used in the tuning circuits of old radios. (*c*) Capacitors wired into a circuit board. (*d*) Capacitors in a printed circuit.

amount of charge for a given potential difference. The equivalent capacitance of two capacitors in parallel is the ratio of the total charge stored to the potential difference:

$$C_{eq} = \frac{Q}{V} = C_1 + C_2 \qquad 21\text{-}20$$

Thus, the equivalent capacitance of two capacitors in parallel equals the sum of the individual capacitances. The same reasoning can be extended to three or more capacitors connected in parallel, as in Figure 21-11:

$$C_{eq} = C_1 + C_2 + C_3 + \cdots \qquad 21\text{-}21$$

Equivalent capacitance for capacitors in parallel

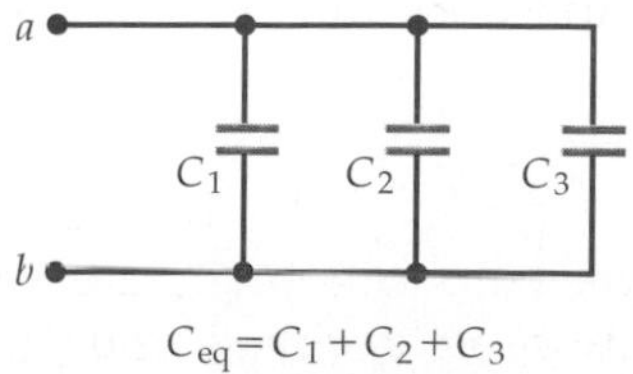

Figure 21-11 Three capacitors in parallel. The effect of adding a parallel capacitor to a circuit is an increase in the equivalent capacitance.

Figure 21-12 shows two **capacitors in series.** When points a and b are connected to the terminals of a battery, there is a potential difference $V = V_a - V_b$ across the two capacitors, but the potential difference across one of the capacitors is not necessarily the same as that across the other. If a charge $+Q$ is placed on the upper plate of the first capacitor, the electric field produced by that charge will induce an equal negative charge $-Q$ on its lower plate. This charge comes from electrons drawn from the upper plate of the second capacitor. Thus, there will be an equal charge $+Q$ on the upper plate of the second capacitor and a corresponding charge $-Q$ on its lower plate. The potential difference across the first capacitor is

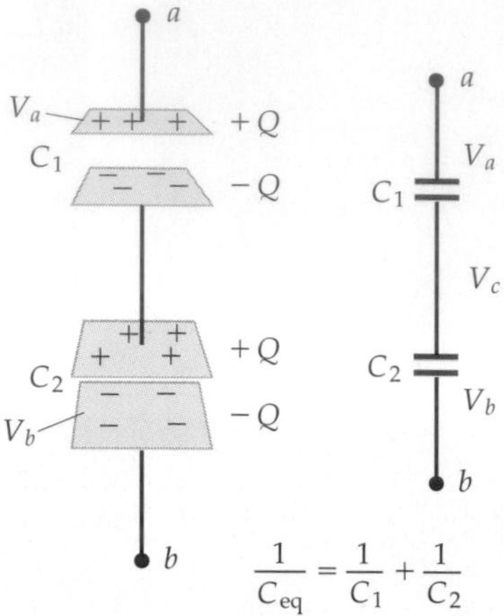

Figure 21-12 Two capacitors in series. The charge is the same on each capacitor. The potential difference across the series combination is the sum of the potential differences across the individual capacitors.

$$V_1 = V_a - V_c = \frac{Q}{C_1}$$

Similarly, the potential difference across the second capacitor is

$$V_2 = V_c - V_b = \frac{Q}{C_2}$$

The potential difference across the two capacitors in series is the sum of these potential differences:

$$\begin{aligned} V = V_a - V_b &= (V_a - V_c) + (V_c - V_b) \\ &= V_1 + V_2 = \frac{Q}{C_1} + \frac{Q}{C_2} \end{aligned}$$

Thus,

$$\begin{aligned} V &= \frac{Q}{C_1} + \frac{Q}{C_2} \\ &= Q\left(\frac{1}{C_1} + \frac{1}{C_2}\right) \end{aligned} \qquad 21\text{-}22$$

The equivalent capacitance of two capacitors in series is that of a single capacitor that could replace the two capacitors and give the potential difference V for the same charge Q. Thus,

$$C_{eq} = \frac{Q}{V} \qquad 21\text{-}23$$

Comparing Equations 21-22 and 21-23, we have

$$\frac{1}{C_{eq}} = \frac{1}{C_1} + \frac{1}{C_2} \qquad 21\text{-}24$$

Equation 21-24 can be generalized to three or more capacitors connected in series:

$$\frac{1}{C_{eq}} = \frac{1}{C_1} + \frac{1}{C_2} + \frac{1}{C_3} + \cdots \qquad 21\text{-}25$$

Equivalent capacitance for capacitors in series

The potential difference across a set of capacitors in series is the sum of the potential differences across the individual capacitors. Note that the addition of a capacitor in series increases $1/C_{eq}$, which means that the equivalent capacitance C_{eq} decreases.

Exercise

Two capacitors have capacitances of 20 μF and 30 μF. Find the equivalent capacitance if the capacitors are connected (*a*) in parallel, (*b*) in series. [Answers: (*a*) 50 μF, (*b*) 12 μF]

Note that in the preceding exercise the equivalent capacitance of the two capacitances in series is less than the capacitance of either capacitor.

There is a maximum voltage that a given capacitor can withstand before breakdown occurs. Suppose we have a power supply of 100 V and two identical capacitors each of which breaks down if the potential across it exceeds 60 V. Either capacitor will break down if connected across the power supply. Breakdown will also occur if we connect the two capacitors in parallel because the potential across each will be 100 V. However, if we connect the two capacitors in series, the potential across each capacitor will be only 50 V, and they will not break down.

Example 21-5

Find the equivalent capacitance of the network of the three capacitors shown in Figure 21-13.

In this network, the 2-μF and 3-μF capacitors are connected in parallel and this parallel combination is connected in series with the 4-μF capacitor. The equivalent capacitance of the two capacitors in parallel is

$$C_{eq} = C_1 + C_2 = 2\ \mu\text{F} + 3\ \mu\text{F} = 5\ \mu\text{F}$$

If we replace the two capacitors in parallel with a single 5-μF capacitor, we have a 5-μF capacitor in series with a 4-μF capacitor. The equivalent capacitance of this series combination is found from

$$\frac{1}{C_{eq}} = \frac{1}{C_1} + \frac{1}{C_2} = \frac{1}{5\ \mu\text{F}} + \frac{1}{4\ \mu\text{F}} = \frac{9}{20\ \mu\text{F}}$$

The equivalent capacitance of the three-capacitor network is thus

$$C_{eq} = \frac{20\ \mu\text{F}}{9} = 2.22\ \mu\text{F}$$

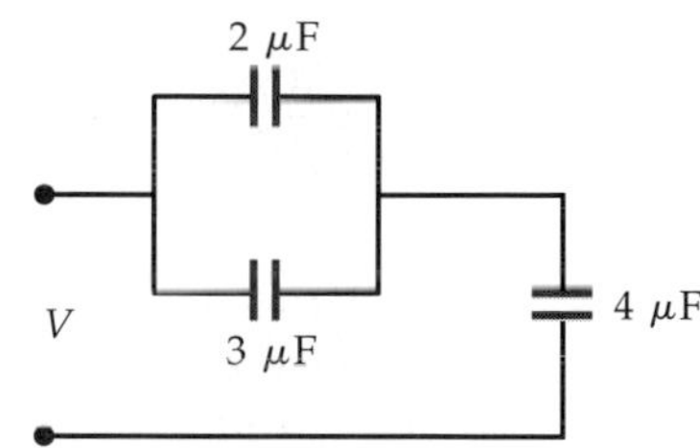

Figure 21-13 Capacitor network for Example 21-5.

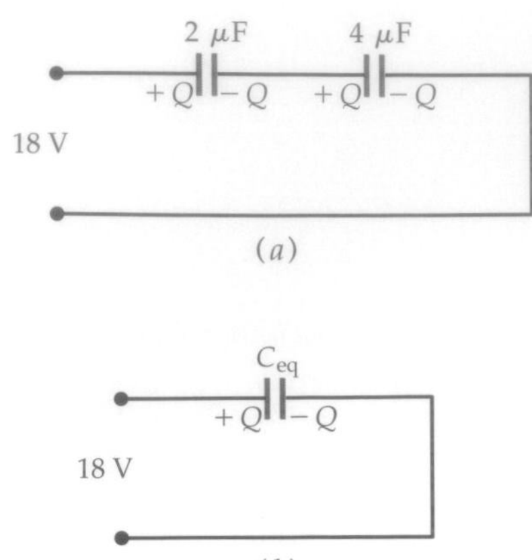

Figure 21-14 (*a*) Two capacitors connected in series across an 18-V battery for Example 21-6. (*b*) The two capacitors in (*a*) can be replaced by an equivalent capacitor.

Example 21-6

A 2-μF capacitor and a 4-μF capacitor are connected in series across an 18-V battery as shown in Figure 21-14*a*. Find the charge on the capacitors and the potential difference across each.

In Figure 21-14*b*, the two capacitors have been replaced by one capacitor with an equivalent capacitance C_{eq}, which is found from

$$\frac{1}{C_{eq}} = \frac{1}{C_1} + \frac{1}{C_2} = \frac{1}{2\ \mu\text{F}} + \frac{1}{4\ \mu\text{F}} = \frac{3}{4\ \mu\text{F}}$$

$$C_{eq} = \tfrac{4}{3}\ \mu\text{F}$$

The charge on either plate of the equivalent capacitor in Figure 21-14*b* is

$$Q = C_{eq}V = (\tfrac{4}{3}\ \mu\text{F})(18\ \text{V}) = 24\ \mu\text{C}$$

This is the magnitude of the charge on each plate of the two original capacitors. The potential difference across the 2-μF capacitor is thus

$$V_1 = \frac{Q}{C_1} = \frac{24\ \mu\text{C}}{2\ \mu\text{F}} = 12\ \text{V}$$

and that across the 4-μF capacitor is

$$V_2 = \frac{Q}{C_2} = \frac{24\ \mu\text{C}}{4\ \mu\text{F}} = 6\ \text{V}$$

Note that the sum of these potential differences is 18 V, as required.

2 μF 4 μF

+24 μC −24 μC +24 μC −24 μC

(*a*)

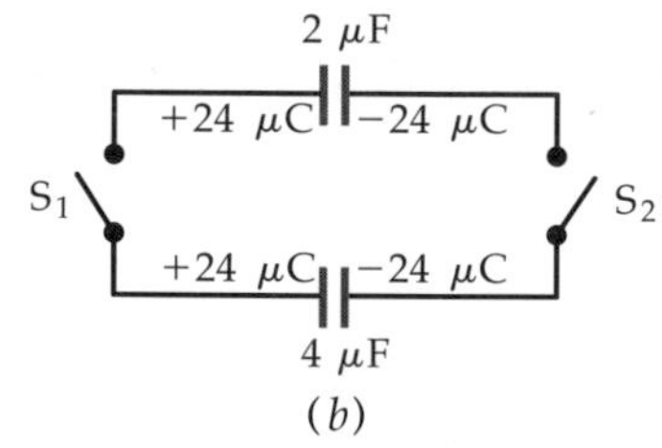

Figure 21-15 (*a*) The two capacitors in Figure 21-14*a* just after they are disconnected from the battery. (*b*) After the switches S_1 and S_2 are closed, the two capacitors are reconnected, positive plate to positive plate and negative plate to negative plate.

Example 21-7

The two capacitors in Example 21-6 are removed from the battery and are carefully disconnected from each other so that the charge on the plates is not disturbed as shown in Figure 21-15*a*. They are then reconnected with positive plate connected to positive plate and negative plate connected to negative plate as shown in Figure 21-15*b*. Find the potential difference across the capacitors and the charge on each capacitor.

After the capacitors are reconnected, the two positive plates form a single conductor as do the two negative plates. The total charge on the positive plates is +48 μC, and that on the negative plates is −48 μC. Furthermore, the potential difference across both capacitors must now be the same. The capacitors are therefore connected in parallel, so the equivalent capacitance is

$$C_{eq} = C_1 + C_2 = 2\ \mu\text{F} + 4\ \mu\text{F} = 6\ \mu\text{F}$$

Since the total charge is 48 μC, the potential difference across the parallel combination is

$$V = \frac{Q}{C_{eq}} = \frac{48\ \mu\text{C}}{6\ \mu\text{F}} = 8\ \text{V}$$

The charge on each capacitor is

$$Q_1 = C_1V = (2\ \mu\text{F})(8\ \text{V}) = 16\ \mu\text{C}$$

and

$$Q_2 = C_2V = (4\ \mu\text{F})(8\ \text{V}) = 32\ \mu\text{C}$$

The total charge adds up to 48 μC, as required.

Example 21-8

Two parallel-plate capacitors, each having a capacitance of 2 μF, are connected in parallel across a 12-V battery. Find (*a*) the charge on each capacitor and (*b*) the total energy stored in the capacitors. The capacitors are disconnected from the battery and a dielectric of constant $\kappa = 3$ is inserted between the plates of one of the capacitors. After the dielectric is inserted, find (*c*) the potential difference across each capacitor, (*d*) the charge on each capacitor, and (*e*) the total energy stored in the capacitors.

(*a*) The original charge on each capacitor is

$$Q = CV = (2\ \mu\text{F})(12\ \text{V}) = 24\ \mu\text{C}$$

(*b*) The energy stored in each capacitor is

$$U = \frac{1}{2}QV = \frac{1}{2}(24\ \mu\text{C})(12\ \text{V}) = 144\ \mu\text{J}$$

so the total energy stored is $2 \times 144\ \mu\text{J} = 288\ \mu\text{J}$.

(*c*) The capacitance of the capacitor with the dielectric is now

$$C' = \kappa C = 3(2\ \mu\text{F}) = 6\ \mu\text{F}$$

Since the capacitors are in parallel, the equivalent capacitance is

$$C_{\text{eq}} = C_1 + C_2 = 6\ \mu\text{F} + 2\ \mu\text{F} = 8\ \mu\text{F}$$

The total charge, which is 48 μC, must remain the same because the battery is disconnected. The potential difference across the parallel combination, which is the potential difference across each capacitor, is

$$V = \frac{Q}{C_{\text{eq}}} = \frac{48\ \mu\text{C}}{8\ \mu\text{F}} = 6\ \text{V}$$

(*d*) When the dielectric is inserted into one of the capacitors, the field is weakened and the potential difference is lowered. Since the two capacitors are connected in parallel, charge must flow from the other capacitor so that the potential difference is the same across both capacitors. Let Q_1 be the charge on the capacitor with the dielectric and Q_2 be that on the other capacitor. Then

$$Q_1 = C_1V = (6\ \mu\text{F})(6\ \text{V}) = 36\ \mu\text{C}$$

and

$$Q_2 = C_2V = (2\ \mu\text{F})(6\ \text{V}) = 12\ \mu\text{C}$$

The total charge is 48 μC as before.

(*e*) The energy in the capacitor with the dielectric is

$$U_1 = \frac{1}{2}Q_1V = \frac{1}{2}(36\ \mu\text{C})(6\ \text{V}) = 108\ \mu\text{J}$$

and the energy in the other capacitor is

$$U_2 = \frac{1}{2}Q_2V = \frac{1}{2}(12\ \mu\text{C})(6\ \text{V}) = 36\ \mu\text{J}$$

The total energy is $U_1 + U_2 = 144\ \mu\text{J}$. This is half the original energy of 288 μJ. Removing the dielectric from the capacitor requires 144 μJ of work which is then stored as electrostatic potential energy.

A mechanical hand is used at Lockheed to install capacitors in a circuit.

Example 21-9

Work parts *c*, *d*, and *e* of Example 21-8, if the dielectric is inserted into one of the capacitors while the battery is still connected.

(*c*) Since the battery is still connected, the potential difference across the capacitors remains 12 V.

(*d*) When the dielectric is inserted into one of the capacitors, additional charge is supplied by the battery to maintain the potential difference of 12 V. Since the new capacitance of the capacitor with the dielectric is 6 μF, the charge on that capacitor is

$$Q_1 = C_1V = (6\ \mu\text{F})(12\ \text{V}) = 72\ \mu\text{C}$$

and the charge on the other capacitor remains the same as before.

(*e*) The energy in the capacitor with the dielectric is

$$U_1 = \frac{1}{2}Q_1V = \frac{1}{2}(72\ \mu\text{C})(12\ \text{V}) = 432\ \mu\text{J}$$

The capacitor without the dielectric has the same charge and potential difference, so it has the same energy:

$$U = \frac{1}{2}QV = \frac{1}{2}(24\ \mu\text{C})(12\ \text{V}) = 144\ \mu\text{J}$$

Therefore, the total energy is 432 μJ + 144 μJ = 576 μJ. In this case, the total energy is increased because the battery supplies more charge.

Summary

1. A capacitor is a device for storing charge and energy. It consists of two conductors, closely spaced but insulated from each other, carrying equal and opposite charges. The capacitance is the ratio of the magnitude of the charge on either conductor Q to the potential difference between the conductors V:

$$C = \frac{Q}{V}$$

Capacitance depends only on the geometrical arrangement of the conductors and not on the charge or the potential difference.

2. The capacitance of a parallel-plate capacitor is proportional to the area of the plates and is inversely proportional to the separation distance:

$$C = \frac{\epsilon_0 A}{s}$$

The capacitance of a cylindrical capacitor is given by

$$C = \frac{2\pi\epsilon_0 L}{\ln (b/a)}$$

where L is the length of the capacitor and a and b are the radii of the inner and outer conductors, respectively.

3. A nonconducting material is called a dielectric. When a dielectric is inserted between the plates of a capacitor, the molecules in the dielectric become polarized and the electric field within the dielectric is weakened. If the field is E_0 without the dielectric, with the dielectric it is

$$E = \frac{E_0}{\kappa}$$

where κ is the dielectric constant. This decrease in electric field leads to an increase in the capacitance by the factor κ:

$$C = \kappa C_0$$

where C_0 is the capacitance without the dielectric. The permittivity of a dielectric ϵ is defined as

$$\epsilon = \kappa \epsilon_0$$

Dielectrics also provide a physical means of separating the plates of a capacitor, and they increase the voltage that can be applied to a capacitor before dielectric breakdown occurs.

4. The electrostatic energy stored in a capacitor of charge Q, potential difference V, and capacitance C is

$$U = \frac{1}{2}\frac{Q^2}{C} = \frac{1}{2}QV = \frac{1}{2}CV^2$$

This energy can be considered to be stored in the electric field between the plates. The energy per unit volume in an electric field E is given by

$$\eta = \frac{\text{energy}}{\text{volume}} = \frac{1}{2}\epsilon E^2$$

5. When two or more capacitors are connected in parallel, the equivalent capacitance of the combination is the sum of the individual capacitances:

$$C_{eq} = C_1 + C_2 + C_3 + \cdots \qquad \text{capacitors in parallel}$$

When two or more capacitors are connected in series, the reciprocal of the equivalent capacitance is found by adding the reciprocals of the individual capacitances:

$$\frac{1}{C_{eq}} = \frac{1}{C_1} + \frac{1}{C_2} + \frac{1}{C_3} + \cdots \qquad \text{capacitors in series}$$

Suggestion for Further Reading

Trotter, Donald M., Jr.: "Capacitors," *Scientific American*, July 1988, p. 86.

Marvelous advances in capacitor miniaturization have been achieved over the past thirty years; without this, the advantages of miniaturizing integrated circuits for radios, computers, and other electronic equipment could not have been realized.

Review

A. Objectives: After studying this chapter, you should:

1. Be able to define capacitance and calculate it for a parallel-plate capacitor and a cylindrical capacitor.

2. Be able to discuss the effects of a dielectric on the capacitance, charge, potential difference, and electric field of a parallel-plate capacitor.

3. Know what is meant by bound charge in a dielectric, and be able to discuss how bound charge arises and what effect it has.

4. Be able to derive the expression $U = \frac{1}{2}QV$ for the energy stored in a charged capacitor.

5. Be able to discuss the concept of electrostatic field energy.

6. Be able to calculate the equivalent capacitance of parallel and series combinations of capacitors.

B. Define, explain, or otherwise identify:

Capacitor
Capacitance
Farad
Dielectric
Dielectric constant
Permittivity of a dielectric
Bound charge
Electrostatic field energy
Energy density
Capacitors in parallel
Equivalent capacitance
Capacitors in series

C. True or false: If the statement is true, explain why it is true. If it is false, give a counterexample.

1. The capacitance of a capacitor is defined as the total amount of charge it can hold.

2. The capacitance of a parallel-plate capacitor depends on the voltage difference between the plates.

3. The capacitance of a parallel-plate capacitor is proportional to the charge on its plates.

4. The equivalent capacitance of two capacitors in parallel equals the sum of the individual capacitances.

5. The equivalent capacitance of two capacitors in series is less than the capacitance of either capacitor alone.

6. A dielectric inserted into a capacitor increases the capacitance.

7. The electrostatic energy per unit volume at some point is proportional to the square of the electric field at that point.

Problems

Level I

21-1 The Parallel-Plate Capacitor

1. (*a*) If a parallel-plate capacitor has a 0.15-mm separation, what must its area be for it to have a capacitance of 1 F? (*b*) If the plates are square, what is the length of their sides?

2. A parallel-plate capacitor has a capacitance of 2.0 μF and a plate separation of 1.6 mm. (*a*) What is the maximum potential difference between the plates such that dielectric breakdown of the air between the plates does not occur? (E_{max} = 3 MV/m) (*b*) How much charge is stored at this maximum potential difference?

3. A parallel-plate capacitor has a charge of 40 μC. The potential difference across the plates is 500 V. What is the capacitance?

4. An electric field of 3×10^6 V/m exists between the plates of a circular parallel-plate capacitor that has a plate separation of 2 mm. (*a*) What is the voltage across the capacitor? (*b*) What plate radius is required if the stored charge is 10 μC?

21-2 The Cylindrical Capacitor

5. A coaxial cable between two cities has an inner radius of 0.8 mm and an outer radius of 6 mm. Its length is 8×10^5 m (about 500 mi). Treat this cable as a cylindrical capacitor and calculate its capacitance.

6. A Geiger tube consists of a wire of radius 0.2 mm and length 12 cm and a coaxial cylindrical shell conductor of the same length and a radius of 1.5 cm. (*a*) Find the capacitance, assuming that the gas in the tube has a dielectric constant of 1. (*b*) Find the charge per unit length on the wire when the potential difference between the wire and shell is 1.2 kV.

21-3 Dielectrics

7. A parallel-plate capacitor is made by placing polyethylene (κ = 2.3) between two sheets of aluminum foil. The area of each sheet is 400 cm^2, and the thickness of the polyethylene is 0.3 mm. Find the capacitance.

8. What is the dielectric constant of a dielectric on which the induced bound charge density is (*a*) 80 percent of the free charge density on the plates of a capacitor filled by the dielectric, (*b*) 20 percent of the free charge density, and (*c*) 98 percent of the free charge density?

9. Two parallel plates have charges Q and $-Q$. When the space between the plates is devoid of matter, the electric field is 2.5×10^5 V/m. When the space is filled with a certain dielectric, the field is reduced to 1.2×10^5 V/m. (*a*) What is the dielectric constant of the dielectric? (*b*) If Q = 10 nC, what is the area of the plates? (*c*) What is the total induced charge on either face of the dielectric?

21-4 The Storage of Electrical Energy

10. (*a*) A 3-μF capacitor is charged to 100 V. How much energy is stored in the capacitor? (*b*) How much additional energy is required to charge the capacitor from 100 to 200 V?

11. How much energy is stored when an isolated spherical conductor of radius 10 cm is charged to 2 kV?

12. A 10-μF capacitor is charged to $Q = 4\ \mu$C. (*a*) How much energy is stored in the capacitor? (*b*) If half the charge is removed, how much energy remains?

13. (*a*) Find the energy stored in a 20-pF capacitor when it is charged to 5 μC. (*b*) How much additional energy is required to increase the charge from 5 to 10 μC?

14. Find the energy per unit volume in an electric field that is equal to the dielectric strength of air (3 MV/m).

15. A parallel-plate capacitor with a plate area of 2 m^2 and a separation of 1.0 mm is charged to 100 V. (*a*) What is the electric field between the plates? (*b*) What is the energy per unit volume in the space between the plates? (*c*) Find the total energy by multiplying your answer to part (*b*) by the total volume between the plates. (*d*) Find the capacitance C. (*e*) Calculate the total energy from $U = \frac{1}{2}CV^2$, and compare your answer with your result for part (*c*).

21-5 Combinations of Capacitors

16. A 10.0-μF capacitor is connected in series with a 20.0-μF capacitor across a 6.0-V battery. (*a*) What is the equivalent capacitance of this combination? (*b*) Find the charge on each capacitor. (*c*) Find the potential difference across each capacitor.

17. A 10.0-μF capacitor and a 20.0-μF capacitor are connected in parallel across a 6.0-V battery. (*a*) What is the equivalent capacitance of this combination? (*b*) What is the potential difference across each capacitor? (*c*) Find the charge on each capacitor.

18. Three capacitors have capacitances of 2.0, 4.0, and 8.0 μF. Find the equivalent capacitance (*a*) if the capacitors are connected in parallel and (*b*) if they are connected in series.

19. A 2.0-μF capacitor is charged to a potential difference of 12.0 V and is then disconnected from the battery. (*a*) How much charge is on the plates? (*b*) When a second capacitor that is initially uncharged is connected in parallel with the first capacitor, the potential difference drops to 4.0 V. What is the capacitance of the second capacitor?

20. (*a*) How many 1.0-μF capacitors connected in parallel would it take to store a total charge of 1 mC with a potential difference of 10 V across each capacitor? (*b*) What would be the potential difference across the combination? (*c*) If the number of 1.0-μF capacitors found in part (*a*) is connected in series and the potential difference across each is 10 V, find the charge on each and the potential difference across the combination.

21. A 1.0-μF capacitor is connected in parallel with a 2.0-μF capacitor, and the combination is connected in series with a 6.0-μF capacitor. What is the equivalent capacitance of this combination?

22. A 3.0-μF capacitor and a 6.0-μF capacitor are connected in series, and the combination is connected in parallel with an 8.0-μF capacitor. What is the equivalent capacitance of this combination?

23. Three capacitors are connected in a triangular network as shown in Figure 21-16. Find the equivalent capacitance across terminals a and c.

Figure 21-16 Problem 23.

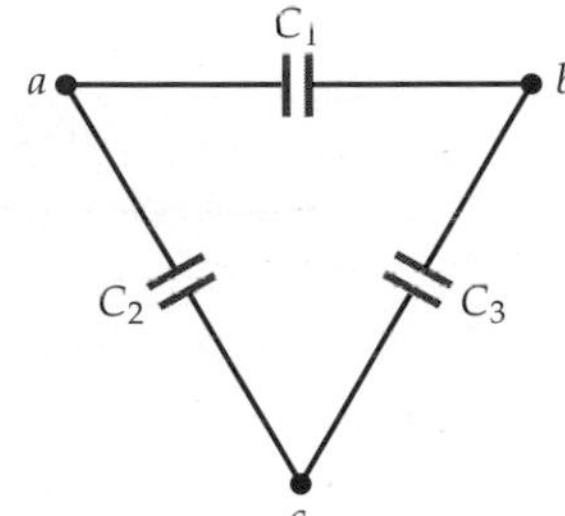

Level II

24. A parallel-plate capacitor has plates of area 600 cm^2 and a separation of 4 mm. It is charged to 100 V and is then disconnected from the battery. (*a*) Find the electric field E_0, the charge density σ, and the electrostatic potential energy U. A dielectric of constant $\kappa = 4$ is then inserted, completely filling the space between the plates. Find (*b*) the new electric field E, (*c*) the potential difference V, and (*d*) the bound charge density.

25. A certain dielectric with a dielectric constant $\kappa = 24$ can withstand an electric field of 4×10^7 V/m. Suppose we want to use this dielectric to construct a 0.1-μF capacitor that can withstand a potential difference of 2000 V. (*a*) What is the minimum plate separation? (*b*) What must the area of the plates be?

26. A parallel-plate capacitor of area A and separation d is charged to a potential difference V and is then disconnected from the charging source. The plates are then pulled apart until the separation is $2d$. Find expressions in terms of A, d, and V for (*a*) the new capacitance, (*b*) the new potential difference, and (*c*) the new stored energy. (*d*) How much work was required to change the plate separation from d to $2d$?

27. A parallel-plate, air-gap capacitor has a capacitance of 0.14 μF. The plates are 0.5 mm apart. (*a*) What is the area of each plate? (*b*) What is the potential difference if there is a charge of 3.2 μC on one plate and a charge of $-3.2\ \mu$C on the other? (*c*) What is the stored energy? (*d*) How much charge can the capacitor carry before dielectric breakdown of the air between the plates occurs?

28. For the circuit shown in Figure 21-17, find (*a*) the total equivalent capacitance between the terminals, (*b*) the charge stored on each capacitor, and (*c*) the total stored energy.

Figure 21-17 Problem 28.

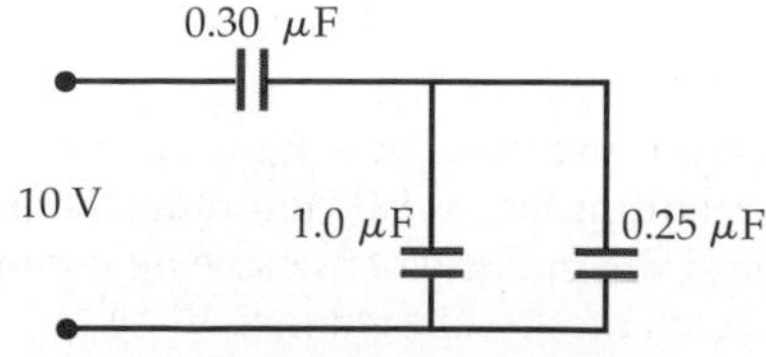

29. For the circuit shown in Figure 21-18, find (*a*) the total equivalent capacitance between the terminals, (*b*) the charge stored on each capacitor, and (*c*) the total stored energy.

Figure 21-18 Problem 29.

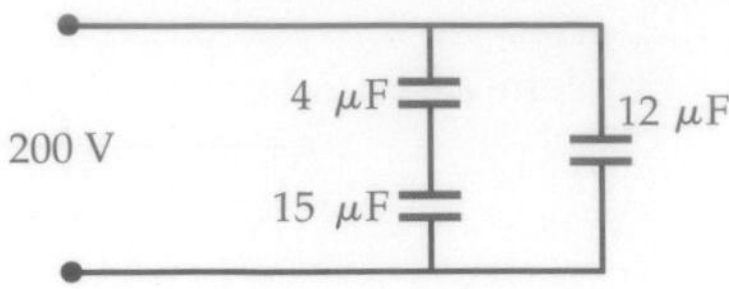

30. (*a*) Show that the equivalent capacitance of two capacitors in series can be written

$$C_{eq} = \frac{C_1C_2}{C_1 + C_2}$$

(*b*) Use this expression to show that $C_{eq} < C_1$ and $C_{eq} < C_2$. (*c*) Show that the correct expression for the equivalent capacitance of three capacitors in series is

$$C_{eq} = \frac{C_1C_2C_3}{C_1C_2 + C_2C_3 + C_1C_3}$$

31. A 20-pF capacitor is charged to 3.0 kV and is then removed from the battery and connected to an uncharged 50-pF capacitor. (*a*) What is the new charge on each capacitor? (*b*) Find the initial energy stored in the 20-pF capacitor and the final energy stored in the two capacitors. Is energy gained or lost when the two capacitors are connected?

32. Find all the different possible equivalent capacitances that can be obtained using a 1.0-, a 2.0-, and a 4.0-μF capacitor in any combination that includes all three or any two of the capacitors.

33. Three identical capacitors are connected so that their maximum equivalent capacitance is 15 μF. (*a*) Describe this combination. (*b*) Find the three other possible combinations using all three capacitors and their equivalent capacitances.

34. Two capacitors $C_1 = 4\ \mu F$ and $C_2 = 12\ \mu F$ are connected in series across a 12-V battery. They are carefully disconnected so that they are not discharged and are reconnected to each other with positive plate to positive plate and negative plate to negative plate. (*a*) Find the potential difference across each capacitor after they are connected. (*b*) Find the initial and final energy stored in the capacitors.

35. Work Problem 34 if the two capacitors are first connected in parallel across the 12-V battery and are then connected, with the positive plate of each capacitor connected to the negative plate of the other.

36. A 100-pF capacitor and a 400-pF capacitor are both charged to 2.0 kV. They are then disconnected from the voltage source and are connected together, positive plate to positive plate and negative plate to negative plate. (*a*) Find the resulting potential difference across each capacitor. (*b*) Find the energy lost when the connections are made.

37. Work Problem 36 if the capacitors are connected with the positive plate of each connected to the negative plate of the other, after they have been charged to 2.0 kV.

38. A parallel-plate capacitor has a capacitance C_0 and a plate separation d. Two dielectric slabs of constants κ_1 and κ_2, each of thickness $\frac{1}{2}d$ and having the same area as the plates, are inserted between the plates as shown in Figure 21-19. When the free charge on the plates is Q, find (*a*) the electric field in each dielectric and (*b*) the potential difference between the plates. (*c*) Show that the new capacitance is given by

$$C = \frac{2\kappa_1\kappa_2}{\kappa_1 + \kappa_2}C_0$$

(*d*) Show that this system can be considered to be two capacitors of thickness $\frac{1}{2}d$ connected in series.

Figure 21-19 Problem 38.

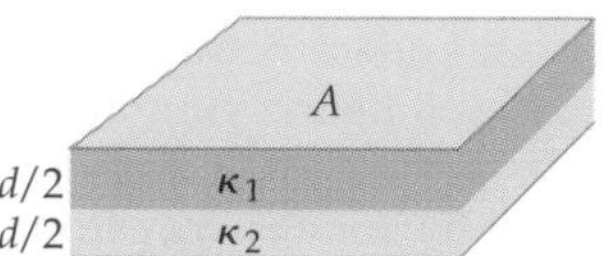

39. The membrane of the axon of a nerve cell is a thin cylindrical shell of radius $r = 10^{-5}$ m, length $L = 0.1$ m, and thickness $d = 10^{-8}$ m. The membrane has a positive charge on one side and a negative charge on the other, and acts as a parallel-plate capacitor of area $A = 2\pi rL$ and separation d. Its dielectric constant is about $\kappa = 3$. (*a*) Find the capacitance of the membrane. If the potential difference across the membrane is 70 mV, find (*b*) the charge on each side of the membrane, and (*c*) the electric field through the membrane.

40. A parallel-plate capacitor has a plate area A and a separation d. A metal slab of thickness t and area A is inserted between the plates. (*a*) Show that the capacitance is given by $C = \epsilon_0 A/(d - t)$, regardless of where the metal slab is placed. (*b*) Show that this arrangement can be considered to be a capacitor of separation a in series with one of separation b, where $a + b + t = d$.

41. In Figure 21-20, $C_1 = 2\ \mu F$, $C_2 = 6\ \mu F$, and $C_3 = 3.5\ \mu F$. (*a*) Find the equivalent capacitance of this combination. (*b*) If the breakdown voltages of the individual capacitors are $V_1 = 100$ V, $V_2 = 50$ V, and $V_3 = 400$ V, what maximum voltage can be placed across points a and b?

Figure 21-20 Problem 41.

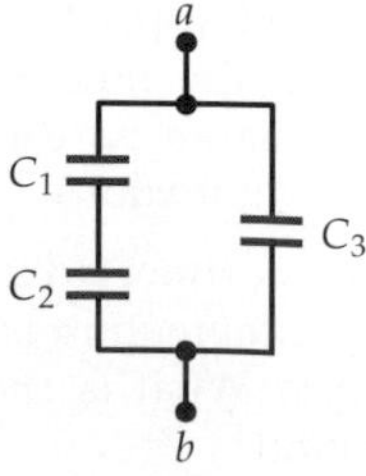

42. A parallel-plate capacitor is filled with two dielectrics of equal size as shown in Figure 21-21. Show (*a*) that this system can be considered to be two capacitors of area $\frac{1}{2}A$ connected in parallel and (*b*) that the capacitance is increased by the factor $(\kappa_1 + \kappa_2)/2$.

Figure 21-21 Problem 42.

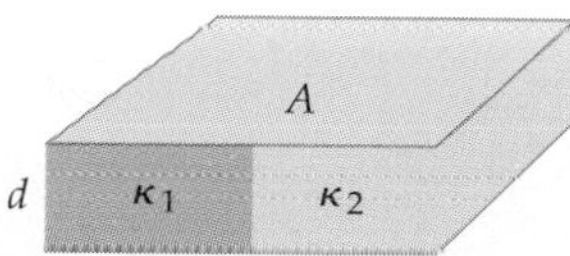

43. A Leyden jar, the earliest type of capacitor, is a glass jar coated inside and outside with metal foil. Suppose that a Leyden jar is a cylinder 40 cm high with 2.0-mm-thick walls and an inner diameter of 8 cm. Ignore any field fringing. (*a*) Find the capacitance of this Leyden jar if the dielectric constant of the glass is 5.0. (*b*) If the dielectric strength of the glass is 15 MV/m, what maximum charge can the Leyden jar carry without undergoing dielectric breakdown?

44. A parallel-plate capacitor of plate area A and separation x is given a charge Q and is then removed from the charging source. (*a*) Find the stored electrostatic energy as a function of x. (*b*) Find the increase in energy dU due to an increase in plate separation dx from $dU = (dU/dx)\,dx$. (*c*) If F is the force exerted by one plate on the other, the work needed to move one plate a distance dx is $F\,dx = dU$. Show that $F = Q^2/2\epsilon_0 A$. (*d*) Show that the force in part (*c*) equals $\frac{1}{2}EQ$, where Q is the charge on one plate and E is the electric field between the plates. Discuss the reason for the factor $\frac{1}{2}$ in this result.

45. Design a network of capacitors that has a capacitance of 2 μF and breakdown voltage of 400 V using as many 2-μF capacitors that have individual breakdown voltages of 100 V as needed.

46. A 1.2-μF capacitor is charged to 30 V. After charging, the capacitor is disconnected from the voltage source and is connected to another uncharged capacitor. The final voltage is 10 V. (*a*) What is the capacitance of the other capacitor? (*b*) How much energy was lost when the connection was made?

47. A rectangular parallel-plate capacitor of length a and width b has a dielectric of width b partially inserted a distance x between the plates as shown in Figure 21-22. (*a*) Find the capacitance as a function of x. Neglect edge effects. (*b*) Show that your answer gives the expected results for $x = 0$ and $x = a$.

Figure 21-22 Problem 47.

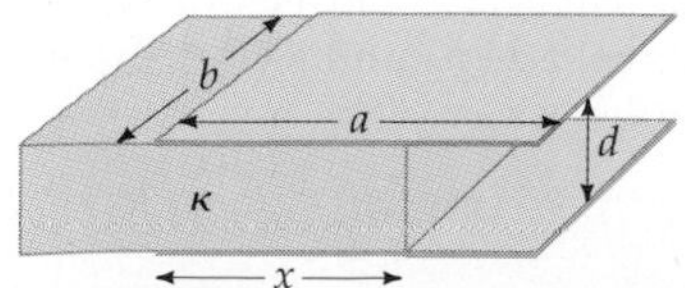

48. Determine the capacitance of each of the networks shown in Figure 21-23.

Figure 21-23 Problem 48.

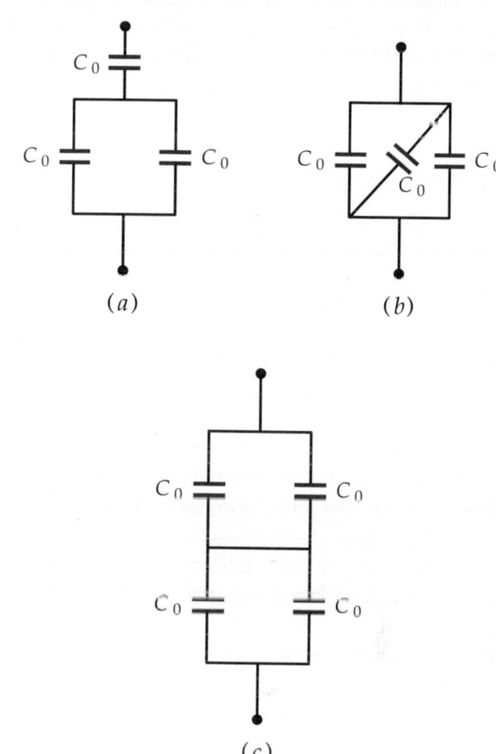

49. Five identical capacitors of capacitance C_0 are connected in a bridge network as shown in Figure 21-24. (*a*) What is the equivalent capacitance between points a and b? (*b*) Find the equivalent capacitance if the capacitance between a and b is changed to $10C_0$.

Figure 21-24 Problem 49.

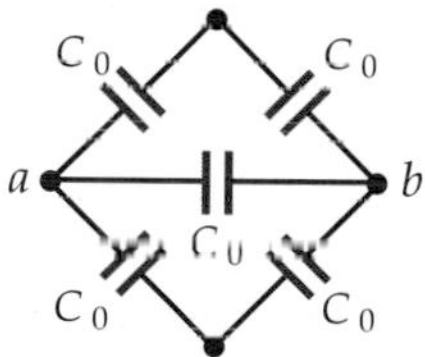

50. A parallel-plate capacitor with plates of area 500 cm^2 is charged to a potential difference V and is then disconnected from the voltage source. When the plates are moved 0.4 cm farther apart, the voltage between the plates increased by 100 V. (*a*) What is the charge Q on the positive plate of the capacitor? (*b*) How much does the energy stored in the capacitor increase due to the movement of the plates?

51. Design a 0.1-μF parallel-plate capacitor with air between the plates that can be charged to a maximum potential difference of 1000 V. (*a*) What is the minimum possible separation between the plates? (*b*) What minimum area must the plates of the capacitor have?

52. Three capacitors, $C_1 = 2\ \mu F$, $C_2 = 4\ \mu F$, and $C_3 = 6\ \mu F$, are connected in parallel and are charged with a 200-V source. They are then disconnected from the source and reconnected positive plates to negative plates as shown in Figure 21-25. (*a*) What is the voltage across each capacitor with switches S_1 and S_2 closed but with switch S_3 open? (*b*) After switch S_3 is closed, what is the final charge on each capacitor? (*c*) Give the voltage across each capacitor after switch S_3 is closed.

Figure 21-25 Problem 52.

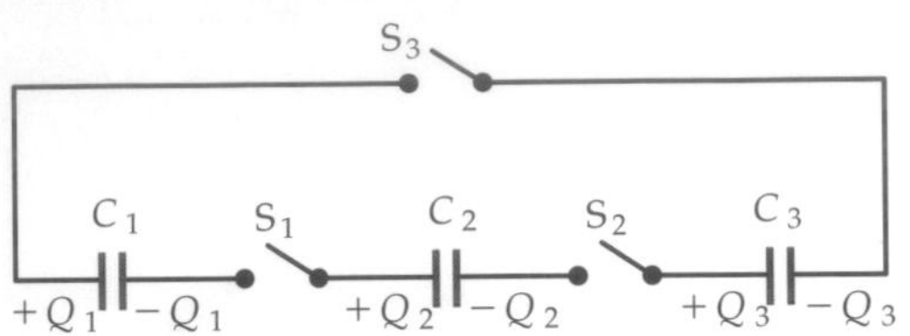

53. A parallel-plate capacitor is constructed from a layer of silicon dioxide of thickness 5×10^{-6} m between two conducting films. The dielectric constant of silicon dioxide is 3.8 and its dielectric strength is 8×10^6 V/m. (*a*) What voltage can be applied across this capacitor without dielectric breakdown? (*b*) What should the surface area of the layer of silicon dioxide be for a 100-pF capacitor? (*c*) Estimate the number of these capacitors that can fit into a square 1 cm by 1 cm.

54. Estimate the electrical energy stored in the atmosphere if the earth's electric field extends upward for 1000 m and has an average magnitude of 200 V/m. *Hint:* You may treat the atmosphere as a rectangular layer with an area equal to the surface area of the earth. Why?

55. You are asked to construct a parallel-plate, air-gap capacitor that will store 100 kJ of energy. (*a*) What minimum volume is required between the plates of the capacitor? (*b*) Suppose you have developed a dielectric that can withstand 3×10^8 V/m and has a dielectric constant of 5. What volume of this dielectric between the plates of the capacitor is required for it to be able to store 100 kJ of energy?

56. Two identical, 4-μF parallel-plate capacitors are connected in series across a 24-V battery. (*a*) What is the charge on each capacitor? (*b*) What is the total stored energy of the capacitors? A dielectric having a dielectric constant of 4.2 is inserted between the plates of one of the capacitors while the battery is still connected. After the dielectric is inserted, (*c*) what is the charge on each capacitor? (*d*) What is the potential difference across each capacitor? (*e*) What is the total stored energy of the capacitors?

57. Consider two parallel-plate capacitors, C_1 and C_2, that are connected in parallel. The capacitors are identical except that C_2 has a dielectric inserted between its plates. A voltage source of 200 V is connected across the capacitors to charge them and is then disconnected. (*a*) What is the charge on each capacitor? (*b*) What is the total stored energy of the capacitors? (*c*) The dielectric is removed from C_2. What is the final stored energy of the capacitors? (*d*) What is the final voltage across the two capacitors?

58. Find the capacitance of the parallel-plate capacitor shown in Figure 21-26.

Figure 21-26 Problem 58.

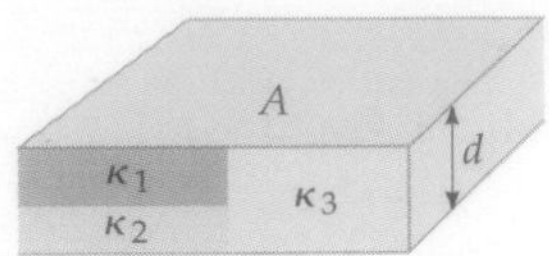

Level III

59. Two parallel-plate capacitors have the same separation and plate area. The capacitance of each is initially 10 μF. When a dielectric is inserted such that it completely fills the space between the plates of one of the capacitors, the capacitance of that capacitor increases to 35 μF. The 35-μF and 10-μF capacitors are connected in parallel and are charged to a potential difference of 100 V. The voltage source is then disconnected. (*a*) What is the stored energy of this system? (*b*) What are the charges on the two capacitors? (*c*) The dielectric is removed from the capacitor. What are the new charges on the plates of the capacitors? (*d*) What is the final stored energy of the system?

60. A parallel-plate capacitor of area A and separation d is charged to a potential difference V and is then removed from the charging source. A dielectric slab of constant $\kappa = 2$, thickness d, and area $\frac{1}{2}A$ is inserted as shown in Figure 21-27. Let σ_1 be the free charge density at the conductor–dielectric surface and σ_2 be the free charge density at the conductor–air surface. (*a*) Why must the electric field have the same value inside the dielectric as in the free space between the plates? (*b*) Show that $\sigma_1 = 2\sigma_2$. (*c*) Show that the new capacitance is $3\epsilon_0 A/2d$ and that the new potential difference is $\frac{2}{3}V$.

Figure 21-27 Problem 60.

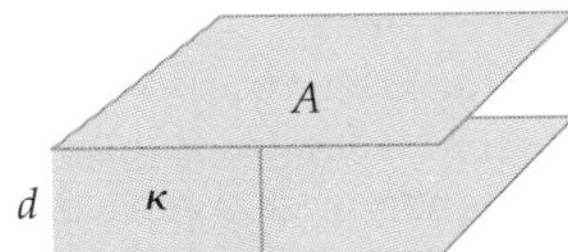

61. Two identical, 10-μF parallel-plate capacitors are given equal charges of 100 μC each and are then removed from the charging source. The charged capacitors are connected by a wire between their positive plates and another wire between their negative plates. (*a*) What is the stored energy of the system? A dielectric having a dielectric constant of 3.2 is inserted between the plates of one of the capacitors such that it completely fills the region between the plates. (*b*) What is the final charge on each capacitor? (*c*) What is the final stored energy of the system?

62. A spherical capacitor consists of two thin concentric spherical shells of radii R_1 and R_2. (*a*) Show that the capacitance is given by

$$C = 4\pi\epsilon_0 R_1 R_2/(R_2 - R_1)$$

(*b*) Show that when the radii of the shells are nearly equal, the capacitance is given approximately by the expression for the capacitance of a parallel-plate capacitor, $C = \epsilon_0 A/d$, where A is the area of the sphere and $d = R_2 - R_1$.

63. A parallel-plate capacitor of plate area 1.0 m^2 and plate separation distance 0.5 cm has a glass plate of the same area that completely fills the space between its plates. The glass has a dielectric constant of 5.0. The capacitor is charged to a potential difference of 12.0 V and is then removed from its charging source. How much work is required to pull the glass plate out of the capacitor?

64. A spherical capacitor has an inner sphere of radius R_1 with a charge of $+Q$ and an outer concentric spherical shell of radius R_2 with a charge of $-Q$. (*a*) Find the electric field and the energy density at any point in space. (*b*) How much energy is there in a spherical shell of radius r, thickness dr, and volume $4\pi r^2\,dr$ between the conductors? (*c*) Integrate your expression from part (*b*) to find the total energy stored in the capacitor, and compare your result with that obtained using $U = \frac{1}{2}QV$.

65. A cylindrical capacitor consists of a long wire of radius R_1 and length L with a charge of $+Q$ and a concentric outer cylindrical shell of radius R_2, length L, and charge $-Q$. (*a*) Find the electric field and the energy density at any point in space. (*b*) How much energy is there in a cylindrical shell of radius r, thickness dr, and volume $2\pi rL\,dr$ between the conductors? (*c*) Integrate your expression from part (*b*) to find the total energy stored in the capacitor, and compare your result with that obtained using $U = \frac{1}{2}CV^2$.

66. A ball of charge of radius R has a uniform charge density ρ and a total charge $Q = \frac{4}{3}\pi R^3\rho$. (*a*) Find the electrostatic energy density at distance r from the center of the ball for $r < R$ and for $r > R$. (*b*) Find the energy in a spherical shell of volume $4\pi r^2\,dr$ for both $r < R$ and $r > R$. (*c*) Compute the total electrostatic energy by integrating your expressions from part (*b*), and show that your result can be written $U = \frac{3}{5}kQ^2/R$. Explain why this result is greater than that for a spherical conductor of radius R carrying a total charge Q.

67. A capacitor is constructed of two concentric cylinders of radii a and b ($b > a$) having a length $L \gg b$. A charge of $+Q$ is on the inner cylinder and a charge of $-Q$ is on the outer cylinder. The region between the two cylinders is filled with a dielectric having a dielectric constant κ. (*a*) Find the potential difference between the cylinders. Find the density of the free charge σ_f on (*b*) the inner cylinder and (*c*) the outer cylinder. Find the bound charge density σ_b on (*d*) the inner cylindrical surface of the dielectric and (*e*) the outer surface of the dielectric. (*f*) Find the total stored electrostatic energy. (*g*) If the dielectric will move without friction, how much mechanical work is required to remove the dielectric cylindrical shell?

68. A conducting sphere of radius R_1 is given a free charge Q. The sphere is surrounded by an uncharged concentric spherical dielectric shell having an inner radius R_1, an outer radius R_2, and a dielectric constant κ. The system is far removed from other objects. (*a*) Find the electric field everywhere in space. (*b*) What is the potential of the conducting sphere relative to $V = 0$ at infinity? (*c*) Find the total electrostatic energy of the system.

69. A parallel-plate capacitor is constructed using a variable dielectric. Let A be the area of the plates and y_0 be their separation. The dielectric constant is given as a function of y according to

$$\kappa = 1 + \frac{3}{y_0}y$$

The bottom plate is at $y = 0$ and the top plate is at $y = y_0$. (*a*) What is the capacitance? (*b*) Find the induced charge density on the surfaces of the dielectric. (*c*) Use Gauss's law to find the induced volume charge density $\rho(y)$ within this dielectric. (*d*) Integrate the expression for the volume charge density found in (*c*) over the dielectric, and show that the total induced bound charge, including that on the surfaces, is zero.

70. A capacitor has rectangular plates of length a and width b. The top plate is inclined at a small angle as shown in Figure 21-28. The plate separation varies from $s = y_0$ at the left to $s = 2y_0$ at the right, where y_0 is much less than a or b. Calculate the capacitance using strips of width dx and length b to approximate differential capacitors of area $b\,dx$ and separation $s = y_0 + (y_0/a)x$ that are connected in parallel.

Figure 21-28 Problem 70.

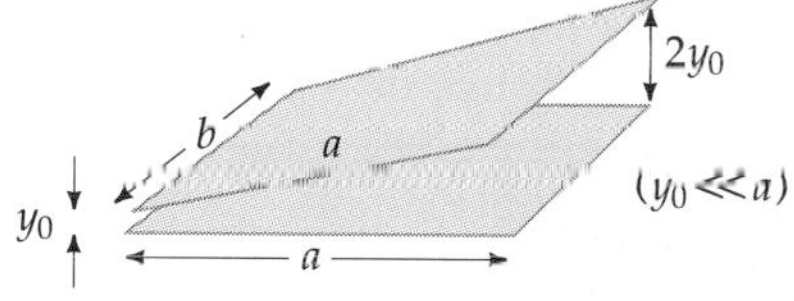

Chapter 22

Electric Current

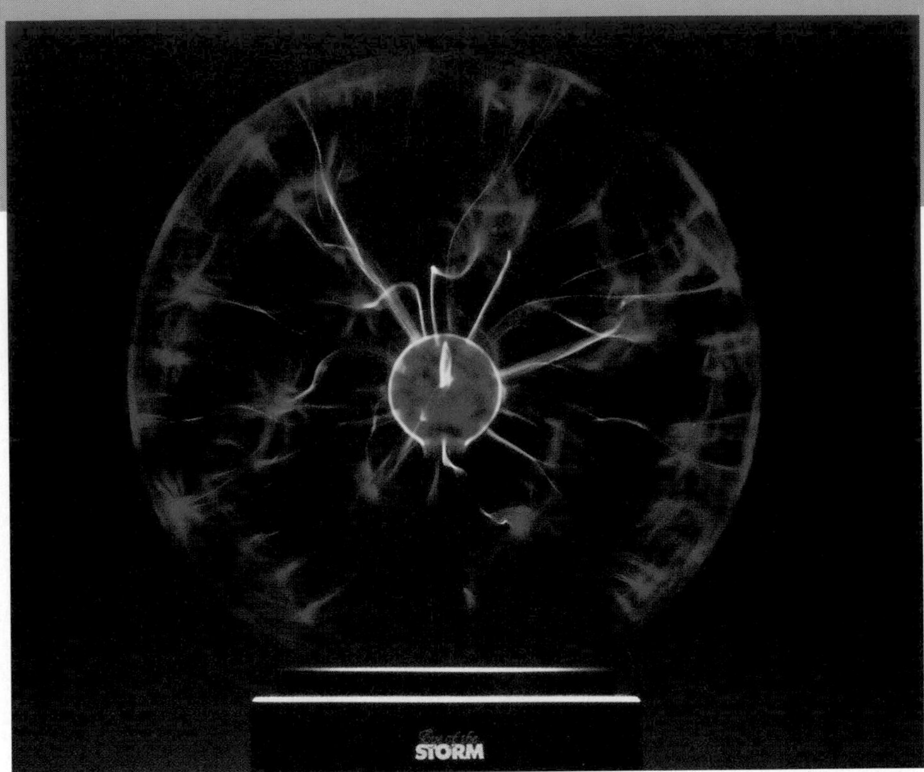

The plasma globe Eye of the Storm. In plasma globes, voltages ranging from 3 kV to 8 kV, at frequencies of between 20 kHz and 50 kHz, are applied between the sphere at the center of the globe and the outer shell, which is at ground. The region in between is filled with a mixture of inert gases. The applied voltage partially ionizes the gases, creating plasma filaments that conduct current. The colored displays are paths along which ionization is occurring.

Inert gases are more easily ionized than many noninert gases—it is easier to achieve dielectric breakdowns in plasma globes than, for instance, in the earth's nitrogen–oxygen atmosphere. The ease of ionization at the voltages applied in Eye of the Storm is apparent in the globe's ability to support multiple simultaneous paths of ionization. Because each of the discharge paths has the same instantaneous polarity, the paths repel one another.

When you touch the outer shell, regions of electrical discharge are drawn to the vicinity of your hand because your body is a much better conductor than the glass outer shell, so it becomes the path of least resistance to the earth for the charge flowing onto the shell. The current passing through your body is low enough to be unnoticeable.

When we turn on a light, we connect the wire filament in the light bulb across a potential difference that causes electric charge to flow through the wire, much like a pressure difference in a garden hose causes water to flow through the hose. The flow of electric charge constitutes an electric current. We usually think of currents as being in conducting wires, but an electric current results from any flow of charge. An example of a current that is not in a conducting wire is the electron beam in a cathode ray tube, such as that in your video monitor, or a beam of charged ions from a particle accelerator. In this chapter, we will define electric current and relate it to the motion of charged particles. After a discussion of electrical resistance and Ohm's law, we will consider the energy aspects of electric currents. Resistors in parallel and in series are then discussed, and the chapter concludes with a brief discussion of the classical microscopic model of electrical conduction.

22-1 Current and Motion of Charges

Electric **current** is defined as the rate of flow of electric charge through a cross-sectional area. Figure 22-1 shows a segment of a current-carrying wire in which charge carriers are moving with some small average velocity. If ΔQ

is the charge that flows through the cross-sectional area A in time Δt, the current is

$$I = \frac{\Delta Q}{\Delta t} \qquad 22\text{-}1$$

Electric current

The SI unit of current is the **ampere** (A):

$$1\ \text{A} = 1\ \text{C/s} \qquad 22\text{-}2$$

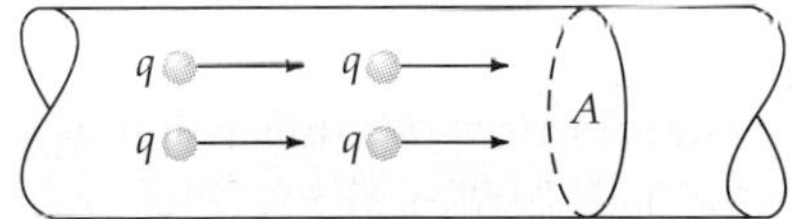

Figure 22-1 A segment of a current-carrying wire. If ΔQ is the amount of charge that flows through the cross-sectional area A in time Δt, the current is $I = \Delta Q/\Delta t$.

By convention, the direction of current is considered to be the direction of flow of positive charge. This convention was established before it was known that free electrons, which are negatively charged, are the particles that are actually moving and thus producing the current in a conducting wire. The motion of negatively charged electrons in one direction is equivalent to a flow of positive charge in the opposite direction. Thus, the electrons move in the direction opposite to the direction of the current. However, not all currents are produced by electrons flowing through a wire. In an accelerator that produces a proton beam, the direction of motion of the positively charged protons is in the direction of the current. In electrolysis, the current is produced by the flow of positive ions in the direction of the current plus the flow of negative ions and electrons opposite the direction of the current. The movement of negative particles in one direction and that of positive particles in the opposite direction both contribute to a current in the same direction. In nearly all applications, the motion of negative charges in one direction is indistinguishable from the motion of positive charges in the opposite direction. We can always think of current as motion of positive charges in the direction of the current and remember (if we need to) that in conducting wires, for example, the electrons are moving in the direction opposite to the current.

The actual motion of the free electrons in a conducting wire is quite complicated. When there is no electric field in the wire, these electrons move in random directions with relatively large speeds due to their thermal energy. Since the velocity vectors of the electrons are randomly oriented, the average velocity due to this thermal energy is zero. When an electric field is applied, for example, by connecting the wire to a battery that applies a potential difference along the wire, the free electrons experience a momentary acceleration due to the force $-e\mathbf{E}$. The electrons acquire a small velocity in the direction opposite the field, but the kinetic energy acquired is quickly dissipated by collisions with the fixed ions in the wire. The electrons are then again accelerated by the field. The net result of this repeated acceleration and dissipation of energy is that the electrons have a small **drift velocity** opposite to the electric field superimposed on their large, random, thermal velocity. The motion of the free electrons in a metal is somewhat similar to that of the molecules of a gas, such as air. In still air, the molecules move with large instantaneous velocities between collisions, but the average velocity is zero. When there is a breeze, the air molecules have a small drift velocity in the direction of the breeze superimposed on their much larger instantaneous velocities. Similarly, when there is no current in a conductor the electrons move about in random directions with very high speeds because of thermal energy. When there is a current, the electrons have a small drift velocity superimposed on their much larger, but random, thermal velocities.

Let us consider a current in a conducting wire of cross-sectional area A. Let n be the number of free charge-carrying particles per unit volume. We will assume that each particle carries a charge q and moves with a drift

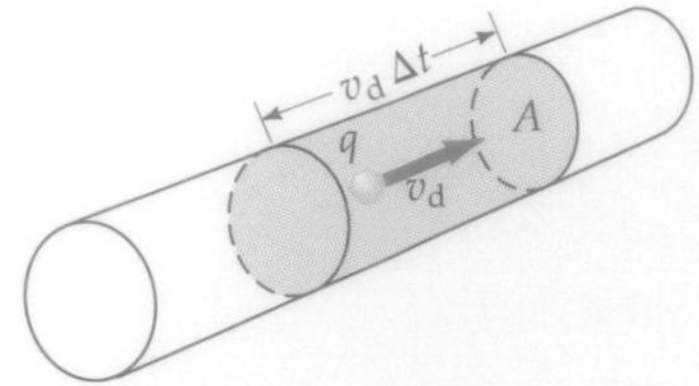

Figure 22-2 In time Δt, all the charges in the shaded volume pass through A. If there are n charge carriers per unit volume, each with charge q, the total charge in this volume is $\Delta Q = nqv_d A\ \Delta t$, where v_d is the drift velocity of the charge carriers. The total current is then $I = \Delta Q/\Delta t = nqv_d A$.

velocity v_d. In a time Δt, all the particles in the volume $Av_d\ \Delta t$, shaded in Figure 22-2, pass through the area element. The number of particles in this volume is $nAv_d\ \Delta t$, and the total charge is

$$\Delta Q = qnAv_d\ \Delta t$$

The current is thus

$$I = \frac{\Delta Q}{\Delta t} = nqAv_d \qquad 22\text{-}3$$

Equation 22-3 can be used to find the current due to the flow of any type of charged particle, simply by replacing the drift velocity v_d with the velocity of the particle.

We can get an idea of the order of magnitude of the drift velocity for electrons in a conducting wire by putting typical magnitudes into Equation 22-3.

Example 22-1

What is the drift velocity of electrons in a typical (14-gauge) copper wire of radius 0.815 mm carrying a current of 1 A?

If we assume one free electron per copper atom, the density of free electrons is the same as the density of atoms n_a, which is related to the mass density ρ, Avogadro's number N_A, and the molar mass M by

$$n_a = \frac{\rho N_A}{M}$$

For copper $\rho = 8.93\ \text{g/cm}^3$ and $M = 63.5$ g/mol. Then

$$n_a = \frac{(8.93\ \text{g/cm}^3)(6.02 \times 10^{23}\ \text{atoms/mol})}{63.5\ \text{g/mol}}$$

$$= 8.47 \times 10^{22}\ \text{atoms/cm}^3$$

The density of electrons is then

$$n = 8.47 \times 10^{22}\ \text{electrons/cm}^3 = 8.47 \times 10^{28}\ \text{electrons/m}^3$$

The drift velocity is therefore

$$v_d = \frac{I}{Ane} = \frac{1\ \text{C/s}}{\pi(0.000815\ \text{m})^2(8.47 \times 10^{28}\ \text{m}^{-3})(1.6 \times 10^{-19}\ \text{C})}$$

$$\approx 3.54 \times 10^{-5}\ \text{m/s}$$

We see that typical drift velocities are of the order of 0.01 mm/s, which is quite small.

Exercise

How long does it take for an electron to drift a distance of 1 m if its drift velocity is 3.54×10^{-5} m/s? (Answer: 7.85 h)

At first, it may seem surprising that an electric light comes on immediately when the switch is thrown since the electrons drift down the wire at such low speeds that it should take them hours to travel from the switch to the light. An analogy with water in a hose may prove useful. When you turn on water into a long, empty hose, it takes quite a few seconds for the water to travel from the faucet to the nozzle. However, if the hose is already full of water, the water emerges from the nozzle almost instantaneously. Because of the water pressure at the faucet, the segment of water there pushes on the water immediately next to it, which pushes on the next segment of water and so on, until the last segment of water is pushed out the nozzle. This pressure wave moves down the hose with the speed of sound in the water (if the hose is rigid), and the water quickly reaches a steady flow rate. The density of the water during a steady flow remains constant in time. The water that flows out of a segment of hose is replaced by an equal amount of water flowing into the segment at the other end. The behavior of wires full of free electrons is similar. When the light switch is turned on, an electric field is propagated down the wire with nearly the speed of light, and the free electrons throughout the wire acquire their drift velocity almost immediately. The charge density within a current-carrying wire remains constant in time. The charge that flows out of a segment of wire is replaced by an equal amount of charge that flows into the segment at the other end. Thus charge starts moving through the filament almost immediately after the light switch is thrown. The transport of a significant amount of charge in a wire is accomplished not by a few charges moving rapidly down the wire, but by a very large number of charges slowly drifting down the wire.

A 1000-A beam of 3-MeV electrons hits a lucite plate. The plate acquires a charge and discharges, producing the beautiful pattern shown. The electron beam, which lasts for about 1 μs, ionizes the air molecules, which give off a bluish glow when they recombine.

Example 22-2

In a certain particle accelerator, a current of 0.5 mA is carried by a 5 MeV proton beam that has a radius of 1.5 mm. (*a*) Find the number of protons per unit volume in the beam. (*b*) If the beam hits a target, how many protons hit the target in one second?

(*a*) From Equation 22-3, we have

$$n = \frac{I}{qAv}$$

where q is the charge on each proton, v is the speed of the protons, and A is the cross-sectional area of the beam. The kinetic energy of each proton is 5 MeV:

$$K_k = \frac{1}{2}mv^2 = 5 \text{ MeV} = 5 \times 10^6 \text{ eV} \times \frac{1.6 \times 10^{-19} \text{ J}}{1 \text{ eV}} = 8 \times 10^{-13} \text{ J}$$

Using $m = 1.67 \times 10^{-27}$ kg for the mass of a proton, we have for the speed of each proton

$$v = \sqrt{\frac{2K_k}{m}} = \sqrt{\frac{(2)(8 \times 10^{-13} \text{ J})}{1.67 \times 10^{-27} \text{ kg}}} = 3.10 \times 10^7 \text{ m/s}$$

The number of protons per unit volume of the beam is then

$$n = \frac{I}{qAv}$$

$$= \frac{0.5 \times 10^{-3} \text{ A}}{(1.6 \times 10^{-19} \text{ C/proton})\pi(1.5 \times 10^{-3} \text{ m})^2(3.10 \times 10^7 \text{ m/s})}$$

$$= 1.43 \times 10^{13} \text{ protons/m}^3$$

(*b*) The number that strikes a target in time Δt is the number in the volume $Av\,\Delta t$, which is $nAv\,\Delta t$. For $\Delta t = 1$ s, this number is

$$N = nAv\,\Delta t$$

$$= (1.43 \times 10^{13} \text{ protons/m}^3)\pi(1.5 \times 10^{-3} \text{ m})^2(3.10 \times 10^7 \text{ m/s})(1 \text{ s})$$

$$= 3.13 \times 10^{15} \text{ protons}$$

We can check this result by noting that, from Equation 22-3,

$$nAv\,\Delta t = \frac{I\,\Delta t}{q} = \frac{Q}{q}$$

where $Q = I\,\Delta t$ is the total charge that strikes the target. Since the current is 0.5 mA, the total charge that strikes the target in 1 s is 0.5 mC. The number of protons is then

$$N = \frac{Q}{q} = \frac{0.5 \times 10^{-3}\text{ C}}{1.6 \times 10^{-19}\text{ C/proton}} = 3.13 \times 10^{15}\text{ protons}$$

22-2 Resistance and Ohm's Law

In our study of conductors in electrostatics, we argued that the electric field inside a conductor in electrostatic equilibrium must be zero. If this were not so, the free charges inside the conductor would move about. We are now considering situations in which the free charge *does* move in a conductor. That is, the conductors are not in electrostatic equilibrium. The current in a conductor is produced by an electric field inside the conductor that exerts a force on the free charges. Since the field **E** is in the direction of the force on a positive charge, and since the direction of the current is the direction of the flow of positive charge, the current is in the direction of the electric field. Figure 22-3 shows a segment of a wire of length ΔL and cross-sectional area A carrying a current I. Since the electric field points from regions of higher potential towards regions of lower potential, the potential at point a is greater than that at point b. Assuming that ΔL is small enough so that we may consider the electric field E to be constant across the segment, the potential difference V between points a and b is*

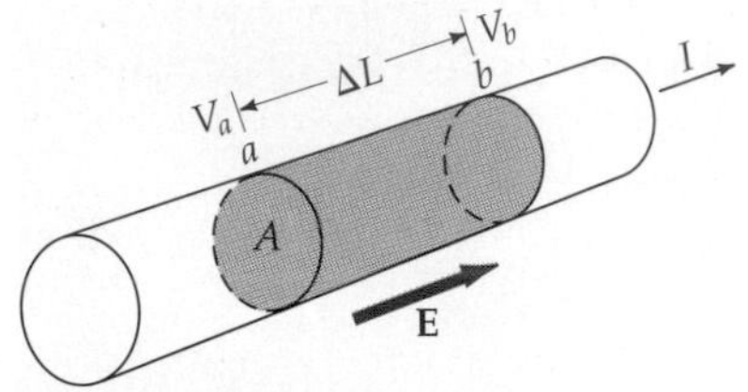

Figure 22-3 A segment of wire carrying a current I. The potential difference is related to the electric field by $V_a - V_b = E\,\Delta L$.

$$V = V_a - V_b = E\,\Delta L \qquad 22\text{-}4$$

For most materials,

The current in a wire segment is proportional to the potential difference across the segment.

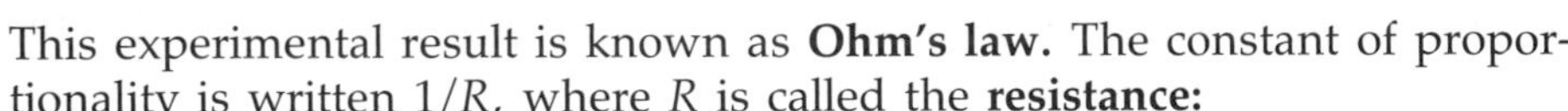

This experimental result is known as **Ohm's law.** The constant of proportionality is written $1/R$, where R is called the **resistance:**

$$I = \left(\frac{1}{R}\right) V$$

or

Resistance defined

$$R = \frac{V}{I} \qquad 22\text{-}5$$

Equation 22-5 gives a general definition of the resistance between two points in terms of the potential drop V between the points. The SI unit of resistance, the volt per ampere, is called an ohm (Ω):

$$1\ \Omega = 1\text{ V/A} \qquad 22\text{-}6$$

The resistance of a material depends on its length, its cross-sectional area, the type of material, and its temperature. For materials obeying Ohm's law,

*Again we use V rather than ΔV for the potential difference (which in this case is a potential *decrease*) to simplify the notation.

the resistance does not depend on the current I; that is, the ratio V/I is independent of I. Such materials, which include most metals, are called **ohmic materials.** For ohmic materials, the potential drop across a segment is proportional to the current:

$$V = IR \qquad R \text{ constant} \tag{22-7}$$

Ohm's law

Equation 22-7, with the qualification that R is constant, gives a mathematical statement of Ohm's law.

For **nonohmic materials,** the ratio V/I depends on the current I, so the current is not proportional to the potential difference. For nonohmic materials, the resistance R, as defined by Equation 22-5, depends on the current I. Figure 22-4 shows the potential difference V versus the current I for ohmic and nonohmic materials. For ohmic materials (the bottom curve), the relation is linear, so $R = V/I$ does not depend on I; but for nonohmic materials (the top curve), the relation is not linear, and $R = V/I$ *does* depend on I. Ohm's law is not a fundamental law of nature like Newton's laws or the laws of thermodynamics but rather is an empirical description of a property shared by many materials.

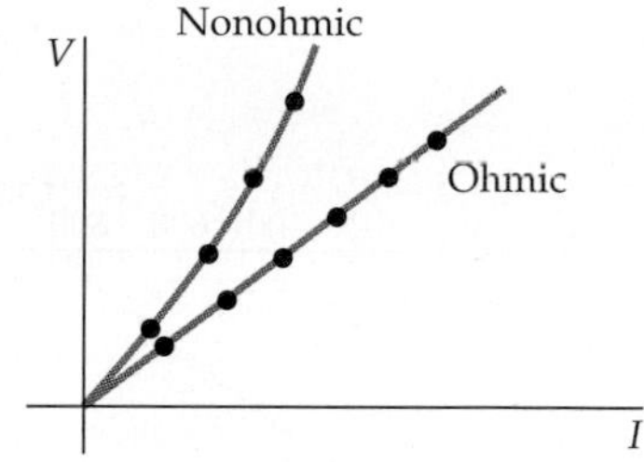

Figure 22-4 Plots of V versus I for ohmic and nonohmic materials. The resistance $R = V/I$ is independent of I for ohmic materials, as is indicated by the constant slope of the line.

Exercise

A wire of resistance 3 Ω carries a current of 1.5 A. What is the potential drop across the wire? (Answer: 4.5 V)

The resistance of a conducting wire is found to be proportional to the length of the wire and inversely proportional to its cross-sectional area:

$$R = \rho \frac{L}{A} \tag{22-8}$$

where the proportionality constant ρ is called the **resistivity** of the conducting material. The unit of resistivity is the ohm-meter (Ω·m).

Example 22-3

A nichrome wire (resistivity 10^{-6} Ω·m) has a radius of 0.65 mm. What length of wire is needed to obtain a resistance of 2.0 Ω?

The cross-sectional area of this wire is

$$A = \pi r^2 = (3.14)(6.5 \times 10^{-4}\ \text{m})^2 = 1.33 \times 10^{-6}\ \text{m}^2$$

From Equation 22-8, we have

$$L = \frac{RA}{\rho} = \frac{(2\ \Omega)(1.33 \times 10^{-6}\ \text{m}^2)}{10^{-6}\ \Omega\cdot\text{m}} = 2.66\ \text{m}$$

We sometimes refer to a wire as a conductor, and at other times we call it a resistor, depending on which property we wish to emphasize. The reciprocal of the resistivity is called the **conductivity** σ*:

$$\sigma = \frac{1}{\rho} \tag{22-9}$$

*The symbols ρ and σ, used here for the resistivity and conductivity, were used in previous chapters for volume charge density and surface charge density. Thus, care must be taken to distinguish what quantities are being referred to when these symbols are used. Usually this will be clear from the context.

Equation 22-8 may be written in terms of the conductivity instead of the resistivity:

$$R = \frac{L}{\sigma A} \tag{22-10}$$

Note that Equations 22-7 and 22-10 for electrical conduction and electrical resistance are of the same form as Equations 16-13 ($\Delta T = IR$) and 16-14 ($R = \Delta x/kA$) for thermal conduction and thermal resistance. For electrical equations, the potential difference V replaces the temperature difference ΔT and the electrical conductivity σ replaces the thermal conductivity k. Ohm was, in fact, led to his law by the similarity between the conduction of electricity and the conduction of heat.

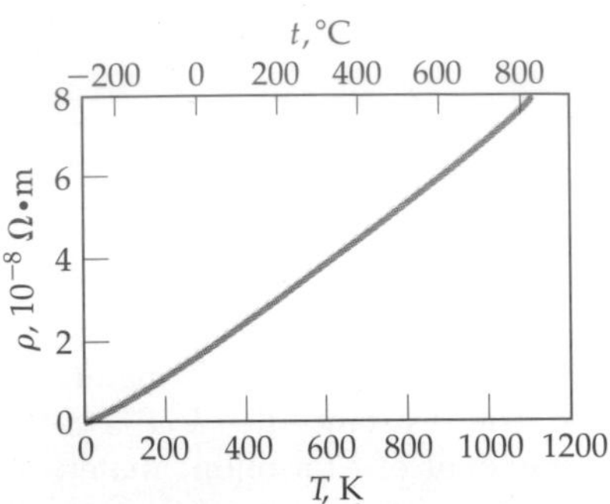

Figure 22-5 Plot of resistivity ρ versus temperature for copper.

The resistivity (and conductivity) of any given metal depends on the temperature. Figure 22-5 shows the temperature dependence of the resistivity of copper. This graph is nearly a straight line, which means that the resistivity varies nearly linearly with temperature. (There is a breakdown in this linearity at very low temperatures, which is not shown on the graph.) In tables, the resistivity is usually given in terms of its value at 20°C (ρ_{20}) along with the **temperature coefficient of resistivity** α, which is the slope of the ρ-versus-T curve. The resistivity at some other Celsius temperature t_C is then given by

$$\rho = \rho_{20}[1 + \alpha(t_C - 20°C)] \tag{22-11}$$

(Since the Celsius and absolute temperatures differ only in the choice of zero, the resistivity has the same slope whether it is plotted against t_C or T.) Table 22-1 gives the resistivity at 20°C and temperature coefficient α for various materials. Note the tremendous difference between the values of ρ for conductors (the metals listed) and those for nonconductors.

Wires used to carry electric current are manufactured in standard sizes. The diameter of the circular cross section is indicated by a "gauge number," with higher numbers corresponding to smaller diameters. Table 22-2 gives the diameters and cross-sectional areas for some common wire gauges. Handbooks typically give ρ/A-R/L in ohms per centimeter or ohms per foot.

Table 22-1 **Resistivities and Temperature Coefficients**

Material	Resistivity ρ at 20°C, Ω·m	Temperature coefficient α at 20°C, K^{-1}
Silver	1.6×10^{-8}	3.8×10^{-3}
Copper	1.7×10^{-8}	3.9×10^{-3}
Aluminum	2.8×10^{-8}	3.9×10^{-3}
Tungsten	5.5×10^{-8}	4.5×10^{-3}
Iron	10×10^{-8}	5.0×10^{-3}
Lead	22×10^{-8}	4.3×10^{-3}
Mercury	96×10^{-8}	0.9×10^{-3}
Nichrome	100×10^{-8}	0.4×10^{-3}
Carbon	3500×10^{-8}	-0.5×10^{-3}
Germanium	0.45	-4.8×10^{-2}
Silicon	640	-7.5×10^{-2}
Wood	$10^{8} - 10^{14}$	
Glass	$10^{10} - 10^{14}$	
Hard rubber	$10^{13} - 10^{16}$	
Amber	5×10^{14}	
Sulfur	1×10^{15}	

Table 22-2 **Wire Diameters and Cross-sectional Areas for Commonly Used Copper Wires**

Gauge number	Diameter at 20°C, mm	Area, mm^2
4	5.189	21.15
6	4.115	13.30
8	3.264	8.366
10	2.588	5.261
12	2.053	3.309
14	1.628	2.081
16	1.291	1.309
18	1.024	0.8235
20	0.8118	0.5176
22	0.6438	0.3255

Example 22-4

Calculate ρ/A in ohms per meter for 14-gauge copper wire, which has a diameter $d = 1.63$ mm.

From Table 22-1, the resistivity of copper is

$$\rho = 1.7 \times 10^{-8}\ \Omega\cdot\text{m}$$

The cross-sectional area of 14-gauge wire is

$$A = \frac{\pi d^2}{4} = \frac{\pi(0.00163\ \text{m})^2}{4} = 2.1 \times 10^{-6}\ \text{m}^2$$

Thus

$$\frac{\rho}{A} = \frac{1.7 \times 10^{-8}\ \Omega\cdot\text{m}}{2.1 \times 10^{-6}\ \text{m}^2} = 8.1 \times 10^{-3}\ \Omega/\text{m}$$

This example shows that the copper connecting wires used in the laboratory have very small resistances.

Resistors for use in the laboratory are often made by winding a fine wire around an insulating tube to get a long wire in a short space. Carbon, which has a relatively high resistivity, is usually used for resistors in electronic equipment. Such a resistor is often painted with colored stripes to indicate the value of its resistance.

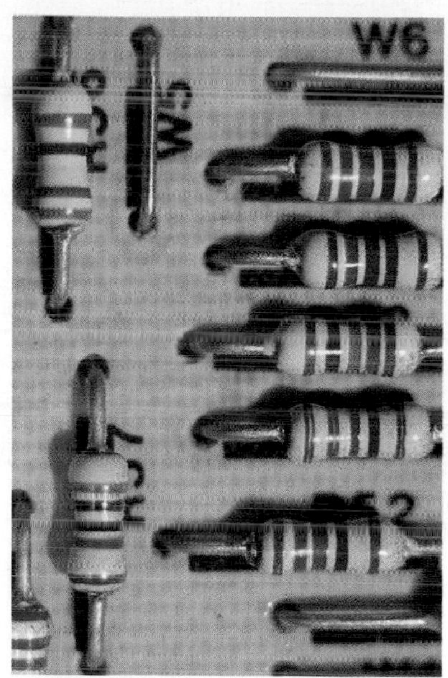

Color-coded carbon resistors on a circuit board.

Example 22-5

Assuming the electric field to be uniform, find its magnitude in a 14-gauge copper wire that is carrying a current of 1 A.

According to Example 22-4, the resistance of a 1-m length of 14-gauge copper wire is $8.1 \times 10^{-3}\ \Omega$. From Ohm's law, the voltage drop across 1 m of this wire is

$$V = IR = (1\ \text{A})(8.1 \times 10^{-3}\ \Omega) = 8.1 \times 10^{-3}\ \text{V}$$

so the electric field is

$$E = \frac{V}{\Delta L} = \frac{8.1 \times 10^{-3}\ \text{V}}{1\ \text{m}} = 8.1 \times 10^{-3}\ \text{V/m}$$

Note that the electric field in a conducting wire is very small.

Example 22-6

By what percentage does the resistance of a copper wire increase when its temperature increases from 20 to 30°C?

From Equation 22-11, the fractional change in the resistivity is

$$\frac{\rho - \rho_{20}}{\rho_{20}} = \alpha(t_C - 20°\text{C})$$

Using $\alpha = 3.9 \times 10^{-3}/\text{K}$ for copper from Table 22-1 and $(t_C - 20°\text{C}) = 10\ \text{C}° = 10$ K, we obtain

$$\frac{\rho - \rho_{20}}{\rho_{20}} = (3.9 \times 10^{-3}/\text{K})(10\ \text{K}) = 3.9 \times 10^{-2}$$

The percentage change is thus 3.9%.

Exercise:

What temperature change will produce a 10% increase in the resistance of an iron bar? (Answer: $\Delta T = 20$ K)

Superconductivity

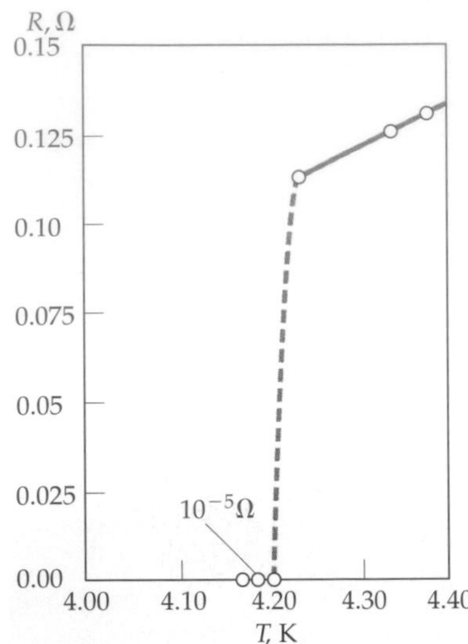

Figure 22-6 Plot by Kamerlingh Onnes of the resistance of mercury versus temperature, showing sudden decrease at the critical temperature $T = 4.2$ K.

There are some materials for which the resistivity is zero below a certain temperature, called the **critical temperature** T_c. This phenomenon, called **superconductivity,** was discovered in 1911 by the Dutch physicist H. Kamerlingh Onnes. Figure 22-6 shows his plot of the resistance of mercury versus temperature. The critical temperature for mercury is 4.2 K. Critical temperatures for other superconducting elements range from less than 0.1 K for hafnium and iridium to 9.2 K for niobium. Many metallic compounds are also superconductors. For example, the superconducting alloy Nb_3Ge, discovered in 1973, has a critical temperature of 23.2 K, which was the highest known until 1986. Despite the cost and inconvenience of refrigeration with expensive liquid helium, which boils at 4.2 K, many superconducting magnets were built using such materials, because such magnets produce no heat. In late 1986 and early 1987, it was discovered that certain ceramic oxides become superconducting at much higher temperatures. For example, the critical temperature for yttrium-barium-copper oxide ($YBa_2Cu_3O_7$) is about 92 K. These discoveries have revolutionized the science of superconductivity because relatively inexpensive liquid nitrogen, which boils at 77 K, can be used to cool them. However, there are many problems, such as the brittleness of ceramics, that make these new superconductors difficult to use. The search continues for new materials that will superconduct at even higher temperatures.

The conductivity of a superconductor cannot be defined since its resistance is zero. There can be a current in a superconductor even when the electric field in the superconductor is zero. Indeed, steady currents have been observed to persist for years without apparent loss in superconducting rings in which there was no electric field. The phenomenon of superconductivity cannot be understood in terms of classical physics. Instead, quantum mechanics, developed in the twentieth century, is needed. We will discuss some of the ideas of quantum mechanics in the latter chapters of this book. The first successful theory of superconductivity was published by John Bardeen, Leon Cooper, and J. Robert Schrieffer in 1957 and is known as the BCS theory. These physicists were awarded the Nobel Prize in Physics in 1972 for their accomplishment. The BCS theory describes the older superconductors well, but it is apparently not sufficient for understanding the newer, higher-temperature superconductors. (We will discuss superconductivity and the BCS theory in more detail in Chapter 39 of the extended version of this book.)

An ohmmeter measures zero resistance across this superconducting coil, which is made of a barium-yttrium-copper-oxide compound and cooled in liquid nitrogen.

Questions

1. Wire *a* and wire *b* have the same electric resistance and are made of the same material. Wire *a* has twice the diameter of wire *b*. How do the lengths of the wires compare?
2. In our study of electrostatics, we concluded that there is no electric field within a conductor in electrostatic equilibrium. How is it that we can now discuss electric fields inside a conductor?

22-3 Energy in Electric Circuits

When there is electric current in a conductor, electrical energy is continually being converted into thermal energy within the conductor. The electric field in the conductor accelerates each free electron for a short time, giving it an increased kinetic energy; but this additional energy is quickly transferred into thermal energy of the conductor by collisions between the electron and the lattice ions of the conductor. Thus, though the electrons continually gain energy from the electric field, this energy is immediately transferred to thermal energy of the conductor, and the electrons maintain a steady drift velocity.

When positive charge flows in a conductor, it flows from high potential to low potential in the direction of the electric field. (The negatively charged electrons move in the opposite direction.) The charge thus loses potential energy. The loss of potential energy appears as kinetic energy of the charge carriers only momentarily before it is transferred to the conducting material by collisions with the fixed ions. The loss of potential energy therefore goes into the increased thermal energy of the conductor.

Consider a segment of wire of length ΔL and cross-sectional area A shown in Figure 22-7. During a time interval Δt, an amount of charge ΔQ passes through area A_1 and enters the segment. If the potential at that point is V_1, the charge has potential energy equal to $\Delta Q\, V_1$. During that time interval, an equal amount of charge leaves the segment passing through area A_2, where the potential is V_2. It has potential energy $\Delta Q\, V_2$, which is less than $\Delta Q\, V_1$. The effect is the same as if the same charge ΔQ entered the segment at a high potential V_1 and left it at a low potential V_2, thereby losing potential energy in the segment given by

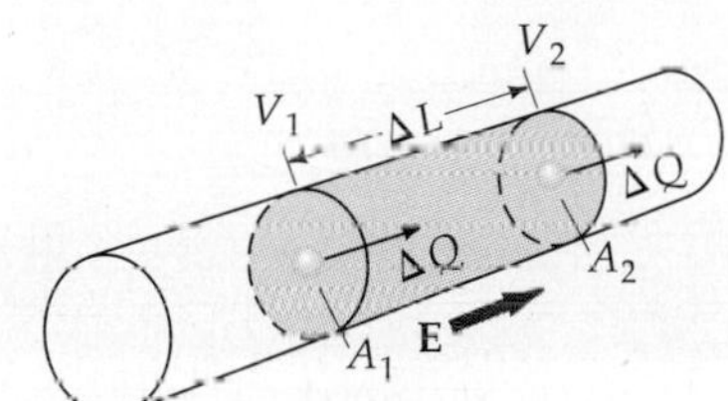

Figure 22-7 During a time Δt, an amount of charge ΔQ passes through area A_1, where the potential is V_1. During the same time interval an equal amount of charge leaves the segment passing through area A_2, where the potential is V_2. The effect is the same as if the same charge ΔQ entered the segment at a high potential V_1 and left it at a low potential V_2, thereby losing potential energy in the segment.

$$\Delta U = \Delta Q\,(V_2 - V_1) = \Delta Q(-V)$$

where $V = V_1 - V_2$ is the potential *decrease* across the segment. The energy lost in this segment of the wire is thus

$$-\Delta U = (\Delta Q)V$$

The rate of energy loss is

$$-\frac{\Delta U}{\Delta t} = \frac{\Delta Q}{\Delta t}V = IV$$

where $I = \Delta Q/\Delta t$ is the current. The energy loss per unit time is the power P dissipated in this conducting segment:

$$P = IV \qquad 22\text{-}12$$

If I is in amperes and V is in volts, the power is in watts. This expression for electric power can be remembered by recalling the definitions of V and I. The voltage drop is the decrease in potential energy per unit charge, and the current is the charge flowing per unit time. The product IV is thus the energy loss per unit time or the power put into the conductor. As we have seen, this power goes into thermal energy in the conductor. Using the definition of resistance, $R = V/I$, we can write Equation 22-12 in other useful forms by eliminating either V or I:

$$P = (IR)I = I^2R \qquad 22\text{-}13$$

or

$$P = \frac{V}{R}V = \frac{V^2}{R} \qquad 22\text{-}14$$

Equations 22-12, 22-13, and 22-14 all contain the same information. The choice of which to use depends on the particular problem. The energy put into a conductor is called **Joule heat.**

(*a*)

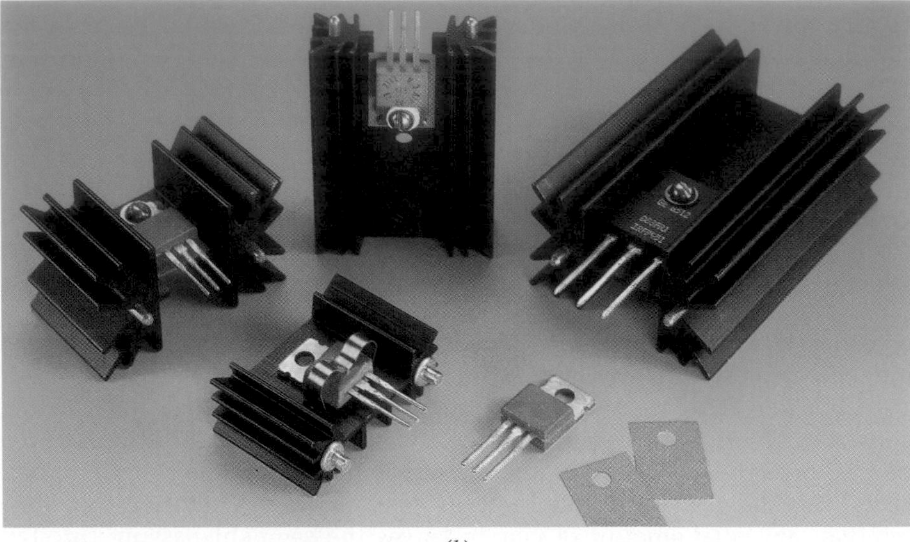

(*b*)

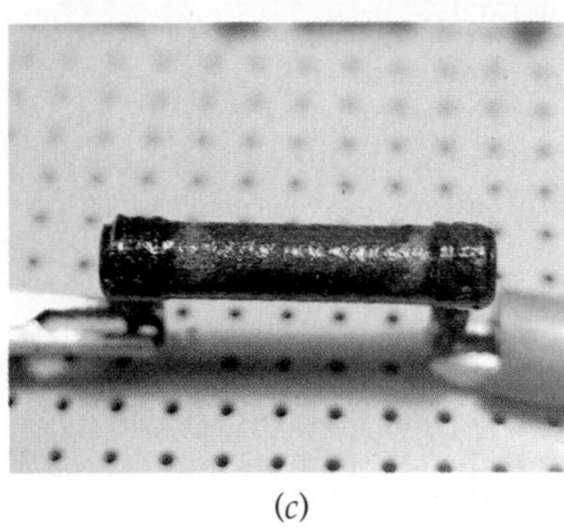

(*c*)

(*a*) Much of the electrical energy dissipated in the filament of this light bulb appears as light energy. (*b*) These rather elaborate cooling fins are used to dissipate the Joule heat developed by transistors and other semiconductor devices. (*c*) The thermal profile of this high-wattage resistor is indicated by the color of the liquid crystal coating.

Example 22-7

A 12-Ω resistor carries a current of 3 A. Find the power dissipated in this resistor.

Since we are given the current and the resistance but not the potential drop, Equation 20-13 is the most convenient to use. We have

$$P = I^2R = (3\ \text{A})^2(12\ \Omega) = 108\ \text{W}$$

Alternatively, we could have first found the potential drop across the resistor from $V = IR = (3\ \text{A})(12\ \Omega) = 36\ \text{V}$, and then used Equation 22-12 to find the power dissipated:

$$P = IV = (3\ \text{A})(36\ \text{V}) = 108\ \text{W}$$

Exercise

A wire of resistance 5 Ω carries a current of 3 A for 6 s. (*a*) How much power is put into the wire? (*b*) How much heat is produced? [Answers: (*a*) 45 W, (*b*) 270 J]

EMF and Batteries

To have a steady current in a conductor, we have to have a constant supply of electrical energy. A device that supplies electrical energy is called a source of **electromotive force** or simply a source of **emf.** It converts chemical, mechanical, or other forms of energy into electrical energy. Examples are a battery, which converts chemical energy into electrical energy, and a generator, which converts mechanical energy into electrical energy. A source of emf does work on the charge passing through it, raising the potential energy of the charge. The work per unit charge is called the emf, $\mathcal{E}$, of the source. When a charge ΔQ flows through a source of emf, its potential energy is increased by the amount $\Delta Q\ \mathcal{E}$. The unit of emf is the volt, the same as the unit of potential difference. An **ideal battery** is a source of emf that maintains a constant potential difference between its two terminals, independent of the rate of flow of charge between them. The potential difference between the terminals of an ideal battery is equal in magnitude to the emf of the battery.

Figure 22-8 shows a simple circuit consisting of a resistance R connected to an ideal battery. In such diagrams a battery is denoted by the symbol ⊣|⊢, the longer line indicating the terminal with the higher potential. The resistance is indicated by the symbol ⩘⩘. The straight lines indicate connecting wires of negligible resistance. Since we are interested only in the potential differences between various points in the circuit, we can choose any point we wish to have zero potential. The source of emf maintains a constant potential difference equal to $\mathcal{E}$ between points a and b, with point a being at the higher potential. There is negligible potential difference between points a and c or between points d and b because the connecting wire is assumed to have negligible resistance. The potential difference between points c and d is therefore equal in magnitude to the emf $\mathcal{E}$, and the current in the resistor is given by $I = \mathcal{E}/R$. The direction of the current in this circuit is clockwise, as shown in the figure. Note that *inside* the source of emf the charge flows from a region of low potential to a region of high potential, so it gains potential energy.* When charge ΔQ flows through the source of emf $\mathcal{E}$, its potential energy is increased by the amount $\Delta Q\ \mathcal{E}$. The charge then flows through the resistor, where this potential energy is converted into thermal energy. The rate at which energy is supplied by the source of emf is the power output:

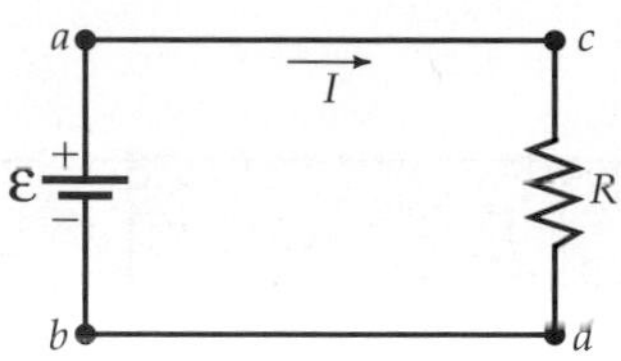

Figure 22-8 A simple circuit consisting of an ideal battery of emf $\mathcal{E}$, a resistance R, and connecting wires that are assumed to be without resistance.

$$P = \frac{\Delta Q\ \mathcal{E}}{\Delta t} = \mathcal{E}I \qquad \text{22-15}$$

In the simple circuit of Figure 22-8, the power put out by the source of emf equals that dissipated in the resistor.

A source of emf can be thought of as a sort of charge pump that pumps the charge from a region of low electrical potential energy to a region of high electrical potential energy, much like a water pump that pumps water from low to high regions of gravitational potential energy. Figure 22-9 shows a mechanical analog to the simple electric circuit just discussed in which marbles of mass m roll down an inclined board with many nails in it. The marbles start at some height h above the bottom and are accelerated between collisions with the nails by the gravitational field. The nails are analogous to the lattice ions in the resistor. During the collisions the marbles transfer the kinetic energy they obtained between collisions to the nails. Because of the many collisions, the marbles move with a small drift velocity toward the bottom. When they reach the bottom, a child picks them up, lifts them to their original height h, and starts them again. The child, who does work mgh on each marble, is analogous to the source of emf. The work per mass is gh, which is analogous to the work per charge done by the source of emf. The energy source in this case is the internal chemical energy of the child.

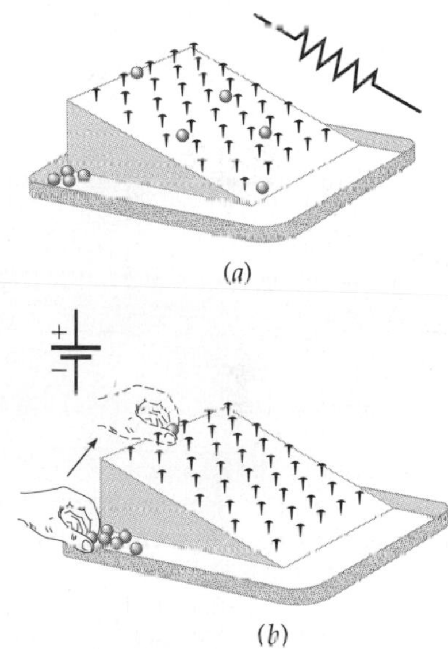

Figure 22-9 A mechanical analogy to a simple circuit consisting of a resistance and source of emf. (*a*) As the marbles roll down the incline, their potential energy is converted into kinetic energy that is quickly converted into heat because of collisions with the nails in the board. (*b*) A child lifts the marbles up from the bottom, where their potential energy is low, to the top, where their potential energy is high, thereby converting her internal chemical energy into potential energy of the marbles.

In a **real battery** the potential difference across the battery terminals, called the **terminal voltage,** is not simply equal to the value of the emf of the battery. Consider the simple circuit consisting of a real battery and a resistor in Figure 22-8. If the current is varied by varying the resistance R and the terminal voltage is measured, the terminal voltage is found to decrease slightly as the current increases, just as if there were a small resistance

*When a battery is being charged by a generator or by another battery, the charge flows from a high potential to a low potential region within the battery being charged, thus losing electrostatic potential energy. The energy lost is stored in the battery being charged.

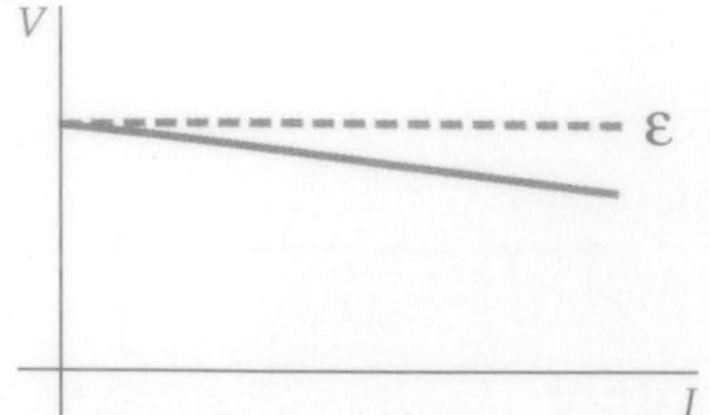

Figure 22-10 Terminal voltage V versus I for a real battery. The dashed line shows the terminal voltage of an ideal battery, which has the same magnitude as $\mathcal{E}$.

within the battery. This is shown graphically in Figure 22-10. Thus we can consider a real battery to consist of an ideal battery of emf $\mathcal{E}$ plus a small resistance r, called the **internal resistance** of the battery. Figure 22-11 shows a simple circuit consisting of a real battery, a resistor, and connecting wires. As before, we can ignore any resistance in the connecting wires. The circuit diagram for this circuit is shown in Figure 22-12. As charge passes from point b to point a, its potential energy is first increased as it passes through the source of emf and is then decreased slightly as it passes through the internal resistance of the battery. (In the actual battery, these energy changes take place concurrently.) If the current in the circuit is I, the potential at point a is related to that at point b by

$$V_a = V_b + \mathcal{E} - Ir$$

The terminal voltage is thus

$$V_a - V_b = \mathcal{E} - Ir \qquad 22\text{-}16$$

The terminal voltage of the battery decreases linearly with current, as we saw in Figure 22-10. The potential drop across the resistor R is IR and is equal to the terminal voltage:

$$IR = V_a - V_b = \mathcal{E} - Ir$$

Solving for the current I, we obtain

$$IR + Ir = \mathcal{E}$$

or

$$I = \frac{\mathcal{E}}{R + r} \qquad 22\text{-}17$$

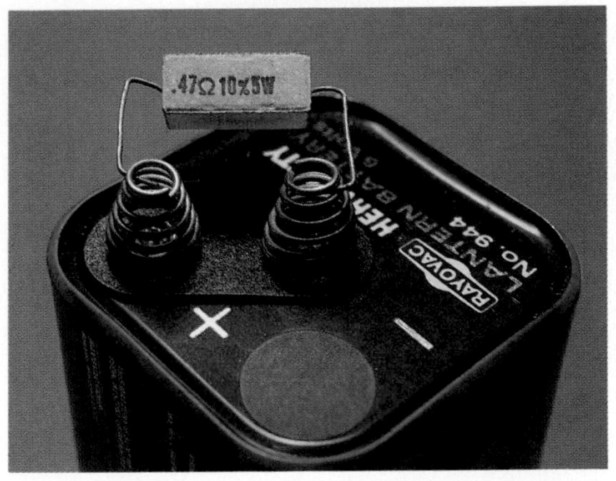

Figure 22-11 A photograph of a simple circuit consisting of a real battery, a resistor, and connecting wires.

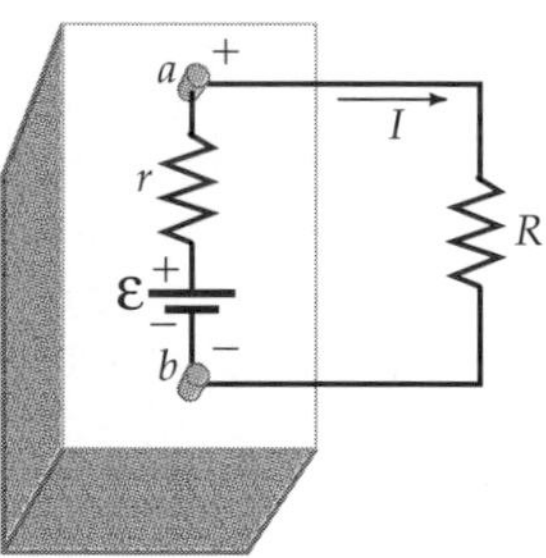

Figure 22-12 Circuit diagram for the circuit in Figure 22-11. A real battery can be represented by an ideal battery of emf $\mathcal{E}$ and a small resistance r.

The terminal voltage given by Equation 22-16 is less than the emf of the battery because of the potential drop across the internal resistance of the battery. Real batteries such as a good car battery usually have an internal resistance of the order of a few hundredths of an ohm, so the terminal voltage is nearly equal to the emf unless the current is very large. One sign of a bad battery is an unusually high internal resistance. If you suspect that your 12-V car battery is bad, and you check the terminal voltage with a voltmeter (discussed in Section 23-3), which draws very little current, the voltmeter may read nearly 12 V, as if the battery were good. However, if you check the terminal voltage while current is being drawn from the battery, such as with the car lights on or while you are trying to start your car, the terminal voltage may drop considerably below 12 V, thereby indicating a high internal resistance and a bad battery.

Example 22-8

An 11-Ω resistance is connected across a battery of emf 6 V and internal resistance 1 Ω.* Find (*a*) the current, (*b*) the terminal voltage of the battery, (*c*) the power delivered by the emf, and (*d*) the power delivered to the external resistance.

(*a*) From Equation 22-17, the current is

$$I = \frac{\mathcal{E}}{R + r} = \frac{6\text{ V}}{11\ \Omega + 1\ \Omega} = 0.5\text{ A}$$

(*b*) The terminal voltage of the battery is

$$V_a - V_b = \mathcal{E} - Ir = 6\text{ V} - (0.5\text{ A})(1\ \Omega) = 5.5\text{ V}$$

(*c*) The power delivered by the source of emf is

$$P = \mathcal{E}I = (6\text{ V})(0.5\text{ A}) = 3\text{ W}$$

(*d*) The power delivered to the external resistance is

$$I^2R = (0.5\text{ A})^2(11\ \Omega) = 2.75\text{ W}$$

This is dissipated as Joule heat. The other 0.25 W of power is dissipated as Joule heat in the internal resistance of the battery.

Example 22-9

For a battery of given emf and internal resistance r, what value of external resistance R should be placed across the terminals to obtain the greatest Joule heating in R?

The external resistance R is sometimes called the load resistance. The power input to R is I^2R, where I is given by Equation 22-17. Thus, the power input is

$$P = I^2R = \frac{\mathcal{E}^2}{(r + R)^2}R = \mathcal{E}^2R(r + R)^{-2}$$

Figure 22-13 shows a sketch of P versus R. At the value of R for which P is maximum, the slope of the curve P versus r is zero. We find this value of R by setting dP/dR equal to zero. We have

$$\frac{dP}{dR} = \mathcal{E}^2(r + R)^{-2} + \mathcal{E}^2R\ (-2)(r + R)^{-3} = 0$$

Multiplying each term by $(r + R)^3/\mathcal{E}^2$, we obtain

$$r + R = 2R$$

or

$$R = r$$

The maximum value of P occurs when $R = r$, that is, when the load resistance equals the internal resistance. A similar result also occurs for more complicated ac circuits and is known as impedance matching.

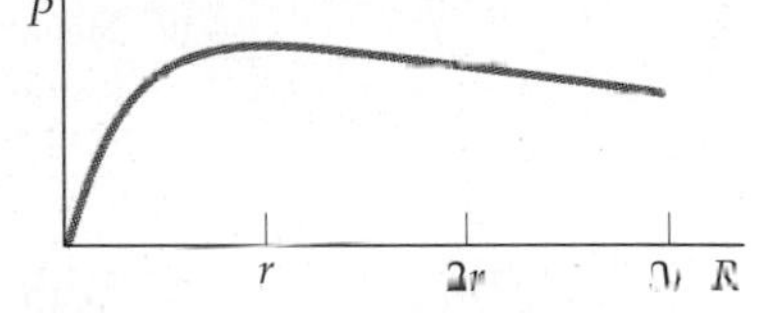

Figure 22-13 Plot of the power input to the external resistor versus R. The power is a maximum when the load resistance R equals the internal resistance of the battery.

*The value of the internal resistance is exaggerated in this example to simplify calculations. In other examples, we may simply ignore the internal resistance.

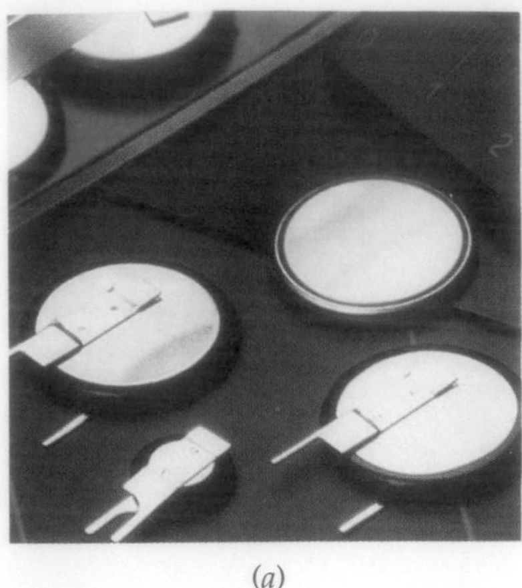
(a)

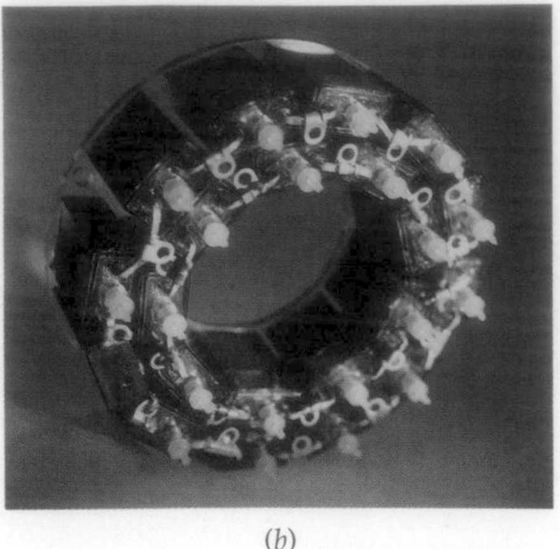
(b)

(c)

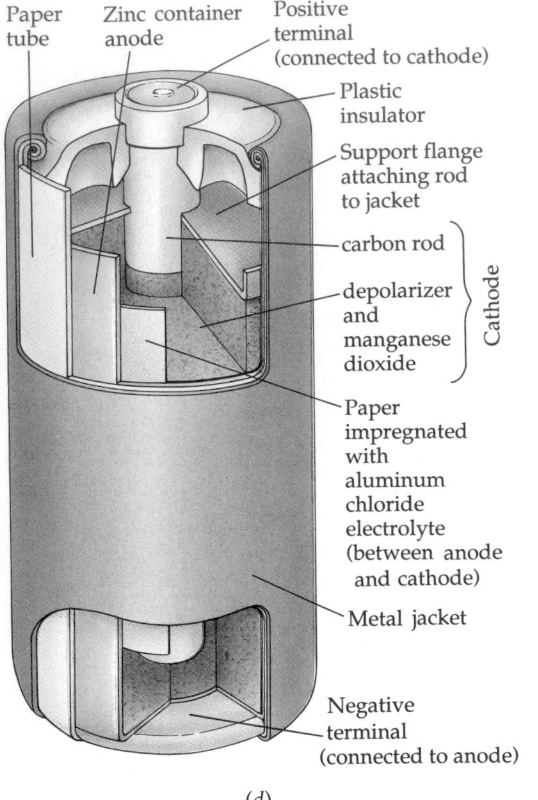

(d)

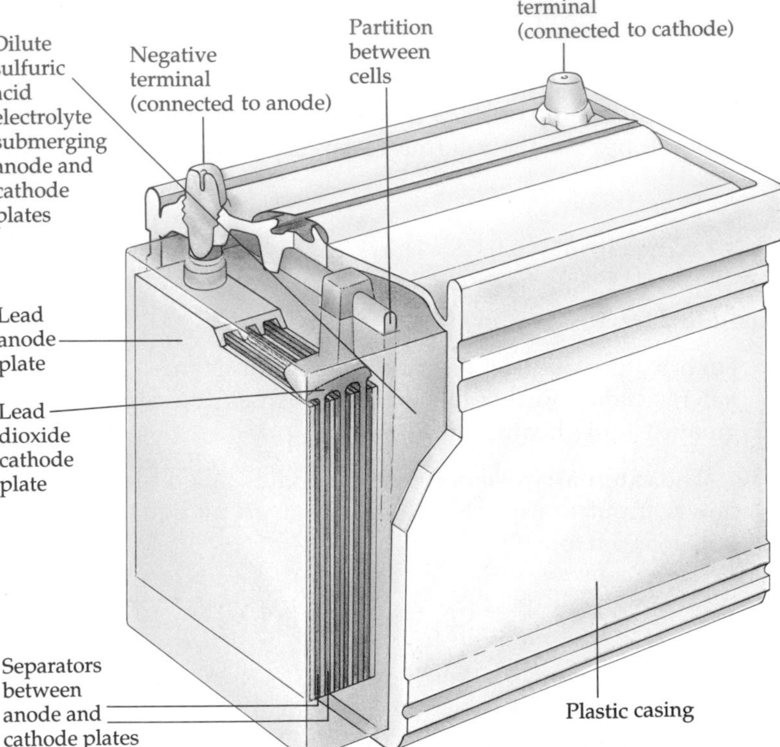

(e)

A battery is a set of chemical cells each of which consists of two metal electrodes immersed in a conducting solution called an electrolyte. Because of chemical reactions between the conductors and the electrolyte, one electrode, the anode, becomes positively charged and the other, the cathode, becomes negatively charged. (*a*) Coin-sized lithium–polymer cells are used in computers to prevent data loss when the power is removed. (*b*) An assembly of high duty silver–zinc cells that is used in satellites. (*c*) This electric car uses eight storage batteries for its power. (*d*) In a dry cell, the electrolyte is a paste of ammonium chloride and other additives. It develops an emf of 1.5 V. (*e*) A 12-V storage battery consists of six cells, which are rechargeable. Each cell has a lead anode, a lead dioxide cathode, and a water solution of sulfuric acid for its electrolyte. (*f*) This giant battery consisting of 200 cells was built in 1870 in the basement of the Royal Institution, London by Humphry Davy.

(f)

Batteries are often rated in ampere-hours (A·h), which is the total charge they can deliver. Since one ampere is one coulomb per second, and one hour contains 3600 seconds, an ampere-hour is equal to 3600 coulombs:

$$1\ \text{A}\cdot\text{h} = 1\frac{\text{C}}{\text{s}}(3600\ \text{s}) = 3600\ \text{C}$$

Questions

3. Name several common sources of emf. What sort of energy is converted into electrical energy in each?
4. In a simple electric circuit like that shown in Figure 22-12, the charge outside the emf flows from positive voltage toward negative voltage, whereas inside the emf it flows from negative voltage to positive voltage. Explain how this is possible.
5. Figure 22-9 illustrates a mechanical analog to the simple electric circuit. Devise another mechanical analog in which the current is represented by a flow of water instead of marbles.
6. A skier is towed up a hill and then skis down with a speed that is constant because of friction. How is this analogous to a simple electric circuit?

22-4 Combinations of Resistors

Series Resistors

Two or more resistors that are connected such that the same charge must flow through each are said to be connected in series. Resistors R_1 and R_2 in Figure 22-14*a* are examples of resistors in series. Since charge does not collect at any point in a wire carrying a steady current, if a charge ΔQ flows into R_1 during some time interval, an equal amount of charge ΔQ must flow out of R_2 during the same interval. The two resistors must therefore carry the same current I. We can often simplify the analysis of a circuit with resistors in series by replacing such resistors with a single equivalent resistance R_{eq} that gives the same total potential drop V when carrying the same current I (see Figure 22-14*b*). The potential drop across R_1 is IR_1 and that across R_2 is IR_2. The potential drop across the two resistors is the sum of the potential drops across the individual resistors:

$$V = IR_1 + IR_2 = I(R_1 + R_2) \qquad \text{22-18}$$

Setting this potential drop equal to IR_{eq}, we obtain

$$R_{eq} = R_1 + R_2$$

Figure 22-14 (*a*) Two resistors in series carry the same current. (*b*) The resistors in (*a*) can be replaced by a single equivalent resistance $R_{eq} = R_1 + R_2$ that gives the same total potential drop when carrying the same current as in (*a*).

Thus, the equivalent resistance for resistors in series is the sum of the original resistances. When there are more than two resistors in series, the equivalent resistance is

$$R_{eq} = R_1 + R_2 + R_3 + \cdots \qquad \text{22-19}$$

Resistors in series

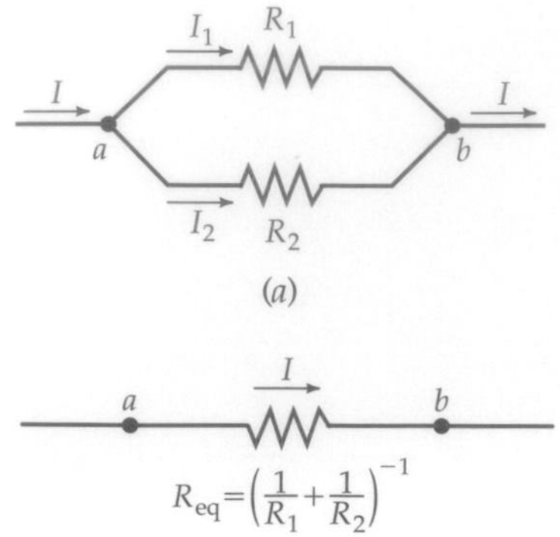

Figure 22-15 (*a*) Two resistors are in parallel when they are connected together at both ends so that the potential drop is the same across each. (*b*) The two resistors in (*a*) can be replaced by an equivalent resistance R_{eq} that is related to R_1 and R_2 by $1/R_{eq} = 1/R_1 + 1/R_2$.

Parallel Resistors

Two resistors that are connected as in Figure 22-15*a* such that they have the same potential difference across them are said to be connected in parallel. Note that the resistors are connected at both ends by wires. Let I be the current from point a to point b. At point a the current splits into two parts, I_1 in resistor R_1 and I_2 in R_2. The total current is the sum of the individual currents:

$$I = I_1 + I_2 \qquad \text{22-20}$$

Let $V = V_a - V_b$ be the potential drop across either resistor. In terms of the currents and resistances,

$$V = I_1R_1 = I_2R_2 \qquad \text{22-21}$$

The equivalent resistance of a combination of parallel resistors is defined to be that resistance R_{eq} for which the same total current I produces the potential drop V (Figure 22-15*b*):

$$R_{eq} = \frac{V}{I}$$

Solving this equation for I and using $I = I_1 + I_2$, we have

$$I = \frac{V}{R_{eq}} = I_1 + I_2 \qquad \text{22-22}$$

But according to Equation 22-21, $I_1 = V/R_1$ and $I_2 = V/R_2$. Equation 22-22 can thus be written

$$I = \frac{V}{R_{eq}} = \frac{V}{R_1} + \frac{V}{R_2}$$

The equivalent resistance for two resistors in parallel is therefore given by

$$\frac{1}{R_{eq}} = \frac{1}{R_1} + \frac{1}{R_2}$$

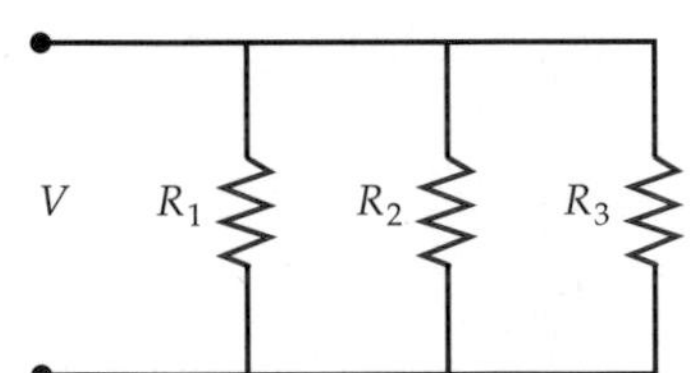

Figure 22-16 Three resistors in parallel.

This result can be generalized for combinations, such as that in Figure 22-16, in which three or more resistors are connected in parallel:

Resistors in parallel

$$\frac{1}{R_{eq}} = \frac{1}{R_1} + \frac{1}{R_2} + \frac{1}{R_3} + \cdots \qquad \text{22-23}$$

Exercise

A 2-Ω and a 4-Ω resistor are connected (*a*) in series and (*b*) in parallel. Find the equivalent resistances. [Answers: (*a*) 6 Ω, (*b*) 1.33 Ω]

Example 22-10

A 4-Ω and a 6-Ω resistor are connected in parallel as shown in Figure 22-17, and a potential difference of 12 V is applied across the combination. Find (*a*) the equivalent resistance, (*b*) the total current, (*c*) the current in each resistor, and (*d*) the power dissipated in each resistor.

(*a*) We first calculate the equivalent resistance from Equation 22-23:

$$\frac{1}{R_{eq}} = \frac{1}{4\ \Omega} + \frac{1}{6\ \Omega} = \frac{3}{12\ \Omega} + \frac{2}{12\ \Omega} = \frac{5}{12\ \Omega}$$

or

$$R_{eq} = \frac{12\ \Omega}{5} = 2.4\ \Omega$$

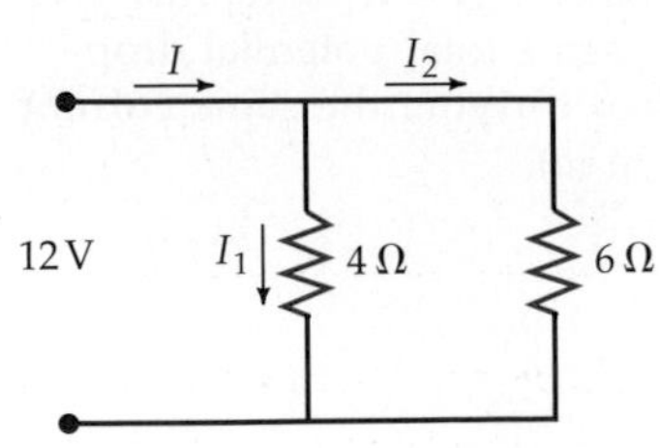

Figure 22-17 Two resistors in parallel across a potential difference of 12 V for Example 22-10.

(*b*) The total current is therefore

$$I = \frac{V}{R_{eq}} = \frac{12\text{ V}}{2.4\ \Omega} = 5\text{ A}$$

(*c*) We obtain the current in each resistor from the fact that the potential drop is 12 V across each resistor (Equation 22-21). Calling the current in the 4-Ω resistor I_1 and that in the 6-Ω resistor I_2, we have

$$V = I_1 R_1 = I_1(4\ \Omega) = 12\text{ V}$$

$$I_1 = \frac{12\text{ V}}{4\ \Omega} = 3.0\text{ A}$$

and

$$I_2 = \frac{12\text{ V}}{6\ \Omega} = 2.0\text{ A}$$

(*d*) The power dissipated in the 4-Ω resistor is

$$P = I_1^2 R_1 = (3.0\text{ A})^2(4\ \Omega) = 36\text{ W}$$

The power dissipated in the 6-Ω resistor is

$$P = (2.0\text{ A})^2(6\ \Omega) = 24\text{ W}$$

This power comes from the source of emf that maintains the 12-V potential difference across the combination of resistors. The power required to deliver 5.0 A at 12 V is

$$P = IV = (5.0\text{ A})(12\text{ V}) = 60\text{ W}$$

which equals the total power dissipated in the two resistors.

Note from Example 22-10 that the equivalent resistance of two parallel resistances is less than the resistance of either resistor alone. This is a general result. Suppose we have a single resistor R_1 carrying current I_1 with potential drop $V = I_1 R_1$. We now add a second resistor R_2 in parallel. If the potential drop is to remain the same, the second resistor must carry additional current $I_2 = V/R_2$ without affecting the original current I_1. The parallel combination thus carries more total current $I = I_1 + I_2$ for the same potential drop, so the ratio of the potential drop to the total current is less. Note also from Example 22-10 that the ratio of the currents in the two parallel resistors equals the inverse ratio of the resistances. This general result follows from Equation 22-21:

$$\frac{I_1}{I_2} = \frac{R_2}{R_1} \qquad \text{parallel resistors} \qquad 22\text{-}24$$

Example 22-11

For the circuit shown in Figure 22-18, find (*a*) the equivalent resistance of the parallel combination of resistors, (*b*) the total current in the source of emf and the current carried by each resistor.

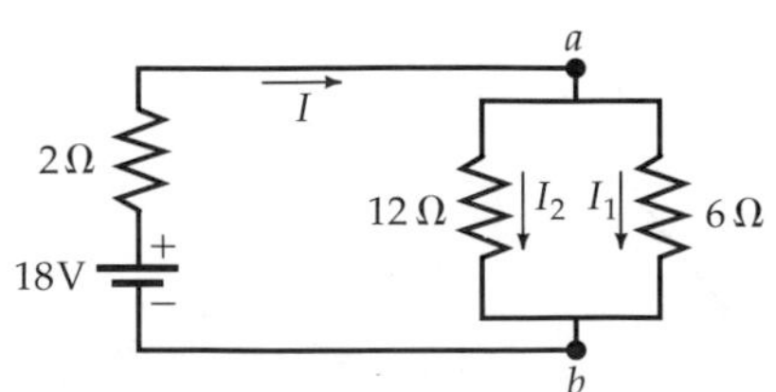

Figure 22-18 Circuit for Example 22-11. The 12- and 6-Ω resistors are in parallel, and this parallel combination is in series with the 2-Ω resistor.

(*a*) The equivalent resistance of the 6- and 12-Ω resistors in parallel is found from

$$\frac{1}{R_{eq}} = \frac{1}{6\ \Omega} + \frac{1}{12\ \Omega} = \frac{3}{12\ \Omega} = \frac{1}{4\ \Omega}$$

$$R_{eq} = 4\ \Omega$$

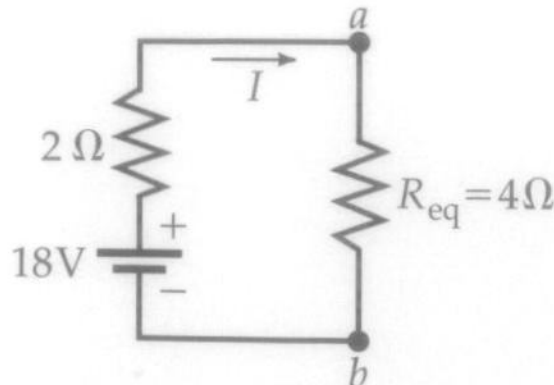

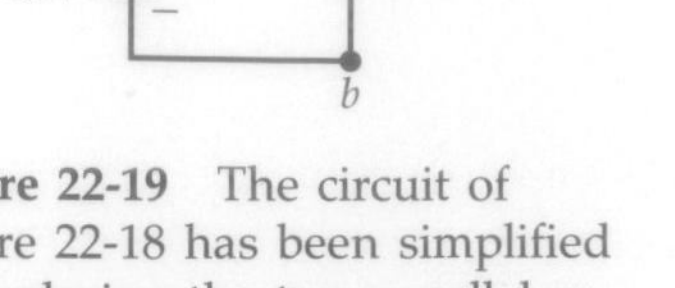
Figure 22-19 The circuit of Figure 22-18 has been simplified by replacing the two parallel resistors with their equivalent resistance.

(*b*) Figure 22-19 shows the circuit with R_{eq} replacing the parallel combination. The resistances $R_{eq} = 4\ \Omega$ and $R = 2\ \Omega$ are in series. The equivalent resistance of this series combination is $R'_{eq} = R_{eq} + R = 6\ \Omega$. The current I in the circuit is therefore

$$I = \frac{\mathcal{E}}{R'_{eq}} = \frac{18\text{ V}}{6\ \Omega} = 3\text{ A}$$

This is the total current in the source of emf. The potential drop from a to b across R_{eq} in Figure 22-19 is $V = IR_{eq} = (3\text{ A})(4\ \Omega) = 12\text{ V}$. The current in the 6-Ω resistor is thus

$$I_1 = \frac{12\text{ V}}{6\ \Omega} = 2\text{ A}$$

and that in the 12-Ω resistor is

$$I_2 = \frac{12\text{ V}}{12\ \Omega} = 1\text{ A}$$

Note that the current in the 6-Ω resistor is twice that in the 12-Ω resistor, as we should expect.

Example 22-12

Find the equivalent resistance between points a and b for the combination of resistors shown in Figure 22-20.

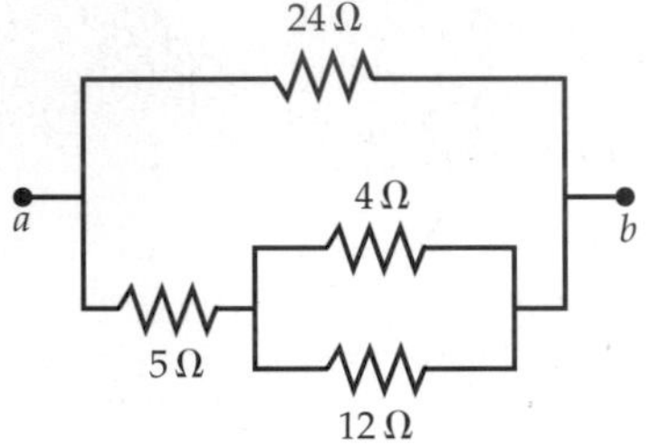

Figure 22-20 Resistor network for Example 22-12.

This combination of resistors may look complicated, but it can be analyzed step by step. The only pair of resistors that are either in series or in parallel are the 4-Ω and 12-Ω resistors, which are in parallel. We therefore first find the equivalent resistance of these resistors. From Equation 22-23, we obtain

$$\frac{1}{R_{eq}} = \frac{1}{4\ \Omega} + \frac{1}{12\ \Omega} = \frac{4}{12\ \Omega} = \frac{1}{3\ \Omega}$$

or

$$R_{eq} = 3\ \Omega$$

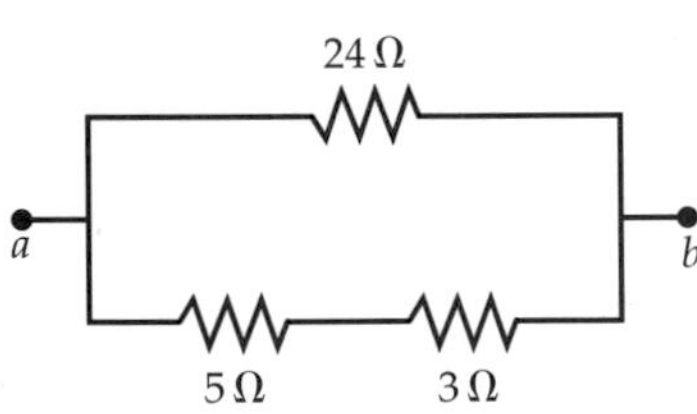

Figure 22-21 Simplification of the resistor network of Figure 22-20. The parallel 4- and 12-Ω resistors in that figure have been replaced by their equivalent resistance of 3 Ω.

In Figure 22-21, the 4-Ω and 12-Ω resistors have been replaced by their equivalent, a 3-Ω resistor. Since this 3-Ω resistor is in series with the 5-Ω resistor, the equivalent resistance of the bottom branch of this combination is 8 Ω. We are now left with an 8-Ω resistor in parallel with a 24-Ω resistor (Figure 22-22). The equivalent resistance of these two parallel resistors is again found from Equation 22-23:

$$\frac{1}{R_{eq}} = \frac{1}{24\ \Omega} + \frac{1}{8\ \Omega} = \frac{4}{24\ \Omega} = \frac{1}{6\ \Omega}$$

$$R_{eq} = 6\ \Omega$$

Thus the equivalent resistance between points a and b is 6 Ω.

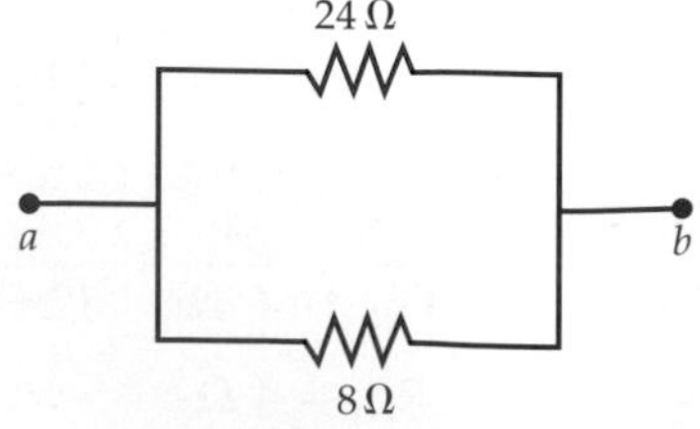

Figure 22-22 Further simplification of the resistor network of Figures 22-20 and 22-21. The 5- and 3-Ω resistors in series in Figure 22-21 are replaced by their equivalent resistance of 8 Ω. The network is now reduced to 24- and 8-Ω resistors in parallel.

Questions

7. Which will produce more heat—a small resistance or a large resistance connected across a source of emf that remains constant?
8. After the introduction of electric lighting, some people were careful to keep bulbs in all the sockets so that the electricity would not leak out. Why is this not necessary?

22-5 A Microscopic Picture of Conduction

A microscopic model of electric conduction was first proposed by P. Drude in 1900 and was developed by Hendrik A. Lorentz about 1909. This model, now called the **classical model of electric conduction,** successfully predicts Ohm's law and relates the conductivity and resistivity of conductors to the motion of free electrons within them. In this model, a metal is pictured as a regular, three-dimensional lattice of ions with a large number of electrons that are free to move throughout the metal. In the absence of an electric field, the free electrons move about the metal much like gas molecules move about in a container. The free electrons are in thermal equilibrium with the lattice ions with which they make collisions. The mean speed of the electrons can be calculated from the equipartition theorem. The result is the same as that for ideal-gas molecules, except that the electron mass m_e is used instead of the mass of a molecule in Equation 15-28. For example, at T = 300 K, the rms speed (which is slightly greater than the mean speed) is

This scanning electron microscope photograph of a 21-μm–wide aluminum conducting strip such as used in computer chips shows voids and clusters of atoms caused by electromigration of aluminum atoms as a result of a current of 0.5 A. The nature of this effect is not well understood, but it adversely affects the reliability of computer chips, which use strips as narrow as 4 μm.

$$v_{rms} = \sqrt{\frac{3kT}{m_e}}$$

$$= \sqrt{\frac{3(1.38 \times 10^{-23}\ \text{J/K})(300\ \text{K})}{9.11 \times 10^{-31}\ \text{kg}}}$$

$$= 1.17 \times 10^5\ \text{m/s} \qquad \text{22-25}$$

This is much greater than the drift velocity that we calculated in Example 22-1.

According to Ohm's law, the current in a conducting wire segment is proportional to the voltage drop across the segment:

$$I = \frac{V}{R}$$

The resistance R is proportional to the length of the wire segment L and inversely proportional to the cross-sectional area A:

$$R = \rho\frac{L}{A}$$

For a uniform electric field E, the voltage across a segment of length L is $V = EL$. Substituting $\rho L/A$ for R, and EL for V, we can write Ohm's law

$$I = \frac{EL}{\rho L/A} = \frac{1}{\rho}EA \qquad \text{22-26}$$

The objective of the classical theory of conduction is to find an expression for ρ in terms of the properties of metals. The current in a wire is related to the number of electrons per unit volume n, the drift velocity v_d, the charge on the electron e, and the cross-sectional area A by Equation 22-3:

$$I = neAv_d$$

In the presence of an electric field, a free electron experiences a force of magnitude eE. If this were the only force acting on the electron, the electron would have an acceleration eE/m_e, and its velocity would steadily increase. However, Ohm's law implies that there is a steady-state situation in which the average velocity of the electron is proportional to the field E, since the current I is proportional to E and also to v_d. In the classical model, it is assumed that a free electron is accelerated for a short time and then makes a collision with a lattice ion. After the collision, the velocity of the electron is completely unrelated to that before the collision. The justification for this assumption is that the drift velocity is very small compared with the random thermal velocity.

Let τ be the average time before an electron, picked at random, makes its next collision. Because the collisions are random, this time does not depend on the time elapsed since the electron's last collision. If we look at an electron immediately after it makes a collision, the average time before its next collision will be τ. Thus τ, called the **collision time**, is the average time between collisions. It is also the average time since the *last* collision of an electron picked at random.*

The drift velocity is the average speed of an electron picked at random. Since the acceleration is eE/m_e, the drift velocity is

$$v_d = \frac{eE}{m_e}\tau \qquad 22\text{-}27$$

Using this result in Equation 22-3, we obtain

$$I = neAv_d = \frac{ne^2\tau}{m_e}EA \qquad 22\text{-}28$$

Using $\rho = EA/I$ from Equation 22-26, we have for the resistivity

$$\rho = \frac{m_e}{ne^2\tau} \qquad 22\text{-}29$$

The average distance the electron travels between collisions is called the **mean free path** λ. It is the product of the mean speed v_{av} and the mean time between collisions τ:

$$\lambda = v_{av}\tau \qquad 22\text{-}30$$

In terms of the mean free path and the mean speed, the resistivity is

$$\rho = \frac{m_e v_{av}}{ne^2\lambda} \qquad 22\text{-}31$$

We can relate the mean free path to the size of the copper ions. Consider one electron moving with speed v through a region of stationary ions (Figure 22-23). If the size of the electron is negligible, it will collide with an ion if it comes within a distance r from the center of the ion, where r is the radius of the ion. In some time t, the electron moves a distance vt and collides with every ion in the cylindrical volume $\pi r^2 vt$. The number of ions in this volume is $n\pi r^2 vt$, where n is the number of ions per unit volume. (After each colli-

*It is tempting but incorrect to think that if τ is the average time between collisions, the average time since its last collision is $\frac{1}{2}\tau$ rather than τ. If you find this confusing, you may take comfort in the fact that Drude used the incorrect result $\frac{1}{2}\tau$ in his original work.

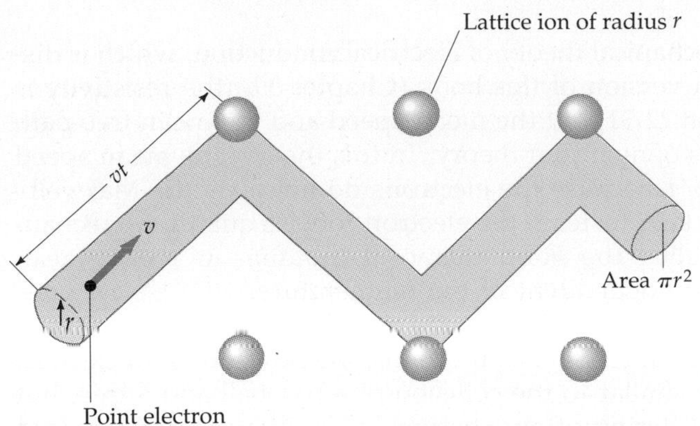

Figure 22-23 Model of an electron moving through the lattice ions in copper. The electron, which is considered to be a point, collides with a lattice ion if it comes within a distance r of the center of the ion, where r is the radius of the ion. If the electron has speed v, it collides with all the ions in the cylindrical volume $\pi r^2 vt$ in time t.

sion, the direction of the electron changes, so the path is really a zigzag one.) The total path length divided by the number of collisions is the mean free path:

$$\lambda = \frac{vt}{n\pi r^2 vt} = \frac{1}{n\pi r^2} \qquad 22\text{-}32$$

Example 22-13

Estimate the mean free path for electrons in copper.

The number of copper ions per unit volume was calculated in Example 22-1 to be 8.47×10^{22} ions/cm^3. Using $r \approx 10^{-10}$ m $= 10^{-8}$ cm for the radius of a copper ion, we obtain for an estimate of the mean free path of the electrons in copper

$$\lambda = \frac{1}{(8.47 \times 10^{22}\ \text{cm}^{-3})\pi(10^{-8}\ \text{cm})^2} \approx 4 \times 10^{-8}\ \text{cm}$$

$$= 4 \times 10^{-10}\ \text{m} = 0.4\ \text{nm}$$

We can use the result of Example 22-13 and $v_{av} \sim 10^5$ m/s from Equation 22-25 to obtain for an estimate of the collision time

$$\tau = \frac{\lambda}{v_{av}} \sim \frac{4 \times 10^{-10}\ \text{m}}{10^5\ \text{m/s}} = 4 \times 10^{-15}\ \text{s}$$

According to Ohm's law, the resistivity is independent of the electric field E. The quantities in Equation 22-31 that might depend on the electric field are the mean speed v_{av} and the mean free path λ. As we have seen, the drift velocity is very much smaller than the mean speed of the electrons in thermal equilibrium with the lattice ions. Thus, the electric field has essentially no effect on the mean speed of the electrons. The mean free path of the electrons depends on the size of the lattice ions and on the density of the ions, neither of which depends on the electric field E. Thus, this model predicts Ohm's law, with the resistivity as given by Equation 22-31.

Although successful in predicting Ohm's law, the classical theory of conduction has several defects. Classical methods for finding the mean free path and the mean speed give a numerical magnitude of the resistivity calculated from Equation 22-31 about six times the measured value at $T = 300$ K, and the temperature dependence is not correct. The temperature dependence of resistivity in Equation 22-31 is given completely by the mean speed v_{av}, which is proportional to $\sqrt{T}$. Thus, this calculation does not give a linear dependence on temperature. Finally, the classical model says nothing about why some materials are conductors, others insulators, and still others semiconductors.

In the quantum-mechanical theory of electrical conduction, which is discussed in the extended version of this book (Chapter 39), the resistivity is again given by Equation 22-31, but the mean speed and the mean free path are interpreted in terms of quantum theory. In this theory, the mean speed is not proportional to $\sqrt{T}$ because the electrons do not obey the Maxwell–Boltzmann distribution law. Instead, the electrons obey a quantum-mechanical distribution law called the Fermi–Dirac distribution in which mean speed is approximately independent of the temperature.

In the quantum-mechanical calculation of the mean free path, the wave nature of the electron (Chapter 35) is important. The collision of an electron with a lattice ion is not similar to the collision of a baseball and a tree, but instead involves the scattering of an electron wave by a regularly spaced lattice. A detailed calculation of the scattering of electron waves in a *perfectly* ordered crystal gives the result that *there is no scattering, and the mean free path is infinite*. Thus, the mean free path is not determined by the size of the lattice ions. At very low temperatures, the ions look like points as far as the scattering of electrons is concerned. The scattering of electron waves arises when the crystal lattice is not perfectly periodic. There are two common causes of deviations from perfect periodicity in a lattice. One is the displacement of the lattice ions due to thermal vibrations. This effect is dominant at ordinary temperatures. The effective area that an ion presents to an electron is proportional to the square of the amplitude of vibration, which is proportional to the energy of vibration, which in turn is proportional to the absolute temperature T. The mean free path therefore varies inversely with T, and ρ is proportional to T as observed experimentally. The other cause of deviations from perfect periodicity is impurities. For example, if some zinc is introduced into pure copper, the previously perfect periodicity is destroyed. At very low temperatures, the resistance of a metal is primarily due to impurities.

Summary

1. Electric current is the rate of flow of charge through a cross-sectional area. By convention, its direction is considered to be that of the flow of positive charge. In a conducting wire, electric current is the result of the slow drift of negatively charged electrons that are accelerated by an electric field in the wire and then quickly collide with the conductor atoms. Typical drift velocities of electrons in wires are of the order of 0.01 mm/s.

2. The resistance of a segment of wire is defined as the ratio of the voltage drop across the segment to the current. In ohmic materials, which include most metals, the resistance is independent of the current, an experimental result known as Ohm's law. For all materials, the potential difference, current, and resistance are related by

$$V = IR$$

3. The resistance of a wire is proportional to its length and inversely proportional to its cross-sectional area:

$$R = \rho\frac{L}{A}$$

where ρ is the resistivity of the material, which depends on its temperature. The reciprocal of the resistivity is called the conductivity σ:

$$\sigma = \frac{1}{\rho}$$

4. The power supplied to a segment of a circuit equals the product of the current and the voltage drop across the segment:

$$P = IV$$

A device that supplies energy to a circuit is called a source of emf. The power supplied by a source of emf is the product of the emf and the current:

$$P = \mathcal{E}I$$

The power dissipated in a resistor is given by

$$P = IV = I^2R = \frac{V^2}{R}$$

An ideal battery is a source of emf that maintains a constant potential difference across its terminals independent of the current. A real battery can be considered to be an ideal battery in series with a small resistance called its internal resistance.

5. The equivalent resistance of a set of resistors in series equals the sum of the resistances:

$$R_{eq} = R_1 + R_2 + R_3 + \cdots \qquad \text{resistors in series}$$

For a set of resistors in parallel, the reciprocal of the equivalent resistance equals the sum of the reciprocals of the individual resistances:

$$\frac{1}{R_{eq}} = \frac{1}{R_1} + \frac{1}{R_2} + \frac{1}{R_3} + \cdots \qquad \text{resistors in parallel}$$

6. In the microscopic model of electrical conduction, the free electrons in a metal are accelerated by the electric field but quickly lose their additional energy in collisions with the lattice ions of the metal. Their drift velocity is proportional to the electric field. The resistivity ρ is related to the mean speed v_{av} and to the mean free path λ (the average distance traveled between collisions) by

$$\rho = \frac{m_e v_{av}}{ne^2\lambda}$$

In the classical model, the mean speed is obtained from the Maxwell–Boltzman speed distribution and is proportional to $\sqrt{T}$, and the mean free path depends on the size of the lattice ions. This model predicts Ohm's law but gives incorrect numerical magnitudes for ρ and an incorrect temperature dependence of ρ. In the modern, quantum-mechanical theory of conductivity, the average speed, which is found from the quantum-mechanical speed distribution, is independent of temperature. The wave nature of electrons is used to determine the mean free path, which turns out to be infinite for a perfectly periodic lattice. Deviations from perfect periodicity arise from impurities and from thermal vibrations of the lattice ions.

Conduction in Nerve Cells

Elizabeth Pflegl Nickles
The Albany College of Pharmacy

During a thunderstorm in 1786, Luigi Galvani touched the muscles of a frog's leg with a metal instrument and noticed that the muscles twitched. He concluded that electricity generated by the storm had been conducted through the frog's nerves, causing the muscles to contract. Nerves *do* transmit impulses from one part of the body to another, but by a mechanism very different from the transmission of electricity through a metallic conductor. For a long time, it was thought that a nerve impulse consisted of a flow of ions (that is, an electric current) along the nerve cell, in a manner analogous to the flow of electrons in a wire. However, the electrical properties of a nerve cell are very different from those of a metallic conductor; for example, neural conduction is much slower and does not vary in strength (it is all-or-nothing conduction).

A nerve cell (neuron) is the basic building block of the nervous system and is specialized to transmit information. It consists of a cell body and one or more branchlike fibers (Figure 1). The fibers are of two types, the dendrites, which along with the cell body receive information in the form of stimuli from sensory receptors or other nerve cells, and the axon, which transmits information to other nerve cells. If the stimuli received on the dendrites or the cell body at any one time are at or above a particular threshold of intensity, a nerve impulse is initiated and propagated along the axon. It flows along the axon away from the cell body toward the terminal branches. The axon is a long, thin cellular extension bounded by a membrane and filled with a viscous intracellular fluid called the axoplasm (Figure 2). Once a nerve impulse reaches the nerve cell's terminal branches, neurotransmitter substances are released that convey the impulse to receptors on the next cell.

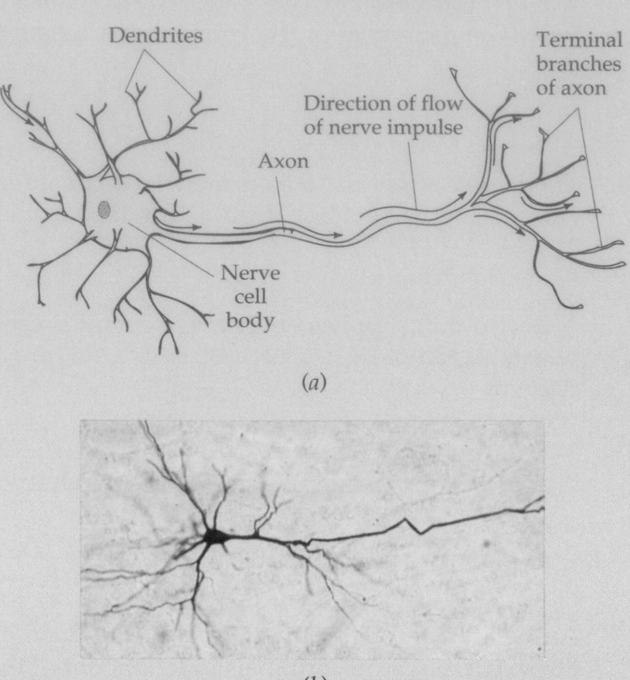

Figure 1 *(a)* A diagram of the structure of a nerve cell. *(b)* A micrograph of a nerve cell from the cerebrum of a cat.

The small cross-sectional area of an axon and the high resistivity ($R = \rho L/A$, Equation 22-8) of the axoplasm combine to yield an extremely high resistance. A piece of nerve axon 1 cm in length has an electrical resistance of about $2.5 \times 10^8\ \Omega$ (comparable to that of wood; see Table 22-1). To understand how nerve impulse transmission is possible through such a seemingly unlikely medium, one must first learn something about the physiology of a nerve cell before it is stimulated, when it is in its "resting state."

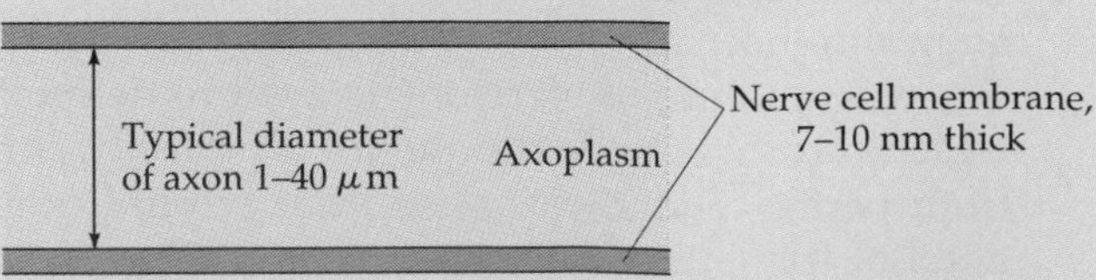

Figure 2 A longitudinal section through an axon.

Elizabeth Pflegl Nickles was born in Catskill, New York and raised on a dairy farm. She earned a B.S. in physics and an M.S. in biology at SUNY Albany, where she is now completing an M.S. in physics. She has taught at SUNY Brockport and at Hudson Valley Community College, and has also held the position of Senior Biophysicist with the New York State Department of Health. Currently, in addition to being a lecturer in physics at the Albany College of Pharmacy, she does research in biological materials using x-ray and backscatter spectra obtained with a scanning microbeam from a linear accelerator.

The Resting Nerve Cell

In nerve cells and other living cells, the cell membrane maintains intracellular conditions that differ from those of the extracellular environment. Critical to the function of the nerve cell is the buildup of a slight excess of negative ions just inside and a slight excess of positive ions just outside the cell membrane (Figure 3).

Electrochemical gradients across the nerve cell membrane are the key to understanding nerve impulse transmission. The concentration of potassium ions (K^+) is 30 times greater in the fluid inside the cell than outside; the concentration of sodium ions (Na^+) is nearly 10 times greater in the fluid outside the cell than inside (see Table 1). Note that anions (particularly chloride, Cl^-) are also unevenly distributed. Like all living cells, nerve cells use both passive diffusion and active transport to maintain these gradients across their cell membranes. The unequal distribution of Na^+ and K^+ is established by an energy-dependent Na^+–K^+ "pump," which moves Na^+ out of the cell and K^+ into the cell. There are also specialized proteins embedded in the nerve cell membrane that function as voltage-dependent channels through which Na^+ and K^+ pass during nerve impulse transmission.

When the nerve cell is in its resting state, the voltage-dependent Na^+ channels are closed, thus maintaining the unequal distribution of Na^+. The resting nerve cell membrane is not permeable to large anions (or to other large negatively charged species, such as proteins) so a slight excess of negative charge builds up immediately inside the nerve cell membrane. The potential difference (Section 20-1) across the membrane is about 70 mV; taking the electric potential outside the cell to be zero, the electric potential inside is therefore −70 mV. This is the nerve cell's *resting potential* (Figure 3).

The arrangement of charge on each surface of the nerve cell membrane resembles that of a charged capacitor. The electric field across a parallel-plate capacitor is uniform (see Section 21-1), so we can calculate the electric field across a cell membrane of 7 nm in thickness:

Table 1 **Ion Concentrations Inside and Outside a Typical Resting Nerve Cell**

	Concentration (mmol/L)	
	Inside	Outside
Na^+	15	145
K^+	150	5
Cl^-	9	120
Other	156	30

$$E = \frac{-dV}{dL} = \frac{-(-70 \times 10^{-3}\ \text{V})}{7.0 \times 10^{-9}\ \text{m}}$$

$$= 1.0 \times 10^{7}\ \text{V/m (inward)}$$

The inward force due to this field on a positive ion would be

$$F = qE = (1.6 \times 10^{-19}\ \text{C})(1.0 \times 10^{7}\ \text{V/m})$$

$$= 1.6 \times 10^{-12}\ \text{N}$$

This force competes against the concentration gradient for K^+, while it supports the effect of the concentration gradient for Na^+

The Stimulated Nerve Cell

The resting potential of a nerve cell can be disturbed by physical or chemical stimuli. A disturbance may only slightly affect the membrane potential at the point of stimulation. In this case, the membrane potential quickly returns to its resting value of −70 mV. The effect of such a subthreshold stimulus, denoted s_1, is illustrated in Figure 4.

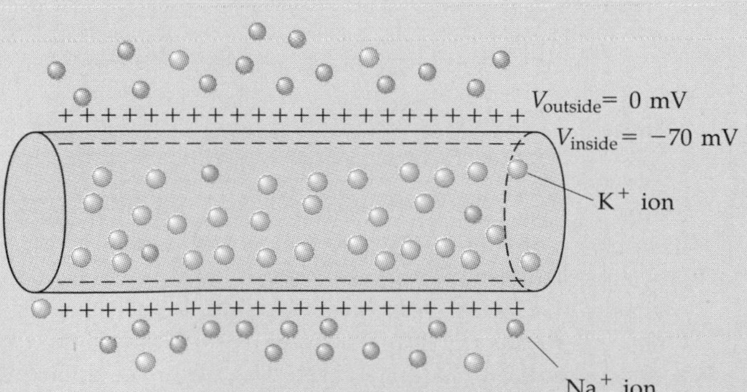

Figure 3 The electric charge distribution along the membrane of a nerve cell in its resting state.

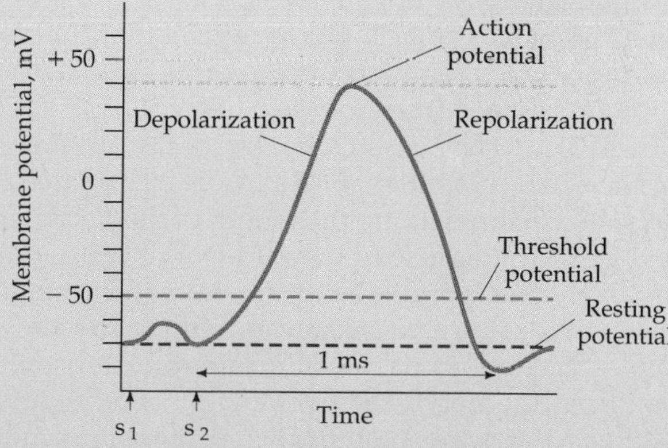

Figure 4 An action potential pulse. s_1 is a subthreshold stimulus; s_2 is a threshold stimulus.

Continued

If, however, the stimulus is strong enough to cause depolarization from the resting potential of -70 mV to around -50 mV, the voltage-dependent Na^+ transmembrane channels open. Favored by both the concentration gradient (see Table 1) and the electric gradient (see the charge distribution in Figure 3), Na^+ ions flow into the cell, creating an electric current ($I = \Delta Q/\Delta t$; see Equation 22-3). The influx of Na^+ causes a local reversal of the electric polarity of the membrane, changing the electric potential to about $+40$ mV (a swing of 110 mV from the resting potential). The effect of such a threshold stimulus, denoted s_2, is also shown in Figure 4.

When the effect of the Na^+ concentration gradient (outside > inside) becomes balanced by that of the electric gradient (due to the membrane potential now having become positive on the *inside*), depolarization is complete at the site of the original stimulus. The Na^+ channels then close again. The K^+ channels respond to the changes in membrane polarity somewhat after the opening of the Na^+ channels, sending K^+ ions flowing out of the cell while the Na^+ ions are flowing in. The movement of the K^+ ions and the slower action of the Na^+–K^+ pump soon restore the concentration gradients and electric gradient to those of the resting state (see repolarization in Figure 4).

After depolarization, the Na^+ channels remain closed for a brief period (of a few milliseconds) during which that portion of the nerve cell membrane cannot again be stimulated. This is called the *refractory period.* The transient change in the electric potential across the membrane is referred to as the *action potential.*

Nerve Impulse Propagation

The preceding paragraphs described a single short electric pulse at the point of stimulation. How does this impulse travel down the axon?

The local depolarization at the site of the original stimulus causes movement—passive diffusion—of ions into areas adjacent to the site of the stimulus (Figure 5). The membrane potential in an adjacent area soon reaches threshold level of -50 mV, so *its* permeability to Na^+ is, in turn, suddenly increased, and the action potential of $+40$ mV is propagated in a wavelike manner along the length of the nerve cell (see Figure 5). Because of the refractory period, during which that portion of the membrane recently depolarized cannot be stimulated again, the nerve impulse can be propagated in one direction only, away from the nerve cell body. The nerve impulse continues along the axon to the terminal branches, where it causes the release of neurotransmitter substances from the nerve cell membrane. These cross the gap (synapse) to the next nerve cell, allowing the process to be repeated.

Note that, during the transmission of a nerve impulse, electric currents flow in and out through the cell membrane *perpendicular* to the direction traveled by the nerve impulse. Furthermore, no matter how long the axon, the impulse never needs amplification; it continues down the axon with each pulse obtaining the same height as that induced by the original stimulus.

Some axons are surrounded by a discontinuous multilayered sheath, the myelin sheath. This is formed when Schwann cells wrap around the axon (Figure 6). Gaps about 1 μm wide in the myelin sheath called nodes of Ranvier occur at regular intervals of 1 to 2 mm along the length of the axon. Propagation of a nerve impulse along a myelinated axon differs somewhat from that seen in unmyelinated axons.

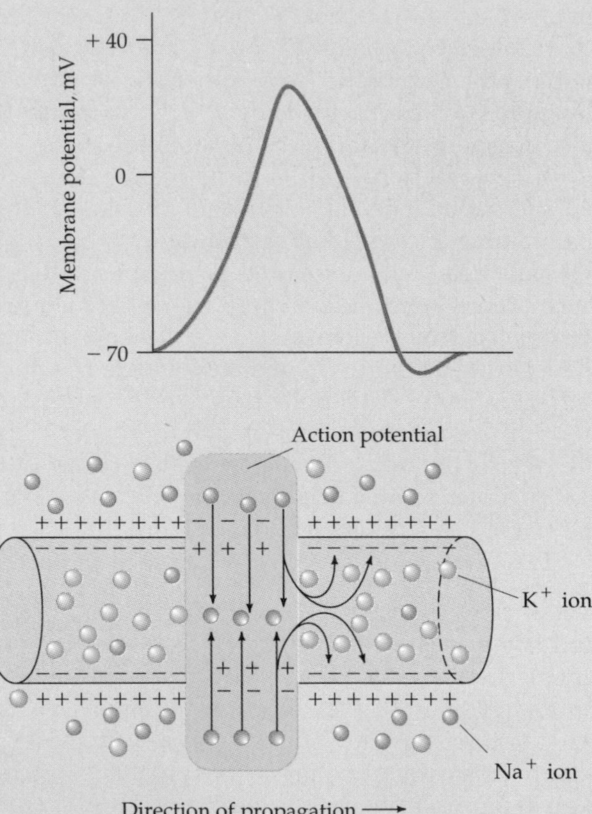

Figure 5 The propagation of an action potential pulse. In advance of the action potential pulse, a small segment of the membrane becomes slightly depolarized owing to the lateral flow of ions inside the membrane. Once this depolarization reaches threshold level, the action potential pulse is generated again in the adjacent segment of the axon. Not shown is the lateral flow of ions to the left, which, because of the refractory period, does not trigger an action potential.

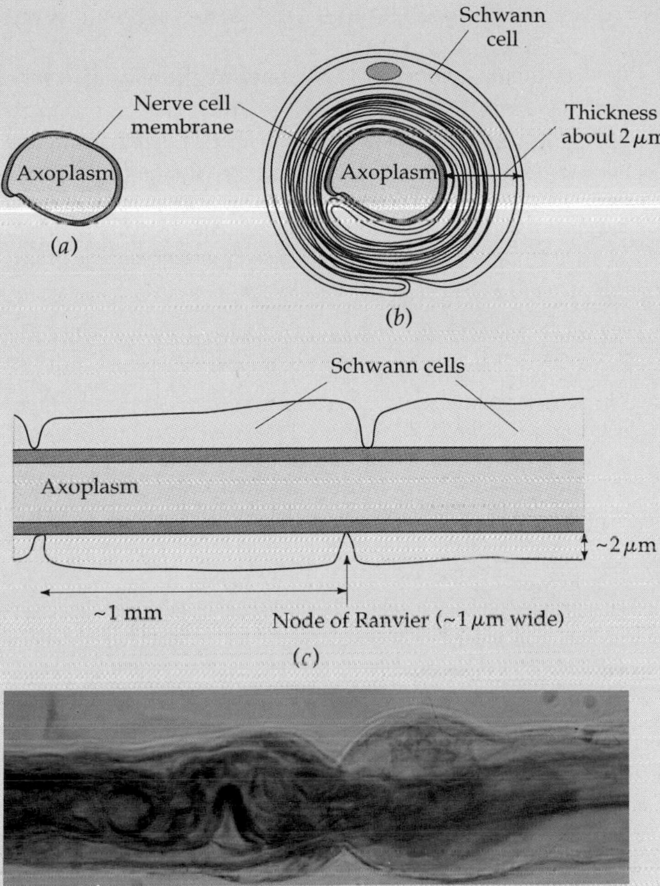

Figure 6 Transverse sections through *(a)* a "bare" (nonmyelinated) nerve axon and *(b)* a nerve axon that has been surrounded by a sheath (myelinated). *(c)* Longitudinal section through the axon of a myelinated nerve cell. *(d)* A photomicrograph of a myelinated axon shows the thickness of the myelin sheath as well as a node of Ranvier in the center of the photo.

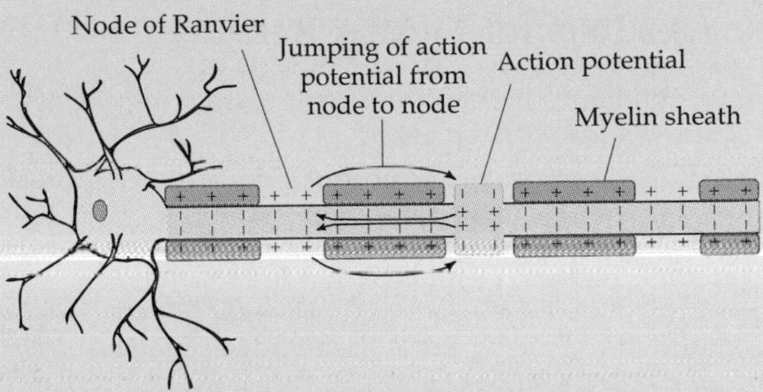

Figure 7 During the propagation of an action potential pulse along a myelinated nerve axon, the impulse jumps from node to node, greatly accelerating nerve impulse conduction. This contrasts with the continuous travel of the action potential along a nonmyelinated axon.

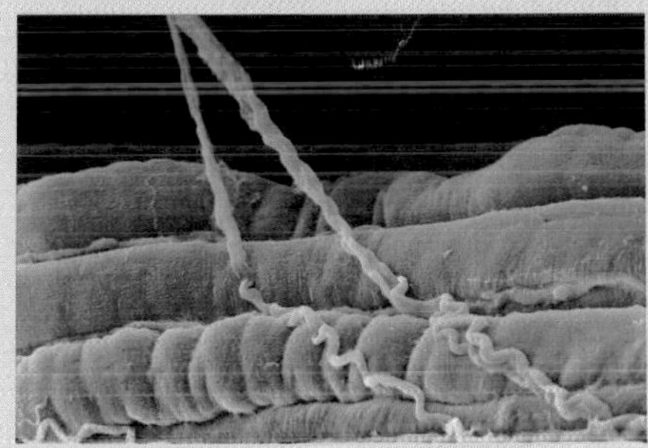

Scanning electron micrograph showing the axons of several nerve cells coming from the upper left of the photo and leading to skeletal muscle fibers lying horizontally in the photo. Electrical impulses traveling down these fibers will result in the release of acetylcholine, a neurotransmitter substance that initiates muscle contraction.

The myelin sheath is a good insulator, so ions cannot flow through it. Electric activity in myelinated nerve cells is confined to the nodes of Ranvier, where there are dense concentrations of voltage-dependent ion channels. Action potentials can be generated only at the nodes of Ranvier and "jump" rapidly from one node to the next along the nerve axon (Figure 7) due to the rapid diffusion of ions through the axoplasm and extracellular fluid. The conduction velocity in a typical myelinated nerve axon is 12 m/s.

Conduction velocity depends on resistivity of the axoplasm and on the membrane capacitance. As resistance is inversely proportional to the cross-sectional area, an axon with a large diameter has a lower resistance and a higher conduction velocity. As was demonstrated for a parallel-plate capacitor (Section 21 1), capacitance is inversely proportional to the separation of the plates. It follows that myelinated axons have a lower capacitance than unmyelinated axons. The lower the membrane capacitance, the smaller the charge, and the less time the membrane will take both to depolarize and to repolarize. This is one explanation for the higher conduction velocity observed in myelinated axons. Measurements of conduction velocities in a wide range of nerve cells have been shown to correlate closely to their calculated resistances and capacitances.

The mechanism of electrical impulse transmission in nerve cells is certainly quite different from electric conduction in metals, but physics also has an important role to play in the understanding of this process.

Suggestions for Further Reading

Allen, Philip B. "Electrical Conductivity," *The Physics Teacher*, vol. 17, 1979, p. 362.

A brief but advanced treatment of the classical and quantum theories of electrical conduction.

Cotterhill, Rodney: "The Busy Electron: Conductors and Insulators," in *The Cambridge Guide to the Material World*, Cambridge University Press, Cambridge, England, 1985.

The distribution of the outer electrons of the atoms making up a material determine whether the material is a conductor, semiconductor, or insulator. This article explains how, with diagrams and without mathematics.

Grundfest, Harry: "Electric Fishes," *Scientific American*, January 1965, p. 82.

Members of many families of fishes are able to produce appreciable voltages outside their bodies using specialized organs containing arrays of "electroplaque membranes" in series or in parallel.

Hamakawa, Yoshihiro: "Photovoltaic Power," *Scientific American*, April 1987, p. 86.

How advances in solar cell design and fabrication are bringing down the cost of photovoltaic electricity.

de Santillana, Giorgio: "Alessandro Volta," *Scientific American*, January 1965, p. 82.

This article describes the argument between Luigi Galvani and Alessandro Volta over whether or not electricity was a living force and a key to the mystery of life or a phenomenon that could manifest itself without requiring the presence of a living being. Volta won the argument in 1800 when he announced his invention of what we would now call a battery.

Schluter, Michael A.: "Unaccountable Conduct," *The Sciences*, May/June 1989, p. 44.

This article presents the background to the discovery of high-temperature superconductivity in 1986, and current attempts to account theoretically for the phenomenon.

Wolsky, Alan M., Robert F. Giese, and Edward J. Daniels: "The New Superconductors: Prospects for Applications," *Scientific American*, February 1989, p. 60.

The authors examine possible applications of the new superconductors and predict that development for practical applications will take at least a decade.

Review

A. Objectives: After studying this chapter you should:

1. Be able to define and discuss the concepts of electric current, drift velocity, resistance, and emf.

2. Be able to state Ohm's law and distinguish between it and the definition of resistance.

3. Be able to define resistivity and describe its temperature dependence.

4. Be able to discuss the simple model of a real battery in terms of a source of emf and an internal resistance, and find the terminal voltage of a battery when it delivers a current I.

5. Be able to give the general relationship between potential difference, current, and power.

6. Be able to determine the equivalent resistances of resistors in series or parallel.

7. Be able to discuss the microscopic model of electrical conduction.

B. Define, explain, or otherwise identify:

Current
Ampere
Drift velocity
Ohm's law
Resistance
Ohmic materials
Nonohmic materials
Resistivity
Conductivity
Temperature coefficient of resistivity
Critical temperature
Superconductivity
Joule heat
Electromotive force
Emf
Ideal battery
Real battery
Terminal voltage
Internal resistance
Resistors in series
Resistors in parallel
Classical model of electric conduction
Collision time
Mean free path

C. True or false: If the statement is true, explain why it is true. If it is false, give a counterexample.

1. Ohm's law is $R = V/I$.

2. Electrons drift in the direction of the current.

3. A source of emf supplies power to an electrical circuit.

4. When the potential drops by V in a segment of a circuit, the power supplied to that segment is IV.

5. The equivalent resistance of two resistors in parallel is always less than the resistance of either resistor alone.

6. The terminal voltage of a battery is the same as its emf.

Problems

Level I

22-1 Current and Motion of Charges

1. A wire carries a steady current of 2.0 A. (*a*) How much charge flows through a cross-sectional area of the wire in 5.0 min? (*b*) How many electrons flow through the cross-sectional area in this time?

2. A 10-gauge copper wire carries a current of 20 A. Assuming one free electron per copper atom, calculate the drift velocity of the electrons.

3. In a fluorescent tube of diameter 3.0 cm, 2.0×10^{18} electrons and 0.5×10^{18} positive ions (with a charge of $+e$) flow through a cross-sectional area each second. What is the current in the tube?

4. In a certain electron beam, there are 5.0×10^6 electrons per cubic centimeter. Suppose the kinetic energy of each electron is 10.0 keV, and the beam is cylindrical, with a diameter of 1.00 mm. (*a*) What is the velocity of an electron? (*b*) Find the beam current.

5. A charge $+q$ moves in a circle of radius r with speed v. (*a*) Express the frequency f with which the charge passes a particular point in terms of r and v. (*b*) Show that the average current is qf and express it in terms of v and r.

6. A ring of radius R with a linear charge density λ rotates about its axis with angular velocity ω. Find an expression for the current.

7. In a certain particle accelerator, a proton beam with a diameter of 2.0 mm constitutes a current of 1.0 mA. The kinetic energy of each proton is 20 MeV. The beam strikes a metal target and is absorbed by it. (*a*) What is the number n of protons per unit volume in the beam? (*b*) How many protons strike the target in 1.0 min? (*c*) If the target is initially uncharged, express the charge of the target as a function of time.

8. A 10-gauge copper wire and a 14-gauge copper wire are welded together end to end. The wires carry a current of 15 A. If there is one free electron per copper atom, find the drift velocity of the electrons in each wire.

22-2 Resistance and Ohm's Law

9. A 10-m–long wire of resistance 0.2 Ω carries a current of 5 A. (*a*) What is the potential difference across the wire? (*b*) What is the magnitude of the electric field in the wire?

10. A potential difference of 100 V produces a current of 3 A in a certain resistor. (*a*) What is its resistance? (*b*) What is the current when the potential difference is 25 V?

11. A copper wire and an iron wire with the same length and diameter carry the same current I. (*a*) Find the potential drop across each wire and the ratio of these drops. (*b*) In which wire is the electric field greater?

12. A block of carbon is 3.0 cm long and has a square cross-sectional area with sides of 0.5 cm. A potential difference of 8.4 V is maintained across its length. (*a*) What is the resistance of the block? (*b*) What is the current in this resistor?

13. A tungsten rod is 50 cm long and has a square cross-sectional area with sides of 1.0 mm. (*a*) What is its resistance at 20°C? (*b*) What is its resistance at 40°C?

14. A carbon rod with a radius of 0.1 mm is used to make a resistor. The resistivity of this material is 3.5×10^{-5} Ω·m. What length of the carbon rod will make a 10-Ω resistor?

15. The third (current-carrying) rail of a subway track is made of steel and has a cross-sectional area of about 55 cm^2. What is the resistance of 10 km of this track?

16. What is the potential difference across one wire of a 30-m extension cord made of 16-gauge copper wire carrying a current of 5.0 A?

17. At what temperature will the resistance of a copper wire be 10 percent greater than it is at 20°C?

18. How long is a 14-gauge copper wire that has a resistance of 2 Ω?

22-3 Energy in Electric Circuits

19. What is the power dissipated in a 10.0-Ω resistor if the potential difference across it is 50 V?

20. Find the power dissipated in a resistor connected across a constant potential difference of 120 V if its resistance is (*a*) 5 Ω and (*b*) 10 Ω.

21. A 10,000-Ω carbon resistor used in electronic circuits is rated at 0.25 W. (*a*) What maximum current can this resistor carry? (*b*) What maximum voltage can be placed across this resistor?

22. A 1-kW heater is designed to operate at 240 V. (*a*) What is its resistance, and what current does it draw? (*b*) What is the power dissipated in this resistor if it operates at 120 V? Assume that its resistance is constant.

23. A 10.0-Ω resistor is rated as being capable of dissipating 5.0 W of power. (*a*) What maximum current can this resistor tolerate? (*b*) What voltage across this resistor will produce this current?

24. If energy costs 9 cents per kilowatt-hour (*a*) how much does it cost to operate an electric toaster for 4 min if the toaster has resistance 11.0 Ω and is connected across 120 V? (*b*) How much does it cost to operate a heater of resistance 5.0 Ω connected across 120 V for 8 h?

25. A battery has an emf of 12.0 V. How much work does it do in 5 s if it delivers a current of 3 A?

26. A battery with 12-V emf has a terminal voltage of 11.4 V when it delivers a current of 20 A to the starter of a car. What is the internal resistance r of the battery?

27. (*a*) How much power is delivered by the emf of the battery in Problem 26 when it delivers a current of 20 A? (*b*) How much of this power is delivered to the starter? (*c*) By how much does the chemical energy of the battery decrease when it delivers a current of 20 A to the starter for 3 min? (*d*) How much heat is developed in the battery when it delivers a current of 20 A for 3 min?

28. A 12-V car battery has an internal resistance of 0.4 Ω. (*a*) What is the current if the battery is shorted momentarily? (*b*) What is the terminal voltage when the battery delivers a current of 20 A to start the car?

29. A physics student runs a 1200-W electric heater constantly in her basement bedroom during the winter time. If electric energy costs 9 cents per kilowatt-hour, how much does this electric heating cost per 30-day month?

22-4 Combinations of Resistors

30. (*a*) Find the equivalent resistance between points *a* and *b* in Figure 22-24. (*b*) If the potential drop between *a* and *b* is 12 V, find the current in each resistor.

Figure 22-24 Problem 30.

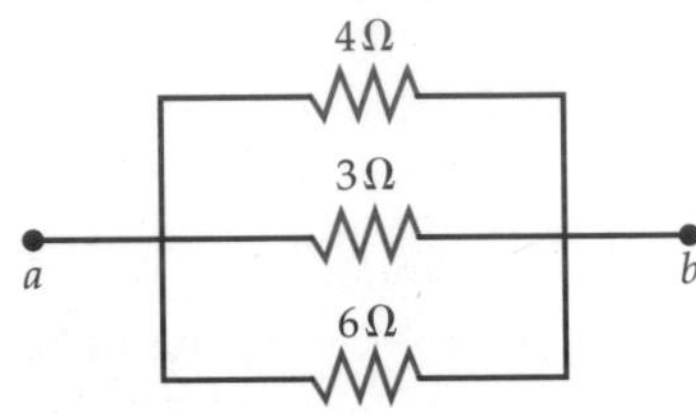

31. Repeat Problem 30 for the resistor network shown in Figure 22-25.

Figure 22-25 Problem 31.

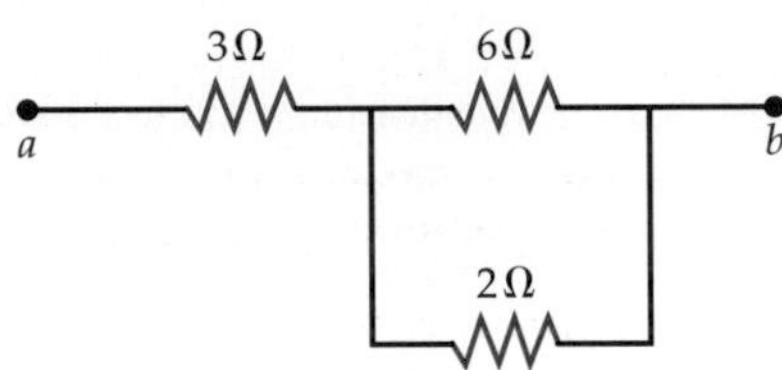

32. Repeat Problem 30 for the resistor network shown in Figure 22-26.

Figure 22-26 Problems 32 and 34.

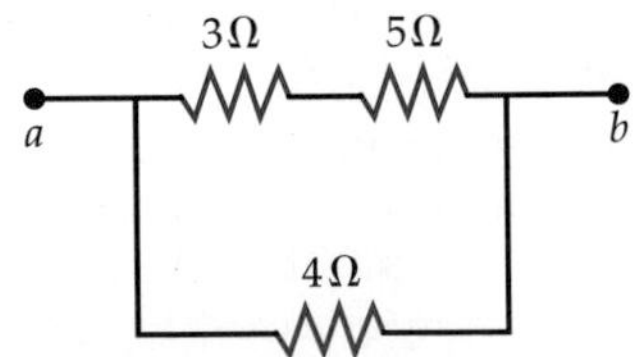

33. Repeat Problem 30 for the resistor network shown in Figure 22-27.

Figure 22-27 Problem 33.

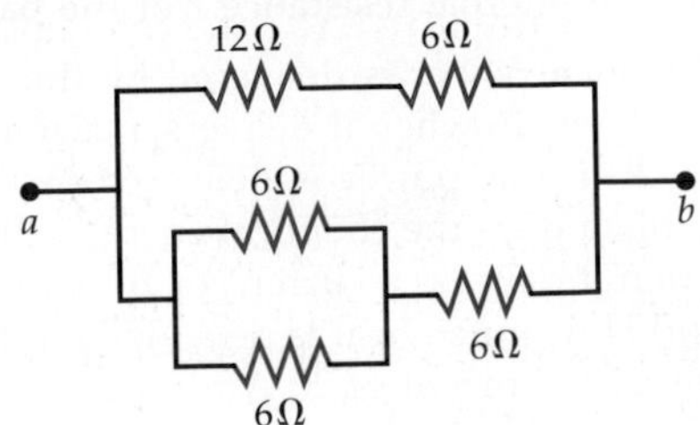

34. In Figure 22-26, the current in the 4-Ω resistor is 4 A. (*a*) What is the potential drop between *a* and *b*? (*b*) What is the current in the 3-Ω resistor?

35. (*a*) Show that the equivalent resistance between points *a* and *b* in Figure 22-28 is *R*. (*b*) What would be the effect of adding a resistance *R* between points *c* and *d*?

Figure 22-28 Problem 35.

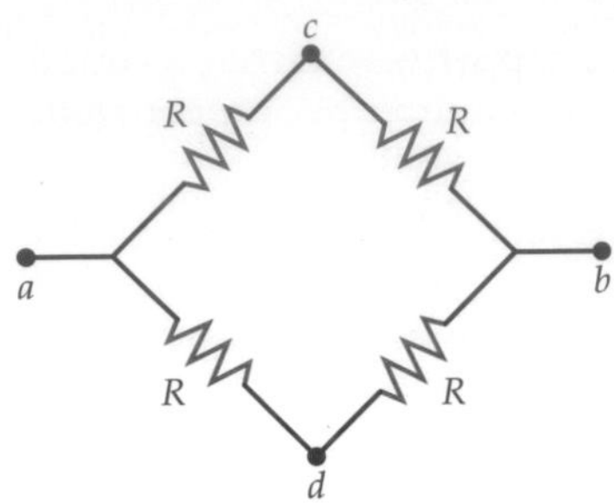

36. Repeat Problem 30 for the resistor network shown in Figure 22-29.

Figure 22-29 Problem 36.

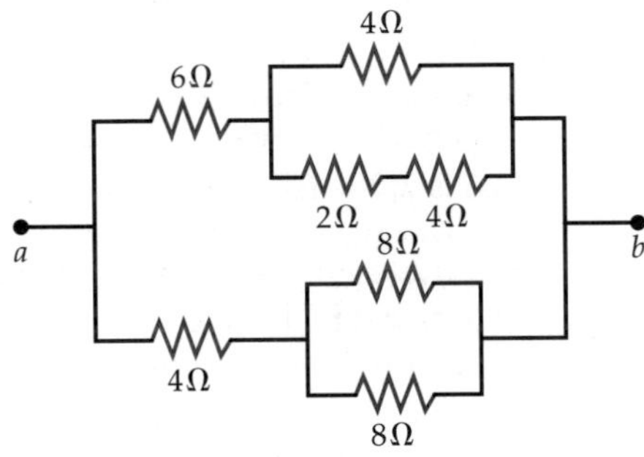

37. The battery in Figure 22-30 has negligible internal resistance. Find (*a*) the current in each resistor and (*b*) the power delivered by the battery.

Figure 22-30 Problem 37.

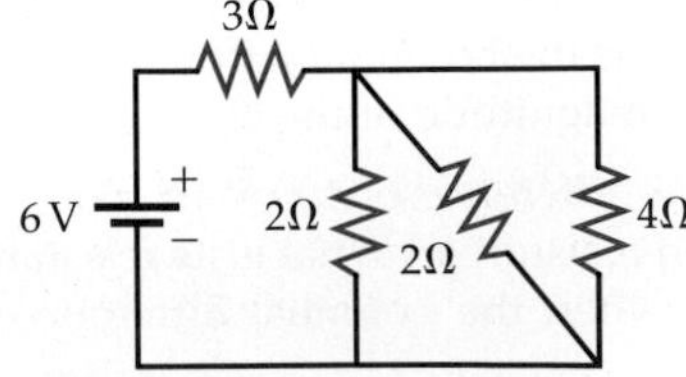

22-5 A Microscopic Picture of Conduction

There are no problems for this section.

Level II

38. A wire of length 1 m has a resistance of 0.3 Ω. It is uniformly stretched to a length of 2 m. What is its new resistance?

39. A cube of copper has sides of 2.0 cm. If it is drawn out to form a 14-gauge wire, what will its resistance be?

40. A 200-W heater is used to heat water in a cup. Assume that 90 percent of the energy produced by the heater goes into heating the water. (*a*) How long does it take to heat 0.25 kg of water from 15 to 100°C? (*b*) How long does it take to boil this water away after it reaches 100°C?

41. Consider the equivalent resistance of two resistors R_1 and R_2 connected in parallel as a function of the ratio $x = R_2/R_1$. (*a*) Show that $R_{eq} = R_1x/(1 + x)$. (*b*) Sketch a plot of R_{eq} as a function of x.

42. A 16-gauge copper wire insulated with rubber can safely carry a maximum current of 6 A. (*a*) How great a potential difference can be applied across 40 m of this wire? (*b*) Find the electric field in the wire when it carries a current of 6 A. (*c*) Find the power dissipated in the wire when it carries a current of 6 A.

43. The belt of a Van de Graaff generator carries a surface charge density of 5 mC/m^2. The belt is 0.5 m wide and moves at 20 m/s. (*a*) What current does it carry? (*b*) If this charge is raised to a potential of 100 kV, what is the minimum power of the motor needed to drive the belt?

44. A coil of nichrome wire is to be used as the heating element in a water boiler that is required to generate 8.0 g of steam per second. The wire has a diameter of 1.80 mm and is connected to a 120-V power supply. Find the length of wire required.

45. An 80.0-m copper wire 1.0 mm in diameter is joined end to end with a 49.0-m iron wire of the same diameter. The current in each is 2.0 A. (*a*) Find the electric field in each wire. (*b*) Find the potential difference across each wire. (*c*) Find the equivalent resistance that would carry a current of 2.0 A at a potential difference equal to the sum of that across the two, and compare it with the sum of the resistances of the two.

46. The current in a wire varies with time according to the relation $I = 20 + 3t^2$, where I is in amperes and t is in seconds. (*a*) How many coulombs are transported by the wire between $t = 0$ s and $t = 10$ s? (*b*) What constant current would transport the same charge in the same time interval?

47. A toaster with a nichrome heating element has a resistance of 80 Ω at 0°C and an initial current of 1.5 A. When the heating element reaches its final temperature, the current is 1.3 A. What is the final temperature of the heating element?

48. A cylinder of glass 1 cm long has a resistivity of 10^{12} Ω·m. How long would a copper wire of the same cross-sectional area need to be to have the same resistance as the glass cylinder?

49. Compact fluorescent light bulbs cost $20 each and have an expected lifetime of 8000 h. These bulbs consume 20 W of power, but produce the illumination equivalent to 75-W incandescent bulbs. Incandescent bulbs cost about 50¢ each and have an expected lifetime of 1200 h. (*a*) If the average household has, on the average, six 75-W incandescent light bulbs on constantly, and if energy costs 9 cents per kilowatt-hour, how much money would a consumer save each year by installing the energy-efficient fluorescent light bulbs? (*b*) At what cost per kilowatt-hour would the total cost of using either type of bulb be the same? (See *Scientific American*, April 1988, p. 56.)

50. The wires in a house must be large enough in diameter so that they do not get hot enough to start a fire. Suppose a certain wire is to carry a current of 20 A, and it is determined that the joule heating of the wire should not exceed 2 W/m. What diameter must a copper wire have to be "safe" for this current?

51. A 12-V automobile battery can deliver a total charge of 160 A·h. (*a*) What is the total stored energy in the battery? (*b*) How long could this battery provide 150 W to a pair of headlights?

52. A lightweight electric car is powered by ten 12-V batteries. At a speed of 80 km/h the average frictional force is 1200 N. (*a*) What must be the power of the electric motor if the car is to travel at a speed of 80 km/h? (*b*) If each battery can deliver a total charge of 160 A·h before recharging, what is the total charge in coulombs that can be delivered by the 10 batteries before charging? (*c*) What is the total electrical energy delivered by the 10 batteries before recharging? (*d*) How far can the car travel at 80 km/h before the batteries must be recharged? (*e*) What is the cost per kilometer if the cost of recharging the batteries is 9 cents per kilowatt-hour?

53. Suppose the bulb in a two-cell flashlight draws 4 W of power. The batteries go dead in 45 min and cost $7.99. (*a*) How many kilowatt-hours of energy can be supplied by the two batteries? (*b*) What is the cost per kilowatt-hour of energy if the batteries cannot be recharged? (*c*) If the batteries can be recharged at a cost of 9 cents per kilowatt hour, what is the cost of recharging them?

54. In a proton supercollider, the proton in a 5-mA beam move with nearly the speed of light. (*a*) How many protons are there per meter of the beam? (*b*) If the cross-sectional area of the beam is 10^{-6} m^2, what is the average separation of the protons?

55. A 10-Ω resistor is wired into an electronic circuit by copper wire of length 50 cm and diameter 0.6 mm. (*a*) What additional resistance does the wire introduce? (*b*) What percentage error is produced by neglecting the resistance of the wiring? (*c*) If the resistor is made from nichrome wire, what change in its temperature would produce a change in its resistance equal to the resistance of the wiring?

56. The filament of a certain lamp has a resistance that increases linearly with temperature. When a constant voltage is switched on, the initial current decreases until the filament reaches its steady-state temperature. The temperature coefficient of resistivity of the filament is 4×10^{-3} K^{-1}. The final current through the filament is one-eighth the initial current. What is the change in temperature of the filament?

57. A rubber tube 1 m long with an inside diameter of 4 mm is filled with a salt solution that has a resistivity of 10^{-3} Ω·m. Metal plugs form electrodes at the ends of the tube. (*a*) What is the resistance of the filled tube? (*b*) What is the resistance of the filled tube if it is uniformly stretched to a length of 2 m?

58. A space heater in an old home draws a 15-A current. A pair of 12-gauge copper wires carries the current from the fuse box to the wall outlet, a distance of 30 m. The voltage at the fuse box is exactly 120 V. (*a*) What is the voltage delivered to the space heater? (*b*) If the fuse will blow at a current of 20 A, how many 60-W bulbs can be supplied by this line when the space heater is on? (Assume that the wires from the wall to the space heater and to the light fixtures have negligible resistance.)

59. An electric space heater has a nichrome-wire heating element with a resistance of 8 Ω at 0°C. When 120 V are applied, the electric current heats the nichrome wire to 1000°C. (*a*) What is the initial current drawn by the cold heating element? (*b*) What is the resistance of the heating element at 1000°C? (*c*) What is the operating wattage of this heater?

60. Currents up to 30 A can be carried by 10-gauge copper wire. (*a*) What is the resistance of 100 m of 10-gauge copper wire? (*b*) What is the electric field in the wire when the current is 30 A? (*c*) How long does it take for an electron to travel 100 m in the wire when the current is 30 A?

61. An automobile jumper cable 3 m long is made of three strands of 12-gauge copper wire twisted together. (*a*) What is the resistance of the jumper cable? (*b*) When the cable is used to start a car, it carries a current of 90 A. What is the potential drop that occurs across the jumper cable? (*c*) How much joule heating occurs in the jumper cable?

Level III

62. A linear accelerator produces a pulsed beam of electrons. The current is 1.6 A for the 0.1-μs duration of each pulse. (*a*) How many electrons are accelerated in each pulse? (*b*) What is the average current of the beam if there are 1000 pulses per second? (*c*) If each electron acquires an energy of 400 MeV, what is the average power output of the accelerator? (*d*) What is the peak power output? (*e*) What fraction of the time is the accelerator actually accelerating electrons? (This is called the *duty factor* of the accelerator.)

63. A wire of cross-sectional area A, length L_1, resistivity ρ_1, and temperature coefficient α_1 is connected end to end to a second wire of the same cross-sectional area, length L_2, resistivity ρ_2, and temperature coefficient α_2, so that the wires carry the same current. (*a*) Show that if $\rho_1 L_1 \alpha_1 + \rho_2 L_2 \alpha_2 = 0$, the total resistance R is independent of temperature for small temperature changes. (*b*) If one wire is made of carbon and the other is copper, find the ratio of their lengths for which R is approximately independent of temperature.

64. A 100-W heater is designed to operate across 120 V. (*a*) What is its resistance, and what current does it draw? (*b*) Show that if the potential difference across the heater changes by a small amount ΔV, the power changes by a small amount ΔP, where $\Delta P/P \approx 2\,\Delta V/V$. *Hint:* Approximate the changes with differentials. (*c*) Find the approximate power dissipated in the heater if the potential difference is decreased to 115 V.

65. Find the resistance between the ends of the half ring shown in Figure 22-31. The resistivity of the material of the ring is ρ.

Figure 22-31 Problem 65.

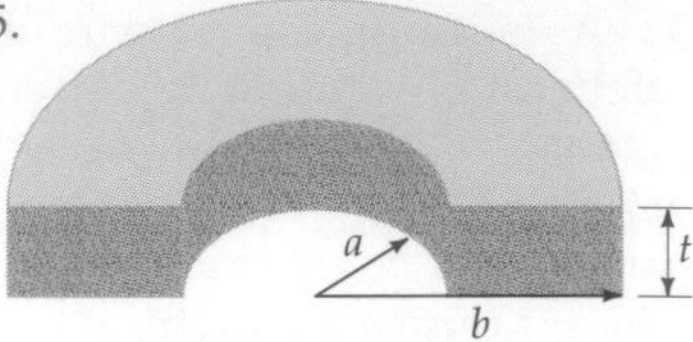

66. The space between two spherical-shell conductors is filled with a material that has a resistivity of 10^9 Ω·m. If the inner shell has a radius of 1.5 cm and the outer shell has a radius of 5 cm, what is the resistance between the conductors? *Hint:* Find the resistance of a spherical-shell element of the material of area $4\pi r^2$ and length dr, and integrate to find the total resistance of the set of shells in series.

67. The space between two metallic coaxial cylinders of length L and radii a and b is completely filled with a material having a resistivity ρ. (*a*) What is the resistance between the two cylinders? (See the hint in Problem 66.) (*b*) Find the current between the two cylinders if $\rho = 30$ Ω·m, $a = 1.5$ cm, $b = 2.5$ cm, $L = 50$ cm, and a potential difference of 10 V is maintained between the two cylinders.

68. A semiconducting diode is a nonlinear device whose current I is related to the voltage V across the diode by

$$I = I_0(e^{eV/kT} - 1)$$

where k is Boltzmann's constant, e is the charge on an electron, and T is the absolute temperature. (*a*) What is the resistance of the diode for $V = 0.5$ V if $I_0 = 10^{-9}$ A? (*b*) What is the resistance for $V = 0.6$ V?

69. The radius of a wire of length L increases linearly along its length according to

$$r = a + \frac{b-a}{L}x$$

where x is the distance from the small end of radius a. What is the resistance of this wire in terms of its resistivity ρ, length L, radius a, and radius b?

Chapter 23

Direct-Current Circuits

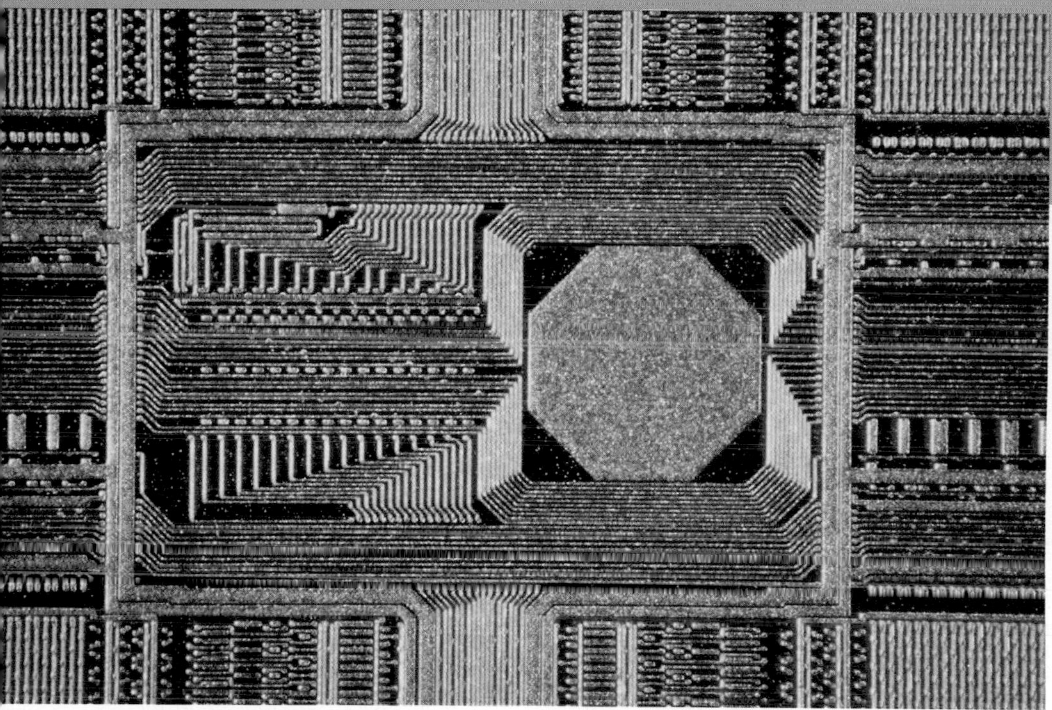

A photomicrograph (about 1000X magnification) of the center portion of a 256K dynamic random-access memory (DRAM) chip that can store 256,000 bits of data. The chip is smaller than a postage stamp and contains over 600,000 electronic components—resistors, capacitors, diodes, and transistors. These components are constructed on a single slice of silicon crystal by layering precisely defined regions of the crystal with impurities. The layering process that constructs the components also creates the wires (100 times finer than a human hair) connecting them. This kind of circuit, called an integrated circuit (IC), functions the same as circuits assembled from discrete components. But ICs operate faster and use less power because, being smaller, currents in them travel shorter distances.

In this chapter, we will analyze some simple circuits consisting of batteries, resistors, and capacitors in various combinations; that is, we will find the values of V and I and quantities derived from them at various points in the circuits. These circuits are called direct-current (dc) circuits because the current in any part of the circuit is always in the same direction. In Chapter 28, we will discuss alternating-current (ac) circuits, which are circuits in which the current in any part alternates in direction.

When a switch is thrown to turn on a circuit, an electric field propagates along the elements of the circuit, causing many complicated changes to occur in the circuit as the current builds up and charge accumulates at various points. However, since the electric field propagates at nearly the speed of light, these changes occur rapidly and equilibrium is quickly established. The time for equilibrium to be established depends on the conductivity of the elements in the circuit, but it is practically instantaneous for most purposes. In equilibrium, charge no longer accumulates at points along the circuit and the current is steady. For circuits containing capacitors, the current may increase or decrease slowly, but appreciable changes occur only over times much longer than the time needed to reach equilibrium. Thus, in circuits of this type each successive value of the current may be considered a quasi-equilibrium state.

23-1 Kirchhoff's Rules

The methods discussed in Chapter 22 for replacing combinations of resistors in series or parallel with their equivalent resistances are very useful for simplifying many combinations of resistors. However, they are not sufficient for the analysis of many simple circuits. Figure 23-1 gives an example of such a circuit. The two resistors R_1 and R_2 in this circuit look as if they might be in parallel, but they are not. The potential drop is not the same across both resistors because of the presence of the emf $\mathcal{E}_2$ in series with R_2. Also, R_1 and R_2 do not carry the same current, so they are not in series.

There are two rules, called **Kirchhoff's rules,** that apply to any circuit in the steady state:

Kirchhoff's rules

1. When any closed circuit loop is traversed, the algebraic sum of the changes in potential must equal zero.
2. At any junction point in a circuit where the current can divide, the sum of the currents into the junction must equal the sum of the currents out of the junction.

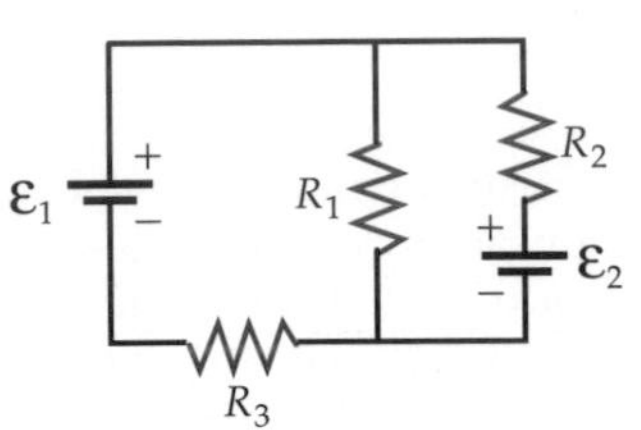

Figure 23-1 An example of a simple circuit that cannot be analyzed by replacing combinations of resistors in series or parallel with their equivalent resistances. The potential drops across R_1 and R_2 are not equal because of the emf $\mathcal{E}_2$, so these resistors are not in parallel. Note also that these resistors are not connected together at both ends. The resistors do not carry the same current, so they are not in series.

Kirchhoff's first rule, called the **loop rule,** follows from the fact that the potential difference between any two points in a circuit in the steady state is constant. In the steady state, the electric field at any point (outside a source of emf) is due to charge that has accumulated on the surfaces of battery terminals, resistors, wires, or other elements of the circuit. Since the electric field is conservative, a potential function exists at any point in space (except inside a source of emf). As we move around a loop of the circuit, the potential may decrease or increase as we pass through a resistor or battery, but when we have completely traversed the loop and have arrived back at our starting point, *the net change in the potential must be zero.* This rule is a direct result of the conservation of energy. If we have a charge q at some point where the potential is V, the potential energy of the charge is qV. As the charge traverses a loop in a circuit, it loses or gains energy as it passes through resistors, batteries, or other devices, but when it arrives back at its starting point, its energy must again be qV.

Exercise

Draw a rectangle representing some arbitrary circuit, and label any six points (such as the four corners and two other points) with the letters a, b, c, d, e, and f. Assign any number you like to each point on your diagram. Begin at point a and add the differences between each pair of numbers as you traverse the circuit, keeping track of the signs of the differences. What is the algebraic sum of these differences for one complete traversal of the circuit? (Answer: 0)

Kirchhoff's second rule, called the **junction rule,** follows from the conservation of charge. It is needed for multiloop circuits containing points where the current can divide. In the steady state, there is no further accumulation of electric charge at any point in a circuit so the amount of charge entering any point must equal the amount of charge leaving that point. Figure 23-2 shows the junction of three wires carrying currents I_1, I_2, and I_3. During a time interval Δt, charge $I_1\,\Delta t$ flows into the junction from the left. During the same time interval, charges $I_2\,\Delta t$ and $I_3\,\Delta t$ flow out of the junction to the right. Since charge does not originate at this point nor continues to accumulate there in the steady state, the conservation of charge implies the junction rule, which for this case gives

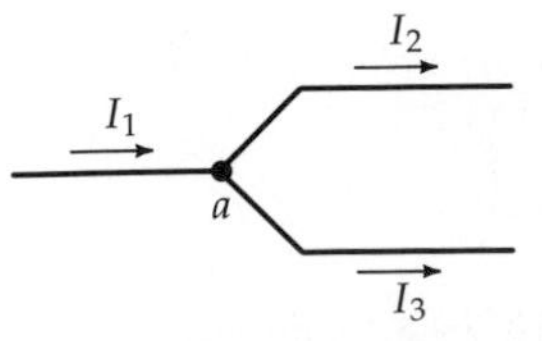

Figure 23-2 Illustration of Kirchhoff's junction rule. The current I_1 into point a equals the sum $I_2 + I_3$ of the currents out of point a.

$$I_1 = I_2 + I_3 \tag{23-1}$$

Figure 23-3 shows a circuit containing two batteries with internal resistances r_1 and r_2 and three external resistors. We wish to find the current in terms of the emfs and resistances, which are assumed to be known. We cannot predict the direction of the current unless we know which battery has the greater emf, but we do not have to know the direction of the current to analyze the circuit. We can assume either direction and solve the problem based on that assumption. If the assumption is incorrect, we will get a negative value for the current, indicating that its actual direction is opposite the direction assumed. Let us assume that I is clockwise, as indicated in the figure, and apply Kirchhoff's loop rule as we traverse the circuit in the assumed direction of the current, beginning at point a. The high- and low-potential sides of the resistors for this choice of current direction are indicated by the plus and minus signs in the figure. The potential decreases and increases are given in Table 23-1. Note that we encounter a potential drop as we traverse the source of emf between points c and d and a potential increase as we traverse the source of emf between f and g. Beginning at point a, we obtain from Kirchhoff's loop rule

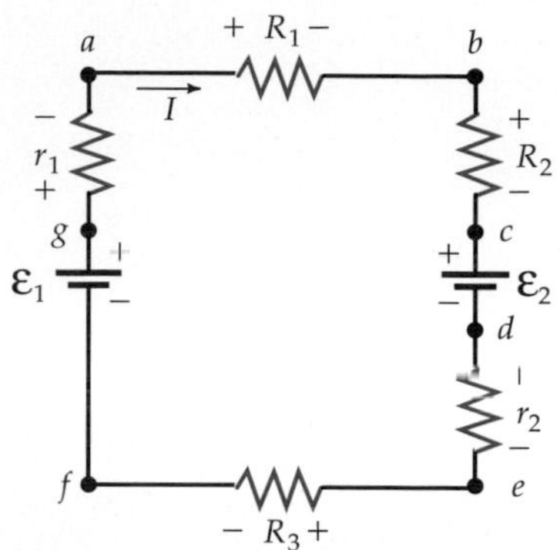

Figure 23-3 Circuit containing two batteries and three external resistors. The plus and minus signs on the resistors are there to help us remember which side of each resistor is at the higher potential for the current direction we have assumed.

$$-IR_1 - IR_2 - \mathcal{E}_2 - Ir_2 - IR_3 + \mathcal{E}_1 - Ir_1 = 0 \tag{23-2}$$

Solving for the current I, we obtain

$$I = \frac{\mathcal{E}_1 - \mathcal{E}_2}{R_1 + R_2 + R_3 + r_1 + r_2} \tag{23-3}$$

Note that if $\mathcal{E}_2$ is greater than $\mathcal{E}_1$, we will get a negative value for the current I, indicating that we have assumed the wrong direction for I. That is, if $\mathcal{E}_2$ is greater than $\mathcal{E}_1$, the current will be in the counterclockwise direction. On the other hand, if $\mathcal{E}_1$ is the greater emf, we will get a positive value for I, which means that the direction we have assumed is correct. Let us assume for this example that $\mathcal{E}_1$ is the greater emf. In battery 2, the charge flows from high potential to low potential. Therefore, a charge ΔQ moving through battery 2 from point c to point d loses energy $\mathcal{E}_2\,\Delta Q$. This electrical energy is converted into chemical energy and stored in the battery, which means that battery 2 is *charging*.

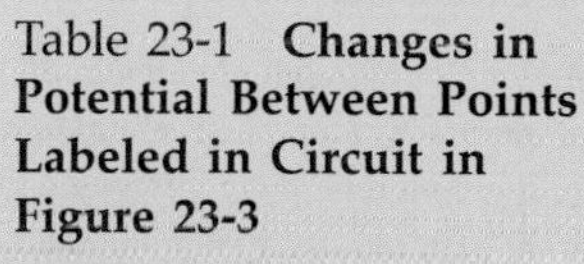

Table 23-1 **Changes in Potential Between Points Labeled in Circuit in Figure 23-3**

$a \to b$	Drop IR_1
$b \to c$	Drop IR_2
$c \to d$	Drop $\mathcal{E}_2$
$d \to e$	Drop Ir_2
$e \to f$	Drop IR_3
$f \to g$	Increase $\mathcal{E}_1$
$g \to a$	Drop Ir_1

We can account for the energy balance in this circuit by rearranging Equation 23-2 and multiplying each term by the current I:

$$\mathcal{E}_1 I = \mathcal{E}_2 I + I^2R_1 + I^2R_2 + I^2R_3 + I^2r_2 + I^2r_1 \tag{23-4}$$

The term $\mathcal{E}_1 I$ is the rate at which battery 1 puts energy into the circuit. This energy comes from the internal chemical energy of the battery. The term $\mathcal{E}_2 I$ is the rate at which electric energy is converted into chemical energy in battery 2. The term I^2R_1 is the rate at which Joule heat is produced in resistor R_1. Similarly, the terms for the other resistances give the rate of Joule heating in them.

Example 23-1

The elements in the circuit in Figure 23-3 have the values $\mathcal{E}_1 = 12$ V, $\mathcal{E}_2 = 4$ V, $r_1 = r_2 = 1\ \Omega$, $R_1 = R_2 = 5\ \Omega$, and $R_3 = 4\ \Omega$ as shown in Figure 23-4. Find the potentials at points a through g in the figure, assuming that the potential at point f is zero, and discuss the energy balance in the circuit.

The analysis of a circuit is usually simplified if we choose one point to be at zero potential and then find the potentials of the other points relative to it. Since only potential differences are important, any point in a circuit can be chosen to have zero potential. In this example, we will choose point f to be at zero potential. This is indicated by the ground symbol ⏚ at point f. As we saw in Section 18-2, the earth can be consid-

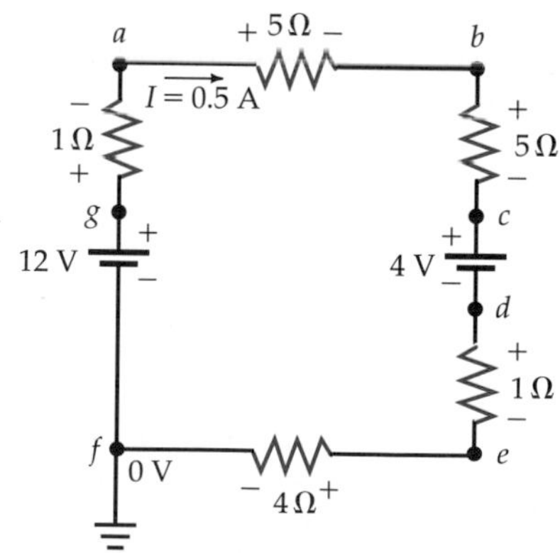

Figure 23-4 The circuit in Figure 23-3 with sample values for Example 23-1. The potential is chosen to be zero at point f. The three horizontal lines near point f indicate that it is grounded.

ered to be a very large conductor with a nearly unlimited supply of charge, which means that the potential of the earth remains essentially constant. Therefore, it is often chosen to be at zero potential. In practice, electrical circuits are often grounded by connecting one point to the earth. The outside metal case of a washing machine, for example, is usually grounded by connecting it by a wire to a water pipe that is in contact with the earth.

We first find the current in the circuit. From Equation 23-3, we have

$$I = \frac{12\text{ V} - 4\text{ V}}{5\ \Omega + 5\ \Omega + 4\ \Omega + 1\ \Omega + 1\ \Omega} = \frac{8\text{ V}}{16\ \Omega} = 0.5\text{ A}$$

We can now find the potentials at points a through g relative to zero potential at point f. Since, by definition, battery 1 maintains a constant potential difference $\mathcal{E}_1 = 12$ V between point g and point f, the potential at point g is 12 V. The potential at point a is less than that at g by the potential drop $Ir_1 = (0.5\text{ A})(1\ \Omega) = 0.5$ V. Thus, the potential at point a is 12 V − 0.5 V = 11.5 V. Similarly, the potential drops across the 5-Ω resistors R_1 and R_2 are each $IR_1 = (0.5\text{ A})(5\ \Omega) = 2.5$ V. The potential at point b is therefore 11.5 V − 2.5 V = 9 V, and that at c is 6.5 V. The potential drop across $\mathcal{E}_2$ is 4 V, so point d is at a potential of 2.5 V. Since the drop across the 1-Ω resistance r_2 is 0.5 V, the potential at e is 2 V. The potential drop across the 4-Ω resistance R_3 is $IR_3 = 2$ V. This gives zero for the potential at f, which is consistent with our original assumption. Figure 23-5 shows the potentials at all the labeled points, beginning and ending with point f.

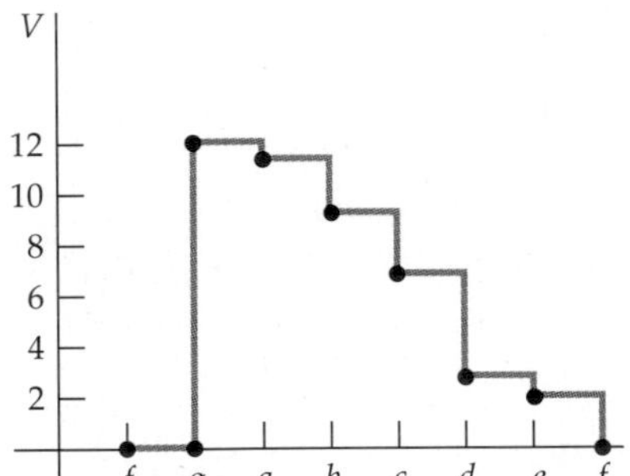

Figure 23-5 The potential at the labeled points of the circuit in Figure 23-4. The potential is zero at f and jumps to 12 V at g. It decreases by IR as we pass through each resistor in the direction of the current. When we get back to f, the potential is again zero.

The power delivered by the emf $\mathcal{E}_1$ is

$$P_{\mathcal{E}_1} = \mathcal{E}_1 I = (12\text{ V})(0.5\text{ A}) = 6.0\text{ W}$$

The power dissipated in the internal resistance of battery 1 is

$$P_{r_1} = I^2 r_1 = (0.5\text{ A})^2(1\ \Omega) = 0.25\text{ W}$$

Thus, the power delivered by battery 1 to the external circuit is 6.0 W − 0.25 W = 5.75 W; this also equals $V_1 I$, where $V_1 = V_a - V_f = 11.5$ V is the terminal voltage of battery 1. The total power dissipated in the external resistances in the circuit is

$$P_R = (0.5\text{ A})^2(5\ \Omega + 5\ \Omega + 4\ \Omega) = 3.5\text{ W}$$

The power going into battery 2, which is being charged, is $(V_c - V_e)I = (6.5\text{ V} - 2\text{ V})(0.5\text{ A}) = 2.25$ W. Part of this power, $P_{r_2} = I^2 r_2 = 0.25$ W, is dissipated in the internal resistance r_2 and part, $P_{\mathcal{E}_2} = \mathcal{E}_2 I = 2$ W, is the rate at which energy is being stored in that battery.

Note that the terminal voltage of the battery that is being charged in Example 23-1 is $V_c - V_e = 4.5$ V, which is greater than the emf of the battery. Because of its internal resistance, a battery is not completely reversible. If the same 4-V battery were to deliver 0.5 A to an external circuit, its terminal voltage would be 3.5 V (again assuming that its internal resistance is 1 Ω). If the internal resistance is very small, the terminal voltage of a battery is nearly equal to its emf, whether the battery is delivering current to an external circuit or is being charged. Some real batteries, such as those used in automobiles, are nearly reversible and can easily be recharged. Other types of batteries are not reversible. If you attempt to recharge one of these by driving current from its positive to its negative terminal, most, if not all, of the energy will go into heat rather than into the chemical energy of the battery, and the battery may explode.

Example 23-2

A good car battery is to be connected by jumper cables to a weak car battery to charge it. (*a*) To which terminal of the weak battery should the positive terminal of the good battery be connected? (*b*) Assume that the good battery has an emf $\mathcal{E}_1 = 12$ V and the weak battery has an emf $\mathcal{E}_2 = 11$ V, that the internal resistances of the batteries are $r_1 = r_2 = 0.02\ \Omega$, and that the resistance of the jumper cables is $R = 0.01\ \Omega$, all of which are typical values. What will the charging current be? (*c*) What will the current be if the batteries are connected incorrectly?

(*a*) Since the weak battery is to be charged, we connect the positive terminal of the good battery to the positive terminal of the weak battery (and the negative terminal of the good battery to the negative terminal of the weak battery) so as to drive charge through the weak battery from the positive terminal to the negative terminal. Figure 23-6 shows the circuit diagram for this process.

(*b*) The charging current is given by

$$I = \frac{\mathcal{E}_1 - \mathcal{E}_2}{R + r_1 + r_2} = \frac{12\ \text{V} - 11\ \text{V}}{0.05\ \Omega} = 20\ \text{A}$$

(*c*) Figure 23-7 shows the circuit diagram for the case in which the batteries are connected incorrectly, positive terminals to negative terminals. In this case, the current would be

$$I = \frac{\mathcal{E}_1 + \mathcal{E}_2}{R + r_1 + r_2} = \frac{12\ \text{V} + 11\ \text{V}}{0.05\ \Omega} = 460\ \text{A}$$

If the batteries are connected in this way, both may explode in a shower of boiling battery acid.

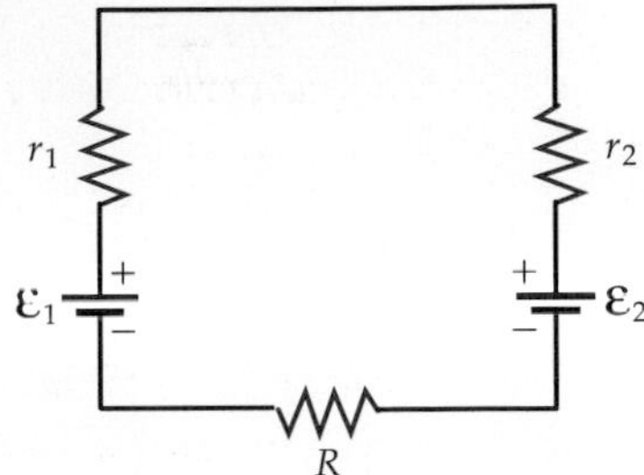

Figure 23-6 Two batteries that are connected such that one is charging the other. Since the sum of the internal resistances of the batteries and the resistance R of the connecting cables is very small, relatively large currents are obtained even when the emfs are nearly equal.

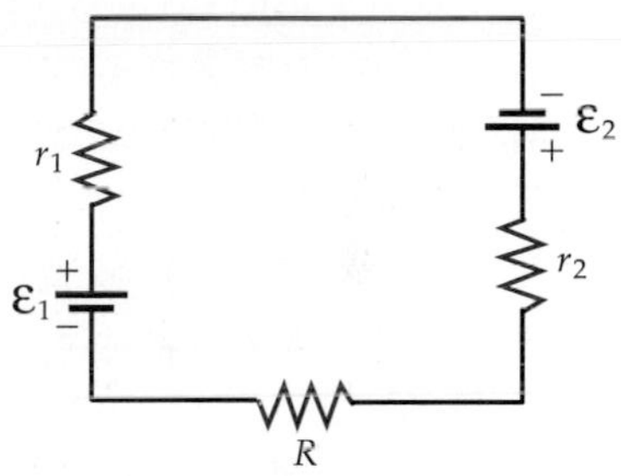

Figure 23-7 Two batteries that are *not* connected correctly for charging. Since the total resistance of the circuit is of the order of hundredths of an ohm, the current is very large, and the batteries could explode.

Multiloop Circuits

We will now consider circuits containing more than one loop. To analyze such circuits, we need to apply Kirchhoff's junction rule at points where the current splits into two or more parts.

Example 23-3

(*a*) Find the current in each part of the circuit shown in Figure 23-8. (*b*) Find the energy dissipated in the 4-Ω resistor in 3 s.

(*a*) This circuit is the same as that in Figure 23-1 with $\mathcal{E}_1 = 12$ V, $\mathcal{E}_2 = 5$ V, $R_1 = 4\ \Omega$, $R_2 = 2\ \Omega$, and $R_3 = 3\ \Omega$. Let I be the current through the 12-V battery in the direction shown in the figure. At point b, this current divides into currents I_1 and I_2. The current directions shown for I_1 and I_2 are merely guesses because we cannot be sure of their directions until after we have analyzed the circuit. For example, we need to know whether point b or point e is at the higher potential in order to know the direction of the current through the 4-Ω resistor. Applying the junction rule to point b, we obtain

$$I = I_1 + I_2$$

Applying the junction rule at point e gives us the same information since the currents I_1 and I_2 join there to form the current I directed towards point f. There are three loops to which the loop rule can be applied, the interior loops *abef* and *bcde* and the outer loop *abcdef*. We need only two more equations to determine the three unknown currents, so equations

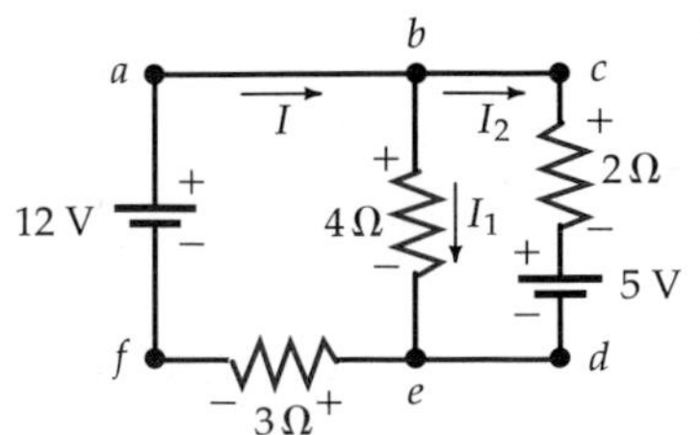

Figure 23-8 Circuit for Example 23-3. The direction of the current I_1 from b to e is not known before the circuit is analyzed. The plus and minus signs on the 4-Ω resistor are for the assumed direction of I_1 from b to e.

for any two of the loops will be sufficient. (The equation for the third loop will give redundant information.) Replacing I with $I_1 + I_2$ and applying the loop rule to the outer loop (*abcdef*), we get

$$12\text{ V} - (2\ \Omega)I_2 - 5\text{ V} - (3\ \Omega)(I_1 + I_2) = 0$$

Simplifying this equation and dividing it by 1 Ω, recalling that 1 V/1 Ω = 1 A, we obtain

$$7\text{ A} - 3I_1 - 5I_2 = 0 \qquad \text{23-5}$$

Similarly, applying the loop rule to the loop on the left (*abef*) gives

$$12\text{ V} - (4\ \Omega)I_1 - (3\ \Omega)(I_1 + I_2) = 0$$

or

$$12\text{ A} - 7I_1 - 3I_2 = 0 \qquad \text{23-6}$$

Equations 23-5 and 23-6 can now be solved for the currents I_1 and I_2. To eliminate I_2, we multiply each term in Equation 23-5 by 3 and each term in Equation 23-6 by 5 and obtain

$$21\text{ A} - 9I_1 - 15I_2 = 0 \qquad \text{23-5}a$$

and

$$60\text{ A} - 35I_1 - 15I_2 = 0 \qquad \text{23-6}a$$

Then, subtracting Equation 23-5*a* from Equation 23-6*a*, we obtain

$$39\text{ A} - 26I_1 = 0$$

or

$$I_1 = \frac{39\text{ A}}{26} = 1.5\text{ A}$$

Substituting this value for I_1 into Equation 23-5, we obtain

$$7\text{ A} - 3(1.5\text{ A}) - 5I_2 = 0$$

$$I_2 = \frac{2.5\text{ A}}{5} = 0.5\text{ A}$$

The total current through the 12-V battery is therefore

$$I = I_1 + I_2 = 1.5\text{ A} + 0.5\text{ A} = 2.0\text{ A}$$

(*b*) Since the current in the 4-Ω resistor is 1.5 A, the power dissipated in this resistor is

$$P = I_1^2R = (1.5\text{ A})^2(4\ \Omega) = 9\text{ W}$$

The total energy dissipated in the 4-Ω resistor in 3 s is then

$$W = Pt = (9\text{ W})(3\text{ s}) = 27\text{ J}$$

Example 23-3 illustrates some general features of the analysis of multiloop circuits. Consider a general circuit containing 1 outer loop and n interior loops. There will be $n - 1$ junction points where the current divides and another $n - 1$ junction points where the currents join. We can solve for the currents by applying the loop rule to any n loops and the junction rule to the $n - 1$ junction points where the current divides. Further application of the junction rule or loop rule will merely lead to redundant information. In Example 23-3, we had two interior loops, so we applied the loop rule to two loops and the junction rule to the one junction point (point b) where the current divided.

Example 23-4

Find the current in each part of the circuit shown in Figure 23-9.

We first note that the 3-Ω and 6-Ω resistors in the loop at the right are in parallel. Our work will be easier if we replace these resistors with the equivalent resistance R_{eq} found from

$$\frac{1}{R_{eq}} = \frac{1}{3\ \Omega} + \frac{1}{6\ \Omega} = \frac{3}{6\ \Omega} = \frac{1}{2\ \Omega}$$

or

$$R_{eq} = 2\ \Omega$$

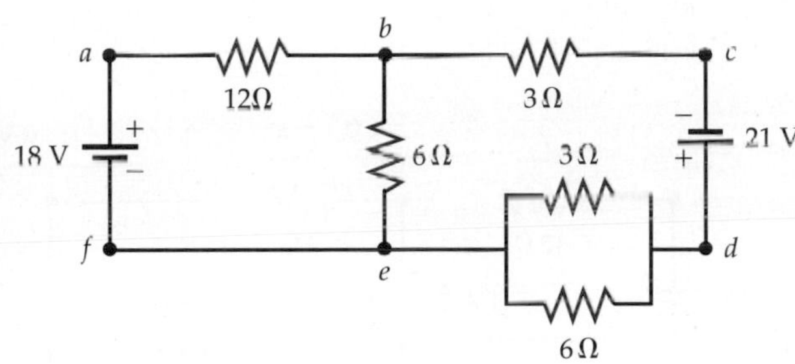

Figure 23-9 Circuit for Example 23-4.

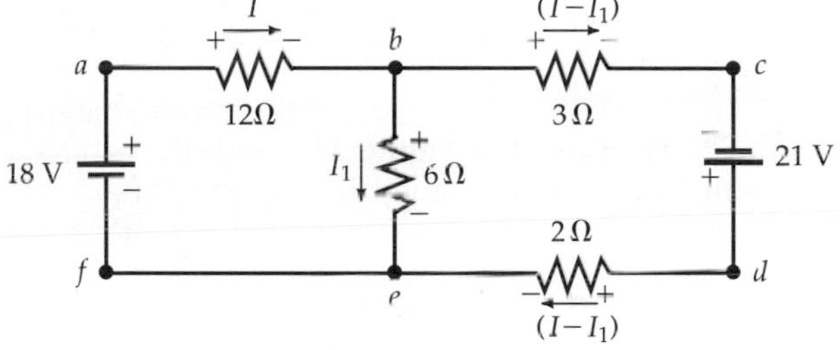

Figure 23-10 Circuit of Figure 23-9 with $R_{eq} = 2\ \Omega$ replacing the parallel combination. The plus and minus signs on the 6-Ω resistor are for the assumed direction of I_1. The junction rule is applied immediately at point b by using $(I - I_1)$ for the current from b to c.

In Figure 23-10, we have made this replacement, and we have chosen directions for the currents. We have called the current through the 18-V battery I and that through the 6-Ω resistor I_1, which we have assumed to be downward. Note that we have applied the junction rule to point b immediately by using $I - I_1$ for the current from point b to point c. Applying Kirchhoff's loop rule in the clockwise direction to the loop *abef* gives

$$18\ \text{V} - (12\ \Omega)I - (6\ \Omega)I_1 = 0$$

Simplifying this equation by dividing each term by 6 Ω and transposing the term $(18\ \text{V})/(6\ \Omega) = 3$ A gives

$$2I + I_1 = 3\ \text{A} \qquad \text{23-7}$$

Applying the loop rule to loop *bcde* gives us another equation:

$$-(3\ \Omega)(I - I_1) + 21\ \text{V} - (2\ \Omega)(I - I_1) + (6\ \Omega)I_1 = 0$$

Note that in moving from e to b we encounter a voltage increase because the assumed direction of current I_1 is opposite the direction in which we are traversing the loop. Combining terms, rearranging, and dividing the terms by 1 Ω, we obtain

$$5I - 11I_1 = 21\ \text{A} \qquad \text{23-8}$$

Solving Equations 23-7 and 23-8 for the unknown currents I and I_1, we obtain

$$I = 2\ \text{A}$$

and

$$I_1 = -1\ \text{A}$$

The negative value for I_1 shows that our original assumption about the direction of I_1 was incorrect. That is, the current through the 6-Ω resistor is actually in the direction from point e to point b. The current through the 21-V battery is therefore

$$I - I_1 = 2\text{ A} - (-1\text{ A}) = 3\text{ A}$$

In the original circuit of Figure 23-9, this current splits just after point d, with 2 A going through the 3-Ω resistor and 1 A going through the 6-Ω resistor. Figure 23-11 shows the complete circuit with the correct magnitudes and directions for the currents. A good way to check the solution of a circuit problem is to assign a potential of 0 to one point in the circuit and use the values of the currents found to determine the potential at the other points. In Figure 23-11 we have chosen point c to be at 0 V. Then point d is at +21 V, point e is at 15 V, point a is at 33 V and point b is at 9 V as indicated.

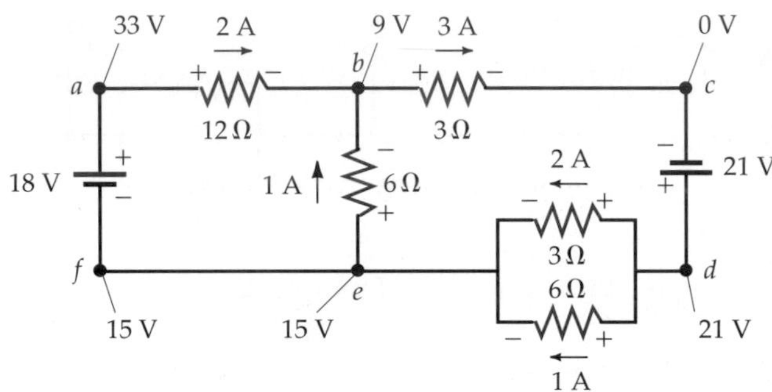

Figure 23-11 Circuit of Figure 23-9 with the correct magnitudes and directions for the currents found in Example 23-4.

We can summarize the general method for solving multiloop circuits:

General methods for solving multiloop circuits

1. Replace any combinations of resistors in series or parallel with their equivalent resistances.
2. Choose a direction for the current in each branch of the circuit, and label the currents in a circuit diagram. Add plus and minus signs to indicate the high and low potential sides of each source of emf, resistor, or capacitor.
3. Apply the junction rule to each junction where the current divides.
4. In a circuit containing n interior loops, apply the loop rule to any n loops.
5. Solve the equations to obtain the values of the unknowns.
6. Check your results by assigning a potential of zero to one point in the circuit and use the values of the currents found to determine the potentials at other points in the circuit.

Example 23-5

Find the current in each part of the circuit shown in Figure 23-12a.

This circuit contains three interior loops, so we will need to apply the loop rule 3 times. In Figure 23-12a, we have assumed a current I to the right through the 3-Ω resistor and have applied the junction rule at the two junctions c and e where the current divides. The current from c to d is labeled I_1, so the current from c to h is $I - I_1$. Similarly, the current from e to f is labeled I_2, so the current from e to h is $I_1 - I_2$. Applying

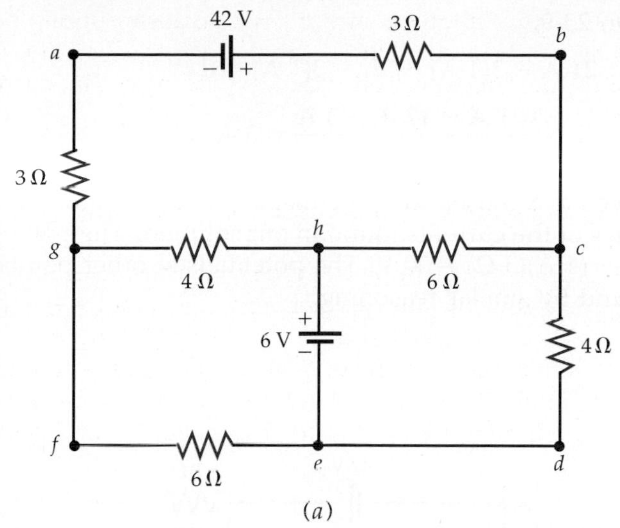

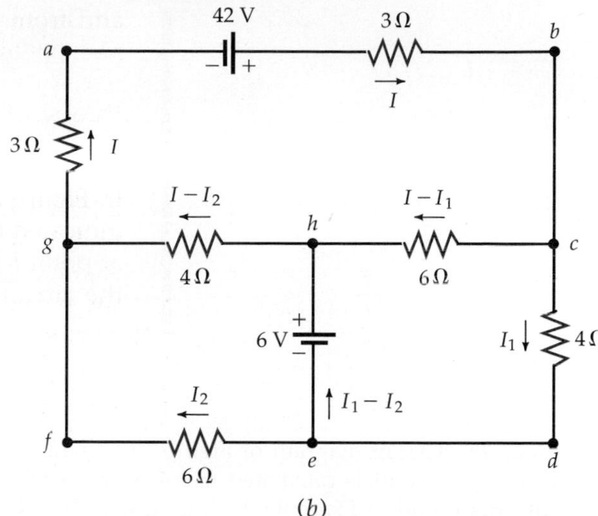

Figure 23-12 (*a*) Circuit containing three interior loops for Example 23-5. (*b*) Circuit diagram with assumed currents. The junction rule has been applied at points *c* and *e* where the current divides.

Kirchhoff's loop rule to the outer loop beginning at point *a*, we have

$$+42\text{ V} - (3\ \Omega)I - (4\ \Omega)I_1 - (6\ \Omega)I_2 - (3\ \Omega)I = 0$$

which can be simplified to

$$21\text{ A} - 3I - 2I_1 - 3I_2 = 0 \tag{23-9}$$

Similarly, applying the loop rule to the upper loop (*abchga*) gives

$$+42\text{ V} - (3\ \Omega)I - (6\ \Omega)(I - I_1) - (4\ \Omega)(I - I_2) - (3\ \Omega)I = 0$$

which simplifies to

$$21\text{ A} - 8I + 3I_1 + 2I_2 = 0 \tag{23-10}$$

For our third loop, we choose the lower left loop (*efghe*). We obtain

$$-(6\ \Omega)I_2 + (4\ \Omega)(I - I_2) - 6\text{ V} = 0$$

which simplifies to

$$2I - 5I_2 - 3\text{ A} = 0 \tag{23-11}$$

We can eliminate I_1 by multiplying each term in Equation 23-9 by 3 and each term in Equation 23-10 by 2 to obtain

$$63\text{ A} - 9I - 6I_1 - 9I_2 = 0$$

$$42\text{ A} - 16I + 6I_1 + 4I_2 = 0$$

Adding the resulting equations gives

$$105\text{ A} - 25I - 5I_2 = 0 \tag{23-12}$$

Substituting $5I_2 = 2I - 3$ A from Equation 23-11 into Equation 23-12, we obtain

$$105\text{ A} - 25I - (2I - 3\text{ A}) = 0$$

$$I = \frac{108\text{ A}}{27} = 4\text{ A}$$

Then from Equation 23-11,

$$5I_2 = 2I - 3\text{ A} = 2(4\text{ A}) - 3\text{ A} = 5\text{ A}$$

$$I_2 = 1\text{ A}$$

and from Equation 23-9,

$$21\text{ A} - 3(4\text{ A}) - 2I_1 - 3(1\text{ A}) = 0$$

$$I_1 = \frac{21\text{ A} - 12\text{ A} - 3\text{ A}}{2} = 3\text{ A}$$

In Figure 23-13, we have chosen the potential to be zero at point a and indicated the values of the currents found in our solution. The potential at point b is 42 V − (4 A)(3 Ω) = 30 V. The potentials at other points in the circuit are found by similar reasoning.

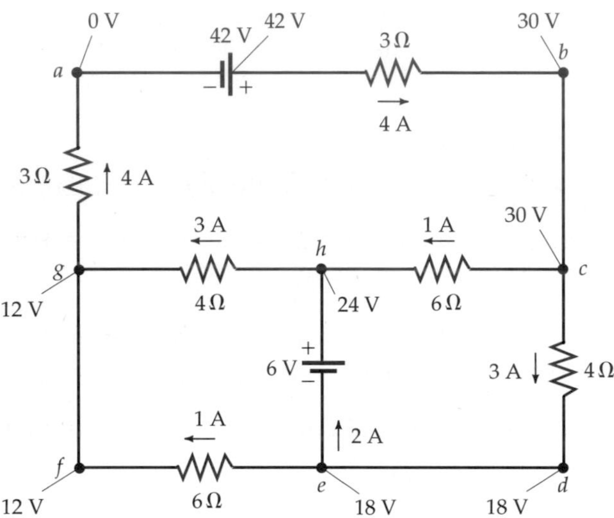

Figure 23-13 Circuit diagram of Figure 23-12 with the values of the potentials calculated from the values of the currents found in Example 23-5, assuming $V = 0$ at point a.

Analysis of Circuits by Symmetry

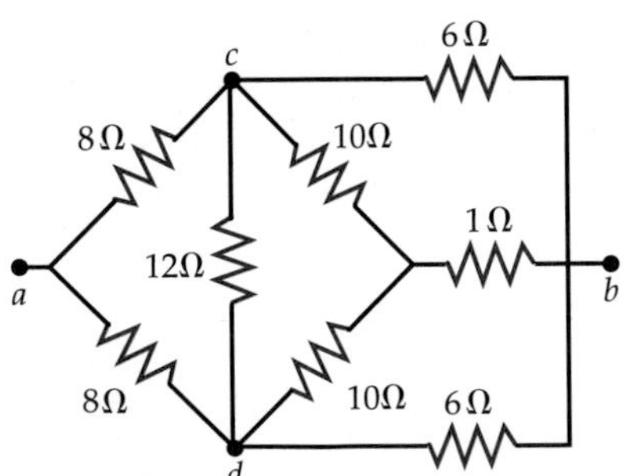

Figure 23-14 A complex, multibranch circuit. The circuit can be simplified by noting that by symmetry points c and d must be at the same potential. Because there is no current through the 12-Ω resistor between c and d, that resistor can be removed without affecting the current or potential anywhere in the circuit. Thus, points c and d can be considered a single point cd.

Some complicated circuits or networks of resistors can be more easily analyzed by using the junction rule and symmetry considerations rather than the loop rule. If two points in a circuit are at the same potential, they can be connected with a wire without causing any changes in the currents or potentials at other points in the circuit. Such points can often be identified by considering the symmetry of the circuit. The circuit can then be simplified by connecting these points and drawing a new circuit diagram.

Consider the circuit in Figure 23-14. We wish to find the current in each part of the circuit when a potential difference V_{ab} is applied between points a and b. Since this circuit has four interior loops, analyzing it by applying the loop rule to each loop becomes quite complicated. However, by symmetry, we can see that the points labeled c and d must be at the same potential, so there is no current in the 12-Ω resistor connecting these points. Thus, if we connect these points by a wire, there will be no current in the wire. In Figure 23-15 the circuit has been redrawn with points c and d connected. We can now find the equivalent resistance R_{eq} between points a and b using the methods for analyzing combinations of resistors in series and parallel discussed in Chapter 22. The total current from a to b is then V_{ab}/R_{eq}, and the current in each part of the circuit can be found by the methods discussed previously in this section.

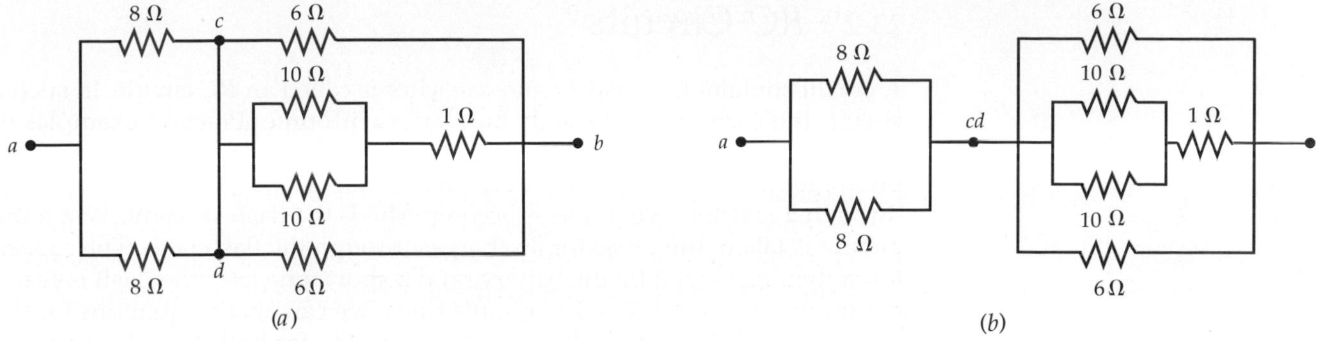

Figure 23-15 (*a*) Simplified diagram for the circuit of Figure 23-14 with points *c* and *d* connected. The two 8-Ω resistors are in parallel, and there are three parallel paths from the point *cd* to point *b*. (*b*) Alternative diagram in which points *c* and *d* are considered to be the same point.

Exercise

Find (*a*) the equivalent resistance between points *a* and *b* in the circuit in Figure 23-15 and (*b*) the current in the 10-Ω resistors if $V_{ab} = 12$ V. [Answer: (*a*) $R_{eq} = 6\ \Omega$, (*b*) $I_{10} = \frac{1}{3}$ A.]

Figure 23-16 shows 12 equal resistors that comprise the edges of a cube. We wish to find the equivalent resistance between the opposite corners of the cube labeled points *a* and *g* in the figure. Analyzing this network without using symmetry would clearly be difficult. By symmetry, we can see that if we apply a potential difference V_{ag} between points *a* and *g*, the points *b*, *d*, and *e* must all be at the same potential. If we connect these points by a wire, we have three equal resistors in parallel between point *a* and the common point *bde*. We can also see that points *c*, *f*, and *h* are at the same potential and can therefore be connected by a wire. There are six parallel paths, each of resistance *R*, between point *bde* and point *cfh* and three parallel paths from point *cfh* to point *g*. The simplified diagram for this circuit is shown in Figure 23-17. The equivalent resistance between *a* and *g* is thus

$$R_{eq} = \tfrac{1}{3}R + \tfrac{1}{6}R + \tfrac{1}{3}R = \tfrac{5}{6}R$$

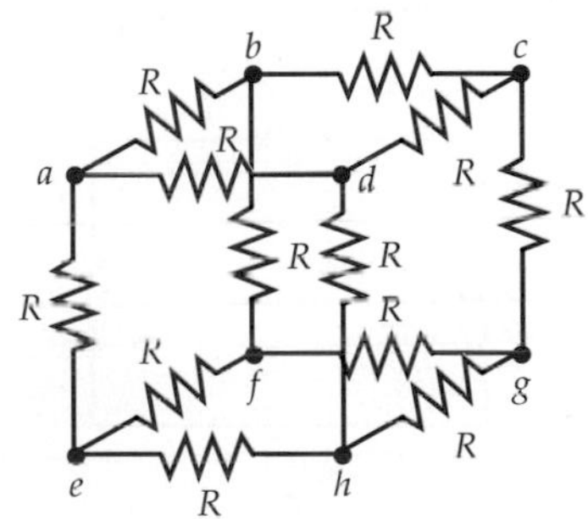

Figure 23-16 Twelve equal resistors comprising the edges of a cube. The circuit diagram can be simplified by noting from the symmetry of the figure that points *b*, *d*, and *e* must be at the same potential and points *c*, *f*, and *h* must be at the same potential.

There is another way to solve this problem. Let *I* be the current from point *a* to point *g* when the potential difference is V_{ag}. By symmetry, the current splits into three equal parts at point *a*, so the current from *a* to *b* is $\frac{1}{3}I$. At point *b*, the current splits into two equal parts since points *f* and *c* are at the same potential. Thus, the current from *b* to *c* is $\frac{1}{6}I$. Finally, since we have three symmetrical paths into point *g*, the current from *c* to *g* is $\frac{1}{3}I$. Therefore, if we move from point *a* to *b* to *c* to *g*, the potential drops are

$$V_{ag} = \tfrac{1}{3}IR + \tfrac{1}{6}IR + \tfrac{1}{3}IR = \tfrac{5}{6}IR = IR_{eq}$$

or

$$R_{eq} = \tfrac{5}{6}R$$

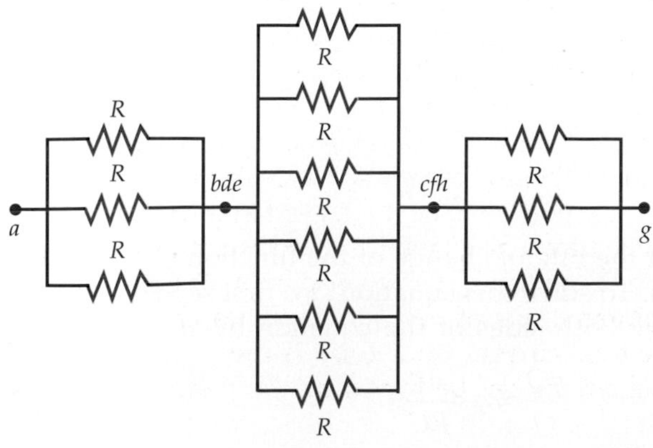

Figure 23-17 Simplified diagram for the circuit in Figure 23-16 with points *b*, *d*, and *e* connected and points *c*, *f*, and *h* connected. The six parallel paths between point *bde* and point *cfh* are the six edges of the cube *bc*, *bf*, *dc*, *dh*, *ef*, and *eh*.

23-2 RC Circuits

A circuit containing a resistor and capacitor is called an ***RC* circuit.** In such a circuit, the current is not steady but varies with time. Practical examples of *RC* circuits are the circuits in the flash attachment of a camera. Before a flash photograph is taken, a battery in the flash attachment charges the capacitor through a resistor. When this is accomplished, the flash is ready. When the picture is taken, the capacitor discharges through the flash bulb. The capacitor is then recharged by the battery, and a short time later the flash is ready for another picture. Using Kirchhoff's rules, we can obtain equations for the charge Q and the current I as functions of time for both the charging and discharging of a capacitor through a resistor.

Discharging a Capacitor

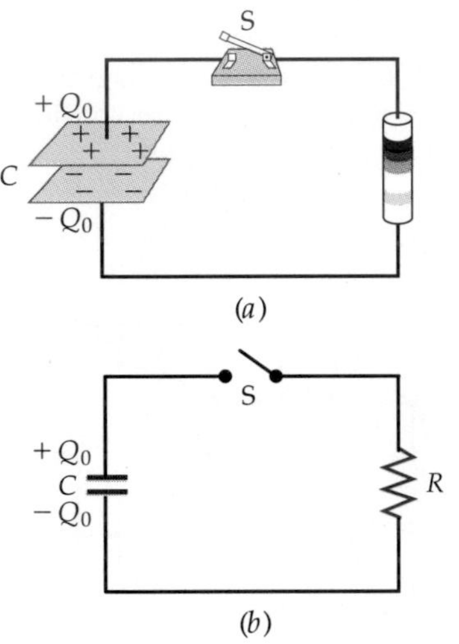

Figure 23-18 (*a*) A parallel-plate capacitor in series with a switch and a resistor R. (*b*) Circuit diagram for (*a*).

Figure 23-18 shows a capacitor with initial charges of $+Q_0$ on the upper plate and $-Q_0$ on the lower plate. It is connected to a resistor R and a switch S, which is open to prevent the charge from flowing through the resistor. The potential difference across the capacitor is initially $V_0 = Q_0/C$, where C is the capacitance. Since there is no current when the switch is open, there is no potential drop across the resistor. Thus, there is also a potential difference V_0 across the switch.

We close the switch at time $t = 0$. Since there is now a potential difference across the resistor, there must be a current in it. The initial current is

$$I_0 = \frac{V_0}{R} = \frac{Q_0}{RC} \qquad \text{23-13}$$

The current is due to the flow of charge from the positive plate of the capacitor to the negative plate through the resistor. After a time, the charge on the capacitor is reduced. Since the charge on the capacitor is *decreasing*, and we are taking the clockwise current to be positive, the current equals the rate of *decrease* of that charge. If Q is the charge on the capacitor at any time, the current at that time is

$$I = -\frac{dQ}{dt} \qquad \text{23-14}$$

Traversing the circuit in the direction of the current, we encounter a potential drop IR across the resistor and a potential increase Q/C across the capacitor. Kirchhoff's loop rule gives

$$\frac{Q}{C} - IR = 0 \qquad \text{23-15}$$

where both Q and I are functions of time and are related by Equation 23-14. Substituting $-dQ/dt$ for I in Equation 23-15, we have

$$\frac{Q}{C} + R\frac{dQ}{dt} = 0$$

or

$$\frac{dQ}{dt} = -\frac{1}{RC}Q \qquad \text{23-16}$$

Equation 23-16 states that the rate of change of the function $Q(t)$ is proportional to the function $Q(t)$. To solve this equation, we first separate the variables Q and t. Multiplying both sides of the equation by dt/Q, we obtain

$$\frac{dQ}{Q} = -\frac{dt}{RC} \qquad \text{23-17}$$

Integrating, we obtain

$$\ln Q = -\frac{t}{RC} + A$$

where A is an arbitrary constant of integration that is determined by the initial conditions. Taking the exponential of both sides of this equation, we obtain

$$e^{\ln Q} = e^{-t/RC+A} = e^{A}e^{-t/RC}$$

or

$$Q = Be^{-t/RC}$$

We obtain the constant $B = e^{A}$ from the initial condition that $Q = Q_0$ at $t = 0$. Then

$$Q(t) = Q_0e^{-t/RC} = Q_0e^{-t/\tau} \qquad 23\text{-}18$$

where τ, which is called the **time constant**, is the time it takes for the charge to decrease to $1/e$ of its original value:

$$\tau = RC \qquad 23\text{-}19$$

Time constant

Figure 23-19 shows the charge on the capacitor in the circuit of Figure 23-18 as a function of time. The dashed line is the initial slope of the charge-versus-time function. If the charge continued to decrease at a constant rate equal to this initial rate, it would reach zero in a time equal to the time constant τ. However, the actual rate of decrease $-dQ/dt$ is not constant but also decreases with time. This is evident from Equation 23-16, which shows that the rate of decrease of the charge is proportional to the charge itself. After a time $t = \tau$, the charge is $Q = Q_0e^{-1} = 0.37Q_0$. After a time $t = 2\tau$, the charge is $Q = Q_0e^{-2} = 0.135Q_0$, and so forth. After a time equal to several time constants, the charge on the capacitor is negligible. This type of decrease, which is called an **exponential decrease,** is very common in nature. It occurs whenever the rate at which a quantity decreases is proportional to the quantity itself. (We encountered exponential decreases in Chapter 11 when we studied the decrease in air pressure with altitude and again in Chapter 12 when we studied the decrease in the energy of a damped oscillator with time.) The decrease in the charge on a capacitor can be likened to the decrease in the amount of water in a bucket that has a small hole in the bottom. The rate at which the water flows out of the bucket is proportional to the pressure of the water, which is in turn proportional to the amount of water still in the bucket.

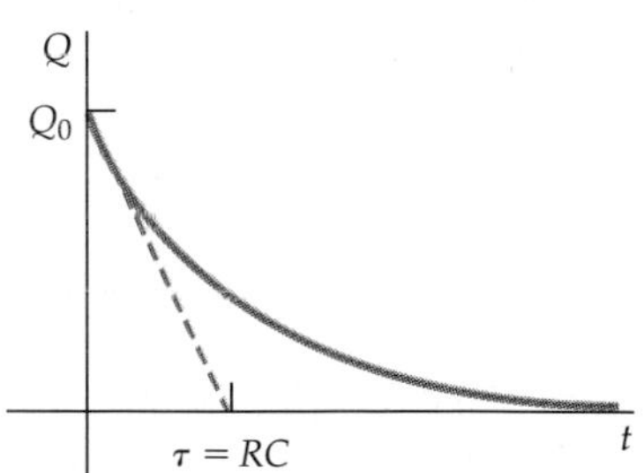

Figure 23-19 Plot of the charge on the capacitor versus time for the circuit in Figure 23-18 when the switch is closed at time $t = 0$. The time constant $\tau = RC$ is the time it takes for the charge to decrease to $e^{-1}Q_0$. After a time equal to two time constants, the charge is $e^{-2}Q_0$. This is an example of an exponential decrease. The time constant is also the time it would take the capacitor to discharge fully if its discharge rate were constant, as indicated by the dashed line.

The current is obtained from Equation 23-14 by differentiating Equation 23-18:

$$I = -\frac{dQ}{dt} = \frac{Q_0}{RC}e^{-t/RC}$$

or

$$I = \frac{V_0}{R}e^{-t/RC} = I_0e^{-t/\tau} \qquad 23\text{-}20$$

where $I_0 = Q_0/RC = V_0/R$ is the initial current. The current as a function of

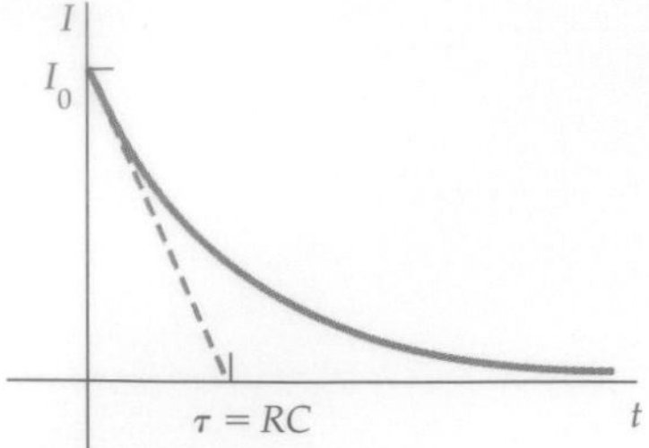

Figure 23-20 Plot of the current versus time for the circuit in Figure 23-18. The curve has the same shape as that in Figure 23-19.

time is shown in Figure 23-20. The current also decreases exponentially with time and falls to $1/e$ of its initial value after a time $t = \tau = RC$.

Example 23-6

A 4-μF capacitor is charged to 24 V and then connected across a 200-Ω resistor. Find (*a*) the initial charge on the capacitor, (*b*) the initial current through the 200-Ω resistor, (*c*) the time constant, and (*d*) the charge on the capacitor after 4 ms.

(*a*) The initial charge is $Q_0 = CV = (4\ \mu\text{F})(24\ \text{V}) = 96\ \mu\text{C}$.
(*b*) The initial current is $I_0 = V_0/R = (24\ \text{V})/(200\ \Omega) = 0.12\ \text{A}$.
(*c*) The time constant is $\tau = RC = (200\ \Omega)(4\ \mu\text{F}) = 800\ \mu\text{s} = 0.8\ \text{ms}$.
(*d*) At $t = 4$ ms, the charge on the capacitor is

$$Q = Q_0 e^{-t/\tau} = (96\ \mu\text{C})e^{-(4\ \text{ms})/(0.8\ \text{ms})} = (96\ \mu\text{C})e^{-5} = 0.647\ \mu\text{C}$$

Exercise

Find the current through the 200-Ω resistor at $t = 4$ ms. (Answer: 0.809 mA)

Charging a Capacitor

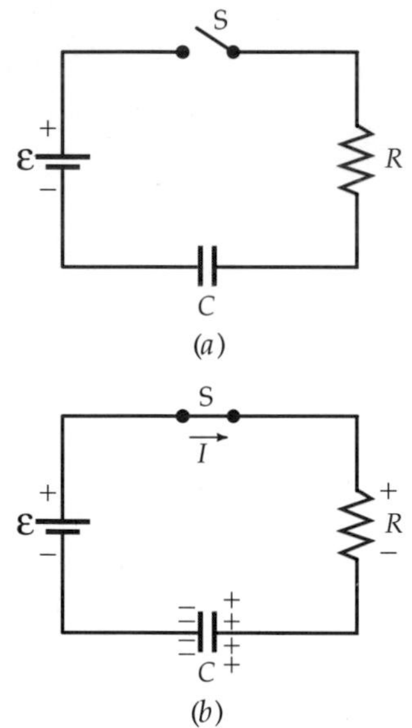

Figure 23-21 (*a*) Circuit for charging a capacitor to a potential difference $\mathcal{E}$. (*b*) After the switch is closed, there is a potential drop across the resistor and a charge on the capacitor.

Figure 23-21*a* shows a circuit for charging a capacitor, which we will assume to be initially uncharged. The switch, originally open, is closed at time $t = 0$. Charge immediately begins to flow through the resistor and onto the positive plate of the capacitor (Figure 23-21*b*). If the charge on the capacitor at some time is Q and the current in the circuit is I, Kirchhoff's loop rule gives

$$\mathcal{E} - V_R - V_C = 0$$

or

$$\mathcal{E} - IR - \frac{Q}{C} = 0 \tag{23-21}$$

In this circuit, the current equals the rate at which the charge on the capacitor is *increasing:*

$$I = +\frac{dQ}{dt}$$

Substituting $+dQ/dt$ for I in Equation 23-21 gives

$$\mathcal{E} = R\frac{dQ}{dt} + \frac{Q}{C} \tag{23-22}$$

At time $t = 0$, the charge on the capacitor is zero and the current is $I_0 = \mathcal{E}/R$. The charge then increases and the current decreases, as can be seen from Equation 23-21. The charge reaches a maximum value of $Q_f = C\mathcal{E}$ when the current I equals zero, as can also be seen from Equation 23-21.

Equation 23-22 is slightly more difficult to solve than was Equation 23-16. Multiplying each term by C and rearranging, we can write Equation 23-22 as

$$RC\frac{dQ}{dt} = C\mathcal{E} - Q$$

We can now separate the variables Q and t by multiplying each side by dt/RC and dividing by $C\mathcal{E} - Q$:

$$\frac{dQ}{C\mathcal{E} - Q} = \frac{dt}{RC} \tag{23-23}$$

Integrating each side of Equation 23-23, we obtain

$$-\ln (C\mathcal{E} - Q) = t/RC + A \qquad 23\text{-}24$$

where A is an arbitrary constant of integration. Taking the exponent of each side of Equation 23-24, we obtain

$$C\mathcal{E} - Q = e^{-A}e^{-t/RC} = Be^{-t/RC}$$

or

$$Q = C\mathcal{E} - Be^{-t/RC} \qquad 23\text{-}25$$

where $B = e^{-A}$ is another constant. The value of B is determined by the initial condition $Q = 0$ at $t = 0$. Setting $t = 0$ and $Q = 0$ in Equation 23-25 gives

$$0 = C\mathcal{E} - B$$

or

$$B = C\mathcal{E}$$

Substituting $B = C\mathcal{E}$ into Equation 23-25, we obtain for the charge

$$Q = C\mathcal{E}(1 - e^{-t/RC}) = Q_f(1 - e^{-t/\tau}) \qquad 23\text{-}26$$

where $Q_f = QE$ is the final charge. The current is obtained by differentiating Equation 23-26:

$$I = \frac{dQ}{dt} = -C\mathcal{E}e^{-t/RC}(-1/RC)$$

or

$$I = \frac{\mathcal{E}}{R}e^{-t/RC} = I_0e^{-t/\tau} \qquad 23\text{-}27$$

Figures 23-22 and 23-23 show the charge and the current as functions of time. Note from Figure 23-22 that the time constant τ is also the time in which the capacitor would become fully charged if the current remained constant at its initial value.

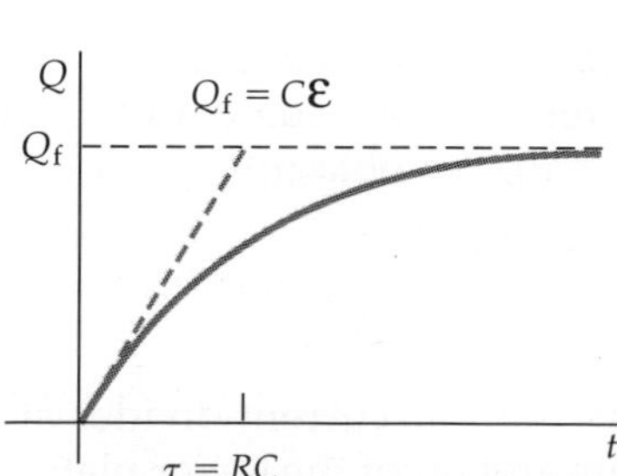

Figure 23-22 Plot of the charge on the capacitor versus time for the charging circuit of Figure 23-21 after the switch is closed at $t = 0$. After a time $t = \tau = RC$, the charge on the capacitor is $0.63C\mathcal{E}$, where $C\mathcal{E}$ is its final charge. If the charging rate were constant, the capacitor would be fully charged after a time $t = \tau$.

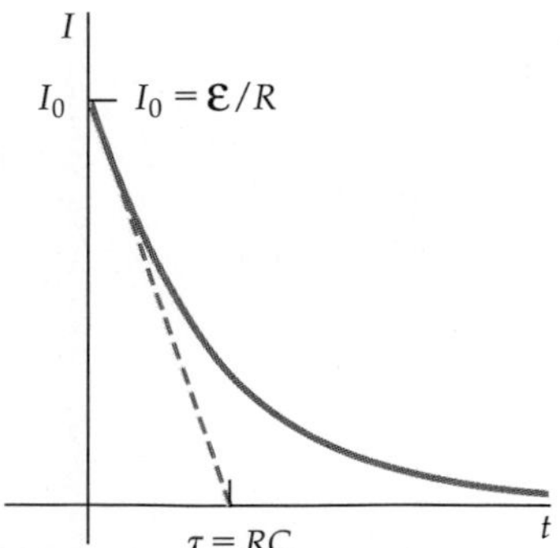

Figure 23-23 Plot of the current versus time for the charging circuit of Figure 23-21. The current is initially $\mathcal{E}/R$, and it decreases exponentially with time.

Exercise

Show that Equation 23-26 does indeed satisfy Equation 23-22 by substituting $Q(t)$ and dQ/dt into Equation 23-22.

Exercise

What fraction of the maximum charge is on the charging capacitor after a time $t = \tau$? (Answer: 0.63)

Example 23-7

A 6-V battery of negligible internal resistance is used to charge a 2-μF capacitor through a 100-Ω resistor. Find (*a*) the initial current, (*b*) the final charge on the capacitor, and (*c*) the time required for the charge to reach 90 percent of its final value.

(*a*) The initial current is

$$I_0 = \mathcal{E}/R = (6\text{ V})/(100\ \Omega) = 0.06\text{ A}$$

(*b*) The final charge on the capacitor is

$$Q_f = \mathcal{E}C = (6\text{ V})(2\ \mu\text{F}) = 12\ \mu\text{C}$$

(*c*) The time constant for this circuit is $\tau = RC = (100\ \Omega)(2\ \mu\text{F}) = 200\ \mu\text{s}$. We should expect the charge to reach 90 percent of its final value in a time of the order of several time constants. We can find the exact solution from Equation 23-26, using $Q = 0.9\mathcal{E}C$:

$$Q = 0.9\mathcal{E}C = \mathcal{E}C(1 - e^{-t/RC})$$

$$0.9 = 1 - e^{-t/RC}$$

$$e^{-t/RC} = 1 - 0.9 = 0.1$$

$$\ln e^{-t/RC} = -\frac{t}{RC} = \ln 0.1 = -2.3$$

Thus,

$$t = 2.3RC = 2.3(200\ \mu\text{s}) = 460\ \mu\text{s}$$

Example 23-8

The capacitor in the circuit shown in Figure 23-24*a* is initially uncharged. Find the current through the battery (*a*) immediately after the switch is closed and (*b*) a long time after the switch is closed.

(*a*) Since the capacitor is initially uncharged, the potential at point *d* is the same as that at point *c* just after the switch is closed. There is thus no initial current through the 8-Ω resistor between *b* and *e*. Applying the loop rule to the outer loop (*abcdefa*), we obtain

$$12\text{ V} - (4\ \Omega)I_0 = 0$$

$$I_0 = 3\text{ A}$$

(*b*) While the capacitor is charging, the current divides at point *b* and charge flows onto the upper plate and off of the lower plate. After a long time, the capacitor becomes fully charged, and no more charge flows onto or off of the plates. Applying the loop rule to the left loop (*abefa*), we obtain

$$12\text{ V} - (4\ \Omega)I_f - (8\ \Omega)I_f = 0$$

$$I_f = 1\text{ A}$$

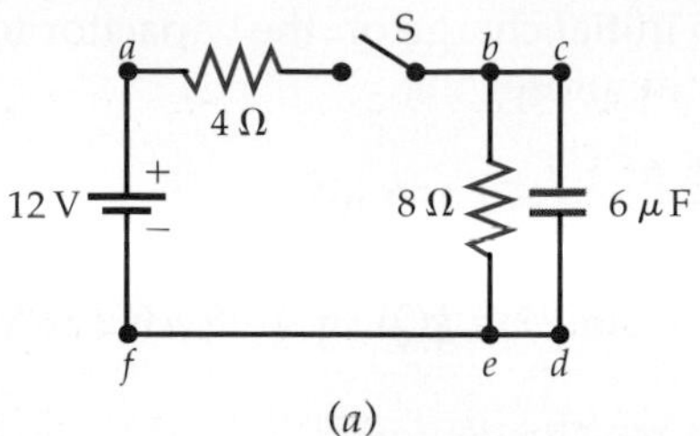

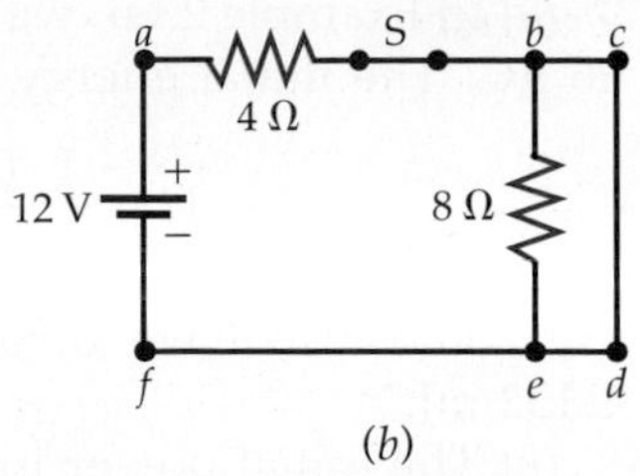

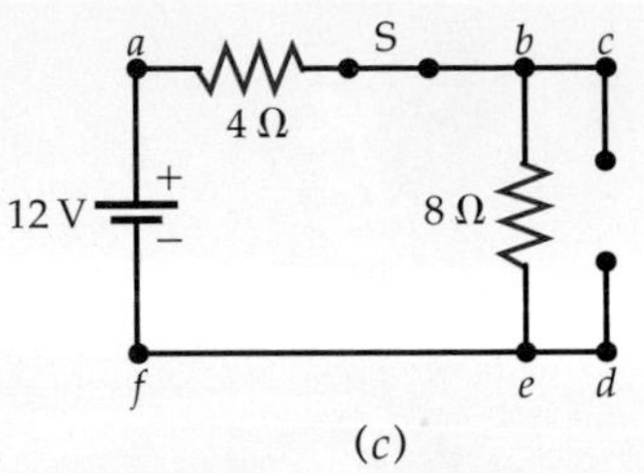

Figure 23-24 (*a*) A capacitor in parallel with a resistor across an emf. The capacitor is assumed to be uncharged before the switch is closed. (*b*) Immediately after the switch is closed, there is no potential drop across the capacitor, so the capacitor can be replaced with a wire of zero resistance. (*c*) A long time after the switch has been closed, the capacitor is fully charged and all the charge entering point *b* flows through the 8-Ω resistor. The capacitor can thus be replaced by an open circuit.

We see that the analysis of this circuit at the extreme times, when the capacitor is either uncharged or fully charged, is simple. When the capacitor is uncharged, it acts like a short circuit between points *c* and *d*, that is, the circuit is the same as shown in Figure 23-24*b*, where we have replaced the capacitor by a wire of zero resistance. When the capacitor is fully charged, it acts like an open circuit as shown in Figure 23-24*c*.

During the charging process, a total charge $Q_f = \mathcal{E}C$ flows through the battery. The battery therefore does work

$$W = Q_f\mathcal{E} = \mathcal{E}^2C$$

Half of this work is stored as energy in the capacitor. From Equation 21-16,

$$U = \tfrac{1}{2}QV = \tfrac{1}{2}Q_f\mathcal{E} = \tfrac{1}{2}\mathcal{E}^2C$$

We now show that the other half of the energy provided by the battery goes into Joule heat in the resistor. The rate of energy put into the resistor is

$$\frac{dW_R}{dt} = I^2R$$

Using Equation 23-27 for the current, we have

$$\frac{dW_R}{dt} = \left(\frac{\mathcal{E}}{R}e^{-t/RC}\right)^2 R = \frac{\mathcal{E}^2}{R}e^{-2t/RC}$$

We find the total Joule heat produced by integrating from $t = 0$ to $t = \infty$:

$$W_R = \int_0^\infty \frac{\mathcal{E}^2}{R}e^{-2t/RC}\,dt$$

The integration can be done by substituting $x = 2t/RC$. Then,

$$dt = \frac{RC}{2}\,dx$$

and

$$W_R = \frac{\mathcal{E}^2}{R}\frac{RC}{2}\int_0^\infty e^{-x}\,dx = \frac{1}{2}\mathcal{E}^2C$$

since the integral is 1. This result is independent of the resistance R. Thus, when a capacitor is charged by a battery with a constant emf, half the energy provided by the battery is stored in the capacitor and half goes into heat, independent of the resistance. The energy that goes into heat includes the energy that goes into the internal resistance of the battery.

Example 23-9

For the discharging capacitor in Example 23-6, find (*a*) the initial energy stored in the capacitor, (*b*) the initial power input to the resistor, and (*c*) the energy stored at $t = 0.8$ ms $= 1\tau$.

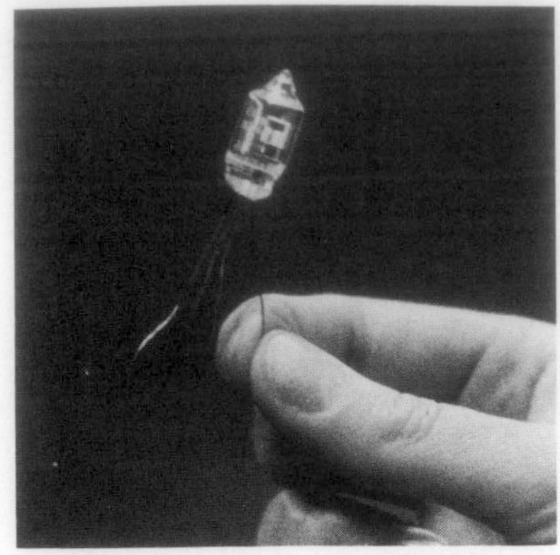

A krytron, shown here, is a high-speed triggering device used to discharge capacitors in the detonation of an atomic bomb, which consists of a uranium core surrounded by a shell of explosives. The capacitor, triggered by the krytron, discharges its energy to detonators that set off the explosive shell. The explosion rapidly and symmetrically compresses the uranium core, which reaches critical mass, undergoes runaway nuclear fission, and explodes. The sale of krytrons and certain specialized capacitors is controlled. In 1985 and 1990, they were the subject of publicized unsuccessful smuggling operations.

(*a*) In Example 23-6, we found the initial charge on the capacitor to be 96 μC. The initial energy stored is therefore

$$U = \frac{1}{2}\frac{Q^2}{C} = \frac{1}{2}\frac{(96\ \mu\text{C})^2}{4\ \mu\text{F}} = 1.152\ \text{mJ}$$

We could also have obtained this from $U = \frac{1}{2}QV = \frac{1}{2}(96\ \mu\text{C})(24\ \text{V}) =$ 1.152 mJ.

(*b*) The initial power input to the resistor is

$$P_0 = I_0^2R = (0.12\ \text{A})^2(200\ \Omega) = 2.88\ \text{W}$$

where we have used $I_0 = 0.12$ A from Example 23-6.

(*c*) The charge on the capacitor after $t = 0.8$ ms $= 1\tau$ is

$$Q = Q_0e^{-t/\tau} = (96\ \mu\text{C})e^{-1} = 35.3\ \mu\text{C}$$

The energy stored at this time is thus

$$U = \frac{1}{2}\frac{Q^2}{C} = \frac{1}{2}\frac{(35.3\ \mu\text{C})^2}{4\ \mu\text{F}} = 0.156\ \text{mJ}$$

Note that we could not have used $U = \frac{1}{2}QV$ unless we first found the potential V, which also decreases as the capacitor discharges.

Example 23-10

Show that the decrease in the energy stored in the capacitor in Example 23-9 from $t = 0$ to $t = 1\tau$ equals the joule heat dissipated in the resistor.

In Example 23-9, we found the energy stored in the capacitor to be 1.152 mJ at $t = 0$ and 0.156 mJ at $t = 1\tau$. The decrease in the energy stored is thus

$$-\Delta U = 1.152\ \text{mJ} - 0.156\ \text{mJ} = 0.996\ \text{mJ}$$

The power input into the resistor is I^2R where I is given by Equation 23-20. Since this power is not constant in time, we must integrate from $t = 0$ to $t = 1\tau$ to find the total energy dissipated in the resistor. We have

$$W = \int_0^\tau I^2R\ dt = \int_0^\tau (I_0e^{-t/\tau})^2R\ dt$$

$$= I_0^2R\int_0^\tau e^{-2t/\tau}\ dt = I_0^2R\left(\frac{\tau}{-2}\right)e^{-2t/\tau}\Big|_0^\tau$$

$$= I_0^2R\left(\frac{\tau}{2}\right)(1 - e^{-2})$$

where we have used

$$\int e^{ax}\ dx = \frac{1}{a}e^{ax}$$

from integral tables. Then, using $I_0^2R = 2.88$ W from Example 23-9 and $\tau = 0.8$ ms from Example 23-6, we have

$$W = (2.88\ \text{W})(0.4\ \text{ms})(1 - e^{-2}) = 0.996\ \text{mJ}$$

which equals the decrease in the energy stored.

Question

1. A swimming pool is filled by siphoning water from a nearby lake. How is this analogous to the charging of a capacitor?

23-3 Ammeters, Voltmeters, and Ohmmeters

We turn now to the consideration of the measurement of electrical quantities in dc circuits. The devices that measure current, potential difference, and resistance are called **ammeters, voltmeters,** and **ohmmeters,** respectively. Often, all three of these meters are included in a single "multimeter" that can be switched from one use to another. You might use a voltmeter to measure the terminal voltage of your car battery and an ohmmeter to measure the resistance between two points in some electrical device at home (such as a toaster) where you suspect a short circuit or a broken wire. Therefore, some knowledge of the basic operation of these devices might prove very useful.

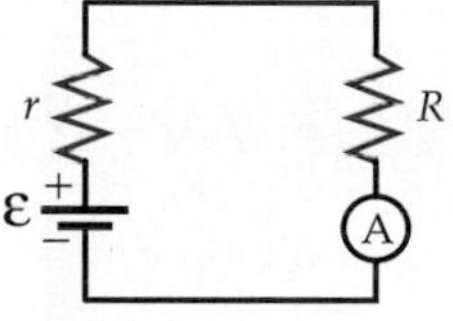

Figure 23-25 To measure the current in a resistor R, an ammeter —Ⓐ— is placed in series with the resistor so that it carries the same current as the resistor.

To measure the current through a resistor in a simple circuit, we place an ammeter in series with the resistor, as shown in Figure 23-25, so that the ammeter and the resistor carry the same current. Since the ammeter has some resistance, the current in the circuit decreases slightly when the ammeter is inserted. Ideally, the ammeter should have a very small resistance so that only a small change is caused in the current to be measured.

The potential difference across a resistor is measured by placing a voltmeter across the resistor in parallel with it, as shown in Figure 23-26, so that the potential drop across the voltmeter is the same as that across the resistor. The voltmeter reduces the resistance between points a and b, thus increasing the total current in the circuit and changing the potential drop across the resistor. A good voltmeter has a very large resistance so that its effect on the circuit is minimal.

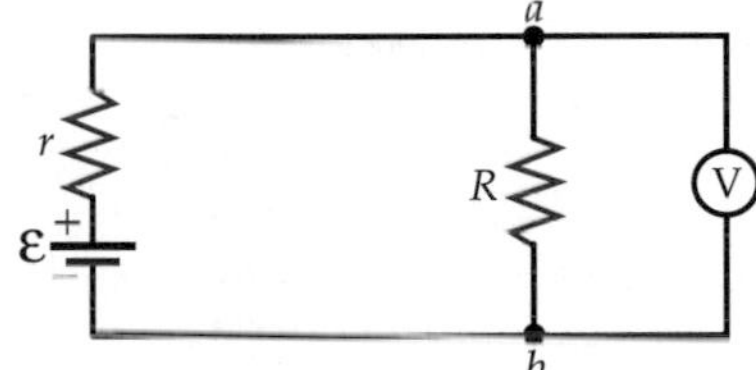

Figure 23-26 To measure the voltage drop across a resistor, a voltmeter —Ⓥ— is placed in parallel with the resistor so that the potential drops across the voltmeter and the resistor are the same.

The principal component of an ammeter and a voltmeter is a **galvanometer,** which is a device that detects a small current that passes through it. The galvanometer is designed so that the scale reading is proportional to the current passing through it. Many meters today have a digital readout rather than an indicator and a scale, but the basic way they operate is similar to that discussed here.

There are two properties of a galvanometer that are important for its use in an ammeter or a voltmeter. These are the resistance of the galvanometer R_g and the current needed to produce a full-scale deflection I_g. Typical values of these for a portable laboratory galvanometer are $R_g = 20\ \Omega$ and $I_g = 0.5$ mA. The voltage drop across such a galvanometer for full-scale deflection is thus

$$V = I_g R_g = (20\ \Omega)(5.0 \times 10^{-4}\ \text{A}) = 10^{-2}\ \text{V}$$

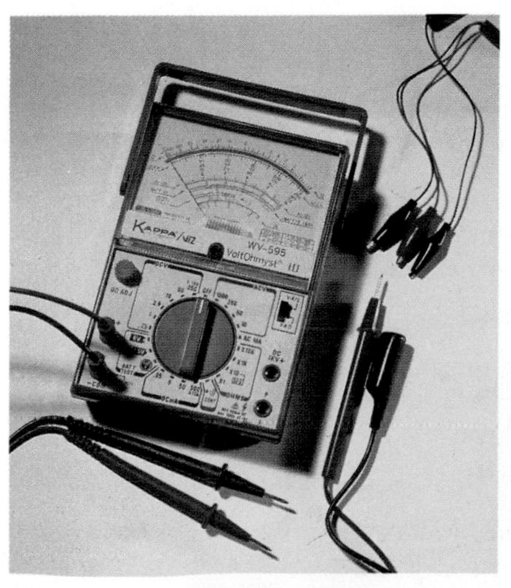

(a)

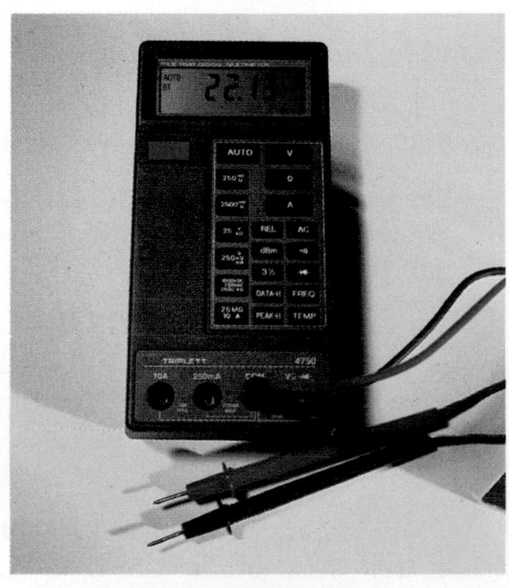

(b)

(a) An analog multimeter. (b) A digital multimeter. Digital multimeters are generally more accurate and more expensive than their analog counterparts.

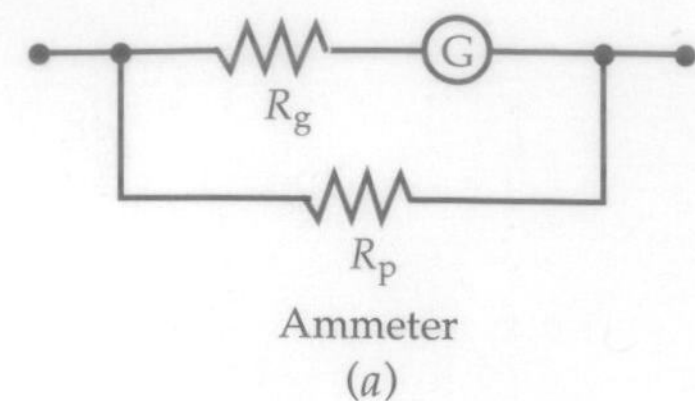

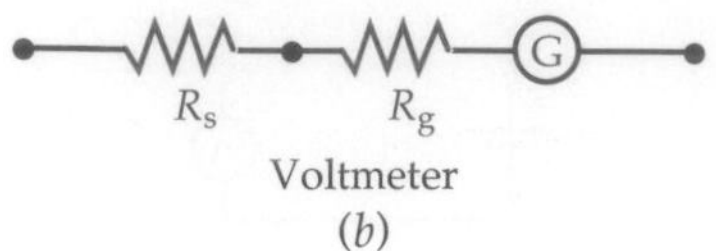

Figure 23-27. (a) An ammeter consists of a galvanometer —Ⓖ— whose resistance is R_g and a small parallel resistance R_p. (b) A voltmeter consists of a galvanometer —Ⓖ— and a large series resistance R_s. In these diagrams, the galvanometer's resistance is shown next to the symbol for the galvanometer.

To construct an ammeter from a galvanometer, we place a small resistor called a **shunt resistor** in parallel with the galvanometer. The shunt resistance is usually much smaller than the resistance of the galvanometer so that most of the current is carried by the shunt resistor and the equivalent resistance of the ammeter is much smaller than that of the galvanometer alone. In fact, the equivalent resistance of the ammeter is approximately equal to the shunt resistance. To construct a voltmeter, we place a resistor with a large resistance in series with the galvanometer so that the equivalent resistance of the voltmeter is much larger than that of the galvanometer alone. Figure 23-27 illustrates the construction of an ammeter and voltmeter from a galvanometer. The resistance of the galvanometer R_g is shown separately in these schematic drawings, but it is actually part of the galvanometer since it is essentially due to the resistance of the galvanometer coil. The choice of appropriate resistors for the construction of an ammeter or voltmeter from a galvanometer is best illustrated by example.

Example 23-11

Using a galvanometer with a resistance of 20 Ω, for which a current of 5×10^{-4} A gives a full-scale deflection, design an ammeter that will give a full-scale deflection when the current is 5 A.

Since the total current through the ammeter must be 5 A when the current through the galvanometer is just 5×10^{-4} A, most of the current must go through the shunt resistor. Let R_p be the shunt resistance and I_p be the current through it. Since the galvanometer and shunt resistor are in parallel, we have

$$I_g R_g = I_p R_p$$

and

$$I_p + I_g = 5 \text{ A}$$

or

$$I_p = 5 \text{ A} - I_g = (5 \text{ A}) - (5 \times 10^{-4} \text{ A}) \approx 5 \text{ A}$$

Thus, the value of the shunt resistor should be

$$R_p = \frac{I_g}{I_p} R_g = \frac{5 \times 10^{-4} \text{ A}}{5 \text{ A}}(20\ \Omega) = 2 \times 10^{-3}\ \Omega$$

Since the resistance of the shunt resistor is so much smaller than the resistance of the galvanometer, the effective resistance of the parallel combination is approximately equal to the shunt resistance.

Example 23-12

Using the same galvanometer as in Example 23-11, design a voltmeter that will give a full-scale deflection for a potential difference of 10 V.

Let R_s be the resistance of the resistor in series with the galvanometer. We want to choose R_s so that a current of $I_g = 5 \times 10^{-4}$ A gives a potential drop of 10 V. Thus,

$$I_g(R_s + R_g) = 10 \text{ V}$$

$$R_s + R_g = \frac{10 \text{ V}}{5 \times 10^{-4} \text{ A}} = 2 \times 10^4\ \Omega$$

$$R_s = 2 \times 10^4\ \Omega - R_g = (2 \times 10^4\ \Omega) - 20\ \Omega$$

$$= 19{,}980\ \Omega \approx 20 \text{ k}\Omega$$

Example 23-13

The resistance of a 100-Ω resistor is to be measured using the circuit shown in Figure 23-28. The resistance of the voltmeter is 2000 Ω and that of the ammeter is 0.002 Ω. What error is made in calculating R from V/I where V is the reading of the voltmeter and I is the reading of the ammeter?

In the circuit shown, the voltmeter measures the voltage drop across the resistor, but the ammeter measures the total current in the circuit, including the current through the voltmeter. The equivalent resistance R'_{eq} of the voltmeter in parallel with the resistor is

$$R'_{eq} = \left(\frac{1}{100\ \Omega} + \frac{1}{2000\ \Omega}\right)^{-1} = 95.238\ \Omega$$

and the equivalent resistance of the entire circuit is

$$R_{eq} = R_a + R'_{eq} = 0.002\ \Omega + 95.238\ \Omega = 95.240\ \Omega$$

The current in the ammeter is

$$I = \frac{12\ \text{V}}{R_{eq}} = \frac{12\ \text{V}}{95.240\ \Omega} = 0.126\ \text{A}$$

If I_1 is the current through the 100-Ω resistor and I_2 is the current through the voltmeter, we have $100I_1 = 2000I_2$ or $I_2 = 0.05I_1$. The current through the 100-Ω resistor is thus

$$I_1 = I - I_2 = I - 0.05I_1$$

or

$$I_1 = \frac{I}{1.05} = \frac{0.126\ \text{A}}{1.05} = 0.120\ \text{A}$$

The voltage drop across the 100-Ω resistor as measured by the voltmeter is thus $RI_1 = (100\ \Omega)(0.120\ \text{A}) = 12.0\ \text{V}$, and the measured value of the resistance is

$$R = \frac{V}{I} = \frac{12.0\ \text{V}}{0.126\ \text{A}} = 95.2\ \Omega$$

which differs from the true value by about 5 percent. This is what is to be expected because the resistance of the voltmeter is only 20 times that of the resistor, so putting the voltmeter and resistor in parallel increases the total current by about 5 percent.

Figure 23-29 shows a better circuit for measuring this resistance with these meters. In this circuit, the ammeter measures the true current in the resistor, but the voltmeter measures the total potential drop across the resistor plus the ammeter. Since the resistance of the ammeter is only 0.002 percent of that of the resistor, the error made with this circuit will be only about 0.002 percent. If the resistance of the resistor were 0.1 Ω instead of 100 Ω, the circuit in Figure 23-28 would be better.

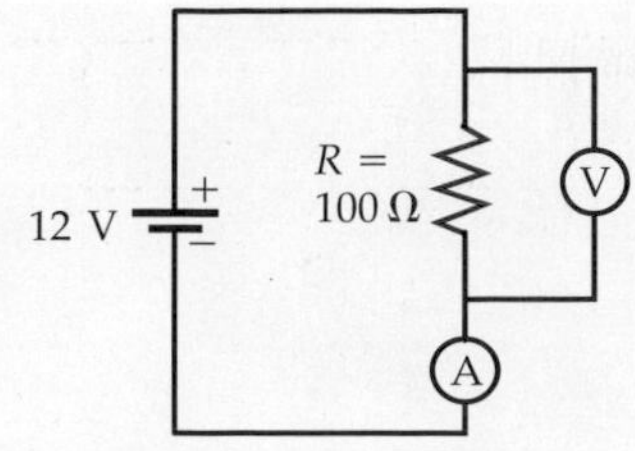

Figure 23-28 A possible circuit for measuring the resistance of a 100 Ω resistor of Example 23-13.

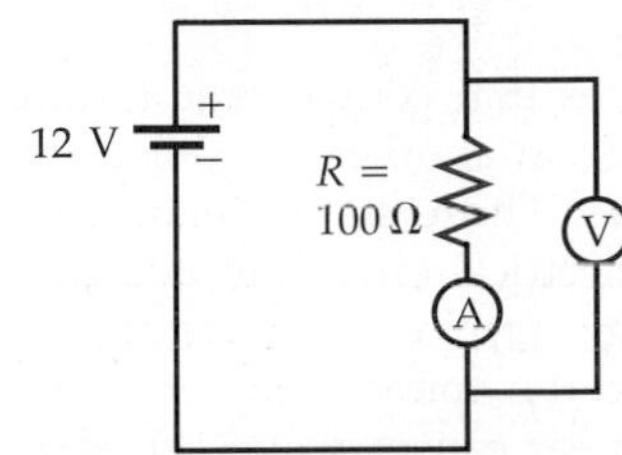

Figure 23-29 A better alternative for measuring the resistance of the 100-Ω resistor with the meters of Example 23-13.

A simple ohmmeter for measuring resistance consists of a battery connected in series with a galvanometer and a resistor, as shown in Figure 23-30a. The resistance R_s is chosen such that when the terminals a and b are shorted (touched together), which means that there is no resistance between them, the current through the galvanometer is I_g. This is the current for which the galvanometer gives a full-scale deflection. Thus, a full-scale deflection indicates no resistance between terminals a and b, and a zero deflection indicates an infinite resistance between the terminals. When the termi-

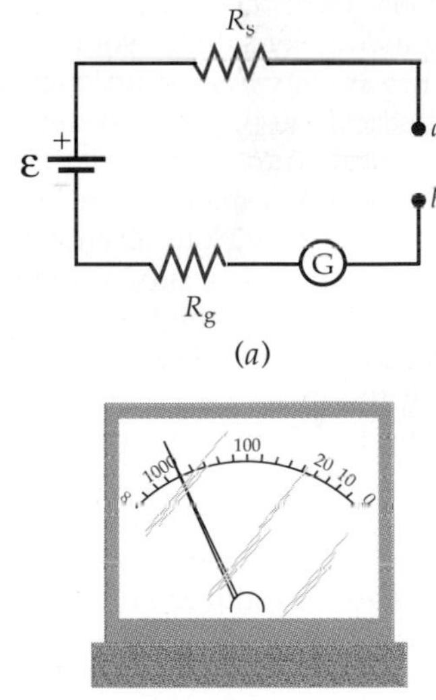

Figure 23-30 (a) An ohmmeter consists of a battery in series with a galvanometer and a resistor R_s, which is chosen such that the galvanometer gives full-scale deflection when points a and b are shorted. (b) The galvanometer scale calibrated to give a readout in ohms.

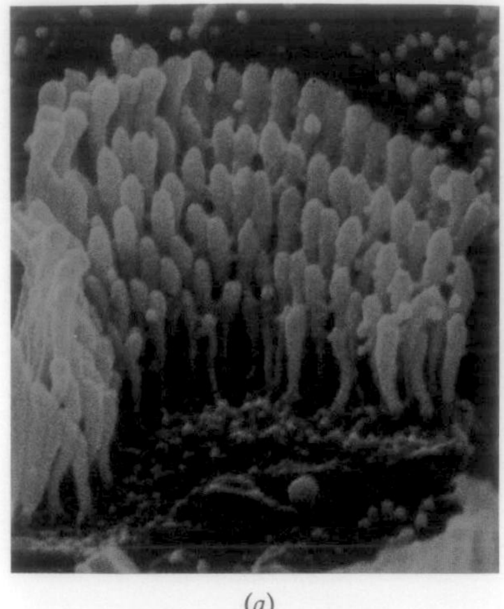

(a)

(b)

Devices that convert one form of energy to another are called transducers. Those shown here convert nonelectrical energy to electrical energy. (a) Sound waves transmitted to the spiral chamber of the inner ear causes watery fluid there to vibrate. Sensory receptor cells (shown here at about 3500X magnification) are attached to the walls of the chamber. The receptor cells are stimulated by the vibrating fluid and cause neurons, connected to their base, to transmit electrical impulses. The impulses, traveling along a chain of neurons, eventually register in the brain as the sensation "sound." Because it converts sound energy to electrical energy, this system is a biological counterpart to a microphone. (b) A microphone converts sound energy to electrical energy. In the kind shown here, a copper ring is attached to a thin plastic membrane. Sound waves hitting the membrane cause it and the ring to vibrate. The ring is mounted in the field of a permanent magnet. Motion back and forth across the field, caused by the vibration, induces an alternating current in the ring. This current causes a secondary alternating magnetic field to arise, which in turn creates a secondary alternating current, this time in a wire coil positioned behind the ring. (c) Photovoltaic cells convert light energy to electrical energy. The cells are composed of two semiconductors. At the junction where they meet, the semiconductors interact to form an electric dipole. Light absorbed near the

nals are connected across an unknown resistance R, the current through the galvanometer is less than I_g. Specifically, it is

$$I = \frac{\mathcal{E}}{R + R_s + R_g} \tag{23-28}$$

Since this current depends on R, the scale can be calibrated to give a direct reading of R, as shown in Figure 23-30*b*. Note that this scale is far from linear. Because the accuracy of the reading depends on the constancy of the emf of the battery, this type of simple ohmmeter is not a highly accurate instrument. However, it is quite useful for making a quick, rough determination of an unknown resistance.

Because an ohmmeter sends a current through the resistance to be measured, some caution must be exercised when using this instrument. For example, consider an ohmmeter constructed from a 1.5-V battery and a galvanometer similar to that in Examples 23-11 and 23-12. The series resistance R_s needed is found from

$$I_g(R_s + R_g) = 1.5\text{ V}$$

or

$$R_s = \frac{1.5\text{ V}}{5 \times 10^{-4}\text{ A}} - R_g = 3000\ \Omega - 20\ \Omega = 2980\ \Omega$$

Suppose we use this ohmmeter to measure the resistance of a very sensitive laboratory galvanometer that gives a full-scale reading when the current through it is 10^{-5} A and has a resistance of about 20 Ω. When the terminals a and b are placed across this sensitive galvanometer, the current sent through it will be just slightly less than 5×10^{-4} A because the total resistance is 3020 Ω, which is just slightly more than 3000 Ω. This current is about 50 times that needed to produce a full-scale deflection. The likely results will be some popping sounds, a puff of smoke, one less sensitive galvanometer, and probably some unkind words from the laboratory instructor.

Questions

2. Under what conditions might it be advantageous to use a galvanometer that is less sensitive than the one discussed in Examples 23-11 and 23-12, that is, one that requires a greater current I_g for a full-scale deflection?
3. When the series resistance R_s is properly chosen for the emf of a particular ohmmeter, any resistance from zero to infinity can be measured. Why, then, do practical ohmmeters have different scales for measuring different ranges of resistance?
4. A none-too-bright student decides to measure the internal resistance of his car battery with an ohmmeter borrowed from his physics laboratory. Why is this a bad idea?

(*c*)

(*d*)

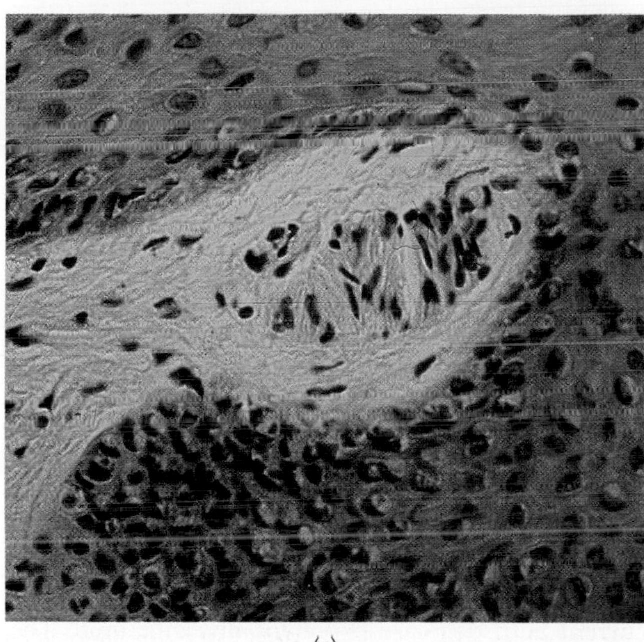

(*e*)

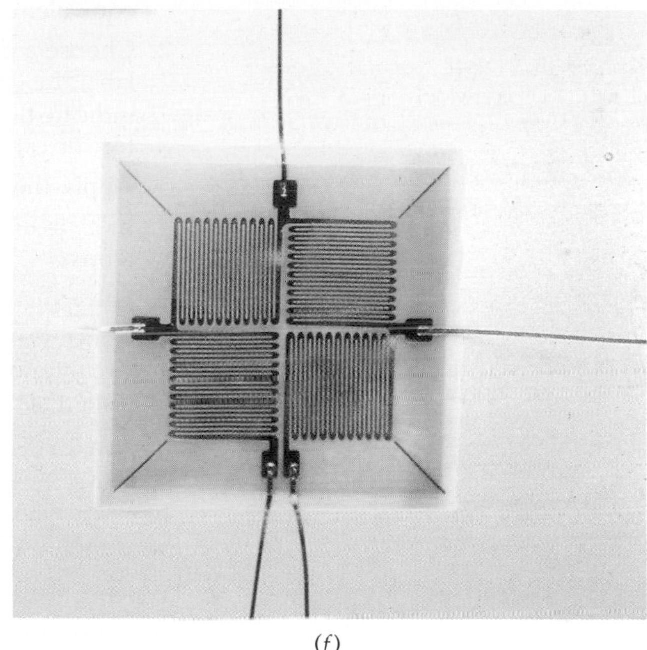

(*f*)

junction creates negative and positive charges which, if they drift into the junction, are swept in opposite directions by the dipole field. This separation of charge creates a voltage difference between the two semiconductors. At present, the best photovoltaic cells convert light energy to electrical energy with an efficiency of between 20 and 30 percent. (*d*) Piezoelectric crystals such as quartz, tourmaline, and topaz become electrically polarized when they are compressed, bent, or otherwise distorted. The polarization creates a voltage drop across the stressed crystal, which can be used to produce an electric current. The process is reversible: voltage applied across a piezoelectric will cause it to distort. Quartz piezoelectrics (shown here) are commonly introduced into oscillating electrical circuits having nearly the same frequency as the natural frequency of vibration of the crystal. The result is that the crystal and the electrical circuit both vibrate at the natural frequency of the crystal. This effect can be used to stabilize the frequency of radio oscillators and to make clocks with errors of less than 0.1 s per year. (*e*) Our sense of touch arises from encapsulated nerve fibers called Meissner's corpuscles that are located directly under the outer layer of skin. When skin overlying them is touched, the corpuscles are deformed, triggering electrical impulses in the nerve fibers. A stronger touch produces greater deformation and increases the frequency of impulses. The system is in some respects a biological counterpart to a strain gauge: in both transducers, the energy of mechanical stress is used to modulate changes in electrical conduction. (*f*) A strain gauge consists of a grid of very fine wire or foil of a substance, such as carbon, which changes its electrical resistance when mechanically stressed. The wire is bonded to a thin insulating backing, which is attached by adhesive to an object. Stresses that distort the object deform the attached strain gauge as well. The degree of deformation is measured by the change in resistance of the gauge. If a fixed voltage is applied across the ends of the gauge wire, a varying resistance will produce variations in the current.

Summary

1. Kirchhoff's rules are

1. When any closed circuit loop is traversed, the algebraic sum of the changes in potential must equal zero.
2. At any junction in a circuit where the current can divide, the sum of the currents into the junction must equal the sum of the currents out of the junction.

2. The general methods for analyzing multiloop circuits can be summarized as follows:

1. Replace any combinations of resistors in series or parallel with their equivalent resistances.
2. Choose a direction for the current in each branch of the circuit, and label the currents in a circuit diagram. Add plus and minus signs to indicate the high- and low-potential sides of each source of emf, resistor, or capacitor.
3. Apply the junction rule to each junction where the current divides.
4. In a circuit containing n interior loops, apply the loop rule to any n loops.
5. Solve the equations to obtain the values of the unknowns.
6. Check your results by assigning a potential of zero to one point in the circuit and using the values of the currents found to determine the potentials at other points in the circuit.

3. Complex circuits or networks of resistors can often be simplified by considering their symmetry. If the potential of two points is equal, the points can be connected with a wire, and a simplified circuit diagram can be drawn.

4. When a capacitor is discharged through a resistor, the charge on the capacitor and the current both decrease exponentially with time. The time constant $\tau = RC$ is the time it takes for either to decrease to $e^{-1} = 0.37$ times its original value. When a capacitor is charged through a resistor, the rate of charging, which equals the current, decreases exponentially with time. After a time $\tau = RC$, the charge on the capacitor has reached 63 percent of its final charge.

5. A galvanometer is a device that detects a small current that passes through it and gives a scale deflection that is proportional to the current. An ammeter is a device for measuring current. It consists of a galvanometer plus a parallel resistor called a shunt resistor. To measure the current through a resistor an ammeter is inserted in series with the resistor. The ammeter has a very small resistance so that it has little effect on the current to be measured. A voltmeter measures potential difference. It consists of a galvanometer plus a large series resistor. To measure the potential drop across a resistor, a voltmeter is placed in parallel with the resistor. The voltmeter has a very large resistance so that it has little effect on the potential drop to be measured. An ohmmeter is a device for measuring resistance. It consists of a galvanometer, a source of emf, and a resistor.

Suggestion for Further Reading

Rosenfeld, L.: "Gustav Robert Kirchhoff," *The Dictionary of Scientific Biography*, vol. 7, Charles C. Gillespie (ed.), Charles Scribner's Sons, New York, 1973, p. 379.

Kirchhoff's famous rules for electrical circuits were only his first of a number of important discoveries.

Review

A. Objectives: After studying this chapter, you should:

1. Be able to state Kirchhoff's rules and use them to analyze various dc circuits.

2. Be able to use symmetry to find the equivalent resistance of complex but symmetrical networks of resistors.

3. Be able to sketch both the charge Q on a capacitor and the current I as functions of time for charging and discharging a capacitor.

4. Be able to calculate the proper series or shunt resistors needed to make an ammeter, voltmeter, or ohmmeter from a given galvanometer and draw the circuit diagrams for these devices.

B. Define, explain, or otherwise identify:

Kirchhoff's rules	Voltmeter
Loop rule	Ammeter
Junction rule	Ohmmeter
RC circuit	Galvanometer
Time constant	Shunt resistor
Exponential decrease	

C. True or false: If the statement is true, explain why it is true. If it is false, give a counterexample.

1. The net change in the potential around a complete circuit in the steady state is zero.

2. The time constant of an *RC* circuit is the time needed to completely discharge the capacitor.

3. To measure the potential drop across a resistor, a voltmeter is placed in series with the resistor.

Problems

Level I

23-1 Kirchhoff's Rules

1. A battery with an emf of 6 V and an internal resistance of 0.3 Ω is connected to a variable resistance R. Find the current and power delivered by the battery when R is (*a*) 5 Ω and (*b*) 10 Ω.

2. A variable resistance R is connected across a potential difference V that remains constant independent of R. When $R = R_1$, the current is 6.0 A. When R is increased to $R_2 = R_1 + 10.0\ \Omega$, the current drops to 2.0 A. Find (*a*) R_1 and (*b*) V.

3. A battery has an emf $\mathcal{E}$ and an internal resistance r. When a 5.0-Ω resistor is connected across the terminals, the current is 0.5 A. When this resistor is replaced by an 11.0-Ω resistor, the current is 0.25 A. Find (*a*) the emf $\mathcal{E}$ and (*b*) the internal resistance r.

4. In Figure 23-31, the emf is 6 V and $R = 0.5\ \Omega$. The rate of Joule heating in R is 8 W. (*a*) What is the current in the circuit? (*b*) What is the potential difference across R? (*c*) What is r?

5. For the circuit in Figure 23-32, find (*a*) the current, (*b*) the power delivered or absorbed by each emf, and (*c*) the rate of Joule heating in each resistor. (Assume that the batteries have negligible internal resistance.)

Figure 23-31 Problem 4.

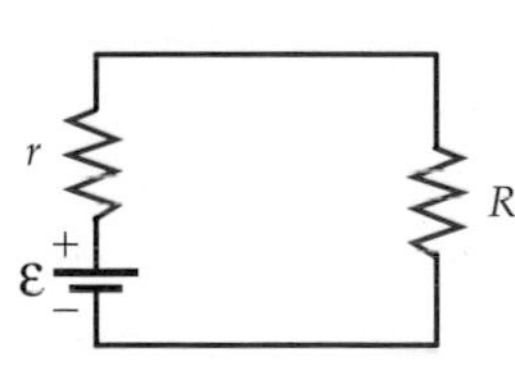

Figure 23-32 Problem 5.

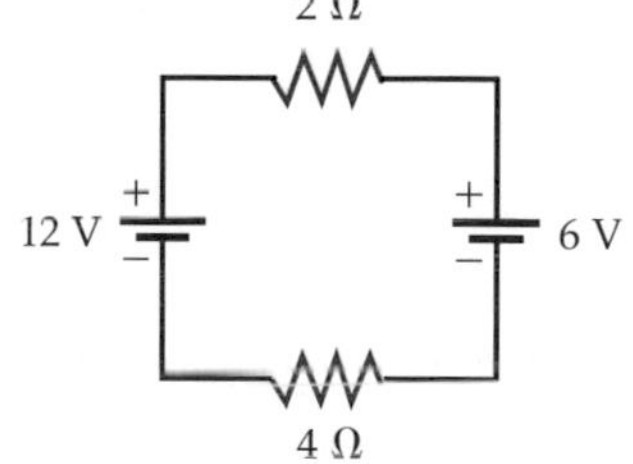

6. In the circuit in Figure 23-33, the batteries have negligible internal resistance, and the ammeter has negligible resistance. (*a*) Find the current through the ammeter. (*b*) Find the energy delivered by the 12-V battery in 3 s. (*c*) Find the total Joule heat produced in 3 s. (*d*) Account for the difference in your answers to parts (*b*) and (*c*).

Figure 23-33 Problem 6.

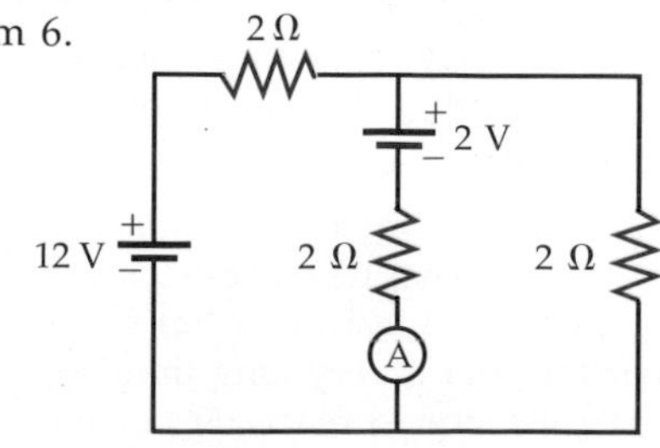

7. In the circuit in Figure 23-34, the batteries have negligible internal resistance. Find (*a*) the current in each resistor, (*b*) the potential difference between points *a* and *b*, and (*c*) the power supplied by each battery.

Figure 23-34 Problem 7.

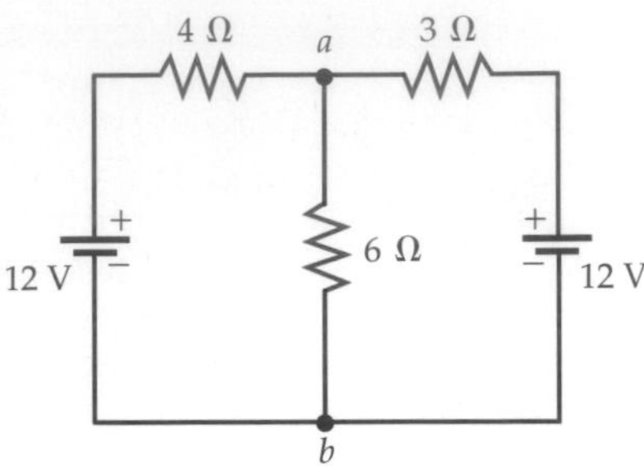

8. Repeat Problem 7 for the circuit in Figure 23-35.

Figure 23-35 Problem 8.

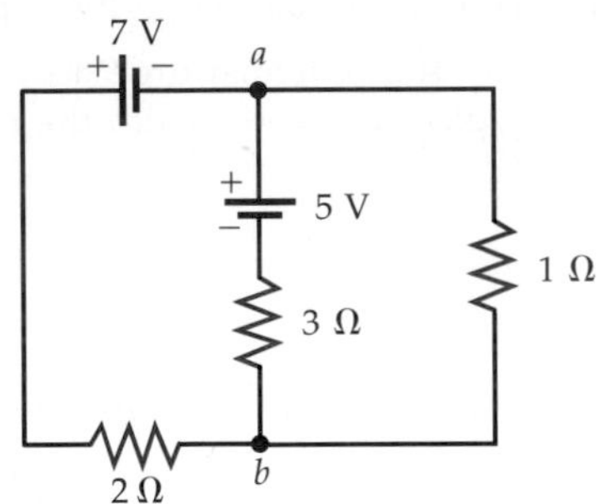

23-2 *RC* Circuits

9. A 6-μF capacitor is charged to 100 V and is then connected across a 500-Ω resistor. (*a*) What is the initial charge on the capacitor? (*b*) What is the initial current just after the capacitor is connected to the resistor? (*c*) What is the time constant of this circuit? (*d*) How much charge is on the capacitor after 6 ms?

10. (*a*) Find the initial energy stored in the capacitor of Problem 9. (*b*) Show that the energy stored in the capacitor is given by $U = U_0 e^{-2t/\tau}$, where U_0 is the initial energy and $\tau = RC$ is the time constant. (*c*) Sketch a plot of the energy U in the capacitor versus time t.

11. A 0.12-μF capacitor is given a charge Q_0. After 4 s, its charge is $\frac{1}{2}Q_0$. What is the effective resistance across this capacitor?

12. A 1.6-μF capacitor, initially uncharged, is connected in series with a 10-kΩ resistor and a 5.0-V battery of negligible internal resistance. (*a*) What is the charge on the capacitor after a very long time? (*b*) How long does it take the capacitor to reach 99 percent of its final charge?

13. A 2-MΩ resistor is connected in series with a 1.5-μF capacitor and a 6.0-V battery of negligible internal resistance. The capacitor is initially uncharged. After a time $t = \tau = RC$, find (*a*) the charge on the capacitor, (*b*) the rate at which the charge is increasing, (*c*) the current, (*d*) the power supplied by the battery, (*e*) the power dissipated in the resistor, and (*f*) the rate at which the energy stored in the capacitor is increasing.

14. Repeat Problem 13 for the time $t = 2\tau$.

23-3 Ammeters, Voltmeters, and Ohmmeters

15. A galvanometer has a resistance of 140 Ω. It requires 1.2 mA to give a full-scale deflection. (*a*) What resistance should be placed in parallel with the galvanometer to make an ammeter that gives a full-scale deflection for a current of 2 A? (*b*) What resistance should be placed in series with the galvanometer to make a voltmeter that gives a full-scale deflection for a potential difference of 5 V?

16. Sensitive galvanometers can detect currents as small as 1 pA. How many electrons per second produce this current?

17. A sensitive galvanometer has a resistance of 120 Ω and requires 1.4 μA of current to produce a full-scale deflection. (*a*) Find the shunt resistance needed to construct an ammeter that gives a full-scale deflection for a current of 1.0 mA. (*b*) What is the resistance of the ammeter? (*c*) What resistance would be required to construct a voltmeter that gives a full-scale deflection for a potential difference of 3.0 V?

18. A galvanometer with a resistance of 90 Ω gives a full-scale deflection when its current is 1.5 mA. It is used to construct an ammeter that gives a full-scale reading for a current of 200 A. (*a*) Find the shunt resistance needed. (*b*) What is the resistance of the ammeter? (*c*) If the shunt resistor consists of a piece of 10-gauge copper wire (diameter 2.59 mm), what should its length be?

19. The galvanometer in Problem 18 is used with a 1.5-V battery of negligible internal resistance to make an ohmmeter. (*a*) What resistance R_s should be placed in series with the galvanometer? (*b*) What resistance R will give a half-scale deflection? (*c*) What resistance R will give a deflection that is one-tenth full scale?

20. For the ohmmeter in Problem 19, show how the galvanometer scale should be calibrated by representing the scale by a straight line of length L, where the end of the line ($x = L$) represents a full-scale reading for $R = 0$. Divide the line into 10 equal parts and give the value of the resistance at each division.

21. A galvanometer with a resistance of 110 Ω gives a full-scale reading when its current is 0.13 mA. It is to be used in the multirange voltmeter shown in Figure 23-36, where the full-scale readings for the ranges are shown by the indicated connections. Determine R_1, R_2, and R_3.

Figure 23-36 Problem 21.

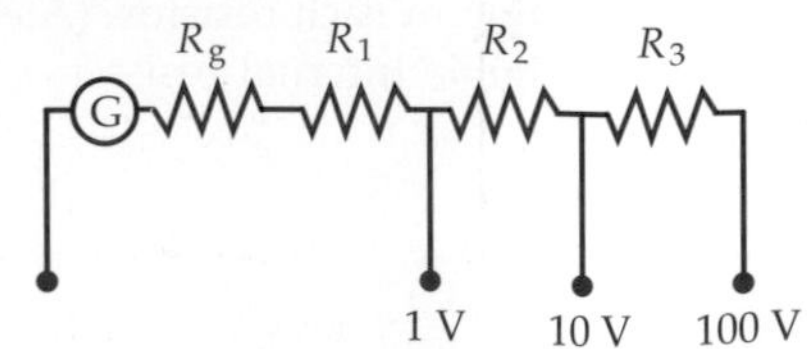

22. The galvanometer of Problem 21 is to be used in the multirange ammeter shown in Figure 23-37. Again, the full-scale reading for the ranges are shown by the indicated connections. Determine R_1, R_2, and R_3.

Figure 23-37 Problem 22.

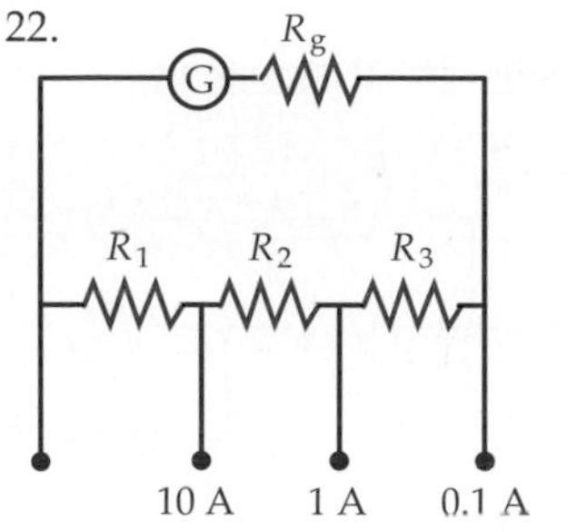

Level II

23. Two identical batteries, each with an emf $\mathcal{E}$ and an internal resistance r, can be connected across a resistance R either in series or in parallel. For which type of connection is the power supplied to R greater when (a) $R < r$ and (b) $R > r$?

24. A sick car battery with an emf of 11.4 V and an internal resistance of 0.01 Ω is connected to a load of 2.0 Ω. To help the ailing battery, a second battery with an emf of 12.6 V and an internal resistance of 0.01 Ω is connected by jumper cables to the terminals of the first battery. (a) Draw a diagram for this circuit. (b) Find the current in each part of the circuit. (c) Find the power delivered by the second battery and discuss where this power goes, assuming that the emfs and internal resistances of both batteries remain constant.

25. For the circuit in Figure 23-38, find (a) the current in each resistor, (b) the power supplied by each emf, and (c) the power dissipated in each resistor.

Figure 23-38 Problem 25.

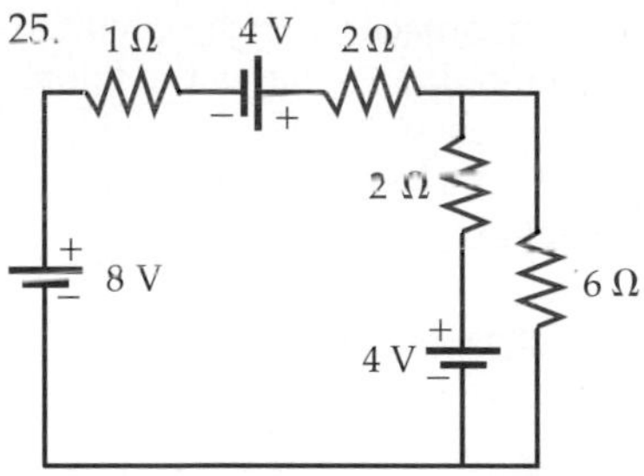

26. For the circuit in Figure 23-39, find the potential difference between points a and b.

Figure 23-39 Problem 26.

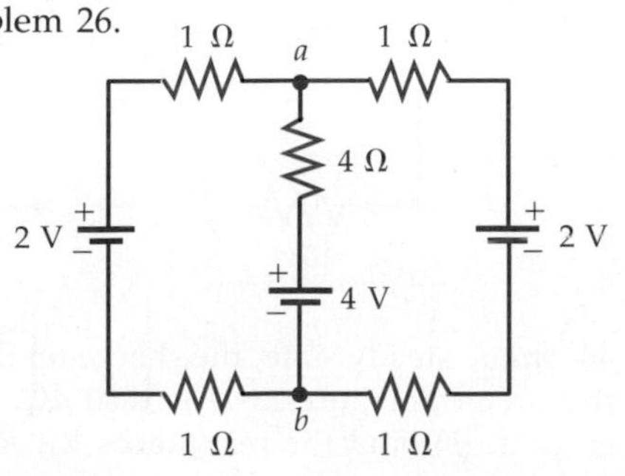

27. The space between the plates of a parallel-plate capacitor is filled with a dielectric of constant κ and resistivity ρ. (a) Show that the time constant for the decrease of charge on the plates is $\tau = \varepsilon_0\kappa\rho$. ($b$) If the dielectric is mica, for which $\kappa = 5.0$ and $\rho = 9 \times 10^{13}$ Ω·m, find the time it takes for the charge to decrease to $1/e^2 \approx 14$ percent of its initial value.

28. The battery in the circuit in Figure 23-40 has an internal resistance of 0.01 Ω. (a) An ammeter with a resistance of 0.01 Ω is inserted at point a. What is the reading of the ammeter? (b) By what percentage is the current changed because of the ammeter? (c) The ammeter is removed and a voltmeter with a resistance of 1 kΩ is connected from a to b. What is the reading of the voltmeter? (d) By what percentage is the voltage drop from a to b changed by the presence of the voltmeter?

Figure 23-40 Problem 28.

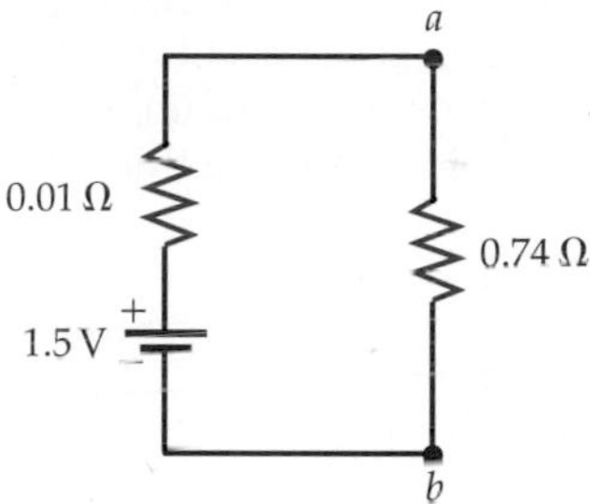

29. You have two batteries, one with $\mathcal{E} = 9.0$ V and $r = 0.8$ Ω and the other with $\mathcal{E} = 3.0$ V and $r = 0.4$ Ω. (a) Show how you would connect the batteries to give the largest current through a resistor R. Find the current for (b) $R = 0.2$ Ω, (c) $R = 0.6$ Ω, (d) $R = 1.0$ Ω, and (e) $R = 1.5$ Ω.

30. In the circuit in Figure 23-41, the reading of the ammeter is the same with both switches open and both closed. Find the resistance R.

Figure 23-41 Problem 30.

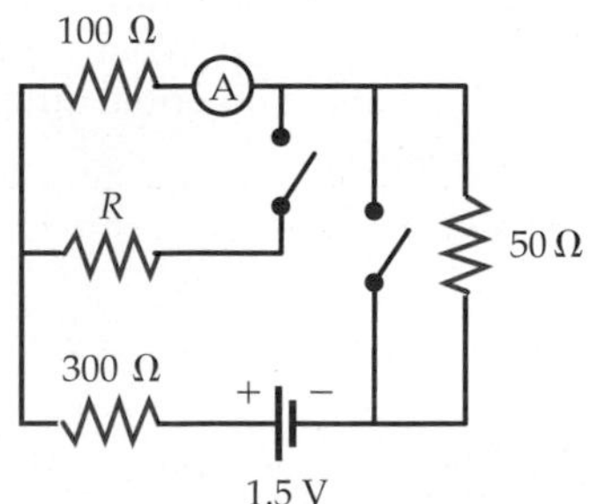

31. A galvanometer gives a full-scale deflection when the voltage across it is 10 mV and the current through it is 50 μA. (a) Design a voltmeter that gives a full-scale reading for a potential difference of 50 V using this galvanometer. (b) Design an ammeter that gives a full-scale reading for a current of 10 A using this galvanometer.

32. (*a*) Find the current in each part of the circuit shown in Figure 23-42. (*b*) Use your results of (*a*) to assign a potential at each indicated point assuming the potential at point *a* is zero.

Figure 23-42 Problem 32.

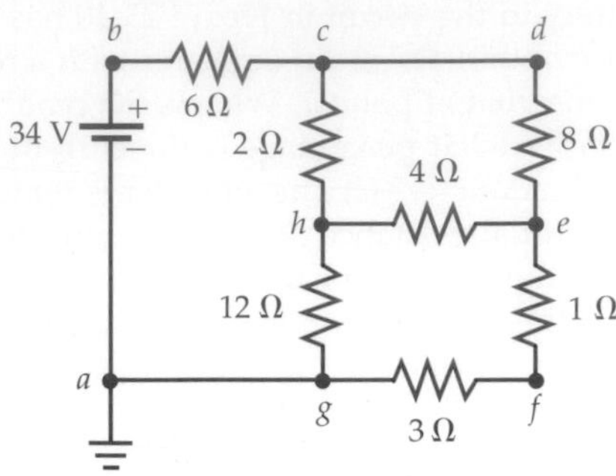

33. (*a*) Find the current in each part of the circuit shown in Figure 23-43. (*b*) Use your results of (*a*) to assign a potential at each indicated point assuming the potential at point *a* is zero.

Figure 23-43 Problem 33.

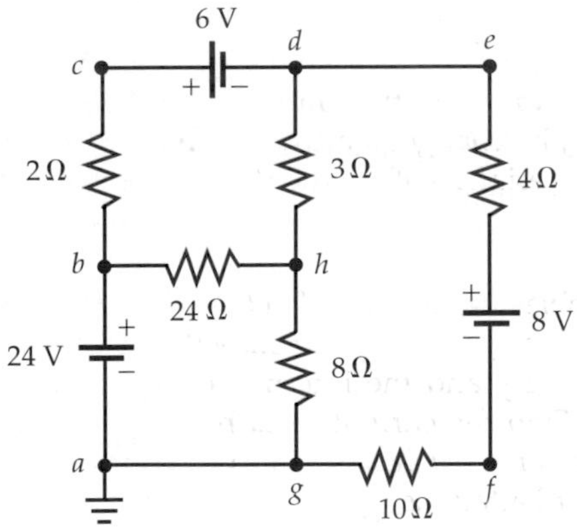

34. (*a*) Use symmetry to find the equivalent resistance of the network in Figure 23-44. (*b*) What is the current in each resistor if *R* is 10 Ω and a potential difference of 80 V is applied between *a* and *b*?

Figure 23-44 Problem 34.

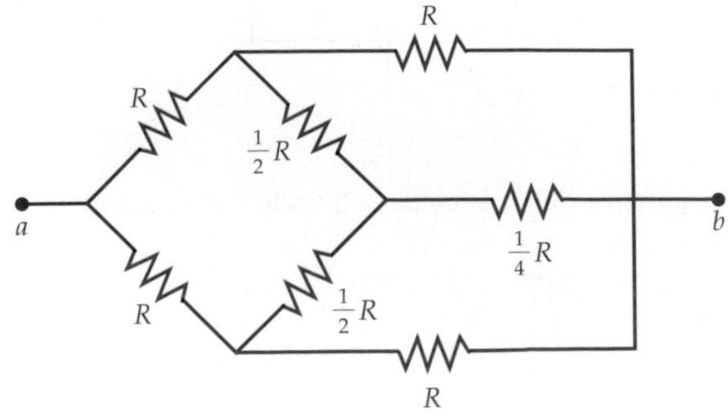

35. Nine 10-Ω resistors are connected as shown in Figure 23-45, and a potential difference of 20 V is applied between points *a* and *b*. (*a*) What is the equivalent resistance of this network? (*b*) Find the current in each of the nine resistors.

Figure 23-45 Problem 35.

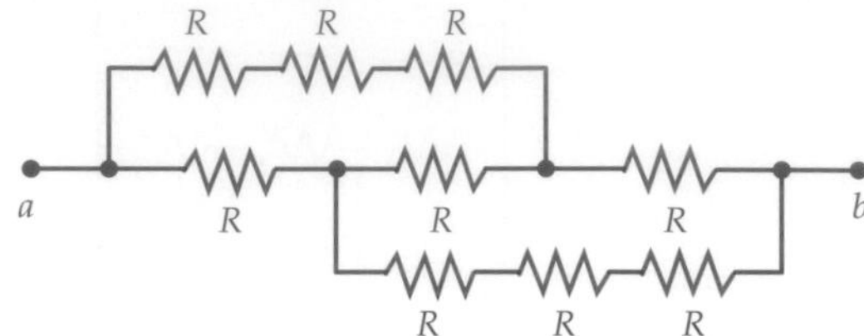

36. A parallel combination of an 8-Ω resistor and an unknown resistor *R* is connected in series with a 16-Ω resistor and a battery. The three resistors are then connected in series with each other and the same battery. In both arrangements, the current through the 8-Ω resistor is the same. What is the unknown resistance *R*?

37. A closed box has two metal terminals *a* and *b*. The inside of the box contains an unknown emf $\mathcal{E}$ in series with a resistance *R*. When a potential difference of 21 V is maintained between *a* and *b*, a current of 1 A enters the box at *a* and exits at *b*. If this potential difference is reversed, a current of 2 A in the reverse direction is observed. Find $\mathcal{E}$ and *R*.

38. A voltmeter has a resistance of 10^5 Ω. A 60-V battery with a 10-Ω internal resistance is connected across a 68-kΩ and a 56-kΩ resistor connected in series. (*a*) What is the voltmeter reading across the 68-kΩ resistor? (*b*) What is the voltmeter reading across the 56-kΩ resistor? (*c*) What is the voltmeter reading across the battery? (*d*) Give the percentage error of each voltage measurement.

39. The capacitors in the circuit in Figure 23-46 are initially uncharged. (*a*) What is the initial value of the battery current when switch S is closed? (*b*) What is the battery current after a long time? (*c*) What are the final charges on the capacitors?

Figure 23-46 Problem 39.

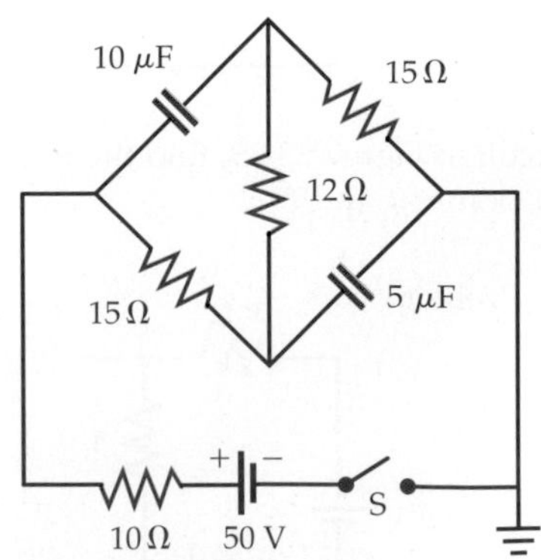

40. In the steady state, the charge on the 5-μF capacitor in the circuit in Figure 23-47 is 1000 μC. (*a*) Find the battery current. (*b*) Find the resistances R_1, R_2, and R_3.

Figure 23-47 Problem 40.

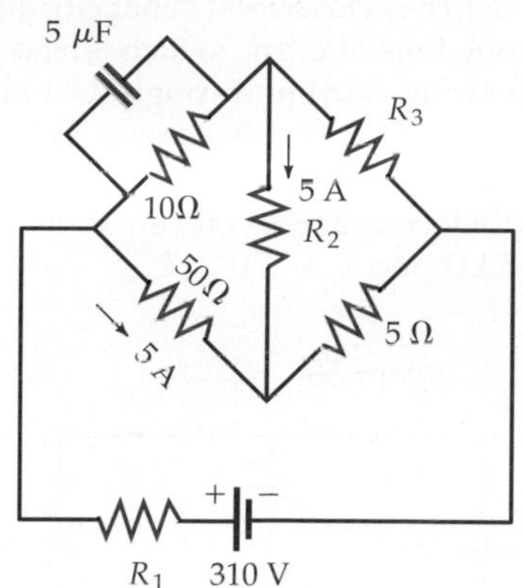

41. Consider the circuit shown in Figure 23-48. From your knowledge of how capacitors behave in circuits find (*a*) the initial current through the battery just after the switch is closed, (*b*) the steady-state current through the battery when the switch has been closed for a long time, and (*c*) the maximum voltage across the capacitor.

Figure 23-48 Problem 41.

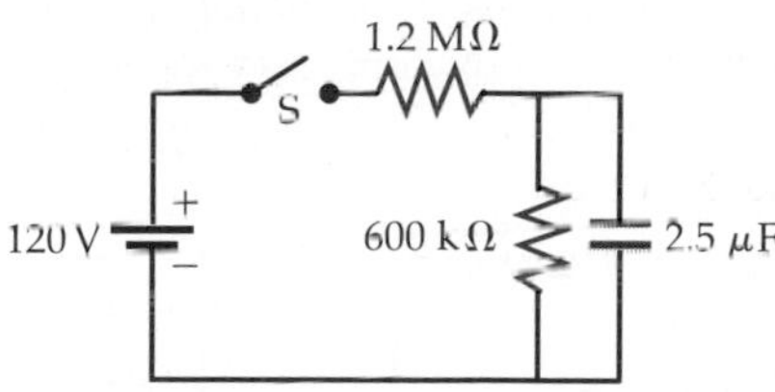

42. (*a*) What is the voltage across the capacitor in the circuit in Figure 23-49? (*b*) If the battery is disconnected, give the capacitor current as a function of time. (*c*) How long does it take the capacitor to discharge until the potential difference across it is 1 V?

Figure 23-49 Problems 42 and 59.

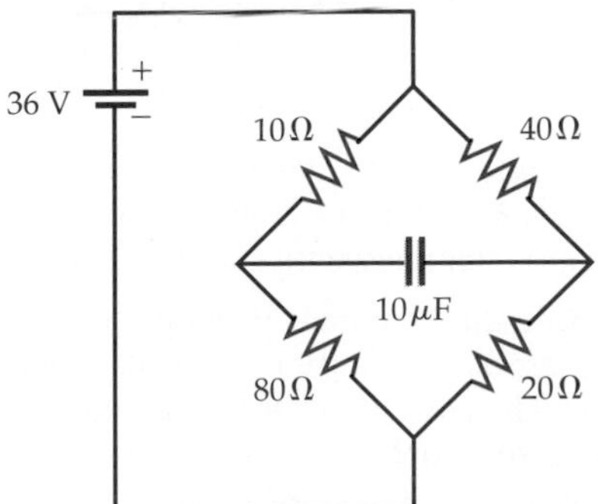

43. The circuit in Figure 23-50 is a slide-type *Wheatstone bridge*. It is used for determining an unknown resistance R_x in terms of the known resistances R_1, R_2, and R_0. The resistances R_1 and R_2 comprise a wire 1 m long. Point *a* is a sliding contact that is moved along the wire to vary these resistances. Resistance R_1 is proportional to the distance from the left end of the wire (0 cm) to point *a*, and R_2 is proportional to the distance from point *a* to the right end of the wire (100 cm). The sum of R_1 and R_2 remains constant. When points *a* and *b* are at the same potential, there is no current in the galvanometer and the bridge is said to be balanced. (Since the galvanometer is used to detect the absence of a current, it is called a *null detector*.) If the fixed resistance $R_0 = 200\ \Omega$, find the unknown resistance R_x if (*a*) the bridge balances at the 18-cm mark, (*b*) the bridge balances at the 60-cm mark, and (*c*) the bridge balances at the 95-cm mark.

Figure 23-50 Problem 43.

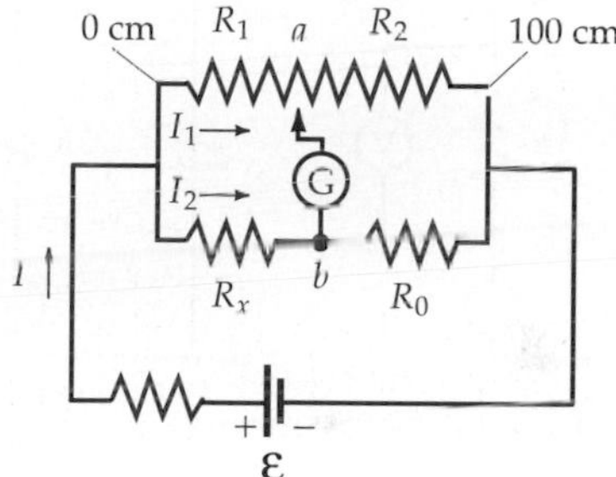

44. For the Wheatstone bridge of Problem 43, the bridge balances at the 98-cm mark when $R_0 = 200\ \Omega$. (*a*) What is the unknown resistance? (*b*) What effect would an error of 2 mm in the location of the balance point have on the measured value of the unknown resistance? (*c*) How should R_0 be changed so that the balance point for this unknown resistor will be nearer the 50-cm mark?

Level III

45. In Problem 24, assume that the emf of the first battery increases at a constant rate of 0.2 V/h while the emf of the second battery and the internal resistances remain constant. (*a*) Find the current in each part of the circuit as a function of time. (*b*) Sketch a graph of the power delivered to the first battery as a function of time.

46. In the ammeter in Figure 23-51, a galvanometer with a resistance of 10 Ω is connected across a 90-Ω resistor. Different ranges of current can be measured by choosing connections *ab*, *ac*, *ad*, or *ae*. (*a*) How should the 90-Ω resistor be divided so that the current that causes full-scale deflection decreases by a factor of 10 for each successive connection? (*b*) What should be the value of the current I_g for which the galvanometer gives a full-scale deflection so that the ranges of this ammeter are 1 A, 100 mA, 10 mA, and 1.0 mA?

Figure 23-51 Problem 46.

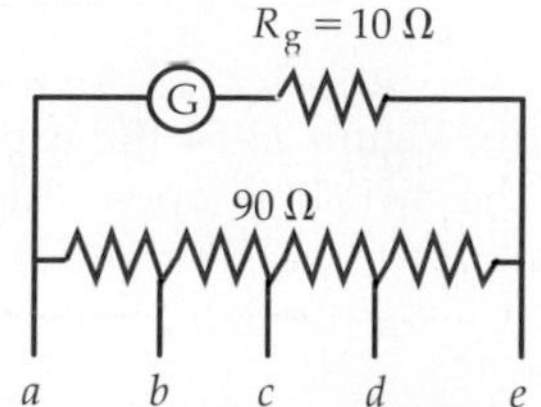

47. Figure 23-52 shows two ways a voltmeter and an ammeter can be used to measure an unknown resistance R. Assume that the internal resistance of the battery is negligible and that the resistance of the voltmeter is 1000 times that of the ammeter, that is, $R_v = 1000R_a$. The calculated value of R is taken to be $R_c = V/I$, where V and I are the readings of the voltmeter and the ammeter. (*a*) Which circuit in the figure should be preferred for values of R in the range from $10R_a$ to $0.9R_v$? Why? Find R_c for each circuit if $R_a = 0.1\ \Omega$, $R_v = 100\ \Omega$, and (*b*) $R = 0.5\ \Omega$, (*c*) $R = 3\ \Omega$, and (*d*) $R = 80\ \Omega$.

Figure 23-52 Problems 47 and 48.

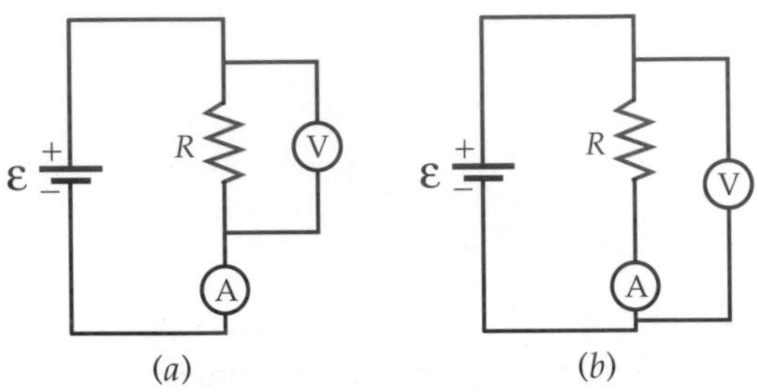

48. (*a*) For the circuits in Figure 23-52, show that $R_c = V/I$ is related to the true value R by $1/R_c = 1/R + 1/R_v$ for circuit *a* and $R_c = R + R_a$ for circuit *b*. (See Problem 47.) If $\mathcal{E} = 1.5$ V, $R_a = 0.01\ \Omega$, and $R_v = 10$ kΩ, for what range of values of R is R_c within 5 percent of R using (*b*) circuit *a* and (*c*) circuit *b*?

49. In the circuit in Figure 23-53, r is the internal resistance of the source of emf and R_a is the resistance of the ammeter. (*a*) Show that the ammeter reading is given by

$$\mathcal{E}\left(R_2 + R_a + r + \frac{R_2 + R_a}{R_1}r\right)^{-1}$$

(*b*) Show that if the ammeter and source of emf are interchanged, the ammeter reading will be

$$\mathcal{E}\left(R_2 + R_a + r + \frac{R_2 + r}{R_1}R_a\right)^{-1}$$

Note that if $R_a = r$ or if both are negligible, the ammeter reading is the same in both cases. (When R_a and r can be neglected, this symmetry can be very useful in analyzing circuits that have only one source of emf. It is not valid for circuits that have more than one source of emf, however.)

Figure 23-53 Problem 49.

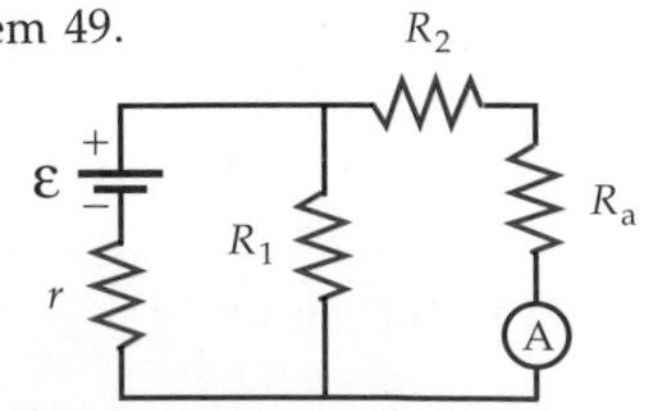

50. In the circuit in Figure 23-54 the capacitor is initially uncharged and the switch is open. At time $t = 0$, the switch is closed. (*a*) What current is supplied by the emf just after the switch is closed? (*b*) What current is supplied by the emf a long time after the switch is closed? (*c*) Derive an expression for the current through the emf at any time after the switch is closed. (*d*) After a long time t', the switch is opened. How long does it take for the charge on the capacitor to decrease to 10 percent of its value at $t = t'$ if $R_1 = R_2 = 5$ kΩ and $C = 1.0\ \mu$F?

Figure 23-54 Problem 50.

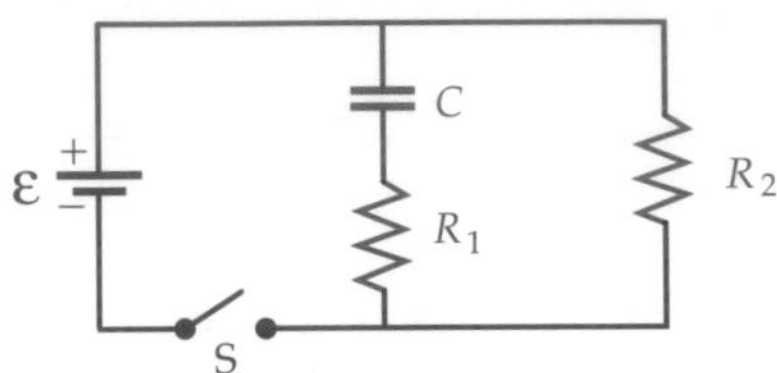

51. Two batteries with emfs $\mathcal{E}_1$ and $\mathcal{E}_2$ and internal resistances r_1 and r_2 are connected in parallel. Prove that the optimal load resistance R (for the delivery of maximum power) connected in parallel with this combination is $R = r_1r_2/(r_1 + r_2)$.

52. Figure 23-55 shows an infinite, two-dimensional, planar network of equal resistors. If the resistance of each resistor is R, find the equivalent resistance between points *a* and *b*.

Figure 23-55 Problem 52.

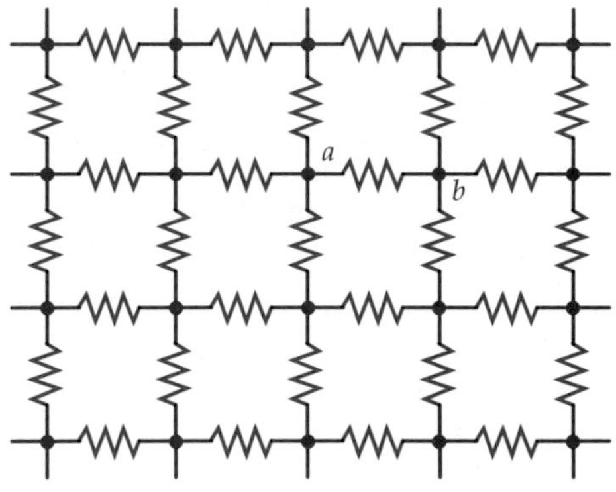

53. Consider an infinite, two-dimensional, periodic, triangular lattice of resistors. If R is the resistance of each resistor, what is the equivalent resistance across any resistor?

54. Consider a three-dimensional, periodic, cubic lattice of resistors that extends to infinity in all directions. If R is the resistance of each resistor, what is the equivalent resistance across any resistor?

55. Each of the six terminals *a*, *b*, *c*, *d*, *e*, and *f* in Figure 23-56 is connected to each other terminal by a wire of resistance R. The wires are insulated so that they make electrical contact only with the terminals. Use symmetry to find the resistance between any two terminals.

Figure 23-56 Problem 55.

56. (*a*) Find the equivalent resistance between points a and b for the 12-resistor cube in Figure 23-16. (*b*) What is the equivalent resistance between a and b if the resistor directly between these two points is removed?

57. An infinite chain of resistors is shown in Figure 23-57. Find the equivalent resistance between points a and b. *Hint:* The resistance R_{ab} is the same as $R_{a'b'}$ if the section to the left of $a'b'$ is removed. The network remains an infinite chain with the same structure.

Figure 23-57 Problem 57.

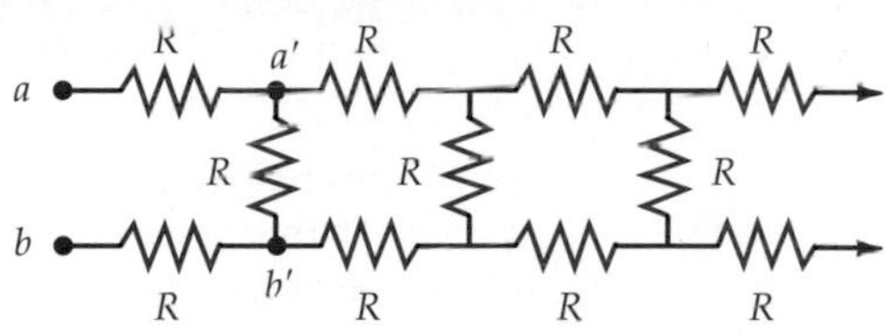

58. An infinite chain of resistors is shown in Figure 23-58. Find the input resistance. (See Problem 57.)

Figure 23-58 Problem 58.

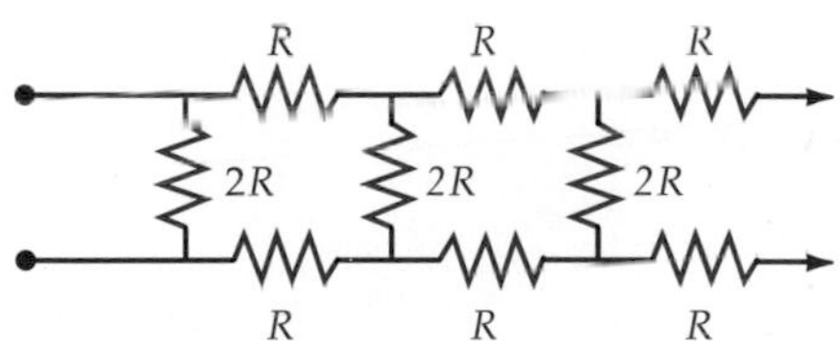

59. If the capacitor in the circuit in Figure 23-49 is replaced by a 30-Ω resistor, what currents flow through the resistors?

60. For the circuit in Figure 23-59, (*a*) what is the initial battery current immediately after switch S is closed? (*b*) What is the battery current a long time after switch S is closed? (*c*) What is the current in the 600-Ω resistor as a function of time?

Figure 23-59 Problem 60.

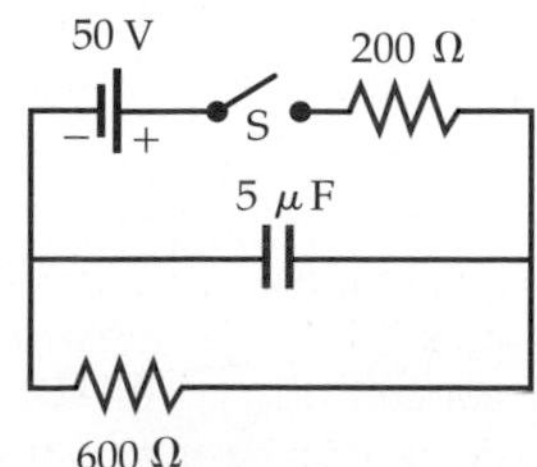

61. For the circuit in Figure 23-60, (*a*) what is the initial battery current immediately after switch S is closed? (*b*) What is the battery current a long time after switch S is closed? (*c*) If the switch has been closed for a long time and is then opened, find the current through the 600-kΩ resistor as a function of time.

Figure 23-60 Problem 61.

62. Capacitors C_1 and C_2 are connected in parallel by a resistor and two switches as shown in Figure 23-61. Capacitor C_1 is initially charged to a voltage V_0, and capacitor C_2 is uncharged. The switches S are then closed. (*a*) What are the final charges on C_1 and C_2? (*b*) Compare the initial and final stored energies of the system. (*c*) What caused the decrease in the capacitor-stored energy?

Figure 23-61 Problem 62.

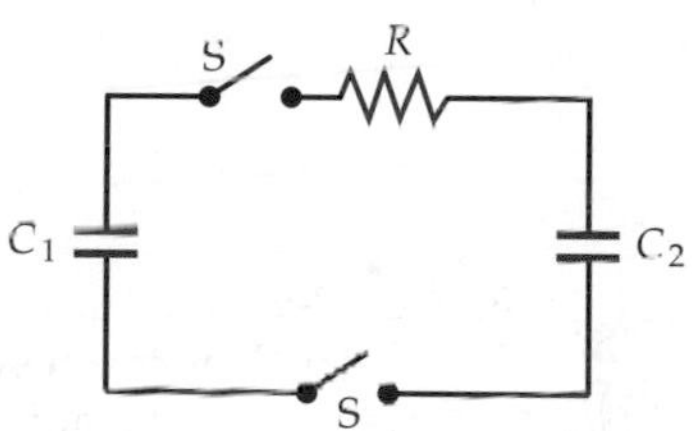

63. (*a*) In Problem 62, find the current through R after the switches S are closed as a function of time. (*b*) Find the energy dissipated in the resistor as a function of time. (*c*) Find the total energy dissipated in the resistor and compare it with the loss of stored energy found in part (*b*) of Problem 62.

64. In the circuit in Figure 23-62, the capacitors are initially uncharged. Switch S_2 is closed and then switch S_1 is closed. (*a*) What is the battery current immediately after S_1 is closed? (*b*) What is the battery current a long time after both switches are closed? (*c*) What is the final voltage across C_1? (*d*) What is the final voltage across C_2? (*e*) Switch S_2 is opened again after a long time. Give the current in the 150-Ω resistor as a function of time.

Figure 23-62 Problem 64.

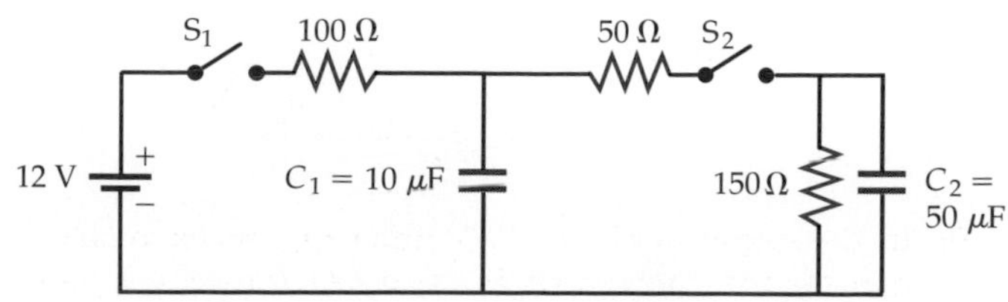

65. In the RC circuit in Figure 23-63 the capacitor is initially uncharged and the switch is closed at time $t = 0$. (*a*) What is the power supplied by the battery as a function of time? (*b*) What is the power dissipated in the resistor as a function of time? (*c*) What is the rate at which energy is stored in the capacitor as a function of time? Plot your answers to parts (*a*), (*b*), and (*c*) versus time on the same graph. (*d*) Find the maximum rate at which energy is stored in the capacitor as a function of the battery voltage $\mathcal{E}$ and the resistance R. At what time does this maximum occur?

Figure 23-63 Problem 65.

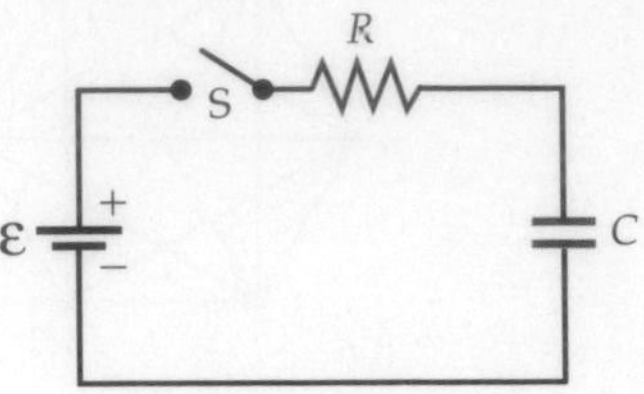

Chapter 24

The Magnetic Field

This experimental Maglev train using magnetic repulsion for levitation, guidance, and propulsion has achieved speeds greater than 300 km/h.

When the existence of magnetism was first noted is not known. However, more than 2000 years ago, the Greeks were aware that a certain type of stone (now called magnetite) attracts pieces of iron, and there are written references to the use of magnets for navigation dating from the twelfth century.

In 1269, Pierre de Maricourt discovered that a needle laid at various positions on a spherical natural magnet orients itself along lines that pass through points at the opposite ends of the sphere. He called these points the poles of the magnet. Subsequently, many experimenters noted that every magnet of whatever shape has two poles, a north pole and a south pole, where the force exerted by the magnet is strongest. It was also noted that the like poles of two magnets repel each other and the unlike poles attract each other.

In 1600, William Gilbert discovered that the earth itself is a natural magnet with magnetic poles near the north and south geographic poles. (Since the north pole of a compass needle points north, what we call the north magnetic pole is actually a south pole, as illustrated in Figure 24-1.) Around 1750, John Michell did a quantitative study of the attraction and repulsion of magnetic poles using a torsion balance. He discovered that the force exerted

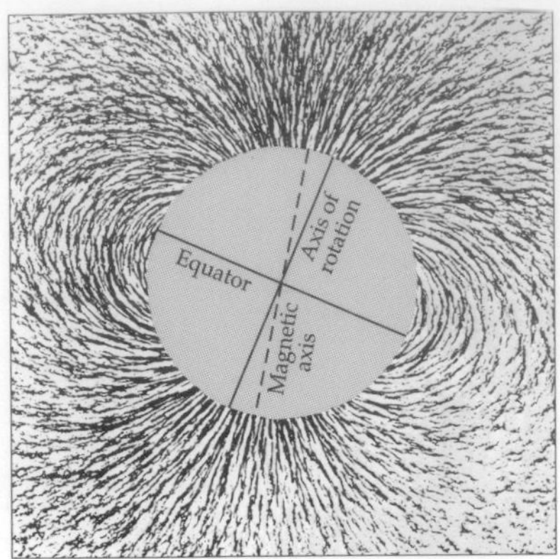

Figure 24-1 Magnetic-field lines of the earth indicated by iron filings around a uniformly magnetized sphere. The lines are somewhat similar to but not exactly the same as those of a bar magnet.

by one pole on another varies inversely with the square of the distance between the poles. These results were confirmed by Coulomb shortly thereafter.

Although the force between two magnetic poles is similar to that between two electric charges, there is an important difference between electric charges and magnetic poles, namely, magnetic poles always occur in pairs. If a magnet is broken in half, there will be equal and opposite poles at either side of the break point; that is, there will be two magnets, each with a north and south pole. There has been much speculation throughout the years as to the existence of an isolated magnetic pole, and in recent years considerable experimental effort has been made to find such an object. Thus far, there seems to be no conclusive evidence that an isolated magnetic pole exists.

The connection between electricity and magnetism was not known until the nineteenth century, when Hans Christian Oersted discovered that an electric current affects the orientation of a compass needle. Subsequent experiments by André-Marie Ampère and others showed that electric currents attract bits of iron and that parallel currents attract each other. Ampère proposed a theoretical model of magnetism that still serves as the basis of the modern theory of magnetism. He speculated that the fundamental source of magnetism is not a magnetic pole but rather an electric current. He further proposed that the magnetism of a permanent magnet is due to the alignment of molecular current loops within the material. Today, we know that these current loops result partly from the motion of electrons within the atom and partly from electron spin, a quantum-mechanical property of the electron. The basic magnetic interaction is the magnetic force one moving charge exerts on another moving charge. This force is in addition to the electric force between the two charges. As with the electric force, we consider the magnetic force to be transmitted by another agent, the magnetic field. The moving charge produces a **magnetic field,** and the field, in turn, exerts a force on the other moving charge. Since a moving charge constitutes an electric current, the magnetic interaction can also be thought of as an interaction between two currents.

In the early 1830s, Michael Faraday and Joseph Henry demonstrated in independent experiments that a changing magnetic field produces an electric field. Some years later (about 1860), James Clerk Maxwell developed a complete theory of electricity and magnetism that showed that a changing electric field produces a magnetic field.

In this chapter, we will consider only the effects of a given magnetic field on moving charges and on wires carrying currents. The sources of magnetic fields will be discussed in Chapter 28.

24-1 The Force Exerted by a Magnetic Field

The existence of a magnetic field* **B** at some point in space can be demonstrated in a straightforward way. We simply place a compass needle at that point and see if it tends to align in a particular direction. If there are no magnets or electric currents nearby, the needle will point in the direction of the magnetic field of the earth. If there are magnets or electric currents nearby, the needle will point in the direction of the net magnetic field due to the earth and the magnets or currents.

*For historical reasons, the magnetic field **B** is sometimes called the *magnetic-induction vector* or the *magnetic flux density*. We will refer to it as the magnetic field.

It is observed experimentally that, when a charge q has velocity $\mathbf{v}$ in a magnetic field, there is a force on it that depends on q and on the magnitude and direction of the velocity. Let us assume that we know the direction of the magnetic field $\mathbf{B}$ at a point in space from a measurement with a compass. Experiments with various charges moving with various velocities at such a point give the following results for the magnetic force:

1. The force is proportional to the charge q. The force on a negative charge is in the direction opposite that on a positive charge with the same velocity.
2. The force is proportional to the speed v.
3. The force is perpendicular to both the magnetic field and the velocity.
4. The force is proportional to $\sin\theta$, where θ is the angle between the velocity $\mathbf{v}$ and the magnetic field $\mathbf{B}$. If $\mathbf{v}$ is parallel or antiparallel to $\mathbf{B}$, the force is zero.

These experimental results can be summarized as follows. When a charge q moves with velocity $\mathbf{v}$ in a magnetic field $\mathbf{B}$, the magnetic force $\mathbf{F}$ on the charge is

$$\mathbf{F} = q\mathbf{v} \times \mathbf{B} \tag{24-1}$$

Magnetic force on a moving charge

Since $\mathbf{F}$ is perpendicular to both $\mathbf{v}$ and $\mathbf{B}$, it is perpendicular to the plane defined by these two vectors. The direction of $\mathbf{F}$ is given by the right-hand rule as $\mathbf{v}$ is rotated into $\mathbf{B}$, as illustrated in Figure 24-2.

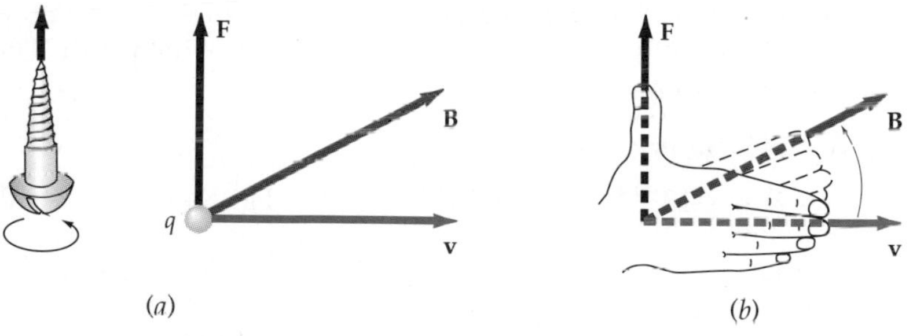

Figure 24-2 Right-hand rule for determining the direction of the magnetic force exerted on a charge moving in a magnetic field. (*a*) The force is perpendicular to both $\mathbf{v}$ and $\mathbf{B}$ and is in the direction in which a right-hand-threaded screw will advance if it is turned in the direction that will rotate $\mathbf{v}$ into $\mathbf{B}$ through the smaller of the two possible angles. (*b*) If the fingers of the right hand are in the direction of $\mathbf{v}$ such that they can be curled into $\mathbf{B}$, the thumb points in the direction of $\mathbf{F}$.

Examples of the direction of the forces exerted on moving charges when the magnetic-field vector $\mathbf{B}$ is in the vertical direction are given in Figure 24-3. Note that the direction of any particular magnetic field $\mathbf{B}$ can be found experimentally by measuring $\mathbf{F}$ and $\mathbf{v}$ and then applying Equation 24-1.

Equation 24-1 defines the **magnetic field B** in terms of the force exerted on a moving charge. The SI unit of magnetic field is the **tesla** (T). A charge of

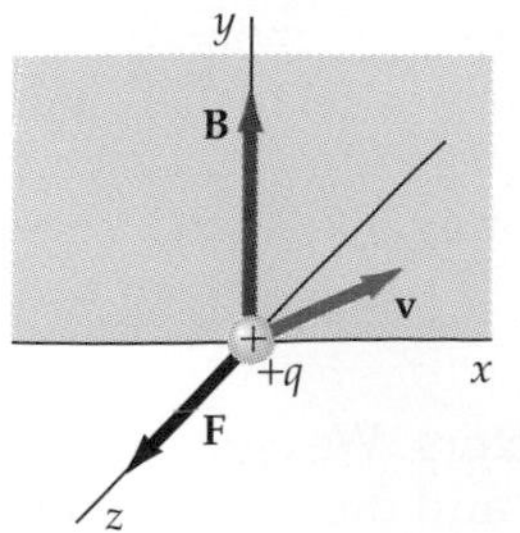

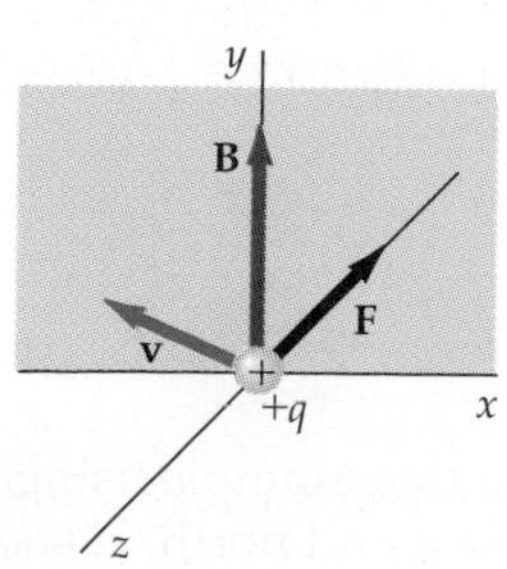

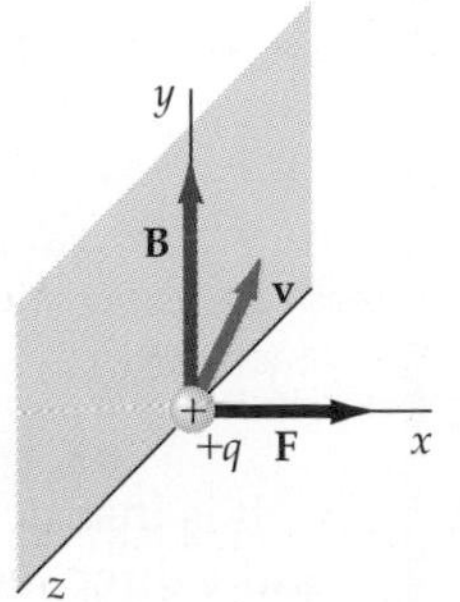

Figure 24-3 Direction of the magnetic force on a charged particle moving with velocity $\mathbf{v}$ in a magnetic field $\mathbf{B}$. The shading indicates the plane of $\mathbf{v}$ and $\mathbf{B}$.

one coulomb moving with a velocity of one meter per second perpendicular to a magnetic field of one tesla experiences a force of one newton:

$$1\ \text{T} = 1\ \frac{\text{N/C}}{\text{m/s}} = 1\ \text{N/A}\cdot\text{m} \qquad 24\text{-}2$$

This unit is rather large. The magnetic field of the earth is slightly less than 10^{-4} T. The magnetic fields near powerful permanent magnets are about 0.1 to 0.5 T, and powerful laboratory and industrial electromagnets produce fields of 1 to 2 T. Fields greater than 10 T are quite difficult to produce because the resulting magnetic forces will tear the magnets apart or crush them. A commonly used unit, derived from the cgs system, is the **gauss** (G), which is related to the tesla as follows:

$$1\ \text{T} = 10^4\ \text{G} \qquad 24\text{-}3$$

Since magnetic fields are often given in gauss, which is not an SI unit, it is important to remember to convert from gauss to teslas when making calculations.

Example 24-1

The magnetic field of the earth has a magnitude of 0.6 G and is directed downward and northward, making an angle of about 70° with the horizontal. (The magnitude and direction of the earth's magnetic field vary from place to place. These data are approximately correct for the central United States.) A proton of charge $q = 1.6 \times 10^{-19}$ C is moving horizontally in the northward direction with speed $v = 10^7$ m/s. Calculate the magnetic force on the proton.

Figure 24-4 shows the directions of the magnetic field **B** and the proton's velocity **v**. The angle between them is $\theta = 70°$. The magnetic force is parallel to **v** × **B**, which is west for a proton moving north. The magnitude of the magnetic force is

$$F = qvB \sin\theta$$
$$= (1.6 \times 10^{-19}\ \text{C})(10^7\ \text{m/s})(0.6 \times 10^{-4}\ \text{T})(0.94) = 9.02 \times 10^{-17}\ \text{N}$$

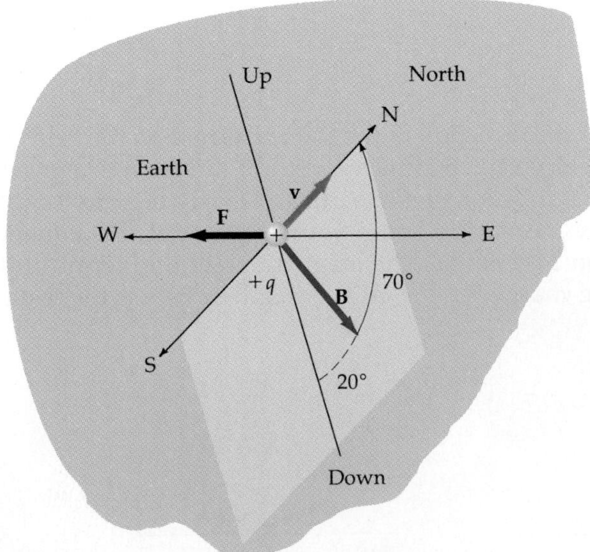

Figure 24-4 Magnetic force on a proton moving north in the magnetic field of the earth, which dips below the horizontal north direction at an angle of 70°, for Example 24-1. The force is directed toward the west.

It is instructive to do this example using unit vectors. We choose the x and y directions to be east and north, respectively, and the z direction to be upwards as shown in Figure 24-5. Then the velocity vector is in the y

direction, and the magnetic field of the earth has the components $B_x = 0$, $B_y = B\cos 70° = (0.6 \times 10^{-4}\text{ T})(0.342) = 2.05 \times 10^{-5}\text{ T}$, and $B_z = -B\sin 70° = (-0.6 \times 10^{-4}\text{ T})(0.940) = -5.64 \times 10^{-5}\text{ T}$. The magnetic-field vector is thus given by

$$\mathbf{B} = 0\,\mathbf{i} + 2.05 \times 10^{-5}\text{ T}\,\mathbf{j} - 5.64 \times 10^{-5}\text{ T}\,\mathbf{k}$$

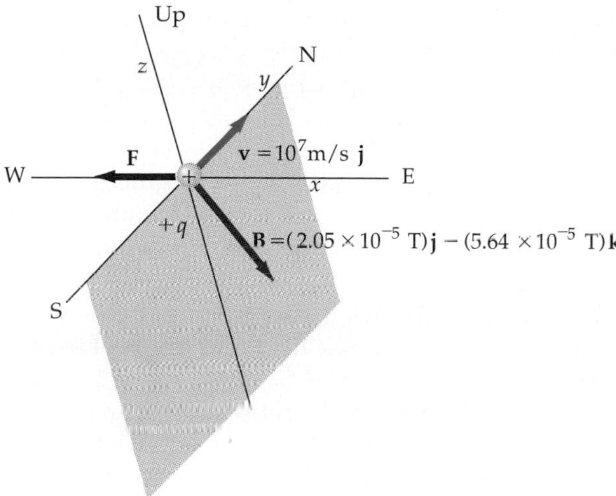

Figure 24-5 Coordinate system for expressing the velocity **v**, magnetic field **B**, and force **F** in terms of the unit vectors **i**, **j**, and **k**.

and the magnetic force on the proton is

$$\mathbf{F} = q\mathbf{v} \times \mathbf{B}$$
$$= (1.6 \times 10^{-19}\text{ C})(10^7\text{ m/s}\,\mathbf{j}) \times (0\,\mathbf{i} + 2.05 \times 10^{-5}\text{ T}\,\mathbf{j} - 5.64 \times 10^{-5}\text{ T}\,\mathbf{k})$$

Since $\mathbf{j} \times \mathbf{j} = 0$ and $\mathbf{j} \times \mathbf{k} = \mathbf{i}$, we have

$$\mathbf{F} = (1.6 \times 10^{-19}\text{ C})(10^7\text{ m/s}\,\mathbf{j}) \times (-5.64 \times 10^{-5}\text{ T}\,\mathbf{k})$$
$$= -9.02 \times 10^{-17}\text{ N}\,\mathbf{i}$$

Exercise

Find the force on a proton moving with velocity $\mathbf{v} = 4 \times 10^6\text{ m/s}\,\mathbf{i}$ in a magnetic field $\mathbf{B} = 2.0\text{ T}\,\mathbf{k}$. (Answer: $-1.28 \times 10^{-12}\text{ N}\,\mathbf{j}$)

When a wire carries a current in a magnetic field, there is a force on the wire that is equal to the sum of the magnetic forces on the charged particles whose motion produces the current. Figure 24-6 shows a short segment of wire of cross-sectional area A and length ℓ carrying a current I. If the wire is in a magnetic field **B**, the magnetic force on each charge is $q\mathbf{v}_d \times \mathbf{B}$, where $\mathbf{v}_d$ is the drift velocity of the charge carriers. The number of charges in the wire segment is the number n per unit volume times the volume $A\ell$. Thus the total force on the wire segment is

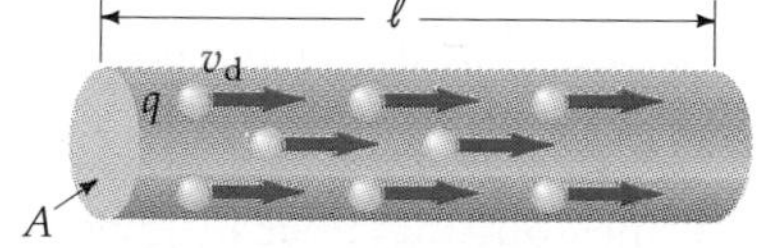

Figure 24-6 Wire segment of length ℓ carrying a current I. If the wire is in a magnetic field, there will be a force on each charge carrier resulting in a force on the wire.

$$\mathbf{F} = (q\mathbf{v}_d \times \mathbf{B})nA\ell$$

From Equation 22-3, the current in the wire is

$$I = nqv_dA$$

Hence the force can be written

$$\mathbf{F} = I\boldsymbol{\ell} \times \mathbf{B} \qquad 24\text{-}4$$

Magnetic force on a segment of current-carrying wire

where $\boldsymbol{\ell}$ is a vector whose magnitude is the length of the wire and whose direction is parallel to $q\mathbf{v}_d$, the direction of the current I. For the current in the positive x direction and the magnetic field in the xy plane shown in Figure 24-7, the force on the wire is directed along the positive z axis, as indicated. In Equation 24-4 it is assumed that the wire segment is straight and that the magnetic field does not vary over its length. The equation can be generalized for an arbitrarily shaped wire in any magnetic field. If we choose a very small wire segment $d\boldsymbol{\ell}$ and write the force on this segment as $d\mathbf{F}$, we have

Magnetic force on a current element

$$d\mathbf{F} = I\, d\boldsymbol{\ell} \times \mathbf{B} \qquad 24\text{-}5$$

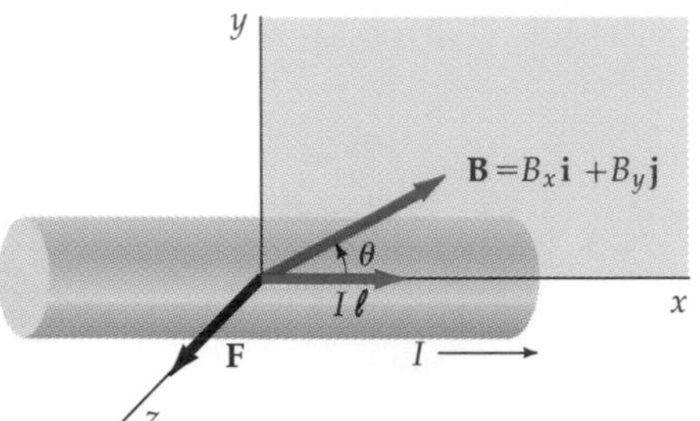

Figure 24-7 Magnetic force on a current-carrying segment of wire in a magnetic field. The current is in the x direction. The magnetic field is in the xy plane and makes an angle θ with the x axis. The force $\mathbf{F}$ is in the z direction, perpendicular to both $\mathbf{B}$ and $I\boldsymbol{\ell}$. Its magnitude is $I\ell B \sin\theta$.

where $\mathbf{B}$ is the magnetic-field vector at the segment. The quantity $I\, d\boldsymbol{\ell}$ is called a **current element.** We find the total force on the wire by summing (integrating) over all the current elements in the wire, using the appropriate field $\mathbf{B}$ at each element.

Equation 24-5 is the same as Equation 24-1 with the current element $I\, d\boldsymbol{\ell}$ replacing $q\mathbf{v}$. It defines the magnetic field $\mathbf{B}$ in terms of the force extended on a current element.

Just as the electric field $\mathbf{E}$ can be represented by electric-field lines, the magnetic field $\mathbf{B}$ can be represented by **magnetic-field lines.** In both cases, the direction of the field is indicated by the direction of the field lines and the magnitude of the field is indicated by their density.

There are, however, two important differences between electric-field lines and magnetic-field lines. The first has to do with the direction of the force the field exerts on a charge. The electric force on a positive charge is in the direction of the electric field and thus of the electric-field lines. The magnetic force on a moving charge, however, is perpendicular to the magnetic field, so magnetic-field lines are *not* in the direction of the magnetic force on a moving charge.

The second difference is that electric-field lines begin on positive charges and end on negative charges. However, because isolated magnetic poles apparently do not exist, there are no points in space where magnetic-field lines begin or end. Instead, they form closed loops. Figure 24-8 shows the magnetic-field lines both inside and outside a bar magnet.

Figure 24-8 (*a*) Magnetic-field lines inside and outside a bar magnet. The lines emerge from the north pole and enter the south pole, but they have no beginning or end. Instead, they form closed loops. (*b*) Magnetic-field lines outside a bar magnet, as indicated by iron filings.

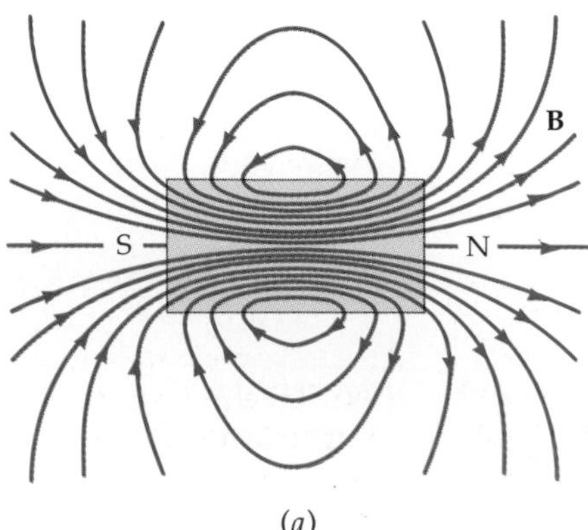

(*a*)

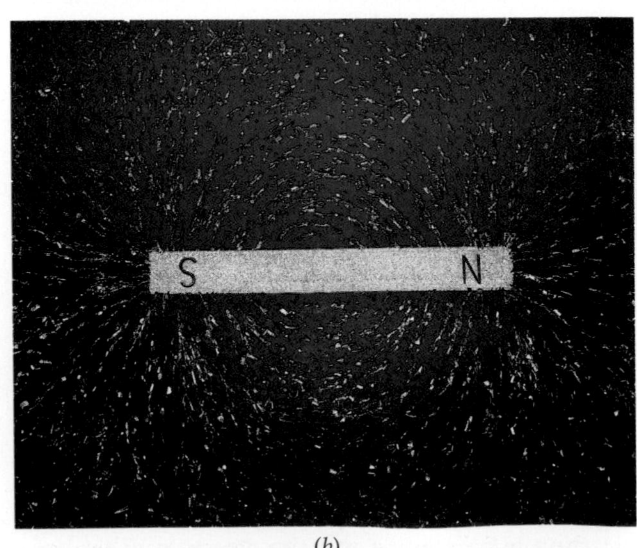

(*b*)

Example 24-2

A wire segment 3 mm long carries a current of 3 A in the x direction. It lies in a magnetic field of magnitude 0.02 T that is in the xy plane and makes an angle of 30° with the x axis as shown in Figure 24-7. What is the magnetic force exerted on the wire segment?

From the figure, we can see that the magnetic force is in the z direction. Its magnitude is given by Equation 24-4:

$$\mathbf{F} = I\boldsymbol{\ell} \times \mathbf{B} = I\ell B \sin 30^\circ \, \mathbf{k}$$

$$= (3.0 \text{ A})(0.003 \text{ m})(0.02 \text{ T})(\sin 30^\circ) \, \mathbf{k}$$

$$= 9 \times 10^{-5} \text{ N } \mathbf{k}$$

The total force on a current-carrying wire is found by adding the forces on each of the segments.

Questions

1. Charge q moves with velocity $\mathbf{v}$ through a magnetic field $\mathbf{B}$. At a particular instant, it experiences a magnetic force $\mathbf{F}$. How would the force differ if the charge had the opposite sign? If the velocity were in the opposite direction? If the magnetic field were in the opposite direction?
2. For what angle between $\mathbf{B}$ and $\mathbf{v}$ is the magnetic force on q greatest? The least?
3. A moving electric charge may experience both electric and magnetic forces. How could you determine whether a force causing a charge to deviate from a straight path is an electric or a magnetic force?
4. How can a charge move through a magnetic field without ever experiencing any magnetic force?
5. Show that the force on a current element is the same in direction and magnitude regardless of whether positive charges, negative charges, or a mixture of positive and negative charges create the current.
6. A current-carrying wire is in a magnetic field, but the wire does not experience any magnetic force. How is this possible?
7. How are magnetic-field lines similar to electric-field lines? How are they different?

24-2 Motion of a Point Charge in a Magnetic Field

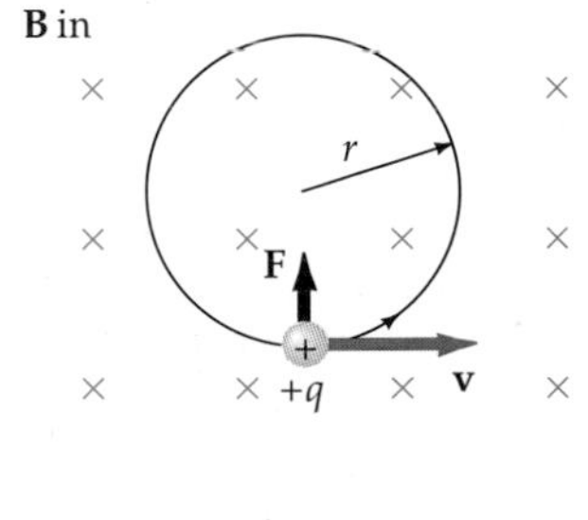

Figure 24-9 Charged particle moving in a plane perpendicular to a uniform magnetic field. The magnetic field is into the page, as indicated by the crosses. (A field out of the plane of the page would be indicated by dots.) The magnetic force is perpendicular to the velocity of the particle, causing it to move in a circular orbit.

An important characteristic of the magnetic force on a charged particle moving through a magnetic field is that the force is always perpendicular to the velocity of the particle. The magnetic force thus changes the direction of the velocity but not its magnitude. It therefore does no work on the particle and does not affect the kinetic energy of the particle.

In the special case where the velocity of a particle is perpendicular to a uniform magnetic field, as shown in Figure 24-9, the particle moves in a circular orbit. The magnetic force provides the centripetal force necessary for circular motion. We can relate the radius of the circle r to the magnetic field B and the speed v of the particle by setting the net force equal to the mass m of the particle times the centripetal acceleration v^2/r in accordance with

Newton's second law. The net force in this case is qvB since **v** and **B** are perpendicular. Thus, Newton's second law gives

$$F = ma$$

$$qvB = \frac{mv^2}{r}$$

or

$$r = \frac{mv}{qB} \quad 24\text{-}6$$

The period of the circular motion is the time it takes the particle to travel once around the circumference of the circle. From Equation 3-23, the period is related to the speed by

$$T = \frac{2\pi r}{v}$$

(*a*) Circular path of electrons moving in the magnetic field produced by two large coils. The electrons ionized the gas in the tube causing it to give off a bluish glow that indicates the path of the beam. (*b*) False-color photograph showing tracks of a 1.6-MeV proton (red) and a 7-MeV α particle (yellow) in a cloud chamber. The radius of curvature is proportional to the momentum and inversely proportional to the charge of the particle. For these energies, the momentum of the α particle, which has twice the charge of the proton, is about four times that of the proton and so its radius of curvature is greater.

(*a*)

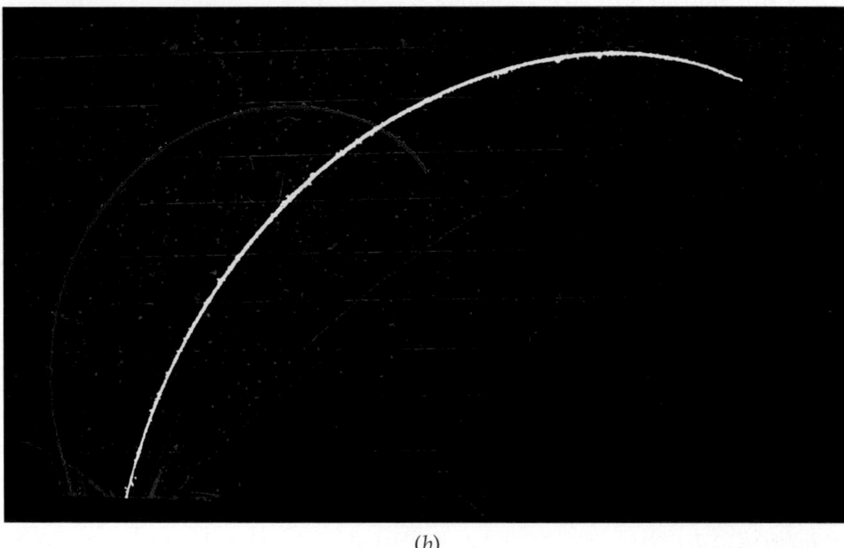

(*b*)

Substituting in $r = mv/qB$ from Equation 24-6, we obtain,

$$T = \frac{2\pi(mv/qB)}{v} = \frac{2\pi m}{qB} \quad 24\text{-}7$$

The frequency of the circular motion is the reciprocal of the period.

Cyclotron frequency

$$f = \frac{1}{T} = \frac{qB}{2\pi m} \quad 24\text{-}8$$

Note that the period and the frequency given by Equations 24-7 and 24-8 do not depend on the radius of the orbit or the velocity of the particle. This period is called the **cyclotron period,** and the frequency the **cyclotron frequency.** Two of the many interesting applications of the circular motion of charged particles in a uniform magnetic field, the mass spectrometer and the cyclotron, will be discussed later in this section.

Example 24-3

A proton of mass $m = 1.67 \times 10^{-27}$ kg and charge $q = e = 1.6 \times 10^{-19}$ C moves in a circle of radius 21 cm perpendicular to a magnetic field $B =$ 4000 G. Find (*a*) the period of the motion and (*b*) the speed of the proton.

(*a*) We do not need to know the radius of the circle to find the period. Converting the magnetic field to SI units (4000 G = 0.4 T) and using Equation 24-7, we have

$$T = \frac{2\pi m}{qB} = \frac{2\pi(1.67 \times 10^{-27}\text{ kg})}{(1.6 \times 10^{-19}\text{ C})(0.4\text{ T})}$$
$$= 1.64 \times 10^{-7}\text{ s}$$

(*b*) The speed v of the proton is related to the radius of the circle by Equation 24-6:

$$v = \frac{rqB}{m} = \frac{(0.21\text{ m})(1.6 \times 10^{-19}\text{ C})(0.4\text{ T})}{1.67 \times 10^{-27}\text{ m}}$$
$$= 8.05 \times 10^{6}\text{ m/s}$$

We can check our results by noting that the product of the speed v and the period T is the circumference of the circle $2\pi r$. Then

$$r = \frac{vT}{2\pi} = \frac{(8.05 \times 10^{6}\text{ m/s})(1.64 \times 10^{-7}\text{ s})}{2\pi}$$
$$= 0.21\text{ m} = 21\text{ cm}$$

In these calculations, we put each quantity in SI units so that the results will be in SI units, namely, seconds for the period and meters per second for the speed.

Note from Equation 24-6 that the radius of the circular motion is proportional to the speed. If we double the speed of the proton in this example, the radius would double, but the period and the frequency would remain unchanged.

Suppose that a charged particle enters a uniform magnetic field with a velocity that is not perpendicular to **B**. We can resolve the velocity of the particle into components $v_{\parallel}$ parallel to **B** and $v_{\perp}$ perpendicular to **B**. The motion due to the perpendicular component is the same as that just discussed. The component of the velocity parallel to **B** is not affected by the magnetic field. It therefore remains constant. The path of the particle is thus a helix, as shown in Figure 24-10.

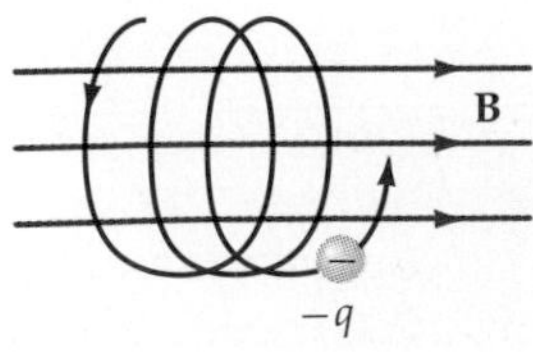

(*a*)

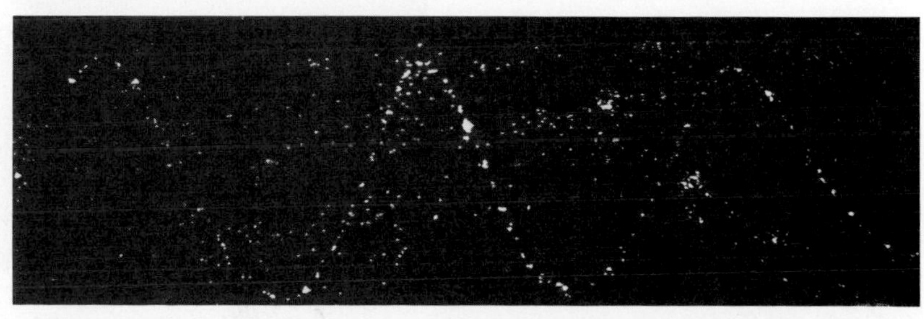
(*b*)

Figure 24-10 (*a*) When a charged particle has a velocity component that is parallel to a magnetic field as well as one that is perpendicular, it moves in a helical path around the field lines. (*b*) Cloud-chamber photograph of the helical path of an electron moving in a magnetic field. The path of the electrons is made visible by the condensation of water droplets in the cloud chamber.

The motion of charged particles in nonuniform magnetic fields is quite complicated. Figure 24-11 shows a **magnetic bottle,** an interesting magnetic-field configuration in which the field is weak at the center and strong at both ends. A detailed analysis of the motion of a charged particle in such a field shows that the particle will spiral around the field lines and become trapped, oscillating back and forth between points P_1 and P_2 in the figure. Such magnetic-field configurations are used to confine dense beams of charged parti-

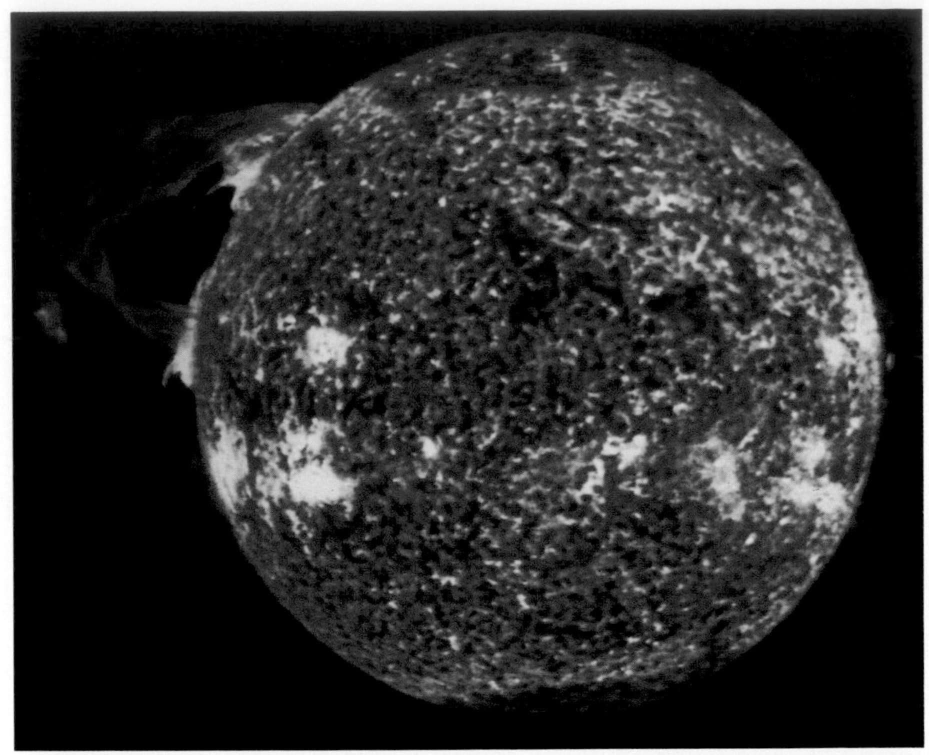

(*a*)

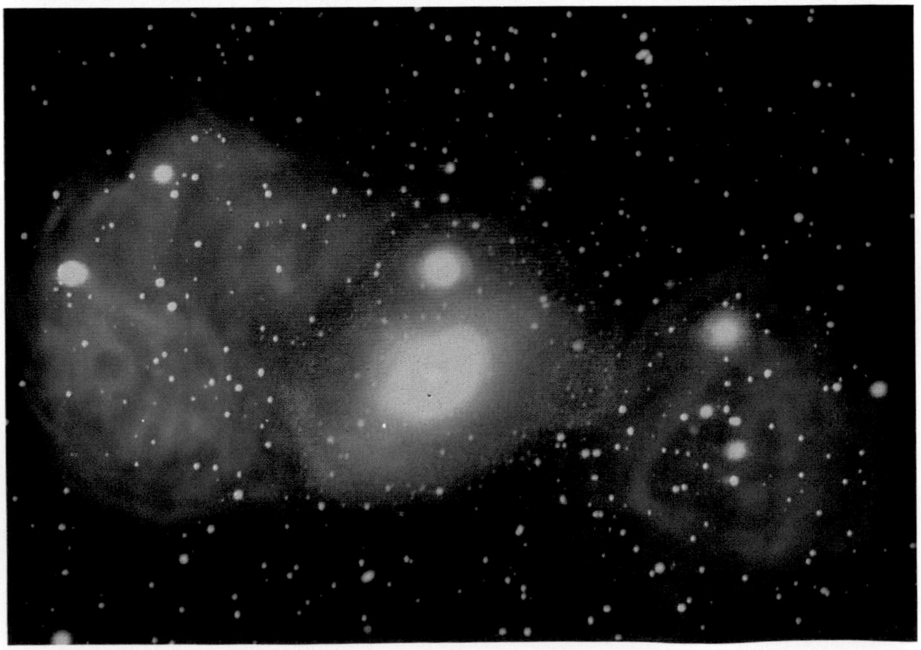

(*b*)

(*a*) The solar flare shown on the left of this skylab-4 photograph consists of charged particles confined by the magnetic field of the sun. (*b*) Interstellar dust grains, aligned by the magnetic fields of distant galaxies, act like polarizing filters. This map of the magnetic field of the galaxy NGC 1316 obtained with a radio telescope shows double-lobed magnetic jets in red. The magnetic map is superimposed on a visible-light photograph of the galaxy.

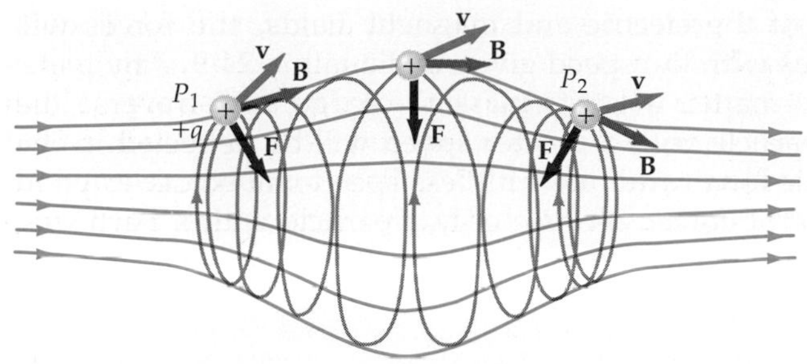

Figure 24-11 Magnetic bottle. When a charged particle moves in such a field, which is strong at both ends and weak in the middle, the particle becomes trapped and moves back and forth, spiraling around the field lines.

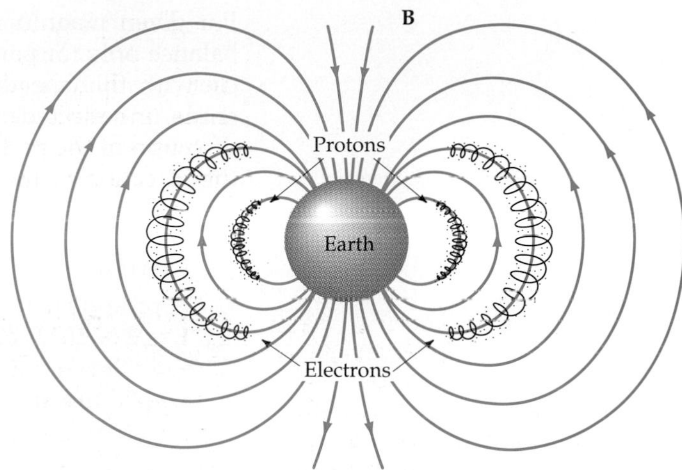

Figure 24-12 Van Allen belts. Protons (the inner belts) and electrons (the outer belts) are trapped in the earth's magnetic field and travel in helical paths around the field lines between the north and south poles.

cles, called *plasmas,* in nuclear fusion research. A similar phenomenon is the oscillation of ions back and forth between the earth's magnetic poles in the Van Allen belts (Figure 24-12).

The Velocity Selector

The magnetic force on a charged particle moving in a uniform magnetic field can be balanced by an electric force if the magnitudes and directions of the magnetic and electric fields are properly chosen. Since the electric force is in the direction of the electric field (for positive particles) and the magnetic force is perpendicular to the magnetic field, the electric and magnetic fields in the region through which the particle is moving must be perpendicular to each other if the forces are to balance. Such a region is said to have **crossed fields.** Figure 24-13 shows a region of space between the plates of a capacitor where there is an electric field and a perpendicular magnetic field (produced by a magnet not shown). Consider a particle of charge q entering this space from the left. If q is positive, the electric force of magnitude qE is down and the magnetic force of magnitude qvB is up. If the charge is negative, each of these forces is reversed. The two forces will balance if $qE = qvB$ or

$$v = \frac{E}{B} \tag{24-9}$$

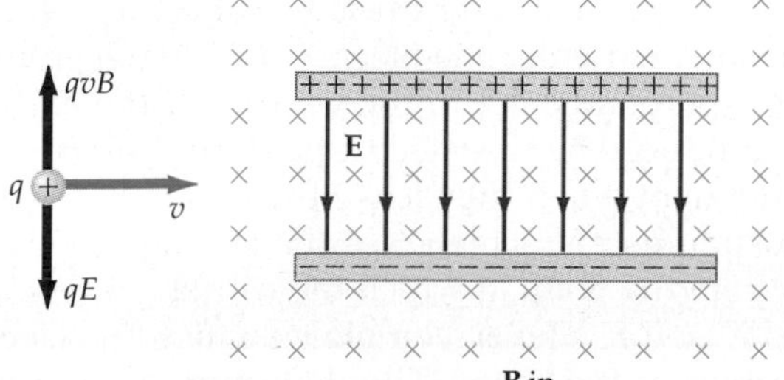

Figure 24-13 Crossed electric and magnetic fields. When a positive particle moves to the right, it experiences a downward electric force qE and an upward magnetic force qvB. These forces balance if the speed of the particle is related to the magnitudes of the fields by $v = E/B$.

For given magnitudes of the electric and magnetic fields, the forces will balance only for particles with the speed given by Equation 24-9. Any particle with this speed, no matter what its mass or charge, will traverse the space undeflected. A particle with a greater speed will be deflected in the direction of the magnetic force, and one with less speed will be deflected in the direction of the electric force. Because only a particle with a particular speed can traverse this arrangement of fields, it is called a **velocity selector.**

Exercise

A proton is moving in the x direction in a region of crossed fields where $\mathbf{E} = 2 \times 10^5$ N/C $\mathbf{k}$ and $\mathbf{B} = -3000$ G $\mathbf{j}$. (*a*) What is the speed of the proton if it is not deflected? (*b*) If the proton moves with twice this speed, in which direction will it be deflected? [Answers: (*a*) 667 km/s; (*b*) in the negative z direction]

Thomson's Measurement of q/m for Electrons

An example of the use of a velocity selector is the famous experiment performed by J. J. Thomson in 1897 in which he showed that the rays of a cathode-ray tube can be deflected by electric and magnetic fields and therefore consist of charged particles. By measuring the deflections of these particles caused by various combinations of electric and magnetic fields, Thomson showed that all the particles have the same charge-to-mass ratio q/m. He also showed that particles with this charge-to-mass ratio can be obtained using any material for the cathode, which means that these particles, now called electrons, are a fundamental constituent of all matter.

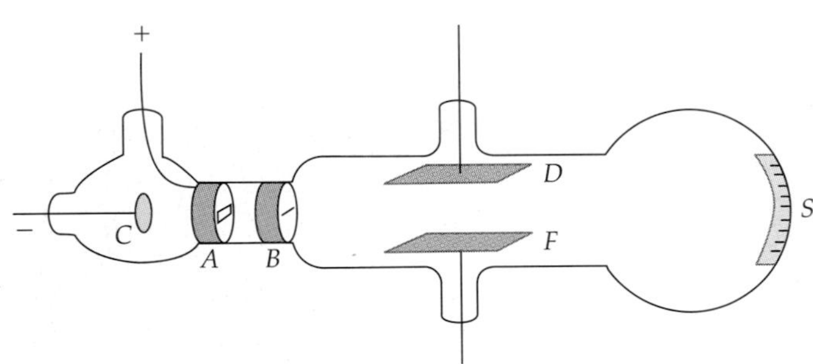

(*a*)

(*b*)

Figure 24-14 (*a*) A schematic diagram of the cathode-ray tube Thomson used to measure q/m for the particles that comprise cathode rays (electrons). Electrons from the cathode C pass through the slits at A and B and strike a phosphorescent screen S. The beam can be deflected by an electric field between plates D and F or by a magnetic field (not shown). (*b*) J. J. Thomson in his laboratory.

Figure 24-14 shows a schematic diagram of the cathode-ray tube Thomson used. Electrons are emitted from the cathode C, which is at a negative potential relative to the slits A and B. An electric field in the direction from A to C accelerates the electrons, and they pass through slits A and B into a field-free region. The electrons then enter the electric field between the capacitor plates D and F that is perpendicular to the velocity of the electrons. Because of the acceleration produced by this electric field, the velocity of the electrons has a vertical component when they leave the region between the plates. They strike the phosphorescent screen S at the far right side of the tube at some deflection Δy from the point at which they strike when there is no field between the plates D and F. The screen glows where the electrons strike it, indicating the location of the beam. The deflection Δy consists of

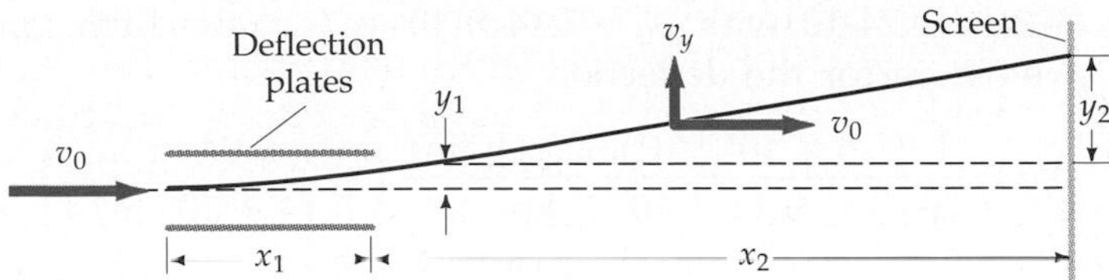

Figure 24-15 The total deflection of the beam in the J. J. Thomson experiments consists of the deflection while the electrons are between the plates, y_1, plus the deflection y_2 that occurs in the field-free region between the plates and the screen.

two parts: the deflection Δy_1, while the electrons are between the plates, and the deflection Δy_2, which occurs after the electrons leave the region between the plates (Figure 24-15).

Let x_1 be the horizontal distance across the deflection plates D and F. If the electron is moving horizontally with speed v_0 when it enters the plates, the time spent between the plates is $t_1 = x_1/v_0$, and the vertical velocity when it leaves the plates is

$$v_y = at_1 = \frac{qE}{m}t_1 = \frac{qE}{m}\frac{x_1}{v_0}$$

where E is the electric field between the plates. The deflection in this region will be

$$\Delta y_1 = \frac{1}{2}at_1^2 = \frac{1}{2}\frac{qE}{m}\left(\frac{x_1}{v_0}\right)^2$$

The electron then travels an additional horizontal distance x_2 in the field-free region from the deflection plates to the screen. Since the velocity of the electron is constant in this region, the time to reach the screen is $t_2 = x_2/v_0$, and the additional vertical deflection is

$$\Delta y_2 = v_y t_2 = \frac{qE}{m}\frac{x_1}{v_0}\frac{x_2}{v_0}$$

The total deflection at the screen is therefore

$$\Delta y = \Delta y_1 + \Delta y_2 = \frac{1}{2}\frac{qE}{m}\left(\frac{x_1}{v_0}\right)^2 + \frac{qE}{m}\frac{x_1 x_2}{v_0^2} \qquad 24\text{-}10$$

The initial speed v_0 is determined by introducing a magnetic field **B** between the plates in a direction that is perpendicular to both the electric field and the initial velocity of the electrons. The magnitude of **B** is then adjusted until the beam is not deflected. The speed is then found from Equation 24-9, and the measured deflection Δy is used to determine the charge-to-mass ratio, q/m, from Equation 24-10.

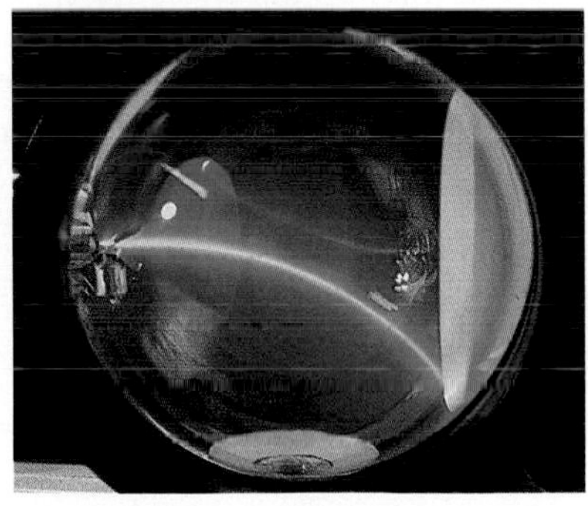

An electron beam moving from left to right is deflected downward by a magnetic field produced by coils. In what direction is the magnetic field?

Example 24-4

Electrons pass undeflected through the plates of Thomson's apparatus when the electric field is 3000 V/m and there is a crossed magnetic field of 1.40 G. If the plates are 4 cm long and the end of the plates is 30 cm from the screen, find the deflection on the screen when the magnetic field is turned off.

In this example, we use the known charge $e = 1.6 \times 10^{-19}$ C and mass $m = 9.11 \times 10^{-31}$ kg of the electron to find the deflection Δy. The initial speed of the electrons is found from Equation 24-9:

$$v_0 = \frac{E}{B} = \frac{3000\ \text{V/m}}{1.40 \times 10^{-4}\ \text{T}} = 2.14 \times 10^7\ \text{m/s}$$

Using Equation 24-10 with $v_0 = 2.14 \times 10^7$ m/s, $x_1 = 4$ cm, and $x_2 =$ 30 cm, we obtain for the deflection

$$\Delta y = \frac{1}{2}\frac{(1.6 \times 10^{-19}\text{ C})(3000\text{ V/m})}{9.11 \times 10^{-31}\text{ kg}}\left(\frac{0.04\text{ m}}{2.14 \times 10^7\text{ m/s}}\right)^2 + \frac{(1.6 \times 10^{-19}\text{ C})(3000\text{ V/m})}{9.11 \times 10^{-31}\text{ kg}}\frac{(0.04\text{ m})(0.30\text{ m})}{(2.14 \times 10^7\text{ m/s})^2}$$

$$= 9.20 \times 10^{-4}\text{ m} + 1.38 \times 10^{-2}\text{ m}$$

$$= 0.92\text{ mm} + 13.8\text{ mm} = 14.7\text{ mm}$$

The Mass Spectrometer

The **mass spectrometer,** first designed by Francis William Aston in 1919 and later improved by Kenneth Bainbridge and others, was developed as a means of measuring the masses of isotopes. Such measurements are an important way of determining both the existence of isotopes and their abundance in nature. For example, natural magnesium has been found to consist of 78.7 percent ^{24}Mg, 10.1 percent ^{25}Mg, and 11.2 percent ^{26}Mg. These isotopes have masses in the approximate ratio 24:25:26.

The mass spectrometer is used to find the mass-to-charge ratio of ions of known charge by measuring the radius of their circular orbits in a uniform magnetic field. Equation 24-6, $r = mv/qB$, gives the radius r for the circular orbit of a particle of mass m and charge q moving with speed v in a magnetic field B that is perpendicular to the velocity of the particle. Figure 24-16 shows a simple schematic drawing of a mass spectrometer. Ions from an ion source are accelerated by an electric field and enter a uniform magnetic field produced by an electromagnet. If the ions start from rest and move through a potential drop ΔV, their kinetic energy when they enter the magnetic field equals their loss in potential energy, $q\,\Delta V$:

$$\tfrac{1}{2}mv^2 = q\,\Delta V \qquad 24\text{-}11$$

The ions move in a semicircle of radius r given by Equation 24-6 and strike a photographic plate at point P_2, a distance $2r$ from the point where they enter the magnet. The speed v can be eliminated from Equations 24-6 and 24-11 to

Figure 24-16 (*a*) Schematic drawing of a mass spectrometer. Ions from an ion source are accelerated though a potential difference ΔV and enter a uniform magnetic field. The magnetic field is out of the plane of the page as indicated by the dots. The ions are bent into circular arcs and strike a photographic plate at P_2. The radius of the circle is proportional to the mass of the ion. (*b*) A modern mass spectrometer used for research at Rockefeller University.

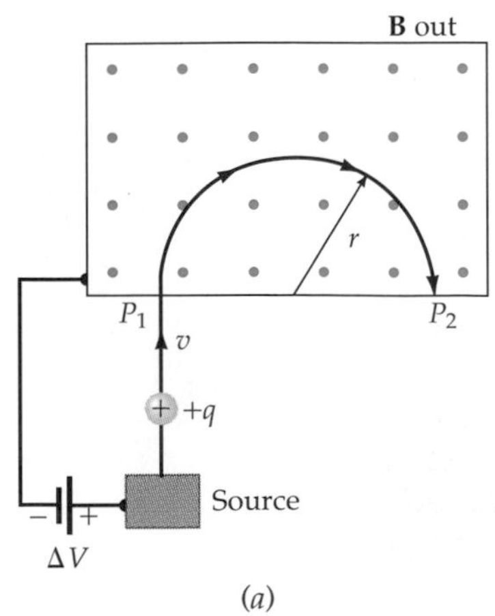

(*a*)

(*b*)

find m/q in terms of the known quantities ΔV, B, and r. We first solve Equation 24-6 for v and square each term, which gives

$$v^2 = \frac{r^2q^2B^2}{m^2}$$

Substituting this expression for v^2 into Equation 24-11, we obtain

$$\frac{1}{2}m\left(\frac{r^2q^2B^2}{m^2}\right) = q\,\Delta V$$

Simplifying this equation and solving it for m/q, we obtain

$$\frac{m}{q} = \frac{B^2r^2}{2\,\Delta V} \qquad 24\text{-}12$$

In Aston's original mass spectrometer, mass differences could be measured to a precision of about 1 part in 10,000. The precision has been improved by introducing a velocity selector between the ion source and the magnet, which makes it possible to limit the range of velocities of the incoming ions and to determine the velocities of the ions more accurately.

Example 24-5

A ^{58}Ni ion of charge $+e$ and mass 9.62×10^{-26} kg is accelerated through a potential difference of 3 kV and deflected in a magnetic field of 0.12 T. (*a*) Find the radius of curvature of the orbit of the ion. (*b*) Find the difference in the radii of curvature of ^{58}Ni ions and ^{60}Ni ions. (Assume that the mass ratio is 58/60.)

(*a*) Using Equation 24-12, we obtain

$$r^2 = \frac{2m\,\Delta V}{qB^2} = \frac{2(9.62 \times 10^{-26}\text{ kg})(3000\text{ V})}{(1.6 \times 10^{-19}\text{ C})(0.12\text{ T})^2} = 0.251\text{ m}^2$$

$$r = \sqrt{0.251\text{ m}^2} = 0.501\text{ m}$$

(*b*) The radius of the orbit of an ion in a given magnetic field is proportional to the square root of its mass, for a given accelerating voltage. If r_1 is the radius of the orbit of the ^{58}Ni ion and r_2 is that of the ^{60}Ni ion, the ratio of the radii is

$$\frac{r_2}{r_1} = \sqrt{\frac{m_2}{m_1}} = \sqrt{\frac{60}{58}} = 1.017$$

Then, the radius of the orbit of the ^{60}Ni ion is

$$r_2 = 1.017\,r_1 = (1.017)(0.501\text{ m}) = 0.510\text{ m}$$

The difference in the radii of the orbits is thus

$$r_2 - r_1 = 0.510\text{ m} - 0.501\text{ m} = 0.009\text{ m} = 9\text{ mm}$$

The Cyclotron

The **cyclotron** was invented by E. O. Lawrence and M. S. Livingston in 1934 to accelerate particles such as protons or deuterons to high kinetic energies. (A deuteron is the nucleus of heavy hydrogen, ^{2}H, which consists of a proton and neutron tightly bound together.) The high-energy particles are then used to bombard atomic nuclei to cause nuclear reactions that are then studied to obtain information about the nucleus. High-energy protons and deuterons are also used to produce radioactive materials and for medical purposes.

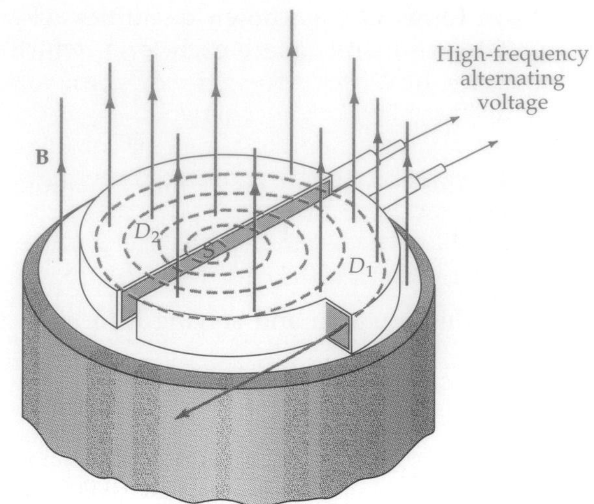

Figure 24-17 Schematic drawing of a cyclotron. The upper portion of the magnet has been omitted. Charged particles such as protons from a source S at the center are accelerated by the potential difference across the gap between the dees. The potential difference across the gap alternates with the cyclotron period of the particle, which is independent of the radius of the circle. Thus, when the particles arrive at the gap again, the potential difference has changed sign, so they are again accelerated across the gap and move in a larger circle.

The operation of the cyclotron is based on the fact that the period of the motion of a charged particle in a uniform magnetic field is independent of the velocity of the particle, as can be seen from Equation 24-7:

$$T = \frac{2\pi m}{qB}$$

Figure 24-17 is a schematic drawing of a cyclotron. The particles move in two semicircular metal containers called *dees* (because of their shape). The dees are housed in a vacuum chamber that is in a uniform magnetic field provided by an electromagnet. (The region in which the particles move must be evacuated so that the particles will not lose energy and be scattered in collisions with air molecules.) Between the dees there is maintained a potential difference ΔV that alternates in time with a period T, which is chosen to be equal to the cyclotron period given by Equation 24-7. This potential difference creates an electric field across the gap between the dees. At the same time, there is no electric field within each dee because of the shielding of the metal dees.

The charged particles are initially injected into dee 1 with a small velocity from an ion source S near the center of the dees. They move in a semicircle in dee 1 and arrive at the gap between dee 1 and dee 2 after a time $\frac{1}{2}T$, where T is the cyclotron period and is also the period with which the potential across the dees is alternated. The alternation of the potential is adjusted

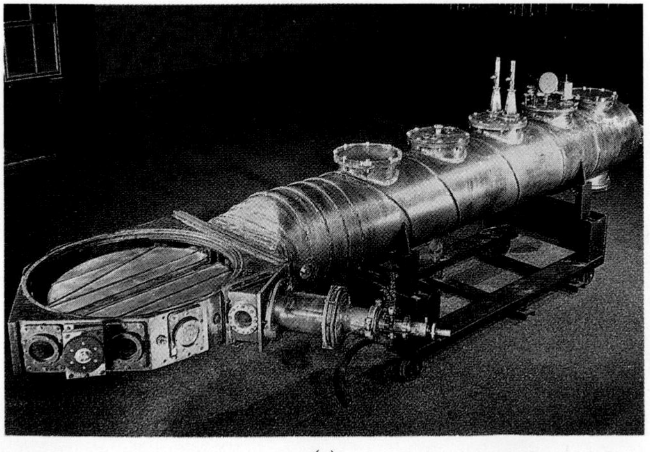

(*a*)

(*b*)

(*a*) The first cyclotron built by E. O. Lawrence in 1932. (*b*) A modern cyclotron used to accelerate protons or deuterons for making short-lived radioactive isotopes for medical diagnosis.

so that dee 1 is at a higher potential than dee 2 when the particles arrive at the gap between them. Each particle is therefore accelerated across the gap by the electric field across the gap and gains kinetic energy equal to $q\ \Delta V$. Because it has more kinetic energy, the particle moves in a semicircle of larger radius in dee 2, and again arrives at the gap after a time $\frac{1}{2}T$. By this time the potential between the dees has been reversed so that dee 2 is now at the higher potential. Once more the particle is accelerated across the gap and gains additional kinetic energy equal to $q\ \Delta V$. Each time the particle arrives at the gap, it is accelerated and gains kinetic energy equal to $q\ \Delta V$. Thus, it moves in larger and larger semicircular orbits until it eventually leaves the magnetic field. In the typical cyclotron, each particle may make 50 to 100 revolutions and exit with energies of up to several hundred MeV.

This kinetic energy of a particle leaving a cyclotron can be calculated by setting r in Equation 24-6 equal to the maximum radius of the dees and solving the equation for v:

$$r = \frac{mv}{qB}$$

$$v = \frac{qBr}{m}$$

Then,

$$K = \frac{1}{2}mv^2 = \frac{1}{2}\left(\frac{q^2B^2}{m}\right)r^2 \qquad \text{24-13}$$

Example 24-6

A cyclotron for accelerating protons has a magnetic field of 1.5 T and a maximum radius of 0.5 m. (*a*) What is the cyclotron frequency? (*b*) Find the kinetic energy of the protons when they emerge.

(*a*) The cyclotron frequency is given by Equation 24-8:

$$f = \frac{qB}{2\pi m} = \frac{(1.6 \times 10^{-19}\ \text{C})(1.5\ \text{T})}{2\pi(1.67 \times 10^{-27}\ \text{kg})} = 2.29 \times 10^7\ \text{Hz} = 22.9\ \text{MHz}$$

(*b*) The kinetic energy of the emerging protons is given by Equation 24-13:

$$K = \frac{1}{2}\left[\frac{(1.6 \times 10^{-19}\ \text{C})^2(1.5\ \text{T})^2}{1.67 \times 10^{-27}\ \text{kg}}\right](0.5\ \text{m})^2$$
$$= 4.31 \times 10^{-12}\ \text{J}$$

The energies of protons and other elementary particles are usually expressed in electron volts. Since 1 eV = 1.6×10^{-19} J, we have

$$K = 4.31 \times 10^{-12}\ \text{J} \times \frac{1\ \text{eV}}{1.6 \times 10^{-19}\ \text{J}} = 26.9\ \text{MeV}$$

Questions

8. How can you determine by observing the path of a deflected particle whether the particle is being deflected by a magnetic field or an electric field?

9. A beam of positively charged particles passes undeflected from left to right through a velocity selector in which the electric field is up. The beam is then reversed so that it travels from right to left. Will the beam now be deflected in the velocity selector? If so, in which direction?

24-3 Torques on Current Loops and Magnets

Figure 24-18 shows a rectangular wire loop of length a and width b carrying a current I in a uniform external magnetic field **B** that is parallel to the plane of the loop. The forces on each segment of the loop are shown in the figure. There are no forces on the top or bottom of the loop because the current in those segments is parallel or antiparallel to the magnetic field **B**, so $I\,d\boldsymbol{\ell} \times \mathbf{B}$ is zero. The forces on the sides of the loop have the magnitude

$$F_1 = F_2 = IaB$$

Since these forces are equal and opposite, they form a couple. The net force is therefore zero, and the torque about any point is independent of the location of the point. Point P is a convenient point about which to compute the torque. The magnitude of the torque is

$$\tau = F_1 b = IabB = IAB$$

where $A = ab$ is the area of the loop. The torque is therefore the product of the current, the area of the loop, and the magnetic field B. This torque tends to twist the loop so that its plane is perpendicular to **B**.

The orientation of the loop can be described conveniently by a unit vector $\hat{\mathbf{n}}$ that is perpendicular to the plane of the loop. The sense of $\hat{\mathbf{n}}$ is chosen to be that given by the right-hand rule applied to the circulating current, as illustrated in Figure 24-19. The torque tends to rotate $\hat{\mathbf{n}}$ into the direction of **B**.

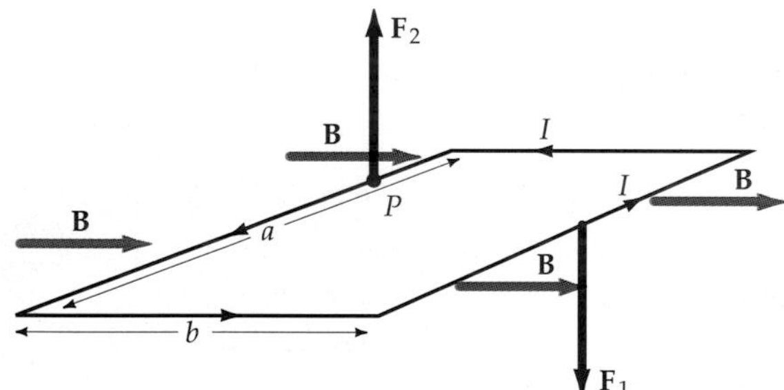

Figure 24-18 Forces exerted on a rectangular current loop in a uniform magnetic field **B** that is parallel to the plane of the loop. The forces produce a torque that tends to twist the loop so that its plane is perpendicular to **B**.

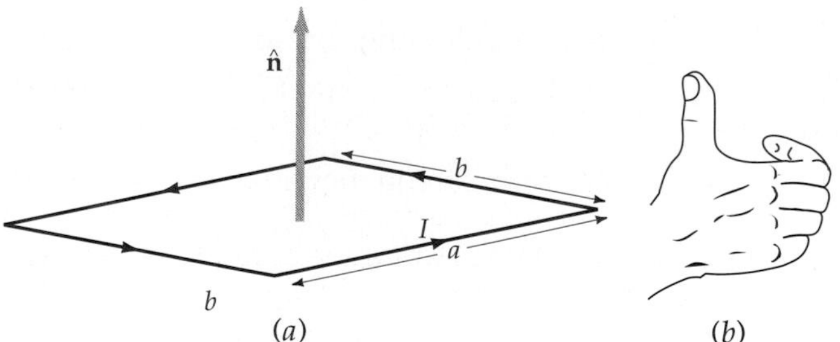

Figure 24-19 (*a*) The orientation of a current loop is described by the unit vector $\hat{\mathbf{n}}$ that is perpendicular to the plane of the loop. (*b*) Right-hand rule for determining the sense of $\hat{\mathbf{n}}$. When the fingers of the right hand curl around the loop, with the fingers pointing in the direction of the current, the thumb points in the direction of $\hat{\mathbf{n}}$.

Figure 24-20 shows the forces exerted by a uniform magnetic field on a rectangular loop whose normal unit vector $\hat{\mathbf{n}}$ makes an angle θ with the magnetic field **B**. Again, the net force on the loop is zero. The torque about any point is the product of the force and the lever arm. For example, the torque about point P is the force $F_2 = IaB$ times the lever arm $b \sin\theta$. The torque thus has the magnitude

$$\tau = IaBb \sin\theta = IAB \sin\theta$$

where again $A = ab$ is the area of the loop. For a loop with N turns, the torque has the magnitude

$$\tau = NIAB \sin\theta$$

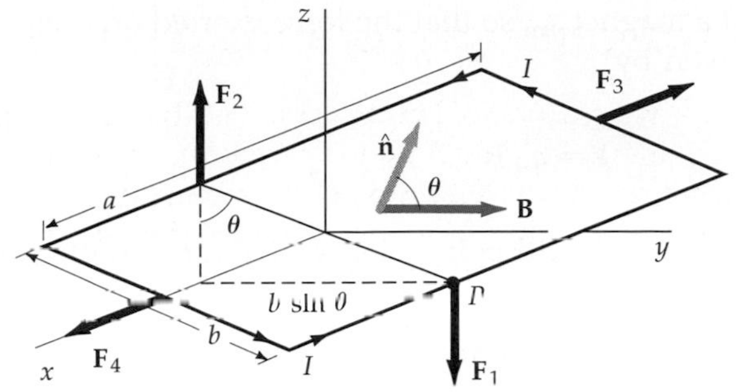

Figure 24-20 Rectangular current loop whose unit normal $\hat{\mathbf{n}}$ makes an angle θ with a uniform magnetic field **B**. The torque on the loop has magnitude $IAB \sin\theta$ and is in the direction such that $\hat{\mathbf{n}}$ tends to rotate into **B**. The torque can be written $\boldsymbol{\tau} = \mathbf{m} \times \mathbf{B}$, where $\mathbf{m} = IA\hat{\mathbf{n}}$ is the magnetic moment of the loop.

The torque can be written conveniently in terms of the **magnetic dipole moment m** (also referred to simply as the **magnetic moment**) of the current loop, which is defined as

$$\mathbf{m} = NIA\,\hat{\mathbf{n}} \tag{24-14}$$

Magnetic dipole moment of a current loop

The SI unit of magnetic moment is the ampere-meter² (A·m²). In terms of the magnetic dipole moment, the torque on the current loop is given by

$$\boldsymbol{\tau} = \mathbf{m} \times \mathbf{B} \tag{24-15}$$

Torque on a current loop

Equation 24-15, which we have derived for a rectangular loop, holds in general for a loop of any shape. The torque on any loop is the cross product of the magnetic moment **m** of the loop and the magnetic field **B**, where the magnetic moment is defined to be a vector that is perpendicular to the area of the loop (Figure 24-21) and has magnitude equal to NIA. Comparing Equation 24-15 with Equation 18-11 for the torque on an electric dipole, we see that a current loop in a magnetic field acts the same as does an electric dipole in an electric field.

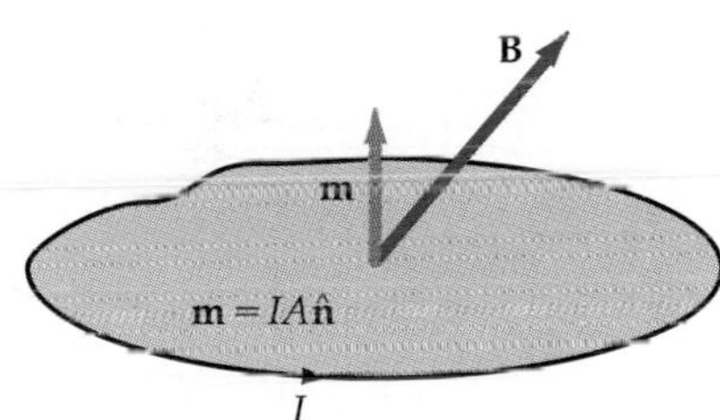

Figure 24-21 The magnetic moment of a current loop of arbitrary shape is $\mathbf{m} = IA\hat{\mathbf{n}}$. In a magnetic field **B**, the loop experiences a torque $\mathbf{m} \times \mathbf{B}$.

When a small permanent magnet such as a compass needle is placed in a magnetic field **B**, it tends to orient itself so that the north pole points in the direction of **B**. This effect also occurs with previously unmagnetized iron filings, which become magnetized in the presence of a **B** field. Figure 24-22 shows a small magnet that makes an angle θ with a magnetic field **B**. There is a force $\mathbf{F}_1$ on the north pole in the direction of **B** and an equal but opposite force $\mathbf{F}_2$ on the south pole. These two forces produce no translational motion because they are equal and opposite, but they do produce a torque that tends to rotate the magnet so that it lines up with the field. A small bar magnet thus behaves like a current loop. The fact that a small magnet behaves like a current loop is not accidental. The origin of the magnetic moment of a bar magnet is, in fact, microscopic current loops that result from the motion of electrons in the atoms of the magnet.

We can use the experimentally observed forces and torque on a bar magnet to define the pole strength and magnetic moment of the magnet. We

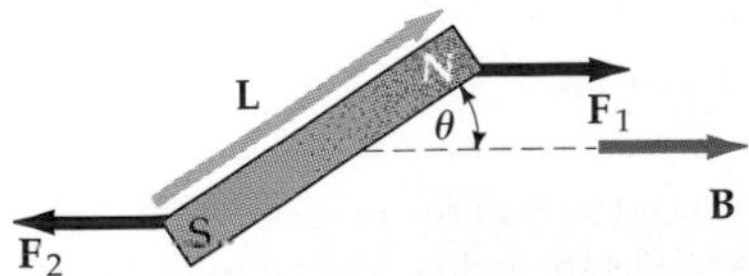

Figure 24-22 A small magnet in a uniform magnetic field experiences a torque that tends to rotate the magnet into the direction of the field. The magnetic moment of the magnet is in the direction of the vector **L** from the south pole to the north pole.

define the pole strength of a magnet q_m so that the force exerted on the pole in a magnetic field **B** is given by*

Magnetic pole strength defined

$$\mathbf{F} = q_m\mathbf{B} \tag{24-16}$$

The pole strength is positive for a north pole and negative for a south pole. The magnetic moment **m** of a magnet is then defined as

$$\mathbf{m} = |q_m|\mathbf{L} \tag{24-17}$$

where **L** is the vector from the south pole to the north pole. The torque exerted on a bar magnet in a magnetic field is then given by $\boldsymbol{\tau} = \mathbf{m} \times \mathbf{B}$, the same as Equation 24-15. Although we have defined the magnetic pole strength to be analogous to electric charge, we should remember that magnetic poles always come in pairs; that is, the fundamental unit of magnetism is the magnetic dipole. Experimentally, it is the magnetic dipole moment **m** of a magnet that is easily measured by placing the magnet in a magnetic field of known strength and measuring the torque. The magnetic pole strength is then found from Equation 24-17 by dividing the magnitude of the magnetic moment by the length of the magnet.

Example 24-7

A circular loop of radius 2 cm has 10 turns of wire and carries a current of 3 A. The axis of the loop makes an angle of 30° with a magnetic field of 8000 G. Find the torque on the loop.

The magnitude of the magnetic moment of the loop is

$$m = NIA = (10)(3\text{ A})\pi(0.02\text{ m})^2 = 3.77 \times 10^{-2}\text{ A}\cdot\text{m}^2$$

The magnitude of the torque is then

$$\tau = mB\sin\theta = (3.77 \times 10^{-2}\text{ A}\cdot\text{m}^2)(0.8\text{ T})(\sin 30°)$$
$$= 1.51 \times 10^{-2}\text{ N}\cdot\text{m}$$

where we have used 8000 G = 0.8 T and 1 T = 1 N/A·m.

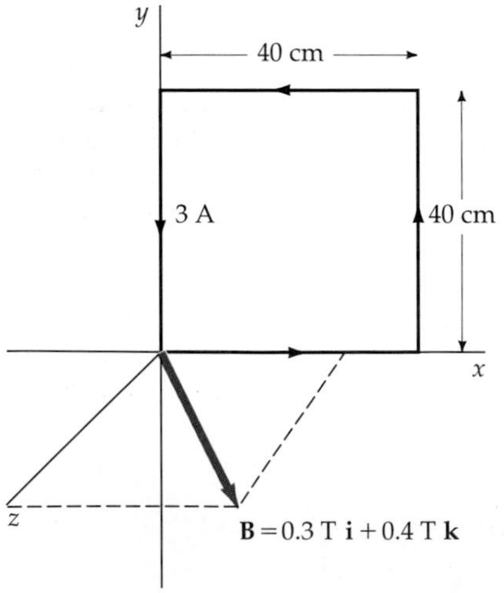

Figure 24-23 Square current loop in the xy plane in a magnetic field **B** = 0.3 T **i** + 0.4 T **k** for Example 24-8.

Example 24-8

A square 12-turn coil with sides of length 40 cm carries a current of 3 A. It lies in the xy plane as shown in Figure 24-23 in a uniform magnetic field **B** = 0.3 T **i** + 0.4 T **k**. Find (*a*) the magnetic moment of the coil and (*b*) the torque exerted on the coil. (*c*) What is the pole strength and orientation of a bar magnet of length 8 cm that has a magnetic moment equal to that of the coil?

(*a*) From Figure 24-23, we see that the magnetic moment of the loop is in the positive z direction. Its magnitude is $m = NIA = (12)(3\text{ A})(0.40\text{ m})^2 = 5.76\text{ A}\cdot\text{m}^2$. The magnetic moment is thus

$$\mathbf{m} = 5.76\text{ A}\cdot\text{m}^2\ \mathbf{k}$$

(*b*) The torque on the current loop is given by Equation 24-15:

$$\boldsymbol{\tau} = \mathbf{m} \times \mathbf{B} = (5.76\text{ A}\cdot\text{m}^2\ \mathbf{k}) \times (0.3\text{ T}\ \mathbf{i} + 0.4\text{ T}\ \mathbf{k}) = 1.73\text{ N}\cdot\text{m}\ \mathbf{j}$$

where we have used $\mathbf{k} \times \mathbf{k} = 0$ and $\mathbf{k} \times \mathbf{i} = \mathbf{j}$.

*The notation for magnetic pole strength q_m is used so that the magnetic equations resemble the corresponding equations for electric charges in electric fields. The subscript m reminds us that q_m denotes a magnetic pole and not an electric charge.

(*c*) A bar magnet with a magnetic moment in the positive z direction must lie along the z axis or be parallel to it with the vector **L** from south to north pole in the positive z direction. For $L = 8$ cm $= 0.08$ m and $m = 5.76$ A·m^2, the pole strength q_m is

$$q_m = \frac{m}{L} = \frac{5.76\ \text{A·m}^2}{0.08\ \text{m}} = 72\ \text{A·m} = 72\ \text{N/T}$$

Question

10. The magnetic moment of a current loop is antiparallel to a uniform magnetic field **B**. What is the torque on the loop? Is this equilibrium stable or unstable?

24-4 The Hall Effect

In Section 24-1, we calculated the force exerted by a magnetic field on a current-carrying wire. This force is transferred to the wire by the forces that bind the electrons to the wire at the surface. Since the charge carriers themselves experience the magnetic force when a current-carrying wire is in an external magnetic field, the carriers are accelerated toward one side of the wire. This results in a separation of charge in the wire called the **Hall effect.** This phenomenon allows us to determine the sign of the charge on the charge carriers and the number of charge carriers per unit volume n in a conductor. It also provides a convenient method for measuring magnetic fields.

Figure 24-24 shows two conducting strips each of which carries a current I to the right because the left sides of the strips are connected to the positive terminal of a battery, and the right sides are connected to the negative terminal. The strips are in a magnetic field that is directed into the paper. Let us assume for the moment that the current consists of positively charged particles moving to the right as shown in Figure 24-24*a*. The magnetic force on these particles is $q\mathbf{v}_d \times \mathbf{B}$ (where $\mathbf{v}_d$ is the drift velocity of the charge carriers). This force is directed upward. The positive particles therefore move up to the top of the strip, leaving the bottom of the strip with an excess negative charge. This separation of charge produces an electrostatic field in the strip that opposes the magnetic force on the charge carriers. When the electrostatic and magnetic forces balance, the charge carriers will no longer move upward. In this equilibrium situation, the upper part of the strip is positively charged, so it is at a greater potential than the negatively charged lower part. If the current consists of negatively charged particles, as shown in Figure 24-24*b*, the charge carriers must move to the left (since the current is still to the right). The magnetic force $q\mathbf{v}_d \times \mathbf{B}$ is again up because the signs of both q

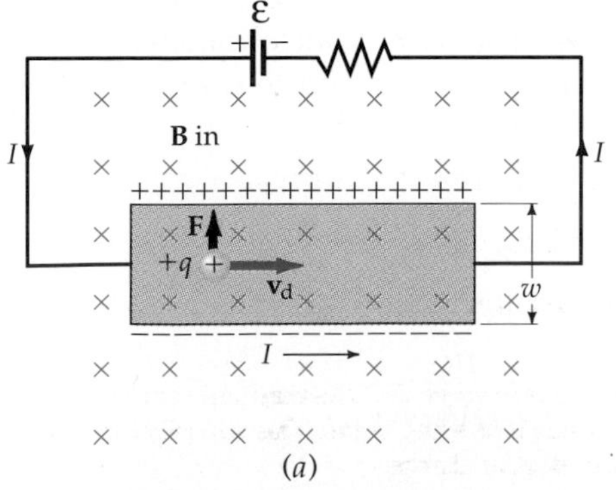

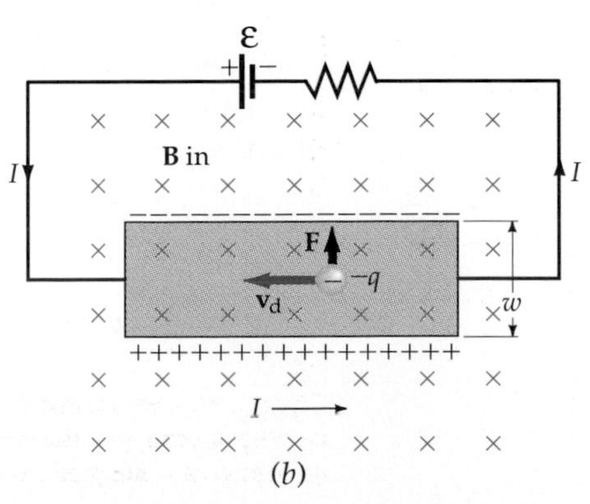

Figure 24-24 The Hall effect. The magnetic field is directed into the plane of the page as indicated by the crosses. The magnetic force on a charged particle is upward for a current to the right whether the current is due to (*a*) positive particles moving to the right or (*b*) negative particles moving to the left.

and $\mathbf{v}_d$ have been changed. Again the carriers are forced to the upper part of the strip, but the upper part of the strip now carries a negative charge (because the charge carriers are negative) and the lower part carries a positive charge.

A measurement of the sign of the potential difference between the upper and lower part of the strip tells us the sign of the charge carriers. For a normal metallic conductor, we find that the upper part of the strip in Figure 24-24 is at a lower potential than the lower part—which means that the upper part must carry a negative charge. It was this type of experiment that led to the discovery that the charge carriers in metallic conductors are negative. Thus Figure 24-24*b* is the correct illustration of the current in a normal conductor.

If we connect the upper and lower portions of the strip with a wire of resistance R, the negative electrons will flow from the upper part of the strip through the wire to the lower part. As electrons leave the upper part of the strip and enter the lower part, the magnitude of the charge separation across the strip is momentarily reduced. As a result, the electrostatic force on the electrons in the strip is momentarily weakened such that it no longer balances the magnetic force on them. The magnetic force will therefore drive more electrons toward the top of the strip, thereby maintaining the potential difference across the strip. The strip is thus a source of emf. The potential difference between the top and bottom of the strip is called the **Hall voltage.**

The magnitude of the Hall voltage is not hard to calculate. The magnitude of the magnetic force on the charge carriers in the strip is qv_dB. This magnetic force is balanced by the electrostatic force of magnitude qE, where E is the electric field due to the charge separation. Thus we have $E = v_dB$. If the width of the strip is w, the potential difference is Ew. The Hall voltage is therefore

$$V_H = Ew = v_dBw \qquad \text{24-18}$$

Exercise

A conducting strip of width $w = 2.0$ cm is placed in a magnetic field of 8000 G. Calculate the Hall voltage if the drift velocity is 4.0×10^{-5} m/s. (Answer: 0.64 μV)

Since the drift velocity for ordinary currents is very small, we can see from Equation 24-18 that the Hall voltage is very small for ordinary-sized strips and magnetic fields. From measurements of the Hall voltage for a strip of a given size carrying a known current in a known magnetic field, we can determine the number of charge carriers per unit volume in the strip. By Equation 22-3 the current is

$$I = nqv_dA$$

where A is the cross-sectional area of the strip. For a strip of width w and thickness t, the cross-sectional area is $A = wt$. Since the charge carriers are electrons, the quantity q is the charge on one electron e. The number density of charge carriers n is thus given by

$$n = \frac{I}{Aqv_d} = \frac{I}{wtev_d} \qquad \text{24-19}$$

Substituting $v_dw = V_H/B$ from Equation 24-18, we have

$$n = \frac{IB}{etV_H} \qquad \text{24-20}$$

Example 24-9

When a silver slab of thickness 1 mm and width 1.5 cm carries a current of 2.5 A in a region in which there is a magnetic field of magnitude 1.25 T perpendicular to the slab, the Hall voltage is measured to be 0.334 μV. (*a*) Calculate the number density of the charge carriers. (*b*) Compare your answer in part (*a*) to the number density of atoms in silver, which has mass density $\rho = 10.5$ g/cm^3 and molar mass $M =$ 107.9 g/mol.

(*a*) From Equation 24-20, we have

$$n = \frac{(2.5\text{ A})(1.25\text{ T})}{(1.6 \times 10^{-19}\text{ C})(0.001\text{ m})(3.34 \times 10^{-7}\text{ V})}$$

$$= 5.85 \times 10^{28}\text{ electrons/m}^3$$

(*b*) The number of atoms per unit volume n_a is given by

$$n_a = \frac{N_A\rho}{M} = \frac{(6.02 \times 10^{23}\text{ atoms/mol})(10.5\text{ g/cm}^3)}{107.9\text{ g/mol}}$$

$$= 5.86 \times 10^{22}\text{ atoms/cm}^3 = 5.86 \times 10^{28}\text{ atoms/m}^3$$

These results indicate that the number of charge carriers in silver is very nearly one per atom.

Although the Hall voltage is ordinarily very small, it provides a convenient method for measuring magnetic fields. If we rearrange Equation 24-20, we can write for the Hall voltage

$$V_H = \frac{I}{net}B \qquad 24\text{-}21$$

A given strip can be calibrated by measuring the Hall voltage for a given current in a known magnetic field. The strength of the magnetic field B of an unknown field can then be measured by placing the strip in the unknown field, sending a current through the strip, and measuring V_H.

The Quantum Hall Effect

According to Equation 24-21, the Hall voltage should increase linearly with magnetic field B for a given current in a given slab. In 1980, while studying the Hall effect in semiconductors at very low temperatures and very large magnetic fields, the German physicist Klaus von Klitzing discovered that a plot of V_H versus B resulted in a series of plateaus, as shown in Figure 24-25,

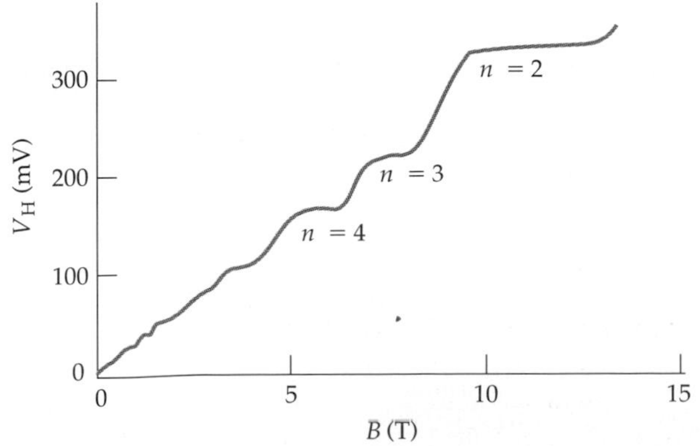

Figure 24-25 A plot of the Hall voltage versus applied magnetic field shows plateaus, indicating that the Hall voltage is quantized. These data were taken at a temperature of 1.39 K with the current I held fixed at 25.52 μA.

rather than a straight line. That is, the Hall voltage is quantized. For this discovery of the **quantum Hall effect,** von Klitzing won the Nobel Prize in Physics in 1985. According to the theory of the quantum Hall effect, the Hall resistance, defined as $R_H = V_H/I$, can take on only the values

$$R_H = \frac{V_H}{I} = \frac{R_K}{n} \qquad n = 1, 2, 3, \ldots \tag{24-22}$$

where n is an integer and R_K, called the **von Klitzing constant,** is related to the fundamental electronic charge e and Planck's constant h by

$$R_K = \frac{h}{e^2} = \frac{6.626 \times 10^{-34}\ \text{J·s}}{(1.602 \times 10^{-19}\ \text{C})^2} = 25\ 813\ \Omega \tag{24-23}$$

Because the von Klitzing constant can be measured to an accuracy of a few parts in 10^9, the quantum Hall effect is now used to define a standard of resistance. As of January 1990, the ohm is now defined so that R_K has the value 25 812.807 Ω exactly.

Recent experiments have shown that under certain special conditions the Hall resistance is given by Equation 24-22 with the integer n replaced by a rational fraction f. At present, the theory of this **fractional quantum Hall effect** is incomplete.

Summary

1. Moving charges interact with each other via the magnetic force. Since electric currents consist of moving charges, they also exert magnetic forces on each other. This force is described by saying that one moving charge or current creates a magnetic field that, in turn, exerts a force on the other moving charge or current. Ultimately, all magnetic fields are caused by charges in motion.

2. When a charge q moves with velocity $\mathbf{v}$ in a magnetic field $\mathbf{B}$, it experiences a force

$$\mathbf{F} = q\mathbf{v} \times \mathbf{B}$$

The force on a current element is given by

$$d\mathbf{F} = I\, d\boldsymbol{\ell} \times \mathbf{B}$$

The SI unit of magnetic field is the tesla (T). A commonly used unit is the gauss (G), which is related to the tesla by

$$1\ \text{T} = 10^4\ \text{G}$$

3. A particle of mass m and charge q moving with speed v in a plane perpendicular to a magnetic field moves in a circular orbit of radius r given by

$$r = \frac{mv}{qB}$$

The period and frequency of this circular motion are independent of the radius of the orbit or of the speed of the particle. The period, called the cyclotron period, is given by

$$T = \frac{2\pi m}{qB}$$

The frequency, called the cyclotron frequency, is given by

$$f = \frac{1}{T} = \frac{qB}{2\pi m}$$

4. A velocity selector consists of crossed electric and magnetic fields such that the electric and magnetic forces balance for a particle whose speed is given by $v = E/B$.

5. The mass-to-charge ratio of an ion of known speed can be determined by measuring the radius of the circular path taken by the ion in a known magnetic field in a mass spectrometer.

6. A current loop in a uniform magnetic field behaves as a magnetic dipole with magnetic moment **m** given by

$$\mathbf{m} = NIA\,\hat{\mathbf{n}}$$

where N is the number of turns, A is the area of the loop, I is the current, and $\hat{\mathbf{n}}$ is a unit vector perpendicular to the plane of the loop in the direction given by the right-hand rule. When a magnetic dipole is in a magnetic field, it experiences a torque given by

$$\boldsymbol{\tau} = \mathbf{m} \times \mathbf{B}$$

tending to align the magnetic moment of the current loop with the external field. The net force on a current loop in a uniform magnetic field is zero.

7. A bar magnet also experiences a torque in a magnetic field. The experimentally measured torque can be used to define the magnetic moment of the bar magnet from $\boldsymbol{\tau} = \mathbf{m} \times \mathbf{B}$. The pole strength of a bar magnet q_m can be defined by writing the force exerted on the pole as $\mathbf{F} = q_m\mathbf{B}$. The north magnetic pole has a positive pole strength and the south pole has a negative pole strength. In terms of the pole strength, the magnetic moment of a bar magnet is $\mathbf{m} = |q_m|\mathbf{L}$, where **L** is the vector from the south pole to the north pole.

8. When a conducting strip carrying a current is placed in a magnetic field, the magnetic force on the charge carriers causes a separation of charge called the Hall effect. This results in a voltage V_H, called the Hall voltage, which is given by

$$V_H = v_d Bw = \frac{I}{nqt}B$$

where v_d is the drift velocity, B is the magnetic field, w is the width of the strip, t is the thickness of the strip, n is the number density of charge carriers, and q is the charge of each carrier. The sign of the charge carriers can be determined from a measurement of the sign of the Hall voltage, and the number of carriers per unit volume can be determined from the magnitude of V_H. Measurements at very low temperatures and very large magnetic fields indicate that the Hall resistance $R_H = V_H/I$ is quantized and can take on values given by

$$R_H = \frac{V_H}{I} = \frac{R_K}{n}$$

where n is an integer and R_K is the von Klitzing constant, which has the value

$$R_K = \frac{h}{e^2} \approx 25\,813\ \Omega$$

Suggestions for Further Reading

Akasofu, Syun-Ichi: "The Dynamic Aurora," *Scientific American*, May 1989, p. 90.

How the "solar wind" of charged particles interacts with the earth's magnetic field to produce the aurora, why the aurora appears to move and change, and where its power comes from.

Nier, Alfred O. C.: "The Mass Spectrometer," *Scientific American*, March 1953, p. 68.

This device, which has made possible great advances in chemistry and other sciences, allows the determination of the composition of a substance based on the principle of the deflection of a beam of charged particles in a magnetic field.

Shamos, Morris H.: "The Electron—J. J. Thomson," in *Great Experiments in Physics*, Henry Holt and Co., New York, 1959. Reprinted by Dover, 1987.

Thomson's account of his discovery of the electron, with editorial annotations for clarity, and a biographical sketch.

Van Allen, James A.: "Interplanetary Particles and Fields," *Scientific American*, September 1975, p. 160.

This article, written by the man after whom the Van Allen radiation belts are named, describes the deflection of the "solar wind" in the earth's magnetic field.

Review

A. Objectives: After studying this chapter, you should:

1. Be able to calculate the magnetic force on a current element and on a moving charge in a given magnetic field.

2. Be able to calculate the magnetic dipole moment of a current loop and the torque exerted on a current loop in a magnetic field.

3. Be able to discuss the experiment in which Thomson measured q/m for electrons.

4. Be able to describe a velocity selector, a mass spectrometer, and a cyclotron.

5. Be able to discuss the Hall effect.

B. Define, explain, or otherwise identify:

Magnetic field
Tesla
Gauss
Current element
Magnetic-field lines
Cyclotron period
Cyclotron frequency
Magnetic bottle
Crossed fields
Velocity selector
Mass spectrometer
Cyclotron
Magnetic dipole moment
Hall effect
Hall voltage
Quantum Hall effect
von Klitzing constant

C. True or false: If the statement is true, explain why it is true. If it is false, give a counterexample.

1. The magnetic force on a moving charged particle is always perpendicular to the velocity of the particle.

2. The torque on a magnet tends to align the magnetic moment in the direction of the magnetic field.

3. A current loop in a uniform magnetic field behaves like a small magnet.

4. The period of a particle moving in a circle in a magnetic field is proportional to the radius of the circle.

5. The drift velocity of electrons in a wire can be determined from the Hall effect.

Problems

Level I

24-1 The Force Exerted by a Magnetic Field

1. Find the magnetic force on a proton moving with velocity 4.46 Mm/s in the positive x direction in a magnetic field of 1.75 T in the positive z direction.

2. A charge $q = -2.64$ nC moves with a velocity of 2.75×10^6 m/s **i**. Find the force on the charge if the magnetic field is (*a*) $\mathbf{B} = 0.48$ T **j**, (*b*) $\mathbf{B} = 0.65$ T **i** + 0.65 T **j**, (*c*) $\mathbf{B} = 0.75$ T **i**, (*d*) $\mathbf{B} = 0.65$ T **i** + 0.65 T **k**.

3. A uniform magnetic field of magnitude 1.28 T is in the positive z direction. Find the force on a proton if its velocity is (*a*) $\mathbf{v} = 3.5$ Mm/s **i**, (*b*) $\mathbf{v} = 2.5$ Mm/s **j**, (*c*) $\mathbf{v} = 6.5$ Mm/s **k**, and (*d*) $\mathbf{v} = 3.0$ Mm/s **i** + 4.0 Mm/s **j**.

4. An electron moves with a velocity of 3.75 Mm/s in the xy plane at an angle of 60° to the x axis and 30° to the y axis. A magnetic field of 0.85 T is in the positive y direction. Find the force on the electron.

5. A straight wire segment 2 m long makes an angle of 30° with a uniform magnetic field of 0.5 T. Find the magnitude of the force on the wire if it carries a current of 2 A.

6. A straight wire segment $I\boldsymbol{\ell} = (2.5\text{ A})(3\text{ cm }\mathbf{i} + 4\text{ cm }\mathbf{j})$ is in a uniform magnetic field $\mathbf{B} = 1.5$ T **i**. Find the force on the wire.

7. A long wire parallel to the x axis carries a current of 8.5 A in the positive x direction. There is a uniform magnetic field $\mathbf{B} = 1.65$ T **j**. Find the force per unit length on the wire.

24-2 Motion of a Point Charge in a Magnetic Field

8. A proton moves in a circular orbit of radius 65 cm perpendicular to a uniform magnetic field of magnitude 0.75 T. (*a*) What is the period for this motion? Find (*b*) the speed of the proton and (*c*) the kinetic energy of the proton.

9. An electron of kinetic energy 45 keV moves in a circular orbit perpendicular to a magnetic field of 0.325 T. (*a*) Find the radius of the orbit. (*b*) Find the frequency and the period of the motion.

10. An alpha particle (charge $+2e$) travels in a circular path of radius 0.5 m in a magnetic field of 1.0 T. Find the (*a*) period, (*b*) speed, and (*c*) kinetic energy (in electron volts) of the alpha particle. Take $m = 6.65 \times 10^{-27}$ kg for the mass of the alpha particle.

11. A beam of protons moves along the x axis in the positive x direction with a speed of 12.4 km/s through a region of crossed fields balanced for zero deflection. (*a*) If there is a magnetic field of magnitude 0.85 T in the positive y direction, find the magnitude and direction of the electric field. (*b*) Would electrons of the same velocity be deflected by these fields? If so, in what direction?

12. A velocity selector has a magnetic field of magnitude 0.28 T perpendicular to an electric field of magnitude 0.46 MV/m. (*a*) What must the speed of a particle be for it to pass through undeflected? What energy must (*b*) protons and (*c*) electrons have to pass through undeflected?

13. An electron from the sun with a speed of 1×10^8 m/s enters the earth's magnetic field high above the equator where the magnetic field is 4×10^{-7} T. The electron moves nearly in a circle except for a small drift along the direction of the earth's magnetic field that will take it toward the north pole. (*a*) What is the radius of the circular motion? (*b*) What is the radius of the circular motion near the north pole where the magnetic field is 2×10^{-5} T?

14. A singly ionized ^{24}Mg ion (mass 3.983×10^{-26} kg) is accelerated through a 2.5-kV potential difference and deflected in a magnetic field of 55.7 mT in a mass spectrometer. (*a*) Find the radius of curvature of the orbit for the ion. (*b*) What is the difference in radius for ^{26}Mg and ^{24}Mg ions? (Assume that their mass ratio is 26/24.)

15. A cyclotron for accelerating protons has a magnetic field of 1.4 T and a radius of 0.7 m. (*a*) What is the cyclotron frequency? (*b*) Find the maximum energy of the protons when they emerge. (*c*) How will your answers change if deuterons, which have the same charge but twice the mass, are used instead of protons?

16. A certain cyclotron with magnetic field 1.8 T is designed to accelerate protons to 25 MeV. (*a*) What is the cyclotron frequency? (*b*) What must the minimum radius of the magnet be to achieve a 25-MeV emergence energy? (*c*) If the alternating potential applied to the dees has a maximum value of 50 kV, how many revolutions must the protons make before emerging with an energy of 25 MeV?

24-3 Torques on Current Loops and Magnets

17. A small circular coil of 20 turns of wire lies on a uniform magnetic field of 0.5 T such that the normal to the plane of the coil makes an angle of 60° with the direction of **B**. The radius of the coil is 4 cm, and it carries a current of 3 A. (*a*) What is the magnitude of the magnetic moment of the coil? (*b*) What is the magnitude of the torque exerted on the coil?

18. The SI unit for the magnetic moment of a current loop is A·m^2. Use this to show that 1 T = 1 N/A·m.

19. What is the maximum torque on a 400-turn circular coil of radius 0.75 cm that carries a current of 1.6 mA and resides in a uniform magnetic field of 0.25 T?

20. The unit of magnetic pole strength, as defined by Equation 24-16, is a newton/tesla (N/T). Show that this unit is also equal to an ampere-meter (A·m).

21. A current-carrying wire is bent into the shape of a square of sides $L = 6$ cm and is placed in the xy plane. It carries a current $I = 2.5$ A. What is the torque on the wire if there is a uniform magnetic field of 0.3 T (*a*) in the z direction, (*b*) in the x direction?

22. Repeat Problem 21 if the wire is bent into an equilateral triangle of sides 8 cm.

23. A small bar magnet of length 8.5 cm and pole strength 25 N/T lies along the x axis in a uniform magnetic field $\mathbf{B} = 1.5\text{ T }\mathbf{i} + 2.5\text{ T }\mathbf{j} + 1.6\text{ T }\mathbf{k}$. (*a*) What is the magnetic moment of the magnet? (*b*) Find the torque exerted on the magnet.

24. A small magnet of length 6.8 cm is placed at an angle of 60° to the direction of a uniform magnetic field of magnitude 0.04 T. The observed torque has the magnitude of 0.10 N·m. (*a*) Find the magnetic moment of the magnet. (*b*) Find the pole strength q_m.

24-4 The Hall Effect

25. A metal strip 2.0 cm wide and 0.1 cm thick carries a current of 20 A in a uniform magnetic field of 2.0 T, as shown in Figure 24-26. The Hall emf is measured to be 4.27 μV. (*a*) Calculate the drift velocity of the electrons in the strip. (*b*) Find the number density of the charge carriers in the strip.

26. (*a*) In Figure 24-26, is point a or b at the higher potential? (*b*) If the metal strip is replaced by a p-type semiconductor in which the charge carriers are positive, which point will be at the higher potential?

Figure 24-26 Problems 25, 26, and 27.

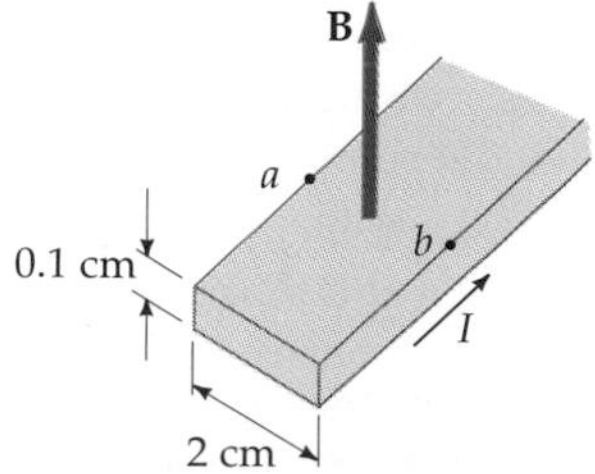

27. The number density of free electrons in copper is 8.47×10^{22} electrons per cubic centimeter. If the metal strip in Figure 24-26 is copper and the current is 10 A, find (*a*) the drift velocity v_d and (*b*) the Hall voltage. (Assume that the magnetic field is 2.0 T.)

28. A copper strip ($n = 8.47 \times 10^{22}$ electrons per cubic centimeter) 2 cm wide and 0.1 cm thick is used to measure the magnitudes of unknown magnetic fields which are perpendicular to the strip. Find the magnitude of B when $I = 20$ A and the Hall voltage is (*a*) 2.00 μV, (*b*) 5.25 μV, and (*c*) 8.00 μV.

29. Blood contains charged ions so that moving blood develops a Hall voltage across the diameter of an artery. A large artery with a diameter of 0.85 cm has a flow speed of 0.6 m/s. If a section of this artery is in a magnetic field of 0.2 T, what is the potential difference across the diameter of the artery?

Level II

30. A beam of ^{6}Li and ^{7}Li ions passes through a velocity selector and enters a magnetic spectrometer. If the diameter of the orbit of the ^{6}Li ions is 15 cm, what is the diameter of that for ^{7}Li ions?

31. The wire segment in Figure 24-27 carries a current of 1.8 A from a to b. There is a magnetic field $\mathbf{B} = 1.2$ T $\mathbf{k}$. Find the total force on the wire and show that it is the same as if the wire were a straight segment from a to b.

Figure 24-27 Problem 31.

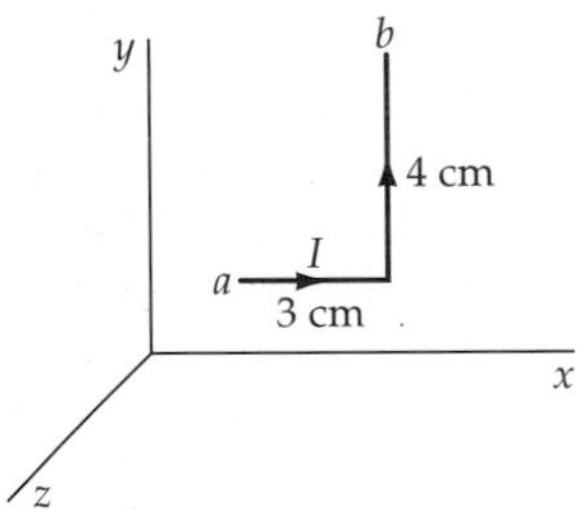

32. A straight, stiff, horizontal wire of length 25 cm and mass 50 g is connected to a source of emf by light, flexible leads. A magnetic field of 1.33 T is horizontal and perpendicular to the wire. Find the current necessary to float the wire, that is, the current such that the magnetic force balances the weight of the wire.

33. The plates of a Thomson q/m apparatus are 6.0 cm long and are separated by 1.2 cm. The end of the plates is 30.0 cm from the tube screen. The kinetic energy of the electrons is 2.8 keV. (*a*) If a potential of 25.0 V is applied across the deflection plates, by how much will the beam deflect? (*b*) Find the magnitude of the crossed magnetic field that will allow the beam to pass through undeflected.

34. A simple gaussmeter for measuring horizontal magnetic fields consists of a stiff 50-cm wire that hangs from a conducting pivot so that its free end makes contact with a pool of mercury in a dish below. The wire has a mass of 5 g and conducts a current downward. (*a*) What is the equilibrium angular displacement of the wire from vertical if the horizontal magnetic field is 0.04 T and the current is 0.20 A? (*b*) If the current is 20 A and a displacement from vertical of 0.5 mm can be detected for the free end, what is the horizontal magnetic field sensitivity of this gaussmeter?

35. A rectangular, 50-turn coil has sides 6.0 and 8.0 cm long and carries a current of 1.75 A. It is oriented as shown in Figure 24-28 and pivoted about the z axis. (*a*) If the wire in the xy plane makes an angle of 37° with the y axis as shown, what angle does the unit normal $\hat{\mathbf{n}}$ make with the x axis? (*b*) Write an expression for $\hat{\mathbf{n}}$ in terms of the unit vectors $\mathbf{i}$ and $\mathbf{j}$. (*c*) What is the magnetic moment of the coil? (*d*) Find the torque on the coil when there is a uniform magnetic field $\mathbf{B} = 1.5$ T $\mathbf{j}$.

Figure 24-28 Problems 35 and 36.

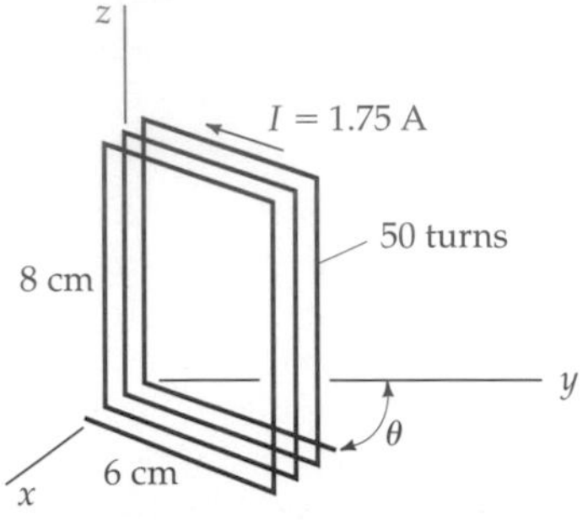

36. The coil in Problem 35 is pivoted about the z axis and held at various positions in a uniform magnetic field $\mathbf{B} = 2.0$ T $\mathbf{j}$. Sketch the position of the coil and find the torque exerted when the unit normal is (*a*) $\hat{\mathbf{n}} = \mathbf{i}$, (*b*) $\hat{\mathbf{n}} = \mathbf{j}$, (*c*) $\hat{\mathbf{n}} = -\mathbf{j}$, and (*d*) $\hat{\mathbf{n}} = (\mathbf{i} + \mathbf{j})/\sqrt{2}$.

37. A particle of charge q and mass M moves in a circle of radius r and with angular velocity ω. (*a*) Show that the average current is $I = q\omega/2\pi$ and that the magnetic moment has the magnitude $m = \frac{1}{2}q\omega r^2$. (*b*) Show that the angular momentum of this particle has the magnitude $L = Mr^2\omega$ and that the magnetic moment and angular momentum vectors are related by $\mathbf{m} = (q/2M)\mathbf{L}$.

38. A particle of charge q and mass m has momentum $p = mv$ and kinetic energy $K = \frac{1}{2}mv^2 = p^2/2m$. If it moves in a circular orbit of radius r perpendicular to a uniform magnetic field B, show that (*a*) $p = Bqr$ and (*b*) $K = B^2q^2r^2/2m$.

39. Protons, deuterons (each with charge $+e$), and alpha particles (with charge $+2e$) of the same kinetic energy enter a uniform magnetic field $\mathbf{B}$ that is perpendicular to their velocities. Let r_p, r_d, and r_α be the radii of their circular orbits. Find the ratios r_d/r_p and r_α/r_p. Assume that $m_\alpha = 2m_d = 4m_p$.

40. Show that the cyclotron frequencies of deuterons and alpha particles are the same and are half that of a proton in the same magnetic field. (See Problem 39.)

41. A proton and an alpha particle move in a uniform magnetic field in circles of the same radii. Compare (*a*) their velocities, (*b*) their kinetic energies, and (*c*) their angular momenta. (See Problem 39.)

42. Beryllium has a density of 1.83 g/cm^3 and a molar mass of 9.01 g/mol. A slab of beryllium of thickness 1.4 mm and width 1.2 cm carries a current of 3.75 A in a region in which there is a magnetic field of magnitude 1.88 T perpendicular to the slab. The Hall voltage is measured to be 0.130 μV. (*a*) Calculate the number density of the charge carriers. (*b*) Calculate the number density of atoms in beryllium. (*c*) How many free electrons are there per atom of beryllium?

43. A rigid, circular loop of radius R and mass M lies in the xy plane on a rough, flat table. The magnetic field is $\mathbf{B} = B_x\,\mathbf{i} + B_y\,\mathbf{j}$. How large must the current I be before one edge of the loop will lift off the table?

44. A moving-coil galvanometer consists of a coil of wire suspended in a radial magnetic field by a thin, highly flexible fiber. As a current I is passed through the coil, a torque is produced that tends to rotate the coil. The fiber, in turn, supplies a restoring torque $\tau = k\theta$ that is proportional to the angle θ through which it has been twisted. The constant k is called the torsion constant. Show that $I = k\theta/(NAB)$, where N is the number of turns of wire in the coil, A is the area of the coil, and B is the magnetic field.

45. A wire of length L is wound into a circular coil of N loops. Show that when this coil carries a current I, its magnetic moment has the magnitude $IL^2/4\pi N$.

46. A metal disk of radius 6 cm is mounted on a frictionless axle. Current can flow through the axle out along the disk to a sliding contact at the rim of the disk. A uniform magnetic field $B = 1.25$ T is parallel to the axis of the disk. When the current is 3 A, the disk rotates with constant angular velocity. What is the frictional force at the rim between the stationary electrical contact and the rotating rim?

47. A particle of mass m and charge q enters a region where there is a uniform magnetic field B along the x axis. The initial velocity of the particle is $\mathbf{v} = v_{0x}\,\mathbf{i} + v_{0y}\,\mathbf{j}$, so the particle moves in a helix. (*a*) Show that the radius of the helix is $r = mv_{0y}/qB$. (*b*) Show that the particle takes a time $t = 2\pi m/qB$ to make one orbit around the helix.

48. Before entering a mass spectrometer, ions pass through a velocity selector consisting of parallel plates separated by 2.0 mm and having a potential difference of 160 V. The magnetic field between the plates is 0.42 T. The magnetic field in the mass spectrometer is 1.2 T. Find (*a*) the speed of the ions entering the mass spectrometer and (*b*) the difference in the diameters of the orbits of singly ionized ^{238}U and ^{235}U. (The mass of a ^{235}U ion is 3.903×10^{-25} kg.)

49. A conducting wire is parallel to the y axis. It moves in the positive x direction with a speed of 20 m/s in a magnetic field $\mathbf{B} = 0.5\text{ T}\,\mathbf{k}$. (*a*) What are the magnitude and direction of the magnetic force on an electron in the conductor? (*b*) Because of this magnetic force, electrons move to one end of the wire leaving the other end positively charged, until the electric field due to this charge separation exerts a force on the electrons that balances the magnetic force. Find the magnitude and direction of this electric field in the steady state. (*c*) Suppose the moving wire is 2 meters long. What is the potential difference between its two ends due to this electric field?

50. A metal crossbar of mass M rides on a pair of long, horizontal conducting rails separated by a distance L and connected to a device that supplies constant current I to the circuit, as shown in Figure 24-29. A uniform magnetic field B is established as shown. (*a*) If there is no friction and the bar starts from rest at $t = 0$, show that at time t the bar has velocity $v = (BIL/M)t$. (*b*) In which direction will the bar move? (*c*) If the coefficient of static friction is μ_s, find the minimum field B necessary to start the bar moving.

Figure 24-29 Problems 50 and 51.

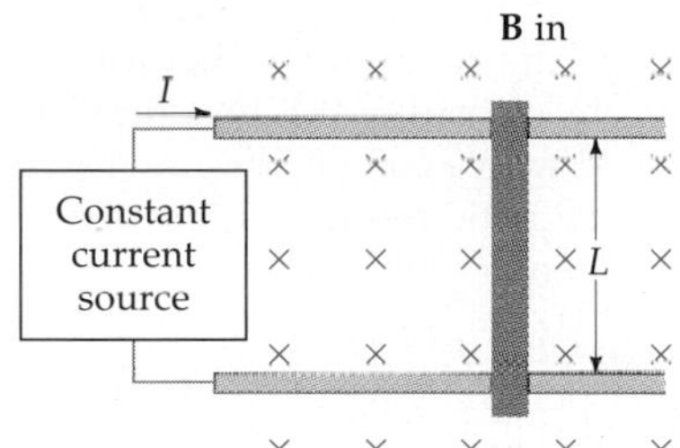

51. Assume that the rails in Figure 24-29 are frictionless but tilted upward so that they make an angle θ with the horizontal. (*a*) What vertical magnetic field B is needed to keep the bar from sliding down the rails? (*b*) What is the acceleration of the bar if B has twice the value found in part (*a*)?

Level III

52. A stiff, straight, horizontal wire of length 25 cm and mass 20 g is supported by electrical contacts at its ends, but is otherwise free to move vertically upward. The wire is in a uniform, horizontal magnetic field of magnitude 0.4 T perpendicular to the wire. A switch connecting the wire to a battery is closed and the wire is shot upward, rising to a maximum height h. The battery delivers a total charge of 2 C during the short time it makes contact with the wire. Find the height h.

53. A circular loop of wire with mass M carries a current I in a uniform magnetic field. It is initially in equilibrium with its magnetic moment vector aligned with the magnetic field. The loop is given a small twist about a diameter and then released. What is the period of the motion? (Assume that the only torque exerted on the loop is due to the magnetic field.)

54. A current-carrying wire is bent into a semicircular loop of radius R, which lies in the xy plane. There is a uniform magnetic field $\mathbf{B} = B\,\mathbf{k}$ perpendicular to the plane of the loop (Figure 24-30). Show that the force acting on the loop is $\mathbf{F} = 2IRB\,\mathbf{j}$.

Figure 24-30 Problem 54.

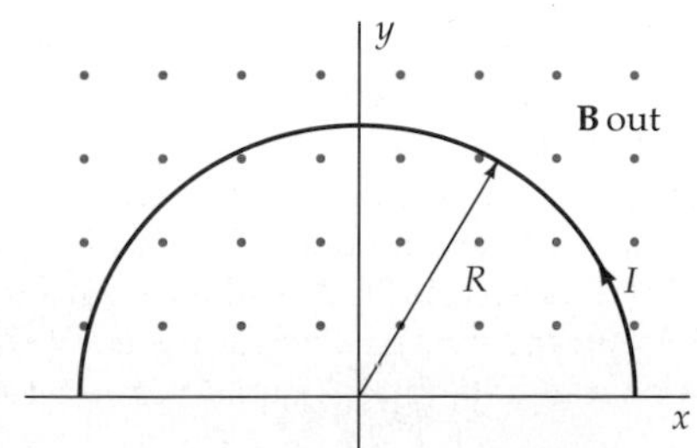

55. Show that the radius of the orbit of a charged particle in a cyclotron is proportional to the square root of the number of orbits completed.

56. A wire bent in some arbitrary shape carries a current I in a uniform magnetic field $\mathbf{B}$. Show that the total force on the part of the wire from some point a to some point b is $\mathbf{F} = I\mathbf{L} \times \mathbf{B}$, where $\mathbf{L}$ is the vector from a to b.

57. If you make a coil of N turns from a wire of fixed length L, the larger the number of turns, the smaller the area enclosed by the wire. Show that for a wire of given length carrying a current I, the maximum magnetic moment is achieved with a coil of just one turn and the magnitude of the magnetic moment is $IL^2/4\pi$. (You need to consider only circular coils. Why?)

58. A nonconducting rod of mass M and length ℓ has a uniform charge per unit length λ and rotates with angular velocity ω about an axis through one end and perpendicular to the rod. (*a*) Consider a small segment of the rod of length dx and charge $dq = \lambda\, dx$ at a distance x from the pivot. Show that the magnetic moment of this segment is $\frac{1}{2}\lambda\omega x^2\, dx$. (*b*) Integrate your result to show that the total magnetic moment of the rod is $m = \frac{1}{6}\lambda\omega\ell^3$. (*c*) Show that the magnetic moment $\mathbf{m}$ and angular momentum $\mathbf{L}$ are related by $\mathbf{m} = (Q/2M)\mathbf{L}$, where Q is the total charge on the rod.

59. A nonconducting disk of mass M and radius R has a surface charge density σ and rotates with angular velocity ω about its axis. (*a*) Consider a ring of radius r and thickness dr. Show that the total current in this ring is $dI = (\omega/2\pi)\, dq = \omega\sigma r\, dr$. (*b*) Show that the magnetic moment of the ring is $dm = \pi\omega\sigma r^3\, dr$. (*c*) Integrate your result for part (*b*) to show that the total magnetic moment of the disk is $m = \frac{1}{4}\pi\omega\sigma R^4$. (*d*) Show that the magnetic moment $\mathbf{m}$ and angular momentum $\mathbf{L}$ are related by $\mathbf{m} = (Q/2M)\mathbf{L}$, where Q is the total charge on the disk.

60. A small magnet of moment $\mathbf{m}$ makes an angle θ with a uniform magnetic field $\mathbf{B}$. (*a*) How much work must be done by an external torque to twist the magnet by a small amount $d\theta$? (*b*) Show that the work required to rotate the magnet until it is perpendicular to the field is $W = mB \cos \theta$. (*c*) Use your result for part (*b*) to show that if the potential energy of the magnet is chosen to be zero when the magnet is perpendicular to the field, the potential energy at angle θ is $U(\theta) = -\mathbf{m}\cdot\mathbf{B}$. (*d*) Would any part of this problem be different if the magnet were replaced by a current-carrying coil with magnetic moment $\mathbf{m}$?

61. A beam of particles with velocity $\mathbf{v}$ enters a region of uniform magnetic field $\mathbf{B}$ that makes a small angle θ with $\mathbf{v}$. Show that after a particle moves a distance $2\pi(m/qB)v \cos \theta$ measured along the direction of $\mathbf{B}$ the velocity of the particle is in the same direction as it was when it entered the field.

62. A small bar magnet has a magnetic moment $\mathbf{m}$ that makes an angle θ with the x axis and lies in a nonuniform magnetic field given by $\mathbf{B} = B_x(x)\,\mathbf{i} + B_y(y)\,\mathbf{j}$. Show that there is a net force on the magnet that is given approximately by

$$\mathbf{F} \approx m_x \frac{\partial B_x}{\partial x}\mathbf{i} + m_y \frac{\partial B_y}{\partial y}\mathbf{j} \qquad \text{24-24}$$

Chapter 25

Sources of the Magnetic Field

These coils at the Kettering Magnetics Laboratory at Oakland University are called Helmholtz coils. They are used to cancel the earth's magnetic field and to provide a uniform magnetic field in a small region of space for studying the magnetic properties of matter.

We now turn to a consideration of the origins of the magnetic field **B**. The earliest known sources of magnetism were permanent magnets. One month after Oersted announced his discovery that a compass needle is deflected by an electric current, Jean Baptiste Biot and Felix Savart announced the results of their measurements of the force on a magnet near a long, current-carrying wire and analyzed these results in terms of the magnetic field produced by each element of the current. André-Marie Ampère extended these experiments and showed that current elements also experience a force in the presence of a magnetic field and that two currents exert forces on each other.

We begin by considering the magnetic field produced by a single moving charge and by the charges in a current element. We then calculate the magnetic fields produced by some common current configurations such as a straight wire segment, a long straight wire, a current loop, and a solenoid. Finally, we discuss Ampère's law, which relates the line integral of the magnetic field around a closed loop to the total current that passes through the loop.

25-1 The Magnetic Field of Moving Point Charges

When a point charge q moves with velocity $\mathbf{v}$, it produces a magnetic field $\mathbf{B}$ in space given by

Magnetic field of a moving charge

$$\mathbf{B} = \frac{\mu_0}{4\pi} \frac{q\mathbf{v} \times \hat{\mathbf{r}}}{r^2} \qquad 25\text{-}1$$

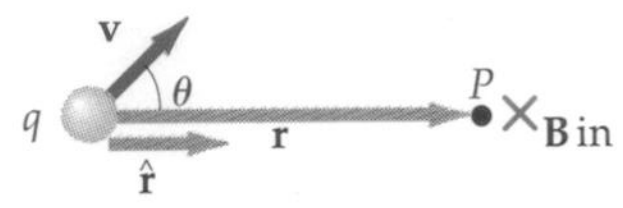

where $\hat{\mathbf{r}}$ is a unit vector that points from the charge q to the field point P (Figure 25-1) and μ_0 is a constant of proportionality called the **permeability of free space,** which has the value

$$\mu_0 = 4\pi \times 10^{-7}\ \text{T}\cdot\text{m/A} = 4\pi \times 10^{-7}\ \text{N/A}^2 \qquad 25\text{-}2$$

Figure 25-1 A point charge q moving with velocity $\mathbf{v}$ produces a magnetic field $\mathbf{B}$ at a field point P that is in the direction $\mathbf{v} \times \hat{\mathbf{r}}$, where $\hat{\mathbf{r}}$ is the unit vector pointing from the charge to the field point. (The blue x at the field point indicates that the direction of the field is into the page.)

The units of μ_0 are such that B is in teslas when q is in coulombs, v is in meters per second, and r is in meters. The unit N/A^2 comes from the fact that $1\ \text{T} = 1\ \text{N/A}\cdot\text{m}$. The constant $1/4\pi$ is arbitrarily included in Equation 25-1 so that the factor 4π will not appear in Ampere's law (Equation 25-15), which we will study in Section 25-4. Equation 25-1 for the magnetic field due to a moving point charge is analogous to Coulomb's law for the electric field due to a point charge:

$$\mathbf{E} = \frac{1}{4\pi\epsilon_0} \frac{kq}{r^2}\hat{\mathbf{r}}$$

We see from Equation 25-1 that the magnetic field of a moving point charge has the following characteristics:

1. The magnitude of $\mathbf{B}$ is proportional to the charge q and to the speed v and varies inversely with the square of the distance from the charge.
2. The magnetic field is zero along the line of motion of the charge. At other points in space, it is proportional to $\sin\theta$, where θ is the angle between the velocity $\mathbf{v}$ and the vector $\mathbf{r}$ from the charge to the field point.
3. The direction of $\mathbf{B}$ is perpendicular to both the velocity $\mathbf{v}$ and the vector $\mathbf{r}$. It is in the direction given by the right-hand rule as $\mathbf{v}$ is rotated into $\mathbf{r}$.

Example 25-1

A point charge of magnitude $q_1 = 4.5$ nC is moving with speed 3.6×10^7 m/s parallel to the x axis along the line $y = 3$ m. Find the magnetic field produced by this charge at the origin when the charge is at the point $x = -4$ m, $y = 3$ m, as shown in Figure 25-2.

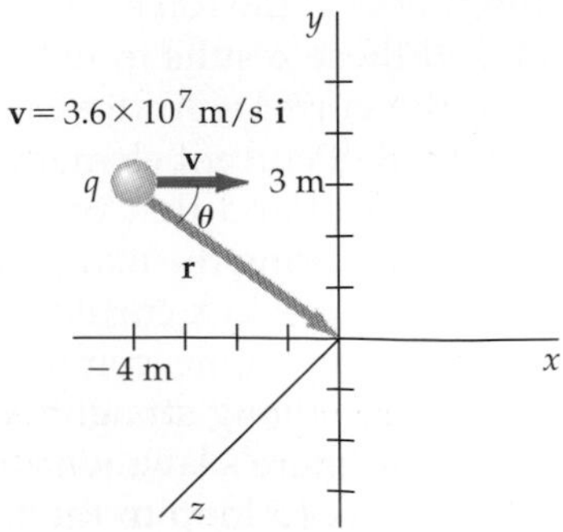

Figure 25-2 Charged particle moving parallel to the x axis for Example 25-1.

The velocity of the charge is $\mathbf{v} = v\mathbf{i} = 3.6 \times 10^7$ m/s $\mathbf{i}$, and the vector from the charge to the origin is given by $\mathbf{r} = 4$ m $\mathbf{i} - 3$ m $\mathbf{j}$. Then $r = 5$ m and the unit vector $\hat{\mathbf{r}}$ is

$$\hat{\mathbf{r}} = \frac{\mathbf{r}}{r} = \frac{4 \text{ m } \mathbf{i} - 3 \text{ m } \mathbf{j}}{5 \text{ m}}$$

$$= 0.8\mathbf{i} - 0.6\mathbf{j}$$

Then

$$\mathbf{v} \times \hat{\mathbf{r}} = (v\mathbf{i}) \times (0.8\mathbf{i} - 0.6\mathbf{j})$$

$$= -\,0.6\, v\mathbf{k}$$

and Equation 25-1 gives

$$\mathbf{B} = \frac{\mu_0}{4\pi} \frac{q\mathbf{v} \times \hat{\mathbf{r}}}{r^2}$$

$$= \frac{\mu_0}{4\pi} \frac{q(-0.6v\mathbf{k})}{r^2}$$

$$= -(10^{-7}\text{ T·m/A}) \frac{(4.5 \times 10^{-9}\text{ C})(0.6)(3.6 \times 10^7\text{ m/s})}{(5\text{ m})^2}\mathbf{k}$$

$$= -3.89 \times 10^{-10}\text{ T }\mathbf{k}$$

We can also find the magnetic field without explicitly finding the unit vector $\hat{\mathbf{r}}$. From the figure we see that $\mathbf{v} \times \hat{\mathbf{r}}$ is in the negative z direction. Since the magnitude of $\mathbf{v} \times \hat{\mathbf{r}}$ is $v \sin \theta$, where $\sin \theta = 3$ m/5 m $= 0.6$, we have

$$\mathbf{v} \times \hat{\mathbf{r}} = v \sin \theta\, (-\mathbf{k}) = -v(0.6)\mathbf{k}$$

which is the same as we found above. We note from this example that the magnetic field due to a moving charge is very small. For comparison, the magnitude of the magnetic field of the earth near its surface is about 10^{-4} T.

The Magnetic Force and Conservation of Momentum

The magnetic force exerted by one moving charge on another is found by combining Equation 24-1 for the force on a moving charge in a magnetic field and Equation 25-1 for the magnetic field of a charge. The force $\mathbf{F}_{12}$ exerted by a charge q_1 moving with velocity $\mathbf{v}_1$ on a charge q_2 moving with velocity $\mathbf{v}_2$ is given by

$$\mathbf{F}_{12} = q_2\mathbf{v}_2 \times \mathbf{B}_1 = q_2\mathbf{v}_2 \times \left(\frac{\mu_0}{4\pi} \frac{q_1\mathbf{v}_1 \times \hat{\mathbf{r}}_{12}}{r_{12}^2}\right) \qquad \text{25-3a}$$

where $\mathbf{B}_1$ is the magnetic field at the position of charge q_2 due to charge q_1, and $\hat{\mathbf{r}}_{12}$ is the unit vector pointing from q_1 to q_2. Similarly, the force $\mathbf{F}_{21}$ exerted by a charge q_2 moving with velocity $\mathbf{v}_2$ on a charge q_1 moving with velocity $\mathbf{v}_1$ is given by

$$\mathbf{F}_{21} = q_1\mathbf{v}_1 \times \mathbf{B}_2 = q_1\mathbf{v}_1 \times \left(\frac{\mu_0}{4\pi} \frac{q_2\mathbf{v}_2 \times \hat{\mathbf{r}}_{21}}{r_{21}^2}\right) \qquad \text{25-3b}$$

These relations are remarkable in that the force exerted by charge q_1 on charge q_2 is generally not equal and opposite to that exerted by charge q_2 on charge q_1. That is, these forces do not obey Newton's third law, as can be

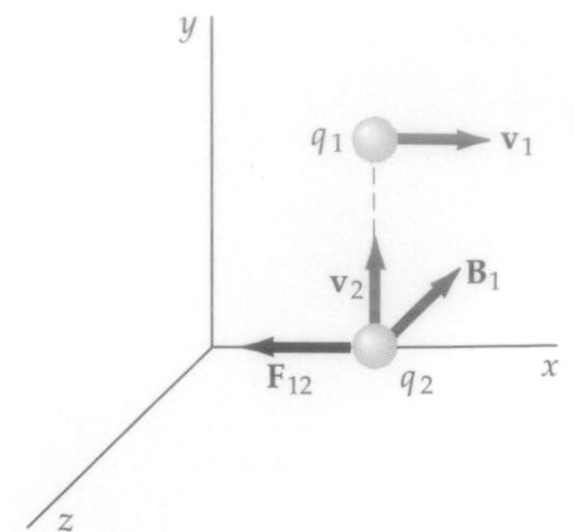

Figure 25-3 The forces exerted by moving charges on each other are not equal and opposite. The magnetic field $\mathbf{B}_1$ at charge q_2 due to charge q_1 is in the negative z direction, so it exerts a force $\mathbf{F}_{12}$ on q_2 to the left in the negative x direction. However, $\mathbf{B}_2$ at charge q_1 due to charge q_2 is zero, so there is no force on q_1.

demonstrated by considering the special case illustrated in Figure 25-3. Here, the magnetic field $\mathbf{B}_1$ due to charge q_1 at charge q_2 is in the negative z direction, and the force on q_2 is to the left in the negative x direction. However, the magnetic field $\mathbf{B}_2$ due to q_2 at q_1 is zero because q_1 lies along the line of motion of q_2. Thus, there is no magnetic force exerted by q_2 on q_1. There is thus a net force $\mathbf{F}_{12}$ acting on the two-charge system. The system will accelerate in the direction of this force, and linear momentum will not be conserved. This apparent violation of the law of conservation of linear momentum results from our treating the force exerted by one charge on another as an action-at-a-distance force and neglecting the momentum carried by the electric and magnetic fields of the moving charges. We saw in Chapter 21 that there is energy associated with an electric field, and we will see later that there is also energy associated with a magnetic field. Advanced treatments of the electric and magnetic fields of moving charges show that there is also momentum associated with these fields. When the charges move, as in Figure 25-3, the linear momentum produced when the system accelerates to the left is balanced by momentum in the opposite direction carried by the fields. Thus, when we include the momentum of the fields, the total momentum of the system is conserved.

Example 25-2

A point charge q_1 is at the point $\mathbf{R} = x\mathbf{i} + y\mathbf{j}$ and is moving parallel to the x axis with velocity $\mathbf{v}_1 = v_1\mathbf{i}$. A second point charge q_2 is at the origin and moving along the x axis with velocity $\mathbf{v}_2 = v_2\mathbf{i}$, as shown in Figure 25-4*a*. Find the magnetic force exerted by each charge on the other.

We first find the force exerted on charge q_1. We note that the vector $\mathbf{r}_{21}$ from charge q_2 to charge q_1 is just $\mathbf{R}$. Using $\hat{\mathbf{r}}_{21} = \mathbf{r}_{21}/r_{21} = \mathbf{R}/R$, we have

$$\frac{\mathbf{v}_2 \times \hat{\mathbf{r}}_{21}}{r_{21}^2} = \frac{\mathbf{v}_2 \times \mathbf{R}}{R^3} = \frac{v_2\mathbf{i} \times (x\mathbf{i} + y\mathbf{j})}{R^3} = \frac{yv_2}{R^3}\mathbf{k}$$

so the magnetic field produced by charge q_2 at the position of charge q_1 is

$$\mathbf{B}_2 = \frac{\mu_0}{4\pi}\frac{q_2 y v_2}{R^3}\mathbf{k}$$

The magnetic force exerted by charge q_2 on charge q_1 is then

$$\mathbf{F}_{21} = q_1\mathbf{v}_1 \times \mathbf{B}_2 = q_1(v_1\mathbf{i}) \times \left(\frac{\mu_0}{4\pi}\frac{q_2 y v_2}{R^3}\mathbf{k}\right)$$

$$= -\frac{\mu_0}{4\pi}\frac{q_1 q_2 v_1 v_2 y}{R^3}\mathbf{j}$$

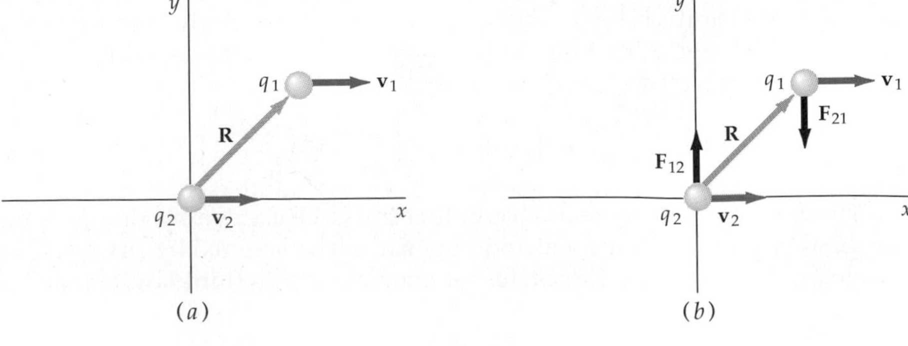

Figure 25-4 (*a*) Two charges moving in parallel directions for Example 25-2. (*b*) The magnetic forces exerted by the charges on each other are equal and opposite, but they are not along the line joining the charges.

To find the magnetic force exerted by charge q_1 on charge q_2, we note that the vector $\mathbf{r}_{12}$ from charge q_1 to charge q_2 is $-\mathbf{R}$. Then,

$$\frac{\mathbf{v}_1 \times \hat{\mathbf{r}}_{12}}{r_{12}^2} = \frac{\mathbf{v}_1 \times (-\mathbf{R})}{R^3} = \frac{v_1\mathbf{i} \times (-x\mathbf{i} - y\mathbf{j})}{R^3} = -\frac{yv_1}{R^3}\mathbf{k}$$

The magnetic force exerted by charge q_1 on charge q_2 is then

$$\mathbf{F}_{12} = q_2\mathbf{v}_2 \times \mathbf{B}_1 = q_2(v_2\mathbf{i}) \times \left(-\frac{\mu_0}{4\pi}\frac{q_1yv_1}{R^3}\mathbf{k}\right) = +\frac{\mu_0}{4\pi}\frac{q_1q_2v_1v_2y}{R^3}\mathbf{j}$$

In this case, the forces are equal and opposite as shown in Figure 25-4*b*, but they are not along the line joining the two particles. The magnetic forces thus exert a torque on the two-particle system. Here, the apparent lack of conservation of *angular* momentum implied by the existence of this torque is resolved by the consideration of the *angular* momentum carried by the electromagnetic field.

25-2 The Magnetic Field of Currents: The Biot–Savart Law

As we did in the previous chapter when we found the force exerted on charges and current elements, we can find the magnetic field $d\mathbf{B}$ produced by a current element $I\,d\boldsymbol{\ell}$ by replacing $q\mathbf{v}$ in Equation 25-1 with the current element $I\,d\boldsymbol{\ell}$. We then have

$$d\mathbf{B} = \frac{\mu_0}{4\pi}\frac{I\,d\boldsymbol{\ell} \times \hat{\mathbf{r}}}{r^2} \qquad 25\text{-}4$$

Biot–Savart law

Equation 25-4, known as the **Biot–Savart law,** was also deduced by Ampère. The Biot–Savart law, like Equation 25-1, is analogous to Coulomb's law for the electric field of a point charge. The source of the magnetic field is a moving charge $q\mathbf{v}$ or a current element $I\,d\boldsymbol{\ell}$, just as the charge q is the source of the electrostatic field. The magnetic field decreases with the square of the distance from the moving charge or current element, just as the electric field decreases with the square of the distance from a point charge. However, the directional aspects of the electric and magnetic fields are quite different.

(*a*)

(*b*)

Oersted's experiment. (*a*) With no current in the wire, the compass needle points north. (*b*) When the wire carries a current, the needle is deflected in the direction of the resultant magnetic field. The current in the wire is directed upwards from left to right. The insulation has been stripped from the wire to improve the contrast of the photograph.

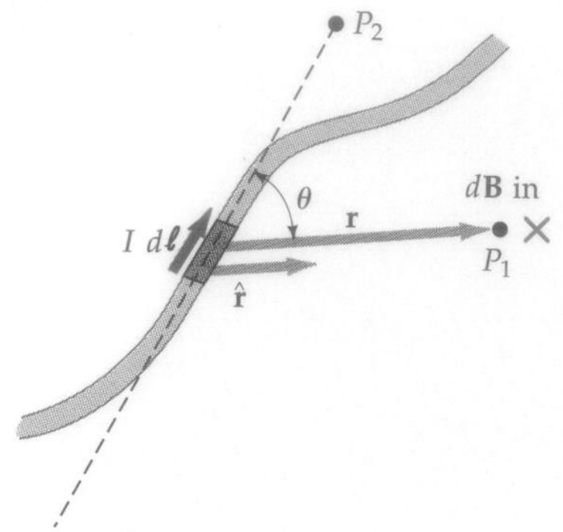

Figure 25-5 The current element $I\,d\boldsymbol{\ell}$ produces a magnetic field at point P_1 that is perpendicular to both $I\,d\boldsymbol{\ell}$ and $\hat{\mathbf{r}}$. It produces no magnetic field at point P_2, which is along the line of $I\,d\boldsymbol{\ell}$.

Whereas the electric field points in the radial direction $\mathbf{r}$ from the point charge to the field point (for a positive charge), the magnetic field is perpendicular both to $\mathbf{r}$ and to the direction of motion of the charges $\mathbf{v}$, which is along the direction of the current element. At a point along the line of a current element, such as point P_2 in Figure 25-5, the magnetic field due to that element is zero, because the angle θ between $I\,d\boldsymbol{\ell}$ and the vector $\mathbf{r}$ to that point is zero. The magnetic field due to the total current in a circuit can be calculated by using the Biot–Savart law to find the field due to each current element and then summing (integrating) over all the current elements in the circuit. This calculation is very difficult for all but the most simple circuit geometries.

B Due to a Current Loop

One calculation that is relatively straightforward is to find the magnetic field at the center of a circular loop. Figure 25-6 shows a current element $I\,d\boldsymbol{\ell}$ of a current loop of radius R and the unit vector $\hat{\mathbf{r}}$ that is directed from the element to the center of the loop. The magnetic field at the center of the loop due to this element is directed along the axis of the loop, and its magnitude is given by

$$dB = \frac{\mu_0}{4\pi}\frac{I\,d\ell\,\sin\theta}{R^2}$$

where θ is the angle between $I\,d\boldsymbol{\ell}$ and $\hat{\mathbf{r}}$, which is 90° for each current element, so $\sin\theta = 1$. The magnetic field due to the entire current is found by integrating over all the current elements in the loop. Since R is the same for all elements, we obtain

$$B = \oint dB = \frac{\mu_0}{4\pi}\frac{I}{R^2}\oint d\ell$$

The integral of $d\ell$ around the complete loop gives the total length $2\pi R$, the circumference of the loop. The magnetic field due to the entire loop is thus

$$B = \frac{\mu_0}{4\pi}\frac{I\,2\pi R}{R^2} = \frac{\mu_0 I}{2R} \qquad \text{(at center of loop)} \qquad 25\text{-}5$$

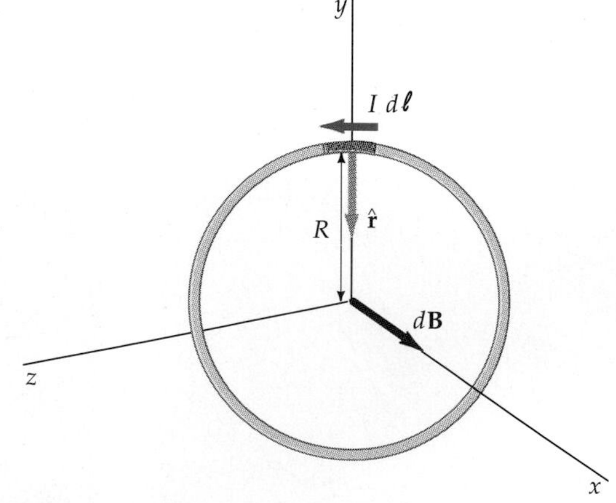

Figure 25-6 Current element for calculating the magnetic field at the center of a circular current loop. Each element produces a magnetic field that is directed along the axis of the loop.

Exercise

Find the current in a circular loop of radius 8 cm that will give a magnetic field of 2 G at the center of a loop. (Answer: 25.5 A)

Figure 25-7 shows the geometry for calculating the magnetic field at a point on the axis of a circular current loop a distance x from its center. We first consider the current element at the top of the loop. Here, as everywhere around the loop, $I\,d\boldsymbol{\ell}$ is tangent to the loop and perpendicular to the vector $\mathbf{r}$ from the current element to the field point P. The magnetic field $d\mathbf{B}$ due to this element is in the direction shown in the figure, perpendicular to $\mathbf{r}$ and also perpendicular to $I\,d\boldsymbol{\ell}$. The magnitude of $d\mathbf{B}$ is

$$|d\mathbf{B}| = \frac{\mu_0}{4\pi}\frac{I|d\boldsymbol{\ell} \times \hat{\mathbf{r}}|}{r^2} = \frac{\mu_0}{4\pi}\frac{I\,d\ell}{x^2 + R^2}$$

where we have used the facts that $r^2 = x^2 + R^2$ and that $d\boldsymbol{\ell}$ and $\hat{\mathbf{r}}$ are perpendicular, so $|d\boldsymbol{\ell} \times \hat{\mathbf{r}}| = d\ell$.

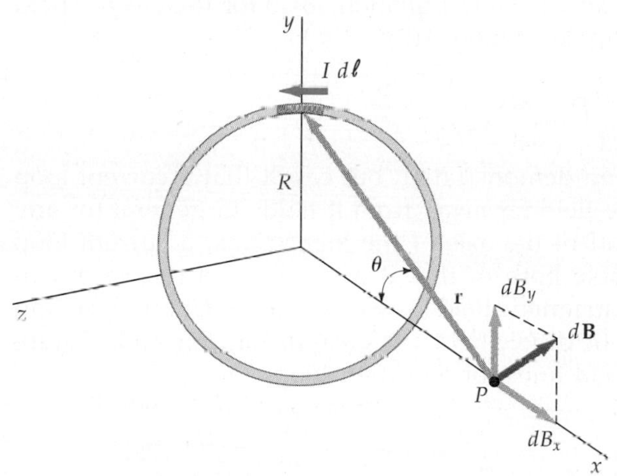

Figure 25-7 Geometry for calculating the magnetic field at a point on the axis of a circular current loop.

When we sum around all the current elements in the loop, the components of $d\mathbf{B}$ perpendicular to the axis of the loop, such as dB_y in the figure, sum to zero, leaving only the components dB_x that are parallel to the axis. We thus compute only the x component of the field. From the figure, we have

$$dB_x = dB \sin\theta = dB\left(\frac{R}{\sqrt{x^2 + R^2}}\right) = \frac{\mu_0}{4\pi}\frac{I\,d\ell}{x^2 + R^2}\frac{R}{\sqrt{x^2 + R^2}}$$

To find the field due to the entire loop of current, we integrate dB_x around the loop:

$$B_x = \oint dB_x = \oint \frac{\mu_0}{4\pi}\frac{IR}{(x^2 + R^2)^{3/2}}\,d\ell$$

Since neither x nor R varies as we sum over the elements in the loop, we can remove these quantities from the integral. Then,

$$B_x = \frac{\mu_0 IR}{4\pi(x^2 + R^2)^{3/2}}\oint d\ell$$

The integral of $d\ell$ around the loop gives $2\pi R$. Thus,

$$B_x = \frac{\mu_0}{4\pi}\frac{IR(2\pi R)}{(x^2 + R^2)^{3/2}} = \frac{\mu_0}{4\pi}\frac{2\pi R^2 I}{(x^2 + R^2)^{3/2}} \qquad \text{25-6}$$

B on the axis of a current loop

Exercise

Show that Equation 25-6 reduces to Equation 25-5 at the center of the loop.

At great distances from the loop, x is much greater than R, so $(x^2 + R^2)^{3/2} \approx (x^2)^{3/2} = x^3$. Then,

$$B_x \longrightarrow \frac{\mu_0}{4\pi}\frac{2I\pi R^2}{x^3}$$

or

Magnetic-dipole field on the axis of the dipole

$$B_x = \frac{\mu_0}{4\pi}\frac{2m}{x^3} \qquad 25\text{-}7$$

where $m = I\pi R^2$ is the magnitude of the magnetic moment of the loop. Note the similarity of this expression to Equation 18-10 for the electric field on the axis of an electric dipole of moment p:

$$E_x \longrightarrow \frac{1}{4\pi\epsilon_0}\frac{2p}{x^3}$$

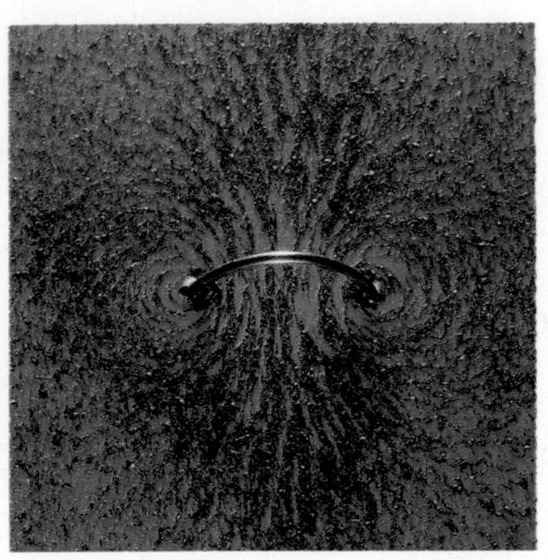

Figure 25-8 The magnetic-field lines of a circular current loop indicated by iron filings.

Although it has not been demonstrated, our result that a current loop produces a magnetic dipole field far away from it holds in general for any point whether it is on or off of the axis of the loop. Thus, a current loop behaves as a magnetic dipole both in that it experiences a torque **m × B** when placed in an external magnetic field (as was shown in Chapter 24) and in that it produces a magnetic dipole field at a great distance from it. Figure 25-8 shows the magnetic-field lines for a current loop.

Example 25-3

A circular loop of radius 5.0 cm has 12 turns and lies in the xy plane. It carries a current of 4 A in the direction such that the magnetic moment of the loop is along the x axis. Find the magnetic field on the x axis at (*a*) $x = 15$ cm and (*b*) $x = 3$ m.

(*a*) Equation 25-6 gives the magnetic field due to the current in a single turn. The magnetic field due to a loop with N turns is N times that due to a single turn. The magnetic field at $x = 15$ cm is thus

$$B_x = \frac{\mu_0}{4\pi}\frac{2\pi R^2 NI}{(x^2 + R^2)^{3/2}}$$

$$= (10^{-7}\ \text{T·m/A})\frac{2\pi(0.05\ \text{m})^2(12)(4\ \text{A})}{[(0.15\ \text{m})^2 + (0.05\ \text{m})^2]^{3/2}}$$

$$= 1.91 \times 10^{-5}\ \text{T}$$

Note that the magnetic field is considerably larger than that due to the single moving charge calculated in Example 25-1.

(*b*) Since 3 m is much greater than the radius 0.05 m, we can use Equation 25-7 for the magnetic field far from the loop. The magnitude of the magnetic dipole moment of a loop with N turns is

$$m = NIA = (12)(4\ \text{A})\pi(0.05\ \text{m})^2$$

$$= 0.377\ \text{A·m}^2$$

The magnetic field at $x = 3$ m is then

$$B_x = \frac{\mu_0}{4\pi}\frac{2m}{x^3} = (10^{-7}\ \text{T·m/A})\frac{2(0.377\ \text{A·m}^2)}{(3\ \text{m})^3} = 2.79 \times 10^{-9}\ \text{T}$$

B Due to a Current in a Solenoid

We now use Equation 25-6 to calculate the magnetic field of a **solenoid,** which is a wire tightly wound into a helix of closely spaced turns, as illustrated in Figure 25-9. A solenoid is used to produce a strong, uniform magnetic field in the region surrounded by its loops. Its role in magnetism is analogous to that of the parallel-plate capacitor in electrostatics, in that the capacitor produces a strong, uniform electric field between its plates. The magnetic field of a solenoid is essentially that of a set of N identical current loops placed side by side. Figure 25-10 shows the magnetic field lines for two such loops. In the region between the loops near their axis, the fields of the individual loops are in the same direction and their magnitudes add, whereas in the region between the loops but at distances from the axis that are large compared with the radius of the loops, the fields tend to cancel. Figure 25-11 shows the magnetic-field lines for a long, tightly wound solenoid. Inside the solenoid, the field lines are approximately parallel to the axis and are closely and uniformly spaced, indicating a strong, uniform magnetic field. Outside the solenoid, the lines are much less dense. They diverge from one end and converge at the other end. Comparing this figure with Figure 24-8, we see that the field lines of a solenoid, both inside and outside, are the same as those of a bar magnet of the same shape as the solenoid.

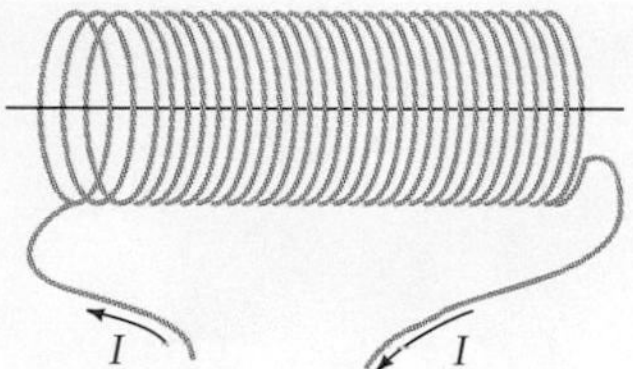

Figure 25-9 A tightly wound solenoid can be considered to be a set of circular current loops placed side by side that carry the same current. It produces a uniform magnetic field inside it.

We will calculate the magnetic field only at a point on the axis of the solenoid and between its ends. We consider a solenoid of length L consisting of N turns of wire carrying a current I. We choose the axis of the solenoid to

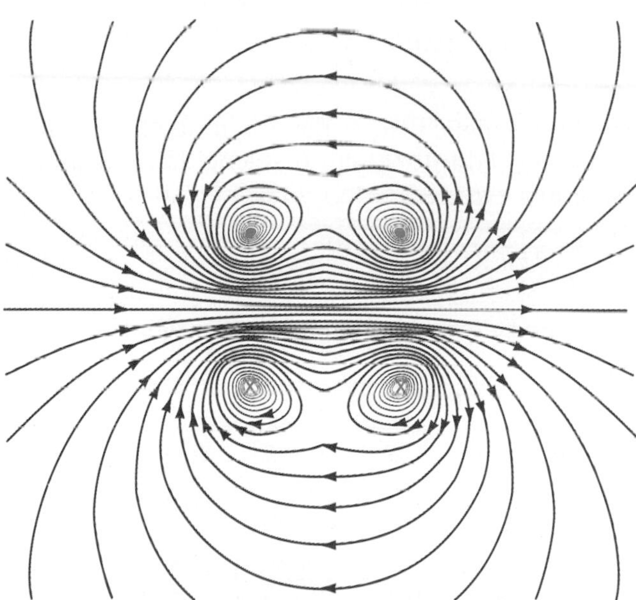

Figure 25-10 Magnetic-field lines due to two loops carrying the same current in the same sense. The points where the loops intersect the plane of the page are marked by a blue x where the current enters and a blue · where the current emerges. In the region between the loops, the magnetic fields of the individual loops add so the resultant field is strong, whereas in the regions away from the loops, the fields subtract so the resultant field is weak.

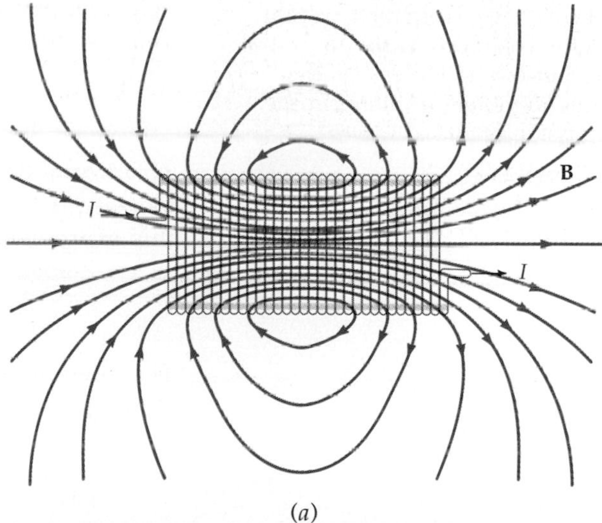

(a)

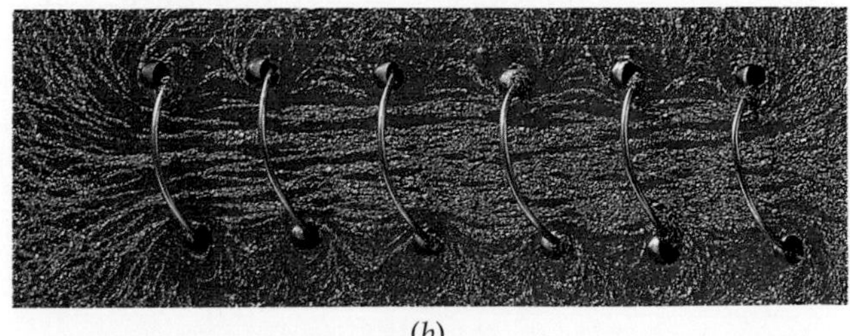

(b)

Figure 25-11 (a) Magnetic-field lines of a solenoid. The lines are identical to those of a bar magnet of the same shape (Figure 24-8). (b) Magnetic-field lines of a solenoid shown by iron filings.

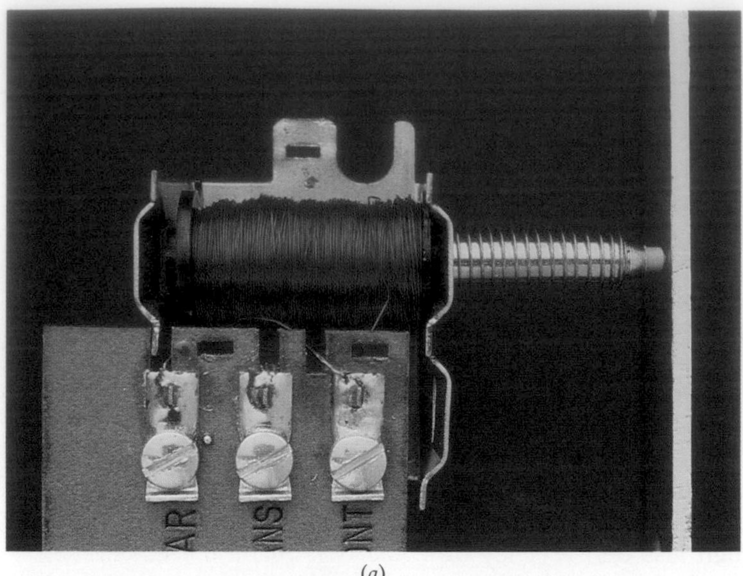

(a)

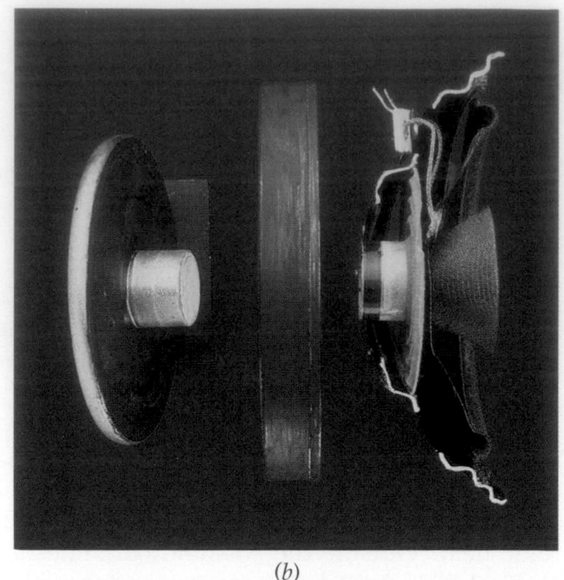

(b)

(*a*) A cross section of a doorbell. When the outer solenoid is energized, its magnetic field causes the inner plunger to strike the bell. (*b*) A cross section of a speaker. The speaker cone is attached to a coil, which is in the magnetic field of a permanent magnet. An alternating current in the coil, such as that from the output of a radio, causes the cone to vibrate resulting in the emission of sound waves.

be along the x axis, with the left end at $x = -a$ and the right end at $x = +b$ as shown in Figure 25-12. We will calculate the magnetic field at the origin. The figure shows an element of the solenoid of length dx at a distance x from the origin. If $n = N/L$ is the number of turns per unit length, there are $n\,dx$ turns of wire in this element, with each turn carrying a current I. The element is thus equivalent to a single loop carrying a current $di = nI\,dx$. The magnetic field at a point x on the axis due to a loop at the origin carrying a current $nI\,dx$ is given by Equation 25-6 with I replaced by $nI\,dx$:

$$dB_x = \frac{\mu_0}{4\pi}\frac{2\pi nIR^2\,dx}{(x^2+R^2)^{3/2}}$$

This expression also gives the magnetic field at the origin due to a current loop at x. We find the magnetic field due to the entire solenoid by integrating this expression from $x = -a$ to $x = b$:

$$B_x = \frac{\mu_0}{4\pi}\,2\pi nIR^2\int_{-a}^{b}\frac{dx}{(x^2+R^2)^{3/2}} \qquad \text{25-8}$$

The integral in Equation 25-8 can be found in standard tables of integrals. Its value is

$$\int_{-a}^{b}\frac{dx}{(x^2+R^2)^{3/2}} = \left.\frac{x}{R^2\sqrt{x^2+R^2}}\right|_{-a}^{b} = \frac{b}{R^2\sqrt{b^2+R^2}} + \frac{a}{R^2\sqrt{a^2+R^2}}$$

Substituting this into Equation 25-8, we obtain

$$B = \frac{1}{2}\mu_0 nI\left(\frac{b}{\sqrt{b^2+R^2}} + \frac{a}{\sqrt{a^2+R^2}}\right) \qquad \text{25-9}$$

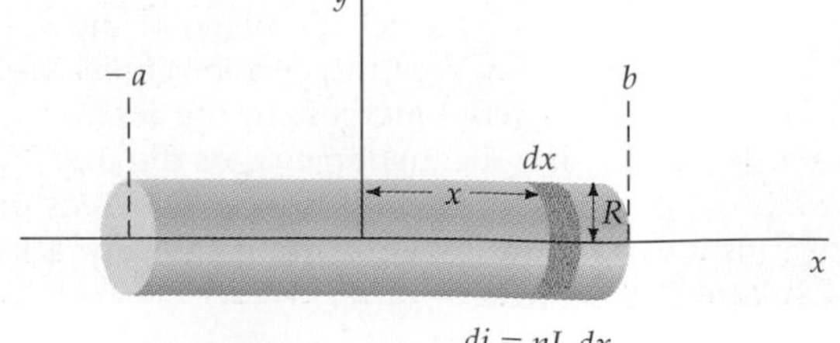

Figure 25-12 Geometry for calculating the magnetic field inside a solenoid on the axis. The number of turns in the element dx is $n\,dx$, where $n = N/L$ is the number of turns per unit length. The element dx is treated as a current loop carrying a current $di = nI\,dx$.

For a long solenoid for which a and b are much larger than R, the two terms in the parentheses each tend toward 1. For this approximation, the magnetic field is

$$B = \mu_0 nI \tag{25-10}$$

B inside a long solenoid

If the origin is at one end of the solenoid, either a or b is zero. Then, if the other end is far away compared with the radius, one of the terms in the parentheses of Equation 25-9 is zero and the other is 1, so $B \approx \frac{1}{2}\mu_0 nI$. Thus, the magnitude of **B** at a point near either end of a long solenoid is about half that at points within the solenoid away from the ends. Figure 25-13 gives a plot of the magnetic field on the axis of a solenoid versus position (with the origin at the center of the solenoid). The approximation that the field is constant independent of the position along the axis is quite good except for very near the ends.

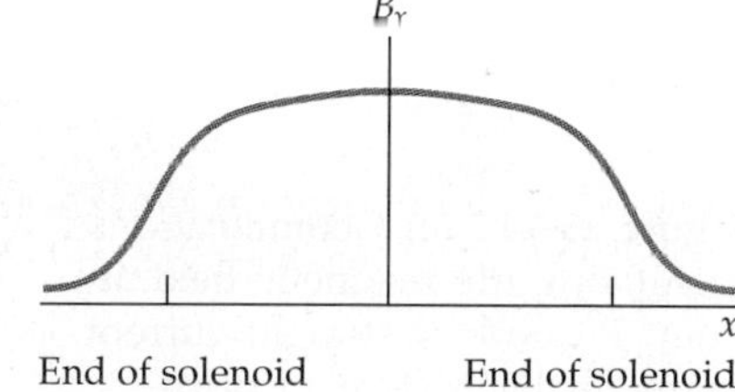

Figure 25-13 Graph of the magnetic field on the axis inside a solenoid versus the position x on the axis. The field inside the solenoid is nearly constant except near the ends.

Example 25-4

Find the magnetic field at the center of a solenoid of length 20 cm, radius 1.4 cm, and 600 turns that carries a current of 4 A.

We will calculate the field exactly using Equation 25-9. For a point at the center of the solenoid, $a = b = 10$ cm. Thus, each term in the parentheses in Equation 25-9 has the value

$$\frac{a}{\sqrt{a^2 + R^2}} = \frac{b}{\sqrt{b^2 + R^2}} = \frac{10 \text{ cm}}{\sqrt{(10 \text{ cm})^2 + (1.4 \text{ cm})^2}} = 0.990$$

Then, from Equation 25-9, the magnetic field at the center of the solenoid is

$$B = \tfrac{1}{2}\mu_0 nI\left(\frac{b}{\sqrt{b^2 + R^2}} + \frac{a}{\sqrt{a^2 + R^2}}\right)$$

$$= (0.5)(4\pi \times 10^{-7} \text{ T·m/A})(600 \text{ turns}/0.2 \text{ m})(4 \text{ A})(0.990 + 0.990)$$

$$= 1.50 \times 10^{-2} \text{ T}$$

Note that the approximation obtained using Equation 25-10 amounts to replacing 0.99 by 1.00, which differs by only 1 percent. Note also that the magnitude of the magnetic field inside this solenoid is fairly large—about 100 times the magnetic field of the earth.

(*a*)

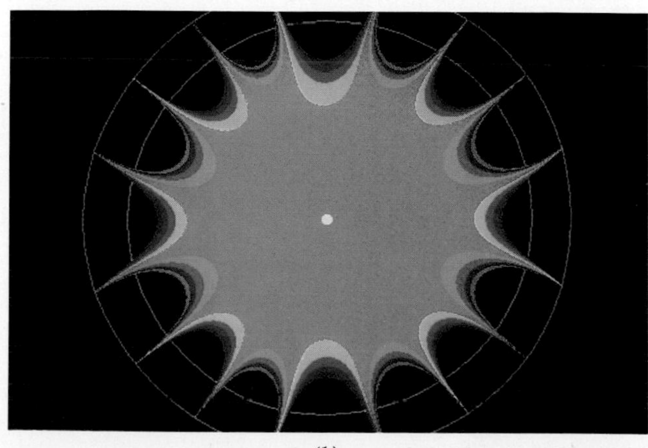

(*b*)

(*a*) A sextupole magnet for focusing beams of charged particles. (*b*) Computer graphic showing the magnetic field distribution in a superconducting magnet used at Brookhaven National Laboratory. The contours show equal deviations of the field from that at the center of the magnet.

B Due to a Current in a Straight Wire

Figure 25-14 shows the geometry for calculating the magnetic field **B** at a point P due to the current in the straight wire segment shown. We choose the wire to be the x axis and point P to be on the y axis. Note that because of the symmetry in this problem, any direction perpendicular to the wire could be chosen for the y axis. A typical current element $I\,d\ell$ at a distance x from the origin is shown. The vector **r** points from the element to the field point P.

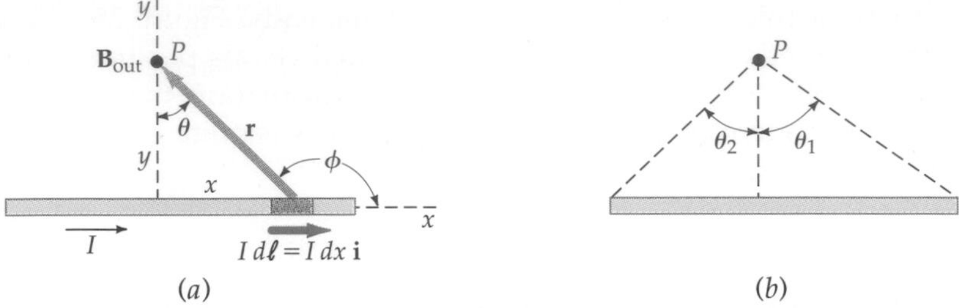

Figure 25-14 (*a*) Geometry for calculating the magnetic field at point P due to a straight current segment. Each element of the segment contributes to the total magnetic field at point P, which is directed out of the paper. (*b*) The result is expressed in terms of the angles θ_1 and θ_2.

The direction of the magnetic field at P due to this element is the direction of $I\,d\boldsymbol{\ell} \times \mathbf{r}$, which is out of the paper. Note that the magnetic fields due to all the current elements of the wire are in this same direction. Thus, we only need to compute the magnitude of the field. The field due to the current element shown has the magnitude

$$dB = \frac{\mu_0}{4\pi}\frac{I\,dx}{r^2}\sin\phi$$

It is more convenient to write this in terms of θ rather than ϕ:

$$dB = \frac{\mu_0}{4\pi}\frac{I\,dx}{r^2}\cos\theta \qquad \text{25-11}$$

To sum over all the current elements, we need to relate the variables θ, r, and x. It turns out to be easiest to express x and r in terms of θ. We have

$$x = y\tan\theta$$

Then,

$$dx = y\sec^2\theta\,d\theta = y\frac{r^2}{y^2}\,d\theta = \frac{r^2}{y}\,d\theta$$

where we have used $\sec\theta = r/y$. Substituting this expression for dx into Equation 25-11, we obtain

$$dB = \frac{\mu_0}{4\pi}\frac{I}{r^2}\frac{r^2\,d\theta}{y}\cos\theta = \frac{\mu_0}{4\pi}\frac{I}{y}\cos\theta\,d\theta$$

Let us first calculate the contribution from the current elements to the right of the point $x = 0$. We sum over these elements by integrating from $\theta = 0$ to $\theta = \theta_1$, where θ_1 is the angle between the line perpendicular to the wire and the line from P to the right end of the wire, as shown in Figure 25-14*b*. For this contribution, we have

$$B_1 = \int_0^{\theta_1}\frac{\mu_0}{4\pi}\frac{I}{y}\cos\theta\,d\theta$$

$$= \frac{\mu_0}{4\pi}\frac{I}{y}\int_0^{\theta_1}\cos\theta\,d\theta = \frac{\mu_0}{4\pi}\frac{I}{y}\sin\theta_1$$

Similarly, the contribution from elements to the left of $x = 0$ is

$$B_2 = \frac{\mu_0}{4\pi}\frac{I}{y}\sin\theta_2$$

The total magnetic field due to the wire segment is the sum of B_1 and B_2. Writing R instead of y for the perpendicular distance from the wire segment to the field point, we obtain

$$B = \frac{\mu_0}{4\pi}\frac{I}{R}(\sin\theta_1 + \sin\theta_2) \qquad 25\text{-}12$$

B due to a straight wire segment

This result gives the magnetic field due to any wire segment in terms of the perpendicular distance R and the angles subtended at the field point by the ends of the wire. If the wire is very long, these angles are nearly 90°. The result for a very long wire is obtained from Equation 25-12 by setting $\theta_1 = \theta_2 = 90°$:

$$B = \frac{\mu_0 I}{2\pi R} = \frac{\mu_0}{4\pi}\frac{2I}{R} \qquad 25\text{-}13$$

B due to a long, straight wire

At any point in space, the magnetic field lines of a long, straight, current-carrying wire is tangent to a circle of radius R about the wire, where R is the perpendicular distance from the wire to the field point. The direction of **B** can be determined by applying the right-hand rule as shown in Figure 25-15*a*. The magnetic-field lines thus encircle the wire as shown in Figure 25-15*b*.

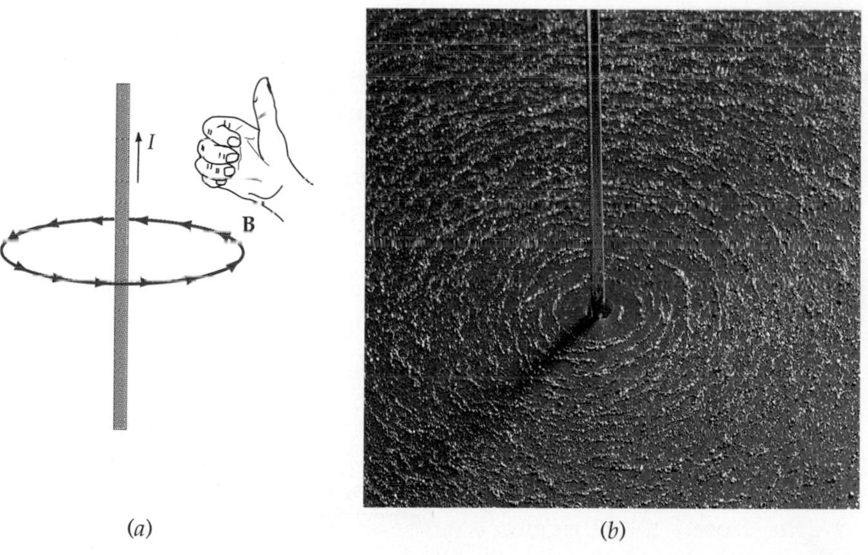

Figure 25-15 (*a*) Right-hand rule for determining the direction of the magnetic field due to a long, straight, current-carrying wire. The magnetic-field lines encircle the wire in the direction of the fingers of the right hand when the thumb points in the direction of the current. (*b*) Magnetic-field lines due to a long wire indicated by iron filings.

The result expressed by Equation 25-13 was found experimentally by Biot and Savart in 1820. From an analysis of it, they were able to discover the expression for the magnetic field due to a current element given in Equation 25-4.

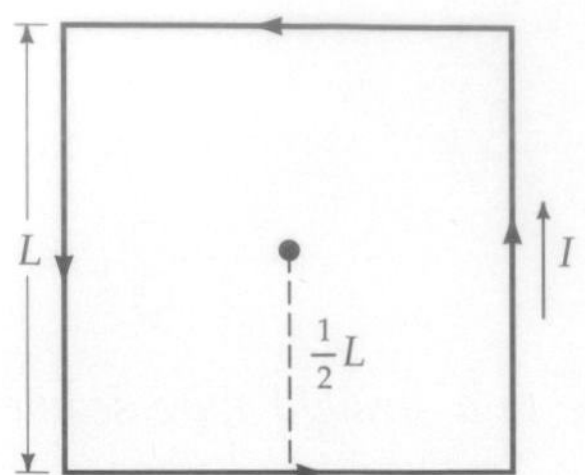

Figure 25-16 Square current loop for Example 25-5.

Example 25-5

Find the magnetic field at the center of a square current loop of side $L = 50$ cm carrying current 1.5 A (Figure 25-16).

From Figure 25-16, we see that each side of the loop contributes a field in the direction out of the paper. Because of the symmetry of this figure, we need only calculate the field due to one side of the loop and multiply by 4. The distance between one side and the field point is $R = \frac{1}{2}L = 0.25$ m. Thus, from Equation 25-12, the field is

$$B = 4\left(\frac{\mu_0}{4\pi}\right)\frac{I}{\frac{1}{2}L}(\sin 45° + \sin 45°) = (4 \times 10^{-7}\ \text{T·m/A})\frac{1.5\ \text{A}}{0.25\ \text{m}}(2 \sin 45°)$$
$$= 3.39 \times 10^{-6}\ \text{T}$$

Exercise

Compare the magnetic field at the center of a circular current loop of radius R with that at the center of a square current loop of side $L = 2R$. Which is larger? (Answer: B is larger for the circle by about 10 percent)

Example 25-6

Find the magnetic field at a distance of 20 cm from a long, straight wire carrying a current of 5 A.

From Equation 25-13, we have

$$B = \frac{\mu_0}{4\pi}\frac{2I}{y} = (10^{-7}\ \text{T·m/A})\frac{2(5\ \text{A})}{0.2\ \text{m}} = 5.00 \times 10^{-6}\ \text{T}$$

We note from this example that the magnetic field near a wire carrying a current of ordinary size is small. In this case, it is only about 1 percent of the magnetic field due to the earth.

Example 25-7

A long, straight wire carrying a current of 1.7 A in the positive z direction lies along the line $x = -3$ cm. A similar wire carrying a current of 1.7 A in the positive z direction lies along the line $x = +3$ cm as shown in Figure 25-17a. Find the magnetic field at a point on the y axis at $y = 6$ cm.

In Figure 25-17b, we have drawn the xy plane in the plane of the page and indicated the directions of the magnetic field $\mathbf{B}_L$ due to the current

Figure 25-17 (a) Two parallel wires carrying currents in the same sense for Example 25-7. (b) Geometry for calculating the resultant magnetic field due to the two currents.

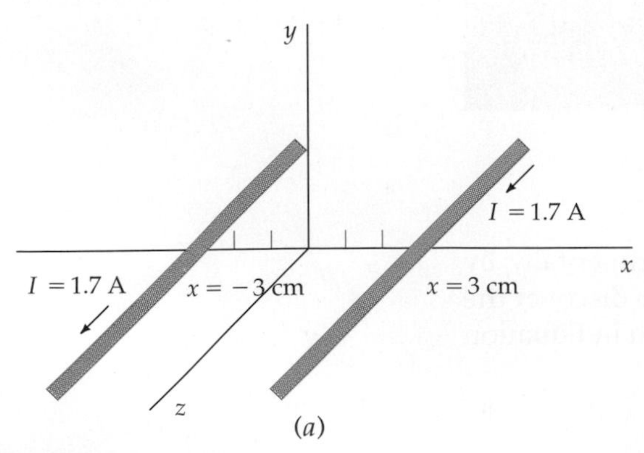

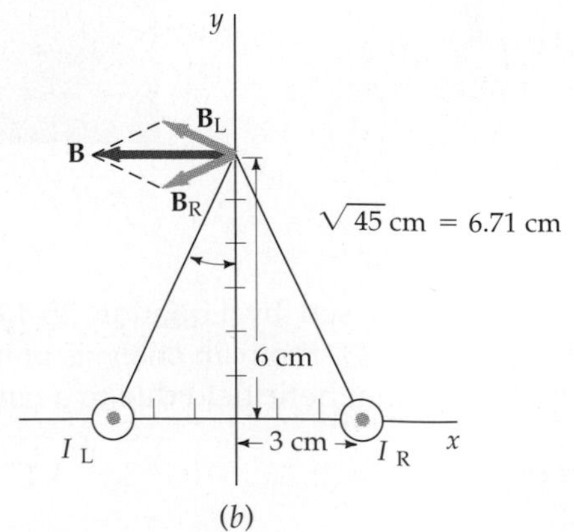

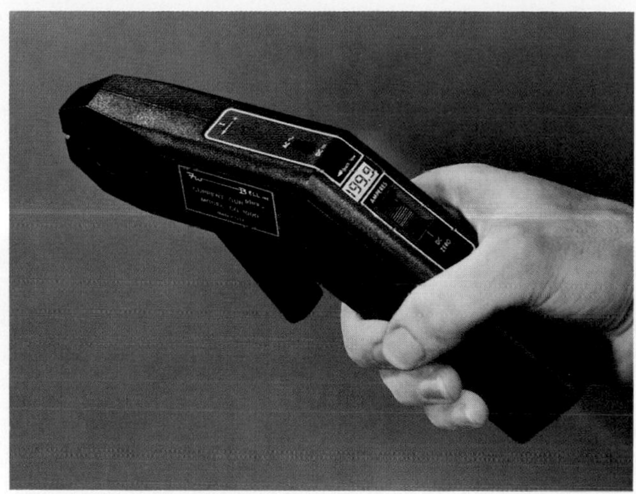

A current gun for measuring electric current. The jaws of the current gun clamp around a current-carrying wire without touching the wire. The magnetic field produced by the wire is measured with a Hall-effect device mounted in the current gun. The Hall-effect device puts out a voltage proportional to the magnetic field, which in turn is proportional to the current in the wire.

on the left and $\mathbf{B}_R$ due to the current on the right. Since the currents have equal magnitudes and are each at a distance $R = \sqrt{(3\text{ cm})^2 + (6\text{ cm})^2} = 6.71$ cm from the field point, these fields have equal magnitudes given by

$$B_R = B_L = \frac{\mu_0}{4\pi}\frac{2I}{R} = (10^{-7}\text{ T·m/A})\frac{2(1.7\text{ A})}{0.0671\text{ m}} = 5.07 \times 10^{-6}\text{ T}$$

From Figure 25-17*b*, we see that the resultant magnetic field is in the negative x direction and has the magnitude $2B_L \cos\theta$, where $\cos\theta = (6\text{ cm})/(6.71\text{ cm}) = 0.894$. The resultant magnetic field is thus

$$\mathbf{B} = -2B_L\cos\theta\,\mathbf{i} = -2(5.07 \times 10^{-6}\text{ T})(0.894)\mathbf{i} = 9.07 \times 10^{-6}\text{ T}\,\mathbf{i}$$

Example 25-8

An infinitely long wire carrying a current of 4.5 A is bent as shown in Figure 25-18. Find the magnetic field at the point $x = 3$ cm, $y = 2$ cm.

We first note that the magnetic field due to each part of the wire is in the positive z direction. The magnitude of the field due to the wire segment along the y axis B_1 is given by Equation 25-12, with $R = 3$ cm, $\theta_1 = 90°$, and $\theta_2 = \alpha = \tan^{-1}(\frac{2}{3}) = 33.7°$:

$$B_1 = \frac{\mu_0}{4\pi}\frac{I}{R}(\sin\theta_1 + \sin\theta_2)$$

$$= (10^{-7}\text{ T·m/A})\frac{4.5\text{ A}}{0.03\text{ m}}(\sin 90° + \sin 33.7°) = 2.33 \times 10^{-5}\text{ T}$$

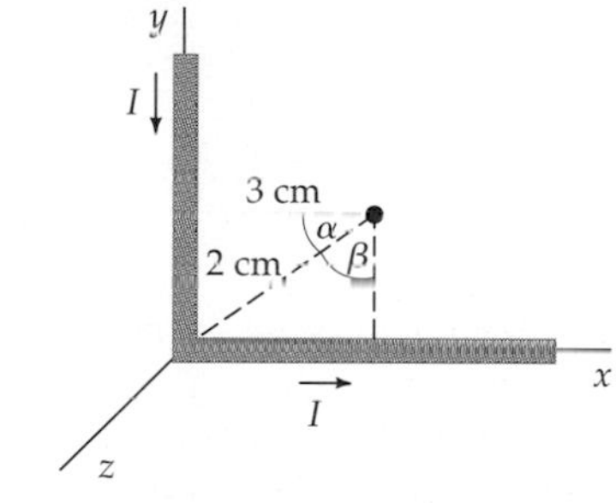

Figure 25-18 Current-carrying wire for Example 25-8.

Similarly, the magnitude of the field B_2 due to the wire segment along the x axis is given by Equation 25-12 with $R = 2$ cm, $\theta_1 = \beta = 90° - 33.7° = 56.3°$, and $\theta_2 = 90°$:

$$B_2 = \frac{\mu_0}{4\pi}\frac{I}{R}(\sin\theta_1 + \sin\theta_2)$$

$$= (10^{-7}\text{ T·m/A})\frac{4.5\text{ A}}{0.02\text{ m}}(\sin 56.3° + \sin 90°) = 4.12 \times 10^{-5}\text{ T}$$

The resultant magnetic field is thus

$$\mathbf{B} = (B_1 + B_2)\mathbf{k} = (2.33 \times 10^{-5}\text{ T} + 4.12 \times 10^{-5}\text{ T})\mathbf{k} = 6.45 \times 10^{-5}\text{ T}\,\mathbf{k}$$

25-3 Definition of the Ampere

We can use Equation 25-13 for the magnetic field due to a long, straight, current-carrying wire and Equation 24-4 for the force exerted by a magnetic field on a segment of a current-carrying wire to find the force exerted by one long, straight current on another. Figure 25-19 shows two long, parallel wires carrying currents in the same direction. We consider the force on a segment $\Delta \ell_2$ carrying current I_2 as shown. The magnetic field $\mathbf{B}_1$ at this segment due to current I_1 is perpendicular to the segment $I_2 \, \Delta \ell_2$ as shown. This is true for all current elements along the wire. The magnetic force on current segment $I_2 \, \Delta \ell_2$ is directed toward current I_1. Similarly, a current segment $I_1 \, \Delta \ell_1$ will experience a magnetic force directed toward current I_2 due to a magnetic field arising from current I_2. Thus, two parallel currents attract each other. If one of the currents is reversed, the force will be reversed. Thus, two antiparallel currents will repel each other. The attraction or repulsion of parallel or antiparallel currents was discovered experimentally by Ampère one week after he heard of Oersted's discovery of the effect of a current on a compass needle.

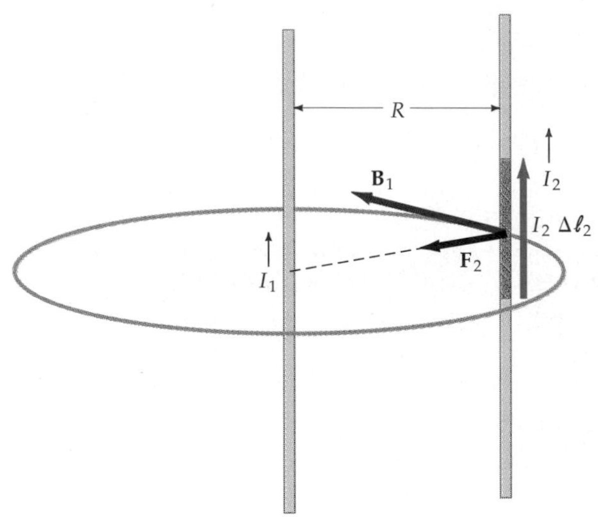

Figure 25-19 Two long, straight wires carrying parallel currents. The magnetic field $\mathbf{B}_1$ due to current I_1 is perpendicular to current I_2. The force on current I_2 is toward current I_1. There is an equal and opposite force exerted by current I_2 on I_1. The currents thus attract each other.

The magnitude of the magnetic force on the segment $I_2 \, \Delta \ell_2$ is

$$F_2 = |I_2 \, \Delta \boldsymbol{\ell}_2 \times \mathbf{B}_1|$$

Since the magnetic field at segment $I_2 \, \Delta \ell_2$ is perpendicular to the current segment, we have

$$F_2 = I_2 \, \Delta \ell_2 \, B_1$$

If the distance R between the wires is much less than their length, the field at $I_2 \, \Delta \ell_2$ due to current I_1 will approximate the field due to an infinitely long, current-carrying wire given by Equation 25-13. The magnitude of the force on the segment $I_2 \, \Delta \ell_2$ is therefore

$$F_2 = I_2 \, \Delta \ell_2 \, \frac{\mu_0 I_1}{2\pi R}$$

The force per unit length is

$$\frac{F_2}{\Delta \ell_2} = \frac{\mu_0}{2\pi} \frac{I_1 I_2}{R} = 2 \frac{\mu_0}{4\pi} \frac{I_1 I_2}{R} \qquad \text{25-14}$$

In Chapter 18, the coulomb was defined in terms of the ampere, but the definition of the ampere was deferred. The ampere is defined as follows:

If two very long parallel wires one meter apart carry equal currents, the current in each is defined to be one ampere when the force per unit length on each wire is 2×10^{-7} N/m.

Ampere defined

This definition of the ampere makes the permeability of free space μ_0 equal to exactly $4\pi \times 10^{-7}$ N/A^2. It also allows the unit of current (and therefore the unit of electric charge) to be determined by a mechanical measurement. In practice, currents much closer together than 1 m are used so that the wires need not be so long and the force is large enough to measure accurately. Figure 25-20 shows a **current balance,** which is a device that can be used to calibrate an ammeter from the definition of the ampere. The upper conductor is free to rotate about the knife edges and is balanced so that the wires (or conducting rods) are a small distance apart. The conductors are connected in series to carry the same current but in opposite directions so that they will repel each other. Weights are placed on the upper conductor until it balances again at the original separation. The force of repulsion is thus determined by measuring the total weight needed to balance the upper conductor.

Figure 25-20 Current balance used in an elementary physics laboratory to calibrate an ammeter. The two parallel rods in front carry equal but oppositely directed currents and therefore repel each other. The force of repulsion is balanced by weights placed on the upper rod, which is part of a rectangle that is balanced on knife edges at the back. The mirror on top is used to reflect a beam of laser light for accurately determining the position of the upper rod.

Example 25-9

Two straight rods 50 cm long and 1.5 mm apart in a current balance carry currents of 15 A each in opposite directions. What mass must be placed on the upper rod to balance the magnetic force of repulsion?

The force exerted by the lower rod on the upper rod of length L has the magnitude

$$F = \frac{\mu_0}{2\pi}\frac{I_1 I_2}{R}L$$

$$= (2 \times 10^{-7}\ \text{N/A}^2)\,\frac{(15\ \text{A})(15\ \text{A})}{0.0015\ \text{m}}\,(0.5\ \text{m}) = 1.5 \times 10^{-2}\ \text{N}$$

This force can be balanced by a weight mg:

$$mg = 1.5 \times 10^{-2}\ \text{N}$$

$$m = \frac{1.5 \times 10^{-2}\ \text{N}}{9.81\ \text{N/kg}} = 1.53 \times 10^{-3}\ \text{kg} = 1.53\ \text{g}$$

We see from this example that the magnetic force between two current-carrying wires is small even for currents as large as 15 A.

25-4 Ampère's Law

In Chapter 24, we noted that no fundamental magnetic charges or "poles" playing a role similar to that of electric charges have ever been observed. Instead, the fundamental source of magnetic fields is electric current. The magnetic fields arising from currents do not originate or end at points in space, but instead form closed loops encircling the current. For instance, we saw in Section 25-2 that the magnetic-field lines due to a long, straight, current-carrying wire encircle the wire. There is thus a definite difference between the characteristic shapes of electric-field lines and magnetic-field lines since electric-field lines begin and end on electric charges.

In Chapter 19, we studied Gauss's law, which relates the normal component of the electric field summed over a closed surface to the net charge inside the surface. There is an analogous equation for the magnetic field called Ampère's law that relates the tangential component of **B** summed around a closed curve C to the current I_C that passes through the curve. In mathematical form, **Ampère's law** is

Ampère's law

$$\oint_C \mathbf{B}\cdot d\boldsymbol{\ell} = \mu_0 I_C \qquad C\text{, any closed curve} \qquad 25\text{-}15$$

where I_C is the net current that penetrates the area bounded by the curve C. Ampère's law holds for any curve C as long as the currents are continuous, that is, they do not begin or end at any finite point. Like Gauss's law, Ampère's law can be used to obtain an expression for the magnetic field in situations that have a high degree of symmetry. If the symmetry is great enough, the line integral $\oint \mathbf{B}\cdot d\boldsymbol{\ell}$ can be written as the product of B and some distance. Then if I_C is known, B can be determined. Also, like Gauss's law, Ampère's law is of no use in finding an expression for the magnetic field if there is no symmetry. It is, however, of considerable theoretical importance.

The simplest application of Ampère's law is to find the magnetic field of an infinitely long, straight, current-carrying wire. Figure 25-21 shows a circular curve around a point on a long wire with its center at the wire. If we assume that we are far from the ends of the wire, we can use symmetry to rule out the possibility of any component of **B** parallel to the wire. We may then assume that the magnetic field is tangent to this circle and has the same magnitude B at any point on the circle. Ampère's law then gives

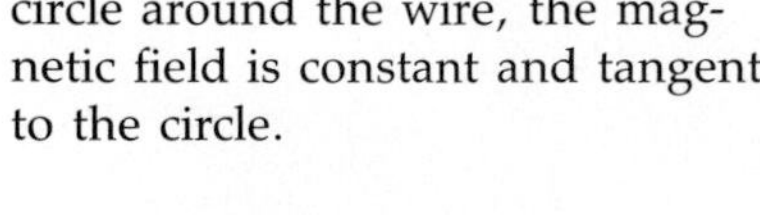

Figure 25-21 Geometry for calculating the magnetic field of a long, straight, current-carrying wire using Ampère's law. On a circle around the wire, the magnetic field is constant and tangent to the circle.

$$\oint_C \mathbf{B}\cdot d\boldsymbol{\ell} = B \oint_C d\ell = \mu_0 I_C$$

where we have taken B out of the integral because it has the same value everywhere on the circle. The integral of $d\ell$ around the circle equals $2\pi r$, the circumference of the circle. The current I_C is the current I in the wire. We thus obtain

$$B(2\pi r) = \mu_0 I$$

$$B = \frac{\mu_0}{2\pi}\frac{I}{r}$$

which is Equation 25-13.

Example 25-10

A long, straight wire of radius a carries a current I that is uniformly distributed over the cross-sectional area of the wire (Figure 25-22). Find the magnetic field both inside and outside the wire.

We can use Ampère's law to calculate **B** because of the high degree of symmetry. At a distance r, we know that **B** is tangent to the circle of radius r about the wire and constant in magnitude everywhere on the circle. Thus,

$$\oint_C \mathbf{B}\cdot d\boldsymbol{\ell} = B\oint_C d\ell = B2\pi r$$

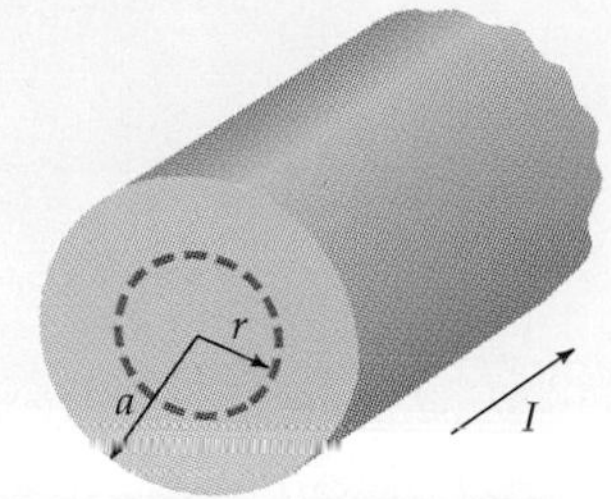

Figure 25-22 Long wire of radius a carrying a current I that is uniformly distributed over its cross-sectional area. Because of the symmetry of the wires, the magnetic field at any distance r can be calculated by applying Ampère's law to a circle of radius r.

The current through C depends on whether r is less than or greater than the radius of the wire a. For r greater than a, the total current I crosses the area bounded by C, and we obtain Equation 25-13 for the magnetic field due to a long, straight, current-carrying wire. Inside the wire, we consider a circle of radius $r < a$. The current passing through this circle is

$$I_C = \frac{\pi r^2}{\pi a^2} I$$

Ampère's law then gives

$$\oint_C \mathbf{B}\cdot d\boldsymbol{\ell} = B2\pi r = \mu_0 \frac{r^2}{a^2} I$$

$$B = \frac{\mu_0}{2\pi}\frac{I}{a^2} r \qquad r < a \qquad \text{25-16}$$

Figure 25-23 gives a graph of B versus r for this example.

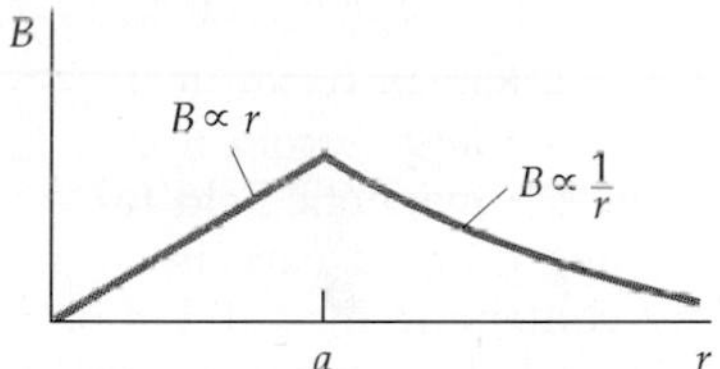

Figure 25-23 Graph of B versus r for a wire of radius a carrying a current uniformly distributed over its cross-sectional area. Inside the wire the magnetic field is proportional to the distance r from the axis of the wire.

For our next application of Ampère's law, we will calculate the magnetic field of a tightly wound **toroid,** which consists of loops of wire wound around a doughnut-shaped form as shown in Figure 25-24. There are N turns of wire, each carrying a current I. To calculate B, we evaluate the line integral $\oint \mathbf{B}\cdot d\boldsymbol{\ell}$ around a circle of radius r centered at the center of the toroid. By symmetry, **B** is tangent to this circle and constant in magnitude at every point on the circle. Then,

$$\oint \mathbf{B}\cdot d\boldsymbol{\ell} = B2\pi r = \mu_0 I_C$$

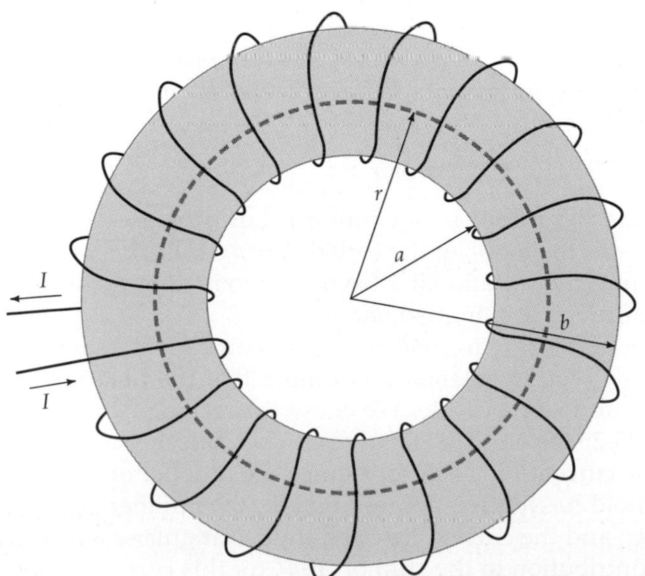

Figure 25-24 A toroid consists of loops of wire wound around a doughnut-shaped form. The magnetic field at any distance r can be found by applying Ampère's law to the circle of radius r.

(a)

(b)

(*a*) The Tokamak fusion test reactor is a large toroid that produces a magnetic field for confining charged particles. Coils containing over 10 km of water-cooled copper wire carry a pulsed current, which has a peak value of 73,000 A and produces a magnetic field of 5.2 T for about 3 s. (*b*) Inspection of the assembly of the Tokamak from inside the toroid.

Let a and b be the inner and outer radii of the toroid, respectively. The total current through the circle of radius r for $a < r < b$ is NI. Ampère's law then gives

$$\oint \mathbf{B} \cdot d\boldsymbol{\ell} = B2\pi r = \mu_0 I_C = \mu_0 NI$$

or

B inside a tightly wound toroid

$$B = \frac{\mu_0 NI}{2\pi r} \qquad a < r < b \tag{25-17}$$

If r is less than a, there is no current through the circle of radius r. If r is greater than b, the total current through r is zero because for each current I into the page in Figure 25-24, at the inner surface of the toroid, there is an equal current I out of the page at the outer surface. Thus, the magnetic field is zero for both $r < a$ and $r > b$:

$$B = 0 \qquad r < a \text{ or } r > b$$

The magnetic field inside the toroid is not uniform but decreases with r. However, if the diameter of the loops of the toroid, $b - a$, is much less than the radius of the doughnut, the variation in r from $r = a$ to $r = b$ is small, and B is approximately uniform as it is in a solenoid.

We can also use Ampère's law to find an expression for the magnetic field inside a long, tightly wound solenoid, assuming that the field is uniform inside the solenoid and zero outside. We chose the rectangle of sides a and b shown in Figure 25-25 for our closed curve C. The current that passes through this curve is the current I in each turn times the number of turns in the length a. If the solenoid has n turns per unit length, the number of turns in the length a will be na, and the current through the rectangular curve will be $I_C = naI$. The only contribution to the sum of $\oint \mathbf{B} \cdot d\boldsymbol{\ell}$ for this curve is along

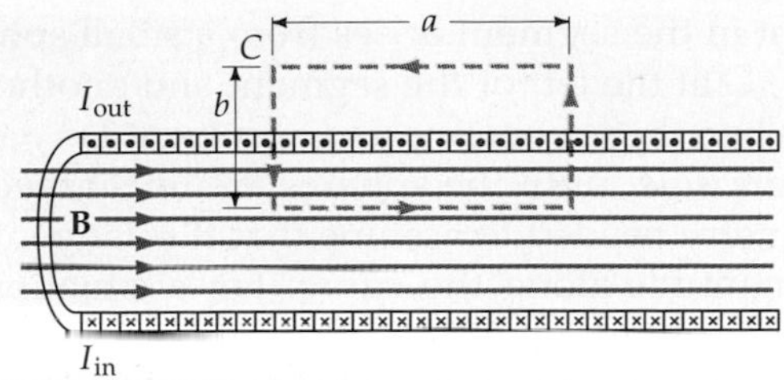

Figure 25-25 The magnetic field inside a solenoid can be calculated by applying Ampère's law to the rectangular curve C. If $\mathbf{B}$ is assumed to be uniform inside the solenoid and zero outside it, the line integral $\oint \mathbf{B}\cdot d\boldsymbol{\ell}$ around the curve C is just Ba.

the long side of the rectangle inside the solenoid, which gives Ba. Ampère's law thus gives

$$\oint \mathbf{B}\cdot d\boldsymbol{\ell} = Ba = \mu_0 I_C = \mu_0 naI$$

The magnetic field inside the solenoid is thus

$$B = \mu_0 nI$$

in agreement with Equation 25-10, which we obtained using the Biot–Savart law.

Limitations of Ampère's law

For an example of a case for which Ampère's law is not useful for calculating the magnetic field, consider the current loop shown in Figure 25-26. We have already calculated the field on the axis of such a loop using the Biot–Savart law. According to Ampère's law, the line integral $\oint \mathbf{B}\cdot d\boldsymbol{\ell}$ around a curve such as curve C in the figure equals μ_0 times the current I in the loop. Although Ampère's law is valid for this curve, the magnetic field $\mathbf{B}$ is not constant along any curve encircling the current, nor is it everywhere tangent to any such curve. Thus, there is not enough symmetry in this situation to allow us to calculate B using Ampère's law.

Figure 25-27 shows a finite current segment of length ℓ. We wish to find the magnetic field at point P, which is equidistant from the ends of the segment and at a distance r from the center of the segment. A direct application of Ampère's law gives

$$B = \frac{\mu_0}{2\pi}\frac{I}{r}$$

This result is the same as for an infinitely long wire, since we have the same symmetry arguments. It does not agree with the result obtained from the Biot–Savart law, which depends on the length of the current segment and which agrees with experiment. If the current segment is just one part of a continuous circuit carrying a current, as shown in Figure 25-28, Ampère's law for curve C is valid, but it cannot be used to find the magnetic field at point P because there is no symmetry.

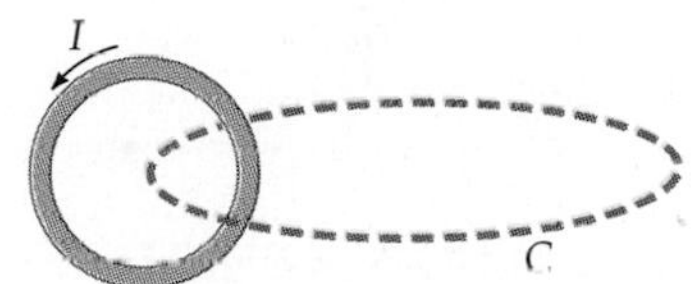

Figure 25-26 Ampère's law holds for the curve C encircling the current in the circular loop, but it is not useful in finding $\mathbf{B}$ because $\mathbf{B}$ is neither constant along the curve nor tangent to it.

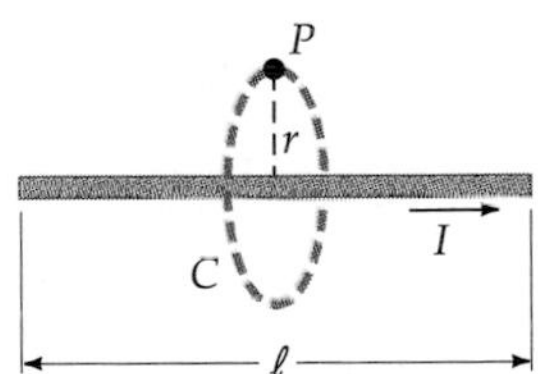

Figure 25-27 The application of Ampère's law to find the magnetic field on the bisector of a finite current segment gives an incorrect result.

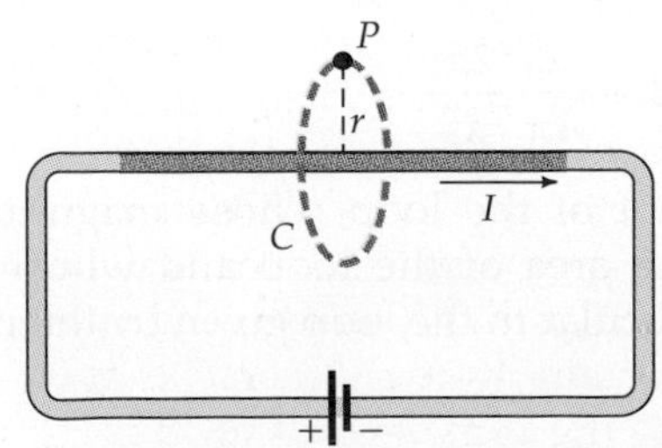

Figure 25-28 If the current segment in Figure 25-27 is part of a complete circuit, Ampère's law for the curve C is valid, but there is not enough symmetry to use it to find the magnetic field at point P.

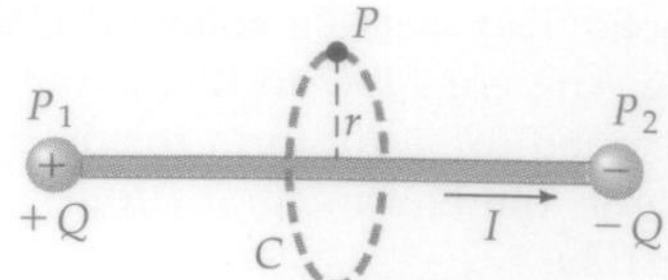

Figure 25-29 If the current segment in Figure 25-27 is due to a momentary flow of charge from a small conductor on the left to the one at the right, there is enough symmetry to use Ampère's law to compute the magnetic field at P, but Ampère's law is not valid because the current is not continuous in space.

In Figure 25-29, the current in the segment arises from a small spherical conductor with initial charge $+Q$ at the left of the segment and another one at the right with charge $-Q$. When they are connected, a current $I = -dQ/dt$ exists in the segment for a short time, until the spheres are uncharged. For this case, we *do* have the symmetry needed to assume that **B** is tangential to the curve and constant in magnitude along the curve. For a situation like this, in which the current is discontinuous in space, Ampère's law is not valid. In Chapter 29, we will see how Maxwell was able to modify Ampère's law so that it holds for all currents. When Maxwell's generalized form of Ampère's law is used to calculate the magnetic field for a current segment, such as that shown in Figure 25-29, the result agrees with that found from the Biot–Savart law.

Summary

1. The magnetic field produced by a point charge q moving with velocity **v** at a point a distance r away is given by

$$\mathbf{B} = \frac{\mu_0}{4\pi}\frac{q\mathbf{v} \times \hat{\mathbf{r}}}{r^2}$$

where $\hat{\mathbf{r}}$ is a unit vector that points from the charge to the field point and μ_0 is a constant, called the permeability of free space, that has the magnitude

$$\mu_0 = 4\pi \times 10^{-7}\ \text{T·m/A} = 4\pi \times 10^{-7}\ \text{N/A}^2$$

2. The magnetic field $d\mathbf{B}$ at a distance r from a current element $I\,d\boldsymbol{\ell}$ is

$$d\mathbf{B} = \frac{\mu_0}{4\pi}\frac{I\,d\boldsymbol{\ell} \times \hat{\mathbf{r}}}{r^2}$$

which is known as the Biot–Savart law. The magnetic field is perpendicular to both the current element and to the vector **r** from the current element to the field point.

3. The magnetic forces between two moving charges do not obey Newton's third law of action and reaction, implying that linear momentum of the two-charge system is not conserved. However, when the momentum associated with t the electromagnetic field is included, the total linear momentum of the two-charge system plus field is conserved.

4. The magnetic field on the axis of a current loop is given by

$$\mathbf{B} = \frac{\mu_0}{4\pi}\frac{2\pi R^2 I}{(x^2 + R^2)^{3/2}}\mathbf{i}$$

where **i** is a unit vector along the axis of the loop. At great distances from the loop, the field is a dipole field:

$$\mathbf{B} = \frac{\mu_0}{4\pi}\frac{2\mathbf{m}}{x^3}$$

where **m** is the dipole moment of the loop whose magnitude is the product of the current and the area of the loop and whose direction points in the direction perpendicular to the loop given by the right-hand rule.

5. Inside a solenoid and far from the ends, the magnetic field is uniform and has the magnitude

$$B = \mu_0 nI$$

where n is the number of turns per unit length of the solenoid.

6. The magnetic field of a straight, current-carrying wire segment is

$$B = \frac{\mu_0}{4\pi}\frac{I}{R}(\sin\theta_1 + \sin\theta_2)$$

where R is the perpendicular distance to the wire and θ_1 and θ_2 are the angles subtended at the field point by the ends of the wire. If the wire is very long, or the field point is very close to the wire, the magnetic field is approximately

$$B = \frac{\mu_0}{4\pi}\frac{2I}{R}$$

The direction of **B** is such that the lines of **B** encircle the wire in the direction of the fingers of the right hand if the thumb points in the direction of the current.

7. The magnetic field inside a tightly wound toroid is given by

$$B = \frac{\mu_0 NI}{2\pi r}$$

where r is the distance from the center of the toroid.

8. The ampere is defined such that two long parallel wires each carrying a current of 1 A and separated by 1 m exert a force of exactly 2×10^{-7} N/m on each other.

9. Ampère's law relates the integral of the tangential component of the magnetic field around a closed curve to the total current I_C passing through the area bounded by the curve:

$$\oint_C \mathbf{B}\cdot d\boldsymbol{\ell} = \mu_0 I_C \qquad C\text{, any closed curve}$$

Ampère's law is valid only if the currents are continuous. It can be used to derive expressions for the magnetic field for situations with a high degree of symmetry, such as a long, straight, current-carrying wire; a tightly wound toroid; and a long, tightly wound solenoid.

Suggestions for Further Reading

Banerjee, Subir K.: "Polar Flip-Flop," *The Sciences*, November/December 1984, p. 24.

The first half of this article is a fine short history of our understanding of the earth's magnetic field. Following this, the modern theory of earth magnetism is discussed, as well as evidence that the field may undergo a north–south reversal over the next 2000 years.

Carrigan, Charles R., and David Gubbins: "The Source of the Earth's Magnetic Field," *Scientific American*, February 1979, p. 118.

The earth may act as a huge dynamo in which electric currents in the molten and flowing metallic core maintain themselves by producing the magnetic field that deflects compass needles on the surface.

Shamos, Morris H.: "Electromagnetism—Hans Christian Oersted," in *Great Experiments in Physics*, Henry Holt and Co., New York, 1959. Reprinted by Dover, 1987.

Oersted's account, in translation, of his discovery of the magnetic field accompanying an electric current, with editorial annotations for clarity and a biographical sketch.

Williams, L. Pearce: "André-Marie Ampère," *Scientific American*, January 1989, p. 90.

A short biography of Ampère, describing his experiments and discoveries in electricity and magnetism, and the rationale behind his methods of investigation.

Review

A. Objectives: After studying this chapter, you should:

1. Be able to state the Biot–Savart law and use it to calculate the magnetic field **B** due to a straight, current-carrying wire and on the axis of a circular current loop.

2. Be able to sketch the magnetic-field lines for a long, straight current; a circular current loop; a solenoid; and a toroid.

3. Be able to state Ampère's law and discuss its uses and limitations.

4. Be able to use Ampère's law to derive expressions for **B** due to an infinite straight current; a toroid; and a long, tightly wound solenoid.

B. Define, explain, or otherwise identify:

Permeability of free space	Current balance
Biot–Savart law	Ampère's law
Solenoid	Toroid

C. True or false: If the statement is true, explain why it is true. If it is false, give a counterexample.

1. The magnetic field due to a current element is parallel to the current element.

2. The magnetic field due to a current element varies inversely with the square of the distance from the element.

3. The magnetic field due to a long wire varies inversely with the square of the distance from the wire.

4. Ampère's law is valid only if there is a high degree of symmetry.

5. Ampère's law is valid only for continuous currents.

Problems

Level I

25-1 The Magnetic Field of Moving Point Charges

1. At a certain instant of time a particle with charge $q = 12\ \mu C$ is located at $x = 0$, $y = 2$ m; its velocity at that time is $\mathbf{v} = 30$ m/s $\mathbf{i}$. Find the magnetic field at (*a*) the origin; (*b*) $x = 0$, $y = 1$ m; (*c*) $x = 0$, $y = 3$ m; and (*d*) $x = 0$, $y = 4$ m.

2. For the particle in Problem 1, find the magnetic field at (*a*) $x = 1$ m, $y = 3$ m; (*b*) $x = 2$ m, $y = 2$ m; and (*c*) $x = 2$ m, $y = 3$ m.

3. A proton (charge $+e$) traveling with a velocity of $\mathbf{v} = 1 \times 10^4$ m/s $\mathbf{i} + 2 \times 10^4$ m/s $\mathbf{j}$ is located at $x = 3$ m, $y = 4$ m at some time. Find the magnetic field at the following positions: (*a*) $x = 2$ m, $y = 2$ m; (*b*) $x = 6$ m, $y = 4$ m; and (*c*) $x = 3$ m, $y = 6$ m.

4. Two equal charges q located at (0, 0, 0) and (0, b, 0) at time zero are moving with speed v in the positive x direction ($v \ll c$). Find the ratio of the magnitudes of the magnetic and electrostatic force on each.

5. An electron orbits a proton at a radius of 5.29×10^{-11} m. What is the magnetic field at the proton due to the orbital motion of the electron?

Section 25-2 The Magnetic Field of Currents: The Biot–Savart Law

6. A small current element $I\,d\boldsymbol{\ell}$, with $d\boldsymbol{\ell} = 2$ mm $\mathbf{k}$ and $I = 2$ A is centered at the origin. Find the magnetic field $d\mathbf{B}$ at the following points: (*a*) on the x axis at $x = 3$ m, (*b*) on the x axis at $x = -6$ m, (*c*) on the z axis at $z = 3$ m, and (*d*) on the y axis at $y = 3$ m.

7. For the current element in Problem 6, find the magnitude and direction of $d\mathbf{B}$ at $x = 0$, $y = 3$ m, $z = 4$ m.

8. For the current element in Problem 6, find the magnitude of $d\mathbf{B}$ and indicate its direction on a diagram at (*a*) $x = 2$ m, $y = 4$, $z = 0$ and (*b*) $x = 2$ m, $y = 0$, $z = 4$ m.

9. A single-turn, circular loop of radius 10.0 cm is to produce a field at its center that will just cancel the earth's magnetic field at the equator, which is 0.7 G directed north. Find the current in the loop and make a sketch showing the orientation of the loop and the current.

10. A solenoid with length 30 cm, radius 1.2 cm, and 300 turns carries a current of 2.6 A. Find B on the axis of the solenoid (*a*) at the center, (*b*) inside the solenoid at a point 10 cm from one end, and (*c*) at one end.

11. A solenoid 2.7 m long has a radius of 0.85 cm and 600 turns. It carries a current I of 2.5 A. What is the approximate magnetic field B on the axis of the solenoid?

12. A single loop of wire of radius 3 cm carries a current of 2.6 A. What is the magnitude of **B** on the axis of the loop at (*a*) the center of the loop, (*b*) 1 cm from the center, (*c*) 2 cm from the center, and (*d*) 35 cm from the center?

13. For the loop of wire in Problem 12, at what point along the axis of the loop is the magnetic field (*a*) 10 percent of the field at the center, (*b*) 1 percent of the field at the center, and (*c*) 0.1 percent of the field at the center?

14. A long, straight wire carries a current of 10 A. Find the magnitude of **B** at (*a*) 10 cm, (*b*) 50 cm, and (*c*) 2 m from the center of the wire.

Problems 15 to 20 refer to Figure 25-30, which shows two long, straight wires in the xy plane and parallel to the x axis. One wire is at $y = -6$ cm and the other is at $y = +6$ cm. The current in each wire is 20 A.

Figure 25-30 Problems 15 through 20.

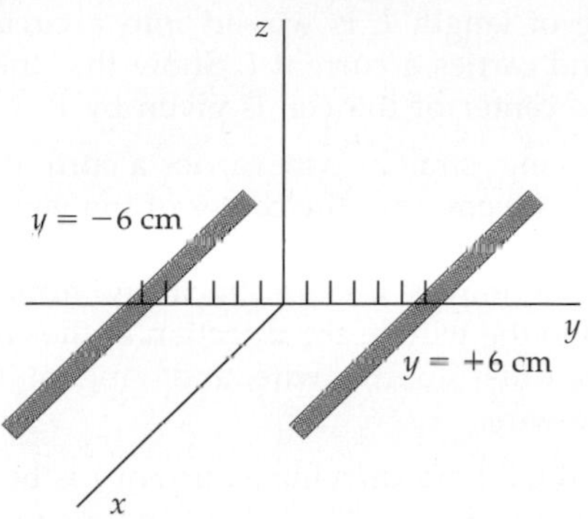

15. If the currents in Figure 25-30 are in the negative x direction, find **B** at the points on the y axis at (*a*) $y = -3$ cm, (*b*) $y = 0$, (*c*) $y = +3$ cm, and (*d*) $y = +9$ cm.

16. Sketch B_z versus y for points on the y axis when both currents are in the negative x direction.

17. Find **B** at points on the y axis as in Problem 15 when the current in the wire at $y = -6$ cm is in the negative x direction and the current in the wire at $y = +6$ cm is in the positive x direction.

18. Sketch B_z versus y for points on the y axis when the directions of the currents are opposite to those in Problem 17.

19. Find **B** on the z axis at $z = +8$ cm if (*a*) the currents are parallel, as in Problem 15 and (*b*) the currents are antiparallel, as in Problem 17.

20. Find the magnitude of the force per unit length exerted by one wire on the other.

21. The current in the wire of Figure 25-31 is 8.0 A. Find **B** at point P due to each wire segment and sum to find the resultant **B**.

Figure 25-31 Problem 21.

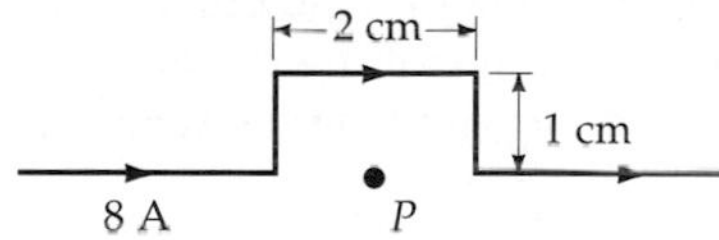

22. Find the magnetic field at point P in Figure 25-32.

Figure 25-32 Problem 22.

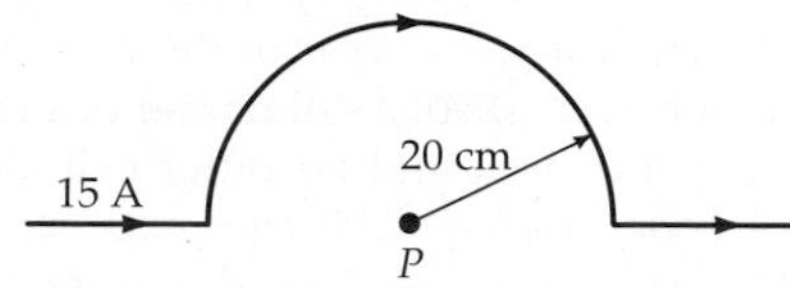

23. A single-turn circular loop of radius 8.5 cm is to produce a field at its center that will just cancel the earth's field of magnitude 0.7 G directed at 70° below the horizontal north direction. Find the current in the loop and make a sketch showing the orientation of the loop and the current.

24. In Figure 25-33, find the magnetic field at point P, which is at the common center of the two semicircular arcs.

Figure 25-33 Problem 24.

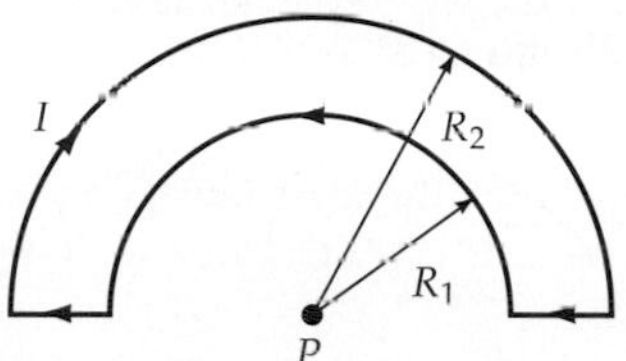

25-3 Definition of the Ampere

25. Two long, straight, parallel wires 8.6 cm apart carry currents of equal magnitude I. They repel each other with a force per unit length of 3.6 nN/m. (*a*) Are the currents parallel or antiparallel? (*b*) Find I.

26. A wire of length 16 cm is suspended by flexible leads above a long, straight wire. Equal but opposite currents are established in the wires such that the 16-cm wire floats 1.5 mm above the long wire with no tension in its suspension leads. If the mass of the 16-cm wire is 14 g, what is the current?

27. In an experiment with a current balance, the upper wire of length 30 cm is pivoted such that with no current it balances at 2 mm above a fixed, parallel wire also 30 cm long. When the wires carry equal but opposite currents I, the upper wire balances at its original position when a mass of 2.4 g is placed on it. What is the current I?

28. Three long, parallel, straight wires pass through the corners of an equilateral triangle of sides 10 cm as shown in Figure 25-34, where a dot means that the current is out of the paper and a cross means that it is into the paper. If each current is 15.0 A, find (*a*) the force per unit length on the upper wire and (*b*) the magnetic field **B** at the upper wire due to the two lower wires. *Hint:* It is easier to find the force per unit length directly from Equation 25-14 and use your result to find **B** than to find **B** and use it to find the force.

Figure 25-34 Problems 28 and 29.

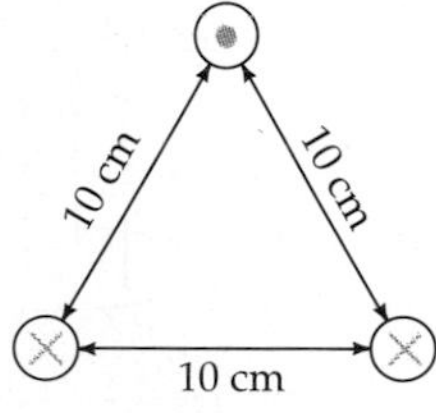

29. Work Problem 28 with the current in the lower right corner of Figure 25-34 reversed.

25-4 Ampère's Law

30. A long, straight, thin-walled, cylindrical shell of radius R carries a current I. Find **B** inside and outside the cylinder.

31. In Figure 25-35, one current is 8 A into the paper; the other current is 8 A out of the paper; and each curve is a circular path. (*a*) Find $\oint \mathbf{B}\cdot d\boldsymbol{\ell}$ for each path indicated. (*b*) Which path, if any, can be used to find **B** at some point due to these currents?

Figure 25-35 Problem 31.

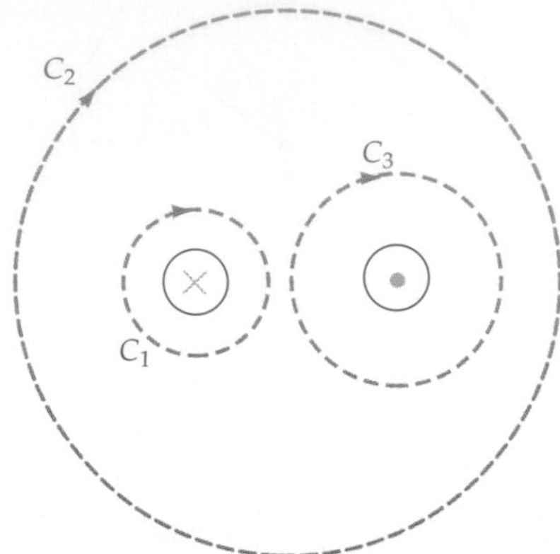

32. A very long coaxial cable consists of an inner wire and a concentric outer cylindrical conducting shell of radius R. At one end, the wire is connected to the shell. At the other end, the wire and shell are connected to opposite terminals of a battery, so there is a current down the wire and back up the shell. Assume that the cable is straight. Find **B** (*a*) at points between the wire and the shell far from the ends, and (*b*) outside the cable.

33. A wire of radius 0.5 cm carries a current of 100 A that is uniformly distributed over its cross-sectional area. Find B (*a*) 0.1 cm from the center of the wire, (*b*) at the surface of the wire, and (*c*) at a point outside the wire 0.2 cm from the surface of the wire. (*d*) Sketch a graph of B versus the distance from the center of the wire.

34. Show that a uniform magnetic field with no fringing field, such as that shown in Figure 25-36, is impossible because it violates Ampère's law. Do this by applying Ampère's law to the rectangular curve shown by the dashed lines.

Figure 25-36 Problem 34.

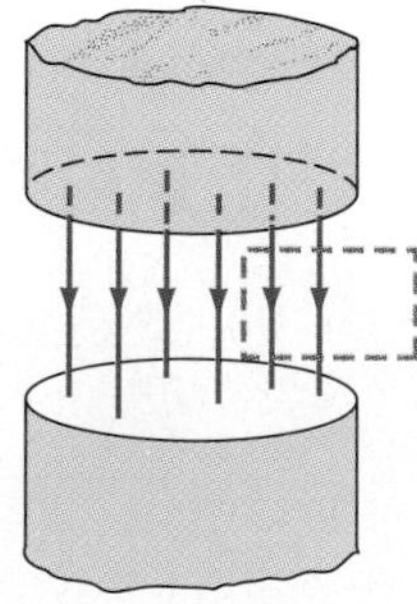

35. A tightly wound toroid of inner radius 1 cm and outer radius 2 cm has 1000 turns of wire and carries a current of 1.5 A. (*a*) What is the magnetic field at a distance of 1.1 cm from the center? (*b*) What is the field 1.5 cm from the center?

Level II

36. A wire of length L is wound into a circular coil of N loops and carries a current I. Show that the magnetic field at the center of the coil is given by $B = \mu_0 \pi N^2 I/L$.

37. A very long, straight wire carries a current of 20.0 A. An electron 1.0 cm from the center of the wire is moving with a speed of 5.0×10^6 m/s. Find the force on the electron when it moves (*a*) directly away from the wire, (*b*) parallel to the wire in the direction of the current, and (*c*) perpendicular to the wire and tangent to a circle around the wire.

38. A very long wire carrying a current I is bent into the shape shown in Figure 25-37. Find the magnetic field at point P.

Figure 25-37 Problem 38.

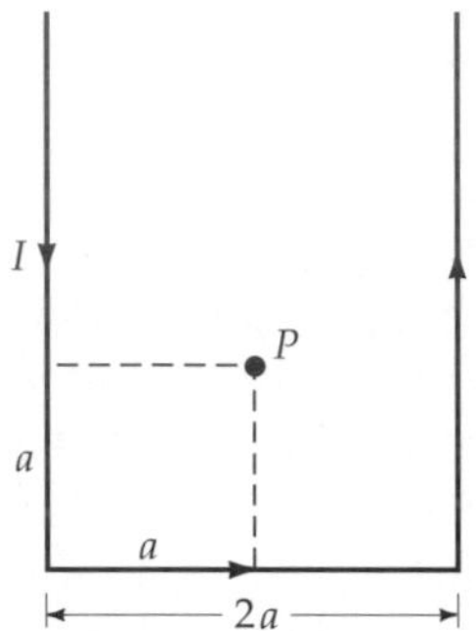

39. A loop of wire of length L carries a current I. Compare the magnetic fields at the center of the loop when it is (*a*) a circle, (*b*) a square, and (*c*) an equilateral triangle. Which field is largest?

40. A power cable carrying 50.0 A is 2.0 m below the earth's surface, but its direction and precise position are unknown. Show how you could locate the cable using a compass. Assume that you are at the equator, where the earth's magnetic field is 0.7 G north.

41. Four long, straight, parallel wires each carry current I. In a plane perpendicular to the wires, the wires are at the corners of a square of side a. Find the force per unit length on one of the wires if (*a*) all the currents are in the same direction and (*b*) the currents in the wires at adjacent corners are oppositely directed.

42. A current I is uniformly distributed over the cross section of a long, straight wire of radius 1.40 mm. At the surface of the wire, the magnitude of the magnetic field is $B = 2.46$ mT. Find the magnitude of the magnetic field at (*a*) 2.10 mm from the axis and (*b*) 0.60 mm from the axis. (*c*) Find the current I.

43. A coaxial cable consists of a solid inner cylindrical conductor of radius 1.00 mm and an outer cylindrical shell conductor of inner radius 2.00 mm and outer radius 3.00 mm. There is a current of 18 A down the inner wire and an equal return current in the outer conductor. The currents are uniform over the cross section of each conductor. Find the numerical value of $\oint \mathbf{B}\cdot d\boldsymbol{\ell}$ for a closed circular path (centered on the axis of the cable and in a plane perpendicular to the axis) which has a radius r for (*a*) $r = 1.50$ mm, (*b*) $r = 2.50$ mm, and (*c*) $r = 3.50$ mm.

44. An infinitely long, insulated wire lies along the x axis and carries current I in the positive x direction. A second infinitely long, insulated wire lies along the y axis and carries current $I/4$ in the positive y direction. Where in the xy plane is the resultant magnetic field zero?

45. A large, 50-turn circular coil of radius 10.0 cm carries a current of 4.0 A. At the center of the large coil is a small 20-turn coil of radius 0.5 cm carrying a current of 1.0 A. The planes of the two coils are perpendicular. Find the torque exerted by the large coil on the small coil. (Neglect any variation in **B** due to the large coil over the region occupied by the small coil.)

46. A relatively inexpensive ammeter called a *tangent galvanometer* can be made using the earth's field. A plane circular coil of N turns and radius R is oriented such that the field B_c it produces in the center of the coil is either east or west. A compass is placed at the center of the coil. When there is no current in the coil, the compass needle points north. When there is a current I, the compass needle points in the direction of the resultant magnetic field **B** at an angle θ to the north. Show that the current I is related to θ and the horizontal component of the earth's field B_e by

$$I = \frac{2RB_e}{\mu_0 N} \tan \theta$$

47. An infinitely long, straight wire is bent as shown in Figure 25-38. The circular portion has a radius of 10 cm with its center a distance r from the straight part. Find r such that the magnetic field at the center of the circular portion is zero.

Figure 25-38 Problem 47.

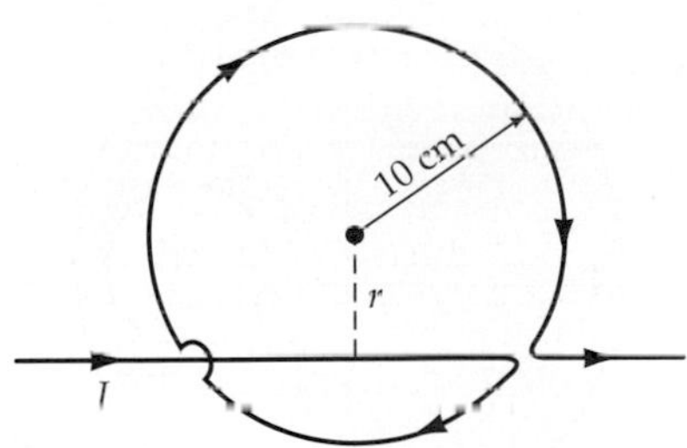

48. An infinitely long, nonconducting cylinder of radius R lies along the z axis. Five long conducting wires are parallel to the cylinder and spaced equally on the upper half of its surface. Each wire carries a current I in the positive z direction. Find the magnetic field on the z axis.

49. An infinitely long wire lies along the z axis and carries a current of 20 A in the positive z direction. A second infinitely long wire is parallel to the z axis at $x = 10$ cm. (*a*) Find the current in the second wire if the magnetic field at $x = 2$ cm is zero. (*b*) What is the magnetic field at $x = 5$ cm?

50. Three very long, parallel wires are at the corners of a square, as shown in Figure 25-39. Find the magnetic field **B** at the unoccupied corner of the square when (*a*) all the currents are into the paper, (*b*) I_1 and I_3 are in and I_2 is out, and (*c*) I_1 and I_2 are in and I_3 is out.

Figure 25-39 Problem 50.

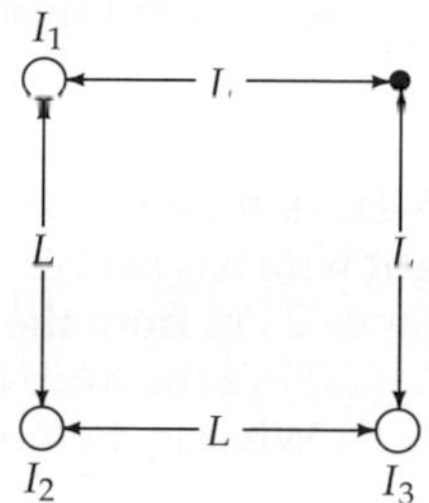

51. (*a*) Find the magnetic field at point P for the wire carrying current I shown in Figure 25-40. (*b*) Use your result from (*a*) to find the field at the center of a polygon of N sides. Show that when N is very large, your result approaches that for the magnetic field at the center of a circle.

Figure 25-40 Problem 51.

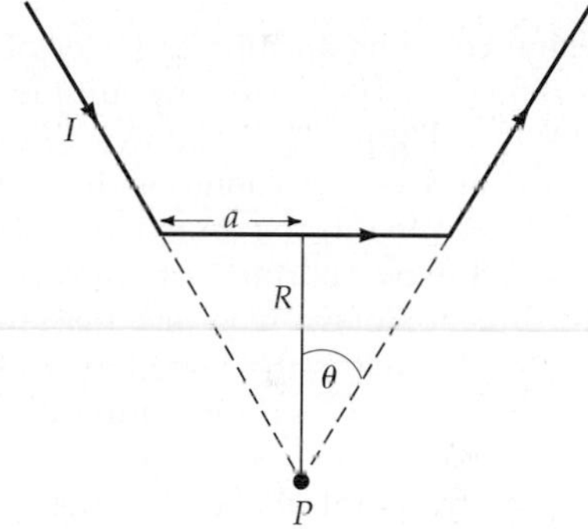

52. A circular current loop of radius R carrying a current I is centered at the origin with its axis along the x axis. Its current is such that it produces a magnetic field in the positive x direction. (*a*) Sketch a graph of B_x versus x for points on the x axis. Include both positive and negative values of x. Compare this graph with that for E_x due to a charged ring of the same size. (*b*) A second, identical current loop, carrying an equal current in the same sense, is in a plane parallel to the yz plane with its center at $x = d$. Sketch graphs of the magnetic field on the x axis due to each loop separately and the resultant field due to the two loops. Show from your sketch that dB_x/dx is zero midway between the two loops.

53. Two coils which are separated by a distance equal to their radius and carry equal currents such that their axial fields add are called **Helmholtz coils.** A feature of Helmholtz coils is that the resultant magnetic field between the coils is very uniform. Let $R = 10$ cm, $I = 20$ A, and $N = 300$ turns for each coil. Place one coil in the yz plane with its center at the origin and the other in a parallel plane at $x = 10$ cm. (*a*) Calculate the resultant field B_x at $x = 5$ cm,

$x = 7$ cm, $x = 9$ cm, and $x = 11$ cm. (*b*) Use your results and the fact that B_x is symmetric about the midpoint of the coils to sketch B_x versus x. (See also Problem 62.)

54. An infinitely long, thick cylindrical shell of inner radius a and outer radius b carries a current I uniformly distributed across a cross section of the shell. Find the magnetic field for (*a*) $r < a$, (*b*) $a < r < b$, and (*c*) $r > b$.

55. A long, straight wire carries a current of 20 A as shown in Figure 25-41. A rectangular coil with two sides parallel to the straight wire has sides 5 cm and 10 cm with the near side a distance 2 cm from the wire. The coil carries a current of 5 A. (*a*) Find the force on each segment of the rectangular coil. (*b*) What is the net force on the coil?

Figure 25-41 Problem 55.

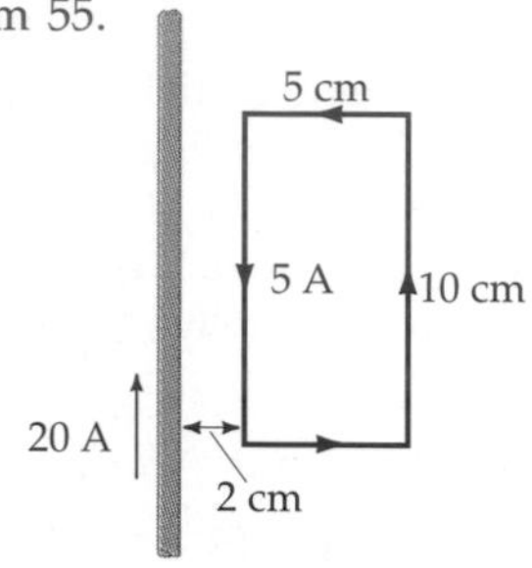

56. The xz plane contains an infinite sheet of current in the positive z direction. The current per unit length (along the x direction) is λ. Figure 25-42*a* shows a point P above the sheet ($y > 0$) and two portions of the current sheet labeled I_1 and I_2. (*a*) What is the direction of the magnetic field **B** at P due to the two portions of the current shown? (*b*) What is the direction of the magnetic field **B** at P due to the entire sheet? (*c*) What is the direction of **B** at a point below the sheet ($y < 0$)? (*d*) Apply Ampère's law to the rectangular curve shown in Figure 25-42*b* to show that the magnetic field at any point above the sheet is given by

$$\mathbf{B} = -\tfrac{1}{2}\mu_0\lambda\mathbf{i}$$

Figure 25-42 Problem 56.

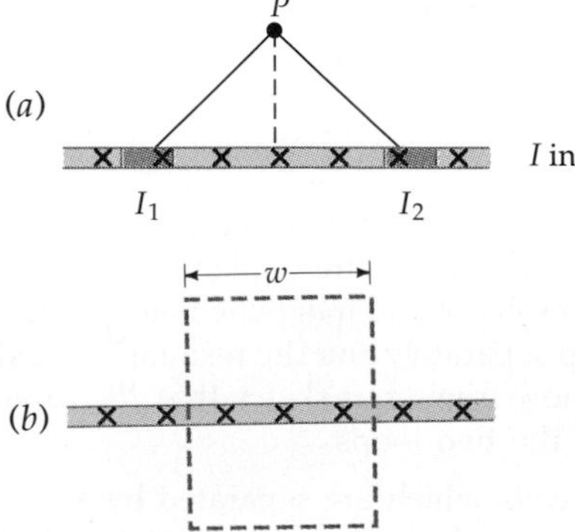

Level III

57. A square loop of side L lies in the yz plane with its center at the origin. It carries a current I. Find the magnetic field **B** at any point on the x axis and show from your expression that for x much larger than L

$$\mathbf{B} \approx \frac{\mu_0}{4\pi}\frac{2\mathbf{m}}{x^3}$$

where $m = IL^2$ is the magnetic moment of the loop.

58. A circular loop carrying current I lies in the yz plane with its axis along the x axis. (*a*) Evaluate the line integral $\oint \mathbf{B}\cdot d\boldsymbol{\ell}$ along the axis of the loop from $x = -L$ to $x = +L$. (*b*) Show that when $L \to \infty$, the line integral approaches $\mu_0 I$. This result can be related to Ampère's law by closing the curve of integration with a semicircle of radius L on which $B \approx 0$ for L very large.

59. A very long, straight conductor with a circular cross section of radius R carries a current I. Inside the conductor, there is a cylindrical hole of radius a whose axis is parallel to the axis of the conductor a distance b from it (Figure 25-43). Let the z axis be the axis of the conductor, and let the axis of the hole be at $x = b$. Find the magnetic field **B** at the point (*a*) on the x axis at $x = 2R$, and (*b*) on the y axis at $y = 2R$. *Hint*: Consider a uniform current distribution throughout the cylinder of radius R plus a current in the opposite direction in the hole.

60. For the cylinder with the hole in Problem 59, show that the magnetic field inside the hole is uniform, and find its magnitude and direction.

Figure 25-43 Problems 59 and 60.

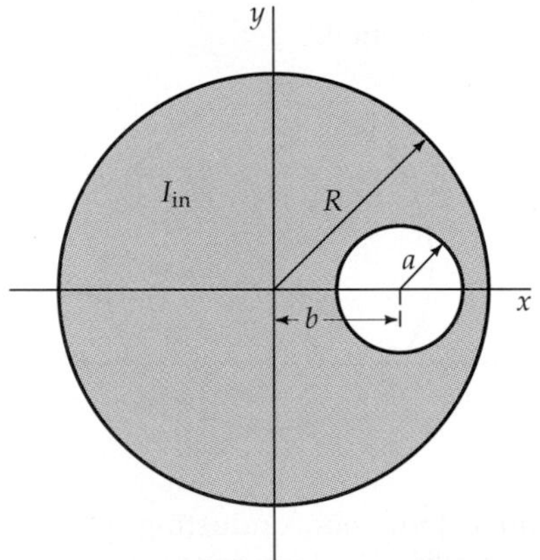

61. A disk of radius R carries a fixed charge density σ and rotates with angular velocity ω. (*a*) Consider a circular strip of radius r and width dr with charge dq. Show that the current produced by this strip $dI = (\omega/2\pi)\,dq = \omega\sigma r\,dr$. (*b*) Use your result from part (*a*) to show that the magnetic field at the center of the disk is $B = \frac{1}{2}\mu_0\sigma\omega R$. (c) Use your result from part (a) to find the magnetic field at a point on the axis of the disk a distance x from the center.

62. Two Helmholtz coils have radii R and their axes along the x axis (see Problem 53). One coil is in the yz plane and the other is in a parallel plane at $x = R$. Show that at the midpoint of the coils ($x = \frac{1}{2}R$), $dB_x/dx = 0$, $d^2B_x/dx^2 = 0$, and $d^3B_x/dx^3 = 0$. This shows that the magnetic field at points near the midpoint is approximately equal to that at the midpoint.

63. A solenoid has n turns per unit length and radius R and carries a current I. Its axis is along the x axis with one end at $x = -\frac{1}{2}L$ and the other end at $x = +\frac{1}{2}L$, where L is the total length of the solenoid. Show that the magnetic field B at a point on the axis outside the solenoid is given by

$$B = \tfrac{1}{2}\mu_0 nI(\cos\theta_1 - \cos\theta_2) \qquad \text{25-18}$$

where

$$\cos\theta_1 = \frac{x + \frac{1}{2}L}{[R^2 + (x + \frac{1}{2}L)^2]^{1/2}}$$

and

$$\cos\theta_2 = \frac{x - \frac{1}{2}L}{[R^2 + (x - \frac{1}{2}L)^2]^{1/2}}$$

64. In Problem 63, a formula for the magnetic field along the axis of a solenoid is given. For $x \gg L$ and $L > R$, the angles θ_1 and θ_2 in Equation 25-18 are very small so the small-angle approximation $\cos\theta \approx 1 - \theta^2/2$ is valid. (a) Draw a diagram and show that

$$\theta_1 \approx \frac{R}{x + \frac{1}{2}L}$$

and

$$\theta_2 \approx \frac{R}{x - \frac{1}{2}L}$$

(b) Show that the magnetic field at a point far from either end of the solenoid can be written

$$B = \frac{\mu_0}{4\pi}\left(\frac{q_m}{r_1^2} - \frac{q_m}{r_2^2}\right) \qquad \text{25-19}$$

where $r_1 = x - \frac{1}{2}L$ is the distance to the near end of the solenoid, $r_2 = x + \frac{1}{2}L$ is the distance to the far end, and $q_m = nI\pi R^2 = m/L$, where $m = NI\pi R^2$ is the magnetic moment of the solenoid.

65. In this problem, you will derive Equation 25-19 by another method. Consider a long, tightly wound solenoid of length L and radius $R \ll L$ lying along the axis with its center at the origin. It has N turns and carries a current I. Consider an element of the solenoid of length dx. (a) What is the magnetic moment of this element? (b) Show that the magnetic field dB due to this element at a point on the x axis x_0 far from the element is given by

$$dB = \frac{\mu_0}{2\pi} nIA \frac{dx}{x'^3}$$

where $A = \pi R^2$ and $x' = x_0 - x$ is the distance from the element to the field point. (c) Integrate this expression from $x = -\frac{1}{2}L$ to $x = +\frac{1}{2}L$ to obtain Equation 25-19.

Chapter 26

Magnetic Induction

A cross section of a pickup for an electric guitar. The pickup consists of a cylindrical permanent magnet wrapped with a wire coil. The guitar string is magnetized by the magnet below it and produces an oscillating magnetic flux through the coil of the pickup when it vibrates. A current of frequency equal to that of the vibrating string is thereby induced in the coil. The current is amplified and sent to a speaker. (For this photo, the guitar string has been placed parallel to six pairs of pickups. When mounted for playing, the strings are perpendicular to the direction shown here with each string crossing over one pair of pickups.)

In the previous chapter, we learned that a current in a wire creates a magnetic field. In the early 1830s, Michael Faraday in England and Joseph Henry in America independently discovered that a magnetic field induces a current in a wire but that this happens only when the magnetic field is *changing*. When you pull the plug of an electric cord from its socket, you sometimes observe a small spark. Before the cord is disconnected, it carries a current, which, as we have seen, produces a magnetic field encircling the current. When the cord is disconnected, the current abruptly ceases and the magnetic field around it collapses. The changing magnetic field produces an emf that tries to maintain the original current, resulting in a spark across the plug. Once the magnetic field has reached zero and is therefore no longer changing, the emf is zero. The emfs and currents caused by changing magnetic fields are called **induced emfs** and **induced currents.** The process itself is referred to as **magnetic induction.**

In the case of an electric cord being disconnected from its power source, the changing magnetic field is caused by a change in the electric currents. However, a changing magnetic field can also be produced by moving a magnet. Figure 26-1 illustrates a simple classroom demonstration of induced currents. The ends of a coil are attached to a galvanometer and a strong magnet is moved toward or away from the coil. The momentary deflection shown by the galvanometer *during* the motion indicates that there is an induced electric current in the coil–galvanometer circuit. A current is also induced if the coil is moved toward or away from the magnet, or if the coil is rotated in a fixed magnetic field. A coil rotating in a magnetic field is the basic element of a generator, which converts mechanical or thermal energy into electrical energy. In a hydrostatic power plant, a river is dammed up and the water is released in a controlled way to turn the coils. In a steam-based power plant, water is heated and converted to steam by energy from burning coal or from nuclear fission. The pressure of the steam is then used to turn the coils.

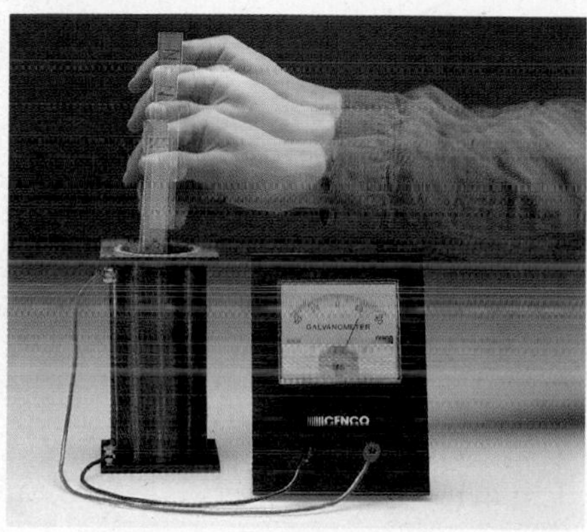

Figure 26-1 Demonstration of induced emf. When the magnet is moving away from the coil, an emf is induced in the coil as shown by the galvanometer's deflection. No deflection is observed when the magnet is stationary.

All of the various methods of magnetic induction can be summarized by a single relation known as Faraday's law, which relates the induced emf in a circuit to the change in magnetic flux through the circuit.

26-1 Magnetic Flux

Magnetic flux is the magnetic analog to electric flux, which we learned about in Section 19-2. It is related to the number of magnetic-field lines that pass through a given area. In Figure 26-2, the magnetic field is perpendicular to the area bounded by a simple circuit consisting of one turn of wire. In this case, the **magnetic flux** ϕ_m is defined as the product of the magnetic field **B** and the area A bounded by the circuit:

$$\phi_m = BA$$

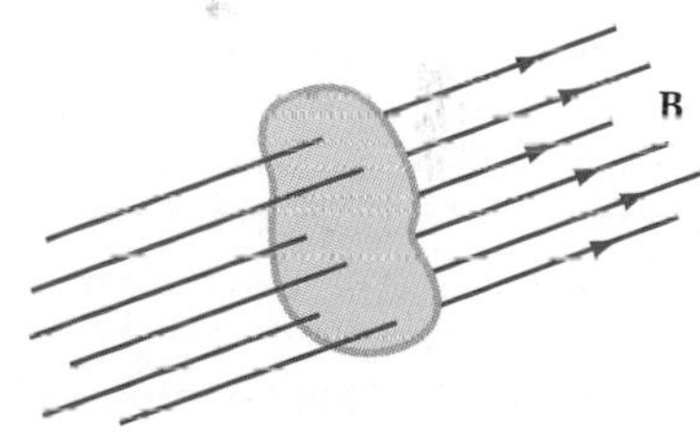

Figure 26-2 When the magnetic field **B** is perpendicular to the area enclosed by a loop, the magnetic flux through the loop is BA.

The unit of magnetic flux is that of magnetic field times area, tesla-meter squared, which is called a **weber** (Wb):

$$1 \text{ Wb} = 1 \text{ T}\cdot\text{m}^2 \qquad 26\text{-}1$$

Since the magnetic field is proportional to the number of magnetic-field lines per unit area, the magnetic flux is proportional to the number of lines through the area.

If the magnetic field is not perpendicular to the surface, as in Figure 26-3, the flux ϕ_m is defined as

$$\phi_m = \mathbf{B}\cdot\hat{\mathbf{n}}A = BA\cos\theta = B_n A \qquad 26\text{-}2$$

where $B_n = \mathbf{B}\cdot\hat{\mathbf{n}}$ is the component of the magnetic-field vector that is perpendicular or normal to the surface.

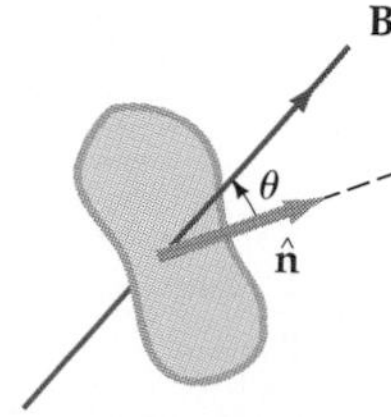

Figure 26-3 When **B** makes an angle θ with the normal to the area of a loop, the flux through the loop is $B\cos\theta\, A$.

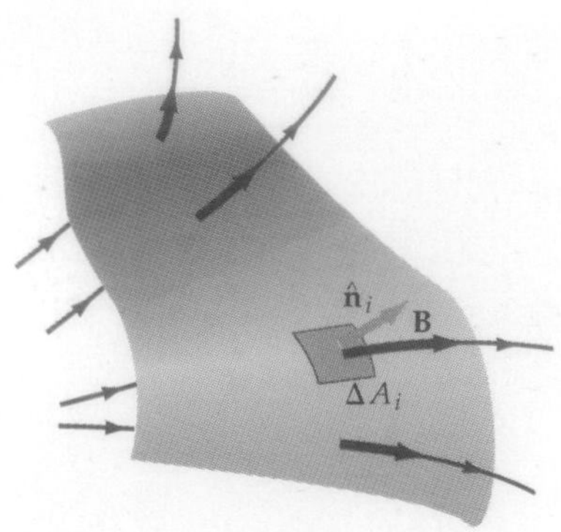

Figure 26-4 When **B** varies in either magnitude or direction, the area is divided into small elements ΔA_i. The flux through the area is computed by summing $B_n \Delta A_i$ over all the area elements.

We can generalize our definition of magnetic flux to curved surfaces over which the magnetic field may vary in magnitude or direction or both by dividing the surface up into a large number of very small area elements. If each element is small enough, it can be considered to be a plane, and the variation of the magnetic field across the element can be neglected. Let $\hat{\mathbf{n}}_i$ be the unit vector perpendicular to such an element and ΔA_i be its area (Figure 26-4). The flux of the magnetic field through this element is

$$\Delta\phi_{mi} = \mathbf{B}\cdot\hat{\mathbf{n}}_i\ \Delta A_i$$

The total flux through the surface is the sum of $\Delta\phi_{mi}$ over all the elements. In the limit as the number of elements approaches infinity and the area of each element approaches zero, this sum becomes an integral. The general definition of magnetic flux is then

$$\phi_m = \lim_{\Delta A_i \to 0} \sum_i \mathbf{B}\cdot\hat{\mathbf{n}}_i\ \Delta A_i = \int_S \mathbf{B}\cdot\hat{\mathbf{n}}\ dA$$

For a coil of N turns, the flux through the coil is N times the flux through a single turn:

$$\phi_m = NBA\cos\theta \qquad 26\text{-}3$$

For the general case in which **B** is not necessarily constant over the area, the flux is

Magnetic flux defined

$$\phi_m = \int_S N\mathbf{B}\cdot\hat{\mathbf{n}}\ dA = \int_S NB_n\ dA \qquad 26\text{-}4$$

Example 26-1

A uniform magnetic field of magnitude 2000 G makes an angle of 30° with the axis of a circular coil of 300 turns and a radius of 4 cm. Find the magnetic flux through the coil.

Since 1 G = 10^{-4} T, this magnetic field in SI units is 0.2 T. The area of the coil is

$$A = \pi r^2 = (3.14)(0.04\ \text{m})^2 = 0.00502\ \text{m}^2$$

The flux through the coil is thus

$$\phi_m = NBA\cos\theta = (300)(0.2\ \text{T})(0.00502\ \text{m}^2)(0.866) = 0.26\ \text{Wb}$$

Example 26-2

Find the magnetic flux through a solenoid that is 40 cm long, has a radius of 2.5 cm, has 600 turns, and carries a current of 7.5 A.

The magnetic field inside the solenoid is given by Equation 25-10:

$$B = \mu_0 nI = (4\pi \times 10^{-7}\ \text{T·m/A})(600\ \text{turns}/0.40\ \text{m})(7.5\ \text{A})$$
$$= 1.41 \times 10^{-2}\ \text{T}$$

Since the magnetic field is essentially constant across the cross-sectional area of the coil the magnetic flux is

$$\phi_m = NBA = (600)(1.41 \times 10^{-2}\ \text{T})\pi(0.025\ \text{m})^2 = 1.66 \times 10^{-2}\ \text{Wb}$$

Note that since $\phi_m = NBA$ and B is proportional to the number of turns N, the flux is proportional to N^2.

26-2 Induced Emf and Faraday's Law

Michael Faraday (1791–1867).

The experiments of Faraday, Henry, and others have shown that if the magnetic flux through a circuit is changed by any means, an emf equal in magnitude to the rate of change of the flux is induced in the circuit. The emf is usually detected by observing a current in the circuit, but it is present even if the circuit is incomplete (not closed) so that no current exists. In our previous discussions, the emf in a circuit has been localized in a specific region of the circuit, such as between the terminals of the battery. However, the emf induced by a changing magnetic flux can be considered to be distributed throughout the circuit.

Consider a single loop of wire in a magnetic field as shown in Figure 26-5. If the flux through the loop is changing, there is an emf induced in the loop. Since emf is the work done per unit charge, there must be a force exerted on the charge associated with the emf. The force per unit charge is the electric field **E**, which in this case is induced by the changing flux. The line integral of the electric field around a complete circuit equals the work done per unit charge, which, by definition, is the emf in the circuit:

$$\mathcal{E} = \oint_C \mathbf{E} \cdot d\boldsymbol{\ell} \qquad 26\text{-}5$$

Emf defined

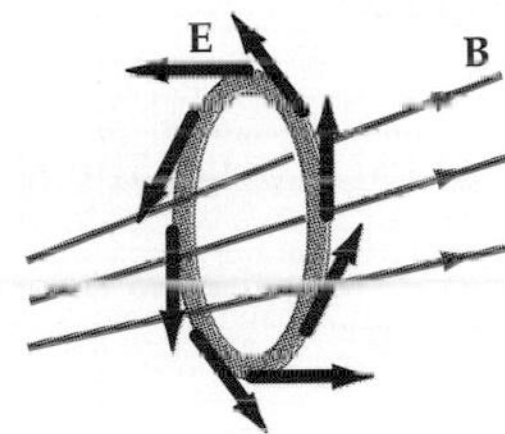

Figure 26-5 When the magnetic flux through the wire loop is changing, an emf is induced in the loop. The emf is distributed throughout the loop and is equivalent to a nonconservative electric field **E** parallel to the wire. In this figure, the direction of **E** corresponds to the case in which the flux through the loop is increasing.

The electric fields that we studied previously resulted from static electric charges. These fields are conservative, meaning that the line integral of the electrostatic field around a closed curve is zero. However, the electric field resulting from a changing magnetic flux is not conservative. The line integral around a closed curve equals the induced emf, which equals the rate of change of the magnetic flux:

$$\mathcal{E} = \oint_C \mathbf{E} \cdot d\boldsymbol{\ell} = -\frac{d\phi_m}{dt} \qquad 26\text{-}6$$

Faraday's law

This result is known as **Faraday's law.** The negative sign in Faraday's law has to do with the direction of the induced emf, which will be discussed shortly.

Exercise

Show that a weber per second is a volt.

The magnetic flux through a circuit can be changed in many different ways. The current producing the magnetic field may be increased or decreased, permanent magnets may be moved toward the circuit or away from it, the circuit itself may be moved toward or away from the source of the flux, the orientation of the circuit may be changed, or the area of the circuit in a fixed magnetic field may be increased or decreased. In every case, an emf is induced in the circuit that is equal in magnitude to the rate of change of the magnetic flux.

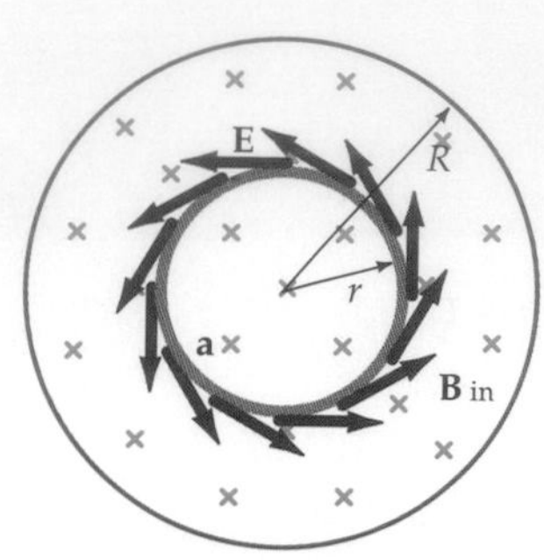

Figure 26-6 Diagram for Example 26-3. The magnetic field **B** is into the page and uniform over a circular region of radius R. When **B** changes, the magnetic flux changes and an emf $\mathcal{E} = \oint \mathbf{E} \cdot d\boldsymbol{\ell}$ is induced around any curve enclosing the flux. The induced electric field **E** at a distance r is tangent to the circle of radius r and is in the direction shown if **B** is increasing.

Example 26-3

A magnetic field **B** is perpendicular to the plane of the page and uniform in a circular region of radius R as shown in Figure 26-6. Outside of the circular region, $\mathbf{B} = 0$. The rate of change of the magnitude of **B** is dB/dt. What is the magnitude of the induced electric field in the plane of the page at a distance r from the center of the circular region?

According to Equation 26-6, the line integral of **E** around a closed curve equals the negative of the rate of change of the magnetic flux through the curve. Since we are interested only in magnitudes, we neglect the negative sign in this equation. Then

$$\oint_C \mathbf{E} \cdot d\boldsymbol{\ell} = \frac{d\phi_{\mathrm{m}}}{dt}$$

In the figure, we have chosen a circular curve of radius $r < R$ to compute the line integral $\oint \mathbf{E} \cdot d\boldsymbol{\ell}$. By symmetry, **E** is tangent to this curve and has the same value at any point on it. Then

$$\oint_C \mathbf{E} \cdot d\boldsymbol{\ell} = E(2\pi r)$$

Since **B** is perpendicular to the plane of this curve, the flux through the curve is

$$\phi_{\mathrm{m}} = BA = B\pi r^2$$

and the rate of change of the flux is

$$\frac{d\phi_{\mathrm{m}}}{dt} = \pi r^2 \frac{dB}{dt}$$

Equation 26-6 then gives

$$2\pi r E = \pi r^2 \frac{dB}{dt}$$

or

$$E = \frac{r}{2}\frac{dB}{dt} \qquad r < R$$

For a circle with $r > R$, the line integral $\oint \mathbf{E} \cdot d\boldsymbol{\ell}$ again gives $2\pi r E$. However, since $\mathbf{B} = 0$ for $r > R$, the flux is now $\pi R^2 B$. For this case, Equation 26-6 gives

$$2\pi r E = \pi R^2 \frac{dB}{dt}$$

$$E = \frac{R^2}{2r}\frac{dB}{dt}$$

This example shows that Faraday's law implies that a changing magnetic field produces an electric field.

Example 26-4

An 80-turn coil has a radius of 5.0 cm and a resistance of 30 Ω. At what rate must a perpendicular magnetic field change to produce a current of 4.0 A in the coil?

The emf in the coil must equal the potential drop across its resistance:

$$\mathcal{E} = IR = (4.0 \text{ A})(30 \ \Omega) = 120 \text{ V}$$

Since the plane of the coil is perpendicular to the field, the flux is

$$\phi_m = NBA = NB\pi r^2$$

According to Faraday's law, the magnitude of the induced emf equals the rate of change of this flux. Thus,

$$\mathcal{E} = 120 \text{ V} = \frac{d\phi_m}{dt} = N\pi r^2 \frac{dB}{dt}$$

$$\frac{dB}{dt} = \frac{120 \text{ V}}{(80)\pi(0.05 \text{ m})^2} = 191 \text{ T/s}$$

Example 26-5

A small coil of N turns has its plane perpendicular to a uniform magnetic field **B** as shown in Figure 26-7. The coil is connected to a current integrator ©, a device designed to measure the total charge passing through it. Find the charge passing through the coil if the coil is rotated through 180° about its diameter.

The flux through the coil is

$$\phi_m = NBA$$

where N is the number of turns and A is the area of the coil. If the coil is rotated through 180°, the flux reverses, so the total change in flux has the magnitude $2NBA$. While the flux is changing, there is an emf and therefore a current in the coil. The current is

$$I = \frac{\mathcal{E}}{R} = \frac{1}{R}\frac{d\phi_m}{dt}$$

where R is the total resistance of the coil. The total charge that passes through the coil is

$$Q = \int I \, dt = \frac{1}{R}\int d\phi_m$$

$$= \frac{\Delta\phi_m}{R} = \frac{2NBA}{R}$$

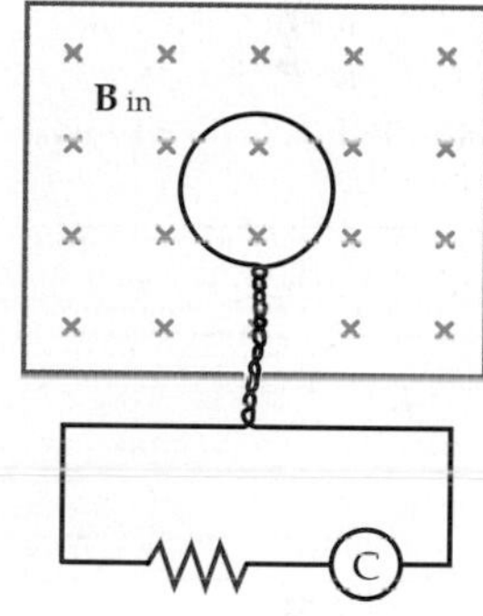

Figure 26-7 Flip-coil circuit for measuring the magnetic field **B**. When the coil is flipped over, the total charge that flows through the current integrator © is proportional to B.

The coil described in Example 26-5 is called a **flip coil.** It is used to measure magnetic fields. If a total charge Q passes through the coil when it is flipped 180°, the magnetic field B can be found from

$$B = \frac{RQ}{2NA} \qquad 26\text{-}7$$

Exercise

A flip coil of 40 turns has a radius of 3 cm and a resistance of 16 Ω. If the coil is turned through 180° in a magnetic field of 5000 G, how much charge passes through it? (Answer: 7.07 mC)

26-3 Lenz's Law

The negative sign in Faraday's law has to do with the direction of the induced emf. The direction of the induced emf and the induced current can be found from a general physical principle known as **Lenz's law:**

Lenz's law

> The induced emf and induced current are in such a direction as to oppose the change that produces them.

This statement of Lenz's law does not specify just what kind of change it is that causes the induced emf and current. It was purposefully left vague to cover a variety of conditions. A few illustrations will clarify this point.

Figure 26-8 shows a bar magnet moving toward a loop that has a resistance R. Since the magnetic field from the bar magnet is to the right, out of the north pole of the magnet, the movement of the magnet toward the loop tends to increase the flux through the loop to the right. (The magnetic field at the loop is stronger when the magnet is closer.) The induced current in the loop produces a magnetic field of its own. This induced current is in the direction shown, so the magnetic flux it produces is opposite that of the magnet. The induced magnetic field tends to *decrease* the flux through the loop. If the magnet were moved away from the loop, which would decrease the flux through the loop due to the magnet, the induced current would be in the opposite direction from that in Figure 26-8. In that case, the current would produce a magnetic field to the right, which would tend to increase the flux through the loop. As we might expect, moving the loop toward or away from the magnet has the same effect as moving the magnet. Only the relative motion is important.

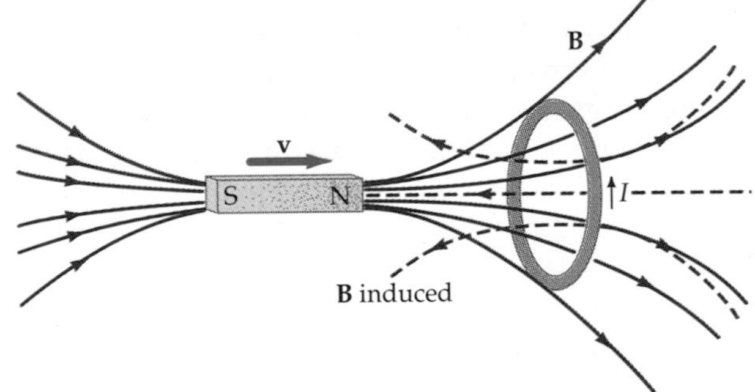

Figure 26-8 When the bar magnet is moving toward the loop, the emf induced in the loop produces a current in the direction shown. The magnetic field due to the induced current in the loop (indicated by the dashed lines) produces a flux that opposes the increase in flux through the loop due to the motion of the magnet.

Figure 26-9 shows the induced magnetic moment of the current loop when the magnet is moving toward it as in Figure 26-8. The loop acts like a small magnet with its north pole to the left and its south pole to the right. Since opposite poles attract and like poles repel, the induced magnetic moment of the loop exerts a force on the bar magnet to the left to oppose its motion toward the loop. Thus, we can express Lenz's law in terms of forces rather than flux. If the bar magnet is moved toward the loop, the induced current must produce a magnetic moment to oppose this change.

Note that Lenz's law is required by the law of conservation of energy. If the current in the loop in Figure 26-9 were opposite the direction shown, the induced magnetic moment of the loop would attract the magnet when it is moving toward the loop and cause it to accelerate toward the loop. If we begin with the magnet a great distance from the loop and give it a very slight push toward the loop, the force due to the induced current would be toward the loop, which would increase the velocity of the magnet. As the speed of the magnet increases, the rate of change of the flux would increase, thereby increasing the induced current. This would further increase the force on the magnet. Hence, the kinetic energy of the magnet and the rate at which Joule heat is produced in the loop (I^2R) would both increase with no source of energy. This would violate the law of conservation of energy.

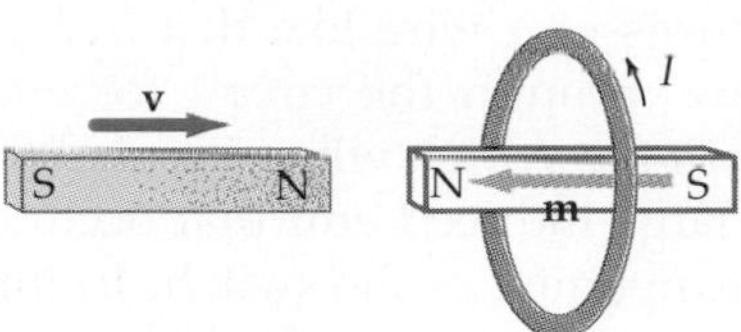

Figure 26-9 The magnetic moment of the loop (indicated by the outline magnet) due to the induced current is such as to oppose the motion of the bar magnet. Here the bar magnet is moving toward the loop, so the induced magnetic moment repels the bar magnet.

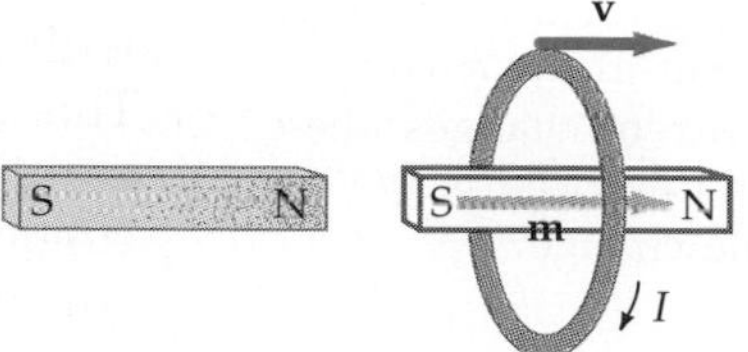

Figure 26-10 When the loop is moving away from the stationary bar magnet, the induced magnetic moment in the loop attracts the bar magnet, again opposing the relative motion.

In Figure 26-10, the bar magnet is at rest, and the loop is moving away from it. The induced current and magnetic moment are shown in the figure. In this case, the magnetic moment of the loop attracts the bar magnet, thus opposing the motion of the loop as required by Lenz's law.

In Figure 26-11, when the current in circuit 1 is changing, there is a changing flux through circuit 2. Suppose that the switch S in circuit 1 is initially open so that there is no current in the circuit (Figure 26-11*a*). When the switch is closed (Figure 26-11*b*), the current in circuit 1 does not reach its steady value $\mathcal{E}_1/R_1$ instantaneously but takes some time to change from zero to this value. During this time, while the current is increasing, the flux through circuit 2 is changing, and there is an induced current in that circuit in the direction shown. When the current in circuit 1 reaches its steady value, the flux through circuit 2 is no longer changing, so there is no induced current in circuit 2. An induced current in circuit 2 in the opposite direction appears momentarily when the switch in circuit 1 is opened (Figure 26-11*c*) and the current is decreasing to zero. It is important to understand that there is an induced emf *only while the flux is changing.* The emf does not depend on the magnitude of the flux, only on its rate of change. If there is a large, steady flux through a circuit, there is no induced emf.

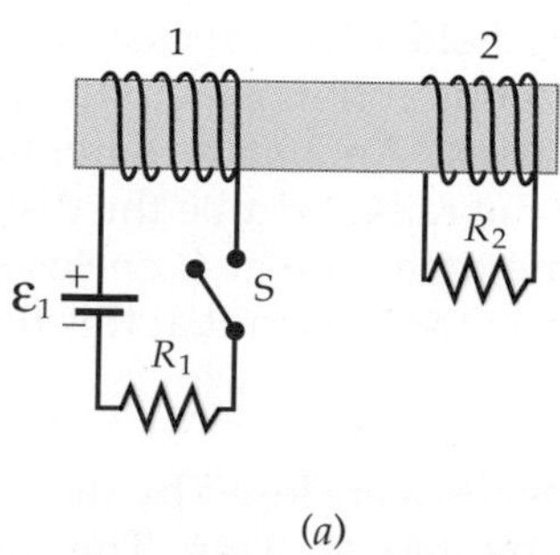

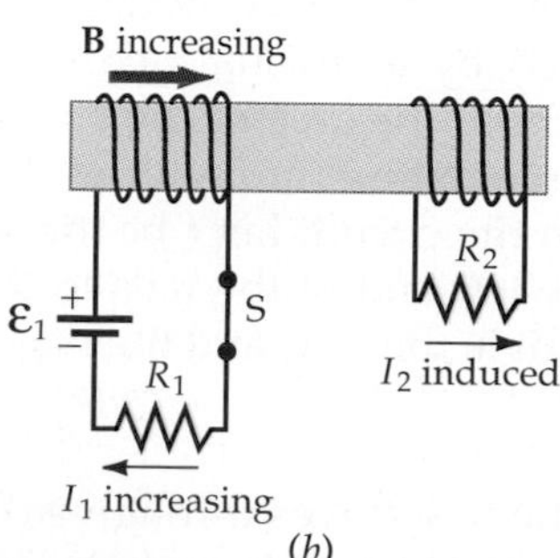

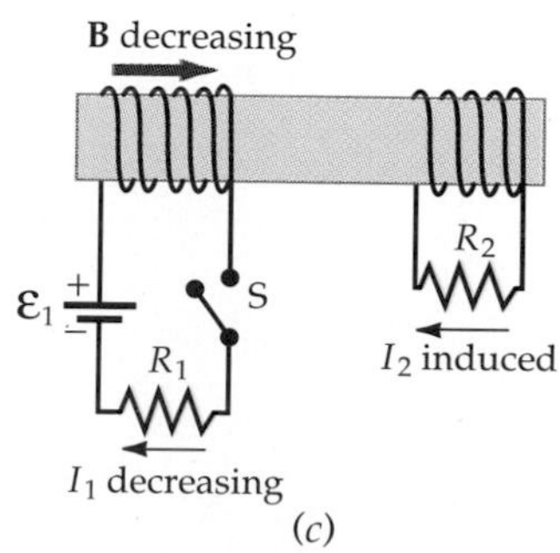

Figure 26-11 (*a*) Two adjacent circuits. (*b*) Just after the switch is closed, I_1 is increasing in the direction shown. The changing flux in circuit 2 induces the current I_2. The flux due to I_2 opposes the increase in flux due to I_1. (*c*) As the switch is opened, I_1 decreases and B decreases. The induced current I_2 then tends to maintain the flux in the circuit, opposing the change.

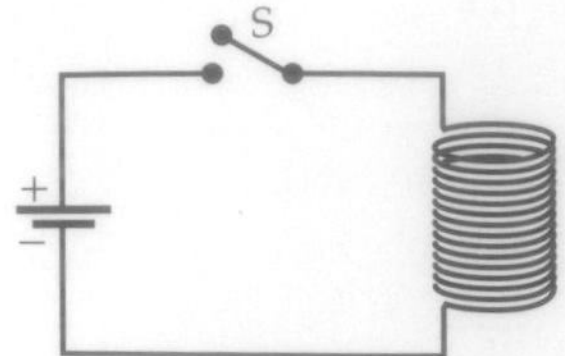

Figure 26-12 The coil with many turns of wire gives a large flux for a given current in the circuit. When the current changes, there is a large emf induced in the coil opposing the change.

For our next example, we consider the single, isolated circuit shown in Figure 26-12. When there is a current in the circuit, there is a magnetic flux through the coil due to its own current. When the current is changing, the flux in the coil is changing and there is an induced emf in the circuit. This *self-induced* emf opposes the change in the current. It is therefore called a **back emf.** Because of this self-induced emf, the current in a circuit cannot jump instantaneously from zero to some finite value or from some finite value to zero. Henry first noticed this effect when he was experimenting with a circuit consisting of many turns of a wire like that in Figure 26-12. This arrangement gives a large flux through the circuit for even a small current. Henry noticed a spark across the switch when he tried to break the circuit. Such a spark is due to the large induced emf that occurs when the current varies rapidly, as during the opening of the switch. In this case, the induced emf tries to maintain the original current. The large induced emf produces a large voltage drop across the switch as it is opened. The electric field between the contacts of the switch is large enough to tear electrons from the air molecules, causing dielectric breakdown. When the molecules in the air dielectric are ionized, the air conducts electric current in the form of a spark.

Question

1. Figure 26-13*a* shows a rectangular loop in a uniform magnetic field into the paper. Indicate the direction of the current induced in loop as it is twisted into the position shown in Figure 26-13*b*.

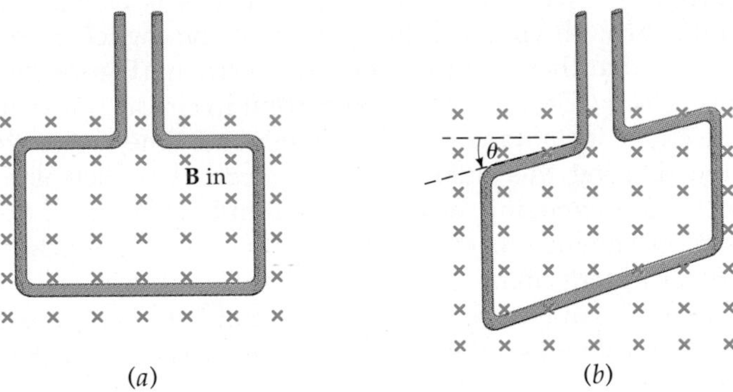

Figure 26-13 (*a*) A rectangular loop whose plane is perpendicular to a magnetic field **B**. (*b*) When the loop is twisted, the flux through it is changed and an emf is induced in it.

26-4 Motional Emf

Figure 26-14 shows a conducting rod sliding along conducting rails that are connected by a resistor. A uniform magnetic field **B** is directed into the paper. Since the area of the circuit increases as the rod moves to the right, the magnetic flux through the circuit is increasing. An emf is therefore induced in the circuit. Let ℓ be the separation of the rails and x be the distance from the left end of the rails to the rod at some time. The area enclosed by the circuit is then ℓx, and the magnetic flux through the circuit at this time is

$$\phi_m = BA = B\ell x$$

When the rod moves through a distance dx, the area enclosed by the circuit changes by $dA = \ell\, dx$ and the flux changes by $d\phi_m = B\ell\, dx$. The rate of change of the flux is

$$\frac{d\phi_m}{dt} = B\ell\frac{dx}{dt} = B\ell v$$

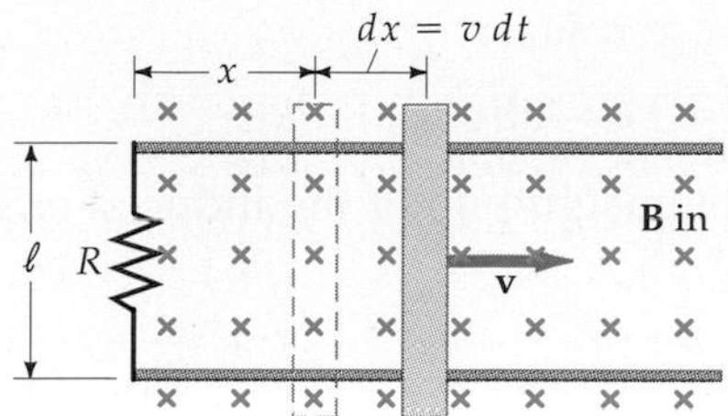

Figure 26-14 A conducting rod sliding on conducting rails in a magnetic field. As the rod moves to the right, the area of the circuit increases, so the magnetic flux through the circuit into the paper increases. An emf of magnitude $B\ell v$ is induced in the circuit, producing a counterclockwise current that produces flux out of the paper opposing the change.

where $v = dx/dt$ is the speed of the rod. The magnitude of the emf induced in this circuit is therefore

$$|\mathcal{E}| = \frac{d\phi_{\rm m}}{dt} = B\ell v$$

The direction of the emf in this case is such as to produce a current in the counterclockwise sense. The flux produced by this induced current is out of the paper, opposing the increase in flux due to the motion of the rod. Because of the induced current, which is upward in the rod, there is a magnetic force on the rod of magnitude $I\ell B$. The direction of this force, obtained from the right-hand rule, is to the left, opposing the motion of the rod. If the rod is given some initial velocity **v** to the right and is then released, the force due to the induced current slows the rod down until it stops. To maintain the motion of the rod, an external force must be exerted on it to the right.

The emf in this case is called **motional emf.** More generally,

Motional emf is any emf induced by the relative motion of a magnetic field and a current path.

Motional emf defined

Motional emf is induced in a conducting rod or wire moving in a magnetic field even when there is no complete circuit and thus no current.

Figure 26-15 shows an electron in a conducting rod that is moving through a uniform magnetic field directed into the paper. Because the electron is moving horizontally with the rod, there is a magnetic force on the electron that has a downward component of magnitude qvB. Because of this magnetic force, free electrons in the rod move downward, producing a net negative charge at the bottom and leaving a net positive charge at the top. The electrons continue to move down until the electric field produced by the separated charges exerts an upward force of magnitude qE on the electrons that balances the magnetic force qvB. In equilibrium, the electric field in the rod is thus

$$E = vB$$

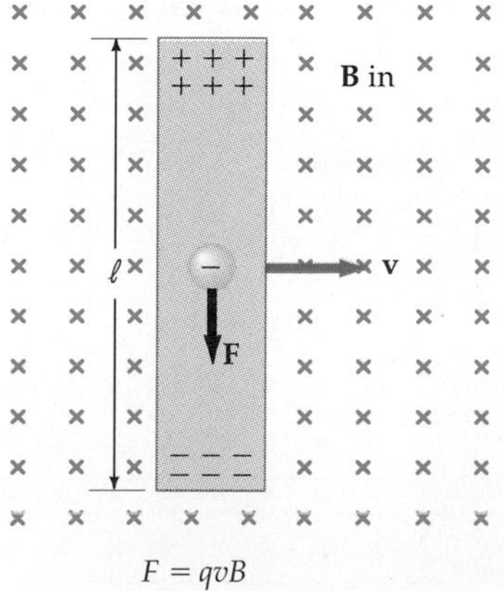

Figure 26-15 An electron in a conducting rod that is moving through a magnetic field experiences a magnetic force that has a downward component. Electrons move to the bottom of the rod, leaving the top of the rod positive. The charge separation produces an electric field of magnitude $E = vB$. The potential at the top of the rod is greater than that at the bottom by $E\ell = vB\ell$.

The potential difference across the rod is

$$\Delta V = E\ell = vB\ell$$

This potential difference equals the magnitude of the induced emf, that is, the motional emf:

Motional emf

$$|\mathcal{E}| = vB\ell \qquad 26\text{-}8$$

Motional emf is an example of Faraday's law in which we can understand the origin of the emf by considering the known forces acting on the electrons in the circuit. Figure 26-16 shows a typical electron in the conducting bar that is moving to the right in a magnetic field directed into the paper. The velocity of a typical electron $\mathbf{v}_e$ makes an angle θ with the rod as shown in the figure. It has a downward vertical component, $v_d = v_e \cos\theta$, the drift velocity, and a horizontal component $v = v_e \sin\theta$ equal to the speed of the rod. The magnetic force $\mathbf{f}_m = -e\mathbf{v}_e \times \mathbf{B}$ is in the plane of the figure and perpendicular to $\mathbf{v}_e$ as shown. It has the magnitude

$$f_m = ev_eB \qquad 26\text{-}9$$

If $\mathbf{f}_m$ were the only force acting on the electron, the electron could not stay in the rod as the rod moves to the right. The rod exerts a horizontal force $\mathbf{f}_r$ on the electron to balance the horizontal component of $\mathbf{f}_m$, which is $f_m \cos\theta$:

$$f_r = f_m \cos\theta \qquad 26\text{-}10$$

Since the magnetic force $\mathbf{f}_m$ is perpendicular to the motion of the electron, it does no work. The work done on the electron is done by the force $\mathbf{f}_r$. As the electron moves down the rod, the rod moves to the right so the electron moves along a diagonal path of length $S = \ell/\cos\theta$ shown in the figure. Since the component of $\mathbf{f}_r$ in the direction of the motion is $f_r \sin\theta$, the work done on the electron as it moves down the complete length of the rod is

$$W = f_r \sin\theta\ S = (f_m \cos\theta)\sin\theta\ S = f_m \sin\theta\ \ell$$

where $\ell = S\cos\theta$ is the length of the rod. Substituting ev_eB for f_m, we obtain

$$W = ev_eB \sin\theta\ \ell$$

But $v_e \sin\theta$ is the velocity of the rod v. The work done on the electron is thus

$$W = eBv\ell \qquad 26\text{-}11$$

The work per unit charge is the emf $B\ell v$, in agreement with our result obtained from Faraday's law.

Since the rod exerts a force f_r on each electron, each electron exerts an

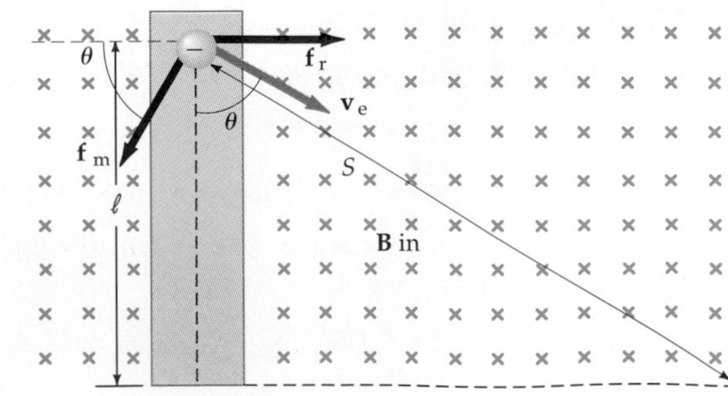

Figure 26-16 Forces on an electron in the rod of Figure 26-14. The electron's velocity $\mathbf{v}_e$ has a horizontal component v, the speed of the rod, and a vertical component v_d, its drift velocity along the rod. The magnetic force $\mathbf{f}_m$ is perpendicular to $\mathbf{v}_e$ and does no work. The rod exerts a horizontal force $\mathbf{f}_r$ on the electron with magnitude $f_m \cos\theta$. This force has a component in the direction of motion of the electron and therefore does work on it. The work per charge is equal to $B\ell v$.

equal and opposite force on the rod. This force is exerted to the left in Figure 26-16. If the rod has a cross-sectional area A and there are n free electrons per unit volume in the rod, the total number of electrons in the rod is $nA\ell$. The total force exerted on the rod is thus

$$F = nA\ell f_r = nA\ell f_m \cos\theta = nA\ell e v_e B \cos\theta \qquad \text{26-12}$$

But $v_e \cos\theta = v_d$, the drift velocity of the electrons, and $nAev_d = I$, the total current in the rod. Substituting I for $nAev_e \cos\theta$ in Equation 26-12, we obtain for the total force exerted by the electrons on the rod

$$F = I\ell B \qquad \text{26-13}$$

which is the same as Equation 24-4 for the magnetic force exerted on a current-carrying segment. To keep the rod moving with a constant speed v, an external force of magnitude $F = I\ell B$ must be exerted to the right. The power input of this force is the force times the speed:

$$P = Fv = I\ell Bv$$

Setting the power equal to the rate of production of Joule heat in the resistor I^2R, we obtain

$$IB\ell v = I^2R$$

or

$$B\ell v = IR$$

Thus, the induced emf $\mathcal{E} = B\ell v$ equals the potential drop across the resistor $\Delta V = IR$.

Exercise

A rod 40 cm long moves at 12 m/s in a plane perpendicular to a magnetic field of 3000 G. Its velocity is perpendicular to its length. Find the emf induced in the rod. (Answer: 1.44 V)

Example 26-6

In Figure 26-14, let $B = 0.6$ T, $v = 8$ m/s, $\ell = 15$ cm, and $R = 25\ \Omega$, and assume that the resistance of the rods and rails is negligible. Find (*a*) the induced emf in the circuit, (*b*) the current in the circuit, (*c*) the force needed to move the rod with constant velocity, and (*d*) the power dissipated in the resistor.

(*a*) The emf is given by Equation 26-8:

$$\mathcal{E} = Bv\ell = (0.6\ \text{T})(8\ \text{m/s})(0.15\ \text{m}) = 0.72\ \text{V}$$

(*b*) Since the total resistance in the circuit is 25 Ω, the current is

$$I = \frac{\mathcal{E}}{R} = \frac{0.72\ \text{V}}{25\ \Omega} = 28.8\ \text{mA}$$

(*c*) The force needed to move the rod with constant velocity is equal and opposite to the force exerted by the magnetic field on the rod. The magnitude of this force is

$$F = IB\ell = (0.0288\ \text{A})(0.6\ \text{T})(0.15\ \text{m}) = 2.59\ \text{mN}$$

(*d*) The power dissipated in the resistor is

$$P = I^2R = (0.0288\ \text{A})^2(25\ \Omega) = 20.7\ \text{mW}$$

We can check our answer in (*d*) by computing the power input of the force found in (*c*):

$$P = Fv = (2.59 \times 10^{-3}\ \text{N})(8\ \text{m/s}) = 2.07 \times 10^{-2}\ \text{W} = 20.7\ \text{mW}$$

Example 26-7

In Figure 26-14, the rod has a mass m. At time $t = 0$, the rod is moving with initial speed v_0, and the external force acting on it is removed. Find the speed of the rod as a function of time.

The current induced in the circuit is $\mathcal{E}/R$, where $\mathcal{E} = B\ell v$ is the induced emf. The magnitude of the magnetic force acting on the rod is therefore

$$F = IB\ell = \frac{\mathcal{E}}{R}B\ell$$

$$= \frac{B\ell v}{R}B\ell = \frac{B^2\ell^2 v}{R}$$

This force is directed opposite the direction of motion. If we take the positive direction as the direction of the initial velocity, the force is negative and Newton's second law for the rod is

$$F = ma = m\frac{dv}{dt}$$

$$-\frac{B^2\ell^2 v}{R} = m\frac{dv}{dt}$$

Separating the variables and integrating, we obtain

$$\frac{dv}{v} = -\frac{B^2\ell^2}{mR}dt$$

$$\ln v = -\frac{B^2\ell^2}{mR}t + C$$

where C is an arbitrary constant of integration. Then

$$v = e^C e^{-(B^2\ell^2/mR)t} = v_0 e^{-(B^2\ell^2/mR)t}$$

where $v_0 = e^C$ is the velocity at time $t = 0$.

26-5 Eddy Currents

In the examples we have discussed, the currents produced by a changing magnetic flux were set up in clearly defined circuits. Often, though, a changing flux sets up circulating currents, called **eddy currents,** in a piece of bulk metal such as the core of a transformer. (We will discuss transformers in Section 28-6.) The heat produced by eddy currents constitutes a power loss in the transformer.

Consider a conducting slab between the pole faces of an electromagnet as shown in Figure 26-17. If the magnetic field **B** between the pole faces is changing with time (as it will be if the current in the magnet windings is alternating current), the flux through any closed loop in the slab will be changing. For example, the flux through the closed loop C in the figure is the product of the magnetic field B and the area enclosed by the loop. If B varies, the flux will vary, and there will be an induced emf around the loop C. Since loop C is in a conductor, there will be a current along it equal to the emf divided by the resistance of the loop. In the figure, just one of the many closed loops that will contain currents if the magnetic field between the pole faces varies has been indicated.

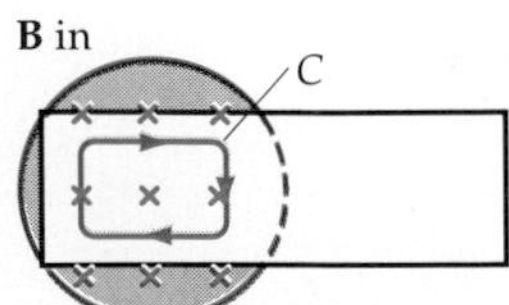

Figure 26-17 Eddy currents. When the magnetic field through a metal slab is changing, an emf is induced in any closed loop in the metal such as loop C. The induced emf causes a current in the loop.

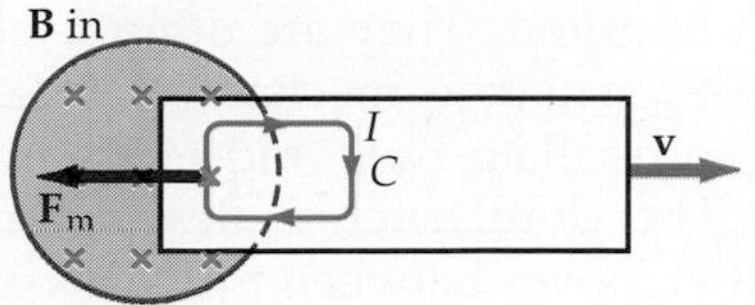

Figure 26-18 Demonstration of eddy currents. When the metal sheet is pulled to the right, there is a magnetic force to the left on the induced current opposing the motion.

The existence of eddy currents can be demonstrated by pulling a copper or aluminum sheet from between the poles of a strong magnet (Figure 26-18). Part of the area enclosed by loop C in this figure is in the magnetic field, and part is outside the field. As the sheet is pulled to the right, the flux through loop C decreases (assuming that the flux into the paper is positive). According to Faraday's law and Lenz's law, a clockwise current will be induced around the loop. Since this current is directed upward in the region between the pole faces, the magnetic field exerts a force on the current to the left, opposing motion of the sheet. You can feel this force on a conducting sheet if you try to pull it quickly through a strong magnetic field.

Eddy currents are usually unwanted because the heat they produce is a power loss. In addition, this heat must be dissipated. The power loss can be reduced by increasing the resistance of the possible paths of the eddy currents. In Figure 26-19, for instance, the conducting slab in Figure 26-17 is laminated; that is, it is made up of small strips of metal glued together. Because of the insulating glue between the strips, the eddy currents are essentially confined to the strips. Thus, the large eddy-current loops are broken up, and the power loss is greatly reduced. Similarly, if the sheet in Figure 26-19 has slots cut in it, as in Figure 26-20, the eddy currents are lessened and the force is greatly reduced. Figure 26-21 shows a common lecture demonstration of a magnetic brake. The apparatus consists of a sheet of metal fixed to the end of a pivoted rod that swings from the pivot like a pendulum. When the sheet swings through the gap between two poles of a magnet, the oscillation rapidly damps out. If the magnetic field is strong enough, the motion is stopped suddenly during the first encounter of the sheet with the magnetic field. If slots are cut in the sheet, as in Figure 26-21*c*, the damping is considerably reduced.

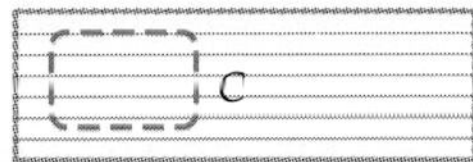

Figure 26-19 Eddy currents in a metal slab can be reduced by constructing the slab from small strips of metal glued together. The resistance of the closed loop C is now larger because of the insulating glue between the strips.

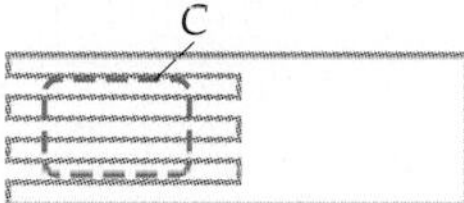

Figure 26-20 If the metal sheet of Figure 26-17 has slots cut, the eddy currents are greatly reduced because of the lack of good conducting paths.

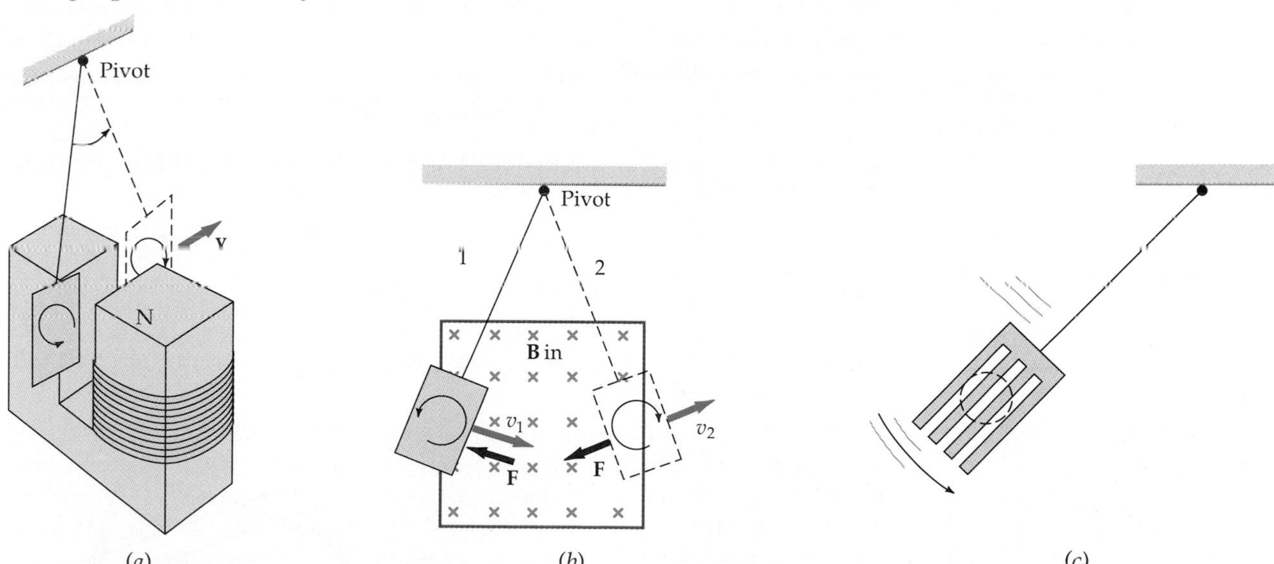

Figure 26-21 Lecture demonstration of a magnetic brake. (*a*) A pendulum with a metallic sheet for a bob is released from a large angle and swings between the poles of an electromagnet. (*b*) The bob is slowed considerably by the magnetic force on the induced eddy currents as the sheet enters or leaves the field. If the magnetic field is strong enough, the bob is stopped between the poles of the magnet. (*c*) If the sheet has slots cut in it, the eddy currents are greatly reduced and the bob swings through the magnetic field with little slowing.

Eddy currents are not always undesired. They are often used to damp unwanted oscillations. For example, sensitive mechanical balance scales used to weigh small masses tend to oscillate back and forth around their equilibrium reading many times. Therefore, such scales are usually designed with a small piece of metal that moves between the poles of a magnet as the scales oscillate. The resulting eddy currents dampen the oscillations so that equilibrium is reached more quickly. Another use of eddy currents is to provide *magnetic braking* for rapid transit cars. A large electromagnet is positioned in the car over the rails. When the magnet is energized by a current in its windings, eddy currents are induced in the rails by the motion of the magnet, and the resulting magnetic forces provide a drag force on the car that stops it.

Question

2. A bar magnet is dropped inside a long vertical tube. If the tube is made of metal, the magnet quickly reaches a terminal velocity, but if the tube is made of cardboard, it does not. Explain.

26-6 Generators and Motors

Most electrical energy used today is produced by electric generators in the form of alternating current (ac). A simple **generator** of alternating current is a coil rotating in a uniform magnetic field as shown in Figure 26-22. The ends of the coil are connected to rings called slip rings that rotate with the coil. Electrical contact is made with the coil by stationary graphite brushes in contact with the rings. When the line perpendicular to the plane of the coil makes an angle θ with a uniform magnetic field **B**, as shown in the figure, the magnetic flux through the coil is

$$\phi_{\mathrm{m}} = NBA \cos \theta \qquad 26\text{-}14$$

where N is the number of turns in the coil and A is the area of the coil. When the coil is mechanically rotated, the flux through it will change, and an emf will be induced in the coil according to Faraday's law. If the initial angle is δ, the angle at some later time t is given by

$$\theta = \omega t + \delta$$

where ω is the angular frequency of rotation. Substituting this expression for θ into Equation 26-14, we obtain

$$\phi_m = NBA \cos (\omega t + \delta) = NBA \cos (2\pi f t + \delta)$$

The emf in the coil will then be

$$\mathcal{E} = -\frac{d\phi_{\mathrm{m}}}{dt} = -NBA \frac{d}{dt} \cos (\omega t + \delta) = +NBA\omega \sin (\omega t + \delta) \qquad 26\text{-}15$$

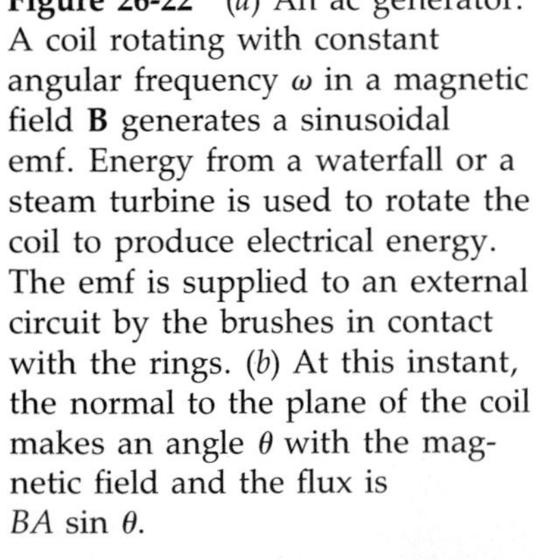

Figure 26-22 (*a*) An ac generator. A coil rotating with constant angular frequency ω in a magnetic field **B** generates a sinusoidal emf. Energy from a waterfall or a steam turbine is used to rotate the coil to produce electrical energy. The emf is supplied to an external circuit by the brushes in contact with the rings. (*b*) At this instant, the normal to the plane of the coil makes an angle θ with the magnetic field and the flux is $BA \sin \theta$.

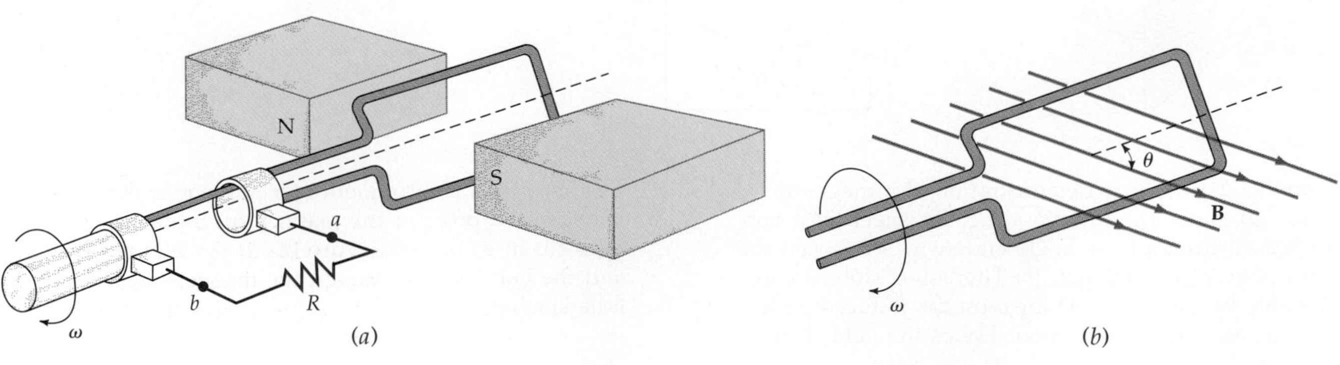

A generator under construction in China. Shown here is the rotor of Turbo Generator #1 at Dahua Power Station being lowered into place.

This can be written

$$\mathcal{E} = \mathcal{E}_{max} \sin (\omega t + \delta) \qquad 26\text{-}16$$

where

$$\mathcal{E}_{max} = NBA\omega \qquad 26\text{-}17$$

is the maximum value of the emf. We can thus produce a sinusoidal emf in a coil by rotating it with constant frequency in a magnetic field. In this source of emf, the mechanical energy of the rotating coil is converted into electric energy. The mechanical energy usually comes from a waterfall or a steam turbine. Although practical generators are considerably more complicated, they work on the same principle that an alternating emf is produced in a coil rotating in a magnetic field, and they are designed so that the emf produced is sinusoidal.

The same coil in a magnetic field that can be used to generate an alternating emf can also be used as an ac **motor.** Instead of mechanically rotating the coil to generate an emf, we apply an alternating current to the coil from another ac generator as shown in Figure 26-23. (In circuit diagrams, an ac generator is represented by the symbol ⏦.) We learned in Chapter 24 that a current loop in a magnetic field experiences a torque that tends to rotate the loop such that its magnetic moment points in the direction of **B** and the plane of the loop is perpendicular to **B**. If direct current were supplied to the coil in Figure 26-23, the torque on the coil would change directions when the coil rotates past its equilibrium position, which is when the plane of the coil is vertical in the figure. The coil would then oscillate about its equilibrium position, eventually coming to rest there with its plane vertical. However, if the direction of the current is reversed just as the coil passes the vertical position, the torque does not change direction but continues to rotate the coil in the same direction. As the coil rotates in the magnetic field, a back emf is generated that tends to counter the emf that supplies the current. When the motor is first turned on, there is no back emf and the current is very large, being limited only by the resistance in the circuit. As the motor begins to rotate, the back emf increases and the current decreases.

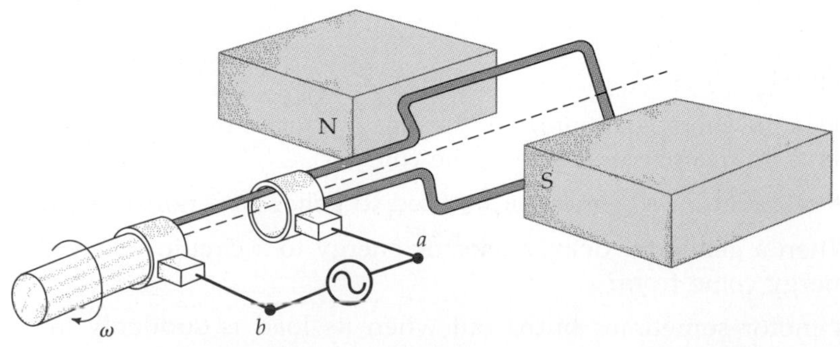

Figure 26-23 When alternating current is supplied to the coil of Figure 26-22, the coil becomes a motor. As the coil rotates, a back emf is generated, limiting the current.

(a)

(b)

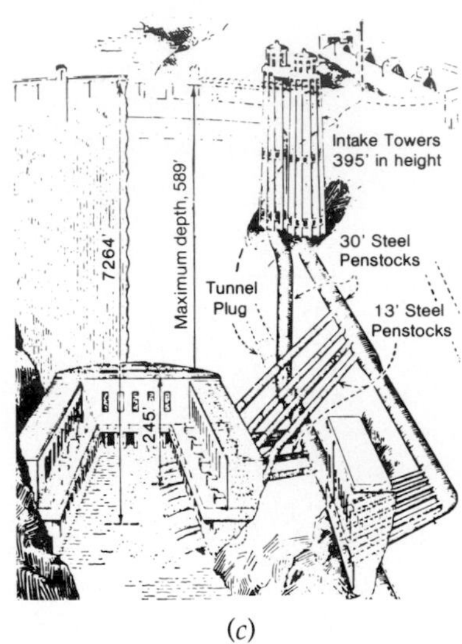

(c)

(*a*) River-level view of Hoover Dam with the Nevada wing of its power plant on the left and the Arizona wing on the right. The mechanical energy of falling water drives the turbines shown (*b*) for the generation of electricity. (*c*) Schematic drawing of Hoover Dam showing the intake towers and pipes (penstocks) that carry the water to the generators below.

Example 26-8

A 250-turn coil has an area of 3 cm^2. If it rotates in a magnetic field of 0.4 T at 60 Hz, what is $\mathcal{E}_{max}$?

From Equation 26-17, we have

$$\mathcal{E}_{max} = NBA\omega = NBA(2\pi f) = (250)(0.4\text{ T})(3 \times 10^{-4}\text{ m}^2)(2\pi)(60\text{ Hz})$$
$$= 11.3\text{ V}$$

Example 26-9

The windings of a dc motor have a resistance of 1.5 Ω. When the motor is connected across 40 V and running at full speed, the current in the windings is 2.0 A. (*a*) What is the back emf when the motor is running at full speed? (*b*) What is the initial current in the windings at start-up when the back emf is negligible?

(*a*) The potential drop across the windings is

$$V = IR = (2.0\text{ A})(1.5\ \Omega) = 3\text{ V}$$

Since the total potential drop across the motor is 40 V, the back emf is 40 V − 3 V = 37 V.

(*b*) When the motor is first turned on, the back emf is negligible. Since the potential drop across the motor is still 40 V, the current is

$$I = \frac{40\text{ V}}{1.5\ \Omega} = 26.7\text{ A}$$

Questions

3. Does the sinusoidal nature of the emf produced by an ac generator depend on the size or shape of the coil?
4. How could an ac generator be used to generate a nonsinusoidal emf?
5. When a generator delivers electric energy to a circuit, where does the energy come from?
6. A motor sometimes burns out when its load is suddenly increased. Why?

26-7 Inductance

Self Inductance

The magnetic flux through a circuit can be related to the current in that circuit and the currents in other, nearby circuits. (We will assume that there are no permanent magnets around.) Consider a coil carrying a current I. The current produces a magnetic field that could, in principle, be calculated from the Biot–Savart law. Since the magnetic field at every point in the neighborhood of the coil is proportional to I, the magnetic flux through the coil is also proportional to I:

$$\phi_m = LI \qquad 26\text{-}18$$

Self inductance defined

where L is a constant called the **self inductance** of the coil. The self inductance depends on the geometric shape of the coil. The SI unit of inductance is the **henry** (H). From Equation 26-18, we can see that the unit of inductance equals the unit of flux, the weber, divided by the unit of current, the ampere:

$$1\text{ H} = 1\frac{\text{Wb}}{\text{A}} = 1\frac{\text{T}\cdot\text{m}^2}{\text{A}}$$

In principle, the self inductance of any coil or circuit can be calculated by assuming a current I, finding the flux ϕ_m, and using $L = \phi_m/I$. In actual practice, the calculation is very difficult. However, there is one case, that of a tightly wound solenoid, for which the self inductance can be calculated directly. The magnetic field inside a tightly wound solenoid of length ℓ carrying a current I is given by Equation 25-10:

$$B = \mu_0 nI$$

where $n = N/\ell$ is the number of turns per unit length. If the solenoid has a cross-sectional area A, the flux through the N turns is

$$\phi_m = NBA = n\ell BA = \mu_0 n^2 A\ell I$$

As expected, the flux is proportional to the current I. The proportionality constant is the self inductance:

$$L = \frac{\phi_m}{I} = \mu_0 n^2 A\ell \qquad 26\text{-}19$$

Self inductance of a solenoid

The self inductance is proportional to the square of the number of turns per unit length n and to the volume $A\ell$. Thus, like capacitance, self inductance depends only on geometric factors. From the dimensions of Equation 26-19, we can see that μ_0 can be expressed in henrys per meter:

$$\mu_0 = 4\pi \times 10^{-7}\text{ H/m}$$

Example 26-10

Find the self inductance of a solenoid of length 10 cm, area 5 cm^2, and 100 turns.

We can calculate the self inductance in henrys from Equation 26-19 if we put all the quantities in SI units. Using $\ell = 0.1$ m, $A = 5 \times 10^{-4}$ m^2, $n = N/\ell = (100 \text{ turns})/(0.1\text{ m}) = 1000$ turns/m, and $\mu_0 = 4\pi \times 10^{-7}$ H/m, we obtain

$$L = \mu_0 n^2 A\ell = (4\pi \times 10^{-7}\text{ H/m})(10^3\text{ turns/m})^2(5 \times 10^{-4}\text{ m}^2)(0.1\text{ m})$$
$$= 6.28 \times 10^{-5}\text{ H}$$

When the current in a circuit is changing, the magnetic flux due to the current is also changing, so an emf is induced in the circuit. Since the self inductance of a circuit is constant, the change in flux is related to the change in current by

$$\frac{d\phi_m}{dt} = \frac{d(LI)}{dt} = L\frac{dI}{dt}$$

According to Faraday's law, we have

$$\mathcal{E} = -\frac{d\phi_m}{dt} = -L\frac{dI}{dt} \qquad 26\text{-}20$$

Thus, the self-induced emf is proportional to the rate of change of the current.

Exercise

At what rate must the current in the solenoid of Example 26-10 change to induce an emf of 20 V? (Answer: 3.18×10^5 A/s)

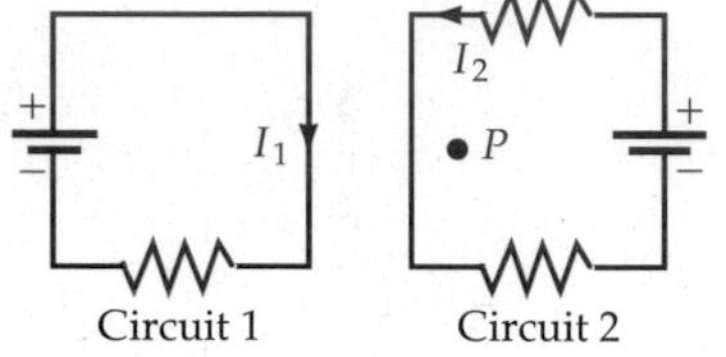

Figure 26-24 Two adjacent circuits. The magnetic field at point P is partly due to current I_1 and partly due to I_2. The flux through either circuit is the sum of two terms, one proportional to I_1 and the other to I_2.

Mutual Inductance

When two or more circuits are close to each other, as in Figure 26-24, the magnetic flux through one circuit depends not only on the current in that circuit but also on the current in the nearby circuits. Let I_1 be the current in circuit 1 on the left in Figure 26-24 and I_2 be that in circuit 2 on the right. The magnetic field at some point P has a part due to I_1 and a part due to I_2. These fields are proportional to the currents that produce them. We can therefore write the flux through circuit 2, ϕ_{m2}, as the sum of two parts, one proportional to the current I_1 and the other proportional to the current I_2:

Mutual inductance defined

$$\phi_{m2} = L_2 I_2 + M_{12} I_1 \qquad 26\text{-}21a$$

where L_2 is the self inductance of circuit 2 and M_{12} is called the **mutual inductance** of the two circuits. The mutual inductance depends on the geometrical arrangement of the two circuits. For instance, if the circuits are far apart, the flux through circuit 2 due to the current I_1 will be small and the mutual inductance will be small. An equation similar to Equation 26-21*a* can be written for the flux through circuit 1:

$$\phi_{m1} = L_1 I_1 + M_{21} I_2 \qquad 26\text{-}21b$$

where L_1 is the self inductance of circuit 1.

Figure 26-25 shows a long, narrow, tightly wound solenoid inside another tightly wound solenoid of equal length but larger radius. For this situation we can actually calculate the mutual inductance of the two solenoids. Let ℓ be the length of both solenoids, and let the inner solenoid have N_1 turns and radius r_1 and the outer solenoid have N_2 turns and radius r_2. We will first calculate the mutual inductance M_{12} by assuming that the inner solenoid carries a current I_1 and finding the magnetic flux ϕ_{m2} due to this current through the outer solenoid. The magnetic field due to the current in the inner solenoid is constant in the space within the solenoid and has magnitude

$$B_1 = \mu_0 n_1 I_1 \qquad 26\text{-}22$$

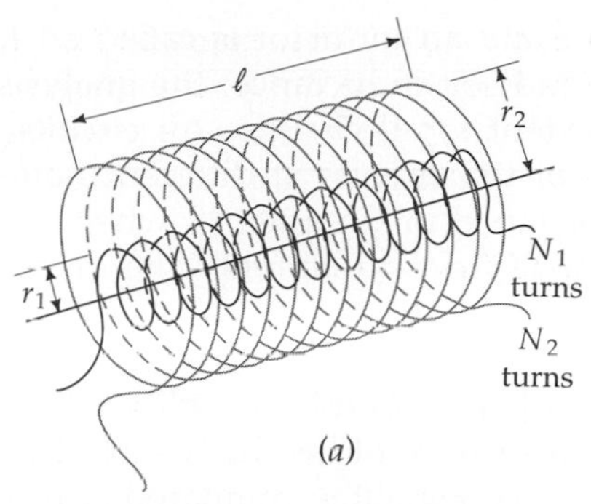

Figure 26-25 (*a*) A long, narrow solenoid inside a second solenoid of the same length. A current in either solenoid produces magnetic flux in the other. (*b*) Tesla coil illustrating the geometry of the wires in part (*a*). Such a device functions as a transformer (Chapter 28). Here, low-voltage alternating current in the outer winding is transformed into a higher-voltage alternating current in the inner winding. Induced alternating voltage from the changing fields is great enough to light the bulb above the coil.

(*b*)

Outside the inner solenoid, the magnetic field is zero. The flux through the outer solenoid due to this magnetic field is therefore

$$\phi_{m2} = N_2 B_1(\pi r_1^2) = n_2 \ell B_1(\pi r_1^2) = \mu_0 n_2 n_1 \ell (\pi r_1^2) I_1$$

Note that the area used to compute the flux through the outer solenoid is not the area of that solenoid, πr_2^2, but rather is the area of the inner solenoid, πr_1^2, because there is no magnetic field outside the inner solenoid. The mutual inductance M_{12} is thus

$$M_{12} = \frac{\phi_{m2}}{I_1} = \mu_0 n_2 n_1 \ell \pi r_1^2 \qquad \text{26-23}$$

We will now calculate M_{21} by finding the magnetic flux through the inner solenoid due to a current I_2 in the outer solenoid. When the outer solenoid carries a current I_2, there is a uniform magnetic field B_2 inside that solenoid given by Equation 26-22 with I_2 replacing I_1 and n_2 replacing n_1:

$$B_2 = \mu_0 n_2 I_2$$

The magnetic flux through the inner solenoid is then

$$\phi_{m1} = N_1 B_2(\pi r_1^2) = n_1 \ell B_2(\pi r_1^2) = \mu_0 n_1 n_2 \ell (\pi r_1^2) I_2$$

The area used here is also πr_1^2 because it is the cross-sectional area of the inner solenoid, and the magnetic field is uniform everywhere inside that solenoid. The mutual inductance M_{21} is thus

$$M_{21} = \frac{\phi_{m1}}{I_2} = \mu_0 n_1 n_2 \ell (\pi r_1^2) \qquad \text{26-24}$$

Note that Equations 26-23 and 26-24 are the same; that is, $M_{21} = M_{12}$. It can be shown that this is a general result. We will therefore drop the subscripts for mutual inductance and simply write M.

Question

7. How would the self inductance of a solenoid be changed if the same length of wire were wound onto a cylinder of the same diameter but twice the length? If twice as much wire were wound onto the same cylinder?

26-8 *LR* Circuits

As we have seen, the self inductance of a circuit prevents the current from rising or falling instantaneously. Circuits containing coils or solenoids of many turns have a large self inductance. Such a coil or solenoid is called an **inductor.** The symbol for an inductor is —ooo—. We can often neglect the self inductance of the rest of the circuit compared with that of an inductor.

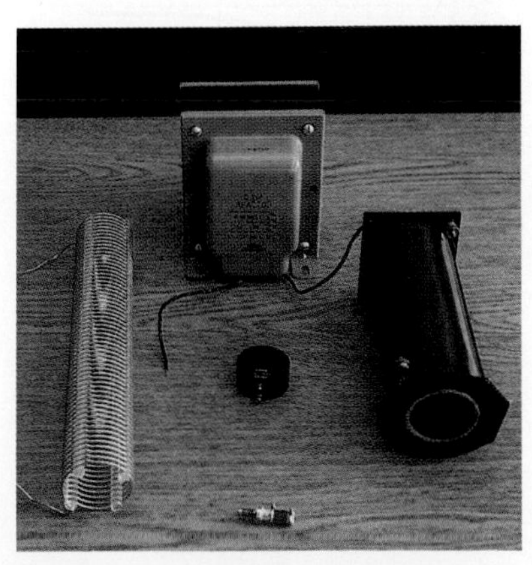
Various inductors.

A circuit containing a resistor and an inductor is called an ***LR* circuit.** Since all circuits have resistance and self inductance, the analysis of an *LR* circuit can be applied to some extent to all circuits. All circuits also have some capacitance between parts of the circuits at different potentials. We will consider the effects of capacitance in Chapter 28 when we study ac circuits. Here we will neglect capacitance to simplify the analysis and to focus on the effects of inductance.

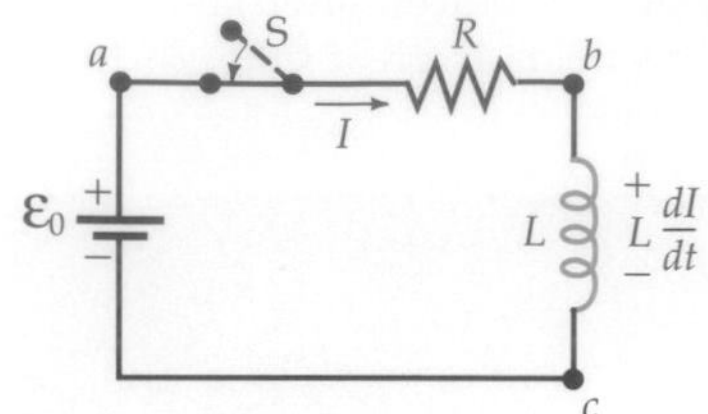

Figure 26-26 A typical *LR* circuit. Just after the switch S is closed, the current begins to increase in the circuit and a back emf of magnitude $L\,dI/dt$ is generated in the inductor. The potential drop across the resistor IR plus the potential drop across the inductor equals the emf of the battery.

Figure 26-26 shows an *LR* circuit in which an inductance L and a resistance R are in series with a battery of emf $\mathcal{E}_0$ and a switch S. We will assume that the resistance R includes the resistance of the inductor coil and that the inductance of the rest of the circuit is negligible compared with that of the inductor. The switch is initially open, so there is no current in the circuit. Just after the switch is closed, the current is still zero, but it is changing at a rate dI/dt, and there is a back emf of magnitude $L\,dI/dt$ in the inductor. In the circuit diagram, plus and minus signs have been placed on the inductor to indicate the direction of the emf when the current is increasing, that is, when dI/dt is positive. A short time after the switch is closed, there is a current I in the circuit and a potential drop IR across the resistor. Applying Kirchhoff's loop rule to this circuit gives

$$\mathcal{E}_0 - IR - L\frac{dI}{dt} = 0 \qquad 26\text{-}25$$

We can understand many of the features of the current in this circuit from this equation without solving the equation. Initially (just after the switch is closed) the current is zero and the back emf $L\,dI/dt$ equals the emf of the battery $\mathcal{E}_0$. The initial rate of change of the current is, from Equation 26-25,

$$\left(\frac{dI}{dt}\right)_0 = \frac{\mathcal{E}_0}{L} \qquad 26\text{-}26$$

As the current increases, the potential drop IR increases, and the rate of change of the current decreases. After a short time, the current has reached a positive value I, and the rate of change of the current is

$$\frac{dI}{dt} = \frac{\mathcal{E}_0}{L} - \frac{IR}{L}$$

At this time the current is still increasing, but its rate of increase is less than at $t = 0$. The final value of the current can be obtained by setting dI/dt equal to zero. The final value of the current is thus

$$I_f = \frac{\mathcal{E}_0}{R} \qquad 26\text{-}27$$

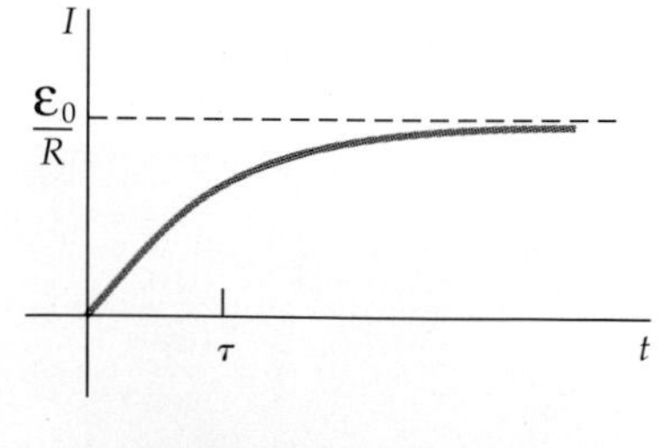

Figure 26-27 shows the current in this circuit as a function of time. This figure is similar to that for the charge on a capacitor when the capacitor is charged in an *RC* circuit (Figure 23-22).

Equation 26-25 is of the same form as Equation 23-22 for the charging of a capacitor and can be solved in the same way. The result is

$$I = \frac{\mathcal{E}_0}{R}(1 - e^{-Rt/L}) = \frac{\mathcal{E}_0}{R}(1 - e^{-t/\tau}) = I_f(1 - e^{-t/\tau}) \qquad 26\text{-}28$$

Figure 26-27 Current versus time in an *LR* circuit. At a time $t = \tau = L/R$, the current is at 63% of its maximum value $\mathcal{E}_0/R$.

where

$$\tau = \frac{L}{R} \qquad 26\text{-}29$$

is the **time constant** of the circuit. The larger the self inductance L or the smaller the resistance R, the longer it takes for the current to build up. Note that the product of the time constant L/R and the initial slope $\mathcal{E}_0/L$ equals the final current $I_f = \mathcal{E}_0/R$. If the rate of increase of the current were constant at its original rate, the current would reach its maximum value in a time $t = \tau$. However, the rate of increase of the current is not constant, but decreases with time. At a time τ, the current is $0.63I_f$.

Example 26-11

A coil of self inductance 5.0 mH and a resistance of 15.0 Ω is placed across the terminals of a 12-V battery of negligible internal resistance. (*a*) What is the final current? (*b*) What is the current after 100 μs?

(*a*) The final current is

$$I_f = \frac{\mathcal{E}_0}{R} = \frac{12\text{ V}}{15\ \Omega} = 0.800\text{ A}$$

(*b*) The time constant for this circuit is

$$\tau = \frac{L}{R} = \frac{5 \times 10^{-3}\text{ H}}{15\ \Omega} = 333\ \mu\text{s}$$

The current after 100 μs is given by Equation 26-28:

$$I = \frac{\mathcal{E}_0}{R}(1 - e^{-t/\tau}) = (0.800\ A)(1 - e^{-100/333}) = (0.800\ A)(1 - 0.741)$$

$$= 0.207\text{ A}$$

In Figure 26-28, the circuit has an additional switch that allows us to remove the battery, and an additional resistor R_1 to protect the battery so that it is not shorted when both switches are momentarily closed. If both switches are initially open, and we close switch S_1, the current builds up in the circuit just as discussed, except that the total resistance is now $R_1 + R$ and the final current is $\mathcal{E}_0/(R + R_1)$. Suppose that this switch has been closed for a long time compared with the time constant so that the current is approximately steady at its final value, which we will call I_0. Switch S_2 is then closed and switch S_1 is opened to remove the battery from consideration completely. Let us choose the time when switch S_2 is closed to be $t = 0$. We now have a circuit with just a resistor and an inductor (loop *abcd*) carrying an initial current I_0. Applying Kirchhoff's loop rule to this circuit gives

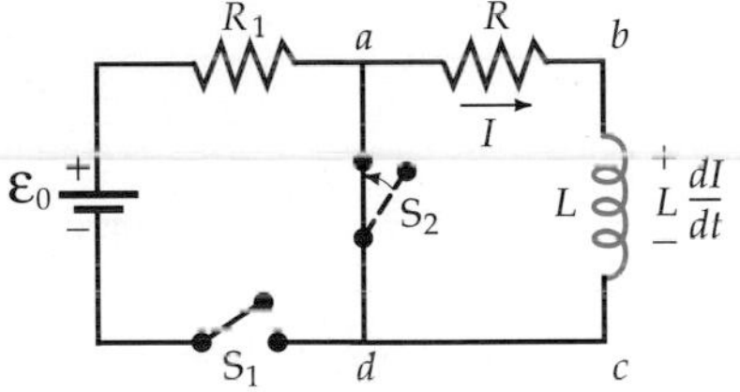

Figure 26-28 An *LR* circuit with two switches so that the battery can be removed from the circuit. After the current in the inductor reaches its maximum value with S_1 closed, S_2 is closed and S_1 is opened. The current then decreases exponentially with time.

$$-IR - L\frac{dI}{dt} = 0$$

or

$$\frac{dI}{dt} = -\frac{R}{L}I \qquad 26\text{-}30$$

Note that to apply Kirchhoff's loop rule, we have assumed a direction for I and assigned plus and minus signs to the inductor in Figure 26-28 for positive dI/dt, as usual. The solution of the equation then gives us the correct algebraic signs for these quantities, which indicate whether or not our assumption was correct. In this case, the direction of I is already known from the initial conditions. Equation 26-30 then tells us that, since I is positive, dI/dt is negative, indicating that the current is decreasing. For a negative dI/dt, the induced emf, $-L\,dI/dt$, is in the direction of the current and opposes the decrease in current in agreement with Lenz's law. Equation 26-30

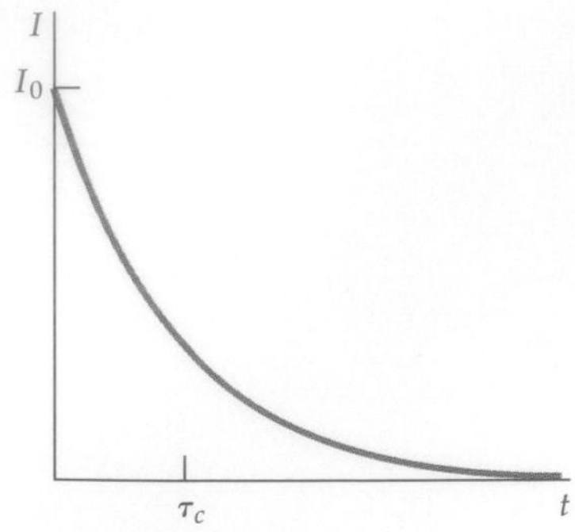

Figure 26-29 Current versus time for the circuit in Figure 26-28. The current decreases exponentially with time.

is of the same form as Equation 23-16 for the discharge of a capacitor. It can be solved by direct integration. We will omit the details and merely state the solution. The current I is given by

$$I = I_0 e^{-Rt/L} = I_0 e^{-t/\tau} \qquad 26\text{-}31$$

where $\tau = L/R$ is the time constant. Figure 26-29 shows the current as a function of time.

Exercise

What is the time constant of a circuit of resistance 85 Ω and inductance 6 mH? (Answer: 70.6 μs)

Example 26-12

Find the total heat produced in the resistor R in Figure 26-28 when the current in the inductor decreases from its initial value of I_0 to 0.

The rate of heat production is

$$P = \frac{dW}{dt} = I^2R$$

where I is given by Equation 26-31. In a time dt, the heat produced is

$$dW = I^2R\,dt$$

The total energy dissipated as heat in the resistor is thus

$$W = \int_0^\infty I^2R\,dt = \int_0^\infty I_0^2 e^{-2Rt/L}R\,dt = I_0^2R \int_0^\infty e^{-2Rt/L}\,dt$$

The integration can be done by substituting $x = 2Rt/L$. Then

$$dt = \frac{L}{2R}\,dx$$

and

$$W = I_0^2R\frac{L}{2R}\int_0^\infty e^{-x}\,dx = \frac{1}{2}LI_0^2$$

since the value of the integral is 1. This energy was originally stored in the inductor. In the next section, we will see that, in general, the energy stored in an inductor carrying a current I is $\frac{1}{2}LI^2$.

26-9 Magnetic Energy

In Section 21-4, we saw that it takes work to charge a capacitor and that a charged capacitor stores energy given by

$$U = \frac{1}{2}QV = \frac{1}{2}CV^2 = \frac{1}{2}\frac{Q^2}{C}$$

where Q is the charge on either plate, V is the potential difference between the plates, and C is the capacitance. We also saw that this energy can be considered to be stored in the electric field between the plates and found that, in general, when there is an electric field E in space, the electric energy density (the electric energy per unit volume) is

$$\eta = \tfrac{1}{2}\epsilon_0 E^2$$

There is a similar expression for the energy in a magnetic field. It takes work to produce a current in an inductor. We can see this by multiplying each term in Equation 26-25 by the current I and then rearranging:

$$\mathcal{E}_0 I = I^2R + LI\frac{dI}{dt} \qquad \text{26-32}$$

The term $\mathcal{E}_0 I$ is the power output of the battery. The term I^2R is the power dissipated as heat in the resistance of the circuit. The term $LI\,dI/dt$ is the rate at which energy is put into the inductor. If U_m is the energy in the inductor, then

$$\frac{dU_m}{dt} = LI\frac{dI}{dt}$$

We can find the total energy in the inductor by integrating this equation from time $t = 0$, when the current is zero, to $t = \infty$, when the current has reached its final value I_f:

$$U_m = \int dU_m = \int_0^{I_f} LI\,dI = \tfrac{1}{2}LI_f^2$$

The energy stored in an inductor carrying a current I is thus given by

$$U_m = \tfrac{1}{2}LI^2 \qquad \text{26-33}$$

Energy stored in an inductor

This is in agreement with the result calculated in Example 26-12 that the heat produced in the resistor is $\frac{1}{2}LI^2$ when the current in the inductor decreases from I to 0.

Exercise

How much energy is stored in the inductor of Example 26-11 when the final current has been attained? (Answer: 1.6×10^{-3} J)

In the process of producing a current in an inductor, a magnetic field is created in the space within the inductor coil. The work done in producing a current in an inductor can be thought of as the work required to create a magnetic field. That is, we can think of the energy stored in an inductor as energy stored in the magnetic field of the inductor. For the special case of a solenoid, the magnetic field is related to the current I and the number of turns per unit length n by

$$B = \mu_0 nI$$

and the self inductance is given by Equation 26-19:

$$L = \mu_0 n^2 A\ell$$

where A is the cross-sectional area and ℓ is the length. Substituting $B/\mu_0 n$ for I and $\mu_0 n^2 A\ell$ for L in Equation 26-33, we obtain

$$U_m = \frac{1}{2}LI^2 = \frac{1}{2}\mu_0 n^2 \ell A\left(\frac{B}{\mu_0 n}\right)^2 = \frac{B^2}{2\mu_0}\ell A$$

The quantity $A\ell$ is the volume of the space within the solenoid containing the magnetic field.

The energy per unit volume is the **magnetic energy density** η_m:

$$\eta_m = \frac{B^2}{2\mu_0} \qquad \text{26-34}$$

Magnetic energy density

Although we have obtained Equation 26-34 by considering the special case of the magnetic field in a solenoid, the result is general. That is, whenever there is a magnetic field in space, the magnetic energy per unit volume is given by Equation 26-34.

Example 26-13

A certain region of space contains a magnetic field of 200 G and an electric field of 2.5×10^6 N/C. Find (*a*) the total energy density and (*b*) the energy in a cubical box of side 12 cm.

(*a*) The electrical energy density is

$$\eta_e = \tfrac{1}{2}\varepsilon_0 E^2 = (0.5)(8.85 \times 10^{-12}\ \mathrm{C^2/N{\cdot}m^2})(2.5 \times 10^6\ \mathrm{N/C})^2 = 27.7\ \mathrm{J/m^3}$$

and the magnetic energy density is

$$\eta_m = \frac{B^2}{2\mu_0} = \frac{(0.02\ \mathrm{T})^2}{2(4\pi \times 10^{-7}\ \mathrm{N/A^2})} = 159\ \mathrm{J/m^3}$$

The total energy density is therefore

$$\eta = \eta_e + \eta_m = 27.7\ \mathrm{J/m^3} + 159\ \mathrm{J/m^3} = 187\ \mathrm{J/m^3}$$

(*b*) The volume in a cube of side 12 cm is

$$V = (0.12\ \mathrm{m})^3 = 1.73 \times 10^{-3}\ \mathrm{m^3}$$

The total energy in this volume is then

$$U = \eta V = (187\ \mathrm{J/m^3})(1.73 \times 10^{-3}\ \mathrm{m^3}) = 0.324\ \mathrm{J}$$

Summary

1. For a magnetic field that is constant in space, the magnetic flux through a coil is the product of the component of the magnetic field that is perpendicular to the plane of the coil and the area of the coil. In general, for a coil of N turns, the magnetic flux through the coil is

$$\phi_m = \int NB_n\, dA$$

The SI unit of magnetic flux is the weber:

$$1\ \mathrm{Wb} = 1\ \mathrm{T{\cdot}m^2}$$

2. When the magnetic flux through a circuit changes, there is an emf induced in the circuit given by Faraday's law

$$\mathcal{E} = \oint_C \mathbf{E}\cdot d\boldsymbol{\ell} = -\frac{d\phi_m}{dt}$$

The induced emf and induced current are in such a direction as to oppose the change that produces them. This is known as Lenz's law.

3. The emf induced in a conducting wire or rod of length ℓ moving with velocity $\mathbf{v}$ perpendicular to a magnetic field $\mathbf{B}$ is called motional emf. Its magnitude is

$$|\mathcal{E}| = \frac{d\phi_m}{dt} = B\ell v$$

4. Circulating currents that are set up in bulk metal by a changing magnetic flux are called eddy currents.

5. A coil rotating with angular frequency ω in a magnetic field generates an alternating emf given by

$$\mathcal{E} = \mathcal{E}_{\max} \sin(\omega t + \delta)$$

where $\mathcal{E}_{\max} = NBA\omega$ is the maximum value of the emf.

6. The magnetic flux through a circuit is related to the current in the circuit by

$$\phi_m = LI$$

where L is the self inductance of the circuit, which depends on the geometrical arrangement of the circuit. The SI unit of inductance is the henry (H):

$$1 \text{ H} = 1 \text{ Wb/A} = 1 \text{ T}\cdot\text{m}^2/\text{A}$$

The self inductance of a tightly wound solenoid of length ℓ and area A with n turns per unit length is given by

$$L = \frac{\phi_m}{I} = \mu_0 n^2 A\ell$$

If there is another circuit nearby carrying current I_2, there is additional flux through the first circuit

$$\phi_m = MI_2$$

where M is the mutual inductance, which depends on the geometrical arrangement of the two circuits.

7. When the current in an inductor is changing, the emf induced in the inductor is given by

$$\mathcal{E} = -\frac{d\phi_m}{dt} = -L\frac{dI}{dt}$$

8. In an LR circuit, which consists of a resistance R, an inductance L, and a battery of emf $\mathcal{E}_0$ in series, the current does not reach its maximum value I instantaneously but rather takes some time to build up. If the current is initially zero, its value at some later time t is given by

$$I = \frac{\mathcal{E}_0}{R}(1 - e^{-Rt/L}) = \frac{\mathcal{E}_0}{R}(1 - e^{-t/\tau})$$

where $\tau = L/R$ is the time constant of the circuit.

9. The energy stored in an inductor carrying a current I is

$$U_m = \tfrac{1}{2}LI^2$$

This energy can be considered to be stored in the magnetic field inside the inductor. In general, the magnetic energy density (the energy per unit volume) of a magnetic field B is given by

$$\eta_m = \frac{B^2}{2\mu_0}$$

The Aurora

Syun-Ichi Akasofu
Geophysical Institute, University of Alaska, Fairbanks

The aurora appears as an awesome but silent light display in the far northern and far southern night skies (Figure 1). It often appears as a faint greenish-white arc, but it is actually a long, shimmering, undulating curtain of glowing bands and rays in several colors. The intensity of auroral light varies. At its brightest, the colors can be dramatic and beautiful. The lower edge of the auroral curtain is located at an altitude of 100 km, and the upper edge can extend to as high as 1000 km above the surface of the earth within two ring-shaped zones lying between 60 and 75 degrees of latitude (Figures 2*a*, 2*b*), each centered on one of the earth's magnetic poles. These ring-shaped belts are called the auroral ovals.

Auroral light was once thought to be sunlight reflected by ice crystals in the sky. However, in 1888 Anders Jonas Ångström showed that auroral light differs from sunlight; many of the wavelengths present in sunlight are absent in auroral light (Figure 3). A spectrum analogous to that of auroral light can result from applying a high voltage to electrodes inserted into a vacuum glass tube containing a gas such as neon. The electrons flow from the negative electrode to the positive one. They collide with neon atoms, exciting them and ultimately causing them to emit light. Similarly, auroral light is the result of an electrical discharge process and is emitted by atoms and molecules in the upper atmosphere (Figure 4) as they are stuck by high-speed electrons.

Syun-Ichi Akasofu, Director of the Geophysical Institute of the University of Alaska, Fairbanks, is the author of hundreds of publications on the aurora. His research work on the aurora has earned him national and international recognition and honors. Since becoming Director of the Geophysical Institute, Professor Akasofu has concentrated his efforts on establishing it as the key research center in the Arctic.

Figure 1 The aurora borealis.

(*a*)

Figure 2 (*a*) The southern aurora, as photographed by astronaut Robert Overmyer using a 35-mm camera.

Figure 3 A comparison of the visible light spectra of the sun and the aurora.

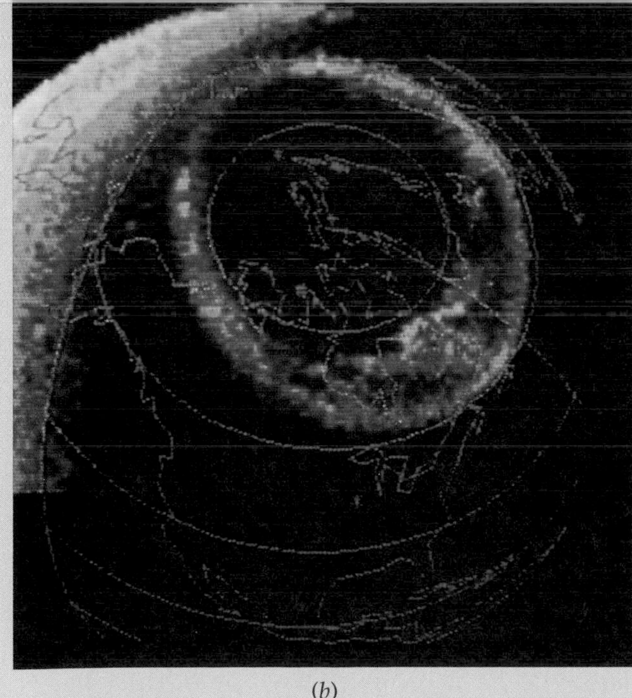

(*b*)

(*b*) A computer-enhanced image of the northern aurora, taken by satellite at a distance of 3 earth radii.

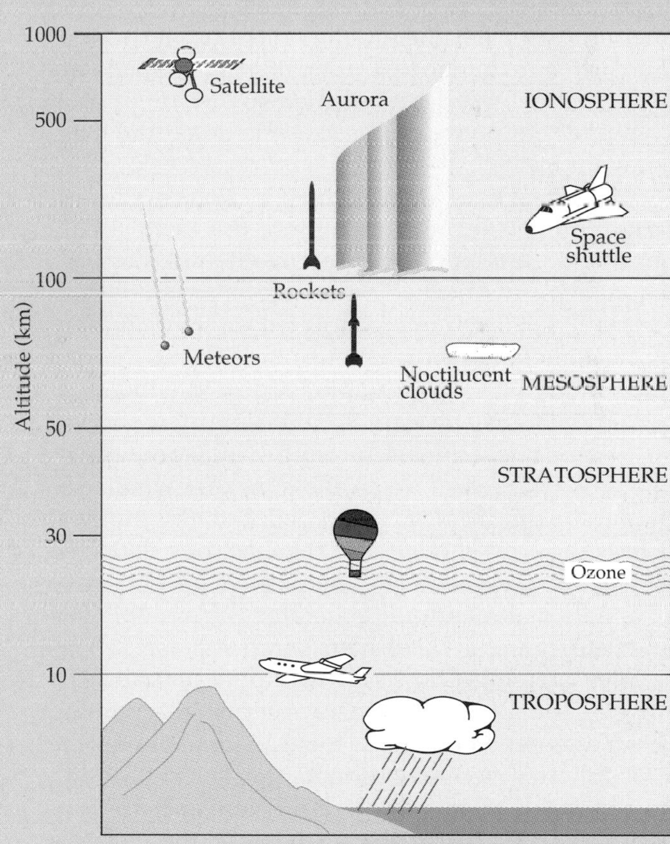

Figure 4 A schematic representation of the earth's atmosphere showing some artificial and some natural features and their altitudes. Auroras form in and above the ionosphere, the layer of the atmosphere that contains many free electrons and ions that have been created by the effects of solar ultraviolet and x-ray emissions.

Continued

In trying to understand the process that powers the auroral discharge, it is useful to consider an analogy with an electric generator. In a generator, an electrical current is produced in a conductor by moving it through a magnetic field. Auroral power is generated in a similar way: A stream of charged particles flowing outward from the sun (called the "solar wind") functions as a conductor, and the earth itself provides the magnetic field (see Figure 5).

The outer layer of the sun's atmosphere, the corona, consists of gas (primarily hydrogen) that is so hot that electrically neutral atoms break down into positive ions (primarily protons) and electrons. The solar wind, streaming out from the corona, is a tenuous hot plasma of these charged particles. Moving at a speed between 300 and 1000 km/s, it travels away from the sun in all directions to the limit of the solar system. Magnetic field lines in the solar wind behave like elastic strings. When the solar wind blows outward, it carries the sun's magnetic field lines and stretches them out. On its way, it confines the earth's magnetic field into a comet-shaped cavity, which is called the magnetosphere (Figure 6). The outer boundary of this cavity is called the magnetopause.

At distances of about 10 earth radii from the surface of our planet, the strength of the earth's magnetic field (30×10^{-5} G) is equal to that of the sun's magnetic field stretched out by the solar wind. Both magnetic fields are interconnected on the boundary of the comet-shaped magnetosphere. Here charged particles in the solar wind blow across the interconnected field. This motion is equivalent to motion of an electrical conductor through a magnetic field. Looking at the earth from the sun, one would see the protons in the solar wind deflected (by the $e\mathbf{v} \times \mathbf{B}$ force) toward the left and the electrons deflected toward the right, creating the positive and negative terminals of the auroral generator (Figure 7*a*). The magnetosphere is filled with tenuous plasma. This enables current to flow between the terminals. The current flows from the positive terminal, spiraling down magnetic field lines, into the ionosphere (the atmosphere's electrically conducting layer), through the ionosphere across the polar region, and up magnetic field lines from the ionosphere to the negative terminal. This is the primary electrical discharge circuit.

On what is called the "morning side" of the magnetosphere, current flows into the inner edge of the auroral oval as part of the primary discharge circuit and is conducted to the outer edge of the oval. Since the region outside the oval is not very conductive, some current finds its way back out along magnetic field lines, giving rise to a parallel, secondary circuit. A corresponding process occurs on the "evening side" (Figure 7*b*). Thus there are a pair of electrical currents (upward and downward) flowing along magnetic field lines in both the morning and evening sides of the magnetosphere. The upward current in both sectors is carried by downward-flowing electrons, which collide with atoms and molecules in the atmosphere, exciting them and causing them to emit light. This is the part of the discharge circuit that produces auroral light (exactly as occurs in a neon tube, as described earlier).

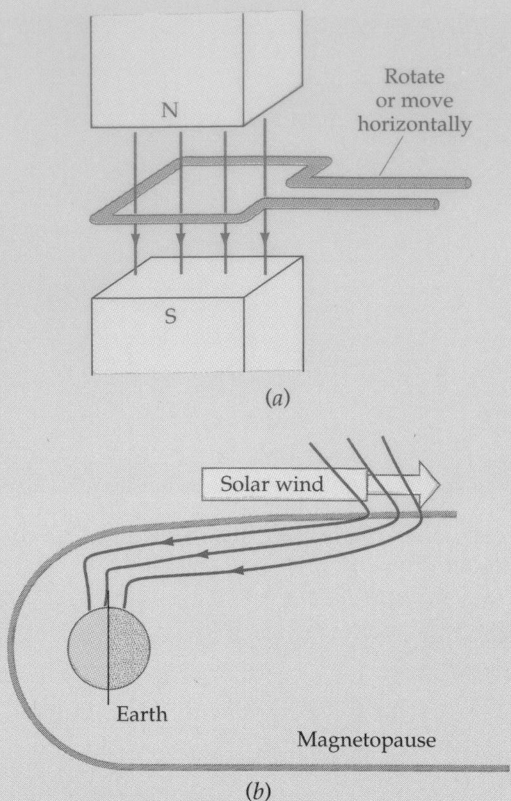

Figure 5 (*a*) Schematic of a conventional generator. (*b*) The interaction of the solar wind with the earth's magnetic field creates a naturally occurring generator.

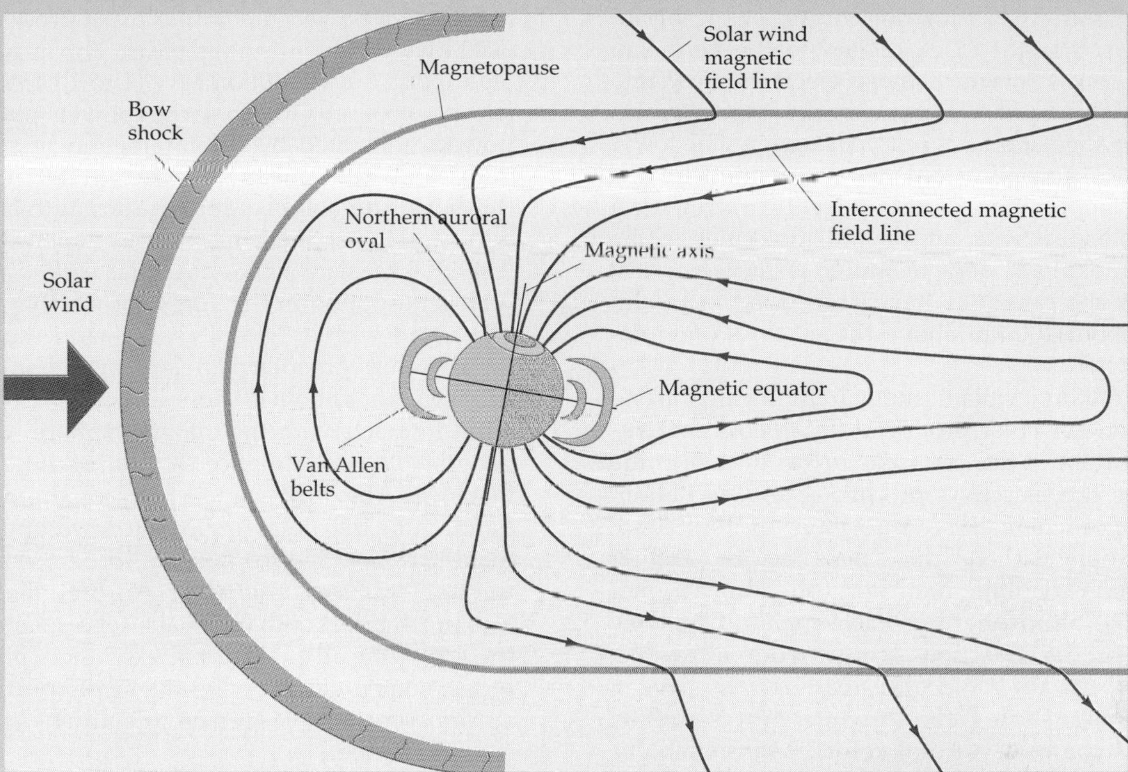

Figure 6 The earth's magnetosphere. The solar wind confines the earth's magnetic field to a comet-shaped zone that has our planet as its nucleus. The distance between the earth and the sunward side of the magnetosphere is about 10 earth radii. The magnetosphere is drawn out into a very long tail (not shown) that stretches for more than 1000 earth radii away from the sun (to the right of this illustration).

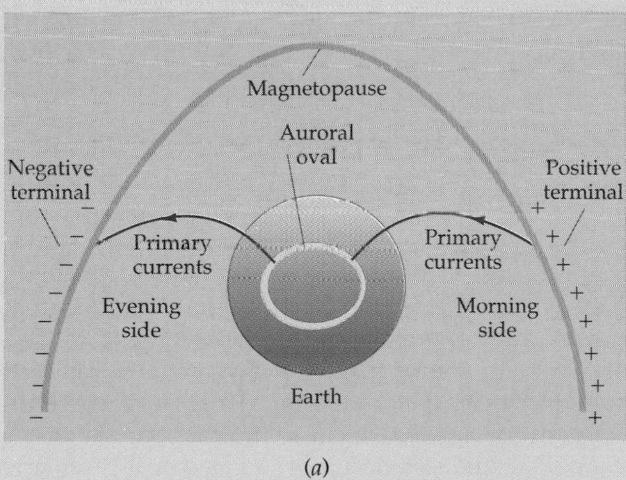

(*a*)

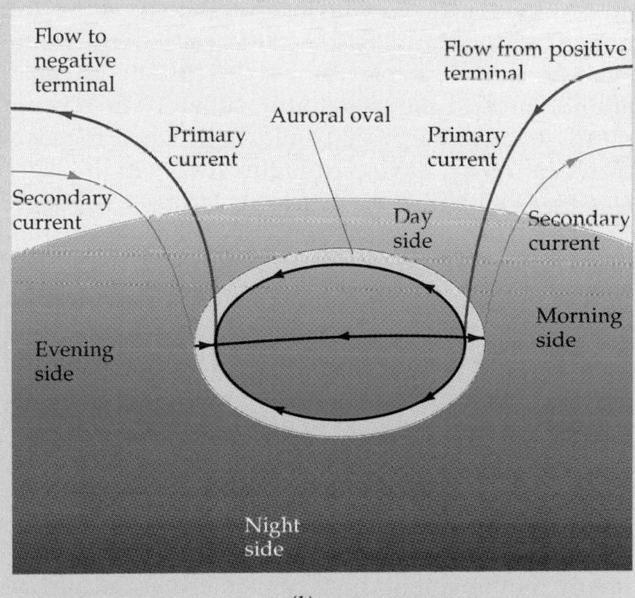

(*b*)

Figure 7 (*a*) View of the magnetosphere and the earth from above. The positive and negative terminals (morning and evening side magnetopause, respectively), together with the primary currents, are shown. (*b*) The primary and secondary electrical discharge circuits. The currents across the polar cap and along the auroral oval depend on the atmosphere's conductivity.

Continued

What is the basis for the curtain shape of auroras? It is thought to be related to the very thin sheetlike forms in which these electrons flow into the upper atmosphere, but the mechanism by which these thin sheets arise is not yet known. The lower limit of the auroral curtain is determined by the depth of penetration of the current-carrying electrons. Collisions with atoms and molecules in the upper atmosphere deplete much of the electrons' energy by the time they have descended to an altitude of about 100 km above the earth, so few descend farther.

Two factors explain the variations in auroral color. First, the color produced by an electrical discharge process varies from gas to gas and with the energy of the electrons producing the excitation. Second, the chemical composition of the atmosphere differs with height. These factors together explain the variations in auroral color. In the ionosphere, the atmosphere consists mainly of atomic oxygen, produced when energy from ultraviolet sunlight splits O_2 molecules apart. When oxygen atoms are excited, a greenish-white light is emitted (the most common auroral color). More energetic electrons penetrating farther into the atmosphere collide with neutral nitrogen molecules, producing auroras with red-violet or pink borders and rippled edges. Ionized nitrogen molecules produce a blue-violet light. Visible light is only a small portion of auroral emissions; x-rays, ultraviolet, and infrared radiation are also produced by the aurora.

To understand the movement observed in most auroral displays, consider an analogy with the image produced on the screen of a cathode-ray tube, such as that in a television set. The screen corresponds here to the upper atmosphere. The coating on the back of the screen emits light when the electron beam from the electron gun strikes it; this light is seen from the front of the screen as the image. Similarly, as just described, the ionosphere fluoresces when it is struck by the current-carrying electron sheets. Just as the impact point of an electron beam in a cathode-ray tube can change, causing movement of the image on the screen, so too can the auroral electron sheet rapidly shift, causing the auroral curtain to move, often violently. In the case of both the cathode-ray tube and the aurora, changes in a magnetic field and/or an electric field modulate the behavior of the electron beam. So, it is changes in the magnetic field, rather than atmospheric motions, that cause movement in the auroral curtain.

A large power-generating plant produces about 1000 MW on a continuous basis. The aurora generates about 1 to 10 million MW (1 to 10 TW), equivalent to 1000 to 10,000 large power plants. This power, generated by the interaction of solar wind with the earth's magnetosphere, fluctuates, sometimes considerably, because the strength of the solar wind and its magnetic field varies depending on the level of solar flare activity. A solar flare is associated with an eruption in the sun's corona that in turn is associated with a "gusty" solar wind that radiates rapidly through interplanetary space and reaches the earth after about 40 hours. As this "gusty" solar wind interacts with the magnetosphere, the power generated can be enhanced a thousandfold. In these instances, the ring-shaped belts of the aurora expand from the polar regions toward the equator, making it possible for auroras to be seen south of the U.S.–Canadian border. Auroral displays are also much brighter than usual after a solar flare, and the top of the auroral curtain extends to greater altitudes, sometimes allowing the upper portion of the aurora borealis to be seen as far south as Mexico and central Europe.

The enhanced electrical discharge currents associated with the gusty solar wind produce intensely fluctuating magnetic fields. When such fields are recorded, we say that a magnetic storm is in progress. The electric currents heat the upper atmosphere, causing upward motion of the lower, denser atmosphere and thus increasing the density at higher altitudes. This, in turn, increases the friction between orbiting satellites and the atmosphere, causing reductions in orbital altitudes. Satellites have even been known to fall out of orbit after major magnetic storms.

We now have a partial understanding of auroras: of the origin of the ring-shaped auroral belts around the geomagnetic poles; of the processes that power the gigantic electrical discharges that create them; of the amount of power generated during a display; of the reasons for the fluctuations in that power, and of the relationship between auroral and solar activity that is manifested by various transient solar processes including solar flares. As the twentieth century draws to a close, the challenge is to advance our understanding of the electrical discharge process that underlies this beautiful phenomenon, a powerful natural generator.

Suggestions for Further Reading

Akasofu, Syun-Ichi: "The Dynamic Aurora," *Scientific American,* May 1989, p. 90.

How the "solar wind" of charged particles interacts with the earth's magnetic field to produce the aurora, why the aurora appears to move and change, and where its power comes from.

Kondo, Herbert: "Michael Faraday," *Scientific American,* October 1953, p. 90.

This article describes Faraday's experiments and his revolutionary concept of the electromagnetic field.

Shamos, Morris H.: "Electromagnetic Induction and Laws of Electrolysis—Michael Faraday," in *Great Experiments in Physics,* Henry Holt and Company, New York, 1959. Reprinted by Dover, 1987.

This is Faraday's account of his discovery of electromagnetic induction, with editorial annotations for clarity and a biographical sketch.

Shamos, Morris H.: "Lenz's Law—Heinrich Lenz," in *Great Experiments in Physics,* Henry Holt and Company, New York, 1959. Reprinted by Dover, 1987.

This is Lenz's account of some experiments on electromagnetic induction resulting in a rule for determining the direction of an induced current, with editorial annotations and a biographical sketch.

Review

A. Objectives: After studying this chapter, you should:

1. Be able to state Faraday's law and use it to find the emf induced by a changing magnetic flux.

2. Be able to state Lenz's law and use it to find the direction of the induced current in various applications of Faraday's law.

3. Be able to discuss eddy currents.

4. Be able to discuss how simple ac generators and motors work.

5. Be able to sketch a graph of current versus time in an LR circuit.

B. Define, explain, or otherwise identify:

Magnetic induction
Magnetic flux
Weber
Faraday's law
Flip coil
Lenz's law
Back emf
Motional emf
Eddy currents
Generator
Motor
Self inductance
Henry
Mutual inductance
Inductor
LR circuit
Time constant of an *LR* circuit
Magnetic energy density

C. True or false: If the statement is true, explain why it is true. If it is false, give a counterexample.

1. The induced emf in a circuit is proportional to the magnetic flux through the circuit.

2. There can be an induced emf at an instant when the flux through the circuit is zero.

3. Lenz's law is related to the conservation of energy.

4. The inductance of a solenoid is proportional to the rate of change of the current in it.

5. The magnetic energy density at some point in space is proportional to the square of the magnetic field at that point.

Problems

Level I

26-1 Magnetic Flux

1. A uniform magnetic field of magnitude 2000 G is parallel to the x axis. A square coil of side 5 cm has a single turn and makes an angle θ with the z axis as shown in Figure 26-30. Find the magnetic flux through the coil when (*a*) $\theta = 0$, (*b*) $\theta = 30°$, (*c*) $\theta = 60°$, and (*d*) $\theta = 90°$.

Figure 26-30 Problem 1.

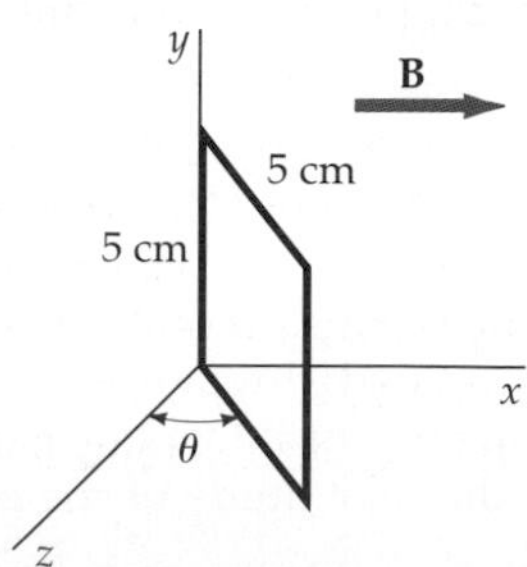

2. A circular coil has 25 turns and a radius of 5 cm. It is at the equator, where the earth's magnetic field is 0.7 G north. Find the magnetic flux through the coil when (*a*) its plane is horizontal, (*b*) its plane is vertical and its axis points north, (*c*) its plane is vertical and its axis points east, and (*d*) its plane is vertical and its axis makes an angle of 30° with north.

3. Find the magnetic flux through a solenoid of length 25 cm, radius 1 cm, and 400 turns that carries a current of 3 A.

4. Work Problem 3 for a solenoid of length 30 cm, radius 2 cm, and 800 turns that carries a current of 2 A.

5. A circular coil of radius 3.0 cm has 6 turns. A magnetic field $B = 5000$ G is perpendicular to the coil. (*a*) Find the magnetic flux through the coil. (*b*) Find the magnetic flux through the coil if the coil makes an angle of 20° with the magnetic field.

6. A magnetic field of 1.2 T is perpendicular to a square coil of 14 turns. The length of each side of the coil is 5 cm. (*a*) Find the magnetic flux through the coil. (*b*) Find the magnetic flux through the coil if the magnetic field makes an angle of 60° with the normal to the plane of the coil.

7. A circular coil of radius 3.0 cm has its plane perpendicular to a magnetic field of 400 G. (*a*) What is the magnetic flux through the coil if it has 75 turns? (*b*) How many turns must the coil have for the flux to be 0.015 Wb?

26-2 Induced Emf and Faraday's Law

8. A uniform magnetic field **B** is established perpendicular to the plane of a loop of radius 5.0 cm, resistance 0.4 Ω, and negligible self inductance. The magnitude of **B** is increasing at a rate of 40 mT/s. Find (*a*) the induced emf in the loop, (*b*) the induced current in the loop, and (*c*) the rate of Joule heating in the loop.

9. A 100-turn coil has a radius of 4.0 cm and a resistance of 25 Ω. At what rate must a perpendicular magnetic field change to produce a current of 4.0 A in the coil?

10. The flux through a loop is given by $\phi_m = (t^2 - 4t) \times 10^{-1}$ T·m^2, where t is in seconds. (*a*) Find the induced emf $\mathcal{E}$ as a function of time. (*b*) Find both ϕ_m and $\mathcal{E}$ at $t = 0$, $t = 2$ s, $t = 4$ s, and $t = 6$ s.

11. (*a*) For the flux given in Problem 10, sketch graphs of ϕ_m and $\mathcal{E}$ versus t. (*b*) At what time is the flux maximum? What is the emf at this time? (*c*) At what times is the flux zero? What is the emf at these times?

12. A 100-turn circular coil has a diameter of 2.0 cm and resistance of 50 Ω. The plane of the coil is perpendicular to a uniform magnetic field of magnitude 1.0 T. The direction of the field is suddenly reversed. (*a*) Find the total charge that passes through the coil. If the reversal takes 0.1 s, find (*b*) the average current in the coil and (*c*) the average emf in the coil.

13. At the equator, a 1000-turn coil with a cross-sectional area of 300 cm^2 and a resistance of 15.0 Ω is aligned with its plane perpendicular to the earth's magnetic field of 0.7 G. If the coil is flipped over, how much charge flows through it?

14. A circular coil of 300 turns and radius 5.0 cm is connected to a current integrator. The total resistance of the circuit is 20 Ω. The plane of the coil is originally aligned perpendicular to the earth's magnetic field at some point. When the coil is rotated through 90°, the charge that passes through the current integrator is measured to be 9.4 μC. Calculate the magnitude of the earth's magnetic field at that point.

15. The magnetic field in Problem 5 is steadily reduced to zero in 1.2 s. Find the emf induced in the coil when (*a*) the magnetic field is perpendicular to the coil and (*b*) the magnetic field makes an angle of 20° with the normal to the coil.

16. The magnetic field in Problem 7 is steadily reduced to zero in 0.8 s. What is the magnitude of the emf induced in the coil of part (*b*)?

17. A solenoid of length 25 cm and radius 0.8 cm with 400 turns is in an external magnetic field of 600 G that makes an angle of 50° with the axis of the solenoid. (*a*) Find the magnetic flux through the solenoid. (*b*) Find the magnitude of the emf induced in the solenoid if the external magnetic field is reduced to zero in 1.4 s.

26-3 Lenz's Law

18. The two loops in Figure 26-31 have their planes parallel to each other. As viewed from A toward B, there is a counterclockwise current in loop A. Give the direction of the current in loop B and state whether the loops attract or repel each other if the current in loop A is (*a*) increasing and (*b*) decreasing.

Figure 26-31 Problem 18.

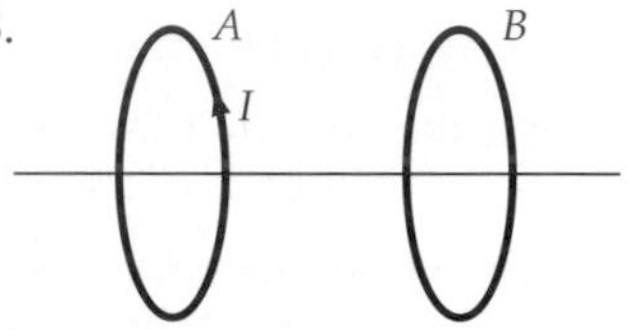

19. A bar magnet moves with constant velocity along the axis of a loop as shown in Figure 26-32. (*a*) Make a qualitative graph of the flux ϕ_m through the loop as a function of time. Indicate the time t_1 when the magnet is halfway through the loop. (*b*) Sketch a graph of the current I in the loop versus time, choosing I to be positive when it is counterclockwise as viewed from the left.

Figure 26-32 Problem 19.

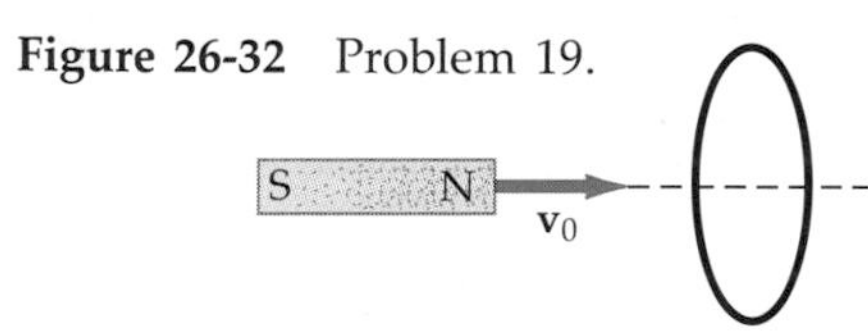

20. Give the direction of the induced current in the circuit on the right in Figure 26-33 when the resistance in the circuit on the left is suddenly (*a*) increased and (*b*) decreased.

Figure 26-33 Problem 20.

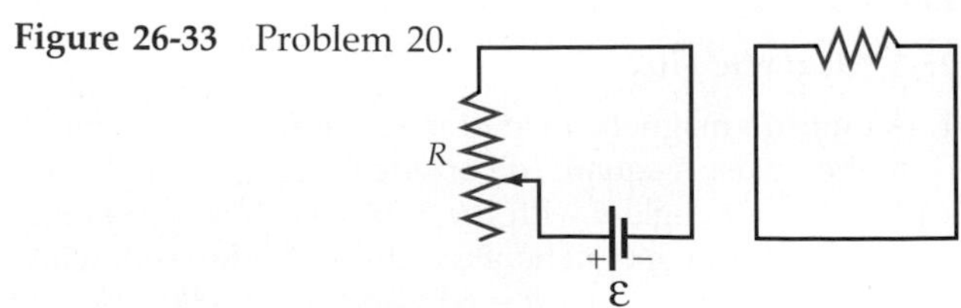

21. A bar magnet is mounted on the end of a coiled spring in such a way that it moves with simple harmonic motion along the axis of a loop as shown in Figure 26-34. (*a*) Make a qualitative graph of the flux ϕ_m through the loop as a function of time. Indicate the time t_1 when the magnet is halfway through the loop. (*b*) Sketch the current I in the loop versus time, choosing I to be positive when it is counterclockwise as viewed from above.

Figure 26-34 Problem 21.

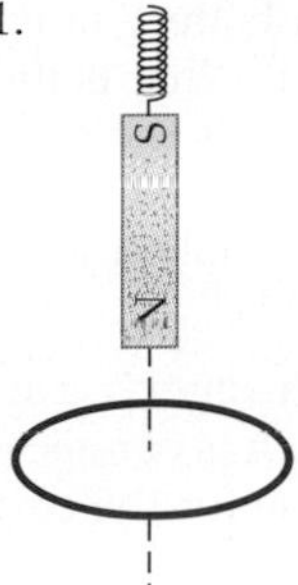

26-4 Motional Emf

22. A rod 30 cm long moves at 8 m/s in a plane perpendicular to a magnetic field of 500 G. The velocity of the rod is perpendicular to its length. Find (*a*) the magnetic force on an electron in the rod, (*b*) the electrostatic field **E** in the rod, and (*c*) the potential difference V between the ends of the rod.

23. Find the speed of the rod in Problem 22 if the potential difference between the ends is 6 V.

24. In Figure 26-14, let B be 0.8 T, v = 10.0 m/s, ℓ = 20 cm, and R = 2 Ω. Find (*a*) the induced emf in the circuit, (*b*) the current in the circuit, and (*c*) the force needed to move the rod with constant velocity assuming negligible friction. Find (*d*) the power input by the force found in part (*c*) and (*e*) the rate of Joule heat production I^2R.

25. Work Problem 24 for B =1.5 T, v = 6 m/s, ℓ = 40 cm, and R = 1.2 Ω.

26-5 Eddy Currents

There are no problems for this section.

26-6 Generators and Motors

26. A 200-turn coil has an area of 4 cm^2. It rotates in a magnetic field of 0.5 T. (*a*) What should the frequency be to generate a maximum emf of 10 V? (*b*) If the coil rotates at 60 Hz, what is the maximum emf?

27. In what magnetic field must the coil of Problem 26 be rotating to generate a maximum emf of 10 V at 60 Hz?

28. A 2-cm by 1.5-cm rectangular coil has 300 turns and rotates in a magnetic field of 4000 G. (*a*) What is the maximum emf generated when the coil rotates at 60 Hz? (*b*) What must its frequency be to generate a maximum emf of 110 V?

29. The coil of Problem 28 rotates at 60 Hz in a magnetic field B. What must the value of B be so that the maximum emf generated is 24 V?

26-7 Inductance

30. A coil with a self inductance of 8.0 H carries a current of 3 A that is changing at a rate of 200 A/s. Find (*a*) the magnetic flux through the coil and (*b*) the induced emf in the coil.

31. A coil with self inductance L carries a current I, given by $I = I_0 \sin 2\pi ft$. Find and graph the flux ϕ_m and the self-induced emf as functions of time.

32. A solenoid has a length of 25 cm, a radius of 1 cm, and 400 turns and carries a 3-A current. Find (*a*) B on the axis at the center of the solenoid; (*b*) the flux through the solenoid, assuming B to be uniform; (*c*) the self inductance of the solenoid; and (*d*) the induced emf in the solenoid when the current changes at 150 A/s.

33. Two solenoids of radii 2 cm and 5 cm are coaxial. They are each 25 cm long and have 300 and 1000 turns, respectively. Find their mutual inductance.

26-8 *LR* Circuits

34. The current in an LR circuit is zero at time t = 0 and increases to half its final value in 4.0 s. (*a*) What is the time constant of this circuit? (*b*) If the total resistance is 5 Ω, what is the self inductance?

35. A coil of resistance 8.0 Ω and self inductance 4.0 H is suddenly connected across a constant potential difference of 100 V. Let t = 0 be the time of connection, at which the current is zero. Find the current I and its rate of change dI/dt at times (*a*) t = 0, (*b*) t = 0.1 s, (*c*) t = 0.5 s, and (*d*) t = 1.0 s.

36. How many time constants must elapse before the current in an LR circuit that is initially zero reaches (*a*) 90 percent, (*b*) 99 percent, and (*c*) 99.9 percent of its final value?

37. The current in a coil with a self inductance of 1 mH is 2.0 A at t = 0, when the coil is shorted through a resistor. The total resistance of the coil plus the resistor is 10.0 Ω. Find the current after (*a*) 0.5 ms and (*b*) 10 ms.

26-9 Magnetic Energy

38. In the circuit of Figure 26-26, let $\mathcal{E}_0$ = 12.0 V, R = 3.0 Ω, and L = 0.6 H. The switch is closed at time t = 0. At time t = 0.5 s, find (*a*) the rate at which the battery supplies power, (*b*) the rate of Joule heating, and (*c*) the rate at which energy is being stored in the inductor.

39. Do Problem 38 for the times t = 1 s and t = 100 s.

40. A coil with a self inductance of 2.0 H and a resistance of 12.0 Ω is connected across a 24-V battery of negligible internal resistance. (*a*) What is the final current? (*b*) How much energy is stored in the inductor when the final current is attained?

41. Find (*a*) the magnetic energy, (*b*) the electric energy, and (*c*) the total energy in a volume of 1.0 m^3 in which there is an electric field of 10^4 V/m and a magnetic field of 1 T.

Level II

42. A circular coil of 15 turns of radius 4 cm is in a uniform magnetic field of 4000 G in the positive x direction. Find the flux through the coil when the unit normal vector to the plane of the coil is (*a*) $\hat{\mathbf{n}} = \mathbf{i}$, (*b*) $\hat{\mathbf{n}} = \mathbf{j}$, (*c*) $\hat{\mathbf{n}} = (\mathbf{i} + \mathbf{j})/\sqrt{2}$, (*d*) $\hat{\mathbf{n}} = \mathbf{k}$, and (*e*) $\hat{\mathbf{n}} = 0.6\mathbf{i} + 0.8\mathbf{j}$.

43. A uniform magnetic field **B** is perpendicular to the base of a hemisphere of radius R. Calculate the magnetic flux through the spherical surface of the hemisphere.

44. An elastic circular conducting loop is expanding at a constant rate so that its radius is given by $R = R_0 + vt$. The loop is in a region of constant magnetic field perpendicular to the loop. What is the emf generated in the expanding loop? Neglect possible effects of self inductance.

45. A solenoid has n turns per unit length, radius R_1, and carries a current I. (*a*) A large circular loop of radius $R_2 > R_1$ and N turns encircles the solenoid at a point far away from the ends of the solenoid. Find the magnetic flux through the loop. (*b*) A small circular loop of N turns and radius $R_3 < R_1$ is completely inside the solenoid, far from its ends, with its axis parallel to that of the solenoid. Find the magnetic flux through the loop.

46. Show that if the flux through each turn of an N-turn coil of resistance R changes from ϕ_{m1} to ϕ_{m2} in any manner, the total charge passing through the coil is given by $Q = N(\phi_{m1} - \phi_{m2})/R$.

47. An ac generator's rectangular loop of dimensions a and b has N turns. The loop is connected to slip rings (Figure 26-35) and rotates with an angular velocity ω in a uniform magnetic field **B**. (*a*) Show that the potential difference between the two slip rings is $\mathcal{E} = NBab\omega \sin \omega t$. (*b*) If $a = 1.0$ cm, $b = 2.0$ cm, $N = 1000$, and $B = 2$ T, at what angular frequency ω must the coil rotate to generate an emf whose maximum value is 110 V?

Figure 26-35 Problem 47.

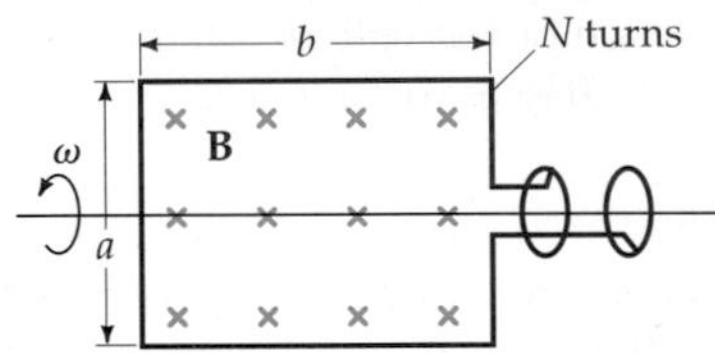

48. A dc motor has coils of resistance 5.5 Ω. When connected to dc power supply of 120 V, it draws 6 A. (*a*) How large is the back emf? (*b*) What is the initial current drawn before it starts to rotate?

49. To limit the current drawn by a motor as it starts up, a resistance is sometimes placed in series with the motor. The resistance is removed when the motor attains operating speed. (*a*) How much resistance should be placed in series with a motor that has a resistance of 0.75 Ω and draws 8 A when operated at 220 V if the current is not to exceed 15 A? (*b*) What is the back emf of this motor when it attains operating speed and the resistance is removed?

50. Compute the initial slope dI/dt at $t = 0$ from Equation 26-31, and show that if the current decreased steadily at this rate, it would be zero after one time constant.

51. An inductance L and resistance R are connected in series with a battery as in Figure 26-28. A long time after switch S_1 is closed, the current is 2.5 A. When the battery is switched out of the circuit by opening switch S_1 and closing S_2, the current drops to 1.5 A in 45 ms. (*a*) What is the time constant for this circuit? (*b*) If $R = 0.4$ Ω, what is L?

52. A coil with inductance 4 mH and resistance 150 Ω is connected across a battery of emf 12 V and negligible internal resistance. (*a*) What is the initial rate of increase of the current? (*b*) What is the rate of increase when the current is half its final value? (*c*) What is the final current? (*d*) How long does it take for the current to reach 99 percent of its final value?

53. A large electromagnet has an inductance of 50 H and a resistance of 8.0 Ω. It is connected to a dc power source of 250 V. Find the time for the current to reach (*a*) 10 A and (*b*) 30 A.

54. When the current in a certain coil is 5.0 A and is increasing at the rate of 10.0 A/s, the potential difference across the coil is 140 V. When the current is 5.0 A and is decreasing at the rate of 10 A/s, the potential difference is 60 V. Find the resistance and self inductance of the coil.

55. In a plane electromagnetic wave such as a light wave, the magnitudes of the electric and magnetic fields are related by $E = cB$, where $c = 1/\sqrt{\epsilon_0\mu_0}$ is the speed of light. Show that in this case the electric and the magnetic energy densities are equal.

56. Show that the effective inductance for two inductors L_1 and L_2 connected in series such that none of the flux from either passes through the other is given by $L_{eff} = L_1 + L_2$.

57. Given the circuit shown in Figure 26-36, assume that switch S has been closed for a long time so that steady currents exist in the circuit and that the inductor L is made of superconducting wire so that its resistance may be considered to be zero. (*a*) Find the battery current, the current in the 100-Ω resistor, and the current through the inductor. (*b*) Find the initial voltage across the inductor when switch S is opened. (*c*) Give the current in the inductor as a function of time measured from the instant of opening switch S.

Figure 26-36 Problem 57.

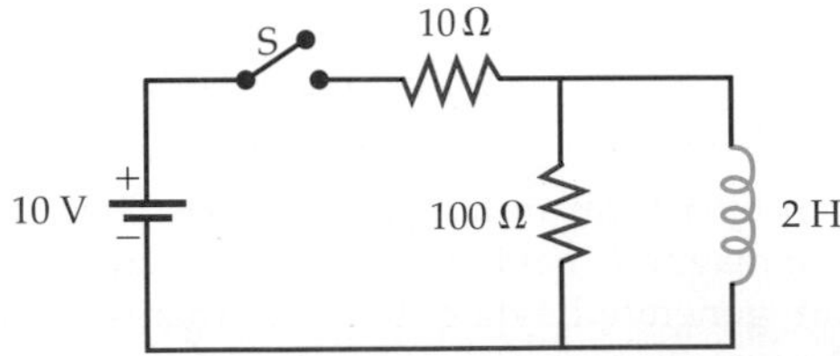

58. Show that the effective inductance for two inductors L_1 and L_2 connected in parallel such that none of the flux from either passes through the other is given by

$$1/L_{eff} = 1/L_1 + 1/L_2$$

59. For the circuit of Figure 26-37, (*a*) find the rate of change of current in each inductor and in the resistor just after the switch is closed. (*b*) What is the final current? (See Problem 58.)

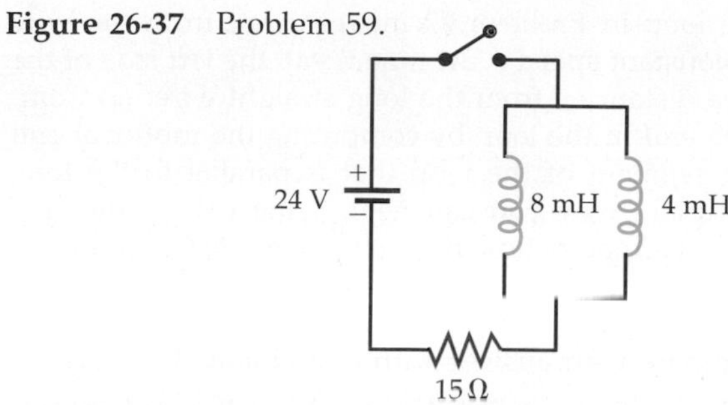

Figure 26-37 Problem 59.

60. For the circuit shown in Figure 26-38, find the currents I_1, I_2, and I_3 (*a*) immediately after switch S is closed and (*b*) a long time after switch S has been closed. After the switch has been closed for a long time, it is opened. Find the three currents (*c*) immediately after switch S is opened and (*d*) a long time after switch S was opened.

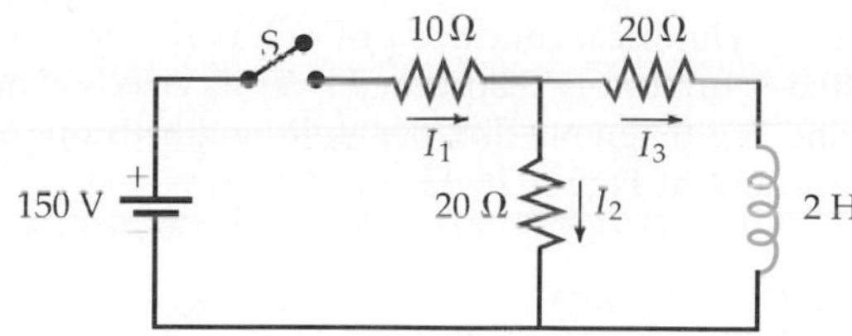

Figure 26-38 Problem 60.

61. A solenoid of 2000 turns, area 4 cm^2, and length 30 cm carries a current of 4.0 A. (*a*) Calculate the magnetic energy stored in the solenoid from $\frac{1}{2}LI^2$. (*b*) Divide your answer in part (*a*) by the volume of the solenoid to find the magnetic energy per unit volume in the solenoid. (*c*) Find B in the solenoid. (*d*) Compute the magnetic energy density from $\eta_m = B^2/2\mu_0$, and compare your answer with your result for part (*b*).

62. A toroid of mean radius 25 cm and a coil radius of 2 cm is wound with a superconducting wire of length 1000 m that carries a current of 400 A. (*a*) What is the number of turns on the coil? (*b*) What is the magnetic field at the mean radius? (*c*) Assuming that B is constant over the area of the coil, calculate the magnetic energy density and the total energy stored in the toroid.

63. A long solenoid has n turns per unit length and carries a current given by $I = I_0 \sin \omega t$. The solenoid has a circular cross section of radius R. Find the induced electric field at a radius r from the axis of the solenoid for (*a*) $r < R$ and (*b*) $r > R$.

64. A uniform magnetic field of magnitude 1.2 T is in the z direction. A conducting rod of length 15 cm lies parallel to the y axis and oscillates in the x direction with displacement given by $x = (2 \text{ cm}) \cos 120 \pi t$. What is the emf induced in the rod?

65. A 10-cm by 5-cm rectangular loop with resistance 2.5 Ω is pulled through a region of uniform magnetic field $B = 1.7$ T (Figure 26-39) with constant speed $v = 2.4$ cm/s. The front of the loop enters the region of the magnetic field at time $t = 0$. (*a*) Find and graph the flux through the loop as a function of time. (*b*) Find and graph the induced emf and the current in the loop as functions of time. Neglect any self inductance of the loop and extend your graphs from $t = 0$ to $t = 16$ s.

Figure 26-39 Problem 65.

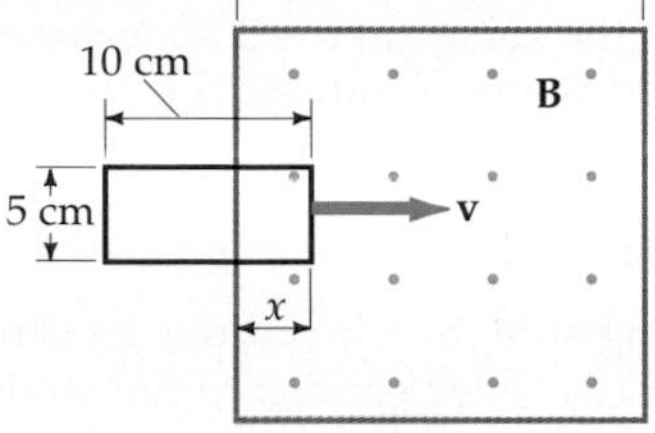

66. In Example 26-7, find the total energy dissipated in the resistance and show that it is equal to $\frac{1}{2}mv_0^2$.

67. In Figure 26-40, the rod has a resistance R and the rails have negligible resistance. A battery of emf $\mathcal{E}$ and negligible internal resistance is connected between points a and b such that the current in the rod is downward. The rod is placed at rest at $t = 0$. (*a*) Find the force on the rod as a function of the speed v and write Newton's second law for the rod when it has speed v. (*b*) Show that the rod reaches a terminal velocity and find an expression for it. (*c*) What is the current when the rod reaches its terminal velocity?

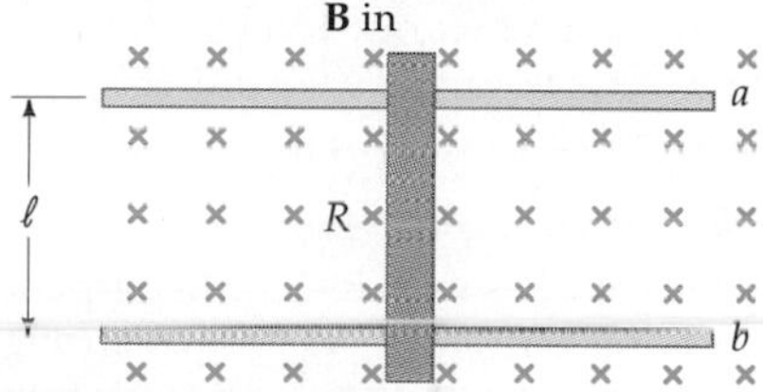

Figure 26-40 Problems 67 and 68.

68. In Figure 26-40, the rod has a resistance R and the rails have negligible resistance. A capacitor with charge Q_0 and capacitance C is connected between points a and b such that the current in the rod is downward. The rod is placed at rest at $t = 0$. (*a*) Write the equation of motion for the rod on the rails. (*b*) Show that the terminal speed of the rod down the rails is related to the final charge on the capacitor.

69. A conducting rod of mass m and resistance R is free to slide without friction along two parallel rails of negligible resistance separated by a distance ℓ and inclined at an angle θ to the horizontal. There is a magnetic field B directed upward. (*a*) Show that there is a retarding force directed up the incline given by

$$F = (B^2\ell^2 v \cos^2 \theta)/R$$

(*b*) Show that the terminal speed of the rod is

$$v_t = (mgR \sin \theta)/(B^2\ell^2 \cos^2 \theta)$$

70. A simple pendulum has a wire of length L supporting a metal ball of mass m. The wire has negligible mass and moves in a uniform horizontal magnetic field B. This pendulum executes simple harmonic motion having angular amplitude θ_0. What is the emf generated along the wire?

71. A wire lies along the z axis and carries current $I = 20$ A in the positive z direction. A small conducting sphere of radius $R = 2$ cm is initially at rest on the y axis at a distance $h = 45$ m above the wire. The sphere is dropped at time $t = 0$. (*a*) What is the electric field at the center of the sphere at $t = 3$ s? Assume that the only magnetic field is that produced by the wire. (*b*) What is the voltage across the sphere at $t = 3$ s?

Level III

72. A long straight wire carries a current I. A rectangular coil with two sides parallel to the straight wire has sides a and b with its near side a distance of d from the wire as shown in Figure 26-41. (*a*) Compute the magnetic flux through the rectangular coil. *Hint:* Calculate the flux through a strip of area $dA = b\,dx$ and integrate from $x = d$ to $x = d + a$. (*b*) Evaluate your answer for $a = 5$ cm, $b = 10$ cm, $d = 2$ cm, and $I = 20$ A.

Figure 26-41 Problems 72 and 74.

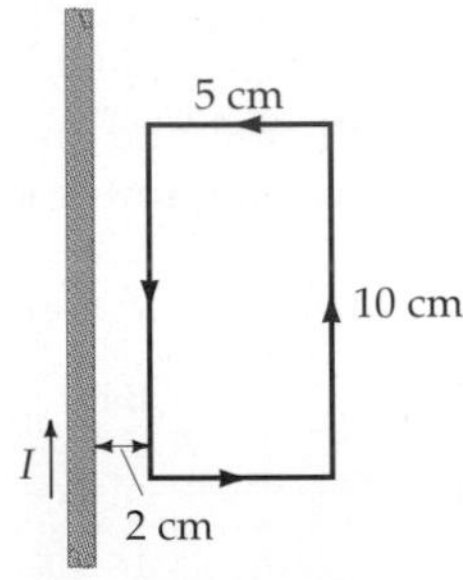

73. A rod of length ℓ lies with its length perpendicular to a long wire carrying current I, as shown in Figure 26-42. The near end of the rod is a distance d away from the wire. The rod moves with a speed v in the direction of the current I. (*a*) Show that the potential difference between the ends of the rod is given by

$$V = \frac{\mu_0 I}{2\pi} v \ln \frac{d + \ell}{d}$$

(*b*) Use Faraday's law to obtain this result by considering the flux through a rectangular area $A = \ell v t$ swept out by the rod.

Figure 26-42 Problem 73.

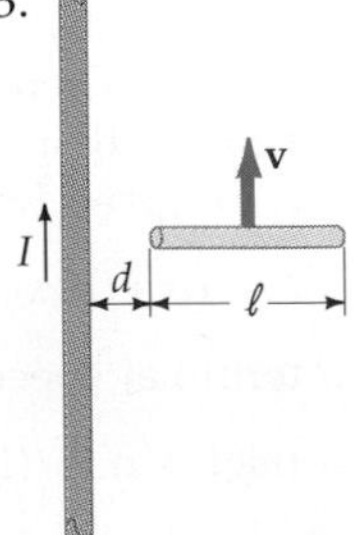

74. The loop in Problem 72 moves away from the wire with a constant speed v. At time $t = 0$, the left side of the loop is a distance d from the long straight wire. (*a*) Compute the emf in the loop by computing the motional emf in each segment of the loop that is parallel to the long wire. Explain why you can neglect the emf in the segments that are perpendicular to the wire. (*b*) Compute the emf in the loop by first computing the flux through the loop as a function of time and then using $\mathcal{E} = -d\phi_m/dt$ and compare your answer with that obtained in part (*a*).

75. A thin-walled hollow wire of radius a lies with its axis along the z axis and carries current I in the positive z direction. A second identical wire is parallel to it with its axis along the line $x = d$. It carries current I in the negative z direction. (*a*) Find the magnetic flux per unit length through the space in the xz plane between the wires. (*b*) If the far ends of the wires are connected together so that the parallel wires form two sides of a loop, find the self inductance per unit length of the loop.

76. A long cylindrical conductor of radius R carries a current I that is uniformly distributed over its cross-sectional area. Find the magnetic flux per unit length through the area indicated in Figure 26-43.

Figure 26-43 Problem 76.

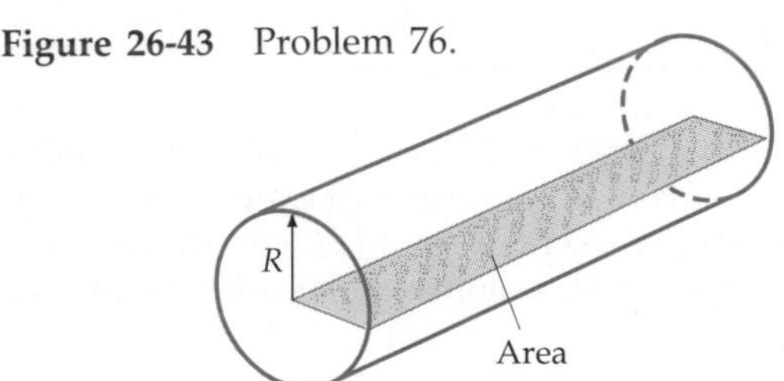

77. A conducting rod of length ℓ rotates at constant angular velocity about one end, in a plane perpendicular to a uniform magnetic field B (Figure 26-44). (*a*) Show that the magnetic force on a charge q at a distance r from the pivot is $Bqr\omega$. (*b*) Show that the potential difference between the ends of the rod is $V = \frac{1}{2}B\omega\ell^2$. (*c*) Draw any radial line in the plane from which to measure $\theta = \omega t$. Show that the area of the pie-shaped region between the reference line and the rod is $A = \frac{1}{2}\ell^2\theta$. Compute the flux through this area, and show that $\mathcal{E} = \frac{1}{2}B\omega\ell^2$ follows when Faraday's law is applied to this area.

Figure 26-44 Problem 77.

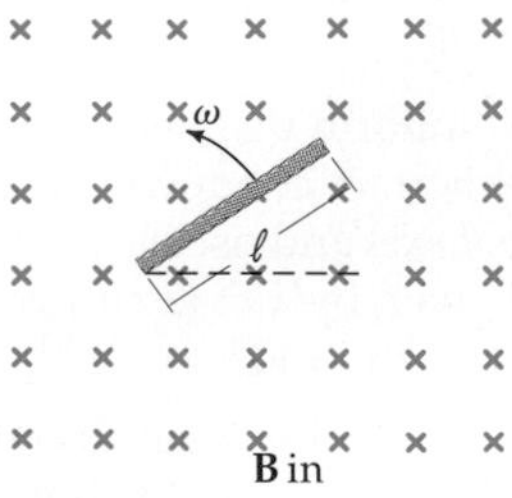

78. In the circuit of Figure 26-26, let $\mathcal{E}_0 = 12.0$ V, $R = 3.0\ \Omega$, and $L = 0.6$ H. The switch is closed at time $t = 0$. From time $t = 0$ to $t = \tau$, find (*a*) the total energy that has been supplied by the battery, (*b*) the total energy that has been dissipated in the resistor, and (*c*) the energy that has been stored in the inductor. *Hint:* Find the rates as functions of time and integrate from $t = 0$ to $t = \tau = L/R$.

79. In Figure 26-45, circuit 2 has a total resistance of $300\ \Omega$. A total charge of 2×10^{-4} C flows through the galvanometer in circuit 2 when switch S in circuit 1 is closed. After a long time, the current in circuit 1 is 5 A. What is the mutual inductance between the two coils?

Figure 26-45 Problem 79.

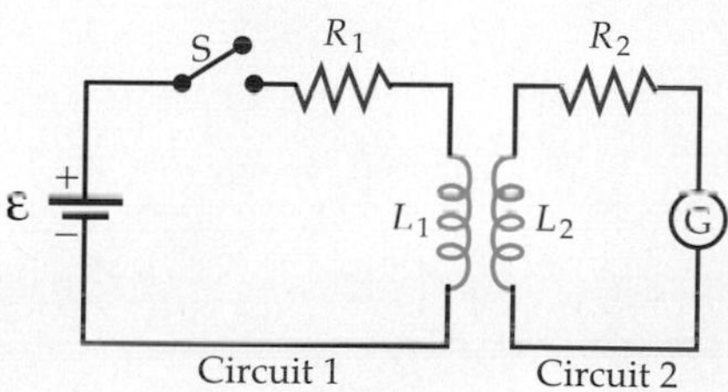

80. A coaxial cable consists of two very thin-walled conducting cylinders of radii r_1 and r_2 (Figure 26-46). Current I goes in one direction down the inner cylinder and in the opposite direction in the outer cylinder. (*a*) Use Ampere's law to find B. Show that $B = 0$ except in the region between the conductors. (*b*) Show that the magnetic energy density in the region between the cylinders is

$$\eta_m = \frac{\mu_0 I^2}{8\pi^2 r^2}$$

(*c*) Find the magnetic energy in a cylindrical shell volume element of length ℓ and volume $d\mathcal{V} = \ell 2\pi r\, dr$, and integrate your result to show that the total magnetic energy in the volume of length ℓ is

$$U_m = \frac{\mu_0}{4\pi} I^2 \ell \ln \frac{r_2}{r_1}$$

(*d*) Use the result in part (*c*) and $U_m = \frac{1}{2}LI^2$ to show that the self inductance per unit length is

$$\frac{L}{\ell} = \frac{\mu_0}{2\pi} \ln \frac{r_2}{r_1}$$

Figure 26-46 Problems 80 and 81.

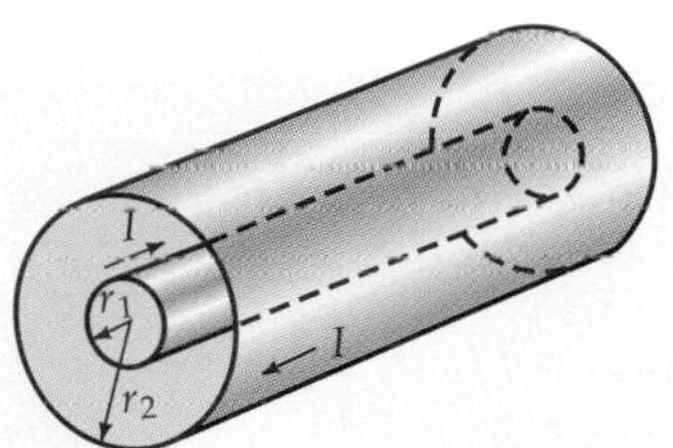

81. In Figure 26-46, compute the flux through a rectangular area of sides ℓ and $r_2 - r_1$ between the conductors. Show that the self inductance per unit length can be found from $\phi_m = LI$ [see part (*d*) of Problem 80].

82. Show that the inductance of a toroid of rectangular cross section as shown in Figure 26-47 is given by

$$L = \frac{\mu_0 N^2 h \ln (b/a)}{2\pi}$$

where N is the total number of turns, a is the inside radius, b is the outside radius, and h is the height of the toroid.

Figure 26-47 Problem 82.

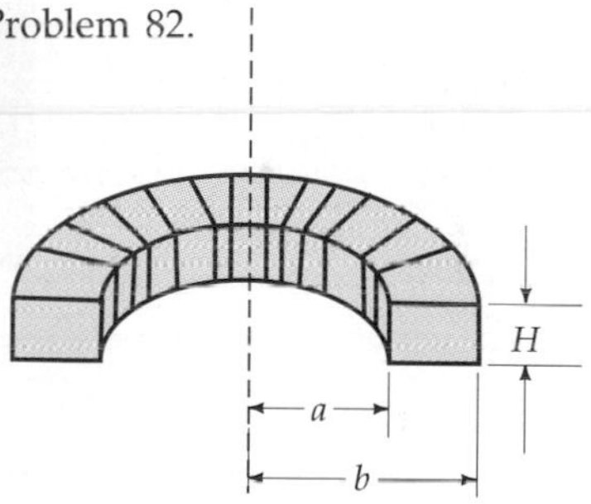

Chapter 27

Magnetism in Matter

Magnetic domains on the surface of an Fe-3%Si crystal—observed when using a scanning electron microscope with polarization analysis. The four colors indicate four possible domain orientations.

In studying electric fields in matter, we found that the electric field is affected by the presence of electric dipoles. Polar molecules have permanent electric dipole moments that are partially aligned by the electric field in the direction of the field. In nonpolar molecules, electric dipole moments are induced by the electric field in the direction of the field. In both cases, the alignment of the dipole moments parallel to the external electric field tends to weaken the field.

Similar but more complicated effects occur in magnetism. Atoms have magnetic dipole moments due to the motion of their electrons. In addition, each electron has an intrinsic magnetic dipole moment associated with its spin. The net magnetic moment of an atom depends on the arrangement of the electrons in the atom. Unlike the situation with electric dipoles, the alignment of magnetic dipoles parallel to an external magnetic field tends to *increase* the field. We can see this difference by comparing the electric-field lines of an electric dipole with the magnetic-field lines of a magnetic dipole, such as a small current loop, as in Figure 27-1. Far from the dipoles, the field lines are identical. However, between the charges of the electric dipole, the electric-field lines are opposite the direction of the dipole moment, whereas inside the current loop, the magnetic-field lines are parallel to the magnetic dipole moment. Thus, inside an electrically polarized material, the electric dipoles create an electric field that is *antiparallel* to their dipole-moment vectors, whereas inside a magnetically polarized material, the magnetic dipoles create a magnetic field that is *parallel* to the magnetic-dipole-moment vectors.

Materials fall into three categories—paramagnetic, diamagnetic, and ferromagnetic—according to the behavior of their molecules in an external magnetic field. Paramagnetic and ferromagnetic materials have molecules with permanent magnetic dipole moments. In paramagnetic materials, the magnetic dipoles do not interact strongly with each other and are normally randomly oriented. In the presence of an external magnetic field, the dipoles are partially aligned in the direction of the field, thereby increasing the field. However, in external magnetic fields of ordinary strength at ordinary temperatures, only a very small fraction of the molecules are aligned because thermal motion tends to randomize their orientation. The increase in the total magnetic field is therefore very small. Ferromagnetism is much more complicated. Because of a strong interaction between neighboring magnetic dipoles, a high degree of alignment occurs even in weak external magnetic fields, causing a very large increase in the total field. Even when there is no external magnetic field, a ferromagnetic material may have its magnetic dipoles aligned, as in permanent magnets. Diamagnetism is observed in materials whose molecules have no permanent magnetic moments. It is the result of an induced magnetic dipole moment opposite in direction to the external field. The induced dipoles thus decrease the total magnetic field. This effect actually occurs in all materials, but because it is very small, it is masked by paramagnetic or ferromagnetic effects when the individual molecules of the material have permanent magnetic dipole moments.

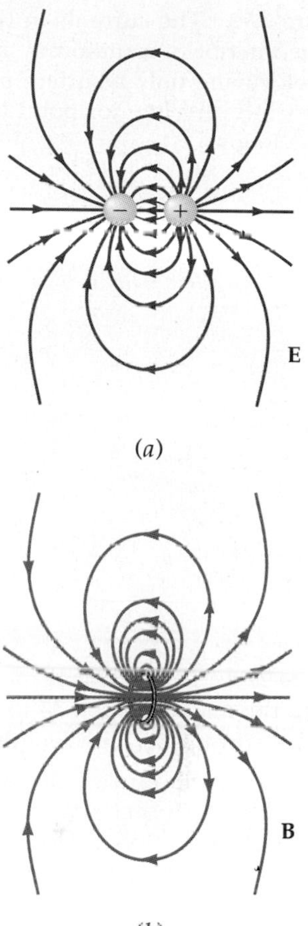

Figure 27-1 (*a*) Electric-field lines of an electric dipole. (*b*) Magnetic-field lines of a magnetic dipole. Far from the dipoles, the field lines are identical. In the region between the charges in (*a*), the electric field is opposite the dipole moment, whereas inside the loop in (*b*), the magnetic field is parallel to the dipole moment.

27-1 Magnetization and Magnetic Susceptibility

In our study of the electrical effects of matter, we placed a nonconducting material (dielectric) in a strong, uniform electric field between the plates of a parallel-plate capacitor. We found that the applied field tends to align the electric dipole moments (either permanent or induced). Similarly, when some material is placed in a strong magnetic field, such as that of a solenoid, the magnetic field of the solenoid tends to align the magnetic dipole moments (either permanent or induced) inside the material, and the material is said to be magnetized. We describe a magnetized material by its **magnetization M,** which is defined as the net magnetic dipole moment per unit volume of the material:

$$\mathbf{M} = \frac{d\mathbf{m}}{dV} \qquad 27\text{-}1$$

Long before we had any understanding of atomic or molecular structure, Ampère proposed a model of magnetism in which the magnetization of materials is due to microscopic current loops inside the magnetized material. We now know that these current loops are the result of the intrinsic motion of atomic charges. Although these motions are very complicated, for Ampère's model, we need only assume that the motions are equivalent to closed-circuit loops. Let us assume that our magnetized material is a cylinder. Figure 27-2 shows atomic current loops in the cylinder aligned with their magnetic moments along the axis of the cylinder. If the material is homogeneous, the net current at any point inside the material is zero because of cancellation of neighboring current loops. However, since there is no cancellation on the surface of the material, the result of these current

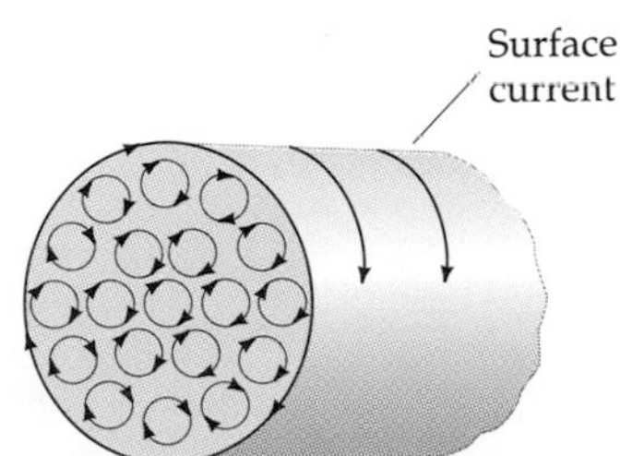

Figure 27-2 A model of atomic current loops in which all the atomic dipoles are parallel to the axis of the cylinder. The net current at any point inside the material is zero due to cancellation of neighboring atoms. The result is a surface current similar to that of a solenoid.

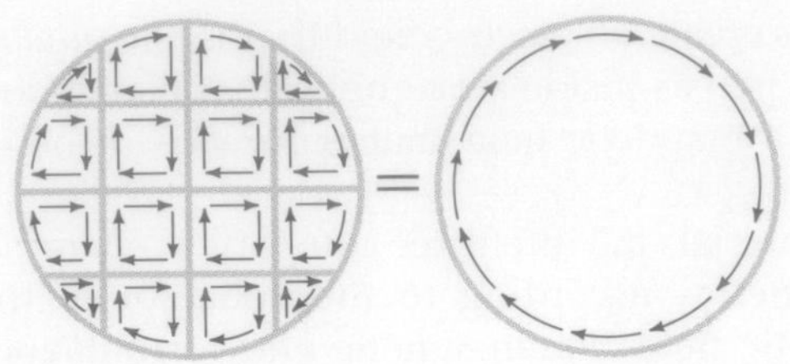

Figure 27-3 The currents in the adjacent current loops in the interior of a uniformly magnetized material cancel leaving only a surface current. This cancellation occurs at every interior point independent of the shape of the loops.

loops is equivalent to a current on the surface of the material (Figure 27-3). This surface current, called an **amperian current,** is similar to the real current in the windings of the solenoid.

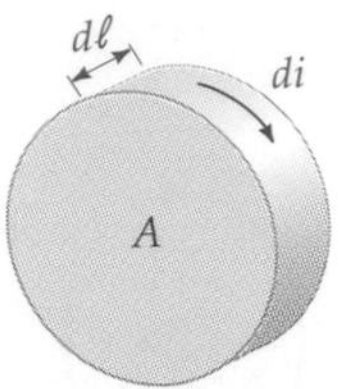

Figure 27-4 Disk element for relating the magnetization M to the surface current per unit length.

Figure 27-4 shows a small disk-shaped section of the cylinder. The disk has a cross-sectional area A, length $d\ell$, and volume $d\mathcal{V} = A\,d\ell$. Let di be the amperian current on the surface of the disk. The magnitude of the magnetic dipole moment dm of the disk is the same as that of a current loop of area A carrying a current di:

$$dm = A\,di$$

The magnetization M of the disk is the magnetic moment per unit volume:

$$M = \frac{dm}{d\mathcal{V}} = \frac{A\,di}{A\,d\ell} = \frac{di}{d\ell} \qquad 27\text{-}2$$

Thus the magnitude of the magnetization vector is the amperian current per unit length along the surface of the magnetized material. We see from this result that the units of M are amperes per meter.

Consider a cylinder that has a uniform magnetization **M** parallel to its axis. As we have just seen, the effect of the magnetization is the same as if the cylinder carried a surface current per unit length of magnitude M. This current is similar to the current carried by a tightly wound solenoid. For a solenoid, the current per unit length is nI where n is the number of turns per unit length and I is the current in each turn. We can calculate the magnetic field produced by the magnetized cylinder in the same way that we calculated the field produced by a tightly wound solenoid. If the cylinder is the same shape as the solenoid and if $M = nI$, the magnetic field produced by the cylinder at any point is exactly the same as that produced by the solenoid. In particular, the magnetic field B inside a solenoid and far from its ends is given by

$$B = \mu_0 nI$$

Then the magnetic field B_m inside and far from the ends of a cylinder with a uniform magnetization M is given by

$$B_m = \mu_0 M \qquad 27\text{-}3$$

Example 27-1

A small, cylindrical bar magnet of radius 0.5 cm and length 12 cm has a magnetic dipole moment of magnitude $m = 1.5\ \text{A}\cdot\text{m}^2$. (*a*) Find the magnetization M, assuming it to be uniform in the magnet. Find the magnetic field (*b*) at the center of the magnet and (*c*) just outside one end of the magnet. (*d*) Find the pole strength q_m of the magnet.

(*a*) The volume of the magnet is $\mathcal{V} = \pi r^2\ell = \pi(0.005\ \text{m})^2(0.12\ \text{m}) = 9.42 \times 10^{-6}\ \text{m}^3$. The magnetization is the magnetic moment per unit volume

$$M = \frac{m}{\mathcal{V}} = \frac{1.5\ \text{A}\cdot\text{m}^2}{9.42 \times 10^{-6}\ \text{m}^3} = 1.59 \times 10^5\ \text{A/m}$$

(*b*) The magnetic field inside a cylindrical magnet is the same as that inside a solenoid with nI replaced by M—the amperian current per unit length on the surface of the magnet. If end effects are neglected, the magnetic field at the center of the magnet is

$$B = \mu_0 M$$
$$= (4\pi \times 10^{-7}\ \text{T·m/A})(1.59 \times 10^5\ \text{A/m}) = 0.200\ \text{T}$$

(*c*) In Chapter 25, we saw that the magnetic field near the end of a solenoid is half that at the center of the solenoid. Thus, the magnetic field near the end of the cylinder is

$$B = \tfrac{1}{2}\mu_0 M = 0.100\ \text{T}$$

(*d*) The magnetic pole strength of the bar magnet equals the magnitude of the magnetic dipole moment divided by length. The pole strength is thus

$$q_m = \frac{m}{\ell} = \frac{1.5\ \text{A·m}^2}{0.12\ \text{m}} = 12.5\ \text{A·m}$$

Consider a long solenoid with n turns per unit length that carries a current I. We will call the magnetic field due to the current in the solenoid the applied field $\mathbf{B}_{app}$. We now place a cylinder of material inside the solenoid. The applied field of the solenoid magnetizes the material so that it has a magnetization **M**. The resultant magnetic field at a point inside the solenoid and far from its ends due to the current in the solenoid plus the magnetized material is

$$\mathbf{B} = \mathbf{B}_{app} + \mu_0 \mathbf{M} \qquad 27\text{-}4$$

For paramagnetic and ferromagnetic materials, **M** is in the same direction as $\mathbf{B}_{app}$; for diamagnetic materials, **M** is opposite to $\mathbf{B}_{app}$. For paramagnetic and diamagnetic materials, the magnetization is found to be proportional to the applied magnetic field that produces the alignment of the magnetic dipoles in the material. We can thus write

$$\mathbf{M} = \chi_m \left(\frac{\mathbf{B}_{app}}{\mu_0}\right) \qquad 27\text{-}5$$

where χ_m is a dimensionless number called the **magnetic susceptibility.** Equation 27-4 is then

$$\mathbf{B} = \mathbf{B}_{app} + \mu_0 \mathbf{M} = \mathbf{B}_{app}\,(1 + \chi_m) \qquad 27\text{-}6$$

For paramagnetic materials, χ_m is a small positive number that depends on temperature. For diamagnetic materials, it is a small negative constant independent of temperature. Table 27-1 lists the magnetic susceptibility of various paramagnetic and diamagnetic materials. We see that the magnetic susceptibility for the solids listed is of the order of 10^{-5}.

Equations 27-5 and 27-6 are not very useful for ferromagnetic materials because χ_m depends on B_{app} and on the previous state of magnetization of the material.

Table 27-1 **Magnetic Susceptibility of Various Materials at 20°C**

Material	χ_m
Aluminum	2.3×10^{-5}
Bismuth	-1.66×10^{-5}
Copper	-0.98×10^{-5}
Diamond	-2.2×10^{-5}
Gold	-3.6×10^{-5}
Magnesium	1.2×10^{-5}
Mercury	-3.2×10^{-5}
Silver	-2.6×10^{-5}
Sodium	-0.24×10^{-5}
Titanium	7.06×10^{-5}
Tungsten	6.8×10^{-5}
Hydrogen (1 atm)	-9.9×10^{-9}
Carbon dioxide (1 atm)	-2.3×10^{-9}
Nitrogen (1 atm)	-5.0×10^{-9}
Oxygen (1 atm)	2090×10^{-9}

Question

1. Why are some values of χ_m in Table 27-1 positive and others negative?

27-2 Atomic Magnetic Moments

The magnetization of a paramagnetic or ferromagnetic material can be related to the permanent magnetic moments of the individual atoms of the material. In general, the magnetic moment of an atom is related to its angular momentum. We can illustrate this by considering a particle of mass m_q and charge q moving with speed v in a circle of radius r as shown in Figure 27-5. (We use m_q here for the mass to avoid confusion with m for the magnetic moment.) The angular momentum of the particle is

$$L = m_q vr \tag{27-7}$$

Figure 27-5 Particle of charge q and mass m_q moving in a circle of radius r. The angular momentum is into the paper and has a magnitude $m_q vr$, and the magnetic moment is into the paper (if q is positive) and has a magnitude $\frac{1}{2}qvr$.

The magnitude of the magnetic moment is the product of the current and the area of the circle:

$$m = IA = I\pi r^2$$

For a charge moving in a circle, the current equals the charge times the frequency:

$$I = qf = \frac{q}{T}$$

where T is the period of the motion. Since the speed times the period equals the circumference of the circle, the period is given by

$$T = \frac{2\pi r}{v}$$

The current is thus

$$I = \frac{q}{T} = \frac{qv}{2\pi r}$$

and the magnetic moment is

$$\begin{aligned} m &= IA \\ &= \frac{qv}{2\pi r}\,\pi r^2 = \frac{1}{2}\,qvr \end{aligned} \tag{27-8}$$

Using $vr = L/m_q$ from Equation 27-7, we have for the magnetic moment

$$m = \frac{q}{2m_q}\,L$$

If the charge q is positive, the angular momentum and magnetic moment are in the same direction. We can therefore write

Magnetic moment and angular momentum

$$\mathbf{m} = \frac{q}{2m_q}\,\mathbf{L} \tag{27-9}$$

Equation 27-9 is the general classical relation between magnetic moment and angular momentum. It also holds in the quantum theory of the atom for orbital angular momentum, but not for the intrinsic spin angular momentum of the electron. For electron spin, the magnetic moment is twice that predicted by this equation. The extra factor of 2 is a result from quantum theory that has no analog in classical mechanics.

In the quantum theory of the atom, angular momentum is quantized; that is, it can have only a certain discrete set of values. The orbital angular momentum must be an integral multiple of $h/2\pi$, and spin angular momentum must be a half-integral multiple of $h/2\pi$, where h is a fundamental constant called Planck's constant* and has the value

$$h = 6.67 \times 10^{-34}\ \text{J·s}$$

The combination $h/2\pi$ occurs often and is designated by $\hbar$ (read "h bar"):

$$\hbar = \frac{h}{2\pi} = 1.05 \times 10^{-34}\ \text{J·s}$$

The magnetic moment of an atom is therefore also quantized. It is convenient to write Equation 27-9 for the magnetic moment as

$$\mathbf{m} = \frac{q\hbar}{2m_q}\frac{\mathbf{L}}{\hbar}$$

For an electron, $m_q = m_e$ and $q = -e$, so the magnetic moment of the electron is

$$\mathbf{m} = -\frac{e\hbar}{2m_e}\frac{\mathbf{L}}{\hbar} = -m_B\frac{\mathbf{L}}{\hbar} \qquad 27\text{-}10$$

where

$$m_B = \frac{e\hbar}{2m_e} = 9.27 \times 10^{-24}\ \text{A·m}^2 = 9.27 \times 10^{-24}\ \text{J/T} \qquad 27\text{-}11$$

Bohr magneton

is called a **Bohr magneton.** The magnetic moment of an electron due to its intrinsic spin angular momentum has a magnitude of 1 Bohr magneton. Although the calculation of the magnetic moment of any atom is a complicated problem in quantum theory, the result for all atoms, according to both theory and experiment, is that the magnetic moment is of the order of a few Bohr magnetons (or zero for atoms with closed-shell electronic structures that have zero angular momentum). See Section 37-6 in the *Extended Version* for discussion of the shell structure of atoms.

If all the atoms or molecules in some material have their magnetic moments aligned, the magnetic moment per unit volume of the material is the product of the number of molecules per unit volume n and the magnetic moment m of each molecule. For this extreme case, the **saturation magnetization** M_s is

$$M_s = nm \qquad 27\text{-}12$$

The number of molecules per unit volume can be found from the molecular mass $\mathcal{M}$, the density of the material ρ, and Avogadro's number N_A:

$$n = \frac{N_A(\text{atoms/mol})}{\mathcal{M}(\text{kg/mol})}\rho(\text{kg/m}^3)$$

Example 27-2

Find the saturation magnetization and the magnetic field it produces for iron, assuming that each iron atom has a magnetic moment of 1 Bohr magneton.

The density of iron is 7.9×10^3 kg/m^3, and its molecular mass is 55.8×10^{-3} kg/mol. The number of iron molecules (atoms) per unit volume is then

$$n = \frac{6.02 \times 10^{23}\ \text{atoms/mol}}{55.8 \times 10^{-3}\ \text{kg/mol}}(7.9 \times 10^3\ \text{kg/m}^3)$$
$$= 8.52 \times 10^{28}\ \text{atoms/m}^3$$

*Quantization and Planck's constant will be discussed in Chapter 35.

Taking the magnetic moment of each atom to be 1 Bohr magneton, we have for the saturation magnetization

$$M_s = (8.52 \times 10^{28}\ \text{atoms/m}^3)(9.27 \times 10^{-24}\ \text{A}\cdot\text{m}^2)$$
$$= 7.90 \times 10^5\ \text{A/m}$$

The magnetic field on the axis inside a long iron cylinder resulting from this maximum magnetization is then (Equation 27-3)

$$B = \mu_0 M_s = (4\pi \times 10^{-7}\ \text{T}\cdot\text{m/A})(7.90 \times 10^5\ \text{A/m})$$
$$= 0.993\ \text{T} \approx 1\ \text{T}$$

The measured saturation magnetic field of annealed iron is about 2.16 T, indicating that the magnetic moment of an iron atom is slightly greater than 2 Bohr magnetons. This magnetic moment is due mainly to the spins of two unpaired electrons in the iron atom.

Questions

2. Can a particle have angular momentum and not have a magnetic moment?
3. Can a particle have a magnetic moment and not have angular momentum?
4. A circular loop of wire carries a current I. Is there angular momentum associated with the magnetic moment of the loop? If so, why is it not noticed?

27-3 Paramagnetism

Paramagnetic materials are those having a very small, positive magnetic susceptibility χ_m. Paramagnetism occurs in materials whose atoms have permanent magnetic moments that interact with each other only very weakly. When there is no external magnetic field, these magnetic moments are randomly oriented. In the presence of an external magnetic field, they tend to line up parallel to the field, but this is counteracted by the tendency for the moments to be randomly oriented due to thermal motion. The fraction of the moments that line up with the field depends on the strength of the field and on the temperature. In strong external magnetic fields at very low temperatures, nearly all the moments are aligned with the field. In this situation, the contribution to the total magnetic field due to the material is very large, as indicated in the numerical estimate in Example 27-2. However, even with the strongest magnetic field obtainable in the laboratory, the temperature must be as low as a few kelvins to obtain a high degree of alignment. At higher temperatures only a small fraction of the moments are aligned with the external field, and the contribution of the material to the total magnetic field is very small. We can state this more quantitatively by comparing the energy of a magnetic moment in an external magnetic field with the thermal energy of an atom of the material, which is of the order of kT, where k is Boltzmann's constant and T is the absolute temperature.

In Chapter 18, we saw that the potential energy of an electric dipole of moment $\mathbf{p}$ in an electric field $\mathbf{E}$ is given by (Equation 18-12)

$$U = -pE\cos\theta = -\mathbf{p}\cdot\mathbf{E}$$

Liquid oxygen, which is paramagnetic, is attracted by the magnetic field of a permanent magnet. In a uniform magnetic field, a magnetic dipole experiences a torque but no net force. In a nonuniform field a dipole experiences a net force that depends on the spatial rate of change or gradient of the field. Here the liquid oxygen collects near the edges of the closely spaced region where the field gradient is maximum.

The potential energy of a magnetic dipole of moment **m** in an external magnetic field **B** is given by a similar equation:

$$U = -mB\cos\theta = -\mathbf{m}\cdot\mathbf{B} \qquad 27\text{-}13$$

The potential energy when the moment is parallel with the field ($\theta = 0$) is thus lower than when it is antiparallel ($\theta = 180°$) by the amount $2mB$ (Figure 27-6). For a typical magnetic moment of 1 Bohr magneton and a typical strong magnetic field of 1 T, the difference in potential energy is

$$\Delta U = 2m_BB = 2(9.27 \times 10^{-24}\ \text{J/T})(1\ \text{T}) = 1.85 \times 10^{-23}\ \text{J}$$

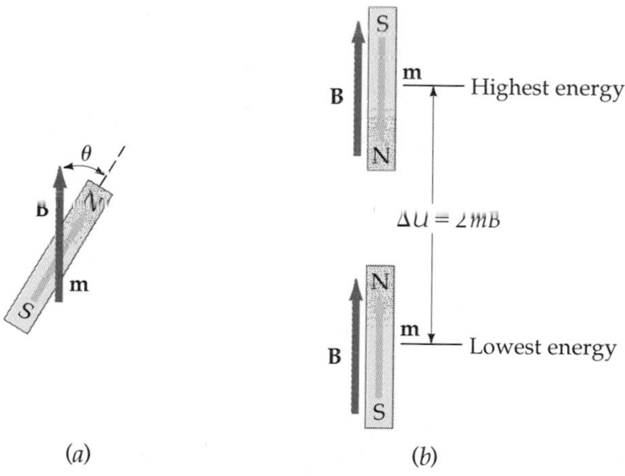

Figure 27-6 (*a*) A magnetic moment that makes an angle θ with a magnetic field **B**. (*b*) The potential energy is greatest when the moment is aligned with the field and least when it is antiparallel to the field. The difference in energy is $2mB$.

At a normal temperature of $T = 300$ K, the typical thermal energy kT is

$$kT = (1.38 \times 10^{-23}\ \text{J/K})(300\ \text{K}) = 4.14 \times 10^{-21}\ \text{J}$$

which is about 200 times greater than $2m_BB$. Thus, even in a strong magnetic field of 1 T, most of the magnetic moments will be randomly oriented because of thermal motions.

Figure 27-7 shows a plot of the magnetization M versus an applied external magnetic field B_{app} at a given temperature. In very strong fields, nearly all the magnetic moments are aligned with the field and $M \approx M_s$. (For magnetic fields attainable in the laboratory, this can only occur for very low temperatures.) When $B_{app} = 0$, $M = 0$, indicating that the orientation of the moments is completely random. In weak fields, the magnetization is approximately proportional to the applied field, as indicated by the orange dashed line in the figure. In this region, the magnetization is given by

Curie's law

$$M = \frac{1}{3}\frac{mB_{app}}{kT} M_s \qquad 27\text{-}14$$

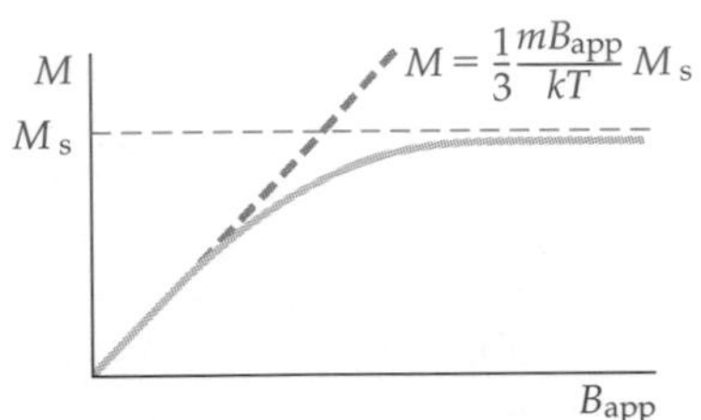

Figure 27-7 Plot of magnetization M versus applied field B_{app}. In very strong fields, the magnetization approaches the saturation value M_s. This can be achieved only at very low temperatures. In weak fields, the magnetization is approximately proportional to B_{app}, a result known as Curie's law.

Note that (mB_{app}/kT) is the ratio of the maximum energy of a dipole in the magnetic field to the characteristic thermal energy and is, therefore, a dimensionless number. The result that the magnetization varies inversely with the absolute temperature was discovered experimentally by Pierre Curie and is known as **Curie's law.**

Example 27-3

If $m = m_B$, at what temperature will the magnetization be 1 percent of the saturation magnetization in an applied magnetic field of 1 T?

From Curie's law, we have

$$M = \frac{1}{3}\frac{mB_{app}}{kT} M_s = 0.01\ M_s$$

Then,

$$T = \frac{mB_{app}}{0.03k} = \frac{(9.27 \times 10^{-24}\ \text{J/T})(1\ \text{T})}{(0.03)(1.38 \times 10^{-23}\ \text{J/K})} = 22.4\ \text{K}$$

From this example, we see that even in a strong applied magnetic field of 1 T the magnetization is less than 1 percent of saturation at temperatures above 22.4 K.

Exercise

If $m = m_B$, what fraction of the saturation magnetization is M at 300 K for an external magnetic field of 15,000 G? (Answer: $M/M_s = 1.12 \times 10^{-3}$)

A chunk of magnetite (loadstone) attracts the needle of a compass.

27-4 Ferromagnetism

Ferromagnetic materials are those having very large, positive values of magnetic susceptibility χ_m (as measured under conditions described below). Ferromagnetism occurs in pure iron, cobalt, and nickel and in alloys of these metals with each other. It also occurs in gadolinium, dysprosium, and a few compounds. In these substances, a small external magnetic field can produce a very large degree of alignment of the atomic magnetic dipole moments. In some cases, this alignment can persist even when the external magnetizing field is removed. This occurs because the magnetic dipole moments of the atoms of these substances exert strong forces on their neighbors so that over a small region of space the moments are aligned with each other even when there is no external field. The region of space over which the magnetic dipole moments are aligned is called a **magnetic domain.** The size of a domain is usually microscopic. Within the domain, all the magnetic moments are aligned, but the direction of alignment varies from domain to domain so that the net magnetic moment of a macroscopic piece of ferromagnetic material is zero in the normal state. Figure 27-8 illustrates this situation. The dipole forces that produce this alignment are predicted by quantum theory but cannot be explained with classical physics. At temperatures above a critical temperature, called the **Curie temperature,** thermal agitation is great enough to break up this alignment, and ferromagnetic materials become paramagnetic.

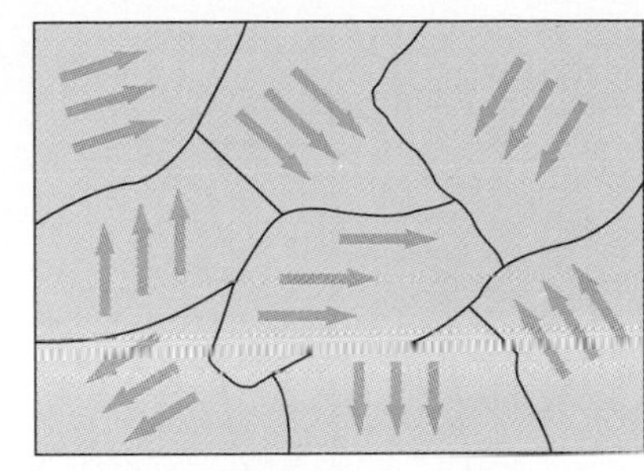

Figure 27-8 Schematic illustration of ferromagnetic domains. Within a domain, the magnetic dipoles are aligned, but the direction of alignment varies from domain to domain so that the net magnetic moment is zero. A small external magnetic field may cause the enlargement of those domains that are aligned parallel to the field, or it may cause the alignment within a domain to rotate. In either case, the result is a net magnetic moment parallel to the field.

When an external magnetic field is applied, the boundaries of the domains may shift or the direction of alignment within a domain may change so that there is a net macroscopic magnetic moment in the direction of the applied field. Since the degree of alignment is large for even a small external field, the magnetic field produced in the material by the dipoles is often much greater than the external field.

Let us consider what happens when we magnetize a long iron rod by placing it inside a solenoid and gradually increase the current in the solenoid windings. We assume that the rod and the solenoid are long enough to permit us to neglect end effects. The magnetic field at the center of the rod is given by Equation 27-4:

$$\mathbf{B} = \mathbf{B}_{\text{app}} + \mu_0\mathbf{M} \qquad 27\text{-}15$$

where

$$B_{\text{app}} = \mu_0 nI$$

In ferromagnetic materials, the magnetic field $\mu_0 M$ due to the magnetic moments is often greater than the magnetizing field B_{app} by a factor of several thousand.

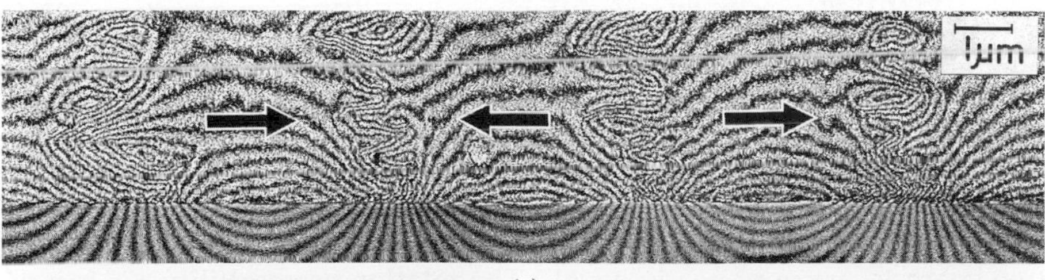

(*a*)

(*b*)

(*c*)

(*a*) Magnetic field lines on a cobalt magnetic recording tape. The solid arrows indicate the encoded magnetic bits. (*b*) Cross section of a magnetic tape recording head. Current from an audio amplifier is sent to wires around a magnetic core in the recording head where it produces a magnetic field. When the tape passes over a gap in the core of the recording head, the fringing magnetic field encodes information on the tape. The information is retrieved when the tape passes over a reading head shown in cross section in (*c*). In this case, the change in flux due to the magnetized tape induces currents in the wires around the core of the reading head.

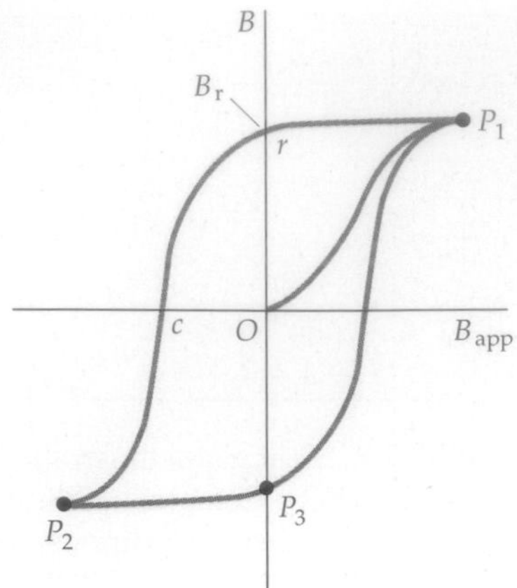

Figure 27-9 Plot of B versus the applied field B_{app}. The outer curve is called a hysteresis curve. The field B_r is called the remnant field. It remains when the applied field returns to zero.

Figure 27-9 shows a plot of B versus the magnetizing field B_{app}. As the current is gradually increased from zero, B increases from zero along the part of the curve from the origin O to point P_1. The flattening of this curve near point P_1 indicates that the magnetization M is approaching its saturation value M_s, at which all the atomic magnetic moments are aligned. Above saturation, B increases only because the magnetizing field $B_{app} = \mu_0 nI$ increases. When B_{app} is gradually decreased from point P_1, there is not a corresponding decrease in the magnetization. The shift of the domains in a ferromagnetic material is not completely reversible, and some magnetization remains even when B_{app} is reduced to zero, as indicated in the figure. This effect is called **hysteresis,** from the Greek word *hysteros* meaning later or behind, and the curve in Figure 27-9 is called a **hysteresis curve.** The value of the magnetic field at point r when B_{app} is zero is called the **remnant field** B_r. At this point, the iron rod is a permanent magnet. If the current in the solenoid is now reversed so that B_{app} is in the opposite direction, the magnetic field B is gradually brought to zero at point c. The remaining part of the hysteresis curve is obtained by further increasing the current in the opposite direction until point P_2 is reached, which corresponds to saturation in the opposite direction, and then decreasing the current to zero at point P_3 and increasing it again in its original direction.

Since the magnetization M depends on the previous history of the material, and it can have a large value even when the applied field is zero, it is not simply related to the applied field B_{app}. However, if we confined ourselves to that part of the magnetization curve from the origin to point P_1 in Figure 27-9, **M** and $\mathbf{B}_{app}$ are parallel and M is zero when B_{app} is zero. We can then define the magnetic susceptibility as in Equation 27-5,

$$M = \chi_m(B_{app}/\mu_0)$$

and

$$B = B_{app} + \mu_0 M = B_{app}(1 + \chi_m) = \mu_0 nI(1 + \chi_m) = \mu nI \qquad 27\text{-}16$$

where

$$\mu = (1 + \chi_m)\mu_0 \qquad 27\text{-}17$$

is called the **permeability** of the material. (The permeability is defined in the same way for paramagnetic and diamagnetic materials, but since χ_m is much less than 1 for these materials, the permeability μ and the permeability of free space μ_0 are very nearly equal.) The **relative permeability** K_m is a dimensionless number that is defined by

$$K_m = \frac{\mu}{\mu_0} = 1 + \chi_m = \frac{B}{B_{app}} \qquad 27\text{-}18$$

Since B does not vary linearly with B_{app}, as can be seen from Figure 27-9, the relative permeability is not constant. The maximum value of K_m occurs at a magnetization that is considerably less than the saturation magnetization. Table 27-2 lists the saturation magnetic field $\mu_0 M_s$ and the maximum values of K_m for some ferromagnetic materials. Note that the maximum values of K_m are much greater than 1.

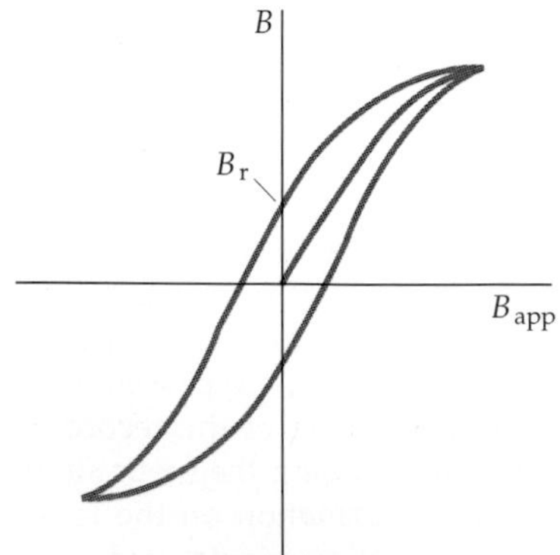

Figure 27-10 Hysteresis curve for a magnetically soft material. The remnant field is very small compared with that for a magnetically hard material such as that in Figure 27-9.

The area enclosed by the hysteresis curve is proportional to the energy dissipated as heat in the irreversible process of magnetizing and demagnetizing. If the hysteresis effect is small, so that the area is small indicating a small energy loss, the material is called **magnetically soft.** Soft iron is an example. The hysteresis curve for a magnetically soft material is shown in Figure 27-10. Here the remnant field B_r is nearly zero, and the energy loss per cycle is small. Magnetically soft materials are used for transformer cores to allow the magnetic field B to change without incurring large energy losses

Table 27-2 **Maximum Values of $\mu_0 M$ and K_m for Some Ferromagnetic Materials**

Material	$\mu_0 M_s$, T	K_m
Iron (annealed)	2.16	5,500
Iron-silicon (96% Fe, 4% Si)	1.95	7,000
Permalloy (55% Fe, 45% Ni)	1.60	25,000
Mu-metal (77% Ni, 16% Fe, 5% Cu, 2% Cr)	0.65	100,000

as the field alternates. On the other hand, a large remnant field is desirable in a permanent magnet. **Magnetically hard** materials, such as carbon steel and the alloy Alnico 5, are used for permanent magnets.

(*a*)

(*b*) 10μm

(*a*) Computer hard disk drive for magnetic storage of information. (*b*) A magnetic test pattern on a hard disk, magnified 2400 times. The light and dark regions correspond to oppositely directed magnetic fields. The smooth region just outside the pattern is a region of the disk that has been erased just prior to writing.

Example 27-4

A long solenoid with 12 turns per centimeter has a core of annealed iron. When the current is 0.50 A, the magnetic field inside the iron core is 1.36 T. Find (*a*) the applied field B_{app}, (*b*) the relative permeability K_m, and (*c*) the magnetization M.

(*a*) The applied field is

$$B_{app} = \mu_0 nI = (4\pi \times 10^{-7}\ \text{T·m/A})(1200\ \text{turns/m})(0.50\ \text{A})$$
$$= 7.54 \times 10^{-4}\ \text{T}$$

Note that the total magnetic field is 1.36 T, so this applied magnetic field is a negligible fraction of the total field.

(*b*) From Equation 27-18, the relative permeability is

$$K_m = \frac{B}{B_{app}} = \frac{1.36\ \text{T}}{7.54 \times 10^{-4}\ \text{T}} = 1.80 \times 10^3 = 1800$$

Note that this is considerably smaller than the maximum value of K_m which is about 5500 (Table 27-2). To the three-place accuracy with which we have calculated K_m, the susceptibility χ_m is equal to the relative permeability:

$$\chi_m = K_m - 1 \approx K_m = 1800$$

(*c*) We can find the magnetization from Equation 27-3 or from Equation 27-6. Using Equation 27-6, we have

$$\mu_0 M = B - B_{app} = 1.36\ \text{T} - 7.54 \times 10^{-4}\ \text{T} \approx B = 1.36\ \text{T}$$

Then

$$M = \frac{B}{\mu_0} = \frac{1.36\ \text{T}}{4\pi \times 10^{-7}\ \text{T·m/A}} = 1.08 \times 10^6\ \text{A/m}$$

Questions

5. In a common lecture demonstration, a long iron rod is held such that its axis is along the magnetic field of the earth and is then struck with a hammer. This makes the rod a permanent magnet. It can be demagnetized if it is held perpendicular to the earth's field and struck again. Explain what happens in the rod.

6. A permanent magnet may lose much of its magnetization if it is dropped or banged against something. Why?

27-5 Diamagnetism

Diamagnetic materials are those having very small, negative values of magnetic susceptibility χ_m. Diamagnetism was discovered by Faraday in 1846 when he found that a piece of bismuth is repelled by either pole of a magnet, indicating that the external field of the magnet induces a magnetic moment in bismuth in the direction opposite the field. We can understand this effect qualitatively using Lenz's law. Figure 27-11 shows two positive charges moving in circular orbits with the same speed but in opposite directions. Their magnetic moments are in opposite directions and therefore cancel. (It is simpler to consider positive charges even though it is the negatively charged electrons that provide the magnetic moments in matter.) Consider now what happens when an external magnetic field **B** is turned on in the direction into the paper. According to Lenz's law, current will be induced to oppose the change in flux. If we assume that the radius of the circle does not change, the charge on the left will be speeded up to increase its flux out of the page, and that on the right will be slowed down to decrease its flux into the page. In each case, the *change* in the magnetic moment of the charges will be in the direction out of the page, opposite that of the external applied field. Since the permanent magnetic moments of the two charges are equal and oppositely directed, they add to zero, leaving only the induced magnetic moments, which are both opposite the direction of the applied magnetic field.

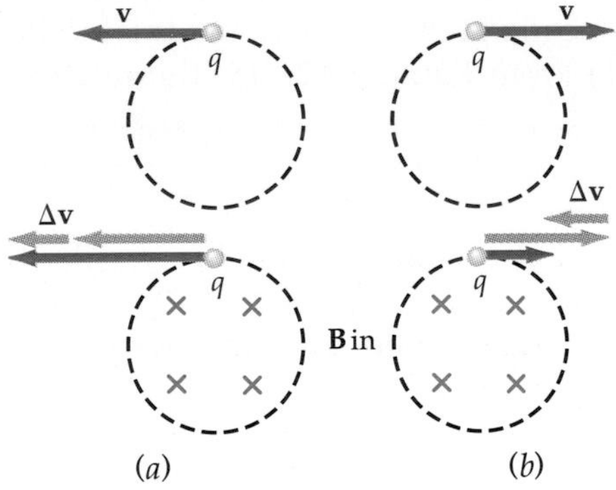

Figure 27-11 (*a*) A positive charge moving counterclockwise in a circle has its magnetic moment outward. When an external, inward magnetic field is turned on, the speed of the particle will increase to oppose the change in flux. The change in the magnetic moment is outward. (*b*) A positive charge moving clockwise in a circle has its magnetic moment inward. When an external, inward magnetic field is turned on, the speed of the particle will decrease to oppose the change in flux. As in (*a*), the change in the magnetic moment is outward.

Atoms that have a closed-shell electronic structure have a net angular momentum of zero and therefore no net permanent magnetic moment. Materials that have such atoms—bismuth, for example—are diamagnetic. As we will see below, the induced magnetic moments that cause diamagnetism have magnitudes of the order of 10^{-5} Bohr magnetons. Since this is much smaller than the permanent magnetic moments of the atoms of paramagnetic or ferromagnetic materials, which do not have closed-shell structure, the diamagnetic effect in these atoms is masked by the alignment of their permanent magnetic moments. However, since this alignment decreases with temperature, all materials are theoretically diamagnetic at sufficiently high temperatures.

A superconductor is a perfect diamagnet, that is, it has a magnetic susceptibility of -1. When a superconductor is placed in an external magnetic field, electric currents are induced on its surface so that the net magnetic field in the superconductor is zero. Consider a superconducting rod inside a solenoid of n turns per unit length. When the solenoid is connected to a source of emf so that it carries a current I, the magnetic field due to the solenoid is $\mu_0 nI$. A surface current of $-nI$ per unit length is induced on the

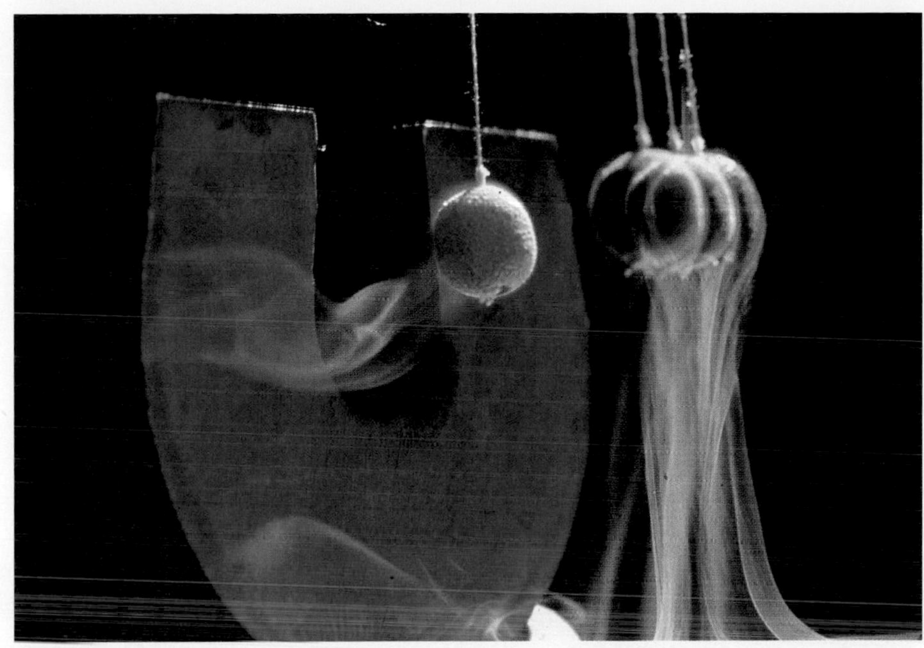

A superconductor is a perfect diamagnet. Here, the superconducting pendulum bob is repelled by the permanent magnet.

superconducting rod that cancels out the field due to the solenoid so that the net field inside the superconductor is zero. From Equation 27-6,

$$\mathbf{B} = \mathbf{B}_{\text{app}}(1 + \chi_{\text{m}}) = 0$$

so

$$\chi_{\text{m}} = -1$$

Estimating Induced Magnetic Moments

We can estimate the magnitude of the induced magnetic moments in diamagnetic materials by relating the change in the speeds of the electrons to the change in the centripetal force due to the external magnetic field. We assume that the radius of the orbit does not change and that the change in speed is small compared with the original speed. Both of these assumptions can be justified. The original centripetal force is provided by the electrostatic force of attraction F of the electron to the nucleus. Setting this force equal to the mass times the acceleration, we obtain

$$F = \frac{m_{\text{e}}v^2}{r} \qquad 27\text{-}19$$

where m_{e} is the mass of the electron. In the presence of an external magnetic field **B**, there is an additional force $q\mathbf{v} \times \mathbf{B}$ on each particle. For the particle on the left in Figure 27-11, this force is directed inward (for a positively charged particle), thus increasing the net inward force, which is necessary because this particle speeds up when the magnetic field is turned on. Similarly, for the particle on the right, the magnetic force is outward, which reduces the net inward force. Again, this is in the correct direction since the particle on the right slows down when the field is turned on.

Since the change in the net inward force is small, we can approximate it by a differential. Differentiating Equation 27-19, we obtain

$$dF = \frac{2m_{\text{e}}v}{r}\,dv \approx \Delta F$$

Setting this change in force equal to the magnetic force qvB, we have

$$qvB = \frac{2m_e v}{r}\,dv$$

or

$$dv = \frac{qrB}{2m_e}$$

This change in speed causes a small change in the magnetic moment of the charge. Using Equation 27-8, we have for the magnetic moment

$$m = \tfrac{1}{2}qvr$$

Then

$$\Delta m \approx dm = \frac{1}{2}qr\,dv = \frac{1}{2}qr\left(\frac{qrB}{2m_e}\right) = \frac{q^2r^2}{4m_e}B \tag{27-20}$$

Using $r = 10^{-10}$ m for a typical atomic radius, $B = 1$ T for a typical strong magnetic field, and the values of q and m_e for the electron, we obtain for the order of magnitude of the induced magnetic moment

$$\Delta m \approx \frac{(1.60 \times 10^{-19}\ \text{C})^2(10^{-10}\ \text{m})^2}{4(9.11 \times 10^{-31}\ \text{kg})}(1\ \text{T}) \approx 7 \times 10^{-29}\ \text{A·m}^2 \sim 10^{-28}\ \text{A·m}^2$$

Comparing this induced magnetic moment with the Bohr magneton ($m_B = 9.27 \times 10^{-24}\ \text{A·m}^2 \sim 10^{-23}\ \text{A·m}^2$), we see that the induced magnetic moments are about 100,000 times smaller than the magnetic moment of an electron due to its intrinsic spin or to its orbital motion in an atom. Thus, materials that have atoms with permanent magnetic dipole moments, which are of the order of a Bohr magneton, are paramagnetic (or ferromagnetic). The resultant magnetic moment in the direction of the applied magnetic field that results from even a very small partial alignment of these permanent magnetic moments at ordinary temperatures is much greater than the induced magnetic dipole moment in the direction opposite the applied magnetic field.

Question

7. Why might you expect a heavy diamagnetic element to have a susceptibility of greater magnitude than a light diamagnetic element?

Summary

1. All materials can be classified as either paramagnetic, ferromagnetic, or diamagnetic.

2. A magnetized material is described by its magnetization vector **M**, which is defined to be the net magnetic dipole moment per unit volume of the material:

$$\mathbf{M} = \frac{d\mathbf{m}}{dV}$$

The magnetic field due to a uniformly magnetized cylinder is the same as if the cylinder carried a current per unit length of magnitude M on its surface. This current, which is due to the intrinsic motion of the atomic charges in the cylinder, is called an amperian current.

3. If a long cylinder of magnetic material is placed inside a solenoid of n turns per unit length carrying a current I, the resultant magnetic field at

a point inside the solenoid and far from its ends due to the current in the solenoid plus the magnetized material is

$$\mathbf{B} = \mathbf{B}_{\text{app}} + \mu_0\mathbf{M}$$

where the applied magnetic field has the magnitude

$$B_{\text{app}} = \mu_0 nI$$

For paramagnetic and ferromagnetic materials, **M** is in the same direction as $\mathbf{B}_{\text{app}}$; for diamagnetic materials, **M** is opposite to $\mathbf{B}_{\text{app}}$.

4. In paramagnetic and diamagnetic materials the magnetization is proportional to the magnetizing field B_{app}:

$$M = \chi_{\text{m}}(B_{\text{app}}/\mu_0)$$

where χ_{m} is called the magnetic susceptibility. For paramagnetic materials, χ_{m} is a small positive number that depends on temperature. For diamagnetic materials (other than superconductors), it is a small negative constant independent of temperature. For superconductors, $\chi_{\text{m}} = -1$. For ferromagnetic materials, the magnetization depends not only on the magnetizing current but also on the past history of the material.

5. The magnetic moment of a particle of charge q and mass m_q is related to its angular momentum **L** by

$$\mathbf{m} = \frac{q}{2m_q}\mathbf{L} = \frac{q\hbar}{2m_q}\frac{\mathbf{L}}{\hbar}$$

where

$$\hbar = \frac{h}{2\pi} = 1.05 \times 10^{-34}\ \text{J·s}$$

is a convenient unit for expressing the angular momentum of electrons and atoms and

$$h = 6.67 \times 10^{-34}\ \text{J·s}$$

is a fundamental constant called Planck's constant. The magnetic moments of electrons and atoms are conveniently expressed in units of the Bohr magneton m_{B}:

$$m_{\text{B}} = \frac{e\hbar}{2m_{\text{e}}} = 9.27 \times 10^{-24}\ \text{A·m}^2 = 9.27 \times 10^{-24}\ \text{J/T}$$

The magnetic moment associated with the spin angular momentum of the electron is 1 Bohr magneton, and the magnetic moment of an atom is of the order of a few Bohr magnetons.

6. Paramagnetic materials have permanent atomic magnetic moments that have random directions in the absence of an external magnetic field. In an external field, some of these dipoles are aligned producing a small contribution to the total field that adds to the external field. The degree of alignment is small except in very strong fields and at very low temperatures. At ordinary temperatures, thermal motion tends to maintain the random directions of the magnetic moments. At low fields, the magnetization is approximately proportional to the external field and is given by Curie's law:

$$M = \frac{1}{3}\frac{mB_{\text{app}}}{kT}M_{\text{s}}$$

where M_{s} is the saturation magnetization that results when all the magnetic dipole moments are aligned.

7. Ferromagnetic materials have small regions of space called magnetic domains in which the permanent atomic magnetic moments are aligned.

When unmagnetized, the direction of alignment in one domain is independent of that in another so that no net magnetic field is produced. When magnetized, the domains of a ferromagnetic material are aligned, producing a very strong contribution to the magnetic field. This alignment can persist even when the external field is removed, thus leading to permanent magnetism.

8. A plot of the magnetic field in a ferromagnetic material versus magnetizing field is called a hysteresis curve. In the upper right quadrant of this curve **M** and $\mathbf{B}_{app}$ are in the same direction and a magnetic susceptibility χ_m can be defined for ferromagnetic materials in the same way as it is defined for paramagnetic and diamagnetic materials. The magnetic field inside a ferromagnetic material in a solenoid carrying current I is then related to the applied field by

$$B = B_{app} + \mu_0 M = B_{app}(1 + \chi_m) = \mu_0 nI(1 + \chi_m) = \mu nI$$

where

$$\mu = (1 + \chi_m)\mu_0$$

is the permeability of the material. The relative permeability K_m is a dimensionless number, which is defined as the ratio of the permeability to the permeability of free space:

$$K_m = \frac{\mu}{\mu_0} = 1 + \chi_m = \frac{B}{B_{app}}$$

The maximum value of K_m is much greater than 1 for ferromagnetic materials.

9. Diamagnetic materials are those in which the magnetic moments of all electrons in each atom cancel, leaving each atom with zero magnetic moment in the absence of an external field. In an external field, a very small magnetic moment is induced that tends to weaken the field. This effect is independent of temperature. Superconductors are diamagnetic with susceptibility equal to -1.

Suggestion for Further Reading

Becker, Joseph J.: "Permanent Magnets," *Scientific American,* December 1970, p. 92.

Magnets made from the new alloys described in this article can be made many times stronger than those made from conventional metals.

Review

A. Objectives: After studying this chapter, you should:

1. Be able to list the three types of magnetism and discuss the origins, directions, and strengths of the magnetic effects in each.

2. Be able to derive the relation between magnetic moment and angular momentum for a charged particle moving in a circle.

3. Know the sign and order of magnitude of the magnetic susceptibility of paramagnetic and diamagnetic materials.

4. Be able to describe the general temperature dependence of the magnetization in paramagnetic materials and explain its origin.

5. Be able to sketch B versus B_{app} for ferromagnetic materials.

B. Define, explain, or otherwise identify:

Magnetization
Amperian current
Magnetic susceptibility
Bohr magneton

Saturation magnetization
Paramagnetic materials
Curie's law
Ferromagnetic materials
Magnetic domain
Curie temperature
Hysteresis
Hysteresis curve
Remnant field
Permeability
Relative permeability
Magnetically soft
Magnetically hard
Diamagnetic materials

C. True or false: If the statement is true, explain why it is true. If it is false, give a counterexample.

1. Diamagnetism occurs in all materials.

2. Diamagnetism is the result of induced magnetic dipole moments.

3. Paramagnetism is the result of the partial alignment of permanent magnetic dipole moments.

4. Hysteresis is associated with a loss in electromagnetic energy.

5. Magnetically hard materials are good for transformer cores.

Problems

Level I

27-1 Magnetization and Magnetic Susceptibility

1. A tightly wound solenoid 20 cm long has 400 turns and carries a current of 4 A such that its axial field is in the z direction. Neglecting end effects, find B and B_{app} at the center when (*a*) there is no core in the solenoid and (*b*) there is an iron core with a magnetization $M = 1.2 \times 10^6$ A/m.

2. Which of the four gases listed in Table 27-1 are diamagnetic and which are paramagnetic?

3. If the solenoid of Problem 1 has an aluminum core, find B_{app}, M, and B at the center, neglecting end effects.

4. Repeat Problem 3 for a tungsten core.

5. A long solenoid is wound around a tungsten core and carries a current. (*a*) If the core is removed while the current is held constant, does the magnetic field inside the solenoid decrease or increase? By what percentage? (*b*) Does the self inductance of the solenoid decrease or increase? By what percentage?

6. When a sample of liquid is inserted into a solenoid carrying a constant current, the magnetic field inside the solenoid decreases by 0.004 percent. What is the magnetic susceptibility of the liquid?

7. A long solenoid carrying a current of 10 A has 50 turns/cm. What is the magnetic field in the interior of the solenoid when the interior is (*a*) a vacuum, (*b*) filled with aluminum, and (*c*) filled with silver?

27-2 Atomic Magnetic Moments

8. Nickel has a density of 8.7 g/cm^3 and molecular mass of 58.7 g/mol. Its saturation magnetization is given by $\mu_0 M_s = 0.61$ T. Calculate the magnetic moment of a nickel atom in Bohr magnetons.

9. Repeat Problem 8 for cobalt, which has a density of 8.9 g/cm^3, a molecular mass of 58.9 g/mol, and a saturation magnetization given by $\mu_0 M_s = 1.79$ T.

27-3 Paramagnetism

10. Show that Curie's law predicts that the magnetic susceptibility of a paramagnetic substance is $\chi_m = m\mu_0 M_s/3kT$.

11. Assume that the magnetic moment of an aluminum atom is 1 Bohr magneton. The density of aluminum is 2.7 g/cm^3, and its molecular mass is 27 g/mol. (*a*) Calculate M_s and $\mu_0 M_s$ for aluminum. (*b*) Use the results of Problem 10 to calculate χ_m at $T = 300$ K. (*c*) Explain why the result for part (*b*) is larger than the value listed in Table 27-1.

27-4 Ferromagnetism

12. The saturation magnetization for annealed iron occurs when $B_{app} = 0.201$ T. Find the permeability μ and the relative permeability K_m of annealed iron at saturation. (See Table 27-2.)

13. For annealed iron, the relative permeability K_m has its maximum value of about 5500 at $B_{app} = 1.57 \times 10^{-4}$ T. Find M and B when K_m is maximum.

14. The coercive force is defined to be the applied magnetic field needed to bring B back to zero along the hysteresis curve (point c in Figure 27-9). For a certain permanent bar magnet, $B_{app} = 5.53 \times 10^{-2}$ T. The bar magnet is to be demagnetized by placing it inside a 15-cm-long solenoid with 600 turns. What minimum current is needed in the solenoid to demagnetize the magnet?

15. A long solenoid with 50 turns/cm carries a current of 2 A. The solenoid is filled with iron, and B is measured to be 1.72 T. (*a*) Neglecting end effects, what is B_{app}? (*b*) What is M? (*c*) What is the relative permeability K_m?

16. When the current in Problem 15 is 0.2 A, the magnetic field is measured to be 1.58 T. (*a*) Neglecting end effects, what is B_{app}? (*b*) What is M? (*c*) What is the relative permeability K_m?

27-5 Diamagnetism

There are no problems for this section.

Level II

17. A long, iron-core solenoid with 2000 turns/m carries a current of 20 mA. At this current, the relative permeability of the iron core is 1200. (*a*) What is the magnetic field within the solenoid? (*b*) With the iron core removed, find the current necessary to produce the same field within the solenoid.

18. The magnetic dipole moment of an iron atom is 2.219 m_B. (*a*) If all the atoms in an iron bar of length 20 cm and cross-sectional area 2 cm^2 have their dipole moments aligned, what is the dipole moment of the bar? (*b*) What torque must be supplied to hold the iron bar perpendicular to a magnetic field of 0.25 T?

19. A small magnetic sample is in the form of a disk. It has a radius of 1.4 cm, a thickness of 0.3 cm, and a uniform magnetization throughout its volume. The magnetic moment of the sample is 1.5×10^{-2} A·m^2. (*a*) What is the magnetization **M** of the sample? (*b*) If this magnetization is due to the alignment of *N* electrons each with a magnetic moment of 1 m_B, what is *N*? (*c*) If the magnetization is along the axis of the disk, what is the magnitude of the amperian surface current?

20. A very long solenoid of length ℓ and cross-sectional area *A* has *n* turns per unit length and carries a current *I*. It is filled with iron of relative permeability K_m. (*a*) Find the self inductance of the solenoid. (*b*) Use $U_m = \frac{1}{2}LI^2$ to find the magnetic energy stored in the solenoid in terms of the magnetic field *B*. (*c*) Show that the energy density in the solenoid is $\eta_m = B^2/(2K_m\mu_0) = B^2/2\mu$.

21. The magnetic moment of the earth is about 9×10^{22} A·m^2. (*a*) If the magnetization of the earth's core were 1.5×10^9 A/m, what is the core volume? (*b*) What is the radius of such a core if it were spherical and centered with the earth?

22. In a simple model of paramagnetism, we can consider that some fraction *f* of the molecules have their magnetic moments aligned with the external magnetic field and that the rest of the molecules are randomly oriented and so do not contribute to the magnetic field. (*a*) Use this model and Curie's law to show that at temperature *T* and external magnetic field *B* the fraction of aligned molecules is $f = mB/3kT$. (*b*) Calculate this fraction for $T = 300$ K, $B = 1$ T, assuming *m* to be 1 Bohr magneton.

23. It is desired to fill a solenoid with a mixture of oxygen and nitrogen at room temperataure and 1 atmosphere pressure such that K_m is exactly 1. Assume that the magnetic dipole moments of the gas molecules are all aligned and that the susceptibility of a gas is proportional to the number density of its molecules. What should the ratio of the number densities of oxygen to nitrogen molecules be so that $K_m = 1$?

24. A cylinder of magnetic material is placed in a long solenoid of *n* turns per unit length and current *I*. Table 27-3 gives the magnetic field *B* versus *nI*. Use these values to plot *B* versus B_{app} and K_m versus *nI*.

Table 27-3 **For Problem 24**

nI, A/m	*B*, T
0	0
50	0.04
100	0.67
150	1.00
200	1.2
500	1.4
1,000	1.6
10,000	1.7

25. A toroid with *N* turns carrying a current *I* has mean radius *R* and cross-sectional radius *r*, where $r < R$ (Figure 27-12). When the toroid is filled with material, it is called a *Rowland ring*. Find $\mathbf{B}_{app}$ and **B** in such a ring, assuming a magnetization **M** everywhere parallel to $\mathbf{B}_{app}$.

Figure 27-12 Problem 25.

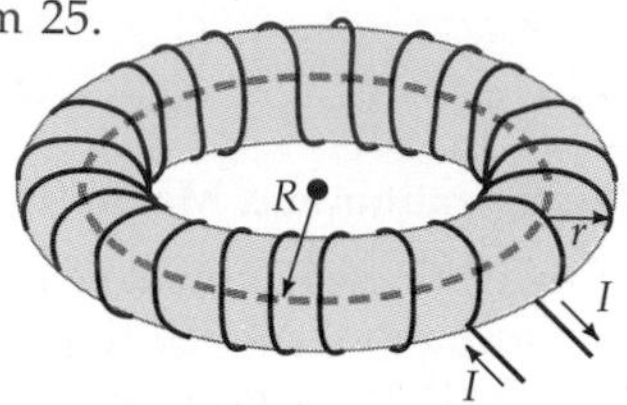

26. A toroid is filled with liquid oxygen that has a susceptibility of 4×10^{-3}. The toroid has 2000 turns and carries a current of 15 A. Its mean radius is 20 cm, and the radius of its cross section is 0.8 cm. (*a*) What is the magnetization *M*? (*b*) What is the magnetic field *B*? (*c*) What is the percentage increase in *B* produced by the liquid oxygen?

27. The toroid of Problem 26 has its core filled with iron. When the current is 10 A, the magnetic field in the toroid is 1.8 T. (*a*) What is the magnetization *M*? (*b*) Find the values for K_m, μ, and χ_m for the iron sample.

28. A toroid has average radius 14 cm and a cross-sectional area of 3 cm^2. It is wound with fine wire with 60 turns/cm measured along its mean circumference that carries a current of 4 A. The core is filled with a paramagnetic material of magnetic susceptibility 2.9×10^{-4}. (*a*) What is the magnitude of the magnetic field within the substance? (*b*) What is the magnitude of the magnetization? (*c*) What would the magnitude of the magnetic field be if there were no paramagnetic core present?

29. What would the result be for Problem 28 if soft iron, having a relative permeability of 500, were substituted for the paramagnetic core?

30. Two long, straight wires 4.0 cm apart are embedded in a uniform insulator having relative permeability of $K_m = 120$. The wires carry 40 A in opposite directions. (*a*) What is the magnetic field at the midpoint of the plane of the wires? (*b*) What is the force per unit length on the wires?

31. A long, narrow bar magnet that has magnetic moment **m** parallel to its long axis is suspended at its center as a frictionless compass needle. When placed in a magnetic field **B**, the needle lines up with the field. If it is displaced by a small angle θ, show that the needle will oscillate about its equilibrium position with frequency $f = (1/2\pi)\sqrt{mB/I}$, where I is the moment of inertia about the point of suspension.

32. Suppose the needle in Problem 31 is a uniformly magnetized iron rod that is 8 cm long and has a cross-sectional area of 3 mm^2. Assume that the magnetic dipole moment for each iron atom is 2.2 m_B and that all the iron atoms have their dipole moments aligned. Calculate the frequency of small oscillations about the equilibrium position when the magnetic field is 0.5 G.

33. The needle of a magnetic compass has a length of 3 cm, a radius of 0.85 mm, and a density of 7.96×10^3 kg/m^3. It is free to rotate in a horizontal plane, where the horizontal component of the earth's magnetic field is 0.6 G. When disturbed slightly, the compass executes simple harmonic motion about its midpoint with a frequency of 1.4 Hz. (*a*) What is the magnetic dipole moment of the needle? (*b*) What is the magnetization M? (*c*) What is the amperian current on the surface of the needle? (See Problem 31.)

34. A long, straight wire with a radius of 1.0 mm is coated with an insulating ferromagnetic material that has a thickness of 3.0 mm and a relative magnetic permeability of $K_m = 400$. The coated wire is in air and the wire itself is nonmagnetic. The wire carries a current of 40 A. (*a*) Find the magnetic field inside the wire as a function of radius r. (*b*) Find the magnetic field inside the ferromagnetic material as a function of radius r. (*c*) Find the magnetic field outside the ferromagnetic material as a function of r. (*d*) What must the magnitudes and directions of the amperian currents be on the surfaces of the ferromagnetic material to account for the magnetic fields observed?

35. In Section 27-5, the change in speed of an electron in an atom is found for the case in which a magnetic field is impressed on the atom. From this result, show that the change in angular frequency is $\Delta\omega = eB/2m$. This frequency is called the *Larmor frequency*.

36. An iron bar of length 1.4 m has a diameter of 2 cm and a uniform magnetization of 1.72×10^6 A/m directed along the bar's length. The bar is stationary in space and is suddenly demagnetized so that its magnetization disappears. What is the rotational angular velocity of the bar if its angular momentum is conserved? (Assume that Equation 27-9 holds where m_q is the mass of an electron and $q = -e$.)

37. A bar magnet has a diameter of 2 cm and has a magnetic field in its center of 0.1 T. If the magnet cracks apart in its middle, the magnetic attraction holds the two pieces together. (*a*) Show that if the pieces move apart a small distance dx, the additional magnetic energy is $dU_m = (B^2/2\mu_0)A\,dx$, where A is the cross-sectional area of the magnet and B is the magnetic field in the gap, which is assumed to be the same as the field inside the magnet. (*b*) Estimate the required force needed to pull apart the two pieces by computing the work done to separate them by a distance dx.

Level III

38. A proton has a magnetic moment **m**, which is parallel to its angular momentum **L**. It is in a uniform magnetic field **B** that makes an angle θ with **m** and **L**. Show that the magnetic moment vector precesses about the magnetic field, and find the angular frequency of precession.

39. Two long conducting strips are each 20 m wide and 0.3 mm thick. The strips are in parallel planes, which are separated by a 1.0-cm ferromagnetic spacer that has a relative permeability of $K_m = 400$. The conducting strips carry a uniform current of 4800 A in opposite directions. In the space between the strips far from their edges find (*a*) B_{app}, (*b*) B, and (*c*) the magnetic energy per unit volume.

40. In our derivation of the magnetic moment induced in an atom, we assumed that the radius of the electron's orbit did not change in the presence of an external magnetic field. (*a*) Show that this assumption of constant radius is justified by showing that, when **B** is applied, there is an impulse that increases or decreases the speed of the electron by just the correct amount, which is given by $\Delta v = qrB/2m$. (*b*) Use Faraday's law to show that the induced electric field is related to the rate of change of the magnetic field by $E = \frac{1}{2}r\,dB/dt$, assuming r to be constant. (*c*) Use Newton's second law to show that the change in speed of the electron dv is related to the change in B by $dv = (qr/2m)\,dB$. Integrate this expression to find Δv.

41. Equation 27-20 gives the induced magnetic moment for a single electron in an orbit that has its plane perpendicular to **B**. If an atom has Z electrons, a reasonable simplifying assumption is that, on the average, one-third have their planes perpendicular to **B**. Show that the diamagnetic susceptibility obtained from Equation 27-20 is then

$$\chi_m = \frac{-nZq^2r^2}{12m_e}\mu_0$$

where n is the number of atoms per unit volume. Use $n \approx 6 \times 10^{28}$ atoms/m^3 and $r \approx 5 \times 10^{-11}$ m to estimate χ_m for $Z \approx 50$.

Chapter 28

Alternating-Current Circuits

Most high-voltage long-distance power transmission is presently accomplished using alternating current. Historically, the difficulty and expense in converting ac to dc at the sending end (such as a hydroelectric power plant) and from dc back to ac at the consuming end have been major disadvantages in transmitting power via high-voltage direct current (HVDC). Because of recent advances in technology, interest in HVDC has been revived. Shown here is a dc to ac conversion station near Boston, linked by HVDC lines to a 2000-MW hydroelectric generator unit in James Bay, Quebec. Converter valves called thyristors are linked via Δ-Y connections to bushings that lead out of the station. This particular station is used to convert direct current to alternating current; but, like many others, it can be run in reverse.

Toward the end of the nineteenth century, there was a heated debate as to whether direct or alternating current should be used to deliver electrical energy to consumers in the United States. Thomas Edison lobbied for the use of direct current while Nikola Tesla and George Westinghouse were proponents of the use of alternating current. In 1893, alternating current was chosen to light the World Columbian Exposition in Chicago, and a contract was awarded to Westinghouse to deliver alternating current generated at Niagara Falls to American homes and factories.

Alternating current has the great advantage that electrical energy can be transported over long distances at very high voltages and low currents to reduce energy losses in the form of Joule heat. It can then be transformed, with almost no energy loss, to lower and safer voltages and correspondingly higher currents for everyday use. The transformer that accomplishes this change in voltage and current works on the basis of magnetic induction. We will study the transformer in Section 28-6.

More than 99 percent of the electrical energy used today is produced by electrical generators in the form of alternating current. In North America, power is delivered by a sinusoidal current of frequency 60 Hz. Devices such as radios, television sets and microwave ovens detect or generate alternating currents of much greater frequencies. Alternating current is easily produced by magnetic induction in an ac generator, as we discussed in Chapter 26. Although practical generators are considerably more complicated than the simple device we studied in Section 26-6, they are designed to put out a sinusoidal emf. We will see that when the generator output is sinusoidal, the current in an inductor, a capacitor, or a resistor is also sinusoidal, though it is generally not in phase with the generator's emf. When the emf and current are both sinusoidal, their maximum values can be simply related. The study of sinusoidal currents is important because even currents that are not sinusoidal can be analyzed in terms of sinusoidal components using Fourier analysis.

We will first look at the behavior of alternating current in resistors, inductors, and capacitors and in some simple circuits containing these elements.

28-1 Alternating Current in a Resistor

In our study of dc circuits in Chapter 23, we noted that Kirchhoff's rules apply to any circuit in a steady state. We also noted that steady states are reached in circuit elements almost immediately after a change in voltage or current is introduced. Since the time it takes to reach steady state is much smaller than the period of oscillation of ac circuits, we can apply Kirchhoff's rules to alternating-current circuits as well as direct-current circuits.

Figure 28-1 shows a simple ac circuit consisting of a generator and a resistor. In the figure, plus and minus signs indicate the higher-potential and lower-potential sides of the source of emf for the assumed direction of the current. Plus and minus signs have also been placed on the resistor to indicate the direction of the potential drop for the assumed direction of the current. Note that the point at which current enters the resistor is at a higher potential than the point at which the current leaves. The voltage drop across the resistor V_R is given by

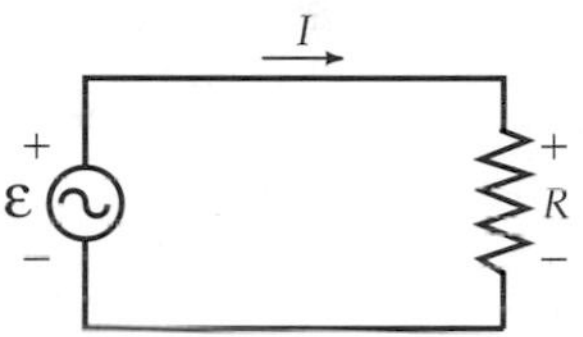

Figure 28-1 An ac generator in series with a resistor R.

$$V_R = V_+ - V_- = IR \qquad 28\text{-}1$$

If $\mathcal{E}$ is the emf supplied by the generator, applying Kirchhoff's loop rule to this circuit gives

$$\mathcal{E} - V_R = 0$$

If the generator produces an emf* given by

$$\mathcal{E} = \mathcal{E}_{max} \cos \omega t$$

we have

$$\mathcal{E}_{max} \cos \omega t - IR - 0 \qquad 28\text{-}2$$

The current in the resistor is

$$I = \frac{\mathcal{E}_{max}}{R} \cos \omega t \qquad 28\text{-}3$$

The maximum value of I occurs when $\cos \omega t$ has its maximum value of 1, in which case

$$I_{max} = \frac{\mathcal{E}_{max}}{R} \qquad 28\text{-}4$$

We may thus write Equation 28-3 as

$$I = I_{max} \cos \omega t \qquad 28\text{-}5$$

*The general equation for the emf of a generator was found in Section 26-6 to be $\mathcal{E} = \mathcal{E}_{max} \sin(\omega t + \delta)$. We are free to choose any convenient phase constant δ since it depends merely on the choice of zero time. For simplicity, we choose $\delta = \pi/2$ so that $\mathcal{E} = \mathcal{E}_{max} \sin(\omega t + \pi/2) = \mathcal{E}_{max} \cos \omega t$.

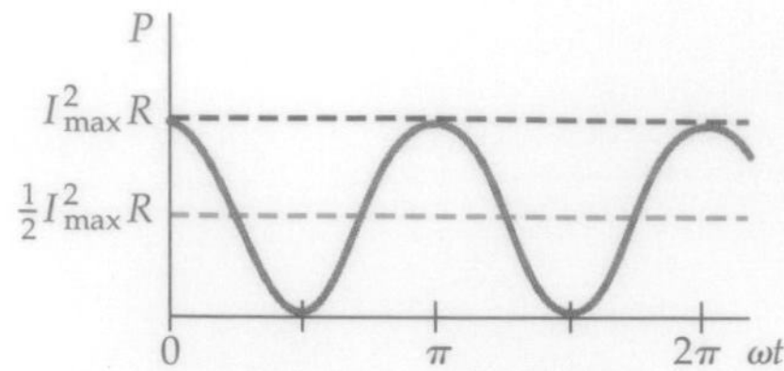

Figure 28-2 Plot of the power dissipated in the resistor in Figure 28-1 versus time. The power varies from zero to a maximum value $I^2_{max}R$. The average power is half the maximum power.

Note that the current through the resistor is in phase with the voltage across the resistor.

The power dissipated in the resistor varies with time. Its instantaneous value is

$$P = I^2R = (I_{max} \cos \omega t)^2R = I^2_{max}R \cos^2 \omega t \qquad 28\text{-}6$$

Figure 28-2 shows the power as a function of time. It varies from zero to its maximum value $I^2_{max}R$ as shown. We are usually interested in the average power over one or more cycles. The energy W_T delivered during the time of one period ($t = T = 2\pi/\omega$) is

$$W_T = \int_0^T P\, dt = \int_0^T I^2_{max}R \cos^2 \omega t\, dt$$

Substituting $\theta = \omega t$, we obtain

$$W_T = \frac{I^2_{max}R}{\omega} \int_0^{2\pi} \cos^2 \theta\, d\theta$$

The integral in this expression can be found in tables. It has the value π. The average power delivered to the resistor during one period is this energy divided by T:

$$P_{av} = \frac{W_T}{T} = \frac{(\pi I^2_{max}R)/\omega}{2\pi/\omega} = \frac{1}{2}I^2_{max}R$$

We could also obtain this result directly from Equation 28-6 by noting that the average value of $\cos^2 \omega t$ over one or more periods is $\frac{1}{2}$. This can be seen from the identity $\cos^2 \omega t + \sin^2 \omega t = 1$. A plot of $\sin^2 \omega t$ looks the same as one of $\cos^2 \omega t$ except that it is shifted by 90°. Both have the same average value over one or more periods, and since their sum is 1, the average value of each must be $\frac{1}{2}$. The average power dissipated in the resistor is thus

$$P_{av} = (I^2R)_{av} = \tfrac{1}{2}I^2_{max}R \qquad 28\text{-}7$$

rms Values

Most ac ammeters and voltmeters are designed to measure **root-mean-square (rms) values** of current and voltage rather than the maximum or peak values. The **rms** value of a current I_{rms} is defined by

rms current defined

$$I_{rms} = \sqrt{(I^2)_{av}} \qquad 28\text{-}8$$

The average value of I^2 is

$$(I^2)_{av} = [(I_{max} \cos \omega t)^2]_{av} = \tfrac{1}{2}I^2_{max}$$

where we have used $(\cos^2 \omega t)_{av} = \frac{1}{2}$. Substituting $\frac{1}{2}I^2_{max}$ for $(I^2)_{av}$ in Equation 28-8, we obtain

$$I_{rms} = \frac{1}{\sqrt{2}}I_{max} \qquad 28\text{-}9$$

The rms value of any quantity that varies sinusoidally equals the maximum value of that quantity divided by $\sqrt{2}$.

Substituting I^2_{rms} for $\frac{1}{2}I^2_{max}$ in Equation 28-7, we obtain for the average power dissipated in the resistor

$$P_{av} = I_{rms}^2 R \qquad 28\text{-}10$$

We can see from Equation 28-10 that the rms current equals the steady dc current that would produce the same Joule heating as the ac current of Equation 28-5.

For this simple circuit in Figure 28-1, the average power delivered by the generator is equal to that dissipated in the resistor:

$$P_{av} = (\mathcal{E}I)_{av} = [(\mathcal{E}_{max}\cos\omega t)(I_{max}\cos\omega t)]_{av} = \mathcal{E}_{max}I_{max}(\cos^2\omega t)_{av}$$

or

$$P_{av} = \tfrac{1}{2}\mathcal{E}_{max}I_{max}$$

Using $I_{rms} = I_{max}/\sqrt{2}$ and $\mathcal{E}_{rms} = \mathcal{E}_{max}/\sqrt{2}$, this can be written

$$P_{av} = \mathcal{E}_{rms}I_{rms} \qquad 28\text{-}11$$

Average power delivered by a generator

The rms current is related to the rms emf in the same way that the maximum current is related to the maximum emf. We can see this by dividing each side of Equation 28-4 by $\sqrt{2}$ and using $I_{rms} = I_{max}/\sqrt{2}$ and $\mathcal{E}_{rms} = \mathcal{E}_{max}/\sqrt{2}$:

$$I_{rms} = \frac{\mathcal{E}_{rms}}{R} \qquad 28\text{-}12$$

Equations 28-10, 28-11, and 28-12 are of the same form as the corresponding equations for direct-current circuits with I replaced by I_{rms} and $\mathcal{E}$ replaced by $\mathcal{E}_{rms}$. Thus, if we use rms values for the current and emf, we can calculate the power input and the heat generated using the same equations as we did for direct current.

Exercise

A 12-Ω resistor is connected across a sinusoidal emf that has a peak value of 48 V. Find (*a*) the rms current, (*b*) the average power, and (*c*) the maximum power. [Answer: (*a*) 2.83 A, (*b*) 96 W, (*c*) 192 W]

The ac power supplied to your house by the power company has a frequency of 60 Hz and a voltage of 120 V rms. (For some high-power appliances, such as an electric clothes dryer or an oven, separate lines carrying power at 240 V are often supplied. For a given power requirement, only half as much current is required at 240 V as at 120 V, but 240 V is much more dangerous than 120 V. A shock at 240 V is much more likely to be fatal than one at 120 V.) If you plug in a 1600-W heater, it will draw a current of

$$I_{rms} = \frac{P_{av}}{\mathcal{E}_{rms}} = \frac{1600\text{ W}}{120\text{ V}} = 13.3\text{ A}$$

The voltage across the outlets is maintained at 120 V, which is independent of the current drawn. Thus, all appliances plugged into the outlets of a single circuit are essentially in parallel. Therefore, if you plug a 500-W toaster into another outlet of the circuit supplying the heater, it will draw a current of 500 W/120 V = 4.17 A, so the total current through the circuit will be more than 17 A. Most household wiring is rated at 15 or 20 A. A current greater than this will overheat the wiring and create a fire hazard. Each circuit is therefore equipped with a circuit breaker (or a fuse in older houses). For a 20-A circuit, the circuit breaker trips (or the fuse blows), opening the circuit, when the current exceeds 20 A. The maximum power load that can be handled by a circuit with a 20-A circuit breaker is

$$P_{\text{av}} = \mathcal{E}_{\text{rms}} I_{\text{rms}} = (120 \text{ V})(20 \text{ A}) = 2.4 \text{ kW}$$

Since most modern houses require considerably more than 2.4 kW of power, many circuits are supplied, each with its own circuit breaker and each having several outlets.

Example 28-1

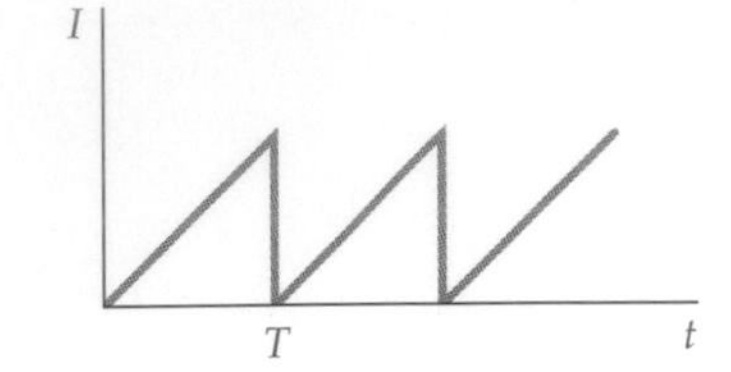

A current with a sawtooth waveform as shown in Figure 28-3 occurs sometimes in electronics. In the region $0 < t < T$, the current is given by $I = (I_0/T)t$. Find (*a*) the average current and (*b*) the rms current for this waveform.

(*a*) The average of any quantity over some interval T is the integral of the quantity over the interval divided by T. The average current is thus

$$I_{\text{av}} = \frac{1}{T}\int_0^T I\,dt = \frac{1}{T}\int_0^T (I_0/T)t\,dt = \frac{I_0}{T^2}\frac{T^2}{2} = \frac{1}{2}I_0$$

The average current is half the maximum current as we would expect.

(*b*) The average squared current is

$$(I^2)_{\text{av}} = \frac{1}{T}\int_0^T I^2\,dt = \frac{1}{T}\int_0^1 (I_0/T)^2 t^2\,dt = \frac{I_0^2}{T^3}\frac{T^3}{3} = \frac{1}{3}I_0^2$$

The rms current is therefore $I_{\text{rms}} = I_0/\sqrt{3}$.

Figure 28-3 Saw-tooth waveform for the current in Example 28-1.

Questions

1. What is the average current in the resistor in Figure 28-1?
2. Is the instantaneous power in the resistor in Figure 28-1 ever negative?

28-2 Alternating Current in Inductors and Capacitors

The behavior of alternating current in inductors and capacitors is very different from that of direct current. For example, when a capacitor is in series in a dc circuit, the current stops completely when the capacitor becomes fully charged. But if the current alternates, charge continually flows onto or off of the plates of the capacitor. We will see that if the frequency of the alternating current is great, a capacitor hardly impedes the current at all. Conversely, an inductor coil usually has a very small resistance and therefore has little effect on direct current. But when the current is changing in an inductor, a back emf is generated that is proportional to the rate of change of the current. The greater the frequency of alternating current in an inductor, the greater the rate of change of current and, therefore, the greater the back emf. Thus, an inductor has just the opposite effect on alternating current as a capacitor. At very low frequencies, an inductor hardly impedes the current at all, but at high frequencies it impedes the current flow greatly because of its back emf.

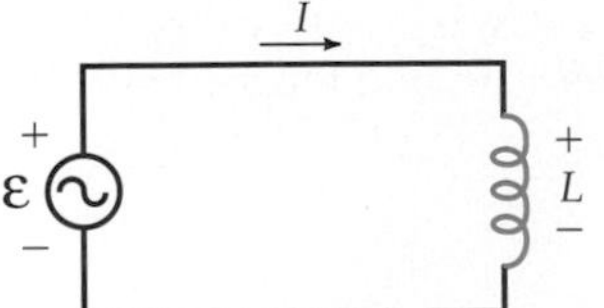

Figure 28-4 An ac generator in series with an inductor L.

Inductors

Figure 28-4 shows an inductor coil across the terminals of an ac generator. When the current increases in the inductor, a back emf of magnitude $L\,dI/dt$ is generated due to the changing flux (Equation 26-20). Usually, the potential drop across the inductor due to this back emf is much greater than the drop IR due to the resistance of the coil. Hence, we can normally neglect any resistance in the coil. In the figure, plus and minus signs have been placed on the inductor to indicate the direction of the potential drop when dI/dt is positive for the assumed direction of the current. Note that for a positive dI/dt, the point at which the current enters the inductor is at a higher potential than the point at which the current leaves. The voltage drop across the inductor V_L is then given by

$$V_L = V_+ - V_- = L\frac{dI}{dt} \qquad 28\text{-}13$$

Applying Kirchhoff's loop rule to this circuit, we obtain

$$\mathcal{E} - V_L = 0$$

where $V_L = L\,dI/dt$ is the magnitude of the emf of the inductor. Setting the emf of the generator equal to $\mathcal{E}_{max} \cos \omega t$, we obtain

$$\mathcal{E} = L\frac{dI}{dt} = \mathcal{E}_{max} \cos \omega t \qquad 28\text{-}14$$

Multiplying both sides by dt and dividing by L, we obtain

$$dI = \frac{\mathcal{E}_{max}}{L} \cos \omega t\, dt \qquad 28\text{-}15$$

We solve for the current I by integrating both sides of Equation 28-15:

$$I = \frac{\mathcal{E}_{max}}{L}\int \cos \omega t\, dt = \frac{\mathcal{E}_{max}}{\omega L} \sin \omega t + C \qquad 28\text{-}16$$

where C is a constant of integration. The constant is the average value of the current, since the average of $\sin \omega t$ over one or more cycles is zero. Choosing the dc component of the current to be zero, we have

$$I = \frac{\mathcal{E}_{max}}{\omega L} \sin \omega t = I_{max} \sin \omega t \qquad 28\text{-}17$$

where

$$I_{max} = \frac{\mathcal{E}_{max}}{\omega L} \qquad 28\text{-}18$$

is the maximum value of the current.

Figure 28-5 shows the current I and the voltage drop across the inductor V_L as functions of time. The current is not in phase with the voltage drop across the inductor, which equals the generator voltage. From the figure, we can see that the maximum value of the voltage occurs 90° or one-fourth period before the corresponding maximum value of the current. Thus, *the voltage drop across an inductor is said to lead the current by 90°*. We can understand this physically. When the current is zero but increasing, its rate of change is at its maximum, so the back emf induced in the inductor is at its maximum. One-quarter cycle later, the current is at its maximum. At this time, dI/dt is zero, so V_L is zero. Using the trigonometric identity $\sin \omega t = \cos (\omega t - \pi/2)$, Equation 28-17 for the current can be written

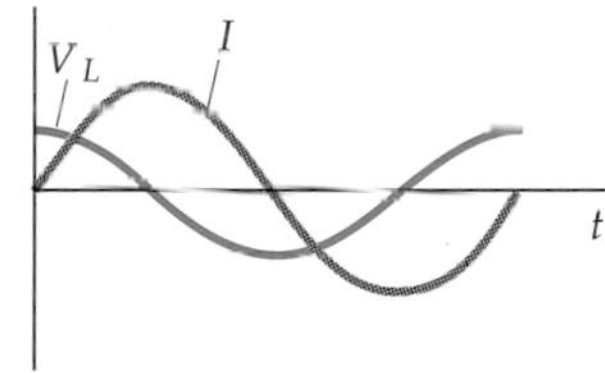

Figure 28-5 Current and voltage across the inductor in Figure 28-4 as functions of time. The maximum voltage occurs one-fourth period before the maximum current. Thus, the voltage is said to lead the current by one-fourth period or 90°.

$$I = I_{max} \cos (\omega t - \pi/2) \qquad 28\text{-}19$$

The relation between the maximum (or rms) current and the maximum (or rms) voltage for an inductor can be written in a form similar to Equation 28-4 for a resistor. From Equation 28-18, we have

$$I_{max} = \frac{\mathcal{E}_{max}}{\omega L} = \frac{\mathcal{E}_{max}}{X_L} \qquad 28\text{-}20$$

where

Inductive reactance

$$X_L = \omega L \tag{28-21}$$

is called the **inductive reactance.** Since $I_{\text{rms}} = I_{\max}/\sqrt{2}$ and $\mathcal{E}_{\text{rms}} = \mathcal{E}_{\max}/\sqrt{2}$, the rms current is given by

$$I_{\text{rms}} = \frac{\mathcal{E}_{\text{rms}}}{\omega L} = \frac{\mathcal{E}_{\text{rms}}}{X_L} \tag{28-22}$$

Like resistance, inductive reactance has units of ohms. As we can see from Equation 28-20, the larger the reactance for a given emf, the smaller the current. Unlike resistance, the inductive reactance depends on frequency of the current—the greater the frequency, the greater the reactance.

The instantaneous power input to the inductor from the generator is

$$P = \mathcal{E}I = (\mathcal{E}_{\max} \cos \omega t)(I_{\max} \sin \omega t)$$
$$= \mathcal{E}_{\max} I_{\max} \cos \omega t \sin \omega t$$

The average power into the inductor is zero. We can see this by using

$$\cos \omega t \sin \omega t = \tfrac{1}{2} \sin 2\omega t$$

The value of this term oscillates twice during each cycle and is negative as often as it is positive. Thus, no energy is dissipated in an inductor. (This is true only if the resistance of the inductor can be neglected.)

Example 28-2

A 40-mH inductor is placed across an ac generator that has a maximum emf of 120 V. Find the inductive reactance and the maximum current when the frequency is 60 Hz and when it is 2000 Hz.

The inductive reactance at 60 Hz is

$$X_{L1} = \omega_1 L = 2\pi f_1 L = (2\pi)(60 \text{ Hz})(40 \times 10^{-3} \text{ H}) = 15.1 \ \Omega$$

and that at 2000 Hz is

$$X_{L2} = \omega_2 L = 2\pi f_2 L = (2\pi)(2000 \text{ Hz})(40 \times 10^{-3} \text{ H}) = 503 \ \Omega$$

The maximum value of the currents for these frequencies are

$$I_{1,\max} = \frac{\mathcal{E}_{\max}}{X_{L1}} = \frac{120 \text{ V}}{15.1 \ \Omega} = 7.95 \text{ A}$$

$$I_{2,\max} = \frac{120 \text{ V}}{503 \ \Omega} = 0.239 \text{ A}$$

Capacitors

Figure 28-6 shows a capacitor connected across the terminals of a generator. For the current direction shown, the current is related to the charge by

$$I = \frac{dQ}{dt}$$

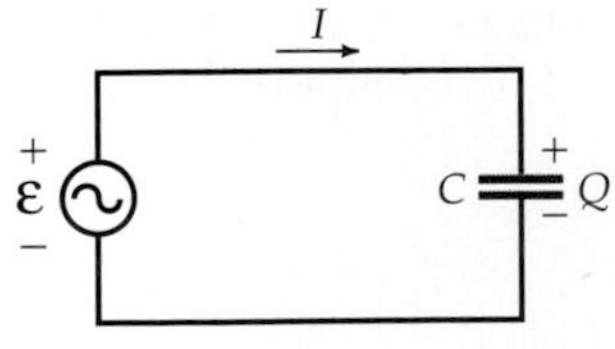

Figure 28-6 An ac generator in series with a capacitor *C*.

Again plus and minus signs have been placed on the capacitor plates showing a positive charge on the plate where the current enters and a negative charge on the plate where the current leaves. The voltage drop across the capacitor is

$$V_C = V_+ - V_- = \frac{Q}{C} \qquad \text{28-23}$$

From Kirchhoff's loop rule, we have

$$\mathcal{E} - V_C = 0$$

or

$$\mathcal{E} = \mathcal{E}_{max} \cos \omega t = \frac{Q}{C}$$

Thus

$$Q = \mathcal{E}_{max} C \cos \omega t$$

The current is

$$I = \frac{dQ}{dt} = -\omega \mathcal{E}_{max} C \sin \omega t$$

The maximum value of I occurs when $\sin \omega t = -1$, in which case

$$I_{max} = \omega \mathcal{E}_{max} C \qquad \text{28-24}$$

The current can thus be written

$$I = -\omega \mathcal{E}_{max} C \sin \omega t = -I_{max} \sin \omega t$$

Using the trigonometric identity $\sin \omega t = -\cos(\omega t + \pi/2)$, we obtain

$$I = -\omega C \mathcal{E}_{max} \sin \omega t = I_{max} \cos(\omega t + \pi/2) \qquad \text{28-25}$$

As with the inductor, the capacitor current is not in phase with the voltage drop across the capacitor, which equals the generator voltage. In Figure 28-7, the current I and the voltage drop across the capacitor $V_C = Q/C$ are plotted as functions of time. We can see that the maximum value of the voltage occurs 90° or one-fourth period *after* the maximum value of the current. Thus, *the voltage drop across a capacitor lags the current by 90°*. Again, we can understand this physically. At $\omega t = 3\pi/2$ in Figure 28-7, the current I is maximum. The maximum rate of charge buildup dQ/dt must occur when the charge Q is zero and, therefore, V_C is zero. As the charge on the capacitor plate increases, the current decreases until the charge is a maximum (so that V_C is a maximum) and the current is zero. The current then becomes negative as the charge flows back in the opposite direction, off the capacitor.

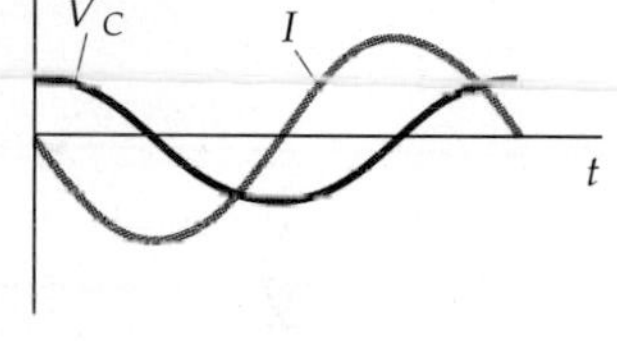

Figure 28-7 Current and voltage across the capacitor in Figure 28-6 versus time. The maximum voltage occurs one-fourth period after the maximum current. Thus, the voltage is said to lag the current by 90°.

Again, the relation between the maximum (or rms) current and the maximum (or rms) voltage for a capacitor can be written in a form similar to Equation 28-4 for a resistor. From Equation 28-24, we have

$$I_{max} = \omega C \mathcal{E}_{max} = \frac{\mathcal{E}_{max}}{1/\omega C} = \frac{\mathcal{E}_{max}}{X_C}$$

and, similarly,

$$I_{rms} = \frac{\mathcal{E}_{rms}}{1/\omega C} = \frac{\mathcal{E}_{rms}}{X_C} \qquad \text{28-26}$$

where

$$X_C = \frac{1}{\omega C} \qquad \text{28-27}$$

Capacitive reactance

is called the **capacitive reactance** of the circuit. Like resistance and inductive reactance, capacitive reactance has units of ohms, and like inductive reactance, capacitive reactance depends on the frequency of the current. In this case, the greater the frequency, the smaller the reactance. As for an inductor, the average power input to a capacitor from an ac generator is zero. This is because the emf is proportional to cos ωt and the current is proportional to sin ωt and $(\cos \omega t \sin \omega t)_{av} = 0$. Thus, like inductors, ideal capacitors dissipate no energy.

Since charge cannot pass across the space between the plates of a capacitor, it may seem strange that there is a continuing alternating current in the circuit of Figure 28-6. Recall, though, that when an uncharged capacitor is placed across the terminals of a dc voltage source (such as a battery), there is a current that decreases exponentially with time until the plates are charged to the same potential as the battery. Consider an initially uncharged capacitor across a source of emf, with the upper plate attached to the positive terminal. Initially, positive charge flows to the upper plate and away from the lower plate. (Of course, it is actually the negative electrons flowing in the opposite direction.) The effect is the same as if the charge actually flows *across* the space between the plates. If the source of emf is an ac generator, the potential difference changes sign every half period, as shown in Figure 28-7. Let us keep the generator emf constant but increase the frequency. During each half cycle, the same charge $\Delta Q = 2C\mathcal{E}_{max}$ is transferred into or out of the capacitor, but the number of cycles per second increases, so the current "through" the capacitor increases in proportion to the frequency. Hence, the greater the frequency, the less the capacitor impedes the flow of charge.

(*a*)

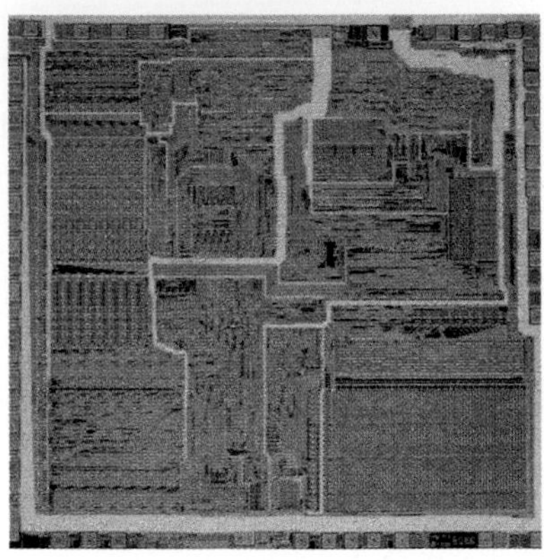

(*b*)

(*a*) The 8086 microprocessor is an integrated circuit containing 29,000 transistors used primarily in personal computers. Much of the 8086 design was drawn by hand on paper, which was used to make cutouts from rubyliths, red cellophanelike sheets as shown in (*b*). Today computers are used to draw microcircuit designs.

Example 28-3

A 20-μF capacitor is placed across a generator that has a maximum emf of 100 V. Find the capacitive reactance and the maximum current when the frequency is 60 Hz and when it is 5000 Hz.

The capacitive reactance at 60 Hz is

$$X_{C1} = \frac{1}{\omega_1 C} = \frac{1}{2\pi f_1 C}$$
$$= [2\pi(60\text{ Hz})(20 \times 10^{-6}\text{ F})]^{-1} = 133\ \Omega$$

and that at 5000 Hz is

$$X_{C2} = \frac{1}{\omega_2 C} = \frac{1}{2\pi f_2 C}$$
$$= [2\pi(5000\text{ Hz})(20 \times 10^{-6}\text{ F})]^{-1} = 1.59\ \Omega$$

The maximum currents are then

$$I_{1,max} = \frac{\mathcal{E}_{max}}{X_{C1}} = \frac{100\text{ V}}{133\ \Omega} = 0.754\text{ A}$$

and

$$I_{2,max} = \frac{100\text{ V}}{1.59\ \Omega} = 62.8\text{ A}$$

The circuits of Figures 28-4 and 28-6 contain only a generator and an inductor or capacitor. In them, the voltage drop across the inductor or capacitor equals the voltage of the generator. In more complicated circuits containing three or more elements, the voltage drop across each element is usually not equal to the generator voltage. It is useful, therefore, to write Equations 28-22 and 28-26 in terms of the voltage drops across the inductor and capacitor, respectively. If $V_{L,\text{rms}}$ is the rms voltage drop across an inductor, the rms current in the inductor is given by

$$I_{\text{rms}} = \frac{V_{L,\text{rms}}}{\omega L} = \frac{V_{L,\text{rms}}}{X_L} \qquad 28\text{-}28$$

The voltage drop across the inductor leads the current by 90°. Similarly, if $V_{C,\text{rms}}$ is the rms voltage across a capacitor, the rms current in the capacitor is given by

$$I_{\text{rms}} = \frac{V_{C,\text{rms}}}{1/\omega C} = \frac{V_{C,\text{rms}}}{X_C} \qquad 28\text{-}29$$

The voltage drop across the capacitor lags the current by 90°. Equations 28-28 and 28-29 can also be written in terms of the maximum voltages and maximum currents.

Questions

3. In a circuit consisting of a generator and an inductor, are there any times when the inductor absorbs power from the generator? Are there any times when the inductor supplies power to the generator?
4. In a circuit consisting of a generator and a capacitor, are there any times when the capacitor absorbs power from the generator? Are there any times when the capacitor supplies power to the generator?

28-3 Phasors

In the previous sections, we saw that the voltage across a resistor is in phase with the current, whereas the voltage across an inductor leads the current by 90° and the voltage across a capacitor lags the current by 90°. These phase relations can be represented by two-dimensional vectors called **phasors.** In Figure 28-8, the voltage across a resistor V_R is represented by a vector $\mathbf{V}_R$ that has magnitude $I_{\text{max}}R$ and makes an angle θ with the x axis. This voltage is in phase with the current. In general, a steady-state current in an ac circuit varies with time as

$$I = I_{\text{max}} \cos \theta = I_{\text{max}} \cos (\omega t - \delta) \qquad 28\text{-}30$$

where ω is the angular frequency and δ is some phase constant. The voltage drop across a resistor is then given by

$$V_R = IR = I_{\text{max}}R \cos (\omega t - \delta) \qquad 28\text{-}31$$

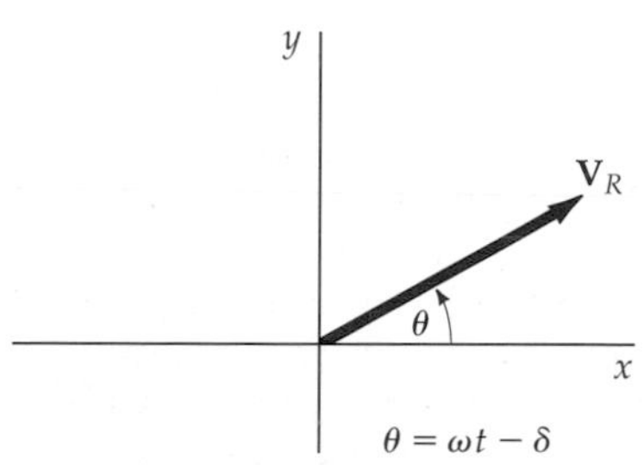

Figure 28-8 The voltage across a resistor can be represented by a vector $\mathbf{V}_R$, called a phasor, that has magnitude $I_{\text{max}}R$ and makes an angle $\theta = \omega t - \delta$ with the x axis. The phasor rotates with an angular frequency ω. The voltage $V_R = IR$ is the x component of $\mathbf{V}_R$.

The instantaneous value of the voltage drop across a resistor is thus equal to the x component of the phasor vector $\mathbf{V}_R$, which rotates counterclockwise with an angular frequency ω. The current I may be written as the x component of a phasor $\mathbf{I}$ having the same orientation as $\mathbf{V}_R$.

When several components are connected together in a series circuit, their voltages add. When they are connected in parallel, their currents add. Adding sines or cosines of different amplitudes and phases algebraically is awkward. It is much easier to do this by vector addition.

Phasors are used as follows. Any ac voltage or current is written in the form $A \cos (\omega t - \delta)$, which in turn is treated as the x component A_x of a

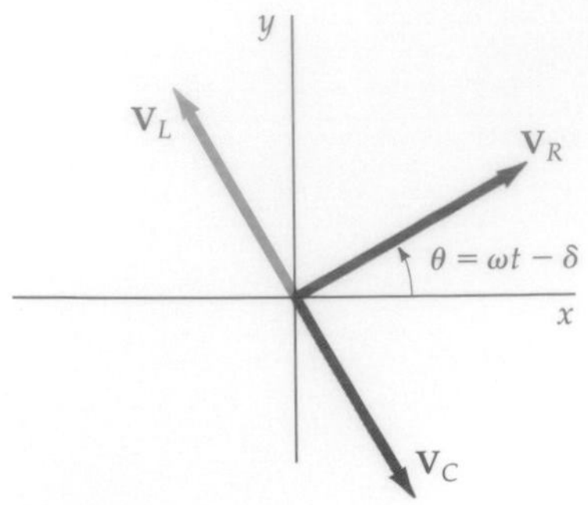

Figure 28-9 Phasor representations of the voltages V_R, V_L, and V_C. Each vector rotates in the counterclockwise direction with an angular frequency ω. At any instant, the voltage across an element equals the x component of the corresponding phasor, and the sum of the voltages equals the x component of the vector sum $\mathbf{V}_R + \mathbf{V}_L + \mathbf{V}_C$.

phasor **A** that makes an angle $(\omega t - \delta)$ with the x axis. Instead of adding two voltages or currents algebraically as $A\cos(\omega t - \delta_1) + B\cos(\omega t - \delta_2)$, we represent these quantities as phasors **A** and **B** and find the phasor sum $\mathbf{C} = \mathbf{A} + \mathbf{B}$ geometrically. The resultant voltage or current is then the x component of the resultant phasor, $C_x = A_x + B_x$. The geometric representation conveniently shows the relative amplitudes and phases of the phasors.

Consider a circuit containing an inductor L, a capacitor C, and a resistor R all connected in series. They all carry the same current, which is represented as the x component of the current phasor **I**. The voltage across the inductor V_L is represented by a phasor $\mathbf{V}_L$ that has magnitude $I_{max}X_L$ and leads the current phasor **I** by 90°. Similarly, the voltage across the capacitor V_C is represented by a phasor $\mathbf{V}_C$ that has magnitude $I_{max}X_C$ and lags the current by 90°. Figure 28-9 shows the three phasors $\mathbf{V}_R$, $\mathbf{V}_L$, and $\mathbf{V}_C$. As time goes on, the three phasors rotate counterclockwise with an angular frequency ω, so the relative positions of the vectors do not change. At any time, the instantaneous value of the voltage drop across any of these elements equals the x component of the corresponding phasor.

28-4 *LC* and *LCR* Circuits without a Generator

In this section, we will study some simple circuits containing inductance, capacitance, and resistance but no generator. We will add a generator to these circuits in the next section. We first look at a simple circuit with inductance and capacitance but no resistance as shown in Figure 28-10. We assume that the capacitor carries an initial charge Q_0 and that the switch is initially open. After the switch is closed at $t = 0$, the charge begins to flow through the inductor. In the figure, the signs of Q on the capacitor and the direction of the current I have been chosen such that

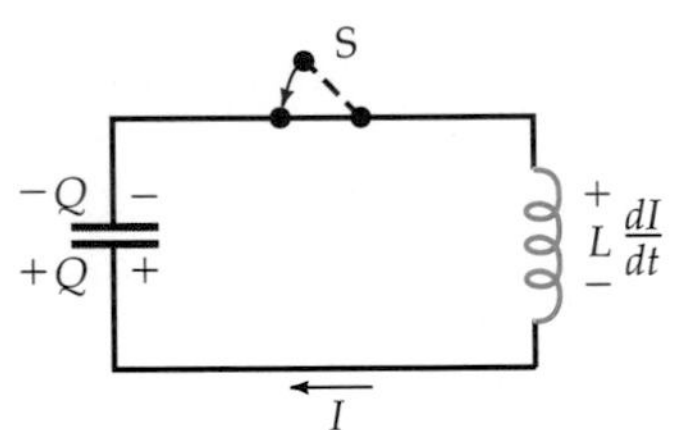

Figure 28-10 An *LC* circuit. When the switch is closed, the initially charged capacitor discharges through the inductor, producing a back emf.

$$I = \frac{dQ}{dt}$$

Applying Kirchhoff's loop rule to the circuit for the assumed signs of Q and I, we have

$$L\frac{dI}{dt} + \frac{Q}{C} = 0 \qquad \text{28-32}$$

Substituting dQ/dt for I in this equation, we obtain

$$L\frac{d^2Q}{dt^2} + \frac{Q}{C} = 0 \qquad \text{28-33}$$

Equation 28-33 is of the same form as the equation for the acceleration of a mass on a spring:

$$m\frac{d^2x}{dt^2} + kx = 0 \qquad \text{28-34}$$

The behavior of an *LC* circuit is thus analogous to that of a mass on a spring, with L analogous to the mass m, Q analogous to the position x, and $1/C$ analogous to the spring constant k. Also, the current I is analogous to the velocity v, since $v = dx/dt$ and $I = dQ/dt$. In mechanics, the mass of an object describes the inertia of the object. The greater the mass, the more difficult it is to change the velocity of the object. Similarly, the inductance L can be thought of as the inertia of an ac circuit. The greater the inductance, the more difficult it is to change the current I.

If we divide each term in Equation 28-33 by L and rearrange, we obtain

$$\frac{d^2Q}{dt^2} = -\frac{1}{LC}Q \qquad 28\text{-}35$$

which is analogous to

$$\frac{d^2x}{dt^2} = -\frac{k}{m}x = -\omega^2 x \qquad 28\text{-}36$$

where $\omega^2 = k/m$. In Chapter 12, we found that we could write the solution of Equation 28-36 for simple harmonic motion in the form

$$x = A\cos(\omega t - \delta)$$

where $\omega = \sqrt{k/m}$ is the angular frequency, A is the amplitude, and δ is the phase constant, which depends on the initial conditions. We can put Equation 28-35 in this same form by writing ω^2 for $1/LC$. Then

$$\frac{d^2Q}{dt^2} = -\omega^2 Q \qquad 28\text{-}37$$

$$\omega = \frac{1}{\sqrt{LC}} \qquad 28\text{-}38$$

The solution of Equation 28-38 is

$$Q = A\cos(\omega t - \delta)$$

The current is found by differentiating this solution:

$$I = \frac{dQ}{dt} = \omega A\sin(\omega t - \delta)$$

If we choose our initial conditions to be $Q = Q_0$ and $I = 0$ at $t = 0$, the phase constant δ is zero and $A = Q_0$. Our solutions are then

$$Q - Q_0\cos\omega t \qquad 28\text{-}39$$

and

$$I = -\omega Q_0\sin\omega t = -I_{max}\sin\omega t \qquad 28\text{-}40$$

where $I_{max} = \omega Q_0$.

Figure 28-11 shows graphs of Q and I versus time. The charge oscillates between the values $+Q_0$ and $-Q_0$ with angular frequency $\omega = \sqrt{1/LC}$. The current oscillates between $+\omega Q_0$ and $-\omega Q_0$ with the same frequency and is 90° out of phase with the charge. The current is maximum when the charge is zero and zero when the charge is maximum.

In our study of the oscillations of a mass on a spring, we found that the total energy is constant but oscillates between potential and kinetic energy. In our LC circuit, we also have two kinds of energy, electric energy and magnetic energy. The electric energy stored in the capacitor is

$$U_e = \frac{1}{2}QV_C = \frac{1}{2}\frac{Q^2}{C}$$

Substituting $Q_0\cos\omega t$ for Q, we have for the electric energy

$$U_e = \frac{Q_0^2}{2C}\cos^2\omega t \qquad 28\text{-}41$$

The electric energy oscillates between its maximum value $Q_0^2/2C$ and zero. The magnetic energy stored in the inductor is

$$U_m = \tfrac{1}{2}LI^2 \qquad 28\text{-}42$$

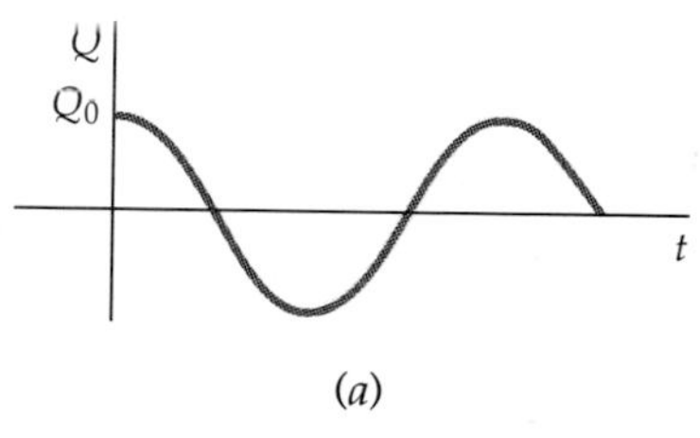

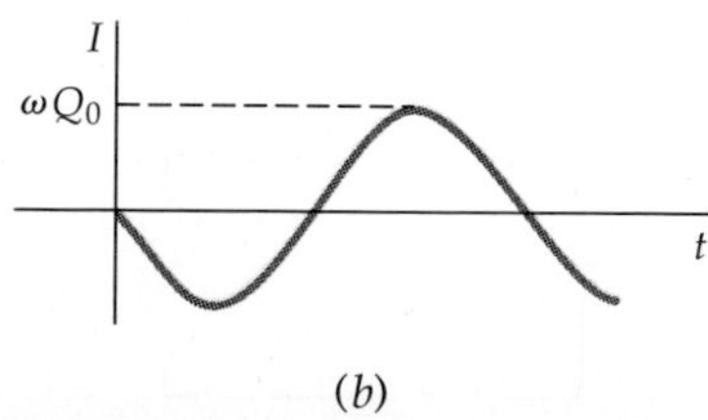

Figure 28-11 Graphs of (*a*) Q versus t and (*b*) I versus t for the LC circuit of Figure 28-10.

Substituting in the value for the current from Equation 28-30, we obtain

$$U_m = \frac{1}{2}LI_{max}^2 \sin^2 \omega t = \frac{1}{2}L\omega^2 Q_0^2 \sin^2 \omega t = \frac{Q_0^2}{2C} \sin^2 \omega t \qquad 28\text{-}43$$

where we have used the fact that $\omega^2 = 1/LC$. The magnetic energy also oscillates between its maximum value of $Q_0^2/2C$ and zero. The sum of the electrostatic and magnetic energies is the total energy, which is constant in time:

$$U_{total} = U_e + U_m = \frac{Q_0^2}{2C} \cos^2 \omega t + \frac{Q_0^2}{2C} \sin^2 \omega t = \frac{Q_0^2}{2C}$$

which is the energy initially stored on the capacitor.

Example 28-4

A 2-μF capacitor is charged to 20 V and is then connected across a 6-μH inductor. (*a*) What is the frequency of oscillation? (*b*) What is the maximum value of the current?

(*a*) The frequency of oscillation depends only on the values of the capacitance and inductance:

$$f = \frac{\omega}{2\pi} = \frac{1}{2\pi\sqrt{LC}}$$

$$= \frac{1}{2\pi\sqrt{(6 \times 10^{-6}\ \text{H})(2 \times 10^{-6}\ \text{F})}} = 4.59 \times 10^4\ \text{Hz}$$

(*b*) According to Equation 28-40, the maximum value of the current is related to the maximum value of the charge by

$$I_{max} = \omega Q_0 = \frac{Q_0}{\sqrt{LC}}$$

The initial charge on the capacitor is

$$Q_0 = CV_0 = (2\ \mu\text{F})(20\ \text{V}) = 40\ \mu\text{C}$$

Thus,

$$I_{max} = \frac{40\ \mu\text{C}}{\sqrt{(6\ \mu\text{H})(2\ \mu\text{F})}} = 11.5\ \text{A}$$

Exercise

A 5-μF capacitor is charged and is then discharged through an inductor. What should the value of the inductance be so that the current oscillates with frequency 8 kHz? (Answer: 79.2 μH)

In Figure 28-12, a resistor is included in series with the capacitor and inductor. Again we assume that the switch is initially open, that the capacitor carries an initial charge Q_0, and that we close the switch at $t = 0$. Since there is now a drop IR across the resistance, Kirchhoff's loop rule gives

$$L\frac{dI}{dt} + \frac{Q}{C} + IR = 0 \qquad 28\text{-}44a$$

or

$$L\frac{d^2Q}{dt^2} + \frac{Q}{C} + R\frac{dQ}{dt} = 0 \qquad 28\text{-}44b$$

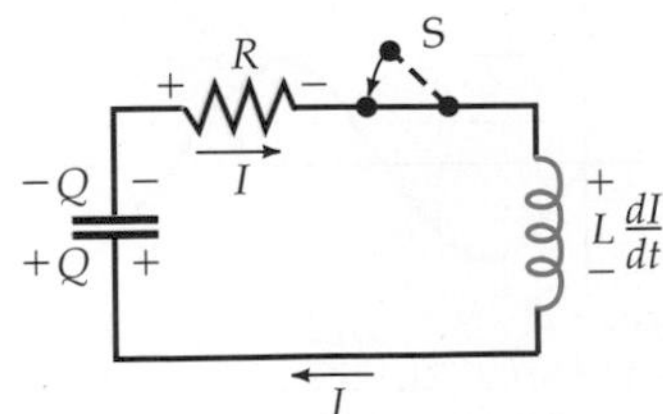

Figure 28-12 An *LCR* circuit.

where we have used $I = dQ/dt$ as before. Equations 28-44*a* and *b* are analogous to the equation for a damped harmonic oscillator (see Equation 12-46):

$$m\frac{d^2x}{dt^2} + kx + b\frac{dx}{dt} = 0$$

The first term, $L\,dI/dt = L\,d^2Q/dt^2$, is analogous to the mass times the acceleration, $m\,dv/dt = m\,d^2x/dt^2$; the second term, Q/C, is analogous to the restoring force kx; and the third term, $IR = R\,dQ/dt$, is analogous to the damping term, $bv = b\,dx/dt$. In the oscillation of a mass on a spring, the damping constant b leads to a dissipation of mechanical energy as heat. In an *LCR* circuit, the resistance R is analogous to the damping constant b and leads to a dissipation of electrical energy as Joule heat.

If the resistance is small, the charge and current oscillate with a frequency that is very nearly equal to $1/\sqrt{LC}$, but the oscillations are damped; that is, the maximum values of the charge and current decrease with each oscillation. We can understand this qualitatively from energy considerations. If we multiply each term in Equation 28-44*a* by the current I, we obtain

$$IL\frac{dI}{dt} + I\frac{Q}{C} + I^2R = 0 \qquad 28\text{-}45$$

The first term in this equation is the current times the voltage across the inductor. This is the rate at which energy is put into the inductor or taken out of it; that is, it is the rate of change of the magnetic energy, $d(\frac{1}{2}LI^2)/dt$, which is positive or negative depending on whether I and dI/dt have the same sign or different signs. Similarly, the second term is the current times the voltage across the capacitor. This is the rate of change of the energy of the capacitor, which may be positive or negative. The last term, I^2R, is the rate at which energy is dissipated in the resistor as Joule heat. This term is always positive. The sum of the electric and magnetic energies is not constant for this circuit because energy is continually dissipated in the resistor. Figure 28-13 shows graphs of Q versus t and I versus t for a small resistance R. If we increase R, the oscillations become more heavily damped until a critical value of R is reached for which there is not even one oscillation. Figure 28-14 shows Q versus t when the value of R is greater than the critical damping value.

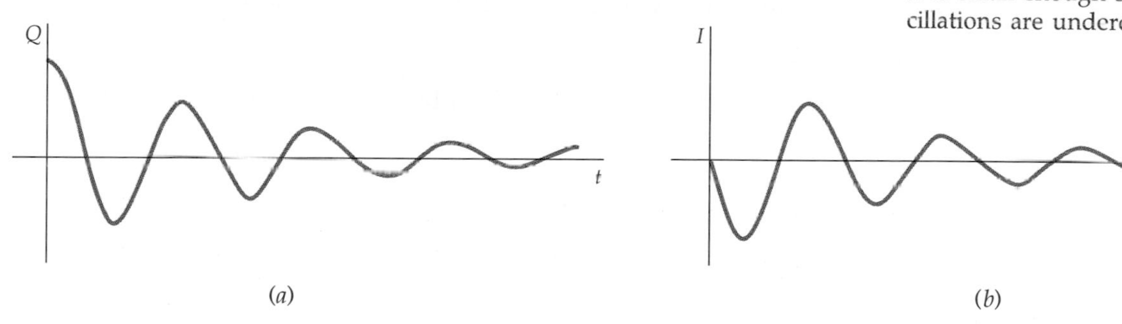

Figure 28-13 Graphs of (*a*) Q versus t and (*b*) I versus t for the *LCR* circuit of Figure 28-12 when R is small enough so that the oscillations are underdamped.

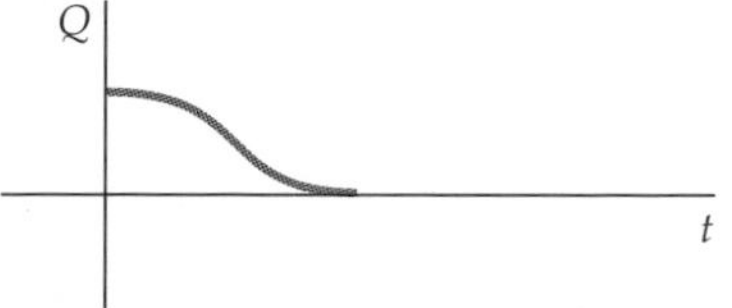

Figure 28-14 Graph of Q versus t for the *LCR* circuit of Figure 28-12 when R is so large that the oscillations are overdamped.

Question

5. It is easy to make *LC* circuits that have frequencies of oscillation of thousands of hertz or more, but it is difficult to make *LC* circuits that have small frequencies. Why?

28-5 *LCR* Circuits with a Generator

Series

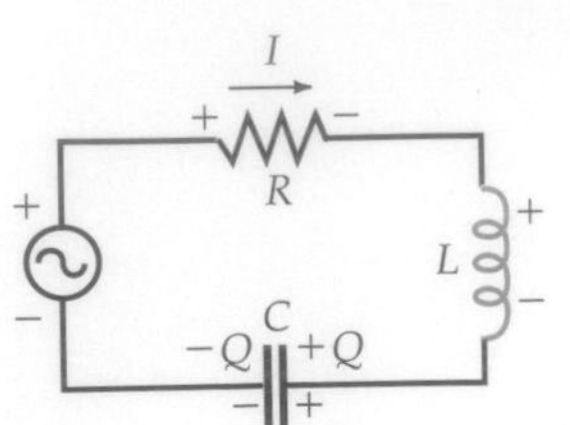

Figure 28-15 A series *LCR* circuit with an ac generator.

An important circuit that has many of the features of most ac circuits is the series *LCR* circuit with a generator shown in Figure 28-15. We assume that the emf of the generator varies with time as $\mathcal{E} = \mathcal{E}_{\text{max}} \cos \omega t$. For this circuit, Kirchhoff's loop rule gives

$$\mathcal{E}_{\text{max}} \cos \omega t - L\frac{dI}{dt} - \frac{Q}{C} - IR = 0$$

Using $I = dQ/dt$ and rearranging, we obtain

$$L\frac{d^2Q}{dt^2} + R\frac{dQ}{dt} + \frac{Q}{C} = \mathcal{E}_{\text{max}} \cos \omega t \qquad \text{28-46}$$

This equation is analogous to Equation 12-63 for the forced oscillation of a mass on a spring:

$$m\frac{d^2x}{dt^2} + b\frac{dx}{dt} + m\omega_0^2 x = F_0 \cos \omega t$$

(In Equation 12-63, the force constant k was written in terms of the mass m and the natural angular frequency ω_0 using $k = m\omega_0^2$. In Equation 28-46, the capacitance could be similarly written in terms of L and the natural angular frequency using $1/C = L\omega_0^2$.)

Rather than solving Equation 28-46, we will discuss its solution qualitatively as we did with Equation 12-63 for the forced oscillator in Chapter 12. The current in the circuit consists of two parts, a transient current and a steady-state current. The transient current depends on the initial conditions, such as the initial phase of the generator and the initial charge on the capacitor. The steady-state current, on the other hand, is independent of the initial conditions. The transient current decreases exponentially with time and is eventually negligible compared with the steady-state current. We will ignore the transient current and concentrate on the steady-state current.

The steady-state current obtained by solving Equation 28-46 is

$$I = I_{\text{max}} \cos (\omega t - \delta) \qquad \text{28-47}$$

where the phase angle δ is given by

$$\tan \delta = \frac{X_L - X_C}{R} \qquad \text{28-48}$$

The maximum current is

Impedance of series LCR circuit

$$I_{\text{max}} = \frac{\mathcal{E}_{\text{max}}}{\sqrt{R^2 + (X_L - X_C)^2}} = \frac{\mathcal{E}_{\text{max}}}{Z} \qquad \text{28-49}$$

where

$$Z = \sqrt{R^2 + (X_L - X_C)^2} \qquad \text{28-50}$$

The quantity $X_L - X_C$ is called the **total reactance,** and the quantity Z is called the **impedance.** Combining these results, we have

$$I = \frac{\mathcal{E}_{max}}{Z}\cos(\omega t - \delta) \qquad 28\text{-}51$$

Equation 28-51 can also be obtained from a simple diagram using the phasor representations discussed in Section 28-3. Figure 28-16 shows the phasors representing the voltage drops across the resistance, the inductance, and the capacitance. The x component of each of these vectors equals the instantaneous voltage drop across the corresponding element. Since the sum of the x components equals the x component of the sum of the vectors, the sum of the x components equals the sum of the voltage drops across these elements, which by Kirchhoff's loop rule equals the instantaneous emf. If we represent the applied emf, $\mathcal{E}_{max}\cos\omega t$, as a phasor $\boldsymbol{\mathcal{E}}$ that has the magnitude $\mathcal{E}_{max}$, we have

$$\boldsymbol{\mathcal{E}} = \mathbf{V}_R + \mathbf{V}_L + \mathbf{V}_C \qquad 28\text{-}52$$

In terms of the magnitudes,

$$\mathcal{E}_{max} = |\mathbf{V}_R + \mathbf{V}_L + \mathbf{V}_C| = \sqrt{V_{R,max}^2 + (V_{L,max} - V_{C,max})^2}$$

But $V_R = I_{max}R$, $V_L = I_{max}X_L$, and $V_C = I_{max}X_C$. Thus,

$$\mathcal{E}_{max} = I_{max}\sqrt{R^2 + (X_L - X_C)^2} = I_{max}Z$$

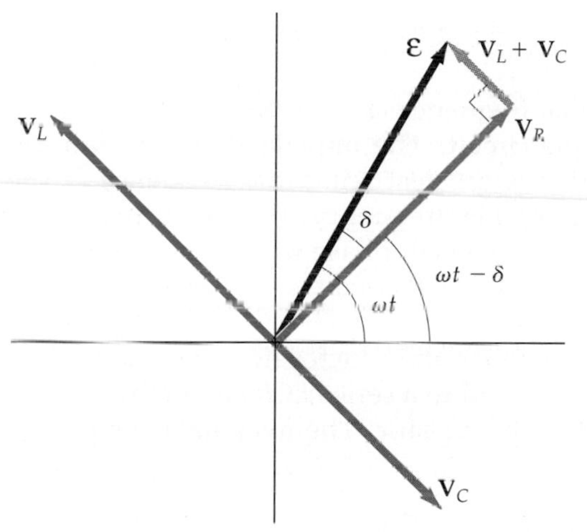

Figure 28-16 Phase relations among voltages in a series *LCR* circuit. The voltage across the resistor is in phase with the current. The voltage across the inductor V_L leads the current by 90°. The voltage across the capacitor lags the current by 90°. The sum of the vectors representing these voltages gives a vector at an angle δ with the current representing the applied emf. For the case shown here, V_L is greater than V_C and the current lags the emf by δ.

The phasor $\boldsymbol{\mathcal{E}}$ makes an angle δ with $\mathbf{V}_R$ as shown in Figure 28-16. From the figure, we can see that

$$\tan\delta = \frac{|\mathbf{V}_L + \mathbf{V}_C|}{|\mathbf{V}_R|} = \frac{I_{max}X_L - I_{max}X_C}{I_{max}R} = \frac{X_L - X_C}{R}$$

in agreement with Equation 28-48. Since $\boldsymbol{\mathcal{E}}$ makes an angle ωt with the x axis, $\mathbf{V}_R$ makes an angle $\omega t - \delta$ with the x axis. This voltage is in phase with the current, which is therefore given by

$$I = I_{max}\cos(\omega t - \delta) = \frac{\mathcal{E}_{max}}{Z}\cos(\omega t - \delta)$$

This is Equation 28-51. The relation between the impedance Z and the resistance R and the total reactance $X_L - X_C$ can be remembered using the right triangle shown in Figure 28-17.

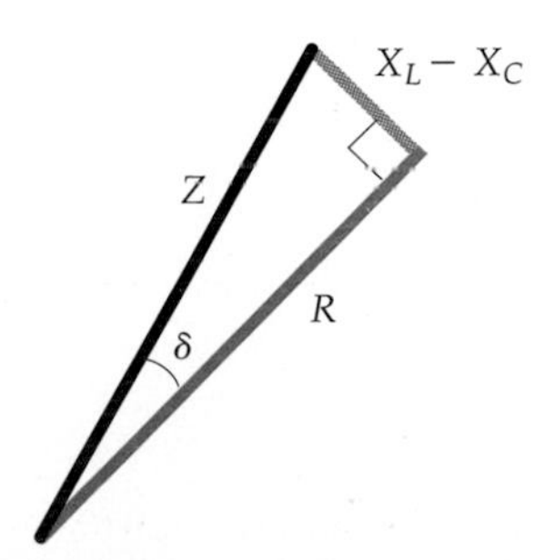

Figure 28-17 Triangle relating capacitive and inductive reactance, resistance, impedance, and the phase angle in an *LCR* circuit.

Resonance

Although Equations 28-50 and 28-51 appear to be complicated, we can use them to learn some simple and important features of the behavior of the circuit in Figure 28-15. Since both the inductive reactance $X_L = \omega L$ and the capacitive reactance $X_C = 1/\omega C$ depend on the frequency of the applied emf, the impedance Z and the maximum current I_{max} do also. At very low frequencies, $X_C = 1/\omega C$ is much greater than $X_L = \omega L$, so the impedance is large and I_{max} is small. The phase angle δ is negative, which means that the current leads the generator voltage. As we increase ω, the inductive reactance increases and the capacitive reactance decreases. When X_L and X_C are equal, the impedance Z has its smallest value, equal to R, and I_{max} has its greatest value. At $X_L = X_C$ the phase angle δ is zero, which means that the current is in phase with the applied emf. As we increase ω further, X_L becomes greater than X_C. The impedance increases and the maximum current decreases. The phase angle is positive, meaning that the current lags the generator voltage.

The value of ω for which X_L and X_C are equal is obtained from

$$X_L = X_C$$

$$\omega L = \frac{1}{\omega C}$$

or

$$\omega = \frac{1}{\sqrt{LC}} = \omega_0 = 2\pi f_0$$

The frequency f_0 (or the angular frequency ω_0) is called the natural frequency or **resonance frequency** of the circuit. The impedance is smallest and the maximum value of the current is greatest when the frequency of the emf equals the natural frequency. At this frequency, the circuit is said to be at **resonance.** At resonance, the current is in phase with the generator voltage. This resonance condition in a driven *LCR* circuit is similar to that in a driven simple harmonic oscillator.

We noted previously that neither an inductor nor a capacitor dissipates energy. The average power delivered to a series *LCR* circuit therefore equals the average power supplied to the resistor. The instantaneous power supplied to the resistor is

$$P = I^2R = [I_{max} \cos (\omega t - \delta)]^2 R$$

Averaging over one or more cycles and using $(\cos^2 \theta)_{av} = \frac{1}{2}$, we obtain for the average power

$$P_{av} = \tfrac{1}{2} I_{max}^2 R$$

which is the same as Equation 28-7. Using $V_R = I_{max}R$, this can be written

$$P_{av} = \tfrac{1}{2} I_{max} V_R$$

From Figure 28-16, we can see that $V_R = \mathcal{E}_{max} \cos \delta$. The average power supplied to the circuit can thus be written

$$P_{av} = \tfrac{1}{2} I_{max} \mathcal{E}_{max} \cos \delta$$

In terms of the rms values, $I_{rms} = I_{max}/\sqrt{2}$ and $\mathcal{E}_{rms} = \mathcal{E}_{max}/\sqrt{2}$, the average power is

$$P_{av} = \mathcal{E}_{rms} I_{rms} \cos \delta \qquad 28\text{-}53$$

The quantity $\cos \delta$ is called the **power factor** of the *LCR* circuit. At resonance, δ is zero, and the power factor is 1.

The power can be expressed as a function of the angular frequency ω. From the triangle in Figure 28-17, we have

$$\cos\delta = \frac{R}{Z} \quad \text{28-54}$$

Using this result and $I_{\text{rms}} = \mathcal{E}_{\text{rms}}/Z$, we have for the average power

$$P_{\text{av}} = \mathcal{E}_{\text{rms}}^2 \frac{R}{Z^2}$$

From the definition of impedance Z, we have

$$Z^2 = (X_L - X_C)^2 + R^2 = \left(\omega L - \frac{1}{\omega C}\right)^2 + R^2$$

$$= \frac{L^2}{\omega^2}\left(\omega^2 - \frac{1}{LC}\right)^2 + R^2$$

$$= \frac{L^2}{\omega^2}(\omega^2 - \omega_0^2)^2 + R^2$$

where we have used $\omega_0^2 = 1/LC$. Using this expression for Z^2, we obtain the average power as a function of ω:

$$P_{\text{av}} = \frac{\mathcal{E}_{\text{rms}}^2 R\omega^2}{L^2(\omega^2 - \omega_0^2)^2 + \omega^2 R^2} \quad \text{28-55}$$

Figure 28-18 shows the average power supplied by the generator to the circuit as a function of generator frequency for two different values of the resistance R. These curves, called **resonance curves,** are the same as the power-versus-frequency curves for a driven damped oscillator (see Section 12-8). The average power is maximum when the generator frequency equals the resonance frequency. When the resistance is small, the resonance curve is narrow; when it is large, the curve is broad. A resonance curve can be characterized by the **resonance width** Δf shown in the figure. This width is the frequency difference between the two points on the curve where the power is half its maximum value. When the width is small compared with the resonance frequency, the resonance is sharp, that is, the resonance curve is narrow.

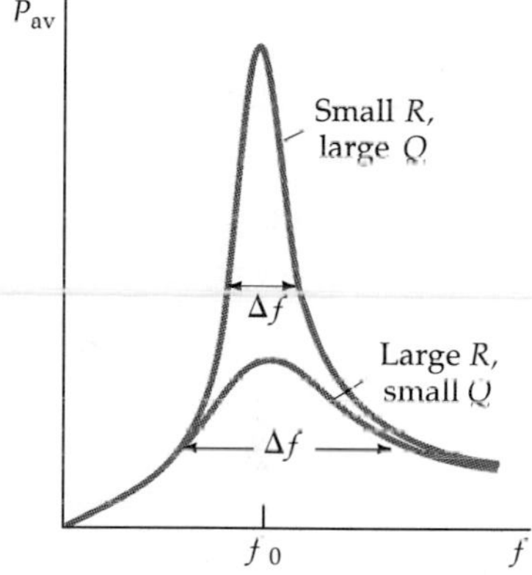

Figure 28-18 Plot of average power versus frequency for a series *LCR* circuit. The power is maximum when the frequency of the generator f equals the natural frequency of the circuit $f_0 = (1/2\pi\sqrt{LC})$. If the resistance is small, the Q factor is large and the resonance is sharp. The resonance width Δf of the curves is measured between points where the power is half its maximum value.

In Chapter 12, the Q factor for a mechanical oscillator was defined as $Q = 2\pi E/|\Delta E|$ (Equation 12-54), where E is the total energy of the system and ΔE is the energy lost in one cycle. We saw that $Q = 2\pi m/bT$, where m is the mass, b is the damping constant, and T is the period. Since $\omega_0 = 2\pi/T$, the Q factor for the damped driven mechanical oscillator is given by

$$Q = \frac{\omega_0 m}{b} \quad \text{28-56}$$

The **Q factor** for an *LCR* circuit can be defined in a similar way. Since L is analogous to the mass m and R is analogous to the damping constant b, the Q factor for an *LCR* circuit is given by

$$Q = \frac{2\pi E}{|\Delta E|} = \frac{\omega_0 L}{R} \quad \text{28-57}$$

When the resonance is reasonably narrow (that is, when Q is greater than about 2 or 3), the Q factor can be approximated by

$$Q = \frac{\omega_0}{\Delta\omega} = \frac{f_0}{\Delta f} \quad \text{28-58}$$

Q factor for an LCR circuit

A shipboard radio, circa 1920. Exposed at the operator's left are the inductance coils and capacitor plates of the tuning circuit.

Resonance circuits are used in radio receivers, where the resonance frequency of the circuit is varied by varying the capacitance. Resonance occurs when the natural frequency of the circuit equals one of the frequencies of the radio waves picked up at the antenna. At resonance, there is a relatively large current in the antenna circuit. If the Q factor of the circuit is sufficiently high, currents due to other station frequencies off resonance will be negligible compared with those due to the station frequency to which the circuit is tuned.

Example 28-5

A series LCR circuit with $L = 2$ H, $C = 2\ \mu$F, and $R = 20\ \Omega$ is driven by a generator with a maximum emf of 100 V and a variable frequency. (*a*) Find the resonance frequency f_0. Find (*b*) the maximum current I_{max} and (*c*) the phase angle δ when the generator frequency is $f = 60$ Hz.

(*a*) The resonance frequency is

$$f_0 = \frac{\omega_0}{2\pi} = \frac{1}{2\pi\sqrt{LC}} = \frac{1}{2\pi\sqrt{(2\text{ H})(2 \times 10^{-6}\text{ F})}} = 79.6\text{ Hz}$$

(*b*) When the generator frequency is 60 Hz, it is well below the resonance frequency. The capacitive and inductive reactances at 60 Hz are

$$X_C = \frac{1}{\omega C} = \frac{1}{(2\pi)(60\text{ Hz})(2 \times 10^{-6}\text{ F})} = 1326\ \Omega$$

and

$$X_L = \omega L = (2\pi)(60\text{ Hz})(2\text{ H}) = 754\ \Omega$$

The total reactance is $X_L - X_C = 754\ \Omega - 1326\ \Omega = -572\ \Omega$. This is of much greater magnitude than the resistance, a result that always holds far from resonance. The total impedance is

$$Z = \sqrt{R^2 + (X_L - X_C)^2} = \sqrt{(20\ \Omega)^2 + (-572\ \Omega)^2} \approx 572\ \Omega$$

since $(20)^2$ is negligible compared with $(572)^2$. The maximum current is then

$$I_{max} = \frac{\mathcal{E}_{max}}{Z} = \frac{100\text{ V}}{572\ \Omega} = 0.175\text{ A}$$

This is small compared with I_{max} at resonance, which is (100 V)/(20 Ω) = 5 A.

(*c*) The phase angle δ is given by

$$\tan \delta = \frac{X_L - X_C}{R} = \frac{-572\ \Omega}{20\ \Omega} = -28.6$$

$$\delta = -88°$$

From Equation 28-51 (or from Figure 28-16), we can see that a negative phase angle means that the current leads the generator voltage.

Example 28-6

Find the average power delivered by the generator in Example 28-5 at 60 Hz.

Since we are given the maximum emf and have calculated the maximum current in Example 28-5, it is convenient to write the average power in terms of these quantities. We have

$$P_{av} = \mathcal{E}_{rms} I_{rms} \cos \delta = \tfrac{1}{2} \mathcal{E}_{max} I_{max} \cos \delta$$
$$= \tfrac{1}{2}(100\ \text{V})(0.175\ \text{A})[\cos(-88°)] = 0.306\ \text{W}$$

As we have noted, this power goes into Joule heat in the resistor. We could also have calculated the average power from

$$P_{av} = I_{rms}^2 R = \tfrac{1}{2} I_{max}^2 R = \tfrac{1}{2}(0.175\ \text{A})^2 (20\ \Omega) = 0.306\ \text{W}$$

Example 28-7

Find (*a*) the *Q* value and (*b*) the resonance width of the circuit in Example 28-5.

(*a*) In Example 28-5, we found the resonance frequency to be $f_0 =$ 79.6 Hz. The *Q* value is therefore

$$Q = \frac{\omega_0 L}{R} = \frac{2\pi(79.6\ \text{Hz})(2\ \text{H})}{20\ \Omega} = 50$$

(*b*) The width of the resonance is

$$\Delta f = \frac{f_0}{Q} = \frac{79.6\ \text{Hz}}{50} = 1.6\ \text{Hz}$$

This is a very sharp resonance. The width is only 1.6 Hz at the resonance frequency of 79.6 Hz.

Example 28-8

Find the maximum voltage across the resistor, the inductor, and the capacitor at resonance for the circuit in Example 28-5.

At resonance, the impedance is just the resistance $R = 20\ \Omega$. Since the maximum emf is 100 V, the maximum current is

$$I_{max} = \frac{\mathcal{E}_{max}}{Z} = \frac{100\ \text{V}}{20\ \Omega} = 5\ \text{A}$$

The maximum voltage across the resistor is thus

$$V_{R,max} = I_{max} R = (5\ \text{A})(20\ \Omega) = 100\ \text{V}$$

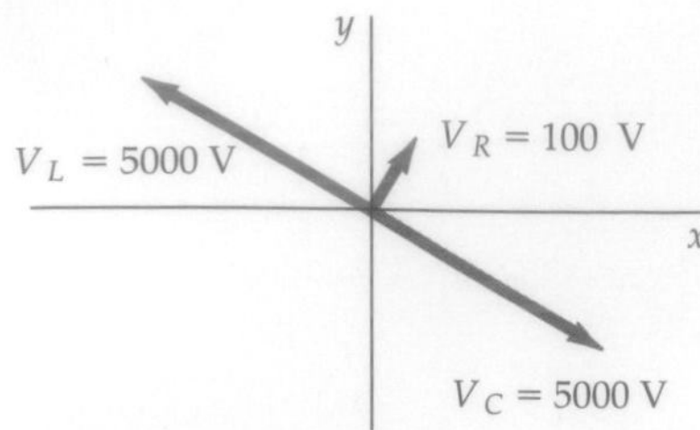

Figure 28-19 Voltages in Example 28-8. The voltages across the inductor and capacitor in a series *LCR* circuit are always 180° out of phase with each other. At resonance, they are equal in magnitude, so they sum to zero, leaving the sum of the voltages across all three elements equal to V_R. In this example, the maximum voltage drop across the resistor is 100 V, whereas the maximum drops across the inductor and the capacitor are 5000 V.

The resonance frequency found in Example 28-4 was $f_0 = 79.6$ Hz. The inductive and capacitive reactances at resonance are

$$X_L = \omega_0 L = (2\pi)(79.6\text{ Hz})(2\text{ H}) = 1000\ \Omega$$

and

$$X_C = \frac{1}{\omega_0 C} = \frac{1}{(2\pi)(79.6\text{ Hz})(2 \times 10^{-6}\text{ F})} = 1000\ \Omega$$

The inductive and capacitive reactances are equal, as we would expect since we found the resonance frequency by setting them equal. The maximum voltage across the inductor is then

$$V_{L,\text{max}} = I_{\text{max}} X_L = (5\text{ A})(1000\ \Omega) = 5000\text{ V}$$

and that across the capacitor is

$$V_{C,\text{max}} = I_{\text{max}} X_C = (5\text{ A})(1000\ \Omega) = 5000\text{ V}$$

Figure 28-19 shows the phasor diagram for these voltages. The maximum voltage across the resistor is a relatively safe 100 V, equal to the maximum emf of the generator. However, the maximum voltages across the inductor and the capacitor are a dangerously high 5000 V. These voltages are 180° out of phase. At resonance, the voltage across the inductor at any instant is the negative of that across the capacitor, so they always sum to zero, leaving the voltage across the resistor equal to the emf in the circuit.

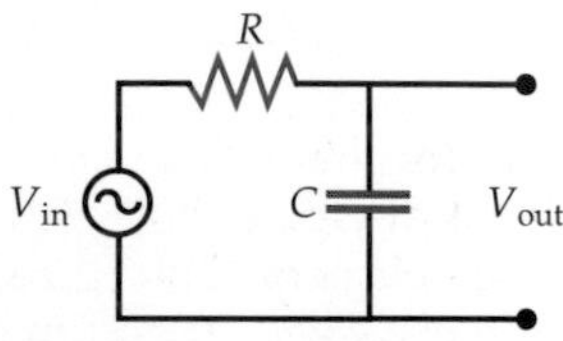

Figure 28-20 Circuit for Example 28-9.

Example 28-9

A resistor *R* and capacitor *C* are in series with a generator, which has a voltage given by $V_{\text{in}} = V_0 \cos \omega t$ as shown in Figure 28-20. Find the voltage across the capacitor as a function of frequency ω.

This circuit is simpler than the ones we have been discussing because there is no inductance. In the figure, we have labeled the generator voltage, the input voltage, and that across the capacitor—the output voltage. Figure 28-21 shows the phasors representing the voltage drops across the resistance and capacitance. The total impedance in the circuit is

$$Z = \sqrt{R^2 + X_C^2}$$

where $X_C = 1/\omega C$. The rms current is then

$$I_{\text{rms}} = \frac{V_{\text{in rms}}}{Z} = \frac{V_{\text{in rms}}}{\sqrt{R^2 + X_C^2}}$$

The rms output voltage across the capacitor is

$$V_{\text{out rms}} = I_{\text{rms}} X_C = \frac{X_C V_{\text{in rms}}}{\sqrt{R^2 + X_C^2}}$$

$$= \frac{(1/\omega C) V_{\text{in rms}}}{\sqrt{R^2 + (1/\omega C)^2}} = \frac{V_{\text{in rms}}}{\sqrt{\omega^2 C^2 R^2 + 1}}$$

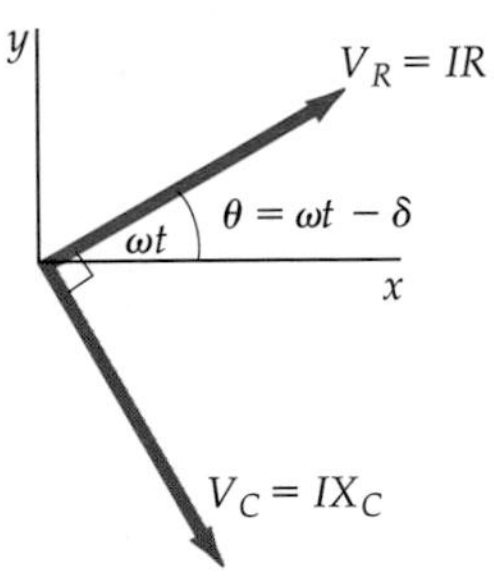

Figure 28-21 Phasor diagram for the voltages across the resistor and capacitor of Figure 28-20.

Figure 28-22 shows the ratio of the output voltage to the input voltage as a function of frequency ω. This circuit is called an *RC* **low-pass filter** because low input frequencies are transmitted with greater amplitude than high input frequencies.

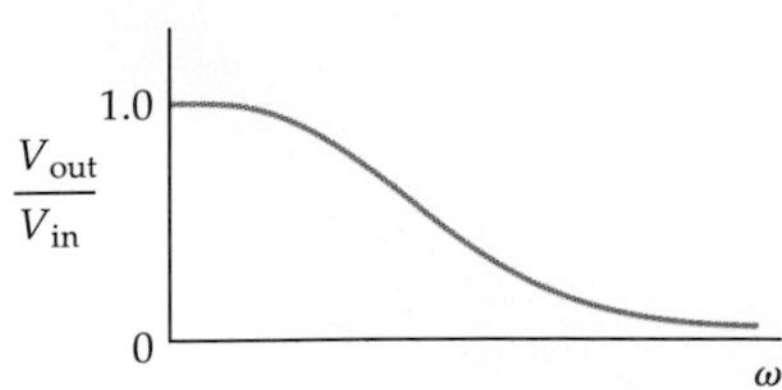

Figure 28-22 Graph of the ratio of output voltage to input voltage for the low-pass filter circuit of Example 28-9.

Parallel

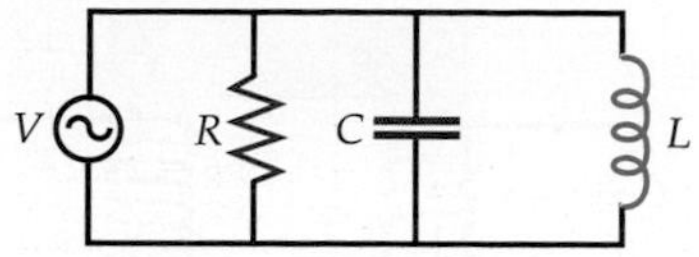

Figure 28-23 shows a resistor R, capacitor C, and an inductor L connected in parallel across an ac generator. The total current I from the generator divides into three currents, the current I_R in the resistor, the current I_C in the capacitor, and the current I_L in the inductor. The instantaneous voltage V is the same across each element. The current in the resistor is in phase with the voltage and has amplitude V/R. Since the voltage drop across an inductor *leads* the current in the inductor by 90°, the current in the inductor *lags* the voltage by 90° and has magnitude V/X_L. Similarly, the current in the capacitor leads the voltage by 90° and has magnitude V/X_C. These currents are represented by phasors in Figure 28-24. The total current I is the x component of the vector sum of the individual currents as shown in the figure. The magnitude of the total current is

Figure 28-23 A parallel *LCR* circuit.

$$I = \sqrt{I_R^2 + (I_L - I_C)^2} = \sqrt{\left(\frac{V}{R}\right)^2 + \left(\frac{V}{X_L} - \frac{V}{X_C}\right)^2} = \frac{V}{Z} \qquad \text{28-59}$$

where the impedance Z is related to the resistance and the capacitive and inductive reactances by

$$\frac{1}{Z} = \sqrt{\left(\frac{1}{R}\right)^2 + \left(\frac{1}{X_L} - \frac{1}{X_C}\right)^2} \qquad \text{28-60}$$

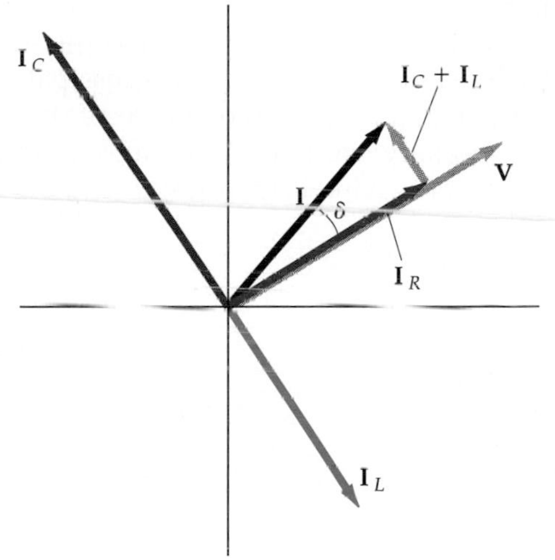

Figure 28-24 Phasor diagram for the voltage and currents in the parallel *LCR* circuit of Figure 28-23. The voltage is the same across each element. The current in the resistor is in phase with the voltage. The current in the capacitor leads the voltage by 90° and that in the inductor lags the voltage by 90°. The phase difference δ between the total current and the voltage depends on the relative magnitudes of the currents, which depend on the values of the resistance and of the capacitive and inductive reactances.

At resonance, the generator frequency ω equals the natural frequency $\omega_0 = 1/\sqrt{LC}$, and the inductive and capacitive reactances are equal. Then from Equation 28-60, we see that $1/Z$ has its minimum value $1/R$, so the impedance Z is maximum and the total current is minimum. We can understand this if we note that at resonance $X_C = X_L$, and the currents in the inductor and capacitor are equal but 180° out of phase, so the total current is just the current in the resistor.

Questions

6. Does the power factor depend on the frequency?
7. Are there any disadvantages in a radio tuning circuit having an extremely large Q factor?
8. What is the power factor for a circuit that has inductance and capacitance but no resistance?

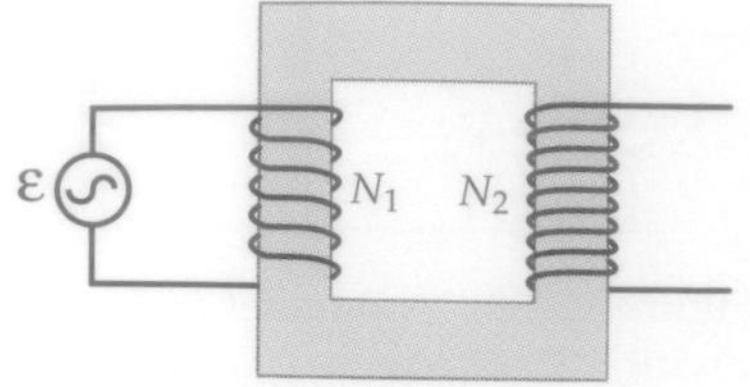

Figure 28-25 Transformer with N_1 turns in the primary and N_2 turns in the secondary.

28-6 The Transformer

A **transformer** is a device for changing alternating voltage and current without an appreciable loss in power. Its operation is based on the fact that an alternating current in one circuit will induce an alternating emf in a nearby circuit because of the mutual inductance of the two circuits. Figure 28-25 shows a simple transformer consisting of two coils of wire around a common iron core. The coil carrying the input power is called the **primary,** and the other coil is called the **secondary.** Either coil of a transformer can be used for the primary or secondary. The function of the iron core is to increase the magnetic field for a given current and to guide it so that nearly all the magnetic flux through one coil goes through the other coil. The iron core is laminated to reduce eddy-current losses. Other power losses arise because of the Joule heating in the small resistances in both coils and from hysteresis in the iron cores. We will neglect these losses and consider an ideal transformer of 100 percent efficiency, for which all of the power supplied to the primary coil appears in the secondary coil. Actual transformers are often 90 to 95 percent efficient.

Consider a transformer with an ac generator of emf $\mathcal{E}$ across the primary of N_1 turns and the circuit of the secondary coil of N_2 turns open. Because of the iron core, there is a large flux through each coil even when the **magnetizing current** I_m in the primary circuit is very small. We can neglect the resistances of the coils compared with their inductive reactances. The primary is then a simple circuit consisting of an ac generator and a pure inductance like that discussed in Section 28-2. The (magnetizing) current and the voltage in the primary are out of phase by 90°, and the average power dissipated in the primary coil is zero. If ϕ_{turn} is the magnetic flux in one turn of the primary coil, the voltage drop across the primary coil is $V_{L1} = N_1\, d\phi_{\text{turn}}/dt$. Applying Kirchhoff's loop rule to the primary circuit then gives

$$\mathcal{E} - N_1 \frac{d\phi_{\text{turn}}}{dt} = 0$$

or

$$\mathcal{E} = N_1 \frac{d\phi_{\text{turn}}}{dt} \qquad 28\text{-}61$$

If there is no flux leakage out of the iron core, the flux through each turn is the same for both coils. Thus, the total flux through the secondary coil is $N_2\, \phi_{\text{turn}}$, and the voltage across the secondary coil is

$$V_2 = -N_2 \frac{d\phi_{\text{turn}}}{dt} \qquad 28\text{-}62$$

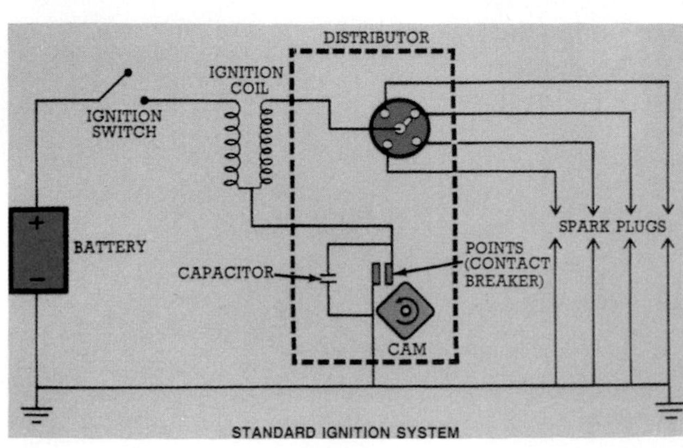

(*a*)

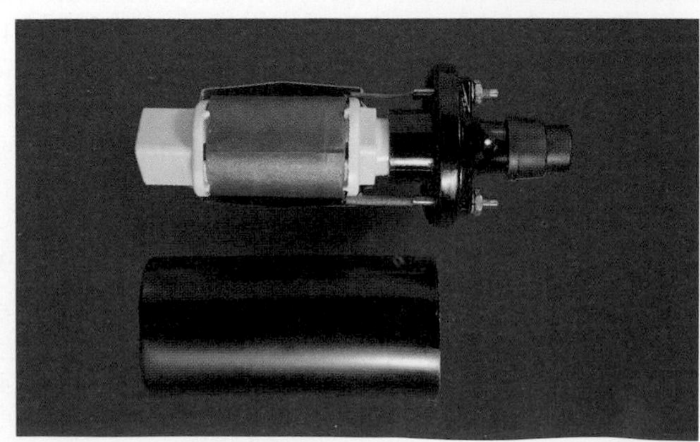

(*b*)

(*a*) Circuit for automobile ignition. The ignition shown in the photo in (*b*) is a transformer with a primary winding of heavy wire and a secondary winding of finer wire. With the ignition switch on and contact points closed, there is a current through the primary of the ignition coil. The rotating cam opens or closes the contact points, resulting in a rapid change in the primary current that produces a voltage of 30 to 40 kV in the secondary coil. This high voltage is distributed through the rotor to the spark plugs at precisely timed intervals, resulting in a spark across the gap in the plugs. The spark ignites the air-fuel mixture in the cylinders.

Comparing these equations, we can see that

$$V_2 = -\frac{N_2}{N_1}\mathcal{E} \qquad \text{28-63}$$

If N_2 is greater than N_1, the voltage in the secondary coil is greater than that in the primary coil, and the transformer is called a **step-up transformer.** If N_2 is less than N_1, the voltage in the secondary coil is less than that in the primary coil, and the transformer is called a **step-down transformer.**

Consider now what happens when we put a resistance R called a **load resistance** across the secondary coil. There will then be a current I_2 in the secondary circuit that is in phase with the voltage V_2 across the resistance. This current will set up an additional flux ϕ'_{turn} through each turn that is proportional to N_2I_2. This flux opposes the original flux set up by the original magnetizing current I_m in the primary. However, the voltage across the primary coil is determined by the generator emf, which is unaffected by the secondary circuit. According to Equation 28-61, the flux in the iron core must change at the original rate; that is, the total flux in the iron core must be the same as when there is no load across the secondary. The primary coil thus draws an additional current I_1 to maintain the original flux ϕ_{turn}. The flux through each turn produced by this additional current is proportional to N_1I_1. Since this flux equals $-\phi'_{\text{turn}}$, the additional current I_1 in the primary is related to the current I_2 in the secondary by

$$N_1I_1 = -N_2I_2 \qquad \text{28-64}$$

These currents are 180° out of phase and produce counteracting fluxes. Since I_2 is in phase with V_2, the additional current I_1 is in phase with the applied emf. The power input from the generator is $\mathcal{E}_{\text{rms}}I_{1,\text{rms}}$, and the power output is $V_{2,\text{rms}}I_{2,\text{rms}}$. (The magnetizing current does not contribute to the power input because it is 90° out of phase with the generator voltage.) If there are no losses,

$$\mathcal{E}_{\text{rms}}I_{1,\text{rms}} = V_{2,\text{rms}}I_{2,\text{rms}} \qquad \text{28-65}$$

In most cases the additional current in the primary I_1 is much greater than the original magnetizing current I_m that is drawn from the generator when there is no load. This can be demonstrated by putting a light bulb in series with the primary coil. The bulb is much brighter when there is a load across the secondary than when the secondary circuit is open. If I_m can be neglected, Equation 28-64 relates the total currents in the primary and secondary circuits.

(*a*)

(*b*)

(*a*) A power box with transformer for stepping down voltage for distribution to homes. (*b*) A suburban power substation where transformers step down voltage from high-voltage transmission lines.

Example 28-10

A doorbell requires 0.4 A at 6 V. It is connected to a transformer whose primary contains 2000 turns that is connected to 120 V ac. (*a*) How many turns should there be in the secondary? (*b*) What is the current in the primary?

(*a*) Since the input voltage is 120 V and the output is 6 V, the turns ratio can be obtained from Equation 28-63:

$$\frac{N_2}{N_1} = \frac{V_2}{\mathcal{E}} = \frac{6\text{ V}}{120\text{ V}}$$

The number of turns in the secondary is thus

$$N_2 = \frac{6}{120}(2000\text{ turns}) = 100\text{ turns}$$

(*b*) Since we are assuming 100 percent efficiency in power transmission, the input and output currents are related by

$$V_2 I_2 = \mathcal{E} I_1$$

The current in the primary is thus

$$I_1 = \frac{V_2}{\mathcal{E}} I_2 = \frac{6}{120}(0.4\text{ A}) = 0.02\text{ A}$$

An important use of transformers is in the transport of electrical power. To minimize the I^2R heat loss in transmission lines, it is economical to use a high voltage and a low current. On the other hand, safety and other considerations, such as insulation, make it necessary to use power at lower voltage and higher current to run motors and other electrical appliances. Suppose, for example, that each person in a city with a population of 50,000 uses 1.2 kW of electric power. (The per capita consumption of power in the United States is actually somewhat higher than this.) At 120 V, the current required for each person would be

$$I = \frac{1200\text{ W}}{120\text{ V}} = 10\text{ A}$$

The total current for 50,000 people would then be 500,000 A. The transport of such a current from a power-plant generator to a city many kilometers away would require wires of enormous size (actually, the wires would probably have to be large copper cylinders), and the I^2R power loss would be substantial. Rather than transmit the power at 120 V, step-up transformers are used at the power plant to step up the voltage to some very large value, such as 600,000 V. The current needed is thus reduced to

$$I = \frac{120\text{ V}}{600{,}000\text{ V}}(500{,}000\text{ A}) = 100\text{ A}$$

To reduce the voltage to a safer level for transport within a city, power stations are located just outside the city to step down the voltage to a safer value, such as 10,000 V. Transformers in boxes attached to the power poles outside each house again step down the voltage to 120 V (or 240 V) for distribution to the house. It is because of the ease of stepping the voltage up or down with transformers that alternating current rather than direct current is in common use.

Example 28-11

A transmission line has a resistance of 0.02 Ω/km. Calculate the I^2R power loss if 200 kW of power is transmitted from a power generator to a city 10 km away at (*a*) 240 V and (*b*) 4.4 kV.

(*a*) The total resistance of 10 km of wire is $R = (0.02\ \Omega/\text{km})(10\ \text{km}) = 0.20\ \Omega$. The current required to transmit 200 kW at 240 V is

$$I = \frac{200{,}000\ \text{W}}{240\ \text{V}} = 833\ \text{A}$$

The power loss is then

$$I^2R = (833\ \text{A})^2(0.20\ \Omega) = 139\ \text{kW}$$

Thus, about 70 percent of the power is wasted through heat loss.

(*b*) If the transmission voltage is 4.4 kV, the current is

$$I = \frac{200\ \text{kW}}{4.4\ \text{kV}} = 45.5\ \text{A}$$

The power loss is then

$$I^2R = (45\ \text{A})^2(0.20\ \Omega) = 414\ \text{W}$$

The energy loss is considerably less.

28-7 Rectification and Amplification

Although alternating current is readily available, direct current is often needed to power such devices as portable radios or calculators. These devices often come with batteries and with ac–dc converters to save the batteries when ac power is available. These converters contain a transformer for stepping down the voltage from 120 V to whatever voltage is needed (typically 9 V) and a circuit for converting from ac to dc. The process of converting alternating current to direct current is called **rectification.** The principal element in a rectifier circuit is a **diode.** The first diodes, developed by John Fleming in 1904, were vacuum tubes containing two main elements, a cathode that emits electrons and an anode, called the plate, that collects them. The important feature of a diode is that it conducts current in one direction and not the other. Most diodes in use today are semiconductor devices, which will be discussed in Chapter 39 of the extended version of this book. The symbol for a diode as a circuit element is →|. The arrow indicates the direction in which current can pass through the diode. (This is opposite the direction of the motion of the electrons.)

A transformer and rectifier for converting 120 V ac from a wall outlet to 9 V dc for use as a battery saver to power a radio or calculator.

Figure 28-26 shows a vacuum-tube diode. When the cathode is heated (by a heating element in a separate circuit), it emits electrons, a process called **thermionic emission** discovered by Thomas Edison in 1883. If the

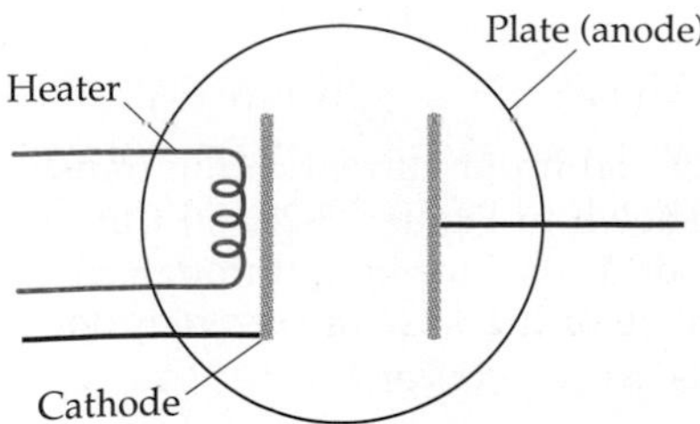

Figure 28-26 Vacuum-tube diode. When the cathode is heated, it emits electrons. The electrons are drawn to the plate when it is at a higher potential than the cathode.

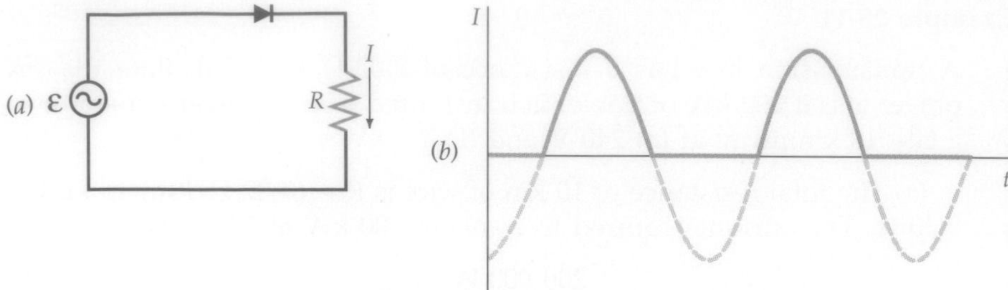

Figure 28-27 (*a*) Simple circuit containing an ac generator, a diode, and a resistor. (*b*) Current versus time in the resistor in the circuit in (*a*). The negative current indicated by the dashed lines does not get through the diode.

plate is at a higher potential than the cathode, it attracts the electrons and the tube conducts a current. The current is called the **plate current.** If the plate is at a lower potential than the cathode, the electrons are repelled and there is no current through the tube.

Figure 28-27*a* shows a simple circuit containing an ac generator, a diode, and a resistor. The current in the resistor is shown in Figure 28-27*b*. The diode is said to be a half-wave rectifier because there is current in the resistor for only half of each cycle of the ac generator. Figure 28-28 shows a circuit that gives full-wave rectification. In it, two diodes are connected to the terminals *a* and *b* of a transformer. The outputs of the diodes are connected together and to a resistor. The other side of the resistor is connected to the midpoint *c* of the transformer. When point *a* is at a higher potential than point *c*, diode 1 conducts the current I_1 to the resistor. One-half cycle later, point *b* is at a higher potential than point *c* and diode 2 conducts the current I_2 to the resistor. The current $I = I_1 + I_2$ through the resistor is shown in Figure 28-29*d*. The undesirable variations in rectifier output are called **ripple.**

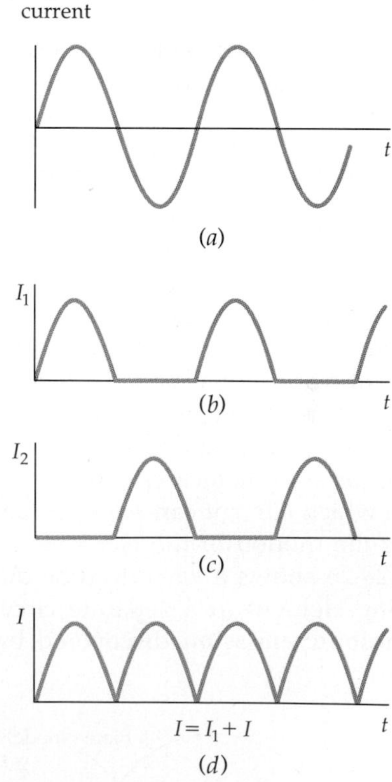

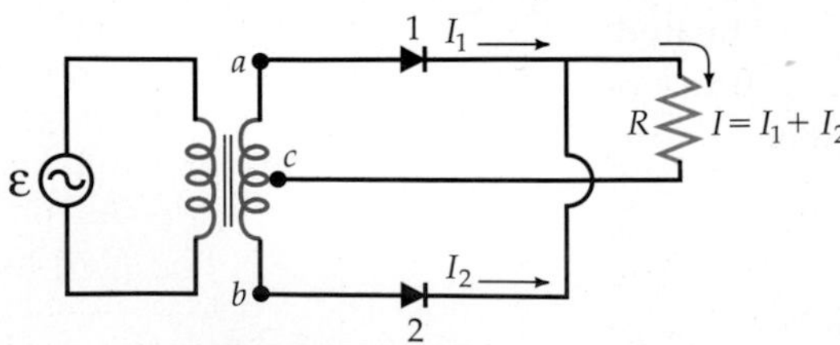

Figure 28-28 Full-wave rectifying circuit. When the potential of point *a* is greater than that of point *c*, the current I_1 passes through diode 1. One-half cycle later, the potential of point *b* is greater than that of point *c*, and the current I_2 passes through diode 2.

Figure 28-29 (*a*) Input current to the transformer in the circuit shown in Figure 28-28. (*b*) Current I_1 through diode 1. (*c*) Current I_2 through diode 2. (*d*) The total current $I = I_1 + I_2$ through the resistor in Figure 28-28.

(a) 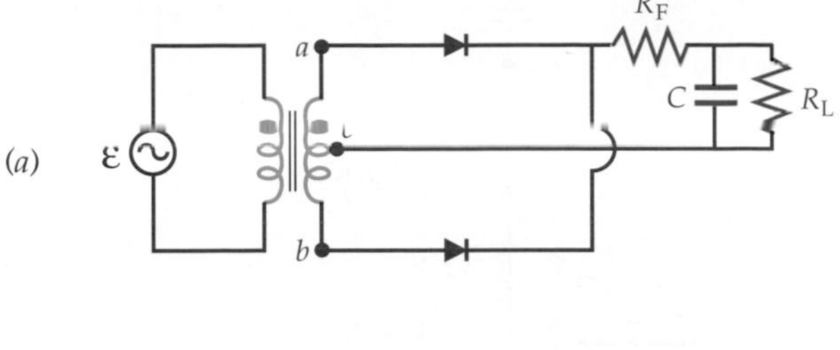

(b) 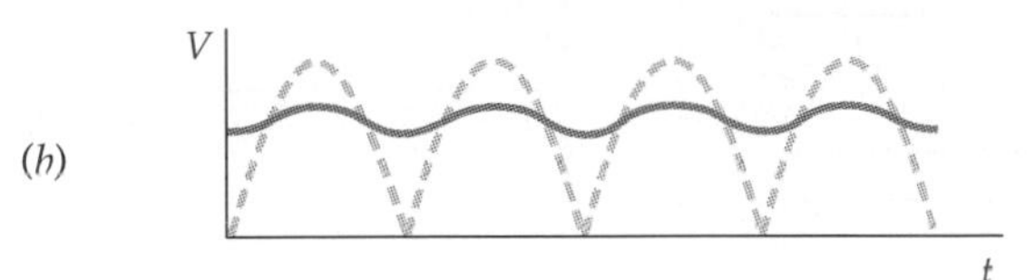

Figure 28-30 (*a*) Full-wave rectifying circuit of Figure 28-28 with a low-pass filter to smooth out some of the ripples in the rectified voltage. (*b*) Input voltage (dashed) and output voltage (solid) of the low-pass filter.

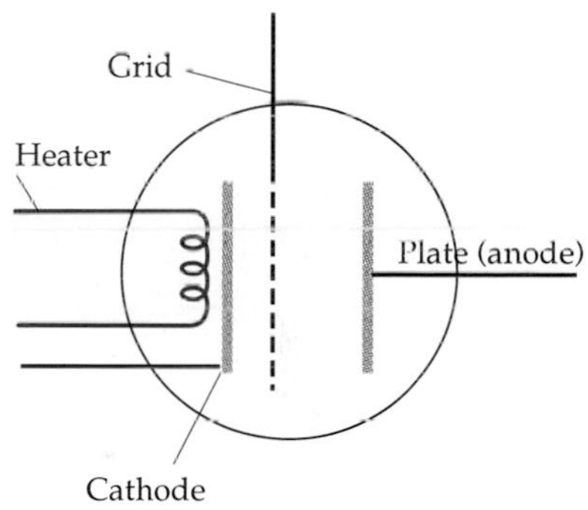

Figure 28-31 Vacuum-tube triode. The grid placed near the cathode controls the plate current. When the grid is negative relative to the cathode, it repels the electrons emitted by the cathode and diminishes the plate current. When the grid is positive relative to the cathode, it attracts the electrons emitted by the cathode and increases the plate current.

In Figure 28-30*a* a low-pass filter consisting of a resistor R_F and a capacitor C have been added between the rectifier and the load resistor R_L. (This filter was analyzed above in Example 28-9.) The resistance R_F is chosen to be much smaller than R_L, so that the dc voltage drop (and hence the power loss) across the filter resistor is small compared to that across the load resistor. The capacitance C is large so that the filter has a time constant R_FC much larger than the period between ripple cycles, so the change of the capacitor charge (and voltage) due to the ripple is very small compared to what it could become if the ripple varied slowly. The waveforms for input and output voltage for the filter are plotted in Figure 28-30*b*.

In 1907, Lee de Forest discovered that the plate current could be greatly modified by small voltage changes on a third electrode inserted between the cathode and the plate. A vacuum-tube **triode** is shown in Figure 28-31. The third electrode is a fine wire mesh called the **grid.** As with the diode, the cathode of the triode is heated and emits electrons that are collected by the plate, which is at a higher potential (typically 100 to 200 V) than the cathode. Because the grid is closer to the cathode than the plate, the potential of the grid relative to the cathode has a large effect on the plate current.

(*a*)

(*b*)

(*a*) The triode vacuum tube, invented by Lee De Forest in 1907. On either side of the cathode (which is hidden from view at the center) is the partially visible control grid, two zigzagging nickel wires. A pair of nickel plates surrounding the grid serves as the anode. The tube (not counting its ceramic base) is 9 cm tall. (*b*) A silicon chip containing six PIN diodes (dark, octagonal elements). PIN diodes function as a resistance that varies according to the voltage applied across them. They are used to switch microwave signals on and off by short-circuiting the waveguides that transmit them.

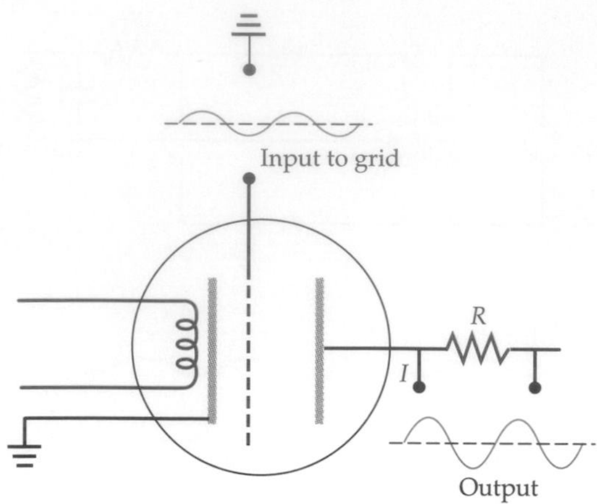

Figure 28-32 Amplification by a triode. A small sinusoidal signal applied to the grid results in a large sinusoidal signal across the resistance R.

When the grid is at the same potential as the cathode, the plate current is essentially unaffected by the grid. When the grid is negative relative to the cathode, the electrons emitted by the cathode are repelled by the grid, and the plate current is greatly diminished. When the grid is positive relative to the cathode, the plate current is increased. Figure 28-32 shows how a triode can be used as an **amplifier.** The input signal is a small sinusoidal voltage that is applied between the grid and the cathode. The output signal is the voltage across the resistance R. The output signal is considerably larger than the input signal because small voltage changes on the grid produce large changes in the plate current. Today, vacuum-tube triodes have been largely replaced by transistors, which will be discussed in Chapter 38 of the extended version of this book.

Question

9. Explain why the rms current in a half-wave rectifier circuit is half that in a full-wave rectifier circuit.

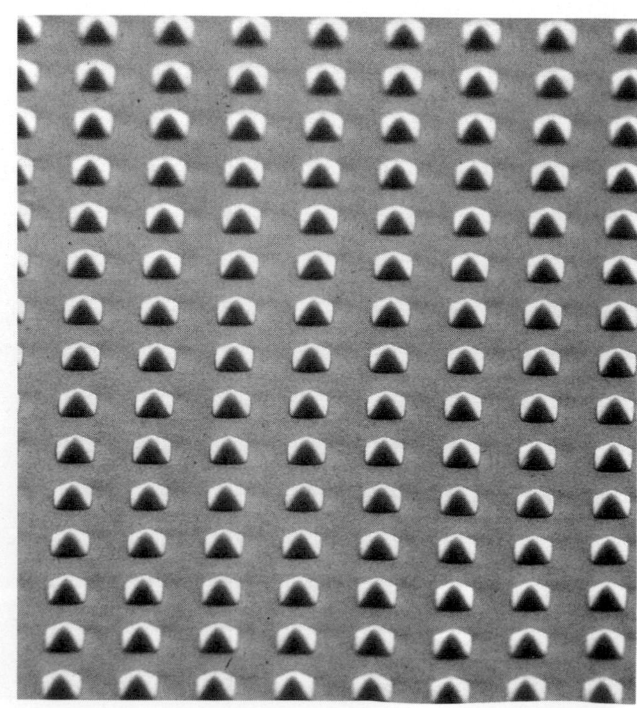

Vacuum tubes can now be constructed in miniaturized forms known as microelectric vacuum devices. Such tubes, presently in the research state, may someday replace the cathode-ray tubes in television sets, making much thinner sets possible. This array of tungsten-clad pyramids etched in a single silicon crystal functions as a replacement for a cathode. The pyramids are 3 μm high and separated by 10 μm. Instead of heat, a strong electric field is applied above the pyramids so that they emit electrons. (At a sufficiently strong field, electrons overcome the forces that bind them to the tungsten surface.) Unlike transistors, microvacuum tubes are insensitive to heat and radiation. Furthermore, at equivalent sizes, microvacuum tubes operate more rapidly than transistors because electrons travel through them without collisions.

Summary

1. The root-mean-square (rms) value of alternating current, I_{rms}, is defined as

$$I_{rms} = \sqrt{(I^2)_{av}}$$

It is related to the maximum current by

$$I_{rms} = \frac{I_{max}}{\sqrt{2}}$$

The average power dissipated in a resistor carrying a sinusoidal current is

$$P_{av} = \tfrac{1}{2}\mathcal{E}_{max}I_{max} = \mathcal{E}_{rms}I_{rms} = I^2_{rms}R$$

2. The voltage across an inductor leads the current by 90°. The rms or maximum current is related to the rms or maximum voltage by

$$I = \frac{V_L}{X_L}$$

where

$$X_L = \omega L$$

is the inductive reactance of the inductor. The average power dissipated in an inductor is zero.

The voltage across a capacitor lags the current by 90°. The rms or maximum current is related to the rms or maximum voltage by

$$I = \frac{V_C}{X_C}$$

where

$$X_C = \frac{1}{\omega C}$$

is the capacitive reactance. The average power dissipated in a capacitor is zero. Like resistance, inductive and capacitive reactances have units of ohms.

3. The phase relations among the voltages across a resistor, a capacitor, and an inductor in an ac circuit can be described graphically by representing the voltages with rotating, two-dimensional vectors called phasors. These phasors rotate in the counterclockwise direction with an angular frequency ω that is equal to the angular frequency of the current. The phasor **I** represents the current. The phasor $\mathbf{V}_R$ representing the voltage across a resistor is in phase with the current. The phasor $\mathbf{V}_L$ representing the voltage across an inductor leads the current by 90°. The phasor $\mathbf{V}_C$ representing the voltage across a capacitor lags the current by 90°. The x component of each phasor equals the magnitude of the current or the corresponding voltage drop at any instant.

4. If a capacitor is discharged through an inductor, the charge and voltage on the capacitor oscillate with angular frequency

$$\omega_0 = \frac{1}{\sqrt{LC}}$$

The current in the inductor oscillates with the same frequency, but it is out of phase with the charge by 90°. The energy oscillates between electric energy in the capacitor and magnetic energy in the inductor. If the circuit also has resistance, the oscillations are damped because energy is dissipated in the resistor.

5. The current in a series *LCR* circuit driven by an ac generator is given by

$$I = \frac{\mathcal{E}_{max}}{Z} \cos (\omega t - \delta)$$

where the impedance Z is

$$Z = \sqrt{R^2 + (X_L - X_C)^2}$$

and the phase angle δ is found from

$$\tan \delta = \frac{X_L - X_C}{R}$$

The average power input to such a circuit depends on the frequency and is given by

$$P_{av} = \mathcal{E}_{rms} I_{rms} \cos \delta$$

where $\cos \delta$ is called the power factor. The average power is maximum at the resonance frequency, which is given by

$$f_0 = \frac{1}{2\pi\sqrt{LC}}$$

At the resonance frequency, the phase angle δ is zero, the power factor is 1, the inductive and capacitive reactances are equal, and the impedance Z equals the resistance R.

6. The sharpness of the resonance is described by the Q factor, which is defined by

$$Q = \frac{\omega_0 L}{R}$$

When the resonance is reasonably narrow, the Q factor can be approximated by

$$Q = \frac{\omega_0}{\Delta\omega} = \frac{f_0}{\Delta f}$$

where Δf is the width of the resonance curve.

7. A transformer is a device for changing alternating voltage and current without an appreciable loss in power. For a transformer with N_1 turns in the primary and N_2 turns in the secondary, the voltage across the secondary coil is related to the generator emf across the primary coil by

$$V_2 = -\frac{N_2}{N_1} \mathcal{E}$$

A transformer is called a step-up transformer if N_2 is greater than N_1 so that the output voltage is greater than the input voltage. If N_2 is less than N_1, it is called a step-down transformer.

8. A diode is a device that conducts current in one direction only. It can be used to convert alternating current to direct current, a process called rectification.

9. Small changes in the voltage of the grid in a triode produce large changes in the plate current, an effect that can be used to amplify ac signals.

Electric Motors

John Dentler
United States Naval Academy

The development of a variety of electric motors has revolutionized our society. Early in the twentieth century, the machines in most large factories were driven by one or two large steam engines via belts and pulleys. Automobiles were started with hand cranks, refrigerators used large blocks of ice for cooling, and sewing machines were powered by foot treadles. Today, electric motors perform these tasks.

The wide variety of applications for electric motors requires many different designs. Electric-clock motors must operate at a precise speed. Automobile starters must deliver a tremendous torque from a standstill. A hand-held hair dryer must be lightweight and operate at several different speeds. Engineers design motors for various applications using models derived from the physical principles discussed in this text. These models are equations that predict a motor's performance for a set of specific applications or loads.

John Dentler graduated from the United States Naval Academy in 1972 with a B.S. in Physics. After completing the Navy Nuclear Power Training Program, he was assigned as Reactor Controls Division Officer on the USS Nimitz. He returned to the Nuclear Power Prototype as an instructor and department head. Following three shipboard tours in Engineering, Weapons and Operations, he attended the Naval Postgraduate School, earning a Master of Science in Electrical Engineering (concentrating in electro-optic, electronic warfare, and radar systems). Assigned as an instructor at the United States Naval Academy, he taught Fundamental Electrical Engineering for all midshipmen, and the Energy Conversion Course in the Electrical Engineering Major Curriculum. He served as Associate Chairman of the Electrical Engineering Department. He is currently assigned as a Power Systems Engineer at the Navy's David Taylor Research Center.

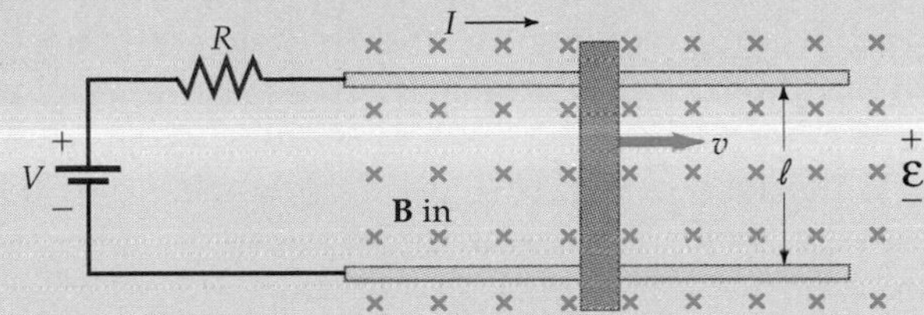

Figure 1 A simple electric motor constructed by adding a battery to the linear machine shown in Figures 26-14 and 26-16.

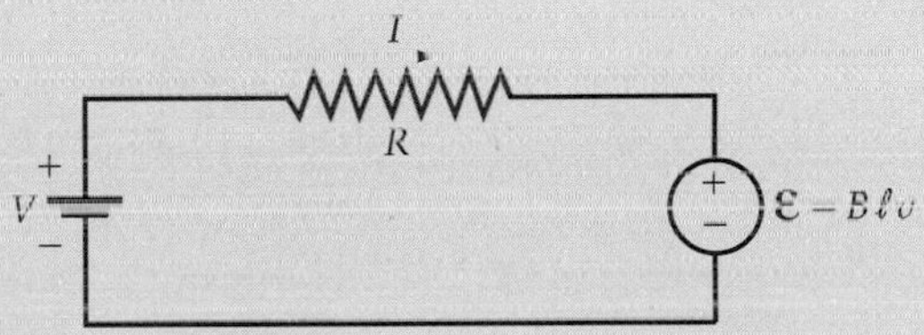

Figure 2 Circuit for the linear electric motor in Figure 1.

The simplest motor to model is derived from the linear machine introduced in Section 26-4. Figure 1 is similar to Figures 26-14 and 26-16 except a battery of voltage V is inserted with its voltage in the same sense as the emf developed across the moving bar. In general, the rails and the bar have resistance and inductance and the battery has internal resistance. For simplicity, we assume that the inductance is negligible and that the total resistance of the system can be considered to be that of a single resistor R. The bar can be modeled as an ideal source of emf with $\mathcal{E} = B\ell v$. The circuit that models this simple linear motor is shown in Figure 2.

Applying Kirchhoff's loop rule to this circuit yields

$$V - IR - \mathcal{E} = 0 \tag{1}$$

Substituting $\mathcal{E} = B\ell v$, we obtain

$$V - IR - B\ell v = 0 \tag{2}$$

The current is thus given by

$$I = -\frac{B\ell}{R}v + \frac{V}{R} \tag{3}$$

The current is thus a linear function of velocity provided that V, B, ℓ, and R are constant. For low velocities, the emf is small and the current is positive (down the bar). For high velocities, the emf is larger

Continued

than the battery voltage V and the current is negative (up the bar). If the velocity is $V/B\ell$, the current is zero.

If the rails in Figure 1 were frictionless and the magnetic field were sufficiently broad, the bar would accelerate to a terminal speed $V/B\ell$, at which time the force, $F = IB\ell$, is zero because the current is zero. If the bar is accelerated to the right by an external force, the current charges the battery as the generator charges the battery in an automobile. If the bar is slowed by an external force directed to the left, the current is driven through the bar by the battery. Sufficient current will be delivered to match the opposing force and the bar will travel at an equilibrium speed that is less than $V/B\ell$.

A motor designer is interested in predicting how a motor will respond (by changing speed) to a load. For this linear motor, the load is an external force on the bar. The equal and opposite force delivered by the motor at the equilibrium speed is $F = IB\ell$. Equation 3 can be rearranged to show the velocity as a linear function of the current:

$$v = -\frac{R}{B\ell}I + \frac{V}{B\ell} \tag{4}$$

Substituting $F/B\ell$ for the current, we obtain

$$v = -\frac{R}{(B\ell)^2}F + \frac{V}{B\ell} \tag{5}$$

Equation 5 relating the velocity to the load is called the *motor performance characteristic*. Figure 3 shows a graph of the velocity versus force. Line 1 represents v versus F for typical values of the battery voltage V and the magnetic field B. Line 2 shows the effect of raising the voltage V. Line 3 shows the effect of lowering the magnetic field B. Line 4 represents a typical (for instance, frictional) load that increases proportionately with speed. The motor operates at the intersection of the load line and the motor performance characteristic. Thus, the operating speed of the motor shown in Figure 1 can be controlled by changing either the voltage or the magnetic field.

The linear motor in Figure 1 is not practical for most applications. Instead, a rotating motor is more appropriate. Figure 4 shows most of the parts of a simple rotating electric motor. Although this motor appears to be quite different from the linear one, the operation of the two motors is similar.

Like the linear motor, the rotating motor has current-carrying conductors that react with an external field. The field, called the *stator field*, is created

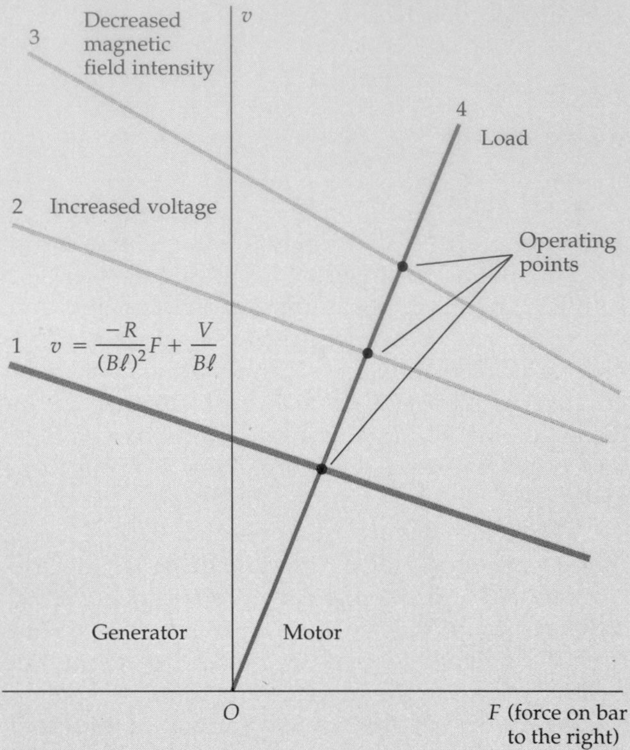

Figure 3 Graph of different values of the motor performance characteristic of the linear electric motor in Figure 1 and a typical load. The load line indicated is typical for a frictional load.

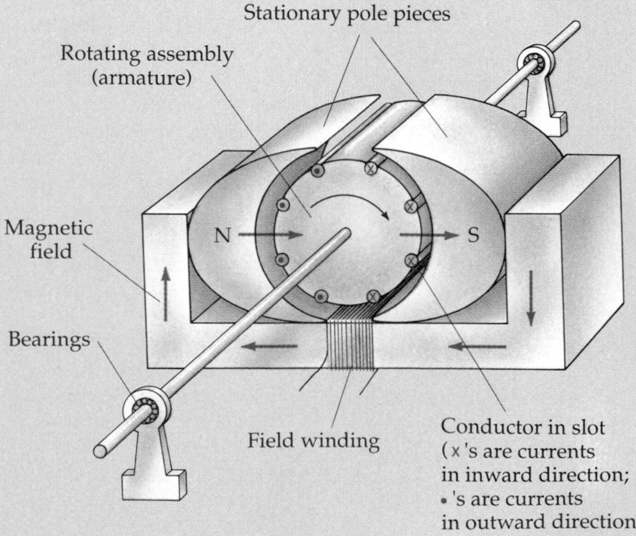

Figure 4 A simple rotating electric motor.

and controlled by the coil of wire at the bottom of Figure 4. Flux from the wire coil passes through the core, creating a north pole on the left and a south pole on the right of the rotating element. The rotating assembly is supported by two bearings, one at the front and one at the back of the motor. The rotating assembly, called the *armature,* consists of an iron cylinder with eight slots. The slots contain conductors, which are similar to the bar in the linear motor. If current can be driven through these conductors in the direction shown (front to back near the south pole and back to front near the north pole), then a net clockwise torque (down on the south-pole side and up on the north-pole side) will be developed to turn the armature. The torque developed in the rotating motor is analogous to the force developed in the linear motor.

Constructing a device that maintains the proper current direction in each conductor as the armature turns is a complicated task. Such devices are called *commutator brush assemblies.* Figure 5 shows the connections of such an assembly for the motor in Figure 4. Figure 6 is a photo of an actual commutator brush assembly used in a car starter motor. The photo shows an armature with many slots and a commutator with many segments. Figure 7 is a photo of a slot-car motor with three discrete slots, each containing many windings and a three-segment commutator (of which only two are visible).

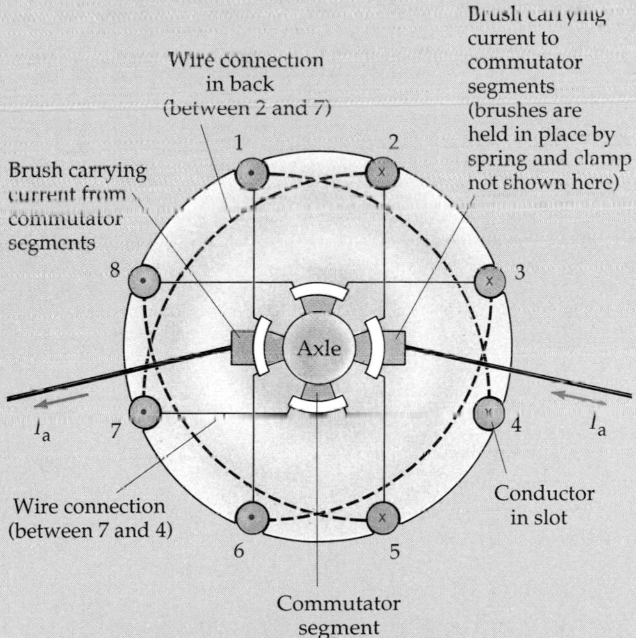

Figure 5 A commutator brush assembly for the motor in Figure 4. A commutator, in its most general sense, is a switching device. The device shown switches current direction through the armature to maintain the clockwise rotation.

The commutator brush assembly shown in Figure 5 consists of four segments that protrude along the motor shaft and two brushes that conduct current from a source to the segments. Each segment is connected to two conductors, which run through the slots of the rotating assembly. The conductors are interconnected through wires in the rear of the rotating assembly and by the commutator segments in the front of the assembly. This method of connection results in two parallel paths between the brushes; thus all of the conductors are used all of the time.

In the commutator shown in Figure 5, current is delivered from the brush on the right. It then follows one of the two parallel paths through the armature. The conductors in slots 2 and 5 both carry the current from the front to the back of the armature. Conductors 2 and 5 are connected to conductors 7 and 8 via wires in the rear of the armature. The current returns to the front through slots 7 and 8, which are connected to slots 3 and 4 via the common connection on the commutator segments. The current is carried to the back along slots 3 and 4 and is then returned to the front through slots 1 and 6, where it is picked off by the brush on the left. The commutator assembly rotates with the armature and moves under the brushes. The brushes are stationary and will contact different commutator segments when the armature has moved 90 degrees. Since the armature is symmetrically wound, the slots on the right will always carry current from front to back and the slots on the left will always carry current from back to front, thus maintaining the clockwise torque.

The total torque turning the motor is the sum of the torques exerted by the conductors in each slot. In any position there are four armature conductors acting on the right and four on the left; therefore, the torque is approximately constant. Similarly, the total emf developed between the brushes is the sum of the emfs on each conductor. At any position there are two parallel paths, each consisting of four conductors. From Section 26-6, the emf developed across a single length of wire can be shown to be

$$\mathcal{E} = B\ell r\omega \sin (\omega t + \delta) \tag{6}$$

where ℓ is the length of the armature (front to back) and r is the radius of the rotor. The total emf across the armature will be the average of that developed

Continued

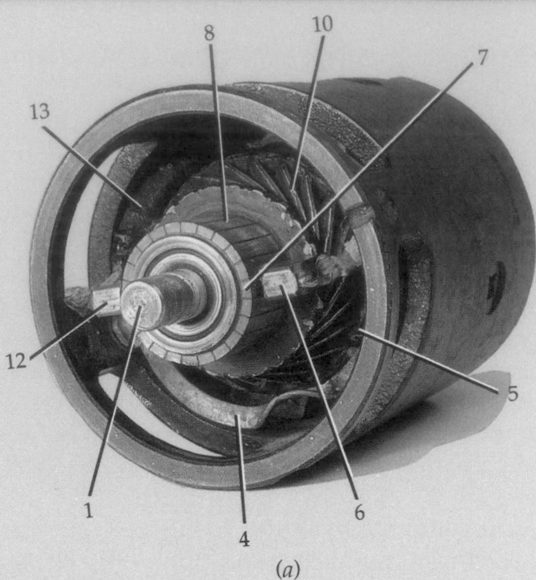

(a)

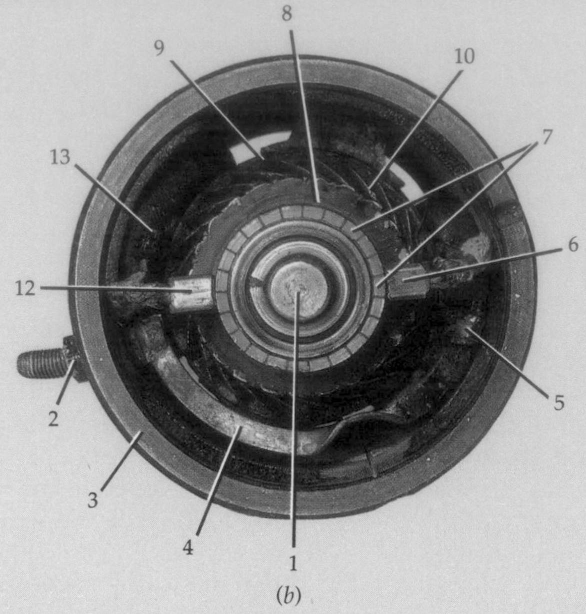

(b)

Figure 6 An automobile starter motor shown from several views. The shaft **(1)** is supported by bearings normally contained in the end bells of the motor housing. (The end bells have been removed.) Current from the battery enters the motor from a wire attached to the bolt **(2)** on the side of the motor. The bolt is insulated from the motor housing **(3)** and is connected through the housing to the field via the metal strip **(4)** running under the shaft. (This extra length is for thermal expansion.) The strip is coiled on the right side to form the winding for the right side of the stator pole **(5).** The strip conductor is then connected to the wire and the soft metal brush **(6)** on the right side. The brush is normally held tight against the commutator **(7)** by a spring clip. (The spring clip has been removed.) Current is conducted through one of the 23 commutator segments **(8)** to the armature. The armature (Figure 6*d*) has 23 slots **(9)** each containing a pair of conductors for a total of 46 conductors **(10)** (one of each pair is close to the axle and hidden, and one is close to the outer circumference of the armature and can be seen). In this motor, the armature is the rotating assembly. These conductors are interconnected in the rear of the motor **(11)** forming two parallel paths, each 23 conductors long, between the brushes. Current is conducted from a commutator segment on the left to the brush on the left **(12),** and to the metal strip that is coiled to form the left side of the stator pole **(13).** Current is conducted to the automobile's "common" or ground via a hard connection between the left stator pole winding and the motor housing. This connection is deep inside the motor and cannot be seen. The magnetic flux path for the stator field runs from the right side, through the armature, to the left side and is completed through the motor housing.

(c)

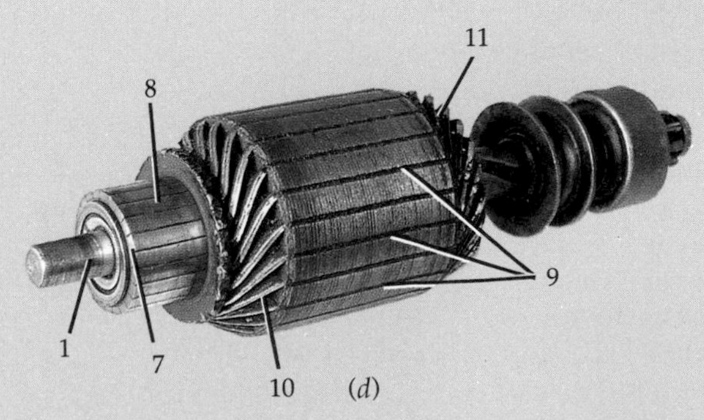

(d)

Figure 7 A common slot car motor with the top half of the housing cut away. The shaft **(1)** is supported by bearings **(2)** on the right and the left. The armature is made of 12 thin plates **(3)** of magnetically permeable material, laminated together, forming three slots and three salient poles. The slots in this armature are very large and contain many turns of wire. The set of turns around one pole is called a winding. The windings are soldered together at their ends **(4)** then connected to the commutator segments **(5)**. The armature is surrounded by black permanent magnets **(6)**, creating a stationary field that reacts with the currents and fields of the armature. A battery connected to the external connections **(7)** drives current through the top hairpin brush **(8)**, to the commutator segment on the top, through the interconnected windings, and back to the battery through the commutator segment on the bottom **(9)**. Notice how the commutator segments are separated **(10)**. The operation of this motor can be thought of in two different ways. Either the current through the windings acts with the stationary field to cause a torque, or current in the windings around the salient armature pieces form poles that alternately repel and attract the stationary pole pieces. Very careful analysis will show that these two approaches are really exactly the same.

across the two parallel paths described above. The slots are separated by only 45 degrees, so the variation in the total emf across the armature as the armature turns will be relatively small. Therefore, the time-varying term of Equation 6, sin $(\omega t + \delta)$, can be discarded and the total emf can be expressed as

$$\mathcal{E}_{\text{total}} = BK\omega \tag{7}$$

where the constant K, the motor constant, includes ℓ and r and the results of the summing and averaging of the total emf across the armature. The validity of Equation 7 improves as more slots and commutator segments are added to the armature.

Equation 7 is similar to $\mathcal{E} = B\ell v$ with K replacing ℓ and ω replacing v. The power delivered to the armature is the product of the emf and the armature current I_a. For a rotating motor, the load is a torque τ applied to the shaft opposing the direction of rotation. The mechanical power delivered to the load is the product of the torque and the angular velocity. At equilibrium, the driving torque from the motor is equal and opposite to the load torque. Thus

$$P = \mathcal{E}I_a = \tau\omega \tag{8}$$

Substituting $BK\omega$ for the emf from Equation 7, we obtain

$$P = BKI_a\omega = \tau\omega \tag{9}$$

Applying the same logic used to develop the model for the linear motor, we can represent the armature by a simple voltage source with an external resistance R_a. The field winding connections of the coil shown at the bottom of Figure 4 can be connected either in series or in parallel (shunt) with the armature. These two methods of connection yield motors with extremely different characteristics.

Parallel or Shunt Connection

Figure 8 shows the circuit for the shunt or parallel field connection. A variable resistance called a rheostat is included to control the field and thereby control the speed of the motor. Applying Kirchhoff's loop rule to this circuit yields

$$V - I_aR_a - BK\omega = 0 \tag{10}$$

which can be rearranged to express the rotational speed ω in terms of the armature current I_a:

$$\omega = -\frac{R_a}{BK}I_a + \frac{V}{BK} \tag{11}$$

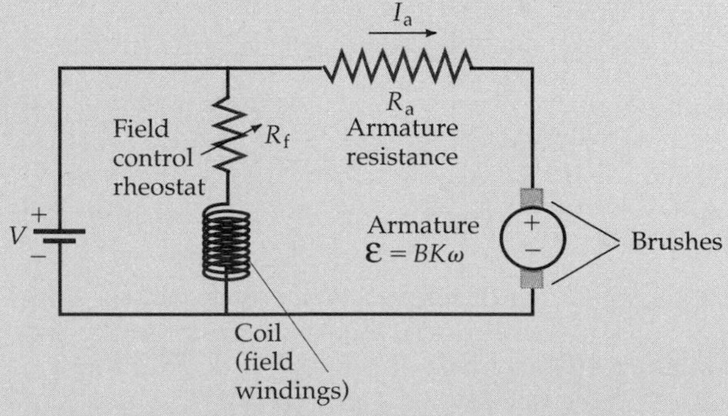

Figure 8 Circuit for a typical dc shunt motor.

Continued

If we substitute τ/BK for the current from Equation 9, the rotational speed is

$$\omega = -\frac{R_a}{(BK)^2}\tau + \frac{V}{BK} \qquad (12)$$

Equation 12 is a linear equation relating the rotational speed to the load. It is analogous to the performance characteristic of the linear motor. The speed can be controlled either by varying the voltage V or, more commonly, by varying the current into the coil with the rheostat.

At high armature currents, the armature core saturates, the voltage drop due to the armature inductance becomes significant, and the relationship between the torque and speed becomes nonlinear. However, for normal loads, Equation 12 accurately describes the motor's operation. Figure 9 shows how the performance characteristic is affected by armature saturation.

Series Connection

In the motor circuit of Figure 10, the coil is connected in series with the armature, so the field strength is a function of the armature current. If the armature current is small and the field does not saturate, the product of the field strength and the motor constant K can be expressed as a linear function of the armature current:

$$BK = CI$$

where C is some constant. Substituting this value of BK in the expressions for the armature emf, power, and torque yields

$$\mathcal{E} = CI\omega \qquad P = CI^2\omega \qquad \text{and} \qquad \tau = CI^2$$

The Kirchhoff loop rule then gives

$$V - IR - CI\omega = 0$$

where R represents the total resistance of the coil and armature, and I represents the only current in the circuit. This gives the following speed-versus-current equation:

$$\omega = \frac{V}{CI} - \frac{R}{C}$$

Substituting $\sqrt{\tau/C}$ for I, we obtain the equation for the speed versus torque, which is the performance characteristic for the series motor:

$$\omega = \frac{V}{(C\tau)^{1/2}} - \frac{R}{C}$$

Figure 11 shows the performance characteristic for the series motor. Comparing this performance characteristic with that for the shunt motor reveals striking differences. At low torques, the series motor runs very fast, almost without limit (the only load is the friction of its bearings and the air around the rotor) whereas the shunt motor is regulated to run close to the speed V/BK. At high torques, the speed of the shunt motor tapers off and the motor stalls, but the series motor delivers its greatest torque when the motor is stopped. A series motor is therefore the best choice to start a car engine, which re-

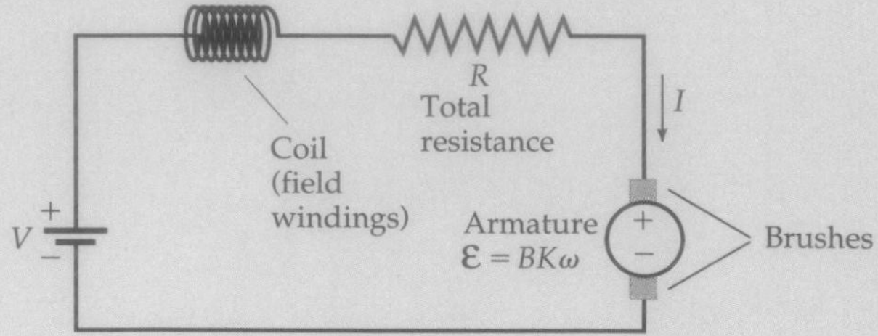

Figure 10 Circuit for a typical dc series motor.

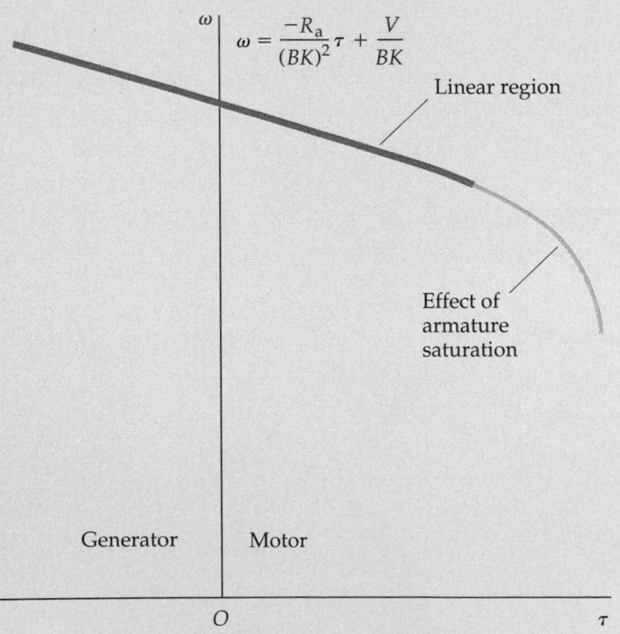

Figure 9 Graph of torque versus rotational speed showing the effect of armature saturation on the performance characteristic of the typical dc shunt motor.

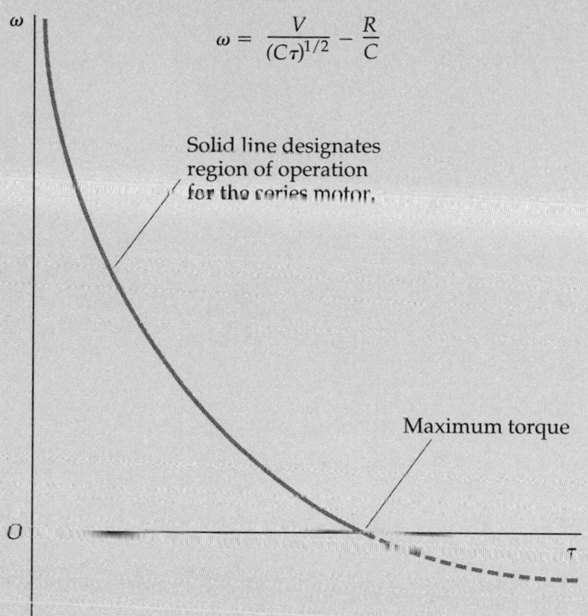

Figure 11 Graph of torque versus rotational speed showing the performance characteristic of the typical dc series motor.

quires a high torque at $\omega = 0$. On the other hand, a shunt motor is the better choice to drive a speed-sensitive load, like a tape recorder.

The linear, shunt, and series motors discussed so far all run on direct current whereas power companies deliver alternating current. With only minor modifications, the principles of dc motor construction and operation also apply to ac motors.

The torque of a series motor is proportional to I^2 and is thus independent of current direction. This is because the same current is in both the stationary field and rotating armature. With this cursory examination, it might be concluded that any series dc motor would run on alternating current. However, an assumption made to simplify the analysis of the dc motor was that inductance could be ignored. Inductance cannot be ignored when driving a motor from an ac source. Inductance has two effects: (1) it acts as a throttle limiting the amount of ac current for a given input voltage and (2) it changes the phase relationship of the current and voltage.

A dc shunt motor typically has field windings with high resistance and an armature with high inductance. Applying alternating current to such a motor would create a phase difference between the field and the armature currents resulting in unsatisfactory performance.

A series motor, like the car starter in Figure 6, has a very tight magnetic circuit with close tolerances to develop a very high torque in a small package. Such a device has a high inductance, thus limiting the ac current drawn by the motor. A series motor designed to run on alternating current must have a relatively low inductance. The low inductance is achieved by limiting the amount of iron used in the pole pieces and the armature. Such a motor is called a *universal motor*. By its nature, it is both lightweight and limited to driving devices with relatively light loads such as vacuum cleaners, food blenders, hair dryers, and sewing machines. Its performance characteristic is similar to that for the dc series motor shown in Figure 11. From this figure, can you explain why a vacuum cleaner motor speeds up when the suction is blocked? (Hint: The load on the vacuum cleaner motor is the air moving through it. If air flow decreases, the load decreases.)

The most common ac motor is the *induction motor*. This motor has a rotating assembly like the one shown in Figure 4, but unlike the dc motor, the commutator and interconnecting wires are replaced with shorting plates, connecting all the slotted conductors, mounted on the front and back. Such a design greatly simplifies construction. The challenge becomes how to make the shorted rotor rotate. The solution is to make the field from the stator appear to rotate. If the field rotates, there will be a relative velocity between the rotor and the stator field. An emf develops across the shorted rotor, driving current through the conductors in the slots. The rotating stator field produces a torque on the induced current in the rotor. Recall that the linear motor tended to move just fast enough to develop an emf that matched the source voltage. The induction motor responds in the same way, but the source voltage to the rotor is zero. To keep the emf close to zero, the rotor moves such as to minimize the relative motion between it and the field. Thus, the rotor turns almost as fast as the rotating stator field.

There are many schemes for creating apparent field rotation. The one shown in Figure 12 is known as the *shaded pole*. The motor is identical to the one in Figure 4 with the exception that the rotor is shorted on either end and the stator pole pieces have been sliced, with a conducting band around the small pieces of each pole. This construction allows the magnetic field to be quickly established through the faces of the large pole pieces and to be delayed through the small faces by the inductance of the conducting band. The phase delay between the field

Continued

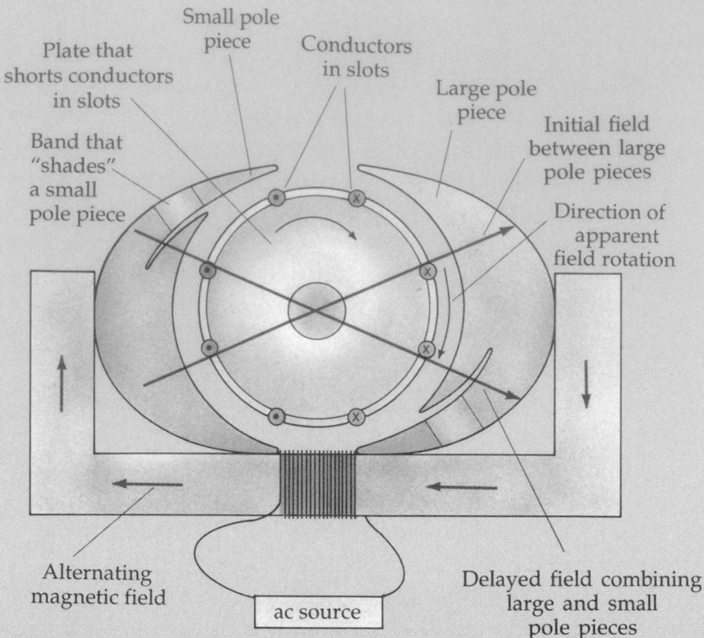

Figure 12 A shaded-pole induction motor.

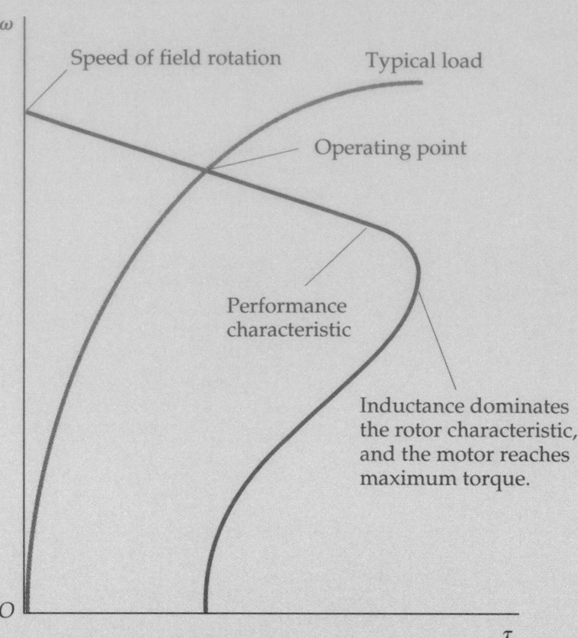

Figure 13 Graph of torque versus speed showing the performance characteristic of a typical induction motor. The load line indicated is typical for a centrifugal pump.

through the large pole faces and the field through the small pole faces creates the appearance of a rotating field.

The performance characteristic of a typical induction motor is shown in Figure 13. Normal operation is close to the speed of field rotation. If the motor in Figure 12 were connected to a 60-Hz source, the rotational speed would be somewhat less than 60 rev/s.

A rotor with conductors surrounding a metal, magnetically permeable core has characteristics of inductance and resistance. The effect of inductance within the rotor is proportional to the frequency of the rotor currents, which in turn is proportional to the difference between the field rotation and the rotor rotation. The maximum torque shown on the performance characteristic curve occurs where the difference between the rotor speed and the field rotation speed is large enough for the effects of rotor inductance to significantly delay rotor currents. The delayed rotor currents cannot interact with the stator field, and so the motor stalls if the load is increased.

Shaded-pole motors are used in devices with light loads such as cooling fans in electrical equipment. More complex schemes for creating field rotation are used in the induction motors for refrigerators and air conditioners. Large industrial induction motors use three-phase electricity to rotate the field. No matter what the size of the motor or the method used to create field rotation, the principles and fundamental operating characteristics for all induction motors are the same.

Whether run off direct or alternating current, all electric motors are based upon the fundamental principles discussed in Chapters 26 through 28 of this text. There is an extraordinary opportunity for engineers to creatively combine these fundamental principles with advances in related fields, such as superconductivity, and design motors meeting the needs of the twenty-first century. Development and improvement of electric cars, trains, and satellites offer just a few of the challenges facing the motor designer.

There are many excellent electrical engineering textbooks devoted to the subject of motors or machines. Two of my favorites are *Electric Machines and Power Systems* by Vincent Deltoro (Prentice Hall, Englewood Cliffs, New Jersey, 1985) and *Electromechanical Motion Devices* by Paul Krause and Oleg Wasynczuk. (McGraw-Hill, New York, 1989).

Suggestion for Further Reading

Coltman, John W.: "The Transformer," *Scientific American*, January 1988, p. 86.

This article describes the nineteenth-century struggle between proponents of alternating current and those of direct current and how the transformer helped to settle the matter. More recent advances in transformer technology are also discussed.

Review

A. Objectives: After studying this chapter, you should:

1. Be able to define rms current and relate it to the maximum current in an ac circuit.

2. Be able to define capacitive reactance, inductive reactance, and impedance.

3. Be able to give the phase relations between the current and the voltage across a resistor, inductor, or capacitor.

4. Be able to draw a phasor diagram for a series *LCR* circuit and from it relate the phase angle δ to the capacitive reactance, inductive reactance, and resistance.

5. Be able to define Q factor and discuss its significance.

6. Be able to state the resonance condition for a series *LCR* circuit with a generator and sketch the power versus frequency for both a high-Q and a low-Q circuit.

7. Be able to describe a step-up and a step-down transformer.

B. Define, explain, or otherwise identify:

Root-mean-square (rms) values
Inductive reactance
Capacitive reactance
Phasors
Total reactance
Impedance
Resonance frequency
Resonance
Power factor
Resonance curves
Resonance width
Q factor
Low-pass filter
Transformer
Primary
Secondary
Magnetizing current
Step-up transformer
Step-down transformer
Load resistance
Rectification
Diode
Thermionic emission
Plate current
Ripple
Triode
Grid
Amplifier

C. True or false: If the statement is true, explain why it is true. If it is false, give a counterexample.

1. Alternating current in a resistance dissipates no power because the current is negative as often as it is positive.

2. At very high frequencies, a capacitor acts like a short circuit.

3. An *LCR* circuit with a high Q factor has a narrow resonance curve.

4. At resonance, the impedance of an *LCR* circuit equals the resistance R.

5. At resonance, the current and generator voltage are in phase.

6. If a transformer increases the current, it must decrease the voltage.

Problems

Level I

28-1 Alternating Current in a Resistor

1. A 100-W light bulb is plugged into a standard 120-V (rms) outlet. Find (*a*) I_{rms}, (*b*) I_{max}, and (*c*) the maximum power.

2. A 3-Ω resistor is placed across a generator having a frequency of 60 Hz and a maximum emf of 12.0 V. (*a*) What is the angular frequency ω of the current? (*b*) Find I_{max} and I_{rms}. What is (*c*) the maximum power into the resistor, (*d*) the minimum power, and (*e*) the average power?

3. A 5.0-kW electric clothes dryer runs on 240 V rms. Find (*a*) I_{rms} and (*b*) I_{max}. (*c*) Find the same quantities for a dryer of the same power that operates at 120 V rms.

4. A circuit breaker is rated for a current of 15 A rms at a voltage of 120 V rms. (*a*) What is the largest value of I_{max} that the breaker can carry? (*b*) What average power can be supplied by this circuit?

28-2 Alternating Current in Inductors and Capacitors

5. What is the reactance of a 1.0-mH inductor at (*a*) 60 Hz, (*b*) 600 Hz, and (*c*) 6 kHz?

6. An inductor has a reactance of 100 Ω at 80 Hz. (*a*) What is its inductance? (*b*) What is its reactance at 160 Hz?

7. At what frequency would the reactance of a 10.0-μF capacitor equal that of a 1.0-mH inductor?

8. Sketch a graph of X_L versus f for $L = 3$ mH.

9. What is the reactance of a 1.0-nF capacitor at (*a*) 60 Hz, (*b*) 6 kHz, and (*c*) 6 MHz?

10. Find the reactance of a 10.0-μF capacitor at (*a*) 60 Hz, (*b*) 6 kHz, and (*c*) 6 MHz.

11. Sketch a graph of X_C versus f for $C = 100$ μF.

12. An emf of 10.0 V maximum and frequency 20 Hz is applied to a 20-μF capacitor. Find (*a*) I_{max} and (*b*) I_{rms}.

13. At what frequency is the reactance of a 10-μF capacitor (*a*) 1 Ω, (*b*) 100 Ω, and (*c*) 0.01 Ω?

28-3 Phasors

14. Draw the resultant phasor diagram for a series *LCR* circuit when $V_L < V_C$. Show on your diagram that the emf will lag the current by the phase angle δ given by

$$\tan \delta = \frac{V_C - V_L}{V_R}$$

28-4 *LC* and *LCR* Circuits without a Generator

15. Show from the definitions of the henry and the farad that $1/\sqrt{LC}$ has the unit s^{-1}.

16. What is the period of oscillation of an *LC* circuit consisting of a 2-mH coil and a 20-μF capacitor?

17. What inductance is needed with an 80-μF capacitor to construct an *LC* circuit that oscillates with a frequency of 60 Hz?

18. An *LC* circuit has capacitance C_1 and inductance L_1. A second circuit has $C_2 = \frac{1}{2}C_1$ and $L_2 = 2L_1$, and a third circuit has $C_3 = 2C_1$ and $L_3 = \frac{1}{2}L_1$. (*a*) Show that each circuit oscillates with the same frequency. (*b*) In which circuit would the maximum current be the greatest if the capacitor in each were charged to the same potential *V*?

19. A 5-μF capacitor is charged to 30 V and is then connected across a 10-mH inductor. (*a*) How much energy is stored in the system? (*b*) What is the frequency of oscillation of the circuit? (*c*) What is the maximum current in the circuit?

28-5 *LCR* Circuits with a Generator

20. A series *LCR* circuit with $L = 10$ mH, $C = 2$ μF, and $R = 5$ Ω is driven by a generator with a maximum emf of 100 V and a variable angular frequency ω. Find (*a*) the resonant frequency ω_0 and (*b*) I_{rms} at resonance. When $\omega = 8000$ rad/s, find (*c*) X_C and X_L, (*d*) Z and I_{rms}, and (*e*) the phase angle δ.

21. For the circuit of Problem 20, let the generator frequency be $f = \omega/2\pi = 1$ kHz. Find (*a*) the resonance frequency $f_0 = \omega_0/2\pi$, (*b*) X_C and X_L, (*c*) the total impedance Z and I_{rms}, and (*d*) the phase angle δ.

22. A series *LCR* circuit in a radio receiver is tuned by a variable capacitor so that it can resonate at frequencies from 500 to 1600 kHz. If $L = 1.0$ μH, find the range of capacitances necessary to cover this range of frequencies.

23. FM radio stations have carrier frequencies that are separated by 0.20 MHz. When the radio is tuned to a station, such as 100.1 MHz, the resonance width of the receiver circuit should be much smaller than 0.2 MHz so that adjacent stations are not received. If $f_0 = 100.1$ MHz and $\Delta f = 0.05$ MHz, what is the *Q* factor for the circuit?

24. (*a*) Find the power factor for the circuit in Example 28-5 when $\omega = 400$ rad/s. (*b*) At what angular frequency is the power factor 0.5?

25. Find (*a*) the *Q* factor and (*b*) the resonance width for the circuit in Problem 20. (*c*) What is the power factor when $\omega = 8000$ rad/s?

26. An ac generator with a maximum emf of 20 V is connected in series with a 20-μF capacitor and an 80-Ω resistor. There is no inductance in the circuit. Find (*a*) the power factor, (*b*) the rms current, and (*c*) the average power if the angular frequency of the generator is 400 rad/s.

27. A coil can be considered to be a resistance and an inductance in series. Assume that $R = 100$ Ω and $L = 0.4$ H. The coil is connected across a 120-V-rms, 60-Hz line. Find (*a*) the power factor, (*b*) the rms current, and (*c*) the average power supplied.

28. Find the power factor and the phase angle δ for the circuit in Problem 20 when the generator frequency is (*a*) 900 Hz, (*b*) 1.1 kHz, and (*c*) 1.3 kHz.

28-6 The Transformer

29. A transformer has 400 turns in the primary and 8 turns in the secondary. (*a*) Is this a step-up or step-down transformer? (*b*) If the primary is connected across 120 V rms, what is the open-circuit voltage across the secondary? (*c*) If the primary current is 0.1 A, what is the secondary current, assuming negligible magnetization current and no power loss?

30. The primary of a step-down transformer has 250 turns and is connected to a 120-V-rms line. The secondary is to supply 20 A at 9 V. Find (*a*) the current in the primary and (*b*) the number of turns in the secondary, assuming 100 percent efficiency.

31. A transformer has 500 turns in its primary, which is connected to 120 V rms. Its secondary coil is tapped at three places to give outputs of 2.5, 7.5, and 9 V. How many turns are needed for each part of the secondary coil?

32. The distribution circuit of a residential power line is operated at 2000 V rms. This voltage must be reduced to 240 V rms for use within the residences. If the secondary side of the transformer has 400 turns, how many turns are in the primary?

28-7 Rectification and Amplification

33. The maximum output current in a half-wave rectifier circuit is 3.5 A. (*a*) Find the rms current. (*b*) Find the rms current if the maximum output current is 3.5 A in a full-wave rectifier circuit.

34. Sketch a graph of the current versus time if a low-pass filter such as that in Figure 28-30*a* is inserted before the load resistor in Figure 28-27.

Level II

35. Figure 28-33 shows the voltage V versus time t for a "square wave" voltage. If $V_0 = 12$ V, (*a*) what is the rms voltage of this waveform? (*b*) If this alternating waveform is rectified so that only the positive voltages remain, what now is the rms voltage of the rectified waveform?

Figure 28-33 Problem 35.

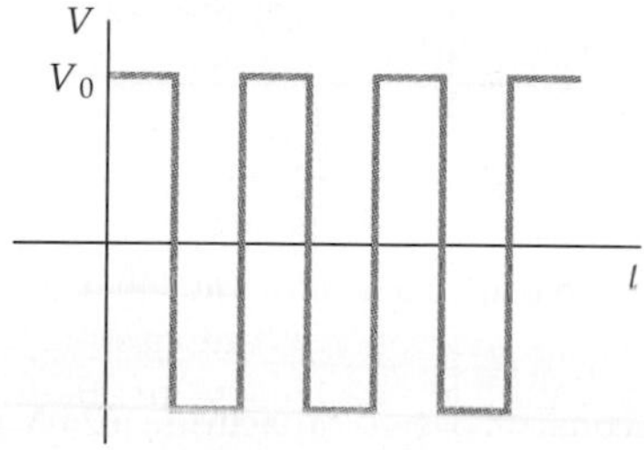

36. A pulsed current has a constant value of 15 A for the first 0.1 s of each second and is then 0 for the next 0.9 s of each second. (*a*) What is the rms value for this current waveform? (*b*) Each current pulse is generated by a 100-V pulse. What is the average power delivered by the pulse generator?

37. A 100-V-rms voltage is applied to a series RC circuit. The voltage across the capacitor is 80 V. What is the voltage across the resistor?

38. Show that the formula $P_{av} = \mathcal{E}^2_{rms}R/Z^2$ gives the correct result for a circuit containing a generator and only (*a*) a resistor, (*b*) a capacitor, and (*c*) an inductor.

39. Sketch the impedance Z versus ω for (*a*) a series LR circuit, (*b*) a series RC circuit, and (*c*) a series LCR circuit.

40. The charge on the capacitor of a series LC circuit is given by $Q = (15\ \mu\text{C}) \cos (1250\, t + \pi/4)$ where t is in seconds. (*a*) Find the current as a function of time. (*b*) Find C if $L = 28$ mH. (*c*) Write expressions for the electrical energy U_e, the magnetic energy U_m, and the total energy U.

41. A resistance R and a 1.4-H inductance are in series across a 60-Hz ac voltage. The voltage across the resistor is 30 V and the voltage across the inductor is 40 V. (*a*) What is the resistance R? (*b*) What is the ac input voltage?

42. A resistance R carries a current $I = (5.0\text{ A}) \sin 120\pi t + (7.0\text{ A}) \sin 240\pi t$. (*a*) What is the rms current? (*b*) If the resistance R is 12 Ω, what is the power dissipated in the resistor? (*c*) What is the rms voltage across the resistor?

43. A coil has a dc resistance of 80 Ω and an impedance of 200 Ω at a frequency of 1 kHz. One may neglect the wiring capacitance of the coil at this frequency. What is the inductance of the coil?

44. Two ac voltage sources are connected in series with a resistor $R = 25$ Ω. One source is given by $V_1 = (5.0\text{ V}) \cos (\omega t - \alpha)$, and the other source is $V_2 = (5.0\text{ V}) \cos (\omega t + \alpha)$, with $\alpha = \pi/6$. (*a*) Find the current in R using a trigonometric identity for the sum of two cosines. (*b*) Use phasor diagrams to find the current in R. (*c*) Find the current in R if $\alpha = \pi/4$ and the amplitude of V_2 is increased from 5.0 V to 7.0 V.

45. Given the circuit shown in Figure 28-34, (*a*) find the power loss in the inductor. (*b*) Find the resistance r of the inductor. (*c*) Find the inductance L.

Figure 28-34 Problem 45.

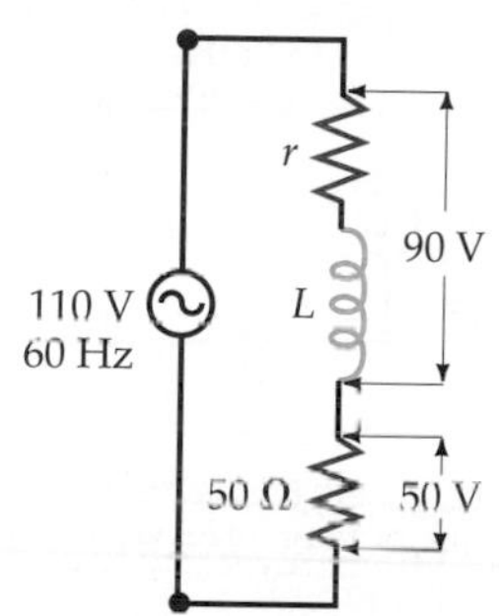

46. A coil of resistance R, inductance L, and negligible capacitance has a power factor of 0.866 at a frequency of 60 Hz. What is the power factor for a frequency of 240 Hz?

47. A coil draws 15 A when connected to a 220-V ac 60-Hz line. When it is in series with a 4-Ω resistor and the combination is connected to a 100-V battery, the battery current after a long time is observed to be 10 A. (*a*) What is the resistance in the coil? (*b*) What is the inductance of the coil?

48. A coil is connected to a 60-Hz, 100-V ac generator. At this frequency the coil has an impedance of 10 Ω and a reactance of 8 Ω. (*a*) What is the current in the coil? (*b*) What is the phase angle between the current and the applied voltage? (*c*) What series capacitance is required so that the current and voltage are in phase? (*d*) What then is the voltage measured across the capacitor?

49. A 0.25 H inductor and a capacitor C are connected in series with a 60-Hz ac generator. An ac voltmeter is used to measure the rms voltages across the inductor and capacitor separately. The rms voltage across the capacitor is 75 V and that across the inductor is 50 V. (*a*) Find the capacitance C and the rms current in the circuit. (*b*) What would be the measured rms voltage across both the capacitor and inductor together?

50. Show that Equation 28-49 can be written as

$$I_{max} = \frac{\omega \mathcal{E}_{max}}{\sqrt{L^2(\omega^2 - \omega_0^2)^2 + \omega^2 R^2}}$$

51. (*a*) Show that Equation 28-48 can be written as

$$\tan \delta = L(\omega^2 - \omega_0^2)/\omega R$$

Find δ approximately at (*b*) very low frequencies and (*c*) very high frequencies.

52. (*a*) Show that in a series *RC* circuit with no inductance, the power factor is given by

$$\cos \delta = \frac{RC\omega}{\sqrt{1 + (RC\omega)^2}}$$

(*b*) Sketch a graph of the power factor versus ω.

53. In the circuit in Figure 28-35, the ac generator produces an rms voltage of 115 V when operated at 60 Hz. What is the rms voltage across points (*a*) *AB*, (*b*) *BC*, (*c*) *CD*, (*d*) *AC*, and (*e*) *BD*?

Figure 28-35 Problem 53.

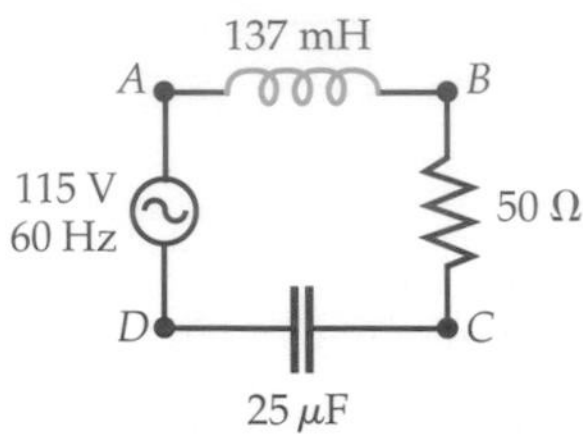

54. A variable-frequency ac generator is connected to a series *LCR* circuit for which $R = 1$ kΩ, $L = 50$ mH, and $C = 2.5$ μF. (*a*) What is the resonance frequency of the circuit? (*b*) What is the *Q* value? (*c*) At what frequencies is the value of the average power delivered by the generator half of its maximum value?

55. A series *LCR* circuit is driven at a frequency of 500 Hz. The phase angle between the applied voltage and current is determined from an oscilloscope measurement to be $\delta = 75°$. If the total resistance is known to be 35 Ω and the inductance is 0.15 H, what is the capacitance of the circuit?

56. A series *LCR* circuit with $R = 400$ Ω, $L = 0.35$ H, and $C = 5$ μF is driven by a generator of variable frequency *f*. (*a*) What is the resonance frequency f_0? Find *f* and f/f_0 when the phase angle δ is (*b*) 60°, and (*c*) −60°.

57. An experimental physicist wishes to design a series *LCR* circuit with a *Q* value of 10 and a resonance frequency of 33 kHz. She has as a 45-mH inductor with negligible resistance. What values for the resistance *R* and capacitance *C* should she use?

58. The generator voltage in Figure 28-36 is given by $V = (100 \text{ V}) \cos \omega t$. (*a*) For each branch, what is the current amplitude and its phase relative to the applied voltage? (*b*) What is the angular frequency ω such that the current in the generator vanishes? (*c*) At this resonance, what is the current in the inductor? What is the current in the capacitor? (*d*) Draw a phasor diagram showing the general relationships between the applied voltage, the generator current, the capacitor current, and the inductor current for the case where the inductive reactance is larger than the capacitive reactance.

Figure 28-36 Problem 58.

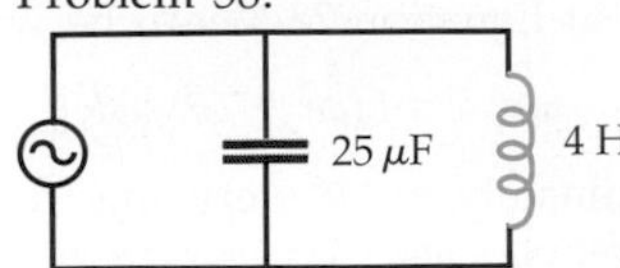

59. The circuit shown in Figure 28-37 is called an *RC* high-pass filter because high input frequencies are transmitted with greater amplitude than low input frequencies. (*a*) If the input voltage is $V_{in} = V_0 \cos \omega t$, show that the output voltage is

$$V_{out} = \frac{V_0}{\sqrt{(1/\omega RC)^2 + 1}}$$

(*b*) At what angular frequency is the output voltage half the input voltage? (*c*) Sketch a graph of V_{out}/V_0 as a function of ω.

Figure 28-37 Problem 59.

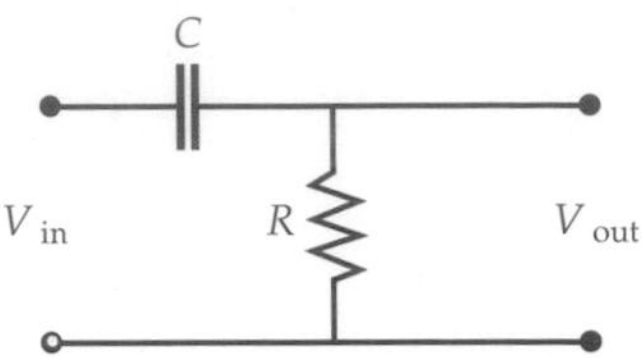

60. A circuit consists of two capacitors, a 24-V battery, and an ac voltage connected as shown in Figure 28-38. The ac voltage is given by (20 V) cos $120\pi t$, where *t* is in seconds. (*a*) Find the charge on each capacitor as a function of time. Assume transient effects have had sufficient time to decay. (*b*) What is the steady-state current? (*c*) What is the maximum energy stored in the capacitors? (*d*) What is the minimum energy stored in the capacitors?

Figure 28-38 Problem 60.

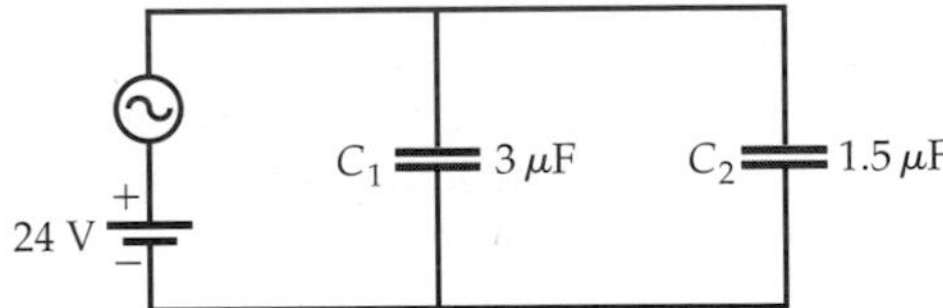

61. A single transmission line carries two voltage signals given by $V_1 = (10 \text{ V}) \cos 100t$ and $V_2 = (10 \text{ V}) \cos 10{,}000t$, where *t* is in seconds. A series inductor of 1 H and a shunting resistor of 1 kΩ is inserted into the transmission line as indicated in Figure 28-39. (*a*) What is the voltage signal observed at the output side of the transmission line? (*b*) What is the ratio of the low-frequency amplitude to the high-frequency amplitude?

Figure 28-39 Problem 61.

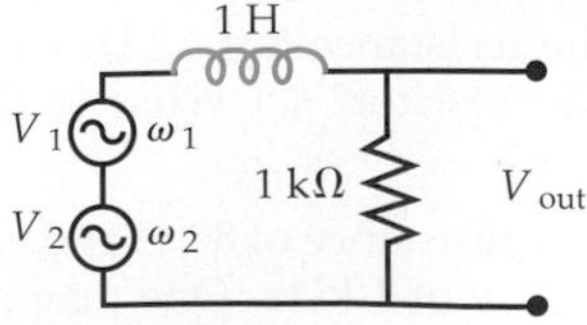

62. A coil with resistance and inductance is connected to a 120-V-rms, 60-Hz line. The average power supplied to the coil is 60 W, and the rms current is 1.5 A. Find (*a*) the power factor, (*b*) the resistance of the coil, and (*c*) the inductance of the coil. (*d*) Does the current lag or lead the voltage? What is the phase angle δ?

63. In a series *LCR* circuit $X_C = 16\ \Omega$ and $X_L = 4\ \Omega$ at some frequency. The resonance frequency is $\omega_0 = 10^4$ rad/s. (*a*) Find *L* and *C*. If $R = 5\ \Omega$ and $\mathcal{E}_{max} = 26$ V, find (*b*) the *Q* factor and (*c*) the maximum current.

64. In a series *LCR* circuit connected to an ac generator whose maximum emf is 200 V, the resistance is 60 Ω and the capacitance is 8.0 μF. The inductance can be varied from 8.0 to 40.0 mH by the insertion of an iron core in the solenoid. The angular frequency of the generator is 2500 rad/s. If the capacitor voltage is not to exceed 150 V, find (*a*) the maximum current and (*b*) the range of inductance that is safe to use.

65. When an *LCR* series circuit is connected to a 120-V-rms, 60-Hz line, the current is $I_{rms} = 11.0$ A and the current leads the voltage by 45°. (*a*) Find the power supplied to the circuit. (*b*) What is the resistance? (*c*) If the inductance $L = 0.05$ H, find the capacitance *C*. (*d*) What capacitance or inductance would you add to make the power factor 1?

Level III

66. Consider the parallel circuit shown in Figure 28-40. (*a*) What is the impedance of each branch? (*b*) For each branch, what is the current amplitude and its phase relative to the applied voltage? (*c*) Give the current phasor diagram, and use it to find the total current and its phase relative to the applied voltage.

Figure 28-40 Problem 66.

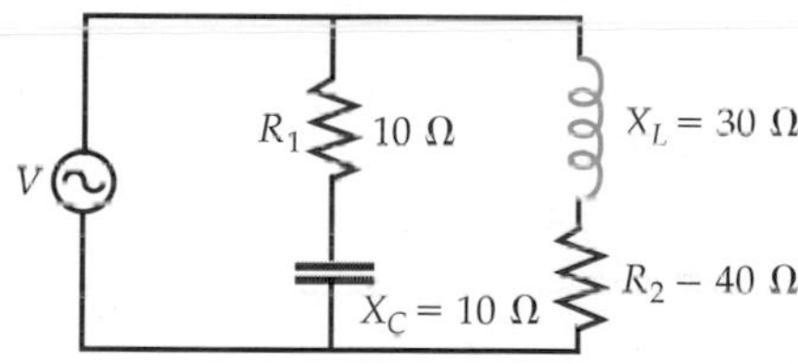

67. (*a*) Show that Equation 28-48 can be written as

$$\tan\delta = \frac{Q(\omega^2 - \omega_0^2)}{\omega\omega_0}$$

(*b*) Show that near resonance

$$\tan\delta \approx \frac{2Q(\omega - \omega_0)}{\omega}$$

(*c*) Sketch a plot of δ versus *x*, where $x = \omega/\omega_0$, for a circuit with high *Q* and for one with low *Q*.

68. Show by direct substitution that the current given by Equation 28-47 with δ and I_{max} given by Equations 28-48 and 28-49, respectively, satisfies Equation 28-46. *Hint:* Use trigonometric identities for the sine and cosine of the sum of two angles, and write the equation in the form

$$A\sin\omega t + B\cos\omega t = 0$$

Since this equation must hold for all times, $A = 0$ and $B = 0$.

69. A certain electrical device draws 10 A rms and has an average power of 720 W when connected to a 120-V-rms, 60-Hz power line. (*a*) What is the impedance of the device? (*b*) What series combination of resistance and reactance is this device equivalent to? (*c*) If the current leads the emf, is the reactance inductive or capacitive?

70. An ac generator is in series with a capacitor and an inductor in a circuit with negligible resistance. (*a*) Show that the charge on the capacitor obeys the equation

$$L\frac{d^2Q}{dt^2} + \frac{Q}{C} = \mathcal{E}_{max}\cos\omega t$$

(*b*) Show by direct substitution that this equation is satisfied by $Q = Q_{max}\cos\omega t$ if

$$Q_{max} = -\frac{\mathcal{E}_{max}}{L(\omega^2 - \omega_0^2)}$$

(*c*) Show that the current can be written as $I = I_{max}\cos(\omega t - \delta)$, where

$$I_{max} = \frac{\omega\mathcal{E}_{max}}{L|\omega^2 - \omega_0^2|} = \frac{\mathcal{E}_{max}}{|X_L - X_C|}$$

and $\delta = -90°$ for $\omega < \omega_0$ and $\delta = 90°$ for $\omega > \omega_0$.

71. A method for measuring inductance is to connect the inductor in series with a known capacitance, a known resistance, an ac ammeter, and a variable-frequency signal generator. The frequency of the signal generator is varied and the emf is kept constant until the current is maximum. (*a*) If $C = 10\ \mu$F, $\mathcal{E}_{max} = 10$ V, $R = 100\ \Omega$, and *I* is maximum at $\omega = 5000$ rad/s, what is *L*? (*b*) What is I_{max}?

72. A resistor and an inductor are connected in parallel across an emf $\mathcal{E} = \mathcal{E}_{max}\cos\omega t$ as shown in Figure 28-41. Show that (*a*) the current in the resistor is $I_R = (\mathcal{E}_{max}/R)\cos\omega t$, (*b*) the current in the inductor is $I_L = (\mathcal{E}_{max}/X_L)\cos(\omega t - 90°)$, and (*c*) $I = I_R + I_L = I_{max}\cos(\omega t - \delta)$, where $\tan\delta = R/X_L$ and $I_{max} = \mathcal{E}_{max}/Z$ with $Z^{-2} = R^{-2} + X_L^{-2}$.

Figure 28-41 Problem 72.

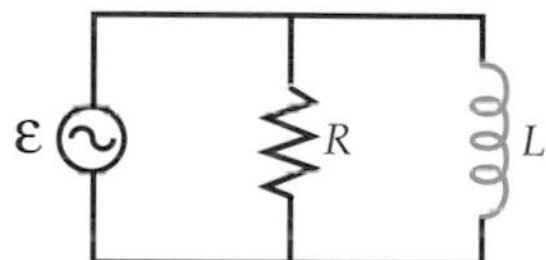

73. A resistor and a capacitor are connected in parallel across a sinusoidal emf $\mathcal{E} = \mathcal{E}_{max}\cos\omega t$ as shown in Figure 28-42. (*a*) Show that the current in the resistor is $I_R = (\mathcal{E}_{max}/R)\cos\omega t$. (*b*) Show that the current in the capacitor branch is $I_C = (\mathcal{E}_{max}/X_C)\cos(\omega t + 90°)$. (*c*) Show that the total current is given by $I = I_R + I_C = I_{max}\cos(\omega t + \delta)$, where $\tan\delta = R/X_C$ and $I_{max} = \mathcal{E}_{max}/Z$ with $Z^{-2} = R^{-2} + X_C^{-2}$.

Figure 28-42 Problem 73.

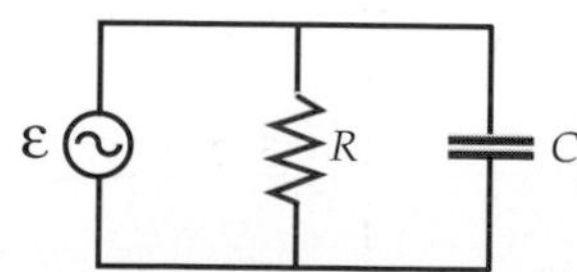

74. Figure 28-18 shows a plot of average power P_{av} versus generator frequency f for an LCR circuit with a generator. The average power P_{av} is given by Equation 28-55, where $\omega = 2\pi f$. The "full width at half maximum" Δf is the width of the resonance curve between the two points where P_{av} is one-half its maximum value. Show that, for a sharply peaked resonance, $\Delta f \approx R/2\pi L = \Delta\omega/2\pi$ and, hence, that $Q \approx \omega_0/\Delta\omega = f_0/\Delta f$ in this case (Equation 28-58). *Hint:* At resonance, the denominator of the expression on the right of Equation 28-55 is ω^2R^2. The half-power points will occur when the denominator is twice the value near resonance, that is, when $L^2(\omega^2 - \omega_0^2)^2 = \omega^2R^2 \approx \omega_0^2R^2$. Let ω_1 and ω_2 be the solutions of this equation. For a sharply peaked resonance, $\omega_1 \approx \omega_0$ and $\omega_2 \approx \omega_0$. Then, using the fact that $\omega + \omega_0 \approx 2\omega_0$, one finds that $\Delta\omega = \omega_2 - \omega_1 \approx R/L$.

75. One use of a transformer is for *impedance matching*. For example, the output impedance of a stereo amplifier is matched to the impedance of a speaker by a transformer. In Equation 28-65, the currents I_1 and I_2 can be related to the impedance Z in the secondary since $I_2 = V_2/Z$. Using Equations 28-63 and 28-64, show that

$$I_1 = \frac{\mathcal{E}}{(N_1/N_2)^2Z}$$

and, therefore, $Z_{eff} = (N_1/N_2)^2Z$.

76. Show by direct substitution that Equation 28-44*b* is satisfied by

$$Q = Q_0e^{-Rt/2L}\cos(\omega' t)$$

where

$$\omega' = \sqrt{(1/LC) - (R/2L)^2}$$

and Q_0 is the charge on the capacitor at $t = 0$.

77. (*a*) Compute the current $I = dQ/dt$ from the solution of Equation 28-44*b* given in Problem 76, and show that

$$I = -I_0\left(\sin\omega' t + \frac{R}{2L\omega'}\cos\omega' t\right)e^{-Rt/2L}$$

where $I_0 = \omega' Q_0$. (*b*) Show that this can be written

$$I = -\frac{I_0}{\cos\delta}(\cos\delta\sin\omega' t + \sin\delta\cos\omega' t)e^{-Rt/2L}$$

$$= -\frac{I_0}{\cos\delta}\sin(\omega' t + \delta)\,e^{-Rt/2L}$$

where $\tan\delta = R/2L\omega'$. When $R/2L\omega'$ is small, $\cos\delta \approx 1$, and

$$I \approx I_0\sin(\omega' t + \delta)e^{-Rt/2L}.$$

Chapter 29

Maxwell's Equations and Electromagnetic Waves

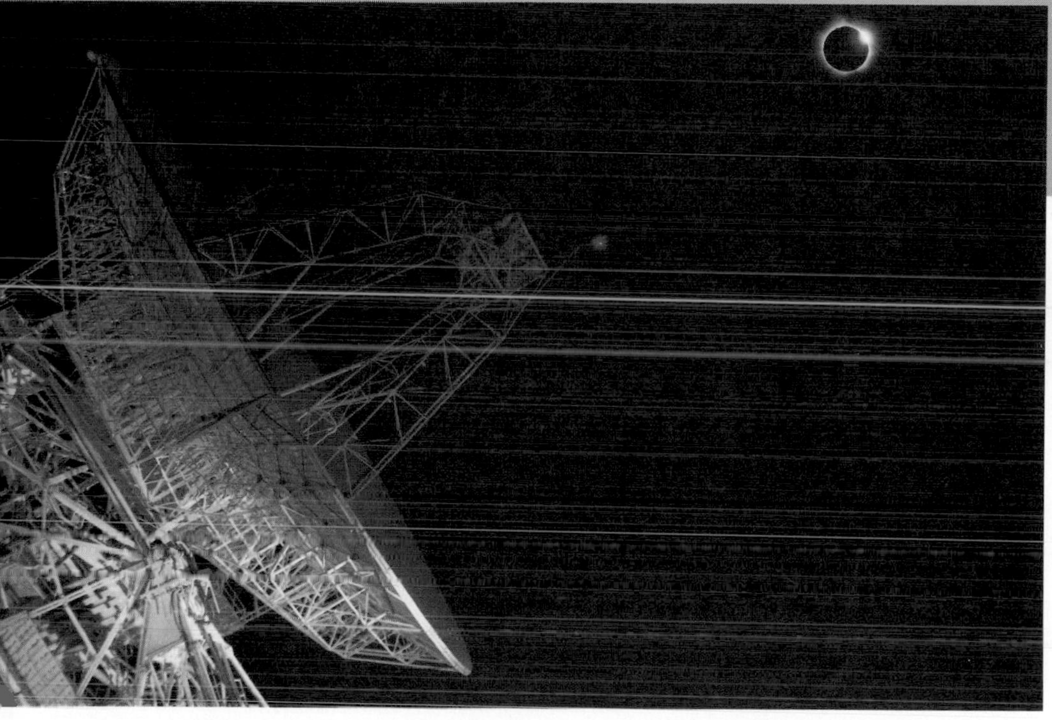

A multiple-exposure view showing the 26 m tracking antenna at Wallops Station, Virginia, and a total solar eclipse. Electromagnetic radiation at radio wavelengths, like that at optical wavelengths, is not readily absorbed by the earth's atmosphere—making it a viable means of communication between two distant points on the ground or between a point on the ground and a plane, satellite, or spacecraft. Objects are tracked by aiming a continuous radar beam at them and receiving the reflected beam.

About 1860, the great Scottish physicist James Clerk Maxwell found that the experimental laws of electricity and magnetism—the laws of Coulomb, Gauss, Biot–Savart, Ampère, and Faraday, which we have studied in Chapters 18 through 28—could be summarized in a concise mathematical form now known as Maxwell's equations. One of the laws, Ampère's law, contained an inconsistency, which Maxwell was able to remove with the invention of the displacement current (Section 29-1). The new consistent set of equations predicts the possibility of electromagnetic waves.

Maxwell's equations relate the electric- and magnetic-field vectors **E** and **B** to their sources, which are electric charges, currents, and changing fields. These equations play a role in classical electromagnetism analogous to that of Newton's laws in classical mechanics. In principle, all problems in classical electricity and magnetism can be solved using Maxwell's equations, just as all problems in classical mechanics can be solved using Newton's laws. Maxwell's equations are considerably more complicated than Newton's laws, however, and their application to most problems involves mathematics beyond the scope of this book. Nevertheless, Maxwell's equations are of great theoretical importance.

Maxwell showed that these equations could be combined to yield a wave equation for the electric- and magnetic-field vectors **E** and **B**. Such electromagnetic waves are caused by accelerating charges, for example, the charges in an alternating current in an antenna. They were first produced in the laboratory by Heinrich Hertz in 1887. Maxwell showed that the speed of electromagnetic waves in free space should be

$$c = \frac{1}{\sqrt{\mu_0 \epsilon_0}} \qquad 29\text{-}1$$

where ϵ_0, the permittivity of free space, is the constant appearing in Coulomb's and Gauss's laws and μ_0, the permeability of free space, is that in the Biot–Savart law and Ampère's law. When the measured value of ϵ_0 and the defined value of μ_0 are put into Equation 29-1, the speed of electromagnetic waves is found to be about 3×10^8 m/s, the same as the measured speed of light. Maxwell noted this "coincidence" with great excitement and correctly surmised that light itself is an electromagnetic wave.

In this chapter we begin by showing that Ampère's law as stated in Chapter 25 does not hold for discontinuous currents. We then show how Maxwell generalized Ampère's law by adding a term now called Maxwell's displacement current. After stating Maxwell's equations and relating them to the laws of electricity and magnetism that we have already studied, we will show that these equations imply that electric and magnetic field vectors obey a wave equation that describes waves that propagate through free space with speed $c = 1/\sqrt{\mu_0 \epsilon_0}$. Finally, we will illustrate how electromagnetic waves carry energy and momentum, and discuss the electromagnetic spectrum.

29-1 Maxwell's Displacement Current

As we studied in Chapter 25, Ampère's law (Equation 25-15) relates the line integral of the magnetic field around some closed curve C to the current that passes through any area bounded by that curve:

$$\oint_C \mathbf{B} \cdot d\boldsymbol{\ell} = \mu_0 I \qquad \text{for any closed curve } C \qquad 29\text{-}2$$

We noted that this equation holds only for continuous currents. We can see that it does not hold for discontinuous currents by considering the charging of a capacitor (Figure 29-1). According to Ampère's law, the line integral of the magnetic field **B** around a closed curve equals μ_0 times the total current through any surface bounded by the curve. Such a surface need not be a plane. Two surfaces bounded by the curve C are indicated in Figure 29-1.

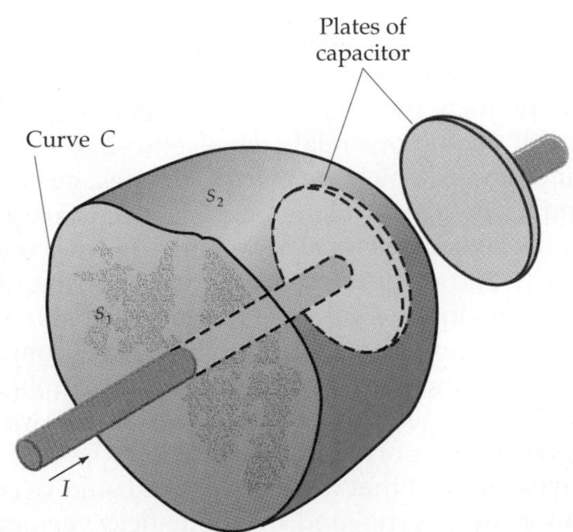

Figure 29-1 Two surfaces S_1 and S_2 bounded by the same curve C. The current I passes through surface S_1 but not S_2. Ampère's law, which relates the line integral of the magnetic field **B** around the curve C to the total current passing through any surface bounded by C, is not valid when the current is not continuous, as when it stops at the capacitor plate here.

The current through surface S_1 is I. There is no current through surface S_2 because the charge stops on the capacitor plate. There is thus ambiguity in the phrase "the current through any surface bounded by the curve." For continuous currents, we get the same current no matter which surface we choose.

Maxwell recognized this flaw in Ampère's law and showed that the law can be generalized to include all situations if the current I in the equation is replaced by the sum of the conduction current I and another term I_d, called **Maxwell's displacement current,** defined as

$$I_d = \epsilon_0 \frac{d\phi_e}{dt} \qquad 29\text{-}3$$

where ϕ_e is the flux of the electric field through the same surface bounded by the curve C. The generalized form of Ampère's law is then

$$\oint_C \mathbf{B} \cdot d\boldsymbol{\ell} = \mu_0(I + I_d) = \mu_0 I + \mu_0 \epsilon_0 \frac{d\phi_e}{dt} \qquad 29\text{-}4$$

We can understand this generalization by considering Figure 29-1 again. Let us call the sum $I + I_d$ the generalized current. According to the argument just stated, the same generalized current must cross any area bounded by the curve C. Thus, there can be no net generalized current into or out of the closed volume. If there is a net true current I into the volume, there must be an equal net displacement current I_d out of the volume. In the volume in the figure, there is a net conduction current I into the volume that increases the charge within the volume:

$$I = \frac{dQ}{dt}$$

The flux of the electric field out of the volume is related to the charge by Gauss's law

$$\phi_{e,\text{net}} = \oint_S E_n \, dA = \frac{1}{\epsilon_0} Q_{\text{inside}}$$

The rate of increase of the charge is thus proportional to the rate of increase of the net flux out of the volume:

$$\epsilon_0 \frac{d\phi_{e,\text{net}}}{dt} = \frac{dQ}{dt} = I_d$$

Thus, the net conduction current into the volume equals the net displacement current out of the volume. The generalized current is always continuous.

It is interesting to compare Equation 29-4 to Faraday's law (Equation 26-6)

$$\mathcal{E} = \oint_C \mathbf{E} \cdot d\boldsymbol{\ell} = -\frac{d\phi_m}{dt} \qquad 29\text{-}5$$

where $\mathcal{E}$ is the induced emf in a circuit and ϕ_m is the magnetic flux through the circuit. According to Faraday's law, a changing magnetic flux produces an electric field whose line integral around a curve is proportional to the rate of change of magnetic flux through the curve. Maxwell's modification of Ampère's law shows that a changing electric flux produces a magnetic field whose line integral around a curve is proportional to the rate of change of the electric flux. We thus have the interesting reciprocal result that a changing magnetic field produces an electric field (Faraday's law) and a changing electric field produces a magnetic field (generalized form of Ampère's law.) Note that there is no magnetic analogue of a conduction current I.

Example 29-1

A parallel-plate capacitor has closely spaced circular plates of radius R. Charge is flowing onto the positive plate and off of the negative plate at the rate $I = dQ/dt = 2.5$ A. Compute the displacement current between the plates.

Since the plates are closely spaced, the electric field between them is uniform in the direction from the positive plate to the negative plate with magnitude $E = \sigma/\epsilon_0$, where σ is the magnitude of the charge per unit area on either plate. Consider any plane between the plates and parallel to them. Since **E** is perpendicular to the plates and therefore to the plane, and is uniform between the plates and zero outside the plates, the electric flux through the plane is

$$\phi_e = \pi R^2 E = (\pi R^2)(\sigma/\epsilon_0) = Q/\epsilon_0$$

where $Q = \pi R^2 \sigma$ is the magnitude of the total charge on either plate. The displacement current is then

$$I_d = \epsilon_0 d\phi_e/dt = dQ/dt = 2.5 \text{ A}$$

Example 29-2

The circular plates in Example 29-1 have a radius of $R = 3.0$ cm. Find the magnetic field at a point between the plates at a distance $r = 2.0$ cm from the axis of the plates when the current into the positive plate is 2.5 A.

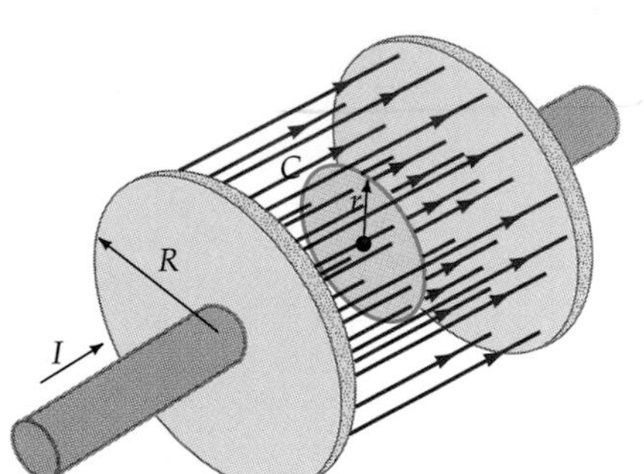

Figure 29-2 Curve C for computing the displacement current in Example 29-2.

We find B from the generalized form of Ampère's law (Equation 29-4). In Figure 29-2, we have chosen a circular path of radius $r = 2.0$ cm about the center line joining the plates to compute $\oint \mathbf{B} \cdot d\boldsymbol{\ell}$. By symmetry, **B** is tangent to this circle and has the same magnitude everywhere on it. Then

$$\oint \mathbf{B} \cdot d\boldsymbol{\ell} = B(2\pi r)$$

The electric flux through the area bounded by this curve is

$$\phi_e = \pi r^2 E = (\pi r^2)\frac{\sigma}{\epsilon_0}$$

$$= (\pi r^2)\frac{Q}{\pi R^2 \epsilon_0} = \frac{r^2 Q}{R^2 \epsilon_0}$$

where we have used $\sigma = Q/\pi R^2$. Since there is no conduction current between the plates of the capacitor, the generalized current is just the displacement current

$$I_d = \epsilon_0 \frac{d\phi_e}{dt} = \frac{r^2}{R^2}\frac{dQ}{dt}$$

$$\oint_C \mathbf{B} \cdot d\boldsymbol{\ell} = \mu_0(I + I_d) = \mu_0 I_d = \mu_0 \epsilon_0 \frac{d\phi_e}{dt}$$

$$B(2\pi r) = \mu_0 \frac{r^2}{R^2}\frac{dQ}{dt}$$

$$B = \frac{\mu_0}{2\pi}\frac{r}{R^2}\frac{dQ}{dt} = (2 \times 10^{-7} \text{ T·m/A})\frac{0.02 \text{ m}}{(0.03 \text{ m})^2}(2.5 \text{ A}) = 1.11 \times 10^{-5} \text{ T}$$

29-2 Maxwell's Equations

Maxwell's equations are

$$\oint_S E_n \, dA = \frac{1}{\epsilon_0} Q_{inside} \quad \text{29-6a}$$

$$\oint_S B_n \, dA = 0 \quad \text{29-6b}$$

Maxwell's equations

$$\oint_C \mathbf{E} \cdot d\boldsymbol{\ell} = -\frac{d}{dt} \int_S B_n \, dA \quad \text{29-6c}$$

$$\oint_C \mathbf{B} \cdot d\boldsymbol{\ell} = \mu_0 I + \mu_0 \epsilon_0 \frac{d}{dt} \int_S E_n \, dA \quad \text{29-6d}$$

Equation 29-6*a* is Gauss's law; it states that the flux of the electric field through any closed surface equals $1/\epsilon_0$ times the net charge inside the surface. As discussed in Chapter 19, Gauss's law implies that the electric field due to a point charge varies inversely as the square of the distance from the charge. This law describes how electric-field lines diverge from a positive charge and converge on a negative charge. Its experimental basis is Coulomb's law.

Equation 29-6*b*, sometimes called Gauss's law for magnetism, states that the flux of the magnetic-field vector **B** is zero through any closed surface. This equation describes the experimental observation that magnetic-field lines do not diverge from any point in space or converge on any point; that is, it implies that isolated magnetic poles do not exist.

Equation 29-6*c* is Faraday's law; it states that the integral of the electric field around any closed curve *C*, which is the emf, equals the (negative) rate of change of the magnetic flux through any surface *S* bounded by the curve. (This is not a closed surface, so the magnetic flux through *S* is not necessarily zero.) Faraday's law describes how electric-field lines encircle any area through which the magnetic flux is changing, and it relates the electric-field vector **E** to the rate of change of the magnetic-field vector **B**.

Equation 29-6*d*, Ampère's law with Maxwell's displacement-current modification, states that the line integral of the magnetic field **B** around any closed curve *C* equals μ_0 times the current through any surface bounded by the curve plus $\mu_0\epsilon_0$ times the rate of change of the electric flux through the surface. This law describes how the magnetic-field lines encircle an area through which a current is passing or the electric flux is changing.

29-3 The Wave Equation for Electromagnetic Waves (Optional)

In Section 13-8, we saw that the harmonic wave functions for waves on a string obey a partial differential equation called the **wave equation:**

$$\frac{\partial^2 y(x, t)}{\partial x^2} = \frac{1}{v^2} \frac{\partial^2 y(x, t)}{\partial t^2} \quad \text{29-7}$$

In this equation, $y(x, t)$ is the wave function, which for string waves is the displacement of the string. The derivatives are partial derivatives because the wave function depends on both x and t. The quantity v is the velocity of the wave, which depends on the medium (and on the frequency if the medium is dispersive). We also saw that the wave equation for string waves can be derived by applying Newton's laws of motion to a string under tension, and we found that the velocity of the waves is $\sqrt{F/\mu}$, where F is the tension and μ the linear mass density. The solutions of this equation are harmonic wave functions of the form

$$y(x, t) = y_0 \sin (kx - \omega t)$$

where $k = 2\pi/\lambda$ is the wave number and $\omega = 2\pi f$ is the angular frequency.

A double rainbow over the radio telescope at Socorro, New Mexico. The telescope consists of a Very Large Array (VLA) of antenna dishes. The direction of incoming radio waves from distant galaxies can be determined by the interference of signals detected in the array.

In this section, we will use Maxwell's equations to derive the wave equation for electromagnetic waves. We will not consider how such waves arise from the motion of charges but merely show that the laws of electricity and magnetism imply a wave equation, which in turn implies the existence of electric and magnetic fields **E** and **B** that propagate through space with the velocity of light c. We will consider only free space, in which there are no charges or currents. We will assume that the electric and magnetic fields **E** and **B** are functions of time and one space coordinate only, which we will take to be the x coordinate. Such a wave is called a **plane wave,** because field quantities are constant across any plane perpendicular to the x axis.

To obtain the wave equation relating the time and space derivatives of either the electric field **E** or the magnetic field **B**, we first relate the space derivative of one of the field vectors to the time derivative of the other. We do this by applying Equations 29-6c and 29-6d to appropriately chosen curves in space. We first relate the space derivative of E_y to the time derivative of B_z by applying Equation 29-6c (which is Faraday's law) to the rectangular curve of sides Δx and Δy lying in the xy plane shown in Figure 29-3. If Δx and Δy are very small, the line integral of **E** around this curve is approximately

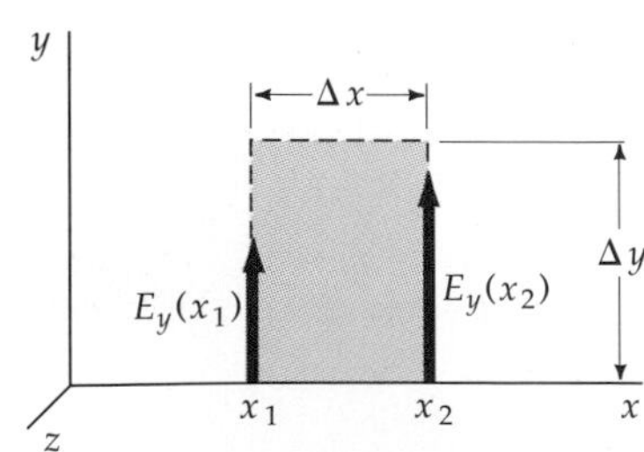

Figure 29-3 A rectangular curve in the xy plane for the derivation of Equation 29-8.

$$\oint \mathbf{E} \cdot d\boldsymbol{\ell} = E_y(x_2)\, \Delta y - E_y(x_1)\, \Delta y$$

where $E_y(x_1)$ is the value of E_y at the point x_1 and $E_y(x_2)$ is the value of E_y at the point x_2. The contributions of the type $E_x\, \Delta x$ from the top and bottom of this curve cancel because we have assumed that **E** does not depend on y (or

z). Because Δx is very small, we can approximate the difference in E_y at the points x_1 and x_2 by

$$E_y(x_2) - E_y(x_1) = \Delta E \approx \frac{\partial E_y}{\partial x} \Delta x$$

Then

$$\oint \mathbf{E} \cdot d\boldsymbol{\ell} \sim \frac{\partial E_y}{\partial x} \Delta x \, \Delta y$$

The flux of the magnetic field through this curve is approximately

$$\int_S B_n \, dA = B_z \, \Delta x \, \Delta y$$

Faraday's law then gives

$$\frac{\partial E_y}{\partial x} \Delta x \, \Delta y = -\frac{\partial B_z}{\partial t} \Delta x \, \Delta y$$

or

$$\frac{\partial E_y}{\partial x} = -\frac{\partial B_z}{\partial t} \qquad \text{29-8}$$

Equation 29-8 implies that if there is a component of the electric field E_y that depends on x, there must be a component of the magnetic field B_z that depends on time or, conversely, that if there is a component of the magnetic field B_z that depends on time, there must be a component of the electric field E_y that depends on x. We can get a similar equation relating the space derivative of the magnetic field B_z to the time derivative of the electric field E_y by applying Equation 29-6*d* to the curve of sides Δx and Δz in the xz plane shown in Figure 29-4. For the case of no conduction currents, Equation 29-6*d* is

$$\oint \mathbf{B} \cdot d\boldsymbol{\ell} = \mu_0 \epsilon_0 \frac{d}{dt} \int_S E_n \, dA$$

The details of this calculation are similar to those for Equation 29-8. The result is

$$\frac{\partial B_z}{\partial x} = -\mu_0 \epsilon_0 \frac{\partial E_y}{\partial t} \qquad \text{29-9}$$

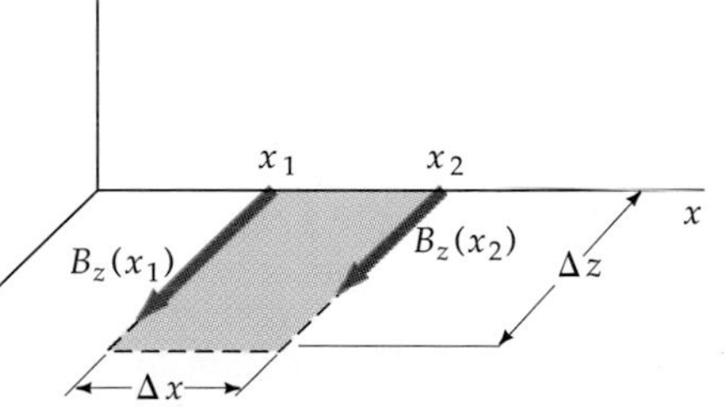

Figure 29-4 A rectangular curve in the xz plane for the derivation of Equation 29-9.

We can eliminate either B_z or E_y from Equations 29-8 and 29-9 by differentiating either equation with respect to x or t. If we differentiate both sides of Equation 29-8 with respect to x, we obtain

$$\frac{\partial}{\partial x}\left(\frac{\partial E_y}{\partial x}\right) = -\frac{\partial}{\partial x}\left(\frac{\partial B_z}{\partial t}\right)$$

or

$$\frac{\partial^2 E_y}{\partial x^2} = -\frac{\partial}{\partial t}\left(\frac{\partial B_z}{\partial x}\right)$$

where the order of the time and space derivatives on the right side have been interchanged. We now use Equation 29-9 for $\partial B_z/\partial x$:

$$\frac{\partial^2 E_y}{\partial x^2} = -\frac{\partial}{\partial t}\left(-\mu_0 \epsilon_0 \frac{\partial E_y}{\partial t}\right)$$

which yields the wave equation

$$\frac{\partial^2 E_y}{\partial x^2} = \mu_0 \epsilon_0 \frac{\partial^2 E_y}{\partial t^2} \qquad \text{29-10}$$

Comparing this equation with Equation 29-7, we see that E_y obeys a wave equation for waves with speed

$$c = \frac{1}{\sqrt{\mu_0 \epsilon_0}}$$

which is Equation 29-1.

If we had instead chosen to eliminate E_y from Equations 29-8 and 29-9 (by differentiating Equation 29-8 with respect to t, for example), we would have obtained an equation identical to Equation 29-10 except with B_z replacing E_y. We can thus see that both the electric field E_y and the magnetic field B_z obey a wave equation for waves traveling with the velocity $1/\sqrt{\mu_0 \epsilon_0}$, which is the velocity of light.

By following the same line of reasoning as used above, we can readily show (as in Problem 29-49) that if Equation 29-6*c* (Faraday's law) is applied to the curve in the xz plane shown in Figure 29-4, the spatial variation in E_z is related to the time variation in B_y by

$$\frac{\partial E_z}{\partial x} = \frac{\partial B_y}{\partial t} \qquad \text{29-11}$$

Similarly, the application of Equation 26-6*d* to the curve in the xy plane of Figure 29-3 gives

$$\frac{\partial B_y}{\partial x} = \mu_0 \epsilon_0 \frac{\partial E_z}{\partial t} \qquad \text{29-12}$$

We can use these results to show that, for a wave propagating in the x direction, the components E_z and B_y also obey the wave equation.

So far we have considered only the y and z components of the electric and magnetic fields. The same type of analysis may be applied to a rectangular loop in the yz plane, similar to the loops in Figure 29-3 and 29-4, to obtain equations analogous to Equations 29-8, 29-9, 29-11, and 29-12 in which the time derivatives of E_x and B_x are proportional to the y and z derivatives of the field quantities. In the plane wave under consideration, the y and z derivatives of the field quantities are zero, so the time derivatives of E_x and B_x must be zero. In an electromagnetic wave, we are interested only in time-varying fields, and so we may subtract out any x components of the fields that are constant in time.

We have shown that in any plane electromagnetic wave traveling parallel to the x axis the x components of the fields are zero, so the vectors **E** and **B** are perpendicular to the x axis. They are also perpendicular to each other and each obeys the wave equation:

Wave equation for **E**

$$\frac{\partial^2 \mathbf{E}}{\partial x^2} = \frac{1}{c^2} \frac{\partial^2 \mathbf{E}}{\partial t^2} \qquad \text{29-13}a$$

Wave equation for **B**

$$\frac{\partial^2 \mathbf{B}}{\partial x^2} = \frac{1}{c^2} \frac{\partial^2 \mathbf{B}}{\partial t^2} \qquad \text{29-13}b$$

As we noted in discussing harmonic waves, a particularly important solution to a wave equation like Equation 29-10 is the harmonic wave function of the form

$$E_y = E_{y0} \sin (kx - \omega t) \qquad 29\text{-}14$$

If we substitute this solution into either Equation 29-8 or 29-9, we can see that the magnetic field B_z is in phase with the electric field E_y. From Equation 29-8, we have

$$\frac{\partial B_z}{\partial t} = -\frac{\partial E_y}{\partial x} = -kE_{y0} \cos (kx - \omega t)$$

Solving for B_z gives

$$B_z = \frac{k}{\omega} E_{y0} \sin (kx - \omega t)$$

$$= B_{z0} \sin (kx - \omega t) \qquad 29\text{-}15$$

where

$$B_{z0} = \frac{k}{\omega} E_{y0} = \frac{E_{y0}}{c}$$

and $c = \omega/k$ is the velocity of the wave.* Since the electric and magnetic fields oscillate in phase with the same frequency, we have the general result that the magnitude of the electric field is c times the magnitude of the magnetic field for an electromagnetic wave:

$$E = cB \qquad 29\text{-}16$$

Suppose the electric field vector **E** is confined to the y direction, as exemplified by Equation 29-14. Then $E_z = 0$, and, according to Equation 29-11, $dB_y/dt = 0$. Thus, if **E** is in the y direction, then the time-varying part (the only part we are interested in) of **B** is in the z direction, as shown in Figure 29-5. Such a wave is said to be **linearly polarized,** because if we plot **E** (or **B**) as a function of time in any plane perpendicular to the x axis, we obtain a straight line.

We see that Maxwell's equations imply wave equations 29-13a and b for the electric and magnetic fields; and that if E_y varies harmonically, as in Equation 29-14, the magnetic field B_z is in phase with E_y and has an amplitude related to the amplitude of E_y by Equation 29-16. The electric and magnetic fields are perpendicular to each other and to the direction of the wave propagation, as shown in Figure 29-5. In general, the direction of propagation of an electromagnetic wave is the direction of the cross product **E** × **B**.

*In obtaining Equation 29-15 by the integration of the previous equation, an arbitrary constant of integration arises. We have omitted this constant magnetic field from Equation 29-15 because it plays no part in the electromagnetic waves we are interested in. Note that if any constant electric field is added to Equation 29-14, the new electric field still satisfies the wave equation.

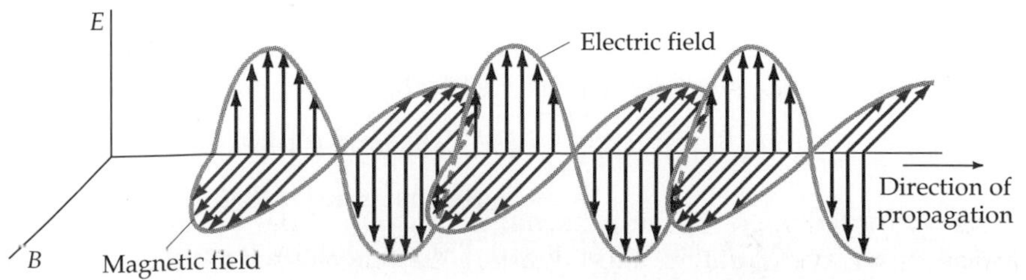

Figure 29-5 The electric- and magnetic-field vectors in a plane-polarized electromagnetic wave. The fields are in phase, perpendicular to each other, and perpendicular to the direction of propagation of the wave.

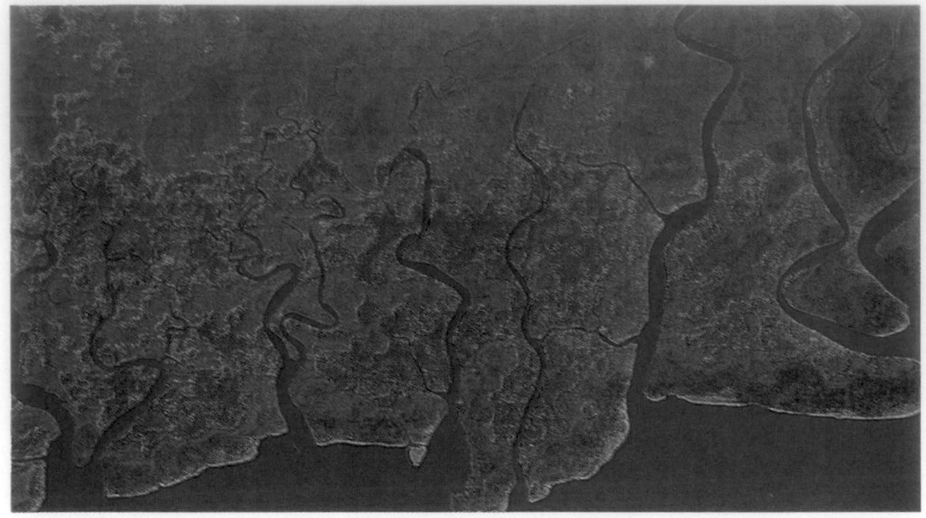

A radar image of the south coast of New Guinea. A radar system operates by transmitting radio waves toward objects, sensing the echo of waves reflected back, and determining the distance to the object from the intervening time interval.

Example 29-3

The electric field vector of an electromagnetic wave is given by $\mathbf{E}(x, t) = E_0 \sin(kx - \omega t)\,\mathbf{j} + E_0 \cos(kx - \omega t)\,\mathbf{k}$. (*a*) Find the corresponding magnetic field. (*b*) Compute $\mathbf{E}\cdot\mathbf{B}$ and $\mathbf{E} \times \mathbf{B}$.

(*a*) We can use either Equation 29-11 or 29-12 to find B_y. From Equation 29-11, we obtain

$$\frac{\partial B_y}{\partial t} = \frac{\partial E_z}{\partial x} = \frac{\partial}{\partial x}[E_0 \cos(kx - \omega t)] = -kE_0 \sin(kx - \omega t)$$

Then, neglecting the arbitrary constant of integration, we obtain

$$B_y = [kE_0 \cos(kx - \omega t)](-1/\omega) = -B_0 \cos(kx - \omega t)$$

where $B_0 = kE_0/\omega = E_0/c$. We can find B_z from either Equation 29-8 or 29-9. Using Equation 29-8, we obtain

$$\frac{\partial B_z}{\partial t} = -\frac{\partial E_y}{\partial x} = -\frac{\partial}{\partial x}[E_0 \sin(kx - \omega t)] = -kE_0 \cos(kx - \omega t)$$

and

$$B_z = [-kE_0 \sin(kx - \omega t)](-1/\omega) = B_0 \sin(kx - \omega t)$$

where again $B_0 = kE_0/\omega = E_0/c$. The magnetic field is thus given by

$$\mathbf{B}(x, t) = -B_0 \cos(kx - \omega t)\,\mathbf{j} + B_0 \sin(kx - \omega t)\,\mathbf{k}.$$

This type of electromagnetic wave is said to be **circularly polarized.** Both **E** and **B** are constant in magnitude, as can be seen by computing $\mathbf{E}\cdot\mathbf{E}$ or $\mathbf{B}\cdot\mathbf{B}$. For example, $\mathbf{E}\cdot\mathbf{E} = E_x^2 + E_y^2 = E_0^2 \sin^2(kx - \omega t) + E_0^2 \cos^2(kx - \omega t) = E_0^2$. At a fixed point x, both vectors rotate in a circle in the plane perpendicular to x with angular frequency ω.

(*b*) Computing $\mathbf{E}\cdot\mathbf{B}$, with $\theta = kx - \omega t$ to simplify the notation, we obtain

$$\begin{aligned}\mathbf{E}\cdot\mathbf{B} &= [E_0 \sin\theta\,\mathbf{j} + E_0 \cos\theta\,\mathbf{k}]\cdot[-B_0 \cos\theta\,\mathbf{j} + B_0 \sin\theta\,\mathbf{k}]\\ &= -E_0B_0 \sin\theta \cos\theta\,\mathbf{j}\cdot\mathbf{j} + E_0B_0 \sin^2\theta\,\mathbf{j}\cdot\mathbf{k}\\ &\qquad - E_0B_0 \cos^2\theta\,\mathbf{k}\cdot\mathbf{j} + E_0B_0 \cos\theta \sin\theta\,\mathbf{k}\cdot\mathbf{k}\\ &= -E_0B_0 \sin\theta \cos\theta + 0 - 0 + E_0B_0 \cos\theta \sin\theta = 0\end{aligned}$$

The electric and magnetic fields are perpendicular to each other as well as to the direction of propagation. Computing $\mathbf{E} \times \mathbf{B}$ and using $\mathbf{j} \times \mathbf{j} = \mathbf{k} \times \mathbf{k} = 0$, $\mathbf{j} \times \mathbf{k} = \mathbf{i}$, and $\mathbf{k} \times \mathbf{j} = -\mathbf{i}$, we obtain

$$\begin{aligned}\mathbf{E} \times \mathbf{B} &= [E_0 \sin \theta\, \mathbf{j} + E_0 \cos \theta\, \mathbf{k}] \times [-B_0 \cos \theta\, \mathbf{j} + B_0 \sin \theta\, \mathbf{k}] \\ &= E_0B_0 \sin^2 \theta\, \mathbf{j} \times \mathbf{k} + (-E_0B_0 \cos^2 \theta\, \mathbf{k} \times \mathbf{j}) \\ &= E_0B_0 \sin^2 \theta\, \mathbf{i} + E_0B_0 \cos^2 \theta\, \mathbf{i} = E_0B_0\mathbf{i}\end{aligned}$$

We note that $\mathbf{E} \times \mathbf{B}$ is in the direction of propagation of the wave.

29-4 Energy and Momentum in an Electromagnetic Wave

In our discussion of the transport of energy by a wave of any kind, we saw that the intensity of the wave (the average energy per unit time per unit area) is equal to the product of the average energy density (the energy per unit volume) and the speed of the wave. The energy density stored in the electric field is (Equation 21-17)

$$\eta_e = \tfrac{1}{2}\epsilon_0 E^2$$

and the energy density stored in the magnetic field is (Equation 26-34)

$$\eta_m = \frac{B^2}{2\mu_0}$$

In an electromagnetic wave in free space, $E = cB$, so we can express the magnetic energy density in terms of the electric field:

$$\eta_m = \frac{B^2}{2\mu_0} = \frac{(E/c)^2}{2\mu_0} = \frac{E^2}{2\mu_0 c^2} = \frac{1}{2}\epsilon_0 E^2$$

where we have used $c^2 = 1/\epsilon_0\mu_0$. Thus, the electric and magnetic energy densities are equal. The total energy density η in the wave is the sum of the electric and magnetic energy densities. Using $E = cB$, we may express the total energy density in several useful ways:

$$\eta = \eta_e + \eta_m = \frac{1}{2}\epsilon_0 E^2 + \frac{1}{2}\epsilon_0 E^2 = \epsilon_0 E^2 = \frac{B^2}{\mu_0} = \frac{EB}{\mu_0 c} \qquad 29\text{-}17$$

Energy density in an electromagnetic wave

In Section 14-3, we saw that the intensity of a wave (the average power flowing across an area per unit area) is equal to the product of the average energy density and the speed of the wave. The instantaneous intensity is the instantaneous power flowing across an area per unit area. It equals the product of the instantaneous energy density and the wave speed. For an electromagnetic wave in free space, the instantaneous intensity is therefore

$$I_{\text{instantaneous}} = \eta c = c\epsilon_0 E^2 = c\frac{B^2}{\mu_0} = \frac{EB}{\mu_0} \qquad 29\text{-}18$$

Equation 29-18 can be generalized to a vector expression:

$$\mathbf{S} = \frac{\mathbf{E} \times \mathbf{B}}{\mu_0} \qquad 29\text{-}19$$

The vector **S** is called the **Poynting vector** after its discoverer Sir John Poynting. Since **E** and **B** are perpendicular in an electromagnetic wave, the magnitude of **S** is the instantaneous intensity of the wave and the direction of **S** is the direction of propagation of the wave.

In a harmonic plane wave of angular frequency ω and wave number k, the instantaneous electric and magnetic fields are given by

$$E = E_0 \sin(kx - \omega t) \qquad \text{and} \qquad B = B_0 \sin(kx - \omega t)$$

Using these results for E and B in Equation 29-17, we obtain for the instantaneous energy density

$$\eta = \frac{EB}{\mu_0 c} = \frac{E_0 B_0 \sin^2(kx - \omega t)}{\mu_0 c}$$

When we average the sine squared function over space or time, we obtain a factor of $\frac{1}{2}$. The average energy density is therefore

$$\eta_{av} = \frac{1}{2}\frac{E_0 B_0}{\mu_0 c} = \frac{E_{rms} B_{rms}}{\mu_0 c} \qquad \text{29-20}$$

where we have used $E_{rms} = E_0/\sqrt{2}$ and $B_{rms} = B_0/\sqrt{2}$. The intensity is thus

Intensity of an electromagnetic wave

$$I = \eta_{av} c = \frac{1}{2}\frac{E_0 B_0}{\mu_0} = \frac{E_{rms} B_{rms}}{\mu_0} = |\mathbf{S}|_{av} \qquad \text{29-21}$$

We will now show by a simple example that an electromagnetic wave carries momentum. In this example, we will calculate the momentum and energy absorbed from the wave by a free charged particle. Consider an electromagnetic wave moving along the x axis with the electric field in the y direction and the magnetic field in the z direction that is incident on a stationary charge on the x axis as shown in Figure 29-6. For simplicity, we will neglect the time dependence of the electric and magnetic fields. The particle experiences a force $q\mathbf{E}$ in the y direction and is thus accelerated by the electric field. At any time t, the velocity in the y direction is

$$v_y = at = \frac{qE}{m}t$$

After a short time t_1 the charge has acquired a velocity in the y direction given by

$$v_y = at_1 = \frac{qE}{m}t_1$$

The energy acquired by the charge after time t_1 is

$$K = \frac{1}{2}mv_y^2 = \frac{1}{2}\frac{mq^2E^2t_1^2}{m^2} = \frac{1}{2}\frac{q^2E^2}{m}t_1^2 \qquad \text{29-22}$$

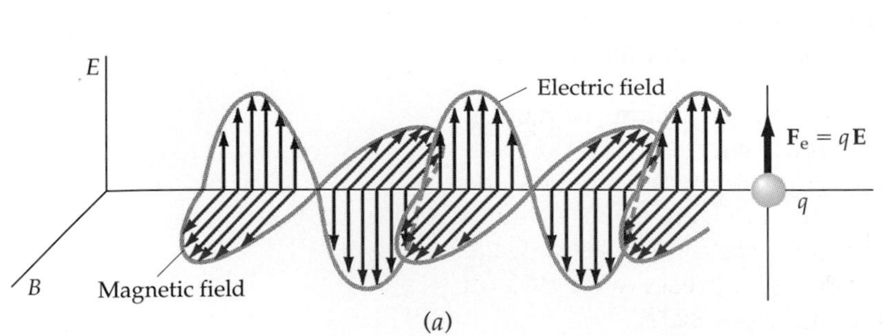

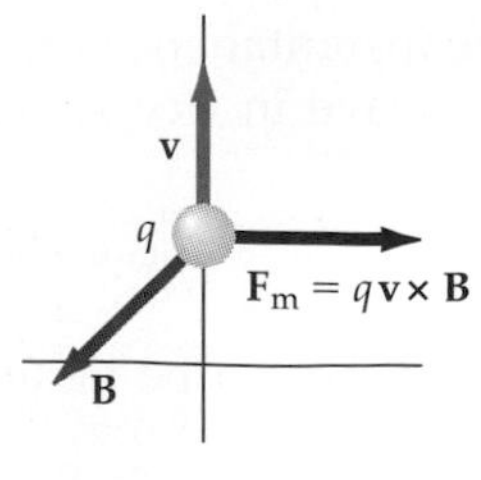

Figure 29-6 An electromagnetic wave incident on a point charge that is initially at rest on the x axis. (a) The electric force $q\mathbf{E}$ accelerates the charge in the upward direction. (b) When the charge has acquired a velocity **v** upward, the magnetic force $q\mathbf{v} \times \mathbf{B}$ accelerates the charge in the direction of the wave.

When the charge is moving in the y direction, it experiences a magnetic force $q\mathbf{v} \times \mathbf{B}$, which is in the positive x direction (the direction of propagation of the wave) when $\mathbf{B}$ is in the z direction. The magnetic force at any time t is

$$F_x = qv_yB = \frac{q^2EB}{m}t$$

The impulse of this force equals the momentum transferred by the wave to the particle. Setting the impulse equal to the momentum p_x, we obtain

$$p_x = \int_0^{t_1} F_x\, dt$$

$$= \int_0^{t_1} \frac{q^2EB}{m}t\, dt = \frac{1}{2}\frac{q^2EB}{m}t_1^2$$

If we use $B = E/c$, this becomes

$$p_x = \frac{1}{c}\left(\frac{1}{2}\frac{q^2E^2}{m}t_1^2\right) \qquad 29\text{-}23$$

Comparing Equations 29-22 and 29-23, we see that the momentum acquired by the charge in the direction of the wave is $1/c$ times the energy. Although our simple calculation was not rigorous, the results are correct. In general,

> The magnitude of the momentum carried by an electromagnetic wave is $1/c$ times the energy carried by the wave:

$$p = \frac{U}{c} \qquad 29\text{-}24$$

Momentum and energy in an electromagnetic wave

Since the intensity of a wave is the energy per unit time per unit area, the intensity divided by c is the momentum carried by the wave per unit time per unit area. The momentum carried per unit time is a force. The intensity divided by c is thus a force per unit area, which is a pressure. This pressure is called **radiation pressure** P_r:

$$P_r = \frac{I}{c} \qquad 29\text{-}25$$

We can relate the radiation pressure to the electric or magnetic fields by using Equations 29-21 and 29-16:

$$P_r = \frac{I}{c} = \frac{E_0B_0}{2\mu_0c} = \frac{E_{rms}B_{rms}}{\mu_0c} = \frac{E_0^2}{2\mu_0c^2} = \frac{B_0^2}{2\mu_0} \qquad 29\text{-}26$$

Radiation pressure

Consider an electromagnetic wave incident normally on some surface. If the surface absorbs energy U from the electromagnetic wave, it also absorbs momentum p given by Equation 29-24, and the pressure exerted on the surface equals the radiation pressure. If the wave is reflected, the momentum transferred is twice the energy incident on the surface because the wave now carries momentum in the opposite direction. The pressure exerted on the surface by the wave is then twice the radiation pressure.

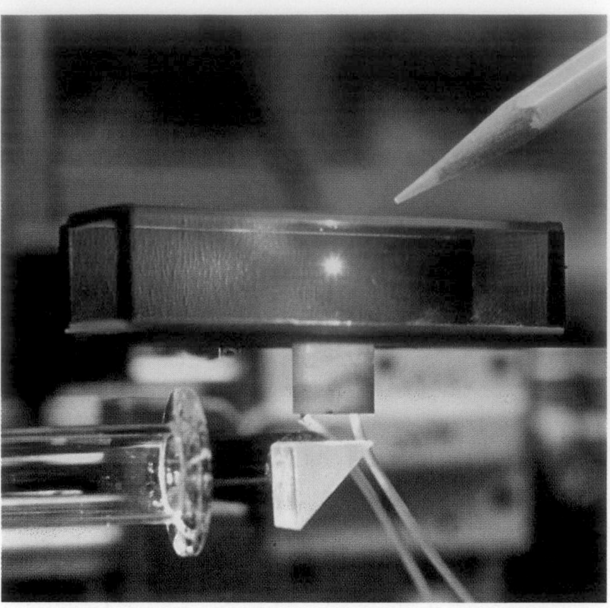

(*a*)

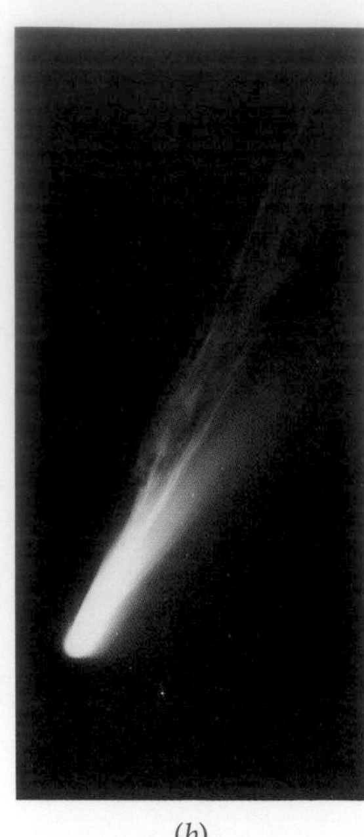

(*b*)

(*a*) A transparent glass sphere, about 25 μm in diameter (and visible here as a starlike speck) is suspended by the radiation pressure of an upward-directed 250-mW laser beam. (*b*) The comet Mrkos photographed in August 1957. The tail is pushed away from the sun and split into two parts by solar radiation pressure and by the solar wind, which is a stream of charged particles emitted by the sun. The splitting occurs because lighter particles in the tail are more readily deflected than heavier ones.

Example 29-4

A 100-W light bulb emits spherical electromagnetic waves uniformly in all directions. Find the intensity, the radiation pressure, and the electric and magnetic fields at a distance of 3 m from the bulb, assuming that 50 W goes into electromagnetic radiation.

At a distance r from the bulb, the energy is spread uniformly over an area $4\pi r^2$. The intensity is therefore

$$I = \frac{50 \text{ W}}{4\pi r^2}$$

At $r = 3$ m, the intensity is

$$I = \frac{50 \text{ W}}{4\pi(3 \text{ m})^2} = 0.442 \text{ W/m}^2$$

The radiation pressure is the intensity divided by the speed of light:

$$P_r = \frac{I}{c} = \frac{0.442 \text{ W/m}^2}{3 \times 10^8 \text{ m/s}} = 1.47 \times 10^{-9} \text{ Pa}$$

This is a very small pressure compared with atmospheric pressure of the order of 10^5 Pa.

The maximum value of the magnetic field is, from Equation 29-26,

$$\begin{aligned} B_0 &= (2\mu_0 P_r)^{1/2} \\ &= [2(4\pi \times 10^{-7})(1.47 \times 10^{-9})]^{1/2} \\ &= 6.08 \times 10^{-8} \text{ T} \end{aligned}$$

The maximum value of the electric field is the speed of light times B_0:

$$E_0 = cB_0 = 18.2 \text{ V/m}$$

The electric and magnetic fields are of the form $E = E_0 \sin(kx - \omega t)$ and $B = B_0 \sin(kx - \omega t)$ with $E_0 = 18.2$ V/m and $B_0 = 6.08 \times 10^{-8}$ T.

Example 29-5

An astronaut is stranded in space a distance of 20 m from her spaceship. She carries a 1-kW laser gun. If the total mass of the astronaut, space suit, and laser gun is 95 kg, how long will it take her to reach the ship if she points the laser gun directly away from it?

The power of the laser gun is the rate at which it emits energy. If we divide the power by c, we obtain the rate at which momentum is emitted. If P is the power of the laser gun, the rate of change of momentum of the system due to the momentum carried away by the radiation is

$$\frac{dp}{dt} = \frac{P}{c}$$

The astronaut must then have an equal and opposite rate of change of momentum. The force on the astronaut is thus

$$F = \frac{dp}{dt} = \frac{P}{c} = ma$$

If the power of the laser gun is constant, the acceleration of the astronaut will be constant and given by

$$a = \frac{P}{mc} = \frac{1000 \text{ W}}{(95 \text{ kg})(3 \times 10^8 \text{ m/s}^2)}$$
$$= 3.51 \times 10^{-8} \text{ m/s}^2$$

This is a very small acceleration. The time required for the astronaut to travel 20 m is found from the constant-acceleration formula:

$$x = \tfrac{1}{2}at^2$$

$$t = \sqrt{\frac{2x}{a}} = \sqrt{\frac{2(20 \text{ m})}{3.51 \times 10^{-8} \text{ m/s}^2}}$$
$$= 3.38 \times 10^4 \text{ s} = 9.38 \text{ h}$$

Dave Cooke and his colleagues at the University of Chicago use a 40.6-cm diameter silver-coated telescope mirror and a tiny cone-shaped sapphire crystal to concentrate sunlight to intensities that are 84,000 times that received normally at the surface of the earth. The world record 72 kW/m^2 exceeds the intensity of light at the surface of the sun itself (63 kW/m^2)—making it the highest concentration of sunlight produced in the solar system. The light is focused to a 1-cm diameter spot by the mirror and is then compressed to a 1-mm diameter spot by the nonimaging sapphire concentrator. The sapphire was chosen for its low absorption and high index of refraction. Sunlight hits the base of the cone, is funnelled through the sapphire, and exits at its tip. The specific shape of the crystal maximizes the concentration of light leaving the tip—rather than providing an accurate image of the source of light (as in conventional lenses). Despite its low absorptivity, the sapphire crystal is often heated by the concentrated sunlight to the point of explosion.

29-5 The Electromagnetic Spectrum

Electromagnetic waves include light, radio waves, x rays, gamma rays, microwaves, and others. The various types of electromagnetic waves differ only in wavelength and frequency, which is related to the wavelength in the usual way:

$$f = \frac{c}{\lambda}$$

Table 29-1 gives the **electromagnetic spectrum** and the names usually associated with the various frequency and wavelength ranges. These ranges are often not well defined and sometimes overlap. For example, electromagnetic waves with wavelengths of about 0.1 nm are usually called x rays, but if they originate from nuclear radioactivity, they are called gamma rays.

The human eye is sensitive to electromagnetic radiation with wavelengths from about 400 to 700 nm, the range called **visible light.** The shortest wavelengths in the visible spectrum correspond to violet light and the longest to red light, with all the colors of the rainbow falling between these extremes. Electromagnetic waves with wavelengths slightly less than those of visible light are called **ultraviolet rays,** and those with wavelengths

Table 29-1 **The Electromagnetic Spectrum**

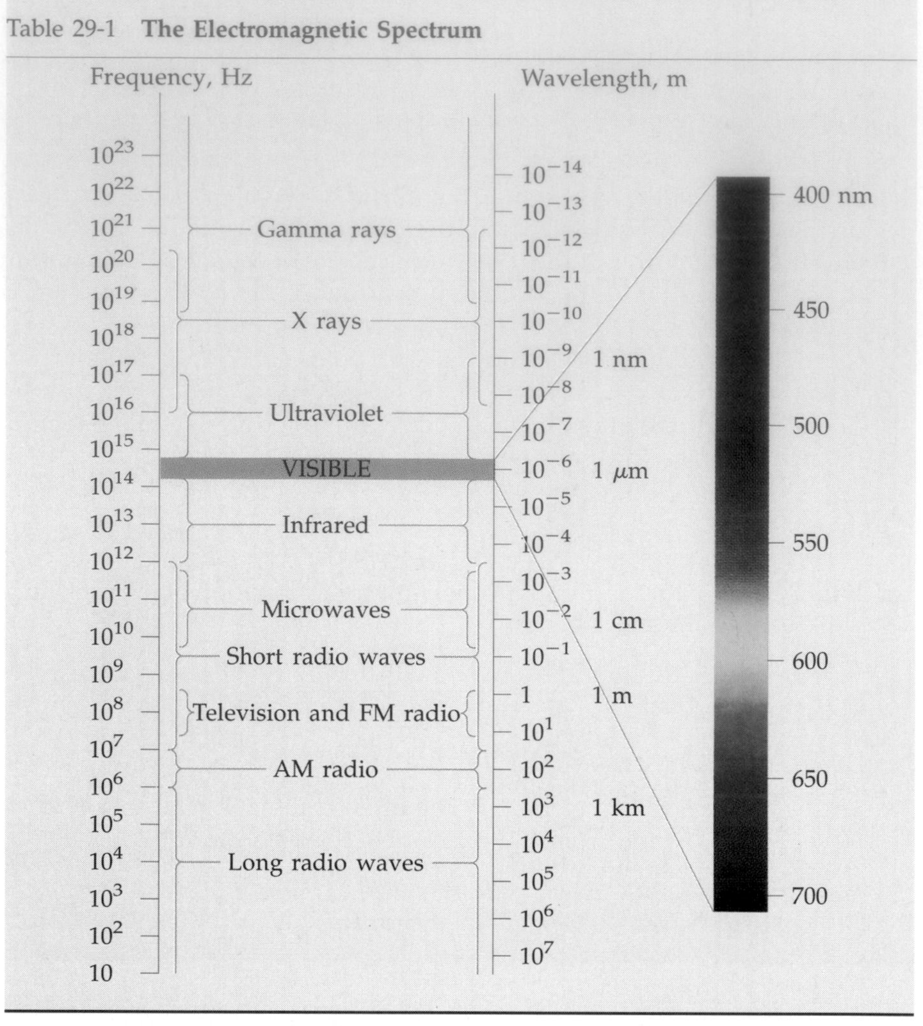

slightly greater than those of visible light are called **infrared waves.** Heat radiation given off by bodies at ordinary temperatures is in the infrared region of the electromagnetic spectrum. There are no limits on the wavelengths of electromagnetic radiation; that is, all wavelengths (or frequencies) are theoretically possible.

The differences in wavelengths of the various kinds of electromagnetic waves are very important. As we know, the behavior of waves depends strongly on the relative sizes of the wavelengths and the physical objects or apertures the waves encounter. Since the wavelengths of light are in the rather narrow range from about 400 to 700 nm, they are much smaller than most obstacles, so the ray approximation is often valid. The wavelength and frequency are also important in determining the kinds of interactions between electromagnetic waves and matter. X rays, for example, which have very short wavelengths and high frequencies, easily penetrate many materials that are opaque to lower-frequency light waves, which are absorbed by the materials. Microwaves have wavelengths of the order of a few centimeters and frequencies that are close to the natural resonance frequencies of water molecules in solids and liquids. Microwaves are therefore readily absorbed by the water molecules in foods, which is the mechanism for heating in microwave ovens.

(a)

(b)

(*a*) Television antennas, shown here, operate at radio frequencies. Messages are transmitted by encoding them as modulations of frequency (FM) or amplitude (AM). The same antenna will function as a transmitter or receiver. However, antennas used for transmission operate at significantly higher power levels than those used for reception. (*b*) The Caltech radio telescope in Owens Valley, California. Stars, galaxies, quasars, and pulsars are all sources of radio waves. Nearly one-fifth of cosmic radio sources are presently "unidentified", that is, are not correlated with any source that has been seen optically.

Electromagnetic waves are produced when electric charges accelerate. When electric charges oscillate they radiate electromagnetic waves whose frequency equals the frequency of oscillation. The wavelength of the waves emitted is therefore determined by the frequency of oscillation of the charges. Radio waves, which have frequencies from about 550 to 1600 kHz for AM radio waves and from about 88 to 108 MHz for FM radio waves, are produced by macroscopic electric currents oscillating in radio antennas. Light waves, which have frequencies of the order of 10^{14} Hz, originate from the motion of atomic charges.

Figure 29-7 is a schematic drawing of an **electric-dipole antenna,** which consists of two bent conducting rods fed by an alternating-current generator. At the time $t = 0$, shown in Figure 29-6*a*, the ends of the rods are charged, and there is an electric field near the rods parallel to the rods. There is also a magnetic field (not shown) encircling the rods due to the current in the rods. The magnetic field is perpendicular to the page. These fields move outward from the rods with the speed of light. After one-fourth period, at $t = \frac{1}{4}T$, the rods are uncharged, and the electric field near the rods is zero, as shown in Figure 29-7*b*. At $t = \frac{1}{2}T$, the rods are again charged, but the charges are opposite to those at $t = 0$, as shown in Figure 29-7*c*. The electric and

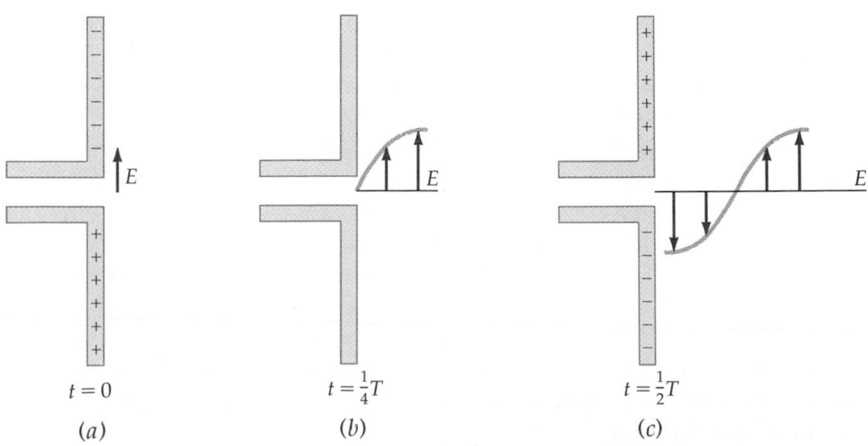

Figure 29-7 An electric-dipole antenna. Alternating current is supplied to the antenna by a generator (not shown). The electric field due to the charges in the antenna propagates outward at the speed of light. There is also a propagating magnetic field (not shown) perpendicular to the paper due to the current in the antenna.

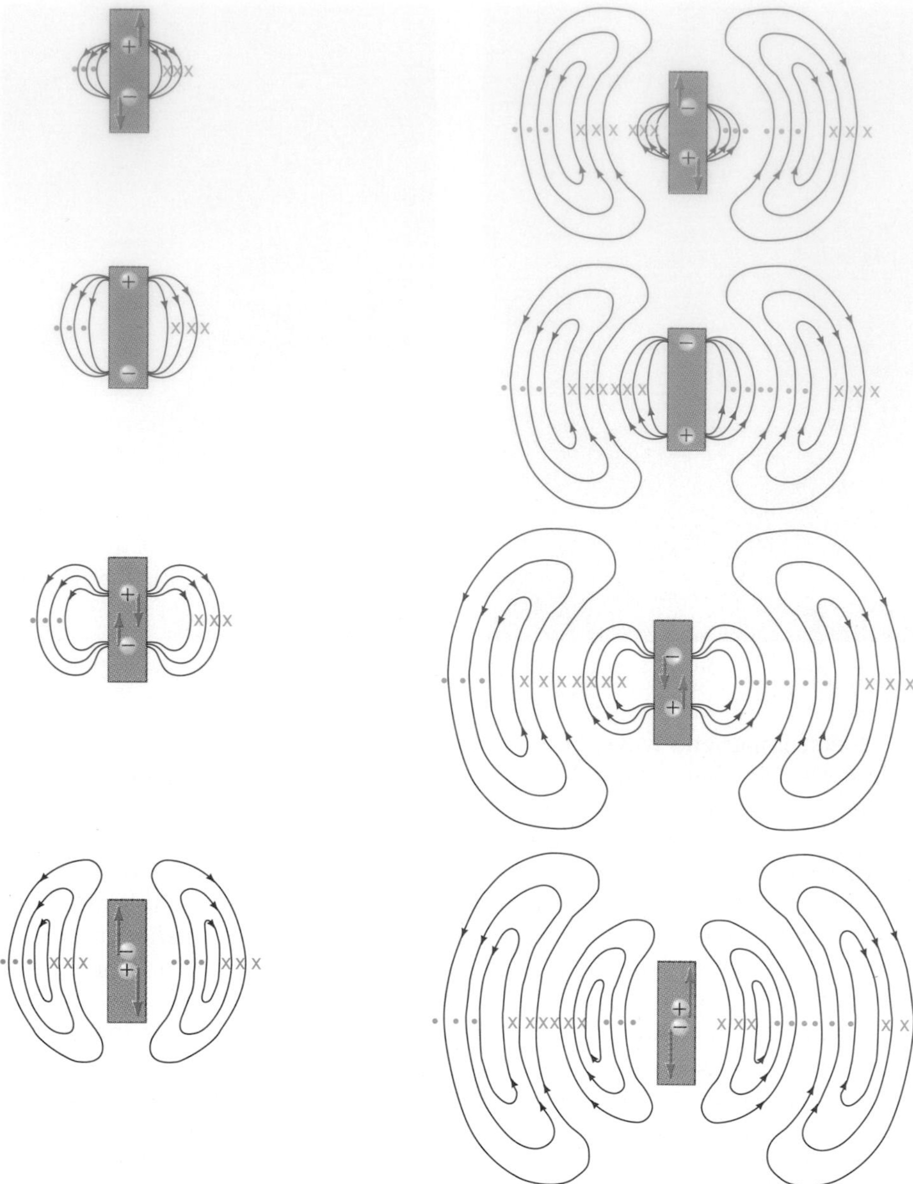

Figure 29-8 Electric- and magnetic-field lines produced by an oscillating electric dipole.

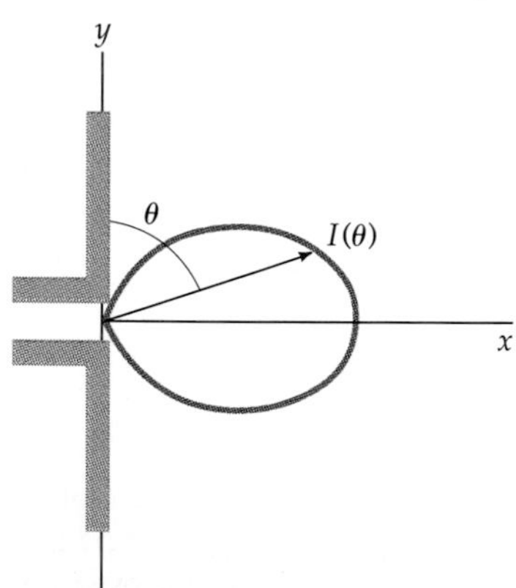

Figure 29-9 Polar plot of the intensity of electromagnetic radiation from an electric-dipole antenna versus angle. The intensity $I(\theta)$ is proportional the length of the arrow. The intensity is maximum perpendicular to the antenna at $\theta = 90°$ and minimum along the antenna at $\theta = 0°$ or $\theta = 180°$.

magnetic fields at a great distance from this transmitting antenna are quite different from the fields near the antenna. Far from the antenna, the electric and magnetic fields oscillate in phase with simple harmonic motion, perpendicular to each other and to the direction of propagation of the wave. The wave is thus a transverse wave. Figure 29-8 shows the electric and magnetic fields far from an electric-dipole antenna.

The radiation from an electric-dipole antenna such as that in Figure 29-7 is called **electric-dipole radiation.** Many electromagnetic waves exhibit the characteristics of electric-dipole radiation. A characteristic that will be important in our discussion of light in Chapter 30 is that the intensity of the electromagnetic waves radiated by a dipole antenna is zero along the axis of the antenna and maximum in the directions perpendicular to the axis. That is, if the dipole is in the z direction with center at the origin, the intensity is zero along the z axis and maximum in the xy plane. In the direction of a line making an angle θ with the z axis, as in Figure 29-9, the intensity is proportional to $\sin^2 \theta$. Far from the antenna, the electric field is parallel to the dipole.

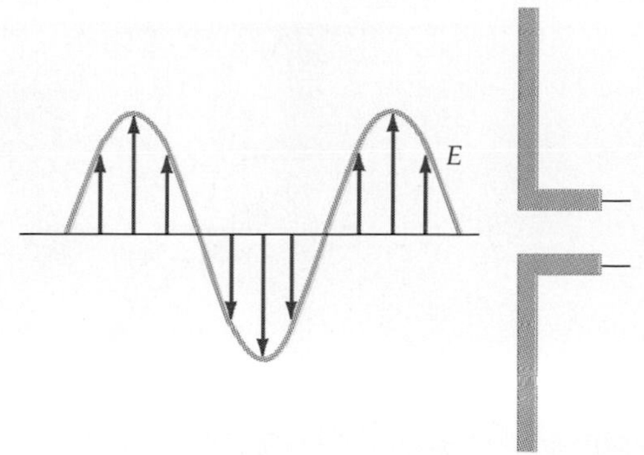

Figure 29-10 An electric-dipole antenna for detecting electromagnetic radiation. The alternating electric field of the radiation produces an alternating current in the antenna.

Electromagnetic waves of radio or television frequencies can be detected by a dipole receiving antenna oriented parallel to the electric field so that it induces an alternating current in the antenna (Figure 29-10). They can also be detected by a loop antenna oriented perpendicular to the magnetic field so that the changing magnetic flux through the loop induces a current in the loop (Figure 29-11). Electromagnetic waves of frequencies in the visible light range can be detected by the eye or by photographic film, both of which are mainly sensitive to the electric field.

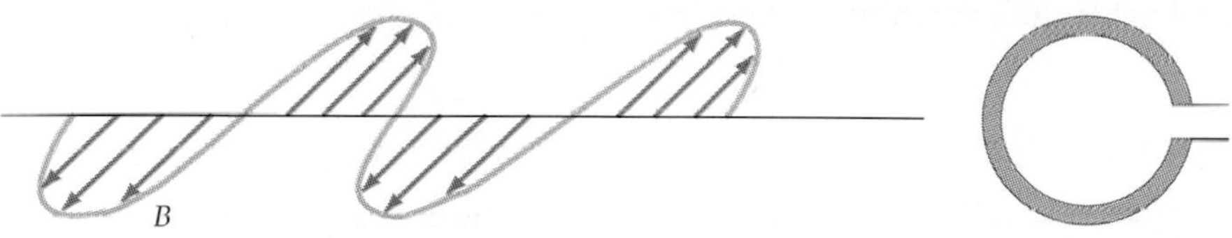

Figure 29-11 A loop antenna for detecting electromagnetic radiation. The alternating magnetic flux in the loop due to the magnetic field of the radiation induces an alternating current in the loop.

Example 29-6

An antenna consisting of a single loop of wire of radius 10 cm is used to detect electromagnetic waves for which $E_{rms} = 0.15$ V/m. Find the rms induced emf in the loop if the wave frequency is (*a*) 600 kHz and (*b*) 600 MHz.

(*a*) From Faraday's law the magnitude of the induced emf is

$$|\mathcal{E}| = \frac{d\phi_m}{dt} = \pi r^2 \frac{dB}{dt}$$

Then $\mathcal{E}_{rms} = \pi r^2 (dB/dt)_{rms}$. If $B = B_0 \sin(kx - \omega t)$, $dB/dt = -\omega B_0 \cos(kx - \omega t)$ and $(dB/dt)_{rms} = \omega B_{rms} = \omega E_{rms}/c = (2\pi f/c)E_{rms}$. The rms induced emf is then

$$\mathcal{E}_{rms} = \pi r^2 (2\pi f/c) E_{rms}$$

At $f = 600$ kHz, the emf is

$$\mathcal{E}_{rms} = \pi(0.1\text{ m})^2 2\pi(6 \times 10^5\text{ Hz})(3 \times 10^8\text{ m/s})^{-1}(0.15\text{ V/m})$$
$$= 5.92 \times 10^{-5}\text{ V}$$

(*b*) The induced emf is proportional to the frequency, so at 600 MHz it will be 1000 times greater than at 600 kHz. Then $\mathcal{E}_{rms} = (10^3)(5.92 \times 10^{-5}\text{ V}) = 0.0592$ V.

(*Left*) Magnetrons are used to generate the radar beams in tracking antennas and, as in the one shown above, to power microwave ovens. The central horizontal cylinder is a cathode that is heated and emits electrons. (The thin strips attached to the central cylinder are cooling fins.) Two disk-shaped magnets at either end provide an axial magnetic field. The emitted electrons are accelerated, creating oscillating electric fields that generate oscillating magnetic fields. The anode is configured so that electromagnetic oscillations at microwave frequencies can be sustained with little energy loss—that is, the magnetron acts as a resonant cavity for standing electromagnetic waves that have wavelengths of the order of a few centimeters. The microwaves exit from their standing pattern via the tube at the right, much as sound waves exit from a clarinet.

(*Right*) A beam of visible light emitted by electrons undergoing acceleration in a synchrotron. Electrons oscillating continuously in a typical radio antenna radiate sinusoidal electromagnetic fields. In a synchrotron, bunches of electrons move in circular paths at speeds near the speed of light and emit radiation that is largely compressed into brief pulses in the direction of motion of the electrons.

Questions

1. Which waves have greater frequencies, light waves or x rays?
2. Are the frequencies of ultraviolet radiation greater or less than those of infrared radiation?
3. What kind of waves have wavelengths of the order of a few meters?

Summary

1. Ampère's law can be generalized to apply to discontinuous currents if the conduction current I is replaced by $I + I_d$, where I_d is called Maxwell's displacement current:

$$I_d = \epsilon_0 \frac{d\phi_e}{dt}$$

2. The laws of electricity and magnetism are summarized by Maxwell's equations, which are

$$\oint_S E_n\, dA = \frac{1}{\epsilon_0} Q_{\text{inside}}$$ Gauss's law

$$\oint_S B_n\, dA = 0$$ Gauss's law for magnetism (isolated magnetic poles do not exist)

$$\oint_C \mathbf{E} \cdot d\boldsymbol{\ell} = -\frac{d}{dt} \int_S B_n\, dA$$ Faraday's law

$$\oint_C \mathbf{B} \cdot d\boldsymbol{\ell} = \mu_0 I + \mu_0 \epsilon_0 \frac{d}{dt} \int_S E_n\, dA$$ Ampère's law modified

3. Maxwell's equations imply that the electric- and magnetic-field vectors in free space obey a wave equation of the form

$$\frac{\partial^2 \mathbf{E}}{\partial x^2} = \frac{1}{c^2}\frac{\partial^2 \mathbf{E}}{\partial t^2}$$

where

$$c = \frac{1}{\sqrt{\mu_0 \epsilon_0}}$$

is the wave speed. The fact that this speed equals the speed of light led Maxwell to surmise correctly that light is an electromagnetic wave.

4. In an electromagnetic wave, the electric and magnetic field vectors are perpendicular to each other and to the direction of propagation. Their magnitudes are related by

$$E = cB$$

5. Electromagnetic waves carry energy and momentum. The average energy density of an electromagnetic wave is

$$\eta_{av} = \frac{1}{2}\frac{E_0 B_0}{\mu_0 c} = \frac{E_{rms} B_{rms}}{\mu_0 c}$$

The intensity of an electromagnetic wave is given by

$$I = \eta_{av} c = \frac{1}{2}\frac{E_0 B_0}{\mu_0} = \frac{1}{2}\frac{E_0^2}{\mu_0 c} = \frac{1}{2}\frac{cB_0^2}{\mu_0} = |\mathbf{S}|_{av}$$

where **S**, called the Poynting vector, describes the transport of electromagnetic energy:

$$\mathbf{S} = \frac{\mathbf{E} \times \mathbf{B}}{\mu_0}$$

6. An electromagnetic wave carries momentum that is equal to $1/c$ times the energy carried by the wave:

$$p = \frac{U}{c}$$

The intensity of an electromagnetic wave divided by c is the momentum carried by the wave per unit time per unit area, which is called the radiation pressure of the wave:

$$P_r = \frac{I}{c}$$

If the wave is incident normally on a surface and completely absorbed, it exerts a pressure equal to its radiation pressure. If it is incident normally and reflected, the pressure exerted is twice the radiation pressure.

7. Electromagnetic waves include light, radio waves, x rays, gamma rays, microwaves, and others. The various types of electromagnetic waves differ only in wavelength and frequency, which is related to the wavelength in the usual way:

$$f = \frac{c}{\lambda}$$

8. Electromagnetic waves are produced when electric charges accelerate. Oscillating charges in an electric-dipole antenna radiate electromagnetic waves with an intensity that is maximum in directions perpendicular to the antenna and zero along the axis of the antenna. Perpendicular to the antenna and far away from it, the electric field of the electromagnetic wave is parallel to the antenna.

James Clerk Maxwell (1831–1879)

C. W. F. Everitt
Stanford University

One day in 1877 a young Scottish undergraduate named Donald MacAlister, afterwards a distinguished physician and academic statesman, wrote home from Cambridge University that he had just had dinner with a professor who was "one of the best of our men, and a thorough old Scotch laird in ways and speech." This description of James Clerk Maxwell was accurate. He was wealthy, an expert swimmer and horseman, proprietor of an estate of 2,000 acres in the southwest of Scotland—few would suspect that this man who seemed to MacAlister charmingly old-fashioned in 1877 was also a scientist whose writings remain astonishingly up to date in the 1990s; that he was the greatest mathematical physicist since Newton; that he had created the electromagnetic theory of light and predicted the existence of radio waves; that he had written the first significant paper on control theory; that he was just then writing a profound article on statistical mechanics, a science he and Ludwig Boltzmann jointly invented; that he had performed with his wife's aid a brilliant series of experiments on color vision and had taken the first color photograph; and that in the remaining 2 years until his death due to cancer in 1879, at the age of 48, he would lay the foundations of another new subject that was to reach fruition in the twentieth century, rarefied gas dynamics.

C.W.F. (Francis) Everitt obtained his doctorate in 1959 from Imperial College, London, in the then new field of paleomagnetic studies of plate tectonics. He decided that, as much as he liked geology, his true métier was physics. So he moved in 1960 to the University of Pennsylvania where he investigated "third sound," a peculiar kind of surface wave propagating in thin films of superfluid helium. He then transferred to Stanford University and with others initiated research on a long-range NASA program (Gravity Probe B) to test Einstein's general theory of relativity using precise gyroscopes in Earth orbit. Everitt's interest in history goes back to his high school days. He has written on the history of precise measurement; on scientific creativity; on the relationship between theory and experiment in physics; on spectroscopy; on science, history and religion; on the dynamics of "big physics"; and on several physicists, including three books about Maxwell.

Maxwell's undergraduate career was unusually protracted. He spent 3 years at Edinburgh University and another $3\frac{1}{4}$ at Cambridge. His student friendships were more with classical scholars than scientists. Unlike Einstein, he enjoyed student life and was fortunate to gain the attention of some outstanding teachers. At Edinburgh he was influenced by two powerful and sharply contrasted men, James David Forbes and Sir William Hamilton, the metaphysician. Forbes was an experimentalist, who had invented the seismometer and done important work on the polarization of infrared radiation and the study of the motion of glaciers. He gave Maxwell the run of his laboratory and with his help began the experiments on color vision that led eventually to Maxwell's own work on the subject. Hamilton, who had a genius for inspiring youth, imparted to Maxwell the ranging philosophic vision that can be seen in the many interesting metaphysical asides in his papers.

In 1850 Maxwell went to Cambridge. By then his mathematical bent was clear and, like many another clever undergraduate before and since, Maxwell worked hard while pretending not to. His tutor was William Hopkins, the founder of modern geophysics and arguably the greatest teacher Cambridge has ever produced. Others who influenced him were G. G. Stokes, the mathematical physicist who held the chair Newton had occupied, and William Whewell.

Maxwell's electromagnetic theory of light was rooted in the work of two men, Michael Faraday and William Thomson. Faraday's invention of the electric motor and his researches on electromagnetic induction, electrochemistry, dielectric and diamagnetic action, and magneto-optical rotation made him in Maxwell's words "the nucleus of everything electric since 1830." His contributions to theory lay in his progressively advancing ideas about lines of electric and magnetic force, in particular the geometrical relations governing electromagnetic phenomena and the idea that magnetic forces might be accounted for not by direct attractions and repulsions between elements of current but by attributing

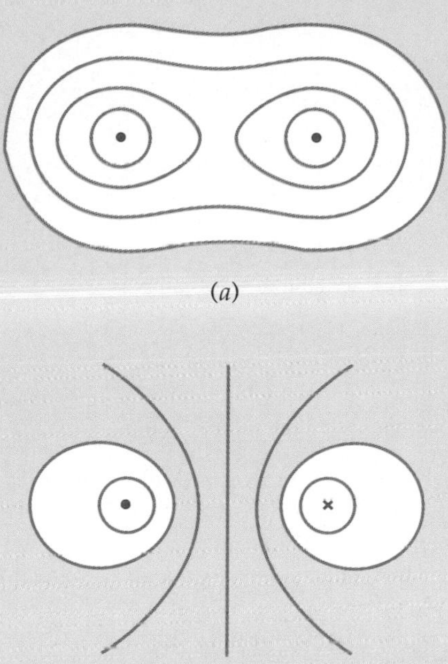

Figure 1 Faraday's explanation of forces between current-carrying wires. The two diagrams show the lines of force observed when currents are flowing in parallel wires. Faraday assumed that the lines of force tend to shorten and repel each other sideways. (*a*) For wires with currents flowing in the same direction, the lines of force pull the two wires together. (*b*) For wires with currents flowing in opposite directions, the lines of force push the wires apart.

to lines of force the property of shortening themselves and repelling each other sideways (Figure 1). Thomson's role was to relate lines of force to existing theories in electrostatics and magnetostatics, to invent a number of highly ingenious analytical techniques for solving electrical problems, and to emphasize the cardinal importance of energy principles in electromagnetism. Maxwell then introduced a series of new concepts: the *electrotonic function* (vector potential), the energy density of the field, the displacement current (see Equation 29-4), and the significance of the operation *curl* in the field equations; he organized the subject into a coherent structure and made in 1861 the momentous discovery of the equivalence between light and electromagnetic waves.

The first part of Maxwell's paper "On Faraday's Lines of Force" (1855–1856) developed an analogy, due in essence to Thomson, between lines of electric and magnetic force and streamlines in a moving incompressible fluid. Maxwell applied this to interpret many of Faraday's observations, prefacing his paper with a luminous discussion of the significance of analogies in physics. Next, still building on Faraday and Thomson, Maxwell extended the discussion to electromagnetism. He formulated a group of equations summarizing the relations of the electric and magnetic fields to the charges and currents producing them—the beginnings of what we now call Maxwell's equations (see Section 29-2). They described the phenomena with elegant precision from a point of view completely different from the then-popular action-at-a-distance theories of André-Marie Ampère and Wilhelm Weber. The central theorem in all this work was one which, following Maxwell, we now call *Stokes' theorem*, which had been published as a question by Stokes in one of the examinations taken by Maxwell in January 1854, while he was still an undergraduate.

After such a brilliant start one might have expected a rush of papers following up the new ideas. But other physicists ignored them, and Maxwell had the habit of investigating different subjects in turn, often with long intervals between successive papers in the same field. Six years elapsed before the appearance of his next paper, "On Physical Lines of Force," published in four parts in 1861–1862. During the interval Maxwell made brilliant contributions to three distinct subjects before returning to electromagnetism: color vision, the theory of Saturn's rings, and the kinetic theory of gases. He left Cambridge, became a professor at Marischal College in Aberdeen, married the daughter of the Principal of the College, and then found himself in the odd position of being forced to retire at age 29 with a life pension after his chair had been abolished when the two universities in Aberdeen were united by Act of Parliament. Fortunately, a suitable post (the chair at King's College in London) had just fallen vacant, and so he went there.

"On Physical Lines of Force" contained Maxwell's extraordinary molecular-vortex model of the electromagnetic field. In order to account for the pattern of stresses associated with lines of force by Faraday, Maxwell investigated the properties of a medium occupying all space in which tiny molecular vortices rotate with their axes parallel to the lines of force. The closer together the lines are, the faster the rotation of the vortices. In a medium of this kind the lines of force do tend to shorten themselves and repel each other sideways, yielding the right forces between currents and magnets. The question is: what makes the vortices rotate? Here Maxwell put forward an idea as ingenious as it was weird. He postulated that an electric current consists in the motion of tiny particles that mesh like gear wheels with the vortices, and that the medium is filled with similar particles between the vortices. Figure 2 gives the picture. Maxwell remarks:

> I do not bring [this hypothesis] forward as a mode of connexion existing in nature . . . [but] I venture to say that anyone who understands [its] provisional and temporary character . . . will find himself helped rather than hindered by it in his search for the true interpretation of [electromagnetic] phenomena.

The question then was how to fit electrostatic phenomena into the model. Maxwell made the medium an elastic one. Thus magnetic forces were accounted for by rotations in the medium, and electric forces by its elastic distortion. Any elastic medium will transmit waves. In Maxwell's medium the velocity of the waves turned out to be related to the ratio of electric to magnetic forces. Putting in numbers from an experiment of 1856 by G. Kohlrausch and W. Weber, Maxwell found to his astonishment that the propagation velocity was equal to the velocity of light. With excitement he wrote "we can scarcely avoid the inference that *light consists in the transverse undulation of the same medium which is the cause of electric and magnetic phenomena*" (see Section 29-3).

Having made the great discovery, Maxwell promptly jettisoned his model. Instead of attempting a more refined mechanical explanation of the phenomena, he formulated a system of electromagnetic equations from which he deduced that waves of electric and magnetic force would propagate through space with the velocity of light. That is why his is called an *electromagnetic* theory of light, in contrast to the theories of the mechanical ether that preceded it. The theory appeared in two papers of 1865 and 1868, and in its most general form in the great *Treatise on Electricity and Magnetism*, published in 1873. This was a work of such scope that Robert Andrews Millikan, author of the famous oil-drop experiment to measure the charge on the electron, ranked it with Newton's *Principia* in considering them the two most influential books in the history of physics, "the one creating our modern mechanical world and the other our modern electrical world."

Equally profound were Maxwell's contributions to statistical and molecular physics. They began with a paper in 1859 on the kinetic theory of gases, in which Maxwell introduced the velocity distribution function and enunciated the *equipartition theorem* (Section 16-7), which in its original form stated that the average translational and rotational energies of large numbers of colliding molecules, whether of the same or different species, are equal. One surprising conclusion, afterwards confirmed experimentally by Maxwell and his wife, was that the viscosity of a gas should over a wide range be independent of its pressure. Another result was Maxwell's estimate of the mean free path of a gas molecule, which Loschmidt in 1865 applied to make the first serious estimates of the diameters of molecules. Later, Maxwell developed the general theory of transport phenomena, from which the Boltzmann equation is derived; invented the concept of ensemble averaging; created rarefied-gas dynamics; and conceived that "very small BUT lively being" the Maxwell demon.

The demon, so named by Kelvin, is one of the earliest examples in physics of a "thought experiment." Maxwell imagined two chambers of gas, A and B, separated by a wall in which there was a trap door guarded by a tiny being with eyesight so acute that it could discern the motion of individual molecules. By opening the door when a fast molecule approached from chamber A or as a slow molecule approached from chamber B, the demon could redistribute the velocities to make B hotter than A without doing any work and thereby defeat the sec-

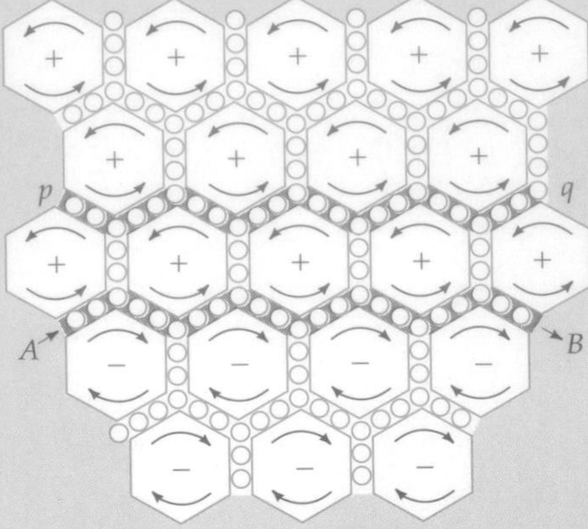

Figure 2 Maxwell's vortex model of the magnetic field. The rotating vortices represent lines of magnetic force. They mesh with small particles that act like gear wheels. In free space the particles are restrained from moving, except for a small elastic reaction (the displacement current), but in a conducting wire they are free to move. Their motion constitutes an electric current, which in turn sets the vortices in rotation, creating the magnetic field around the wire. *A* and *B* represent current through a wire, and *p* and *q* represent an induced current in an adjacent wire. (Redrawn from The Scientific Papers of James Clerk Maxwell, Vol. I, fig. 2 after p. 488.)

James Clerk Maxwell (1831–1878) with his wife, Katherine Mary, and dog.

ond law of thermodynamics. Maxwell's point was to demonstrate that the second law of thermodynamics is inherently a statistical law and not a dynamical one.

The work by Maxwell and Boltzmann on statistical mechanics had profound implications for modern physics. Brilliant as its successes were, the failures were—as Maxwell saw—in some ways even more striking. The equipartition theorem gave an answer for the ratio of the specific heats of gases that disagreed with experiment, while some of Boltzmann's theorems "proved too much" because they would apply to the properties of solids and liquids as well as gases. These questions remained shrouded in mystery until the emergence in 1900 of Planck's quantum hypothesis (see Section 35-1). Writing about them in 1877, Maxwell confessed his bewilderment and stated that nothing remained but to adopt the attitude of "thoroughly conscious ignorance that is the prelude to every real advance in knowledge."

Maxwell was an unusually sensitive man, with strong religious feeling and a fascinating and astonishing sense of humor. Many of his letters reveal a delightfully sly irony. He also had some talent for writing poetry, usually light, but occasionally touching a deeper note. The last stanza of one poem to his wife, written in 1867, was

All powers of mind, all force of will
May lie in dust when we are dead,
But love is ours, and shall be still
When earth and seas are fled.

Suggestions for Further Reading

Campbell, L., and W. Garnett: *The Life of James Clerk Maxwell,* Johnson Reprint Corporation, Harcourt, Brace & Jovanovich, New York, 1970 (reprint of Oxford 1882 edition).

Everitt, C. W. F.: *James Clerk Maxwell: Physicist and Natural Philosopher,* Scribner, New York, 1975.

These books offer more details on Maxwell's fascinating life and work than found in this chapter's essay.

Mulligan, Joseph F.: "Heinrich Hertz and the Development of Physics," *Physics Today,* March 1989, p. 50.

This article describes Hertz's life and work, including not only his experiments on electromagnetism, but also his discovery of the photoelectric effect and his work with cathode rays.

Shamos, Morris H: "The Electromagnetic Field—James Clerk Maxwell," in *Great Experiments in Physics,* Henry Holt and Company, New York, 1959. Reprinted by Dover, 1987.

In annotated excerpts from his paper of 1865, "A Dynamical Theory of the Electromagnetic Field," Maxwell discusses his conception of the electromagnetic field and introduces his equations. The excerpts are introduced by a brief biography.

Shamos, Morris H.: "Electromagnetic Waves—Heinrich Hertz," in *Great Experiments in Physics,* Henry Holt and Company, New York, 1959. Reprinted by Dover, 1987.

Hertz's account of the set of experiments in which he showed the existence of electromagnetic waves of much greater wavelength than light waves. He was able to demonstrate in these experiments on reflection, refraction, and polarization that these waves behave the same as light. The chapter also includes a brief biography.

Review

A. Objectives: After studying this chapter, you should:

1. Be able to write Maxwell's equations and discuss the experimental basis of each.

2. Be able to state the expression for the speed of an electromagnetic wave in terms of the fundamental constants μ_0 and ϵ_0.

3. Be able to give the expression for the Poynting vector and discuss its significance.

4. Be able to state the relationships between the Poynting vector, the intensity of an electromagnetic wave, and radiation pressure.

5. Be able to calculate the radiation pressure and the maximum values of E and B from the intensity of an electromagnetic wave.

B. Define, explain, or otherwise identify:

Maxwell's displacement current
Maxwell's equations
Wave equation
Plane wave
Linearly polarized wave
Circularly polarized wave
Poynting vector
Radiation pressure
Electromagnetic spectrum
Visible light
Ultraviolet rays
Infrared waves
Electric-dipole antenna
Electric-dipole radiation

C. True or false: If the statement is true, explain why it is true. If it is false, give a counterexample.

1. Maxwell's equations apply only to fields that are constant in time.

2. The wave equation can be derived from Maxwell's equations.

3. Electromagnetic waves are transverse waves.

4. In an electromagnetic wave, the electric and magnetic fields are in phase.

5. In an electromagnetic wave, the electric- and magnetic-field vectors **E** and **B** are equal in magnitude.

6. In an electromagnetic wave, the electric and magnetic energy densities are equal.

Problems

Level I

29-1 Maxwell's Displacement Current

1. A parallel-plate capacitor in air has circular plates of radius 2.3 cm separated by 1.1 mm. Charge is flowing onto the upper plate and off the lower plate at a rate of 5 A. (*a*) Find the time rate of change of the electric field between the plates. (*b*) Compute the displacement current between the plates and show that it equals 5 A.

2. In a region of space, the electric field varies according to

$$E = (0.05 \text{ N/C}) \sin 2000t$$

where t is in seconds. Find the maximum displacement-current through a 1 m^3 area perpendicular to **E**.

29-2 Maxwell's Equations

There are no problems for this section.

29-3 The Wave Equation for Electromagnetic Waves

3. Show by direct substitution that the wave function

$$E_y = E_0 \sin (kx - \omega t) = E_0 \sin k(x - ct)$$

where $c = \omega/k$, satisfies Equation 29-10.

4. Use the known values of μ_0 and ϵ_0 in SI units to compute

$$c = \frac{1}{\sqrt{\mu_0 \epsilon_0}}$$

and show that it is approximately 3×10^8 m/s.

29-4 Energy and Momentum in an Electromagnetic Wave

5. An electromagnetic wave has an intensity of 100 W/m^2. Find (*a*) the radiation pressure P_r, (*b*) E_{rms}, and (*c*) B_{rms}.

6. The amplitude of an electromagnetic wave is $E_0 =$ 400 V/m. Find (*a*) E_{rms}, (*b*) B_{rms}, (*c*) the intensity I, and (*d*) the radiation pressure P_r.

7. (*a*) Show that if E is in volts per meter and B is in teslas, the units of the Poynting vector $\mathbf{S} = (\mathbf{E} \times \mathbf{B})/\mu_0$ are watts per square meter. (*b*) Show that if the intensity I is in watts per square meter, the units of radiation pressure $P_r = I/c$ are newtons per square meter.

8. (*a*) An electromagnetic wave of intensity 200 W/m^2 is incident normally on a rectangular black card with sides of 20 and 30 cm that absorbs all the radiation. Find the force exerted on the card by the radiation. (*b*) Find the force exerted by the same wave if the card reflects all the radiation incident on it.

9. Find the force exerted by the electromagnetic wave on the reflecting card in part (*b*) of Problem 8 if the radiation is incident at an angle of 30° to the normal.

10. The rms value of the electric field in an electromagnetic wave is $E_{rms} = 400$ V/m. (*a*) Find B_{rms}, (*b*) the average energy density, and (*c*) the intensity.

11. Show that the units of $E = cB$ are consistent; that is, show that when B is in teslas and c is in meters per second, the units of cB are volts per meter or newtons per coulomb.

12. The root-mean-square value of the magnitude of the magnetic field in an electromagnetic wave is $B_{rms} = 0.245\ \mu T$. Find (*a*) E_{rms}, (*b*) the average energy density, and (*c*) the intensity.

29-5 The Electromagnetic Spectrum

13. Find the wavelength for (*a*) a typical AM radio wave with a frequency of 1000 kHz and (*b*) a typical FM radio wave of 100 MHz.

14. What is the frequency of a 3-cm microwave?

15. What is the frequency of an x ray with a wavelength of 0.1 nm?

Level II

16. For Problem 1, show that at a distance r from the axis of the plates the magnetic field between the plates is given by $B = (1.89 \times 10^{-3}\ T/m)r$ if r is less than the radius of the plates.

17. (*a*) Show that for a parallel-plate capacitor the displacement current is given by $I_d = C\,dV/dt$, where C is the capacitance and V the voltage across the capacitor. (*b*) A parallel plate capacitor $C = 5$ nF is connected to an emf $\mathcal{E} = \mathcal{E}_0 \cos \omega t$, where $\mathcal{E}_0 = 3$ V and $\omega = 500\ \pi$. Find the displacement current between the plates as a function of time. Neglect any resistance in the circuit.

18. Current of 10 A flows into a capacitor having plates with areas of 0.5 m^2. (*a*) What is the displacement current between the plates? (*b*) What is dE/dt between the plates for this current? (*c*) What is the line integral of $\mathbf{B}\cdot d\boldsymbol{\ell}$ around a circle of radius 10 cm that lies within and parallel to the plates?

19. The intensity of radiation from an electric dipole is proportional to $(\sin^2 \theta)/r^2$, where θ is the angle between the electric dipole moment and the position vector $\mathbf{r}$. A radiating electric dipole lies along the z axis (its dipole moment is in the z direction). Let I be the intensity of the radiation at a distance $r = 10$ m and at angle $\theta = 90°$. Find the intensity (in terms of I) at (*a*) $r = 30$ m, $\theta = 90°$; (*b*) $r = 10$ m, $\theta = 45°$; and (*c*) $r = 20$ m, $\theta = 30°$.

20. (*a*) For the situation described in Problem 19, at what angle is the intensity at $r = 5$ m equal to I? (*b*) At what distance is the intensity equal to I at $\theta = 45°$?

21. A typical AM radio station radiates an isotropic sinusoidal wave with an average power of 50 kW. What are the amplitudes of E_{max} and B_{max} at a distance of (*a*) 500 m, (*b*) 5 km, and (*c*) 50 km?

22. The intensity of sunlight striking the earth's upper atmosphere is called the solar constant and is 1.35 kW/m^2. (*a*) Find E_{rms} and B_{rms} due to the sun at the upper atmosphere of the earth. (*b*) Find the average power output of the sun. (*c*) Find the intensity and the radiation pressure at the surface of the sun.

23. At the surface of the earth, there is an approximate average solar flux of 0.75 kW/m^2. A family wishes to construct a solar energy conversion system to power their home. If the conversion system is 30 percent efficient and the family needs a maximum of 25 kW, what effective surface area is needed for perfectly absorbing collectors?

24. Instead of sending power by a 750-kV, 1000-A transmission line, one desires to beam this energy via an electromagnetic wave. The beam has a uniform intensity within a cross-sectional area of 50 m^2. What are the rms values of the electric and the magnetic fields?

25. A demonstration laser has an average output power of 0.9 mW and a beam diameter of 1.2 mm. What is the force exerted by the laser beam on (*a*) a 100 percent absorbing black surface? (*b*) a 100 percent reflecting surface?

26. A laser beam has a diameter of 1.0 mm and average power of 1.5 mW. Find (*a*) the intensity of the beam, (*b*) E_{rms}, (*c*) B_{rms}, and (*d*) the radiation pressure.

27. A laser pulse has an energy of 20 J and a beam radius of 2 mm. The pulse duration is 10 ns and the energy density is constant within the pulse. (*a*) What is the spatial length of the pulse? (*b*) What is the energy density within the pulse? (*c*) Find the electric and magnetic amplitudes of the laser pulse.

28. An electromagnetic wave has a frequency of 100 MHz and is traveling in a vacuum. The magnetic field is given by

$$\mathbf{B}(z, t) = (10^{-8}\ \mathrm{T}) \cos (kz - \omega t)\ \mathbf{i}$$

(*a*) Find the frequency, wavelength, and the direction of propagation of this wave. (*b*) Find the electric vector $\mathbf{E}(z, t)$. (*c*) Give Poynting's vector, and find the intensity of this wave.

29. The electric field of an electromagnetic wave oscillates in the y direction and the Poynting vector is given by

$$\mathbf{S}(x, t) = (100\ \mathrm{W/m^2}) \cos^2 [10x - (3 \times 10^9)t]\ \mathbf{i}$$

where x is in meters and t is in seconds. (*a*) What is the direction of propagation of the wave? (*b*) Find the wavelength and the frequency. (*c*) Find the electric and magnetic fields.

30. A pulsed laser fires a 1000-MW pulse of 200 ns duration at a small object of mass 10 mg suspended by a fine fiber 4 cm long. If the radiation is completely absorbed without other effects, what is the maximum angle of deflection of this pendulum?

31. A very long wire of radius 4 mm is heated to 1000 K. The surface of the wire is an ideal blackbody radiator. (*a*) What is the total power radiated per unit length? Find (*b*) the Poynting vector S, (*c*) E_{rms}, and (*d*) B_{rms} at a distance of 25 cm from the wire.

32. A blackbody sphere of radius R is a distance 2×10^{11} m from the sun. The effective area of the body for absorption of energy from the sun is πR^2, but the area for radiation by the object is $4\pi R^2$. The power output of the sun is 3.83×10^{26} W. What is the temperature of the sphere?

33. (*a*) If the earth were an ideal blackbody with infinite thermal conductivity and no atmosphere, what would be the temperature of the earth? (*b*) If 40 percent of the incident sun's energy were reflected, what then would be the temperature of the earth? (See Problem 32.)

34. Two harmonic waves of angular frequency ω_1 and ω_2 have their electric fields given by $\mathbf{E}_1 = E_{10} \cos (k_1 x - \omega_1 t)\, \mathbf{j}$ and $\mathbf{E}_2 = E_{20} \cos (k_2 x - \omega_2 t + \delta)\, \mathbf{j}$. Find (*a*) the instantaneous Poynting vector for the resultant wave motion and (*b*) the time-average Poynting vector. If $\mathbf{E}_2 = E_{20} \cos (k_2 x + \omega t + \delta)\, \mathbf{j}$ find (*c*) the instantaneous Poynting vector for the resultant wave motion and (*d*) the time-average Poynting vector.

35. A 10- by 15-cm card has a mass of 2 g and is perfectly reflecting. The card hangs in a vertical plane and is free to rotate about a horizontal axis through one edge. The card is illuminated uniformly by an intense light that causes the card to make an angle of 1° with the vertical. Find the intensity of the light.

36. A valuable 0.08-kg diamond gem and a 105-kg spaceperson are separated by 95 m. Both objects are initially at rest. The spaceperson has a 1.5-kW laser that can be used as a photon rocket motor to propel the person towards the diamond. How long would it take the spaceperson to move 95 m using the laser rocket propulsion?

37. A circular loop of wire can be used to detect electromagnetic waves. Suppose a 100-MHz FM station radiates 50 kW uniformly in all directions. What is the maximum rms voltage induced in a loop of radius 30 cm at a distance of 10^5 m from the station?

38. Suppose one has an excellent radio capable of detecting a signal as weak as 10^{-14} W/m^2. This radio has a 2000-turn coil antenna having a radius of 1 cm wound on an iron core of permeability 200. The ratio frequency is 140 KHz. (*a*) What is the amplitude of the magnetic field in this wave? (*b*) What is the emf induced in the antenna? (*c*) What would be the emf induced in a 2-m wire oriented in the direction of the electric field?

39. The electric field from a radio station some distance from the transmitter is given by

$$E = (10^{-4}\ \text{N/C}) \cos 10^6 t$$

where t is in seconds. (*a*) What voltage is picked up on a 50-cm wire oriented along the electric field direction? (*b*) What voltage can be induced in a loop of radius 20 cm?

40. A parallel-plate capacitor with circular plates is given a charge Q_0. Between the plates is a leaky dielectric having a dielectric constant of κ and a resistivity ρ. (*a*) Find the conduction current between the plates as a function of time. (*b*) Find the displacement current between the plates as a function of time. What is the total (conduction plus displacement) current? (*c*) Find the magnetic field as a function of time between the plates produced by the leakage discharge current. (*d*) Find the magnetic field as a function of time between the plates produced by the displacement current. (*e*) What is the total magnetic field between the plates during discharge of the capacitor?

41. The leaky capacitor of Problem 40 is charged such that the voltage across the capacitor is given by $V(t) = 10^{-2} t$. (*a*) Find the conduction current as a function of time. (*b*) Find the displacement current. (*c*) Find the time for which the displacement current is equal to the conduction current.

42. The space between the plates of a capacitor is filled with a material of resistivity ρ of 10^4 Ω·m and dielectric constant $\kappa = 2.5$. The parallel plates are circular with a radius of 20 cm and are separated by 1 mm. The voltage across the plates is given by $V_0 \cos \omega t$, with $V_0 = 40$ V and $\omega = 120\pi$ rad/s. (*a*) What is the displacement current density? (*b*) What is the conduction current between the plates? (*c*) At what angular frequency is the total current 45° out of phase with the applied voltage?

43. A circular capacitor of radius R has a thin resistor connecting the centers of the two plates. A voltage $V_0 \sin \omega t$ is applied between the plates. (*a*) What is the current drawn by this capacitor? (*b*) What is the magnetic field as a function of radial distance r from the centerline within the plates of this capacitor? (*c*) What is the phase angle between current and applied voltage?

44. Show that the normal component of the magnetic field $\mathbf{B}$ is continuous across a surface. Do this by applying Gauss's law for $\mathbf{B}$ ($\oint B_n\, dA = 0$) to a pillbox gaussian surface that has a face on each side of the surface.

Level III

The following two problems do not concern waves but illustrate the use of the Poynting vector to describe the flow of electromagnetic energy.

45. A long, cylindrical conductor of length L, radius a, and resistivity ρ carries a steady current I that is uniformly distributed over its cross-sectional area. (*a*) Use Ohm's law to relate the electric field E in the conductor to I, ρ, and a. (*b*) Find the magnetic field B just outside the conductor. (*c*) Use the results for parts (*a*) and (*b*) to compute the Poynting vector $\mathbf{S} = \mathbf{E} \times \mathbf{B}/\mu_0$ at $r = a$ (the edge of the conductor). In what direction is $\mathbf{S}$? (*d*) Find the flux $\oint S_n\, dA$ through the surface of the conductor into the conductor, and show that the rate of energy flow into the conductor equals I^2R, where R is the resistance. (Here S_n is the *inward* component of $\mathbf{S}$ perpendicular to the surface of the conductor.)

46. A long solenoid of n turns per unit length has a current that slowly increases with time. The solenoid has radius r, and the current in the windings has the form $I(t) = at$. (*a*) Find the induced electric field at a distance $r < R$ from the solenoid axis. (*b*) Find the magnitude and direction of the Poynting vector $\mathbf{S}$ at the cylindrical surface $r = R$ just inside the solenoid windings. (*c*) Calculate the flux $\oint S_n\, dA$ into the solenoid, and show that it equals the rate of increase of the magnetic energy inside the solenoid. (Here S_n is the *inward* component of $\mathbf{S}$ perpendicular to the surface of the solenoid.)

47. Small particles might be blown out of solar systems by the radiation pressure of sunlight. Assume the particles are spherical with a radius r and a density of 1 g/cm^3 and absorb all the radiation in a cross-sectional area of πr^2. They are a distance R from the sun, which has a power output of 3.83×10^{26} W. What is the radius r for which the radiation force of repulsion just balances the gravitational force of attraction to the sun?

48. In this problem, you are to show that the generalized form of Ampère's law (Equation 29-4) and the Biot–Savart law give the same result in a situation in which they both can be used. Figure 29-12 shows two charges $+Q$ and $-Q$

Figure 29-12 Problem 48.

on the x axis at $x = -a$ and $x = +a$, with a current $I = -dQ/dt$ along the line between them. Point P is on the y axis at $y = R$. (a) Use Equation 25-12, obtained from the Biot–Savart law, to show that the magnitude of B at point P is

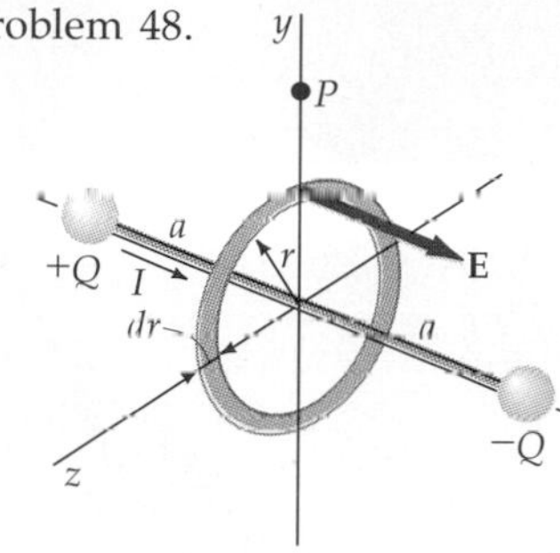

$$B = \frac{\mu i a}{2\pi R}\frac{1}{\sqrt{R^2 + a^2}}$$

(b) Consider a circular strip of radius r and width dr in the yz plane with its center at the origin. Show that the flux of the electric field through this strip is

$$E_x\, dA = (Q/\epsilon_0)a(r^2 + a^2)^{-3/2}\, r\, dr$$

(c) Use your result for part (b) to find the total flux ϕ_e through a circular area of radius R. Show that

$$\epsilon_0\phi_e = Q(1 - a/\sqrt{a^2 + R^2})$$

(d) Find the displacement current I_d, and show that

$$I + I_d = I\frac{a}{\sqrt{a^2 + R^2}}$$

(e) Then show that Equation 29-4 gives the same result for B as that found in part (a).

49. (a) Using arguments similar to those given in the text, show that for a plane wave, in which **E** and **B** are independent of y and z,

$$\frac{\partial E_z}{\partial x} = \frac{\partial B_y}{\partial t}$$

and

$$\frac{\partial B_y}{\partial x} = \mu_0\epsilon_0\frac{\partial E_z}{\partial t}$$

(b) Show that E_z and B_y also satisfy the wave equation.

50. Some science fiction writers have used solar sails to propel interstellar spaceships. Imagine a giant sail erected on a spacecraft subjected to the solar radiation pressure. (a) Show that the spacecraft's acceleration is given by

$$a = \frac{P_s A}{4\pi r^2 cm}$$

where P_s is the power output of the sun and is equal to 3.8×10^{26} W, A is the surface area of the sail, m is the total mass of the spacecraft, r is the distance from the sun, and c is the speed of light. (b) Show that the velocity of the spacecraft at a distance r from the sun is found from

$$v^2 = v_0^2 + \left(\frac{P_s A}{2\pi mc}\right)\left(\frac{1}{r_0} - \frac{1}{r}\right)$$

where v_0 is the initial velocity at r_0. (c) Compare the relative accelerations due to the radiation pressure and the gravitational force. Use reasonable values for A and m. Will such a system work?

51. Novelty stores sell a device called a radiometer shown in Figure 29-13, in which a balanced vane spins rapidly. A card is mounted on each arm of the vane. One side of each card is white and the other is black. Assume that the mass of each card is 2 g, that the light-collecting area for each card is 1 cm^2, and that each arm of the vane has a length of 2 cm. (a) If a 100-W light bulb produces 50 W of electromagnetic energy and the bulb is 50 cm from the radiometer, find the maximum angular acceleration of the vane. (Estimate the moment of inertia of the vane by assuming all the mass of each card is at the end of the arms.) (b) How long will it take for the vane to accelerate to 10 rev/min if it starts from rest and is subject to the maximum angular acceleration at all times? (c) Can the radiation pressure account for the rapid motion of the radiometer? (The radiometer actually spins in the opposite direction from what would be expected if the force were due to radiation pressure. The reason is that the air near the black side is warmer than that near the white side, so the air molecules hitting the black side have greater energy than those hitting the white side.)

Figure 29-13 A radiometer. See Problem 51.

Part 5 Optics

Stress patterns around a crack in a sheet of transparent plastic are revealed by polarized light. The stress is perpendicular to the crack. Two smaller cracks have propagated from the lower end of the large one, creating additional patterns of stress. Smaller circular flaws surround the upper end of the large crack.

Chapter 30

Light

A bright primary rainbow and the fainter secondary rainbow in a sheet of rain over Lake Michigan. The primary bow is formed by light rays that enter spherical drops of water and are reflected once internally before leaving the drops. The secondary bow results from rays that experience two internal reflections before leaving the drops.

Light has intrigued mankind for centuries. Early theories considered light to be something that emanated *from* the eye. Later it was realized that light must come from the objects seen and enter the eye causing the sensation of vision. The question of whether light consists of a beam of particles or is some kind of wave motion is one of the most interesting in the history of science. The most influential proponent of the **particle theory of light** was Newton. Using it, he was able to explain the laws of reflection and refraction. However, his derivation of the law of refraction depends on the assumption that light travels faster in water or glass than in air, an assumption later shown to be false. The chief proponents of the **wave theory of light** were Christian Huygens and Robert Hooke. Using his own theory of wave propagation, Huygens was able to explain reflection and refraction by assuming that light travels more slowly in glass or water than in air. Newton understood the virtues of the wave theory of light, particularly as it explained the colors formed by thin films, which he studied extensively. However, he rejected the wave theory because of the apparent rectilinear propagation of light. Diffraction, the bending of a light beam around obstacles, had not been observed in his time. Because of Newton's great reputation and authority, his reluctant rejection of the wave theory of light was strictly adhered to by his followers. Even after evidence of diffraction was available,

Newton's followers sought to explain it as the scattering of light particles from the edges of slits.

Newton's particle theory of light was accepted for more than a century. Then, in 1801, Thomas Young revived the wave theory of light. He was one of the first to introduce the idea of interference as a wave phenomenon that occurs in both light and sound. His observation of interference with light was a clear demonstration of the wave nature of light. However, Young's work went unnoticed by the scientific community for more than a decade. Perhaps the greatest advance in the general acceptance of the wave theory of light was due to the French physicist Augustin Fresnel (1788–1827), who performed extensive experiments on interference and diffraction and put the wave theory on a mathematical basis. He showed, for example, that the observed rectilinear propagation of light is a result of the very short wavelengths of visible light. In 1850, Jean Foucault measured the speed of light in water and showed that it is less than that in air, thus ruling out Newton's particle theory. In 1860, James Clerk Maxwell published his mathematical theory of electromagnetism, which predicted the existence of electromagnetic waves that propagate with a speed calculated from the laws of electricity and magnetism to be 3×10^8 m/s, the same as the speed of light. Maxwell's theory was confirmed in 1887 by Hertz, who used a tuned electrical circuit to generate the waves and another, similar circuit to detect them. In the latter half of the nineteenth century, Kirchhoff and others applied Maxwell's equations to explain the interference and diffraction of light and other electromagnetic waves and put Huygens' empirical methods of construction on a firm mathematical basis.

Although the wave theory is generally correct in describing the propagation of light (and other electromagnetic waves), it fails to account for all properties of light, particularly the interaction of light with matter. In his famous experiment of 1887 that confirmed Maxwell's wave theory, Hertz also discovered the photoelectric effect, which will be discussed in detail in Chapter 35. This effect can be explained only by a particle model of light, as Einstein showed only a few years later. Thus, a particle model of light was reintroduced. Light particles are called **photons.** The energy of a photon E is related to the frequency f of the associated light wave by the famous Einstein relation $E = hf$, where h is a constant called *Planck's constant.* Complete understanding of the dual nature of light did not come until the 1920s, when experiments by C. J. Davisson and L. Germer and by G. P. Thompson showed that electrons (and other "particles") also have a dual nature and exhibit the wave properties of interference and diffraction in addition to their well-known particle properties. (We will discuss the dual nature of light and electrons in Chapter 35.)

The development of the quantum theory of atoms and molecules by Rutherford, Bohr, Schrödinger, and others in the twentieth century led to an understanding of the emission and absorption of light by matter. Light emitted or absorbed by atoms is now known to be the result of energy changes of the outermost electrons in the atom. Because these energy changes are quantized rather than being continuous, the photons emitted have discrete energies resulting in light waves with a discrete set of frequencies and wavelengths similar to the set of frequencies and wavelengths observed in standing sound waves. Viewed through a spectroscope with a narrow-slit aperture, the light emitted by an atom appears as a discrete set of lines of different colors or wavelengths, with the spacing and intensities of the lines being characteristic of the element.

The technological developments in the second half of the twentieth century have led to a renewed interest in both theoretical and applied optics. The advent of high-speed computers has brought vast improvement in the design of complex optical systems. Optical fibers are rapidly replacing elec-

(*a*)

(*b*)

(*a*) Since its invention in 1960, the laser has found many uses. (*b*) A carbon dioxide laser takes just two minutes to cut out a steel saw blade. (*c*) The supermarket laser scanner is now a familiar sight. (*d*) A helium-neon laser mounted on a truck is used to precisely measure the clearance in a tunnel. (*e*) A liquid dye laser beam transmitted by an optical fiber is used here in eye surgery.

trical wires for the transmission of data. The laser, invented in 1960, has led to the discovery of a number of new optical effects. Today, lasers are used to scan labels in supermarkets, to perform surgery in hospitals, to set type for newspapers, and to read compact discs for home audio systems. The wavefront reconstruction technique known as holography, developed in the late 1940s, is now used in nondestructive testing and in data storage.

In this chapter, we will begin by discussing some of the early measurements of the speed of light. We will then discuss the basic phenomena of reflection, refraction, dispersion, and polarization. These can all be adequately understood by using rays to describe the straight-line propagation of light and neglecting interference and diffraction effects. As discussed in Chapter 14, the ray approximation is valid for the propagation of any wave motion if the wavelength is small compared with any apertures or obstacles. This is often the situation in optics because the wavelengths of visible light range from about 400 nm (1 nm $= 10^{-9}$ m) for violet light to about 700 nm for red light.

30-1 The Speed of Light

The first effort to measure the speed of light was made by Galileo. He and a partner stood on hilltops about a kilometer apart, each with a lantern and a shutter to cover it. Galileo proposed to measure the time it took for light to travel back and forth between the experimenters. First, one would uncover his lantern, and when the other saw the light, he would uncover his. The time between the first partner's uncovering his lantern and his seeing the light from the other lantern would be the time it took for light to travel back and forth between the experimenters. Though this method is sound in principle, the speed of light is so great that the time interval to be measured is much smaller than fluctuations in human response time, so Galileo was unable to obtain any value for the speed of light.

The first indication of the true magnitude of the speed of light came from astronomical observations of the period of Io, one of the moons of Jupiter.

(c)

(d)

(e)

This period is determined by measuring the time between eclipses (when the moon Io disappears behind Jupiter). The eclipse period is about 42.5 hours, but measurements made when the earth is moving away from Jupiter along path *ABC* in Figure 30-1 give a greater time for this period than do measurements made when the earth is moving toward Jupiter along path *CDA* in the figure. Since these measurements differ from the average value by only about 15 seconds, the discrepancies were difficult to measure accurately. In 1675, the astronomer Ole Römer attributed these discrepancies to the fact that the speed of light is not infinite. During the 42.5 hours between eclipses of Jupiter's moon, the distance between the earth and Jupiter changes, making the path for the light longer or shorter. Römer devised the following method for measuring the cumulative effect of these discrepancies. Because Jupiter moves much more slowly than earth, we can neglect its motion. When the earth is at point *A*, nearest to Jupiter, the distance between the earth and Jupiter is changing negligibly. The period of Io's eclipse is measured, providing the time between the beginnings of successive eclipses. Based on this measurement, the number of eclipses in 6 months is computed, and the time when an eclipse should begin a half-year later when the earth is at point *C* is predicted. When the earth is actually at *C*, the observed beginning of the eclipse is about 16.6 minutes later than predicted. This is the time it takes light to travel a distance equal to the diameter of the earth's orbit.

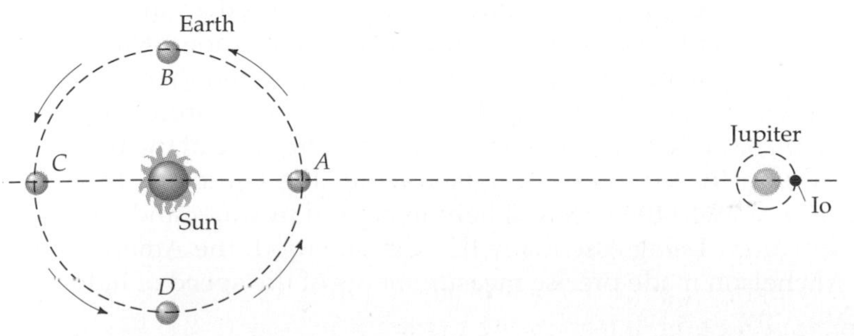

Figure 30-1 Römer's method of measuring the speed of light. The time between eclipses of Jupiter's moon Io appears to be greater when the earth is moving along path *ABC* than when it is moving along path *CDA*. The difference is due to the time it takes light to travel the distance traveled by the earth along the line of sight during one period of Io. (The distance traveled by Jupiter in one earth year is negligible.)

Example 30-1

The diameter of the earth's orbit is 3.00×10^{11} m. If light takes 16.6 min to travel this distance, what is the speed of light?

The number of seconds in 16.6 min is (16.6 min) × (60 s/min) = 996 s. The measured speed of light is thus

$$c = \frac{\Delta x}{\Delta t} = \frac{3.00 \times 10^{11}\ \text{m}}{996\ \text{s}} = 3.01 \times 10^{8}\ \text{m/s}$$

Römer obtained a considerably smaller value for c because he used 22 min for Δt.

The first nonastronomical measurement of the speed of light was made by the French physicist Fizeau in 1849. On a hill in Paris, Fizeau placed a light source and a system of lenses arranged such that the light reflected from a semitransparent mirror was focused on a gap in a toothed wheel as shown in Figure 30-2. On a distant hill (about 8.63 km away), he placed a mirror to reflect the light back to be viewed by an observer as shown. The toothed wheel was rotated, and the speed of rotation was varied. At low speeds of rotation, no light was visible because the reflected light was obstructed by the teeth of the rotating wheel. The speed of rotation was then increased. The light suddenly became visible when the rotation speed was such that the reflected light passed through the next gap in the wheel.

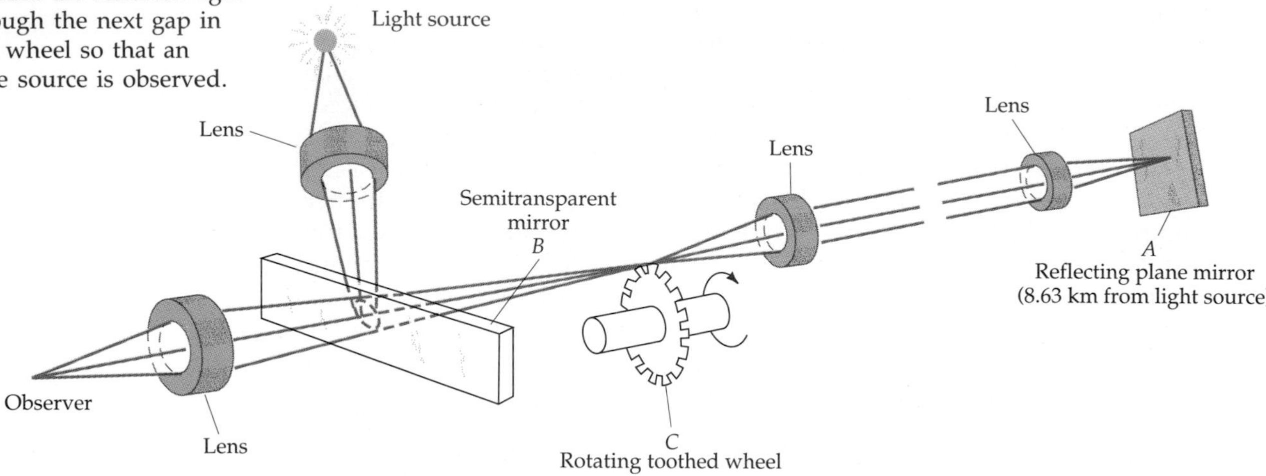

Figure 30-2 Fizeau's method of measuring the speed of light. Light from the source is reflected by mirror B and is transmitted through a gap in the toothed wheel to mirror A. The speed of light is determined by measuring the angular speed of the wheel that will permit the reflected light to pass through the next gap in the toothed wheel so that an image of the source is observed.

Fizeau's method was improved upon by Foucault, who replaced the toothed wheel with an eight-sided rotating mirror as shown in Figure 30-3. Light strikes one face of the mirror and is reflected from a distant fixed mirror to another face and then to an observing telescope. (The actual experimental arrangement was somewhat more complicated than the simple diagram shown in Figure 30-3.) When the mirror rotates through one-eighth of a turn (or $n/8$ turns, where n is an integer) another face of the mirror is in the right position for the reflected light to enter the telescope. In about 1850, Foucault measured the speed of light in air and in water and showed that it is less in water. Using essentially the same method, the American physicist A. A. Michelson made precise measurements of the speed of light from 1880 to 1930.

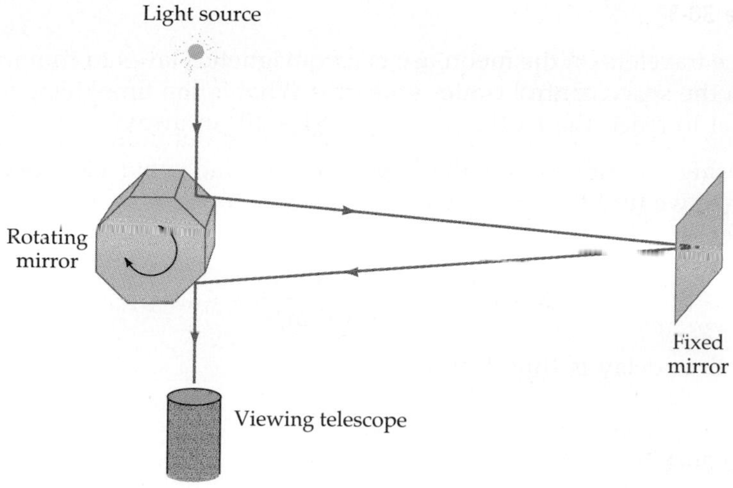

Figure 30-3 Simplified drawing of Foucault's method of measuring the speed of light. Essentially, Fizeau's rotating toothed wheel is replaced by a rotating octagonal mirror. When the mirror makes an eighth of a revolution during the time it takes for the light to travel to the fixed mirror and back, the next face of the mirror is in the proper position to reflect the light into the telescope.

Another method of determining the speed of light involves the measurement of the electrical constant ϵ_0 to determine c from

$$c = \frac{1}{\sqrt{\epsilon_0 \mu_0}}$$

The constant ϵ_0 can be obtained by measuring the capacitance of a parallel-plate capacitor. The constant μ_0 is defined in terms of the definition of the ampere, which in turn determines the coulomb.

The various methods we have discussed for measuring the speed of light are all in general agreement. Today, the speed of light is defined to be exactly

$$c = 299\ 792\ 457 \text{ m/s}$$

and the standard unit of length, the meter, is defined in terms of this speed. A measurement of the speed of light is therefore now a measurement of the size of the meter, which is the distance light travels in (1/299 792 457) s. The value 3×10^8 m/s for the speed of light is accurate enough for nearly all calculations. The speed of radio waves and all other electromagnetic waves (in a vacuum) is the same as the speed of light.

Example 30-2

In Fizeau's experiment, the wheel had 720 teeth, and light was observed when the wheel rotated at 25.2 revolutions per second. If the distance from the wheel to the distant mirror was 8.63 km, what was Fizeau's value for the speed of light?

The total distance traveled by the light from the gap in the wheel to the mirror and back was 2×8.63 km = 17.3 km. The reflected light passed through the next gap in the wheel, so the wheel made 1/720 revolution. Since the wheel made 25.3 revolutions in 1 s, the time it took to make 1/720 revolution was

$$\Delta t = \frac{1 \text{ s}}{25.3 \text{ rev}}\left(\frac{1}{720} \text{ rev}\right) = 5.49 \times 10^{-5} \text{ s}$$

The value for the speed of light from this experiment is then

$$c = \frac{\Delta x}{\Delta t} = \frac{17.3 \times 10^3 \text{ m}}{5.49 \times 10^{-5} \text{ s}} = 3.15 \times 10^8 \text{ m/s}$$

which is about 5 percent too high.

Example 30-3

Space travelers on the moon use electromagnetic waves to communicate with the space control center on earth. What is the time delay for their signal to reach the earth, which is 3.84×10^8 m away?

Using 3×10^8 m/s for the speed of light and other electromagnetic waves, we find the time it takes for a signal to travel from the moon to the earth to be

$$\Delta t = \frac{\Delta x}{c} = \frac{3.84 \times 10^8 \text{ m}}{3 \times 10^8 \text{ m/s}} = 1.28 \text{ s}$$

The time delay is thus 1.28 s each way.

Example 30-4

The sun is 1.50×10^{11} m from the earth. How long does it take for the sun's light to reach the earth?

$$\Delta t = \frac{\Delta x}{c} = \frac{1.50 \times 10^{11} \text{ m}}{3 \times 10^8 \text{ m/s}} = 500 \text{ s} = 8.33 \text{ min}$$

Large distances are often given in terms of the time it takes for light to travel those distances. For example, the distance to the sun is 8.33 light-minutes, written 8.33 c-min. A light-year is the distance light travels in one year. We can easily find a conversion factor between light-years and meters. The number of seconds in one year is

$$1 \text{ y} = 1 \text{ y} \times \frac{365.24 \text{ d}}{1 \text{ y}} \times \frac{24 \text{ h}}{1 \text{ d}} \times \frac{3600 \text{ s}}{1 \text{ h}} = 3.156 \times 10^7 \text{ s}$$

The number of meters in one light-year is thus

$$1 \; c\text{-year} = (2.998 \times 10^8 \text{ m/s})(3.156 \times 10^7 \text{ s}) = 9.46 \times 10^{15} \text{ m}$$

Question

1. Estimate the time required for light to make the round trip in Galileo's experiment to determine the speed of light.

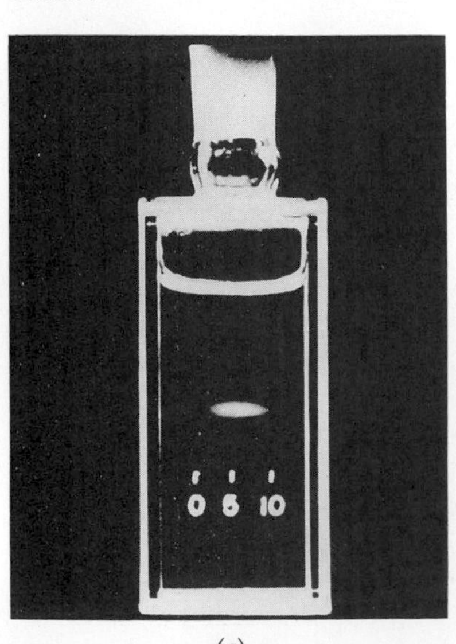

(*a*)

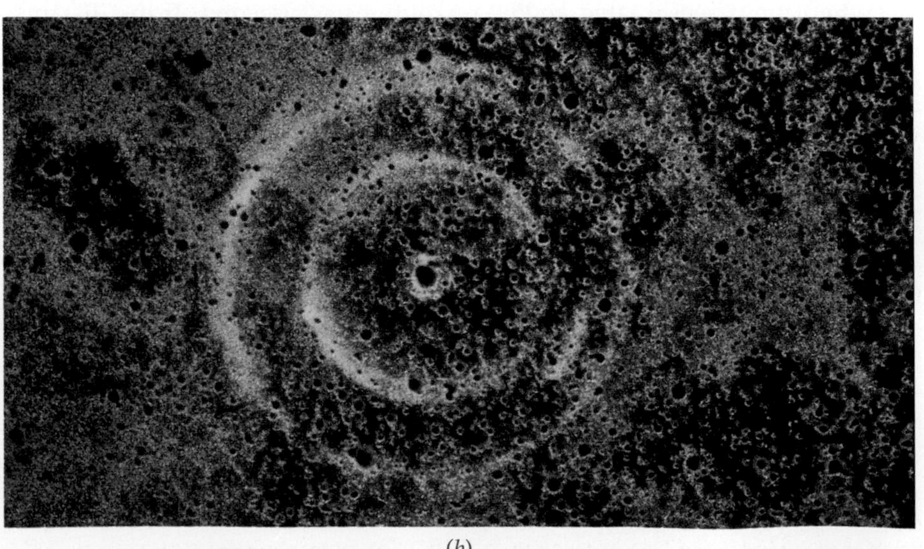

(*b*)

(*a*) A 10-ps pulse of light from a neodymium-doped glass laser passing through a water cell calibrated in millimeters. The pulse length is about 2.2 mm. (*b*) Light echoes from a supernova explosion that occurred 170,000 years ago. Direct light from the explosion was first seen on earth in February 1987. Light reflected from dust clouds near the explosion arrives at the earth months or years later. Comparison of photograph plates exposed in 1987 and 1989 show the outer rings that result from reflection from dust clouds.

30-2 The Propagation of Light: Huygens' Principle

Figure 30-4 shows a portion of a spherical wavefront emanating from a point source. The wavefront is the locus of points of constant phase. If the radius of the wavefront is r at time t, its radius at time $t + \Delta t$ is $r + c\,\Delta t$, where c is the speed of the wave. However, if a part of the wave is blocked by some obstacle or if the wave passes through a different medium, as in Figure 30-5, the determination of the new wavefront at time $t + \Delta t$ is much more difficult.

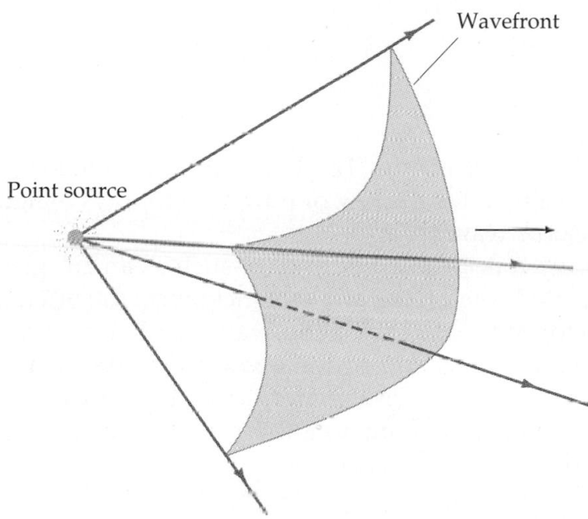

Figure 30-4 Spherical wavefront from a point source.

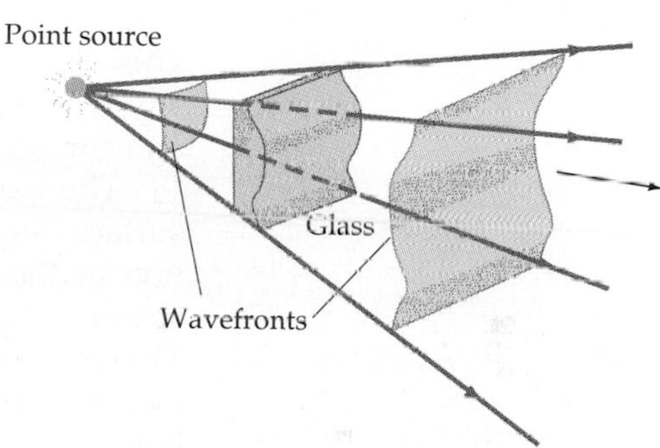

Figure 30-5 Wavefront from a point source before and after passing through an irregularly shaped piece of glass.

The propagation of any wave through space can be described using a geometric method discovered by Christian Huygens about 1678 that is now known as **Huygens' principle** or **Huygens' construction:**

> Each point on a primary wavefront serves as the source of spherical secondary wavelets that advance with a speed and frequency equal to those of the primary wave. The primary wavefront at some later time is the envelope of these wavelets.

Figure 30-6 shows the application of Huygens' principle to the propagation of a plane wave and of a spherical wave. Of course, if each point on a wavefront were really a point source, there would be waves in the backward direction as well. Huygens ignored these back waves.

Huygens' principle was later modified by Fresnel so that the new wavefront was calculated from the old wavefront by superposition of the wavelets considering their relative amplitudes and phases. Kirchhoff later showed that the Huygens–Fresnel principle was a consequence of the wave equation, thus putting it on a firm mathematical basis. Kirchhoff showed that the intensity of the wavelets depends on the angle and is zero in the backward direction.

In this chapter, we will use Huygens' principle to derive the laws of reflection and refraction. In Chapter 33, we will apply Huygens' principle with Fresnel's modification to calculate the diffraction pattern of a single slit.

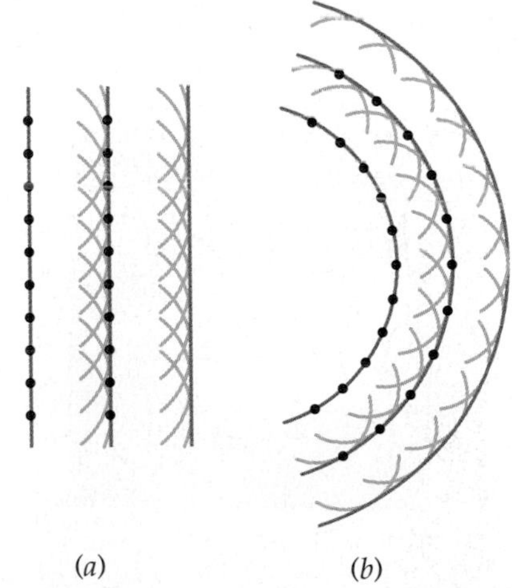

Figure 30-6 Huygens' construction for the propagation to the right of (*a*) a plane wave and (*b*) an outgoing spherical, or circular, wave.

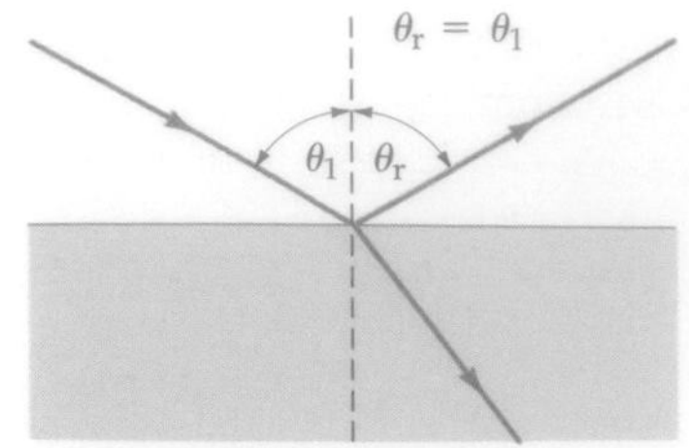

Figure 30-7 The angle of reflection θ_r equals the angle of incidence θ_1.

30-3 Reflection

When waves of any type strike a plane barrier such as a mirror, new waves are generated that move away from the barrier. This phenomenon is called **reflection.** Reflection occurs at a boundary between two different media such as an air–glass surface, in which case part of the incident energy is reflected and part is transmitted. Figure 30-7 shows a light ray striking a smooth air–glass surface. The angle θ_1 between the incident ray and the normal (the line perpendicular to the surface) is called the **angle of incidence,** and the plane defined by these two lines is called the **plane of incidence.** The reflected ray lies in the plane of incidence and makes an angle θ_r with the normal that is equal to the angle of incidence as shown in the figure:

Law of reflection

$$\theta_r = \theta_1 \tag{30-1}$$

This result is known as the **law of reflection.** The law of reflection holds for any type of wave. Figure 30-8 illustrates the law of reflection for rays of light and for wavefronts of ultrasonic waves.

The fraction of light energy reflected at a boundary such as an air–glass surface depends in a complicated way on the angle of incidence, the orientation of the electric-field vector associated with the wave, and the relative speed of light in the first medium (air) and in the second medium (glass). The speed of light in a medium such as glass, water, or air is characterized by the **index of refraction** n, which is defined as the ratio of the speed of light in a vacuum c to the speed in the medium v:

Index of refraction

$$n = \frac{c}{v} \tag{30-2}$$

For the special case of normal incidence ($\theta_1 = \theta_r = 0°$), the reflected intensity can be shown to be

$$I = \left(\frac{n_1 - n_2}{n_1 + n_2}\right)^2 I_0 \tag{30-3}$$

where I_0 is the incident intensity and n_1 and n_2 are the indexes of refraction of the two media. For a typical case of reflection from an air–glass surface for

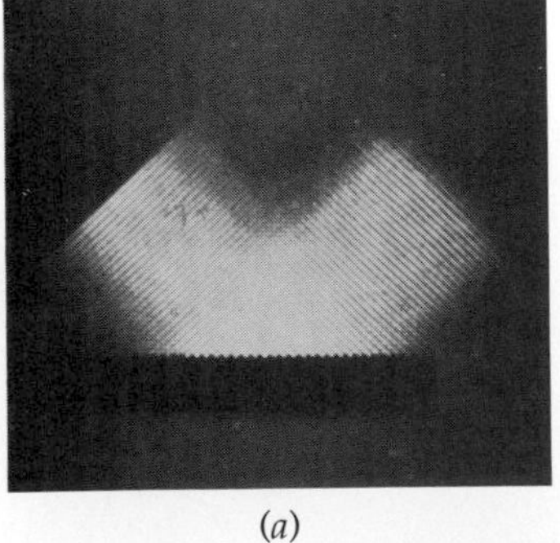

(a)

(b)

Figure 30-8 (a) Ultrasonic plane waves in water reflecting from a steel plate. (b) Light rays reflecting from an air–glass surface showing equal angles of incidence and reflection.

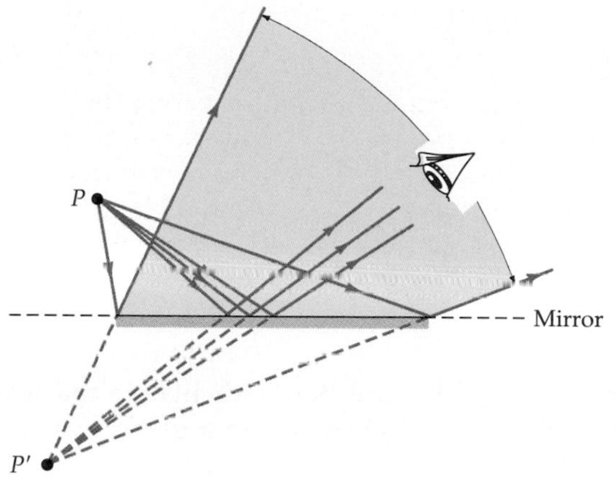

Figure 30-9 Rays from a source P reflected by a mirror into the eye appear to come from the image point P' behind the mirror. The image can be seen when the eye is anywhere in the shaded region.

which $n_1 = 1$ and $n_2 = 1.5$, Equation 30-3 gives $I = I_0/25$. That is, only about 4 percent of the energy is reflected; the rest is transmitted.

Figure 30-9 shows a narrow bundle of light rays from a point source P that are reflected from a flat surface. After reflection, the rays diverge exactly as if they came from a point P' behind the surface. The point P' is called the **image** of point P. When these rays enter the eye, they cannot be distinguished from rays diverging from a source at P' as though no reflecting surface were present. (We will study the formation of images by reflecting and refracting surfaces in the next chapter.) Reflection from a smooth surface is called **specular reflection.** It differs from **diffuse reflection,** which is illustrated in Figure 30-10. Here, because the surface is rough, the rays enter the eye after reflecting from many different points on the surface, so there is no image. The reflection of light from the page of this book is diffuse reflection. Ground glass is sometimes used in a picture frame to give diffuse reflection and thereby cut out the glare from the light used to illuminate the picture. It is diffuse reflection from the surface of the road that allows you to see the road when you are driving at night because some of the light from your headlights reflects back toward you.

The physical mechanism of the reflection of light can be understood in terms of the absorption and reradiation of the light by the atoms in the reflecting medium. When light traveling in air strikes a glass surface, the atoms in the glass absorb the light and reradiate it at the same frequency in all directions. The waves radiated backwards by the glass atoms interfere constructively at an angle equal to the angle of incidence to produce the reflected wave.

The law of reflection can be derived from Huygens' principle. Figure 30-11 shows a plane wavefront AA' striking a mirror at point A. As can be seen from the figure, the angle ϕ_1 between the wavefront and the mirror is the same as the angle of incidence θ_1, which is the angle between the per-

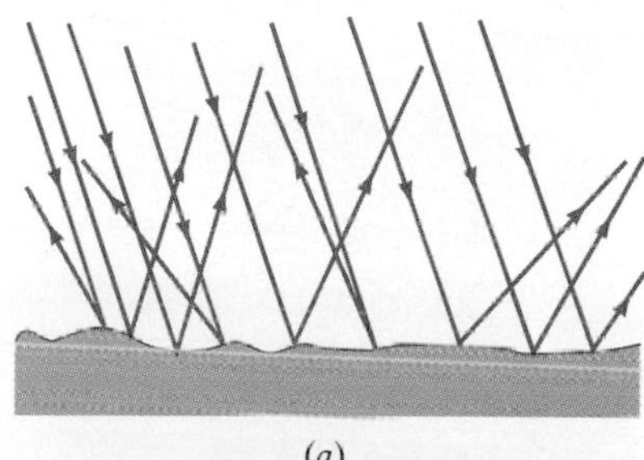

(a)

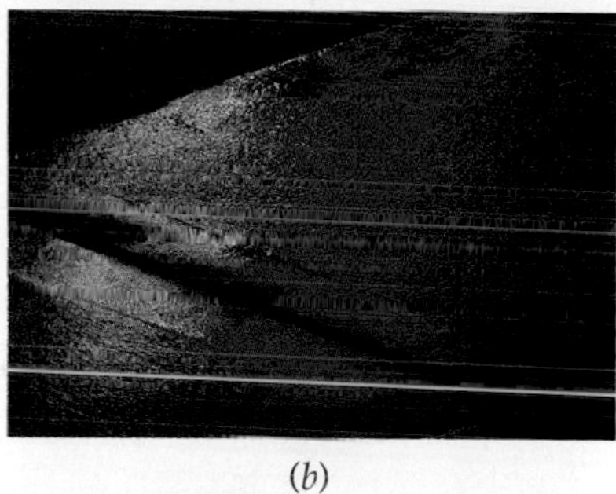

(b)

Figure 30-10 (a) Diffuse reflection from a rough surface. (b) Photograph of diffuse reflection of colored lights from a sidewalk.

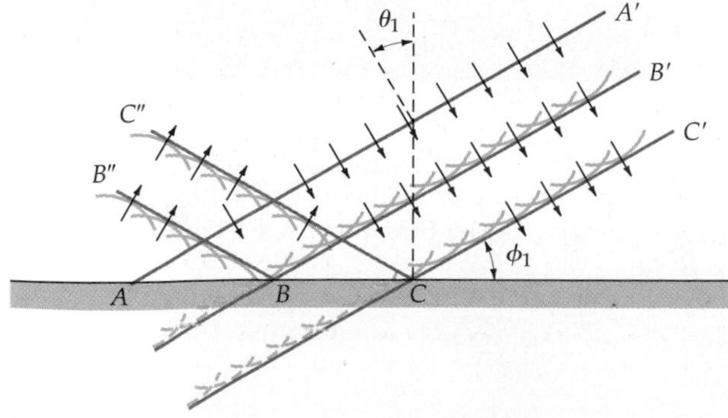

Figure 30-11 Plane wave reflected at a plane mirror. The angle θ_1 between the incident ray and the normal to the mirror is the angle of incidence. It is equal to the angle ϕ_1 between the incident wavefront and the mirror.

Figure 30-12 Geometry of Huygens' construction for the calculation of the law of reflection. The wavefront AP initially strikes the mirror at point A. After a time t, the Huygens' wavelet from P strikes the mirror at point B, and the one from A reaches point B''.

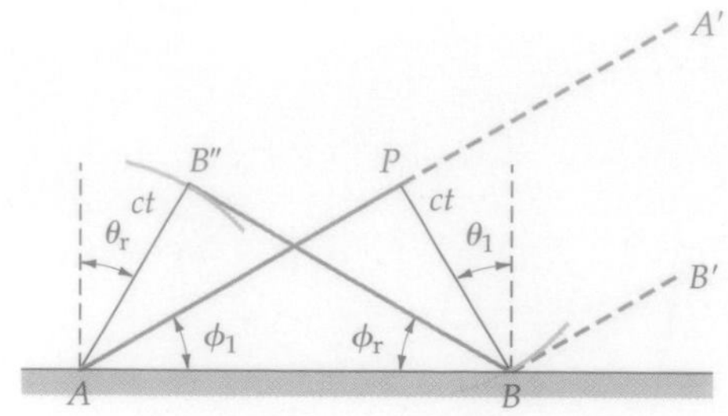

pendicular to the mirror and the rays that are perpendicular to the wavefronts. According to Huygens' principle, each point on a given wavefront can be considered to be a point source of secondary wavelets. The position of the wavefront after a time t is found by constructing wavelets of radius ct with their centers on the wavefront AA'. Wavelets that do not strike the mirror form the portion of the new wavefront BB'. Wavelets that do strike the mirror are reflected and form the portion of the new wavefront BB''. By a similar construction, the wavefront $C''CC'$ is obtained from the Huygens' wavelets originating on the wavefront $B''BB'$. Figure 30-12 is an enlargement of a portion of Figure 30-11 showing the part of the original wavefront AP that strikes the mirror during time t. In this time, the wavelet from point P reaches the mirror at point B, and the wavelet from point A reaches point B''. The reflected wave BB'' makes an angle ϕ_r with the mirror that is equal to the angle of reflection θ_r between the reflected ray and the normal to the mirror. The triangles ABP and BAB'' are both right triangles with a common side AB and equal sides $AB'' = BP = ct$. Hence, these triangles are congruent, and the angles ϕ_1 and ϕ_r are equal, implying that the angle of reflection θ_r equals the angle of incidence θ_1.

Question

2. How does a thin layer of water on the road affect the light you see reflected off the road from your own headlights? How does it affect the light you see reflected from the headlights of an oncoming car?

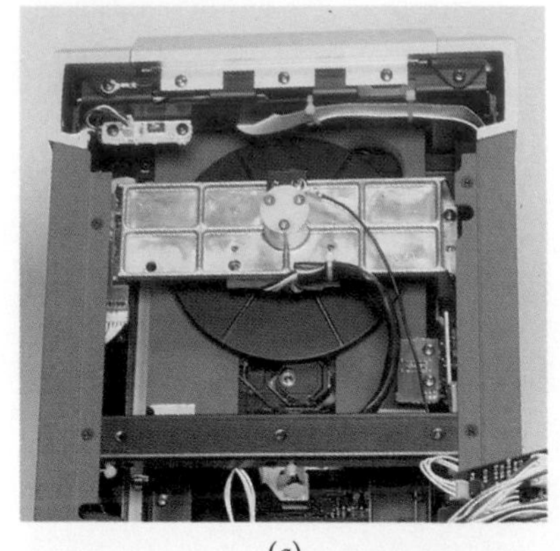

(*a*)

(*b*)

(*c*)

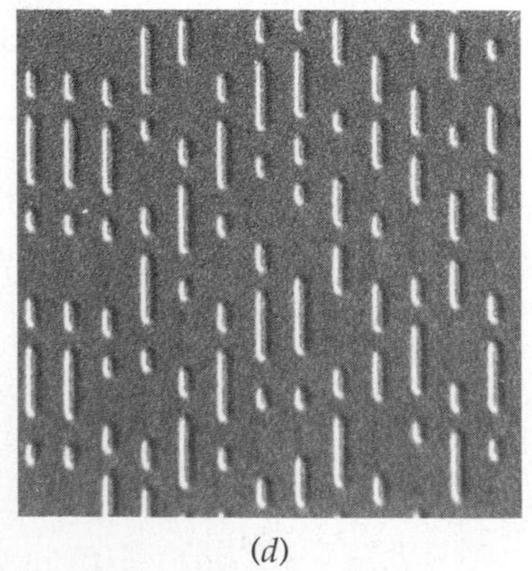

(*d*)

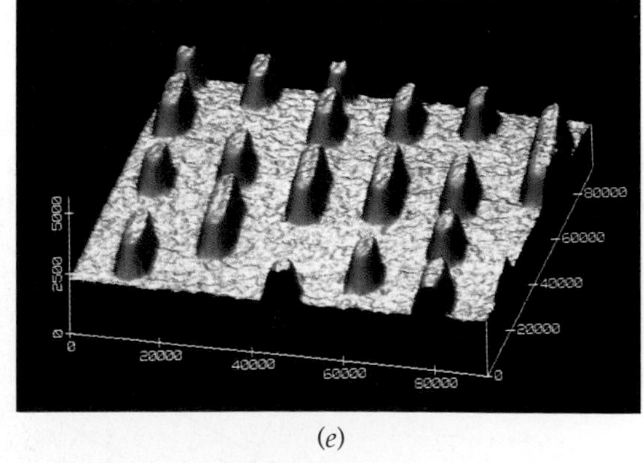

(*e*)

The reflection of laser light is used to read compact discs. (*a*) In a compact disc player, the laser is less than 2 mm from the bottom of the disc as it spins. (*b*) The laser head used to read a compact disc. (*c*) A compact disc has a diameter of 12 cm and is 1.2 mm thick. (*d*) Information is stored digitally on a compact disc in these pits, which are 0.12 μm deep, 0.6 μm wide, and 0.9 to 3.3 μm long. The leading and trailing edges of a pit represent a binary 1, whereas intervening areas represent binary 0's. (*e*) Stamper used to produce the pits in (*d*).

30-4 Refraction

When a beam of light strikes a boundary surface separating two different media, such as an air–glass surface, part of the light energy is reflected and part enters the second medium. The change in direction of the transmitted ray is called **refraction.**

The transmitted wave is the result of the interference of the incident wave and the wave produced by the absorption and reradiation of light energy by the atoms in the medium. For light entering glass from air, there is a phase lag between the reradiated wave and the incident wave. There is therefore also a phase lag between the resultant wave and the incident wave. This phase lag means that the position of a wave crest of the transmitted wave is retarded relative to the position of a wave crest of the incident wave in the medium. Therefore, in a given time, the transmitted wave does not travel as far in the medium as the original incident wave; that is, the velocity of the transmitted wave is less than that of the incident wave. The index of refraction, which is the ratio of the speed of light in a vacuum to that in the medium, is therefore greater than 1. For example, the speed of light in glass is about two-thirds of the speed of light in free space. The index of refraction of glass is therefore about $n = c/v = 3/2$.

Since the frequency of the light in the second medium is the same as that of the incident light the atoms absorb and reradiate the light at the same frequency—but the wave speed is different, the wavelength of the transmitted light is different from that of the incident light. If λ is the wavelength of light in a vacuum, the wavelength λ' in a medium of index of refraction n is

$$\lambda' = \frac{v}{f} = \frac{c/n}{f} = \frac{\lambda}{n} \qquad 30\text{-}4$$

Exercise

Sodium light has a wavelength of 589 nm in a vacuum. Find the wavelength of sodium light (*a*) in water, for which $n = 1.33$, and (*b*) in glass, for which $n = 1.50$. [Answers: (*a*) 443 nm, (*b*) 393 nm]

Figure 30-13 shows light striking a flat air–glass surface. The ray that enters the glass is called the refracted ray, and the angle θ_2 is called the angle of refraction. The angle of refraction is less than the angle of incidence θ_1, as shown in the figure; that is, the refracted ray is bent toward the normal. If, on the other hand, the light beam originates in the glass and is refracted into

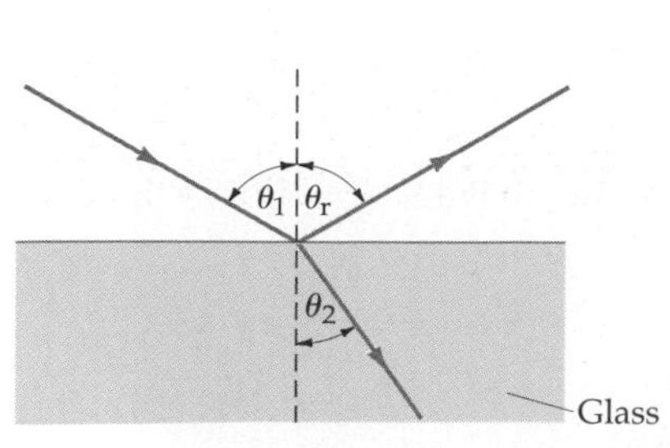

(*a*)

(*b*)

Figure 30-13 (*a*) Incident, reflected, and refracted rays for light striking an air–glass surface. The angle of refraction θ_2 is less than the angle of incidence θ_1. (*b*) Reflection and refraction of a beam of light incident on a glass slab. The refracted beam is partially reflected and partially refracted at the bottom glass–air surface.

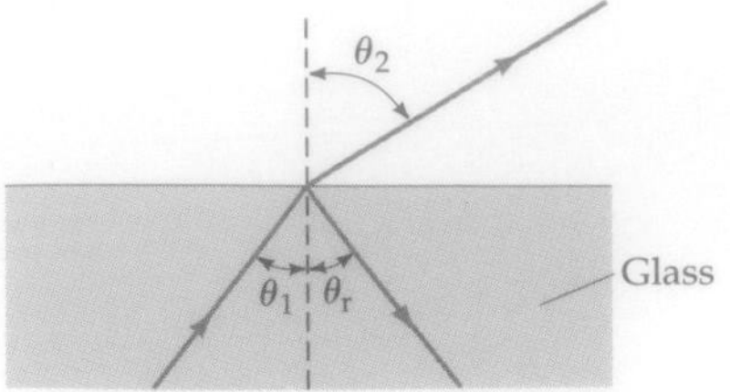

Figure 30-14 Refraction from a dense medium to a less dense medium. Here the angle of refraction is greater than the angle of incidence. The light ray is bent away from the normal.

the air, the angle of refraction is greater than the angle of incidence, and the refracted ray is bent away from the normal as shown in Figure 30-14. We can relate the angle of refraction θ_2 to the indexes of refraction of the two media n_1 and n_2 and to the angle of incidence θ_1 using Huygens' principle. Figure 30-15 shows a plane wave incident on an air–glass surface. We apply Huygens' construction to find the wavefront of the transmitted wave. Line *AP* indicates a portion of the wavefront in medium 1 that strikes the glass surface at an angle of incidence θ_1. In time t the wavelet from *P* travels the distance $v_1 t$ and reaches the point *B* on the line *AB* separating the two media, while the wavelet from point *A* travels a shorter distance $v_2 t$ into the second medium. The new wavefront BB' is not parallel to the original wavefront *AP* because the speeds v_1 and v_2 are different. From the triangle *APB*,

$$\sin\phi_1 = \frac{v_1 t}{AB}$$

or

$$AB = \frac{v_1 t}{\sin\phi_1} = \frac{v_1 t}{\sin\theta_1}$$

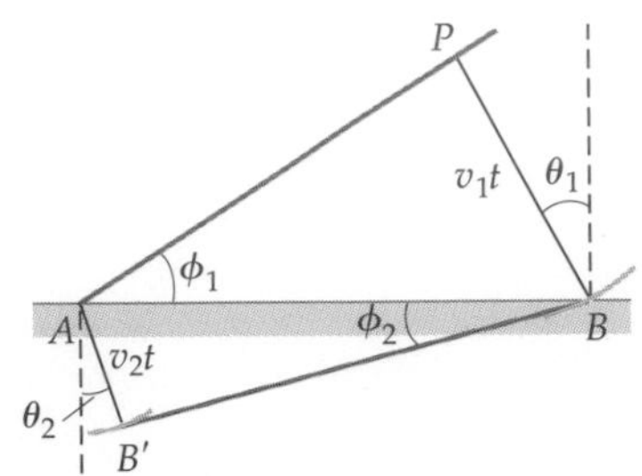

Figure 30-15 Application of Huygens' principle to the refraction of plane waves at the surface separating a medium in which the wave speed is v_1 from a medium in which the wave speed v_2 is less than v_1. The angle of refraction in this case is less than the angle of incidence.

using the fact that the angle ϕ_1 equals the angle of incidence θ_1. Similarly, from triangle $AB'B$,

$$\sin\phi_2 = \frac{v_2 t}{AB}$$

or

$$AB = \frac{v_2 t}{\sin\phi_2} = \frac{v_2 t}{\sin\theta_2}$$

where $\theta_2 = \phi_2$ is the angle of refraction. Equating the two values for *AB*, we obtain

$$\frac{\sin\theta_1}{v_1} = \frac{\sin\theta_2}{v_2} \qquad \text{30-5}$$

Substituting $v_1 = c/n_1$ and $v_2 = c/n_2$ in this equation and multiplying by c, we obtain

Snell's law of refraction

$$n_1 \sin\theta_1 = n_2 \sin\theta_2 \qquad \text{30-6}$$

This result was discovered experimentally in 1621 by Willebrod Snell, a Dutch scientist, and is known as **Snell's law** or the **law of refraction.** It was independently discovered a few years later by René Descartes. Equation 30-6 holds for the refraction of any kind of wave incident on a boundary surface separating two media. Figure 30-16 shows the refraction of plane water waves at a boundary at which the wave speed changes because the depth of the water changes.

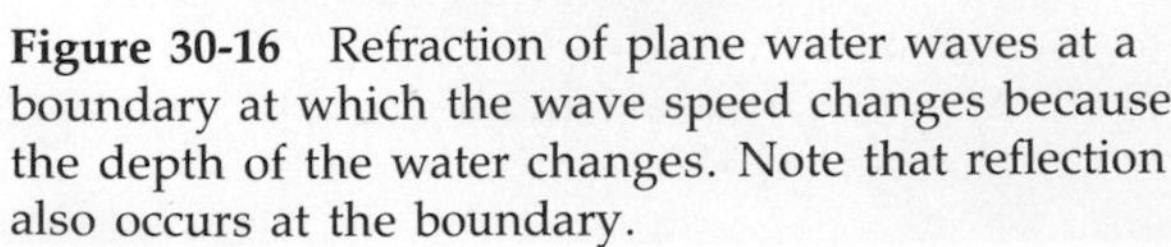

Figure 30-16 Refraction of plane water waves at a boundary at which the wave speed changes because the depth of the water changes. Note that reflection also occurs at the boundary.

Example 30-5

Light traveling in air enters water with an angle of incidence of 45°. If the index of refraction of water is 1.33, what is the angle of refraction?

Taking $n = 1$ for air, we obtain from Equation 30-6

$$(1.00) \sin 45^\circ = (1.33) \sin \theta_2$$

$$\sin \theta_2 = \frac{(1.00) \sin 45^\circ}{1.33} = \frac{(1.00)(0.707)}{1.33} = 0.53$$

The angle whose sine is 0.53 is 32°.

Figure 30-17 shows a point source in glass with rays striking the glass–air surface at various angles. All the rays are bent away from the normal. As the angle of incidence is increased, the angle of refraction increases until a critical angle of incidence θ_c is reached for which the angle of refraction is 90°. For incident angles greater than this critical angle, there is no refracted ray. All the energy is reflected. This phenomenon is called **total internal reflection.** The critical angle can be found in terms of the indexes of refraction of the two media by solving Equation 30-6 for $\sin \theta_1$ and setting θ_2 equal to 90°:

$$\sin \theta_1 = (n_2/n_1) \sin \theta_2$$

Setting $\theta_2 = 90^\circ$, we obtain

$$\sin \theta_c = \frac{n_2}{n_1} \qquad 30\text{-}7$$

Critical angle for total internal reflection

Note that total internal refraction occurs only when the light is originally in the medium with the higher index of refraction. Mathematically, if n_2 is greater than n_1, Equation 30-7 cannot be satisfied because there is no real angle whose sine is greater than 1.

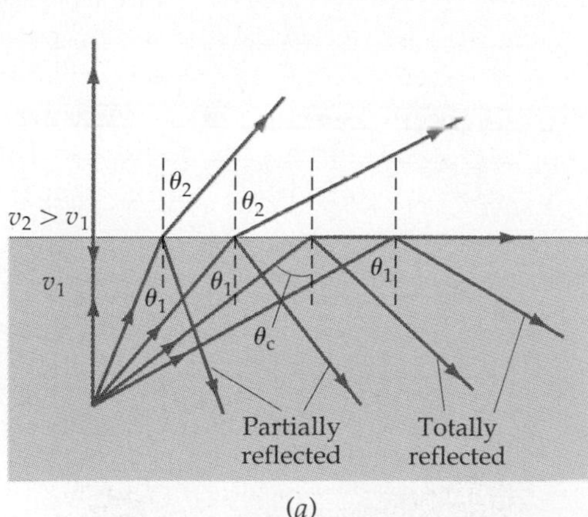

(a)

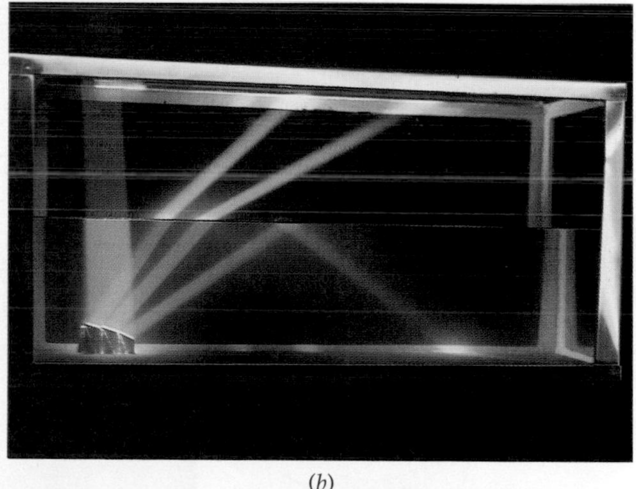

(b)

Figure 30-17 (*a*) Total internal reflection. For refraction from glass or water to air, the refracted ray is bent away from the normal. As the angle of incidence is increased, the angle of refraction is increased until, at a critical angle of incidence θ_c, the angle of refraction is 90°. For angles of incidence greater than the critical angle, there is no refracted ray. All the energy is reflected. (*b*) Photograph of refraction and total internal reflection from a water–air surface.

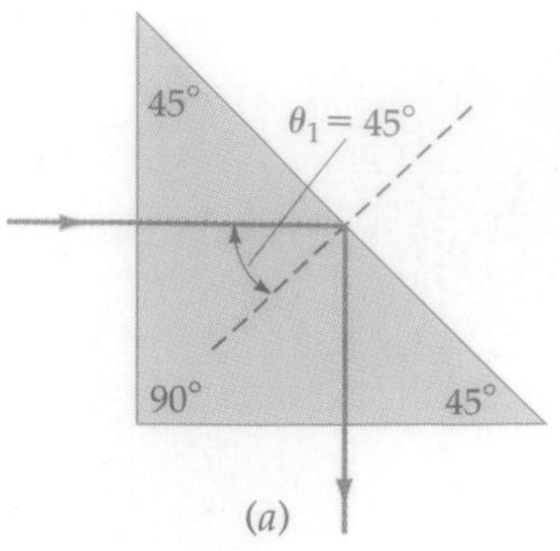

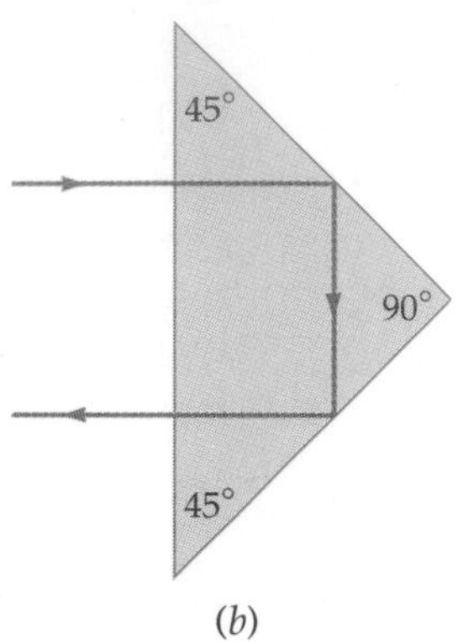

Figure 30-18 (*a*) Light entering through one of the short sides of a 45-45-90° glass prism is totally reflected in the prism and emerges through the other short side at 90° to the direction of the incident light. (*b*) Light entering through the long side of the prism is totally reflected twice and emerges in the direction opposite to that of the incident light.

Example 30-6

A particular glass has an index of refraction of $n = 1.50$. What is the critical angle for total internal reflection for light leaving this glass and entering air, for which $n = 1.00$?

From Equation 30-7, we obtain for the critical angle

$$\sin \theta_c = \frac{1.00}{1.50} = 0.667$$

The angle whose sine is 0.667 is 42°.

Figure 30-18*a* shows light incident normally on one of the short sides of a 45-45-90° glass prism. If the index of refraction of the prism is 1.5, the critical angle for total internal reflection is 42°, as we found in Example 30-6. Since the angle of incidence of the ray on the glass–air surface is 45°, the light will be totally reflected and will exit perpendicular to the other face of the prism as shown. In Figure 30-18*b*, the light is incident perpendicular to the hypotenuse of the prism and is totally reflected twice such that it emerges at 180° to its original direction. Prisms are used to change the direction of a light ray. In binoculars, four prisms are used to reinvert the image that was inverted by the binocular lens. Diamonds have a very high index of refraction ($n \approx 2.4$), so nearly all the light that enters a diamond is eventually reflected back out, giving the diamond its sparkle.

An interesting application of total internal reflection is the transmission of a beam of light down a long, narrow, transparent glass fiber (Figure 30-19). If the beam begins approximately parallel to the axis of the fiber, it will strike the walls of the fiber at angles greater than the critical angle (if the bends in the fiber are not too sharp) and no light energy will be lost through the walls of the fiber.

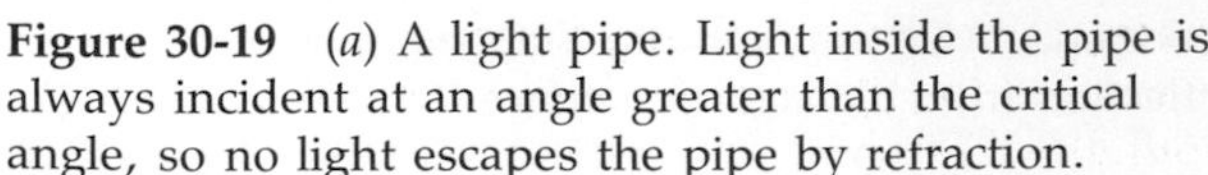

Figure 30-19 (*a*) A light pipe. Light inside the pipe is always incident at an angle greater than the critical angle, so no light escapes the pipe by refraction. (*b*) These laser beams are totally reflected three times at the sides of the glass bar before exiting the end.

(a)

(b)

(a) Light emerging from a loose bundle of glass fibers. (b) In this demonstration at the Naval Research Laboratory, a combination of laser sources generates different colors that excite adjacent fiber sensor elements leading to a separation of the information as indicated by the separation of the colors.

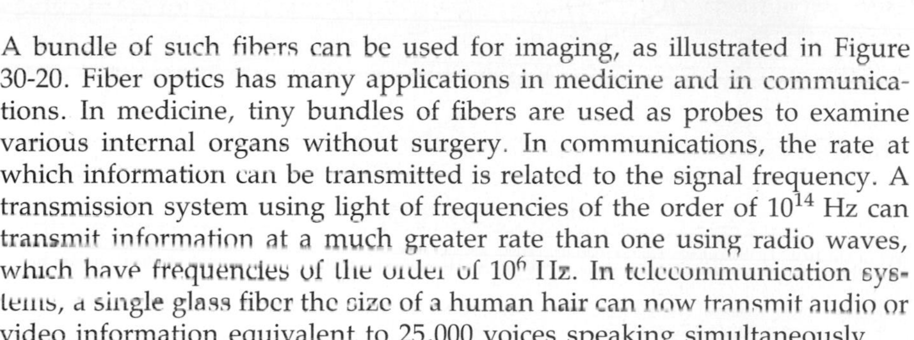

A bundle of such fibers can be used for imaging, as illustrated in Figure 30-20. Fiber optics has many applications in medicine and in communications. In medicine, tiny bundles of fibers are used as probes to examine various internal organs without surgery. In communications, the rate at which information can be transmitted is related to the signal frequency. A transmission system using light of frequencies of the order of 10^{14} Hz can transmit information at a much greater rate than one using radio waves, which have frequencies of the order of 10^6 Hz. In telecommunication systems, a single glass fiber the size of a human hair can now transmit audio or video information equivalent to 25,000 voices speaking simultaneously.

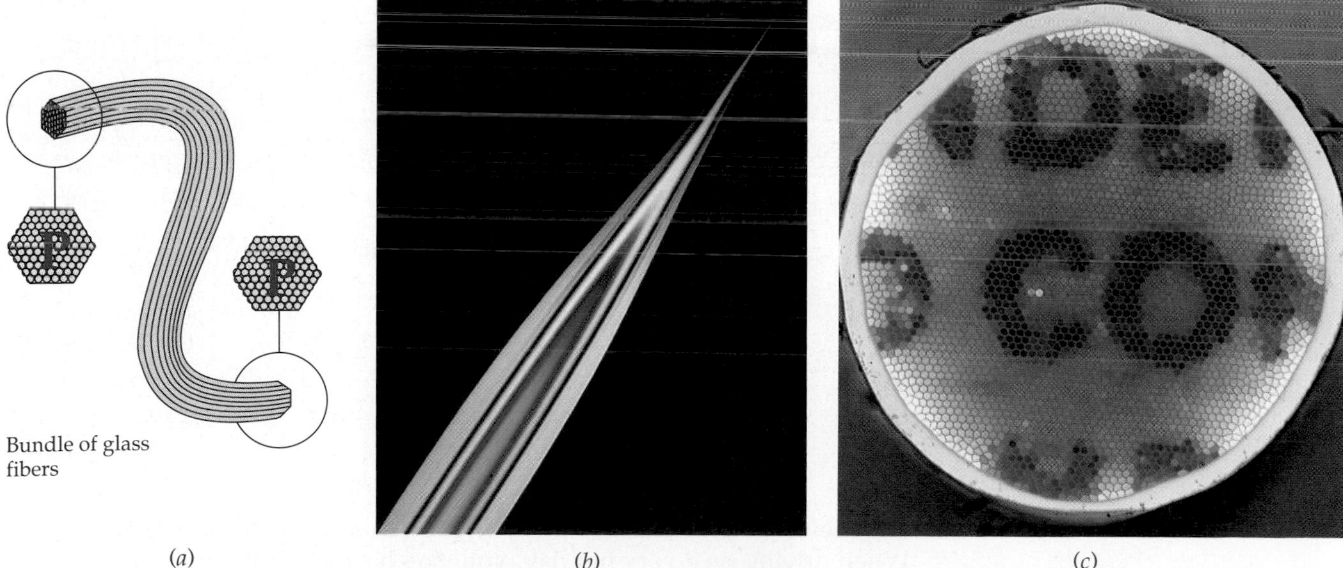

(a) (b) (c)

Figure 30-20 (a) Light from the object is transported by a bundle of glass fibers to form an image of the object at the other end of the pipe. (b) The tip of a lightguide preform is softened by heat and drawn into a long tiny fiber. The colors in the preform indicate a layered structure of differing compositions, which is retained in the fiber. (c) An image transmitted through a bundle of fibers.

When the index of refraction of a medium changes gradually, the refraction is continuous, leading to a gradual bending of the light. An interesting example of this is the formation of a mirage. On a hot day, there is often a layer of very hot air near the ground. This air is warmer and is therefore less dense than the air just above it. The speed of light is slightly greater in this less dense layer, so a light beam passing from the cooler layer into the warmer layer is bent. Figure 30-21*a* shows the light from a tree when all the air is at the same temperature. The wavefronts are spherical, and the rays are straight lines. In Figure 30-21*b*, the air near the ground is warmer, resulting in a greater speed of light there. The portions of the wavefronts near the ground travel faster and get ahead of the higher portions, creating a non-spherical wavefront and causing a curving of the rays. Thus, the ray shown initially heading for the ground is bent upwards. As a result, the viewer sees an image of the tree and thinks the light has been reflected from the ground.

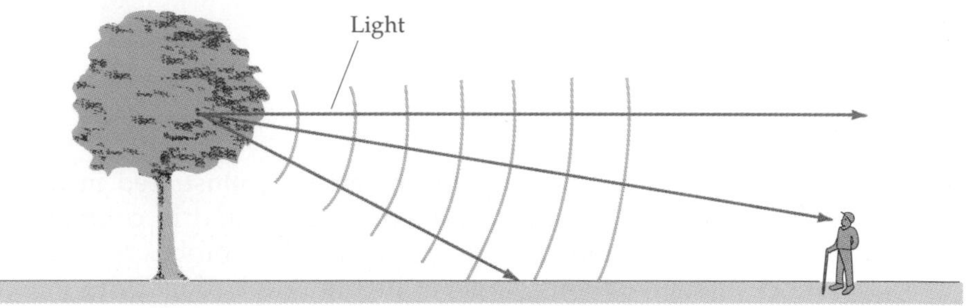

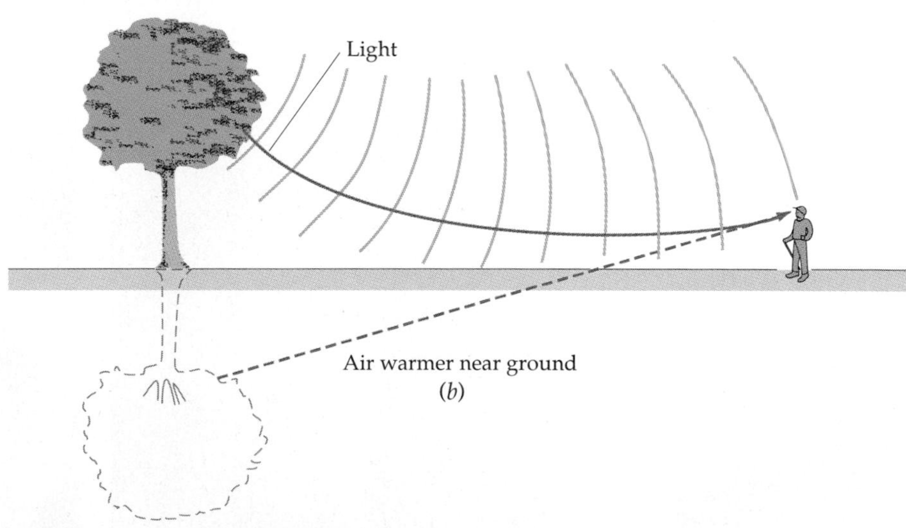

Figure 30-21 A mirage. (*a*) When the air is at a uniform temperature, the wavefronts of the light from the tree are spherical. (*b*) When the air near the ground is warmer, the wavefronts are not spherical and the light from the tree is continuously refracted into a curved path. Because an image of the tree is seen, the viewer may think that there is a reflecting body of water in front of the tree. (*c*) Photograph of apparent reflections of motorcycles and cars on a hot road.

The viewer often attributes this reflection to a water surface near the tree. When driving on a very hot day, you may have noticed apparent wet spots on the highway that disappear when you get to them. This is due to the refraction of light from a hot air layer near the pavement.

Table 30-1 **Index of Refraction for Yellow Sodium Light (λ = 589 nm)**

Substance	Index of refraction	Substance	Index of refraction
Solids		Liquids at 20°C	
Ice (H_2O)	1.309	Methyl alcohol (CH_3OH)	1.329
Fluorite (CaF_2)	1.434	Water (H_2O)	1.333
Rock salt (NaCl)	1.544	Ethyl alcohol (C_2H_5OH)	1.36
Quartz (SiO_2)	1.544	Carbon tetrachloride (CCl_4)	1.460
Zircon ($ZrO_2 \cdot SiO_2$)	1.923	Turpentine	1.472
Diamond (C)	2.417	Glycerine	1.473
Glasses (typical values)		Benzene	1.501
Crown	1.52	Carbon disulfide (CS_2)	1.628
Light flint	1.58		
Medium flint	1.62		
Dense flint	1.66		

Dispersion

Table 30-1 lists the indexes of refraction for sodium light of wavelength 589 nm for various transparent materials. The index of refraction of a material has a slight dependence on wavelength. Figure 30-22 shows this dependence for several materials. We can see that the indexes of refraction for these materials decrease slightly as the wavelength increases. This dependence of the index of refraction on wavelength (and therefore on frequency) is called **dispersion.** When a beam of white light is incident at some angle on the surface of a glass prism, the angle of refraction for the shorter wavelengths toward the violet end of the visible spectrum is slightly larger than that for the longer wavelengths toward the red end of the spectrum. The light of shorter wavelength is therefore bent more than that of longer wavelength. The beam of white light is thus spread out or dispersed into its component colors or wavelengths (Figure 30-23).

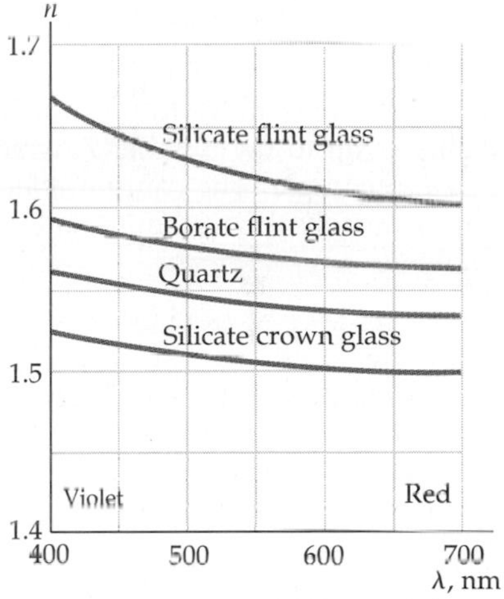

Figure 30-22 The graph of the index of refraction versus wavelength for various materials.

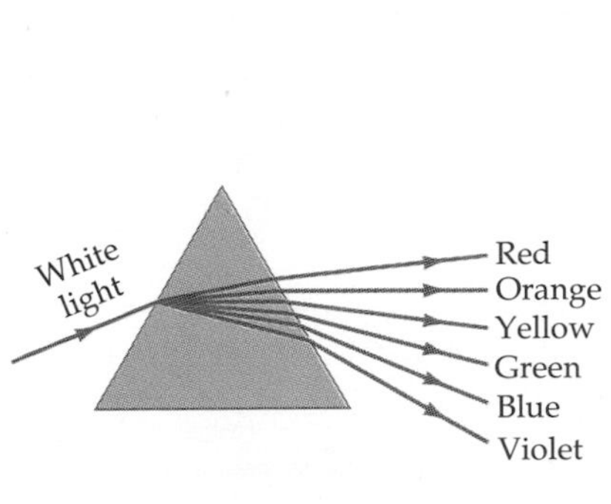

(*b*)

Figure 30-23 (*a*) A beam of white light incident on a glass prism is dispersed into its component colors. The index of refraction decreases as the wavelength increases so that the longer wavelengths (red) are bent less than the shorter wavelengths (blue). (*b*) Photograph of the dispersion of light by a glass prism.

Rainbows

The formation of a rainbow is a familiar example of the dispersion of sunlight by refraction in water drops. Figure 30-24 is a diagram originally drawn by Descartes showing parallel rays of light from the sun entering a spherical water drop. First, the rays are refracted as they enter the drop. They are then reflected from the back water–air surface and finally are refracted again as they leave the drop. Ray 1 enters the drop along a diameter (with an angle of incidence of zero) and is reflected back along its incident path. Ray 2 enters slightly above the diameter and emerges below the diameter at a small angle with it. The rays entering farther and farther away from the diameter emerge at greater and greater angles up to ray 7, shown as the heavy line.

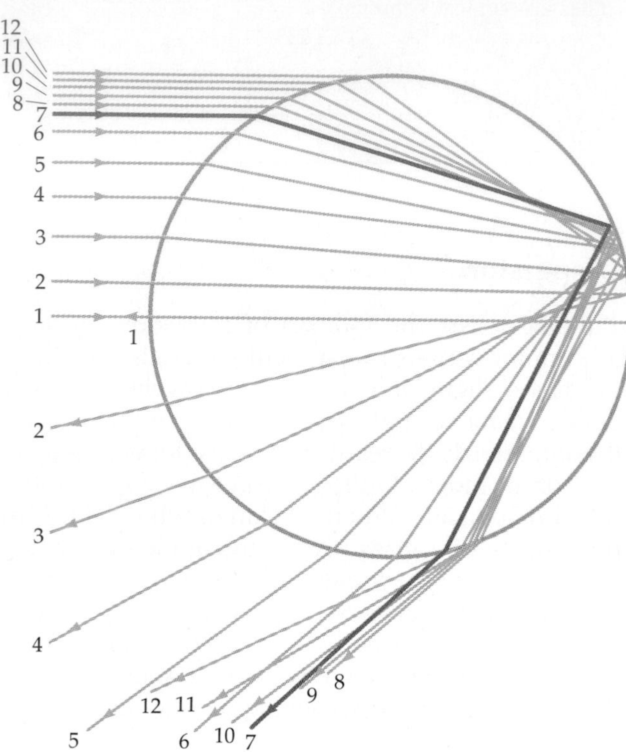

Figure 30-24 Descartes' construction of parallel rays of light entering a spherical water drop. The rays are refracted at the first surface, reflected from the back surface, and refracted again as they leave the drop. The angle between the emerging ray and the diameter increases as we move to rays further away from the diameter up to ray number 7, which emerges at the maximum angle. The concentration of rays emerging at approximately the maximum angle gives rise to the rainbow.

Rays entering above ray 7 emerge at smaller and smaller angles with the diameter. We can see from the diagram that a concentration of rays emerges at angles near the maximum angle. This concentration of rays near the maximum angle gives rise to the rainbow. By construction (using the law of refraction), Descartes showed that the maximum angle is about 42°. To observe a rainbow, we must therefore look at the water drops at an angle of 42° relative to the line back to the sun as shown in Figure 30-25. The angular radius of the rainbow is therefore 42°.

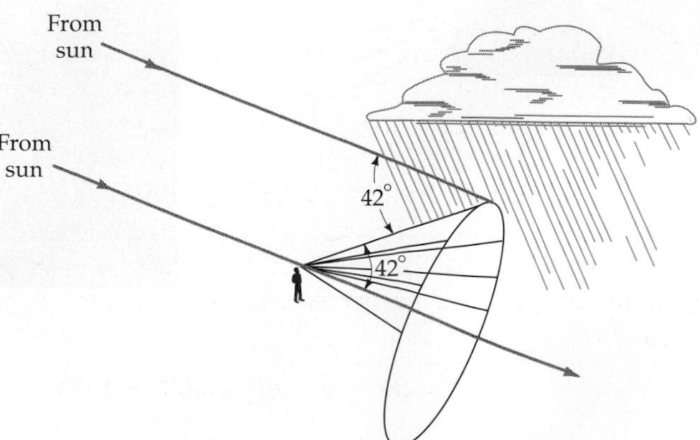

Figure 30-25 A rainbow is viewed at an angle of 42° from the line to the sun, as predicted by Descartes' construction in Figure 30-24.

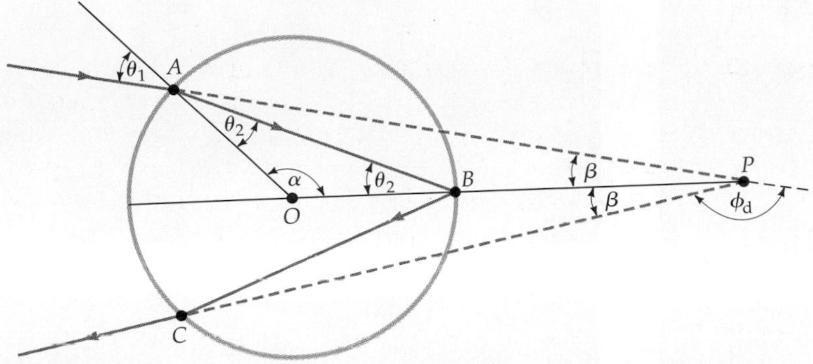

Figure 30-26 Light ray incident on a spherical water drop. The ray is refracted at point A, reflected at point B, and again refracted at point C, where it leaves the drop. The line of the incident ray intersects that of the emerging ray at point P. The angle ϕ_d is called the angle of deviation of the ray.

We can calculate the angular radius of the rainbow from the laws of reflection and refraction. Figure 30-26 shows a ray of light incident on a spherical water droplet at point A. The angle of refraction θ_2 is related to the angle of incidence θ_1 by Snell's law:

$$n_{air} \sin \theta_1 = n_{water} \sin \theta_2 \qquad \text{30-8}$$

The refracted ray strikes the back of the water droplet at point B. It makes an angle θ_2 with the radial line OB and is reflected at an equal angle. The ray is refracted again at point C, where it leaves the droplet. Point P is the intersection of the line of the incident ray and the line of the emerging ray. The angle ϕ_d is called the angle of deviation of the ray. It is related to the angle β by

$$\phi_d + 2\beta = \pi \qquad \text{30-9}$$

The angle 2β is the angular radius of the rainbow. We wish to relate the angle of deviation ϕ_d to the angle of incidence θ_1. From the triangle AOB, we have

$$2\theta_2 + \alpha = \pi \qquad \text{30-10}$$

Similarly, from the triangle AOP, we have

$$\theta_1 + \beta + \alpha = \pi \qquad \text{30-11}$$

Eliminating α from Equations 30-10 and 30-11 and solving for β, we obtain

$$\beta = \pi - \theta_1 - \alpha = \pi - \theta_1 - (\pi - 2\theta_2) = 2\theta_2 - \theta_1$$

Substituting this value for β into Equation 30-9, we obtain for the angle of deviation

$$\phi_d = \pi - 2\beta = \pi - 4\theta_2 + 2\theta_1 \qquad \text{30-12}$$

Equation 30-12 can be combined with Snell's law (Equation 30-8) to eliminate θ_2 and give the angle of deviation ϕ_d in terms of the angle of incidence θ_1:

$$\phi_d = \pi + 2\theta_1 - 4 \arcsin\left(\frac{n_{air} \sin \theta_1}{n_{water}}\right) \qquad \text{30-13}$$

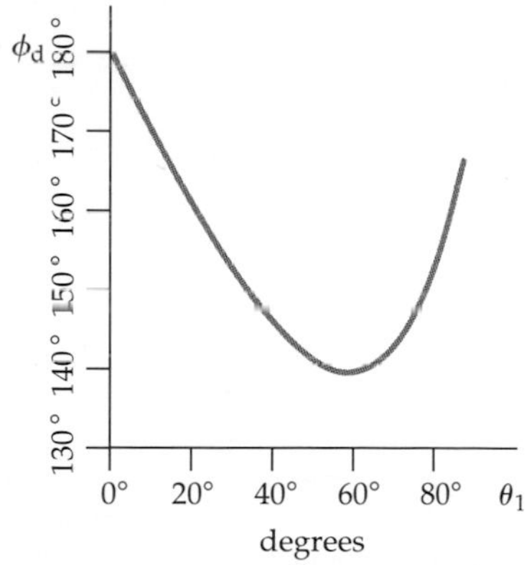

Figure 30-27 Plot of the angle of deviation ϕ_d as a function of incident angle θ_1. The angle of deviation has its minimum value of 138° when the angle of incidence is 60°. Since $d\phi_d/d\theta_1 = 0$ at minimum deviation, the deviation of rays with incident angles slightly less or slightly greater than 60° will be approximately the same.

(The arcsine of a quantity is the angle whose sine is that quantity, that is, the arcsin x is the angle whose sine is x.) Figure 30-27 shows a plot of ϕ_d versus θ_1. The angle of deviation ϕ_d has its minimum value when $\theta_1 = 60°$. At this angle of incidence, the angle of deviation is $\phi_d = 138°$. This angle is called the **angle of minimum deviation.** At incident angles that are slightly greater or slightly smaller than 60°, the angle of deviation is approximately the same. Therefore, the light reflected by the water droplet will be concentrated near the angle of minimum deviation. The angular radius of the rainbow is thus

$$2\beta = \pi - \phi_d = 180° - 138° = 42°$$

(a)

(b)

(*a*) This 22° halo around the sun results from reflection and refraction from hexagonal ice crystals that are randomly oriented in the atmosphere. (*b*) When the ice crystals are not randomly oriented but are falling with their flat bases horizontal, only parts of the halo on each side of the sun, called "sun dogs," are seen.

The separation of the colors in the rainbow results from the fact that the index of refraction of water depends slightly on the wavelength of light. The angle of minimum deviation and the angular radius of the bow will therefore depend slightly on the wavelength of the light. The observed rainbow is made up of light rays from many different droplets of water as shown in Figure 30-28. The color seen at a particular angular radius corresponds to the wavelength of light that has an angle of minimum deviation that allows the light to reach the eye from the droplets at that angular radius. For example, since n_{water} is smaller for red light than for blue light, the arcsine in Equation 30-13 is greater for red light than for blue light. Then ϕ_d is smaller for red light than for blue light and β is greater for red light than for blue light. Consequently, the red part of the rainbow is at a slightly greater angular radius than the blue part of the rainbow.

When a light ray strikes a surface separating water and air, part of the light is reflected and part is refracted. A secondary rainbow results from the light rays that are reflected twice within a droplet (Figure 30-29). The secondary bow has an angular radius of 51°, and its color sequence is the reverse of that of the primary bow; that is, the violet is on the outside in the secondary bow. Because of the small fraction of light reflected from a water–air surface, the secondary bow is considerably fainter than the primary bow.

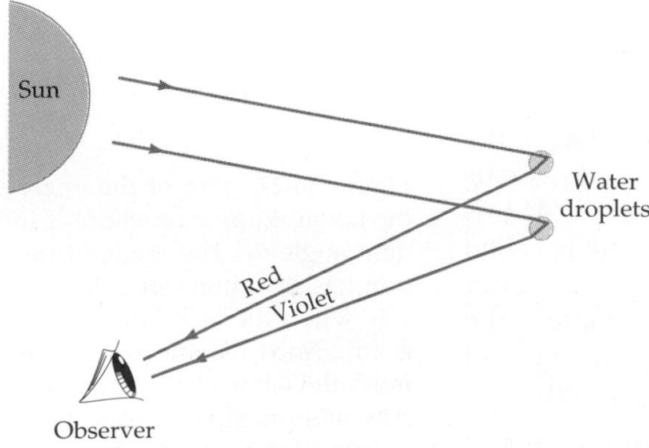

Figure 30-28 The rainbow results from light from many different water droplets.

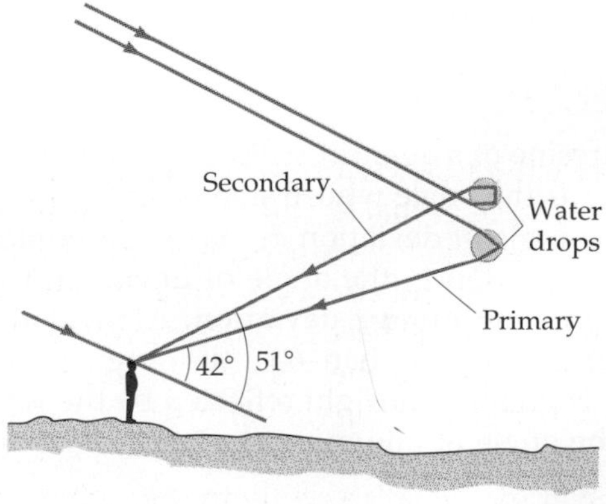

Figure 30-29 The secondary rainbow results from light rays that are reflected twice within a water droplet.

30-5 Fermat's Principle

We have seen that the propagation of light and other waves can be described by Huygens' principle. The propagation of light can also be described by **Fermat's principle** first enunciated in the seventeenth century by the French mathematician Pierre de Fermat:

> The path taken by light in traveling from one point to another is such that the time of travel is a minimum.

Fermat's principle

This statement of Fermat's principle does not cover all cases. The time of travel for the path taken by light is not always a minimum. Sometimes it is a maximum. A more complete and general formulation of Fermat's principle is

> The path taken by light in traveling from one point to another is such that the time of travel is stationary with respect to variations in that path.

Fermat's principle; general form

If the time t is expressed in terms of some parameter x, the path taken by light will be such that $dt/dx = 0$; that is, t is either a minimum, a maximum, or a point of inflection in the t-versus-x curve. The important characteristic of a stationary path is that the time taken along nearby paths will be approximately the same as that along the true path.

In this section, we will use Fermat's principle for alternative derivations of the laws of reflection and refraction.

Reflection

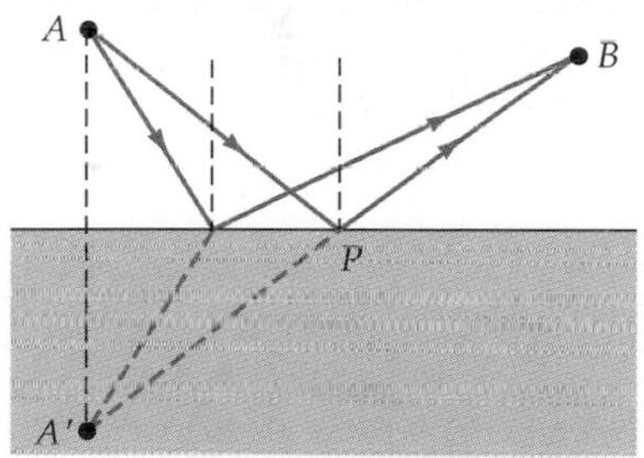

Figure 30-30 Geometry for deriving the law of reflection from Fermat's principle. The time it takes for the light to travel from point A to point B is a minimum when the light strikes the surface at point P.

In Figure 30-30, we assume that light leaves point A, strikes the plane surface, which we can consider to be a mirror, and travels to point B. We wish to find the path taken by the light. The problem for the application of Fermat's principle to reflection can be stated as follows: At what point P in Figure 30-30 must the light strike the mirror so that it will travel from point A to point B in the least time? Since the light is traveling in the same medium for this problem, the time will be minimum when the distance is minimum. In Figure 30-30, the distance APB is the same as the distance $A'PB$, where A' is the image of the source A. Point A' lies along the perpendicular from A to the mirror and is equidistant behind the mirror. Obviously, as we vary point P, the distance $A'PB$ is least when the points A', P, and B lie on a straight line. We can see from the figure that this occurs when the angle of incidence equals the angle of reflection.

Refraction

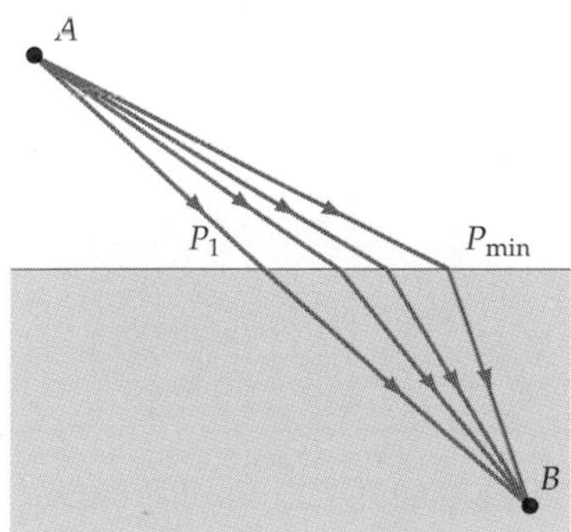

Figure 30-31 Geometry for deriving Snell's law of refraction from Fermat's principle. The point P_{min} is the point at which light must strike the glass for the travel time from A to B to be a minimum.

The derivation of Snell's law of refraction from Fermat's principle is more complicated than that for the law of reflection. Figure 30-31 shows the possible paths for light traveling from point A in air to point B in glass. Point P_1 is on the straight line between A and B, but this path is not the one for the shortest travel time because light travels with a smaller speed in the glass. If we move slightly to the right of P_1, the total path length is greater, but the distance traveled in the slower medium is less than for the path through P_1. It is not apparent from the figure which path is that of least time, but it is not surprising that a path slightly to the right of the straight-line path takes less time because the time gained by traveling a shorter distance in the glass more than compensates for the time lost traveling a longer distance in the air. As we move the point of intersection of the possible path to the right of point P_1, the total time of travel from A to B decreases until we reach a minimum at point P_{min}. Beyond this point, the time saved by traveling a shorter distance in the glass does not compensate for the greater time required for the greater distance traveled in the air.

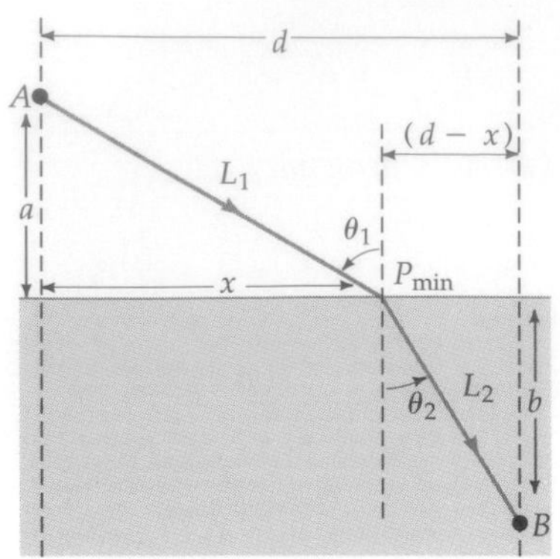

Figure 30-32 Geometry for calculating the minimum time in the derivation of Snell's law from Fermat's principle.

Figure 30-32 shows the geometry for finding the path of least time. If L_1 is the distance traveled in medium 1 with index of refraction n_1 and L_2 is the distance traveled in medium 2 with index of refraction n_2, the time for light to travel the total path AB is

$$t = \frac{L_1}{v_1} + \frac{L_2}{v_2} = \frac{L_1}{c/n_1} + \frac{L_2}{c/n_2} = \frac{n_1 L_1}{c} + \frac{n_2 L_2}{c} \qquad 30\text{-}14$$

We wish to find the point P_{min} for which this time is a minimum. We do this by expressing the time in terms of a single parameter indicating the position of point P_{min}. In terms of the distance x in Figure 30-32, we have

$$L_1^2 = a^2 + x^2 \quad \text{and} \quad L_2^2 = b^2 + (d - x)^2 \qquad 30\text{-}15$$

Figure 30-33 shows the time t as a function of x. At the value of x for which the time is a minimum, the slope of this graph is zero:

$$\frac{dt}{dx} = 0$$

Differentiating each term in Equation 30-14 with respect to x, we obtain

$$\frac{dt}{dx} = \frac{1}{c}\left(n_1 \frac{dL_1}{dx} + n_2 \frac{dL_2}{dx}\right)$$

Setting $dt/dx = 0$, we obtain

$$n_1 \frac{dL_1}{dx} + n_2 \frac{dL_2}{dx} = 0 \qquad 30\text{-}16$$

We can compute these derivatives from Equations 30-15. We have

$$2L_1 \frac{dL_1}{dx} = 2x$$

or

$$\frac{dL_1}{dx} = \frac{x}{L_1}$$

But x/L_1 is just $\sin\theta_1$, where θ_1 is the angle of incidence. Thus,

$$\frac{dL_1}{dx} = \sin\theta_1$$

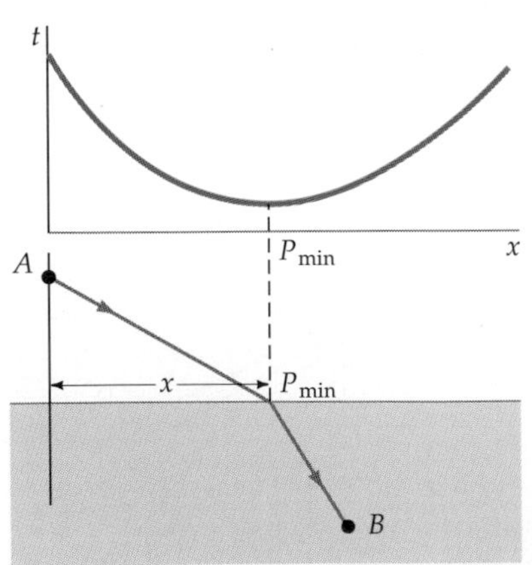

Figure 30-33 Graph of the time it takes for light to travel from A to B versus x, measured along the refracting surface. The time is a minimum at the point at which the angles of incidence and refraction obey Snell's law.

Similarly,

$$2L_2 \frac{dL_2}{dx} = 2(d - x)(-1)$$

or

$$\frac{dL_2}{dx} = -\frac{d - x}{L_2} = -\sin\theta_2$$

where θ_2 is the angle of refraction. Hence, Equation 30-16 is

$$n_1 \sin\theta_1 + n_2(-\sin\theta_2) = 0$$

or

$$n_1 \sin\theta_1 = n_2 \sin\theta_2$$

which is Snell's law.

30-6 Polarization

In any transverse wave, the vibration is perpendicular to the direction of propagation of the wave. For example, in waves moving down a string, the elements of the string move in a plane perpendicular to the string. Similarly, in a light wave traveling in the z direction, the electric field is perpendicular to the z direction. (The magnetic field of a light wave is also perpendicular to the z direction.) If the vibration of a transverse wave remains parallel to a fixed line in space, the wave is said to be **linearly polarized.** We can visualize polarization most easily by considering mechanical waves on a string. If one end is moved up and down, the resulting waves on the string are linearly polarized with each element of the string vibrating in the vertical direction. Similarly, if one end is moved along a horizontal line (perpendicular to the string), the displacements of the string are linearly polarized in the horizontal direction. If one end of the string is moved with constant speed in a circle, the resulting wave is said to be **circularly polarized.** In this case, each element of the string moves in a circle. Unpolarized waves can be produced by moving the end of the string vertically and horizontally in a random way. Then, if the string itself is in the z direction, the vibrations will have both x and y components that vary in a random way.

Most waves produced by a single source are polarized. For example, string waves produced by the regular vibration of one end of a string or electromagnetic waves produced by a single atom or by a single antenna are polarized. Waves produced by many sources are usually unpolarized. A typical light source, for example, contains millions of atoms acting independently. The electric field for such a wave can be resolved into x and y components that vary randomly because there is no correlation between the individual atoms producing the light.

There are four phenomena that produce polarized light from unpolarized light: (1) absorption, (2) scattering, (3) reflection, and (4) birefringence (also called double refraction).

Polarization by Absorption

Several naturally occurring crystals, when cut into appropriate shapes, absorb and transmit light differently depending on the polarization of the light. These crystals can be used to produce linearly polarized light. In 1938, E. H. Land invented a simple commercial polarizing film called Polaroid. This material contains long-chain hydrocarbon molecules that are aligned when the sheet is stretched in one direction during the manufacturing process. These chains become conducting at optical frequencies when the sheet is dipped in a solution containing iodine. When light is incident with its electric-field vector parallel to the chains, electric currents are set up along the chains, and the light energy is absorbed. If the electric field is perpendicular to the chains, the light is transmitted. The direction perpendicular to the chains is called the **transmission axis.** We will make the simplifying assumption that all the light is transmitted when the electric field is parallel to the transmission axis and all is absorbed when it is perpendicular to the transmission axis.

Consider an unpolarized light beam traveling in the z direction incident on a polarizing film with its transmission axis in the y direction. On the average, half of the incident light has its electric field in the y direction and half has it in the x direction. Thus, half the intensity is transmitted, and the transmitted light is linearly polarized with its electric field in the y direction.

Suppose we have a second polarizing film whose transmission axis makes an angle θ with that of the first as shown in Figure 30-34. If $\mathbf{E}$ is the

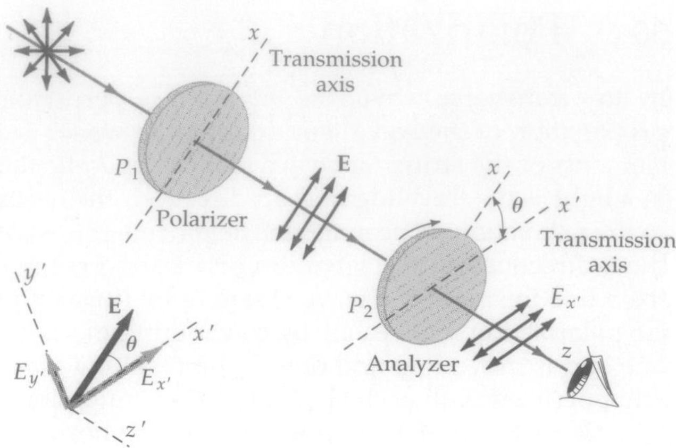

Figure 30-34 Two polarizing films with their transmission axes making an angle θ with each other. Only the component $E \cos \theta$ is transmitted through the second film. If the intensity between the films is I_0, that transmitted by both films is $I_0 \cos^2 \theta$.

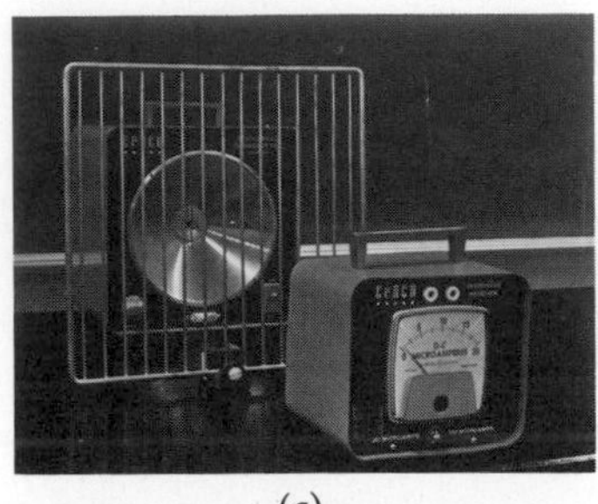

(*a*)

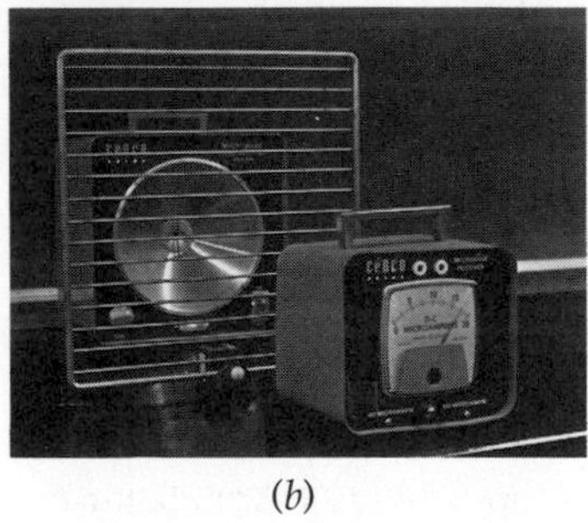

(*b*)

Figure 30-35 Demonstration showing the polarization of microwaves. The electric field of the microwaves is vertical, parallel to the vertical dipole antenna. (*a*) When the metal wires of the absorber are vertical, electric currents are set up in the wires and energy is absorbed, as indicated by the low reading on the detector. (*b*) When the wires are horizontal, no currents are set up, and the microwaves are transmitted, as indicated by the high reading on the detector. The transmission axis for the wire screen is perpendicular to the wires.

electric field between the films, its component along the direction of the transmission axis of the second film is $E \cos \theta$. Since the intensity of light is proportional to E^2, the intensity of light transmitted by both films will be given by

$$I = I_0 \cos^2 \theta \tag{30-17}$$

where I_0 is the intensity incident on the second film and is, of course, half the intensity incident on the first film. When two polarizing elements are placed in succession in a beam of light as described here, the first is called the **polarizer** and the second is called the **analyzer.** If the polarizer and analyzer are crossed, that is, if their transmission axes are perpendicular to each other, no light gets through. Equation 30-17 is known as **Malus's law** after its discoverer E. L. Malus (1775–1812). It applies to any two polarizing elements whose transmission axes make an angle θ with each other.

The polarization of electromagnetic waves can be demonstrated with microwaves, which have wavelengths on the order of centimeters. In a typical microwave generator, polarized waves are radiated by a dipole antenna. In Figure 30-35, the dipole antenna is vertical, so the electric-field vector **E** of the radiated waves is vertical. An absorber can be made of a screen of parallel straight wires. When the wires are vertical, as in Figure 30-35*a*, the electric field parallel to the wires sets up currents in the wires and energy is absorbed. When the wires are horizontal and so perpendicular to **E**, as in Figure 30-35*b*, no currents are set up and the waves are transmitted. The transmission axis of the wire screen is thus perpendicular to the wires.

Example 30-7

Unpolarized light of intensity 3.0 W/m^2 is incident on two polarizing films whose transmission axes make an angle of 60°. What is the intensity of light transmitted by both films?

The intensity of the light that is transmitted by the first film is half the incident intensity, or 1.5 W/m^2. Calling this intensity I_0, that transmitted by the second film is

$$\begin{aligned} I = I_0 \cos^2 \theta &= (1.5\ \text{W/m}^2) \cos^2 60^\circ \\ &= (1.5\ \text{W/m}^2)(0.500)^2 \\ &= 0.375\ \text{W/m}^2 \end{aligned}$$

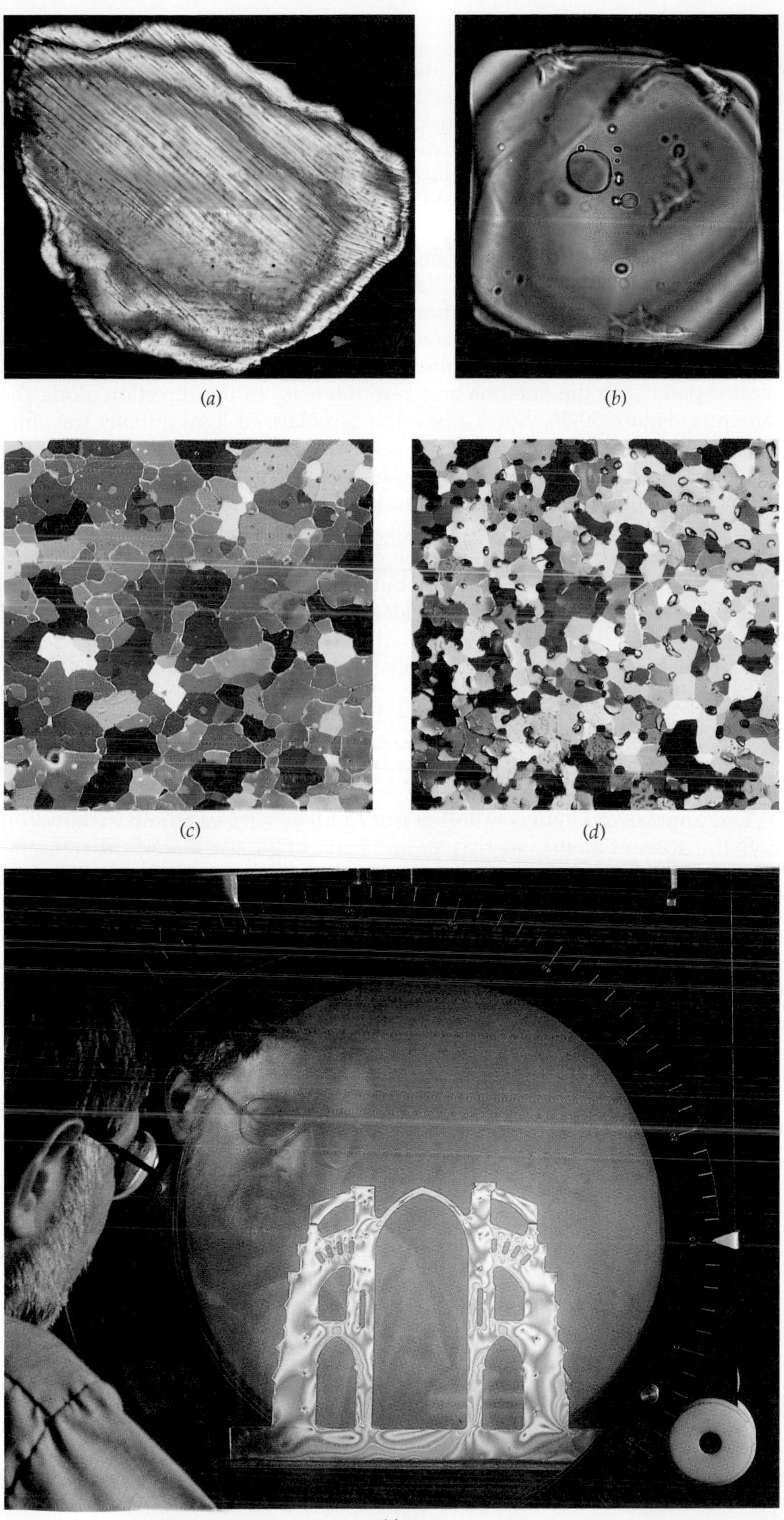

When the transmission axes of two polarizing films are perpendicular, the polarizers are said to be crossed and no light is transmitted. However, many materials are birefringent (page 1001) or become so under stress. Such materials rotate the direction of polarization of the light so that light of a particular wavelength is transmitted through both polarizers. When a birefringent material is viewed between crossed polarizers, information about its internal structure is revealed. (*a*) A thin section of a shocked quartz grain from the site of a meteorite crater. The layered structure, evidenced by the parallel lines, arises from the shock of the impact of the meteor. (*b*) A grain of quartz typically found in silicic volcanic rocks. No shock lines are seen. (*c*) Thin sections of ice core from the antarctic ice sheet reveal bubbles of trapped CO_2, which appear amber-colored. This sample was taken from a depth of 194 meters corresponding to air trapped 1600 years ago, whereas that in (*d*) is from a depth of 56 meters corresponding to air trapped 450 years ago. Ice core measurements have replaced the less reliable technique of analyzing carbon in tree rings to compare current atmospheric CO_2 levels with those of the recent past. (*e*) Robert Mark of the Princeton School of Architecture examines the stress patterns in a plastic model of the nave structure of Chartres Cathedral.

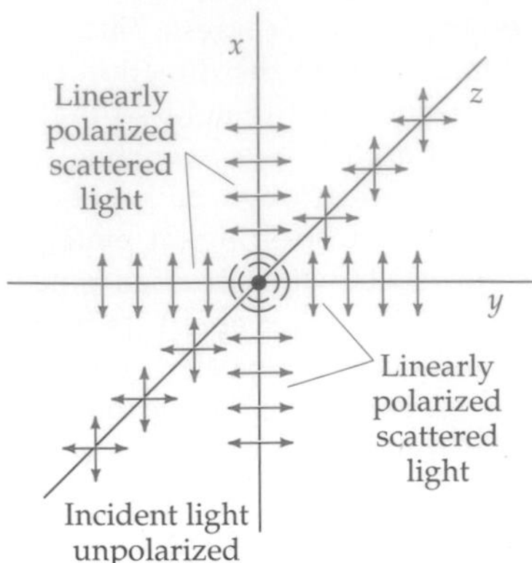

Figure 30-36 Polarization by scattering. Unpolarized light propagating in the z direction is incident on a scattering center at the origin. The light scattered in the x direction is polarized in the y direction and that scattered in the y direction is polarized in the x direction.

Polarization by Scattering

The phenomenon of absorption and reradiation is called **scattering.** Scattering can be demonstrated by passing a light beam through a container of water to which a small amount of powdered milk has been added. The milk particles absorb light and reradiate it, making the light beam visible. Similarly, laser beams can be made visible by introducing chalk or smoke particles into the air to scatter the light. A familiar example of light scattering is that from clusters of air molecules (due to random fluctuations in the density of air), which tend to scatter short wavelengths more than long wavelengths, thereby giving the sky its blue color.

We can understand polarization by scattering if we think of an absorbing molecule as an electric-dipole antenna that radiates waves with a maximum intensity in the direction perpendicular to the antenna with the electric-field vector parallel to the antenna and zero intensity in the direction along the antenna. Figure 30-36 shows a beam of unpolarized light initially traveling along the z axis striking a scattering center at the origin. The electric field in the light beam has components in both the x and y directions perpendicular to the direction of motion of the light beam. These fields set up oscillations of the scattering center in both the x and y directions, but there is no oscillation in the z direction. The oscillation of the scattering center in the x direction produces light along the y axis but not along the x axis, which is along the line of oscillation. The light radiated along the y axis is thus polarized in the x direction. Similarly, the light radiated along the x axis is polarized in the y direction. This can be seen easily by examining the scattered light with a piece of polarizing film.

Polarization by Reflection

When unpolarized light is reflected from a plane surface boundary between two transparent media, such as air and glass or air and water, the reflected light is partially polarized. The degree of polarization depends on the angle of incidence and the indexes of refraction of the two media. When the angle of incidence is such that the reflected and refracted rays are perpendicular to each other, the reflected light is completely polarized. This result was discovered experimentally by Sir David Brewster in 1812.

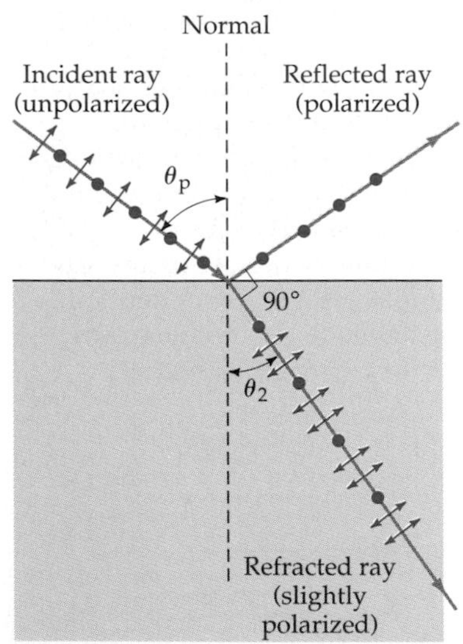

Figure 30-37 Polarization by reflection. The incident wave is unpolarized and has components of the electric field parallel to the plane of incidence (arrows) and components perpendicular to this plane (dots). For incidence at the polarizing angle, the reflected wave is completely polarized, with its electric field perpendicular to the plane of incidence.

Figure 30-37 shows light incident at the polarizing angle θ_p for which the reflected light is completely polarized. The electric field of the incident light can be resolved into components parallel and perpendicular to the plane of incidence. The reflected light is completely polarized with its electric field perpendicular to the plane of incidence. We can relate the polarizing angle to the indexes of refraction of the media using Snell's law. If n_1 is the index of refraction of the first medium and n_2 is that of the second medium, Snell's law gives

$$n_1 \sin \theta_p = n_2 \sin \theta_2$$

where θ_2 is the angle of refraction. From Figure 30-37, we can see that the sum of the angle of reflection and the angle of refraction is 90°. Since the angle of reflection equals the angle of incidence, we have

$$\theta_2 = 90° - \theta_p$$

Then

$$n_1 \sin \theta_p = n_2 \sin (90° - \theta_p)$$
$$= n_2 \cos \theta_p$$

or

$$\tan \theta_p = \frac{n_2}{n_1} \qquad 30\text{-}18$$

Brewster's law

Equation 30-18 is known as **Brewster's law.** Although the reflected light is completely polarized when the incident angle is θ_p, the transmitted light is only partially polarized because only a small fraction of the incident light is reflected. If the incident light itself is polarized with **E** in the plane of incidence, there is no reflected light when the angle of incidence is θ_p. We can understand this qualitatively from Figure 30-38. If we consider the molecules in the second medium to be oscillating parallel to the electric field of the refracted ray, there can be no reflected ray because no energy is radiated along the line of oscillation.

Because of the polarization of reflected light, sunglasses made of polarizing material can be very effective in cutting out glare. If light is reflected from a horizontal surface such as a lake or snow on the ground, the plane of incidence will be vertical and the electric field of the reflected light will be predominantly horizontal. Polarized sunglasses with a vertical transmission axis will then reduce glare by absorbing much of the reflected light. If you have polarized sunglasses, you can observe this by looking through them at such reflected light and then rotating the glasses 90°, so that much more of the light is transmitted.

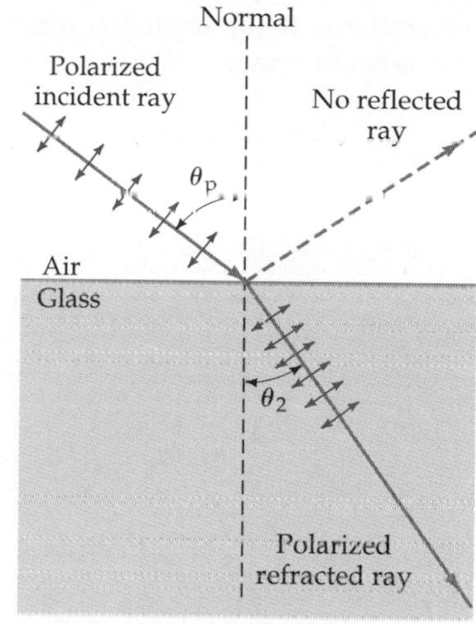

Figure 30-38 Polarized light incident at the polarizing angle. When the light is polarized with **E** in the plane of incidence there is no reflected ray.

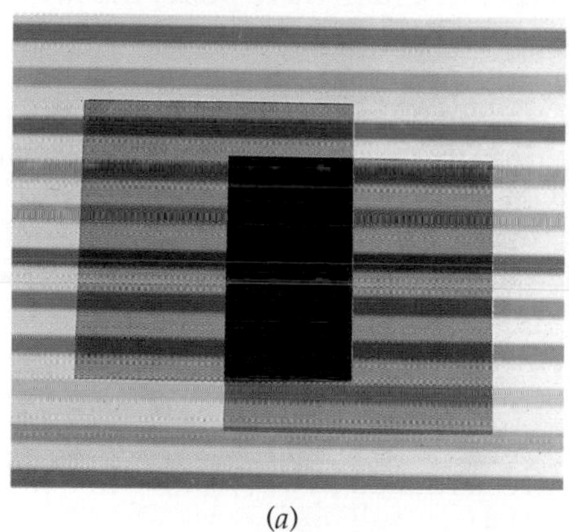

(*a*)

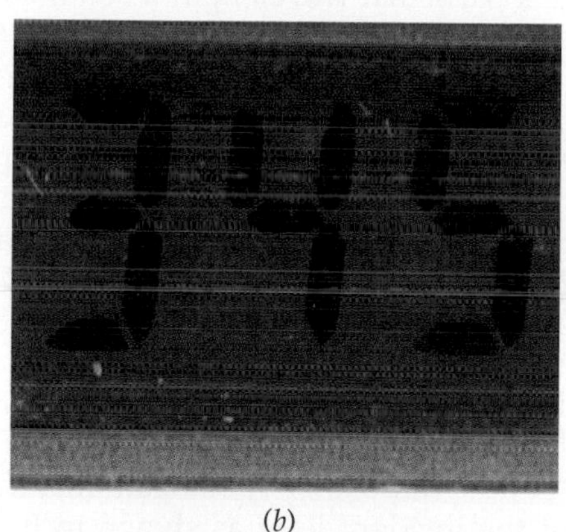

(*b*)

(*a*) Cross polarizers block out all of the light. (*b*) In a liquid crystal display, the crystal is between crossed polarizers. Light incident on the crystal is transmitted because the crystal rotates the direction of polarization of the light 90°. The light is reflected back out through the crystal by a mirror behind the crystal, and a uniform background is seen. When a voltage is applied across a small segment of the crystal, the polarization is not rotated, so no light is transmitted and the segment appears black.

Polarization by Birefringence

Birefringence, or **double refraction,** is a complicated phenomenon that occurs in calcite and other noncubic crystals and in some stressed plastics such as cellophane. In most materials, the speed of light is the same in all directions. Such materials are **isotropic.** Because of their atomic structure, birefringent materials are **anisotropic.** The speed of light depends on its direction of propagation through the material. When a light ray is incident on such materials, it may be separated into two rays called the *ordinary ray* and the *extraordinary ray*. These rays are polarized in mutually perpendicular directions, and they travel with different speeds. Depending on the relative orientation of the material and the incident light, the rays may also travel in different directions.

There is one particular direction in a birefringent material in which both rays propagate with the same speed. This direction is called the **optic axis** of the material. (The optic axis is actually a *direction* rather than a line in the material.) Nothing unusual happens when light travels along the optic axis.

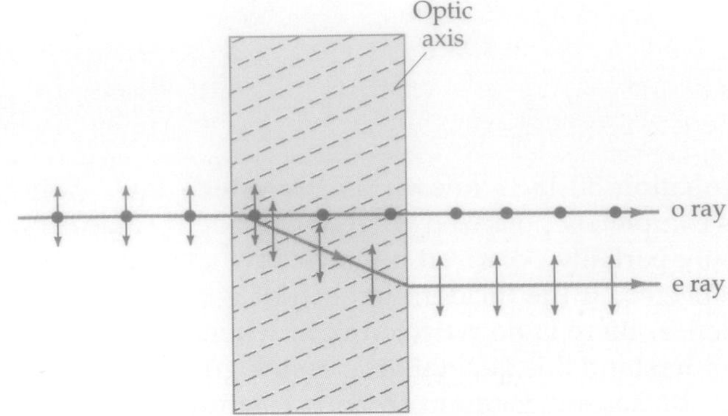

Figure 30-39 A narrow beam of light incident on a birefringent crystal such as calcite is split into two beams, called the ordinary ray (o ray) and the extraordinary ray (e ray) that have mutually perpendicular polarizations. If the crystal is rotated, the extraordinary ray rotates in space.

However, when light is incident at an angle to the optic axis as shown in Figure 30-39, the rays travel in different directions and emerge separated in space. If the material is rotated, the extraordinary ray (the e ray in the figure) rotates in space.

If light is incident on a birefringent plate perpendicular to its crystal face and perpendicular to the optic axis, the two rays travel in the same direction but at different speeds. The number of wavelengths in the two rays in the plate is different because the wavelengths ($\lambda = v/f$) of the rays differ. The rays emerge with a phase difference that depends on the thickness of the plate and on the wavelength of the incident light. In a **quarter-wave plate,** the thickness is such that there is a 90° phase difference between the waves of a particular wavelength when they emerge. In a **half-wave plate,** the rays emerge with a phase difference of 180°.

Suppose that the incident light is linearly polarized such that the electric-field vector is at 45° to the optic axis, as illustrated in Figure 30-40. The ordinary and extraordinary rays start out in phase and have equal amplitudes. With a quarter-wave plate, the waves emerge with a phase difference of 90°, so the resultant electric field has components $E_x = E_0 \sin \omega t$ and $E_y = E_0 \sin (\omega t + 90°) = E_0 \cos \omega t$. The electric-field vector thus rotates in a circle and the wave is circularly polarized.

With a half-wave plate, the waves emerge with a phase difference of 180°, so the resultant electric field is linearly polarized with components $E_x = E_0 \sin \omega t$ and $E_y = E_0 \sin (\omega t + 180°) = -E_0 \sin \omega t$. The net effect is that the direction of polarization of the wave is rotated by 90° relative to that of the incident light as shown in Figure 30-41.

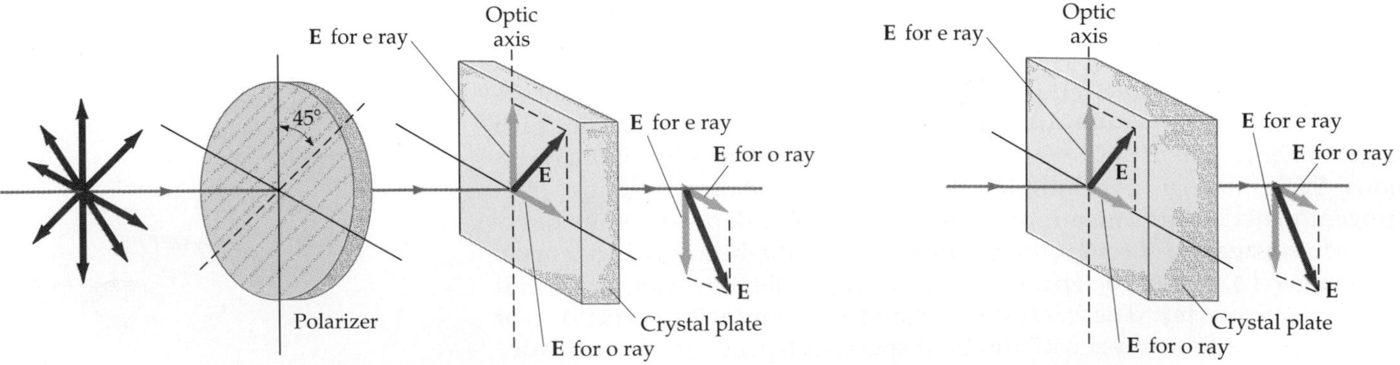

Figure 30-40 Polarized light emerging from the polarizer is incident on a birefringent crystal such that the electric-field vector makes a 45° angle with the optic axis, which is perpendicular to the light beam. The ordinary and extraordinary rays travel in the same direction but at different speeds.

Figure 30-41 When the birefringent crystal in Figure 30-40 is a half-wave plate, the emerging light has its direction of polarization rotated.

Interesting and beautiful patterns, like those on page 999, can be observed by placing birefringent materials, such as cellophane or a piece of stressed plastic, between two polarizing sheets with their transmission axes perpendicular to each other (crossed Polaroids). Ordinarily, no light is transmitted through crossed polarizing sheets because the polarization direction of light transmitted by the first sheet is perpendicular to the axis of the second. However, if we place a birefringent material between the crossed Polaroids, the material acts as a half-wave plate for light of a certain color depending on the material's thickness. The direction of polarization is rotated and some light gets through both films. Various glasses and plastics become birefringent when under stress. The stress patterns can be observed when the material is placed between crossed polarizing sheets.

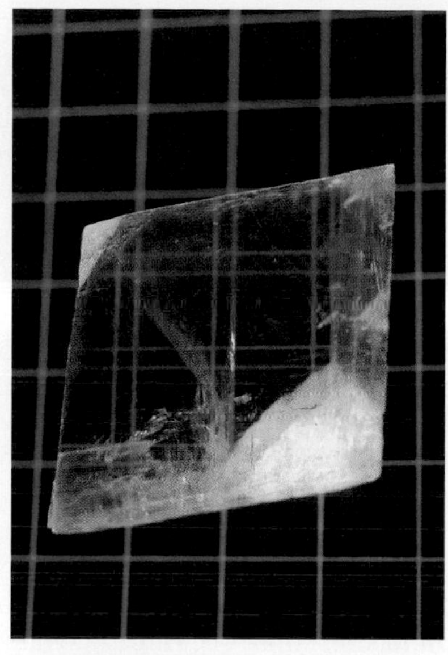

A double image of the cross hatching is produced by this birefringent crystal of calcium carbonate.

Summary

1. When light is incident on a surface separating two media in which the speed of light differs, part of the light energy is transmitted and part is reflected. The angle of reflection equals the angle of incidence:

$$\theta_r = \theta_1$$

The angle of refraction depends on the angle of incidence and on the indexes of refraction of the two media and is given by Snell's law of refraction:

$$n_1 \sin \theta_1 = n_2 \sin \theta_2$$

where the index of refraction of a medium n is the ratio of the speed of light in a vacuum c to that in the medium v:

$$n = \frac{c}{v}$$

2. When light is traveling in a medium with an index of refraction n_1 and is incident on the boundary of a second medium with a lower index of refraction $n_2 < n_1$, the light is totally reflected if the angle of incidence is greater than the critical angle θ_c given by

$$\sin \theta_c = \frac{n_2}{n_1}$$

3. The speed of light in a medium and therefore the index of refraction of that medium depends on the wavelength of light, a phenomenon known as dispersion. Because of dispersion, a beam of white light incident on a refracting prism is dispersed into its component colors. Similarly, the reflection and refraction of sunlight by raindrops produces a rainbow.

4. When two polarizers have their transmission axes at an angle θ, the intensity transmitted by the second polarizer is reduced by the factor $\cos^2 \theta$, a result known as Malus's law. If I_0 is the intensity of the light between the polarizers, the intensity transmitted by the second polarizer is

$$I = I_0 \cos^2 \theta$$

5. The four phenomena that produce polarized light from unpolarized light are (1) absorption, (2) scattering, (3) reflection, and (4) birefringence.

Beyond the (Visible) Rainbow*

Robert Greenler
University of Wisconsin, Milwaukee

Sometimes science is a very personal activity. Ever since I was a small boy, I have been excited by the beauty and grandeur of the rainbow. This was my reaction long before I had acquired the tools of the scientist with which to understand the origin of this marvelous arch of color.

A professional interest in which I have invested a considerable amount of energy over the past three decades is the understanding of the structure of molecules that become attached (adsorbed) to the surface of a solid material. This understanding is important for such a diverse assortment of phenomena as the functioning of catalysts, the electrical properties of small integrated circuits, the separation of ores, and the processes that take place within a fusion reactor.

It might seem that this interest would have nothing to do with rainbows, but not so. I have developed a technique for deducing the structure of molecules adsorbed on a metal surface using infrared radiation. So, some understanding of the nature of infrared radiation is one of the tools of my scientific trade.

These two different strands of my experience came together one day while I was sitting at my desk woolgathering rather than addressing the task at hand. The question occurred to me: I wonder if there is an infrared rainbow in the sky?

How does one explore such a question? Here is the process I went through. For there to be an infrared rainbow, a number of conditions must be met. *First*, the source of light must emit infrared radiation (the sun emits light over the entire electromagnetic spectrum from x-rays to radio waves; see Section 29-5). *Second*, the infrared radiation must pass through the Earth's atmosphere (water vapor and carbon dioxide in the atmosphere absorb some infrared wavelengths, as discussed in the essay on global warming, but others pass through unimpeded). The rainbow is caused by light rays that enter a droplet of water and are reflected internally before emerging from the drop (see Figure 30-26 and Section 30-4). For there to be an infrared rainbow, the *third* requirement is that the infrared rays would have to pass through a water droplet. This is a serious consideration. Just because a droplet of water appears transparent in visible light, we cannot assume that it is transparent to infrared "light"; indeed, liquid water does absorb over a broad range of infrared wavelengths. However, the measured spectral transmittance of water shows that water drops should be quite transparent from the visible region out to an infrared wavelength of about 1300 nm. *Finally*, after emerging from the raindrop, the infrared rays that have survived all these losses must again pass through air to the (unseeing) eye of the would-be observer.

The Search

This line of reasoning produced a tentative answer to the question that prompted the speculation: Yes, there should be an infrared rainbow in the sky and it should lie in a band just outside of the red of the visible rainbow.

I decided to try to photograph this invisible bow using a film that is sensitive to a portion of the infrared spectrum. Figure 1 shows the curve of the sensitivity of the film. The figure also shows a curve of the sensitivity of the human eye, as a way of defining the limits of the visible spectral region (extending from about 400 nm at the violet end of the spectrum to 700 nm at the red end). Note that the infrared film has a sensitivity extending out to about 930 nm.

*This essay is adapted from an article that first appeared in *Optic News*, published by the Optical Society of America, in November 1988.

Robert Greenler has been professor of physics at the University of Wisconsin–Milwaukee since 1962, where he has been instrumental in the development of the Laboratory for Surface Studies at Milwaukee. His research interests include the study of the structure of molecules adsorbed on solid surfaces, optical effects of the sky, and the understanding of the iridescent colors seen in many biological organisms.

The pursuit of rainbows (and other scientific interests) has taken him to the University of East Anglia in Norwich, England; the Fritz Haber Institute in West Berlin, Germany; the Institute Teknologi MARA in Shah Alam, Malaysia; and the U.S. Antarctic research station at the South Pole.

He served as President of the Optical Society of America in 1987 and, in 1988, he received the Millikan Lecture Award of the American Association of Physics Teachers for "notable and creative contributions to the teaching of physics."

The problem in using this film to record an infrared scene is that the film is not only sensitive to the infrared but throughout the visible region (it is *very* sensitive to blue light). If we were to look at the black-and-white image produced by such a film, we would have no way of knowing which parts of the image resulted from exposure to infrared and which parts from exposure to visible radiation. This problem was solved by using a filter that appears to be an opaque sheet of black plastic. The material is opaque to visible light and transmits only wavelengths longer than about 800 nm. As can be seen from Figure 1, this combination of film and filter will permit the recording of only those wavelengths in a band between 800 and 930 nm, well removed from the visible spectral region.

The Capture

Anyone who has tried to photograph rainbows knows that they usually occur when a camera is not at hand and fade just before one is located. I decided to first try an easier subject—that of a rainbow in a water spray that I could turn on at my convenience in my backyard. Figure 2 shows one of the first photographic results. A garden hose with many holes was wrapped back and forth across a board resting on top of the ladder. And in the spray of the hose—the infrared rainbow! You can also see the fainter, secondary rainbow outside the brighter, primary bow. This corresponds to the secondary bow seen rather commonly in visible light and it results from rays that enter a water droplet and experience two internal reflections before leaving the drop (see Figure 30-29).

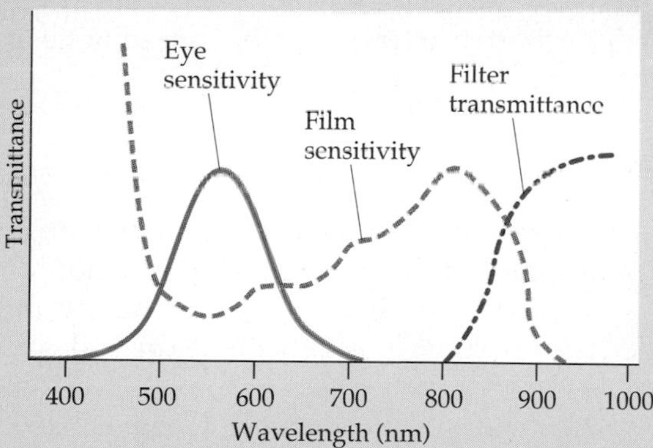

Figure 1 The infrared film (Eastman Kodak infrared film IR 135) has a sensitivity extending throughout the visible and into the near infrared region. The filter (Eastman Kodak 87C infrared transmitting filter) is opaque to visible light but transmits in the infrared for wavelengths longer than 800 nm. The combination of film and filter records an image with wavelengths between 800 and 930 nm, well outside the visible spectrum.

Figure 2 An infrared rainbow photographed in the water spray of a garden hose. The fainter secondary bow is shown outside (to the left of) the primary bow. The fringes seen inside (to the right of) the primary bow are caused by interference effects. A rare interference fringe outside the secondary bow is visible on the original photograph but may be difficult to pick out on the reproduction.

There is another interesting feature in this infrared photograph: immediately *inside* (to the right side of) the bright primary bow there is another bright band—or perhaps two bands. Such fringes, sometimes seen inside a visible bow, are called supernumerary bows and result from the interference of light waves (see Chapter 33).

Close inspection of the negative of Figure 2 reveals yet another feature, which is difficult to reproduce in a printed picture. There is a faint fringe just *outside* the secondary bow. A process similar to the one that produces supernumerary bows inside the primary should, in theory, produce a similar set of fringes outside the secondary. I have never seen any of those fringes associated with any rainbow or with any rainbow photograph, but they are visible in the original of this photograph. They are even visible when I project the slide onto a screen. For the first attempt, that was quite an exciting collection of effects.

Other Effects in the Infrared Photos

Some other features of these infrared photographs are worth considering. If the only radiation that produced these photographic images is invisible, infrared radiation, is it surprising that we can see the ladder, trees, and grass? One should not be too surprised. These objects absorb some wavelengths and reflect or scatter others. Objects that absorb the infrared appear dark in the photos and those that scatter it strongly appear bright. To make it clear just what these photos show, we need to understand the difference between reflected (or scattered) radiation and emitted radiation.

Continued

Normally when you look at objects in your landscape, you see them only by the light they scatter. However, if the temperature of an object is high enough, it emits light (see discussion of radiation in Section 16-3). If it is very hot—you might call it "white hot"—it emits a broad spectrum of wavelengths with the peak of the emission curve in the visible spectrum. If the object cools down a bit, the peak in the emission curve moves to longer wavelengths. The result is that there is more red light than blue light being emitted, and the appropriate description for its temperature is "red hot." At a lower temperature, you might see a dull red glow. At this point, the peak of emission curve is in the infrared with just a small amount of emission in the red end of the visible region. At a slightly lower temperature, the object appears dark; the emission peak has moved further into the infrared, so no visible radiation can be seen. If the object cools to where it is warm to the touch, its emission peak is far out in the infrared—perhaps at 10,000 nm—and it is emitting almost nothing in the visible or in the near infrared region to which photographic films are sensitive.

If, however, you could produce a picture with 10,000-nm radiation, objects slightly warmer than their surroundings would appear to be bright—they would be glowing with emitted infrared radiation. There *are* ways to produce such pictures; they are used to show sources of heat loss in homes or to record relatively warm spots on a human body (thermograms) that may indicate the site of some physical disorder. These pictures are usually described as infrared pictures, but they are quite different from the photographs taken with infrared-sensitive film. This film is sensitive only to the near infrared, but the "heat pictures" result from emitted radiation in the far infrared. So the infrared photographs shown here show only the infrared radiation from the sun that is scattered by the leaves or ladder or transformed by raindrop spheres into an invisible rainbow.

Another interesting feature of the photographs is the darkness of the clear sky background. We see light in the clear, clean sky, away from the sun, due to scattering from the molecules of gases in the air. These small scattering particles (much smaller than the wavelength of the light) scatter the shorter waves more effectively than the longer waves. Thus, more blue light is scattered than red light, giving the sky its blue color. This same effect, which makes the sky darker in red light than in blue, makes it even darker in the infrared light sampled by these photographs.

After taking the initial photographs, made with the leaky hose, I waited to capture Nature's own natural, infrared rainbow. It was four years before I saw a natural rainbow when I had at hand my camera, infrared film, the filter, *and* time enough to put them together for the photographs shown in Figures 3 and 4.

Figure 3 A natural infrared rainbow. This photograph shows the clouds to be brighter inside the bow than outside, a common feature of visible rainbows.

Figure 4 A photograph of Nature's own invisible rainbow, showing the primary bow, secondary bow, and a series of interference fringes (supernumerary bows) inside the primary.

Public Response

I received an interesting collection of letters in response to a brief published note describing the infrared rainbow. Some were from people who had a "scientific interest" in the matter; others from friends, with whom ties had been stretched by distance and neglect, saying: I'm glad to see you're still at it. Other letters represented unique interests, such as the psychologist studying color blindness, wondering whether such a disability might be a reason for a person named Greenler to be interested in invisible light—or the person from Belgium Television wanting photographs of the infrared rainbow for a show they were producing, insisting that they be *in color*. But most of the letters were from people who shared with me the fascination of "seeing" for the first time this bow, whose undetected presence in the sky predated that of a human consciousness on this planet.

Suggestions for Further Reading

Boyle, W. S.: "Light-Wave Communications," *Scientific American*, August 1977, p. 40.

The physics and technology of a telephone system that transmits signals via light pulses carried along optical fibers are discussed in this article.

Greenler, Robert: *Rainbows, Halos, and Glories*, Cambridge University Press, Cambridge, 1980.

In this book, the author of this chapter's essay ["Beyond the (Visible) Rainbow"] discusses rainbows as well as reflection and refraction effects due to atmospheric ice crystals and refraction through a nonuniform atmosphere.

Katzir, Abraham: "Optical Fibers in Medicine," *Scientific American*, May 1989, p. 120.

Optical fibers can carry light into and out of the body for diagnostic purposes (imaging, blood flow measurement) and treatment (laser surgery). This article examines instruments presently used and those likely to be developed in the future.

Können, G. P.: *Polarized Light in Nature*, Cambridge University Press, Cambridge, 1985.

This unusual book is a kind of field guide for observing the polarization of light from diverse objects, including the sky, clouds, rainbows, plants, ice sheets, beetles, and minerals. It includes many color photographs and diagrams.

"Light," *Scientific American* special issue, September 1968.

How light interacts with both living and nonliving matter, how images are formed, vision, and laser light are some of the topics covered in this issue.

Sobel, Michael I: *Light*, University of Chicago Press, Chicago, 1987.

This book presents light as a central concept of the natural sciences, one that ties together nonmathematical discussions of x rays and radio waves, cosmic background radiation, fiber-optic communication, light-sensitive pigments in the eye, and much more.

Walker, Jearl: "The Amateur Scientist: Studying Polarized Light with Quarter-Wave and Half-Wave Plates of One's Own Making," *Scientific American*, December 1977, p. 172.

Walker, Jearl: "The Amateur Scientist: More about Polarizers and How to Use Them, Particularly for Studying Polarized Sky Light," *Scientific American*, January 1978, p. 132.

These two articles are instructive even if one doesn't choose to repeat the experiments.

Wehner, Rüdiger: "Polarized-Light Navigation by Insects," *Scientific American*, July 1976, p. 106.

This article describes how ants and bees are able to use the natural polarization of sky light as an aid to navigation.

Review

A. Objectives: After studying this chapter, you should:

1. Be able to state Huygens' principle and use it to derive the law of reflection and Snell's law of refraction.

2. Be able to state Fermat's principle and use it to derive the law of reflection and Snell's law of refraction.

3. Be able to derive an expression relating the critical angle for total internal reflection to the index of refraction of a substance.

4. Be able to describe how a rainbow is formed, and explain qualitatively why the primary bow is seen at an angular radius of 42°.

5. Be able to list the four means of producing polarized light from unpolarized light.

6. Be able to state Malus's law and use it in problems involving the transmission of light through a polarizer.

7. Be able to derive Brewster's law using Snell's law and the fact that at the polarizing angle the reflected and refracted rays are perpendicular.

B. Define, explain, or otherwise identify:

Particle theory of light
Wave theory of light
Photons
Huygens' principle
Huygens' construction
Reflection
Angle of incidence
Plane of incidence
Law of reflection
Index of refraction
Image
Specular reflection
Diffuse reflection
Refraction
Snell's law of refraction
Total internal reflection
Dispersion
Angle of minimum deviation
Fermat's principle
Polarization
Linearly polarized
Circularly polarized
Transmission axis
Polarizer
Analyzer
Malus's law
Scattering
Brewster's law
Birefringence
Double refraction
Isotropic materials
Anisotropic materials
Optic axis
Quarter-wave plate
Half-wave plate

C. True or false: If the statement is true, explain why it is true. If it is false, give a counterexample.

1. Light and radio waves travel with the same speed through a vacuum.

2. Most of the light incident normally on an air–glass surface is reflected.

3. The angle of refraction of light is always less than the angle of incidence.

4. The index of refraction of water is the same for all wavelengths in the visible spectrum.

5. Longitudinal waves cannot be polarized.

Problems

Level I

30-1 The Speed of Light

1. The spiral galaxy in the Andromeda constellation is about 2×10^{19} km away from us. How many light-years is this?

2. On a spacecraft sent to Mars to take pictures, the camera is triggered by radio waves, which like all electromagnetic waves travel with the speed of light. What is the time delay between sending the signal from the earth and receiving it at Mars? (Take the distance to Mars to be 9.7×10^{10} m.)

3. The distance from a point on the surface of the earth to one on the surface of the moon is measured by aiming a laser light beam at a reflector on the surface of the moon and measuring the time required for the light to make a round trip. The uncertainty in the measured distance Δx is related to the uncertainty in the time Δt by $\Delta x = c\,\Delta t$. If the time intervals can be measured to ± 1.0 ns, find the uncertainty in the distance in meters.

30-2 The Propagation of Light: Huygens' Principle

There are no problems for this section.

30-3 Reflection

4. Calculate the fraction of light energy reflected from an air–water surface at normal incidence ($n = 1.33$ for water).

5. Light is incident normally on a slab of glass with an index of refraction $n = 1.5$. Reflection occurs at both surfaces of the slab. About what percentage of the incident light energy is transmitted by the slab?

30-4 Refraction

6. The index of refraction of water is 1.33. Find the angle of refraction of a beam of light in air that hits a water surface at an angle with the normal of (*a*) 20°, (*b*) 30°, (*c*) 45°, and (*d*) 60°. Show these rays on a diagram.

7. Repeat Problem 6 for a beam of light initially in water that is incident on a water–air surface.

8. What is the critical angle for total internal reflection for light traveling initially in water ($n = 1.33$) that is incident on a water–air surface?

9. Find the speed of light in water ($n = 1.33$) and in glass ($n = 1.5$).

10. A beam of monochromatic red light with a wavelength of 700 nm in air travels in water. (*a*) What is the wavelength in water? (*b*) Does a swimmer underwater observe the same color or a different color for this light?

11. A slab of glass with an index of refraction of 1.5 is submerged in water with an index of refraction of 1.33. Light in the water is incident on the glass. Find the angle of refraction if the angle of incidence is (*a*) 60°, (*b*) 45°, and (*c*) 30°.

12. Repeat Problem 11 for a beam of light initially in the glass that is incident on the glass–water surface at the same angles.

13. A glass surface ($n = 1.50$) has a layer of water ($n = 1.33$) on it. Light in the glass is incident on the glass–water surface. Find the critical angle for total internal reflection.

14. The index of refraction for silicate flint glass is 1.66 for light with a wavelength of 400 nm and 1.61 for light with a

wavelength of 700 nm. Find the angles of refraction for light of these wavelengths that is incident on this glass at an angle of 45°.

30-5 Fermat's Principle

15. A physics student playing pocket billiards wants to strike her cue ball such that it hits a cushion and then hits the eight ball squarely. She chooses several points on the cushion and for each point measures the distance from it to the cue ball and to the eight ball. She aims at the point for which the sum of these distances is least. (*a*) Will her cue ball hit the eight ball? (*b*) How is her method related to Fermat's principle?

16. A swimmer at *S* in Figure 30-42 develops a leg cramp while swimming near the shore of a calm lake and calls for help. A lifeguard at *L* hears the call. The lifeguard can run 9 m/s and swim 3 m/s. He knows physics and chooses a path that will take the least time to reach the swimmer. Which of the paths shown in Figure 30-42 does he take?

Figure 30-42 Problem 16.

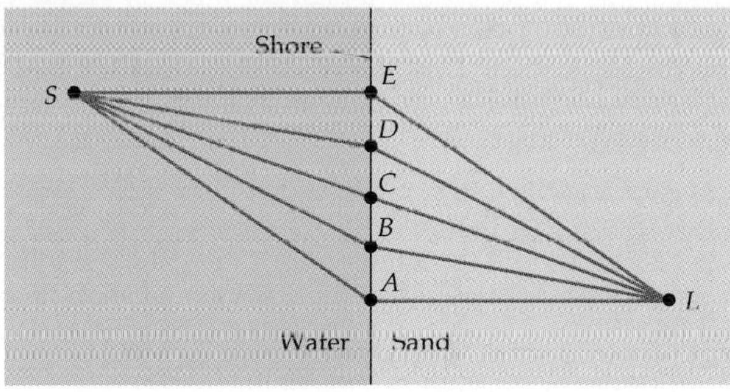

30-6 Polarization

17. Two polarizing sheets have their transmission axes crossed so that no light gets through. A third sheet is inserted between the first two such that its transmission axis makes an angle θ with that of the first sheet. Unpolarized light of intensity I_0 is incident on the first sheet. Find the intensity of the light transmitted through all three sheets if (*a*) $\theta = 45°$ and (*b*) $\theta = 30°$.

18. The polarizing angle for a certain substance is 60°. (*a*) What is the angle of refraction of light incident at this angle? (*b*) What is the index of refraction of this substance?

19. The critical angle for total internal reflection for a substance is 45°. What is the polarizing angle for this substance?

20. What is the polarizing angle for (*a*) water with $n = 1.33$ and (*b*) glass with $n = 1.5$?

Level II

21. A point source of light is 5 cm above a plane reflecting surface (such as a mirror). Draw a ray from the source that strikes the surface at an angle of incidence of 45° and two more rays that strike the surface at angles slightly less than 45°, and draw the reflected ray for each. The reflected rays appear to diverge from a point called the image of the light source. Draw dotted lines extending the reflected rays back until they meet at a point behind the surface to locate the image point.

22. A silver coin sits on the bottom of a swimming pool that is 4 m deep. A beam of light reflected from the coin emerges from the pool making an angle of 20° with respect to the water's surface and enters the eye of an observer. Draw a ray from the coin to the eye of the observer. Extend this ray, which goes from the water–air surface to the eye, straight back until it intersects with the vertical line drawn through the coin. What is the apparent depth of the swimming pool to this observer?

23. Two affluent students decide to improve on Galileo's experiment to measure the speed of light. One student goes to London and calls the other in New York on the telephone. The telephone signals are transmitted by reflecting electromagnetic waves from a satellite that is 37.9 Mm above the earth's surface. If the distance between London and New York is neglected, the distance traveled is twice this distance. One student claps his hands, and when the other student hears the sound over the phone, she claps her hands. The first student measures the time between his clap and his hearing the second one. Calculate this time lapse, neglecting the students' response times. Do you think this experiment would be successful? What improvements for measuring this time interval would you suggest? (Time delays in the electronic circuits that are greater than those due to the light traveling to the satellite and back make this experiment not feasible.)

24. In Galileo's attempt to determine the speed of light, he and his assistant were located on hilltops about 3 km apart. Galileo flashed a light and received a return flash from his assistant. (*a*) If his assistant had an instant reaction, what time difference would Galileo need to be able to measure for this method to be successful? (*b*) How does this time compare with human reaction time, which is about 0.2 s?

25. A point source of light is located 5 m below the surface of a large pool of water. Find the area of the largest circle on the pool's surface through which light coming directly from the source can emerge.

26. A swimmer at the bottom of a pool 3 m deep looks up and sees a circle of light. If the index of refraction of the water in the pool is 1.33, find the radius of the circle.

27. Show that when a mirror is rotated through an angle θ, the reflected beam of light is rotated through 2θ.

28. Light is incident normally on the largest face of an isosceles–right-triangle prism. What is the speed of light in this prism if the prism is just barely able to produce total internal reflection?

29. Show that the transmitted intensity through a glass slab with an index of refraction of n for normally incident light is approximately given by

$$I_T = I_0 \left[\frac{4n}{(n+1)^2} \right]^2$$

30. A ray of light begins at the point $x = -2$ m, $y = 2$ m, strikes a mirror in the yz plane at some point x, and reflects through the point $x = 2$ m, $y = 6$ m. (*a*) Find the value of x that makes the total distance traveled by the ray a minimum. (*b*) What is the angle of incidence on the reflecting plane? What is the angle of reflection?

31. Light passes symmetrically through a prism having an apex angle of α as shown in Figure 30-43. (*a*) Show that the angle of deviation δ is given by

$$\sin \frac{\alpha + \delta}{2} = n \sin \frac{\alpha}{2}$$

(*b*) If the refractive index for red light is 1.48 and for violet light is 1.52, what is the angular separation of visible light for a prism with an apex angle of 60°?

32. A beam of light strikes the plane surface of silicate flint glass at an angle of incidence of 45°. The index of refraction of the glass varies with wavelength as shown in the graph in Figure 30-22. How much smaller is the angle of refraction for violet light of wavelength 400 nm than that for red light of wavelength 700 nm?

33. Repeat Problem 32 for quartz.

34. Use Figure 30-22 to calculate the critical angles for total internal reflection for light initially in silicate flint glass that is incident on a glass–air surface if the light is (*a*) violet light of wavelength 400 nm, and (*b*) red light of wavelength 700 nm.

35. (*a*) For a light ray inside a transparent medium having a planar interface with a vacuum, show that the polarizing angle and the critical angle for internal reflection satisfy $\tan \theta_p = \sin \theta_c$. (*b*) Which angle is the larger?

36. Light is incident from air on a transparent substance at an angle of 58.0° with the normal. The reflected and refracted rays are observed to be mutually perpendicular. (*a*) What is the index of refraction of the transparent substance? (*b*) What is the critical angle for total internal reflection in this substance?

37. Two polarizing sheets have their transmission axes crossed and a third sheet is inserted so that its transmission axis makes an angle θ with that of the first sheet as in Problem 17. Show that the intensity transmitted through all three sheets is maximum when $\theta = 45°$.

38. If the middle polarizing sheet in Problem 37 is rotating at an angular velocity ω about an axis parallel to the light beam, find the intensity transmitted through all three sheets as a function of time. Assume that $\theta = 0$ at time $t = 0$.

39. Given a stack of $N + 1$ ideal polarizing sheets with each sheet rotated by an angle of $\pi/2N$ rad with respect to the preceding sheet. A plane linearly polarized light wave of intensity I_0 is incident normally on the stack. The incident light is polarized along the transmission axis of the first sheet and therefore normal to the transmission axis of the last sheet in the stack. (*a*) What is the transmitted intensity through the stack? (*b*) For 3 sheets ($N = 2$), what is the transmitted intensity? (*c*) For 101 sheets, what is the transmitted intensity? (*d*) What is the direction of polarization of the transmitted beam in each case?

40. A point source of light is located at the bottom of a steel tank, and an opaque circular card of radius 6.0 cm is placed over it. A transparent fluid is gently added to the tank such that the card floats on the surface with its center directly above the light source. No light is seen by an observer above the surface until the fluid is 5 cm deep. What is the index of refraction of the fluid?

41. A light ray in dense flint glass with index of refraction 1.655 is incident on the glass interface. An unknown liquid condenses on the surface of the glass. Total internal reflection on the glass–liquid surface occurs for an angle of incidence on the glass–liquid interface of 53.7°. (*a*) What is the refractive index of the unknown liquid? (*b*) If the liquid is removed, what is the angle of incidence for total internal reflection? (*c*) For the angle of incidence found in part (*b*), what is the angle of refraction of the ray into the liquid film? Does a ray emerge from the liquid film into the air above? Assume the glass and liquid have perfect planar surfaces.

42. Given that the index of refraction for red light in water is 1.3318 and that the index of refraction of blue light in water is 1.3435, find the angular separation of these colors in the primary rainbow. (Use the equation given in Problem 51.)

43. In Figure 30-44, light is initially in a medium (such as air) of index of refraction n_1. It is incident at angle θ_1 on the surface of a liquid (such as water) of index of refraction n_2. The light passes through the layer of water and enters glass of index of refraction n_3. If θ_3 is the angle of refraction in the glass, show that $n_1 \sin \theta_1 = n_3 \sin \theta_3$. That is, show that the second medium can be neglected when finding the angle of refraction in the third medium.

44. A ray of light falls on a rectangular glass block ($n =$ 1.5) that is almost completely submerged in water ($n =$ 1.33) as shown in Figure 30-45. (*a*) Find the angle θ for which total internal reflection just occurs at point P. (*b*) Would total internal reflection occur at point P for the value of θ found in part (*a*) if the water were removed? Explain.

Figure 30-43 Problem 31.

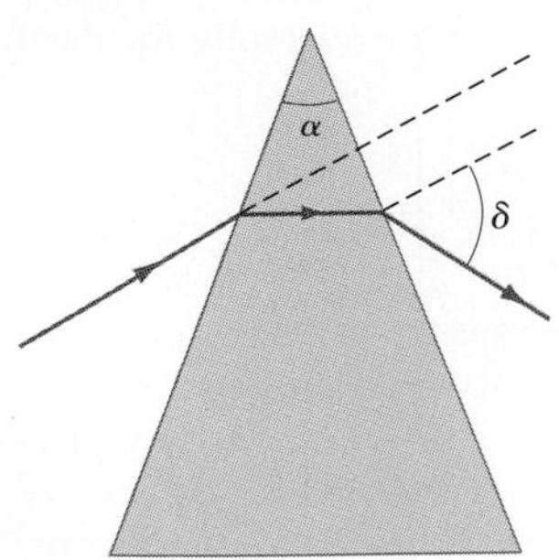

Figure 30-44 Problem 43.

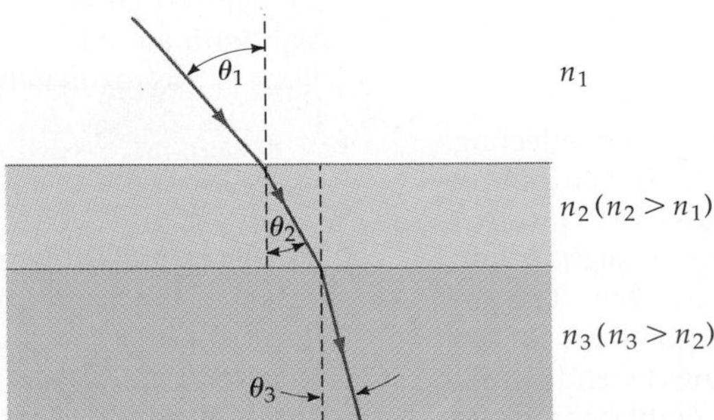

Figure 30-45 Problem 44.

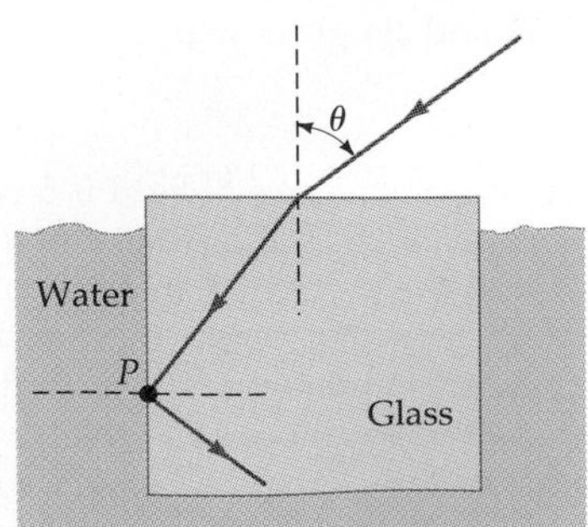

45. Light of wavelength λ in air is incident on a slab of calcite such that the ordinary and extraordinary rays travel in the same direction as shown in Figure 30-46. Show that the phase difference between these rays after they traverse a thickness t of calcite is

$$\delta = \frac{2\pi}{\lambda}(n_o - n_e)t$$

46. (*a*) Use the result for Problem 29 to find the ratio of the transmitted intensity to the incident intensity through N parallel slabs of glass for light of normal incidence. (*b*) Find this ratio for three slabs of glass with $n = 1.5$. (*c*) How many slabs of glass with $n = 1.5$ will reduce the intensity to 10 percent of the incident intensity?

47. Light is incident on a slab of transparent material at an angle θ_1 as shown in Figure 30-47. The slab has a thickness t and an index of refraction n. Show that

$$n = \frac{\sin \theta_1}{\sin [\arctan(d/t)]}$$

where d is the distance shown in the figure and $\arctan(d/t)$ is the angle whose tangent is d/t.

48. Suppose rain falls vertically from a stationary cloud 10,000 m above a confused marathoner running in a circle with constant speed of 4 m/s. The rain has a terminal speed of 9 m/s. (*a*) What is the angle that the rain appears to make with the vertical to the marathoner? (*b*) What is the apparent motion of the cloud as observed by the marathoner? (*c*) A star on the axis of the earth's orbit appears to have a circular orbit of angular diameter of 41.2 seconds of arc. How is this angle related to the earth's speed in its orbit and the velocity of photons falling from this distant star? (*d*) What is the speed of light using this method?

49. This problem is a refraction analogy. A band is marching down a football field with a constant speed v_1. About midfield, the band comes to a section of muddy ground that has a sharp boundary making an angle of 30° with the 50-yd line as shown in Figure 30-48. In the mud, the marchers move with speed $v_2 = \frac{1}{2}v_1$. Diagram how each line of marchers is bent as it encounters the muddy section of the field so that the band is eventually marching in a different direction. Indicate the original direction by a ray, the final direction by a second ray, and find the angles between the rays and the line perpendicular to the boundary. Is their direction of motion bent toward the perpendicular to the boundary or away from it?

Level III

50. Light is incident normally upon one face of a prism of glass with an index of refraction n (Figure 30-49). The light is totally reflected at the right side. (*a*) What is the minimum value n can have? (*b*) When the prism is immersed in a liquid whose index of refraction is 1.15, there is still total reflection, but when it is immersed in water whose index of refraction is 1.33, there is no longer total reflection. Use this information to establish limits for possible values of n.

51. Equation 30-13 gives the relation between the angle of deviation ϕ_d of a light ray incident on a spherical drop of water in terms of the incident angle θ_1 and the index of refraction of water. (*a*) Assume that $n_{air} = 1$, and differentiate ϕ_d with respect to θ_1. [Hint: If $y = \arcsin x$, $dy/dx = (1 - x^2)^{-1/2}$.] (*b*) Set $d\phi_d/d\theta_1 = 0$ and show that the angle of incidence θ_{1m} for minimum deviation is given by

$$\cos \theta_{1m} = \sqrt{\frac{n^2 - 1}{3}}$$

and find θ_{1m} for water, where the index of refraction for water is 1.33.

52. Investigate how a thin film of water on a glass surface affects the critical angle for total reflection. Take $n = 1.5$ for glass and $n = 1.33$ for water. (*a*) What is the critical angle for total internal reflection at the glass–water surface? (*b*) Is there any range of incident angles that are greater than θ_c for glass-to-air refraction, and for which light rays will leave the glass and the water and pass into the air?

53. A laser beam is incident on a plate of glass of thickness 3 cm. The glass has an index of refraction of 1.5 and the angle of incidence is 40°. The top and bottom surfaces of the glass are parallel and both produce reflected beams of nearly the same intensity. What is the perpendicular distance d between the two adjacent reflected beams?

54. (*a*) Show that a light ray transmitted through a glass slab emerges parallel to the incident ray but displaced from it. (*b*) For an incident angle of 60°, glass of index of refraction $n = 1.5$, and a slab of thickness 10 cm, find the displacement measured perpendicularly from the incident ray.

55. An isotropic point source of light is placed below the surface of a large pool of liquid having an index of refraction n. What fraction of light energy leaves the surface directly?

Figure 30-46 Problem 45. **Figure 30-47** Problem 47.

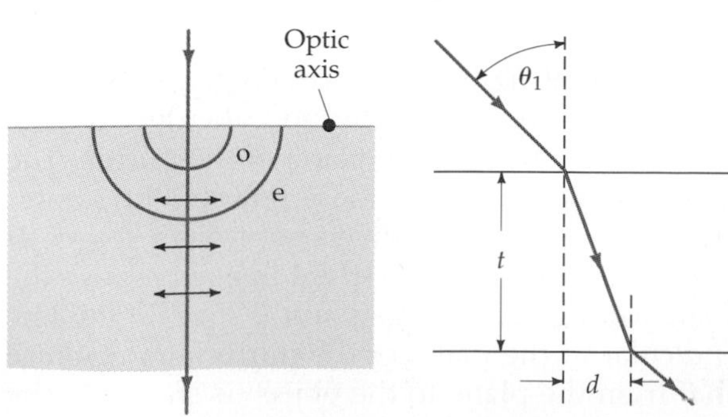

Figure 30-48 Problem 49. **Figure 30-49** Problem 50.

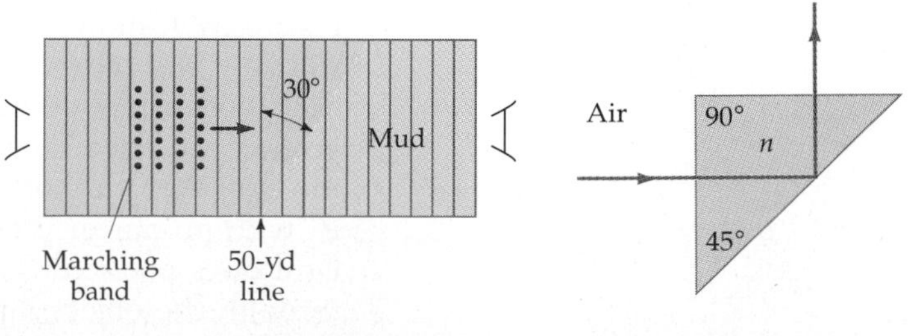

Chapter 31

Geometrical Optics

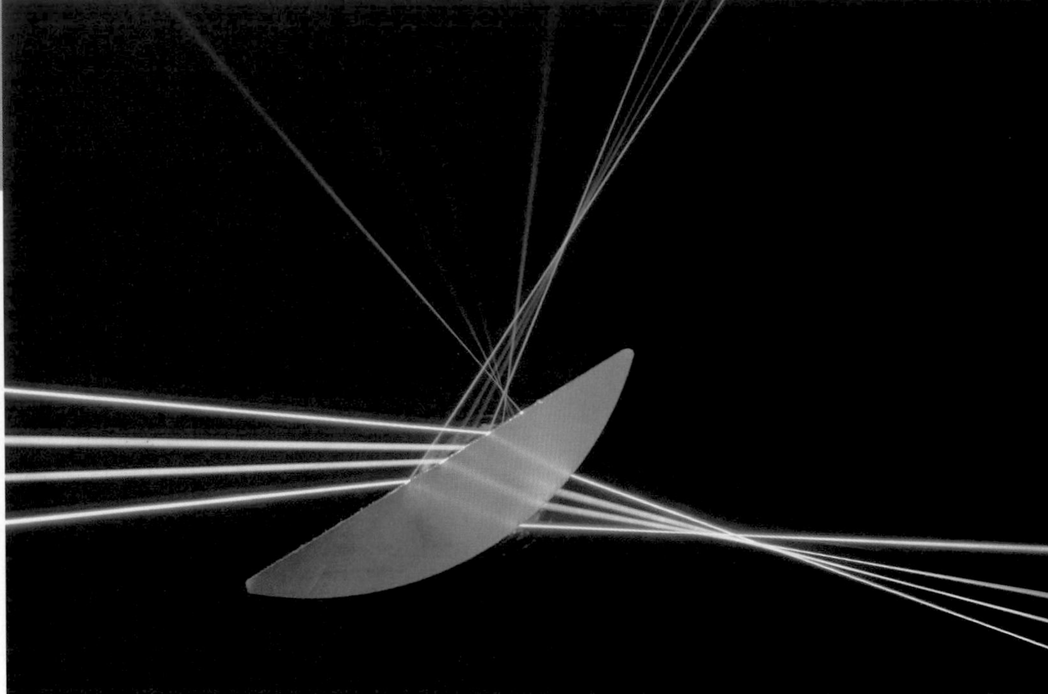

The focusing of rays by reflection and by refraction is illustrated by these laser beams incident on a glass lens.

The wavelength of light is very small compared with most obstacles and openings. Because of this, diffraction—the bending of waves around corners—is often negligible, and the ray approximation, in which waves are considered to propagate in straight lines, is valid. **Geometrical optics** is the study of those phenomena in which the ray approximation is valid. In this chapter, we will apply the laws of reflection and refraction to the formation of images by mirrors and lenses.

31-1 Plane Mirrors

Figure 31-1 shows a narrow bundle of light rays emanating from a point source P and reflected from a plane mirror. After reflection, the rays diverge exactly as if they came from a point P' behind the plane of the mirror. The point P' is called the **image** of the object P. When these reflected rays enter the eye, they cannot be distinguished from rays diverging from a source at P' with no mirror present. This image is called a **virtual image** because the light does not actually emanate from it. The image point P' lies on the line through the object P perpendicular to the plane of the mirror at a distance behind the plane equal to that from the plane to the object as shown in the

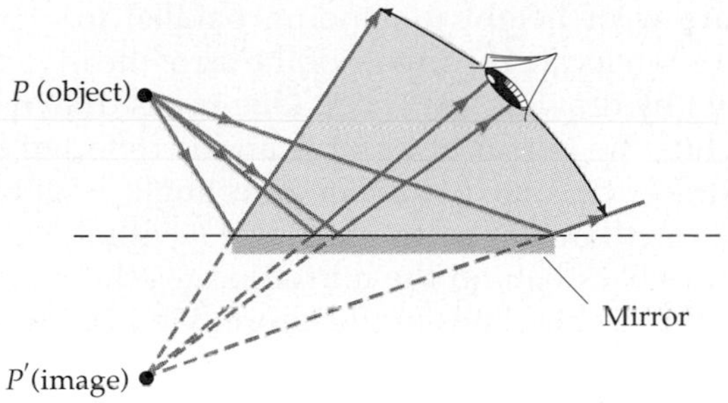

Figure 31-1 Image formed by a plane mirror. The rays from point P that strike the mirror and enter the eye appear to come from the image point P' behind the mirror. The image can be seen by the eye anywhere in the shaded region.

figure. (Figure 31-1 was produced by geometric construction using the law of reflection.) The image can be seen by an eye anywhere in the shaded region indicated in which a straight line from the image to the eye passes through the mirror. The object need not be directly in front of the mirror. The image can be seen as long as the object is not behind the plane of the mirror.

The image you see if you hold the palm of your right hand up to a plane mirror is shown in Figure 31-2. The image is the same size as the object, but the image is not the same as what would be seen by someone facing you or what you would see if you looked at the palm of your right hand. The image of a right hand in a mirror is a left hand. This right-to-left inversion is a result of **depth inversion;** that is, the hand is transformed from a right hand to a left hand because the front and back of the hand are reversed by the mirror. Depth inversion is also illustrated in Figure 31-3, which shows a person lying down with his feet touching a plane mirror. The image of a simple rectangular coordinate system that has its x and y axes parallel to the plane of the mirror is shown in Figure 31-4. The images of the arrows along the x and y axes are parallel to the object arrows. But the image of the arrow along the z axis is directed opposite to the object arrow along the z axis. The mirror transforms a right-handed coordinate system for which $\mathbf{i} \times \mathbf{j} = \mathbf{k}$, where $\mathbf{i}$, $\mathbf{j}$, and $\mathbf{k}$ are the unit vectors along the x, y, and z axes, respectively, into a left-handed coordinate system for which $\mathbf{i} \times \mathbf{j} = -\mathbf{k}$.

Figure 31-2 The image of a right hand in a plane mirror is a left hand. This right-to-left reversal is a result of depth inversion.

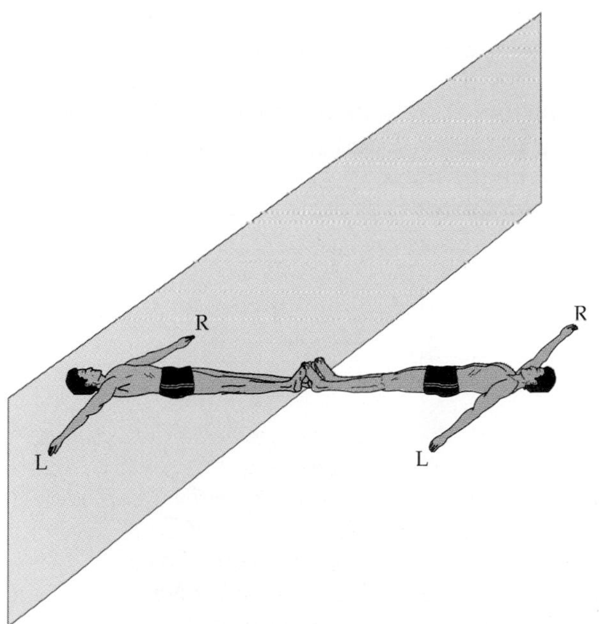

Figure 31-3 A person lying down with his feet against the mirror. The image is depth inverted.

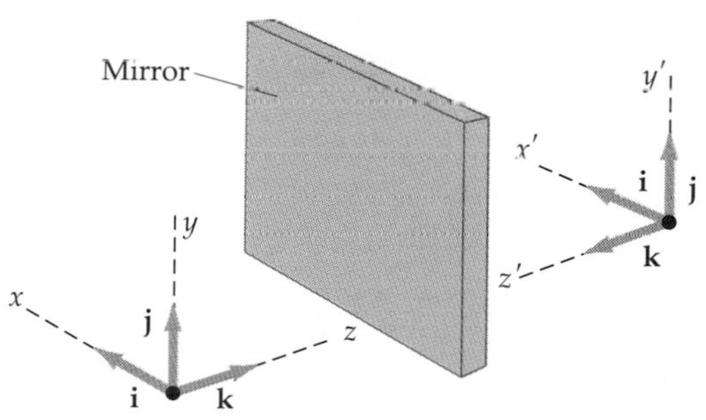

Figure 31-4 Image of a rectangular coordinate system in a plane mirror. The arrows along the x and y axes, which are parallel to the plane of the mirror, are in the same directions in the image as in the object. The direction of the arrow along the z axis is reversed in the image. The image of the original right-handed coordinate system, for which $\mathbf{i} \times \mathbf{j} = \mathbf{k}$, is a left-handed coordinate system, for which $\mathbf{i} \times \mathbf{j} = -\mathbf{k}$.

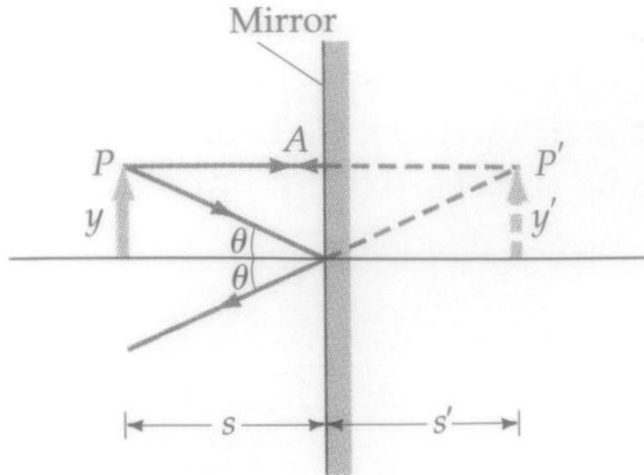

Figure 31-5 Ray diagram for locating the image of an arrow in a plane mirror.

Figure 31-5 shows an arrow of height y standing parallel to a plane mirror a distance s from it. We can locate the image of the arrowhead (and of any other point on the arrow) by drawing two rays. One ray is drawn perpendicular to the mirror. It hits the mirror at point A and is reflected back onto itself. The other ray strikes the mirror, making an angle θ with the normal to the mirror. It is reflected, making an equal angle θ with the x axis. The extension of these two rays back behind the mirror locates the image of the arrowhead, as shown by the dashed lines in the figure. We can see from this figure that the image is the same distance behind the mirror as the object is in front of the mirror and that the image is erect and the same size as the object.

The formation of multiple images by two plane mirrors making an angle with each other is illustrated in Figure 31-6. We frequently see this phenomenon in clothing stores that provide adjacent mirrors. Light reflected from mirror 1 strikes mirror 2 just as if it came from the image point P_1'. The image P_1' is called the object for mirror 2. Its image is at point $P_{1,2}''$. This image will be formed whenever the image point P_1' is in front of the plane of mirror 2.

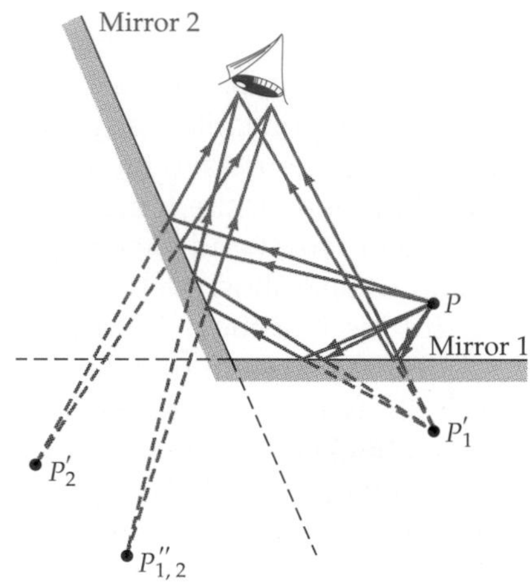

Figure 31-6 Images formed by two plane mirrors. P_1' is the image of the object P in mirror 1, and P_2' is the image of the object in mirror 2. Point $P_{1,2}''$ is the image of P_1' in mirror 2 seen when light rays from the object reflect first from mirror 1 and then from mirror 2. The image P_2' does not have an image in mirror 1 because it is behind that mirror.

The image at point P_2' is due to rays from the object that reflect directly from mirror 2. Since P_2' is behind the plane of mirror 1, it cannot serve as an object point for a further image in mirror 1. The number of multiple images formed by two mirrors depends on the angle between the mirrors and the position of the object.

Figure 31-7 shows two mirrors at right angles to each other. Rays from the object to the eye that strike mirror 1 and then mirror 2 are shown in Figure 31-7*a*. In this case, the image point $P_{1,2}''$ is the same as that for rays that strike mirror 2 first and then mirror 1, as can be seen from Figure 31-7*b*.

Figure 31-7 Two plane mirrors at right angles to each other. (*a*) Rays that strike mirror 1 first and then mirror 2. The image of P_1' in mirror 2 is $P_{1,2}''$. (*b*) Rays that strike mirror 2 first and then mirror 1. The image of P_2' in mirror 1 is $P_{2,1}''$, which coincides with $P_{1,2}''$ for perpendicular mirrors.

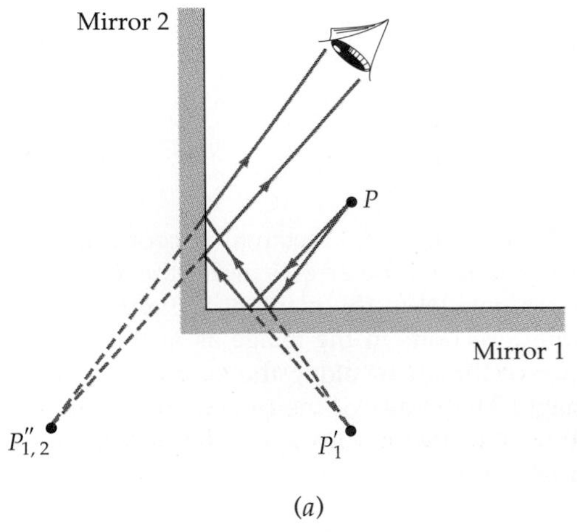

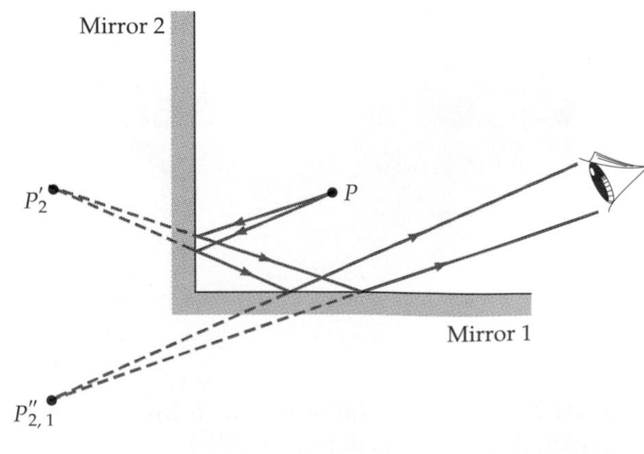

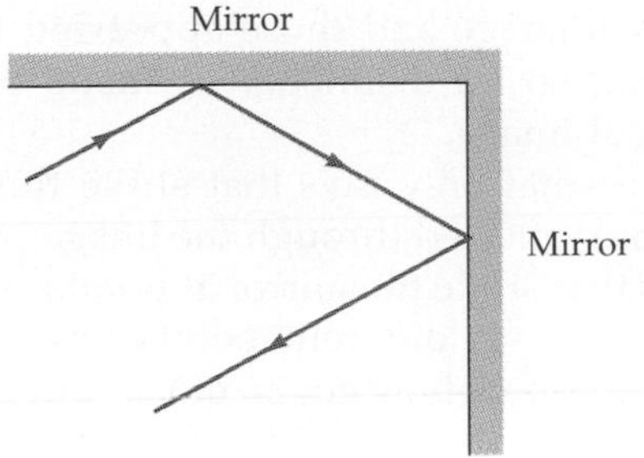

Figure 31-8 A ray striking one of two perpendicular plane mirrors is reflected from the second mirror in the direction opposite the original direction for any angle of incidence.

If you stand in front of two vertical mirrors that are perpendicular to each other, such as in the corner of a room, the image you see *is* the same as that seen by others who are facing you because depth inversion occurs twice, once in each mirror. Figure 31-8 illustrates the fact that a horizontal ray reflected from two perpendicular vertical mirrors is exactly reversed in direction no matter what angle the ray makes with the mirrors. If three mirrors are placed perpendicular to each other like the sides of an inside corner of a box, any ray incident on any of the mirrors from any direction is exactly reversed. A set of mirrors of this type (actually a set of reflecting prisms) was placed on the moon facing the earth. A laser beam from earth directed at the mirrors is reflected back to the same place on the earth. Such a beam has been used to measure the distance to the mirrors to within a few centimeters by measuring the time it takes for the light to reach the mirrors and return.

This reflecting pool serves as a plane mirror producing a virtual image of the Taj Mahal.

Questions

1. Can a virtual image be photographed?
2. Suppose each axis of a coordinate system like the one in Figure 31-4 is painted a different color. One photograph is take of the coordinate system and another is taken of its image in a plane mirror. Is it possible to tell that one of the photographs is of a mirror image rather than both being photographs of the real coordinate system from different angles?

31-2 Spherical Mirrors

Figure 31-9 shows a bundle of rays from a point source P on the axis of a concave spherical mirror reflecting from the mirror and converging at point P'. The rays then diverge from this point just as if there were an object at that point. This image is called a **real image** because light actually does emanate from the image point. The image can be seen by an eye at the left of the image looking into the mirror. It could also be observed on a ground glass viewing screen or photographic film placed at the image point. A virtual image, such as that formed by a plane mirror as discussed in the previous section, cannot be observed on a screen at the image point because there is no light there. Despite this distinction between real and virtual images,

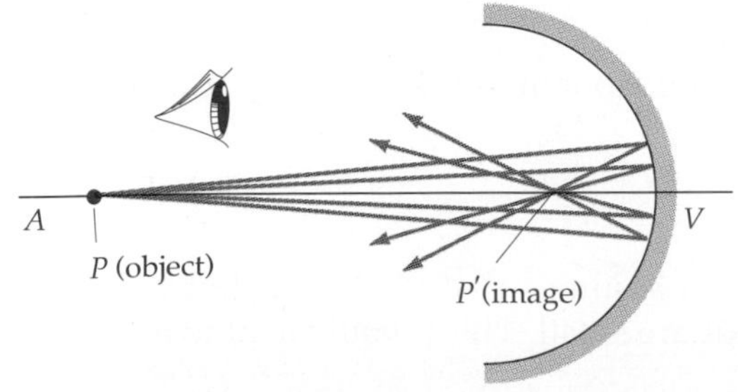

Figure 31-9 Rays from a point object P on the axis AV of a concave spherical mirror form an image at P'. The image is sharp if the rays strike the mirror near the axis.

the light rays diverging from a real image and those appearing to diverge from a virtual image are identical, so no distinction is made by the eye between viewing a real or a virtual image.

From Figure 31-10, we can see that only rays that strike the spherical mirror at points near the axis AV are reflected through the image point. Such rays are called **paraxial rays.** Rays that strike the mirror at points far from the axis, called *nonparaxial rays*, converge to different points near the image point. Such rays cause the image to appear blurred, an effect called **spherical aberration.** The image can be sharpened by blocking off all but the central part of the mirror so that nonparaxial rays do not strike it. Although the image is then sharper, its brightness is reduced because less light is reflected to the image point.

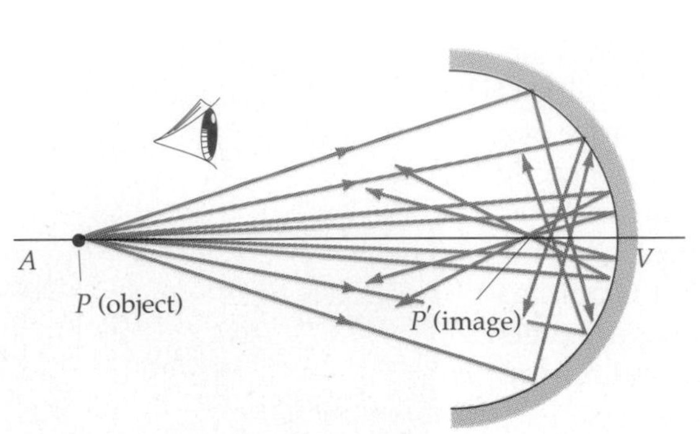

Figure 31-10 Spherical aberration. Nonparaxial rays that strike the mirror at points far from the axis AV are not reflected through the image point P'. These rays blur the image.

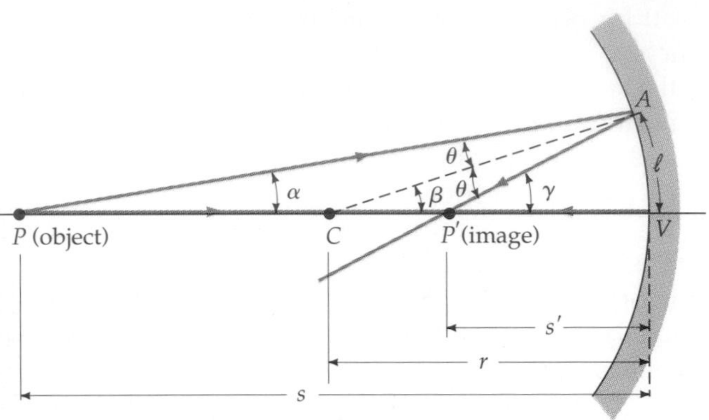

Figure 31-11 Geometry for calculating the image distance s' from the object distance s and the radius of curvature r.

The image distance from the vertex of the mirror V to P' can be related to the object distance from the vertex V to point P and the radius of curvature of the mirror by elementary geometry. Figure 31-11 shows a ray from an object point P reflecting off the mirror and passing through the image point P'. Point C is the center of curvature of the mirror. The incident and reflected rays make equal angles with the radial line CA, which is perpendicular to the surface of the mirror. Let s be the object distance, s' be the image distance, and r be the radius of curvature of the mirror. The angle β is an exterior angle to the triangle PAC and is therefore equal to $\alpha + \theta$:

$$\beta = \alpha + \theta \tag{31-1}$$

Similarly, from the triangle PAP',

$$\gamma = \alpha + 2\theta \tag{31-2}$$

Eliminating θ from these equations gives

$$2\theta = \gamma - \alpha = 2\beta - 2\alpha$$

or

$$2\beta = \alpha + \gamma \tag{31-3}$$

Using the small angle approximations $\alpha \approx \ell/s$, $\beta \approx \ell/r$, and $\gamma \approx \ell/s'$, we have

$$\frac{1}{s} + \frac{1}{s'} = \frac{2}{r} \tag{31-4}$$

The derivation of this equation assumes that angles made by the incident and reflected rays with the axis are small. This is equivalent to assuming that the rays are paraxial.

When the object distance is large compared with the radius of curvature of the mirror, the term $1/s$ in Equation 31-4 is much smaller than $2/r$ and can be neglected. For $s = \infty$, the image distance is $s' = \frac{1}{2}r$. This distance is called the **focal length** f of the mirror:

$$f = \tfrac{1}{2}r \tag{31-5}$$

Focal length for mirror

In terms of the focal length f, the mirror equation is

$$\frac{1}{s} + \frac{1}{s'} = \frac{1}{f} \tag{31-6}$$

Mirror equation

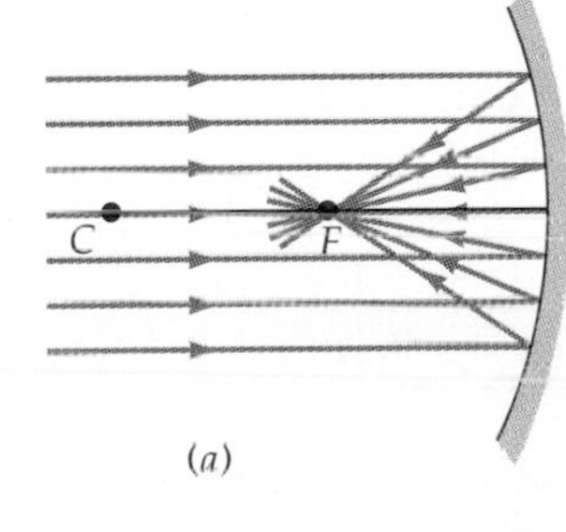

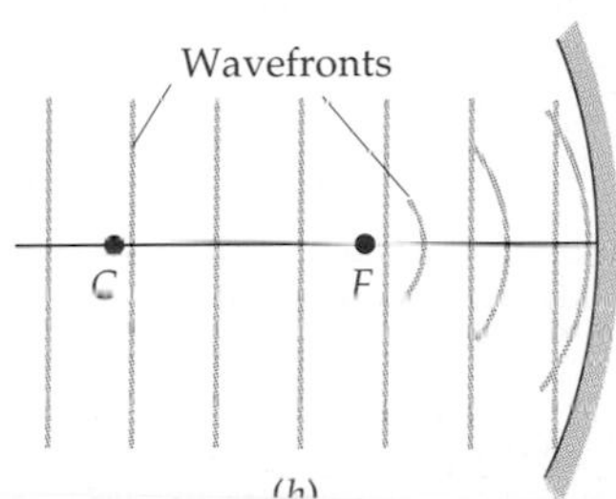

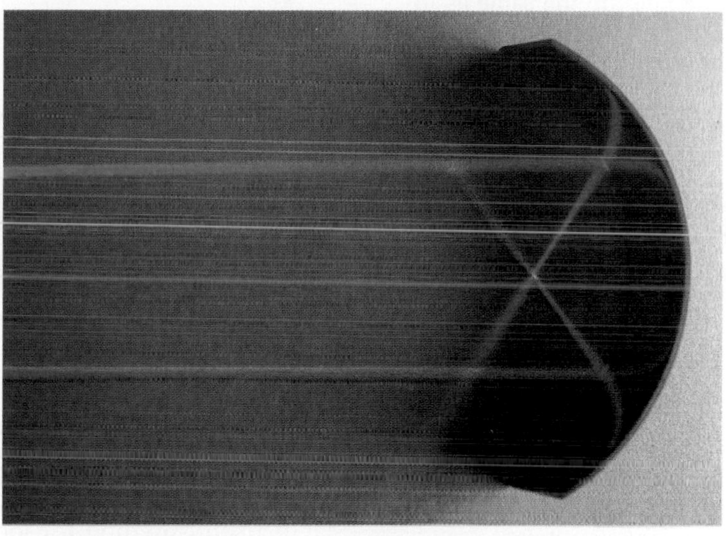

Figure 31-12 *(a)* Parallel rays strike a concave mirror and are reflected through the focal point F at a distance $r/2$. *(b)* The incoming wavefronts are plane waves; upon reflection they become spherical waves that converge at the focal point. *(c)* Photograph of parallel rays focused by a concave mirror.

The **focal point** F (the image point) is the point at which parallel rays incident on the mirror are focused, as illustrated in Figure 31-12*a*. (Again, only paraxial rays are focused at a single point.)

When an object is very far from the mirror, the wavefronts are approximately planes, as shown in Figure 31-12*b*, and the rays are parallel. In Figure 31-12*b*, note how the edges of the wavefront hit the concave mirror surface before the central portion near the axis, resulting in a spherical wavefront upon reflection. Figure 31-13 shows the wavefronts and rays for plane waves striking a convex mirror. In this case, the central part of the wavefront strikes the mirror first, and the reflected waves appear to come from the focal point behind the mirror.

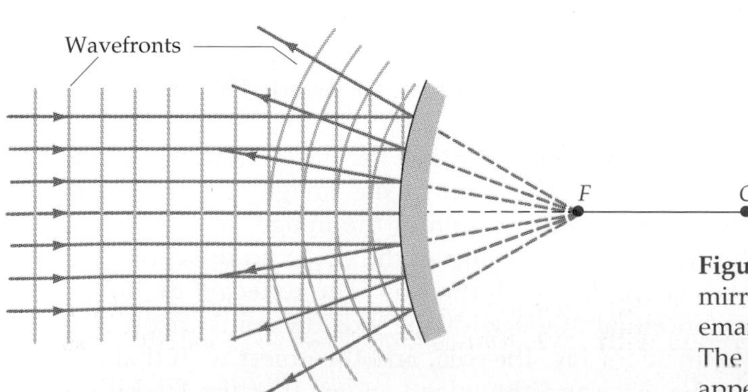

Figure 31-13 Reflection of plane waves from a convex mirror. The outgoing wavefronts are spherical as if emanating from the focal point F behind the mirror. The rays are perpendicular to the wavefronts, and appear to diverge from F.

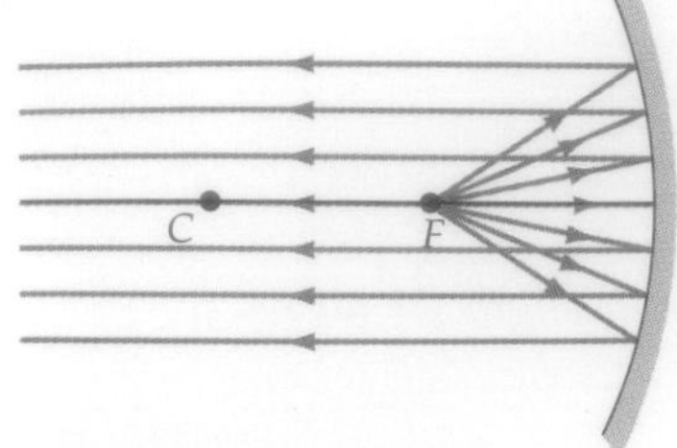

Figure 31-14 Illustration of reversibility. Rays diverging from a point source at the focal point of a concave mirror are reflected from the mirror as parallel rays. The rays are the same as in Figure 31-12*a* but in the reverse direction.

In Figure 31-14, rays from a point source at the focal point strike a concave mirror and are reflected parallel to the axis. This illustrates a property of waves called **reversibility.** If we reverse the direction of a reflected ray, the law of reflection assures us that the reflected ray will be along the original incoming ray but in the opposite direction. (Reversibility holds also for refracted rays, which are discussed in later sections.) Thus, if we have a real image of an object formed by a reflecting (or refracting) surface, we can place an object at the image point and a new image will be formed at the position of the original object.

Example 31-1

An object is 12 cm from a concave mirror with a radius of curvature of 6 cm. Find the focal length of the mirror and the image distance.

From Equation 31-5, the focal length is

$$f = \tfrac{1}{2}r = \tfrac{1}{2}(6 \text{ cm}) = 3 \text{ cm}$$

Equation 31-6 then gives

$$\frac{1}{12 \text{ cm}} + \frac{1}{s'} = \frac{1}{3 \text{ cm}}$$

$$\frac{1}{s'} = \frac{4}{12 \text{ cm}} - \frac{1}{12 \text{ cm}} = \frac{3}{12 \text{ cm}}$$

$$s' = 4 \text{ cm}$$

Ray Diagrams for Mirrors

A useful method for locating images is by geometric construction of a **ray diagram.** This is illustrated in Figure 31-15, where the object is a human figure perpendicular to the axis a distance s from the mirror. By the judicious choice of rays from the head of the figure, we can quickly locate the image. There are four **principal rays** that are convenient to use:

Principal rays for a mirror

1. The **parallel ray,** drawn parallel to the axis. This ray is reflected through the focal point.
2. The **focal ray,** drawn through the focal point. This ray is reflected parallel to the axis.
3. The **radial ray,** drawn through the center of curvature. This ray strikes the mirror perpendicular to its surface and is thus reflected back on itself.
4. The **central ray,** drawn to the vertex of the mirror. It reflects at an equal angle to the axis.

The first three of these rays are shown in Figure 31-15. The intersection of any two rays locates the image point of the head. The third ray can be used to provide a check.

We can see from Figure 31-15 that the image is inverted and is not the same size as the object. The ratio of the image size to the object size is defined as the **lateral magnification** of the image. In Figure 31-16 we have drawn the central ray from the top of the object to the center of the mirror. This ray makes an angle θ with the axis. The reflected ray to the top of the image makes an equal angle with the axis. A comparison of the triangle formed by the incident ray, the axis, and the object with that formed by the reflected ray, the axis, and the image shows that the lateral magnification

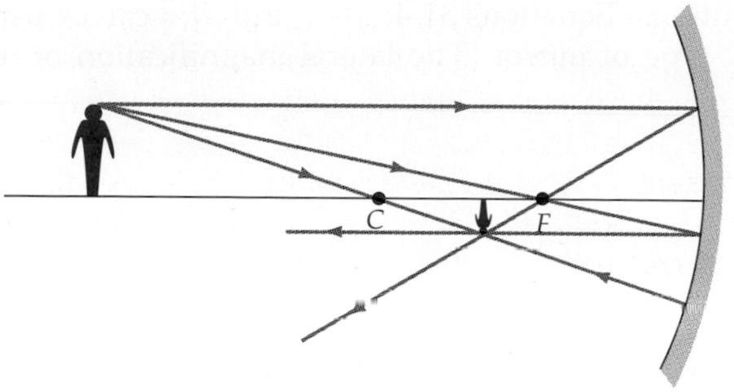

Figure 31-15 Ray diagram for the location of the image by geometric construction.

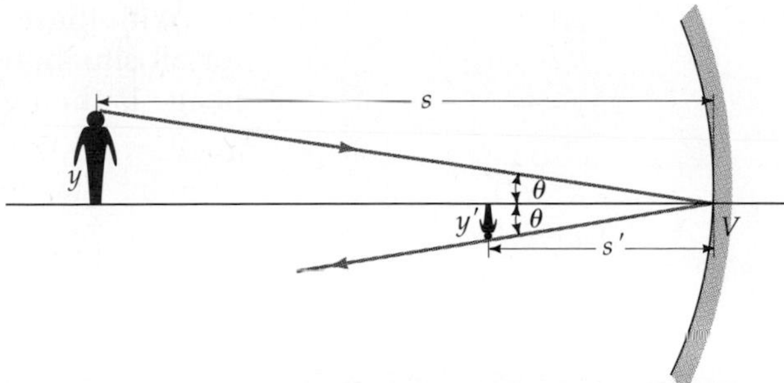

Figure 31-16 Geometry for finding the magnification of a concave spherical mirror.

y'/y equals the ratio of the distances s'/s. Ray diagrams are easier to draw if the mirror is replaced by a straight line that extends as far as necessary to intercept the rays as shown in Figure 31-17.

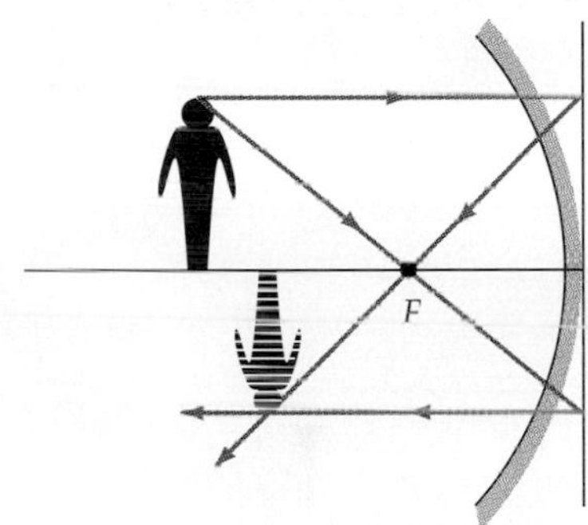

Figure 31-17 Ray diagrams are easier to construct if the curved surface is replaced by a plane surface.

When the object is between the mirror and its focal point, the rays reflected from the mirror do not converge but appear to diverge from a point behind the mirror as illustrated in Figure 31-18. In this case the image is virtual and erect. ("Erect" means not inverted relative to the object.) For an object between the mirror and the focal point, s is less than $\frac{1}{2}r$, so the image distance s' calculated from Equation 31-4 turns out to be negative. We can apply Equations 31-4, 31-5 and 31-6 to this case and to convex mirrors if we adopt a convenient sign convention. Whether the mirror is convex or concave, real images can be formed only in front of the mirror, that is, on the same side of the mirror as the object. Virtual images are formed behind the mirror where there are no actual light rays. Our sign convention is as follows:

Sign conventions for reflection

s	+ if the object is in front of the mirror (real object)
	− if the object is behind the mirror (virtual object)*
s'	+ if the image is in front of the mirror (real image)
	− if the image is behind the mirror (virtual image)
r, f	+ if the center of curvature is in front of the mirror (concave mirror)
	− if the center of curvature is behind the mirror (convex mirror)

*You may wonder how an object can be behind a mirror. This occurs when there is a lens in front of the mirror and the rays to the image of the lens are intercepted by the mirror. The image of the lens is then never formed, but the distance to the unformed image behind the mirror is taken as the object distance for the mirror, and the object is called a virtual object. We will discuss examples of this in Section 31-4 when we discuss lenses.

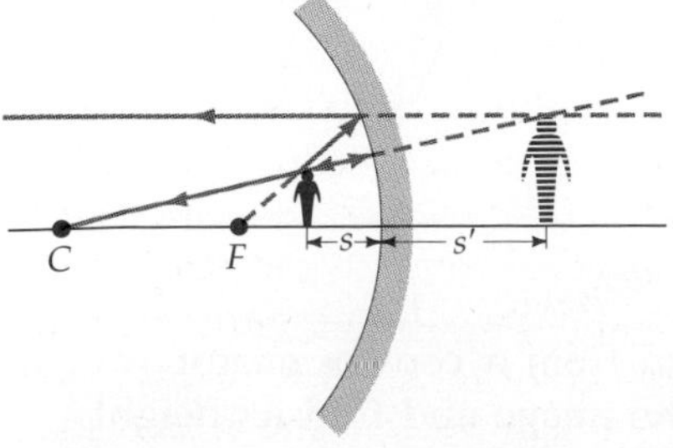

Figure 31-18 A virtual image formed by a concave mirror. The image is located by the radial ray, which is reflected back on itself, and the focal ray, which is reflected parallel to the axis. These two rays appear to diverge from a point behind the mirror found by extending them. A third ray (not shown) could be drawn from the object parallel to the axis. It would be reflected through the focal point, and its extension would intersect the other two rays at the image point.

With these sign conventions, Equations 31-4, 31-5, and 31-6 can be used for all situations with any type of mirror. The lateral magnification of the image is then given by

Lateral magnification

$$m = \frac{y'}{y} = -\frac{s'}{s} \quad \text{31-7}$$

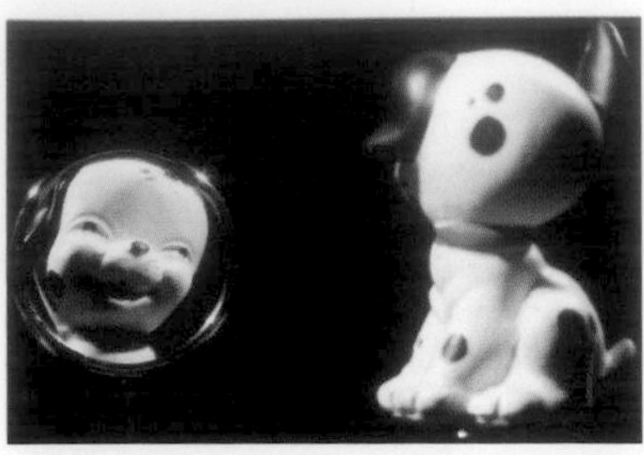

(a)

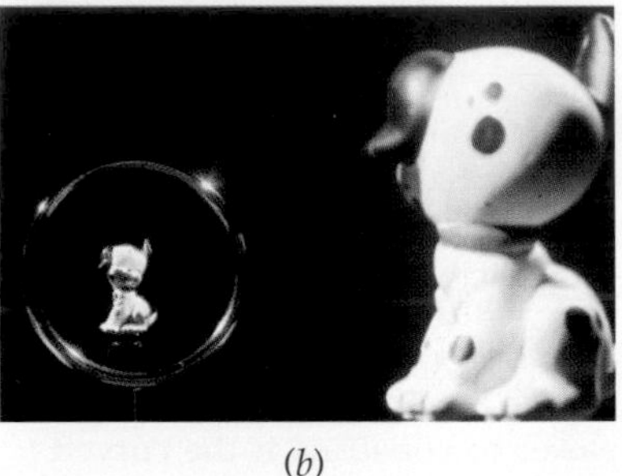

(b)

Reflection in (a) a concave mirror and (b) a convex mirror.

A negative magnification, which occurs when both s and s' are positive, indicates that the image is inverted.

Exercise

A concave mirror has a focal length of 4 cm. (a) What is its radius of curvature? (b) Find the image distance for an object 2 cm from the mirror. Draw a ray diagram for this situation. Is the image erect or inverted? [Answers: (a) 8 cm; (b) $s' = -4$ cm, erect]

For plane mirrors, discussed in the previous section, the radius of curvature is infinite. The focal length given by Equation 31-5 is then also infinite. Equation 31-6 then gives $s' = -s$, indicating that the image is behind the mirror at a distance equal to the object distance. The magnification given by Equation 31-7 is then +1, indicating that the image is erect and the same size as the object.

Although the preceding equations coupled with our sign conventions are relatively easy to use, we often need to know only whether an image is real or virtual, whether it is erect or inverted, and its approximate location and magnification. This knowledge is usually easiest to obtain by constructing a ray diagram. However, it is always a good idea to use both the graphical method and the algebraic method to locate an image so that one method serves as a check on the results of the other.

Convex Mirrors

Figure 31-19 shows a ray diagram for an object in front of a convex mirror. The central ray heading toward the center of curvature C is perpendicular to the mirror and is reflected back on itself. The parallel ray is reflected as if it came from the focal point F behind the mirror. The focal ray (not shown) would be drawn towards the focal point and would be reflected parallel to the axis. We can see from the figure that the image is behind the mirror and is therefore virtual. It is also erect and smaller than the object.

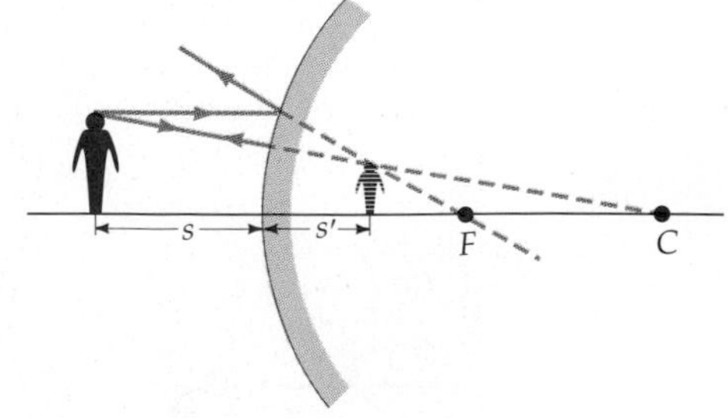

Figure 31-19 Ray diagram for a convex mirror. The parallel ray is reflected as if it came from the focal point behind the mirror, and the radial ray is reflected back on itself. These rays appear to diverge from a point behind the mirror. A third ray (not shown) could be drawn toward the focal point. It would be reflected parallel to the axis and its extension would intersect the other two rays at the image point.

Example 31-2

An object 2 cm high is 10 cm from a convex mirror with a radius of curvature of 10 cm. Locate the image and find its height.

Since the center of curvature of a convex mirror is behind the mirror, the radius and the focal length are negative:

$$f = \tfrac{1}{2}r = \tfrac{1}{2}(-10\text{ cm}) = -5\text{ cm}$$

Using Equation 31-6 to find the image distance, we obtain

$$\frac{1}{10\text{ cm}} + \frac{1}{s'} = \frac{1}{f} = -\frac{1}{5\text{ cm}}$$

$$\frac{1}{s'} = -\frac{2}{10\text{ cm}} - \frac{1}{10\text{ cm}} = -\frac{3}{10\text{ cm}}$$

$$s' = -3.33\text{ cm}$$

The image distance is negative, indicating a virtual image behind the mirror. The magnification is

$$m = -\frac{s'}{s} = -\frac{-3.33\text{ cm}}{10\text{ cm}} = +0.333$$

Thus, the image is erect and is one-third the size of the object. Since the object height is 2 cm, the image height is 2/3 cm. The ray diagram for this example is similar to Figure 31-19.

Exercise

Find the image distance and magnification for an object 5 cm away from the mirror in Example 31-2. Draw a ray diagram. (Answers: $s' = -2.5$ cm, $m = +0.5$)

Questions

3. Under what condition will a concave mirror produce an erect image? A virtual image? An image smaller than the object? An image larger than the object?
4. Answer Question 3 for a convex mirror.
5. Convex mirrors are often used for rear-view mirrors on cars and trucks to give a wide-angle view. Below the mirror is written, "Warning, objects are closer than they appear." Yet, according to a ray diagram such as Figure 31-19, the image distance for distant objects is much smaller than the object distance. Why then do they appear more distant?

(a)

(b)

(a) A convex mirror resting on paper with equally spaced parallel stripes. Note the large number of lines imaged in a small space and the reduction in size and distortion in shape of the image. *(b)* A convex mirror is used for security in a store.

31-3 Images Formed by Refraction

The formation of an image by refraction at a spherical surface separating two media with indexes of refraction n_1 and n_2 is illustrated in Figure 31-20. In this figure, n_2 is greater than n_1, so the waves travel more slowly in the second medium. Again, only paraxial rays converge to one point. An equation relating the image distance to the object distance, the radius of curvature, and the indexes of refraction can be derived by applying Snell's law of refraction to these rays and using small-angle approximations. The geometry is shown in Figure 31-21. The angles θ_1 and θ_2 are related by Snell's law:

$$n_1 \sin \theta_1 = n_2 \sin \theta_2$$

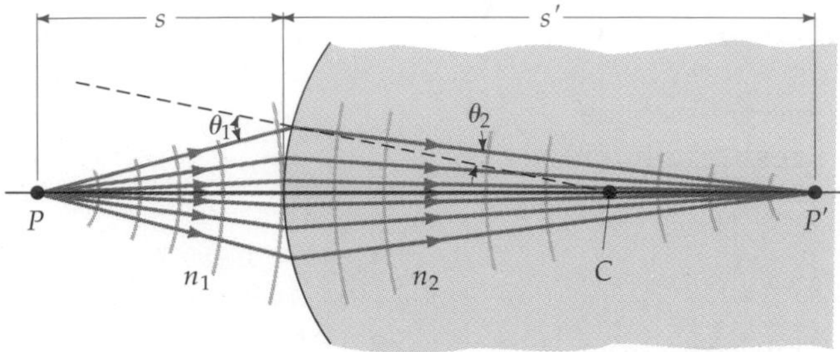

Figure 31-20 Image formed by refraction at a spherical surface between two media where the waves move slower in the second medium.

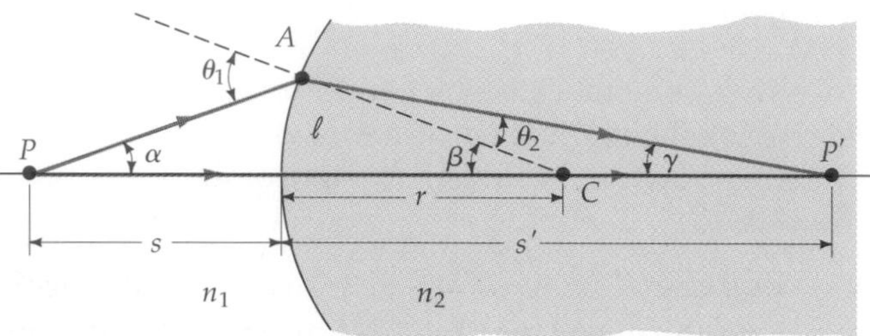

Figure 31-21 Geometry for relating the image position to the object position for refraction at a single spherical surface. Snell's law is applied to the ray incident at point A, and the small-angle approximation is used.

Using the small-angle approximation $\sin \theta \approx \theta$, we have

$$n_1\theta_1 = n_2\theta_2 \quad \text{31-8}$$

From triangle ACP', we have

$$\beta = \theta_2 + \gamma = \frac{n_1}{n_2}\theta_1 + \gamma \quad \text{31-9}$$

We can obtain another relation for θ_1 from triangle PAC:

$$\theta_1 = \alpha + \beta \quad \text{31-10}$$

Eliminating θ_1 from Equations 31-9 and 31-10, we obtain

$$n_1\alpha + n_1\beta + n_2\gamma = n_2\beta$$

or

$$n_1\alpha + n_2\gamma = (n_2 - n_1)\beta \quad \text{31-11}$$

Using the small-angle approximations $\alpha \approx \ell/s$, $\beta \approx \ell/r$, and $\gamma \approx \ell/s'$, we obtain

$$\frac{n_1}{s} + \frac{n_2}{s'} = \frac{n_2 - n_1}{r} \quad \text{31-12}$$

In refraction, real images are formed in back of the surface, which we will call the transmission side, whereas virtual images occur on the incident side in front of the surface. The sign conventions we use for refraction are similar to those for reflection:

s	+ (real object) for objects in front of the surface (incident side) − (virtual object) for objects in back of the surface (transmission side)
s'	+ (real image) for images in back of the surface (transmission side) − (virtual image) for images in front of the surface (incident side)
r, f	+ if the center of curvature is on the transmission side − if the center of curvature is on the incident side

Sign conventions for refraction

If we compare these sign conventions with those for reflection, we see that s' is positive and the image is real when the image is on the side of the surface traversed by the reflected or refracted light. For reflection, this side is in front of the mirror, whereas for refraction, it is behind the refracting surface. Similarly, r and f are positive when the center of curvature is on the side traversed by the reflected or refracted light.

We can obtain an expression for the magnification of an image formed by a refracting surface by considering Figure 31-22, which shows a ray from the top of the object to the top of the image. The ray is bent toward the normal as it crosses the surface, so θ_2 is less than θ_1. These angles are related by Snell's law:

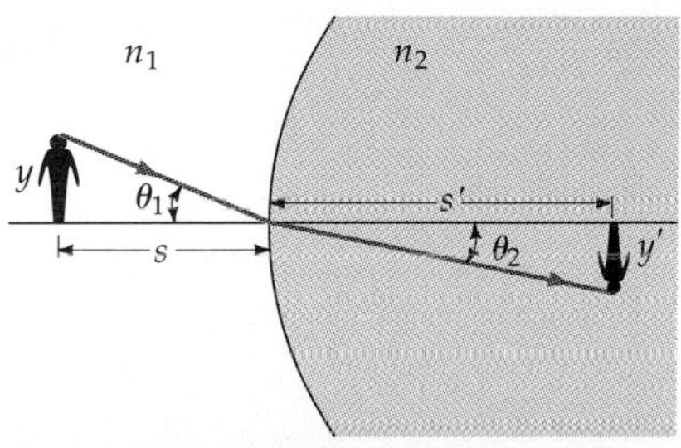

Figure 31-22 Geometry for finding the lateral magnification of an image formed by refraction from a single spherical surface.

$$n_1 \sin \theta_1 = n_2 \sin \theta_2$$

The object and image sizes are related to the angles by

$$\tan \theta_1 = \frac{y}{s}$$

$$\tan \theta_2 = -\frac{y'}{s'}$$

where the minus sign arises because y' is negative. Since we are considering only paraxial rays for which the angles are small, the sine of an angle is approximately equal to its tangent. With this approximation, Snell's law becomes

$$n_1 \frac{y}{s} = n_2 \frac{-y'}{s'}$$

The magnification is thus

$$m = \frac{y'}{y} = -\frac{n_1 s'}{n_2 s} \qquad \text{31-13}$$

Example 31-3

A fish is in a spherical bowl of water of index of refraction 1.33. The radius of the bowl is 15 cm. The fish looks through the bowl and sees a cat sitting on the table with its nose 10 cm from the bowl. Where is the image of the cat's nose, and what is its magnification? Neglect any effect of the thin glass wall of the bowl.

The object distance between the cat and the bowl is 10 cm. The indexes of refraction are $n_1 = 1$ and $n_2 = 1.33$. The radius of curvature is +15 cm. Equation 31-12 then gives for the image distance

$$\frac{1.00}{10 \text{ cm}} + \frac{1.33}{s'} = \frac{1.33 - 1.00}{15 \text{ cm}}$$

Solving for s', we obtain

$$s' = -17.1 \text{ cm}$$

The negative image distance means that the image is virtual and in front

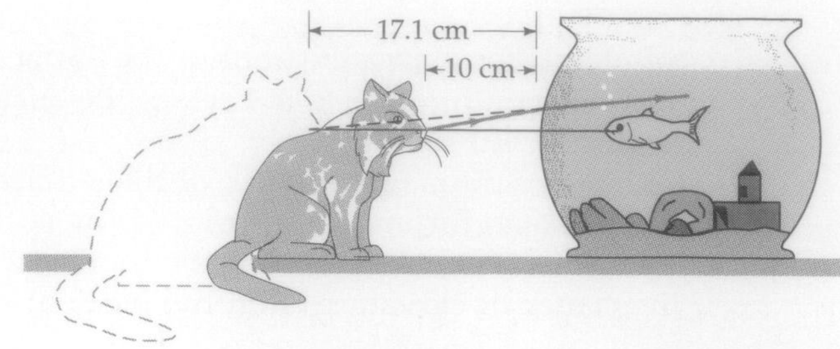

Figure 31-23 Fish looking at cat for Example 31-3. Because of refraction at the spherical surface, the cat appears farther away and slightly larger.

of the refracting surface, on the same side as the object, as shown in Figure 31-23. The magnification of the image is

$$m = -\frac{n_1 s'}{n_2 s} = -\frac{-17.1 \text{ cm}}{1.33(10 \text{ cm})} = 1.29$$

Thus, the cat appears to be farther away and slightly larger.

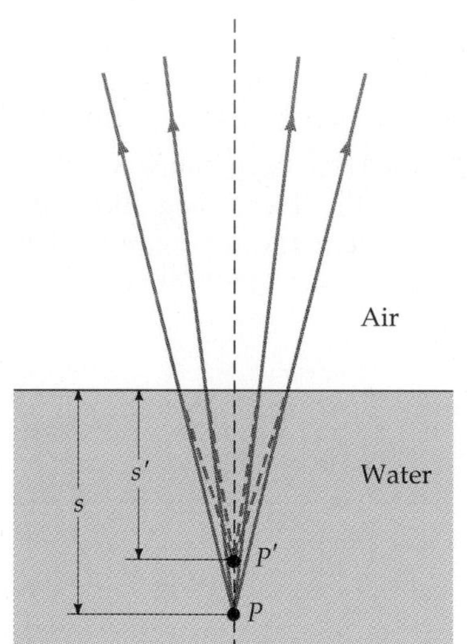

Figure 31-24 Ray diagram for the image of an object in water as viewed from directly overhead. The depth of the image is less than the depth of the object. The apparent depth equals the real depth divided by the index of refraction of water.

We can use Equation 31-12 to find the **apparent depth** of an object under water when it is viewed from directly overhead. For this case, the surface is a plane surface, so the radius of curvature is infinite. The image and object distances are related by

$$\frac{n_1}{s} + \frac{n_2}{s'} = 0$$

where n_1 is the index of refraction of the first medium (water) and n_2 is that of the second medium (air). The apparent depth is therefore

$$s' = -\frac{n_2}{n_1} s \qquad 31\text{-}14$$

The negative sign indicates that the image is virtual and on the same side of the refracting surface as the object, as shown in the ray diagram in Figure 31-24. The magnification is

$$m = -\frac{n_1 s'}{n_2 s} = +1$$

Since $n_2 = 1$ for air, we see from Equation 31-14 that the apparent depth equals the real depth divided by the index of refraction of water.

Example 31-4

Find the apparent depth of a fish resting 1 m below the surface of water that has an index of refraction $n = 4/3$.

Using $n_1 = 4/3$ and $n_2 = 1$ in Equation 31-14, we obtain

$$s' = -\frac{1}{(4/3)}(1 \text{ m}) = -\frac{3}{4}(1 \text{ m}) = -0.75 \text{ m}$$

The apparent depth is three-fourths the actual depth, so the fish appears to be 75 cm below the surface. Note that this result holds only when the object is viewed from directly overhead so that the rays are paraxial.

Questions

6. If a fish under water is viewed from a point not directly above it, is its apparent depth greater or less than three-fourths its actual depth? (Draw rays from the fish to the eye for large angles to help answer this question.)

7. A bird on a limb above the water is viewed by a scuba diver submerged beneath the water's surface. Does the bird appear to the diver to be closer to or father from the surface than it actually is?

Because of refraction, the apparent depth of the submerged portion of the straw is less than the real depth. Consequently, the straw appears to be bent. A reflected image of the straw is also seen.

31-4 Thin Lenses

The most important application of Equation 31-12 is finding the position of the image formed by a lens. This is done by considering the refraction at each surface separately to derive an equation relating the image distance to the object distance, the radius of curvature of each surface of the lens, and the index of refraction of the lens.

We will consider a very thin lens of index of refraction n with air on both sides. Let the radii of curvature of the surfaces of the lens be r_1 and r_2. If an object is at a distance s from the first surface (and therefore from the lens), the distance s'_1 of the image due to refraction at the first surface can be found using Equation 31-12:

$$\frac{1}{s} + \frac{n}{s'_1} = \frac{n-1}{r_1} \qquad 31\text{-}15$$

This image is not formed because the light is again refracted at the second surface. Figure 31-25 shows the case when the image distance s'_1 for the first surface is negative, indicating a virtual image to the left of the surface. Rays in the glass refracted from the first surface diverge as if they came from the image point P'_1. They strike the second surface at the same angles as if there were an object at this image point. The image for the first surface therefore becomes the object for the second surface. Since the lens is of negligible thickness, the object distance is equal in magnitude to s'_1, but since object distances in front of the surface are positive whereas image distances are negative there, the object distance for the second surface is $s_2 = -s'_1$. (If s'_1 were positive, the rays would converge as they strike the second surface. The object for the second surface would then be to the right of the surface. This object would be a **virtual object.** Again, $s_2 = -s'_1$.) We now write Equation 31-12 for the second surface with $n_1 = n$, $n_2 = 1$, and $s = -s'_1$. The image distance for the second surface is the final image distance s' for the lens.

$$\frac{n}{-s'_1} + \frac{1}{s'} = \frac{1-n}{r_2} \qquad 31\text{-}16$$

We can eliminate the image distance for the first surface s'_1 by adding Equations 31-15 and 31-16. We obtain

$$\frac{1}{s} + \frac{1}{s'} = (n-1)\left(\frac{1}{r_1} - \frac{1}{r_2}\right) \qquad 31\text{-}17$$

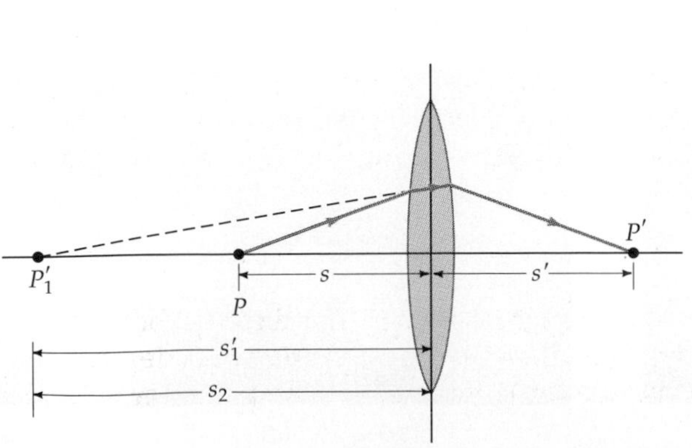

Figure 31-25 Refraction occurs at both surfaces of a lens. Here, the refraction at the first surface leads to a virtual image at P'_1. The rays strike the second surface as if they came from P'_1. Since image distances are negative when the image is on the incident side of the surface whereas object distances are positive for objects there, $s_2 = -s'_1$ is the object distance for the second surface of the lens.

Equation 31-17 gives the image distance s' in terms of the object distance s and the properties of the thin lens—r_1,r_2, and the index of refraction n. As with mirrors, the focal length of a thin lens is defined as the image distance when the object distance is infinite. Setting s equal to infinity and writing f for the image distance s', we obtain

Lens-maker's equation

$$\frac{1}{f} = (n - 1)\left(\frac{1}{r_1} - \frac{1}{r_2}\right) \tag{31-18}$$

Equation 31-18 is called the **lens-makers' equation;** it gives the focal length of a thin lens in terms of the properties of the lens. Substituting $1/f$ for the right side of Equation 31-17, we obtain

Thin-lens equation

$$\frac{1}{s} + \frac{1}{s'} = \frac{1}{f} \tag{31-19}$$

Equation 31-19 is called the **thin-lens equation.** Note that it is the same as the mirror equation (Equation 31-6). Recall, however, that the sign conventions for refraction are somewhat different from those for reflection. For lenses, the image distance s' is positive when the image is on transmission side of the lens, that is, when it is on the side opposite the side on which light is incident. The sign convention for r in Equation 31-18 is the same as that for refraction at a single surface. The radius is positive if the center of curvature is on the transmission side of the lens and negative if it is on the incident side.

Figure 31-26*a* shows the wavefronts of plane waves incident on a double convex lens. The central part of the wavefront strikes the lens first. Since the wave speed in the lens is less than that in air (assuming $n > 1$), the central

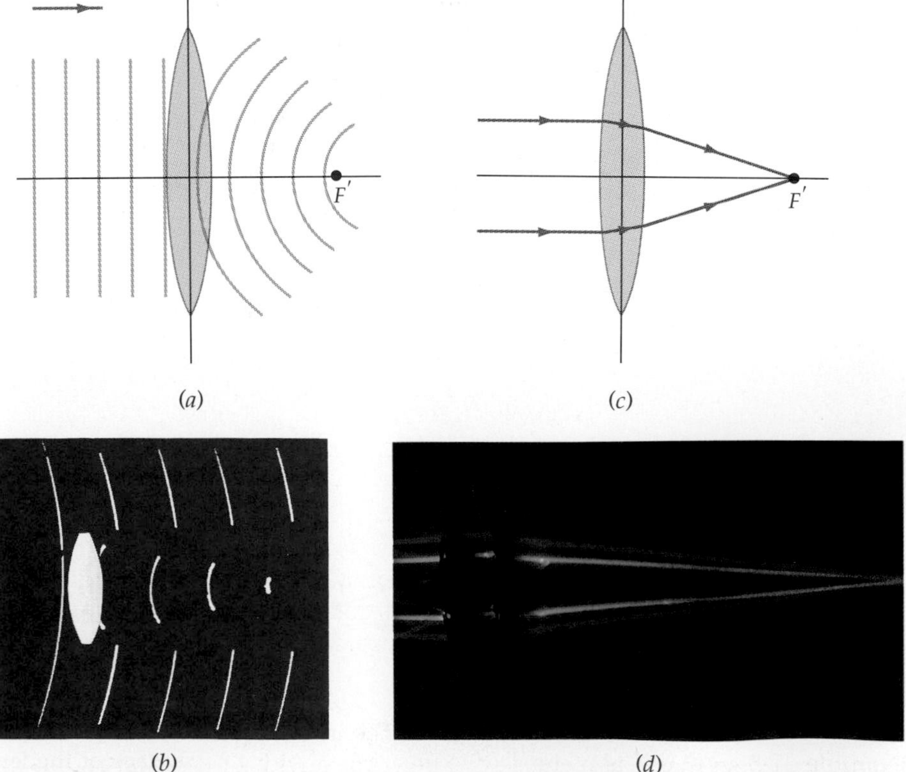

Figure 31-26 *(a)* Wavefronts for plane waves striking a converging lens. The central part of the wavefront is retarded more by the lens than the outer part, resulting in a spherical wave that converges at the focal point F'. *(b)* Wavefronts passing through a lens, shown by a photographic technique called *light-in-flight-recording* that uses a pulsed laser to make a hologram of the wavefronts of light. *(c)* Rays for plane waves striking a converging lens. The rays are bent at each surface and converge at the focal point. *(d)* Photograph of rays focused by a converging lens.

part of the wavefront lags behind the outer parts, resulting in a spherical wave that converges at the focal point F'. The rays for this situation are shown in Figure 31-26*c*. Such a lens is called a **converging lens.** Since its focal length as calculated from Equation 31-18 is positive, it is also called a **positive lens.** Any lens that is thicker in the middle than at the edges is a converging lens (providing that the index of refraction of the lens is greater than that of the surrounding medium). Figures 31-27*a* and 31-27*b* show the wavefronts and rays for plane waves incident on a double concave lens. In this case, the outer part of the wavefronts lag behind the central parts, resulting in outgoing spherical waves that diverge from a focal point on the incident side of the lens. The focal length of this lens is negative. Any lens (with index of refraction greater than that of the surrounding medium) that is thinner in the middle than at the edges is a **diverging, or negative, lens.**

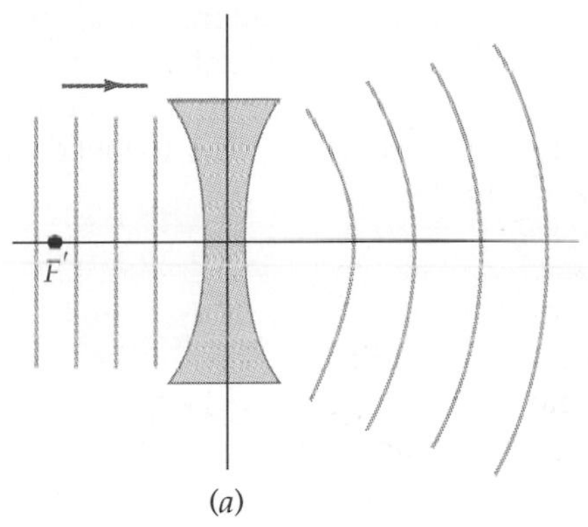

(*a*)

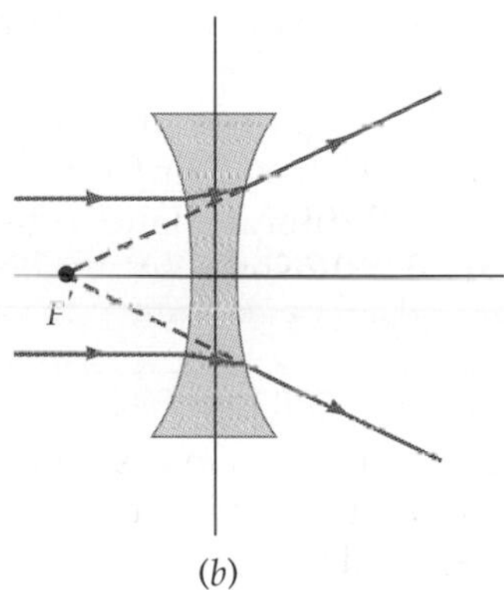

(*b*)

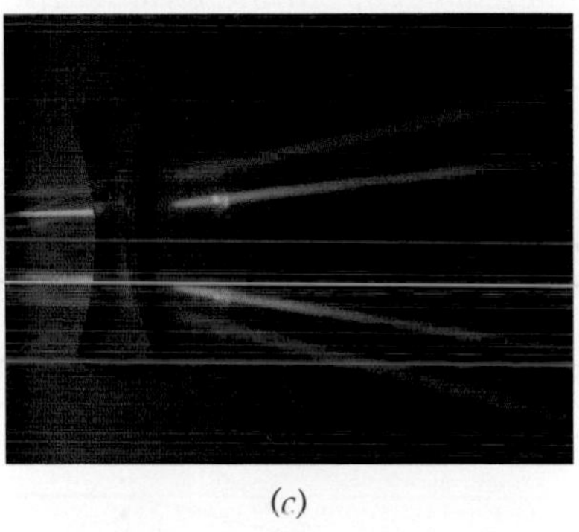
(*c*)

Figure 31-27 (*a*) Wavefronts for plane waves striking a diverging lens. Here, the outer parts of the wavefronts are retarded more than the central part, resulting in a spherical wave that diverges as it moves out as if it came from the focal point F' in front of the lens. (*b*) Rays for plane waves striking the same diverging lens. The rays are bent outward and diverge as if they came from the focal point F'. (*c*) Photograph of rays passing through a diverging lens.

Example 31-5

A double convex thin lens made of glass of index of refraction $n = 1.5$ has radii of curvature of magnitude 10 cm and 15 cm as shown in Figure 31-28. Find its focal length.

We assume that the light is incident on the surface with the smaller radius of curvature (the left surface in Figure 31-28). The center of curvature of the first surface, C_1, is on the transmission side of the lens, so r_1 is positive and equal to +10 cm. The center of curvature of the second surface, C_2, is on the incident side, so r_2 is negative and equal to −15 cm. Equation 31-18 is then

$$\frac{1}{f} = (1.5 - 1)\left(\frac{1}{+10\text{ cm}} - \frac{1}{-15\text{ cm}}\right)$$

$$= 0.5\left(\frac{3}{30\text{ cm}} + \frac{2}{30\text{ cm}}\right) = 0.5\left(\frac{1}{6\text{ cm}}\right)$$

$$f = 12\text{ cm}$$

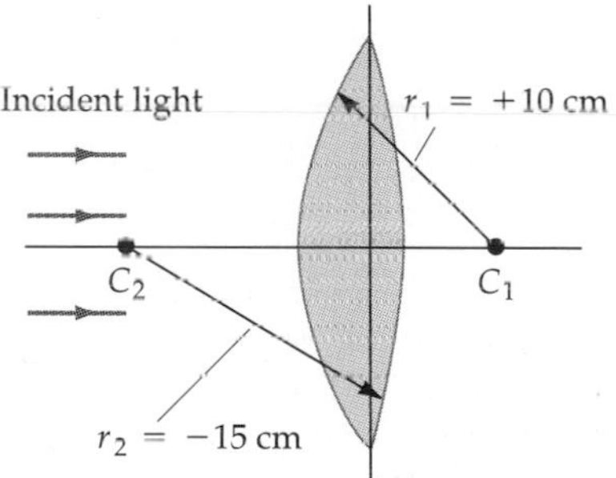

Figure 31-28 Double convex lens with radii of curvature of magnitude 15 cm and 10 cm for Example 31-5. The center of curvature of the first surface is on the transmission side of the lens, so r is positive for this surface. The center of curvature for the second surface is on the incident side of the lens, so r for it is negative. Both surfaces tend to converge the light rays and contribute to a positive focal length for the lens.

Exercise

A double convex thin lens has an index of refraction $n = 1.6$ and radii of curvature of equal magnitude. If its focal length is 15 cm, what is the magnitude of the radius of curvature of each surface? (Answer: 18 cm)

Note that if we reverse the direction of the incoming light for the lens in Example 31-5 so that it is incident on the surface with the greater radius of

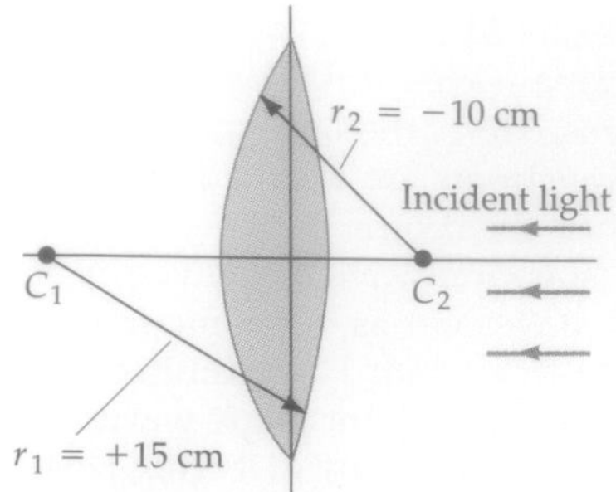

Figure 31-29 The same lens as in Figure 31-28 with the light incident from the other side. The order of the surfaces and the signs of the radii of curvature are interchanged, but the focal length is the same.

curvature (from the right as in Figure 31-29), the order of the surfaces is interchanged. The radius of the first surface has a magnitude of 15 cm and is positive because the center of curvature of that surface is on the transmission side, whereas the center of curvature of the surface with the 10-cm radius is on the incident side, so $r_2 = -10$ cm. Using these values in Equation 31-18 gives the same result, $f = 12$ cm, for the focal length. Thus, the focal length of a lens is the same for light incident on either side. If parallel light strikes the lens of Example 31-5 from the left, it is focused at a point 12 cm to the right of the lens, whereas if parallel light strikes the lens from the right, it is focused at 12 cm to the left of the lens. Both of these points are focal points of the lens. Using the reversibility property of light rays, we can see that light diverging from a focal point and striking the lens will leave the lens as a parallel beam as shown in Figure 31-30. In a particular lens problem in which the direction of the incident light is specified, the object point for which light emerges as a parallel beam is called the **first focal point** F and the point at which parallel light is focused is called the **second focal point** F'. For a positive lens, the first focal point is on the incident side and the second focal point is on the transmission side. If parallel light is incident on the lens at a small angle with the axis as in Figure 31-31, it is focused at a point in the **focal plane** a distance f from the lens.

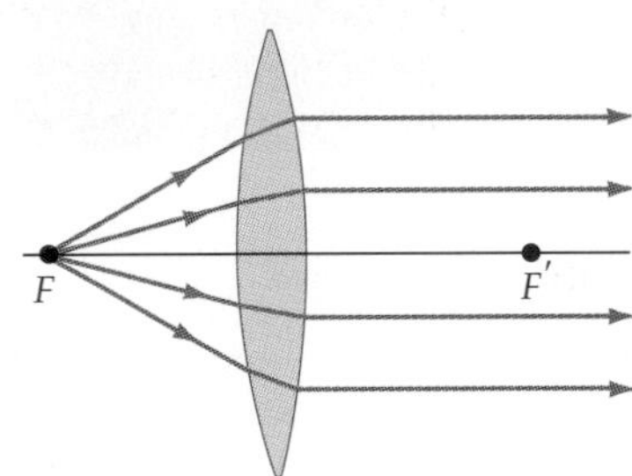

Figure 31-30 Light rays diverging from the focal point of a positive lens emerge parallel to the axis. This point is called the first focal point F. The point at which parallel light is converged by the lens is called the second focal point F'.

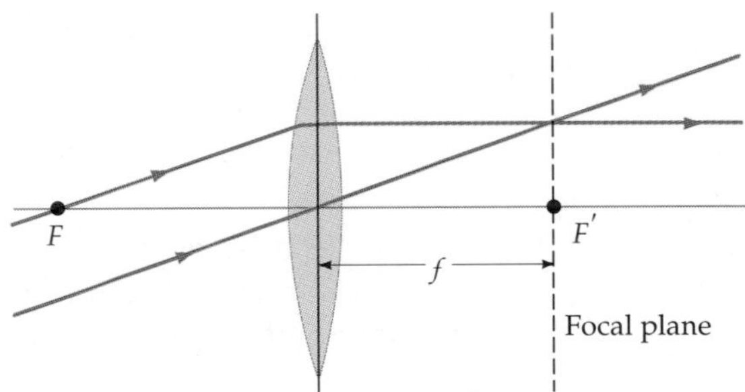

Figure 31-31 Parallel rays incident on the lens at an angle to its axis are focused at a point in the focal plane of the lens.

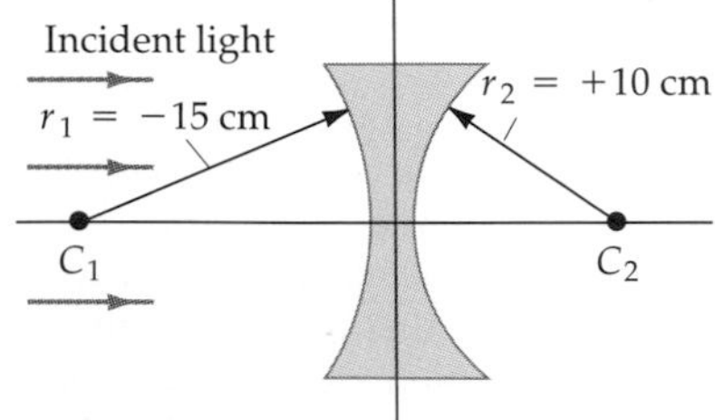

Figure 31-32 Double concave lens with radii of magnitude 15 cm and 10 cm. The center of curvature of the first surface is on the incident side of the lens and that of the second surface is on the transmission side, so r_1 is negative and r_2 is positive. Both surfaces tend to diverge light rays and contribute to a negative focal length.

Example 31-6

A double concave lens has an index of refraction of 1.5 and radii of curvature of magnitude 10 cm and 15 cm. Find its focal length.

For the orientation of the lens relative to the incident light shown in Figure 31-32, the radius of curvature of the first surface is $r_1 = -15$ cm and that of the second surface is $r_2 = +10$ cm. The lens-maker's equation (Equation 31-18) then gives

$$\frac{1}{f} = (1.5 - 1.0)\left(\frac{1}{-15\text{ cm}} - \frac{1}{+10\text{ cm}}\right)$$

Solving for f, we obtain $f = -12$ cm. Again, we obtain the same result no matter which surface the light strikes first.

In laboratory experiments involving lenses, it is usually much easier to measure the focal length rather than to calculate it from the radii of curvature of the surfaces.

Ray Diagrams for Lenses

As with images formed by mirrors, it is convenient to locate the images of lenses by graphical methods. Figure 31-33 illustrates the graphical method for a converging lens. We use three principal rays. For the sake of simplicity, we consider the rays to bend at the plane through the center of the lens. For a positive lens, the principal rays are

1. The **parallel ray,** drawn parallel to the axis. This ray is bent through the second focal point of the lens.
2. The **central ray,** drawn through the center (the vertex) of the lens. This ray is undeflected. (The faces of the lens are parallel at this point, so the ray emerges in the same direction but displaced slightly. Since the lens is thin, the displacement is negligible.)
3. The **focal ray,** drawn through the first focal point. This ray emerges parallel to the axis.

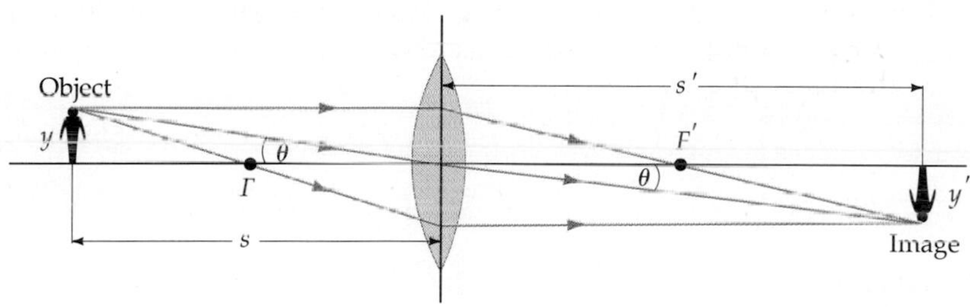

Figure 31-33 Ray diagram for a thin converging lens. For the sake of simplicity, we assume that all the bending takes place at the central plane. The ray through the center is undeflected because the lens surfaces there are parallel and are close together.

These three rays converge to the image point, as shown in the figure. In this case, the image is real and inverted. From Figure 31-33, we have $\tan\theta = y/s = -y'/s'$. The lateral magnification is then

$$m = \frac{y'}{y} = -\frac{s'}{s}$$

This expression is the same as that for mirrors. Again, a negative magnification indicates that the image is inverted.

The principal rays for a negative, or diverging, lens are

Principal rays for a negative lens

1. The **parallel ray,** drawn parallel to the axis. This ray diverges from the lens as if it came from the second focal point.
2. The **central ray,** drawn through the center (the vertex) of the lens. This ray is undeflected.
3. The **focal ray,** drawn toward the first focal point. This ray emerges parallel to the axis.

The ray diagram for a diverging lens is shown in Figure 31-34.

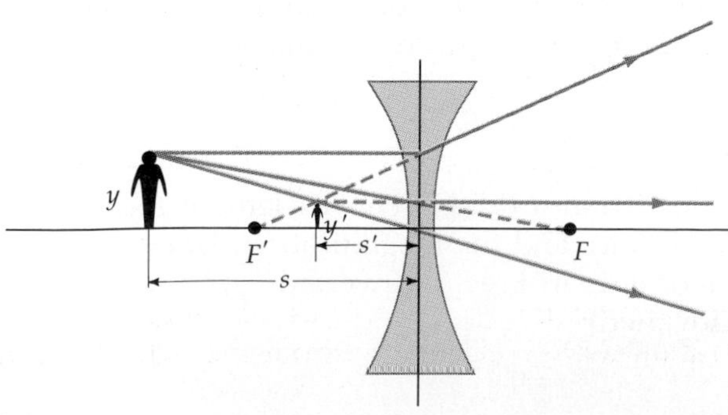

Figure 31-34 Ray diagram for a diverging lens. The parallel ray is bent away from the axis as if it came from the second focal point F'. The ray toward the first focal point F emerges parallel to the axis. The central ray is undeflected. The three rays appear to diverge from the image point.

Example 31-7

An object 1.2 cm high is placed 4 cm from the double convex lens of Example 31-5. Locate the image, state whether it is real or virtual, and find its height.

In Example 31-5, the focal length for this lens was found to be $f = 12$ cm. Figure 31-35 shows the ray diagram for an object placed 4 cm in front of a positive lens of focal length 12 cm. The parallel ray is bent through the second focal point, and the central ray is undeflected. These rays are diverging on the transmission side of the lens. The image is located by extending the rays back until they meet. These two rays are sufficient to locate the image. (As a check, we could draw the third ray, the focal ray, along the line from the first focal point F on the incident side of the lens. This ray would leave the lens parallel to the axis.) We can see immediately from the figure that the image is virtual, erect, and enlarged. It is on the same side of the lens as the object and is slightly farther away from the lens. Since it is quite easy to make an error when calculating the image distance using Equation 31-19, it is always a good idea to check your result with a ray diagram.

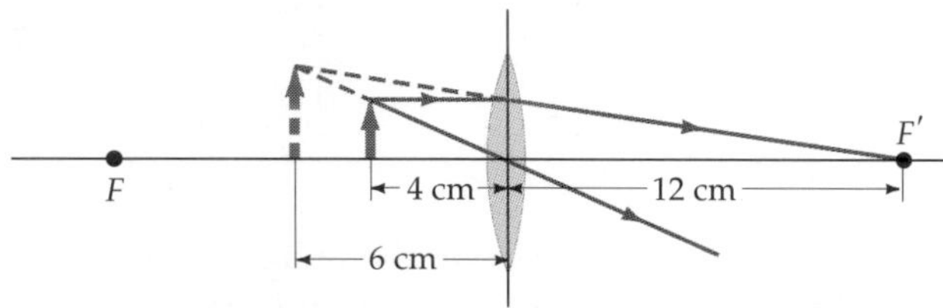

Figure 31-35 Ray diagram for Example 31-7. When the object is between the first focal point and the converging lens, the image is virtual and erect.

The image distance is found algebraically using Equation 31-19:

$$\frac{1}{4\text{ cm}} + \frac{1}{s'} = \frac{1}{12\text{ cm}}$$

$$\frac{1}{s'} = \frac{1}{12\text{ cm}} - \frac{1}{4\text{ cm}} = \frac{1}{12\text{ cm}} - \frac{3}{12\text{ cm}} = -\frac{1}{6\text{ cm}}$$

$$s' = -6\text{ cm}$$

The image distance is negative, indicating that the image is virtual and on the incident side of the lens. The magnification is

$$m = -\frac{s'}{s} = -\frac{-6\text{ cm}}{4\text{ cm}} = +1.5$$

The image is thus 1.5 times as large as the object and is erect. Since the height of the object is 1.2 cm, the height of the image is 1.8 cm.

Exercise

An object is placed 15 cm from a double convex lens of focal length 10 cm. Find the image distance and the magnification. Draw a ray diagram. Is the image real or virtual? Erect or inverted? (Answers: $s' =$ 30 cm, $m = -2$, real, inverted)

Exercise

Work the previous exercise for an object placed 5 cm from a lens with a focal length of 10 cm. (Answers: $s' = -10$ cm, $m = 1$, virtual, erect)

Multiple Lenses

If we have two or more thin lenses, we can find the final image produced by the system by finding the image distance for the first lens and using it along with the distance between lenses to find the object distance for the second lens. That is, we consider each image, whether it is real or virtual and whether it is formed or not, as the object for the next lens.

Example 31-8

A second lens of focal length +6 cm is placed 12 cm to the right of the lens in Example 31-7. Locate the final image.

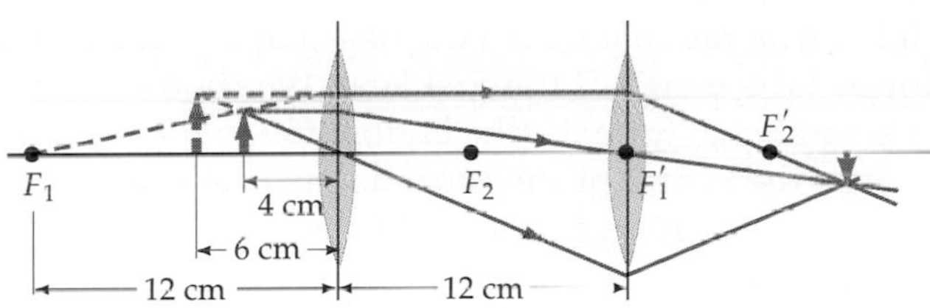

Figure 31-36 Ray diagram for Example 31-8. The image of the first lens acts as the object for the second lens. The final image is located by drawing two rays from the first image through the second lens. In this case, one of the rays used to locate the first image happens to be the central ray for the second lens. A second ray parallel to the axis from the first image locates the final image.

Figure 31-36 shows the ray diagram for this example. The rays used to locate the image of the first lens will not necessarily be the principal rays for the second lens. If they are not, we merely draw additional rays from the first image that are principal rays for the second lens, such as a ray from the image parallel to the axis and one from the image through the first focal point of the second lens or one through the vertex of the second lens. In this case, two of the principal rays for the first lens are also principal rays for the second lens. The parallel ray for the first lens turns out to be the central ray for the second lens. Also, the focal ray for the first lens emerges parallel to the axis and is therefore refracted through the focal point of the second lens. (In the figure, we have extended the central ray for the first lens so that it passes through the image found from the other two rays.) We can see that the final image is real, inverted, and just outside the second focal point of the second lens. We can locate its position algebraically by noting that the virtual image of the first lens is 6 cm to the left of that lens and is therefore 18 cm to the left of the second lens. Using $s_2 = 18$ cm and $f_2 = 6$ cm, we have

$$\frac{1}{s_2} + \frac{1}{s_2'} = \frac{1}{f_2}$$

$$\frac{1}{18\text{ cm}} + \frac{1}{s_2'} = \frac{1}{6\text{ cm}}$$

giving

$$s_2' = 9\text{ cm}$$

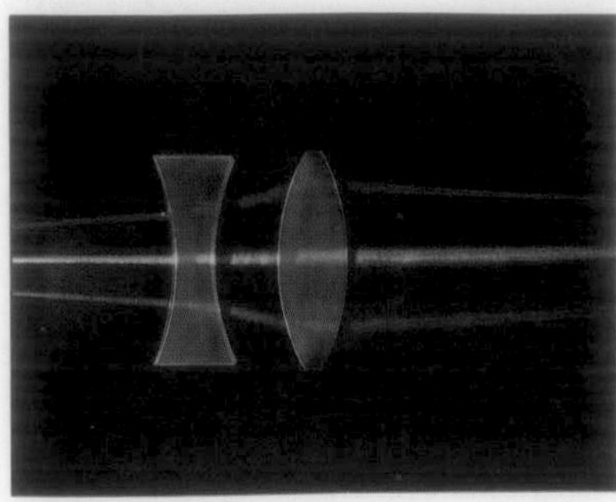

Light rays passing through a diverging lens followed by a converging lens, which is placed at the first focal point of the diverging lens.

Example 31-9

Two lenses, each of focal length 10 cm, are 15 cm apart. Find the final image of an object 15 cm from one of the lenses.

In the ray diagram of Figure 31-37, the image of the first lens would be 30 cm to the right of the lens if the second lens were not there. We calculate this using $s_1 = 15$ cm and $f_1 = 10$ cm in the thin-lens equation:

$$\frac{1}{s_1} + \frac{1}{s_1'} = \frac{1}{f_1}$$

$$\frac{1}{15\text{ cm}} + \frac{1}{s_1'} = \frac{1}{10\text{ cm}}$$

Solving for s_1', we obtain

$$s_1' = 30\text{ cm}$$

This image is not formed because the light rays strike the second lens before they reach the image position. We can locate the final image graphically by choosing rays that are heading toward the unformed image when they strike the lens. These rays need not be the principal rays for the first lens. Any ray that leaves the object and strikes the first lens is directed toward the image of the first lens. We choose a ray that leaves the first lens parallel to the axis (the bottom ray in the figure) and one that goes through the center of the second lens (the top ray in the figure). We see that the final image is between the second lens and its

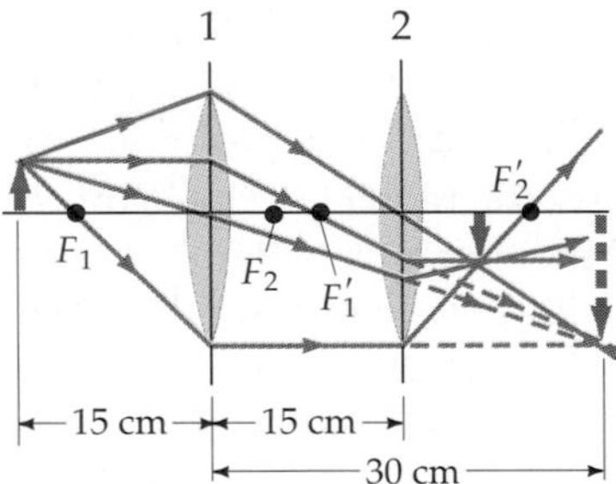

Figure 31-37 Ray diagram for Example 31-9. The image of the first lens is to the right of the second lens. This image is not formed because the rays are refracted by the second lens before they get to the first image. Nevertheless, this unformed image acts as a virtual object for the second lens. The final image is found by drawing rays toward the first image as shown. A ray through the center of the second lens and a ray parallel to the axis as it strikes the second lens are used.

focal point. (In the figure, we have extended the middle two rays used to locate the first, unformed image so that they pass through the final image.) The final image can be located algebraically by using the first image as the object for the second lens. Since this unformed image is on the transmission side of the second lens, it is a virtual object. Since it is 15 cm from the second lens, the object distance is $s_2 = -15$ cm. Then

$$\frac{1}{-15\text{ cm}} + \frac{1}{s_2'} = \frac{1}{f_2} = \frac{1}{10\text{ cm}}$$

Solving for s_2', we obtain

$$s_2' = 6\text{ cm}$$

Example 31-10

Two thin lenses of focal lengths f_1 and f_2 are placed together. Show that the equivalent focal length of the combination f is given by

$$\frac{1}{f} = \frac{1}{f_1} + \frac{1}{f_2} \qquad \text{31-20}$$

Let s be the object distance for the first lens (and, therefore, for the lens combination) and s'_1 be the image distance. Applying the thin-lens equation to the first lens, we have

$$\frac{1}{s} + \frac{1}{s'_1} = \frac{1}{f_1}$$

Since the lenses are together, the object distance for the second lens is the negative of the image distance for the first lens, so $s_2 = -s'_1$. Calling the final image distance s', we have for the second lens

$$\frac{1}{-s'_1} + \frac{1}{s'} = \frac{1}{f_2}$$

Adding these two equations to eliminate s'_1, we obtain

$$\frac{1}{s} + \frac{1}{s'} = \frac{1}{f_1} + \frac{1}{f_2} = \frac{1}{f}$$

The weight and bulk of a large diameter lens can be reduced by constructing the lens from annular segments at different angles such that light from a point is refracted by the segments into a parallel beam. Such an arrangement is called a Fresnel lens. Several Fresnel lenses are used in this lighthouse to produce intense parallel beams of light from a source at the focal point of the lenses.

Example 31-10 gives us the important result that when two lens are placed in contact (or very close together), the reciprocals of their focal lengths add. The reciprocal of the focal length is called the **power of a lens.** When the focal length is expressed in meters, the power is given in reciprocal meters called **diopters** (D):

$$P = \frac{1}{f} \quad \text{diopters} \qquad 31\text{-}21$$

The power of a lens measures its ability to focus parallel light at a short distance from the lens. The shorter the focal length, the greater the power. For example, a lens with a focal length of 25 cm = 0.25 m has a power of 4.0 diopters. A lens with a focal length of 10 cm = 0.10 m has a power of 10 diopters. Since the focal length of a diverging lens is negative, its power is also negative.

Example 31-11

A lens has a power of −2.5 diopters. What is its focal length?

Solving Equation 31-21 for the focal length, we obtain

$$f = \frac{1}{P} = \frac{1}{-2.5\text{ D}} = -0.40\text{ m} = -40\text{ cm}$$

where we have used the fact that a diopter is the same as a reciprocal meter, that is, $1\text{ D} = 1\text{ m}^{-1}$.

The results of Example 31-10 can be simply stated in terms of the power of a lens. When two lenses are in contact, the power of the combination equals the sum of the powers of the lenses:

$$P = P_1 + P_2 \quad \text{two lenses in contact} \qquad 31\text{-}22$$

Questions

8. Under what conditions will the focal length of a thin lens be positive? Negative?
9. The focal length of a simple lens is different for different colors of light. Why?

31-5 Aberrations

When all the rays from a point object are not focused at a single image point, the resulting blurring of the image is called **aberration.** Figure 31-38 shows rays from a point source on the axis traversing a thin lens with spherical surfaces. Rays that strike the lens far from the axis are bent much more than those near the axis, with the result that not all the rays are focused at a single point. Instead, the image appears as a circular disk. The **circle of least confusion** is at point C, where the diameter is minimum. This type of aberration is called **spherical aberration.** It is the same as the spherical aberration of mirrors discussed in Section 31-2. Similar but more complicated aberrations called *coma* (for the comet-shaped image) and *astigmatism* occur when objects are off axis. The aberration in the shape of the image of an extended object due to the fact that the magnification depends on the distance of the object point from the axis is called **distortion.** We will not discuss these aberrations further except to point out that they do not arise from any defect in the lens or mirror but instead result from the application of the laws of refraction and reflection to spherical surfaces. They are not evident in our simple equations because we used small-angle approximations in the derivation of these equations.

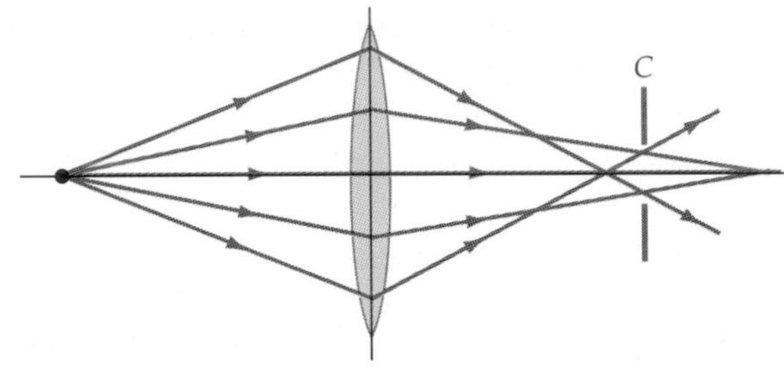

Figure 31-38 Spherical aberration. Rays from a point object on the axis are not focused at a point. The image is thus a circle about the axis rather than a point. At point C, the circle has its least diameter. This is called the circle of least confusion. Spherical aberration can be reduced by blocking off the outer parts of the lens and thereby reducing the diameter of the circle of least confusion, but this also reduces the amount of light reaching the image.

Some aberrations can be eliminated or partially corrected by using nonspherical surfaces for mirrors or lenses, but nonspherical surfaces are usually much more difficult and costly to produce than spherical surfaces. One example of a nonspherical reflecting surface is the parabolic mirror illustrated in Figure 31-39. Rays that are parallel to the axis of a parabolic surface are reflected and focused at a common point no matter how far they are from the axis. Parabolic reflecting surfaces are commonly used in large astronomical telescopes, which need a large reflecting surface to gather as much light as possible to make the image as intense as possible. A parabolic surface can also be used in a searchlight to produce a parallel beam of light from a small source placed at the focal point of the surface.

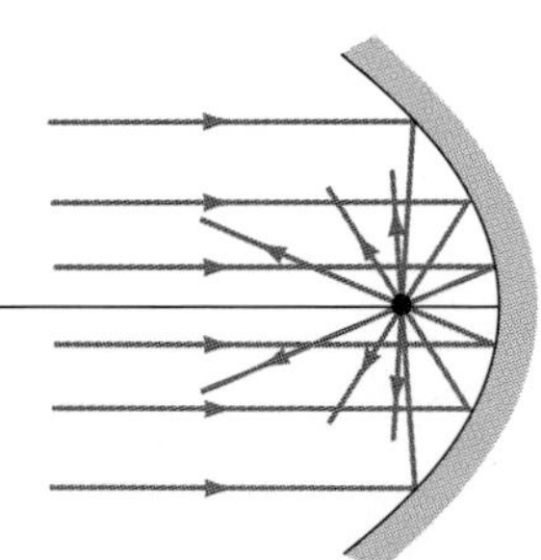

Figure 31-39 A parabolic mirror focuses all rays parallel to the axis to a single point with no spherical aberration. Parabolic surfaces are more costly to produce than spherical surfaces.

An important aberration found with lenses but not with mirrors is **chromatic aberration,** which is due to variations in the index of refraction with wavelength. From Equation 31-18, we can see that the focal length of a lens depends on its index of refraction and is therefore different for different wavelengths. Since n is slightly greater for blue light than for red light (see Figure 30-22), the focal length for blue light will be shorter than that for red light. Since chromatic aberration does not occur for mirrors, many large telescopes use mirrors rather than lenses.

Chromatic and other aberrations can be partially corrected by using combinations of lenses instead of a single lens. For example, a positive lens and a negative lens of greater focal length can be used together to produce a converging lens system that has much less chromatic aberration than a single lens of the same focal length. The lens of a good camera typically contains six elements to correct for the various aberrations that are present.

Summary

1. The image formed by a spherical mirror or by a lens is at a distance s', which is related to the object distance s by

$$\frac{1}{s}+\frac{1}{s'}=\frac{1}{f}$$

where f is the focal length, which is the image distance when $s = \infty$. For a mirror, the focal length is equal to half the radius of curvature. For a thin lens in the air, the focal length is related to the index of refraction n and radii of curvature of the two sides r_1 and r_2 by

$$\frac{1}{f}=(n-1)\left(\frac{1}{r_1}-\frac{1}{r_2}\right)$$

In these equations s, s', r, r_1, and r_2 are taken to be positive when the object, image, or center of curvature lies on the real side of the element. For mirrors, the real side is the incident side. For lenses, the real side is the incident side for objects and the transmission side for images and centers of curvature. When s' is positive, the image is real, meaning that light rays actually diverge from the image point. Real images can be seen on a ground-glass viewing screen or photographic film placed at the image point. When s' is negative, the image is virtual, meaning that no light actually diverges from the image point.

2. The lateral magnification of the image is given by

$$m=\frac{y'}{y}=\frac{-s'}{s}$$

where y is the object size and y' is the image size. A negative magnification means that the image is inverted.

3. For a plane mirror, r and f are infinite, $s' = -s$, and the image is virtual, erect, and the same size as the object.

4. Images can be conveniently located by a ray diagram using any two principal rays. The point from which these rays diverge or appear to diverge is the image point. For spherical mirrors, there are four principal rays: the parallel ray, parallel to the axis; the focal ray, through the focal point; the radial ray, through the center of curvature of the mirror; and the central ray, toward the vertex of the mirror. For a lens, there are three principal rays: the parallel ray, parallel to the axis; the focal ray, through the second focal point; and the central ray, through the center of the lens.

5. A positive or converging lens is one that is thicker at the middle than at the edges. Parallel light incident on a positive lens is focused at the second focal point, which is on the transmission side of the lens. A negative or diverging lens is one that is thicker at the edges than at the

middle. Parallel light incident on a negative lens emerges as if it originated from the second focal point, which is on the incident side of the lens.

6. The power of a lens equals the reciprocal of the focal length. When the focal length is in meters, the power is in diopters (D):

$$P = \frac{1}{f} \qquad \text{diopters}$$

$$1\ \text{D} = 1\ \text{m}^{-1}$$

7. The image distance s' for refraction at a single spherical surface of radius r is related to the object distance s and the radius of curvature of the surface r by

$$\frac{n_1}{s} + \frac{n_2}{s'} = \frac{n_2 - n_1}{r}$$

where n_1 is the index of refraction of the medium on the incident side of the surface and n_2 is the index of refraction of the medium on the transmission side. The magnification of the image due to refraction at a single surface is

$$m = -\frac{n_1 s'}{n_2 s}$$

8. The blurring of the image of a single object point is known as aberration. Spherical aberration results from the fact that a spherical surface focuses only paraxial rays (those that travel close to the axis) at a single point. Nonparaxial rays are focused at nearby points depending on the angle made with the axis. Spherical aberration can be reduced by reducing the size of the spherical surface, which also reduces the amount of light reaching the image. Chromatic aberration, which occurs with lenses but not mirrors, results from the variation in the index of refraction with wavelength. Lens aberrations are most commonly reduced by using a series of lens elements.

Review

A. Objectives: After studying this chapter, you should:

1. Be able to draw simple ray diagrams for mirrors and lenses to locate images and determine whether they are real or virtual, erect or inverted, and enlarged or reduced.

2. Be able to determine algebraically the location of the image formed by a mirror or by a thin lens and calculate the magnification of the image.

3. Be able to use the lens-maker's equation to determine the focal length of a lens from the radii of curvature of the surfaces.

4. Be able to discuss spherical aberration and chromatic aberration.

B. Define, explain, or otherwise identify:

Geometrical optics
Image
Virtual image
Depth inversion
Real image
Paraxial ray
Spherical aberration
Focal length
Focal point
Reversibility
Ray diagram
Principal ray
Lateral magnification
Apparent depth
Virtual object
Lens-maker's equation
Converging lens
Positive lens
Diverging lens
Negative lens
Focal plane
Virtual object
Power of a lens
Diopter
Chromatic aberration

C. True or false: If the statement is true, explain why it is true. If it false, give a counterexample.

1. A virtual image cannot be displayed on a screen.

2. Aberrations occur only for real images.

3. A negative image distance implies that the image is virtual.

4. All rays parallel to the axis of a spherical mirror are reflected through a single point.

5. A diverging lens cannot form a real image from a real object.

6. The image distance for a positive lens is always positive.

7. Chromatic aberration does not occur with mirrors.

Suggestions for Further Reading

Walker, Jearl: "The Amateur Scientist: The Kaleidoscope Now Comes Equipped with Flashing Diodes and Focusing Lenses," *Scientific American*, December 1985, p. 134.

Multiple reflections created by arrangements of different numbers of plane mirrors are investigated.

Walker, Jearl: "The Amateur Scientist: What Is a Fish's View of a Fisherman and the Fly He Has Cast on the Water?" *Scientific American*, March 1984, p. 138.

Refraction causes the world above water to appear to a fish the way things appear to us when viewed through a "fish-eye" lens.

Walker, Jearl: "The Amateur Scientist: Shadows Cast on the Bottom of a Pool Are Not Like Other Shadows. Why?" *Scientific American*, July 1988, p. 116.

This article describes simple experiments in refraction from curved surfaces to try the next time you take a bath!

Problems

Use $n = 1.33$ for the index of refraction of water unless otherwise specified.

Level I

31-1 Plane Mirrors

1. The image of the point object P in Figure 31-40 is viewed by an eye as shown. Draw a bundle of rays from the object that reflect from the mirror and enter the eye. For this object position and mirror, indicate the region of space in which the eye can see the image.

Figure 31-40 Problem 1.

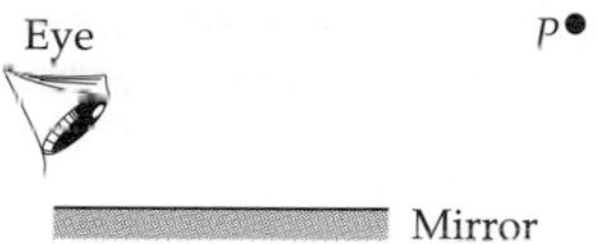

2. When two plane mirrors are parallel, such as on opposite walls in a barber shop, multiple images arise because each image in one mirror serves as an object for the other mirror. A point object is placed between parallel mirrors separated by 30 cm. The object is 10 cm in front of the left mirror and 20 cm in front of the right mirror. (*a*) Find the distance from the left mirror to the first four images in that mirror. (*b*) Find the distance from the right mirror to the first four images in that mirror.

3. A person 1.62 m tall wants to be able to see her full image in a plane mirror. (*a*) What must be the minimum height of the mirror? (*b*) How far above the floor should it be placed, assuming that the top of the person's head is 15 cm above her eye level? Draw a ray diagram.

4. Two plane mirrors make an angle of 90°. Show by considering various object positions that there are three images for any position of an object. Draw appropriate bundles of rays from the object to the eye for viewing each image.

5. (*a*) Two plane mirrors make an angle of 60° with each other. Show on a sketch the location of all the images formed of a point object on the bisector of the angle between the mirrors. (*b*) Repeat for an angle of 120°.

31-2 Spherical Mirrors

6. A concave spherical mirror has a radius of curvature of 40 cm. Draw ray diagrams to locate the image (if one is formed) for an object at a distance of (*a*) 100 cm, (*b*) 40 cm, (*c*) 20 cm, and (*d*) 10 cm from the mirror. For each case, state whether the image is real or virtual; erect or inverted; and enlarged, reduced, or the same size as the object.

7. Use the mirror equation to locate and describe the images for the object distances and mirror of Problem 6.

8. Repeat Problem 6 for a convex mirror with the same radius of curvature.

9. Repeat Problem 7 for the convex mirror in Problem 8.

10. Show that a convex mirror cannot form a real image of a real object, no matter where the object is placed, by showing that s' is always negative for a positive s.

11. Convex mirrors are used in stores to provide a wide angle of surveillance for a reasonable mirror size. The mirror shown in Figure 31-41 allows a clerk 5 m away from the mirror to survey the entire store. It has a radius of curvature of 1.2 m. (*a*) If a customer is 10 m from the mirror, how far from the mirror surface is his image? (*b*) Is the image in front of or behind the mirror? (*c*) If the customer is 2 m tall, how high is his image?

Figure 31-41 Problem 11.

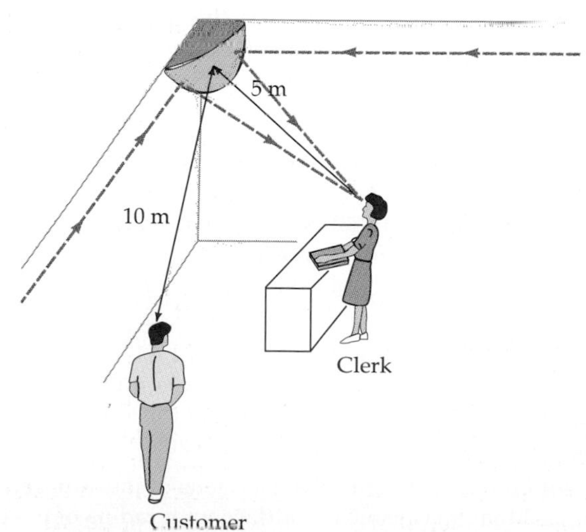

12. A certain telescope uses a concave spherical mirror of radius 8 m. Find the location and diameter of the image of the moon formed by this mirror. The moon has a diameter of 3.5×10^6 m and is 3.8×10^8 m from the earth.

13. A dentist wants a small mirror that will produce an upright image with a magnification of 5.5 when the mirror is located 2.1 cm from a tooth. (*a*) What should the radius of the mirror be? (*b*) Should it be concave or convex?

31-3 Images Formed by Refraction

14. A sheet of paper with writing on it is protected by a thick glass plate having an index of refraction of 1.5. If the plate is 2 cm thick, at what distance beneath the top of the plate does the writing appear when it is viewed from directly overhead?

15. A very long glass rod has one end ground to a convex hemispherical surface of radius 5 cm. Its index of refraction is 1.5. (*a*) A point object in air is on the axis of the rod 20 cm from the surface. Find the image and state whether it is real or virtual. Repeat for (*b*) an object 5 cm from the surface and (*c*) an object very far from the surface. Draw a ray diagram for each case.

16. At what distance from the rod of Problem 15 should the object be placed so that the light rays in the rod are parallel? Draw a ray diagram for this situation.

17. Repeat Problem 15 for a glass rod with a concave hemispherical surface of radius −5 cm.

18. Repeat Problem 15 when the glass rod and objects are immersed in water.

19. Repeat Problem 15 for a glass rod with a concave hemispherical surface of radius −5 cm when it and the objects are immersed in water.

20. A fish is 10 cm from the front surface of a fish bowl of radius 20 cm. (*a*) Where does the fish appear to be to someone in air viewing it from in front of the bowl? (*b*) Where does the fish appear to be when it is 30 cm from the front surface of the bowl?

21. A scuba diver wears a diving mask with a face plate that bulges outward with a radius of curvature of 0.5 m. There is thus a convex spherical surface between the water and the air in the mask. A fish is 2.5 m in front of the diving mask. (*a*) Where does the fish appear to be? (*b*) What is the magnification of the image of the fish?

31-4 Thin Lenses

22. The following thin lenses are made of glass with an index of refraction of 1.5. Make a sketch of each lens, and find its focal length in air: (*a*) double convex, $r_1 = 10$ cm and $r_2 = -21$ cm; (*b*) plano-convex, $r_1 = \infty$ and $r_2 = -10$ cm; (*c*) double concave, $r_1 = -10$ cm and $r_2 = +10$ cm; (*d*) plano-concave, $r_1 = \infty$ and $r_2 = +20$ cm.

23. Glass with an index of refraction of 1.6 is used to make a thin lens that has radii of equal magnitude. Find the radii of curvature and make a sketch of the lens if the focal length in air is (*a*) +5 cm and (*b*) −5 cm.

24. Find the focal length of a glass lens of index of refraction 1.62 that has a concave surface with radius of magnitude 100 cm and a convex surface with a radius of magnitude 40 cm.

25. A double-concave lens of index of refraction 1.45 has radii of magnitude 30 cm and 25 cm. An object is located 80 cm to the left of the lens. Find (*a*) the focal length of the lens, (*b*) the location of the image, and (*c*) the magnification of the image. (*d*) Is the image real or virtual? Upright or inverted?

26. A converging lens made of polystyrene (index of refraction, 1.59) has a focal length of 50 cm. One surface is convex with radius of magnitude 50 cm. Find the radius of the second surface. Is it convex or concave?

27. The following thin lenses are made of glass of index of refraction 1.6. Make a sketch of each lens, and find its focal length in air: (*a*) $r_1 = 20$ cm, $r_2 = 10$ cm; (*b*) $r_1 = 10$ cm, $r_2 = 20$ cm; (*c*) $r_1 = -10$ cm, $r_2 = -20$ cm.

28. Show that a diverging lens can never form a real image from a real object. (*Hint:* Show that s' is always negative.)

29. For the following object distances and focal lengths of thin lenses in air, find the image distance and the magnification and state whether the image is real or virtual and erect or inverted: (*a*) $s = 40$ cm, $f = 20$ cm; (*b*) $s = 10$ cm, $f = 20$ cm; (*c*) $s = 40$ cm, $f = -30$ cm; (*d*) $s = 10$ cm, $f = -30$ cm.

30. An object 3.0 cm high is placed 20 cm in front of a thin lens of power 20 diopters. Draw a precise ray diagram to find the position and size of the image and check your results using the thin-lens equation.

31. Repeat Problem 30 for an object 1.0 cm high placed 10 cm in front of a thin lens of power 20 diopters.

32. Repeat Problem 30 for an object 1.0 cm high placed 10 cm in front of a thin lens whose power is −20 diopters.

33. A thin converging lens of focal length 10 cm is used to obtain an image that is twice as large as a small object. Find the object and image distances if (*a*) the image is to be erect and (*b*) the image is to be inverted. Draw a ray diagram for each case.

34. (*a*) What is meant by a negative object distance? How can it occur? Find the image distance and magnification and state whether the image is virtual or real and erect or inverted for a thin lens in air when (*b*) $s = -20$ cm, $f = +20$ cm and (*c*) $s = -10$ cm, $f = -30$ cm. Draw a ray diagram for each of these cases.

35. Two converging lenses, each of focal length 10 cm, are separated by 35 cm. An object is 20 cm to the left of the first lens. (*a*) Find the position of the final image using both a ray diagram and the thin-lens equation. (*b*) Is the image real or virtual? Erect or inverted? (*c*) What is the overall lateral magnification of the image?

36. Work Problem 35 for a second lens that is a diverging lens of focal length −15 cm.

31-5 Aberrations

37. A double-convex lens of radii $r_1 = +10$ cm and $r_2 = -10$ cm is made from glass with indexes of refraction of 1.53 for blue light and 1.47 for red light. Find the focal length of this lens for (*a*) red light and (*b*) blue light.

Level II

38. A concave spherical mirror has a radius of curvature of 6.0 cm. A point object is on the axis 9 cm from the mirror. Construct a precise ray diagram showing rays from the object that make angles of 5°, 10°, 30°, and 60° with the axis, strike the mirror, and are reflected back across the axis. (Use a compass to draw the mirror, and use a protractor to measure the angles needed to find the reflected rays.) What is the spread along the axis of the image for these rays?

39. An object placed 8 cm from a concave spherical mirror produces a virtual image 10 cm behind the mirror. (*a*) If the object is moved back to 25 cm from the mirror, where is the image located? (*b*) Is it real or virtual?

40. A concave mirror has a radius of curvature 6.0 cm. Draw rays parallel to the axis at 0.5, 1.0, 2.0, and 4.0 cm above the axis and find the points at which the reflected rays cross the axis. (Use a compass to draw the mirror and a protractor to find the angle of reflection for each ray.) (*a*) What is the spread Δx of the points where these rays cross the axis? (*b*) By what percentage could this spread be reduced if the edge of the mirror were blocked off so that parallel rays more than 2.0 cm from the axis could not strike the mirror?

41. An object located 100 cm from a concave mirror forms a real image 75 cm from the mirror. The mirror is then turned around so that its convex side faces the object. The mirror is moved so that the image is now 75 cm behind the mirror. How far should the mirror be moved? Should it be moved toward or away from the object?

42. (*a*) Show that if f is the focal length of a thin lens in air, its focal length in water is f' given by

$$f' = \frac{n_w(n-1)}{n_w - n}$$

where n_w is the index of refraction of water and n is that of the lens. (*b*) Calculate the focal length in air and in water of a double-concave lens of index of refraction $n = 1.5$ that has radii of magnitude 30 cm and 35 cm.

43. You wish to see an image of your face for applying makeup or shaving. If you want the image to be upright, virtual, and magnified 1.5 times when your face is 30 cm from the mirror, what kind of mirror should you use, convex or concave, and what should its focal length be?

44. A glass rod 96 cm long with an index of refraction of 1.6 has its ends ground to convex spherical surfaces of radii 8 cm and 16 cm. A point object is in air on the axis of the rod 20 cm from the end with the 8-cm radius. (*a*) Find the image distance due to refraction at the first surface. (*b*) Find the final image due to refraction at both surfaces. (*c*) Is the final image real or virtual?

45. Repeat Problem 44 for a point object in air on the axis of the rod 20 cm from the end with the 16-cm radius.

46. (*a*) Find the focal length of a *thick* double-convex lens with an index of refraction of 1.5, a thickness of 4 cm, and radii of +20 cm and −20 cm. (*b*) Find the focal length of this lens in water.

47. A thin lens of index of refraction 1.5 has one convex side with a radius of magnitude 20 cm. When an object 1-cm in height is placed 50 cm from this lens, an upright image 2.15 cm in height is formed. (*a*) Calculate the radius of the second side of the lens. Is it concave or convex? (*b*) Draw a sketch of the lens.

48. (*a*) Show that to obtain a magnification of magnitude m with a converging thin lens of focal length f, the object distance must be given by

$$s = \frac{m+1}{m} f$$

(*b*) A camera lens with 50-mm focal length is used to take a picture of a person 1.75 m tall. How far from the camera should the person stand so that the image size is 24 mm?

49. A 2-cm thick layer of water ($n = 1.33$) floats on top of a 4-cm thick layer of carbon tetrachloride ($n = 1.46$) in a tank. How far below the top surface of the water does the bottom of the tank appear to be to an observer looking from above at normal incidence?

50. While sitting in your car, you see a jogger in your side mirror, which is convex with a radius of curvature of magnitude 2 m. The jogger is 5 m from the mirror and is approaching at 3.5 m/s. How fast does the jogger appear to be running when viewed in the mirror?

51. Parallel light from a distant object strikes the large mirror in Figure 31-42 ($r = 5$ m) and is reflected by the small mirror that is 2 m from the large mirror and is actually spherical, not planar as shown. The light is focused at the vertex of the large mirror. (*a*) What is the radius of curvature of the small mirror? (*b*) Is it convex or concave?

Figure 31-42 Problem 51.

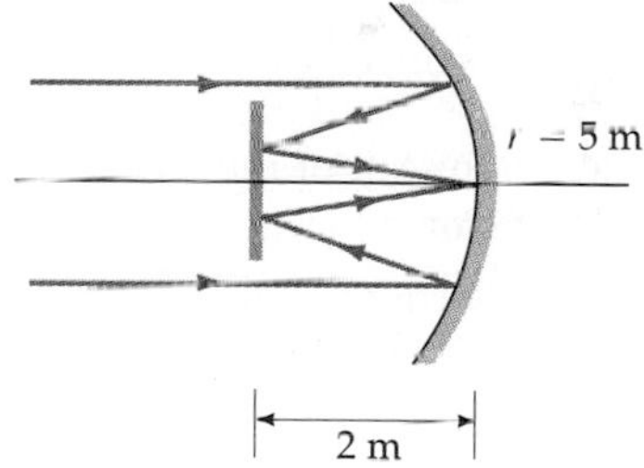

52. A small object is 20 cm from a thin positive lens of focal length 10 cm. To the right of the lens is a plane mirror that crosses the axis at the second focal point of the lens and is tilted so that the reflected rays do not go back through the lens (see Figure 31-43). (*a*) Find the position of the final image. (*b*) Is this image real or virtual? (*c*) Sketch a ray diagram showing the final image.

Figure 31-43 Problem 52.

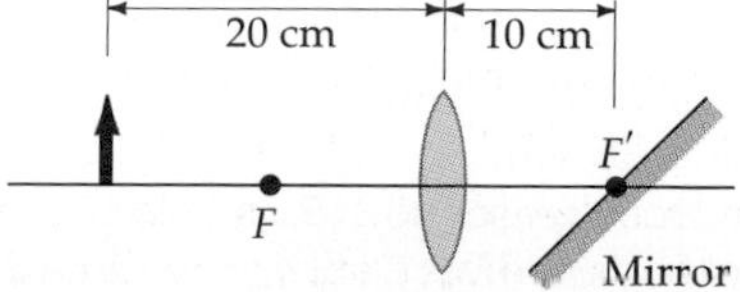

53. An object is placed 12 cm to the left of a lens of focal length 10 cm. A second lens of focal length 12.5 cm is placed 20 cm to the right of the first lens. (*a*) Find the position of the final image. (*b*) What is the magnification of the image? (*c*) Sketch a ray diagram showing the final image.

54. An object is 15 cm in front of a positive lens of focal length 15 cm. A second positive lens of focal length 15 cm is 20 cm from the first lens. Find the final image and draw a ray diagram.

55. Work Problem 54 for a second lens with a focal length of −15 cm.

56. In the seventeenth century, Antonie van Leeuwenhoek, the first great microscopist, used simple spherical lenses made first of water droplets and then of glass for his first instruments. He made staggering discoveries with these simple lenses. Consider a glass sphere of radius 2.0 mm with an index of refraction of 1.50. Find the focal length of this lens. *Hint:* Use the equation for refraction at a single spherical surface to find the image distance for an infinite object distance for the first surface. Then use this image point as the object point for the second surface.

Level III

57. An object is 15 cm to the left of a thin convex lens of focal length 10 cm. A concave mirror of radius 10 cm is 25 cm to the right of the lens. (*a*) Find the position of the final image formed by the mirror and lens. (*b*) Is the image real or virtual? Erect or inverted? (*c*) Show on a diagram where your eye must be to see this image.

58. Find the final image for the situation in Problem 52 when the mirror is not tilted. Assume that the image is viewed by an eye to the left of the object looking through the lens into the mirror.

59. A women uses a concave makeup mirror of radius 1.5 m. How far from the mirror should her face be for the image to be 80 cm from her face?

60. When a bright light source is placed 30 cm in front of a lens, there is an erect image 7.5 cm from the lens. There is also a faint inverted image 6 cm in front of the lens due to reflection from the front surface of the lens. When the lens is turned around, this weaker, inverted image is 10 cm in front of the lens. Find the index of refraction of the lens.

61. A horizontal concave mirror with radius of curvature of 50 cm holds a layer of water with an index of refraction of 1.33 and a maximum depth of 1 cm. At what height above the mirror must an object be placed so that its image is at the same position as the object?

62. A lens with one concave side with a radius of magnitude 17 cm and one convex side with a radius of magnitude 8 cm has a focal length in air of 27.5 cm. When placed in a liquid with an unknown index of refraction, the focal length increases to 109 cm. What is the index of refraction of the liquid?

Figure 31-44 Problem 63.

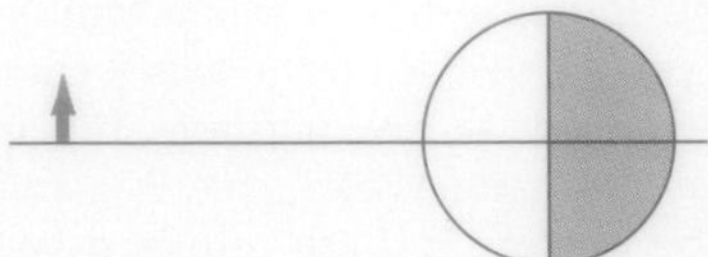

63. A glass ball of radius 10 cm has an index of refraction of 1.5. The back half of the ball is silvered so that it acts as a concave mirror (Figure 31-44). Find the position of the final image seen by an eye to the left of the object and ball for an object at (*a*) 30 cm and (*b*) 20 cm to the left of the front surface of the ball.

64. In a convenient form of the thin-lens equation used by Newton, the object and image distances are measured from the focal points. Show that if $x = s - f$ and $x' = s' - f$, the thin-lens equation can be written as $xx' = f^2$, and the lateral magnification is given by $m = -x'/f = -f/x$. Indicate x and x' on a sketch of a lens.

65. An object is placed 2.4 m from a screen, and a lens of focal length f is placed between the object and the screen so that a real image of the object is formed on the screen. When the lens is moved 1.2 m toward the screen, another real image of the object is formed on the screen. (*a*) Where was the lens located before it was moved? (*b*) What is the focal length of the lens?

66. An object is 17.5 cm to the left of a lens of focal length 8.5 cm. A second lens of focal length −30 cm is 5 cm to the right of the first lens. (*a*) Find the distance between the object and the final image formed by the second lens. (*b*) What is the overall magnification? (*c*) Is the final image real or virtual? Upright or inverted?

67. (*a*) Show that a small change dn in the index of refraction of a lens material produces a small change in the focal length df given approximately by

$$\frac{df}{f} = \frac{-dn}{n-1}$$

(*b*) Use this result to find the focal length of a thin lens for blue light, for which $n = 1.53$, if the focal length for red light, for which $n = 1.47$, is 20 cm.

68. The lateral magnification of a spherical mirror or a thin lens is given by $m = -s'/s$. Show that for objects of small horizontal extent, the longitudinal magnification is approximately $-m^2$. *Hint:* Show that $ds'/ds = s'^2/s^2$.

69. A thin double-convex lens has radii r_1 and r_2 and an index of refraction n_L. The surface of radius r_1 is in contact with a liquid of index of refraction n_1, and the surface of radius r_2 is in contact with a liquid of index of refraction n_2. Show that the thin-lens equation for this situation can be expressed as

$$\frac{n_1}{s} + \frac{n_2}{s'} = \frac{n_2}{f}$$

where the focal length is given by

$$\frac{1}{f} = \frac{n_L - n_1}{n_1 r_1} - \frac{n_L - n_2}{n_2 r_2}$$

Chapter 32

Optical Instruments

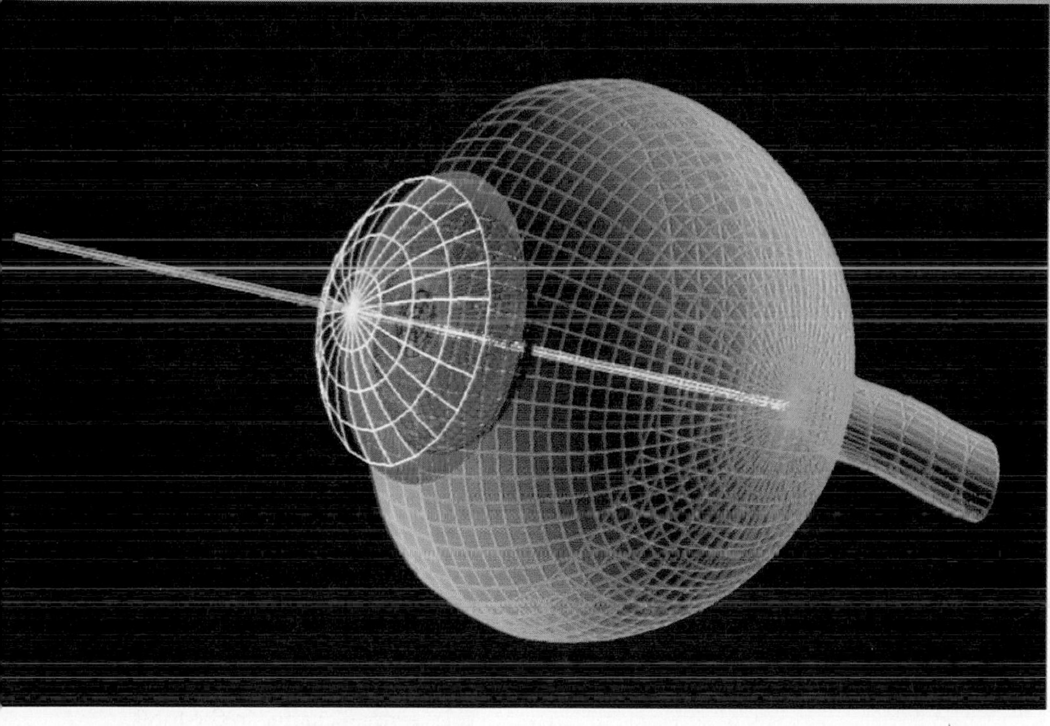

A computer-modeled image of the human eye.

In this chapter, we will use what we have learned about mirrors and lenses to examine the workings of various optical instruments, such as the camera, the simple magnifier, the microscope, and the telescope. The most important optical instrument is the eye, which we will study first. Many optical instruments used today are quite complicated. The basic principles behind their workings are often quite simple, but imaginative applications of these principles have revolutionized our capacity to see and understand the world around us.

32-1 The Eye

The optical system of prime importance is the eye, shown in Figure 32-1. Light enters the eye through a variable aperture, the *pupil*, and is focused by the *cornea–lens system* on the *retina*, a film of nerve fibers covering the back surface. The retina contains tiny light-sensing structures called *rods* and *cones* that receive the image and transmit the information along the optic nerve to the brain. The shape of the crystalline lens can be altered slightly by the action of the *ciliary muscle*. When the eye is focused on an object far away, the muscle is relaxed and the cornea–lens system has its maximum

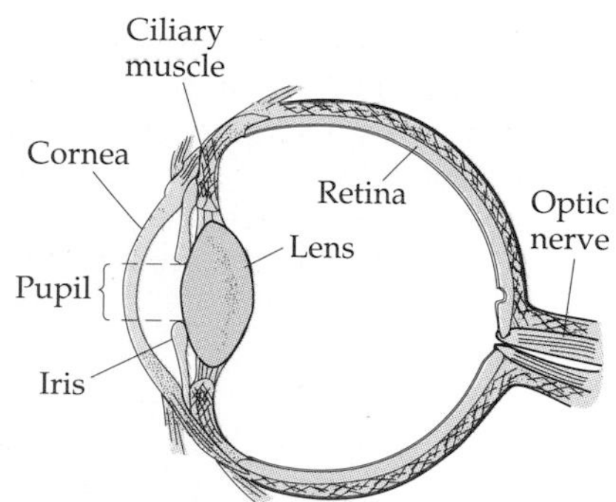

Figure 32-1 Cutaway view of the human eye. The amount of light entering the eye is controlled by the iris, which regulates the size of the pupil. The lens thickness is controlled by the ciliary muscle.

focal length, about 2.5 cm, the distance from the cornea to the retina. When the object is brought closer to the eye, the ciliary muscle increases the curvature of the lens slightly, thereby decreasing its focal length so that the image is again focused on the retina. This process is called **accommodation.** If the object is too close to the eye, the lens cannot focus the light on the retina and the image is blurred. The closest point for which the lens can focus the image on the retina is called the **near point.** The distance from the eye to the near point varies greatly from one person to another and changes with age. At the age of 10 years, the near point may be as close as 7 cm, whereas by 60 years it may have receded to 200 cm because of the loss of flexibility of the lens. The standard value taken for the near point is 25 cm.

If the eye underconverges, resulting in the images being focused behind the retina, the person is said to be **farsighted.** A farsighted person can see distant objects, for which little convergence is required, but has trouble seeing close objects clearly. Farsightedness is corrected with a converging (positive) lens (Figure 32-2).

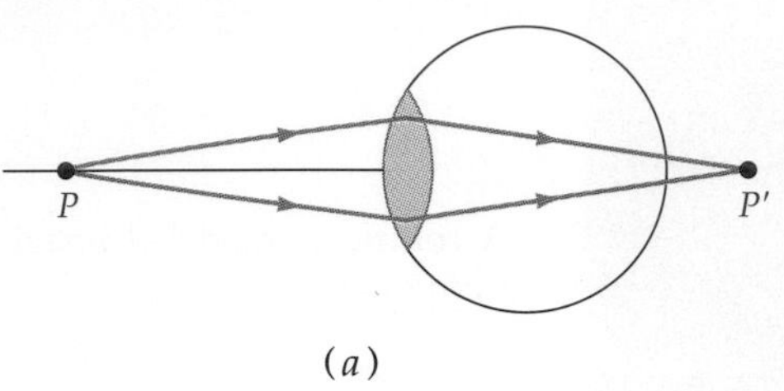

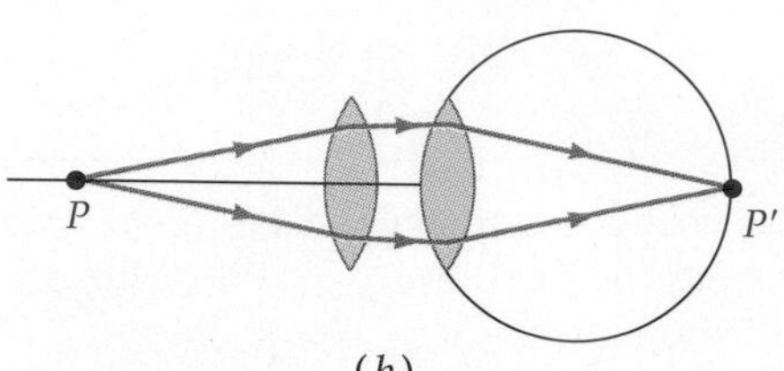

Figure 32-2 (*a*) A farsighted eye focuses rays from a nearby object P to a point P' behind the retina. (*b*) A converging lens corrects this defect by bringing the image onto the retina. These diagrams and those following are drawn as if all the focusing of the eye is done at the lens; in fact, the cornea–lens system acts more like a spherical refracting surface than a thin lens.

On the other hand, the eye of a **nearsighted** person overconverges and focuses light from distant objects in front of the retina. A nearsighted person can see nearby objects, for which the widely diverging incident rays can be focused on the retina, but has trouble focusing on distant objects. Nearsightedness is corrected with a diverging (negative) lens (Figure 32-3).

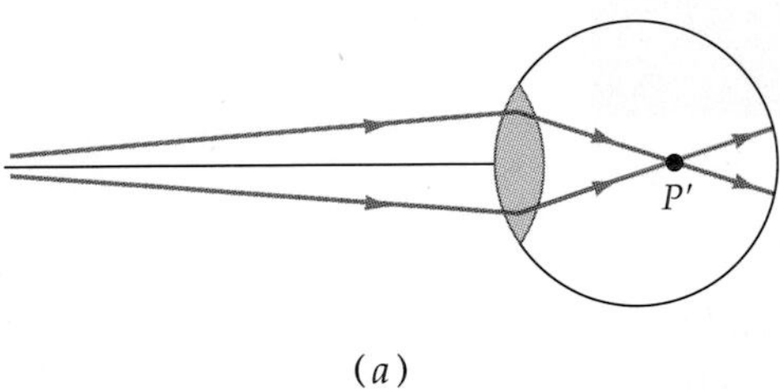

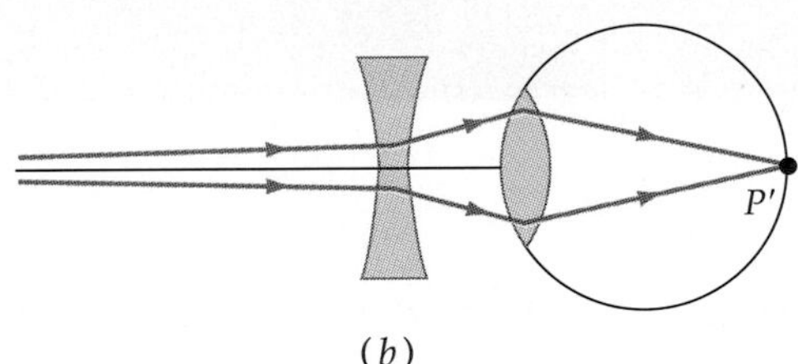

Figure 32-3 (*a*) A nearsighted eye focuses rays from a distant object to a point P' in front of the retina. (*b*) A diverging lens corrects this defect.

Another common defect of vision is *astigmatism,* which is caused by the cornea being not quite spherical but having a different curvature in one plane than in another. This results in the blurring of the image of a point object into a short line. Astigmatism is corrected by glasses with lenses of cylindrical rather than spherical shape.

Example 32-1

By how much must the focal length of the cornea–lens system of the eye change when an object is moved from infinity to the near point at 25 cm? Assume that the distance from the cornea to the retina is 2.5 cm.

When the object is at infinity, the rays from the object are parallel and are focused by the eye on the retina, giving a focal length for the cornea–lens system of 2.5 cm. When the object is at 25 cm, the focal length f must be such that the image distance is 2.5 cm. Using $s = 25$ cm for the object distance and $s' = 2.5$ cm for the image distance in the thin-lens equation (Equation 31-19), we have

$$\frac{1}{25\text{ cm}} + \frac{1}{2.5\text{ cm}} = \frac{1}{f}$$

$$\frac{1}{f} = \frac{1}{25\text{ cm}} + \frac{10}{25\text{ cm}} = \frac{11}{25\text{ cm}}$$

$$f = \frac{25\text{ cm}}{11} = 2.27\text{ cm}$$

The focal length must therefore decrease by 0.23 cm. In terms of the power of the cornea–lens system, when the focal length is 2.5 cm = 0.025 m for distant objects, the power is $P = 1/f = 40$ diopters. When the focal length is 2.27 cm, the power is 44 diopters.

Exercise

Find the change in the focal length of the eye when an object originally at 4 m is brought to 40 cm from the eye. (Assume the distance from the cornea to the retina is 2.5 cm.) (Answer: 0.13 cm)

The apparent size of an object is determined by the size of the image on the retina. The larger the image on the retina, the greater the number of rods and cones activated. From Figures 32-4*a* and *b*, we can see that the size of the image on the retina is greater when the object is close than it is when the object is far away. Thus, even though the actual size of the object does not change, its apparent size is greater when it is brought closer to the eye. A convenient measure of the size of the image on the retina is the angle θ subtended by the object at the eye as shown in Figure 32-4. From Figure 32-4*c*, we can see that the angle θ is related to the image size y' by

$$\theta = \frac{y'}{2.5\text{ cm}} \qquad 32\text{-}1$$

The image size is therefore directly proportional to the angle subtended by the object. From Figure 32-4*a* or *b*, we can see that the angle θ is related to the object size y and object distance s by

$$\tan\theta = \frac{y}{s}$$

For small angles, we can use the approximation $\tan\theta \approx \theta$ and write

$$\theta \approx \frac{y}{s} \qquad 32\text{-}2$$

Combining Equations 32-1 and 32-2, we obtain

$$y' = (2.5\text{ cm})\theta \approx (2.5\text{ cm})\frac{y}{s} \qquad 32\text{-}3$$

Thus, the size of the image on the retina is proportional to the size of the object and inversely proportional to the distance between the object and the eye.

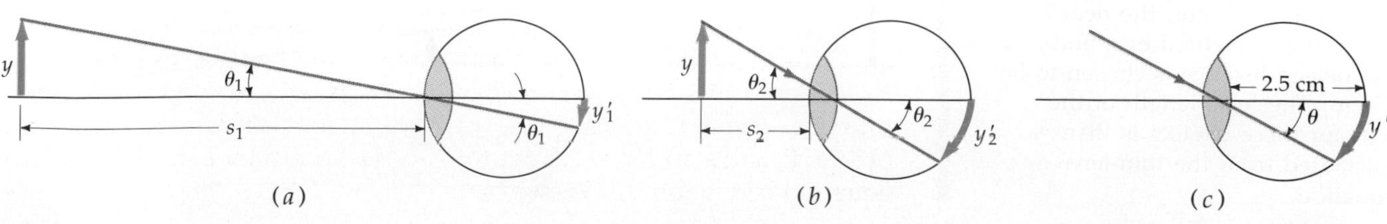

Figure 32-4 (*a*) A distant object of height y looks small because the image on the retina is small. (*b*) When the same object is closer, it looks larger because the image on the retina is larger. The size of the image on the retina is proportional to the angle θ subtended by the object, which in turn is inversely proportional to the object distance. (*c*) The angle subtended is $\theta = y'/(2.5\text{ cm})$.

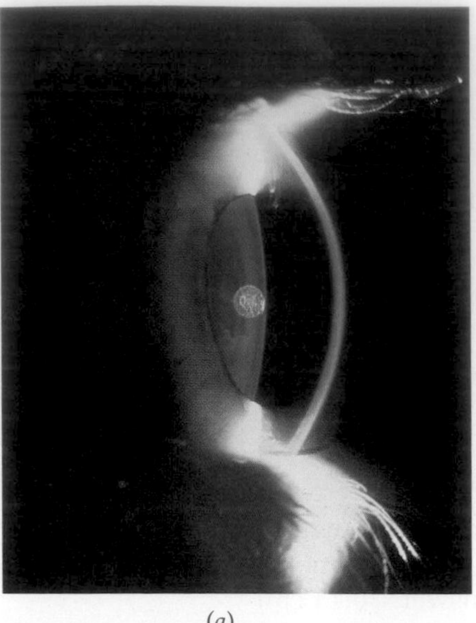
(a)

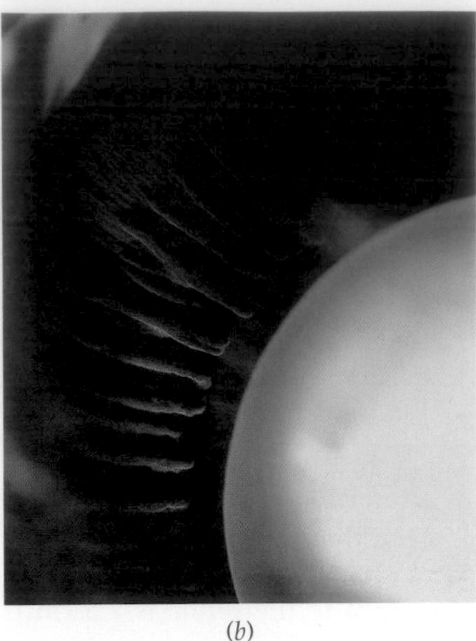
(b)

(a) The human eye in profile. (b) The lens of the eye is kept in place by ciliary muscle fibers shown here in the upper left. When the fibers contract, the tension on the lens is reduced and the lens, which is made of elastic tissue, tends to balloon outward. The greater lens curvature enables the eye to focus on nearby objects. (c) Some of the 120 million rods and 7 million cones in the eye magnified approximately 5000 times. The rods (the more slender of the two) are more sensitive in dim light, whereas the cones are more sensitive to color. The rods and cones form the bottom layer of the retina and are covered by nerve cells, blood vessels, and supporting cells. Most of the light entering the eye is reflected or absorbed before reaching the rods and cones. The light that does reach them triggers electrical impulses along nerve fibers that ultimately reach the brain. (d) A neural net used in the vision system of certain robots. Loosely modeled on the human eye, it contains 1920 sensors.

Example 32-2

Assume that the near point of your eye is 75 cm. What power reading glasses should you use to bring your near point to 25 cm?

If your near point is 75 cm, you are farsighted. To read a book you must hold the book at least 75 cm from your eye so that you can focus on the print. The image of the print on your retina is then very small. A converging lens, which is used in reading glasses, allows the book to be brought closer to the eye so that the image of the print is larger. When the book is 25 cm from the eye, we want the image formed by the converging lens to be 75 cm from the eye. Recall that a converging lens forms a virtual, erect image when the object is between the lens and the focal point. We therefore expect the focal length of the lens to be greater than 25 cm.

Figure 32-5 shows a diagram of an object 25 cm from a converging lens that produces virtual, erect image at $s' = -75$ cm. Using the thin-lens equation with $s = 25$ cm and $s' = -75$ cm, we obtain

$$\frac{1}{25\text{ cm}} + \frac{1}{-75\text{ cm}} = \frac{1}{f}$$

$$\frac{1}{f} = \frac{2}{75\text{ cm}} = \frac{1}{0.375\text{ m}} = 2.67\text{ diopters}$$

$$f = \frac{1}{2.67\text{ m}^{-1}} = 0.375\text{ m} = 37.5\text{ cm}$$

Thus, the power of your reading glasses should be 2.67 diopters, which is obtained when the focal length of the glasses is 37.5 cm.

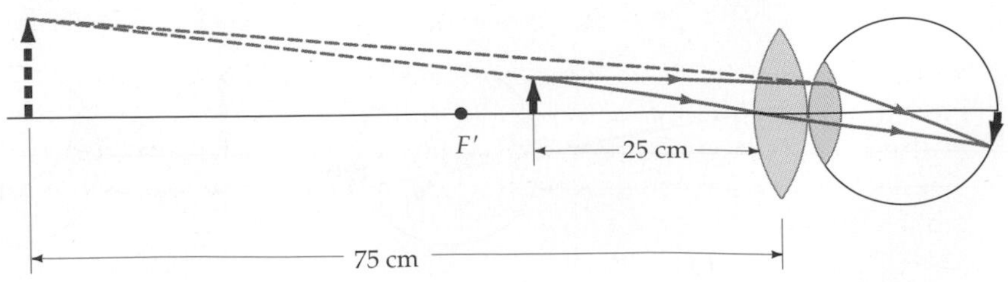

Figure 32-5 Ray diagram for Example 32-2. When the object is placed just inside the first focal point of a converging lens, the image is virtual, erect, enlarged, and far from the lens. In this example, the image distance is chosen to be 75 cm, the near point of a farsighted eye, and the object distance is chosen to be 25 cm. The focal length of the lens for these choices is then calculated from the thin-lens equation.

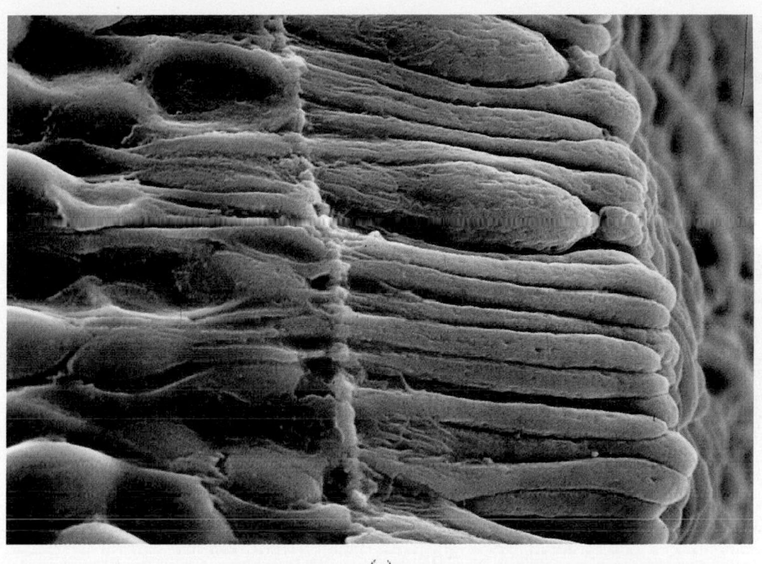
(c)

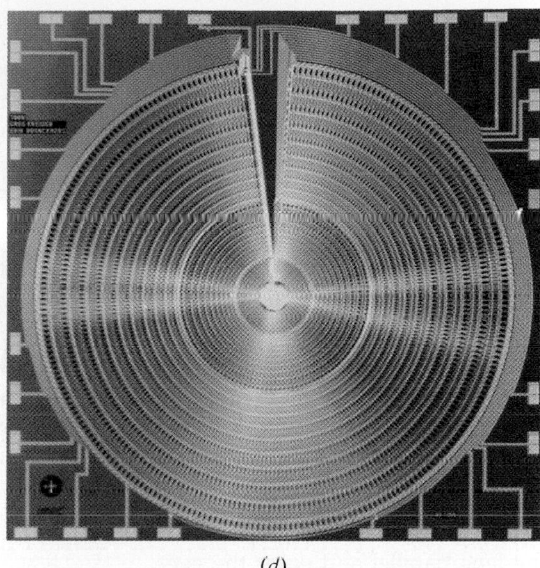
(d)

There is an alternative way to solve this problem that uses the result we found in Section 31-4 that the power of two lenses in contact is the sum of the powers of the individual lenses. Without reading glasses, when the object is at 75 cm = 0.75 m from the eye, the image is at 2.5 cm = 0.025 m, the distance from the eye lens to the retina. The focal length of the eye can then be determined from the thin-lens equation:

$$\frac{1}{0.75\text{ m}} + \frac{1}{0.025\text{ m}} = \frac{1}{f_e}$$

The power of the eye lens is then

$$P_e = \frac{1}{f_e} = 1.33\text{ m}^{-1} + 40.00\text{ m}^{-1} = 41.33\text{ diopters}$$

When the reading-glasses lens is used, the image distance of the combination should be 0.025 m when the object is at a distance of 25 cm. If f_c is the focal length of the combination and $P_c = 1/f_c$ is the power of the combination, we have

$$\frac{1}{0.25\text{ m}} + \frac{1}{0.025\text{ m}} = \frac{1}{f_c} = P_c$$

or

$$P_c = 4.00\text{ m}^{-1} + 40\text{ m}^{-1} = 44.0\text{ diopters}$$

The power of the combination equals the sum of the power of the eye lens and that of reading-glasses lens:

$$P_c = P_e + P_g$$

The power of the reading-glasses lens is thus

$$P_g = P_c - P_e = 44.0\text{ diopters} - 41.33\text{ diopters} = 2.67\text{ diopters}$$

which is in agreement with our first calculation. In both calculations, we have assumed that the lens of the reading glasses is in contact with that of the eye. Our results therefore apply for contact lenses. For reading glasses that are a short distance in front of the eye, the results are somewhat different.

Question

1. Glasses with a power of −2 diopters are prescribed for a certain person. Is that person nearsighted or farsighted?

32-2 The Simple Magnifier

We saw in Example 32-2 that the apparent size of an object can be increased by using a converging lens to allow the object to be brought closer to the eye and thereby increase the size of the image on the retina. Such a converging lens is called a **simple magnifier.** In Figure 32-6*a*, a small object of height y is at the near point of the eye at a distance x_{np}. As we discussed previously, the size of the image on the retina is proportional to the angle θ_o subtended by the object at the eye. In this case, θ_o is given approximately by

$$\theta_o = \frac{y}{x_{np}}$$

In Figure 32-6*b*, a converging lens of focal length f, which is less than x_{np}, has been placed in front of the eye, and the object has been placed at the focal point of the lens. The rays emerge from the lens parallel, indicating that the image is an infinite distance in front of the lens. The parallel rays are focused by the relaxed eye on the retina. If the lens is in close contact with the eye, the angle subtended by the object is now approximately

$$\theta = \frac{y}{f}$$

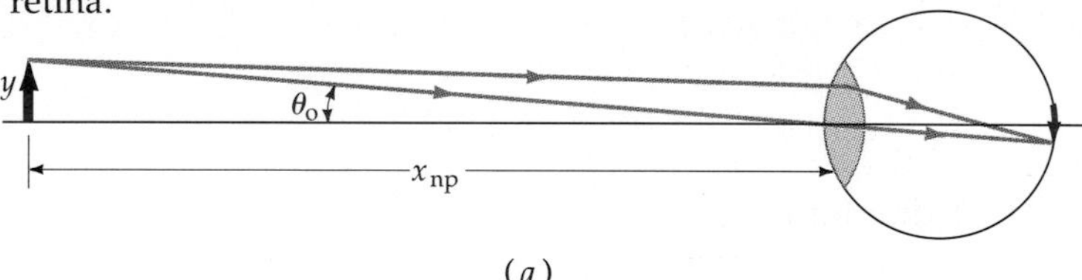

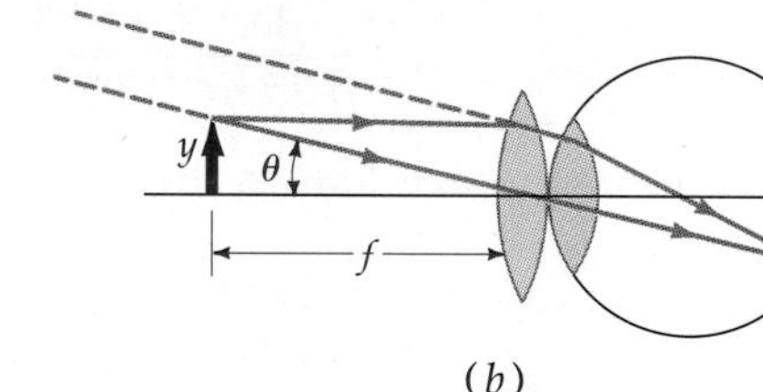

Figure 32-6 (*a*) An object at the near point subtends an angle θ_o at the eye. (*b*) When the object is at the focal point of the converging lens, the rays emerge from the lens parallel and enter the eye as if they came from an object a very large distance away. The image can thus be viewed at infinity by the relaxed eye. When f is less than the near point, the converging lens allows the object to be brought closer to the eye, increasing the angle subtended by the object to θ and thereby increasing the size of the image on the retina.

The ratio θ/θ_o is called the **angular magnification** or **magnifying power** M of the lens:

Magnifying power of a lens

$$M = \frac{\theta}{\theta_o} = \frac{x_{np}}{f} \qquad 32\text{-}4$$

Example 32-3

A person with a near point of 25 cm uses a 40-diopter lens as a simple magnifier. What magnification is obtained?

The focal length of a 40-diopter lens is

$$f = \frac{1}{P} = \frac{1}{40\ \text{m}^{-1}} = 0.025\ \text{m} = 2.5\ \text{cm}$$

Using $x_{np} = 25$ cm and $f = 2.5$ cm in Equation 32-4, we obtain for the magnification

$$M = \frac{25\ \text{cm}}{2.5\ \text{cm}} = 10$$

The object looks 10 times larger because it can be placed at 2.5 cm rather than at 25 cm from the eye, so the size of the image on the retina is increased tenfold. The additional converging power of the magnifier is used to focus the very divergent rays from the very close object.

Exercise

What is the magnification in Example 32-3 if the near point of the person is 30 cm rather than 25 cm? (Answer: $M = 12$)

It is possible to increase this magnification slightly by moving the object closer to the magnifier. When the object is inside the focal point of the magnifier, the image is virtual and erect as in Figure 32-5. As the object is moved toward the magnifier, the image moves closer to the eye, and the angle subtended increases slightly. The largest usable magnification occurs when the image is at the near point of the eye, as it was in Example 32-2. Calculation shows that the magnification in this case is just 1 greater than that with the image at infinity (see Problem 26). For example, if the near point distance is 25 cm, a lens with focal length of 2.5 cm would give a magnification of 11 with the image at 25 cm rather than a magnification of 10 with the image at infinity. In Example 32-2, we found that a lens of focal length 37.5 cm produces an image at 75 cm from the eye when the object at 25 cm from the eye. Since the angle subtended by the image on the retina varies inversely as the distance of the object (or image) from the eye, the magnification was (75 cm)/(25 cm) = 3 in that example. If we were to move the object back from 25 cm to the focal point of the lens at 37.5 cm, the image could be viewed with a relaxed eye and the magnification would be 2 instead of 3. In practice, the gain in magnification obtained by viewing the image at the near point, rather than at infinity with a relaxed eye, is not worth the additional eye strain, so we will use Equation 32-4 for the magnification of a simple magnifier.

Simple magnifiers are used as eyepieces (called *oculars*) in compound microscopes and telescopes to view the image formed by another lens or lens system. To correct aberrations, combinations of lenses with a resulting short positive focal length are often used instead of a single lens, but the principle is the same as that of the simple magnifier.

32-3 The Camera

The basic camera consists of a positive lens, a variable aperture, a shutter that can be opened for a short time that can be varied, a light-tight box, and a film (Figure 32-7). Unlike the eye, which has a lens of variable focal length, the focal length of a camera lens is fixed. Typically, the focal length for the lens in a 35-mm camera is 50 mm. (The 35-mm refers to the width of film.) Focusing is accomplished by varying the distance from the lens to the film by moving the lens closer to or farther from the film.

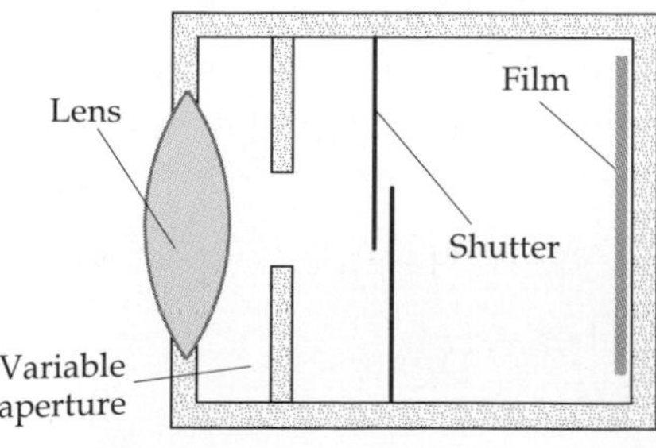

Figure 32-7 Schematic diagram of a camera. The positive lens focuses the light on the film. The variable aperture limits the amount of light entering the camera and the area of the lens used. The shutter speed can be varied to vary the exposure time.

Example 32-4

The focal length of a camera lens is 50 mm. By how much must the lens be moved to change from focusing on an object far away to one 2 m away?

When the object is far away, the image of the lens is at the focal length of the lens, so the film should be 50 mm from the lens. When the object is 2 m away, the image distance is s', given by

$$\frac{1}{2\text{ m}} + \frac{1}{s'} = \frac{1}{f} = \frac{1}{50\text{ mm}}$$

$$\frac{1}{s'} = \frac{1}{50\text{ mm}} - \frac{1}{2000\text{ mm}}$$

$$= \frac{40}{2000\text{ mm}} - \frac{1}{2000\text{ mm}}$$

$$s' = \frac{2000\text{ mm}}{39} = 51.3\text{ mm}$$

The lens must therefore be moved 1.3 mm farther away from the film.

Exercise

In Example 32-4, how far must the lens be moved to change from focusing on an object far away to one that is 1 m away? (Answer: 2.6 mm)

The amount of light that strikes the film can be controlled by varying the time that the shutter is open and by varying the size of the aperture. For a given film type, there is an optimum amount of light that will give a good picture of proper contrast. Too little light results in a dark picture. Too much light results in a washed-out picture with too little contrast. The amount of light needed for proper contrast is related to the film's "speed," which is rated by an ASA number. The higher the ASA number, the faster the film and the smaller the amount of light needed. A film with a high ASA number, such as ASA 400 or ASA 1000, is good for taking pictures indoors where there is little available light. With high-speed films, some reduction in picture quality (the sharpness of the image or the trueness of the color reproduction) usually occurs; so for outdoor photographs taken with plenty of available light, it is usually preferable to use a lower-speed film, such as ASA 100 or ASA 64. Lower-speed film may also be needed in bright-light situations if the shutter speed of the camera has limited variability. Shutter speeds on good cameras can often be varied from exposures of several seconds for low-light photography to 1/1000 of a second for stop-action photography. For a hand-held camera, exposure times of more than about 1/60 of a second often result in a blurring of the image because of camera motion.

The maximum size of the aperture is limited by the size of the lens, which is in turn limited by the various lens aberrations we discussed in Section 31-5. (Although we treat the camera lens as a single positive lens, the optical systems in good cameras are combinations of lenses designed to reduce chromatic, spherical, and other aberrations.) The size of the aperture is given by the ***f*-number,** which is the ratio of the focal length to the diameter of the aperture:

f-number

$$f\text{-number} = \frac{f}{D} \qquad 32\text{-}5$$

Cutaway view of a 35-mm camera.

The maximum aperture is the *f*-number of the lens. For example, an $f/2.8$ lens with a focal length of 50 mm has a maximum usable diameter given by

$$D = \frac{f}{f\text{-number}} = \frac{50 \text{ mm}}{2.8} = 17.9 \text{ mm}$$

Lenses with large diameters (small *f*-numbers) are costly to make because of the expense of correcting for aberrations. Aperture settings on a camera are usually marked $f/22$, $f/16$, $f/11$, $f/8$, $f/5.6$, $f/4$, $f/2.8$, $f/2.0$, $f/1.4$, $f/1.0$ and on down to the lowest *f*-number, which corresponds to the largest usable diameter of the lens. Note that for each successive setting, called an *f*-stop, the diameter of the aperture is $\sqrt{2} = 1.4$ times that of the previous one. Since the amount of light that enters the camera is proportional to the area of the lens, which is in turn proportional to the square of the diameter, opening up the aperture by one *f*-stop, for example, from $f/2.0$ to $f/1.4$, increases the area and therefore the amount of light entering in a given time by a factor of 2. Increasing the aperture, however, reduces the depth of focus—the range of distances to objects that are in sharp focus on the film.

Exercise

What is the *f*-number of a lens whose focal length is 50 mm and whose maximum usable diameter is 8.93 mm? (Answer: $f/5.6$)

Example 32-5

The instructions for a certain film say to set the aperture at $f/11$ and the shutter speed at 1/250 s to take pictures on bright sunny days. What should the shutter speed be if the aperture is set at $f/5.6$?

Since $11^2 = 121$ and $5.6^2 = 31.4$, the area of the aperture at $f/5.6$ is $121/(31.4) = 3.85$ or approximately 4 times greater than that at $f/11$. Thus, to get the desired amount of light, you should use a shutter speed 4 times faster or 1/1000 s.

A 35-mm camera with a lens having a focal length of 50 mm gives a field of view that is approximately the same as that of ordinary vision, about 45°.

Wide-angle lens photograph of the skating rink at Rockefeller Center in New York. Note the distortion at the top of the photograph.

To increase the field of view, a wide-angle lens with a smaller focal length, for example, 24 mm, is used. When the object distance is much greater than the focal length, which is usually the case for a camera, the image distance s' is approximately equal to f. Since the lateral magnification of a lens is $m = (-)s'/s$, the size of the image on the film is approximately proportional to the focal length. A wide-angle lens thus gives a smaller image on the film than does a normal lens for a given object size. A telephoto lens has a large focal length to increase the size of the image on the film and thus make the object seem closer. A telephoto lens with a focal length of 200 mm would give a magnification approximately 4 times that of an ordinary lens with a focal length of 50 mm.

Questions

2. What are the advantages of having a camera with a fast shutter?
3. Why is an $f/1.0$ lens more expensive than an $f/2.8$ lens?

32-4 The Compound Microscope

The compound microscope (Figure 32-8) is used to look at very small objects at short distances. In its simplest form, it consists of two converging lenses. The lens nearest the object, called the **objective,** forms a real image of the object. This image is enlarged and inverted. The lens nearest the eye, called the **eyepiece** or **ocular,** is used as a simple magnifier to view the image formed by the objective. The eyepiece is placed such that the image formed by the objective falls at the first focal point of the eyepiece. The light thus emerges from the eyepiece as a parallel beam as if it were coming from a point a great distance in front of the lens. (This is commonly called "viewing the image at infinity.") As we discussed in Section 32-2, the function of a simple magnifier (the eyepiece in this case) is to allow the object (the image formed by the objective in this case) to be brought closer to the eye than the near point. Since a simple magnifier produces a virtual image that is erect, the final image produced by the two lenses is inverted.

(*a*)

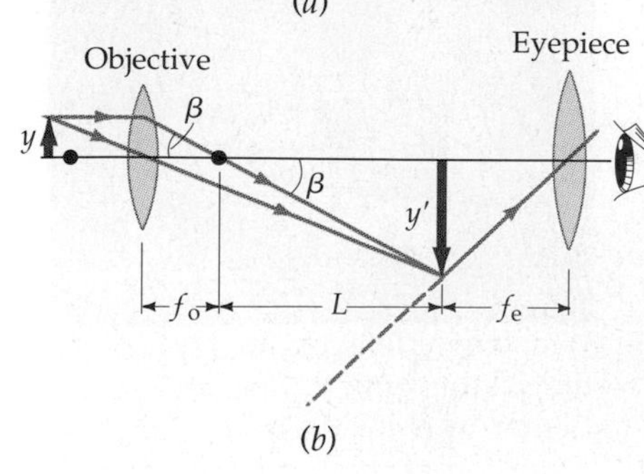

(*b*)

Figure 32-8 (*a*) William Marin at the Brookhaven National Laboratory chooses an objective lens for viewing a sample of radioactive waste with a modern optical microscope. (*b*) Schematic diagram of a compound microscope consisting of two positive lenses, the objective of focal length f_o and the ocular, or eyepiece, of focal length f_e. The real image of the object formed by the objective is viewed by the eyepiece, which acts as a simple magnifier. The final image is at infinity.

The distance between the second focal point of the objective and the first focal point of the eyepiece is called the **tube length** L. It is typically fixed at 16 cm. The object is placed just outside the first focal point of the objective so that an enlarged image is formed at the first focal point of the eyepiece a distance $L + f_o$ from the objective, where f_o is the focal length of the objective. From Figure 32-8, $\tan \beta = y/f_o = -y'/L$. The lateral magnification of the objective is therefore

$$m_o = \frac{y'}{y} = -\frac{L}{f_o} \qquad 32\text{-}6$$

The angular magnification of the eyepiece is

$$M_e = \frac{x_{np}}{f_e}$$

where x_{np} is the near point of the viewer and f_e is the focal length of the eyepiece. As we discussed in Section 32-2, a slightly greater angular magnification can be obtained by placing the object (image formed by the objective) at a point just inside the first focal point of the eyepiece so that the final image is at the near point. The slight gain in the angular magnification of the eyepiece is usually not worth the eye strain caused by viewing the image at the near point rather than viewing it at infinity with a relaxed eye. The magnifying power of the compound microscope is the product of the lateral magnification of the objective and the angular magnification of the eyepiece:

$$M = m_o M_e = -\frac{L}{f_o}\frac{x_{np}}{f_e} \qquad 32\text{-}7$$

Magnifying power of a microscope

Example 32-6

A microscope has an objective lens of focal length 1.2 cm and an eyepiece of focal length 2.0 cm separated by 20 cm. (*a*) Find the magnifying power if the near point of the viewer is 25 cm. (*b*) Where should the object be placed if the final image is to be viewed at infinity?

(*a*) The distance between the second focal point of the objective and the first focal point of the eyepiece is 20 cm − 2 cm − 1.2 cm = 16.8 cm. The magnifying power is given by Equation 32-7 with $L = 16.8$ cm, $f_o = 1.2$ cm, $f_e = 2.0$ cm, and $x_{np} = 25$ cm:

$$M = -\frac{16.8\text{ cm}}{1.2\text{ cm}}\frac{25\text{ cm}}{2\text{ cm}} = -175$$

The negative sign indicates that the final image is inverted.

(*b*) We can calculate the object distance between the original object and the objective from the thin-lens equation. From Figure 32-8, we can see that the image distance is

$$s' = f_o + L = 1.2\text{ cm} + 16.8\text{ cm} = 18\text{ cm}$$

The object distance is then found from

$$\frac{1}{s} + \frac{1}{s'} = \frac{1}{f}$$

$$\frac{1}{s} + \frac{1}{18\text{ cm}} = \frac{1}{1.2\text{ cm}}$$

Solving for s, we obtain $s = 1.29$ cm. The object should thus be placed at 1.29 cm from the objective or 0.09 cm outside its first focal point.

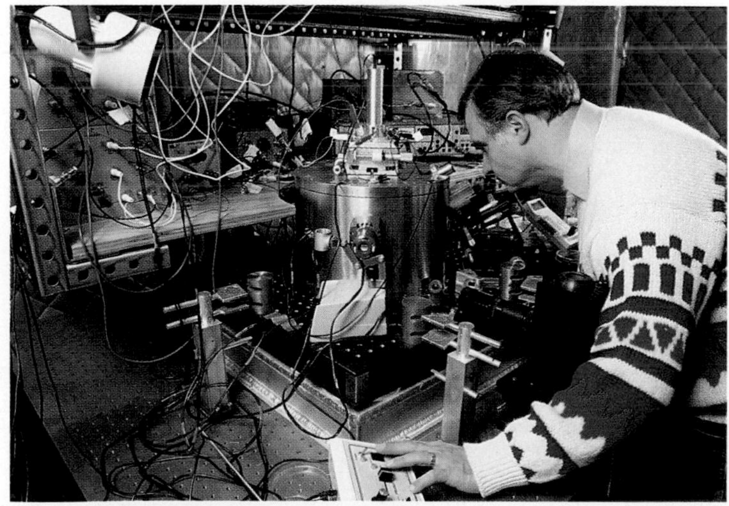

(*a*)

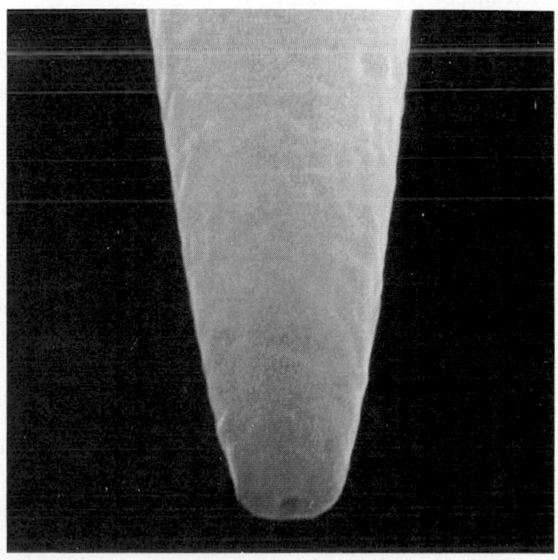

(*b*)

The shorter the wavelength of light used to image a specimen, the smaller the specimen that can be imaged. Since shorter wavelength light carries more energy, there is a trade-off between resolving power of the scope and potential damage to the illuminated specimen. (*a*) A state-of-the-art near-field microscope. (*b*) The aperture of the microscope is a 50-nm wide hole at the tip of a glass pipette. The specimen is positioned so near the aperture that light encounters it before the light has a chance to diverge. The beam is swept over the specimen's surface, partially transmitted through the specimen, and collected by a photodetector. Line by line an image is eventually built up on a video monitor. Near-field microscopes using greenish-yellow light of 500-nm wavelength have achieved a resolution of about 40 nm. This resolution is an order of magnitude better than that possible with a conventional optical microscope.

32-5 The Telescope

A telescope is used to view objects that are far away and often large. Its purpose is to bring the image of the object closer, that is, to increase the angle subtended by the image so that the object appears larger. The astronomical telescope, illustrated schematically in Figure 32-9, consists of two positive lenses—an objective lens that forms a real, inverted image and an eyepiece that is used as a simple magnifier to view that image. Since the object is very far away, the image of the objective lies at the focal point of the objective, and the image distance equals the focal length f_o. Since the object distance is much larger than the focal length of the objective, the image formed by the objective is much smaller than the object. For example, if we are looking at the moon, the image of the moon formed by the objective is much smaller than the moon itself. The purpose of the objective is not to magnify the object, but to produce an image that is close so it can be viewed by the eyepiece. Since this image is at the second focal point of the objective and at the first focal point of the ocular, the objective and ocular must be separated by the sum of the focal lengths of the objective and eyepiece, $f_o + f_e$, where f_e is the focal length of the eyepiece.

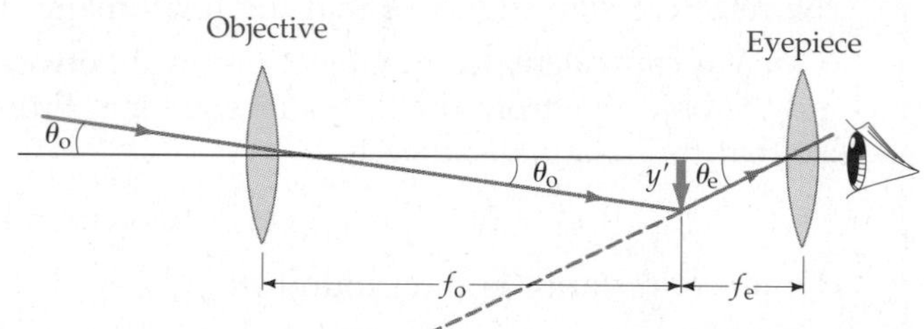

Figure 32-9 Schematic diagram of an astronomical telescope. The objective forms a real image of a distant object near its second focal point, which coincides with the first focal point of the eyepiece. The eyepiece serves as a simple magnifier for viewing the image.

The magnifying power of the telescope is the angular magnification θ_e/θ_o, where θ_e is the angle subtended by the final image as viewed through the eyepiece and θ_o is the angle subtended by the object when it is viewed directly by the unaided eye. The angle θ_o is the same as that subtended by the object at the objective shown in Figure 32-9. (The distance from a distant object, such as the moon, to the objective is essentially the same as the distance to the eye.) From this figure, we can see that

$$\tan \theta_o = -\frac{y'}{f_o} \approx \theta_o$$

where we have used the small-angle approximation $\tan \theta \approx \theta$ and have introduced a negative sign to make θ_o positive when y' is negative. The angle θ_e in the figure is that subtended by the final image:

$$\tan \theta_e = \frac{y'}{f_e} \approx \theta_e$$

Since y' is negative, θ_e is negative, indicating that the image is inverted. The magnifying power of the telescope is then

Magnifying power of a telescope

$$M = \frac{\theta_e}{\theta_o} = -\frac{f_o}{f_e} \qquad 32\text{-}8$$

From Equation 32-8, we can see that a large magnifying power is obtained with an objective of large focal length and an eyepiece of short focal length.

(a) (b) (c) (d) (e)

Astronomy at optical wavelengths began with Galileo approximately 400 years ago. In this century, astronomers have begun to explore the electromagnetic spectrum at other wavelengths beginning with radio astronomy in the 1940s, satellite based x-ray astronomy in the early 1960s, and more recently ultraviolet, infrared, and gamma-ray astronomy. (*a*) Galileo's seventeenth-century telescope with which he discovered mountains on the moon, sunspots, Saturn's rings, and the bands and moons of Jupiter. (*b*) An engraving of the reflector telescope built in the 1780s, which was used by the great astronomer Hershel who was the first to observe galaxies outside our own. (*c*) Because of the difficulty of making large, flaw-free lenses, refractor telescopes as this 91.4-cm refractor telescope at Lick Observatory, have been superseded in light-gathering power by reflector telescopes. (*d*) The great astronomer Hubble, who discovered the apparent expansion of the universe, is shown seated in the observer's cage of the 5.08-m Hale reflecting telescope, which is large enough for the observer to sit at the prime focus itself. (*e*) This 10-meter optical reflector at the Whipple Observatory in southern Arizona is the largest instrument designed exclusively for use in gamma-ray astronomy. High-energy gamma rays of unknown origin strike the upper atmosphere and create cascades of particles, such as high-energy electrons which emit Čherenkov radiation that can be observed from the ground. According to one hypothesis, high-energy gamma rays are emitted when matter is accelerated towards ultradense, rotating stars called pulsars. (For an image of Čherenkov radiation, see p. 469; for an image of a pulsar, see p. 237.)

(a)

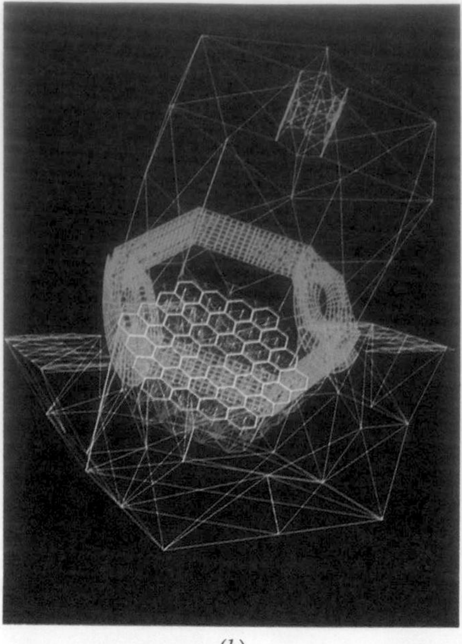

(b)

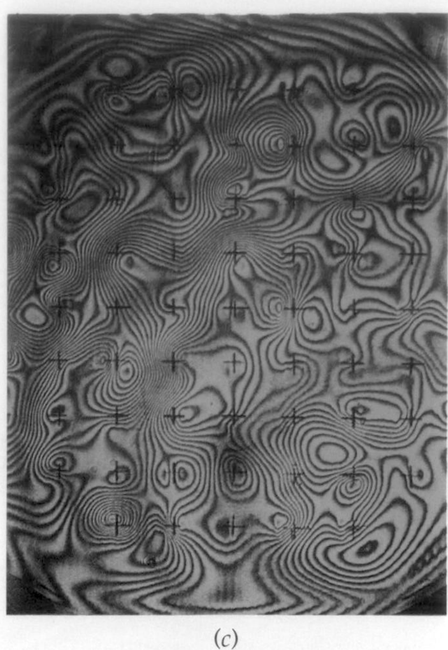

(c)

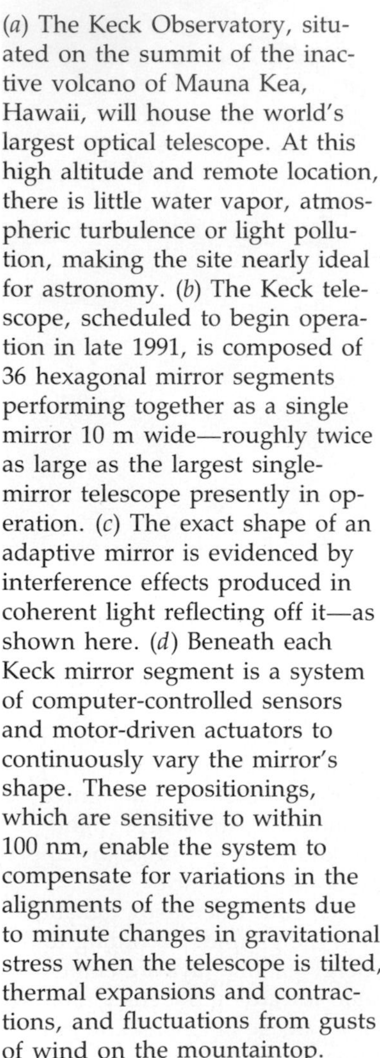

(*a*) The Keck Observatory, situated on the summit of the inactive volcano of Mauna Kea, Hawaii, will house the world's largest optical telescope. At this high altitude and remote location, there is little water vapor, atmospheric turbulence or light pollution, making the site nearly ideal for astronomy. (*b*) The Keck telescope, scheduled to begin operation in late 1991, is composed of 36 hexagonal mirror segments performing together as a single mirror 10 m wide—roughly twice as large as the largest single-mirror telescope presently in operation. (*c*) The exact shape of an adaptive mirror is evidenced by interference effects produced in coherent light reflecting off it—as shown here. (*d*) Beneath each Keck mirror segment is a system of computer-controlled sensors and motor-driven actuators to continuously vary the mirror's shape. These repositionings, which are sensitive to within 100 nm, enable the system to compensate for variations in the alignments of the segments due to minute changes in gravitational stress when the telescope is tilted, thermal expansions and contractions, and fluctuations from gusts of wind on the mountaintop.

(d)

Example 32-7

The world's largest refracting telescope is at the Yerkes Observatory of the University of Chicago at Williams Bay, Wisconsin. The objective has a diameter of 102 cm and a focal length of 19.5 m. The focal length of the eyepiece is 10 cm. What is the magnifying power of this telescope?

From Equation 32-8, we have

$$M = -\frac{f_o}{f_e} = -\frac{19.5\text{ m}}{0.10\text{ m}} = -195$$

The main consideration with an astronomical telescope is not its magnifying power but its light-gathering power, which depends on the size of the objective. The larger the objective, the brighter the image. Very large lenses without aberrations are difficult to produce. In addition, there are mechanical problems in supporting very large, heavy lenses by their edges. A reflecting telescope uses a concave mirror instead of a lens for its objective. This offers several advantages. For one, a mirror does not produce chromatic aberration. In addition, mechanical support is much simpler since the mirror weighs far less than a lens of equivalent optical quality and can be supported over its entire back surface.

One problem with the reflecting telescope is that the image of the objective mirror must be viewed in the region of the incoming rays (Figure 32-10). In very large reflecting telescopes, such as the 200-in (5.1 m) diameter telescope at Mt. Palomar, California, the viewer sits in an observer's cage near the focal point of the mirror. To obstruct as little light as possible, the cage is very small and cramped, so there is little space for auxiliary instruments such as spectrographs. In smaller telescopes, the fraction of light obstructed by such an arrangement would be too great. One method of reducing the amount of light obstructed is to use a second, smaller mirror to reflect the rays through a small hole in the center of the objective as shown in Figure 32-11. This has the further advantages of making the viewing area more accessible and providing more room for auxiliary instruments.

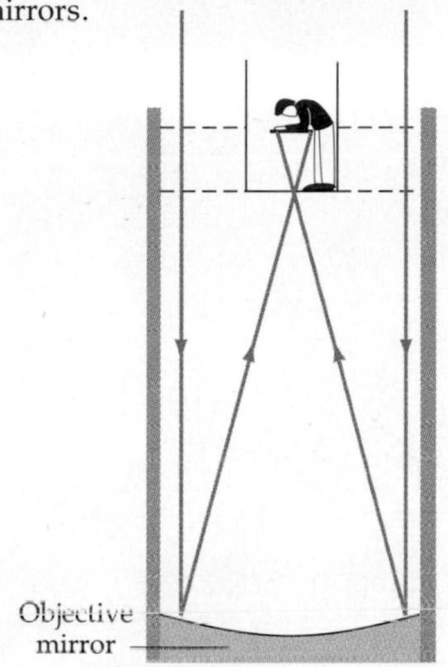

Figure 32-10 A reflecting telescope uses a mirror for its objective. Because the viewer compartment blocks off some of the incoming light, the arrangement shown here is used only in telescopes with very large objective mirrors.

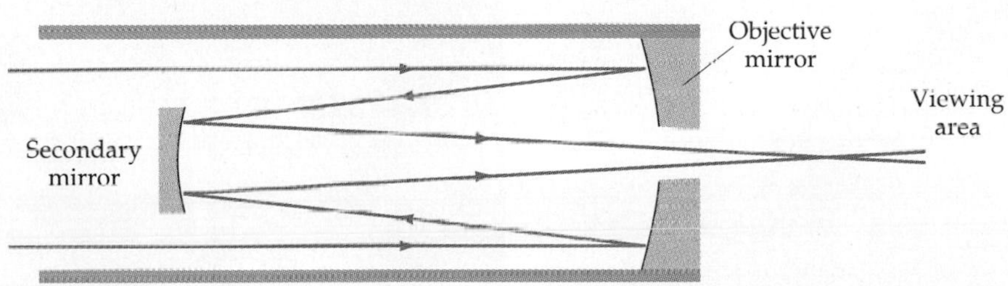

Figure 32-11 Reflecting telescope with secondary mirror to redirect the light through a small hole in the objective mirror. This arrangement has a further advantage over that of Figure 32-10 in that there is more room for auxiliary instruments in the viewing region.

The fact that the final image is inverted in a simple telescope is not a disadvantage when viewing astronomical objects such as stars and planets, but it is when viewing terrestrial objects. Binoculars use two 45-45-90° prisms in each side to provide a second inversion of the image so that the final image is upright. Figure 32-12*a* shows a 45-45-90° prism with its hypotenuse horizontal and its reflecting sides vertical. Light entering through the long face is reflected twice and emerges back through the long face in the opposite direction. Horizontal images are inverted, but vertical ones are not. In Figure 32-12*b*, a second prism with its hypotenuse vertical redirects the light back into its original direction and inverts vertical images without changing horizontal ones. The multiple reflections of the prisms also in-

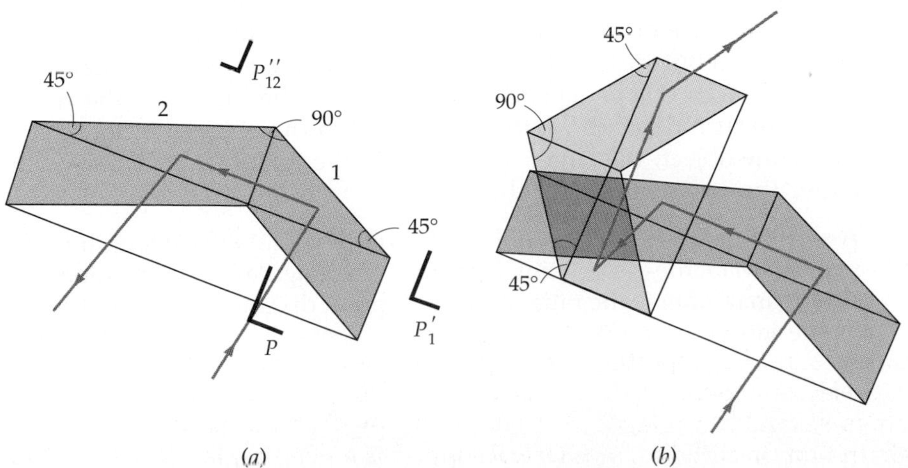

Figure 32-12 (*a*) A 45-45-90° prism with its hypotenuse horizontal and the reflecting faces 1 and 2 vertical. The image P''_{12} due to reflection from surfaces 1 and 2 is inverted in the horizontal direction but not in the vertical direction. (*b*) If the light enters a second, identical prism oriented with its hypotenuse vertical, the image will be inverted in the vertical direction. After passing through both prisms, the light emerges in its original direction with the image inverted in both the vertical and horizontal directions.

(a)

(b)

The Hubble Space Telescope (*a*) during and (*b*) after its deployment from the cargo bay of the space shuttle using a long mechanical arm. The telescope orbits nearly 615 km above the earth's surface, high above atmospheric turbulence that limits the ability of ground-based telescopes to resolve images at optical wavelengths. Due to an error in the device used to measure the shape of the Hubble's main reflecting mirror, the telescope suffers from spherical aberration. At present it is not clear to what extent this can be remedied. (*c*) A false-color image of 30 Doradus, a star cluster in the Large Magellanic Cloud from a 2.2-m ground-based telescope. (*d*) The same region as imaged by the Hubble with a sixfold improvement in resolution. (*e*) A computer-processed version of the Hubble image, removing the haze attributed to spherical aberrations. Such processing may work for bright sources, but is unlikely to enable the Hubble to image dim ones from very distant galaxies as originally hoped.

crease the path length of the light so that a relative long focal length for the objective can be used in a relatively short space.

Summary

1. The cornea–lens system of the eye focuses light on the retina, where it is sensed by the rods and cones, which transmit the information along the optic nerve to the brain. When the eye is relaxed, the focal length of the cornea–lens system is about 2.5 cm, the distance from the cornea to the retina. When objects are brought near the eye, the shape of the lens changes slightly to decrease the overall focal length so that the image is again focused on the retina. The closest distance for which the lens can focus the image on the retina is called the near point, which is typically about 25 cm but varies with age and from person to person. The apparent size of an object depends on the size of the image on the retina, which in turn depends on the distance from the object to the eye. The closer the object, the larger the image on the retina and, therefore, the larger the apparent size of the object.

2. A simple magnifier consists of a lens with a positive focal length that is smaller than the near point distance. The angular magnification of a simple magnifier is the ratio of the near-point distance to the focal length of the lens:

$$M = \frac{x_{np}}{f}$$

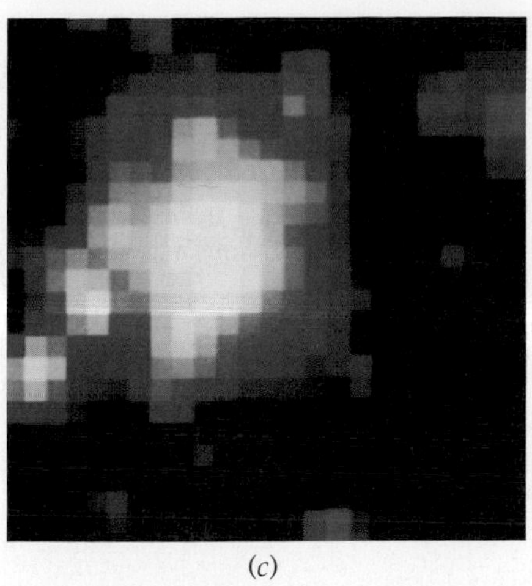

(c)

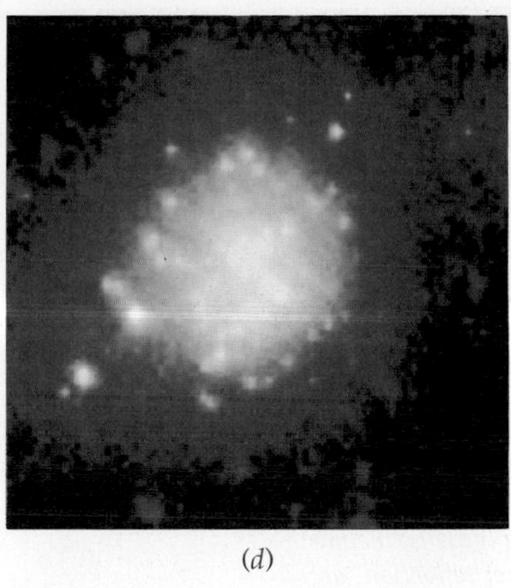

(d)

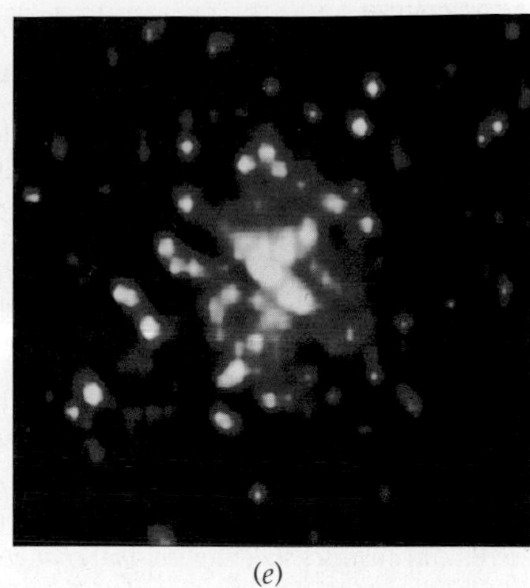

(e)

3. A basic camera consists of a lens, a variable aperture, a shutter, a light-tight box, and a film. Since the focal length of the lens is fixed, focusing is accomplished by moving the lens toward or away from the film. The *f*-number of the aperture is the ratio of the focal length to the diameter of the lens:

$$f\text{-number} = \frac{f}{D}$$

The focal length of a typical lens in a 35-mm camera is about 50 mm. A telephoto lens has a larger focal length, which gives a larger image on the film but a narrower field of view. A wide-angle lens has a smaller focal length, which gives a smaller image on the film but a wider field of view.

4. The compound microscope is used to look at very small objects at short distances. In its simplest form, it consists of two lenses—an objective and an ocular or eyepiece. The object to be viewed is placed just outside the focal point of the objective, which forms an enlarged image of the object at the focal point of the eyepiece. The eyepiece acts as a simple magnifier for viewing the final image. The magnifying power of the microscope is the product of the lateral magnification of the objective and the angular magnification of the eyepiece:

$$M = m_o M_e = -\frac{L}{f_o}\frac{x_{np}}{f_e}$$

where L is the tube length, which is the distance between the second focal point of the objective and the first focal point of the eyepiece.

5. The telescope is used to view objects that are far away. The objective of the telescope forms a real image that is much smaller than the object but much closer. The eyepiece is then used as a simple magnifier for viewing the image. A reflecting telescope uses a mirror for its objective. The magnifying power of a telescope equals the (negative) ratio of the focal length of the objective to the focal length of the eyepiece:

$$M = -\frac{f_o}{f_e}$$

The most important feature of an astronomical telescope is its light-gathering power, which is proportional to the area of the objective.

Suggestions for Further Reading

Everhart, Thomas E., and Thomas L. Hayes: "The Scanning Electron Microscope," *Scientific American*, January 1972, p. 54.

This article describes how the interaction between a beam of high-energy electrons and matter is used by the scanning electron microscope to create an image of three-dimensional appearance.

Koretz, Jane F., and George H. Handelman: "How the Human Eye Focuses," *Scientific American*, July 1988, p. 92.

Throughout a person's lifetime, the shortest distance at which the eye can focus gradually increases. This article describes measurements and analyses that suggest that changes in the lens, including increasing thickness and decreasing index of refraction, are responsible.

Land, Michael F.: "Animal Eyes with Mirror Optics," *Scientific American*, December 1978, p. 126.

Several sea creatures, described in this article, have been found to use reflection rather than refraction to form images of their surroundings.

Price, William H.: "The Photographic Lens," *Scientific American*, August 1976, p. 72.

A history of lens design, including a description of the defects and aberrations one wishes to minimize in a lens, and a discussion of modern computer-aided design are presented in this article.

Review

A. Objectives: After studying this chapter, you should:

1. Be able to discuss how the eye works.

2. Be able to show with a simple diagram why an object appears larger when it is brought closer to the eye.

3. Be able to describe how a simple magnifier works and calculate its angular magnification.

4. Be able to discuss how a camera works.

5. Be able to describe with diagrams and equations how a microscope and a telescope work.

B. Define, explain, or otherwise identify:

Accommodation	Magnifying power
Near point	*f*-number
Farsighted	Objective
Nearsighted	Eyepiece
Simple magnifier	Ocular
Angular magnification	Tube length

C. True or false: If the statement is true, explain why it is true. If it is false, give a counterexample.

1. The lens of the eye forms a real image.

2. A simple magnifier should have a short focal length.

3. A simple magnifier forms a virtual image.

4. The lens of a camera forms a real image.

5. The area of a camera's aperture is proportional to the *f*-number.

6. The focal length of a telephoto lens is greater than that of a wide-angle lens.

7. The image formed by the objective of a microscope is inverted and larger than the object.

8. The image formed by the objective of a telescope is inverted and larger than the object.

9. A reflecting telescope uses a mirror for its ocular.

Problems

Level I

32-1 The Eye

In the following problems, take the distance from the cornea–lens system of the eye to the retina to be 2.5 cm and assume that the lenses of corrective glasses are in contact with the eye unless otherwise stated.

1. Suppose the eye were designed like a camera with a lens of fixed focal length $f = 2.5$ cm that could move toward or away from the retina. Approximately how far would the lens have to move to focus the image of an object 25 cm from the eye onto the retina? *Hint:* Find the distance from the retina to the image behind it for an object at 25 cm.

2. Find the change in the focal length of the eye when an object originally at 3 m is brought to 30 cm from the eye.

3. A farsighted person needs to read from a computer screen that is 45 cm from her eye. Her near point is at 80 cm. (*a*) Find the focal length of the lenses in reading glasses that will produce an image of the screen at 80 cm from her eye. (*b*) What is the power of the lenses?

4. Find (*a*) the focal length and (*b*) the power of a lens that will produce an image at 80 cm from the eye of a book that is 30 cm from the eye.

5. A farsighted person requires lenses with a power of 1.75 diopters to read comfortably from a book that is 25 cm from his eye. What is his near point without the lenses?

6. A nearsighted person cannot focus clearly on objects more distant that 225 cm from her eye. What power lenses are required for her to see distant objects clearly?

7. Since the index of refraction of the lens of the eye is not very different from that of the surrounding material, most of the refraction takes place at the cornea, where n changes abruptly from 1.0 in air to about 1.4. Assuming the cornea to be a homogeneous sphere with an index of refraction of 1.4, calculate its radius if it focuses parallel light on the retina a distance 2.5 cm away. Do you expect your result to be larger or smaller than the actual radius of the cornea?

8. If two point objects that are close together are to be seen as two distinct objects, their images must fall on the retina on two different cones that are not adjacent. That is, there must be an unactivated cone between them. The separation of the cones is about 1 μm. (*a*) What is the smallest angle the two points can subtend? (See Figure 32-13.) (*b*) How close together can two points be if they are 20 m from the eye?

Figure 32-13 Problem 8. The two points will look like two separate points only if their images fall on two different, nonadjacent cones of the retina.

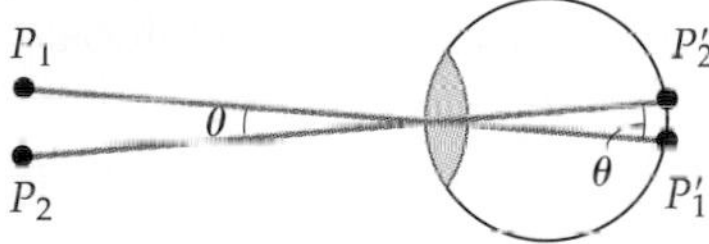

32-2 The Simple Magnifier

9. A person with a near-point distance of 30 cm uses a simple magnifier of power 20 diopters. What is the magnification obtained if the final image is at infinity?

10. A person with a near-point distance of 25 cm wishes to obtain a magnifying power of 5 with a simple magnifier. What should be the focal length of the lens used?

11. What is the magnifying power of a lens of focal length 7 cm when the image is viewed at infinity by a person whose near point is at 35 cm?

12. A lens of focal length 6 cm is used as a simple magnifier with the image at infinity by one person whose near point is at 25 cm and by another whose near point is at 40 cm. (*a*) What is the effective magnifying power of the lens for each person? (*b*) Compare the size of the image on the retina when each looks at the same object with the magnifier.

32-3 The Camera

13. What is the diameter of an $f/1.4$ lens if its focal length is 50 mm?

14. A lens has a usable diameter of 2.5 cm. What is its f-number if its focal length is 50 mm?

15. A telephoto lens has a focal length of 200 mm. By how much must it be moved to change from focusing on an object at infinity to one at a distance of 30 m?

16. A wide-angle lens has a focal length of 28 mm. By how much must it be moved to change from focusing on an object at infinity to one at a distance of 5 m?

17. A camera produces a proper exposure for an aperture stop of $f/16$ at 1/30 s. What shutter speed should be used at (*a*) $f/11$, (*b*) $f/8$, (*c*) $f/5.6$, (*d*) $f/4$, and (*e*) $f/2.8$?

18. Light conditions on a certain day for a certain film call for an aperture stop of $f/8$ at 1/250 s. (*a*) If you wish to take a picture of a humming bird at 1/1000 s, what f/number should you use? (*b*) If you set the aperture at $f/22$ for another picture, what shutter speed should you use?

32-4 The Compound Microscope

19. A microscope objective has a focal length of 0.5 cm. It forms an image at 16 cm from its second focal point. What is the magnifying power for a person whose near point is at 25 cm if the focal length of the eyepiece is 3 cm?

20. A microscope has an objective of focal length 16 mm and an eyepiece that gives an angular magnification of 5 for a person whose near point is 25 cm. The tube length is 18 cm. (*a*) What is the lateral magnification of the objective? (*b*) What is the magnifying power of the microscope?

21. A crude, symmetric, hand-held microscope consists of two converging 20-diopter lenses fastened in the ends of a tube 30 cm long. (*a*) What is the tube length of this microscope? (*b*) What is the lateral magnification of the objective? (*c*) What is the magnifying power of the microscope? (*d*) How far from the objective should the object be placed?

22. Repeat Problem 21 for the same two lenses separated by 40 cm.

32-5 The Telescope

23. A simple telescope has an objective of focal length 100 cm and an eyepiece of focal length 5 cm. It is used to look at the moon, which subtends an angle of about 0.009 rad. (*a*) What is the diameter of the image formed by the objective? (*b*) What angle is subtended by the final image at infinity? (*c*) What is the magnifying power of the telescope?

24. The objective lens of the refracting telescope at the Yerkes Observatory has a focal length of 19.5 m. It is used to look at the moon, which subtends an angle of about 0.009 rad. What is the diameter of the image of the moon formed by the objective?

25. The 200-in reflecting telescope at Mt. Palomar has a mirror with a diameter of 200 in = 5.1 m and a focal length of 1.68 m. (*a*) By what factor is the light-gathering power increased over the 40-in (1.02-m) diameter refract-

ing lens of the telescope at the Yerkes Observatory? (*b*) If the focal length of the eyepiece is 1.25 cm, what is the magnifying power of this telescope?

Level II

26. (*a*) Show that if the final image of a simple magnifier is at the near point of the eye rather than at infinity, the angular magnification is given by

$$M = \frac{x_{np}}{f} + 1$$

(*b*) Find the magnification of a 20-diopter lens for a person with a near point of 30 cm if the final image is at the near point. Draw a ray diagram for this situation.

27. A botanist examines a leaf using a convex lens of power 12 diopters as a simple magnifier. What is the expected angular magnification (*a*) if the final image is at infinity and (*b*) if the final image is at 25 cm?

28. Show that when the image of a simple magnifier is viewed at the near point, the lateral and angular magnification of the magnifier are equal.

29. A 35-mm camera has a picture size of 24 mm by 36 mm. It is to be used to take a picture of a person 175 cm tall in which the image of the person just fills the height (24 mm) of the film. How far should the person stand from the camera if the focal length of the lens is 50 mm?

30. A 35-mm camera with interchangeable lenses is used to take a picture of a hawk that has a wingspan of 2 m. The hawk is 30 m away. What would be the ideal focal length for the lens so that the image of the wings just fills the width of the film, which is 36 mm?

31. An astronomical telescope has a magnifying power of 7. The two lenses are 32 cm apart. Find the focal length of each lens.

32. The near point of a certain person is 80 cm. Reading glasses are prescribed so that he can read a book at 25 cm from his eye. The glasses are 2 cm from the eye. What power lenses should be used in the glasses?

33. A disadvantage of the astronomical telescope for terrestrial use (for example, at a football game) is that the image is inverted. A galilean telescope uses a converging lens as its objective but a diverging lens as its eyepiece. The image formed by the objective is behind the eyepiece at its focal point so that the final image is virtual, erect, and at infinity. (*a*) Show that the magnifying power is $M = -f_o/f_e$, where f_o is the focal length of the objective and f_e is that of the eyepiece (which is negative). (*b*) Draw a ray diagram to show that the final image is indeed virtual, erect, and at infinity.

34. A galilean telescope (see Problem 33) is designed so that the final image is at the near point, which is at 25 cm, rather than at infinity. The focal length of the objective is 100 cm and that of the eyepiece is −5 cm. (*a*) If the object distance is 30 m, where is the image of the objective? (*b*) What is the object distance for the eyepiece for which the final image is at the near point? (*c*) How far apart are the lenses? (*d*) If the object height is 1.5 m, what is the height of the final image? (*e*) What is the angular magnification of the image?

35. A compound microscope has an objective with a power of 45 diopter and an eyepiece with a power of 80 diopter. The lenses are separated by 28 cm. Assuming that the final image is formed 25 cm from the eye, what is the magnifying power?

Level III

36. At age 45, a person was fitted for reading glasses of power 2.1 diopters so she could read a newspaper at 25 cm. Now at age 55, she has discovered she must hold her newspaper at a distance of 40 cm in order to see it clearly with her glasses on. (*a*) Where was her near point at age 45? (*b*) Where is her near point now? (*c*) What power is now required for the lenses of her reading glasses so that she can again read at 25 cm? (Assume the glasses are 2.2 cm from her eyes.)

37. If you look into the wrong end of a telescope, that is, into the objective, you will see a distant object that is reduced in size. For a refracting telescope with an objective of focal length 2.25 m and an eyepiece of focal length 1.5 cm, by what factor is the angular size of the object reduced?

38. An aging physics professor discovers that he can see objects clearly only between 0.75 m and 2.5 m, so he decides he needs bifocals. The upper part of the lens allows him to see objects clearly at infinity, and the lower part allows him to see objects clearly at 25 cm. Assume that the lens is 2 cm from his eye. (*a*) Calculate the power of the lens required for the upper part his bifocals. (*b*) Calculate the power of the lens required for the lower part of his bifocals. (*c*) Is there a range of distances over which he cannot see objects clearly no matter which part of the bifocals he looks through? If so, what is that range? (*d*) Is there a range of distances over which he cannot see objects clearly whether or not he is wearing his bifocals? If so, what is that range?

39. A microscope has a magnifying power of −600 and an eyepiece with an angular magnification of 15. The objective lens is 22 cm from the eyepiece. Without making any approximations, calculate (*a*) the focal length of the eyepiece, (*b*) the location of the object at which it is in focus for a normal relaxed eye, and (*c*) the focal length of the objective lens.

40. A hunter lost in the mountains tries to make a telescope from two lenses, one of power 2.0 diopters and the other of power 6.5 diopters, and a cardboard tube. (*a*) What is the maximum possible magnifying power? (*b*) How long must the tube be? (*c*) Which lens should be used as the eyepiece? Why?

Chapter 33

Interference and Diffraction

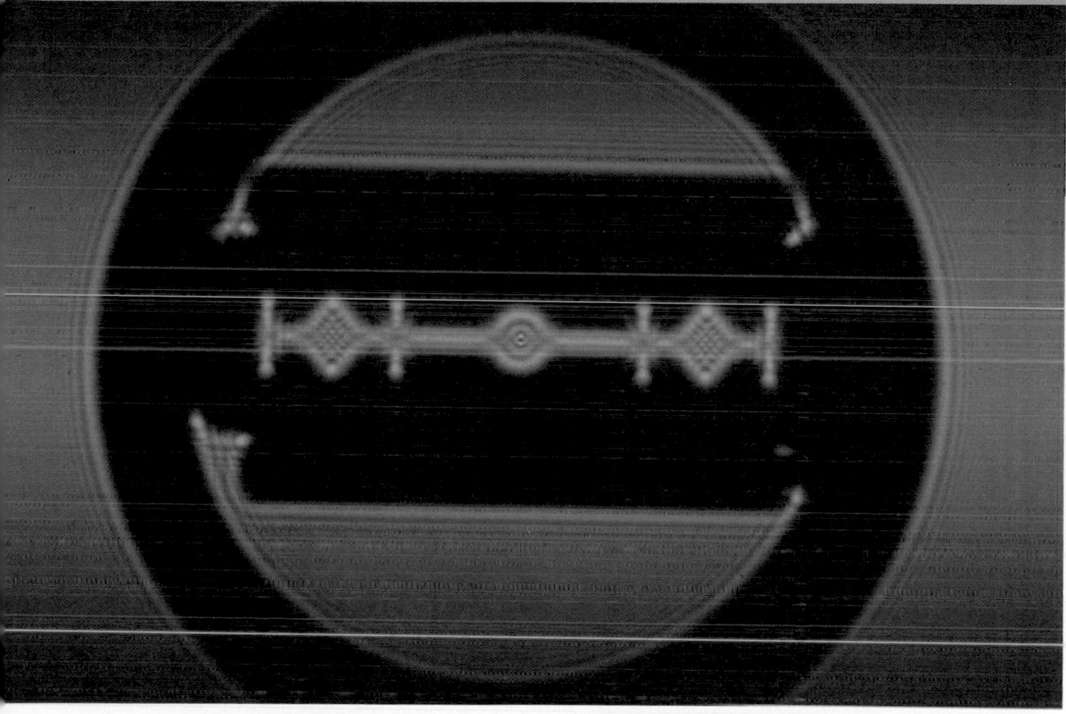

The diffraction of light incident on a razor blade.

Interference and diffraction are the important phenomena that distinguish waves from particles. Interference is the combining by superposition of two or more waves that meet at one point in space. Diffraction is the bending of waves around corners that occurs when a portion of a wavefront is cut off by a barrier or obstacle. The pattern of the resulting wave can be calculated by treating each point on the original wavefront as a point source according to Huygens' principle and calculating the interference pattern resulting from these sources.

In Chapter 14, we discussed the interference of sound waves from two point sources and the diffraction of sound qualitatively. Since the analytical treatment of interference and diffraction is the same for all waves whether they are sound waves, waves on strings, water waves, or electromagnetic waves, you should review Chapter 14 before you begin this chapter.

33-1 Phase Difference and Coherence

When two harmonic waves of the same frequency and wavelength but differing in phase combine, the resultant wave is a harmonic wave whose amplitude depends on the phase difference. If the phase difference is 0 or an integer times 360°, the waves are in phase and interfere constructively. The resultant amplitude equals the sum of the individual amplitudes, and the intensity (which is proportional to the square of the amplitude) is maximum. If the phase difference is 180° (π radians) or any odd integer times 180°, the waves are out of phase and interfere destructively. The resultant amplitude is then the difference between the individual amplitudes, and the intensity is a minimum. If the amplitudes are equal, the maximum intensity is 4 times that of either source and the minimum intensity is zero.

A phase difference between two waves is often the result of a difference in path length traveled by the two waves. A path difference of one wavelength produces a phase difference of 360°, which is equivalent to no phase difference at all. A path difference of one-half wavelength produces a 180° phase difference. In general, a path difference of Δr contributes a phase difference δ given by

$$\delta = \frac{\Delta r}{\lambda}2\pi = \frac{\Delta r}{\lambda}360° \qquad 33\text{-}1$$

Another cause of phase difference is the 180° phase change a wave sometimes undergoes upon reflection from a boundary surface. This phase change is analogous to the inversion of a pulse on a string when it reflects from a point where the density suddenly increases, such as when a light string is attached to a heavier string or rope. The inversion of the reflected pulse is equivalent to a phase change of 180° for a sinusoidal wave, which can be thought of as a series of pulses. When light traveling in air strikes the surface of a medium in which light travels more slowly, such as glass or water, there is a 180° phase change in the reflected light. When light is originally traveling in glass or water, there is no phase change in the light reflected from the glass–air or water–air surface. This is analogous to the reflection without inversion of a pulse on a heavy string at a point where the heavy string is attached to a lighter string.

As we saw in Chapter 14, interference of waves from two sources is not observed unless the sources are coherent, that is, unless the phase difference between the waves is constant in time. Because a light beam is usually the result of millions of atoms radiating independently, two light sources are usually not coherent. Indeed, the phase difference between the waves from such sources fluctuates randomly many times per second. Coherence in optics is often achieved by splitting the light beam from a single source into two or more beams, which can then be combined to produce an interference pattern. This splitting can be achieved by reflecting the light from the two closely spaced surfaces of a thin film (Section 33-2); the simultaneous reflection from and transmission through a half-silvered mirror, as in the Michelson interferometer (Section 33-3); or the diffraction of the beam by two small openings or slits in an opaque barrier (Section 33-4). Coherent sources can also be obtained by using a single point source and its image in a plane mirror for the two sources, an arrangement called *Lloyd's mirror*. Today, lasers are the most important sources of coherent light in the laboratory. Lasers have the property that all the atoms within a laser radiate in phase with one another, which leads to strong collimation of the light radiated.

Example 33-1

(*a*) What is the minimum path difference that will produce a phase difference of 180° for light of wavelength 800 nm? (*b*) What phase difference will that path difference produce in light of wavelength 700 nm?

(*a*) From Equation 33-1, we have

$$\delta = \frac{\Delta r}{\lambda}360° = 180°$$

$$\Delta r = \tfrac{1}{2}\lambda = \tfrac{1}{2}(800 \text{ nm}) = 400 \text{ nm}$$

(*b*) For $\lambda = 700$ nm and $\Delta r = 400$ nm, we have

$$\delta = \frac{\Delta r}{\lambda}360° = \frac{400 \text{ nm}}{700 \text{ nm}}360° = 206° = 3.59 \text{ rad}$$

33-2 Interference in Thin Films

Undoubtedly, you have noticed the colored bands in a soap bubble or in an oil film on a water-covered street. These bands are due to the interference of light reflected from top and bottom surfaces of the film. The different colors arise because of variations in the thickness of the film, causing interference for different wavelengths at different points.

Consider a thin film of water (such as a small section of a soap bubble) of uniform thickness viewed at small angles with the normal as shown in Figure 33-1. Part of the light is reflected from the upper, air–water surface. Since light travels more slowly in water than in air, there is a 180° phase change in this reflected light. Some of the light enters the film and is partially reflected by the bottom water–air surface. There is no phase change in this reflected light. If the light is nearly perpendicular to the surfaces, both the ray reflected from the top surface and the one reflected from the bottom surface can enter the eye at point P in the figure. The path difference between these two rays is $2t$, where t is the thickness of the film. This path difference produces a phase difference of $(2t/\lambda')360°$, where λ' is the wavelength of the light in the film. The wavelength in the film is related to the wavelength in air λ by $\lambda' = \lambda/n$ (Equation 30-4), where n is the index of refraction of the film. The total phase difference between these two rays is thus 180° plus that due to the path difference. Destructive interference occurs when the path difference $2t$ is zero or a whole number of wavelengths λ' (in the film). Constructive interference occurs when the path difference is an odd number of half wavelengths.

Interference of light rays from the front and back surface of a thin soap film. At the top where the film is very thin, the rays from the front surface of the film (that undergo a 180° phase change) and the rays from the back surface of the film (that do not change phase) interfere destructively and the film appears dark. At other parts of the film, the interference is destructive or constructive depending on the wavelength and on the film thickness.

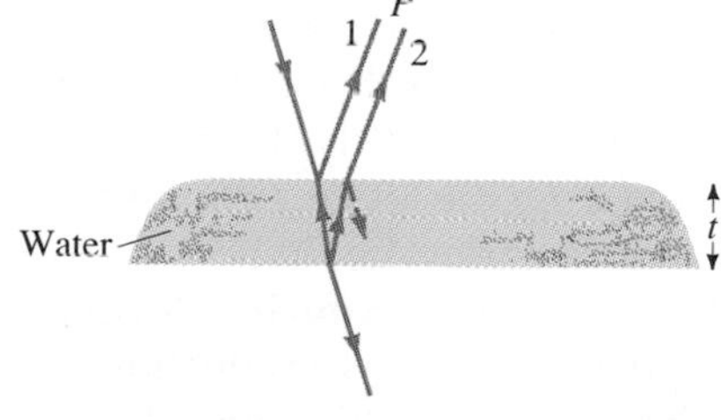

Figure 33-1 Light rays reflected from the top and bottom surfaces of a thin film are coherent because both rays come from the same source. If the light is incident nearly normally, the two reflected rays will be very close to each other and will produce interference.

We can express these conditions mathematically. When there is one phase change of 180° due to reflection, the conditions for interference are

$$\frac{2t}{\lambda'} = m \qquad m = 0, 1, 2, 3, \ldots \quad \text{(destructive)} \tag{33-2a}$$

$$\frac{2t}{\lambda'} = m + \frac{1}{2} \qquad m = 0, 1, 2, 3, \ldots \quad \text{(constructive)} \tag{33-2b}$$

Conditions for interference with one 180° phase change

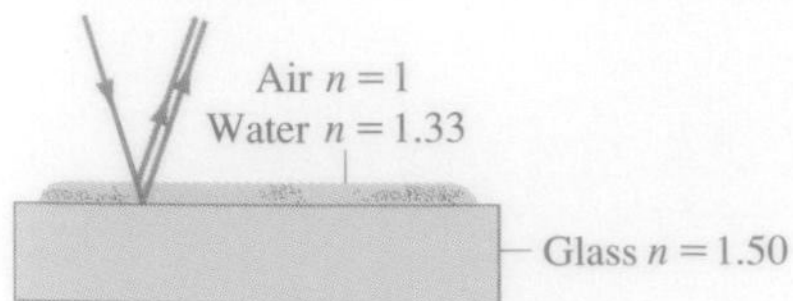

Figure 33-2 Interference of light reflected from a thin film of water resting on a glass surface. In this case, both rays undergo a change in phase of 180° upon reflection.

When a thin water film lies on a glass surface as in Figure 33-2, the ray that reflects from the lower, water–glass surface also undergoes a 180° phase change because the index of refraction of glass (about 1.5) is greater than that of water (about 1.33). Thus, both the rays shown in the figure have undergone a 180° phase change upon reflection. The phase difference between these rays is due solely to the path difference and is given by $\delta = (2t/\lambda')360°$. When there are two 180° phase changes upon reflection (or if there are no phase changes), the conditions for interference are

Conditions for interference with two 180° phase changes

$$\frac{2t}{\lambda'} = m \qquad m = 0, 1, 2, 3, \ldots \quad \text{(constructive)} \qquad \text{33-3}a$$

$$\frac{2t}{\lambda'} = m + \frac{1}{2} \qquad m = 0, 1, 2, 3, \ldots \quad \text{(destructive)} \qquad \text{33-3}b$$

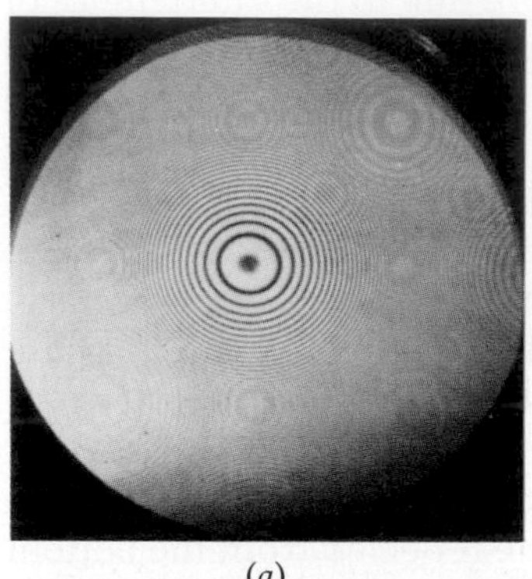
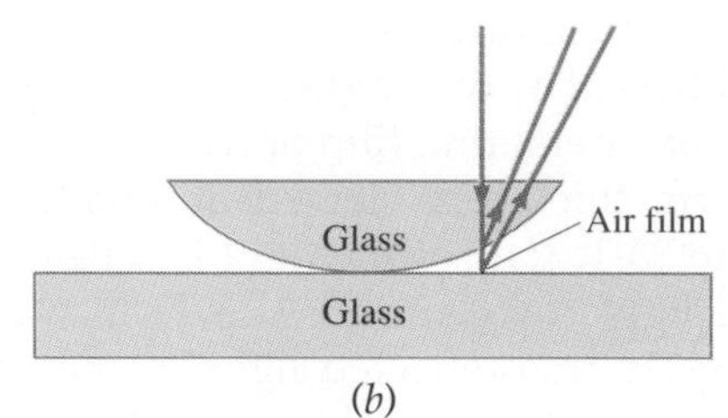

(*a*) (*b*)

Figure 33-3 (*a*) Newton's rings observed with light reflected from a thin film of air between a plane glass surface and a spherical glass surface. At the center, the thickness of the air film is negligible and the interference is destructive because of the phase change of one of the rays. (*b*) Glass surfaces for the observation of Newton's rings shown in part (*a*). The thin film in this case is the film of air between the glass surfaces.

When a thin film of varying thickness is viewed with monochromatic light, such as the yellow light from a sodium lamp, alternating bright and dark bands or lines called **fringes** are observed. The distance between a bright fringe and a dark fringe is that distance over which the film's thickness changes such that the path difference $2t$ is $\lambda'/2$. Figure 33-3*a* shows the interference pattern observed when light is reflected from an air film between a spherical glass surface and a plane glass surface in contact. These circular interference fringes are known as **Newton's rings.** Typical rays reflected at the top and bottom of the air film are shown in Figure 33-3*b*. Near the point of contact of the surfaces, where the path difference between the ray reflected from the upper, glass–air surface and the ray reflected from the lower, air–glass surface is essentially zero or is at least small compared with the wavelength of light, the interference is perfectly destructive because of the 180° phase shift of the ray reflected from the lower air–glass surface. This central region in Figure 33-3*a* is therefore dark. The first bright fringe occurs at the radius at which the path difference is $\lambda/2$, which contributes a phase difference of 180°. This adds to the phase shift due to reflection to produce a total phase difference of 360°, which is equivalent to a zero phase difference. The second dark region occurs at the radius at which the path difference is λ, and so on.

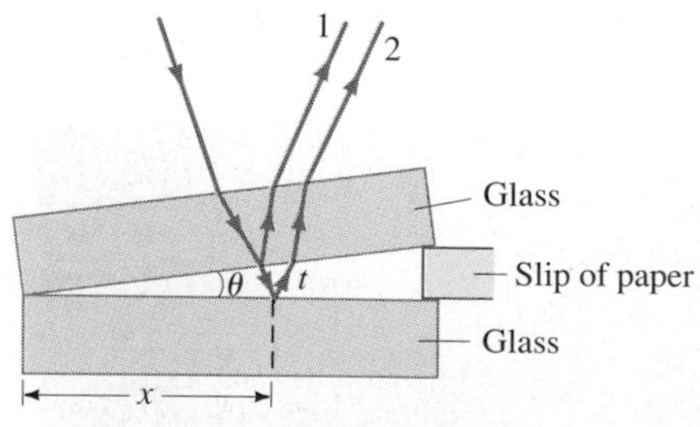

Figure 33-4 Light incident normally on a wedge-shaped film of air between two glass plates. The path difference $2t$ is proportional to x. When viewed from above, alternate bright and dark bands or fringes are seen.

Example 33-2

A wedge-shaped film of air is made by placing a small slip of paper between the edges of two flat pieces of glass as shown in Figure 33-4. Light of wavelength 500 nm is incident normally on the glass plates, and interference fringes are observed by reflection. If the angle θ made by the plates is 3×10^{-4} rad, how many interference fringes per centimeter are observed?

Because of the 180° phase shift in the ray reflected from the bottom plate, the first fringe near the point of contact (where the path difference is zero) will be dark. Let x be the horizontal distance to the mth dark fringe, where the plate separation is t as shown in the figure. Since the angle θ is very small, it is given approximately by

$$\theta = \frac{t}{x}$$

Using Equation 33-2a for m, we have

$$m = \frac{2t}{\lambda'} = \frac{2t}{\lambda}$$

since the film is an air film. Substituting $t = x\theta$ gives

$$m = \frac{2x\theta}{\lambda}$$

or

$$\frac{m}{x} = \frac{2\theta}{\lambda} = \frac{2(3 \times 10^{-4})}{5 \times 10^{-5}\ \text{cm}} = 12\ \text{cm}^{-1}$$

where we have used $\lambda = 5 \times 10^{-7}$ m $= 5 \times 10^{-5}$ cm. We therefore observe 12 dark fringes per centimeter. In practice, the number of fringes per centimeter, which is easy to count, can be used to determine the angle.

Note that if the angle of the wedge is increased, the fringes become more closely spaced. The distance along the glass between adjacent dark (or adjacent bright) fringes is that distance that results in an additional path difference equal to the wavelength of the light in the film. If the angle of the wedge is increased, this distance is decreased.

Exercise

How many fringes per centimeter are observed in Example 33-2 if light of wavelength 650 nm is used? (Answer: $9.2\ \text{cm}^{-1}$)

Figure 33-5a shows interference fringes produced by a wedge-shaped air film between two flat glass plates as in Example 33-2. The straightness of the fringes indicates the flatness of the glass plates. Such plates are said to be **optically flat.** A similar wedge-shaped air film formed by two ordinary glass plates yields the irregular fringe pattern in Figure 33-5b, which indicates that these plates are not optically flat.

(*a*)

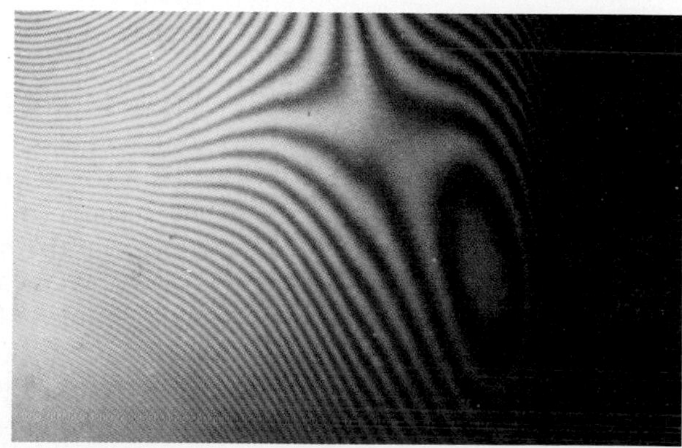

(*b*)

Figure 33-5 (*a*) Straight-line fringes from a wedge-shaped film of air like that in Figure 33-4. The straightness of the fringes indicates the flatness of the glass plates. (*b*) Fringes from a wedge-shaped film of air between glass plates that are not optically flat.

(a)

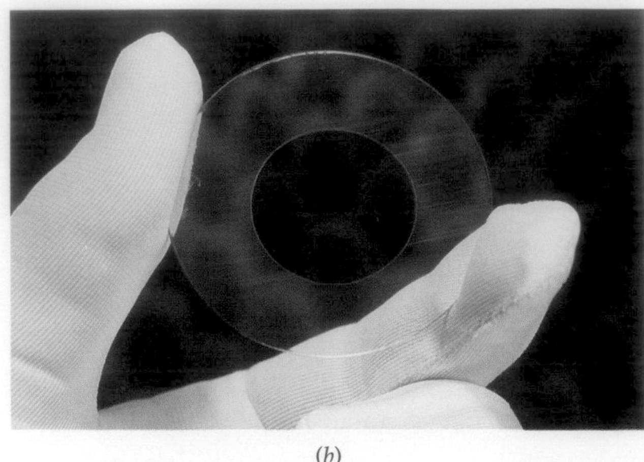

(b)

(*a*) A narrow-angle lens (*left*) and a wide-angle lens (*right*) prior to assembly in the cameras of Voyager 2. Each lens has two types of thin-film coatings: a low thermal emittance coating that reflects thermal wavelengths to prevent heat loss and keep the telescope elements warm and an antireflecting coating at optical wavelengths. (*b*) The antireflecting film used on the lenses in (*a*) is demonstrated here on the central region of a clear, polycarbonate disk.

A nonreflecting lens is made by covering the lens with a thin film of a material that has an index of refraction of about 1.22, which is between that of glass and that of air, so that the intensities of the light reflected from the top and bottom surfaces of the film are approximately equal. Since both rays undergo a 180° phase change, there is no phase difference between the rays due to reflection. The thickness of the film is chosen to be $\lambda'/4$, where $\lambda' = \lambda/n$ and λ is in the middle of the visible spectrum, so that there is a phase change of 180° due to the path difference of $\lambda'/2$. Reflection from the coated surface is thus minimized.

Questions

1. Why must a film used to observe interference colors be thin?
2. If the angle of a wedge-shaped air film such as that in Example 33-2 is too large, fringes are not observed. Why?
3. The spacing between Newton's rings decreases rapidly as the diameter of the rings increases. Explain qualitatively why this occurs.

33-3 The Michelson Interferometer

An **interferometer** is a device that uses interference fringes to make precise measurements of distance. Figure 33-6 is a schematic diagram of a Michelson interferometer. Light from a broad source strikes plate A, a beam splitter that is partially silvered so that the light is partially reflected and partially transmitted. The reflected beam travels to mirror M_2 and is reflected back toward the eye at O. The transmitted beam travels through a compensating plate B, which is of the same thickness as plate A, to mirror M_1 and is reflected back to plate A and then to the eye at O. The purpose of the compensating plate B is to make both beams pass through the same thickness of glass. Mirror M_1 is fixed, but mirror M_2 can be moved back and forth with fine and accurately calibrated screw adjustments. The two beams combine at O and form an interference pattern. This pattern is most easily understood by considering the mirror M_2 and the image of mirror M_1 produced by the mirror in the beam splitter A. This image is labeled M_1' in the diagram. If the mirrors M_1 and M_2 are exactly perpendicular to each other and are equidistant from the beam splitter, the image M_1' will coincide with M_2. If not, M_1' will be slightly displaced and will make a small angle with M_2, as shown in the diagram. The interference pattern at O will then be that of a thin, wedge-shaped film of air between M_1' and M_2 similar to that discussed in Example 33-2. If mirror M_2 is now moved, the fringe pattern will shift. If, for

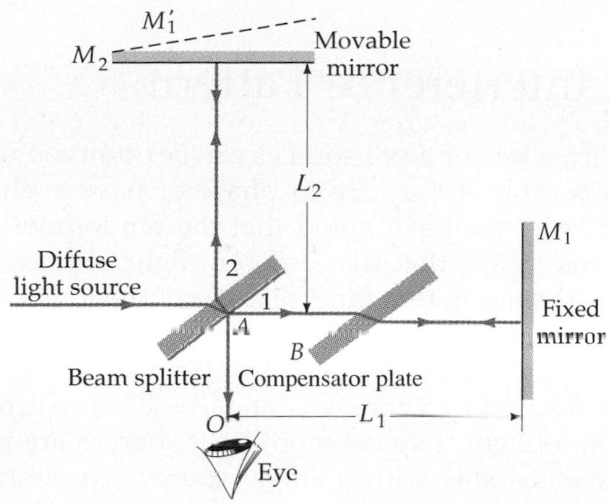

Figure 33-6 Michelson interferometer. The dashed line M_1' is the image of mirror M_1 in the mirror A. The interference fringes observed are those of a small wedge-shaped film of air formed by the sources M_1' and M_2. As M_2 is moved, the fringes move across the field of view.

example, the mirror M_2 is moved toward the splitter by a distance equal to $\frac{1}{4}\lambda$, the thickness of the wedge will increase by $\frac{1}{4}\lambda$ at each point. This will introduce an additional path difference of $\frac{1}{2}\lambda$ everywhere in the wedge (because the light traverses the wedge twice). The fringe pattern will move over by one-half fringe; that is, a previously dark fringe will now be a bright fringe, and so forth. If the distance mirror M_2 is moved is known, the wavelength of the light can be determined. Michelson used such an interferometer to measure the wavelength of a spectral line of light emitted by krypton 86 in terms of the standard meter bar. This measurement was then used to redefine the standard meter in terms of this wavelength. (The definition of the standard meter has since been changed. It is now defined in terms of the speed of light.)

Another use for the Michelson interferometer is to measure the index of refraction of air (or of some other gas). One of the beams from plate A is made to pass through a container that can be evacuated. The wavelength of the light in air λ' is related to that in a vacuum by $\lambda' = \lambda/n$, where n is the index of refraction of air (about 1.0003). When the container is evacuated, the wavelength of the light passing through it increases so that there are fewer waves in the length of the container. This causes a shift in the fringe pattern. By measuring the shift, the index of refraction can be determined (see Problem 10).

Michelson also used his interferometer in a famous experiment with Edward W. Morley in 1887 in which they attempted to measure the difference between the speed of light relative to the earth in the direction of motion of the earth and in a perpendicular direction. This experiment will be described in some detail in the next chapter.

Example 33-3

A film of index of refraction 1.33 and thickness 12 μm is inserted in one arm of a Michelson interferometer. The light used has a wavelength of 589 nm in air. By how many fringes is the interference pattern shifted?

The number of waves in the thin film N_f is the thickness $2t$ divided by the wavelength $\lambda' = \lambda/n$:

$$N_f = \frac{2t}{\lambda'} = \frac{2nt}{\lambda} = \frac{2(1.33)(12 \times 10^{-6}\ \text{m})}{589 \times 10^{-9}\ \text{m}} = 54.2 \text{ waves}$$

The number of waves originally in this space when it was occupied by air is

$$N_a = \frac{2t}{\lambda} = \frac{2(12 \times 10^{-6}\ \text{m})}{589 \times 10^{-9}\ \text{m}} = 40.8 \text{ waves}$$

There are thus $54.2 - 40.8 = 13.4$ more waves in one arm of the interferometer, so the interference pattern will be shifted by 13.4 fringes.

Figure 33-7 Plane water waves in a ripple tank encountering a barrier with a small opening. The waves to the right of the barrier are circular waves that are concentric about the opening just as if there were a point source at the opening.

33-4 The Two-Slit Interference Pattern

Interference patterns of light from two or more sources can be observed only if sources are coherent, that is, only if they are in phase or have a phase difference that is constant in time. We have noted that the randomness of the emissions of light by atoms means that two different light sources are generally incoherent. The interference in thin films discussed previously can be observed because the two beams come from the same light source but are separated by reflection.

In the famous experiment devised by Thomas Young in 1801 in which he demonstrated the wave nature of light, two coherent light sources are produced by illuminating two parallel slits with a single source. We assume here that each slit is very narrow. (We will treat the general case in Section 33-8.) We saw in Chapter 14 that when a wave encounters a barrier with a very small opening, the opening acts as a point source of waves (Figure 33-7). In Young's experiment each slit acts as a line source, which is equivalent to a point source in two dimensions. The interference pattern is observed on a screen far from the slits, which are separated by a distance d. At very large distances from the slits, the lines from the two slits to some point P on the screen are approximately parallel, and the path difference is approximately $d \sin \theta$, as shown in Figure 33-8*c*. We thus have interference maxima at an angle given by

Two-slit interference maxima

$$d \sin \theta = m\lambda \qquad m = 0, 1, 2, \ldots \tag{33-4}$$

The interference minima occur at

Two-slit interference minima

$$d \sin \theta = (m + \tfrac{1}{2})\lambda \qquad m = 0, 1, 2, \ldots \tag{33-5}$$

The phase difference δ at a point P is $2\pi/\lambda$ times the path difference $d \sin \theta$:

$$\delta = \frac{2\pi}{\lambda} d \sin \theta \tag{33-6}$$

The distance y_m measured along the screen from the central point to the mth bright fringe (see Figure 33-8*b*) is related to the angle θ by

$$\tan \theta = \frac{y_m}{L}$$

where L is the distance from the slits to the screen. For small θ, we have

$$\sin \theta \approx \tan \theta = \frac{y_m}{L}$$

so $d \sin \theta$ is given approximately by

$$d \sin \theta \approx d \frac{y_m}{L}$$

Substituting this into Equation 33-4, we obtain

$$d \frac{y_m}{L} = m\lambda$$

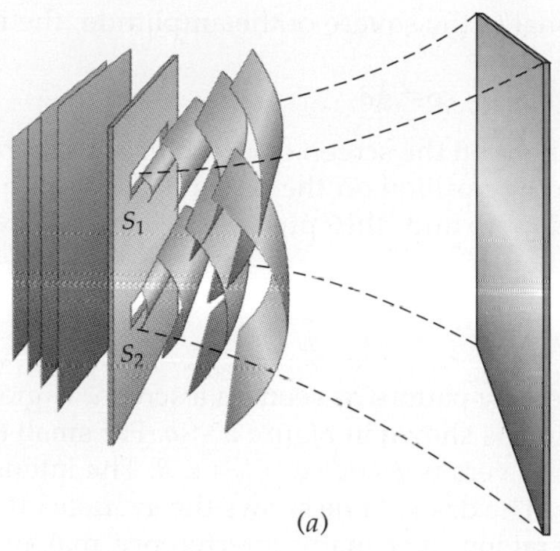

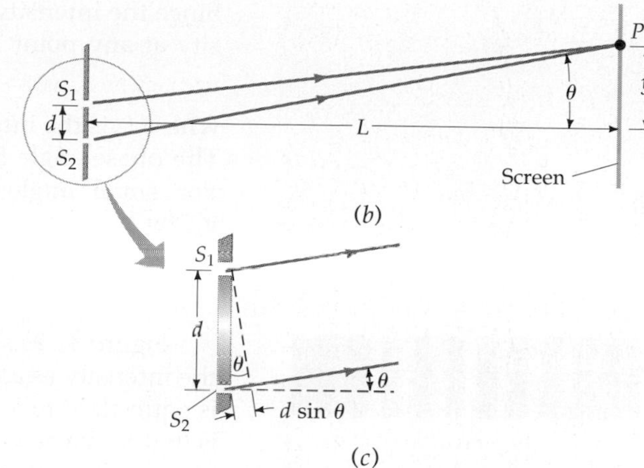

Figure 33-8 (*a*) Two slits act as coherent sources of light for the observation of interference in Young's experiment. Cylindrical waves from the slits overlap and produce an interference pattern on a screen far away. (*b*) Geometry for relating the distance y measured along the screen to L and θ. (*c*) Because the screen is very far away compared with the slit separation, the rays from the slits to a point on the screen are approximately parallel, and the path difference between the two rays is $d \sin \theta$.

Thus, for small angles (which is nearly always the case), the distance measured along the screen to the mth bright fringe is given by

$$y_m = m\frac{\lambda L}{d} \tag{33-7}$$

Note from this result that the fringes are equally spaced on the screen, with the distance between two successive bright fringes given by

$$\Delta y = \frac{\lambda L}{d}$$

To calculate the intensity of the light on the screen at a general point P, we need to add two harmonic wave functions that differ in phase, as we did in Chapter 14 where we discussed variations in intensity due to the interference of sound waves. The wave functions for electromagnetic waves are the electric-field vectors. Let E_1 be the electric field at some point P on the screen due to the waves from slit 1, and let E_2 be the electric field at that point due to waves from slit 2. Since the angles of interest are small, we can assume that these fields are parallel and consider only their magnitudes. Both electric fields oscillate with the same frequency (since they result from a single source that illuminates both slits) and they have the same amplitude. (The path difference is only of the order of a few wavelengths of light at most.) They have a phase difference δ given by Equation 33-6. If we represent these wave functions by

$$E_1 = A_0 \sin \omega t \qquad \text{and} \qquad E_2 = A_0 \sin(\omega t + \delta)$$

the resultant wave function is

$$E = E_1 + E_2 = A_0 \sin \omega t + A_0 \sin(\omega t + \delta) \tag{33-8}$$

We use the following trigonometric identity for the two sine functions:

$$\sin \alpha + \sin \beta = 2 \cos \tfrac{1}{2}(\alpha - \beta) \sin \tfrac{1}{2}(\alpha + \beta) \tag{33-9}$$

Equation 33-8 then becomes

$$E = 2A_0 \cos \tfrac{1}{2}\delta \sin(\omega t + \tfrac{1}{2}\delta) \tag{33-10}$$

The amplitude of the resultant wave is thus $2A_0 \cos \frac{1}{2}\delta$. It has its maximum value of $2A_0$ when the waves are in phase ($\delta = 0$ or an integer times 2π) and is zero when they are 180° out of phase ($\delta = \pi$ or an odd integer times π).

Since the intensity is proportional to the square of the amplitude, the intensity at any point P is

$$I = 4I_0 \cos^2 \tfrac{1}{2}\delta \qquad \text{33-11}$$

where I_0 is the intensity of the light on the screen from either slit separately. The phase angle δ is related to the position on the screen by Equation 33-6. For small angles, $d \sin\theta \approx dy_m/L$ and the phase angle is related to y_m by

$$\delta = \frac{2\pi}{\lambda} d \sin\theta \approx \frac{2\pi}{\lambda}\frac{y_m d}{L} \qquad \text{33-12}$$

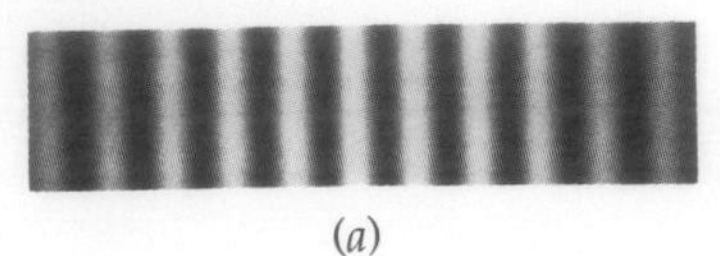

(a)

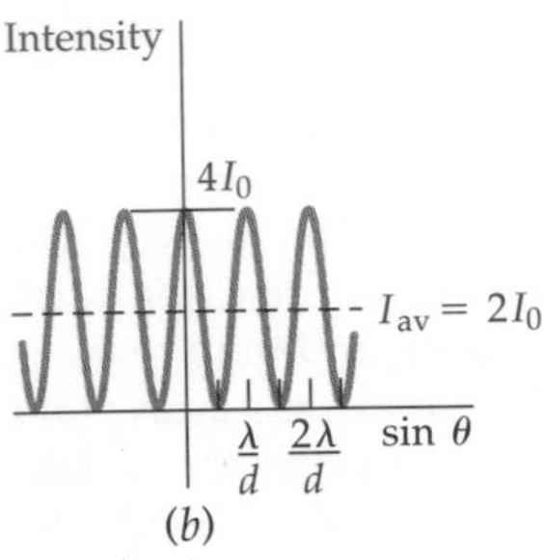

(b)

Figure 33-9 (*a*) The interference pattern observed on a screen far away from the two slits of Figure 33-8. (*b*) Plot of intensity versus $\sin\theta$. The maximum intensity is $4I_0$, where I_0 is the intensity due to each slit separately. The average intensity (dashed line) is $2I_0$. For small θ, this is also a plot of intensity versus the distance y measured along the screen because $y = L\tan\theta \approx L\sin\theta$.

Figure 33-9*a* shows the intensity pattern as seen on a screen. A graph of the intensity as a function of $\sin\theta$ is shown in Figure 33-9*b*. For small θ, this is equivalent to a plot of intensity versus y since $y \approx L\sin\theta$. The intensity I_0 is that from each slit separately. The dashed line shows the average intensity $2I_0$, which is the result of averaging over many interference maxima and minima. This is the intensity that would arise from the two sources if they acted independently without interference. In other words, it is the intensity we would observe if the sources were incoherent because there would then be an additional phase difference between them that would fluctuate randomly so that only the average intensity would be observed.

Example 33-4

Two narrow slits separated by 1.5 mm are illuminated by sodium light of wavelength 589 nm. Interference fringes are observed on a screen 3 m away. Find the spacing of the fringes on the screen.

The distance y_m measured along the screen to the mth bright fringe is given by Equation 33-7, with $L = 3$ m, $d = 1.5$ mm, and $\lambda = 589$ nm. The spacing of the fringes is this distance divided by the number of fringes, or y_m/m. Solving Equation 33-7 for y_m/m and substituting in the given values, we obtain

$$\frac{y_m}{m} = \lambda\frac{L}{d} = \frac{(589 \times 10^{-9}\ \text{m})(3\ \text{m})}{0.0015\ \text{m}} = 1.18 \times 10^{-3}\ \text{m} = 1.18\ \text{mm}$$

The fringes are thus 1.18 mm apart.

Figure 33-10 shows another method of producing the two-slit interference pattern, an arrangement known as **Lloyd's mirror.** A single slit is placed at a distance $\frac{1}{2}d$ above the plane of a mirror. Light striking the screen directly from the source interferes with that reflected from the mirror. The reflected light can be considered to come from the virtual image of the slit formed by the mirror. Because of the 180° change in phase upon reflection at the mirror, the interference pattern is that of two coherent line sources that differ in phase by 180°. The pattern is the same as that shown in Figure 33-9 for two slits except that the maxima and minima are interchanged. The central fringe just above the mirror at a point equidistant from the sources is

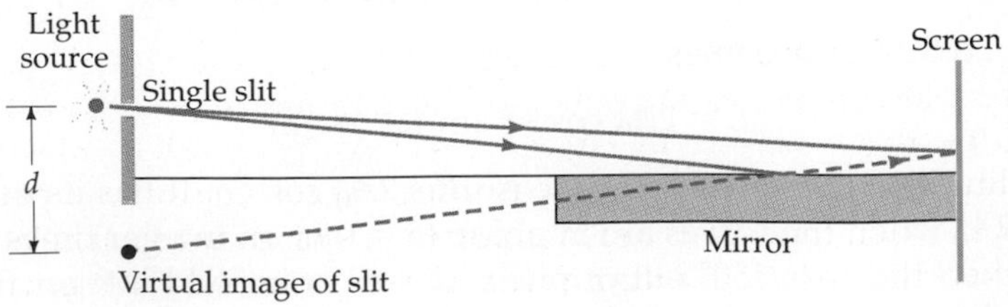

Figure 33-10 Lloyd's mirror for producing a two-slit interference pattern. The two sources (the slit and its image) are coherent and are 180° out of phase. The central interference band at the point equidistant from the sources is dark.

dark. Constructive interference occurs at points for which the path difference is a half wavelength or any odd number of half wavelengths. At these points, the 180° phase difference due to the path difference combines with the 180° phase difference of the sources to produce constructive interference.

Exercise

A point source of light (λ = 589 nm) is placed 0.4 mm above a mirror. Interference fringes are observed on a screen 6 m away. Find the spacing of the fringes on the screen. (Answer: 4.42 mm)

Question

4. When destructive interference occurs, what happens to the energy in the light waves?

33-5 The Addition of Harmonic Waves Using Phasors

To calculate the interference pattern produced by three, four, or more coherent light sources and to calculate the diffraction pattern of a single slit, we need to combine several harmonic waves of the same frequency that differ in phase. A simple geometric interpretation of harmonic wave functions leads to a method of adding harmonic waves of the same frequency by geometric construction. This method lets us find the sum of two or more harmonic waves geometrically without having to remember the trigonometric identity given in Equation 33-9. It is useful even if the amplitudes of the waves are different. The method is based on the fact that the y (or x) component of the resultant of two vectors equals the sum of the y (or x) components of the vectors.

Let

$$E_1 = A_1 \sin \omega t \qquad \text{and} \qquad E_2 = A_2 \sin (\omega t + \delta)$$

be the wave functions of the two waves at some point. (We have chosen our time t so that $E_1 = 0$ at $t = 0$.) We can simplify our notation by writing α for the quantity ωt. Our problem is then to find the sum

$$E_1 + E_2 = A_1 \sin \alpha + A_2 \sin (\alpha + \delta)$$

Consider a vector in the xy plane of magnitude A_1 that makes an angle α with the x axis (Figure 33-11). The y component of this vector is $A_1 \sin \alpha$, which is the wave function E_1. As the time varies, this vector rotates in the xy plane with angular frequency ω. Such a vector is called a **phasor.** We encountered phasors in our study of ac circuits in Section 28-3. The wave function $E_2 = A_2 \sin (\alpha + \delta)$ is the y component of a phasor of magnitude A_2 that makes an angle $\alpha + \delta$ with the x axis. By the laws of vector addition, the sum of these components equals the y component of the resultant vector $\mathbf{A}'$, as shown in Figure 33-11. The y component of the resultant vector, $A' \sin (\alpha + \delta')$, is a harmonic wave function that is the sum of the two original wave functions:

$$A_1 \sin \alpha + A_2 \sin (\alpha + \delta) = A' \sin (\alpha + \delta') \qquad 33\text{-}13$$

where A' (the amplitude of the resultant wave) and δ' (the phase of the resultant wave relative to the first wave) are found by adding the phasors representing the waves as in Figure 33-11. As time varies, α varies. The phasors representing the two wave functions and the resultant phasor representing the resultant wave function rotate in space, but their relative positions do not change because they all rotate with the same angular velocity ω.

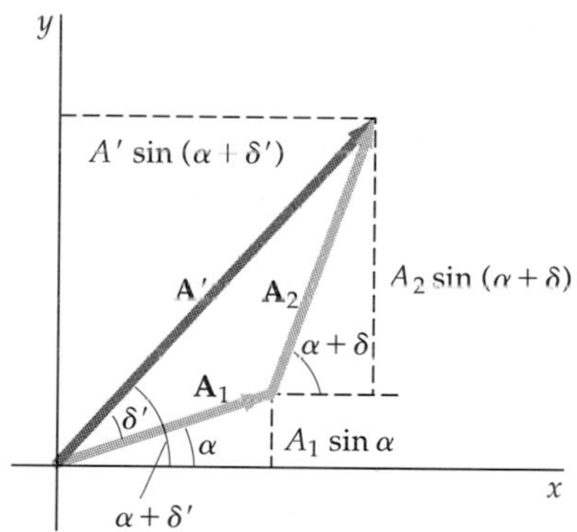

Figure 33-11 The wave function $A_1 \sin \alpha$ is the y component of the vector $\mathbf{A}_1$, which makes an angle α with the x axis. The wave function $A_2 \sin (\alpha + \delta)$ is the y component of the vector $\mathbf{A}_2$, which makes an angle $\alpha + \delta$ with the x axis. The sum of these wave functions is $A' \sin (\alpha + \delta')$, which is the y component of the resultant vector $\mathbf{A}' = \mathbf{A}_1 + \mathbf{A}_2$.

Example 33-5

Use the phasor method of addition to derive Equation 33-10 for the superposition of two waves of the same amplitude.

Figure 33-12 shows the phasors representing two waves of equal amplitude A_0 and the resultant wave of amplitude A'. These three phasors form an isosceles triangle in which the two equal angles are δ'. Since the sum of these angles equals the exterior angle δ, we have

$$\delta' = \tfrac{1}{2}\delta$$

The amplitude A' can be found from the right triangles formed by bisecting the resultant phasor, as shown in Figure 33-12*b*. From these triangles, we have

$$\cos \tfrac{1}{2}\delta = \frac{\tfrac{1}{2}A'}{A_0}$$

Therefore, the amplitude is given by $A' = 2A_0 \cos \tfrac{1}{2}\delta$, and the resultant wave is

$$A' \sin (\alpha + \delta') = 2A_0 \cos \tfrac{1}{2}\delta \sin (\alpha + \tfrac{1}{2}\delta)$$

which is in agreement with Equation 33-10 when $\alpha = \omega t$.

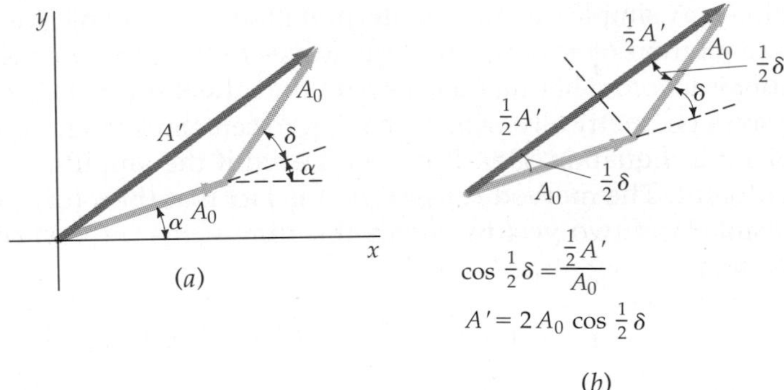

Figure 33-12 The phasor addition of two waves having equal amplitudes A_0 and a phase difference of δ. (*a*) The phasors at a particular time at which $\alpha = \omega t$. (*b*) Geometric construction for finding the amplitude A' of the resultant wave.

Example 33-6

Find the resultant of the two waves

$$E_1 = 4 \sin (\omega t) \qquad \text{and} \qquad E_2 = 3 \sin (\omega t + 90^\circ)$$

Figure 33-13 shows the phasor diagram for this addition. The phasors make an angle of 90° with each other. The resultant of these two phasors has a magnitude of 5 and makes an angle of 37° with the first phasor, as shown in the figure. The sum of these two waves is therefore

$$E_1 + E_2 = 5 \sin (\omega t + 37^\circ)$$

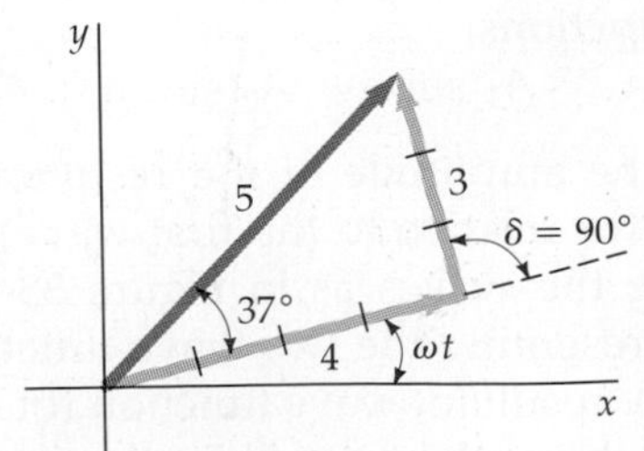

Figure 33-13. Phasor diagram for the addition of the two waves in Example 33-6.

33-6 Interference Pattern of Three or More Equally Spaced Sources

If we have three or more sources that are equally spaced and in phase with each other, the intensity pattern on a screen far away is similar to that due to two sources, but there are important differences. The positions on the screen of the intensity maxima are the same no matter how many sources we have, but these maxima have much greater intensities and are much sharper if there are many sources. We can calculate the intensity pattern for the interference of three or more equally spaced sources using the phasor method of adding harmonic waves discussed in the previous section. We will be most interested in the points of perfectly constructive interference and perfectly destructive interference, that is, in the interference maxima and minima.

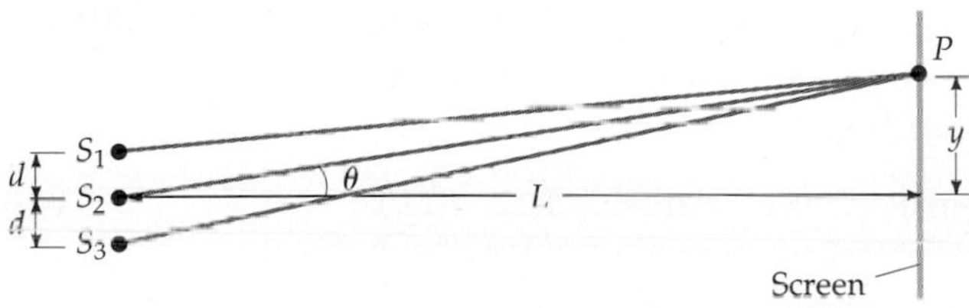

Figure 33-14 Geometry for calculating the intensity pattern far from three equally spaced sources that are in phase.

Figure 33-14 shows the situation for the case of three sources. The geometry is the same as for two sources. At a great distance from the sources, the rays from the sources to a point P on the screen are approximately parallel. The path difference between the first and second source is then $d \sin \theta$, as before, and that between the first and third source is $2d \sin \theta$. The wave at point P is the sum of three waves. Let $\alpha = \omega t$ be the phase of the first wave at point P. We thus have the problem of adding three waves of the form

$$E_1 = A_0 \sin \alpha$$
$$E_2 = A_0 \sin (\alpha + \delta) \qquad 33\text{-}14$$
$$E_3 = A_0 \sin (\alpha + 2\delta)$$

where

$$\delta = \frac{2\pi}{\lambda} d \sin \theta \approx \frac{2\pi}{\lambda} \frac{yd}{L} \qquad 33\text{-}15$$

as in the two-slit problem.

It is easiest to analyze the resulting pattern in terms of the phase angle δ between the first and second sources or between the second and third sources instead of directly in terms of the space angle θ. If we know the resultant amplitude due to the three waves at some point P corresponding to a particular phase angle δ, we can relate this phase angle to the space angle θ by using Equation 33-15.

At $\theta = 0$, the phase angle δ is zero; that is, all the waves are in phase. The amplitude of the resultant wave is 3 times that of each individual wave. Since the intensity is proportional to the square of the amplitude, the intensity at $\delta = 0$ is 9 times that due to each source acting separately. As the angle δ increases from $\theta = 0$, the phase angle δ increases and the intensity decreases. The position $\theta = 0$ is thus a position of maximum intensity.

Figure 33-15 shows the phasor addition of three waves for a phase angle δ of about 30° = $\pi/6$ rad. (This corresponds to a point P on the screen for which θ is given by $\sin \theta = \lambda\delta/2\pi d = \lambda/12d$.) The resultant amplitude is considerably less than 3 times that of each source. As the phase angle δ

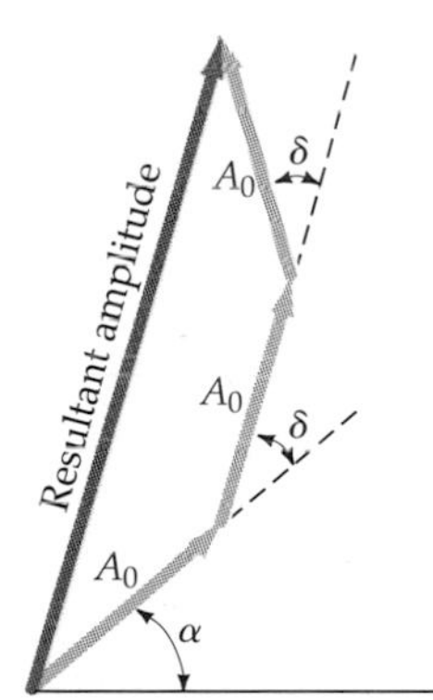

Figure 33-15 Phasor diagram for determining the resultant amplitude due to three waves, each of amplitude A_0, that have phase differences of δ and 2δ due to path differences of $d \sin \theta$ and $2d \sin \theta$. The angle $\alpha = \omega t$ varies with time but does not affect the calculation of the resultant amplitude.

increases, the resultant amplitude decreases until the amplitude is zero at $\delta = 120°$. For this phase difference, the three phasors form an equilateral triangle (Figure 33-16). This first interference minimum for three sources occurs at a smaller phase angle (and therefore at a smaller space angle θ) than it does for only two sources (for which the first minimum occurs at $\delta = 180°$). As δ increases from 120°, the resultant amplitude increases, reaching a **secondary maximum** near $\delta = 180°$. At the phase angle $\delta = 180°$, the amplitude is the same as that from a single source since the waves from the first two sources cancel each other, leaving only the third. The intensity of the secondary maximum is one-ninth that of the maximum at $\theta = 0$. As δ increases beyond 180°, the amplitude again decreases and is zero at $\delta = 180° + 60° = 240°$. For δ greater than 240°, the amplitude increases and is again 3 times that of each source when $\delta = 360°$. This phase angle corresponds to a path difference of 1 wavelength for the waves from the first two sources and 2 wavelengths for the waves from the first and third sources. Hence, the three waves are in phase at this point. The largest maxima, called the **principal maxima,** are at the same positions as for just two sources, which are those points corresponding to the angles θ given by

Principal interference maxima

$$d \sin \theta = m\lambda \qquad m = 0, 1, 2, \ldots \tag{33-16}$$

These maxima are stronger and narrower than those for two sources. They occur at points for which the path difference between adjacent sources is zero or an integral number of wavelengths.

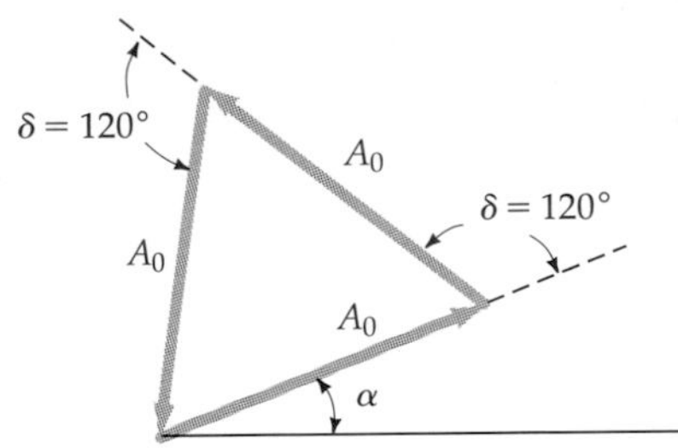

Figure 33-16 The resultant amplitude for the waves from three sources is zero when δ is 120°. This interference minimum occurs at a smaller angle θ than does the first minimum for two sources, which occurs when δ is 180°.

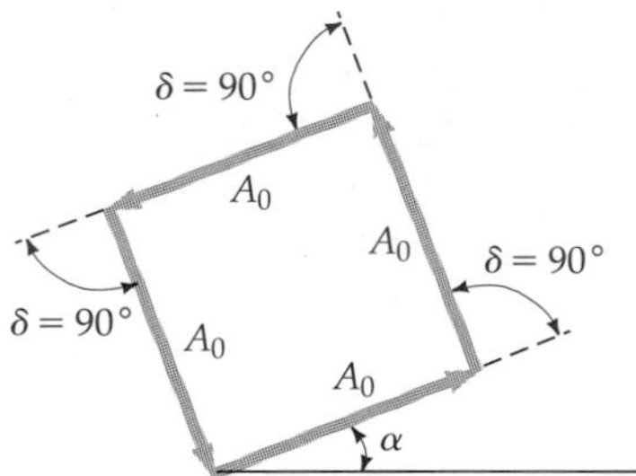

Figure 33-17 Phasor diagram for the first minimum for four equally spaced in-phase sources. The amplitude is zero when the phase difference of the waves from adjacent sources is 90°.

These results can be generalized to more than three sources. For example, if we have four equally spaced sources that are in phase, the principal interference maxima are again given by Equation 33-16, but these maxima are even more intense and narrower and there are two small secondary maxima between each pair of principal maxima. At $\theta = 0$, the intensity is 16 times that due to a single source. The first interference minimum occurs when δ is 90°, as can be seen from the phasor diagram of Figure 33-17. The first secondary maximum is near $\delta = 120°$, where the waves from three of the sources cancel, leaving only the wave from the fourth source. The intensity of the secondary maximum is approximately one-sixteenth that of the central maximum. There is another minimum at $\delta = 180°$, another secondary maximum near $\delta = 240°$, and another minimum at $\delta = 270°$ before the next principal maximum at $\delta = 360°$.

Figures 33-18*a–c* show the intensity patterns for two equally spaced sources, three equally spaced sources, and four equally spaced sources. In

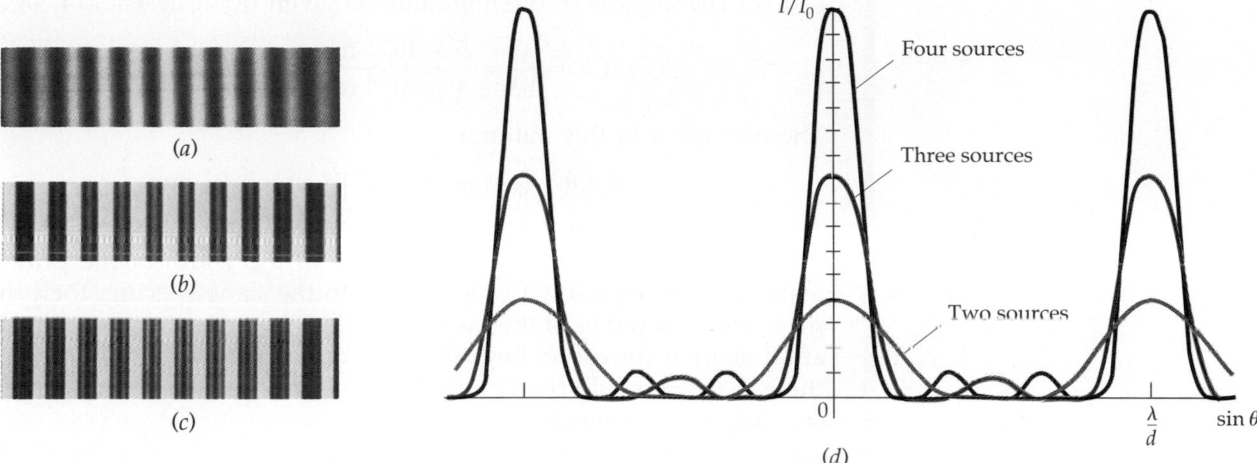

Figure 33-18 Intensity patterns for (a) two, (b) three, and (c) four equally spaced coherent sources. There is a secondary maximum between each pair of principal maxima for three sources, and two secondary maxima for four sources. (d) Plot of intensity versus sin θ for two, three, and four equally spaced coherent sources.

Figure 33-18*d*, I_0 is the intensity due to each source acting separately. For three sources, there is a very small secondary maximum between each pair of principal maxima, and the principal maxima are sharper and more intense than those due to just two sources. For four sources, there are two small secondary maxima between each pair of principal maxima, and the principal maxima are even more narrow and intense.

From this discussion, we can see that as we increase the number of sources, the intensity becomes more and more concentrated in the principal maxima given by Equation 33-16, and these maxima become narrower. For N sources, the intensity of the principal maxima is N^2 times that due to a single source. The first minimum occurs at a phase angle of $\delta = 360°/N$, for which the N phasors form a closed polygon of N sides. There are $N - 2$ secondary maxima between each pair of principal maxima. These secondary maxima are very weak compared with the principal maxima. As the number of sources is increased, the principal maxima become sharper and more intense, and the intensities of the secondary maxima become negligible compared to those of the principal maxima.

Example 33-7

Four equally spaced coherent light sources with a wavelength of 500 nm are separated by a distance $d = 0.1$ mm. The interference pattern is viewed on a screen at a distance of 1.4 m. Find the positions of the principal interference maxima and compare their width with that for just two sources with the same spacing.

According to Equation 33-16, the maxima are at angles given by

$$\sin\theta = m\frac{\lambda}{d} = m\frac{5 \times 10^{-7}\text{ m}}{1 \times 10^{-4}\text{ m}} = (5 \times 10^{-3})m$$

where $m = 0, 1, 2, 3, \ldots$. Since θ is small, we can approximate $\sin\theta \approx \tan\theta \approx \theta$. The distance y measured along the screen from the central maximum is related to θ by

$$y = L\tan\theta \approx L\theta$$

The position of the mth principal maximum is thus

$$y_m = L\theta_m = m(1.4\text{ m})(5 \times 10^{-3}) = m(7.0\text{ mm})$$

The principal maxima are thus separated by 7.0 mm on the screen.

The first minimum occurs when the phase difference between two adjacent sources is $\delta = 90° = \pi/2$. This corresponds to a path difference of $\lambda/4$. The angle θ of this minimum is given by $d \sin\theta = \lambda/4$, or

$$\sin\theta = \frac{\lambda}{4d} = \frac{5 \times 10^{-7}\ \text{m}}{4 \times 10^{-4}\ \text{m}} = 1.25 \times 10^{-3}$$

The position y of this minimum is

$$y = L\theta = (1.4\ \text{m})(1.25 \times 10^{-3}) = 1.75\ \text{mm}$$

The distance between the first minima on either side of the maximum at $\theta = 0$ is $2y = 3.5$ mm. This is a measure of the width of the principal maxima. If we had only two sources with the same spacing, the principal maxima would be at the same points but the first minimum would be at an angle θ corresponding to a path difference of $\lambda/2$. The width of these maxima would therefore be twice as great as when there are four sources, as is shown in Figure 33-18.

Question

5. How many secondary maxima would there be between the main maxima in the interference pattern produced by five equally spaced sources? Why would they be difficult to see?

33-7 Diffraction Pattern of a Single Slit

Qualitative

In our discussion of the interference patterns produced by two or more slits, we assumed the slits were very narrow so that we could consider them to be line sources of cylindrical waves, which in our two-dimensional diagrams are point sources of circular waves. We could therefore assume that the intensity due to one slit acting alone was the same (I_0) at any point P on the screen independent of the angle θ made between the ray to point P and the normal line between the slit and the screen. When the slit is not narrow, the intensity on a screen far away is not independent of angle but decreases as the angle increases. Let us consider a slit of width a. Figure 33-19 shows

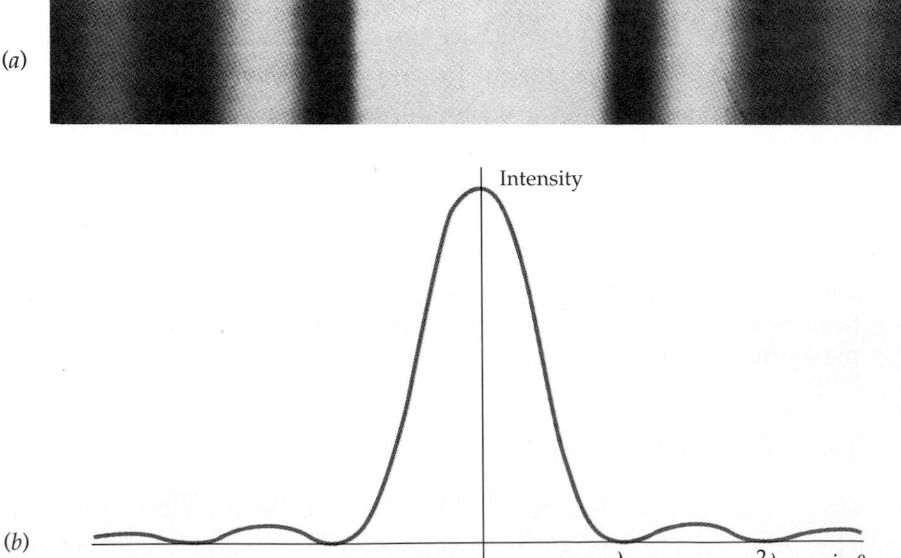

Figure 33-19 (*a*) Diffraction pattern of a single slit as observed on a screen far away. (*b*) Plot of intensity versus $\sin\theta$ for the pattern in (*a*).

the intensity pattern on a screen far away from the slit of width a as a function of $\sin\theta$. We can see that the intensity is maximum in the forward direction ($\sin\theta = 0$) and decreases to zero at an angle that depends on the slit width a and the wavelength λ. Most of the light intensity is concentrated in the broad **central diffraction maximum,** though there are minor secondary maxima bands on either side of the central maximum. The first zeroes in the intensity occur at angles given by

$$\sin\theta = \lambda/a \tag{33-17}$$

Note that for a given wavelength λ, the width of the central maximum varies inversely with the width of the slit. That is, if we *increase* the slit width a, the angle θ at which the intensity first becomes zero *decreases,* giving a more narrow central diffraction maximum. Conversely, if we *decrease* the slit width, the angle of the first zero *increases,* giving a wider central diffraction maximum. When a is very small, there are no points of zero intensity in the pattern, and the slit acts as a line source (a point source in two dimensions), radiating light energy essentially equally in all directions.

We can write Equation 33-17 slightly differently. Multiplying both sides by a, we obtain

$$a\sin\theta = \lambda \tag{33-18}$$

The quantity $a\sin\theta$ is the path difference between a light ray leaving the top of the slit and one leaving the bottom of the slit. We see that the first diffraction minimum occurs when these two rays are in phase, that is, when their path difference equals 1 wavelength. We can understand this result by considering each point on a wavefront to be a point source of light in accordance with Huygens' principle. In Figure 33-20, we have placed a line of dots on the wavefront at the slit to represent these point sources schematically. Suppose, for example, that we have 100 such dots and that we look at an angle θ for which $a\sin\theta = \lambda$, that is, the angle for which the waves from the top and bottom of the slit are in phase. Let us consider the slit to be divided into two regions, with the first 50 sources in the first, upper region and sources 51 through 100 in the second, lower region. When the path difference between the top and bottom of the slit equals one wavelength, the path difference between source 1 (the first source in the upper region) and source 51 (the first source in the lower region) is $\frac{1}{2}$ wavelength. The waves from these two sources will be out of phase by 180° and will thus cancel. Similarly, waves from the second source in each region (source 2 and source 52) will cancel. Continuing this argument, we can see that the waves from each pair of sources separated by $a/2$ will cancel. Thus, there will be no light energy at this angle. We can extend this argument to the second and third minima in the diffraction pattern of Figure 33-19. At an angle such that $a\sin\theta = 2\lambda$, we can divide the slit into four regions, two for the top half and two for the bottom half. Using this same argument, the light intensity from the top half is zero because of the cancellation of pairs of sources, and, similarly, that from the bottom half is zero. The general expression for the points of zero intensity in the diffraction pattern of a single slit is thus

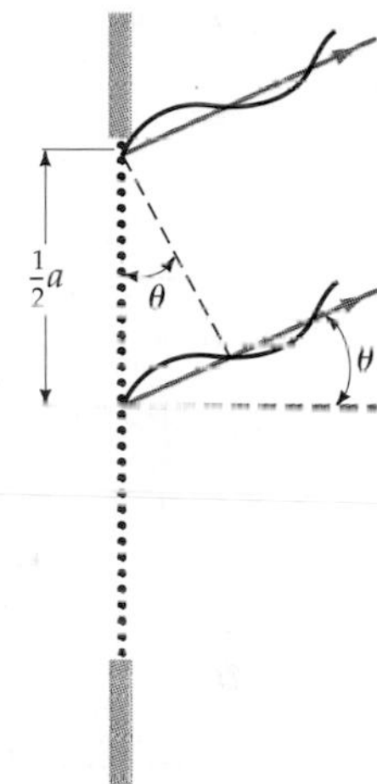

Figure 33-20 A single slit is represented by a large number of point sources of equal amplitude. At the first diffraction minimum of a single slit, the waves from the source near the top and those from the source just below the middle of the slit are 180° out of phase and cancel, as do all other pairs of sources.

$$a\sin\theta = m\lambda \qquad m = 1, 2, 3, \ldots \tag{33-19}$$

Points of zero intensity for a single-slit diffraction pattern

Usually, we are just interested in the first occurrence of a minimum in the light intensity because nearly all of the light energy is contained in the central diffraction maximum.

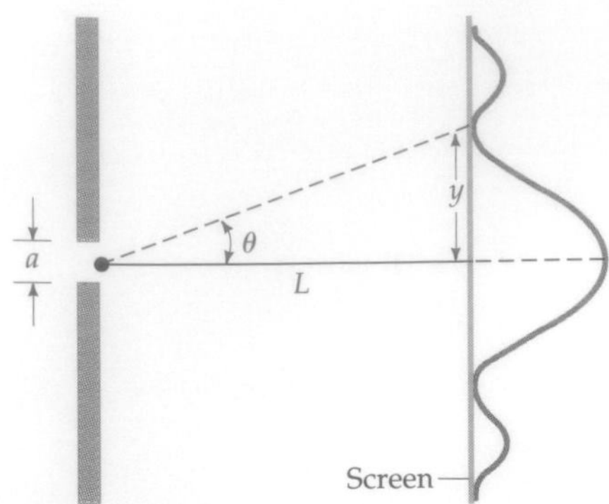

Figure 33-21 The distance y measured along the screen from the central maximum to the first diffraction minimum is related to the angle θ by $\tan\theta = y/L$, where L is the distance to the screen. Since the angle is very small, $\tan\theta \approx \sin\theta$. Then $y = L\tan\theta \approx L\sin\theta = L\lambda/a$.

In Figure 33-21, the distance y from the central maximum to the first diffraction minimum is related to the angle θ and the distance L from the slit to the screen by

$$\tan\theta = \frac{y}{L}$$

Since this angle is very small, $\tan\theta \approx \sin\theta$. Then, according to Equation 33-17, we have

$$\sin\theta = \frac{\lambda}{a} \approx \frac{y}{L}$$

or

$$y = \frac{L\lambda}{a} \qquad \text{33-20}$$

Example 33-8

In a lecture demonstration of single-slit diffraction, a laser beam of wavelength 700 nm passes through a vertical slit 0.2 mm wide and hits a screen 6 m away. Find the width of the central diffraction maximum on the screen, that is, the distance between the first minimum on the left and the first minimum on the right of the central maximum.

The width of the central diffraction maximum in Figure 33-21 is $2y$. From Equation 33-20, we have

$$y = \frac{L\lambda}{a} = \frac{(6\text{ m})(700\times10^{-9}\text{ m})}{0.0002\text{ m}} = 2.1\times10^{-2}\text{ m} = 2.1\text{ cm}$$

The width of the central maximum is thus $2y = 4.2$ cm.

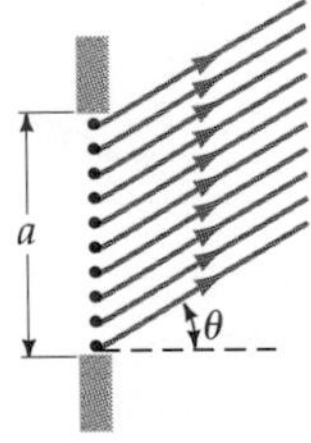

Figure 33-22 Diagram for calculating the diffraction pattern far away from a narrow slit. The slit width a is assumed to contain a large number of in-phase point sources separated by a distance d. The rays from these sources to a point very far away are approximately parallel. The path difference for the waves from adjacent sources is $d\sin\theta$.

Quantitative

We will now use the phasor method of addition of harmonic waves to calculate the intensity pattern shown in Figure 33-19. We assume that the slit of width a is divided into N equal intervals and that there is a point source of waves at the midpoint of each interval. If d is the distance between two adjacent sources and a is the width of the opening, we have $d = a/N$. Since the screen on which we are calculating the intensity is very far from the sources, the rays from the sources to a point P on the screen are approximately parallel. The path difference between any two adjacent sources is $d\sin\theta$ (Figure 33-22), and the phase difference is

$$\delta = \frac{2\pi}{\lambda} d\sin\theta$$

If A_0 is the amplitude due to a single source, the amplitude at the central maximum, where $\theta = 0$ and all the waves are in phase, is $A_{max} = NA_0$ (Figure 33-23). We can find the amplitude at some other point at an angle θ by using the phasor method for the addition of harmonic waves. As in the addition of two, three, or four waves, the intensity is zero at any point

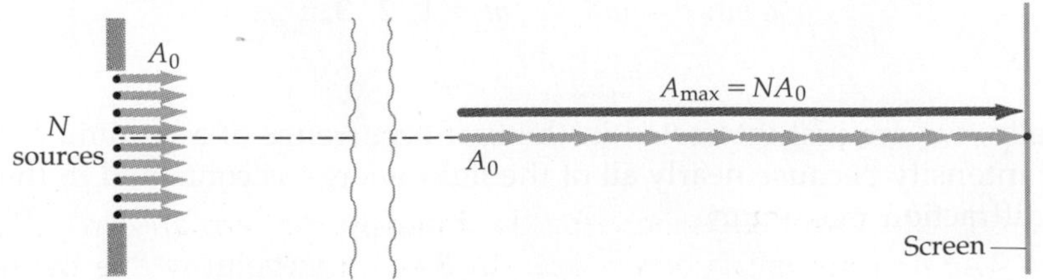

Figure 33-23 A single slit is represented by N sources, each of amplitude A_0. At the central maximum point at $\theta = 0$, the waves from the sources add in phase, giving a resultant amplitude $A_{max} = NA_0$.

where the phasors representing the waves form a closed polygon. In this case the polygon has N sides (Figure 33-24). At the first minimum, the wave from the first source just below the top of the opening and that from the source just below the middle of the opening are 180° out of phase. In this case, the waves from the source near the top of the opening differ from those from the bottom by nearly 360°. (The phase difference is, in fact, 360° − (360°)/N.) Thus, if the number of sources is very large, we get complete cancellation when the waves from the first and last sources are out of phase by 360°, corresponding to a path difference of 1 wavelength in agreement with Equation 33-18.

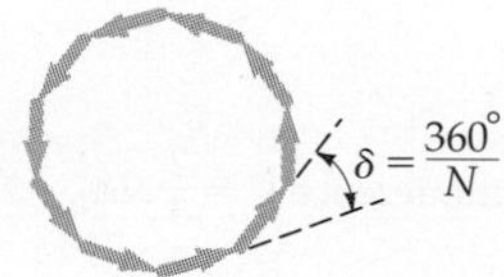

Figure 33-24 Phasor diagram for calculating the first minimum in the single-slit diffraction pattern. When the waves from the N sources completely cancel, the N phasors form a closed polygon. The phase difference between the waves from adjacent sources is then $\delta = 360°/N$. When N is very large, the waves from the first and last sources are approximately in phase.

We will now calculate the amplitude at a general point at which the waves from two adjacent sources differ in phase by δ. Figure 33-25 shows the phasor diagram for the addition of N waves where the subsequent waves differ in phase from the first wave by $\delta, 2\delta, \ldots, (N-1)\delta$. When N is very large and δ is very small, the phasor diagram approximates the arc of a circle. The resultant amplitude A is the length of the chord of this arc. We will calculate this resultant amplitude in terms of the phase difference ϕ between the first wave and the last wave. From Figure 33-25, we have

$$\sin \tfrac{1}{2}\phi = \frac{A/2}{r}$$

or

$$A = 2r \sin \tfrac{1}{2}\phi \qquad 33\text{-}21$$

where r is the radius of the arc. Since the length of the arc is $A_{max} = NA_0$ and the angle subtended is ϕ, we have

$$\phi = \frac{A_{max}}{r} \qquad 33\text{-}22$$

or

$$r = \frac{A_{max}}{\phi}$$

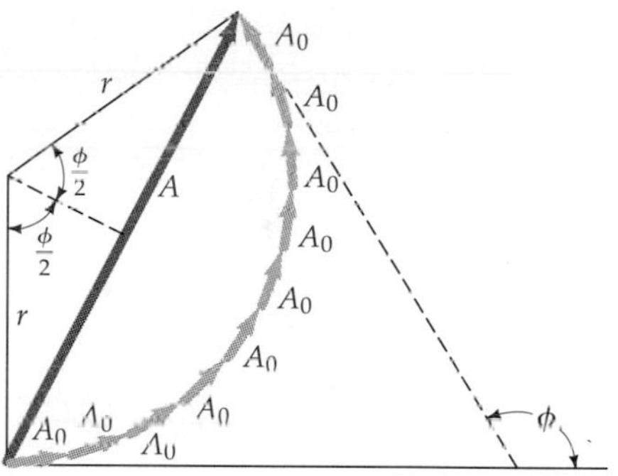

Figure 33-25 Phasor diagram for calculating the resultant amplitude due to the waves from N sources in terms of the phase difference ϕ between the wave from the first source just below the top of the slit and that from the last source just above the bottom of the slit. When N is very large, the resultant amplitude A is the chord of a circular arc of length $NA_0 = A_{max}$.

Substituting this into Equation 33-21 gives

$$A = \frac{2A_{max}}{\phi} \sin \tfrac{1}{2}\phi = A_{max} \frac{\sin \frac{1}{2}\phi}{\frac{1}{2}\phi}$$

Since the amplitude at the center of the central maximum ($\theta = 0$) is A_{max}, the ratio of the intensity at any other point to that at the center of the central maximum is

$$\frac{I}{I_0} = \frac{A^2}{A_{max}^2} = \left(\frac{\sin \frac{1}{2}\phi}{\frac{1}{2}\phi}\right)^2$$

or

$$I = I_0 \left(\frac{\sin \frac{1}{2}\phi}{\frac{1}{2}\phi}\right)^2 \qquad 33\text{-}23$$

Intensity for a single-slit diffraction pattern

The phase difference ϕ between the first and last waves is $2\pi/\lambda$ times the path difference $a \sin \theta$ between the top and bottom of the opening:

$$\phi = \frac{2\pi}{\lambda} a \sin \theta \qquad 33\text{-}24$$

Equations 33-23 and 33-24 describe the intensity pattern shown in Figure 33-19. The first minimum occurs at $a \sin \theta = \lambda$, the point where the waves

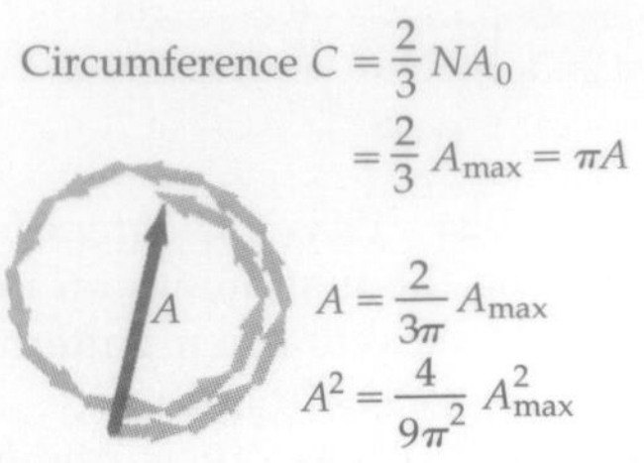

Circumference $C = \frac{2}{3}NA_0 = \frac{2}{3}A_{max} = \pi A$

$A = \frac{2}{3\pi}A_{max}$

$A^2 = \frac{4}{9\pi^2}A^2_{max}$

Figure 33-26 Phasor diagram for calculating the approximate amplitude of the first secondary maximum of the single-slit diffraction pattern. This secondary maximum occurs near the midpoint between the first and second minima when the N phasors complete $1\frac{1}{2}$ circles.

from the top and bottom of the opening have a path difference of λ and are in phase. The second minimum occurs at $a \sin\theta = 2\lambda$, where the waves from the top and bottom of the opening have a path difference of 2λ.

There is a secondary maximum approximately midway between the first and second minima at $a \sin\theta \approx \frac{3}{2}\lambda$. Figure 33-26 shows the phasor diagram for determining the approximate intensity of this secondary maximum. The phase difference between the first and last waves is approximately 360° + 180°. The phasors thus complete $1\frac{1}{2}$ circles. The resultant amplitude is the diameter of a circle with a circumference that is two-thirds the total length A_{max}. If $C = \frac{2}{3}A_{max}$ is the circumference, the diameter A is

$$A = \frac{C}{\pi} = \frac{\frac{2}{3}A_{max}}{\pi} = \frac{2}{3\pi}A_{max}$$

and

$$A^2 = \frac{4}{9\pi^2}A^2_{max}$$

The intensity at this point is

$$I = \frac{4}{9\pi^2}I_0 = \frac{1}{22.2}I_0 \qquad 33\text{-}25$$

Question

6. As the width of a slit producing a single-slit diffraction pattern is slowly and steadily reduced, how will the diffraction pattern change?

33-8 Interference–Diffraction Pattern of Two Slits

When there are two or more slits, the intensity pattern on a screen far away is a combination of the single-slit diffraction pattern and the multiple-slit interference pattern we have studied. Figure 33-27 shows the intensity pattern on a screen far from two slits whose separation d is 10 times the width a of each slit. The pattern is the same as the two-slit pattern with very narrow slits (Figure 33-9) except that it is modulated by the single-slit diffraction pattern; that is, the intensity due to each slit separately is now not constant but decreases with angle as shown in Figure 33-27*b*. The intensity can be calculated from the two-slit pattern (Equation 33-11) with the intensity of each slit (I_0 in that equation) replaced by the diffraction-pattern intensity due to each slit, I, given by Equation 33-23. The intensity for the two-slit interference–diffraction pattern is thus

Interference–diffraction intensity for two slits

$$I = 4I_0\left(\frac{\sin\frac{1}{2}\phi}{\frac{1}{2}\phi}\right)^2 \cos^2\tfrac{1}{2}\delta \qquad 33\text{-}26$$

where ϕ is the difference in phase between rays from the top and bottom of each slit, which is related to the width of each slit by

$$\phi = \frac{2\pi}{\lambda}a\sin\theta$$

and δ is the difference in phase between rays from the centers of two adjacent slits, which is related to the slit separation by

$$\delta = \frac{2\pi}{\lambda}d\sin\theta$$

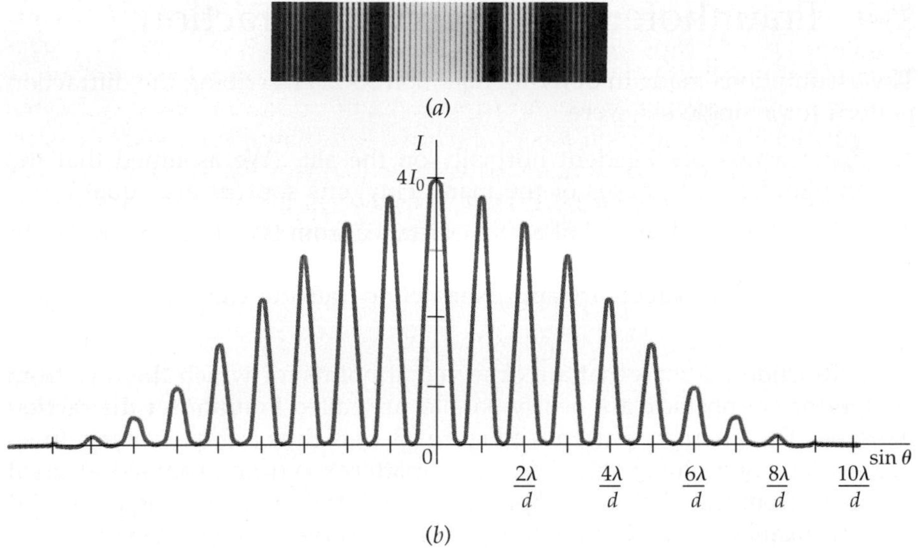

Figure 33-27 (*a*) Interference–diffraction pattern for two slits whose separation d is equal to 10 times their width a. The tenth interference maximum on either side of the central interference maximum is missing because it falls at the first diffraction minimum. (*b*) Plot of intensity versus $\sin\theta$ for the central band of the pattern in (*a*).

In Equation 33-26, the intensity I_0 is the intensity at $\theta = 0$ due to one slit alone. Note that in Figure 33-27 the central diffraction maximum contains 19 interference maxima the central interference maximum and 9 maxima on either side. The tenth interference maximum on either side of the central one is at the angle θ given by $\sin\theta = 10\lambda/d = \lambda/a$ since $d = 10a$. This coincides with the position of the first diffraction minimum, so this interference maximum is not seen. At these points, the light from the two slits would be in phase and would interfere constructively, but there is no light from either slit because the points are diffraction minima.

Example 33-9

Two slits of width $a = 0.015$ mm are separated by a distance $d = 0.06$ mm and illuminated by light of wavelength $\lambda = 650$ nm. How many bright fringes are seen in the central diffraction maximum?

The number of bright fringes in the central diffraction maximum does not depend on the wavelength of light, but only on the ratio of the slit separation to the slit width:

$$\frac{d}{a} = \frac{0.06\text{ mm}}{0.015\text{ mm}} = 4$$

The angle of the first diffraction minimum is given by

$$\sin\theta = \lambda/a$$

Because $a = d/4$, this can be written

$$\sin\theta = 4\lambda/d$$

Thus the position of the fourth interference maximum coincides with the position of the first diffraction minimum, so there will be 3 interference maxima on either side of the central interference maximum for total of 7 bright fringes in the central diffraction maximum.

Question

7. How many interference maxima will be contained in the central diffraction maximum in the diffraction–interference pattern of two slits if the separation of the slits d is 5 times their width a? How many will there be if $d = Na$ for any value of N?

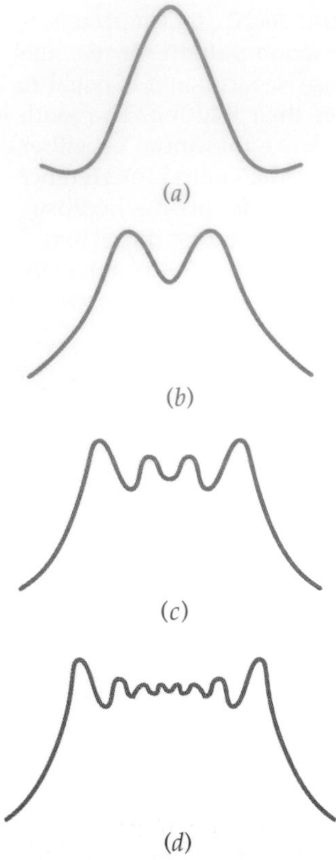

Figure 33-28 Diffraction patterns for a single slit at various screen distances. As the screen is moved closer to the slit, the Fraunhofer pattern (*a*) observed far from the slit gradually changes into the Fresnel pattern (*d*) observed near the slit.

33-9 Fraunhofer and Fresnel Diffraction

The assumptions made in deriving Equation 33-23 describing the diffraction pattern for a single slit were

1. Plane waves are incident normally on the slit. (We assumed that the amplitudes and phases of the many Huygens sources are equal.)
2. The pattern is observed at a great distance from the slit compared with the size of the openings. (We assumed that the rays from the sources to a point on the screen are approximately parallel to simplify the geometry.)

Diffraction patterns that are observed at points for which the rays from an aperture or obstacle are nearly parallel are called **Fraunhofer diffraction patterns.** The intensity pattern of Figure 33-19 is thus a Fraunhofer diffraction pattern of a single slit. Fraunhofer patterns can be observed at great distances from the obstacle or aperture so that the rays reaching any point are approximately parallel, or they can be observed using a lens to focus parallel rays on a viewing screen placed in the focal plane of the lens. If a slit is many wavelengths wide, the Fraunhofer pattern will not be observed because the angle of the first minimum will be very small. For example, if $a = 1000\lambda$, the first minimum will occur at an angle θ given by $\sin\theta = 1/1000 \approx \theta$. This small angle is not much different from the angle made by the rays from the top and bottom of the slit to the central maximum, rays which we assumed to be parallel in our derivation.

When the diffraction pattern is observed near an aperture or obstacle, it is called a **Fresnel diffraction pattern.** Because of the geometry, this pattern is much more difficult to analyze. Figure 33-28 illustrates the difference between the Fresnel and Fraunhofer patterns for a single slit.*

Figure 33-29*a* shows the Fresnel diffraction pattern of an opaque disk illuminated by light from a source on its axis. Note the bright spot at the center of the pattern caused by the constructive interference of the light waves diffracted from edge of the disk. This pattern is of some historical interest. In an attempt to discredit Fresnel's wave theory of light, Poisson applied it to this situation and considered the prediction of the bright spot at the center of the shadow to be a ridiculous contradiction of fact. However, Fresnel immediately demonstrated experimentally that such a spot does, in fact, exist. This demonstration convinced many doubters of the validity of the wave theory of light. The Fresnel diffraction pattern of a circular aperture is shown in Figure 33-29*b*. Comparing this with the pattern of the opaque disk in Figure 33-29*a*, we can see that the two patterns are complements of each other.

*See Richard E. Haskel, "A Simple Experiment on Fresnel Diffraction," *American Journal of Physics*, vol. 38, 1970, p. 1039.

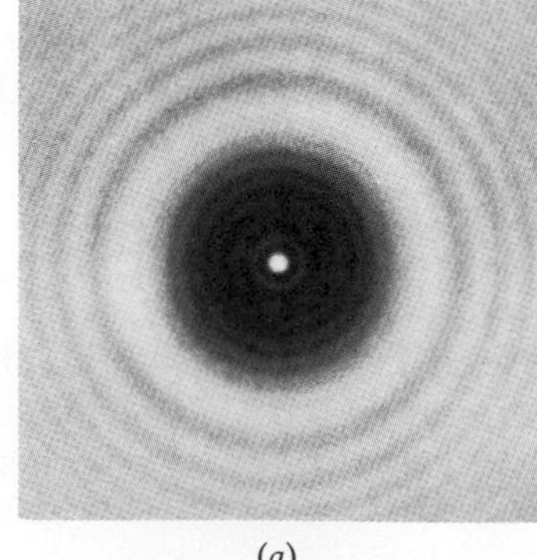

(*a*)

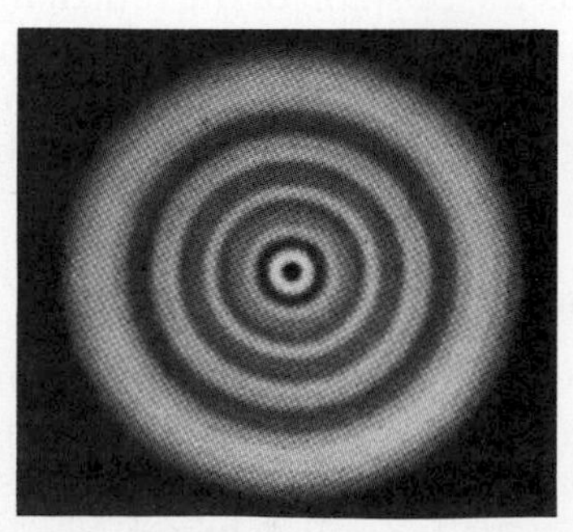

(*b*)

Figure 33-29 (*a*) Fresnel diffraction pattern of an opaque disk. At the center of the shadow, the light waves diffracted from the edge of the disk are in phase and produce a bright spot called the *Poisson spot*. (*b*) Fresnel diffraction pattern of a circular aperture. Compare this with part (*a*).

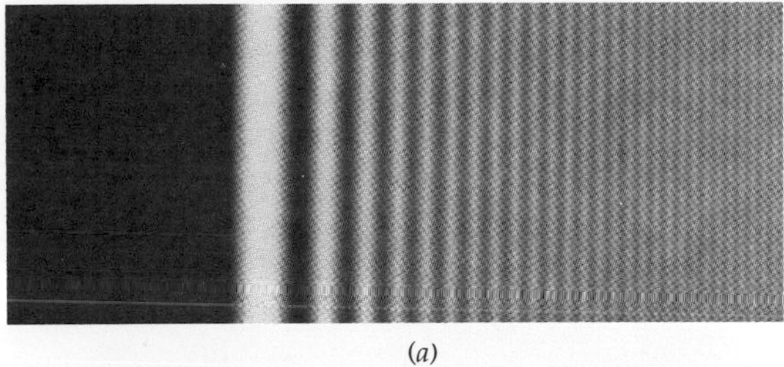

(a)

Intensity

Geometric shadow

Edge

Distance

(b)

Figure 33-30 (a) Fresnel diffraction of a straightedge. (b) Intensity versus distance along a line perpendicular to the edge.

Figure 33-30*a* shows the Fresnel diffraction pattern of a straightedge illuminated by light from a point source. A graph of the intensity versus distance (measured along a line perpendicular to the edge) is shown in Figure 33-30*b*. The light intensity does not fall abruptly to zero in the geometric shadow, but it decreases rapidly and is negligible within a few wavelengths of the edge. The Fresnel diffraction pattern of a rectangular aperture is shown in Figure 33-31. These patterns cannot be seen with broad light sources like an ordinary light bulb because the dark fringes of the pattern produced by light from one point on the source overlap the bright fringes of the pattern produced by light from another point.

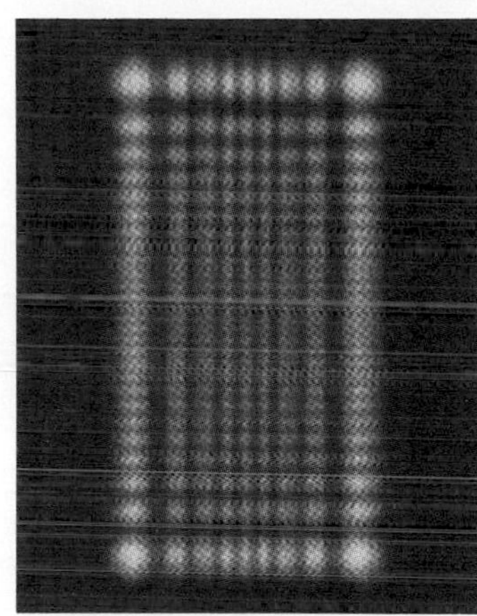

Figure 33-31 Fresnel diffraction of a rectangular aperture.

33-10 Diffraction and Resolution

Figure 33-32 shows the Fraunhofer diffraction pattern of a circular aperture. This pattern has important applications to the resolution of many optical instruments. The angle θ subtended by the first diffraction minimum is related to the wavelength and the diameter of the opening D by

$$\sin\theta = 1.22\frac{\lambda}{D} \qquad 33\text{-}27$$

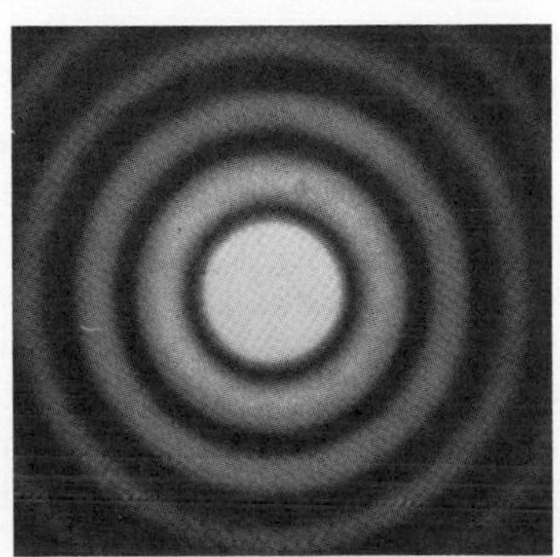

Figure 33-32 Fraunhofer diffraction pattern of a circular aperture.

Equation 33-27 is similar to Equation 33-17 except for the factor 1.22. This factor arises from the mathematical analysis, which is similar to that for a single slit but more complicated because of the circular geometry. In many applications, the angle θ is small, so $\sin\theta$ can be replaced by θ. The first diffraction minimum is then at an angle θ given by

$$\theta \approx 1.22\frac{\lambda}{D} \qquad 33\text{-}28$$

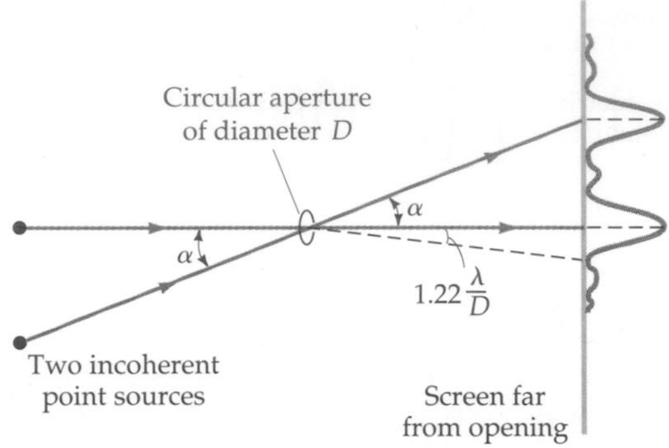

Figure 33-33 Two distant sources that subtend an angle α. If α is much greater than $1.22\ \lambda/D$, where λ is the wavelength of light and D is the diameter of the aperture, the diffraction patterns have little overlap and the sources are easily seen as two sources. If α is not much greater than $1.22\ \lambda/D$, the overlap of the diffraction patterns makes it difficult to distinguish two sources from one.

Figure 33-33 shows two point sources that subtend an angle α at a circular aperture far from the sources. The Fraunhofer diffraction patterns are also included in this figure. If α is much greater than $1.22\lambda/D$, they will be seen as two sources. However, as α is decreased, the overlap of the diffraction patterns increases, and it becomes difficult to distinguish the two sources from one source. At the critical angular separation α_c given by

$$\alpha_c = 1.22\frac{\lambda}{D} \qquad 33\text{-}29$$

the first minimum of the diffraction pattern of one source falls at the central maximum of the other source. These objects are said to be just resolved by **Rayleigh's criterion for resolution.** Figure 33-34 shows the diffraction patterns for two sources when α is greater than the critical angle for resolution and when α is just equal to the critical angle for resolution.

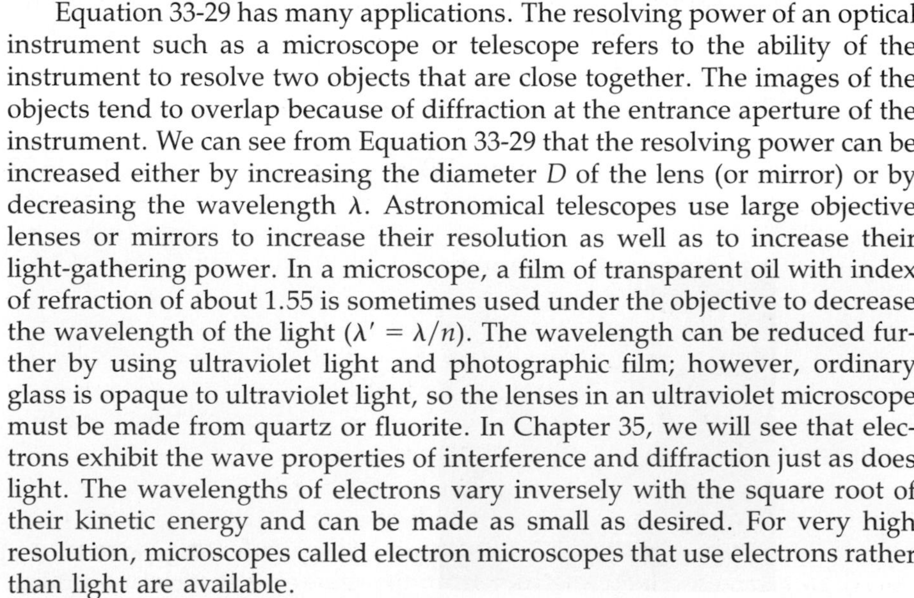

Equation 33-29 has many applications. The resolving power of an optical instrument such as a microscope or telescope refers to the ability of the instrument to resolve two objects that are close together. The images of the objects tend to overlap because of diffraction at the entrance aperture of the instrument. We can see from Equation 33-29 that the resolving power can be increased either by increasing the diameter D of the lens (or mirror) or by decreasing the wavelength λ. Astronomical telescopes use large objective lenses or mirrors to increase their resolution as well as to increase their light-gathering power. In a microscope, a film of transparent oil with index of refraction of about 1.55 is sometimes used under the objective to decrease the wavelength of the light ($\lambda' = \lambda/n$). The wavelength can be reduced further by using ultraviolet light and photographic film; however, ordinary glass is opaque to ultraviolet light, so the lenses in an ultraviolet microscope must be made from quartz or fluorite. In Chapter 35, we will see that electrons exhibit the wave properties of interference and diffraction just as does light. The wavelengths of electrons vary inversely with the square root of their kinetic energy and can be made as small as desired. For very high resolution, microscopes called electron microscopes that use electrons rather than light are available.

(*a*)

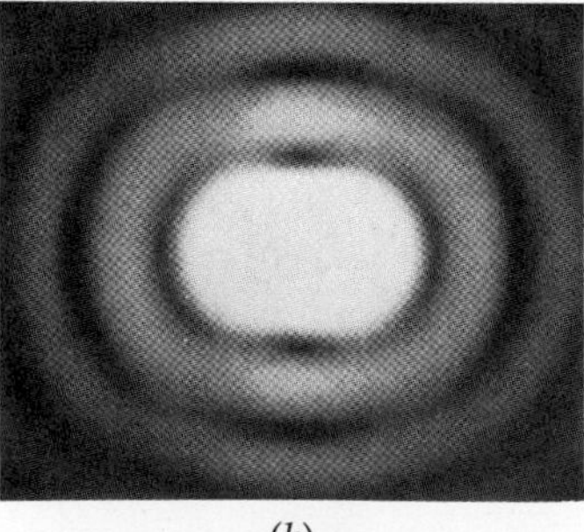

(*b*)

Figure 33-34 Diffraction patterns for a circular aperture and two incoherent point sources when (*a*) α is much greater than $1.22\ \lambda/D$ and (*b*) when α is at the limit of resolution, $\alpha_c = 1.22\ \lambda/D$.

Example 33-10

What minimum angular separation must two point objects have if they are to be just resolved by the eye? How far apart must they be if they are 100 m away? Assume that the diameter of the pupil of the eye is 5 mm and that the wavelength of the light is 600 nm.

Using Equation 33-29 with $D = 5$ mm and $\lambda = 600$ nm, we have for the minimum angular separation

$$\alpha_c = 1.22\frac{6 \times 10^{-7}\ \text{m}}{5 \times 10^{-3}\ \text{m}} = 1.46 \times 10^{-4}\ \text{rad}$$

If the objects are separated by a distance y and are 100 m away, they will be just barely resolved if $\tan \alpha_c = y/(100\ \text{m})$. Then

$$y = (100\ \text{m}) \tan \alpha_c \approx (100\ \text{m})\ \alpha_c = 1.46 \times 10^{-2}\ \text{m} = 1.46\ \text{cm}$$

where we have used the small-angle approximation $\tan \alpha_c \approx \alpha_c$.

It is instructive to compare the limitation on resolution of the eye due to diffraction as seen in Example 33-10 with that due to the separation of the receptors (cones) on the retina. To be seen as two distinct objects, the images of the objects must fall on the retina on two nonadjacent cones. (See Problem 8 in Chapter 32.) Because the retina is about 2.5 cm from the eye lens, the distance y on the retina corresponding to an angular separation of 1.5×10^{-4} rad is found from

$$\alpha_c = 1.5 \times 10^{-4}\ \text{rad} = \frac{y}{2.5\ \text{cm}}$$

or

$$y \approx 4 \times 10^{-4}\ \text{cm} = 4 \times 10^{-6}\ \text{m} = 4\ \mu\text{m}$$

The actual separation of the cones in the fovea centralis, where the cones are the most tightly packed, is about 1 μm. Outside this region, they are about 3 to 5 μm apart.

Exercise

Two objects are 4 cm apart. How far away from them can you be and still resolve them with your eye if $\lambda = 600$ nm and the diameter of the pupil of your eye is 5 mm? (Answer: 274 m)

33-11 Diffraction Gratings

A useful tool for measuring the wavelength of light is the **diffraction grating,** which consists of a large number of equally spaced lines or slits on a flat surface. Such a grating can be made by cutting parallel, equally spaced grooves on a glass or metal plate with a precision ruling machine. With a reflection grating, light is reflected from the ridges between the lines. A phonograph record exhibits some of the properties of a reflection grating. In a transmission grating, the light passes through the clear gaps between the rulings. Inexpensive plastic gratings with 10,000 or more slits per centimeter are not uncommon. The spacing of the slits in a grating with 10,000 slits per centimeter is $d = (1\ \text{cm})/10{,}000 = 10^{-4}$ cm.

Consider a plane light wave incident normally on a transmission grating (Figure 33-35) and assume that the width of each slit is very small so that each slit produces a widely diffracted beam. The interference pattern produced on a screen a large distance from the grating is that due to a large

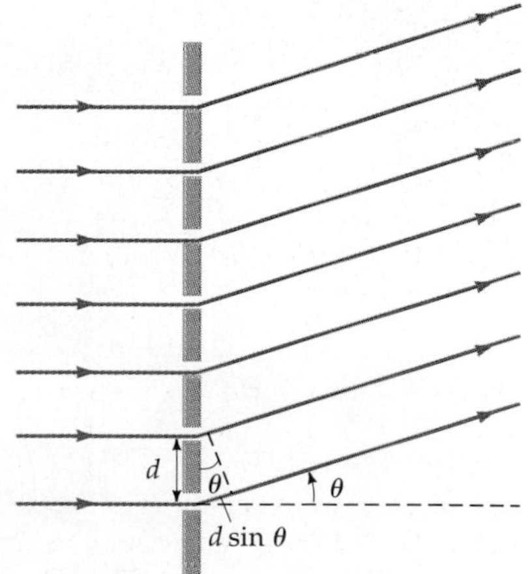

Figure 33-35 Light incident normally on a diffraction grating. At an angle θ, the path difference between rays from adjacent slits is $d \sin \theta$.

number of equally spaced light sources. The interference maxima are at angles θ given by

$$d \sin \theta = m\lambda \qquad m = 0, 1, 2, \ldots \tag{33-30}$$

where m is called the **order number.** The position of an interference maximum does not depend on the number of sources, but the more sources there are, the sharper and more intense the maximum will be, as was illustrated in Figure 33-18.

Figure 33-36b shows a typical spectroscope, which uses a diffraction grating to analyze light from a source such as a tube containing atoms of a gas, for example, helium or sodium vapor. The gas atoms are excited because of bombardment by electrons that are accelerated by the high voltage across the tube. The light emitted by such a source does not consist of a continuous spectrum. Instead the spectrum contains only certain wavelengths that are characteristic of the atoms in the source. Light from the source passes through a narrow collimating slit and is made parallel by a lens. Parallel light from the lens is incident on the grating. Instead of falling on a screen a large distance away, the parallel light from the grating is focused by a telescope and viewed by the eye. The telescope is mounted on a rotating platform that has been calibrated so that the angle θ can be measured. In the forward direction ($\theta = 0$), the central maximum for all wavelengths is seen. If light of a particular wavelength λ is emitted by the source, the first interference maximum is seen at the angle θ given by Equation 33-30 with $m = 1$. Each wavelength emitted by the source produces a separate image of the collimating slit in the spectroscope called a **spectral line.** The set of lines corresponding to $m = 1$ is called the **first-order spectrum.** The **second-order spectrum** corresponds to $m = 2$ for each wavelength. Higher orders may be seen if the angle θ given by Equation 33-30 is less than 90°. Depending on the wavelengths and the spacing of the slits in the grating, the orders may be mixed; that is, the third-order line for one wavelength may occur before the second-order line for another wavelength. If the spacing of the slits in the grating is known, the wavelengths emitted by the source can be determined by measuring the angle θ.

Figure 33-36 (a) A late nineteenth century spectroscope belonging to Gustav Kirchhoff used a prism rather than a diffraction grating to disperse the light. (b) Typical student spectroscope. Light from a collimating slit near the source is made parallel by a lens and falls on a grating. The diffracted light is viewed with a telescope at an angle that can be accurately measured.

Example 33-11

Sodium light is incident on a diffraction grating with 10,000 lines per centimeter. At what angles will the two yellow lines of wavelengths 589.00 nm and 589.59 nm be seen in the first order?

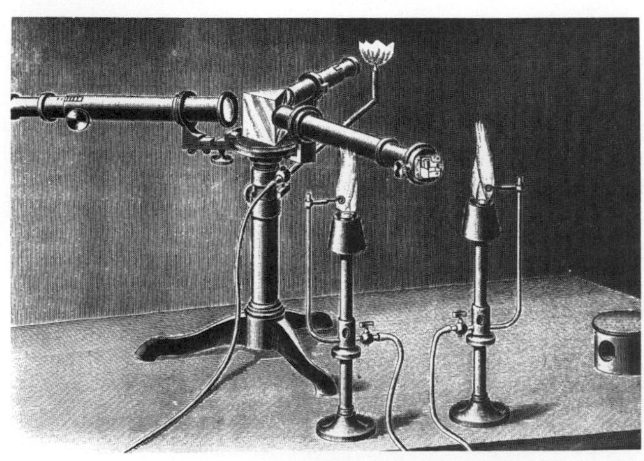

(a)

(b)

Using $m = 1$ and $d = 10^{-4}$ cm $= 10^{-6}$ m in Equation 33-30, we have for $\lambda = 589 \times 10^{-9}$ m

$$\sin\theta = \frac{\lambda}{d} = \frac{589 \times 10^{-9}\ \text{m}}{10^{-6}\ \text{m}} = 0.589$$

$$\theta = 36.09°$$

For $\lambda = 589.59$ nm, a similar calculation gives $\sin\theta = 0.58959$, resulting in $\theta = 36.13°$.

An important feature of a spectroscope is its ability to measure light of two nearly equal wavelengths λ_1 and λ_2. For example, the two prominent yellow lines in the spectrum of sodium have wavelengths 589.00 and 589.59 nm, which can be seen as two separate wavelengths if their interference maxima do not overlap. According to Rayleigh's criterion for resolution, these wavelengths are resolved if the angular separation of their interference maxima is greater than the angular separation between one interference maximum and the first interference minimum on either side of it. The **resolving power** of a diffraction grating is defined to be $\lambda/|\Delta\lambda|$, where $|\Delta\lambda|$ is the smallest difference between two nearby wavelengths, each approximately equal to λ, that may be resolved. The resolving power is proportional to the number of slits illuminated because the more slits illuminated, the sharper the interference maxima. The resolving power R can be shown to be

$$R = \frac{\lambda}{|\Delta\lambda|} = mN \qquad \text{33-31}$$

where N is the number of slits and m is the order number (see Problem 73). We can see from Equation 33-31 that to resolve the two yellow lines in the sodium spectrum the resolving power must be

$$R = \frac{589.00\ \text{nm}}{589.59 - 589.00\ \text{nm}} = 998$$

Thus, to resolve the two yellow sodium lines in the first order ($m = 1$), we need a grating containing about 1000 slits in the area illuminated by the light.

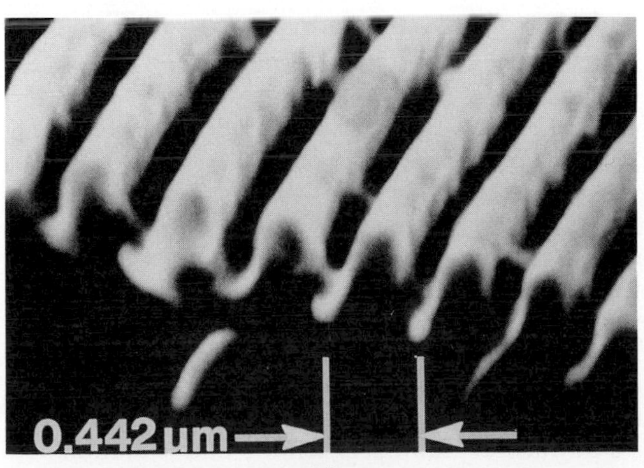

(a)

(b)

(*a*) Photomicrograph of the ridged pattern on the surface of a diffraction grating. (*b*) Aerial view of the very large array (VLA) radio telescope in New Mexico. Radio signals from distant galaxies add constructively when Equation 33-30 is satisfied, where d is the distance between two adjacent telescopes.

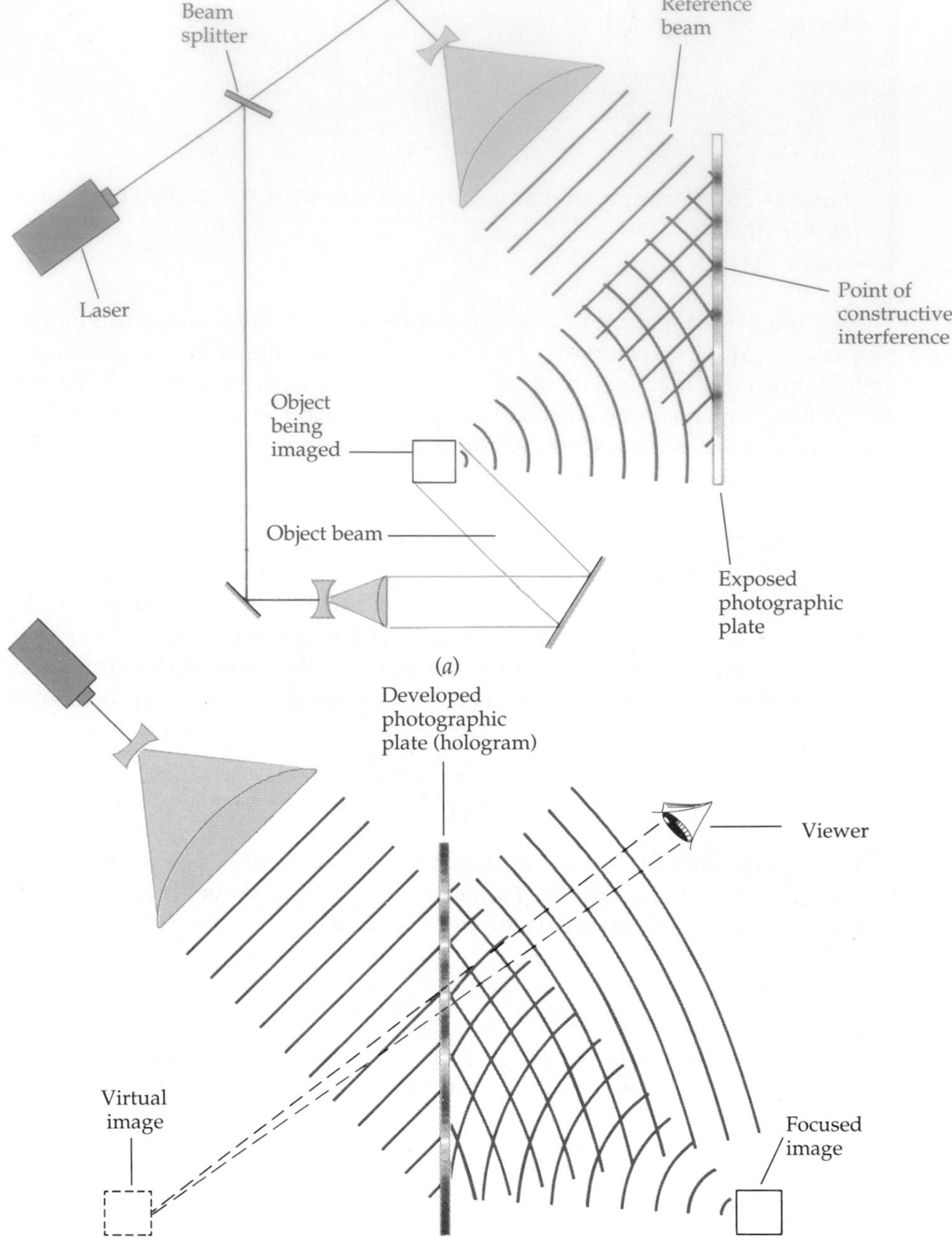

(*a*) The production of a hologram. The interference pattern produced by the reference beam and object beam is recorded on a photographic film. (*b*) When the film is developed and illuminated by coherent laser light, a three-dimensional image is seen. Holograms that you see on credit cards or postage stamps, called rainbow holograms, are more complicated. A horizontal strip of the original hologram is used to make a second hologram. The three-dimensional image can be seen as the viewer moves from side to side, but if viewed with laser light, the image disappears when the viewer's eyes move above or below the slit image. When viewed with white light, the image is seen in different colors as the viewer moves in the vertical direction.

Holograms

An interesting application of diffraction gratings is the production of a three-dimensional photograph called a **hologram.** In an ordinary photograph, the intensity of reflected light from an object is recorded on a film. When the film is viewed by transmitted light, a two-dimensional image is produced. In a hologram, a beam from a laser is split into two beams, a reference beam and an object beam. The object beam reflects from the object to be photographed and the interference pattern between it and the reference beam is recorded on a photographic film. This can be done because the laser beam is coherent so that the relative phase difference between the reference beam and object beam can be kept constant during the exposure. The interference fringes on the film act as a diffraction grating. When the film is illuminated with a laser, a three-dimensional replica of the object is produced.

(*a*)

(*a*) A technician produces a hologram of a statuette at the University of Strasbourg. When the glass plate is later illuminated by laser light, the statuette appears as a three-dimensional image. (*b*) and (*c*) Two views of the hologram "Digital." Note that different parts of the circuit board appear behind the front magnifying lens. (*d*) A holograph emulsion magnified 100 times. (*e*) A "head-up display" uses a holographic projection of important information from the airplane's control panel, so the pilot can view the runway and the control panel at the same time.

(*b*)

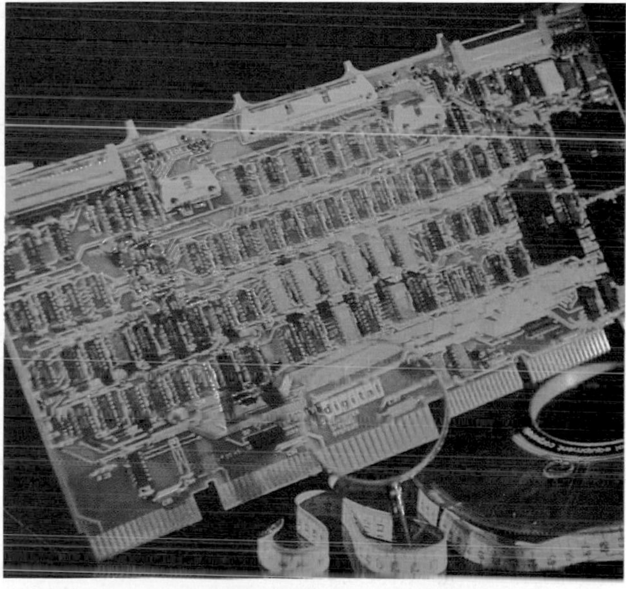

(*c*)

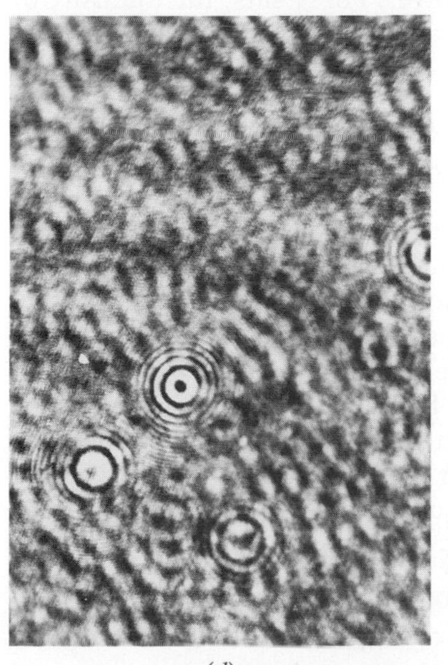

(*d*)

(*e*)

Summary

1. Two light rays interfere constructively if their phase difference is zero or an integer times 360°. They interfere destructively if their phase difference is 180° or an odd integer times 180°. A common source of phase difference is a path difference. A path difference Δr introduces a phase difference δ given by

$$\delta = \frac{\Delta r}{\lambda} 2\pi = \frac{\Delta r}{\lambda} 360°$$

A phase difference of 180° is introduced when a light wave is reflected from a boundary between two media for which the wave speed is greater in the original medium, such as the one between air and glass.

2. The interference of light rays reflected from the top and bottom surfaces of a thin film produces colored bands or fringes commonly observed in soap films or oil films. The difference in phase between the two rays results from the path difference of twice the thickness of the film plus any phase change due to reflection of one or both of the rays.

3. The Michelson interferometer uses interference to measure small distances such as the wavelength of light, or to measure a small difference in the index of refraction such as that between air and a vacuum.

4. The path difference at an angle θ on a screen far away from two narrow slits separated by a distance d is $d \sin \theta$. When this path difference is an integer times the wavelength, the interference is constructive and the intensity is maximum. When the path difference is an odd integer times $\lambda/2$, the interference is destructive, resulting in minimum intensity.

$$d \sin \theta = m\lambda \qquad m = 0, 1, 2, \ldots \qquad \text{maxima}$$

$$d \sin \theta = (m + \tfrac{1}{2})\lambda \qquad m = 0, 1, 2, \ldots \qquad \text{minima}$$

If the intensity due to each slit separately is I_0, the intensity at points of constructive interference is $4I_0$ and that at points of destructive interference is 0. When there are many equally spaced slits, the principal interference maxima occur at the same points as for two slits, but the maxima are much more intense and much narrower. For N slits, the intensity of the principal maxima is N^2I_0, and there are $N - 2$ secondary maxima between each pair of principal maxima.

5. Diffraction occurs whenever a portion of a wavefront is limited by an obstacle or aperture. The intensity of light at any point in space can be computed using Huygens' principle by taking each point on the wavefront to be a point source and computing the resulting interference pattern. Fraunhofer patterns are observed at great distances from the obstacle or aperture so that the rays reaching any point are approximately parallel, or they can be observed using a lens to focus parallel rays on a viewing screen placed in the focal plane of the lens. Fresnel patterns are observed at points close to the source. Diffraction of light is often difficult to observe because the wavelength is so small or because the light intensity is not great enough. Except for the Fraunhofer pattern of a long narrow slit, diffraction patterns are usually difficult to analyse.

6. When light is incident on a single slit of width a, the intensity pattern on a screen far away shows a broad central diffraction maximum that de-

creases to zero at an angle θ given by

$$a \sin \theta = \lambda$$

The width of the central maximum is inversely proportional to the width of the slit. Other zeros in the single-slit diffraction pattern occur at angles given by

$$\sin \theta = m\frac{\lambda}{a} \qquad m = 1, 2, 3 \ldots$$

On each side of the central maximum are secondary maxima of much smaller intensity.

7. The Fraunhofer interference–diffraction pattern of two slits is the same as the interference pattern for two narrow slits modulated by the single-slit diffraction pattern.

8. When light from two point sources that are close together passes through an aperture, the diffraction patterns of the sources may overlap. If the overlap is too great the two sources cannot be resolved as two separate sources. When the central diffraction maximum of one source falls at the diffraction minimum of the other source, the two sources are said to be just resolved by Rayleigh's criterion for resolution. For a circular aperture, the critical angular separation of two sources for resolution by Rayleigh's criterion is

$$\alpha_c = 1.22\frac{\lambda}{D}$$

where D is the diameter of the aperture.

9. A diffraction grating consisting of a large number of closely spaced lines or slits is used to measure the wavelength of light emitted by a source. The positions of the interference maxima from a grating are at angles given by

$$d \sin \theta = m\lambda \qquad m = 0, 1, 2, \ldots$$

where m is the order number. The resolving power of a grating is

$$R = \frac{\lambda}{|\Delta\lambda|} = mN$$

where N is the number of slits of the grating that are illuminated and m is the order number.

Suggestions for Further Reading

Baumeister, Philip, and Gerald Pincus: "Optical Interference Coatings," *Scientific American*, December 1970, p. 58.

Color television cameras, lasers, projector bulbs, and lenses of various types employ thin films to reflect or transmit light of certain wavelengths.

Nassau, Kurt: "Color Involving Geometrical and Physical Optics," *The Physics and Chemistry of Color: The Fifteen Causes of Color*, Part IV, John Wiley & Sons, New York, 1983.

Contains a good discussion of color production by thin films.

Walker, Jearl: "The Amateur Scientist: A Ball Bearing Aids in the Study of Light and Also Serves as a Lens," *Scientific American*, November 1984, p. 186.

This is a report of an unusual investigation into the properties of the diffraction pattern of a ball bearing placed in the beam of a laser.

Review

A. Objectives: After studying this chapter, you should:

1. Be able to work problems involving interference in thin films.

2. Be able to describe the Michelson interferometer.

3. Be able to sketch the two-slit interference intensity pattern and calculate the positions of the interference maxima and minima.

4. Be able to use the phasor method to find the sum of several harmonic waves.

5. Be able to sketch the interference pattern of three or more equally spaced slits.

6. Be able to sketch the single-slit diffraction pattern and calculate the position of the first diffraction minimum.

7. Be able to sketch the combined interference–diffraction pattern for several slits.

8. Be able to state Rayleigh's criterion for resolution and use it to investigate the conditions for the resolution of two nearby objects.

9. Be able to discuss the use of diffraction gratings and find the resolving power of a grating.

B. Define, explain, or otherwise identify:

Fringes
Newton's rings
Optically flat
Interferometer
Lloyd's mirror
Phasor
Secondary maxima
Principal maxima
Central diffraction maximum
Fraunhofer diffraction pattern
Fresnel diffraction pattern
Rayleigh's criterion for resolution
Diffraction grating
Order number
Spectral line
Resolving power
Hologram

C. True or false: If the statement is true, explain why it is true. If it is false, give a counterexample.

1. When waves interfere destructively, the energy is converted into heat energy.

2. Interference is observed only for waves from coherent sources.

3. In the Fraunhofer diffraction pattern for a single slit, the narrower the slit, the wider the central maximum of the diffraction pattern.

4. A circular aperture can produce both a Fraunhofer and a Fresnel diffraction pattern.

5. The ability to resolve two point sources depends on the wavelength of the light.

Problems

Level I

33-1 Phase Difference and Coherence

1. Which of the following pairs of light sources are coherent? (*a*) Two candles. (*b*) One candle and its image in a plane mirror. (*c*) Two pinholes uniformly illuminated by the same source. (*d*) Two headlights of a car. (*e*) Two images of a candle due to reflection from the front and back surfaces of a glass window pane.

2. (*a*) What minimum path difference is needed to introduce a phase shift of 180° in light of wavelength 600 nm? (*b*) What phase shift will that path difference introduce in light of wavelength 800 nm?

3. Two coherent microwave sources that produce waves of wavelength 1.5 cm are in the xy plane, one on the y axis at $y = 15$ cm and the other at $x = 3$ cm, $y = 14$ cm. If the sources are in phase, find the difference in phase between the two waves from these sources at the origin.

4. Light of wavelength 500 nm is incident normally on a film of water 10^{-4} cm thick. The index of refraction of water is 1.33. (*a*) What is the wavelength of the light in the water? (*b*) How many wavelengths are contained in the distance $2t$, where t is the thickness of the film? (*c*) What is the phase difference between the wave reflected from the top of the film and the one reflected from the bottom after it has traveled this distance?

33-2 Interference in Thin Films

5. A loop of wire is dipped in soapy water and held so that the soap film is vertical. (*a*) Viewed by reflection with white light, the top of the film appears black. Explain why. (*b*) Below the black region are colored bands. Is the first band red or violet? (*c*) Describe the appearance of the film when it is viewed by *transmitted* light.

6. A wedge-shaped film of air is made by placing a small slip of paper between the edges of two flat plates of glass. Light of wavelength 700 nm is incident normally on the glass plates, and interference bands are observed by reflection. (*a*) Is the first band near the point of contact of the plates dark or bright? Why? (*b*) If there are five dark bands per centimeter, what is the angle of the wedge?

7. A thin layer of a transparent material with an index of refraction of 1.30 is used as a nonreflective coating on the surface of glass with an index of refraction of 1.50. What

should the thickness of the material be for it to be nonreflecting for light of wavelength 600 nm (in a vacuum)?

8. The diameters of fine wires can be accurately measured using interference patterns. Two optically flat pieces of glass of length L are arranged with the wire between them as shown in Figure 33-37. The setup is illuminated by monochromatic light, and the resulting interference fringes are detected. Suppose $L = 20$ cm and yellow sodium light ($\lambda \approx 590$ nm) is used for illumination. If 19 bright fringes are seen along this 20 cm distance, what are the limits on the diameter of the wire? *Hint:* The nineteenth fringe might not be right at the end, but you do not see a twentieth fringe at all.

Figure 33-37 Problem 8.

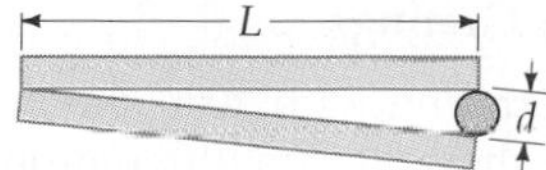

33-3 The Michelson Interferometer

9. A thin film of index of refraction $n = 1.5$ for light of wavelength 600 nm is inserted in one arm of a Michelson interferometer. (*a*) If a fringe shift of 12 fringes occurs, what is the thickness of this film? (*b*) If the illuminating light is changed to 400 nm, the fringe shift when this film is inserted is 16 fringes. What is the index of refraction of this film for light of wavelength 400 nm?

10. A hollow cell of length 5 cm with glass windows is inserted into one arm of a Michelson interferometer. The air is pumped out of the cell and the mirrors are adjusted to give a bright fringe at the center. As the air is gradually let back into the cell, there is a shift of 49.6 fringes when light of wavelength 589.29 nm is used. (*a*) How many waves are there in the 5.0 cm long cell when it is evacuated? (*b*) How many waves are there in the cell when it contains air? (*c*) What is the index of refraction of air as determined by this experiment?

33-4 The Two-Slit Interference Pattern

11. Two narrow slits separated by 1 mm are illuminated by light of wavelength 600 nm, and the interference pattern is viewed on a screen 2 m away. Calculate the number of bright fringes per centimeter on the screen.

12. Using a conventional two-slit apparatus with light of wavelength 589 nm, 28 bright fringes per centimeter are observed on a screen 3 m away. What is the slit separation?

13. Two narrow slits are separated by a distance d. Their interference pattern is to be observed on a screen a large distance L away. (*a*) Calculate the spacing y of the maxima on the screen for light of wavelength 500 nm when $L =$ 1 m and $d = 1$ cm. (*b*) Would you expect to observe the interference of light on the screen for this situation? (*c*) How close together should the slits be placed for the maxima to be separated by 1 mm for this wavelength and screen distance?

14. A long, narrow, horizontal slit lies 1 mm above a plane mirror, which is in the horizontal plane. The interference pattern produced by the slit and its image is viewed on a screen 1 m from the slit. The wavelength of the light is 600 nm. (*a*) Find the distance from the mirror to the first maximum. (*b*) How many dark bands per centimeter are seen on the screen?

15. In a lecture demonstration, laser light is used to illuminate two slits separated by 0.5 mm, and the interference pattern is observed on a screen 5 m away. The distance on the screen from the centerline to the thirty-seventh bright fringe is 25.7 cm. What is the wavelength of the light?

16. Light of wavelength 633 nm from a helium–neon laser is shone normally on a plane containing two slits. The first interference maximum is 82 cm from the central maximum on a screen 12 m away. (*a*) Find the separation of the slits. (*b*) How many interference maxima can be observed?

33-5 The Addition of Harmonic Waves Using Phasors

17. Find the resultant of the two waves

$$E_1 = 2 \sin \omega t \quad \text{and} \quad E_2 = 3 \sin (\omega t + 270°)$$

18. Find the resultant of the two waves

$$E_1 = 4 \sin \omega t \quad \text{and} \quad E_2 = 3 \sin (\omega t + 60°)$$

33-6 Interference Pattern of Three or More Equally Spaced Sources

19. Three waves have electric fields given by

$$E_1 = E_0 \sin \omega t \; E_2 = E_0 \sin (\omega t + \delta) \; E_3 = E_0 \sin (\omega t + 2\delta)$$

Using the phasor method, draw the appropriate phasors. Calculate the amplitude (in terms of E_0) and phase of the resultant wave for (*a*) $\delta = 30°$, (*b*) $\delta = 60°$, (*c*) $\delta = 90°$, and (*d*) $\delta = 240°$.

20. Three equally spaced slits separated by 0.1 mm are uniformly illuminated by light of wavelength 600 nm. The interference pattern is viewed on a screen 2 m away. Find the positions of the interference maxima and minima.

21. Five equally spaced slits are uniformly illuminated, and the slit spacing is large enough that the small angle approximation $\sin \theta \approx \theta$ is good for the first few maxima. (*a*) Find the angle θ_1 between the first principal interference maximum and the central maximum ($\theta = 0$), and compare this angle with the angle to the first minimum. (*b*) Sketch the interference pattern.

22. Repeat Problem 21 for six equally spaced slits.

33-7 Diffraction Pattern of a Single Slit

23. Equation 33-4, $d \sin \theta = m\lambda$, and Equation 33-19, $a \sin \theta = m\lambda$, are sometimes confused. For each equation, define the symbols and explain the equation's application.

24. Light of wavelength 600 nm is incident on a long, narrow slit. Find the angle of the first diffraction minimum if the width of the slit is (*a*) 1 mm, (*b*) 0.1 mm, and (*c*) 0.01 mm.

25. The single-slit diffraction pattern of light is observed on a screen a large distance L from the slit. Note from Equation 33-20 that the width $2y$ of the central maximum varies inversely with the width a of the slit. Calculate the width $2y$ for $L = 2$ m, $\lambda = 500$ nm, and (*a*) $a = 0.1$ mm, (*b*) $a = 0.01$ mm, and (*c*) $a = 0.001$ mm.

26. In a lecture demonstration of diffraction, a laser beam of wavelength 700 nm passes through a vertical slit 0.5 mm wide and hits a screen 6 m away. Find the horizontal length of the principal diffraction maximum on the screen; that is, find the distance between the first minimum on the left and the first minimum on the right of the central maximum.

27. Plane microwaves are incident on a long, narrow metal slit of width 5 cm. The first diffraction minimum is observed at $\theta = 37°$. What is the wavelength of the microwaves?

33-8 Interference–Diffraction Pattern of Two Slits

28. A two-slit Fraunhofer interference–diffraction pattern is observed with light of wavelength 500 nm. The slits have a separation of 0.1 mm and a width of a. (*a*) Find the width a if the fifth interference maximum is at the same angle as the first diffraction minimum. (*b*) For this case, how many bright interference fringes will be seen in the central diffraction maximum?

29. A two-slit Fraunhofer interference-diffraction pattern is observed with light of wavelength 700 nm. The slits have widths of 0.01 mm and are separated by 0.2 mm. How many bright fringes will be seen in the central diffraction maximum?

30. Suppose that the *central* diffraction maximum for two slits contains 17 interference fringes for some wavelength of light. How many interference fringes would you expect in the first *secondary* diffraction maximum?

33-9 Fraunhofer and Fresnel Diffraction

There are no problems for this section.

33-10 Diffraction and Resolution

31. Light of wavelength 700 nm is incident on a pinhole of diameter 0.1 mm. (*a*) What is the angle between the central maximum and the first diffraction minimum for a Fraunhofer diffraction pattern? (*b*) What is the distance between the central maximum and the first diffraction minimum on a screen 8 m away?

32. Two sources of light of wavelength 700 nm are 10 m away from the pinhole of Problem 31. How far apart must the sources be for their diffraction patterns to be resolved by Rayleigh's criterion?

33. (*a*) How far apart must two objects be on the moon to be resolved by the eye? Take the diameter of the pupil of the eye to be 5 mm, the wavelength of the light to be 600 nm, and the distance to the moon to be 380,000 km. (*b*) How far apart must the objects on the moon be to be resolved by a telescope that has a 5-m-diameter mirror?

34. Two sources of light of wavelength 700 nm are separated by a horizontal distance x. They are 5 m from a vertical slit of width 0.5 mm. What is the least value of x for which the diffraction pattern of the sources can be resolved by Rayleigh's criterion?

35. What minimum aperture, in millimeters, is required for opera glasses (binoculars) if an observer is to be able to distinguish the soprano's individual eyelashes (separated by 0.5 mm) at an observation distance of 25 m? Assume the effective wavelength of the light to be 550 nm.

36. The headlights on a small car are separated by 112 cm. At what maximum distance could you resolve them if the diameter of your pupil is 5 mm and the effective wavelength of the light is 550 nm?

37. You are told not to shoot until you see the whites of their eyes. If their eyes are separated by 6.5 cm and the diameter of your pupil is 5 mm, at what distance can you resolve the two eyes using light of wavelength 550 nm?

33-11 Diffraction Gratings

38. A diffraction grating with 2000 slits per centimeter is used to measure the wavelengths emitted by hydrogen gas. At what angles θ in the first-order spectrum would you expect to find the two violet lines of wavelengths 434 and 410 nm?

39. With the grating used in Problem 38, two other lines in the first-order hydrogen spectrum are found at angles $\theta_1 = 9.72 \times 10^{-2}$ rad and $\theta_2 = 1.32 \times 10^{-1}$ rad. Find the wavelengths of these lines.

40. Repeat Problem 38 for a diffraction grating with 15,000 slits per centimeter.

41. A diffraction grating of 2000 slits per centimeter is used to analyze the spectrum of mercury. (*a*) Find the angular separation in the first-order spectrum of the two lines of wavelength 579.0 and 577.0 nm. (*b*) How wide must the beam on the grating be for these lines to be resolved?

42. What is the longest wavelength that can be observed in the fifth-order spectrum using a diffraction grating with 4000 slits per centimeter?

Level II

43. Laser light falls normally on three evenly spaced, very narrow slits. When one of the side slits is covered, the first-order maximum is at 0.60° from the normal. If the center slit is covered and the other two are open, find (*a*) the angle of the first-order maximum and (*b*) the order number of the maximum that now occurs at the same angle as the fourth-order maximum did before.

44. The ceiling of your lecture hall is probably covered with acoustic tile, which has small holes separated by about 6 mm. (*a*) Using light with a wavelength of 500 nm, how far could you be from this tile and still resolve these holes? The diameter of the pupil of your eye is about 5 mm. (*b*) Could you resolve these holes better with red light or with violet light?

45. The telescope on Mount Palomar has a diameter of 200 inches. Suppose a double star were 4 light-years away. Under ideal conditions, what must be the minimum separation of the two stars for their images to be resolved?

46. A mica sheet 1.2 μm thick is suspended in air. In reflected light, there are gaps in the visible spectrum at 421, 474, 542, and 633 nm. Find the index of refraction of the mica sheet.

47. A thin film having an index of refraction of 1.5 is surrounded by air. It is illuminated normally by white light and is viewed by reflection. Analysis of the resulting reflected light shows that the wavelengths 360, 450, and 602 nm are the only missing wavelengths in or near the visible portion of the spectrum. That is, for these wavelengths, there is destructive interference. (*a*) What is the thickness of the film? (*b*) What visible wavelengths are brightest in the reflected interference pattern? (*c*) If this film were resting on glass with an index of refraction of 1.6, what wavelengths in the visible spectrum would be missing from the reflected light?

48. For a ruby laser of wavelength 694 nm, the end of the ruby crystal is the aperture that determines the diameter of the light beam emitted. If the diameter is 2 cm and the laser is aimed at the moon, 380,000 km away, find the approximate diameter of the light beam when it reaches the moon, assuming the spread is due solely to diffraction.

49. Sodium light of wavelength 589 nm falls normally on a 2-cm-square diffraction grating ruled with 4000 lines per centimeter. The Fraunhofer diffraction pattern is projected onto a screen at 1.5 m by a lens of focal length 1.5 m placed immediately in front of the grating. Find (*a*) the positions of the first two intensity maxima on one side of the central maximum, (*b*) the width of the central maximum, and (*c*) the resolution in the first order.

50. At the second secondary maximum of the diffraction pattern of a single slit, the phase difference between the waves from the top and bottom of the slit is approximately 5π. The phasors used to calculate the amplitude at this point complete 2.5 circles. If I_0 is the intensity at the central maximum, find the intensity I at this second secondary maximum.

51. A camera lens is made of glass with an index of refraction of 1.6. This lens is coated with a magnesium fluoride film ($n = 1.38$) to enhance its light transmission. This film is to produce zero reflection for light of wavelength 540 nm. Treat the lens surface as a flat plane and the film as a uniformly thick flat film. (*a*) How thick must the film be to accomplish its objective in the first order? (*b*) Would there be destructive interference for any other visible wavelengths? (*c*) By what factor would the reflection for light of wavelengths 400 and 700 nm be reduced by this film? Neglect the variation in the reflected light amplitudes from the two surfaces.

52. (*a*) Show that the positions of the interference minima on a screen a large distance L away from three equally spaced sources (spacing d, with $d >> \lambda$) are given approximately by

$$y = \frac{n\lambda L}{3d} \qquad \text{where } n = 1, 2, 4, 5, 7, 8, 10, \ldots$$

that is, n is not a multiple of 3. (*b*) For $L = 1$ m, $\lambda = 5 \times 10^{-7}$ m, and $d = 0.1$ mm, calculate the width of the principal interference maxima (the distance between successive minima) for three sources.

53. (*a*) Show that the positions of the interference minima on a screen a large distance L away from four equally spaced sources (spacing d, with $d >> \lambda$) are given approximately by

$$y = \frac{n\lambda L}{4d} \qquad \text{where } n = 1, 2, 3, 5, 6, 7, 9, 10, \ldots$$

that is, n is not a multiple of 4. (*b*) For $L = 2$ m, $\lambda = 6 \times 10^{-7}$ m, and $d = 0.1$ mm, calculate the width of the principal interference maxima (the distance between successive minima) for four sources. Compare this width with that for two sources with the same spacing.

54. In a pinhole camera, the image is fuzzy because of geometrical optics effects of the pinhole and because of diffraction. As the pinhole is made smaller, the fuzziness due to its size (that is, due to rays arriving at the film through different parts of the pinhole) is reduced, but the fuzziness due to diffraction is increased. The optimum size of the pinhole for the sharpest possible image occurs when the spread due to diffraction equals that due to geometrical optics effects of the pinhole. Estimate the optimum size of the pinhole if the distance from it to the film is 10 cm and the wavelength of the light is 550 nm.

55. Light of wavelength 480 nm falls normally on four slits. Each slit is 2 μm wide and is separated from the next by 6 μm. (*a*) Find the angle from the center to the first point of zero intensity of the single-slit diffraction pattern on a distant screen. (*b*) Find the angles of any bright interference maxima that lie inside the central diffraction maximum. (*c*) Find the angular spread between the central interference maximum and the first interference minimum on either side of it. (*d*) Sketch the intensity as a function of angle.

56. A drop of oil ($n = 1.22$) floats on water ($n = 1.33$). When reflected light is observed from above as shown in Figure 33-38, what is the thickness of the drop at the point where the second red fringe, counting from the edge of the drop, is observed? Assume red light has a wavelength of 650 nm.

Figure 33-38 Problem 56.

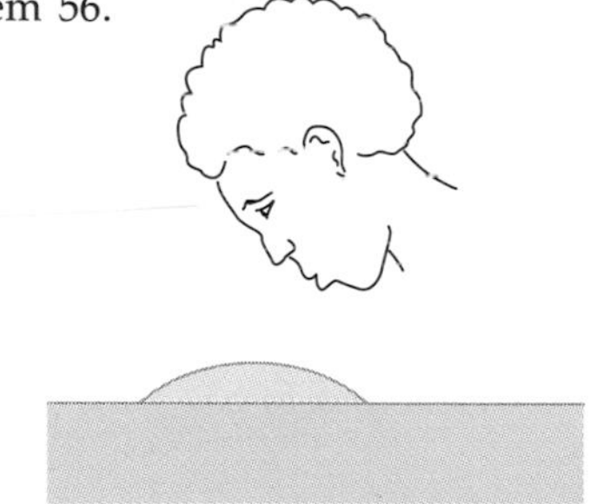

57. The Impressionist painter Georges Seurat used a technique called "pointillism," in which his paintings are composed of small, closely spaced dots of pure color, each about 2 mm in diameter. The illusion of the colors blending together smoothly is produced in the eye of the viewer by diffraction effects. Calculate the minimum viewing distance for this effect to work properly. Use the wavelength of visible light that requires the *greatest* distance, so that you're sure the effect will work for *all* visible wavelengths. Assume the pupil of the eye has a diameter of 5 mm.

58. Light of wavelength 600 nm is used to illuminate normally two glass plates 22 cm in length that touch at one end and are separated at the other end by a wire of radius 0.025 mm. How many bright fringes appear along the total length of the plates?

59. A square diffraction grating with an area of 25 cm^2 has a resolution of 22,000 in the fourth order. At what angle should you look to see a wavelength of 510 nm in the fourth order?

60. Light of wavelength 550 nm illuminates two slits of width 0.03 mm and separation 0.15 mm. (*a*) How many interference maxima fall within the full width of the central diffraction maximum? (*b*) What is the ratio of the intensity of the third interference maximum to the side of the centerline (not counting the center interference maximum) to the intensity of the center interference maximum?

61. Light is incident at an angle ϕ with the normal to a vertical plane containing two slits of separation d (Figure 33-39). Show that the interference maxima are located at angles θ given by $\sin\theta + \sin\phi = m\lambda/d$.

Figure 33-39 Problem 61.

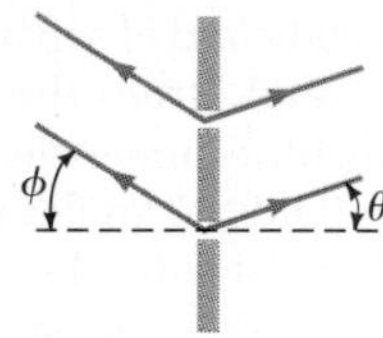

62. White light falls at an angle of 30° to the normal of a plane containing a pair of slits separated by 2.5 μm. What visible wavelengths give a bright interference maximum in the transmitted light in the direction normal to the plane? (See Problem 61.)

Level III

63. A Newton's-ring apparatus consists of a glass lens with radius of curvature R that rests on a flat glass plate as shown in Figure 33-40. The thin film is air of variable thickness. The pattern is viewed by reflected light. (*a*) Show that for a thickness t the condition for a bright (constructive) interference ring is

$$t = (m + \tfrac{1}{2})\frac{\lambda}{2} \qquad m = 0, 1, 2, \ldots$$

Figure 33-40 Problem 63.

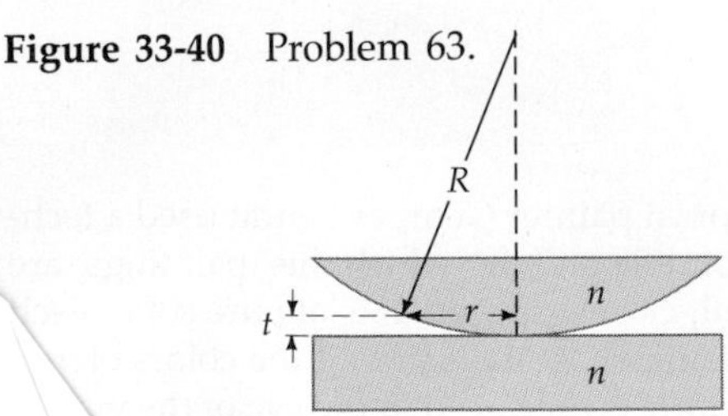

that as long as $t/R << 1$, the radius r of a bright …ge is given by

$$\sqrt{(m + \tfrac{1}{2})\lambda R} \qquad m = 0, 1, 2, \ldots$$

(*c*) How would the transmitted pattern look in comparison with the reflected one? (*d*) Use $R = 10$ m and a diameter of 4 cm for the lens. How many bright fringes would you see if the apparatus were illuminated by yellow sodium light ($\lambda \approx 590$ nm) and were viewed by reflection? (*e*) What would be the diameter of the sixth bright fringe? (*f*) If the glass used in the apparatus has an index of refraction $n = 1.5$ and water is placed between the two pieces of glass, what change will take place in the bright fringes?

64. A *Jamin refractometer* is a device for measuring or comparing the indexes of refraction of fluids. A beam of monochromatic light is split into two parts, each of which is directed along the axis of a separate cylindrical tube before being recombined into a single beam that is viewed through a telescope. Suppose that each tube is 0.4 m long and that sodium light of wavelength 589 nm is used. Both tubes are initially evacuated, and constructive interference is observed in the center of the field of view. As air is slowly allowed to enter one of the tubes, the central field of view changes to dark and back to bright a total of 198 times. (*a*) What is the index of refraction of air? (*b*) If the fringes can be counted to ±0.25 fringe, where one fringe is equivalent to one complete cycle of intensity variation at the center of the field of view, to what accuracy can the index of refraction of air be determined by this experiment?

65. With a diffraction grating, we are interested not only in its resolving power R, which is the ability of the grating to separate two close wavelengths, but also in the dispersion D of the grating. This is defined by $D = \Delta\theta_m/\Delta\lambda$ in the mth order. (*a*) Show that D can be written

$$D = \frac{m}{\sqrt{d^2 - m^2\lambda^2}}$$

where d is the slit spacing. (*b*) If a diffraction grating with 2000 slits per centimeter is to resolve the two yellow sodium lines in the second order (wavelengths 589.0 and 589.6 nm), how many slits must be illuminated by the beam? (*c*) What would the separation be between these resolved yellow lines if the pattern were viewed on a screen 4 m from the grating?

66. Light of wavelength λ is diffracted through a single slit of width a, and the resulting pattern is viewed on a screen a long distance L away from the slit. (*a*) Show that the width of the central maximum on the screen is approximately $2L\lambda/a$. (*b*) If a slit of width $2L\lambda/a$ is cut in the screen and is illuminated, show that the width of its central diffraction maximum at the same distance L is a to the same approximation.

67. A double-slit experiment uses a helium–neon laser with a wavelength of 633 nm and a slit separation of 0.12 mm. When a thin sheet of plastic is placed in front of one of the slits, the interference pattern shifts by 5.5 fringes. When the experiment is repeated under water, the shift is 3.5 fringes. Calculate (*a*) the thickness of the plastic sheet and (*b*) the index of refraction of the plastic sheet.

68. Three slits, each separated from its neighbor by 0.06 mm, are illuminated by a coherent light source of wavelength 550 nm. The slits are extremely narrow, so you may ignore the effects of diffraction. A screen is located 2.5 m from the slits. The intensity on the centerline is 0.05 W/m^2. Consider a location 1.72 cm from the centerline. (*a*) Draw the phasors, according to the phasor model for the addition of harmonic waves, appropriate for this location. (*b*) From the phasor diagram, calculate the intensity of light at this location.

69. Two coherent sources are located on the y axis at $+\lambda/4$ and $-\lambda/4$. They emit waves of wavelength λ and intensity I_0. (*a*) Calculate the net intensity I as a function of the angle θ measured from the $+x$ axis. (*b*) Make a polar plot of $I(\theta)$.

70. Repeat Problem 69 for four sources that are located on the y axis at $+3\lambda/4$, $+\lambda/4$, $-\lambda/4$, and $-3\lambda/4$.

71. For single-slit diffraction, calculate the first three values of ϕ (the total phase difference between rays from each edge of the slit) that produce subsidiary maxima by (*a*) using the phasor model and (*b*) by setting $dI/d\phi = 0$, where I is given by Equation 33-26.

72. For a diffraction grating in which all the surfaces are normal to the incident radiation, most of the energy goes into the zeroth order, which is useless from a spectroscopic point of view since the various wavelengths are not dispersed in angle. Therefore, modern gratings have shaped, or *blazed*, grooves as shown in Figure 33-41. This shifts the specular reflection, which contains most of the energy, from the zeroth order to some higher order. (*a*) Calculate the blaze angle ϕ in terms of a (the groove separation), λ, (the wavelength), and m (the order in which specular reflection is to occur). (*b*) Calculate the proper blaze angle for the specular reflection to occur in the second order for light of wavelength 450 nm incident on a grating with 10,000 lines per centimeter.

73. In this problem you will derive Equation 33-31 for the resolving power of a diffraction grating containing N slits separated by a distance d. To do this you will calculate the angular separation between the maximum and minimum for some wavelength λ and set it equal to the angular separation of the mth-order maximum for two nearby wavelengths. (*a*) Show that the phase difference ϕ between the light from two adjacent slits is given by

$$\phi = \frac{2\pi d}{\lambda}\sin\theta$$

(*b*) Differentiate this expression to show that a small change in angle $d\theta$ results in a change in phase of $d\phi$ given by

$$d\phi = \frac{2\pi d}{\lambda}\cos\theta\, d\theta$$

(*c*) For N slits, the angular separation between an interference maximum and interference minimum corresponds to a phase change of $d\phi = 2\pi/N$. Use this to show that the angular separation $d\theta$ between the maximum and minimum for some wavelength λ is given by

$$d\theta = \frac{\lambda}{Nd\cos\theta} \qquad \text{33-32}$$

(*d*) The angle of the mth-order interference maximum for wavelength λ is given by Equation 33-30. Compute the differential of each side of this equation to show that angular separation of the mth-order maximum for two nearly equal wavelengths differing by $d\lambda$ is given by

$$d\theta \approx \frac{m\, d\lambda}{d\cos\theta} \qquad \text{33-33}$$

(*e*) According to Rayleigh's criterion, two wavelengths will be resolved in the mth-order if the angular separation of the wavelengths given by Equation 33-33 equals the angular separation of the interference maximum and interference minimum given by Equation 33-32. Use this to derive Equation 33-31 for the resolving power of a grating.

Figure 33-41 Problem 72.

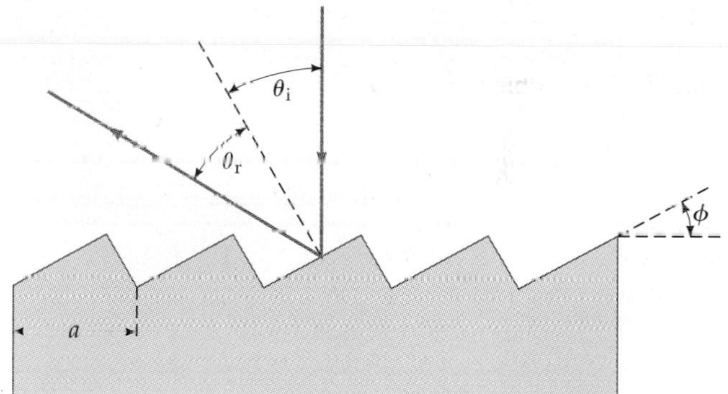

Part 6 Modern Physics

The surface of silicon, an important ingredient in many semiconductor devices, is shown here at a magnification of about 1,000,000,000. Individual atoms of a silicon crystal are seen as hills in this micrograph, which was obtained with a scanning tunneling microscope. The digital information collected by the microscope is plotted by a computer, which assigns false colors to accentuate the crystalline structure.

Chapter 34

Relativity

Albert Einstein in 1916.

Near the end of the nineteenth century, many physicists thought that all the important laws of physics had been discovered and that there was little left for them to do other than work out the remaining details. Newton's laws of motion and gravity seemed to describe all known motion on earth as well as that of the planets and other heavenly bodies, whereas Maxwell's equations of electricity and magnetism seemed to give a complete description of electromagnetic phenomena. Even as evidence of the microscopic world of molecules and atoms began to accumulate, it was assumed that these new phenomena would be adequately described by the theories of Newton and Maxwell. However, the discovery of radioactivity by Becquerel in 1896, the theoretical papers of Planck in 1897 and Einstein in 1905, and the work of Rutherford, Millikan, Bohr, de Broglie, Schrödinger, Heisenberg, and others in the early twentieth century led to two completely new theories: relativity and quantum mechanics. These theories revolutionized the world of science and became the foundation for new technologies that have changed the face of civilization.

In this chapter we will study relativity. The theory of relativity consists of two rather different theories, the special theory and the general theory.

The special theory, developed by Einstein and others in 1905, concerns the comparison of measurements made in different inertial reference frames moving with constant velocity relative to one another. Its consequences, which can be derived with a minimum of mathematics, are applicable in a wide variety of situations encountered in physics and engineering. On the other hand, the general theory, also developed by Einstein and others around 1916, is concerned with accelerated reference frames and gravity. A thorough understanding of the general theory requires sophisticated mathematics, and the applications of this theory are chiefly in the area of gravitation. It is of great importance in cosmology, but it is rarely encountered in other areas of physics or in engineering. We will therefore concentrate on the special theory (often referred to as *special relativity*) and discuss the general theory only briefly in the last section of this chapter.

34-1 Newtonian Relativity

Newton's first law does not distinguish between a particle at rest and one moving with constant velocity. If there is no net external force acting, the particle will remain in its initial state—either at rest or moving with its initial velocity. Consider a particle at rest relative to you with no forces acting on it. According to Newton's first law, it will remain at rest. Now consider the same particle from the point of view of a second observer who is moving with constant velocity relative to you. From this observer's "frame of reference," both you and the particle are moving with constant velocity. Newton's first law holds also for him. (Note that if the second observer were accelerating relative to you, he would see the particle accelerating relative to him with no external forces acting on it. Thus Newton's first law would not hold for him.) How might we distinguish whether you and the particle are at rest and the second observer is moving with constant velocity, or the second observer is at rest and you and the particle are moving?

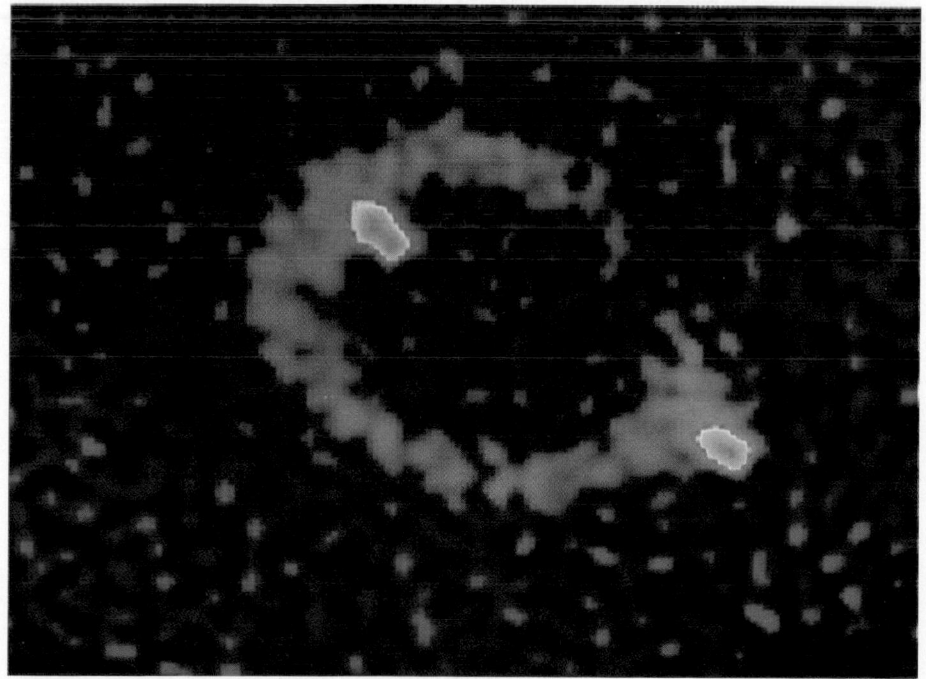

This ringlike structure of the radio source MG1131 + 0456 is thought to be due to "gravitational lensing," first proposed by Einstein in 1936, in which a source is imaged into a ring by a large massive object in the foreground.

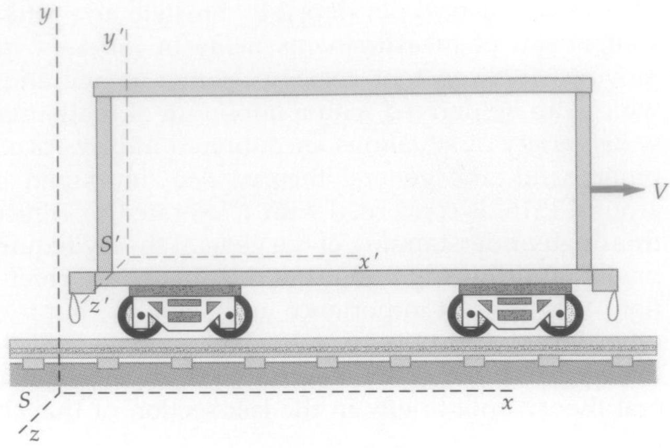

Figure 34-1 A boxcar moving with constant velocity along a straight track. The reference frame S' is at rest relative to the car and is moving with speed V relative to S, which is at rest relative to the track. It is impossible to tell by doing mechanics experiments inside the car whether the car is moving to the right with speed V or the track is moving to the left with speed V.

Let us consider some simple experiments. Suppose we have a train moving along a straight, flat track with a constant velocity V. (We assume there are no bumps or shakes in the motion.) Let us choose a coordinate system xyz with its x axis along the track as shown in Figure 34-1. It doesn't matter where along the track we choose the origin. For different choices, the position (relative to the origin) of the train and its contents will differ, but their velocities will be the same. A set of coordinate systems at rest relative to each other is called a **reference frame.** We will call the reference frame at rest relative to the track frame S. We now consider doing various mechanics experiments in a closed boxcar of the train. For these we choose a coordinate system at rest relative to the train. This coordinate system is in reference frame S', which is moving to the right with speed V relative to frame S. We note that a ball at rest in the train remains at rest. If we drop the ball, it falls straight down in frame S' with an acceleration g due to gravity. (Of course, when viewed in frame S, the ball moves along a parabolic path because it has an initial velocity V to the right.) No mechanics experiment that we can do—measuring the period of a pendulum or a body on a spring, observing the collisions between two bodies, or whatever—will tell us whether the train is moving and the track is at rest or the track is moving and the train is at rest. Newton's laws hold for reference frame S' as well as reference frame S.

A reference frame in which Newton's laws hold is called an **inertial reference frame.**

All reference frames moving at constant velocity relative to an inertial reference frame are also inertial reference frames.

If we have two inertial reference frames moving with constant velocity relative to each other, such as S and S', there are no mechanics experiments that can tell us which is at rest and which is moving or if they are both moving. This result is known as the principle of **newtonian relativity:**

Absolute motion cannot be detected.

This principle was well known by Galileo, Newton, and others in the seventeenth century. By the late nineteenth century, however, this view had changed. It was then generally thought that newtonian relativity was not valid and that absolute motion could be detected in principle by a measurement of the speed of light.

34-2 The Michelson–Morley Experiment

From our study of wave motion, we know that all mechanical waves require a medium for their propagation, and that the speed of such waves depends only on the properties of the medium. For example, the speed of sound waves in air depends on the temperature of the air. This speed is relative to still air. Motion relative to still air can indeed be detected. If you are moving relative to still air, you feel a wind.

It was therefore natural to expect that some kind of medium supports the propagation of light and other electromagnetic waves. This proposed medium was called the **ether.** As proposed, the ether had to have unusual properties. For example, it had to have great rigidity to support waves of such high velocity. (Recall that the velocity of waves on a string depends on the tension of the string and that of longitudinal sound waves in a solid depends on the bulk modulus of the solid.) Yet the ether could introduce no drag force on the planets, as their motion is fully accounted for by the law of gravitation. It was suspected that the ether was at rest relative to the distant stars, but this was considered to be an open question. It was therefore of considerable interest to determine the velocity of the earth relative to the ether. Experiments to do this were undertaken by Albert Michelson, first in 1881 and then again with Edward Morley in 1887 with greater precision. It was thought that a measurement of the speed of light relative to some reference frame moving through the ether would yield a result greater or less than c by an amount that depended on the speed of the frame relative to the ether and the direction of motion relative to the direction of the light beam. Thus, in 1881 Michelson set out to measure the speed of light relative to the earth and from this measurement to determine the velocity of the earth relative to the ether.

According to Maxwell's theory of electromagnetism, the speed of light and other electromagnetic waves is

$$c = \frac{1}{\sqrt{\epsilon_0 \mu_0}} = 3 \times 10^8 \text{ m/s}$$

where ϵ_0 and μ_0 are, respectively, the permittivity and permeability of free space. There is nothing in Maxwell's equations that tells us in what reference frame the speed of light will have this value, but the expectation was that this was the speed of light relative to its natural medium, the ether.

In the usual measurements of the speed of light (Section 30-1), the time it takes for a light pulse to travel to and from a mirror is determined. Figure 34-2 shows a light source and a mirror a distance L apart. If we assume that both are moving with speed v through the ether, classical theory predicts that the light will travel toward the mirror with speed $c - v$ and back with speed $c + v$ (both speeds being relative to the mirror and the light source). The time for the total trip will be

$$t_1 = \frac{L}{c - v} + \frac{L}{c + v} = 2c \frac{L}{c^2 - v^2} = \frac{2L}{c}\left(1 - \frac{v^2}{c^2}\right)^{-1} \qquad 34\text{-}1$$

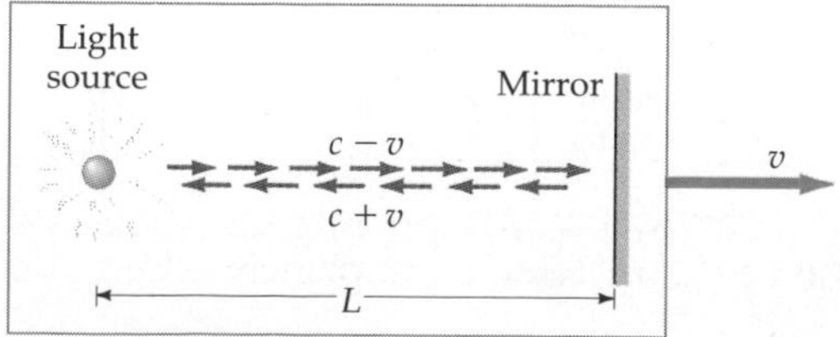

Figure 34-2 A light source and mirror moving with speed v relative to the "ether." According to classical theory, the speed of light relative to the source and mirror is $c - v$ toward the mirror and $c + v$ away from the mirror.

We can see that this differs from the time $2L/c$ by the factor $(1 - v^2/c^2)^{-1}$, which is very nearly equal to 1 if v is much less than c. We can simplify this expression for small values of v/c by using the binomial expansion

$$(1 + x)^n = 1 + nx + n(n - 1)\frac{x^2}{2} + \cdots \approx 1 + nx \qquad 34\text{-}2$$

when x is much less than 1. Using $n = -1$ and $x = v^2/c^2$, Equation 34-1 becomes

$$t_1 \approx \frac{2L}{c}\left(1 + \frac{v^2}{c^2}\right) \qquad 34\text{-}3$$

The orbital speed of the earth about the sun is about 3×10^4 m/s. If we take this for an estimate of v we have $v = 3 \times 10^4$ m/s, $v/c = (3 \times 10^4 \text{ m/s})/(3 \times 10^8 \text{ m/s}) = 10^{-4}$, and $v^2/c^2 = 10^{-8}$. Thus the correction for the earth's motion is small indeed.

Michelson realized that although this effect is too small to be measured directly, it should be possible to determine v^2/c^2 by a difference measurement. For this measurement, he used the Michelson interferometer, which we discussed in Section 33-3. In this experiment, one beam of light moves along the direction of the earth's motion and another moves perpendicular to that direction (Figure 34-3). The difference between the round-trip times of these beams depends on the speed of the earth and can be determined by an interference measurement. Let us assume that the interferometer is oriented such that the beam that strikes mirror M_1 is in the direction of the assumed motion of the earth. Equation 34-3 then gives the classical result for the round-trip time t_1 for the transmitted beam. The beam that reflects from the beam splitter and strikes mirror M_2 travels with some velocity **u** (relative to the earth) perpendicular to the earth's velocity. Relative to the ether, it travels with velocity **c** as shown in Figure 34-4. The velocity **u** (according to classical theory) is then the vector difference $\mathbf{u} = \mathbf{c} - \mathbf{v}$, as shown in the figure. The magnitude of **u** is $\sqrt{c^2 - v^2}$, so the round-trip time t_2 for this beam is

$$t_2 = \frac{2L}{\sqrt{c^2 - v^2}} = \frac{2L}{c}(1 - v^2/c^2)^{-1/2} \qquad 34\text{-}4$$

Again using the binomial expansion, we obtain

$$t_2 \approx \frac{2L}{c}\left(1 + \frac{1}{2}\frac{v^2}{c^2}\right) \qquad 34\text{-}5$$

This expression is slightly different from that given for t_1 in Equation 34-3.

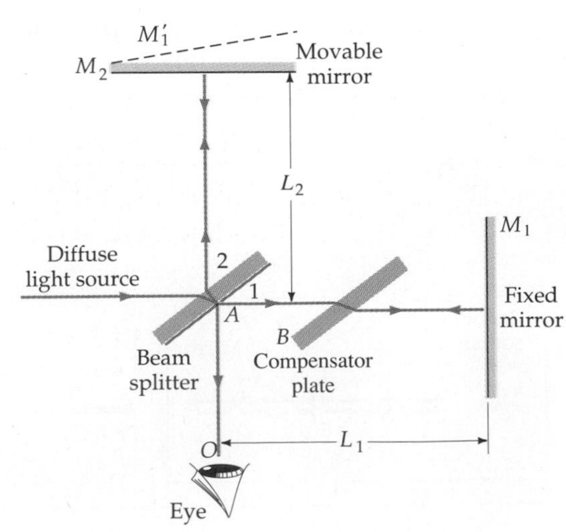

Figure 34-3 Michelson interferometer. The dashed line M_1' is the image of mirror M_1 in mirror A. The interference fringes observed are those of a small wedge-shaped film of air formed between the sources M_2 and M_1'. Assume that the light beam reflecting off mirror M_1 is parallel to the motion of the earth and that reflecting off mirror M_2 is perpendicular to the earth's motion. The interference of the two beams depends on the relative number of waves in each path, which depends in turn on the speed of the light beams relative to the earth. If the speed of light along the parallel path is different from that along the perpendicular path, the interference fringe pattern will shift when the interferometer is rotated through 90°.

The difference in these two times is

$$\Delta t = t_1 - t_2 \approx \frac{L}{c}\frac{v^2}{c^2} \qquad 34\text{-}6$$

This time difference is to be detected by observing the interference of the two beams of light.

Because of the difficulty of making the two paths of equal length to the precision required, the interference pattern of the two beams is observed and the whole apparatus is then rotated 90°. The rotation produces a time difference given by Equation 34-6 for each beam. The total time difference of $2\ \Delta t$ results in a phase difference $\Delta\phi$ between the two beams, where

$$\Delta\phi = 2\pi\frac{2c\ \Delta t}{\lambda}$$

and λ is the wavelength of the light. The interference fringes observed in the first orientation should thus shift by a number of fringes ΔN given by

$$\Delta N = \frac{\Delta\phi}{2\pi} = \frac{2c\ \Delta t}{\lambda} = \frac{2L}{\lambda}\frac{v^2}{c^2} \qquad 34\text{-}7$$

In Michelson's first attempt in 1881, L was about 1.2 m and λ was 590 nm. For $v^2/c^2 = 10^{-8}$, ΔN was expected to be 0.04 fringes. However, no shift was observed. In case the earth just happened to be at rest relative to the ether when the experiment was performed, the experiment was repeated six months later when the motion of the earth relative to the sun was in the opposite direction. Even though the experimental uncertainties were estimated to be of about the same magnitude as the expected fringe shift, Michelson reported the observation of no fringe shift as evidence that the earth did not move relative to the ether. In 1887, when he repeated the experiment with Edward W. Morley, he used an improved system for rotating the apparatus without introducing a fringe shift due to mechanical strains, and he increased the effective path length L to about 11 m by a series of multiple reflections. Figure 34-5 shows the configuration of the Michelson–Morley apparatus. For this attempt, ΔN was expected to be 0.4 fringes, about 20 to 40 times the minimum that could be observed. Once again, no shift was observed. The experiment has since been repeated under various conditions by a number of people, and no shift has ever been found.

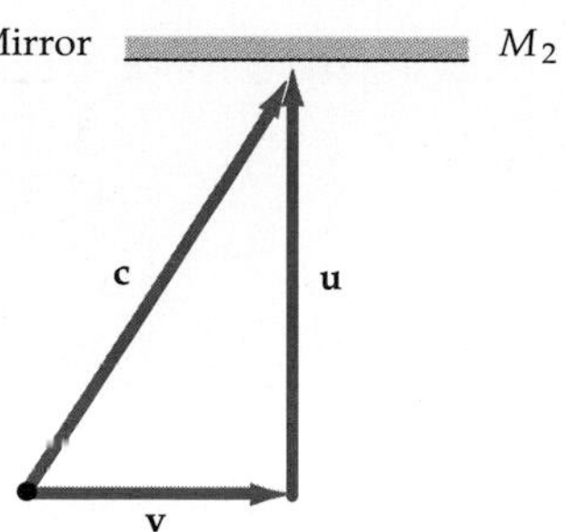

Figure 34-4 The light beam reflected from the beam splitter in a Michelson interferometer. The interferometer is moving to the right relative to the ether with velocity **v**, and the light beam moves perpendicular to mirror M_2 with velocity **u**. The velocity of light is **c** in the frame of the ether. Relative to the earth, in which the interferometer is fixed, the velocity of light is $\mathbf{u} = \mathbf{c} - \mathbf{v}$. The speed of light relative to the earth according to classical theory is then $u = (c^2 - v^2)^{1/2} = c(1 - v^2/c^2)^{1/2}$.

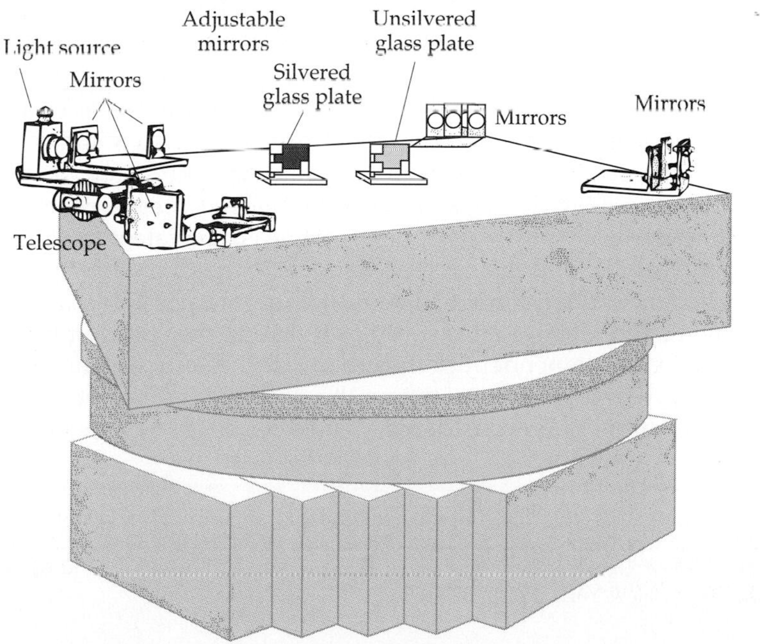

Figure 34-5 Drawing of Michelson and Morley's apparatus for their 1887 experiment. The optical parts were mounted on a sandstone slab 5 ft square, which was floated in mercury to reduce the strains and vibrations that had affected the earlier experiments. Observations could be made in all directions by rotating the apparatus in the horizontal plane.

In 1905, at the age of 26, Albert Einstein published a paper on the electrodynamics of moving bodies.* In this paper, he postulated that absolute motion cannot be detected by any experiment. (We will discuss the Einstein postulates in detail in the next section.) The null result of the Michelson–Morley experiment is therefore expected. We can consider the whole apparatus and the earth to be at rest. Thus no fringe shift is expected when the apparatus is rotated 90° since all directions are equivalent. Einstein did not set out to explain the results of the Michelson–Morley experiment. His theory arose from his considerations of the theory of electricity and magnetism and the unusual property of electromagnetic waves that they propagate in a vacuum. In his first paper, which contains the complete theory of special relativity, he made only a passing reference to the Michelson–Morley experiment, and in later years he could not recall whether he was aware of the details of this experiment before he published his theory.

34-3 Einstein's Postulates

The theory of special relativity can be derived from two postulates proposed by Einstein in his original paper in 1905. Simply stated, these postulates are

Einstein's postulates

> Postulate 1. Absolute, uniform motion cannot be detected.
>
> Postulate 2. The speed of light is independent of the motion of the source.

Postulate 1 is merely an extension of the newtonian principle of relativity to include all types of physical measurements (not just those that are mechanical). Postulate 2 describes a common property of all waves. For example, the speed of sound waves does not depend on the motion of the sound source. When an approaching car sounds its horn, the frequency heard increases according to the doppler effect we studied in Section 14-6, but the speed of the waves traveling through the air does not depend on the speed of the car. The speed of the waves depends only on the properties of the air, such as its temperature.

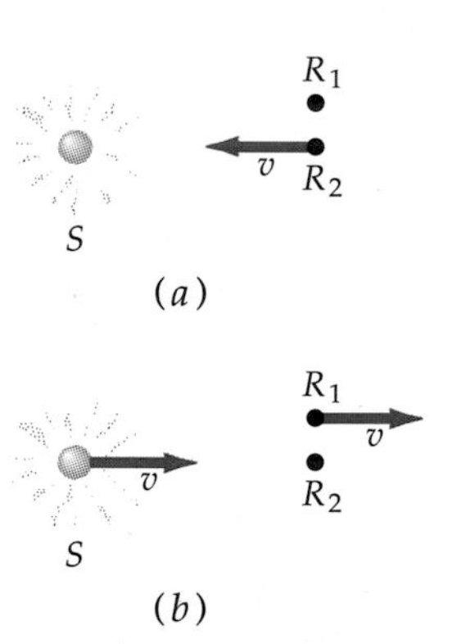

Figure 34-6 (*a*) A stationary light source S and a stationary observer R_1, with a second observer R_2 moving toward the source with speed v. (*b*) In the reference frame in which the observer R_2 is at rest, the light source S and observer R_1 move to the right with speed v. If absolute motion cannot be detected, the two views are equivalent. Since the speed of light does not depend on the motion of the source, observer R_2 measures the same value for that speed as observer R_1.

Although each postulate seems quite reasonable, many of the implications of the two together are quite surprising and contradict what is often called common sense. For example, one important implication of these postulates is that every observer measures the same value for the speed of light independent of the relative motion of the source and the observer. Consider a light source S and two observers, R_1 at rest relative to S and R_2 moving toward S with speed v, as shown in Figure 34-6*a*. The speed of light measured by R_1 is $c = 3 \times 10^8$ m/s. What is the speed measured by R_2? The answer is *not* $c + v$. By postulate 1, Figure 34-6*a* is equivalent to Figure 34-6*b*, in which R_2 is at rest and the source S and R_1 are moving with speed v. That is, since absolute motion cannot be detected, it is not possible to say which is really moving and which is at rest. By postulate 2, the speed of light from a moving source is independent of the motion of the source. Thus, looking at Figure 34-6*b*, we see that R_2 measures the speed of light to be c, just as R_1 does. This result is often considered as an alternative to Einstein's second postulate:

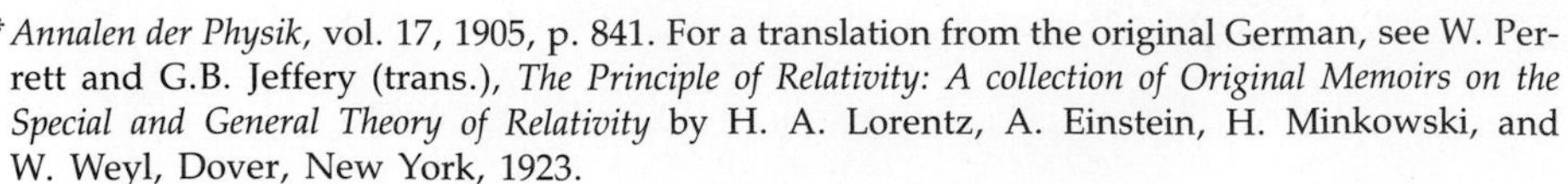
**Annalen der Physik*, vol. 17, 1905, p. 841. For a translation from the original German, see W. Perrett and G.B. Jeffery (trans.), *The Principle of Relativity: A collection of Original Memoirs on the Special and General Theory of Relativity* by H. A. Lorentz, A. Einstein, H. Minkowski, and W. Weyl, Dover, New York, 1923.

Postulate 2 (Alternate). Every observer measures the same value c for the speed of light.

This result contradicts our intuitive ideas about relative velocities. If a car moves at 50 km/h away from an observer and another car moves at 80 km/h in the same direction, the velocity of the second car relative to the first car is 30 km/h. This result is easily measured and conforms to our intuition. However, according to Einstein's postulates, if a light beam is moving in the direction of the cars, observers in both cars will measure the same speed for the light beam. Our intuitive ideas about the combination of velocities are approximations that hold only when the speeds are very small compared with the speed of light. Even in an airplane moving with the speed of sound, it is not possible to measure the speed of light accurately enough to distinguish the difference between the results c and $c + v$, where v is the speed of the plane. In order to make such a distinction, we must either move with a very great velocity (much greater than that of sound) or make extremely accurate measurements, as in the Michelson–Morley experiment.

34-4 The Lorentz Transformation

Einstein's postulates have important consequences for measuring time intervals and space intervals as well as relative velocities. Throughout this chapter we will be comparing measurements of the positions and times of events (such as lightning flashes) made by observers who are moving relative to each other. We will use a rectangular coordinate system xyz with origin O, called the S reference frame, and another system $x'y'z'$ with origin O', called the S' frame, that is moving with a constant velocity $\mathbf{V}$ relative to the S frame. Relative the S' frame, the S frame is moving with a constant velocity $-\mathbf{V}$. For simplicity, we will consider the S' frame to be moving with speed V along the x axis in the positive x direction relative to S. Then, relative to S', the S frame is moving with speed V in the negative x' direction along the x' axis. In each frame, we will assume that there are as many observers as are needed who are equipped with measuring devices, such as clocks and metersticks, that are identical when compared at rest (see Figure 34-7).

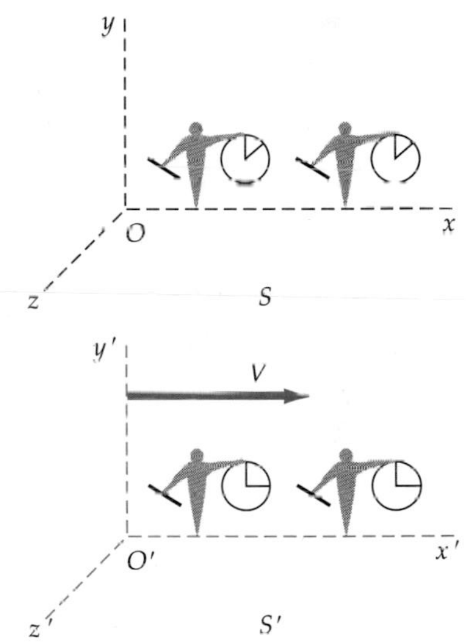

Figure 34-7 Coordinate reference frames S and S' moving with relative speed V. In each frame, there are observers with metersticks and clocks that are identical when compared at rest.

We need many observers, for example, to determine the times of events. If one observer is distant from an event, then his time observation can be thrown off by the time it takes for the information about the event to travel to his location (such as the transit time for a light pulse). The observer can avoid such problems by recording only events *local* to him, and leaving other events to other observers at those locations. It's like having one official at the beginning of a racetrack and another at the end.

We will use Einstein's postulates to find the general relation between the coordinates x, y, and z and the time t of an event as seen in reference frame S and the coordinates x', y', and z' and the time t' of the same event as seen in reference frame S', which is moving with uniform velocity relative to S. We will consider only the simple case in which the origins are coincident at time $t = t' = 0$. The classical relation, called the **galilean transformation,** is

$$x = x' + Vt' \qquad y = y' \qquad z = z' \qquad t = t' \qquad \text{34-8}a$$

Galilean transformation

The inverse transformation is

$$x' = x - Vt \qquad y' = y \qquad z' = z \qquad t' = t \qquad \text{34-8}b$$

These equations are consistent with experimental observations as long as V is much less than c. They lead to the familiar classical addition law for velocities. If a particle has velocity $u_x = dx/dt$ in frame S, its velocity in frame S' is

$$u'_x = \frac{dx'}{dt'} = \frac{dx'}{dt} = \frac{dx}{dt} - V = u_x - V \tag{34-9}$$

If we differentiate this equation again, we find that the acceleration of the particle is the same in both frames:

$$a_x = du_x/dt = du'_x/dt' = a'_x$$

It should be clear that the galilean transformation is not consistent with Einstein's postulates of special relativity. If light moves along the x axis with speed c in S, these equations imply that the speed in S' is $u'_x = c - V$; rather than $u'_x = c$, which is consistent with Einstein's postulates and with experiment. The classical transformation equations must therefore be modified to make them consistent with Einstein's postulates. We will give a brief outline of one method of obtaining the relativistic transformation.

We assume that the relativistic transformation equation for x is the same as the classical equation (Equation 34-8*a*) except for a constant multiplier on the right side. That is, we assume the equation is of the form

$$x = \gamma(x' + Vt') \tag{34-10}$$

where γ is a constant that can depend on V and c but not on the coordinates. The inverse transformation must look the same except for the sign of the velocity:

$$x' = \gamma(x - Vt) \tag{34-11}$$

Let us consider a light pulse that starts at the origin of S at $t = 0$. Since we have assumed that the origins are coincident at $t = t' = 0$, the pulse also starts at the origin of S' at $t' = 0$. Einstein's postulates require that the equation for the x component of the wavefront of the light pulse is $x = ct$ in frame S and $x' = ct'$ in frame S'. Substituting ct for x and ct' for x' in Equations 34-10 and 34-11, we obtain

$$ct = \gamma(ct' + Vt') = \gamma(c + V)t' \tag{34-12}$$

and

$$ct' = \gamma(ct - Vt) = \gamma(c - V)t \tag{34-13}$$

We can eliminate either t' or t from these two equations and determine γ. We get

$$\gamma^2 = (1 - V^2/c^2)^{-1}$$

$$\gamma = \frac{1}{\sqrt{1 - V^2/c^2}} \tag{34-14}$$

(It is important to note that γ is always greater than 1 and that when V is much less than c, $\gamma \approx 1$.) The relativistic transformation for x and x' is therefore given by Equations 34-10 and 34-11 with γ given by Equation 34-14. We can obtain equations for t and t' by combining Equation 34-10 with the inverse transformation given by Equation 34-11. Substituting $x = \gamma(x' + Vt')$ for x in Equation 34-11, we obtain

$$x' = \gamma[\gamma(x' + Vt') - Vt] \tag{34-15}$$

which can be solved for t in terms of x' and t'. The complete relativistic transformation is

$$x = \gamma(x' + Vt') \qquad y = y' \qquad z = z' \tag{34-16}$$

$$t = \gamma\left(t' + \frac{Vx'}{c^2}\right) \tag{34-17}$$

Lorentz transformation

The inverse is

$$x' = \gamma(x - Vt) \qquad y' = y \qquad z' = z \tag{34-18}$$

$$t' = \gamma\left(t - \frac{Vx}{c^2}\right) \tag{34-19}$$

The transformation described by Equations 34-16 through 34-19 is called the **Lorentz transformation.** It relates the space and time coordinates x, y, z, and t of an event in frame S to the coordinates x', y', z', and t' of the same event as seen in frame S', which is moving along the x axis with speed V relative to frame S.

We will now look at some applications of the Lorentz transformation.

Time Dilation

An important consequence of Einstein's postulates and the Lorentz transformation is that the time interval between two events that occur at the same place in some reference frame is always less than the time interval between the same events that is measured in another reference frame in which the events occur at different places. Consider two events that occur at x'_0 at times t'_1 and t'_2 in frame S'. We can find the times t_1 and t_2 for these events in S from Equation 34-17. We have

$$t_1 = \gamma\left(t'_1 + \frac{Vx'_0}{c^2}\right)$$

and

$$t_2 = \gamma\left(t'_2 + \frac{Vx'_0}{c^2}\right)$$

so

$$t_2 - t_1 = \gamma(t'_2 - t'_1)$$

The time between events that happen at the *same place* in a reference frame is called **proper time** t_p. In this case, the time interval $\Delta t_p = t'_2 - t'_1$ measured in frame S' is proper time. The time interval Δt measured in any other reference frame is always longer than the proper time. This expansion is called **time dilation:**

$$\Delta t = \gamma\, \Delta t_p \tag{34-20}$$

Time dilation

Example 34-1

Two events occur at the same point x'_0 at times t'_1 and t'_2 in frame S', which is traveling at speed V relative to frame S. What is the spatial separation of these events in frame S?

From Equation 34-16, we have

$$x_1 = \gamma(x_0' + Vt_1')$$

and

$$x_2 = \gamma(x_0' + Vt_2')$$

Then

$$x_2 - x_1 = \gamma V(t_2' - t_1')$$
$$= V(t_2 - t_1)$$

The spatial separation of these events in S is the distance a single point, such as x_0' in S', moves in S during the time interval between the events.

We can understand time dilation directly from Einstein's postulates without using the Lorentz transformation. Figure 34-8*a* shows an observer A' a distance D from a mirror. The observer and the mirror are in a spaceship that is at rest in frame S'. He explodes a flash gun and measures the time interval $\Delta t'$ between the original flash and his seeing the return flash from the mirror. Since light travels with speed c, this time is

$$\Delta t' = \frac{2D}{c}$$

We now consider these same two events, the original flash of light and the receiving of the return flash, as observed in reference frame S, in which observer A' and the mirror are moving to the right with speed V as shown in Figure 34-8*b*. The events happen at two different places x_1 and x_2 in frame S. During the time interval Δt (as measured in S) between the original flash and the return flash, observer A' and his spaceship have moved a horizontal distance $V\,\Delta t$. In Figure 34-8*b*, we can see that the path traveled by the light is longer in S than in S'. However, by Einstein's postulates, light travels with the same speed c in frame S as it does in frame S'. Since it travels farther in S at the same speed, it takes longer in S to reach the mirror and return. The time interval in S is thus longer than it is in S'. From the triangle in Figure 34-8*c*, we have

$$\left(\frac{c\,\Delta t}{2}\right)^2 = D^2 + \left(\frac{V\,\Delta t}{2}\right)^2$$

or

$$\Delta t = \frac{2D}{\sqrt{c^2 - V^2}} = \frac{2D}{c}\frac{1}{\sqrt{1 - V^2/c^2}}$$

Figure 34-8 (*a*) Observer A' and the mirror are in a spaceship at rest in frame S'. The time it takes for the light pulse to reach the mirror and return is measured by A' to be $2D/c$. (*b*) In frame S, the spaceship is moving to the right with speed V. If the speed of light is the same in both frames, the time it takes for the light to reach the mirror and return is longer than $2D/c$ in S because the distance traveled is greater than $2D$. (*c*) A right triangle for computing the time Δt in frame S.

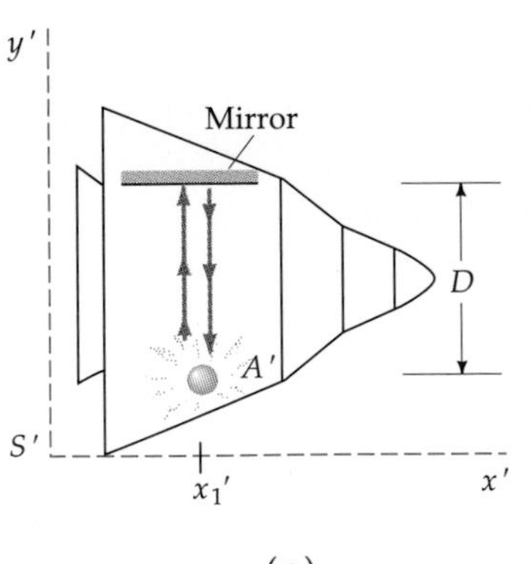

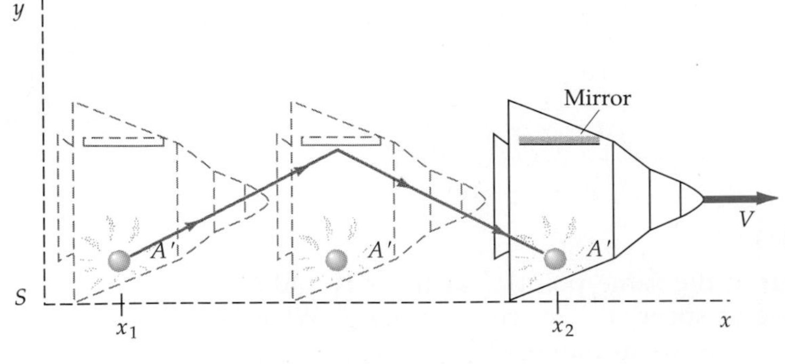

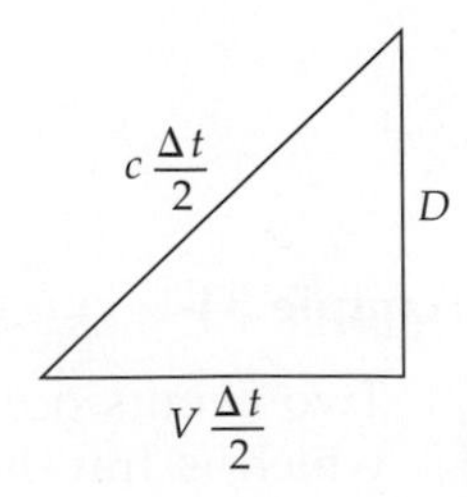

Using $\Delta t' = 2D/c$, we obtain

$$\Delta t = \frac{\Delta t'}{\sqrt{1 - V^2/c^2}} = \gamma\, \Delta t'$$

Example 34-2

Astronauts in a spaceship traveling away from the earth at $V = 0.6c$ sign off from space control, saying that they are going to nap for 1 hour and then call back. How long does their nap last as measured on earth?

Since the astronauts go to sleep and wake up at the same place in their reference frame, the time interval for their nap of 1 hour as measured by them is proper time. In the earth's reference frame, they move a considerable distance between these two events. The time interval measured in the earth's frame (using two clocks located at those events) is longer by the factor γ. With $V = 0.6c$, we have

$$1 - \frac{V^2}{c^2} = 1 - (0.6)^2 = 0.64$$

Then γ is

$$\gamma = \frac{1}{\sqrt{1 - V^2/c^2}} = \frac{1}{\sqrt{0.64}} = \frac{1}{0.8} = 1.25$$

The nap thus lasts for 1.25 hours as measured on earth.

Exercise

If the spaceship in Example 34-2 is moving at $V = 0.8c$, how long would a 1-hour nap last as measured on earth? (Answer: 1.67 h)

Length Contraction

A phenomenon closely related to time dilation is **length contraction.** The length of an object measured in the reference frame in which the object is at rest is called its **proper length** L_p. In a reference frame in which the object is moving, the measured length is shorter than its proper length. Consider a rod at rest in frame S' with one end at x_2' and the other end at x_1'. The length of the rod in this frame is its proper length $L_p = x_2' - x_1'$. Some care must be taken to find the length of the rod in frame S. In this frame, the rod is moving to the right with speed V, the speed of frame S'. The length of the rod in frame S is *defined* as $L = x_2 - x_1$, where x_2 is the position of one end at some time t_2, and x_1 is the position of the other end *at the same time* $t_1 = t_2$ as measured in frame S. Equation 34-18 is convenient to use to calculate $x_2 - x_1$ at some time t because it relates x, x' and t, whereas Equation 34-16 is not convenient because it relates x, x', and t':

$$x_2' = \gamma(x_2 - Vt_2)$$

and

$$x_1' = \gamma(x_1 - Vt_1)$$

Since $t_2 = t_1$, we obtain

$$x_2' - x_1' = \gamma(x_2 - x_1)$$

$$x_2 - x_1 = \frac{1}{\gamma}(x_2' - x_1') = \sqrt{1 - V^2/c^2}\,(x_2' - x_1')$$

or

Length contraction

$$L = \frac{1}{\gamma} L_p = \sqrt{1 - V^2/c^2}\, L_p \qquad 34\text{-}21$$

Thus the length of a rod is smaller when it is measured in a frame in which it is moving. Before Einstein's paper was published, Lorentz and FitzGerald tried to explain the null result of the Michelson–Morley experiment by assuming that distances in the direction of motion contracted by the amount given in Equation 34-21. This contraction is now known as the **Lorentz–FitzGerald contraction.**

Example 34-3

A stick that has a proper length of 1 m moves in a direction along its length with speed V relative to you. The length of the stick as measured by you is 0.914 m. What is the speed V?

The length of the stick measured in a frame in which it is moving with speed V is related to its proper length by Equation 34-21:

$$L = \frac{L_p}{\gamma}$$

Then

$$\gamma = \frac{L_p}{L} = \frac{1\text{ m}}{0.914\text{ m}} = \frac{1}{\sqrt{1 - V^2/c^2}} = 1.094$$

$$\sqrt{1 - V^2/c^2} = 0.914$$

$$1 - \frac{V^2}{c^2} = (0.914)^2 = 0.835$$

$$\frac{V^2}{c^2} = 1 - 0.835 = 0.165$$

$$V = 0.406c$$

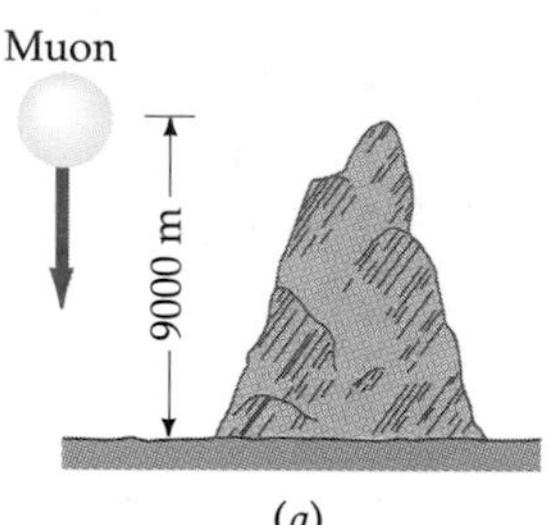

(a)

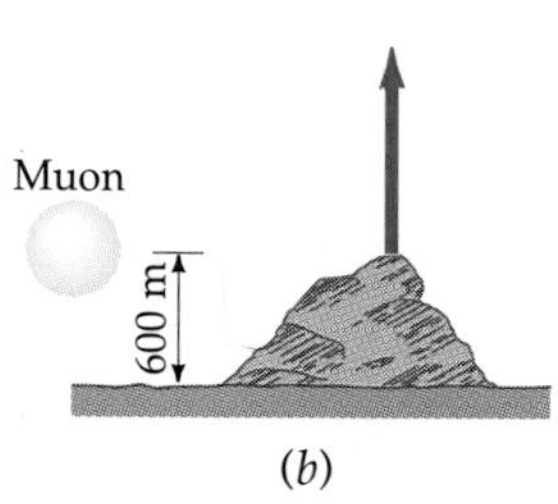

(b)

Figure 34-9 Although muons are created high above the earth and their mean lifetime is only about 2 μs when at rest, many appear at the earth's surface. (a) In the earth's reference frame, a typical muon moving at $0.998c$ has a mean lifetime of 30 μs and travels 9000 m in this time. (b) In the reference frame of the muon, the distance traveled by the earth is only 600 m in the muon's lifetime of 2 μs.

An interesting example of time dilation or length contraction is afforded by the appearance of muons as secondary radiation from cosmic rays. Muons decay according to the statistical law of radioactivity:

$$N(t) = N_0\, e^{-t/\tau} \qquad 34\text{-}22$$

where N_0 is the original number of muons at time $t = 0$, $N(t)$ is the number remaining at time t, and τ is the mean lifetime, which is about 2 μs for muons at rest. Since muons are created (from the decay of pions) high in the atmosphere, usually several thousand meters above sea level, few muons should reach sea level. A typical muon moving with speed $0.998c$ would travel only about 600 m in 2 μs. However, the lifetime of the muon measured in the earth's reference frame is increased by the factor $1/\sqrt{1 - V^2/c^2}$, which is 15 for this particular speed. The mean lifetime measured in the earth's reference frame is therefore 30 μs, and a muon with speed $0.998c$ travels about 9000 m in this time. From the muon's point of view, it lives only 2 μs, but the atmosphere is rushing past it with a speed of $0.998c$. The distance of 9000 m in the earth's frame is thus contracted to only 600 m in the muon's frame as indicated in Figure 34-9.

It is easy to distinguish experimentally between the classical and relativistic predictions of the observation of muons at sea level. Suppose that we observe 10^8 muons at an altitude of 9000 m in some time interval with a muon detector. How many would we expect to observe at sea level in the same time interval? According to the nonrelativistic prediction, the time it takes for these muons to travel 9000 m is (9000 m)/0.998$c \approx 30\ \mu$s, which is 15 lifetimes. Substituting $N_0 = 10^8$ and $t = 15\tau$ into Equation 34-22, we obtain

$$N = 10^8\, e^{-15} = 30.6$$

We would thus expect all but about 31 of the original 100 million muons to decay before reaching sea level.

According to the relativistic prediction, the earth must travel only the contracted distance of 600 m in the rest frame of the muon. This takes only $2\ \mu\text{s} = 1\tau$. Therefore the number of muons expected at sea level is

$$N = 10^8\, e^{-1} = 3.68 \times 10^7$$

Thus relativity predicts that we would observe 36.8 million muons in the same time interval. Experiments of this type have confirmed the relativistic predictions.

Question

1. You are standing on a corner and a friend is driving past in an automobile. Both of you note the times when the car passes two different intersections and determine from your watch readings the time that elapses between the two events. Which of you has determined the proper time interval?

34-5 Clock Synchronization and Simultaneity

We saw in Section 34-4 that proper time is the time interval between two events that occur at the same point in some reference frame. It can therefore be measured on a single clock. However, in another reference frame moving relative to the first, the same two events occur at different places, so two clocks are needed to record the times. The time of each event is measured on a different clock, and the interval is found by subtraction. This procedure requires that the clocks be **synchronized.** We will show in this section that

> Two clocks that are synchronized in one reference frame are not synchronized in any other frame moving relative to the first frame.

A corollary to this result is

> Two events that are simultaneous in one reference frame are not simultaneous in another frame moving relative to the first.

(This is true unless the events and clocks are in the same plane and are perpendicular to the relative motion). Comprehension of these facts usually resolves all relativity paradoxes. Unfortunately, the intuitive (and incorrect) belief that simultaneity is an absolute relation is difficult to overcome.

Suppose we have two clocks at rest at points A and B a distance L apart in frame S. How can we synchronize these two clocks? If an observer at A looks at the clock at B and sets her clock to read the same time, the clocks will not be synchronized because of the time L/c it takes light to travel from one clock to another. To synchronize the clocks, the observer at A must set her clock ahead by the time L/c. Then she will see that the clock at B reads a time that is L/c behind the time on her clock, but she will calculate that the clocks are synchronized when she allows for the time L/c for the light to reach her. Any other observers except those equidistant from the clocks will see the clocks reading different times, but they will also calculate that the clocks are synchronized when they correct for the time it takes the light to reach them. An equivalent method for synchronizing two clocks would be for an observer C at a point midway between the clocks to send a light signal and for observers at A and B to set their clocks to some prearranged time when they receive the signal.

We now examine the question of **simultaneity.** Suppose A and B agree to explode flashguns at t_0 (having previously synchronized their clocks). Observer C will see the light from the two flashes at the same time, and since he is equidistant from A and B, he will conclude that the flashes were simultaneous. Other observers in frame S will see the light from A or B first, depending on their location, but after correcting for the time the light takes to reach them, they also will conclude that the flashes were simultaneous. We can thus define simultaneity as follows:

> Two events in a reference frame are simultaneous if light signals from the events reach an observer halfway between the events at the same time.

To show that two events that are simultaneous in frame S are not simultaneous in another frame S' moving relative to S, we will use an example introduced by Einstein. A train is moving with speed V past a station platform. We will consider the train to be at rest in S' and the platform to be at rest in S. We have observers A', B', and C' at the front, back, and middle of the train. We now suppose that the train and platform are struck by lightning at the front and back of the train and that the lightning bolts are simultaneous in the frame of the platform (S) (Figure 34-10). That is, an observer C on the platform halfway between the positions A and B, where the lightning strikes, sees the two flashes at the same time. It is convenient to suppose that the lightning scorches the train and platform so that the events can be easily located in each reference frame. Since C' is in the middle of the train, halfway between the places on the train that are scorched, the events are simultaneous in S' only if C' sees the flashes at the same time. However, the flash from the front of the train is seen by C' before the flash from the back of the train. We can understand this by considering the motion of C' as seen in frame S (Figure 34-11). By the time the light from the front flash reaches C', C' has moved some distance toward the front flash and some distance away from the back flash. Thus the light from the back flash has not yet reached C' as indicated in the figure. Observer C' must therefore conclude that the events are not simultaneous and that the front of the train was struck before the back. Furthermore, all observers in S' on the train will agree with C' when they have corrected for the time it takes the light to reach them.

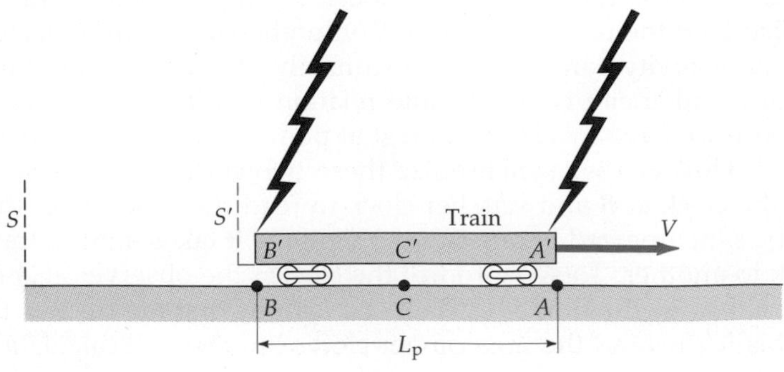

Figure 34-10 Simultaneous lightning bolts strike the ends of a train traveling with speed V in frame S attached to the platform. The light from these simultaneous events reaches observer C midway between the events at the same time. The distance between the bolts is $L_{p,\text{platform}}$.

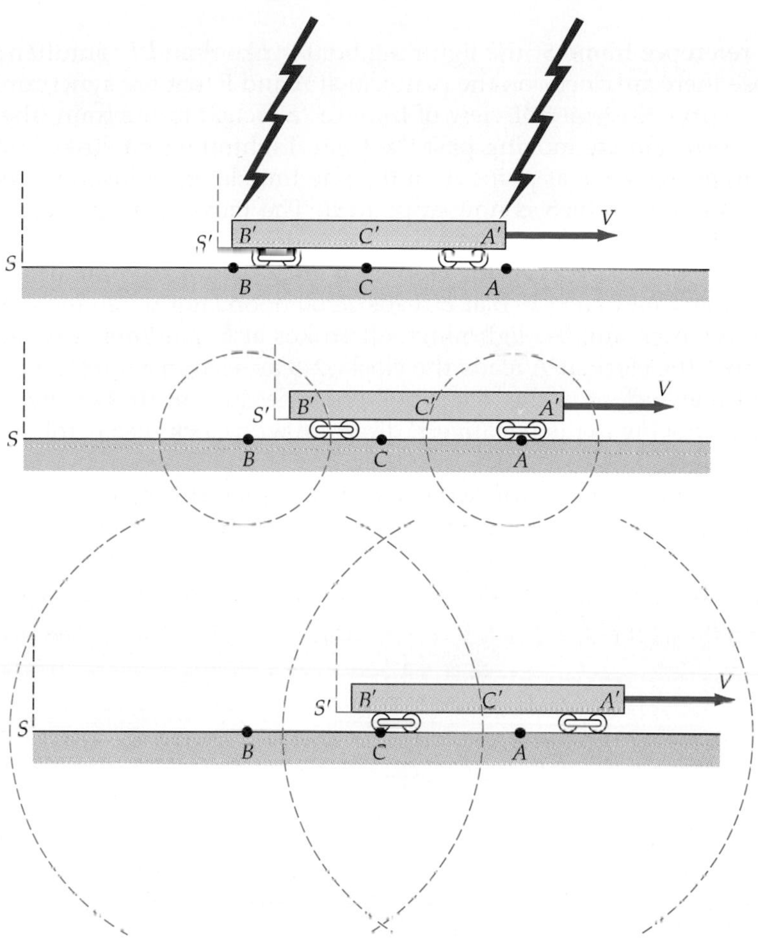

Figure 34-11 The light from the lightning bolt at the front of the train reaches observer C' at the middle of the train before that from the bolt at the back of the train. Since C' is midway between the events (which occur at the front and rear of the train), these events are not simultaneous for him.

Let $L_{\text{p,train}}$ be the proper length of the train, that is, the length of the train as measured in S' in which it is at rest. Also, let $L_{\text{p,platform}}$ be the proper length of the platform, that is, the distance between the scorch marks as seen in S. Since the scorch marks on the platform coincide with the front and back of the train at the instant (in S) that the lightning bolts strike, the distance between the scorch marks $L_{\text{p,platform}}$ equals the length of the train L_{T} as measured in frame S in which it is moving. This length is smaller than the proper length of the train because of length contraction; that is, $L_{\text{T}} = L_{\text{p,platform}} < L_{\text{p,train}}$.

Figure 34-12 shows the events of the lightning bolts as seen in the reference frame of the train (S') in which the train is at rest and the platform is moving. In this frame, the distance between the burns on the platform is contracted, so the platform is shorter than it is in S, and the train is at rest, so the train is longer than it is in S. When the lightning bolt strikes the front of the train at A', the front of the train is at point A, and the back of the train has not yet reached point B. Later, when the lightning bolt strikes the back of the train at B', the back has reached point B on the platform.

Figure 34-12 The lightning bolts of Figure 34-10 as seen in frame S' of the train. In this frame, the distance between A and B on the platform is less than $L_{\text{p,platform}}$, and the proper length of the train $L_{\text{p,train}}$ is longer than $L_{\text{p,platform}}$. The first lightning bolt strikes the front of the train when A' and A are coincident. The second bolt strikes the rear of the train when B' and B are coincident.

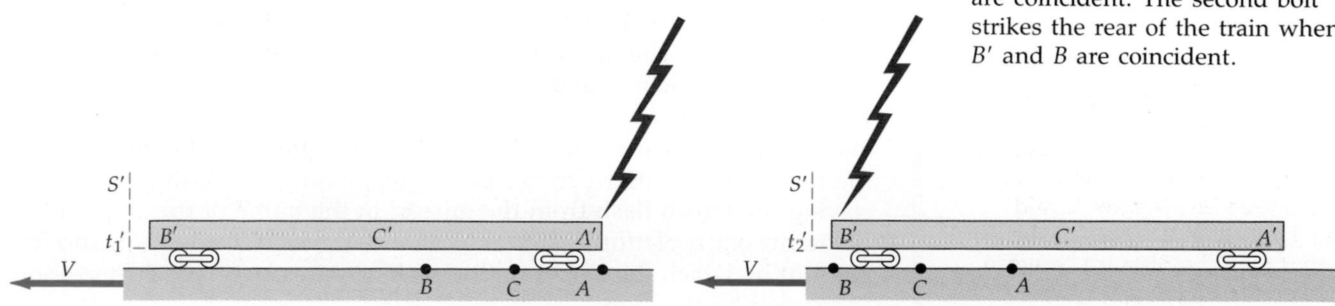

In reference frame S, the lightning bolts strike A and B simultaneously. Suppose there are clocks on the platform at A and B that are synchronized in frame S. From the point of view of frame S' attached to the train, the clocks and the platform are moving past the train. Lightning first strikes the front of the train, which is at point A, and some time later, lightning strikes the back of the train, which is now at point B. The moving clocks are thus not synchronized as seen from frame S'. If the clock at A reads 12:00 noon when the lightning bolt strikes A, the clock at B must read some time before 12:00 noon at that time. The clock at B reads 12:00 noon later when it reaches the back of the train and the lightning bolt strikes at B. Another way of saying this is that the clock at A leads the clock at B as seen in S'. In frame S', we will call the clock at A the "chasing" clock because in that frame the two clocks are moving in the negative x' direction with clock A at x_2 following, or chasing, clock B at x_1.

The time discrepancy of two clocks that are synchronized in frame S as seen in frame S' can be found from the Lorentz transformation equations. Suppose we have clocks at points x_1 and x_2 that are synchronized in S. What are the times t_1 and t_2 on these clocks as observed from frame S' at a time t_0'? From Equation 34-19, we have

$$t_0' = \gamma\left(t_1 - \frac{Vx_1}{c^2}\right)$$

and

$$t_0' = \gamma\left(t_2 - \frac{Vx_2}{c^2}\right)$$

Then

$$t_2 - t_1 = \frac{V}{c^2}(x_2 - x_1) \tag{34-23}$$

Note that the chasing clock (at x_2) leads the other (at x_1) by an amount that is proportional to their proper separation $x_2 - x_1$.

If two clocks are synchronized in the frame in which they are at rest, they will be out of synchronization in another frame. In the frame in which they are moving, the chasing clock leads (shows a later time) by an amount

$$\Delta t_s = L_p \frac{V}{c^2}$$

where L_p is the proper distance between the clocks.

A numerical example should help clarify time dilation, clock synchronization, and the internal consistency of these results.

Example 34-4

An observer in a spaceship has a flash gun and a mirror (as in our time dilation example in Figure 34-8). The distance from the gun to the mirror is 15 light-minutes (written $15c\cdot$min) and the spaceship travels with speed $V = 0.8c$. The spaceship travels past a very long space platform that has two synchronized clocks, one at the position of the spaceship when the observer explodes the flash gun and the other at the position of the spaceship when the light returns to the gun from the mirror. Find the time intervals between the events (exploding the flash gun and receiving the return flash from the mirror) in the frame of the ship and in the frame of the platform. Find the distance traveled by the ship, and the amount by which the clocks on the platform are out of synchronization as viewed by the ship.

We will call the frame of the spaceship S' and that of the platform S. In the spaceship, the light travels from the gun to the mirror and back, a total distance $D = 30c\cdot\text{min}$. The time it takes for light to travel $30c\cdot\text{min}$ is

$$\Delta t' = \frac{D}{c} = \frac{(30c\cdot\text{min})}{c} = 30 \text{ min}$$

Since these events happen at the same place in the spaceship, the time interval is proper time:

$$\Delta t_{\text{P}} = D/c = 30 \text{ min}$$

During this time, the platform travels backwards past the ship a distance equal to the distance L' between the platform clocks measured in frame S':

$$L' = \Delta x' = V\,\Delta t' = (0.8c)(30 \text{ min}) = 24c\cdot\text{min}$$

In frame S, the time between the events is longer by the factor γ. Since $V/c = 0.8$, $1 - V^2/c^2 = 1 - 0.64 = 0.36$. The factor γ is thus

$$\gamma = \frac{1}{\sqrt{1 - V^2/c^2}} = \frac{1}{\sqrt{0.36}} = \frac{1}{0.6} = \frac{5}{3}$$

The time between the events as observed in frame S is therefore

$$\Delta t = \gamma\,\Delta t_{\text{P}} = \frac{5}{3}(30 \text{ min}) = 50 \text{ min}$$

During this time, the spaceship travels a distance in frame S equal to the proper distance between the platform clocks:

$$L_{\text{P}} = \Delta x = V\,\Delta t = (0.8c)(50 \text{ min}) = 40c\cdot\text{min}$$

Note that this distance is longer than the contracted distance between the clocks that is measured by observers in frame S' of the spaceship.

Observers on the platform would say that the spaceship's clock is running slow because it records a time of only 30 min between the events, whereas the time measured on the platform is 50 min.

Figure 34-13 shows the situation viewed from the spaceship in S'. The platform is traveling past the ship with speed $0.8c$. There is a clock at point x_1, which coincides with the ship when the flash gun is exploded, and another at point x_2, which coincides with the ship when the return flash is received from the mirror. We assume that the clock at x_1 reads 12:00 noon at the time of the light flash. The clocks at x_1 and x_2 are synchronized in S but not in S'. In S', the clock at x_2, which is chasing the one at x_1, leads by

$$\frac{L_{\text{P}}V}{c^2} = \frac{(40c\cdot\text{min})(0.8c)}{c^2} = 32 \text{ min}$$

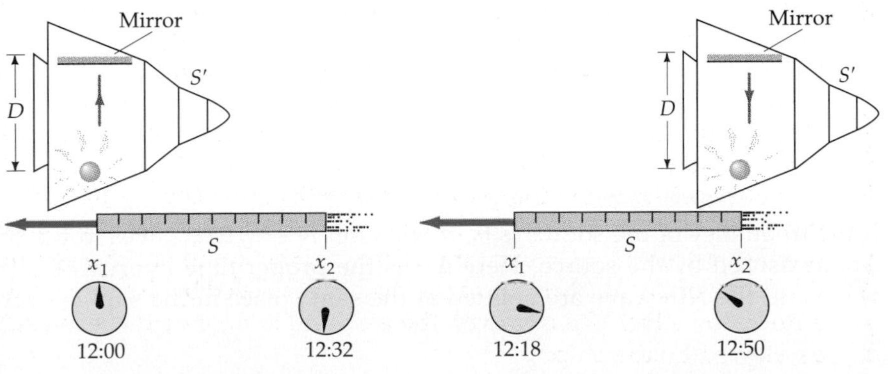

Figure 34-13 Example 34-4. Clocks on a platform as observed from the spaceship's frame of reference S'. During the time $\Delta t' =$ 30 min it takes for the platform to pass the spaceship, the clocks on the platform run slow and tick off $(30 \text{ min})/\gamma = 18$ min. But the clocks are unsynchronized, with the chasing clock leading by $L_{\text{P}}V/c^2$, which for this case is 32 min. The time it takes for the spaceship to pass as measured on the platform is therefore 32 min + 18 min = 50 min.

When the spaceship coincides with x_2, the clock there reads 12:50. The time between the events is therefore 50 min in S. Note that according to observers in S', this clock ticks off 50 min − 32 min = 18 min for a trip that takes 30 min in S'. Thus, observers in S' see this clock run slow by the factor 30/18 = 5/3.

Every observer in one frame sees the clocks in the other frame run slow. According to observers in S, who measure 50 min for the time interval, the time interval in S' (30 min) is too small, so they see the single clock in S' run too slow by the factor 5/3. According to the observers in S', the observers in S measure a time that is too *long* despite the fact that their clocks run too slow because the clocks in S are out of synchronization. The clocks tick off only 18 min, but the second one leads the first by 32 min, so the time interval is 50 min.

Questions

2. Two observers are in relative motion. In what circumstances can they agree on the simultaneity of two different events?
3. If event A occurs before event B in some frame, might it be possible for there to be a reference frame in which event B occurs before event A?
4. Two events are simultaneous in a frame in which they also occur at the same point in space. Are they simultaneous in other reference frames?

34-6 The Doppler Effect

In our discussion of the doppler effect for sound (Section 14-9), we found that the change in frequency for a given velocity V depends on whether it is the source or the receiver that is moving with that speed. Such a distinction is possible for sound because there is a medium (the air) relative to which the motion takes place, so it is not surprising that the motion of the source or the receiver relative to the still air can be distinguished. Such a distinction between motion of the source or receiver cannot be made for light or other electromagnetic waves in a vacuum. Therefore, the expressions we have derived for the doppler effect cannot be correct for light. We will now derive the relativistic doppler-effect equations that are correct for light.

We will consider a source moving toward a receiver with velocity V, and we will work in the frame of the receiver. Let the source emit N electromagnetic waves. The first wave will travel a distance $c\,\Delta t_R$ and the source will travel a distance $V\,\Delta t_R$ in the time Δt_R measured in the frame of the receiver. The wavelength will be

$$\lambda' = \frac{(c\,\Delta t_R - V\,\Delta t_R)}{N}$$

The frequency f' observed by the receiver will therefore be

$$f' = \frac{c}{\lambda'} = \frac{c}{c - V}\frac{N}{\Delta t_R}$$

$$= \frac{1}{1 - V/c}\frac{N}{\Delta t_R}$$

If the frequency of the source is f_0, it will emit $N = f_0\,\Delta t_S$ waves in the time Δt_S measured by the source. Here Δt_S is the proper time interval (the first wave and the Nth wave are emitted at the same place in the source's reference frame). Times Δt_S and Δt_R are related by Equation 34-20 for time dilation, so $\Delta t_R = \gamma\, \Delta t_S$. Thus, when the source and receiver are moving toward one another we obtain

$$f' = \frac{1}{1 - V/c} \frac{f_0\, \Delta t_S}{\Delta t_R} = \frac{f_0}{1 - V/c} \frac{1}{\gamma}$$

or

$$f' = \frac{\sqrt{1 - V^2/c^2}}{1 - V/c} f_0 = \sqrt{\frac{1 + V/c}{1 - V/c}}\, f_0 \qquad \text{approaching} \qquad 34\text{-}24a$$

This differs from our classical equation only in the time-dilation factor. When the source and receiver are moving away from one another, the same analysis shows that the observed frequency is given by

$$f' = \frac{\sqrt{1 - V^2/c^2}}{1 + V/c} f_0 = \sqrt{\frac{1 - V/c}{1 + V/c}}\, f_0 \qquad \text{receding} \qquad 34\text{-}24b$$

It is left as a problem (Problem 34-64) for you to show that the same results are obtained if the calculations are done in the reference frame of the source.

Example 34-5

The longest wavelength of light emitted by hydrogen in the Balmer series (see Chapter 35) has a wavelength of $\lambda_0 = 656$ nm. In light from a distant galaxy, this wavelength is measured to be $\lambda' = 1458$ nm. Find the speed at which the distant galaxy is receding from the earth.

If we substitute $f' = c/\lambda'$ and $f_0 = c/\lambda_0$ into Equation 34-24b, we obtain

$$\sqrt{\frac{1 - V/c}{1 + V/c}} = \frac{f'}{f_0} = \frac{\lambda_0}{\lambda'}$$

This equation is somewhat simplified if we use $\beta = V/c$. Then squaring the above equation and taking the reciprocal of each side, we obtain

$$\frac{1 + \beta}{1 - \beta} = \left(\frac{\lambda'}{\lambda_0}\right)^2 = \left(\frac{1458 \text{ nm}}{656 \text{ nm}}\right)^2 = 4.94$$

so

$$1 + \beta = 4.94 - 4.94\,\beta$$

$$\beta = \frac{4.94 - 1}{4.94 + 1} = 0.663 = \frac{V}{c}$$

The galaxy is thus receding at a speed of $V = 0.663c$. The shift towards longer wavelengths of light from distant galaxies that are receding from us is called the **redshift.**

34-7 The Twin Paradox

Homer and Ulysses are identical twins. Ulysses travels at high speed to a planet beyond the solar system and returns while Homer remains at home. When they are together again, which twin is older, or are they the same age? The correct answer is that Homer, the twin who stays at home, is older. This problem, with variations, has been subject of spirited debate for decades, though there are very few who disagree with the answer.* The problem is a paradox because of the seemingly symmetric roles played by the twins with the asymmetric result in their aging. The paradox is resolved when the asymmetry of the twins' roles is noted. The relativistic result conflicts with common sense based on our strong but incorrect belief in absolute simultaneity. We will consider a particular case with some numerical magnitudes that, though impractical, make the calculations easy.

Let planet P and Homer on earth be at rest in reference frame S a distance L_p apart, as illustrated in Figure 34-14. We neglect the motion of the earth. Reference frames S' and S'' are moving with speed V toward and away from the planet, respectively. Ulysses quickly accelerates to speed V, then coasts in S' until he reaches the planet, where he stops and is momentarily at rest in S. To return he quickly accelerates to speed V toward earth and then coasts in S'' until he reaches earth, where he stops. We can assume that the acceleration times are negligible compared with the coasting times. We use the following values for illustration: $L_p = 8$ light-years and $V = 0.8c$. Then $\sqrt{1 - V^2/c^2} = 3/5$ and $\gamma = 5/3$.

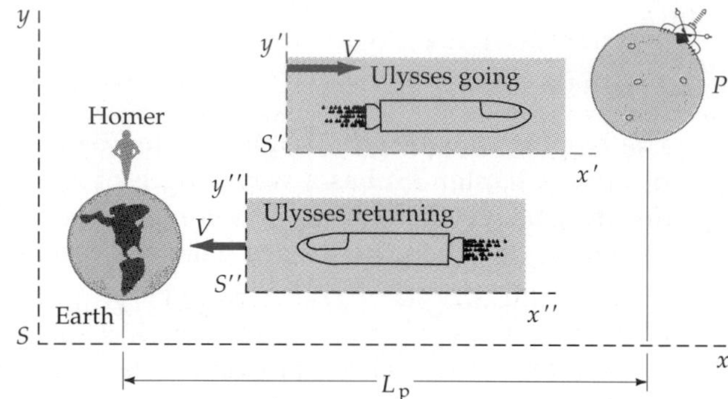

Figure 34-14 The twin paradox. The earth and a distant planet are fixed in frame S. Ulysses coasts in frame S' to the planet and then coasts back in frame S''. His twin Homer stays on earth. When Ulysses returns, he is younger than his twin. The roles played by the twins are not symmetric. Homer remains in one inertial reference frame, but Ulysses must accelerate if he is to return home.

It is easy to analyze the problem from Homer's point of view on earth. According to Homer's clock, Ulysses coasts in S' for a time $L_p/V = 10$ y and in S'' for an equal time. Thus Homer is 20 y older when Ulysses returns. The time interval in S' between Ulysses' leaving earth and his arriving at the planet is shorter because it is proper time. The time it takes to reach the planet by Ulysses' clock is

$$\Delta t' = \frac{\Delta t}{\gamma} = \frac{10\text{ y}}{5/3} = 6\text{ y}$$

Since the same time is required for the return trip, Ulysses will have recorded 12 y for the round trip and will be 8 y younger than Homer upon his return.

From Ulysses' point of view, the distance from the earth to the planet is contracted and is only

$$L' = \frac{L_p}{\gamma} = \frac{8\text{ light-years}}{5/3} = 4.8\text{ light-years}$$

At $V = 0.8c$, it takes only 6 y each way.

The real difficulty in this problem is for Ulysses to understand why his twin aged 20 y during his absence. If we consider Ulysses as being at rest and Homer as moving away, Homer's clock should run slow and measure

*A collection of some important papers concerning this paradox can be found in *Special Relativity Theory, Selected Reprints,* American Association of Physics Teachers, New York, 1963.

only 3/5(6) = 3.6 y. Then why shouldn't Homer age only 7.2 y during the round trip? This, of course, is the paradox. The difficulty with the analysis from the point of view of Ulysses is that he does not remain in an inertial frame. What happens while Ulysses is stopping and starting? To investigate this problem in detail, we would need to treat accelerated reference frames, a subject dealt with in the study of general relativity and beyond the scope of this book. However, we can get some insight into the problem by having the twins send regular signals to each other so that they can record the other's age continuously. If they arrange to send a signal once a year, each can determine the age of the other merely by counting the signals received. The arrival frequency of the signals will not be 1 per year because of the doppler shift. The frequency observed will be given by Equations 34-24*a* and 34-24*b*. Using $V/c = 0.8$ and $V^2/c^2 = 0.64$, we have for the case in which the twins are receding from each other

$$f' = \frac{\sqrt{1 - V^2/c^2}}{1 + V/c} f_0 = \frac{\sqrt{1 - 0.64}}{1 + 0.8} f_0 = \frac{1}{3} f_0$$

When they are approaching, Equation 34-24 gives $f' = 3f_0$.

Consider the situation first from the point of view of Ulysses. During the 6 y it takes him to reach the planet (remember that the distance is contracted in his frame), he receives signals at the rate of $\frac{1}{3}$ signal per year, and so he receives 2 signals. As soon as he turns around and starts back to earth, he begins to receive 3 signals per year. In the 6 y it takes him to return he receives 18 signals, giving a total of 20 for the trip. He accordingly expects his twin to have aged 20 years.

We now consider the situation from Homer's point of view. He receives signals at the rate of $\frac{1}{3}$ signal per year not only for the 10 y it takes Ulysses to reach the planet but also for the time it takes for the last signal sent by Ulysses before he turns around to get back to earth. (He cannot know that Ulysses has turned around until the signals begin reaching him with increased frequency.) Since the planet is 8 light-years away, there is an additional 8 y of receiving signals at the rate of $\frac{1}{3}$ signal per year. During the first 18 y, Homer receives 6 signals. In the final 2 y before Ulysses arrives, Homer receives 6 signals, or 3 per year. (The first signal sent after Ulysses turns around takes 8 y to reach earth, whereas Ulysses, traveling at $0.8c$, takes 10 y to return and therefore arrives just 2 y after Homer begins to receive signals at the faster rate.) Thus Homer expects Ulysses to have aged 12 y. In this analysis, the asymmetry of the twins' roles is apparent. When they are together again, both twins agree that the one who has been accelerated will be younger than the one who stayed home.

The predictions of the special theory of relativity concerning the twin paradox have been tested using small particles that can be accelerated to such large speeds that γ is appreciably greater than 1. Unstable particles can be accelerated and trapped in circular orbits in a magnetic field, for example, and their lifetimes can then be compared with those of identical particles at rest. In all such experiments, the accelerated particles live longer on the average than those at rest, as predicted. These predictions have also been confirmed by the results of an experiment in which high-precision atomic clocks were flown around the world in commercial airplanes, but the analysis of this experiment is complicated due to the necessity of including gravitational effects treated in the general theory of relativity.*

*The details of this experiment can be found in J. C. Hafele and Richard E. Keating, "Around-the-world Atomic Clocks: Predicted Relativistic Time Gains" and "Around-the-world Atomic Clocks: Observed Relativistic Time Gains," *Science*, July 14, 1972, p. 166.

34-8 The Velocity Transformation

We can find how velocities transform from one reference frame to another by differentiating the Lorentz transformation equations. Suppose a particle has velocity $u'_x = dx'/dt'$ in frame S', which is moving to the right with speed V relative to frame S. Its velocity in frame S is

$$u_x = \frac{dx}{dt}$$

From the Lorentz transformation equations (Equations 34-16 and 34-17), we have

$$dx = \gamma(dx' + V\,dt')$$

and

$$dt = \gamma\left(dt' + \frac{V\,dx'}{c^2}\right)$$

The velocity in S is thus

$$u_x = \frac{dx}{dt} = \frac{\gamma(dx' + V\,dt')}{\gamma\left(dt' + \dfrac{V\,dx'}{c^2}\right)} = \frac{\dfrac{dx'}{dt'} + V}{1 + \dfrac{V}{c^2}\dfrac{dx'}{dt'}} = \frac{u'_x + V}{1 + Vu'_x/c^2}$$

If a particle has components of velocity along the y or z axes, we can use the same relation between dt and dt', with $dy = dy'$ and $dz = dz'$, to obtain

$$u_y = \frac{dy}{dt} = \frac{dy'}{\gamma\left(dt' + \dfrac{V\,dx'}{c^2}\right)} = \frac{dy'/dt'}{\gamma\left(1 + \dfrac{V}{c^2}\dfrac{dx'}{dt'}\right)} = \frac{u'_y}{\gamma\left(1 + \dfrac{Vu'_x}{c^2}\right)}$$

and

$$u_z = \frac{u'_z}{\gamma\left(1 + \dfrac{Vu'_x}{c^2}\right)}$$

The complete relativistic velocity transformation is

Relativistic velocity transformation

$$u_x = \frac{u'_x + V}{1 + Vu'_x/c^2} \qquad \text{34-25}a$$

$$u_y = \frac{u'_y}{\gamma(1 + Vu'_x/c^2)} \qquad \text{34-25}b$$

$$u_z = \frac{u'_z}{\gamma(1 + Vu'_x/c^2)} \qquad \text{34-25}c$$

The inverse velocity transformation equations are

$$u'_x = \frac{u_x - V}{1 - Vu_x/c^2} \qquad \text{34-26}a$$

$$u'_y = \frac{u_y}{\gamma(1 - Vu_x/c^2)} \qquad \text{34-26}b$$

$$u'_z = \frac{u_z}{\gamma(1 - Vu_x/c^2)} \qquad \text{34-26}c$$

These equations differ from the classical and intuitive result $u_x = u'_x + V$, $u_y = u'_y$ and $u_z = u'_z$ because the denominators in Equations 34-25 and 34-26 are not equal to 1. When V and u'_x are small compared with the speed of light c, $\gamma \approx 1$ and $Vu'_x/c^2 \ll 1$. Then the relativistic and classical expressions are the same.

Example 34-6

A supersonic plane moves with speed 1000 m/s (about 3 times the speed of sound) along the x axis relative to you. Another plane moves along the x axis at speed 500 m/s relative to the first plane. How fast is the second plane moving relative to you?

According to the classical formula for combining velocities, the speed of the second plane relative to you is 1000 m/s + 500 m/s = 1500 m/s. If we assume that you are at rest in the S reference frame and that the first plane is at rest in the S' frame, which is moving at $V = 1000$ m/s relative to S, the second plane has velocity $u'_x = 500$ m/s in S'. The correction term for u_x in the denominator of Equation 34-25*a* is then

$$\frac{Vu'_x}{c^2} = \frac{(1000)(500)}{(3 \times 10^8)^2} \approx 5 \times 10^{-12}$$

This correction term is so small that the classical and relativistic results are essentially the same.

Example 34-7

Work Example 34-6 if the first plane moves with speed $V = 0.8c$ relative to you and the second plane moves with the same speed $0.8c$ relative to the first plane.

In this case, the correction term is

$$\frac{Vu'_x}{c^2} = \frac{(0.8c)(0.8c)}{c^2} = 0.64$$

The speed of the second plane in frame S is then

$$u_x = \frac{0.8c + 0.8c}{1 + 0.64} = 0.98c$$

This is quite different from the classically expected result of $0.8c + 0.8c = 1.6c$. In fact, it can be shown from Equation 34-25 that if the speed of an object is less than c in one frame, it is less than c in all other frames moving relative to that frame with a speed less than c. We will see in Section 34-10 that it takes an infinite amount of energy to accelerate a particle to the speed of light. The speed of light c is thus an upper, unattainable limit for the speed of a particle having mass. (Massless particles, such as photons, always move at the speed of light.)

Example 34-8

Light moves along the x axis with speed $u_x = c$. What is its speed in S'?

From Equation 34-26a, we have

$$u'_x = \frac{c - V}{1 - Vc/c^2} = \frac{c(1 - V/c)}{1 - V/c} = c$$

as required by Einstein's postulates.

Question

5. The Lorentz transformation for y and z is the same as the classical result: $y = y'$ and $z = z'$. Yet the relativistic velocity transformation does not give the classical result $u_y = u'_y$ and $u_z = u'_z$. Explain.

34-9 Relativistic Momentum

We have seen in previous sections that Einstein's postulates require important modifications in our ideas of simultaneity and in our measurements of time and length. Perhaps more importantly, they also require modifications in our concepts of mass, momentum, and energy. In classical mechanics, the momentum of a particle is defined as the product of its mass and its velocity, $\mathbf{p} = m\mathbf{u}$, where $\mathbf{u}$ is the velocity. In an isolated system of particles, with no net force acting on the system, the total momentum of the system remains constant.

In this section we will see from a simple thought experiment that the classical expression for momentum, $\mathbf{p} = m\mathbf{u}$, is just an approximation. That is, this quantity is not conserved in an isolated system. We consider two observers: observer A in reference frame S and observer B in frame S'. Each observer has a ball of mass m. The two balls are identical when compared at rest. Each observer throws his ball vertically with a speed u_0 such that it travels a distance L, makes an elastic collision with the other ball, and returns. Figure 34-15 shows how the collision looks in each reference frame. Classically, each ball has vertical momentum of magnitude mu_0. Since the vertical components of the momenta are equal and opposite, the total vertical component of momentum is zero before the collision. The collision merely reverses the momentum of each ball, so the total vertical momentum is zero after the collision.

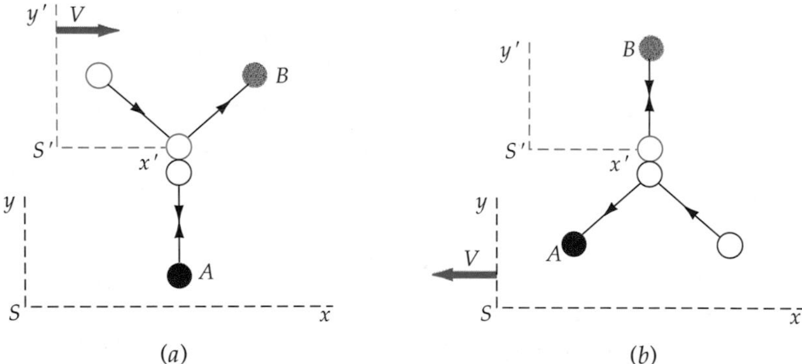

Figure 34-15 (*a*) Elastic collision of two identical balls as seen in frame S. The vertical component of the velocity of ball B is u_0/γ in S if it is u_0 in S'. (*b*) The same collision as seen in S'. In this frame, ball A has vertical component of velocity equal to u_0/γ.

Relativistically, however, the vertical components of the velocities of the two balls as seen by either observer are not equal and opposite. Thus, when they are reversed by the collision, classical momentum is not conserved. Consider the collision as seen by A in frame S. The velocity of his ball is $u_{Ay} = +u_0$. Since the velocity of B's ball in frame S' is $u'_{Bx} = 0$, $u'_{By} = -u_0$, the y component of the velocity of B's ball in frame S is (Equation 34-25*b*) $u_{By} = -u_0/\gamma$. Thus, if the classical expression for momentum $\mathbf{p} = m\mathbf{u}$ is used, the vertical components of momentum of the two balls are not equal and opposite as seen by observer A. Since the balls are reversed by the collision, momentum is not conserved. Of course, the same result is observed by B. In the classical limit, when u is much less than c, γ is approximately 1, and the momentum of the system is conserved as seen by either observer.

The reason that the total momentum of a system is important in classical mechanics is that it is conserved when there are no external forces acting on the system, as is the case in collisions. We now see that the quantity $\Sigma m\mathbf{u}$ is conserved only in the approximation that $u \ll c$. We will define the relativistic momentum $\mathbf{p}$ of a particle to have the following properties:

1. In collisions, $\mathbf{p}$ is conserved.
2. As u/c approaches zero, $\mathbf{p}$ approaches $m\mathbf{u}$.

We will now show that the quantity

$$\mathbf{p} = \frac{m\mathbf{u}}{\sqrt{1 - u^2/c^2}} \tag{34-27}$$

is conserved in the elastic collision shown in Figure 34-15. Since this quantity also approaches $m\mathbf{u}$ as u/c approaches zero, we take this equation for the definition of the **relativistic momentum** of a particle.

We will compute the y component of the relativistic momentum of each particle in reference frame S and show that the y component of the total relativistic momentum is zero. The speed of ball A in S is u_0, so the y component of its relativistic momentum is

$$p_{Ay} = \frac{mu_0}{\sqrt{1 - u_0^2/c^2}}$$

The speed of ball B in S is more complicated. Its x component is V and its y component is $-u_0/\gamma$. Thus

$$u_B^2 = u_{Bx}^2 + u_{By}^2 = V^2 + (-u_0\sqrt{1 - V^2/c^2})^2 = V^2 + u_0^2 - \frac{u_0^2V^2}{c^2}$$

Using this result to compute $\sqrt{1 - u_B^2/c^2}$, we obtain

$$1 - \frac{u_B^2}{c^2} = 1 - \frac{V^2}{c^2} - \frac{u_0^2}{c^2} + \frac{u_0^2V^2}{c^4} = (1 - V^2/c^2)(1 - u_0^2/c^2)$$

and

$$\sqrt{1 - u_B^2/c^2} = \sqrt{1 - V^2/c^2}\sqrt{1 - u_0^2/c^2} = \frac{1}{\gamma}\sqrt{1 - u_0^2/c^2}$$

The y component of the relativistic momentum of ball B as seen in S is therefore

$$p_{By} = \frac{mu_{By}}{\sqrt{1 - u_B^2/c^2}} = \frac{-mu_0/\gamma}{(1/\gamma)\sqrt{1 - u_0^2/c^2}} = \frac{-mu_0}{\sqrt{1 - u_0^2/c^2}}$$

Since $p_{By} = -p_{Ay}$, the y component of the total momentum of the two balls is zero. If the speed of each ball is reversed by the collision, the total momentum will remain zero and momentum will be conserved.

One interpretation of Equation 34-27 is that the mass of an object increases with speed. The quantity $m/\sqrt{1 - u^2/c^2}$ is called the **relativistic mass** of a particle. The mass of a particle when it is at rest in some reference frame is called its **rest mass** m_0. The mass thus increases from m_0 at rest to $m_r = m_0/\sqrt{1 - u^2/c^2}$ when it is moving at speed u. To avoid confusion, we will label the rest mass m_0 and the relativistic mass $m_0/\sqrt{1 - u^2/c^2}$ and not use m for either. The rest mass of a particle is the same in all reference frames. Using this notation, the relativistic momentum of a particle is then

$$\mathbf{p} = \frac{m_0\mathbf{u}}{\sqrt{1 - u^2/c^2}} \tag{34-28}$$

Relativistic momentum

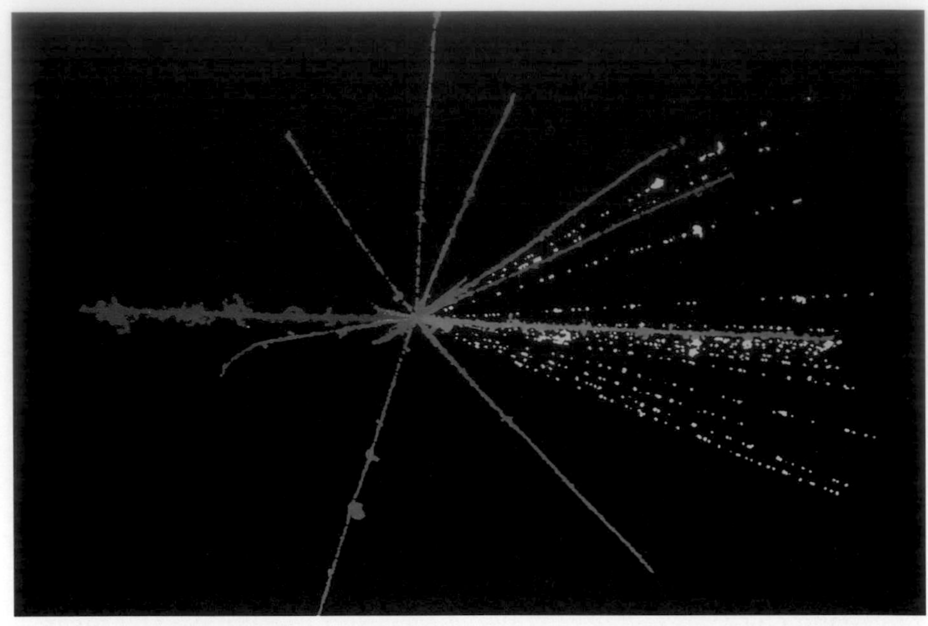

The creation of elementary particles demonstrates the conversion of kinetic energy to rest energy. In this 1950 photograph of a cosmic ray shower, a high-energy sulfur nucleus (red) collides in a photographic emulsion and produces a spray of particles, including a fluorine nucleus (green), other nuclear fragments (blue), and about 16 pions (yellow).

34-10 Relativistic Energy

In classical mechanics, the work done by an unbalanced force acting on a particle equals the change in the kinetic energy of the particle. In relativistic mechanics, we equate the unbalanced force to the rate of change of the relativistic momentum. The work done by such a force can then be calculated and set equal to the change in kinetic energy. As in classical mechanics, we will define kinetic energy as the work done by an unbalanced force in accelerating a particle from rest to some velocity. Considering one dimension only, we have

$$K = \int_{u=0}^{u} \sum F\, ds = \int_0^u \frac{dp}{dt}\, ds = \int_0^u u\, dp = \int_0^u u\, d\left(\frac{m_0 u}{\sqrt{1 - u^2/c^2}}\right) \qquad \text{34-29}$$

where we have used $u = ds/dt$. It is left as a problem (Problem 70) for you to show that

$$d\left(\frac{m_0 u}{\sqrt{1 - u^2/c^2}}\right) = m_0\left(1 - \frac{u^2}{c^2}\right)^{-3/2} du$$

If we substitute this expression into the integrand in Equation 34-29, we obtain

$$K = \int_0^u u\, d\left(\frac{m_0 u}{\sqrt{1 - u^2/c^2}}\right) = \int_0^u m_0\left(1 - \frac{u^2}{c^2}\right)^{-3/2} u\, du$$

$$= m_0 c^2\left(\frac{1}{\sqrt{1 - u^2/c^2}} - 1\right)$$

or

$$K = \frac{m_0 c^2}{\sqrt{1 - u^2/c^2}} - m_0 c^2 \qquad \text{34-30}$$

The expression for kinetic energy consists of two terms. The first term depends on the speed of the particle. The second, m_0c^2, is independent of the speed. The quantity m_0c^2 is called the **rest energy** of the particle E_0. The rest energy is the product of the rest mass and c^2:

$$E_0 = m_0c^2 \tag{34-31}$$

Rest energy

The total **relativistic energy** E is then defined to be the sum of the kinetic energy and the rest energy:

$$E = K + m_0c^2 = \frac{m_0c^2}{\sqrt{1 - u^2/c^2}} \tag{34-32}$$

Relativistic energy

Thus, the work done by an unbalanced force increases the energy from the rest energy m_0c^2 to the final energy $m_0c^2/\sqrt{1 - u^2/c^2} = m_rc^2$, where $m_r = m_0/\sqrt{1 - u^2/c^2}$ is the relativistic mass. We can obtain a useful expression for the velocity of a particle by multiplying Equation 34-28 for the relativistic momentum by c^2 and comparing the result with Equation 34-32 for the relativistic energy. We have

$$pc^2 = \frac{m_0c^2u}{\sqrt{1 - u^2/c^2}} = Eu$$

or

$$\frac{u}{c} = \frac{pc}{E} \tag{34-33}$$

Example 34-9

An electron with rest energy 0.511 MeV moves with speed $u = 0.8c$. Find its total energy, kinetic energy, and momentum.

We first calculate the factor $1/\sqrt{1 - u^2/c^2}$.

$$\frac{1}{\sqrt{1 - u^2/c^2}} = \frac{1}{\sqrt{1 - 0.64}} = \frac{5}{3} = 1.67$$

The total energy is then

$$E = \frac{m_0c^2}{\sqrt{1 - u^2/c^2}} = 1.67(0.511 \text{ MeV}) = 0.853 \text{ MeV}$$

The kinetic energy is the total energy minus the rest energy:

$$K = E - m_0c^2 = 0.853 \text{ MeV} - 0.511 \text{ MeV} = 0.342 \text{ MeV}$$

The magnitude of the momentum is

$$p = \frac{m_0u}{\sqrt{1 - u^2/c^2}} = (1.67)m_0(0.8c) = \frac{1.33m_0c^2}{c}$$

$$= \frac{(1.33)(0.511 \text{ MeV})}{c} = 0.680 \text{ MeV}/c$$

The unit MeV/c is a convenient unit for momentum.

The expression for kinetic energy given by Equation 34-30 doesn't look much like the classical expression $\frac{1}{2}m_0u^2$. However, when u is much less than c, we can approximate $1/\sqrt{1 - u^2/c^2}$ using the binomial expansion (Equation 34-2):

$$\frac{1}{\sqrt{1 - u^2/c^2}} = \left(1 - \frac{u^2}{c^2}\right)^{-1/2}$$

$$\approx 1 + \frac{1}{2}\frac{u^2}{c^2}$$

From this result, when u is much less than c, the expression for relativistic kinetic energy becomes

$$K = m_0c^2\left(\frac{1}{\sqrt{1 - u^2/c^2}} - 1\right)$$

$$\approx m_0c^2\left(1 + \frac{1}{2}\frac{u^2}{c^2} - 1\right)$$

$$= \frac{1}{2}m_0u^2$$

Thus at low speeds, the relativistic expression is the same as the classical expression.

We note from Equation 34-32 that as the speed u approaches the speed of light c, the energy of the particle becomes very large because $1/\sqrt{1 - u^2/c^2}$ becomes very large. At $u = c$, the energy becomes infinite. For u greater than c, $\sqrt{1 - u^2/c^2}$ is the square root of a negative number and is therefore imaginary. A simple interpretation of the result that it takes an infinite amount of energy to accelerate a particle to the speed of light is that no particle that is ever at rest in any inertial reference frame can travel as fast or faster than the speed of light c. As we noted in Example 34-7, if the speed of a particle is less than c in one reference frame, it is less than c in all other reference frames moving relative to that frame at speeds less than c.

In practical applications, the momentum or energy of a particle is often known rather than the speed. Equation 34-28 for the relativistic momentum and Equation 34-32 for the relativistic energy can be combined to eliminate the speed u. (See Problem 48.) The result is

Relation for total energy, momentum, and rest energy

$$E^2 = p^2c^2 + (m_0c^2)^2 \qquad \text{34-34}$$

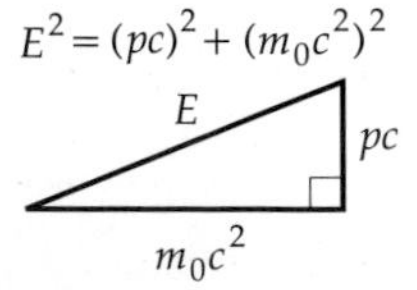

This useful equation can be conveniently remembered from the right triangle shown in Figure 34-16. If the energy of a particle is much greater than its rest energy mc^2, the second term on the right of Equation 34-34 can be neglected, giving the useful approximation

$$E \approx pc \qquad \text{for } E \gg m_0c^2 \qquad \text{34-35}$$

Equation 34-35 is an exact relation between energy and momentum for particles with no rest mass, such as photons and neutrinos.

$E^2 = (pc)^2 + (m_0c^2)^2$

E pc m_0c^2

Figure 34-16 Right triangle for remembering Equation 34-34.

Exercise

A proton with a rest mass of 938 MeV/c^2 has a total energy of 1400 MeV. Find (*a*) $1/\sqrt{1 - u^2/c^2}$, (*b*) its momentum, and (*c*) its speed u. [Answers: (*a*) 1.49, (*b*) $p = 1040$ MeV/c, (*c*) $u = 0.74c$]

able difference in mass. Some very heavy nuclei, such as radium, are radioactive and decay into a lighter nucleus plus an alpha particle. In this case, the original nucleus has a rest energy greater than that of the decay particles. The excess energy appears as the kinetic energy of the decay products.

Example 34-11

In a typical nuclear fusion reaction, a tritium nucleus (^{3}H) and a deuterium nucleus (^{2}H) fuse together to form a helium nucleus (^{4}He) plus a neutron. How much energy is released in this fusion reaction?

The reaction is written

$$^2\text{H} + {^3\text{H}} \longrightarrow {^4\text{He}} + n$$

From Table 34-1, we see that the rest energy of the deuterium plus tritium nuclei is 1875.628 MeV + 2808.944 MeV = 4684.572 MeV. The rest energy of the helium nucleus plus the neutron is 3727.409 + 939.573 = 4666.982 MeV. This is less than that of the deuterium plus tritium by 4684.572 − 4666.982 = 17.59 MeV. The energy released in this reaction is thus 17.59 MeV. This and other fusion reactions occur in the sun and are responsible for the energy supplied to the earth. As the sun gives off energy, its rest mass continually decreases.

Example 34-12

A hydrogen atom consisting of a proton and an electron has a binding energy of 13.6 eV. By what percentage is the mass of the proton plus the electron greater than that of the hydrogen atom?

The rest energy of a proton plus that of an electron is 938.28 MeV + 0.511 eV = 938.791 MeV. The sum of the masses of these two particles is 938.791 MeV/c^2. The mass of the hydrogen atom is less than this by 13.6 eV/c^2. The percentage difference is

$$\frac{13.6\ \text{eV}/c^2}{938.791 \times 10^6\ \text{eV}/c^2} = 1.45 \times 10^{-8} = 1.45 \times 10^{-6}\%$$

This mass difference is so small as to be hardly measurable.

Example 34-13

A particle of rest mass 2 MeV/c^2 and kinetic energy 3 MeV collides with a stationary particle of rest mass 4 MeV/c^2. After the collision, the two particles stick together. Find (*a*) the initial momentum of the system, (*b*) the final velocity of the two-particle system, and (*c*) the rest mass of the two-particle system.

(*a*) Since the moving particle has kinetic energy of 3 MeV and rest energy of 2 MeV, its total energy is $E_1 = 5$ MeV. We obtain its momentum from Equation 34-34,

$$pc = \sqrt{E_1^2 - (m_0^2c^2)^2} = \sqrt{(5\ \text{MeV})^2 - (2\ \text{MeV})^2} = \sqrt{21}\ \text{MeV}$$

or

$$p = 4.58\ \text{MeV}/c$$

Since the other particle is rest, this is the total momentum of the system.

(*b*) We can find the final velocity of the two-particle system from its total energy E and its momentum p using Equation 34-33. By the conservation of total energy, the final energy of the system equals the initial total energy of the two particles:

$$E_f = E_i = E_1 + E_2 = 5\text{ MeV} + 4\text{ MeV} = 9\text{ MeV}$$

By the conservation of momentum, the final momentum of the two-particle system equals the initial momentum, $p = 4.58\text{ MeV}/c$. The velocity of the two-particle system is thus given by

$$\frac{u}{c} = \frac{pc}{E} = \frac{4.58\text{ MeV}}{9\text{ MeV}} = 0.509$$

(*c*) We can find the rest mass of the final two-particle system from Equation 34-34 using $pc = 4.58$ MeV and $E = 9$ MeV. We have

$$E^2 = (pc)^2 + (M_0c^2)^2$$

$$(9\text{ MeV})^2 = (4.58\text{ MeV})^2 + (M_0c^2)^2$$

$$M_0c^2 = \sqrt{81 - 21}\text{ MeV} = 7.75\text{ MeV}$$

$$M_0 = 7.75\text{ MeV}/c^2$$

It is instructive to check our answers by computing the initial and final kinetic energies. The initial kinetic energy is $K_i = 3$ MeV. The final kinetic energy is

$$K_f = E - M_0c^2 = 9\text{ MeV} - 7.75\text{ MeV} = 1.25\text{ MeV}$$

The loss in kinetic energy is

$$K_i - K_f = 3\text{ MeV} - 1.25\text{ MeV} = 1.75\text{ MeV}$$

Since the initial rest energy is 2 MeV + 4 MeV = 6 MeV and the final rest energy is $M_0c^2 = 7.75$ MeV, the gain in rest energy is 7.75 MeV − 6 MeV = 1.75 MeV.

34-11 General Relativity

The generalization of the theory of relativity to noninertial reference frames by Einstein in 1916 is known as the general theory of relativity. It is much more difficult mathematically than the special theory of relativity, and there are fewer situations in which it can be tested. Nevertheless, its importance calls for a brief qualitative discussion.

The basis of the general theory of relativity is the **principle of equivalence:**

A homogeneous gravitational field is completely equivalent to a uniformly accelerated reference frame.

This principle arises in newtonian mechanics because of the apparent identity of gravitational mass and inertial mass. In a uniform gravitational field, all objects fall with the same acceleration **g** independent of their mass because the gravitational force is proportional to the (gravitational) mass, whereas the acceleration varies inversely with the (inertial) mass. Consider a compartment in space far from any matter and undergoing a uniform acceleration **a**, as shown in Figure 34-18*a*. No mechanics experiment can be performed *inside* the compartment that will distinguish whether the compartment is actually accelerating in space or is at rest (or is moving with uniform

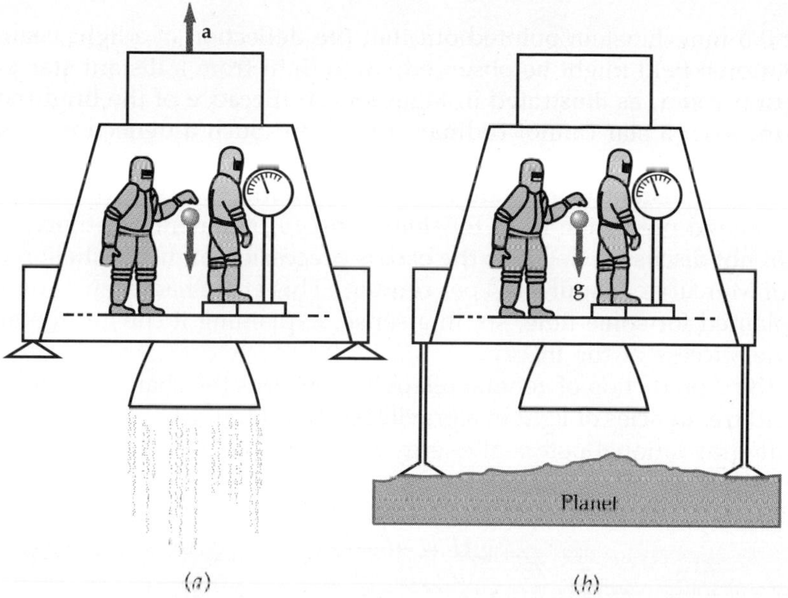

Figure 34-18 The results of experiments in a uniformly accelerated reference frame (*a*) cannot be distinguished from those in a uniform gravitational field (*b*) if the acceleration **a** and the gravitational field **g** have the same magnitude.

velocity) in the presence of a uniform gravitational field $\mathbf{g} = -\mathbf{a}$, as shown in Figure 34-18*b*. If objects are dropped in the compartment, they will fall to the "floor" with an acceleration $\mathbf{g} = -\mathbf{a}$. If people stand on a spring scale, it will read their "weight" of magnitude ma.

Einstein assumed that the principle of equivalence applies to all physics and not just to mechanics. In effect, he assumed that there is no experiment of any kind that can distinguish uniformly accelerated motion from the presence of a gravitational field. We will look qualitatively at a few of the consequences of this assumption.

The first consequence of the principle of equivalence we will discuss, the deflection of a light beam in a gravitational field, was one of the first to be tested experimentally. Figure 34-19 shows a beam of light entering a compartment that is accelerating. Successive positions of the compartment at equal time intervals are shown in Figure 34-19*a*. Because the compartment is accelerating, the distance it moves in each time interval increases with time. The path of the beam of light as observed from inside the compartment is therefore a parabola, as shown in Figure 34-19*b*. But according to the principle of equivalence, there is no way to distinguish between an accelerating compartment and one moving with uniform velocity in a uniform gravitational field. We conclude, therefore, that a beam of light, like objects having mass, will accelerate in a gravitational field. For example, near the surface of the earth, light will fall with an acceleration of 9.81 m/s^2. This is difficult to observe because of the enormous speed of light. For example, in a distance of 3000 km, which takes about 0.01 s to traverse, a beam of light should fall

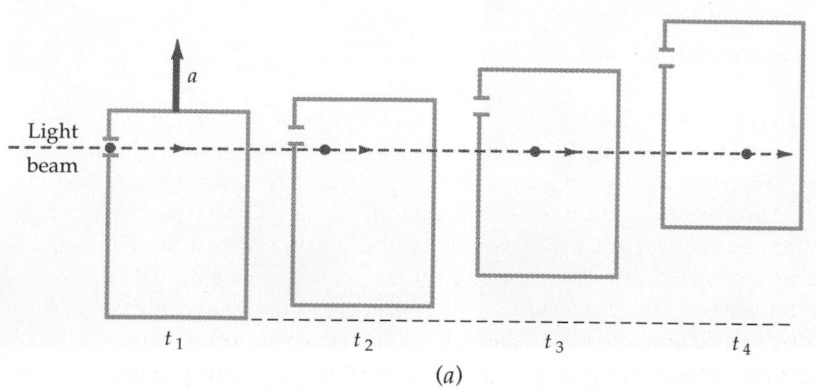

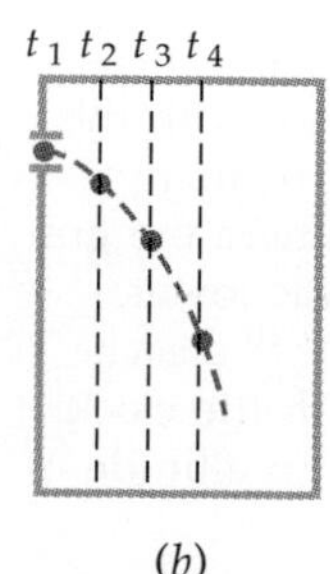

Figure 34-19 (*a*) A light beam moving in a straight line through a compartment that is undergoing uniform acceleration. The position of the beam is shown at equally spaced times t_1, t_2, t_3, and t_4. (*b*) In the reference frame of the compartment, the light travels in a parabolic path as a ball would if it were projected horizontally. The vertical displacements are greatly exaggerated in both (*a*) and (*b*) for emphasis.

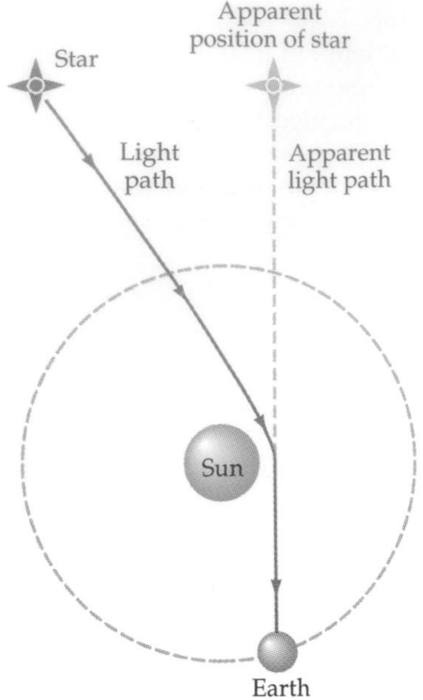

Figure 34-20 The deflection (greatly exaggerated) of a beam of light due to the gravitational attraction of the sun.

about 0.5 mm. Einstein pointed out that the deflection of a light beam in a gravitational field might be observed when light from a distant star passes close to the sun, as illustrated in Figure 34-20. Because of the brightness of the sun, such a star cannot ordinarily be seen. Such a deflection was first observed in 1919 during an eclipse of the sun. This well-publicized observation brought instant worldwide fame to Einstein.

A second prediction from Einstein's theory of general relativity, which we will not discuss in detail, is the excess precession of the perihelion of the orbit of Mercury of about 0.01° per century. This effect had been known and unexplained for some time, so, in a sense, explaining it constituted an immediate success of the theory.

A third prediction of general relativity concerns the change in time intervals and frequencies of light in a gravitational field. In Chapter 10, we found that the gravitational potential energy between two masses M and m a distance r apart is

$$U = -\frac{GMm}{r}$$

where G is the universal gravitational constant, and the point of zero potential energy has been chosen to be when the separation of the masses is infinite. The potential energy per unit mass near a mass M is called the *gravitational potential* ϕ:

$$\phi = -\frac{GM}{r} \qquad 34\text{-}38$$

According to the general theory of relativity, clocks run more slowly in regions of low gravitational potential. (Since the gravitational potential is negative, as can be seen from Equation 34-38, low gravitational potential occurs near the mass where the *magnitude* of the potential is large.) If Δt_1 is a time interval between two events measured by a clock where the gravitational potential is ϕ_1 and Δt_2 is the interval between the same events as measured by a clock where the gravitational potential is ϕ_2, general relativity predicts

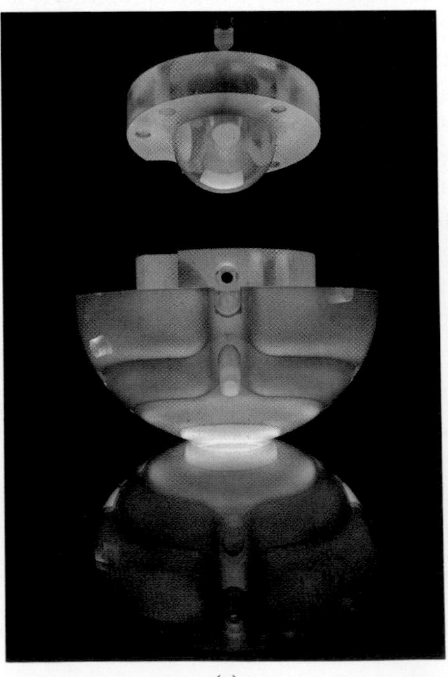

(*a*)

(*b*)

(*a*) This quartz sphere in the top part of the container is probably the world's most perfectly round object. It is designed to spin as a gyroscope in a satellite orbiting the earth. General relativity predicts that the rotation of the earth will cause the axis of rotation of the gyroscope to precess in a circle at a rate of about 1 revolution in 100,000 years. (*b*) This extremely accurate hydrogen maser clock was launched in a satellite in 1976, and its time was compared to that of an identical clock on earth. In accordance with the prediction of general relativity, the clock on earth, where the gravitational potential was lower, "lost" about 4.3×10^{-10} s each second compared with the clock orbiting the earth at an altitude of about 10,000 km.

that the fractional difference between these times will be approximately

$$\frac{\Delta t_2 - \Delta t_1}{\Delta t} = \frac{1}{c^2}(\phi_2 - \phi_1) \qquad 34\text{-}39$$

(Since this shift is usually very small, it does not matter by which interval we divide on the left side of the equation.) A clock in a region of low gravitational potential will therefore run slower than one in a region of high potential. Since a vibrating atom can be considered to be a clock, the frequency of vibration of an atom in a region of low potential, such as near the sun, will be lower than that of the same atom on earth. This shift toward a lower frequency and therefore a longer wavelength is called the **gravitational redshift.**

As our final example of the predictions of general relativity, we mention **black holes,** which were first predicted by Oppenheimer and Snyder in 1939. According to the general theory of relativity, if the density of an object such as a star is great enough, its gravitational attraction will be so great that once inside a critical radius, nothing can escape, not even light or other electromagnetic radiation. (The effect of a black hole on objects outside the critical radius is the same as that of any other mass.) A remarkable property of such an object is that nothing that happens inside it can be communicated to the outside. As sometimes occurs in physics, a simple but incorrect calculation gives the correct results for the relation between the mass and the critical radius of a black hole. In newtonian mechanics, the speed needed for a particle to escape from the surface of a planet or star of mass M and radius R is given by Equation 10-21:

$$v_e = \sqrt{\frac{2GM}{R}}$$

If we set the escape speed equal to the speed of light and solve for the radius, we obtain the critical radius R_S, called the **Schwarzschild radius:**

$$R_S = \frac{2GM}{c^2} \qquad 34\text{-}40$$

For an object with a mass equal to that of our sun to be a black hole, its radius would have to be about 3 km. Since no radiation is emitted from a black hole and its radius is expected to be small, the detection of a black hole is not easy. The best chance of detection would occur if a black hole were a companion to a normal star in a binary star system. The black hole would affect a number of properties of its visible companion. Measurements of the doppler shift of the light from the normal star, for example, might allow a computation of the mass of the unseen companion to determine whether it is great enough to be a black hole. At present there are several excellent candidates—one in the constellation Cygnus, one in the Small Magellanic Cloud, and perhaps one in our own galaxy—but the evidence is not conclusive.

This antenna, consisting of a 1400-kg aluminum cylinder freely suspended by a steel cable, was built by Joseph Webber, David Zippy, and Robert Foward at the University of Maryland to detect gravitational waves. In theory, the antenna should vibrate as the gravity waves pass through it.

Summary

1. The special theory of relativity is based on two postulates of Albert Einstein:
   Postulate 1. Absolute, uniform motion cannot be detected.

Postulate 2. The speed of light is independent of the motion of the source.

An important implication of these postulates is

Postulate 2 (Alternate). Every observer measures the same value for the speed of light independent of the relative motion of the sources and observer.

All of the results of special relativity can be derived from these postulates.

2. The Michelson–Morley experiment was an attempt to measure the absolute velocity of the earth by comparing the speed of light in the direction of motion of the earth with that in a direction perpendicular to that motion. Their null result for the difference in these speeds is consistent with Einstein's postulates.

3. The Lorentz transformation relates the coordinates x, y, and z and the time t of an event seen in frame S to the coordinates x', y', and z' and the time t' of the same event as seen in frame S', which is moving with speed V relative to S:

$$x = \gamma(x' + Vt') \qquad y = y' \qquad z = z'$$

$$t = \gamma\left(t' + \frac{Vx'}{c^2}\right)$$

where

$$\gamma = \frac{1}{\sqrt{1 - V^2/c^2}}$$

The inverse transformation is

$$x' = \gamma(x - Vt) \qquad y' = y \qquad z' = z$$

$$t' = \gamma\left(t - \frac{Vx}{c^2}\right)$$

The transformation equations for velocities are

$$u_x = \frac{u'_x + V}{1 + Vu'_x/c^2}$$

$$u_y = \frac{u'_y}{\gamma(1 + Vu'_x/c^2)}$$

$$u_z = \frac{u'_z}{\gamma(1 + Vu'_x/c^2)}$$

The inverse velocity transformation equations are

$$u'_x = \frac{u_x - V}{1 - Vu_x/c^2}$$

$$u'_y = \frac{u_y}{\gamma(1 - Vu_x/c^2)}$$

$$u'_z = \frac{u_z}{\gamma(1 - Vu_x/c^2)}$$

4. The time interval measured between two events that occur at the same point in space in some reference frame is called the proper time. In another reference frame in which the events occur at different places,

the time interval between the events is longer by the factor γ. This result is known as time dilation. A related phenomenon is length contraction. The length of an object measured in a frame in which it is at rest is called its proper length L_p. When measured in another reference frame, the length of the object is L_p/γ.

5. Two events that are simultaneous in one reference frame are not simultaneous in another frame that is moving relative to the first. If two clocks are synchronized in the frame in which they are at rest, they will be out of synchronization in another frame. In the frame in which they are moving, the "chasing" clock leads by an amount $\Delta t_s = L_p V/c^2$, where L_p is the proper distance between the clocks.

6. The relativistic momentum of a particle is related to its mass and velocity by

$$\mathbf{p} = \frac{m_0 \mathbf{u}}{\sqrt{1 - u^2/c^2}}$$

where m_0 is the rest mass of the particle.

7. The kinetic energy of a particle is given by

$$K = \frac{m_0 c^2}{\sqrt{1 - u^2/c^2}} - m_0 c^2 = \frac{m_0 c^2}{\sqrt{1 - u^2/c^2}} - E_0$$

where

$$E_0 = m_0 c^2$$

is the rest energy. The total energy is

$$E = K + E_0 = \frac{m_0 c^2}{\sqrt{1 - u^2/c^2}}$$

The speed of a particle is related to its momentum and its total energy by

$$\frac{u}{c} = \frac{pc}{E}$$

The total energy is related to the momentum and rest energy by

$$E^2 = p^2c^2 + (m_0 c^2)^2$$

For particles with energies much greater than their rest energies, a useful approximation is

$$E \approx pc \qquad \text{for } E \gg m_0 c^2$$

This is an exact equation for particles of zero rest mass such as photons.

8. The total rest mass of bound systems of particles, such as nuclei or atoms, is less than the sum of the rest masses of the particles making up the system. The difference in mass times c^2 equals the binding energy of the system. The binding energy is the energy that must be added to break up the system into its parts. The binding energies of electrons in atoms are of the order of eV or keV, leading to a negligible difference in rest mass. The binding energies of nuclei are of the order of several MeV, and the difference in rest mass is noticeable.

9. The basis of the general theory of relativity is the principle of equivalence: A homogeneous gravitational field is completely equivalent to a uniformly accelerated reference frame. Important consequences of general relativity include the bending of light in a gravitational field, the prediction of the precession of the perihelion of the orbit of Mercury, the gravitational redshift, and probably the existence of black holes.

Suggestions for Further Reading

Bondi, Hermann: *Relativity and Common Sense: A New Approach to Einstein*, Doubleday, Garden City, New York, 1964.

This book uses familiar phenomena to help show how logical and easy it is to understand the ideas of special relativity.

Chaffee, Frederick H., Jr.: "The Discovery of a Gravitational Lens," *Scientific American*, November 1980, p. 70.

General relativity predicts that light should be deflected by concentrations of matter. This article describes how an elliptical galaxy can act as a giant lens in space.

Gamow, George: "Gravity," *Scientific American*, March 1961, p. 94.

Einstein's general theory of relativity is explained in an entertaining and nonmathematical fashion.

Goldberg, Stanley: *Understanding Relativity: Origin and Impact of a Scientific Revolution*, Birkhaeuser, Boston, 1984.

This book examines the intellectual and social context from which Einstein's special theory grew and the theory's early reception by communities of scientists in four countries.

MacKeown, P. K.: "Gravity is Geometry," *The Physics Teacher*, vol. 22, 1984, p. 557.

This article is an excellent, brief exposition of the ideas of general relativity.

Marder, L.: *Time and the Space Traveller*, George Allen & Unwin, Ltd., London, 1971.

This book presents some of the arguments which have been made in the long and colorful debate over the twin paradox. It also examines some practical limitations of space travel, the implications of time dilation for the long-distance space traveller, and the nature of living clocks.

Mook, Delo E., and Thomas Vargish: *Inside Relativity*, Princeton University Press, Princeton, 1987.

This is a book for nonscientists written by two scholars, one working in the physical sciences and the other in the humanities. The book provides a historical and scientific context for Einstein's work, and explains the special and general theories with the aid of drawings and graphs but no mathematics.

Schwinger, Julian: *Einstein's Legacy: The Unity of Space and Time*, Scientific American Books, Inc., New York, 1986.

A modern and well-illustrated exposition of the special and general theories of relativity and some of their consequences.

Shankland, R. S.: "The Michelson–Morley Experiment," *Scientific American*, November 1964, p. 107.

This article sets the experiment in its historical context and considers its influence on the development of the theory of relativity.

Will, Clifford M.: *Was Einstein Right?: Putting General Relativity to the Test*, Basic Books, Inc., New York, 1986.

Starting around 1960, new discoveries in astronomy motivated a renewed interest in experimentally testing predictions of general relativity. This book, written by a physicist who began his career during this relativity "renaissance," describes the tests with great enthusiasm.

Review

A. Objectives: After studying this chapter, you should:

1. Be able to discuss the results and significance of the Michelson–Morley experiment.

2. Be able to state the Einstein postulates of special relativity.

3. Be able to use the Lorentz transformation to derive expressions for time dilation and length contraction and to solve problems in which time and space intervals in different reference frames are compared.

4. Be able to discuss the lack of synchronization of clocks in moving reference frames.

5. Be able to discuss the twin paradox.

6. Be able to state the definition of relativistic momentum and the equations relating to kinetic energy and total energy of a particle to its speed.

7. Be able to discuss the relation between mass and energy in special relativity and compute the binding energy of various systems from the known rest masses of their constituents.

8. Be able to state the principle of equivalence and discuss three predictions derived from it.

B. Define, explain, or otherwise identify:

Reference frame
Inertial reference frame
Newtonian relativity
Ether
Michelson–Morley experiment
Einstein's postulates
Galilean transformation
Lorentz transformation
Proper time

Time dilation
Length contraction
Proper length
Lorentz–FitzGerald contraction
Synchronized clocks
Simultaneity
Redshift
Twin paradox
Relativistic momentum
Relativistic mass
Rest mass
Rest energy
Relativistic energy
Binding energy
Principle of equivalence
Gravitational redshift
Black hole
Schwarzschild radius

C. True or false: If the statement is true, explain why it is true. If it is false, give a counterexample.

1. The speed of light is the same in all reference frames.

2. Proper time is the shortest time interval between two events.

3. Absolute motion can be determined by means of length contraction.

4. The light-year is a unit of distance.

5. Simultaneous events must occur at the same place.

6. If two events are not simultaneous in one frame, they cannot be simultaneous in any other frame.

7. If two particles are tightly bound together by strong attractive forces, the rest mass of the system is less than the sum of the masses of the individual particles when separated.

Problems

Level I

34-1 Newtonian Relativity

There are no problems for this section.

34-2 The Michelson–Morley Experiment

1. In one series of measurements of the speed of light, Michelson used a path length L of 27.4 km (22 mi). (*a*) What is the time needed for light to make the round-trip distance of $2L$? (*b*) What is the classical correction term in seconds in Equation 34-1, assuming earth's speed is $v = 10^{-4}c$? (*c*) From about 1600 measurements, Michelson quoted the result for the speed of light as 299,796 ± 4 km/s. Is this experiment accurate enough to be sensitive to the correction term in Equation 34-1?

2. An airplane flies with speed u relative to still air from point A to point B and returns. Compare the time required for the round trip when the wind blows from A to B with speed v with that when the wind blows perpendicularly to the line AB with speed v.

34-3 Einstein's Postulates

There are no problems for this section.

34-4 The Lorentz Transformation

3. The proper mean lifetime of pions is 2.6×10^{-8} s. If a beam of pions has a speed of $0.85c$, (*a*) what would their mean lifetime be as measured in the laboratory? (*b*) How far would they travel, on the average, before they decay? (*c*) What would your answer be to part (*b*) if you neglect time dilation?

4. (*a*) In the reference frame of the pion in Problem 3, how far does the laboratory travel in a typical lifetime of 2.6×10^{-8} s? (*b*) What is this distance in the laboratory's frame?

5. The proper mean lifetime of a muon is 2 μs. Muons in a beam are traveling at $0.999c$. (*a*) What is their mean lifetime as measured in the laboratory? (*b*) How far do they travel, on the average, before they decay?

6. (*a*) In the reference frame of the muon in Problem 5, how far does the laboratory travel in a typical lifetime of 2 μs? (*b*) What is this distance in the laboratory's frame?

7. A spaceship of proper length 100 m passes you at a high speed. You measure the length of the spaceship to be 85 m. What was the speed of the spaceship?

8. A spaceship departs from earth for the star Alpha Centauri, which is 4 light-years away. The spaceship travels at $0.75c$. How long does it take to get there (*a*) as measured on earth and (*b*) as measured by a passenger on the spaceship?

9. A spaceship travels to a star 95 light-years away at a speed of 2.2×10^8 m/s. How long does it take to get there (*a*) as measured on earth and (*b*) as measured by a passenger on the spaceship?

10. The mean lifetime of a pion traveling at high speed is measured to be 7.5×10^{-8} s. Its lifetime when measured at rest is 2.6×10^{-8} s. How fast is the pion traveling?

11. How fast must a muon travel so that its mean lifetime is 46 μs if its mean lifetime at rest is 2 μs?

12. A meterstick moves with speed $V = 0.8c$ relative to you in the direction parallel to the stick. (*a*) Find the length of the stick as measured by you. (*b*) How long does it take for the stick to pass you?

13. How fast must a meterstick travel relative to you in the direction parallel to the stick so that its length as measured by you is 50 cm?

14. Use the binomial expansion (Equation 34-2) to derive the following results for the case when V is much less than c, and use the results when applicable in the following problems:

(*a*) $\gamma \approx 1 + \frac{1}{2}\frac{V^2}{c^2}$ (*b*) $\frac{1}{\gamma} \approx 1 - \frac{1}{2}\frac{V^2}{c^2}$

(*c*) $\gamma - 1 \approx 1 - \frac{1}{\gamma} \approx \frac{1}{2}\frac{V^2}{c^2}$

15. Supersonic jets achieve maximum speeds of about $(3 \times 10^{-6})c$. (*a*) By what percentage would you see a jet traveling at this speed contracted in length? (*b*) During a time of 1 y = 3.15×10^7 s on your clock, how much time would elapse on the pilot's clock? How many minutes are lost by the pilot's clock in 1 y of your time?

16. How great must the relative speed of two observers be for the time-interval measurements to differ by 1 percent? (See Problem 14.)

34-5 Clock Synchronization and Simultaneity

Problems 17 through 21 refer to the following situation: An observer in S′ lays out a distance L′ = 100c·min between points A′ and B′ and places a flashbulb at the midpoint C′. She arranges for the bulb to flash and for clocks at A′ and B′ to be started at zero when the light from the flash reaches them (see Figure 34-21). Frame S′ is moving to the right with speed 0.6c relative to an observer C in S who is at the midpoint between A′ and B′ when the bulb flashes. At the instant he sees the flash, observer C sets his clock to zero.

Figure 34-21 Problems 17 through 21.

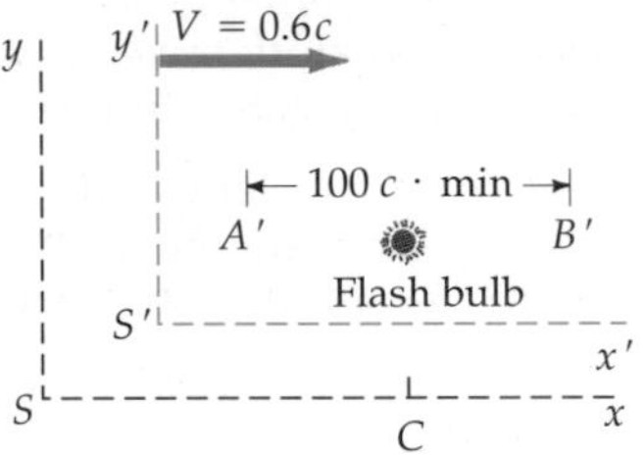

17. What is the separation distance between clocks A' and B' according to the observer in S?

18. As the light pulse from the flashbulb travels toward A' with speed c, A' travels toward C with speed $0.6c$. Show that the clock in S reads 25 min when the flash reaches A'. (*Hint*: In time t, the light travels a distance ct and A' travels $0.6ct$. The sum of these distances must equal the distance between A' and the flashbulb as seen in S.

19. Show that the clock in S reads 100 min when the light flash reaches B', which is traveling away from C with speed $0.6c$. (See the hint for Problem 18.)

20. The time interval between the reception of the flashes at A' and B' in Problems 18 and 19 is 75 min according to the observer in S. How much time does he expect to have elapsed on the clock at A' during this 75-min interval?

21. The time interval calculated in Problem 20 is the amount that the clock at A' leads that at B' according to the observer in S. Compare this result with $L_p V/c^2$.

34-6 The Doppler Effect

22. How fast must you be moving toward a red light (λ = 650 nm) for it to appear green (λ = 525 nm)?

23. A distant galaxy is moving away from us at a speed of 1.85×10^7 m/s. Calculate the fractional redshift $(\lambda' - \lambda_0)/\lambda_0$ in the light from this galaxy.

24. Show that if V is much less than c, the doppler shift is given approximately by $\Delta f/f \approx \pm V/c$.

25. A distant galaxy is moving away from the earth with a speed that results in each wavelength received on earth being shifted such that $\lambda' = 2\lambda_0$. Find the speed of the galaxy relative to the earth.

26. Sodium light of wavelength 589 nm is emitted by a source that is moving toward the earth with speed V. The wavelength measured in the frame of the earth is 620 nm. Find V.

27. A student on earth hears a tune on her radio that seems to be coming from a record that is being played too fast. She has a 33-rev/min record of that tune and determines that the tune sounds the same as when her record is played at 78 rev/min, that is, the frequencies are all too high by a factor of 78/33. If the tune is being played correctly, but is being broadcast by a spaceship that is approaching the earth at speed V, determine V.

34-7 The Twin Paradox

28. A friend of yours who is the same age as you travels at $0.999c$ to a star 15 light-years away. She spends 10 y on one of the star's planets and then returns at $0.999c$. How long has she been away (*a*) as measured by you and (*b*) as measured by her?

35-8 The Velocity Transformation

29. Two spaceships are approaching each other. (*a*) If the speed of each is $0.6c$ relative to the earth, what is the speed of one relative to the other? (*b*) If the speed if each relative to the earth is 30,000 m/s (about 100 times the speed of sound), what is the speed of one relative to the other?

30. A light beam moves along the y' axis with speed c in frame S', which is moving to the right with speed V relative to frame S. (*a*) Find the x and y components of the velocity of the light beam in frame S. (*b*) Show that the magnitude of the velocity of the light beam in S is c.

31. A spaceship is moving east at speed $0.90c$ relative to the earth. A second spaceship is moving west at speed $0.90c$ relative to the earth. What is the speed of one spaceship relative to the other?

32. A particle moves with speed $0.8c$ along the x'' axis of frame S'', which moves with speed $0.8c$ along the x' axis relative to frame S'. Frame S' moves with speed $0.8c$ along the x axis relative to frame S. (*a*) Find the speed of the particle relative to frame S'. (*b*) Find the speed of the particle relative to frame S.

34-9 Relativistic Momentum; 34-10 Relativistic Energy

33. How much rest mass must be converted into energy (*a*) to produce 1 J and (*b*) to keep a 100-W light bulb burning for 10 years?

34. Sketch a graph of the momentum p of a particle versus its speed u.

35. (*a*) Calculate the rest energy in 1 g of dirt. (*b*) If you could convert this energy into electrical energy and sell it for 10 cents per kilowatt-hour, how much money would you get? (*c*) If you could power a 100-W light bulb with this energy, for how long could you keep the bulb lit?

36. Find the ratio of the total energy to the rest energy of a particle of rest mass m_0 moving with speed (*a*) $0.1c$, (*b*) $0.5c$, (*c*) $0.8c$, and (*d*) $0.99c$.

37. An electron with rest energy of 0.511 MeV moves with speed $u = 0.2c$. Find its total energy, kinetic energy, and momentum.

38. A muon has a rest energy of 105.7 MeV. Calculate its rest mass in kilograms.

39. A proton with rest energy of 938 MeV has a total energy of 1400 MeV. (*a*) What is its speed? (*b*) What is its momentum?

40. The total energy of a particle is twice its rest energy. (*a*) Find u/c for the particle. (*b*) Show that its momentum is given by $p = \sqrt{3}m_0c$.

41. For the fusion reaction of Example 34-11, calculate the number of reactions per second that are necessary to generate 1 kW of power.

42. Using Table 34-1 to find how much energy is needed to remove one neutron from ^{4}He, leaving ^{3}He plus a neutron.

43. A free neutron decays into a proton plus an electron:

$$n \rightarrow p + e$$

Use Table 34-1 to calculate the energy released in this reaction.

44. How much energy would be required to accelerate a particle of mass m_0 from rest to (*a*) $0.5c$, (*b*) $0.9c$, and (*c*) $0.99c$? Express your answers as multiples of the rest energy.

45. If the kinetic energy of a particle equals its rest energy, what error is made by using $p = m_0u$ for its momentum?

46. In another nuclear fusion reaction, ^{2}H nuclei are combined to produce ^{4}He. (*a*) How much energy is released in this reaction? (*b*) How many such reactions must take place per second to produce 1 kW of power?

34-11 General Relativity

There are no problems for this section.

Level II

47. A friend of yours who is the same age as you travels to the star Alpha Centauri, which is 4 light-years away and returns immediately. He claims that the entire trip took just 6 y. How fast did he travel?

48. Use Equations 34-28 and 34-32 to derive the equation $E^2 = p^2c^2 + m_0^2c^4$.

49. If a plane flies at a speed of 2000 km/h, for how long must it fly before its clock loses 1 s because of time dilation?

50. Use the binomial expansion (Equation 34-2) and Equation 34-34 to show that when $pc \ll m_0c^2$, the total energy is given approximately by

$$E \approx m_0c^2 + \frac{p^2}{2m_0}$$

51. A clock is placed in a satellite that orbits the earth with a period of 90 min. By what time interval will this clock differ from an identical clock on earth after 1 y? (Assume that special relativity applies.)

52. *A* and *B* are twins. *A* travels at $0.6c$ to Alpha Centauri (which is $4c$·years from earth as measured in the reference frame of the earth) and returns immediately. Each twin sends the other a light signal every 0.01 year as measured in her own reference frame. (*a*) At what rate does *B* receive signals as *A* is moving away from her? (*b*) How many signals does *B* receive at this rate? (*c*) How many total signals are received by *B* before *A* has returned? (*d*) At what rate does *A* receive signals as *B* is receding from her? (*e*) How many signals does *A* receive at this rate? (*f*) How many total signals are received by *A*? (*g*) Which twin is younger at the end of the trip, and by how many years?

53. In frame *S*, event *B* occurs 2 μs after event *A*, which occurs at $\Delta x = 1.5$ km from event *A*. How fast must an observer be moving along the $+x$ axis so that events *A* and *B* occur simultaneously? Is it possible for event *B* to precede event *A* for some observer?

54. Observers in reference frame *S* see an explosion located at $x_1 = 480$ m. A second explosion occurs 5 μs later at $x_2 = 1200$ m. In reference frame S', which is moving along the $+x$ axis at speed *V*, the explosions occur at the same point in space. What is the separation in time between the two explosions as measured in S'?

55. An interstellar spaceship travels from the earth to a distant star system $12c$·years away (as measured in the earth's frame). The trip takes 15 years as measured on the ship. (*a*) What is the speed of the ship relative to the earth? (*b*) When the ship arrives, it sends a signal to the earth. How long after the ship leaves the earth will it be before the earth receives the signal?

56. Show that the speed *u* of a particle of mass m_0 and total energy *E* is given by

$$\frac{u}{c} = 1 - \left[\frac{(m_0c^2)^2}{E^2}\right]^{1/2}$$

and that when *E* is much greater than m_0c^2, this can be approximated by

$$\frac{u}{c} \approx 1 - \frac{(m_0c^2)^2}{2E^2}$$

Find the speed of an electron with kinetic energy of (*b*) 0.51 MeV and (*c*) 10 MeV.

57. Two spaceships, each 100 m long when measured at rest, travel toward each other with speeds of $0.85c$ relative to the earth. (*a*) How long is each ship as measured by someone on earth? (*b*) How fast is each ship traveling as measured by an observer on the other? (*c*) How long is one ship when measured by an observer on the other? (*d*) At time $t = 0$ on earth, the fronts of the ships are together as they just begin to pass each other. At what time on earth are their ends together? (*e*) Sketch diagrams in the frame of one of the ships showing the passing of the other ship.

58. In the Stanford linear collider, small bundles of electrons and positrons are fired at each other. In the laboratory's frame of reference, each bundle is about 1 cm long and 10 μm in diameter. In the collision region, each particle has an energy of 50 GeV, and the electrons and positrons are moving in opposite directions. (*a*) How long and how wide is each bundle in its own reference frame? (*b*) What must be the minimum proper length of the accelerator for a bundle to have both its ends simultaneously in the accelerator in its own reference frame? (The actual length of the accelerator is less than 1000 m.) (*c*) What is the length of a positron bundle in the reference frame of the electron bundle?

59. An electron with rest energy of 0.511 MeV has a total energy of 5 MeV. (*a*) Find its momentum in units of MeV/c from Equation 34-34. (*b*) Find the ratio of its speed u to the speed of light.

60. The rest energy of a proton is about 938 MeV. If its kinetic energy is also 938 MeV, find (*a*) its momentum and (*b*) its speed.

61. What percent error is made in using $\frac{1}{2}m_0u^2$ for the kinetic energy of a particle if its speed is (*a*) $0.1c$ and (*b*) $0.9c$?

62. A rocket with a proper length of 1000 m moves in the $+x$ direction at $0.6c$ with respect to an observer on the ground. An astronaut stands at the rear of the rocket and fires a bullet toward the front of the rocket at $0.8c$ relative to the rocket. How long does it take the bullet to reach the front of the rocket (*a*) as measured in the frame of the rocket, (*b*) as measured in the frame of the ground, and (*c*) as measured in the frame of the bullet?

63. A rocket with a proper length of 700 m is moving to the right at a speed of $0.9c$. It has two clocks, one in the nose and one in the tail, that have been synchronized in the frame of the rocket. A clock on the ground and the nose clock on the rocket both read $t = 0$ as they pass. (*a*) At $t = 0$, what does the tail clock on the rocket read as seen by an observer on the ground? When the tail clock on the rocket passes the ground clock, (*b*) what does the tail clock read as seen by an observer on the ground, (*c*) what does the nose clock read as seen by an observer on the ground, and (*d*) what does the nose clock read as seen by an observer on the rocket? (*e*) At $t = 1$ h, as measured on the rocket, a light signal is sent from the nose of the rocket to an observer standing by the ground clock. What does the ground clock read when the observer receives this signal? (*f*) When the observer on the ground receives the signal, he sends a return signal to the nose of the rocket. When is this signal received at the nose of the rocket as seen on the rocket?

64. Derive Equation 34-24*a* for the frequency received by an observer moving with speed V toward a stationary source of electromagnetic waves.

65. Frames S and S' are moving relative to each other along the x and x' axis. They set their clocks to $t = 0$ when their origins coincide. In frame S, event 1 occurs at $x_1 = 1.0c\cdot$year and $t_1 = 1$ y and event 2 occurs at $x_2 = 2.0c\cdot$year and $t_2 = 0.5$ y. These events occur simultaneously in frame S'. (*a*) Find the magnitude and direction of the velocity of S' relative to S. (*b*) At what time do both these events occur as measured in S'?

66. An observer in frame S standing at the origin observes two flashes of colored light separated spatially by $\Delta x = 2400$ m. A blue flash occurs first, followed by a red flash 5 μs later. An observer in S' moving along the x axis at speed V relative to S also observes the flashes 5 μs apart and with a separation of 2400 m, but the red flash is observed first. Find the magnitude and direction of V.

67. The sun radiates energy at the rate of about 4×10^{26} W. Assume that this energy is produced by a reaction whose net result is the fusion of 4 H nuclei to form 1 He nucleus, with the release of 25 MeV for each He nucleus formed. Calculate the sun's loss of rest mass per day.

68. A spaceship of mass 10^6 kg is coasting through space when it suddenly becomes necessary to accelerate. The ship ejects 10^3 kg of fuel in a very short time at a speed of $c/2$ relative to the ship. (*a*) Neglecting any change in the rest mass of the system, calculate the speed of the ship in the frame in which it was initially at rest. (*b*) Calculate the speed of the ship using classical, newtonian mechanics. (*c*) Use your results from (*a*) to estimate the change in the rest mass of the system.

69. Reference frame S' is moving along the x' axis at $0.6c$ relative to frame S. A particle that is originally at $x' = 10$ m at $t_1' = 0$ is suddenly accelerated and then moves at a constant speed of $c/3$ in the $-x'$ direction until time $t_2' = 60$ m/c, when it is suddenly brought to rest. As observed in frame S, find (*a*) the speed of the particle, (*b*) the distance and direction the particle traveled from t_1' to t_2', and (*c*) the time the particle traveled.

70. Show that

$$d\left(\frac{m_0u}{\sqrt{1 - u^2/c^2}}\right) = m_0\left(1 - \frac{u^2}{c^2}\right)^{-3/2} du$$

71. Two protons approach each other head on at $0.5c$ relative to reference frame S'. (*a*) Calculate the total kinetic energy of the two protons as seen in frame S'. (*b*) Calculate the total kinetic energy of the protons as seen in reference frame S, which is moving with speed $0.5c$ relative to S' such that one of the protons is at rest.

72. A particle of rest mass 1 MeV/c^2 and kinetic energy 2 MeV collides with a stationary particle of rest mass 2 MeV/c^2. After the collision, the particles stick together. Find (*a*) the speed of the first particle before the collision, (*b*) the total energy of the first particle before the collision, (*c*) the initial total momentum of the system, (*d*) the total kinetic energy after the collision, and (*e*) the rest mass of the system after the collision.

73. The radius of the orbit of a charged particle in a magnetic field is related to the momentum of the particle by

$$p = BqR \qquad 34\text{-}41$$

This equation holds classically for $p = mu$ and relativistically for $p = m_0u/\sqrt{1 - u^2/c^2}$. An electron with kinetic energy of 1.50 MeV moves in a circular orbit perpendicular to a uniform magnetic field $B = 5 \times 10^{-3}$ T. (*a*) Find the radius of the orbit. (*b*) What result would you obtain if you used the classical relations $p = mu$ and $K = p^2/2m$?

74. Oblivious to economics and politics, physicists propose building a circular accelerator around the earth's circumference using bending magnets that provide a magnetic field of magnitude 1.5 T. (*a*) What would be the kinetic energy of protons orbiting in this field in a circle of radius R_E? (See Problem 73.) (*b*) What would be the period of rotation of these protons?

75. In a simple thought experiment, Einstein showed that there is mass associated with electromagnetic radiation. Consider a box of length L and mass M resting on a frictionless surface. At the left wall of the box is a light source that emits radiation of energy E, which is absorbed at the right wall of the box. According to classical electromagnetic theory, this radiation carries momentum of magnitude $p = E/c$ (Equation 29-24). (*a*) Find the recoil velocity of the box such that momentum is conserved when the light is emitted. (Since p is small and M is large, you may use classical mechanics.) (*b*) When the light is absorbed at the right wall of the box, the box stops, so the total momentum remains zero. If we neglect the very small velocity of the box, the time it takes for the radiation to travel across the box is $\Delta t = L/c$. Find the distance moved by the box in this time. (*c*) Show that if the center of mass of the system is to remain at the same place, the radiation must carry mass $m = E/c^2$.

76. An antiproton $\bar{p}$ has the same rest energy at a proton. It is created in the reaction $p + p \rightarrow p + p + p + \bar{p}$. In an experiment, protons at rest in the laboratory are bombarded with protons of kinetic energy K_L, which must be great enough so that kinetic energy equal to $2m_0c^2$ can be converted into the rest energy of the two particles. In the frame of the laboratory, the total kinetic energy cannot be converted into rest energy because of conservation of momentum. However, in the zero-momentum reference frame in which the two initial protons are moving toward each other with equal speed u, the total kinetic energy can be converted into rest energy. (*a*) Find the speed of each proton u such that the total kinetic energy in the zero-momentum frame is $2m_0c^2$. (*b*) Transform to the laboratory's frame in which one proton is at rest, and find the speed u' of the other proton. (*c*) Show that the kinetic energy of the moving proton in the laboboratory's frame is $K_L = 6m_0c^2$.

Level III

77. A stick of proper length L_p makes an angle θ with the x axis in frame S. Show that the angle θ' made with the x' axis in frame S', which is moving along the $+x$ axis with speed V, is given by $\tan\theta' = \gamma \tan\theta$ and that the length of the stick in S' is

$$L' = L_p\left[\frac{1}{\gamma^2}\cos^2\theta + \sin^2\theta\right]^{1/2}$$

78. Show that if a particle moves at an angle θ with the x axis with speed u in frame S, it moves at an angle θ' with the x' axis in S' given by

$$\tan\theta' = \frac{\sin\theta}{\gamma(\cos\theta - V/u)}$$

where the frame S' is moving with speed V relative to S.

79. For the special case of a particle moving with speed u along the y axis in frame S, show that its momentum and energy in frame S' are related to its momentum and energy in S by the transformation equations

$$p_x' = \gamma\left(p_x - \frac{VE}{c^2}\right) \qquad p_y' = p_y \qquad p_z' = p_z$$

$$\frac{E'}{c} = \gamma\left(\frac{E}{c} - \frac{Vp_x}{c^2}\right)$$

Compare these equations with the Lorentz transformation for x', y', z', and t'. These equations show that the quantities p_x, p_y, p_z, and E/c transform in the same way as do x, y, z, and ct.

80. The equation for the spherical wavefront of a light pulse that begins at the origin at time $t = 0$ is $x^2 + y^2 + z^2 - (ct)^2 = 0$. Using the Lorentz transformation, show that such a light pulse also has a spherical wavefront in frame S' by showing that $x'^2 + y'^2 + z'^2 - (ct')^2 = 0$ in S'.

81. In Problem 80, you showed that the quantity $x^2 + y^2 + z^2 - (ct)^2$ has the same value (0) in both S and S'. Such a quantity is called an *invariant*. From the results of Problem 79, the quantity $p_x^2 + p_y^2 + p_z^2 - (E/c)^2$ must also be an invariant. Show that this quantity has the value $-m_0c^2$ in both the S and S' reference frames.

82. Two events in S are separated by a distance $D = x_2 - x_1$ and a time $T = t_2 - t_1$. (*a*) Use the Lorentz transformation to show that in frame S', which is moving with speed V relative to S, the time separation is $t_2' - t_1' = \gamma(T - VD/c^2)$. (*b*) Show that the events can be simultaneous in frame S' only if D is greater than cT. (*c*) If one of the events is the *cause* of the other, the separation D must be less than cT since D/c is the smallest time that a signal can take to travel from x_1 to x_2 in frame S. Show that if D is less than cT, t_2' is greater than t_1' in all reference frames. This shows that if the cause precedes the effect in one frame, it must precede it in all reference frames. (*d*) Suppose that a signal could be sent with speed $c' > c$ so that in frame S the cause precedes the effect by the time $T = D/c'$. Show that there is then a reference frame moving with speed V less than c in which the effect precedes the cause.

83. Two identical particles of rest mass m_0 are each moving toward the other with speed u in frame S. The particles collide inelastically with a spring that locks shut (Figure 34-22) and come to rest in S, and their initial kinetic energy is transformed into potential energy. In this problem you are going to show that the conservation of momentum in reference frame S', in which one of the particles is initially at rest, requires that the total rest mass of the system after the collision be $2m_0/\sqrt{1-u^2/c^2}$. (*a*) Show that the speed of the particle not at rest in frame S' is

$$u' = \frac{2u}{1+u^2/c^2}$$

and use this result to show that

$$\sqrt{1-\frac{u'^2}{c^2}} = \frac{1-u^2/c^2}{1+u^2/c^2}$$

(*b*) Show that the initial momentum in frame S' is $p' = 2m_0u/(1-u^2/c^2)$. (*c*) After the collision, the composite particle moves with speed u in S' (since it is at rest in S). Write the total momentum after the collision in terms of the final rest M_0, and show that the conservation of momentum implies that $M_0 = 2m_0/\sqrt{1-u^2/c^2}$. (*d*) Show that the total energy is conserved in each reference frame.

84. A horizontal turntable rotates with angular speed ω. There is a clock at the center of the turntable and one at a distance r from the center. In an inertial reference frame, the clock at distance r is moving with speed $u = r\omega$. (*a*) Show that from time dilation in special relativity, time intervals Δt_0 for the clock at rest and Δt_r for the moving clock are related by

$$\frac{\Delta t_r - \Delta t_0}{\Delta t_0} \approx -\frac{r^2\omega^2}{2c^2} \qquad \text{if } r\omega \ll c$$

(*b*) In a reference frame rotating with the table, both clocks are at rest. Show that the clock at distance r experiences a pseudoforce $F_r = mr\omega^2$ in this accelerated frame and that this is equivalent to a difference in gravitational potential between r and the origin of $\phi_r - \phi_0 = \frac{1}{2}r^2\omega^2$. Use this potential difference in Equation 34-39 to show that in this frame the difference in time intervals is the same as in the inertial frame.

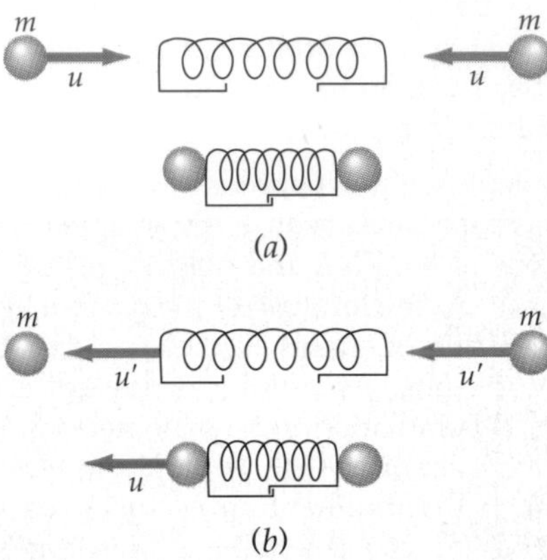

Figure 34-22 Problem 83. An inelastic collision between two identical objects (*a*) in the zero-momentum reference frame S and (*b*) in frame S', which is moving to the right with speed $V = u$ relative to frame S such that one of the particles is initially at rest. The spring, which is assumed to be massless, is merely a device for visualizing the storage of potential energy.

Chapter 35

The Origins of Quantum Theory

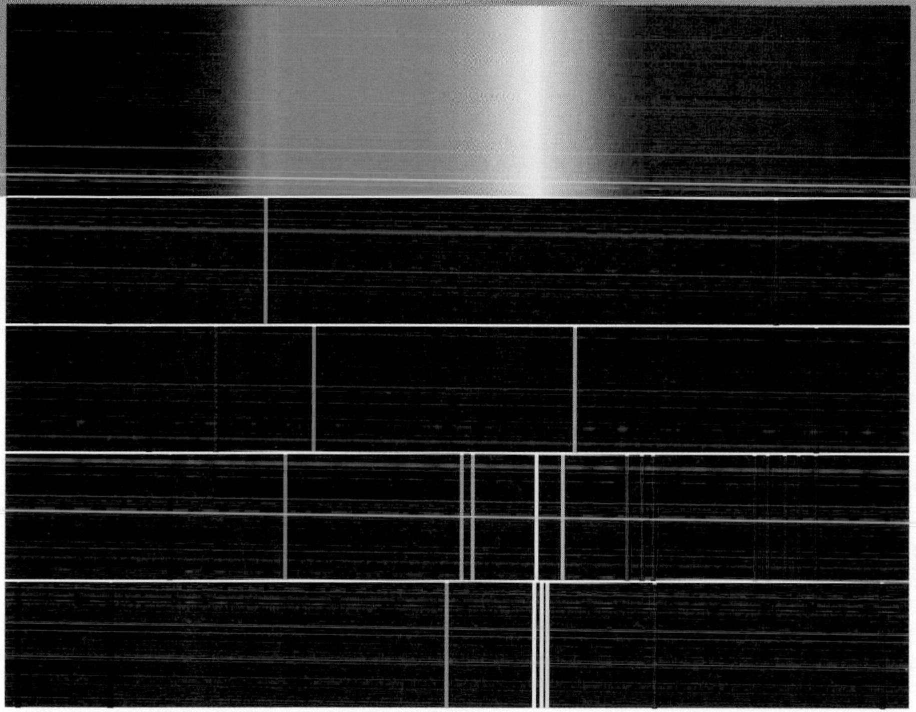

The continuous visible spectrum (*top*) along with the characteristic optical line spectrum (*from top to bottom*) emitted by hydrogen, helium, sodium, mercury, and barium. Niels Bohr's model of the hydrogen atom, which could be used to predict the wavelengths of the observed spectrum of hydrogen, was one of the great triumphs of twentieth century physics and was an important step in the development of our understanding of the microscopic world.

In Chapter 34, we saw that Newton's laws must be modified when they are applied to objects that move at speeds comparable to the speed of light. In the last 20 years of the nineteenth century and the first 30 years of the twentieth century, many startling discoveries, both experimental and theoretical, demonstrated that the laws of classical physics also break down when they are applied to microscopic systems, such as the particles within an atom. This failure is as drastic as the failure of newtonian mechanics at high speeds. The interior of the atom can be described only in terms of *quantum theory* (sometimes called *quantum mechanics* or *wave mechanics*), which requires the modification of some of our fundamental ideas about the relationships between physical theory and the physical world. Table 35-1 lists the approximate dates of some of the important experiments performed and theories proposed between 1881 and 1932.

The development of quantum theory was very different from that of the theory of relativity. In a sense, special relativity was presented as a complete theory in 1905 (and the general theory, in 1916) by a single scientist, Albert Einstein. Quantum theory, on the other hand, was developed over a long period by many different people. Many of the discoveries initially seemed unrelated, and it wasn't until the late 1920s that any consistent theory emerged. This theory is now the basis of our understanding of the micro-

Table 35-1 **Approximate Dates of Some Important Experiments and Theories, 1881–1932**

1881	Michelson obtains null result for absolute velocity of earth
1884	Balmer finds empirical formula for spectral lines of hydrogen
1887	Hertz produces electromagnetic waves, verifying Maxwell's theory and accidently discovering photoelectric effect
1887	Michelson repeats his experiment with Morley, again obtaining null result
1895	Röntgen discovers x rays
1896	Becquerel discovers nuclear radioactivity
1897	J. J. Thomson measures e/m for cathode rays, showing that electrons are fundamental constituents of atoms
1900	Planck explains blackbody radiation using energy quantization involving new constant h
1900	Lenard investigates photoelectric effect and finds energy of electrons independent of light intensity
1905	Einstein proposes special theory of relativity
1905	Einstein explains photoelectric effect by suggesting quantization of radiation
1907	Einstein applies energy quantization to explain temperature dependence of heat capacities of solids
1908	Rydberg and Ritz generalize Balmer's formula to fit spectra of many elements
1909	Millikan's oil-drop experiment shows quantization of electric charge
1911	Rutherford proposes nuclear model of atom based on alpha-particle scattering experiments of Geiger and Marsden
1912	Friedrich and Knipping and von Laue demonstrate diffraction of x rays by crystals showing that x rays are waves and that crystals are regular arrays
1913	Bohr proposes model of hydrogen atom
1914	Moseley analyzes x-ray spectra using Bohr model to explain periodic table in terms of atomic number
1914	Franck and Hertz demonstrate atomic energy quantization
1915	Duane and Hunt show that the short-wavelength limit of x rays is determined from quantum theory
1916	Wilson and Sommerfeld propose rules for quantization of periodic systems
1916	Millikan verifies Einstein's photoelectric equation
1923	Compton explains x-ray scattering by electrons as collision of photon and electron and verifies results experimentally
1924	De Broglie proposes electron waves of wavelength h/p
1925	Schrödinger develops mathematics of electron wave mechanics
1925	Heisenberg invents matrix mechanics
1925	Pauli states exclusion principle
1927	Heisenberg formulates uncertainty principle
1927	Davisson and Germer observe electron wave diffraction by single crystal
1927	G. P. Thomson observes electron wave diffraction in metal foil
1928	Gamow and Condon and Gurney apply quantum mechanics to explain alpha-decay lifetimes
1928	Dirac develops relativistic quantum mechanics and predicts existence of positron
1932	Chadwick discovers neutron
1932	Anderson discovers positron

scopic world. It is extremely successful, yet there is still debate about its philosophical interpretations. As with special relativity, quantum theory reduces to classical physics when it is applied to macroscopic (large-scale) systems, that is, to objects in our familiar, everyday world.

The origins of quantum theory were not, strangely enough, in the discoveries of radioactivity or x rays or atomic spectra but in thermodynamics. In his study of the radiation spectrum of a blackbody, Max Planck found that he could reconcile theory and experiment if he assumed that radiant energy was emitted and absorbed not continuously but in discrete lumps or quanta. It was Einstein who first recognized that this quantization of radiant energy was not just a calculational device but a general property of radiation. Niels Bohr then applied Einstein's ideas of energy quantization to the energy of an atom and proposed a model of the hydrogen atom that was spectacularly successful in calculations of the wavelengths of the radiation emitted by hydrogen. In this chapter, we will look qualitatively at the origins of the idea of energy quantization.

35-1 The Origin of the Quantum Constant: Blackbody Radiation

One of the most puzzling phenomena studied near the end of the nineteenth century was the spectral distribution of blackbody radiation.* A blackbody is an ideal system that absorbs all the radiation incident on it. It can be approximated by a cavity with a very small opening, as illustrated in Figure 35-1. The characteristics of the radiation in such a cavity depend only on the temperature of the walls. At ordinary temperatures (below about 600°C), the thermal radiation emitted by a blackbody is not visible because the energy is concentrated in the infrared region of the electromagnetic spectrum. As the body is heated, the amount of energy radiated increases (according to the Stefan–Boltzmann law, Equation 16-21), and the concentration of energy moves to shorter wavelengths. Between about 600 and 700°C, there is enough energy in the visible spectrum for the body to glow a dull red. At higher temperatures, it becomes bright red or even "white hot."

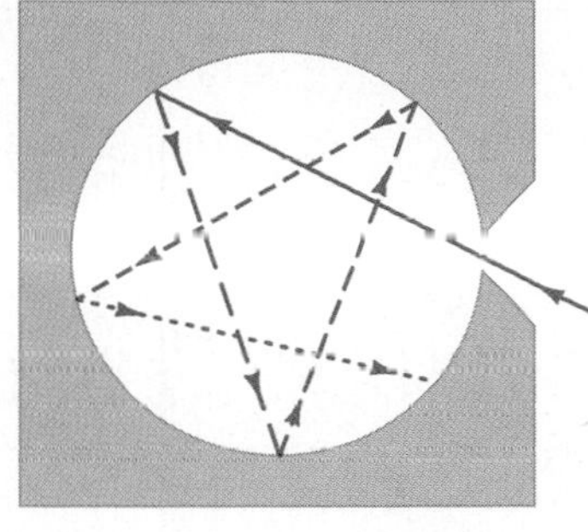

Figure 35-1 Cavity approximating an ideal blackbody. Radiation entering the cavity has little chance of leaving. It is usually completely absorbed.

Figure 35-2 shows the power radiated by a blackbody as a function of wavelength for three different temperatures. These curves are known as spectral distribution curves. The quantity P in this figure is the power radiated per unit wavelength. It is a function of both the wavelength λ and the temperature T and is called the spectral distribution function. This function $P(\lambda,T)$ has a maximum at a wavelength λ_{max} that varies inversely with temperature according to Wien's displacement law (Equation 16-25) discussed in Section 16-3:

$$\lambda_{max} = \frac{2.898 \text{ mm·K}}{T}$$

The spectral distribution function $P(\lambda,T)$ can be calculated from classical thermodynamics in a straightforward way, and the result can be compared with the experimentally obtained curves of Figure 35-2. The result of this classical calculation, known as the **Rayleigh–Jeans law,** is

$$P(\lambda,T) = 8\pi kT\lambda^{-4} \qquad 35\text{-}1$$

where k is Boltzmann's constant. This result agrees with experimental results in the region of long wavelengths, but it disagrees violently at short wavelengths. As λ approaches zero, the experimentally determined $P(\lambda,T)$

Portrait of Max Planck (1858–1947).

*Blackbodies were discussed briefly in Section 16-3. Some of that discussion is repeated here.

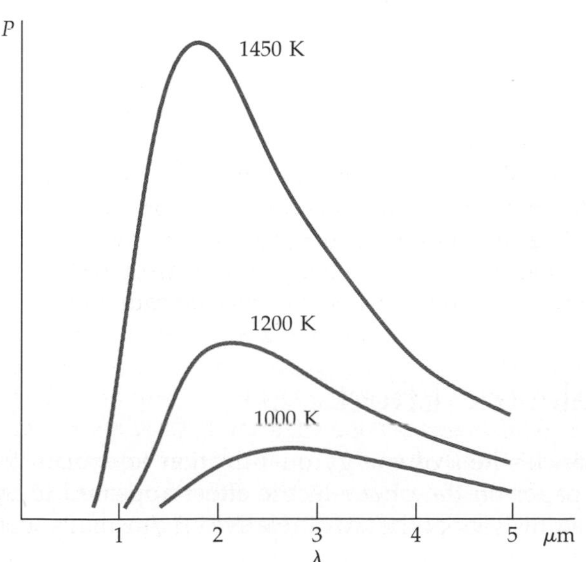

Figure 35-2 Spectral distribution of radiation from a blackbody for three different temperatures.

also approaches zero, but the calculated function approaches infinity because it is proportional to λ^{-4}. Thus, according to the classical calculation, blackbodies radiate an infinite amount of energy concentrated in the very short wavelengths. This result was known as the **ultraviolet catastrophe.**

In 1900, the German physicist Max Planck announced that by making a strange modification in the classical calculation he could derive a function $P(\lambda,T)$ that agreed with the experimental data at all wavelengths. Planck's result is shown in Figure 35-3 along with experimental data and the Rayleigh–Jeans law. Planck first found an empirical function that fit the data and then searched for a way to modify the usual calculation. He found that he could "derive" this function if he made the unusual assumption that the energy emitted and absorbed by the blackbody was not continuous but was instead emitted or absorbed in discrete packets or **quanta.** Planck found that the size of an energy quantum is proportional to the frequency of the radiation:

Quantization of energy of radiation

$$E = hf \tag{35-2}$$

where h is the proportionality constant now known as **Planck's constant.** The value of h was determined by Planck by fitting his function to the experimentally obtained data. The accepted value of this constant is now

$$h = 6.626 \times 10^{-34}\ \text{J·s} = 4.136 \times 10^{-15}\ \text{eV·s} \tag{35-3}$$

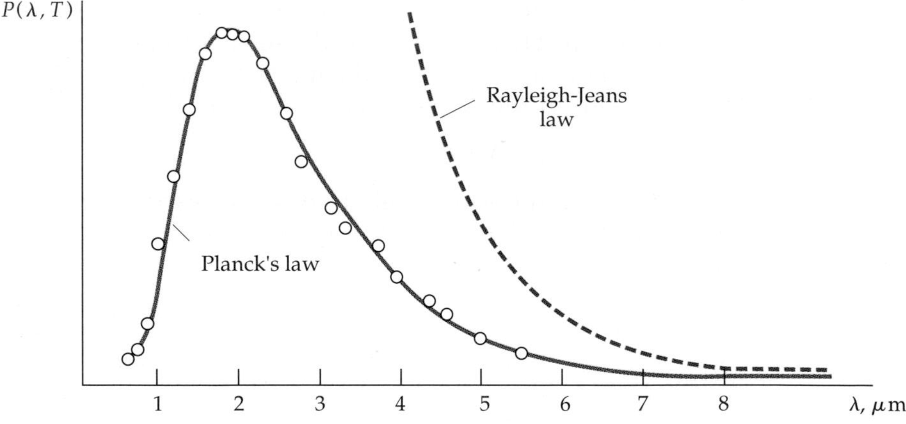

Figure 35-3 Spectral distribution of blackbody radiation versus wavelength at T = 1600 K. The classical theoretical calculation gives the Rayleigh–Jeans law, which agrees with experimental results at very long wavelengths but disagrees with them violently at short wavelengths.

Planck was unable to fit the constant h into the framework of classical physics. The fundamental importance of his assumption of energy quantization, implied by Equation 35-2, was not generally appreciated until Einstein applied similar ideas to explain the photoelectric effect and suggested that quantization is a fundamental property of electromagnetic radiation.

35-2 The Photoelectric Effect

In 1905, Einstein used Planck's idea of energy quantization to explain the photoelectric effect. (His paper on the photoelectric effect appeared in the same journal that contained his special theory of relativity.) Einstein's work marked the beginning of quantum theory, and for it he received the Nobel prize for physics. Whereas Planck looked at energy quantization in his blackbody-radiation theory as a calculational device, Einstein made the bold suggestion that energy quantization is a fundamental property of electromagnetic energy. Three years later, he applied the idea of energy quantization to molecular energies to clear up another puzzle in physics—the discrepancy between the specific heats calculated from the equipartition theorem and those observed experimentally at low temperatures. Later, the ideas of energy quantization were applied to atomic energies by Niels Bohr in the first explanation of atomic spectra.

The photoelectric effect was discovered by Hertz in 1887 and was studied by Lenard in 1900. Figure 35-4 shows a schematic diagram of the basic apparatus. When light is incident on the clean metal surface of the cathode C, electrons are emitted. If some of these electrons strike the anode A, there is a current in the external circuit. The number of emitted electrons that reach the anode can be increased or decreased by making the anode positive or negative with respect to the cathode. Let V be the difference in potential between the cathode and the anode. Figure 35-5 shows the current versus V for two values of the intensity of the light incident on the cathode. When V is positive, the electrons are attracted to the anode. At sufficiently large values of V, all the emitted electrons reach the anode, and the current is at its maximum value. A further increase in V does not affect the current. Lenard observed that the maximum current is proportional to the light intensity. When V is negative, the electrons are repelled from the anode. Only electrons with initial kinetic energies $\frac{1}{2}mv^2$ that are greater than $|eV|$ can then reach the anode. From Figure 35-5, we can see that if V is less than $-V_0$, no electrons reach the anode. The potential V_0 is called the **stopping potential.** It is related to the maximum kinetic energy of the emitted electrons by

$$(\tfrac{1}{2}mv^2)_{\max} = eV_0$$

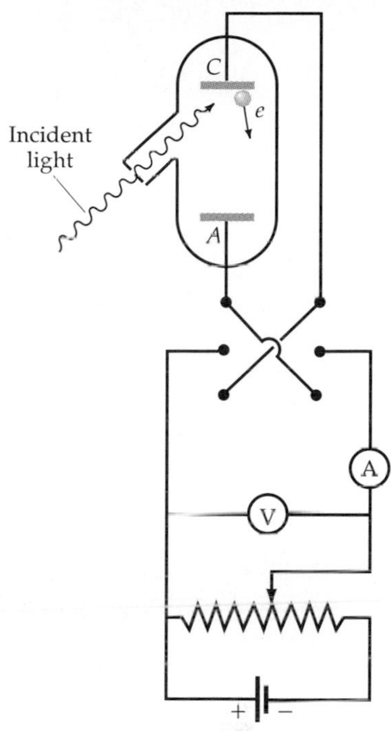

Figure 35-4 Schematic drawing of the apparatus for studying the photoelectric effect. Light strikes the cathode C and ejects electrons. The number of electrons that reach the anode A is measured by the current in the ammeter. The anode can be made positive or negative with respect to the cathode to attract or repel the electrons.

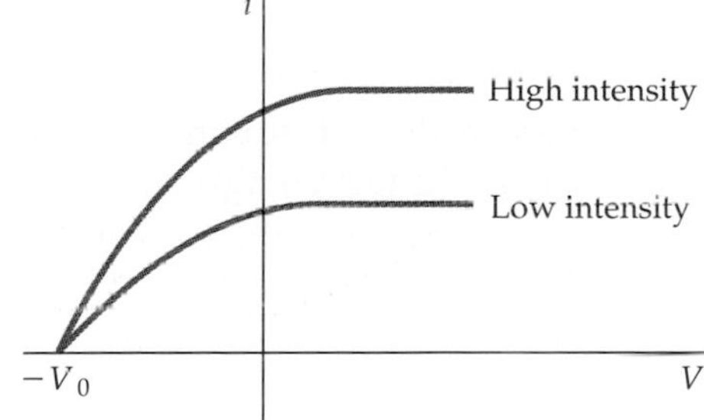

Figure 35-5 Photoelectric current i versus voltage V for two values of light intensity. There is no current when V is less than $-V_0$. The saturation current observed for large values of V is proportional to the intensity of the incident light.

The experimental result that V_0 is independent of the intensity of the incident light was surprising. Classically, increasing the rate of light energy falling on the cathode should increase the energy absorbed by an electron and should therefore increase the maximum kinetic energy of the electrons emitted. Apparently, this is not what happens. In 1905, Einstein demonstrated that this experimental result can be explained if light energy is not distributed continuously in space but rather is quantized in small bundles called **photons.** The energy of each photon is hf, where f is the frequency and h is Planck's constant. An electron emitted from a metal surface exposed to light receives its energy from a single photon. When the intensity of light of a given frequency is increased, more photons fall on the surface in unit time, but the energy absorbed by each electron is unchanged. If ϕ is the energy necessary to remove an electron from a metal surface, the maximum kinetic energy of the electrons emitted will be

Einstein's photoelectric equation

$$(\tfrac{1}{2}mv^2)_{max} = eV_0 = hf - \phi \tag{35-4}$$

The quantity ϕ, called the **work function,** is a characteristic of the particular metal. Some electrons will have kinetic energies less than $hf - \phi$ because of the loss of energy from traveling through the metal. Equation 35-4 is known as **Einstein's photoelectric equation.** From it, we can see that the slope of V_0 versus f should equal h/e.

Einstein's photoelectric equation was a bold prediction, for at the time it was made there was no evidence that Planck's constant had any applicability outside of blackbody radiation, and there were no experimental data on the stopping potential V_0 as a function of frequency. Experimental verification of Einstein's theory was quite difficult. Careful experiments by R. C. Millikan, reported first in 1914 and then in more detail in 1916, showed that Einstein's equation was correct and that measurements of h agreed with the value found by Planck. Figure 35-6 shows a plot of Millikan's data.

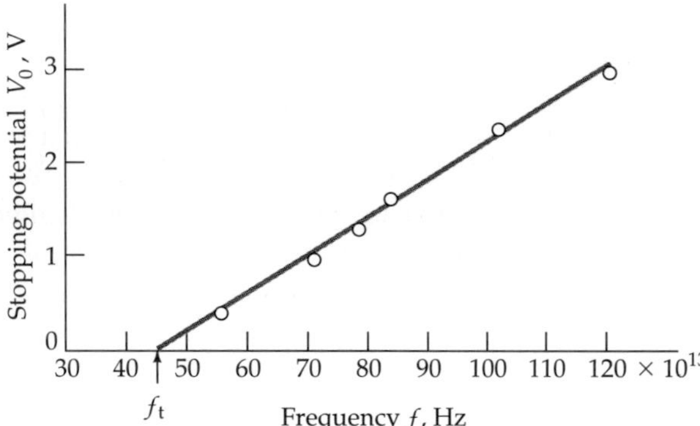

Figure 35-6 Millikan's data for the stopping potential V_0 versus frequency f for the photoelectric effect. The data fall on a straight line that has a slope h/e, as predicted by Einstein a decade before the experiment was performed.

Photons with frequencies less than a **threshold frequency** f_t, and therefore with wavelengths greater than a **threshold wavelength** λ_t, do not have enough energy to eject an electron from a particular metal. The threshold frequency and the corresponding threshold wavelength can be related to the work function ϕ by setting the maximum kinetic energy of the electrons equal to zero in Equation 35-4. Then

$$\phi = hf_t = \frac{hc}{\lambda_t} \tag{35-5}$$

Work functions for metals are typically a few electron volts. Since wavelengths are usually given in nanometers and energies in electron volts, it is useful to have the value of hc in electron volt–nanometers:

$$hc = (4.14 \times 10^{-15} \text{ eV·s})(3 \times 10^8 \text{ m/s}) = 1.24 \times 10^{-6} \text{ eV·m}$$

or

$$hc = 1240 \text{ eV·nm} \tag{35-6}$$

Example 35-1

Calculate the photon energy for light of wavelengths 400 nm (violet) and 700 nm (red). These are approximately the extreme wavelengths in the visible spectrum.

A collection of photomultiplier tubes used to detect very weak light. The face of each tube is a photosensitive area that emits electrons, via the photoelectric effect, when struck by photons. Each electron is accelerated and strikes a metal electrode, resulting in the emission of several more electrons—which are in turn accelerated and strike other electrodes. The electron beam cascades down the tube until it strikes the anode and produces a measurable electric current.

Using Equation 35-2, we have

$$E = hf = \frac{hc}{\lambda} = \frac{1240\ \text{eV}\cdot\text{nm}}{400\ \text{nm}} = 3.1\ \text{eV}$$

for $\lambda = 400$ nm. For $\lambda = 700$ nm, the photon energy is 4/7 that for $\lambda = 400$ nm or 1.77 eV. We can see from these calculations that visible light contains photons with energies that range from about 1.8 to 3.0 eV.

Example 35-2

The intensity of sunlight at the earth's surface is approximately 1400 W/m^2. Assuming the average photon energy is 2 eV (corresponding to a wavelength of about 600 nm), calculate the number of photons that strike an area of 1 cm^2 in one second.

Since 1 watt is 1 joule per second, the energy striking the earth's surface in one second is 1400 J/m^2. The energy per second per square centimeter is then

$$\frac{1400\ \text{J}}{\text{m}^2} \times \frac{1\ \text{m}^2}{(100\ \text{cm})^2} = 0.14\ \text{J/cm}^2$$

If N is the number of 2-eV photons that together have a total energy of 0.14 J, we have

$$N(2\ \text{eV}) = 0.14\ \text{J}$$

$$N = \frac{0.14\ \text{J}}{2\ \text{eV}} \times \frac{1\ \text{eV}}{1.6 \times 10^{-19}\ \text{J}}$$

$$= 4.38 \times 10^{17}\ \text{photons}$$

This is an enormous number. In most everyday situations, the number of photons is so great that a few more or less make no difference. That is, quantization is not noticed.

Example 35-3

The threshold wavelength for potassium is 564 nm. (*a*) What is the work function for potassium? (*b*) What is the stopping potential when light of wavelength 400 nm is incident on potassium?

(*a*) From Equation 35-5, we have for the work function

$$\phi = hf_t = \frac{hc}{\lambda_t} = \frac{1240 \text{ eV}\cdot\text{nm}}{564 \text{ nm}} = 2.20 \text{ eV}$$

(*b*) The energy of a photon with a wavelength of 400 nm was calculated in Example 35-1 to be 3.1 eV. The maximum kinetic energy of the emitted electrons is then

$$(\tfrac{1}{2}mv^2)_{\max} = eV_0 = hf - \phi = 3.10 \text{ eV} - 2.20 \text{ eV} = 0.90 \text{ eV}$$

The stopping potential is therefore 0.90 V.

Exercise

Find the energy of a photon corresponding to electromagnetic radiation in the FM radio band of wavelength 3 m. (Answer: 4.13×10^{-7} eV)

Exercise

Find the wavelength of a photon whose energy is (*a*) 0.1 eV, (*b*) 1 keV, and (*c*) 1 MeV. [Answers: (*a*) 12.4 μm, (*b*) 1.24 nm, (*c*) 1.24 pm]

Another interesting feature of the photoelectric effect is the absence of any lag between the time the light first strikes the metal and the time the electrons appear. In the classical theory, given the intensity (the power per unit area), the time it takes for enough energy to fall on the area of an atom to eject an electron can be calculated. However, even when the intensity is so small that such a calculation gives a time lag of hours, essentially no time lag is observed. The explanation of this result is simple. When the intensity is low, the number of photons hitting the metal per unit time is very small, but each photon has enough energy to eject an electron. There is therefore a good chance that one photon will be absorbed immediately. The classical calculation gives the correct *average* number of electrons ejected per unit time.

35-3 X Rays

While working with a cathode-ray tube in 1895, W. Röntgen discovered that "rays" from the tube could pass through materials that were opaque to light and activate a fluorescent screen or photographic film. These rays originated from a point where the electrons in the tube hit a target within the tube or the glass tube itself. Röntgen was not able to deflect these rays in a magnetic field, as would be expected if they were charged particles, nor was he able to observe diffraction or interference, as would be expected if they were waves. He therefore gave the rays the somewhat mysterious name **x rays.** Röntgen investigated these rays extensively and found that all materials were transparent to them to some degree and that the degree of transparency decreased with increasing density of the material. This fact led to the medical use of x rays within months after Röntgen's first paper. Röntgen was the first recipient of the Nobel Prize for physics in 1901.

Since classical electromagnetic theory predicts that electric charges will radiate electromagnetic waves when they are accelerated (or decelerated), it

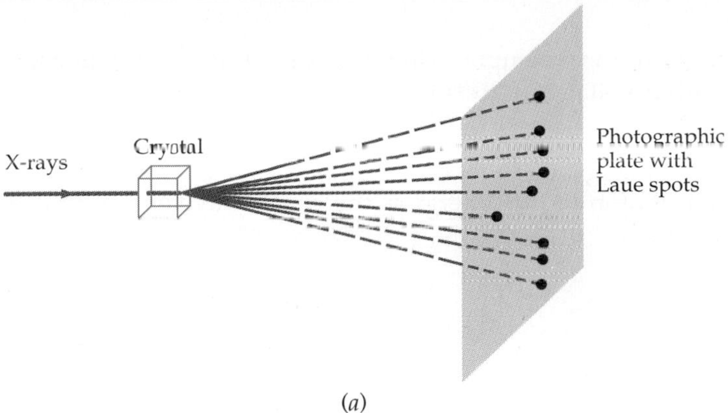

(a)

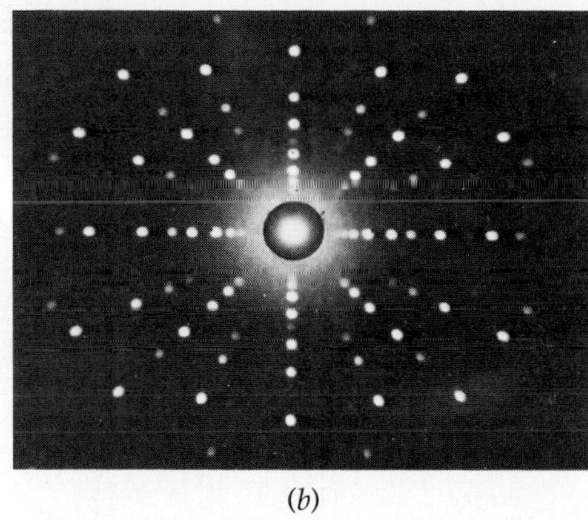

(b)

Figure 35-7 (a) Schematic diagram of the Laue experiment. The crystal acts as a three-dimensional grating that diffracts the x-ray beam and produces a regular array of spots called a *Laue pattern* on a photographic plate. (b) A modern Laue x-ray diffraction pattern from a niobium diboride crystal using 20-kV x rays from a molybdenum target.

was natural to assume that x rays are electromagnetic waves produced when electrons decelerate as they are stopped by a target. A few years later, a slight broadening of an x-ray beam after it passed through slits a few thousandths of a millimeter wide was observed. This was assumed to be due to diffraction, and the wavelength of x rays was estimated to be about 0.1 nm. In 1912, M. Laue suggested that, since the wavelengths of x rays were of the same order of magnitude as the spacing of the atoms in a crystal, the regular array of atoms in a crystal might act as a three-dimensional grating for the diffraction of x rays. Acting on this suggestion, W. Friedrich and P. Knipping allowed a collimated beam of x rays to pass through a crystal and strike a photographic plate (Figure 35-7a). In addition to the central beam, they observed a regular array of spots like those shown in Figure 35-7b. From an analysis of the positions of the spots, they were able to calculate that their x-ray beam had wavelengths ranging from about 0.01 to 0.05 nm. This important experiment confirmed two important assumptions: (1) x rays are electromagnetic radiation and (2) the atoms in crystals are arranged in a regular array.

Figure 35-8 shows a plot of intensity versus wavelength for the spectrum emitted from a typical x-ray tube, in which a target (molybdenum in this case) is bombarded with electrons. The spectrum consists of a series of sharp lines called the **characteristic spectrum** superimposed on a continuous spectrum called a **bremsstrahlung spectrum** (from the German for "braking radiation"). The line spectrum is characteristic of the target material and varies from element to element. It is similar to the optical spectrum of the elements except the x-ray spectrum involves transitions of the inner atomic electrons whereas the optical spectrum involves transitions of the outer atomic electrons. We will discuss both the optical spectrum and the characteristic x-ray spectrum in Chapter 37 in the extended version of this book. The continuous bremsstrahlung spectrum is produced by the rapid deceleration of the bombarding electrons when they crash into the target. If the voltage across the x-ray tube is V, the maximum kinetic energy of the electrons is eV when they hit the target. Often several photons are emitted as an electron slows down. However, sometimes just one photon with the maximum energy eV is emitted. Since the wavelength of a photon varies inversely with its energy ($\lambda = hc/hf = hc/E$), the minimum wavelength in the bremsstrahlung spectrum corresponds to a photon with the maximum energy eV. The minimum wavelength is called the **cutoff wavelength** and is labeled λ_m in the figure. The cutoff wavelength is related to the voltage of the x-ray tube by

$$\lambda_m = \frac{hc}{E} = \frac{hc}{eV} \qquad 35\text{-}7$$

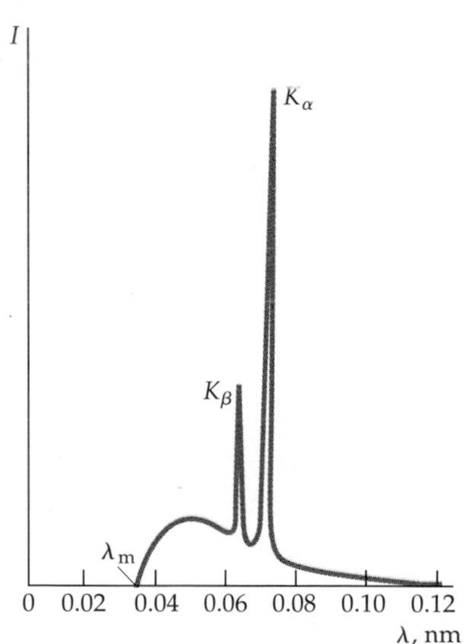

Figure 35-8 X-ray spectrum of molybdenum. The sharp peaks labeled K_α and K_β are characteristic of the target element. The cutoff wavelength λ_m is independent of the target element and is related to the voltage of the x-ray tube by $\lambda_m = hc/eV$.

Example 35-4

What is the minimum wavelength of the x rays emitted by a television picture tube with a voltage of 2000 V?

The maximum kinetic energy of the electrons is 2000 eV, so this will be the maximum energy of the photons in the x-ray spectrum. The wavelength of a photon of this energy is the cutoff wavelength, which from Equation 35-7 is

$$\lambda_m = \frac{hc}{E} = \frac{1240 \text{ eV}\cdot\text{nm}}{2000 \text{ eV}} = 0.62 \text{ nm}$$

Exercise

An x-ray tube operates at a potential of 30 kV. What is the minimum wavelength of the continuous x-ray spectrum from this tube? (Answer: 0.041 nm)

35-4 Compton Scattering

Further evidence of the correctness of the photon concept was furnished by Arthur H. Compton, who measured the scattering of x rays by free electrons. According to classical theory, when an electromagnetic wave of frequency f_1 is incident on material containing charges, the charges will oscillate with this frequency and will reradiate electromagnetic waves of the same frequency. Compton pointed out that if this interaction were described as a scattering process involving a collision between a photon and an electron, the electron would recoil and thus absorb energy. The scattered photon would then have less energy and therefore a lower frequency than the incident photon.

According to classical theory, the energy and momentum of an electromagnetic wave are related by

$$E = pc \qquad 35\text{-}8$$

This result is consistent with the relativistic expression relating the energy and momentum of a particle (Equation 34-34),

$$E^2 = p^2c^2 + (mc^2)^2$$

if the mass m of the photon is assumed to be zero. Figure 35-9 shows the geometry of a collision between a photon of wavelength λ_1 and an electron initially at rest. Compton related the scattering angle θ to the incident and scattered wavelengths λ_1 and λ_2 by treating the scattering as a relativistic-mechanics problem and using the conservation of energy and momentum. Let $\mathbf{p}_1$ be the momentum of the incident photon, $\mathbf{p}_2$ be that of the scattered photon, and $\mathbf{p}_e$ be that of the recoiling electron. The conservation of momentum gives

$$\mathbf{p}_1 = \mathbf{p}_2 + \mathbf{p}_e \qquad 35\text{-}9$$

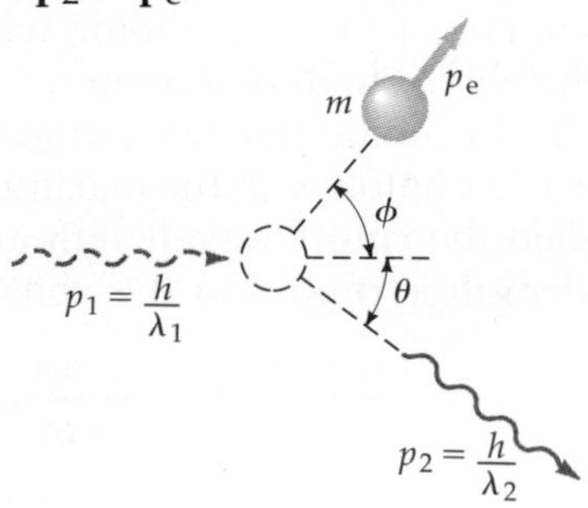

Figure 35-9 Compton scattering of an x ray by an electron. The scattered photon has less energy and therefore a greater wavelength than the incident photon because of the recoil energy of the electron. The change in wavelength is found from conservation of energy and momentum.

or

$$\mathbf{p}_e = \mathbf{p}_1 - \mathbf{p}_2$$

Taking the dot product of each side with itself, we obtain

$$p_e^2 = p_1^2 + p_2^2 - 2\mathbf{p}_1 \cdot \mathbf{p}_2$$

or

$$p_e^2 = p_1^2 + p_2^2 - 2p_1p_2 \cos \theta \qquad 35\text{-}10$$

The energy before the collision is $p_1c + mc^2$, where mc^2 is the rest energy of the electron. After the collision, the electron has energy $\sqrt{(mc^2)^2 + p_e^2c^2}$. Conservation of energy then gives

$$p_1c + mc^2 = p_2c + \sqrt{(mc^2)^2 + p_e^2c^2} \qquad 35\text{-}11$$

Compton eliminated the electron momentum p_e from Equations 35-10 and 35-11 and expressed the photon momenta in terms of the wavelengths to obtain an equation relating the incident and scattered wavelengths λ_1 and λ_2 and the angle θ. The algebraic details are left as a problem (see Problem 60). Compton's result is

$$\lambda_2 - \lambda_1 = \frac{h}{mc}(1 - \cos \theta) \qquad 35\text{-}12$$

The change in wavelengths is independent of the original wavelength. The quantity h/mc depends only on the mass of the electron. It has the dimension of length and is called the **Compton wavelength.** Its value is

$$\lambda_C = \frac{h}{mc} = \frac{hc}{mc^2} = \frac{1240 \text{ eV}\cdot\text{nm}}{5.11 \times 10^5 \text{ eV}} = 2.43 \times 10^{-12} \text{ m} = 2.43 \text{ pm} \qquad 35\text{-}13$$

Because $\lambda_2 - \lambda_1$ is small, it is difficult to observe unless λ_1 is so small that the fractional change $(\lambda_2 - \lambda_1)/\lambda_1$ is appreciable. Compton used x rays of wavelength 71.1 pm. The energy of a photon of this wavelength is $E = hc/\lambda = (1240 \text{ eV}\cdot\text{nm})/(0.0711 \text{ nm}) = 17.4$ keV. Since this is much greater than the binding energy of the valence electrons in atoms (which is of the order of a few electron volts), these electrons can be considered to be essentially free. Compton's experimental results for $\lambda_2 - \lambda_1$ as a function of the scattering angle θ agreed with Equation 35-12, thereby confirming the correctness of the photon concept.

Example 35-5

Calculate the percentage change in wavelength observed in the Compton scattering of 20-keV photons at $\theta = 60°$.

The change in wavelength at $\theta = 60°$ is given by Equation 35-12:

$$\lambda_2 - \lambda_1 = \lambda_C(1 - \cos \theta) = (2.43 \text{ pm})(1 - \cos 60°) = 1.22 \text{ pm}$$

The wavelength of the incident 20-keV photons is

$$\lambda_1 = \frac{1240 \text{ eV}\cdot\text{nm}}{20{,}000 \text{ eV}} = 0.062 \text{ nm} = 62 \text{ pm}$$

The percentage change in wavelength is thus

$$\frac{\lambda_2 - \lambda_1}{\lambda_1} = \frac{1.22 \text{ pm}}{62 \text{ pm}} \times 100\% = 1.97\%$$

35-5 Quantization of Atomic Energies: The Bohr Model

The most famous application of energy quantization to microscopic systems was that of Niels Bohr. In 1913, Bohr proposed a model of the hydrogen atom that had spectacular success in calculations of the wavelengths of the lines in the known hydrogen spectrum and in predicting new lines (later found experimentally) in the infrared and ultraviolet spectra.

Near the turn of the century, much data was collected on the emission of light by atoms in a gas when they are excited by an electric discharge. Viewed through a spectroscope with a narrow-slit aperture, this light appears as a discrete set of lines of different colors or wavelengths; the spacing and intensities of the lines are characteristic of the element. It was possible to determine the wavelengths of these lines accurately, and much effort went into finding regularities in the spectra. In 1884, a Swiss schoolteacher, Johann Balmer, found that the wavelengths of some of the lines in the spectrum of hydrogen can be represented by the formula

$$\lambda = (364.6 \text{ nm}) \frac{m^2}{m^2 - 4} \qquad 35\text{-}14$$

where m is an integer that takes on the values $m = 3, 4, 5, \ldots$. Figure 35-10 shows the set of spectral lines of hydrogen, now known as the **Balmer series,** whose wavelengths are given by Equation 35-14.

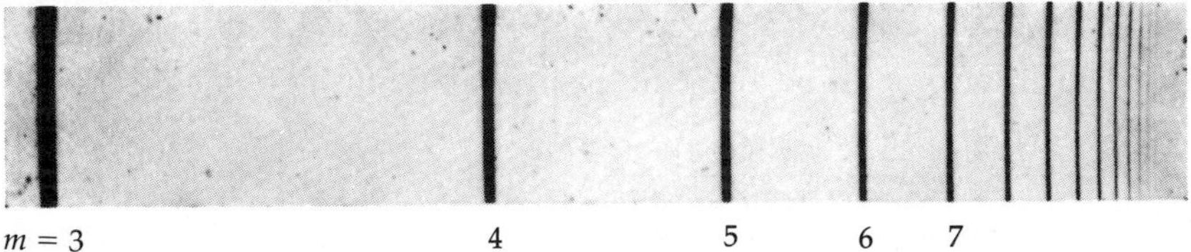

Figure 35-10 The Balmer series for light emitted from the hydrogen atom. The wavelengths of these lines are given by Equation 35-14 for different values of the integer m.

Balmer suggested that his formula might be a special case of a more general expression that would be applicable to the spectra of other elements. Such an expression, found by Johannes R. Rydberg and Walter Ritz and known as the **Rydberg–Ritz formula,** gives the reciprocal wavelength as

$$\frac{1}{\lambda} = RZ^2\left(\frac{1}{n_2^2} - \frac{1}{n_1^2}\right) \qquad n_1 > n_2 \qquad 35\text{-}15$$

This formula is valid not only for hydrogen, with atomic number $Z = 1$, but also for heavier atoms of nuclear charge Ze from which all electrons but one have been removed. R, called the **Rydberg constant** or simply the **Rydberg,** is the same for all spectral series of the same element and varies only slightly in a regular way from element to element. For very massive elements, R approaches the value

$$R_\infty = 10.97373 \ \mu\text{m}^{-1} \qquad 35\text{-}16$$

If we take the reciprocal of Equation 35-14 for the Balmer series, we obtain

$$\frac{1}{\lambda} = \frac{1}{364.6 \text{ nm}}\left(\frac{m^2 - 4}{m^2}\right) = \frac{1}{364.6 \text{ nm}}\left(\frac{1}{1} - \frac{4}{m^2}\right)$$
$$= \frac{4}{364.6 \text{ nm}}\left(\frac{1}{4} - \frac{1}{m^2}\right) = 10.97 \ \mu\text{m}^{-1}\left(\frac{1}{2^2} - \frac{1}{m^2}\right)$$

We can thus see that the Balmer formula is indeed a special case of the Rydberg–Ritz formula (Equation 35-15) for hydrogen with $n_2 = 2$ and $n_1 =$

m. The Rydberg–Ritz formula and various modifications of it were very successful in predicting other spectra. For example, hydrogen lines outside the visible spectrum were predicted and found. Setting $n_2 = 1$ in Equation 35-15 leads to a series in the ultraviolet region called the *Lyman series,* whereas setting $n_2 = 3$ leads to the *Paschen series* in the infrared region.

Many attempts were made to construct a model of the atom that would yield these formulas for its radiation spectrum. The most popular early model, developed by J. J. Thomson, considered electrons to be embedded in various arrangements in some kind of fluid that contained most of the mass of the atom and had enough positive charge to make the atom electrically neutral. Thomson's model, called the "plum pudding" model, is illustrated in Figure 35-11. Since classical electromagnetic theory predicted that a charge oscillating with a frequency f would radiate light of the same frequency, Thomson searched for configurations of electrons that were stable and had normal modes of vibration with frequencies equal to those of the spectrum of the atom. A difficulty with this model and all others was that electric forces alone cannot produce stable equilibrium. Thomson was unable to find a configuration of electrons that predicted the observed frequencies for any atom.

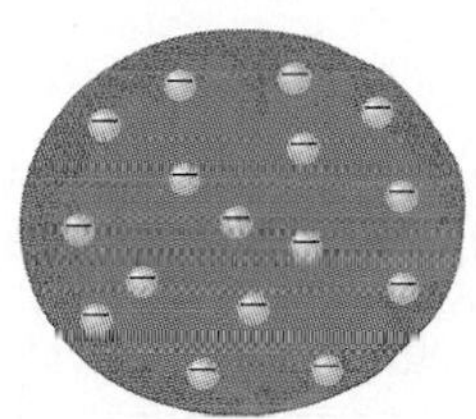

Figure 35-11 J. J. Thomson's "plum pudding" model of the atom. In this model, the negative electrons are embedded in a fluid of positive charge. For a given configuration of electrons in such a system, the resonance frequencies of the oscillations of the electrons can be calculated. According to classical theory, the atom should radiate light with a frequency equal to the frequency of oscillation of the electrons. Thomson could not find any configuration of electrons that would give frequencies that agreed with the measured frequencies of the spectrum of any atom.

The Thomson model was essentially ruled out by a set of experiments performed by H. W. Geiger and E. Marsden under the supervision of E. Rutherford around 1911 in which alpha particles from radioactive radium were scattered by atoms in a gold foil. Rutherford showed that the number of alpha particles scattered at large angles could not be accounted for by an atom in which the positive charge was distributed throughout the atomic volume (known to be about 0.1 nm in diameter). The findings required that the positive charge and most of the mass of the atom be concentrated in a very small region, now called the nucleus, with a diameter of the order of 10^{-6} nm = 1 fm. (Before the establishment of the SI, the femtometer, 1 fm = 10^{-15} m, was called a *fermi* after the Italian physicist Enrico Fermi.)

Niels Bohr, who was working in the Rutherford laboratory at the time, proposed a model of the hydrogen atom that combined the work of Planck, Einstein, and Rutherford and successfully predicted the observed spectra. Bohr assumed that the electron in the hydrogen atom moved under the influence of the Coulomb attraction between it and the positive nucleus according to classical mechanics, which predicts circular or elliptical orbits with the force center at one focus, as in the motion of the planets around the sun. For simplicity he chose a circular orbit as shown in Figure 35-12. Although mechanical stability is achieved because the Coulomb attractive force provides the centripetal force necessary for the electron to remain in orbit, such an atom would be unstable electrically according to classical theory because the electron must accelerate when moving in a circle and must therefore radiate electromagnetic energy of a frequency equal to that of its motion. According to classical electromagnetic theory, such an atom would quickly collapse because the electron would spiral into the nucleus as it radiates away its energy.

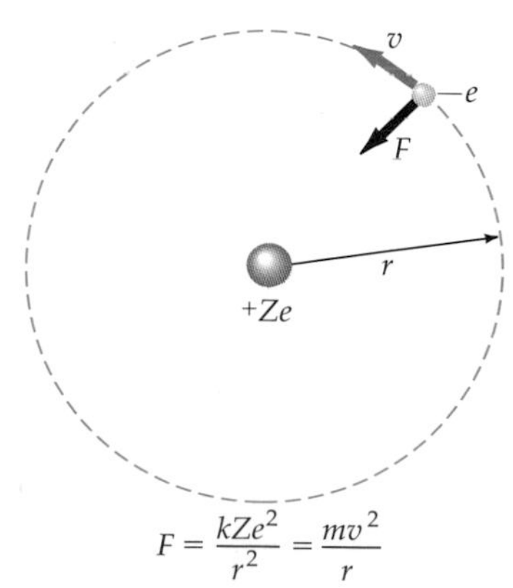

Figure 35-12 An electron of charge $-e$ traveling in a circular orbit of radius r around the nuclear charge $+Ze$. The attractive electrical force kZe^2/r^2 provides the centripetal force to hold the electron in its orbit.

Bohr's first postulate: nonradiating orbits

Bohr "solved" this difficulty by modifying the laws of electromagnetism and *postulating* that the electron could only move in certain nonradiating orbits. This idea is referred to as Bohr's first postulate. He called these stable orbits **stationary states.** The atom radiates only when the electron somehow makes a transition from one stationary state to another. The frequency of the radiation is not the frequency of the electron's motion in either stable orbit. Instead, it is related to the energies of the orbits by

Bohr's second postulate: photon frequency from energy conservation

$$f = \frac{E_i - E_f}{h} \qquad 35\text{-}17$$

where h is Planck's constant and E_i and E_f are the total energies in the initial and final orbits, respectively. This assumption, which is equivalent to the assumption of conservation of energy with the emission of a photon, is a key one in the Bohr theory because it deviates from classical theory, which requires the frequency of radiation to be that of the motion of the charged particle.

If the nuclear charge is $+Ze$ and there is only one electron of charge $-e$, the potential energy at a distance r (see Equation 20-8) is

$$U = -\frac{kZe^2}{r}$$

where k is the Coulomb constant. (For hydrogen, $Z = 1$, but it is convenient not to specify Z at this time so that the results can be applied to other hydrogen-like atoms.) The total energy of the electron moving in a circular orbit with a speed v is then

$$E = \frac{1}{2}mv^2 + U = \frac{1}{2}mv^2 - \frac{kZe^2}{r}$$

The kinetic energy can be obtained as a function of r by using Newton's second law $F = ma$. Setting the Coulomb attractive force equal to the mass times the centripetal acceleration, we obtain

$$\frac{kZe^2}{r^2} = m\frac{v^2}{r}$$

or

$$\frac{1}{2}mv^2 = \frac{1}{2}\frac{kZe^2}{r} \qquad 35\text{-}18$$

For circular orbits, the kinetic energy equals half the magnitude of the potential energy, a result that holds for circular motion in any inverse-square force field. The total energy is then

$$E = \frac{1}{2}\frac{kZe^2}{r} - \frac{kZe^2}{r} = -\frac{1}{2}\frac{kZe^2}{r} \qquad 35\text{-}19$$

Using Equation 35-17 for the frequency of radiation emitted when the electron changes from orbit 1 of radius r_1 to orbit 2 of radius r_2, we obtain

$$f = \frac{E_1 - E_2}{h} = \frac{1}{2}\frac{kZe^2}{h}\left(\frac{1}{r_2} - \frac{1}{r_1}\right) \qquad 35\text{-}20$$

To obtain the Rydberg–Ritz formula, $f = c/\lambda = cR(1/n_2^2 - 1/n_1^2)$, it is evident that the radii of stable orbits must be proportional to the squares of integers. Bohr searched for a quantum condition for the radii of the stable orbits that would yield this result. After much trial and error, he found that he could obtain it if he postulated that the angular momentum of the electron in a stable orbit equals an integer times Planck's constant divided by 2π. Since the angular momentum of a circular orbit is just mvr, this postulate is

$$mvr = \frac{nh}{2\pi} = n\hbar \qquad 35\text{-}21$$

Bohr's third postulate: quantized angular momentum

where

$$\hbar = \frac{h}{2\pi} = 1.05 \times 10^{-34}\ \text{J·s}$$

(The constant $\hbar = h/2\pi$, read "h bar", is often more convenient to use than h itself, just as the angular frequency $\omega = 2\pi f$ is often more convenient to use than the frequency f.) We can determine r by eliminating v between Equations 35-18 and 35-21. Solving Equation 35-21 for v and squaring, and using Equation 35-18, we obtain

$$v^2 = n^2\frac{\hbar^2}{m^2r^2} = \frac{kZe^2}{mr}$$

Solving for r we obtain

$$r = n^2\frac{\hbar^2}{mkZe^2} = n^2\frac{a_0}{Z} \qquad 35\text{-}22$$

where

$$a_0 = \frac{\hbar^2}{mke^2} \approx 0.0529\ \text{nm} \qquad 35\text{-}23$$

Bohr radius

is called the first **Bohr radius.** Combining Equations 35-22 and 35-20, we get

$$f = Z^2\frac{mk^2e^4}{4\pi\hbar^3}\left(\frac{1}{n_2^2} - \frac{1}{n_1^2}\right) \qquad 35\text{-}24$$

If we compare this expression for $f = c/\lambda$ with the empirical Rydberg–Ritz formula (Equation 35-15), we obtain for the Rydberg constant

$$R = \frac{mk^2e^4}{4\pi c\hbar^3} \qquad 35\text{-}25$$

Using the values of m, e, and $\hbar$ known in 1913, Bohr calculated R and found his result to agree (within the limits of the uncertainties of the constants) with the value obtained from spectroscopy. Figure 35-13 illustrates the Bohr model of the hydrogen atom.

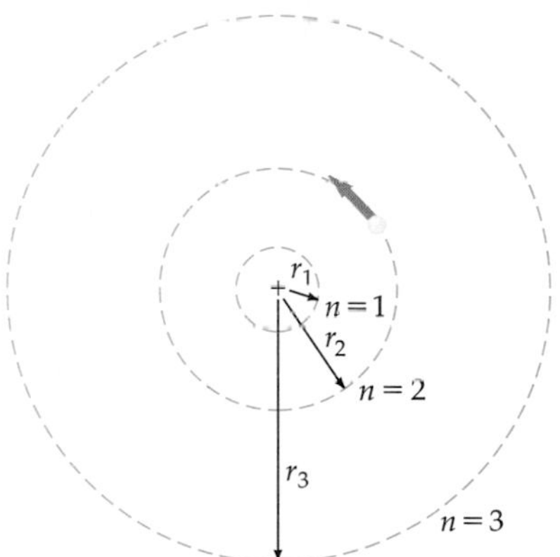

Figure 35-13 Stable orbits in the Bohr model of the hydrogen atom. The radii of the stable orbits are given by $r_n = n^2a_0$, where n is an integer and a_0 is the smallest radius.

The possible values of the energy of the hydrogen atom predicted by the Bohr model and given by Equation 35-19, when r is given by Equation 35-22, are

$$E_n = -\frac{k^2e^4m}{2\hbar^2}\frac{Z^2}{n^2} = -Z^2\frac{E_0}{n^2} \qquad 35\text{-}26$$

Energy levels

where

$$E_0 = \frac{k^2e^4m}{2\hbar^2} \approx 13.6\ \text{eV} \qquad 35\text{-}27$$

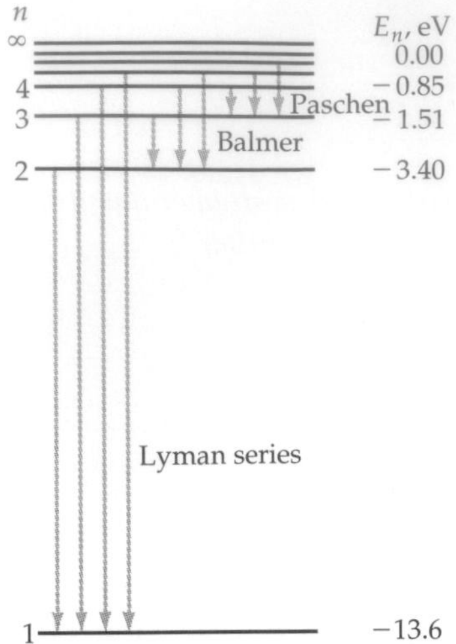

Figure 35-14 An energy-level diagram for hydrogen showing a few of the transitions in the Lyman, Balmer, and Paschen series. The energies of the levels are given by Equation 35-26.

It is sometimes convenient to represent these energies in an **energy-level diagram,** as in Figure 35-14. The lowest energy level is called the ground state. The energy of the hydrogen atom in the ground state is −13.6 eV. The highest energy state E_n is reached as $n \to \infty$ or $r \to \infty$, as can be seen from Equations 35-26 and 35-22. This process of removing the electron from an atom is termed **ionization.** The ionization energy of the hydrogen atom has been measured as 13.6 eV. This is thus the energy with which the electron is bound to the atom.

Various series of radiative transitions are indicated in Figure 35-14 by vertical arrows between the energy levels. The frequency of the light emitted in one of these transitions is the energy difference divided by h (Equation 35-17). At the time of Bohr's paper (1913), the Balmer series, corresponding to $n_2 = 2$ and $n_1 = 3, 4, 5, \ldots$, and the Paschen series, corresponding to $n_2 = 3$ and $n_1 = 4, 5, 6, \ldots$, were known. In 1916, T. Lyman found the series corresponding to $n_2 = 1$, and in 1922 and 1924, F. Brackett and H. A. Pfund, respectively, found series corresponding to $n_2 = 4$ and $n_2 = 5$. As can be determined by computing the wavelengths of these series, only the Balmer series lies in the visible portion of the electromagnetic spectrum.

In our derivations, we have assumed that the electron revolves around a stationary nucleus. This is equivalent to assuming that the nucleus has infinite mass. Since the mass of the hydrogen nucleus is not infinite but about 2000 times that of the electron, a correction must be made for the motion of the nucleus. This correction leads to a very slight dependence of the Rydberg constant, as given in Equation 35-25, on the nuclear mass, in precise agreement with the observed variation.

Example 35-6

Find the energy and wavelength of the line with the longest wavelength in the Lyman series.

From Figure 35-14, we can see that the Lyman series corresponds to transitions ending at the ground-state energy, $E_f = E_1 = -13.6$ eV. Since λ varies inversely with energy, the transition with the longest wavelength is the transition with the lowest energy, which is that from the first excited state $n = 2$ to the ground state $n = 1$. The energy of the first excited state is $E_2 = (-13.6 \text{ eV})/4 = -3.40$ eV. Since this is 10.2 eV above the ground-state energy, the energy of the photon emitted is 10.2 eV. The wavelength of this photon is

$$\lambda = \frac{hc}{\Delta E} = \frac{1240 \text{ eV}\cdot\text{nm}}{10.2 \text{ eV}} = 121.6 \text{ nm}$$

This photon is outside the visible spectrum and in the ultraviolet region. Since all the other lines in the Lyman series have even greater energies and shorter wavelengths, the Lyman series is completely in the ultraviolet region.

Exercise

Find the shortest wavelength for a line in the Lyman series. (Answer: 91.2 nm)

Questions

1. If an electron moves to a larger orbit, does its total energy increase or decrease? Does its kinetic energy increase or decrease?
2. How does the spacing of adjacent energy levels change as n increases?
3. What is the energy of the photon with the shortest wavelength emitted by the hydrogen atom?

35-6 Electron Waves and Quantum Theory

In 1924, a French student, L. de Broglie, suggested in his dissertation that electrons may have wave properties. His reasoning was based on the symmetry of nature. Since light was known to have both wave and particle properties, perhaps matter—especially electrons—might also have both wave and particle characteristics. This suggestion was highly speculative since there was no evidence at that time of any wave aspects of electrons. For the frequency and wavelength of electron waves, de Broglie chose the equations

$$f = \frac{E}{h} \qquad 35\text{-}28$$

$$\lambda = \frac{h}{p} \qquad 35\text{-}29$$

where p is the momentum and E is the energy of the electron. Equation 35-28 is the same as the Planck–Einstein equation for the energy of a photon. Equation 35-29 also holds for photons, as can be seen from

$$\lambda = \frac{c}{f} = \frac{hc}{hf} = \frac{hc}{E}$$

Since the momentum of a photon is related to its energy by $E = pc$, we have

$$\lambda = \frac{hc}{pc} = \frac{h}{p}$$

De Broglie's equations are thought to apply to all matter. However, for macroscopic objects, the wavelengths calculated from Equation 35-29 are so small that it is impossible to observe the usual wave properties of interference or diffraction. Even a particle as small as 1 μg is much too massive for any wave characteristics to be noticed, as we will see in the following example.

Example 35-7

Find the de Broglie wavelength of a particle of mass 10^{-6} g moving with a speed of 10^{-6} m/s.

From Equation 35-29, we have

$$\lambda = \frac{h}{p} = \frac{h}{mv} = \frac{6.63 \times 10^{-34}\ \text{J·s}}{(10^{-9}\ \text{kg})(10^{-6}\ \text{m/s})}$$

$$= 6.63 \times 10^{-19}\ \text{m}$$

Since the wavelength found in Example 35-7 is much smaller than any possible apertures or obstacles (the diameter of the nucleus of an atom is about 10^{-15} m, roughly 10,000 times this wavelength), diffraction or interference of such waves cannot be observed. As we have discussed, the propagation of waves of very small wavelength is indistinguishable from the propagation of particles. Note that the momentum in Example 35-7 is extremely small. A macroscopic particle with a greater momentum would have an even smaller de Broglie wavelength. We therefore do not observe the wave properties of such macroscopic objects as baseballs or billiard balls.

Exercise

Find the de Broglie wavelength of a baseball of mass 0.17 kg moving at 100 km/h. (Answer: 1.4×10^{-34} m)

The situation is different for low-energy electrons. Consider an electron with kinetic energy K. If the electron is nonrelativistic, its momentum is found from

$$K = \frac{p^2}{2m}$$

or

$$p = \sqrt{2mK}$$

Its wavelength is then

$$\lambda = \frac{h}{p} = \frac{h}{\sqrt{2mK}} = \frac{hc}{\sqrt{2mc^2K}}$$

Using $hc = 1240$ eV·nm and $mc^2 = 0.511$ MeV, we obtain

$$\lambda = \frac{1240 \text{ eV·nm}}{\sqrt{2(0.511 \times 10^6 \text{ eV})K}}$$

or

$$\lambda = \frac{1.226}{\sqrt{K}} \text{ nm} \qquad K \text{ in electron volts} \qquad 35\text{-}30$$

From Equation 35-30, we can see that electrons with energies of the order of tens of electron volts have de Broglie wavelengths of the order of nanometers. This is the order of magnitude of the size of the atom and the spacing of atoms in a crystal. Thus, when electrons with energies of the order of 10 eV are incident on a crystal, they are scattered in much the same way as are x rays of the same wavelength.

Exercise

Find the wavelength of an electron whose kinetic energy is 10 eV. (Answer: 0.388 nm)

The crucial test for the existence of wave properties of electrons was the observation of diffraction and interference of electron waves. This was first accomplished accidentally in 1927 by C. J. Davisson and L. H. Germer as they were studying electron scattering from a nickel target at the Bell Telephone Laboratories. After heating the target to remove an oxide coating that had accumulated during an accidental break in the vacuum system, Davisson and Germer found that the intensity of the scattered electrons as a function of the scattering angle showed maxima and minima. Their target had crystallized, and by accident they had observed electron diffraction. They then prepared a target consisting of a single crystal of nickel and investigated this phenomenon extensively. Figure 35-15 illustrates their experiment. Electrons from an electron gun are directed at a crystal and are detected at some ϕ that can be varied. Figure 35-16 shows a typical pattern observed. There is a strong scattering maximum at an angle of 50°. The angle for maximum intensity of scattering of waves from a crystal depends on the wavelength of the waves and the spacing of the atoms in the crystal. Using the known spacing of atoms in their crystal, Davisson and Germer calculated the wavelength that could produce such a maximum and found that it agreed with the de Broglie equation (Equation 35-29) for the electron energy they were using. By varying the energy of the incident electrons, they could vary the electron wavelengths and produce maxima and minima at different locations in the diffraction patterns. In all cases, the measured wavelengths agreed with de Broglie's hypothesis.

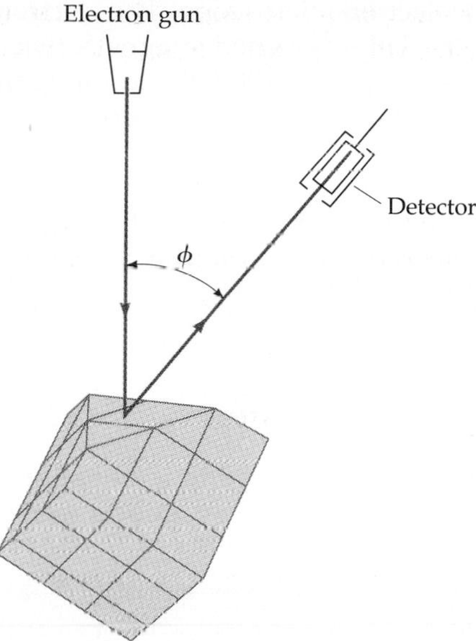

Figure 35-15 The Davisson–Germer experiment. Electrons from the electron gun incident on a crystal are scattered into a detector at some angle ϕ that can be varied.

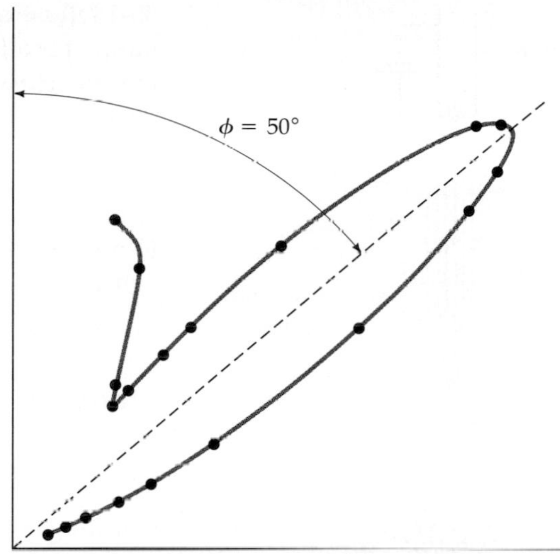

Figure 35-16 Plot of intensity versus angle for the scattered electron in the Davisson–Germer experiment. If this pattern is assumed to be a diffraction–interference pattern, the wavelength of the electrons can be calculated from the known spacing of the atoms in the crystal and the position of the maximum. The result agrees with the de Broglie hypothesis for the wavelength of an electron.

In the same year G. P. Thomson (son of J. J. Thomson) also observed electron diffraction in the transmission of electrons through thin metal foils. A metal foil consists of tiny, randomly oriented crystals. The diffraction pattern resulting from such a foil is a set of concentric circles. Since Thomson performed his experiment, diffraction has been observed for neutrons, protons, and other particles. Figure 35-17*a* to *c* shows the diffraction patterns of x rays, electrons, and neutrons of similar wavelength transmitted through thin metal foils. Figure 35-17*d* shows a diffraction pattern produced by electrons incident on two narrow slits. This experiment is equivalent to Young's famous double-slit diffraction–interference experiment with light. The pattern is identical to that observed with photons of the same wavelength.

Shortly after the wave properties of the electron were demonstrated, it was suggested that electrons rather than light might be used to "see" small objects. Today, the electron microscope is an important research tool. Figure

(*a*)

(*b*)

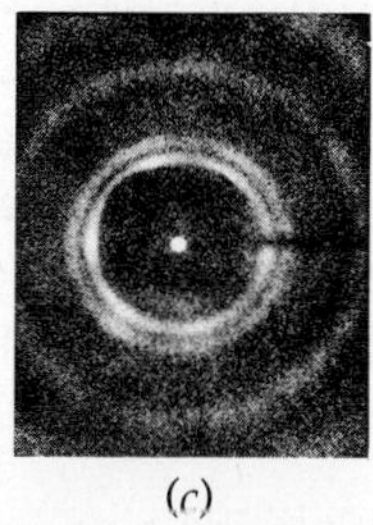

(*c*)

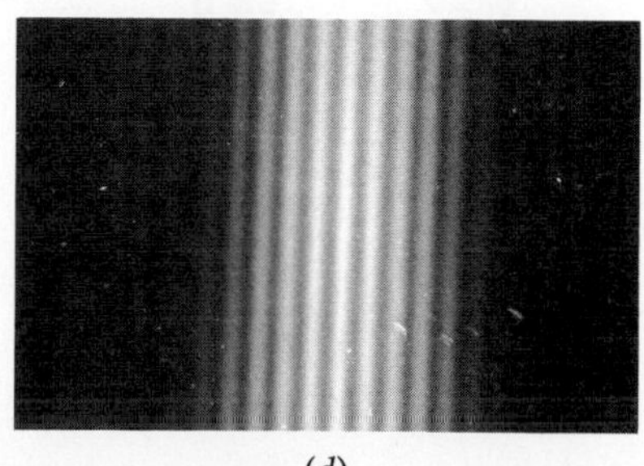

(*d*)

Figure 35-17 Diffraction pattern produced by (*a*) x rays and (*b*) electrons incident on an aluminum foil target and (*c*) neutrons incident on a target of polycrystalline copper. Note the similarity in the patterns produced. (*d*) A two-slit electron diffraction–interference pattern. This pattern is the same as that obtained with photons.

35-18 illustrates the features of an electron microscope. The electron beam is made parallel and focused by specially designed magnets that serve as lenses. The energy of the electrons is typically 100 keV, resulting in a wavelength of about 0.004 nm. The target specimen must be very thin so that the transmitted beam will not be slowed down or scattered too much. The final image is projected onto a fluorescent screen or film. Various distortions resulting from focusing problems with the magnetic lenses limit the resolution to a few tenths of a nanometer, which is about a thousand times better than can be achieved with visible light.

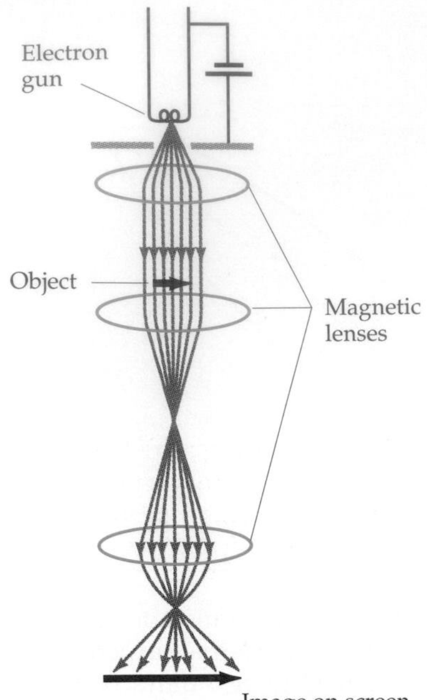

Figure 35-18 Electron microscope. Electrons from a heated filament (the electron gun) are accelerated by a large voltage difference. The electron beam is made parallel by a magnetic focusing lens. The electrons strike a thin target and are then focused by a second magnetic lens that is equivalent to the objective lens in an ordinary microscope. The third magnetic lens takes the place of the eyepiece in a microscope. It projects the electron beam onto a fluorescent screen for viewing the image.

Standing Waves and Energy Quantization

De Broglie pointed out that the Bohr quantum condition (Equation 35-21) for the angular momentum of the electron in a hydrogen atom is equivalent to a standing-wave condition. This condition states that

$$mvr = n\frac{h}{2\pi}$$

Substituting h/λ for the momentum mv gives

$$\frac{h}{\lambda}r = n\frac{h}{2\pi}$$

or

$$n\lambda = 2\pi r = C \qquad 35\text{-}31$$

where C is the circumference of the Bohr orbit. Thus, Bohr's quantum condition is equivalent to saying that an integral number of electron waves must fit into the circumference of the circular orbit as shown in Figure 35-19.

Example 35-8

The kinetic energy of the electron in the ground (lowest energy) state of the hydrogen atom is 13.6 eV. (Its potential energy is −27.2 eV and its total energy is −13.6 eV, leading to a binding energy of 13.6 eV.) Find the de Broglie wavelength for this electron.

Using $K = 13.6$ eV in Equation 35-30, we have

$$\lambda = \frac{1.226}{\sqrt{13.6}}\text{ nm} = 0.332\text{ nm} = 2\pi(0.0529\text{ nm})$$

This is the circumference of the first Bohr orbit in the hydrogen atom.

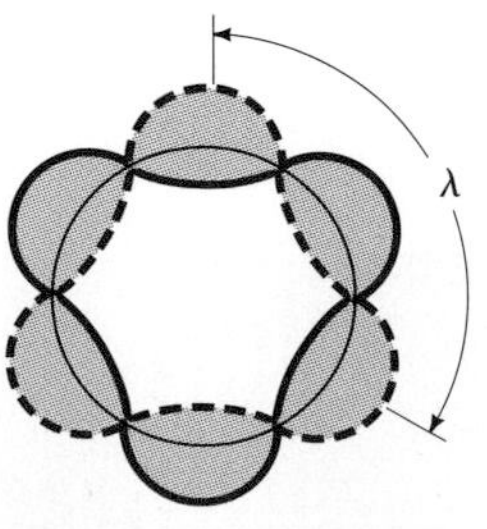

Figure 35-19 Standing waves around the circumference of a circle.

The fitting of an integral number of electron waves into the circumference of a Bohr orbit is similar to the fitting of an integral number of half wavelengths into the length of a string or organ pipe, as in standing waves on strings or standing sound waves. In classical wave theory, standing waves lead to a quantization of frequency. For example, for standing waves on a string of length L fixed at both ends (Figure 35-20), the standing wave condition is

$$n\frac{\lambda}{2} = L$$

For waves traveling with a speed v, the frequency of such standing waves on a string is then given by

$$f = \frac{v}{\lambda} = n\frac{v}{2L}$$

If energy is associated with the frequency of a standing wave, as in Equation 35-28, then standing waves imply quantized energies.

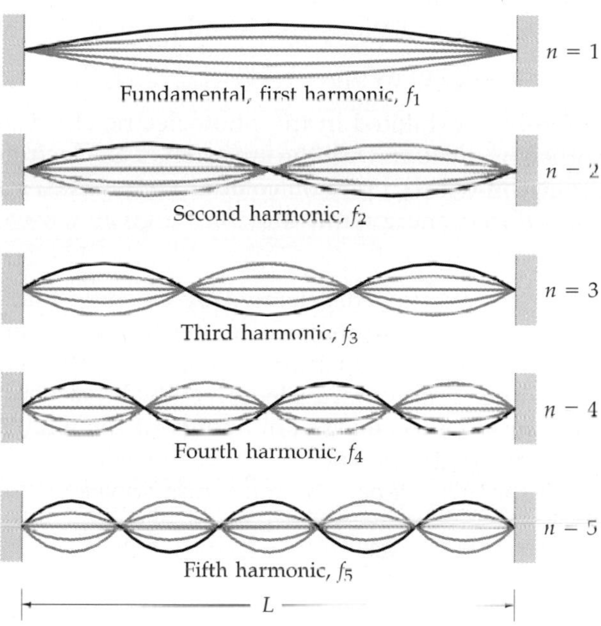

Figure 35-20 Standing waves on a string fixed at both ends. The frequencies of these waves are quantized; that is, they may have only certain values given by $f_n = nf_1$, where f_1 is the fundamental frequency.

The idea of explaining the discrete energy states of matter by standing waves led to the development of a detailed mathematical theory by Erwin Schrödinger and others in 1928. In this theory, known as **quantum theory, quantum mechanics,** or **wave mechanics,** the electron is described by a wave function ψ that obeys a wave equation that is somewhat similar to the classical wave equations for sound and light waves. The frequency and wavelength of electron waves are related to the energy and momentum of the electron just as the frequency and wavelength of light waves are related to the energy and momentum of photons. Schrödinger solved the standing-wave problem for the hydrogen atom, the simple harmonic oscillator, and other systems of interest. He found that the allowed frequencies combined with the de Broglie relation $E = hf$ led to the set of energy levels for the hydrogen atom found by Bohr (Equation 35-26), thereby demonstrating that quantum theory provides a general method of finding the quantized energy levels for a given system. Quantum theory is the basis for our understanding of the modern world, from the inner workings of the atomic nucleus to the radiation spectrum of distant galaxies in cosmology.

Summary

1. The energy in electromagnetic radiation is not continuous but comes in quanta with energies given by

$$E = hf = \frac{hc}{\lambda}$$

where f is the frequency, λ is the wavelength, and h is Planck's constant, which has the value

$$h = 6.626 \times 10^{-34}\ \text{J·s} = 4.136 \times 10^{-15}\ \text{eV·s}$$

The quantity hc occurs often in calculations and has the value

$$hc = 1240\ \text{eV·nm}$$

The quantum nature of light is exhibited in the photoelectric effect, in which a photon is absorbed by an atom with the emission of an electron, and in Compton scattering, in which a photon collides with a free electron and emerges with reduced energy and therefore a greater wavelength.

2. X rays are emitted when electrons are decelerated by crashing into a target in an x-ray tube. An x-ray spectrum consists of a series of sharp lines called the characteristic spectrum superimposed on the continuous bremsstrahlung spectrum. The minimum wavelength in the bremsstrahlung spectrum λ_m corresponds to the maximum energy of the emitted photon, which equals the maximum kinetic energy of the electrons eV, where V is the voltage of the x-ray tube. The minimum wavelength is then given by

$$\lambda_m = \frac{hc}{eV}$$

3. The wavelengths of x rays are typically a few nanometers, which is also approximately equal to the spacing of atoms in a crystal. Diffraction maxima are observed when x rays are scattered from a crystal, indicating that x rays are electromagnetic waves and that the atoms in crystals are arranged in a regular array.

4. In order to derive the Balmer formula for the spectrum of the hydrogen atom, Bohr proposed the following postulates:

Postulate 1: The electron in the hydrogen atom can move only in certain nonradiating circular orbits called stationary states.

Postulate 2: The atom radiates a photon when the electron makes a transition from one stationary orbit to another. The frequency of the photon is given by

$$f = \frac{E_i - E_f}{h}$$

where E_i and E_f are the initial and final energies of the atom.

Postulate 3: The radius (and hence the energy) of a stationary state orbit is determined by classical physics together with the quantum condition that the angular momentum of the electron must equal an integer times Planck's constant divided by 2π:

$$mvr = \frac{nh}{2\pi} = n\hbar$$

where $\hbar = h/2\pi = 1.05 \times 10^{-34}$ J·s.

These postulates lead to allowed energy levels in the hydrogen atom given by

$$E_n = -\frac{k^2e^4m}{2\hbar^2}\frac{Z^2}{n^2} = -Z^2\frac{E_0}{n^2}$$

where n is an integer and

$$E_0 = \frac{k^2e^4m}{2\hbar^2} \approx 13.6\text{ eV}$$

The radii of the stationary orbits are given by

$$r = n^2\frac{\hbar^2}{mkZe^2} = n^2\frac{a_0}{Z}$$

where

$$a_0 = \frac{\hbar^2}{mke^2} \approx 0.0529\text{ nm}$$

is the first Bohr radius.

5. The wave nature of electrons was first suggested by de Broglie, who postulated the equations

$$f = \frac{E}{h} \quad \text{and} \quad \lambda = \frac{h}{p}$$

for the frequency and wavelength of electron waves. With these equations, the Bohr quantum condition can be understood as a standing-wave condition. The wave nature of electrons was observed experimentally first by Davisson and Germer and later by G. P. Thomson, who measured the diffraction and interference of electrons.

6. The mathematical theory of the wave nature of matter is known as quantum theory. In this theory, the electron is described by a wave function that obeys a wave equation. Energy quantization arises from standing-wave conditions applied to electrons in various systems. Quantum theory is the basis for our understanding of the physical nature of the modern world.

Suggestions for Further Reading

Feinberg, Gerald: "Light," *Scientific American*, September 1968, p. 50.

This article presents an introduction to our present understanding of light as a phenomenon of both wavelike and particlelike properties, as manifested by diffraction, two-slit interference, the photoelectric effect, and blackbody radiation.

Moran, Paul R., R. Jerome Nickles, and James A. Zagzebski: "The Physics of Medical Imaging," *Physics Today*, vol. 36, no. 7, 1983, p. 36.

This article briefly describes such new medical imaging techniques as digital subtraction angiography, computed tomography (CAT), nmr imaging (MRI), positron-emission tomography (PET), and ultrasound imaging.

Wheeler, John Archibald: "Niels Bohr, the Man," *Physics Today*, vol. 38, no. 10, 1985, p. 66.

Bohr's very personal approach to science is recounted by a former collaborator, who is himself a highly respected physicist. This article appears as part of a special issue commemorating the centennial of Bohr's birth.

Review

A. Objectives: After studying this chapter you should:

1. Be able to sketch the spectral distribution curve for blackbody radiation and the curve predicted by the Rayleigh–Jeans law.

2. Be able to discuss the photoelectric effect and state the Einstein equation describing it.

3. Be able to discuss how the photon concept explains all the features of the photoelectric effect and the Compton scattering of x rays.

4. Be able to sketch a typical x-ray spectrum and relate the minimum wavelength of the spectrum to the voltage of the x-ray tube.

5. Be able to state the Bohr postulates and describe the Bohr model of the hydrogen atom.

6. Be able to draw an energy-level diagram for hydrogen, indicate on it transitions involving the emission of a photon, and use it to calculate the wavelengths of the emitted photons.

7. Be able to state the de Broglie relations for the frequency and wavelength of electron waves and use them and the standing-wave condition to derive the Bohr condition for the quantization of angular momentum in the hydrogen atom.

8. Be able to discuss the experimental evidence for the existence of electron waves.

B. Define, explain, or otherwise identify:

Blackbody radiation
Rayleigh–Jeans law
Ultraviolet catastrophe
Quanta
Planck's constant
Photoelectric effect
Stopping potential
Photons
Work function
Einstein's photoelectric equation
Threshold frequency
Threshold wavelength
X rays
Characteristic spectrum
Bremsstrahlung spectrum
Cutoff wavelength
Compton wavelength
Balmer series
Rydberg–Ritz formula
Rydberg
Stationary states
Bohr radius
Energy-level diagram
Ionization
Quantum theory
Quantum mechanics
Wave mechanics

C. True or false: If the statement is true, explain why it is true. If it is false, give a counterexample.

1. The spectral distribution of radiation in a blackbody depends only on the temperature of the body.

2. In the photoelectric effect, the maximum current is proportional to the intensity of the incident light.

3. The work function of a metal depends on the frequency of the incident light.

4. The maximum kinetic energy of electrons emitted in the photoelectric effect varies linearly with the frequency of the incident light.

5. The energy of a photon is proportional to its frequency.

6. One of Bohr's assumptions is that atoms never radiate light.

7. In the Bohr model, the energy of a hydrogen atom is quantized.

8. In the ground state of the hydrogen atom, the potential energy is -27.2 eV.

9. The de Broglie wavelength of an electron varies inversely with its momentum.

10. Electrons can be diffracted.

11. Neutrons can be diffracted.

12. An electron microscope is used to look at electrons.

Problems

Level I

35-1 The Origin of the Quantum Constant: Blackbody Radiation

There are no problems for this section

35-2 The Photoelectric Effect

1. Find the photon energy in joules and in electron volts for an electromagnetic wave in the FM radio band of frequency 100 MHz.

2. Repeat Problem 1 for an electromagnetic wave in the AM radio band of frequency 900 kHz.

3. What is the frequency of a photon of energy (*a*) 1 eV, (*b*) 1 keV, and (*c*) 1 MeV?

4. Find the photon energy for light of wavelength (*a*) 450 nm, (*b*) 550 nm, and (*c*) 650 nm.

5. Find the range of photon energies in the visible spectrum, which ranges from wavelengths of 400 to 700 nm.

6. Find the photon energy if the wavelength is (*a*) 0.1 nm (about 1 atomic diameter) and (*b*) 1 fm (1 fm = 10^{-15} m, about 1 nuclear diameter).

7. The work function for tungsten is 4.58 eV. (*a*) Find the threshold frequency and wavelength for the photoelectric effect. Find the stopping potential if the wavelength of the incident light is (*b*) 200 nm and (*c*) 250 nm.

8. When light of wavelength 300 nm is incident on potassium, the emitted electrons have maximum kinetic energy of 2.03 eV. (*a*) What is the energy of the incident photon?

(*b*) What is the work function for potassium? (*c*) What would be the stopping potential if the incident light had a wavelength of 430 nm? (*d*) What is the threshold wavelength for the photoelectric effect with potassium?

9. The threshold wavelength for the photoelectric effect for silver is 262 nm. (*a*) Find the work function for silver. (*b*) Find the stopping potential if the incident radiation has a wavelength of 175 nm.

10. The work function for cesium is 1.9 eV. (*a*) Find the threshold frequency and wavelength for the photoelectric effect. Find the stopping potential if the wavelength of the incident light is (*b*) 250 nm and (*c*) 350 nm.

11. A light beam of wavelength 400 nm has an intensity of 100 W/m^2. (*a*) What is the energy of each photon in the beam? (*b*) How much energy strikes an area of 1 cm^2 perpendicular to the beam in 1 s? (*c*) How many photons strike this area in 1 s?

35-3 X Rays

12. An x-ray tube operates at a potential of 460 kV. What is the minimum wavelength of the continuous x-ray spectrum from this tube?

13. The minimum wavelength in the continuous x-ray spectrum from a television tube is 0.134 nm. What is the voltage of the tube?

14. What is the minimum wavelength of the continuous x-ray spectrum from a television tube operating at 2500 V?

35-4 Compton Scattering

15. Find the shift in wavelength of photons scattered at $\theta = 60°$.

16. When photons are scattered by electrons in carbon, the shift in wavelength is 0.33 pm. Find the scattering angle.

17. Find the momentum of a photon in eV/c and in kg · m/s if the wavelength is (*a*) 400 nm, (*b*) 2 nm, (*c*) 0.1 nm, and (*d*) 3 cm.

18. The wavelength of Compton-scattered photons is measured at $\theta = 90°$. If $\Delta\lambda/\lambda$ is to be 1.5 percent, what should the wavelength of the incident photons be?

19. Compton used photons of wavelength 0.0711 nm. (*a*) What is the energy of these photons? (*b*) What is the wavelength of the photon scattered at $\theta = 180°$? (*c*) What is the energy of the photon scattered at this angle?

20. For the photons used by Compton, find the momentum of the incident photon and that of the photon scattered at 180°, and use momentum conservation to find the momentum of the recoil electron in this experiment (see Problem 19).

35-5 Quantization of Atomic Energies: The Bohr Model

21. Use the known values of the constants in Equation 35-22 to show that a_0 is approximately 0.0529 nm.

22. The wavelength of the longest wavelength of the Lyman series was calculated in Example 35-6. Find the wavelengths for the transitions (*a*) $n_1 = 3$ to $n_2 = 1$ and (*b*) $n_1 = 4$ to $n_2 = 1$. (*c*) Find the shortest wavelength in the Lyman series.

23. Find the photon energy for the three longest wavelengths in the Balmer series and calculate the wavelengths.

24. (*a*) Find the photon energy and wavelength for the series limit (shortest wavelength) in the Paschen series ($n_2 = 3$). (*b*) Calculate the wavelengths for the three longest wavelengths in this series and indicate their positions on a horizontal linear scale.

25. Repeat Problem 24 for the Brackett series ($n_2 = 4$).

26. A hydrogen atom is in its tenth excited state according to the Bohr model ($n = 11$). (*a*) What is the radius of the Bohr orbit? (*b*) What is the angular momentum of the electron? (*c*) What is the electron's kinetic energy? (*d*) What is the electron's potential energy? (*e*) What is the electron's total energy?

35-6 Electron Waves and Quantum Theory

27. Use Equation 35-30 to calculate the de Broglie wavelength for an electron of kinetic energy (*a*) 2.5 eV, (*b*) 250 eV, (*c*) 2.5 keV, and (*d*) 25 keV.

28. An electron is moving at $v = 2.5 \times 10^5$ m/s. Find its de Broglie wavelength.

29. An electron has a wavelength of 200 nm. Find (*a*) its momentum and (*b*) its kinetic energy.

30. Through what potential must an electron be accelerated so that its de Broglie wavelength is (*a*) 5 nm and (*b*) 0.01 nm?

31. A thermal neutron in a reactor has kinetic energy of about 0.02 eV. Calculate the de Broglie wavelength of this neutron from

$$\lambda = \frac{hc}{\sqrt{2mc^2K}}$$

where $mc^2 = 940$ MeV is the rest energy of the neutron.

32. Find the de Broglie wavelength of a proton (rest energy $mc^2 = 938$ MeV) that has a kinetic energy of 2 MeV. (See Problem 31.)

33. A proton is moving at $v = 0.003c$, where c is the speed of light. Find its de Broglie wavelength.

34. What is the kinetic energy of a proton whose de Broglie wavelength is (*a*) 1 nm and (*b*) 1 fm?

35. Find the de Broglie wavelength of a baseball of mass 0.145 kg moving at 30 m/s.

36. The energy of the electron beam in Davisson and Germer's experiment was 54 eV. Calculate the wavelength for these electrons.

37. The distance between Li^+ and Cl^- ions in a LiCl crystal is 0.257 nm. Find the energy of electrons that have wavelengths equal to this spacing.

38. An electron microscope uses electrons of energy 70 keV. Find the wavelength of these electrons.

Level II

39. An x ray undergoes Compton scattering and emerges with a wavelength of 0.20 nm at a scattering angle of 100°. What was the initial energy of the x-ray photon?

40. When the kinetic energy of an electron is much greater than its rest energy, the relativistic approximation $E \approx pc$ is good. (*a*) Show that in this case photons and electrons of the same energy have the same wavelength. (*b*) Find the de Broglie wavelength of an electron of energy 200 MeV.

41. Suppose that a 100-W source radiates light of wavelength 600 nm uniformly in all directions and that the eye can detect this light if only 20 photons per second enter a dark-adapted eye having a 7-mm diameter pupil. How far from the source can the light be detected under these rather extreme conditions?

42. Data for stopping potential versus wavelength for the photoelectric effect using sodium are

λ, nm	200	300	400	500	600
V_0, V	4.20	2.06	1.05	0.41	0.03

Plot these data so as to obtain a straight line and from your plot find (*a*) the work function, (*b*) the threshold frequency, and (*c*) the ratio h/e.

43. The diameter of the pupil of the eye is about 5 mm. (It can vary from about 1 mm to 8 mm). Find the intensity of light of wavelength 600 nm such that 1 photon per second passes through the pupil.

44. Show that the speed of an electron in the nth Bohr orbit of hydrogen is given by $v_n = e^2/2\epsilon_0 hn$.

45. A light bulb radiates 90 W uniformly in all directions. (*a*) Find the intensity at a distance of 1.5 m. (*b*) If the wavelength is 650 nm, find the number of photons per second that strike a 1-cm^2 area oriented so that its normal is along the line to the bulb.

46. How many head-on Compton scattering events are necessary to double the wavelength of a photon having initial wavelength 200 pm?

47. An x-ray photon of wavelength 6 pm makes a head-on collision with an electron so that it is scattered by an angle of 180°. (*a*) What is the change in wavelength of the photon? (*b*) What is the energy lost by the photon? (*c*) What is the kinetic energy of the scattered electron?

48. A 0.200-pm photon scatters from a free electron that is initially at rest. For what photon scattering angle will the kinetic energy of the recoiling electron equal the energy of the scattered photon?

49. The binding energy of an electron is the minimum energy required to remove the electron from its ground state to a large distance from the nucleus. (*a*) What is the binding energy for the hydrogen atom? (*b*) What is the binding energy for He^+? (*c*) What is the binding energy for Li^{2+}?

50. A hydrogen atom has its electron in the $n = 2$ state. The electron makes a transition to the ground state. (*a*) What is the energy of the photon according to the Bohr model? (*b*) If angular momentum is conserved, what is the angular momentum of the photon? (*c*) The linear momentum of the emitted photon is E/c. If we assume conservation of linear momentum, what is the recoil velocity of the atom? (*d*) Find the recoil kinetic energy of the atom in electron volts. By what percent must the energy of the photon calculated in part (*a*) be corrected to account for this recoil energy?

51. A particle of mass m moves in a one-dimensional box of length L. (Take the potential energy of the particle in the box to be zero so that its total energy is its kinetic energy $p^2/2m$). Its energy is quantized by the condition $n(\lambda/2) = L$, where λ is the de Broglie wavelength of the particle and n is an integer. (*a*) Show that the allowed energies are given by

$$E_n = n^2E_1 \qquad \text{where } E_1 = h^2/8mL^2$$

(*b*) Evaluate E_n for an electron in a box of size $L = 0.1$ nm and make an energy-level diagram for the states from $n = 1$ to $n = 5$. Use Bohr's second postulate $f = \Delta E/h$ to calculate the wavelength of electromagnetic radiation emitted when the electron makes a transition from (*c*) $n = 2$ to $n = 1$, (*d*) $n = 3$ to $n = 2$, and (*e*) $n = 5$ to $n = 1$.

52. (*a*) Use the results of Problem 51 to find the energy of the ground state ($n = 1$) and the first two excited states of a proton in a one-dimensional box of length $L = 10^{-15}$ m = 1 fm. (These are of the order of magnitude of nuclear energies.) Calculate the wavelength of electromagnetic radiation emitted when the proton makes a transition from (*b*) $n = 2$ to $n = 1$, (*c*) $n = 3$ to $n = 2$, and (*d*) $n = 3$ to $n = 1$.

53. (*a*) Use the results of Problem 51 to find the energy of the ground state ($n = 1$) and the first two excited states of a proton in a one-dimensional box of length 0.2 nm (about the diameter of a H_2 molecule). Calculate the wavelength of electromagnetic radiation emitted when the proton makes a transition from (*b*) $n = 2$ to $n = 1$, (*c*) $n = 3$ to $n = 2$, and (*d*) $n = 3$ to $n = 1$.

54. (*a*) Use the results of Problem 51 to find the energy of the ground state ($n = 1$) and the first two excited states of a small particle of mass 1 μg confined to a one-dimensional box of length 1 cm. (*b*) If the particle moves with a speed of 1 mm/s, calculate its kinetic energy and find the approximate value of the quantum number n.

55. In the center-of-mass reference frame of the electron and the nucleus of an atom, the electron and nucleus have equal and opposite momenta of magnitude p. (*a*) Show that the total kinetic energy of the electron and nucleus can be written

$$K = \frac{p^2}{2\mu}$$

where

$$\mu = \frac{m_e M}{m_e + M} = \frac{m_e}{1 + m_e/M}$$

is called the reduced mass, m_e is the mass of the electron, and M is the mass of the nucleus. It can be shown that the motion of the nucleus can be accounted for by replacing the mass of the electron by the reduced mass. (*b*) Use Equation 35-25 with m replaced by μ to calculate the Rydberg for hydrogen ($M = m_p$) and for a very massive nucleus ($M = \infty$). (*c*) Find the percentage correction for the ground-state energy of the hydrogen atom due to the motion of the proton.

56. The kinetic energy of rotation of a diatomic molecule can be written $K = L^2/2I$, where L is its angular momentum and I is the moment of inertia. (*a*) Assuming that the angular momentum is quantized as in the Bohr model of the hydrogen atom, show that the energy is given by $K_n = n^2 K_1$, where $K_1 = \hbar^2/2I$. (*b*) Make an energy-level diagram for such a molecule. (*c*) Estimate K_1 for the hydrogen molecule assuming the separation of the atoms to be $r = 0.1$ nm and considering rotation about an axis through the center of mass and perpendicular to the line joining the atoms. Express your answer in electron volts. (*d*) When K_1 is greater than kT (where k is Boltzmann's constant), molecular collisions do not result in rotation and so rotational energy does not contribute to the internal energy of the gas. Use your result of part (*c*) to find the critical temperature $T_c = K_1/k$.

Level III

57. This problem is one of estimating the time lag (expected classically but not observed) in the photoelectric effect. Let the intensity of the incident radiation be 0.01 W/m^2. (*a*) If the area of the atom is 0.01 nm^2, find the energy per second falling on an atom. (*b*) If the work function is 2 eV, how long would it take classically for this much energy to fall on one atom?

58. A photon cannot transfer all of its energy to a single free electron. Prove this by considering the problem of conservation of energy and momentum.

59. An electron and positron are moving towards each other with equal speeds of 3×10^6 m/s. The two particles annihilate each other and produce two photons of equal energy. (*a*) What were the de Broglie wavelengths of the electron and positron? Find the (*b*) energy, (*c*) momentum, and (*d*) wavelength of each photon.

60. (*a*) Solve Equation 35-11 for $p_e{}^2$ to obtain $p_e^2 = p_1^2 + p_2^2 - 2p_1p_2 + 2mc(p_1 - p_2)$. (*b*) Eliminate p_e^2 from your result in part (*a*) and Equation 35-10 to obtain $mc(p_1 - p_2) = p_1p_2(1 - \cos\theta)$. (*c*) Multiply both sides of your result in part (*b*) by h/mcp_1p_2 and use the de Broglie relation $h/p = \lambda$ to obtain the Compton formula (Equation 35-12).

61. The total energy density of radiation in a blackbody is given by

$$\eta = \int f(\lambda, T)\, d\lambda$$

where $f(\lambda, T)$ is given by the Planck formula

$$f(\lambda, T) = \frac{8\pi hc\lambda^{-5}}{e^{hc/\lambda kT} - 1}$$

Change the variable to $x = hc/\lambda kT$ and show that the total energy density can be written

$$\eta = \left(\frac{kT}{hc}\right)^4 8\pi hc \int_0^\infty \frac{x^3}{e^x - 1} dx = \alpha T^4$$

where α is some constant independent of T. This shows that the energy density in a blackbody is proportional to T^4.

62. The frequency of revolution of an electron in a circular orbit of radius r is $f_{rev} = v/2\pi r$, where v is the speed. (*a*) Show that in the nth stationary state

$$f_{rev} = \frac{k^2Z^2e^4m}{2\pi h^3}\frac{1}{n^3}$$

(*b*) Show that when $n_1 = n$, $n_2 = n - 1$, and n is much greater than 1,

$$\frac{1}{n_2^2} - \frac{1}{n_1^2} \approx \frac{2}{n^3}$$

(*c*) Use your result in part (*b*) in Equation 35-24 to show that in this case the frequency of radiation emitted equals the frequency of motion. This result is an example of Bohr's correspondence principle: when n is large, so that the energy difference between adjacent states is a small fraction of the total energy, classical and quantum physics must give the same results.

Appendix A

Review of Mathematics

In this appendix, we will review some of the basic results of algebra, geometry, trigonometry, and calculus. In many cases, we will merely state results without proof. Table A-1 lists some mathematical symbols.

Equations

The following operations can be performed on mathematical equations to facilitate their solution:

1. The same quantity can be added to or subtracted from each side of the equation.
2. Each side of the equation can be multiplied or divided by the same quantity.
3. Each side of the equation can be raised to the same power.

It is important to understand that the preceding rules apply to each *side* of the equation and not to each *term* in the equation.

Table A-1 **Mathematical Symbols**

$=$	is equal to
$\neq$	is not equal to
$\approx$	is approximately equal to
$\sim$	is of the order of
$\propto$	is proportional to
$>$	is greater than
$\geq$	is greater than or equal to
$\gg$	is much greater than
$<$	is less than
$\leq$	is less than or equal to
$\ll$	is much less than
Δx	change in x
$\lvert x \rvert$	absolute value of x
$n!$	$n(n-1)(n-2)\cdots 1$
Σ	sum
lim	limit
$\Delta t \rightarrow 0$	Δt approaches zero
$\dfrac{dx}{dt}$	derivative of x with respect to t
$\dfrac{\partial x}{\partial t}$	partial derivative of x with respect to t
$\int$	integral

Example A-1

Solve the following equation for x:

$$(x-3)^2+7=23$$

We first subtract 7 from each side of the equation to obtain $(x-3)^2 = 16$. Taking the square root of each side, which is the same as raising each side to the $\frac{1}{2}$ power, we obtain $\pm(x-3) = \pm 4$. We have included the plus-or-minus signs because either $(+4)^2 = 16$ or $(-4)^2 = 16$. We do not need to write $\pm$ on both sides of the equation as all the possibilities are included in $x - 3 = \pm 4$. We can now solve for x by adding 3 to each side. There are two solutions: $x = 4 + 3 = 7$ and $x = -4 + 3 = -1$. These values can be checked by substituting them into the original equation.

Example A-2

Solve the following equation for x:

$$\frac{1}{x}+\frac{1}{4}=\frac{1}{3}$$

This type of equation occurs both in geometric optics and in analyses of electric circuits. Although it is easy to solve, errors are often made. We solve it by first subtracting $\frac{1}{4}$ from each side to obtain

$$\frac{1}{x}=\frac{1}{3}-\frac{1}{4}=\frac{4}{12}-\frac{3}{12}=\frac{1}{12}$$

We then multiply each side by $12x$ to obtain $x = 12$. Note that this is equivalent to taking the reciprocal of each side of the equation. A typical mistake in handling this type of equation is to take the reciprocal of each term first to obtain $x + 4 = 3$. This operation is not allowed; it changes the relative values of each side of the equation and leads to incorrect results.

Direct and Inverse Proportion

The relationships of direct proportion and inverse proportion are so important in physics that they deserve special consideration. Often much algebraic manipulation can be avoided through a simple knowledge of these relationships. Suppose, for example, that you work for 5 days at a certain pay rate and earn \$400. How much would you earn at the same pay rate if you worked 8 days? In this problem, the money earned is *directly proportional* to the time worked. We can write an equation relating the money earned M to the time worked t using a constant of proportionality R:

$$M = Rt$$

The constant of proportionality in this case is the pay rate. We can express R in dollars per day. Since \$400 was earned in 5 d, the value of R is \$400/(5 d) = \$80/d. In 8 d, the amount earned is therefore

$$M = (\$80/\text{d})(8\text{ d}) = \$640$$

However, we do not have to find the pay rate explicitly to work the problem. Since the amount earned in 8 d is $\frac{8}{5}$ times that earned in 5 d, this amount is

$$M = \tfrac{8}{5}(\$400) = \$640$$

We can use a similar example to illustrate inverse proportion. If you get a 25 percent raise, how long would you need to work to earn $400? Here we consider R to be a variable and we wish to solve for t:

$$t = \frac{M}{R}$$

In this equation, the time t is *inversely proportional* to the pay rate R. Thus, if the new rate is $\frac{5}{4}$ times the old rate, the new time will be $\frac{4}{5}$ times the old time or 4 d.

There are some situations in which one quantity varies as the square or some other power of another quantity where the ideas of proportionality are also very useful. Suppose, for example, that a 10-in diameter pizza costs $8.50. How much would you expect a 12-in diameter pizza to cost? We expect the cost of a pizza to be approximately proportional to the amount of its contents, which is proportional to the area of the pizza. Since the area is in turn proportional to the square of the diameter, the cost should be proportional to the square of the diameter. If we increase the diameter by a factor of 12/10, the area increases by a factor of $(12/10)^2 = 1.44$, so we should expect the cost to be (1.44)($8.50) = $12.24.

Example A-3

The intensity of light from a point source varies inversely with the square of the distance from the source. If the intensity is 3.20 W/m^2 at 5 m from a source, what is it at 6 m from the source?

The equation expressing the fact that the intensity varies inversely with the square of the distance can be written

$$I = \frac{C}{r^2}$$

where C is some constant. Then, if $I_1 = 3.20\ W/m^2$ at $r_1 = 5$ m and I_2 is the unknown intensity at $r_2 = 6$ m, we have

$$\frac{I_2}{I_1} = \frac{C/r_2^2}{C/r_1^2} = \frac{r_1^2}{r_2^2} = \left(\frac{5}{6}\right)^2 = 0.694$$

The intensity at 6 m from the source is thus

$$I_2 = 0.694(3.20\ W/m^2) = 2.22\ W/m^2$$

Linear Equations

An equation in which the variables occur only to the first power is said to be linear. A linear equation relating y and x can always be put into the standard form

$$y = mx + b \qquad \text{A-1}$$

where m and b are constants that may be either positive or negative. Figure A-1 shows a graph of the values of x and y that satisfy Equation A-1. The constant b, called the **intercept**, is the value of y at $x = 0$. The constant m is the **slope** of the line, which equals the ratio of the change in y to the corresponding change in x. In the figure, we have indicated two points on the line, x_1, y_1 and x_2, y_2, and the changes $\Delta x = x_2 - x_1$ and $\Delta y = y_2 - y_1$. The slope m is then

$$m = \frac{y_2 - y_1}{x_2 - x_1} = \frac{\Delta y}{\Delta x}$$

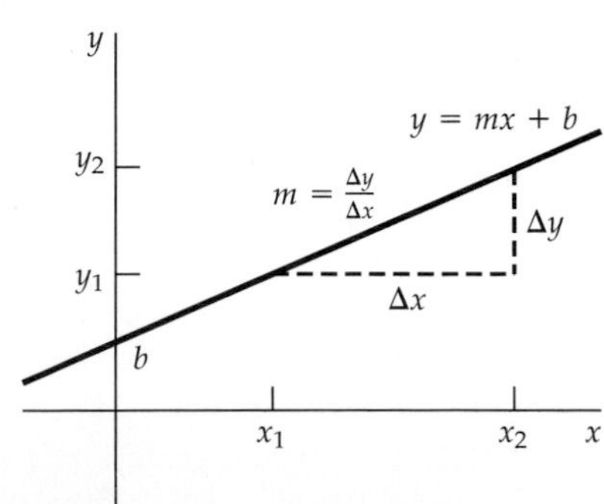

Figure A-1 Graph of the linear equation $y = mx + b$, where b is the intercept and $m = \Delta y/\Delta x$ is the slope.

If x and y are both unknown, there is no unique solution for their values. Any pair of values x_1, y_1 on the line in Figure A-1 will satisfy the equation. If we have two equations, each with the same two unknowns x and y, the equations can be solved simultaneously for the unknowns.

Example A-4

Find the values of x and y that satisfy

$$3x - 2y = 8 \tag{A-2}$$

and

$$y - x = 2 \tag{A-3}$$

Figure A-2 shows a graph of each of these equations. At the point where the lines intersect, the values of x and y satisfy both equations. We can solve two simultaneous equations by first solving either equation for one variable in terms of the other variable and then substituting the result into the other equation. From Equation A-3, we have

$$y = x + 2$$

Substituting this value for y in Equation A-2, we obtain

$$3x - 2(x + 2) = 8$$

$$3x - 2x - 4 = 8$$
$$x = 12$$

Then

$$y = x + 2 = 14$$

An alternative method that is sometimes easier is to multiply one equation by a constant such that one of the unknown terms is eliminated

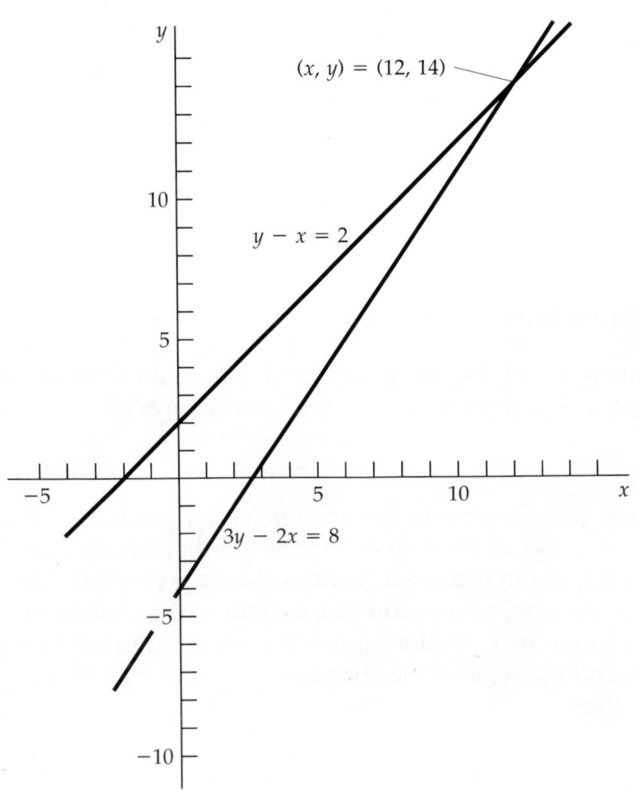

Figure A-2 Graph of Equations A-2 and A-3. At the point where the lines intersect, the values of x and y satisfy both equations.

when the equations are added or subtracted. If we multiply Equation A-3 by 2, we can add the resulting equation to Equation A-2 and eliminate y:

$$3x - 2y = 8$$

$$2y - 2x = 4$$

Adding, we obtain $3x - 2x = 12$ or $x = 12$, as before.

Factoring

Equations can often be simplified by factoring. Three important examples are

1. Common factor: $2ax + 3ay = a(2x + 3y)$
2. Perfect square: $x^2 \pm 2xy + y^2 = (x \pm y)^2$
3. Difference of squares: $x^2 - y^2 = (x + y)(x - y)$

The Quadratic Formula

An equation that contains a variable to the second power is called a *quadratic equation*. The standard form for a quadratic equation is

$$ax^2 + bx + c = 0 \qquad \text{A-4}$$

where a, b, and c are constants. The general solution of this equation is

$$x = -\frac{b}{2a} \pm \frac{1}{2a}\sqrt{b^2 - 4ac} \qquad \text{A-5}$$

When b^2 is greater than $4ac$, there are two solutions corresponding to the + and − signs. Figure A-3 shows a graph of y versus x where $y = ax^2 + bx + c$. The curve, called a **parabola,** crosses the x axis twice. The values of x for which $y = 0$ are the solutions to Equation A-4. When $b^2 < 4ac$, the graph of y versus x does not intersect the x axis, as is shown in Figure A-4, and there are no real solutions to Equation A-4. When $b^2 = 4ac$, the graph of y versus x is tangent to the x axis at the point $x = -b/2a$.

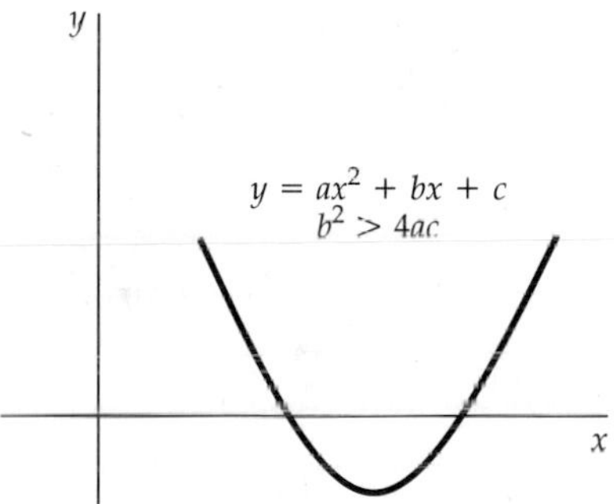

Figure A-3 Graph of y versus x when $y = ax^2 + bx + c$ for the case $b^2 > 4ac$. The two values of x for which $y = 0$ satisfy the quadratic equation (Equation A-4).

Exponents

The notation x^n stands for the quantity obtained by multiplying x times itself n times. For example, $x^2 = x \cdot x$ and $x^3 = x \cdot x \cdot x$. The quantity n is called the **power,** or the **exponent,** of x. When two powers of x are multiplied, the exponents are added:

$$(x^m)(x^n) = x^{m+n} \qquad \text{A-6}$$

This can be readily seen from an example:

$$x^2x^3 = (x \cdot x)(x \cdot x \cdot x) = x^5$$

Any number raised to the 0 power is defined to be 1:

$$x^0 = 1 \qquad \text{A-7}$$

Then

$$x^n x^{-n} = x^0 = 1$$

$$x^{-n} = \frac{1}{x^n} \qquad \text{A-8}$$

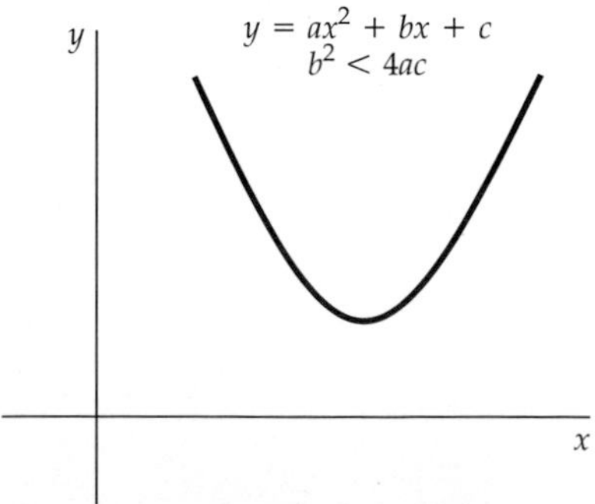

Figure A-4 Graph of y versus x when $y = ax^2 + bx + c$ for the case $b^2 < 4ac$. In this case, there are no (real) values of x for which $y = 0$.

When two powers are divided, the exponents are subtracted:

$$\frac{x^n}{x^m} = x^n x^{-m} = n^{n-m} \qquad \text{A-9}$$

Using these rules, we have

$$x^{1/2} \cdot x^{1/2} = x$$

so

$$x^{1/2} = \sqrt{x}$$

When a power is raised to another power, the exponents are multiplied:

$$(x^n)^m = x^{nm} \qquad \text{A-10}$$

Logarithms

When y is related to x by $y = a^x$, the number x is said to be the logarithm of y to the base a and is written

$$x = \log_a y$$

If $y_1 = a^n$ and $y_2 = a^m$, then

$$y_1 y_2 = a^n a^m = a^{n+m}$$

and

$$\log_a y_1 y_2 = n + m = \log_a y_2 + \log_a y_1 \qquad \text{A-11}$$

It then follows that

$$\log_a y^n = n \log_a y \qquad \text{A-12}$$

Since $a^1 = a$ and $a^0 = 1$,

$$\log_a a = 1 \qquad \text{A-13}$$

and

$$\log_a 1 = 0 \qquad \text{A-14}$$

There are two bases in common use: base 10, called **common logarithms,** and base e ($e = 2.728 \ldots$), called **natural logarithms.** When no base is specified, the base is understood to be 10. Thus, $\log 100 = \log_{10} 100 = 2$ since $100 = 10^2$.

The symbol ln is used for natural logarithms. Thus,

$$y = \ln x \qquad \text{A-15}$$

implies

$$x = e^y \qquad \text{A-16}$$

Logarithms can be changed from one base to another. Suppose that

$$z = \log x \qquad \text{A-17}$$

Then

$$10^z = x \qquad \text{A-18}$$

Taking the natural logarithm of both sides of Equation A-18, we obtain

$$z \ln 10 = \ln x$$

or

$$\ln x = (\ln 10) \log x \qquad \text{A-19}$$

The Exponential Function

When the rate of change of a quantity is proportional to the quantity itself, the quantity increases or decreases exponentially. An example of *exponential decrease* is nuclear decay. If N is the number of radioactive nuclei at some time, then the change dN in some very small time interval dt will be proportional to N and to dt:

$$dN = -\lambda N\, dt$$

where the constant of proportionality λ is the decay rate. The function N satisfying this equation is

$$N = N_0\, e^{-\lambda t} \qquad \text{A-20}$$

where N_0 is the number at time $t = 0$. Figure A-5 shows N versus t. A characteristic of exponential decay is that N decreases by a constant factor in a given time interval. The time interval for N to decrease to half its original value is its half-life $t_{1/2}$, which is related to the decay rate by

$$t_{1/2} = \frac{\ln 2}{\lambda} = \frac{0.693}{\lambda} \qquad \text{A-21}$$

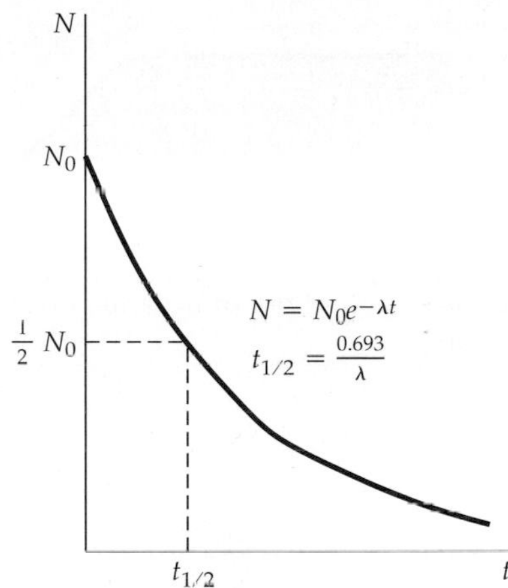

Figure A-5 Graph of N versus t when N decreases exponentially. The time $t_{1/2}$ is the time it takes for N to decrease by one-half.

An example of *exponential increase* is population growth. If the number of organisms is N, the change in N after a small time interval dt is given by

$$dN = +\lambda N\, dt$$

where λ is a constant that characterizes the rate of increase. The function N satisfying this equation is

$$N = N_0\, e^{\lambda t} \qquad \text{A-22}$$

A graph of this function is shown in Figure A-6. An exponential increase is characterized by a doubling time T_2, which is related to λ by

$$T_2 = \frac{\ln 2}{\lambda} = \frac{0.693}{\lambda} \qquad \text{A-23}$$

If the rate of increase λ is expressed as a percentage, $r = \lambda/100\%$, the doubling time is

$$T_2 = \frac{69.3}{r} \qquad \text{A-24}$$

For example, if the population increases by 2 percent per year, the population will double every $69.3/2 \approx 35$ years. Table A-2 lists some useful relations for exponential and logarithmic functions.

Table A-2 Exponential and Logarithmic Functions

$e = 2.71828 \qquad e^0 = 1$

If $y = e^x$, then $x = \ln y$

$e^{\ln x} = x$

$e^x e^y = e^{(x+y)}$

$(e^x)^y = e^{xy} = (e^y)^x$

$\ln e = 1 \qquad \ln 1 = 0$

$\ln xy = \ln x + \ln y$

$\ln \frac{x}{y} = \ln x - \ln y$

$\ln e^x = x \qquad \ln a^x = x \ln a$

$\ln x = (\ln 10) \log x$
$\quad = 2.3026 \log x$

$\log x = \log e \ln x = 0.43429 \ln x$

$e^x = 1 + x + \frac{x^2}{2!} + \frac{x^3}{3!} + \cdots$

$\ln (1 + x) = x - \frac{x^2}{2} + \frac{x^3}{3} - \frac{x^4}{4} + \cdots$

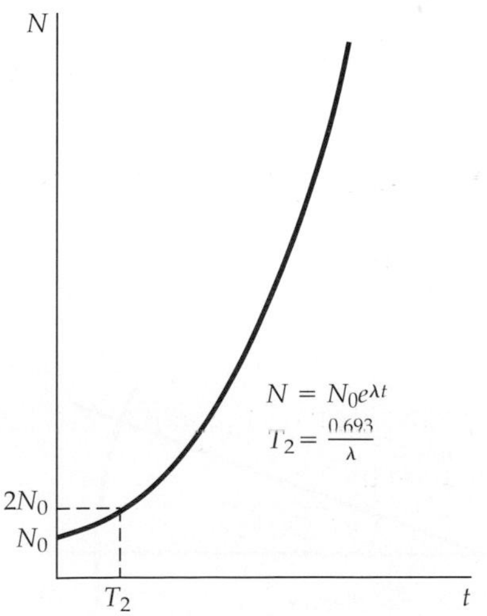

Figure A-6 Graph of N versus t when N increases exponentially. The time T_2 is the time it takes for N to double.

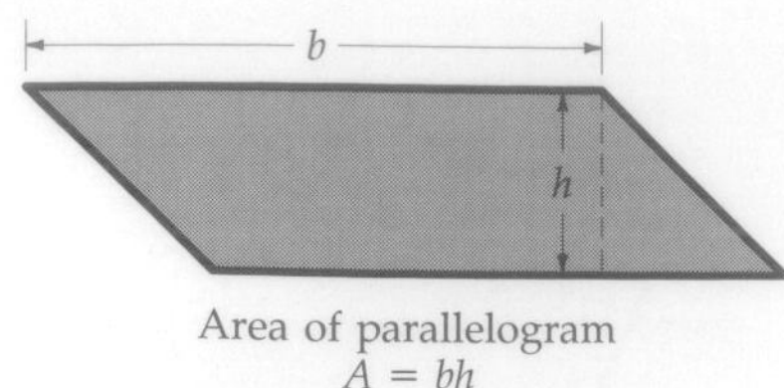

Figure A-7 Area of a parallelogram.

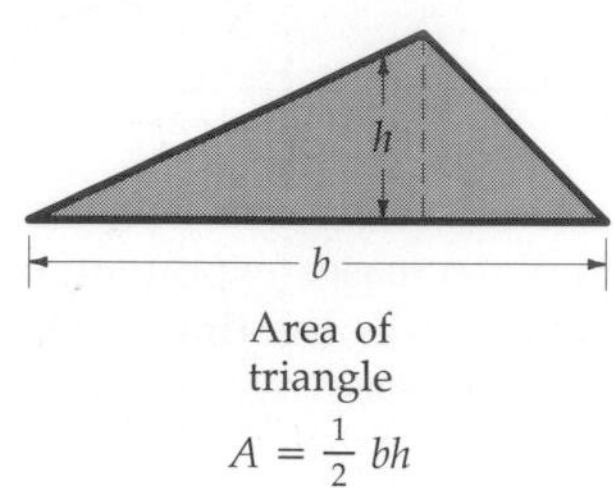

Figure A-8 Area of a triangle.

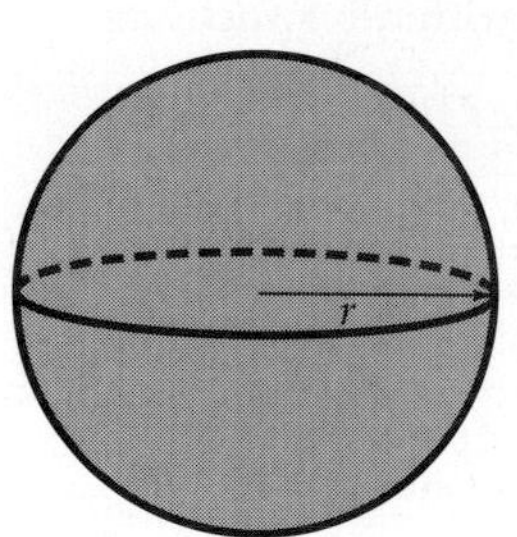

Figure A-9 Surface area and volume of a sphere.

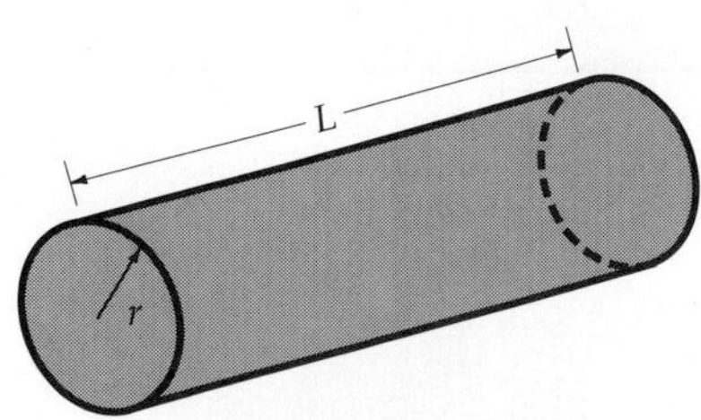

Figure A-10 Surface area (not including the end faces) and volume of a cylinder.

Geometry

The ratio of the circumference of a circle to its diameter is a natural number π, which has the approximate value

$$\pi = 3.141592$$

The circumference of a circle C is thus related to its diameter d and its radius r by

$$C = \pi d = 2\pi r \qquad \text{circumference of circle} \qquad \text{A-25}$$

The area of a circle is

$$A = \pi r^2 \qquad \text{area of circle} \qquad \text{A-26}$$

The area of a parallelogram is the base b times the height h (Figure A-7) and that of a triangle is one-half the base times the height (Figure A-8). A sphere of radius r (Figure A-9) has a surface area given by

$$A = 4\pi r^2 \qquad \text{spherical surface area} \qquad \text{A-27}$$

and a volume given by

$$V = \tfrac{4}{3}\pi r^3 \qquad \text{spherical volume} \qquad \text{A-28}$$

A cylinder of radius r and length L (Figure A-10) has surface area (not including the end faces) of

$$A = 2\pi rL \qquad \text{cylindrical surface} \qquad \text{A-29}$$

and volume of

$$V = \pi r^2 L \qquad \text{cylindrical volume} \qquad \text{A-30}$$

Trigonometry

The angle between two intersecting straight lines is measured as follows. A circle is drawn with its center at the intersection of the lines, and the circular arc is divided into 360 parts called **degrees.** The number of degrees in the arc between the lines is the measure of angle between the lines. For very small angles, the degree is divided into minutes (′) and seconds (″) with $1' = 1°/60$ and $1'' = 1'/60 = 1°/3600$. For scientific work, a more useful measure of an angle is the radian (rad), which is defined as the length of the circular arc between the lines divided by the radius of the circle (Figure A-11). If s is the arc length and r is the radius of the circle, the angle θ measured in radians is

$$\theta = \frac{s}{r} \qquad \text{A-31}$$

Since the angle measured in radians is the ratio of two lengths, it is dimensionless. The relation between radians and degrees is

$$360° = 2\pi \text{ rad}$$

or

$$1 \text{ rad} = \frac{360°}{2\pi} = 57.3° \qquad \text{A-32}$$

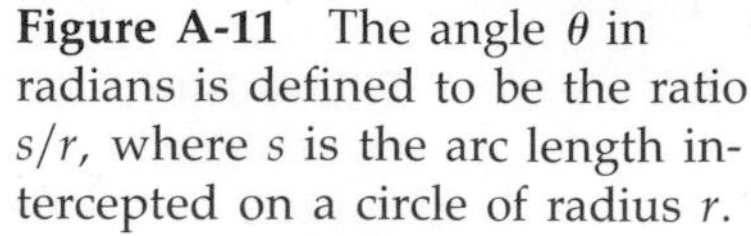

Figure A-11 The angle θ in radians is defined to be the ratio s/r, where s is the arc length intercepted on a circle of radius r.

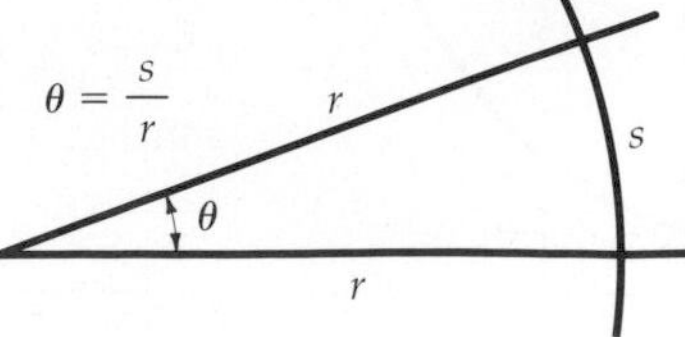

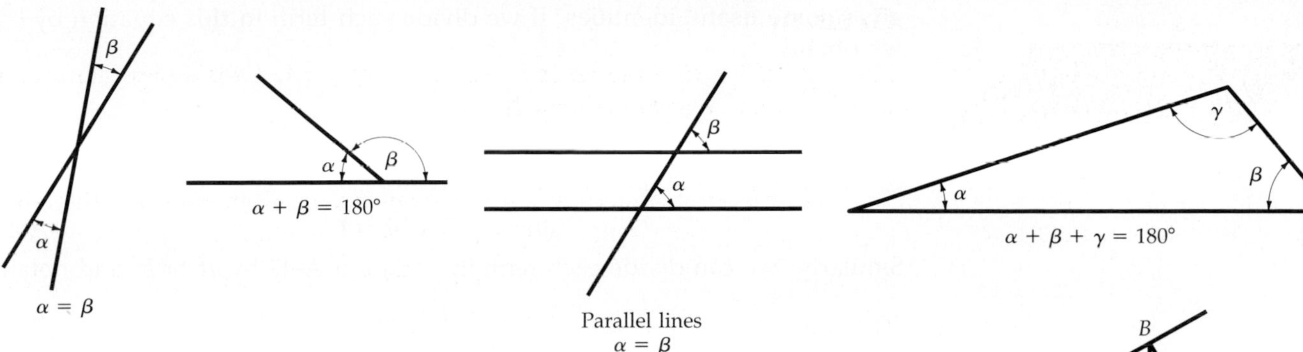

Figure A-12 shows some useful relations for angles.

Figure A-13 shows a right triangle formed by drawing the line BC perpendicular to AC. The lengths of the sides are labeled a, b, and c. The trigonometric functions $\sin\theta$, $\cos\theta$, and $\tan\theta$ for an acute angle θ are defined as

$$\sin\theta = \frac{a}{c} = \frac{\text{opposite side}}{\text{hypotenuse}} \qquad \text{A-33}$$

$$\cos\theta = \frac{b}{c} = \frac{\text{adjacent side}}{\text{hypotenuse}} \qquad \text{A-34}$$

$$\tan\theta = \frac{a}{b} = \frac{\text{opposite side}}{\text{adjacent side}} = \frac{\sin\theta}{\cos\theta} \qquad \text{A-35}$$

Three other trigonometric functions are the reciprocals of these functions and are defined as

$$\sec\theta = \frac{c}{b} = \frac{1}{\cos\theta} \qquad \text{A-36}$$

$$\csc\theta = \frac{c}{a} = \frac{1}{\sin\theta} \qquad \text{A-37}$$

$$\cot\theta = \frac{b}{a} = \frac{1}{\tan\theta} = \frac{\cos\theta}{\sin\theta} \qquad \text{A-38}$$

The angle θ whose sine is x is called the arcsine of x, and is written $\sin^{-1}x$. That is, if

$$\sin\theta = x$$

then

$$\theta = \arcsin x = \sin^{-1} x \qquad \text{A-39}$$

The arcsine is the inverse of the sine. The inverse of the cosine and tangent are defined similarly. The angle whose cosine is y is the arccosine of y. That is, if

$$\cos\theta = y$$

then

$$\theta = \arccos y = \cos^{-1} y \qquad \text{A-40}$$

The angle whose tangent is z is the arctangent of z. That is, if

$$\tan\theta = z$$

$$\theta = \arctan z = \tan^{-1} z \qquad \text{A-41}$$

The pythagorean theorem

$$a^2 + b^2 = c^2 \qquad \text{A-42}$$

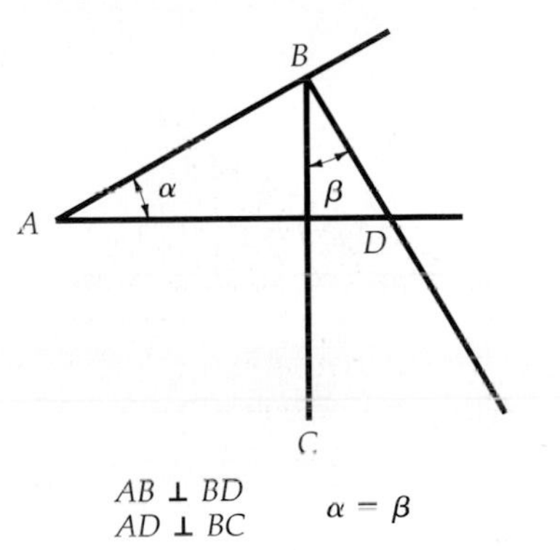

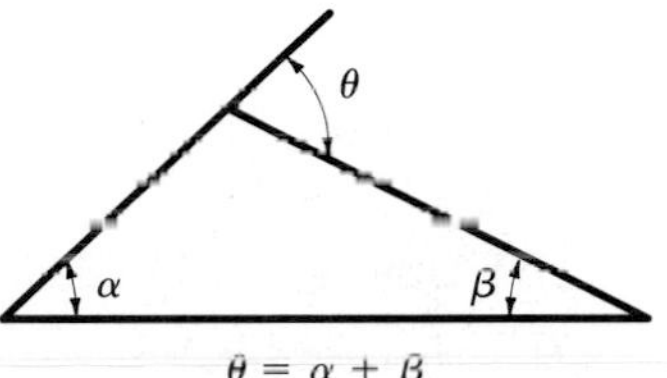

Figure A-12 Some useful relations for angles.

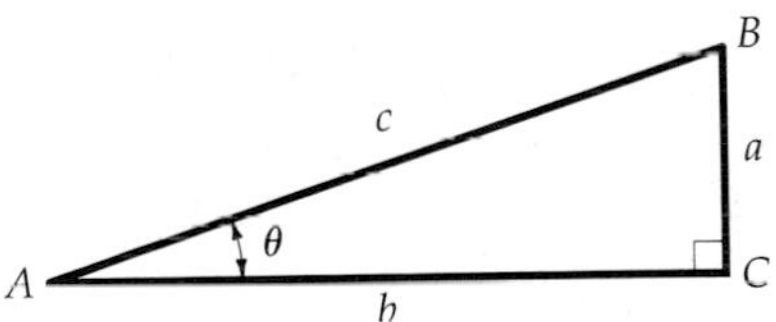

Figure A-13 A right triangle with sides of length a and b and a hypotenuse of length c.

gives some useful identities. If we divide each term in this equation by c^2, we obtain

$$\frac{a^2}{c^2} + \frac{b^2}{c^2} = 1$$

or, from the definitions of sin θ and cos θ,

$$\sin^2 \theta + \cos^2 \theta = 1 \qquad \text{A-43}$$

Similarly, we can divide each term in Equation A-42 by a^2 or b^2 and obtain

$$1 + \cot^2 \theta = \csc^2 \theta \qquad \text{A-44}$$

and

$$1 + \tan^2 \theta = \sec^2 \theta \qquad \text{A-45}$$

These and other useful trigonometric formulas are listed in Table A-3.

Table A-3 **Trigonometric Formulas**

$\sin^2 \theta + \cos^2 \theta = 1 \qquad \sec^2 \theta - \tan^2 \theta = 1 \qquad \csc^2 \theta - \cot^2 \theta = 1$

$\sin 2\theta = 2 \sin \theta \cos \theta$

$\cos 2\theta = \cos^2 \theta - \sin^2 \theta = 2 \cos^2 \theta - 1 = 1 - 2 \sin^2 \theta$

$$\tan 2\theta = \frac{2 \tan \theta}{1 - \tan^2 \theta}$$

$$\sin \frac{1}{2} \theta = \sqrt{\frac{1 - \cos \theta}{2}} \qquad \cos \frac{1}{2} \theta = \sqrt{\frac{1 + \cos \theta}{2}} \qquad \tan \frac{1}{2} \theta = \sqrt{\frac{1 - \cos \theta}{1 + \cos \theta}}$$

$\sin (A \pm B) = \sin A \cos B \pm \cos A \sin B$

$\cos (A \pm B) = \cos A \cos B \mp \sin A \sin B$

$$\tan (A \pm B) = \frac{\tan A \pm \tan B}{1 \mp \tan A \tan B}$$

$\sin A \pm \sin B = 2 \sin [\tfrac{1}{2}(A \pm B)] \cos [\tfrac{1}{2}(A \mp B)]$

$\cos A + \cos B = 2 \cos [\tfrac{1}{2}(A + B)] \cos [\tfrac{1}{2}(A - B)]$

$\cos A - \cos B = 2 \sin [\tfrac{1}{2}(A + B)] \sin [\tfrac{1}{2}(B - A)]$

$$\tan A \pm \tan B = \frac{\sin (A \pm B)}{\cos A \cos B}$$

Example A-5

Use the isosceles right triangle shown in Figure A-14 to find the sine, cosine, and tangent of 45°.

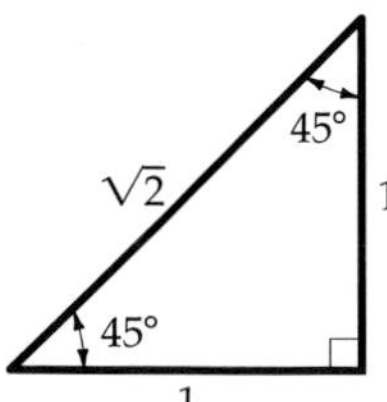

Figure A-14 An isosceles right triangle for Example A-5.

It is clear from the figure that the two acute angles of this triangle are equal. Since the sum of the three angles in a triangle must equal 180° and the right angle is 90°, each acute angle must be 45°. If we multiply each side of any triangle by a common factor, we obtain a similar triangle with the same angles as the first. Since the trigonometric functions involve the ratios of only two sides of a triangle, we can choose any convenient length for one side. Let the equal sides of this triangle have a length of 1 unit. The length of the hypotenuse can then be found from the pythagorean theorem:

$$c = \sqrt{a^2 + b^2} = \sqrt{1^2 + 1^2} = \sqrt{2} \text{ units}$$

The sine, cosine, and tangent for the angle 45° are then given by Equations A-33, A-34, and A-35, respectively:

$$\sin 45° = \frac{1}{\sqrt{2}} = 0.707 \qquad \cos 45° = \frac{1}{\sqrt{2}} = 0.707 \qquad \tan 45° = \frac{1}{1} = 1$$

Example A-6

The sine of 30° is exactly $\frac{1}{2}$. Find the ratios of the sides of a 30–60° right triangle.

This common triangle is shown in Figure A-15. We choose a length of 1 unit for the side opposite the 30° angle. The hypotenuse is then obtained from the fact that sin 30° = 0.5:

$$c = \frac{a}{\sin 30°} = \frac{1}{0.5} = 2$$

The length of the side opposite the 60° angle is found from the pythagorean theorem:

$$b = \sqrt{c^2 - a^2} = \sqrt{2^2 - 1^2} = \sqrt{3}$$

From these results, we can obtain the following trigonometric functions for the angles 30° and 60°:

$$\cos 30° = \frac{b}{c} = \frac{\sqrt{3}}{2} = 0.866$$

$$\tan 30° = \frac{a}{b} = \frac{1}{\sqrt{3}} = 0.577$$

$$\sin 60° = \frac{b}{c} = \cos 30° = \frac{\sqrt{3}}{2} = 0.866$$

$$\cos 60° = \frac{a}{c} = \sin 30° = 0.500$$

$$\tan 60° = \frac{b}{a} = \frac{\sqrt{3}}{1} = 1.732$$

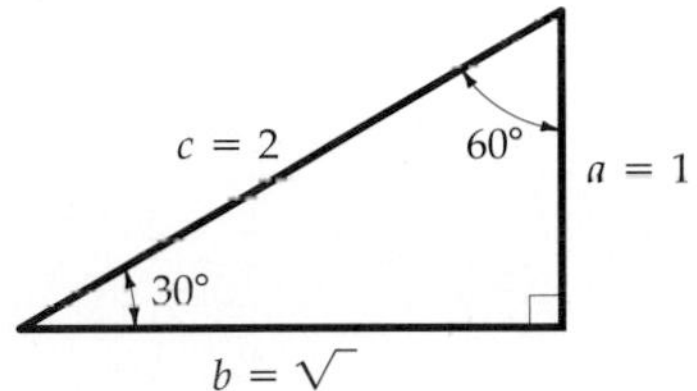

Figure A-15 A 30–60° right triangle for Example A-6.

For small angles, the length a is nearly equal to the arc length s, as can be seen in Figure A-16. The angle $\theta = s/c$ is therefore nearly equal to $\sin \theta = a/c$:

$$\sin \theta \approx \theta \qquad \text{for small values of } \theta \tag{A-46}$$

Similarly, the lengths c and b are nearly equal, so $\tan \theta = a/b$ is nearly equal to both θ and $\sin \theta$ for small values of θ:

$$\tan \theta \approx \sin \theta \approx \theta \qquad (\text{for small values of } \theta) \tag{A-47}$$

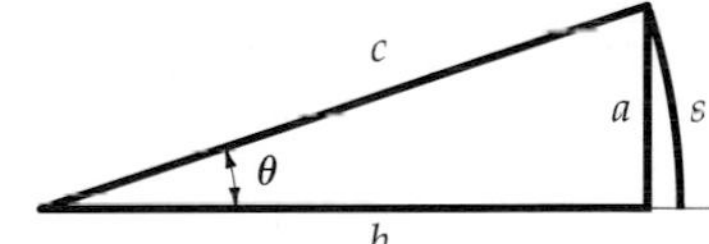

Figure A-16 For small angles, $\sin \theta = a/c$, $\tan \theta = a/b$, and the angle $\theta = s/c$ are all approximately equal.

Equations A-46 and A-47 hold only if θ is measured in radians. Since $\cos \theta = b/c$ and these lengths are nearly equal for small values of θ, we have

$$\cos \theta \approx 1 \qquad \text{for small values of } \theta \tag{A-48}$$

Example A-7

By how much do $\sin \theta$, $\tan \theta$, and θ differ when $\theta = 15°$?

This angle in radians is

$$\theta = 15° \frac{2\pi \text{ rad}}{360°} = 0.262 \text{ rad}$$

Using a calculator or a table of trigonometric functions, we obtain

$$\sin 15° = 0.259$$

and

$$\tan 15° = 0.268$$

Thus, $\sin \theta$ and θ (in radians) differ by 0.003 or about 1 percent, and $\tan \theta$ and θ differ by 0.006 or about 2 percent. For small angles, the approximation $\theta \approx \sin \theta \approx \tan \theta$ is even more accurate.

Example A-7 shows that if accuracy of a few percent is needed, small angle approximations can be used only for angles of about 15° or less. Figure A-17 shows graphs of θ, $\sin \theta$, and $\tan \theta$ versus θ, for small values of θ.

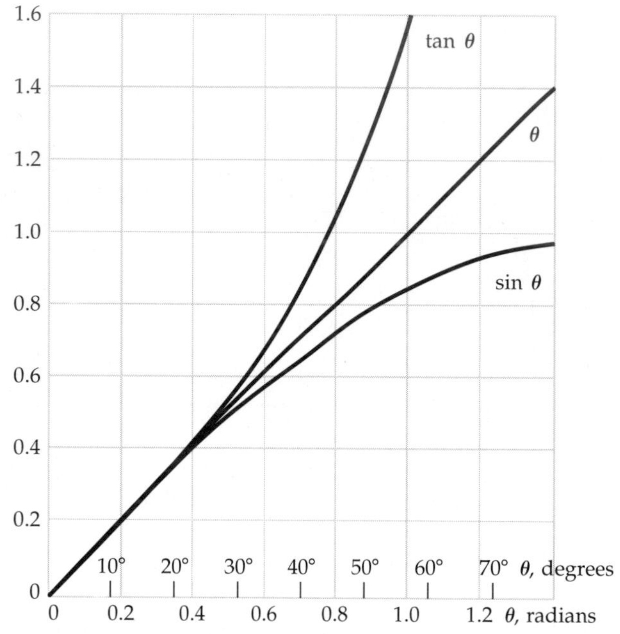

Figure A-17 Graphs of $\tan \theta$, θ, and $\sin \theta$ versus θ for small values of θ.

Figure A-18 shows an obtuse angle with its vertex at the origin and one side along the x axis. The trigonometric functions for a general angle such as this are defined by

$$\sin \theta = \frac{y}{c} \qquad \text{A-49}$$

$$\cos \theta = \frac{x}{c} \qquad \text{A-50}$$

$$\tan \theta = \frac{y}{x} \qquad \text{A-51}$$

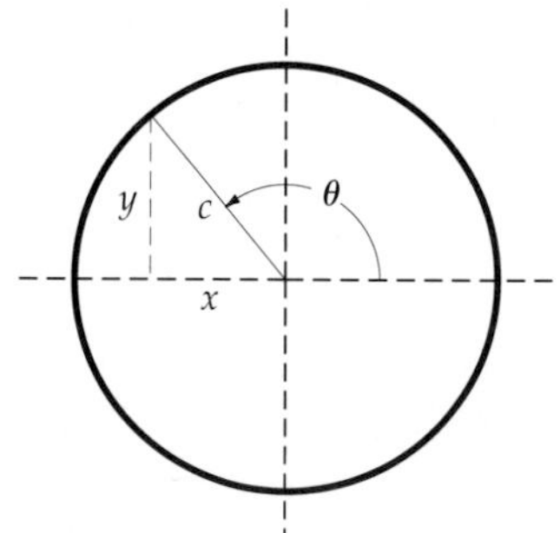

Figure A-18 Diagram for defining the trigonometric functions for an obtuse angle.

Figure A-19 shows plots of these functions versus θ. All trigonometric functions have a period of 2π. That is, when an angle changes by 2π rad, the function returns to its original value. Thus, $\sin(\theta + 2\pi) = \sin \theta$ and so forth. Some other useful relations are

$$\sin(\pi - \theta) = \sin \theta \qquad \text{A-52}$$

$$\cos(\pi - \theta) = -\cos \theta \qquad \text{A-53}$$

$$\sin(\pi/2 - \theta) = \cos \theta \qquad \text{A-54}$$

$$\cos(\pi/2 - \theta) = \sin \theta \qquad \text{A-55}$$

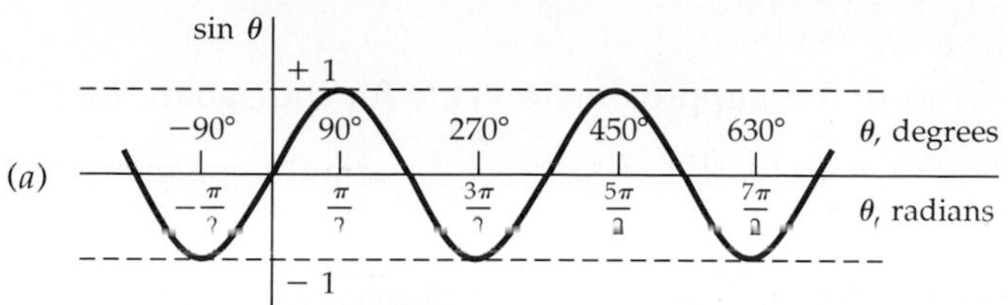

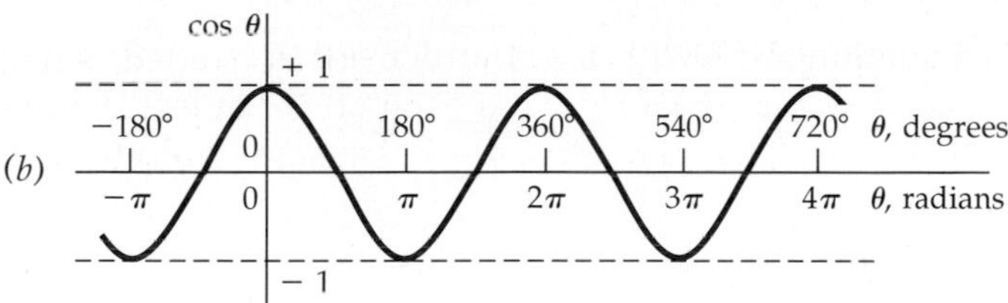

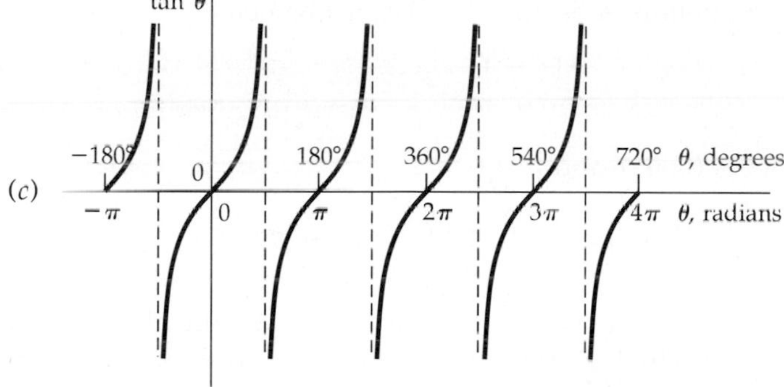

Figure A-19 The trigonometric functions sin θ, cos θ, and tan θ versus θ.

The trigonometric functions can be expressed as power series in θ. The series for sin θ and cos θ are

$$\sin \theta = \theta - \frac{\theta^3}{3!} + \frac{\theta^5}{5!} - \frac{\theta^7}{7!} + \cdots \qquad \text{A-56}$$

$$\cos \theta = 1 - \frac{\theta^2}{2!} + \frac{\theta^4}{4!} - \frac{\theta^6}{6!} + \cdots \qquad \text{A-57}$$

When θ is small, good approximations are obtained using only the first few terms in the series.

The Binomial Expansion

The binomial theorem is very useful for making approximations. One form of this theorem is

$$(1 + x)^n = 1 + nx + \frac{n(n-1)}{2!}x^2 + \frac{n(n-1)(n-2)}{3!}x^3 + \frac{n(n-1)(n-2)(n-3)}{4!}x^4 + \cdots \qquad \text{A-58}$$

If n is a positive integer, there are $n + 1$ terms in this series. If n is a real number other than a positive integer, there are an infinite number of terms. The series is valid for any value of n if x^2 is less than 1. It is also valid for $x^2 = 1$ if n is positive. The series is particularly useful if $|x|$ is much less than 1. Then each term is much smaller than the previous term and we can drop all but the first two or three terms in the equation. If $|x|$ is much less than 1, we have

$$(1 + x)^n \approx 1 + nx \qquad |x| \ll 1 \qquad \text{A-59}$$

Example A-8

Use Equation A-59 to find an approximate value for the square root of 101.

We first restate the problem in the form $(1 + x)^n$ with x much less than 1:

$$(101)^{1/2} = (100 + 1)^{1/2} = (100)^{1/2} (1 + 0.01)^{1/2} = 10(1 + 0.01)^{1/2}$$

We can now use Equation A-59 with $n = \frac{1}{2}$ and $x = 0.01$:

$$(1 + 0.01)^{1/2} \approx 1 + \tfrac{1}{2}(0.01) = 1.005$$

Then

$$(101)^{1/2} \approx 10(1.005) = 10.05$$

We can get an idea of the accuracy of this approximation by looking at the first term in Equation A-58 that has been neglected. This term is

$$\frac{n(n-1)}{2} x^2$$

Since x is 0.01, x^2 is 0.0001 and

$$\frac{n(n-1)}{2} x^2 = \frac{\frac{1}{2}(-\frac{1}{2})}{2} (0.0001) = -\frac{0.0001}{8} \approx -0.00001 = -0.001\%$$

We therefore expect our answer to be correct to within about 0.001 percent. The value of $(101)^{1/2}$ to eight significant figures is 10.049876, which differs from our approximate value of 10.05 by 0.000124 or about 0.001% of 10.05.

Complex Numbers

A general complex number z can be written

$$z = a + bi \qquad \text{A-60}$$

where a and b are real numbers and $i = \sqrt{-1}$. The quantity a is called the real part and the quantity ib is called the imaginary part of z. We can represent a complex number in a plane as shown in Figure A-20, where the x axis is the real axis and the y axis is the imaginary axis. We can use the relations $a = r \cos\theta$ and $b = r \sin\theta$ from Figure A-20 to write the complex number z in polar coordinates:

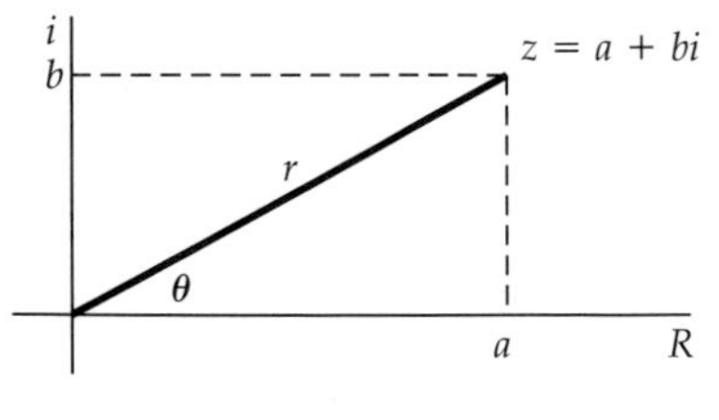

$$\begin{aligned} z &= a + bi \\ &= r\cos\theta + (r\sin\theta)i \\ &= r(\cos\theta + i\sin\theta) \end{aligned}$$

Figure A-20 Representation of a complex number in a plane. The real part of the complex number is plotted along the horizontal axis, and the imaginary part is plotted along the vertical axis.

$$z = r\cos\theta + (r\sin\theta)\, i \qquad \text{A-61}$$

where $r = \sqrt{a^2 + b^2}$ is called the magnitude of z.

When complex numbers are added or subtracted, the real and imaginary parts are added or subtracted separately:

$$z_1 + z_2 = (a_1 + b_1 i) + (a_2 + b_2 i) = (a_1 + a_2) + (b_1 + b_2)i \qquad \text{A-62}$$

However, when two complex numbers are multiplied, each part of one number is multiplied by each part of the other number:

$$\begin{aligned} z_1 z_2 = (a_1 + b_1 i)(a_2 + b_2 i) &= a_1 a_2 + b_1 b_2 i^2 + (a_1 b_2 + a_2 b_1)i \qquad \text{A-63} \\ &= a_1 a_2 - b_1 b_2 + (a_1 b_2 + a_2 b_1)i \end{aligned}$$

where we have used $i^2 = -1$.

The complex conjugate of a complex number z^* is that number obtained by replacing i with $-i$:

$$z^* = (a + bi)^* = a - bi \qquad \text{A-64}$$

The product of a complex number and its complex conjugate equals the square of the magnitude of the number:

$$zz^* = (a + bi)(a - bi) = a^2 + b^2 \qquad \text{A-65}$$

A particularly useful function of a complex number is the exponential $e^{i\theta}$. Using the expansion for e^x given in Table A-2, we have

$$e^{i\theta} = 1 + i\theta + \frac{(i\theta)^2}{2!} + \frac{(i\theta)^3}{3!} + \frac{(i\theta)^4}{4!} + \cdots$$

Using $i^2 = -1$, $i^3 = -i$, $i^4 = +1$, and so forth and separating the real parts from the imaginary parts, this expansion can be written

$$e^{i\theta} = 1 - \frac{\theta^2}{2!} + \frac{\theta^4}{4!} + \cdots + i(\theta - \frac{\theta^3}{3!} + \cdots)$$

Comparing this result with Equations A-56 and A-57, we can see that

$$e^{i\theta} = \cos\theta + i\sin\theta \qquad \text{A-66}$$

Using this result, we can express a general complex number as an exponential:

$$z = a + bi = r\cos\theta + (r\sin\theta)i = re^{i\theta} \qquad \text{A-67}$$

where $r = \sqrt{a^2 + b^2}$.

Differential Calculus

When we say that x is a function of t, we mean that for each value of t there is a corresponding value of x. An example is $x = At^2$, where A is a constant. To indicate that x is a function of t, we sometimes write $x(t)$ for x. Figure A-21 is a graph of x versus t for a typical function $x(t)$. At a particular value $t = t_1$, x has the value of x_1 as indicated. At another value t_2, x has the value x_2. The change in t, $t_2 - t_1$, is written $\Delta t = t_2 - t_1$ and the corresponding change in x is written $\Delta x = x_2 - x_1$. The ratio $\Delta x/\Delta t$ is the slope of the straight line connecting (x_1, t_1) and (x_2, t_2). If we make Δt smaller and smaller, the line connecting (x_1, t_1) and (x_2, t_2) approaches the line that is tangent to the curve at the point (x_1, t_1). The slope of this tangent line is called the derivative of x with respect to t and is written dx/dt:

$$\frac{dx}{dt} = \lim_{\Delta t \to 0} \frac{\Delta x}{\Delta t} \qquad \text{A-68}$$

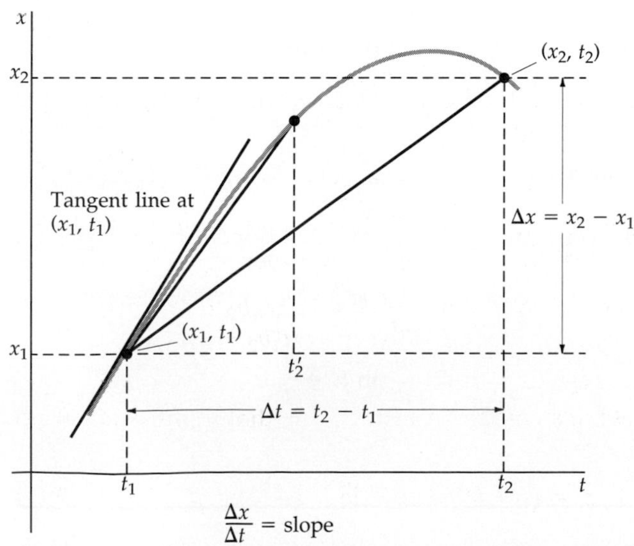

Figure A-21 Graph of a typical function $x(t)$. The points (x_1, t_1) and (x_2, t_2) are connected by a straight line. The slope of this line is $\Delta x/\Delta t$. As the time interval beginning at t_1 is decreased, the slope for that interval approaches the slope of the line tangent to the curve at time t_1, which is the derivative of x with respect to t.

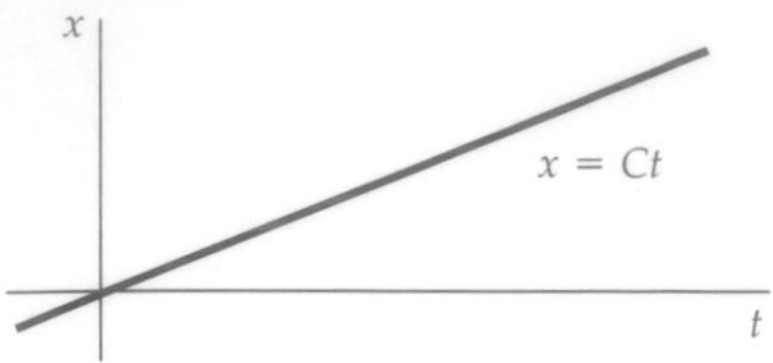

Figure A-22 Graph of the linear function $x = Ct$. This function has a constant slope C.

The derivative of a function of t is another function of t. If x is a constant, the graph of x versus t is a horizontal line with zero slope. The derivative of a constant is thus zero. In Figure A-22, x is proportional to t:

$$x = Ct$$

This function has a constant slope equal to C. Thus the derivative of Ct is C. Table A-4 lists some properties of derivatives and the derivatives of some particular functions that occur often in physics. It is followed by comments aimed at making these properties and rules clearer. More detailed discussion can be found in most calculus books.

Table A-4 **Properties of Derivatives and Derivatives of Particular Functions**

Linearity

1. The derivative of a constant times a function equals the constant times the derivative of the function:

$$\frac{d}{dt}[Cf(t)] = C\frac{df(t)}{dt}$$

2. The derivative of a sum of functions equals the sum of the derivatives of the functions:

$$\frac{d}{dt}[f(t) + g(t)] = \frac{df(t)}{dt} + \frac{dg(t)}{dt}$$

Chain rule

3. If f is a function of x and x is in turn a function of t, the derivative of f with respect to t equals the product of the derivative of f with respect to x and the derivative of x with respect to t:

$$\frac{d}{dt}f(x) = \frac{df}{dx}\frac{dx}{dt}$$

Derivative of a product

4. The derivative of a product of functions $f(t)g(t)$ equals the first function times the derivative of the second plus the second function times the derivative of the first:

$$\frac{d}{dt}[f(t)g(t)] = f(t)\frac{dg(t)}{dt} + \frac{df(t)}{dt}g(t)$$

Reciprocal derivative

5. The derivative of t with respect to x is the reciprocal of the derivative of x with respect to t, assuming that neither derivative is zero:

$$\frac{dx}{dt} = \left(\frac{dt}{dx}\right)^{-1} \quad \text{if} \quad \frac{dt}{dx} \neq 0$$

Derivatives of particular functions

6. $\frac{dC}{dt} = 0$ where C is a constant

7. $\frac{d(t^n)}{dt} = nt^{n-1}$

8. $\frac{d}{dt}\sin\omega t = \omega\cos\omega t$

9. $\frac{d}{dt}\cos\omega t = -\omega\sin\omega t$

10. $\frac{d}{dt}\tan\omega t = \omega\sec^2\omega t$

11. $\frac{d}{dt}e^{bt} = be^{bt}$

12. $\frac{d}{dt}\ln bt = \frac{1}{t}$

Example A-9

Find the derivative of $x = at^2 + bt + c$, where a, b, and c are constants.

From rule 2, we can differentiate each term separately and add the results. Using rules 1 and 7, we have

$$\frac{d(at^2)}{dt} = 2at^1 - 2at$$

Similarly, $d(bt)/dt = b$ and $dc/dt = 0$. Adding these results, we obtain

$$\frac{dx}{dt} = 2at + b$$

Comments on Rules 1 through 5

Rules 1 and 2 follow from the fact that the limiting process is linear. We can understand rule 3, the chain rule, by multiplying $\Delta f/\Delta t$ by $\Delta x/\Delta x$ and noting that, since x is a function of t, both Δx and Δf approach zero as Δt approaches zero. Since the limit of a product of two functions equals the product of their limits, we have

$$\lim_{\Delta t \to 0} \frac{\Delta f}{\Delta t} = \lim_{\Delta t \to 0} \frac{\Delta f}{\Delta x}\frac{\Delta x}{\Delta t} = \left(\lim_{\Delta x \to 0} \frac{\Delta f}{\Delta x}\right)\left(\lim_{\Delta t \to 0} \frac{\Delta x}{\Delta t}\right) = \frac{df}{dx}\frac{dx}{dt}$$

Rule 4 is not immediately apparent. The derivative of a product of functions is the limit of the ratio

$$\frac{f(t + \Delta t)g(t + \Delta t) - f(t)g(t)}{\Delta t}$$

If we add and subtract the quantity $f(t + \Delta t)g(t)$ in the numerator, we can write this ratio as

$$\frac{f(t + \Delta t)g(t + \Delta t) - f(t + \Delta t)g(t) + f(t + \Delta t)g(t) - f(t)g(t)}{\Delta t}$$

$$= f(t + \Delta t)\left[\frac{g(t + \Delta t) - g(t)}{\Delta t}\right] + g(t)\left[\frac{f(t + \Delta t) - f(t)}{\Delta t}\right]$$

As Δt approaches zero, the terms in brackets become $dg(t)/dt$ and $df(t)/dt$, respectively, and the limit of the expression is

$$f(t)\,\frac{dg(t)}{dt} + g(t)\,\frac{df(t)}{dt}$$

Rule 5 follows directly from the definition:

$$\frac{dx}{dt} = \lim_{\Delta t \to 0} \frac{\Delta x}{\Delta t} = \lim_{\Delta t \to 0} \left(\frac{\Delta t}{\Delta x}\right)^{-1} = \lim_{\Delta x \to 0} \left(\frac{\Delta t}{\Delta x}\right)^{-1} = \left(\lim_{\Delta x \to 0} \frac{\Delta t}{\Delta x}\right)^{-1}$$

Comments on Rule 7

We can obtain this important result using the binomial expansion. We have

$$f(t) = t^n$$

$$f(t + \Delta t) = (t + \Delta t)^n = t^n\left(1 + \frac{\Delta t}{t}\right)^n$$

$$= t^n\left[1 + n\,\frac{\Delta t}{t} + \frac{n(n-1)}{2!}\left(\frac{\Delta t}{t}\right)^2 + \frac{n(n-1)(n-2)}{3!}\left(\frac{\Delta t}{t}\right)^3 + \cdots\right]$$

Then

$$f(t-\Delta t)-f(t)=t^n\left[n\,\frac{\Delta t}{t}+\frac{n(n-1)}{2!}\left(\frac{\Delta t}{t}\right)^2+\cdots\right]$$

and

$$\frac{f(t-\Delta t)-f(t)}{\Delta t}=nt^{n-1}+\frac{n(n-1)}{2!}\,t^{n-2}\,\Delta t+\cdots$$

The next term omitted from the last sum is proportional to $(\Delta t)^2$, the following to $(\Delta t)^3$, and so on. Each term except the first approaches zero as Δt approaches zero. Thus

$$\frac{df}{dt}=\lim_{\Delta x\to 0}\frac{f(t+\Delta t)+f(t)}{\Delta t}=nt^{n-1}$$

Comments on Rules 8 to 10

We first write $\sin\omega t=\sin\theta$ with $\theta=\omega t$ and use the chain rule,

$$\frac{d\sin\theta}{dt}=\frac{d\sin\theta}{d\theta}\,\frac{d\theta}{dt}=\omega\,\frac{d\sin\theta}{d\theta}$$

We then use the trigonometric formula for the sine of the sum of two angles θ and $\Delta\theta$:

$$\sin(\theta+\Delta\theta)=\sin\Delta\theta\cos\theta+\cos\Delta\theta\sin\theta$$

Since $\Delta\theta$ is to approach zero, we can use the small-angle approximations

$$\sin\Delta\theta\approx\Delta\theta\qquad\text{and}\qquad\cos\Delta\theta\approx 1$$

Then

$$\sin(\theta+\Delta\theta)\approx\Delta\theta\cos\theta+\sin\theta$$

and

$$\frac{\sin(\theta+\Delta\theta)-\sin\theta}{\Delta\theta}\approx\cos\theta$$

Similar reasoning can be applied to the cosine function to obtain rule 9.

Rule 10 is obtained by writing $\tan\theta=\sin\theta/\cos\theta$ and applying Rule 4 along with Rules 8 and 9.

$$\begin{aligned}\frac{d}{dt}(\tan\theta)&=\frac{d}{dt}(\sin\theta)(\cos\theta)^{-1}=\sin\theta\,\frac{d}{dt}(\cos\theta)^{-1}+\frac{d\,(\sin\theta)}{dt}(\cos\theta)^{-1}\\&=\sin\theta\,(-1)(\cos\theta)^{-2}(-\sin\theta)+(\cos\theta)(\cos\theta)^{-1}\\&=\frac{\sin^2\theta}{\cos^2\theta}+1=\tan^2\theta+1=\sec^2\theta\end{aligned}$$

Comments on Rule 11

Again we use the chain rule

$$\frac{de^\theta}{dt}=b\,\frac{de^\theta}{d\theta}\qquad\text{with}\qquad\theta=bt$$

and the series expansion for the exponential function:

$$e^{\theta+\Delta\theta}=e^\theta e^{\Delta\theta}=e^\theta\left[1+\Delta\theta+\frac{(\Delta\theta)^2}{2!}+\frac{(\Delta\theta)^3}{3!}+\cdots\right]$$

Then

$$\frac{e^{\theta+\Delta\theta}-e^\theta}{\Delta\theta}=e^\theta+e^\theta\,\frac{\Delta\theta}{2!}+e^\theta\,\frac{(\Delta\theta)^2}{3!}+\cdots$$

As $\Delta\theta$ approaches zero, the right side of the equation above approaches e^θ.

Comments on Rule 12

Let

$$y = \ln bt$$

Then

$$e^y = bt \qquad \text{and} \qquad \frac{dl}{dy} = \frac{1}{b} e^y = t$$

Then using rule 5, we obtain

$$\frac{dy}{dt} = \left(\frac{dt}{dy}\right)^{-1} = \frac{1}{t}$$

Integral Calculus

Integration is related to the problem of finding the area under a curve. It is also the inverse of differentiation. Figure A-23 shows a function $f(t)$. The area of the shaded element is approximately $f_i \, \Delta t_i$, where f_i is evaluated anywhere in the interval Δt_i. This approximation improves if Δt_i is very small. The total area from t_1 to t_2 is found by summing all the area elements from t_1 to t_2 and taking the limit as each Δt_i approaches zero. This limit is called the integral of f over t and is written

$$\int_{t_1}^{t_2} f \, dt = Area = \lim_{\Delta t_i \to 0} \sum_i f_i \, \Delta t_i$$

If we integrate some function $f(t)$ from t_1 to some general value of t, we obtain another function of t. Let us call this function y:

$$y = \int_{t_1}^{t} f \, dt$$

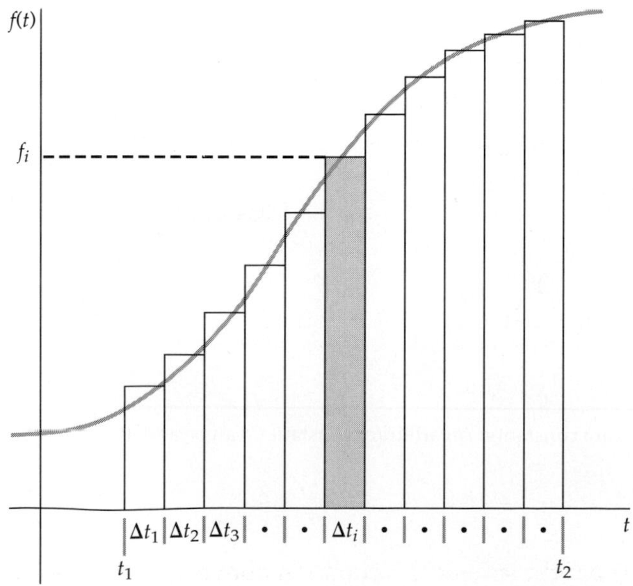

Figure A-23 A general function $f(t)$. The area of the shaded element is approximately $f_i \, \Delta t_i$, where f_i is evaluated anywhere in the interval.

The function y is the area under the f-versus-t curve from t_1 to a general value t. For a small interval Δt, the change in the area Δy is approximately $f\,\Delta t$.

$$\Delta y \approx f\,\Delta t$$

$$f \approx \frac{\Delta y}{\Delta t}$$

If we take the limit as Δt approaches 0, we can see that f is the derivative of y:

$$f = \frac{dy}{dt}$$

The relation between y and f is often written

$$y = \int f\,dt$$

where $\int f\,dt$ is called an **indefinite integral.** To evaluate an indefinite integral, we find the function y whose derivative is f. The definite integral of f from t_1 to t_2 is $y(t_1) - y(t_2)$, where $df/dt = y$:

$$\int_{t_1}^{t_2} f\,dt = y(t_2) - y(t_1)$$

Example A-10

Find the indefinite integral of $f(t) = t$.

The function whose derivative is t is $\frac{1}{2}t^2$ plus any constant. Thus,

$$\int t\,dt = \tfrac{1}{2}t^2 + C$$

where C is any constant.

Table A-5 lists some important integration formulas. More extensive lists of differentiation and integration formulas can be found in handbooks such as Herbert Dwight's *Tables of Integrals and Other Mathematical Data,* fourth edition, Macmillan Publishing Company, Inc., New York, 1961.

Table A-5 **Integration Formulas†**

1. $\int A\,dt = At$	5. $\int e^{bt}\,dt = \frac{1}{b}e^{bt}$
2. $\int At\,dt = \frac{1}{2}At^2$	6. $\int \cos\omega t\,dt = \frac{1}{\omega}\sin\omega t$
3. $\int At^n\,dt = A\frac{t^{n+1}}{n+1}$ $\quad n \neq -1$	7. $\int \sin\omega t\,dt = -\frac{1}{\omega}\cos\omega t$
4. $\int At^{-1}\,dt = A\ln t$	

†In these formulas, A, b, and ω are constants. An arbitrary constant C can be added to the right side of each equation.

Appendix B

SI Units

Basic Units	
Length	The *meter* (m) is the distance traveled by light in a vacuum in 1/299,792,458 s
Time	The *second* (s) is the duration of 9,192,631,770 periods of the radiation corresponding to the transition between the two hyperfine levels of the ground state of the ^{133}Cs atom
Mass	The *kilogram* (kg) is the mass of the international standard body preserved at Sèvres, France
Current	The *ampere* (A) is that current in two very long parallel wires 1 m apart that gives rise to a magnetic force per unit length of 2×10^{-7} N/m
Temperature	The *kelvin* (K) is 1/273.16 of the thermodynamic temperature of the triple point of water
Luminous intensity	The *candela* (cd) is the luminous intensity, in the perpendicular direction, of a surface of area 1/600,000 m^2 of a blackbody at the temperature of freezing platinum at a pressure of 1 atm

Derived Units		
Force	newton (N)	$1\ N = 1\ kg{\cdot}m/s^2$
Work, energy	joule (J)	$1\ J = 1\ N{\cdot}m$
Power	watt (W)	$1\ W = 1\ J/s$
Frequency	hertz (Hz)	$1\ Hz = s^{-1}$
Charge	coulomb (C)	$1\ C = 1\ A{\cdot}s$
Potential	volt (V)	$1\ V = 1\ J/C$
Resistance	ohm (Ω)	$1\ \Omega = 1\ V/A$
Capacitance	farad (F)	$1\ F = 1\ C/V$
Magnetic field	tesla (T)	$1\ T = 1\ N/A{\cdot}m$
Magnetic flux	weber (Wb)	$1\ Wb = 1\ T{\cdot}m^2$
Inductance	henry (H)	$1\ H = 1\ J/A^2$

Appendix C

Numerical Data

Terrestrial Data	
Acceleration of gravity g	
Standard value	9.80665 m/s^2
	32.1740 ft/s^2
At sea level, at equator†	9.7804 m/s^2
At sea level, at poles†	9.8322 m/s^2
Mass of earth M_E	5.98×10^{24} kg
Radius of earth R_E, mean	6.37×10^6 m
	3960 mi
Escape speed $\sqrt{2R_E g}$	1.12×10^4 m/s
	6.95 mi/s
Solar constant‡	1.35 kW/m^2
Standard temperature and pressure (STP):	
Temperature	273.15 K
Pressure	101.325 kPa
	1.00 atm
Molar mass of air	28.97 g/mol
Density of air (STP), ρ_{air}	1.293 kg/m^3
Speed of sound (STP)	331 m/s
Heat of fusion of H_2O (0°C, 1 atm)	333.5 kJ/kg
Heat of vaporization of H_2O (100°C, 1 atm)	2.257 MJ/kg

†Measured relative to the earth's surface.
‡Average power incident normally on 1 m^2 outside the earth's atmosphere at the mean distance from the earth to the sun.

Astronomical Data	
Earth	
Distance to moon†	3.844×10^8 m
	2.389×10^5 mi
Distance to sun, mean†	1.496×10^{11} m
	9.30×10^7 mi
	1.00 AU
Orbital speed, mean	2.98×10^4 m/s
Moon	
Mass	7.35×10^{22} kg
Radius	1.738×10^6 m
Period	27.32 d
Acceleration of gravity at surface	1.62 m/s^2
Sun	
Mass	1.99×10^{30} kg
Radius	6.96×10^8 m

†Center to center.

Physical Constants		
Gravitational constant	G	$6.672\,6 \times 10^{-11}$ N·m²/kg²
Speed of light	c	$2.997\,924\,58 \times 10^{8}$ m/s
Electron charge	e	$1.602\,177 \times 10^{-19}$ C
Avogadro's number	N_A	$6.022\,137 \times 10^{23}$ particles/mol
Gas constant	R	8.314 51 J/mol·K
		1.987 22 cal/mol·K
		$8.205\,78 \times 10^{-2}$ L·atm/mol·K
Boltzmann's constant	$k = R/N_A$	$1.380\,658 \times 10^{-23}$ J/K
		$8.617\,385 \times 10^{-5}$ eV/K
Unified mass unit	$u = (1/N_A)$ g	$1.660\,540 \times 10^{-24}$ g
Coulomb constant	$k = 1/4\pi\epsilon_0$	$8.987\,551\,788 \times 10^{9}$ N·m²/C²
Permittivity of free space	ϵ_0	$8.854\,187\,817 \times 10^{-12}$ C²/N·m²
Permeability of free space	μ_0	$4\pi \times 10^{-7}$ N/A²
		$1.256\,637 \times 10^{-6}$ N/A²
Planck's constant	h	$6.626\,076 \times 10^{-34}$ J·s
		$4.135\,669 \times 10^{-15}$ eV·s
	$\hbar = h/2\pi$	$1.054\,573 \times 10^{-34}$ J·s
		$6.582\,122 \times 10^{-16}$ eV·s
Mass of electron	m_e	$9.109\,390 \times 10^{-31}$ kg
		510.999 1 keV/c^2
Mass of proton	m_p	$1.672\,623 \times 10^{-27}$ kg
		938.272 3 MeV/c^2
Mass of neutron	m_n	$1.674\,929 \times 10^{-27}$ kg
		939.565 6 MeV/c^2
Bohr magneton	$m_B = e\hbar/2m_e$	$9.274\,015\,4 \times 10^{-24}$ J/T
		$5.788\,382\,63 \times 10^{-5}$ eV/T
Nuclear magneton	$m_n = e\hbar/2m_p$	$5.050\,786\,6 \times 10^{-27}$ J/T
		$3.152\,451\,66 \times 10^{-8}$ eV/T
Magnetic flux quantum	$\phi_0 = h/2e$	$2.067\,834\,6 \times 10^{-15}$ T·m²
Quantized Hall resistance	$R_K = h/e^2$	$2.581\,280\,7 \times 10^{4}$ Ω
Rydberg constant	R_H	$1.097\,373\,153\,4 \times 10^{7}$ m⁻¹
Josephson frequency–voltage quotient	$2e/h$	$4.835\,979 \times 10^{14}$ Hz/V
Compton wavelength	$\lambda_C = h/m_e c$	$2.426\,310\,58 \times 10^{-12}$ m

For additional data, see the front and back endpapers and the following tables in the text

Table 1-1 Prefixes for powers of 10, p. 3

Table 1-2 Order of magnitude of some lengths, p. 9

Table 1-3 Order of magnitude of some masses, p. 9

Table 1-4 Order of magnitude of some time intervals, p. 10

Table 3-1 Properties of vectors, p. 52

Table 5-1 Approximate values of frictional coefficients, p. 110

Table 8-1	Moments of inertia of uniform bodies of various shapes, p. 233
Table 8-2	Comparison of linear motion and rotational motion, p. 245
Table 10-1	Measurements of *G*, p. 306
Table 11-1	Densities of selected substances, p. 332
Table 11-2	Young's modulus *Y* and strengths of various materials, p. 334
Table 11-3	Approximate values of the shear modulus M_s of various materials, p. 336
Table 11-4	Approximate values of the bulk modulus *B* of various materials, p. 337
Table 11-5	Coefficients of viscosity of various fluids, p. 352
Table 14-1	Intensity and intensity level of some common sounds, $I_0 = 10^{-12}$ W/m^2, p. 447
Table 15-1	Approximate values of the coefficients of thermal expansion for various substances, p. 493
Table 15-2	Vapor pressure of water versus temperature, p. 507
Table 15-3	Critical temperatures T_c for various substances, p. 507
Table 16-1	Specific heat and molar heat capacities for various solids and liquids at 20°C, p. 519
Table 16-2	Normal melting point (MP), latent heat of fusion L_f, normal boiling point (BP), and latent heat of vaporization L_v for various substances at 1 atm, p. 523
Table 16-3	Thermal conductivities *k* for various materials, p. 526
Table 16-4	*R* factors $\Delta x/k$ for various building materials, p. 527
Table 16-5	Molar heat capacities in J/mol·K of various gases at 25°C, p. 543
Table 18-1	Some Electric Fields in Nature, p. 608
Table 21-1	Dielectric Constants and Strengths of Various Materials, p. 697
Table 22-1	Resistivities and Temperature Coefficients, p. 722
Table 22-2	Wire Diameters and Cross-sectional Areas for Commonly Used Copper Wires, p. 722
Table 27-1	Magnetic Susceptibility of Various Materials at 20° C, p. 881
Table 27-2	Maximum Values of $\mu_0 M$ and K_m for Some Ferromagnetic Materials, p. 889
Table 29-1	The Electromagnetic Spectrum, p. 958
Table 30-1	Index of Refraction for Yellow Sodium Light (λ = 589 nm), p. 991
Table 34-1	Rest Energies of Some Elementary Particles and Light Nuclei, p. 1130
Table 35-1	Approximate Dates of Some Important Experiments and Theories, 1881–1932, p. 1146

Appendix D

Conversion Factors

Conversion factors are written as equations for simplicity; relations marked with an asterisk are exact.

Length

1 km = 0.6215 mi

1 mi = 1.609 km

1 m = 1.0936 yd = 3.281 ft = 39.37 in

*1 in = 2.54 cm

*1 ft = 12 in = 30.48 cm

*1 yd = 3 ft = 91.44 cm

1 lightyear = 1 c·y = 9.461×10^{15} m

*1 Å=0.1 nm

Area

*1 $m^2 = 10^4$ cm^2

1 km^2 = 0.3861 mi^2 = 247.1 acres

*1 in^2 = 6.4516 cm^2

1 $ft^2 = 9.29 \times 10^{-2}$ m^2

1 m^2 = 10.76 ft^2

*1 acre = 43,560 ft^2

1 mi^2 = 640 acres = 2.590 km^2

Volume

*1 $m^3 = 10^6$ cm^3

*1 L = 1000 $cm^3 = 10^{-3}$ m^3

1 gal = 3.786 L

1 gal = 4 qt = 8 pt = 128 oz = 231 in^3

1 in^3 = 16.39 cm^3

1 ft^3 = 1728 in^3 = 28.32 L = 2.832×10^4 cm^3

Time

*1 h = 60 min = 3.6 ks

*1 d = 24 h = 1440 min = 86.4 ks

1 y = 365.24 d = 31.56 Ms

Speed

1 km/h = 0.2778 m/s = 0.6215 mi/h

1 mi/h = 0.4470 m/s = 1.609 km/h

1 mi/h = 1.467 ft/s

Angle and Angular Speed

*π rad = 180°

1 rad = 57.30°

1° = 1.745×10^{-2} rad

1 rev/min = 0.1047 rad/s

1 rad/s = 9.549 rev/min

Mass

*1 kg = 1000 g

*1 tonne = 1000 kg – 1 Mg

1 u = 1.6606×10^{-27} kg

1 kg = 6.022×10^{23} u

1 slug = 14.59 kg

1 kg = 6.852×10^{-2} slug

1 u = 931.50 MeV/c^2

Density

*1 g/cm^3 = 1000 kg/m^3 = 1 kg/L

(1 g/cm^3)g = 62.4 lb/ft^3

Force

1 N = 0.2248 lb = 10^5 dyn

1 lb = 4.4482 N

(1 kg)g = 2.2046 lb

Pressure

*1 Pa = 1 N/m^2

*1 atm = 101.325 kPa = 1.01325 bars

1 atm = 14.7 lb/in^2 = 760 mmHg

= 29.9 inHg = 33.8 ftH_2O

1 lb/in^2 = 6.895 kPa

1 torr = 1 mmHg = 133.32 Pa

1 bar = 100 kPa

Energy

*1 kW·h = 3.6 MJ

*1 cal = 4.1840 J

1 ft·lb = 1.356 J = 1.286 × 10^{-3} Btu

*1 L·atm = 101.325 J

1 L·atm = 24.217 cal

1 Btu = 778 ft·lb = 252 cal = 1054.35 J

1 eV = 1.602 × 10^{-19} J

1 u·c^2 = 931.50 MeV

*1 erg = 10^{-7} J

Power

1 horsepower = 550 ft·lb/s = 745.7 W

1 Btu/min = 17.58 W

1 W = 1.341 × 10^{-3} horsepower

= 0.7376 ft·lb/s

Magnetic Field

*1 G = 10^{-4} T

*1 T = 10^4 G

Thermal Conductivity

1 W/m·K = 6.938 Btu·in/h·ft^2·F°

1 Btu·in/h·ft^2·F° = 0.1441 W/m·K

Appendix E Periodic Table

1	2	3	4	5	6	7	8	9	10	11	12	13	14	15	16	17	18
1 **H** 1.00797																	2 **He** 4.003
3 **Li** 6.941	4 **Be** 9.012											5 **B** 10.81	6 **C** 12.011	7 **N** 14.007	8 **O** 15.9994	9 **F** 19.00	10 **Ne** 20.179
11 **Na** 22.990	12 **Mg** 24.31											13 **Al** 26.98	14 **Si** 28.09	15 **P** 30.974	16 **S** 32.064	17 **Cl** 35.453	18 **Ar** 39.948
19 **K** 39.102	20 **Ca** 40.08	21 **Sc** 44.96	22 **Ti** 47.88	23 **V** 50.94	24 **Cr** 52.00	25 **Mn** 54.94	26 **Fe** 55.85	27 **Co** 58.93	28 **Ni** 58.69	29 **Cu** 63.55	30 **Zn** 65.38	31 **Ga** 69.72	32 **Ge** 72.59	33 **As** 74.92	34 **Se** 78.96	35 **Br** 79.9[illegible]	36 **Kr** 83.80
37 **Rb** 85.47	38 **Sr** 87.62	39 **Y** 88.906	40 **Zr** 91.22	41 **Nb** 92.91	42 **Mo** 95.94	43 **Tc** (98)	44 **Ru** 101.1	45 **Rh** 102.905	46 **Pd** 106.4	47 **Ag** 107.870	48 **Cd** 112.41	49 **In** 114.82	50 **Sn** 118.69	51 **Sb** 121.75	52 **Te** 127.60	53 **I** 126.90	54 **Xe** 131.29
55 **Cs** 132.905	56 **Ba** 137.33	57–71 **Rare** **Earths**	72 **Hf** 178.49	73 **Ta** 180.95	74 **W** 183.85	75 **Re** 186.2	76 **Os** 190.2	77 **Ir** 192.2	78 **Pt** 195.09	79 **Au** 196.97	80 **Hg** 200.59	81 **Tl** 204.37	82 **Pb** 207.19	83 **Bi** 208.98	84 **Po** (210)	85 **At** (210)	86 **Rn** (222)
87 **Fr** (223)	88 **Ra** (226)	89–103 **Actinides**	104 **Rf** (261)	105 **Ha** (260)	106 (263)	107 (262)	108 (265)	109 (266)									

Rare Earths (Lanthanides)	57 **La** 138.91	58 **Ce** 140.12	59 **Pr** 140.91	60 **Nd** 144.24	61 **Pm** (147)	62 **Sm** 150.36	63 **Eu** 152.0	64 **Gd** 157.25	65 **Tb** 158.92	66 **Dy** 162.50	67 **Ho** 164.93	68 **Er** 167.26	69 **Tm** 168.93	70 **Yb** 173.04	71 **Lu** 174.97
Actinides	89 **Ac** 227.03	90 **Th** 232.04	91 **Pa** 231.04	92 **U** 238.03	93 **Np** 237.05	94 **Pu** (244)	95 **Am** (243)	96 **Cm** (247)	97 **Bk** (247)	98 **Cf** (251)	99 **Es** (252)	100 **Fm** (257)	101 **Md** (258)	102 **No** (259)	103 **Lr** (260)

The 1–18 group designation has been recommended by the International Union of Pure and Applied Chemistry (IUPAC).

Atomic Numbers and Atomic Masses

Element	Symbol	Atomic Number	Atomic Mass
Actinium	Ac	89	227.03
Aluminum	Al	13	26.98
Americium	Am	95	(243)
Antimony	Sb	51	121.75
Argon	Ar	18	39.948
Arsenic	As	33	74.92
Astatine	At	85	(210)
Barium	Ba	56	137.3
Berkelium	Bk	97	(247)
Beryllium	Be	4	9.012
Bismuth	Bi	83	208.98
Boron	B	5	10.81
Bromine	Br	35	79.90
Cadmium	Cd	48	112.41
Calcium	Ca	20	40.08
Californium	Cf	98	(251)
Carbon	C	6	12.011
Cerium	Ce	58	140.12
Cesium	Cs	55	132.905
Chlorine	Cl	17	35.453
Chromium	Cr	24	52.00
Cobalt	Co	27	58.93
Copper	Cu	29	63.55
Curium	Cm	96	(247)
Dysprosium	Dy	66	162.50
Einsteinium	Es	99	(252)
Erbium	Er	68	167.26
Europium	Eu	63	152.0
Fermium	Fm	100	(257)
Fluorine	F	9	19.00
Francium	Fr	87	(223)
Gadolinium	Gd	64	157.25
Gallium	Ga	31	69.72
Germanium	Ge	32	72.59
Gold	Au	79	196.97
Hafnium	Hf	72	178.49
Hahnium	Ha	105	(260)
Helium	He	2	4.003
Holmium	Ho	67	164.93
Hydrogen	H	1	1.00797
Indium	In	49	114.82
Iodine	I	53	126.90
Iridium	Ir	77	192.2
Iron	Fe	26	55.85
Krypton	Kr	36	83.80
Lanthanum	La	57	138.91
Lawrencium	Lr	103	(260)
Lead	Pb	82	207.19
Lithium	Li	3	6.941
Lutetium	Lu	71	174.97
Magnesium	Mg	12	24.31
Manganese	Mn	25	54.94
Mendelevium	Md	101	(258)
Mercury	Hg	80	200.59
Molybdenum	Mo	42	95.94
Neodymium	Nd	60	144.24
Neon	Ne	10	20.179
Neptunium	Np	93	237.05
Nickel	Ni	28	58.69
Niobium	Nb	41	92.91
Nitrogen	N	7	14.007
Nobelium	No	102	(259)
Osmium	Os	76	190.2
Oxygen	O	8	15.9994
Palladium	Pd	46	106.4
Phosphorus	P	15	30.974
Platinum	Pt	78	195.09
Plutonium	Pu	94	(244)
Polonium	Po	84	(210)
Potassium	K	19	39.102
Praseodymium	Pr	59	140.91
Promethium	Pm	61	(147)
Protactinium	Pa	91	231.04
Radium	Ra	88	(226)
Radon	Rn	86	(222)
Rhenium	Re	75	186.2
Rhodium	Rh	45	102.905
Rubidium	Rb	37	85.47
Ruthenium	Ru	44	101.1
Rutherfordium	Rf	104	(261)
Samarium	Sm	62	150.36
Scandium	Sc	21	44.96
Selenium	Se	34	78.96
Silicon	Si	14	28.09
Silver	Ag	47	107.870
Sodium	Na	11	22.990
Strontium	Sr	38	87.62
Sulfur	S	16	32.064
Tantalum	Ta	73	180.95
Technetium	Tc	43	(98)
Tellurium	Te	52	127.60
Terbium	Tb	65	158.92
Thallium	Tl	81	204.37
Thorium	Th	90	232.04
Thulium	Tm	69	168.93
Tin	Sn	50	118.69
Titanium	Ti	22	47.88
Tungsten	W	74	183.85
Uranium	U	92	238.03
Vanadium	V	23	50.94
Xenon	Xe	54	131.29
Ytterbium	Yb	70	173.04
Yttrium	Y	39	88.906
Zinc	Zn	30	65.38
Zirconium	Zr	40	91.22

Illustration Credits

Part Openers
Part 1 p. xix, 18 © James Sugar/Black Star; **p. 19** © D. Cavagnaro/Peter Arnold, Inc.
Part 2 p. xxii, 366–367 © Michael Freeman.
Part 3 p. xxiii, 484–485 Sandia National Laboratories.
Part 4 p. xxiv, 596–597 © 1988 Richard Megna/Fundamental Photos.
Part 5 p. xxvii, 972–973 Courtesy AT&T Archives.
Part 6 p. xxviii, 1098–1099 Courtesy AT&T Archives.

Chapter 1
Opener p. 1 The Granger Collection; **p. 3 (a)** The Granger Collection; **(b)** Frequency Electronics, Inc.; **p. 4 (a)** McDonald Observatory; **(b)** Bruce Coleman; **p. 5** Owen Franken/Stock, Boston; **p. 9 (a)** IBM Almaden Research Center; **(b)** © Lennart Nilsson, from *The Body Victorious,* Delacourt Press, 1985; **(c)** Kent and Donnan Dannen/Photo Researchers; **(d)** N.A.S.A. (68-HC-74); **(e)** Smithsonian Institution; **p. 11** B. A. Watkinson; **p. 12** Fermi Institute, University of Chicago; **p. 14** AIP Niels Bohr Library.

Chapter 2
Opener p. 20 © George D. Lepp/Comstock; **p. 24** © Gunter Ziesler/Peter Arnold, Inc., **p. 29** Estate of Harold E. Edgerton/Courtesy Palm Press, Inc.; **p. 30** © Sidney Harris; **p. 34 (a)** Courtesy Stanford Linear Accelerator Center, U.S. Department of Energy; **(b)** John Seeman, SLAC; **p. 35** Courtesy General Motors Corp.

Chapter 3
Opener p. 47 © Lewis Portnoy/Spectra-Action, Inc.; **p. 60** © 1968 Fundamental Photographs; **p. 63** Scala/Art Resource; **p. 64** *PSSC Physics,* 2nd ed., 1965. D. C. Heath & Co. and Education Development Center, Inc., Newton, Massachusetts; **p. 65** N.A.S.A. (82-HC-78); **Figure 3-27** AIP Niels Bohr Library; **p. 68** N.A.S.A. (81-HC-408); **p. 75** Courtesy New York Public Library.

Chapter 4
Opener p. 77 The Granger Collection; **p. 79** (*left*) © George Hall/Woodfin Camp and Assoc.; (*right*) © Arthur Tilley/FPG International; **p. 83** © Berenice Abbott/Commerce Graphics, Ltd., Inc.; **p. 84** N.A.S.A. (76-HC-6); **p. 87** Culver Pictures; **p. 88 (a)** © Cotton Coulson/Woodfin Camp and Assoc.; **(b)** © 1972 Gary Ladd; **p. 89 (c)** Los Alamos Scientific Laboratory; **(d)** © Science Photo Library/Photo Researchers; **p. 90** © Fundamental Photographs; **Figure 4-7 (b)** © Michael Abbey/Photo Researchers; **p. 95** Museum of Modern Art, New York; **p. 105** © David de Lossy/The Image Bank.

Chapter 5
Opener p. 107 Courtesy Center for Engineering Design, University of Utah, Salt Lake City; **p. 108 (a)** F. P. Bowden and D. Tabor, *Friction and Lubrication of Solids,* Oxford University Press, 1950. Reprinted by permission of the publisher; **(b)** Uzi Landman, W. David Luedtke, Georgia Institute of Technology; **Figure 5-5** © Jean-Claude Lejeune/Stock, Boston; **Figure 5-10 (b)** © Malraux Photography/The Image Bank; **p. 115** © Robert Alexander/Photo Researchers; **p. 116** (*top*) © Fundamental Photographs; **Figure 5-12 (b)** © Tom Walker/Stock, Boston; **p. 117** © **(a)** N.A.S.A. (81-HC-945); **(b)** © Jonathan Blair/Woodfin Camp and Assoc.; **p. 123** Sandia National Laboratories; **p. 124** N.A.S.A. (85-HC-345); **p. 131** © P. Thomann/The Image Bank.

Chapter 6
Opener p. 135 © 1990 Estate of Harold E. Edgerton/Courtesy Palm Press, Inc.; **p. 139** Naples National Museum/Scala/Art Resource; **p. 142** © L. Grant/FPG International; **p. 148** © Steve Leonard/Black Star; **Figure 6-12 (b)** © T. Alvez/FPG International; **p. 156** © M. C. Escher/Cordon Art, B. V. De Baarn, Holland; **p. 157** © 1973 Berenice Abbott/Photo Researchers; **p. 167** (*left*) USGS; (*right*) New York State Commerce Department; **p. 168 (a)** © Stan Sholik/FPG International; **(b)** © Leicester University/Science Photo Library/Photo Researchers; **p. 169** © Visual Horizons/FPG International.

Chapter 7
Opener p. 182 © David Parker/Science Photo Library/Photo Researchers; **Figure 7-1** © Estate of Harold E. Edgerton/Courtesy Palm Press, Inc.; **p. 198** © Estate of Harold E. Edgerton/Courtesy Palm Press, Inc.; **Figure 7-23** © Estate of Harold E. Edgerton/Courtesy Palm Press, Inc.; **Figure 7-24 (b)** © Estate of Harold E. Edgerton/Courtesy Palm Press, Inc.; **p. 205** Courtesy of Mercedes-Benz of North America, Montvale, New Jersey; **p. 209 (a)** © 1973 Berenice Abbott/Photo Researchers; **(b)** Brookhaven National Laboratory; **(c) (1–4)** Alar and Juri Toomre, from *Cambridge*

Atlas of Astronomy, Cambridge University Press; **(d)** Kitt Peak National Observatory, from *Cambridge Atlas of Astronomy*, Cambridge University Press, 1985; **p. 211** © Dr. David Jones/Science Photo Library/Photo Researchers; **p. 215 (a)** N.A.S.A. (393-479)/Shostal Superstock; **(b)** © Bill Curtsinger/Photo Researchers; **p. 220** Courtesy Central Scientific Company.

Chapter 8

Opener p. 227 © David F. Malin/Anglo-Australian Telescope Board; **p. 229** Courtesy Center for Engineering Design, University of Utah, Salt Lake City; **p. 232** Culver Pictures; **p. 236** PAR/NYC, Inc. Archives; **p. 237 (a)** and **(b)** © David Malin, Anglo-Australian Telescope Board; **Figure 8-21** Focus on Sports; **Figure 8-22** © Estate of Harold E. Edgerton/Courtesy Palm Press, Inc.; **p. 247** © Dick Luria/FPG International; **p. 248 (a)** and **(b)** N.A.S.A. (84-HC-582); **p. 254** National Oceanographic and Atmospheric Administration; **p. 257** © Dick Luria/Photo Researchers; **p. 261** (*left*) Richard Minnix; (*right*) Courtesy Central Scientific Company.

Chapter 9

Opener p. 275 © Tom McHugh/Photo Researchers; **p. 276** Manitowoc Engineering Co., Manitowoc, Wisconsin; **p. 277** © Mike and Carol Werner/Comstock; **p. 278** © Malcolm Hanes/Bruce Coleman; **p. 281** © Marc Romanelli/The Image Bank; **p. 286** © Rick Rickman/Black Star; **p. 287** (*left*) © Yulsman/The Image Bank; (*right*) Demetrios Zangos.

Chapter 10

Opener p. 295 Collection of Historical Scientific Instruments, Harvard University; **p. 296** © Erich Lessing/Magnum; **p. 298 (a)** N.A.S.A.; **(b)** © Dennis di Cicco/Peter Arnold, Inc.; **p. 302** © Erich Lessing, Culture and Fine Arts Archive, Vienna; **Figure 10-8 (a)** Courtesy Deutsches Museum, Munich; **p. 306** Courtesy Central Scientific Company; **p. 308 (a)** *Physics Today*, May 1988; **(b)** N.A.S.A.; **p. 313** N.A.S.A. (69-HC-905); **p. 317 (b)** N.A.S.A. (80-HC-627); **Figure 1 p. 322** By permission of the Syndics of Cambridge University Library, Additional MS 3975, p. 15; **Figure 2 p. 323** From Voltaire, *The Elements of Sir Isaac Newton's Philosophy*, 1738, New York: Frank Cass, 1967, p. 97; **Figure 3 p. 323** From Newton, *Opticks*, 1730; **p. 327** Art Resource.

Chapter 11

Opener p. 331 Dr. Tai Il Mah, from *Popular Science*, June 1990; **p. 332** © Daemmrich/Stock, Boston; **p. 335 (a)** Scala/Art Resource; **(b)** © Jonathan Watts/Science Photo Library/Photo Researchers; **p. 341 (a)** © David Burnett/Woodfin Camp and Assoc.; **(b)** PAR/NYC, Inc. Archives; **p. 342** © Chuck O'Rear/Woodfin Camp and Assoc.; **p. 344 (a)** and **(b)** © Estate of Harold E. Edgerton/Courtesy Palm Press, Inc.; **p. 346 (a)** © Jerry Howard/Stock, Boston; **(b)** © Alan Oddie/Photo Edit; **Figure 11-16 (a)** © Richard Megna/Fundamental Photographs; **(b)** © Estate of Harold E. Edgerton/Courtesy Palm Press, Inc.; **p. 348** Picker International; **p. 350 (a)** and **(b)** Office National d'Etudes et de Rechèrches Aérospatiales; **p. 353** Dr. Owen M. Griffin, Naval Research Laboratory; **Figure 3 p. 359** © Jack Zohrt/FPG International; **Figure 4 p. 359** © Dave Custer, 1986.

Chapter 12

Opener p. 368 © Fundamental Photographs; **p. 371** N.A.S.A. (73-HC-787); **p. 376** Institute for Marine Dynamics; **p. 383** © Berenice Abbott/Commerce Graphics Ltd., Inc.; **p. 388** © Estate of Harold E. Edgerton/Courtesy Palm Press, Inc.; **p. 390** Courtesy of Monroe Auto Equipment, Monroe, Michigan; **p. 392** Courtesy Citicorp Building, New York City; **p. 394** Royal Swedish Academy of Music, courtesy Prof. Thomas D. Rossing, Northern Illinois University, DeKalb; **p. 395** Courtesy Central Scientific Company **p. 402** (*top*) from H.-O. Peitgen, P. H. Richter, *The Beauty of Fractals*, Springer-Verlag, New York, 1986; (*bottom*) Prof. Julio M. Ottino, University of Massachusetts, Amherst.

Chapter 13

Opener p. 409 (c) © Hashi; **Figure 13-1 (b)** *PSSC Physics*, 2nd ed., 1965. D. C. Heath & Co. and Education Development Center, Inc., Newton, Massachusetts; **Figure 13-3 (b)** and **(d)** *PSSC Physics* 2nd ed., 1965. D. C. Heath & Co. and Education Development Center Inc., Newton, Massachusetts; **Figure 13-5 (a)** and **(b)** © Berenice Abbott/Photo Researchers; **p. 414 (a)** and **(b)** © Four By Five, Inc.; **p. 421 (a,b,c,d)** *PSSC Physics*, 2nd ed., 1965. D. C. Heath & Co. and Education Development Center, Inc., Newton, Massachusetts; **p. 422** © David Yost/Courtesy Steinway & Sons; **p. 424 (a,b,c)** University of Washington; **p. 428** Prof. Thomas D. Rossing, Northern Illinois University, DeKalb.

Chapter 14

Opener p. 439 From Winston E. Kock, *Lasers and Holography*, Dover Publications, New York, 1981; **Figure 14-4** *PSSC Physics*, 2nd ed., 1965. D. C. Heath & Co. and Education Development Center, Inc., Newton, Massachusetts; **Figure 14-10 (a)** © 1973 Berenice Abbott/Photo Researchers; **p. 459 (a)** from Adolphe Ganot, *Elementary Treatise on Physics*, William Wood & Publishers Ltd., 1910; **(b)** © 1989 Mike Chesser, Yamaha BBDO Worldwide Network, Los Angeles; **p. 460** Davis Hathaway/N.A.S.A./Courtesy of National Optical Astronomy Observatories; **p. 462 (a)** Courtesy Naval Research Laboratory; **(b)** Courtesy San Francisco Symphony; **Figure 14-28** and **Figure 14-30** *PSSC Physics*, 2nd ed., 1965. D. C. Heath & Co. and Education Development Center, Inc., Newton, Massachusetts; **p. 463** Dr. John S. Shelton; **p. 464 (a)** © 1986 Howard Sochurek/Medichrome; **(b)** N.A.S.A. (81-HC-670); **Figure 14-31 (a)** *PSSC Physics*, 2nd ed., 1965. D. C. Heath & Co. and Education Development Center, Inc., Newton, Massachusetts; **Figure 14-33 (b)** *PSSC Physics*, 2nd ed., 1965. D. C. Heath & Co. and Education

Development Center, Inc., Newton, Massachusetts; **p. 469 (a)** Sandia National Laboratories; **(b)** © Robert de Gast/Photo Researchers; **(c)** © Estate of Harold E. Edgerton/Courtesy Palm Press, Inc.; **(d)** Sandia National Laboratories, **Figure 7 p. 476** New Scientist/IPC Magazines Limited/World Press Network 1990.

Chapter 15
Opener p. 487 Lockheed Corporation; **Figure 15-2** © Damien Lovegrove/Science Photo Library/Photo Researchers; **p. 488 (a)** © Michael Melford/The Image Bank; **(b)** Courtesy Central Scientific Company; **(c)** Instituto e Museo della Storia Scienza; **Figure 15-5 (b)** © Demetrios Zangos; **p. 491** Dr. William Mangum/National Bureau of Standards; **p. 494** © Richard Choy/Peter Arnold, Inc.; **p. 505 (a)** N.A.S.A. (78-HC-575); **(b)** N.A.S.A./Jet Propulsion Laboratory; **p. 507 (a,b,c)** Courtesy Central Scientific Company; **p. 510 (a)** © Alison Wright/Image Works; **(b)** © Eric Neurath/Stock, Boston.

Chapter 16
Opener p. 517 Lana Berkovich. Photo caption by kind cooperation of Robert Serlin, Phoenix Pipe and Tube Co., Phoenixville, Pennsylvania; **p. 522 (a)** Electro Scan Corporation, Wilmington, Massachusetts; **(b)** © Manfred Kage/Peter Arnold, Inc.; **p. 523 (c)** Dr. Mary Neubert, Liquid Crystal Institute, Kent State University; **(d)** © Don Gray/f/Stop Pictures; **(e)** © Ken Karp; **p. 529** Dr. Al Bartlett, University of Colorado, Boulder; **p. 530** From *Proceedings of Third International Heat Transfer Conference,* The Science Press, Ephrata, Pennsylvania, 1966; **p. 531 (a)** © Dr. R. P. Clark and M. Goff/Science Photo Library/Photo Researchers; **(b)** N.A.S.A. (75-HC-620); **p. 533** © 1990 John W. Warden. Photo caption by kind cooperation of David A. Jensen, Alyeska Pipeline Service Co.; **p. 534** The Royal Society; **Figure 16-7 (b)** Science Museum, London; **Figure 16-10 (b)** © David F. Malin/Anglo Australian Telescope Board; **p. 549** © Will and Deni McIntyre/Photo Researchers; **p. 556** N.A.S.A. (83-HC-227).

Chapter 17
Opener p. 563 Sandia National Laboratories; **p. 566–567** R. F. Sawyer, University of California, Berkeley; **p. 571** © J. M. Mejuto/FPG International; **p. 574 (a)** © Michael Collier/Stock Boston; **(b)** © Jean-Pierre Horlin/The Image Bank; **p. 575 (c)** Sandia National Laboratories; **(d)** © Peter Miller/The Image Bank; **(e)** Sandia National Laboratories; **p. 583** © Sidney Harris.

Chapter 18
Opener p. 598 Lawrence Berkeley Laboratory/Science Photo Library/Photo Researchers; **Figure 18-1** *PSSC Physics,* 2nd ed., 1965. D. C. Heath & Co. and Education Development Center, Inc., Newton, Massachusetts; **p. 600** (*top*) Courtesy National Institute of Standards and Technology; (*margin*) Bruce Terris/IBM Almaden Research Center; **p. 603** (*top left*) © Grant Heilman; (*top right*) Ann Ronan Picture Library; (*margin*) Burndy Library, Norwalk, Connecticut; **Figure 18-16 (b)** Harold M. Waage; **Figure 18-17 (b)** Harold M. Waage, **Figure 18-18 (b)** Harold M. Waage, **p. 615** Courtesy of Hulon Forrester/Video Display Corporation, Tucker, Georgia; **Figure 18-25** Courtesy Tripos-Evans and Sutherland Corporation.

Chapter 19
Opener p. 624 © Dagmar Hailer-Hamann/Peter Arnold, Inc.; **Figure 19-4 (b)** Ben Damsky Electric Power Research Institute; **Figure 19-22 (b)** Runk/Schoenberger from Grant Heilman; **p. 645** Harold M. Waage.

Chapter 20
Opener p. 656 © 1990 Richard Megna/Fundamental Photographs; **p. 659** © Mark Antman/The Image Works; **Figure 20-19 (b)** © Karen R. Preuss; **(c)** © Hank Morgan/Rainbow; **p. 678 (a)** Stanford Linear Accelerator Center/Science Photo Library/Photo Researchers; **(b)** © 1986 Wetmore/Photo Researchers; **Figure 20-20 (a)** Harold M. Waage; **p. 683** Courtesy Ohio Edison Company; **p. 684** Courtesy Xerox Corporation.

Chapter 21
Opener p. 690 Courtesy Tusonix, Tucson, Arizona; **Figure 21-2 (b)** Harold M. Waage; **p. 693** © Mark Antman/The Image Works; **p. 697 (a,b)** © Bruce Iverson; **(c)** © Manfred Kage/Peter Arnold, Inc.; **p. 700** © Lawrence Livermore National Laboratory; **p. 703 (a)** © Picturepoint Ltd., London; **(b)** © Paul Brierly; **(c)** © Lenman Connell/The Stock Market; **(d)** © Robert Essel/The Stock Market; **p. 707** Courtesy Lockheed Corporation.

Chapter 22
Opener p. 716 © Rabbit System, Santa Monica, California. Photo caption by kind cooperation of Larry Albright; **p. 719** Courtesy Sandia National Laboratories; **p. 723** © Chris Rogers/The Stock Market; **p. 724** Courtesy AT&T Archives; **p. 726 (a)** © 1985 Bob O'Shaughnessy/The Stock Market; **(b)** Courtesy EG&G Wakefield; **(c)** Courtesy Liquid Crystal Applications, Inc.; **Figure 22-11** © Paul Silverman/Fundamental Photographs; **p. 730 (a)** © *Popular Science,* June 1990; **(b)** © Paul Brierly; **(c)** © Coco McCoy/Rainbow; **(f)** Ann Ronan Picture Library; **p. 735** Donald Calabrese/Rome Air Development Center, Grittiss AFB, New York; **p. 740 Figure 1 (b)** Carolina Biological Supply Company; **p. 743 Figure 6 (d)** and (*right*) © Lennart Nilsson.

Chapter 23
Opener p. 749 Courtesy AT&T Archives; **p. 766** UPI/The Bettmann Archive; **p. 767 (a,b)** Courtesy Central Scientific Company; **p. 770 (a)** © Lennart Nilsson; **(b)** © Bruce Iverson; **p. 771 (c)** Courtesy Sandia National Laboratories; **(d)** © Joel E. Arem;

(e) © Lennart Nilsson; **(f)** Courtesy of Omega Engineering, Inc.

Chapter 24
Opener p. 781 © Stadler GmbH, Courtesy Transrapid International, Munich; **Figure 24-1** Fred Weiss; **Figure 24-8 (b)** © 1990 Richard Megna/Fundamental Photographs; **p. 788 (a)** Larry Langrill; **(b)** © Lawrence Berkeley Laboratory/Science Photo Library; **Figure 24-10 (b)** Carl E. Nielsen; **p. 790 (a)** © N.A.S.A. (74-HC-260); **(b)** Courtesy NRAO/AUI; **Figure 24-14 (b)** Courtesy Cavendish Laboratory, University of Cambridge; **p. 793** Courtesy Central Scientific Company; **Figure 24-16 (b)** © Stephen Cohen, Rockefeller University; **p. 796 (a)** © 1980 Smithsonian Institution; **(b)** Courtesy Scanditronix Inc., Essex, Massachusetts.

Chapter 25
Opener p. 811 Bob Williamson, Oakland University, Rochester, Michigan; **p. 815 (a,b)** © 1990 Richard Megna/Fundamental Photographs; **Figure 25-8** © 1990 Richard Megna/Fundamental Photographs; **Figure 25-11 (b)** © 1990 Richard Megna/Fundamental Photographs; **p. 820 (a,b)** © Bruce Iverson; **p. 821 (a)** Courtesy CERN; **(b)** Courtesy Brookhaven National Laboratory; **Figure 25-15 (b)** © 1990 Richard Megna/Fundamental Photographs; **p. 825** Courtesy F. W. Bell Co.; **Figure 25-20** Clarence Bennett, Oakland University, Rochester, Michigan; **p. 830 (a,b)** Courtesy Princeton University Plasma Physics Laboratory.

Chapter 26
Opener p. 840 © Bruce Iverson; **Figure 26-1** © 1990 Richard Megna/Fundamental Photographs; **p. 843** Courtesy National Portrait Gallery, London; **p. 855** Sun Yihons, China Reconstructs, October, 1985; **p. 856 (a,c)** Courtesy U.S. Dept. of the Interior, Dept. of Reclamation; **(b)** © Lee Langum/Photo Researchers; **Figure 26-25 (b)** © Michael Holford, Collection of the Science Museum, London; **p. 859** (*bottom*) Clarence Bennett, Oakland University, Rochester, Michigan; **p. 866** (*bottom left*) Y. Yoshida; **Figure 1** Prof. Neal Brown, Geophysical Institute, University of Alaska, Fairbanks; **Figure 2 (a)** N.A.S.A. (85-HC-148); **Figure 3 p. 867** Courtesy Prof. Syun-Ichi Akasofu; **Figure 2 (b)** Courtesy Prof. Lou Frank, University of Iowa.

Chapter 27
Opener p. 878 Robert J. Celotta, National Institute of Standards and Technology; **p. 885** J. F. Allen, St. Andrews University, Scotland; **p. 886** © Paul Silverman/Fundamental Photographs; **p. 887 (a)** © Akira Tonomura, Hitachi Advanced Research Laboratory, Hatoyama, Japan; **(b,c)** © Bruce Iverson; **p. 889 (a)** © Seagate Technologies; **(b)** John Mamin, IBM Almaden Research Center; **p. 891** © Bill Pierce/Time.

Chapter 28
Opener p. 898 © 1990 C. J. Allen/New England Hydro Transmission Corporation; **p. 906 (a)** Courtesy Intel Corporation; **(b)** Courtesy Sperry Corporation; **p. 916** © George A. Clark Radioana Collection-Archive Center, National Museum of American History; **p. 920 (a)** Courtesy Champion Spark Plug Co; **(b)** © Bruce Iverson; **p. 921 (a)** © Yoav/Phototake; **(b)** © Daniel S. Brody/Stock, Boston; **p. 923** Lana Berkovich; **p. 925 (a)** Photo by Jolin Petit/Courtesy J. W. Petit, The Vestal Press Ltd.; **(b)** © Bruce Iverson; **p. 926** Courtesy Dr. H. H. Busta, Amoco Research Center, Naperville, Illinois; **Figure 6, p. 932 (a-d)** © Lloyd Wolf; **Figure 7, p. 933** © Bruce Iverson.

Chapter 29
Opener p. 943 N.A.S.A. (70-HC-220); **p. 948** © Doug Johnson/SPL/Science Source; **p. 952** NOAA-N.A.S.A./JPL; **p. 955** Dave Cooke, University of Chicago; **p. 956 (a)** Courtesy AT&T Archives; **(b)** Hale Observatories; **p. 959 (a)** © 1984 Frank Zullo; **(b)** © Galen Powell/Peter Arnold, Inc.; **p. 962** (*left*) Alistair Steyn-Ross, University of Waikato, New Zealand/Courtesy *The Physics Teacher;* (*right*) Courtesy Brookhaven National Laboratory; **p. 967** Courtesy Cavendish Laboratory, University of Cambridge; **Figure 29-13** Courtesy Central Scientific Company.

Chapter 30
Opener p. 974 Robert Greenler; **p. 976 (a)** Courtesy Optical Coating Laboratory, Inc. (OCLI); **(b)** © Chuck O'Rear/West Light; **p. 977 (c)** © William James Warren/West Light; **(d,e)** © Chuck O'Rear/West Light; **p. 980 (a)** M. A. Duguay and A. T. Mattick, Bell Telephone Laboratories, Murray Hill, New Jersey; **(b)** © 1989 Anglo-Australian Telescope Board; **Figure 30-8 (a)** Courtesy Battelle-Northwest Laboratories; **(b)** © 1987 Ken Kay/Fundamental Photographs; **Figure 30-10 (b)** © 1987 Pete Saloutos/The Stock Market; **p. 984 (a)** Courtesy *Audio* Magazine; **(b)** Demetrios Zangos, with kind cooperation of Photographic Sciences Corporation, Webster, New York; **(c)** © Leonard Lessin/Peter Arnold, Inc.; **(d)** Courtesy University of Miami, Music Engineering; **(e)** Courtesy Digital Instruments, Santa Barbara, California; **Figure 30-13 (b)** © 1990 Richard Megna/Fundamental Photographs; **Figure 30-16** *PSSC Physics,* 2nd ed., 1965. D. C. Heath & Co. and Education Development Center, Inc., Newton, Massachusetts; **Figure 30-17 (b)** © 1987 Ken Kay/Fundamental Photographs; **Figure 30-19 (b)** © 1987 Ken Kay/Fundamental Photographs; **p. 989 (a)** © Lawrence Manning/West Light; **(b)** © Dan Boyd/Courtesy Naval Research Laboratory; **Figure 30-20 (b)** Courtesy AT&T Archives; **(c)** © 1983 C. Falco/Photo Researchers; **Figure 30-21 (c)** © Robert Greenler; **Figure 30-23 (b)** © Paul Silverman/Fundamental Photographs; **p. 994 (a)** © Robert Greenler; **(b)** © Giovanni DeAmici; **Figure**

30-35 (a,b) Larry Langrill; **p. 999 (a,b)** Glen A. Izett, U.S. Geological Survey, Denver, Colorado; **(c,d)** Dr. Anthony J. Gow, Cold Regions Research and Engineering Laboratory, Hanover, New Hampshire; **(e)** © Sepp Seitz/Woodfin Camp and Associates, Inc.; **p. 1001 (a)** © 1970 Fundamental Photographs; **(b)** © 1990 PAR/NYC, Inc./Photo by Elizabeth Algieri; **p. 1003** © 1987 Paul Silverman/Fundamental Photographs; **Figure 2 p. 1005** Robert Greenler; **Figures 3,4 p. 1006** Robert Greenler.

Chapter 31
Opener p. 1012 © Dagmar Hailer-Hamann/Peter Arnold, Inc.; **Figure 31-2** Demetrios Zangos; **p. 1015** © 1982 Jon Brenneis Photo; **Figure 31-12 (c)** © 1990 Richard Megna/Fundamental Photographs; **p. 1020 (a,b)** A.G.E. FotoStock/© Peter Arnold, Inc.; **p. 1021 (a,b)** © 1990 Richard Megna/Fundamental Photographs; **p. 1025** © 1990 Richard Megna/Fundamental Photographs; **Figure 31-26 (b)** Nils Abramson; **(d)** © 1974 Fundamental Photographs; **Figure 31-27 (c)** © Fundamental Photographs; **p. 1032** © 1990 Richard Megna/Fundamental Photographs; **p. 1033** © Bohdan Hrynewych/Stock, Boston.

Chapter 32
Opener p. 1041 Robert Parshall, Biomedical Visualization Laboratories, University of Illinois at Chicago; **p. 1044 (a,b)** © Lennart Nilsson; **p. 1045 (c)** © Lennart Nilsson; **(d)** Courtesy IMEC and University of Pennsylvania, Dept. of Electrical Engineering; **p. 1049** (*top*) Courtesy Nikon, Inc.; (*margin*) © Chuck O'Rear/West Light; **Figure 32-8 (a)** © 1983 Jonathan Levine; **p. 1051 (a)** Cornell University; **(b)** M. Issacson, M. Barshatsky, J. Cline/Cornell University; **p. 1053 (a)** AIP Niels Bohr Library; **(b)** © Royal Astronomical Society Library; **(c)** © Lick Observatory, U. C. Regents; **(d)** Hale Observatories; **(e)** © 1980 Gary Ladd; **p. 1054 (a,b,d)** © California Association for Research in Astronomy; **(c)** © Ray Ellis/Photo Researchers, Inc.; **p. 1056 (a)** N.A.S.A. (531-76-026); **(b)** N.A.S.A. (531-76-0390); **p. 1057 (c,d,e)** N.A.S.A. (90-HC-508).

Chapter 33
Opener p. 1061 © Ken Kay/Fundamental Photographs; **p. 1063** © 1990 Richard Megna/Fundamental Photographs; **Figure 33-3 (a)** Courtesy Bausch & Lomb; **Figure 33-5 (a,b)** Courtesy T. A. Wiggins; **p. 1066 (a,b)** Optical Coating Laboratory, Inc. (OCLI); **Figure 33-7** *PSSC Physics*, 2nd ed., 1965. D. C. Heath & Co. and Education Development Center, Newton, Massachusetts; **Figure 33-9 (a)** Courtesy Michel Cagnet; **Figure 33-18 (a,b,c)** Courtesy Michel Cagnet; **Figure 33-19 (a)** Courtesy Michel Cagnet; **Figure 33-27 (a)** Courtesy Michel Cagnet; **Figure 33-29 (a,b)** M. Cagnet, M. Françon, J. C. Thrierr, *Atlas of Optical Phenomena;* **Figure 33-30 (a)** Courtesy Battelle-Northwest Laboratories; **Figure 33-31** Courtesy Michel Cagnet; **Figure 33-32** Courtesy Michel Cagnet; **Figure 33-34 (a,b)** Courtesy Michel Cagnet; **Figure 33-36 (a)** The Bettmann Archive; **(b)** Clarence Bennett, Oakland University, Rochester, Michigan; **(c)** Courtesy Holotek Ltd., Rochester, New York; **(d)** NRAO/AUI/Science Photo Library; **p. 1088** Courtesy Lawrence Livermore National Laboratory, Laser Program; **p. 1089 (a)** © Philippe Plailly/Science Photo Library; **(b,c)** © Ronald R. Erickson, 1981. Hologram by Nicklaus Phillips, 1978, for Digital Equipment Corporation; **(d)** © 1983 Ronald R. Erickson; **(e)** © Chuck O'Rear/West Light.

Chapter 34
Opener p. 1100 The Hebrew University of Jerusalem/Courtesy AIP Niels Bohr Library; **p. 1101** Courtesy NRAO/AUI; **p. 1126** C. Powell, P. Fowler & D. Perkins/Science Photo Library/Photo Researchers; **p. 1134 (a)** © Michael Freeman; **(b)** N.A.S.A. (76-HC-612); **p. 1135** © Michael Freeman.

Chapter 35
Opener p. 1145 Adapted from Eastman Kodak and Wabash Instrument Corporation; **p. 1147** Max Planck Institute, Berlin; **p. 1151** Courtesy Thorn EMI Electron Tubes Ltd.; **Figure 35-7 (b)** Courtesy General Electric Company; **Figure 35-10** From G. Herzberg, *Annalen de Physick*, Vol. 84, p. 565, 1927; **Figure 35-17 (a,b)** *PSSC Physics*, 2nd ed., 1965. D. C. Heath & Co. and Education Development Center, Newton, Massachusetts; **(c)** © C. G. Shull; **(d)** © Claus Jonsson.

Answers

These answers are calculated using $g = 9.81\ m/s^2$ unless otherwise specified in the exercise or problem. The results are usually rounded to three significant figures. Differences in the last figure can easily result from differences in rounding the input data and are not important.

Chapter 1

True or False

1. True
2. False; for example, $x = vt$ where the speed v and the time t have different dimensions
3. True

Problems

1. (*a*) 1 MW (*b*) 2 mg (*c*) 3 μm (*d*) 30 ks
3. (*a*) 1 picoboo (*b*) 1 gigalow (*c*) 1 microscope (*d*) 1 attoboy (*e*) 1 megaphone (*f*) 1 nanogoat (*g*) 1 terabull
5. (*a*) C_1, ft; C_2, ft/s (*b*) C_1, ft/s^2 (*c*) C_1, ft/s^2 (*d*) C_1, ft; C_2, s^{-1} (*e*) C_1, ft/s; C_2, s^{-1}
7. (*a*) 62.1 mi/h (*b*) 23.6 in (*c*) 91.4 m
9. (1.61 km/h)/(1 mi/h)
11. (*a*) 36.00 km/h·s (*b*) 10.00 m/s^2 (*c*) 88 ft/s (*d*) 27 m/s
13. 4046.9 m^2
15. (*a*) C_1, L; C_2, L/T (*b*) C_1, L/T^2 (*c*) C_1, L/T^2 (*d*) C_1, L; C_2, T^{-1} (*e*) C_1, L/T; C_2, T^{-1}
17. $F = mv^2/r$
19. (*a*) 30,000 (*b*) 0.0062 (*c*) 0.000004 (*d*) 217,000
21. (*a*) 1.14×10^5 (*b*) 2.25×10^{-8} (*c*) 8.27×10^3 (*d*) 6.27×10^2
23. 1 in = 3.63×10^6 cell membranes
25. (*a*) 1690 (*b*) 5 (*c*) 5.6 (*d*) 10
27. 3.51 Mm
29. (*a*) 6000 km (*b*) 8760 tankers/y (*c*) 43.8 billion dollars
31. (*a*) 1.41×10^{17} kg/m^3 (*b*) 216 m
33. (*a*) 4.92×10^9 m (*b*) 12.8 r_{e-m}
35. (*a*) $T = Cm^n$, with $n = 0.5$ and $C = 1.81\ s/kg^{1/2}$. (*b*) Those points with the greatest deviation are at T = 1.05 and 1.75 s.

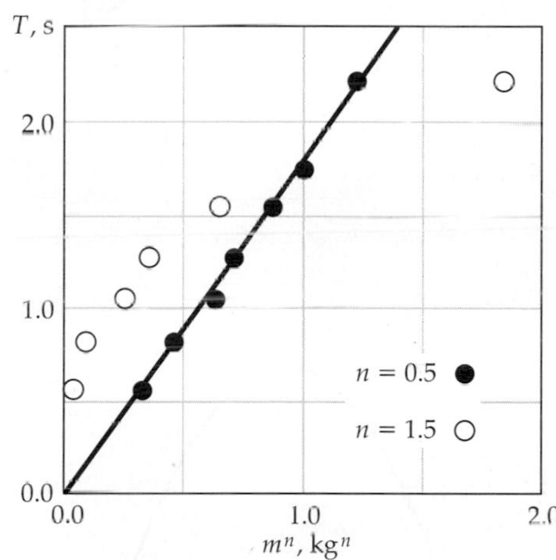

37. (*a*) $T = C\sqrt{L/g}$ (*b*) example results: L = 1 m, T = 2 s; L = 0.5 m, T = 1.4 s (*c*) $C = 2\pi$
39. (*a*) R increases with increasing H and with increasing v (*b*) $R = v\sqrt{2H/g}$
41. (*a*) 8×10^8 bits = 1×10^8 bytes (*b*) $\approx$ 500 books

Chapter 2

True or False **1.** False; this equation holds only for constant acceleration **2.** False; a ball at the top of its flight is momentarily at rest but has acceleration g **3.** False; for example, motion with constant velocity **4.** True **5.** True, by definition of average velocity

Problems

1. (*a*) 24 km/h (*b*) −12 km/h (*c*) 0 km/h (*d*) 16 km/h
3. (*a*) −2 m/s (*b*) 2.25 m/s (*c*) −0.3 m/s
5. (*a*) 260 km (*b*) 65 km/h
7. (*a*) 1.82 h (*b*) 4.85 h (*c*) 945 km/h (*d*) 621 km/h
9. (*a*) 8.33 min (*b*) 1.28 s (*c*) 1 lightyear = 9.47×10^{12} km = 5.88×10^{12} mi
11. $66\frac{2}{3}$ km/h
13. (*a*) 1 m/s (*b*) 2 m/s
15. (*a*) 1 m/s (*b*) 0.7 m/s (*c*) 8 s
17. −2 m/s^2
19. (*a*) 80 m/s (*b*) 400 m (*c*) 40 m/s
21. 15.6 m/s^2
23. t = 4.47 s, v = 44.7 m/s, v_{av} = 22.4 m/s
25. 4.59 km
27. (*a*) 25.1 s (*b*) 3.08 km
29. (*a*) The area under the curve in the indicated time interval is −36 m.

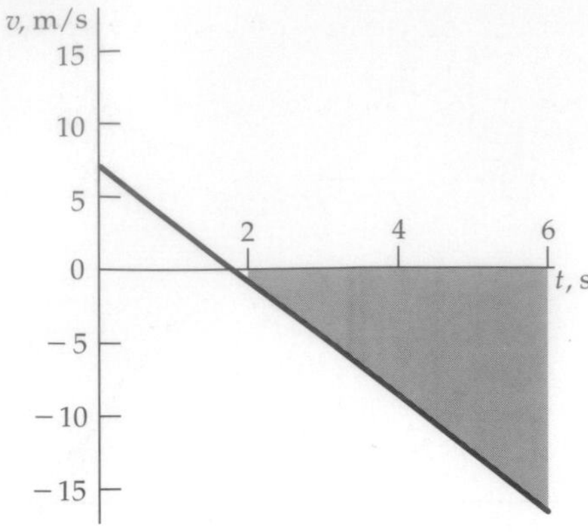

(*b*) $x(t) = 7t - 2t^2 + c$, $\Delta x = -36$ m (*c*) −9 m/s
31. (*a*) 0 m/s (*b*) $\frac{1}{3}$ m/s (*c*) −2 m/s (*d*) 1 m/s
33. (*a*)

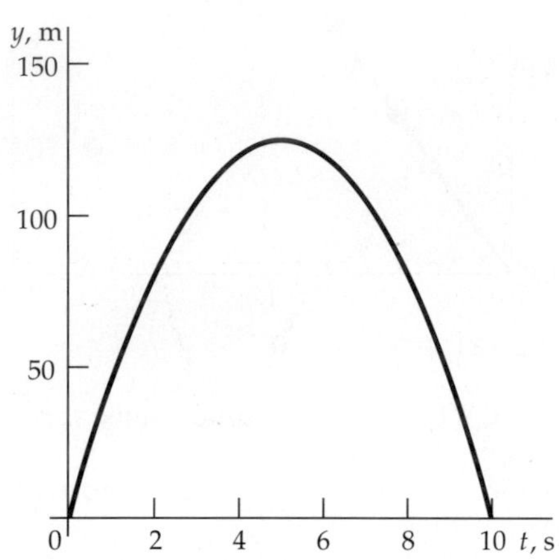

(*b*) $t = 0$ to 1 s, $v_{av} = 45$ m/s;
$t = 1$ to 2 s, $v_{av} = 35$ m/s;
$t = 2$ to 3 s, $v_{av} = 25$ m/s; $t = 3$ to 4 s, $v_{av} = 15$ m/s;
$t = 4$ to 5 s, $v_{av} = 5$ m/s; $t = 5$ to 6 s, $v_{av} = -5$ m/s;
$t = 6$ to 7 s, $v_{av} = -15$ m/s;
$t = 7$ to 8 s, $v_{av} = -25$ m/s;
$t = 8$ to 9 s, $v_{av} = -35$ m/s;
$t = 9$ to 10 s, $v_{av} = -45$ m/s.

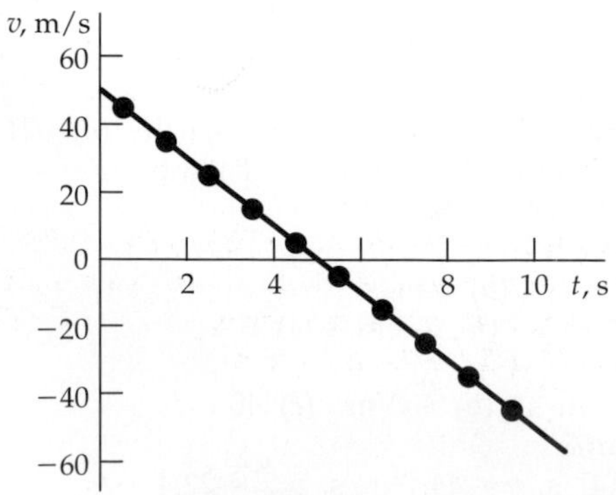

(*c*) $v(t) = -10t + 50$

35. (*a*) Velocity is greatest for $t \approx 2$ to 3 s

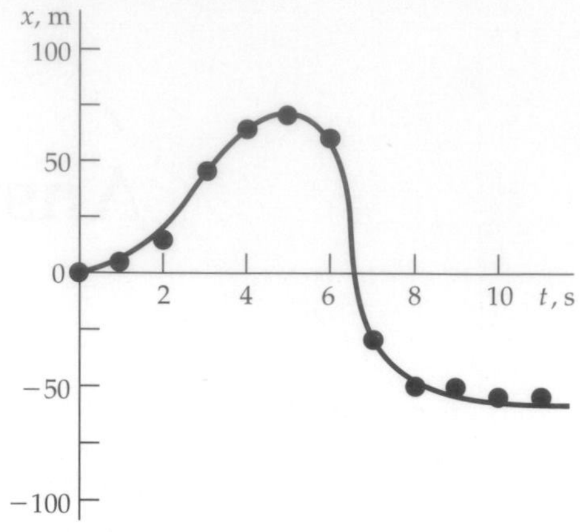

(*b*) Speed is least near $t = 5$ s, and for $t \approx$ 8 s and $t \approx 11$ s (*c*) Same as part (*b*). (*d*) Speed is constant for $t > 10$ s (*e*) Acceleration is positive for $t \approx 0$ s to 3 s and for $t \approx 7$ s to 10 s. (*f*) Acceleration is negative for $t \approx 3$ s to 7 s.
37. (*a*) −7.67 m/s (*b*) $a_{av} = 6.26$ m/s (*c*) 697 m/s^2 upward
39. (*a*) $x(t) = (C/6)t^3 + Dt + E$, where E is a constant in meters. (*b*) At $t = 5$ s, we find $v = 37.5$ m/s and $x =$ 62.5 m.
41. $\Delta x = 4\frac{1}{3}$ m. The average velocity is not equal to the mean of the initial and final values when the acceleration is not constant, as is the case here.
43.

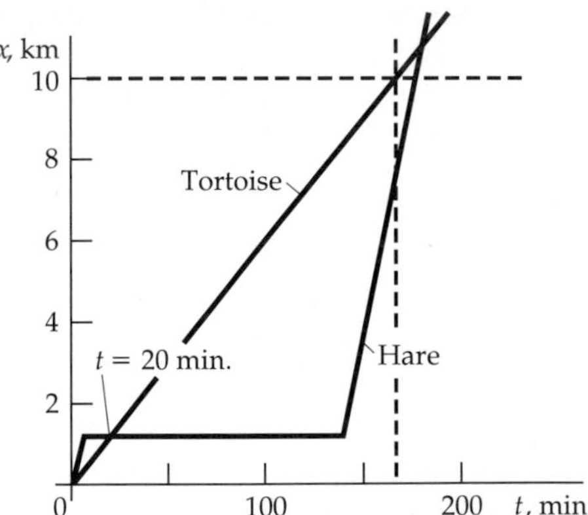

Tortoise passes Hare at $t = 20$ min. When Tortoise finishes, Hare is 2.4 km behind. Hare could nap up to 125 min without losing the race.
45. (*a*) t_0, t_1 (*b*) t_3, t_4, t_6, t_7 (*c*) t_2, t_5 (*d*) t_4 (*e*) t_2, t_6 (*f*) t_0, t_1, t_3, t_5, t_7
47. For Figure 2–22*a*, (*a*) $a > 0$ for $t \approx 3$ to 6 s; $a < 0$ for $t \approx 1$ to 3 s; $a = 0$ near $t = 3$ and for $t = 6$ s to 7 s. (*b*) $a =$ constant where the curve is straight. (*c*) $v = 0$ at $t \approx 8.75$ s. For Figure 2–22*b*, (*a*) $a > 0$ for $t \approx 5$ to $t \approx 7$ s; $a < 0$ for $t \approx 0$ to 3 s and $t > 7$ s; $a = 0$ for $t \approx 3$ to 5 s (*b*) $a =$ constant on the parabolic and straight-line segments of the curve. (*c*) $v = 0$ near $t = 2$, 6, and 8 s.

49. (*a*) 34.7 s (*b*) 1.21 km
(*c*)

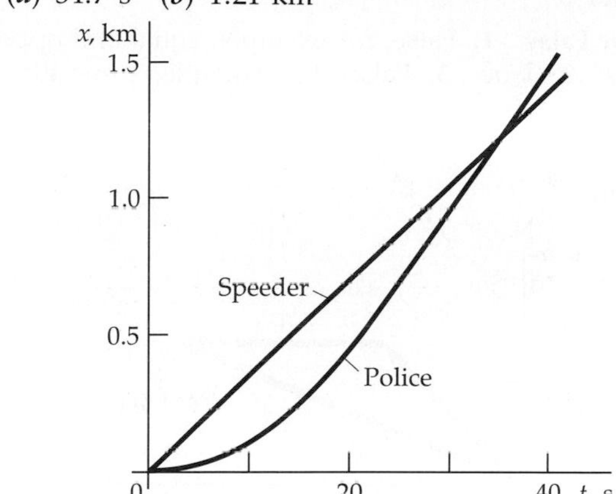

51. $v_0 = 40$ cm/s; $a = -6.88$ cm/s^2
53. (*a*) $6\frac{2}{3}$ s (*b*) $6\frac{2}{3}$ s
55. 18.4 m
57. (*a*) 26.1 m (*b*) 23.4 m/s
59.

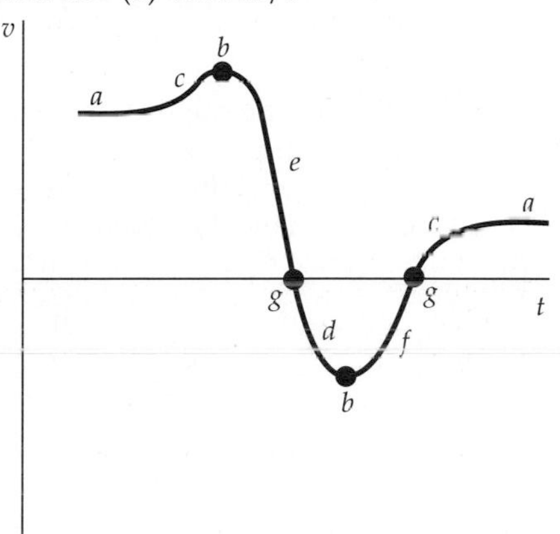

61. (*a*) Yes
(*b*)

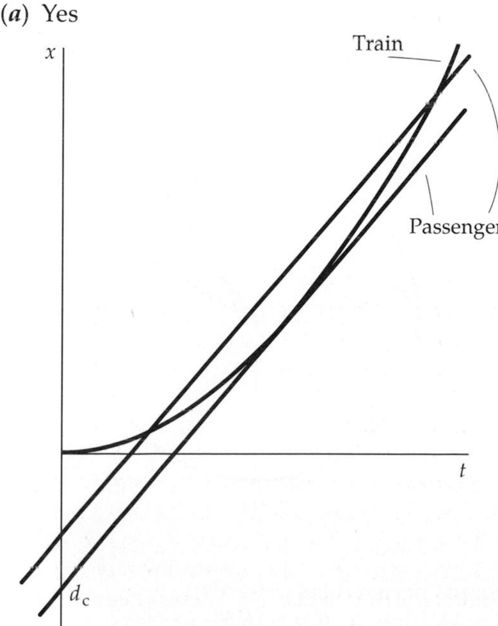

(*c*) The train's speed is 8 m/s, its average speed is 4 m/s and $d_c = 32$ m.
63. The balls collide two-thirds of the way to the top of the building.
65. (*a*)

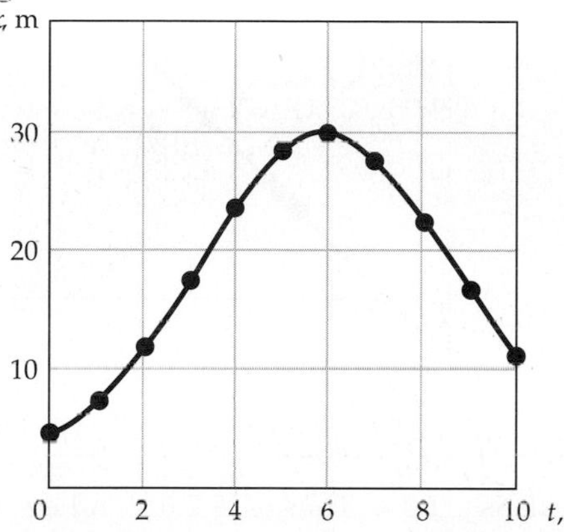

(*b*)

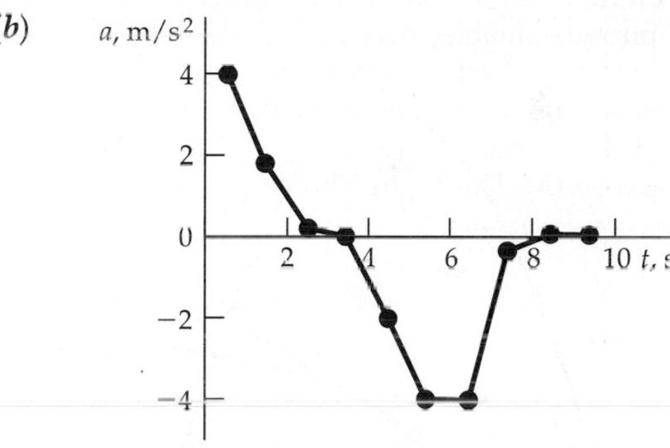

67. (*a*) 0.875 cm/s

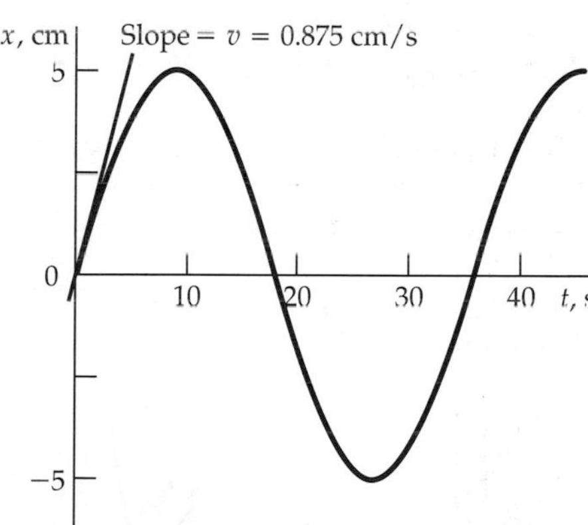

(*b*) The average velocities are 0.723, 0.835, 0.857, 0.871, 0.874, and 0.875 cm/s for the time intervals ending at 6, 3, 2, 1, 0.5, and 0.25 s, respectively. (*c*) $dx/dt = A\omega \cos \omega t = 0.875$ cm/s at $t = 0$ s.

69. (*a*)

(*b*) $x = \frac{1}{2}v_0T + aT(t - T)$, or $x = v_0(t - \frac{1}{2}T)$ (*c*) The slope is ≈ 11.6 m/s $= v_0$, and the intercept with the time axis is ~ 1.3 s. Thus, $T = 2.6$ s and $a = 4.46$ m/s^2 (*d*) The sprint is run at top speed. For longer races, top speed cannot be maintained and a runner must run at a slower, but sustainable, pace
71. (*a*) Initially the particle accelerates with the acceleration of gravity, but as its velocity increases, its acceleration diminishes. When $v = g/b$, the acceleration becomes zero. (*b*) The figure shows a rough sketch of v versus t

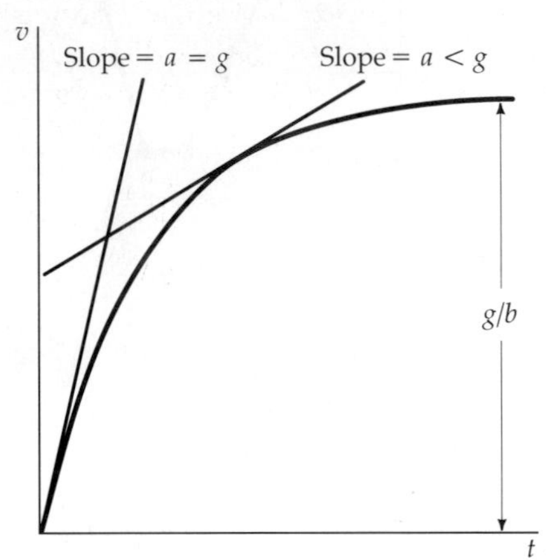

73. $x(t) = x_0e^{A(t-t_0)}$, where $A = 1$ s^{-1}
75. (*a*)

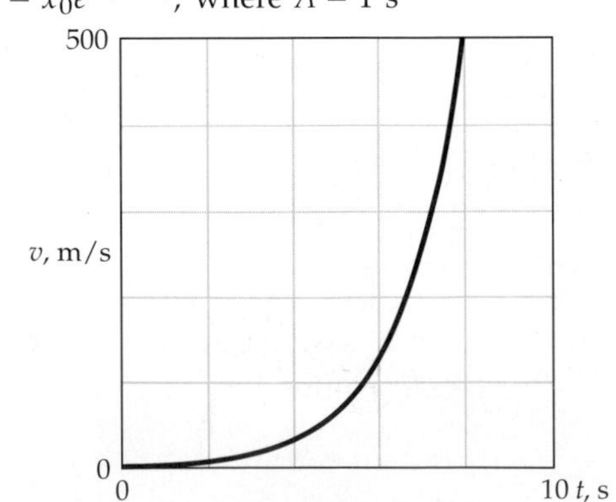

(*b*) 295 m/s

Chapter 3

True or False **1.** False; for example, equal and opposite vectors **2.** True **3.** False; for example, projectile motion **4.** False; for example, circular motion **5.** True **6.** True **7.** True

Problems

1.

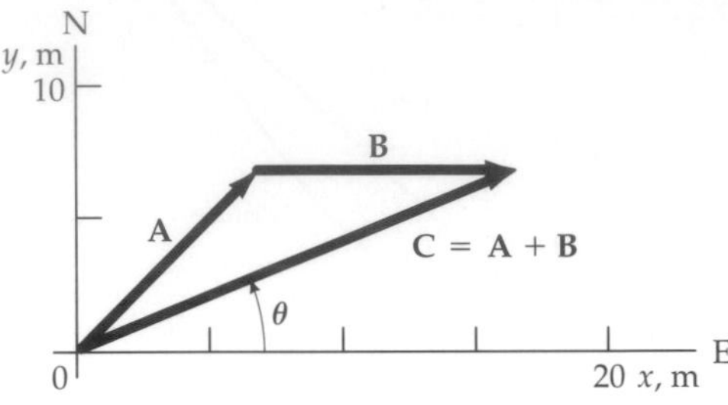

18.5 m, 22.5° north of east.
3. (*a*) $t = 15$ s, $\Delta r = 14.1$ m, and $\theta = 45°$; $t = 30$ s, $\Delta r = 20$ m, and $\theta = 90°$; $t = 45$ s, $\Delta r = 14.1$ m, and $\theta = 135°$; $t = 60$ s, $\Delta r = 0$ m (*b*) $\Delta r = 14.1$ m, $\theta = 45°$; $\Delta r = 14.1$ m, $\theta = 135°$; $\Delta r = 14.1$ m, $\theta = 225°$; $\Delta r = 14.1$ m, $\theta = 315°$ (*c*) same magnitude, but rotated by 90° (*d*) same magnitude but in the opposite direction
5. (*a*) 1 km (*b*) 57.3° north of east (*c*) 1/5
7. (*a*) $A_x = 8.66$ m, $A_y = 5$ m (*b*) $A_x = 3.54$ m, $A_y = 3.54$ m (*c*) $A_x = 3.5$ km, $A_y = 6.06$ km (*d*) $A_x = 0$ km, $A_y = 5$ km (*e*) $A_x = -13$ km/s, $A_y = 7.5$ km/s (*f*) $A_x = -5$ m/s, $A_y = -8.66$ m/s (*g*) $A_x = 0$ m/s^2, $A_y = -8$ m/s^2
9. (*a*) $A_x = 1.41$ m $= A_y$, $B_x = 1.73$ m, $B_y = -1$ m (*b*) $C_x = 3.15$ m, $C_y = 0.414$ m, $C = 3.17$ m, $\theta = 7.5°$ (*c*) $C_x = -0.318$ m, $C_y = 2.41$ m, $C = 2.44$ m, $\theta = 97.5°$
11. (*a*) $A = 8.06$, $\theta = 240°$; $B = 3.61$, $\theta = -33.7°$; $C = 9.06$, $\theta = 264°$ (*b*) $A = 4.12$, $\theta = -76°$; $B = 6.32$, $\theta = 71.6°$; $C = 3.61$, $\theta = 33.7°$
13. (*a*) $\mathbf{v} = (5$ m/s$)\mathbf{i} + (8.66$ m/s$)\mathbf{j}$ (*b*) $\mathbf{A} = (-3.54$ m$)(\mathbf{i} + \mathbf{j})$ (*c*) $\Delta\mathbf{r} = (14$ m$)\mathbf{i} + (-6$ m$)\mathbf{j}$
15. $A_y/A_x = \pm B_y/B_x$
17.

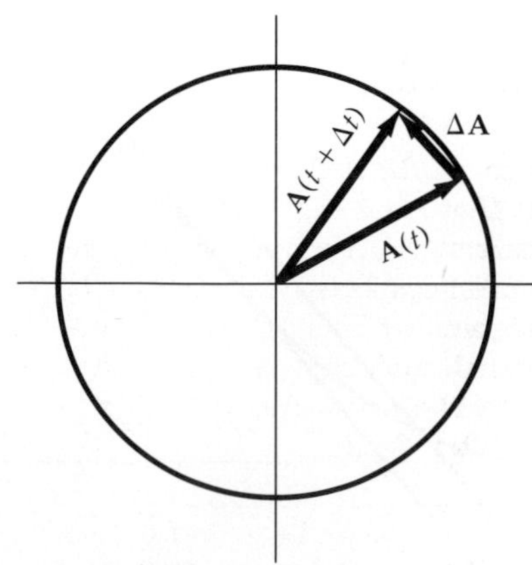

$\Delta\mathbf{A}$ becomes perpendicular to $\mathbf{A}(t)$.
19. $v_{av} = 14.7$ km/h, $\theta = -16.3°$

21. (*a*)

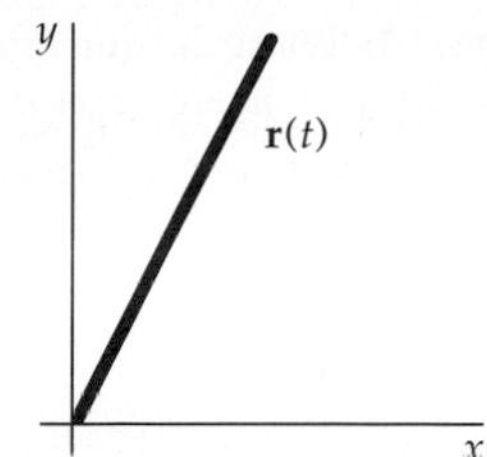

(*b*) $\mathbf{v} = d\mathbf{r}/dt = 5\mathbf{i} + 10\mathbf{j}$; $v = 11.2$ m/s

23. (*a*) *AB*, north; *BC*, northeast; *CD*, east; *DE*, southeast; *EF*, south (*b*) *AB*, north; *BC*, southeast; *CD*, acceleration = 0; *DE*, southwest; *EF*, north (*c*) The acceleration is greater on segment *DE*

25. (*a*) $\mathbf{v}_{av} = 33\frac{1}{3}\mathbf{i} + 26\frac{2}{3}\mathbf{j}$ (*b*) $\mathbf{a}_{av} = (-3\mathbf{i} - 1.77\mathbf{j})$ m/s^2

27. (*a*) west of north by $\theta = 13.1°$ (*b*) 300 km/h

29. 0.533 s

31. (*a*) 2294 m (*b*) 43.2 s (*c*) 9174 m

33. $\mathbf{v} = (15$ m/s$)\mathbf{i}$, where $\mathbf{i}$ is taken to be horizontal; $a = 9.81$ m/s^2 downward

35. 45 m/s^2

37. (*a*) $a_r \rightarrow 4a_r$ (*b*) $a_r \rightarrow \frac{1}{2}a_r$ (*c*) A perfectly sharp turn implies $r = 0$, which would require an infinitely large acceleration for any finite speed.

39. 29.9 rev/min

41. (*a*) $\mathbf{a} = 2$ m/s^2 $\mathbf{i} - 3.5$ m/s^2 $\mathbf{j}$ (*b*) $\mathbf{v} = (1 + 2t)$ m/s $\mathbf{i} + (1 - 3.5t)$ m/s $\mathbf{j}$ (*c*) $\mathbf{r} = (4 + t + t^2)$ m $\mathbf{i} + (3 + t - 1.75t^2)$ m $\mathbf{j}$

43. (*a*) 31.6 m/s^2 (*b*) 26.1 m/s^2 (*c*) 23.9 m/s^2

45. $\mathbf{v} = 30\mathbf{i} + (40 - 10t)\mathbf{j}$, $\mathbf{a} = d\mathbf{v}/dt = -10\mathbf{j}$

47. (*a*) 8.14 m/s (*b*) 23.2 m/s

49. The projectile covers a horizontal distance $x =$ 408 m

51. (*a*) $(-120\mathbf{i} + 4\mathbf{j})$ m (*b*) $(-20\mathbf{i} - 12\mathbf{j})$ m/s (*c*) $-2\mathbf{j}$ m/s^2

53. 3.20 m/s in the direction of the thrower

55. 14.2 m/s = 31.7 mi/h

57. 4.29 m

59. 0.785 m

61. (*a*) 2 s (*b*) π m/s (*c*) $-\mathbf{j}$ (*d*) $a_r = \pi^2$ m/s^2, $a_t = \pi/2$ m/s^2 (*e*) $a = 9.99$ m/s^2, $\theta = 189°$

63. 34.6 m/s

65. (*a*) 20.0 m/s (*b*) 2.87 m

67. (*a*) Note that $r = 4$ m for all time; thus, the path is a circle of radius 4 m centered on the origin. (*b*) $\mathbf{v} = (8\pi$ m/s$)(\cos 2\pi t\ \mathbf{i} - \sin 2\pi t\ \mathbf{j})$ (*c*) $\mathbf{a} = (-16\pi^2$ m/s$^2)(\sin 2\pi t\ \mathbf{i} + \cos 2\pi t\ \mathbf{j}) = (-4\pi^2$ s$^{-2})\mathbf{r}$; hence $\mathbf{a}$ is radial, and in particular, is radially inward as indicated by the minus sign.

69. (*a*) $H = W \tan\theta$ (*b*)

$$v_0 = \frac{W}{\cos\theta}\sqrt{\frac{g}{2(W\tan\theta - H)}}$$

71. (*a*) 54.3 m/s (*b*) 269 m (*c*) 6.56 s

73. (*a*) $\mathbf{v} = (10 - 10 \sin 2t)\mathbf{i} - (10 \cos 2t)\mathbf{j}$ (*b*) The particle is in contact with the surface at $5\pi/2$, $25\pi/2$, . . . = 7.85, 39.2, ...

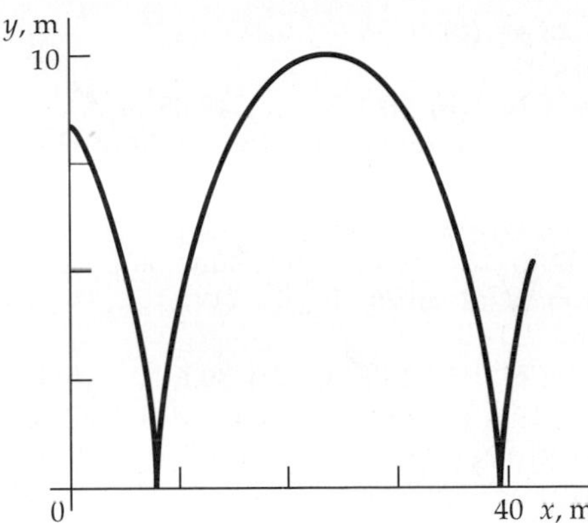

(*c*) $\mathbf{a} = -20 \cos 2t\ \mathbf{i} + 20 \sin 2t\ \mathbf{j}$ (*d*) $\mathbf{v} = 0$ when $t = (4n + 1)\pi/4$, where $n = 0, 1, 2,$ . . . . The point in contact with the ground is instantaneously at rest.

Chapter 4

True or False 1. True 2. False; the forces acting sum to zero 3. False; the *acceleration* is in the direction of the unbalanced force 4. True 5. False 6. False

Problems

1. (*a*) 8 m/s^2 (*b*) $m_2 = \frac{1}{2}m_1$ (*c*) 8/3 m/s^2

3. (*a*) $a = 7.07$ m/s^2 in a direction bisecting the angle between the two forces (*b*) $a = 14.0$ m/s^2 in a direction 14.6° away from $2\mathbf{F}_0$ toward $\mathbf{F}_0$

5. 12 m/s^2

7. (*a*) 6 m/s (*b*) 9 m

9. 10 kg

11.

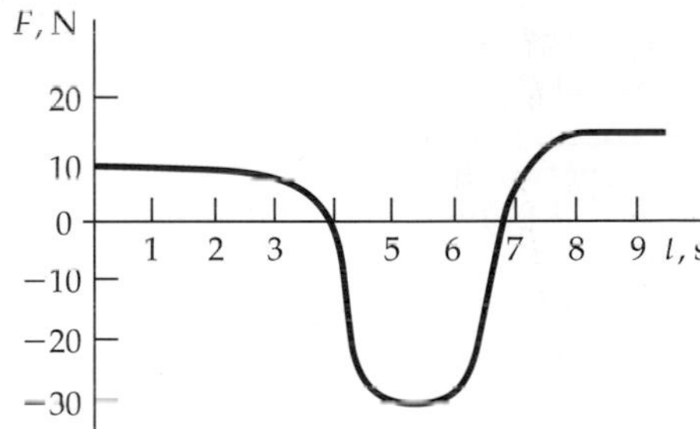

13. *AB*, force vertically upward; *BC*, force radial toward center of circular arc; *CD*, no force; *DE*, force is toward center of circular arc; *EF*, force acts vertically upward

15. (*a*) 79.4 kg (*b*) 7.94×10^4 g

17. (*a*) $W = 784.8$ N = 176 lb (*b*) $W = 716$ N = 161 lb (*c*) $m = 80$ kg = 5.48 slugs

19.

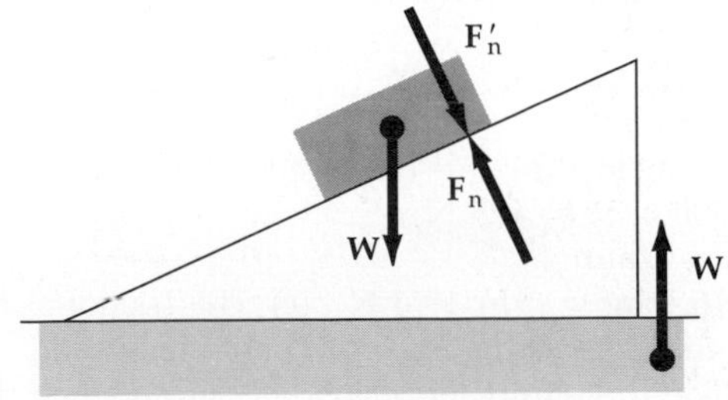

21. (*a*) 1.28 s (*b*) 8.7 y
23. $5\frac{1}{3}$ m/s^2
25. (*a*) 98.1 N (*b*) 98.1 N (*c*) 49.05 N (*d*) (98.1 N) sin θ, where θ is the angle of elevation of the inclined plane, 56.3°
27. (*a*) 8.77 m/s^2 (*b*) 26.3 N
29. (*a*) 132 lb (*b*) 105 lb (*c*) Same as part (*b*)
31. (*a*) $T = W/(2 \sin\theta)$, $T_{min} = \frac{1}{2}W$, $T_{max} \to \infty$ (*b*) 19.6 N
33. (*a*) 19.6 N (*b*) 19.6 N (*c*) 39.6 N (*d*) For $t = 0$ to 2 s, $T = 19.6$ N; for $t = 2$ to 4 s, $T = 9.62$ N
35. (*a*) 11,810 N (*b*) 9810 N (*c*) T = 7810 N
37. (*a*) 3821 N (*b*) 5732 N
39. 500 N
41. (*a*) 149 N (*b*) 3.41 m/s^2
43. **F** = 329 N = 73.9 lb directed toward the man.
45. (*a*) 17 m/s^2 (*b*) If a is made greater, the block will accelerate up the wedge.
47. (*a*) 398 N (*b*) 368 N
49. (*a*) The direction of deflection is opposite to the direction of acceleration. (*c*) $\theta = 9.31°$ in the forward direction.
51. $\theta = 51.6°$

Chapter 5

True or False **1.** False **2.** False; it opposes the relative motion of the surfaces **3.** True **4.** True **5.** False; round-off errors increase and may be important

Problems

1. 0.417
3. (*a*) Since $\mu_s > \mu_k$ in general, the car's acceleration is greater when the wheels do not spin. (*b*) 0.212
5. (*a*) 0.289 (*b*) 600 N
7. 4.57°
9. (*a*) M/T, kg/s (*b*) M/L, kg/m
11. 2.79×10^{-4} kg/m
13. 205 km/h
15. (*a*) $a = 2.45$ m/s^2, $T = 36.8$ N (*b*) $T = m_1m_2g(1 + \sin\theta)/(m_1 + m_2)$
17. Consider first the mass m_2. If we take radially inward as positive, then $T_1 = 4\pi^2[m_1L_1 + m_2(L_1 + L_2)]/T^2$ and $T_2 = m_2v_2^2/(L_1 + L_2)$.
19. (*a*) In the inertial frame, the object moves with constant speed. In the boxcar frame, the object has an acceleration $a = -5$ m/s^2. (*b*) 4 s
21. In the inertial frame, the object slows with time, then sticks to the floor and accelerates with the boxcar. In the boxcar frame, the object slows, then comes to rest and remains at rest.
23. 6.61 cm
25. 24.5 m/s
27. 83.9 m
29. (*a*) 10 m/s^2 (*b*) $T = (100 - 8x)$ N
31. (*a*) 400 N (*b*) 625 N
33. 23.6 rev/min
35. (*a*) 16.4 m/s^2 (*b*) 19.6 N (*c*) The frictional force will not increase if the acceleration is increased.
37. (*a*) The force initially decreases with θ since the vertical component of **F** will decrease the normal force $\mathbf{F}_n$ exerted by the table. As θ increases further, however, the horizontal component of **F** decreases so much that **F** must become larger again in order to move the box (*b*) $F = \mu_s mg/(\cos\theta + \mu_s \sin\theta)$; F is a minimum near 31°.

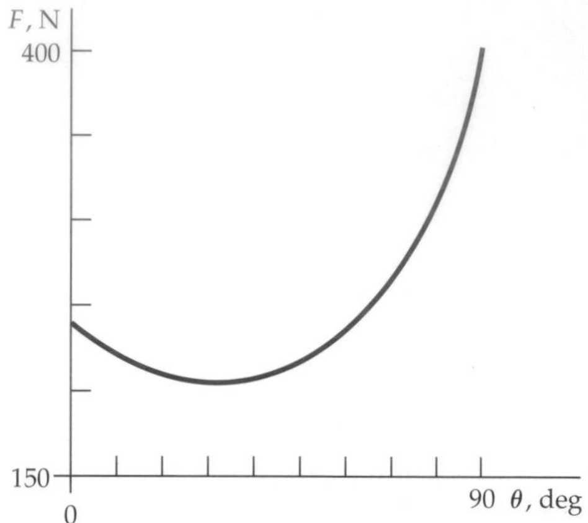

39. (*a*) 51.7 km/h (*b*) 27.6 km/h
41. (*a*) 80 N (*b*) $F_{net} = 600$ N, $F = 680$ N (*c*) 6.8 m/s^2
43. (*a*) $a = \frac{1}{2}$ m/s^2, $F_c = 2$ N (*b*) $F_c = m_2F/(m_1 + m_2)$
45. (*a*) The maximum m is 3.39 kg; the minimum m is 0.614 kg. (*b*) 9.81 N
47. (*a*) 17.7 N (*b*) $a = 1.47$ m/s^2, $f = 2.94$ N (*c*) $a_2 = 1.96$ m/s^2, $a_3 = 6.87$ m/s^2, $T = 88.3$ N
49. (*a*) $F(\theta) = \mu_s mg/(\cos\theta + \mu_s \sin\theta)$ (*b*) $dF/d\theta = \mu_s mg(\sin\theta - \mu_s\cos\theta)/(\cos\theta + \mu_s\sin\theta)^2$, $\theta = \tan^{-1}(0.6) = 31°$
51. (*a*) 7.25 m/s (*b*) 0.536
53. $v(t) = v_0/[1 + (\mu_k/r)v_0t]$

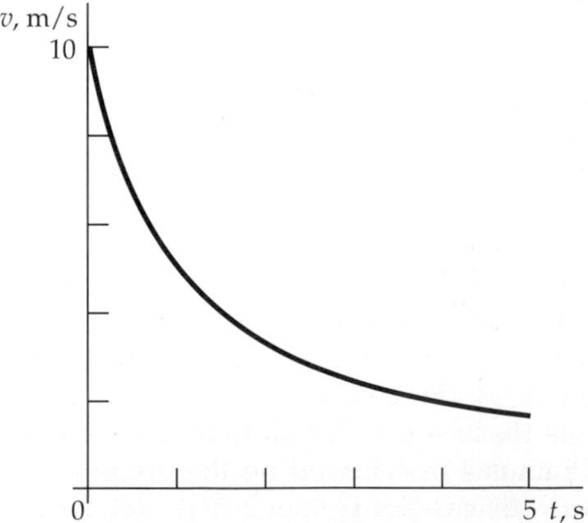

55. $v_{max} = 56.0$ km/h, $v_{min} = 20.1$ km/h
57. $a = -1.12$ m/s^2, $T = 44.7$ N
59. (*a*) $F_n = 4\pi^2 mrB^2/3600$ (*b*) Relative to the compartment, the mass falls to the floor. From the point of view of the rotating frame of reference, this is caused by a pseudoforce of magnitude mv^2/r acting radially outward. (*c*) In an inertial frame, no forces act on the mass; therefore, it travels in a straight line with the same velocity it had when dropped. This straight-line path intercepts the circular path of the compartment floor; thus, the mass is seen to drop to the floor.

Chapter 6

True or False **1.** False **2.** True **3.** False **4.** True **5.** False; it is a unit of energy **6.** False **7.** False **8.** True

Problems

1. (*a*) 7200 J (*b*) 1800 J (*c*) 28,800 J
3. 1 J = 0.738 ft·lb
5. (*a*) 9 J (*b*) 12 J (*c*) 4.58 m/s
7. $\frac{81}{4}C$
9. (*a*) $F = 327$, 245, and 196 N (*b*) 981 J, independent of L (*c*) The work is the same in each case, but the force required is less as L becomes greater
11. 18 m^2
13. (*a*) 142° (*b*) 101° (*c*) 90°
15. (*a*) $\mathbf{A}\cdot\mathbf{i} = A_x(\mathbf{i}\cdot\mathbf{i}) + A_y(\mathbf{j}\cdot\mathbf{i}) + A_z(\mathbf{k}\cdot\mathbf{i}) = A_x$ (*b*) $\mathbf{A}/\sqrt{A_x^2 + A_y^2 + A_z^2}$ (*c*) 2
17. (*a*) nonconservative (*b*) conservative (*c*) conservative (*d*) nonconservative
19. (*a*) 392 J (*b*) $x = 2.45$ m, $v = 4.91$ m/s (*c*) $K = 24.1$ J, $U = 368$ J (*d*) $K = 392$ J, $v = 19.8$ m/s
21. (*a*) $U(x) - U(x_0) = -4(x - x_0)$ (*b*) $U(x) = -4x + 24$ (*c*) $U(x) = -4x + 36$
23. (*a*) A, negative; B, zero; C, positive; D, zero; E, negative; F, zero (*b*) Point A would appear to have the force with greatest magnitude, followed closely by point E. (*c*) Points B, D, and F are equilibrium points. B is unstable, D is stable, and F is neutral.
25.

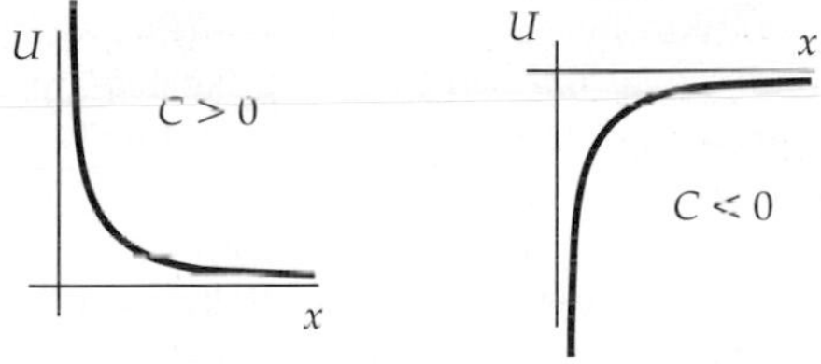

(*a*) C/x^2 (*b*) The force points away from the origin when $C > 0$. (*c*) U decreases as x increases for $C > 0$. (*d*) With $C < 0$, the force is toward the origin and U increases with increasing x.
27. (*a*) The work done by the woman is 1.22 J. The work done by the spring is −1.22 J. (*b*) 1.10 m/s
29. (*a*) 0.858 m. (*b*) After coming to rest, the object is accelerated by the spring. The kinetic energy gained by the object is just enough to allow it to rise again to the original height of 5 m. Assuming no nonconservative forces act on the object, it will continue to oscillate back and forth indefinitely.
31. (*a*) The object stops 2.5 m below the point Q, slides back to the left, rising 2.5 m above the point P, and continues this motion. (*b*) The object passes the point Q with a speed of 2.10 m/s. (*c*) 9.90 m/s
33. 5.05 m
35. (*a*) 104 J (*b*) −94.2 J (*c*) $\Delta K = 9.75$ J (*d*) 1.56 m/s
37. (*a*) 7.67 m/s (*b*) −58.9 J (*c*) 1/3
39. (*a*) 3.14×10^{16} J (*b*) 7.47 megatons TNT
41. (*a*) No work is done on the skater (*b*) 560 J (*c*) The kinetic energy of the skater results from the conversion of chemical energy in the muscles
43. (*a*) 1.67 m/s (*b*) 15 J
45. 1.37×10^6 kW
47. (*a*) 8.64×10^6 J (*b*) 2065 kcal
49. (*a*) 60° (*b*) $v^2 = \frac{2}{3}gL(2 - \cos\theta_0)$
51. (*a*) −11 J, −10 J, −7 J, −3 J, 0 J, 1 J, 0 J, 2 J, and −3 J
(*b*)

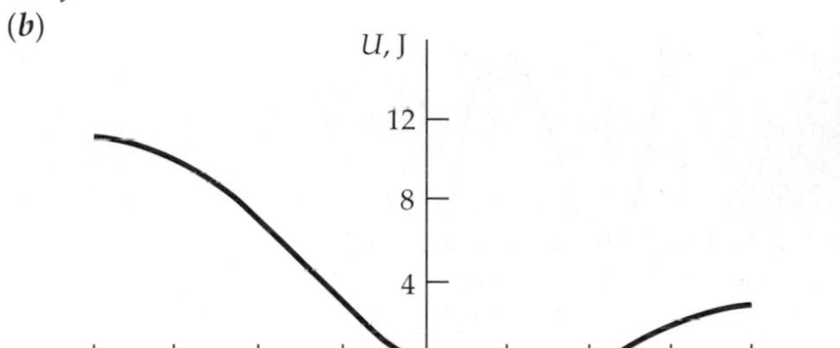

53. (*a*) 11/4 J (*b*) 6.13 J (*c*) 2.02 m/s (*d*) 14.25/4 J (*e*) 2.15 m/s
55. (*a*) 57.6 J (*b*) $U = 14.4$ J, $K = 43.2$ J (*c*) 175 J (*d*) 41.9 m/s
57. $U(x) = \frac{1}{3}ax^3$

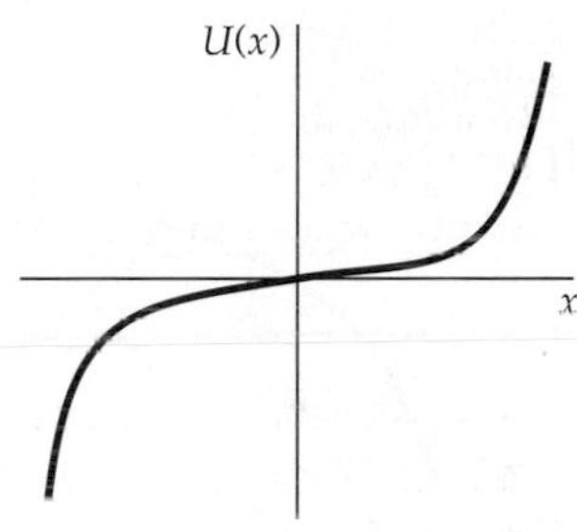

59 (*a*) 0.989 m (*b*) 0.783 m (*c*) 1.54 m
61. (*a*) $M(y) = 40 - \frac{1}{2}y$ (*b*) 1.18×10^4 J
63. (*a*) The forces and the work done by them are: gravity, $-mgL \sin 60°$; friction, $-\mu_k(mg \cos 60°)L$; the normal force, no work. (*b*) 0.451 m (*c*) The work done by gravity, friction, and the normal force is $mgL \sin 60°$, $-\mu_k mgL \cos 60°$, and zero, respectively. (*d*) 2.52 m/s
65. (*a*) $W_{nc} = -\frac{3}{4}(\frac{1}{2}mv_0^2)$ (*b*) $\mu_k = -W_{nc}/(2\pi rmg)$ (*c*) $\frac{1}{3}$
67. y_0
69. 16.0 hp
71. (*a*) 0 (*b*) 78 J
73. (*a*) $U(y) = 2Mg[(d^2 + y^2)^{1/2} - d] - mgy$
(*b*)

$$y_0 = d\frac{\left(\dfrac{m}{2M}\right)}{\sqrt{1 - \dfrac{m^2}{4M^2}}}$$

75. 1.45 hp
77. (*a*) $mg(h - 2R)$ (*b*) $2g(h - 2R)/R$ (*c*) 2.5R
79. $T_{rest} = mg - K/L$, $T_{bottom} = mg + 2K/L = T_{rest} + 3K/L$
81. (*a*)

$$v(\theta) = \sqrt{v_0^2 + 2gR(1 - \cos\theta)}$$

(*b*) $\cos\theta = \frac{2}{3} + v_0^2/(3gR)$

83. (*a*) $x = 0$ m, $x = 2$ m, and $x > 3$ m (*b*) A plot of U versus x is presented below

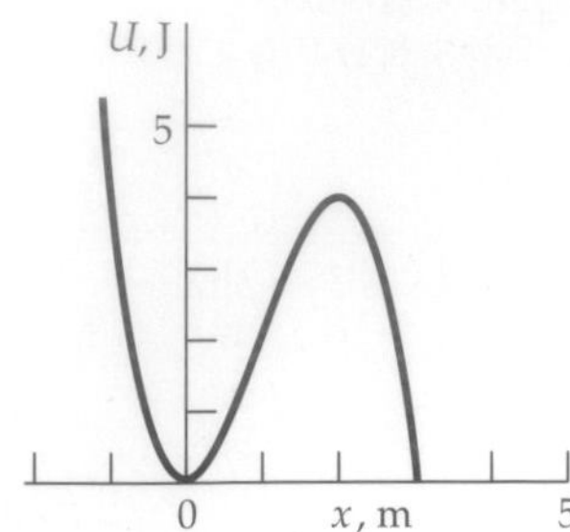

(*c*) The equilibria at $x = 0$ m, $x = 2$ m, and $x > 3$ m are stable, unstable, and neutral, respectively (*d*) 2 m/s
85. For the path from $x = 4$ m, $y = 1$ m to $x = 4$ m, $y = 4$ m, $W = 36A$. For a path consisting of three parts—horizontal from $x = 4$ m to $x = 5$ m with $y = 1$ m; vertical from $y = 1$ m to $y = 4$ m with $x = 5$ m; horizontal from $x = 5$ m to $x = 4$ m with $y = 4$ m—$W = 45A$
87. (*a*)

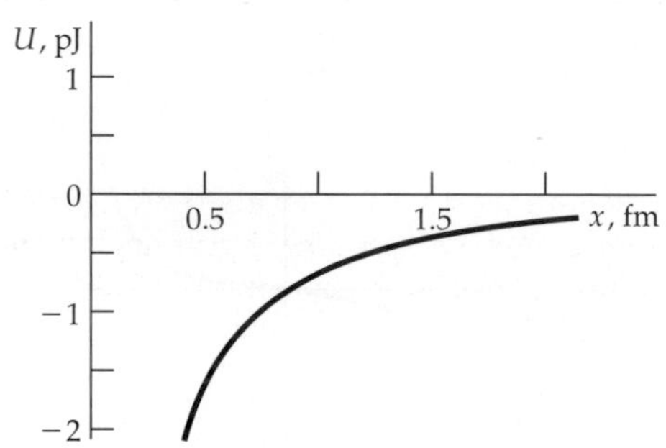

(*b*)

$$F_x = -\frac{dU}{dx} = -U_0 e^{-x/a}\left[\frac{a}{x^2} + \frac{1}{x}\right]$$

(*c*) $F_x(x = 2a)/F_x(x = a) = 0.138$
(*d*) $F_x(x = 5a)/F_x(x = a) = 0.00220$
89. (*a*)

$$\frac{dx}{\sqrt{E - U}} = \sqrt{\frac{2}{m}}\,dt$$

(*b*)

$$\frac{dx}{\sqrt{(A^2 - x^2)}} = \sqrt{\frac{k}{m}}\,dt$$

(*c*)

$$x = A\cos\left(\sqrt{\frac{k}{m}}t\right)$$

Chapter 7

True or False **1.** True **2.** False; it is true only in the center-of-mass reference frame **3.** True **4.** True **6.** True

Problems

1. 0.233 m
3. $X_{cm} = 6\frac{2}{3}$m, $Y_{cm} = 6\frac{2}{3}$m
5. $X_{cm} = 52.5$ cm, $Y_{cm} = 22.5$ cm
7. $X_{cm} = 2$ cm, $Y_{cm} = 1.4$ cm
9. 4 m/s to the east
11. (*a*) 11.8 m/s **i** (*b*) 32.9 m
13. 1.5 m/s, in the direction opposite the girl's motion
15. 0.192 m/s
17. (*a*) The ball moves in pure translational motion (*b*) The ball rotates about the center of mass
19. (*a*) 43.5 J (*b*) 1.5 m/s (*c*) $u_1 = 3.5$ m/s, $u_2 = -3.5$ m/s, taking motion to the right to be positive. (*d*) 36.75 J (*e*) $K_{tot} - K_{rel} = 6.75$ J. For the center of mass, $K_{cm} = \frac{1}{2}MV_{cm}^2 = 6.75$ J.
21. 0.652 m/s
23. 3.13 m/s
25. (*a*) $v_1 = v_2 = 3.75$ m/s (*b*) $v_{1f} = 2.5$ m/s, $v_{2f} = 4.5$ m/s
27. (*a*) 0.588 m/s (*b*) The collision is inelastic
29. (*a*) One-fifth of the mechanical energy is lost with each bounce. (*b*) 0.894
31. (*a*) The cue ball moves at an angle of 60° on the opposite side of the original direction from the 8 ball. (*b*) $v_c = 2.5$ m/s, $v_8 = 4.33$ m/s
33. (*a*) 10.75 kg·m/s (*b*) 1.34×10^3 N
35. (*a*) 6 kg·m/s (*b*) 4.62×10^3 N
37. 3290 N
39. 1.2×10^6 N
41. (*a*) 4.83 km/s (*b*) 6.91 km/s (*c*) 13.8 km/s
43. 450 m/s
45. (*a*) 212 m from the launch site (*b*) $K_i = 3600$ J, $K_f = \frac{1}{2}(2\text{ kg})(34.6\text{ m/s})^2 + \frac{1}{2}(4\text{ kg})(69.2\text{ m/s})^2 = 10{,}800$ J. The explosion released at least $K_f - K_i = 7200$ J of energy
47. $v_{1f} = -1.2$ m/s; $v_{2f} = 0.8$ m/s
49. (*a*) 1700 km below the earth's surface (*b*) The predominant external force is that exerted by the sun, though the other planets and every other body in the universe also act on the system. (*c*) To a very good approximation, the center of mass accelerates directly toward the sun. (*d*) 9340 km radially every 14 days
51. $r_{cm} = \frac{r}{14}$, where positive is taken as the direction away from the hole
53. (*a*) 1.22 m/s (*b*) 1.17 m/s
55. (*a*) 58.8 N (*b*) Yes. The weight of the ball is $w = 1.47$ N, only about 2.5 percent of the force exerted during the throw
57.

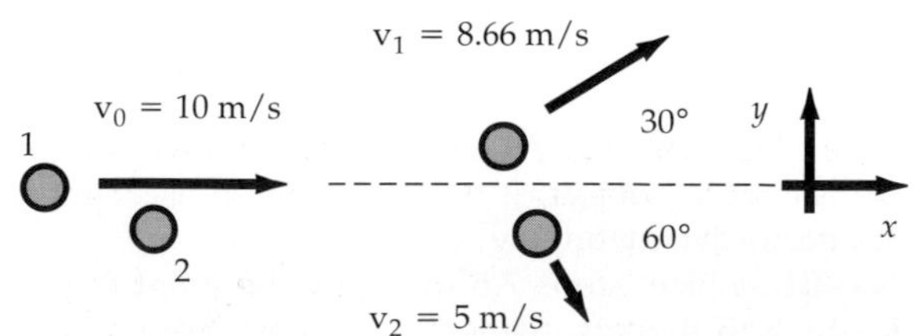

$v_1 = 8.66$ m/s; $v_2 = 5$ m/s
59. (*a*) 4.8 kg·m/s (*b*) 1600 N (*c*) 2.4 kg·m/s (*d*) 19.2 N
61. 36.4 m/s
63. $m_1^2v^2/(8gm_2^2)$
65. (*a*) 3.6×10^5 N (*b*) 120 s (*c*) 1720 m/s
67. (*a*) 0.0165 N (*b*) The weight of a single water droplet is $mg = 0.0981$ N
69. 2252 m/s

71. $v_{1f} = v_{2f} = 0.693v_{ci}$, $v_{cf} = -0.2v_{ci}$. The cue ball returns along its original direction of motion; the other two balls move at 30° above and below the original motion of the cue ball.

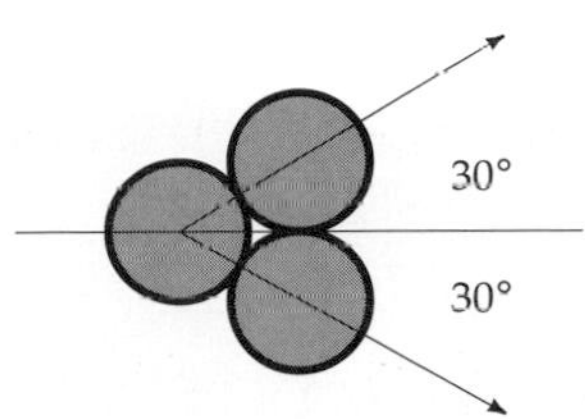

73.

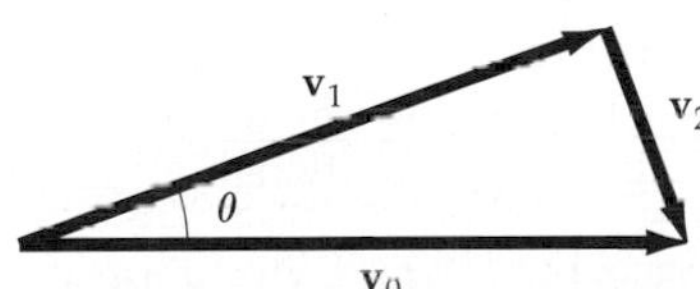

$E_2 = (\sin^2 \theta)\, E_0$
75. $(X_{cm}, Y_{cm}) = (6.67\text{ m}, 3.33\text{ m})$
77. (*a*) 9.66 km/s (*b*) 8.68 km/s (*c*) The rocket rises ≈ 400 km. Since this is only ≈ 7 percent of the earth's radius, the variation of g can be safely neglected.

Chapter 8

True or False **1.** False **2.** True **3.** True **4.** True **5.** False **6.** False; if the net torque is zero, the *rate of change* of the angular momentum is zero **7.** False; it is true for a rigid body.

Problems

1. (*a*) 0.2 rad/s (*b*) 0.955 rev
3. (*a*) −0.0291 rad/s² (*b*) 1.75 rad/s (*c*) $33\frac{1}{3}$ rev
5. 0.625 s
7. (*a*) 0.628 rad/s² (*b*) 6.28 rad/s (*c*) 11.3 m
9. (*a*) −7.33 rad/s² (*b*) −0.0359 N·m
11. (*a*) $\tau = FR = 1.5$ N·m, $\alpha = 100$ rad/s² (*b*) $\omega = 400$ rad/s
13. (*a*) 56 kg·m² (*b*) 112 J
15. (*a*) 0.136 J (*b*) 356 rev/min
17. 28 kg·m²
19. (*a*) $I_x = 28$ kg·m² (*b*) $I_y = 28$ kg·m² (*c*) $I_z = 56$ kg·m²
21. $\frac{1}{4}MR^2$
23. (*a*) 60 kg·m²/s (*b*) 75 kg·m² (*c*) 0.8 rad/s
25. (*a*) L is doubled (*b*) L is doubled
27. (*a*) zero (*b*) $v_1/v_2 = r_2/r_1$
29. (*a*) K_{rot} is 28.6 percent of the total kinetic energy (*b*) K_{rot} is 33.3 percent of the total kinetic energy (*c*) K_{rot} is 50 percent of the total kinetic energy
31. (*a*) $\frac{7}{5}g \sin\theta$ (*b*) $(2/7)\, mg \sin\theta$ (*c*) $\tan\theta = (7/2)\mu_s$
33. $\mathbf{F} = -F\mathbf{i}$; $\mathbf{r} = R\mathbf{j}$; $\boldsymbol{\tau} = FR\mathbf{k}$
35. (*a*) 36**k** (*b*) −36**j** (*c*) 12**k**
37. (*a*) (−25.8 kg·m²/s)**k** (*b*) (12.9 N·m)**k**
39. (*a*) 20.8 kg·m²/s (*b*) 0.520 rad/s (*c*) 12.1 s (*d*) 0.172 kg·m²/s, upward (downward) when the angular momentum due to the spin of the wheel is away (toward) the pivot
41. (*a*) 2.5 N (*b*) 25 cm from the axis in a direction away from the center of the disk
43. (*a*) 2.4 N·m (*b*) 66.7 rad/s² (*c*) 200 rad/s (*d*) 720 J (*e*) 7.2 kg·m²/s (*f*) 300 rad (*g*) The work done by the torque is $W = 720$ J
45. (*a*) 283 kg·m²/s (*b*) 9.43 N·m (*c*) 9.43 N·m (*d*) 31.4 N
47. (*a*) 1.33×10^{-5} kg·m²/s (*b*) 1.33×10^{-5} kg·m²/s (*c*) 1.33×10^{-5} kg·m²/s (*d*) $L_{cm} = 8.83 \times 10^{-5}$ kg·m²/s or 6.18×10^{-5} kg·m²/s, depending on the direction of $\mathbf{v}_{cm}$.
49. $K_{rot} = 2.60 \times 10^{29}$ J; $K_{orb} = 2.67 \times 10^{33}$ J
51. (*a*) 13.6 kg·m² (*b*) 7.14 N·m
53. 1.95 m
55. (*a*) 5.42 rad/s (*b*) $(5/2)mg$
57. $a = 0.2$ m/s²; $v = at$; $x = x_0 + \frac{1}{2}at^2$; $\alpha = 3.33$ rad/s²; $\omega = \alpha t$
59. (*a*)

$$a = \frac{g}{\left(1 + \dfrac{2}{5}\dfrac{M}{m}\right)}$$

(*b*) $T = \frac{2}{5}Ma$
61. (*a*) 2.7R (*b*) 2.5R
63. (*b*) $T = \frac{1}{3}Mg$
65. (*a*) 5 rev/s (*b*) 592 J (*c*) The additional energy is supplied by the man's muscles
67. (*a*) 5.36 rad/s (*b*) 9.99 J
69. (*a*) 552 N (*b*) 1.77 rad/s² (*c*) 2.17 rad/s
71. 1.13×10^{-4} s
73. (*a*) 0.0918 m/s² (*b*) $T_1 = 1.952$ N; $T_2 = 4.955$ N; $T_2 - T_1 = 2.37 \times 10^{-3}$ N (*c*) If you neglect the motion of the pulley, that is, if you set $a = -0.0971$ m/s², $T_1 = T_2 = 4.95$ N
75. (*b*)

$$W = -\frac{L_0^2}{m}\int_{r_0}^{r_f} \frac{1}{r^3}\,dr = \frac{L_0^2}{2m}(r_f^{-2} - r_0^{-2})$$

(*c*) $v_f = L_0/mr_f$, $\Delta K = L_0^2(r_f^{-2} - r_0^{-2})/2m = W$ (*d*) 0.282 m
77. The day would lengthen by 0.554 s
79. (*a*) $I = 0.04$ kg·m² (*b*) $I = 0.0415$ kg·m², about 3.5 percent greater than the approximate result
81. (*a*) $s_1 = (12/49)v_0^2/\mu_k g$; $t_1 = (2/7)v_0/\mu_k g$; $v_1 = (5/7)v_0$ (*b*) $s_1 = 3.99$ m; $t_1 = 0.583$ s; $v_1 = 5.71$ m/s
83. (*a*) $v_{cm} = (L/2)\omega = (3/2)\, P_0 x/ML$ (*b*) $P = P_0(3x/2L - 1)$
85. (*a*) 0.0443 m/s² (*b*) 2.21 rad/s² (*c*) 3.63 J (*d*) 399 J
87. (*a*) 200 m/s (*b*) 8000 rad/s (*c*) 257 m/s (*d*) 11.7 s
89.

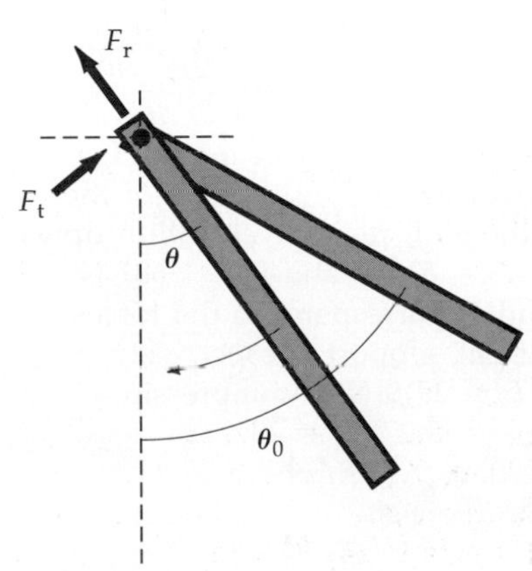

$F_r = \frac{1}{2}Mg(5\cos\theta - 3\cos\theta_0)$; $F_t = \frac{1}{4}Mg\sin\theta$
91. (*a*) $(11/7)v_0$ (*b*) $4v_0/(7\mu_k g)$ (*c*) $(36/49)v_0^2/\mu_k g$
93. 54.0°

Chapter 9

True or False **1.** False; the resultant torque must also be zero **2.** True **3.** True **4.** False **5.** False

Problems

1. 1.4 m away from the center
3. $F_B = 417$ N; $F_J = 270$ N
5. $F_1 = 117$ N; $F_2 = 333$ N
7. (*b*) Part (*a*) gives the maximum angle the leg may make with the vertical before the foot slips. (*c*) On ice μ_s is very small; thus the maximum θ is also small. This implies that short steps must be taken
9. The U-shaped figure has its center of gravity on the vertical line about which the shape is symmetric. The height of the center of gravity above the base of the "U" is $Y_{cg} = 4L/5$. For the L-shaped wire, choosing the origin at the bend in the "L" gives $X_{cg} = L/6$, $Y_{cg} = 2L/3$. For the triangular shape, the center of gravity is on the vertical line of symmetry, and $Y_{cg} = \sqrt{15}L/5$
11. 99.0 cm from the man's feet
13. Let F_1 and F_2 denote the forces exerted by the left and right supports, respectively. $F_1 = 245$ N; $F_2 = 736$ N
15. (*a*) 5400 N (*b*) 9
17. $\frac{1}{2}F(\sqrt{3}b - a)$
19. (*a*) $\tau = Fa$ (*b*) The effective point of application of the normal force is a distance $a/6$ in from the corner of the cube. (*c*) $\frac{1}{2}Mg$
21. $x_{cg} = 0.944a$, $y_{cg} = a$
23. Let F_1 and F_2 be the forces exerted on the cylinder by the left and right planes, respectively. $F_1 = 0.866W$; $F_2 = 0.5W$
25. (*a*) 42.4 N (*b*) 57.0 N, along a direction 7.13° above the strut
27. 26.6°
29. (*a*) 450 N (*b*) 241 N (*c*) The tension decreases as the brace is moved downward
31.

$$\mathbf{F} = \frac{Mg\sqrt{h(2R-h)}}{(R-h)}$$

33. 9 m
35. (*a*)

$$\mathbf{N} = M\mathbf{g} - \mathbf{F}\sqrt{\frac{(2R-h)}{h}}$$

(*b*) $-\mathbf{F}$
(*c*)

$$F_{cy} = F\sqrt{\frac{(2R-h)}{h}}$$

37. (*a*) 49.1 N (*b*) 73.6 N·m (*c*) Let F_1 and F_2 refer to the forces exerted on the end of the rod and 0.1 m in from the end, respectively. With upward taken as positive, $F_2 = 736$ N and $F_1 = -687$ N
39. F_1 and F_2 correspond to the forces exerted by the left and right supports, respectively. $F_1 = -1938$ N, a tension; $F_2 = 2919$ N, a compression
41. $a = g/3 = 3.27$ m/s^2
43. $x_{cg} = 0$, $y_{cg} = 2R/\pi$
45. 566 N
47. 61.6°
49. (*a*) 0.361mg (*b*) 0.313mg (*c*) 0.820mg

Chapter 10

True or False **1.** False **2.** True **3.** True

Problems

1. 2.92 AU
3. 10^4 AU
5. 248 y
7. 2.63×10^6 m/s
9. (*a*) 1.16×10^6 s (*b*) 8.79×10^{25} kg
11. 6.02×10^{24} kg
13. 5.97×10^{24} kg
15. 10 times as much
17. (*a*) 4.80×10^{-9} N (*b*) 8.64×10^{-10} N·m
19. (*a*) 5.77 kg (*b*) gravitational
21. 5330 mi/h
23. (*a*) -6.25×10^9 J (*b*) -3.13×10^9 J (*c*) 7.92×10^3 m/s
25. (*a*) 0 (*b*) 0 (*c*) 3.20×10^{-9} m/s^2
27.

$$W = \frac{GMm_0}{R}$$

29. (*a*) $GM_1m/(3a)^2 + GM_2m/(2.2a)^2$ (*b*) $GM_1m/(1.1a)^2$ (*c*) 0
31. $\sqrt{3}v_e = 1.94 \times 10^4$ m/s
33. 51.2 km/s
35. $h = 0.147\,R_E = 935$ km
37. (*a*) 3.31 y (*b*) 1.49×10^{29} kg (*c*) Planet 2. Planet 2 also has the greater total energy. (*d*) $v_A = v_P/1.8$
39. (*b*) $\mathbf{g} = \mathbf{F}/m_0 = [-2Gmx/(x^2 + a^2)^{3/2}]\mathbf{i}$
41. 9.66×10^{-8} N
43. (*a*) 7.38 m (*b*) 0.0319 mm
45. (*a*) 2.74×10^{-3} N (*b*) 0.480 N, a value 175 times that due to the moon (*c*) −0.06 percent
47. 1.88 N
49.

$$\frac{GMm}{a(a+L)}$$

51. (*a*) $F_r = GMm_0/r$ for $r > R$ and $F_r = 0$ for $r < R$. (*b*) $U(r) = -GMm_0/r$ (*c*) Since $dU = -F_r\,dr = 0$ everywhere inside the shell, it follows that U is constant in value there. From the results of part (*c*), we conclude that $U(r) = -GMm_0/R$ for $r \leq R$
(*d*)

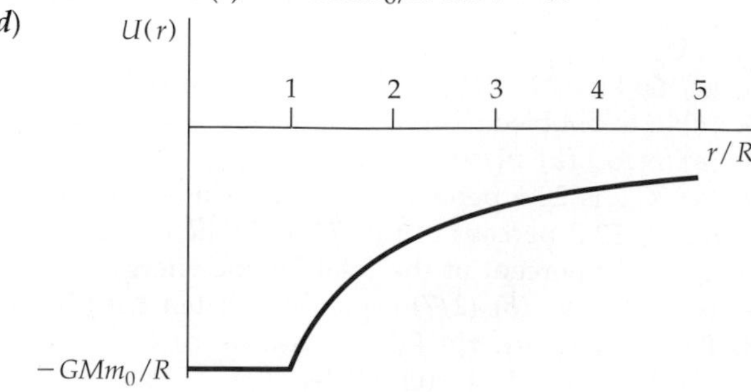

53. (*b*) $U(x) = -(GMm_0/L)\ln[(x + L/2)/(x - L/2)]$ for $x > \frac{1}{2}L$. (*c*) $F_x = -GMm_0/(x^2 - L^2/4)$
55.

$$g = F/m_0 = GM\left[\frac{1}{x^2} - \frac{1}{8(x - R/2)^2}\right]$$

57. $\omega = (GM/R^3)^{1/2} = (\frac{4}{3}G\pi\rho_0)^{1/2}$

59. (*a*) $F_s/F_m = (M_s/M_m)(r_m/r_s)^2$. Substituting the appropriate numerical values yields 169 for this ratio, showing that the sun exerts by far the greater force. (*c*) Numerically, this is equal to 0.422, showing that the moon has the greater effect in causing the tides in earth's oceans

61. (*a*) $\mathbf{F} = -(GMm/d^2)[1 - \frac{1}{4}d^3/(d^2 + R^2/4)^{3/2}]\mathbf{i}$
(*b*) $\mathbf{F} = -0.821(GMm/R^2)\mathbf{i}$

Chapter 11

True or False 1. False 2. True 3. False; it can float because of surface tension 4. True

Problems

1. 0.676 kg
3. 103 kg
5. 54 kg
7. 0.976 mm
9. 5.01°
11. (*a*) 10.3 m (*b*) 0.757 m
13. $F = 6.46 \times 10^4$ N The table doesn't collapse because air pressure also exerts an upward force on the bottom surface of the card table
15. 230 N
17. (*a*) $F = 8.97 \times 10^4$ N (*b*) $F = 5.02 \times 10^4$ N (*c*) If a window is rolled down the pressure on the two sides of the door will equalize, and it may then be opened
19. 1060 kg/m^3
21. 15.8 cm
23. 4.36 N
25. (*a*) 11.1×10^3 kg/m^3 (*b*) lead
27. 0.0233 N/m
29. (*a*) 4.58 L/min (*b*) 763 cm^2
31. (*a*) 12 m/s (*b*) 133 kPa (*c*) The flow rates are identical in the two sections
33. $F = 1.31 \times 10^5$ N = 29,400 lb
35. 1.43 mm
37. 1.49 ft
39. 183 m^3
41. $2[h(H - h)]^{1/2}$
43. 0.441 Pa
45. 3.89 kg
47. The readings on the upper and lower scales are 12.4 N and 36.7 N, respectively
49. 0.25×10^3 kg/m^3
51. (*a*) 21.2 kg (*b*) 636 N·s
(*c*)

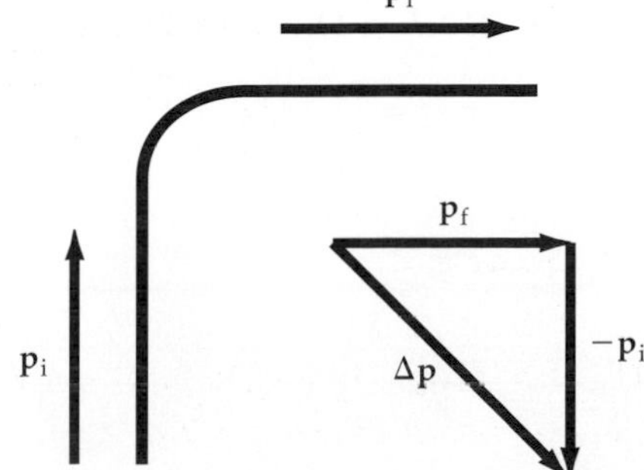

$\Delta\mathbf{p} = 900$ N·s, $F = 900$ N

53. (*a*) $dF = (\rho g y)(w\,dy)$
(*b*)

$$F = \tfrac{1}{2}\rho g w d^2 = 9.20 \times 10^7 \text{ N}$$

(*c*) The reason atmospheric pressure may be neglected is that it acts equally on both sides of the dam and thus gives rise to no net force

55. (*a*) 34.7 cm (*b*) 7.76×10^{-2} J
57. (*a*) 14.7 g (*b*) 1.03×10^3 kg/m^3
59. (*b*) 1.12 mm
61. 5.02 h
63. (*c*)

$$t = \frac{1}{\sqrt{2g}}\frac{2A_1}{A_2}(\sqrt{H} - \sqrt{h})$$

$$\text{or } h = (\sqrt{H} - A_2 t/\sqrt{2}\,A_1)^2$$

(*d*) $t = (2H/g)^{1/2}(A_1/A_2) = 1.77$ h

Chapter 12

True or False 1. False; it is independent of the amplitude 2. True 3. True 4. False; it is simple harmonic only for small angular displacements 5. True 6. True 7. True 8. True 9. True

Problems

1. (*a*) 7.96 Hz (*b*) 0.126 s (*c*) 0.1 m (*d*) 5 m/s (*e*) 250 m/s^2 (*f*) $t = 0.0314$ s; $x = 0$
3. (*a*) 474 N/m (*b*) 0.5 s (*c*) 1.26 m/s (*d*) 15.8 m/s^2
5. 0.801 Hz
7. (*a*) 1.49 kg (*b*) 5.84 Hz (*c*) 0.171 s
9. (*a*) 62.8 cm/s (*b*) 790 cm/s^2
11. (*a*) $f = 0.477$ Hz, $T = 2.09$ s (*b*) 0.283 m (*c*) −0.262 m
13. (*a*) $x(t) = (0.25 \text{ m}) \cos (4\pi/3)t$ (*b*) $v = -(\pi/3 \text{ m/s}) \sin (4\pi/3)t$ (*c*) $a = -(4\pi^2/9 \text{ m/s}^2) \cos (4\pi/3)t$
15. (*a*) 1.05 m/s (*b*) 4.39 m/s^2 (*c*) 1.13 s
17. (*a*) 0.314 m/s (*b*) $2\pi/3$ rad/s (*c*) $x = (0.15 \text{ m}) \cos 2\pi t/3$
19. 1.92 J
21. (*a*) 3 cm (*b*) 77.5 cm/s
23. (*a*) 0.270 J (*b*) −0.736 J (*c*) 1.01 J (*d*) 0.270 J
25. (*a*) 1.90 cm (*b*) 0.054 J (*c*) 0.223 J (*d*) 0.277 J
27. 12.3 s
29. 11.7 s
31. 1.10 s
33. 0.504 kg·m^2
35. (*a*) 0.444 s (*b*) 0.180 J (*c*) $Q = 628$; $b = 0.045$ kg/s
37. (*a*) The energy decreases by 10 percent each cycle (*b*) 30 s (*c*) 62.8
39. (*a*) 314 (*b*) 6 rad/s
41. 0.420 J

43. (*a*)

(*b*) 2.93 cm, 7.07 cm, 7.07 cm, 2.93 cm
45. 2.62 s
47. 12.9 Hz
49. (*a*) 70 J (*b*) 22 J (*c*) 7.33 W
51. (*a*) 0.314 (*b*) 0.000313 percent
53. (*a*) 1.57 percent (*b*) $E_n = (0.984)^n E_0$
(*c*) $(0.430)E_0$
55. (*a*) 7.85 cm (*b*) 3.92 cm (*c*) 0.397 s
(*d*) 62.0 cm/s; $t = 0.0993$ s
57. 9.10 Hz
59. (*a*) 5.23×10^4 N/m (*c*) 2×10^{11} N/m
61. (*a*) $A = 0.141$ m; $T = 0.444$ s (*b*) $A = 0.231$ m; $T = 0.363$ s (*c*) For the inelastic collision, we have $x(t) = (0.141\text{ m}) \cos [(14.1\text{ rad/s})t - \pi/2]$; $I_2 = 4$ kg·m/s. For the elastic collision, $x(t) = (0.231\text{ m}) \cos [(17.3\text{ rad/s})t - \pi/2]$; $I_2 = 8$ kg·m/s.
63. (*a*) 1.55 cm (*b*) 1.72 m/s (*c*) 16.6 cm
65. (*a*) $F_r = -(GM_E m/R_E^3)r = -kr$, so $F_r \sin\theta = -kr \sin\theta = -kx$. The motion is simple harmonic, with a force constant $k = GM_E m/R_E^3$.
(*b*) $T = 2\pi(m/k)^{1/2} = 2\pi(R_E/g)^{1/2} = 84.4$ min
67. (*a*) 29.4 (*b*) 3.53 s (*c*) 6 J
69. (*c*) The error is 0.008 percent. For the error to be 1 percent the requirement is $r = 22.4$ cm
71. answer given in problem
73. answer given in problem

Chapter 13

True or False **1.** True **2.** False **3.** False; even though $v = f\lambda$, the speed depends only on the tension and mass density. When λ changes, f also changes so that v remains the same. **4.** True **5.** True

Problems

1.

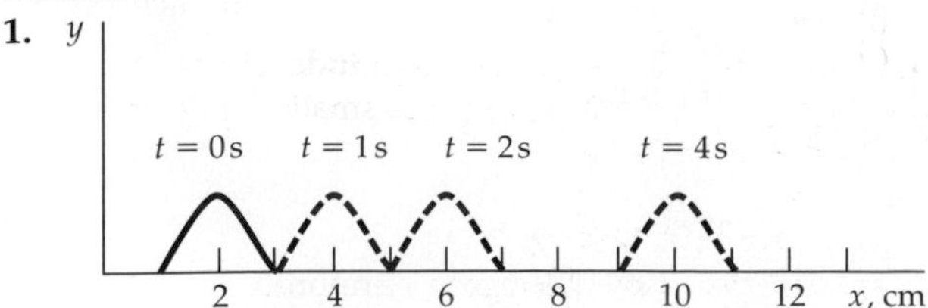

3.

5.

7. (*a*)

(*b*) $y(x, t) = (0.12\text{ m}^3)/[(2.00)^2 + (x - 10t)^2]\text{ m}^2$
(*c*) $y(x, t) = (0.12\text{ m}^3)/[(2.00)^2 + (x + 10t)^2]\text{ m}^2$
9. 251 m/s
11. (*a*) 20 m/s (*b*) 28.3 m/s (*c*) 14.1 m/s
13. 0.252 s
15. (*a*) 66.7 m/s (*b*) 889 N
17. (*a*) $y = y_0 \sin k(x - vt)$ (*b*) $y = y_0 \sin 2\pi(x/\lambda - ft)$
(*c*) $y = y_0 \sin 2\pi(x/\lambda - t/T)$ (*d*) $y = y_0 \sin (2\pi/\lambda)(x - vt)$ (*e*) $y = y_0 \sin 2\pi f(x/v - t)$
19. $y = (0.02\text{ m}) \sin 12\pi(x - 10t)$
21. $y = (0.5\text{ m}) \cos (2\pi/50)(x - 10t)$
23. 9.87 W
25. (*a*) 3.86 cm (*b*) 3.46 cm
27. $A_{resul} = 2A \cos \frac{1}{2}\delta = 2(0.05\text{ m}) \cos \pi/4 = 7.07$ cm
29. (*a*) $\lambda = 2$ m, $f = 25$ Hz (*b*) $y = (4 \times 10^{-3}\text{ m}) \cos 50\pi t \sin \pi x$
31. (*a*) 1.25 Hz (*b*) 3.75 Hz (*c*) 6.25 Hz
33. 141 Hz
35. (*a*) $\lambda = 31.4$ cm, $f = 47.7$ Hz (*b*) 1500 cm/s
(*c*) 62.8 cm

37. (*a*)

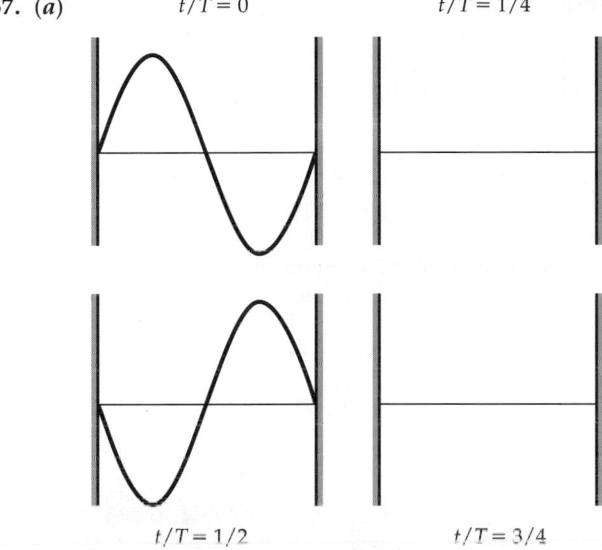

(*b*) 0.0126 s (*c*) The energy of the wave is kinetic
39. (*a*) 2.66 m (*b*) 0.666 m (*c*) 160 m/s
41. (*a*) 5/3, 7/5 (*b*) Since these are successive frequencies, it is clear that the even harmonics are missing, as is the case for a string fixed on one end only. (*c*) 25 Hz (*d*) The frequencies 75, 125, and 175 Hz are the third, fifth, and seventh harmonics, respectively. (*e*) 4 m
43. For $y(x, t) = A \sin kx \cos \omega t$, it is clear that $\partial^2 y/\partial t^2 = -\omega^2 y$ and $\partial^2 y/\partial x^2 = -k^2 y$. Therefore, $\partial^2 y/\partial x^2 = (k^2/\omega^2)\partial^2 y/\partial t^2 = (1/\omega^2)\partial^2 y/\partial t^2$, and thus $y(x,t)$ satisfies the wave equation
45. (*a*) 75 Hz (*b*) 375 Hz is the fifth harmonic, and 450 Hz is the sixth harmonic. (*c*) 2 m
47. (*a*) $f = 400$ Hz, $T = 2.5 \times 10^{-3}$ s (*b*) 316 m/s (*c*) $\lambda = 0.791$ m, $k = 7.95\ \text{m}^{-1}$ (*d*) $(0.5 \times 10^{-3}\ \text{m}) \sin 2\pi(x/\lambda - ft)$ (*e*) $v_{max} = 1.26$ m/s, $a_{max} = 3.16 \times 10^3\ \text{m/s}^2$ (*f*) 2.50 W
49. (*a*) $v = -(7.54\ \text{m/s}) \sin 2.36x \sin 377t$
(*b*) The maximum speed occurs at the point where $\sin 2.36x = 1$ or $x = 0.666$ m; $v_{max} = 7.54$ m/s (*c*) $a = -(2.84 \times 10^3\ \text{m/s}^2) \sin 2.36x \cos 377t$ (*d*) The greatest acceleration also occurs at $x = 0.666$ m; $a_{max} = 2.84 \times 10^3\ \text{m/s}^2$
51. (*a*) 1.11 W (*b*) The amplitude must be increased by a factor of 10 or the frequency must be increased by a factor of 10, which implies that the tension must be increased by 10,000 times its original value. (*c*) The frequency of the power source is probably easiest to change
53. (*a*)

$$K = \frac{1}{2}\mu n^2 \omega_1^2 A^2 \sin^2 n\omega_1 t \int_0^L \sin^2 \frac{n\pi x}{L}\, dx$$

(*b*) $K = \frac{1}{4}\mu n^2 \omega_1^2 A_n^2 L$ (*c*) $y(x, t) = 0$ for all x (*d*) $K = (\frac{1}{2}\mu\omega_1^2 L)(n^2 A_n^2)$
55. (*a*) $df_n/f_n = \frac{1}{2}dF/F$ (*b*) A 1.54 percent increase in the tension is required
57. (*a*) $y(x, t) = (0.01\ \text{m}) \sin (2\pi x/3 - 160\pi t)$
(*b*) 80 Hz (*c*) 0.0125 s (*d*) 5.03 m/s
59. (*a*) 0.8 m (*b*) 480 N (*c*) The finger should be placed 9.23 cm from the end
61. (*a*) 60 cm (*b*) $2\pi/5$ (*c*) 24 m/s
63. (*a*)

$$v = -A_1\omega_1 \sin \omega_1 t \sin k_1 x - A_2\omega_2 \sin \omega_2 t \sin k_2 x$$

(*b*)

$$dK = \tfrac{1}{2}\mu v^2\, dx = \tfrac{1}{2}\mu[A_1^2\omega_1^2 \sin^2 \omega_1 t \sin^2 k_1 x + A_2\omega_2^2 \sin^2 \omega_2 t \sin^2 k_2 x + 2A_1A_2\omega_1\omega_2 \sin \omega_1 t \sin \omega_2 t \sin k_1 x \sin k_2 x]\, dx$$

(*c*) $\frac{1}{4}\mu L\omega_1[(n_1A_1)^2 + (n_2A_2)^2]$
65. (*a*)

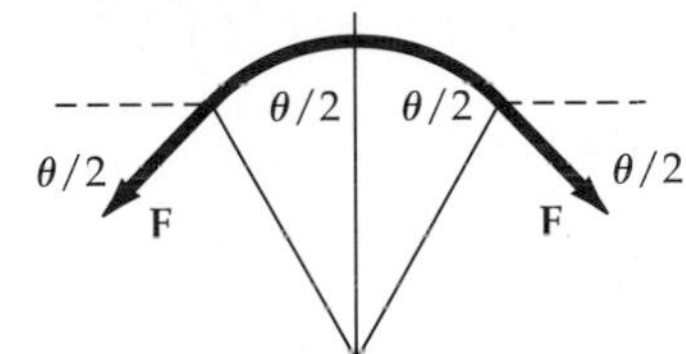

$2F \sin \frac{\theta}{2} \approx F\theta = mv_0^2/R = \mu R\theta v_0^2/R$, so $F = \mu v_0^2$. (*b*) v_0
(*c*) A forward-moving pulse always remains in contact with the ground (where it was generated), whereas a backward-moving pulse appears to make two revolutions for one rotation of the loop
67. (*b*) 2.21 s
69. (*a*) 4 m/s (*b*) 0.00424 m (*c*) $y(x, t) = (0.00424\ \text{m}) \cos (12.5x - 50t)$

Chapter 14

True or False **1.** False; sound waves are longitudinal **2.** False **3.** False; it has 1000 times the intensity **4.** False **5.** False **6.** True **7.** False **8.** True **9.** True **10.** True **11.** False

Problems
1. $[(B/\rho)^{1/2}] = [(\text{kg/m}\cdot\text{s}^2)/(\text{kg/m}^3)]^{1/2} = (\text{m/s})$
3. 5.09×10^3 m/s
5. $2.70 \times 10^{10}\ \text{N/m}^2$
7. (*a*) 1.30 m (*b*) 0.649 m
9. 8.27×10^{-2} Pa
11. (*a*) zero (*b*) 3.67×10^{-6} m
13. (*a*) 138 Pa (*b*) 21.6 W/m^2 (*c*) 0.216 W
15. (*a*) $10^{-11}\ \text{W/m}^2$ (*b*) $2 \times 10^{-12}\ \text{W/m}^2$
(*c*) For $\beta = 3$ dB, $p_0 = 4.18 \times 10^{-5}$ Pa; for $\beta = 10$ dB, $p_0 = 9.37 \times 10^{-5}$ Pa
17. 99/100 of the acoustic power must be eliminated to reduce β by 20 dB
19. (*a*) 90° (*b*) $\sqrt{2}A$
21. (*a*) 85 Hz; 255 Hz (*b*) Some sound will reflect off walls and other surfaces, thus giving a variety of path differences
23. (*a*) zero (*b*) $2I_0$ (*c*) $4I_0$
25. (*a*) $\lambda/4$ (*b*) The smallest magnitude of $r_2 - r_1$ occurs for $r_2 - r_1 = -\lambda/4$, and the smallest positive value is $r_2 - r_1 = 3\lambda/4$
27. 437 Hz
29. (*a*) 17 Hz (*b*) 8.5 Hz
31. (*a*) 2267 Hz (*b*) The eighth harmonic
33. (*a*) 3400 Hz, 10,200 Hz and 17,000 Hz
(*b*) We might expect the ear to be more sensitive near the above resonant frequencies, since these special

frequencies produce standing waves within the ear canal

35. (*a*) $N \approx \Delta t/T = f_0\,\Delta t$ (*b*) $\lambda \approx \Delta x/N$ (*c*) $k = 2\pi/\lambda \approx 2\pi N/\Delta x$ (*d*) N is uncertain because the waveform dies out gradually, rather than stopping abruptly at some time. (*e*) $\Delta k = (2\pi/\Delta x)\,\Delta N$

37. (*a*) $\lambda = 15$ m and $f = 20$ pies/min (*b*) $\lambda = 13.5$ m; $f = 22.2$ pies/min (*c*) $\lambda = 15$ m; $f = 22$ pies/min

39. (*a*) 1.3 m (*b*) 262 Hz

41. (*a*) 2.1 m (*b*) 162 Hz

43. (*a*) 80 m/s (*b*) 420 m/s (*c*) 1.7 m (*d*) 247 Hz

45. 529 Hz; 474 Hz

47. 3.42 m/s

49. (*a*) $I_1/I_2 = (M_1/M_2)^{1/2} = 4$ (*b*) $I_1/I_2 = s_{01}/s_{02} = \frac{1}{4}$ (*c*) $s_{01}/s_{02} = [\rho_2 v_2/\rho_1 v_1]^{1/2} = \frac{1}{2}$; $p_{01}/p_{02} = \rho_1 v_1 s_{01}/\rho_2 v_2 s_{02} = 2$

51. A reasonable time for Δt might be $\frac{1}{4}$ to $\frac{1}{2}$ s, giving a distance to the wall of 20 to 40 m

53. (*a*) At a node, the powder is not vibrated; hence, it collects there. (*b*) $v = \lambda f = 2Sf$ (*c*) $S = 5$ cm, $f = v/\lambda = 3400$ Hz (*d*) Suppose the tube is 1 m long. The lowest frequency in air is $f = 85$ Hz, the highest frequency is 8500 Hz. For the same tube, the appropriate frequencies in helium would be 35 to 3500 Hz

55. $v = 338$ m/s. This method lacks accuracy since the antinode generally does not occur exactly at the end of an open pipe

57. (*a*) $dv/dT = \frac{1}{2}(\gamma R/MT)^{1/2} = \frac{1}{2}(v/T)$ (*b*) 4.95 percent (*c*) $v = 347.4$ m/s. An exact calculation gives $v = (331 \text{ m/s})[(273 + 27)/273]^{1/2} = 347$ m/s, showing the approximate result to be quite accurate

59. (*a*) 0 dB (*b*) 66 dB (*c*) 63 dB

61. (*a*) $I_1 = 1.99 \times 10^{-5}$ W/m^2; $I_2 = 0.884 \times 10^{-5}$ W/m^2 (*b*) $[(I_1)^{1/2} + (I_2)^{1/2}]^2 = 5.53 \times 10^{-5}$ W/m^2 (*c*) $[(I_1)^{1/2} - (I_2)^{1/2}] = 0.221 \times 10^{-5}$ W/m^2 (*d*) $I_1 + I_2 = 2.87 \times 10^{-5}$ W/m^2

63. (*a*) 4.5 m/s (*b*) 22.1 beats/s (*c*) 22 beats/s

65. (*a*) 55.1 N/m^2 (*b*) 3.46 W/m^2 (*c*) 0.245 W

67. 0.773 m/s

69. (*a*) $r = 100$ m (*b*) 0.126 W

71. (*a*) 100 dB (*b*) 25.1 W (*c*) 2 m (*d*) 96.5 dB

73. 3.97×10^{-5} W

75. (*a*) 26 percent increase per year. This seems to be an unreasonably large increase. (*b*) 3.01 years

77. 87.8 dB

79. answer given in problem

81. (*a*) 0.279 m (*b*) 1.22×10^3 Hz (*c*) $\theta_3 = 24.7°$; $\theta_4 = 33.9°$; $\theta_5 = 44.2°$; $\theta_6 = 56.9°$; $\theta_7 = 77.6°$ (*d*) 4°

83. answer given in problem

85. (*a*) (80, 0) m; (75.2, ±27.2) m; (58.7, ±54.4) m (*b*) (78.8, ±13.6) m; (68.8, ±40.8) m

87. answer given in problem

Chapter 15

True or False **1.** False **2.** False **3.** False; the degree size is also different **4.** True **5.** False **6.** False; it could remain the same or even decrease depending on the change in the volume **7.** True **8.** True **9.** True

Problems

1. 10.4°F to 19.4°F

3. 56.7°C and −62.2°C

5. (*a*) 8.4 cm (*b*) 107°C

7. −320.44°F

9. (*a*) 54.9 torr (*b*) 3704 K

11. 30.0264 cm

13. 2.10 km

15. $V_2 = 1.16V_1$

17. 1.79 mol; 1.07×10^{24} molecules

19. (*a*) 3.66×10^3 moles (*b*) 59.9 moles

21. 152 J

23. $v_{rms} = 4.99 \times 10^5$ m/s; $K_{av} = \frac{3}{2}kT = 2.07 \times 10^{-16}$ J

25.

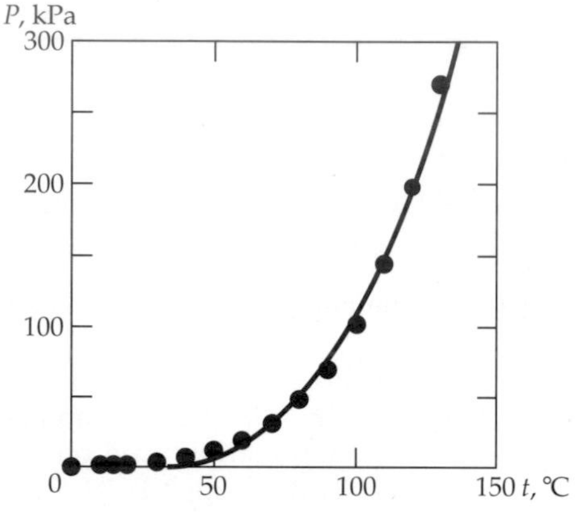

(*a*) 88.2°C (*b*) 80.4°C (*c*) 173 kPa

27. 3.39 kPa

29. 70.8

31. (*a*) 1846 m/s (*b*) 461 m/s (*c*) 393 m/s (*d*) One-sixth of the escape speed on Mars is 833 m/s; hence, O_2 and CO_2 should still be found on Mars, but not H_2

33. If $t_F = -40°$F, then $t_C = -40°$C.

35. (*a*) 74.98 cm (*b*) 75.02 cm

37. (*a*) $L_B' - L_A' = L_B - L_A$ if $\alpha_B L_B = \alpha_A L_A$ or $L_A/L_B = \alpha_B/\alpha_A$ (*b*) 182 cm

39. $V = 3.94 \times 10^{-23}$ cm^3; $r = 2.11$ Å

41. (*a*) 231 kPa (*b*) 201 kPa

43. 0.78 L

45. 375 K

47. $K_{av} = 6.21 \times 10^{-21}$ J. The change in gravitational potential energy is $mgh = 7.82 \times 10^{-26}$ J

49. (*a*) 4.7 m/s (*b*) 5.09 m/s

51.

$$\int_0^\infty f(v)\,dv = \frac{4}{\sqrt{\pi}}\left(\frac{m}{2kT}\right)^{3/2}\frac{\sqrt{\pi}}{4}\left(\frac{m}{2kT}\right)^{-3/2} = 1$$

53. $dP/dV = -nRT/(V - nb)^2 + 2an^2/V^3$ and $d^2P/dV^2 = 2nRT/(V - nb)^3 - 6an^2/V^4$. Setting these two derivatives to zero yields $V_c = 3nb$, which is $3b$ per mole

55. (*a*) $B = 3.95 \times 10^3$ K, $R_0 = 3.89 \times 10^{-3}$ ohms (*b*) 1320 ohms (*c*) For the ice point and the steam point, dR/dT is −393 ohms/K and −4.37 ohms/K, respectively. The thermistor is more sensitive at the ice point since the resistance changes more there for a given change in temperature

57. 1.34×10^5 N

59. 400.5 K

61. (*a*) 1.61×10^5 K (*b*) 1.01×10^4 K (*c*) For H_2, $f(v) \approx 10^{-9}$; for O_2, $f(v) \approx 10^{-106}$ (*d*) On the moon, the escape speed is $v_m = 2.38 \times 10^3$ m/s. O_2 and H_2 have an rms speed equal to v_m when the temperature is 7.28×10^3 K and 455 K, respectively

Chapter 16

True or False **1.** True **2.** False **3.** True **4.** False; it is proportional to the fourth power of T **5.** False **6.** False **7.** True **8.** True **9.** False **10.** True

Problems

1. (*a*) 1.046×10^7 J (*b*) 121 W
3. 0.0924 kcal/kg·K
5. (*a*) 0°C (*b*) 125 g
7. 99.8 g
9. 2073 Btu/h
11. (*a*) $I_{Al} = 569$ W, $I_{Cu} = 962$ W (*b*) $I_{total} = I_{Al} + I_{Cu} = 1531$ W (*c*) 0.0522 K/W
13. 9470 nm
15. 874 kJ
17. (*a*) 0.117 K (*b*) 1.74 K
19. (*a*)

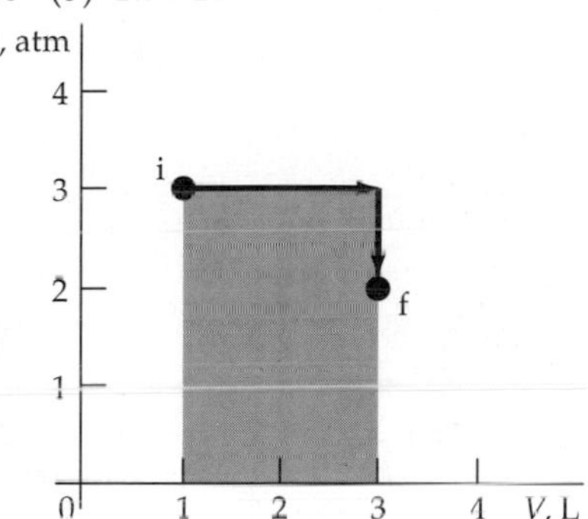

$W = 608$ J (*b*) 254 cal
21. (*a*)

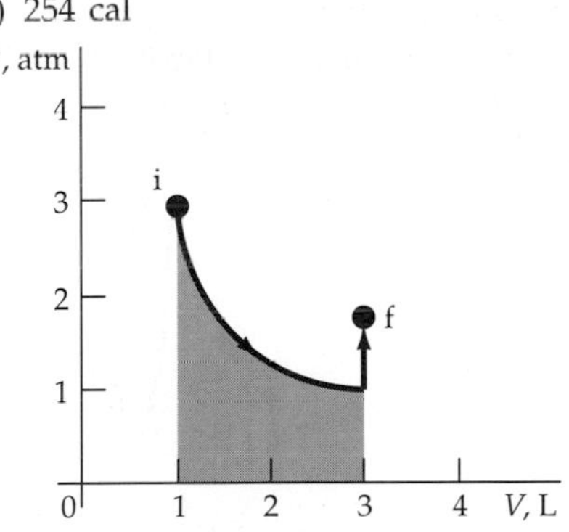

$W = 3.30$ L·atm (*b*) 189 cal
23. (*a*) 1.57 kJ (*b*) 1.57 kJ
25. (*a*) 0.713 J/g·K (*b*) 252 J/L
27. (*a*) $\Delta U = 6.24$ kJ; $W = 0$; $Q = 6.24$ kJ (*b*) $\Delta U = 6.24$ kJ; $Q = 8.73$ kJ; $W = 2.49$ kJ (*c*) $W = 2.49$ kJ
29. (*a*) $V_i = 2.24$ L; $V_f = 5.89$ L (*b*) $T_f = 143$ K (*c*) 1.62 kJ
31. (*a*) 28.5°C (*b*) 15.5°C
33. (*a*) 0.0920 J/kg·K (*b*) 0.0584 J/kg
35. (*a*) 34.0 km (*b*) yes (*c*) The potential energy released is the same with or without air resistance, so there should be little net effect
37. (*a*) 2.99°C (*b*) 199.85 (*c*) no
39. 18
41. 101.3°C
43. 47.8 cm
45. 96.6 g
47. 0.127 g
49. (*a*) For an ideal gas, U depends only on T; hence, for a given ΔT there is a unique ΔU. (*b*) $\Delta U - Q - W = (C_p - nR)\,\Delta T = C_v\,\Delta T$
51. (*a*) 3.40 kJ (*b*) $U_f = 3.70$ kJ, $W = 200$ J (*c*) $U_f = 3.90$ kJ, $W = 0$
53. (*a*) 263 K (*b*) 10.8 L (*c*) 1.48 kJ (*d*) −1.48 kJ
55. (*a*)

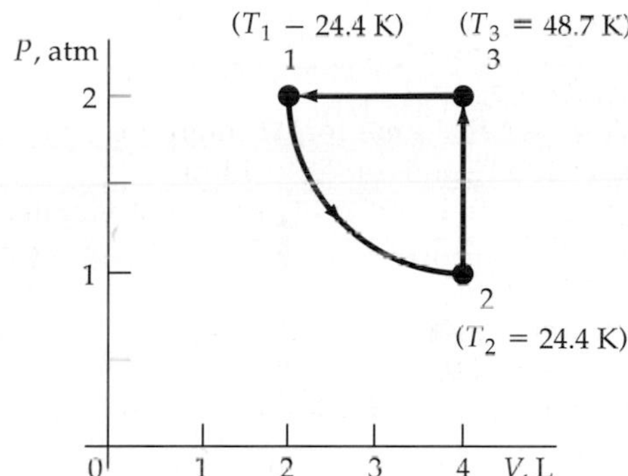

(*b*) For 1 to 2, $W - 2.77$ L·atm, $Q - 2.77$ L·atm. For 2 to 3, $W = 0$; $Q = 6.0$ L·atm. For 3 to 1, $W = -4$ L·atm; $Q = -10.0$ L·atm. (*c*) $T_1 = 24.4$ K; $T_2 = T_1$; $T_3 = 48.7$ K
57. 626°C = 1159°F
59. 577 J
61. (*a*) $\frac{1}{2}P_0$ (*b*) diatomic (*c*) No change occurs during the isothermal expansion; during the adiabatic compression, the kinetic energy increases
63. answer given in problem
65. answer given in problem
67. (*a*)

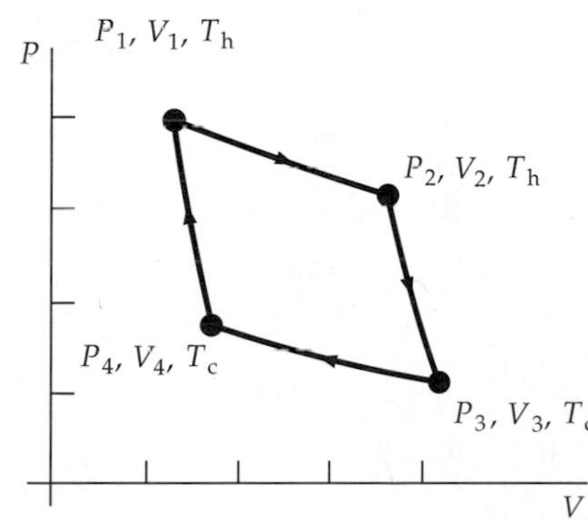

Chapter 17

True or False **1.** False; it can be in a noncyclic process **2.** False **3.** False **4.** False **5.** False **6.** True **7.** False **8.** True

Problems

1. (*a*) 500 J (*b*) 400 J
3. (*a*) 0.4 (*b*) 80 W

5. (*a*) 1.67 (*b*) 0.375
7. answer given in problem
9. answer given in problem
11. (*a*) $\frac{1}{3}$ (*b*) 33.3 J (*c*) 66.7 J (*d*) 2
13. (*a*) 13.7 (*b*) 8.77
15. (*a*) 0.51 (*b*) 102 kJ (*c*) 98 kJ
17. (*a*) 5.26 (*b*) 3.19 kW (*c*) 5.32 kW
19. (*a*) 303 kJ (*b*) 212 kJ
21. (*a*) 11.5 J/K (*b*) zero
23. (*a*) 50 J (*b*) 0.167 J/K (*c*) zero (*d*) The answers to parts (*a*) and (*b*) stay the same; the answer to part (*c*) is $\Delta S_u > 0$
25. (*a*) 11.5 J/K (*b*) 11.5 J/K
27. 0.417 J/K
29. 6.05 kJ/K
31. (*a*) Process 1 is more wasteful. (*b*) $\Delta S_1 =$ 1.67 J/K, $\Delta S_2 = 0.833$ J/K
33.

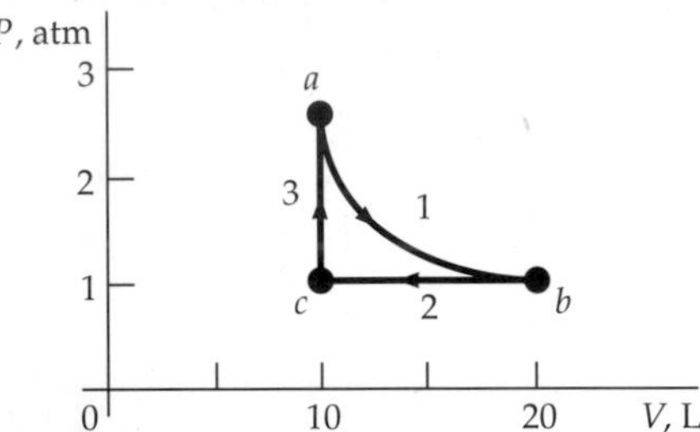

$\epsilon = W/Q_3 = 0.146$
35. (*a*) The maximum possible efficiency is 40.5 percent. (*b*) 1.68×10^9 J
37. (*a*)

(*b*) $V_3 = 2.30$ L; $T_3 = 345$ K (*c*) 5.15 L·atm (*d*) 0.232
39. (*a*) 10.1°C (*b*) 22 J/K
41. 10.7 J/K
43. 1.97×10^3 J/K
45. (*a*) 0.2 (*b*) $\Delta S_{engine} = 0$; $\Delta S_h = -2.5$ J/K; $\Delta S_c =$ 4.0 J/K; $\Delta S_u = 1.5$ J/K (*c*) 0.5 (*d*) 500 J
47. 93.9 W
49. (*a*) The heat input is $Q_h = C_v\,(T_c - T_b)$, and the magnitude of the heat output is $|Q_c| = C_v(T_d - T_a)/(T_c - T_b)$. (*c*) 0.565 (*d*) In a real engine, the processes are not truly adiabatic, nor are they quasi-static
51. $\epsilon = (T_h - T_c)/\{T_h + c_v(T_h - T_c)/[R \ln (V_a/V_b)]\}$
53. answer given in problem
55. (*a*) $T_1 = 301$ K; $T_2 = T_3 = 601$ K (*b*) $Q_{12} = 3.75$ kJ; $Q_{23} = 3.47$ kJ; $Q_{31} = -6.25$ kJ (*c*) 0.134
57. answer given in problem
59. answer given in problem

Chapter 18

True or False 1. False; it points toward a negative charge 2. True (except for the charges on quarks which are $e/3$ or $2e/3$, but isolated quarks have not been found) 3. False; they diverge from positive point charges 4. True 5. True

Problems
1. 5×10^{12} electrons
3. 4.82×10^7 C
5. (*a*)

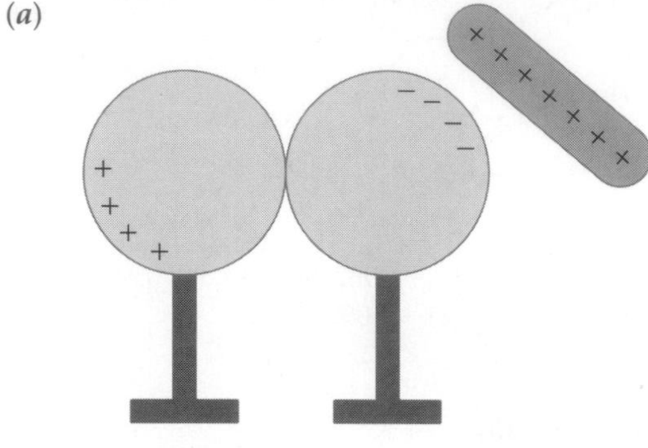

(*b*)

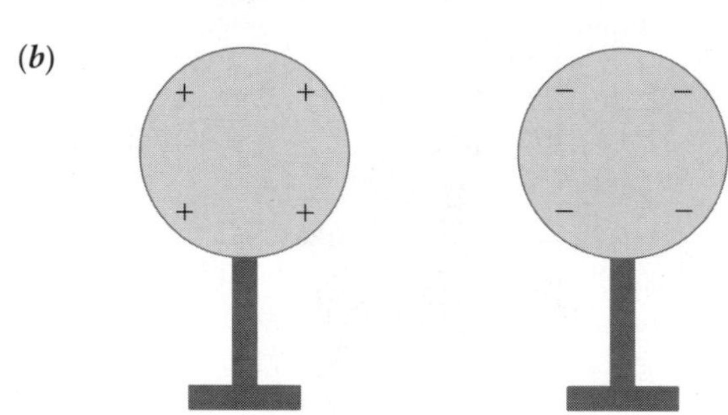

7. $(1.50 \times 10^{-2}$ N)**i**
9. 2.09×10^{-5} N along the diagonal, away from the −3-nC charge
11. (*a*) (999 N/C)**i** (*b*) (−360 N/C)**i**

(*c*)

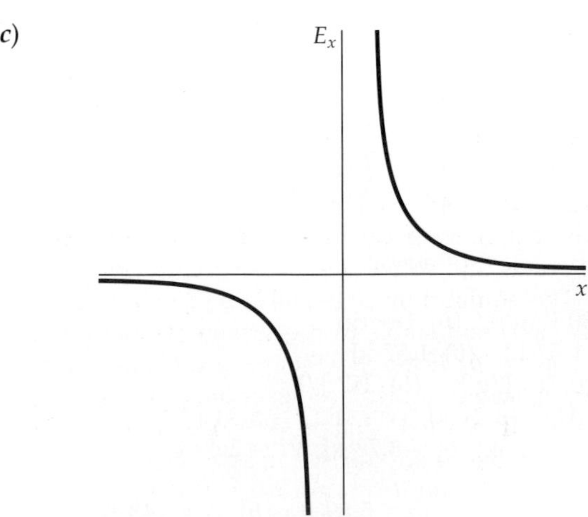

13. (*a*) $(3.45 \times 10^4$ N/C)**i** (*b*) $(6.90 \times 10^{-5}$ N)**i**
15. 8.18×10^5 N/C, upward

17. (*a*) The particle on the left has the greater charge by a factor of 4 (*b*) The particles on the left and right are positive and negative, respectively (*c*) The field is strong above and below the particle on the left; the field is weak to the right and to the left of the two particles

19.

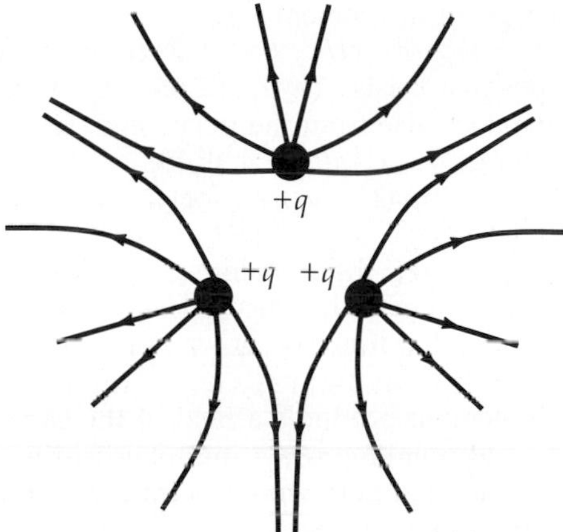

21. (*a*) 1.76×10^{11} C/kg (*b*) 1.76×10^{13} m/s², in the direction opposite **E** (*c*) 0.171 μs (*d*) 25.6 cm
23. (*a*) 7.03×10^{13} m/s² (*b*) 5×10^{-8} s (*c*) 8.78 cm in the negative *y* direction
25. (*a*) 8×10^{-18} C·m

(*b*)

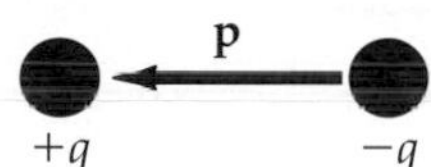

27. (*a*) 3.3×10^{-7} percent (*b*) 32.4 N
29. (*a*) E = 1.90×10^3 N/C, $\theta = 235°$ (*b*) F = 3.04×10^{-16} N, $\theta = 235°$
31. (*a*) 3.21×10^3 N/C (*b*) -5.88×10^6 N/C
33. (*a*) 6.4 mm below the tube axis (*b*) 17.7° below the tube axis (*c*) 4.48 cm below the tube axis
35. (*a*) 4 μC and 2 μC (*b*) +7.12 μC and −1.12 μC
37. (*c*) For large values of *x*, the system is essentially the same as one with a charge 2*q* located at the origin
39. $\mathbf{E} = -\dfrac{2kqa}{y(y^2+a^2)}\mathbf{i} \to -\dfrac{kp}{y^3}\mathbf{i}$
41. (*b*) 0.241 μC
43. (*a*) The equilibrium is unstable to displacements along the *x* axis and stable to displacements along the *y* axis (*b*) The equilibrium is unstable to displacements along the *y* axis and stable to displacements along the *x* axis (*c*) $-q/4$ (*d*) If the $+q$ charges are fixed in place, the system is stable to displacements along the *y* axis, as in part (*b*). If all three charges are free to move, the system is unstable to any displacement
45. (*a*) For $+q$, $\mathbf{F} = (q)C(x_1 + a)\mathbf{i}$, for $-q$, $\mathbf{F} = (-q)C(x_1 - a)\mathbf{i}$
47. (*a*) $\mathbf{E} = (-3kqa^2/x^4)\mathbf{i}$ (*b*) $\mathbf{E} = (6kq/y^4)\mathbf{j}$

Chapter 19

True or False **1.** False; the flux through the surface must be zero, but **E** need not be zero everywhere **2.** False; it holds for any charge distribution, but is useful for finding **E** only for symmetric distributions **3.** True **4.** True **5.** False **6.** False; it can be positive in some regions and negative in others **7.** False; for example, **E** is continuous at the edge of a spherical volume charge. **E** is discontinuous at points where there is a surface charge density σ **8.** True

Problems

1. (*a*) 17.5 nC (*b*) 26.2 N/C (*c*) 4.37 N/C (*d*) 2.57×10^{-3} N/C (*e*) For a point charge $E_x = 2.52 \times 10^{-3}$ N/C, about 2 percent lower than the correct result for a line charge
3. (*a*) 4.69×10^5 N/C (*b*) 1.13×10^6 N/C (*c*) 1.54×10^3 N/C (*d*) 1.55×10^3 N/C, about 0.07 percent greater than for the ring of charge
5. (*a*) 2.00×10^5 N/C (*b*) 2.54 N/C
7. $a/(3)^{1/2}$
9. (*a*) $(0.804)(2\pi k\sigma)$ (*b*) $(0.553)(2\pi k\sigma)$ (*c*) $(0.427)(2\pi k\sigma)$ (*d*) $(0.293)(2\pi k\sigma)$ (*e*) $(0.106)(2\pi k\sigma)$ (*f*)

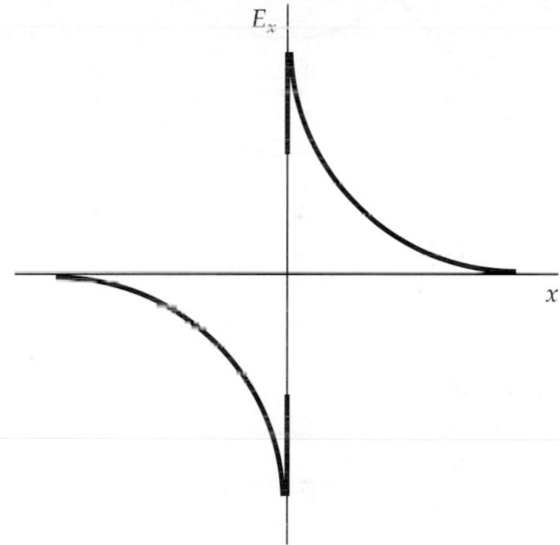

11. (*a*)

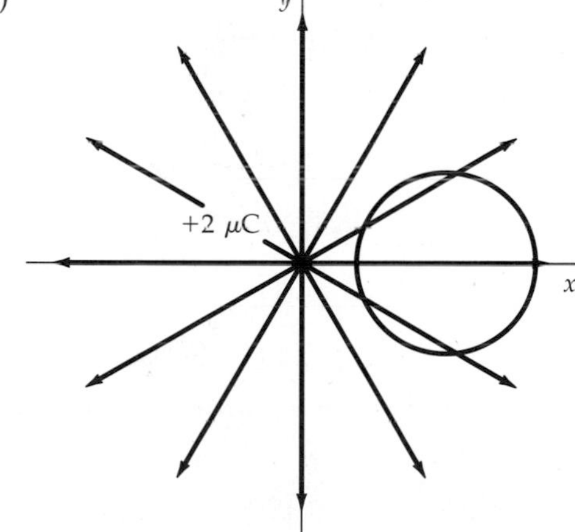

There are lines entering and leaving the surface. (*b*) zero (*c*) zero
13. (*a*) N (*b*) N/6 (*c*) q/ϵ_0 (*d*) $q/6\epsilon_0$ (*e*) Parts (*b*) and (*d*) would change
15. (*a*) 3.14 m² (*b*) 7.19×10^4 N/C (*c*) 2.26×10^5 N·m²/C (*d*) No (*e*) 2.26×10^5 N·m²/C
17. (*a*) 0.407 nC (*b*) 0 (*c*) 0 (*d*) 984 N/C (*e*) 366 N/C

19. (*a*) $Q = 40.7$ nC (*b*) $E_r = 0$ (*c*) $E_r = 0$ (*d*) $E_r =$ 999 N/C (*e*) $E_r = 610$ N/C
21. (*a*) $E = 0$ for $r < R_1$, $E = kq_1/r^2$ for $R_1 < r < R_2$, $E = k(q_1 + q_2)/r^2$ for $r > R_2$ (*b*) $|q_1/q_2| = 1$, and the signs of q_1 and q_2 are opposite (*c*) The electric-field lines corresponding to part (*b*) for $q_1 > 0$ are

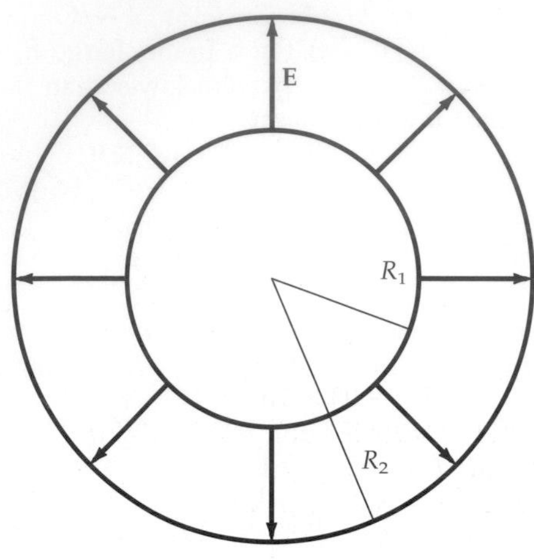

23. 1.15×10^5 N/C
25. 9.41×10^3 N/C
27. (*a*) For $r < a$, $E = kq/r^2$; for $a < r < b$, $E = 0$; for $r > b$, $E = kq/r^2$ (*b*) The electric-field lines are

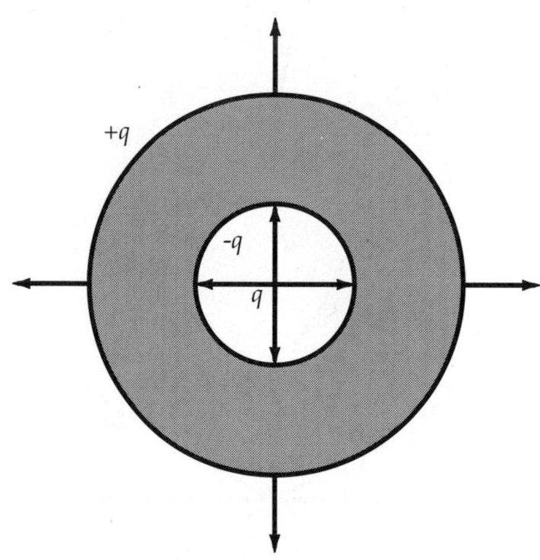

(*c*) On the inner surface, $\sigma = -q/4\pi a^2$; on the outer surface, $\sigma = q/4\pi b^2$
29. -1.18×10^{-12} C/m^3
31. (*a*) $\mathbf{E} = (9.41 \times 10^4\text{ N/C})\mathbf{i}$ (*b*) $\mathbf{E} = (3.36 \times 10^4\text{ N/C})\mathbf{j}$ (*c*) $\mathbf{E} = (1.56 \times 10^4\text{ N/C})(2\mathbf{i} - 3\mathbf{j})/\sqrt{13}$
33. (*a*) At the center of the electron sphere of charge (*b*) The equilibrium position is a distance $d = E_0R^3/kZe$ from the center of the electron sphere (*c*) E_0R^3/k
35. (*a*) For the inner surface the induced charge is -2.5 μC, and the surface charge density is -0.553 μC/m^2; for the outer surface, the induced charge is 2.5 μC, and the surface charge density is 0.246 μC/m^2 (*b*) For $r < r_1$, $E = kq_1/r^2$; for $r_1 < r < r_2$, $E = 0$; for $r > r_2$, $E = kq_1/r^2$ (*c*) The results for the inner surface are unchanged. For the outer surface, the total charge is 6 μC, and the surface charge density is 0.589 μC/m^2; for $r < r_1$, $E = kq_1/r^2$; for $r_1 < r < r_2$, $E = 0$; for $r > r_2$, $E = k(q_1 + q_2)/r^2$
37. answer given in problem
39. (*a*) $E = 2.04 \times 10^5$ N/C, $\theta = 56.3°$ counterclockwise from the positive x axis (*b*) $E = 2.63 \times 10^5$ N/C, $\theta =$ 153° counterclockwise from the positive x axis
41. Total charge $q = \rho[\frac{4}{3}\pi(b^3 - a^3)]$; for $r < a$, $E = 0$; for $a < r < b$, $E = k(\frac{4}{3}\pi\rho)(r^3 - a^3)/r^2$; for $r > b$, $E = kq/r^2 = k(\frac{4}{3}\pi\rho)(b^3 - a^3)/r^2$
43. (*a*) $q_2/q_1 = r_2/r_1$; the element s_1 gives the greater field (*b*) Each element produces a field pointing away from it, along a line from its center to the point P; the total field points away from s_1 (*c*) 0 (*d*) $q_2/q_1 = r_2^2/r_1^2$; each element produces a field of the same magnitude and pointing away; the total field is zero; for $E \propto 1/r$, the total field would point away from s_2
45. $E_x = -k\lambda/y$, $E_y = k\lambda/y$
47. For $r < a$, $E = 0$; for $a < r < b$, $E = 2\pi\rho k(r^2 - a^2)/r$; for $r > b$, $E = 2\pi\rho k(b^2 - a^2)/r$
49. (*a*) $Q = 2\pi BR^2$ (*b*) for $r < R$, $E_r = 2\pi kB$; for $r \geq R$, $E_r = 2\pi kBR^2/r^2 = kQ/r^2$

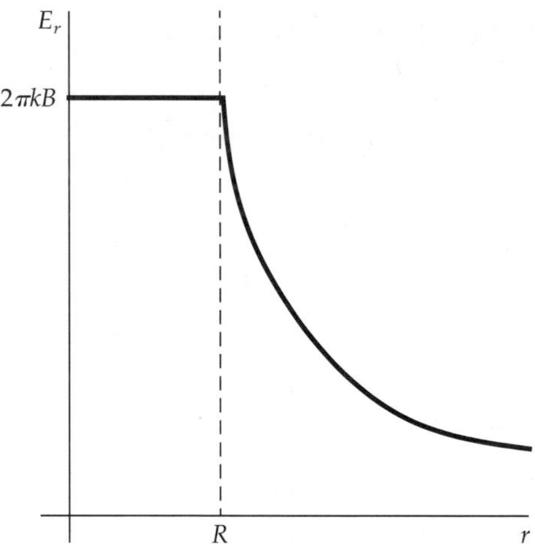

51. $F = kQq/[R(R + d)]$
53. $E = kQx/[(x^2 + L^2/4)(x^2 + L^2/2)^{1/2}]$, where $Q = 4L\lambda$ is the total charge on the square; for a ring of radius $r = L/2$, $E = kQx/(x^2 + L^2/4)^{3/2}$
55. (*b*) Half of the field just outside a conductor is due to the charge in the area ΔA, and the other half is due to all other charges; it is only this latter half that contributes to the force (*c*) 14.3 N/m^2

Chapter 20

True or False **1.** False; if $\mathbf{E} = 0$ in some region, V is constant in that region, but not necessarily zero. **2.** True; if $V =$ constant, $-dV/dx = 0$ **3.** False; $\mathbf{E}$ depends on the spatial rate of change of V, not on its value at any point **4.** True **5.** True **6.** True **7.** False; dielectric breakdown depends on the magnitude of the electric field E, not on the magnitude of the potential. It occurs in air when $E = 3$ MV/m

Problems

1. (*a*) 2.4×10^{-2} J (*b*) -2.4×10^{-2} J (*c*) -8000 V (*d*) $(-2$ kV/m$)x$ (*e*) 4000 V $-$ (2 kV/m)x (*f*) 2000 V $-$ (2 kV/m)x
3. (*a*) Positive (*b*) 25,000 V/m
5. (*a*) N/C·m = kg/C·s^2 = V/m^2 (*b*) $q_0ax^2/2$ (*c*) $V(x) = -ax^2/2$
7. (*a*) 1.29×10^4 V (*b*) 7.55×10^3 V (*c*) 4.44×10^3 V
9. (*a*) 2.68×10^5 V (*b*) 1.91×10^5 V
11. (*a*) 0.0487 J (*b*) 0 J (*c*) -0.0232 J (*d*) -0.0127 J
13. (*a*) 0.190 J (*b*) -0.0634 J (*c*) -0.0634 J
15. (*a*) Just outside the shell, $E = 6.24 \times 10^3$ V/m; just inside the shell, $E = 0$ (*b*) $V = 749$ V, both just inside and just outside the shell (*c*) $V = 749$ V, $E = 0$
17. (*a*) 6.02×10^3 V (*b*) -1.27×10^4 V (*c*) -4.23×10^4 V
19. (*a*) At $x = 3$ m, $V(x) = 8.99 \times 10^3$ V; at $x = 3.01$ m, $V(x) = 8.96 \times 10^3$ V (*b*) The potential decreases as x increases; $-\Delta V/\Delta x = 2.97 \times 10^3$ V/m (*c*) $E = 2.997 \times 10^3$ V/m (*d*) At $x = 3$ m, $y = 0.01$ m, $V = 8.99 \times 10^3$ V; V is nearly the same for the two points in question because they are approximately on an equipotential surface
21. (*a*) -3000 V/m (*b*) -3000 V/m (*c*) 3000 V/m (*d*) Zero
23. 0.506 mm
25. (*a*) ± 8.54 μC (*b*) $\pm 4.80 \times 10^5$ V
27. 26.6 μC/m^2
29. 250 W
31. (*a*) $E_x = 2\sqrt{2}kq/a^2$, $E_y = 0$ (*b*) $3\sqrt{2}kq/a$ (*c*) $v = q(6\sqrt{2}k/ma)^{1/2}$
33. $V(x = 2 \text{ m}) - V(x = 1 \text{ m}) = -7500$ V
35. (*a*) 3.10×10^7 m/s (*b*) 2.5×10^6 V/m
37. (*a*) 234 MeV (*b*) 2.67×10^{16} fissions per second
39. (*a*) 30,000 eV (*b*) 4.8×10^{-15} J (*c*) 1.03×10^8 m/s
41. $kq(1/a - 1/b)$
43. $V_a - V_b = (2kq/L) \ln (b/a)$
45. (*a*)

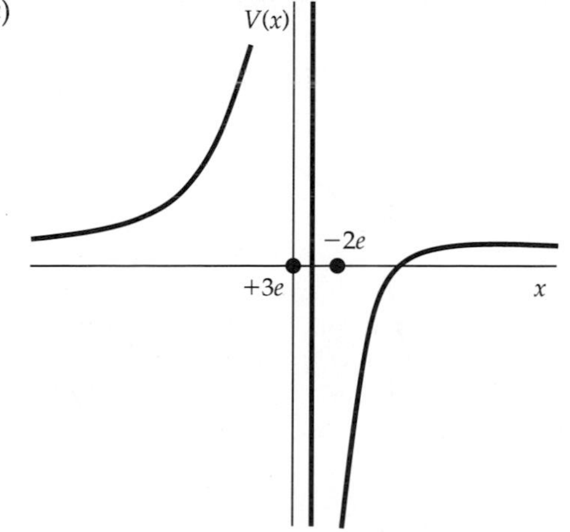

(*b*) $x = 3a/5$, $x = 3a$, $x = +\infty$, $x = -\infty$ (*c*) $2ke^2/a$
47. (*a*) $V(x) = kq(1/|x| - 3/|x - 1|)$ (*b*) $x = -0.5$ m, $x = 0.25$ m, $x = +\infty$, $x = -\infty$ (*c*) At $x = -0.5$ m, $E_x = -8kq/3$; at $x = 0.25$ m, $E_x = 64kq/3$; $E_x = 0$ at $x = \pm\infty$

(*d*)

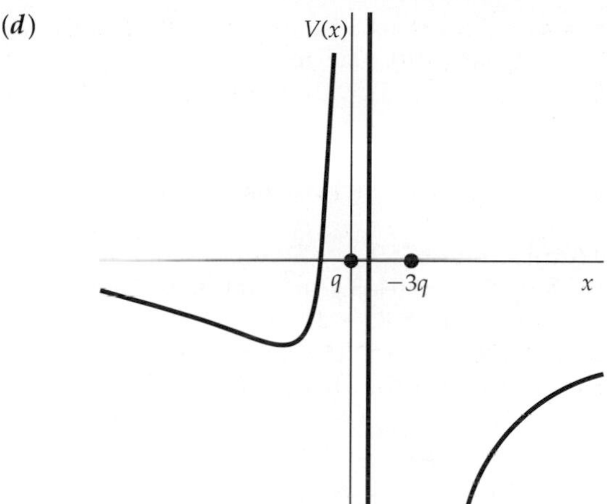

49. 1.45×10^{-7} J $= 9.03 \times 10^{11}$ eV
51. (*a*) $kQ^2(4 + \sqrt{2})/2L$ (*b*) $kQ^2(2 + \sqrt{2})/2L$ (*c*) kQ^2/L (*d*) 0
53. $\sigma_t = 9$ μC/m^2, $\sigma_b = 3$ μC/m^2
55. (*a*) $V(x) = kQ/(x^2 + a^2)^{1/2} + kQ'/|x - 2a|$ (*b*) For $x < 2a$, $E_x = kQx/(x^2 + a^2)^{3/2} - kQ'/(x - 2a)^2$; for $x > 2a$, $E_x = kQx/(x^2 + a^2)^{3/2} + kQ'/(x - 2a)^2$
57. $E_x = -8$ V/m, $E_y = -2$ V/m, $E_z = -1$ V/m
59. (*a*) $V(r) = kQ/r$ (*b*) $V(r) = (kQ/2R)(3 - r^2/R^2)$ (*c*) $3kQ/2R$ (*d*)

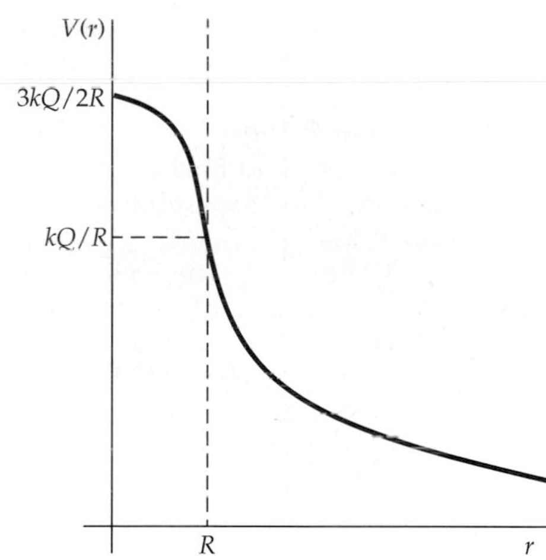

61. (*b*) $E_x = 3kpzx/r^5$, $E_y = 3kpzy/r^5$, $E_z = -kp/r^3 + 3kpz^2/r^5$
63. (*a*) $V(a) = kQ(1/b - 1/c) = V(b)$, $V(c) = 0$ (*b*) $V(a) = V(c) = 0$, $V(b) = -kQ_a(b - a)/ba = kQ(c - b)(b - a)/[(c - a)b^2]$; $Q_a = -Q(a/b)(c - b)/(c - a)$, $Q_c = -Q(c/b)(b - a)/(c - a)$; $Q_b = Q$
65. (*a*) $v(x) = [(kQ^2/2m)(1/x - 1/a)]^{1/2}$ (*b*) $t = (\pi/2)(2ma^3/kQ^2)^{1/2}$

Chapter 21

True or False **1.** False; C is the ratio of the charge to the voltage **2.** False; it depends only on the area and separation of the plates **3.** False; although $C = Q/V$, V is proportional to Q, so the ratio does not depend on Q **4.** True **5.** True **6.** True **7.** True

Problems

1. (*a*) 1.69×10^7 m^2 (*b*) 4117 m or 2.56 mi
3. 8×10^{-8} F
5. 22.1 μF
7. 2.71 nF
9. (*a*) 2.08 (*b*) 45.2 cm^2 (*c*) 5.2 nC
11. 2.22×10^{-5} J
13. (*a*) 0.625 J (*b*) 1.875 J
15. (*a*) 10^5 V/m (*b*) 0.0443 J/m^3 (*c*) 8.85×10^{-5} J (*d*) 1.77×10^{-8} F (*e*) 8.85×10^{-5} J
17. (*a*) 30 μF (*b*) 6 V (*c*) The charge on the 10-μF capacitor is 60 μC, and the charge on the 20-μF capacitor is 120 μC
19. (*a*) 24 μC (*b*) 4 μF
21. 2 μF
23. $C_{eq} = (C_1C_2 + C_2C_3 + C_1C_3)/(C_1 + C_3)$
25. (*a*) 0.05 mm (*b*) 235 cm^2
27. (*a*) 7.91 m^2 (*b*) 22.9 V (*c*) 3.66×10^{-5} J (*d*) 210 μC
29. (*a*) 15.2 μF (*b*) The 12-μF capacitor has a charge of 2400 μC; both the 4-μF and the 15-μF capacitors have charges of 632 μC (*c*) 0.303 J
31. (*a*) The charge on the 20-pF capacitor is 1.71×10^{-8} C; the charge on the 50-pF capacitor is 4.29×10^{-8} C (*b*) The initial energy is 9×10^{-5} J; the final energy is 2.57×10^{-5} J, so energy is lost when the capacitors are connected
33. (*a*) The maximum equivalent capacitance occurs when the capacitors are connected in parallel, so the appropriate combination is three 5-μF capacitors in parallel (*b*) The other possible equivalent capacitances are 10/3 μF, 7.5 μF, and 5/3 μF
35. (*a*) 6 V (*b*) The initial and final energies are 1.15×10^{-3} J and 2.88×10^{-4} J, respectively
37. (*a*) 1200 V (*b*) 6.4×10^{-4} J
39. (*a*) 1.67×10^{-8} F (*b*) 1.17×10^{-9} C (*c*) 7×10^6 V/m
41. (*a*) 5 μF (*b*) 133 V
43. (*a*) 2.28×10^{-9} F (*b*) 6.67×10^{-5} C
45. Connect four of the capacitors in series for an equivalent capacitance of 0.5 μF and a breakdown voltage of 400 V; now connect four such groups of four capacitors in parallel for a final capacitance of 2 μF
47. (*a*) $C_{eq} = \epsilon_0 b[(\kappa - 1)x + a]/d$ (*b*) For $x = 0$, $C_{eq} = \epsilon_0 ba/d$; for $x = a$, $C_{eq} = \kappa\epsilon_0 ba/d$
49. (*a*) $2C_0$ (*b*) $11C_0$
51. (*a*) 3.33×10^{-4} m (*b*) 3.77 m^2
53. (*a*) 40 V (*b*) 1.49×10^{-5} m^2 (*c*) 6
55. (*a*) 2.51×10^3 m^3 (*b*) 5.02×10^{-2} m^3
57. (*a*) Let the magnitude of C_1 and C_2 in farads be c; then the charges in coulombs are $Q_1 = 200c$ and $Q_2 = 200\kappa c$ (*b*) The energy in joules is $U_i = 20{,}000(1 + \kappa)c$ (*c*) $U_f = 10{,}000(1 + \kappa)^2 c$ (*d*) $V = (100 \text{ V})(1 + \kappa)$
59. (*a*) 0.255 J (*b*) The capacitor without a dielectric has a charge of 10^{-3} C, and the capacitor with the dielectric has a charge of 3.5×10^{-3} C (*c*) Both capacitors now have the same charge, 2.25×10^{-3} C (*d*) 0.506 J
61. (*a*) 0.001 J (*b*) The capacitor without a dielectric has the charge 47.6 μC; the capacitor with the dielectric has the charge 152 μC (*c*) 4.76×10^{-4} J
63. 2.55×10^{-6} J
65. (*a*) $E = 2kQ/rL$, $\eta = 2\epsilon_0 k^2Q^2/r^2L^2$ (*b*) $(kQ^2/rL)\,dr$ (*c*) $(kQ^2/L)\ln(R_2/R_1)$
67. (*a*) $(Q/2\pi\epsilon_0\kappa L)\ln(b/a)$ (*b*) $Q/2\pi aL$ (*c*) $-Q/2\pi aL$ (*d*) $-Q(\kappa - 1)/2\pi aL\kappa$ (*e*) $Q(\kappa - 1)/2\pi aL\kappa$ (*f*) $(kQ^2/L\kappa)\ln(b/a)$ (*g*) $(kQ^2/L\kappa)(\kappa - 1)\ln(b/a)$
69. (*a*) $3\epsilon_0 A/(y_0 \ln 4)$ (*b*) $\sigma_{top} = -3\sigma/4$, $\sigma_{bot} = 0$ (*c*) $(3\sigma/y_0)(1 + 3y/y_0)^{-2}$

Chapter 22

True or False 1. False; $R = V/I$ is the definition of resistance. Ohm's Law states that R is independent of I or V 2. False; they drift in the opposite direction 3. True 4. True 5. True 6. False

Problems

1. (*a*) 600 C (*b*) 3.75×10^{21}
3. 0.4 A
5. (*a*) $v/2\pi r$ (*b*) $vq/2\pi r$
7. (*a*) 3.21×10^{13} protons/m^3 (*b*) 3.75×10^{17} (*c*) $q = It = (10^{-3} \text{ C/s})t$
9. (*a*) 1 V (*b*) 0.1 V/m
11. (*a*) $V_{Cu} = IL\rho_{Cu}/A$, $V_{Fe} = IL\rho_{Fe}/A$, $V_{Fe}/V_{Cu} = 5.88$ (*b*) **E** is greater in the iron wire
13. (*a*) 0.0275 Ω (*b*) 0.030 Ω
15. 0.182 Ω
17. 45.6°C
19. 250 W
21. (*a*) 5 mA (*b*) 50 V
23. (*a*) 0.707 A (*b*) 7.07 V
25. 180 J
27. (*a*) 240 W (*b*) 228 W (*c*) 4.32×10^4 J (*d*) 2160 J
29. \$77.76
31. (*a*) 4.5 Ω (*b*) $I_3 = 2.67$ A, $I_2 = 2$ A, $I_6 = 2/3$ A
33. (*a*) 6 Ω (*b*) The top two resistors both have a current of 2/3 A; the two 6-Ω resistors in parallel each have a current of 2/3 A, and the single 6-Ω resistor on the bottom has a current of 4/3 A
35. (*b*) No effect
37. (*a*) $I_3 = 30/19$ A, $I_2 = 12/19$ A, $I_4 = 6/19$ A (*b*) 9.47 W
39. 0.0314 Ω
41. (*b*)

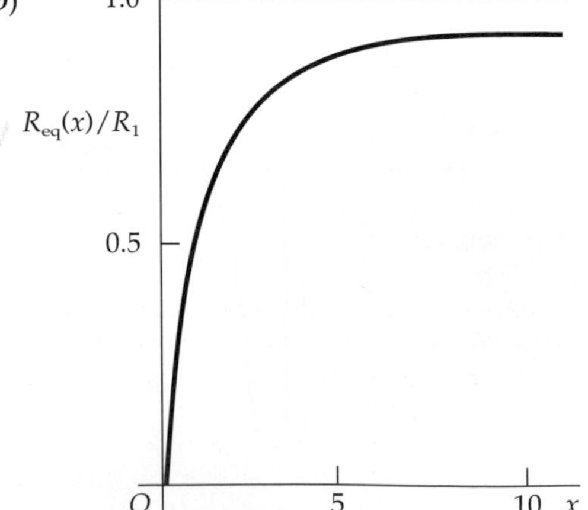

43. (*a*) 0.05 A (*b*) 5000 W
45. (*a*) $E_{Cu} = 0.0433$ V/m, $E_{Fe} = 0.255$ V/m (*b*) $V_{Cu} = 3.46$ V, $V_{Fe} = 12.5$ V (*c*) $R_{eq} = 7.97$ Ω, $R_{Cu} = 1.73$ Ω, $R_{Fe} = 6.24$ Ω

47. 382°C
49. (*a*) \$150.77 (*b*) 3.79 cents/kW·h
51. (*a*) 6.91×10^6 J (*b*) 12.8 h
53. (*a*) 3×10^{-3} kW·h (*b*) \$2663 per kW·h (*c*) 0.027 cents
55. (*a*) 0.03 Ω (*b*) 0.3 percent (*c*) 7.5C°
57. (*a*) 79.6 Ω (*b*) 318 Ω
59. (*a*) 15 A (*b*) 11.2 Ω (*c*) 1.28 kW
61. (*a*) 5.14×10^{-3} Ω (*b*) 0.462 V (*c*) 41.6 W
63. (*b*) $L_{Cu} = 264L_C$
65. $R = \rho\pi/[t \ln(b/a)]$
67. (*a*) $R = (\rho/2\pi L)\ln(b/a)$ (*b*) 2.05 A
69. $R = \rho L/\pi ab$

Chapter 23

True or False 1. True 2. False; it is the time for the charge to be reduced by a factor of e^{-1} 3. False; the voltmeter is placed in parallel with the resistor

Problems

1. (*a*) $I = 1.13$ A, $P = 6.79$ W (*b*) $I = 0.583$ A, $P = 3.50$ W
3. (*a*) 3 V (*b*) 1 Ω
5. (*a*) 1 A (*b*) 12 W for the emf on the left; 6 W for the emf on the right (*c*) 2 W for the 2-Ω resistor; 4 W for the 4-Ω resistor
7. (*a*) $I_4 = 2/3$ A, $I_3 = 8/9$ A, $I_6 = 14/9$ A (*b*) $V_b - V_a = -28/3$ V (*c*) 8 W supplied by the left-side emf; 32/3 W by the right side emf
9. (*a*) 6×10^{-4} C (*b*) 0.2 A (*c*) 3×10^{-3} s (*d*) 8.12×10^{-5} C
11. 4.81×10^7 Ω
13. (*a*) 5.69 μC (*b*) 1.10 μC/s (*c*) 1.10 μA (*d*) 6.62×10^{-6} W (*e*) 2.44×10^{-6} W (*f*) 4.19×10^{-6} J/s
15. (*a*) 0.0841 Ω (*b*) 4027 Ω
17. (*a*) 0.168 Ω (*b*) 0.168 Ω (*c*) 2.14×10^6 Ω
19. (*a*) 910 Ω (*b*) 1000 Ω (*c*) 9000 Ω
21. $R_1 = 7582$ Ω, $R_2 = 69{,}231$ Ω, $R_3 = 692{,}308$ Ω
23. (*a*) Parallel connection supplies greater power (*b*) Series connection supplies greater power
25. (*a*) The currents in the 1-Ω and 6-Ω resistors are 2 A and 1 A, respectively; the current in the horizontal 2-Ω resistor is 2 A; the current in the vertical 2-Ω resistor is 1 A (*b*) The 8-V emf supplies 16 W, the upper 4-V emf supplies 8 W, and the lower 4-V emf absorbs 4 W (*c*) The power dissipated in the resistors is 4 W, 6 W, 8 W, and 2 W, where the order is the same as in part (*a*)
27. (*b*) 9.65 h
29. (*a*) Connecting the batteries in parallel gives the largest current for small *R*; connecting them in series works best for large *R* (*b*) 10.7 A (*c*) 6.67 A (*d*) 5.45 A (*e*) 4.44 A
31. (*a*) Connect the galvanometer in series with a resistor $R = 999{,}800$ Ω (*b*) Connect the galvanometer in parallel with a resistor $R = 10^{-3}$ Ω
33. (*a*) $I_{a\to b} = 2$ A; $I_{b\to c} = 1.5$ A; $I_{b\to h} = 0.5$ A; $I_{d\to h} = 1.5$ A; $I_{c\to f} = 0.5$ A; $I_{h\to g} = 1.5$ A (*b*) $V_a = 0$ V, $V_b = 24$ V, $V_c = 21$ V, $V_d = 15$ V, $V_e = 15$ V, $V_f = 5$ V, $V_g = 0$ V, $V_h = 12$ V
35. (*a*) 50/3 Ω (*b*) For the top three and lower three resistors, each has a current of 2/5 A; for the middle three resistors, the first and last have a current of 4/5 A, and the current in the middle resistor is 2/5 A
37. $\mathcal{E} = 7$ V, $R = 14$ Ω
39. (*a*) 3.42 A (*b*) 0.962 A (*c*) $Q_{10} = 260$ μC, $Q_5 = 130$ μC
41. (*a*) 10^{-4} A (*b*) 6.67×10^{-5} A (*c*) 40 V
43. (*a*) 43.9 Ω (*b*) 300 Ω (*c*) 3800 Ω
45. (*a*) The current in the car battery is -57.0 A + (10 A/h)*t*, where the minus sign means that current is entering the battery; the current in the second battery is 63.0 A − (10 A/h)*t* (*b*)

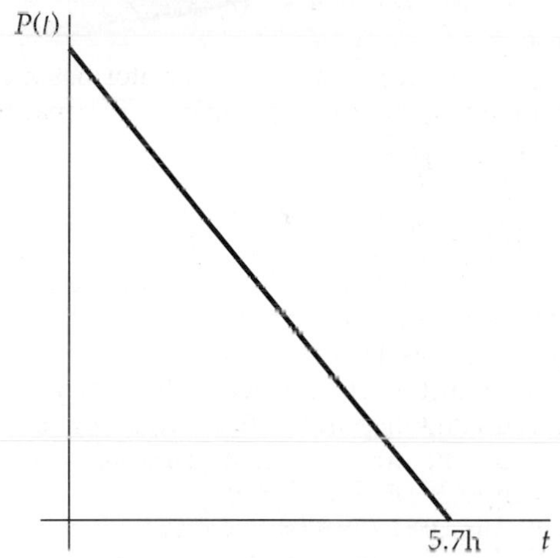

47. (*a*) Case *a* is preferred for small *R*, case *b* for large *R*; case *a* is the correct configuration for an ideal voltmeter with R_V infinitely large; if *R* is comparable to R_V then case *b* compensates for the fact that a finite current flows through the voltmeter (*b*) Case *a*, $R = 0.498$ Ω; case *b*, $R = 0.6$ Ω (*c*) Case *a*, $R = 2.91$ Ω; case *b*, $R = 3.10$ Ω (*d*) Case *a*, $R = 44.4$ Ω; case *b*, $R = 80.1$ Ω
49. answers given in problem
51. answer given in problem
53. $R_{eq} = \frac{1}{3}R$
55. $\frac{1}{3}R$
57. $R_{eq} = (1 + \sqrt{3})R$
59. $I_{10} = 104.4/141$ A, $I_{40} = 66.6/141$ A, $I_{30} = 54/141$ A, $I_{80} = 50.4/141$ A, $I_{20} = 120.6/141$ A
61. (*a*) 4.17×10^{-5} A (*b*) 2.78×10^{-5} A (*c*) $I(t) = (2.78 \times 10^{-5}\text{ A})\, e^{-t/(1.5\text{ s})}$
63. (*a*) $I(t) = (V_0/R)\, e^{-t/RC}$, where $C = C_{eq} = C_1C_2/(C_1 + C_2)$ (*b*) $P(t) = (V_0^2/R)\, e^{-2t/RC}$ (*c*) $U = \frac{1}{2}C_{eq}V_0^2$

65. (*a*) $\mathcal{E}I(t) = (\mathcal{E}^2/R)\, e^{-t/RC}$ (*b*) $I(t)^2R = (\mathcal{E}^2/R)\, e^{-2t/RC}$ (*c*) $dU/dt = (\mathcal{E}^2/R)\, e^{-t/RC} - (\mathcal{E}^2/R)\, e^{-2t/RC}$

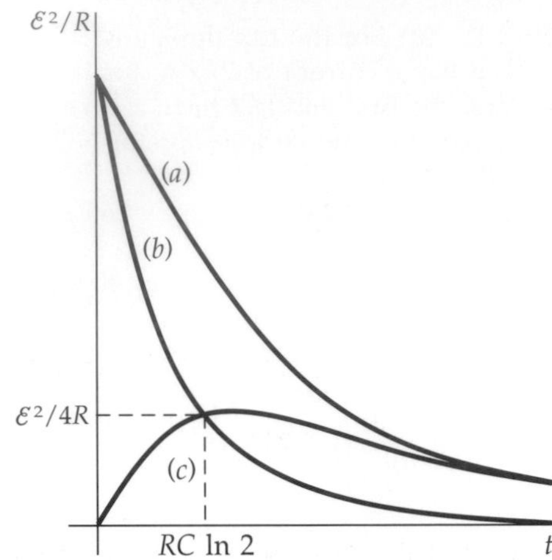

(*d*) $(dU/dt)_{\max} = \mathcal{E}^2/4R,\ t = RC \ln 2$

Chapter 24

True or False **1.** True **2.** True **3.** True **4.** False; it is independent of the radius **5.** True

Problems

1. -1.25×10^{-12} N **j**
3. (*a*) -7.17×10^{-13} N **j** (*b*) 5.12×10^{-13} N **i** (*c*) 0 (*d*) 8.19×10^{-13} N **i** $-$ 6.14×10^{-13} N **j**
5. 1 N
7. 14.0 N/m **k**
9. (*a*) 2.20 mm (*b*) $f = 9.08 \times 10^9\ \text{s}^{-1}$, $T = 1.10 \times 10^{-10}$ s
11. (*a*) -1.05×10^4 N/C **k** (*b*) No
13. (*a*) 1.42 km (*b*) 28.5 m
15. (*a*) $2.13 \times 10^{-7}\ \text{s}^{-1}$ (*b*) 46.0 MeV (*c*) Both the frequency and the kinetic energy will be reduced by a factor of 2
17. (*a*) 0.302 A·m^2 (*b*) 0.131 N·m
19. 2.83×10^{-5} N·m
21. (*a*) 0 (*b*) 2.7×10^{-3} N·m
23. (*a*) 2.125 N·m/T **i** (*b*) -3.40 N·m **j** + 5.31 N·m **k**
25. (*a*) 1.07×10^{-4} m/s (*b*) 5.85×10^{28} electrons/m^3
27. (*a*) 3.69×10^{-5} m/s (*b*) 1.48 μV
29. 1.02×10^{-3} V
31. 0.0864 N **i** $-$ 0.0648 N **j**
33. (*a*) 7.35 mm (*b*) 6.64×10^{-5} T
35. (*a*) The normal points 37° below the x axis (*b*) $\hat{\mathbf{n}} = 0.799\,\mathbf{i} - 0.602\,\mathbf{j}$ (*c*) $\mathbf{m} = 0.335$ A·m^2**i** $-$ 0.253 A·m^2**j** (*d*) 0.503 N·m **k**
37. answers given in problem
39. $r_d/r_p = \sqrt{2}$, $r_\alpha/r_p = 1$
41. (*a*) $v_p/v_\alpha = 2$ (*b*) $K_p/K_\alpha = 1$ (*c*) $L_p/L_\alpha = \frac{1}{2}$
43. $I = Mg/\pi RB_x$
45. answer given in problem
47. answers given in problem
49. (*a*) 1.6×10^{-18} N **j** (*b*) 10 V/m **j** (*c*) 20 V
51. (*a*) $B = (Mg/IL)\tan\theta$ (*b*) $a = g\sin\theta$, uphill
53. $T = 2\pi(M/\pi IB)^{1/2}$
55. answer given in problem
57. answer given in problem
59. answers given in problem
61. answer given in problem

Chapter 25

True or False **1.** False **2.** True **3.** False; it varies inversely with distance **4.** False; it is useful in finding **B** only if there is symmetry, but it is valid for any continuous current **5.** True

Problems

1. (*a*) $\mathbf{B} = -9 \times 10^{-12}$ T **k** (*b*) $\mathbf{B} = -3.6 \times 10^{-11}$ T **k** (*c*) $\mathbf{B} = 3.6 \times 10^{-11}$ T **k** (*d*) $\mathbf{B} = 9 \times 10^{-12}$ T **k**
3. (*a*) 0 (*b*) -3.56×10^{-23} T **k** (*c*) 4×10^{-23} T **k**
5. 12.5 T
7. -9.6×10^{-12} T **i**
9. 11.1 A

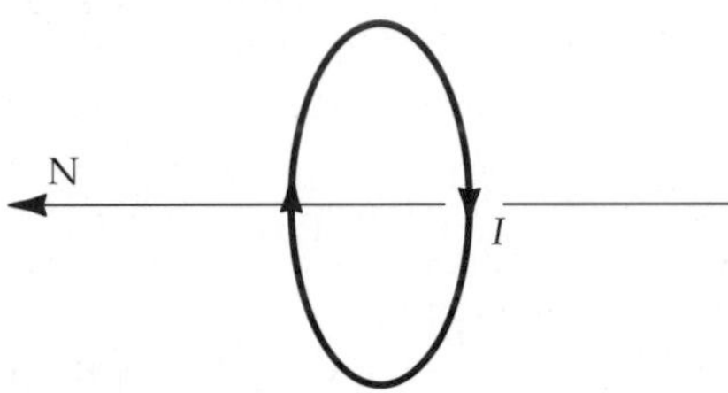

11. 6.98×10^{-4} T
13. (*a*) $x = \pm 5.72$ cm (*b*) $x = \pm 13.6$ cm (*c*) $x = \pm 29.8$ cm
15. (*a*) -8.89×10^{-5} T **k** (*b*) 0 (*c*) 8.89×10^{-5} T **k** (*d*) -1.6×10^{-4} T **k**
17. (*a*) -1.78×10^{-4} T **k** (*b*) -1.33×10^{-4} T **k** (*c*) -1.78×10^{-4} T **k** (*d*) 1.07×10^{-4} T **k**
19. (*a*) 6.4×10^{-5} T **j** (*b*) -4.8×10^{-5} T **k**
21. The fields caused by the wire segments, going from left to right, are 0, 56.6 μT, 113 μT, 56.6 μT, and 0; all of the fields are into the page; the total field is 226 μT into the page
23. 9.47 A

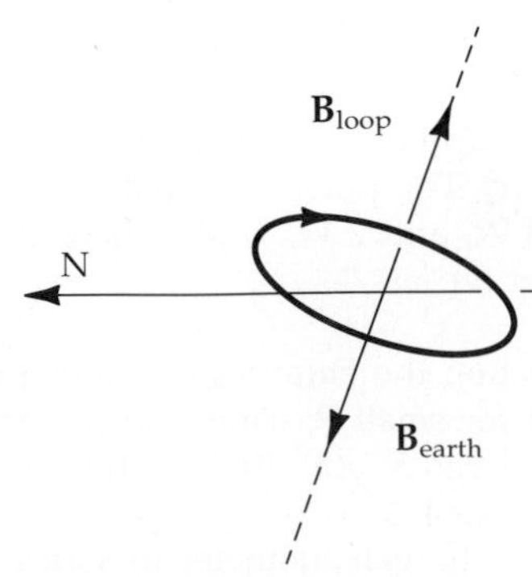

25. (*a*) Antiparallel (*b*) 39.3 mA
27. 28 A
29. (*a*) 4.5×10^{-4} N/m to the right (*b*) 30 μT down
31. (*a*) C_1, (8 A)μ_0; C_2, 0; C_3, (-8 A)μ_0 (*b*) None of them

33. (*a*) 8×10^{-4} T (*b*) 4×10^{-3} T (*c*) 2.86×10^{-3} T (*d*)

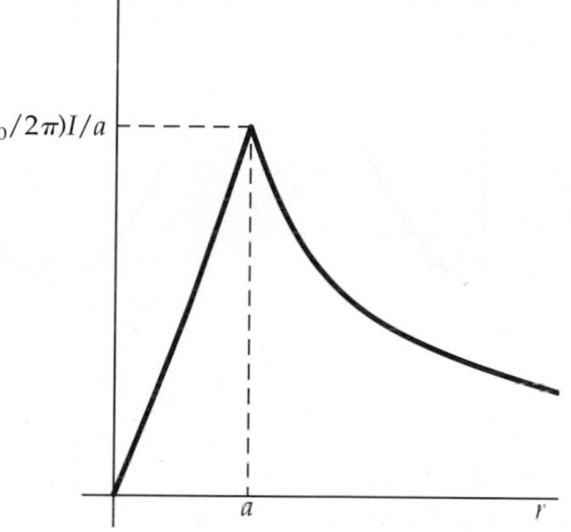

35. (*a*) 0.0273 T (*b*) 0.0200 T
37. (*a*) 3.2×10^{-16} N, in the direction opposite to the current (*b*) 3.2×10^{-16} N, away from the wire (*c*) 0
39. (*a*) $\pi(\mu_0 I/L) = 3.14(\mu_0 I/L)$ (*b*) $(8\sqrt{2}/\pi)(\mu_0 I/L) = 3.60(\mu_0 I/L)$ (*c*) $(27/2\pi)(\mu_0 I/L) = 4.30(\mu_0 I/L)$
41. (*a*) $3\sqrt{2}\mu_0 I^2/4\pi a$, along the diagonal toward the opposite corner (*b*) $\sqrt{2}\mu_0 I^2/4\pi a$, along the diagonal away from the opposite corner
43. (*a*) 2.26×10^{-5} T·m (*b*) 1.13×10^{-5} T·m (*c*) 0
45. $\tau = \pi r_2^2 \mu_0 N_1 N_2 I_1 I_2 / 2r_1 = 1.97 \times 10^{-6}$ N·m, where the subscripts 1 and 2 refer to the large and small coil, respectively
47. 3.18 cm
49. (*a*) 80 A, in the positive z direction. (*b*) -2.4×10^{-4} T $\mathbf{j}$
51. (*a*)

$$B = \frac{\mu_0 I}{2\pi R} \sin\theta,$$

where R is the perpendicular distance from the point P to the wire (*b*) For a polygon of N sides,

$$B = \frac{N\mu_0 I}{2\pi R} \sin(\pi/N);$$

for large values of N, the field approaches $\mu_0 I/2R$
53. (*a*) $x = 5$ cm, $B = 0.0540$ T; $x = 7$ cm, $B = 0.0539$; $x = 9$ cm, $B = 0.0526$ T; $x = 11$ cm, $B = 0.0486$ T (*b*)

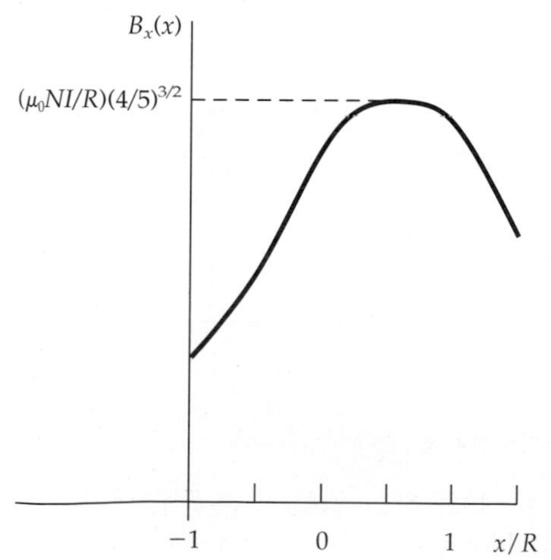

55. (*a*) The force on each of the horizontal segments is 0.501×10^{-4} N, down on the upper segment and up on the lower segment; the force on the left vertical segment is 2×10^{-4} N to the right, and the force on the right vertical segment is 0.571×10^{-4} N to the left (*b*) 1.43×10^{-4} N to the right
57. $B_x = (\mu_0 I L^2/2\pi x^3)(1 + L^2/4x^2)^{-1}(1 + 2L^2/4x^2)^{-1/2}$
59. (*a*) $B_y = \dfrac{\mu_0 I}{2\pi(R^2 - a^2)}\left[\dfrac{R}{2} - \dfrac{a^2}{2R - b}\right]$
(*b*) $B_x = \dfrac{\mu_0 I}{\pi(R^2 - a^2)}\left[\dfrac{a^2 R}{4R^2 + b^2} - \dfrac{R}{4}\right]$
$B_y = \dfrac{\mu_0 I a^2 b}{2\pi(R^2 - a^2)(4R^2 + b^2)}$
61. (*c*) $B_x = \frac{1}{2}\mu_0\sigma\omega[(R^2 + 2x^2)/(x^2 + R^2)^{1/2} - 2x]$
63. answer given in problem
65. (*a*) $dm = (N/L)I\pi R^2\,dx = nIA\,dx$

Chapter 26

True or False **1.** False; it depends only on the rate of change of the flux **2.** True **3.** True **4.** False **5.** True

Problems

1. (*a*) 5×10^{-4} Wb (*b*) 4.33×10^{-4} Wb (*c*) 2.5×10^{-4} Wb (*d*) 0
3. 7.58×10^{-4} Wb
5. (*a*) 8.48×10^{-3} Wb (*b*) 7.97×10^{-3} Wb
7. (*a*) 8.48×10^{-3} Wb (*b*) 133 turns
9. 199 T/s
11. (*a*)

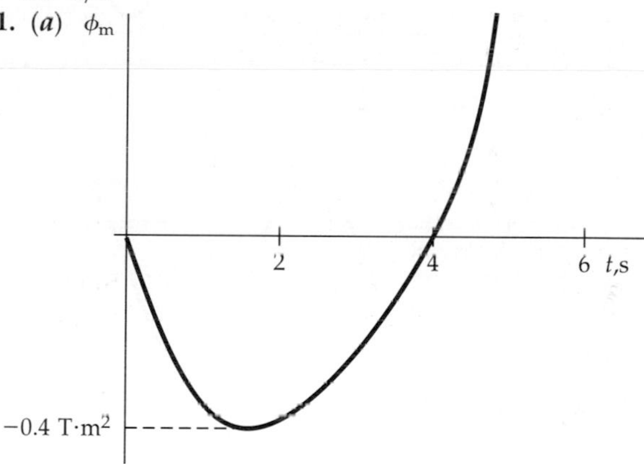

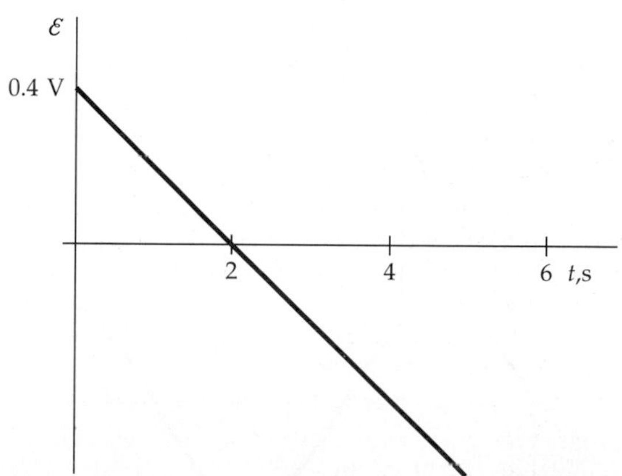

(*b*) At $t = 2$ s, ϕ_m has its maximum negative value; ϕ_m increases indefinitely as t goes to infinity (*c*) $\phi_m = 0$ at $t = 0$ and $t = 4$ s; at $t = 0$ s, $\mathcal{E} = 0.4$ V, and at $t = 4$ s, $\mathcal{E} = -0.4$ V

13. 2.8×10^{-4} C
15. (*a*) 7.07×10^{-3} V (*b*) 6.64×10^{-3} V
17. (*a*) 3.1×10^{-3} Wb (*b*) 2.21×10^{-3} V
19. (*a*)

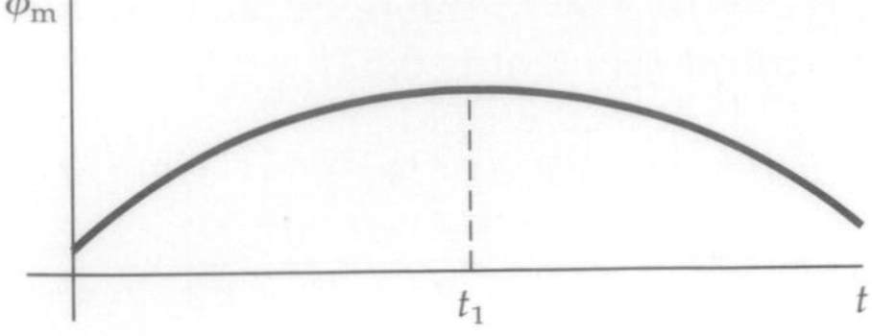

(*b*) $\mathcal{E}$ t_1 t

21. (*a*)

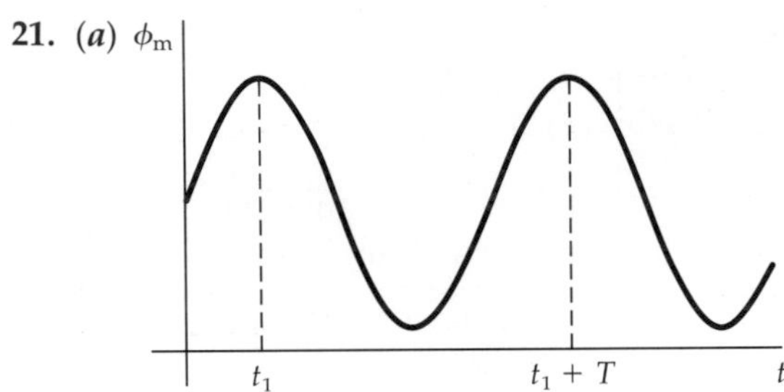

(*b*) I t_1 $t_1 + T$ t

23. 400 m/s
25. (*a*) 3.6 V (*b*) 3 A (*c*) 1.8 N (*d*) 10.8 W (*e*) 10.8 W
27. 0.332 T
29. 0.707 T
31. $\phi_m = LI_0 \sin 2\pi ft$

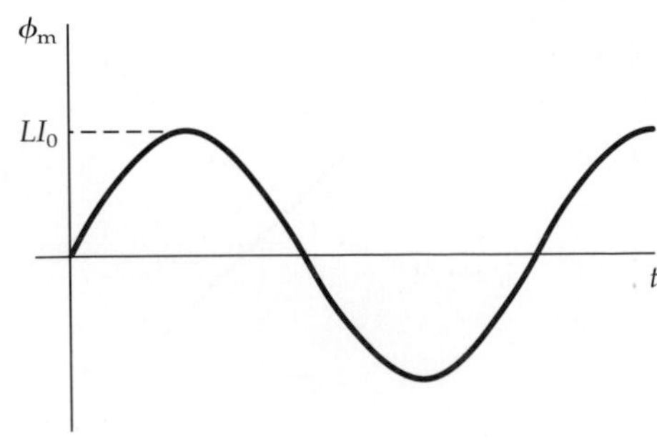

$\mathcal{E} = -LI_0 2\pi f \cos 2\pi ft$

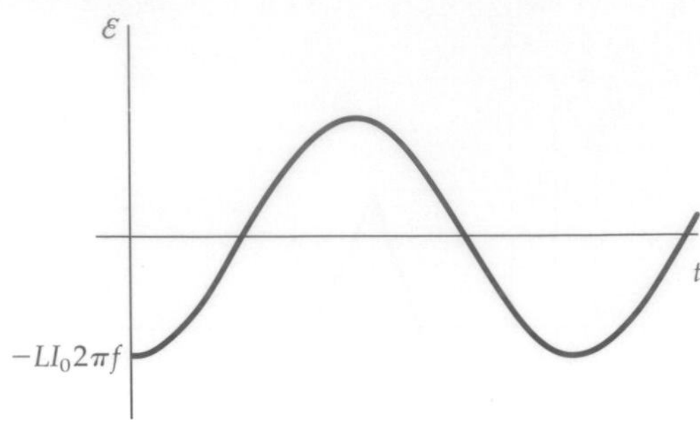

33. 1.89 mH
35. (*a*) $I = 0$ A, $dI/dt = 25$ A/s (*b*) $I = 2.27$ A, $dI/dt = 20.5$ A/s (*c*) $I = 7.90$ A, $dI/dt = 9.20$ A/s (*d*) $I = 10.8$ A, $dI/dt = 3.38$ A/s
37. (*a*) 13.5 mA (*b*) 7.44×10^{-44} A
39. (*a*) $t = 0$ s, $P = 47.7$ W; $t = 100$ s, $P = 48.0$ W (*b*) $t = 0$ s, $I^2R = 47.4$ W; $t = 100$ s, $I^2R = 48.0$ W (*c*) $t = 0$ s, $dU_m/dt = 0.321$ W; $t = 100$ s, $dU_m/dt = 0$ W
41. (*a*) 3.98×10^5 J (*b*) 4.43×10^{-4} J (*c*) 3.98×10^5 J
43. $B\pi R^2$
45. (*a*) $\mu_0 nIN\pi R_1^2$ (*b*) $\mu_0 nIN\pi R_3^2$
47. (*b*) 275 rad/s
49. (*a*) 13.9 Ω (*b*) 214 V
51. (*a*) 88.1 ms (*b*) 35.2 mH
53. (*a*) 2.41 s (*b*) 20.1 s
55. answer given in problem
57. (*a*) $I_B = I_{10} = I_L = 1$ A, $I_{100} = 0$ A (*b*) $I(t) = (1\text{ A})\, e^{-50t}$
59. (*a*) For the resistor, $dI/dt = 9000$ A/s; for the inductors, $dI_1/dt = 3000$ A/s and $dI_2/dt = 6000$ A/s, where 1 and 2 refer to the 8-mH and 4-mH inductors, respectively (*b*) 1.6 A
61. (*a*) 0.0536 J (*b*) 447 J/m^3 (*c*) 0.0335 T (*d*) 447 J/m^3
63. (*a*) $E = \frac{1}{2}\mu_0 nrI_0\omega \cos \omega t$ (*b*) $E = \frac{1}{2}\mu_0 n(R^2/r)I_0\omega \cos \omega t$
65. (*a*) $0\text{ s} \le t \le 4.17$ s, $\phi_m = (2.04 \times 10^{-3}\text{ Wb/s})t$; $4.17\text{ s} \le t \le 8.33$ s, $\phi_m = 8.5 \times 10^{-3}$ Wb; $8.33\text{ s} \le t \le 12.5$ s, $\phi_m = 8.5 \times 10^{-3}\text{ Wb} - (2.04 \times 10^{-3}\text{ Wb/s})(t - 8.33)$; $t > 12.5$ s, $\phi_m = 0$

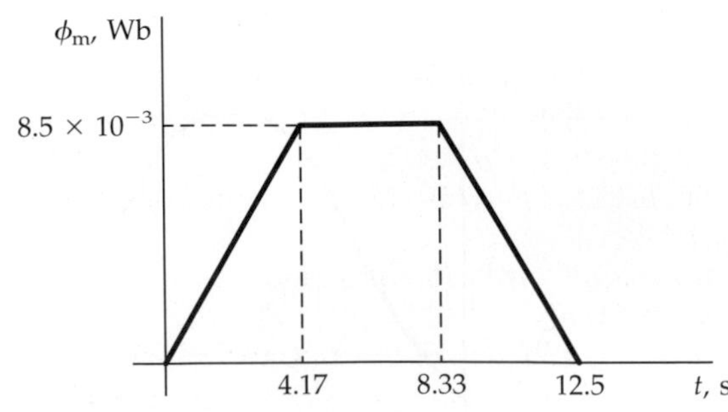

(*b*) $0 \text{ s} \le t \le 4.17$ s, $\mathcal{E} = -2.04 \times 10^{-3}$ V; $4.17 \text{ s} \le t \le 8.33$ s, $\mathcal{E} = 0$ V; $8.33 \text{ s} \le t \le 12.5$ s, $\mathcal{E} = 2.04 \times 10^{-3}$ V; $t > 12.5$ s, $\mathcal{E} = 0$

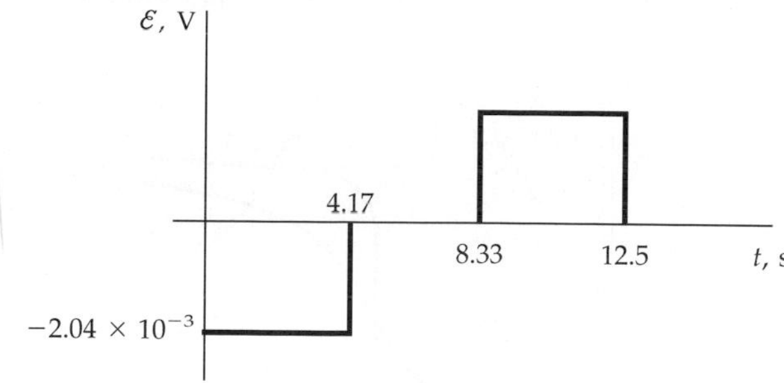

67. (*a*) $\Gamma = (\mathcal{E} - B\ell v)B\ell/R - m\,dv/dt$ (*b*) $v_t = \mathcal{E}/B\ell$ (*c*) 0
69. answers given in problem
71. (*a*) 1.38×10^{-4} V/m (*b*) 5.51×10^{-6} V
73. answers given in problem
75. (*a*) $(\mu_0 I/\pi) \ln [(d-a)/a]$ (*b*) $(\mu_0/\pi) \ln [(d-a)/a]$
77. (*c*) $\frac{1}{2}\ell^2 B\theta$
79. 12 mH
81. $(\mu_0 I\ell/2\pi) \ln (r_2/r_1)$

Chapter 27

True or False **1.** True theoretically, but it is masked by paramagnetism or ferromagnetism in materials whose molecules have permanent dipole moments **2.** True **3.** False **4.** True **5.** False

Problems

1. (*a*) $B_{app} = 0.0101$ T, $B = 0.0101$ T (*b*) $B_{app} = 0.0101$ T, $B = 1.52$ T
3. $B_{app} = 0.0101$ T, $M = 0.183$ A/m, $B = 0.0101$ T
5. (*a*) The field decreases by 6.8×10^{-3} percent (*b*) The self-inductance also decreases by 6.8×10^{-3} percent
7. (*a*) 0.0628 T (*b*) 0.0628 T to three places, though greater than (*a*) by 2.3×10^{-3} percent (*c*) 0.0628 T to three places, though less than (*a*) by 2.6×10^{-3} percent
9. $m = 1.69\, m_B$
11. (*a*) $M_s = 5.58 \times 10^5$ A/m, $\mu_0 M_s = 0.701$ T (*b*) 5.23×10^{-4} (*c*) Diamagnetic effects have been neglected, and these effects tend to reduce the susceptibility
13. $M = 6.87 \times 10^5$ A/m, $B = 0.864$ T
15. (*a*) 0.0126 T (*b*) 1.36×10^6 A/m (*c*) 137
17. (*a*) 0.0603 T (*b*) 24 A
19. (*a*) 8.12×10^3 A/m (*b*) 1.62×10^{21} electrons (*c*) 24.4 A
21. (*a*) 6×10^{13} m^3 (*b*) 24.3 km
23. $n_N = 418 n_O$
25. $B_{app} = \mu_0 NI/2\pi R$, $B = \mu_0(NI/2\pi R + M)$
27. (*a*) 1.42×10^6 A/m (*b*) $K_m = 90$, $\mu = 90\mu_0 = 1.13 \times 10^{-4}$ T·m/A, $\chi_m = 89$
29. (*a*) 15.1 T (*b*) 1.2×10^7 A/m (*c*) 0.0302 T
31. answer given in problem
33. (*a*) 0.0524 A·m^2 (*b*) 7.70×10^5 A/m (*c*) 2.31×10^4 A
35. answer given in problem
37. (*b*) 1.25 N
39. (*a*) 3.02×10^{-4} T (*b*) 0.121 T (*c*) 14.5 J/m^3
41. -2.21×10^{-5}

Chapter 28

True or False **1.** False; the power dissipated is proportional to I^2 **2.** True **3.** True **4.** True **5.** True **6.** True

Problems

1. (*a*) 0.833 A (*b*) 1.18 A (*c*) 200 W
3. (*a*) 20.8 A (*b*) 29.5 A (*c*) $I_{rms} = 41.7$ A, $I_{max} = 58.9$ A
5. (*a*) 0.377 Ω (*b*) 3.77 Ω (*c*) 37.7 Ω
7. 1.59 kHz
9. (*a*) 2.65 MΩ (*b*) 26.5 kΩ (*c*) 26.5 Ω
11.

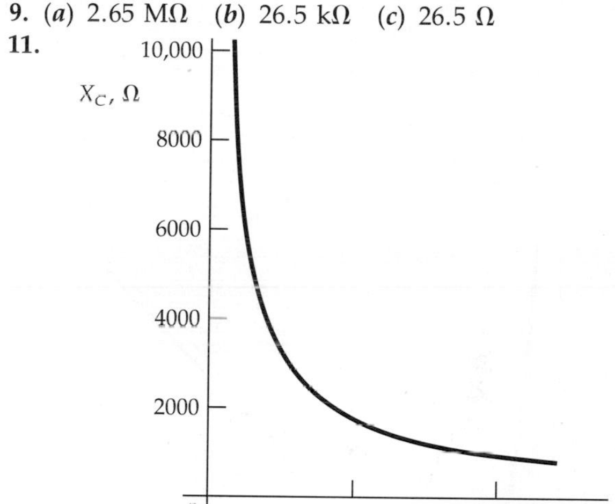

13. (*a*) 15.9 kHz (*b*) 159 Hz (*c*) 1.59 MHz
15. answer given in problem
17. 88 mH
19. (*a*) 2.25 mJ (*b*) 712 Hz (*c*) 0.671 A
21. (*a*) 1.13 kHz (*b*) $X_C = 79.6$ Ω, $X_L = 62.8$ Ω (*c*) $Z = 17.5$ Ω, $I_{rms} = 4.04$ A (*d*) $-73.4°$
23. 2002
25. (*a*) 14.1 (*b*) 79.6 Hz (*c*) 0.275
27. (*a*) 0.553 (*b*) 0.663 A (*c*) 44 W
29. (*a*) A step-down transformer (*b*) 2.4 V rms (*c*) 5 A
31. 10.4 turns for 2.5 V; 31.3 turns for 7.5 V; 37.5 turns for 9 V
33. (*a*) 1.75 A (*b*) 2.47 A
35. (*a*) 12 V (*b*) 8.49 V
37. 60 V
39. (*a*)

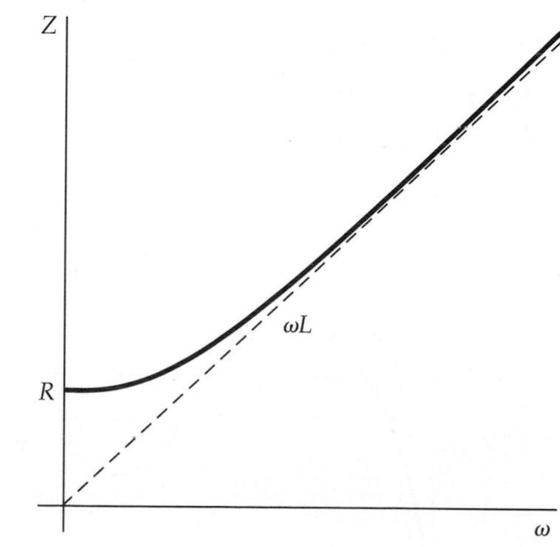

(*b*)

Z

R

ω

(*c*)

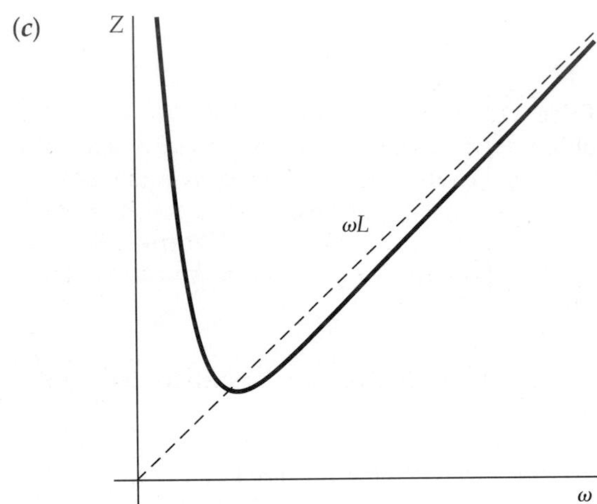

41. (*a*) 396 Ω (*b*) 50 V
43. 29.2 mH
45. (*a*) 15 W (*b*) 15 Ω (*c*) 0.235 H
47. (*a*) 6 Ω (*b*) 35.5 mH
49. (*a*) $C = 18.8\ \mu\text{F}$, $I_{rms} = 0.531$ A (*b*) 25 V
51. (*b*) $\delta = -\pi/2 + \omega RC$ (*c*) $\delta = \pi/2 - R/L\omega$
53. (*a*) 80.3 V (*b*) 77.8 V (*c*) 165 V (*d*) 112 V (*e*) 182 V
55. 0.935 μF
57. $R = 933\ \Omega$, $C = 0.517$ pF
59. (*b*) $\omega = 1/\sqrt{3}RC$ (*c*)

V_{out}/V_0

1

ω

61. (*a*) $V_{out} = (9.95\text{ V})\cos 100t + (0.995\text{ V})\cos 10{,}000t$ (*b*) 10
63. (*a*) $L = 0.8$ mH, $C = 12.5\ \mu\text{F}$ (*b*) 1.6 (*c*) 2 A
65. (*a*) 933 W (*b*) 7.71 Ω (*c*) 99.8 μF (*d*) Add a capacitance of 40.9 μF
67. (*c*)

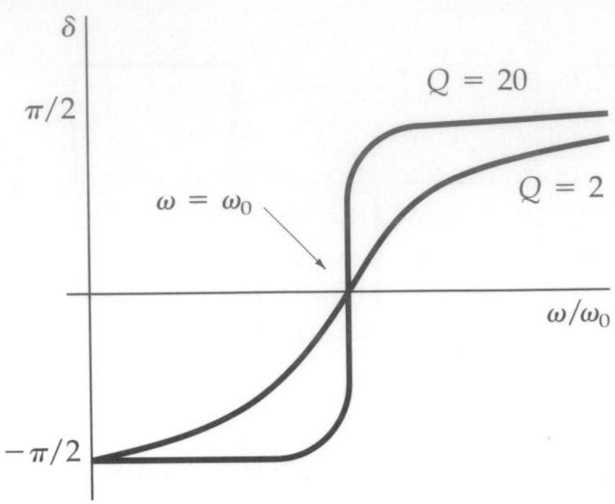

69. (*a*) 12 Ω (*b*) $R = 7.2\ \Omega$, $X = 9.6\ \Omega$ (*c*) Capacitive
71. (*a*) 4 mH (*b*) 0.1 A
73. answers given in problem
75. answer given in problem
77. answers given in problem

Chapter 29

True or False **1.** False **2.** True **3.** True **4.** True **5.** False **6.** True

Problems

1. (*a*) 3.4×10^{14} V/m·s (*b*) 5 A
3. answer given in problem
5. (*a*) $3.33 \times 10^{-7}\ \text{N/m}^2$ (*b*) 194 V/m (*c*) 647 nT
7. answers given in problem
9. 6.93×10^{-8} N
11. answer given in problem
13. (*a*) 300 m (*b*) 3 m
15. 3×10^{18} Hz
17. (*b*) $(-2.36 \times 10^{-5}\text{ A})\sin 500\pi t$
19. (*a*) $I/9$ (*b*) $I/2$ (*c*) $I/16$
21. (*a*) $E_{max} = 12$ V/m, $B_{max} = 4 \times 10^{-8}$ T (*b*) $E_{max} = 0.12$ V/m, $B_{max} = 4 \times 10^{-10}$ T (*c*) $E_{max} = 1.2 \times 10^{-3}$ V/m, $B_{max} = 4 \times 10^{-12}$ T
23. 111 m^2 or 10.5 by 10.5 m
25. (*a*) 2×10^{-12} N (*b*) 6×10^{-12} N
27. (*a*) 3 m (*b*) $5.31 \times 10^5\ \text{J/m}^3$ (*c*) $E_{max} = 3.46 \times 10^8$ V/m, $B_{max} = 1.15$ T
29. (*a*) Positive *x* direction (*b*) $\lambda = 0.628$ m, $f = 4.77 \times 10^8$ Hz (*c*) $E = (194\text{ V/m})\cos[10x - (3 \times 10^9)t]$, $B = (0.647 \times 10^{-6}\text{ T})\cos[10x - (3 \times 10^9)t]$
31. (*a*) 1417 W/m (*b*) 902 W/m^2 (*c*) $E_{rms} = 583$ V/m (*d*) 1.94×10^{-6} T
33. (*a*) 279 K (*b*) 245 K
35. $3.42 \times 10^6\ \text{W/m}^2$
37. 7.25×10^{-3} V
39. (*a*) $(5 \times 10^{-5}\text{ V})\cos 10^6 t$ (*b*) $(4.19 \times 10^{-8}\text{ V})\sin 10^6 t$
41. (*a*) $At/100\rho d$, where *d* is the separation between the plates (*b*) $\kappa\epsilon_0 A/100d$ (*c*) $\kappa\epsilon_0\rho$
43. (*a*) $V_0[(1/R)\sin\omega t + (\epsilon_0 A\omega/d)\cos\omega t]$ (*b*) $(\mu_0/2\pi)[(V_0/rR)\sin\omega t + (\epsilon_0\omega\pi V_0 r/d)\cos\omega t]$ (*c*) $\tan\delta = \epsilon_0 A\omega R/d$

45. (*a*) $\rho I/\pi a^2$ (*b*) $\mu_0 I/2\pi a$ (*c*) $S = \rho I^2/2\pi^2 a^3$, radially inward (*d*) $S(2\pi aL) = I^2(\rho L/A) = I^2R$
47. 5.74×10^{-7} m
49. answer given in problem
51. (*a*) 3.12×10^{-6} rad/s^2 (*b*) 3.88 days (*c*) no

Chapter 30

True or False **1.** True **2.** False **3.** False; it is greater when the second medium has a lower index of refraction—for example, for refraction from water to air **4.** False; if it were, there would be no rainbow **5.** True

Problems

1. 2.11×10^6 *c*-year
3. ± 0.3 m
5. 92 percent
7. (*a*) 27.1° (*b*) 41.7° (*c*) 70.1° (*d*) Totally reflected
9. $v_{water} = 2.26 \times 10^8$ m/s, $v_{glass} = 2 \times 10^8$ m/s
11. (*a*) 50.2° (*b*) 38.8° (*c*) 26.3°
13. 62.5°
15. (*a*) Yes (*b*) Since the ball moves with constant speed, least distance implies least time
17. (*a*) $I_0/8$ (*b*) $3I_0/32$
19. 35.3°
21.

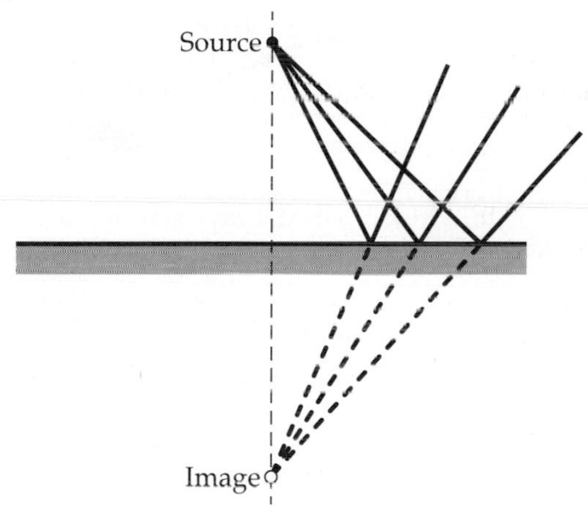

23. $\Delta t = 0.505$ s
25. 238 m^2
27. answer given in problem
29. answer given in problem
31. (*b*) 3.47°
33. $\theta_{violet} = 27.0°$, $\theta_{red} = 27.3°$
35. (*b*) The critical angle is greater
37. answer given in problem
39. (*a*) $I_0(\cos \pi/2N)^{2N}$ (*b*) $I_0/4$ (*c*) $0.976I_0$ (*d*) Perpendicular to initial polarization
41. (*a*) 1.33 (*b*) 37.2° (*c*) $\theta_r = 48.7°$; no, the refraction into the liquid doesn't change the outcome; the final beam is parallel to the liquid surface
43. answer given in problem
45. answer given in problem
47. answer given in problem
49. $\theta_r = 14.5°$; the refracted ray is bent toward the perpendicular

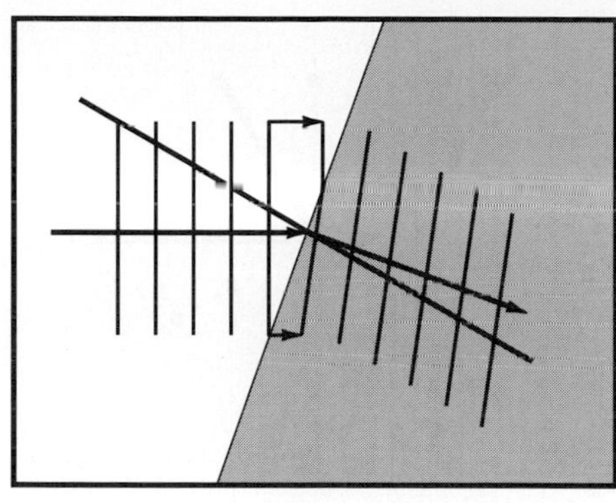

51. (*a*) $d\theta_d/d\theta_1 = 2 - (4 \cos \theta_1)/(n^2 - \sin^2 \theta_1)^{1/2}$
53. 2.18 cm
55. $\frac{1}{2}[1 - (1 - 1/n^2)^{1/2}]$

Chapter 31

True or False **1.** True **2.** False **3.** False; it is true for positive object distances. An example of a real image with a negative image distance occurs when converging light is reflected by a plane mirror **4.** False; spherical aberration occurs for rays far from the axis of the mirror **5.** True **6.** False; for example, the image distance is negative for a simple magnifier **7.** True

Problems

1. The eye can see the image from anywhere between rays 1 and 2

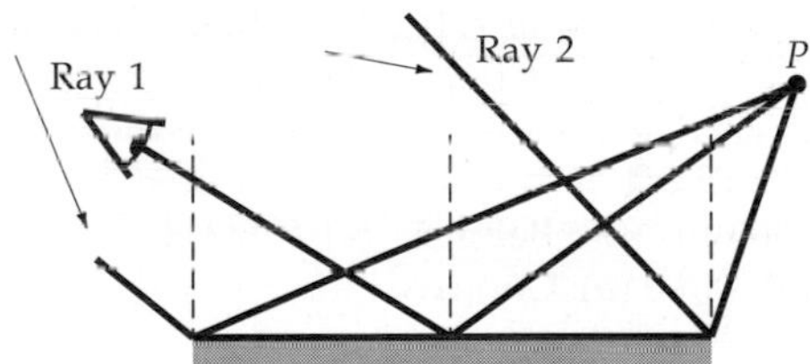

3. (*a*) 0.81 m (*b*) The bottom of the mirror should be 0.735 m above the floor

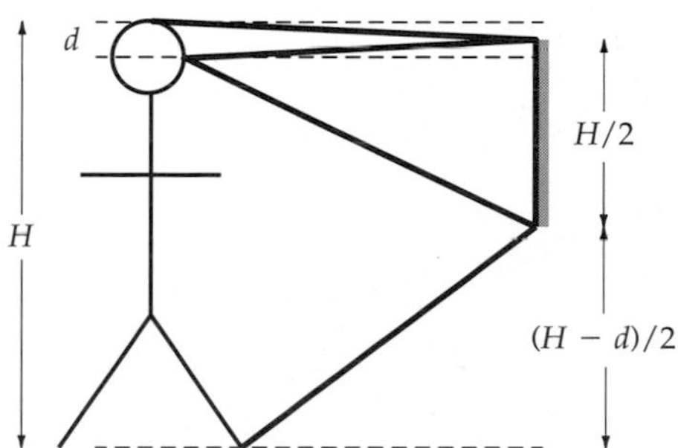

5. (*a*)

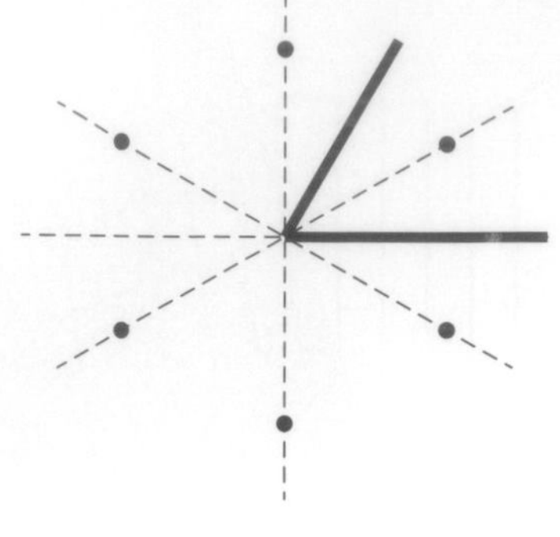

(*b*)

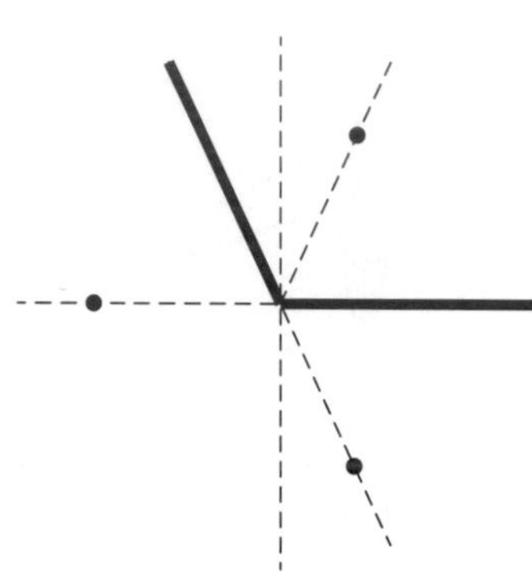

7. (*a*) $s' = 25$ cm, $m = -0.25$, real, inverted, reduced (*b*) $s' = 40$ cm, $m = -1$, real, inverted, same size (*c*) $s' = \infty$, $m = -\infty$, real, inverted, enlarged (*d*) $s' = -20$ cm, $m = 2$, virtual, erect, enlarged
9. (*a*) $s' = -16.7$ cm, $m = 0.167$, virtual, erect, reduced (*b*) $s' = -13.3$ cm, $m = 0.333$, virtual, erect, reduced (*c*) $s' = 10$ cm, $m = 0.5$, virtual, erect, reduced (*d*) $s' = -6.67$ cm, $m = 0.667$, virtual, erect, reduced
11. (*a*) 0.566 m (*b*) Behind (*c*) 0.113 m
13. (*a*) 5.13 cm (*b*) Concave
15. (*a*) $s' = 30$ cm, real

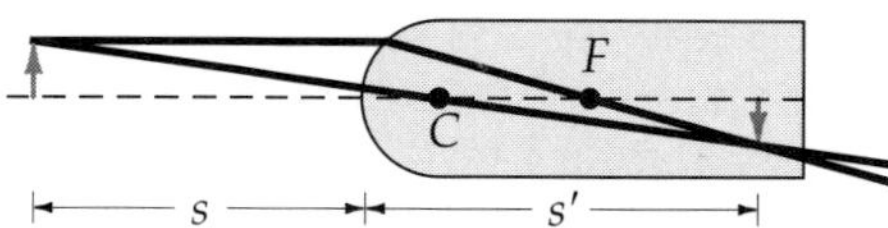

(*b*) $s' = -15$ cm, virtual

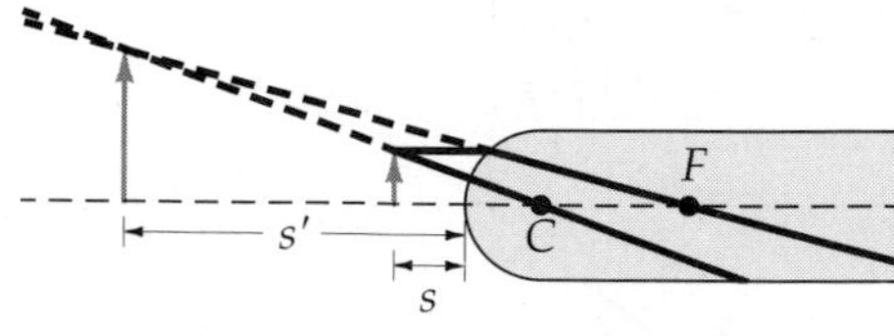

(*c*) $s' = 15$ cm, real; image is of zero size and located at *F*

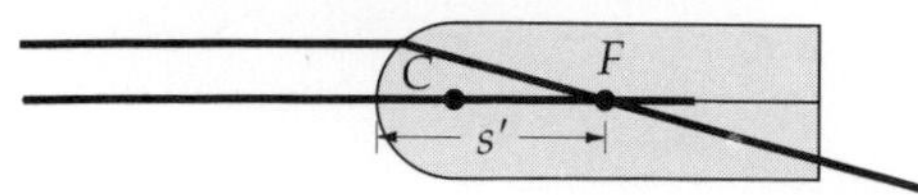

17. (*a*) $s' = -10$ cm, virtual

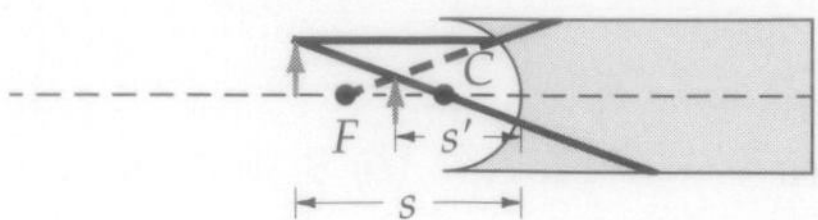

(*b*) $s' = -5$ cm, virtual; paraxial rays starting at *C* are undeflected, thus the image and object are identical

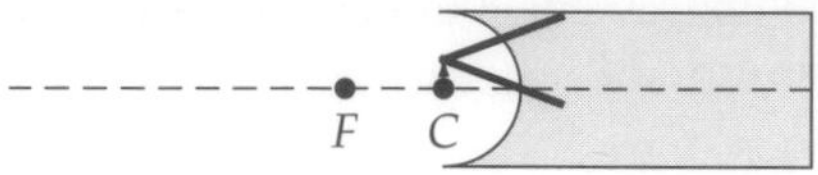

(*c*) $s' = -15$ cm, virtual; image is of zero size and located at *F*

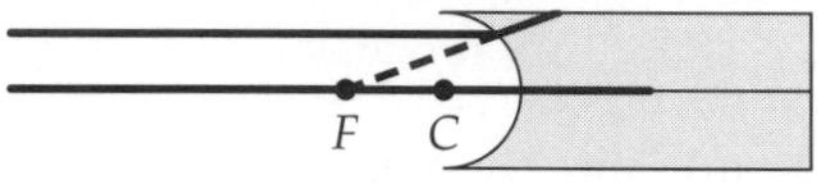

19. (*a*) $s' = -14.9$ cm, virtual

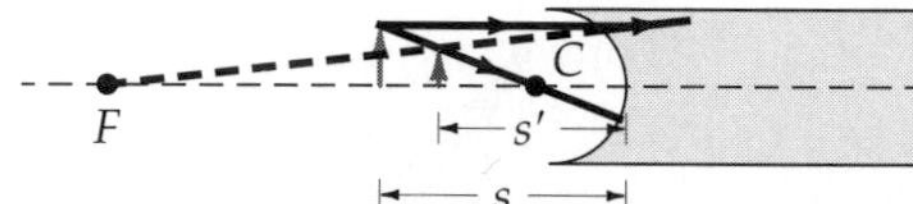

(*b*) $s' = -5$ cm, virtual; paraxial rays starting at *C* are undeflected, thus the image and object are identical

(*c*) $s' = -44.1$ cm, virtual; image is of zero size and located at *F*

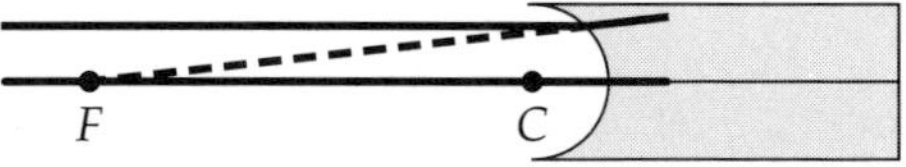

21. (*a*) −0.839 m (*b*) 0.336
23. (*a*) 6 cm

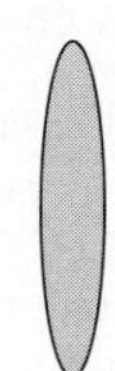

(*b*) −6 cm

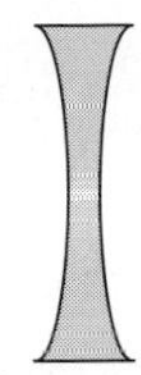

25. (*a*) −30.3 cm (*b*) −22.0 cm (*c*) 0.275
(*d*) Virtual, upright
27. (*a*) −33.3 cm

(*b*) 33.3 cm

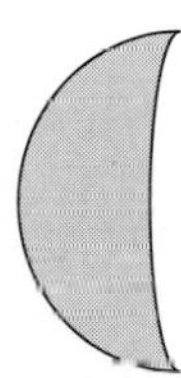

(*c*) −33.3 cm

29. (*a*) $s' = 40$ cm, $m = -1$, real, inverted (*b*) $s' = 20$ cm, $m = 2$, real, erect (*c*) $s' = -17.1$ cm, $m = 0.429$, virtual, erect (*d*) $s' = -7.5$ cm, $m = 0.75$, virtual, erect
31. $s' = 10$ cm, $m = -1$

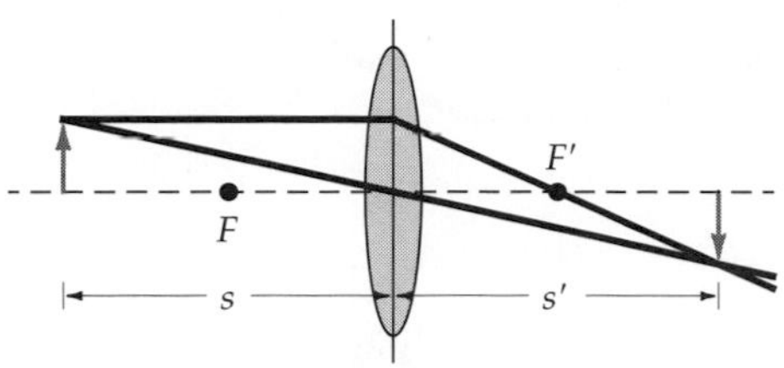

33. (*a*) $s = 5$ cm, $s' = -10$ cm

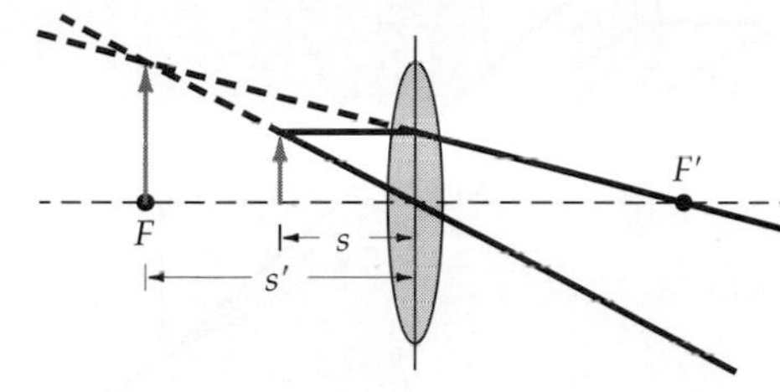

(*b*) $s = 15$ cm, $s' = 30$ cm

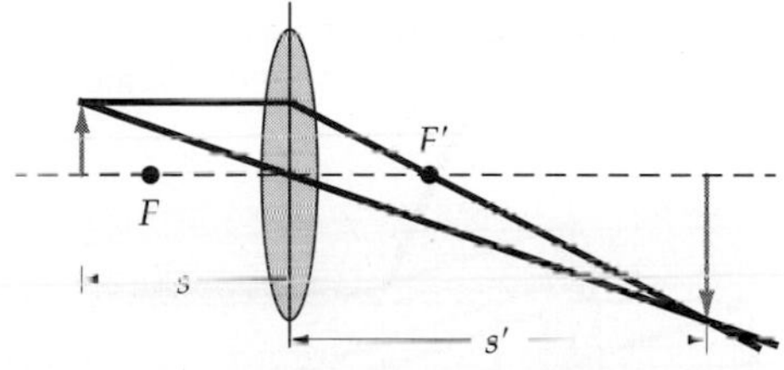

35. (*a*) 30 cm to the far side of the second lens

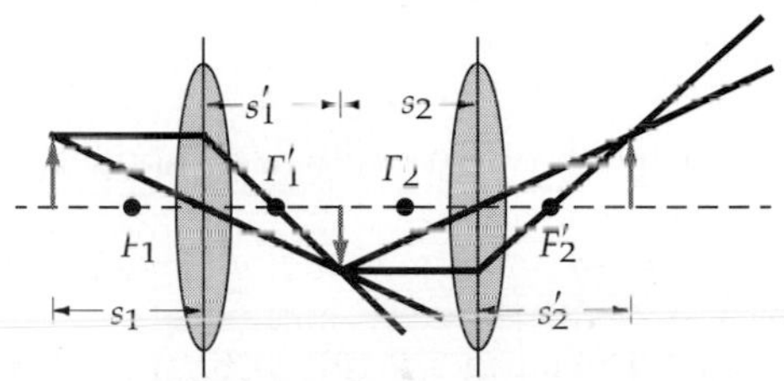

(*b*) Real, erect (*c*) 2
37. (*a*) 10.6 cm (*b*) 9.43 cm
39. (*a*) −66.7 cm (*b*) Virtual
41. The mirror should be moved 91 cm away from the object
43. Concave, $f = 90$ cm
45. (*a*) −128 cm (*b*) 14.7 cm (*c*) Real
47. (*a*) $r_2 = 35.0$ cm, concave

(*b*)

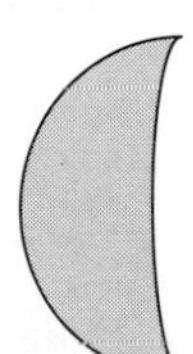

49. 4.10 cm
51. (*a*) −1.33 m (*b*) Convex
53. (*a*) 9.52 cm (*b*) −1.19

(*c*)

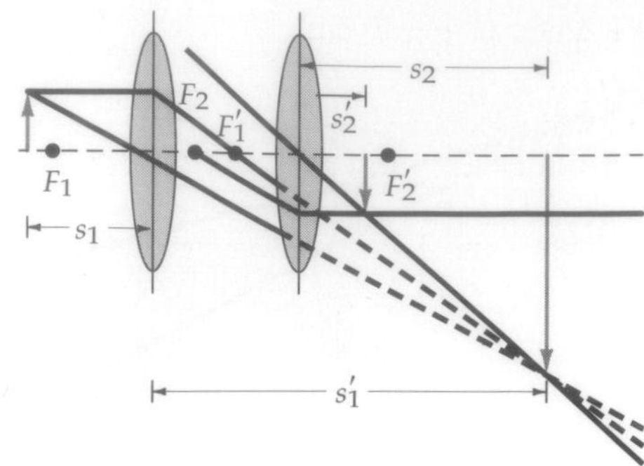

55. The final image is at the left focal point of the second lens; the image is erect and the same size as the object

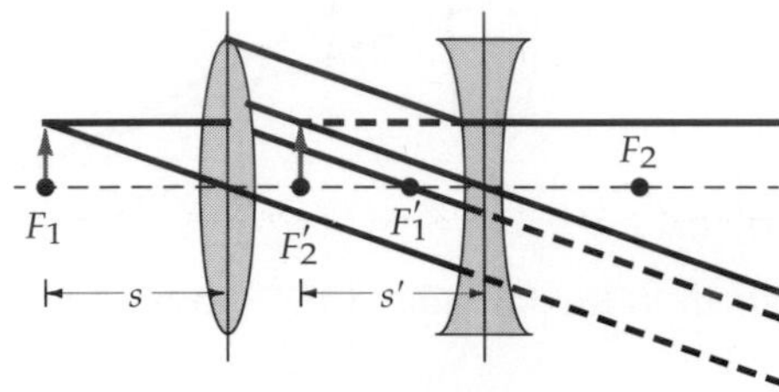

57. (*a*) 18 cm to left of lens (*b*) Real, upright

(*c*)

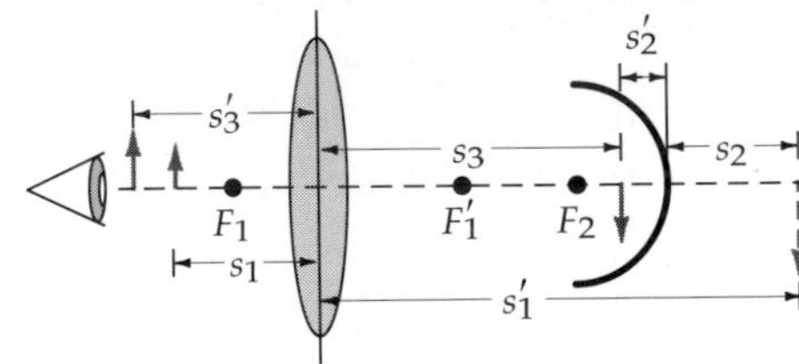

59. 200 cm
61. 43.5 cm
63. (*a*) The final image is 0.9 cm behind the back surface (*b*) The final image is on the back surface
65. (*a*) 1.8 m from screen (*b*) 45 cm
67. (*b*) 17.5 cm
69. answer given in problem

Chapter 32

True or False **1.** True **2.** True **3.** True **4.** True **5.** False; it varies inversely with the square of the *f*-number **6.** True **7.** True **8.** False; it is inverted and smaller than the object **9.** False; it uses a mirror for its objective

Problems

1. 0.278 cm
3. (*a*) 103 cm (*b*) 0.972 diopters
5. 44.4 cm
7. 0.714 cm; the actual radius should be smaller
9. 6
11. 5
13. 35.7 mm
15. 1.3 mm
17. (*a*) ≈1/64 s (*b*) ≈1/120 s (*c*) ≈1/250 s (*d*) ≈1/500 s (*e*) ≈1/1000 s
19. −267
21. (*a*) 20 cm (*b*) −4 (*c*) −20 (*d*) 6.25 cm
23. (*a*) 0.9 cm (*b*) 0.18 rad (*c*) −20
25. (*a*) 25 (*b*) −134
27. (*a*) 3 (*b*) 4
29. 3.7 m
31. $f_e = 4$ cm, $f_o = 28$ cm
33.

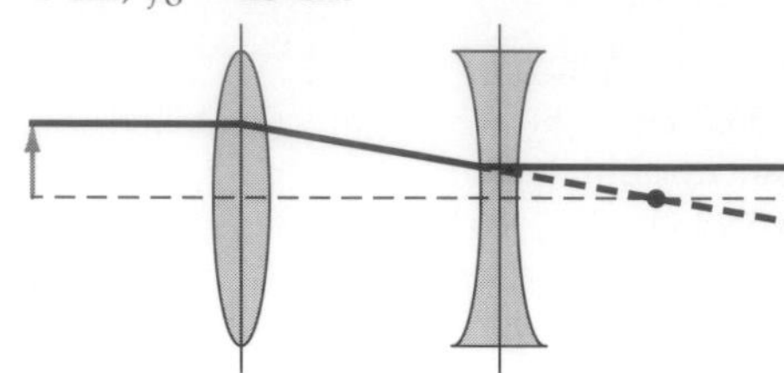

35. −232
37. 0.00667
39. (*a*) 1.67 cm (*b*) 0.508 cm in front of the objective (*c*) 0.496 cm

Chapter 33

True or False **1.** False **2.** True **3.** True **4.** True **5.** True

Problems

1. (*a*) Incoherent (*b*) Coherent (*c*) Coherent (*d*) Incoherent (*e*) Coherent
3. 164°
5. (*a*) The top of the film approaches zero thickness, so the phase difference approaches 180° (*b*) Violet (*c*) The top of the film is white, the color of the first band is red
7. 115 nm
9. (*a*) 7.2 μm (*b*) 1.44
11. 8.33 fringes/cm
13. (*a*) 50 μm (*b*) No (*c*) 0.5 mm
15. 695 nm
17. $E = 3.61 \sin(\omega t - 56.3°)$
19. (*a*) $A' = 2.73\,E_0$, $\delta' = 30°$

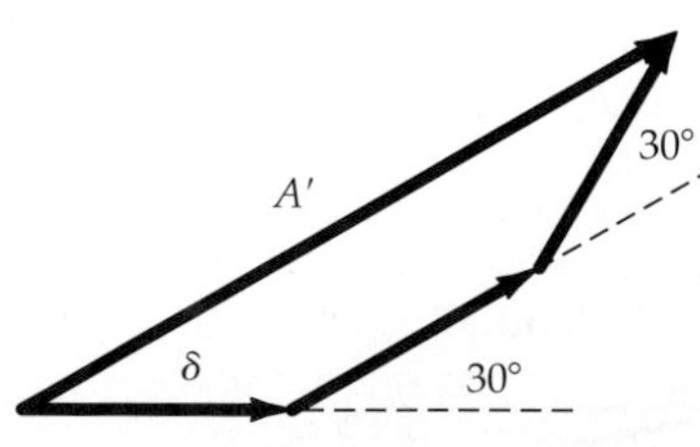

(*b*) $A' = 2E_0$, $\delta' = 60°$

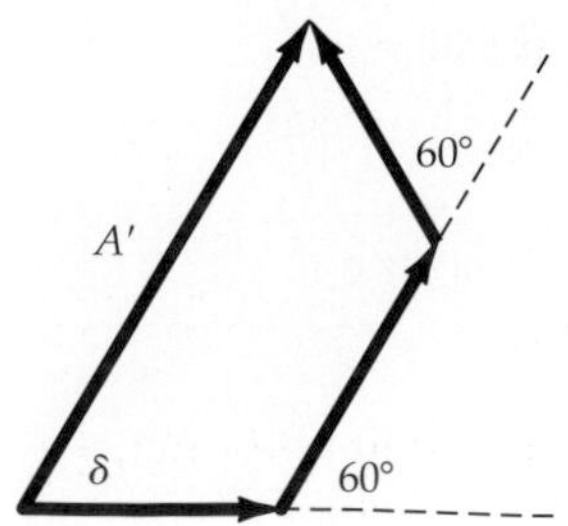

(*c*) $A' = E_0$, $\delta' = 90°$

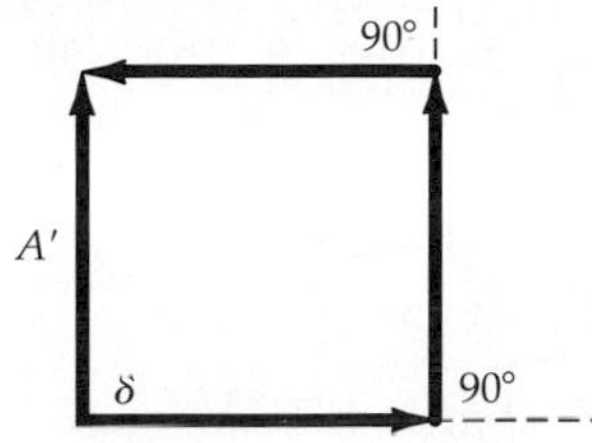

(*d*) $A' = 0$

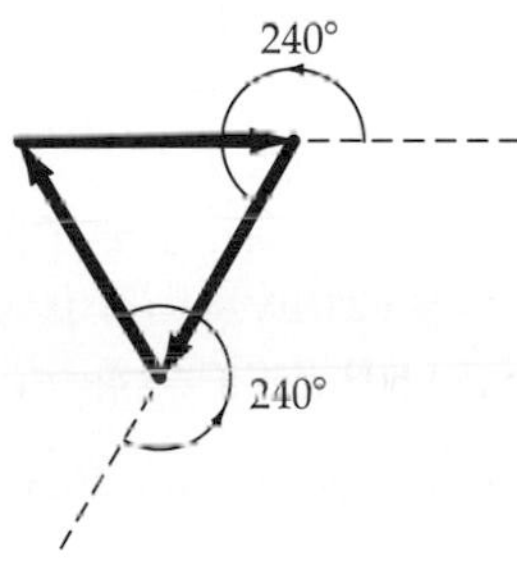

21. (*a*) $\theta_1 = \lambda/d$, $\theta_{min} = \lambda/5d$

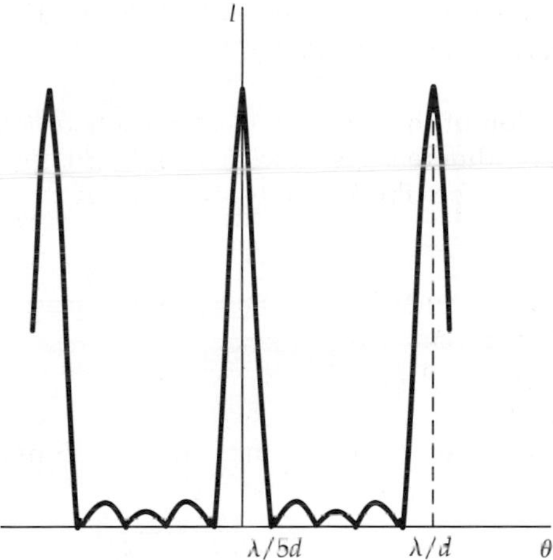

23. The separation between the slits is d, and the condition for an interference maximum is $d \sin\theta = m\lambda$; the width of an individual slit is a, and the condition for a diffraction minimum is $a \sin\theta = m\lambda$
25. (*a*) 2 cm (*b*) 20 cm (*c*) 2.31 m
27. 3.01 cm
29. 39 fringes
31. (*a*) 8.54×10^{-3} rad (*b*) 6.83 cm
33. (*a*) 55.6 km (*b*) 55.6 m
35. 33.6 mm
37. 484 m
39. 486 nm, 660 nm
41. (*a*) 0.0231° (*b*) 0.145 cm
43. (*a*) 0.30° (*b*) 8
45. 4.5×10^6 km
47. (*a*) 0.6 μm (*b*) 400 nm, 514 nm, 720 nm (*c*) 400 nm, 514 nm, 720 nm
49. (*a*) 0.530 m, 0.883 m (*b*) 0.707 m (*c*) 8000
51. (*a*) 97.8 nm (*b*) No (*c*) $I_{400} = 0.273 I_{max}$, $I_{700} = 0.124 I_{max}$
53. (*b*) The width of the principal interference maximum is 6 mm for four sources, 12 mm for two sources
55. (*a*) 0.242 rad (*b*) 0.08 rad, 0.161 rad (*c*) 0.04 rad (*d*)

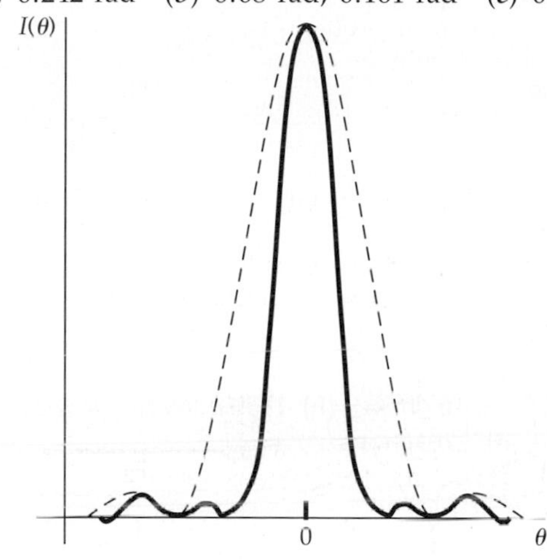

57. 20.5 m
59. 13.0°
61. answer given in problem
63. (*c*) Reversed (*d*) 67 fringes (*e*) 1.14 cm (*f*) The fringes will be closer together
65. (*b*) 491 (*c*) 1 mm
67. (*a*) 1.92 μm (*b*) 1.91
69. (*a*) $I = I_{max} \cos^2 [(\pi/2) \sin\theta]$

(*b*)

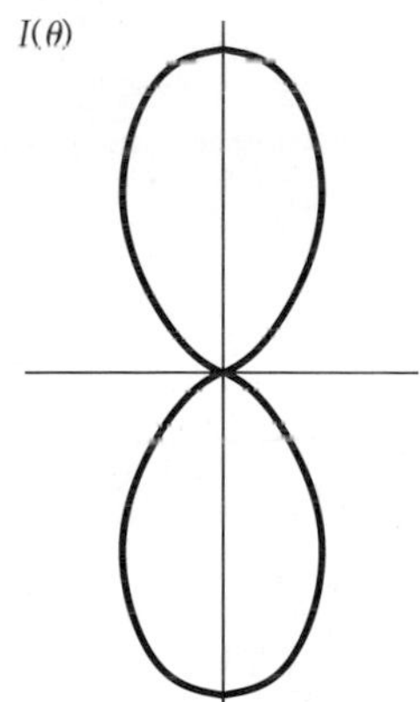

71. (*a*) $\phi = \pm 2.86\pi, \pm 4.92\pi, \pm 6.94\pi$ (*b*) Same as part (*a*)
73. Answer given in problem

Chapter 34

True or False **1.** True **2.** True **3.** False **4.** True **5.** False **6.** False **7.** True

Problems

1. (*a*) 0.183 ms (*b*) 1.83×10^{-12} s (*c*) No
3. (*a*) 4.94×10^{-8} s (*b*) 12.6 m (*c*) 6.63 m
5. (*a*) 44.7 μs (*b*) 13.4 km
7. 0.527*c*
9. (*a*) 130 y (*b*) 88.1 y

11. $0.9991c$
13. 2.60×10^8 m/s
15. (*a*) 4.5×10^{-10} percent (*b*) The time elapsed on the pilot's clock is 3.15×10^7 s $- 1.42 \times 10^{-4}$ s; the time lost in minutes is 2.37×10^{-6} min
17. 80 c·min
19. answer given in problem
21. $L_p V/c^2 = 60$ min
23. 0.0637
25. $0.6c$
27. $0.696c$
29. (*a*) $-0.882c$ (*b*) $-60{,}000$ m/s $+ 6 \times 10^{-4}$ m
31. $-0.994c$
33. (*a*) 1.11×10^{-17} kg (*b*) 0.351 mg
35. (*a*) 9×10^{13} J (*b*) \$2.5 million (*c*) 28,571 y
37. $E = 0.522$ MeV, $K = 1.05 \times 10^{-2}$ MeV, $p = 0.104$ MeV/c
39. (*a*) 2.23×10^8 m/s (*b*) 1039 MeV/c
41. 3.55×10^{14} reactions/s
43. 0.782 MeV
45. 50 percent
47. $0.8c$
49. 1.85×10^4 y
51. 9.61 ms
53. The required speed is $0.4c$; event B precedes event A for an observer moving with a speed $0.4c < v < c$
55. (*a*) $0.66c$ (*b*) 31.3 y
57. (*a*) 52.7 m (*b*) $-0.987c$ (*c*) 16.1 m (*d*) 2.07×10^{-7} s (*e*)

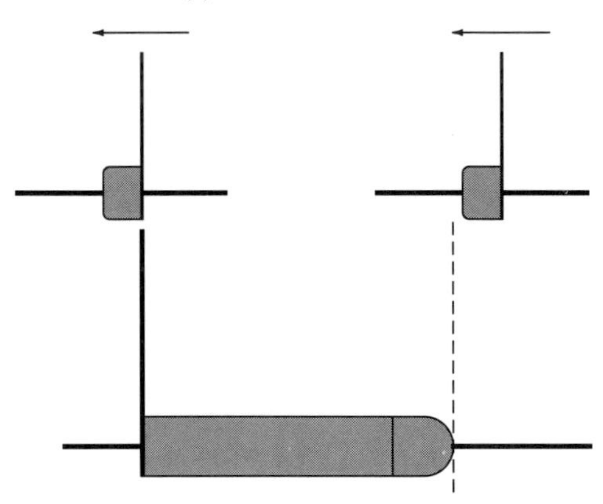

59. (*a*) 4.97 MeV/c (*b*) $0.995c$
61. (*a*) 0.75 percent (*b*) 68.7 percent
63. (*a*) 630 m/c (*b*) 777 m/c (*c*) 148 m/c (*d*) [illegible] m/c (*e*) 4.36 h (*f*) 19 h
65. (*a*) To the left with speed $0.5c$ (*b*) 1.73 y
67. 3.84×10^{14} kg/day
69. (*a*) $0.333c$ (*b*) 20 m in the $+x$ direction (*c*) 60 m/c
71. (*a*) 290 MeV (*b*) 625 MeV
73. (*a*) 1.30 m (*b*) 0.825 m
75. (*a*) E/Mc (*b*) EL/Mc^2
77. answer given in problem
79. answer given in problem
81. answer given in problem
83. answers given in problem

Chapter 35

True or False **1.** True **2.** True **3.** False **4.** True **5.** True **6.** False **7.** True **8.** True **9.** True **10.** True **11.** True **12.** False

Problems

1. $E = 6.626 \times 10^{-26}$ J $= 4.14 \times 10^{-7}$ eV
3. (*a*) 2.42×10^{14} Hz (*b*) 2.42×10^{17} Hz (*c*) 2.42×10^{20} Hz
5. $E_{400} = 3.10$ eV, $E_{700} = 1.77$ eV
7. (*a*) $f_t = 1.11 \times 10^{15}$ Hz, $\lambda_t = 271$ nm (*b*) 1.63 V (*c*) 0.39 V
9. (*a*) 4.74 eV (*b*) 1.68 V
11. (*a*) 4.97×10^{-19} J (*b*) 0.01 J (*c*) 2.01×10^{16} photons/s
13. 9.27×10^3 V
15. 1.215 pm
17. (*a*) $p = 1.66 \times 10^{-27}$ kg·m/s $= 3.11$ eV/c (*b*) $p = 3.32 \times 10^{-25}$ kg·m/s $= 621$ eV/c (*c*) $p = 6.63 \times 10^{-24}$ kg·m/s $= 12.4$ keV/c (*d*) $p = 2.21 \times 10^{-32}$ kg·m/s $= 4.14 \times 10^{-5}$ eV/c
19. (*a*) 17.5 keV (*b*) 76.0 pm (*c*) 16.3 keV
21. answer given in problem
23. $\Delta E_{3\to 2} = 1.89$ eV, $\lambda_{3\to 2} = 656$ nm; $\Delta E_{4\to 2} = 2.55$ eV, $\lambda_{4\to 2} = 486$ nm; $\Delta E_{5\to 2} = 2.86$ eV, $\lambda_{5\to 2} = 434$ nm
25. (*a*) $\Delta E_{\infty\to 4} = 0.850$ eV, $\lambda_{\infty\to 4} = 1459$ nm (*b*) $\lambda_{5\to 4} = 4052$ nm, $\lambda_{6\to 4} = 2627$ nm, $\lambda_{7\to 4} = 2168$ nm

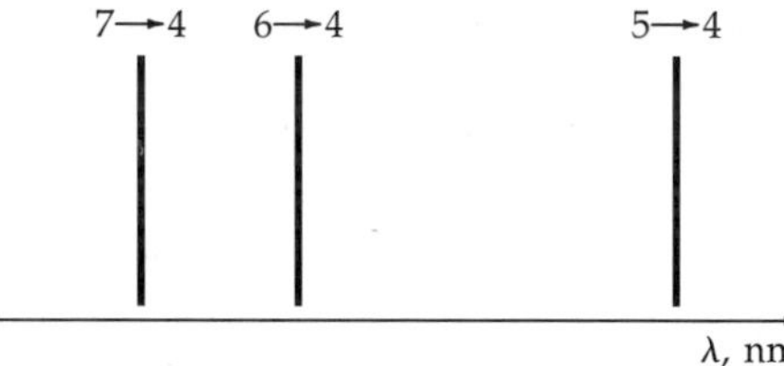

27. (*a*) 0.775 nm (*b*) 0.0775 nm (*c*) 0.0245 nm (*d*) 0.00775 nm
29. (*a*) 3.313×10^{-27} kg·m/s (*b*) 6.024×10^{-24} J
31. 0.203 nm
33. 4.40×10^{-13} m
35. 1.52×10^{-34} m
37. 22.8 eV
39. 6.11 keV
41. 4.86×10^6 m $= 3018$ mi
43. 1.69×10^{-14} W/m^2
45. (*a*) 3.18 W/m^2 (*b*) 1.04×10^{15} photons/s
47. (*a*) 4.86 pm (*b*) 92.7 keV (*c*) 92.7 keV
49. (*a*) 13.6 eV (*b*) 54.4 eV (*c*) 122.4 eV

51. (*b*) $E_n = n^2(37.6\ \text{eV})$

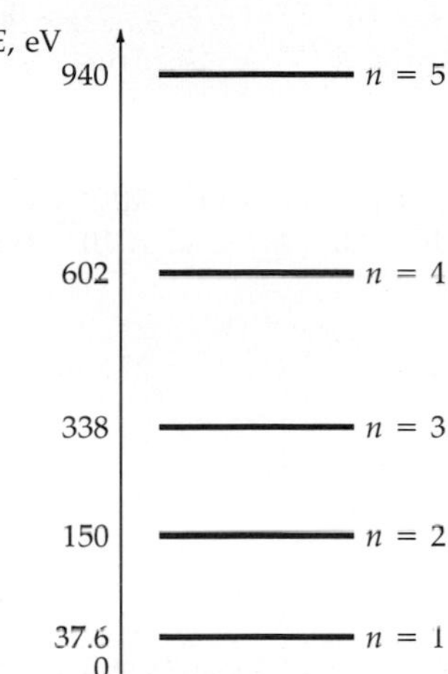

(*c*) 11.0 nm (*d*) 6.60 nm (*e*) 1.37 nm
53. (*a*) $E_1 = 5.13 \times 10^{-3}$ eV, $E_2 = 2.05 \times 10^{-2}$ eV, $E_3 = 4.61 \times 10^{-2}$ eV (*b*) 80.8 μm (*c*) 48.5 μm (*d*) 30.3 μm
55. (*b*) $R_\infty = 10.97373\ \mu\text{m}^{-1}$; $R_H = 10.96776\ \mu\text{m}^{-1}$ (*c*) 0.0545 percent
57. (*a*) 10^{-22} W (*b*) 53 min
59. (*a*) 2.42×10^{-10} m (*b*) 0.512 MeV (*c*) 2.73×10^{-22} kg·m/s (*d*) 2.42 pm
61. answer given in problem

Index

Aberration
astigmatism, 1034
chromatic, 1034–1035
coma, 1034
correction of, 1034
distortion, 1034
spherical, 1016, 1034
Absolute pressure, 339
Absolute temperature scale, 491–492, 496
average kinetic energy and, 500–501
Acceleration
average, 28–29, 56
centripetal, 65–67
charge in electric field, 614
constant, 30–36, 61
due to gravity (g), 28–29, 30, 58–62, 83–84, 383
instantaneous, 28–29, 56
vector, 56–57
AC circuits (*see* Alternating current)
Accommodation, 1042
Action and reaction, 86–87 (*see also* Newton's third law of motion)
Action at a distance, 89, 93, 607
Adhesive force, 345
Adiabatic process, 547–549, 572
Aerodynamics of bicycling, *essay*, 356–359
Akasofu, Syun-Ichi, 866–870
Alternating current, 898–936
advantages of, 898, 922
in capacitors, 904–907
in inductors, 902–904
LCR circuits, 908–919
rectification and amplification, 923–926
in resistors, 899–902
rms values, 900–902
transformer, 920–923
Ammeter, 767–769, 900
calibration, 827
Ampère, André-Marie, 782, 811, 815, 826, 879, 965
Ampere (A), unit of current, 3, 600, 717
definition, 826–827
Ampère's law, 811, 812, 828–832
generalized form, 945, 947
inconsistency in, 943–945
limitations of, 831
for straight current-carrying wire, 828–829
symmetry requirements, 828
for tightly wound solenoid, 830–831
for tightly wound toroid, 829–830
Amperian current, 880
Amplification, 926
Amplitude (A), 370–375, 377–379, 416, 446, 951, 1173
Andromeda galaxy, *photo*, 9
Ångström, Anders Jonas, 866
Angular acceleration
definition, 229
tangential acceleration versus, 230
units of, 229
Angular displacement, 228, 230
Angular frequency, 370–373, 416–417, 442
dimensions of, 370
force constant and, 371
units of, 370
Angular magnification, 1046–1047, 1050, 1052
Angular momentum, 242–249
center of mass and, 252
conservation of, 244–246
definition, 242, 257
kinetic energy and, 246–247
magnetic moments and, 882
orbital, 882
of a particle, 242, 257
quantized, 1159, 1161
of rotating object, 242–243
spin, 882–883
total, 242, 258
as vector cross product, 257–258
Angular speed, 229
Angular velocity
analog of linear velocity, 230
definition, 228
units of, 229
Anisotropic materials, 1001
Antenna
electric dipole, 961
for gravity waves, *photo*, 1135
loop, 961
television, *photo*, 959
tracking, *photo*, 943
Antinode, 420–421, 425, 457
Antireflecting film, *photo*, 1066
Aphelion, 298
Apparent depth, 1024, *photo*, 1025
Arc discharge, 678, *photo*, 678
Archimedes, 341
Archimedes' principle, 340–343
Astigmatism, 1034, 1042
Aston, Francis William, 794–795
Astronomical data, *table*, AP-22
Astronomical unit (AU), 299
Atmosphere
of Earth, *photo*, 556
escape speed, 308, 312–313, 315, 504, 1135
pressure variation, 340
of Venus and Jupiter, *photo*, 505
Atmosphere (atm), unit of pressure, 336–337
Atomic magnetic moments
angular momentum and, 882
electron spin, 882
period of the motion, 882
Atomic masses, AP-28
Atomic numbers (Z), 599, AP-28
Atomizer, 351
Atwood's machine, 160
Aurora, *essay*, 866–870
magnetosphere, 868
solar flare, 870
solar wind, 868, 870
Avogadro's number, 497
Axis of rotation, 228, 232–233, 254

Back emf, 848, 855–856, 902–903
Bainbridge, Kenneth, 794
Ballistic pendulum, 204
Balmer, Johann, 1156
Balmer formula, 1156
Balmer series, 1119, 1156, 1160
Banking of roads, 115
Bar, unit of pressure, 339
Bardeen, John, 724
Barometer, 339
Bartlett, Albert A., 586
Batteries
dry cell, *schematic*, 730
emf and, 726–731
ideal, 728
internal resistance, 728–729, 752
jumper cable connections, 753
Kirchhoff's rules and, 751–753
lithium-polymer, *photo*, 730
load resistance, 729
power output, 727, 729
rating in ampere-hours, 731
real, 727–728
silver-zinc, *photo*, 730
storage, *schematic*, 730
terminal voltage, 727–729, 752
BCS theory of superconductivity, 724

Beat frequency, 452–453
Beats, 452–453, 460
Becquerel, Antoine Henri, 218, 1100
Bender, Paul A., 398
Benzene, *photo*, 9
Bernoulli's equation, 349–351
Beta decay, 89, 218–220
Bicycling, aerodynamics of, *essay*, 356–359
Binding energy, 312, 314, 1130–1131
Binoculars, 988
Biot, Jean Baptiste, 811, 823
Biot–Savart law, 815, 831–832
Birefringence, 1001–1003
  extraordinary ray, 1001–1002
  half-wave plate, 1002–1003
  optic axis, 1001
  ordinary ray, 1001–1002
  quarter-wave plate, 1002
Birefringent crystal, *photo*, 1003
Blackbody radiation, 531–533, 1147–1148
Black holes, 1135
Bohr, Niels, 218–219, 975, 1100, 1146, 1149, 1157–1159, 1165
Bohr magneton, 883, 885, 892
Bohr model of atom, 1156–1160
Bohr orbit, 1164
Bohr radius, 1159
Bohr's postulates, 1158–1159
Boiling point (*see* Normal boiling point)
Boltzmann, Ludwig, 530, 964, 967
Boltzmann's constant, 497, 544, 1147
Boundary conditions for standing-wave functions, 426
Bound charge, 695–696
Boyle, Robert, 496
Boyle's law, 496
Brackett, F., 1160
Brahe, Tycho, 1, 297, 299
Bremsstrahlung spectrum, 1153
Brewster, David, 1000
Brewster's law, 1001
Btu (British thermal unit), 519
Bulk modulus (B)
  adiabatic, 548
  and sound waves, 440, 442, 444
  table, 337
Buoyant force, 340–342 (*see also* Archimedes' principle)

Calorie, unit of heat energy, 518–519
Calorimetry, 519
Camera, 1047–1050
  ASA number, 1048
  f-number, 1048–1049
  focal length of lens, 1047–1050
  photo, 1049
Candela (cd), unit of luminous intensity, 3
Capacitance
  definition, 691
  dielectrics and, 694–696
  equivalent, 702–706
  of cylindrical capacitor, 693
  of parallel-plate capacitor, 692–693
  SI unit of, 691
Capacitive reactance, 905–906, 914, 918–919
Capacitors
  alternating current in, 902, 904–907
  capacitive reactance, 905–906
  ceramic, *photos*, 690, 697
  charging, 762–766
  current versus frequency, 906
  cylindrical, 693–694
  discharging, 760–762, 765–766
  ideal, 906
  Leyden jar, *photo*, 703
  in parallel, 702–703, 706–707
  in printed circuit, *photo*, 703
  parallel-plate, 691–693
  in series, 704–706
  time constant, 761, 763–764
  uses of, 690
  variable air gap, *photo*, 703
  voltage drop versus current, 905, 907
Capillarity (capillary action), 344–346
Carlson, Charles, 683
Carnot, Sadi, 569
Carnot cycle, 571
Carnot efficiency, 571–573
  second-law efficiency, 573
Carnot engine, 569–574, 581
Carnot theorem, 569–570
Cathode ray tube, 792, *photo*, 615
Cavendish, Henry, 300, 303, 305, 336
Cavendish experiment, 603
Celsius temperature scale, 488–489
  Celsius-absolute conversion, 492
  Celsius-Fahrenheit conversion, 489
Center of gravity, 278–279 (*see also* Center of mass)
Center of mass, 182, 183–187 (*see also* Center of gravity)
  calculation of, 185–187
  coordinates, 183–184
  definition for continuous systems, 184
  definition for discrete systems, 184
  experimental method for finding, 185, 385
  kinetic energy of, 196
  motion of, 187–190
  pseudowork, 197
  reference frame, 194–195, 200–202
Centigrade scale (*see* Celsius temperature scale)
Central force, 300, 309
Centrifugal force, 122–123
Centripetal acceleration, 65–67, 96, 130, 415
Centripetal force, 96
  friction and, 114
  road banking and, 115
Čerenkov radiation, 468
Chaos, *essay*, 397–402
Characteristic spectrum, 1153
Charge, electric (*see* Electric charge)
Charge separation, 676, 801
Charge sharing, 676
Charging by induction, 601–602
Charles, Jacques, 496
Chromatic aberration, 1034–1035
Chromosome, *photo*, 9
Ciliary muscle, 1041–1042
  photo, 1044
Circuit breaker, 901–902
Circuits (*see* Electric circuits)
Circular motion, 56, 65–69
  centripetal acceleration, 65–67
  with constant speed, 66–67
  period T, 67
  satellite motion, 65–66, 68
  simple harmonic motion and, 376
  with varying speed, 69
Clausius statement of second law of thermodynamics, 568
Clock synchronization, 1113–1118
Coaxial cable
  capacitance of, 693–694
  photo, 693
Coefficient of friction, *table*, 110
Coefficient of kinetic friction, 109–110
Coefficient of performance, 575–576
Coefficient of restitution, 205–206
Coefficient of rolling friction, 109–110
Coefficient of static friction, 108–110, 114
Coefficient of thermal expansion, *table*, 493
Coefficient of viscosity, *table*, 352
Coherence, 1061–1063
Coherent sources, 451, 1062
Cohesive force, 345
Collisions
  elastic, 198–203
  inelastic, 198, 203–206
  in one dimension, 198–206
  in three dimensions, 206–209
  of two galaxies, *photo*, 209
Collision time, 736–737
Coma, 1034
Comet Mrkos, *photo*, 956
Compact disc, *photos*, 984
Compound microscope, 1050–1051, 1084
  angular magnification, 1050
  eyepiece or ocular, 1050
  lateral magnification, 1050
  objective lens, 1050–1051
  tube length, 1050
Compressibility, 337
Compressive strain, 337
Compressive strength, 334, 336
Compressive stress, 335, 337
Compton, Arthur H., 1054
Compton scattering, 1154–1155
Compton wavelength, 1155
Computer hard-disk drive, *photo*, 889
Conduction, electric (*see* Electric conduction)
Conduction, thermal, 525–530
  thermal conductivities, *table*, 525–526
Conduction in nerve cells, *essay*, 740–743
  action potential of nerve, 742
  nerve impulse propagation, 742–743
  resting potential of nerve, 741

Conductivity, 721–722, 735
Conductor, electric, 601–602
  charge and field at surface, 644–648
  charging by induction, 601–602
  earth as, 602, 752
  in electrostatic equilibrium, 644–648
  free electrons in, 601–602, 644, 717–719
Cones and rods, 1041, 1085
Conservation of angular momentum, 244–246
Conservation of charge, 600, 750
Conservation of energy, 155–161, 166–168, 218–220, 750, 1129, 1158 (*see also* Conservation of mechanical energy)
  first law of thermodynamics and, 518, 534
  Lenz's law and, 846–847
Conservation of mechanical energy, 155–161
  applications of, 156–161
Conservation of momentum, 182, 190–194
  magnetic force and, 813–815
Conservative force
  definition, 148
  electric force, 656
  examples, 149–151, 161–165
  potential energy and, 148–151, 152–153
Constraints, 94
Continuity equation for fluids, 347
Convection, 525, 530
  forced, 533
Conversion factors, 5, AP-25–26
Cooling fins, *photos*, 726
Cooper, Leon, 724
Copernicus, 296
Coriolis force, 123–124
Cornea, 1041–1042
Corona discharge, *photo*, 628
Cosmic ray, *photo*, 1126
Cosmology, 1101
Coulomb, Charles Augustin de, 305, 603, 676, 782
Coulomb (C), unit of charge, 600, 826
Coulomb constant, value of, 604
Coulomb's law, 305, 603–607, 625–631, 634, 673, 947
Couple, 284, 798
Cowan, Clyde, 220
Critical angle, 987–988
Critical damping, 391, 911
Critical point, 507
Critical temperature (T_C)
  for liquefying a gas, 507
  for superconductivity, 724
  table, 508
Cross product, 255–260
  definition, 256
  properties, 256–257
Crossed fields, 791
Crossed polarizers, 972, 999, 1000
Curie, Pierre, 886
Curie's law, 886
Curie temperature, 887
Current (*see also* Alternating current; Direct-current circuits; Magnetic field)
  amperian, 880
  displacement, 943–946, 965
  eddy, 852–854, 920
  electric (*see* Electric current)
  generalized, 945–946
  induced, 840, 846–847
  magnetizing, 920–921
  rms, 900–903, 905
Current balance, 827, *photo*, 827
Current element, 786
Current gun, *photo*, 825
Current integrator, 845
Cyclotron, 788, 795–797, *photo*, 796
Cyclotron frequency, 788, 797
Cyclotron period, 788, 796

Dalton's law, 509
Damped oscillations, 388–393
  critical damping, 391
  exponential decrease, 388
  in *LCR* circuit, 910–911
  overdamping, 391
  Q factor, 390–391, 393
  time constant, 390
Davisson, C. J., 975, 1162
DC circuits (*see* Direct-current circuits)
De Broglie, Louis Victor, 1100, 1161, 1164
De Broglie wavelength, 1161–1162, 1164
Deceleration, 34 (*see also* Acceleration)
Decibels (db), unit of intensity level, 447
Decimal system, 3
  prefixes for powers of 10, *table*, 3
De Forest, Lee, 925
Degrees of freedom, 544–545
Density
  definition, 332
  table, 333
  units of, 332
  weight density, 332–333
Density waves, 439, 443
Dentler, John, 929–936
Depth inversion, 1013, 1015
Derivative, AP-15–19
  chain rule for, 143
  definition, 25
  partial, 429
Descartes, René, 986, 992
Deuteron, 1130
Dew point, 510
Diamagnetism, 879, 881, 890–892
  estimating induced magnetic moments, 891–892
  superconductors, 890–891
Dielectric breakdown, 677–679, 697, 848
Dielectric constant, 694–695
  table, 697
Dielectrics, 694–698
  electric field inside, 695
  functions of, 696
  permittivity of, 695
Dielectric strength, 677–679, [illegible]–697
  table, 697
Differential calculus, AP-15–19
Differentiation, 37 (*see also* Derivative)
Diffraction, 974–975, 1076–1089
  central maximum, 1077–1079, 1081
  of circular aperture, *photo*, 1083, 1084
  first minimum, 1079, 1081, 1084
  Fraunhofer, 1082–1084
  Fresnel, 1082–1083, *photos*, 1082, 1083
  gratings, 1085–1089 (*see also* Gratings, diffraction), *photo*, 1087
  intensity for single-slit, 1079–1080
  and interference for two slits, 1080–1081
  of light, *photo*, 1061
  points of zero intensity, 1077–1079
  and resolution, 1083–1085
  secondary maximum, 1080
  single-slit, 1076–1080, *photo*, 1076
  of sound waves, 462
Diffuse reflection, 983, *photo*, 983
Dimensions, 5–6
Diode
  semiconductor, 923
  vacuum tubes, 923
Diopter (D), unit of power of lens, 1033, 1043–1045
Dipole moment (*see also* Magnetic moment)
  electric, 611, 616–617
  induced, 616
  magnetic, 799
  unit for, 617
Direct current, high voltage, 898
Direct-current circuits, 749–771
  Kirchhoff's rules, 750–759
  RC circuits, 760–766
Dispersion, 410, 460–461, 991
  photo, 991
  rainbow, 992–994
Displacement, 21, 23, 25
Displacement antinode, 455
Displacement current, 943–946, 965
Displacement node, 454, 457
Displacement vector, 48–49, 51
Displacement wave, 443
Dissociation energy, 154
Distortion, 1034
Doorbell, *photo*, 820
Doppler effect, 453, 464–469, 1106
  applications, 467
  for electromagnetic waves, 468
  relativistic, 1118–1119, 1121
  for sound waves, 464–469
Dot product, 144–147
Drag forces, 116–117, 351
  damping, 389
  effect of speed, 116
  terminal speed, 116–117
Drift velocity, 717–718, 725, 735–736, 785, 802, 851
Driven oscillations, 392–395
  in *LCR* circuit, 912–919
  resonance, 393–395
  transient and steady-state solutions, 393–394
Drude, P., 735
Dulong–Petit law, 545–546

Earth
electric conductor, 602, 752
natural magnet, 781–782, 784
Earthquake magnitudes and frequencies, *table*, 472
Eddy currents, 852–854, 920
Edison, Thomas, 898, 923
Efficiency
Carnot, 572
of heat engines, 566
second law efficiency, 573
Einstein, Albert, 218, 307, 468, 975, 1100, 1106, 1129, 1157
energy quantization, 1146, 1148–1149
general relativity, 1100–1101, 1132–1135, 1145
photo, 1100
photoelectric effect, 1148–1152
special relativity, 1100–1101, 1106, 1145
train illustration of simultaneity, 1114–1116
Einstein relation, 975, 1148
Einstein's photoelectric equation, 1150
Einstein's postulates of special relativity, 1106–1110, 1123–1124
Elastic collisions, 198–203
center-of-mass reference frame, 200–202
Elastic limit, 333
Electric charge, 599–601
attraction and repulsion, 599
conservation of, 600
coulomb, SI unit of, 600
fundamental unit *e*, 599–600
motion and electric current, 716–720
motion in electric field, 614–615
motion in magnetic field, 787–798
pair production, 600
quantization of, 599
Electric circuits (*see also* Alternating current; Direct-current circuits)
energy in, 725–731
LC, 908–910
LCR, 910–919
LR, 859–862
power loss, 725
RC, 760–766, 918–919
Electric conduction
classical model, 735–737
microscopic picture of, 735–738
quantum mechanical theory, 738
Electric current, 716–748 (*see also* Alternating current; Direct-current circuits)
definition, 716
direction of, 717
formula for, 718
free electrons in conducting wire, 717–719
motion of charges and, 716–720
resistance and Ohm's law, 720–724
SI unit, ampere, 717
Electric dipole, 611, 616–617, 632 (*see also* Dipole moment)
electric field due to, 611
in electric fields, 616–617
potential of, 656
torque on dipole, 616–617
Electric-dipole antenna, 959
Electric dipole moment, 611, 616–617
Electric-dipole radiation, 960
Electric field (**E**), 607–611, 624–655, 625–631
on axis of a disk charge, 629–630
on axis of finite line charge, 625–631
on axis of ring charge, 629
of charged solid sphere, 642–643
at conductor surfaces, 644–648
for continuous charge distributions, 624–649
from Coulomb's law, 625–631
of cylindrical shell of charge, 638–639
definition, 607
of a dipole, 611, 613
direction of, 612–613
discontinuity of, 643–644
electric potential and, 671–673
from Gauss's law, 635–644
inside a conductor, 644, 674
lines, 612–614, 658, 671
in nature, *table*, 608
near infinite line charge, 627–628, 637–638
near infinite plane of charge, 630–631, 636–637
outside a conductor, 646–647
on perpendicular bisector of finite line charge, 626–627
of a point charge, 609, 635–636
rules for drawing electric field lines, 613
SI unit of, 608, 658
of solid cylinder of charge, 639–640
of spherical shell of charge, 640–642
superposition of, 608, 662
of system of point charges, 609–611
work done by, 661
Electric flux
definition, 632–633
Gauss's law and, 634
net, through closed surface, 633–634
units of, 632
Electric force, 604–607, 617 (*see also* Coulomb's law)
action at a distance, 607
conservative, 656
electric field and, 608
superposition of, 606
Electric guitar, pickup, *photo*, 840
Electric potential, 656–689, 721, 725, 726, 731 (*see also* Voltage)
on axis of charged disk, 666–667
on axis of charged ring, 665–666
continuity of *V*, 667–669
for continuous charge distributions, 665–671
due to spherical shell of charge, 669–670
due to system of point charges, 660–664
electric field and, 671–673
near infinite line charge, 670–671
near infinite plane of charge, 667–669
potential difference and, 657–660
zero at infinity, 660–661
Electric shocks, 678, 901
Electrical discharge, *photo*, 624
Electrolysis, 717
Electromagnetic force, 88 (*see also* Electric force; Magnetic force)
Electromagnetic spectrum, 957–962
Electromagnetic waves
circularly polarized, 952
direction of propagation, 951
energy density in, 953–954
intensity of, 953–956
linearly polarized, 951
momentum in, 954–957
radiation pressure, 955–956
speed of, 944, 950
wave equation for, 947–953
Electromigration, *photo*, 735
Electromotive force (*see* Emf)
Electron
charge-to-mass ratio, 614, 792–794
diffraction, 1162–1163
free, in conductor, 601–602, 644, 717–719
magnetic moment, 878, 883
microscope, 1084, 1164
motion in electric field, 614–615
wavelengths, 1162
Electron diffraction, 1162–1163
Electron microscope, 1084, 1164
Electron spin, 782, 878, 882–883
Electron volt (eV), unit of energy, 659–660, 1129
Electron waves, 1161–1165
Electroscope, 601
Electrostatic equilibrium, 644–645
Electrostatic field energy, 701
Electrostatic induction, 602 (*see also* Charging by induction)
Electrostatic potential energy, 658, 664–665, 698–702, 707–708
Electrostatics and xerography, *essay*, 682–684
Elevator problem, 98
Emf
back, 848
and batteries, 726–731
definition, 843
Hall voltage, 802–803
induced, 840, 843–845
motional, 848–852
rms, 901, 903, 905
self-induced, 848
unit of, 726
Emissivity, 530, 533–534
Energy, 135–181
binding, 312, 314, 1130–1131
in capacitor, 698–702
conservation of, 136, 155–161, 166–168, 750, 846–847, 1129, 1158
conversion of, 166–168

dissociation, 154
distribution function, 504–505
in electric circuits, 725–731, 909–910
electrostatic, 658, 664–665, 698–702, 707–708
gravitational, 149–150, 310–315
internal, 518, 536–538
kinetic, 136–147, 195–197, 236–237, 246–247, 500–501, 1126–1132
magnetic, 862–864
mechanical, 155–161
particle systems, 148–151
potential, 136, 148–151, 148–154, 310, 313–315
quantization, 1146, 1148–1149, 1164–1165
relativistic, 1126–1132
rest, 1127–1132
in simple harmonic motion, 377–379
thermal, 136, 139, 166, 525, 530–534, 884–885
transmitted by waves, 417–418, 427
unit of, 136, 169
work-energy theorem, 138, 161–165, 167, 348–349
Energy density
in electromagnetic waves, 953–954
of electrostatic field, 701, 862, 953
magnetic, 863–864, 953
Energy-level diagram, 1159–1160
Entropy
and disorder, 576–582
entropy change, 577–581
and probability, 582–584
statement of second law, 580
Equation of state, 498, 539, 547
Equilibrium
conditions for, 152, 275–278
examples of, 280–283
general motion near, 386–387
neutral, 153, 286
potential energy and, 152–154
of rigid bodies, 275–294
stable, 152–153, 285, 387
static, 92, 275–294
unstable, 153, 285, 387
Equipartition theorem, 544–546, 735, 966–967, 1149
Equipotential surface, 671, 674, 679
Equipotential volume, 674–675
Equivalence principle, 307, 1132–1133
Escape speed, 308, 312–313, 315, 504, 1135
Ether, 1103
Euler's method, 125
Everitt, C.W.F., 964–967
Exponential decrease, 340, 388–389, 761
Exponential growth and decay, *essay*, 586–589
Extraordinary ray, 1001–1002
Eye, 1041–1046
accommodation, 1042
astigmatism, 1042
farsightedness, 1042, 1044
focal length, 1042–1043
near point, 1042, 1044
nearsightedness, 1042
optic nerve, 1041
parts of, 1041
photo, 1044
reading glasses, 1044–1045
resolution of, 1085
rods and cones, 1041, 1085, *photo,* 1044
visible light, 957
Eyepiece, 1050, 1052

Fahrenheit temperature scale, 488–489
Faraday, Michael, 691, 843
dielectrics in capacitors, 694
discovery of diamagnetism, 890
induced emf, 782, 840, 843
influence on Maxwell, 964
Faraday's law, 841, 843–845, 858, 945
eddy currents, 853
generator, 854
Maxwell's equations and, 947
motional emf, 850
wave equation and, 948–950
Farad (F), unit of capacitance, 691
Farsightedness, 1042, 1044
Fermat, Pierre de, 995
Fermat's principle, 995–996
law of reflection, 995
law of refraction, 995–996
Fermi, Enrico, 10, 11–14, 219–220, 1057
Fermi, unit of length, 1057
Fermi-Dirac distribution, 738
Fermi questions, 10, 11–14
Ferromagnetism, 879, 881, 886–889, 892
Curie temperature, 887
definition, 886
hysteresis, 888
magnetically hard material, 889
magnetically soft material, 888
remnant field, 888–889
Fiber optics, 975, 988–989, *photos,* 989
Fictitious forces (*see* Pseudoforces)
Field, concept of, 89
Field, electric (*see* Electric field)
Field emission, 645
Field point, 609
First harmonic (*see* Fundamental mode)
First law of thermodynamics, 518, 534–537
Joule's experiments, 534–535
FitzGerald, George Francis, 1112
Fizeau, A.H.L., 978–979
Fleming, John, 923
Flinner, Jack L., 472
Flip coil, 845
Fluids
Archimedes' principle, 340–343
Bernoulli's equation, 349–350
bulk modulus, 337
capillarity, 344–346
compressibility, 337
continuity equation, 347
density, 331–333
nonviscous, 346
Pascal's principle, 338
Poiseuille's law, 352–353
pressure, 336–340
resistance to flow, 352
Reynolds number, 353–354
surface tension, 344–346
Torricelli's law, 349
turbulence, 346, 347, 353
Venturi effect, 350
viscous flow, 351–354
volume flow rate, 347, 352
Focal length
of camera lens, 1047–1050
of eye lens, 1042–1043
for spherical mirrors, 1017
for thin lens, 1026–1028
Focal plane for thin lens, 1028
Focal point for spherical mirrors, 1017
Foot, definition, 3
Foot-pound (ft.lb), unit of energy, 136
Forbes, James David, 964
Force
action-at-a-distance, 89, 93, 607
adhesive, 345
buoyant, 340–342
central, 300, 309
centrifugal, 122–123
centripetal, 96, 114–115
cohesive, 345
conservative, 148–153, 161–165
contact, 90–92, 93
Coriolis, 123–124
definition, 80
drag, 116–117, 389
due to gravity, 83–85, 88
electric, 604–607, 617, 656
electromagnetic, 88
electromotive (*see* Emf)
frictional, 92 (*see also* Friction)
by magnetic field, 782–787, 813–815
mass and, 80–82
Newton's second law and, 80–82
nonconservative, 151, 162–165, 166
normal, 91
pseudoforce, 121–124
restoring, 90
strong nuclear, 88–89
tension (T), 93, 96–97
tidal, 317
time average of, 110–112
units, 3, 81, 85
vector sum, 80
weak nuclear, 88–89
Force constant, 90, 92
Forced convection, 533
Forced oscillations (*see* Driven oscillations)
Foucault, Jean, 975, 978
Fourier, Jean Baptiste, 458
Fourier analysis, 458, 899 (*see also* Harmonic analysis)
Franklin, Benjamin, 599, 690
Franklin panes, 690
Fraunhofer diffraction, 1082–1084
Free-body diagram, 93, 117

Free expansion, 538, 576, 579, 583
Free fall, 85
French, A. P., 322
Frequency (f)
  beat, 393, 452–453
  cyclotron, 788, 797
  in Einstein relation, 975, 1148
  fundamental, 420–425, 454–455
  LC circuit, 909–910
  resonance, 393, 420, 423–425, 457, 914–918
  for simple harmonic motion, 369–374, 416
  unit of, 369
Fresnel, Augustin, 975, 981, 1082
Fresnel diffraction, 1082–1083
Fresnel lens, *photo*, 1033
Friction, 34, 78, 87, 107–115
  coefficients of, *table*, 110
  damped oscillations and, 388
  kinetic (f_k), 108–110
  rolling, 113
  source of, 107
  static (f_s), 107–112
Friedrich, W., 1153
Fringes, 1064–1071, 1081
Fundamental frequency, 420–425, 454
Fundamental mode, 420–425
Fundamental unit of charge (e), 599–600
Fusion, 830
Fusion, nuclear, 791
Fusion, phase change, 522

Galaxy
  Andromeda, *photo*, 9
  collisions of, *photo*, 209
Galilean transformation, 1107–1108
Galileo, Galilei, 63–64, 78, 307, 324, 517, 976, 1102
Galvani, Luigi, 740
Galvanometer, 767–768, 840
Gamma rays, 957
Gauge number of wire, *table*, 722
Gauss (G), unit of magnetic field, 784
Gaussian surface, 635, 645–646
Gauss's law, 631–635, 828, 945, 947
  calculation of electric field from, 635–644
  Coulomb's law and, 635
  mathematical derivation, 649
  qualitative statement of, 632
Gay-Lussac, Joseph Louis, 496
Geiger, H. W., 1157
Generalized current, 945–946
General relativity, 307, 1100–1101, 1132–1135, 1145
  black holes, 1135
  deflection of light in gravitational field, 1133–1134
  frequencies of light in gravitational field, 1134–1135
  perihelion of orbit of Mercury, 1134
Generator, 841, 854–855, 868, 899, 901, 904
  photos, 855, 856
  Van de Graaff, *photo*, 677
Geometrical optics, 1012–1035
  aberrations, 1034–1035
  images formed by reflection, 983
  images formed by refraction, 1022–1025
  plane mirrors, 1012–1015
  spherical mirrors, 1015–1021
  thin lenses, 1025–1034
Geosynchronous orbit, 304
Germer, L., 975, 1162
Gilbert, William, 781
Global warming, *essay*, 553–556
Gradient, 671
Gram (g), unit of mass, 3
Gratings, diffraction, 1085–1089
  holograms, 1088
  order number, 1086
  resolving power, 1087–1088
  spectroscope, 1086–1088
Gravitation, law, 82
Gravitational constant (G), 299–306
  measurement of, 305–306, *table*, 306
Gravitational field
  of earth, 84, 303, 308
  of solid sphere, 317
  of spherical shell, 316–317, 318–320
Gravitational lensing, *photo*, 1101
Gravitational mass, 307
Gravitational potential, 1134
Gravitational red shift, 1135
Gravity
  acceleration due to (g), 28–29, 58–62, 83–84, 383
  center of, 278–279
  force due to, 83–85, 88
  gravitational field, 84, 303, 308, 316–320
  gravitational potential energy, 149–150, 310–311, 313–315
  Newton's law of, 82, 295, 299–304
"Greenhouse effect," 553–556
Greenler, Robert, 1004–1007
Grid, 925–926
Ground, electrical, 602
Group velocity, 461
Gyroscope, 260–262, *photo*, 1134

Hadron, 88
Hadronic force, 88–89
Half-wave plate, 1002–1003
Hall effect, 801–804, 825
Halley, Edmund, 325
Halley's comet, *photo*, 298
Hall voltage, 802–803
Halo, *photo*, 994
Harmonic analysis, 458–459
Harmonic series, 423
Harmonic sound waves, 442–445
Harmonic synthesis, 458–459
Harmonic waves, 416–417
  addition by phasors, 1071–1075, 1078–1080
  superposition, 419–420
Heat
  caloric theory of, 517–518
  definition, 517
  energy Q, 518–525
  friction and, 517–518
  of fusion, 522–523, *table*, 523
  transfer of thermal energy, 525–534
  of vaporization, 522–523, *table*, 523
Heat capacity, 518–521 (*see also* Specific heat)
  at constant pressure or constant volume, 519
  and equipartition theorem, 542–547
  molar, 519
  table, 519, 543
Heat engines
  efficiency of, 566
  internal-combustion engine, 565–566
  purpose, 566
  and second law of thermodynamics, 564–567
  working substance, 566
Heat pump, 575–576
Heat reservoir, 566
Heisenberg, Werner, 1100
Helmholtz, Herman von, 218
Helmloltz coils, *photos*, 811, 838
Henry, Joseph, 782, 840, 843, 848
Henry (H), unit of inductance, 857
Hertz, Heinrich, 944, 975, 1149
Hertz (Hz), unit of frequency, 369
Holograms, 976, 1088, *photos*, 1088–1089
Hooke, Robert, 974
Hooke's law, 90, 141
  mass on a spring, 369
  strain versus stress, 334
  for torsional stress, 336
Hoover Dam, *photo*, 856
Hopkins, William, 964
Horsepower (hp), unit of power, 169
Hubble Space Telescope, *photo*, 1056
Humidity
  dew point, 510
  relative, 509–510
Hunt, Robert G., 356
Huygens, Christiaan, 974
Huygens' construction (*see* Huygens' principle)
Huygens-Fresnel principle, 981
Huygens' principle, 981, 983–984, 986, 995, 1061
Huygens' wavelets, 984
Hydraulic lift, 338
Hydrogen atom
  Bohr model, 1156–1160
  energy levels, 1159–1160, 1165
  standing-wave problem, 1165
Hydrogen spectrum, 1156
  Balmer series, 1119, 1156, 1160
  Lyman series, 1157, 1160
  Paschen series, 1157, 1160
Hydrostatic paradox, 339
Hysteresis, 888, 920
  curve, 888

- Ice-point temperature, 488
- Ideal gas
  - definition, 497
  - internal energy of, 537
  - isotherms for, 498
  - law, 496–499
  - temperature scale, 491–492
- Ideal-gas law, 496–499
- Ignition coil, *photo*, 920
- Image
  - real, 1015–1016, 1019–1020, 1022
  - by reflection, 983
  - by refraction, 1022–1025
  - virtual, 1012, 1015–1016, 1019–1022
- Image point, 1012–1017
- Imbalance, static and dynamic, 262–263
- Impact parameter, 207
- Impedance, 912–919
  - matching, 729
- Impulse (I), 110–112, 441
  - definition, 110
  - units of, 110
- Incoherent sources, 451
- Index of refraction, 982–983, 985–987, 1034
  - dispersion, 991
  - measurement by interferometer, 1067
  - *table*, 991
- Induced currents, 840, 846–847
- Induced emfs, 840, 843–845, 848
- Inductance
  - mutual inductance, 858–859
  - self inductance, 857–858
  - SI unit of, henry, 857
  - in transformer, 920
- Induction
  - electrostatic, 602
  - magnetic, 840–870
- Inductive reactance, 904, 914, 918–920
- Inductor, 859, 862
  - alternating current in, 902–904
  - energy stored in, 863
  - inductive reactance, 904
  - *photo*, 859
  - power from generator, 904
  - voltage drop versus current, 903, 907
- Inelastic collisions, 198, 203–206
  - center of mass and, 198
  - perfectly, 198, 203
- Inertia, 78 (*see also* Newton's first law)
- Inertial mass, 307
- Inertial reference frame, 80, 121, 1102
- Infrared rainbow, *essay*, 1004–1007
- Infrared rays, 958
- Initial conditions, 37
- Initial-value problems, 37
- Insulator
  - electric, 601
  - thermal, 526–529
- Integrated circuit, *photo*, 907
- Integration, 37–39, AP-19–20
  - numerical, 124–127
- Intensity
  - amplitude and, 446, 1173
  - definition, 445
  - diffraction, 1077–1080
  - of electromagnetic waves, 953–956, 1062
  - intensity level (decibels), 447
  - of reflected light, 982
  - of sounds, *table*, 447
  - threshold of hearing, 447–448
  - units of, 445
- Intensity level
  - pain threshold, 447–448
  - *table*, 447
- Interference (*see also* Superposition)
  - beats, 452–453
  - conditions for, 1063–1064
  - constructive, 419, 449–450, 452, 1062–1063, 1073
  - destructive, 420, 449, 452–453, 1062–1063, 1073
  - and diffraction for two slits, 1080–1081
  - of electromagnetic waves, 975, 1061–1076, 1080–1081
  - fringes, 1064–1067
  - of harmonic waves, 419–420
  - intensity of wave, 1062
  - Newton's rings, 1064
  - optically flat plates, 1065
  - phase difference and coherence, 1061–1063
  - principal maxima, 1074–1076
  - secondary maxima, 1074–1076
  - of sound waves, 449–453
  - in thin films, 1063–1066
  - of three or more sources, 1073–1076
  - two-slit pattern, 1068–1071
- Interferometer (*see* Michelson interferometer)
- Internal energy, 518, 536–538 (*see also* Thermal energy)
  - heat capacity and, 542–546
  - of ideal gas, 537–538
- Internal resistance of battery, 728–729
- International Bureau of Weights and Measures, 81, 491
- Inverse-square laws, 604
- Io, eclipse of, 977
- Ion, 601
- Ionization, 1160
- Irreversible process, 563–564, 569, 573, 576–577, 580, 583
- Isobaric expansion, 540
- Isotherm, 498, 506–507
- Isothermal expansion, 540
- Isotopes, measurement of masses, 794
- Isotropic materials, 1001

- Jet propulsion, 213–216
  - rocket equation, 213–214
  - thrust of the rocket, 213–215
- Joule, James, 518
- Joule heat, 726, 729, 751, 765–766, 847, 851, 898, 901, 917, 920
- Joule (J), unit of energy, 136
- Junction rule, 750, 753–755, 758

- Kelvin, Lord (*see* Thomson, William)
- Kelvin (K), unit of temperature, 2, 491
- Kelvin-Planck statement of second law of thermodynamics, 567
- Kelvin temperature scale, 491–492
- Kepler, Johannes, 297, 299, 302
- Kepler's laws, 295–299, 300–301
- Kilogram (kg), unit of mass, 2–3
- Kilowatt-hour (kW-h), unit of energy, 169
- Kinetic energy (*K*)
  - absolute temperature and, 500–501
  - angular momentum and, 246–247
  - definition, 137, 1126
  - motion in one dimension, 136–140
  - motion in three dimensions, 143–147
  - relativity and, 1126–1132
  - of rotation, 236–237
  - of a system of particles, 195–197
  - work and, 136–143, 147
- Kinetic friction (f_k), 107–109
  - coefficient of, 109–110
  - nonconservative force, 162–165
- Kinetic theory of gases
  - assumptions, 500
  - distribution of molecular speeds, 502–505
  - kinetic energy and absolute temperature, 501
  - microscopic theory, 500
- Kirchhoff, Gustav Robert, 975, 981
- Kirchhoff's rules, 750–759
  - analysis of circuits by symmetry, 758–759
  - junction rule, 750, 753–755, 758
  - loop rule, 750–751, 754–757, 760, 762, 764, 860–861, 899, 903, 908–913, 920
  - multiloop circuits, 753–758
- Knipping, P., 1153
- Kohlrausch, G., 966
- Krenz, Jerrold H., 553
- Krytron, *photo*, 766

- Laminar flow, 353
- Land, E. H., 997
- Laser, 976, 1062, 1088, *photos*, 976–977
- Latent heat
  - of fusion (L_f), *table*, 523
  - of vaporization (L_v), *table*, 523
- Lateral magnification, 1018–1020, 1050
- Lattice, 601
- Laue, M., 1153
- Law of equal areas, 297–298, 301 (*see also* Kepler's laws)
- Law of reflection, 982, 995, 1013
- Law of refraction (*see* Snell's law)
- Lawrence, E. O., 795
- *LC* circuit
  - analogy of mass on spring, 908–909
  - electric and magnetic energy, 909–910
  - frequency, 909–910

LCR circuit
analogy of damped harmonic oscillator, 911
analogy of forced oscillator, 912
critical damping value, 911
with generator, 912–919
impedance, 912–919
parallel, 919
phasor diagram, 913
power factor, 914–915
Q factor, 915–917
resonance, 914–918
series, 912–913, 916
total reactance, 912–913
without generator, 910–911
Lenard, P. E., 1149
Length contraction, 1111–1113, 1115
Lenses
converging, 1027, 1029, 1044, 1046
diverging, 1027, 1029
of eyes, 1041–1045, 1085
lens-maker's equation, 1026, 1029
negative, 1027
positive, 1027
power of, diopter, 1031
principal rays for negative, 1029
principal rays for positive, 1029
ray diagrams for, 1029–1031
thin, 1025–1034
thin-lens equation, 1026, 1044–1045
Lens-maker's equation, 1026, 1029
Lenz's law, 846–848, 853, 861, 890
Lever arm, 231
Levitation, *photo*, 597, 781
Leyden jar, 690 (*see also* Capacitors)
Light, 974–1089 (*see also* Diffraction; Geometrical optics; Interference; Optical instruments)
diffraction, 974–975, 1076–1089
electromagnetic spectrum, 958
Huygens' principle, 981
interference, 975, 1061–1076, 1080–1081
polarization, 997–1003
propagation of, 981
reflection, 982–984
scattering, 1000
speed of, 2, 944, 950, 975–980, 1102–1103
wave versus particle, 974–975
Light-minute (c-min), unit of distance, 980
Lightning, *photo*, 678
Lightning rod, 679, *photo*, 603
Light-year (c-year), unit of distance, 980
Linear charge density, 625
Linear motion, comparison with rotational, *table*, 245
Line spectrum, *photos*, 1145
Lines of force
electric, 612–614, 658, 671
gravitational, 614
Lines of magnetic field, 786, 818–819
Liquid crystal
coating on resistor, *photo*, 726
display, *photo*, 1001
Liquid crystal display, *photo*, 1001
Liter (L), unit of volume, 332
Livingston, M. S., 795
Llewellyn, Ralph A., 218
Lloyd's mirror, 1062, 1070
Load resistance, 729, 920
Loadstone, *photo*, 886
Longitudinal waves, 412, 439
Loop antenna, 961
Loop rule, 750–751, 754–757, 760, 762, 764, 860–861, 899, 903, 908–913, 920
Lorentz, Hendrik A., 735, 1112
Lorentz-FitzGerald contraction, 1112
Lorentz transformation, 1107–1113, 1116, 1122
length contraction, 1111–1113, 1115
time dilation, 1109–1111, 1116
Loudness, 447–448
Low pass filter, 918, 925
LR circuits, 859–862
heat production, 862
time constant, 860–862
Lyman, T., 1160
Lyman series, 1157, 1160

Mach number, 468
Macroscopic variables, 500
Maglev train, *photo*, 781
Magnetically hard material, 889
Magnetic bottle, 790
Magnetic brake, 853–854
Magnetic dipole moment (*see* Magnetic moment)
Magnetic domain, 886–887, *photo*, 878
Magnetic energy, 862–864
density, 863–864, 953
stored in inductor, 863
Magnetic field (**B**), 781–804
cgs unit, gauss, 784
of currents, 815–825
cyclotron frequency, 788, 797
cyclotron period, 788–789, 796
definition, 783
directional aspects, 815–816
due to a current loop, 816–88
due to current in a solenoid, 819–821
due to current in straight wire, 822–825
of earth, 781–782, 784–785
field lines, 786
force exerted by, 782–787
Hall effect, 801–804
magnetic-dipole field, 818
magnetic force and conservation of momentum, 813–815
motion of point charge in, 787–798
of moving point charges, 812–815
nonuniform, 790–791
right-hand rule, 783, 823
SI unit, tesla, 783–784
sources of, 811–832
torque on current loops and magnets, 798–801
velocity selector, 791–795
Magnetic flux
definition, 841–842
Faraday's law and, 843–845
Lenz's law and, 846–848
unit of, weber, 841
Magnetic force
on current-carrying wire, 785–787
on a current element, 786
on moving charge, 783, 813–815
Magnetic induction, 840–870
eddy currents, 852–854
Faraday's law, 841, 843–845, 850, 853, 858
generators and motors, 854–856
inductance, 857–859
Lenz's law, 846–848
LR circuits, 859–862
magnetic energy, 862–864
magnetic flux, 841–842
motional emf, 848–852
Magnetic moment (*see also* Magnetization)
angular momentum and, 882
atomic, 878, 882–884 (*see also* Atomic magnetic moments)
of current loop, 799, 818
definition, 799
of electron, 883
induced, 846, 891–892
of a magnet, 799–801
SI unit of, 799
torque and, 799, 818
Magnetic pickup, *photo*, 840
Magnetic poles, 781–782, 786
Magnetic pole strength, 799–800
Magnetic recording, *photo*, 887
Magnetic susceptibility, 881
diamagnetic materials, 890
ferromagnetic materials, 886
paramagnetic materials, 884
table, 881
Magnetism, theoretical model of, 782, 879
Magnetite, 781, *photo*, 886
Magnetization (**M**), 879–881, 886, 888–889
Magnetizing current, 920–921
Magnetron, *photo*, 962
Magnification, 1021, 1023–1024
angular, 1046–1047, 1050, 1052
lateral, 1018–1020, 1050
simple magnifier, 1046–1047
Magnifying power
of lens, 1046
of microscope, 1050
of telescope, 1052
Malus, E. L., 998
Malus's law, 998
Manometer, open-tube, 339
Maricourt, Pierre de, 781
Marsden, E., 1157
Mass density, 410, 414–415
Mass (*m*)
center of, 182–187
definition, 81
energy and, 1129
gravitational versus inertial, 1132
Newton's second law and, 80–82

of a planet, calculation, 302–303
relativistic, 1125
standard of, 2, 81
units of, 85
Mass spectrometer, 788, 794–795, *photo*, 794
Mathematical symbols, *table*, AP-1
Mathematics, review of, AP-1–20
Maxwell, James Clerk, 782, 832, 943 (*see also* Maxwell's equations)
equipartition theorem, 966–967
essay, 964–967
portrait, 967
Treatise on Electricity and Magnetism, 966
wife as co-experimenter, 966
Maxwell, Katherine Mary, 966–967
Maxwell–Boltzmann distribution, 738, 967
definition, 503
energy distribution function, 504–505
most probable speed, 503
Maxwell demon, 966
Maxwell's displacement current, 943–946, 965
Maxwell's equations, 943, 947, 965, 975, 1103
May, Robert, 398
Mean free path, 736–738
Mean speed, 738
Mechanical energy (*see also* Energy)
conservation of, 155–161
total, 155
Mechanical equivalence of heat, 535
Mega electron volt (MeV), unit of energy, 1129
Melting point (*see* Normal melting point)
Meter (m), definition, 2, 979
Metric system, 3
Michell, John, 305, 781
Michelson, Albert, 978, 1067, 1103–1105
Michelson interferometer, 1062, 1066–1067, 1104
Michelson–Morley experiment, 1103–1106, 1107, 1112
Microphone, *photo*, 770
Microprocessor, *photo*, 906
Microscope (*see also* Compound microscope)
electron, 1084, 1164
near-field, *photo*, 1051
ultraviolet, 1084
Microwaves, 957–958, 962
microwave oven, 617
Millibar, unit of pressure, 339
Millikan, Robert A., 966, 1100, 1150
Mirage, 990, *photo*, 990
Mirror equation, 1016–1017
Mirrors (*see also* Plane mirrors; Spherical mirrors)
lateral magnification, 1018–1020
parabolic, 1034
principal rays for, 1018
ray diagrams for, 1018–1020
sign conventions for reflection, 1019–1020
Moiré pattern, 453
Molar heat capacitites
for gases, *table*, 543
for solids and liquids, *table*, 519
Molar mass, 498
Molecular mass, 498
Molecular weight, 498
Mole (mol), 2, 497
Moment of inertia (I), 232–235, 243
calculations, 238–242
for physical pendulum, 385
table, 233
Momentum, 182
conservation, 182, 190–194
definition, 182
in electromagnetic waves, 954–957
relativistic, 1124–1125, 1127–1128, 1131–1132
total, of a system, 191
Moon, measurement of distance to, 1015
Morley, Edward M., 1067, 1103, 1105
Motion
comparison of linear and rotational, *table*, 245
with constant acceleration, 30–36
in one dimension, 20–46
projectile, 58–65
in two and three dimensions, 47–76
Motional emf, 848–852
definition, 849
electron in moving rod, 849–851
Faraday's law and, 850
Motors, electric, 855–856
armature, 931, 933–934
commutator brush assemblies, 931
dc and ac, 935
essay, 929–936
induction motor, 935
linear machine, 929–930, 935
performance characteristic, 930, 934, 936
rotating motor, 930–931
shaded-pole motors, 935–936
universal motor, 935
velocity-versus-load equation, 930, 934
Multimeter, 767, *photo*, 767
Muon, 1112–1113
Mutual inductance, 858–859

Natural frequency, 392–395, 423, 914 (*see also* Resonance frequency)
Near point, 1042, 1044
Nearsightedness, 1042
Neutrino, 1128
discovery of, *essay*, 218–220
Neutron diffraction, 1163
Newton, Isaac, 65, 77, 299–303, 307, 517, 1102
development of calculus, 316
essay, 322–325
Opticks, 323, 325
particle theory of light, 974–975
Principia, 323–325
System of the World, 68
Newtonian relativity, 1101–1102, 1106
Newton (N), unit of force, 3, 81, 85
Newton's first law of motion, law of inertia, 77, 78–80, 1101
reference frame, 79–80
Newton's law of cooling, 533
Newton's law of gravity, 82, 295, 299–304, 604
Newton's laws of motion, applications to problem solving, 92–99
Newton's rings, 1064
Newton's second law of motion, 78, 80–83, 120, 137, 500, 1158
force and, 80–82
mass and, 80–82
pseudoforces and, 121–123
for rotation, 233, 243, 259
for a system, 188, 197
Newton's third law of motion, law of interaction, 78, 86–87, 500
magnetic force and, 813–815
Nickles, Elizabeth Pflegl, 740–743
Nobel prize, 724, 804, 1149, 1152
Node, 420–421, 451
Nonconservative force, 151, 161–165
deformations, 166
friction, 162–165, 166
Nondispersive medium, 461
Noninertial reference frame, 121–124
Nonohmic materials, 721
Nonparaxial rays, 1016
Nonpolar molecules, 616
Normal boiling point (BP), *table*, 523
of water, 488, 507, 524 (*see also* Steam-point temperature)
Normal force, 91
friction and, 108
Normal freezing point of water (*see* Ice-point temperature)
Normal melting point (MP), *table*, 523
Nuclear fusion, 791
Nucleus, 1057
Numerical methods, 124–127
Euler's method, 125
numerical integration, 125–127
Nutation, 262

Objective lens, 1050–1052, 1054
Ocular, 1047, 1050
Oersted, Hans Christian, 782, 811, 826
Oersted's experiment, *photo*, 815
Ohm, Georg Simon, 722
Ohm (Ω), unit of resistance, 720, 804
Ohmic materials, 721
Ohmmeter, 767, 769–770
Ohm's law, 720–723, 735, 737
Oil-drop experiment, 966
Onnes, H. Kamerlingh, 724
Oppenheimer, J. Robert, 1135
Optical fibers (*see* Fiber optics)
Optical instruments, 1041–1056 (*see also* Eye)
camera, 1047–1050
compound microscope, 1050–1051

resolving power of, 1084
simple magnifier, 1046–1047
telescope, 1052–1056
Optics
geometrical (*see* Geometrical optics)
physical (*see* Diffraction; Interference)
Order number, 1086
Orders of magnitude
of lengths, *table*, 9
of masses, *table*, 9
of time intervals, *table*, 10
Ordinary ray, 1001–1002
Organ pipe, 456–457
Oscillations, 368–408
damped, 388–391
driven, 392–395
in *LC* circuits, 908–910
in *LCR* circuits, 910–919
mass on spring, 369–375
motion near equilibrium, 386–387
physical pendulum, 384–385
resonance, 392–395
simple harmonic motion, 369–379
simple pendulum, 382–384
torsional pendulum, 385–386
vertical spring, 380–381
Oscilloscope, 614
Overtone, 423

Pair production, 598
Parabolic mirror, 1034
Parallel-axis theorem, 240–241
Parallel resistors, 732–734
Paramagnetism, 879, 881, 884–886, 887, 892
Curie's law, 886
definition, 884
temperature dependence, 884–886
Paraxial rays, 1016
Partial pressures, 509
Particle theory of light, 974–975
Pascal, Blaise, 338
Pascal's principle, 338
Paschen series, 1157, 1160
Pauli, Wolfgang, 219–220
Pendulum
acceleration due to gravity and, 383
angular displacement, 383
ballistic, 204
Newton's experiments, 307
period, 383–386
physical, 384–385
simple, 382–384
torsional, 385–386
Performance, coefficient of, 575–576
Perihelion, 298, 1134
Periodic table, AP-27
Period (*T*)
for circular motion, 67, 882
of cyclotron, 788, 796
of pendulum, 383–386
of planets, 297–299, 301–302
for simple harmonic motion, 369–374, 416–417
Permeability, magnetic, 888 (*see also* Relative permeability, magnetic)
Permeability of free space, 812, 827, 944, 1103
Permittivity of a dielectric, 695
Permittivity of free space, 634–635, 692, 944, 979, 1103
Perpendicular-axis theorem, 241–242
Pfund, H. A., 1160
Phase, of simple harmonic motion, 370–372
Phase change, 522–524
Phase constant, 370–372, 375
Phase diagrams, 508–509
Phase difference
coherence and, 1061–1063
for sound waves, 449–451
for waves on a string, 419–420
Phase lag, 985
Phase velocity, 461
Phasors, 907–908
addition of harmonic waves, 1071–1075, 1078–1080
diagram for *LCR* circuits, 913, 918–919
diagram for *RC* circuit, 919
Photoconductivity, 683–684
Photoelectric effect, 975, 1148–1152
Einstein's equation, 1150, 1152
Millikan experiments, 1150
stopping potential, 1149, 1152
threshold frequency, 1150
threshold wavelength, 1150, 1152
work function, 1150, 1152
Photomultiplier, *photos*, 1151
Photon, 975, 1128, 1150
Compton scattering, 1154–1155
energy, 1149, 1151–1152
speed of light, 1123
Photo-voltaic cells, *photo*, 771
Physical constants, *table*, AP-23
Physical pendulum, 384–385
PIN diode, *photo*, 925
Pitch, 458
Planck, Max, 532, 1100, 1146, 1148–1149, 1157, *photo*, 1147
Planck's constant, 883, 975, 1148, 1158
Plane mirrors, 1012–1015, 1020
depth inversion, 1013, 1015
image point, 1012–1017
multiple images, 1014–1015
virtual image, 1012, 1015–1016, 1019–1022
Plane wave, 461, 948
Plasma, 791
Plasma globe, *photo*, 719
Plate current, 924
"Plum pudding" model of atom, 1157
Poise, unit of viscosity, 352
Poiseuille's law, 352–353
Poisson, S. D., 1082
Polarization of dielectric, 694, 697
Polarization of light, 997–1003
by absorption, 997–999
analyzer, 998
by birefringence, 1001–1003
circular, 997
linear, 997
polarizer, 998
by reflection, 1000–1001
by scattering, 1000
transmission axis, 997
Polar molecules, 616–617
Polaroid, 997
Pole strength (*see* Magnetic pole strength)
Pollution, 575
Potential, electric (*see* Electric potential)
Potential difference, 667
definition, 657–658, 698
electric potential and, 657–660
SI unit for, 658
Potential energy (*U*), 148–151
conservative force and, 148–151, 152–153
definition, 148–149
electrostatic, 658, 664–665, 698–702, 707–708
equilibrium in one dimension and, 152–154
gravitational, 149–150, 310, 313–315
Pound (lb), unit of force, 3
definition, 85
Power, 169–170
average, delivered by generator, 901
definition, 169
dissipated in resistor, 900
of *LCR* circuit, 914–915
of lens, 1033
loss in electric circuit, 725, 733
for rotation, 236
transmitted by waves, 418
units of, 3, 169
Power conversion, *photo*, 898
Powers of 10, prefixes for, *table*, 3
Poynting, John, 954
Poynting vector, 954
Precession, 261–262
Pressure
absolute, 339
altitude versus, 340
atmospheric, 339
bulk modulus, 337
definition, 336
gauge pressure, 339
partial, 509
Pascal's principle, 338
radiation, 955–956
units of, 336, 339
Pressure waves, 439, 443, 454
Printed circuit, *photo*, 749
Prism, 988
Probability and entropy, 582–584
Projectile motion, 58–65
optimum angle of projection, 63
parabolic path, 58–64
range, 60–63
Proper length, 115, 1111–1112
Proper time, 1109, 1111, 1113
Proton diffraction, 1163
Pseudoforces, 121–124
centrifugal force, 122–123
Coriolis force, 123–124

noninertial frames, 122
Pseudowork, 197
Ptolemy, 296
Pupil, of eye, 1041–1043
PV isothermal curves, 506–507, 539–541

Q factor
damped oscillations, 390–391, 393
for *LCR* circuit, 915–917
resonance and, 393
Quality factor (*see Q* factor)
Quanta, 1148
Quantization
of angular momentum, 1159, 1164
of atomic energies, 1156–1160
of electric charge, 599
of energy, 532, 1146, 1148–1149, 1164–1165
standing-wave condition, 1164–1165
Quantum Hall effect, 803–804 (*see also* Hall effect)
fractional, 804
Quantum mechanics (*see* Quantum theory)
Quantum theory, 1100
blackbody radiation, 1147–1148
Bohr model, 1156–1160
Compton scattering, 1154–1155
dates of experiments and theories, *table*, 1146
of electric conduction, 738
electron waves, 1161–1165
emission and absorption of light, 975
photoelectric effect, 975, 1148–1152
quantization of atomic energies, 1156–1160
superconductivity, 724
Quarter-wave plate, 1002
Quasi-static process, 539
adiabatic compression of a gas, 549, 572
adiabatic expansion of a gas, 547–549, 572

Radar image, *photo*, 952
Radiation, thermal, 525, 530–534
blackbody radiation, 531–533, 1147–1148
power radiated, 531
Radiation pressure, 955–956, *photo*, 956
Radioactivity, 1100
Radio, *photo*, 916
Radio receiver, 916
Radio telescope, *photo*, 943, 959
Radio waves, 957, 959
Rainbow, 992–994
angle of minimum deviation, 993–994
angular radius of, 993–994
"Beyond the Visible Rainbow", *essay*, 1004–1007
infrared, 1004–1007
photo, 974
secondary rainbow, 994
supernumerary bows, 1005
Ray approximation, 463, 976, 1012
Ray diagrams
for lenses, 1029–1031
for mirrors, 1018–1020
Rayleigh–Jeans law, 1147–1148
Rayleigh's criterion for resolution, 1084, 1087
Rays, 445
RC circuits, 760–766
charging a capacitor, 762–766
discharging a capacitor, 760–762, 765–766
low pass filter, 918
phasor diagram for, 919
time constant, 761, 763–764
Real image, 1015–1016, 1019–1020, 1022
Rectification, 923–925
ac-dc converters, 923
full-wave, 924
half-wave, 924
ripple, 925
Red shift, 1119
gravitational, 1135
Reference frame, 79–80
center-of-mass, 194–195, 200–202
inertial, 80, 121, 1102
noninertial, 121–122
rotating, 122–124
zero-momentum, 194–195
Reflection
angle of incidence, 982
diffuse, 983, *photo*, 983
Fermat's principle, 995
Huygens' principle, 983–984
law of, 982–983
of light, 982–984
plane of incidence, 982
reflected intensity, 982
reflecting telescope, 1054–1055
of sound waves, 461, 463
specular, 983, *photo*, 982, 985, 1012, 1013, 1015, 1017, 1020, 1021
total internal, 987–989, *photo*, 987, 988, 989
Refraction
angle of, 985–987
angle of incidence, 985
apparent depth of object, 1024
definition, 985
dispersion, 991
double (*see* Birefringence)
Fermat's principle, 995–996
images formed by, 1022–1025
index of, 982–983, 985–987, 991
mirage, 990
photo, 985, 1012
rainbow, 992–994, 1004–1007
refracted ray, 985
refracting telescope, 1054
Snell's law, 986
of sound waves, 461
Refrigerators and second law of thermodynamics, 568
Reines, Frederick, 220
Relative permeability, magnetic, 888–889
table, 889
Relativistic energy, 1126–1132
Relativistic mass, 1125
Relativistic momentum, 1124–1125, 1127–1128, 1131–1132
Relativistic transformation equation, 1108–1109 (*see also* Lorentz transformation)
Relativity, 1100–1135
clock synchronization and simultaneity, 1113–1118
doppler effect for light, 1118–1119, 1121
Einstein's postulates, 1106–1110
general theory, 1100–1101, 1132–1135
Lorentz transformation, 1107–1113, 1116
Michelson-Morley experiment, 1103–1107, 1112
Newtonian, 1101–1102
relativistic energy, 1126–1132
relativistic momentum, 1124–1125, 1127–1128, 1131–1132
special theory, 1100–1101, 1106, 1121
twin paradox, 1119–1121
velocity transformation, 1122–1124
Remnant field, 888–889
Resistance (*R*)
electrical, definition, 720
equivalent for parallel resistors, 732–734, 769
equivalent for series resistors, 731
to flow, 352
load, 920
SI unit, ohm, 720, 804
thermal, 526–529
Resistivity, 721–723, 735–738
temperature coefficients, *table*, 722
Resistors, 723
alternating current in, 899–902
on circuit board, *photo*, 723
color coded, *photo*, 723
in parallel, 732–734
in series, 731
shunt, 768
Resolution
diffraction and, 1083–1085
of the eye, 1085
Rayleigh's criterion, 1084, 1087
Resonance, 393–395
condition, 423, 914
curves, 393, 915
frequency, 393, 420, 423–425, 457, 914–918 (*see also* Natural frequency)
in *LCR* circuits, 914–918, 919
Q factor and, 393
standing waves and, 422–423
width, 915, 917
Resonance condition, 423, 914
Resonance frequency, 393, 420, 423–425, 457, 914–918 (*see also* Natural frequency)

Rest energy, 1127–1132
  table, 1130
Restitution, coefficient of, 205–206
Rest mass, 1125, 1129, 1131
Restoring force, 90
Retarding force, 116
Retina, 1041–1043, 1085
Reversibility of waves, 1018
Reversible process, 569–570, 579
Reynolds number (N_R), 353–354, 357
R factor, *table*, 527 (*see also* Thermal resistance)
Right-hand rule, 255, 783, 823
Rigid body, 228
Ripple, 925
Ripple tank, 461
Ritz, Walter, 1056
Rms values
  rms current, 900–903, 905
  rms emf, 901, 903, 905
  rms speed, 502–504
Rocket equation, 213–214
Rods and cones, 1041, 1085
Rolling condition, 249–254
Rolling friction, coefficient of, 113
Rolling objects, 249–254
  ball down an incline, 252–254
  billiards, 251–252
  bowling ball, 250–251
Römer, Ole, 977–978
Röntgen, W., 1152
Root mean square values (*see* Rms values)
Rotating reference frame, 122–124
Rotation, 227–274
  angular momentum, 242–249
  angular velocity and acceleration, 228–231
  centripetal acceleration, 230
  comparison with linear motion, *table*, 245
  equations, *table*, 245
  gyroscope, 260–262
  imbalance, static and dynamic, 262–263
  kinetic energy, 236–237
  moment of inertia, 231–235, 238–242
  Newton's second law for, 233, 243, 259
  power, 236
  rolling objects, 249–254
  torque, 231–235
  vector nature of, 254–260
Rutherford, Ernest, 218, 975, 1100,1157
Rydberg, Johannes R., 1056
Rydberg constant, 1056, 1059–1060
Rydberg–Ritz formula, 1056–1059

Satellite motion, 65–66, 68, 304, 314–315
Saturation magnetic field, 888
  table, 889
Saturation magnetization, 883–884
Saturn, rings of, *photo*, 65, 317
Savart, Felix, 811, 823
Scalar, 47
Scalar product, 144
Scattering of light, 1000
Schrieffer, Robert, 724
Schrödinger, Erwin, 975, 1100, 1165
Schwarzschild radius, 1135
Scientific notation, 6–7
Second-law efficiency, 573
Second law of thermodynamics, 563–584
  Carnot engine, 569–574
  Clausius statement of, 568
  entropy and disorder, 576–582
  entropy and probability, 582–584
  heat engines and, 564–567
  Kelvin-Planck statement of, 567
  refrigerators and, 568
Second (s), unit of time, 2
Seismic waves, *essay*, 472–476
  reflection seismology, 474–476
Self-induced emf, 848
Self inductance, 857–858
  definition, 857
  of a solenoid, 857
Semiconductor, 923
Sensory receptor cells, *photos*, 770–771
Series resistors, 731
Sextupole magnet, *photo*, 821
Shear force, 336
Shear modulus (M_S), 336
  table, 336
Shear strain, 336
Shear stress, 336
Sherwin, Chalmers, 220
Shock waves, 468
  photo, 469
Shunt resistor, 768
Sign conventions
  for reflection, 1019–1020
  for refraction, 1023
Significant figures, 7–8
Silicon, *photo*, 1098
Simple harmonic motion
  amplitude, 370–375, 377–379
  angular frequency, 370–373
  circular motion and, 376
  conditions for, 369, 383
  definition, 369–370
  energy in, 377–379
  frequency, 369–374
  mass on a spring, 369–375
  period, 369–374, 416–417
  phase, 370
  sinusoidal curve, 369
Simple magnifier, 1046–1047
Simple pendulum, 382–384
Simultaneity, 1113–1118
SI units, 2–3, 21, 85, AP-21
Sliding friction (*see* Kinetic friction)
Slope, 23–28, 30, 33
Slug, unit of mass, 85
Snell, Willebrod, 986
Snell's law, 986, 993, 995–996, 1000, 1022–1023
Snyder, 1135
Solar flare, *photo*, 790
Solenoid
  in doorbell, 820
  energy stored in, 863
  magnetic field of, 819–821, 830–831
  magnetic flux through, 842
  photo, 859
  self inductance, 857
Solid angle, 649
Solids, 331–336
  bulk modulus, *table*, 337
  density, 331–333
  shear modulus, *table*, 336
  stress and strain, 331–336
  Young's modulus, *table*, 334
Sonogram, 463
Sound waves
  beats, 452–453
  coherent sources, 451
  diffraction, 462–463
  doppler effect, 464–469
  harmonic, 442–445
  harmonic analysis and synthesis, 458–459
  incoherent sources, 451
  intensity, 445–448
  intensity level, *table*, 447
  interference, 449–451
  loudness, 447–448
  reflection, 461–462
  refraction, 461
  shock, 468, *photo*, 469
  speed of, 440–442, 456
  standing, 454–457
  in three dimensions, 445–448
  waveform, 458
  wave functions for, 439
  wave packets and dispersion, 460–461
Space shuttle, *photo*, 68, 117
Speaker, *photo*, 820
Special theory of relativity, 1100–1101, 1106, 1121, 1145
Specific gravity, 332, 341
Specific heat, 518–521 (*see also* Heat capacity)
  at constant pressure or constant volume, 519
  table, 519
Spectral distribution curves, 1147
Spectral distribution function, 1147
Spectral line, 1087
Spectroscope, 1086–1088, 1156, *photos*, 1086
Spectrum, line, *photos*, 1145
Specular reflection, 983
Speed (*see also* Velocity)
  angular, 229
  average, 21
  of electromagnetic waves, 944, 950, 1026
  escape, 308, 312–313, 315, 504, 1135
  instantaneous, 25
  of light, 2, 944, 950, 975–980, 1102–1103
  mean, 738
  of molecules, distribution, 502–505

root mean square (rms), 502–505
sound waves, 440–442, 456
terminal, 116–117
units of, 2, 21
of waves on a string, 414–415
Spherical aberration, 1016, 1034
Spherical mirrors, 1015–1021
concave, 1015–1017
convex, 1017, 1020–1021
focal length, 1017
mirror equation, 1016–1017
nonparaxial rays, 1016
paraxial rays, 1016
real image, 1015–1016, 1019–1020
spherical aberration, 1016
Springs, 90–92, 369–375, 380–381
potential energy, 377–379
total energy, 377–379
vertical, 380–381
Square-wave generator, 459
Standard conditions, 332, 498
Standing waves, 420–427
condition for, 421–422, 425, 426, 455
energy quantization and, 1164–1165
functions, 426–427
fundamental frequency, 420–425, 454–455
for hydrogen atom, 1165
nodes, antinodes, 420–421, 425
resonance condition, 423
resonance frequencies, 420, 423–425, 457
sound, 454–457
superposition, 427–428
State of the system, 537, 540
Static equilibrium, 92
Static friction (f_s), 107–112, 120
coefficient of, 108–110, 114
maximum, 108
normal force and, 108
Stationary states, 1158
Statistical mechanics, 503
Steady-state solution, 393
Steam-point temperature, 488 (*see also* Normal boiling point, of water)
Stefan, Josef, 530
Stefan-Boltzmann law, 530, 1147
Stefan's constant, 530, 553
Steradian, unit of solid angle, 649
Stevin, Simon, 307
Stokes, G. G., 964–965
Stokes' theorem, 964–965
Stopping distance, 33–35
Stopping potential, 1149, 1152
Strain, 333–336
Strain gauge, *photo,* 771
Strengths of materials and Young's modulus, *table,* 334
Stress, 333–336
Stress pattern, *photo,* 90, 972, 999
Strong nuclear force, 88–89
Sublimation, 508, 522
Sun dog, *photo,* 994
Superconducting magnet, *photo,* 821
Superconductivity
BCS theory of, 724
critical temperature, 724
diamagnetism and, 890–891
high-temperature superconductors, 724
quantum mechanics and, 724
Superconductor, *photo,* 597, 724, 891
Supernova explosion, 980
Superposition, 413, 431
of electric field, 608, 662
of electric forces, 606
of harmonic waves, 419–420 (*see also* Interference)
of standing waves, 427–428
Surface charge density, 624
Surface tension, 344–346
Symmetry axis, 258
Synchronization of clocks, 1113–1118
Synchrotron radiation, *photo,* 962
Systems of particles, 182–226

Tables, list of, AP-24
Tangential acceleration, 230
Telescope, 1052–1056, *photos,* 1053, 1054
astronomical, 1052–1056, 1084
eyepiece, 1052
Hubble Space Telescope, *photo,* 1056
Keck, *photo,* 1054
light-gathering power, 1054
Mt. Palomar, 1055
objective lens, 1052, 1054
reflecting, 1054–1055
refracting, 1054
resolving power, 1084
Yerkes Observatory, 1054
Television antennas, *photo,* 959
Television picture tube, 614
Temperature, 486–516
absolute or Kelvin scale, 491–492, 572
Celsius and Fahrenheit scales, 487–489
critical, 506–508, *table,* 508
ideal-gas law, 496–499
ideal-gas scale, 491–492, 572
molecular interpretation of, 500–505
Temperature gradient, 525
Tensile strength, 334, 336
Tensile stress (*see* Stress)
Tension (T), 93, 96–97, 111–112, 118–119, 382, 414–415
Terminal speed, 116–117
Terminal voltage, 727–729, 752
Terrestrial data, *table,* AP-22
Tesla, Nikola, 898
Tesla (T), unit of magnetic field, 783–784
Tesla coil, *photo,* 859
Test charge, 607–608
Thermal conductivity (k), 525–530 (*see also* R factor)
table, 526
Thermal energy, 487, 518, 884–885 (*see also* Internal energy)
conduction, 525–530
convection, 525, 530
radiation, 525, 530–534
transfer of, 525–534
Thermal equilibrium, 487, 531
Thermal expansion
coefficient of linear, 493
coefficient of volume, 493
table, 493
water as anomaly, 494
Thermal resistance (R factor), 526
in parallel, 529
in series, 527–528
table, 527
Thermal velocity, 717, 736
Thermionic emission, 923
Thermodynamics, 486 (*see also* Temperature)
first law, 518, 534–537
free expansion, 538, 576, 579, 583
quasi-static adiabatic expansion of a gas, 547–550
second law, 563–584
work and PV diagram, 539–541
zeroth law, 487
Thermograph, 532
Thermometer
bimetallic strip, 489
calibration points, 488, 490
constant-volume gas, 490–492
mercury, 487–488
Thermometric property, 487
Thermostat, 489
Thin films, interference in, 1063–1066
Thin-film coatings, *photo,* 1066
Thin-lens equation, 1026, 1044–1045
Thin lenses, 1025–1034
equation, 1026, 1044–1045
first focal point, 1028
focal length, 1026–1028, 1028
focal plane, 1028
image distance, 1030–1031
lens-maker's equation, 1026, 1029
multiple lenses, 1031–1034
second focal point, 1028
virtual object, 1025, 1032
Thompson, Benjamin (Count Rumford), 517–518
Thompson, G. P., 975, 1163
Thomson, J. J., 614, 792–793, 1157, *photo,* 792
Thomson, William, 964–966
Threshold frequency for photoelectric effect, 1150
Threshold wavelength for photoelectric effect, 1150, 1152
Thrust of a rocket, 213–215
Tidal force, 317
Tides, 88
Time average of a force, 110–112
Time constant
capacitors, 761, 763–764
damped oscillations, 390
in LR circuits, 860–862
in RC circuits, 761, 763–764

Time dilation, 1109–1111, 1116
Tokamak, *photo*, 830
Tone quality, 458
Toroid, 829–830, *photo*, 830
Torque, 231–235, 243, 259
  clockwise and counterclockwise, 277–278
  couple, 284, 798
  on current loops, 798–799
  on electric dipole, 616–617
  equilibrium and, 276–278
  on magnets, 799–800
Torr, unit of pressure, 339
Torricelli, Evangelista, 339
Torricelli's law, 349
Torsional pendulum, 385–386
Torsion balance, 305, 336, 386, 603
Torsion constant, 305, 386
Torsion modulus (*see* Shear modulus)
Total internal reflection, 987–989
  fiber optics, 988–989
  photos, 987, 988, 989
Total reactance, 912–913
Tracking antenna, *photo*, 943
Transducers
  piezoelectric crystals, *photo*, 771
  photovoltaic cells, *photo*, 771
  strain gauge, *photo*, 771
  microphone, *photo*, 770
  auditory receptor cells, *photo*, 770
  Meissner corpuscle, *photo*, 771
Transformer, 852, 898, 920–923, *photos*, 921
  magnetizing current, 920–921
  primary and secondary coils, 920–922
  rectifier and, *photo*, 923
  step-down, 920, 922
  step-up, 920, 922
  use in transmission lines, 922–923
Transient solution, 393
Transverse waves, 411
Triode, 925–926
Triode vacuum tube, *photo*, 925
Triple point, 491, 508–509
Turbulence, 346, 347, 353
Twin paradox, 1119–1121
Two-body problems, 117–120
Two-slit interference pattern, 1068–1071
  fringes, 1069–1071
  Lloyd's mirror, 1070
  maxima, 1068–1070
  minima, 1068–1070

Ultrasonic waves, 463, *photo*, 982
Ultraviolet catastrophe, 1148
Ultraviolet rays, 957
Units, 2–4
  cgs system, 3
  conversion of, 4–5
  mks system, 2
  SI, 2–3, 21, 85, AP-21
  U.S. customary system, 3, 21, 85
Unit vector, 53
Universal gas constant (R), 497
Universal gravitational constant (*see* Gravitational constant)

Vacuum tubes
  miniaturized, *photo*, 926
  triode, *photo*, 925
Van Allen belts, 791
Van de Graaff generator, 677, 679, *photo*, 641, 677
Van der Waals equation, 506–507
Vaporization, 522
Vapor pressure
  definition, 507
  of water at various temperatures, *table*, 507
Vector, 47
  acceleration, 56–57
  addition by components, 50–54
  addition by parallelogram method, 50
  cross product, 255–260
  displacement, 48–49
  dot product, 144–147
  equality, 49
  graphic addition, 48
  magnitude of, 48
  multiplication by scalars, 53
  position, 54
  Poynting, 954
  properties of, *table*, 52
  for rotational motion, 254–260
  scalar product, 144
  unit, 53
  velocity, 54–55
Vector product (*see* Cross product)
Velocity, 21 (*see also* Speed)
  average, 22–27, 29, 36, 39, 54–55
  instantaneous, 24–27, 54–55
  relative, 57–58, 199
  selector, 791–795
  transformation, 1122–1124
  vector, 54–55
Velocity selector, 791–795
Velocity transformation, 1122–1124
Venturi effect, 350
Very Large Array (VLA), *photo*, 948, 1087
Vibrations (*see* Oscillations)
Virtual image, 1012, 1015–1016, 1019–1022, 1030–1031
Viscosity, coefficient of, *table*, 352
Viscous forces, 351
Visible light, 957
Voltage, 657–658 (*see also* Electric potential)
Voltmeter, 767–769, 900
Volt (V), unit of electric potential, 656, 658, 720, 725, 726, 843
Volume charge density, 624
Volume flow rate, 347, 352
Von Baeyer, Hans Christian, 11
Von Klitzing, Klaus, 803–804
Von Klitzing constant, 804
Vortex shedding, 353
Walker, James S., 397
Water molecule, *model*, 617
Watt (W), unit of power, 3, 169, 725
Wave equation
  for electromagnetic waves, 947–953
  for string waves, 429–431, 947–948
Waveform, 458
Wavefront, 445, 982, 986
Wave function, 413, 416, 426, 429, 439
Wavelength, 416–417
  Compton, 1155
  cutoff, for x-ray spectrum, 1153–1154
  de Broglie, 1161–1162, 1164
  electron, 1162
  threshold, for photoelectric effect, 1150, 1152
Wave mechanics (*see* Quantum theory)
Wave number, 416, 442
Wave packet, 460–461 (*see also* Wave pulse)
Wave pulse, 410–413, 460 (*see also* Wave packet)
Waves (*see also* Sound waves)
  diffraction, 462–463, 1076–1089
  dispersion, 410
  electromagnetic, 944, 947–957
  electron, 1161–1165
  energy, 417–418, 427
  equation, 429–431, 947–953
  harmonic, 416–417
  interference, 419–420, 975, 1061–1076, 1080–1081
  longitudinal, 412, 439
  plane, 461, 948
  power, 418
  propagation of disturbance, 410–411, 439
  reflection, 461–462, 982–984, 987–989, 1013, 1019–1020
  refraction, 461, 982–987, 991–996, 1022–1025
  reversibility, 1018
  shock, 468, *photo*, 469
  speed of, 414–415
  standing, 420–427, 454–455
  on a string, 409–438
  superposition, 413, 419–420, 427–428
  transverse, 411
  ultrasonic, 463
  wave function, 413, 416, 426, 429, 439
Wave speed, 1026
Wave theory of light, 974–975, 1068, 1082
Wave train, 416
Weak nuclear force, 88–89
Weber, Wilhelm, 965–966
Weber (Wb), unit of magnetic flux, 841, 843
Weight (w), 83–85
  apparent weight, 84
  units of, 85
  weightlessness, 84
Westinghouse, George, 898
Whewell, William, 964

Wien's displacement law, 532, 553, 1147
Work, 135–181
  to charge a capacitor, 698–701
  by constant force, 136
  definition, 135, 136, 143
  dimensions of, 136
  first law of thermodynamics, 534–537, 539–541
  kinetic energy and, 136–143, 147
  motion in one dimension, 136–140
  motion in three dimensions, 143–147
  particle systems, 148–151
  pseudowork, 197
  units of, 136
  by variable force, 140–142
Work-energy theorem, 138, 167
  applied to fluids, 348–349
  generalized, 161–165
Work function, 1150, 1152
World Columbian Exposition, 898

Xerography (*see* Electrostatics and xerography)
X-ray diffraction, 1153
X rays, 957–958, 1152–1154
  Compton scattering, 1154–1155
  diffraction, 1153
X-ray spectrum
  bremsstrahlung spectrum, 1153
  characteristic spectrum, 1153
  cutoff wavelength, 1153–1154

Young, Thomas
  two-slit interference experiment, 1068, 1163
  wave theory of light, 975, 1068
Young's modulus (Y)
  for compressive stress, 335, 495
  definition, 334
  and sound waves, 440
  table, 334
  units of, 334

Zallen, Richard, 682–684
Zero-momentum reference frame, 194–195
Zeroth law of thermodynamics, 487

Some Conversion Factors

1 m = 39.37 in = 3.281 ft = 1.094 yd

1 m = 10^{15} fm = 10^{10} Å = 10^9 nm

1 km = 0.6215 mi

1 mi = 5280 ft = 1.609 km

1 lightyear = 1 $c \cdot$y = 9.461×10^{15} m

1 in = 2.540 cm

1 L = 10^3 cm^3 = 10^{-3} m^3 = 1.057 qt

1 h = 3.6 ks

1 y = 365.24 d = 3.156×10^7 s

1 km/h = 0.278 m/s = 0.6215 mi/h

1 ft/s = 0.3048 m/s = 0.6818 mi/h

1 rev = 2π rad = 360°

1 rad = 57.30°

1 rev/min = 0.1047 rad/s

1 slug = 14.59 kg

1 tonne = 10^3 kg = 1 Mg

1 atm = 101.3 kPa = 1.013 bar = 76.00 cmHg = 14.70 lb/in^2

1 N = 10^5 dyn = 0.2248 lb

1 lb = 4.448 N

1 Pa·s = 10 poise

1 J = 10^7 erg = 0.7373 ft·lb = 9.869×10^{-3} L·atm

1 kW·h = 3.6 MJ

1 cal = 4.184 J = 4.129×10^{-2} L·atm

1 L·atm = 101.3 J = 24.22 cal

1 eV = 1.602×10^{-19} J

1 Btu = 778 ft·lb = 252 cal = 1054 J

1 horsepower = 550 ft·lb/s = 746 W

1 W/m·K = 6.938 Btu·in/h·ft^2·°F

1 T = 10^4 G

1 kg weighs about 2.205 lb